With fondest memories
of our times together
and faith in our
continuing relationship
*

Mel Lewis
9/30/83

*and Szechuan

MATRAS

The end papers for this book were designed and drawn by Mr. Gene Matras, a wildlife artist from New Hampshire who specializes in detailed pen and ink drawings of farm animals and scenes from nature. The artist has depicted for *Developmental-Behavioral Pediatrics* various scenarios in the lives of geese. An array of behavioral and developmental challenges and pleasures confronting these waterfowl is portrayed.

DEVELOPMENTAL-BEHAVIORAL PEDIATRICS

MELVIN D. LEVINE, M.D.
Associate Professor of Pediatrics
Harvard Medical School
Chief, Division of Ambulatory Pediatrics
The Children's Hospital
Boston, Massachusetts

WILLIAM B. CAREY, M.D.
Clinical Associate Professor of Pediatrics
University of Pennsylvania School of Medicine
Private Practice
Media, Pennsylvania

ALLEN C. CROCKER, M.D.
Associate Professor of Pediatrics
Harvard Medical School
Director, Developmental Evaluation Clinic
The Children's Hospital
Lecturer on Maternal and Child Health
Harvard School of Public Health
Boston, Massachusetts

RUTH T. GROSS, M.D.
Katharine Dexter and Stanley McCormick
Professor of Pediatrics
Stanford University School of Medicine
Director, Division of General Pediatrics
Stanford University Hospital
Director, Stanford–Children's Ambulatory Care Center
Palo Alto, California

1983
W. B. SAUNDERS COMPANY
Philadelphia London Toronto Mexico City Rio de Janeiro Sydney Tokyo

W. B. Saunders Company: West Washington Square
Philadelphia, PA   19105

1 St. Anne's Road
Eastbourne, East Sussex BN21 3UN, England

1 Goldthorne Avenue
Toronto, Ontario M8Z 5T9, Canada

Apartado 26370—Cedro 512
Mexico 4, D.F., Mexico

Rua Coronel Cabrita, 8
Sao Cristovao Caixa Postal 21176
Rio de Janeiro, Brazil

9 Waltham Street
Artarmon, N.S.W. 2064, Australia

Ichibancho, Central Bldg., 22-1 Ichibancho
Chiyoda-Ku, Tokyo 102, Japan

Library of Congress Cataloging in Publication Data

Main entry under title:

Developmental-behavioral pediatrics.

1. Pediatrics.   2. Pediatrics—Psychological aspects.
    I. Levine, Melvin D. [DNLM: 1. Child behavior disorders.
    2. Child development disorders.   3. Child development.
    4. Child behavior. WS 105 D48912]

RJ47.D48 1983   618.92′8   82–40187

ISBN 0–7216–5744–3   AACR2

Developmental-Behavioral Pediatrics   ISBN   0-7216-5744-3

Last digit is the print number: 9   8   7   6   5   4   3   2   1

CONTRIBUTORS

Israel F. Abroms, M.D.
Professor of Pediatrics and Neurology, University of Massachusetts Medical School. Chief, Pediatric Neurology, University of Massachusetts Medical Center, Worcester, Massachusetts.

Thomas F. Anders, M.D.
Professor of Psychiatry and Behavioral Sciences, Professor of Pediatrics, and Director, Division of Child Psychiatry and Child Development, Stanford University Medical Center. Chief of Psychiatry, Children's Hospital at Stanford, Stanford, California.

Ronald G. Barr, M.D., M.A., C.M.
Assistant Professor of Pediatrics, McGill University. Director, Developmental Medicine, Department of Ambulatory Services, Montreal Children's Hospital, Montreal, Quebec, Canada.

Anthony S. Bashir, Ph.D.
Lecturer in Pediatrics, Harvard Medical School. Senior Speech and Language Pathologist, Division of Hearing and Speech, The Children's Hospital, Boston, Massachusetts.

Myron L. Belfer, M.D.
Acting Chief and Director, Child and Adolescent Services, Department of Psychiatry, Cambridge Hospital, Cambridge, Massachusetts.

Stanley J. Berman, Ph.D.
Instructor of Psychology, Department of Psychiatry, Harvard Medical School. Staff Psychologist, Sidney Farber Cancer Institute. Assistant in Psychology, The Children's Hospital, Boston, Massachusetts.

William G. Bithoney, M.D.
Instructor in Pediatrics, Harvard Medical School. Director, Comprehensive Child Health Program, The Children's Hospital, Boston, Massachusetts.

Gilbert J. Botvin, Ph.D.
Assistant Professor of Public Health and Clinical Assistant Professor of Psychiatry, Cornell University Medical College. Assistant Attending Psychologist, The New York Hospital–Cornell Medical Center, New York, New York.

M. Patricia Boyle, Ph.D.
Associate Director of Child Psychiatry and Director, Division of Child Psychology, Department of Psychiatry, Cambridge Hospital, Cambridge, Massachusetts.

Michael J. Bresnan, M.D.
Associate Professor of Neurology, Harvard Medical School. Associate Chief for Clinical Services, Department of Neurology, The Children's Hospital, Boston, Massachusetts.

Robert B. Brooks, Ph.D.
Instructor, Department of Psychiatry, Harvard Medical School, Boston. Director, Outpatient Child Psychology and Psychoeducation Services, Hall-Mercer Children's Center, McLean Hospital, Belmont, Massachusetts.

Christina Browne, M.S.N., R.N.C.
Director, Nurse Practitioner–Physician Assistant Program, Wrentham Project, The Children's Hospital, Boston, Massachusetts.

Hilde Bruch, M.D.
Professor Emeritus of Psychiatry, Baylor College of Medicine, Houston, Texas.

William B. Carey, M.D.
Clinical Associate Professor of Pedi-

atrics, University of Pennsylvania School of Medicine. Private Practice, Media, Pennsylvania.

Robert W. Chamberlin, M.D., M.P.H.
Coordinator, Maternal Child Health Program, Indian Health Service, Billings, Montana.

Stella Chess, M.D.
Professor of Child Psychiatry, New York University Medical Center. Attending Psychiatrist, University Hospital and Bellevue Hospital, New York, New york.

Walter P. Christian, Ph.D.
Assistant Professor of Neurology, Tufts University School of Medicine. Consultant in Autism, Developmental Evaluation Clinic, The Children's Hospital, Boston. Adjunct Associate Professor of Psychology, University of Massachusetts, Worcester. Executive Director, The May Institute, Chatham, Massachusetts. Adjunct Associate Professor, Rehabilitation Institute, Southern Illinois University, Carbondale, Illinois. Adjunct Assistant Professor of Human Development, University of Kansas, Kansas City, Kansas.

Edward R. Christophersen, Ph.D.
Professor and Chief, Behavioral Sciences Section, Department of Pediatrics, University of Kansas Medical Center, Kansas City, Kansas.

Donald J. Cohen, M.D.
Professor of Pediatrics, Psychiatry, and Psychology, Yale University School of Medicine and Child Study Center. Clinical Staff, Yale–New Haven Hospital, New Haven, Connecticut.

Thomas E. Cone, Jr., M.D.
Clinical Professor of Pediatrics Emeritus, Harvard Medical School. Senior Associate in Clinical Genetics and Medicine, The Children's Hospital, Boston, Massachusetts.

Allen C. Crocker, M.D.
Associate Professor of Pediatrics, Harvard Medical School. Lecturer on Maternal and Child Health, Harvard School of Public Health. Director, Developmental Evaluation Clinic, The Children's Hospital, Boston, Massachusetts.

Christine E. Cronck, Sc.D.
Associate Dean, Nesbitt College of Design, Nutrition, Human Behavior, and Home Economics, Drexel University, Philadelphia, Pennsylvania.

Marie M. Cullinane, R.N., M.S.
Adjunct Assistant Professor of Maternal-Child Nursing, Boston College Graduate School. Nurse Specialist in Developmental Disabilities and Mental Retardation and Director of Interdisciplinary Training, Developmental Evaluation Clinic, The Children's Hospital, Boston, Massachusetts.

Bruce Cushna, Ph.D.
Principal Associate in Pediatrics (Psychology), Harvard Medical School. Adjunct Assistant Professor of Special Education, Boston College. Senior Associate in Psychology and Associate Director, Developmental Evaluation Clinic, The Children's Hospital, Boston, Massachusetts.

Philip W. Davidson, Ph.D
Associate Professor of Pediatrics, Psychiatry (Psychology), and Psychology, University of Rochester School of Medicine and Dentistry. Director, University Affiliated Diagnostic Clinic for Developmental Disorders, and Senior Associate Psychologist, Strong Memorial Hospital. Consultant, Monroe Developmental Center, Rochester, New York.

Edith G. DeAngelis, Ed.D., Ed.M.
Professor, Department of Physical Education and Recreation, University of Massachusetts at Boston, Boston, Massachusetts.

Sanford M. Dornbusch, Ph.D.
Reed-Hodgson Professor of Human Biology and Professor of Sociology and Education, Stanford University, Stanford, California.

Paula M. Duke, M.D.
Assistant Professor of Pediatrics, Stanford University School of Medicine, Stanford, California.

Paul H. Dworkin, M.D.
Associate Professor of Pediatrics, University of Connecticut School of Medicine. Director of Ambulatory Care, Department of Pediatrics, University of Connecticut Health Center, Farmington, Connecticut.

Bernice T. Eiduson, Ph.D.
Professor, Department of Psychiatry and Biobehavioral Sciences, University of California at Los Angeles School of Medicine, Los Angeles, California.

Constantine John Falliers, M.D.
Associate Clinical Professor of Pediatrics and Medicine, University of Colorado School of Medicine. Attending Allergist, National Jewish Hospital and National Asthma Center, Denver, Colorado.

Richard Famularo, M.D.
Clinical Instructor in Child Psychiatry, Harvard Medical School. Clinical Staff, The Children's Hospital, Boston, Massachusetts.

Marianne E. Felice, M.D.
Assistant Professor of Pediatrics, University of California, San Diego. Chief, Division of Adolescent Medicine, University of California, San Diego, Medical Center, San Diego, California.

Deborah A. Frank, M.D.
Assistant Clinical Professor of Pediatrics, Boston University School of Medicine. Pediatrician, Child Development Unit, Boston City Hospital, Boston, Massachusetts.

William K. Frankenburg, M.D.
Professor of Pediatrics and Preventive Medicine and Director, Rocky Mountain Child Development Center, University of Colorado School of Medicine, Denver, Colorado.

Park S. Gerald, M.D.
Professor of Pediatrics, Harvard Medical School. Senior Associate in Medicine, The Children's Hospital, Boston, Massachusetts.

Edmund W. Gordon, Ed.D., M.S., M.A.
Professor of Psychology and Afro-American Studies, Yale College and Graduate School. Professor of Child Psychology, Yale University School of Medicine, New Haven, Connecticut.

Marvin I. Gottlieb, Ph.D., M.D.
Professor of Pediatrics, College of Medicine and Dentistry of New Jersey–Rutgers Medical School, Piscataway. Director, Institute for Child Development, Hackensack Medical Center, Hackensack, New Jersey.

John W. Graef, M.D.
Assistant Clinical Professor, Harvard Medical School. Associate in Medicine and Director, Lead-Toxicology Clinic, The Children's Hospital, Boston, Massachusetts.

John M. Graham, Jr., M.D., Sc.D.
Assistant Professor, Department of Maternal and Child Health, Dartmouth Medical School. Attending Pediatrician and Dysmorphologist, Dartmouth-Hitchcock Medical Center, Hanover, New Hampshire.

Morris Green, M.D.
Perry W. Lesh Professor and Chairman, Department of Pediatrics, Indiana University School of Medicine. Physician-in-Chief, James Whitcomb Riley Hospital for Children, Indianapolis, Indiana.

Ruth T. Gross, M.D.
Katharine Dexter and Stanley McCormick Professor of Pediatrics, Stanford University School of Medicine. Director, Division of General Pediatrics, Stanford University Hospital. Di-

rector, Stanford–Children's Ambulatory Care Center, Palo Alto, California.

Michael J. Guralnick, Ph.D.
Professor of Communication and Psychology and Director, The Nisonger Center, Ohio State University, Columbus, Ohio.

Beatrix A. Hamburg, M.D.
Professor of Psychiatry in Pediatrics, Mt. Sinai School of Medicine, Mt. Sinai Medical Center, New York, New York.

Robert Hampton, Ph.D.
Associate Professor of Sociology, Connecticut College, New London, Connecticut. Research Associate, Family Development Study, and Project Director, "Hospitals as Gatekeepers" Study, The Children's Hospital, Boston, Massachusetts.

Alfred Healy, M.D.
Professor of Pediatrics and Special Education, University of Iowa. Chairman, Division of Developmental Disabilities, Department of Pediatrics, University of Iowa Hospitals and Clinics. Director, Iowa University Affiliated Program, University Hospital School, Iowa City, Iowa.

Herman A. Hein, M.D.
Professor of Pediatrics, University of Iowa. Director, Iowa Statewide Perinatal Care Program and Iowa High-Risk Infant Follow-up Program, Iowa City, Iowa.

Anne Henderson, Ph.D., O.T.R.
Associate Professor of Occupational Therapy, Sargent College of Allied Health Professions, Boston University, Boston, Massachusetts.

Elaine M. Hicks, M.B., M.R.C.P.
Clinical Fellow in Neurology, Harvard Medical School. Clinical Fellow in Neurophysiology, The Children's Hospital, Boston, Massachusetts.

Debra L. Olenick Hirsch, Ph.D.
Psychologist, Pediatric Oncology, Tom Baker Cancer Centre, Calgary, Alberta, Canada.

E. Lawrence Hoder, M.D.
Assistant Professor, Department of Pediatrics and Child Study Center, Yale University School of Medicine. Clinical Staff, Yale–New Haven Hospital, New Haven, Connecticut.

Rosanne B. Howard, R.D., M.P.H.
Visiting Lecturer in Human Nutrition, University of Massachusetts, Amherst. Director of Nutrition, Developmental Evaluation Clinic, The Children's Hospital, Boston, Massachusetts.

Carol Nagy Jacklin, Ph.D.
Senior Research Associate, Department of Psychology, Stanford University, Stanford, California.

Juel M. Janis, Ph.D.
Assistant Dean and Adjunct Associate Professor of Public Health, University of California at Los Angeles, Los Angeles, California.

Marcia A. Keener, Ph.D.
Research Assistant, Infant Sleep Laboratory, Division of Child Psychiatry and Child Development, Stanford University School of Medicine, Stanford, California.

John H. Kennell, M.D.
Professor of Pediatrics, Case Western Reserve University School of Medicine. Associate Pediatrician, Rainbow Babies and Children's Hospital and Cleveland Metropolitan General Hospital, Cleveland, Ohio.

William Kessen, Ph.D.
Eugene Higgins Professor of Psychology and Professor of Pediatrics, Yale University School of Medicine, New Haven, Connecticut.

William E. Kiernan, Ph.D.
Clinical Associate, Sargent College of Allied Health Professionals, Boston University. Director of Rehabilitation, Developmental Evaluation Clinic, The Children's Hospital, Boston, Massachusetts.

Marshall H. Klaus, M.D.
Professor and Chairman, Department of Pediatrics, Michigan State University School of Medicine, East Lansing, Michigan.

Gerald P. Koocher, Ph.D.
Assistant Professor of Psychology, Harvard Medical School. Director of Training and Senior Associate in Psychology, The Children's Hospital, Boston, Massachusetts.

Jules G. Leroy, M.D., Ph.D.
Professor of Genetics and Medical Genetics, Antwerp University Medical School. Head, Division of Medical Genetics, Academic Hospital, Antwerp University Medical School, Wilryk, Belgium.

Melvin D. Levine, M.D.
Associate Professor of Pediatrics, Harvard Medical School. Chief, Division of Ambulatory Pediatrics, The Children's Hospital, Boston, Massachusetts.

Gregory S. Liptak, M.D., M.P.H.
Assistant Professor of Pediatrics, University of Rochester School of Medicine and Dentistry. Associate Pediatrician, Strong Memorial Hospital, Rochester, New York.

Iris F. Litt, M.D.
Associate Professor of Pediatrics and Director, Division of Adolescent Medicine, Stanford University School of Medicine, Stanford, California.

Pauline F. Lukens, M.Ed., M.S.W.
Lecturer, Department of Psychiatry, and Staff Social Worker, University of Massachusetts Medical Center, Worcester, Massachusetts.

Eleanor E. Maccoby, Ph.D.
Barbara Kimball Browning Professor of Psychology, Department of Psychology, Stanford University, Stanford, California.

John A. Martin, Ph.D.
Research Associate, Department of Developmental Psychology, Stanford University School of Medicine, Stanford, California.

Wendy S. Matthews, Ph.D.
Adjunct Assistant Professor of Pediatrics (Psychology), Department of Pediatrics, College of Medicine and Dentistry of New Jersey–Rutgers Medical School, Piscataway, New Jersey.

Lynn J. Meltzer, Ph.D.
Instructor in Psychology, Department of Psychiatry, Harvard Medical School. Director, Psychoeducational Services, Division of Ambulatory Pediatrics, The Children's Hospital, Boston, Massachusetts.

David L. Meryash, M.D.
Instructor in Pediatrics, Harvard Medical School. Assistant in Medicine, The Children's Hospital, Boston, Massachusetts.

Robert B. Millman, M.D.
Professor of Clinical Public Health and Associate Professor of Clinical Psychiatry, Cornell University Medical College. Director, Drug and Alcohol Abuse Service, New York Hospital–Payne Whitney Psychiatric Clinic, New York, New York.

Ann P. Murphy, M.S.W., A.C.S.W.
Clinical Assistant Professor, Boston University School of Social Work. Director of Social Work, Developmental Evaluation Clinic, The Children's Hospital, Boston, Massachusetts.

Philip R. Nader, M.D.
Professor of Pediatrics and Director of General Pediatrics, Department of Pediatrics, University of California at San Diego, San Diego, California.

Richard P. Nelson, M.D.
Assistant Professor of Pediatrics, University of Minnesota Medical School, Minneapolis. Director, Developmental Disabilities Program, Gillette Children's Hospital, St. Paul, Minnesota.

Charlotte G. Neumann, M.D., M.P.H.
Professor of Public Health and Pediatrics, University of California at Los Angeles Schools of Public Health and Medicine. Attending Pediatrician, University of California at Los Angeles Hospital, Los Angeles, California.

Eli H. Newberger, M.D.
Assistant Professor of Pediatrics, Harvard Medical School. Director, Family Development Study, and Senior Associate in Medicine, The Children's Hospital, Boston, Massachusetts.

David R. Offord, M.D., F.R.C.P.(C.)
Professor, Department of Psychiatry, McMaster University. Research Director, Child and Family Centre, Chedoke-McMaster Hospitals, Chedoke Division, Hamilton, Ontario, Canada.

Judith S. Palfrey, M.D.
Clinical Instructor in Pediatrics, Harvard Medical School. Associate in Medicine, The Children's Hospital, Boston, Massachusetts.

Charles W. Popper, M.D.
Clinical Instructor in Psychiatry, Harvard Medical School, Boston. Assistant Child Psychiatrist, McLean Hospital, Belmont, Massachusetts.

Siegfried M. Pueschel, M.D., M.P.H.
Associate Professor of Pediatrics, Brown University Program in Medicine. Physician and Director, Child Development Center, Rhode Island Hospital, Providence, Rhode Island.

Joaquim Puig-Antich, M.D.
Assistant Professor of Clinical Psychiatry, Columbia University College of Physicians and Surgeons. Director, Child and Adolescent Depression Clinic, New York State Psychiatric Institute. Assistant Attending Psychiatrist, Presbyterian Hospital and Medical Center, New York, New York.

Harris Rabinovich, M.D.
Instructor of Clinical Psychiatry, Columbia University College of Physicians and Surgeons. Research Psychiatrist, Child and Adolescent Depression Clinic, New York Psychiatric Institute. Assistant Attending Psychiatrist, Presbyterian Hospital and Hospital Center, New York, New York.

Jennifer M. Rathbun, M.D., M.A.
Instructor in Psychiatry, Harvard Medical School. Assistant in Psychiatry, The Children's Hospital. Clinical Associate in Psychiatry and Director, Failure to Thrive Team and Growth Development Clinic, Massachusetts General Hospital, Boston, Massachusetts.

H. Burtt Richardson, Jr., M.D.
Adjunct Associate Professor of Child Health and Development, George Washington University School of Medicine and Health Sciences, Washington, D. C. Pediatrician, Winthrop Area Medical Center, Winthrop, Maine.

Julius B. Richmond, M.D.
Professor of Health Policy, Harvard Medical School. Advisor, Child Health Policy, The Children's Hospital, Boston, Massachusetts.

I. Leslie Rubin, M.B., B.Ch.
Instructor in Pediatrics, Harvard Medical School. Associate in Medicine and Pediatrician, Developmental Evaluation Clinic, The Children's Hospital, Boston, Massachusetts.

Dennis C. Russo, Ph.D.
Assistant Professor of Psychology, Department of Psychiatry, Harvard Medical School. Director, Behavioral Medicine Program, The Children's Hospital, Boston, Massachusetts.

Neil L. Schechter, M.D.
Assistant Clinical Professor of Pediatrics, University of Connecticut School of Medicine, Farmington. Director, Section of Behavioral and Developmental Pediatrics, St. Francis Hospital and Medical Center. Pediatric Devel-

opmental Consultant, Hartford Public Schools, Hartford, Connecticut.

Albert P. Scheiner, M.D.
Professor of Pediatrics and Co-Director, Child Development Service, University of Massachusetts Medical School, Worcester, Massachusetts.

Barton D. Schmitt, M.D.
Associate Professor of Pediatrics, University of Colorado School of Medicine, Denver, Colorado.

Richard R. Schnell, Ph.D.
Associate in Pediatrics (Psychology), Harvard Medical School. Director of Psychology, Developmental Evaluation Clinic, The Children's Hospital, Boston, Massachusetts.

Lisbeth Bamberger Schorr, B.A.
Adjunct Professor of Maternal and Child Health Policy, University of North Carolina School of Public Health, Chapel Hill, North Carolina.

Daniel S. P. Schubert, M.D., Ph.D.
Associate Professor of Psychiatry and Clinical Associate Professor of Psychology, Case Western Reserve University School of Medicine. Director of Research, Department of Psychiatry, Cleveland Metropolitan General Hospital, Cleveland, Ohio.

Herman J. P. Schubert, Ph.D.
Research Psychologist, State University College at Buffalo, Buffalo, New York.

David Scott, Ph.D.
Assistant Professor of Pediatrics, Yale University School of Medicine, New Haven, Connecticut.

Bennett A. Shaywitz, M.D.
Associate Professor of Pediatrics and Neurology and Director, Pediatric Neurology, Yale University School of Medicine. Attending Physician, Yale–New Haven Hospital, New Haven, Connecticut.

Sally E. Shaywitz, M.D.
Associate Professor of Pediatrics and Director, Learning Disorders Unit, Yale University School of Medicine. Attending Physician, Yale–New Haven Hospital, New Haven, Connecticut.

Alice M. Shea, P.T., M.P.H.
Adjunct Assistant Professor, Sargent College of Allied Health Professionals, Boston University. Teaching Fellow, Harvard School of Public Health. Director of Physical Therapy, Developmental Evaluation Clinic, The Children's Hospital, Boston, Massachusetts.

Eunice Shishmanian, R.N., M.S.
Adjunct Assistant Clinical Professor, Department of Parent-Child Nursing, Boston University. Adjunct Instructor, Boston College Graduate School of Nursing. Director of Nursing, Developmental Evaluation Clinic, The Children's Hospital, Boston, Massachusetts.

Jack P. Shonkoff, M.D.
Assistant Professor of Pediatrics, University of Massachusetts Medical School. Co-Director, Child Development Service, University of Massachusetts Medical Center, Worcester, Massachusetts.

Maarten S. Sibinga, M.D.
Professor of Pediatrics, Temple University School of Medicine. Chief, Section of Gastroenterology, Department of Pediatrics, St. Christopher's Hospital for Children, Philadelphia, Pennsylvania.

Jane C. Snyder, Ph.D.
Lecturer in Pediatrics, Harvard Medical School. Research Associate and Fellow in Psychology, The Children's Hospital, Boston, Massachusetts.

Burton Z. Sokoloff, M.D.
Assistant Clinical Professor of Pediatrics, University of California at Los Angeles School of Medicine. Attend-

ing Physician, Cedars–Sinai Medical Center, Los Angeles. Attending Pediatrician, West Hills Hospital, Canoga Park, California.

Rachel E. Stark, Ph.D.
Associate Professor, Department of Neurology, The Johns Hopkins University School of Medicine. Director, Division of Hearing and Speech, John F. Kennedy Institute, Baltimore, Maryland.

George Storm, M.D.
Instructor in Pediatrics, Harvard Medical School. Assistant in Medicine, The Children's Hospital, Boston, Massachusetts. Clinical Staff, Department of Pediatrics, Exeter Hospital, Exeter, New Hampshire.

Ludwik S. Szymanski, M.D.
Assistant Professor of Psychiatry, Harvard Medical School. Director of Psychiatry, Developmental Evaluation Clinic, and Associate in Psychiatry, The Children's Hospital, Boston, Massachusetts.

Lawrence T. Taft, M.D.
Professor and Chairman, Department of Pediatrics, College of Medicine and Dentistry of New Jersey–Rutgers Medical School, Piscataway, New Jersey.

Alexander Thomas, M.D.
Professor of Psychiatry, New York University Medical Center. Attending Psychiatrist, University Hospital and Bellevue Hospital, New York, New York.

Victor C. Vaughan, III, M.D.
Professor in Pediatrics, Temple University School of Medicine. Senior Medical Evaluation Officer, National Board of Medical Examiners. Attending Physician, St. Christopher's Hospital for Children, Philadelphia, Pennsylvania.

Mazie Earle Wagner, Ph.D.
Psychologist and Director of Counseling Emeritus, State University College at Buffalo, Buffalo, New York.

Judith S. Wallerstein, Ph.D.
Senior Lecturer, School of Social Welfare and School of Law, University of California, Berkeley. Executive Director, Center for the Family in Transition, Corte Madera, California.

Charles Walton, M.D.
Clinical Professor of Psychiatry, Stanford University School of Medicine. Attending Psychiatrist, Children's Hospital at Stanford, Stanford, California.

Brent G. H. Waters, M.B., Dip. Psych., F.R.C.P.(C.), M.R.A.N.Z.C.P.
Staff Psychiatrist, Department of Child and Family Psychiatry, Royal Alexandra Hospital for Children, Camperdown, New South Wales, Australia.

Esther H. Wender, M.D.
Associate Professor of Pediatrics and Director, Division of Behavioral Pediatrics, Albert Einstein College of Medicine. Attending Pediatrician, North Central Bronx Hospital, Bronx Municipal Hospital Center, and Hospital of the Albert Einstein College of Medicine, Bronx. Attending Pediatrician, Montefiore Hospital and Medical Center, New York, New York.

Kathleen M. White, Ed.D.
Associate Professor of Psychology, Boston University. Project Director, Developmental Consequences of Severe Inflicted Violence, The Children's Hospital, Boston, Massachusetts.

Gregg F. Wright, M.Ed., M.D.
Assistant Professor of Pediatrics, University of Texas Medical Branch, Galveston, Texas.

Alayne Yates, M.D.
Associate Professor of Psychiatry and Pediatrics and Chief of Child Psychiatry, Department of Psychiatry, University of Arizona Health Sciences Center, Tucson, Arizona.

Jean M. Zadig, Ph.D.
Assistant Professor of Education, Bos-

ton College. Director of Special Education, Developmental Evaluation Clinic, The Children's Hospital, Boston, Massachusetts.

Philip R. Ziring, M.D.
Associate Clinical Professor of Pediatrics, Columbia University College of Physicians and Surgeons. Associate Attending Pediatrician, Babies Hospital, Columbia Presbyterian Medical Center, New York, New York. Chairman, Department of Pediatrics, Morristown Memorial Hospital, Morristown, New Jersey.

Barry S. Zuckerman, M.D.
Assistant Professor of Pediatrics, Boston University School of Medicine. Director, Child Development Unit, Boston City Hospital, Boston, Massachusetts.

PREFACE
Toward a Better Armful

THE ARMFUL

For every parcel I stoop down to seize
I lose some other off my arms and knees
And the whole pile is slipping, bottles, buns,
Extremes too hard to comprehend at once,
Yet nothing I should care to leave behind.
With all I have to hold with, hand and mind
And heart, if need be, I will do my best
To keep their building balanced at my breast.
I crouch down to prevent them as they fall;
Then sit down in the middle of them all.
I had to drop the armful in the road
And try to stack them in a better load.

Robert Frost
From "West-Running Brook"*

The quest for insight—its acquisition, modification, and reclamation—is the essence of professional growth. Those who support the health and life quality of children have struggled to deliver an ever weightier stack of theories and practices. *Developmental-Behavioral Pediatrics* arose out of a need to rearrange and recombine the relevant hypotheses, discoveries, and methods.

The Hyphen

The hyphenated title of this book symbolizes one of its principal missions, namely, the joining of two frequently divided efforts within child health. Developmental pediatrics has concerned itself largely with cognitive competence and the associated physical and mental disabilities that constrain function in childhood. Behavioral pediatrics has emphasized the prevention and treatment of disorders of personality and the effects of family function and social adaptation. *Developmental-Behavioral Pediatrics* splices these strands so as to emphasize their shared themes, their compatible missions, and their complementary contributions to general pediatrics and other disciplines concerned with health and function in childhood.

This book, then, is an outgrowth of the recognition that children with disabilities or handicapping conditions bring forth inevitable "behavioral" concerns, while those with "psychosocial problems" commonly harbor underlying developmental or constitutional traits that render them exquisitely vulnerable to their environmental stresses. The interplay makes it unjust to classify as exclusively behavioral or developmental such conditions as attention deficit, bed wetting, colic, teenage pregnancy, and mental retardation. Moreover, sound preventive pediatric practice covers the wide spectrum of developmental and behavioral health.

Linkages

This volume constitutes a forum for disciplines beyond as well as within pediatrics. Multiple child oriented professions are represented among its authors. Among those included arè child psychiatry, adolescent medicine, pediatric neurology, psychology (with its subspecialities), education, social work, speech and language pathology, recreation therapy, and nursing. While the unique perspective of each profession is preserved, it is evident that these related disciplines share many concepts, diagnostic procedures, and therapeutic techniques.

Although *Developmental-Behavioral Pediatrics* was edited by four pediatricians and written largely with the needs of health care providers in mind, its content is relevant for all who work toward sustaining healthy child development. Consequently we anticipate that its readership will include individuals from mental health professions, education, and other related pursuits.

Normalcy, Variation, and Deviation

Developmental-Behavioral Pediatrics encourages its readers to explore and perhaps rearrange their ideas of normalcy, variation, and deviation. Throughout this text an effort has been made to accentuate the heterogeneity of children and to depict the network of pathways of normal development, while portraying common and sometimes stressful forces that influence development and behavior in all children. A continuum emerges; it extends from the most common patterns of function to less universal themes and variations, to subtle dysfunctions, and then to some more prominent and less common clinical disorders. Underlying such a model is the acknowledgment that "line drawing" is difficult and possibly injurious. If we make rigid distinctions between variation and deviation, we sometimes run the risk of superimposing observer biases that mislead and burden a child with self-fulfilling prophecy and consequent constrained opportunity.

Whenever possible, the use of restrictive labels has been avoided. Some contributors have followed this course more than others, and certain subject areas are far more conducive to a descriptive, nonlabeling approach.

Theory, Knowledge, and Practice

In planning this book the editors sought to achieve a balance among theory, documentation from existing literature, clinical wisdom based on experience, and applicable advice. The text was not intended to be a compilation of detailed reviews of the literature. In keeping with traditional formats of textbooks, contributors were asked to select "key" references or suggested readings with a focus on theoretical constructs and current research findings of clinical relevance. On the other hand, it would be inappropriate and irresponsible to produce an anthology of anecdotal accounts and home remedies for the disturbances of childhood. A fusion of theoretical constructs and current research findings with the insights of experienced clinicians has resulted.

The Plan

Developmental-Behavioral Pediatrics is subdivided into eight parts. The first of these provides a historical backdrop, a review of earlier efforts in the pediatrics of development and behavior. The eight chapters in Part II are concerned with "patterns of variation over time," or common themes and influences that characterize normal developmental change from the prenatal period until the end of adolescence. Included are a review of general issues and models of development as well as discussions of the sources and effects of physical maturation, gender identity, and behavioral style. There also is a chapter on parent development. Integrated throughout Part II of *Developmental-Behavioral Pediatrics* are recommended roles of health professionals as children age.

Parts III and IV describe a range of inputs during childhood. Part III deals with "milieux and circumstances," describing vital forces that either benefit or thwart development and behavior. These seven chapters concentrate on the effects of environmental or extrinsic factors. Part IV, on the other hand, enumerates "biological intrusions" or health related stresses that sometimes threaten optimum development and behavioral adaptation.

As Parts III and IV deal with extrinsic and intrinsic inputs that may alter the course of development, Part V comprises fifteen chapters covering problematic outcomes, difficult states, variations, dysfunctions, and clinical disorders frequently brought to the attention of clinicians. Within each chapter there is an attempt to clarify some of their causes and manifestations. Although some management suggestions are provided, more definitive discussions of intervention are presented in a later part of the book.

Part VI concerns techniques for assessment. The emphasis is on description of childhood function, with the implicit theme of collaboration and the use of multiple diagnostic approaches.

Part VII concentrates on treatment, including counseling, behavioral management, special education, and other related services. There is a chapter on psychopharmacology. Another chapter reviews the pros and cons of various alternative therapies, some of which are controversial. This separate part on intervention is included in recognition of the pronounced overlap in goals and approaches to various developmental and behavioral problems.

Part VIII explores four general areas that will continue to promote and shape developmental and behavioral pediatrics, namely, relevant legal issues, research and methodology, professional training, and ethical considerations.

This volume will be informative, and it will enrich the professional care of children by demonstrating that health and function, nature and nurture, and variation and deviation are all part of the same "armful."

Like the narrator in Robert Frost's poem, the editors of this book have "sat down in the middle" of a vast array of promising parcels for children and families. We have felt the need to "drop the armful in the road" and thence to gather it up in a manageable format. The result, we believe, is a subject that has been stacked "in a better load."

CONTENTS

I

The Background

In the Preface, the editors have summarized some of the justifications for Developmental-Behavioral Pediatrics. There is also an explanation of the book's overall structure and organization that may help the reader to understand the particular arrangement of the chapters herein.

The first two chapters of this book provide some relevant background information that is likely to have a bearing on the reader's understanding of its rationale and historical context. Chapter 1, on early concepts of behavior and development within pediatrics, presents the changing views of variation and deviation from the sixteenth century up through the second decade of the twentieth. This is followed by a more recent perspective, focusing on changes in the field of pediatrics and the "coming of age" of development and behavior within the context of child health care.

1

Historical Trends: Evolving Pediatric Concepts of Variation and Deviation Up to 1920

> . . . Children have always been, and still are, a mirror to us—
> ourselves writ small, so to speak.
>
> ROBERT COLES (1979)

Nothing is more difficult in a historical analysis of child study than to follow the emergence of the child from his older place as an ill-formed adult at the fringes of society to his present place as a cultural hero. Through the centuries, with the exception of a few bright spots, the child was hardly considered a worthy enough member of the human family to warrant much interest as a legitimate object of scientific curiosity. Philippe Ariès (1962) has brilliantly supported the thesis that childhood is an invention of modern times and, as a phenomenon unto itself, needs special institutions for its protection. He dates the emergence of the modern family—i.e., one bound by love rather than authority—at about 1700. Ariès wrote:

In medieval society the idea of childhood did not exist; this is not to suggest that children were neglected, forsaken, or despised. The idea of childhood is not to be confused with affection for children: it corresponds to an awareness of the particular nature of childhood, that particular nature which distinguishes the child from the adult, even the young adult. In medieval society this awareness was lacking. That is why, as soon as the child could live without the constant solicitude of his mother, his nanny, or his cradle-rocker, he belonged to adult society. . . . Language did not give the word "child" the restricted meaning we give it today: people said "child" much as we say "lad" in everyday speech. The absence of definition extended to every sort of social activity: games, crafts, arms.

Accustomed as we now are to the centrality of the child in the family and in the exhaustive current studies of the motives, thoughts, and behavior of children, the lack of importance attached to childhood in earlier centuries seems at first acquaintance astonishing. Perhaps it becomes less astonishing when one remembers how brief the total span of life—particularly of infant life—could be in all the centuries prior to the present.

SIXTEENTH CENTURY

In *The Booke of Children* (1545), the first book on pediatrics ever to have been written by an Englishman, Thomas Phaire listed 40 of "the manye grevous and perilous diseases" that commonly "vexed and greved" children of his day (Fig. 1–1). Among the diseases in his list were "aposteme of the brayne" (? meningitis), "consumpcion" (tuberculosis), "fallying evill" (epilepsy), "gogle eyes" (squint), "sacer ignis" (erysipelas), "swelling of the head" (hydrocephalus), and "the stone." For none of these were there satisfactory therapeutic methods available to Phaire or to any of his contemporaries. For example, to prevent subsequent attacks of epilepsy, Phaire recommended the following: "the stone that is founde in the bellye of a yonge swallow being the first brood of the dame . . . hanged about the necke of the child saveth and preserveth it from the sayd sickness [epilepsy]."

It was not until the sixteenth century that pediatric textbooks appeared, pointing out for the first time an awareness of the differences between the child and the adult. However, even as late as the end of the sixteenth century, paintings of the great Renaissance artists failed to show the differences in body proportions between the child and the adult (Cone, 1962). Perhaps even harder for us to understand is why a statistic so mundane as

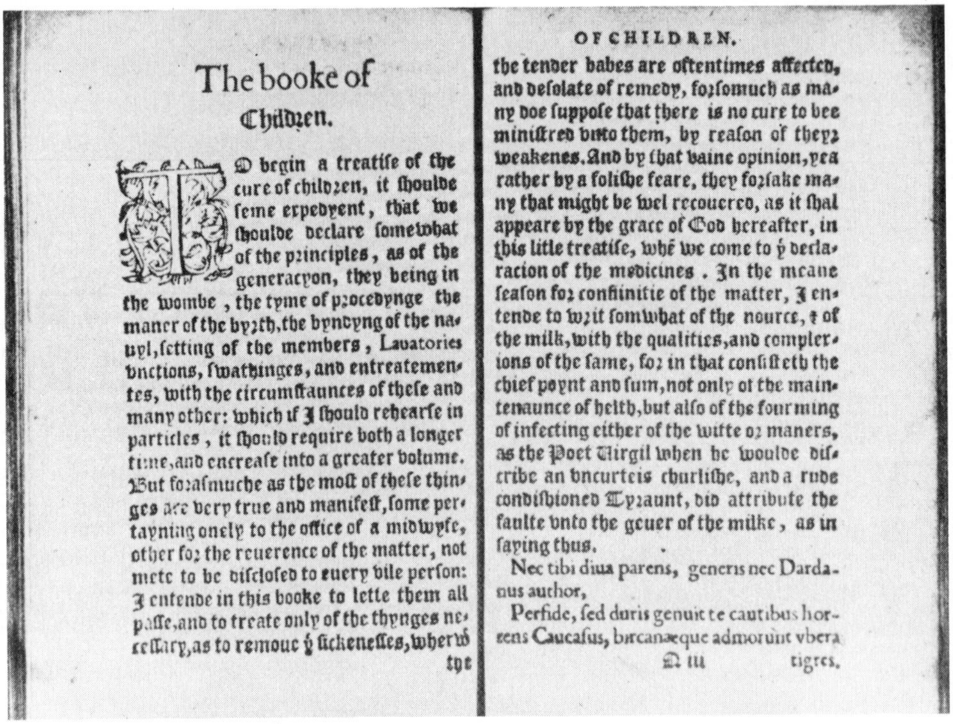

The booke of
Chyldren.

TO begin a treatise of the cure of chyldren, it shoulde seme expedyent, that we shoulde declare somewhat of the principles, as of the generacyon, they being in the wombe, the tyme of procedynge the maner of the byrth, the byndyng of the nauyl, setting of the members, Lavatories vnctions, swathinges, and entreatementes, with the circumstaunces of these and many other: which if I should reherse in particles, it should require both a longer tyme, and encrease into a greater volume. But forasmuche as the most of these thinges are very true and manifest, some pertaynyng onely to the office of a midwyfe, other for the reuerence of the matter, not mete to be disclosed to euery vile person: I entende in this booke to lette them all passe, and to treate only of the thynges necessary, as to remoue ỹ sickenesses, wherw the

OF CHILDREN.

the tender babes are oftentimes affected, and desolate of remedy, forsomuch as many doe suppose that there is no cure to bee ministred vnto them, by reason of theyr weakenes. And by that vaine opinion, yea rather by a folishe feare, they forsake many that might be wel recouered, as it shal appeare by the grace of God hereafter, in this litle treatise, whẽ we come to ỹ declaracion of the medicines. In the meane season for continuitie of the matter, I entende to wryt somwhat of the nource, & of the milk, with the qualities, and complexions of the same, for in that consisteth the chief poynt and sum, not only of the maintenaunce of helth, but also of the fourming of infecting either of the witte or maners, as the Poet Virgil when he woulde discribe an vncurteis churlishe, and a rude condishioned Tyraunt, did attribute the faulte vnto the geuer of the milke, as in saying thus.

Nec tibi diua parens, generis nec Dardanus author,
Perfide, sed duris genuit te cautibus horrens Caucasus, hircanæque admorunt vbera tigres.

Q iii

Figure 1–1. The opening pages of Thomas Phaire's *The Booke of Children* (1545).

the infant's birth weight was not mentioned by anyone prior to 1694, when Mauriceau, the greatest French obstetrician of his day, claimed that the infant weighed 14 to 15 pounds at birth—an erroneous figure even so (Cone, 1961). Obviously, the infant and child were not objects of intense interest or study before the eighteenth century.

Typical of the attitude shown toward children in pre-Restoration England was the tone taken by Hugh Latimer, Bishop of Worcester, in the course of a sermon he preached in Lincolnshire in 1552:

> I exhort you, in God's behalf, to consider the matter, ye parents: suffer not your children to let, or tell false tales. When you hear one of your children to make a lie, take him up, and give him three or four good stripes and tell him that it is naught; and when he maketh another lie, give him six or eight stripes and I am sure when you serve him so, he will leave it (Pinchbeck and Hewitt, 1969).

Adolescent girls were treated with equal severity. For example, Lady Jane Grey (1570) told her tutor* when she was 15 years old:

> When I am in the presence of either father or mother, whether I speake, kepe silence, sit, stand or go, eate, drinke, be merrie or sad, be sewying, playing, dancing, or doing anie thing els, I must do it, as it were, in such weight, mesure and number, even so perfitelie as God made the world, or els I am so sharply taunted, so cruelie threatened yea presentlie some tymes with pinches, and nippes and bobbes, and other waies . . . I will not name for the honour I bear them . . . that I think myself in hell (Pinchbeck and Hewitt, 1969).

SEVENTEENTH CENTURY

Toward the end of the seventeenth century, a change—albeit a frail one—occurred in the way society looked at children and childhood, and there arose the first flicker of curiosity about the development and behavior of children. However, as the century began, practitioners of medicine were still mumbling the shibboleths of Hippocrates and Galen, and pathology was still tethered to the humoral theory. But before the end of the century, medicine would be revolutionized by Harvey's discovery of the circulation of blood and by Sydenham's first-hand accounts of malarial fevers, measles, chorea, and scarlet fever. Yet, as a general rule, the seventeenth century still remained the great age of the whip. Violence done to a child by parents and teachers was socially acceptable; the rod kept its place of primacy as a weapon of punishment.

Inevitably the continued and terrifyingly high rates of mortality among seventeenth century children militated against the individual child's being the focus and principal object of parental interest and affection—and certainly not that of the physician. Indeed, the appalling infant mortality rates

*The tutor was Roger Ascham, author of *The Scholemaster* (1570) and also tutor to the future Queen Elizabeth I.

sustained the centuries-old belief that many children were conceived so that a few might be preserved—a view that is shocking in light of our present-day sensibility. This outlook bred a curiously detached attitude toward the death of a child. "I have lost two or three children in their infancy," wrote Montaigne, "not without regret, but without great sorrow."

Brevity of life of both parents and children, dramatic as it appears in contrast to the longevity that characterizes the twentieth century, only partially explains why children and childhood were of so little importance almost until modern times. Of profound significance too was the conception of society in which the family, not the individual, was the essential unit of social organization. Promotion of family ambition and advancement of family interest, not the realization of private ambition or the achievement of personal success, were seen as the common, all-important task. Hence, unlike modern society, in which personal happiness and fulfillment are regarded as the *sine qua non* of a successful marriage and family life, the personal affections of husband and wife, parents and children were of minor concern in seventeenth century England (Pinchbeck and Hewitt, 1969) and America (Cone, 1979). In a society in which the romantic concept of marriage played so small a role, it is hardly to be expected that the parent-child relationship would be sentimentalized.

A seventeenth century American child had neither voice nor rights and was under the complete control of the father. According to John Robinson, a Puritan leader, fathers governed "by their greater wisdom and authority." A mother's role was to bear, nurse, and care for infants, but after the first few months, fathers should take over and "by their severity" correct "the fruits of their mother's indulgence" (Cable, 1975).

Seventeenth and eighteenth century American parents, particularly those living in New England, believed that family government would fail and authority collapse unless parental will was obeyed invariably. Many parents from the earliest months of their children's life through the years of childhood acted upon the assumption that parental authority was unlimited and incontrovertible. As Greven (1977) has noted, "Total power of parents, total dependency and obedience of children—this was the persistent polarity. For this reason the parent-child relationship was shaped by a stark and sharply defined gulf between generations—an enormous and unbridgeable distance between parents and children, which implicitly denied to children any rights to their own desires, needs or wishes that might be at odds with those designed for them by their parents." Further, children were expected to be grateful to their parents for having

given them birth and apologetic for the trouble they caused by simply being children.

CHILD CARE AND CHILD REARING IN COLONIAL AMERICA

Although colonial life was woven of many strands, such as English, Scotch-Irish, Dutch, French, and German, all the new groups, whatever their ethnic differences, shared the common belief that the family was, in Benjamin Franklin's phrase, the "sacred cement of all societies." As Cotton Mather put it in 1693, "Families are the Nurseries of all Societies; and the First Combinations of Mankind. Well-Ordered *Families* naturally produced a Good Order in other Societies." Not only was the family thus decreed by custom and by religion as the basis for sound community life, but also in the New World the tradition of family solidarity was made the more urgent by economic need.

The nature of childhood in the colonies is largely hidden in obscurity. Unfortunately letters and diaries contain little mention of the child except for records of their births, deaths, and illnesses. Day-to-day observations of children at play, at school, or in their joys or griefs are absent, and we read with sadness that there were no inventories of toys in Pilgrim families (Cone, 1979).

Childhood as such—as we have seen—was barely recognized during the seventeenth century. There was little sense in colonial America that children might somehow be a special group with their own needs and interests and capacities. Instead they were viewed largely as miniature adults: the boy was a little model of his father, likewise the girl of her mother. In many seventeenth century colonial families children 7 or 8 years of age were already veteran workers. For example, the Reverend Francis Higginson wrote in a letter to friends in England in 1629 that "little children here by setting of corn may earn much more than their own maintenance" (Cone, 1979).

Mortality among colonial infants was appalling. Putrid fevers, epidemic influenzas, bloody fluxes, malignant sore throats, and smallpox carried off thousands of the children who survived baptism. Perhaps for this reason parents gave their children names that had deep significance, were appropriate to conditions, or might have a profound influence—presumably on the child's life. Abigail, meaning father's joy, was frequently given, as was Hannah, meaning grace. Names such as Comfort, Deliverance, Temperance, Peace, Hope, Patience, Charity, Faith, Love, Submit, and Rejoice were not uncommon. Some children of Roger Clap were named Experience, Waitsill, Preserved, Hopstill, Wait, Thanks, Desire, Unite, and Supply (Cone,

1979). One wonders what effect these names had on the children who received them!

EIGHTEENTH CENTURY

Eighteenth century concepts of child rearing were immensely influenced by the writings of Jean Jacques Rousseau and John Locke. Rousseau vividly set forth the belief in the innate morality of children and their goodness in conflict with an immoral society. He believed that children, uninhibited by adult rules and conduct, would develop into blameless adults who would organize socially through the formulation of social contracts. The emphasis, according to Grotberg (1976), was not so much on development—and certainly not sequential development—as on the right of children to develop their innate goodness without the restrictions of the adult world. Personality was seen as the result of the unfolding of predetermined and innate characteristics. With the publication of Rousseau's *Émile*, the child became a unique part of society. Even today, Rousseauian ideas are subscribed to by some educators, child development researchers, and psychologists who study children and design programs for them in which adults play a minimal or merely supportive role. Child centered education, including the educational philosophy of John Dewey, is based on the capacity of children to develop and relate to others through innate abilities developed without adult interference.

John Locke, founder of the English empiricist tradition, viewed the child as passive in the hands of environmental forces; i.e., the child's behavior is molded by experience. On the other hand, the Continental tradition, represented by Rousseau, regarded the child as actively manipulating and engaging his environment, trying to mold it to his skills and attempting to understand it in terms of the existing knowledge (Kagan, 1971). Later, behaviorism and psychoanalysis would regard the child as passive, while Piagetian theory viewed the child as active.

Locke, although himself a bachelor, gave detailed advice about the feeding, bathing, clothing, and toilet training of young children in his *Thoughts on Education* (1692/3). For example, he wrote, "I will also advise his [the child's] feet be washed every day in cold water, and to have his shoes so thin that they might leak and let in water, whenever he comes near it."

To toilet train the child, Locke recommended that every small child, daily after breakfast, "be set upon the stool, as if disburdening were as much in his power as filling his belly; and let not him or his maid know anything to the contrary . . . and if he be forced to endeavour, by being hindered from his play or eating again till he has been effectually at stool, or at least done his utmost, I doubt not but in a little while it will be natural to him."

At a time when purging was almost an act of faith, Locke wrote, "Perhaps it will be expected from me that I should give some Directions of Physick to prevent Diseases: for which I have only this one very sacredly to be observ'd, never to give Children any Physick for Prevention."

His thoughts about the training of the child's character, manners, and intellect display an understanding of the psychology of childhood that leaves one wondering how and when he acquired it, for other than directing the education of two or three individual children, he seems to have had no special opportunities to study them (Still, 1931).

James Nelson, an English apothecary, wrote a widely known and thoughtful book, *An Essay on the Government of Children under Three General Heads, Health, Manners, and Education* (1753). This book, less philosophical than Locke's *Thoughts on Education*, was conceived in the same spirit, but looks at child behavior and development more from the standpoint of the every day practitioner. In his introduction he writes, with wit and truth:

> Were none to engage in a State of Wedlock in order to become parents till their abilities to train up their little offspring were try'd and approved, I am of Opinion the number of Marriage Licences would be greatly abrig'd.

Of his three headings, he stressed manners:

> Manners . . . is the Grand Point I aim at; everything else is secondary to that. . . . By manners I do not mean that external Shew of good Breeding, which consists only in a Bow or a Curtsy. . .tho' this too is of Importance; but I mean such a uniform Deportment, such a ready engaging Behaviour, and such a Propensity to what is right as testify as happy Disposition of the Mind and Heart.

Nelson wrote that the training of character must begin in infancy, the first year of life, and that parents must be consistent in their training. "In the government of Children, he wrote, "Parents should be obstinately good, that is, set out upon right Principles and then pursue them. . . . The first Rule Parents are to lay down to themselves is never to deceive their Children. . . . The next Rule is to avoid the Practice of Bribes. . . . The influence of both Father and Mother should if possible be equal; at least it is necessary that Parents go hand in hand and not counteract one another in the Government of them."

NINETEENTH CENTURY

As the nineteenth century began, the Calvinist belief that a playful, pleasure-loving child was on the road to perdition still remained strong, at least in America (Fig. 1–2). But by 1830 many influential

Figure 1–2. An American Calvinist family at prayer, from *Evenings at Home*, circa 1820, by John Aiken, M.D., and his sister, Anna Letitia Barbauld.

American theologians were questioning the strict Calvinist notion of infant depravity and the accompanying stress on conversion as the sole aim of religious education and child rearing. Although infants were liable to be punished, this did not invariably mean that they would be sent to Hell. Lyman Beecher, for example, wrote that no exact knowledge of who was saved or damned could be gained, since this was known only to God (Wishy, 1968).

At about the same time, two nationally read and influential women writers, Mrs. Lydia Sigourney and Catherine Beecher, began to publish books and pamphlets that were to have a great impact on American mothers. Mrs. Sigourney's classic, *Letters to Mothers* (1838), stated that the pivotal problem of child nurture was "how the harp might be so tuned as not to injure its tender and intricate harmony." She described the "waxen state of children's minds," of the body as "a miniature temple." The mother with her "mission so sacred" should learn to "feel with Rousseau" that "the greatest respect is due to children" for the "ark of the nation as well as the child's soul were in her hands."

By the Civil War most writers involved in the emerging public child-nurture debate offered parents two choices: the child was either God's or Satan's, either a glory to the republic or its shame and corruption. The argument was greatly hampered because there was no sizable body of facts about the physical and emotional development of children to provide support for any particular view (Wishy, 1968).

A famous American nurture book, published in 1871 by Jacob Abbott, entitled *Gentle Measures in the Management and Training of the Young*, stressed two principal themes: authority "without violence or anger" and "right development . . . in harmony with the structure and characteristics of the juvenile mind." Abbott emphasized that moral training must take into account physiological and neurological conditions. This was a novel, previously unappreciated concept. As for the old question of innate depravity versus the effects of environment, "bad tendencies" in the child were blamed not on a supernatural essence but on heredity; "bad habits of action" were not inevitable consequences of a partial spiritual corruption but came from improper training. He also recognized that the growth of the child's moral life is molded more by the person the child wishes to resemble than by what the child is told or thinks he ought to do. The parent was advised to be tolerant, patient, and sympathetic as the child's moral growth took place (Wishy, 1968); (Fig. 1–3).

After the 1870's and Abbott's book, there was a plethora of manuals and journals on child nurture giving evidence of how rapidly the study of the child came under professional scrutiny and control. However, the growth of so-called expertise in the last quarter of the century remained eclectic, uncoordinated, and widely scattered over two continents. Unlike today, there were as yet no specialty journals, conferences, or "institutes" to evaluate and "synthesize" the latest reports.

Among the European and American child development experts whose work became known to the public between 1880 and 1900 were John Dewey, Alfred Binet, William James, E. L. Thorndike, and G. Stanley Hall. These men published books and monographs and helped to establish systematic studies of the child at universities. At Clark University in Worcester, Massachusetts, largely sponsored by Hall, years of child study culminated in the founding of the Children's Institute in 1909.

HABIT FORMATION AND CHILD DEVELOPMENT

Catherine Beecher, an extraordinarily popular mid-nineteenth century writer of books and articles about proper child-rearing practices, wrote

Figure 1–3.   Portrait of the Reverend John Atwood Family, by Henry Darby (1845). (Courtesy of Museum of Fine Arts, Boston.)

extensively on the development of good habits in children—and thus nondeviant ones (see later). Her writings, typical of those of other writers of this period, were directed to mothers because she saw them as having "a superior influence in all questions relating to morals or manners." She wrote:

> Do not allow a child to form such habits, that it will not be quiet, unless tended and amused. A healthy child should be accustomed to lie or sit in its cradle, much of the time; but it should occasionally be taken up, and tossed, or carried about, for exercise and amusement. . . . In regard to forming habits of obedience, there have been two extremes, both of which need to be shunned. One is, a stern and unsympathizing maintenance of parental authority, demanding perfect and constant obedience, without any attempt to convince the child of the propriety and benevolence of the requisitions, and without any sympathy and tenderness for the pain and difficulties which are to be met. . . . In shunning this danger, other parents pass to the opposite extreme. They put themselves too much on the footing of equals with their children, as if little were due to superiority of relation, age and experience. . . . This system produces a most pernicious influence. Children soon perceive this position, thus allowed them, and take every advantage of it. They soon learn to dispute parental requirements. . . . But children can be very early taught, that their happiness, both now and hereafter, depends on the formation of *habits of submission, self-denial* and *benevolence*. . .; whenever their wishes are crossed, or their wills subdued, they can be taught, that all is done, not merely to please the parent, or to secure some good to themselves or to others; but as a part of that merciful training, which is designed to form such a character, and such habits, that they can hereafter find their chief happiness. . .in living to do good to others, instead of living merely to please themselves.

CHARACTER DEVELOPMENT AND CHILD-REARING PRACTICES

The emerging interest shown by American mothers in problems of child rearing toward the end of the nineteenth century is evident from the many articles dealing with physical and character development of infants published in such popular women's magazines as the *Ladies' Home Journal* and *Good Housekeeping*. A review of these articles indicates that the mother of 1890 was primarily concerned with the development of good moral character; the child's character, not personality, was the focal point of her interest. Mothers were told that good character was formed through the influence of the home. The mother was still viewed as the essential person in the formation of the child's character. As one writer put it, "The roots of all pure love, of piety and honor must spring from the home. . . . No honor can be higher than to know that she [the mother] has built such a home. . . . To preside there with such skill that husband and children will rise up and call her blessed is nobler than to rule an empire. . ." (Stendler, 1950).

Twenty years later, mothers were still interested primarily in good character development, but the means to a good character was no longer through a shower of love; discipline became more important. Mothers were admonished to insist upon obedience at all times, and if temper tantrums developed, they should be ignored. Stimulation of any sort, mothers were told, would lead to precocity in the older child and dullness in the man.

TWENTIETH CENTURY

Our attitudes toward children today are related to and determined by attitudes in the past. Why has no one dared go back beyond the twentieth century? Is it that the heartlessness and cruelty toward children which peer through related subjects as found in historical documents have proved unbearable to the modern student of behavior?

J. LOUISE DESPERT (1965)

With the beginning of this century the focus slowly shifted to an interest in how children develop. As the century progressed, this interest would eventually be granted equal importance to the study of the diseases afflicting children.

Arnold Gesell, founder of the Clinic for Child Development at Yale in 1911, was the first to study children over time by using scientific procedures; he approached child development from a biological and behavioral point of view, as did Freud. But Gesell's studies described age-related behavior by years and emphasized as major developmental thrusts the divergent and convergent aspects of development. Of interest is that Gesell was the first to use the words *child development* in describing major events of childhood. His clinic was the first visible evidence of a program to study the development of children rather than their separate health problems, educational needs, or welfare status.

Freud's *Infantile Sexuality* (1910), built on a developmental framework, outlined the biologically determined stages of human development—as he saw them—and then described the characteristics of each stage. They comprised the oral, anal, phallic, latent, and heterosexual stages. With publication of this book, Freud became a major influence on child development studies in his country. Alfred Kazin (1957) has written that "the greatest and most beautiful effect of Freudianism is the increasing awareness of childhood as the most important single influence on human development." Further evidence of concern with the development of children and youth was the establishment of the Judge Baker Foundation, in 1917, in Boston, Massachusetts, to study psychological development and problems of juvenile delinquents; in 1921, a Child Guidance Clinic was also established in Boston and was concerned with guiding the development of children (Grotberg, 1976).

During the first two decades of this century there emerged four distinct trends in studying the behavior of young children: (1) longitudinal studies, (2) statistical studies, (3) study of conditioned reflexes in early learning, and (4) interpretation of behavior in terms of individual motivation. At the end of the first quarter of this century, the sporadic interest in child development finally gave way to a general desire to include "everyday problems of the everyday child" in the practice and teaching of pediatricians (Kanner, 1948).

BEGINNING STUDIES OF CHILDREN'S BEHAVIOR AND DEVELOPMENT

The behavior and psychologic development of the child was first approached on an educational basis, and from this flowed the first attempts to observe and record the sequential development of the child. Johann Heinrich Pestalozzi, the Swiss educator, stressed the importance of childhood as a separate learning phase. In 1774, he recorded some early concepts of child development by publishing the first well documented developmental biography of a child—his own son—which may be considered the first scientific record of this kind (Kessen, 1965). A few years later (1787) Dieterich Tiedemann* traced sensory-motor and language development during the first 2½ years of an infant's life.

However, it took Charles Darwin in his book *On the Origin of Species* (1859) to formulate concepts that have provided the basis for a major scientific perspective on the development of humans and indeed all life. Darwin's greatest contribution to the study of children was in his assignment of scientific value to childhood per se. Kessen (1965) deftly describes the effect of Darwin's writings on child development studies by reminding us that with their publication "there was a riot of parallel-drawing between animal and child, between early human history and child. The developing human being was seen as a natural museum of human physiology and history; by careful observation of the infant and child, one could see the descent of man." With the chapter comparing "the mental powers of man and the lower animals" in *The Descent of Man* (1871), Darwin invented the field of comparative psychology.

Darwin, a biologist, and Preyer, a physiologist, each kept a diary of the behavior of his infant son (Darwin, 1877; Preyer, 1890), and each compared the behavior of the human infant with that of other animal species. Their studies, essentially repeated, cumulative observations on the same subject, were the forerunners of the twentieth century longitudinal approach to studying large groups of children.

---

*Apparently, no complete English translation exists of Tiedemann's *Record of an Infant's Life* (1787).

In 1872 Darwin published *The Expression of the Emotions in Man and Animals*, a book that stressed the concept of evolving growth and behavior. In this book he examined the causes, physiologic and psychologic, of all the fundamental emotions in man and animals. He concluded that "the chief expressive actions exhibited by man and by lower animals are now innate or inherited," and that most of the infant's movements of expression must have been gradually acquired. His theory of the evolutionary thrust of development within life forms and across time, combined with new views of scientific study, contributed to clearer formulations of the nature and scope of human development. His "macroview" of evolution set the conceptual framework for a "microview" of human development (Grotberg, 1976).

Darwin deepened and widened public curiosity about human origins and development; he also provided scientists with a fruitful basis for a systematic "unified understanding of the vast disparate problems of the body and the mind" that were under scientific investigation by the middle of the nineteenth century.

Darwinian theories of evolution were largely biologic and influenced the thinking not only of those trained in biology but also of those interested in personality development and human behavior. Sigmund Freud was one such person. He was the first to introduce the notion that stages of development are sequential and arise in a biologically predetermined order. He further taught that the experiences and conflict the developing human faced with respect to social standards and mores determined at what phase of development a personality might fixate or what phase would predominate and be recognizable in adult personality traits and behaviors. Freud wrote, "As early as 1896, I had already emphasized the significance of childhood for the origin of certain important phenomena connected with the sexual life, and since then I have not ceased to put into the foreground the importance of the infantile factor for sexuality."

EVOLVING CONCEPTS OF VARIATION AND DEVIATION

The modern reader may not realize that "the good child" was a concept of little concern to parents in the past, because there were no overriding rules or special standards for children that were not also applicable to the rest of society. An interest in good children, namely those who were obedient and innocent and who showed no signs of deviancy, blossomed fitfully with the Puritanical reform of the late seventeenth century. Good children were expected to be pious, obedient, diligent, literate, and self-reliant.

The "bad child," as seen by our Puritan ancestors, was in abundance because, as I have mentioned, eighteenth century American parents were told to be suspicious of the appearance of innocence in children, since, without following God, such "innocence" was but a mask of the devil. As Jonathan Edwards warned, "As innocent as children seem to be to us . . . they are young vipers, and are infinitely more hateful than vipers, and are in a most miserable condition. . . . Why should we conceal the truth from them?" (Wishy, 1968).

Bad children could be spotted easily: disobedience, the original sin, was the first and most obvious offense to "the representatives of God," their parents. They could also be identified by other outward appearances or marks of the evil inner self.

Susanna Wesley, mother of 19 children, one of whom became the founder of Methodism, raised her children in the belief that food could be used to both nourish and keep them from falling into "sin." She wrote:

As soon as they were grown pretty strong they were confined to three meals a day. At dinner their little table and chairs were set by ours, where they could be overlooked; and they were suffered to eat and drink (small beer) as much as they would; but not to call for anything. If they wanted aught, they used to whisper to the maid that attended them, who came and spake to me; and as soon as they could handle a knife and fork they were set to our table. They were never suffered to choose their meat, but always made to eat such things as were provided for the family.

Morning they had always spoon-meat*; sometimes at nights. But whatever they had, they were never permitted to eat at those meals, of more than one thing; and of that sparingly enough. Drinking or eating between meals was never allowed, unless in case of sickness; which seldom happened. Nor were they suffered to go into the kitchen to ask anything of the servants, when they were at meat; if it was known they did, they were certainly beat, and the servants severely reprimanded. . . .

They were so constantly used to eat and drink what was given them, that when any of them was ill, there was no difficulty in making them take the most unpleasant medicine: for they durst not refuse it, though some of them would presently throw it up. This I mention to show that a person may be taught to take anything, though it be never so much against his stomach.

In the eighteenth century an *idle child*, even more so than a disobedient one, was considered a *bad child*, because idleness led to "self-pollution" and eventually to a life of poverty, beggarliness, vagrancy, and thievery. Poverty was thought to be the main cause of "badness," or as the social reformer viewed it, the idleness of poverty was the sin that created the bad child. To save these poor and idle children, "humanitarian" efforts were instituted to put them to work by apprenticing them to work on farms or as domestic servants. England solved the problem by sending thousands of idle or abandoned children to the colonies as indentured servants.

---

*Soft or liquid food taken with a spoon.

For most of the nineteenth century, at least in America, the idle child was still branded as "bad." He or she would likely have been shiftless, poor, illiterate, and mischievous. Children read that God and not economics or the greed of the mercantile class was responsible for the sorting out of the haves and have nots. Toward the century's end, the irresponsible poor were now said to be "bad" because they were slothful and therefore ignorant, and above all they did not fear God (Schorsch, 1979).

The virtue of personal industry was taught to thousands of nineteenth century American children in their school readers, one of which was written by William Holmes McGuffey (Fig. 1–4). An excellent example is shown in Figure 1–5 as it appeared in the third of the McGuffey eclectic readers, published in 1857.

Until this century, most physicians would not have felt that a child's deviant behavior was any concern of theirs unless a medical diagnosis could be pinned on the child. The father's belt would suffice in coping with minor infringements. For more serious problems, jails or juvenile reforma-tories were places where appropriate "punish-ment" would be meted out.

Eighteenth and nineteenth century attempts to explain deviant behavior led to the widely ac-cepted assumption that there was a "disease" called "moral imbecility" or "moral insanity." This was thought to be the result of a selective "ethical defect"—a loss, as Kanner (1948) has put it, of the "power of conscience" with normal development of all other faculties. Terms such as "ethical de-generation" and "faulty development of the moral fiber system" were favorite nineteenth century explanations of deviant behavior. Lombroso's (1876) concept of "the born criminal" added im-petus to the existence of moral insanity (Kanner, 1948).

Even after the notion of moral insanity had been discarded, parents, teachers, and educators con-tinued to look for some inherent common factor that "predisposed" children toward deviancy in behavior. Some invoked heredity; others tried to find an explanation by means of anthropometric measurements. When heredity and anthropome-try failed to develop convincing proof, psychom-

Figure 1–4. William Holmes McGuffey (1800–1873), from an 1836 portrait by Horace Harding. (Courtesy of McGuffey Museum, Oxford, Ohio.)

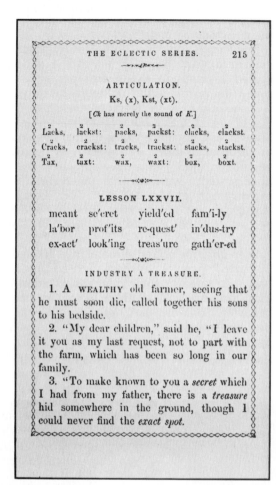

Figure 1–5. "Industry a Treasure," a lesson from McGuffey's *Third Reader* (1857). It serves as one of the most concise illustrations of a virtue regularly emphasized throughout the McGuffey Readers.

etry came forward, followed in about 1920 by a gradual awareness of the value of psychology and psychiatry.

EXAMPLES OF SUPPOSEDLY DEVIANT BEHAVIOR

"He sits on the floor, crosses his thighs tightly, and rocks backward and forward. I tell him he will go crazy if he continues to do so."

"He embarrasses me so because he plays with himself in public."

The attitudes expressed in these two statements referring to *masturbation* are less commonly held by parents than was so a generation ago. But, even today, such expressions are occasionally still heard during interviews with parents. They do not derive from folklore, however, as do so many other incorrect notions. Kanner (1948) reminds us that they were "brought into circulation by medical men of high repute and presented to the public as irrefutable 'scientific' facts."

Simon Tissot* published a book in 1760 on the "diseases produced by masturbation." In it, the worst possible physical and mental ailments were predicted for the unfortunate offenders. Parents were told that if their children should masturbate, locomotor ataxia and early insanity could be anticipated. His dire warnings were accepted by other prominent physicians, among them Claude François Lallemand and Christoph Hufeland.† In 1875, Richard von Krafft-Ebing described cases of "insanity through masturbation." Hypochondriasis, hysteria, epilepsy, depression, and even schizophrenia were said by various nineteenth century authors to be brought on by masturbation (Kanner, 1948).

In the first edition (1897) of L. Emmett Holt's superb textbook, *The Diseases of Infancy and Childhood,* which had an enormous influence on Amer-

---

*Simon A. Tissot (1728–1797) wrote a tract on medicine for the lay public entitled *Avis au Peuple sur la Santé;* it ran through several editions and was translated into all European languages.

†Christoph W. Hufeland (1762–1836) wrote *Makrobiotik* in 1797, and it became one of the most popular books of its time on personal hygiene. Hufeland was court physician at Weimar.

ican pediatrics, the author claimed that children who masturbate "are pale and anaemic; they have dark rings under the eyes; they sleep poorly, are easily fatigued and frequently complain of headaches." He further wrote that "they avoid the society of other children, and lose all animation and all interest in out-of-door amusements."

Like many of his contemporaries, Holt believed that masturbation could lead to insanity and epilepsy. He described a boy of 7 years "who was having from six to ten epileptic seizures a week, in whose case masturbation appeared to be the principal cause."

By 1914 Holt still believed that "masturbation is the most injurious of all the bad habits, and should be broken up just as early as possible." For treatment he suggested putting "parents and nurses on their guard, and the first suspicion should be reported and the child carefully watched until all doubt is removed." Mechanical restraints to immobilize the child's hand were strongly recommended. And the "child should be removed from all vicious companions." Hypnotism was said to have "been employed with excellent results." Finally, "circumcision should be done if phimosis exists, and even where it is not; the moral effect of the operation is sometimes of great value."

Eduard Henoch, a leader in nineteenth century German pediatrics, wrote in 1889 that masturbation could cause enuresis, hysteria, and migraine.

Two other deviant behaviors, *nail-biting* and *thumb-sucking* were, according to Holt, less important than masturbation. Contrary to the need for active treatment in the latter condition, these activities were said to be self-limiting habits.

As late as 1915, in a popular book written for parents, Shannon labeled *self-pollution*, the term often used in the lay literature to refer to onanism or masturbation, as "by far the worst form of venereal indulgence" because "it impairs the intellectual and moral faculties and debases the mind in the greatest degree, and causes the most deep and lasting regret which sometimes rises to the most pungent remorse and despair."

TRENDS IN INFANT CARE

An excellent source of information about the main trends in infant care in America is the series of bulletins put out by the United States Children's Bureau entitled *Infant Care*. Since its first appearance in 1914, this publication has undergone several drastic revisions. By the time the 10th edition was published in 1955, over 48 million copies had been distributed for use by American parents.

Review of the many editions vividly demonstrates the marked temporal changes in emphasis with respect to the severity or mildness with which the child's impulses—as manifested in the areas of thumb-sucking, masturbation, and bowel and bladder training—have been handled (Wolfenstein, 1953). At first, the danger presented by the child's autoerotic impulses was emphatically impressed on the parents. If not promptly and rigorously prevented, thumb-sucking and masturbation would become uncontrollable and could permanently damage the child. While in bed, the child was to be bound down, hand and foot, so that it would not be possible to suck the thumb, touch the genitals, or rub the thighs together.

In the 1914 edition of *Infant Care*, masturbation is described as an "injurious practice" that "must be eradicated" through the use of "mechanical restraints." By the 1921 revision, it is viewed a bit less ominously; "a common habit" that "grows worse if left uncontrolled." The mechanical restraints are less severe; the nightgown sleeves must still be pinned down, but it is no longer specified (as it was in 1914) that the child's legs should be tied to opposite sides of the crib. In 1929, the atmosphere became much more relaxed: this "early period of what may be called sex awareness will pass away unless it is emphasized by unwise treatment on the part of adults." Physical restraints were by this time considered of little value.

Although the alarm about thumb-sucking in the 1914 edition is somewhat less extreme than that about masturbation, it persists longer. In the period from 1914 to 1921, mothers are cautioned that thumb-sucking deforms the mouth and causes constant drooling. "Thumb or finger must be persistently and constantly removed from the mouth and the baby's attention diverted to something else." Thus, diversion, a relatively mild technique not yet considered in the case of masturbation, is said to be partially effective against thumb-sucking. However, it is not enough; sleeves should also be "pinned or sewed down over the offending hand for several days and nights or the hand put in a cotton mitten." In case the mother's efforts to keep the child's hand inaccessible is too zealous, this caution is added: "The baby's hands should be set free now and then, especially if he is old enough to use his hands for his toys, and at meal time to save as much unnecessary strain on his nerves as possible, but with the approach of sleeping time the hands must be covered." Use of a pacifier is regarded as a "disgusting habit," the introduction of which is blamed on adults. The pacifier "must be destroyed." "Thumb and finger sucking babies will rebel fiercely at being deprived of this comfort when they are going to sleep, but this must be done if the habit is to be broken up." Thus, in this period of open struggle against the baby's oral pleasures the ferocity of this drive is nonetheless fully acknowledged (Wolfenstein, 1953).

Bowel training was to be begun "by the third

month or even earlier," according to the 1914 edition. The mother was advised to use "the utmost gentleness. . . . Scolding and punishment will serve only to frighten the child and to destroy the natural impulses, while laughter will tend to relax the muscles and to promote an easy movement." The chamber was to be presented "persistently each day" at the same hour. The mother was told that establishing bowel regularity would be a great saving of her time and would be "of untold value to the child, not only in babyhood, but throughout the whole of life." Unlike the relaxation of guidelines seen with thumb-sucking and masturbation, the trend in bowel training was characterized by an increasingly severe approach, as the 1921 revision demanded an earlier attempt at training: "As early as the end of the first month and as soon as the mother takes charge of the baby after her confinement she should begin upon this task." The time for the completion of bowel training was now stipulated: "Almost any baby can be trained so that there are no more soiled diapers after the end of the first year." The time that the baby is to be placed on the chamber each day is specified more rigidly, "not varying the time by five minutes." Gentleness and laughter are no longer mentioned, and the warning against scolding and punishment is deleted. Whereas in 1914 this training required "much time and patience" from the mother, it now takes "unlimited patience." The value of establishing and maintaining bowel regularity is that it prevents "endless misery from constipation in the adult."

Bladder training was not correlated exactly with that of bowel training. In 1929, when severity in bowel training was at its peak, severity in bladder training was decreasing. Intolerance toward wetting was most intense in 1921, but from 1929 on, the attitude became steadily gentler (Wolfenstein, 1953).

EPILOGUE

Ariès reminds us that nothing escapes history and culture—not even the essential elements of life itself—a man, a woman, a child. This historical review, in providing a sense of how others in the past struggled, often faltered, but more often succeeded in rearing their children, demonstrates the resilience with which most children have managed to grow up successfully in spite of the wide vari-ations in child-rearing practices carried out in different countries and from time to time within the same country.

THOMAS E. CONE, JR.

REFERENCES

Abbott, J.: Gentle Measures in the Management and Training of the Young. New York, Harper & Brothers, 1871.

Ariès, P.: Centuries of Childhood. New York, Alfred A. Knopf, 1962.

Beecher, C. E.: A Treatise on Domestic Economy. New York, Harper & Brothers, 1848.

Cable, M.: The Little Darlings. New York, Charles Scribner's Sons, 1975.

Cone, T. E., Jr.: De pondere infantum recens natorum. The history of weighing the newborn infant. Pediatrics 28:490, 1961.

Cone, T. E., Jr.: The emerging awareness of the artist in the true proportion of the human infant. Clin. Pediat. 1:176, 1962.

Cone, T. E., Jr.: History of American Pediatrics. Boston, Little, Brown and Co., 1979.

Darwin, C.: On the Origin of Species by Means of Natural Selection. London, J. Murray, 1859.

Darwin, C.: Descent of Man, and Selection in Relation to Sex (2 vols.). London, J. Murray, 1871.

Darwin, C.: The Expression of the Emotions in Man and Animals. London, J. Murray, 1872.

Darwin, C.: A biographical sketch of an infant. Mind 2:285, 1877.

Despert, J. L.: The Emotionally Disturbed Child—Then and Now. New York, Vantage Press, 1965.

Greven, P.: The Protestant Temperament. New York, Alfred A. Knopf, 1977.

Grotberg, E. H.: Child development. In Grotberg, E. H. (ed.): 200 Years of Children. Washington, D.C., U. S. Department of Health, Education, and Welfare, 1976.

Henoch, E.: Lectures on Children's Diseases. London, The New Sydenham Society, 1889.

Holt, L. E.: Diseases of Infancy and Childhood. New York, Appleton, 1897.

Kagan, J.: Personality development. In Talbot, N. B., Kagan, J., and Eisenberg, L. (eds.): Behavioral Science in Pediatric Practice. Philadelphia, W. B. Saunders Co., 1971.

Kanner, L.: Child Psychiatry, ed. 2. Springfield, Illinois, Charles C Thomas, 1948.

Kazin, A.: The Freudian revolution analyzed. In Nelson, B. (ed.): Freud and the Twentieth Century. Cleveland, World Publishing, 1957.

Kessen, W. (ed.): Darwin and the beginnings of child psychology. In Kessen, W. (ed.): The Child. New York, John Wiley, 1965.

Parker, P. L. (ed.): The Heart of John Wesley's Journal. New York, Fleming H. Revell Co., ca. 1902.

Pinchbeck, I., and Hewitt, M.: Children in English Society (vol. 1). London, Routledge & Kegan Paul, 1969.

Preyer, W.: The Mind of the Child. Part I: The Senses and the Will. New York, Appleton, 1890.

Schorsch, A.: Images of Childhood. New York, Mayflower Books, 1979.

Shannon, T. W.: Eugenics or the Laws of Sex Life and Heredity. Marietta, Ohio, The S. A. Mullikin Co., 1917.

Sigourney, L. H.: Letters to Mothers. Hartford, Hudson and Skinner, 1838.

Stendler, C. B.: Sixty years of child training practices. J. Pediat. 36:122, 1950.

Still, G. F.: The History of Paediatrics. London, Oxford University Press, 1931.

Wishy, B.: The Child and The Republic. Philadelphia, University of Pennsylvania Press, 1968.

Wolfenstein, M.: Trends in infant care. Am. J. Orthopsychiat. 23:120, 1953.

2

Ripeness Is All: The Coming of Age of Behavioral Pediatrics

The adoption of new ideas and programs has always been associated with long incubation periods. In general, it takes 20 to 30 years for new social developments to take hold from the time they were first proposed. Incorporating ideas about what is now often generically referred to as developmental or behavioral pediatrics into the education of the pediatrician has been no exception.

Although pediatrics as a discipline has always emphasized the importance of treating the whole child, for most of this century efforts to examine the social and psychologic factors affecting growth and development have been viewed as being peripheral to the central core of teaching pediatrics. Within the past two decades, however, there has been a growing movement to ensure that teaching and research in areas relating to child development and behavior are integrated into the mainstream of academic pediatrics. Recognition of the importance of incorporating issues relating to developmental pediatrics may be attributed to several factors:

1. The greatly improved health of children. In 1900, 870 of every 100,000 children between the ages of one and 14 died annually. By 1977, that figure was down to 43 (Surgeon General's Report, 1979). This significant change in the health status of children has allowed pediatricians to redirect their efforts from a predominant concern with life-threatening diseases and to incorporate the growing knowledge of psychologic, social, cultural, and environmental issues that contribute to children's health and well-being into practice. This shift has been from the predominant study of disease to a more dynamic study of processes: growth, development, and pathogenesis.

2. The general movement within medicine away from considerations of biologic processes and disease exclusively to a broader consideration of the many factors that can influence behavior and in turn have biologic effects. This trend has been conceptualized as a biopsychosocial approach to medicine—and particularly medical education—by Engel (1977). This movement within medicine

is, of course, closely tied to the improving health status of people throughout the developed countries. Within this context, the frequently quoted World Health Organization's definition of health as being a "state of complete physical, mental and social well-being and not merely the absence of disease" reflects the increasing emphasis by health professionals of all kinds on the prevention of illness, including as a major component an understanding of the behavioral processes that contribute to healthy growth and development.

3. The existence of a significant body of scientific literature in the area of child growth and development that is so clearly applicable to a pediatrician's work. Unfortunately, until recent years, this literature was little known to pediatricians, and they made few contributions to it. Within the past decade, this has been changing rapidly as a corps of academic pediatricians has become active in research on psychologic and social issues. This trend has been catalyzed by the support provided by several foundations and public agencies for such work.

Today it is probably fair to say that a mastery of developmental pediatrics in pediatric training is an idea whose time has arrived. However, its arrival has been not quite as soon as some would have liked and not without considerable resistance. To understand why it has taken so long for pediatrics to incorporate developmental pediatrics as a part of the major body of its research and teaching efforts, we must examine the way pediatrics developed as a discipline.

OVERVIEW—THE DEVELOPMENT OF PEDIATRICS

The State of the Art in 1900

The historical development of pediatrics from 1900 to the present can be conceptualized in terms of a time line, with basically four different eras. As shown in Figure 2–1, the period prior to 1900 may be referred to as the Prescientific Era, a period rich in contributions to our understanding of child

15

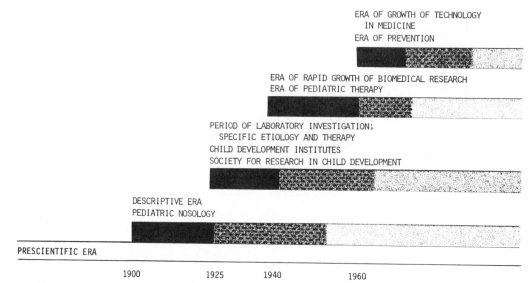

ERA OF GROWTH OF TECHNOLOGY
IN MEDICINE
ERA OF PREVENTION

ERA OF RAPID GROWTH OF BIOMEDICAL RESEARCH
ERA OF PEDIATRIC THERAPY

PERIOD OF LABORATORY INVESTIGATION;
SPECIFIC ETIOLOGY AND THERAPY
CHILD DEVELOPMENT INSTITUTES
SOCIETY FOR RESEARCH IN CHILD DEVELOPMENT

DESCRIPTIVE ERA
PEDIATRIC NOSOLOGY

PRESCIENTIFIC ERA

1900   1925   1940   1960

Figure 2–1. Periods in the historical development of pediatrics. Each period extends into a lighter area, indicating that there is no end point; the developments of each period continue into the new eras. (Adapted from Richmond, J. B.: Child development: a basic science for pediatrics. Pediatrics, *39*:650, 1967.)

development. Pediatrics as a discipline originated at the turn of the century and grew out of a concern for infant welfare—not exclusively health—as evidenced in the newly developing infant health and welfare stations. Attention was focused on feeding and other infant and child care practices as well as on the living conditions of the family and their effects on the development of the child. Unfortunately this concern with the child's total welfare was soon relegated to a somewhat secondary role as pediatrics began to develop as a separate specialty within medicine, and it did not resurface until more than half a century later, in the mid-1960's, when issues relating to a child's growth and development were seen to be more clearly linked to many of the circumstances in his family.

1900 to 1925: The Descriptive Era and Pediatric Nosology

With the growth of the natural sciences, interests and priorities in medicine became oriented toward the new knowledge being acquired. Rapid advances in microbiology, immunology, biochemistry, physiology, pathology, and pharmacology, among many other sciences, led to a reorganization of medical education and established a solid base of knowledge on which new practices could be developed. The dramatic reduction in infant and child morbidity and mortality rates from 1900 to the present testifies to the effectiveness of the public health and clinical practices that emerged from this scientific revolution.

Clinical descriptions and classifications were called for to allow more precise etiologic definitions of disease. The unique aspects of disease in children became apparent, and pediatrics was es-

tablished as a discipline. Textbooks of pediatrics from this era contain the most lucid descriptions of the natural history of disease in childhood.

Although this period resulted in an elaboration of pediatric nosology, it would be incorrect to suggest that the discipline was confined to descriptions of clinical conditions. In fact, during these years, many public health and therapeutic advances were taking place. The pasteurization of milk, the use of cod liver oil and orange juice to prevent rickets and scurvy on an empirical basis, and the introduction of the technique of parenteral fluid administration had much to do with lowering morbidity and mortality in infants and children.

1925 to 1940: Laboratory Investigation—Child Development Institutes and The Society for Research in Child Development

After about 1925, there was an increase in laboratory investigations into specific etiologies and therapies. During this period, elegant and elaborate studies of infant and child metabolism greatly expanded our understanding of nutrition and specific food factors and provided the basis for modern parenteral replacement therapy. The elucidation of specific infectious agents and concomitant advances in the field of immunology led to the early successes of antibacterial therapy and the development of immunizing procedures that are still evolving. Interactions between endocrine secretions and growth were recognized, hastening the establishment of endocrinology as a subspecialty.

Moreover, there occurred a renewal of interest in child growth and development as a result of financial support for activities in this area provided by the Laura Spelman Rockefeller Fund. In particular, these funds made possible two significant developments:

1. Establishment of a number of child research institutes. While most of these developed in universities and were largely outside the mainstream of medical centers, a number of child development research programs were carried out under medical auspices, including those of Dr. Alfred Washburn at the Child Research Council in Denver, Dr. Harold Stuart at the Harvard School of Public Health, Dr. Lester Sontag at the Fels Research Institute at Yellow Springs, and Dr. Milton Senn, first at the Cornell Medical Center and then at the Yale Child Study Center. It is also worth noting that Dr. Arnold Gesell, also a pediatrician, was conducting observations on the development of children at the Yale Child Study Center during this period.

2. Organization of the Society for Research in Child Development. This group provided the only scientific forum for an exchange of information among workers in the interdisciplinary field of child development and allowed pediatricians to become more involved in this general area. Over the years this organization has stimulated teaching and research in growth and development while the more clinically oriented disciplines concentrated on understanding and managing disease. Pediatricians were slow to respond to the advantages of these interdisciplinary activities until the 1960's, when their interest and research in child development began to grow.

1940 to 1960: The Era of Pediatric Therapy and the Rapid Growth of Biomedical Research

This era, described as the golden era of pediatric therapy, overlaps the preceding period. It might also be called the period of the rise of biomedical research, because of the remarkable growth of the National Institutes of Health, which channeled federal funds for research to universities and research institutes across the nation. The revolution in biology was well under way.

Dr. Samuel Z. Levine, in his 1950 presidential address before the American Pediatric Society, summarized this period as follows:

From 1920 to 1950, many of us lived through or participated in what might be called the golden age of curative pediatrics—prophylactic immunization, the widespread use of vitamins, nutritional knowledge, water and electrolyte metabolism, discovery and availability of antibiotics, isolation and synthesis of hormones, tranquilizers and diuretics, plus an ever growing body of scientific knowledge, led to radical changes in both the scope and direction of modern pediatrics (Levine, 1950).

In his address, Dr. Levine also alluded to the restlessness of some of the "elder statesmen" of pediatrics regarding the status quo:

"While a major revolution has taken place in community and individual child health needs during the past couple of decades, pediatric education would appear to be still plugging along in the pattern of the late 1920's or, at best, the early 1930's.

In fact, it was not until almost a decade later, in the 1960's, that pediatrics began to acknowledge that the conventional classifications of morbidity and mortality would no longer suffice and to recognize that the new pediatric nosology should also be concerned with child ecology and child rearing practices and their effects on the development of the child.

1960 to the Present: The Era of Child Development, of Prevention, and of the Growth of Medical Technology

As might be expected, this era in pediatrics corresponded with the renewed interest in the issue of preventive medicine and an increasing willingness to adopt information from other disciplines when applicable. Not surprisingly, several pioneers foresaw these changes. In 1951 Dr. Senn wrote:

In formulating a treatment program or one of prevention, the physician and nurse need to know the potentialities for growth and change which are inherent within the human being, and which need to be mobilized for overcoming ill health of any kind in maintaining good health. Under the stimulation of teachers who have an understanding of the dynamic concepts of personality development and of behavior, students in a meaningful way may acquire knowledge of biological and psychological patterning, and of the developmental characteristics of the human organism. It is clear, however, that one cannot consider behavior, growth trends, or developmental traits in the abstract or in isolation as separate items from others, either within the person or outside himself. To give the student a simple understanding of things so complex is difficult, but it may be attempted through focusing on the family as a psychological and social unit, and particularly on the infant and child in relationship first to his mother, and then in relation to other persons in and outside the family as he grows, changes and differentiates himself constantly.

In the same year the American Board of Pediatrics, recognizing the need to expand the curriculum in the training of pediatricians to include more information about growth and development of the well child, published the following statement:

The Board does not believe that pediatricians should be less adequately trained in the care of the sick infant and child. It does believe, however, that a study of growth and developmental processes can be advantageously incorporated into such training. One of the outstanding defects of current candidates is the fact that on the basis of examination requirements they have didactically mastered a certain number of facts relating to growth and development without understanding the practical implications of these facts.

It becomes obvious, therefore, that pediatric training centers must increasingly assume the responsibility for the day-to-day teaching of growth and development in a clinical setting. This means a practical evaluation of the total development of each patient and its meaning to parents and physicians. It also means that some opportunity to observe presumably well infants and children becomes part of the training program where it is not at present. Intensive study of relatively few patients longitudinally provides an opportunity to teach not only growth and development but the importance of continuity of pediatric care, which is an important element of pediatric practice.

It is to be emphasized that the Board recognizes that only pediatricians themselves adequately trained in growth and development can teach effectively. While other personnel (nutritionists, anthropologists, psychologists, psychiatrists) may participate advantageously at various times, only the pediatrician is in a position to teach by example. At the resident level this is probably the most effective teaching technique.

This statement is presented, not with the idea of making something different of pediatricians or pediatric education. Rather it is intended to strengthen both, and to render both more adequate for their responsibility of caring for the total health of children (American Board of Pediatrics, 1951).

Dr. Gesell (1951) approved of this developing emphasis in a letter to the journal *Pediatrics*:

The excellent statement of the American Board of Pediatrics in the March issue of *Pediatrics* has stirred me to communicate with your editorial department. . . . The statement, however, deserves more than acquiescence and should initiate new undertakings to meet the training requirements so clearly indicated. The Board does not assume the task of outlining specific training centers, devoted to special aspects of infant and child development. If the approach is soundly developmental, the danger of overspecialization is small; because the development is an inclusive, integrative concept which embraces the "total health." The concept applies with equal force to the well child and to the handicapped. It applies equally and conjointly to physical and to mental health. In fact, health can now be defined as that condition which permits and promotes optimal development. These considerations must govern any adequate training program. The clinical implications are far-reaching. . . .

The demands for a developmental type of pediatrics are rising in volume and penetration. The American Board of Pediatrics has issued a timely challenge.

RECENT DEVELOPMENTS IN PEDIATRICS

The preceding statements were all made 30 years ago; now it is appropriate to review recent trends that reveal considerable progress in the field of pediatrics.

Community Pressures for Child Health Services

In the past two decades, organized consumers and public officials have become more aware of the desirability of high quality child health and child care services. Along with the professionals' increasing knowledge of growth and development have come greater literacy and sophistication among citizens in general throughout the country.

These developments were reflected in programs started in the 1960's, such as Head Start, Maternal and Infant Care, Children and Youth Programs,

and EPSDT (Early Periodic Screening, Diagnosis, and Treatment) for children eligible for Medicaid. In the 1970's there was the WIC (Special Supplemental Food Program for Women, Infants and Children), the Education for All Handicapped Children Act (PL 94–142), and special legislation that provided amendments to the Community Mental Health Services Act to create a new program for children's mental health services (Part F of CMHS Act). In addition to these measures, which focused exclusively on women and children, other health programs were begun such as the community health centers, migrant health programs, and the National Health Service Corps, which also served the health needs of mothers and children.

Rechanneling of Pediatric Time and Effort

Much has been written in recent years concerning shortages of pediatric health care professionals. By now, the health manpower shortages of the 1940's and 1950's appear to have been largely overcome, although the problem of distribution of care has not. Haggerty (1972) has pointed out that there is relatively little reason to believe that there is a significant shortage in terms of pediatrician-child population ratios.

Although a report submitted in 1980 by the Graduate Medical Education National Advisory Committee (GEMENAC) to the Secretary of Health, Education, and Welfare (now the Department of Health and Human Services) projected a surplus of 7500 physicians in general pediatrics by 1990, it also cautioned that this estimate might be due to a statistical error and consequently recommended that medical school graduates be encouraged "to enter training and practice in general pediatrics" (GEMENAC, 1980). Nevertheless, the American Academy of Pediatrics deemed this report an "inadequate basis" for developing policies and strategies for dealing with child health care needs (Newsletter, 1981). The GEMENAC report noted the "positive" role of nonphysician health care providers (i.e., nurse practitioners, physician assistants, and nurse midwives), and observed that the potential increase in the number of these new health professionals should augment the time that a pediatrician can devote to the issues of growth and development.

Along with the pediatrician's responsibilities in monitoring the child's physical growth, there should be an attempt to ensure psychologic and social growth as well. This demands a transformation of training programs from those limited to care of the sick child to child health programs in the broadest sense. Programs to prepare faculty members to teach child development and comprehensive care in pediatric educational centers have emerged as manifestations of this shift.

Increasing Knowledge of Child Development and Advances in Research Methodology

After World War II, Freud's theories and observations concerning emotional development became accepted and were further explored. Erikson's extension of Freud's work and his formulation of the "eight stages of man" are particularly helpful to clinicians. The work of Piaget concerning cognitive development has been widely recognized, albeit somewhat belatedly, and has stimulated many investigators to become interested in the early years in a child's development. Skinner's work interested psychologists in behavior modification, which in turn has influenced pediatricians. Pediatricians have also been stimulated by the leadership of Arnold Gesell and Milton Senn and more recently Albert Solnit at the Yale Child Study Center, Charles Janeway at Harvard, Morris Green at Indiana University, Henry Kempe at the University of Colorado, and many others as well as by workers in the related field of child psychiatry. In child psychiatry there were many efforts to stimulate the interest and competence of pediatricians, and it is not possible to list them all. The pioneering efforts of Dr. Leo Kanner in developing a child psychiatry unit at the Johns Hopkins Hospital under the auspices of the Department of Pediatrics and the work of his successor, Dr. Leon Eisenberg, were especially noteworthy.

Increasing interest and methodologic competence in research in the social sciences have spawned a large group of investigators (far too numerous to mention) in all aspects of child development in recent years. Thus, child development has increasingly become a basic science for pediatricians and all other child caring professions, as evidenced by increased membership in the Society for Research in Child Development, new journals on child development, establishment of the Section on Child Development of the American Academy of Pediatrics, and to a considerable extent the phenomenal growth of the Ambulatory Pediatrics Association.

PROBLEMS IN TEACHING BEHAVIORAL PEDIATRICS

Although each of the preceding developments has drawn the field of child development into the mainstream of pediatrics, several factors have complicated the teaching of behavioral pediatrics:

Defining Pediatrics—What Are Its Boundaries? With the broadening range of issues involving child health, it may be difficult at times to define what is truly relevant for the pediatrician and what is not. Although most would agree that psychosocial considerations are important in pediatrics, the use that the pediatrician may make of this information differs from that of child psychiatrists or other behavioral scientists—but how? In short, what is the scope of the pediatrician's concern in relation to that of other health professionals?

Institutionalization of Pediatric Education. In the past, early educational concepts and programs in pediatrics tended to be rigidly institutionalized, inhibiting their development. It has been said that no institution is as resistant to change as one that has once attained excellence. But excellent institutions can deteriorate. There is always the danger of concentrating entirely on disease oriented research rather than allocating a fair share of attention and resources to developmental biology, which should become pervasive in pediatric education.

John Gardner has eloquently stated the need for renewal and change:

> In my earlier book, *Excellence*, I stress the importance of high standards, but high standards are not enough. There are kinds of excellence—very important kinds—that are not necessarily associated with the capacity for self-renewal. A society that has reached heights of excellence may already be caught up on the rigidities that will bring it down. An institution may hold itself to the highest standards and yet already be entombed in the complacency that will eventually spell its decline.
>
> We are beginning to understand the processes of growth and decline in societies. We understand better than ever before how and why an aging society loses its adaptiveness and stifles creativity in its members. And we are beginning to comprehend the conditions under which a society may renew itself. Renewal is not just innovation and change. It is also the process of bringing the results of change into line with our purposes.

In pediatrics, failure to recognize the need for change can result in practicing pediatricians being inadequately prepared for and thus frustrated by the demands of their practice. This frustration is compounded by the fact that many trainees are initially attracted to pediatrics based on the model of a hospital setting. Unfortunately, this model bears little relationship to the typical daily demands of a community based pediatric practice.

"Hard" Versus "Soft" Data. One of the more subtle forms of resistance to research and training in child development has been the inference by many pediatric investigators that biologic research provides "hard data" while research in the social sciences provides "soft data." This position is both unfortunate and misguided, since it has nothing in common with good scientific practice. The scientific method demands excellence in experimental design, data collection, and data analysis. The criterion for determining quality is excellence—not hardness or softness.

Dependence on a Psychiatric Model. Another factor that has inhibited the effective teaching of behavioral pediatrics has been the conceptual confusion arising from pediatrics' early dependence on models developed in child psychiatry. That is, as pediatricians came to realize the need for competence in guiding the psychosocial development of children in the period immediately after World War II, they turned predominantly to the field of child psychiatry for help. At that time, however, the field of child psychiatry was still young and was ill equipped to provide pediatricians with the core conceptual and methodologic background and skills they needed. Why was this so?

Child psychiatry grew out of a concern for the disturbed child and adult, although a few child psychiatrists who left the clinical setting contributed some original observations concerning normal development. As a consequence, the field was relatively rich in theory (drawn largely from retrospective reconstructions during treatment of psychiatric patients) but relatively deficient in terms of empirical observations of the development of children. This is not to suggest that the concerns of child psychiatry were or are inappropriate; the study of psychopathology and the management of disturbed children is a proper and socially necessary function. But pediatricians have grown up in a culture based on empirical observation. They are concerned predominantly with the developmental process and prevention as well as with treatment. Their frame of reference includes the dynamic development and individualization of behavior patterns; the observation of child rearing practices and their consequences; the emergence of curiosity, learning patterns, coping behavior, and personality; and the capacity of children and families to overcome adversity. Pediatricians can observe this process in their daily work, whereas child psychiatrists generally must leave their settings to gain opportunities for such studies. Considering these opportunities and challenges, it is indeed surprising that pediatricians have, until recently, so neglected research in child development.

Because they were heavily dependent on child psychiatry for the teaching of psychosocial development, pediatricians also relied on concepts of development that were related primarily to psychiatry and thus tended to neglect the considerable body of literature concerning child development in general.

Closely related to this issue is the neglect of research opportunities available in the pediatric setting. The clinical discipline of child psychiatry is still trying to establish a strong research tradition and has thus provided a weak model for investigative orientation. As a result, foster care, institutional care, the impact of illness, and handicapping conditions as a focus for study in pediatrics were largely overlooked until relatively recently.

THE COMING OF AGE OF DEVELOPMENTAL PEDIATRICS

The 1960's were a time of self-examination and reassessment for many disciplines in our society, including pediatrics. This decade renewed our awareness of the relationship between poverty and illness, demonstrated the need to approach pediatric problems from a multidisciplinary perspective, and emphasized the importance of programs for disabled children and their families. In particular, the President's Panel on Mental Retardation awakened national interest in the causes of retardation; other reports appeared that focused on the special needs of the handicapped child; and a body of literature developed dealing with the emotional care of the hospitalized child.

AMERICAN ACADEMY OF PEDIATRICS

In the mid-1960's the American Academy of Pediatrics (AAP) issued a special report on *Standards of Child Health Care*, which noted that the dramatic and "complex changes during the past three decades have resulted in a need for an up-to-date comprehensive definition of the practice of pediatrics." Acknowledging that the primary concern in pediatric practice was no longer with infections and nutritional disorders, the Academy reported that the emphasis was now on "preventive pediatrics." This meant that a significant amount of time should be spent in providing "anticipatory guidance" and in managing "behavioral and emotional disorders" (AAP, 1967).

Five years after the release of this report, the Academy issued a revision that included a more detailed discussion of sex education in the schools and of drug abuse among adolescents (AAP, 1972). A third edition appeared in 1975, which incorporated a discussion of new legislation affecting children and noted the changes taking place in "concepts as to when, where, how and by whom care should be delivered to children" (AAP, 1975).

Regarding this last point, as government supported programs proliferated in the late 1960's and during the 1970's, the pediatrician's role expanded to include consultation and services to children in different settings and circumstances. As a consequence, the sphere of interests and concerns relevant to the practice of pediatrics became much wider.

TASK FORCE ON PEDIATRIC EDUCATION

In 1976 there was growing concern that the health needs of infants, children, and adolescents were not being met as effectively as they should be. A Task Force on Pediatric Education (made up of several organizations concerned with pediatric education) was therefore formed to identify these needs and draw up educational strategies to help pediatricians meet them.

In preparing their report, the Task Force conducted a survey of pediatricians asking them to evaluate the adequacy of their residency experience. Significantly, 54 per cent felt that their residency experience did not sufficiently prepare them to deal with the psychosocial or behavioral problems of their patients; 64 per cent felt they were not well trained to deal with school health issues; and 73 per cent felt insufficiently trained for work involving community programs such as custodial institutions, nursery schools, juvenile courts, and programs for disabled children.

The report of the Task Force (issued in 1979) therefore strongly supported the need for future pediatric training to "emphasize the process of human growth and development." It indicated that the area of "biosocial and developmental aspects of pediatrics" had been greatly underemphasized in pediatric education in the past and stressed the belief of the Task Force members that "growth and development are the bases of the specialty of pediatrics." The Task Force report defined biosocial* problems as "those health problems which are socially induced or complicated by social and environmental factors."

One of the specific components cited by the Task Force as being essential to undergraduate pediatric education was the student's ability to display an understanding of:

— A systematic approach to the identification of common behavioral disorders and their initial management, providing parental guidance and counseling for children and adolescents when appropriate.
— The importance of health maintenance, life style, preventive care, and anticipatory guidance.
— The recognition, etiology, and management of child abuse and neglect.

The tone of the Task Force Report clearly reflects the profound changes in social problems and demography that took place in the 1960's. Noting the "dramatic increase in our recognition of child health problems associated with poverty, a deteriorating physical environment, changing family structure, and other social factors," the Report advocated the need for pediatricians to be able to manage a wide range of circumstances:

Residents should learn to manage such family crises as death and bereavement, suicide attempts, sexual assault, accidents,

*The term "biosocial" is used by the Task Force in preference to such terms as "psychosocial" or "behavioral."

child abuse, birth of a defective child, separation, divorce, abortion, and a wide range of common behavioral disorders. Furthermore, they should be able to work with the family to resolve problems in parenting, well child care, adoption/foster care, school management, and learning. They should be familiar with the role of the pediatrician in the management of disease states in which psychological elements play an etiologic or contributory role.

These charges are surely more comprehensive than most pediatricians would believe legitimate or even feasible. Nevertheless they point out that the leaders in pediatrics have significantly revised their thinking regarding the education of pediatricians. In truth, we have come full circle, since this report may be seen as an elaboration of the statement made by the American Board of Pediatrics in 1951 (see p. 17).

ADDITIONAL RECOGNITION

This increase in the responsibilities of the pediatrician was further endorsed by two subsequent reports. The 1979 report of the President's Commission on Mental Health recommended that the Department of Health, Education, and Welfare provide funding to primary health care givers and students "for education in mental health principles, psychiatric evaluation, and treatment." A year later the Select Panel for the Promotion of Child Health, a national panel made up of child health professionals, issued a report advocating training programs to provide those involved in child health care with a "firm foundation in human growth and development, and an understanding of the influences of genetic, familial, environmental, and social factors on the health status of children and mothers." The Panel also endorsed the recommendations of the Task Force on Pediatric Education as well as the efforts of the Robert Wood Johnson Foundation and the W.T. Grant Foundation to provide support for academic pediatric departments in order to encourage the teaching of behavioral pediatrics.

CURRENT TRENDS

One of the most significant indicators that behavioral pediatrics has "come of age" is the proliferation of current textbooks in this field, especially between 1978 and 1980. In the previous 12 years, only a few general textbooks were available on this subject (Bakwin and Bakwin, 1972; Talbot et al., 1971). In addition, between 1978 and 1980 two new journals devoted to behavioral pediatrics began publication—*Monographs in Developmental Pediatrics* and *Journal of Developmental and Behavioral Pediatrics*. Proposed future topics for the *Monographs* include early language development and the pediatrician, the multiply handicapped child and the physician, developmental assessment of the preschool child, and infant behavioral scales.

Textbooks for pediatricians dealing with psychosocial and developmental issues can be considered in two categories: (1) those that provide a general overview of behavioral issues (Allmond et al., 1979; Camp, 1980; Cava et al., 1979; Friedman and Hoekelman, 1980; Gellert, 1978; Kenny and Clemmens, 1980; Magrab, 1978; Sahler and McAnerney, 1981; Prugh, 1982; Toback, 1980; Williams et al., 1981) and (2) those that examine specific areas in depth (Khan, 1979; Levine et al., 1980; Palmer, 1978; Thomas and Chess, 1980; Werry, 1978). (A complete reference list is provided at the end of this chapter.) Interestingly, although the titles may appear similar in some cases, there are significant differences in the subjects covered and in the general emphasis each provides. Some are oriented toward the pediatrician's role in relation to the family (Allmond et al.,1979); others focus on the problems of physically ill children, children with chronic illness, or other handicapping conditions (Magrab, 1978; Gellert, 1978); and still others adopt a prevention oriented approach (Cava et al., 1979). One of the more eclectic texts presents a general summary of theories of child growth and development and also deals with issues such as interviewing and psychologic testing and counseling (Kenny and Clemmens, 1980). From the standpoint of the changing nature of child health problems, it is somewhat puzzling to note that only one text deals in any depth with issues such as the response of children and adolescents to illness or injury, adolescent suicide, or the effects of television on the developing child (Camp, 1980).

In 1975 the senior author of this chapter observed, "From a perspective of thirty years of attempting to teach child development in pediatrics, it is unfortunately necessary to record that we still have a long way to go to realize the objective that the American Board of Pediatrics expounded in 1951:[6] . . .that pediatric training centers must increasingly assume the responsibility for the day to day teaching of growth and development in a clinical setting" (Richmond, 1975). Surely texts such as those mentioned as well as the present text represent a significant move forward in helping pediatric training centers to realize this objective. The growing number of faculty members in departments of pediatrics who are well trained in child development is perhaps the best evidence that this long term goal is being achieved.

In conclusion, it is important to note that the efforts to incorporate psychologic principles of growth and development in a pediatrician's education should not be viewed as an attempt to encourage pediatricians to become either psychologists or psychiatrists. Rather, the intent is to broaden the range and depth of diagnosis available to the practicing pediatrician. Such training should help the pediatrician to become more comfortable with well-child care and to recognize those problems that require referral as well as those that can be appropriately cared for in the pediatric setting.

EPILOGUE

Progress is achieved through the efforts of people with visions of the future and the courage to pursue these visions. We would therefore like to pay tribute to two such pioneers: Dr. Milton J. E. Senn, who provided a helpful model for the study of child development for all of us to follow, and Dr. Charles A. Janeway, who provided moral support for those interested in venturing into a new, uncharted area. Both succeeded in spawning a generation of pediatricians who are moving rapidly to realize their vision—to build child development into the mainstream of pediatric research, teaching, and practice.

JULIUS B. RICHMOND

JUEL M. JANIS

REFERENCES

Allmond, B. W., Buckman, W., and Gofman, H. F.: The Family is the Patient: An Approach to Behavioral Pediatrics for the Clinician. St. Louis, The C. V. Mosby Co., 1979.

American Academy of Pediatrics: Standards of Child Health Care. Evanston, Illinois, 1967, 1972, 1975.

American Academy of Pediatrics Newsletter, August 1981.

American Board of Pediatrics: Statement on training requirements in growth and development. Pediatrics, 7:430, 1951.

Block, R., and Rash, F.: Handbook of Behavioral Pediatrics. Chicago, Year Book Medical Publishers, Inc., 1981.

Bakwin, H., and Bakwin, R. M.: Clinical Management of Behavior Disorders in Children. Ed. 4. Philadelphia, W. B. Saunders Company 1972.

Camp, B. (Editor): Advances in Behavioral Pediatrics. Greenwich, Connecticut, Jai Press, Inc., 1980.

Cava, E., et al.: A Pediatrician's Guide to Child Behavior Problems. New York, Masson Publishing USA, Inc., 1979.

Engel, G. L.: The need for a new medical model: a challenge for biomedicine. Science, 196:129–136, 1977.

Friedman, S. B., and Hoekelman, R. A. (Editors): Behavioral Pediatrics. New York, McGraw-Hill Book Co., 1980.

Gellert, E. (Editor): Psychosocial Aspects of Pediatric Care. New York, Grune & Stratton, Inc., 1978.

Gesell, A.: The pediatrician and the public (letter to the editor). Pediatrics, 8:734, 1951.

Graduate Medical Education National Advisory Committee (GEMENAC): Summary Report to the Secretary, Department of Health and Human Services, Vol. I, September 30, 1980.

Haggerty, R. J.: Commentary: do we really need more pediatricians? Pediatrics, 50:681, 1972.

Kenny, T. J., and Clemmens, R. L.: Behavioral Pediatrics in Child Development. Baltimore, The Williams & Wilkins Co., 1980.

Khan, A.: Psychiatric Emergencies in Pediatrics. Chicago, Year Book Medical Publishers, Inc., 1979.

Levine, M. D., Brooks, R., and Shonkoff, J. P.: A Pediatric Approach to Learning Disorders. New York, John Wiley & Sons, Inc., 1980.

Levine, S. Z.: Pediatric education at the crossroads. (Presidential address, American Pediatric Society.) Am. J. Dis. Child., *100*:651, 1950.

Magrab, P. (Editor): Psychosocial Management of Pediatric Problems. Baltimore, University Park Press, 1978, Vols. I and II.

Palmer, S.: Pediatric Nutrition and Developmental Disorders. Springfield, Illinois, Charles C Thomas, 1978.

Prugh, D. G.: The Psychosocial Aspects of Pediatrics. Philadelphia, Lea & Febiger, 1982.

Report of the Select Panel for the Promotion of Child Health: 1980. Better health for our children: A national strategy. Washington, D.C., U.S., Department of Health and Human Services, 1980.

Richmond, J. B.: An idea whose time has arrived. Pediat. Clin. N. Am., *22*:517, 1975.

Sahler, O. J., and McAnerney, E. R.: The child from 3 to 18. St. Louis, The C. V. Mosby Co., 1981.

Senn, M. J. E.: The contribution of psychiatry to child health services. Am. J. Orthopsychiatry, *21*:138, 1958.

Stone, F. H.: Psychiatry and the Paediatrician. London, Butterworth & Co. (Publishers) Ltd., 1976.

Surgeon General's Report on Health Promotion and Disease Prevention: Healthy People. Public Health Service, DHEW (PHS) Publication 79-55071. Washington, D.C., U.S. Department of Health, Education, and Welfare, 1979.

Talbot, N., Kagan, J., and Eisenberg, L.: Behavioral Science in Pediatric Medicine. Philadelphia, W. B. Saunders Company, 1971.

Task Force on Pediatric Education: The Future of Pediatric Education. Evanston, Illinois, 1979.

Thomas, A., and Chess, S.: The Dynamics of Psychological Development. New York, Brunner-Mazel, Inc., 1980.

Toback, C.: Pediatrician's Psychological Handbook. Chicago, Medical Examination Publishing Co., Inc., 1980.

Werry, J. S. (Editor): Pediatric Psychopharmacology. New York, Brunner-Mazel, Inc., 1978.

Williams, B. J., Foreyt, J. P., and Goodrick, G. K. (Editors): Pediatric Behavioral Medicine. New York, Praeger Publishers, 1981.

II
Patterns of Variation Over Time

This part of the book is intended to provide the reader with a conceptual scaffolding against which other portions of this book will be considered. Emphasis is placed on the common or normal progression of development during childhood. Chapter 3 considers various conceptual models of behavior and development and the ways in which professional thought has evolved and continues to do so. The balance of chapters deals with patterns of variation over time, covering a range of issues and influences that pertain to the development of all children. Chapter 4 considers the ways in which parents change as their children develop. Since parenting plays such a critical role in development, the pediatrician's understanding of the developmental stages that parents pass through would seem to be crucial.

Each of the five following chapters covers a specific period in the life of children, beginning with pregnancy and concluding at the end of adolescence. Various normal events are reviewed and their implications for health care and anticipatory guidance explored.

Chapter 10 deals with evolving individuality; in the first section, issues such as the interactional model, childhood plasticity, temperament, and the evolution of coping styles and self-esteem are discussed—issues that, of course, transcend the age-related considerations of the preceding five chapters. A second recurring theme—gender differentiation over time—is considered in the second section of Chapter 10.

11

Patterns of Variation Over Time

3

The Development of Behavior: Problems, Theories, and Findings

The curious mixture of stability and variety in human development—the texture of change and continuity—has fascinated parents and scholars at least since Aristotle. In a sense, this book is a contemporary statement of that fascination. One part of the continuing mystery has been psychological and psychosocial (misleadingly but conventionally called "behavioral") development, i.e., the course of changes over time in perception, learning, thinking, language, and personality. The vast domain of behavioral development has been worked over intensely for the last century, chiefly by biologists, pediatricians, and psychologists. Of necessity, any single-chapter survey of the domain today will be incomplete, selective, and crude; our attempt will meet those requirements. In the pages that follow, we have attempted to explore four aspects of the study of behavioral development. These include (1) a brief historical statement of the several lines of thought and work, descending almost entirely from Darwin, that have organized the developmental study of behavior over the last hundred years or so; (2) an exposition of the core prejudicial or conceptual issues that have defined—and divided—the field; (3) a summary statement of five theoretical positions that have been influential in the study of child development; and (4) the primary methods used and exemplary findings in three of the major empirical fields of developmental study.

A SKETCH HISTORY OF DEVELOPMENTAL STUDY, 1850–1980

Unlike babies, fields of study do not have rules for establishing birth dates; however, if a conventional opening date for the systematic study of human behavioral development is required, the best candidate is early in 1877 when Charles Darwin published his observations on his first child's first months, "A Biographical Sketch of an Infant," in *Mind*. And, whether or not a specific birthday for developmental studies is accepted, there can be no doubt that Darwin's observations about

human change—especially in *The Descent of Man* (1871)—both forecast and began the flood of theoretical and empirical studies about child development that bridge Darwin's time to ours. In fact, a look at Darwin's intellectual lineage will form a sound introduction to contemporary research in psychological and psychosocial development; one of several possible genealogies is shown in Figure 3–1.

The first line of descent in the Darwinian pedigree called on the analogy between the development of children and the development of species (represented most famously in Ernst Haeckel's cry, "Ontogeny recapitulates phylogeny" [Haeckel, 1874]). The *child-animal analogy* has since appeared in many guises; ludicrous to us now is the turn-of-the-century assignment of children to fish stages, ape stages, and the like (Chamberlain, 1901; Romanes, 1889). The founder of American child psychology, G. Stanley Hall (1923), was not as extravagant in his claims for the parallel between species and person, but his confessed "intoxication" with evolutionism marked the early years of child study in the United States. Far more sophisticated applications of the child-animal analogy derived from the work of Jakob von Uexküll (1909), who understood better than any of his contemporaries the social context of animal behavior, and Nikolaas Tinbergen (1951), from whom productive speculations were derived about the nature of the attachment between parent and child. In the last years of the twentieth century, the child-animal analogy appeared with new virulence in the work of E. O. Wilson (1978), who enlarged it to form an account of human social behavior.

Less productive of empirical work but richly influential in theory has been the allied analogy between embryological development and postnatal behavioral development. Laid out explicitly by Arnold Gesell (1928, 1945) in all his well-regarded *opera*, the embryological image of all human change returned to child development with the revival of the work of Jean Piaget. In more focused form, C. H. Waddington's (1957) notion of "can-

27

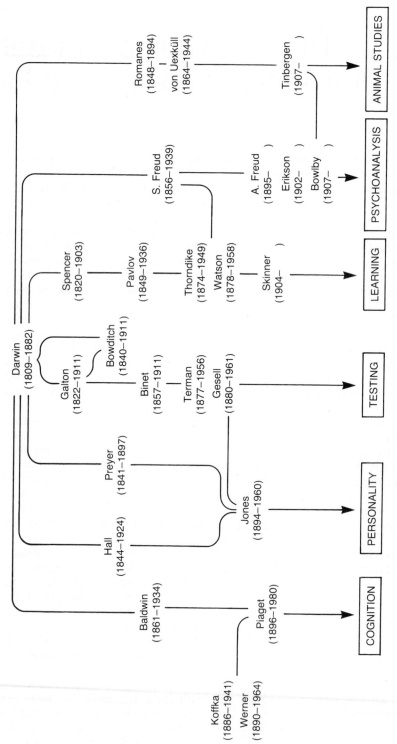

Figure 3–1. "Pedigree" for the study of human behavioral development.

alization" offered developmentalists a possible escape from the vise formed by nature opposing nurture.

Never far from the child-animal association was the line of influence labeled "Learning" in Figure 3–1. Even before Darwin's major work, Hippolyte Taine and Herbert Spencer had suggested the powerful idea that human behavior was selected by the action of the environment. Ivan Pavlov (1927) and John Watson (1928), with one model of learning, and Edward Thorndike (1932), with another, provided the learning-theoretical fuel that was to keep American child psychology running for many years. The simplicity of the learning vision was its major strength—development could be understood as the action of rewards and punishments on behavioral atoms—and it was left to B. F. Skinner (1961) to forge the most relentless and powerful version of the argument for development as learning.

Darwin's concern with variation also led steadily from his cousin Francis Galton's (1889) early studies of the inheritance of talent to the still flourishing field of individual testing. Galton's central purpose was to demonstrate the genetic superiority of the English ruling classes, but the possibility of systematically measuring individual variation had educational implications as well. Alfred Binet (1905–1911) in France and, somewhat later, Gesell and Lewis Terman (1916) in the United States set in motion a line of research on and application of the testing of intelligence, aptitude, achievement, and personality that has probably produced a larger literature than any other segment of developmental study over the last century. It is telling that Galton's invention, "intelligence," should return to the center of theoretical debate in the late twentieth century, exactly in the terms he first used—an argument about the priority of nature over nurture. Two less dramatic consequences of the interest in individual variation have persisted in scholarly attention—the close study of physical growth and the assessment of the developmental status of newborn and young infants.

It is difficult to imagine Francis Galton and Sigmund Freud (1905, 1949) maintaining a conversation much beyond initial stiff greetings, and it is a measure of the enormous power of the Darwinian vision that they are intellectual joint heirs. Freud's biologicism, his commitment to development, and his emphasis on conflict are part of the heritage he shared with and derived from Darwin. We will say more about psychoanalysis later on, but Freud's place in the genealogy—uneasily between the animal folk and the students of cognition—is secure. In the second half of the twentieth century, Freud's flag has been carried by his daughter Anna (1966), the only true descendant in the pedigree, and by the deviationists Erik Erikson (1950) and John Bowlby (1969, 1973).

Developmental study through much of the last century has been dominated by the opinions and observations of Jean Piaget (1947, 1967), who brought together the strands of European interest in the philosophy of human thought and the evolutionary models of phylogenetic and ontogenetic development. Drawing on the work of the brilliant and neglected American, James Mark Baldwin (1906–1911), Piaget built a unique form of the evolutionist argument. Still, his persistent use of physiological analogies, his insistence on the construction of new thought from the interaction of organism and world, and his certainty that a small set of orderly rules could encompass the explanation of human development set him surely in the line of intellectual march we have been sketching.

As might be expected, some parts of the developmental story do not fit easily into the structure of our family tree. The most important aberrant group are those researchers who studied the development of human individuality across more or less wide stretches of age. Allied to the sturdy Darwinian, William Preyer (a physiologist who wrote two volumes [1881] on his observations of babies and commentary thereon), the so-called "baby biographers" were succeeded in the 1930's by industrious investigators at several centers (at Berkeley, Fels, Denver, and Harvard, among others) devoted to collecting as much information as could be found about the changes over time in the lives of samples of American children. Their new empiricism left little room for the play of theory.

Because they fit even less neatly, some significant chapters of the developmental story cannot be properly drawn from our Darwinian display. The Gestalt movement, which contributed three important theorists concerned with children— Kurt Koffka (1924), Kurt Lewin (1935), and Heinz Werner (1948)—drew its strength from Continental sources in philosophy and physiology. Similarly, the interest in human thought and language that engaged American philosophers such as William James and Charles Pierce did not draw heavily on evolutionist dogma. The presence of these divergent lines—the cognitionists—is crucial in any account of developmental study; as this chapter is being written, the wave of interest in the emergence of language, thought, and problem-solving in the child is at its flood-peak.

Weaving through the story of these scholars and their relation to one another is the more important story of the critical issues that have defined and divided the field of developmental study. In keeping with our convictions about the complexity and variety of the human mind, we find that people and issues do not always smoothly dovetail. The closer we get to children, the less secure become our historical and conceptual simplifications.

PERSISTENT CONCEPTUAL ISSUES IN DEVELOPMENTAL STUDY

A neat antithesis—nature versus nurture or whole child versus separated functions—is often misleading, chiefly because the wisest theorists and scholars in a field of study find adroit and innovative ways to evade it. Nonetheless, in our truncated staking out of the territory, a discussion of several dimensions of disagreement among developmentalists will serve to create a conceptual space in which we can locate theorists and research studies. Recognizing with our readers that serious workers rarely hang out at either end of our dichotomies, we will use the following contrasts to present the most persistent conceptual issues in developmental study.

LAWS OF DEVELOPMENT: UNDERSTANDING INDIVIDUAL VARIETY

Folk knowledge as well as somewhat sturdier empirical evidence suggests that people favor either a search for the simple general rules governing a domain of study or an exploration of the domain's marvelous intricacy. For the developmentalist, this contrast can be represented in the difference between an interest in *the child* and an interest in *this child*. In the larger definition, the contrast is between those of us who consider development a characteristic of the species and those who consider it a characteristic of particular individual children. Many disputes in developmental study have persisted because the contestants were arguing at cross purposes; to put the point in statistical terms, the phenomenon of special interest for the "individualist" constitutes experimental error for the "law seeker." Cronbach (1957) has presented an elegant analysis of how, over the years, this difference in attitude has produced striking variations in research method and analysis as well as in theoretical stance.

The group of developmental researchers least ambivalent in their attitude toward the law-and-variety dimension are the experimental psychologists concerned with perception, learning, and cognition. When a study was made of infants' responses to voice-onset-time (i.e., the acoustic parameter that permits adults to tell the difference between the sounds *b* and *p*, for example), the baby subjects certainly did not all behave in the same way. As a matter of fact, some of them would not play in the experiment at all. But the empirical question posed—"Do infants perceive voice-onset-time in the manner of adults?"—required the researchers to treat the resulting creative individuality as noise in the observations. A great number of studies of children, particularly in the years from 1950 to 1970, fell near the law end of the law-and-variety dimension. Of course, the law-seeking developmentalists know as well as anyone else that children are not alike. But their research problems are chosen and their data-analysis procedures are devised under the guidance of an overarching search for general principles.

But the devotion of the law-seeker can be more subtly revealed. When Darwin described the behavior of his son Doddy or when Preyer described, in a much longer account, the behavior of his son Axel, their paternal pride may have been aroused by the uniqueness of their offspring, but their scientific interest was in Doddy and Axel as particular children representing general human characteristics. To the degree that the children were different from other children, to that degree were they false witnesses to developmental laws. Later, Piaget (1936, 1937, 1945), on the basis of his observations of his three children, elaborated a theory of infancy—a statement that can also stand as a general theory of human development. His strategies were the same as Darwin's. They all believed that beneath the surface variations of style and pace and personality, one can see in operation—if one has cut nature at the joints—general, unexceptionable principles of human development.

Galton, both because of his interest in variety and because of his life-long fascination with statistics, carried the law-variety debate a long way forward by describing individual variation as being distributed about some central tendency of the group. The central tendency—the middle place in the array of variation—stood as the description of the species, while the scatter around the center—shorter people, dumber people, stronger people — was understood as departures from the middle. Binet had no fondness for the statistical simplicities of Galton; ironically, when the Binet "tests" had been laundered by Terman and Gesell, most human variation came to be seen as statistical deviation from "the norm."

The Galtonian interpretation of variety remains the most powerful idea used nowadays in attempts to resolve the tension of our first dichotomy; the laws of species development can be described by placing the person in the appropriate position within the general distribution of values obtained from many observations. In their reach to understand variation among infants, Gesell and his successors (Griffiths [1954] and Brazelton [1973], to name only two) follow in the same tradition. A closer look at Gesell's writings suggests that however much he wanted to fulfill his clinical obligation to describe individuality accurately, the heart of his scientific task was the statement of universal laws of human development.

Who, then, speaks for the special character of each of us? We will not consider here the sentimentalists who are satisfied to declare the wonder of human uniqueness and who would attack all systematic attempts to understand it. Three strategies of research and theory remain: First are the *trait theorists*, who postulate a spectrum or—if the metaphor will survive—several intersecting spectra of personal characteristics. We are each of us defined by the assembly of several, or many, basic human qualities. As a waggish statistician has noticed, one needs only eight traits at about 15 different levels of intensity to designate uniquely every human being in the world. The task for the trait theorists, thus far unresolved, has been to detect the correct eight traits. The second group, with even less certain supportive research, are the *personal-history theorists*, who propose that human uniqueness depends on variations in our personal histories and circumstances (Mischel, 1968), those subtle and sometimes not-so-subtle turns in our lives that determine our present position. We will meet this idea later; it represents so inchoate a strategy of explanation of human variation that we can only note its recent emergence. Then, of course, there is the *interactionist strategy*, which would, sensibly enough, account for human variation by adducing a version of trait theory to work in harness with a version of personal-history theory. The proposal is so balanced and so plausible that it can only be faulted by the absence of any significant empirical demonstration of its accuracy.

As we noted at the outset, the dichotomy of development defined by species and development defined by person is rationally based and thus artificial. Its appeal—with the possible exception of dissention concerning internal and external causation (see below)—is that it is the dimension of theoretical stance most useful for determining what students of children believe about the primary tasks of developmental study.

NORMAL: EXCEPTIONAL

Floating through the foregoing discussion has been an unstated distinction between those of us most concerned with empirical research and those most concerned with caring for sick and needful children. Even before the relation between research and practice can be clarified, however, there is the linking issue of whether one's primary interest is in the normal or usual or "garden-variety" child or, by contrast, in the exceptional or unusual or deviant child. In fact, the two issues can scarcely be separated. Everyone is aware of the correlation between research interest and normal behavior on the one hand and clinical interest and the exceptional child on the other. Of course, significant empirical work on exceptional children has been carried out (the rich research literature on mental retardation stands as a measure), but, by and large, the center of scientific developmental study has been on the normal or typical paths of development. Only psychoanalysis, among the major visions of development, includes a systematic theory of deviance.

Not surprisingly, several important consequences have derived from the long standing difference between studying the abnormal and studying the normal and its correlated distinction between empirical research and clinical practice. The most obvious consequence—which we will meet again in different forms—is that different research methods, different avenues of publication, and different networks of colleagues have been established for the two strategies of study. Thus, it is not the case that some people were interested in normal behavior while others were interested in deviant behavior and all shared a common set of theories, methods, and colleagues. Rather, two cultures of study emerged, each with its own sense of scientific value and professional standing. Over the years, it has become increasingly difficult to establish a group of shared problems in which members of the two cultures could bring their best skills to bear jointly on a research issue. The need for productive interaction is now critical to the lives of children; although a glimmering of such an exchange has recently appeared in the disparate fields of cognitive-behavior therapy, behavioral medicine, and the study of at-risk infants, we are still far from the cooperation that is called for.

A less often noted consequence of the separation of cultures between the normalists and the abnormalists has been the common but by no means universal tendency for developmental deviation—whatever its character—to be seen as a function of personality differences. Reasons for this curious outcome are not hard to find. Empirical scientists (particularly those whose work space is the laboratory) who study memory and problem-solving rarely become interested in memory disturbances or eccentric problem-solving. Laboratory students of simple learning rarely become concerned with educational problems in the classroom. Therefore, when a child at home or in school is judged to need attention on the basis of unusual behavior, the professional person most likely to be called in is someone trained in one of the several schools of trait or personality variation rather than an expert on the mental structure underlying the disturbance. Again, however, particularly in the recent application of behavioral and biobehavioral procedures, there are signs of a broader definition of the sources of developmental deviation.

Thus, in our survey of the disciplinary space that contains developmental study we can detect a significant defining characteristic of investigators in the relative emphasis given to the normal or to the deviant.

INNER FORCES: EXTERNAL DETERMINATION

The war (no weaker word will do) between those scholars who see human development and behavior as directed by internal organizing forces (Nature) and those who see development and behavior as determined by personal history and experience (Nurture) has been going on for thousands of years. Despite valiant attempts at compromise—"all behavior is 100% genetic and 100% environmental," "development occurs in the interaction of nature and nurture," "biology sets limits from within which personal experience selects"—there is little sign of a truce. Even the people who are reasonable on the issue (and all of us *believe* that we are) are "reasonable" on one side or the other. The debate that took its first formal statement from the fifth century B. C. reached a new frenzy of polemical rhetoric in twentieth century America. No one has ever matched the soaring promise of the behaviorist John Watson:

Give me a dozen healthy infants, well formed, and my own specified world to bring them up in and I'll guarantee you to take any one at random and train him to become any type of specialist I might select—doctor, lawyer, artist, merchant, chief and, yes, even beggarman and thief, regardless of his talents, penchants, tendencies, abilities, vocations, and race of his ancestors (Watson, 1925).

And, in continuation of the everlasting debate, Watson was answered in his own time and has been answered repeatedly in ours:

It is doubtful whether the basic temperamental qualities of infants can be measurably altered by environmental influence. Training and hygiene may exert [a] very palpable and important influence on the organization of the personality without necessarily altering the underlying nature or habitus (Gesell, 1928).

The positions of genes having indirect effects on the most complex forms of behavior will soon be mapped on the human chromosomes. These genes are unlikely to prescribe particular patterns of behavior. . . .The behavioral genes more probably influence the ranges of the form and intensity of emotional responses, the thresholds of arousals, the readiness to learn certain stimuli as opposed to others. . . (Wilson, 1978).

When a disagreement persists for so long and with such vehemence among intellectuals who are committed to the steady use of mind and method to solve problems rationally, one has to suspect (1) that the problem has been wrongly stated, (2) that there is not yet sufficient research information, or (3) that the conflict is fired by ethical and political beliefs outside the usual canons of science. In this case, all three conditions obtain.

For friends of children, tension between the belief in internal forces and the belief in external determination has been especially painful over the last 40 years of this century. Reforms of the 1960's, best exemplified by the enactment of the Head Start program, were based on a commitment made by the political establishment, and supported by academic developmentalists, that there was great space in the lives of children for the impact of environmental intervention. After a few tentative years, the balance shifted sharply with the renewed debate over the genetic basis of intelligence. Since 1970, both the national policy and the scientific establishment have been divided and ill at ease about the chances that society could be transformed by manipulating changes in the lives of the children of deprived and dispossessed members of our culture.

Two advances in the terms of the antique division between Nature and Nurture are worth noting; both address the issue of our common ignorance. First, theoretical models for elucidating the action of genes have become enriched in their complexity and thereby more ambiguous with regard to their ultimate effect on behavior and research methods; both in biochemical embryology and in behavioral genetics, the biological models are now more ingenious and more revealing of the variety of human development. If the study of behavioral development follows its traditional imitation of biological advances, the complications of genetic theory will be mirrored in new complexities of behavioral explanation.

The second advance has been slower in coming but is no less important for our ideas about development. During the recent period of prodigious change in our understanding of genetic mechanisms, there has been no corresponding advance in our ability to talk about and research the remarkable variety of environments in which children find themselves. In brief, we have not had theories of environment with power comparable to that of our theories of gene action. This imbalance is slowly being redressed, chiefly in the recognition that earlier conceptions of the environment were too narrowly drawn; fewer developmental psychologists nowadays are committed to a view of the environment as an assembly of atomistic stimuli, and fewer still believe that the child's social surroundings are fully captured by describing characteristics of a mother. Recent work in cross-cultural psychology has shown anew that the child's world comprises cultural values, economic constraints, and networks of adult interaction that defy simplistic analysis. If environmental theory over the next years advances at half the pace that gene theory advanced between 1960 and 1980, our conceptions of child development will be radically transformed.

Despite the visible improvement in our comprehension of the relation between genetic influence and environmental influences on human development, the acrimony of the Nature-Nurture debate will not be significantly reduced as long as the conflict is tied to our varying visions of the just society.

DEVELOPMENTAL CHANGE: SUDDEN OR SLOW

It is a long descent from the tortuous and politically vital tensions of the Nature-Nurture debate to the academic feuds over whether development is continuous or proceeds in sudden spurts, usually called "stages." But, for reasons rooted in philosophical traditions since Aristotle, developmentalists have been testy concerning their colleagues' position on whether development leaps or slinks (Brainerd, 1978).

At the grosser level of theory, the issue of *stage* divides the positions of Freud and Piaget, dependent as they both are on radical and relatively sudden transformations in mental structure, from the positions of the learning theorists and most of the cognitive developmentalists, who prefer to see changes in the child as gradual and continuous. However, the disagreement goes deeper than a theorist's epistemological prejudice; it reveals a sharp distinction in the meaning of developmental change.

For stage theorists, Piaget in particular (and it is his theory that has provoked the most intense debates), to say that the child has entered a new stage of cognitive development is to say that a fundamental change has taken place in the way the child understands and uses the environment. In a useful analogy to computer simulation, a new stage requires a new program. For Piaget, once a reorganization has taken place in the child's mental structures, even memory will be transformed; the past, like the present, will be seen according to new rules and a new logic—a new set of principles of knowledge—will be at work in the life of the child. Although psychoanalytic theorists have not fussed about stages as often as the Piagetian folk have, it is plausible to see the transformation from one psychosexual stage to another as representing a saltatory shift in the way impulses and prohibitions are organized.

The other side of the argument derives from the traditions of nineteenth-century biology and physics. Nature does not proceed by leaps. Rather, if we have secure control of the underlying reasons for change—the relevant variables—we could display a family of smoothly changing functions. In the most extreme statement of the position, age itself is only a parameter in forming the most elegant mathematical statement of a continuous function.

Controversy over the rate and nature of developmental change cooled somewhat in the 1970's, but a scholar's opinion on the stage issue remains a revealing way of finding out his position in the larger developmental domain.

WHOLE CHILD: REDUCED CHILD

Advances in knowledge seem inevitably linked to increasingly refined and necessarily narrower analysis. Thus, the marvelous intricacy of a newborn is compressed into a bilirubin level, the attractive troubled second-grader becomes a case of learning disability, and the particularity and social complexity of the adolescent are screened into our heads as a heroin addict. The tendency to see only parts of the child—the "organ systems" of our special interest—is surely not confined to the laboratory developmentalist. Dramatic examples of simplification can be advanced by looking in the laboratory—the assessment of reaction times as a measure of problem-solving or the use of adrenalin levels to measure anxiety—but the clinician seems just about as likely as the bench scientist to reduce the fullness of a child to the comprehensible smaller package of a diagnosis.

We must analyze, then, in order to understand. Is there a defensible scholarly and humane alternative to that ancient strategy, a way of re-assembling the child to make a whole recognizable being?

With cyclical regularity, and in all the developmental sciences, a call has been made for a new look at the *whole child*. Whether in child psychology or in pediatrics, there is a recurrent professional declaration that the child is not just a laboratory report or an aptitude test score, that a new try must be made to broaden our vision of the growing infant and child. The book you are now reading is, in fact, a careful and systematic characterization of the currently perceived need in pediatrics to take into account the social and behavioral aspects of human development that have often been underappreciated by clinicians and scholars alike. Unhappily for our joint attempt here, as for the whole-child argument more generally conceived, developmental workers have been notoriously deficient in their ability to devise theories of reconstruction, ways of sewing back the pieces of the child that were cut apart in order to study their weave and structure more closely.

The human and scientific problem is compounded by the apparent fact that the dichotomy between advocacy of the whole child and advocacy of the reduced child is not a division among schools or philosophies; rather, it is a division that occurs chiefly *within* concerned developmentalists. That is, each of us, as scholar-researcher, is necessarily committed to one form or another of the strategies of methodical analysis; each of us, as a student of the larger issues of development (not to speak of our roles as parents and citizens), wants to get back to the child entire. Unlike the earlier contrasting views on conceptual issues, which tend to divide groups, the search for the whole child divides each developmentalist. The solution to this dilemma, at least for the foreseeable future, is to recognize that the reconstruction of the whole child will not soon become part of the scientific enterprise as it is usually defined; rather, the cyclical call for a broader vision must be seen as a philosophical and ethical caution to the analytic scientist in us: Do the best science you can, but remember that you are working on a tiny fragment of the entire mosaic and, above all, do not claim that your fragment is the whole.

EMOTION: COGNITION

Developmentalists differ on many specific questions about content of the field and about what critical problems deserve close study. In our examination of the general topography of developmental study we cannot pause to describe all the diverging paths; however, one among them has been so enduring and so divisive that it reaches the status of an epistemological separation. The dividing question has been, Shall we conceive of behavioral development as turning centrally around emotion (or motivation) or as based on cognition (or thought)?

In the United States the early years of child study did not elaborate the dichotomy between feeling and knowledge. G. Stanley Hall's vision of the child was so aggregative and far-reaching that he was able (in his questionnaire studies of development) to range from the investigation of children's fears to studies of language to inquiries about the nature of dolls. However, the general psychology of the time was heavily biased toward the study of thought and language and perception and consciousness with almost no place left for the systematic study of the conative or emotional side of the human experience.

The standard emphasis was severely disturbed when Freud visited the United States in 1909, and it is convenient to date from that event the slow but steady growth of interest among developmentalists in the emotional life of the child. The Freudian image, or at least the American version

of this image, had a couple of surprising effects. One, directly influencing Watson, led to the laboratory study of infantile emotion and the postulation of primitive stages of rage, fear, and love. The second and far more extended effect was the introduction of the notion of *drive* or *motivation* into the powerful emerging field of learning studies (Woodworth, 1918). From the early 1920's until Dollard and Miller's statement of the intersection between psychoanalysis and learning theory (1950), the mainstream of American psychology (and its deposits in developmental study) saw the energetic and emotional side of the child's experience chiefly as a matter of primary and learned "drives" (hunger, comfort, desire for learned rewards such as praise, and the like). But the psychoanalytic emphasis on the emotional life of children did not modify the field overnight; rather, the study of emotion seeped into American developmental study and came to the surface most often in studies of personality variation and in the great longitudinal studies of the 1930's and 1940's.

The domination of American developmental studies by learning theory—motivation and all— and by students of personality development received a toppling challenge with the reappearance of the work of Piaget in the United States. Although Piaget had been anticipated by Baldwin and had a mild youthful success among American psychologists between 1925 and 1930, his impact on the larger pattern of developmental study was felt with decisive force in the years between 1950 and 1975. Piaget presented a rich and persuasive vision of the child (as we shall see in the next section) that was organized around the idea that development was primarily cognitive; human change was a matter of the child's epistemological development, the child's increasingly rational theory of the way the world worked. The Piagetian revolution was supported by the concurrent and vigorous increase in American interest in computer simulation of human mental structures and by the renewed empirical study of thought and language. For 15 years, American developmental psychology and allied fields were devoted to an examination and extension of Piaget's image of the child as thoughtful or on the way to being thoughtful.

At the moment of this writing, there is no central cohesive prejudice in developmental studies about the role of mind and emotion. Even the subdomain of language studies, for so long fully in the possession of the cognitivists, shows signs of invasion by those observers of children who see speech and meaning as matters of social or personality development rather than as changes in the rational organization of language. But, as has been the case in each of the dichotomies proposed here, scholars of human growth can be perceived, however ambiguously, as biased toward either cognition or emotion as the main fuel for human development.

SUMMARY

We have proposed six rough dimensions along which to locate the predispositions and attitudes of developmental scholars in the late twentieth century. The dimensions are neither linear nor orthogonal, but they will serve to set some boundaries on the conversations among serious researchers of children in our time. The fully sensible reader would, we suppose, choose to take a middling position on all the seesaws, claiming the sixfold fulcrum. However balanced such a solution seems, the message of the foregoing paragraphs is that no one can maintain that calm neutrality; we will be able to understand the major developmental theories about to be discussed if we understand first that all of us bring to these theories several biasing commitments that are made intellectually unhurtful only by being made visible.

FIVE IMAGES OF DEVELOPMENT

Perhaps the first discriminating characteristic of all major theories of development is that they are social entities as well as conceptual ones. Put in the baldest terms, developmental theories are separated not only by their methods, applications, and core theoretical ideas; they are, as noted earlier, separated as professional associations. Theories are collections of people who agree on the main terms of the child's nature and who gather together—at conventions, in departments, and through journals—to debate their residual minor disagreements. Such chasmic separation of different points of view has two consequences: (1) the members of each group are protected, to some degree, from doubt about their convictions because they are surrounded by other convinced people, and (2) more sadly, exigent problems arising from the lives of children can rarely be addressed from several sides because the group definition—the theoretical purity—must be protected against leakage. No more will be said here about the cult character of developmental theory; the formation of the social clusters of developmental science has been insufficiently studied by historians and demographers of academic disciplines.

Setting the social meaning of theories aside, there are a number of other dimensions along which we can compare different systematic ways of seeing children. For each of the five theoretical attitudes to be discussed in the following pages,

we will select a prototypical theory and then present four summary statements about each position: (1) a word about the history of the theory; (2) the major theoretical ideas tied to the position (specifically, the structural unit of development and the major mechanism of change); (3) the theory's favored methods and favored empirical problem; and (4) some implications of the theory for the people, professionals or parents, who care for children.

DEVELOPMENT AS MATURATION

History

The body grows; behavior grows. The infant is a growing action system. He comes by his mind in the same way that he comes by his body, through the processes of development. As the nervous system undergoes growth differentiations, the forms of behavior also differentiate (Gesell and Amatruda, 1947).

The idea that the child becomes more sensible and more competent in much the same way that he becomes taller and stronger is surely the most ancient folk-theory of development. A million-year-old phylogenetic heritage has laid down regularities of mind just as it has laid down regularities of tibial elongation. Unexamined commitment to the maturationalist attitude was behind Darwin's willingness to present his son Doddy as a model of mankind; internal forces of growth required the unfolding of predictable patterns of behavior.

Arnold Gesell remains the smartest and best informed champion of the conviction that development is maturation. He collected miles of film about babies and wrote numerous books to demonstrate the thesis that human behavior is as lawful in its changes with age as any other aspect of human physiology. Moreover, Gesell formulated a series of principles—the principles of developmental direction, reciprocal interweaving, functional asymmetry, and self-regulating fluctuation—that were meant to carry the maturational theory of grandmother onto a scientific plane. Not insensitive to the variety of human growth, Gesell suggested that many differences among children could be best understood as early, perhaps genetic, biases of temperament, but he knew that children were different from one another, and he spent a good part of his career devising procedures to describe the differences in precise ways.

Structure and Change. For Gesell, the structural unit of development, i.e., what develops, is behavior. Although the maturationalists are persuaded that there are biological structures underlying all visible changes in the actions of children, it is one of the ironies of history that Gesell shared with his intellectual enemy, Watson, the commitment to the notion that what had to be explained in development were changes over age in observable acts. Like less thoughtful maturationalists, Gesell saw age as the primary mechanism of change. Of course, age was a countable marker for more fundamental biological processes, but the central variable in human development is, for maturational theorists, level of maturity.

Method and Problem.   In the hands of the baby-biographers, natural observation was the primary method for finding out about developmental change. Gesell and other later scholars constrained the range of observations and looked at babies and children either in controlled environmental situations or more specifically through the use of standard tests.

Gesell was particularly concerned that his observations be made in as rigorous and reliable a way as possible; he introduced the use of film as a method of data collection, and he organized his observations into what came to be known as developmental scales, a layering of test performance that would indicate each child's relation to the expected performance for children of that age. A number of developmental tests (especially aimed at infants) have been proposed since Gesell's early observations; most widely used in the 1970's has been the Brazelton Infant Assessment Scale, a test that combines many earlier devices.

Implications.   For parents, the maturationalist position is quietly reassuring. The child will develop at his individual pace according to the program that inheres in biological makeup; the task of the parents is to provide the care and sustenance that good gardeners provide for their plants. And, in the popular literature of the maturationalist school, specific measures are given so that parents and professional practitioners may determine whether the baby is on his expected maturational course or how far (in weeks or months or years) the child departs from the predicted species pattern (Gesell and Ilg, 1943, 1946).

DEVELOPMENT AS LEARNING

History.   Opposite to maturation on the theoretical seesaw has been the equally ancient and equally exaggerated notion that human development could be understood by an examination of the individual child's personal history, environment, and social interactions. The idea of development as learning was forcefully stated in the late seventeenth century by John Locke; it was refined in the work of the associationistic British philosophers; and it reached its apogean claims by the midtwentieth century in the proposals of the American psychologists, John Watson and B. F. Skinner and their students. As maturationalist theory depends on evolutionist ideas of regularity

and species differentiation, so learning theory depends on evolutionist ideas about selection. In the course of each individual life, particular ways of behaving and thinking are chosen by the exigencies of environmental variation (in most readings, by rewards and punishments) to shape the character and personality of each child. In the statement by Watson, which appeared earlier in this chapter, we find the most arrogant presentation of the learning position.

Structure and Change.   What changes during development? The answer from the learning theorists has been remarkably consistent—"the responses of the child." Because Watson and Skinner reflect an epistemological position as well as a developmental position, it has been an integral part of learning theories of development to see the "structural unit" (a term none of them would use!) of change as the aggregation of learned responses, of more or less generality, that are acquired over a lifetime. The only data that can be admitted into the scientific canon are those which can be seen and described by several objective observers. Subdermal processes such as images or thoughts or ideas or dreams need not be called on to understand the development of the child.

Proposals about the mechanisms of change in children have been relatively straightforward as well. The most general answer is environmental contingencies, those varied ways in which the inanimate and social environments respond to the behavior of the child. Conditioning in the Pavlovian manner has been called on as a specific mechanism of change, and quite subtle conceptions of stimulus differentiation and response generalization have been used in explaining behavioral change, but the most persistent theoretical mechanism used by learning theories of development to comprehend variation in behavior has been reinforcement, a notion that in earlier folk theories was encompassed by the ideas of reward and punishment.

Less radical learning theorists than Watson and Skinner, such as Dollard and Miller, earlier on, and Bandura, more recently, have proposed less adamant versions of the learning model; however, throughout the emphasis has remained on simplicity of behavior description and elegance of explanatory paradigms.

Method and Problem.   The method of choice for the student of children's learning has been the laboratory experiment; literally thousands of studies have been performed in which infants and children were seen in a carefully monitored setting where tight controls were maintained over environmental change, where the child's responses were measured with care and precision, and where experimental effects could be unambiguously assigned to some variation of circumstances designed by the experimenter. It is somewhat more difficult to specify the prototypical problem of the learning theorists. To be sure, studies of simple learning have predominated (for example, left-right choices guided by reward or the tendency of a child to aggress as a function of the behavior of a valued adult), but a great diversity of phenomena has been brought under the explanatory umbrella of learning theory. The discriminating marks of the learning approach have been explanatory simplicity, close definition of the experimental procedures, and an unyielding insistence on the efficacy of definable environmental contingencies.

Implications.   The messages for practice of the learning ideology are mixed. On one hand is the almost boundless optimism of a theory that maintains that if you control the environmental contingencies, you control all development. Like Archimedes, with sufficient leverage, parents and practitioners can move the developmental world. On the other hand, there is the more hidden message of assigned responsibility; if the child does not grow well and fruitfully, then one or another of the caregivers was inadequate or ill-informed. The underside of an Archimedean theory of human development is that failure of normal development must be assigned, not to some recessed and unobservable physiological or genetic source, but rather to the adult manipulators of the environmental rewards and punishments.

Since the 1890's at least, one form or another of the "learning" vision has dominated American culture. Its optimism, its assignment of personal responsibility, and its theoretical simplicity have been the critical signs of American behavioral science in the twentieth century.

DEVELOPMENT AS RESOLUTION OF CONFLICT

History.   The proposition that life and mind are results of conflict—the resolution of antitheses—is as old as philosophy itself, and the idea of development as the resolution of conflict had a lively effluence in the speculations of a number of nineteenth century scholars. Perhaps a derivative of that tradition, perhaps another expression of Darwin's implied phylogenetic struggle, Sigmund Freud, in the first years of the twentieth century, put forth a textured and elaborate theory of the human mind that, first, leaned heavily on a conflict theory of change and, second, irreversibly transformed our conceptions of children. The working out of the implications for child development of the general theory of psychoanalysis was left largely to Freud's disciples and to his vaguely ambivalent intellectual successors. However conveyed, the psychoanalytic image of children has become (in curiously naturalized forms) almost unconsciously part of the American folk theory of children.

Structure and Change.   Freud's proposals about the nature of the human mind and, by implication, of human development were the most intricate and conceptually demanding of any psychological theory prior to the arrival of the chip-based computer. Therefore, the easy simplicities of Gesell and Skinner are barred to us. Nonetheless, it is not a distortion of Freud's vision to see the primary structural unit of development as the child's theory of impulse and prohibition and the primary mechanism of change as the resolution of personal conflict. In the expression of the operation of his major theoretical ideas, Freud depended as well on a maturational premise about developmental changes in sensitive areas of the body. His unique attempt to reconcile the demands of physiology and the demands of society made Freud's image of the child the most pervasive and influential theory of the twentieth century.

Largely because of neuroanatomical development, the child's focus of pleasure-seeking interest was held by Freud to move from sucking to defecation to stimulation of genitals over the first years of life. Such a progression poses parallel problems in social development—the ambivalent and continually significant relation with the child's mother, the conflict that arises from the demands of adults for the child to exercise personal control, and the far more complicated prohibitions on the expression of unlimited sensual gratification. Freud emphasized a critical conflict between the child's impulse and the requirements of his parents (standing in for the larger society) for graded limitation of pleasure. Some stable resolution of the resulting conflicts was established in most children by age six or so and, healthy or unhealthy, the resolution marked the rest of their lives.

Freud explored all the forbidden areas of conventional child development in his time—dreams, wishes, terrors, craziness, and defensive maneuvers—and he presented to a shocked and disbelieving world a version of the child that was discrepant from all the contemporaneous standard stories of innocent peaceful youth. In spite of the grim and pessimistic tone of Freud's speculations, American interpreters were able (thanks to the wildly optimistic temper of our society when Freud was worrying his theory through) to see the psychoanalytic vision as a liberating and uplifting message: children had only to be allowed to indulge in their impulsive, sensuous urges in order to become healthy and fully formed. Someday the story of a marvelous irony will be told: while Freud's ideas were being shredded into a form acceptable to the American temper, i.e., all freedom and a slack rein on childish urges, Watson was simultaneously advocating and persuading parents to accept rigidly controlled procedures for child rearing, devoid of pacifiers, late toilet training, or masturbation. Here we find yet another demonstration of the capacity of the American temper to find support for individualism, self-control, and progressivism no matter what the intentions of the theorist.

Freud's emphasis on the importance of conflict in development brought to general child development theory a sensitivity to human motivation and intention. Also, observers became concerned about the pathology of children; the almost unbroken commitment of American psychology to the rational and the ordinary was severely tested by Freud's delving into the hidden and the strange in childish behavior.

Method and Problem. Despite Freud's dream of a physiological psychology and despite his early training in laboratory science, psychoanalysis evolved a method of test and proof that was, and still is, eccentric to the usual canons of empirical science. The critical problem for psychoanalytic child psychology is the child's attempt to reconcile the demands of impulse, danger, and social restriction. Elaborate mental structures are erected to monitor and control the child's interpretation of what is felt in desire and what is required by the parental surround. To plumb the mysteries of this set of interpretations, the primary method of study for Freud came to be the therapeutic encounter between child and analyst. Early on, the difficulties of applying the required therapeutic style to children led to a substitution of speculation and natural observation as ways to the child's mind. It is a measure of the complexities and difficulties of the psychoanalytic research mode that, three-quarters of a century after Freud's observations on Little Hans, there exists no established and defensible empirical literature on the early psychodynamic lives of children.

Implications. Because of the general seepage of psychoanalytic ideas into the American culture, alluded to earlier, it is not easy to specify the major implications of psychoanalytic child psychology for the lives of practitioners and parents. Without doubt, Freud taught us that the child, like the adult, is a far more conflicted, troubled, and vulnerable being than other theories had maintained. He also drove our attention to the importance for the child of early social encounters, particularly those encounters having to do with the continuing struggle between the push of ancient urges on one side and the controlling modifications of civilization on the other. Finally, by elaborating a theory of pathological variation, of alternative endings to the struggle, Freud presented conventional developmental study with the task of understanding why some of the variations in human growth were frightening and beyond ordinary experience.

DEVELOPMENT AS COGNITIVE CHANGE

History. The line of nineteenth century philosophy that led from Schopenhauer to Nietzsche to Freud was contrasted by a line that ran from Kant to Piaget. Barely touched by questions of human drives and urges, the cognitive line had as its main intention understanding the ways in which human beings came to know, that is, how we are able to use the regularities of mind and the regularities of personal history to build the traditional epistemological notions of time, space, object, and cause. Although American developmental scholars were concerned with problems of human knowledge early in the twentieth century—found in particular in the brilliant work of James Mark Baldwin—questions about the development of memory, thought, and inference were off base for the usual American student of children until the revival of interest in the work of Jean Piaget in the 1950's. For the three ensuing decades, the cognitive vision of children dominated academic child study; thousands of studies were published that extended, refuted, modified, or commented on the work of the great Genevan theorist. And, even as we now write, when enthusiasm for Piaget's ideas has begun to fade, an even older tradition of American interest in cognitive issues—now clothed in the language of information processing and artificial intelligence—has appeared in new strength and with new interest in the lives of children.

Structure and Change. Like psychoanalytic theory, Piaget's cognitive theory saw the child's growth, in the largest view, as a series of quantal changes—from the hand and eye construction of the world (sensorimotor) to a primitively theoretical way of dealing with intellectual problems in an immediately relevant here-and-now fashion (concrete operations) to a fully abstract understanding of the structure of the world (logical operations). Each state in the child's developing comprehension of his universe was a newly organized theory of the world and self; the core structural unit of development for Piaget was the scheme, an integrated theory, correct or not by adult standards, that permitted the child to receive information from the world, make sense of it, and predict the future. Development is the increase in scope and elegance of the child's schemes, the growth of intellectual sophistication. What is sometimes difficult to comprehend for adults—even professionally trained adults—is Piaget's recognition that the child's world-view differs from ours. The child's theory of the world is not an incomplete adult view; rather, it is a different theoretical organization of presented evidence.

As a mechanism for change, Piaget proposed an analogy with two biological processes of adaptation—*assimilation* and *accommodation*. In a parallel with ingestion and growth, the process of cognitive change is seen as a taking in of information from the world and making it useful for the child's present mental structures (e.g., a wooden block may stand for a cat in play) balanced against a changing of mental structures in order to adapt to the requirements of the environment (e.g., the child modifies his gestural repertoire to conform with the practices of his parents). Piaget saw human development as truly constructive, depending neither on maturation nor on learning for its working out. Rather, on each problem the child encounters in the world, a theory is brought to bear (at whatever level of generality and elegance). If confirmed, the theory is strengthened; if not, the child will adapt, or attempt to adapt, to the new demands of the environment. Thus, through the processes of assimilation and accommodation the child becomes more effectively adapted to the world, both inanimate and social, that surrounds him.

Following on the work of Piaget and the much older work of the psychology of adult thought, there has been in the late twentieth century a sudden growth of interest in understanding the child's thinking as a system of information coding, storage, organization, and retrieval. Backed up by the technology of digital computing and often using the computer as a powerful metaphor, the newer developmental psychology of thought has concentrated its attention on memory, language studies, and problem solving.

Method and Problem. The study of cognitive change has drawn on just about every method in the developmentalist's kitbag. Piaget presented his seminal account of the first three years of intellectual life as commentaries on observations he made on his own three children. In his work with older children, Piaget used a method of conversational inquiry with children in which he probed, gently and sympathetically, their reasons for believing in their theories of physics or chemistry or morality. Finally, both Piaget and his imitators have used the laboratory experiment extensively to test particular propositions about the variables that influence cognitive change. Through all the variation in method, the core problem for students of the development of thought has been the child's conception of the world, the ways in which evidence from the world is organized and used to understand how things (and people) work. Thus, the emphasis for Piaget and other cognizers shifted

from behavior as the focus of attention to mental structures as the focus of attention. The similarity to Freud's strategies is clear, but there should be no confusion about the fact that Freud and Piaget were in pursuit of far different domains of mental structure.

Implications. The image of the thinking child that Piaget brought so dramatically to the attention of developmental scholars has been widely influential in shaping the character of academic research and, to a lesser degree, in the design of school curricula and teaching procedures. In contrast, the view of the child as a largely cognitive being has had relatively little direct impact on the behavior of parents and of clinical practitioners concerned with children. If there is an effect to be detected, it is from the general cultural diffusion of the conviction that intelligence and academic competence are the major markers of a successful modern life.

DEVELOPMENT AS CULTURAL (ECOLOGICAL) ADAPTATION

History. The scholarly interest in variation that sprouted from Darwin's proposals and produced systematic studies of the development of children also produced the anthropological interest in variation from one cultural group to another. A number of early explorations of cultural variation had, as part of their agenda, the demonstration of the superiority of Western European forms; however, over the years, the emphasis on deficiency of cultures has been supplanted in anthropological studies by an emphasis on understanding how cultures come to vary, without assignment of evaluative labels (Jahoda, 1980).

Strange to say, after an initial flurry of attention, American students of children did not draw on cultural variation in patterns of development to amplify their understanding of children. The search for uniformity, the desire for simple descriptions of growth and development, was uncongenial to attempts to account for the nonuniform and the unusual. Then, a bubble of interest in the 1950's was followed by an explosion of research on cultural variation in the development of children. Moreover, the focus of study was no longer the exotic ranges of isolated and nonliterate groups alone, but also on the subdivisions of American society itself—for example, ethnic variation and social class variation in the rearing of children.

Structure and Change. The revived enthusiasm for the comparative study of human development has not brought in its train a well worked-out theory of structure and a statement of mechanisms for change. However, commentaries on cultural variation tend to lean toward a cognitive interpretation of the child's initiation into the peculiarities of his culture. Ritual, social patterns, power structures, caretaking arrangements, theories of the inanimate world—all are seen as problems posed for the child's solution. Becoming a member of a particular society is, in large measure, a matter of assimilating and accommodating to the theories of action and intention that adult members of the culture share. Therefore, the child in middle-class white American culture can be understood as differing from the peasant child in a Chinese commune primarily in terms of their varying theories of social structure and the natural order. From such a point of view, developmentalists are only beginning to formulate systematic theories of cultural differences in child development.

Method and Problem. The problem of cultural, or ecological, developmental study is easy enough to state superficially—the examination of the nature and sources of variation in the mind of the child from one culture to another—but, just as in the case of theory, investigators have only preliminary and borrowed ways of addressing the problem. What is required is a meeting of anthropological sensitivities and historical knowledge with the rigors and care of empirical research in psychology. Signs of such a meeting can now be detected (Cole and Scribner, 1974).

Implications. Just as early anthropologists sought the better and the worse in cultural patterns, so developmentalists early interested in cultural variation in child development sought to define the ways in which one pattern of child care or education was preferable to another. An evaluative stance about variation was notably present in the first attempts by social scientists to understand variation within American culture—the differences between poor black and poor white families, for instance. Happily, derogatory evaluation has usually given way to the recognition that, in child rearing as in other aspects of cultural variety, different does not mean deficient. It is too early to say how the new appreciation among academic scholars of cultural variety will affect the lives of American parents and their children.

The five theoretical positions just presented are a selection and an abstraction of a vast range of attitudes and beliefs that have been expressed concerning the lives of children over the last century. All are attempts to encompass the richness—or the jumble—of children's variety and, in a way, to provide systematic answers to the core conceptual questions with which we began the chapter. Now it is time to turn from the grander statements of developmental theory to a more down-to-earth consideration—again highly selective—of the em-

pirical findings about human development that pass muster as reliable, interesting, or important.

THREE DOMAINS OF EMPIRICAL RESEARCH

MOTOR DEVELOPMENT

Descriptive and Normative Approaches

During the 1920's and 1930's, the motor development of infants and children received close attention on both coasts of the United States; in Connecticut, Arnold Gesell and his colleagues at the Yale Clinic of Child Development compiled narratives of mental and motor development, while Nancy Bayley in California and Psyche Cattell in Boston undertook rather more psychometric approaches.

Gesell. The Yale group conducted studies of infantile movement patterns with an old Pathé camera and a relatively new technique, which they termed "cinemanalysis." The studies were designed to provide information about the regularities and patterns of child development; Gesell's intention was to improve the "developmental diagnosis" of various types of mental and motor dysfunction (Gesell and Amatruda, 1947).

The early normative studies made an important clinical contribution by permitting more reliable identification of children with developmental lags, but the bookfuls of age-related "norms" did not lead to much theory construction or hypothesis testing. Motor development was portrayed largely as a gradual unfolding or elaboration of simple inborn movements into more complex motor patterns, and in some instances all motor development was subsumed under the general rubric of "posture." In this rather atheoretical, maturationist approach, sitting, for example, was described as "a transitional stage between supine and standing postures."

The Gesell group did reiterate and codify a number of regularities of motor development. A general cephalocaudal progression of extensor tone was reported. The early standing and automatic walking reflexes were noted and described. The increasingly "automatic" quality of motor skills was also noted: As a new skill was being mastered, it seemed to occupy the attention of the child, making it very difficult for the child to conduct some other activity simultaneously; after mastery, however, the new motor skill seemed to occupy no appreciable space in the child's consciousness. Yet little attention was given to the implication that learning and overlearning (in addition to maturation) might therefore be involved in the acquisition of motor skills. Likewise, observed sex differences in (say) throwing a ball—with boys throwing farther, more accurately, and with a relatively more "mature" style—were carefully described with no mention of the discrepant training patterns that had no doubt preceded these skill assessments. Oddly enough, however, the Gesell group implicated learning processes in perceptual development; depth perception, for example, was attributed to "the gradual integration of visual and proprioceptive cues, through the process of trial and error."

From Binet to Bayley. The clearest case of attempts to relate the work of Gesell to the older work of Binet was the Developmental and Intelligence Scale developed by Psyche Cattell at Harvard during the 1930's and published by the Psychological Corporation in 1940. Cattell tried to restandardize items from Gesell and others based on a sample of 274 Boston children from largely working class households. Many of the gross-motor items were dropped along the way, however, so that most of the remaining motor items examined fine-motor performance.

The Cattell IQ was computed in the same manner as the original Stanford-Binet IQ, i.e., the ratio of mental age to chronological age. Not surprisingly, the resulting IQ's were remarkably unstable over the first three years of life. Also, given the demographic limitations of the sample on which the age grading had been based, the Cattell test was (and remains) of indeterminate value in evaluating infants from culturally varied backgrounds.

The Griffiths Mental Development Scale for Testing Babies from Birth to Two Years was developed by the late British psychologist Ruth Griffiths during the postwar years and was first made available for distribution in 1954. The Griffiths scales comprised 260 items that were evenly classified into five subscales: Locomotor, Personal-Social, Hearing and Speech, Eye and Hand, and Performance. The Griffiths test also followed a format similar to those of the Stanford-Binet and the Cattell, yielding a mental age and a profile of scores for each of the five subscales. A "GQ" (general intelligence quotient) was obtained, again by using the ratio method.

The Griffiths standardization research was conducted on 571 infants in day-care nurseries in one area of London. Although an attempt was made to control for the social ranking of the father's occupation, it is still not clear to what extent the resulting sample might or might not have reflected the postwar British population in general. In addition, subsequent research called into question Griffiths' five-subscale structure (Munro, 1968; Bayley, 1969). Factor analyses of test-item variation suggested that only two or three factors could be discerned—a language factor and one or two motor factors.

The Bayley Scales of Infant Development (BSID) were published by Nancy Bayley and the Psychological Corporation in 1969 as the culmination of 40 years of Bayley's experience and an estimated 4500 infant examinations. There were several stages in the development of the instrument, beginning with the Berkeley Growth Study from 1928 to 1933 and the subsequent publication of the California Infant Scale of Motor Development (Bayley, 1936) and two scales for assessing mental development. These instruments underwent further revision before the final standardization, resulting in a 163-item Bayley Mental Scale and an 81-item Bayley Motor Scale. The Motor Scale contains both gross- and fine-motor items, but many of the items on the Mental Scale require fine-motor skill as well as more traditionally "cognitive" abilities.

The cardinal virtue of the BSID lies in its ambitious standardization; it was standardized on a stratified national sample that was constructed to be representative of the population of the United States as measured by the 1960 census. The standardization comprised 1262 "normal" children, with nearly 100 children at each of 14 ages ranging from 2 to 30 months. The sample was controlled for infant's gender, race, type of community (rural vs. urban), and education of the head of household.

The Bayley test uses a deviation-score method to calculate the standardized mental and motor indices rather than the ratio method. The mental and motor indices, like IQs, have been normalized to have a mean of 100 and a standard deviation of 16 in the normal population.

The BSID represents a significant advance in the fledgling field of infant assessment. Although the long-term predictive validity of the BSID mental index is quite modest in the normal population, it predicts considerably better the performance of certain high-risk populations; infants with unusually low BSID indices are more likely to become children with low IQ's or motor dysfunction. The validity of the Motor Scale requires a great deal of further study, however. Fortunately, the renewed interest in motor development in fields such as pediatrics, neurology, and physical therapy (e.g., Drillien, 1977; Bobath, 1980) may at least permit construction of improved motor instruments for later childhood with which to assess the predictive validity of the Bayley Motor Scale.

The Physical Therapy Models

Motor development of the human infant has been studied in some detail by psychiatrists and physical therapists. A number of accounts of normal motor development have arisen from the clinical study of normal infants and of infants with a variety of pathological conditions, such as cerebral palsy (Peiper, 1963; André-Thomas and Aut-gaerden, 1966; Bobath, 1966, 1980; Fiorentino, 1973).

Many of these accounts of motor development attribute to the infant's "primitive reflexes" a prime role in the progression toward motor maturity. These primitive reflexes are reliably elicited early in the infant's development, but they subsequently disappear in the normal infant as they are "integrated" into more complex motor patterns, which are taken to be more clearly under cortical control. This approach to motor development certainly has a strong affiliation with biomedical approaches, although its biological model is often the functional neurology of the early decades of this century (von Uexküll, 1904; Sherrington, 1913; Magnus, 1926). Nevertheless, despite the biomedical language and the presupposition of a developing nervous system that is becoming ever more differentiated, great emphasis is also placed on sensorimotor experience and on the "learning of movements" (Fiorentino, 1972). Indeed, the physical therapy approach to motor development has stressed afferent pathways that initiate and guide motor patterns, because it is by taking advantage of these pathways that therapists hope to alter the motor output patterns of children with motor dysfunction.

Bobath, for example, has applied to physical therapy the axiom that posture produces proprioceptive feedback that guides and attenuates the motor reflexes of the central nervous system. Over several decades Bobath and his colleagues have compiled a group of empirically derived techniques in which manipulation of posture permits the control and inhibition of abnormal reflexes and the consequent expression of more normal alternative movements.

Conceptual Models in Motor Development

It would seem that motor development has been relatively short-changed by the theorists of behavior and development. As is often the case in many other disciplines, the special, narrow theories or frameworks have preceded the general, unifying ones. Several decades have been required to develop the conceptual tools with which to describe the phenomena of motor development. One is reminded of the state of affairs in biology before Darwin: there are taxonomies, competing schemes of classification, and special theories of limited scope.

The normative school of motor development has concerned itself principally with the rate of development, but also, to a lesser extent, with patterns.While this work has been a prerequisite to further progress, its explanatory power is limited.

The psychometric refinements of the normative approach have improved its clinical utility by making possible more quantitative interpretations. Clinicians may now determine with various levels of confidence whether one score or group of scores is reliably different from another score or group of scores.

The more biomedical approach of Bobath and others seems to offer somewhat greater explanatory promise. Movement patterns are guided and constrained by the gross-anatomical architecture of the organism, which reflects the cumulative history of phylogenetic influences. Yet, wholly contained within the lifespan of the individual are a host of ontogenetic factors that can either enhance or obstruct the individual's motor potential. Many of the guiding ideas in posture theory have taken for their model the biological investigations of earlier decades. It seems likely that this cross-disciplinary traffic in conceptual models will grow and enhance our ultimate understanding of motor development.

SENSORY AND PERCEPTUAL DEVELOPMENT

Two forces governed the study of perception at the beginnings of developmental science—the traditions of nineteenth century physiology and the drive to establish phylogenetic parallels between animals and human beings. In 1880, Wilhelm Preyer left his embryological laboratory for a few years to survey the literature on "the senses" in animals and infants, expecting to find evolutionary links. His report heads a line of research on infantile vision, audition, olfaction, taste, and touch that continues to our own time. The century-long research activity on perceptual development can be divided, roughly but conveniently, into two epochs: the studies and theoretical attitudes prior to about 1950 and those that came thereafter. The first epoch, summarized best by Peiper (1963), can be crudely called the era of infantile incompetence. Much of the work details the animal-like character of the sensory systems, the neuroanatomical or neurophysiological insufficiency of the infant, and the incompetence of the baby's perceptual organization (Preyer went so far as to assert that the infant was deaf at birth). The image of development as a form of phylogenetic or embryological recapitulation led researchers to see the baby as perceptually feeble.

In contrast, more recent lines of research have emphasized the wisdom of babies. Surprising, but documentable, claims have been made for the ability of very young infants to receive and organize the information displayed in the world about them. We suspect that the shift of attitude was pushed by the fading of the evolutionary analogy and pulled by the interests of investigators in the infant as a legitimate research preparation in its own right, as a problem-solver and processor of information. In any case, a collection of research papers 20 years into the new era was given the revealing title, *The Competent Infant.* Because the research of the second epoch is technically of a higher order than that which preceded it, the summary notes that follow will draw on illustrations from work done after 1950. Another limitation will be evident: most of the examples to be shown are from the field of vision research, largely because of the density of the literature. Work in audition and in taste, although on the upswing, has not yet achieved the richness and breadth of the work on the infant and child as looker.

Infantile Looking. In the late 1950's, Robert Fantz took advantage of an observation made long ago by lovers—that the human cornea is a mirror reflecting whatever the eye looks at. Fantz wisely deduced that the cornea-as-mirror could be used to study the visual orientation and interest of very young infants (Fantz, Fagan, and Miranda, 1975). Subsequent development of keenly sensitive television cameras and computer-based control systems has made it possible to examine the eye movements of infants and young children in remarkable detail and with remarkable accuracy (Haith, 1980).

An abundant literature on infantile looking warrants the following general conclusions: First, the human infant is an active processor of his surroundings. From birth, the child actively scans his visual environment, pausing to examine parts of the field (lines, bright objects, peripheral events) in a systematic fashion (Cohen and Salapatek, 1975); the image of the infant as a passive receptor surface is indefensible. Second, the visual behavior of the child is rule-governed and predictable. Marshall Haith has made the most economical and elegant statement of the rules that permit an understanding of early visual behavior, and his summary, whether or not it turns out to be correct in detail, is a demonstration of the organizing activity of the infant. Third, although the infant is far more competent visually in the first days of life than we had believed before 1950, there are regular and significant changes in visual ability over the first six months of life. Salapatek and Banks (1978) have described orderly developments during the first half-year in acuity, accommodation, color sensitivity, and binocularity. Fourth, the cornea-as-mirror procedure can be used in a wide variety of settings, not just to study visual functioning narrowly defined, but as a tool for investigating social responsiveness (when does the baby reliably differentiate his primary caretakers?) and, as we shall see, for studying higher-order cognitive development. The cornea has truly become a window on the mind of the child.

Habituation. When an infant or young child is shown the same display a number of times, he becomes less and less visually interested in it. When, as the child's attention slackens, a new display is presented, attention to the stimulus is renewed and a new cycle begins. This simple phenomenon, the habituation of looking time (or any other systematic measure) with repeated presentation of a stimulus, and the revival of interest with a new display, has become a second window on the world of the child. Two illustrations of its use—one on the study of color perception and one on the study of perceptual categories—will illustrate the flexibility and power of the habituation procedure.

Habituation and Color. A lively debate has been carried out among anatomists, psychologists, and anthropologists for many years on the question of the baby's ability to see and appreciate color. Conventional wisdom has maintained that the baby is deficient in color perception until well into the first year of life, and in some anthropological readings, the very categories of color (what we usually call blue, green, yellow, and red) are dependent on the child's learning the language and color conventions of his culture. Bornstein (Collins, 1981) has used the habituation paradigm to demonstrate that, no later than four months of age, the child divides the spectral range in almost exactly the same fashion as does the adult (of any culture, incidentally).

Briefly summarized, the procedure is as follows: The child watches small spots of colored light. A stimulus at the edge of the category called "blue" by adults (480 nm.) is presented 15 times; the baby's interest flags. Then two new stimuli are presented—one that is also called "blue" by adults (450 nm.) and one, equidistant in physical units from the habituated stimulus, that is usually called "green" (510 nm.). The baby behaves as though the two blues were the same (no recovery of interest to the 450 nm. stimulus) and as though the green were different (recovery of interest to the 510 nm. stimulus). Bornstein tested 4-month-old infants across all the conventional color boundaries and found the regularities characteristic of adult color perception, as shown in Figure 3-2. At least for infants 4 months of age, the old debate seems resolved; the child organizes color in essentially the same way as does the adult.

Habituation and Categories. The case of color perception is interesting for what it tells us about the child's perceptual competence, but we present the example in order to illustrate the use of the habituation procedure as a method of detecting the character and complexity of the infantile mind.

Another study may serve to indicate the range

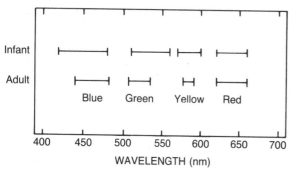

Figure 3–2. Equivalence classes of hue in infant and adult. Data for the infant have been summarized from Bornstein, Kessen, and Weiskopf (1976); those for adults are from Boynton and Gordon (1965) and represent regions of the wavelength spectrum that adults name exclusively blue, green, yellow, or red 70 per cent or more of the time. (Adapted from Bornstein. In Collins, 1981.)

of application of habituation studies. Cohen (1977), working with infants between 5 and 8 months of age, displayed pictures of a female face. After habituation of the child's looking time, Cohen presented several different test pictures that varied in novelty of content (a different female face) or in novelty of orientation (full face instead of side orientation). An examination of the results justified the following conclusion (among others): By 30 weeks of age, children can elaborate the general notion of "faceness." As McClusky (Bloom, 1981) writes in a summary of the study, "By 30 weeks of age the human infant can discern invariant aspects of facial stimuli and use this information to establish concepts into which similar stimuli can be incorporated."

Our all-too-brief sweep over recent research in early perceptual development was intended to propose and to illustrate the central conclusion that the human infant moves from sensation to cognition (Cohen and Salapatek, 1975) during the first year of life in a systematic fashion that can be studied with reliable methods. It is a long stretch from the newborn's attention to lines and lights to the 8-month-old's formation of perceptual categories. Although many issues remain to be resolved, the image of the helpless and incompetent baby can happily be abandoned. From the moment of birth (and probably even earlier) the human being is an active investigator and organizer of his world.

COGNITIVE DEVELOPMENT

Thinking

Over some six decades, Jean Piaget maintained a program of writing and research that established the empirical and theoretical framework within which the development of the child's thought is studied. He saw the development of knowledge

as a complex joint function of maturation, learning, social influence, and, most importantly, the mind's search for equilibrium between problems posed by the world and mental structures that have worked in the past. All the relevant factors are displayed in the major succession of stages that Piaget perceived in the child's development: sensorimotor (birth to about 18 months), preoperational (about 2 to about 5 years of age), concrete operational (5 or 6 years to about 11 or 12), and logical operational (from about age 11 on). We can give only the barest sketch of what Piaget saw as the tasks of each period; the literature, from Geneva and from other laboratories, now runs to thousands of articles on each level (Flavell, 1977; Ginsberg and Opper, 1969).

Sensorimotor Stage. The task for the child in the preverbal months is to construct the fundamental notions of stable object, of space, and of the dimension of time. One of Piaget's most provocative and persistent contributions was his demonstration that, for the young infant, tables and bowls and people are not fixed entities that exist independently of the baby's perception. Rather, the object as a permanent continuing thing must be constructed as the child matures and encounters stabilities in the world. In a classic and oft-repeated demonstration, Piaget showed that when a 5-month-old child watches an object being hidden, first under pillow A and then moved, while the child watches, to be hidden under pillow B, the baby will immediately search for the hidden object under pillow A. Ingenious natural experiments of this kind pepper Piaget's trilogy on infancy, and by and large, they have stood the test of replication. The young infant clearly lives and moves in a world that does not have the fixed values of the adult world. Not that the child is inactive or incompetent; rather, Piaget has shown us that the evidence of eyes and ears and touch is assembled by babies into a construction of reality different from ours.

Preoperational Stage. In the years before school, even though a relatively stable world of objects in space has been established, the child's solution of intellectual problems has a hit-and-miss character, an unregulated and irregular strategy. The archetypal observation, first made by Piaget, is on the young child's conception of number. A display of objects (for example, flowers and vases [Fig. 3–3]) is shown to the child, and a question is posed: Do the two rows contain the same or a different number of objects? Until near the end of the preschool period, most children will not have an invariant one-to-one correspondence rule for deciding whether the rows are equal or not. Thus, the array in Figure 3–3B would be judged by the child to contain more vases than flowers. In a sense, overall length of row or width

Figure 3–3. Problem posed in the study of the young child's understanding of number. Children under 5 years of age often see *B* as containing more pots than flowers. (Adapted from descriptions by Piaget.)

of intervening spaces intrudes on the evaluation of "pure" number. Although the particular case of number judgment has been subjected to close scrutiny and to some correction of Piaget's interpretations (Gelman and Gallistel, 1978), the burden of repeated studies has confirmed Piaget's conclusion that the child under five is not governed in his thinking by the operation of general rules (see, for example, Siegler, 1978). The child has difficulty in ordering a series of objects by size; the child often cannot sort mixed picture collections of (say) running rabbits and black rabbits into nonoverlapping categories; in short, the

preoperational child, however adept in managing the everyday world, usually does not possess systematic principles of inference.

Operational Throught—From Concrete to Logical. Piaget's overall contention about school-age children is that, in the early part of the school years, the child solves most problems, academic and naturally appearing, in a strongly empirical and concrete fashion. That is, the young scholar may solve quite demanding problems, but this solution tends to be here-and-now, tied to the problems presented, and not susceptible to being cast in abstract form.

Take the case, one among many such demonstration cases, of predicting the course of a billiard-like carom. Figure 3–4 shows an apparatus used by Inhelder and Piaget (1958) to determine the child's tactics and strategies of problem solving. Targets are placed on the table, and the child is asked to aim the plunger so as to hit the target (a bounce off the side wall is always necessary) and then to report what happened. Inhelder and Piaget detected three levels of thought in the movement from concrete toward more abstract solution and statement of the problem.

Wirt (age 5 years and 5 months) shows a typical early concrete-operational solution:

Wirt: "It came out here and it went over there. . . . I'm sure to make it," etc. He succeeds occasionally but describes the trajectories with his finger only in the form of curves not

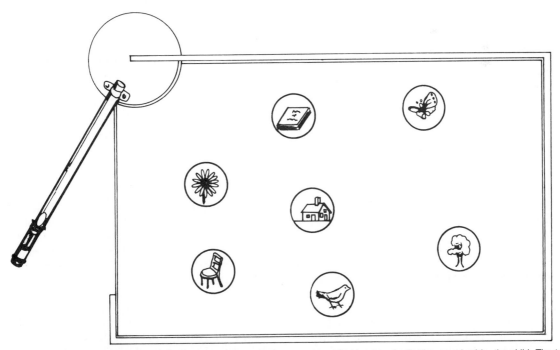

Figure 3–4. The principle of a billiard game is used to demonstrate the angles of incidence and reflection perceived by the child. The tubular plunger can be pivoted and aimed. Balls are launched from this plunger against the projection wall and rebound to the interior of the apparatus. The circled drawings represent targets that are placed successively at different points. (Adapted from Inhelder, B., and Piaget, J.: The Growth of Logical Thinking from Childhood to Adolescence: An Essay on the Construction of Formal Operational Structures. New York, Basic Books, Inc., 1958. Copyright © 1958 by Basic Books, Inc. Used by permission.)

touching the walls of the apparatus; he considers only the goal as if there were no rebounds.

At the second level of solution, the child almost always succeeds and has formulated the argument in the correct terms of straight lines and the angle of impact on the wall, but the case is still particular and not symbolized. Nic is 9 years and 4 months old:

Nic: "You have to move the plunger according to the location of the target; the ball has to make a slanting line with the target."

When the highest level of logical operations is reached, the problem is described by the child in terms satisfactory to a physicist (at least in Geneva). Pom is 15 years and 5 months old:

Pom: "I look a bit at an angle. . . . The higher up you want to aim, the wider the angle has to be." [Pom experiments with several angles for the shooter.] "You have to have two angles: the inclination of the lever equals the angle that the trajectory of the ball makes."

Across the school years, there is a steady change in strategy from concentration on a single problem and its practical solution toward an ability to state the abstract characteristics of a presented problem. Less research has been devoted to the examination of changes in cognitive strategies among school-age children and adolescents than to the preoperational period, but, in general, Piaget's observations have been confirmed for Western European children in schools. The emergence of logical operations, however, seems to be heavily dependent on tuition (Cole, in press), and the expected Piagetian sequence cannot be seen in all cultures, at least not in forms usual for children schooled in the West.

Although Piaget's thought has dominated the study of cognitive development in the late twentieth century, and although many of his original observations have withstood repeated testing, several new interpretations of developmental data have begun to appear. In general, the new cognitive developmental psychology stresses three considerations: (1) the child's increasingly sophisticated use of rules for problem-solving, (2) the importance of what has come to be called "metacognition," and (3) the embedding of thought in social context.

The notion of children's thought as a rule-governed activity has been particularly attractive because developmental changes in the strategies and tactics of thought can be understood as changes in a few operating rules of encoding or inference rather than requiring, as Piaget's theory does, a massive change of scale in the nature of thought. Moreover, our sensitivity to the rule-governance of thought makes possible computer simulation of some developmental changes and adds a powerful new procedure to the study of

cognition (Klahr and Wallace, 1976). Finally, the idea that the child's thought can be conceived as the operation of a group of specifiable rules leads almost inevitably to what Piaget called "the American question"—How can we speed up development? Perhaps, some educators hope, we can systematically teach children effective rules for thinking.

Metacognition reflects the interest of scholars in "the thinker's knowledge, control, and coordination of his own cognitions" (Brown and Deloache, 1978). Fairly early on, the child begins to talk to himself, either aloud or silently, about the procedures being used to solve a problem. The consequent awareness and explicit statement of strategies turns out to be a major component of the child's increasing ability to acquire and organize knowledge (Flavell and Wellman, 1976). One side of metacognition is seen in Pom's analysis of the billiard-table problem; the task is described in explicit terms that can be manipulated, as though on a mental blackboard, and subjected to higher-order rules of evidence and inference.

The embedding of thought in social context was effectively described by the Russian, Lev Vygotsky, in the 1930's and has undergone a renaissance in the work of Michael Cole and Urie Bronfenbrenner (1977). Instead of an emphasis on the inevitable unfolding of developmental sequences, researchers studying social context see development largely as the child's taking on the rules of the culture for the examination and solution of problems. As Cole and Scribner have observed, "Traditional cultures in transition would thus seem to offer an important natural laboratory in which to explore . . . the developmental factors . . . which contribute to specific cognitive organizations" (1974). The effects of schooling, the restraints of particular rituals, the intellectual place assigned to children in a culture—all provide rarely examined opportunities to study the variety of cognitive development.

Language

Between 1920 and 1955, most developmental psychologists were concened with issues of simple behavior, learning, and personality; little intellectual room was left for the study of general cognitive development or for the study of its most interesting and resistant aspect, language acquisition and development. But, in the late 1940's, almost simultaneously with the revival of interest in Piaget's work and with the emergence of information processing as a way of thinking about adult thought, two books appeared that recreated and redefined research on language development—Chomsky's *Syntactic Structures* (1957) and Brown's *Words and Things* (1958). Over the ensuing quarter-century, the study of the child's evolving use of language forms became one of the core problems of the field, spawning journals, books by the score, and literature of great breadth and detail.

The Development of Grammar. In part because of the traditions of linguistics and in part because of the power of Chomsky's intelligence, the first part of the language puzzle that captured the attention of the developmental psychologist was the study of grammar. How did the child come to use the elaborate transformations of sentences that permit adults to understand passive voice, nested phrases, inversions, and the various eccentricities of English (in particular), represented in exaggerated form in the sentence, "The gun the boy the girl loved bought misfired"? Psychologists accepted Chromsky's distinction between the surface structure of language (the sentences that we, in fact, say and hear) and its deep structure (how we reduce the heard sentence to its basic syntactical form) and went in search of children's grammar. The search produced a great deal of new information about early language, including the child's tendency to generalize grammatical rules even when the generalization leads to error, the regularity of the child's acquisition of certain forms (pluralization rules, intonation, the use of what-where-who questions, and the like), and the existence of individually distinctive styles of early language. However, no simple, singular theory of childish grammar emerged. As research results accumulated, it became clear that if we were to understand early language, we would have to look beyond grammar.

The Development of Meaning. Since the early 1970's, the core problem in developmental language study has been the acquisition of meaning, the development of the child's ability to use a conventional communication system appropriately. Attempts to understand childish communication in terms of finished adult forms were pretty much given up, and the phenomena to be explained were substantially broadened. At the present time, the field is in a yeasty productive turmoil that defies summary, but several research questions are shaping the domain of inquiry:

1. How shall we describe the child's knowledge at any moment? The question recognizes that, behind any particular childish sentence, there lies a developing world-knowledge. When the 2-year-old, standing at the window, says with rising inflection, "Daddy home," a remarkable amount of knowledge about the workings of the world is represented. The question about knowledge achieves the further benefit of tying language study tightly to the larger enterprise of understanding cognitive development.

2. What are the rules by which the child maps his knowledge (including social knowledge and desires) into the conventional language of the culture? The question is a variant of the older questions about grammar but with some striking differences. The child's language is now seen as dependent on the context of communication (as, by the way, is an adult's language). The appropriate response to "Can you pass the salt?" at the dinner table is not "Yes" but the action of passing the salt. The question about rules has also directed attention to individual differences. It turns out that some children tend to concentrate their use of language on describing the world about them ("What's that?" is the defining question), whereas others concentrate their linguistic attention on the social and emotional aspects of their lives. There may be, in fact, a number of different strategies, or sets of rules, for making speech out of knowledge.

3. In what settings and with what assistance do children learn language? The answers have ranged from "Language is innate" to "Language is learned by conditioning," but these two easy extremes cannot be taken seriously nowadays. The search for a plausible middle ground has led to research attention to the particular circumstances of early language learning and to an examination of the special ways adults talk to language-learning children. "Motherese"—the adult teaching language consisting of exaggerated intonation, simplified grammar, and age-appropriate vocabulary together with a sound diagnosis of the child's world-knowledge—is the summary concept to express the textured social interchange that is necessary for the child to learn the subtleties of any natural language.

A fourth question regarding language acquisition—What are the disorders and retardations of language use?—is addressed in Chapter 40.

Intelligence

For centuries, philosophers and scientists have looked for simple answers to complicated questions; by and large, the scientists have done better than the philosophers—although neither has done well—and the urge persists. When Alfred Binet was asked by the Minister of Education in 1904 to devise an examination to identify children who were likely to have trouble in school, he did not know that he stood at the head of a contentious research lineage—the use of testing to measure intelligence. Americans leapt at the idea of a simple means of predicting success in school and, for some additional, unattractive reasons (racism, ethnocentrism, and school tracking), glommed onto the Binet test.

In several chapters of this book, the uses and abuses of assessment will be discussed. Within our small compass, we can hope only to sketch out some new directions in intelligence testing and interpretation. Between 1970 and 1980, three issues arose that will persist at least until the end of the century.

Most notorious has been the debate about the genetic contribution to tested intelligence and, close by but muted, the debate about the sources of racial differences in IQ (intelligence quotient) scores. The latter issue can be readily vacated; there is no believable evidence of genetic sources for racial differences in IQ. The former issue is more ambiguous. To be sure, identical twins are more alike in IQ scores than are fraternal twins or siblings of different ages. The interpretation of the facts, however, spreads from Jensen (1969), who finds the evidence of genetic influence convincing, to Kamin (1974), who remains profoundly skeptical about the genetic argument.

Second, there has been a resurgence of interest in the dissection of the intelligence test. Whatever its predictive possibilities, which is an issue addressed by the National Research Council Committee on Testing (1982), the intelligence test remains a conceptual puzzle. The new wave, represented best by the work of Sternberg (1980), is trying to pull the IQ test apart in order to examine the components—the atoms—of intelligence testing.

Finally, and closely related to our earlier concern with contextual considerations in language and cognitive development, there is abroad now an active interest in understanding the practical applications of intelligence in natural settings. Cole (in press) has presented the best case for defining intelligence, not as a test score, but as a mode of solving problems in everyday settings. Drawing on cross-cultural data, he has raised reasonable doubts about the utility of IQ tests as more than stand-ins for school success.

The study of cognition continues to dominate developmental study and is likely to do so for some time, particularly in its extensions into artificial intelligence and cognitive science (a field that aims to join together linguistics, cognitive psychology, anthropology, and philosophy). The enormous intellectual structure that Piaget, Chomsky, and Brown built in the middle of the century still contains many unexplored rooms.

THE REMAINDER OF DEVELOPMENTAL STUDY

Motor development, perceptual development, and cognitive development make up only a part—although a big part—of the "pie" of developmental study. In our account, there have been two whopping omissions—social development and

pathological development. Fortunately, discussions of both themes can be found within several other chapters in this book. Our survey here must be seen, not as an accurate map of a bounded terrain, but as a collection of quick sketches of a broad landscape, unkempt perhaps, but flowering.

WILLIAM KESSEN

DAVID SCOTT

REFERENCES*

Baldwin, J. M.: Thought and Things: Genetic Logic. 3 Vols. New York, The Macmillan Co., 1906–1911.

Bandura, A.: Social Learning Theory. Englewood Cliffs, N.J., Prentice-Hall, 1977.

Binet, A., and Simon, T.: The Development of Intelligence in Children. (Translation of papers in L'Année Psychologique, 1905–1911). Baltimore, Williams and Wilkins, 1916.

Bowlby, J.: Attachment and Loss. New York, Basic Books, 1969, 1973, and 1980.

Brainerd, C. J., et al.: The stage question in cognitive-developmental theory. Behavioral Brain Sci. 2:173, 1978.

Brazelton, T. B.: Neonatal Behavioral Assessment Scale. Philadelphia, J. B. Lippincott, 1973.

Bronfenbrenner, U.: Toward an experimental ecology of human development. Am. Psychologist 32:513, 1977.

Cattell, R. B.: Personality: A Systematic, Theoretical, and Factual Study. New York, McGraw-Hill Book Co., 1950.

Chamberlain, A. F.: The Child: A Study in the Evolution of Man. London, Walter Scott, 1901.

Cole, M., and Scribner, S.: Culture and Thought. New York, John Wiley and Sons, 1974.

Cronbach, L. J.: The two disciplines of scientific psychology. Am. Psychologist 12:671, 1957.

Darwin, C. R.: The Descent of Man, and Selection in Relation to Sex. New York, Appleton, 1871.

Darwin, C. R.: A biographical sketch of an infant. Mind 2:286, 1877.

Dollard, J., and Miller, N. E.: Personality and Psychotherapy. New York, McGraw-Hill Book Co., 1950.

Erikson, E. H.: Childhood and Society. New York, W. W. Norton, 1950.

Flavell, J. H.: Cognitive Development. Englewood Cliffs, N.J., Prentice-Hall, 1977.

Freud, A.: Normality and Pathology in Childhood. London, Hogarth Press, 1966.

Freud, S.: Three contributions to the theory of sex (German edition, 1905). In Brill, A. A. (ed.): Basic Writings of Sigmund Freud. New York, Modern Library, 1938.

Freud, S.: An Outline of Psychoanalysis. New York, W. W. Norton, 1949.

Galton, F.: Natural Inheritance. New York, The Macmillan Co., 1889.

Gesell, A.: Infancy and Human Growth. New York, The Macmillan Co., 1928.

Gesell, A.: The Embryology of Behavior. The Beginnings of the Human Mind. New York, Harper Brothers, 1945.

Gesell, A., and Amatruda, C. S.: Developmental Diagnosis: Normal and Abnormal Child Development. 2nd ed. New York, Hoeber, 1947.

Gesell, A., and Ilg, F. L.: Infant and Child in the Culture of Today: The Guidance of Development. New York, Harper Brothers, 1943.

Gesell, A., and Ilg, F. L.: The Child from Five to Ten. New York, Harper Brothers, 1946.

Griffiths, R.: The Ability of Babies: A Study in Mental Measurement. New York, McGraw-Hill Book Co., 1954.

Haeckel, E.: Anthropogenie oder Entwicklungs-Geschichte des Menschens. Leipzig, Engelmann, 1874.

Hall, G. S.: Life and Confessions of a Psychologist. New York, Appleton, 1923.

Jahoda, G.: Theoretical and systematic approaches in cross-cultural psychology. In Triandis, H. C., and Lambert, W. W. (eds.): Hand-

---

*A complete list of the works referred to in the chapter can be obtained by writing to William Kessen, Psychology, Box 11A, Yale Station, New Haven, CT 06520.

book of Cross-Cultural Psychology: Perspectives. Boston, Allyn and Bacon, 1980, pp. 69–141.

Koffka, K.: The Growth of the Mind. London, Kegan Paul, 1924.

Lewin, K.: A Dynamic Theory of Personality. New York, McGraw-Hill Book Co., 1935.

Mischel, W.: Personality and Assessment. New York, John Wiley and Sons, 1968.

Pavlov, I. P.: Conditioned Reflexes. London, Oxford University Press, 1927.

Piaget, J.: The Psychology of Intelligence (French edition, 1947). London, Routledge and Kegan Paul, 1950.

Piaget, J.: Play, Dreams, and Imitation in Childhood (French edition, 1945). New York, W. W. Norton, 1951.

Piaget, J.: The Origins of Intelligence in Children (French edition, 1936). New York, International University Press, 1952.

Piaget, J.: The Construction of Reality in the Child (French edition, 1937). New York, Basic Books, 1954.

Piaget, J.: Biology and Knowledge (French edition, 1967). Chicago, University of Chicago Press, 1971.

Preyer, W. T.: The Mind of the Child (German edition, 1881). New York, Appleton, 1888–1889.

Romanes, G. J.: Mental Evolution in Man. Origins of Human Faculty. New York, Appleton, 1889.

Siegler, R. S.: Children's Thinking: What Develops? Hillsdale, N.J., Lawrence Erlbaum Associates, 1978.

Skinner, B. F.: Cumulative Record (enlarged edition). New York, Appleton-Century-Crofts, 1961.

Terman, L. J.: The Measurement of Intelligence. Boston, Houghton-Mifflin, 1916.

Thorndike, E. L.: The Fundamentals of Learning. New York, Teachers College, 1932.

Tinbergen, N.: The Study of Instinct. Oxford, Clarendon Press, 1951.

von Uexküll, J. J.: Umwelt and Innerwelt der Tiere. Berlin, Springer Verlag, 1909.

Waddington, C. H.: The Strategy of the Genes. London, Allen and Unwin, 1957.

Watson, J. B.: Psychological Care of Infant and Child. New York, W. W. Norton, 1928.

Werner, H.: Comparative Psychology of Mental Development. New York, Follett Publishing Co., 1948.

Wilson, E. O.: On Human Nature. Cambridge, Harvard University Press, 1978.

4

Evolving Parenthood:
A Developmental Perspective

As the study and application of behavioral pediatrics have advanced, pediatricians have become increasingly aware that the demands of a contemporary practice are multidimensional. Along with the child's physical health and well-being, many other aspects of growth and development have properly become areas for concern. With the need to know the child more fully, pediatricians also find that they need to learn more about the family. They recognize, too, the degree to which compliance and overall treatment success are fostered by the development of a strong parental alliance. Clearly, such an alliance depends largely upon the pediatricians' ability to understand parents and to convey to them a sense of that understanding and an assurance of support. Physicians therefore seek to develop a heightened sensitivity to the complexities and predicaments of parenting.

Guarding against the popular view that being a perfect parent is a realistic or attainable goal, many pediatricians must also resist the tendency to become aligned exclusively with the interests of the child. This may be a difficult task, since so much of the current literature represents the child as the center of the universe and the parents as existing primarily to maintain that position! While supporting the cause of children is a wise and prudent course for any society, it is critical today that the dilemmas of parenthood be addressed carefully. Commonly, discussions that purport to be concerned with the subject of parenting actually focus on the developmental issues and needs of the child.

In recent years, many theorists have promulgated the concept of an adult life cycle (Erikson, 1950; Levinson, 1978; Lidz, 1976; Vaillant, 1977). In his discussion of the "eight ages of man," Erikson posited a stage theory of development extending from birth to old age. Much more recently, Levinson elaborated specific stages in adult development, each of which occurs in an orderly progression and has its own developmental tasks. Like Erikson, Levinson proposes a life-long sequence that is fixed and interdependent. Problems or issues unresolved in earlier stages of childhood

or adulthood may significantly hamper the resolution of an ongoing period. Although Levinson's work was limited to the study of adult males, his conviction regarding the existence of an adult structure of development is not gender-specific. He states, "We are now prepared to maintain that everyone lives through the same developmental periods in adulthood, just as in childhood, though people go through them in radically different ways."

Subscribing to a life-cycle principle of adult development is of particular value to the pediatrician because it offers a theoretical paradigm for improving counseling skills and other intervention techniques. Being able to consider parents as engaged in identifiable developmental issues of their own can increase the pediatrician's empathic response and lead to a stronger bond between physician and parent. It can also contribute to clarifying diagnoses and enlisting parental cooperation.

Since this chapter will examine the parent-child relationship from the parents' perspective, it is critical to recognize that parents, subject to adult imperatives, are themselves faced with unavoidable developmental tasks throughout their lives. While meeting the changing and complex demands of the parenting role, husbands and wives must strive to evolve as individuals in their own right and to secure a deep and lasting relationship with each other. Although it is well established that the child's growth and advancement are shaped and determined by the parents' genetic endowment, familial background, and child-rearing practices, the extent to which the parents' lives are changed and affected by their developing offspring remains relatively unexplored. It should be noted that the degree and kinds of change experienced by parents are relative to the investment they make—both knowingly and unknowingly—in raising their children. Societal pressure for "perfect" parents with "perfect" children suggests that for many parents the investment may be heavy indeed.

There is a tendency for all cultures to maintain

some idealized view of their social patterns and structures as real and certain. In the same way that the belief in idealistic goals and methods of parenting has evolved today, we are inclined also to assume the existence of a model family unit. Families that qualify for inclusion in this category are those who have chosen a rather conventional approach to child rearing. Conventional parents are those young married couples who have a first child before or around the age of 30, after they have achieved a reasonably successful integration as adults. Having given up a dependent relationship with their own parents, they have acquired a sense of individual identity and of independence. By marrying, they have elected to forfeit some of their hard-won independence so as to gain emotional stability and satisfaction with each other. In deciding to have a child they have embarked on a course of change and self-denial, the impact of which is seldom perceived at the outset.

The image of a nuclear family, comprising two natural parents living in the home with one or more children born of their union, is less and less reflective of the norm in present-day society (see Chapter 12). While there continues to be a large representation of families that meet the aforementioned two-parent standards, recognition of the direction of current social trends calls into question the permanence of many such units. There are today a growing number of children being raised by single parents, many of whom are teenage mothers, and children living in restructured families with siblings who were born to one of their parents in another marriage. With the rising rate of divorce and remarriage, unconventional parental styles seem in a sense to represent more realistically our changing social patterns.

However, in the present exploration of the dynamics of parenthood, it is not possible to examine the countless parental permutations as they currently exist. There is, moreover, a commonality of basic parental issues extending across the widest experiential gaps to which this presentation is properly addressed. Because conventional child rearing continues to provide the most familiar and meaningful structure for organizing concepts about the interrelatedness of parenting and adult development, it will be the subject for most of our discussion.

There are, however, a number of individuals, both conventional and nonconventional in their parental styles, for whom the course of parenting is precipitous, delayed, interrupted, problematic, and, for all—in varying degrees—distressful. For some, the alteration or adjustment to an unplanned pregnancy either within or outside of marriage causes disruption in their lives and in the lives of all those with whom they are closely related. For others the decision to have a child, while readily made, may not be so readily realized. Frequent miscarriages, stillbirths, or failure to conceive at all weakens the resolve of some couples planning a family. Pregnancies, threatened by or ending in premature births, also create inordinate stress for parents. The ability to parent may be painfully challenged for both husbands and wives in those instances in which a child is produced who is biologically defective or physically at risk in the neonatal period. The occurrence of serious or fatal illness at any time in the child's life is a major source of pain and disturbance for all family members.

Regardless of their age, parents are ill prepared for any of these crises. The ways in which they respond to such disruption and disappointment may be attributed to a number of factors: (a) the availability and timeliness of meaningful support from within the marital relationship, extended family, and appropriate professionals (among whom the pediatrician may be the key figure); (b) the style, philosophy, and emotional resources of individual parents; and (c) the mother's and father's own developmental levels and stages of maturity. Although a comprehensive exploration of all these factors is clearly beyond the scope of this chapter, one such extraordinary situation will be presented in order to illustrate the management issues that arise with a complex problem.

The overall plan of this chapter is to focus attention on how the transition to and transitions within "ordinary parenthood" are made, how the ease and success of transitions are facilitated, and, finally, in what ways being a parent contributes to the critical development of the adult. There is no question that the birth of a child, particularly the first child, produces marked change in the new parents' lives. However, Lidz's statement, "The arrival of the first child transforms spouses into parents and turns a marriage into a family," implies a magic process that, in the beginning, is rarely more than nominal (Lidz, 1976). In order to appreciate how truly complex and confounding the parenting process can be, the following propositions, found to have application to almost all parental experience, will be examined individually:

Proposition One: Parenthood is one of a series of developmental tasks of adult life wherein the parent as well as the child is modified and changed. It is unique in its irrevocability and in its power to transform and transcend all other aspects of development.

Proposition Two: The interactional effects between parent and child are stage-dependent. Each stage of the child's development presents its own goals and challenges, which are faced by parents of widely divergent experiences and developmental stages.

Proposition Three: Parents are frequently caught up in patterns of behavior with their children wherein they re-experience and re-enact an earlier conflict of their own as a child reaches the developmental stage in which the parent's conflict had originated. Such patterns can be, and often are, negative and circular in their functioning.

Proposition Four: Parents bring from their own life experiences expectations and aspirations for themselves and their children of which they are not always aware. Yet these often unconscious motivations significantly influence the child-rearing process, with both positive and negative effects.

Proposition Five: A major challenge for parents is to love and bestow affection and caring on significant family members from two distinct generations: their parents and their children. They must also prepare to yield the intensity of their attachments as the parents grow old and the children grow away.

While a detailed presentation of adult life-cycle theory is not a realistic goal for this chapter, some elaboration of the concepts of adult development is essential to any study of parent-child relationships and the adult parent role. Although Levinson's conceptualization of the developmental periods of adulthood is, as noted earlier, somewhat limited in its application to the present discussion, certain of his delineative constructs, appropriate to both men and women, are relevant. A major finding of Levinson's study is that the evolution of life structure in adults follows an orderly sequence that consists of alternating stable "structure-building" periods and transitional "structure-changing" periods. As explained by Levinson:

> The developmental tasks are crucial to the evolution of the periods. The specific character of a period derives from the nature of its tasks. A period begins when its major tasks become predominant in a man's life. A period ends when its tasks lose their primacy and new tasks emerge to initiate a new period. The orderly progression of periods stems from the recurrent change in tasks. The most fundamental tasks of a stable period are to make firm choices, rebuild the life structure and enhance one's life within it. Those of a transitional period are to question and reappraise the existing structure, to search for new possibilities in self and world and to modify the present structure enough so that a new one is formed.

Learning to apply a theory of adult development can broaden our understanding of the myriad responses of parents to their children. To achieve some sense of the value of this knowledge, Levinson's conception of adult development will be examined relative to the pre-parenting experience—the period of pregnancy—perceived by some authors as the initial stage of actual parenting (Lidz, 1968; Rapoport et al., 1980). The effects of conceiving a child and anticipating parenthood are as varied as are the ages and circumstances of the prospective parents.

Levinson's Novice Phase, extending from 17 to 35 years of age, is composed of three distinct periods: Early Adult Transition, Entering the Adult World, and Age Thirty Transition. Each has its own developmental tasks while sharing the major task of emerging from adolescence to enter adulthood. The Novice Phase has four ultimate goals: forming a dream and giving it a place in the life structure; forming mentor relationships; form-

ing an occupation; and forming love relationships, marriage, and family.

The Early Adult Transition phase (age 17 to 22) presents two tasks: to "leave the world of childhood" (to separate from the family) and to "form a basis for living in the adult world before becoming fully a part of it." Entering the Adult World, a more stable stage, begins around age 22 and lasts about six years. The two major developmental tasks of this stage are exploring and examining options and considering possibilities, and creating a stable structure. There exists in all adult developmental tasks a paradoxical character, exemplified in the foregoing statement. The need for young adults to explore freely is countered by societal pressures on them to make serious choices and commitments. The Age Thirty Transition phase, "a remarkable gift and a burden," begins around age 28 and ends most often at 33. Levinson conceives this period as an "opportunity to work out the flaws and limitations of the first adult life structure, and to create the basis for a more satisfactory structure with which to complete the era of early adulthood." From here, the transition is to the Settling Down Stage, in which the tasks are to become truly adult and acquire seniority in the self-defined dimensions of one's world.

In commenting on the fourth developmental task of the Novice Phase, Levinson believes that it makes a great difference when, in the evolution of the developmental period, a marriage takes place. If it occurs at the beginning of a period, the earlier character of the marriage may be closely tied to the struggles of that phase; coming more optimally at the end of a developmental period, it may reflect a satisfactory formulation of life goals and the successful entry into adulthood. This observation applies also to the initial phase of establishing a family—the one irreversible step in the life of the individual, which, perhaps more than any other, precipitates the assumption of an adult role.

PROPOSITION ONE

Parenthood is one of a series of developmental tasks of adult life wherein the parent as well as the child is modified and changed. It is unique in its irrevocability and in its power to transform and transcend all other aspects of development.

According to Erikson, "Parenthood is, for most, the first, and for many, the prime generative encounter. . . ." In his essay "Human Strength and the Cycle of Generations," he states, "The love of young adulthood is, above all, a *chosen*, an *active* love. . . . The problem is one of transferring the experience of being cared for in a parental setting, in which one happened to grow up, to a new, an adult affiliation which is actively chosen and cultivated as mutual concern." Care is "the quality essential for psychosocial evolution, . . . the widening concern for what has been generated by love, necessity or accident; it overcomes the ambivalence adhering to irreversible obligation" (Erikson, 1964). The concept and application of the words "generativity" and "generate" were first defined by Erikson as "concern in establishing and guiding the next generation," a concern of far greater complexity than biological reproduction, since, as he sees it, "the mere act of having, or even wanting children. . .does not 'achieve' generativity" (Erikson, 1964). Less well known but of significant value are the works of LeMasters (1957) and Rossi (1968). One of the first to posit parenthood as a development crisis, LeMasters believed that parenthood and not marriage denotes the final transition to adult responsibility in our society. "The arrival of the first child forces young married couples to take the last painful step into the adult world." Interested in the effect of parenthood on adults and in parent development, Rossi examined the social aspects of the transition to parenthood, focusing on the parent more than on the child. She, too, felt that the first pregnancy, rather than marriage, is the major transition point, particularly for women. Her formulations are somewhat limited in application, since they offer little about the father's perspective.

While it is generally agreed that the timing of the first pregnancy is of great significance, the decision to have a baby is not always well considered. Some babies are conceived after much careful thought, some after very little, and an increasingly large number with no thought at all. The most thoughtfully arranged pregnancies are those of the nonconventional, older husbands and wives (whose numbers seem to be growing) who have accorded priority to business or professional careers and postponed parenting until they are well into the later Settling Down Period (ages 32 to 40) or even just after entering the Midlife Transition of 40 to 45 years of age. Many younger couples briefly defer having a first child while they resolve concerns related to housing, finances, or advanced educational goals. Other young married adults manifest a willingness or perhaps an ambivalence relative to becoming parents by the degree of casualness or inefficiency with which they attend to contraception. Finally, and lamentably, there is a growing population of young—even very young—girls who, having become pregnant, decide to keep their babies with or without outside support from either the baby's father or the mother's family.

Romanticization of the experience of parenting and failure to anticipate the impact of having a child on one's life are nowhere more apparent than among adolescent girls trying to be mothers. Yet the many unexpected stresses present in the early stages of parenting are universally felt. Even when children are planned and desired, first-time parents of any age are seldom prepared for the relentless demands of child rearing on their energy and resources and the endless constraints on their hitherto unencumbered lives. What is lacking in contemporary culture is a more realistic understanding of the commitment and self-denial involved in parenthood. Practical programs are needed to teach prospective parents early caretaking skills and child development and to prepare them for the ways in which their relationship and their lives may soon be changed. Finally, the vulnerability of the period of pregnancy should be addressed and support should be available for those assuming the role of parents, particularly for the first time.

The need for anticipatory guidance and programmatic training has come about as a result of the tensions created by rapidly changing social mores and the uncertainty of world events within the context of a diminishing sense of community. By the act of conception, couples expecting their first child are precipitated into a transitional stage for which they are frequently unprepared. Often lacking the proximity of extended families and feeling somewhat at a loss, they may seek support by becoming organized and task-oriented. They may, for example, contact in advance the physician who will care for their child. In recognition of the couple's needs, many behavioral pediatricians and family practitioners schedule an extended prenatal visit in order to become acquainted, answer questions, and allay fears. In this way the physician is offered a rare opportunity to learn about prospective parents as individuals while they are free of the anxiety and confusion common to the postpartum period. The need for intervention can also be considered and implemented wherever necessary.

With careful interviewing it is often possible to ascertain the individual level of maturity of the husband and wife, to identify potential trouble spots, and to begin to develop an alliance. The ages of the parents will suggest where they might be in the developmental scheme of their lives. Open-ended questions and background data can provide additional information. Often, showing an interest in early familial experiences can be of great value in uncovering myths and legends from one or the other's extended family: "My mother always told me I'd have trouble delivering a child," or "My dad was too nervous to drive my mother to the hospital when she had me." Such remarks, which give important clues to underlying concerns, might have been made by a somewhat troubled couple who arranged to meet with a pediatrician prior to the birth of their child:

The 28-year-old husband, a computer expert with a national corporation, has recently been promoted and transferred to the area; he enthusiastically relates his satisfaction with his job and the brightness of his future. Somewhat more reservedly, his 27-year-old wife described a high school courtship and a marriage of six years. She speaks wistfully of her parents' disappointment at having their only child move so far away, particularly at the time of her pregnancy. She wonders if she will continue in her own academic pursuits, since she has recently returned to college after working for three and a half years while her husband studied part-time for his MBA. She goes on to say that, although she hasn't made any friends yet, she'll get along because soon all her time will be taken up caring for the baby.

Observing the husband's growing restlessness and seeming impatience with this line of discussion, the pediatrician reflected inwardly on the source of some of the discomfort. It had been at the husband's insistence that he and his wife leave a newly established apartment in their home town to move a distance of 600 miles. The decision to have a child just prior to his promotion had been more in response to his wish to "settle down" than to the needs of his wife, who had set aside her own career explorations while contributing to her husband's educational dream. Dismissing her troubled feelings with a shrug and an "Oh well," the young wife (also seemingly aware of her husband's unease) turned to a practical discussion of rooming-in and breast feeding. The pediatrician, deciding not to probe further, sought to reassure them and to respond to their stated questions and their unspoken concerns.

Later, in reviewing this session, the pediatrician recognized the presence of stress but did not minimize the strength of the marital relationship and the warmth and caring the couple demonstrated for each other. Her analysis of this preparental couple is relatively straightforward. She recognizes the husband's developmental status as being comfortably stable in the adult world as he approaches the Age Thirty Transition described by Levinson. Having made two very serious commitments (to work and to being a parent), he will now have the opportunity to consolidate, to continue to grow, and to prepare for a more mature adult life period. His wife, on the other hand, replacing one dream with another, must now defer completion of her own education and, together with her husband, create a stable family structure.

In counseling them, the pediatrician recommends that they visit their families over the next several months whenever it is practical and that they arrange that their parents (particularly the wife's) be present at the time of delivery. They are assured that while making friends will take some additional time, becoming members of the community will foster feelings of security and a new sense of belonging for them. This is followed by a brief discussion of the location and curriculum of the state university and the time for a new mother to take courses. Convinced of her husband's continued support of her return to school, the wife seems more comfortable with the interruption in her education and the likelihood that it will not be too prolonged. In this eventful meeting the pediatrician is able to offer guidance and to gain insight and understanding that will affect the course of the family's development and the quality and effectiveness of pediatric care.

Far more disquieting for the pediatrician was an earlier interview with a young, unmarried adolescent girl and her parents as they struggled with their feelings about the daughter's decision to have and to keep her baby. It is apparent that the pregnant teenager (16 years of age) is at great risk. Not having fully negotiated adolescence, she has yet to begin the transition to early adulthood. She must now, in a sense, attempt to "skip" several developmental stages to achieve a maturity beyond her years or experience. The pediatrician perceives the need to assume an active role. Recommending and conferring with an obstetrician is an obvious objective, as is preparing the family for a referral to a mental health professional experienced with young, unmarried, prospective mothers. The pediatrician can provide immediate support and assistance during this period of initial stress. Exploring and resolving feelings is a task for all those involved, as is the making of practical decisions and arrangements. Planning for and learning about the processes of pregnancy, childbirth, and child care should be left for the young patient to explore with a counselor. Also deferred are issues related to the father of the baby. These are not broached by the pediatrician, who recognizes the extreme sensitivity of this subject. Continuing in or returning to school often depends upon parental guidance and peer support. While parents frequently find themselves capable and willing to assume an active caretaking role with a grandchild, the decision to do so may require radical adjustment and professional counseling.

For the physician who is committed to providing comprehensive pediatric care to the child and family, the foregoing situation exemplifies the ways in which a need may be recognized and met. Having become a trusted friend over many years, the pediatrician has an opportunity to listen, support, mediate, and intervene. Knowing the parents' current status is also of inestimable value. Very well established in a law firm, the father is financially capable of assuming the added expenses, which could be considerable. The mother, newly returned to work, has sufficient interest and investment outside her home to help her to maintain the family's equilibrium in the difficult months to come. The pediatrician reflected with irony that the husband and wife had each been struggling with anxieties related to their own mid-lives and were only recently resolved in their decision to continue their marriage. The effect of their daughter's pregnancy on this tenuous adjustment is a critical element in the present crisis. This issue, together with the question of the interrelatedness of the two disruptions, is best left to family counseling. At this juncture, the presence, interest, and support of the pediatrician have served to hold the family and have significantly ameliorated a difficult family problem.

In a third interview, the pediatrician met with a married couple in their mid-30's who were mutually overjoyed about the coming birth of their first child but were becoming apprehensive as the wife's pregnancy neared term. At the outset, the pediatrician's role in this situation, in which there were few *stated* concerns, was far less apparent. During this first interview the couple referred in some detail to the many obstacles they had had to overcome: delayed conception, concern about fertility, and subsequent struggle with the question of normalcy. The decision for amniocentesis had implied a willingness to terminate the pregnancy that also required resolution. They view this last trimester as a plateau to be thoroughly enjoyed. Yet, as they spoke, they betrayed an anxiety inconsistent with their identified purpose of coming merely to introduce themselves. Encouraged to talk more freely about their expectations regarding the baby, they expressed their underlying concerns. Because of fear that amniocentesis would damage the fetus, they had canceled the scheduled procedure. Recently, a close friend had delivered a baby with Down's syndrome, revealing dramatically how much could still go awry for them too. Overcome with fearfulness, they had agreed not to seek reassurance from anyone outside and to avoid further discussions between themselves. Yet they found this increasingly impossible to do; the last two months loomed now as a time to be endured rather than enjoyed.

Following this hastily delivered narrative, the wife subsided into quiet sobs while her husband distractedly patted her shoulder. The pediatrician conceded how painfully they had been reminded of their vulnerability and sought to support and comfort them. Recognizing the futility of trying to dissuade people from their fears, she commented on how difficult the waiting must be. She spoke of the couples long-standing good health and the absence of birth defects in their family histories, seeking to reassure them in a gentle, reasonable manner. While offering to meet with them again, she suggested that they discuss their distress with their obstetrician. They asked instead that another appointment be arranged. The couple came a few times thereafter for supportive counseling and were able to recapture many of their earlier, secure feelings of anticipation. When the child was born, the parents' fears were allayed, and their beginning relationship with their child reflected the joy they had temporarily lost. A major facilitative factor in this instance was the couple's developmental stability in the middle of their early adult lives.

For the pediatrician, her ability in each of the three instances to develop an alliance and to plan successful intervention derived from her dedication, training, and experience. An understanding of human development provided the theoretical foundation upon which she had developed her approach and skills.

PROPOSITION TWO

The interactional effects between parent and child are stage-dependent. Each stage of the child's development presents its own goals and challenges, which are faced by parents of widely divergent experiences and developmental stages.

With the arrival of the first child, some of the life changes parents will experience are immediately apparent. Yet the degree to which their lives will be modified and the child's power to transform and transcend all other aspects of their development may be blissfully obscured. Until very recently the initial impact has been felt more directly by the mother and, by way of her response, indirectly by the father. The trend toward fathers' participating more actively in preparing for childbirth and child care has contributed to their feeling less isolated and excluded from the early birthing process.

Because of biological design, the intimate awareness of the child growing within her provides the mother with a head start in adjusting to the idea of her baby as a real person. She has, in fact, an actual opportunity to practice being a parent. An expectant mother may begin almost immediately to provide care for her unborn child by caring for herself, by tolerating discomfort, and by giving up certain pleasures such as smoking or drinking coffee. Having gradually detached herself somewhat from daily concerns in order to "plan for the baby," she pursues a state of psychological preparedness directed toward providing nurturance and fostering reciprocity.

Without the physical sensations and discomforts, but also deprived of the positive emotional and psychological experiences, the father will, in a sense, always feel some sense of exclusion. Yet during the pregnancy, he too struggles with his own psychological upset. Concerns about a wife's successful negotiation of pregnancy and delivery, financial stress created by adding another family member and, perhaps, loss of a wife's income, or the need to work out child care arrangements are some of the issues he may face. Increasingly deprived of the physical closeness of his wife as the pregnancy progresses, he may perceive the infant as divisive and begin to dread its arrival. The prospect of future deprivation created by his wife's need to attend to their dependent infant has enough validity to create increasing tensions for him. Very often the father is caught in the dilemma of wanting the child born immediately and not wanting the child at all. Unable to admit the latter, he seeks to repress it along with other unresolved concerns.

The child's arrival can almost as readily stress parents and threaten their closeness as it can enhance their relationship. Clearly challenged is their ability to set aside the sense of totality they may have previously known: the totality of privacy, of freedom, and of preoccupation with their own goals and their concerns for each other. While putting first and making way for the wondrous product of their common bond can become a new totality for parents, it does not happen immediately. As in so many other aspects of adult life, the opportunity to grow by meeting responsibilities has more appeal in theory and outcome than in practice.

During the early neonatal period, the responsibilities may seem inordinate. Anxieties deriving from lack of experience, the tedium and demands of providing almost constant care, the exhaustion due to sleep deprivation, worry about feeding problems, and a cranky or colicky infant may make this period, in retrospect, one of only blurred recollection. Unable to create a comfortable environment or to soothe an infant, some mothers lose their sense of confidence and adequacy. Fathers who share caretaking responsibilities often feel similar frustration. Others, who had been involved deeply in their work, and are now left somewhat on their own, may become driven. Feeling deprived and abandoned, some fathers become angry and demanding, contributing further to the stress within the family. Any unusual crisis from outside, from work or extended family, adds further to the distress some parents experience.

Even under optimal circumstances, the developmental task of the neonatal period may seem to be one more of survival than of transition. Although with the birth of subsequent children the parents have the added concern of sibling adjustment and rivalry, they are increasingly expert and sanguine. Self-assured and purposeful parents are the most accomplished and successful throughout the parenting years. Regrettably, this expertise is frequently achieved only through hard-won struggles and too often at the expense of the first-born child.

Since all first-born children are recognized to be at some risk, pediatricians should be alerted to the significance of problems in the first neonatal period. Frequent or anxious telephone calls and emergency room or office visits may indicate the presence of a real problem such as postpartum depression or other serious parent-child disruption. Considering the capability of each individual to accept and to fill the parenting role, one realizes the degree to which his or her own experience and maturity contribute to the ability to parent. The young mother or father whose early childhood was one of emotional deprivation and neglect will require generous increments of nurturance and support; either parent may, in addition, require some professional help.

The ability to parent well is also affected by the parents' ages and developmental levels. A 22-year-old mother who had always longed for a family of her own was sufficiently overjoyed with the arrival of her first child that she could easily tolerate most stressful experiences in the neonatal period. Well adjusted and well informed about the demands a newborn makes by way of its total dependency, this young woman convinced her husband of their mutual readiness to become parents. Having already made a first career commitment in computer technology, the husband at 25 years is willing to take on further responsibility. By consolidating this couple's assumption of adult roles, the birth of a first child added stability to their lives. A couple in their early 30's found it similarly manageable and rewarding to provide care for their newborn. They were far more comfortable and relaxed parents of their third child by virtue of feeling experienced and successful as parents and more settled in their own development.

Since the baby's beginning emotional security rests upon the parents' ability to understand and to meet the earliest, nonverbal needs, the mother and father must be readily available and invested. The developmental task of the first stage of life is to achieve a sense of mutuality and reliability— the attainment of basic trust as Erikson so well defined it, wherein the meeting of all one's needs by the first caretakers is the primary condition from which evolves trust in one's self and trustworthiness in others (Erikson, 1950).

Unhappily, not all parents and babies begin their relationships under optimal circumstances, regardless of the positive, even eager attitudes parents may have. Some babies are born with physiological sensitivities, some with difficult or uneven temperaments; some may simply not live up to parental expectation. There are parents who respond well to placid, smiling, sleepy babies and parents who like their babies active, alert, and more wakeful. Most effective early parenting is achieved by those mothers and fathers who are able to adjust to the baby's temperament and needs, to tolerate disappointment, and to delay gratification. Pediatricians benefit from understanding that early difficulties rarely arise from a conscious decision on the part of parents and that the most helpful response to distressed and troubled parents is to accept and support them and to avoid making critical judgments.

It is not at all uncommon, for example, for a pediatrician to receive a frantic telephone call from an agitated parent relating that the baby has not slept in two days. Having become alerted to other signs of family tension in recent weeks, the pediatrician, after attending to the sleep problem, suggests an office visit with both parents within the week. Prior to the visit he or she reviews the family history and recalls that the young father has recently changed jobs, that the loss of the wife's income has created additional financial worry, and that both parents are struggling with a question about the wife's returning to work. Because of their inexperience and recent entry into a transitional period in their own development, the couple is manifesting normal, predictable stress. By providing them with ample time to describe the baby's irregular sleep patterns and making some suggestions about management, the pediatrician confirms his or her interest and support. Encouraging the young couple to discuss how things are going may, or may not, elicit a detailed commentary about present woes. Whether the pediatrician is offered an opportunity to address the problems directly may not be as critical as are the concern and availability implicit in his or her response. An alliance of this nature can create a sense of comfort and support for a husband and wife that will prove meaningful to their development as parents.

PROPOSITION THREE

Parents are frequently caught up in patterns of behavior with their children wherein they re-experience and re-enact an earlier conflict of their own as a child reaches the developmental stage in which the parent's conflict had originated. Such patterns can be, and often are, negative and circular in their functioning.

Having survived the neonatal period, many parents experience the remainder of their child's infancy as calm and pleasurable. They tend, under ordinary circumstances, to begin reinvesting some of their energies in earlier interests and considerations. A husband who has been feeling distracted and fragmented becomes more stimulated and productive in his work and more actively involved in projects related to the home. With the lessening of some of the baby's demands on her time and her body, a wife rediscovers herself as a separate person and as a wife. Perhaps either or both new parents, having felt at times regressed and infantile during the procreative period, become once again more fully aware of how far they have actually advanced in establishing themselves as mature adults. Separated by the demands and anxieties of producing and providing care to a newborn, they may now enjoy reunion, realizing again the happiness they find in each other.

Yet, as Fraiberg tells us, there are ghosts in every nursery. The ghosts to which she refers are the unresolved conflicts from a parent's own past that may "break through the magic circle in an unguarded moment, and a parent and his child may find themselves re-enacting a moment from another time with another set of characters" (Fraiberg, 1976). There is some pain and irresolution in all our childhoods; it is one reason why we not only forget details but are sometimes unable to recall entire periods from our early lives. Application of this defensive process is known clinically as "repression." According to Fraiberg, it is a form of repression, together with the mechanism of isolation, that children most commonly use to bury painful feelings. Both repression and isolation are unconscious activities that serve to banish anxiety and other painful affects from the child's consciousness. Fraiberg's theory holds that parents who have blocked the early sensations of pain associated with certain childhood experiences tend to identify more closely with the aggressive adult who inflicted the pain and to re-enact with their children the unpleasant and even cruel behaviors they had experienced. An extreme example is the father who, having been beaten by his own father, physically abuses his child.

Pediatricians may often encounter and be greatly puzzled by this phenomenon operating in a much milder form. Consider the inexplicable behavior of an irate mother who attributes consciously malevolent motivation to her smiling, cherubic 6-month-old daughter: "As soon as my husband and I sit down to dinner, we hear her screaming in her crib; she is always trying to keep us apart." Or another, who complains, "The baby wakes me up night after night; I think he's out to get me."

While it is relatively easy to recognize and understand that such statements represent distortion, it is difficult to help a parent realize how inaccurate or inappropriate his or her perception is. In fact, attempting to create instant insight by confronting directly can cause such acute distress that the parents decide immediately to change pediatricians. Even though the matter seems simple and straightforward to the pediatrician, it may be related to so sensitive and critical an area that a direct approach is the least felicitous. Forcing awareness should be guarded against, since it is seldom, if ever, productive and often jeopardizes an alliance. In almost all instances in which the disruption is mild, the pediatrician should begin by offering behavioral management techniques that may be effective in smoothing the way for improved parent-child interaction. If the problem appears intractable or seems to escalate, consultation with a mental health professional for the pediatrician or directly with the parent is indicated.

A situation with which pediatricians are most familiar often arises when a toddler begins to assert himself. A mother whose own development of autonomy was thwarted and whose relationship with *her* mother has been one of long-standing tension and conflict is particularly vulnerable to a child's emerging strong-mindedness. With increasing conviction she describes her interaction with her child as a "battlefield," failing to recognize, perhaps, her own overriding need to be victorious. Since the establishment of a continuing sense of self and a beginning sense of self-control is the developmental task of this stage for a young child, he must be prepared to defend his autonomy at any cost. In fact, it is a manifestation of psychological health for him to do so. Moreover, it provides the climate for future dyadic conflict, if left unresolved.

Parents report skirmishes around all the early basic functions that present potential conflicts: eating, sleeping, and toileting. As the narrative continues, it often appears that the negative interaction has begun to develop a life of its own. The parent is unable to withdraw from the battle or change the stakes; the child by this time is far too entrenched and perhaps, in a way, too gratified to yield. When the child's negativism becomes generalized, as it tends to do, the parent and child may even—like two adults—forget what the original fight was about. Interrupting such circular patterning is a difficult process; the pediatrician must rely on tact; on the strength of his or her alliances with both the parents and the child, respectively; and on a knowledge of innovative child management techniques. While avoiding direct reference to the parent's own underlying conflict (of which the parent is likely to be unaware), the pediatrician may succeed in clarifying what is happening in the relationship and in supporting parental efforts to reverse negative trends.

Clearly this pediatric problem exemplifies the sooner-the-better principle; if it is left unresolved

for too long, outside intervention may be necessary, such as enrollment in a special play group or nursery school or referral for counseling. Often, however, escalation can be avoided by a prompt response on the part of a pediatrician, who with empathy and sometimes humor can be truly effective by encouraging the parent's recollection of similar childhood frustration and thereby promoting self-evaluation. If unsuccessful in these efforts, the pediatrician should be equally prompt in recommending further intervention when indicated.

It should not be inferred that negative displacement by a parent always produces opposing behavior in the child, or that the pattern automatically becomes circular. Furthermore, such projective identification by a parent can occur at any time in the child's development with varying intensity and outcomes. Affected directly by the interrelatedness of a parent's and child's respective life-cycle stages (see Proposition Two above), it may be moderated or heightened accordingly. As the last child to leave her family, a young mother who continues to struggle with her own separation issues will have far more difficulty when her child first boards the school bus than will a woman in her mid-30's for whom the same landmark represents freedom to take courses or return to work. A 40-year-old father who has successfully established himself in the business world, thereby completing the Becoming One's Own Man phase (Levinson, 1978), may have less need to prove himself through his child's academic or athletic achievement than will perhaps a 26-year-old father who is dissatisfied with his own progression up the corporate ladder.

PROPOSITION FOUR

Parents bring from their own life experiences expectations and aspirations for themselves and their children of which they are not always aware. Yet these often unconscious motivations significantly influence the child-rearing process, with both positive and negative effects.

Family physicians and pediatricians may feel dismayed by much of the foregoing discussion if it seems to suggest that they must be all things to all parents. It is not likely, however, that they have reacted with surprise or unfamiliarity to the family interactional patterns and problems discussed so far. While physicians are perhaps not systematically attentive to such occurrences in daily practice, the complexity of the parent-child interaction is consistently apparent as a major determinant in the quality of pediatric care they are able to provide.

During the early years, the pediatrician hears most frequently about the conflict between mother and child. Since the mother serves as principal caretaker, rule-maker, and disciplinarian, it is with her that the infant, toddler, or preschooler naturally has the greatest opportunity to begin to resolve the critical issues of attachment, individuation, and separation and to acquire autonomy, independence, and a sense of competence. In two-parent families with young children in which the mother works outside the home, she most often works part-time or is involved in educational or training pursuits. Mothers who are actively engaged in full-time professional or business careers may have regular caretaking help; in these instances, too, husbands and wives have worked out special arrangements whereby fathers are far more available to share parenting responsibilities equally. Regardless of child-care provisions in the former cases, mother and child interact quite similarly to at-home dyads. However, the latter families, in which mothers have full time careers, tend to differ in that the children are more inclined to negotiate and resolve issues with either parent—whoever is there at the time or with whomever the issue and the temperamental match provide an impetus. Yet, in general, during the earlier stages of a child's development, pediatricians only rarely have an opportunity to meet with fathers unless an interview is specifically arranged.

In the majority of families, fathers at this stage of their development and the development of the family spend a considerable part of each day out of the house investing their energies in making a living. Often preoccupied with being productive, proving themselves, and attaining gratification from work, they seek to settle themselves in a stable position vocationally yet are open to discovering other options for making the best possible decisions and advancements. Many men feel a sense of comfort when they enter management level positions, while for others a move upward is accompanied by a renewed sense of urgency. Some fathers in the Age Thirty Transition phase described by Levinson (1978) find themselves moving smoothly along an "occupational path" and feeling freer to appreciate family and friends. Also, they are often the fathers who avow that they are just beginning to enjoy their school-age children who are now old enough to express ideas or to understand and share their interests. They fail to realize the degree to which they themselves are more available and the positive effects of that availability.

It may be that present-day societal demands are creating added stress for many fathers who are expected to share child-care responsibilities and be actively involved in the daily experiences of their offspring. To some extent in many households fathers can accept this role. In families where mothers are employed outside the home, paternal involvement often develops more naturally and perhaps, more readily, out of necessity. Dropping a toddler off at day care, shopping for groceries, and folding laundry are reasonable tasks for fathers in many families. Other fathers with less well-defined familial responsibilities may be somewhat remiss in attending to their children's early lives. Yet paternal investment does not derive solely from external pressure. Many fathers have their own reasons for wishing to be actively involved with their very young children. Early positive familial experience inspires a father to nurture and support his family readily and well. If the father's own childhood was harsh and deprived, he may resolve to prevent his children from having a similar experience. For a father whose parents structured and stimulated him toward academic achievement, athletic excellence, and a strong sense of self, his creation of similar child-rearing conditions is a naturally evolving process. Another father who thinks of himself as a victim of ignorance or disinterest on the part of his parents, which led to school failure and inadequate vocational training, is determined to provide his children with "every opportunity." Parental motivation derives from many sources: from a marriage partner; from society; from necessity; and from one's own expectations, experiences, and ideals. It varies according to the child's developmental stage and the parents' developmental tasks and conditions, e.g., familial, political, and economic. Having aspirations and expectations for one's children is natural and universal. Many parents, aware of their child-directed goals, accurately perceive their motivation and seek, in a reasonable fashion, to help their children to develop and to achieve accordingly (as in the examples just provided).

Less consciously and with too little restraint, some parents are engaged with their children in strivings and maneuvers related to their own unmet needs from earlier periods of development. There is generally a further lack of awareness that the pursuit of certain qualities or accomplishments for the child reflects the parent's wish to undo or assuage old hurts, humiliations, and feelings of inadequacy. In some such situations, the essence and true potential of a child are obscured because of the endless adjusting and readjusting of a parental template. The degree of success achieved by a parent in this pursuit varies widely, since it depends, of course, upon many factors: the natural endowment of the child, the cooperation or interference of the other parent, the available resources and other environmental conditions, the strength of the parent's determination, and the intensity of the need from which it arises.

A bright preschooler may be taught to read and to memorize many facts. A sturdy kindergarten youngster can begin skating lessons at age 5 and go on to play hockey by age 7. Piano lessons or French tutoring for a talented first grader may be within the child's level of ability and energy. Yet a mother who plans to have her offspring begin school as a reader may suffer and create suffering for her intellectually average child when each experiences frustration and failure in pursuing this goal. Some children are poorly coordinated, and learn to skate only after long hours of practice at an age when they are sufficiently mature physically and motivated personally to do so. Musical ability and language facility are not inherent in all children. In their absence, the achievement of modest expertise depends upon the child's effort and investment.

Predictably, a child's entrance into school precipitates the emergence of overdetermined aspirations in many families hitherto uninvolved in "pushing" their children. For both children and their parents, school entry symbolizes a formal coming of age. Going to first grade, perhaps with older siblings and certainly with older schoolmates, contributes to children's awareness that they will now be judged on their own merits. While many have attended nursery school and most, kindergarten, entering elementary school represents the opening phase of a complicated evaluative process whereby the children and their parents will be measured by an independent agency for many years to come. The children, alone and on their own, must measure up to an additional and even sterner "agency"—a peer group.

Since it is fairly commonplace for children today to enter day care as infants and toddlers and have ongoing collective experiences throughout their early years, such demarcation at grade one may seem an overstatement. There are, however, specific, well-recognized tasks for first graders with which parents are identified. Acquiring basic learning skills, beginning competitive sports, getting along with age-mates, and behaving well in the classroom are perceived by many parents as the start of serious business; for some, it is even the beginning of college and career preparation. Setting inordinately high standards and responding harshly to mediocrity or failure is extreme but not uncommon parental behavior. It can be quite difficult for many parents to discover just the right approach—of feeling interested, supportive, and helpful while avoiding overinvestment, nagging, anger, and control. It is even more difficult to maintain such an attitude when a child lacks motivation, refuses to study, demonstrates a learning disability, or has trouble socializing.

Other problems arise between the parents and the school deriving from lack of communication and misunderstanding. At times, both school personnel and parents may displace anger, each on the other, as a result of their own sense of frustration and failure. Pediatricians who have ongoing relationships with schools—public, private, and special—are often able to mediate and to be effective in having children evaluated, referred for remediation, and counseled. They often serve another very meaningful, mediating role in helping parents to understand the child's perspective. Intervening effectively in this regard may be relatively simple and straightforward; it may, on the other hand, require patience and an extended time period.

Effects of sex typing, documented by Kagan (1964) continue to have an impact on children both at home and at school (see also Chapter 10). Many mothers and teachers maintain a double standard for school-age children, expecting girls to be quiet, alert, and compliant while allowing their male age-mates to be at least slightly noisy, distracted, and obstreperous.

Consider the daughter for whom parental goals are that she be assertive, competitive, and academically superior. In many typical classroom settings, this daughter is destined to be somewhat at odds with her teachers and her peers. She may, however, press forward, seemingly unaware or unconcerned with the "different" way in which she is perceived. Closely identified with her parents' values and intellectually talented, she is determined to excel. Her parents, invested both in their own professional careers and in each other, make the time and the effort to support her. As her parents work hard and enjoy greater achievement, their daughter's motivation increases. In many ways these parents are fully cognizant of their goals for their only child and are convinced that they are directing her wisely. Coming themselves from similar familial and educational backgrounds, theirs is a narrow, rather rigid perspective. They are, moreover, so accustomed to their daughter's complete cooperation that they have never considered the possibility of her having a developmental disruption. It is, therefore, not remarkable that they fail to attend to some changes in her eating habits or to note that she has lost 10 pounds. It is noted, however, by her pediatrician on a routine checkup. Both he and they become alarmed by her "dieting," by her denying the presence of a problem, and by the many ways in which she meets the "good girl" criteria of the anorexic. The pediatrician recommends a psychological evaluation to which the parents immediately agree, as they do with a subsequent referral for family therapy. Diagnosed and treated at an early stage, the physical symptoms of the pre-anorexic condition abate fairly quickly. The underlying conflicts, reflective of the daughter's adolescent turmoil, respond somewhat more slowly to family therapy, since this requires significant alterations and shifts in the family structure.

The preceding example is relevant to the practice of pediatrics because it illustrates the need for flexibility and sensitivity in deciding about intervention. Knowing well the parents' determination and single-mindedness, the daughter's potential for developmental difficulty, and the absence of actual symptomatology, the pediatrician has long refrained from exceeding the role assigned to him by the parents. By scheduling more frequent "regular" visits, he was able to follow his young patient's progress, alert to any signals of disruption. When the need arose, his approach was direct and straightforward and his recommendations were put forth firmly and promptly.

PROPOSITION FIVE

A major challenge for parents is to love and bestow affection and caring on significant family members from two distinct generations: their parents and their children. They must also prepare to yield the intensity of their attachments as the parents grow old and the children grow away.

This final proposition relates in part to the many problems that arise as parents and children confront and are confronted by the crisis of adolescence. Although reminiscent of the child's very early attempts as a toddler to separate and to establish autonomy and independence, normal adolescence differs in that it is experienced as an upheaval with far more frequency than are the earlier phases of parenting. It has, in fact, been stated that the absence of conflict between parent and child at this time may signal the potential for very real difficulty (as was illustrated in the previous example). Many more parents seek support and guidance at this stage of parenthood than at any other. As children change in size and contour, they reveal new interests and longings. Eschewing parental advice and assistance, they seem to immerse themselves in the values and standards of the peer group. A change in their cognitive abilities leads to a qualitative change in the way they are able to hypothesize and to think conceptually. Acquisition of this cognitive stage of *formal operations*, identified and described by Inhelder and Piaget (1958), can lead to the adolescents' overvaluing their cognitive solutions, believing that they have all the answers and needing to adhere with tenacity to a single point of view.

Parents, well able to describe adolescent behavior, are at a loss as to how it should be managed.

They consult the family pediatrician and are encouraged by his interest and his explanation about the healthy need of adolescents to separate and to become independent. Stressing the fact that the course of adolescence need not be smooth helps parents acknowledge their own negative feelings of hurt, anger, and disappointment. They are helped also to understand that the adolescent child, struggling with ambivalence about growing up, uses conflict and disruption as a means of moving away from parents so as to become autonomous. Encouraged to continue to set limits, to enforce rules, and to express their anger, parents should be advised of the need to provide continued caring, investment, and protection even though adolescent children seem to reject, defy, and disown them. Often tempted or goaded to abdicate their responsibility during this period, parents must realize how critical it is that they remain stable—even stolid at times—in the face of such unpleasant and turbulent interaction.

A far more challenging task for the pediatrician is to help parents become aware of those aspects of the conflict to which they are contributing. This awareness is often relatively inaccessible, since it approaches areas of vulnerability that parents are unwilling and unprepared to address. Taking refuge in their feelings of anger and disappointment with the child's behavior is a poignantly successful way to avoid the true issues. In fact, the more actively parents struggle, and the more energy they unknowingly invest in maintaining a state of conflict, the less sadness and pain they have to experience about the child's growing up and eventually leaving home. For parents, being needed and needing are closely related, as are separating and being separated from. Trying to prevent children from making the same mistakes they made, keeping them secure and safe, and monitoring their motives and whereabouts are necessary activities with which parents protect themselves from deeper and more searing pain. Asking a parent to affirm and to encourage the child's wish for independence may be unconsciously experienced as being asked to participate in the amputation of one's own limb.

As painful and conflicted as they may be, parents' experiences with a child's adolescence are heightened further by other circumstances and conditions. It is certainly unfortunate that adolescent turmoil occurs coincidentally with the time when parents reach a critical period in their own lives. Recognition of the many aspects of the Mid-life Transition has illuminated the stresses that arise for a parent who is also parenting an adolescent child. Levinson (1978) places the Mid-life Transition phase of the adult male between ages 40 and 45 years. Both he and Elliott Jaques (1965) addressed themselves to this period of the middle years. Jaques, the first theorist to posit the concept of a mid-life "crisis," differs from Levinson in the age span of the transition, which he identifies as starting in the late 30's and continuing for several years. Like Levinson, he sees the core of the crisis to be the experiencing of one's own mortality. Levinson, moreover, is rigorous in his view that while a modest decline in functioning may occur with some frequency, it is not developmentally normal. Rapoport (1980) hypothesizes a different time determinant for the "mid-life phase," fixed not by chronology but by the period between the first and last child's reaching adolescence.

Regardless of the exact age when it occurs, all parents reach a time when they realize their mortality, and a time when their children reach adolescence. In addition, the parents' position often becomes a mid-generational one, in which the mortality they must confront first is that of their own parents. There is a time when they must perceive and accept *their* parents' decline and assume responsibility for their care. It is critical to recognize how rapidly the patterns shift for parents of an adolescent at this juncture. Adjusting to, and accepting, the psychological loss of aging parents represents a final yielding of dependency, with the sad and reluctant knowledge that one's parents are ill or growing old and that a time will come when they will no longer be there. Caught between the loss of support from their parents and the beginning loss of their adolescent children, parents may be unable to find solace in their relationship with each other. While often financially more secure than at any other time in their lives, parents are faced with new responsibilities and demands at a period anticipated as a respite. Many marriages reflect the stresses of this period of transition and are vulnerable to disruption as well. For women, recent changes in occupational role patterns have created a new philosophy and new opportunities that may relieve some of the familial tension of this period. On the other hand, in some marriages, the entry of a wife into an interesting and engaging job may create added dissatisfactions for the husband.

There are endless variations in the human life-cycle drama of adults sharing their lives with the lives of their aging parents and their growing children. Stage settings and scene changes are often beyond the control of the members of the cast. Happily, many parents achieve career success, acquire status, gain community respect, and grow in wisdom and perspective. Unhappily, there are those who lose jobs, fail to meet self-actualizing goals, suffer ill health, and experience marital disruption and dissolution. Given the trials and tensions of negotiating the many stages of human development, the degree of success achieved by so very many parents is impressive and reassuring.

With the many shifts and changes that occur during parenthood, mothers and fathers often encounter disappointments. However, these are offset by the many satisfactions and rewards children bring to their lives. As the children grow older, they cease to be a major responsibility but continue as a source of interest, affection, and caring. Parents look forward to their children assuming adult status and to their establishing homes and families of their own. The arrival of a grandchild confirms the parents' lives and closes the circle of creativity to which they have made a lasting contribution.

As a developmental stage, parenthood presents a series of challenges for adults. When successfully resolved, these challenges contribute to their psychological integration and maturity. The adjustments and adaptations of the pre-parenting phase exemplify a potential for maturation as the psychobiological factors (conception and pregnancy) interact with the new, reality experiences of a specific period (being pregnant and preparing for birth). In examining the effects of different stages of development upon the parental role, it is readily apparent that parents differ in their interactions and experiences with their children by virtue of their own adult development. Parents change as they negotiate adult transitions and stages. A young father, just beginning to establish a home and a career, will interact very differently with his first child than he will, nine years later, with his third child, when he is enjoying some security and success. He may vary in the amount of time he invests, the extent of his demands and expectations, and the intensity of his emotional investment.

Throughout the life cycle, progressive and regressive aspects of developmental experiences occur for all adults. Challenges arise for them when a critical issue in a child's development revives a developmental conflict in the parent. It has been noted that this often presents an opportunity for parents to resolve an earlier conflict as they help the child negotiate developmental tasks. In addition, many dreams for both parent and offspring may be realized by a highly motivated and talented child. There are advantages for siblings, too, when other family members succeed; many benefits are, therefore, accrued reciprocally. Similarly, families suffer stress when any one member is in difficulty. The resolution of such stress contributes very often to the growth of all family members.

Finally, and often much later in the life cycle, parents seek to resolve a major issue from their own early development—that of separation. In some place and time, all adults must continue to negotiate the one task with which they have struggled all their lives. As they accept and resign themselves to being left, finally, by their parents, they are required to foster and encourage in their children the acquisition of independence, thereby helping them also to grow up and to leave. When this occurs, the compensation for adult parents can be the arrival of grandchildren to complete for them the circle of their generativity.

M. PATRICIA BOYLE

REFERENCES

Erikson, E.: Childhood and Society. New York, W.W. Norton, 1950.

Erikson, E.: Insight and Responsibility. New York, W.W. Norton, 1964.

Fraiberg, S.: Clinical Studies in Infant Mental Health. New York, Basic Books, 1980.

Inhelder, B., and Piaget, J.: The Growth of Logical Thinking from Childhood to Adolescence. New York, Basic Books, 1958.

Jaques, E.: Death and the midlife crisis. Intern. J. Psychoanalysis 46:203, 1965.

Kagan, J.: Acquisition and significance of sex-typing and sex-role identity. In Hoffman, L. W., and Hoffman, M. L. (eds.): Review of Child Development Research. New York, Russell Sage, 1964.

LeMasters, E. E.: Parenthood as crisis. Marriage Family Living 19:352, 1957.

Levinson, D., et al.: The Seasons of a Man's Life. New York, Alfred A. Knopf, 1978.

Lidz, T.: The Person. New York, Basic Books, 1976. (Revised edition, 1968.)

Rapoport, R., Rapoport, N., and Shelitz, Z.: Fathers, Mothers and Society. New York, Random House, 1980.

Rossi, A.: Transition to parenthood. J. Marriage Family 30:26, 1968.

Vaillant, G.: Adaptation to Life. Boston, Little, Brown and Co., 1977.

5

Pregnancy, Birth, and the First Days of Life

Mental growth and development of an infant is often greatly affected by the pregnancy and birth experience and by events during the days and weeks shortly after birth. This chapter will explore what is now known about this time in the life of the infant, the parents, and the family and will emphasize what interventions aid the maturation of the family and what alters its development.

PREGNANCY

Pregnancy for a woman has been considered a process of maturation, with a series of adaptive tasks, each dependent upon successful completion of the preceding one. There are two general time periods during the pregnancy and another in the neonatal period during which a wide range of stressful factors may profoundly influence a woman's subsequent mothering behavior and ultimately the developmental outcome of her child. Behavioral changes in the mother during pregnancy have been described in detail.

Many mothers are initially disturbed by feelings of grief and anger when they become pregnant, because of factors ranging from economic and/or housing hardships to interpersonal difficulties. However, by the end of the first trimester, most women who initially rejected pregnancy come to accept it. Characteristically, in this initial stage the mother identifies the growing fetus as an integral part of herself.

The second stage consists of a growing perception of the fetus as a separate individual, usually beginning with her awareness of fetal movement. After quickening, a woman will generally begin to have some fantasies about what the baby may be like, attributing to him or her some human personality characteristics and developing a sense of attachment and value. At this time, further acceptance of the pregnancy and marked changes in attitude toward the fetus may be observed. Unplanned, unwanted infants may now seem more acceptable. Objectively, the health worker will usually find some outward evidence of the moth-

er's preparation by such actions as the purchase of clothes or a crib, selecting a name, and arranging a space for the baby.

It has been difficult to assess which factors determine the parenting behavior of an adult human who has lived for 20 to 30 years. A mother's and father's actions and responses toward their infant derive from a complex combination of their own genetic endowment, the way the baby responds to them, a long history of interpersonal relations with their own families and with each other, past experiences with this or previous pregnancies, the absorption of the practices and values of their cultures, and—probably most importantly—how each was raised by his or her own mother and father. The mothering or fathering behavior of each woman and man, his or her ability to tolerate stresses, and his or her need for special attention and support differ greatly and depend upon a mixture of these factors.

Figure 5–1 illustrates the major influences on parental behavior and the resulting disturbances that we hypothesize may arise from them. Included under parental background are the following:

1. Parent's care by his or her own mother
2. Endowment or genetics of parents
3. Practices of the culture
4. Relationships within the family
5. Experiences with previous pregnancies
6. Planning, course, and events during pregnancy

Included under care practices are the following:

1. Vulnerable child syndrome
2. Disturbed parent-child relationships
3. Some developmental and emotional problems in high-risk infants

By the time the infant is conceived, some of these factors are already established in the parents—parenting they received as infants, the practices of their culture, their endowments, and their relationships with their own families. Although the effects of these particular determinants were once thought to be fixed and unchangeable, it has been observed that their impact may be altered, both favorably and unfavorably, during the crisis

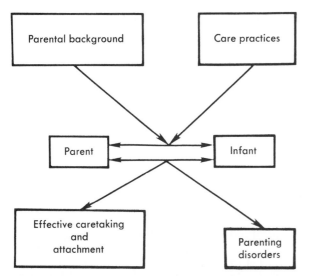

Figure 5–1. Major influences on parent-infant attachment and the resulting outcomes. (From Klaus, M. H., and Kennell, J. H.: Parent-Infant Bonding. St. Louis, The C. V. Mosby Co., 1982, p. 37.)

of birth. Parenting behavior and the parent-child relationship are also influenced by such factors as the attitudes, statements, and practices of the nurses and physicians in the hospital; whether the mother is alone for short periods during her labor; whether or not there is separation from the infant in the first days of life; the nature and temperament of the infant and whether he or she is healthy, sick, or malformed.

The actual process of parent-to-infant attachment or bond formation is not yet completely understood, but through a diversity of observations we are beginning to piece together some of the various phases. Events that are important to the formation of a mother's bond to her infant are

Prior to pregnancy: Planning the pregnancy
During pregnancy: Confirming the pregnancy
  Accepting the pregnancy
  Fetal movement
  Accepting the fetus as an individual
After birth: Birth
  Seeing the baby
  Touching the baby
  Giving care to the baby

Cohen (1966) suggests the following questions to learn the special needs of each mother:

1. How long have you lived in this immediate area and where does most of your family live?

2. How often do you see your mother or other close relatives?

3. Has anything happened to you in the past (or do you currently have any condition) that causes you to worry about the pregnancy or the baby?

4. What was your husband's reaction to your becoming pregnant?

5. What other responsibilities do you have outside the family?

It is important to inquire about how the pregnant woman was mothered: Did she have a neglected and deprived infancy and childhood, or grow up with a warm and intact family life?

Australian primigravid women were interviewed in an effort to explore their feelings about the pregnancy. At 8 to 12 weeks' gestation, 70 per cent of these women said they couldn't imagine the fetus or believe that it was really there; to them, the fetus was not a real person. A minority (30 per cent) did think of the fetus as a person but one about whom they expressed anxieties, such as the possibility of abnormality or fear of miscarriage. They predicted severe grief in the event of a miscarriage. Four times as many women talked about the fetus as a sort of animal. After the interview, the mothers were asked to draw an image of the fetus. During the 8 to 12 week gestation period, the fetus was generally represented as shapeless and formless, but as the pregnancies progressed, the images developed a more human form. Those mothers who showed early feelings of attachment usually came from larger families and were women who had worked in nursing or teaching. However, in cases where there were severe symptoms of pregnancy, or when the woman's husband was not interested in the fetus or did not provide his wife with emotional support, the process was inhibited.

Recent information from Japan has now revealed that in the uterus the fetus is responsive to light and sound as well as to some of the mother's extreme changes in emotion. In late gestation, if a mother is concerned that the infant is not moving, a small transistor radio placed on the abdomen will usually make the infant move. The environment during gestation also affects the response after delivery. If an infant gestates in utero for the first six months near the noisy Osaka airport, after birth noisy planes landing over their heads will awaken only 5 per cent of the infants. If they begin living near the Osaka airport in the sixth month of gestation, after birth 55 per cent will awaken and cry when a plane flies over their home.

LABOR: THE IMPORTANCE OF CONTINUING SUPPORT

Of the 150 human cultures studied by anthropologists, in all but one a family member or friend, usually a woman, remained with a mother during labor and birth. In 98 per cent of the cultures studied there was a tradition of postpartum confinement of the mother and baby together. In our own culture, before childbirth moved from the

Table 5–1. **PERINATAL PROBLEMS FOR MOTHER-NEWBORN PAIRS IN FIRST STUDY**

	No Doula		Doula	
	%	*No. of Mothers*	%	*No. of Mothers*
Cesarean birth	27.3	26	18.7	6
Meconium staining	25.3	24	9.4	3
Asphyxia	5.3	5	0.0	0
Subtotal	57.9	55 (p <0.01)	28.1	9
Drugs and forceps	21.0	20 (p <0.30)	9.3	3
Total	78.9	75 (p <0.001)	37.4	12

home to the hospital, family members frequently provided active support to the laboring mother, often with the assistance of a trained or untrained midwife. In the past 10 years in the United States, more fathers, relatives, and friends have been allowed into labor and delivery rooms, yet a considerable number of mothers still give birth without the support or presence of family members or close friends. In 1962, Newton and Newton reported that mothers who were relaxed and had good emotional relations with their attendants during labor and birth were more pleased with the first sight of their babies, but there has been little systematic study of this issue since then.

Recent observations by our research group support and extend these observations and suggest that there may be major perinatal benefits from constant human support during labor. We studied the effects of a supportive lay woman ("doula") on the length of labor and on mother-infant interaction after delivery in healthy Guatemalan primigravid women. Initial assignment of mothers to the experimental (doula) or control group was random, but controls showed a higher rate (p <0.001) of subsequent perinatal problems (e.g., cesarean section and meconium staining) (Table 5–1). It was necessary to admit 103 mothers to the control group and 33 to the experimental group to obtain 20 in each group with uncomplicated deliveries. In the final sample, the length of time from admission to delivery was shorter in the experimental group (8.8 versus 19.3 hours, p <0.001). Mothers who had a doula present during labor

were awake more after delivery (p <0.02) and stroked (p <0.001), smiled at, and talked to their babies more than the control mothers.

Prompted by the results of this first study on the short-term benefits of a doula, we designed a second study to explore the long-term advantages of having a doula during labor and birth. Four hundred seventeen primiparous, full-term, healthy Guatemalan mothers in early labor were randomly assigned either to the doula or the control group. Mothers who gave birth without perinatal problems were then randomly assigned to early (first hour) or late (2 to 3 hours) suckling. Research workers who did not know the previous experience of the mothers made home visits to the women in both groups at one, three, and six months to assess feeding practices, to evaluate the infant's health status, and to make anthropometric measurements of the infant. Table 5–2 demonstrates that significantly more perinatal problems (cesarean birth, meconium staining, fetal distress) occurred in the control group (59.1 versus 26.8 per cent, p <0.001). This replicates the findings of the first study. The presence of the doula or early suckling did not affect the length of breastfeeding or anthropometric measurements.

These findings point to the importance of human companionship during labor and birth. The quality of care a woman receives during this vulnerable period is crucial to her subsequent evaluation of the experience, to her later maternal behavior, and to her self-concept. Benefits of supportive care include significantly shortened labor,

Table 5–2. **PERINATAL PROBLEMS FOR MOTHER-NEWBORN PAIRS IN SECOND STUDY**

	No Doula		Doula	
	%	*No. of Mothers*	%	*No. of Mothers*
Cesarean birth	17.3	43 p<.002	6.5	11
Meconium staining	17.7	44	12.5	21
Asphyxia	3.2	8	2.4	4
Subtotal	38.2	95 p<.001	21.4	36
Pitocin	13.3	33 p<.001	2.4	4
Analgesia	4.0	10	1.2	2
Forceps	2.8	7	1.2	2
Other	0.8	2	0.6	1
Subtotal	20.9	52 p<.001	5.4	9
Total	59.1	147 p<.001	26.8	45

reduced perinatal problems, and heightened aspects of maternal behavior in the first hours of life.

Results of these investigations suggest that further studies of any intervention must ensure that the groups involved receive the same amount of attention and support from nursing and medical personnel. All too often, as a response to the many inhibiting aspects of a hospital environment, a woman may experience false or dramatically slowed labor. In our Guatemalan study we found that an untrained female companion (doula) provided comforting and friendly support at this time. Similar or even greater benefits might be expected when a family member (especially the father) or friend remains with the mother throughout labor and birth. This type of low cost intervention is a simple way of reducing length of labor and potential perinatal problems.

More studies on the effects of various delivery procedures and anesthetic interventions are required for us to better understand the attitude and behavior of mothers and the mother-infant interaction in the first hours and days of life. Most anesthetics clearly do depress infant responsivity and so may directly influence the first interchanges between mother and child. In contrast to countries in which little or no maternal analgesia or anesthesia is used, neonates in the United States, delivered under minimal analgesia and conduction anesthesia, need to be treated as postsurgical (or postanesthesia) patients for many hours.

FIRST WEEK OF LIFE

In vigorous high Apgar score term infants at the end of labor, the fetal heart rate fluctuates around a baseline rate with a rapid return to this baseline after either tachycardia or bradycardia. These wide swings mirror the intensity of the input reaching the fetus at the end of labor and the rapidity of his response to these stimuli. Following delivery there is an abrupt increase in heart rate. For a short time oscillations occur around a higher baseline, and then the rate begins to fall irregularly.

In infants with a suboptimal response to delivery and low one-minute Apgar scores, heart rate may remain at an extremely low or extremely high level and does not return promptly to baseline levels after the wide oscillations, indicating autonomic imbalance.

On delivery, if the infant is vigorous and reactive to the experience of being born, a characteristic series of changes in vital signs and clinical appearance takes place. These include a first period of reactivity, a relatively unresponsive interval (or sleep period), and another period of reactivity (Fig. 5–2).

During the first 15 to 30 minutes after birth (first period of reactivity) the normal infant with an Apgar score of 7 to 10 will be vigorous and highly responsive due to the numerous stimuli he has been subjected to during the labor and delivery process. During the first 60 minutes of life, he spends up to 40 minutes in a quiet alert state in which he is unusually responsive. This is often the longest period of this state during the first four days. In this state, he can turn his head toward sounds, follow a face, and be taught to mimic.

The infant exhibits changes that are at first predominantly sympathetic, including tachycardia (mean peak heart rate is 180 beats per minute at 3 minutes of age), with some lability of the heart

Figure 5–2. Summary of physical findings in normal transition as seen in the first 10 hours of extrauterine life of a representative high–Apgar-score infant delivered under spinal anesthesia without premedication. (From Klaus, M. H., and Fanaroff, A. A.: Care of the High Risk Neonate. Ed. 2. Philadelphia, W. B. Saunders Company, 1979, p. 55.)

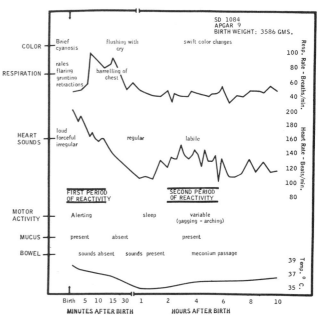

rate. An active cry initiates filling and expansion of the lungs. Lung fluid remaining after the "squeeze" through the birth canal is rapidly reabsorbed. As a result, rapid and irregular respiration (ranging from 60 to 90 per minute), transient rales, grunting, flaring of alae nasi, and retraction of the chest give evidence of the establishment of pulmonary respiration in the neonate. There is a falling body temperature with increased activity and increased muscle tone and alerting exploratory behavior. Characteristic reactions and responses with such behavior include nasal flaring or "sniffing" unrelated to respiratory activity; movements of the head from side to side; spontaneous startles and the Moro reflex; grimacing; sucking; chewing and swallowing; pursing and smacking of the lips; tremors of the extremities and mandible; opening and closing of the eyelids; short, rapid, jerky movements of the eyeball; outcries and sudden onset and cessation of crying. Bowel sounds become evident as the parasympathetic nervous system activates peristaltic activity in the intestine and the bowel begins to inflate as the baby swallows air. Saliva production is also increased by parasympathetic stimulation, resulting in increased mucus in the mouth. Brief periods of apnea and sternal retraction are not unusual during this period. This massive reaction dissipates rapidly, and after this first period of reactivity (usually between 10 and 60 minutes of age), heart and respiratory rates decline. The diffuse, apparently purposeless motor activity reaches a peak and then diminishes, and the infant passes into an unresponsive or sleep period.

With return of tonus to normal and diminished responsiveness, color should be excellent. Rapid respirations without dyspnea should not be alarming at this time. An increase in the anteroposterior diameter of the chest (barreling) may be noted accompanying the periods of shallow, rapid respiration. The chest, however, is not fixed in this position; the barreling disappears promptly with any change in respiratory pattern if the infant is handled or begins to cry spontaneously. It recurs with resumption of rapid, shallow, synchronous breathing. The abdomen should be rounded and bowel sounds audible.

Peristaltic waves beginning in the left upper quadrant of the abdomen and moving from left to right (gastric peristalsis) may occasionally be visible during periods of quiet activity or sleep. Small amounts of watery mucus may be visible at the lips. General responsiveness declines and the infant sleeps. Heart rate at this time (average, 120 to 140 beats per minute) is relatively unresponsive. During this sleep period, spontaneous jerks and twitches are common, but the infant quickly returns to rest.

After this sleep period, the infant enters a second period of reactivity (between 2 and 6 hours of age). Responsiveness returns and may become exaggerated. This is not unlike the reactivity of a patient following anesthesia. The infant again exhibits tachycardia, brief periods of rapid respiration, and abrupt changes in tonus, color, and bowel sounds. Oral mucus may again become prominent, and gagging and vomiting are not unusual; the infant becomes more responsive to exogenous and endogenous stimuli, and heart rate becomes labile. The bowel is cleared of meconium. Some infants exhibit waves of heightened autonomic activity. Wide swings in heart rate (bradycardia to tachycardia) are seen, along with the passage of meconium stools and the clearing of mucus, vasomotor instability, and irregular respiration with apneic pauses. This period may be brief or may last several hours. As it diminishes, the infant appears to be relatively stable and ready for feeding.

The sequence of clinical behavior just described is common to all newborns after birth regardless of gestational age or route of delivery. However, the time sequence of changes is altered in infants who are immature or who have demonstrated difficulty in establishing respiration promptly on delivery (low Apgar score infants).

CLINICAL EVALUATION

The state of quiet alertness is necessary to obtain optimal response. An important aspect of clinical evaluation and care of the neonate is to observe and consider the behavioral state of the infant. The "state" or "pattern" of behavior refers to the infant's overall level of functioning at any given time—ranging from deep sleep to awakeness, activity, and crying.

Peter Wolff designed the first descriptive rating scale for the term infant and defined six states, while Heinz Prechtl omitted the drowsy state, regarding it as a transition between states, and defined five states (Table 5–3). Others describe coma as a state. Any of these behavioral classifications can be used clinically for describing the infant. States can also be differentiated by record-

Table 5–3. STATES OF ALERTNESS*

Prechtl's	Wolff's
1. *Deep sleep:* Eyes closed, regular respiration, no movements	Same
2. *Active sleep (REM):* Eyes closed, irregular respiration, small movements	Same
3. *Quiet alert:* Eyes open, no movements	Drowsy
4. *Active alert:* Eyes open, gross movements, no crying	Quiet alert: No movements
5. *Crying (vocalization):* Eyes open	Active alert: Gross movements, no crying
6. - - - - - - -	Crying (vocalization)

*Adapted from Wolff (1959) and Prechtl (1977).

Table 5–4. RESPONSE TO STIMULATION IN DIFFERENT STATES OF ALERTNESS

Stimulus (Function)	Deep Sleep	Active Sleep (REM)	Quiet Alert
Visual attention and pursuit	−	−	+ + +
Auditory (hearing)	+	+ +	+ + +
Reflexes:			
Knee	+ +	−	+ +
Ankle clonus	+ +	−	+ +
Moro	+ +	−	+ +
Palmar grasp	−	+	+ +
Plantar grasp	−	+	+ +
Babinski	+	+	+

ing physiological measurements including respiration, heart rate, eye movements (EOG), and electromyography (EMG).

It is important to stress that although the states seem like a continuing spectrum, they differ qualitatively, with distinct types of organization and brain center control. They are also relatively stable and, with the exception of short transitional periods, recur in regular cycles during the day and night.

Practically every behavior and body function of the newborn depends on state and the stability and control of this state. Important examples are autonomic functions, such as heart rate and respiration. Even the presence of simple reflexes during a routine neurological examination will depend on whether the infant is awake or asleep and what type of sleep he is in (Table 5–4).

Cognitive function in the newborn period can best be assessed in the quiet alert state. For a sick newborn, the state of quiet alertness cannot always be achieved.

The Brazelton Neonatal Behavioral Assessment Scale is useful for testing the whole spectrum of state and state control as well as the infant's individual characteristics. In the exam, a graded series of procedures are carried out starting with a sleeping infant and slowly arousing him; 20 reflexes and 26 behavioral responses are tested over 30 to 40 minutes, and these are rated on a 9-point scale. The infant is always rated on his best performance. This assessment scale is being increasingly utilized clinically not only to assess the infant's cognitive and neurological status but to

Table 5–5. SENSORY-PERCEPTUAL ABILITIES OF NORMAL NEWBORN

1. Identify mother's voice (change temporal pattern of suckling to hear mother's voice)
2. Follow object or face with eyes
3. Select visual pattern
4. Turn head to bell, to voice
5. Imitate facial expressions
6. Identify smell of mother
7. Movements entrained to adult speech
8. Intermodel matching
9. Discriminate taste, prefer sweet

demonstrate his marvelous sensory abilities to his parents (Table 5–5).

PARENT-TO-INFANT ATTACHMENT

Bonding is the term used for the process by which a unique relationship is established between two people. It is important to avoid any implication that the speed of this reaction resembles the epoxy bonding of inorganic materials. Although it is difficult to define this human relationship operationally, certain attachment behaviors—fondling, kissing, cuddling, and prolonged gazing—serve both to indicate its presence and to maintain close contact between the individuals involved.

The noun "bond" refers generally to the tie from parent to infant while "attachment" labels the tie in the opposite direction, from infant to parent though these words may be used interchangeably. The parents' bond to their child may be the strongest of all human relations, characterized as it is by gestation within the mother's body and the utter dependence of the infant after birth. This parent-infant tie is crucial to the survival and development of the infant. It is the major source for subsequent attachments and the formative relationship from which he draws a sense of himself. Finally, the power of this attachment enables parents to make the sacrifices necessary in the day-to-day care of their child.

Since the human infant is wholly dependent on his mother or caregiver to meet all his physical and emotional needs, the strength and durability of this attachment may well determine whether or not he will survive and develop optimally. We will begin to piece together components of the affectional bond between a human mother and her infant and to try to determine the factors that may alter or distort its formation.

FATHERS

The role of the father in early infancy has been largely ignored by American culture and its theorists. Recently, however, changes in birthing procedures, particularly a renewed interest in home births, have been accompanied by changing views of and a sharper focus on the father's role.

The powerful impact of a newborn on the father is best described by the term "engrossment"— that is, absorption, preoccupation, and interest. As the father's bond to his new child develops, so may his perception of the infant as "perfect." During this period many fathers also experience extreme elation and an increased sense of self-esteem.

Investigators taking photographs every second of a father holding his naked infant 15 minutes after birth noted an orderly progression of behavior. To begin with, the father touched the infant's extremities, then proceeded to fondle the baby with his fingertips, then to use his palms, and finally the dorsal side of his fingers. Father and infant demonstrated increased eye-to-eye contact. When compared to early contacts between mothers and infants, this father-infant behavior appeared to be surprisingly similar.

These findings are borne out by Parke (1974), who observed parents in three different situations at 2 to 4 days of age: the mother alone with the infant; the father alone with the infant; the father, mother, and infant together in the mother's hospital room (i.e., triadic interaction). His studies have revealed few significant behavioral differences between fathers alone with their infants and mothers alone with their infants.

Parke's studies show that fathers are just as responsive as mothers to infant cues; for example, both parents increase their rate of vocalization following sound from the infant, although they differ in behaviors used in this response. Mothers are more likely to touch the infant when he vocalizes, while fathers are more apt to talk rapidly. According to Parke, "these data indicate that fathers and mothers both react to the newborn infant's cues in a contingent and functional manner even though they differ in their specific response patterns." Fathers and mothers demonstrated similar sensitivity to the infant and, surprisingly, were equally successful in bottle-feeding the baby, judging from the amount of milk consumed.

In the triadic situation, the father plays a more active role than the mother—a contrast to the cultural stereotype of the father as a passive participant. In fact, in these situations where both parents were present, the mother's overall interaction actually declines, although she tends to smile at the infant more than does the father. All but one of the fathers studied by Parke were present for labor and birth, and this could be expected to produce a strong degree of father-to-infant attachment. In another study of similar design but in which fathers rarely participated in labor and birth, they still played a more active and

dominant role with increased holding, vocalizing, and touching. These facts indicate that the father is much more interested in, and responsive toward, his infant than our culture has previously acknowledged. As a result, Parke believes that the father ought to have extensive early exposure to his infant in the hospital where the initial bond is formed.

Studies in Sweden have contributed other observations including increased paternal care in the first three months of life in cases in which infants had an hour of eye-to-eye contact with their fathers and had been twice undressed by them in the first three days of life. These investigations have also taken advantage of different policies following cesarean birth in two maternity hospitals in Göteborg, Sweden, to study the effect of allowing fathers to handle newborns immediately after birth. The fathers in question were compared with another group who were permitted only to look at their infants. Three months later, in a play situation, the early-contact fathers demonstrated much more touching behavior toward their infants than did the noncontact group. They also held their babies more frequently *en face*. Although the fact that the noncontact babies were kept in incubators may have had a contributory effect, these findings suggest that early contact might have altered paternal behavior.

As a result of changes in our society and the subsequently modified expectations of young parents, there is a temptation to consider the roles of the father and the mother as increasingly the same in relation to their newborns. However, we tend to agree with Donald Winnicott (1964) that each parent has a separate and unique role. He noted that fathers "can provide a space in which the woman has elbow room." When so protected, the mother does not have to deal with her surroundings just at the time when she wants to be "concerned with the inside of the circle which she can make with her arms in the center of which is the baby." This period does not last long, but "the mother's bond with the baby is very powerful at the beginning and we must do all we can to enable her to be preoccupied with her baby at this time— the natural time."

Several studies have shown that when the father was more supportive of the mother, she evaluated her maternal skills more positively and was more effective in feeding the baby. It might also be the case that competent mothers generally elicit more positive evaluations from their husbands. It is important in this time of change that we exercise caution in not fitting the data to our prejudices. However, the facts seem to indicate that increased paternal contact and involvement at the time of early infancy provides important benefits to the newborn, the mother, and to the father himself.

EARLY POSTNATAL PERIOD

Fifteen to 20 years ago, the staffs of intensive care nurseries observed that small premature infants who were sent home intact and thriving sometimes returned to emergency rooms having failed to thrive or battered by their parents. These observations became a major impetus to the study of parent-infant bonding. Since that time, careful studies have documented an increase in the incidence of child abuse and failure to thrive without organic disease—a syndrome in which the infant does not grow, gain weight, or show motor or behavioral progress at a normal rate during the first months at home but then shows rapid gains in all aspects of development when given warm, affectionate care during hospitalization (see Chapter 30). The relationship between these facts and early-contact practices is of great importance to the whole subject of parent-infant bonding.

Studies of a Sensitive Period

The question of whether additional time for close contact of the mother and full-term infant in the first minutes and hours of life alters the quality of the maternal-infant bond over time has been the subject of 17 separate studies. Because hospital practices have recently been altered on the basis of these studies, it is essential to explore in depth their design, ecology, outcome measures, and general strengths and weaknesses. Figure 5–3 illustrates the timing of the contact in these studies. In three studies (group A) the extra time was added not only during the first 2 hours but also during the next three days of life. In one study (group B), contact was added on days 1 and 2. In 13 studies (group C), the additional mother-infant contact occurred only in the first hour of life.

Group A Studies. In the first study, the investigators observed that a group of poor, primarily single, primiparous, inner-city mothers who had 16 hours of extra contact with their infants in the first three days of life, fed their infants with more

affection prior to discharge. At one month, these mothers were more supportive and affectionate when the infant cried during a stressful office visit than were control mothers who had only received their infants for 20 minutes every 4 hours in the first days. During an office visit at one year similar differences were again noted between the two groups of mothers. In a 2-year follow-up study, the extra-contact group of mothers talked to their infants differently. Follow-up studies at 5 years showed a strikingly close correlation between the child's language and cognitive development and his or her mother's speech to the infant at 2 years in the experimental group only; no such correlation was seen in the control group.

Earl Siegel and associates assessed 202 patients in a beautifully designed investigation. Their aim was to explore the effect on maternal attachment of early and extended contact as well as the influence of home visits by well-trained paraprofessionals. The home-visit interventions had no significant effects, but early- and extended-contact mothers showed differences in attachment variables, such as acceptance of the infant and consolation of the crying infant at four months as noted by the home visitor. At 12 months, infants in this same group showed a significant increase in positive versus negative behaviors in the home.

This study makes an important contribution. Investigators calculated that 2.5 to 3.0 per cent of the recorded variance could be explained by early and extended contact, whereas 10 to 22 per cent of the variance could be attributed to background variables such as the mother's economic status, race, housing, education, parity, and age. These were low-income mothers who were primarily multiparous and lived in a rural area of North Carolina. Finally—and importantly—there was no difference between the experimental and control groups in the reports of child abuse and neglect in the first year of life.

Group B Studies. It is useful to compare this single study by O'Connor and associates (1980) with the large study by Siegel et al. already

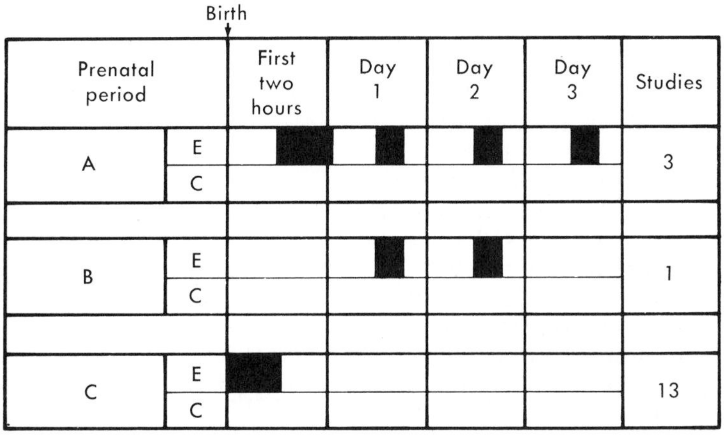

Figure 5–3. Time patterns and number of three types of randomized controlled studies in which one group of mothers had additional contact with their infants (E) compared to another group with routine contact (C). (From Klaus, M. H., and Kennell, J. H.: Parent-Infant Bonding. St. Louis, The C. V. Mosby Co., 1982, p. 41.)

Table 5–6. CHILD ABUSE OR NEGLECT IN THE FIRST YEAR OF LIFE

Study	Total	Abuse or Neglect
O'Connor et al.		
Extended contact	134	2
Control	143	10*
Siegel et al.		
Extended contact	97	7
Control	105	10

*p<0.05.

described (Table 5–6). In the two groups he studied, Siegel noted no differences in parenting disorders, child abuse, neglect, abandonment, and nonorganic failure to thrive. O'Connor, on the other hand, noted that infants of mothers who were allowed 12 additional hours of contact in the first two days had significantly lower hospital admission rates as well as fewer accidents and poisonings. Thus there is disagreement over whether additional early contact prevents or alters parenting failures.

Group C Studies. Thirteen separate studies have looked at the effect of additional mother-infant contact in the first hour of life, with contact after this period being similar in both the experimental and the control groups. In nine of the studies, differences in the behavior of the mother or infant were noted in the experimental group.

Effect of Early Contact on Breastfeeding

Breastfeeding according to six out of nine studies continued for a significantly longer period for those mothers who had contact that involved suckling their babies in the first hour after birth (Table 5–7). It is difficult to know whether it was the early contact or, more specifically, the suckling that altered the length of time that these mothers continued to breastfeed. It may be argued that the length of breastfeeding is not a valid assessment of the strength of the mother-infant bond, since it

is culture-bound. In a study currently in progress in Guatemala, mothers had their babies in the first hour but one group did not suckle and the other did. There was then a brief period of separation for both groups, after which mothers and babies were united 2 to 4 hours after birth. To date, there appears to be no difference in the length of breastfeeding between these two groups.

It should be noted that in all the studies with length of breastfeeding as an end point, both the control and experimental groups of mothers came from the same population except for one study in which there were no significant differences.

Effect of Early Contact on Maternal and Infant Behavior

De Chateau and Wiberg observed interesting behavioral differences in the home three months after birth. At this time early-contact mothers kissed, looked *en face* significantly more, and cleaned their babies less during free play. These same infants smiled and laughed more and cried less. An interesting confirmation of these findings has been reported by Ali and Lowry in Jamaica, in which at three months, early-contact mothers were more likely to rise and follow when their baby was taken from them, they looked at the baby more frequently during feeding, and they were more likely to talk to their baby. On the other hand, the early-contact infants were less likely to cry or be restless during interviews at 6 and 12 weeks.

In other studies, Carlsson and associates noted differences on the second and fourth day after early contact, but not at 6 weeks. In the Campbell and Taylor study, in which differences were not noted at 3 days or at 1 month following early contact, it should be noted that the mothers in the control group who received routine care had their infants for 5 minutes in the first hour of life. The same was true for the control group of mothers in the study of Svejda and associates. In these two studies it is reasonable to question whether 5 minutes in the first hour was enough contact to

Table 5–7. PERCENTAGE OF MOTHERS BREASTFEEDING 2 OR 3 MONTHS POST PARTUM AFTER ADDITIONAL EARLY MOTHER-INFANT CONTACT

Study	No. of Mothers	Hospital Routine	Early Contact
Johnson, 1976 (USA)	12	16	100
Sosa et al., 1976 (Roosevelt I, Guatemala)	60	92	74*
Sosa et al., 1976 (Roosevelt II, Guatemala)	68	58	70*
Sosa et al., 1976 (Social Security Hospital, Guatemala)	40	59	85
Sousa 1974 (Brazil)	200	27	77
de Chateau and Wilberg, 1977 (Sweden)	40	26	58
Ali and Lowry, 1981 (Jamaica)	100	27	57
Salariya et al., 1978 (England)	108	50	60*
Thomson et al., 1979 (Canada)	30	20	60

*No significant difference.

result in similar behavior in both the experimental and control groups.

Fragility of the Sensitive Period Phenomenon

There are at present no reliable studies to tell us the length of time required in the first hour(s) and days after birth to produce an effect on the behavior of the mother or child in the subsequent days and weeks of life. It is worth reemphasizing Pat Bateson's admonition about copying the conditions in which a sensitive period was first described: "Even small changes can cause the evidence to evaporate." Could 5 minutes be long enough to affect the mother's later behavior with her infant?

When Does Love Begin?

The first feelings of love for the infant are not necessarily instantaneous with the initial contact. Many mothers have shared with us their distress and disappointment when they did not experience feelings of love for their baby in the first minutes or hours after birth. It should be reassuring for them and others like them to learn about two studies of normal, healthy mothers in England.

Aidan MacFarlane and associates asked 97 Oxford mothers, "When did you first feel love for your baby?" The replies were as follows: during pregnancy, 41%; at birth, 24%; first week, 27%; and after the first week, 8%.

In a study of two groups of primiparous mothers (n = 112 and n = 41), 40 recalled that their predominant emotional reaction when holding their babies for the first time was one of indifference. Twenty-five percent of 40 multiparous mothers marked a similar response. Forty per cent of both groups felt immediate affection. But within the first week, most of the mothers in both groups had developed affection for their babies.

Primary Maternal Preoccupation

In 1957, Donald Winnicott made remarkably perceptive observations that appear to describe what we call the sensitive period. He proposed that a healthy mother goes through a period of "Primary Maternal Preoccupation," which "gradually develops and becomes a state of heightened sensitivity during, and especially toward the end of, the pregnancy. It lasts for a few weeks after the birth of the child." He continues "I do not believe that it is possible to understand the functioning of the mother at the very beginning of the infant's life without seeing that she must be able to reach this state of heightened sensitivity, and to recover from it." Interestingly, the heightened sensitivity of mothers at this time is sometimes misinterpreted by physicians and nurses as excessive anxiety.

According to Winnicott, "The mother who develops this state . . . provides a setting for the infant's constitution to begin to make itself evident . . . and for the infant to . . . become the owner of the sensations that are appropriate to this early phase of life." He notes that "only if a mother is sensitized in the way I am describing can she feel herself into her infant's place, and so meet the infant's needs."

Since both the Primary Maternal Preoccupation and the maternal sensitive period were determined independently using extremely different techniques, it is interesting to note that the timing and course of each should be so similar. We suspect that endocrine changes play a significant role in starting and enhancing both these processes. Are they perhaps initiated by the rise in estradiol and drop in progesterone in the five weeks prior to delivery? Studies of other cultures are important to the understanding of an early sensitive period because in most societies the mother and baby are placed together with support, protection, and isolation for at least seven days after birth. The provision of food and water and a private time for the mother and infant to get to know each other are common in other cultures.

The effects of increased mother-infant contact in the 17 studies described earlier may be ascribed in part to normal human behavior—something that has been present in our genetic makeup for centuries. The remarkable effects gained by increasing mother-infant contact may in part also make up for the marked deprivation that tends to be a part of present-day hospital routines.

THE BEGINNINGS OF MOTHER-INFANT INTERACTION

Figure 5–4 depicts a common situation—a mother feeding her infant in the first hour of life. The scene, however, is not as simple as it appears at first, as one can see in the surrounding diagram that presents multiple, simultaneous interactions between mother and child. Each is intimately involved with the other on a number of sensory levels. Their behaviors complement each other and serve to lock the pair together. The infant elicits behaviors from the mother which in turn are satisfying to him, and vice versa. For example, the infant's hard crying is likely to bring the mother near and trigger her to pick him up. When she picks him up, he is likely to be quiet, open his eyes, and follow. Looking at the process in the opposite direction, when the mother touches the infant's cheek, he is likely to turn his head, bringing him into contact with her nipple, on which he will suck. His sucking in turn is pleasurable to

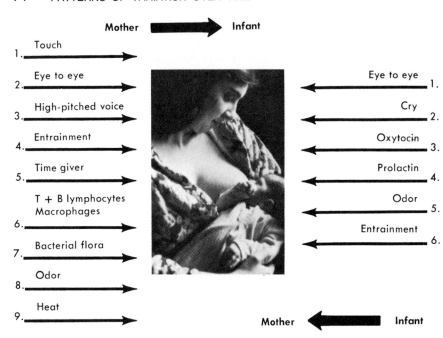

Figure 5–4. Mother-to-infant and infant-to-mother interactions that can occur simultaneously in the first days of life. (From Klaus, M. H., and Kennell, J. H.: Parent-Infant Bonding. St. Louis, The C. V. Mosby Co., 1982, p. 71.)

both of them. Actually, this is a necessarily over-simplified description of these interactions, for these behaviors do not occur in a chainlike sequence but rather each behavior triggers several others.

What we see is an immensely complex "fail-safe" system designed to ensure the proximity of mother and child. Keeping the mother and baby together soon after birth is likely to initiate and enhance the operation of known sensory, hormonal, physiological, immunological, and behavioral mechanisms that probably help lock the parent to the infant.

In summary, although there is increasing evidence from many studies of a sensitive period that is significant to bonding, this does not imply that every mother and father develops a close tie to their infant within a few minutes of the first contact. Each parent does not react in a standard or predictable manner to the complex environmental influences that occur in this brief period. We do not consider this evidence against the concept of a sensitive period but rather as representative of the multiple individual differences of mothers and fathers. (See Parts II and III of this volume for subsequent factors affecting formation and alteration of the parent-child relationship.)

CLINICAL CONSIDERATIONS

PRACTICAL OBSERVATIONS AND RECOMMENDATIONS BEFORE LABOR

1. It is important for the physician and nurse, during the pregnancy, to gather facts about the father and mother concerning diseases in the family, prior abortions, stillbirths, neonatal deaths or illnesses, and experiences with previous pregnancies and deliveries. Information should also be obtained about the parenting these people experienced in their first years of life. Questions should be asked about their major concerns in this new situation. There is much evidence that the loss of a parent in the first 11 years of life is a major psychological risk factor for new fathers and mothers, so information about this can be very helpful to the nurse or pediatrician who may note unexplained parental anxiety and unreasonable concerns about a newborn infant.

2. Major changes in the ecology of the family, such as moving to a new community or the death of a close relative, can have a devastating effect on the pregnancy, early maternal caretaking, and affectionate interaction with the young infant.

3. It is necessary to modify hospital practices to permit and encourage strong family support when the pregnant woman must be hospitalized for prolonged periods because of toxemia, diabetes, hypertension, and intrauterine growth retardation. On a number of occasions, women in these situations have reported that everyone involved seemed primarily interested in the high-risk pregnancy but had little concern for them or their babies as individuals. Changes in this approach should take into account visiting policies for young siblings-to-be, extra beds for fathers-to-be to stay overnight with their wives, special dining rooms so the family can eat together, and any other alterations that might help to make the hospital situation less alien. Such revisions will not come about easily, since they do not fit with the usual

medical model. With this in mind, other systems for care of the high-risk mother might be developed, such as homelike surroundings adjacent to the hospital.

4. Interventions in the perinatal period probably should begin *during* the pregnancy rather than after the mother has taken her newborn home. We encourage pediatricians to meet and talk with mothers prior to the birth of the child for a pediatric prenatal visit. In some beautifully executed observations, Larson reveals that when home health caretakers made contact with mothers before delivery and again 48 hours after birth, rather than starting six weeks post partum, there was a significant change in the mother's ability to parent.

Inevitably questions arise about the impact of the separation between physical facilities and caretakers for the entire process of birth. Often the pregnancy is managed by one physician or midwife, and the infant is cared for by another health care worker, with little exchange of information between all those involved. We should keep in mind how recently it was that the change was made from a system in which there was one continuous caretaker, who was available to the mother during her pregnancy, labor, and delivery and who provided care for both mother and baby during the postpartum period.

5. In cases of psychiatrically disturbed women, evidence from a small number of individual case observations suggests that close and continuing support beginning in the third trimester of pregnancy and continuing through delivery and into the early postnatal period may have a powerful and beneficial effect on the later parenting behavior and family life.

6. We strongly support the many prenatal classes in which parents can share with each other as well as with other couples their many individual experiences, concerns, needs, and questions. These forums allow all involved to recognize and appreciate how many of their worries are simply a normal part of pregnancy.

LABOR, DELIVERY, AND THE FIRST DAYS

Until 80 years ago, events surrounding the delivery had changed little over the centuries. Elaborate societal customs helped parents through this time. Has the enormous improvement in medical management contributed to a waning concern about the many other problems a mother faces during pregnancy? In 1959, Bibring wrote, "What was once a crisis with carefully worked-out traditional customs of giving support to the woman passing through this crisis period has become at this time a crisis with no mechanisms within the society for helping the woman involved in this profound change of conflict-solutions and adjustive tasks." This deficiency may account for the development of the many support systems in our society, such as the wide assortment of childbirth classes that attempt to continue previous customs. These groups help the mother through the delivery period as well as aiding her in later infant and child care. We, therefore, believe that these courses have a valuable supportive role during pregnancy.

In addition, the following principles have been helpful:

1. The less anxiety the mother experiences during delivery, the better will be her immediate relationship with her baby. Therefore, she and her husband should visit the maternity unit to see where labor and delivery will take place. She should also learn about the anesthetic (if she is to receive one); delivery routines and procedures; and medication she will receive before, during, and after delivery. By reducing the possibility of surprise, such advance preparation will increase confidence during labor and delivery. For an adult, just as for a child entering the hospital for surgery, the more meticulously every step and event is detailed in advance, the less the subsequent anxiety.

2. The mother ought to have one person (her husband, mother, a friend, a midwife, a nurse, or an obstetrician) with her throughout labor and birth for guidance and reassurance. The idea of a supportive companion—a feature of human childbirth that was universal until this century—does not fit the medical model of care. For this reason, there is a real danger that this type of intervention may be considered unscientific and so less important than more strictly medical ones. However, because of its proven clinical advantages (shorter labor, decreased perinatal problems, and increased affectionate maternal-infant interaction in the first hour), we feel it is crucial that no woman go through labor or give birth without the presence of a supportive companion.

3. In an effort to reduce the amount of tension on the mother, she should labor and deliver in the same room, preventing the necessity of rushing to a delivery room in the last minutes of labor. Once the delivery is completed and the mother has had a quick glance at the infant, it is important for her to have a few seconds to regain her composure before she proceeds to the next task—taking on the infant. It has been our experience that it is best not to give a mother her baby until a short but complete examination of the infant reveals that it is completely normal and she indicates that she is ready to take it on. It should be her decision.

4. In many hospitals it is customary to put the

baby on the mother's chest for one or two minutes shortly after delivery. This is helpful, but the lack of privacy, the narrow table and the short time period do not allow sufficient opportunity for the mother to touch and explore her baby. Although it is a reasonable procedure, it is not sufficient to optimize maternal attachment.

5. After delivery, it is extremely helpful for the father, mother, and baby to have a period alone in either the delivery room or an adjacent room (a recovery room). This might be called the family time. Obviously, this is only possible if the infant is normal and the mother is well. The mother should have the infant with her on the bed so she can hold him; he should not be off in a bassinet where she can see only his face. She should be given the baby naked and allowed to examine him completely. We have found it valuable to encourage the mother to move over in her regular hospital bed, leaving the other half for her partially dressed or naked infant. A heat panel can be used to maintain or, if need be, increase the infant's body temperature. Several mothers have told us of the unforgettable experience of holding their nude baby against their own bare chest. The father sits or stands at the side of the bed by the infant, allowing both parents to become acquainted with their newborn. Because the eyes are so important for both the parents and baby, we withhold the application of silver nitrate to the eyes until after this rendezvous.

6. We have found it valuable for the mother, father, and infant to be together for at least an hour. After 30 to 45 minutes, the mother and baby often fall asleep. The mother and father usually never forget this significant and stimulating shared experience. It helps some parents to begin to attach to the real infant. We must emphasize that this should be a private session and that many normal parents take many days to fall in love with their infants.

7. The mother and infant should be kept together continuously or have long periods together in the days after the birth. The average mother in the United States today has had so little previous experience with infants that she may worry unnecessarily that many normal features and behaviors are signs of disease or abnormality. The postpartum period should be a time when the mother interacts with her infant, becoming acquainted with him, learning about his needs, and gaining confidence in her own ability to meet these needs before she decides on the time of discharge. We suggest that the infant stay in a small bassinet at his mother's side for a minimum of five hours per day. During this time the mother should have control over the care of the newborn while the nurse acts as consultant. Reversing the usual pattern of "experts" is a step in the direction of establishing the new mother's competence and confidence.

8. We believe it is essential that parents be involved in the many decisions associated with labor and birth—the birth environment, procedures such as shaving the perineum, use and choice of drugs, use of the electronic fetal monitor, ambulation, food and drink during labor, the position for labor and birth, and who should be present.

9. Whenever possible, we suggest that infants requiring additional heat in an incubator be allowed to remain with the mother. We also recommend that phototherapy for hyperbilirubinemia take place in the mother's room.

10. Breastfeeding mothers should be encouraged to feed on demand. This can mean between 8 and 18 feedings in a 24-hour period. However, studies by DeCarvalho reveal that women who feed this frequently in the first 14 days have a larger milk output at 30 days, minimal nipple soreness, and infants with significantly lower bilirubin levels than those women who nursed less than 8 times a day in the first two weeks.

11. The mother needs the emotional support of close contact with her husband or chosen companion as well as with any other of her children during this period. Not only may separation have both a severe immediate effect and long-term effects on siblings, especially those under 3 years of age, but sadness about separation from husband and children often compels a woman to leave the hospital before she is physically ready.

12. We strongly recommend that nurses, physicians, and other maternity staff be optimistic and avoid criticism in their interaction with new mothers. In the postpartum period even a perfectly normal woman may be extremely sensitive to opinions and statements coming from staff members.

13. If the baby must be moved to a hospital with an intensive care unit, we have found it helpful to give the mother a chance to see and touch her infant, even if he has respiratory distress and is in an oxygen hood. The house officer or the attending physician stops in the mother's room with the transport incubator and encourages her to touch her baby and look at him at close hand. A comment about the baby's strength and healthy features may be long remembered and appreciated.

14. We encourage the father to follow the transport team to our hospital so he can see what is happening with his baby. He uses his own transportation so that he can stay in the neonatal intensive care unit for three to four hours. This extra time allows him to get to know the nurses and physicians in the unit, to find out how the infant is being treated, and to talk with the physicians about what we expect will happen with the baby in the succeeding days. We allow him to come into the nursery, explaining in detail everything that is going on with the infant. We also ask

him to help act as a link between us and his family by carrying information back to his wife, requesting that he come to our unit before he visits his wife so that he can let her know how the baby is doing. We suggest that he take a Polaroid picture, even if the infant is on a respirator, so that he can show and describe to his wife in detail how the baby is being cared for. Mothers often tell us how valuable the picture is in keeping some contact with their infant while they are physically separated.

CARE OF THE HIGH-RISK MOTHER

It is useful to try to detect in advance those mothers most likely to have special difficulties in relating to their infants. In our own experience, a high incidence of severe mothering difficulties is found among women with one of the following characteristics: (1) previous loss of a newborn infant, including miscarriage and induced abortion; (2) a fertility problem, with no living children; (3) a previous seriously ill newborn infant; (4) primiparity if younger than 17 or older than 38 years; (5) a medical problem with which the infant may be affected, such as Rh disease, toxemia, or diabetes; and (6) unmarried. Certain principles of management for obstetricians and pediatricians apply to all these situations:

1. In almost all high-risk situations, the odds are heavily in favor of the birth of a live baby who will ultimately be healthy and normal, so it is reasonable to stress the positive aspects and be optimistic. This is essential for the mother's later relationship with her baby, which in turn is extremely important for his development.

2. The obstetrician should bring in the pediatrician early and continue to involve him or her in decisions and plans for the management of the mother and baby.

3. The mother should be prepared for the anticipated aspects of care for her newborn.

4. It is important to communicate with the mother about her condition and about the baby's condition before, during, and after the birth. At times, these explanations will be brief and incomplete, but establishing communication is essential.

CARE OF MOTHER AND HIGH-RISK INFANT AFTER DELIVERY

Clinically, we have been impressed and disturbed by the devastating and lasting untoward effects on the mothering capacity of women who have been frightened by the physician's pessimistic outlook about the chance of survival and nor-

mal development of an infant. If there is a close and firm bond between the mother and infant (which occurs after an infant has been home for several months), there is no reason for the physician to withhold concern. However, while the ties of affection are still forming, these can be easily retarded, altered, or permanently damaged. (See Chapter 26 re: vulnerable child syndrome.)

During the past several years, we have made many changes in the physical arrangements for mothers and our approach to them. We find it best to describe what the infant looks like to us and how the infant will appear physically to the mother. We do not talk about chances or survival rates or percentages, but stress that most babies survive in spite of early and often worrisome problems. We do not emphasize problems that may occur in the future. We do try to anticipate common developments (e.g., the need for bilirubin reduction lights for jaundice in small premature infants). The following guidelines may be helpful:

1. A mother's room arrangements should be adjusted to her needs. Mothers are often best able to express themselves and work out their problems when they are alone.

2. If at all possible, mother and infant should be kept near each other in the same hospital, ideally on the same floor.

3. It is useful to talk with the mother and father together whenever possible. When this is not possible, it is often wise to talk with one parent on the phone in the presence of the other. At least once a day we discuss with the parents how the child is doing; we talk with them at least twice a day if the child is critically ill. It is necessary to find out what the mother believes is going to happen or what she has read about the problem. We move at her pace during any discussion.

4. The physician should not relieve his or her own anxiety by adding his worries to those of the parents. If the physician is worried about a slightly high bilirubin, it is not necessary to discuss kernicterus. Once mentioned, the possibility of death or brain damage can never be completely erased.

5. Before the mother comes to the neonatal unit, the physician should describe in detail what the baby and the equipment will look like. During her first visit, she may become distressed when she looks at her infant. We always have a stool nearby so that she can sit down, and a nurse stays at her side during most of the visit, describing in detail the procedures being carried out, such as the monitoring of respiration and heart rate.

6. The nurse should describe in detail all the equipment surrounding the infant. She should be available to answer questions and give support during this difficult period when the mother is first seeing her infant.

7. It is important to remember that feelings of

love for the baby are often elicited through eye-to-eye contact. Therefore, we turn off the lights and remove the eye patches from an infant under bilirubin lights, so that the mother and infant can see each other.

8. Extended visiting for the mother of a normal full-term infant when the mother is able to handle and completely care for her infant from 1 P.M. to 7 P.M. has been a useful practice. As soon as possible we describe to both the father and the mother the value of touching the infant in helping them get to know him, in reducing the number of apneic episodes (if this is a problem), in increasing weight gain, and in hastening his discharge from the unit. This encourages them to visit the baby frequently for extended periods.

9. The nursery should keep a record of all phone calls and visits by parents. Our data reveal that fewer than three phone calls or visits within a two-week period often predicts subsequent severe mothering disorders.

10. Nurses should feel at ease in reporting any worries or problems they have about a father and mother's behavior. To accomplish this, a good working relationship must exist between the physician and nurses. Meetings with the nursery staff in the intensive care unit should be held every two weeks to provide an opportunity to express their concerns and problems.

11. It may be possible to enhance normal attachment behavior in the mother several days or weeks following birth by permitting a special, short nesting period of close physical contact, with privacy and virtual isolation during which the mother provides complete care for her small infant with help and nursing support readily available nearby. If the safety and feasibility of early discharge of premature infants are fully confirmed, we hypothesize that early discharge combined with a period of isolated physical contact with

caretaking may help to normalize mothering behavior toward infants discharged from intensive care nurseries.

CONGENITAL MALFORMATIONS

Birth of an infant with a congenital malformation presents complex challenges to the physician who will care for the affected child and his family. Despite the relatively large number of infants with congenital anomalies, our understanding of how parents develop an attachment to a malformed child remains incomplete. Although previous investigators agree that the child's birth often precipitates major family stress, relatively few have described the process of family adaptation during the infant's first year of life. Solnit and Stark's (1961) conceptualization of parental reactions emphasized that a significant aspect of adaptation is that parents must mourn the loss of the normal child they had expected. Other observers have noted pathological aspects of family reactions, including the chronic sorrow that envelops the family of a defective child. Less attention has been given to the more adaptive aspects of parental attachment to children with malformations.

Parental reactions to the birth of a child with a congenital malformation appear to follow a predictable course. For most parents, initial shock, disbelief, and a period of intense emotional upset (including sadness, anger, and anxiety) are followed by a period of gradual adaptation, which is marked by a lessening of intense anxiety and emotional reaction (Fig. 5–5). This adaptation is characterized by an increased satisfaction with and ability to care for the baby. These stages in parental reactions are similar to those reported in other crisis situations, such as terminally ill children. The shock, disbelief, and denial reported by many

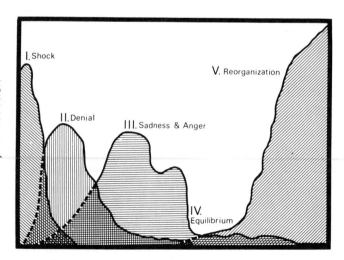

Figure 5–5. Hypothetical model of normal sequence of parental reactions to the birth of a malformed infant. (Adapted from Drotar, D., et al.: Adaptation of parents to birth of an infant with a congenital malformation. Pediatrics, 51:710, 1975.)

parents seem to be an understandable attempt to escape the traumatic news of the baby's malformation, news so much at variance with their expectations that it is impossible to register except gradually.

The intense emotional turmoil described by parents who have produced a child with a congenital malformation corresponds to a period of crisis (defined as "upset in a state of equilibrium caused by a hazardous event which creates a threat, a loss, or a challenge for the individual"). A crisis includes a period of impact, a rise in tension associated with stress, and finally a return to equilibrium. During such crisis periods, a person is at least temporarily unable to respond with his usual problem-solving activities to solve the crisis.

Solnit and Stark have likened the crisis of the birth of a child with a malformation to the emotional crisis following the death of a child, in that the mother must mourn the loss of her expected normal infant. However, the sequence of parental reactions to the birth of a baby with a malformation differs from that following the death of a child in another respect. The mother must become attached to her living but damaged child. Because of the complex issues raised by the continuation of the child's life and hence the demands of his physical care, the parents' sadness, which is initially important in their relationship with the child, diminishes in most instances once they take over the physical care. Most parents reach a point at which they are able to care adequately for their children and cope effectively with disrupting feelings of sadness and anger. The mother's initiation of the relationship with her child is a major step in the reduction of anxiety and emotional upset associated with the trauma of the birth. As with normal children, the parents' initial experience with their infant seems to release positive feelings that aid the mother-child relationship following the stresses associated with the news of the child's anomaly and, in many instances, the separation of mother and child in the hospital.

PRACTICAL SUGGESTIONS FOR PARENTS OF MALFORMED INFANTS

1. We have come to believe that, if medically feasible, it is far better to leave the infant with the mother for the first two to three days of discharge home. If the child is rushed to the hospital where special surgery will eventually be done, the mother will not have a sufficient opportunity to become attached to him. Even if immediate surgery is necessary, as in the case of bowel obstruction, it is best to bring the baby to the mother first, allowing her to touch and handle him and pointing out to her how normal he is in all other respects.

2. The parents' mental picture of the anomaly may often be far more alarming than the actual problem. Any delay greatly heightens their anxiety and causes their imaginations to run wild. Therefore, we suggest bringing the baby to both parents when they are together as soon after delivery as possible.

3. We believe that parents should not be given tranquilizers, which tend to blunt their responses and slow their adaptation to the problem. However, a sedative at night is sometimes helpful.

4. Parents who are adapting reasonably well often ask many questions and indeed at times appear to be almost overinvolved in clinical care. We are pleased by this and are more concerned about the parents who ask few questions and who appear stunned or overwhelmed by the problem. Parents who become involved in trying to find out what the best procedures are and who ask many questions about care are sometimes annoying but often adapt best in the end.

5. Many anomalies are frustrating to both physicians and nurses. The physician may be tempted to withdraw from parents who ask many questions and then appear to forget and ask the same questions over and over.

6. We have found it best to move at the parents' pace. If we move too quickly, we run the risk of losing the parents along the way. It is beneficial to ask parents how they view their infant.

7. Each parent may move through the process of shock, denial, anger, guilt, and adaptation at a different pace. If they are unable to talk with each other about the baby, their own relationship may be disrupted. Therefore, we use the process of early crisis intervention and meet several times with the parents. During these discussions, we ask the mother how she is doing, how she feels her husband is doing, and how he feels about the infant. We then reverse the questions and ask the father how he is doing and how he thinks his wife is progressing. The hope is that they will think not only about their own reactions but will begin to consider each other's as well.

SUMMARY

In assessing the needs of the family during pregnancy, labor, and birth, it becomes apparent that current practices and physical facilities available in the United States are often inadequate for meeting these needs. Postpartum arrangements that bring a newborn to the mother for only 20 to 30 minutes out of every four-hour period are probably insufficient for some women to develop a close attachment with their infants. Furthermore, the small amount of time that the average American mother has with her infant in the hospital does not allow her to discover his reactions and to gain confidence in her ability to care for him properly.

In spite of the lack of early contact experienced by parents in hospital births in the past 20 to 30 years, almost all these parents became bonded to their babies. Humans are highly adaptable, and there are many fail-safe routes to attachment. Sadly, some of those who missed the early bonding experience have felt that all was lost for their future relationship. This response, although incorrect, has impelled us to speak more moderately about our convictions concerning the long-term significance of this early bonding experience. Unfortunately, this has, in turn, caused some skeptics to eliminate the practice of early contact or to make of it a brief charade in which many essential factors were missing.

Evidence would suggest that at least 30 to 60 minutes of early contact in privacy should be provided for every parent and infant in order to enhance the bonding experience. Studies have not clarified how much of the effect may be apportioned to the first hours and how much to the first days; for some parents, one period may be more important than another. However, it appears that additional contact at both times will be beneficial.

We also urge that, whenever possible, the infant remain with the mother as long as she wishes throughout the hospital stay. If the health of either the mother or the infant makes this impossible, then discussion, support and reassurance should help the parents realize that they can still become as completely attached to their child as if they had the usual early bonding experience. In the near future, placement of babies in the large central nursery will probably be phased out. Allowing the newborn to be with the mother, and the father, will mean that parents can have longer periods in which to learn about their baby and to develop a strong tie in the first week of life.

Although first days after discharge are described by many mothers as "hellish," or as the most difficult days of their lives, this need not be the case. By ensuring that the infant remains in close contact with the mother during this period we provide an opportunity for her to prepare for caring for the newborn confidently after discharge. Existing labor and birthing facilities and routines in the United States can easily be adapted (if outmoded state regulations are brought up to date) to provide optimal conditions, in the first few days

of life, for the development of both parents and infants and the formation of the important bond they share.

MARSHALL H. KLAUS
JOHN H. KENNELL

REFERENCES

Bibring, G. L., Dwyer, T. F., Huntington, D. S., and Valenstein, A. F.: A study of the psychological processes in pregnancy and of the earliest mother-child relationship. I. Some propositions and comments. Psychoanal. Study Child 16:9, 1961.

Brazelton, T. B.: The early mother-infant adjustment. Pediatrics 32:931, 1963.

Cohen, R. L.: Some maladaptive syndromes of pregnancy and the puerperium. Obstet. Gynecol. 27:562, 1966.

Drotar, D., Baskiewicz, A., Irwin, N., Kennell, J. H., and Klaus, M. H.: The adaptation of parents to the birth of an infant with a congenital malformation: A hypothetical model. Pediatrics 56:710, 1975.

Hales, D. J., et al.: Defining the limits of the Maternal Sensitive Period. Dev. Med. Child Neurol. 19(4):454, 1977.

Klaus, M. H., and Kennell, J. H.: Parent-Infant Bonding, St. Louis, The C. V. Mosby Co., 1982.

Klaus, M. H., et al.: Maternal attachment: Importance of the first postpartum days. N. Engl. J. Med. 286(9):460, 1972.

Klein, M., and Stern, L.: Low birth weight and the battered child syndrome. Am. J. Dis. Child 122:15, 1971.

Lozoff, B., Brittenham, G. M., Trause, M. A., Kennell, J. H., and Klaus, M. H.: The mother-infant relationship: Limits of adaptability. J. Pediat. 91:1, 1977.

Lumley, J.: The image of the fetus in the first trimester. Birth Fam. J. 7:5, 1980.

Newton, N., and Newton, M.: Mothers' reactions to their newborn babies. JAMA 181:206, 1962.

O'Connor, S., et al.: Reduced incidence of parenting inadequacy following rooming-in. Pediatrics 66:176, 1980.

Parke, R.: Father-infant interaction. In Klaus, M. H., Leger, T., and Trause, M. A. (eds.): Maternal Attachment and Mothering Disorders: A Round Table, Sausalito, Calif., Johnson & Johnson Co., 1974.

Prechtl, H.: The Neurological Examination of the Full-Term Newborn Infant. 2nd ed. Clinics in Developmental Medicine, Number 63. London, Heinemann, 1977.

Solnit, A. J., and Stark, M. H.: Mourning and the birth of a defective child. Psychoanal. Study Child 16:523, 1961.

Sosa, R., Klaus, M. H., Kennell, J. H., and Urrutia, J. J.: The effect of early mother-infant contact on breastfeeding, infection and growth. In Breastfeeding and the Mother, Ciba Foundation Symposium 45 (new series). Amsterdam, Elsevier Publishing Co., 1976.

Sosa, R., Kennell, J. H., Klaus, M. H., et al.: The effect of a supportive companion on perinatal problems, length of labor, and mother-infant interaction. N. Engl. J. Med. 303:597, 1980.

Winnicott, D. W.: The child, the family, and the outside world. New York, Penguin Books, 1964.

Wolff, P. H.: Observations on newborn infants. Psychosom. Med. 21:110, 1959.

Yogman, M. W.: Development of the father-infant relationship. In Fitzgerald, H., et al. (eds.): Theory and Research in Behavioral Pediatrics. Vol I. New York, Plenum Press, 1980.

Editor's Note: Works cited in this chapter but not found in this list may be found in Klaus, M. H., Kennell, J. H.: Parent-Infant Bonding, St. Louis, The C. V. Mosby Co., 1982.

6

Infancy

The health professional who cares for children enjoys an exceptional opportunity to monitor and enhance the development of infants and their families. However, no single theory of behavior and development provides a practical framework for dealing with the diverse developmental issues that arise in the everyday practice of pediatrics. We will propose a transactional model that incorporates data and theory from several fields: neurology, developmental psychology, child psychiatry, and pediatrics. While not all-inclusive, this model facilitates informed clinical supervision of children's behavior and development from birth to age 2. As outlined in Chapters 3 and 10, the transactional model of development assumes that infants and caretakers together determine the child's developmental and behavioral outcome. This differs from other models in which either the child or the environment can unilaterally determine outcome.

The approach to clinical supervision of infant behavior and development presented here has three goals: (1) to nurture the child's internal security and self-control, (2) to decrease parent-child conflicts, and (3) to identify remediable disabilities and problems. The clinician is aided in this endeavor by using the transactional model to clarify the interactions of children and their parents.

In this chapter, we discuss the multiple developmental processes that take place during the first two years of life. Social and emotional development, sensory and motor maturation, cognitive development, language acquisition, and physical growth are outlined. For each developmental process, we describe possible *normal variations and indications for clinical concern*. Drawing on the transactional model, we offer suggestions for clinical supervision and interventions on the basis of both the child's behavioral style and the parents' characteristics. (See also Chapters 10 and 13.)

SOCIAL-EMOTIONAL DEVELOPMENT

Three theoretical constructs—attachment, separation, and autonomy and mastery—will provide a framework for understanding of social and emotional development on a clinical level.

ATTACHMENT

"Attachment" describes the enduring and specific affective bond that develops over time between children and caregivers. Although infant behaviors that create and maintain this bond vary from one developmental stage to the next, their goal—to maintain the child's internal security—remains constant.

The process of attachment begins in utero. With quickening, parents begin to perceive the fetus as a separate individual and enter into an intense relationship with the imagined child to be (see Chapter 5). After birth, parents modify the expectations, hopes, and fears that evolved during the pregnancy as they become acquainted with the real baby.

The infant's actual appearance and behavior become major determinants of the attachment process. For example, an alert infant who reacts readily to parents' faces and responds promptly to consoling maneuvers enhances parents' positive feelings and sense of competence. On the other hand, a drowsy, relatively hypotonic infant who provides less satisfying feedback may disappoint parents when they attempt to gain emotional satisfaction from their infant.

The process of parent-child attachment is not instantaneous. Most families require several months before they feel they know their infant. During this initial adjustment period, parents strive to understand their infant's needs. Through trial and error, they gradually learn effective responses to the infant's needs for food, rest, or social interaction. Parents begin to demonstrate an intuitive understanding of how to enhance their child's social responsiveness. For example, in face-to-face interactions, parents exaggerate their facial expressions and slow their vocalizations in response to the infant's limited ability to process social information. Eyebrows go up, mouths open wide; conversation consists of "aahs" and "oohs." In response to such maneuvers, neonates widen their eyelids, dilate their pupils, and round their mouths. These signs of social interest occur long before the development of responsive smiling at six to eight weeks.

By 3 months, the child and parents achieve social synchrony, manifested by reciprocal vocal and affective exchanges. Parental displays of pleasure are followed by a build-up of smiling, cooing, and movement in the infant. When the excitement peaks, the infant transiently disengages to reorganize for another cycle of excitement. Infants also initiate these pleasurable exchanges. Occurrence of this mutually satisfying synchrony signals the end of the early adjustment period.

The next important step in the attachment process is the development of a clear preference for primary caretakers. By 3 to 5 months, a baby stops crying more readily for familiar caretakers than for strangers. Babies smile sooner and more brightly for their parents, and this clear behavioral preference enhances parents' formation of positive emotional ties to their infants. As recall memory for absent objects emerges between 7 and 9 months, the infant's preference for primary caretakers produces the well-recognized phenomena of separation protest and stranger anxiety.

The goal of the infant's attachment behaviors is to maintain close contact with preferred caretakers. The forms of these behaviors vary with the child's cognitive and motor abilities and the degree of distress. Crying, the first proximity-seeking behavior available to the human infant, continues to serve as a signal of the need for comfort throughout the first years of life. Bell and Ainsworth (1972) have shown that infants whose mothers respond promptly to their cries in the first months of life cry less at one year. Parents sometimes require reassurance that responding to their child's cries will not "spoil" the infant but rather may help to establish a sense of internal security that will facilitate later independence.

With maturation, the child's repertoire of attachment behaviors becomes increasingly elaborate. While infants seek prolonged body-to-body contact, toddlers, sustained by only brief visual or physical contact with caretakers, can happily investigate new people and new places. If the caretaker leaves, however, the child's exploration and playfulness in an unfamiliar setting abruptly decrease (Ainsworth, 1979). With increased stress, such as the approach of a doctor or nurse, the toddler, like the infant, usually desires closer physical contact with the caretaker.

There are two behavioral hallmarks of a secure attachment relationship in the first two years: (1) the child's ability to seek and obtain comfort from familiar caretakers and (2) the child's willingness to explore and master the environment when supported by a caretaker's presence (Sroufe, 1979).

The creation of such a secure attachment relationship requires consistent availability of adults who are affectionate and responsive to the child's physical and emotional needs. Children given the opportunity to develop such a relationship possess a foundation on which to build positive relationships with peers and unrelated adults (Waters et al., 1979).

Normal Variability and Clinical Supervision

The transactional model provides a context for examining variability in the attachment process as a function of both the infant's behavior and the parents' responses to that behavior. The clinician may facilitate the evolution of secure attachment relationships by being available to counsel caretakers during periods of difficulty. For example, during the early adjustment period, when parents have difficulty interpreting the needs of an inattentive or biologically irregular infant, the clinician may decrease their sense of frustration by pointing out that the child's intrinsic behavioral variability is contributing to the confusion. Parents of such an infant especially require support from clinicians to sustain them in their caretaking efforts until the infant matures sufficiently to provide more satisfying feedback.

While some infants appear intrinsically more difficult to nurture than others, parental handling from the earliest months of life appears to be a major determinant of the quality of the attachment relationship (Ainsworth, 1979). Inconsistent, inappropriate, or punitive responses to a child's physical or emotional needs may threaten the development of a secure parent-child attachment. Parents burdened by illness, psychiatric impairment, drug abuse, or other crises may find it particularly difficult to respond warmly and consistently to the infant's frequent demands. Infants and toddlers who have experienced chronically inconsistent nurturing may appear uninterested in exploring the surrounding world, even in the caretakers' presence. Some such children appear unusually clingy in the absence of obvious stress. Others appear actively angry and distrustful of their primary caretakers, ignoring or resisting caretakers' efforts to comfort them after brief separations or other stress (Sroufe, 1979).

An even more serious disturbance of the attachment process should be suspected when youngsters between the ages of 9 months and 2 years fail to demonstrate a behavioral preference for familiar caretakers in response to stress. Lack of discriminant attachment behaviors toward familiar caretakers may be an ominous sign requiring the clinician to search for developmental delay in the child, serious family dysfunction, neglect, or abuse (Gaensbauer and Sands, 1979). Long separation from parents and disorganized patterns of multiple caretaking—conditions that occur in many prolonged hospitalizations—may also produce indiscriminant attachment behaviors. When a child exhibits indiscriminant, avoidant, or resistant attachment behaviors, the clinician should encourage caretakers' continuous availability, warmth, and responsiveness whether the child is at home, in day care, or in the hospital. Hospital and day-care personnel should provide the child with one or two consistently assigned nurses or teachers to augment parents' efforts to restore the child's sense of internal security. If the child's avoidant, resistant, or indiscriminant attachment behaviors persist, mental health referral for the family is indicated.

SEPARATION

Negotiation of separation, both psychological and physical, poses a continuous challenge to parent and child. In psychiatric theory, best outlined by Mahler et al. (1975), separation refers to the internal processes by which the child evolves a satisfying identity as an individual distinct from the parents. Depending on the psychological context, actual physical separations may enhance or impede the child's ability to develop a comfortable individuality. During infancy, as the reliable physical availability of responsive caretakers encourages the infant's efforts toward independence, a complementary process of acceptance of the child's internal separation must take place within the parents. Some parents readily accept an infant's total dependence but have difficulty tolerating a toddler's striving for an independent identity.

Normal Variability and Clinical Supervision—Bedtime, Day Care, and Hospitalization

The child's and parents' responses to everyday experiences of physical separation, such as bedtime, day care, parents' vacations, or hospitalization of parent or child, may vary widely. In assessing the developmental progress of the child's internal process of separation, the clinician should anticipate that both parent and child will show mixed feelings about separations. Appropriate management of a physical separation depends on its duration and context as well as on the developmental readiness of both parent and child.

Brief, predictable physical separations from the parent facilitate successful psychological separation for young children. The first such separation occurs when the infant is put to bed alone at night. The next occurs the first time parents leave their new infant with a relative or babysitter. Most parents are uncomfortable with these first separations. When parents express apparently disproportionate anxiety about their child's well-being during routine separations, they are often expressing ambivalence about the child's evolution of independence. Explicit discussion of the parents' feelings about internal and external separations may be more effective than reassurance about the ostensible concern. For example, the clinician might suggest, "It's not easy for parents to be away from their babies," rather than saying, "Crying doesn't make children sick."

Initial difficulties with separation subside only to become acute again when, at 7 to 9 months of age, children begin to show separation distress by crying whenever the caretaker leaves their presence. As described in subsequent sections, clinicians can help parents recognize that the separation distress that results from normal cognitive phenomena will diminish as the child learns from multiple brief separations and reunions that parents reliably return.

As the psychological process of separation proceeds, the child develops the ability to form relationships with caretakers other than the parents. Parents can facilitate the formation of these new relationships by their physical availability to the child as the relationship is first formed. A new babysitter should be introduced with the mother present for at least one day. When the child begins at a new day-care center, a parent should stay at the center with the child for the first five to seven days, leaving for successively longer periods each day. Parents should expect that the child will initially protest their departure. The child's distress diminishes as familiarity with the new caretakers increases and experience teaches that the parent's return is assured.

Overwhelming stress, such as physical illness and the painful experiences entailed in hospitalization, exceeds any infant's capacity to tolerate physical separation from his or her parents. When the child is tired or ill or has recently sustained a prolonged separation from caretakers, the physical presence of the parent paradoxically supports the process of internal separation by preventing the child from becoming overwhelmed by internal or external stress. The clinician should recommend that a parent or other familiar caretaker remain with a young child during hospitalization. (See also Hospitalization, in Chapter 26).

Since the separation process is mutual, clinicians should be alert to parental issues that may unintentionally sabotage the child's establishment of a separate identity. This process is particularly in jeopardy when the parents perceive the child as unusually "vulnerable" because of past illness or other factors that make a child special (only boy, only girl, last child, and so on). A recent loss in the parents' lives, such as a death or divorce, may also threaten the separation process.

When parents regard their child as uniquely susceptible to harm, they become overprotective. Such overprotection becomes manifest in separation difficulties, insufficient setting of limits, somatic concerns, and overutilization of the health care system. Overprotected, "vulnerable" youngsters have difficulty with the process of internal separation; they become overly dependent, provocative, and defiant. A clinician who encourages parents to discuss their real or imagined losses and their ambivalence about separation may liberate both parent and child. The process of internal separation does not end in infancy for parent or child but must be negotiated repeatedly throughout the life cycle. (See also the Vulnerable Child Syndrome, Chapter 26.)

AUTONOMY AND MASTERY

The infant's intrinsic need for autonomy and mastery drives developmental progress from the earliest weeks of life. *Autonomy* refers to the achievement of behavioral independence. *Mastery* describes the child's quest for ever-increasing competence. These complementary processes require that caretakers and infants continually renegotiate control of the infant's bodily functions and social interactions.

Self-consoling behavior marks the beginning of autonomy. From the earliest days after birth, a crying infant will try to bring the hand to the mouth. Once the hand is inserted, the infant begins to suck and stops crying.

Several studies have demonstrated that sucking facilitates the infant's ability to regulate his or her level of arousal. For example, Brazelton (1962) found that during the first three months of life, infants who engage in frequent hand-sucking cry less than other infants. Cohen (1965) reported that infants given a pacifier to suck during circumcision showed shorter periods of tachycardia and tachypnea than infants not given a pacifier.

As the infant matures, the repertoire for self-consolation expands to include rhythmic behaviors such as body rocking (20 per cent of all children) and head-banging or rolling (6 per cent of all children); these behaviors usually begin between 6 and 10 months of age. In the second year, toddlers employ favored possessions like blankets (transitional objects) and repetitive rituals (e.g., saying goodnight to stuffed animals in a fixed order) to cope autonomously with bedtime and other stressful situations. (See also Chapter 33.)

The infant's drive to master the environment serves as an important motivating force in itself, independent of the need for food, warmth, sleep, and social approval. This intense striving for competence and independence may lead to struggles with caretakers over feeding, sleep, toileting, and exploration. For instance, many 9-month-olds are so intent on practicing new fine-motor skills that they will insist on feeding themselves with their fingers, refusing to allow parents to feed them. Like the legendary explorer, an 18-month-old will repetitively scale a forbidden sofa "because it's there." These behavioral manifestations of the child's drive for autonomy and mastery exasperate caretakers but are essential to the infant's successful development.

Normal Variability and Clinical Supervision—Thumb Sucking, Tantrums, and Toilet Training

As the transactional model predicts, the child's struggles for autonomy and mastery produce varying degrees of discord depending on the temperamental style of the child and the attitudes of the caretakers. For example, temperamentally persistent youngsters delight parents by working at a new task until they have mastered it. However, such persistent children may also infuriate parents by refusing to abandon unsafe explorations of the kitchen stove. Clinicians must often help parents understand these exasperating behaviors as products of the child's temperament and age-appropriate efforts to achieve autonomy, rather than as signs of malevolent defiance.

Parental concerns about thumb sucking, temper tantrums, and toilet training provide three common clinical examples of autonomy issues. *Sucking*, the first organized behavior under the infant's control, is used both to obtain nutrition and to achieve self-regulation by sucking on a pacifier, a hand, or on nothing. Parents unaware of the self-regulatory function of non-nutritive sucking may interpret it as a sign of hunger and inadvertently overfeed the infant. With increasing age, non-nutritive sucking on fingers or pacifiers decreases, becoming a selective coping response to fatigue, illness, or distress. At least 50 per cent of all children suck thumbs or fingers during infancy (Surzon, 1974).

If parents ignore this harmless self-regulating behavior, most children spontaneously relinquish it between the ages of 4 and 5, as other strategies for coping develop. However, if parents try to discourage finger sucking through criticism or restraint, a positive coping mechanism becomes an occasion for a negative struggle over who controls the child's body. To assert their autonomy, children will then stubbornly persist in thumb sucking longer than they would otherwise. Clinicians can help parents perceive the positive functions of non-nutritive sucking and alleviate unnecessary anxiety about orthodontic problems which arise only if thumb sucking persists past the age at which permanent teeth erupt (Surzon, 1974).

Tantrums, common in the second year of life, arise from children's efforts to exercise mastery and autonomy. Clinicians and caretakers can better devise appropriate management if they understand the developmental issues that give rise to tantrums. For example, some tantrums result from the child's frustration at failing to master a task. Distracting the child and permitting success in a more manageable activity can be a helpful maneuver to alleviate this type of tantrum. Most toddlers respond with tantrums as parents impose limits that restrict their autonomy. Parental response to such tantrums should encourage self-control. Young children may need to be held so that they can regain control. Older children should be left alone in a safe place until they have calmed themselves. In using a "time out" procedure, parents should not attempt to inflict a fixed number of minutes of isolation. The goal should be to help the children develop self-regulation. As soon as the tantrum subsides, isolation should end and the child should receive praise for the quieter state. (See also Chapter 55.)

Certain temperamental characteristics make children particularly prone to tantrums. Active, intensely responding children exhibit the most dramatic tantrums—yelling, screaming, and flinging themselves around. A clinician who identifies an intensely responding young infant should, by the time the infant reaches 6 months of age, provide anticipatory guidance concerning the child's probable reaction to limit-setting. Parents will then understand that their child's tantrums arise from a generally intense style of response and not from perversity. The clinician should also emphasize to parents the positive (intensely happy) aspects of the child's style of response.

An appropriate balance between necessary limits and support for independence requires frequent renegotiation as the child's level of development changes. In general, successful limits are firm, consistent, explicit, and selective. Children get confused when parental prohibitions are inconsistent. Limit-setting should include praise for desired behavior as well as disapproval for undesired behavior. Parents of toddlers often need help in choosing which issues are worth a battle. Breaking the child's will should never become an end in itself. Constant tantrum behavior indicates that both family and child have lost control. Such families may benefit from mental health referral.

Toilet training proceeds optimally when parents appreciate the child's need for autonomy and mastery. Anticipatory guidance around toilet training should begin toward the end of the first year, since many parents plan to initiate toilet training on the first birthday. Toilet training begun on an arbitrary schedule, before the child has shown an interest in mastery of this skill, may create unnecessary tension between parent and child. Between the ages of 2 and 3, most children begin spontaneously to imitate the toileting behaviors of siblings and parents. At first, youngsters may just want to sit on the toilet without producing urine or stool. Parents should permit the child to pace the process, as they would the learning of any other new skill (such as riding a tricycle). The drive for autonomous control of body functions and the desire to master socially approved tasks enable children to train themselves within a few days when they choose the time. Toilet training need not be a charged issue. By respecting the child's autonomy and pride in mastery, parents can make toilet training an occasion for growth rather than conflict.

NEUROMATURATION IN INFANCY

Maturing sensory and motor abilities progessively refine the quality of information available to the growing infant. To learn about the social and inanimate world, the baby must actively coordinate the three systems that result in (1) regulation of state (i.e., level of arousal), (2) reception and processing of sensory stimuli, and (3) voluntary control of fine- and gross-motor movements. Normal development of these systems requires both an intact central nervous system and a responsive environment. Preexisting neurological patterns mature and differentiate at varying rates, depending on environmental support. The clinician must be familiar with the normal progression of state control, sensory abilities, and motor skills to distinguish normal variability from pathologic conditions.

STATE CONTROL

The newborn's level of arousal creates six organized clusters of behaviors, called states. There are two sleep states (quiet and active), a drowsy state, an alert, responsive state, a fussy state, and a state characterized by vigorous crying. These six states provide the neurophysiological foundation for the infant's motor and sensory responses. For example, muscle tone will appear hypertonic in crying infants and hypotonic in sleeping ones. During

Table 6–1. STATE CHANGES IN EARLY INFANT*

Behavior	2 Weeks	6 Weeks	12 Weeks	16 Weeks
Longest awake period	2.61	NA	3.41	3.56
Crying	1.75	2.75	1.00	NA
Longest sleep period	4.41	NA	7.67	8.48
Total sleep	16.25	NA	15.11	14.87

*Time = hours per day.
NA = Information not available.
Adapted from Parmelee et al. (1974) and Brazelton (1962).

sleep, infants will gradually decrease responses to repetitive loud noises or bright lights. The same stimuli presented to infants in drowsy or alert states may increase their alertness and activity (Brazelton, 1973).

Most learning occurs when the baby is alert but not moving, since newborns cannot use their large muscles and attend simultaneously. As the baby becomes active and fussy, orienting behavior toward sights and sounds ceases. Periods of quiet alertness are fleeting in the early weeks of life. The ability of some awake, fussy newborns to interact with caretakers may be improved by swaddling, which inhibits movement and facilitates a quiet, alert state.

During the first three months of life, neurophysiological changes (doubling of quiet sleep and diminishing latency of the visual evoked potential) and neuroanatomical changes (rapid myelinization and increased dendritic branching) progressively permit the infant to regulate arousal. This improved regulation of arousal produces increased sustained alertness, decreased crying, and longer periods of uninterrupted sleep. These changes in duration of state are summarized in Table 6–1. Sleep and wakefulness progressively organize into clear day-night cycles. Development of this diurnal pattern depends both on the maturing central nervous system and on signals from the environment. For example, Sander and colleagues (1970) compared sleep-wake patterns during the first 10 days of life of infants cared for in hospital nurseries with those of infants rooming-in with their mothers. The rooming-in infants showed greater duration of both wakefulness and sleep, sleeping more at night than during the day. These findings may reflect the different responses infants in each group receive from adults. Caretaking in the nursery follows a schedule convenient to the staff. In contrast, rooming-in mothers can respond in an individualized fashion to their infants' signals. This contingent caretaking may entrain the sleep cycles of rooming-in infants into more mature patterns.

Normal Variations and Clinical Supervision—Colic and Sleep

The infant's first social responses consist of attaining or maintaining an alert state in response to caretaking maneuvers. The crying baby who quiets down when picked up or the drowsy one who suddenly becomes wide-eyed and attentive at the sound of mother's voice delights caretakers.

There are wide individual variations in infants' control of state, responsiveness to environmental input, and sleep-wake patterns. Some babies spontaneously rouse from active sleep into quiet alertness. Others move directly from sleep to crying, becoming alert only after being consoled. Once roused, many infants independently inhibit their movements in order to attend to an interesting sound or sight. However, some youngsters cannot sustain an alert, receptive state unless assisted by an adult who swaddles them or gently restrains their hands. These inattentive infants can be frustrating to caretakers. The clinician can suggest various methods that will help such infants maintain alertness and can also alleviate parents' distress by pointing out that the baby's inattentiveness reflects immaturity and will improve with time.

Like the inattentive infant, the colicky infant suffers from a disorder of state control that will improve with maturation. *Colic,* or "paroxysmal fussiness," tends to occur in infants with low sensory thresholds. Parents who try to soothe the inexplicably crying infant may inadvertently further overstimulate the baby and prolong the crying bout. This frequently happens when the crying has been of long duration and the infant has not had enough sleep. An inconsolable infant who is not hungry, cold, or wet will most readily move from fussiness to sleep if he or she is swaddled and left alone (Zuckerman, 1981). (See also Colic, in Chapter 28.)

Clinical supervision of *sleep disorders* requires an understanding of the normal variability in children's sleep patterns. From 4 months to 2 years of age, infants sleep 15 to 16 hours a day (Parmelee, 1974). Moore and Ucko (1957) found that by age 9 months, 90 per cent of infants sleep without interruption between midnight and 5 A.M. Traisman et al. (1966) showed that between 6 and 8 months, most infants consolidate their sleep into 12 hours at night with two daytime naps. However, between 9 and 12 months, 25 to 50 per cent of infants resume night-waking (Moore and Ucko, 1957; Carey, 1974). Regular night-waking persists between the ages of 1 and 2 in 20 per cent of children (Richman, 1981).

Duration of sleep depends both on the maturation of the child's central nervous system and on parental handling. Infants with an adverse perinatal history are at increased risk for sleep disruption throughout the first two years of life (Moore and Ucko, 1957; Richman, 1981). Biologically irregular infants or infants with low sensory thresholds also show relatively frequent night-waking (Carey, 1974; Richman, 1981).

While recognizing possible neurological contributions to sleep disturbances, the clinician can help parents devise strategies that may gradually mold the infant's innate biological rhythms into more socially convenient patterns. For example, some new parents require help with distinguishing active sleep from wakefulness. Active sleep occurs every 50 to 60 minutes during a sleep cycle. If parents rush to check or feed the infant at every rustle or moan made during active sleep, the development of sustained sleep will be delayed. The clinician who suspects this to be the case can advise parents to wait until their infant seems fully awake before picking him or her up. Infants who sleep during the day for prolonged periods and remain awake at night may be receiving messages from parents that sustain this day-night reversal. Parents overwhelmed during the day by older children may find a chance to play with their infants only at night. These parents can learn to play in an animated way with their baby after a daytime feeding and to soothe the infant after a night feeding. Early introduction of solids, a traditional remedy for night-waking, does not seem to be reliably effective (Parmelee et al., 1974). (See also Chapter 32.)

SENSORY ABILITIES

Sensory abilities of the infant mature rapidly during the first year of life. By their behavior newborns show preferences for sensory stimuli provided by other people. For example, infants respond poorly to pure tones but respond readily to multifrequency, patterned sounds in the range of human speech, particularly in the higher-pitched range of the female voice. Babies smile in response to sounds weeks before they smile at visual stimuli.

Although immature, the infant's innate visual capacities are also preset to select socially relevant stimuli. The baby's visual response to faces follows a clear developmental sequence. From birth, infants fix their gaze on highly contoured, curved patterns, such as the human face. At first, the infant scans the outlines of the face. By two months, as visual preference shifts toward the internal features of patterns, the baby will reliably look at the caretaker's eyes. Mutual gaze now becomes a preferred form of social interaction. Once infants discriminate the face as an organized whole, they begin to differentiate facial expressions. Smiling faces attract and still faces upset the 4- to 6-month-old. By 6 to 7 months of age, babies make fine sensory discriminations between faces of the same sex and age (Fagan, 1976).

Because visual abilities are immature at birth, not all sensory information is immediately available to the infant. The newborn's visual field is relatively narrow, and only objects at the fixed focal distance of 19 cm (approximately 12 inches) are perceived clearly. Infants ignore visual stimuli that are too close or too distant (Cohen et al., 1979). Thus, the mother's face is seen more clearly by the newborn than are his or her own hands. By 2 to 3 months, visual accommodation matures. The baby then discovers hands and other near objects. Between 6 months and one year, the infant achieves visual acuity and a visual field functionally similar to an adult's, although measured visual acuity does not approach adult values until the preschool years (Cohen et al., 1979).

Very young infants modify their behavior in response to information gathered by smell and taste. By 7 days of age, infants reliably discriminate their own mother's breast pads from those of other nursing mothers (MacFarlane, 1975). Infants vary their sucking patterns in response to the taste of breast milk, formula, and salty or sweet liquid. Infants suck less on an unsweetened liquid, such as breast milk, after they have tasted a sweet solution (Lipsitt, 1979).

Normal Variation and Clinical Supervision—Sensory Threshold and Sensory Deficit

Parents subliminally monitor their infant's responses to sensory input and then modulate that input to enhance the infant's responsiveness. A mother, for example, moves her head slowly back and forth until the infant's expression signals that her face is now in focus. When the baby is startled by the father's deep voice, the father switches to falsetto.

Although caretakers usually make these adjustments automatically, clinicians may need to provide explicit guidance for families whose infants are unusually hypersensitive or unresponsive. Some premature or small-for-date infants have low sensory thresholds. Sounds and sights that are attractive to most infants may be aversive to these hypersensitive babies. For example, while most infants prefer to track a moving face that is making sounds, hypersensitive infants may avert their gaze, vomit, or startle when confronted with this simultaneous visual and auditory stimulation. Clinicians can assist parents in sustaining social interaction with these infants by suggesting that stimulation be offered to only one of the baby's senses at a time while extraneous stimuli, such as bright lights and loud radios, be decreased.

Knowledge of the normal infant's sensory abilities facilitates screening for sensory deficit. Infants may not respond at all to traditional aversive methods of sensory testing such as loud hand claps or bright lights. However, healthy infants should turn to voices and track faces with their eyes. Parents are exquisitely sensitive to their infant's responses. When parents express concern that their infant does not seem to hear or see, that infant should be formally assessed. No child is too young for audiologic testing.

There are many other indications for audiologic testing besides parental concern. Indications for yearly hearing assessment include prematurity, low birthweight, hyperbilirubinemia, perinatal asphyxia, congenital infection, previous treatment with ototoxic drugs, an abnormally formed outer ear, and a family history of hearing impairment. Early detection of hearing loss is crucial, since hearing-impaired children who receive hearing aids before the age of 3 have the best prognosis for acqusition of normal language. (See also Hearing Impairment, in Chapter 39.)

MOTOR DEVELOPMENT

Fine-Motor Development

The evolution of fine-motor skills increases the infant's capacity to explore and change his or her surroundings. At first, the eyes' fixed focal distance limits the newborn's exploratory range. Objects far enough away to be seen are too far to be felt, and objects close enough to be touched cannot be seen clearly. Thus, grasp first develops without visual guidance. During the first three months of life, infants can suck grasped objects but not look at them.

Between 2 and 3 months, decline of the asymmetrical tonic neck reflex and expansion of accommodative abilities permit infants to look at their hands and touch one hand with the other. By furnishing simultaneous information to the senses of vision and touch, this mutual hand grasp provides a foundation for later visual-motor skills.

During the third month, as the near world comes into focus, infants begin swiping at objects with loosely fisted hands. At this stage, infants swipe with one hand only at objects in front of one shoulder or the other. By six months, they reach persistently toward objects in the midline, at first with both hands and then with one. At this age, if one hand is restrained, the infant will cross the midline with the other to reach the desired goal (Provine and Westerman, 1979).

Between 3 and 6 months, the coordination of grasping and reaching gradually comes under visual guidance and voluntary control, as White (1971) has outlined. During early reaching efforts, grasping may occur, but only after the hand has contacted the object. After 3 to 4 months of age, infants begin to shape their hands for grasping in the horizontal or vertical plane of the desired object immediately before touching it. By 9 months, hand-shaping occurs before the object is reached. One-year-olds orient the hand in the appropriate plane when starting to reach for an object (Twitchell, 1965).

Once the infant can reliably obtain an object, clumsy, whole-hand grasping becomes progressively refined. At 4 months, the child holds an object between fingers and palm; at 5 months, the thumb becomes involved. By 7 months, thumb and fingers can grasp and retain an object without resting on the palm at all. At this time, the child uses a raking motion between the thumb and several fingers to scoop up small objects. By 9 months, the child manipulates small objects with a neat pincer grasp, using thumb and forefinger perpendicular to the surface. Every nook and cranny is now accessible to the child's exploration. During the second year, toddlers develop a palmar grasp and wrist supination that permits them to use tools such as spoons and pencils.

Gross-Motor Development

Parents anxiously anticipate their infant's ability to sit up, crawl, and walk. The child health professional must provide families with a sense of the wide range of normal variation in gross-motor development, and must detect early signs of true motor impairment.

In the relatively weightless world of the amniotic sac, the fetus practices many movements that become impossible to perform against the unfamiliar force of gravity encountered after birth. Three processes enable the infant to attain upright posture and the ability to move limbs across the body's midline: (1) balance of flexor and extensor tone, (2) decline of obligatory primary reflexes, and (3) evolution of protective and equilibrium responses.

First, the infant's muscle tone progresses from the neonatal state of predominant flexion to a balance in the tone of flexor and extensor muscles. As this balance develops, the flexed newborn posture gradually unfolds until, by six months, babies can extend their legs so far that they can put their toes in their mouths.

Second, the decline of obligatory primary reflexes (such as the Moro or asymmetrical tonic neck reflex) permits the infant more flexible movement. A one-month-old cannot look to one side or the other without assuming the fencing posture of the asymmetrical tonic neck reflex. Until this reflex dissolves, the child's arm position is determined by the head's orientation. As this reflex disappears, the infant develops the ability to bring his or her hands toward the midline.

Third, in order to sit and walk, the child must establish equilibrium and protective responses. These responses are the automatic changes in trunk and extremity positions that the baby uses to balance and keep from falling. The familiar

Table 6–2. MEDIAN AGE AND RANGE IN ACQUISITION OF MOTOR SKILLS

Motor Skill	Age in Months	
	Median	*Range**
Transfers objects hand to hand	5.5	4 to 8
Sits alone 30 seconds or more	6.0	5 to 8
Rolls from back to stomach	6.4	4 to 10
Has neat pincer grasp	8.9	7 to 12
Stands alone	11.0	9 to 16
Holds crayon adaptively	11.2	8 to 15
Walks alone	11.7	9 to 17
Walks up stairs with help	16.1	12 to 23
Walks up stairs both feet on each step	25.8	19 to 30

*5th to 95th percentile.
Adapted from Bayley (1969).

parachute response of 9-month-olds who extend arms or legs to catch themselves when dropped toward the ground is an example of such protective reactions. Table 6–2 summarizes the age ranges for acquisition of selected milestones in motor development.

As Table 6–2 demonstrates, the age range for normal development of gross-motor skills is wide. Parents and clinicians should not focus on any rigid timetable of discrete motor milestones but should appreciate the ongoing process. In general, infants learn to maintain new positions weeks to months before they can attain them voluntarily. For example, many infants at six months will sit unsupported if placed, but they cannot get themselves into a sitting position until 8 months. Coordinated motion from a new posture takes even longer to develop. Most children cannot walk independently until four to five months after they have learned to pull themselves up to a standing position.

Normal Variation and Clinical Supervision of Gross- and Fine-Motor Development

Both innate variability and environmental input influence the rate of gross-motor development. While the results are not always consistent, it appears that black infants (in the U.S., West Indies, and Africa, compared to their white counterparts), infants from families of lower socioeconomic status (SES) (compared to higher SES infants), and first-born children (compared to later born siblings) attain one or more gross motor milestones earlier than the average norms (Hindley, 1968; Neligan and Prudham, 1969; Super, 1976). The mechanisms of these differences are not know.

The developmental route to walking varies with the child's tone and temperament. Mildly hypotonic (loose-limbed) youngsters sit at the usual age but are late in rolling from back to front or walking.

These hypotonic infants scoot or hitch on their buttocks rather than crawl. By age 3, these mildly hypotonic youngsters have developed a normal gait. Temperamentally inactive or slowly adapting children may not attempt independent walking until long after they are neurologically able to do so. Conversely, very active babies start taking steps as soon as they can stand. During the second year, these active infants rarely walk if they can run.

Parents should be reassured that special efforts to "help" an infant to walk by exercise, walkers, or hard-soled shoes are unnecessary. The motion used to propel a walker differs from that of unsupported walking. Thus, time spent in a walker does not "teach" and may actually *delay* independent walking. Infants' shoes should be flexible, not stiff, since the baby starting to walk uses his or her toes to push off at each step. Parents are often relieved to know that, within the wide range of normal variation, there is no correlation between intelligence and the age at which gross motor skills are acquired.

No single motor skill can be used as an indicator of neurological integrity or dysfunction. In general, the clinician should investigate when delayed motor milestones are associated with global delays, opisthotonic posturing, persistent fisting of the hands, consistent disuse of a limb or side of the body, obligatory and prolonged infantile reflexes, or failure to develop a neat pincer grasp by the first birthday. The early diagnosis of cerebral palsy and other motor disabilities is described in Chapter 39.

COGNITION, PLAY, AND LANGUAGE

COGNITION

Dramatic transformations in infants' cognitive abilities may proceed almost unrecognized until they produce changes in social behavior. For example, parents may complain that a 9-month-old is becoming spoiled because he or she cries when they leave the room. The clinician with knowledge of cognitive development can reassure the mother by suggesting that her baby's behavior reflects a stage in cognitive development. A clinician with this knowledge can thus provide families with a positive model of the child as a thinker and learner. Moreover, knowledge of cognitive development is important for accurate assessment of development as a whole.

The developmental theories of Jean Piaget, as outlined by Ginsberg and Opper (1979), provide the most useful clinical framework for understanding infant cognitive growth. Although subtle modifications of Piaget's findings have recently been suggested on the basis of extensive testing of infants in specially equipped laboratories, his developmental outline of infant behavior remains the most powerful one available to the practicing clinician. Since Piaget derived much of his knowledge of infancy from watching his own children at home, his observations can be readily duplicated by both clinicians and parents.

According to Piaget, infant cognitive development proceeds as a fixed sequence of orderly changes in the child's ability to process information from the environment. These changes occur through two simultaneous processes: (1) *assimilation*, when the child incorporates environmental input into already existing patterns of thought or behavior, and (2) *accommodation*, when the child must modify an earlier pattern of thought or behavior to deal with the environment. Piaget hypothesizes that the process of learning is intrinsically satisfying to the infant. By actively accommodating and assimilating information from the inanimate and social world, the infant learns that objects exist when they are not perceptually present (object permanency), that events have causes and objects have uses (causality), and that one object can represent another (symbolic play).

Piaget organizes cognitive development during the sensorimotor period (birth to 2 years) into six stages. Each stage represents a temporary equilibrium between the infant's skills and the environment's challenges. Infants spend their energy on tasks that activate but do not overwhelm their capacities for assimilation and accommodation. A too-familiar toy no longer engages the toddler, who prefers the greater challenges posed by the contents of the kitchen cupboard. Conversely, a completely insoluble problem (such as use of a crayon presents to a 9-month-old) does not hold the infant's interest. Cognitive development requires opportunities for exploration and manipulation that are neither too easy nor too hard.

An infant is not merely an uninformed adult. Rather, an infant's capacity for handling available information differs from the adult's. During the first year of life, infants process information more slowly than do adults, since, according to Piaget's findings, infants must deal with the cognitive or perceptual features of an object sequentially rather than simultaneously. Infants apparently think egocentrically, unable to realize that objects have an existence independent of the infants' actions upon them. As egocentric thinking declines and simultaneous processing of information increases, the infant's memory gradually transforms from an ability to recognize the familiar to an ability to recall people and events over ever-increasing spans of time and distance. Although infants make rapid gains in their capacity to process information during the first two years of life, cognitive processes do not fully mature until adolescence. (See also Chapter 3.)

Object Permanence

The newborn behaves as though the world consists of shifting images that cease to exist when they are no longer perceived. "Out of sight, out of mind" is a literal description of the infant's world during the first stage of sensorimotor development (stage I). Gradually, infants develop stable mental images of absent objects and people. By 2 months, a baby continues to look expectantly at a person's empty hand after an object has been dropped from sight (stage II). Between 4 and 8 months, the infant will locate a partly hidden object and visually track objects through a vertical trajectory. If they see someone hide an object, however, they will not search for it (stage III). Between 9 and 12 months, infants can find an object that they see hidden (stage IV). However, infants at this age cannot retrieve an object that is moved in plain view from one hiding place to another. By 18 months, babies reliably find objects after multiple changes of position as long as those changes are observed, but they cannot deduce the whereabouts of an object if they do not see it being moved (stage V). Finally, by age 2, the toddler has sufficient symbolic abilities to infer a hidden object's position from other cues without actually observing it being moved to that position (stage VI). People and things now reliably exist for the child as stable entities whether or not they are perceptually present. Behaviors characteristic of each stage of the child's understanding of object permanence are outlined in Table 6–3.

Causality

Piaget observed an orderly sequence of changes in the child's understanding of causal relationships over the first two years of life. First, the infant learns to recreate satisfying bodily sensations by maneuvers such as thumb-sucking (primary circular reaction). At about 3 months of age, the child begins to use causal behaviors to recreate accidentally discovered, interesting effects (secondary circular reaction). For example, babies at this age will repeatedly kick their mattress once they have discovered by chance that this behavior will set in motion a mobile above the bed. The infant's understanding of cause and effect gradually leads to increasingly specific behavior patterns aimed at particular environmental effects. During the second year, the infant becomes more of an experimenter, intent on causing novel events rather than reinstituting familiar ones (tertiary circular reactions). At the same time, the child begins to comprehend that apparently unrelated behaviors may be combined to create a desired effect. For example, by age 2 a child will spontaneously wind up a toy to make it move.

Table 6–3. COGNITION, PLAY, AND LANGUAGE

Piagetian Stage	Age	Object Permanence	Causality	Play	Receptive Language	Expressive Language
I	Birth to 1 month	Shifting images	Generalization of reflexes		Turns to voice	Range of cries (hunger, pain)
II	1 to 4 months	Stares at spot where object disappeared (looks at hand after yarn drops)	Primary circular reactions (thumb sucking)		Searches for speaker with eyes	Cooing Vocal contagion
III	4 to 8 months	Visually follows dropped object through vertical trajectory (tracks dropped yarn to floor)	Secondary circular reactions (recreates accidentally discovered environmental effects, e.g., kicks mattress to shake mobile)	Same behavioral repertoire for all objects (bangs, shakes, puts in mouth, drops)	Responds to own name and to tones of voice	Babbling Four distinct syllables
IV	9 to 12 months	Finds an object after watching it hidden	Coordination of secondary circular reactions	Visual motor inspection of objects Peek-a-boo	Listens selectively to familiar words Responds to "no" and other verbal requests	First real word Jargoning Symbolic gestures (shakes head no)
V	12 to 18 months	Recovers hidden object after multiple visible changes of position	Tertiary circular reactions (deliberately varies behavior to create novel effects)	Awareness of social function of objects Symbolic play centered on own body (drinks from toy cup)	Can bring familiar object from another room Points to parts of body	Many single words—uses words to express needs Acquires 10 words by 18 months
VI	18 months to 2 years	Recovers hidden object after invisible changes in position	Spontaneously uses nondirect causal mechanisms (uses key to move wind-up toy)	Symbolic play directed toward doll (gives doll a drink)	Follows series of two or three commands Points to pictures when named	Telegraphic two-word sentences

PLAY

As Rosenblatt (1977) describes, a child's play provides external evidence of the internal processes of cognitive development. Peek-a-boo signals the emergence of object permanence. An elaborate detour to retrieve a ball rolled under the couch shows that the child understands invisible displacements.

The infant's handling of objects also reflects his or her progressive understanding of the world. At 5 to 6 months of age, an infant can reliably reach and grasp attractive objects. At this stage, the baby subjects all toys to the same behavioral repertoire, regardless of their particular properties. A toy car, a bell, or a spoon are all mouthed, shaken, banged, and dropped. By 9 months, the baby systematically manipulates the object to inspect it with eyes and hands in all orientations, thus demonstrating the cognitive ability to process information simultaneously instead of sequentially.

By the first birthday, the baby demonstrates understanding of the socially assigned function of objects. A toy car is pushed on its wheels; a bell is rung. Next, early representational play, reflecting a stable concept of objects, appears. At first such *symbolic play* centers on the child's own body as the youngster "drinks" from a toy cup or puts a toy telephone to his or her ear. Between 17 and 24 months, the child's thought and play become less egocentric. Now the doll is offered a drink. When the child becomes facile in the use of symbols (between 24 and 30 months), truly imaginative play begins. In such play the child uses one object to represent another (such as putting bits of paper on a plate to symbolize food). Table 6–3 outlines the concurrent development of object permanency, causality, and play.

Normal Variation and Clinical Supervision—Separation Protest, Stranger Awareness, and Cognitive Assessment

Each cognitive transformation alters infants' social behavior. While babies as young as 2 to 3 months of age can recognize their parents, they have no *recall memory*—i.e., no internal symbolic representation of their parents—until they attain stage IV object permanence (the stage at which they will search for a completely hidden object). The child's recall memory for parents evolves prior to memory for inanimate objects (Bell, 1970). The child's experience of "missing" the parent after separation results from the discrepancy between the recalled image of the parent and the parent's absence from the child's perceptual field. For the 4-month-old, the parent does not exist when not seen. The 10-month-old knows that parents still exist when they are absent, but cannot imagine where they might be. No wonder the child vigorously protests separation and anxiously tracks parents even into the bathroom. Not until he or she attains stage V object permanence (at 15 to 18 months of age) can the child predict the position of an object from a series of unseen displacements. Once this cognitive capacity emerges, the child has the ability to infer parents' whereabouts in their absence, and separation protest diminishes.

At the same time that the child achieves stage IV object permanence, the ability to deal simultaneously with several pieces of information develops. Stranger awareness results, as the child can now actively compare unfamiliar to familiar people. The 4-month-old smiles at any smiling adult. The 7- to 9-month old glances warily from parent to stranger to parent, and howls.

The child health professional can reduce parents' bewilderment at separation protest and stranger anxiety by predicting these behaviors as positive signs of normal cognitive development rather than as results of inexplicable emotional disturbance.

While most children show some degree of separation protest and stranger awareness, temperamental differences create variable behavioral patterns in these expressions of cognitive abilities. Persistent children will search longer for hidden objects and cry longer when parents leave than will their less persistent peers. Intensely responding children will seem to be in great distress after separation. By pointing out that the intensely responding child tends to express both happy and angry feelings at a high pitch, the clinician can reassure parents that their child's reaction to separation is both age-appropriate and a usual style of response.

Children progress through the stages of cognitive development in the same sequence but at different rates. A clinician can assess cognitive level in spite of language or motor impairment by determining the child's understanding of object permanence and causality from the child's social behavior and play with objects. Such assessment may prevent the misdiagnosis of general developmental delay in children with other handicaps.

LANGUAGE

Like other developmental phenomena, the infant's acquisition of language follows a predictable sequence. However, the rate and quality of the infant's progression through linguistic development may be more sensitive to caretaking practices than are other sensorimotor skills. Infants can acquire language only through interaction with responsive sources. Television and radio have negligible effects on infant language learning.

Parents and infants begin to construct the basis for later language acquisition long before the baby can understand or produce a single word. Through games and caretaking rituals during the first year, children learn communicative turn-taking. With vocalization and nonverbal cues, caretakers and infants learn to direct each other's attention to interesting environmental events, to signal needs and feelings, and to interpret each other's intentions. During the second year, the child begins to extend the rules of communication learned in action to the use of spoken words.

The actual production of meaningful speech is the result of cognitive, oral-motor, and social processes. By 1 month, the infant has a range of cries that parents associate with hunger, pain, and the like. Between 1 and 3 months of age, the infant develops a range of nondistress vocalizations, onomatopoetically described as "cooing." When the caretaker imitates the baby, the infant's production of sound is prolonged. The caretaker's contingent responses to these early vocalizations shape the infant's vocalizations into conversation-like patterns. Adult speech both elicits and reinforces infant speech (Bloom, 1977).

By 3 to 4 months, infants can produce babbling repetition of all vowel sounds and some consonants. At least two distinct syllables are produced. As babbling matures, the infant starts to produce repetitive two-syllable combinations, such as mama and dada, although as yet these combinations have no symbolic reference. The range of possible syllabic combinations then enlarges, so that most children produce at least four distinct syllables by the age of 8 months.

The emergence of actual words, or sounds used as symbols, depends on the infant's attainment of rudimentary object permanence. Before a person or object can be named, it must have a stable existence in the infant's mind. Not surprisingly, the first true words usually refer to parents and other family members, since the concept of object permanence occurs for people before inanimate objects. Between 10 and 15 months, infants speak their first real words. During this time infants begin also to use symbolic gestures, such as shaking the head to indicate no. Jargoning, long utterances that sound like statements or questions but contain no real words, also occurs initially around the first birthday.

Receptive language ability precedes expression during the toddler years. When asked to do so, children can point to pictures or objects before they can name them. Most 1-year-old infants will respond to simple commands, such as "bye-bye" or "no-no," and most 18-month-olds can point to one or two body parts.

During the second and third year, expressive vocabulary expands exponentially. Infants have acquired a mean of 10 words by age 18 months and about 1000 words by age 3. At the same time, the child begins to construct two-word telegraphic sentences, first to comment on his or her own needs ("more cookie") and then to comment on events in the immediate environment ("mommy go"). By conversing with children and expanding statements ("mommy's going out"), caretakers help children to become generally competent speakers of their native language by the age of 5.

Normal Variability and Clinical Supervision

During physical exams, clinicians can provide a model for "talking to" even the youngest infant. In addition, clinicians should help parents differentiate between immaturity and pathology in children's language. Pronunciation should not be a focus of concern for children under the age of 2. At this age, children frequently make sound substitutions and omit final consonants. Parents should provide a model but not demand correct speech. Clinicians must distinguish difficulties with speech (sound production) from difficulties with language (use of symbols). Ability to use symbols can be easily observed in children's play. Isolated speech difficulties from either anatomical or neurological abnormalities may be seen in children with normal symbolic skills. Such children often have difficulty with other oral motor behaviors, such as eating or blowing kisses, or with other fine-motor skills. They are not necessarily cognitively or emotionally impaired. If a child's speech is unintelligible by age 3, referral to a speech pathologist and audiologist for evaluation is in order.

True delay in acquisition of language constitutes a serious developmental dysfunction. An 18-month-old who uses no single words other than mama or dada, a 24-month-old without multiple real words, or a 30-month-old without two-word phrases should undergo evaluation. Language delays of this magnitude do not result from spoiling or laziness, as parents sometimes suggest (e.g., "He never has to ask for anything"). When evaluating a child with delayed expressive language,

the clinician should assess receptive language abilities and search for evidence of global developmental delay, impaired hearing, autism, or an extremely deprived environment. Full evaluation and treatment of language impairments are described in Chapter 40.

NUTRITION AND GROWTH

Caretakers must provide infants with nutrients that will sustain the rapid growth of body and brain, which is greater during the first two years than at any other time in postnatal life.

An infant's own feeding behavior also affects the intake of adequate nutrition. At birth, an infant has a rooting reflex that helps in locating the nipple. An extrusion reflex pushes out solids to prevent ingestion of inappropriate foods. The infant's small, elongated mouth, combined with forward-and-backward movements of the tongue, squeeze the nipple so that milk is suckled.

Extrusion and rooting reflexes disappear after four months. By 3 months, as the mouth enlarges, neuromaturation of the cheek and tongue allows the infant to become progressively efficient at true sucking, which employs negative pressure to obtain milk from the nipple. By 6 to 8 months, infants begin chewing motions and are able to close their lips over the rim of a cup and drink. By 9 to 12 months, the development of the pincer grasp permits finger-feeding. During the second year, infants acquire the ability to use a spoon and hold a cup or bottle. Parental concerns about messiness, decreased appetite, and selective tastes emerge at this time.

Infant nutrition affects both concurrent and future growth patterns. Both over- and underfeeding may jeopardize the infant's later well-being. During this period of rapid growth, the brain is uniquely vulnerable to nutritional insult. Seventy per cent of adult brain weight is attained by age 2.

Although growth of the brain is the most critical organic achievement of infancy, most adult attention focuses on the rapid growth of the baby's body. Birthweight should double by age 5 months, triple by 1 year, and quadruple by 2 years. Length at birth increases 50 per cent in the first year and doubles by age 4.

Bodily proportions also change. Compared to adults, infants are top-heavy, with a relatively larger head (one-fourth vs. one-seventh of total height) and smaller lower limbs (one-third vs. one-half of total height). These "cute" infantile proportions create a high center of gravity, inhibiting early independent movement; on the other hand, they appear to elicit caretaking from most adults.

The rate of growth and consequent caloric re-

Table 6–4. FEEDING, NUTRITION, AND GROWTH IN INFANCY

Age	Feeding Behaviors	Parental Concerns	Recommended Nutrition	Growth Rate
0 to 3 months	Reflexes: Rooting Sucking Extrusion	Adequacy of intake Discomfort with breast feeding	—115 to 130 cal./kg./day —Breast milk with 400 I.U. vitamin D and 0.25 mg. fluoride or iron fortified formula	Weight = 30 gm./day (1 oz./day) Length = 3.5 cm./mo. Head circumference = 2 cm./mo.
4 to 6 months	Reflexes: Extrusion and rooting gone Skills: Sucking Chewing	Introduction of solids	—100 to 110 cal./kg./day —Breast milk with 400 I.U. vitamin D and 0.25 mg. fluoride or iron fortified formula	Weight = 20 gm./day (⅔ oz./day) Length = 2 cm./mo. Head circumference = 1 cm./mo.
6 to 12 months	Skills: Sucking Chewing Cup drinking (with help) Finger feeding	Messiness Control over feeding	—100 to 110 cal./kg./day —Solids 50% of total calories —Limit formula intake to 30 oz./day	Weight = 15 gm./day (½ oz./day) Length = 1.5 cm./mo. Head circumference = 0.5 cm./mo.
12 to 24 months	Skills: Spoon Cup drinking Fork	Messiness Selective tastes Decreased appetite	—90 to 100 cal./kg./day —Limit milk to 24 oz./day —No foods that can be aspirated	Weight = 220 gm./mo. (7 oz./mo.) Length = 1 cm./mo. Head circumference = 0.27 cm./mo. (12 to 18 months) Head circumference = 0.15 cm./mo. (18 to 24 months)

quirement per unit of body weight decline gradually over the first two years. As caloric needs change during maturation, so does the infant's capacity to ingest and digest an increasing variety of foods. Table 6–4 summarizes feeding behaviors, nutritional requirements, growth characteristics, and common parental concerns about feeding during infancy.

Normal Variation and Clinical Supervision

Male newborns are larger than their female counterparts at birth, and they continue to grow at a faster rate during the first three to six months of life. After six months, there are no sex differences in infant growth rate. In well-nourished populations, racial differences in growth are slight during infancy. A decreased rate of weight gain in lower-class children during the first two years of life should be investigated, since it usually reflects inadequate nutrition or complicating illness, not immutable genetic potential (Smith, 1977).

The rate of increase in length in the first two years of life is variable. Two-thirds of normal infants cross percentiles (Smith, 1977). An infant's genotype for height is expressed by 2 years of age. At this age, children's height can be evaluated, by means of standard charts, as a function of midparental height.

Premature infants and infants who are small for gestational age have distinct growth patterns. Premature infants catch up to their term counterparts in head circumference by 18 months, in weight by 24 months, and in height by 40 months after birth (Brandt, 1978). Small-for-gestational-age infants show less predictable growth patterns, depending on the timing, severity, and etiology of their intrauterine failure to grow; those who show growth acceleration in the first six months of life have the best prognosis for later outcome.

Neuromaturation, cognitive and social development, and temperament all influence feeding during infancy. Lethargic infants are difficult to feed. Management should focus on maintaining alertness. A hyperresponsive extrusion reflex (tongue-thrusting) and/or dyskinetic tongue movements result in difficult and prolonged feeding. These responses warrant further evaluation, since they are often associated with dysfunction of the central nervous system.

Anticipatory guidance at the six- and nine-month visits should address potential feeding conflicts. Children's need for autonomy may result in their refusing food from parents. In this case, parents can present finger-foods to facilitate independent feeding. Parents should also be warned that most children at this age explore by banging and dropping. Most parents do not mind when infants repetitively drop toys but may become annoyed when food is dropped. Parents can be told that this behavior is not intentionally provocative but reflects the child's need to practice new cognitive and motor abilities.

Some feeding problems need to be seen in the contexts of development and temperament, since newly acquired skills can present difficulties. For example, parents should anticipate that new gross-motor abilities will make eating less interesting to most toddlers. The clinician can help parents by pointing out the contribution of the child's temperament to feeding behavior. For example, a child with a high activity level may have difficulty sitting long enough to complete a meal. Similarly, children who are distractable and nonpersistent are also unlikely to finish a meal. Children with a withdrawal response are often unwilling to try unfamiliar foods. Intensely responding children will scream if forced to accept new foods. Slowly adapting toddlers will have a selective diet because they do not readily learn to like new foods. The clinician can work with parents to devise specific strategies for dealing with such feeding problems.

CONCLUSION

An understanding of infant development and behavior provides a framework for child care during the first two years of life. This knowledge can be applied to a transactional model for child development, which stresses the contributions of both child and caretaker to developmental outcome. By adopting a developmental framework and the transactional model, the clinician both accepts a satisfying intellectual challenge and gains an opportunity to be uniquely helpful to families. Usually, the child health professional is the only professional involved with families of young children, and he or she may thus be able to prevent unnecessary parental concerns or parent-child conflicts that may contribute to later behavior disturbances. When developmental delays, sensory deficits, or serious behavior problems already exist, the clinician can minimize their long-term impact by early identification and appropriate management and referral.

A developmental framework and the concept of temperament can be useful to the clinician striving to understand the parent-child adaptation and parent-child relationship. The clinician must elicit parents' attitudes, concerns, and expectations and must investigate their behavioral responses to their child. In addition, identifying the child's temperament seems to be especially helpful, in that it provides a nonjudgmental way of viewing and talking about the child's individuality and impact on the family.

Support of families, which is implicit in the approach outlined here, can become an integral part of preventive well child care.

Acknowledgment

Support for our research has been provided by the William T. Grant Foundation. We would like to thank T. Berry Brazelton, M.D., for his excellent teaching and support; Jennifer Rathbun, M.D., David Barrett, Ph.D., and Trudy Norman, M.S., for their helpful suggestions; and Jane Freeman and Susan Simon for their help in preparing the manuscript.

BARRY S. ZUCKERMAN
DEBORAH A. FRANK

REFERENCES

Ainsworth, M. D. S.: Infant-mother attachment. Am. Psychologist 34:932, 1979.

Bayley, N.: Manual for the Bayley Scales of Infant Development. New York, Psychological Corporation, 1969.

Bell, S. M.: The development of the concept of object as related to infant-mother attachment. Child Devel. 41:291, 1970.

Bell, S. M., and Ainsworth, M. D. S.: Infant crying and maternal responsiveness. Child. Devel., 43:1171, 1972.

Bloom, K.: Patterning of infant vocal behavior. J. Child. Psychol. 23:367, 1977.

Brandt, I.: Growth dynamic of low birthweight infants with emphasis on the perinatal period. In Falkner, F., and Tanner, J. (eds.): Human Growth: Postnatal Growth. New York, Plenum Press, 1978.

Brazelton, T. B.: Crying in infancy. Pediatrics 29:579, 1962.

Brazelton, T. B.: Neonatal Behavioral Assessment Scale. Philadelphia, J. B. Lippincott, 1973.

Carey, W. B.: Night waking and temperament in infancy. J. Pediatr., 84:750, 1974.

Cohen, L. B.: Observing responses, visual preferences, and habituation to visual stimuli in infants. Doctoral Dissertation, University of California at Los Angeles (UCLA), 1965.

Cohen, L. B., DeLoach, S., and Strauss, M. S.: Infant visual perception. In Osofsky, J. D. (ed.): Handbook of Infant Development. New York, John Wiley and Sons, 1979, pp. 393–438.

Fagan, J. F.: Infants' recognition of variant features of faces. Child Devel. 47:627, 1976.

Gaensbauer, T., and Sands, K.: Distorted affective communications in abused, neglected infants and their potential impact on caretakers. J. Am. Acad. Child Psychiatr. 18:236, 1979.

Ginsberg H., and Opper, S.: Piaget's Theory of Intellectual Development. Englewood Cliffs, N.J., Prentice-Hall, 1979.

Hindley, C.: Growing up in five countries. Devel. Med. Child Neurol. 10:715, 1968.

Lipsitt, L. P.: The newborn as informant. In Kearsley, R. B., and Sigel, I. E. (eds.): Infants at Risk. Assessment of Cognitive Functioning. Hillsdale, N.J., Lawrence Erlbaum Associates, 1979.

MacFarlane, J. A.: Olfaction in the development of social preferences in the human neonate. In Parent-Infant Interactions. CIBA Foundation Symposium 33. New York, Elsevier, 1975, pp. 103–113.

Mahler, M., Pine, F., and Bergman, A.: The Psychological Birth of the Human Infant. New York, Basic Books, 1975.

Moore, T., and Ucko, C.: Night waking in early infancy: Part I. Arch. Dis. Child. 32:333, 1957.

Neligan, G., and Prudham, D.: Norms for four standard developmental milestones by sex, social class and place in family. Devel. Med. Child Neurol. 11:413, 1969.

Parmelee, A.: Remarks on receiving the C. Anderson Aldrich Award. Pediatrics 59:389, 1977.

Parmelee, A., Weiner, W., and Schulz, H.: Infant sleep patterns: from birth to 16 weeks of age. J. Pediatrics 65:576, 1974.

Provine, R., and Westerman, J.: Crossing the midline: Limits of early eye-hand behavior. Child Devel. 50:437, 1979.

Richman, N.: A community survey of characteristics of one- to two-year-olds with sleep disruptions. J. Am. Acad. Child Psychiatry 20:281, 1981.

Rosenblatt, D.: Developmental trends in infant play. In Tizard, B., and Harvey, D. (eds.): Biology of Play. Philadelphia, J. B. Lippincott, 1977.

Sander, L. W., Julia, H., and Stechler, G.: Regulation and organization in early infant-caretaker interaction. In Robinson, R. J. (ed.): Brain and Early Behavior. London, Academic Press, 1969.

Smith, D. W.: Growth and Its Disorders. Philadelphia, W. B. Saunders Co., 1977.

Sroufe, L. A.: The coherence of individual development: Early care, attachment, and subsequent developmental issues Am. Psychologist 34:834, 1979.

Super, C.: Environmental effects on motor development: The case of African infant precocity. Dev. Med. Child. Neurol. 18:561, 1976.

Surzon, M. E.: Dental implications of thumbsucking. Pediatrics 54:196, 1974.

Traisman, A. S., Traisman, H. S., and Getti, R. A.: The well baby care of 530 infants. A study of immunization, feeding, and behavioral and sleep habits. J. Pediatrics 68:608, 1966.

Twitchell, T.: The automatic grasping responses of infants. Neuropsychologia 3:247, 1965.

Waters, E., Wippman, J., and Sroufe, L. A.: Attachment, positive affect and competence in the peer group: Two studies in construct validation. Child Devel. 50:821, 1979.

White, B.: Human Infants: Experience and Psychological Development. Englewood Cliffs, N.J., Prentice-Hall, 1971.

Zuckerman, B.: Crying and colic. In Gabel, S. (ed.): Behavioral Problems in Childhood. New York, Grune and Stratton, 1981, pp. 257–272.

OVERVIEW

Children between 2 and 5 years of age can be enchanting to observe or an overwhelming problem to manage. Their dramatic transformations during this period may promote parental pleasure and pride as well as frustration and concern.

The accomplishments of preschool youngsters are truly remarkable. They leap from a relatively limited action-bound sensorimotor understanding of the world to more sophisticated levels of knowledge based upon the use of symbols and the elaboration of fantasy. Their growing physical prowess converts coordinated motor skill from an end in itself to a means toward more effective exploration and independence. The toddler who is just beginning to put words together becomes the articulate storyteller and verbal manipulator who has mastered virtually all the major rules of the language before entering first grade. The 2-year-old who struggles with physical separation from his primary caregiver is transformed into a youngster who is ready to enter the world of his peers with a personal code of moral values and a growing sense of himself as an individual (Fraiberg, 1959; Gesell, 1940).

The major developmental and behavioral issues of the preschool years provide a useful framework within which variations and individual differences can be examined and understood by the primary care physician. The child between 2 and 5 years of age is developing a greater sense of autonomy at the same time he is becoming more aware of the perspectives of others. He is developing complex skills and competencies that represent major qualitative advances over his earlier abilities. He is very much influenced by the values of his family at a time when he is increasingly confronted with the demands of a rule-governed society. The way in which these and other developmental challenges are negotiated is influenced by the transactions among a multitude of factors—neurobiological, constitutional, familial, sociocultural, and political. Multiple patterns of adaptation are observed, some more functional than others. Although developmental tasks are the same for all children, their resolution may be as varied as the individual differences among youngsters and families. The boundaries of normal variation are broad. The challenge for the physician is to combine a sophisticated understanding of the developmental-behavioral goals of this period with both a sensitivity to variability within the healthy population and a commitment to those who are truly disabled.

THE EMERGING PRESCHOOLER

PHYSICAL GROWTH AND NEUROMOTOR MATURATION

Growth in size and physical competence provide an obviously important biological substrate for the developmental and behavioral changes that characterize the preschool years. Although the rate of increase in height and weight is slower than in the first two years, the average child during this period gains approximately 2.0 kilograms per year and grows 6 to 8 centimeters per year, attaining half his adult height sometime between 2 and 3 years of age. Postural changes between 3 and 4 years are characterized by a gradual disappearance of the exaggerated lordosis and protuberant abdomen of the typical toddler. With the usual decrease in appetite and increase in activity level, the average preschooler loses a great deal of his "baby fat" and assumes a leaner body habitus and more mature facial features.

As the child gets bigger, he also becomes more competent physically. Through the influences of both neuromaturation and practice, gross-motor coordination and manual dexterity are substantially advanced. On average, 2-year-old youngsters can run and climb well, walk up and down stairs one step at a time, and build a tower of four to six cubes. By 3 years most children can stand on one foot, walk upstairs with an alternating pattern, pedal a tricycle, and copy a circle. Four-year-olds can usually hop on one foot, throw a ball overhand, use scissors adaptively, and draw a person with two to four parts (Table 7-1).

The range of competencies in the motor area is considerably broad. Some children are obviously well endowed, in a genetic or constitutional sense, with natural grace and agility. Others will "pass"

Table 7–1. NEUROMOTOR ACCOMPLISHMENTS

Gross-Motor Skills	Fine-Motor Skills
Age 2 Runs well	Builds tower of six cubes
Kicks ball	Imitates vertical crayon
Climbs well	stroke
Goes up and down stairs	Turns book pages singly
(one step at a time)	
Age 3 Goes up stairs (alternat-	Copies circle
ing feet)	Copies cross
Jumps from bottom step	
Pedals tricycle	
Stands on one foot mo-	
mentarily	
Age 4 Hops on one foot	Copies square
Goes down stairs (alter-	Draws person with two
nating feet)	to four parts
Stands on one foot (5 sec-	Uses scissors
onds)	
Throws ball overhand	
Age 5 Stands on one foot (10	Copies triangle
seconds)	Draws person with body
May be able to skip	Prints some letters

a standardized motor test, but the quality of their performance may be poor. One youngster may develop considerable athletic skills through extensive practice guided by a supportive parent. Another may perform below his or her potential ability level because of excessive pressure from an overambitious caregiver.

Almost all children eventually master basic skills for ambulation and physical manipulation. These gross- and fine-motor abilities are progressively refined and contribute to the emerging independence of the preschool years. Although the development of motor competence seems to be less sensitive to environmental factors than are other areas of development, qualitative differences are significant, and the degree to which relative deficits are perceived as abnormal is frequently determined more by familial and cultural values than by biological fiat.

INCREASING CAPACITY FOR SYMBOLIC FUNCTIONING

Most contemporary theorists of child development support the notion of developmental stages characterized by qualitative change. That is to say, as children mature, they periodically graduate to levels of competence at which their abilities are different, not just better. Perhaps at no time is that phenomenon more dramatic than it is for the child who passes from the sensorimotor period to the age of language and pre-operational thinking.

Language Development

The emergence of linguistic abilities in the preschool child is awesome. It provides a mechanism for describing events, asking questions, expressing desires, and sharing feelings. It facilitates the development of richer and more complex interpersonal relationships and provides a vehicle for creativity and further mastery over one's environment.

Language development has been the object of intense study over the past 20 years. Most psycholinguists focus on three aspects for investigation—phonology, syntax, and semantics. *Phonology* refers to the analysis of the sounds (phonemes) that characterize a given language. Within this context, one can study a child's ability to recognize the various sounds of his native tongue (auditory discrimination) as well as the ability to produce them (articulation). *Syntax* is concerned with the way in which words are combined to form phrases and sentences. It deals with the rules of grammar that govern the structure of a language in its written or spoken form. *Semantics* refers to the meaning of language and therefore considers the relationship between words and their referents.

Although fierce debate continues to rage, most scholars now believe that the human organism has an innate ability to discover the rules of language that is not determined by simple imitation and positive reinforcement. In this regard, it is interesting to note that most children have mastered most of these basic rules by the time they enter elementary school, even though parents usually correct their preschooler's speech for errors in semantics and not syntax.

By 2 years of age, most children have a vocabulary of several dozen words and have begun to put together two or three word sentences. They can usually point to several body parts, can express negation ("no milk"), and may ask questions by making a statement in a rising intonation ("Daddy bye-bye?") (Table 7–2).

The 2½- to 3-year-old usually has a vocabulary of several hundred words that seems to be continually expanding. He can consistently speak in short sentences, although frequent grammatical errors are typical. Most youngsters can give their full name by 2½ years and their age and sex by age 3.

By 4 years of age, most children have mastered many of the grammatical rules but speak in rather simple sentences. Over the next year, their expressive language becomes more complex, and they often combine two or more ideas in a single sentence. At this point, many youngsters are highly competent storytellers, who will produce elaborate narratives for the receptive listener. Most of the articulation errors during this period are related to immaturity and will self-correct. Stuttering, although not completely understood, is a common transient phenomenon that may persist as a problem in the self-conscious child affected by parental anxiety and pressure.

Table 7–2. SELECTED LANGUAGE GUIDEPOSTS

Skill	Age 2	Age 3	Age 4	Age 5
Comprehension	Follows simple commands Identifies body parts Points to common objects	Understands spatial relationships (in, on, under) Knows functions of common objects	Follows two-part commands Understands same/different	Recalls parts of a story Understands number concepts (3–4–5–6) Follows three-part commands
Expression	Labels common objects Uses two or three word sentences Uses minimal jargon	Uses three to four word sentences Uses regular plurals Uses pronouns (I, me, you) Can count three objects Can tell age, sex, and full name	Speaks four to five word sentences Can tell story Uses past tense Names one color Can count four objects	Speaks sentences of ≥ five words Uses future tense Names four colors Can count 10 or more objects
Speech	Intelligible to strangers 25% of time	Intelligible to strangers 75% of time	Normal dysfluency (stuttering)	Dysfluencies resolved

Editor's note: The reader should be aware that mean ages for some milestones vary slightly from one authority to another.

Assessment of language development during the preschool years may present major challenges. Many normative data have been derived from relatively small samples of children. Recent evidence suggests that variation within the normal population may be more common than was previously suspected—not only in the rate at which language skills emerge but also in the pattern of progression. In fact, the possibility of several alternative pathways to ultimate linguistic competence has been suggested by some researchers, but present data provide limited insight into this issue.

The influences of sex and socioeconomic class differences on language development have also been studied. Although some data support the assertion that girls are more precocious than boys, these differences seem to be most striking in the first two years of life.

Reported social class and ethnic differences in language development have been particularly controversial. Several studies have suggested that the poor test performance of children in lower socioeconomic classes is a function of their linguistically "restricted" environment, as compared to the "elaborated" language of the typical middle-class home. In the same spirit, lower scores among black children used to be considered a reflection of their linguistic "inferiority." Subsequent research, however, has demonstrated that "Black English" has its own unique syntax, which is no less complex than that of so-called "Standard English." This has led to current concepts of language "differences" rather than "deficits" regarding ethnically distinct dialects (see also Chapter 40).

Language and Thought

The relationship between cognitive and linguistic development in the preschool years is complex. Some theorists suggest that language and thought develop independently, others postulate that language determines thought once a child passes beyond the sensorimotor period, and a third model contends that language and thought initially develop along parallel but independent pathways that merge after 2 years of age.

Much of what we understand about the cognitive achievements of the preschool period is based upon the monumental work of Jean Piaget. His astute observations and brilliant conceptualizations provide a rich framework for understanding the emerging abilities of what he called the pre-operational stage.

The pre-operational child lives in a world of magic. He believes that cartoon characters are real people who live inside the television, and he may have an imaginary friend. His thinking is pre-logical. He confuses coincidence with causation and is able to center on only one feature of an experience at any given time. This inability to consider more than one aspect of a phenomenon at a time explains the failure of the preschooler to understand the simple principle of conservation, as demonstrated by Piaget's classic experiment in which children who could focus only on height did not realize that a tall, thin glass and a short, fat glass contained the same amount of water. The development of primitive classification skills that fall short of the ability to group objects by more than one simple dimension (e.g., color, shape) at a single time is another illustration of how perception dominates reasoning during this period (Table 7–3).

Learning to distinguish the boundaries between fantasy and reality is one of the major cognitive

Table 7–3. COGNITION IN THE PRESCHOOL PERIOD

Magical thinking
Pre-logic
Egocentrism
Perceptual domination

tasks of the pre-operational years. Achievement of this goal is accompanied by a gradual shift from egocentrism to an increasing ability to understand things from the perspective of others. With the completion of these "milestones," the youngster is ready to move on to the logical thinking of the school years (see also Chapter 3).

Evolution of Play

The crowning achievement of the preschool child is his ability to use symbols. These may take the form of words or mental images and are dramatically demonstrated in the rich fantasy that characterizes much of the play of this age period (Garvey, 1977). The use of transitional objects (e.g., teddy bears and special blankets) in order to work through the challenges of separation and individuation, and of dramatic play (e.g., playing "doctor" and "house") in order to facilitate the processes of early sexual and adult role identification are familiar themes in the preschool repertoire. A fascination with monsters and "super heroes" is yet another way in which play is used by the youngster to master typical feelings of vulnerability that accompany the increasing self-awareness of the age.

The concept of play also describes activities that provide a vehicle for mastering a range of specific abilities. Acquisition of new skills in the perceptual and gross- and fine-motor domains, for example, is facilitated by the practice opportunities provided in the context of play.

Another phenomenon that has been well described is the evolution from solitary and parallel to interactive and cooperative play. Its progression in the preschool period provides important preparation for the rule-governed, competitive play that becomes so critically important in the elementary school years (Table 7–4).

IDENTIFICATION AND EXPANDING SOCIAL VALUES

As the typical 2-year-old is engaged in the classic battle over independence from his primary caregiver, the average 5-year-old should be prepared to enter the world of his peers with a developing sense of himself as an individual. The process by which one attains this personal identification and redefines his attachment to the significant people

in his life constitutes another task of the preschool period.

During these early years, children begin to develop a sense of themselves as individuals in a social system, with conventional guidelines related to such issues as sex roles and moral values regarding the difference between right and wrong. A variety of theories have attempted to explain the basis for this psychosocial transition. The psychoanalytical interpretation emphasizes the power of sexual fantasies and associated guilt, which are ultimately mastered through resolution of the Oedipus or Electra complex, as characterized by identification with the parent of the same sex and adoption of a strict code of moral values (the "super-ego"). Most behaviorists attribute the assumption of sex roles and social values to the effect of positive and negative reinforcement during this period. Social learning theorists emphasize the susceptibility of youngsters to the influences of their parents' behavior; cognitive theorists present an alternative explanation for the development of a moral code at this age, emphasizing the role of maturing cognitive abilities as the determinant of a youngster's ability to understand the reasoning that underlies social values. Regardless of the theoretical position one may favor, developmental-behavioral tasks themselves have been well described and consistently observed. Fantasy play of the preschool child is rich in themes related to sex roles and adult values. Many parents of creative preschoolers have been exposed (and often embarrassed) by the revealing domestic dramas frequently acted out by their children. (See also Chapter 36 concerning socialization.)

In recent years, the degree to which stereotypical sex roles reflect biological imperatives or cultural inventions has been a matter of great debate. Although a considerable body of data supports the notion of greater inherent aggressiveness in boys than girls, differential preferences for activities such as doll play or gross-motor sports are significantly influenced by social reinforcement. Individual families that have attempted to provide their sons or daughters with alternative, nontraditional sex role models often find themselves competing with the impact of television, neighbors, and other youngsters. (See also Chapter 10 concerning gender differentiation.)

BEHAVIORAL STYLE AND INDIVIDUAL DIFFERENCES

Role models and values in the emerging self-concept of the preschooler are always incorporated into a personality matrix unique to each individual child. That is to say, each youngster has his or her own behavioral style within which universal

Table 7–4. PROGRESSION OF PLAY

Age 4	Cooperates with other children	Increasing elaboration of symbolism and fantasy
Age 3	Understands turn-taking	
Age 2	Mostly parallel play	

developmental issues will be negotiated (Thomas and Chess, 1977).

Individual differences among youngsters can be appreciated in several important social dimensions. Manifestations of aggression, for example, provide a rich focus for study. Despite widely divergent theoretical explanations regarding its significance, most would agree that aggression is, to some extent, biologically determined. The intensity with which it is discharged, however, varies among children from the earliest years. In virtually every culture in which this behavior has been studied, boys have been found to be more aggressive than girls. Yet, the extent to which these sex differences are socially influenced remains a matter of debate. Social learning theorists have produced data that suggest that levels of aggression in children are related to the behavior of their parents, especially in the face of excessive permissiveness or harsh punishment. Correlations between degree of aggressiveness and exposure to television violence have also been demonstrated. However, the inability to replicate all these findings consistently relates to differences in the extent to which individual youngsters are susceptible to common environmental influences. For example, some children seem to have large reservoirs of aggression that are easily tapped, while others demonstrate considerably less intensity.

Studies of coping behaviors in preschool children have also illustrated variability in adaptive styles. When dealing with adult supervision, for example, some youngsters automatically defer to authority, others respond with defiance, while others attempt to create a mutual relationship through talking or social interaction. The child's approach to increasing autonomy can also take many different forms. Some youngsters cling more persistently and demonstrate less initiative in their drive toward independence while others are more aggressive in their separation (Murphy, 1962).

Behavioral "health" ultimately depends upon how a child feels about himself. Positive self-esteem most easily flows from a comfortable match between a youngster's style and a caregiver's preferences. Because no single temperamental model has been shown to be most effective for all children, no single style can be arbitrarily designated as "normal." (See also Chapter 10 concerning behavioral style.)

ENVIRONMENTAL INFLUENCES

FAMILY VALUES AND CHILD-REARING STYLE

The American family has been characterized as changing, vanishing, threatened, and surviving.

In varying circumstances, it could be classified as nuclear, extended, or communal. It may be headed by a single parent, or two parents conforming to a matriarchal, patriarchal, or egalitarian model. Whatever form it may take, the family has a powerful influence over the behaviors and developing value system of its children. The preschooler's perception of his parents as infallible makes him particularly impressionable during this period of life.

Children clearly develop most of their concepts of masculinity and femininity from the behaviors and attitudes of their parents. Their ultimate identification with the parent of the same sex is a major task of this period. For youngsters who are raised in single-parent families, the absent model is frequently sought in relatives or family friends. Despite their power, families with strong feelings about the promotion or discouragement of sexist attitudes are often humbled by the realization that societal values may modify the most determined parental influences.

The overall impact of child-rearing styles on emerging values and behaviors has received a great deal of attention. Some accumulated data have suggested that overly strict and authoritarian parents produce less happy but more compliant youngsters, while children of extremely permissive parents tend to be more aggressive and lacking in self-discipline. Such sweeping generalizations have been challenged by contradictory studies that show great variability in behavioral outcomes. Most contemporary observers of family interaction conclude that parental styles of child-rearing have significant impact, but the effects may be modified by differences in age, sex, and temperament of the child as well as by the cultural milieu in which the family lives. In some cases, the influence of the child may be paramount. The primary care physician must avoid generalizations and arbitrary value judgments about specific philosophies of child care. The healthiest and most adaptive outcomes are generally found in those situations where both love and limits are transmitted from parent to child in a manner that most comfortably fits with the style of each (Chamberlin, 1978).

ORGANIZED PRESCHOOL EXPERIENCES

Strictly speaking, the designation of the years from 2 to 5 as "the preschool period" has become a misnomer. In fact, for many youngsters, "school" begins well before 6 years of age, and in American society this trend is clearly increasing.

The proliferation of preschool programs has been fueled by many factors. Much of the demand for alternative child care arrangements is related to the rapid rise in the number of mothers of young children who are entering the work force. In some cases, this phenomenon is related to climbing divorce rates and increasing numbers of single-parent families. In others, it reflects the growth of two-career families in which both parents are employed because of economic necessity or commitment to personal career development. Child-centered motivations also contribute to the demand for preschool programs, as parents increasingly subscribe to the belief that organized peer activities provide their youngsters with additional opportunities for learning and social growth.

A number of alternative models are available. These vary from informal arrangements, such as unlicensed family day care or parent-cooperative play groups, to highly organized and professionally administered day care or nursery school programs. Specific mandates are quite variable. Programs such as Head Start are designed to provide health, education, and socialization experiences for children from families in the lower socioeconomic classes in order to increase their chances for later success in school. Specialized preschool programs for developmentally disabled youngsters combine specific therapeutic and educational interventions for the children with supportive services for their families. Some nursery schools in upper-middle class communities are viewed by overly ambitious parents as an early start in the world of formal instruction and peer competition. Some parents enroll their children in nursery schools or play groups simply to offer them a chance to play with other children their own age.

A number of studies have measured the impact of preschool programs on the development and behavior of young children. Particular attention has been directed to the influences of these programs on the vicissitudes of attachment and separation, intellectual skills, and social development. Although many questions remain unanswered, a number of generalizations can now be considered valid. Preschool interventions for socioeconomically disadvantaged youngsters have been demonstrated to correlate with improved long-term school performance as reflected in lower rates of grade retention and a decreased need for special education services. Early fears that young children in day care programs would experience a higher rate of emotional problems related to attachment and separation difficulties have not been substantiated. In general, most data suggest that the developmental-behavioral effects of day care and other organized preschool programs are quite variable and are ultimately a function of the quality of the program and the age, characteristics, and family circumstances of the child (Rutter, 1981, Zigler and Valentine, 1979).

IMPACT OF SOCIOCULTURAL INFLUENCES

Although the basic developmental tasks of the preschool years are the same for all children, the arena in which these normal "crises" are mastered can vary considerably. Since societal expectations and values dictate the definitions of developmental and behavioral normality, an understanding of variations among different groups is imperative. A physician who assumes responsibility for assessment and guidance regarding developmental-behavioral issues must be cognizant of the cultural milieu in which families live and must carefully avoid ethnocentric or arbitrary judgments and prescriptions.

Ethnic and social differences can appear in a variety of forms. Separation-individuation issues provide one interesting area of potential variation. The goal of increasing autonomy is certainly relevant for all young children, but the road toward independence is not uniform across all cultures. Some ethnic groups, for example, place a great deal of emphasis on early toilet training, while others adopt a laissez-faire attitude. The commonly practiced "late" weaning (from breast, bottle, or pacifier) of one culture may be characterized as "infantilization" by another. A suburban mother who is reluctant to let her 4-year-old play outside the home may be "overprotective," while an inner-city mother who forbids her 5-year-old to play unsupervised in an abandoned tenement building may be appropriately cautious.

Style of discipline is another area in which sociocultural influences on family function may be dramatic. While some societies consider physical punishment to be a sign of lost control, others readily endorse its appropriateness and efficacy. Not all cultural groups support the importance of verbal reasoning with young children, especially if the authoritarian role of the benevolent dictator has been the normative parental model for successive generations.

Sociocultural conflicts in child-rearing philosophy require a reflective response. The physician who shares the value system of his patients is obviously in the best position to be helpful. When differences do exist, it is the responsibility of the clinician to learn about the influences of the family's cultural milieu before any intervention attempt is made. Many cultural differences do not require modification. The developmental and behavioral health of the youngster is a better standard of measurement than an arbitrary formula for child rearing. Only when dysfunction or maladaptive behaviors are apparent may intervention be indicated. In such cases, resolution of cultural conflict is best engineered by considering the individual characteristics of the child and the parents as well as the values and demands of the society within which the youngster will be growing up. Constructive problem-solving that both legitimizes the family's cultural heritage and acknowledges the value system of the "dominant society" will maximize the chances for a successful intervention. (See also Chapters 11 and 12.)

CLINICAL SUPERVISION ISSUES

EVOLUTION OF THE DOCTOR-CHILD-PARENT RELATIONSHIP

The physician who provides care for families with preschool children has a unique opportunity to offer support and guidance through these vital years. For most parents, the child's growing autonomy and individuation engenders a feeling of pride related to the youngster's maturation as well as ambivalence about his diminished dependency. The desire for another infant and separation difficulties are common parental themes during this period. The youngster's identification with adult role models often prompts his caregivers to become more introspective about their own values and behavior. Most families survive these normal crises without assistance. Some can be significantly helped by the sensitive counsel of a physician who has come to know the family well. The doctor who demonstrates empathy and sophistication regarding developmental and behavioral issues can have a far-reaching impact on the lives of many children and parents..

The years from 2 to 5 also provide an important opportunity for the primary care physician to develop a special relationship with the child. Particular attention to this goal not only reinforces the youngster's increasing autonomy in the eyes of the parents but also builds a foundation for an independent relationship that will grow over the years and may pay important dividends during adolescence.

COMMON PARENTAL CONCERNS

Routine developmental and behavioral issues provide a focus for many of the problems that parents of preschoolers bring to the attention of their pediatrician. The clinician who listens carefully and who diligently looks for hidden agendas can help resolve many common concerns before they become magnified.

Regulation of Body Functions

Eating, sleeping, and elimination are major issues in the early infant-caregiver relationship that assume variable importance for families in the later preschool years. Although it is rare for a healthy child not to eat enough if food is made available to him, the range of appetites is broad. Thus, the use of growth curves is a much better indication of the sufficiency of a youngster's diet than parental criteria for adequate intake. Most "picky eaters" thrive; "skinny kids" are usually not malnourished; and relatively few obese youngsters have "gland problems." A complete history and physical examination will usually identify any underlying pathologic condition. For most preschool eating problems, sympathetic exploration of parental concern and an explanation of the range of normal eating patterns will bring about a simple resolution.

Sleep disturbances in the preschool years are extremely common. The monsters who typically roam the minds of these youngsters are particularly threatening when the sun goes down and the lights go out. This is an age when the boundaries between fantasy and reality are frequently blurred, and many children need repeated assistance in defining the distinctions. Night lights, transitional objects, and patient parental reassurance are usually sufficient for managing the child who has difficulty at bedtime or upon awakening in the middle of the night. Sympathetic understanding and a firm avoidance of nonsolutions, such as bringing the youngster into the parents' bed, will generally lead to a satisfactory outcome. (Diagnosis and managment of more serious sleep disorders are discussed in Chapter 32.)

The task of toilet training can be a simple game or one of the toughest battlegrounds for parents and their child. The age at which it is most appropriately begun is very much culturally influenced and ultimately dictated by the neurophysiological readiness and temperamental receptivity of the youngster. Daytime bladder control usually precedes bowel regulation, and girls tend to master both functions earlier than boys. Many American families today do not initiate toilet training until after 2 years of age, and it is not unusual for a normal, emotionally healthy youngster to be in diapers until almost 3 years. A helpful physician can assist parents resist their own peer pressure to begin the training process until the child is motivated to acquire this skill. A casual and nonconfronting approach when the youngster is ready to master this hallmark of autonomy should result in a relatively painless process for everyone.

It is important for the pediatrician to recognize that eating and elimination are highly charged issues for some adults. Their particular salience may be related to unresolved early life experiences that can seriously interfere with rational parental behavior. A supportive physician can help parents understand the impact of such influences and can provide guidelines for action that will minimize parent-child confrontation in areas where the parent cannot expect to win.

Discipline

One of the most important developmental tasks for the preschool child is to learn the limits beyond which certain behaviors are unacceptable and to develop a functional capacity for self-discipline. Confrontations between preschoolers and their parents are inevitable and frequent. The wise mother or father chooses battles carefully and avoids clashes over issues that are not important. When discipline or limit-setting is required, the manner in which it is carried out will vary among different cultural groups as well as according to individual differences in parents and children.

Debates about the relative merits and consequences of physical versus emotional discipline have continued for generations. Arguments generally focus on relative efficacy as well as on the long-term price that one may pay for immediate compliance. Most agree that the purpose of discipline is not to punish but to teach and help a youngster develop his or her own autonomous control. In this regard, positive reinforcement for appropriate behavior has been shown to be more effective than punishment for bad behavior. Expectations should be commensurate with a child's physical and cognitive abilities, and the reasons for specific expectations should be made clear.

The need for consistency is probably the most overworked cliché used by professionals who offer behavioral management advice to parents. On the one hand, it is obviously true that mixed messages and unpredictable responses will complicate a youngster's mastery of the criteria for appropriate behavior. Practically speaking, however, consistent behavior is much easier to recommend than to live up to, and no parents can completely avoid inconsistent responses to their children's challenges. The constructive physician will help parents understand that the ultimate goal is to aim for consistency and to shift the locus of responsibility from themselves to their child. The youngster who has been well-disciplined is not one who is submissive but rather one who exhibits self-control and positive self-esteem. Thus, successful discipline considers the individual characteristics of parents and children as well as their family circumstances and cultural environment. (See also Chapter 36 concerning socialization.)

ANTICIPATORY GUIDANCE

The pediatrician who has a supportive and trusting relationship with a family can be in a strategic position not only to respond to parental concerns but also to provide guidance in an anticipatory manner. Issues related to accident prevention and behavioral management are two areas that are particularly ripe for such an approach.

Accident Prevention

Accidents are by far the leading cause of death in childhood. Automobile-related fatalities make up almost half the total, with the remainder related to drowning, fire, suffocation, firearms, falls, and poisoning. Most studies have demonstrated a much higher incidence of accidental injuries for boys than girls. Over one-third of all accidents occur in the home.

The preschool child is particularly vulnerable because of his potentially lethal combination of increased mobility and independence with relatively immature judgment. Most of a youngster's learning during this period is through experience, and minor accidents are inevitable. The risk of serious injury, however, can be lowered. Preventive strategies fall into three categories. The first includes legislative or institutionalized efforts that reduce the likelihood or severity of childhood injury, regardless of caregiver behavior. Regulations regarding the flammability of children's pajamas, child-proof medicine bottles, and lower maximum temperatures for home hot-water heaters are examples of such actions. The second category of prevention includes measures that are well-publicized, but the efficacy of which requires individual caregiver implementation. A classic example would be the use of automobile seatbelts. The third category is less well-defined and relates to the general level of organization, supervision, and accident consciousness of the youngster's environment.

Guidance concerning accident prevention is a critical aspect of preschool pediatric health care. Some parents need assistance in achieving an appropriate balance between safety-conscious supervision and overprotective restriction of their youngster's play. Identification of the child who is particularly at risk for injury is an important primary care task. Research findings on accident proneness emphasize the contributions of both the child and his or her social circumstances. Some studies suggest that youngsters who are impulsive, highly active, and emotionally labile are especially vulnerable. Other investigations have noted that some agile and well-coordinated "daredevils" repeatedly survive dangerous situations unharmed. However, an increasing body of data suggests that children at greatest risk are those whose families are burdened by significant stress and disorganization. In such circumstances, a more active and directive primary care role may be indicated.

Perhaps in no other aspect of preschool health care is the traditional pediatric emphasis on prevention more compelling than in the area of accidents. The primary care physician must not only offer routine counseling but also provide vigorous anticipatory guidance for those families whose poor organization and limited resources for coping with stress may contribute to an environment where the risks of accidental injury are increased (Feldman, 1980).

Behavioral Understanding and Management

Behavioral styles are frequent matters of concern for parents. The emerging personality of the preschool youngster may create conflict at home as well as arouse parental anxiety about how the child's behavior outside the family is a reflection of his or her upbringing. Anticipatory guidance in the hands of a sensitive clinician can reduce the intensity, if not the actual incidence, of many such difficulties.

The range of individual differences in temperament or behavioral style in the normal population is well accepted. The relative influence of constitutional and experiential factors may vary, but there is nearly universal agreement that the youngster and his caregiver each contribute to the final product. The extent to which a child may be categorized as having a behavioral problem is largely related to the degree of dissonance between his or her particular characteristics and the standards established by the significant people in his or her life. In fact, problematic behaviors themselves are often the secondary symptoms and self-fulfilling prophecies that emerge from such adult-child conflicts.

The so-called "hyperactive" child provides perhaps the most common example of the dilemma of diagnosing preschool behavioral disorders. Activity levels in humans vary as widely as preferences for individual personality tempos. Thus, a youngster who is constantly reprimanded and labeled by his parents as "hyperactive" might, in another family, be repeatedly praised and affectionately characterized as energetic, curious, and exuberant. Conversely, what one preschool teacher classifies as "boys will be boys" behavior, another might label as a disruptive conduct disorder that requires repeated disciplinary action.

The primary care physician who is attuned to both the behavioral style of the growing child and the temperament and values of the parents can identify potential conflicts in their early stages.

When a question of "hyperactivity" is raised, the pediatrician can explain about temperamental mismatches and focus on issues such as attention, cognitive style, and the quality of the youngster's accomplishments rather than on the pace at which the child operates. Parental concerns about "withdrawn" and "passive" behavior in a preschooler can be similarly analyzed in the context of normal ranges and individual differences.

Sibling rivalries present another area of potential behavioral conflict the intensity of which will vary. Although some children demonstrate considerable regression after the birth of a new brother or sister, others seem to take on the new competition with arrogant indifference. The inevitable clashes between two autonomy-seeking preschoolers in the same family may present as minor annoyances or tax the mediating skills of the most patient parents.

The developmental tasks of the preschool period provide a useful standard against which the functional aspects of behavioral differences can be measured. If a child receives frequent negative feedback from adults and peers, and his accomplishments and self-esteem are diminished, active intervention is clearly indicated. If, however, a parent expresses concern about the behavior of a youngster who is developing a sense of increasing independence and competence, gets along well with peers, and feels good about himself, the problem may rest primarily with the parent's perceptions or attitudes.

Each child is born with an individual personality. That inborn style may certainly be modified by life experiences, but the boundaries within which change will take place seem to have constitutional limits. Each adult has individual preferences for childhood behavior. Most transactions between parents and their children involve mutual accommodations. When an unresolved conflict arises, the primary care physician who knows the family well can help distinguish between maladaptive behavioral disturbances that require treatment and individual differences in healthy children that require respect and acceptance.

ADVICE REGARDING PRESCHOOL PROGRAMS

Pediatricians are often asked for guidance on the advisability and selection of preschool programs. Although these matters have been omitted from traditional pediatric training programs, an understanding of the developmental needs of preschool children and a willingness to learn about local resources provide a sufficient data base for the physician to respond constructively to these questions.

One must first learn more about the parents' motivations, attitudes, and needs. Working mothers may feel guilty about abandoning their children at too young an age. Those who are forced to work because of economic pressures may feel that both they and their children are being victimized by the "system," while mothers who choose to work (full- or part-time) for personal fulfillment may feel ambivalent about "compromising" their children's needs for their own "selfish" desires. The pediatrician can help by reassuring parents that available data show that children of working parents who are provided with good alternative care suffer no ill effects and, in fact, may be more independent, responsible, and peer-oriented. It has been reasonably argued that a high quality day care or nursery program is much better for a youngster than remaining at home with an unfulfilled, frustrated parent. The rapid increase in the number of working mothers of young children as well as the rise in shared child care by mothers and fathers has advanced much faster than the growth of institutionalized supports within the American economic and social system. Working mothers must therefore deal with an unsympathetic and nonflexible work situation as well as the burden of responsibility for multiple roles (each of which alone could be a full-time job). When a working mother is supported by other family members and her employment is fulfilling, the experience can be beneficial for her children. Enlightened support from the primary care physician can provide additional reassurance in the process of securing child care arrangements (Zambrana et al., 1979).

For mothers who are home full-time with their preschool children, the consideration of nursery school may still be an important issue. Here again, the pediatrician's advice can be helpful. A high quality nursery school experience provides a potentially valuable opportunity for peer interaction and socialization. For parents who have particular difficulty separating from their youngsters, the astute physician might suggest that a nursery program could be a helpful influence during this critical period of individuation.

Many physicians are asked for advice regarding the selection of a specific program in the community. It is important to emphasize to parents that no prototype has been demonstrated to be clearly superior for all children. Generally speaking, a good program should have a warm, supportive, and noncompetitive atmosphere. It should provide opportunities for varied peer interactions as well as individualized adult attention. Differences among children may dictate more specific recommendations. A shy, slow-to-warm-up youngster, for example, would be better off in a smaller program with a higher ratio of staff to children. A highly active, somewhat disorganized child would probably do best in a more structured setting. Whatever model one selects, the qualities of the staff are obviously of paramount importance. Parents should be urged to talk with teachers personally as well as observe them on-the-job before making a final choice.

DEVELOPMENTAL ASSESSMENT AND PRESCHOOL SCREENING

Developmental evaluation of young children and screening for school readiness have become part of a "growth industry" in this country. Increased emphasis on early intervention for developmental disabilities and a rising interest in identifying precursors of specific learning disorders in the preschool years have stimulated a proliferation of screening programs. Despite the inadequacies of formal pediatric training, the primary care pediatrician is often asked to play a role in the assessment process. Some accept this assignment with enthusiasm, while others defer to developmental specialists. Specific options available to the practicing physician are discussed in detail in Chapter 51.

Regardless of the role chosen by the primary care physician, developmental assessment and preschool screening present a number of important challenges. As the professional with the longest—and usually the most intimate—relationship with the child and family, the pediatrician can make a unique contribution. He or she can construct a life profile that incorporates biological, sociofamilial, and developmental-behavioral data. The physician can elucidate the relevance of medical data and explain how they facilitate or constrain a youngster's performance. No other professional has access to all this information nor can anyone else authoritatively evaluate its importance.

For the pediatrician with sophistication in psychoeducational assessment techniques, the ability to serve an integrating function can be most useful. As a generalist with insight into the range of neuromaturational, physical, psychological, educational, and social factors contributing to a youngster's developmental status, the primary care physician can make sure that an appropriate mix of disciplines is included in an evaluation and that no single perspective is allowed to have a disproportionate influence on ultimate formulations or recommendations. Because of his status in the community, it is likely that the pediatrician who

Table 7–5. CAVEATS REGARDING DEVELOPMENTAL ASSESSMENT

Limitations of evaluation tools
Vague boundaries of normality
Uncertain natural histories of subtle dysfunction
Power of a "medical" diagnosis

chooses to make diagnostic statements or therapeutic recommendations will prevail over other dissenting professionals. Those physicians who choose to exercise this power actively have a responsibility to develop and maintain a high level of multidisciplinary expertise in this rapidly changing field.

A final caution regarding developmental assessment and preschool screening relates to the issue of diagnosis (Table 7-5). Moderate to severe handicaps can be reliably identified in the preschool years. Subtle dysfunction, on the other hand, is extremely difficult to label with any reasonable degree of predictive validity. The technology for evaluation has serious limitations. The natural histories of specific learning disorders and subtle developmental deficits have not been well documented. Distinctions between transient neuromaturational delays, mild disabilities, and normal variations are not always clear. With these caveats in mind, the primary care physician can play a vital role in preventing premature or inappropriate labeling of a youngster by emphasizing the distinction between risks and diagnoses. Unlike most areas of medical practice, developmental-behavioral intervention does not require a definitive diagnosis in order to determine a plan of management. The enlightened pediatrician can avoid the damaging effects of labeling by deemphasizing diagnosis during this extremely fluid period and focusing instead on descriptive analyses of the assets, liabilities, and needs of children and their families. (See Chapter 52 concerning diagnostic formulation.)

CONCLUDING REMARKS

The preschool years begin with the emerging individuation of the toddler within his or her family. They culminate with the entry into society of a little person with a sense of himself as an individual and a set of values that reflects the morality of his caregivers and culture. Under the best of circumstances, the preschooler experiences a sense of increasing competence and positive self-esteem. For parents, the realization that their child is developing an independent life of his own is both exciting and somewhat sad.

The primary care physician may have the privilege to share in this human drama and to help promote its healthy passage. As a clinician, he or she must learn to appreciate the richness and diversity of child development and not succumb to a diagnostic epidemic of developmental and behavioral disorders. When faced with problems, the successful physician is equally able to deal appropriately with disability and comfortably support normality.

JACK P. SHONKOFF

REFERENCES

Chamberlin, R. W.: Relationships between child-rearing styles and child behavior over time. Am. J. Dis. Child. *132*:155, 1978.

Feldman, K. W.: Prevention of childhood accidents: recent progress. Pediatrics in Review. 2:75, 1980.

Fraiberg, S.: The Magic Years. New York, Charles Scribner's Sons, 1959.

Garvey, C.: Play. Cambridge, Harvard University Press, 1977.

Gesell, A.L The First Five Years of Life. New York, Harper and Row, 1940.

Murphy, L.: The Widening World of Childhood. New York, Basic Books, 1962.

Rutter, M.: Social-emotional consequences of day care for preschool children. Am. J. Orthopsychiat. *51*:4, 1981.

Thomas, A., and Chess, S.: Temperament and Development. New York, Brunner/Mazel, 1977.

Zambrana, R. E., Hurst, M., and Hite, R. L.: The working mother in contemporary perspective: A review of the literature. Pediatrics *64*:862, 1979.

Zigler, E., and Valentine, J. (eds.): Project Head Start—A Legacy of the War on Poverty. New York, The Free Press, 1979.

8
Middle Childhood

Childhood's middle years are not a latent period. They are characterized by activation and change—a time of earnest searching, of goal directed exploration, of increasingly sophisticated decision making. This is a period for preparation and rehearsal. It also is an era of debuts and refined performances and of trials and errors. Diverse challenges and constraints entice and imperil the school child. To begin with, this chapter will explore some of the principal missions that await youngsters entering this age span. It will then survey some developmental acquisitions that facilitate the pursuit of these missions. Finally, there will be consideration of the health professional and the ways in which such an adult relates, forms alliances, and collaborates in the missions of middle childhood.

DEVELOPMENTAL MISSIONS

Children in this age group, motivated by sociocultural and endogenous forces, undertake unspoken missions in their quest of growth and fulfillment. Ongoing self-assessment and feedback either reinforce or inhibit the missions, which may be elucidated and classified according to various schema. In the following section, 12 such missions are described briefly as a way of offering insights into the life scenario of this age group.

Mission One: To sustain self-esteem

Feeling good about oneself is a prime requisite for mental health. For the school child, daily life brings forth exposures and experiences that have the potential for strengthening, maintaining, or reducing self-esteem (Coopersmith, 1967).

To explain the quest for self-esteem, certain generalizations can be helpful. Many factors influence a child's feelings of worth. In this age group, these include feedback (i.e., either praise or criticism) from peers, reinforcement from respected

adults, a record of success or mastery, an accommodation with social standards, a sense of having met endogenous or personal needs, confidence in the ability to recover from anxiety, a feeling of being in control, and optimism regarding future challenges or stresses.

Influences on self-esteem are multiple and diffuse (Coopersmith, 1967). Patterns of nurturance, early life experiences, successes and failures, and innate abilities may interact with an internal system of self-assessment to produce either feelings of worthiness or predominantly negative sentiments and reduced self-esteem.

Self-esteem is not a fixed attribute. As children progress through their school years, it is likely to fluctuate from day to day and even hour to hour. Although a general level can be described, it is normal for the developing school child to engage in a cyclical process of questioning and reconstructing feelings about self-worth. Occasional crises or influential encounters may necessitate drastic re-evaluations. Often such momentary turmoil is not evident to adults. Children may disguise anxieties or deny personal feelings of inadequacy, perhaps believing that an appearance of confidence—a veneer of "macho"—is nearly as good as feeling good about oneself. In fact, some youngsters who seriously and chronically doubt their own worth tend to act in a manner that will obscure their dwindling self-esteem. They may be too aggressive, boastful, or controlling as part of an unconscious or planned campaign to conceal the embarrassment of self-doubt.

It should be emphasized that it is healthy for school children to struggle somewhat with issues of self-esteem. It is appropriate that they explore different techniques and measures to sustain self-esteem, to test and to prove for themselves their own merit. On the other hand, some children may fail to achieve reasonable self-esteem. Typically, this is the end product of a series of setbacks, losses, nonreinforcing life events, and failures to meet personal standards and those set by peers and adults. Loss of self-esteem is a common component of childhood success deprivation and depression (see Chapter 41).

Mission Two: To find social acceptance

The school years are a time of coming out—a challenge to the child to "make it" beyond the protective shroud of his or her family. Primarily, this mission entails a quest for social acceptance and admission into one's peer group. For most school children this is a matter of utmost concern, and it commonly becomes a strong preoccupation. For some youngsters peer acceptance is gained with ease, while for others it exacts inordinate effort and the expenditure of great energy and anxiety. On the list of a child's priorities, popularity is likely to rank near the top. It is not unusual in this age group to observe the sacrifice of autonomy and free will for the cause of social success.

The acquisition of social skills will be described later in this chapter. Children are variably prepared to confront the testy challenges set up by peer groups. School becomes a critical arena in which one interacts and makes comparisons between oneself and one's peers. When faced with the choice of pleasing the teacher or appealing to a peer group, many school age children feel constrained to select the latter!

As children grow older, their social effectiveness is measured by their status among peers in general and by the richness of their close relationships. A child's social status may be influenced by many factors, such as physical appearance, the presence or absence of acceptable mannerisms, patterns of behavior, competence in specific performance areas, personality, somatic maturation, and the child's own repertoire of social skills. Some children who are lacking in one or more of these areas develop a most recalcitrant condition, namely, a bad reputation. Those who are popular and esteemed by their peers actually gain in social status by being particularly creative in the ways in which they "put down" socially inept youngsters. This aggravates the plight of the child with a bad reputation. Some socially successful children appear to need to prey upon youngsters who are unpopular. It can be frustrating for an unpopular child to try to climb the social ladder. Withdrawal, maladaptive patterns of interaction, and diminution of self-esteem may be the result.

It is possible for a child to receive too much gratification from his or her social standing. This phenomenon of "too much too soon" may eclipse other priorities for the child. For example, a youngster may underachieve academically as a result of excessive peer adulation. Such a child may also be deprived of the opportunity to develop face-saving strategies and coping skills because these have never been necessary. In a sense, part of the youngster's social education has been forsaken. Obviously, an appropriate balance between downright deprivation of social success and an overdose of such gratification is most desirable.

Mission Three: To reconcile individuality with conformity

The school age child becomes increasingly aware of his unique preferences, strengths, and styles. These then find expression in a range of social contexts. The growing child faces enormous pressures to conform both to adult expectations and to the stringent standards of taste and performance mandated by peers. A child's unique characteristics or interests may differ from those of the peer group, sometimes engendering ostracism and isolation. For most children there is a process of transaction that allows them to modify but not sacrifice totally natural inclinations to yield authentic but socially acceptable patterns of behavior and taste. Eccentricity in this age group is rare. Many cultural-social forces and traditions coerce children toward conformity and uniformity. These include contemporary fads or fashions in the community, religious values, current notions of gender differentiation (see Chapter 10), and competitive pressures and prevailing ideas about success.

A child who likes to collect insects while most others in the neighborhood opt for sports, one who prefers to stay indoors and read rather than pleasing parents and peers by being outside on a beautiful day, a child who disdains the "cool" vernacular used by peers, a youngster who does not wish to accept contemporary dress codes and fashions—all such children may have to struggle to sustain these individual preferences. For most, a conflict between individuality and the pressure to conform may not be as dramatic; nevertheless, the need to reconcile innate preferences with exogenously conditioned expectations is a constant tension, one that may lead to confusion over a child's authentic needs and tastes. The impact of this process increases during this age span. Children of late elementary and junior high age are particularly susceptible to such coercion. In fact, the subculture between ages 11 and 14 may resemble the harshest of totalitarian governments, the despot being the peer group. These young citizens may sacrifice freedom of speech, follow a stiff behavioral code, act in accordance with preset standards (i.e., "coolness," "macho"), and adorn themselves in uniforms that adhere meticulously to current modes. Further elaboration of the social scene in early adolescence can be found in Chapter 9.

Mission Four: To appropriate role models

Children in this age group often seek out and try out models. Various attributes and interests may be borrowed from peers, older children, and adults. In a sense, the school child embarks on an expedition, studying those he esteems, sometimes exaggerating or fantasizing about their virtues, and commonly adapting or attempting to incorporate their qualities. Models may include a parent, an older sibling, other children, a television star, a baseball player, a teacher, or a physician (or all the above).

The process of modeling and identification is a helpful one, and it is universal. At times parents may feel that a child is emulating an inappropriate prototype or is excessive or even obsessive about it. Sometimes entire communities of children extract from the same model—perhaps a television personality or musical sensation. Commercial enterprises may, in fact, exploit the propensity by issuing T-shirts and other paraphernalia that can induce and energize imitation and adulation.

Role models within a family are particularly influential. To some extent, all children identify with mothers or fathers. One parent may be a particularly compelling influence. It is not unusual for a child to try on attributes of a favorite and close grandparent, aunt, or uncle. Modeling can influence academic achievement. Children who have never seen their parents read a book or write a letter may experience more trouble in acquiring academic skills than those who have witnessed these practices throughout early childhood. The drive to imitate is an important part of modeling. In a sense, the school age child often rehearses or tries out an assortment of practices and modes of behavior observed in the adults he respects. Some imaginary play and even athletic efforts of children are based upon this desire to assume the identity and functions of adult role models, to move in and, in a sense, become that model. A strong sense of fulfillment is bestowed upon a child who watches his favorite football star on television and hours later finds himself on the playing field going through the identical motions and perhaps fantasizing that he is that very star.

The modeling processes that are so important in school age development can become either positive or self-destructive influences. Therefore, an important part of developmental health maintenance in this age group is the promotion of healthy types of identification.

Mission Five: To examine values

Emergence from the security of home and the sometimes abrupt entry into an outside world entail a series of cognitive and behavioral adjust-ments for the school child. One of these involves the reconciliation of disparate values. Expectations at home may contrast with those at school. A child's personal and family beliefs may be diametrically opposed to those of a neighborhood friend and his parents. Standards of discipline may differ from those of friends and from friend to friend, and economic values may be sharply contrasting. For the school age child encountering such discrepancies, there is a constant reawakening and sometimes a chain of disconcerting realizations.

Before entering kindergarten, the child assumes that values assimilated at home are absolute and universal. Social exposures in school and the neighborhood begin to introduce *relativism,* or the notion that standards are dependent upon contexts and backgrounds. Monetary values depend in part upon how much money you have. Religious beliefs are founded both on one's creed and on the extent to which it is observed. These awakenings may lead a child to question personal norms; it is not unusual for the seeds of rebellion to be sown as an older school child comes to recognize not only that there exist other possibilities out there, but also that at least some may be superior to those of his own family. The child who begins during early school years assuming that the way things are done at home is universally acceptable soon discovers that there may be more than one set of criteria; ultimately, in preadolescence, he recognizes that at least some of the cherished values at home may be abnormal, arbitrary, or inappropriate. This scrutiny may persist through adolescence. The critical re-examination of previously accepted values and a healthy level of skepticism (and sometimes cynicism) may be important precursors of a late adolescent reaffirmation and commitment. It is wise not to thwart fundamental questioning but to allow children to become introspective and to inspect family values and pursue the processes of comparing and contrasting with others.

Mission Six: To "make it" in a family

Although "breaking away" from earlier dependence upon the family is an important accomplishment of the school years, home remains a critical arena for development throughout this period. Children need to feel as successful at home as they are in their other performance domains. In this age group, youngsters return home each day like warriors back from the plains of battle. They would like to feel and be perceived as triumphant. They would like to get credit at home for what they have achieved afield. In this age group, youngsters have a profound appetite for well-earned positive feedback from family members.

When this fails to occur regularly, either because such reinforcement is unavailable or because there is actual success deprivation and failure in the world beyond the domicile, some of the impetus for healthy development may be lost.

The need to "measure up" at home is a compelling one. Probably the most convenient and most accessible gauges for such measurement are siblings. Brothers and sisters form an influential part of a child's environment, and the standards they set must be taken seriously. They serve as models and competitors, as playmates and adversaries, as gratifiers and humiliators. Therefore, sibling rivalry is an expected phenomenon. Intense competition, conflict, and resentment may be intermittent or constant between siblings during school years. Indulgence in such intense crossfire may actually serve as useful training and practice in the art of social conflict resolution. At an appropriate level, sibling rivalry may also facilitate a child's social education. Parents may need help in their efforts to contain but not suppress such transactions and to insure that no one child is overly oppressed by this developmentally normal warfare.

Being able to "make it" with parents is critical. School age children need to know not only that they are loved by their elders but also that they are respected. There may be a constant drive to impress parents, to solicit their approval, to meet or surpass their expectations.

When youngsters are thwarted excessively in their efforts to gain parental approval, they may tactically change course and become rebellious. If they feel too criticized, if there is a sense that parents are impossible to please, if adult disapproval far outweighs praise or positive reinforcement, a child may be constrained to seek a more appreciative audience outside the family, sometimes from peers who are similarly alienated from their mothers or fathers. Sometimes parents may be unaware of how important it is to demonstrate their recognition of a youngster's strengths, to praise and acknowledge triumphs, to balance compliments with criticisms, and to prioritize and set realistic goals. They need help in communicating approval and should understand that school age children harbor an intense desire for their parents to perceive them as successful and worthy. If children feel this mission is unlikely to succeed, they will abandon this most important and healthy drive, and the toll on self-esteem may be heavy.

Mission Seven: To explore autonomy and its limits

It is not unusual for the school age child to experiment with the constraints imposed upon him by the adult world. He may try repeatedly to determine what he can "get away with." A child may simultaneously take comfort in the regulations promulgated at home and in school while deriving satisfaction from violating them! In particular, older school age children often commit minor social taboos (such as using vulgar language, lying, and stealing insignificant objects). In moderation, these experiments may represent normal testing behaviors during the school years. Various manifestations of defiance may also be within normal limits. Parents may be concerned about minor infractions, about a child's tendency to exaggerate or make up stories, or about his unwillingness to accept discipline or assume responsibility. In fact, it can be hard for them to draw the line between a normal testing of limits and truly difficult behavior.

As school age children progress toward early adolescence, their drive for autonomy intensifies. There may be internal conflict between an intense desire for greater freedom, on the one hand, and lingering feelings of dependency, on the other. In late elementary school this precarious balance engenders awkward situations. A child may want to stay out and play later but be equally desirous of having his parents available when he returns. It should be emphasized that when children seek autonomy, not only are they experimenting with the constraints imposed by parents and society, they are also—perhaps more importantly—testing themselves. There may be an unconscious rehearsal process, an attempt to discover whether they can survive without the help of adults, an effort to find out what it will be like to be grown up. Cigarette smoking, use of adult language, membership in cliques, and increasing adventures after dark may all represent auditions for the biggest show—adult autonomy.

Mission Eight: To acquire knowledge and skill

To observe oneself becoming increasingly competent and masterful is a reinforcing experience for a school age child. Children in this age group need to feel that they are growing more able, that they are mastering skills. School is of course the principal arena for this effort; however, skill and knowledge acquisition are not confined to the classroom. Children in this age group also learn from one another, from their parents, and from a wide range of avocational and recreational exposures. There is greatly expanded knowledge of how things work. As will be discussed later in this chapter, there is a growing ability to generalize, to understand processes, to think in abstract terms. Knowledge of the universe expands. Children strive for physical mastery. For the school age child, the process of learning may be a constant source of gratification and reward, or it may constitute a hazard to self-esteem. School performance thus plays a critical role in his development. A youngster in a classroom is aware of the accomplishments of his peer group. When he feels that he does not "measure up," maladaptive strategies may be appropriated to save face. The school child has little tolerance for failure. When knowledge and skill are not accruing rapidly enough, he may give up rather than face the humiliation of having tried and failed.

Beginning at about age seven, school children are preoccupied with whether or not they are "smart." Derogatory terms such as "retard" are unfortunately quite common and reflect a concern for at least a modicum of intellectual prowess. Many children would rather be regarded as "bad" than "dumb." They would prefer to get into trouble for not completing an assignment in school than to endure the humiliation of bad grades and pages streaked with a teacher's critical red marks.

During the school years, a drive emerges that is likely to operate throughout life, namely, the compulsion to feel successful. For school children, acquiring and applying new knowledge and skill, and recognition of this by peers and adults, constitute one process through which success is perceived. Triumphs in the classroom, on the playground, in the neighborhood, or at home are fortifying. Parents and other adults who want to ensure the optimal development of a school age child may need to program successful experience into such a youngster's domain. They need to be certain that the child is not suffering from success deprivation. Too much failure during these years may constitute a developmental emergency!

Mission Nine: To live within one's body

An important component of the development of the schoolage child is the progressive realization of the extent to which one is entrapped in a body. Various anatomic and physiologic advantages and constraints become an important part of experience. The school child is a student of bodies—his or her own, those of peers, and those of adults—but may feel uncomfortable talking about personal somatic concerns. Becoming reconciled with one's own body is a critical and ongoing developmental challenge. Much experimentation occurs during these years, as children set out to investigate bodily functions and limitations. A child evaluates himself to determine whether his physical appearance, body build, and coordination compare favorably with those of his peers. Sporting events can be one mechanism for this, because they permit the child to explore a variety of corporal performance frontiers. This is but one of many physical challenges. The school child must also learn to deal with pain and discomfort (see Chapter 28) and to tolerate illness.

Acquistion of secondary sexual characteristics becomes an increasing interest as children approach early adolescence, and it is not unusual for youngsters throughout this age group to become curious about the bodies of other children. A common occurrence, one that frequently alarms parents, is when two or more youngsters unclothe (to varying degrees) and "play doctor." Experiments and explorations such as this allow youngsters to pursue one of their major interests, namely, bodily similarities and differences. In a sense, they are acquiring norms. They also experiment with their own bodies. Various forms of masturbation, for example, may be developmental learning experiences. It is only when these become obsessive or extreme or when adults over-react to such activities that they become clinically problematic. It is probably fortunate that most children who undertake bodily investigations never get caught! For those who are surprised in the act, it is most helpful to provide reassurance of normalcy and to alleviate needless anguish or guilt.

It is not unusual for school age children to believe that their bodies are not right, that one or more anatomic parts are of the wrong size or shape, that something does not work right, that the soma is defective in some manner. Usually they do not discuss such concerns with parents. With encouragement, they may be willing to confide in a physician. It can be helpful for a school child to learn of the universality of this phenomenon. Youngsters in this age range may not be aware of the wide normal variation in bodies with respect to looks and functions or of how common it is to worry about one's body.

School age children express somatic concerns in a variety of manners. Some conceal the blemishes about which they are most concerned. Extreme modesty about undressing may be evident. Other youngsters may refrain from activities that expose self-perceived shortcomings of anatomy or body performance. A child with nocturnal enuresis may refrain from sleeping at a friend's house (see Chapter 31). A youngster who is obese may decline an invitation to go swimming. Certain children may conveniently forget their gym suits when they are likely to be humiliated over their poor gross-motor function on the playing fields. It is a common phenomenon in many communities for children to be required to take a shower in physical education classes beginning at about age 12 or 13. Word may spread throughout town, and youngsters may fret for months over the prospects of this threatening exposure. It is important to reassure children about both the uniqueness and the sameness of their bodies while at the same time respecting their privacy and allowing them to appropriate protective strategies. The important interaction between physical maturation at puberty and behavior and school adjustment is considered in detail in Chapter 9.

Mission Ten: To deal with fears

A young child approaches the school years with a bevy of fears, many of which were established considerably earlier in life. There is of course wide variation among youngsters in terms of the targets of their apprehension. In early elementary school, fears of the dark, concerns about monsters and ghosts, idiosyncratic phobias, squeamishness about certain animals (such as snakes, mice, and spiders), and preoccupations with death and injury are typical. Particularly common are fears involving food intake. A child may like peas and enjoy mashed potatoes but become quite anxious when the two come in contact with each other! Some children fear lumps in oatmeal, while others are reluctant to swallow pills, firmly believing somehow that they are likely to choke on them.

No six or seven year old is entirely liberated from such fears. An important mission during school years is the attempt to explore and modify, minimize, and perhaps entirely eliminate targeted fears. Children may do so by becoming increasingly educated about the foci of their concern. Intellectualization and knowledge can be used to overcome anxiety. Through education and experience, a youngster may come to realize and confirm that ghosts do not exist. Usually this requires more than parental reassurance; somehow a child must arrive at a state of readiness to accept and assimilate such an accommodation.

School age children are likely to employ many mechanisms to deal with fears. They may explore these through play. A child with an inordinate fear of injury may actually enjoy playing imaginary dangerous games. Some youngsters with concerns about monsters take great pleasure in reading or observing horror stories on television. In recent years, a game called "Dungeons and Dragons"* has helped some of them to deal with a variety of supernatural threats by actually creating these hazards themselves. Some youngsters with many apprehensions compensate by showing a great deal of bravery and "macho," trying somehow to acquire equanimity by pretending to be tough. There is a sense that one can become fear-free by acting fearless.

Fear of the future is common in school age children. They may be apprehensive about what comes next, about future challenges, or environmental disasters, about the possibility of failure and humiliation. Life transitions may be particularly difficult. A fear of future losses (of relatives, of friends, of possessions) also may be prevalent.

Another common concern in the fear of losing control. Recently the concept of "locus of control" has become recognized (Lefcourt, 1976). Many youngsters may be preoccupied with losing this, with being stripped of their decision-making efficacy, with seeming to be weak or weakened. They may feel that their fate is not within their control or that they will lose control over themselves as well as over others (see Chapter 41). Such apprehensions are common. They become abnormal when they are excessive and interfere with daily function and with other missions of development.

Fears present critical developmental challenges to school children. The process of understanding and controlling or overcoming them is perpetual and becomes an essential form of preparation for adult life. Most youngsters accomplish this with little permanent scarring. Others may require more support in confronting and containing fearfulness.

Mission Eleven: To deal with appetites and drives

Learning how to obtain satisfaction is an important mission for school age children. As the school child experiences appetites and desire, there is a drive toward fulfillment and satisfaction. As is discussed in Chapter 38, some youngsters with attention deficits suffer because of their insatiability. They appear to be in a steady state of hunger. As school children mature, they become increasingly sophisticated at identifying and finding ways to satisfy or curb appetites. Multiple hungers come forth, including desires for specific foods, sexual drives, quests for praise and success, want of material things, searches for pleasure, and the wish for the attention of others. Each of these is represented to varying degrees at various times in the development of school age children. Any one or more may become excessive (even obsessive).

An important mission thus involves finding the tools to limit and control such appetites and drives. The means of doing so may be painful and yet in the long run more rewarding than total capitulation to desire. Included are such processes as compromising with others when there is a conflict over desires, sharing, settling for fulfillment in less than anticipated dosages, delaying gratification, and accepting substitutes or replacements. The process of maturing for the school age child entails, in part, the acquisition and assimilation of these means of control. When this does not occur, children become increasingly frustrated

---

*Dungeons and Dragons, TSR, Lake Genera, Wisconsin. This popular noncompetitive game encourages youngsters to formulate their own "plots" and rules as they proceed to assume the identity of various fictional characters who differ in their attributes and capacities to cope with dangers.

and anxious. There is enormous variation with regard to the capacity to satisfy or extinguish "burning desires."

Mission Twelve: To refine self-awareness

The age-old admonition, "Know thyself," characterizes another mission. Throughout elementary school a youngster is acquiring increasing insight into himself or herself. Formal education is one source of such knowledge, while studying of peers and their attributes is also informative. Observations and feedback regarding one's own performance and limitations further shape the child's sense of identity. Finally, a process of introspection is applied to this effort. The school age child progressively develops a sense of what he or she is "good at" and what he or she is "bad at." There is a growing capacity to articulate and confess specific strengths and weaknesses. Children between ages of six and eight are more apt to deny deficits or to refuse to think or talk about them. By age 12, youngsters are more prepared to reckon with these, although commonly there will still be reluctant to discuss the ways in which they might be different from peers.

The first eleven missions described in this section play roles in shaping a child's identity. As school age children accumulate experience, they become increasingly perceptive about themselves. They observe their own patterns of behavior, ways of reacting, and targets of fear and confidence. They integrate these as they climb toward higher levels of self-awareness. In doing so, they are liable to pass through brief or prolonged periods of confusion. Occasional outpourings of despair may be reported by parents. Children may use a variety of indirect means to discover more about themselves, such as making provocative comments like "Nobody likes me" or "I'm not good at anything." The responses to these experimental self-effacing statements can be important in helping a child see himself as others see him. As children grow older they learn more about themselves by talking about other youngsters. In gossiping with friends or being "catty" about someone else, a child, in reality, may be analyzing himself or herself. Some older elementary school children become very interested in reading biographies. They become close to a particular friend and like to engage in intimate discussions partly as a means of increasing self-awareness. By finding someone to whom they can confess, someone whom they feel they can trust, they may be able to define their own identity simply by describing it.

As children grow in their self-awareness, several issues become increasingly important: To what extent is the child willing and able to accept the way he or she is? How realistic are the youngster's self-perceptions? How prepared does the child feel to meet the challenges of adolescence and adulthood? Those who live with and help children in this age group need to be sensitive to the child's own sense of identity. To the degree that school children develop an authentic or viable version of who and what they are, they are likely to find fulfillment and gratification during adolescence.

DEVELOPMENTAL ACQUISITIONS

Success in the school years is facilitated through the acquisition of new abilities, strategies, and insights that become the means for discovery and accomplishment, the implements for carrying out the aforementioned missions. Psychologists, developmentalists, and other students of childhood have advanced various conceptual models relevant to development in this age group. The reader is referred to the works of Freud (1965), Erikson (1959), Piaget (1968a), and Gesell (1946) (see also Chapter 31). Many possible schemes can be suggested to account for change over the school years. In the present section, a framework has been constructed to be particularly relevant to developmental pediatrics and to the understanding of variation and deviation in this age group. Some basic developmental acquisitions will be described:

- Attention, persistence, and goal-directedness
- Orientation and perception
- Storage and retrieval
- Interpretation and generalization
- Expression and production
- Social reception and interaction
- Protective resiliency and strategy formation
- Other acquisitions

The following sections elaborate upon each of these eight areas of developmental acquisition.

ATTENTION, PERSISTENCE, AND GOAL-DIRECTEDNESS

During the early elementary school years, a progressive honing of selective attention is a major accomplishment. Children in this age group increasingly are able to concentrate on the most informative stimuli in their environments for greater periods of time. A third grade student can focus attention in a classroom and pursue to completion a prolonged exercise more effectively than he could in kindergarten. To do so requires greater tenacity or persistence and a growing ability to filter out what is irrelevant. Both behavior and learning become increasingly goal-directed and progressively less random during the elementary school years.

A number of acquired traits contribute to the strengthening of selective attention and the lengthening of task persistence in this age group. Important is the child's growing ability to delay gratification. A toddler or preschool youngster commonly demands immediate rewards. His goals generally are of short range. Desires and felt needs call for immediate satisfaction. Delayed consequences and events in the distant future have far less motivating power for preschool children. As youngsters enter the school years, the very foundations of education emphasize the concept of preparation (a form of delay). Children become increasingly receptive to the notion of "getting ready" for later life. Enormous pressure is brought to bear, coercing them to look ahead, to sacrifice immediate pleasure for something better later on. Even the Boy Scout troup leader keeps reiterating his faithful motto, "Be Prepared"!

Most school age children encounter little difficulty in sustaining delays of gratification. As they proceed through the elementary school years, successful youngsters realize increasingly the rewards of work and sacrifice. Earlier hedonistic inclinations are tempered, as they come to recognize that completing homework, delivering newspapers, performing family chores, or enduring dull sermons in a place of worship may portend later and greater benefits than those attained through instant gratification. By being able to postpone their reinforcement, children become increasingly able to persist at tasks, to find their way through immediately unrewarding detail, and to engage in activities whose short term motivational content is minimal. These capacities clearly are strengthened when there is a likelihood that children can reap some intermediate profits along the way; those whose efforts consistently meet with failure may have little reason to invest their persistence and patience.

Closely related to the delay of gratification is the capacity to obtain satisfaction. Later in this book (Chapter 38) a group of children will be described who have great difficulty in feeling satisfied. Their insatiability becomes a distraction. They seem never able to arrive at a state of gratification, despite the fact that their appetites and desires are felt so keenly. They may want a particular object, but as soon as they obtain it, it appears to lose its previously attractive qualities. Such youngsters find themselves in a "steady state of hunger." The ability to curb appetites, to crave and then to feel at least temporarily fulfilled represents an important developmental acquisition in this age group. The process of wanting, seeking, and deriving sustained satisfaction energizes task persistence.

A growing capacity for satisfaction may be accompanied by the progressive development of strong areas of interest that can be pursued in depth. Some youngsters, on the other hand, cannot proceed beyond an initial stage of romantic attraction to subject matter. Most children, however, grow in their ability to approach a hobby, an area of interest, or a sport and continue over time to become increasingly involved and adept. This capacity is closed related to a child's ability to derive some satisfaction during the act of pursuit.

Linked to the enhancement of selective attention is the child's improved function as a planner, one who grows steadily in the capacity to reflect on potential actions before undertaking them. Children learn to predict social and personal consequences. Increasingly they are able to inhibit counterproductive impulses, such as those that are apt to lead to inevitable punishment or ridicule. Competent school children become increasingly proficient at "editing" the many possibilities for action that present themselves. Some plans are facilitated, while others are aborted or modified. Multiple factors are integrated as part of this process. Cultural standards, moral values, and social skills play a role. Selective inhibtion of impulses and the development of reflective behaviors also have a profound impact on learning and school performance (see Chapter 38).

Another conditioner of persistent attention is the child's capacity to stave off fatigue, to remain aroused and alert during prolonged purposeful activity or concentration. This too is a progressive acquisition during school years. The preschool child may have a limited ability to pursue a task without becoming excessively tired and encountering the ever-present hazard of boredom. This is seen particularly with regard to cognitive effort. As children mature through the school years, they increase their ability to remain alert while probing, problem solving, and monitoring their own output, even during relatively low motivational activities.

It can be seen therefore that for the school age child the ability to become increasingly selective and goal oriented, to persist, and to be reflective are most useful acquisitions. By the age of 12, a child is able to exercise more discrimination over stimuli that enter the central nervous system while implementing purposefully his own activity level and direction.

ORIENTATION AND PERCEPTION

The school child continues the exploratory work of earlier life and becomes increasingly knowledgeable about his environment. There is a continuing stabilization of orientation, one that enables the child to perceive himself relevant to various ordinate points in daily life. Three principal forms of orientation can be cited: body awareness, appreciation of spatial attributes, and awareness of time relationships. In their daily activities and in the process of learning, school children gain sophistication in these three dimensions.

The school child's perception of his own body evolves steadily. During late preschool and early elementary years, he becomes increasingly proficient at naming body parts. Fantasies about the purposes of various anatomic components give way to knowledge and demystification, while children become adept at managing their bodies independently (see Chapter 35). They are able to respond appropriately to bodily needs and become more autonomous in managing discomfort and other inconveniences incurred by the soma. As they progress through school years, youngsters take increasing pride in their appearance. They may agonize over blemishes, unwanted protuberances, conspicuous dimensions, and the like.

Body position sense and gross-motor function improve, so that children can acquire feelings of mastery over physical space. This is a requisite for effective participation in sports and also for artistic, writing, and craft skills. By age seven, most children are able to discriminate left from right. They then perform increasingly complex operations regarding left-right distinctions, including the ability to identify right and left parts on other people or objects facing them, a sign of enhanced sophistication with regard to body awareness and "outer space."

Spatial perception is another area of development in this age group. It entails the capacity to perceive physical relationships beyond one's own body. The child's interpretation of visual, haptic (i.e., touch), proprioceptive, and kinesthetic data becomes increasingly strengthened. Children are able to perceive more complex and more detailed configurations in the spatial environment. Standarized tests requiring them to copy geometric forms or match complicated designs demonstrate a growing ability to attend simultaneously to multiple aspects of a pattern and to see their interrelationships. Thus, a preschool child may be able to interpret and copy only a simple right triangle, while a nine year old can perceive and utilize a triangle that is at a 45 degree angle on the page and has a second triangle embedded within it. A 12 year old can add to this array the attributes of three-dimensionality. Such progressively sophisticated spatial orientation and perception is important in learning to read and write.

Subtle discriminations in the visual symbol system are facilitated as the child's spatial orientation and perceptual abilities improve. As school age children grow older, in fact, actual perceptions and visual analyses become increasingly less im-portant insofar as they are able to interpret incoming data with more reliance on memory and conceptualization. Thus, a word may be decoded not exclusively by its spatial attributes but more by the overall semantic context and its nearly instant recognition through visual memory and attached verbal associations.

Orientation in time and the capacity to assimilate and make good use of temporal relationships are critical themes of development in this age group. Closely linked is the ability to understand and store sequences of information when the serial order of their components is critical to their meaning. Arranging words in the correct order to form a grammatically acceptable sentence, remembering the sequence of letters in a word, processing three and four step instructions, and completing long division problems are examples of these sequential operations. As is discussed in Chapter 38, children with sequencing problems often experience academic lags. Time and sequential awareness also contribute to a child's organizational abilities. The school years are characterized by an increasing ability to retain, retrieve, and deploy sequences of information. Between ages six and 13, a youngster is able to engage in activities that depend upon increasingly complex sequences. A five year old may be able to follow only a three step instruction, while at age 12 he masters six or seven step directions. The ability to tell time, to learn the days of the week and months of the year, and to master time-related vocabulary (e.g., before and after, now and later, yesterday and tomorrow) may be markers of this aspect of development.

STORAGE AND RETRIEVAL

During the school years, the facility for storage and easy retrieval of information undergoes steady expansion. The capacity to preserve and then find previously encountered data is critical for learning, for the interpretation of new experience, and for the acquistion of skills.

Many trends during the school years produce a generalized enhancement of memory. A growing proportion of stored fact and routine becomes easy to recall quickly. This progressive automatization is a prime facilitator of learning and productivity. A five or six year old child may struggle to recall the "blueprints" for constructing certain letters. As he progresses through the school years, visual retrieval of symbolic configurations becomes increasingly efficient, unconsicious, and rapid. Such facilitated memory enables the child to concentrate on more sophisticated ideas and processes rather than on the mechanics of letter formation. Progressive automatization, the expanding mass of skill and knowledge that can be drawn upon effortlessly, is a most accelerated aspect of memory growth in this age group.

Another important trend with regard to memory is the emergence of a series of strategies for retaining new information. Youngsters acquire a repertoire of "tricks" needed to store relevant data more effectively. These include mnemonics, the ability to associate newly presented information with previous experience, and the use of highly specific memory strengthening techniques. For example, as a child approaches age seven or eight, the use of verbal cuing or labeling is increasingly applied. If a youngster needs later to recall something he is seeing now, he may strengthen the "memory trace" by whispering under his breath (so-called subvocalizing) to reinforce verbally the stimulus set that is entering through his eyes. By so doing, he acquires a multisensory input; i.e., the same general information passes through visual and verbal channels, and this allows information to be registered twice. Morever, a child may find that by whispering to himself while reading and memorizing visually presented information he concentrates more intensely on the data.

The school child's cognitive development also facilitates retention and retrieval. For example, the capacity to classify incoming data makes such information easier to store. An older school age child may memorize a list of words by subgrouping them into categories (e.g., plants, animals, verbs). A five or six year old would be unlikely to use such a device to enhance memory. The school child's retentive capacity is further enhanced by the use of rules and generalizations. Increasing familiarity with these during school years greatly expands memory storage space. Cognitive acquisitions further strengthen memory by improving comprehension and the understanding of phenomena (see next section). The more effectively a child is able to discern logical relationships in what he sees and hears, the easier it is to store and retrieve such data. By understanding, categorizing, relating new data to previous knowledge, and employing various "rehearsal strategies," children can preserve increasingly large chunks of information as they progress through the school years. Their growing facility in this area contributes generously to several of their developmental missions, including the acquisition of skills and knowledge, the sustenance of self-esteem, and the growth of self-awareness.

INTERPRETATION AND GENERALIZATION

The rapid growth of language and cognitive ability in the school age child has been studied and written about extensively. The preadolescent child displays competence in thought processes and in comprehension that surpasses dramatically that of the prekindergarten youngster. Greater facility with language and the ability to understand phenomena on a increasingly abstract level characterize these revolutionary processes.

Over the school years, the child becomes adept at generalizing from experience. He readily applies rules to situations and discerns common denominators that link diverse experiences and stimuli. He becomes increasingly adept at detecting recurring themes and at storing inputs according to their concrete properties and, as he ages, according to more subtle or less directly visible attributes. A preschool child given a so-called "object sorting task," an exercise in which he is asked to group things that are similar to each other, might classify the array according to the materials of which objects are made. Wooden items would be placed in one pile, plastic ones in another, and metal in still another. An 11 year old youngster, on the other hand, should be more likely to classify them according to less concrete or blatant features. He might, for example, infer a function for each object and classify each according to its application, placing writing utensils in one pile, toys in another, and so on.

Perhaps the most widely accepted conceptual model of cognitive development in this age group stems from the work of Piaget (see also Chapter 3). The Swiss epistemologist thought of the school years as representing a period in which children master so-called concrete operations (Piaget, 1968b). The phenomenon is illustrated vividly in what may be the most well known experiment in developmental psychology. Piaget and his collaborator, Inhelder, presented children with two identical beakers, each of which contained the same quantity of liquid, such as water. Each child was asked whether both containers possessed the same amount of water. Necessary adjustments were made until children agreed that this was the case. Subsequently, with the child watching, the investigator poured the liquid from one beaker into a third receptacle, one that was taller and thinner than either of the other containers. Not surprisingly, the water rose to a higher livel in the taller and thinner beaker. The children were then asked, "Does this (i.e., the tall thin) beaker contain as much water as the other (i.e., the initial) beaker, does it contain more water, or does it contain less?" The researchers were interested in finding out whether children would understand that the amount of liquid is "conserved" regardless of the apparent change of height of the liquid in the taller and thinner beaker. They wanted to determine whether children could "conserve continuous quantities." This would provide some evidence that they were able to transcend their own direct sensory observations and appeal to a higher order of logic or reasoning.

In this experiment, four year old children commonly believed that the total amount of water had changed. They attended to only one component of what they saw, justifying their responses by saying such things as "I know there is more; it's higher" or "The glass is bigger, so there is more." Such responses were not changed, even when the experimenter pointed out that no liquid had been added or taken away. Children age five or six tended to give intermediate responses. Often they were not sure. Some were able to conserve the liquid when a difference in appearance was only slight but had more difficulty when it was great. Sometimes they could foresee what would happen (i.e., "When you pour the water, it will stay the same"), but they actually answered incorrectly when confronted with the dramatic visual difference. In other words, they were not yet prepared to overrule sensory data with logical thought.

Beginning at about age seven or eight, genuine conservation was detectable. Children were absolutely sure that the amount of liquid had not changed. Three basic concepts were used to justify their answers, and these were characterized as "compensation," "identity," and "reversal." Children invoking compensation noted that although the liquid was higher, the container was thinner and one made up for the other. Those citing identity pointed out that there could be no difference, since nothing was added or taken away; while those invoking reversal asserted that if one pours the liquid back into the original container, one can prove that the same amount was there all along. As youngsters gain in sophistication they become less exclusively reliant on concrete or sensory data. A child may state, "You know, I really didn't need to look when you were pouring, since I knew you were just pouring. I knew that just by pouring you can't change the amount of water."

It was Piaget's conviction that subjects in the middle childhood years made extensive use of these concepts of compensation, reversability, and identity to make sense of a multitude of phenomena that went beyond continuous quantitites (such as the water in the experiment). Included were discontinuous quantities (such as the conservation of length, of area, of solid substance, and of number). As noted, Piaget characterized this phase of development as concrete operations. By operations he meant those mental activities (i.e., efforts carried out in one's head) through which a youngster derives a more lucid understanding of time, space, number, amount, and other conceptual areas. They are "concrete" because they are applied to the physical entities of everyday life, substances such as pecan pie, blocks, and money. It is not until adolescence, during a stage Piaget referred to as "formal operations," that these processes are applied on a more abstract level to symbols, words, and numbers in addition to observable objects (see Chapter 9).

Thus, it can be seen that the cognitive development of the school age child entails the increasing ability to make inferences, build upon previous experience, and see relationships between newly acquired data and data encountered previously. Toward the end of middle childhood, youngsters become increasingly adept at interpeting analogies and metaphors. Their inferential powers grow, and they are able to detect irony, paradox, double entendre, and hidden meanings. Further development of these functions, however awaits the adolescent years.

Closely linked to a child's conceptual development is increasing sophistication in the understanding of language. Words and sentences become important tools—methods of characterizing relationships, rules, and generalizations. As noted earlier, language skills reinforce memory and attention. Thus they facilitate the mission of skill and knowledge acquisition. Language is also critical for social development.

By age five, most children have acquired a large vocabulary and have mastered the basic grammar and syntax of their language. Between age five and the onset of adolescence, they add a relatively small number of syntactical formations that were lacking in kindergarten (Smith et al., 1970). They become increasingly adept at using and understanding embedded clauses. Carol Chomsky (1969) and others provide interesting examples of this aspect of development. She cites the child's ability to interpret sentences such as, "The monkey promises the dog to jump off." In this kind of sentence a complement verb (i.e., "to jump") occurs, and the challenge is to assign the correct subject (i.e., who will jump, the monkey or the dog?). At age five, a child will select the noun phrase that is closest in the sentence to the complement verb. Therefore, most early elementary school students respond that it is the dog who will jump off. The capacity to deal with exceptions to this rule for certain verbs (such as "promises") is not developed until close to age 10.

With time, the child's ability to judge or explain the structure of sentences also improves, aided by the acquisition of writing and reading skills. The latter fortify the propensity to examine language in a conscious way and to interpret sentences on the basis of their structure. Growing awareness and understanding of more complicated sentence construction then become the major thrust of receptive language development in the school age child.

Other language acquisitions are also relevant. The child becomes sensitive to figurative language, to metaphor and simile. During the school years, children are more discerning of figures of speech that equate two elements from different realms of experience (Winner et al., 1976). In one study, early elementary school students were asked to interpret the sentence, "My friend John is a real tiger." Most believed that John really was a tiger. By about age eight, youngsters surmised that this sentence was really nonsense, and they attempted to alter it to make sense. For example, some tried to change the sentence into a story, stating, "John is friendly with a tiger." It was not until age 10 or 11 that the metaphorical interpretation of this sentence was possible. To accomplish this, the children had to be sensitive to multiple meanings and the ways in which various classes of words can inter-relate. School age children become progressively skilled at penetrating beyond literal meanings, proceeding toward flexibility and even creativity of interpretation.

Other components of language processing and use represent gains for the school child. There is a marked increase in receptive vocabulary. Greater numbers of words can be understood readily. Children begin to derive meaning from the structure of words within their own language and can infer from subcomponents, appreciating prefixes, suffixes, rules of tense, and the like. They also become increasingly sensitive to intonation and language rhythm, deriving greater meaning from attention to these auxiliary aspects of spoken communication.

As understanding and cognitive function flow, school age children are better able to construct generalizations from incomplete verbal contexts. An older school child need not attend to every single word in a sentence to understand it. He is able to extrapolate, to grasp "the general idea," to fill in the parts he may not have heard. This capacity, sometimes referred to as *closure,* represents still another aspect of the expanding horizon of inference from actually perceived data. At times, this new acqustion may infuriate teachers and parents, as a youngster sits in a classroom and stares out the window, missing much of what is being said by the teacher. However, when asked a question, the child gets the correct answer! His ability to generalize from context, to infer, and to extrapolate rescues him from the consequences of daydreaming.

It should be emphasized that language progression is variable. Although psycholinguists are able to characterize stages in the acquisition of good syntactical comprehension, one cannot list unequivocably specific chronologic ages at which particular elements of competence are registered. Many factors contribute to this variation. Language acquisition is particularly vulnerable to cultural effects. Even before entering school, many youngsters have mastered aspects of language relating to specific dialects. The home environment, the child's innate language abilities, the status of cognitive development, and the kinds of cultural and educational exposures a youngster has encountered all influence the rate of language development.

EXPRESSION AND PRODUCTION

Development of the school age child is conditioned to some extent by exogenous expectations and challenges dispatched from the adult world. An example of this is the transition from a predominantly decoding experience in the early elementary grades to the expectation for high efficiency encoding (Levine et al., 1981). Youngsters in kindergarten and first grade need to learn to recognize and interpret various symbols and operations. While they may be required to perform some writing, to "show and tell," and to solve some arithmetic problems, the volume and complexity of these outputs are relatively low. At ages six to eight, children are mastering the alphabet and the number system. They are learning to recognize words and to associate these with sounds and meanings. Beginning around fourth grade, there is a fairly rapid shift, and they are expected to become increasingly productive, to synthesize and express ideas of their own, to write more lengthy assignments, to take tests under timed conditions, to complete long term projects, and to solve multistep mathematics problems. All these activities have in common their emphasis on output, on efficient productivity, on encoding more than decoding, and on expressive skills more than receptive ones.

Fortunately school age children grow in their productivity. There are many reasons for this. As attention and concentration improve, more sustained and more focused effort is feasible. Enhanced abilities in the language and motor areas (see below) further facilitate output, as do the growing abilities to integrate data from multiple sources, to retrieve stored information rapidly and with relatively little conscious effort, and to tap functions of memory and cognition simultaneously. The school child becomes increasingly efficient and organized. Thus, most youngsters are able to satisfy demands for effective output.

Much of the impetus for productivity during the school years is contingent upon positive reinforcement. It is likely that all children need to feel that their products are admirable, that they can elicit boasting from parents and other adults, and that they also can impress peers and siblings. For purposes of analysis, one can describe four common conduits of output: verbal expression (and its most sophisticated medium, writing), gross-motor output, fine-motor function, and what might be termed socialization. The last of these will be described in the next section of this chapter.

A child's verbal expressive ability undergoes a metamorphosis that parallels in many ways the acquisition of receptive language skills described in the previous section. Fluency and word finding improve markedly. The size of the child's active vocabulary grows exponentially under the influence of an educational setting and a home environment. The child has available increasing options for the use of language as he or she becomes adept for the first time at figurative speech. The latter does not become well developed until late in adolescence (and often beyond). The school child grows in the conscious awareness of the uses of language. For example, words and phrases can be mobilized to win friends and influence people. During school years the child begins to acquire the skills to employ writing as an alternative language mode. Those who have not as yet mastered oral speech may encounter particular difficulties with fluent written expression. As mentioned earlier, during the school years the child improves in his ability to use language to reinforce memory, to think, to imagine, and to create. Words also can be used to help a child gain social acceptance.

Most notable is the school age child's expanding lexicon of "off-color" or morally tainted expressions. These can serve a useful purpose in making a youngster seem "cool," fashionable, and sufficiently defiant of adult standards to prove the existence of some autonomy to his peers. The school age child learns to utilize words to signal friends that he is not "out of it," that he is indeed in step with his times. This is particularly true during the late elementary school years, when fashionable language acknowledges conformity with peers. The "in" adjectives and verbs change so quickly with the years and are so dependent upon geographic variations that examples will not be cited here, out of fear of dating this chapter and rendering it too parochial!

School children become increasingly able to exploit language to initiate, modify, and sustain relationships with peers and adults. Subtle regulations of voice tone, word usage, and even rhythm can connote social nuances that lubricate relationships. The growing awareness of this aids in the socialization of the school age child.

Gross-motor skills represent another conduit for output. Physical mastery over space is a valued aspiration for many school children. Maturation and changes in body size facilitate the process. Growing muscle strength adds to physical efficacy. During school years, youngsters become increasingly able to interpret incoming information from their eyes and from proprioceptive-kinesthetic fibers in muscles and joints. These data can be used

to program effective motor activities. They become better able to organize complex motor operations, to master and retrieve multistep motor processes and sequential motor patterns. Complicated athletic pursuits become possible. School children become better able to plan a motor activity, to inhibit extraneous muscle function and to become motor-efficient. Young children show an abundance of associated movements (such as synkinesias), i.e., muscle activities not relevant to the task at hand. By age seven or eight these tend to disappear, and increasing motor efficiency is noted.

Availability of motor memory increases. School children are able to store and retrieve rapidly the blueprints for various forms of motor output, such as particular athletic skills or dance steps. Children differ markedly in their capacity to store and retrieve these motor plans.

Most likely there is a close relationship between gross-motor function and confidence and self-esteem (Cratty, 1979). A child who is well coordinated, one who succeeds in sports and other motor enterprises, is likely to be bolstered in fulfilling the mission of sustaining self-esteem. Motor success can also help youngsters gain social acceptance and come to terms with their own bodies. Children who are delayed or deficient in gross-motor abilities may encounter ostracism by peers. They may have to resign themselves to being selected last for teams, to being excluded from certain activities, to getting derided during physical education classes. In the past, this was considered mainly a boy's problem, but now female adroitness is increasingly valued.

Much of what has been stated regarding gross-motor output is applicable to the area of fine-motor function. During the school years, youngsters become increasingly adept at planning, coordinating, monitoring, and remembering a range of activities requiring effective finger manipulation. Arts and crafts and the act of writing depend heavily upon the ability to take in data from visual and proprioceptive-kinesthetic pathways and then program appropriate fine-motor responses. Increases in skill in eye-hand coordination, in the rate of fine-motor output, and in overall precision become important developmental acquisitions for school age children. A kindergarten or first grade child approaching a fine-motor task may be highly impulsive and lack any systematic approach. The youngster may not be able to carry through a sequence of fine-motor steps in the correct order but may have to struggle inordinately to interpret the visual-spatial data to execute the task. Constructional praxis, or the ability to assemble parts into a whole, may be effective only for relatively simple tasks. The ability to deal with increasingly complex fine-motor demands and to integrate these with a wide range of inputs and memory sources is a significant dimension of development is school age children. Once again, there is likely to be considerable variation in the acquisition of these skills.

It is essential that a school age child feel effective in one or more of the output conduits just delineated. Strengths in one may compensate for weaknesses in another. If all four of these outputs are blocked or are chronically deficient, the mission to sustain self-esteem is likely to be abortive. Some youngsters with developmental output failure (see Chapter 38) become depressed and amotivational because they are unable to keep up with output demands. Their overall working efficiency and development in these areas may fall short of meeting their own and adults' expectations. Such a predicament may be as close as one can come to a true "developmental emergency." These youngsters need to be "rescued," since children must find gratification in the products of their output in order for development to proceed in a healthy manner.

SOCIAL RECEPTION AND INTERACTION

It was pointed out earlier that the quest for social acceptance is an important mission of the school years. A youngster emerges from the protective custody of the home into a social milieu dominated by peers. A wide range of developmental acquisitions during this period facilitate socialization (Flavell and Ross, 1981). Social perceptual and conceptual skills develop under the influence of culture and experience. They are bolstered by innate sensitivities or instincts and further conditioned by ongoing feedback. Through successful social encounters, a child gains in other missions also. Socialization helps a child in selecting and conforming to models. It is a way both of studying and assimilating values or cultural influences and of learning. Interactions with other children help youngsters increase their self-awareness as they "study" the attributes and behaviors of others with whom they are intimate. Peers become allies in the exploration of autonomy and its limits. They collaborate in the satisfaction of appetites and drives. They serve as sources of comparisons that even help a child understand, appreciate, or criticize his own body. Finally, through peer interaction, youngsters come to understand their individuality and to strike a balance between uniqueness and conformity. A number of important developmental attainments seem to catalyze these processes.

Social Perception. With increasing age, experience, and sophistication, school children become adept at interpreting social feedback. They are able to "read" faces and determine whether they convey approval, disdain, or some other response. They become equally adept at perceiving verbal cues. Voice intonation and semantic connotations carry hints about the success or failure of social actions. With such cues, they can tell whether they are saying the right thing, doing the right thing, influencing appropriately, and gaining sufficient esteem in the eyes of others. Conversely, they also are able to detect their own faux pas before they have done too much damage. If they have said something inappropriate or committed acts that are offensive to other persons, they are likely to perceive the responses of a companion and self-correct or counterbalance the indiscretion. Some youngsters have particular problems with this. They are said to be socially imperceptive. They appear oblivious to social feedback and are therefore unable to "titrate" relationships effectively.

Social Cognition. The school age child steadily develops an ability to conceptualize aspects of relating. There is a growing understanding of the differences between kinds of relationships. In the early elementary years, interactions tend to be almost totally spontaneous. They do not involve much forethought, planning, and review. Interacting is not a particularly self-conscious act for a six or seven old. Relating has an "automatic" quality to it. It is more like a bodily function than a cognitive interactive process. As children progress through school years, social relationships become more recognizable phenomena. By age 10 or 11 youngsters are able to identify and discuss the differences among a range of relationships they maintain. They become increasingly aware of a dichotomy between friends and foes. They can talk about a best friend, a pal, a "chum." The latter phenomenon becomes particularly germane between the ages of 10 and 12, at which time boys and girls commonly seek out one or a very few "best friends." These become potent sources of pleasure, feedback, pressure to conform, and self-esteem. At this time they may form surrogate families or highly exclusive cliques. Ongoing approval is particularly critical, and children begin to prepare or even rehearse for relationships. They begin to dress up, to invite selectively, and to schedule and arrange social agendas.

It is in late elementary school that many children become increasingly "other directed" (Riesman et al., 1950). Many of their personal values and actions are shaped by external, socially induced criteria. They are apt to say and do things that are most likely to be "salable" or appealing to their peers. In the minority of instances, children emerge as more "inner directed." Their actions and words may reflect predominantly intrinsic values. They are said to have a kind of inner "gyroscope" based on tradition and moral values, which somehow over-rides the pressure to conform and become socially acceptable.

In acquiring social cognition, a youngster grows aware of the dynamics of relationships. Late elementary school students talk about making friends. They are interested in what it means to be popular, becoming preoccupied in many cases with physical appearance (such as looks and clothing) and likely to offer highly sterotyped discourses on "taste." While the early elementary school student is likely to interact on a less self-conscious level with little planning or forethought, for the older child social relationships constitute a matter of interest. From the study of socialization, the child begins to refine his basic social skills. In particular, he learns to sublimate immediate desires in the interest of long term relationships. He learns to share with others, to praise someone else, to balance competition with cooperation, to say and do things that are likely to please a friend, and to inhibit actions and words that could be damaging to relationships.

In late elementary school, the child becomes a keen observer of the social scene. Able to select friends judiciously, he emerges as increasingly astute at labeling peers and can tell who is "cool," who is a "fag," who is a "wimp," and who is "out of it." He comes to realize that certain youngsters are seen as social "lepers" and that it is best not to be seen or associated with them. Their reputations are bad, and they threaten one with contamination or contagion. There are other peers whose social approval and acceptance can perform miracles in elevating one's own sociopolitical status. For many youngsters in late elementary and junior high school, social life emerges as a major topic of reflection and conversation. It is probably the subject that most commonly jams telephone wires each evening. It is a popular agenda item on the street corner, in the school bus, and elsewhere. Social comparisons, candid critiques, and even calculated scapegoating and group ostracism become vehicles of social education and experimentation during a period of life when such activities at times become nearly obsessive and all-consuming.

Social Prediction.   Planning social interactions depends upon the ability to predict the social consequences of one's statements or actions. This form of reflective behavior contributes to a child's social success. During elementary school, youngsters become good social forecasters. Prognostications condition their actions. They are able to "size up" present conditions and the current cast of characters to determine potential impacts and to minimize social failure. Some youngsters who are particularly impulsive or have other cognitive handicaps seem to experience inordinate difficulty

in predicting such outcomes. As a result, they commit one faux pas after another. They become stigmatized and unpopular and develop a bad reputation, which in many cases may seriously erode self-esteem.

Although the comments in this section have focused largely on peer interaction, many of the same developmental acquisitions can affect a child's relationships with individuals of other ages. During elementary school most children are concerned with children of their own sex and age group. Youngsters who for one reason or another are unable to succeed with their peers may become more involved with older children, adults, the opposite sex, the very young, or even animals. Children's social styles are influenced by observations of adults. In particular, parents and older siblings have a profound effect on the ways in which children form and nurture relationships. Role modeling contributes to social education, as children emulate the interactional styles of those whom they admire. Teachers may become social prototypes; as youngsters proceed through elementary school they become increasingly judgmental about their teachers —not so much for their pedagogic talents (or lack thereof) but for their personality traits (or lack thereof). Thus, all adults who are in contact with the youngsters in this age range must be aware that they are likely to be judged with keen social scrutiny.

Protective Resiliency and Strategy Formation

Children hate to feel hurt. In this regard, of course, they are not different from infants or adults. During school years, they amass an arsenal of defensive materiel to fortify themselves against the threat of humiliation, physical harm, frustration, fear, and failure. They learn to cope and to apply strategies to prevent such injuries. They develop a resiliency that enables them to recover promptly after setbacks. They acquire some immunity from prolonged feelings of despair and find protection through coping, which enables them to succeed at the following life tasks:

Managing Success and Failure.   An important development acquisition is a repertoire for dealing with failure and, analogously, a method of handling success. Too much of either may be problematic. A variety of different styles are appropriated. Children may practice *denial*. After a defeat, they may pretend that they won or that "It wasn't important." *Rationalization* may be incorporated into such a style.

As children pass through elementary school, they sense a need not to become complacent or "cocky" about their successes. They may show a tendency to tone these down, to apply modesty, or to proclaim, for example, "I wasn't that good today" or "It was too easy." This serves as a good social strategy, and, at the same time, it shields the child from some deleterious effects of excessive gratification.

Saving Face. Children develop face-saving techniques to insulate themselves from humiliation. A youngster who strikes out every time he is at bat in baseball may proclaim that he is not really trying or, alternatively, he may clown at home plate. Another way of coping with potential or actual humiliation is *avoidance*—strategies to evade intolerable situations. He conveniently may forget his gym suit when the class is playing baseball or may feign illness on the day of a science examination. A girl may deny that she is concerned about her weight or physical appearance in general, as a way of concealing concern over her inability to "keep up" in the quest for physical attractiveness. Most youngsters acquire a battery of face-saving skills during the elementary school years. Various styles and strategies are apt to be tried. It is important for the adult world to acknowledge and recognize these but not to try to discourage face-saving techniques without, at least, suggesting better ones.

Managing Frustration, Disappointment, and Loss. Losses and disappointments are not unusual at any stage of life. The preschool child may have a difficult time coping with such occurrences. Temper tantrums, violent outbursts, and prolonged tearfulness may be characteristic of the "terrible twos." While sadness and grief are common and appropriate, children need mechanisms for recovery within a reasonable amount of time. There is wide variation in the extent to which this occurs, and a range of acceptable devices may be appropriated.

Some children deal with frustration and loss through denial and rationalization: "I didn't really like that bicycle anyway" or "I don't care if I failed that test. They can't really do anything to me." At other times and in other cases, youngsters can displace or dislodge a source of frustration. They can gain resiliency by seeking alternative routes after a disappointment. A youngster may be so disillusioned with the way his mother or father copes that he develops a totally opposite style. If a parent constantly "holds everything in," to try to appear in full control, and never forsakes equanimity, the child may become so frustrated in observing this behavior that he assumes a low threshold for demonstrative outbursts. Coping styles undergo steady modification in adolescence and adult life. Notwithstanding, some recurring patterns reveal themselves by age nine or 10, and these are apt to be recognizable throughout life.

Accommodating Discomfort. The capacity to deal with pain and discomfort becomes increasingly strengthened during the school years. Pain tolerance develops in some unique ways (see Chapter 28). Closely linked to this aspect of coping is the process of delaying gratification, referred to earlier in the context of task persistence. Youngsters may be better able to endure boredom, drudgery, and inconvenience, knowing that there is a "light at the end of the tunnel." They increasingly postpone gratification and recognize that immediate discomfort may lead to greater pleasure later on. A clearer picture of the need to sacrifice some comfort in the present as a way of enhancing future pleasure becomes evident during this period. The ability to keep future gains in sight may make current pain or boredom easier to tolerate. One may have to endure a long car trip to arrive at an amusement park. One may need to sit through many a dull class session to get a good grade in a course. One may be more apt to accept the inflicted pain of a hypodermic needle by acknowledging that it will prevent the even greater threat and pain of tetanus. Such "trade-offs" become increasingly negotiable in this age group. This capacity to sacrifice today as an "investment" for tomorrow is an important developmental acquisition. There is a vast difference between the ability of a kindergarten child and that of a normal 12 year old to discern benefits in transient discomfort.

Dealing with Feeling. The school child becomes increasingly adept at talking about and coping with personal feelings. Many styles are encountered as one surveys how youngsters experience and react to their moods. Most become more skilled in expressing feelings, in trying to control those that appear to be wasteful and self-destructive, and in modifying inappropriate emotions. The range of coping mechanisms includes "letting it all out," denial, and displacement.

Channeling Appetites. During the school years, the child must cope with a whole range of emerging appetites that demand satisfaction. One example is burgeoning sexual awareness, in which various forms of *sublimation* are appropriate. The older school child may struggle to reach a balance between repression of desires and their outward, judicious expression. In this effort, children may need to use adults as resources. Considerable guilt can be associated with appetites, and to some extent this is normal. Growth of sexual awareness and various methods of coping with it are elaborated in Chapter 35. In addition, appetites for food, attention, and pleasure in general also need to be dealt with. Once again, youngsters appropriate a range of coping styles to pursue, contain, deny, and channel their cravings.

Designing and Implementing Strategies. Some youngsters are master strategists and rational problem-solvers—a skill likely to be of great value for coping with a wide range of setbacks and potentially troublesome challenges. Other children are less adept at masterminding. Strategy formation pervades social and cognitive realms. In acquiring skill and knowledge, some children are particularly talented at developing good learning techniques, being able to bypass intrinsic weaknesses and apply their strengths effectively.

A child may be very good at using rehearsal strategies to enhance memory. He may be good with mnemonics, with "short cuts," with using simplifying rules and generalizations. Similar problem-solving methods may also be applied to solve personal troubles or dilemmas. Some children may be particularly "slick" at techniques of evasion and persuasion that help to ease tension in the family, in the neighborhood, and at school. Most youngsters develop rapidly in this area during the school years. This capacity to formulate effective strategies also enhances their social success rate, which, as noted earlier, requires conscious planning and organization. It also entails an ability to build upon previous experience and to assimilate lessons from past experience in such a way that they can be readily applied to new situations or confrontations. As school children progress toward puberty, the capacity to study, learn, and invent strategies and to apply them grows. This progression is energized through contributions from all the components of development elucidated in this chapter. Moreover, strategy formation in turn makes its own critical contribution to the pursuit of all the missions we have suggested for this age group.

OTHER ACQUISITIONS

Although particular school age acquisitions have been singled out for elaboration in this chapter, one could cite many others that facilitate the pursuit of childhood missions. Several are particularly worthy of mention.

In recent years there has been considerable interest in *moral development*. School children may show a logical progression in this area. Much of the work in this field (some of it controversial) has been carried out by Kohlberg. He elaborates six stages of "moral reasoning" (Kohlberg, 1963), which appear to resonate with the stages of cognitive development depicted by Piaget and his coworkers.

During the earliest stage, moral choices are made to avoid punishment, to acknowledge power, and to "stay out of trouble." At a second stage, doing the right thing entails doing what basically satisfies personal needs and desires, although there may also be some elements of reciprocity and fair play. The latter tend to be pragmatic (i.e., "If you help me, I'll help you"). With the third stage of moral development, the child tries to be a "good" boy or girl, striving to make choices that will conform to what the majority of people want. An action becomes acceptable if it is well intended. In the fourth stage, there is said to be a "law and order" fixation. The child develops a rather inflexible attitude toward authority, laws, and the stabilization of society. The child does his duty; motives become irrelevant. At stage five, he begins to base his actions on his conscience. Correct acts are defined in terms of general individual rights and criteria agreed upon by everyone in society. Also, at this stage, values become somewhat more relative—the law can be changed if it violates principles. Circumstances sometimes justify breaking the law. In the sixth stage, actions are based on personal ethical principles that have been arrived at over time. These are based on rather universal notions of justice, fair play, equality and respect. Although as children progress through the various stages of moral development they may not necessarily agree in their moral decision-making, there is a tendency for them to base their choices on thought processes or standards appropriate for their stage of development.

Kohlberg believes that the stages occur in the same order across cultures and populations. Considerable support for this theory has come from research projects that usually consist of stories or moral dilemmas that individuals at varying ages are asked to solve. The methods of decision-making are scrutinized carefully. Results suggest that during middle childhood children are said to be "preconventional"; i.e., they are operating predominantly in the first two levels of moral development. They arrive at level three beginning at about age 13. Only about 50 per cent of older adolescents have attained the "principle level." The early elementary school child is very much bound to cultural standards of right and wrong, having a strong punishment and obedience orientation. He relies heavily on the consequences of actions as determinants of their goodness or badness. As children progress through these years, their orientation becomes more relativistic. They are able to differentiate the interpersonal aspects of a decision from its purely consequential attributes. A young child therefore may not be able to assimilate the golden rule or its equivalent in his social repertoire. On the other hand, he may recognize during late elementary school that if he is decent to other children, they will reward him in kind.

Another developmental acquisition concerns the *growth of a sense of humor*. One can learn a great deal by studying the ways in which children joke and interpret the humor of others, and much research has been done in this area (Wolfenstein, 1954). Cultural, cognitive, and affective issues have an impact on this particular acquisition. Certain characteristic patterns are evident. For example, during the early elementary years, children are said to be interested in issues of competence, becoming particularly fascinated with riddles and games concerning "morons." They are attracted to stories about people who do something "dumb."

In early elementary school, children are able to memorize jokes for the first time. They are preoccupied with "getting them right." As they progress toward late elementary school, they are able to appreciate increasingly subtle humor. In particular, the element of irony becomes accessible. Children can use humor to work through painful issues, as joke-telling becomes another medium of experimentation and exploration. Troublesome themes may be incorporated in joking in a relatively nonthreatening manner as a way of demystifying them for the child. This is seen particularly in the older elementary school youngster's preoccupation with "dirty jokes," which may be employed as a means of discovery. It is not unusual, for example, for a 10 year old youngster to tell such jokes without really understanding them. Audience reaction can be used as a form of education and confirmation. Joke-telling and humor responsiveness can aid in a number of school age missions. Their social applications, their usefulness for acquiring knowledge, and their deployment as strategies to save face and overcome frustration may be invaluable—especially to youngsters adept in their use.

A closely related developmental acquisition concerns the increasing *sophistication and application of play* (Erikson, 1977). Children can use their play for a variety of purposes. Playing can help with coping. According to Erikson, "The child's play is the infantile form of the human ability to deal with experience by creating model situations and to master reality by experiment and planning." Play obviously has other purposes. It can help a child to build skills—in the motor area, in visual-spatial orientation, in rule compliance, and in language. Play can be used as a medium to explore consequences of actions. Imaginary play affords an opportunity to alter outcomes and plots and is also a way of testing fate without experiencing true pain or risk.

School age children ultimately lose their inclination or option to participate in imaginary games. As they progress in elementary school, such activities are perceived as "babyish." More time is spent in competitive recreation and in activities that enhance skill, focus interests, and build knowledge. To some extent, this progression is culturally determined. Some youngsters may not wish to give up imaginary play but are forced to do so because this activity is perceived as inappropriate after a certain age. The game "Dungeons and Dragons" may represent one example of the older youngster's hunger to continue to play imaginary games—in some socially acceptable form. Other children may seek such escapes in television, private fantasies, comic books, or other, more passive pursuits.

During school years another developmental acquisition involves changing notions about *gender identity*. This too is highly conditioned by culture and by the expectations of peers and adults. As a result, there can be considerable variation in its evolution. This vital component of development in the school child is covered in Chapter 10.

THE HEALTH CARE PROFESSIONAL AND THE SCHOOL CHILD

Health care professionals, like other adults who play a role in the lives of school age children, can have a profound influence on the outcomes of the various missions of this period. The next two sections explore some issues and procedures relevant to what might be termed the "developmental health maintenance" of the school age child.

ON ORCHESTRATING A VISIT

When there is sufficient time, a health visit for a school child can be rewarding. A well child examination or a specific consultation can become an important event in his or her development. Parents can also find such an encounter informative, reassuring, and influential. The flow of events during a visit can facilitate the process.

Whenever possible (and at least occasionally), both parents should accompany the child. Each child should come for a separate visit (i.e., without siblings), because this approach communicates a highly individualized focused interest in that child as a person.

Often it is helpful for the physician to begin a visit alone with one or both parents. Several key questions can be framed to elicit issues regarding the developmental missions of this age group. Open ended questions can be asked. Usually the parents' feelings about the child and about the job they are doing can be inferred from the comments. By spending time alone with parents, one is in a better position to evoke concerns or "hidden agendas" that may not emerge when the youngster is present.

If they are seeing the physician for a specific complaint, some important matters can be explored: Why are they coming in now? What are their real concerns? What is it they are most worried about? Are there any "skeletons in the closet"? A child may be brought to the pediatrician because he seems to be "hyperactive." The physician may wonder why they are coming in now—at the age of nine—when the child has had problems like this all along. With some inquiry, it may turn out that a neighbor's son, now a teenager, had a similar problem and was recently indicted for car theft and possession of drugs. The association between that child and their own may haunt the parents. It is important for the pediatrician to be aware of this real but perhaps concealed concern.

The physician can also use this time alone with the parents to get a sense of how they perceive the visit, what they expect the outcome to be, and what they think they need. To pursue the previous example, parents who bring in a child who is thought to be "hyperactive" may expect that some blood tests will be done and he will be given a special diet. In other instances, they may want a referral to a child psychiatrist, some assistance in formulating an individualized educational plan at school, or advice on day-to-day management. Some parents want and expect an electroencephalogram, while others just crave support and a trusting relationship with the physician. Still others approach the physician to gather influence as they request services in school. Identifying these expectations can be very helpful for the physician.

After meeting with parents, the physician should spend time alone with the school child. Pediatricians accustomed to dealing with preschoolers and infants may not feel inclined to be alone with children. On the other hand, this is an important part of the developmental health maintenance of the school child. The physician needs to construct a positive "one-to-one," trusting relationship with the child, whether the visit is for a specific problem or for an annual physical examination. Time spent talking with the youngster and forming an alliance can be profitable (see next section).

For any child over age six, it is wise to perform the physical examination without parents in the room. Many school age children never get undressed in front of their parents. They are likely to be more self-conscious before a mother or father than they are to be barely clothed before their physician. Often they are more embarrassed about being embarrassed than they are embarrassed! Therefore, they may not suggest that their parents leave the room but are grateful if the physician excludes the parents.

The physical examination can be a rich source of information and an excellent medium for education and counseling. The physician should carry on a dialogue during the physical assessment, reassuring the youngster about any somatic concerns. Sometimes the physician may need to be aggressive in eliciting these. For example, if a child is obese, one may want to comment upon this and state that many children who are overweight are concerned about their appearance. When eliciting potential somatic concerns, it is important for the physician to universalize or generalize the problem, pointing out that many other youngsters face the same anxieties, since children of this age group seem to fear that their problems are unique. The obese 11 year old boy may be self-conscious because he feels that his genitalia are too small. In reality, they are embedded in a pre-pubic fat pad. It might be important to reveal this to the youngster, while stating that many overweight boys harbor this concern but that they are all very normal. Frequently a youngster is too embarrassed or threatened by such a problem to bring it up himself, so that the physician may need to take the initiative. When dealing with potentially humiliating somatic issues, the physician must, of course, promise the youngster complete confidentiality.

After some discussion and a physical examination, the physician should reconvene the entire group—parents and child. It would be inappropriate to see the parents alone at this time. If the physician closes the door and excludes a child following a physical examination, it is likely that the youngster will fantasize about this. "What is he telling them? Does he think I'm a 'mental case'? Did he find out that I have cancer? How come he has to see them alone after he promised he would not tell them anything I said?" Meeting alone with the parents following a physical examination could thus be perceived as a betrayal and may disrupt the physician-child alliance. Instead, it is appropriate for all to meet together. If for some reason the physician needs to speak to the parents alone, this should be done by telephone or on a subsequent visit. In an atmosphere of openness, the physician can share his or her findings, offer guidance, and jointly plan for the future. The child can also use this forum to reassure himself that the physician has not failed to honor the pledge of confidentiality. Forging this kind of relationship among the physician, the patient, and the family during middle childhood can be extremely important later during the adolescent years.

COLLABORATING WITH PARENTS

The pediatric role in counseling, advice-giving, anticipatory guidance, and parent education is a familiar one (see also Chapter 53). Traditionally these activities tend to be better informed and more vigorous with regard to younger rather than school age children. Some pediatricians may feel more comfortable counseling parents about an infant feeding problem than they do advising about how to manage a rivalry between 10 and 12 year old siblings. Notwithstanding, the demand for advice on school age issues is intense and may often remain unfulfilled (Shonkoff et al., 1979; Levine, 1982). Certain aspects of developmental education and management are particularly relevant and therefore form the core of pediatric developmental health maintenance.

Elucidating the Missions and Attainments. It is difficult to separate the responsibilities of the health care professional from those of a health and developmental educator. The physician plays an important role in acquainting parents with the missions and developmental events of this age group, particularly when parents do not seem to understand some of the struggles of their own children.

Asking the Age-Relevant Questions. In providing primary care for school children, it is helpful to inquire about the most relevant areas of function. Certain performance arenas become particularly germane. One wants to gain a sense of how the child is functioning at home, in the neighborhood, and at school. Specific inquiries may yield information about the success or failure of the child's developmental missions. Is he succeeding socially? Is he feeling good about himself? Is he successful at acquiring skills and knowledge? Does he seem to understand his strengths and limitations? Is he excessively fearful? Is he gaining any motor mastery? How well does he deal with discipline and limit-setting? What characteristic styles of coping and behavior are emergent? These and other questions may help to uncover both genuine issues and discrepant perceptions on the part of the parents.

Anticipating Next Steps. A physician can be helpful by sensitizing parents to the "coming attractions" of development. Simultaneously, one can assess and encourage preparedness. For example, the physician might mention to a parent that the youngster is reaching an age of increasing sexual awareness and may inquire about anticipated methods of sex education. An older school age child may be showing early signs of experimenting with autonomy, and parents may need to be clued in about the need to develop ways of handling the child's intensifying yearning for independence. Anticipatory guidance can be offered regarding a wide range of normal events, such as entering school, going away to summer camp, dealing with peer abuse, socializing with the opposite sex, and developing good study habits. By having parents anticipate common decision-making points and possible conflictual issues, one may greatly facilitate parenting and ease some of the common strains in development during middle childhood. As with other age groups, such anticipatory education is particularly important for first-time parents.

Collaborating to Induce Mastery. The physician can help parents induce success in their children. An important question to ask about

school age children during a well child visit concerns their level of success: "When was the last time he had a real triumph?" or "What are the activities or endeavors that bring him consistent rewards and praise?" In this age group, *success deprivation*, a feeling of not being masterful at anything, is devastating. True success (in one or more areas) is like a developmental vitamin; without it serious complications are likely to set in. The physician can help parents to recognize the importance of helping a child taste success. In some cases, this will require a careful effort to identify the youngster's unique strengths and to select those pathways most likely to lead to mastery. Parents need to understand that there is a real difference between actual success and false praise. Their children are more likely to thrive on the former!

Praising, Idolizing, and Criticizing. Parents may need to be helped to achieve a good balance between praise and criticism, between boasting and condemnation, between adulation and approbation. Sometimes it is helpful for a physician to suggest that a child's parent keep some sort of mental ledger to sustain these balances. If one is spending too much time criticizing and nit-picking, a conscious effort may be called for to provide positive reinforcement for the youngster. Parents may need to "catch him doing something right."

It is not unusual for older elementary school students to appear to have their "heads in the clouds." Many have difficulty assuming responsibility, responding to discipline, and getting their lives organized. In early elementary school, it is common for children to be absent-minded and careless about their possessions. Parents may need to be made aware of these normal shortcomings. Although it is important to try to keep these difficulties under control, there is the ever-present danger of being overly critical. It may be possible to follow an 11 year old youngster all day and uncover a lapse or careless act at least once every 60 seconds, so that one might be tempted to spend the entire day criticizing the child. If they are consistently on the negative side of the praise-criticism balance sheet, parents must make a determined effort to increase the dose of praise and positive reinforcement.

Differentiating Normalcy from Extreme Normalcy and Deviancy. The health care professional can be an important source of information with regard to the appropriateness of various problems and occurrences in the lives of school age children. It is common for parents to want to know what is normal. One set of parents may report that their 10 year old son has stolen money from them. Another may lament that their nine year old seems to lie around too much, or that a 12 year old daughter is too concerned about her appearance. Still another may have discovered their 12 year old masturbating. Some may fret because their daughter came home intoxicated or stayed out too late and may have taken drugs.

During middle childhood, parents may be haunted by fears of the kind of adolescent their youngster will become. The parents of a six year old may call their physician and anxiously relate the fact that they caught their son or daughter undressed and playing doctor with the little girl or boy next door. There may be an underlying concern that he or she is becoming a "sex fiend" or is somehow making an abnormal psychosexual adjustment. Even greater panic may ensue when such activities transpire between children of the same sex. The physician will find himself in the position of having to look at the whole child. Is the reported event or concern an isolated finding in an otherwise well adjusted child? Is it extreme or chronic or recurrent? Or is this a single or "experimental" event? What are the parents' concerns? What are their worst fantasies about this problem? How realistic are they?

Sometimes parental reaction to perceived abnormal behavior can be most damaging to a child. The physician can help temper parents' responses, while remaining vigilant for early signs of seriously troubled behavior. Reassurance can be helpful, although there also is some danger in overreliance on the old adage, "He'll outgrow it." Physicians have a natural sense of normal variation in behavior and development. The primary care doctor sees a vast number of normal children. It can be redemptive for parents to learn that a particular struggle they perceive as unique is also occurring in many other families.

Helping to Prioritize.   Sometimes parents overwhelm their children, expecting too much too soon. They may be trying to superimpose rigid adult values on the often fluid and amorphous configurations of the school child's life. In such cases, the physician can help parents prioritize and decide what is really important. An older elementary school child, for example, may be heavily preoccupied with outside interests (such as sports, dance, or music). Life is further encumbered with academic concerns, coercive social pressures from peers, and anxieties about incipient puberty. If, amid all this disquietude, the youngster must come home each day to parents who "hassle" over the cleanliness of a bedroom, dirt under fingernails, a C-minus on the last report card, coming home late last Saturday night, and the low caliber of chosen friends, life can become increasingly confused and frustrating.

Parents may have to recognize that they cannot have everything their way. They must determine what is most important to them. This may necessitate ignoring dirt lodged under fingernails or allowing a child to continue to live in the substand-

ard conditions of a condemned bedroom! They may need help to decide that developing better study habits and coming home on time at night are the most important things to them, and that they will overlook more trivial infractions. The physician can play an important role by encouraging parents to think through their priorities rather than insisting upon perfection or compliance in all areas. Throughout the process, it is important to help parents set realistic goals, objectives that can be achieved reasonably by a particular youngster. It is not up to the clinician to set these priorities but rather to encourage families to do so.

Helping to Cope.   Sometimes parents primarily need advice about how to handle their own reactions rather than the actions of the child. They themselves need to evolve a coping style, a way of managing the frustration of cohabiting with one or more school children. They need to program a time for themselves. They may seek advice about how to channel their own anger, despair, frustration, or disagreements about child-rearing. It may have been easier for them to raise infants than school children, and the physician may need to assist parents to identify and perhaps modify their own coping strategies when they are very angry at their youngster, when they feel like hurting him, or when things seem out of control. There may be a need to rework these strategies and to support parents.

Listening and Reinforcing.   The physician can be a trusted and valuable "sounding board." At times it is appropriate merely to lend an ear, since parents may need to talk about their children and the way they manage them. At other times, they may want to boast about a child or about the success of their own rearing techniques. The physician can do a great deal to offer parents positive reinforcement. By listening and supporting, the health care professional can also alleviate guilt and anxiety. Whenever possible, the physician should help parents feel good about themselves and the job they are doing to develop confidence in their own values, judgments, and practices. Parenting can become increasingly effective to the extent that it is so perceived by competent and confident parents. The physician should always seek, find, and point out to the caretakers areas in which they have shown good practice and judgment.

Managing Crises.   All families have their quota of crises and critical life events. A trusted physician, one who has established a good alliance with parents and their children, can be an invaluable ally during critical times of decision-making. A parent may approach a physician for advice about how to handle the child when there is an impending separation or divorce, how to present a terminal illness or death, or how to prepare a youngster for hospitalization or surgery. Through experience, training, and reading appropriate material, health care professionals can become skilled at educating parents about the proper management of commonly occurring life crises in the school age child. In particular, with this age group, mothers and fathers may need help in learning to be honest and open, to find the right words, to share family problems with children, and to balance their candor with warmth and reassurance. They may need help in recognizing that it is difficult and potentially dangerous to deceive, to conceal, or to deny problems with youngsters of this age group. School children have extraordinary insights of their own and can actually help parents meet a crisis or solve a problem in some cases.

Acknowledging and Accepting Individuality. Parents sometimes must be reminded that no two children are alike. It is important that they define the unique attributes and needs of their individual progeny and perceive a separate profile of strengths and weaknesses for each. They may need help in allowing a child to expand his assets, to express his unique style, to engage in or to avoid activities according to unique inclinations. There is the potential for conflict between a parent's expectations and a child's individuality. A parent may be determined to have a child become an outstanding athlete, although that youngster's gross-motor abilities are limited, and he displays potential for musical or artistic talent. A young girl may be inclined toward computers rather than ballet lessons. The physician can help parents tease out the individuality of the child and pursue rearing strategies consistent with this. Parents can be encouraged not to force square pegs into round holes. At times, they need to be encouraged not to try too hard to mold offspring after their own images or to model them after some immutable preconceived ideals. The child's very special array of traits, his personal temperament or behavioral style, his intrinsic assets and deficits all need to be taken into account in understanding, managing, and planning for the youngster. The physician may need to caution parents who appear to be coercing, redesigning, and even dreaming for a child, whom they perceive more as a mannequin in a shop window than a one-of-a-kind being, replete with dandruff, quirks, deficiencies, and unanticipated attributes!

COLLABORATING WITH THE SCHOOL AGE CHILD

As noted earlier in this section, the physician has a unique opportunity to form a strong supportive alliance with a school child. To some extent this depends on establishing the routine practice of spending time alone with children in this age group. Whether a youngster is being seen in consultation for a specific problem or followed for routine medical care, a one-to-one interaction can be highly complimentary and informative for the child.

Privacy and Confidentiality. The child in the early elementary school grades may find it relatively easy to relate to an adult authority figure, such as a physician. The youngster may anticipate visits eagerly and enjoy sharing experiences and thoughts with the clinician. As children progress through the elementary school years, they may become more wary and, at times, even suspicious of the doctor-patient relationship. They may come to view the visit to a physician as an intrusion upon their privacy. The doctor may be perceived as an agent for the parents rather than a confidante and ally of the child. For this reason it is especially important to reassure an older school child regarding confidentiality. Sometimes it is even appropriate to state initially in front of the parents that everyone should understand that the child and the doctor may want to keep a few secrets—about very personal things. The youngster may seek periodic reassurance regarding this. The physician can say something like "I want to keep you as my patient and have you trust me. If I tell anyone else what you have said to me, I'm afraid you won't want to come back here anymore. I'm going to be very careful not to let that happen."

To what extent should a physician invade a child's traditionally private life? Should one ask personal questions? Or, alternatively, should inquiries be open-ended (such as, "Are there any matters I can help you with?")? The physician should individualize his approach to these issues. Certain youngsters who appear to be harboring problems or about whom there are concerns may benefit from a more "invasive" approach. Direct questions should be asked in an effort to uncover underlying concerns, particularly if there are problems within the family, if the child is in difficulty at home or in school, or if one has the clinical impression that the youngster is anxious about his health, development, or other matters. By spending some time alone with the parents prior to seeing the youngster, one sometimes may use their perceptions to formulate leading questions to ask the child. In "invading" various private domains of a school child, care must be taken in the manner in which questions are asked.

How to Ask. A visit to a physician contains many constraints, not the least of which are limitations of time. This may necessitate a direct approach, without an opportunity to "break the ice." Nevertheless, every attempt should be made to avoid embarrassing the youngster or making him feel too different or "singled out." As suggested earlier, one way to avoid pitfalls is to state questions in such a way that they highlight the universality of issues. For example, if one suspects that a child may be concerned about not growing fast enough, one might initiate the discussion in the following manner: "So many kids who come in here at your age seem to have questions about the way they're growing. Some of them think they're growing too fast, and others are worried that they're not growing fast enough! There are so many different normal ways to grow at your age that almost everyone wonders if he is okay or not. Do you ever have any thoughts about this?" By generalizing issues to all children in that age group, some of the possible pathologic implications are removed, making it easier for the child to express concerns that might not otherwise be discussed with any adults. This technique can be used to deal with such potentially charged issues as smoking, substance abuse, alcohol consumption, masturbation, sexual promiscuity, and other highly personal yet common sources of anxiety for older school children. Periodic reassurance about confidentiality may be needed to sustain the dialogue.

Sometimes a physician may not suspect that a child has any problems at all. Open-ended questions could therefore conceivably uncover unsuspected anxieties. First the physician should obtain a sense of how the usual developmental missions are progressing for the child. Questions can be framed in a nonthreatening manner to pursue areas of socialization, academic success, self-esteem, family life, and so on. Concrete questions are more likely to yield results. For example, if one wants to derive a sense of how a child is doing with peers, one may want to depict a particular scenario: For most school children, critical sites of social interaction include the bus stop, the bus, the corridors at school, the gymnasium, the cafeteria, and the playground. One might ask specifically about one or more of these. "How are things on the playground at your school? Is there a lot of fighting and arguing? Do kids ever call each other names? What do you usually do on the playground? Are there certain kids you spend most of your time with? Have you ever gotten called any names on the playground?"

In this way one progresses from a general discussion of the anthropology of a playground to a more specific inquiry about the youngster's own status therein. The physician can use certain common occurrences as a way of signaling to the child that he understands what things are really like. By choosing questions appropriately, the physician can demonstrate to the child that he sympathizes with him and takes seriously the issues that are confronted in daily life. Some youngsters may be reluctant to mention concerns they have because they feel that these are not viewed as legitimate by the adult world. The following are examples of some good "lead-ins":

"I bet your sister really makes you mad sometimes."

"A lot of kids think that their parents hassle them too much. Have you ever had this problem?"

"I bet there are some teachers at school that really make you mad. Do you find this sometimes?"

The physical examination, as noted earlier in this section, can be a good source of information about a child's body image. The physician can use this experience to elicit concerns or feelings about growth, about physical attractiveness, and about feelings and anxieties over somatic vulnerability of one sort or another. Certain neutral questions can be asked, such as: "Do you get belly aches or headaches?" One of the most common complaints of school age boys is pain in the legs. The physician may want to ask specifically: "Do you ever have leg pains? What do you think causes them?" Girls of 10 frequently are preoccupied about localized or general obesity.

Targeted inquiries can help the child recognize that discussions of bodily concerns are appropriate during a physical examination. More specific questions may need to be raised as a result of direct observations. If the child appears to be unduly self-conscious about physical appearance, the physician may want to try to elicit information about the sources of such anxieties. Certain obvious findings, such as obesity, acne, short stature, a birth mark, or adolescent gynecomastia should be commented upon or tactfully asked about.

Sometimes a physical examination can be a good way to facilitate dialogue with a school age child. During a direct interview, the youngster may indulge only in monosyllables and remain relatively unresponsive. However, after a physical examination, the child may "open up" and become more responsive. Somehow the laying on of the hands may help build rapport and trust. Therefore, in many cases it may be better to postpone discussion of how things are going until after the physical examination has been completed. In some cases, a youngster may be worried about the check-up and may feel much more relaxed once it has been completed.

How to Listen. Despite time constraints, the physician should become a listener. Many important insights are likely to come forth. School children should feel comfortable using a physician as a sounding board. The latter in turn may respond by taking concerns seriously and by revealing a sincere interest in the child's enterprises and accomplishments. Children should be encouraged to gather trophies, report cards, or artwork to show their doctor. The physician can become an influential and valued participant-observer in the development of a school child. The youngster can look forward eagerly to periodic visits, in part to have a chance to boast and to harvest praise. Children with problems may feel more comfortable confiding in a physician who has also shared their triumphs.

School children are often masters of understatement. Many of their comments and responses may be designed to test the physician, to sample his possible responses. A youngster may have some fear about the possible consequences of answers he gives the physician. In response to the doctor's question, "I understand that you have been having pretty bad stomach aches," the child may say "Not that much" or "Not all the time." For many a school age child, such a phrase represents a resoundingly affirmative answer. Subtle nuances and understatements need to be appreciated by the physician. Although the child may want to say a simple "yes," he may fear that this will bring forth hypodermic needles or scalpels. The more tentative or muted response enables the nine or 10 year old youngster to "test the waters."

Over time a physician can acquire good norms about a child's language development. By listening to children speak and express themselves in the office, one can get a sense of the complexity of sentences and syntax, of word-finding ability, and of a youngster's capacity to use language to express feelings and to interact socially. Some youngsters are quiet and nonverbal in a physician's office, which may indicate anxiety, reduced language skills, problems with socialization, or a poor relationship with the physician. On the other hand, some youngsters are just not talkative in any setting, and these "strong silent types" need not be viewed as deviant.

A physician can also spend time watching and listening as the child interacts with parents. At the end of a visit, when all participants are called together for feedback or discussion, one may gather some notion of the quality of communication between the child and parents as well as the style of communication between the two parents, if both are present.

How to Advise and Educate. The school child may solicit advice from a pediatrician. The physician should not minimize or try to render trivial the concerns that a youngster brings to him and should try to achieve a balance between reassurance and over-reaction. A sympathetic response is important—even when the child's anxieties really are unfounded. The physician must resist the temptation to moralize, preach, or devalue an articulated concern. Whenever possible, the child should be given an opportunity to suggest solutions of his own. When the physician wishes to offer advice, presenting several choices will make the child feel like more of a collaborator. It also conveys respect for the youngster by implying that he has the ability to make important decisions.

School children often need demystification. This is particularly true when they are ill, when they are likely to harbor fantasies about symptoms or specific diseases. Even at a young age, children are apt to hear and read about cancer and may fear this disease when they have a swollen gland or are given a blood test. It is therefore important that the physician spend time explaining a child's illness to the youngster. He may need to learn about how a particular medicine works and why he is taking it. The physician should attempt to uncover any fantasies the youngster may have about an illness or treatment program. Nontechnical language and an abundance of examples or analogies are helpful. Sometimes it is useful to draw pictures or work from diagrams to demystify a problem for a child. An example of this is presented in the discussion of encopresis in Chapter 31.

Availability and Accessibility. The physician should communicate to a school child that he is aware of and interested in the various missions of the child's age group. Children might want the freedom to telephone the physician. They should have the sense that the doctor is available for advice and education. The child should know that, in a confidential and private way, the physician can serve as a source of normative feedback, since so many youngsters are concerned with whether they are normal. Parents, too, should feel comfortable asking for advice and presenting concerns about their school child's development. They may seek help from the physician as a diagnostician, educator, advisor, and advocate. When well informed and conscientiously fulfilled, these roles can strengthen the formative developmental acquisitions of the school years. Along with parents, other family members, and other professionals in the community, the physician can co-conspire actively in the processes through which the missions pursued become missions accomplished.

MELVIN D. LEVINE

REFERENCES

Chomsky, C.: The Acquisition of Syntax in Children from Five to Ten. Cambridge, Massachusetts, MIT Press, 1969.
Coopersmith, S.: Antecedents of Self-Esteem. San Francisco, W. H. Freeman, 1967.

Cratty, B. J.: Perceptual and Motor Development in Infants and Children. Englewood Cliffs, New Jersey, Prentice-Hall, 1979.

Erikson, E. H.: Identify and the Life Cycle. New York, International Universities Press, 1959.

Erikson, E. H.: Toys and Reasons. New York, W. W. Norton, 1977.

Flavell, J. H., and Ross, L. (Editors): Social Cognitive Development. Cambridge, Cambridge University Press, 1981.

Freud, S.: New Introductory Lessons on Psychoanalysis. New York, W. W. Norton, 1965.

Gesell, A., and Ilg, F.: The Child from Five to Ten. New York, Harper and Brothers, 1946.

Kohlberg, L.: Development of children's orientation towards a moral order. I. Sequence in the development of moral thought. Vita Humana, 6:11, 1963.

Lefcourt, H. M.: Locus of Control: Current Trends in Theory and Research. Hillsdale, New Jersey, Lawrence Erlbaum Associates, Inc., 1976.

Levine, M. D.: The school child with school problems: an analysis of physician participation. Except. Child., 48: 296, 1982.

Levine, M. D., Oberklaid, F., and Meltzer, L.: Developmental output failure—a study of low productivity in school aged children. Pediatrics, 67:18, 1981.

Piaget, J.: Quantification, conservation, and nativism. Science, 162:976, 1968a.

Piaget, J., and Inhelder, B.: The Psychology of the Child. New York, Basic Books, 1968b.

Riesman, D., Glazer, N., and Denney, R.: The Lonely Crowd. New Haven, Yale University Press, 1950.

Shonkoff, J., Dworkin, P., Leviton, A., and Levine, M.D.: Primary care approaches to developmental disabilities. Pediatrics, 64:506, 1979.

Smith, E. B., Goodman, K. S., and Meredith, R.: Language and Thinking in the Elementary School. New York, Holt, Rinehart, and Winston, 1970, pp. 9-29.

Winner, E., Rosenstiel, A., and Gardner, H.: The development of metaphorical understanding. Devel. Psychol., 12:289, 1976.

Wolfenstein, M.: Children's Humor. Glencoe, Illinois, Free Press, 1954.

Zigler, E., Levine, J., and Gould, L.: Cognitive processes in the development of children's appreciation of humor. Child Devel., 37:507, 1966.

9
Adolescence

Our youth now seem to love luxury. They have bad manners and contempt for authority. They show disrespect for adults and spend their time hanging around places gossiping with one another. They are ready to contradict their parents, monopolize their conversation in company, eat gluttonously, and tyrannize their teachers.

Socrates

9A General Considerations

During adolescence, the individual experiences quantum leaps in physical, psychological, social, cognitive, and moral growth. This exciting stage of development was first scrutinized in the early 1900's, when the term "adolescence" was coined by G. S. Hall. Most historians of the family believe that the present concept of adolescence is a by-product of the social forces and industrial conditions of the late 19th and early 20th centuries. However, Mead's description of girls growing up in Samoa and Socrates' observations above indicate that all young people seem to undergo some universal developmental changes.

The term *adolescence* generally refers to psychosocial growth and development, whereas the term *pubescence* refers to physiological growth and development. Psychosocial attributes of the adolescent years are usually first noticed shortly after the onset of puberty. The ending of adolescence is less clear and may extend well beyond the legal age of 21 years. The transition to adulthood is completed when the physically and intellectually mature individual is able to formulate a distinct individual identity and develop the ability to respond to internal and external conflicts with a consistent and realistic value system.

Adolescence may be classified into three different phases of development: early (10 to 13 years), middle (14 to 16), and late (17 and older). These age ranges are arbitrary and approximate and may overlap or vary within different subcultures. Some authors prefer different terminology, such as pre-adolescent, adolescent, and youth. Regardless of the vocabulary used, the concept is similar. For example, some 15-year-olds may be undergoing early adolescent development, while others of the same age may be in late adolescence. Hence, developmental age may be at variance with chronological age. This concept applies to psychological development just as it has been shown to apply to physical development. In addition, the three phases of adolescence emphasize the different social needs of young adolescents and old adolescents.

PUBESCENCE AND PHYSICAL GROWTH

Girls experience puberty about two years earlier than most boys. The first sign of puberty in girls is usually the development of breast buds or pubic hair, which appear in most girls between the ages of 8 and 10 years. Further breast enlargement, pubic hair growth, a height spurt, and menarche then occur in a well described pattern (Tanner, 1962; Petersen, 1979; Barnes, 1975). Puberty in boys is clinically signaled by darkening of the scrotal skin and lengthening of the penis between the ages of 10 and 12. The proliferation of pubic hair, a height spurt, and additional enlargement of the genitalia then follow over the next two to six years (Tanner, 1962; Barnes, 1975). Acne, axillary hair, deepening of the voice, and chest hair are also characteristics of pubertal growth, but these vary from one youngster to another, depending upon genetic and cultural factors. The most dramatic changes of puberty commonly occur early in adolescence, but many teenagers, particularly boys, continue to grow taller into their early 20's. (See also Chapter 35.)

PSYCHOSOCIAL GROWTH TASKS OF ADOLESCENCE

The psychosocial growth tasks of adolescence have been described from various perspectives. Erikson characterizes adolescence in terms of identity formation; Anna Freud marks adolescence as a time of battle between a relatively strong id and a relatively weak ego; and Blos writes of adolescence as a second individuation process. Table 9–1 summarizes the developmental tasks commonly attributed to the adolescent age group (Felice, 1982).

PSYCHOLOGICAL DEVELOPMENT

During adolescence, teenagers gradually establish independence from parents, find a satisfying body image, and learn to control and express their sexual drives. Each of these psychological issues is discussed separately.

Gradual Development as an Independent Individual

It is extremely important for adolescents to establish independence. This process begins in early adolescence as the young teen begins to separate psychologically from his parents in an effort to formulate his own identity. Prior to adolescence, most school children identify strongly with their families and look to one or both parents as role models. During adolescence, teenagers may shun parental viewpoints, take issue with parental opinions, and test parental values, all in an attempt to establish themselves as individuals separate from their parents. For the adolescent, it is important to be known as "Tommy Burns" and not just the "Burns' boy."

In past years, it was believed that establishing independence resulted in much adolescent rebel-

Table 9–1. DEVELOPMENTAL GROWTH TASKS OF ADOLESCENCE

1. Gradual development as an independent individual
2. Mental evolvement of a satisfying, realistic body image
3. Harnessing appropriate control and expression of sexual drives
4. Expansion of relationships outside the home
5. Implementation of a realistic plan to achieve social and economic stability
6. Transition from concrete to abstract conceptualization
7. Integration of a value system applicable to life events

lion. It is now thought that, in most families, this separation takes place quietly without open rebellion (Offer and Offer, 1975); for some parents, however, it may still pose problems. For example, a mother may complain that her young teenage daughter has become secretive, whereas before "she used to tell me everything." Another parent may be angered because his teenage son "contradicts everything I say." Such parents need reassurance that this is normal separation behavior.

In midadolescence, teenagers may be ambivalent about the separation process as they expose themselves to unfamiliar situations. Sometimes these situations are frightening, and the teenager finds himself retreating to the comfort of his family but feels anger and self-pity at his inability to stand on his own. This ambivalence may be expressed as hostility or bravado in some youth.

By late adolescence, most young people are comfortable away from home. In this stage of development, many older adolescents are able to return to their parents and seek advice and counsel without feeling threatened or ashamed.

Mental Evolvement of a Satisfying, Realistic Body Image

All adolescents need to develop a satisfying and appropriate body image. In early adolescence, the young teen must adjust to the dramatic changes of pubescence. Not only is the young person suddenly taller, but he also experiences a shift in distribution of body fat and musculature, hair grows where it never grew before, breasts and genitalia enlarge, and a once smooth skin erupts with blemishes. It is not surprising that early adolescents may spend hours before the mirror or in the bathroom becoming acquainted with a body that seems to be changing before their eyes. Young adolescents are not only aware of their own body's changes but are also acutely conscious of changes in their friends. They constantly make comparisons and worry that their development may be either too fast or too slow. In the next section, the consequences of early and late maturation are discussed.

Most midadolescents have already experienced puberty, but they may not yet be comfortable with the results. Both boys and girls spend much time, money, and energy trying to improve their faces and figures. Girls experiment with make-up, and both sexes experiment with clothing styles, "trying on" different images to find the real self. Again, this is an attempt to become comfortable with one's body.

This focusing of attention on body development and body changes in early and midadolescence is one factor contributing to the self-centeredness of teenagers. Most young people are very self-conscious. Since they are spending so much time looking at themselves and thinking about themselves, they presume that others are always looking at them, thinking about them, or talking about them. This preoccupation with self may seem to border on paranoia but is completely normal. One example of this self-centered behavior can be observed by watching a group of same sex adolescents in a public place. Each teenager acts, and feels, that all attention is on him or her. That may account for the hair-combing, clothes-straightening, and make-up repairing activities that continually take place in any gathering of teens..

By late adolescence, most young people have experienced most of their linear growth and most of their pubertal changes and have attained a realistic body image. (Some boys may continue height growth into their 20's and many men do not develop chest hair until early adulthood. Some women may experience continuing breast changes past the teenage years.) Body image problems are usually not a concern for most normal older adolescents but remain unresolved for those with chronic illness, obesity, and anorexia nervosa (see Chapter 34).

Handling Sexual Drives

Sexual and aggressive drives may be stronger during adolescence than at any other time of life. Learning to express and control these drives is a major and formidable task of the teenage years, a time when the individual may seem least equipped to master them (Godenne, 1974). In other words, the teenager must learn to be comfortable with his own sexuality.

Early adolescence is marked by sexual curiosity, and masturbation among young adolescents is common. Midadolescents increase their sexual experimentation, and by age 19, more than 50 per cent of single females have become sexually active. Not all teenagers experience sexual intercourse, however, and the clinician must not presume this to be so for individual patients. A characteristic of midadolescent sexuality is that the opposite sex is viewed as a sex object, and both boys and girls may see the dating relationship as an opportunity for social gain. Late adolescence is distinguished by the ability to begin intimacy, that is, to care deeply for another person without the need for exploitation. Adolescent sexuality is discussed in more depth in Chapter 35.

SOCIAL GROWTH

Adolescents experience marked social growth during the teen years.

Expansion of Relationships Outside the Home

As adolescents emotionally move away from parents, they turn to relationships outside the home, including a peer group as well as other adults. For young adolescents, the peer group generally consists of members of the same sex. Usually boys belong to a "gang" from childhood and girls have two or three "best friends." This unisexual peer group provides a psychological shelter in which youngsters can test out ideas. Members of the peer group conform to certain group standards; hence, they dress alike, wear their hair similarly, and may even have group "rituals", such as wearing the same clothing on certain days of the week, or meeting at the same time at the same place every day. It is thought by some that having close peer relationships in early adolescence is a prerequisite to the development of sensitive and intimate relationships in late adolescence or early adulthood (Mannarino, 1979).

It is also common for young adolescents to develop friendships with adults outside the home. Sometimes the relationship may take the form of a "crush" as the young person idolizes a favorite teacher, movie star, sports hero, or even the mother or father of one of his friends. The youngster may prefer the company of this other adult to that of his own parents. Often, parents may be hurt or bewildered by this behavior and need reassurance that, for most young teens, this is normal.

Midadolescents expand close peer-group relationships to include heterosexual friendships. For most teens, this marks the beginning of boy-girl dating patterns, discussed in Chapter 35.

Midadolescents continue to turn to adults outside the home as role models. This exposes the teenager to family structures, religious beliefs, and lifestyles different from his own. Such exposure often serves as an impetus for the teenager to "try on" different styles and philosophies. Many parents may find this behavior distasteful, or confusing. Usually it is reassuring for them to be told that if they can be patient and tolerant during this experimental phase, they can expect their sons and daughters to "psychologically return" to the family fold as young adults (Felice and Friedman, 1982).

For late adolescents, individual relationships assume more importance than the peer-group relationship. Friendships become more intense; issues are discussed in more depth. The superficiality of previous years is dissipated. These bonds are. particularly strong among individuals working toward a common goal or for a common task, such as college roommates, sports team members, and military recruits.

Implementation of a Realistic Plan to Achieve Social and Economic Stability

Adolescents must decide what they want to do as adults in order to support themselves financially and socially. The young adolescent may have vague and unrealistic plans for a seemingly far-off future. Although he may be eager to obtain a job to earn spending money, this reflects a need more for independence than for career planning.

Midadolescents more seriously tackle the problem of eventual career choice, but their thoughts may still be unrealistic. Often the 16-year-old views the future job prospect as an escape from home or the chance to be glamorous. Reality testing may be hampered by the feeling that "I can do anything" or "the more glamorous the better."

Late adolescents are confronted with hard decisions, particularly as seniors in high school: Should I go to college, join the work force, marry and have a family? Continuing education prolongs economic dependence, since most college youth are at least in part financially dependent on parents. Indeed, graduate school and postgraduate school further delay this growth task and may present a source of conflict or concern for many youth. Erikson notes that occupational identity may be a serious problem for many young people (Erikson, 1968). For some youth, it may be appropriate to "drop out" of the usual track (e.g., college), find out what feels comfortable to him, and then tackle the career or job he finally selects.

Some teenagers find the final career choice so difficult that they avoid all decision-making and simply go along with decisions made for them by parents or teachers. Eventually, however, they pay an emotional price and may end up resenting the adults who made the decision for them. Although there are no data to support this view, clinical experience suggests that the adolescent who struggles with this decision-making process and eventually does what *he* wants to do rather than what somebody else wants him to do is more likely to achieve satisfaction.

COGNITIVE GROWTH

During adolescence a transition takes place from concrete to abstract conceptualization. Cognitive development is clearly differentiated in the various phases of adolescence. The young adolescent thinks concretely, with limited abilities for abstraction. This stage is known as concrete operations in the Piagetian scheme of intellectual development (Piaget, 1969). The implications of this type of cognition are numerous. For example, questions are answered literally, not figuratively. Health professionals often do not receive the answers

they seek from youngsters because they may not be asking the right questions. For example, if a clinician wishes to find out if a young teenage girl has been sexually active, it may not be wise to ask "have you ever slept with a boy?" The answer, yes or no, may have nothing to do with sexual intercourse, just sleeping!

By midadolescence, the young person has developed greater capacity for abstraction. Midteens are usually capable of introspection and can reflect on their own thought processes as "objective" (Elkind, 1967). This stage is known as formal operational thinking (Piaget, 1969). This is a giant step in mental development, and the teenager may become fascinated with his newfound intellectual tool. This aspect of adolescence may be another factor contributing to the self-centeredness of midadolescents. These young people now have the ability to utilize symbols (e.g., algebra) and can think through "suppose if " situations (Piaget, 1969). For example, if one began to propose a problem with the statement, "Let's suppose the sky is green and the grass is blue. . .", many children and some young adolescents would begin arguing with the examiner that the sky is blue and the grass is green and not be able to proceed to the heart of the problem. Most mid to late adolescents would be able to proceed to the problem and think through a set of solutions. It is interesting to note that some adults never attain this degree of formal operational reasoning (Kohlberg and Gilligan, 1972).

The late adolescent is capable of stretching his mental faculties immensely. Indeed, he often thinks through solutions to multiple problems, although he may still be limited by a rigid value system. Some authorities believe that creative achievement may be quite remarkable in this age group.

The social implications of this stage of cognitive development are many. Older adolescents may be avid conversationalists with opinions on every issue. In addition, adolescents can now see a host of alternatives to parents' directions (Elkind, 1967) and may promptly point these out to a beleaguered mother or father.

MORAL GROWTH

Adolescents must proceed through various stages of moral development and move beyond a child's concept of mortality to adult morality; Kohlberg divides moral growth into three major levels: pre-conventional, conventional, and postconventional or autonomous (Kohlberg and Gilligan, 1972). Children age 4 to 10 years are usually in the preconventional stage in which "good behavior" results in rewards and misbehavior results in punishment. Hence, good or bad is determined solely by physical consequences, a "child's morality." The second level, a conventional level, is marked by the need to meet the expectations or follow the rules of one's family, peer group, or nation. Maintaining the rules of the group is a value in itself. Early adolescents are usually in this stage of moral development. The postconventional level consists of a major thrust toward autonomous moral principles that have validity apart from the authority of the group or individual who holds them, an "adult morality." This last level begins in mid to late adolescence.

In developing this value system, the adolescent inevitably struggles with his conscience, or the superego. As the young person attempts to formulate his own moral code, he must also adjust to the fact that society's moral standards change. This becomes a source of more confusion and more conflict. The adolescent struggles with his conscience, which may have served him well in childhood but may not be satisfactory for answering the questions he now raises. An example of this struggle is sexuality. The child is taught that sexual activity is forbidden but as an adolescent he experiences strong sexual drives. Obviously, his sexual feelings must be freed from childhood restraints for mature sexuality, but the attainment of that freedom may create a struggle with his conscience.

In early adolescence, it is not unusual for youngsters to experience a temporary decline in the superego. This may occur for several reasons. As discussed previously, the conscience of childhood is largely formed by adult input: "You may not do this!" "That is wrong." Indeed, in childhood, parental presence serves as a continual reminder of right versus wrong actions. In the teenage years, the youngster is away from the watchful eye of the parent and in the company of the peer group. The "collective conscience" of the peer group may be at odds with a teen's parental standards. In addition, the young person may feel a strong need to test his parents' moral code. An example of the decline in the superego in early adolescence may be the stealing of hubcaps in response to a group dare or as a group activity. Under ordinary circumstances, individual youngsters might never consider stealing, but under group pressure they may feel forced to do so. If such youngsters are caught in this activity, they are embarrassed and ashamed about their involvement.

Midadolescence is marked by a narcissistic value system: "What is right is what makes me feel good." "What is right is what I want." This partially explains the sexual exploitation described previously. Hence, much activity in adolescence may be impulsive with little thought about consequences.

This self-serving behavior can be frightening and anxiety-provoking to the adolescent. If his impulses have no check, he may feel out of control. To guard against this outcome, the teenager then becomes severe in his moral standards with rigid concepts of right and wrong. This is particularly true of the older adolescent. Asceticism and idealism are common. The young person becomes other-oriented and causes are embraced with zeal. Teenagers join movements (e.g., religious groups, Peace Corps). Issues are viewed in black-and-white terms, often with self-righteous indignation. Thus the young person consolidates moral growth by developing self-imposed restrictions and prohibitions. In fact, although "justice" and "rightness" are championed by the adolescent, there is little tolerance for different or opposing points of view. Indeed, at one level, the transition to adulthood occurs when an individual finds that there are suddenly more gray issues in life than black and white ones.

CHARACTERISTICS OF EACH PHASE OF DEVELOPMENT

The growth tasks just described can be summarized by the adolescent's responses to the following questions: Who am I? Where am I going? How am I getting there? Progression through all these tasks is necessary for healthy adulthood and emotional maturity. Although adolescents seem to grapple to some extent with all the tasks concomitantly, and growth in one area probably influences growth in another, each phase of adolescence seems to concentrate on different aspects of development. Table 9–2 outlines the differences in the growth tasks of each developmental phase.

Early Adolescence (10 to 13 Years)

The major developmental task of young adolescents is establishing independence from their parents. As this process takes place, they turn to the peer group, who are usually members of the same sex. In addition, it is not unusual for young teens to develop "crushes" on adults outside the home.

Early adolescents are usually in the throes of puberty and must adjust to a rapidly changing body image. Although the young teen is curious and fascinated with sexuality, he generally has not begun to enter into sexual relationships (see Chapter 35).

Young adolescents think concretely and may have vague and even unrealistic plans for a future career. There is some testing of the parents' value

Table 9–2. GROWTH TASKS BY DEVELOPMENTAL PHASE

Tasks	Early: 10 to 13 Years	Mid: 14 to 16 Years	Late: 17 Years and Older
1. Independence	Emotionally breaks from parents and prefers friends to family	Ambivalence about separation	Integration of independence issues
2. Body image	Adjustment to pubescent changes	"Trying on" different images to find real self	Integration of a satisfying body image with personality
3. Sexual drives	Sexual curiosity; occasional masturbation	Sexual experimentation; opposite sex viewed as sex object	Beginning of intimacy and caring
4. Relationships	Unisexual peer group; adult crushes	Begin heterosexual peer group; multiple adult role models	Individual relationships more important than peer group
5. Career plans	Vague and even unrealistic plans ――――――――――――――――→		Specific goals and specific steps to implement them
6. Conceptualization	Concrete thinking ―――――――――――――――――――→ Fascinated by new capacity for thinking		Ability to abstract
7. Value system	Drop in superego; testing of moral system of parents	Self-centered ――――――――→	Idealism; rigid concepts of right and wrong. Other-oriented; asceticism

system as the young adolescent struggles to develop his own moral code. In short, early adolescence is marked by a unisexual peer group, concerns about puberty, and the active establishment of independence.

Midadolescence (14 to 16 years)

The major developmental task of midadolescence is sexual identity, that is, becoming comfortable with one's own sexuality. This includes the need to become comfortable with one's body image. Midadolescents generally begin heterosexual dating patterns.

Midadolescents begin to grapple with issues of morality as their cognitive functions expand with the capacity and capability for abstraction. Career plans begin to take some shape but may not be definite.

Late Adolescence (17 Years and Beyond)

The primary focus of the late adolescent is planning his career or his contribution to society as a responsible adult. This planning is accompanied by high idealism, rigid concepts of right and wrong, and the ability to think through problems with various alternatives. In addition, late adolescents can generally shed the strong need for a peer group in favor of a close, intimate, and caring relationship. For many young people, finding a partner or significant other becomes a major search; this is the typical developmental age for falling in love. In previous generations, being in love automatically led to marriage. In today's society, many young people choose to live together,

or cohabitate. This lifestyle may be a "trial marriage" to test compatibility or may be a tactic to delay the legal bonds of marriage for economic or psychological reasons (Davidoff, 1977).

THE EFFECT OF PARENTING AN ADOLESCENT

As their children enter adolescence, parents can expect to witness changes in themselves and the family as well as in the teenager. They may grow weary of having their ideas challenged by a son or daughter who previously respected their authority. They may feel hurt at their child's seeming rejection of them in favor of others. The youngster who was always thoughtful and altruistic as a child may have become self-centered. Other family members may take sides "for" or "against" the adolescent, as if it were a contest. Parents may find themselves re-examining marital issues with respect to sexual and aggressive behavior.

Parenting an adolescent typically evokes or reawakens in the parent specific issues or conflicts. These contentions should be faced and resolved

Table 9–3. COMMONLY EXPERIENCED CONFLICTS IN PARENTS OF ADOLESCENTS

Challenged authority
Unresolved issues from parent's adolescence
Unresolved issues from adolescent's infancy or childhood
Sexual attractiveness of adolescent
Vicarious satisfaction in adolescent's activities
Recognition of self-aging

for healthy growth of both the parent and the adolescent. Table 9–3 lists the common conflicts faced by most parents of adolescents.

The family of early adolescents may enter a temporary period of disequilibrium which begins to resolve during midadolescence (Ravenscroft, 1974). This conflict usually revolves around the adolescent and occurs as the entire family adjusts to new roles as a result of having an adolescent family member. For some families, this may be a time of crisis; for most, it is a time of growth.

Challenged Authority

As the adolescent becomes more autonomous, establishes independence, and grapples with his own emerging value system, it is natural for him to begin to challenge parental authority. This may be very threatening to parents, particularly in rigid or authoritarian families; at the least, it is unsettling—in all families. Parents may erroneously respond to this challenge in either of two ways: with fear, setting up unrealistic rules and regulations for the teenager, which leads to more authority conflicts; or with frustration, abandoning all attempts to set limits, which the adolescent may interpret as noncaring. Obviously, neither approach is helpful to the parent or to the teenager.

Parents may need help in differentiating between limit-setting and a power struggle. Limit-setting generally refers to giving the adolescent rules and regulations to follow concerning certain behaviors. A power struggle occurs when authority is at stake regardless of the issue being discussed (Aten and McAnarney, 1981). Limit-setting in families is necessary and is actually helpful and reassuring to the adolescent, since it establishes some protective boundaries. Since it consists of rules, these may need to be modified as the adolescent matures and situations change. Adolescents and parents should be encouraged to communicate as these changing needs arise.

A power struggle should be avoided if possible, since it is a no-win situation. Adolescents can rarely back out of a power struggle, since their newly gained autonomy is at stake. In addition, power struggles usually escalate, resulting in a complete breakdown in communication. Parents can help avoid power struggles by not setting up meaningless regulations or arguing about trivial details. This does not mean that they should always agree with the adolescent or give in to his wishes, but they should be willing to listen and accept his right to hold views that are different from theirs. Mostly, it means keeping communication lines open in a spirit of acceptance and affection.

Unresolved Issues from the Parents' Adolescence

In general, adults tend to repress the feelings and conflicts of their own adolescent years. However, having an adolescent son or daughter may reawaken forgotten conflicts or fears. This situation should be suspected if a parent seems to be overreacting to a specific aspect of an adolescent's normal behavior. For example, the mother who is overly strict with her 16-year-old daughter may be fearful that her teenager may become pregnant because she herself became pregnant at that age. In such cases, the parent may need assistance in sorting out her own past from her child's adolescence.

Unresolved Issues from the Adolescent's Infancy or Childhood

Some parent-adolescent conflicts may actually have roots in the youngster's earlier years or previous developmental phases. For example, a mother with strong needs to be nurtured herself may foist this need upon her infant or child, who tolerates the demand until adolescence when he rebels and resents it (GAP, 1978).

Another example is the mother who has never faced the fact that her child was unwanted. As a cute, cuddly baby or as a docile, charming child, the youngster could sway the mother's ambivalence in his favor. When he enters adolescence, this same child may evoke his mother's negative feelings. Quarrels and arguments may be heightened by the mother's conscious or unconscious rejecting feelings and subsequent guilt. She may really be saying, "See, I was right not to want you. I knew you would cause me nothing but trouble."

Sexual Attractiveness of the Adolescent

As adolescents enter puberty and are transformed from a child's physique to the figure of a man or woman, parents may find themselves in the uncomfortable position of being physically attracted to their sons or daughters. These feelings may be conscious or unconscious on the part of the parent; they may be quite anxiety-provoking and even threatening, particularly if the parent does not realize that the feelings are normal so long as they are not acted upon. In some instances, parents may find themselves avoiding all overt expressions of affection with the attractive teenager, as a defense maneuver (GAP, 1978). In certain families the incest barrier may be weak and result in the incestuous form of sexual abuse (see Chapter 14).

In still other families, the presence of an attractive teenager may provoke a sexual rivalry between the same-sex parent and the teenager, a recreation of the Oedipus complex. Anthony notes that the extreme example of this situation is the menopausal-menarche syndrome when the "mother's waning reproductive life is confronted with the flowering sexuality of the girl" (Anthony, 1975). In healthy marriages, this conflict can usually be resolved by the husband and wife, but in others, it may be a bitter conflict that leads to discord and even divorce.

Vicarious Satisfaction in the Adolescent's Activities

It is obviously normal for parents to have pride in their child's accomplishments; in some instances, however, parents may unconsciously take vicarious pleasure in the misbehavior of adolescents. For example, a mother may tell her daughter to be a "good girl" on dates and yet buy her erotic or revealing clothing with the clear message that she would sanction promiscuity. A father may sternly warn his son about having sex with girls, yet brag to his friends about what a "stud" Johnny is when he hears tales of his sexual escapades. In still other examples, an adolescent's rebellion against the system may fulfill the parent's revenge against the system. Parents should be encouraged to take stock of their true feelings about a teenager's activities.

Recognition of Self-Aging

Having one's child become an adolescent is a sharp reminder that one is no longer as young as one used to be. Most parents can accept this fact gracefully but some find it difficult to accept their own aging process and may begin to compete with their children. For example, a father may challenge his son to a strenuous athletic competition in order to prove that he is still as good as he used to be. A mother may begin to dress girlishly even to the point of being inappropriate. Parents may envy their own child's youthfulness and vigor and may express this envy through derision or criticism (Anthony, 1975). Such parents usually need help in adjusting to a new developmental phase of their own. (See also Chapter 4.)

HEALTH CARE DELIVERY TO THE ADOLESCENT

Adolescents receive medical or mental health care services from professionals of many disciplines, among whom are doctors, nurses, social workers, psychologists, nutritionists, teachers, and counselors. In addition, physicians representings various subspecialties—pediatrics, internal medicine, family practice, gynecology, and psychiatry—all interact with adolescents on a regular basis. The following section will offer some general guidelines for providing health care to adolescents. Since the guidelines are broad, they are applicable to various disciplines and types of practitioners.

THE CLINICAL SETTING

Clinics.   Since the 1960's, various clinics have been established for adolescents, including: special adolescent clinics in teaching hospitals with adolescent medicine programs; college health suites; "free" clinics catering to young people living on their own, runaways, or drug addicts; and community clinics designed for specific groups of teenagers in a given locale. These clinical settings may be quite diverse and range from university medical center sites to detention centers to storefront rooms. The entire program, including the physical plant and selection of personnel, may be tailored to the needs of adolescents. Many of these programs have contributed to the training of adolescent health service personnel (Wallace, 1975).

Private Offices.   Adolescents are also seen in private office settings. In some instances, it is necessary for practitioners who accept adolescent patients to make adjustments in their office site or office management to accommodate the teenager. For example, pediatricians whose waiting room and exam rooms are tailored for toddlers may need larger furniture and space. All physicians who provide primary care services to teenagers should have equipment and skills to perform an adequate pelvic examination.

Clinicians who routinely schedule patients every 10 or 15 minutes may find it more practical to schedule double sessions for adolescents. Pediatricians in private practice commonly face the following logistical obstacles to providing health care to adolescents: insufficient time allotted for appointments, inadequate plans for parental interactions, and inappropriately low fee schedules (Stephenson, 1970). Some of these problems may be allayed by scheduling adolescent patients during after-school or evening hours, booking double or triple sessions to accommodate interviews with the adolescent as well as his parents, and charging a fee according to the time spent with the patient and his family.

Multidisciplinary Care.   It is time-consuming to see adolescent patients, and teenagers with multiple problems may be emotionally draining. For these reasons, some clinicians avoid adolescent patients, thus depriving themselves and the teenagers of a potentially rewarding experience. To circumvent these problems as well as to improve the quality of the services provided, many physicians, whether in a clinic or an office setting, find it helpful to utilize or rely upon the assistance and expertise of professionals of other disciplines. Most adolescent clinics routinely provide multidisciplinary health care. For example, most successful prenatal programs for pregnant teenagers include obstetricians, pediatricians, nurses, social workers, and nutritionists on the staff as regular health care providers (McAnarney, 1978; Felice et al., 1981; Hollingsworth and Kreutner, 1980). Some physicians incorporate professionals from other disciplines into their private practices on a regular full-time or part-time basis. When this is not feasible or practical, the clinician should become familiar with private and public resources for adolescents in the community.

Use of Questionnaires

Some health professionals have found that health information questionnaires can be helpful in the initial evaluation of the adolescent patient. Since this questionnaire is brief, the teenager can usually complete it while sitting in the waiting room. (See Table 9–4 for one example.)

This format can be useful to both physicians and teenagers. It saves time, since the clinician can quickly scan the adolescent's responses and then focus on the pertinent positive answers. For some teenagers, it may be easier to write one's concerns than to verbalize them. The questionnaire may also be a source of reassurance, since the checklist implies that these worries are common to other teenagers.

These history forms, however, are not appropriate for every teenager and every clinical setting, nor for every cultural group. Hostile, suspicious, or angry adolescents may resent a questionnaire, particularly if they have been brought to the clinician against their wishes. Lastly, a questionnaire is not helpful if the physician uses it to replace rather than augment communication with the adolescent.

In addition to (or instead of) the health information questionnaire, some clinicians find it helpful to use a standardized History and Physical Exam Form to record data obtained during the initial patient-doctor interaction. These forms include data specifically relevant to the adolescent, both in the psychosocial history and in the physical examination. They are used in many university hospital adolescent clinics and may also be useful

Table 9–4. EXCERPT FROM QUESTIONNAIRE*

Below are listed a number of common problems reported to us by other teenagers. Check yes or no for each, so that we may be in a better position to help you.

	Yes	No
1. Trouble falling asleep	___	___
2. Awakening during the night	___	___
3. Being very tired during the day	___	___
4. Occasionally wetting the bed	___	___
5. Pain with menstrual periods	___	___
6. Bothered by headaches	___	___
7. Bothered by stomach aches	___	___
8. Bothered by dizzy spells	___	___
9. Bothered by leg pains	___	___
10. Worrying about health	___	___
11. Concerned that I am too short	___	___
12. Concerned that I am too tall	___	___
13. Concerned that I am too thin	___	___
14. Concerned that I am too fat	___	___
15. Concerned that my breasts are too small	___	___
16. Concerned that my penis is too small	___	___
17. Worried that I might become pregnant before I am ready	___	___
18. Worried that I might make someone pregnant	___	___
19. Worried that I might not be able to get pregnant	___	___
20. Not yet ready for sex, but feel pressured	___	___
21. Worried about my parents' relationship	___	___
22. Would you like to change something in your relationship with your parents?	___	___
23. Do you have a friend you can talk to about anything at all?	___	___
24. Trouble getting to school	___	___
25. Worried about school	___	___
26. Troubled about future plans	___	___
27. Sometimes I'm so sad that I think about dying	___	___
28. Have other personal problems that I would like to discuss with the doctor but would rather not write down	___	___

*From Stanford University Youth Clinic Medical History Form.

to physicians who see adolescent patients infrequently, serving as a reminder of the type of information that should be obtained on all adolescent patients.

Ambience

Regardless of the nature of the clinical setting, a certain ambience should pervade a facility that hopes to serve teenagers. The teenagers should feel welcomed in the clinical site. This attitude should be projected not only by the physician and nurse but by the entire staff. Physical examinations should be completed in privacy, in a room separate from other patients. Examining gowns and drapes should be used to respect the adolescent's need for privacy. This is particularly important during early adolescence, when youngsters are especially sensitive about their bodies. Chaperones should be present during the pelvic examination and in some instances during the physical examination. (This applies to male doctors with female patients as well as female doctors with male patients.) The presence of a chaperone decreases the intensity and intimacy of the doctor-patient relationship, which may be desirable in some cases. Traditionally, nursing staff have served as chaperones in the clinical setting. One recent study suggests that for some groups of adolescents, other arrangements may be preferred (Phillips et al., 1981).

STAFF-ADOLESCENT INTERACTIONS

The most important feature for the effective delivery of adolescent health care is staff-adolescent interactions. Communicating with adolescents in the clinical setting includes multiple issues such as interviewing techniques, maintaining confidentiality, and being sensitive to the dynamics of transference and counter-transference.

Guidelines for Interviewing Adolescents

Basic interviewing techniques are discussed elsewhere (Chapter 43) and apply to teenagers as well as all other patients. This section highlights specific aspects of the interview process as it relates to the adolescent patient or client. Table 9–5 is a summary of the guidelines to be recalled when interviewing young people. These guidelines apply regardless of the purpose of the interview, whether to seek medical information, to give counsel, or simply to establish rapport (Hammar and Holterman, 1970; Felice and Friedman, 1982).

Table 9–5. GUIDELINES FOR INTERVIEWING ADOLESCENTS

Interview adolescents separately from parents
Create an accepting atmosphere
Ask open-ended questions
Remain objective
Reflect and summarize

Interview Adolescents Separately from Parents. By the time a young person reaches adolescence, he should be interviewed privately and separately from his parents. A private interview with the teenager enables the physician to ask the teenager questions that he may not be willing to answer in front of his parents, including questions regarding sex or drugs. In addition, the separate interview emphasizes the independence of the adolescent to both the parent and the young person. Finally, it enables the clinician to identify himself as the teenager's physician, not the parents'.

There is no set procedure for the interview, and different styles have been recorded. Some clinicians find it effective to greet the adolescent and the parent in the waiting room and then announce, "I'd like to talk to Tommy first, and then I'll talk to you, Mrs. Jones." In this approach, most of the initial interview is spent with the adolescent (Felice and Friedman, 1982) followed by a brief interview with the parent. Others find it useful to interview the parents at the first visit and then interview the teenager at a subsequent visit (Felice and Friedman, 1978). If this technique is used, the clinician may say to the young person, "Your parents have asked me to see you because of [...]. What do you think?" Both styles are appropriate, and in both cases the practitioner is able to establish rapport with the young patient and his parents.

Create an Accepting Atmosphere. To most adolescents, physicians are authority figures, hence the teenager may be anxious, frightened, wary, or even hostile. To allay these feelings, the clinician should strive to create an atmosphere conducive to the interview. If the physician is continually interrupted by phone calls, if he flips through charts or writes continuously, he sends a clear message of disinterest to the patient. If the practitioner is seated, relaxed, and attentive, he conveys interest and warmth.

Ask Open-Ended Questions. When interviewing adolescents, it is best to ask open-ended questions so that the teenager can respond in whatever manner he feels comfortable. It is probably better to ask, "What does the pain feel like?" rather than "Is it sharp or dull?" This technique is usually not threatening and yet productive in obtaining information.

There are times when adolescents talk too much. In some instances, the young person may keep chatting in an effort to keep the physician at a distance and avoid talking about the problem. In other cases, teenagers feel obligated to report grandiose escapades and wild tales. In such situations, the young person may not understand the nature of the therapeutic relationship and may believe that he is obligated to "entertain" the interviewer. Finally, some adolescents feel compelled to report embarrassing information or sordid details about themselves or the family. At times, this is an attempt to test the examiner to see if he is still accepting. If this occurs early in the relationship, the teenager may be too embarrassed or ashamed to return for future appointments. Experienced examiners generally recommend that in the initial interview the adolescent be encouraged to defer a discussion of embarrassing material until a later date.

Since it is more time-consuming to ask open-ended questions rather than a list of yes/no answers, it is often difficult to obtain all the information in one interview. Two or three visits of 40 to 45 minutes each may be required to gather all the initial data on a new patient.

Remain Objective. Some adolescents may be difficult to interview. They may be hostile, angry, or noncommunicative. Usually this is a frightened youngster who chooses to hide his fear behind hostility or silence. If the clinician can remain objective and learn to interpret the adolescent's words or actions as a defense maneuver, there is a chance to develop a therapeutic alliance and help the adolescent. One appropriate response to the hostile teen may be, "It seems that you are pretty angry about being here. Did someone make you come here?"

Remaining objective also means not assuming the role of the adolescent's parent. Sometimes teenagers remind clinicians of their own children, and the physician is tempted to treat the adolescent as his own son or daughter. This is unwise. Usually the teenager needs a physician, not another parent.

Reflect and Summarize. While talking with adolescents, it is helpful to pause and reflect upon the adolescent's statements to make certain that you understand the focus of the discussion. This technique also helps the adolescent to understand the discussion. In some clinical situations, one may say, "It seems that you feel that your mother doesn't trust you. Did I understand you correctly?"

Summarizing the interview at the end of the allotted time helps the teenager feel some progress and accomplishment and lends necessary minimal structure to the interview process.

Issues of Confidentiality

Adolescents should be assured that the information shared with the interviewer is held confidential. In general, information should not be divulged or shared without the adolescent's knowledge or permission. Since some information must be shared with parents, it is usually appropriate to say to the teenager, "When I talk to your mother today, I'd like to tell her about [...]. Is that all right with you?" In this manner, the teenager learns to trust the clinician. If it is necessary to contact other physicians, schools, or insitutions regarding the patient, it is important to inform the teenager in advance and in some cases it is appropriate to obtain the written permission from the adolescent as well as the guardian.

In many states, adolescents may receive medical assistance for specific problems without the parents' knowledge (e.g., for venereal disease, contraception, pregnancy, abortion, or drug usage). These laws are designed to help the adolescent receive necessary medical treatment quickly without fear of parental reprisal. Physicians should be familiar with the laws pertinent to their community (Holder, 1977).

Adolescents should also be informed that there are two situations in which the physician is obligated to inform their parents: if the young person plans to harm himself, or if he plans to harm someone else. Most adolescents understand this reasoning and accept it readily. In fact, these clarifying statements rarely deter the adolescent from telling the physician about plans to run away or commit suicide.

The definition of "emancipated minor" varies from state to state, but the term generally refers to adolescents less than 18 years of age who are no longer subject to parental control or regulations (Holder, 1977). This usually means that the teenager is self-supporting and not living at home. Some adolescents are automatically considered emancipated, for example, minors in the military or married minors. Emancipated minors may give consent for medical treatment for themselves or for their own children.

Transference and Counter-transference

Health professionals who work with teenagers must be sensitive to issues of transference and counter-transference. Transference refers to the projection of feelings about another person upon the physician. For example, the adolescent who is angry with his father may "transfer" this anger to the male clinician caring for him. Counter-transference refers to the transfer of feelings appropriate to someone else onto the patient. For example, the physician may have paternal feelings toward a young male patient. These feelings may be constructive or destructive—constructive if the physician recognizes the feelings yet remains objective and uses them appropriately to help the teenager progress, destructive if the physician denies or ignores the feelings either in himself or the patient. In the latter case, it will be extremely difficult, if not impossible to counsel or guide the younger person objectively, and the therapeutic relationship will probably fail.

ANTICIPATORY GUIDANCE AND PATIENT EDUCATION

As the teenager progressses through the various phases of adolescence, it is appropriate to offer him or his parents counsel in anticipation of predictable issues of concern. Some of the common issues are discussed briefly. Rarely is it effective to lecture the teenager on any of these issues. It is more productive to engage the adolescent in dialogue about a given topic and then build the discussion on the adolescent's level of knowledge. Sometimes the same topic may be discussed more than once. Often the discussion can be augmented by written materials the teenager can read and keep.

PHYSICAL DEVELOPMENT

All adolescents worry about their physical appearance. Early adolescents are particularly sensitive to issues of physical development. Since puberty usually progresses in a predictable manner within a given time-frame, it is reassuring and instructive to review the various components of puberty with the young adolescent: breast development, genitalia growth, pubic hair, linear growth, and menstruation. It is helpful to many young people to be given information about the opposite sex as well as their own. Many 10- to 12-year-olds will be shy about asking for this information, but they are usually quite grateful if the adult takes the initiative. (See the succeeding section of this chapter for further discussion of early and late developers.)

ISSUES OF SEXUAL ACTIVITY

By midadolescence, most teenagers have begun to display an interest in the opposite sex (see Chapter 35). Even if a teenager has not become sexually active, it is appropriate to educate him about responsible sexual activity. This should be done in such a way that the adolescent is comfortable whether or not he has begun sexual experimentation. That is, if the teenager has admitted to having sexual intercourse, the counselor should not give information in a punitive tone to scare him into stopping the activity. On the other hand, if a teenager has elected not to begin sexual intercourse, he should not feel that the counselor is telling him to go out and have sex. Many young people welcome the opportunity to discuss sexual issues, including how not to have sex if one wishes to refrain from or delay sexual activity. It is important to be supportive of the sexually inactive teen while permitting the option of sexual activity at another time.

Venereal Disease

The incidence of venereal disease is increasing among youth. In the last 20 years, the incidence of gonorrhea among 15-19 year olds has risen by 200 per cent. Adolescents should be informed about the different types of venereal diseases and instructed to recognize the signs and symptoms in themselves or their sexual partner.

Birth Control

The words "birth control" have different connotations for different adolescents. To some it simply means the birth control pill. When an adolescent girl requests birth control, it may signify that she has already been sexually active without protection for several months (Zelnik, 1979). It is usually helpful to describe the various forms of birth control—foam, condoms, diaphragms, intrauterine devices, and oral contraception—to both adolescent girls and boys. Some clinicians, counselors, and patient educators have a prepared birth control "kit" containing samples of each type of contraceptive that can easily be reviewed with the young person. (See Chapter 35 for problems of compliance with contraceptive use among sexually active adolescents.)

Sexual Intercourse

It is not unusual for mid to late adolescents to have questions about the sexual act, although they may be embarrassed about making the inquiry or unsure about terminology. Both boys and girls may have questions about their own behavior or that of the opposite sex. There are many myths about sex, and movies often portray sexual intercourse as smooth and glamorous. Young people, unskilled in love-making, may not know whether their experiences are normal. They may simply need information about anatomy and physiology, or they may actually require counseling concerning sexual dysfunction. In order to discuss sexual intercourse with adolescents, the interviewer must be comfortable with the topic and his own sexuality.

Teenage Pregnancy

One-fifth of all births in the United States are to young women between the ages of 10 and 19, resulting in approximately 600,000 births to teenagers each year (Alan Guttmacher Institute, 1976; National Center for Health Statistics, 1976). Indeed, approximately one in 10 adolescent girls becomes pregnant each year; two-thirds deliver an infant and one-third either miscarry or choose to terminate the pregnancy (Alan Guttmacher Institute, 1976). Recent statistics indicate that the incidence of births to older adolescents (15 to 19 years of age) has decreased in the past 10 years while the incidence of births to younger adolescents (less than 15 years of age) is still rising (Baldwin, 1976).

All professionals who work with adolescents should be alerted to the possibility of pregnancy in their patients, clients, or students. If a girl is found to be pregnant, it should be recognized that, for her, this a crisis. Time should be spent with the young woman reviewing the options and alternatives available to her, respecting her wishes and her decision concerning the pregnancy. Professionals should not coerce any adolescent into either having an abortion or continuing the pregnancy. This decision should not be made *for* the adolescent but rather *by* the adolescent.

In some instances, the adolescent will ask the professional to tell her parents about the pregnancy for her. It is probably more constructive to decline that responsibility and instead offer to assist the teen to tell them herself. This assistance may take the form of role-playing, so that the teenager can "practice" the interaction or the professional can invite the teenager to bring her parents to the office, where she can inform her parents about the pregnancy in the clinician's presence. In this way, the professional can offer support to both the teenager and the parents. This same approach may be helpful when the teenager tells the father of the baby about the pregnancy.

Pregnancy in adolescents has medical, psychological, social, and educational implications. Past reports indicate that pregnant teenagers are at high risk for maternal and infant mortality (Alan Guttmacher Institute, 1976), for hypertension or toxemia (Coates, 1970), and for low birth weight infants (McAnarney, 1978). Most recent national studies have concluded that adolescents who receive appropriate prenatal care have good obstetric outcome despite their young age (Duenhoelter, 1975; Felice, 1981; Perkins, 1978).

The most commonly reported negative consequence of teenage pregnancy that affects the young mother is school drop-out. Some authors estimate that nine of 10 girls who become pregnant at age 15 or less quit school (Alan Guttmacher Institute, 1976). Some girls may become pregnant in order to leave school (Cattanach, 1976). Furstenberg's six-year follow-up of a large group of pregnant Baltimore teenagers indicates that the younger the individual when she first becomes a parent, the less the amount of formal education obtained (Furstenberg, 1976). This seems to be true for both boys and girls. Pregnant girls who drop out of school are at high risk of becoming pregnant again while still a teenager (Trussell and Menken, 1978). If she has not completed her schooling, the teenage girl has a difficult time finding employment. Such young women may become bored and lonely and may attempt to have

another baby to alleviate their own misery. In postnatal care, attention must be paid to contraceptive counseling. However, the health provider should not become discouraged if contraceptive advice is not followed. Decisions concerning childbearing still rest with the individuals involved and not with the physician.

DRUGS, CIGARETTES, AND ALCOHOL

Substance use and misuse is discussed in detail in Chapter 37. Many adolescents experiment with drugs and alcohol. Adolescents may be reticent to admit using substances if an adult authority figure asks about it directly. Sometimes it is more productive to ask initially about the peer group's use of substances rather than the individual teen's use. For example, a clinician may say, "I know that a lot of young people are smoking marijuana these days; what are most of your friends doing about smoking grass?" This technique places the subject in the third person and may be less threatening.

In spite of the information available on the risks of heavy cigarette smoking, 20 per cent of high school seniors smoke 10 or more cigarettes per day (Johnson et al., 1977), and the proportion of young women who smoke has been steadily rising (National Center for Health Statistics, 1976). Daily smoking begins at age 12 or 13 for more than half of young smokers and appears to be related to peer pressure. An effective peer education technique has been successfully used to reduce smoking among young adolescents (McAlister et al., 1979) and may be applied to other unhealthy activities, such as drug and alcohol use (Heit, 1977).

ACCIDENT PREVENTION

Accidents are the leading cause of death in adolescence. Early adolescents ride bicycles and skateboards; midadolescents obtain driver's licenses for the first time; and some late adolescents ride motorcycles. One cannot write a "prescription" for safety precautions for young people, but the clinician can review facts and figures about accidents with teenagers and counsel them to wear helmets, use seatbelts, and drive responsibly. The possibility of masked depression should always be considered in the accident-prone adolescent.

GOOD HEALTH HABITS

Self-Examination

Adolescence is a natural time to reinforce good health habits in young people. Learning to be responsible for one's health by keeping track of medications, making doctor and dental appointments, and routinely performing self-examinations is an appropriate task for midadolescents. During this time frame, girls should be taught how to examine their breasts, and boys should be instructed in how to examine their testicles.

Young women in mid to late adolescence are usually receptive to a teaching pelvic examination by a practitioner who utilizes a mirror to demonstrate and explain the various parts of the anatomy.

Nutrition

Adolescents should be taught basic information about nutrition and encouraged to develop good dietary habits. The nutritional requirements of the teenage years have been described elsewhere (Heald et al., 1980). Unfortunately, adolescence is also the age of eating fast-foods, junk foods, and food with high salt content, which may result in poor nourishment. Rather than "lecture" the adolescent on good nutrition, many clinicians find it helpful to plan their remarks around a given medical situation. For example, if a young person is found to have lactase deficiency, it is an ideal opportunity to discuss nutrition by listing ways to obtain calcium other than in milk. In counseling about menstruation, the health professional can point out that girls should maintain iron stores to replace the iron lost in menstrual flow. When treating the teenager with acne, the clinician can both dispel rumors about foods causing acne as well as impart knowledge about nutrition. Since adolescents are always concerned about their bodies, teaching them nutrition in this manner may be effective. Many clinics and doctor's offices have pamphlets or posters in waiting rooms to further the education process.

Obesity and being overweight are common problems, and many teens go on "crash" or fad diets. In general, parents need not be concerned about fad diets and crash diets if they are short-lived, as most of them are. But some fad diets may be unsafe, particularly if maintained for any length of time; this is especially true of low protein or rapid weight loss diets. For active teens, the latter diet may result in fatigue, and the former diet may result in poor growth if pursued over time. Obese teens should be counseled to lose weight slowly but steadily (about 1 or 2 pounds a week) with balanced diets restricted in calories and a regular exercise program. A teenager who insists on dieting continually should be evaluated for nutritional problems (such as anemia) or emotional problems (such as anorexia nervosa). (See Chapter 29 for further discussion of eating problems in adolescents.)

Stress

Stress is common in modern adolescence. Tension headaches, anxiety attacks, and psychogenic abdominal pain are frequent presenting complaints. Adolescents may be helped to recognize and manage stress in themselves. Relaxation techniques, physical exercise, and adequate rest are some suggestions for relieving stress. (See Chapter 55 for further discussion of behavioral management of stress.)

ADOLESCENT PARENTS

By late adolescence, many young people have already become parents, with or without marriage. Teenage parents may need more support from health professionals than do older mothers and fathers. In some instances, young couples try to raise a child with little assistance from their own families. They may drop out of school, marry, and set up housekeeping. In most cases these marriages fail, since the unrealistic expectations of the young man and woman are not met (Furstenberg, 1976).

Often, young teenage mothers who do not marry live with their own parents, who help raise the child. Some grandmothers assume complete responsibility for rearing the infant. At times, this places the adolescent in competition with her own child for her mother's attention. Other adolescents enter into a rivalry with their mothers about the proper care of the new infant (Felice, 1982). The physician's office may become the scene of the battle as mother and daughter both demand that the medical team choose which one is right. In order to assist with these problems additional time must be allotted for these visits. Every opportunity should be taken to help the adolescent mother and father learn about infant and child development.

In recent years, much has been written about child abuse and teenage parents. There is no evidence that teenagers abuse or neglect their children more than does any other parental age-group. However, these younger parents are at high risk for stress, which is associated with abusing situations. By maintaining close contact with young parents and introducing them to multiple community facilities, the health professional can significantly augment the resources available to them.

TRANSFER TO "ADULT" DOCTOR

Pediatricians who care for adolescents should help the adolescent transfer to an "adult" type of health care. This process should be explored with the late adolescent at an appropriate visit. Sometimes this transition takes place at natural intervals: leaving home for college, leaving town for a first job. In those cases, adolescents should be instructed about how to find appropriate health care in a strange city (e.g., teaching hospitals, student health centers). For adolescents staying in town, the pediatrician can prepare a list of local physicians who are interested in caring for young adults. Transferring out of the pediatrician's care should not be done precipitously, because the teenager may feel abandoned. Rather, it should be done gradually over two or three visits, so that the young person will feel a sense of independence and pride in assuming responsibility for himself.

MENTAL HEALTH SCREENING

ROUTINE PSYCHOSOCIAL SCREENING

Normal adolescence includes a wide range of behavior. It is sometimes difficult to distinguish a normal from a disturbed adolescent. Since all teenagers spend most of their time in one of three spheres—in school, with family, or with friends—much information can be gained by evaluating the adolescent's success in each of the areas.

School

Inquiring about the adolescent's school functioning should include questions about his daily schedule at school, grades, his feelings about the school, favorite teachers, favorite courses, the size of the school, the nature of homework, and participating in extracurricular activities. Most teenagers will readily share this information. In some instances, it is appropriate to call or visit the school (with the teenager's and parent's knowledge) to confirm or clarify issues.

Family

The interviewer should strive to find out how the adolescent fits into the family structure and how well he gets along with various family members. In addition, the clinician should ask the young person about living arrangements, the family constellation, family activities, favorite siblings, closest parent, how anger is handled, who makes decisions, and personal chores around the house. Parents should be asked their perception of the adolescent's role in the family.

Peer Group

Questions about the peer group should include the age and gender of friends, activities shared, length of time of friendships, and the characteristics of the friends whom the young teen finds attractive. Does the adolescent have a best friend? What clubs has he joined and why?

Some teenagers may have temporary difficulty in one of these three areas, and this can often be resolved by short-term counseling. Teenagers having difficulty in two or three of the spheres may need to be referred to a mental health professional experienced in adolescent care (Sarles and Friedman, 1978).

ADOLESCENTS IN NEED OF HELP

Depression and Suicide

Depression in children and young adolescents is frequently missed because symptoms may be masked by hyperactivity, school failure, truancy, accident-proneness, running away, or psychosomatic complaints. Gentle, careful interviewing may be required to elicit the adolescent's feelings of hopelessness and helplessness. Three groups of adolescents are high-risk candidates for depression: those with chronic illness, those under extreme stress, and those who feel unwanted (Felice and Friedman, 1978).

The adolescent with a chronic illness often becomes depressed when he finally realizes the significance or impact of his illness. He may feel thwarted in his attempts to become independent, be aware that he is physically immature, be embarrassed by a physical deformity, or have a poor self-image and no confidence. These factors all lead to frustration, anger, and depression (Leichtman and Friedman, 1975).

The adolescent under pressure to perform in order to please his parents may experience extreme stress. Emphasis may be placed on grades, sports, or other activities. Overachievers are particularly vulnerable. As his fear of failure mounts, depression may become acute.

The adolescent who feels unwanted is typified by the teenager whose parents divorce during his adolescence. Often the young person may feel responsible for causing the divorce or may blame himself for not being able to keep his parents together. If he must choose his subsequent living arrangement, his hopelessness and helplessness may be heightened. (See Chapter 13D.)

All depressed adolescents should be evaluated for suicidal tendencies. Suicide is the fourth leading cause of death among adolescents, after accidents, homicides, and malignancy (Felice, 1980). Suicidal teens often present to health-care professionals prior to making the suicide attempt, but physicians may not recognize the adolescent's duress (Teicher and Jacobs, 1966). If a professional suspects that a teen is suicidal, he should ask the adolescent about it in a direct way. Asking the adolescent whether he is suicidal does not *make* him suicidal. On the contrary, the teen is generally relieved that someone has realized his plight and may be willing to help. (For further discussion of adolescent depression, see Chapter 41.)

Psychosis

For many psychotic individuals, mental illness begins to become evident during adolescence. Since normal adolescents commonly have broad mood swings, have episodes of paranoia, and occasionally exercise poor reality testing and judgment, it may be difficult to recognize the disturbed teenager. One study found that early adolescence is a relatively dormant period of emerging psychosis, but a large number of psychotic adolescents are noted between the ages of 15 and 18 (Thomas and Chess, 1976).

Table 9–6 is a brief list of some characteristics found among disturbed adolescents. Primary care physicians who examine a psychotic adolescent may not learn of the presence of loose thought associations nor be told of the patient's auditory hallucinations. However, these youngsters may be extremely anxious during the interview, and all usual attempts to relax them may be unsuccessful. The interviewer may find that the teenager is a loner with no friends (or an occasional younger playmate) and that he has a poor school record in recent months. On occasion, such a teenager may present with bizarre somatic delusions. (See Chapter 41.)

When to Refer Adolescents to Mental Health Professionals

It is not always clear when to refer adolescents to psychiatrists or other experienced mental health professionals. Certain adolescents should always be referred: the self-destructive adolescent, the homicidal youth, the adolescent addicted to substances, and the psychotic adolescent. Other cases may not be as clear-cut. Certainly adolescents who are having difficulty in all three spheres of activity should be referred at least for consultation. Adolescents having difficulty in two of the three spheres may need consultation (Sarles and Friedman, 1978). And, of course, if a youngster is not responsive to the treatment of the primary care

physician, referral should be instituted. (See also Chapter 54.)

In summary, adolescence is a developmental stage marked by rapid physical, psychological, cognitive, and moral growth. Adolescence is usually divided into three phases, each having specific characteristics. Health care may be successfully delivered in many settings as long as the staff is sensitive to the issues that confront this age group.

MARIANNE E. FELICE

REFERENCES

Alan Guttmacher Institute: 11 Million Teenagers; What Can Be Done About the Epidemic of Adolescent Pregnancies in the United States. New York, Planned Parenthood Federation of America, 1976.

Anthony, E. J.: The reactions of adults to adolescents and their behavior. *In* Esman, A. H. (ed.): The Psychology of Adolescence: Essential Readings. New York, International University Press, 1975.

Aten, M. J., and McAnarney, E. R.: A Behavioral Approach to the Care of Adolescents. St. Louis, The C. V. Mosby Co., 1981.

Baldwin, W.: Adolescent pregnancy and childbearing—growing concerns for Americans. Population Bulletin, Vol. 31, No. 2. Washington, D. C., Population Reference Bureau, Inc., 1976.

Barnes, H. V.: Physical growth and development during puberty. Med. Clin. N. Am. 59:1305, 1975.

Cattanach, T. J.: Coping with intentional pregnancies among unmarried teenagers. School Counselor. January, 1976, pp. 211-215.

Coates, J. B.: Obstetrics in the very young adolescent. Am. J. Obstet. Gynecol. 108:68, 1970.

Davidoff, I. F.: "Living together" as a developmental phase: A holistic view. J. Marriage Fam. Counsel. 3:67, 1977.

Duenhoelter, J. H., Jimenez, J. M., and Bauman, G.: Pregnancy performance of patients under 15 years of age. Obstet. Gynecol. 46:49, 1975.

Erikson, E. H.: Identity: Youth and Crisis. New York, W. W. Norton Co., 1968.

Elkind, D.: Cognitive structure and adolescent experience. Adolescence 2:427, 1967.

Felice, M. E.: Teenage pregnancy. *In* Hollingsworth, C. (ed.): Coping with Pediatric Illness: The Child, the Family, and the Caregivers. New York, Spectrum, 1982.

Felice, M. E.: Teenage suicide. Position paper for American Academy of Pediatrics, Committee on Adolescence. Pediatrics 66:144, 1980.

Felice, M. E., and Friedman, S. B.: The adolescent as a patient. J. Cont. Educ. Pediatr. 20:15, 1978.

Felice, M. E., and Friedman, S. B.: Behavioral considerations in the health care of adolescents. Ped. Clin. N. Am. 29:399, 1982.

Felice, M. E., Granados, J. L., Ances, I. G., Hebel, R., Roeder, L. M., and Heald, F. P.: The young pregnant teenager: Impact of comprehensive prenatal care. J. Adol. Health Care 1:193, 1981.

Freud, A.: Adolescence as a developmental disturbance. *In* Caplan, G., and Lebovici, S. (eds.): Adolescence: Psychosocial Perspectives. New York, Basic Books, 1969.

Furstenberg, F. F.: Unplanned Parenthood—The Social Consequences of Teenage Childbearing. New York, The Free Press, 1976.

Godenne, G.: From childhood to adulthood: a challenging sailing. Ann. Am. Soc. Adol. Psychiat. 3:118, 1974.

Group for the Advancement of Psychiatry (GAP): Normal Adolescence: Its Dynamics and Impact. New York, Charles Scribner's Sons, 1968.

Group for the Advancement of Psychiatry (GAP): Power and Authority in Adolescence: The Origins and Resolutions of Intergenerational Conflict. New York, Mental Health Materials Center, 1978.

Hammar, S. L., and Holterman, V.: Interviewing and counselling adolescent patients. Clin. Pediat. 9:47, 1970.

Heald, F. P., Rosebrough, R. H., and Jacobson, M. S.: Nutrition and the adolescent: An update. J. Adol. Health Care 1:142, 1980.

Heit, P.: Initiating a high school health-related student peer group program. J. School Health 47:541, 1977.

Holder, A. R.: Legal Issues in Pediatrics and Adolescent Medicine. New York, John Wiley and Sons, 1977.

Hollingsworth, D. R., and Kreutner, A. K. K.: Teenage pregnancy: Solutions are evolving. N. Engl. J. Med. 303:516, 1980.

Table 9–6. COMMON CHARACTERISTICS OF DISTURBED ADOLESCENTS

Somatic delusions
Severe anxiety
Loose thought associations
Social isolation
Poor school performance
Auditory hallucinations

Johnson, L. D., Bachman, J. G., and O'Malley, P. M.: Drug use among American high school students 1975–1977. Rockville, Maryland, National Institute on Drug Abuse, 1977.

Kohlberg, L., and Gilligan, C.: The adolescent as a philosopher: The discovery of the self in a post-conventional world. *In* Kagan. J., and Coles, R. (eds.): 12 to 16: Early adolescence. New York, W. W. Norton, 1972, pp. 144-179.

Leichtman, S. R., and Friedman, S. B.: Social and psychological development of adolescents and the relationship to chronic illness. Med. Clin. N. Am. *59*:1319, 1975.

Mannarino, A. P.: The inter-actional process in pre-adolescent friendships. Psychiatry *42*:280, 1979.

McAlister, A. L., Perry C., and Maccoby, N.: Adolescent smoking: Onset and prevention. Pediatrics *63*:650, 1979.

McAnarney, E. R., Roghmann, K. J., Adams, B. N., Tatelbaum, R. C., Kash, C., Coulter, M., Plume, M., and Charney, E.: Obstetrics, neonatal and psychosocial outcome of pregnant adolescents. Pediatrics *61*:199, 1978.

Mitchell, J. J.: Moral growth during adolescence. Adolescence *10*:211, 1975.

Nadelson, C. C., Notman, M. T., and Gillon, J. W.: Sexual knowledge and attitudes of adolescents: Relationship to contraceptive use. Obstet. Gynecol. *55*:340, 1980.

National Center for Health Statistics, DHEW (NCHS), Vital Statistics of the United States—1974, "Natality." Washington, D. C., U. S. Government Printing Office, 1976.

Offer, D., and Offer, J. B.: From Teenage to Young Manhood. New York, Basic Books, 1975.

Perkins, R. P., Nakashima, I. I., Mullin, M., Dubansky, L. S., and Chin, M. L.: Intensive care in adolescent pregnancy. Obstet. Gynecol. *52*:179, 1978.

Peterson, A. C.: Female pubertal development. *In* Sugar, M. (ed.): Female Adolescent Development. New York, Brunner-Mazel, 1979.

Phillips, S., Friedman, S. B., Seidenberg, M., and Heald, F. P.: Teenagers' preferences regarding the presence of family members, peers, and chaperones during examination of the genitalia. Pediatrics *68*:665, 1981.

Piaget, J.: The intellectual development of the adolescent. *In* Caplan, G., and Lebovici, S. (eds.): Adolescence: Psychosocial Perspectives. New York, Basic Books, 1969.

Ravenscroft, K.: Normal family regression at adolescence. Am.J. Psychiat. *131*:31, 1974.

Sarles, R. M., and Friedman, S. B.: The processes of consultation and referral. *In* Gellert, E. (ed.): Psychosocial Aspects of Pediatric Care. New York, Grune & Stratton, 1978.

Stephenson, J. R.: Adolescent medicine in a private practice. Group Pract. *19*:14, 1970.

Tanner, J. M.: Growth at Adolescence. Oxford, Blackwell Scientific Publishers, 1962.

Teenage Smoking, National Patterns of Cigarette Smoking, Ages 12 Through 18, in 1972 and 1974. Publication #(NIH) 76–931. Bethesda, U.S. Dept. of Health, Education, and Welfare, 1976.

Teicher, J. D., and Jacobs, J.: The physician and the adolescent suicide attempter. J. School Health *36*:406, 1966.

Thomas, A., and Chess, S.: Evolution of behavior disorders into adolescence. Am. J. Psychiat. *133*:539, 1976.

Trussell, J. and Menken, J.: Early childbearing and subsequent fertility. Family Plan. Perspect. *10*:209, 1978.

Wallace, H. M.: The training of adolescent health service personnel. J. School Health *45*:535, 1975.

Zelnik, M., Kim, Y. J., and Kantner, J. F.: Probabilities of intercourse and conception among U.S. teen-age women, 1971 and 1976. Fam. Planning Perspect. *11*:177, 1979.

9B Effects of Early Versus Late Physical Maturation on Adolescent Behavior

Common lore acknowledges that the behavior of adolescents will change as they begin to undergo pubertal development. Self-consciousness, emotional lability, and a burgeoning interest in the opposite sex are attributes of the stereotypical adolescent. It is important to realize, however, that the onset of puberty may occur in normal subjects as early as age 9 or as late as age 14. Youth at the same chronological age may be at very different levels of sexual maturation. In any normal population there will be adolescent boys and girls who differ strikingly from the majority of their peers because they are either early or late developers. This variation in the timing of normal maturation may produce additional important behavioral consequences. This section will review these behavioral responses and highlight issues in which a knowledgeable clinician may be especially helpful.

THE ONSET OF PUBERTY

The onset of puberty is triggered in the hypothalamus. A series of mediators and receptors in the hypothalamic-pituitary-gonadal axis leads to the elaboration of the sex hormones from the gonads. Along with steroid hormones from the adrenal gland, they stimulate the development of secondary sexual characteristics. In the physiological sense, sexual maturation begins during fetal life, since the hypothalamic-pituitary-gonadal axis is active at that time and remains so until early infancy. The mechanism of negative feedback, in which minute levels of sex hormones cause the hypothalamus to turn off the process, becomes complete by two years of age, at the latest, and remains so until one or two years before the external signs of puberty emerge. If puberty in fact begins with release of this negative feedback mechanism, it is not known at this writing exactly how that action occurs. Many factors, however, have been implicated in the process.

DETERMINING FACTORS

Nutrition has long been assumed to play a role in determining the onset of puberty. The secular trend that has led to a lowering·of the age of menarche in the United States by approximately three years during the past century now appears to have ceased, presumably related to the improved nutritional status of the population. Better nutrition has also been considered the major reason for the earlier occurrence of menarche in the higher socioeconomic classes, in individuals from small families, and in first and only compared to later born children.

On the other hand, racial differences in menarchal age suggest other possible explanations. In the United States, for example, blacks mature earlier than whites, suggesting a difference in genetic control. The familial trend also supports the role of genetics. Mothers and daughters seldom differ by more than one year in their age of menarche. Early maturing mothers have early maturing offspring, both male and female.

Fatter women also mature earlier. Their children, both boys and girls, are fatter and taller and mature earlier than do offspring of less fat women (Garn, 1980). On the average, however, girls tend to begin maturing earlier than boys in all groups. The possibility of a critical weight associated with onset of menarche has been proposed but not definitively established (Frisch, 1974). A relationship between body weight and menarche is further suggested by the fact that female athletes have later menarche than their nonathletic peers. Whether this is the effect of athletics or selection of asthenic young women into these sports is unknown (Malina, 1979). Frisch (1980) has also described delayed menarche in thin, hard-working ballet dancers.

Another influence is photostimulation; blind individuals mature earlier than those with visual access to light. Finally, chronic illness is a deterrent to maturation.

MARKERS FOR THE ONSET AND SEQUENCE OF EVENTS

Despite the many factors that appear to influence the onset of puberty, the sequence of events occurs in an orderly and predictable fashion once the process has begun. A variety of indices have been employed as the hallmarks of puberty (see also Chapter 35). With longitudinal data, the maximum height velocity is a reliable marker; however, it occurs earlier in the pubertal process in females than in males. Bone age has been used effectively in a number of large samples in the past but should not be determined routinely because of the possible radiation hazard. Age of menarche is a well-defined marker of maturation in females, although it is a relatively late event.

Staging of secondary sexual characteristics by Tanner's method is highly reliable and reproducible among observers, can be well correlated with other parameters, and provides measures for identifying the initiation as well as for following the process of puberty (Tanner, 1962). The need to undress the subject in order to make this determination should present no difficulty for the physician in the office, who should be routinely assessing physical development. For the researcher in the school or other settings, the issue of undressing may be eliminated, since adolescents have been shown to be capable of determining their own stage of sexual maturation if shown the Tanner photographs (Duke, 1980).

THE IMPACT OF EARLY VERSUS LATE MATURATION

The importance of the timing of pubertal events and the impact of early versus late maturation on adolescent development are evident in a number of physiological as well as behavioral characteristics. Among the physical developmental characteristics more closely related to maturational age than to chronological age during adolescence are bone age, serum gonadotropins, serum alkaline phosphatase, hematocrit, serum uric acid levels, and sleep latency periods.

The hypothesis that social and cognitive development should be more closely related to physical maturation than to age has been studied to only a limited degree, despite its intuitive appeal. Longitudinal studies initiated at the Institute of Human Development in Berkeley in the 1920's carefully followed a group of individuals from birth to maturity (Jones, 1971). Assessing maturation by bone age, they identified a small group of early maturing and late maturing adolescents, at the extremes, and compared them on a number of behavioral parameters. For males, early maturation was judged to be desirable. The early maturing boys were perceived by adults as more poised, relaxed, good-natured, unaffected, and less rebellious toward their parents than the late maturers. Peer ratings described the early maturers as more attractive and popular with less need to strive for status. From this group arose the outstanding athletes and the student body officers. Late maturers were seen by their peers as less good-looking, less grown-up, more often seeking attention, bossy, restless, and talkative. Follow-up of this group as adults indicated that, although a number of the differences were no longer apparent, the best predictor of adult male social participation was the rate of skeletal development during adolescence.

In Kinsey's cross-sectional study in 1948, early maturing males were characterized by more intense sexual activity, and in adult life the frequency of sexual activity continued to be highest among early maturers regardless of marital status or educational level, although early maturers married earlier.

Interactions with social class and body build have been pointed out by Clausen, who notes that early maturers are most valued in the working class and that even within the latter group, mesomorphs are viewed as more poised and more assertive than ectomorphs or endomorphs (Clausen, 1975).

The important influence of the cultural context was demonstrated by the Berkeley group in their studies of Italian boys in Italy and boys in Boston born of Italian parents. Unlike the American boys studied earlier in Berkeley, early maturing boys in Italy did not have more positive self-concepts, presumably because Italians place much less value on physical size and strength than do Americans. The Italian-American early maturers resembled the Americans in self-confidence but were rebellious toward their parents. The negative attitudes toward parental control may have resulted from exposure to conflicting cultural values between the general American milieu and the parents' Italian mores.

In females, comparisons of early and late maturers have led to somewhat divergent conclusions. In the Berkeley studies it appeared that, in contrast to the boys, it was the early maturing girls who were below average in prestige, popularity, sociability, and leadership, whereas late maturers excelled in these characteristics as well as in personal appearance and attractiveness. Subsequent analyses, however, suggested that early maturation was viewed positively in middle-class girls and negatively in working-class girls. A later study of girls in the sixth to ninth grades indicated that in the sixth grade it was more prestigious to be developmentally in phase, namely prepubertal, whereas during junior high school years, being ahead of the grade developmentally seemed to be an advantage (Faust, 1960). In general, fewer long-term predictive data are available for girls than for boys. Kinsey failed to find differences in sexual activity among females based on the age of menarche.

Recently, Simmons and coworkers have reviewed their five-year longitudinal study of all sixth graders in 18 randomly sampled schools in Milwaukee (Simmons, 1979). Early developing girls were more popular with the opposite sex; dated more; and were more concerned about their height, weight, and figure. Late developers saw themselves as smarter and better at school work; they scored higher in reading and math achievement tests in the sixth and seventh grades and had higher GPAs. The environmental context, however, was extremely important in their studies. Girls who experienced the lowest self-esteem were those who had reached puberty early, had embarked early on dating behavior, and had experienced a major environmental change by moving into junior high school.

Longitudinal studies in Great Britain followed three large cohorts of subjects born in the 1940's (Douglas, 1977). Although superiority was noted in both intellectual behavior and measured ability for early maturing girls, these characteristics were attributed less to their physical maturity than to the fact that a large proportion of the early maturers were from small families. However, in five groups of girls classified by age at menarche, the lowest mean score in intelligence tests was obtained by girls whose menarche occurred after the 15th birthday.

Waber relates sex differences in mental ability to differences in the organization of cortical functioning, resulting in turn from differential rates of physical maturation. Based upon evidence that late maturing individuals of both sexes perform better than early maturers in tests of spatial ability, she postulates that the late maturation of males compared to females explains the superiority of males in tests of spatial visualization. Although females excelled in verbal fluency, there was no correlation between this function and early maturation (Waber, 1977).

NATIONAL HEALTH EXAMINATION SURVEY DATA

The foregoing brief reviews of several studies suggest that correlations exist between maturation and behavior but that they may differ from one cohort to another, between males and females, and among cultures. Spurious differences may result from the use of unreliable measures or small sample size; real differences may stem from heterogeneity in ethnic background, geographical origin, and social class.

In order to explore further the relationship between maturation and behavior, our group* has analyzed data derived from a probability sample of the noninstitutionalized youth of the entire United States drawn by the National Center for Health Statistics and stratified for these and other variables. In Cycle III of the National Health Examination Survey, data were collected between 1966 and 1970 on 3,514 males and 3,196 females, ages 12 through 17, representing the target population of 23 million noninstitutionalized youth. Extensive examinations on each subject included health and behavior histories, obtained separately from parents and youth; detailed physical examinations, including assessments of sexual maturation using Tanner's method of staging; and teachers' ratings and test scores of intellectual ability and achievement.

*J. Merrill Carlsmith, Sanford M. Dornbusch, Paula M. Duke, Ruth T. Gross, Philip Ritter, and Bryna Siegel-Gorelick, from the Departments of Psychology, Sociology, and Pediatrics, and the Center for the Study of Youth Development at Stanford University.

Utilizing the computerized data from this survey, we have examined a variety of behavioral characteristics in relation to the stage of sexual maturation of the youth, and, specifically, to whether or not they were early or late maturers. The categories of early maturation, late maturation, and mid-maturation (neither early nor late) were derived from the frequency of the sex maturation stages, rated from 1 (the lowest) to 5 (the most mature), for each sex at each age from 12 through 17. The highest 20 per cent in each age group were designated the early maturers, the lowest 20 per cent the late maturers, and the remaining 60 per cent the mid-maturers (Fig. 9–1).

Education-Related Variables

In the realm of intellectual ability and academic achievement, some dramatic relationships to early and late development were derived from the National Health Examination Survey data. It was possible to examine the youths' and their parents' expectations and aspirations regarding their future educational achievement; teachers' assessments of their intellectual ability and academic achievement; and test scores for cognitive ability and achievement.

In all these education-related variables, the late maturing males fared worse than the mid and early maturers. They were less likely to want to complete college and less frequently expected to do so. Their parents concurred in their lower expectations and aspirations. The teachers less often characterized late maturing males as above average in intellectual ability and less often in the upper third of their class in terms of academic achievement.

Mean scores on the Wechsler Intelligence Scale for Children (WISC) and the Wide Range Achievement Test (WRAT) were also lower in this late maturation group at all ages except for 12-year-olds. These differences were also consistent within the subsets, vocabulary and block design, in the WISC and reading and mathematics in the WRAT. Waber's finding of superiority in spatial ability for late maturers was not confirmed in these studies. In addition to these overall differences, the negative relationship of late maturation to the educational variables was even more pronounced among the older males (Fig. 9–2). Although the findings were less striking, the early maturing males did consistently better than the rest of the group in all these parameters (Duke, 1980).

It is well known that socioeconomic status, family size, and birth order are related to educational achievement. According to these studies, individuals in lower socioeconomic classes, from large

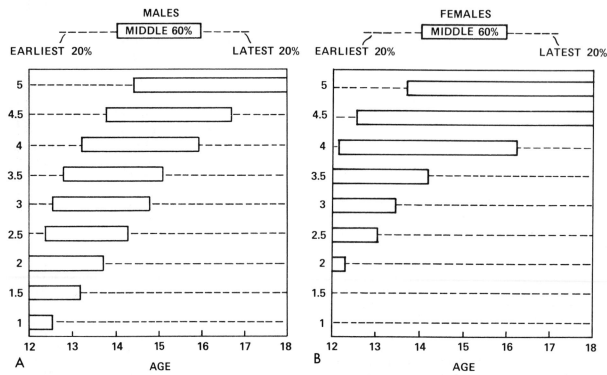

Figure 9–1. Definition of early, late, and mid maturational groups by stage of sexual maturation and chronological age for white males (A) and white females (B). Sex maturity stage is on the y axis; age is on the x axis. (From Duke, P. M., et al.: Educational correlates of early and late sexual maturation in adolescence. J. Pediat., 100:633, 1982.)

INTELLECTUAL DEVELOPMENT

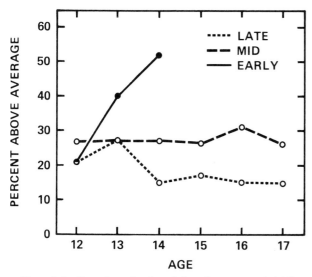

Figure 9–2. Percentage of males rated as above average in intellectual development for the three maturational groups: late, mid, and early.

families, and late in the birth order tended to do less well. Within each of these categories, however, the relationship between the education-related variables and early or late maturation persisted, with one exception. Within the high socioeconomic status group, in which there are already high expectations for education, rate of pubertal development did not alter these expectations.

It was also possible to examine some aspects of the intellectual performance of these youth at an earlier age. The National Health Examination Survey Cycle II produced a probability sample of children ages 6 through 11 gathered a few years earlier in the same geographical regions as Cycle III. Approximately 2,000 subjects were sampled both in Cycle II, at ages 8 through 11, and later in Cycle III, at ages 12 through 16. Identifying the early and late maturers from their stage of sexual maturation in Cycle III, it was possible to look at their test scores prior to puberty. At ages 8, 9, 10, or 11 the mean test scores of those destined to be late maturing males were lower than those of the mid and early maturers. Thus, before there were any physical signs of puberty, the late maturing males were at a disadvantage intellectually.

These observations that late maturing males do less well in a variety of intellectual and educational parameters and that intellectual differences exist prior to the appearance of secondary sexual characteristics raise new questions regarding the academic difficulties encountered by late maturers. The possibility of a link between maturation and cognitive maturation is an intriguing but unresolved issue. Factors common to both pubertal onset and academic achievement, such as social class, family size, and ordinal position must be further explored.

In the same national study, the females did not demonstrate the clear associations between early and late maturation and educational and intellectual achievement seen among the males. It may indeed be true that the relationships observed in males do not pertain to females; alternatively, it is possible that the population of early maturing females in the National Health Examination Survey was under-ascertained because the sample began with 12-year-olds.

Body Build and Self-Image

Concomitant with sexual maturation there is an increase in weight in adolescent males and females. The National Health Examination Survey data provided the opportunity to compute mean weight at each stage of sexual maturation for males and females (Table 9–7). Quetelet's index (weight divided by height squared) was greater at each successive stage of sexual maturation. Furthermore, adiposity, or fatness, could be differentiated from heaviness by measurements of skinfold thickness. These measurements showed that as girls become sexually mature they become fatter, whereas fatness remains relatively constant in boys throughout adolescence. By contrast, the estimate of muscle mass indicated that boys become increasingly muscular whereas girls become relatively less so with maturation (Figs. 9–3 and 9–4). Thus the weight gain in adolescence is indicative of fatness is girls, whereas it represents muscularity in boys (Dornbusch, in preparation).

The relationship between sexual maturation and fatness, however, is independent both of age and of timing of sexual maturation. Early maturing girls simply begin to get fatter at a relatively younger age than their peers. These and other studies indicate that boys, on the average, are satisfied with these changes but that the majority

Table 9–7. HEAVINESS OF BODY BUILD*

Stage of Sexual Maturation	Males	Females
1.0	0.0180	0.0152
1.5	0.0184	0.0168
2.0	0.0188	0.0177
2.5	0.0186	0.0182
3.0	0.0193	0.0188
3.5	0.0194	0.0193
4.0	0.0199	0.0199
4.5	0.0206	0.0210
5.0	0.0218	0.0227

*Quetelet's index: Weight in grams/height in centimeters.²

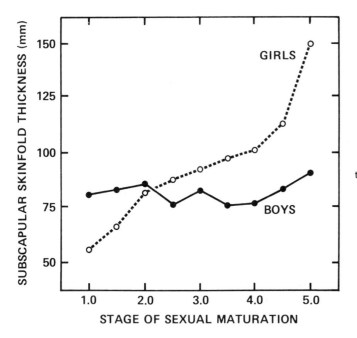

Figure 9–3. Fatness, as indicated by subscapular skinfold thickness, at each stage of sexual maturation.

of girls at each stage of sexual maturation wish to be thinner.

The impact on these attitudes of social class as well as the actual level of fatness showed striking sex differences. Females from the higher social groups were more likely to want to be thinner at each successive stage of sexual maturation, whereas there was no comparable impact of maturation on males, regardless of social class. These differences between males and females were dramatically evident when the real level of fatness was taken into account. Even among the girls who were actually thin, a number in the middle and the higher social groups wished to be thinner (Fig. 9–5A). Among males (Fig. 9–5B), there was no interest in becoming thinner in any social group until they exceeded the 50th percentile for fatness; the real increase in the desire to be thinner occurred in the group who were at the highest percentile for fatness, irrespective of social class. The negative value that society places on a normal developmental event causes many girls to face a difficult adjustment in puberty related to their weight gain. Early maturing girls have the additional burden of coping with this issue two or more years sooner than their age-mates.

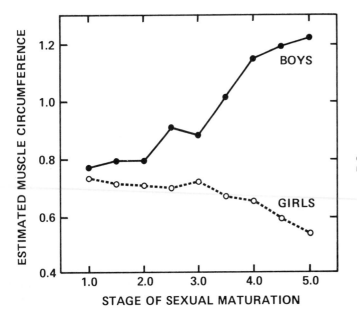

Figure 9–4. Estimated muscle circumference (arm girth in cm./π × triceps skinfold thickness in mm.) at each stage of sexual maturation.

CLINICAL IMPLICATIONS

Data from several cohorts and two continents indicate that important interrelationships exist between early versus late sexual maturation and adolescent development, especially in the areas of intellectual aspiration and achievement, and body build and self-image. These findings have salient clinical implications for physicians and other counselors of adolescents. They should be especially careful, in dealing with males, to monitor the school performance of the late maturer and be alert to the need for encouragement and specific scholastic support. They should inquire into the future academic plans of these late maturers, aware that both the youth and their parents may be negative in their expectations and aspirations. For younger boys experiencing academic difficulties, the clinician should follow them carefully as they reach adolescence, mindful that they may be destined to be late sexual maturers with greater intellectual handicaps.

Physicians are familiar with the reluctance of some adolescents to undress and participate in gym classes. Affected youth are those who perceive themselves as deviant from their peers, even though their physical development may fall within the upper or lower 20 per cent of the normal range. Whether or not these youth should be excused from gym for a time or can be made comfortable through counseling and information about their maturation will be an individual choice. It is very important that their concerns be anticipated and avoided, or at the very least, recognized and addressed so that they do not lead to school avoidance.

For the male would-be athlete, his stage of physical maturation will greatly influence his potential skills. Regardless of body size, his strength and endurance will not increase greatly nor will his muscle mass respond to weight training and dietary regimens until he is well into puberty, at Tanner stage 4 or 5. The late maturing young adolescent may be at increased risk for injuries and may be shunned as an inadequate performer. He may benefit by avoiding active participation in

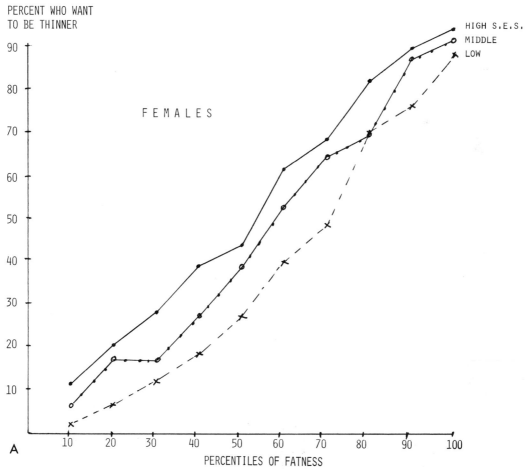

Figure 9–5. Comparison of percentage of females *(A)* and of males *(B)* who want to be thinner by deciles of fatness in low, middle, and high socioeconomic status groups.

Illustration continued on following page

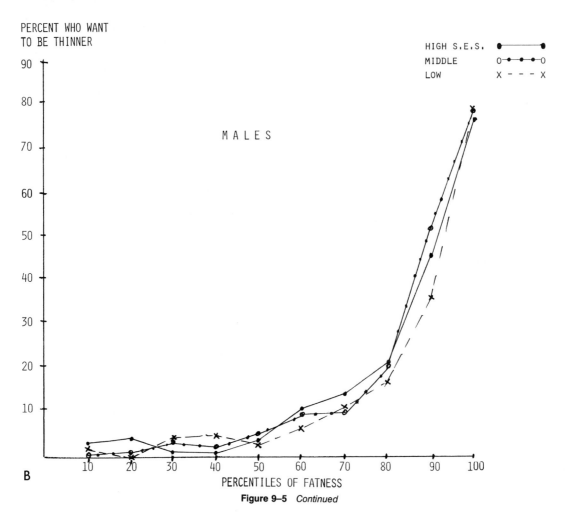

Figure 9–5 *Continued*

contact and collision sports, seeking instead activities in which speed and skill are more important than strength and endurance.

Likewise, the early maturer needs to realize that his advantage in strength and size in comparison to teammates and opponents may be only temporary. At a later age, when his peers mature, he may fail to satisfy his own as well as his parents' and coaches' athletic expectations. Smith suggests that the early maturers be directed to participate with and compete against individuals of comparable maturational stage, irrespective of chronological age (Smith, 1981). As females become increasingly involved in competitive sports, they will need similar developmental guidelines. The relationship of endurance and physical strength to maturational status in the athletic female is an important area for study.

Both the pre-adolescent female and her parents should be counseled about the physical changes that will take place with puberty and her possible reactions to them. If girls realize that the increased weight and the changes in body build are the physiological result of normal maturation, they may make more realistic plans to deal with these changes. For many girls this may involve acceptance of their body build and the employment of regular exercise and dietary moderation. For those who continue to feel uncomfortable about their weight, more intensive diet and exercise regimens can be prescribed. In addition, commonly held misconceptions about body build should be corrected. Some girls and their parents are convinced that excessive carbohydrate intake is responsible for adolescent weight gain. Others, fearing unsightly muscle development, abandon their participation in sports. Prior to the onset of puberty, these issues can be incorporated into a general discussion of the girl's individual rate of maturation, weight, height, menarche, and development of secondary sexual characteristics.

Utilizing what is already known and keeping abreast of future studies of early and late development, physicians and other counselors can offer important information to pre-adolescent and adolescent youths and their parents.

RUTH T. GROSS
PAULA M. DUKE

REFERENCES

Clausen, J. A.: The social meaning of differential physical and sexual maturation. *In* Dragstin, S. E., and Elder, G. H., Jr. (eds.): Adolescence in the Life Cycle. New York, Halsted, 1975.

Dornbusch, S. M., Carlsmith, J. M., Duke, P. M., Gross, R. T., Martin, J. A., Ritter, P., and Siegel-Gorelick, B.: Testing thinness as conspicuous consumption (*submitted for publication*).

Douglas, J. W. B., Kiernan, K. E., and Wadsworth, M. E. T.: A longitudinal study of health and behavior. Proc. Roy. Soc. Med. *70*:530, 1977.

Duke, P. M., Carlsmith, J. M., Jennings, D., Martin, J. A., Dornbusch, S. M., Siegel-Gorelick, B., and Gross, R. T.: Educational correlates of early and late sexual maturation in adolescence. J. Pediat. *100*:633, 1982.

Duke, P. M., Litt, I. F., and Gross, R. T.: Adolescents' self assessment of sexual maturation. Pediatrics *66*:918, 1980.

Faust, M.: Developmental maturity as a determinant of prestige in adolescent girls. Child Dev. *31*:173, 1960.

Frisch, R. E.: Critical weight at menarche, initiation of adolescent growth spurt and control of puberty. *In* Grumbach, M. M., Grave, G. D., and Mayer, F. E. (eds.): Control of the Onset of Puberty. New York, John Wiley and Sons, 1974.

Frisch, R. E., Wyshak, G., and Vincent, L.: Delayed menarche and amenorrhea in ballet dancers. N. Engl. J. Med. *303*:17, 1980.

Garn, S. M.: Continuities and change in maturational timing. *In* Brim, O. G., Jr., and Kagan, J. (eds.): Constancy and Change in Human Development. Cambridge, Harvard University Press, 1980.

Jones, M., Bayley, N., Macfarlane, J., et al. (eds.): The Course of Human Development. Toronto, John Wiley and Sons, 1971.

Malina, R. M., Bouchard, C., Shoup, R. F., Demerjian, A., and Lariviere, G. : Age at menarche, family size and birth order in athletes at Montreal Olympic games, 1976. Med. Sci. in Sports *11*:354, 1979.

Simmons, R. G., Blyth, D. A., Van Cleave, E. F., and Bush, D. M.: Entry into early adolescence: The impact of school structure, puberty and early dating on self-esteem. Am. Sociol. Rev. *44*:948, 1979.

Smith, N. J.: Medical issues in sports medicine. Pediat. Rev. *2*:229, 1981.

Tanner, J. M.: Growth at Adolescence. Oxford, Blackwell Scientific Publications, 1962.

Waber, E.: Sex differences in mental abilities, hemispheric lateralization, and rate of physical growth at adolescence. Develop. Psychol. *13*:29, 1977.

10

Individuality

10A Dynamics of Individual Behavioral Development _____

CONCEPTS OF DEVELOPMENT

Early Models

In past centuries, philosophers and scientists debated two opposite views of the primary force dictating the child's psychological development. In one concept, the newborn infant was considered a *homunculus*, an adult in miniature who already possessed the physical and psychological attributes that would characterize him as an adult. In the other view, the neonate was a *tabula rasa*, as John Locke put it—a clean slate on which the environment would progressively inscribe its influence until the adult personality was etched to completion. (See also Chapter 3.)

The homunculus concept led to such labels as *constitutional inferior* and *constitutional psychopath*—still influential in the 1920's—in which all kinds of complex patterns of deviant behavior were thought to be already present in the newborn infant. A more idealized romantic view was Rousseau's vision of the child as a *noble savage*, endowed with an "innate moral sense" with intuitive knowledge of what is right and wrong "but thwarted by restrictions imposed on him by society." (Freud also considered the infant a savage, dominated by "id impulses," but hardly "noble.") This concept emphasized the influence of heredity and led to such popular aphorisms as "like father, like son" or "the apple does not fall far from the tree."

This mechanical constitutionalist view held a dominant position in psychology and psychiatry in the nineteenth century and into the early decades of the present century. Gradually, it was discredited, largely owing to the work of Freud and Pavlov, who demonstrated from very different vantage points how much of behavior that had been labeled as preformed and predetermined actually arose out of the child's life experiences. Psychodynamic-psychoanalytical and behaviorist studies expanded and deepened our knowledge of the profound significance of the child's environment in shaping his psychological development. With

the excessive swing of the pendulum that so often occurs, however, the rejection of the one-sided constitutionalist model led to an increasing trend to one-sided environmentalist concepts. By the 1950's, this view was firmly in the ascendency in this country, and all types of deviant child behavior, whether minor variations from the average, or severe psychopathology such as autism, schizophrenia, neurosis, or juvenile delinquency, were ascribed in the main to presumed harmful influences of the parents, and in some cases to other intra- and extrafamilial influences. These two views corresponded to the nature versus nurture formulations that were debated in the field of biology.

These opposing formulations, the *homunculus* versus the *tabula rasa*, not only were of theoretical interest to researchers. They also had highly significant practical implications. If the infant was a homunculus, then proper training and development required strict discipline to curb and tame potentially dangerous inborn characteristics. As the phrase had it, "spare the rod and spoil the child." If the child was a tabula rasa, then difficulties in sleeping or toilet-training or the development of more complex behavior disorders led to a search for parental hostility, rejection, or other pathogenic attitudes. If such harmful parental characteristics were not evident on the surface, so much the worse. This meant they were unconscious and therefore presumably more damaging.

In the 1960's and 1970's, however, this one-sided environmentalist view came to be increasingly questioned. Sophisticated developmental research demonstrated the impressive perceptual, learning, and social competencies of the neonate and very young infant (Thomas and Chess, 1980). Identification of significant individual variations in the responses of infants to child-care practices, differences which did not seem to be environmentally determined, indicated that the child's own characteristics also played an important role in the developmental process (Thomas and Chess, 1977). In addition, the findings of a number of longitudinal studies, in which the course of individual behavioral development in different samples was followed from early life to adulthood, were consistent in demonstrating that outcome could not be predicated from the characteristics of the parents or other early environmental influences alone (Chess, 1979).

This did not mean a return to the previously discredited homunculus position. Parents, siblings, other family members, peers, the school—all exerted some influence, often considerable, on the child's development. But so did the child, with his characteristics. A different formulation, the interactive model, attempted to overcome the unidimensional constitutionalist and environmentalist views by creating a two-dimensional structure of an additive nature. Good constitution plus good environment leads to a good outcome; poor constitution plus poor environment leads to a poor outcome. Intermediate outcomes are the result of a good-poor combination. But, as Sameroff, a leading developmental theorist points out, this model "is insufficient to facilitate our understanding of the actual mechanisms leading to later outcomes. The major reason behind the inadequacy of this model is that neither constitution nor environment is necessarily constant over time. At each moment, month, or year the characteristics of both the child and his environment change in important ways. Moreover, these differences are interdependent and change as a function of their mutual influence on one another" (1975). Thus, the simple, static interactive model also proved inadequate to explain the course of a child's development.

Interactional (Transactional) Model

The alternative model, which has by now been formulated and accepted in biology as well as in developmental psychology and psychiatry, can be characterized as interactional, or transactional, the latter term being preferred by some researchers. In this view, behavioral as well as biological attributes must at all times be considered in their reciprocal relationship with other characteristics of the organism and is their interaction with environmental opportunities, demands, and expectations. Consequences of this "interactional" process may modify or change selective features of behavior, and the new behavior may in turn alter recurrent or new environmental influences. In addition, new environmental features may emerge independently or as the result of the previous or ongoing organism-environment interactional process. The same process may modify or change abilities, motives, behavioral style, or psychodynamic defenses. Development thus becomes a fluid dynamic process which may reinforce, modify, or change specific psychological patterns at all age periods.

This interactionist position began to be formulated by a small minority of developmental theorists in biology and psychology in the 1920's to 1930's, was emphasized in our own longitudinal studies from their inception in the mid-1950's, and has finally come to hold a dominant position in the past 10 years. This view has a number of important theoretical and practical implications, which can be cited only briefly here. (A more detailed exposition can be found in our 1980 volume.)

Early life experience, while important, is not decisive for later development. Psychological development is characterized by both continuity and change, and in any individual case linear, one-to-one prediction from early to later life is unreliable. Development does not stop at age 5, or at adolescence, but continues through all age periods. No one pathogenic factor, whether it be the child's characteristics, the parents and family, or the larger social environment, can be labeled as the sole cause of disturbed psychological development. All possible factors and their interaction must be considered in the analysis of the ontogenesis and evolution of an individual's behavior disorder. Such an approach avoids what Mischel describes as "the shortcomings of all simplistic theories that view behavior as the exclusive result of any narrow set of determinants, whether these are habits, traits, drives, reinforcers, constructs, instincts, or genes, and whether they are exclusively inside or outside the person" (1977).

In analyzing the nature of the organism-environment interactional process, we have found the concept of "goodness of fit" and the related ideas of consonance and dissonance to be very useful. Goodness of fit results when the properties of the environment and its expectations and demands are in accord with the organism's own capacities, motivations, and style of behavior. When this consonance between organism and environment is present, optimal development in a progressive direction is possible. Conversely, poorness of fit involves discrepancies and dissonances between environmental opportunities and demands and the capacities and characteristics of the organism, so that distorted development and maladaptive functioning occur. Consonance is never an abstraction but is always goodness of fit in terms of the values and demands of a given culture or socioeconomic group (Fig. 10–1).

It should be emphasized that goodness of fit does not imply an absence of stress and conflict. Quite the contrary. These are inevitable concomitants of the developmental process, in which new expectations and demands for change and pro-

GOODNESS OF FIT

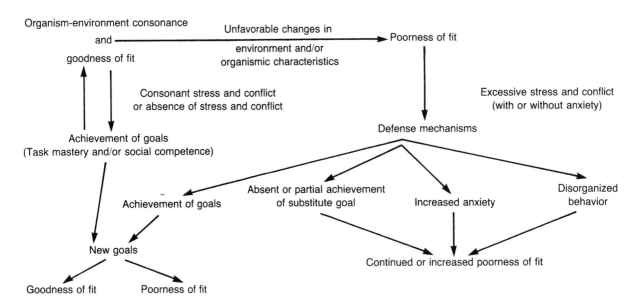

POORNESS OF FIT

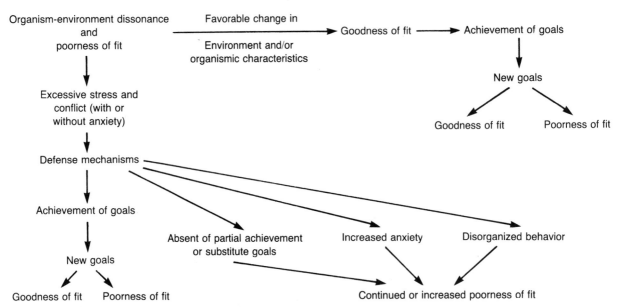

Figure 10–1. The goodness of fit model and the roles of stress, conflict, and development are shown. All kinds of permutations and combinations are possible in the course of any one child's development. Goodness or poorness of fit is rarely global. At any age period for any individual, certain environmental demands and expectations may be consonant with his capacities and coping mechanisms, while others may be dissonant. These consonances and dissonances may even shift with succeeding age periods.

gressively higher levels of functioning occur continuously as the child grows older. Demands, stresses, and conflicts, when in keeping with the child's developmental potentials and capacities for mastery, may have constructive consequences and should not be considered an inevitable cause of behavioral disturbance. The issue involved in dis-

turbed behavioral functioning is rather one of *excessive* stress resulting from poorness of fit between environmental expectation and demands and the capacities of the child at a particular level of development.

The goodness of fit concept does not in any fashion imply some modification of the basic interactionist position. Rather, it is a formulation that facilitates the application of the interactionist conceptual model to specific counseling, early intervention, and treatment situations. The formulation structures a strategy of intervention that includes an assessment of the individual's motivations, abilities, and temperament; his behavioral patterns and their consequences; and the expectations, demands, and limitations of the environment. The specific potential or actual dissonance between individual and environment can then be proportioned. Thus, for example, if consultation is requested for a girl who stands passively at the periphery of a group, and if assessment reveals a slow-to-warm-up temperamental pattern, then attention can be focused on whether the parents and teachers are making a demand for quick, active group involvement. If a boy disrupts his class with bizarre behavior, the assessment may show a severe reading difficulty, with defensive avoidance behavior. If we know that the handicapped youngster may have special difficulties in mastering new complex demands and expectations in adolescence, this can provide a guide to preventive intervention to ensure continued goodness of fit. (See also Chapter 52 concerning diagnostic formulation.)

The interactionist and goodness of fit model can be illustrated by a brief case summary from one of our research studies:

Nancy was a temperamentally difficult child from early infancy onward. She was biologically irregular in feeding, sleeping, and elimination; had intense negative reactions to most new situations and demands; and made positive adaptations slowly. The father was highly critical of her behavior, rigid in his expectations for quick, positive adaptation, which the girl could not live up to, and punitive when Nancy did not respond to his demands. The mother was intimidated by both husband and daughter and was vacillating and anxious in the handling of her child.

By age 6, Nancy had developed a severe behavior disorder with many symptoms. This grew worse over the next few years. Her parents' behavior only made her symptoms worse, exacerbating her father's critical attitude and her mother's fear and vacillation, which in turn increased Nancy's disturbance.

At 9 years of age, however, she began to show evidence of musical and dramatic talent, which led to increasingly favorable attention from her teachers. This talent also ranked high in her parents' own hierarchy of desirable attributes, and her father now began to see his daughter's intense reactions, not as a sign of a "rotten kid" as earlier, but as evidence of a budding artist of whom he could be proud. With this new view of Nancy and her temperament, the mother was able to relax and relate positively to her daughter. Nancy was permitted to adapt at her own pace, the positive aspects of her temperament had a chance to become evident, and her self-image progressively improved. By adolescence all evidence of her neurotic symptoms and functioning had disappeared, a recovery that has continued into her early adult life.

However, Nancy's dramatic change for the better had the opposite effect on her younger sister, Olga. Olga also had a temperamentally difficult pattern from early childhood, although not as intense as that of her older sister. During the years when her parents were preoccupied with Nancy's difficult behavior and adaptations, their attitudes toward Olga were benign and tolerant, since she was less of a problem to them. When, however, they began to view Nancy in a positive light, their critical attention shifted to Olga, who now became the family scapegoat. Olga then began to develop a severe behavior disorder, which continued unabated and even grew worse in adolescence.

Thus, in this family the developmental course of the two daughters could be understood only by applying the interactionist concept. Nancy's behavior disorder arose not because of her own characteristics alone or because of her parents' attitudes or handling alone, but as a result of the interaction of the two factors, with each reinforcing the other over time. The introduction of a new element, her budding talents, transformed her parents' attitudes and behavior and radically changed the nature and direction of her developmental process. For her sister Olga, Nancy's emerging talents also served to change qualitatively the nature of her interaction with her parents, but in this case for the worse instead of the better.

Nancy's development would have been different if her temperament had been quiet and easily adaptable, or if her parents had coped with her difficult temperament as they did at first with Olga. Furthermore, Nancy's development would have been different if her dramatic or muscial talents had not emerged, or if her parents had been lukewarm or negative about these talents. Olga's development would also have been different if her temperament had been quiet and easily adaptable, or if Nancy's course had not changed and shifted so radically at age 9. The two girls showed both consistency and change. Predictability of outcome depended on continuity and consistency in the parent-child interactional process and not on fixation of personality in early childhood.

PLASTICITY OF DEVELOPMENT

The dramatic changes in the developmental courses of the two sisters described above, or in many other subjects whom we have now followed from infancy to adult life, were made possible by highly significant attributes of the human brain, namely flexibility and plasticity.

There are many different pathways to optimal learning and effective social functioning. In human development there are many roads to Rome, and many Romes. The developmental plasticity of behavior must derive from the biological characteristics of the brain, as expressed in interaction with the variety of opportunities and expectations that the social environment provides. The brain's potential in this regard is dramatically evident in the developmental course of children with severe physical handicaps. As we observe these children we realize the futility and destructiveness of the notion that there is only one norm, with deviations being therefore unsatisfactory and inferior.

We have had a special opportunity to follow the developmental course of a group of children with a wide range of severe physical handicaps. These are the children in our congenital rubella longitudinal study, who have been followed from 2 to 4 years of age into early adolescence. The most frequent handicap has been hearing loss (in 65 cases as a single defect and in 112 combined with other defects). Visual loss has been next in frequency, followed closely by neuromuscular disturbances and cardiac abnormalities. Multiple defects are common.

The Deaf Child

Deaf children can see the world but cannot hear it. There is no delay in motor skill acquisition. These children crawl or walk at the expected ages in pursuit of objects or people that catch their attention. They grab just as do the hearing children—and can equally get burned or cut or run over. Rules must be given visually, by touch, or by punishment, since the deaf child cannot be warned of danger from afar unless he is already looking at the warner. Cause and effect cannot be explained, at least not in oral language. They cannot ask questions; they could not hear the answers. Rules for deaf children tend to be more restrictive as a necessary substitution for complex descriptions. Changes cannot be explained. Deaf children cannot anticipate events unless they have already experienced a similar happening that can be used as a concrete reference point. (See also Chapter 39 concerning hearing impairment.)

Yet, the deaf children as a group in our rubella sample showed a significant improvement in level of intellectual test performance between the preschool and middle childhood years. Approximately 50 per cent of those classified as retarded in the preschool period moved either to a lesser degree of retardation or into the normal range. A similar upward trend was strikingly evident in those scoring in the average, dull normal, or borderline range at 3 to 4 years of age.

How do we explain the rising adaptation of these deaf children? The answer appears to lie in the development of communication skills that make language, thought, and conceptualization possible. Most deaf children are deprived of the possibility of learning to communicate in early childhood through words as the hearing child does, and this is reflected in retarded cognitive and adaptive functioning. But soon the child develops a system of communication through visual cues such as gestures, sign language, and lip reading—what is called "Total Communication."

Once the child makes up in this way for his or her earlier inability to communicate and consequent retardation in learning, the deaf child can move ahead rapidly in cognitive and adaptive functioning.

Until very recently, most schools for the deaf used only oral training, that is lip reading and voicing of words. This approach rested on the assumption that there was only one road to the development of language and abstract thinking—that taken by the normal hearing child. Sign language was labeled as inferior and incapable of stimulating cognitive development. Parents were instructed in the most authoritative terms to eliminate all gestures of any kind in communicating with their deaf youngsters, in order to insure the success of "oral" training. But the evidence accumulated indicated that this approach made it most difficult, if not impossible, for deaf children to master the use of language. Schlesinger and Meadows (1972) have shown that deaf children of deaf parents who communicated by means of sign language and natural gestures from infancy onward showed greater academic and language competence than did congenitally deaf children with hearing parents who were not skilled in sign language.

In our own rubella sample of congenitally deaf children, oral speech training did not significantly enhance speaking ability. The use of sentences at age 8 or 9 varied inversely with the severity of hearing loss. Fewer than 15 per cent of those with profound hearing disability used any kind of sentences—long, short, or telegraphic. Of those children who had only oral training in school and had acquired no single words in early childhood, only 7 per cent were able to use sentences by middle childhood. On the other hand, the use of Total Communication led quickly to accelerated academic and social functioning.

Sign language is a true language, with all the implications for learning and abstract thinking, thanks to the plastic potential of the brain. By the use of total communication, the deaf child's visual and motor abilities can be harnessed to compensate for his auditory deficit in learning language. The brain's capacity for plasticity of development makes possible utilization of these alternative pathways to mastering language for the deaf child. He can argue, push limits, understand why and when safety rules are necessary, learn social necessities, express emotions clearly, explore ideas, and master abstraction and symbolization.

The Blind Child

Congenitally blind children show delays in achieving a number of motor developmental landmarks in infancy. They do not raise up their heads to see the world around them as other infants do; they can hear quite well both prone and supine. Sitting, standing, and walking occur later than in their seeing peers, since visually beckoning objects cannot stimulate the blind child to move into a better position to see or to come closer to grasp. But as they mature, these children catch up in neuromuscular skills and cognitive capacity, often in different sequences from sighted babies. For example, in a careful, detailed study of the development of 10 babies blind from birth, Fraiberg (1977) reports that these infants showed the ability to search for a hidden object when they were 8 to 11 months of age. This is the same age when sighted babies show this particular ability; however, sighted babies come to master this search for a hidden object with only a sound cue after having learned at an earlier age to locate and grasp a visually apparent object—a task the blind infant has not mastered. (See also Chapter 39 concerning visual impairment.)

Infants first learn to grasp objects with wide raking motions. As integration of visual and kinesthetic sensation develops, children with normal vision go on to develop precise pincer hand movements for grasping. The blind child continues to use raking motions—the same kind of sweeping movement any one of us would use in trying to locate an object in a pitch-dark room.

The blind child does not respond to the caretaker's face and voice by smiling. The sighted baby who fails to respond to these stimuli with a smile may very well be showing the beginning of a disorder of affective and social development, such as autism. All too often the same interpretation is given to the blind baby's failure to smile, with a judgment that pathology of affective functioning exists. Those who work with blind children are quite familiar with the tendency of psychiatrists and psychologists to overdiagnose autism in such children.

This misjudgment is bolstered by a lack of understanding of the adaptive significance of typical habitual behaviors of blind children. These include pressing the eyeballs, weaving the head from side to side, holding the face down, gazing at a light source and waving the hand in front of the eyes, smelling foods before eating them, and smelling objects and people. These behaviors are so frequently found in blind individuals that they are categorized as "blindisms." Similar habitual behaviors are also found in autistic children and are called "autistic rituals." However, the functional significance of these behaviors is very different in the two groups. The blind children uses these behaviors to increase sensory stimulation and to substitute the use of an intact sense, such as smell,

for the visual defect. Once such a child is otherwise stimulated, such as by someone talking to him, he readily stops the eyeball-pressing or head- or hand-waving. He can also be taught that smelling food or people is socially undesirable. The autistic child's behavioral rituals, in contrast, appear to reflect pathological self-preoccupation and cannot be altered or extinguished in the same manner as in the blind child.

Just as a deaf child is capable of normal language development, so a blind child is capable of normal affective and social development. Such a child cannot see, but he can hear, smell, taste, feel, and be aware of kinesthetic stimuli. He can grasp, learn to cuddle and kiss and to play games identifying parts of the body, and then go on to more complex types of play and social activities that do not require vision.

Although Fraiberg describes the social developmental sequences and patterns of blind children and emphasizes their adaptive value, she fails to see that her own data confirm the key concept that the plasticity of the brain makes it possible for the child to follow more than one normal adaptive sequence of development. Committed to the theoretical position that there is only one normal biological program for behavioral adaptation, she comments that "what we have seen in this typical profile of a blind baby at five months of age is a biological program that has been derailed and for which adaptive solutions have not yet been found. . . . In the biological program it is 'intended' that vision and prehension evolve in synchrony" (1977).

If we choose to assume that there is only one biological program that is "intended," the different developmental sequence of the blind infant does represent a "derailment." If one views the data without such an a priori assumption, the behavioral patterns of the blind child that enable him to develop affectively and cognitively and cope with his environment become a dramatic tribute to the plasticity of the human brain, which makes possible a host of alternative adaptive developmental pathways, depending on the characteristics of the child and the nature of the environment.

The Child with Motor Handicaps

The infant with a motor handicap suffers from diminished sensorimotor experience. Depending upon the nature of the disability—absence or partial failure of limb growth, paresis, amyotonia, spasticity, athetosis—a corresponding limitation on the exploration of the world will be present. The handicapped infant may not be able to touch or hold objects with ease, move them from hand to hand, or take them from or give them to nurturing adults. (See also Chapter 39.)

In Piaget's theory of cognitive development the first stage is that of sensorimotor intelligence in infancy. In this period, the infant learns to coordinate sensory data and motor experiences, leading to an awareness of the external world as a permanent place, with objects having properties independent of one's own perceptions. Piaget designates the development of the basic psychological units as *schemata*. One primary scheme is prehension, which evolves from progressive modification of the grasping reflex by the infant's contact with different shapes, textures, temperatures, and weights as he grasps and handles one object after another.

For the infant with a severe motor handicap, difficulties in grasping, holding, and handling objects preclude the usual sequence of the formation of primary schemata. Piaget considers this sequence the essential first stage in the progressive development of cognition, as indeed it is for the nonhandicapped child. Yet, it is clear that the person born with a handicap that seriously limits sensorimotor experience finds alternative pathways to normal cognitive development. Again, the host of such individuals who attain a superior intellectual level which they use productively and creatively are a vivid testimonial to the inherent capacity of the human brain for plasticity of development.

Implications

The deaf child, the blind child, and the motor-handicapped child—each can find a developmental pathway consonant with his capacities and limitations, thanks to the plasticity of the brain. By the same token, the environmentally handicapped child is not inevitably doomed to an inferior and abnormal psychological developmental course. Whether the handicap comes from social ideology, poverty, a pathological family environment, or special stressful life experiences, the plastic potential of the brain offers promise for positive and corrective change. Although this promise is not a guarantee, professional and political authorities who formulate educational, remedial, and therapeutic programs for these youngsters can translate this promise into reality. To do so requires the abandonment of hierarchical judgments about a single normal sequence of development and a single standard for measuring social and intellectual growth. The assumption that the handicapped child's development must duplicate that of the nonhandicapped child, otherwise it is inferior, inevitably leads to self-fulfilling prophecies of actual inferior outcome.

TEMPERAMENT

Pediatricians, baby nurses, and experienced nurses have always known that young infants and even newborns exhibit marked individual differences in both their spontaneous and reactive behavior. Also, at least some of these characteristics appeared to persist as the child grew older and to influence the way he responds to handling by the parents and to new situations and events.

For psychiatrists and psychologists, however, the concept of the infant as a tabula rasa as well as the abandonment of the mechanical constitutionalist views of the nineteenth century led them increasingly to ignore this phenomenon of individual differences in the infant during the 1950's. The role or organismic characteristics received some attention, mainly through the view that developmental level was a fundamental factor in structuring the child's reactions to his environment. However, the concept of developmental level in the main referred to general laws of responsiveness and to the time sequences in which universals in personality organization were achieved, but not to the issue of individuality or uniqueness of functioning.

It is true that by the 1950's a number of leading theoreticians and researchers, such as Freud, Gesell, and Shirley, had commented on significant individual differences in the behavioral characteristics of infants, and that a number of studies had described observations of individual differences in infants and young children in specific, discrete areas of functioning (see Thomas and Chess, 1977, Ch. 1, for a summary of these studies). However, no long-term investigations had been reported that defined these differences systematically and comprehensively or studied the relationship between these findings in early life and the later course of healthy and deviant psychological development.

Our own interest in exploring the nature and significance of these individual differences in children derived from several considerations:

1. Like many parents, we were struck by the clearly evident individual differences in our children, even in the first weeks of life, and noticed the same phenomenon in the young children of our relatives and friends.

2. As clinicians, we were repeatedly impressed by our inability to make a direct correlation between environmental influences, such as parental attitudes and practices, and the child's psychological development. There was no question, of course, that these influences played an important part in the child's life, and we, like other clinicians, devoted much effort to trying to persuade parents and others to provide a healthier environment for children. However, we saw many, many instances

in which psychopathology in a child occurred even with good parents, or in which a child's development pursued a consistently healthy direction, even into adult life, in the face of severe parental disturbance, family disorganization, or social stress.

3. As mental health professionals, we became increasingly concerned about the dominant professional ideology of the time (the 1950's), in which the cause of all a child's deviation from the accepted behavioral norms was laid at the doorstep of the mother. The guilt and anxiety this created in mothers whose children had even minor behavioral difficulties was enormous. Time after time it was clear that pointing the finger of blame at the mother had devastating psychological effects on her. And yet, in so many cases we could see little or no evidence that the mother was completely or even partially responsible for her child's difficulties.

4. Finally, a review of the research literature revealed that there was a growing body of reports reflecting considerable skepticism of this exclusively environmentalist view (Thomas and Chess, 1977).

With these considerations and concerns in mind we began in 1956 the New York Longitudinal Study (NYLS) to identify, categorize, and rate individual differences in children and to explore their significance for the developmental process. In the NYLS we have followed the behavioral development of a sample of 133 subjects from early infancy to early adult life, using a variety of data-gathering and analytical procedures. Later, several other longitudinal studies of special populations were added to supplement the data obtained from the NYLS. Our methods and findings have been reported in a number of publications over the years and are detailed in our latest volumes (Thomas and Chess, 1977; Thomas and Chess, 1980).

The characteristics of behavioral individuality that we have studied we have subsumed under the term *temperament*, already utilized by some other workers to designate similar behavioral phenomena.

Temperament may best be viewed as a general term referring to the *how* of behavior. It differs from ability, which is concerned with the *what* and *how well* of behaving, and from motivation, which accounts for *why* a person does what he is doing. Temperament, by contrast, concerns the *way* in which an individual behaves. Two children may dress themselves with equal skillfulness or ride a bicycle with the same dexterity and have the same motives for engaging in these activities. Two adolescents may display similar learning ability and intellectual interests and their academic goals may

coincide. Two adults may show the same technical expertness in their work and have the same reason for devoting themselves to their jobs. Yet, these two children, adolescents, or adults may differ significantly with regard to the quickness with which they move; the ease with which they approach a new physical environment, social situation, or task; the intensity and character of their mood expression; and the effort required by others to distract them when they are absorbed in an activity.

In this definition, temperament is a categorical term, with no implications as to etiology or immutability. On the contrary, like any other characteristic of the organism—wehther it be height, weight, intellectual competence, or perceptual skills—temperament is influenced by environmental factors in its expression and even in its nature as development proceeds.

TEMPERAMENTAL CHARACTERISTICS

From the first group of infant behavioral records in the NYLS obtained by parental interviews, nine categories of temperament were established by an inductive content analysis:

1. *Activity Level.* The motor component present in a given child's functioning and the diurnal proportion of active and inactive periods. Data on motility during bathing, eating, playing, dressing, and handling as well as information concerning the sleep-wake cycle, reaching, crawling, and walking are used in scoring this category.

2. *Rhythmicity (Regularity).* The predictability and/or unpredictability in time of any function. It can be analyzed in relation to the sleep-wake cycle, hunger, feeding pattern, and elimination schedule.

3. *Approach or Withdrawal.* The nature of the initial response to a new stimulus, be it a new food, new toy, or new person. Approach responses are positive, whether dispalyed by mood expression (smiling, verbalizations) or motor activity (swallowing a new food, reaching for a new toy, active play). Withdrawal reactions are negative, whether displayed by mood expression (crying, fussing, grimacing, verbalization) or motor activity (moving away, spitting new food out, pushing new toy away).

4. *Adaptability.* Responses to new or altered situations. One is not concerned with the nature of the initial response but with the ease with which they are modified in a desired direction.

5. *Threshold of Responsiveness.* The intensity level of stimulation necessary to evoke a discernible response, irrespective of the specific form that the response may take or the sensory modality af-

fected. Behaviors utilized concern reactions to sensory stimuli, environmental objects, and social contacts.

6. *Intensity of Reaction.* The energy level of response, irrespective of its quality or direction.

7. *Quality of Mood.* The amount of pleasant, joyful, and friendly behavior as contrasted with unpleasant, crying, and unfriendly behavior.

8. *Distractability.* The effectiveness of extraneous environmental stimuli in interfering with or altering the direction of the ongoing behavior.

9. *Attention Span and Persistence.* Two related categories. Attention span concerns the length of time a particular activity is pursued by the child. Persistence refers to the continuation of an activity in the face of obstacles to maintain the activity direction.

Temperamental Constellations

Three temperamental constellations of functional significance have been defined by qualitative analysis of the data and factor analysis. The first group is characterized by regularity, positive approach responses to new stimuli, high adaptability to change, and mild or moderately intense mood that is preponderantly positive. These children quickly develop regular sleep and feeding schedules, take to most new foods easily, smile at strangers, adapt easily to a new school, accept most frustration with little fuss, and accept the rules of new games with no trouble. Such a youngster is aptly called the "easy child" and is usually a joy to his parents, pediatricians, and teachers. This group comprises about 40 per cent of our NYLS sample.

At the opposite end of the temperamental spectrum is the group with irregularity in biological functions, negative withdrawal responses to new stimuli, nonadaptability or slow adaptability to change, and intense mood expressions that are frequently negative. These children show irregular sleep and feeding schedules; slow acceptance of new foods; prolonged adjustment periods to new routines, people, or situations; and relatively frequent and loud periods of crying. Laughter, also, is characteristically loud. Frustration typically produces a violent tantrum. This is the "difficult child," and mothers and pediatricians find such youngsters difficult indeed. This group comprises about 10 per cent of our NYLS sample.

The third noteworthy temperamental constellation is marked by a combination of negative responses of mild intensity to new stimuli with slow adaptability after repeated contact. In contrast to the difficult children, these youngsters are characterized by mild intensity of reactions, whether positive or negative, and by less tendency to show irregularity of biological functions. The negative

mild responses to new stimuli can be seen in the first encounter with the bath, a new food, a stranger, a new place, or a new school situation. If given the opportunity to reexperience such new situations over time and without pressure, such a child gradually comes to show quiet and positive interest and involvement. A youngster with this characteristic sequence of response is referred to as the "slow-to-warm-up" child, an apt if inelegant designation. About 15 per cent of our NYLS sample falls into this category.

As can be seen from the above percentages, not all children fit into one of these three temperamental groups. This results from the varying and different combinations of temperamental traits manifested by individual children. Also, among those children who do fit one of these three patterns, there is a wide range in degree of manifestation. Some are extremely easy children in practically all situations; others are relatively easy but not always so. A few children are extremely difficult with all new situations and demands; others show only some of these characteristics and relatively mildly. For some children it is highly predictable that they will warm up slowly in any new situation; others warm up slowly with certain types of new stimuli or demands but warm up quickly in others.

It should be emphasized that the various temperamental constellations all represent variations within normal limits, i.e., any child may be easy, difficult, or slow to warm up temperamentally, have a high or low activity level, be easily distracted or be persistent; a relatively extreme rating score may be seen in a sample of children for any specific temperamental attribute. However, such an amodal rating is not a criterion of psychopathology but rather an indication of the wide range of behavioral styles exhibited by normal children.

It has been possible to identify each of the nine categories of temperament in each child at different age periods in all our longitudinal samples. In addition, these temperamental characteristics have been identified in a number of populations studied by investigators at many other centers in this country and abroad. It is clear, therefore, that these behavioral traits are ubiquitous among children and can be categorized systematically. Now we are also able to rate our NYLS subjects on these same nine categories in adolescence and early adult life.

Several parental questionnaires for rating temperament at various age periods in childhood as well as self-rating questionnaires for adolescents and young adults have been developed. The childhood questionnaires developed by William Carey and his coworkers have been especially widely used in a large number of studies here and abroad. (See Chapter 44.)

Other workers have suggested some modifications in our scheme of categorization of temperament as well as devised several alternative models, although their functional significance for the developmental process remains to be demonstrated (Thomas and Chess, 1980). It can be expected that future investigators will identify additional categories and determine their functional importance.

TEMPERAMENT AND BEHAVIOR DISORDERS

Studies carried out in our own research unit and by other investigators utilizing our formulations of temperament have indicated its significant role in normal and deviant psychological development. These findings, detailed in our recent volumes (Thomas and Chess, 1977; Thomas and Chess, 1980), will be summarized briefly here. Children with the difficult child pattern are most vulnerable to the development of behavior problems in early and middle childhood. Their intense negative withdrawal reactions to new situations and slow adaptability together with their biological irregularity make the demands of early socialization especially stressful for these children. Seventy per cent of this group in the NYLS developed clinically evident behavior disorders (a mild reactive behavior disorder in most cases) before 10 years of age. With parent counseling and other therapeutic measures when indicated, the great majority recovered or improved markedly by adolescence.

In children with physical handicaps or mild mental retardation, the difficult child group is at even greater risk for behavior problem development than are nonhandicapped children. Children with this temperamental pattern are also vulnerable to psychiatric disorders if they have a mentally ill parent. Infants with colic are also more likely to show the difficult child pattern (Carey, 1972).

However, a child with any temperamental pattern can develop behavior disorders if demands for change and adaptation are dissonant with the particular child's capacities and therefore excessively stressful. Thus, the distractable child is put under excessive stress if expected to concentrate without interruption for long periods of time, the persistent child if his absorption in an activity is prematurely and abruptly terminated, and the high-activity child if restricted in his possibilities for constructive activity. Teachers may underestimate the intelligence of the slow-to-warm-up child or the low-activity child, with unfavorable consequences for the learning situation.

These findings do not imply that temperament is always a significant factor in the development of every behavior disorder. This is also true of

motivations, abilities, or specific environmental influences. In any specific instance, the pattern of interaction of factors responsible for disturbed function cannot be assumed a priori, inasmuch as it may vary qualitatively from case to case.

IMPORTANCE OF TEMPERAMENT TO THE PEDIATRICIAN

The pediatrician's role with regard to the child's temperamental individuality can be considered under a number of headings: advice on child-care practices, reassurance to parents, routine examinations and procedures, management of minor problems, evaluation of the acutely ill child, management of the chronically ill or handicapped child, and preventive counseling (Table 10-1).

In carrying out these activities it is necessary, of course, for the pediatrician to be able to assess the child's temperamental characteristics with a reasonable expenditure of time. To a large extent this can be accomplished through the information the pediatrician obtains from the parents and from his own observations of the child in the course of ongoing care and management. In our own contact with a number of pediatricians, as with parents and teachers, we have been impressed by their ability to rate a child's temperamental qualities, once they were familiar with the categories and the behavioral criteria for their rating. Furthermore, the pediatric examination, whether in the office, clinic, or at home, provides an opportunity to observe the child's behavior in a number of situations. More systematic data on temperament can be obtained quickly by asking the mother to fill out the Carey questionnaire for the appropriate age period (see Chapter 44). Finally, the pediatrician can include questions on temperament as part of the clinical history taking. (A model for the latter procedure is included as Appendix D in our 1977 volume.)

Advice on Child Care

Defining the infant's temperamental characteristics by the pediatrician and parent can be enormously useful for both parent and pediatrician in developing an optimal approach to early child care. More than any other professional, the pediatrician can offer vitally important reassurance to anxious, insecure mothers. For the inexperienced mother who has been bombarded with pat formulations warning her that every move or expression may affect her baby's future mental health, any difficulty or problem in her infant's management may appear ominous indeed. If her child fusses a lot during the night, does not take to new foods or feeding schedules easily, cries when ap-

Table 10–1. ISSUES AND SITUATIONS IN WHICH CONSIDERATION OF TEMPERAMENT MAY BE SIGNIFICANT FOR THE CLINICIAN

1. Reassurance of parents that child's deviation from culturally desirable norm does not mean pathology in child or bad parenting. Especially true with difficult or slow-to-warm-up child.
2. Child-care advice specified in terms of child's temperament, such as approach to weaning, toilet-training, and the like.
3. Evaluation of severity of acute physical illness by estimating deviation of child's behavior from usual temperament. Also, temperament may affect reaction to illness.
4. Evaluation and management of specific symptoms such as colic, night-awakening, or "hyperactivity" as partially influenced by temperament.
5. Child's adaptation to beginning nursery school or day-care center as influenced by reactions to new situation and speed of adaptation.
6. Ease or difficulty of child's establishment of peer relations.
7. School functioning — optimal style of classwork and homework schedule in relation to degree of persistence and distractability.
8. In behavior disorders, identification of influence of temperament and the specific pattern of "poorness of fit."
9. Special influences of temperament in the physically handicapped and the mentally retarded.

proached by strangers, resists toilet-training or ignores her safety rules, she may all too frequently interpret such difficulties as proof that she is a "bad mother." The pediatrician can step in, evaluate the issue, identify the aspects of the child's temperament that are relevant, and advise the mother appropriately. By contrast, simply brushing aside the mother's concern with responses such as "Don't worry, everything will be all right" or "Be patient, your baby will outgrow this" will most often have limited and temporary value.

Routine Examinations and Procedures

An infant's responses to new people and procedures, especially if associated with any discomfort or pain, is strongly influenced by his temperamental characteristics. When brought to the pediatrician's office for the first time or at subsequent irregular intervals, the infant is confronted with a strange place, a number of unfamiliar persons, and perhaps unusual sounds, and is then subjected to a physical examination and vaccinations that are restraining, discomforting, and sometimes painful. Depending on his temperament, the child may fuss quietly and briefly, squirm a bit, and then be immediately cheerful once the procedures are completed. Or he may howl loudly from the moment he enters the pediatrician's office, struggle violently during the

physical examination and inoculation, take up to several hours to calm down, and respond even more intensely at the next visit. The child with a low-activity level will sit quietly in the waiting room, whereas the high-activity youngster will fidget, jump around, try to poke into drawers and closets, and make a nuisance of himself if he has to wait a long time to see the doctor.

The pediatrician who evaluates the temperament of his patient and understands its implications will be a good position to minimize the distress or trauma of each child's office visit. With the easy child, no special management approach is necessary. However, even such an adaptable child may be overwhelmed if the busy doctor rushes him too quickly through the office procedures. The slow-to-warm-up child should be given time to get used to the waiting room and more time in the examining room before one proceeds with the physical examination. The highly active child should be seen quickly. If he has to wait, he should be given toys and space to move around in without disturbing others. The difficult child should also be given a warm-up period, but will probably still respond to the examination or inoculation with loud and long protests. The painfulness or discomfort of any of the office procedures should certainly be minimized for all patients, but this is of special concern for the child with a low sensory threshold.

Management of Minor Problems

Minor problems of a young child brought to the pediatrician's attention are in many cases significantly related to temperamental issues (Carey, 1972). Frequent night waking, as reported by Carey (1974), is more likely to occur in an infant with a low sensory threshold. He suggests two possible explanations for this relationship: the low-threshold child's greater responsiveness to daytime stimulation makes him more arousable at night, or the infant is more responsive to internal and external stimuli at night as well. Carey recommends dealing with the problem by reducing environmental stimulation before bedtime and, if necessary, during the night.

The prolonged loud crying spells of the infant with colic present a difficult and disturbing management problem to the parents. In our own experience, this syndrome appeared to be related to the difficult child temperament and not to pathological maternal attitudes, as was so often assumed in the past. Our impression is confirmed by Carey's work, which also suggests a relationship to low sensory threshold as well as to the difficult child pattern. The pediatrician can provide much needed reassurance and support to the parents in such cases, in addition to any other specific therapeutic measures he utilizes. (See also Chapter 28 concerning colic.)

We have seen a sleep problem develop in several easy children after a minor acute illness. Such children, who previously slept well through the night, awoke frequently when ill, crying with the discomfort of their acute symptoms. Parental care naturally involved picking the child up each time, soothing him, and alleviating the discomfort. Following recovery from the illness, the child continued to awaken crying each night. The parents continued to pick him up and soothe him in the same way, now interpreting the awakening as a sign of anxiety. Upon evaluation, there was no other evidence of anxiety and the night awakening undoubtedly reflected the quick adaptability of the child to the new night-time behavior established during his acute illness. Because he was an easy child, one might predict that the nocturnal crying would quickly disappear if the child were no longer picked up. Indeed, this was the outcome when the parents followed this recommendation. (See also Chapter 32.)

The response of a child to the irritation of a skin rash can be significantly influenced by temperamental traits. The intense reactor may cry loudly and complain vigorously even if treatment is proceeding effectively, and this reaction should not lead doctor or parent into a misjudgment of the treatment procedures. The child with a low sensory threshold may have great difficulty in obeying the injunctions not to rub or scratch the irritated skin, and special measures may be necessary to reduce the sensory stimulation from the rash.

The above instances illustrate—but by no means exhaust—the many ways in which the child's temperament influences the response to physical symptoms and the management of minor ailments. Every pediatrician can document this relationship extensively from his own clinical experience.

Evaluation of Acutely Ill Child

A major professional responsibility of the pediatrician is to evaluate the gravity of symptoms in an acutely ill child. Is the sudden high fever the first sign of a serious illness or of a minor upper respiratory infection? Is lethargy an indication of severe toxicity or a reaction to the discomfort and malaise of a more benign condition? Are acute restlessness and thrashing about an ominous sign or a reaction to acute pain? Certainly these judgments depend primarily on careful diagnostic clinical evaluation. But there are many occasions in which knowledge of the child's temperamental characteristics can be helpful in making these judgments. All other things being equal (which

they often are not!), if the child's behavior when acutely ill is qualitatively different from his usual temperamental style, this is a more serious indication than if his behavior is similar though exaggerated. Mothers are aware of this issue pragmatically when they express their concern to the doctor because the acutely ill child "is not behaving like his usual self."

The high-activity, high-intensity child who is restless, thrashes around vigorously, and complains loudly when taken ill is exhibiting a quantitative exaggeration of his usual temperamental traits. The same behavior in a low-activity, low-intensity child represents a qualitative change from his normal temperament. By and large, the qualitative change is more likely to reflect a more severe acute physical change than is the simple quantitative exaggeration of the child's usual temperament. The same considerations apply in reverse, that is, listlessness in a normally high-activity child is a more ominous sign than an increase in activity. In a similar vein, the complaint of intense pain in a high-threshold child is likely to have substantial significance, while the same complaint of a low-threshold child may be significant or may represent the reaction to a relatively minor irritation.

While these and similar temperamental phenomena are not the decisive issues in the evaluation of an acutely ill child, their consideration has often been of significant value to doctors and nurses in making diagnostic judgments.

Complicating the above considerations, however, is the phenomenon some mothers have reported to us that their child's behavioral pattern is different when he is ill but is also typical of him. The child seems to express one kind of temperament when healthy and another when acutely ill. Thus, manifestations of intense negative responses during illness may be the expression of extreme discomfort, in contrast to the vigor of positive intense mood expression at other times. Here, too, both doctor and parent must be alert to the significance of the acutely ill youngster's behavior.

Management of the Chronically Ill or Handicapped Child

In the management of the more chronically ill or physically handicapped child, temperamental issues are not infrequently of major importance. The ease with which a child can follow a regimen of restricted physical activity and engage in substitute sedentary occupations or adapt to the stringent demand of prolonged bed rest is strongly influenced by his temperamental characteristics. Low-activity and highly adaptive children will find such regimens much easier to tolerate than high-activity or slowly adaptive youngster. The child who is persistent and has a long attention span can be content once his interest is engaged in a sedentary activity. On the other hand, if it is necessary to terminate a more physically exhausting activity in which a persistent youngster has become absorbed, this may be difficult and stressful. The distractable child will be much more easily diverted from an undesirable activity, or one that should be time-limited, but it may be difficult to keep him interested in any single sedentary occupation for more than a short time. When hospitalization is required, the need to adapt to a host of new people and situations may be especially stressful to the difficult or slow-to-warm-up child.

It is important for the physician, together with the parents and nurses, to formulate schedules and routines for the physically restricted child that will be most consonant with the youngster's temperament. This will ensure the maximum cooperation of the child and provide the best assurance that the necessary management procedures will be carried through. It will also minimize the danger that the stress imposed by the illness or handicap will precipitate a reactive behavior disorder.

Certainly the child's temperament is not a crucial or even important factor in every pediatric consultation and treatment. The same is true for the mental health practitioner. As with any other variable that may significantly influence a child's physical or psychological functioning, the importance of temperament may be considerable in certain pediatric practice situations, moderate or modest in others, and negligible in still others. However, it is only when the potential influence of temperament is appreciated that its actual significance can be evaluated in any specific situation with an individual child.

GOALS OF DEVELOPMENT

A fundamental characteristic of all living organisms, including the human being, is that their functioning is goal-directed. This is a qualitative distinction from the nature of the physicochemical laws governing inanimate matter. The student of biology must ask the question "What for?" This question reflects human-centered teleology and is forbidden to the nonbiological scientist, who is allowed to ask only "How?"—a question the biologist must also ask.

Since the beginning of the psychoanalytical movement, consideration of the goals and motives of human behavior has been a primary concern. However, given the primitive state of behavioral research and knowledge regarding the human infant 100 years ago, Freud could rely only on the animal models of instincts and drive state satisfaction and frustration of his time. In recent decades, these concepts have been found to be increasingly inadequate not only for human developmental theory but for animal behavior studies as well, leading many psychoanalysts to propose modifications or a radical revision of libido theory. However, no consensus has as yet been achieved in the psychoanalytical movement for a revision of its traditional formulations of the goals of behavior in light of current research findings (Thomas and Chess, 1980).

In contrast, behaviorism avoided the issue of human goals by focusing on the stimulus-response paradigm, as typified by the conditioned reflex, with the judgment that goals and motives are hidden in the mind's scientifically unknowable "black box." Those behaviorists who have incorporated intrapsychic factors into their conceptual frameworks have moved to the social learning field but have not developed a systematic theoretical model for considering the goals of behavior developmentally.

An alternative approach is possible, based on the weight of recent research involving infants. The goals of human behavior, starting at birth, can be conceptualized as social competence and task mastery. Both are highly developed in the human being, with his unique capacity for learning. Both proceed developmentally, as the individual's capacities mature, as learning takes place, and as the environment makes successive new demands and presents new opportunities. Both proceed by a constant reciprocal interaction. Task mastery facilitates social relationships, and increased social competence promotes the capacity to master the environment. Most activities, such as play, school functioning, sex and athletics, contain both social and task features.

A number of developmental psychologists have emphasized this concept of the central role of social competence and task mastery for goal-directed functioning. Bruner, for example, puts it that "the forms of early competence can be divided into those which regulate interaction with other members of the species and those involved in mastery over objects, tools, and spatially and temporally ordered sequences of events. Obviously, the two cannot be fully separated, as can be seen in the importance of imitation and modeling in the mastery of "thing skills" (Bruner, 1973).

Recent research has provided an impressive body of data on the capacities of the neonate and very young infant. The newborn can recognize visual patterns and shows preferential attention to complexity, movement, three-dimensionality, and representations of the human face. Sound can be localized at birth, and there are data suggesting a spatially relevant and functional relationship between vision and hearing. The neonate also shows a wide range of capacities with regard to neurobehavioral organization, including orienting responses, habituation reactions, and correlation between intensity of auditory stimulation and direction of eye movements. Learning, as demonstrated by the formation of conditioned reflexes, starts actively at birth. Learning by imitation has been demonstrated in the first week of life, and the infant can differentiate between two live female faces during the first weeks of life and can discriminate voices. The newborn is also capable of active social communication—the most basic element of social exchange. Studies have demonstrated precise and synchronous movement of the neonate with adult speech and in breast feeding. Manipulative-exploratory behavior of increasing complexity in the first weeks of life has also been described by a number of investigators (see Chapter 5).

Thus, the neonate and young infant are biologically equipped for the pursuit of two basic adaptive goals—the development of social relations (social competence) and the acquisition of skills (task mastery). Achievement of these goals sequentially with increasing levels of complexity and maturation is a source of satisfaction and gratification and a crucial factor in the progressive development of a positive self-concept.

Whether it is walking, weaning, self-feeding, toilet-training, self-dressing, or the acquisition of language, the normal child is highly motivated at the appropriate developmental levels to engage in these tasks and carry them through to completion. It is true that many humans carry out the process of mastery through stress and tension, whether it is the 10-month-old infant learning to walk and drink from a cup or the adult artist or scientist struggling with a painting or laboratory experiment. To see such stress as undesirable and, if inevitable, regrettable is to misinterpret profoundly the dynamics of healthy psychological development. It is only when demands are made for a level and quality of performance that are excessive and inappropriate for the individual that the stress and tension may have unfavorable consequences. If the demands are not excessive and the stress is resolved by mastery, the effects are positive. Actually, unfavorable consequences may result not only from excessive demands but also from misguided efforts by parents to "protect" their child from stress and tension.

ADAPTIVE AND COPING STYLES

In responding to the demands and expectations of parents, teachers, and peers, and in the pursuit of social competence and task mastery, the child's most effective and positive coping strategy is direct engagement with successful outcome. Even when the child is only partially able to cope directly with a new demand or expectation, mastery and positive adaptation are still often possible. These are the situations in which the individual's direct engagement with the issue results in sufficient modification of the environmental demand or growth of his capacity, or both, so that direct mastery and positive adaptation can occur. For example, a boy who is clumsy because of delayed neuromuscular development may not be able to meet the expectations of his peer group as a ball player. However, he may stick to the group in spite of their initial criticisms and teasings, practice assiduously on his own with a parent or older sibling, improve gradually, and gain the respect of his peers as they begin to appreciate his determined efforts and improvement, until he finally becomes an accepted and modestly competent member of the team. Successful achievement also enhances self-esteem and the child's confidence that successful direct struggle is possible even when the stress is substantial.

However, if the new demand is highly excessive for the child, then direct mastery is not possible unless the level of expectation is modified to bring it within the scope of the individual's capacities. If not modified, the excessive stress and impossibility of direct mastery may cause the child to resort to one defensive strategy or another. This may happen for many reasons: dysphasia, dyslexia, temperament-environment dissonance, academic demands that are beyond an individual's intellectual abilities, overwhelming environmental stress, or severe distortions of brain functioning. On many occasions, and for various reasons, an individual may have the capacity to cope directly and effectively with the new demand yet fail to do so, turning instead to some defensive strategy. Certain past experiences may have created a conditioned response that any stress is dangerous. Or the demand may be presented by parent, teacher, or employer in an ambiguous or confusing form. As another example, the kind of effort required to master the demand may appear, rightly or wrongly, to alienate the individual from his peer group.

Defensive strategies, or defense mechanisms as they are usually called, can be defined operationally as behavioral strategies that attempt to cope with demands or conflicts the individual cannot or will not master directly. This definition does not assume, as Freud did, that defense mechanisms are necessarily unconscious, nor does it assume any a priori theoretical formulations of the causes of stress and conflict. Thus, although derived originally from psychoanalytic theory and practice, the concept of defense mechanisms, as defined here, can be used in the analysis of behavioral dynamics independently of the psychoanalytical framework.

Specific defense mechanisms can also be defined operationally and identified by simple inference from empirical data, again, without commitment to any one theoretical scheme. *Suppression-repression* represents the attempt to extinguish feelings or ideas; *denial*, the posture that the stressful environmental demands are really insignificant and need not be confronted; *avoidance*, the detachment from the stressful situation; *reaction formation*, the attempt to cope by transforming the motivation into its opposite; *rationalization*, ascribing a socially acceptable motive to behavior that has other motivations; *displacement*, the involvement with a less meaningful but less stressful object or situation than the one at issue; *projection*, ascribing motives, feelings, or ideas to others that are really one's own; and *fantasy*, the retreat into intrapsychic thoughts and feelings as a substitute for confronting external reality. *Sublimation*, or avoidance through a socially desirable activity, and *humor* are generally more constructive defense strategies. *Delusional thinking*, the reconstruction of reality to eliminate stress and threat, usually occurs in severe mental illness. Fantasy and humor are not always defensive maneuvers; they may even be prominent features of healthy and creative functioning.

The use of defense mechanisms is by no means always undesirable. An individual may at one time or another be faced with excessively stressful demands or conflicts that cannot be mastered directly. This may evoke the temporary utilization of a defensive strategy, which then gives the person the opportunity to resolve the stress positively. In these situations it would appear that the use of the defense mechanism is necessary to organize the strengths and capacities needed for successful mastery. For example, an adolescent with the slow-to-warm-up temperament may experience his typically uncomfortable, shy response when joining a new peer group. He may rationalize his initial peripheral and outwardly detached involvement to the group by making various excuses (he is worried about a friend who is ill, he has a sprained ankle, he has to get home early) and gain the time necessary to make his gradual positive adaptation. A young girl may be aware that the severe and frequent arguments between her parents threaten to break up her home and family. She may deal with the potential anxiety this creates by denying the seriousness of the situation and displacing her anxiety by worrying over a pet dog's well-being. Although the denial and displacement are unrealistic, they enable her to pursue her own life activities without crippling anxiety.

At times, the price paid for the achievement of psychological equilibrium through a defensive strategy is so high that it must be challenged at all costs. Thus, a child with a school phobia may avoid the anxiety of attending school by elaborating various somatic symptoms. The child may be completely comfortable if not attending school, with no physical complaints and with active pursuit of friendships and other activities, but may be incapable of returning to school without severe acute anxiety. In such a case, this pattern of adaptation has so many serious consequences that it must be treated as a psychiatric emergency.

Finally, there are instances in which a defense mechanism, no matter how strongly and decisively it is utilized, fails to achieve or sustain a state of psychological equilibrium. The resulting disequilibrium may find expression in a panic state, acute dissociative state, severe depression, or behavioral disorganization (see Chapter 41).

Defense mechanisms vary tremendously in their influence on the developmental course of different individuals. These mechanisms may have little or no importance, may be of transient significance at one or another time, or may be intermittent or constitute a continuous and highly important factor. In this regard, they are similar to other factors that influence the course of development—temperament, intellectual level, specific abilities or defects, parental practices and attitudes, other family influences, special events, and the social environment—all of which vary greatly in their impact on different individuals.

ACHIEVING SELF-ESTEEM

The uniqueness of the human mind is perhaps most exquisitely manifested in the self-concept or sense of identity. It is true that a young chimpanzee can recognize and identify its reflection in a mirror, indicating at least a crude sense of "self." However, there is no evidence that the development and differentiation of self in the chimpanzee or other nonhuman species proceed beyond this simple level. In contrast, humans are consciously aware of self—a sense of our own identity as separate, unique persons. We change over time, as we cope with one life experience after another, yet always a central core of our psychological being appears to remain intact and to endure. While the 60-year-old knows he is vastly different from the young adult he was at 20, he still feels deeply that he is the same as he was at 20. The child or adult experiences himself as a unique person who is at the same time similar to others. He has a sense of autonomy yet really knows himself only through his social interactions with other human beings. The sense of self is the quintessence of the subjective yet shows itself continously to others in objective actions and communications. The concept of self is the province of the poet, the philosopher, the social and behavioral scientist, and the clinician; its highly abstract and general nature makes its systematic study difficult. But, because of its central place in the individual's psyche, it cannot be ignored in developmental research, psychological theory, or clinical practice.

The terms *self*, *self-concept*, and *identity* are used loosely in the professional literature, not surprisingly for such general and abstract categories. The term *self* can be used to denote the actual existing psychological structure of the individual, above and beyond the various separate and specific aspects of his behavioral characteristics, abilities, values, beliefs, goals, and so on. The terms *self-concept* and *identity* are often used synonymously to denote the individual's consciousness of self, his evaluative self-assessment.

Empirically, we can include as significant components of the self those attributes which are both enduring and influential in determining the course of psychological development and functioning. Thus, although a severe acute illness may have substantial psychological consequences, it is itself transient and not part of the self. A chronic physical illness or handicap that affects the individual behaviorally because it is enduring does become part of the self.

Societal values and demands clearly play a vital role in determining significant components of the self. In a society such as ours, in which hierarchical value judgments are made on the basis of sex, color, religion, national origin, and socioeconomic class, these aspects of an individual's identity become important attributes of the self. In a society where such value judgments are not made, these items might constitute useful demographic data for statistical purposes but would not be considered significant aspects of the self. In a simple agricultural community, intellectual competence, except for extremes, might be of little importance, while physical strength, endurance, and dexterity might be highly treasured assets and constitute meaningful attributes of the self. In a technologically advanced society, these value judgments are reversed, and cognitive capacity becomes a significant aspect of the self.

Development of the self, therefore, will reflect the existence and influence of enduring and functionally important attributes. The meaning of any one attribute may change with time and lead to a corresponding change in the self. As an example, one young woman in our NYLS group is intellectually superior and completing undergraduate studies with academic honors at a prestigious institution. When she expressed interest in becoming a social worker, she was sharply attacked by her friends for yielding to the stereotype of suitable careers for a woman. With her academic abilities, they insisted, she should choose a career that will be more demanding. The significance of her intellectual capacities for her self is now quite different than it would have been 50 or even 25 years ago.

Development of Self-Concept and Self-Esteem

The unique human capacity for sense of self or self-concept can be either a profoundly positive asset or a source of the greatest distress, unhappiness, and disturbance in functioning. The child who develops a positive self-concept—a strong sense of self-esteem—can approach the challenges, problems, and opportunities of adolescence and adult life with self-confidence and can expect to be successful in coping with new demands and expectations, even if these are unanticipated and complex, and can acquire an inner assuredness of positive responses from other people. By contrast, the youngster with a downgraded self-image—a lack of self-esteem—will face adolescence and adult life with uneasiness and even anxiety, will assume that others will have the same negative image of himself that he has, and will be constantly pressured from within to resort to defensive strategies instead of coping directly with new challenges and opportunities.

Once a positive or negative self-image becomes firmly established within the child, this in itself becomes a highly influential determinant of adaptive strategies and task mastery and social competence during successive age periods. Self-fulfilling prophecies will result unless the environmental influences change radically. The self-confident child who approaches new situations with assurance and direct utilization of his abilities and experiences can consistently expect to succeed, with enhancement of his self-esteem. The unsure, self-doubting youngster who approaches the new with anxiety and self-defeating defensive strategies will all too often fail to live up to his capacities and the expectations of others, which can only further intensify his negative self-image.

What can we say about the development of the self-concept and the formation of a healthy sense of self-esteem?

The young infant's social transactions and individual way of mastering experience can be assumed to lead to the beginnings of self-differentiation soon after birth. This assumption is confirmed by observations of young infants. By the third month or even earlier, infants will spend many minutes at a time gazing at their hands, turning their hands at the wrists and looking intently as they do so. At the same age, or perhaps a few weeks later, infants left alone in a crib will often not only babble for long periods but at times keep repeating the same sound. The 3-month-old child also shows a clear awareness of the responses of others to his behavior, showing pleasure, for example, if his smile produces an active positive reaction.

In all these instances the young infant is initiating behavior that has consequences (movement of his hand, a sound, actions of other people). This is certainly the beginning of that basic constituent of positive self-concept, that "I" can produce changes in the external world. "I" must be a separate entity from the outside world if "I" can accomplish this.

Many a 10-month-old infant will already insist on trying to accomplish many tasks by herself, such as holding the bottle, using a spoon, or putting on her shoes. These efforts must reflect a further development of self-concept, that "I" have the ability to attempt these actions which previously only the caretakers could accomplish. Failure often accompanies these first efforts, as it will for more complex tasks at later ages, like writing the alphabet and riding a bicycle. But persistence and eventual success are fundamental building blocks for a firm structure of self-esteem.

Growth of a positive self-concept in later infancy takes other forms as well. The 18-month-old who says "shoes" to a visitor and thrusts her foot forward with a beaming face to have her new shoes inspected is conveying a sense of self-pride. The 2-year-old who scribbles on a paper and brings it to the parent with a smile is expressing some level of awareness of her worth to her parent.

From its beginnings in infancy, the self and self-concept rapidly develop through the preschool and middle childhood years. As the child's world expands in all areas, the elements of the formation and differentiation of the self come together in a mutually influential interactive process. If there is consonance between child and environment, the foundation for a healthy self-concept and stable self-esteem is laid. If there is dissonance, a negative, denigrated self-evaluation begins to crystallize. Development of self-esteem is always a social process. There are no a priori built-in mechanisms by which the child can assess by himself the worthiness of his achievements, the appropriateness of his efforts at task mastery, and the competence of his social functioning. Communication and judgments from others provide the initial standards and the basis for developing his own standards. The parent who approves uncritically of everything the child does is doing a great disservice to the youngster, as is the parent who criticizes every imperfection of performance.

Special Stress and Self-Esteem

The physically handicapped child is frequently subject to special stress in his efforts to achieve task mastery and social competence. Even his best efforts may leave him lagging significantly behind his nonhandicapped peer at home, in school, in play, in social settings, or in all situations, depending upon the type and severity of the handicap. This may make it harder, but by no means impossible, for such a child to develop a positive self-evaluation. Over and over again, we have been impressed—even awed—by the struggles of our handicapped congenital rubella children to reach a level of functioning that will gain the acceptance and respect of their parents, siblings, peers, and teachers. Those who succeed gain enormously in self-esteem. Those who fail—because the handicap is too severe, the demands too excessive, or the defense mechanisms too self-defeating—suffer great injury to their self-assessments.

The child subjected to discrimination because of race, color, or religion suffers special threats to the development of healthy self-esteem. But it is by no means true, as some mental health professionals would have it, that all such children must inevitably be crippled psychologically. Although some children may be overwhelmed, many others become aware of the sources of the threat to their integrity, develop appropriate anger at the injustices they suffer, and reject the prejudiced judgment that they are inferior (Thomas and Chess, 1980).

Autonomy and Dependency

The human infant begins life completely dependent on his caretakers. In the course of physical and psychological development this dependency is gradually and progressively replaced by both personal autonomy and social interdependence. With each step in task mastery, whether it is self-feeding and control of elimination in the infant, the effective acquisition on language in the toddler, or the sequential steps in learning physical skills and the accumulated wisdom of his culture in the older child, the youngster acquires an increasing sense of autonomy, a sense of self-direction and self-control. At the same time, with each expansion of social competence, the growing child's sense of interconnection with other people, his need for them and their need for him, is reaffirmed and strengthened. When these two processes of personal autonomy and social interdependence proceed in consonance and harmony, healthy psychological development is assured. Disruption of this harmony to produce either a personally autonomous child who is socially isolated or a socially connected youngster who remains helplessly dependent on others signals a seriously pathological developmental course that requires prompt intervention and acute treatment.

Performance and Satisfaction

From a different vantage point, development of a healthy self-concept can also be characterized by a progression of performance and satisfaction in consonance and harmony. The child who successfully performs tasks and meets demands set by his society for regular feeding and sleep schedules and toilet-training in infancy, plays cooperatively with peers and adapts to routines and issues of discipline in the home during the toddler stage, meets school demands and expectations in middle childhood, and faces the host of new issues and expectations of adolescence will likely achieve a sense of increasing self-confidence and autonomy. However, this positive developmental sequence will be diluted or even distorted if successful performance is not accompanied by subjective satisfaction with these achievements. This sense of satisfaction does not come to the child automatically or mystically but through feedback from important people— parents, other family members, peers, and teachers. Such feedback must communicate to the child that his performance is indeed positive and competent. Failure to receive this feedback, because of problems of social relatedness in the child or unrealistic and arbitrary standards set by one or more of these important people, can cancel out or even turn to a sense of failure even the most competent performance by the child.

CONCLUSION

In the past, various developmental theories have conjectured that the growth of positive self-esteem and overall healthy psychological growth are determined by hypothetical "unconscious" or subtle interchanges between child and parent, or by the mother's doling out of sufficient quantities of "tender loving care" and "empathy." In other cases, subjective states of child and parent have been ignored as unknowable "black boxes," and prescriptions for objective training and discipline have been formulated. Of course, parental love and sensitivity and routine training procedures are important, but recent research has broadened our perspective and understanding. We can now formulate the process of subjective expansion of the self and self-concept operationally and objectively in terms of goodness of fit between the child's temperament and other characteristics and the demands and expections of the environment, and the factors involved in the achievement of personal autonomy, social interdependence, successful performances, and subjective satisfaction from these successes. This developmental process also proceeds by a continual, sequential, mutually influential interplay between child and environment, an interplay which may make for continuity at some age periods and dramatic spontaneous changes at other times. The child with unfavorable experiences and excessive stresses at one age is not doomed, the parent's early mistakes are not irrevocable, and our preventive and therapeutic intervention can make a difference at all age periods.

STELLA CHESS
ALEXANDER THOMAS

REFERENCES

Bruner, J.: Organization of early skilled action. Child Devel. 44:1, 1973.

Carey, W. B.: Clinical applications of infant temperament measurements. J. Pediat. 81:823, 1972.

Carey, W. B.: Night wakening and temperament in infancy. J. Pediat. 84:756, 1974.

Chess, S.: Developmental theory revisited: Findings of longitudinal study. Canad. J. Psychiat. 24:101, 1979.

Fraiberg, S.: Insights from the Blind. New York, Basic Books, 1977.

Mischel, W.: On the future of personality measurement. Am. Psychologist. 32:246, 1977.

Sameroff, A. J.: Early influences on development: Fact or fancy? Merrill Palmer Quart. 20:275, 1975.

Schlesinger, H., and Meadows, K.: Sound and Sign: Childhood Deafness and Mental Health. Berkeley, University of California Press, 1972.

Thomas, A., and Chess, S.: Temperament and Development. New York, Brunner-Mazel, 1977.

Thomas, A., and Chess, S.: Dynamics of Psychological Development. New York, Brunner-Mazel, 1980.

10B Issues of Gender Differentiation

This discussion of issues in gender differentiation has been organized according to age. In the first section we consider perinatal biological processes, neonatal behavior, and sex-stereotyped expectations and perceptions on the part of parents. The second section concerns the early years. Temperamental characteristics often alleged to be sex-differentiated will be considered as well as the hypothesis that girls mature faster than boys. Parental behavior toward young sons and daughters will be discussed, with particular reference to the difficulties of studying the family unit. The third section focuses on the preschool-age child, for whom peer interaction becomes an important part of gender differentiation. The differential behavior of preschool teachers toward boys and girls will also be addressed. Finally, sex differences and similarities in cognitive performance during the school years and possible causes for any differences will be examined and behavior of both teachers and parents toward school-age boys and girls will be reviewed.

BIRTH

Birth Processes That Are Sex Differentiated

Conception results in more male than female zygotes. Conservative judgments of this primary sex ratio estimate 120 males conceived for every 100 females. Fetal and neonatal death is more common among males than among females, thus reducing the sex ratio at birth. However, this secondary sex ratio still yields a slight preponderance of males. Although these mortality rates have been dropping over the last decades, and the secondary sex ratio has been lowering as well, these changes are largely limited to white populations (see also Chapter 21).

The birth process itself is somewhat different for the average boy compared to the average girl, with mean length of labor about an hour longer for male than for female infants. Labor length at different hospitals, by sex and ordinal positions, is presented in Table 10-2. (Details of this study are reported elsewhere [Jacklin and Maccoby, 1982.]) The implications of these findings are not yet clear. Although there is evidence that length of labor is related to children's subsequent development, in most studies long labor is confounded by greater usage of perinatal medication, so that the effect of either labor length or drugs alone is difficult to determine.

The greater vulnerability of the male to birth injury and congenital malformations not resulting in death may have subtler effects on results of research in behavioral development. Procedures

Table 10-2. MEAN LENGTH OF LABOR IN MINUTES

	Male Infants	Female Infants	Sexes Combined
First borns	535.91 (79)	453.69 (70)	497.28 (149)
Later borns	304.14 (76)	290.05 (95)	296.32 (171)
Totals	422.27 (155)	359.47 (165)	389.89 (320)

( ) = Number of cases studied (Jacklin and Maccoby, 1982).

for selecting infants to be studied may involve sex bias in one direction or another. If males are included in a sample in proportion to their birth, and the sample is randomly drawn, it will include more male children at risk. On the other hand, if very strict selection criteria are used at birth, a more protected sample of males will be drawn and fewer sex differences may emerge in later behavioral testing.

Sex Steroid Hormones at Birth

One area of study of physiological sexual differentiation is hormone concentrations. Considerable research is under way on the course of some of the sex steroid hormones during development. In particular testosterone has been extensively studied. Briefly, we know that when the testes first become active about six weeks after conception, males have higher concentrations of testosterone than do females, although females do have some. Sex differences during the third and fourth gestational month are greater than they are at birth (Faiman et al., 1974). The critical period for the development of genitalia appears to be the time of greatest activity of the testes (Gandy, 1977).

The critical period of genital differentiation may not be the critical period for other dimorphisms between the sexes. In studies with lower mammals in which testosterone has been injected perinatally, late hormonal interventions do affect sex-dimorphic characteristics (see Reinish, 1974, for a review).

The sex steroid hormones of the neonate are somewhat different in males and females. In the Stanford Longitudinal Study (Maccoby et al., 1979), five sex hormones—testosterone, androstenedione, estrone, estradiol, and progesterone—were assayed from samples of umbilical cord blood in three groups of infants. Testosterone concentrations were significantly greater in the males while the other four hormones did not differ significantly between the sexes. Of interest was the size of the mean sex difference. As shown in Figure 10-2, there was considerable overlap in the male and female distributions of testosterone, the one hormone shown to be at different levels in the two sexes.

This overlap in the distributions among male and female humans contrasts with the differentiation of sex hormones in the neonatal nonhuman primate (Resko, 1974). Sex differences in sex steroid hormones are much larger at birth in nonhuman subjects. Progesterone levels, which do not differ in human neonates (Maccoby et al, 1979), are twice as high in newborn female rhesus monkeys as in the males (Hagemenas and Kittinger, 1972). Similarly, estradiol levels in umbilical cord plasma are twice as high in newborn female

TESTOSTERONE IN UMBILICAL CORD BLOOD

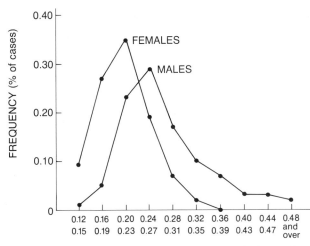

Figure 10–2. Sex similarities in testosterone concentration of human umbilical cord plasma.

rhesus as in males, but there are no sex differences for this hormone in human neonates. The sex difference for testosterone in humans is much smaller than in rhesus, with the male/female ratio of testosterone in the umbilical plasma of rhesus being about 2:1 (Resko, 1974).

Because of the discrepancy in the magnitude of sex differences between human and nonhuman hormone concentrations, simple generalizations about sex differentiation in animals should be applied to humans only with great caution.

Behavior at Birth

At birth, boys and girls seem very alike. Nonetheless, investigators have carried out numerous tests to detect any behavioral differences. Bell and his colleagues (1971) performed 60 or so different behavioral assessments on the first four days of life; of these, 12 revealed stable characteristics (that is, scores correlated significantly from day to day for individual infants during the first four days of life). Of these, two showed sex differences: prone head reaction, showing the clearest sex difference, and tactile threshold, in which a sex difference was found in a breast-fed but not a bottle-fed sample. Lipsitt and Levy (1959) also reported electrodermal thresholds to be lower in female newborns.

The following investigators have not found sex differences in tactile sensitivity at birth: Birns, who measured motoric response to the application of a cold disc; Turkewitz, Moreau, Birch, and Crystal, who measured head turning to the touch of a light brush on the mouth; Yang and Douthitt, who measured reaction to an air jet to the abdomen; Rosenblith and DeLucia, who measured reaction

to cotton and cellophane placed lightly over the infant's nose and mouth; and Jacklin, Snow, and Maccoby (1981), who measured aesthesiometer thresholds in three independent samples.

The discrepancies among these studies are so far unexplained. They do not appear to be a function simply of the measures employed, since studies using the same methods have resulted in conflicting findings. Bell has suggested that chubbiness may be a confounding variable accounting for some of the discrepancies in investigations of tactile sensitivity, but this hypothesis has not been supported in other studies (see Jacklin et al. 1981, for references).

Muscle strength has been reported to differ by sex at birth, with the means for boys being higher. Two measures of muscle strength used for newborns are the prone head reaction and grip strength. The prone head reaction measures the height and duration of the head lift when the infant is placed in a prone position. Bell, Weller, and Waldrop (1971) found boys to have higher scores on this measure. Rosenblith and DeLucia (1963) found no sex difference in prone head reaction; however, they did find grip strength to be greater in newborn boys than girls. These investigators also found a positive relation between weight of the infant and grip strength. They reported that the greater mean weight of their male infants accounted for the observed sex difference in grip strength.

In a study testing three independent samples of newborns (Jacklin et al., 1981), prone head reaction was higher for boys than for girls in all three samples. Two measures of grip strength (responsive and passive) showed somewhat different results. Although the responsive grip strength measure was significantly higher for boys when all three samples were combined, the passive grip strength was not. Both weight and chubbiness were measured and were shown not to account for the difference in responsive grip strength.

Although there are significant sex differences in muscle strength, it must be emphasized that the magnitude of this sex difference is very small. The frequency distributions for boys and girls are almost completely overlapping. Put another way, the correlation between sex and prone head reaction is $r = 0.166$; for responsive grip strength, $r = 0.161$; and for passive grip strength, $r = 0.111$. Knowing the sex of a child does not help predict the child's strength.

Activity level at birth has received considerable research attention. Both the vigor and frequency of movements have been studied (Korner et al., 1981). Interestingly, the amplitude (or vigor) of noncrying movements has higher day-to-day stability than frequency of movements, with correlations of 70 to 80 per cent. Even though individual stability is quite high, no sex differences have been found in any of the measures of activity (Korner et al., 1981).

Parents' Expectations and Beliefs Before They Gain Experience with Their Children

In terms of behavior, boys and girls are very similar at birth. The one difference that has been established, muscle strength, is a difference that parents would probably not be aware of. How parents do see their children at birth has been investigated. When asked to rate their children, parents describe many more sex differences than researchers have been able to document.

In a sample of newborn infants, one study (Rubin et al., 1974) found that parents described their infants in sex-stereotyped ways. Although the infants did not differ in height or weight, for example, parents rated their girl infants as small and delicate and their boy infants as big and strong. Mothers had held and looked at their infants before filling out the rating scales, while fathers were only able to see their infants through the window of the nursery before they produced their ratings. Interestingly, fathers gave more strongly stereotyped ratings than did mothers. Perhaps stereotyping is most likely when little information is available.

Consistent with this view, stereotyping by parents appears to be less evident for older infants. A modified version of Rubin's checklist was given to parents of two cohorts in the Stanford Longitudinal Study. One group of parents had children 6 months old, and the other had children 33 months old. At both ages, only minimal sex differences were noted in the responses to checklist items.

FIRST YEARS

Maturation Issues

It is often said that girls mature faster than boys and that this difference in maturation rate accounts for many sex differences (Garai and Scheninfeld, 1968). Quite clearly there is a sex difference in the onset of the prepubertal growth spurt and in bone development (see Roche, 1979, for a review). Whether a pervasive difference in maturation rate exists between boys and girls in many areas of functioning has not been established.

One area of function that shows clear development with age is sleep. *Sleep patterns* change dramatically and systematically during the first few years of life. Total sleep decreases from 17 hours in each 24-hour period at birth to about 15 hours at 3 months, about 14 hours at 1 year, and about 13 hours at 2 years. Periods of uninterrupted sleep and uninterrupted wakefulness increase with age; concomitantly, the number of transitions from sleep to wakefulness for each child decreases.

Individual children vary considerably in their sleep patterns, but there is evidence for individual stability of these patterns. By six months some children are consistently sleeping more than others; some consistently sleep through the night, while others do not (Jacklin, et al., 1980). (See also Chapter 32.)

No sex differences have been demonstrated in the rate of development of sleep pattern (Moss and Robson, 1970). In the Jacklin et al. study 1980), only one sex difference emerged: boys were found to be more stable in sleep-wake transitions than were girls. This finding would contradict the hypothesis that girls mature faster than boys. Although children of a given age do differ in the overall maturity of their sleep pattern, these differences are not related to sex.

Developmental milestones are commonly used as measures of maturity. Children are compared based on the ages at which they hold up their heads, sit up, and so on (Bayley, 1956). Two socially important milestones, age of walking and age of toilet-training, were measured in the Stanford Longitudinal Study, but no differences were found between boys and girls regarding the age at which they took their first steps, were secure in walking, or did their first running. This finding is consistent with Bayley (1956), who did not find the sexes to differ in developmental milestones. However, age of toilet training in the Stanford sample shows a clear sex difference. On the average, girls are toilet-trained earlier than boys, a finding that may or may not be related to neurological development.

Another instance in which the alleged greater maturity of girls has been used to explain an alleged sex difference is verbal ability. Sex differences in verbal ability among older children and adults tend to be small and elusive (Maccoby and Jacklin, 1974; Macaulay, 1977). With younger children, finding differences is largely a function of the methodology used. Mean length of utterances is longer in younger toddler girls.

In a test of the hypothesis that girls are more verbal than boys, 35 measures of vocalization were made in the Stanford Longitudinal Study. These measures were taken from 6 to 45 months of age and involved three successive samples of children. Children were assessed in both free play and structured testing situations, and measures included both frequency of vocalization and verbal comprehension. Of the 35 measures, only one showed a significant difference, and one would expect to see one difference in 20 by chance. Parents' vocalization to their children was tested in 33 separate measures to determine whether parents talked more to boys or girls (again in both free play and structured situations from 6 to 45 months). Of the 33 measures, two reached the 0.05 level of significance, but in one case mothers talked more to girls and in the other they talked more to boys.

At this time, it is reasonable to conclude that girls are not more verbal than boys in the early years. As suggested earlier, however, the issue of sampling is important here. Since boys are more likely to suffer from developmental anomalies, including speech defects, means from random samples of boys and girls will depend on how many of the small group of boys at the lower end of the distribution happened to be included in a particular study.

In sum, maturation is a complex phenomenon. On the average, girls do show more rapid bone development and are toilet-trained earlier. However, there do not seem to be sex differences in the maturity of sleep patterns, in reaching developmental milestones of a sensory-motor type, nor in tested or spontaneous verbal ability.

Sex Differences in Temperament

Emotional states, particularly negative emotional states, are believed to be stable temperamental characteristics. Most of the work measuring emotional stability of young children has measured irritability, sometimes as a naturally occurring state. For example, Moss measured crying or fussiness in the absence of discernible stimulation. Others have measured fussiness as a reaction to a wide range of unpleasant stimulations from mild to somewhat harsh. Although Moss did find boys crying more than girls, when hours asleep were equated in his sample the sex difference disappeared. Other studies do not tend to find sex differences in crying or irritability (see Maccoby and Jacklin, 1974, for a review). (See also earlier section on temperament, page 164.)

Timidity is commonly believed to be more intense, more frequent, or elicited upon weaker stimulation in girls than in boys. Although many studies using both observational measures and parent reports do not find differences by sex, when differences are found, girls are reported to show more fear (see Maccoby and Jacklin, 1974, for a review). In the Stanford Longitudinal Study, sex differences are borderline: girls more often have higher scores on observed timidity, but differences are small and usually not significant. In sum, boys and girls do not differ in irritability and do not differ very much in levels of timidity. However, when differences do occur, girls show slightly higher levels of timidity.

"Activity level" includes a variety of behaviors, and as is often true in development, the specific behaviors that form the basis of a behavioral cluster change with age. Research in this area is mixed with respect to sex differences. Although it is commonly believed that boys are more active than girls, most studies do not find differences in the first three years. But when differences are found, boys are likely to be more active than girls (see Maccoby and Jacklin, 1974, for a review). Situational factors may play a role in the inconsistency of findings. At least three situational factors relate to activity level: the size of the space in which activity level is measured, the presence or absence of rough-and-tumble play, and the size of the group of peers with whom the child is interacting.

If children are tested in small spaces, sex differences are unlikely, but when children are observed in large play yards, these differences are more likely to emerge, since there are greater opportunities for rough play, an activity more common among boys (see Peer Interaction below.). The issue of group size relative to activity level is a controversial one. Halverson and Waldrop (1973) report no sex difference in activity level for children playing alone, but found boys' activity level increased with group size while girls' did not. However, in an unsuccessful attempt to replicate this finding, no sex differences in activity level as a function of group size were found in two samples of the Stanford Longitudinal Study.

An additional factor may be how activity is organized. Boys are more frequently diagnosed to be "hyperactive" than girls (Safer and Allen, 1976), but when actual activity is measured, "hyperactive" children do not move from place to place more than normal children. However, they do show more restless, squirmy movements when staying in one place. Thus, sex differences in activity level may prove to be differences in *kind* of activity rather than in amount; however, this issue is not yet settled.

Sex differences and similarities of the young child are summarized in Table 10–3.

Early Sex Differences in Parental Treatment

As noted earlier, parents describe their newborn infants in sex-stereotyped ways but are less likely to do this after they have had some experience with their children. But what of their behavior toward their sons and daughters?

Studying the interaction between a parent and a child (or any other two people) is complex. It is difficult to determine which individual is determining the nature of the interaction. Is a child treated in a certain way because of stereotyped views a parent has about that child, or because the child is acting in a way that elicits a particular behavior from the parent? For many years parent-child interactions were assumed to consist of en-

Table 10–3. SEX DIFFERENCES AND SIMILARITIES IN EARLY BEHAVIOR

	Sex Difference	Comments
Behavior at Birth		
Tactile sensitivity	None	Relationship to chubbiness not replicated
Prone head reaction	Boys higher	Differences so small studies must be aggregated to find significance
Grip strength	Boys higher	
Activity level	None	
Temperamental Characteristics		
Irritability	None	
Timidity	Girls higher	Small differences
Activity	Boys higher	Small differences only in gross-motor activities
Peer interaction	None	Some sex-of-partner effects

counters in which parents trained and influenced their children. The evidence given for this influence was the correlation between parent and child behaviors. Bell (1968) and others have pointed out that the child could just as easily be "shaping" the parent. Given only correlational evidence, it is not clear who is influencing whom.

Even with the difficulties involved in dyadic research, if one examines the evidence for differential treatment by parents of boys and girls in the first three or four years of life, one finds that boys and girls are treated very similarly. For example, in such areas as warmth, nurturance, acceptance, restrictiveness, allowing dependency, and allowing aggression, very young boys and girls do not seem to be treated differently by their parents (Maccoby and Jacklin, 1974). Boys do seem to receive more physical punishment. This may be a case in which boys and girls are eliciting different behavior from parents because of different initial behavior, particularly aggressive behavior (Maccoby and Jacklin, 1980).

In most areas of parent-child interaction there appears to be little difference in the first few years in the way boys and girls are socialized, although there is an exception. In the area of sex-role stereotyped play, one can document differential socialization. Fathers offer more sex-appropriate toys to sons and daughters as early as 12 months of age (Snow et al., in press) and react negatively to sex-inappropriate toy play as early as three years of age (Langlois and Downs, 1980). Rough-and-tumble play and high-arousal play is more likely to be initiated by fathers toward their sons than toward their daughters (Jacklin et al., 1981.)

PRESCHOOL

Peer Interaction

In young children playing with peers, some differences in *play behavior* are seen among pairs of boys and girls. In one study of children as young as 33 months, there was slightly more aggression, hitting, and threatening in the boy-boy pairs and more offers of toys in the girl-girl pairs, although, on the whole, differences were outweighed by similarities. Both boy-boy and girl-girl pairs were socially active (Jacklin and Maccoby, 1978). In striking contrast was the play of mixed-sex pairs; in these pairings there was very little interaction. This mutual wariness in mixed-sex pairs at 33 months presages a familiar phenomenon in preschool: although, of course, there is some mixed-sex play, children most often select same-sex partners in free play (Strayer, 1977).

Another difference in early peer behavior may be the *size of the groups* in which boys and girls are most likely to play. In one laboratory nursery school it was found that the average size for the boys' groups was five, while the girls' groups were likely to have two or three members (Waldrop and Halverson, 1975). A cross-national study with somewhat older children confirmed a difference in group size (Omark et al., 1980). However, in other work (Bell et al., 1971) and in two samples of nursery school children from the Stanford Longitudinal Study, no differences were found in the size of group in which boys and girls played. More work needs to be done before we will be able to decide how general the sex difference is relative to group size.

An interesting average difference between boys and girls (and men and women) has been documented in the ability to report accurately on the content of *nonverbal communication*. Children have been studied as early as age 3, and even at that age there is a small (but greater than chance) advantage for girls in correctly labeling nonverbal cues. Surprisingly, this result is not contingent on the age or sex of the stimulus subject. Furthermore the female advantage does not increase *or* decrease with age (see Hall, 1978, for a comprehensive review).

Aggression is consistently found to be a small sex difference (Maccoby and Jacklin, 1980). Boys are more likely than girls to display peer-directed aggression. Further, boys' aggression is most often displayed in the presence of male partners.

Rough-and-tumble play is a behavior that is consistently found to be more frequent in boys than in girls. Sex differences are found in American preschoolers (DiPietro, 1981) and in a variety of diverse cultures (Whiting and Edwards, 1973). The sexes differ fairly consistently, then, in rough-and-tumble play. It may be instructive to illustrate the size of the sex difference in a typical study. Figure

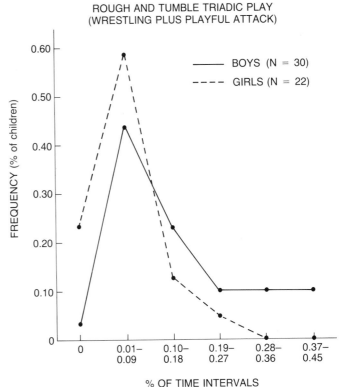

Figure 10–3. Sex similarities in rough and tumble play in preschool children.

10–3 displays the data from DiPietro (1981) in percentage form. As can be seen, the children engaging in the most intervals of rough-and-tumble play are boys. However, the majority of boys and girls are quite similar in the amounts of rough-and-tumble play they display. The overlap and similarities between the sexes should be kept in mind even in a behavior such as rough-and-tumble play, which consistently shows significant average sex differences.

Teachers' Differential Behavior by Sex—Preschool

Recent research in tracing differential socialization of boys and girls in the schools is promising and will be described here for two age groups. Data pertaining to the preschool child will be discussed first followed in a later section by a discussion about the grade school age child.

In the preschool class, boys and girls receive positive reinforcement depending on different circumstances. Girls are more likely to receive positive reinforcement from their teachers when they stand close to their teachers. For boys, receiving positive reinforcement from the teachers is unrelated to their distance from their teachers (Serbin et al., 1973). When teachers change their reinforcement pattern, girls adjust their distance from the teacher, so that we may conclude that the differ-

ential reinforcement pattern by the teachers is at least partly responsible for the girls' greater closeness. Where the child stands with respect to the teacher is also an important determinant of what activities the child engages in, since the preschool teacher tends to be involved in the areas of "fine-motor skill" activities (e.g., bead-stringing, play with clay). If teachers move to other activity areas, the girls follow them (Serbin, 1977). The girls' play experience is thus shaped by the contingencies involved in the teacher's close presence. Play experience with different types of toys has been shown to correlate with the child's tested cognitive abilities (Connor and Serbin, 1978).

We do not wish to exaggerate the teacher's role in generating sex differences. Teachers do not appear to respond differentially to aggression in the two sexes, for example. Boys may be reprimanded more often for fighting, since this occurs more often among boys, but the teacher's response is similar regardless of the sex of the child (Barrett, 1979).

SCHOOL YEARS

Cognitive Sex Differences

Girls, on the average, receive better grades during the grade school years, but this seems to be more a matter of work habits than abilities, since tests of abilities show boys and girls to be very similar at this age. Beginning at about junior high school, there are two or perhaps three sex-related differences in cognitive abilities. (For a more complete review, see Maccoby and Jacklin, 1974.)

At this age, girls receive higher average scores on tests of *verbal ability*. Both boys' and girls' verbal abilities improve throughout high school, but girls' rate of improvement seems to be somewhat higher. However, the average difference between the sexes is very small. At the upper levels of test scores, there does not seem to be much difference in the numbers between boys and girls (Zahalkova et al., 1972). Among high school students taking college board exams, differences in verbal scores have diminished in recent years (Educational Testing Service Report, 1980).

Boys receive higher average scores on tests of *spatial-visualization*, again starting around the time of junior high school. These tests measure the mental ability to rotate objects of two and three dimensions. This ability increases in both sexes throughout the high school years, but male rates of increase are greater. Average differences in spatial-visualization are larger than average differences in verbal abilities. In addition there is a sex-related difference in *mathematical ability*, which may or may not be separable from visual-spatial abilities (Fennema and Sherman, 1977).

Current research into the causes of these sex-related differences include the following (see also Wittig and Petersen, 1979; Jacklin, in press):

1. *Differential brain organization—laterality differences.* When one side of the brain solves problems more quickly or more accurately than the other side, the brain is said to be "lateralized" for that type of problem. There are some sex-related differences in brain laterality, with girls being better able to carry out a variety of functions in either hemisphere and boys being more fully lateralized. However, these differences have *not* been directly related to differences in cognitive abilities.

2. *Maturity rate.* How quickly or slowly children mature may relate to their cognitive abilities. As indicated earlier, there are clear sex-related differences in maturation rate at puberty.

3. *Genetic sex-linkage.* Correlations between parents and children in scores on spatial ability tests had suggested a possible sex-related genetic factor. However, recent data (and reevaluation of older data) have not supported this position.

4. *Hormones.* At puberty, hormone differences between the sexes are increased. Some researchers are examining the relationship between hormones and cognitive abilities.

5. *Socialization.* Pressure on children to adopt behaviors, interests, and values appropriate to their sex may be especially strong just at the time when sex-related differences in cognitive abilities begin to emerge (i.e., junior high school and high school).

Differential Socialization in the Classroom

Boys receive somewhat more praise in class than do girls, and recent research has replicated this finding. Perhaps more importantly, a different pattern of feedback is given to boys and girls. Girls tend to receive negative feedback for the content of their academic work, while boys receive most of their negative feedback for noncontent aspects of their work, such as lack of neatness or not trying hard enough. These feedback patterns lead boys and girls to different attributions about the reasons for their own poor performance when they do fail: boys tend to believe the problem is lack of effort; girls more often believe they failed because of lack of ability. These self-attributions relate, in turn, to how hard a child persists at a task after he or she experiences failure. Girls give up more easily after academic failure, but if the feedback contingencies are experimentally changed, "learned helplessness" can be reversed. (We have greatly simplified this vigorous area of research; interested readers should also see Parsons, 1981.)

Researchers have also documented differential treatment of boys and girls by mathematics teachers. For example, math teachers provide more feedback to males. Perhaps even more importantly, "males were more likely to get feedback from the teacher that sustained the interaction when they gave a partially correct, incorrect, or no response answer." Teachers are also more likely to initiate verbal contact with male students than with female students. Most of the research in this area has been limited to frequency counts of teacher behavior in direct instruction. One study also documented differential behavior by mathematics teachers in less formal contacts in the classroom: 70 per cent of all encouragement in academic abilities and pursuits by the teachers was directed to males. Although active discouragement was rare, female students received 90 per cent of the discouraging comments from teachers. These findings come from a variety of schools and involve teachers of both sexes and all ages. In sum, boys and girls do receive differential treatment in the classroom.

Parental Treatment

As discussed above, it is difficult to document differential treatment of boys and girls by their parents in the early years. Except for the area of sex-typed toys, both mothers and fathers react similarly to their young boys and girls. When children get older, some areas of differential socialization can be seen. Parents are more likely to know where their girls are after school or on the weekend. In a longitudinal study in England, Newson and Newson (1968) did not find any sex differences at age 4 but by age 7, daughters were receiving more of what the Newsons call "chaperonage" than were the sons.

Knowing where a child is and/or allowing the child to roam may be partly a function of the task a child is asked to do in a household. In our culture, and in others as well, girls are more likely to be assigned household chores and child care. Why girls are so often given these tasks is an interesting question, but without information about that point, we can still ask what is the consequence of being given certain tasks. Whiting and her students have done considerable work on this point (Ember, 1973). Task assignment, particularly child care, does have an effect on a child's behavior. In the relatively few cases in which this task is assigned to boys, it seems to increase their nurturance and decrease their aggression outside the child-care situation (Ember, 1973).

CONCLUSION

Earlier we listed possible explanations for sex differences in cognitive functions. Any of these explanations—either biological or social—could, by itself, suffice to explain the very small sex differences that exist. At this point we cannot determine which explanation, or what part of all the explanations, correctly accounts for sex differences.

The most important point is that there is very little to explain. Recent publications concentrate not on whether a sex difference exists but on how large a difference really exists. These analyses are sobering if we consider the social implications of the work. It is completely impossible to predict an individual's abilities on the basis of his or her sex. Large numbers of both sexes score at the high end of distributions for all cognitive tests—enough to fill society's need for persons of exceptional skill.

Despite the biological importance of sex, many aspects of human functioning are not particularly dimorphic with respect to sex. In regard to psychological functions, we should focus our attention on the similarities rather than on the differences.

CAROL NAGY JACKLIN
ELEANOR E. MACCOBY

REFERENCES

Barrett, D. E.: A naturalistic study of sex differences in children's aggression. Merrill-Palmer Quart. 26:193, 1979.

Bayley, N.: Individual patterns of development. Child Devel. 27:45, 1956.

Bell, R. Q.: A reinterpretation of the direction of effects in studies of socialization. Psychol. Rev. 75:81, 1968.

Bell, R. Q., Weller, G. M., and Waldrop, M. F.: Newborn and preschooler: Organization of behavior and relations between periods. Monographs of the Society for Research in Child Development 36:1–145, 1971.

Connor, J. M., and Serbin, L. A.: Behaviorally-based masculine and feminine activity preference scales for preschoolers: Correlates with other classroom behaviors and cognitive tests. Child Devel. 48:1411, 1978.

DiPietro, J. A.: Rough and tumble play: A function of gender. Devel. Psychol. 17:50, 1981.

Ember, C. R.: Feminine task assignment and the social behavior of boys. Ethos, 1:424, 1973.

Faiman, C., Reyes, F. R., and Winter, J. S. D.: Serum gonadotropin patterns during the perinatal period in man and chimpanzee. International Symposium on Sexual Endocrinology of the Perinatal Period. INSERM (Paris) 32:281, 1974.

Fennema, E., and Sherman, J.: Sex-related differences in mathematics achievement, spatial visualization and affective factors. Am. Educ. Res. J. 14:51, 1977.

Gandy, H. M.: Androgens. In Fuchs, F., and Klopper, A. (eds.): Endocrinology of Pregnancy, 2nd ed. New York, Harper and Row, 1977.

Garai, J. E., and Scheninfeld, A.: Sex differences in mental and behavioral traits. Genet. Psychol. Monographs 77:169, 1968.

Hagemenas, F. C., and Kittinger, G. W.: The influence of fetal sex on plasma progesterone levels. Endocrinology 91:253, 1972.

Hall, J. A.: Gender effects in decoding nonverbal cues. Psychol. Bull. 85:845, 1978.

Halverson, C. F., and Waldrop, M. F.: The relations of mechanically recorded activity level to varieties of preschool play behavior. Child Devel. 44:678, 1973.

Jacklin, C. N.: Social factors in sex-related differences in spatial-mathematical abilities. Intern. J. Behav. Devel. In press.

Jacklin, C. N., and Maccoby, E. E.: Social behavior at thirty-months in same-sex and mixed-sex dyads. Child Devel. 49:557, 1978.

Jacklin, C. N., and Maccoby, E. E.: Length of labor differences depending on the sex of the newborn. J. Pediat. Psychol. In press, 1982.

Jacklin, C. N., DiPietro, J. A., and Maccoby, E. E. Sex-typing behavior and sex-typing pressure in child/parent interaction. Arch. Sexual Behavior. In press.

Jacklin, C. N., Snow, M. E., Gahart, M., and Maccoby, E. E.: Sleep pattern development from 6 through 33 months. J. Pediat. Psychol. 5:295, 1980.

Jacklin, C. N., Snow, M. E., and Maccoby, E. E.: Tactile sensitivity and strength in newborn boys and girls. Infant Behav. Devel. 4:285, 1981.

Korner, A. F., Hutchinson, C. A., Koperski, J. A., Kraemer, H. C., and Schneider, P. A.: Stability of individual differences of neonatal motor and crying patterns. Child Devel. 52:83, 1981.

Langlois, J., and Downs, C.: Mothers, fathers and peers as socialization agents of sex-typed play behavior in young children. Child Devel. 51:1237, 1980.

Lipsitt, L. P., and Levy, N.: Electroactual threshold in the human neonate. Child Devel. 30:547, 1959.

Macaulay, R. K. S.: The myth of female superiority in language. J. Child Language 5:353, 1978.

Maccoby, E. E., and Jacklin, C. N.: The Psychology of Sex Differences. Stanford, California, Stanford University Press, 1974.

Maccoby, E. E., and Jacklin, C. N.: Sex difference in aggression: A rejoinder and reprise. Child Devel. 51:964, 1980.

Maccoby, E. E., Doering, C. H., Jacklin, C. N., and Kramer, H.: Concentrations of sex hormones in umbilical cord blood: Their relation to sex and birth order of infants. Child Devel. 50:632, 1979.

Moss, H., and Robson, K.: The relation between the amount of time infants spend at various states and the development of visual behavior. Child Devel. 41:509, 1970.

Newson, J., and Newson, E.: Four year olds in an urban community. Harmondworth, England, Pelican Books, 1968 (and personal communication).

Omark, D. R., Strayer, F. F., and Freedman, D. G.: Dominance Relations: An Ethological View of Human Conflict and Social Interaction. New York, Garland STPM Press, 1980.

Parsons, J. E.: Attributions, learned helplessness and sex differences in achievement. *In* Yussen, S. R., (ed.): The Development of Reflection. New York, Academic Press, 1981.

Reinisch, J. M.: Fetal hormones, the brain, and human sex differences: A heuristic, integrative review of the recent literature. Arch. Sexual Behavior, *3*:51, 1974.

Resko, J. A.: Sex steroids in the circulation of the fetal and neonatal rhesus monkey: A comparison between male and female fetuses. International Symposium on Sexual Endocrinology of the Perinatal Period. INSERM (Paris) *32*:195, 1974.

Roche, A. F.: Secular trends in human growth, maturation, and development. Monographs of the Society for Research in Child Development, Vol. 44, Nos. 3 and 4, 1979.

Rosenblith, J. F., and DeLucia, L. A.: Tactile sensitivity and muscular strength in the neonate. Biologia Neonatorum *5*:266, 1963.

Rubin, J. S., Provenzano, F. J., and Luria, Z.: The eye of the beholder: Parents' views on sex of newborns. Am. J. Orthopsychiat. *5*:353, 1974.

Safer, D. J., and Allen, R. P.: Hyperactive Children. Baltimore, University Park Press, 1976.

Serbin, L. A.: Sex sterotyped play in the preschool classroom: Effects of teacher presence and modeling. Paper presented at the biennial meeting of the Society for Research in Child Development, New Orleans, April 1977.

Serbin, L. A., O'Leary, K. D., Kent, R. N., and Tonick, I. J.: A comparison of teacher response to the preacademic and problem behavior of boys and girls. Child Devel. *44*:796, 1973.

Snow, M. E., Jacklin, C. N., and Maccoby, E. E.: Sex-of-child differences in father-child interaction at one year of age. Child Devel. In press.

Strayer, F. F.: Peer attachment and affiliative subgroups. *In* Strayer, F. F. (ed.): Ethological perspectives on preschool social organization. Research memo #5, University of Quebec (Montreal) Department of Psychology April, 1977.

Waldrop, M. F., and Halverson, C. F., Jr.: Intensive and extensive peer behavior: Longitudinal and cross-sectional analysis. Child Devel. *46*:19, 1975.

Whiting, B., and Edwards, D.: A cross-cultural analysis of sex differences in the behavior of children aged three through eleven. J. Social Psychol. *91*:177, 1973.

Wittig, W. A., and Petersen, A. C.: Sex-Related Differences in Cognitive Functioning. New York, Academic Press, 1979.

Zahalkova, M., Vrzal, A., and Kloboukova, E.: Genetical investigations in dyslexia. J. Med. Genet. *9*:48, 1972

III

Milieux and Circumstances

In this part we will review some typical and atypical backdrops for growing up and some things that happen to children, i.e., primarily exogenous influences. A range of circumstances and life events are explored. All have in common their potential for shaping (or misshaping) developmental and behavioral outcomes during childhood and thereafter.

Chapters 11 and 12 analyze the potential impacts of cultural background, ethnicity, and family lifestyles. These are followed by a four-section chapter that describes some common variations within families, including descriptions of parenting and its problems, the impacts of various sibling constellations, the effects of divorce, and issues in adoption and foster care. Chapter 14 examines the extremes of maladapted family function, including the most injurious input, namely child abuse. The final three chapters in this part explore other environmental influences, including school, critical events, and miscellaneous possible deterrents to development.

11

Culture and Ethnicity

Modern developmentalists no longer give serious attention to the old nature/nurture debate concerning the genesis of organized behavioral development. We no longer ask whether heredity or environment determines the course of an individual's development. With the increased acceptance of a broader perspective regarding development, the focus is now on the nature of the interaction between genetic and environmental forces. However, considerable attention has been given to attributing a "percentage of causation" to each of these factors (Jensen, 1965). Although this line of investigation has received a fair degree of exposure, efforts directed toward describing and analyzing the nature of the genetic-environmental interaction seem to offer greater promise for improving our understanding of both behavioral-developmental processes and intervention procedures. With respect to living organisms, short of alterations in *genotype*, i.e., the genetic potential of the organism, it is an individual's *phenotype*, i.e., the structural or behavioral manifestation of the organism, that is the product of this interaction and that can be modified. In our attempt to elucidate this interaction, it is the organism's environmental encounters that provide our primary data. No aspects of the child's encounters are more important than those that are a function of the cultural and ethnic contexts in which they occur. (See also Chapters 3 and 10.)

Most human beings are raised in families with specific cultural and ethnic identities. Even persons raised in nonfamily-centered institutions are assigned an ethnic identity and are socialized to the culture that is dominant in that institution. Thus, mastery of such primary developmental tasks as acquiring a sense of trust, autonomy, and ultimately identity (Erikson, 1950) occurs within the context of the cultural and ethnic group characteristics of the child's primary care providers. The influence of these early encounters and the ubiquitous character of their continuing presence on the developing person are probably second only to gender socialization in the shaping of attitudes and behavior.

DEFINITIONS

The concept of *culture* has a wide variety of definitions. It is an abstraction. Its interpretation varies widely, perhaps because it constitutes collective customs inferred from behavioral patterns. The fact that these patterns are transmitted by symbols renders language and the communicative structure of a social unit of paramount importance to our understanding of culture.

Writers such as Berry (1951) treat culture as acquired and learned behavior:

> We are born, ignorant and helpless, into a group. . . . We proceed immediately to imitate and acquire these 'group habits' of thought, feeling, and behavior; and the members of the group, at the same time, set about to indoctrinate us with those behavior patterns which they regard as right, proper, and natural.

Some authors distinguish between "material" and "nonmaterial" aspects of a culture. Artifacts, structures, and concrete products of a culture are examples of the material culture, while belief systems, attitudes, attributions, and skills are examples of the nonmaterial culture. In his indices of culture, Tylor (1891) includes "knowledge, beliefs, art, law, custom and any other capabilities and habits acquired by man as a member of [a] society."

If we consider these perceptions together, we can conceptualize five basic dimensions or levels of culture:

1. The *judgmental or normative dimension*, which reflects social standards and values, i.e., those behavior patterns which, according to Berry, people regard as "right, proper, and natural."

2. The *cognitive dimension*, which relates to social perceptions, conceptions, and attributions, all of which may be thought of as categories of mentation expressed through the medium of language.

This dimension therefore involves the communicative functions and structure of a social unit and is exemplified by what Berry describes as group habits of thought.

3. The *affective dimension*, the emotional structure of a social unit, including its common feelings, sources of motivation, joy and sorrow, and sense of value—Berry's group habits of feeling.

4. The *skills dimension*, signifying those special capabilities people develop to meet the demands of their social and technoeconomic environment (Ogbu, 1978).

5. Remaining is the *technological dimension*, the notion of culture as accumulated artifacts, instrumentation, and techniques, which includes things made and used as well as the manner in which they are used.

So far our definition emphasizes those characteristics by which a culture may be identified or by which the culture of a group may be characterized. But culture is not only a descriptive concept or, as Harrington (1982) has cautioned, not simply a "product of human action: observe the action and you can label the culture." Culture also influences human action and must therefore be regarded as an explanatory as well as a descriptive concept. This dual nature of culture not only is important to our understanding of it as a phenomenon but is crucial to use of information concerning culture to inform knowledge production and knowledge utilization.

When we use cultural information to describe and identify, we are being sensitive to culture as a *status phenomenon*. Such information can help us determine a person's position, place, and role in the social order and, to some extent, how the person or the group is perceived. From such information we might even predict how the person or the group is likely to behave. However, in order to make more accurate predictions and to understand behavior, we must use information relative to culture as an explanatory concept, as a *functional phenomenon*. We want to know what this information tells us not simply about status but about the consequences of the phenomenon for the functioning of the person. How does a particular aspect of culture influence the behavior of the person? What does the symbol or the tool enable the person to do? What societal capabilities are provided by the existence of the language? How do specific belief systems influence the patterns of social organization? Questions like these explore the ways in which culture functions to shape individual and group behaviors rather than to describe the culture or the status of specific members. Kluckholn (1965) has captured the dual nature of culture in the following statement:

Culture consists of patterns, explicit and implicit, of and for behavior acquired and transmitted by symbols, constituting the distinctive achievement of human groups. . . . Culture systems may, on the one hand be considered as products of action, on the other as influences upon actions.

Culture is commonly associated with a related but distinct indicator of a social group—*ethnicity*, the dictionary definition of which is "a community of physical and mental traits possessed by members of a group as a product of their common heredity and cultural traditions." According to Isajiw (1974), the most common attributes characterizing an ethnic group are (a) common ancestral origin, (b) same culture, (c) same religion, (d) same language, and (e) same race. Earlier, we noted that belief systems and languages are aspects of culture. In addition, our reference to "same race" and "common ancestral origin" seems redundant. These considerations led Isajiw to use the term ethnicity to define

. . . an involuntary group of people who share the same culture or to descendants of such people who identify themselves and/ or are identified by others as belonging to the same involuntary group.

Ethnicity, then, refers to one's belonging to and identification with a group with common attributes referable to cultural traditions, belief systems, genetic history, and language and sometimes with common physical characteristics and identification systems. Although ethnicity is often used as a synonym for *race*, such use does not necessarily specify biological race but may indicate membership in a group that draws from the same gene pool. An example of this is the ethnic group we call Puerto Ricans, in which two racial groups or gene pools (Caucasian and Negro) are drawn upon in varying degrees of concentration or mixture. Another example is found in Nigeria, where the two ethnic groups Ibo and Hausa draw from the same race. In this country, the most widely encountered example is of English, French, and German ethnic groups all drawn from the Caucasian race. There are three *races* of mankind: Negro, Mongolian, and Caucasian; the *ethnicities* of mankind number in the hundreds or even thousands.

Because of tendencies toward within-group mating and procreational insulation, ethnicity has come to reflect biocultural kinship systems rooted in the common, acquired and/or inherited characteristics and identities of the members of these systems. A child may be born or adopted into such a system early in life, although adoptive ethnicity can be limited by the presence of tell-tale "alien" characteristics such as physiognomy. At the same time, ethnicity is a function of societal perception of and attribution from the fact of group differences. These perceptions and attributions then form the basis for the choice or assignment of subgroup membership. In the latter case, ethnic *identification* (the group to which I am assigned) may differ from ethnic *identity* (the group to which I feel that I belong). Goffman (1971) has used "social identity" and "personal identity" to distinguish between the identificatory perceptions of others and those of the individual, respectively.

Like the status/function distinction discussed in reference to culture, one's ethnicity in part defines one's status. This is particularly likely in societies in which positions in the hierarchy are determined by ethnic identification. However, one's ethnic identity influences the manner in which one functions, the way in which one responds to environmental encounters. Thus, ethnicity can determine the way in which a society treats the individual or group, as in permitting access to and control over resources and power. In this case it is a *status characteristic*. Ethnicity can also determine how one utilizes the resources and power available, and thus it operates as a *functional characteristic*. For example, being black in the United States automatically relegates one to a lower status in the social order. At the same time, identifying oneself as black is likely to influence the manner in which one behaves.

PRACTICAL IMPORTANCE OF CULTURE AND ETHNICITY FOR DEVELOPMENT AND BEHAVIOR

In terms of the relevance of culture and ethnicity to research, practice, and policy in developmental-behavioral pediatrics, this distinction between status and function is of crucial importance. In fact, it may have greater practical importance than the distinctions between culture and ethnicity, since both these social-group indicators influence the manner in which persons are treated by the social order and define the way a person functions within the social order.

All societies, certainly all technologically developed societies, are organized hierarchically in terms of the diverse groups of which they are composed. Since none of these societies has achieved social justice in its allocation of power or its distribution of resources, position within the hierarchy greatly influences an individual's or a group's access to the power (social influence and control) and utilization of the resources (means of subsistence, human services, and means of production) intrinsic to the social order. Since we tend to develop stereotyped assumptions concerning the functional characteristics of status groups, a status designation not only defines the relation to power and resources but also signals (1) what society expects of the person, (2) the nature of available opportunities, and (3) the degree of participation in the reward structure. For example,

we anticipate limited intellectual behavior from a low-status group like blacks despite the fact that many blacks show high levels of intellectual productivity. It is common knowledge that in the United States employment opportunities for black youths are 30 to 40 per cent lower than those for white youths. Ogbu (1978) reports that a black youth may expect to be paid about 25 per cent less than a white youth for the same job requiring the same amount of education.

This relationship between status and expectation, opportunity and reward exists for other groups as well. Often, one's status defines the manner in which one will be treated. Persons whose ethnic "status" is low but whose social position is high may experience dramatic reversals in the way others approach them if they do not "fit the stereotype." Even in the absence of superior achievement, such generalized expectations are more likely than not to be in error. Status signaled by one's culture or ethnicity can be helpful but can also be misleading when distorted by stereotypes or thoughtlessly applied.

To supplement information concerning status, one can examine these two group indicators— culture and ethnicity—as vehicles of self-definition and identification. They provide clues to an individual's affiliation and serve as frames of reference as we define our orientation, make judgments, and set standards. They often provide a source of aspiration or behavior models. They also transmit technique, grant or deny legitimacy, and convey social meaning.

Since most socialization experiences are perceived in reference to one's gender role, culture, and ethnicity, these three contexts for development are powerful forces in the definition of self and the development of personal identity. Even the structure of one's gender role is influenced by the culture in which it is developed and the ethnic group to which an individual belongs. Relative degrees of aggression, assertiveness, independence, nurturance, and the like tend to vary in different cultures and ethnic groups. As such, different groups encourage and support varying degrees of these characteristics in gender-role socialization. Likewise, parental role, the nature of childhood, role as breadwinner, and socialization to work and to leisure are influenced by these contexts. Thus, the emerging personal identity and elements of self-definition are rooted in cultural and ethnic experiences. Similarly, after family, which cannot be separated from culture and ethnicity, these social-group indicators are primary reference points for affiliation. The individual's tie to his or her ethnic group and identification with the indigenous culture are among the most deeply entrenched human associations. So strong is this identity that when one is isolated from one's cultural or ethnic group, there is the tendency to use symbols of cultural or ethnic affiliation and for members of the same group to seek each other out in alien environments. Witness the previously unacquainted "Americans" who greet each other as long-lost relatives on a chance encounter in Accra.

PROBLEMS IN CROSS-CULTURAL AND CROSS-ETHNIC PROFESSIONAL INTERACTION

LANGUAGE BARRIER—MEANINGS

The salience, strength, and ubiquity of culture and ethnicity as forces shaping the behavior of individuals and groups may either facilitate or complicate social adaptation. In isolated or insulated cultures the potential for provincialism and recalcitrance is great and precludes the adoption of new models, values, and behaviors. On the other hand, in situations of cultural and ethnic heterogeneity and pluralism, we confront problems in communication and evaluation and in trying to reconcile contradictions. We are accustomed to dual-language situations, in which we may not recognize or know the meanings of the word symbols a patient uses, and even to some instances where a word is familiar but its meaning is different; We are less accustomed to the situation in which a person's customs, ritual behaviors, and underlying values are different, not understood, and not appreciated. Communication under these circumstances is, more often than not, *miscommunication*. The intent and significance of a message may be missed because it contradicts our own indigenous customs or values. Under such circumstances, we try to reduce this dissonance but may, in the process, distort what is being communicated. For example, introducing infant formula to feed babies in cultures where food processing and handling routines are quite different or less technologically advanced may have deleterious effects on these babies' health because of errors in their measurement or use. These errors often result when traditional practices are applied by the mother to whom these materials and procedures are alien and not fully understood. Similarly, in less affluent and medically uninformed families, medications may be diluted or the dosage reduced in order to make them last longer because the user does not appreciate the need for a certain level of dosage for the drug to be effective. Such frugality may be appropriate to the indigenous culture and circumstances but inappropriate when applied to carefully prescribed and calibrated medication.

LANGUAGE BARRIER—FUNCTIONAL SIGNIFICANCE

It is not only form and meaning of language but *function* of language that may vary from one culture or ethnic group to another. Miscommunication is more likely when a message used in the culture of the sender to pass on information or instructions is interpreted in the culture of the receiver as an indication of status or authority. Under such circumstances, as much attention may need to be given to the one who conveys the message as to the content of the message. As the responsibilities of paramedical personnel increase, we should be mindful that a patient may disregard a directive or prescriptive procedure unless it comes from the doctor. Similarly, with our increased effort to have patients assume greater responsibility for their own health, our efforts at providing options from which patients may choose may result, in certain populations, in the judgment that systematic intervention may not be necessary, since the doctor has left so much to the discretion of the patient or the family.

Thus, since specific cultures may attach different functions as well as different meanings to the language, it is imperative that the impact of culture on language be considered in cross-cultural professional contacts. My non-English speaking patient may know enough English to understand the meaning of the words but may misunderstand the intent of the language or the function I intended the language to serve. Since in health care we depend so heavily on the ability of the patient or the family to follow through, problems in cross-cultural and cross-ethnic communication loom large unless care is taken to ensure that transposition as well as translation of the language is accurate.

SOCIAL STEREOTYPES

Another source of communicative error rests with our tendency to develop stereotyped notions concerning what other people are like and what they mean as a result of extensive or intensive contact. We may then inappropriately generalize our insights or impressions and apply them to an entire group of people, as if they were homogeneous masses. Instead, we must be aware of the diversity within racial or ethnic groups. For example, although Puerto Ricans and Cubans may share some elements of language and culture, they strongly distinguish themselves from one another, sometimes to the extent of showing open hostility toward the others and toward those who lump them together. Students of cultural and ethnic groups increasingly assert that differences *within* groups are as important as differences *between* them. Consequently, if one is to be sensitive to cultural and ethnic variety, one must take the time to identify and understand the specific identity held by the person or persons rather than the general identity assigned by others to the person.

DIFFERENCES IN DIALECT

Occasionally we may incorrectly assume that people share the same culture and language when they in fact use different dialects or have different cultural styles (variations on the same culture). One may tend to consider the dominant culture as the norm and to interpret the language or behavior according to the meaning ascribed to it by those in the dominant culture. However, this approach often leads to a misunderstanding about what is being expressed. One striking illustration of this is the use by young black males of the verbal exchange called the "dozens." Both the language and its expression have special meanings which may be interpreted by an outsider as hostile and belligerent but to the initiated is recognized as friendly competition, seldom if ever used between enemies. In this case the dominant culture norm actually contradicts the meaning of this "specialized" behavior.

PROBLEMS IN EVALUATION

Problems in miscommunication are also reflected in problems of evaluation. When we judge members of a culture or ethnic group other than our own, we confront several sources of distortion.

Evaluative Bias. Banks (1975) noted that when we are called upon to evaluate a person's behavior that is judged to be negative, we tend to look for external causative factors if that person is a member of the same group. If the individual comes from an alien group, however, we tend to look for internal causative factors—within the person. Thus, when the lower-class black mother is late for or misses an appointment the tendency is to conclude that "these people are like that, irresponsible and chronically late." When the middle class, Caucasian mother shows the same behavior, we are likely to conclude that something has prevented her from arriving on time. Similarly, the popularly held notion is that low-status outgroup parents are more likely to be child abusers while higher status ingroup parents have problems that may sometimes drive them to abuse or neglect their children.

Chauvinistic Models. All of us tend to suffer from provincialism, which makes us chauvinistic. The models we know and have experienced are usually the models we look toward for ourselves and others. When called upon to make judgments about behavior, we rely on the models that we are familiar with. Since few of us are genuinely bicultural or multicultural, we refer to models from the culture with which we identify. Thus, our conceptions of and judgments about, say, good parenting or wholesome development, responsible patients or adequate males, are likely to be biased in favor of idiosyncratic or chauvinistic models that bear little relationship to the realities of the lives of people from another culture.

Idiosyncratic Values. Evaluative judgments made across cultural and ethnic boundaries may be invalid unless the values that guide a person's behavior and shape the purposes and goals of that behavior are recognized and understood. In one culture, longevity may be valued more than the richness of contemporary life experience. In some African cultures the number of children in a family may be valued higher than the survival of a single child, since having many children is in part a function of the parents' expectation that some of their offspring will not survive. Similarly, in cultures where disability is viewed as a handicap for the society as well as for the individual, investment in the survival of a disabled person is not highly valued. Cultural groups not faced with the demands of technology and geared to traditions of experimental validation may ascribe less value to precision in measurement, use of language, and observance of time. These and other differences in values idiosyncratic to specific groups significantly affect the behavior and attitudes of culture and ethnic group members. The skillful clinician will probe and observe carefully to discover and appreciate these differences in order to adapt prescriptions and treatments to the individual.

Pluralistic Criteria. In the clinician's process of adaptation, problems of pluralistic criteria and the reconciliation of contradictions are crucial factors. Obviously, in some circumstances, a standard criterion must be set. Certain conditions are essential to survival. Practices that greatly increase the risk of ill health, permanent damage, or death must be prevented. Under such circumstances one cannot accept multiple criteria. However, in some cases, a number of different routes will reach the same end. An example would be nutrition, since, within limits, pluralistic criteria are permitted to assure an adequate diet that takes into account cultural differences. In other situations where there is greater tolerance for variation, the manner in which the task is done, the frequency with which the practice is engaged, the precision demanded, as well as the choice of ends may well be left pluralistic so as to accommodate the lifestyles and preferences of the people served. After all, in health matters, which must be self-monitored, general compliance is often better than no compliance at all, and self-directed compliance is thought to be a function of personal identification with and acceptance of the procedure.

RECONCILING CONTRADICTIONS

We face a more difficult problem in having to reconcile contradictions born of cultural differences. Here the responsible clinician has fewer degrees of freedom. Again, when the contradiction reflects preference, style, or aesthetic values, the conflict should be resolved in favor of the patient's bias. However, when the issues involve life-threatening or permanent negative sequelae, we turn first to the informed judgments of parents and ultimately to our own responsibilities to respect and advance human development. This may require the imposition of an alien standard, goal, or practice upon a reluctant patient. In these cases, I ask myself the following questions when making professional decisions about a patient's care:

1. Does my decision represent the best informed and most honest judgment that I can make under the circumstances?

2. Does my decision increase the alternatives for wholesome development rather than reduce them for this person?

EDMUND W. GORDON

REFERENCES

Banks, C.: Delay of gratification in black adolescent boys. IRCD Bulletin Vol. XII, No. 3, 1975.

Berry, B.: Race and Ethnic Relations. Boston, Houghton-Mifflin Co., 1951.

Erikson, E.: Childhood and Society. New York, Harper Brothers, 1951.

Goffman, E.: Relations in Public. New York, Basic Books, 1971.

Harrington, C.: Culture as a manifestation of human diversity. In Gordon, E. W. (ed.): Human Diversity and Pedagogy. Westport, Connecticut, 1982.

Isajiw, W. W.: Definitions of Ethnicity. Ethnicity 1:111, 1974.

Jensen, A.: Harvard Education Review, 39:1–123, 1969.

Kluckholn, C.: Culture and Behavior. New York, The Free Press, 1965.

Ogbu, J.: Minority Education and Caste: The American System in Cross Cultural Perspective. New York, Academic Press, 1978.

Tylor, E. B.: Primitive Culture. Vol. I. 3rd ed. London, John Murray, 1891, p. 1. (First edition, 1871.)

12

Traditional and Alternative Family Life Styles

CHANGES IN THE AMERICAN FAMILY

In the last decade dramatic changes have taken place in the composition and structure of the American family. In the recent past the modal family was readily described as a two-parent, two-child unit; the children were cared for mainly by the mother, who stayed at home, while the father was the exclusive wage-earner. For most Americans, this was the ideal family.*

By 1970, however, the American family had assumed many forms. It was a pluralistic unit, with less than 20 per cent of these units fitting the once conventional definition. Other kinds of families became more prevalent: over 32 per cent of family households consist of married couples without children at home—either as childless couples or couples whose grown progeny live elsewhere; 28 per cent of families have one or more children at home and both parents are employed; individuals living alone constitute another 22 per cent. These include one-third of all women over 65 years of age.

A variety of other types of units are also seen: at least 7.5 per cent of American families consist of one parent and at least one child at home; in 9 of every 10 of these families the mother heads the household. Separated parents who live apart from spouses and children and reside instead with other relatives make up 5.3 per cent of households. About 6 per cent comprise unrelated, unmarried persons who live together as in living groups or in social-contract or cohabitant families.

These compelling modifications of the family picture, first visible during the 1970's, are proj-

ected to continue through the 1980's (Masnick and Bane, 1980). At the most, slightly more than one-fourth of all American households are expected to be conventional ones, i.e., with mother, father, and young children. At least 13 separate types of households will eclipse this conventional family in incidence, examples being female-headed families, widows with children, or divorced males with no children. Households in which there will be no children under 15 may number as many as 60 million—nearly equaling the total number of American households that existed in 1970. This follows the decline of approximately one million households with children during the period from 1970 to 1980. Husband-wife households in which only one spouse works will account for 14 per cent of all households compared to 43 per cent in the year 1960; wives will contribute about 40 per cent of all family income, compared to 25 per cent at the present time. As more wives shift from part- to full-time work, those living in families having only one wage-earner will be in danger of becoming an economic "under-class."

More than one-third of the couples who married for the first time in the 1970's will be divorced, with more than one-third of all children born during the 1970's spending some part of childhood years living with only one parent. At the beginning of 1980, 19 per cent of all families with a child at home were maintained by a single parent, a unit that arose either through divorce, premarital birth, or, to a lesser extent, death of a parent.

FAMILY CHANGE AS A CHOICE

In the past, departures from the traditional family occurred primarily by default. Death and desertion accounted for non-two-parent families more commonly than did divorce. Today's divorce rate, which is close to 41 per cent, suggests that people are choosing to resolve conflicts by changing their traditional family status and opting for another life style. This has made the "career" of a family less regular and predictable. Formerly one

*Many of the findings in this chapter have been drawn from an intensive longitudinal study conducted by a multidisciplinary research team that has been following children from birth through the first seven years of life. These children were born and reared in one of the following alternative life styles: single mother households headed by never-married women; cohabiting unmarried couples who form a family and have children, referred to as social contract families; and living group or communal families. A comparison group is comprised of conventionally married two-parent nuclear families. The study is directed at documenting how children are socialized in alternative families, and assessing the implications for the child's intellectual, social and emotional development.

193

could anticipate a fairly consistent pattern: marriage and coupleness was followed by parenting of young children, then by the stage of independence for both parents and children as the latter move through adolescence, and finally by the "empty nest" with parents left alone and the children on their own, preparing to start a new family. Today a family's career is often more variable—at any stage the original marriage might dissolve and the family might be restructured. Thus, for the health professional, following a family may involve more than one set of adults and more than one set of children.

Some adults are consciously choosing nonconventional family life styles in order to live in new forms of households that are more likely to express their attitudes and values than does the two-parent nuclear unit in which they were reared. They want family styles that meet their desires for more humanistically oriented, more emotionally close, and less materialistically driven lives (Eiduson et al., 1973). They value families comprising capable, caring adults so children can rely on persons other than their own parents, and the attachments and nurturance of early mother-child bonding can be extended to others. They seek families who see their organic roots in the environment and who try to live close to nature. They question the conventional family's dependence on technology and science; its materialistic orientation; and the way it prizes objectivity and rationality at the expense of personalized, nonobjective, and sensory output. Instead of seeking the goals of success, money, status, and prestige, they hope to reinstitute achievement in the service of creativity and personal self-fulfillment. Instead of the planning and saving for tomorrow, as espoused by the traditional family, new families put a premium on maximizing enjoyment and pleasure in the "here and now." Young adults want to take more control of their lives, to become more qualified to help themselves and less dependent on societal institutions. Many desire to regain the sense of commitment and personal responsibility identified with early America.

Illustrative of the new alternative families are the cohabitant or social-contract relationships in which adults live together as man and wife but are not legally married. This relationship has been adopted by some who consider that their emotional involvement with each other may be lessened when it rests on a marriage certificate and thus is taken for granted. The built-in opportunities for shared child caretaking have also led to living-group, or communal, or intentional-community families, in which people live together who share certain interests, needs, and religious or social benefits to maximize the psychological, emotional, and economic benefits that come from joint activities, living quarters, and responsibilities.

CHILD-REARING IN TRADITIONAL AND ALTERNATIVE FAMILIES

It is remarkable how rapidly the values of persons who have adopted alternative life styles have diffused into traditional families. The growing interest in natural food, the greater variability in household roles, and the lessening of sexual repression and its greater expression in the mass media are dramatic examples, as is the rise of alternative classrooms and alternative schools. The desire for a more humanistic environment when a child is born and the use of the Lamaze and Bradley methods of delivery also represent a change from traditional practices.

Yet important differences between alternative and traditional families still remain with respect to the growing child (Eiduson et al., 1982); (Table 12–1). The traditional child is more likely to be born into a financially advantaged family than is a child in another family form. This differential in financial status remains fairly stable; while the income of the traditional family increases slowly and regularly over time, that in alterantive families may increase but at a less rapid and less consistent rate.

Traditional mothers and fathers are in higher level occupations and more of them work on a full-time basis compared with parents in alternative families. The work model presented by traditional fathers to their children is of an achiever who is the main wage-earner in the family during the child's first few years and who works consistently toward conventional goals and attaining material success in ways our society deems desirable. Parents in many alternative families are directed more toward psychologically fulfilling goals. They value leisure and family time and often opt for vocations that permit more freedom and flexibility in their schedules. To them the pursuit of interests that express their personal styles and creative urges is conducive to a more well-rounded personal and family life.

In line with status ambitions and financial resources, traditionally married families generally live in residences that are larger than alternatives, with more areas designated for particular functions and more private areas not shared by all family members. Their homes are more frequently detached dwellings, while alternative families, often restricted by their means, are directed toward duplexes. Traditionally married families take pride in ownership: cars, savings accounts, insurance, stereos, boats. Alternatives may want to acquire these too, when these serve to promote

Table 12–1. DIFFERENCES IN CHILD-REARING PRACTICES*†

Traditional Families	Alternative Families	No Differences
Use of anesthetics and instruments	Preference for and greater incidence of home births	Ways of comforting child when fussy
Circumcision at birth	Choice of unconventional first names for child	Sleep problems shown by child (6 months and 1 year)
Mate present during birth	Friends and family present at birth	Mother as primary caretaker from 6 months through 3½ years
Time limit placed on feeding	Demand feeding emphasized	Extent of breast feeding at 6 months
Bottle feeding introduced (6 months)	Breast feeding extended over longer period	Attentiveness of caretaking (observed at 6 months)
Introduction of solid foods (6 months)	Greater use of "natural" foods, beginning at one year	Intensity of stimulation of child: visual, verbal, auditory, tactile
Common use of prepared baby foods (6 months and 1 year)	Extent of social stimulation for child greater, with greater variety of people present in household during early years of life	Number of others in household at 1 year
Child given vitamins (6 months, 1 year, 18 months)		Use of vitamins at 2 years
Child given fluorides (1 year, 2 years)	More interest in father caretaking	Frequency of peer group contact (18 months and 2 years)
Child fed vegetables and animal protein (1 year) and more meat in diet (18 months)	Greater extent of parent-child interaction at age 2	Responsiveness of caretaker (as observed at 18 months)
Use of infant seat and stroller	More peer contact by age 2	Mother's perception of child as difficult, shy, confident, or independent: factors rated separately at 18 months and 2 and 3 years
Regular medical visits	Earlier bowel training	
Child sleeps in own room and own bed	Greater tendency to feed vegetarian diet beginning at age 3	Family eats dinner together (2 years)
Child left at others' houses by 1 year	Attendance at school with more male caretakers	Behaviors for which child is disciplined (18 months and 2 years)
Greater frequency of grandparent contact	Parents bathe with child with greater frequency (age 2 and following)	Modes of discipline used
More exposure to T.V. and radio, and to educational T.V.	Child plays in the nude with other children and adults around	Exposure to different child caretakers (2 years)
More toys owned		Extent of consistency among caretakers (2 years)
Greater variability in home stimulation		Nudity of parents in front of child
Child has favorite blanket (18 months)		Exposure of child to adult sexuality
Mother shows more controlled caretaking styles in home observation (18 months)		Parental stress on sex-role typed toys
Mother "hovers" more over child (18 months, home observation)		Favorite activities of child
More no-no's for which child is disciplined		Favorite toys
Child more likely to be in day care or nursery school (age 3)		Frequency of peer contacts at age 3½
		Family travel with child
		Fathers' caretaking role
		Use of alternative schools (4 years)
		Use of dental care (4 years)
		Use of traditional medical services
		Incidence of problem behaviors in child (1 to 3 years)

*Differences cited reflect group differences in those variables (or groups of variables) that have been summarized into macrovariables and shown to be statistically significantly different on *t*-tests when probability levels of 0.05 or 0.01 are applied.

†Adapted from Eiduson, B. T., et al.: Comparative socialization practices in alternative family settings. In Lamb, M. (ed.): Nontraditional Families. New York, Plenum Press, 1982.

their interests or permit them to explore new "worlds."

In general, traditionally married families are more stable than alternatives. They move less and experience fewer changes in terms of people moving in or out of the house. Despite the high divorce rates reported, traditional married couples are found to stay together more frequently during a child's early years than do alternatives. Usually, they see their own parental family as their model for family stability. Also, traditional households tend to be physically more orderly.

Traditional marrieds are significantly more likely to have planned the birth of their child, and the mothers have a history of fewer abortions than do alternatives. However, many young expectant mothers from both traditional and alternative groups take courses to prepare them for pregnancy and seek appropriate prenatal care. Even patterns of nutrition and drug use during pregnancy are similar in traditional and alternative families. By and large, alternative families have had a history of significantly greater drug use prior to pregnancy, including social use of marijuana and experimental and social use of psychedelic drugs; yet drug use declines by the last trimester of the pregnancy in most young Caucasian alternative parents. Alternatives are significantly more likely to choose natural childbirth and home births, while the traditionals are more likely to have instrument-assisted deliveries and cesarean sections. More children in the traditional families are bottle-fed earlier than are alternative children, with breast feeding extending longer in alternative populations, in a few cases as long as three years.

However, there are no differences among families in their use of conventional health care services, the number of well-child doctor visits, or the relative number of doctor visits by type of illness. Likewise, frequency of immunizations is comparable. However, children in the traditionally married population seem to be taken to the doctor more frequently for minor respiratory illnesses such as colds than children raised according to the alternative life style, reflecting the alternatives' greater tendency to deal with minor illnesses through home remedies.*

Few differences in the psychological aspects of caretaking between traditional and alternative groups can be cited. Mutuality of affect, observable visual, vocal, and sensory interchanges, tactile and physical stimulations, display of warmth, attentiveness toward and concern for the child, and appropriate supervision are parent-child behaviors that are found to the same extent in all family styles. Ways of playing with the child, toilet training, and adult responses to a child waking at night are similar. Alternatives are more likely to feed their children natural-organic and homemade foods rather than store-bought products. Traditional marrieds give vitamin supplements more often and are more likely to use fluorides. Alternative parents like to take their infants along wherever they go and carry them physically close, using slings. Traditional marrieds are more likely to use "infant seats" and strollers.

Traditional and alternative caretakers are similar in their outlook toward discipline. Parents discipline in response to the same behavior, but traditional married parents appear to discipline for more types of behaviors than some alternative families. All families use a variety of modes of discipline, with reasoning and verbal restrictions being the mode of choice. Traditional marrieds place higher value on teaching their children obedience and respect for authority and are often more directive in caretaking than are some alternatives.

CHILD-REARING IN ONE-PARENT FAMILIES

Looking more specifically at child-rearing in some of the most common of today's pluralistic family forms, we find that one-parent families constitute 19 per cent of families in the United States when widowed, divorced, maritally sepa-

rated, and never-married units are considered together. According to 1980 figures, these family groups have 20 per cent of the children under the age of 16. Projecting to the end of the century, 40 per cent of children under age 16 in the United States will spend some part of their childhood years in a one-parent family. At present, the one-parent family has an average of 1.9 children, about equal to the 1.98 in two-parent households.

The category of single-parent families comprises the never-married; alternative single-mother unit; and the single-parent family of the divorced, widowed, or deserted parent. The widowed, deserted, and divorced single-parent units, which were initially conventionally married families, resulted when families were separated for long periods as by war, when fathers were absent or unavailable, or, in the case of family dissolution, through divorce. Today, divorce far outweighs death or desertion as the cause of single-parent families.

The unwed mother, a growing phenomenon in America, accounts for one out of six births. Customarily, out-of-wedlock pregnancies have been thought of as a teenage phenomenon, with one-third of the children born to white teenagers and five-sixths of those born to black adolescents accounting for a large percentage. However, added to this group, are the elective single mothers, who are usually Caucasian women who grew up in middle-class families of origin and who decide to start a family unit of their own even though they are not married. This elective single mother is, on the average, 25 years of age, comes from a middle-class or stable working class background, and has more schooling (approximately 13.9 years) than does the adolescent unwed mother. In the UCLA sample, only 14 per cent of the 50 elective single mothers lacked a high school diploma, and 24 per cent had a college degree.

Even within the elective single-mother group one finds diversity. One typology has identified three categories:

Nest-builders: Women who had consciously planned to become pregnant and frequently selected the man who might be a suitable father. They were the women who were highly educated, the most vocationally competent and experienced, and the most career-oriented. They lived by themselves, had the highest income, and were the most economically, socially and psychologically self sufficient (Kornfein et al., 1976).

Post hoc adapters: These women had not intended to become pregnant; however, once pregnant, they did not avail themselves of the possibility of abortion because they felt happy about their situation and felt able to adapt to and enjoy the circumstances. These women lived alone or in rooms or apartments with other female friends or with relatives, such as siblings or aunts. Work experience had primarily been in administrative,

---

*These data are derived from a UCLA study in which economic differences in alternative and traditional families did not account for the difference in usage patterns, since the project paid for pediatric care.

business, clerical, or skilled jobs, because education had been more limited than in the case of the nest-builders and vocational goals less specific.

Unwed mothers: This group's background and attitudes about their "fate" so strongly resembled those of the unwed pregnant teenager that the label "unwed mother" seems most apt. This group was minimally competent compared to the other elective single mothers. Work prior to pregnancy had been clerical, skilled, or semi-skilled, and annual income was at least $1000 lower than that of the post hoc group. Some still resided with their own parents. Pregnancy was unhappily anticipated, but abortion was not elected because of the cost-effective benefits (probably both financial and psychological) of getting pregnant. One such strong motivation (as suggested also for the black adolescent single-mother population) seemed to outweigh the benefits of not being pregnant (Luker, 1977).

One salient feature regarding alternative Caucasian single mothers was that despite the fact that they chose to have a child out of wedlock, they were ambivalent about adopting alternative values. As a group, they appear at times more like the traditional married families and at other times more like their alternative cohorts (Eiduson, 1981). Like traditional parents, single mothers may place a high value on conventional achievement, but they are often less concerned with materialism and, like many alternatives, are drawn to sensory and intuitive modes of thinking as a useful supplement to scientific and rational thought.

The problem of a marginal income level for the single-mother family has been well documented. During the first three years of their child's life, the mean income of the single mothers in the UCLA study was about $5,000 a year; during their child's infancy, only 10 per cent of the single mothers were in high socioeconomic status groups, 35 per cent were in the middle, and over 50 per cent were in the lowest groups. This is in line with the national figures on single-mother status: the income of the single mother is one-third that of married couples. Two million of the more than five million of single mothers with children under age 18 have incomes below the poverty level; 1979 figures cite $7,000 as the median annual income for single mothers, except for those whose children are all under age 6, in which case the median is about $4,500.

Few single mothers work when their children are infants, and in the UCLA study, approximately 51 per cent were not working when the children were 3 years of age. When single mothers did work, they were more likely than other mothers to work full time.

Of all the alternative groups, the single mothers made significantly more residential moves. In one study the single mother had a higher number of total changes (moves, life-style changes, separations from child) than did any other family group during the child's first 18 months. From 18 months to 3 years, and again from 3 to 4½ years, the single mother made significantly more changes than did the traditional married group.

Single mothers report more abortions than traditional marrieds, and many report not planning the baby. Therefore, calling them "elective" single mothers refers to their decision to have a baby once they have become pregnant, not their attitudes toward planning the pregnancy (with the exception of the "nest-builders").

Like the traditional marrieds, almost all single mothers (94 per cent) opted for a hospital delivery with a medical doctor present; 74 per cent agreed to an anesthetic, significantly more than did other alternatives. Generally, they had long labors—in fact significantly longer than traditional married mothers. While these mothers had fewer males present to lend support at birth, they had a greater number of female and other adult friends present to compensate.

Infancy and early childhood patterns of caretaking did not distinguish this group from the other alternative groups (Eiduson et al., 1982). Few differences were significant in terms of feeding or sleep arrangements or the psychological patterns such as mother/child attentiveness or involvement. In terms of extent or kind of caretaker responsiveness to child-initiated contacts, affective interchanges, or verbalizations or contact initiated by the mother, home observations at six months revealed no significant differences between the single-mother sample and any other group. However, some differences were noted in summary scores of intensity and variability of social and nonsocial stimulation with the single mother scoring low.

There were no significant differences based on life style in the ways parents perceived children. However, it is noteworthy—although not significant—that single mothers tended to have more children in the "difficult" and "shy" categories than did the other groups (Eiduson, 1981). Similarly, in a study using a self-report measure in which mothers reported development of their children in a number of areas, such as self-help, comprehension, perceptual motor development, and the like, the sons of single mothers emerged in the borderline or problem areas more often than did children from other groups, although overall, very few children were so rated in the total sample.

The single-mother population does not differ in use of traditional medical care, number of well-child visits to the doctor, or in relative number of visits by type of illness. Single mothers did tend to take their children to the doctor more frequently for psychological or behavioral problems than parents in other alternative life styles. This tendency may not indicate a greater frequency of problems but rather may reflect the single mothers' lack of practical and psychological support within their own households, and perhaps attribution of their own psychological difficulties to their children.

A greater percentage (75 per cent) of single-parent children attended day care or preschool than did children in any other alternative group by age 3½ years. Such exposure to outside inputs was in line with earlier findings that the single-mother's child had greater exposure to radio, television, and other forms of mass media than other alternative groups. Again, the single mother was similar to the traditional mother population and different from alternatives to a significant degree.

Child-rearing by single mothers as a group was distinctive in terms of conscious desire on the parent's part not to sex-stereotype children's toys, activities, or personalities. This is particularly interesting in light of the voiced preference of many elective single mothers for daughters.

By the time the child reaches one year of age, the single mother has usually become very dissatisfied with her fate (Eiduson, 1979). This attitude may be related to low economic status, more stressful experiences, and the problem of "overload" because she is solely responsible for a large variety of economic, psychological, and social tasks. The single-mother group, from the child's infancy on, seems to be the family in which the mother is responsible for more household and child-care tasks by herself than are women in any other group. Also, legal recourse seems generally inaccessible to single parents; less than 4 per cent of divorced women collect alimony and approximately 35 per cent collect child support.

An additional problem confronting the mother who brings up her child alone is social isolation. She has less freedom to maintain previous social contacts because of her work and child-care responsibilities. Even more devastating seem to be the negative attitudes expressed toward a single parent and the loss of status in the immediate and extended community. These concerns apply to the divorced woman as well as to the elective single mother.

As the divorced parent moves from married to single status, areas involving the greatest change and stress are those related to handling everyday reality problems; conflicts concerning self-concept and personal identity; and changes in relationships with others, including the child. The need to adjust to a reduced standard of living is reported to be a major cause of depression, anger, and feelings of incompetence and insecurity among divorced parents. Their daily lives are more dis-organized, and patterns of time spent with their children are more erratic—reverberations of the parent's own feelings of dismay and helplessness. The swing from initial apprehension and depression to feelings of freedom and ebullience that come just when feelings of rejection have abated is often mirrored in the parent-child involvement (Hetherington et al., 1977; Wallerstein, 1977).

In child-rearing, divorced parents seem to make fewer maturity demands than do married controls. They communicate less well, are less affectionate, and are more inconsistent in discipline. Sex of child and parent operates here: for example, divorced mothers communicate less well, are less consistent, and use more negative sanctions with sons than daughters. Poor parenting is most marked for the first years after the divorce, with the resurgence of competence and control over the children by two years. Investigators point out that children give divorced parents a difficult time, too, exhibiting negative behavior, particularly during the first post-divorce year. Subsequently there is more compliance and less opposition; however, hostility and opposition exhibited by boys are not ameliorated as easily as the whining and complaining more typical of girls. These findings support clinical data suggesting that the upset over parental disagreement and subsequent family disruption has long-term effects on children. While not necessarily as overtly displayed as in the first two years after the divorce, children remain anxious, angry, and lonely, and feel deprived and often guilty.

Availability of the male parent who does not have custody seems critical for the child's effective school functioning, especially for boys. While parents, siblings, and close friends may be effective substitutes, they do not compare as supporting figures with the presence of a biological parent (see also Chapter 13).

UNWED COUPLES

The contemporary cohabiting couples or marriages by social (rather than legal) contract are similar but not exactly comparable to the common-law marriage, which took place previously when people experienced social and economic constraints that interfered with legal marriage. Today's social contracts are based on the bond of love and trust that holds a couple together, which is considered more important and stronger than the legal bond authorized by church and state. The "marriage" rests on the couple's personal commitment to each other and will remain viable only so long as their relationship is vital.

This group represents the extreme alternative

family type in terms of nonconventional beliefs and values. As a family group, they are the least oriented toward conventional treatment and the most likely to seek "here-and-now" gratifications. On testing, they identified with using intuitive and sensory modes of input in problem-solving and showed the greatest preference for the natural-organic orientation in environmental concerns, foods, and the like. They also appear to be the most egalitarian-oriented family group, with fathers expressing interest in being intimately involved with the caretaking of the infant and the young child.

Paternal interest in child caretaking is effected because of the more relaxed attitude toward full-time work. In one study, social-contract families had the lowest number of fathers working full-time at each period, from the birth of the child through age 4 years. Vocational level of competency as well as personal preference may play a role in the employment status of social-contract men: their level of educational and vocational competencies was somewhat lower than that found in other family groups during the first years of a child's life; fewer fathers returned to school or sought subsequent college or professional training when data showed that men in other alternative families were returning to these activities.

In this family style, mothers also tend to stay home with the child (Eiduson, 1978). Few seem to return to work when the child is younger than 3 years old, and they seem to prefer part-time employment. As a result, this family group is a low-income group. In part, they have voluntarily reduced their socioeconomic status, in line with their antimaterialistic stance. They are "voluntary poor," taking seasonal or casual jobs in order not to be constrained by their employment and to be available for their family and to pursue other interests. Conventional indices of success are not important to them.

Many of these unwed households are found in rural areas or small towns to which the adults immigrated, feeling that they could thus live closer to nature and have more direct control over how they wanted to spend their lives. They often dwell in hand-hewn houses, which, although small and primitive in terms of middle-class amenities, are often colorful and creative.

The social cohabiting families also tend to be the most extreme alternative family form in terms of infant child-rearing practices (Eiduson et al., 1973). They deliver their babies at home when feasible. Most breast feed their children and continue this practice over a longer period of time than do traditional or single mothers; their infants are fed significantly more often, and their parents are particularly oriented toward demand feeding. For many, the breast was regarded not only as the main nutritional source but also as a pacifier. Natural-organic foods are commonly chosen, with many households restricting the use of meat and the use of fluorides.

In this life style, children are exposed to most family experiences, without much attempt by the parent to differentiate the child's experience from that of the adults. The children are reportedly placed in play pens less often than most other children.

Social-contract households' unconventionality is interesting when discipline is an issue. These parents do not stress obedience and respect for authority and are generally lenient and permissive in their exercise of discipline; yet, they consider it important that even a 3-year-old child be given a reason to mind.

Their more permissive life style extends to their attitudes toward sexuality. Social-contract mothers and fathers bathe more frequently with their children, fathers go nude more frequently in front of children, and children are allowed to be nude in front of other children more frequently than in other family groups.

Social-contract families value school for their children, despite their tendency to disapprove of institutions. However, they are attracted to schools offering alternative programs, especially those which stress an appreciation of nature, social adjustment, and happiness (Eiduson et al., 1982).

LIVING-GROUP FAMILIES

In an effort to replicate the extended family of yesteryear, some families band together to intensify bonds and share aspects of their lives with people of like minds. One group of such families have been called domestic living groups, i.e., people who have banded together because they like each other, want to reside closely so they can share meals and experiences, thus solidifying their friendships (Weisner and Martin, 1979). Triads (two men and one woman, or two women and a man) are considered the smallest living group; the largest may contain hundreds of members. Usually adults bring their children into the extended group with them.

Living arrangements for living groups show an enormous variability, including adapted motels, rural acreage with permanent and/or temporary dwellings, buses, hand-hewn houses, large homes, and mansions. All provide combinations of public and private spaces and are not necessarily under one roof. Private spaces are smaller in square footage than in abodes of most other families.

Other living-group families are group organized on the basis of adherence to a common religion or social philosophy or to a charismatic leader. Examples of such communities in historical America—the Oneida community, the Hutterites, and the Amish—are well known. These social arrangements have been distinguished by philosophies that make for important differences in child-rearing when put into practice. In today's living groups, some of the Eastern religions wear distinctive garb and their daily regimens may be shaped by religious practices in which all members participate. Others are less obviously identifiable, but their adaptation of customary family functions is evident in such practices as pooling personal income, sharing personal possessions, exchanging child care, or delegating child-care roles to certain members.

Living groups differ in the extent of their alternativeness. In some, women assume traditional roles; in others, especially in the domestic groups, more personally determined roles are adopted. Some living groups, despite their unconventional life style, have a highly ordered and authority-oriented structure; others are more casual, laissez-faire, and experimental. In some, members work in jobs outside the family, but others try to absorb all family adults in their own activities, so that the unit is self-sufficient. Their variability seems to rest on the extent to which members are committed to group goals, the precedence of these goals over the goals of any single individual or within-group family units, and the ways and means available to them for surviving economically.

Living groups seem advantageous for particular periods in a family's career, such as when they have small children who need care and supervision. However, incompatibility with regard to parental preferences about child-rearing can lead to tensions and unhappiness. Parents may report that too many adults provide input; in other cases, there is some unevenness in or a lack of the expected sharing of responsibility and caretaking.

The contemporary living group has introduced some changes into traditional child-rearing practices. First, some espouse a noninterventionist philosophy insofar as the parents' role and responsibility for the child are concerned, as contrasted with the modern mainstream position that regards the parent and parenting as formative influences on the young child (Cohen and Eiduson, 1975). The noninterventionist view regards the fate and experiences of an individual child as determined by other than human forces—"the stars" or "the Lord"—thus justifying a "hands-off" policy toward the child and his developmental course.

Also of interest are those living groups in which a more conscious and decisive role by the parent in shaping or "intervening" in the child's development results in child-rearing models that deliberately emphasize discontinuity in the parenting process and among parenting persons in order to reduce intense bonding with one person, usually the biological parent. This multiple caretaking approach is rationalized as a way of enhancing the child's capacity to trust others. With this thrust have come radical child-rearing strategies. Examples include geographical separation of the child from the parent by age 2, as in religious communities in which male children are sent to school to be trained for later roles in the group or the church, and approaches in which group care replaces family-based rearing. In these cases, female parent surrogates act as caretakers and/or teachers, and often parents and children are encouraged not to interact in order to reduce one-to-one attachments. The rationale for mitigating the intensive mother-child bond also rests on a desire to provide the child with an enriched cognitive milieu and a more extensive social environment.

To what extent do the child's needs rather than the group's preferences determine such practices? Does multiple caretaking mainly relieve parental responsibility and allow for more diversity in the adults' activities? The question of neglect has been raised in regard to some "laid-back," casual communal groups who, in "over-relaxing" conventional age- and sex-graded relationships, leave children to their own devices—to fend for themselves or to assume that peers will act as adequate companions, confidants, and allies. Some data obtained through home observations in the more laissez-faire group arrangements have indicated these children remain only peripherally involved in family activities and relationships—an ambience suggestive of those lower-class families in which there is little supervision and little attention paid to the individual child's needs. Just as this latter child-rearing milieu is associated with "street-wise" children, so some of these group children have been described as "family-wise" and "life-wise."

Of course, this laissez-faire attitude is not the case in all multiple-caretaking situations. On the contrary, in some groups child behaviors are carefully prescribed and supervised because it is thought that such education will best prepare the child to be competent in his group (Eiduson, 1978).

Systematic studies at UCLA of 50 living-group families have shown that within every living group there resides a nuclear mother-child and often father-mother-child unit during infancy. Interestingly, these living groups, who based on their overt life styles appear to be anti-mainstream, are usually not so unconventional where the child is concerned. They adopt pre-birth and anticipatory socialization practices that are not easily distinguished from those adopted by traditional families. Although their preference was for home deliveries and natural births, in actuality, parents enrolled in self-preparatory birth classes, all received prenatal care, and 80 per cent have had hospital deliveries. Commune children are breast-fed over longer periods than in traditional nuclear families. However, despite journalistic reports to the contrary, few breasts are "exchanged" among new mothers, and the rituals surrounding births early in the counterculture movement, which aroused anthropological interest, have diminished. The infant's needs for biological and psychological dependency seem to be a strong determinant of how he is reared—even in the modern communes.

At six months, mothers are the main caretakers, but communes did have more fathers and other women as secondary and tertiary caretakers during noontime feeding. With the larger number of persons around and less private space per family, the living-group child tends to be exposed to more social stimulation. Children see their fathers more, for in many groups weekday contacts between father and children occur because fathers work close by or in the complex. However, on weekends, when traditional fathers are usually at home, the commune father is usually more occupied with the activities in which he engages the other days of the week. In communities in which a person is substituted as caretaker for the biological parents, it has been found that the child perceives less of a distinction between parents and other adults in terms of their roles as disciplinarian, teacher, or caretaker. This seems in line with a report from the Hutterite community, which stresses egalitarianism and an absence of visible status symbols; children seem to have a global, undifferentiated perception of social roles.

Children growing up in a group that tries to develop high internal cohesiveness and a sense of within-group responsibility for all aspects of life tend to develop a sense of remoteness from the outside world. Usually relationships within the group-extended family become a substitute for all aspects of the more conventional extended kin network, thus reinforcing the child's social isolation from the surrounding community. If schooling is handled within the group by employing outside tutors or by using group members as teachers, there seems to be a marked sense of isolation from the outside world. For the child this means that he has few contacts outside his community and depends heavily on other living-group children for social exchanges. Since most communities seem to have a wide age spread, with children dispersed in age, a little red schoolhouse approach may be inferred.

However, it is interesting to note that many parents who themselves want to be separate from the outside community because of their disenchantment with conventional values may be eager for their child to become affiliated with small community groups to further ensure some private time for themselves.

Since the stimulation potential of the home has been implicated as a critical variable for early learning, it is noteworthy that commune children have about the same number of toys and playthings as other children but of less variety (Weisner and Martin, 1979). In a few communal families some toys are shared, but all children have some things of their own. Some communal children have little access to television or magazines, especially in creedal groups, but others have the television on 24 hours a day. Interests are often predetermined by the interests of the total group. In one politically oriented domestic group, children were politicized in nursery school, attending rallies and having political poster decorations. In a religious community where teaching of mathematics, science, history, or physical education is forbidden, English, Sanskrit, and ideological texts are substituted. When nontraditional training for the child is adhered to, the question is raised as to how successfully a child will be able to function on the outside. While such training seems counterproductive for effective functioning within the framework of conventional institutions, it does ensure that the individual will remain within the group and thus sustain it.

Although in most groups children reside with their parents, in some group settings the child and mother are geographically separated by choice of the leader or because the mother must play some other role; occasionally a group decides to separate mothers and children who become too "tied" to each other for the good of the group. However, even in this instance all children know who their parents are and differentiate them from the other adults.

As noted, the communal living-group child has the most nonbiologically related peers in the household but the fewest social contacts outside the home. Children are significantly less often left at someone's house outside the group. Also, visits by grandparents are less frequent—one index of the commune's insularity. Only one third to one half the children attend a school that is not an in-house school (Eiduson et al., 1982).

The creedal living-group child is likely to be disciplined in line with the group's needs for conformity to cooperate and with their demand that the child respect authority. Unlike the domestic living-group members, who do not emphasize obedience and who "reason" with their children, creedal group members feel it is paramount to teach obedience and respect for authority. They discipline for more types of child behaviors than do domestic-group parents, although they use the same modes of discipline in about the same proportion: primarily verbal reprimand and secondarily verbal and restrictive modes. As in most large families, today's extended family depends on cooperation, loyalty, and particularly compliance in order to make it work. Even the children must incorporate group goals over individual preference and individual aspirations.

CHILD DEVELOPMENT IN TRADITIONAL AND ALTERNATIVE FAMILIES

Previous studies of the effects of growing up in a one-parent family or in an intentional community had linked deviance to father absence or multiple caretaking. Negatively affected were intellectual performance and academic attainment, with boys whose fathers were not available from an early age being particularly disadvantaged. Analytical thinking patterns in boys without a father tended to approximate the patterns found in girls (high verbal, low mathematics). Also associated with father absence were increased introversion, depression, and ambiguity in direction toward a career goal. Problems of sexual identity seemed to increase. In addition, moral judgments were affected, which seemed substantiated by these children's involvement in delinquency.

These studies ignored confounding factors that might have influenced results, such as the reason for the father's absence, availability of a male substitute for the father, birth order of the child, presence of other siblings, mother's reaction to father's absence, arrangements made by child caretaking, and many others. Lowering of the family's socioeconomic status that often accompanies a change from a two-parent to a one-parent home was often not taken into account. Thus, methodological problems have raised questions about the validity of these findings.

The best documented living-group or extended-family arrangements have been on kibbutzim in Israel. Children reared on the kibbutz show few differences in cognitive characteristics when compared with home-reared children. However, extended immaturity is reported in kibbutz elementary school age children and adolescents and is related to the reduced interaction with their parents. Kibbutz children seem attached to parents early, perceive parents as nurturing, and show no diminution of their capacity for intimacy; yet identification with parents appears more diffuse, with less ambivalence toward and less rivalry with the parent. The child in the group situation seems to have the capacity to use adults other than parents as models; teachers set aspirations and limits for the children and provide personal ties. The cooperation necessitated by the demands of group living seems to foster the child's role-taking ability, ease of interaction with peers, and sensitivity to the needs of others. The few followup studies available provide a picture of group-reared children as well-functioning, adequate, productive adults.

PHYSICAL DEVELOPMENT

The longitudinal UCLA study of alternative parents and children has followed children from birth through 6 years of age. When some of the major trends in the data were singled out, it appeared that from birth, maturation of the total sample fell within the normal range. Mothers had reported a high involvement with drugs prior to the pregnancy. Alternatives were heavier users than were traditionals, but use among both groups was reported throughout the period of study. However, during pregnancy, drug use dramatically declined in all families and remained low. No children showed features at birth that could be attributable to parental drug or alcohol use.

Study of the obstetrical complications revealed few problems for each life style, suggesting that births were generally uneventful. Neurological examinations also indicated that these newborns were average or above average. No differences attributable to life style were significant on these measures; while single mothers' infants were lowest in rank, there was in all only a small difference between the lowest ranking (social contract) infants and the highest ranking (living-group) infants. The 22 at-risk children (11 per cent) were randomly dispersed over all life styles, including the traditionally married comparison group. On the obstetrical complications scale, families were generally similar to each other (Eiduson, 1978).

At one year, a physical examination was conducted, weight and height were checked, and a 24-hour dietary recall was conducted with the mother. Only 18 children of 209 children fell more than one standard deviation below the normal for weight—eight from traditional marriages and 10 from alternative groups. The 16 children who were shorter compared to males and females from standard populations were all from alternative families. Head circumference measurements were within the normal range for all children studied. Standard pediatric examinations detected heart murmurs in three cases, one instance of upper respiratory infection, one case of acute gastroenteritis, and one child in whom polyethylene ear tubes had been placed. These children were from traditional as well as alternative groups.

Nutritional studies at one year suggested that

fewer children in traditional families were given diets that were low in two or more nutrients (i.e., providing less than two-thirds of the RDA) compared to alternative children, but by and large, diets were at least normally adequate in all cases during infancy. Alternatives as a group breast-fed longer than did traditionals. Traditionals fed their children solid foods earlier, tended to use meats more regularly in the diet, and gave their children more fluorides—practices that continued throughout the preschool years. By age 4, only one child was on a macrobiotic diet, although a few additional ones restricted meat in favor of eggs and/or cheese. However, there were no differences between alternative and traditional family groups in the use of vitamins as supplements, in giving children immunizations, or in medical usage patterns. Close to 95 per cent of the providers of services to children in alternative families were physicians, with few nontraditional therapists employed, despite the intellectual interest of some alternative parents in these resources for themselves.

Children averaged 9.5 medical visits in the first six years of life, with no significant differences between alternative and traditional family youngsters. Of the total of 1758 visits, 938 (53.4 per cent) were for ongoing evaluation or supervision, immunization and tests, or routine dental checkups. About three-fourths of the remaining 820 visits involved illnesses: acute problems, accidents, and emergencies. In 4 per cent of these cases, the child was hospitalized.

By age 1, the number of medical visits was dramatically reduced. The number of nonillness visits was highest from birth to one year, dropping by 46 per cent by 3 years of age and by another 11 per cent in the 3- to 6-year period. In contrast, visits for illness rose to a peak in the 1- to 3-year period, the number of visits being 23 per cent higher than in the first year of the child's life. In the 3- to 6-year period, the number of medical visits for illness again dropped significantly below the number during infancy. Traditional families were more likely to seek medical care for their children than were some alternative families, but this was the case only for 3- to 6-year-olds and not at earlier time periods.

In every one of the three age periods under study (birth to 1 year, 1 year to 3 years, and 3 to 6 years) respiratory illnesses, the primary referral problem, accounted for more than one-third of the contacts during the first years of life, and again in the 3- to 6-year period; and for more than half the contacts in the 12- to 36-month period. Dermatological problems and gastrointestinal disturbances were second and third most frequent problems, respectively; in each period after the first year, these incidences steadily decreased. Less than 10 per cent of the children sought help during any single time period for neuromuscular, allergic, immunological, genitourinary, hematological, cardiovascular, endocrine-metabolic, collagen, or bone-joint difficulties.

There were some small but significant differences in illness patterns among alternative and traditional families: for example, a higher proportion of girls in traditional rather than alternative families were referred for respiratory problems, although this was not the case for boys; the incidence of neuromuscular or gastrointestinal problems was significantly lower among children from traditional families, with girls in single-mother families showing the highest incidence of gastrointestinal disturbances. However, by and large, neither alternative nor traditional children in this sample were particularly prone to illness, although it had been hypothsized that there might be differences due to the major differences in home milieu, number of persons with whom the child had close contacts, and other factors (Eiduson and Forsythe, in press).

INTELLECTUAL GROWTH

At six months of age, when the child's range of affective behaviors, ability to initiate responsiveness to caretaker-initiated behaviors, smiling, and crying behaviors were studied, all babies showed a normal range of behaviors, and no notable differences were found between groups with regard to such maturational parameters. Children on the whole achieved average scores on the Bayley Scales of Infant Development at 8 months of age. Few life-style differences have been noted favoring traditional married over alternative children on mental developmental measures in the early years. When differences did appear, they were not maintained consistently over time. For example, in development focusing on motor behavior, single-mother children scored significantly higher than did living-group children at 8 months of age, but these differences disappeared by one year.

When children reached 3 years of age, a Stanford-Binet intelligence test (IQ) was used to assess general functioning level (Eiduson et al., 1982). On this test, which is heavily loaded in favor of verbal material, the entire sample again scored as normal, and a comparison of children by life style revealed only minimal differences between groups. Children from both social-contract and traditional married families had identical mean IQs (103.0), slightly surpassed by living-group children (104.2), who were, incidentally, the most diverse group (SD $p < 0.09$). Single-mother children showed the lowest mean IQ score (101.1) but this difference was not significant.

Major differences in environments of children reared according to different life styles were expected to affect early language development and recognition and labeling of objects. However, this did not prove to be the case. On the Peabody Picture Vocabulary Test, for example, scores for the total sample were average (102.4, SD 16.6) and, again, life style differences were not significant. However, single-mother children had the lowest mean score (98.2), while traditional married children scored highest (104.5) and social-contract and living-group children were close to traditional-group children. Rural-urban differences, differences in parents' education level or socioeconomic status did not appear to discriminate family groups. Hypotheses relating verbal competency to large family size (and therefore more verbal interaction) were not confirmed, although children from one-parent families had slightly lower scores. An "intellectual risk" score, which summarized performance on both the Peabody Vocabulary Test and the Stanford-Binet scale, identified a total of 18 children (9 per cent) as at risk; the three alternative life style groups under study and the traditionally married two-parent nuclear family contributed almost evenly to this group. Three children from the traditional married group appeared to have possible "minimal brain damage," although only one had evidence of difficulty at birth.

A number of other cognitive-related abilities were also tested at three years: all children showed typical articulation skills for their age, although living-group children tended to surpass single-mother children. A global assessment of creativity, evaluating amount and kind of originality showed in fantasy play on a semi-structured test as well as nonconventional responses on one projective test, suggested that traditional married and social-contract children were more free to express their fantasy compared to the more "stifled" living-group children.

A child's persistence, a characteristic related to the kind of task orientation demanded at school, was measured through a variety of experimental tasks. There was considerable variance within groups but no significant difference by life style. Traditional and single-mother children scored highest, however, and social-contract children who scored lower on receptive vocabulary were more apt to refuse these verbal tasks.

SOCIO-EMOTIONAL DEVELOPMENT

A number of socio-emotional behaviors were assessed at one and three years. For age 1, the child's reaction to separation from his mother was observed in the Strange Situation and Mother Attachment Procedure developed by Ainsworth. Using Ainsworth's classification, attachment is of three kinds:

1. *Group A (avoidant)*, in which there is little or no tendency to seek proximity to or interaction or contact with the mother, even in the reunion episodes. In general, either the baby is not distressed during separation or the distress seems to be due to being left alone rather than to his mother's absence. In this study, 15% of the children were rated A.

2. *Group B (attached)* consists of children who want either proximity and contact with the mother or interaction with her and actively seek it, especially in the reunion episodes. Of this sample, 76% were rated B.

3. *Group C (anxious resistant)* identifies the child who might display generally "maladaptive" behavior in the strange situation. Either he tends to be more angry than infants in other groups or he may be conspicuously passive. Only 9% of the sample were rated C.

Children in each life style showed a comparable range of behaviors described in other studies. Few differences in alternatives versus traditionals emerged, thus supporting the position that distress reactions to separation at this early point may be maturational rather than family environment–dependent. Although not statistically significant, traditional children seem to have a differentially strong need for "mothering," as revealed by their wanting to be held more, exhibiting less exploratory or manipulative toy play, and refusing to be soothed following separation.

At age 3, the separation anxiety elicited at one year when the mother leaves is expected to be minimal. Only eight children (4%) were rated as anxious, resistant ("C"), and none of these had been so rated at one year. Four were from the traditional married sample, three were single-mother, and one was a social-contract child. Examination of environmental data for these children showed that all but one of the distressed children had recently been subjected to marked stresses such as parental divorce, separation, or hospitalization. Yet other children experiencing such stresses were rated in accordance with the total sample, in the A (56%) and B (39%) categories, both of which suggest a child's ability to experience separation without visible distress.

Test of the child's fearfulness and/or anxiety showed no significant difference as a function of life style. Results showed a tendency for children of traditional marrieds to be the least fearful and those of the living-group to be most fearful.

Aggression, rated on the basis of extent of motor or verbal expression during play, did not differ significantly by family group. However, levels were high for unwed-couple children and were low for living-group children.

Measures of cooperativeness showed that living-group and single-mother children tended to be more cooperative, while social-contract children were less, although differences were minimal. Thus there is suggestion of compliance in the living-group sample's children, a seemingly desirable characteristic for adaptation in this large and extensive family unit, and we find some tendency toward noncompliance among the social-contract families, the most permissive and alternative group.

Also of interest considering the alternative families' orientation toward the "here and now" was the tolerance for frustration or stress among children of this group. Single-mother children tended to be most tolerant and social-contract children least. No life style differences were significant in activity level, but unwed-couple children were more apt to approach a rating of "hyperactive" in contrast to living-group children. Extent of attentiveness again revealed no significant differences between life styles, but single-mother children were rated highest and social-contract children lowest.

An additional variable, adjustment, was measured in a number of ways: the child's ability to play during separation in the Stranger Test at 3 years of age; the ability to perform a task when confronted with a frustration barrier; attempts at finding a solution to an insoluble task; and the personal resources shown on a projective test. Single-mother children tended to be rated highest and social-contract lowest, but, again, the overlap among children of various life styles suggests that the family unit in which a child is growing up is itself not critical for his or her adjustment in early years. Summary assessments of a child's social, intellectual, and emotional competence scores at age 3 also supported these findings; no differences were evident based on life style in the first three years of life.

Results of these measures seem to be showing more dispersion as the children reach 6 years of age. The competencies adaptive for adjusting in each of the life styles seem to be sufficiently different by 6 years, so that differences in their characteristics are becoming apparent. The implications for the child's later adjustment in the school and community will become follow-on data in this longitudinal study.

IMPLICATIONS FOR PEDIATRICIANS

What do the new life styles and pluralism in family styles mean for the pediatrician who eval-

uates and treats the child and works with the family?

1. *It is important to note that values and practices associated with following a nonconventional life style are not necessarily being expressed only by alternatives.* Values and perspectives of parents—and even of their children—are changing rapidly, so that within nuclear families who live as a two-parent unit, many traditional values are being questioned: marriage and religion as useful institutions, planning and saving for tomorrow, the Puritan ethic of diligence and hard work as the only road to success, and the willingness to put work ahead of leisure and family. Furthermore, values that were once considered "alternative" are beginning to diffuse into traditional society: the lack of sexual repression in the mass media, the interest in natural organic foods and health stores, and the growth of alternative classes in public schools are cases in point.

Similarly, every person who practices an alternative life style does not necessarily share all the values and perspectives associated with alternativeness. Some alternatives identify with certain values, such as lack of interest in material possessions, but not with others; in fact some families living alternatively are quite traditional in some ways; for example, they may adhere to conventional sex-role stereotyping. Today, because of the considerable overlap between alternatives and traditionals in their values and ways of living, it is risky to generalize about values and attitudes according to an individual's life style and vice versa (Eiduson, 1979). Rather, in order to understand those aspects of a child's health problems or of the family's compliance behavior that are relevant to pediatric care, values, attitudes, and practices need to be explored with each family.

2. Another noteworthy change is the rethinking of the parental role (Cohen and Eiduson, 1975). Some parents question a commitment to parenting as their primary or exclusive role. For contemporary adults, the role of parent is regarded as one of many adult roles. The family may resort to ingenious maneuvering in order to try to adapt and interdigitate children's needs with those of the parents without interfering with the child's development. In some cases, this perspective derives from the parents' awareness of the enormous frustrations of their own parents as caretakers, which often led to neurotic relationships with their own children. Other parents question how much "shaping" of children they can actually do, since they recognize powerful differences among their children at birth that seem to have a strong influence on later growth. In addition, strong social and economic forces are fostering changes in the family and, in turn, in parenting roles. *Thus, the needs of parents as expressed in the way the family*

functions and the roles parents play must be considered along with the needs of the child when the clinician makes recommendations and plans for the child.

3. How do changes in values influence what goes on in a family? Alternative children may come to the pediatrician's office looking dirty or dressed so casually in homespun garments that they appear to be neglected. Or they may be brought in by fathers or other caretakers who give support and assistance to the mother. This does not necessarily mean that a child's mother is not concerned, loving, and interested in what is happening to her child. Alternative families operate differently—an attractive feature to many adults. They can offer equivalent parenting in a number of different ways, which have as yet not been proved to be inferior to or less successful than the parenting provided by biological parents. In fact, some alternative arrangements seem preferable. The mother in a living group, for example, has fewer tasks than the traditional mother and has more help with tasks. If mother and father both work outside the home, as is the case for almost half of American families with children under 18, the living group can provide competent and consistent parent-substitutes in the area where the child lives—just as grandparents or other relatives did when the rural family consisted of parents and adult children who continued to live together in an extended arrangement. For some men and women today, alternative life styles seem to offer them ways to live that are personally gratifying and adaptable to the needs of both parent and child. Thus, the traditional way of living must not be seen as the only or the best approach to child care. Although some alternatives have different values and different priorities, the amount or quality of attachments and relationships between child and parent or between parents is not necessarily reduced. *It would be erroneous to conclude that because a life style is externally different, it is also psychologically different and inferior.*

4. As a follow-on, one characteristic of alternative life styles seems to be to seek solutions for conflictual, stressful, or difficult situations through change (Eiduson and Forsythe, in press). Alternatives change location and mates more than do traditionals. Separations between mother and preschool child for more than two weeks are more likely to occur in an alternative rather than a traditional household. Professionals have customarily regarded change—environmental manipulation, as clinicians call it—as indicative of instability and a reluctance to resolve conflict through self-scrutiny and effecting personal psychological change. Adjustment of a child exposed to many moves and changes in household personnel is thought to be jeopardized. However, for many alternatives, change appears to be ego-syntonic

and even adaptive. Some seem to make life style and geographical changes in a trial-and-error way, without showing the expected signs of instability and withdrawal or evasiveness. Furthermore, although mobility and change can pose some risk for an alternative child, this mode of living has so far not proved to be of consequence. In fact, some children develop effective coping skills when faced with stresses, developing early an ability to adapt to change. Our customary assumptions about what is good or bad for a child are therefore undergoing reevaluation. *Until we have conclusive and long-term findings concerning the effects of life style on child development, we would be wise not to consider traditional ways of child-rearing as the only or a better approach.*

5. Thus far, research suggests that alternative children may be developing differently from traditional children, showing some competencies that their families value and have fostered as well as some limitations. If alternative parents are less driven to intellectual achievement and more interested in humanism and social facilitation and the enhanced humanistic relationships these promise, children may develop strong social sensibilities and an appreciation of cooperation and sharing. At the same time, they may be less motivated toward school achievement and less oriented toward academic goals. If parents put a premium on creativity, their children may think differently, look for interesting avenues and novelty in play, and be less interested in accepted or popular ways of problem-solving. Or, if group decision-making is the mode by which the child's family airs and resolves problems, the traditional push for self-reliance and assertiveness may emerge, with the child feeling more comfortable with group activities and relying on peers as allies and joint decision-makers at an earlier age. The early independence of the single-parent child or the lessons in cooperation and sharing with others gained by the child in a living group may foster certain strengths that are not as visible in the high-striving, compliant child of the two-parent traditional unit, whose family values and expectations, on the other hand, may prepare him for doing well and making an acceptable adjustment in school. Thus, each family style seems to offer the child some "trade-offs." Also, no specific data suggest that deficit or deviant development in a child can be attributed to whether or not the mother is married or whether a father is at home. *No one type of family style is consistently associated with particular problems or dysfunctions in child growth and development.* Instead, there seem to be competent and responsible parents in every family style and some less competent and less committed ones as well (Eiduson, 1978).

6. Although the implications of changing values

and life styles are not yet definitely known, a great many institutions in society are changing to serve better the families with whom they interact, hence the adaptations of labor-room practices to facilitate the Lamaze and Bradley training of pregnant women, the growth of women's clinics, the development of holistic medicine, and the publication of self-help medical "texts" for families with young children. Many of these new practices seem a throwback to practices found in America during its preindustrialized eras. They speak to the desire for more personal control of one's life and for more humanistically oriented services, especially when "the body is concerned." The results seem to be a wider spectrum of care and services available to a pluralistic society, with more variety and diversity in techniques and therapeutic regimens, and a closer involvement of the family in the care of its members. This trend can also result in a lessening of quality of care but not necessarily.

Interesting in this regard is the finding that some contemporary parents avail themselves of traditional services provided in traditional medical settings by traditional physicians for their children, while they continue to seek out less conventional therapies and service providers for themselves. This "split" in parental behavior shows that when parents are anxious about their offspring, they try to assure that their children receive tested and true medical attention and the same benefits of quality and experience that they themselves had as children. It also reflects their ambivalence about nontraditional medical practices, for all their consonance with the new value perspectives. In actuality, few alternative parents have forsaken traditional medicine completely. Of 150 alternative families studied who had at least one child, slightly more than 5 per cent sought nontraditional medical care. *Therefore, living alternatively does not necessarily mean preferring alternative medical care.* Alternatives flirt with nontraditional practices and practitioners, but many return to what they know.

7. *A major task for the pediatrician is determining when the pluralism and diversity in family values and attitudes need to be taken into account and when firm rules and procedures should be preserved.* Some experts act in a rigid way when parents look or act in an unconventional way or express unusual perspectives; others seem to abrogate their responsibility as physicians and lean overboard to accommodate differences. For all their questioning of authority, most parents regard pediatricians as knowledgeable, competent, and concerned individuals who can be helpful in time of illness and stress. All parents want sufficient information to alleviate their anxieties and concerns or at least to put these in some realistic perspective. They also want to understand the child's illness behavior as

clearly as possible. In particular, they welcome those physicians who understand and appreciate their desire to become knowledgeable so that as parents they can be better informed and play a more active role in decision-making about their child. Especially meaningful to them are pediatricians whom they can regard as part of their "support-network," i.e., persons who provide service, education, and understanding while at the same time showing a genuine interest in the new ways that they as families are living and relating to children.

Pediatricians, like all clinicians, have to appreciate that for a long time the middle-class Caucasian child has been the norm. Yet tomorrow's child may be listening to a "different drummer." It would perhaps be comfortable if all children and all parents returned to the old tunes. However, who knows how suitable the old tunes and the old ways of adjustment are for the world of tomorrow? The child in the office today is growing up in a fast-paced, overpopulated, stressful world. Hopefully, in such a world, some of the changes in outlook and practices in contemporary families have the potential for providing a child with the competencies and strengths that lead to effectiveness, adaptation, and adjustment tomorrow.

Acknowledgment

This work was supported in part by National Institute of Mental Health Research Scientist Career Award No. 5K05 MH 70541–10 and by United States Public Health Service Grant No. 1 R08 MH 24947 and Carnegie Corporation Grant B3970.

BERNICE T. EIDUSON

REFERENCES

Cohen, J., and Eiduson, B. T.: Changing patterns of child rearing in alternative life styles: Implications for development. *In* Davids, A. (ed.): Child Personality and Psychopathology: Current Topics. New York, John Wiley and Sons, 1975.

Eiduson, B. T.: Emerging families of the 1970's: Values, practices, and impact on children. *In* Reiss, D., and Hoffman, H. (eds.): The Family: Dying or Developing. New York, Plenum Press, 1978.

Eiduson, B. T.: Alternative families: Implications for policy. Paper commissioned by the National Forum for Children, Youth and Families. Cornell University and the National P.T.A., Chicago, August, 1979.

Eiduson, B. T.: Contemporary single mothers and parents. *In* Katz, L. G. (ed.): Current Topics in Early Childhood Education. Vol. 3. Norwood, N.J., Ablex, 1981.

Eiduson, B. T., Cohen, J., and Alexander, J.: Alternatives in child rearing in the 1970's. Am. J. Orthopsychiatry 43:721, 1973.

Eiduson, B. T., and Forsythe, A.: Life change events in alternative family styles. *In* Callahan, E. J., and McClusky, K. A. (eds.): Life-Span Development Psychology: Non-normative Life Events. New York, Academic Press (in press).

Eiduson, B. T., Kornfein, M., Zimmerman, I. L., and Weisner, T. S.: Comparative socialization practices in alternative family settings. *In* Lamb, M. (ed.): Nontraditional Families. New York, Plenum Press, 1982.

Hetherington, E. M., Cox, M., and Cox, R.: The aftermath of divorce. *In* Stevens, J. H., and Mathews, M. (eds.): Mother/Child, Father/Child Relations. Washington, D.C., National Association for Education of Young Children, 1977.

Kornfein, M., Weisner, T. S., and Martin, J. C.: Women into mothers: Experimental lifestyles. *In* Chapman, J. R., and Gates, M. J. (eds.): Women into Wives: Sage Annual of Women's Policy Studies. Vol. 2. Beverly Hills, California, Sage, 1976.

Luker, K.: Contraceptive risk-taking and abortion: Results and implications of a San Francisco Bay Area Study. Studies in Family Planning, *8:*190, 1977.

Masnick, G., and Bane, M. J.: The Nation's Families: 1960–1990. Cambridge, Massachusetts, Joint Center for Urban Studies of MIT and Harvard University, 1980.

Wallerstein, J. S.: Response of the preschool child to divorce: Those who cope. *In* MacMillan, M., and Henao, S. S. (eds.): Child Psychiatry: Treatment and Research. New York, Brunner-Mazel, 1977.

Weisner, T. S., and Martin, J. C.: Learning environments for infants: Communes and conventionally married families in California. Altern. Lifestyles, 2:201, 1979.

13

Family Variation

13A Parenting and Its Problems

CONCEPTUALIZATIONS OF PARENTING

Parenting, mothering, fathering—what do these technical "reductions" of the human experience mean? Is parenting a biologically based instinct, a trait formed as a result of socialization practices, a learned skill—or something else entirely? Should we think of parenting as a characteristic inhering in mothers and fathers, or as a characteristic of a relationship?

Such questions are not merely academic; how we define a construct such as parenting has implications for both research and practice. For example, researchers who conceptualize mothering and fathering as traits or abilities of parents may fail to examine the contribution of the child's characteristics to the parent-child relationship; those who consider mothering to be an innate and instinctual process may fail to recognize cases of child abuse. Conversely, practitioners who adopt an ethological perspective may assume that close parent-infant bonds will never develop if normal attachment processes are disrupted during the neonatal period. Or, to take another example, clinicians who believe that parenting practices are the crystallized products of a long-completed socialization history may assume that the best solution to "bad parenting" is to remove children from the custody of bad parents.

Investigators from different disciplines have conceptualized parenting, mothering, and fathering in different—and often inconsistent—ways. A variety of conceptual frameworks has resulted in a variety of definitions. In this review we therefore begin with a consideration of definition and conceptualization, followed by our own perspective on parenting and an attempt to present briefly the empirical and conceptual literature relevant to the parenting process.

Perspectives on Fathering

The 1964 edition of Webster's *New World Dictionary of the American Language* offered four meanings for the transitive verb "to father" and three meanings of the transitive verb "to mother"; parent was listed as a noun and an adjective but not as a verb. Thus, in 1964, it was correct to speak of fathering a child and mothering a child but not of parenting a child. Moreover, important differences in meaning accompanied the two parental terms. "To father" meant to beget, protect, originate, and take the responsibility for but not "to look after or care for as a mother does."

Even today people commonly use the term "to father" to describe the father's role in procreation. Certainly, one would not be surprised if the question "Did he father that child?" was interpreted as a query about the biological paternity of the child. On the other hand, "Did she mother the child?" would probably be interpreted as a question about the extent to which a woman nurtured, cared for, and looked after the well-being of the child. These different connotations probably represent traditional stereotypes about the roles of mothers and fathers within the family—stereotypes that enable us to make quick sense out of statements such as "Mr. Jones does a better job of mothering the Jones' children than his wife does."

While "parent" had not been approved by lexicographers as a verb form in 1964, it is often used today to describe child care activities in which both mothers and fathers may engage. The emergence of this sexually neutral term reflects changing notions about the potential participation of the father in the family. According to Fein (1978), it is possible to identify three perspectives in the literature on fathering—traditional, modern, and emergent. Because all three perspectives can be found in the literature on parent-child relations, it seems useful to review them briefly here.

In the *traditional perspective*, the father is seen as an aloof and distant figure who cares for his children primarily through his role as breadwinner. Psychoanalytical thinkers operating within the traditional perspective (e.g., Bowlby, 1969) generally see no direct caring role for fathers when their children are small, although these fathers could be indirectly involved by providing companionship and emotional support to their wives. Sociologists subscribing to a traditional point of view have emphasized the complementary family roles fulfilled by instrumental fathers (who served as heads of households and providers) and by expressive mothers (who were the emotional, available caretakers of children). Fein believes that this traditional perspective on fathering did not exist in a vacuum, divorced from social reality; on the contrary, it generally conformed to social ideals and realities of the late 1940's and 1950's. Some authors still conceptualize mothering and fathering in ways consistent with this perspective; moreover, there undoubtedly are parents in today's society who behave in very traditional ways.

In what Fein calls the *modern perspective* on fathering, fathers have been seen as contributing directly to the psychosocial development of their children, especially their sons. Building on research comparing children from father-present and father-absent homes, proponents of the modern perspective argue that fathers play an important part in preparing their sons and daughters to grow into normative adult roles as men and women. Important research is currently under way that is designed to determine both the extent to which fathers interact differently with sons and daughters and the implications of differential interactions for the child's subsequent development.

Finally, Fein notes that the *emergent perspective* on fathering proceeds from the notion that men are psychologically able to participate in a full range of parenting behaviors. "The emergent perspective on fathering is androgynous in assuming that the only parenting behaviors from which men are necessarily excluded by virtue of their gender are gestation and lactation." Recognizing that not all fathers behave in this androgynous fashion, proponents of the emergent perspective often express the value judgment that it may be *good* for both parents and children if men do take active roles in child care and child rearing. Like the traditional and modern perspectives, this emergent perspective on fathering also seems to be reflected in the behavior of at least some fathers. As Lynn (1974) says of contemporary men, "Many fathers are more genuinely motivated to meet the needs of their children, are less likely to impose their own ambitions and hopes on their children, are less authoritarian and arbitrary, and are much less austere and unapproachable than fathers of the recent past."

Side-by-side with the emergent perspective on fathering, controversy continues as to whether or not males are biologically unsuited to include child care activities in their parent role. Fein notes that arguments that men are inherently limited in child-rearing capacities have generally drawn on studies of infrahuman animal species to support the position that parenting (caretaking) behaviors by males are rare. However, after reviewing the relevant data, Howells (1971) concludes that "the main lesson to be found from the study of care given to young animals is that nature is flexible."

In a more recent paper on the question of why men's family roles have been so limited, an even stronger rejection of the biological-evolutionary argument can be found. Reviews of several studies relevant to the biological-incapacity hypothesis in other animal species reveal considerable diversity within the male primates in the role of child rearing and suggest that parental behavior does develop in males simply with exposure to newborns. Furthermore, human fathers in controlled situations feel and act toward their infants in ways hard to distinguish from the ways of mothers. Thus, in general, it is more plausible to say that man's low level of child care occurs *despite* his innate capacity to nurture infants and not because of some biological incapacity (Corfman, 1979).

As this discussion suggests, conceptions of fathering have varied as a function of changing social realities and the particular types of father-child interactions on which investigators have focused as well as their own theoretical frameworks.

Parenting and Parent-Child Bonding

Rossi (1977) believes that the process of bonding with children may come more "naturally" to women than to men because of biologically based differences between the genders rooted in evolution. This notion that women are biologically better prepared than men to mother (in the sense of looking after or caring for children "as a mother does") has been endorsed by others as well.

Not all investigators, however, conceptualize parenting as a biologically based process with an evolutionary history that may—or may not—favor the female gender. Proponents of some perspectives think of parenting simply as a set of behaviors by which the family socializes children into the gender, cultural, and economic roles that the parents and/or society deem appropriate.

While such conceptions avoid biological-evolutionary presumptions and make the question of possible parenting differences between mothers and fathers an empirical one, they are also subject to shortcomings of their own. Definitions of parenting as a set of behaviors omit at least two elements that seem important to our understanding of parent-child relationships. First is the matter of *affective bonds* between parents and children. Workers in institutions, and even many parents, may perform the minimum necessary caretaking activities (e.g., feeding, washing, changing) for the children in their care, and yet we might say that these behaviors lack the "mothering" quality that seems to be a major component of "true parenting." The idea that there is more to family relationships than patterns of behavior is not a new one. In the literature on attachment, a distinction has been made between the attachment bond and attachment behaviors (Rutter, 1974). From this perspective, commonly studied infant behaviors such as smiling, greeting, and protest on being left are seen as signs of a specific and persistent bond but not as constituting the bond. Attached infants may show a number of predictable behaviors when their mothers leave and rejoin them, but the actual attachment relationship goes beyond these behaviors to include feelings that can only be inferred from overt behavior. Similarly, we believe that the conception of parenting should include some acknowledgment of the emotional components of the parent-child relationship.

A second shortcoming of conceptions in which parenting is defined as a set of parent behaviors is that the extent of variation among parental behaviors toward different children is ignored. Parents may "parent" in very different ways when they are dealing with their second child as compared with their first, or with a boy as compared with a girl, or with a sickly child as compared with a healthy one.

Rutter (1974) summarized a number of prevailing opinions to formulate a conception of parenting with the following components: (a) existence of emotional bonds and relationships with children; (b) presence of a secure base from which children can explore (and return to at times of distress); (c) availability of models of behaviors and attitudes; and (d) establishment of communication networks by which children can set standards, establish norms, develop experiences and let their ideas grow. While this view of parenting goes well beyond notions of parenting as a set of behaviors, it leaves unstated the powerful role of a number of variables that shape the form of the parenting process. As Klaus and Kennell remark, "A mother's and father's actions and responses toward their infant are derived from a complex combination of their own genetic endowment, the way the baby responds to them, a long history of interpersonal relations with their own families and with each other, past experience with this or previous pregnancies, the absorption of the practices and values of their cultures, and perhaps most importantly how each was raised by his own mother and father. The mothering or fathering behavior of each woman and man, his or her ability to tolerate stresses, and his or her need for special attention differ greatly and depend upon a mixture of these factors." (See also Chapter 5.)

The major perspective underlying our treatment of parenting is *transactional*—a perspective that is now apparent in and helping to synthesize the work of social scientists from areas such as psychology, sociology, and medicine. One aim of this emerging synthesis is a fuller understanding of individuals and their relationships, stresses, and health and illnesses as these vary across environments and the life span. According to this perspective, children are not passive creatures whose personalities and behaviors are simply the product, for better or worse, of parental acts of loving or hurting. Nor are parental practices seen as the crystallized and static products of parental histories or instincts. Instead, both parents and children are seen as dynamic and interacting individuals whose mutual influence on each other is affected by complex forces, including their own characteristics, their relationships with each other, and the broader contexts (e.g., social, economic, and historical) in which they interact.

Family Interaction

Although a transactional perspective is being applied in investigations of a number of issues, family interaction has been given considerable attention both as a source of influence on particular aspects of development (e.g., cognition) and as an important area of research and theory in its own right. A number of investigators support the notion that the family is the crucible within which individuals learn to behave toward others—in both caring and hurtful ways. While there has long been recognition within both the psychological/sociological academic community and the general public that families are important to individual development, social scientists have, until recently, lacked adequate sets of conceptual and methodological tools for analyzing family functioning.

In a similar vein, Lerner and Spanier (1978) note that the reciprocal dependency of family development on individual development has, until recently, been largely unrecognized. Our own recommendation that individual and family developmental processes be analyzed simultaneously is consistent with their contention that not only is the effect of the person on others moderated by his developmental level but this effect and the feedback received are influenced by the developmental levels of the other people in the social relationship.

TOWARD A MODEL OF PARENTING

Building on a transactional view of human functioning, our model of parenting is multidimensional and multidirectional. We see parenting as a process, an ongoing relationship that is affected by characteristics and experiences of all parties to the relationship. It is possible, empirically, to study mothering and fathering separately, but even if the focus is on mothering, one must consider not only the characteristics and behaviors of mother and child but also the mother's relationship to and interactions with the father (and with other children if there are any in the family). Moreover, a full understanding of the mothering (or fathering) process also demands a consideration of the interpersonal, social, historical, and economic contexts in which the relationship is embedded.

What, then, are the essential components of the parenting process? The parent's own characteristics and experiences certainly contribute one vital component to the parenting process. For example, a father's capacity to nurture and protect his child may be influenced by these aspects of his own upbringing. Similarly, a mother's capacity to foster the healthy development of her child may be influenced by such characteristics as her self-esteem, level of depression, and knowledge of developmental norms. Parents bring such characteristics as level of maturity, personal histories, habitual modes of interrelating, attitudes toward child rearing, and personalities to their relationships, and it would be surprising if such attributes had no effect on those relationships.

Although much of the research on family relationships has focused on dyads (e.g., mother-child, father-child, husband-wife), behavior within dyads is undoubtedly influenced by each member's other relationships. A mother may interact differently with her daughter after the father loses his job and starts drinking heavily. The relationship between father and son may change when the mother faces an identity crisis and decides to start her life over elsewhere. Relationships with extended family members may eliminate or intensify problems of interaction between father and daughter.

Children also, however innocently, bring their own characteristics to their relationships. However nurturant a personality a mother may possess, she may not be able to respond as warmly and nurturantly to a handicapped infant as she would to a normal child. And however much a father may have looked forward to the birth of his baby, he may not be able to respond as positively and enthusiastically to a premature as to a full-term child. Moreover, infants vary widely in temperament and such temperamental differences may have an impact on the parenting process. Even the child's gender may influence the way the child is handled—physically and psychologically—by the parents.

Family relationships do not exist in a vacuum, and interactions among family members may vary in content or meaning as a function of such variables as historical era, geographical location, and social-cultural values. Rearing children in a friendly neighborhood may be qualitatively different from rearing children in a hostile neighborhood. Democratic child-rearing practices may be applauded and supported by some groups but rejected in favor of autocratic methods in other groups.

Finally, parent-child relationships are not fixed and static but may change over time. A little boy may be no more energetic at age 4 than he was at age 2, but his mother may find herself more exhausted from their interactions when he is 4 than when he was 2. A handicapped child may be no more handicapped at age 2 than at birth, but just when the parents believe they have accepted their child's limitations, new challenges may arise that affect the quality of parenting. From the perspective of a professional given to making such judgments, the same mother may be a "good mother" when her child is 3 and a "bad mother" when her child is 5 for any number of reasons— changes in her relationship with her spouse, economic or personal setbacks, a move from a friendly suburb to a seemingly hostile inner-city neighborhood. Within every component of the parenting process there is an opportunity for the relationship to go awry. To some extent, the factors that place children "at risk" (e.g., for maltreatment) are factors that are inherent in the nature of the parenting process—in characteristics of parents, of children, of their interactions, and of their environments.

REVIEW OF LITERATURE

PARENTAL CHARACTERISTICS

As already indicated, a number of investigators assume that parents bring characteristics of their own to their interactions with their children and that these characteristics have an impact on the relationship. A number of assumptions seem like "common sense," for example, that parents may behave differently toward wanted than toward unwanted babies; that men and women who were maltreated during childhood by their parents will have difficulty nurturing their own children; that some people have kind, warm, patient personalities that will enable them to be kind, warm, and patient toward their offspring; and that young inexperienced parents will be less successful in their interactions with their children than older more experienced parents.

While such assumptions may seem highly logical, there is surprisingly little in the way of clear research data to support them. Moreover, because fathers traditionally have been seen as secondary parents at best, there are even less data concerning paternal than maternal characteristics with regard to how these variables affect the parenting relationship. Because most investigators have studied either mothers *or* fathers, and rarely both, we will also consider the literature on mothers and fathers separately, before reviewing the ways in which their interactions with each other can affect their relationship with their babies.

Maternal Characteristics

Attitude-Behavior Relationship.   The literature on mothering is replete with the common-sense assumption that a mother's attitudes toward a child will affect her mothering behavior. What evidence supports this attitude-behavior relationship? Moss (1967) found that mothers' responsiveness to 3-week-old infants was positively related to their "acceptance of the nurturant role" as assessed through an interview completed prior to the baby's birth. Robson, Pedersen, and Moss (1969) found that the frequency of mother-infant gazing at one month post partum was positively related to the mother's "interest in affectionate contact" as measured before the baby's birth. On the other hand, Thomas and others report that child-rearing attitudes expressed before the child's birth were not predictive of the mother's ability to adjust to a baby with a difficult temperament.

Several other studies using self-reports of parenting attitudes indicate some limited evidence of a relationship between certain maternal attitudes and certain maternal behaviors, at least among middle-class mothers. (The relationship of social class to maternal attitudes and behavior will be considered in more detail later.) A reasonable question to ask about this relationship is which comes first, attitudes or behaviors—or do they evolve together?

A longitudinal study by Clarke-Stewart (1973) sheds some light on this question. In a nine-month investigation of 36 first-born children from relatively poor families, both black and white, 12 visits enabled the investigators to make repeat observations of mother-infant interactions; interview mothers about themselves and their babies; and perform tests of the babies' language competence, cognitive development, and social and play behaviors. From the wealth of data gathered on maternal characteristics, Clarke-Stewart was able to determine not only which attitudes and behaviors were related but also the extent to which direction of influence was detectable. Through sophisticated statistical analyses, she found that an initially positive attitude toward infants caused mothers to be less directive toward their own children later on, and to engage in a great deal of physical contact with the baby. On the other hand, the behaviors of playing with the child and providing

stimulation through objects led to a more positive attitude toward the child later. "This suggests one interactional pattern wherein maternal attitudes affect maternal behaviors which the mother imposes on the child, but at the same time, maternal attitudes are influenced by the mother's playful and stimulating contact with the child. This demonstrates the complex reciprocity of mother-infant interaction." Despite her small and relatively disadvantaged sample (and the limits on generalizability that it imposes), Clarke-Stewart's study provides one of the richest pictures available of the ways maternal and infant characteristics interact in the mother-infant relationship. (Data on infant characteristics and the ways these contribute to the relationship will be considered later.)

Another rich and significant set of data comes from a longitudinal study of 84 relatively advantaged couples (mostly middle and upper-middle class) followed by Grossman, Eichler, and Winickoff (1980) from the first trimester of pregnancy through one year after the birth of the child. In this study of adaptation to pregnancy, birth, and parenthood, Grossman and her research team assessed families five times during an 18-month period on a variety of interview, questionnaire, test, and observational measures. The findings of this ambitious and valuable study are relevant to many dimensions of parenting to be reviewed in this chapter. Of particular interest here is the finding that women's antepartum feelings about pregnancy and infants were *not* related to their interactions with their babies as observed at two months post partum. However, by one year post partum, there was some evidence that quality of mothering at one point in time could be predicted from earlier measures of the mother's psychological functioning as well as her age. However, Grossman et al. also point out that "it seems most likely that each aspect of the system—mother, baby and couple—both enhances and is enhanced by the other aspects. Conversely, when one aspect, such as the marriage or the baby's temperament, is somewhat troublesome, all other aspects of the system are affected." This hypothesis is a nice statement of a transactional perspective, and all the aspects mentioned will be considered later in this review.

Maturity and Knowledge of Child Development.   Another type of characteristic assumed to be related to parenting behavior is knowledge of and expectations concerning normal child development. In a longitudinal study of 37 young couples who married in high school, DeLissovoy (1973) collected data over a three-year period on a number of marriage and parenting issues. After almost all the subjects had had their first child, they completed a questionnaire on developmental norms. According to DeLissovoy, "The responses of both parents revealed that knowledge of basic norms was sadly lacking. The answers were skewed to an unrealistically early expectation of development." Moreover, "these young mothers with a few notable exceptions were impatient and intolerant of their children." These observations were consistent with the finding of Sears et al. (1957) that younger mothers tended to be somewhat more severe in their treatment of young children, more irritable and quick to punish. Both DeLissovoy's and Sears' findings are consistent with Grossman's observation that among first-time mothers, younger women were making less successful adaptations to parenthood.

Additional evidence concerning maternal age and lack of knowledge of developmental norms comes from a study by Epstein (1979) indicating that teenage mothers are deficient in their knowledge of child development and that this lack of knowledge may affect their interactions with their children.

Effect of Mother's Own Childhood. When we speak of what a mother brings to her relationship with her baby, clearly one of the variables that has seemed most important is the mother's own childhood experience. However, Rutter and Madge conclude from a review of the literature that "what little evidence there is suggests that, within the normal range of child rearing, there is only slight direct intergenerational continuity" (1976). Despite this conclusion, it seems unlikely that the assumption that childhood experiences with one's parents will affect one's own parenting will be given up easily—and, indeed, despite the absence of strong findings there may be a relationship (Aldous, 1978).

Assessing Predictors of Parenting Processes. If we had adequate measures of psychological maturity, we might find powerful predictors of parenting processes. According to the Group for the Advancement of Psychiatry (GAP) (1973), "When the individual is physically mature enough to have a child and psychologically mature enough to gratify both the child's needs and his own, he is ready to become a parent." Clearly not everyone who actually has a child has reached such a level of maturity, yet the notion of psychological maturity as a useful antecedent to parenting makes sense—as does the GAP assertion that parenthood itself is a developmental process and that the child has an influence on the personality development of the parent.

Unfortunately, there is little empirical evidence for the relationship between any index of parental maturity (other than chronological age) and parents' relationships with their children. One promising line of research comes from the work of Carolyn Newberger (1980), who has developed a theory of Parental Awareness and an interview approach to a broad range of parental tasks that discerns a series of stages of parental development. Responses to questions on such issues as establishing and maintaining communication and trust, discipline and authority, and learning and evaluating parenting reveal parental conceptions that can be ordered hierarchically into four increasingly comprehensive levels, i.e., an egoistic orientation, a conventional orientation, a subjective-individualistic orientation, and a process (or interactional) orientation. As part of an effort to investigate the relationship between levels of parental awareness and actual parent behavior, Newberger has compared the interview responses of parents with a recent history of child abuse or neglect with responses of a matched group of parents with no history of dysfunction in the parent-child relationship. Her data provide useful insights into the potential interaction of social stress and parental awareness as factors contributing to child maltreatment and are discussed in Chapter 14.

Overall, what can we conclude about the impact of maternal characteristics on the mother's role in parenting? There is some evidence for relationships between maternal attitudes and behaviors, although it should be noted that all the evidence reviewed was basically derived after the fact, that is, none of the particular attitude-behavior relationships that emerged through statistical analysis had been predicted in advance of data collection.

Data on maternal personality variables related to mothering are even more limited than the attitude-behavior data. There is some evidence that mothers who scored themselves high on characteristics considered indicative of control and extroversion view their babies differently and interact with them differently than do mothers who scored themselves low on these dimensions. Unfortunately, one shortcoming of most investigations aimed at identifying relationships between maternal personality characteristics and mother-child interaction is that they have not disentangled characteristics of mothers from characteristics of children and of the mother-child interaction. As we shall see in a later section, maternal characteristics may be modified by infant characteristics, and both mothers and babies may undergo changes as their relationship evolves.

Although only a few studies addressed the issue directly, there is some evidence that a lack of knowledge about child rearing and child development is related to maternal restrictiveness and punitiveness. However, further research would be necessary to separate the effects of lack of knowledge from youth (in most studies the less knowledgeable mothers were also the younger mothers) and other variables related to the low social class of most of the women in the samples.

Finally, as for the impact on mothering of the

mother's childhood experiences with her own parents, there is more conviction than evidence for the notion that mothers treat their children the way their parents treated them. After reviewing studies of "normal" and "abusing" parents, Rutter and Madge (1976) concluded that intergenerational continuity seems to be least strong within the range of normal parenting relationships and most strong when parenting is seriously abnormal. However, Rutter and Madge also note that more data are needed for a valid assessment of intergenerational continuity in parenting behaviors.

Paternal Characteristics

Although a growing literature provides evidence concerning father influences on child development and the father-child interaction, few studies directly examine the relationship between characteristics of fathers and the nature of their parenting activities. The paucity of research in this area is probably related both to historical realities (lesser involvement of fathers than mothers in child care, especially during infancy and early childhood) and to the assumptions of researchers (that personal characteristics of fathers, such as a concern with masculinity, are among the very factors that have kept fathers from engaging more actively in early child care). Almost no studies provide data about the relationship between the attitudes and personalities of fathers and their parenting behavior—although there has been considerable speculation about such relationships. Starting with the observation that fathers spend much less time than do mothers in child care activities, especially during the child's infancy, investigators have asked "What are the characteristics of fathers that have prevented them from participating more in the care of their babies?" While acknowledging the culturally prescribed and approved involvement of most men in full-time employment, some authors have suggested that parenting is incompatible with the masculine image in America.

Other authors have argued that "good" parenting is facilitated by an androgynous identity, one in which individuals view themselves as high both in traditionally "masculine" attributes such as initiative and independence and in traditionally "feminine" attributes such as nurturance. Even within this perspective, men are viewed as having characteristics and attributes that place them at a disadvantage when it comes to parenting. According to Weinraub (1978), "Since the characteristics required of the ideal parent are androgynous, rigidly sex-typed individuals of either sex may have difficulties adapting successfully to the parental role. However, the average woman can adapt to the parental role simply by adding some

"masculine" characteristics without threatening her essential femininity. . . . For the average male, this may not be so easy. . . . Just as women in our society have few models of successful career women, so do men in our society have few models of successful modern-day male parents."

No matter how limited the actual studies, it seems clear that there are individual differences among fathers in the extent of their involvement with their children. As noted above, one widespread explanation for variability in fathering focuses on the issue of masculine identity. Russell (1978) examined the relationship between the father's sex-role identity and his involvement with his children. Noting that "little study has been made of the critical factors that are associated with whether and how a father will interact with his children," Russell interviewed 43 Australian mothers and fathers participating in a large parent-child interaction study. With this sample of Australian families, patterns of father-child interaction were identified that were similar to those reported by Kotelchuck for families in the United States. That is, on the average, fathers spent less than an hour a week on each of seven tasks (feeding, dressing, changing, bathing, attending at night, reading stories, and helping with schoolwork) but more than eight hours a week on play.

In Russell's study, individual differences in father involvement in these activities were related to the father's sex-role identity. Using Bem's Sex Role Attitude questionnaire, Russell found that fathers classified as androgynous and feminine were much higher on both day-to-day child care and on other interactions (play, story reading, and helping with schoolwork) than were fathers classified as masculine or undifferentiated. Although Russell's sample is small and select, and the methodology limited in the reliance on self-report measures, the study provides valuable support for the notion that fathers' involvement in parenting may vary as a function of self-perceptions concerning characteristics related to the sex role. The finding of Grossman et al. (1980) that men who described themselves as having more feminine traits tended to be more comfortable with themselves two months after the birth of their children is consistent with Russell's formulation.

Tasch (1952) provides some self-report data on fathers' perceptions of their sons and daughters as well as their interactions with them. In this study, fathers described themselves as viewing their daughters as fragile and delicate. Consistent with this perception, the fathers also reported worrying more about their daughters' safety and engaging more actively in developing the motor abilities, skills, and interests of their sons. After the children reached the age of six, fathers reported engaging in much more rough-and-tumble play with boys than with girls. Clearly, more current investigations that consider actual behavior as well as self-reported behavior are needed to clarify the relationship between fathers' perceptions of their sons and daughters and their behavioral interactions with them—especially since, as we shall see, recent studies of father-child interaction are inconsistent on the issue of differential father interactions with sons as compared with daughters.

While there appear to be few objective and reliable studies available on the attitude-behavior relationship in fathers, there is some evidence to support the idea that the father's own childhood history and relationship with his parents contributes to his fathering. For example, Grossman et al. (1980) found that men who more strongly identified with a mother perceived as nurturant reported feeling more comfortable about themselves and their role as fathers in the postpartum period. Moreover, those so identified were also judged by observers as doing better in their interactions with their infants.

Similarly, Burlingham (1973) concludes from both observational and clinical data that "We have found that whatever handling the father has experienced from his own father, whether it has been loving, unfeeling, secure, lenient, understanding or inconsistent, this affects his own attitudes and behavior towards the child."

To summarize the material so far, there are only limited data on the ways in which characteristics brought by mothers to their interactions with their children influence these interactions. There is still less solid information on the ways father characteristics affect the parenting process. It is not even clear whether our ignorance concerning fathering is a necessary byproduct of a comparative lack of father involvement in parenting (until recent times) or simply a failure of investigators to recognize and study the correlates of important father-child interactions that were taking place.

INFANT CHARACTERISTICS

Since the seminal work of Bell (1968 and 1971), it has become widely recognized that infants have characteristics that contribute to their own personality development and to the relationships in which they participate. In response to the question "Mother/child: who influences whom?", Stevens and Mathews (1978) note that infants have characteristics and behaviors that appear to influence their caregivers. In particular, these investigators believe that infant characteristics such as adaptability and consolability seem to contribute significantly to maternal feelings of competence. Moreover, infant "readability" (that is, the extent to which the baby's behavior is differentiated and predictable) may influence the quality of mother-child interactions. Unfortunately, while many investigators acknowledge the child's active involvement in its own development and in the parenting relationship, both conceptual and methodological issues have complicated the effort to pinpoint salient infant characteristics and their effects (Blank, 1976).

Part of the difficulty in studying the contribution of infant characteristics to parent-infant interactions is that infants are undergoing rapid changes, some of which may be related to the ongoing development of the parent-infant relationship. Indeed, the effort to identify causal relationships between infant characteristics and parent-infant interactions may be misguided as well as doomed to failure. Clearly infants are born with many characteristics that are likely to have an impact on caretakers, e.g., the ability to cry, to suck, to wave their fists, to show gross forms of affect, to cling, to gaze. On the other hand, all these may be malleable, and the extent to which babies laugh or cry, share mutual gazes, or avert their eyes may vary as a function of this experience with their caretakers.

Babies do have some characteristics that are stable and that may be related in consistent ways both to their interactions with caretakers and to their own sociopersonality development. One such characteristic is gender. Moreover, there is some evidence that characteristics related to gestational status (premature or full-term) may have a powerful effect on the subsequent development of the mother-infant relationship.

Gender.   Whether boys and girls bring different characteristics into the world with them by virtue of their gender is an ongoing debate (Maccoby and Jacklin, 1974). Of particular interest to researchers concerned with the family have been the questions of whether parents respond differentially to their children as a function of the child's gender and whether they reinforce different patterns of behavior in boys than in girls. Although not all investigators who have looked for differential parental behavior as a function of child's gender have found it (e.g., Grossman et al., 1980), there does seem to be considerable evidence that both parents handle girls differently from boys at least some of the time. Sex differences in infants have been noted in the earliest months of life. The question is whether the sex differences noted are entirely a function of differential treatment of the sexes by caregivers or whether differential parental behaviors are also in response to subtle sex differences in the infants (Korner, 1978). Whatever the direction of influence, it seems clear that infant gender is a relevant component of the parenting equation. (The reader is referred to Chapter 10 of this volume for a detailed discussion of this subject.)

Gestational Status. Some investigators (e.g., Lamb, 1978) have suggested that premature infants are "at risk"—not just for further medical difficulties but for impaired relationships with parents that can lead to or include abuse. While we do not wish to focus on the possible relationship between prematurity and child abuse in this chapter, we do want to consider briefly some of the evidence concerning differences between premature and full-term infants and the implications of such differences for the parenting process. (See Chapter 14 in this book for data on the relationship between prematurity and abuse.)

Following up on a suggestion by Lamb (1978) that the cries of premature infants may be perceived as even more aversive than are ordinary infant cries, Frodi et al. (1978) showed videotapes of infants who were, in turn quiescent, crying, and quiescent to 64 parents. In accord with the experimental design, half the parents viewed a full-term newborn while the other half viewed a premature infant. Moreover, sound tracks were dubbed so that half the full-term and half the premature infants appeared to emit the cry of a full-term infant while the other half appeared to be emitting the cry of a premature infant. Neurophysiological and self-report responses of all parents to the videotapes were collected. Frodi found that the cry of the premature infant elicited greater autonomic arousal and was perceived as more aversive than the cry of the full-term infant. This effect was particularly strong when the cry of the premature infant was paired with the face of a premature infant. There were no sex differences on physiological measures, but responses to the mood adjective checklist revealed that fathers felt more sympathy overall for the crying infants than did mothers. Fathers reported feeling most sympathetic to the full-term baby with its own cry and least sympathetic to the infants with incongruous characteristics, with the premature infant with its own cry falling in the middle. Mothers, by contrast, were most sympathetic toward the full-term baby with a normal cry and significantly less toward all others, especially the premature baby with the premature cry. Frodi et al. argue that through a process of generalization, some premature babies may become aversive to parents because of the aversiveness of their crying. It seems likely that if babies do come to be seen as aversive to their parents, the whole parenting process may be "at risk."

In a direct observational study, Field (1977) provides provocative data concerning a potentially maladaptive feeding interaction between multiparous mothers and their premature infants. Field's emphasis on the necessity of considering the conjunction of parental and infant behavior is important to understanding not only the potential consequences of prematurity but all parent-child relationships.

Also important is an analysis of the interaction over time: infants and parents change and so do their relationships. Rice (1977) provides evidence that when mothers of premature infants are trained to administer tactile-kinesthetic stimulation to their infants at home, their infants may show significant gains in neurological development, weight gain, and mental development over a period of just one month. Although it was beyond the scope of the study to analyze mother-infant interaction, Rice speculates as follows: "It seems reasonable to predict that a premature infant who achieves a more robust development would elicit responses of more confident coping behavior from his/her mother. This could set up cyclic interactions of stimulus-response patterns that could enhance the mothering behavior of the caretaker and the developmental behavior of the infant."

PARENT-CHILD INTERACTIONS

Our rather circumscribed discussion of infant characteristics reflects the aforementioned difficulty of separating attributes that infants bring to the parent-child interaction from those that develop within those interactions. Recognizing that relationships can have dynamic characteristics of their own that go beyond those of individual participants, investigators who focus on parent-infant interactions are frequently operating within a transactional, or *general systems,* framework. Because such perspectives provide a useful step beyond the notions of static individual characteristics and static relationships, we believe it is useful to give considerable attention to the conceptualizations themselves as well as presenting relevant data.

Thoman and Freese (1980), working within the framework of general systems theory, note that a major assumption of their model of mother-infant interaction is that it constitutes a communication system from the time of the baby's birth. The use of the term "systems" to describe the mother-infant interaction reflects the belief that mother and infant mutually influence and provide feedback to one another. Thoman and Freese argue that traditional notions of causality (e.g., mother attitude A causes infant behavior B) are inapplicable in the systems model, although other noncausal forms of determinism are not precluded. Mother and infant can influence each other, but the influences are bidirectional (if not multidirectional) and change over time. They (like Sander,

1969) see the developmental process as essentially integrative, a process of synthesis of highly complex determinants.

Parental Responsiveness. Proponents of a transactional (or systems) model of parenting often employ the concept of *sensitive mothering* or *maternal responsiveness* to refer to optimal expressions of the parenting process. For example, Hinde (1979) states: "Appropriate sensitivity to the baby's needs to interact or to withdraw from interacting are an important aspect of mothering—especially as the baby's needs change from day to day. . . . But natural responsiveness is neither ubiquitous nor indiscriminate. As mentioned already, babies differ markedly in the stimuli they provide and in their responsiveness to adults, and adults show similar divergence (e.g., Bennett, 1971; Sander, 1969, 1977)."

While acknowledging the constant feedback and change built into parent-child interactions, researchers have often tried to identify the *direction* of influence for particular sets of parent and infant behaviors. For example, in a study of individual differences in infant responses to separations from and reunion with their mothers during the fourth quarter of the first year of life, Stayton and Ainsworth (1973) focused on the extent to which the infant's separation protest was associated with maternal unresponsiveness to crying and to maternal insensitivity to infant signals. They noted via a statistical pattern of cross-correlations among measures that the extent to which the mother was unresponsive to the baby's crying had a greater effect on the extent to which the baby subsequently cried than vice versa. "It seems likely that a baby's confidence in his mother's accessibility is built up in the course of the first year largely through his mother's consistency and promptness in responding to his signals, including his crying, and that infants who are chronically anxious about their mother's whereabouts are those whose crying signals have often fallen on deaf ears."

While Stayton and Ainsworth focus on the mother-infant dyad, parental responsiveness is a construct that can be applied to fathers as well. Reporting on the father's sensitivity to an auditory signal in the feeding context, Parke, Power and Fisher (1980) noted that fathers are as sensitive and responsive as mothers to infant cues. They also noted that fathers are just as responsive as mothers to other newborn infant cues such as vocalizations, concluding that "interactions between fathers and infants—even in the newborn period—is clearly bidirectional in quality; both parents and infants mutually regulate each other's behavior in the course of interaction."

Eye contact, smiling, and expression of positive or negative affect (e.g., pleasure or distress) seem to be important components of the early parent-infant relationship. In the previously discussed longitudinal study by Clarke-Stewart (1973), the infant's expressions of happiness were found to be most clearly related to the mother's expression of positive emotion. Thus, happy, loving mothers had children who were affectionate and smiling. The child's expression of attachment for the mother—e.g., smiling at her, looking at her, following her, giving to her—showed a high correlation with the frequency of her own social behaviors with the child—e.g., looking, talking, playing.

Interested in the direction of influences, Clarke-Stewart analyzed child and maternal behaviors over time with the cross-lagged correlation technique. She found that the child's looking, smiling, and vocalizing to mother when mother was in the room led to mother and child spending more time together. The relationship between maternal responsiveness to distress and infant irritability could not be assigned a causal direction in the Clarke-Stewart study. However, as noted earlier, Stayton and Ainsworth (1973) have provided evidence that maternal responsiveness to distress reduces the child's tendency to cry.

These studies thus provide evidence that some maternal tendencies have a determining influence on some child behaviors and some child tendencies have a determining influence on some maternal behaviors. Others have noted parent and child behaviors that seem consistently bidirectional in influence. Evidence for such bidirectional influences comes from a study by Robson (1967) of eye-to-eye contact as a two-way process of communication. Certain infants from the beginning vigorously sought out their mother's eyes, while others avoided them. The baby's looking behavior appeared to be an important eliciter of maternal responses. However, mothers also differed in the quantity of their visual activity, independent of the infant's behavior. Consequently, the resulting patterns of eye contact were established by both mother and infant behaviors.

Robson and Moss (1970) focus particularly on infant propensities in identifying circumstances in which early parent-infant interactions may go awry. "It is likely that a baby's infantile characteristics, such as his size, helplessness, fussing and crying, help initiate caretaking behavior in his parents. . . . But if, after the first four to six weeks, he fails to smile and look at his mother, this need-fulfilling interaction does little to reinforce her attachment to him." They go on to say that if the helplessness of infancy continues without the countering influence of eye contact and smiling, the mother's love and pity may turn to disenchantment and anger. Such an outcome seems possible even if certain characteristics of the mother (e.g., anxiety, depression) contribute to the infant's avoidance of eye contact and lack of smiling.

One thing that seems clear from a review of the literature is that successful or unsuccessful parenting relationships can never be predicted solely on the basis of either parental or child characteristics. Unfortunately, once a negative interaction is set up, it may have long-term consequences that are harmful to both parents and children.

A number of investigators (e.g., Korn et al., 1978) have reported that a temperamentally difficult child may produce disruptions in the parent-child relationship and in marital and family functioning generally. (See also Chapter 10.) Werner and Smith in a longitudinal study of 660 children on the island of Kauai, Hawaii, found that many mothers who described their infants as having difficult temperaments also reported having major concerns over child rearing. However, at every data point in this long-term longitudinal study there were mothers who reported having difficult children but no major concerns or adjustments as well as mothers who described their children as having average temperaments but who nevertheless saw themselves as having major concerns and adjustments to make. Moreover, Werner and Smith found a tendency over time for many infants to change from difficult to average in temperament, and vice versa, and an even greater instability over time in maternal concern and family adjustment. They warn clinicians not to prognosticate about temperamental difficulties in infants early in life and suggest that it may be more valuable to identify mothers who have major concerns despite the fact that their infants are not classified as difficult. The complexity of the issue of infant temperament is discussed in detail in Chapter 10.

Interactive Play.   Opportunities for play provide an important context for the development of parent-infant interactions. Arguing that mother-infant games constitute an interactive sequence with a developmental course of its own, Crawley et al. (1978) videotaped mothers interacting with infants ages 4, 6, and 8 months.

They interpret their data as providing support for the notion that mothers attempt to encourage the optimal level of infant participation during play. "The four-month-olds' lack of motoric skill restricts their form of participating to smiling, gazing, and vocalizing. The mothers of four-month-olds in the present study, therefore, used games that had a high probability of eliciting positive affective responses while requiring little infant motoric skill. The older infants' ability to coordinate motor activity, in contrast, enabled them to participate in play involving imitation and individual performance of a motor behavior." Also

of interest was the finding that mothers who played a greater number of games with their 8-month-old infants had infants with higher scores on two measures of social responsiveness (smiles and visual regard).

What of the father's participation in playful interactions with his child? Yogman (1980) believes that the ways fathers play with children have important implications for the child's development. On the basis of his own and other studies, Yogman argues that there are qualitative differences between mothers and fathers in the way they interact with their children and, specifically, that fathers engage in more vigorous and exciting play.

He concludes that interactions between fathers and infants commonly involve physical play and that games with mothers and with fathers provide differential experiences for infants. "Conventional games, more common with mothers, may allow the establishment and consolidation of rules of interchange that provide the foundation for later language development while more arousing physical games, more common with fathers, may differentiate into alternative forms of social play, eventually incorporating objects and later leading into further instrumental activities. . . ."

Generally, playful interaction may further the developing relationships of parents and children; in some cases, however, playful interactions may fail to occur or may become aversive because of characteristics of infant, caretaker, or both. Yogman believes that the father's typical style of playful interaction, so positive for the normal infant, may lead into a different cycle of interaction with premature infants. "Given the difficulties of social interaction with a premature infant whose state and motor organization may be labile (Goldberg, 1978; Field, 1977) and whose cry seems more aversive to father and mother, the tendency of many fathers to excite, play with, and vigorously stimulate their infant may stress an already vulnerable infant, interfere with social interaction, and lead a father to withdraw."

Beyond the Parent-Child Dyad.   Most of the studies reviewed in this section are valuable in their emphasis on the bidirectionality of influence within parent-child dyads and the ways that both parent and child characteristics can develop and be modified over time as a function of the parent-child interaction. While such an emphasis represents an important advance in conceptualizing about parenting processes, it omits important contextual elements. First, parent-child dyads typically are embedded in larger family networks, and second, these networks themselves function within particular experiential, environmental, and sociocultural contexts. After considering the various triadic influences of mothers, fathers, and infants in their relationships with one another, we will consider broader environmental and sociocultural influences on family functioning.

MOTHER-FATHER INTERACTIONS

Even before researchers began recognizing that fathers were parents too and that at least some of them might be developing relationships with their children, it was widely assumed that fathers had at least an indirect effect on their offspring through their interactions with their wives. This notion continues to be a popular one. That is, many people assume that if a mother is caught in an angry, conflictual relationship with her husband, she will have difficulty being loving and nurturant with her children. Conversely, it is assumed that warmth and support from a caring husband will make it easier for mothers to adapt to the challenge of child-rearing.

While it seems reasonable to assume that positive husband-wife relationships will facilitate relationships with children and negative ones will impair many aspects of family functioning, the actual data are somewhat inconsistent. Pedersen (reported in Bronfenbrenner, 1979) found that when the father was supportive of the mother, the mother was more effective in feeding the baby. Conversely, marital conflict was associated with inept feeding on the part of the mother. On the other hand, in a study of lower-class mothers' behaviors and attitudes toward child-rearing, Zunich (1971) found no support for the notion that marital conflict is reflected in the mother's interaction with her children. Unfortunately, Zunich's observations of mother-infant interaction were made in a laboratory setting, and it is difficult to know whether these interactions were a valid representation of the range of interactions that took place in the home.

Any consideration of triadic interactions among fathers, mothers, and infants should give some recognition to the impact of the infant on the husband-wife relationship. Dyer (1963) found that the birth of the first child initiated a "crisis" period for couples and that the degree of crisis was related to differences in the strength of the couple's relationship before the child was born. Yalom (1968) reported that two-thirds of all postparturitional women experience some sort of postpartum blues during the 10 days following delivery.

The notion that the father has an influence on mother-infant interaction from early in life has received some other direct empirical support in a number of studies. Feiring and Taylor (cited by Parke et al., 1979) found that mother-infant involvement was positively related to the mother's

support from a secondary parent; in this study, 67 per cent of the "secondary parents" were the children's fathers. Moreover, at least for adolescent mothers, it may be particularly important that the secondary parent be the child's father and not another man. In comparison to those who married the father of the child, adolescent mothers who married other men were less confident in their parenting role, had more behavior problems with their children, and were more critical of their children (Furstenberg, 1976).

A review of a number of relevant studies by Parke, Power, and Gottman (1979) supports the conclusion that "consensus in childbearing attitudes, the father's perception of the mother's caretaking competence, and other qualities of the husband-wife relationship are all related to maternal involvement or competence."

While there has been considerable interest in the father's impact on the mother and the mother-child relationship, some investigators have recognized, wisely, that women are not necessarily passive creatures with no reciprocal influences on their husbands. Moreover, whatever the husband-wife relationship may be like before the birth of the child, the arrival of a baby is bound to have some impact. For example, a whole new area of potential spousal agreement/disagreement is opened up. As Knox (1979) says, "Because parents are different ages, have been reared differently themselves, and have been exposed to different people and environments, they view the parental role differently."

In a thoughtful study of the interrelation of parental and spousal behaviors, Belsky (1979) conducted two 2-hour observations of 40 middle-class families with infants 15 months of age. Observations were made of two classes of behavior: (1) *parental*—i.e., talking to, verbally responding to, teaching, stimulating, restricting, playing with, and holding the child; and (2) *spousal*—i.e., talking about baby-related and nonbaby-related matters and sharing pleasure (e.g., smiling) with regard to the baby's activities. After the systematic observations, both father and mother were rated on several dimensions: (1) intensity of positive affect displayed in interacting with the child, (2) cognitive stimulation to the child, (3) spousal harmony, and (4) facilitation of the three-person interaction (i.e., the extent to which either husband-wife interactions were organized to include the baby or parent-infant interactions were organized to include the other parent).

Belsky's findings suggest that wives may be more influential in involving their husbands in parenting than vice versa. Wives who talked a lot about the baby had husbands who engaged the baby in cognitively stimulating verbal interaction, physical contact during play, and object-mediated play—all while in the presence of their wives. Husbands and wives who conversed frequently about their child were also rated high in intensity of positive affection and on cognitive stimulation. Belsky also found that in some families, involvement of the mother and father in discussion of nonbaby-related topics seemed to preclude active involvement with their babies. In other families, attention directed toward the baby provided the basis for a pleasurable interaction between husband and wife.

On the basis of these findings, Belsky proposed a conceptual model that is quite consistent with the general transactional model being used to organize the current review of literature. In this model, "(1) the transition to parenthood influences the spousal relation which (2) in turn influences and is influenced by the parent-infant relation which (3) itself influences and is influenced by the infant's development and which (4) coming full circle, influences and is influenced by the husband-wife relationship."

Although few studies provide data relevant to more than one or two elements in this conceptual model, the Grossman, Eichler, and Winickoff study (1980) of pregnancy, birth, and parenthood does have evidence relevant to almost every step of the model. On the basis of interview and questionnaire data obtained two months post partum, Grossman et al. report that marital adjustment seems to be "jostled considerably" by a first pregnancy and birth but not by a later pregnancy and birth. "For first-time mothers . . . the role of parent is a new one and the style and quality of that role seems to depend more strongly on the psychological and experiential resources (or lack of them) with which the woman and her husband approach it. Of the primiparous women, those who were older, those with greater marital satisfaction in early pregnancy, and those with greater motivation for the pregnancy at eight months reported themselves happier with their husbands at this time."

Fathers in the Grossman study also provided evidence at two months post partum on the impact of the baby's arrival. The most common source of stress reported was the lack of time they had alone with their wives—a problem handled better by some couples than others. Among the predictors of men's marital satisfaction at two months post partum was their marital satisfaction early in the pregnancy and the degree of difficulty their wives had during labor and delivery.

Block and coworkers (1981) provide evidence that the influence of husbands on the parenting process begins even before the birth of the child. Describing husbands as the "gatekeepers" of childbirth, they note that it is generally the husband who determines whether or not his wife takes a childbirth preparation class, the kind of class, and the degree to which she uses this education during labor and delivery. This role is important because the husband's willingness to participate turns out to be critical in decisions affecting the physical comfort and satisfaction of his wife's birth experience.

The reader is referred to Chapter 4 of this book for further consideration of the development of the parenting role.

FAMILY CONTEXTS

As already noted, family interactions do not take place in a vacuum. Bronfenbrenner (1977) has argued that an understanding of human development requires examination of multiperson systems interacting at a number of different ecological levels, including (a) the immediate setting in which individuals are being studied (e.g., home, laboratory); (b) the relationships among the major settings in which the individual participates (home, school, work, hospital); (c) the major institutions of society (e.g., the neighborhood, the mass media, agencies of government); and (d) the overarching patterns of culture or subculture (e.g., economic, social, educational, and legal systems). To be more concrete, we would suggest the following specific application of Bronfenbrenner's position: the way in which Mrs. Jones performs routine caretaking functions with her new daughter is influenced by *all* the following as well as additional factors: (a) whether she is observed at home or in a hospital or research setting; (b) whether there is adequate income to support a "livable" dwelling; (c) whether the neighborhood is safe or unsafe, friendly or alienating, convenient to public transportation or isolated; (d) whether government agencies involved with the family are seen as helpful or threatening, sympathetic or rejecting; and (e) whether the family is part of the dominant culture or a "subculture," whether society is seen as offering advantages or disadvantages, whether the powers that be are seen as for or against the family.

Clearly, it is rare that any research study encompasses even a small proportion of the interrelated influences. Researchers often seem to assume that Mother A will demonstrate Attitude A and Behavior A toward her infant regardless of whether or not her husband loses his job, her mother has a stroke, the family must move to a new neighborhood, an older child becomes a target for racial discrimination, and TV ads offer spectacles of plenty that seem increasingly unattainable. Even studies that stratify samples by race and/or social class generally do little to elucidate the experiences that actually affect the parenting process. Bronfenbrenner found it easier to find studies that illustrated by default the types of contextual considerations that were essential to ecologically valid research than studies that were exemplary for their consideration of variables beyond the isolated individual or isolated dyad. We will consider briefly some of the studies that provide useful starting points for more sophisticated multivariate research and the implications of some of these studies for thoughtful research and practice.

In Block's study of husband "gatekeeping" around the birth process, race and social class proved to be highly predictive of husband involvement in birth preparation and labor and delivery. Block's results suggested that "the greater power, accessibility to resources, and available time of higher status whites are at least as important [to involvement in the birth] as value and attitude differences." This study raises important questions; for example, why are higher-status white fathers more likely to participate in childbirth preparation and assistance than working-class and black fathers? Are attitudes toward education or assumptions about sex-appropriate behavior the most important influences? Or does the problem lie in prior exposure to the attitudes and behaviors displayed by members of the establishment who impart various kinds of education? And what are the implications of the childbirth experience for future mother-child, father-child, and mother-father interactions?

Other studes giving some attention to social-class variables related to parenting processes also raise more questions than they answer. Tulkin and Cohler (1973) report fewer correlations between parenting attitudes and behaviors in working-class than in middle-class mothers. They speculate that the working-class mothers see themselves as powerless with and overwhelmed by their children. These investigators conclude that greater consideration should be given to personality variables that may be mediating between maternal attitudes and behavior. We would argue that greater consideration should be given to the differential *meaning* that observers in the home may apply to working-class and middle-class mothers as well as to the validity of self-report attitude measures across social-class groups.

Sameroff and Chandler (1975) report the results of research indicating that in advantaged families infants who had suffered perinatal complications generally showed no significant or only small residual effects at follow-up. In lower social-class homes, on the other hand, many infants with identical histories of complications showed significant retardations in later functioning. What variables account for such social-class differences in outcomes for infants with perinatal complications? Is it actual or perceived accessibility of services? Is

it the availability of monetary or interpersonal resources? Is it a history of learned helplessness reinforced by association with institutions where "blaming the victim" is common? Answers to such questions are essential to our understanding of family functioning.

Studies of ethnic differences in parenting also raise major questions. In Clarke-Stewart's (1973) longitudinal study of mother-infant interaction, black and white mothers were equivalent in age, education, occupation, and the stimulation and variety of their homes. Moreover, the black and white mothers reported equally positive attitudes toward their infants in initial interviews and appeared equally accepting of their children during observations. Nevertheless, "there were, however, marked—and increasing—differences in what mothers and children were likely to be doing when they were together": white mothers looked at their children, talked to them, played with them more, and were more openly affectionate; black mothers, by contrast, were more restrictive and spent more time caring for their children's physical needs. Judging by children's immediate responses to maternal behaviors, black mothers' physical and social behaviors also were more effective than were those of white mothers. A difference in the philosophy of child rearing may be reflected by varying behavior in the two groups: black mothers emphasized the physical aspects of child care, white mothers emphasized the educational aspects. If there are consistent race-related differences in child-rearing attitudes, one can ask what experiences give rise to and maintain such differences, and what are the consequences of the differences for the developing parent-child relationship and for family functioning in general.

The impact on parenting of "nontraditional" family situations—e.g., where mothers are single or working—is only beginning to be studied. Moving beyond a simple comparison of employed mothers and homemakers, Hoffman reported differences in child-rearing behaviors between satisfied and dissatisfied working mothers. Mothers who enjoyed working were relatively high in positive affect toward their children, used mild discipline, and tended to avoid assigning household tasks to their children. In contrast, mothers who disliked working appeared less involved with their children, who were relatively more assertive and hostile (Hoffman, 1961).

In a more recent study of the impact of maternal employment on parenting, Henggeler and Bordvin (1981) studied 28 white mother-son pairs from two types of intact families—those where mother was employed outside the home and those where she was not. Children ranged in age from 4.00 to 4.92 and had attended a high-quality preschool for at least two years. All mothers rated themselves as moderately satisfied or extremely satisfied with their present role status. Henggeler and Bordvin (1981) observed mothers and sons on two occasions, once at home and once in a laboratory, partly in free play and partly executing experimental instructions. Maternal behavior, child behavior, maternal psychosocial functioning, and child psychosocial functioning were assessed. They found no significant main effects or interactive effects for maternal working status on any of the maternal or child behaviors across settings and tasks. Working mothers did not evidence lower levels of positive affect or more directive control strategies. Similarly, children of working mothers were as compliant, affectionate, and independent as children of nonworking mothers. No differences were found in maternal or child psychosocial functioning.

Further studies of alternative life styles are presented in Chapter 12.

Most studies of family functioning in "broken homes" are subject to the same limitations as the classic parent-child studies—that is, unidirectional influences are assumed and examined. Typically, children from broken homes are compared with children from intact homes on some "outcome" measure, and it is assumed that it is the relative intactness of the family that is responsible for differences in outcome between the groups. Few investigators have looked directly at family relationships or have considered the complex interactions across the relationships. One exception to this rule is Hetherington's work on the aftermath of divorce (1976 and 1978).

Hetherington found that if after divorce the mother's role as parent was supported by the father, the mother's relationship with the child was more harmonious and the child's adjustment relatively unimpaired. By contrast, if divorced mothers and fathers continued a pattern of conflict, the mother-child relationship was also disrupted and the child's adjustment was less successful. (See the last section of this chapter for further consideration of the impact of divorce.)

Few investigators have considered the impact on parenting of variables that extend beyond parent-child dyads or triads. In a short-term longitudinal study of black mothers (both teenagers and adults) and their infants (both preterm and full-term), Field et al. (1980) concluded the following: "Despite the teenage mothers' less optimal perceptions and attitudes, their offspring did not differ from those of adult mothers on developmental assessments. Any expected developmental differences may have been attenuated by family support systems, the greater availability of substitute caregiving (e.g., by the unemployed grandmothers), and the infants' exposure to a wide range of playmates."

CONCLUSIONS

While there has been considerable research relevant to aspects of parent-child relationships, our understanding of the parenting process is far from complete. There is good reason to believe that the process is complex and that influences are multidirectional and multidimensional.

Researchers and practitioners alike will do well to consider a multitude of points for strength and strain: parental history, current parental characteristics, the spousal relationship, characteristics of the child, the history of parent-child interaction, direct and indirect influences of one parent on another, current life circumstances, and the environments in which family functioning is embedded.

KATHLEEN M. WHITE
ELI H. NEWBERGER

REFERENCES

Aldous, J.: Family Careers: Developmental Change in Families. New York, John Wiley and Sons, 1978.

Bell, R. Q.: A reinterpretation of the direction of effects in studies of socialization. Psychol. Rev. 75:81, 1968.

Bell, R. Q.: Stimulus control of parent or caretaker behavior by offspring. Devel. Psychology 4:63, 1971.

Belsky, J.: The interrelation of parental and spousal behavior during infancy in traditional nuclear families: An exploratory analysis. J. Marriage Family 41:749, 1979.

Blank, M.: The mother's role in infant development. A review. In Rexford, E. N., Sander, L. W., and Shapiro, T. (eds.): Infant Psychiatry: A New Synthesis. New Haven, Yale University, 1976.

Block, C. R., Norr, K. L., Meyering, S., and Norr, J. L.: Husband gatekeeping in childbirth. Family Relations 30:197, 1981.

Bowlby, J.: Attachment and Loss. Vol. I. Attachment. London, Hogarth, 1969.

Bronfenbrenner, U.: Developmental research, public policy, and the ecology of childhood. Child Devel. 45:1, 1974.

Bronfenbrenner, U.: Toward an experimental ecology of human development. Am. Psychologist 32:513, 1977.

Bronfenbrenner, U.: Contexts of child rearing: Problems and contexts. Am. Psychologist 34:844, 1979.

Burlingham, D.: The preoedipal infant-father relationship. Psychoanal. Study Child 28:23, 1973.

Clarke-Stewart, K. A.: Interactions between mothers and their young children: Characteristics and consequences. Monogr. SRCD 38:6 (Serial No. 153), 1973.

Corfman, E.: Introduction and overview. Families Today: A Research Sampler on Families and Children. Bethesda, Md., DHEW Publ. No. (ADM) 79–815, 1979.

Corfman, E.: Married men: Work and family. Families Today: A Research Sampler on Families and Children. Bethesda, Md., DHEW Publ. No. (ADM) 79–815, 1979.

Crawley, S., Rogers, P., Friedman, S., Iacobbo, M., Criticos, A., Richardson, C., and Thompson, M.: Developmental changes in the structure of mother-infant play. Devel. Psychology 14:30, 1978.

DeLissovoy, V.: High school marriages: A longitudinal study. J. Marriage Family 35:245, 1973.

Dyer, E. D.: Parenthood as crisis: A re-study. Marriage Family Living 25:196, 1963.

Epstein, A. S.: Pregnant teenagers' knowledge of infant development. Paper presented at the biennial meeting of the Society for Research in Child Development, San Francisco, March, 1979.

Fagot, B. I.: Sex differences in toddlers' behavior and parental reaction. Dev. Psychology 10:554, 1974.

Fein, R. A.: Research on fathering: Social policy and an emergent perspective. J. Social Issues 34:122, 1978.

Field, T. M.: Maternal stimulation during infant feeding. Dev. Psychology 13:539, 1977.

Field, T. M., Widmayer, S. M., Stringer, S., and Ignnatoff, E.: Teenage, low-class, black mothers and their preterm infants: An intervention and developmental follow-up. Child Devel. 51:426, 1980.

Frodi, A. M., Lamb, M. E., Leavitt, L. A., Donovan, W. C., Meff, C., and Sherry, D.: Fathers' and mothers' responses to the faces and cries of normal and premature infants. Devel. Psychology 14:490, 1978.

Furstenberg, F. F., Jr.: Unplanned Parenthood. New York, The Free Press, 1976.

Goldberg, S.: Prematurity: Effects on parent-infant interaction. J. Pediat. Psychol. 3:137, 1978.

Grossman, F. K., Eichler, L. S., and Winickoff, S. A.: Pregnancy, Birth, and Parenthood. San Francisco, Jossey-Bass, 1980.

Group for the Advancement of Psychiatry: The Joys and Sorrows of Parenthood. New York, Charles Scribner's Sons, 1973.

Henggeler, S. W., and Bordvin, C. M.: Satisfied working mothers and their preschool sons: Interaction and psychosocial adjustment. J. Family Issues 2:322, 1981.

Hetherington, E. M., Cox, M., and Cox, R.: Divorced fathers. Family Coordinator 25:417, 1976.

Hetherington, E. M., Cox, M., and Cox, R.: The aftermath of divorce. In Stevens, J. H., Jr., and Mathews, M. (eds.): Mother-Child, Father-Child Relations. Washington, D.C., National Association for the Education of Young Children, 1978.

Hinde, R. A.: Toward Understanding Relationships. London, Academic Press, 1979.

Hoffman, C. W.: The father's role in the family and the child's peer group adjustment. Merrill-Palmer Quart. 7:97, 1961.

Howells, J. G.: Fathering. In Howells, J. G. (ed.): Modern Perspectives in Child Psychiatry. New York, Brunner-Mazel, 1971.

Knox, D.: Exploring Marriage and the Family. Glencoe, Ill., Scott, Foresman, 1979.

Korn, S. J., Chess, S., and Fernandez, P.: The impact of children's physical handicaps on marital quality and family interaction. In Lerner, R. M., and Spanier, G. B. (eds.): Child Influences on Marital and Family Interaction. New York, Academic Press, 1978.

Korner, A. F.: The effect of the infant's sex on the caregiver. In Bee, H. (ed.): Social Issues in Developmental Psychology. 2nd ed. New York, Harper and Row, 1978.

Lamb, M. E.: Fathers: Forgotten contributors to child development. Hum. Devel. 18:245, 1975.

Lamb, M. E.: Influence of the child on marital quality and family interaction during the prenatal, perinatal, and infancy periods. In Lerner, R. M., and Spanier, G. B. (eds.): Child Influences on Marital and Family Interaction. New York, Academic Press, 1978.

Lamb, M. E.: Paternal influence and the father's role: A personal perspective. Am. Psychologist 34:938, 1979.

Lerner, R. M., and Spanier, G. B. (eds.): Child Influences on Marital and Family Interaction: A Life-Span Perspective. New York, Academic Press, 1978.

Lynn, D. B.: The Father: His Role in Child Development. Monterey, California, Brooks/Cole, 1974.

Maccoby, E. E., and Jacklin, C.: The Psychology of Sex Differences. Stanford, Stanford University, 1974.

Moss, H. A.: Sex, age, and state as determinants of mother-infant interaction. Merrill-Palmer Quart. 13:19, 1967.

Nash, J.: Historical and social changes in the perception of the role of the father. In Lamb, M. E. (ed.): The Role of the Father in Child Development. New York, John Wiley and Sons, 1976.

Newberger, C. M.: The cognitive structure of parenthood: Designing a descriptive method. In Selman, R. and Yando, R. (eds.): New Directions for Child Development. San Francisco, Jossey-Bass, 1980, pp. 45–67.

Newson, E., and Newson, J.: Influences on Parent Behavior. London, Tavistock, 1963.

Parke, R. D., Power, T. G., and Fisher, T.: The adolescent father's impact on the mother and child. J. Soc. Issues, 36:88–106, 1980.

Parke, R. D., Power, T. G., and Gottman, J.: Conceptualizing and quantifying influence patterns in the family triad. In Lamb, M. E., Suomi, S. J., and Stephenson, G. R. (eds.): Social Interaction Analysis: Methodological Issues. Madison, University of Wisconsin Press, 1979.

Pedersen, F. A.: Mother, father and infant as an interaction system. Paper presented at the meeting of the American Psychological Association, Washington, D.C., September, 1976.

Peterson, G. B., Hey, R. N., and Peterson, C. R.: Intersection of family development and moral stage frameworks: Implications for theory and research. J. Marriage Family 41:229, 1979.

Rice, R. D.: Neurophysiological development in premature infants following stimulation. Devel. Psychology 13:69, 1977.

Robson, K. S.: The role of eye-to-eye contact in maternal infant attachment. J. Child Psychology Psychiatry 8:13, 1967.

Robson, K. S., and Moss, H. A.: Patterns and determinants of maternal attachment. J. Pediatrics 77:976, 1970.

Robson, K. S., Pedersen, F. A., and Moss, H. A.: Developmental observations of diadic gazing in relation to fear of strangers and social approach behavior. Child Devel. 40:619, 1969.

Rossi, A.: A biosocial perspective on parenting. Daedalus 106:1, 1977.

Russell, G.: The father role and its relation to masculinity, femininity, and androgyny. Child Devel. 49:1174, 1978.

Rutter, M.: Dimensions of parenthood: Some myths and some suggestions. In Department of Health and Social Security Report: The Family in Society: Dimensions of Parenthood. London, HMSO, 1974.

Rutter, M.: Maternal deprivation: 1972–1978. New findings, new concepts, new approaches. Child Devel. 50:283, 1979.

Rutter, M., and Madge, N.: Cycles of Disadvantage: A Review of Research. London, Heinemann, 1976.

Sameroff, A. J., and Chandler, M. J.: Reproductive risk and the continuum of caretaking casualty. In Horowitz, F. D., Hetherington, M., Scarr-Salopatek, S., and Siegal, G. (eds.): Review of Child Development Research. Chicago, University of Chicago Press, 1975, Vol. 4, pp. 187–244.

Sander, L. W.: Comments on regulation and organization in the early infant-caretaker system. In Robinson, R. J. (ed.): Brain and Early Behavior. London, Academic Press, 1969.

Sears, R. R., Maccoby, E. E., and Levin, H.: Patterns of Child Rearing. Evanston, Ill., Row and Peterson, 1957.

Stayton, D. J., and Ainsworth, M. D. S.: Individual differences in infant responses to brief, everyday separations as related to other infant and maternal behaviors. Devel. Psychology 9:226, 1973.

Stevens, J. H., Jr., and Mathews, M.: Mother/Child, Father/Child Relationships. Washington, D.C., National Association for the Education of Young Children, 1978.

Tasch, R. J.: The role of the father in the family. J. Exper. Educ. 20:319, 1952.

Thoman, E. B., and Freese, M. P.: A model for the study of early mother-infant communication. In Bell, R. W., and Smotherman, W. P. (eds.): Maternal Influences and Early Behavior. New York, Spectrum, 1980.

Tulkin, S. R., and Cohler, B. J.: Child rearing attitudes and mother-child interactions in the first year of life. Merrill-Palmer Quart. 19:95, 1973.

Weinraub, M.: Fatherhood: The myth of the second-class parent. In Stevens, J. E., Jr., and Mathews, M. (eds.): Mother/Child, Father/Child Relationships. Washington, D.C., National Association for the Education of Young Children, 1978.

Yalom, I. D.: Postpartum blues syndrome. Arch. Gen. Psychiatry 28:16, 1968.

Yogman, M. W.: Development of the father-infant relationship. In Fitzgerald, H., Lester, B., and Yogman, M. W. (eds.): Theory and Research in Behavioral Pediatrics. Vol. I. New York, Plenum Press, 1980.

Zunich, M.: Lower class mothers' behavior and attitudes toward child rearing. Psychol. Repr. 29:1051, 1971.

13B Effects of Sibling Constellations

Knowledge of the effects of a child's position in the family and the size and position of that family in society can have an impact on family and population planning as well as on child-rearing techniques. For example, if parents know that children of small families are more intelligent and taller than their counterparts in large families, the parents may want to plan for fewer children. Likewise, if parents know that lower IQs and delinquent and maladjusted children are more frequent among the later-born of large families, they may plan for a smaller family. Parents who are aware that closely spaced like-sexed children as well as twins are at risk of developing identity problems unless treated as individuals from infancy may modify their child-rearing attitudes to allow for more individualized experiences.

In this section, characteristics of first-borns are contrasted with those of later-borns, and the differences between only-borns and eldest children as well as between intermediate and youngest children are discussed. These characteristics include (a) intelligence, achievement, and success; (b) orientation and social achievement; and (c) personality disorders and substance abuse. Sibship size, or the number of siblings in a family, and its effects, together with the ordinal position of the children and the socioeconomic status of the family, are also considered as well as the effects of space between siblings and of being a twin. We conclude by examining sibling variables as they affect pediatric practice. The physician should be aware of the effects of sibship variables not only for understanding pediatric behavior and counseling parents but for possible modification of treatment plans.

The applied aims of any research on sibship variables are to maximize the positive effects of each child's position in the family and to reduce the negative ones. The research reported here is limited to those studies providing relatively conclusive results. To obtain the source citations for the majority of these studies, the reader is referred to our published review article (Wagner et al., 1979).

THE FIRST-BORN VERSUS LATER-BORNS

First-borns (including the only child) differ from later-borns in that they show higher achievement, greater conformity, and greater neuroticism with increased incidence of physical disorders frequently labeled psychosomatic. These characteristics will be described in greater detail and related to differential parenting characteristics that may be the cause of the differential first-born traits.

Greater Achievement Among First-Borns. For at least a half century there has been extensive documentation that first-borns tend to attend and achieve at college, achieve as scientists and scholars, and become eminent in their chosen fields. As one may expect, they tend to be upwardly mobile with regard to social class. In specific areas they tend to show greater creativity, such as multiple-answer problem-solving.

Parents tend to put greater pressure to achieve on the first-born than on later-borns. Specific studies dealing with each of these characteristics, as well as the characteristics of the first-borns are presented in our literature review (Wagner et al., 1979). Mothers of first-born sons are more demanding and exacting, express more disappointment over unsatisfactory behavior, and less often speak of love for their sons. The first two of these especially would tend to increase the pressure to achieve.

On the more positive side, parents tend to show more interactive vocalization and spend more time of a social, affectionate, and caretaking nature with their first-borns in contrast to later-borns. Also first-borns tend to be more encouraged than later-borns if they are from higher social classes, while this effect is less distinct among lower social classes.

Greater Conformity of First-Borns. In a number of studies, first-borns are found to be more conforming to authority. First-borns tend to identify more with their parents and other authority figures, and, perhaps as a consequence, they are less popular with their peers. Also, parents tend to pressure the first-born to conform to a greater extent than later-borns. First-borns are more dominated and receive more of the caretaker's attention, and the mothers of first-born sons are more cautious. These parental actions may relate to conformity.

LATER-BORN ORDINAL POSITIONS

Frequently later-borns have been found to be the reverse of the first-borns with whom they are compared. The other end of the continuum will be briefly sketched in, emphasized only when additional features or dimensions are relevant. Later-borns include, of course, the youngest and the middle-borns. The youngest have been studied more than the middle-borns. In most cases the youngest and middle-borns are described as similar or the same. Therefore, middle-borns will be distinguished here only when they have been found to differ from last-borns.

Achievement and Intelligence. On the whole, later-borns are simply less achieving and school-oriented than are first-borns. They are described as having lower verbal IQs in particular but generally lower IQ scores. They are less upwardly mobile in a social-class sense and less academically motivated, attending college and continuing education after high school less frequently than first-borns. There is increasingly more mental retardation in later birth orders. One study of youngest children showed that if there was a wide space in years prior to the birth of the youngest that this person's IQ might be higher than the rest of the family of siblings. However, last-borns, including the only child, may be disadvantaged in IQ because they have no younger siblings to teach.

Social Orientation and Social Achievement. Later-borns are more popular with their peers than are first-borns. Youngest children are described as gregarious, while middle-borns are described as socially active, having more social skills, and wanting to work with people. Later-borns are found to be both more affectionate and more physically aggressive toward their peers. Youngests are described as being less conforming, less susceptible to social pressure, and less responsible, while middle-borns are described as independent and, again, identify less with parents. The youngest is described as a person who tends to take more risks, including physical risks, and is activity-oriented in contrast to firstborns; middle-borns have not been studied in this regard.

Personality Disorders and Substance Abuse. Later-borns tend to be found more frequently among delinquents and show more personality disorder traits, particularly antisocial personality disorder types. They also tend to be more accident-prone, especially if they are in the latter half of the sibship. Later-borns use alcohol, heroin, and street drugs to a greater extent than do first-borns.

SIBSHIP SIZE AND ITS EFFECT

Family size has been extensively reviewed by Terhune (1974). His monograph should be consulted for its documentation as well as for an annotated bibliography of reports in this area.

Sibship is defined as family size excluding parents. In contrast to large sibships, small sibships have members who are on the average more achieving intellectually and better adjusted and have a better self-concept. There are fewer suicides and fewer juvenile delinquents, school problems, antisocial problems, and alcoholics in this group. Individuals from small sibships take fewer risks.

Members also tend to be more dominant and have less need for approval.

In family relations, small sibship members have parents who are more interested in them. The family of a small sibship is described as having less tension, resentment, and discouragement when the children are in their youth. In the larger sibships, the father becomes more involved, particularly with the eldest. In large families, activities tend to be family socializing, whereas the smaller sibship family may tend to go as a group to a restaurant or theater or on a trip.

The large family or sibship is described as more family-centered. Rules, however, become more stringent in large families, with discipline more punitive and of the "unconcerned" type, exposure to which may lead to lack of spontaneity and to giving up quickly. In a large family, there is more authoritarianism and more power assertion, criticism, and control, and siblings tend to be treated more alike despite their differences. (See also Wagner et al., 1979, for research relating to these specific findings). Children born earlier in large families frequently have heavy household chores, and in certain economic situations, eldest boys may be withdrawn from school to help support the family. With a given income, larger families are obviously less well-off financially. Some writers go so far as to say that large families are a "common cause of poverty." As sibship size increases, there is increased perinatal mortality and morbidity as well as an increased incidence of infectious disease.

With regard to achievement and motivation, members of small sibships are similar to first-borns in that they show greater frequency of college and graduate school attendance and tend to have higher scores on intelligence testing.

Since several characteristics other than achievement and motivation seem similar between small sibships and first-borns, the question arises whether there is a common explanation for these similarities.

RELATIONSHIP BETWEEN BIRTH ORDER AND FAMILY SIZE

Belmont, Stein, and Susser (1975) have related birth order and family size to both intelligence and height for their large sample of 18-year-old Netherlands men. The relationship is similar in that in the sibship of two, intelligence and height were greatest; sibships of one and three shared second place, and progressive decreases in both intelligence and height were seen with sibships of four, five, and six.

However, when their data were divided by ordinal position, a decrement in intelligence was seen for each ordinal rank as well. First-borns for each sibship size are less intelligent as family size increases, although for each family size first-borns are consistently more intelligent than second-borns, who in turn are more intelligent than third-borns, and so on. Here both family size *and* ordinal rank seem to play a role in determining intelligence level.

For height, however, these authors found that family size alone plays a role; for each family size, there was little difference by ordinal rank. Those of sibships of two are about equally tall and are taller than all those in sibships of three, all of whom are of similar height. Members of sibships of four, averaging much the same, are shorter than sibships of three, and taller than sibships of five, with a similar decrement for sibships of six. It has been suggested that this decrease in height as family size increases may be due to less attention by the mother to the eating behavior of each child with the advent of the next as well as to poorer nutrition with increase of number of children per family among the economically disadvantaged.

Not fitting into these progressions in the data is the only child. For height and intelligence, only children are about equal to the eldest child in sibships of three and four.

A study by Page and Grandon (1979) of 22,000 high-school graduates in 1972 in the United States relates mental ability and socioeconomic status (SES) to family size. They found for the highest 25 per cent in SES, there is no relation between family size and mental ability; for the middle 50 per cent in SES, mental ability decreases moderately with each increment in family size; and for the lowest 25 per cent in SES, mental ability decreases considerably with an increase in family size. When the relation is calculated for the total sample, i.e., when summated across all socioeconomic classes, the relationship is strong and negative. Further, they showed that the lowest 25 per cent in SES has a significantly larger proportion of sibships of five, six, and seven members. Belmont, Wittes, and Stein (1977) in a later publication report similar differences by socioeconomic status. For the higher social class nonmanual workers, the effect of family size was weaker than for the manual workers from lower-class homes, for which it was considerable. For the nonmanual sample, ordinal rank had considerable effect on abilities. Ordinal rank had less effect on ability in lower-class groups. Thus, Belmont's data (1977) would seem to be consistent with that of Page and Grandon (1979).

In a recommended book about 11th-grade children, Blau (1981) concluded that socioeconomic level is the strongest predictor of IQ and achievement score, while larger family size also has a definite negative effect. However, she emphasizes that the mother's ambition for her children with a heavy investment of resources (time, interest, effort, and money, all frequently diverted from other desired areas), along with reduced family size, results in high intellectual ability.

Blau believes a strategy of high investment of resources and low use of aversive authoritarian discipline optimize scholastic achievement. High investment with high aversive discipline are associated with lower achievement, while high aversive control with low investment of resources produces children with the lowest achievement levels.

In conclusion, greater sibship size has a definite negative effect on both intellect and physical stature in general. Such effects are not found in those homes with high levels of economic resources, good child-rearing practices, and ample time for each child. Other factors that affect intellect over and above family size are socioeconomic level, type of discipline, and cultural ideology.

These studies show the highly complex interactive effects of family constellation and related variables on such factors as height and intelligence. For example, summating across family size can lead to a totally erroneous conclusion regarding the relation of birth order to height.

SIBLING SPACING

In general, the older child seems to benefit most when there is a long interval before the next sibling is born. A long interval is usually three to four years, as defined by various authors. Short intervals were associated with less verbal ability and lower college attendance in the older child. These results on ability were controlled for family size and social class (Breland, 1974).

With two closely spaced children (less than 18 months apart) the mother was found to talk to the children as a unit and at the level of the younger child, so that the older child was at a disadvantage.

In cross-sexed, closely spaced siblings, the older child had a greater effect on the younger child with regard to sex-role characteristics. For example, a boy with an older sister is likely to be more feminine if his sister is 1½ years older than if she is 4 years older.

TWINS

Twins seem to resemble most later-borns of large families. They tend to have lower IQs than single-borns. Twins have increased perinatal mortality and increased congenital anomalies. Second-born of twins is more disadvantaged than the first.

Generally twins are at a mental and physical disadvantage, perhaps especially verbally. Triplets and other multiple births are even more disadvantaged.

With age, monozygotic twins seem to increase in similarity, whereas dizygotic twins tend to decrease in similarity. Monozygotic twins seem to be more disadvantaged than dizygotic twins.

Koch (1966) reviewed the literature on twins from a psychosocial point of view and studied 6-year-old twins attending school in Chicago. She found, as others have, that twins were closer to each other than were other sibling pairs. Koch did not find that this closeness was detrimental to the social relations, but other studies suggest that it may be difficult for them to achieve identity in light of such closeness.

DISCUSSION

Perhaps the most substantial finding reported in the area of sibling studies is that first-borns tend to excel academically. Although there is great overlap, and the relatively lower academic achievement by later-borns may be small, studies have taken into account enough interrelated variables to suggest that this finding is fairly consistent among middle- and upper-class cultures in urban America. Birth order, family size, and birth interval all tend to be related to one another, and each interrelates with social class (Schooler, 1972). With this in mind, one may say that first-borns tend to be more conforming, while later-borns tend to have more social skills and desire to interact with others. Studies indicate that later-borns are also more often found to have antisocial personality disorders such as rebellion against authority, promiscuity, alcoholism, drug abuse, and delinquency. The individual from a large family has some of these characteristics that have been described for the later-born and also lower levels of academic achievement.

Birth interval tends to decrease as family size increases, so that what is said of large families may also apply to the elder with a closely spaced younger sibling. Verbal fluency increases with length of the birth interval. An elder cross-sexed sibling seems to influence the younger to adopt more of the cross-sexed role if the interval is small than if the interval is large. Mothers tend to interact verbally with two closely spaced siblings simultaneously and at the level of the younger sibling. This seems to be to the verbal detriment of the older closely spaced sibling. Closely spaced siblings, including twins, tend to be treated as a unit, in general, rather than differentially, which seems to have a detrimental effect on the elder of the two and perhaps on both twins. Twins raised singly had higher verbal reasoning scores than those raised together, suggesting that they may have been called on for less verbal interaction when dealt with as a unit of two.

OVERVIEW OF IMPLICATIONS FOR PEDIATRIC PRACTICE

Pediatricians and family physicians are frequently called upon to counsel parents on child-rearing practices. They may effectively share with parents the implications of family constellation effects. Possible needs of later-borns and members of large sibships are parental increase in expressed affection; increase in attention and encouragement; and increase in active vocalizations with more social, affective, and caretaking time. Running the family like a military operation with spanking, physical punishment, and authoritarian rules, as found in many large families, may distance the children. Closely spaced siblings and twins need especially to have their individual characteristics noticed and their needs met. In the case of twins who are physically quite similar, the parents may be encouraged to dress and treat them differently in order to respond to them less as a unit. The elder closely spaced sibling may benefit from increased parental interaction of an age-appropriate verbal nature, especially if the mother is responding to the elder and younger closely spaced sibling as a unit at the level of the younger. The elder in a close or moderately spaced interval from the next younger sibling may benefit from encouragement by the parents with increased warmth and attention. With very wide spacings of greater than five years, the younger sibling also may benefit from increased attention.

Since the inferential links in these observations are not always strong, they should be presented to parents merely as suggestions for possible application of some conclusions from research. From this relatively new area of investigation further studies are sure to emerge.

DANIEL S. P. SCHUBERT
MAZIE EARLE WAGNER
HERMAN J. P. SCHUBERT

REFERENCES

Belmont, L., Stein, Z. A., and Susser, M. W.: Comparison of associations of birth order with intelligence test score and height. Nature (London) 255:54, 1975.

Belmont, L., Wittes, J., and Stein, Z. A.: Relationship of birth order, family size, and social class to psychological function. Perceptual Motor Skills 45:1107, 1977.

Blau, Z. S.: Black Children/White Children: Competence, Socialization and Social Structure. New York, The Free Press, 1981.

Breland, H. M.: Birth order, family configuration, and verbal achievement. Child Devel. 43:1011, 1974.

Koch, H. L.: Twins and Their Relations. Chicago, University of Chicago Press, 1966.

Page, E. B., and Grandon, G. M.: Family configuration and mental ability: Two theories contradicted with U.S. data. Am. Educ. Res. J. 16:257, 1979.

Schooler, C.: Birth-order effects: Not here, not now! Psychol. Bull. 78:161, 1972.

Terhune, K. W.: A review of actual and expected consequences of family size. Washington, D.C., U.S. Public Health Service Publications NIH 75-779, 1974.

Wagner, M. E., Schubert, H. J. P., and Schubert, D. S. P.: Sibship constellation effects on psychosocial development, creativity, and health. Adv. Child Devel. Behavior 14:58, 1979.

Zajonc, R. B.: Family configuration and intelligence: Variations in scholastic aptitude scores parallel trends in family size and the spacing of children. Science 192:227, 1976.

Zajonc, R. B., and Markus, G. B.: Birth order and intellectual development. Psychol. Rev. 82:74, 1975.

13C Adoption and Foster Care

A section on adoption and foster care is included in this chapter because these are significant variations from the usual biologic family structure. Most children raised in these situations do well, at least as well as biologic children in comparable circumstances. However, there are some unique differences potentially affecting development and behavior with which a primary care clinician should be familiar. The professional person advising these parents and children is in a position to play an extremely helpful role (Wessel, 1960).

After a brief historical account of adoption this section will be concerned with a description of these special features of adoption and foster care and with their clinical management. For a discussion of broader legal and policy issues the reader is referred to other sources (American Academy of Pediatrics, 1973; Mech, 1973).

HISTORY

The practice of adoption has persisted from the earliest times of record and mythology. It has been the most universal alternative method utilized by various societies in all ages to insure the continuity of the family. Primitive tribes still practice rites described in mythology wherein the adoptive mother clasps the child to her body and allows him to fall to the ground in imitation of a true birth. Adoption was practiced by ancient Egyptians, Babylonians, Assyrians, Greeks, and Romans, usually for the purpose of acquiring an heir or successor. The Bible refers to the most famous example: the adoption of Moses by the daughter of the Pharaoh. In China, an ancient custom that persisted until the 1800's allowed a childless male to claim the first-born child of any of his younger brothers, because "to die without leaving a male posterity to care for his ashes and to decorate his grave, thereby pacifying his wandering spirit in purgatory" was considered "one of the greatest calamities to be apprehended" by a Chinese individual.

There is reference to adoption in many ancient laws. The first truly structured laws concerning this subject were found in very old Roman manuscripts. Roman law and other ancient codes served the primary purpose of providing and ensuring continuity of the adoptee's family; the individual's rights were strictly controlled by the head of the family unit.

Great Britain provided a form of nonlegal adoption for centuries; basically it provided for the nurturing, protection, education, and training of children. Inheritance and perpetuation of the family line were not factors in these situations. In early America these same practices persisted; children were "adopted" principally for the purpose of obtaining additional labor or farmhands.

In the mid-1850's American laws began to be formulated for the control of adoption practices. The substance of this legislation was to protect the welfare of the adoptee rather than the needs and rights of the adoptive parents. Reformers of this era sought better care for the many children who had been placed in unwholesome, unhealthy, or noncaring homes. For nearly a century adoption laws were passed with the theme of "the best interests of the child" by situating the homeless adoptee in a home as a complete and equal member of the family unit, to include all the rights and privileges given biologic children (American Academy of Pediatrics, 1973).

ADOPTION TODAY

Adoption as a means of creating families is widely accepted in our society. A concept of the family unit based on ties of psychologic needs and affection (not only on biologic relationships) has evolved. This has made it possible for adoptive parents and their children to enjoy the satisfactions of true family life. Adoption is now considered desirable not only for infants but any child who has no family of which he is a part. Communities and social agencies are taking greater responsibility for finding families that can give children the wholesome benefits of adoption. This is especially true for children with special needs, including older children, children with handicaps, and children of minority or mixed racial background. In addition, intercountry adoption has been increasing, based mutually on the needs of these children as well as the desires of the parents seeking presently "hard to find" adoptees.

The availability of adoptees has vastly decreased in the last decade because of the increasing use of contraceptive techniques, the legal and inexpensive availability of abortions, and the fact that more unmarried parents are keeping their children. An increasing proportion of adoptions are now private, rather than being undertaken through agencies. With the State of California as an example, and comparing the year 1966 with 1976, the following statistics are available: The number of independent adoptions decreased 50 per cent, placement by agencies decreased 66 per cent, and the number of intercountry adoptions increased 300 per cent (Los Angeles County Department of Adoptions, 1979).

The United States now has approximately 2,500,000 adoptees under the age of 18—representing nearly 2 per cent of the children and youths in that age group. About half of these have been adopted by relatives. Adoptive families in this country now include nearly 11,000,000 people (adoptees, siblings, and adoptive parents).

UNIQUE FEATURES OF ADOPTIVE FAMILY LIFE

Adoption is a situation engendering unique experiences for all three constituents of the adoptive triad—the birth parents, the adoptee, and the adoptive parents. In the adoptive procedure a mother has given up a child she has borne; parents usually adopt the child of an unfamiliar background, and the child knows that he was not born to the parents who have reared him. Furthermore, the adoptive mother does not have time to prepare herself emotionally during a fixed nine month gestation period, as does the biologic mother.

A re-evaluation of these practices has indicated that there has been a failure by both professional persons and consumers to understand and identify the psychologic complexities inherent in the adoption process. It is imperative that the adoption be viewed as a lifelong process in which the continuing changing needs of all involved persons receive appropriate study and concern. Only in this context is it possible to offer meaningful support and provide adequate intervention (Sorosky et al., 1980).

ADOPTIVE PARENTS

In most cases couples choose to adopt because of infertility. For many the discovery of sterility is extremely traumatic, with psychologic reactions characteristic of grief and mourning. It is imperative that the accompanying feelings of anger, envy, self-incrimination, and mutual blame be resolved before adoption is considered. If they are not, the adoptive parents will have difficulty in helping their adopted child deal with the feelings of loss and abandonment created by the adoption process. Counseling of the potential parents by a knowledgeable physician or social worker becomes appropriate at the time when the decision to adopt is first made.

Children (natural or adoptive) presently in the home should be included in family discussions of possible adoption. Their opinions must be heard. As one such child said, "This is important talk— Mom and Dad aren't just going to bring home a puppy."

In many respects the adoptive parents are placed in a double bind by the advice from the agency: "Make the child your own, but tell him he isn't." Yet if they can accept themselves as being different from the biologic parents rather than denying that a difference exists, they will be better able to communicate openly with their children regarding their adoptive status and deal more effectively and more supportively with problems that may arise (Kirk et al., 1966).

THE ADOPTED CHILD

The adopted child grows up in a family to which he is not biologically related (unless adopted by a relative). How does this affect his development and behavior?

Compared with children who remain with their single mothers, adopted children incur substantially less educational failure and fewer psychiatric disorders. From this point of view, adoption is a significant force for positive mental health (Wolff, 1974).

The majority of adopted children make a normal adjustment to family life and compare favorably with their nonadopted peers. However, various studies of child psychiatric populations have revealed about 4 per cent adopted children among them, a disproportionate representation of the roughly 1 to 2 per cent in the general population (Mech, 1973). It has been impossible to determine whether this indicates a higher incidence of problems in adopted children or an overutilization of these facilities by adoptive parents. Even population surveys have not settled the issue, one in Sweden indicating a higher rate of disturbance

(Bohman, 1972) and one in Great Britain finding no significant difference from controls (Seglow et al., 1972).

If there is a significant incidence of behavioral problems in adopted children, several factors should be considered. One should not generalize about genetic risks because of the varying backgrounds of children made available for adoption; the clinician should examine the known family tree of the individual child. Although the emotional distress of the woman carrying an unwelcome pregnancy may in some way affect the behavior of the child in utero, there is no evidence that such effects persist beyond the newborn period. No greater incidence of difficult temperament in infants was found in the only study exploring this possibility in a randomly selected population (Carey et al., 1974). Until there is clearer evidence of an excess of congenital risks in the development or behavior of adopted children, explanations for any problems in these areas are more likely to be found in the life experience of adopted children—feelings of being different and of having parents who may be older than usual, who have experienced the frustration of infertility, and who are usually poorly prepared for the unique features of adoptive parenting.

Despite the common assumption that children adopted beyond infancy turn out less well, the evidence is that the age of placement by itself usually makes little difference in later adjustment of the child (Eldred et al., 1976; Kadushin, 1970).

Some older adoptees have been moved from place to place prior to adoption, with the subsequent development of recurrent feelings of rejection. These children may be subject to special emotional, learning, or behavioral difficulties. Careful selection of adoptive parents and provision of clear consistent preadoptive advice about problems that may arise will help to lessen the possibility of the development of such disorders.

Another important determination to be made initially is whether a couple actually wants and accepts the child offered to them at the time. If they have been waiting for a newborn, and suddenly are offered a healthy but stubborn two or four year old, are they prepared for the temper tantrums and such behavior that may immediately place obstacles in their attachment to the child? Can they bond themselves to a colicky three month old baby if they have been hoping and longing for a sleepy little newborn to nurture? And will a dark Indochinese fill their needs to be parents instead of the blonde baby they hoped for? This should be openly discussed before and at the time the adoptee becomes available to them.

Counseling in these instances should include some assurance that the parents have sufficient motivation in adopting an older child and are not merely accepting this child in lieu of a much preferred healthy infant. It is likewise critical that the parents accept this child for what he already is rather than for the changes that they think will occur. Realization of these factors, understanding of the child's feelings of possible deprivation and past rejection, and the seeking of counseling for the problems of coping and later limit-setting will help lead to a satisfactory adoption. An adoptee who finally gets his own family will usually fare much better with the love and security he can finally sense, especially if there is no competition from others of his age at home.

RELATIONSHIP WITH BIRTH PARENTS

Despite the probable addition of another emotional burden, the adoptive parents must know before the adoption that there may no longer be a guarantee that the identity of the birth parents will remain confidential. An increasing number of states are establishing legal processes whereby mature adoptees are permitted to obtain information with which to identify, locate, and even meet their birth parents. The couple should be advised of the importance of attempting to obtain and record information about the birth parents so that they can answer the adoptee's questions about his birth parents and thus minimize the chances that the adoptee will later resort to excessive fantasizing to fill an identity gap. Preadoptive counseling of this nature by the physician and social worker will enable the couple to comprehend, understand, and cope with the adoptive process, as well as to adjust to the changing characteristics of the maturing family unit (Carey, 1974).

CLINICAL MANAGEMENT

MEDICAL EVALUATION OF THE NEWBORN ADOPTEE

When the waiting couple is notified that their prospective adoptive child is available for them, the physician should investigate the family history in as great detail as possible—genetic illnesses, age, health, alcohol or drug use by the birth parents, difficulties reported during the pregnancy and delivery, and the health of the child. These concerns deserve the honest and thorough evaluation of all available information. Medical problems should be objectively interpreted for the parents to insure their complete understanding of present and future implications. All available information should be written down for future reference by parents, adoptee, and physician.

BONDING AND ATTACHMENT OF THE ADOPTEE

Adoption of children allows for no neonatal bonding. However, bonding, like adoption, is a lifelong process and can occur at later times (Ward, 1981). Potential parents should be aware that some children are "hard to get attached to" whether adopted or natural, because of variables in their temperament and personality. A child in the stage of the "terrible twos" is obviously going to present difficulties in the bonding process to the adoptive parents. The adoptive parents should be alerted to the fact that biologic parents also find this stage a difficult one to endure.

IS BREAST FEEDING POSSIBLE?

On occasion, adoptive mothers may express a strong determination to breast feed the new child. Induced lactation by adoptive mothers has been described enthusiastically in the lay press. Medical literature does not seem to substantiate this claim that completely successful lactation is possible among these mothers. The use of Lact-Aid and other similar devices can help produce small quantities of a thin milk, but rarely in sufficient quantities to fully breast feed the adoptee (Bose et al., 1981). An adoptive mother may run the risk of another failure of physical function and additional loss of self-esteem. The time and energy required for attempts at breast feeding would best be spent on other vital areas of adjustment (Carey, 1981).

CONTINUING COUNSELING DURING FURTHER OFFICE VISITS

Further counseling of the adoptive family continues as a component of future office visits. A feeling of the impact of the adoption can be obtained as the new mother discusses routine child care problems. Overprotectiveness and uncertainty at this time are normal early reactions that can be helped by basic reassurance. On the other hand, the adoptive parent who requests that the fact that the child is adopted remain confidential and not known to the office staff, or perhaps to the child himself, is signaling a need for further exploration of her feelings and anxieties. Early discussion of the appropriate problem areas is essential because the adoptive parents may be reacting to emotional pressures to dissociate themselves from the child's genealogic identity or to "deny the difference"(Kirk, 1964).

WHEN AND HOW TO TELL THE CHILD OF HIS ADOPTION

The adoptive parents should be advised that an adoptee should learn the truth of his adoption when he is so young that he will not in later years remember a time when he did not know the truth about his life.

Most experienced professionals in the adoption field believe that the revelation should take place at any early age, from two to four years, to avoid the child's learning of the adoption first from outside sources. The depth of the discussions about adoption should always be geared to the child's maturity level. It is a mistake to assume that there is a lack of interest if the child asks no questions about his adoptive status; the passive avoidance of the subject may merely represent a fear of hurting the adoptive parents (Sorosky et al., 1981).

It is of enormous importance for the child to know in whose "tummy" he was and from which "tummy" he came out. If the child grows up with the false belief that his adoptive mother has borne him, there will come a day when this belief will be shattered, giving rise to bitter disillusion. This occurs not infrequently as adoptees learn of their adoption from others instead of from their parents.

Revealing all "life facts" not only those concerning adoption is not a single act; it should be repeated at different stages of the child's maturation, since his ability to comprehend the process develops throughout childhood (Brodzinsky et al., 1981).

Care should be taken by the parents in describing the adoptive process. To inform the child that he was "chosen" because he was the "prettiest" or "smartest" might put him under undue stress to live up to the expectations of others.

During the preschool office visits, discussions concerning adoption, even brief ones, may serve to elicit a reaction connoting problems, anxiety, or even denial. The adoptee's level of awareness of his adoption becomes an important concern.

Even more important than parents' providing specific information at certain times is their readiness to answer their children's questions honestly and appropriately when questions arise spontaneously.

USING ADOPTION AS A "WEAPON"

Physicians may alert adoptive parents in advance that the adoptee may at times use adoption as a weapon: "I am leaving; I will run away from here to my other home, to my real mother." The adoptive parents may be assured that biologic children use the same type of weapon. It is not a true threat unless there is an excessively unstable relationship. Overpermissiveness may result from threats of this nature in a home that is overly insecure or unduly anxious. This symptom may erupt in a physician's office; the depth of this problem should be explored by an open discussion with the parents.

An adoptive parent may use words as a weapon to the adoptee. "I wish I never chose you" or "I may just send you back to your real mother" are remarks that may or may not reflect the true feelings of the parents. These words are shock waves threatening a possible second rejection to the child and never should be used. These "weapons" may also reflect a true statement of displeasure, perhaps not only momentary, in having adopted this youngster.

This is another example of how an adoptive family differs from other families: there can be additional conscious or subconscious feelings of having made a wrong decision by adopting.

ADOLESCENCE

Later—near and during adolescence—as the adoptee becomes increasingly concerned about his own identity, he may inquire about unknown facts ranging from the color of his birth parents' eyes, their stature, their present situation in life (marriage, other children, occupation) to questions of familial diseases such as diabetes, epilepsy, and Tay-Sachs disease (of importance now that he or she is nearing the age of possible parenthood). This reflects a normal concern typical of his age group about his own health and body image. It is essential that the adoptive parents be convinced that the more open and complete the communication about all adoption related matters, the less likely the adolescent will be to resort to excessive concern, fantasizing, or acting out of his uncertainties.

Adolescence is normally a turbulent age in which the teen-ager seeks his own identity. At the same time he is in the midst of a dependent-independent crisis with himself and his parents. This is viewed as a second stage of individualization from the parents. Adoptees now have two sets of parents from which to separate (albeit one set is not immediately present). This phase of life is always difficult; an adoptee and his adoptive parents may have more special problems. Too few adoptees have been provided with enough background information to incorporate it into their developing ego and sense of identity. Some ado-

lescents even may have composed a mental picture of their birth parents as being an immature, confused, and unwed couple. A misconception may have developed because of scanty information provided at the time of the adoption procedure and little or no follow-up of confirmation concerning the birth parents. Some adoptees are hesitant to probe into genealogic questions if they sense their adoptive parents' insecurity in this area.

Acceptance by peers becomes increasingly important as the adolescent gradually transfers his interests and emotional attachments from the family to the outside world. Some are ashamed of their adoptive status and fear being exposed; others compensate for feelings of inadequacy by wearing their adoption as a badge and telling everyone. The dating experience can be affected: Some are reluctant to get too close emotionally to a companion because of an unconscious fear of rejection; others are concerned about a negative reaction from their date's parents; and a few may actually harbor the terrifying fear of establishing an incestuous liaison with an unknown biologic relative.

SEARCH FOR BIRTH RECORDS OR BIRTH PARENTS

Most adoptive parents have warm and loving relationships with their adopted children. At appropriate times most try to pass on to them as much of the birth information as they know and are able to provide. Most adoptees have a warm, loving, and close relationship with their adoptive parents. In spite of this, and regardless of their attachment to the adoptive parents, some adoptees have a compelling desire or need to learn of their birth parents.

Usually this search is not related to the quality of the adoptive relationship. This is not to deny that there are adoptees who do have an obsessive need to search for their birth parents because of a neurotic problem or secondary to an emotionally barren relationship with their adoptive parents (Eldred et al., 1976). It has been documented that many adoptees who press the search have been informed of their adoption late in life; some have learned of their adoption from people other than adoptive parents (Sorosky et al., 1978). Attempts at searching for birth parents, however, should be discouraged during adolescence, since the immature adoptee cannot put the entire experience into a proper and healthy perspective. The adolescent period is complicated enough without the child's being confronted with two sets of parents (Sokoloff, 1979).

The adoption process in the United States traditionally has been designed to safeguard the rights of the adoptive parents, ensure the solidar-

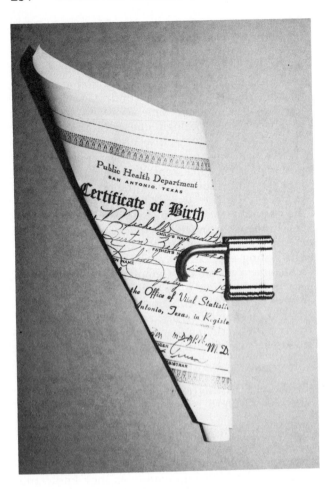

Figure 13–1. In a few states, birth records can be examined by the court for "just cause"—the definition of which varies from state to state.

ity of the adoptive family, and preserve the anonymity of the birth parents. When adoption is finalized, the original birth certificate is "sealed" and a new certificate is issued in the name of the adoptive parents. Once it is sealed, the laws of most states specify that the original record can be opened only by court order and for "just cause." A few states have provisions for opening of the records on demand of the adoptees when they become adults. This provision frequently exists in theory but not in practice, and the definition of "just cause" has varied considerably from court to court (Fig. 13–1).

Many adult adoptees and adoption specialists see this search as being essential to the establishment of a sense of identity. Most reports of reunions indicate that adoptees have been pleased with the meeting and their their ties to their adoptive parents have been strengthened thereby. In addition, there is the growing body of law that has spoken to the right of people to know the content of various personal records.

In Great Britain the Children's Act of 1975 provided for a relatively simple process whereby mature adoptees could obtain the name and oc-

cupation of their birth mother (and possibly the father) following compulsory counseling by a social worker. Evaluation of these experiences discloses that a relatively small number of adoptees attempted to see their birth records; a considerably smaller number of adoptees actually traced their birth parents. Studies of reunions of adoptees and their parents in the United States show that frequently the adoptee developed a greater sense of identity, the relationship between the adoptee and the adoptive parents was improved, and the birth parents ended their deep concern about the welfare of the child they relinquished. The degree of satisfaction obtained usually depended on the nature of the initial psychologic need for the search. Frequently the end of the search occurred in discovering the details of the past rather than in an actual reunion (Leeding, 1977).

The primary physician is in a unique position to assume the role of the child's advocate. He may serve as a clearinghouse for information, such as birth and adoption history. He may advise parents and potential adoptive parents about anticipated problems, particularly the possibility that their adoptive child may later want to learn his biologic identity, and that their cooperation in this effort may improve rather than break down the relationship with their adoptive child.

The physician should encourage the adoptive parents to obtain as much medical and background history of the birth parents as possible at the time of adoption. Many present day adoptions are private rather than being conducted through agencies. The involved physicians and attorneys must help compile a complete written summary of the medical history of both birth parents and their families with a copy for the adopting parents. This information is needed to enable the adoptive parents to provide essential answers as the maturing adoptee inquires about his background.

Later, open access to information and the possibility that the mature adoptee will consider contact with his birth parents could create a more wholesome environment for both parents and child. The more open the communication concerning all adoption related matters between the child and his parents, the less likely will be identity problems for the adoptee (Committee on Adoption and Dependent Care, 1981).

Counseling during this period should include knowledge that the adoptive parents may fear rejection or loss of their child. Furthermore they are probably suffering from a resurgence of old preadoptive feelings of failure and loss. They must be convinced that the adoptee's search for genealogic information about his birth parents is a personal need that cannot be accurately comprehended by a nonadopted individual. They should be reminded that it is always important to attempt to satisfy their child's search with all of the identifying information they can obtain. It is important to guide the parents as they realize that they are the adoptee's true psychologic family and to encourage them not to swerve from normal standards of permissiveness, indulgence, and overprotection. Possible uncertainties burdening the adoptive parents may frequently result in either undesirable extreme—overprotection and rigidity in regard to academic and other achievements by the adoptee, or complete disinclination of the parents to exert effective discipline.

Attention should be devoted to the adolescent adoptee during his critical period of conscious and subconscious "searching." He will welcome reassurances that his feelings of a need to search for identity and the difficulties that he is subsequently encountering are normal. He can be advised that his feelings should be discussed openly with his adoptive parents; this open communication is beneficial to the entire family unit. Sometimes aid is needed to enable the teenager to handle any feelings of betrayal by his adoptive parents, his fears of meeting his birth parents and possibly again facing rejection, and his own fantasizing of what may have caused the birth parents to relinquish him (Sokoloff, 1979b).

THE ADULT ADOPTEE

The psychologic needs of the adult adoptee are often seriously neglected (Eldred et al., 1976). Sometimes, as prospects of marriage emerge, such people become overly concerned about rejection (again) because of their adoptive status. Pregnancies can cause apprehension about unknown hereditary illness and about the complications of delivery and birth. The delivery of their own newborn is often the first opportunity for adoptees to experience a relationship with blood relatives. Adult adoptees are often frustrated by not being able to transmit genealogic information to their offspring; consequently they may be guilty of the same lack of openness about adoption with their own children that they themselves may have been subject to earlier in life.

ADOLESCENT PREGNANCY AND MOTHERHOOD

There are 1,100,000 pregnancies annually among teen-agers (see also Chapter 35). Physicians are increasingly involved in the care of these "children of children." Many—but not all—adolescent parents relinquish their children for adoption. Although the birth parents have relinquished all their right and responsibilities to the child, their memories and feelings of grief and loss do not disappear. Many continue to carry a sense of guilt and experience anniversary reactions of sorrow on the date of the child's birth.

The teen-age birth father is often emotionally and legally involved in the adoption process. It has been shown that he is frequently quite concerned about his child's welfare. Both the adolescent birth parents require counseling and reassurance during the decision making process throughout the pregnancy and after delivery of the baby. If unmarried, they must be told that it has become more socially acceptable for young unmarried women to keep their children; the decision to relinquish the newborn for adoption is, in itself, not automatic. This adds to the importance of the physician's guidance.

CONCLUSION

The need for counseling and support arises frequently during the maturation of the adoptive family. Psychologic aspects of the adoptee probably cannot be fully comprehended by anyone who is not adopted. The adoptive parents react to the adoption procedure because of sensitization and factors that exist prior to the adoptive procedure itself. They continue to respond to the unique intrinsic characteristics of the maturing adoptee. Adoption is truly a life-long process. The physician therefore should be aware of problems inherent in various stages of the maturing adoptive family. It is desirable to maintain open communication on this matter with all members of the involved family. In this manner the physician can more readily recognize possible problems, even in the face of denial, and be able to give appropriate support or counsel. At times referral to psychiatrists or trained adoption agency personnel may still be required.

CONTEMPORARY ISSUES IN ADOPTION

ADOPTION OF THE "HARD TO PLACE" CHILD

Physically and developmentally disabled children are adoptable. The adoption of these children is a placement that is not familiar to many physicians, and it is one that merits attention. Placement of children with special needs (those who are older or emotionally or physically handicapped) has become the successful focus of adoptive activities in recent years. A recent statement by the American Academy of Pediatrics declares that:

Many families are eager to take these children into their homes. These families are not saints, nor are they emotionally unbalanced. They are just people who feel that they can accept a disabled person into their family and provide the kind of stability, support, nurturance and guidance that is so greatly needed.

These families often seek the advice of physicians, lawyers, and clergymen in helping them make their final decision. These professionals must know the facts and present them to inquiring families in a manner that is as free from personal bias as possible.

The facts are:

1. An increasing number of developmentally disabled children are being adopted annually.

2. Research has shown that there is a lower incidence of disruption (failed placements) with this population than with other children with "special needs."

3. The level of functioning of the child is not a determining factor in disruption.

4. Children who live within a stable and permanent family environment tend to maximize their potential more fully than those who live in institutional settings.

5. The adoptive siblings of these children are

accepting of them if the introduction of the child into the family is handled with openness and sensitivity.

6. There are parent support groups that assist adoptive families on a continuing basis as well as during stressful times.

In the United States all 50 states and the District of Columbia have some form of subsidized adoption or financial assistance available to couples who adopt handicapped children. Public Law 96–272, passed in 1980, allows states to continue Medicaid coverage for a child with special needs who is placed in a adoptive family. With this assistance, families need not assume a potentially large financial burden if the child requires special medical attention or other types of care and treatment.

In order for all children to be given the right to permanence and all the accompanying advantages, it is incumbent upon members of all professions concerned with the welfare of children to learn the facts. The fact is that developmentally disabled children are adoptable (Committee on Adoption and Dependent Care, 1981).

TRANSRACIAL ADOPTION

Transracial adoption is the adoption of a child whose race is different from that of the parents, primarily black children adopted by white parents. This form of adoption, now on the decline, reached a peak in the United States in the late 1960's, partly in response to an increase in social idealism and partly because of the large number of black children waiting for adoption.

Although it is generally preferable, other things being equal, to place children in families of their own racial background, transracial adoption can be viewed as a viable alternative. Transracially adopting parents must be special in their abilities to cope with the pressures on them and their children.

White parents contemplating adopting a nonwhite child must first be secure in their marriage. Both adoptive parents must be in unequivocal agreement that they can accept this child as a member of their family, not only during infancy but as he grows into adolescence and adulthood. They must be able to share in his cultural heritage and in the development of his identity. Parents should be made aware of the resources available in their area for counseling, if problems arise, and also for general group fellowship in organizations formed by parents who have adopted children of different races.

The black child knows early that he is in an adoptive home of white parents. There should be early and continued effort to teach the child what is known of his biologic parents and to help him develop pride in his ancestry. Although family ties may be strong, the child must leave the confines of the home and become exposed to a community of his peers and their parents who may not share the same views about race as the family with whom he lives. School is often the source of the first continuous pressure the child faces. If the pupils at school are of several different races and nationalities, the child will adjust more easily than if he is the only one of a different race. Adolescence is often a difficult period for families and their biologic children. The problems of identity and the quest for maturity, as well as increased social interactions, are accentuated in the family whose parents and children differ racially. The child's selection of friends and associates is made from available acquaintances, which may be few. White parents who adopt black children often do not approve of inter-racial dating.

In all adoptions the welfare of the child must receive highest priority. The parents' unspoken attitudes toward black people may be the critical component in determining how the black child in a white family develops his identity as he matures into adulthood. The pressures on a black child in an all white family and community are great; unfaltering love and true concern will help to prevent the development of feelings of inferiority and hostility.

INTERCOUNTRY ADOPTION

Intercountry adoption is the adoption of a child who is a native of one country by residents of another country and the subsequent removal of the child from his native country to the second country. International assistance should first support services that strengthen the biologic family and child welfare within the native country. In addition to this assistance, there are occasions when the interest of the child may be met by intercountry adoption. Most recent intercountry adoptees have arrived as refugees from South Asia. A much smaller number of adoptees have been available in India and South or Central America.

The motivation of parents seeking to adopt children from other countries should be examined, as should the motivation of any prospective adoptive parents. "Humanitarian" motivations as the sole reason for adoption do not provide a sound foundation for beginning a parent-child relationship.

The overwhelming psychologic problem seen in certain groups has been culture shock—the absence of the familiar and the presence of totally different food, clothing, speech, and a multitude of habits and customs. Many children from poverty or war-torn areas may be malnourished and underdeveloped. Considerable "catch-up" in development and growth may occur, especially if these children are placed before they are three years old. The degree of early malnutrition and its effects vary among refugee groups. Disruption of the development of speech is an important factor in behavioral and emotional problems seen among adopted refugee children.

Specific disease entities indigenous to the part of the world in which the child was born should be a consideration. Common problems of Asian children include intestinal parasites, tuberculosis, anemia, malaria (including congenital malaria), and hepatitis. *Pneumocystis carinii* infections may

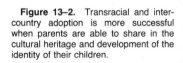

Figure 13–2. Transracial and intercountry adoption is more successful when parents are able to share in the cultural heritage and development of the identity of their children.

occur in malnourished infants. The history and physical examination may suggest other rare disorders. The physician should be reminded that children have occasionally arrived in the United States with obvious severe health problems not noted in the medical report that preceded the child.

Studies have shown that adopted Indochinese refugees have demonstrated a great ability for developmental "catch-up." They have shown the potential for excelling in school after a period of adaptation. A majority of the 1975 "baby-lift" era orphans have done well in all respects; in general there has been a better than anticipated adjustment resulting in wholesome, healthy, and happy adoptive family situations. Numerous psychologic problems, shown, for example, by excessive episodes of night terrors, have usually responded quite well in the new family environment.

Preplacement and postplacement social and advisory services are essential to successful outcomes. The complexities of intercountry adoptions and the difficulties inherent in them may not be fully understood at any time. We do know that the services of volunteer adoptive parent groups prior to the adoption have provided both parents and pediatricians with invaluable aid. They are essential in preparing the adoptive family for the difficulties they are likely to encounter early in the child's placement into their home. Advance preparation is essential in coping with and correcting such problems (Fig. 13–2).

More detailed information concerning intercountry adoption may be found in publications provided by the United States Office of Human Development (Committee on Adoption and Dependent Care, 1981).

ADOPTION OF AMERICAN INDIANS

An alarmingly high percentage of American Indian children have been separated from their families and tribes, having been placed in non-Indian foster or adoptive homes. Poverty, poor housing, lack of modern plumbing, and overcrowding often have been cited by social workers as proof of parental neglect and used as grounds for beginning custody hearings. Some children have been removed from their families on such vague grounds as "social deprivation." Indian communities often have been shocked to learn that people they considered excellent parents had been judged unfit by non-Indian social workers. Eighty-five per cent of all Indian children in foster care reside in non-Indian homes. In some states one of every eight Indian children under age 18 is living in white adoptive homes; in 1971–1972 nearly one in every four Indian infants under one

year of age was adopted by a non-Indian family—many in states other than the home state. In several states the ratio of Indian children placed in foster or adoptive care was (at a per capita rate) five to 13 times greater than in non-Indian children.

Physicians should be aware of the provisions of the Indian Welfare Act of November 8, 1978 (P.L. 95–608), which "responds to the pleas of Indian tribes to stop the 'unwarranted' removal of Indian children from their families and communities." The law established federal standards for the removal of Indian children from their families and the placement of such children in foster or adoptive homes that reflect the values of the local Indian culture.

SINGLE PARENT ADOPTION

The one parent family as a lifestyle is increasingly accepted by society. As a result, community reaction to one parent adoptions has been increasingly positive. Because there are currently fewer babies or young children available for adoptive placement, agencies have reached out to meet the needs of some of the thousands of foster children, age six or older, who need permanent homes. Some of these children have medical or emotional problems that are not best met within two parent families, most of which still continue to adopt children under six.

A second reason agencies place older children with single adoptive parents is because many of these children require an intense one to one relationship owing to early deprivation, abandonment, abuse, rejection, or neglect. Thus, the single adoptive parent family is often the preferred situation, because it gives the child the best opportunity to grow and develop into mature adulthood. The goal is to arrive at a plan that will best meet the specific child's needs (Los Angeles County Department of Adoptions, 1979).

The traits sought in single parents are the same as those looked for in any person who wishes to adopt an older child. Single adoptive parents should be sensitive to children's needs and capable of giving a child a sense of "belonging." Single parents find that they need a high energy level, along with a feeling of confidence, flexibility, and, above all, understanding and patience. Ideally single adoptive parents are mature, stable people. They are willing to wait for love and gratification, while they steadfastly nurture and guide their children. In the "study" process, single parent applicants go through such careful self-examination that all but the most committed people weed themselves out. Single applicants choosing to be adoptive parents have usually attained some security in their work. They are willing to take risks and have shown skills in solving their own problems.

Although single adoptive parents call upon their own inner resources in order to cope with the stresses of instant parenthood, it is believed that they will profit from the added involvement and emotional support of family and friends (Los Angeles County Department of Adoptions, 1979).

"OPEN ADOPTION"

Annually thousands of single birth parents are now reluctant to give up all contacts with their children and refuse to relinquish them entirely. Increasing numbers of parents are practicing "open adoption," wherein the expectant birth mother meets the adoptive parents and participates in the adoption process. At times both sets of parents may actually participate together in a "natural childbirth" situation. The birth mother then retains her right to continue contact and knowledge of the child's whereabouts and welfare, but relinquishes all legal, moral, and nurturing rights to the child. It is too early to accumulate adequate case histories to comment upon this form of adoption.

ARTIFICIAL INSEMINATION

When the husband has been proven to be infertile secondary to azospermia or oligospermia, the couple may choose artificial insemination by a donor (AID). The prevalence has increased markedly since the 1950's; it was estimated in the 1970's that 10,000 to 20,000 such children were being born annually in the United States.

Anonymity has been the established norm for the AID procedure. The donors are often medical students or graduate students who are performing the procedure as a means of defraying educational expenses.

In the case of the AID child there is no legal adoption procedure to clarify the father's status. The birth certificate is falsified, creating a situation of possible perjury if the attending physician is aware of the artificial conception. As a result, many inseminators advise their patients to consult another obstetrician who is unaware of the true nature of the pregnancy and therefore not aware of his own deception when filling out the birth certificate. AID mothers report a fear of impulsively blurting out the secret. Although the "secret" may serve as a bond holding the couple together, it also affects meaningful open communication and intimacy because of the pressures to maintain the secrecy from the outside world.

Many couples attempt never to bring up the subject of AID again.

Whether AID should be treated like adoption, with careful records and opportunity for the child to trace his genetic parents has become an issue for consideration. Although the traumatic effects of the "secret" on the family system are apparent, the revelation of AID origin to the child could also be traumatic and confusing. Opinions at this time indicate that it is probably better not to disclose this fact to the child (Sorosky et al., 1981).

It is essential that potential AID parents attempt to ascertain that the donor has been screened meticulously for genetic diseases. Careful psychologic screening methods must be devised to select donors who will be able to accept their part in the procedure years later, and parents who will be able to handle the controversial aspects of artificial insemination without its affecting their role as parents to the children so conceived (Sorosky et al., 1981).

FOSTER CHILD CARE

There are more than 500,000 children in foster care services in the United States. Most reside among 150,000 foster families during any given year; the remainder are in larger "institutionalized" care centers. A majority of them have been placed in care because of severe personal and social problems that have afflicted their parents and not as a result of the child's own problems. Such factors include parental neglect, other forms of abuse, abandonment, mental illness, or other coping impairments. Only 20 per cent of recent placements were for reasons relating to the manifest difficulties of the children themselves, as behavioral problems beyond the control of their parents. These statistics continue: the average placement has been longer than five years; 68 per cent of the placements range in duration from four to eight years, with frequent movement of the child from one family setting to another (Sokoloff, 1979b).

Many social and legal problems persist in delaying either the return of the children to their parents after appropriate remedying of the home conditions or the eventual adoption of the children.

The termination of parental rights and the description of unfit parents present legal problems that vary among the states. In some states children cannot be relinquished for adoption for more than five years, if the father has terminated his rights to the child but the mother refused to give up her rights—even if she has not attempted to see the child during that length of time. Recent federal legislation (P.L. 95–266) seeks to "promote the healthy development of children who would benefit from adoption by facilitating their placement in adoptive homes" (and out of the prolonged foster environment). It is to be hoped that this legislation will improve the situation and that the foster child will more rapidly obtain a sense of security and family identity through earlier adoption.

It has become apparent that parental neglect of foster children has been replaced by community neglect. These children represent a population at high risk.

At the time of placement most of these children have had only erratic physical and emotional care and little or no medical care. One recent study revealed handicaps in 40 per cent of the children monitored; 15 per cent of these children had multiple handicaps and 33 per cent had various physical ills. Twenty per cent had not even been evaluated.

The health care needs in the emergency, short term placement of a healthy child are very different from those of a long term placement of a handcapped child. However, the Academy of Pediatrics strongly recommends, as a minimum, the following guideline:

> The adequate provision for safeguarding and promoting the health of children in routine foster care should include periodic health supervision examination, appropriate medical care for the ill child or child with special health problems, and dental care.

It is urged that foster families having access to adequate continuing medical care for themselves and their children incorporate the child into their family health care system. By using the same health services as the foster family, the child would not be singled out for different treatment and would become more integrated into the family life. When this is not possible, basic medical services should be provided through the agency or other resources whose services are coordinated with a total plan for the child, thus providing continuity of medical care.

Ideal health services for the child should include preplacement examinations and frequent medical examinations for appraisal of physical growth and development, health status, and the effect of emotional and social factors on the child. These children require extremely close evaluation for learning disabilities and emergence of other psychosocial problems. Immunizations and administration of routine diagnostic laboratory procedures must be kept current.

Inadequate medical records usually result in inadequate and inappropriate medical care for foster children. It is imperative that immunization records be maintained to ensure that proper immunizations are not erroneously omitted or duplicated. Well kept records will maintain a complete current medical history in the hands of the physician—a factor that is so extremely important in such cases as a suspicious abdomen in a child with a scar that may or may not represent an earlier appendectomy. It is frequently beneficial to have two identical sets of medical records—one kept by the family keeping the foster child and the other in the office of the social worker.

The pediatrician should encourage the foster parents to maintain a scrapbook of pictures and other such memories of the child. Some foster care agencies, as in Los Angeles, have been using a "scrapbook experience" to help the child recall his childhood events while growing up in one home

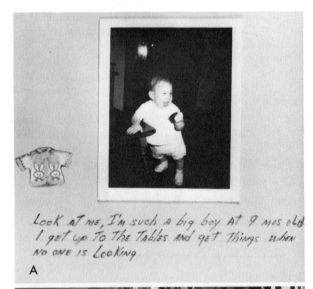

Look at me, I'm such a big boy at 9 mos old. I get up to the tables and get things when no one is looking.

A

B

Figure 13–3. *A,* A page from a scrapbook kept by a foster parent to help maintain an identity for a foster child. This book was kept until the child was adopted at nearly one year of age. *B,* The recipient of the scrapbook.

and then another. Later, especially if there is an adoption, this book serves as a continuing source of identity for the child to look back upon himself while other family members do the same with their own albums. Thoughtful foster parents and social workers maintain the book with photographs of the child during infancy and childhood. Later, other personal items of significance, as perhaps hair from a first haircut or pictures of pets and of other foster family members, may be added to the memories. At any later time the child will then be capable of reviewing his background and can actively contribute to the scrapbook articles of significance (Fig. 13–3).

BURTON Z. SOKOLOFF

REFERENCES

American Academy of Pediatrics: Adoption of Children. Ed. 3. Evanston, Illinois, American Academy of Pediatrics, 1973.

Bohman, M.: A study of adopted children, their background, environment and adjustment. Acta Pediat. Scand., 61:90, 1972.

Bose, C. L., Ercoles, A. J., Lester, A. G., Hunter, R. S., and Barrett, J. R.: Relactation by mothers of sick and premature infants. Pediatrics, 67:565, 1981.

Brodzinsky, D. M., Pappas, L., Singer, L. M. and Braff, A. M.: Children's conception of adoption. J. Ped. Psychol., 6:177, 1981.

Carey, W B.: Adopting children: the medical aspects. Children Today, Jan-Feb., 1974.

Carey, W. B.: Induced lactation. Am. J. Dis. Child., 135:973, 1981.

Carey, W. B., Lipton, W. L., and Myers, R. A.: Temperament in adopted and foster babies. Child Welfare, 53:352, 1974.

Committee on Adoption and Dependent Care: The role of the pediatrician in adoption with reference to "the right to know": an update. Pediatrics, 67:305, 1981.

Committee on Adoption and Dependent Care: Adoption of the hard-to-place child. Pediatrics, 68:598, 1981.

Committee on Adoption and Dependent Care: Intercountry adoption. Pediatrics, 68:596, 1981.

Eldred, C. A., Rosenthal, D., Wender, P. H., Kety, S. S., Schulsinger, F., Welner, J., and Jacobsen, B.: Some aspects of adoption in selected samples of adult adoptees. Am. J. Orthopsychiat., 46:279, 1976.

Kadushin, A.: Adopting Older Children. New York, Columbia University Press, 1970.

Kirk, H. D., Jonasson, K., and Fish, A. D.: Are adopted children especially vulnerable to stress? Arch. Gen. Psychiatry, 14:291, 1966.

Leeding, A.: Access to Birth Records. A Report to the Association of British Adoption and Fostering Agencies. London, October 1977.

Los Angeles County Department of Adoptions: Single parent adoption—a fact sheet. Los Angeles, 1979.

Mech, E. V.: Adoption: a policy perspective. In Caldwell, B. M., and Ricciuti, H.: Review of Child Development Research. Chicago, University of Chicago Press, 1973.

Seglow, Å, Pringle, M. K., and Wedge, P.: Growing Up Adopted. Windsor, England, National Foundation for Educational Research, 1972.

Sokoloff, B.: Adoptive families—needs for counseling. Clin. Pediatr., 18:184, 1979a.

Sokoloff, B.: Adoption and foster care—the pediatrician's role. Peds. Rev., 1:57, 1979b.

Sorosky, A. D., Baran, A., and Pannor, R.: The Adoption Triangle. New York, Doubleday Publishing Co., 1978.

Sorosky, A. D., Baran, A., and Pannor, R.: Adoption. In Kaplan, H. I., Freedman, A. M., and Sadock, B. J. (Editors): Comprehensive Textbook of Psychiatry III. Baltimore, Maryland, The Williams & Wilkins Co., 1980.

Sorosky, A. D., Baran, A., and Pannor, R.: Infertility, adoption, and artificial insemination. In Pasnau, R. O. (Editors): Psychosocial Aspects of Medical Practice. Menlo Park, Addison-Wesley Press, 1981, Vol. II.

Ward, M.: Parental bonding in older child adoptions. Child Welfare, 60:24, 1981.

Wessel, M. A.: The pediatrician and adoption. N. Engl. J. Med., 262:446, 1960.

Wolff, S.: The fate of the adopted child. Arch. Dis. Child., 49:165, 1974.

BOOKS TO RECOMMEND TO ADOPTIVE FAMILIES

Dywasuk, C. T.: Adoption—Is It For You? New York, Harper and Row, 1973.

Kirk, H. D.: Shared Fate: A Theory of Adoption and Mental Health. New York, Free Press of Glencoe, 1964.

Raymond, L.: Adoption and After. New York, Harper and Row, 1974.

Sorosky, A. D., Baran, A., and Pannor, R.: The Adoption Triangle. New York, Doubleday Publishing Co., 1978.

Tizard, B.: Adoption—A Second Chance. New York, Free Press, 1976.

BOOK TO RECOMMEND TO ADOPTED CHILDREN

Livingston, C.: Why Was I Adopted? Secaucus, New Jersey, Lyle Stuart, Inc., 1978.

13D Separation, Divorce, and Remarriage

Changes in family relationships wrought by divorce and remarriage have significantly altered the average expectable experience of growing up. As so often happens, these changes become commonplace before we have had a chance to assess the complex balance of losses and benefits to the individual families and to society.

Inevitably, research in social and behavioral science lags behind changes that occur in the social system. This is especially true for issues regarding children, because the longitudinal perspective is crucial to our understanding of the effects of any event or series of events on the child's development. There is reason to believe that the full effect on a child of stress such as divorce or loss of a parent cannot be truly assessed until that child grows into adulthood and in turn becomes a marital partner and a parent (Kulka and Weingarten, 1979; Wallerstein, in press).

The rise in divorce and remarriage also has an impact on children beyond those whose lives are directly affected. Children whose parents are not divorcing are increasingly fearful about this possibility within their own families. Teachers have reported children at school who ask anxiously, "Will my mom and dad divorce now? They had a big fight last night.'' Another effect, more difficult to assess, is whether the adolescent or the young adult of today is more burdened by fears of commitment to a relationship that has some substantial likelihood of dissolving. Perhaps the recognition that marital relationships may be unreliable affects behaviors in many spheres of our society. Divorce and marriage are indisputably intertwined, and the way society deals with divorce reflects—and in turn influences—the institutions of marriage and family.

Certain consequences of these changes are relevant to the physician's task and role. Divorce-engendered stress and the changes in the family-child relationship bear directly on psychological health and development at the time of the marital rupture and during the several years that follow.

Perhaps most relevant to the physician is the finding that the divorcing family with children is likely to need guidance at the time of the marital rupture and during the extended aftermath. Similarly, the remarried family is likely to need advice, especially during the first years of the remarriage. By and large, adults in divorce and remarriage lack both role models and experience for dealing with the many attendant problems. All too often they feel alone, facing complex difficulties with no place to turn for help at a critical time in their own lives and in the lives of their children.

As a consequence physicians and members of other disciplines whose work brings them into significant contact with these troubled families can expect to be called upon to provide information and counsel to the divorcing parents, to the children, to the adolescents, and to other members of the extended family. The counselor who is knowledgeable in these matters and who welcomes inquiries is in a position to make a major, lasting contribution to the children's psychological health.

There have also been reports that the probability of divorce increases when the family is under stress from other sources, such as following the diagnosis of a severe or fatal illness in a child or other family member (Kaplan et al., 1976). The decision to divorce is not necessarily related to marital unhappiness but may reflect a response to stress in the family from various sources (Wallerstein and Kelly, 1980b). In such families, children are at especially high risk because family supports collapse at the time when they are greatly needed, adding to the cumulative stress affecting the child. The physician whose task it is to convey to the family a message of serious illness in a child should keep in mind the potential far-reaching effects of the tragic message. Efforts should be made to discuss the child's condition with both parents and to prepare them for the reverberations within the family. It may well be indicated for the physician to meet with the parents on several occasions and to meet with the children present

as well, in order to support and bolster the family as a protective structure.

RESEARCH ABOUT CHILDREN IN DIVORCING FAMILIES

The observations reported here are largely the result of an extensive clinical investigation of 131 children and parents from 60 predominantly white, middle-class families in Northern California followed over a five-year period from the decisive marital separation. This study has become known as the California Children of Divorce Project. The results to date are described in the bibliographical citations of Wallerstein, Wallerstein and Kelly, Kelly and Wallerstein. Many of the observations have been corroborated by Hetherington and her coworkers in the only other published longitudinal study of children in divorced families, a Virginia-based study of 36 preschool children and their divorcing parents and 36 controls followed for a two-year period after the legal divorce (Hetherington et al., 1979a and b).

Knowledge regarding the impact of family rupture on children is in a beginning stage. The topic has, for many reasons, suffered neglect in social and behavioral science research. Traditionally divorce has been regarded by social scientists as a single disruptive event in an overtly conflicted marriage that was likely to eventuate in a single parent, mostly father-absent family. Only in recent years has professional awareness come to include the many dimensions of the divorce experience— its extended timetable, the wide range of family structures and relationship patterns that evolve after the divorce, and the far-reaching changes in the relation between child and *both* parents that often ensue.

INCIDENCE OF DIVORCE AND REMARRIAGE IN THE GENERAL POPULATION

The rise in the marital dissolutions that began in the late 1960's in the United States has leveled off but not diminished. Since 1973 one million new children a year have experienced their parents' divorce. The startling prediction made in the mid-1970's (Bane, 1976) that 30 to 40 per cent of children born in the 1970's will experience the divorce of their parents has in fact become a reality. The number of divorces occurring in 1980 to 1981, 1.2 million, will probably remain at that same figure for the next several years.

Over the past decade, the number of divorces has increased across all age groups but most dramatically among young adults. Estimates in 1981 suggest that marriages last an average of 6.6 years before dissolving. As a result, the children in divorcing families are younger, in the infant, toddler, and preschool age groups.

The number of remarriages has also risen. Clearly failure of the first marriage has not discouraged interest in trying again. Forty-one per cent of all marriages contracted in 1980 involved at least one partner who had been previously married (Glick, 1980). About three-fourths of all women and five-sixths of all men reenter marriage after a divorce (Furstenberg, 1980). By 1980 about 10 per cent of children below 18 resided with a parent and a step-parent. Within the next decade we may expect that 30 per cent of children born in 1981 will spend some time in their growing up years within a remarried family.

Practitioners in clinics and private practice alike have in recent years reported a substantial increase within the patient population of children from divorced and remarried families. According to a 1975 national survey of children between 7 and 12 years of age in divorced families, the number of children in need of psychological help was significantly higher than among their counterparts in intact families, even after allowing for socioeconomic and educational differences (Zill, 1978). The same report also noted that parent-child relationships in divorced families are likely to be burdened by parental difficulties and tensions. Reports from clinics have placed the children of divorce at between 50 and 80 per cent of the total number of children receiving psychological or psychiatric treatment (Kalter, 1977).

CUSTODIAL ARRANGEMENTS

Surprisingly, there is no source of national information on the actual awarding of custody of children other than family composition data from census surveys. Contrary to popular expectation, custody arrangements have changed very little throughout the nation over the past quarter-century. Approximately 90 per cent of children of divorce remain in their mother's custody. Although this proportion has not changed in the past 25 years, the number of divorces has increased and therefore the number of fathers with custody of their children has risen. Additionally, a small percentage of parents have opted for joint custody in those states in which this is permitted by law.

Understanding the relative merits and disadvantages of the various custody arrangements for children of different sexes and different ages has been hampered by a dearth of systematic or longitudinal research. Much of the often impassioned discussion of custody has occurred in the political arena.

Most divorcing families expect that the children will remain in the mother's custody. Children in the father's custody are likely to have been placed there either following the mother's abandonment or as a result of litigation. However, there appears to be little hard evidence for preferring mother custody over father custody when all other things are equal. One small study has suggested that school-age boys do somewhat better in father-custody homes, whereas girls do somewhat better in mother-custody homes (Santrock and Warshak, 1979). The findings of this limited study are insufficient to justify a shift in public policy. By and large, judges have continued to award custody in favor of mothers, especially for young children. Evidence is accumulating that the single parent, whether male or female, faces similar problems in raising children without the other parent and that the similarities are greater than the differences. Unquestionably, an extensive longitudinal comparison of father and mother custody is needed, especially with regard to boys and girls at different ages.

The movement for *joint custody* reflects the growing interchangeability of roles for men and women in the work place and in the family. This movement has gained momentum from the findings of the California Children of Divorce Project regarding the continuing psychological importance of the father to the child in the post-divorce family, from Steinman's (1981) report of successful joint custody, and from organizations of fathers who have complained about their limited access and secondary status in relationships with their children following divorce (Roman and Haddad, 1978).

Joint custody is designed to permit both parents to continue to be influential figures in their children's lives. Various living arrangements are possible for the child and the parents. A child may stay alternately with each parent on a nightly, weekly, monthly, or annual basis. The core of the joint-custody arrangement reflects a commitment to two homes for the child and to continuing cooperation and mutual decision making between the parents rather than a commitment to a specific division of the child's time. Within the next few years, it is likely that the number of children in the custody of their fathers and in joint custody will increase.

Under the following conditions joint custody appears to be a helpful arrangement for children:

— When both parents assign high priority to their parenting roles and are willing to make important life decisions in accord with this commitment.

— When both parents are sensitive observers of the child's changing needs and are respectful of the child's wishes and willing to accommodate living arrangements in response to the child's needs and priorities.

— When both parents respect each other as parents and are able to communicate effectively with each other about the children.

— When both parents can live with the ambiguities and differences that inevitably arise.

— When the children can go back and forth between the two homes without disruption of their own psychological adjustment or their activities. School-age children seem better able to tolerate these changes. Preschool children and adolescents are more likely to find the arrangement disruptive to their development. Individual differences among children are salient in the success or failure of joint custody.

ECONOMIC CONSEQUENCES OF DIVORCE

After a divorce, many families face problems that involve diminished financial resources, unemployment, child care arrangements, and social isolation—difficulties similar to those encountered by single-parent families. One major problem related to divorce that has been approached uneasily and with mixed success in different legal jurisdictions is how to enforce the collection of child-support payments from resistant, unreliable, or absent parents. Delinquency in the regular payment of child support is widespread. In a careful summary of available research, Weitzman (1980) reports that "after one year, less than half of the men (studied) are still paying support at all for their children." Moreover, Weitzman and Dixon (1979) found that the average amount paid for child support in a sample studied in California in 1972 and 1977 provided significantly less than one-half the cost of raising children during those years.

Levels of income and education are generally much higher among parents in two-parent families than among parents in families with only the mother present. Even in middle-class families, the decline in the standard of living for divorced mothers and their children is striking and occurs soon after the marital breakup.

THE PSYCHOLOGICAL CHARACTERISTICS OF THE DIVORCING POPULATION

Although a great many persons come to divorce without a history of psychological disorder, the representation of psychological difficulty in the divorcing population is high. Divorce and mental illness may also occur in conjunction (Blumenthal, 1967; Briscoe et al., 1973). Available data suggest that suicide, accidental death, and psychological illnesses such as serious depression are consider-

ably higher among divorced adults (Carter and Glick, 1976). According to a review of eleven studies (Gove, 1972a and b), the rate of mental illness among divorced men was more than five times higher than that for married men and nearly three times higher for divorced women than for married women. Although these data are correlational and do not establish causal links, a higher incidence of psychological illnesses among divorced parents will place children of divorce at greater risk of developing psychiatric and social problems than children whose families are intact.

In a comprehensive review of the psychological, psychiatric, and medical literature relevant to the adult divorced population, using a range of indices including admission to mental hospitals, Bloom and his coworkers (1979) concluded that divorce is associated with a wide variety of severe physical and emotional disorders. A significant number of divorcing people in the California Children of Divorce Project were troubled with psychological problems during the marriage. The most common problem reported by the women was chronic depression. Fewer men than women showed depression, but alcoholism, physical violence accompanying the alcoholism, severe emotional constriction, and social withdrawal were found in a large percentage of divorcing men in that study.

In many families in which there was a high degree of psychopathology, the more troubled person sought the divorce and was awarded custody of the children in the absence of litigation. It is of additional concern that many adults who are already vulnerable deteriorate further under the intense stress of the divorce experience.

A wide range of marriages comes to divorce. Contrary to popular belief, not all divorcing families are conflict-ridden, and many children consider themselves content within these families. Divorcing families include adults who are lonely and isolated from each other as well as those who share a rich history of common recreational interests and happy family holidays. In some families conflicts are explosive and highly visible; in others the differences are muted and concealed. Many couples who divorce are neither bleak nor conflict-ridden and appear outwardly to be well-functioning, close-knit families regarded as "good enough" by the children, the friends, and the neighbors. All these patterns of behavior occur in significant numbers, and no model predominates.

THE DYNAMICS OF THE DIVORCE

THE DECISIVE SEPARATION

The central event of the divorce from the standpoint of the child is the decisive separation of the parents and the departure of one parent from the home. The legal divorce is of little moment either to children or to most adolescents. The child's acute response to the family rupture, as well as early efforts to cope, commences with the parents' announcement of the impending divorce and one parent's departure. Thus, parents who announce their decision to separate but continue to reside together are apt to confuse their children needlessly, burdening them with an ambiguous situation that they have great difficulty comprehending. Such parents should be advised to separate if indeed divorce is their intention.

THE AMBIANCE OF THE DIVORCE PERIOD

There has been insufficient recognition of the disabling impact of divorce itself on the psychological functioning of the adult and particularly on the adult's capacity to carry on with his or her expected roles or responsibilities during the months immediately preceding the marital rupture, the height of the divorce crisis, and the year or more that follows the separation. Many adults who can live alongside each other for years without open anger become openly hostile as the marriage ends. Others who suffer from chronic or moderate depression may become severely depressed or agitated. In the majority of households, feelings of bitterness and scenes of conflict increase. Much of the anger and depression among parents reflects the fact that in a family with children, the decision to divorce is rarely mutual. Most often the divorce is sought by one member of the couple and opposed or reluctantly accepted by the other. This difference between the adults and the humiliation engendered by the one-sided decision set the stage for the interaction at the time of the divorce and during the years that follow.

Perhaps centrally important from the child's perspective is the bizarre, although short-lived, behavior frequently exhibited by the parents, such as verbal accusations and threats as well as rage accompanied by violence and depression, and possibly suicidal ideation. A marriage that has been humdrum for years is likely to spring to life with the decision to divorce, and the unrestrained expression of aggressive sexual impulses and intense feelings tends to dominate households that were accustomed to a more circumspect way of life prior to the divorce decision.

DIMINISHED CAPACITY TO PARENT

The acute phase of the divorce is characterized by diminished physical and psychological availa-

bility of the parents. Newly employed parents are likely to leave children alone after school or for long periods with new sitters in a strange setting. Sometimes the burden of work for the parent is such that very young children prepare their own lunches, get themselves to school, and put themselves to bed. The practical changes of life take an immense toll in reducing the time and attention available for children and so increase a child's anxiety and anger.

In addition, new relationships become important and demand considerable time. The flurry of social or sexual activity that often immediately follows the marital rupture is likely to absorb weekend time and evenings and to take custodial and noncustodial parents out of the home or to bring in new sexual partners.

At the outset, the visiting and noncustodial parent's house is often inadequate for extended visits, and early visits tend to be uneasy, tense, or on the run. Also, new partners, some of whom have children, are likely to be present during these early visits, and children may feel aggrieved because their access to the parent has been blocked.

The household disorder that prevails in the aftermath of divorce, the rising tempers in both mother and child—especially between mothers and small sons—seem to eventuate in a sense of reduced competence and a greater sense of helplessness in the mother, provoking a continuing cycle of mutually interactive destructive behaviors (Hetherington et al., 1979). Further cause for household deterioration can be found in mothers' fear of rejection by their children. Wishing to avoid reproach for the divorce, and fearful of invoking the children's anger, she may retreat from requiring certain standards of behavior or meeting ordinary household standards.

These problems, added to the preoccupation with his or her own decisions and concerns, cause the custodial parent to be less competent as a parent and less able to maintain the structure of the household. As a result, there is mounting disorder, less discipline, less caretaking, and a sense among the children that the divorce has led to the loss of not one but both parents.

EXPERIENCES OF THE CHILD AND ADOLESCENT IN DIVORCE

The separation and its aftermath are remembered by many children and adolescents as the most stressful period of their lives. Family rupture evokes an acute sense of shock, intense fear, and grieving that children find overwhelming. A significant number of children have no prior knowledge of the divorce and did not understand that their parents were unhappy in the marriage. In many divorcing families children enjoy a close relationship with one parent, perhaps because of the absence of a good marital relationship. One can only imagine the terrible distress of the child who has spent every weekend in the company of his father and is now told that his parents are divorcing and that the father is leaving home.

Knowing beforehand of his parents' marital conflict does not necessarily mute the child's response to the divorce nor does it distinguish such a child from children who are not consciously aware of their parents' marital unhappiness. There is little evidence that long-standing foreknowledge and ignorance represent critical variables that determine the child's response to family rupture. Perhaps some events are too overwhelmingly stressful to master far in advance. For the child who has not yet reached adolescence, family breakup may fall into this category.

THE ABSENCE OF SUPPORT AT THE SEPARATION

One of the more striking findings of the California study was the loneliness of the child in the divorcing family. Over half the children felt that their father was entirely insensitive to their distress at this critical time, and one-third of the children felt that their mother was entirely unaware of their distress. Children age 9 and older were acutely aware of the lapses in parenting and felt neglected and aggrieved, whereas the younger children felt in danger of imminent abandonment.

The children experienced an extraordinary lack of support outside the family. Only 25 per cent were helped by grandparents or other members of the extended family. Those children who did have the support of grandparents appeared to benefit considerably from special concern and attention. Outside of the school, few institutions touched these children's lives, according to the California findings. Fewer than 5 per cent were approached by a member of a church or synagogue, although half the families in the study were active members of religious groups. All in all, less than 10 per cent of the children received any help or support from adults outside the family. It is particularly relevant that although most of the children were regularly under the care of pediatricians, none of these physicians was contacted at the time of the crisis and none talked to the children.

It is often useful to encourage parents to seek help from members of the extended family, including grandparents. Although grandparents often are unavailable, there is evidence that a significant number of grandparents are hesitant about intruding at the time of crisis and, if called upon, would happily make themselves available as a resource for their grandchildren. Grandparents and members of the extended family can be especially helpful in visiting regularly, providing a special treat, babysitting, and other forms of support. In addition, parents should be encouraged to take their concerns about the children directly to the school and to enlist the teacher's help during the crisis, requesting perhaps some special attention for the child or some special words of encouragement or suggesting an assignment related to the child's interest that might provide both pleasure and recognition at this difficult time.

The physician who knows the child can convey to him directly a compassionate recognition of his many feelings, fears, and concerns. Such acknowledgment of the stress resulting from family rupture can be beneficial in allaying the child's sense of loneliness and of diminished adult support and availability, and the child's fears of being lost or overlooked in the welter of sudden family changes.

TELLING THE CHILD ABOUT THE DIVORCE

Children who are told of the divorce before the parent's departure from the household and who are also assured that they will continue to see the departing parent are significantly calmer than those who must confront the divorce without any preparation. It is essential to inform the child and to allow several discussions for the child to believe and understand the impending breakup of his family.

Parents often experience difficulty in telling their children about the decision to divorce, not knowing how much to reveal about their marital intimacy. They may be confused about details, when or where to tell their children and how far ahead of one parent's departure from the household, or whether they should tell the children together or separately. Moreover, parents are apprehensive that their children may be unhappy, frightened, or angered by their decision and, feeling somewhat battered and depleted by their own ordeal, are often reluctant to take on this issue. As a result, children—especially young children—are often not made aware of the final rupture and may even wake up one morning to find that one parent has left.

Ideally, informing the child about the divorce is not a single, separate announcement but part of a supportive process over time, which the parents provide for the child and which enables the child to understand the divorce and thus begin to cope with the family changes. The purpose of the initial discussion is to begin to explain the divorce so that it appears as a rational step to the child and, at the same time, to prepare the child for the changes that lie ahead. Because this is only the first step in the process of overall support for the child, communication about the divorce should be kept open and the parent(s) should expect and encourage continued questions and repetitions in the normal course of the household routines and especially during times of particular intimacy between parent and child.

WHAT SHOULD CHILDREN KNOW?

Children should understand what divorce means, what the family structure will be like in the immediate future, and what immediate changes they can expect in their living arrangements and daily routines. They need to understand that they will continue to be cared for in the present and into the future and that their needs will be considered and their wishes given some priority in the new post-divorce family. Finally, they need the explicit assurance that they will not get "lost in the shuffle."

Moreover, children need to believe that their relationship will endure with each parent and that they will not be abandoned by either parent. The pattern of visiting that will be established with the noncustodial parent should be explained to them. They need to feel confident that the father has not disappeared, and they need to be protected from their worry that he has no place to sleep and no one who cares for him. (It is helpful for children to see where the father is residing in order to help allay specific worries about him.) The frequency of the visiting should be based on the child's needs and the father-child relationship and should not be determined by the degree of conflict between mother and father.

Children should be apprised of the reasons for the divorce and assured that their parents are rational people who thought carefully before making such a weighty decision that would so powerfully affect all their lives. Children need to know that their parents have tried and exhausted every resource before coming to the divorce. They should be offered an explanation appropriate to their age and level of understanding. Young children can be told that the parents are unhappy and fighting, and that the purpose of the divorce is to bring an end to both the fighting and the unhappiness. Older children should be informed of the various attempts the parents made to resolve their conflicts and the parents' disappointment and sorrow with the marital failure. Details of sexual infidelity are not helpful to children. On the other hand, if the child is aware of an extramarital relationship, it should not be misrepresented but should be represented as symptomatic of the unhappiness within the marriage.

Finally, children need to understand clearly that they did not cause the divorce, that their efforts cannot mend the broken marriage, and that the divorce represents the parents' decision and is entirely separate from the children's involvement. They need to be assured that neither parent expects them to take sides against the other. They need permission to love both parents and to experience fully their feelings of sadness, anger, and disappointment. They need to understand that their family structure has not been destroyed and that they can expect a return to order and familiar routine in their lives following the transitional instability and difficulty of the immediate future.

THE IMPACT OF SEPARATION AND DIVORCE ON CHILDREN AND ADOLESCENTS

It is important to distinguish the initial responses to and the impact of divorce from its long-lasting effects. Initial responses are, of course, more observable, whereas the long-lasting impact on development is complex and impossible to predict. Developmental factors are critical in the response of children and adolescents at the time of the marital rupture. Even though there may be significant differences among individuals, the child's age and developmental stage appear to be the most important factors governing the initial response. The child's stage of development profoundly influences his need for and expectation of the parents, the perception and understanding of the divorce, and his available armamentarium of defensive and coping strategies.

One of the major findings of research on divorce has been the recognition of patterns of response within different age groups. The age groups identified were preschool (2½ to 5 years of age), early school-age (6 to 8 years), later school-age (9 to 12 years), and adolescents. These groupings do not reflect a priori categories but instead the unexpected commonalities that were discovered in the children's responses.

PRESCHOOL CHILDREN

Preschool children are most likely to show regression following the decisive separation. This regression usually affects the most recent developmental achievement of the child, e.g., in toilet training, going to nursery school, venturing into carpools unattended by the parent, playing with peers, or remaining at nursery school unattended by mother. Youngsters may return to thumb-sucking, to security blankets and to increased masturbatory activity. Intensified fears are evoked by routine separations from the custodial parent during the day and at bedtime. Sleep disturbances are particularly frequent and appear to be linked to the young child's terrifying preoccupation with the thought of awakening to an empty household and his fear of abandonment by both parents. Indeed, the central concerns of many young children are abandonment and starvation. Many play out elaborate play scenes which portray adults caring for children and children caring for children. These children are also likely to become irritable and demanding with parents and to behave aggressively with younger siblings and with their peers (Wallerstein and Kelly, 1975).

The physician can help by providing the parent with direct guidance. Whenever possible, household routines should be maintained and the child's life should be stabilized to the extent that the parent is able to do so. In addition, the young child should be assured repeatedly that the parent loves the child and will continue to care for him. At each separation the child should be told when the parent will return. At bedtime the parent is well advised to spend additional time with bath rituals and bedtime stories. The parent should clarify where she will be when the child is asleep and when he awakens.

It is of utmost importance that the child visit with the departed parent whenever possible. Very young children fear that disasters have overtaken the parent whom they do not see daily. Children at a very young age are more likely to feel that they have caused the divorce. The child should be told explicitly that this is not so and that the parents decided to divorce to improve the quality of life for the entire family.

The young child is easier to reassure and to comfort directly than are other siblings. Often the new symptoms will disappear in a few days or weeks with direct assurance by the parent and the reinstating of stable routines that enhance the child's sense of predictability and order in the household. If the child is fearful about leaving the home or going to nursery school, the parent would be well advised either to go with him for a few days and then gradually leave or to permit the child to remain at home temporarily in recognition of his fear of loss. All such parental behavior should be accompanied by verbal assurance and an explanation that the child's fear is related directly to the marital separation.

CHILDREN AGES 5 TO 8

Children in this age group are likely to show open grieving, including sighing and sobbing. They are preoccupied with feelings of rejection, with longing for the departed parent, and with the fear that they will never see him or her again. They share the terrifying and humiliating fantasy of being replaced, "Will my Daddy get a new dog? a new Mommy? a new little boy?" Children are likely to weave fantasies that can be conceptualized as "Madame Butterfly" fantasies, asserting that the departed father will some day return—"he loved me the best"—and unable to believe that the divorce will endure. Children in this group feel torn by conflicting loyalties and by guilt if they turn toward one parent. In the California study, about half the children in this age group suffered a precipitous decline in their level of school work and reported difficulty in concentrating and worry about their parents. A few children at this age appeared untroubled and were able to maintain their usual composure and their full range of activities (Kelly and Wallerstein, 1976).

During the visit with the noncustodial parent, the children are likely to be well-behaved out of fear that the visiting parent may not reappear; upon their return home, they may be irritable with the custodial parent as a reaction to having been on such good behavior during the visit. On the other hand, they may become fearful of expressing anger with either parent. Boys are likely to feel especially vulnerable to the mother's anger and may be fearful of being "thrown out." They are likely to assume that the divorce occurred because of a fight in which the father was ejected from the household, and they fear that the same fate will befall them if they disobey their mother.

The physician can be helpful to parents of children in this age group by requesting to see both parents in order to explain the importance of allowing the child to love both parents without feeling that he is called upon to make choices in the marital conflict. Parents should be urged to make it easy for the child to cross back and forth between the parents and, in fact, should try to help the child with these different transitions. In addition, each parent should make an effort to reduce the child's worry about himself and about the parents. The child should be repeatedly and simply reassured by each parent that the conditions of life will improve and that the child should not worry about either of them but should try to concentrate on his customary activities and on school.

The physician can also meet the child to hear his concerns directly and to help him diminish his worry about the present and the future. The physician can directly address the child's loyalty conflict and assure the child that love *for* both parents continues after divorce.

CHILDREN AGES 9 TO 12

Children in this age group are frightened by the divorce and often worried about entering puberty and adolescence without the supportive structure of the intact family. They may feign nonchalance or disinterest in the family events. School performances and peer relationships are likely to deteriorate during the year following the separation.

More characteristic of this age, however, is intense anger at one or both parents for the divorce decision. As these children mourn the loss of the intact family and struggle with anxiety, loneliness, and a sense of their own powerlessness, they often cast one parent in the role of the "good" parent and the other as "bad." They are vulnerable to the blandishments of one or another of the parents who are actively engaged in fighting with the other and are easily co-opted as allies in strategies designed to harass and humiliate the other parent. Children in this group are also often helpful to a troubled parent and can show enhanced maturity and greater compassion as a response to the divorce.

It may be difficult to speak directly with youngsters in this age group, since they are likely to deny discomfort or to remain silent. Nevertheless, it is very helpful for the physician to express a continuing interest in and specific concern for the youngster who may feel lonely and beleaguered; this will also indicate to the child that the door is open for future conversation, which the youngster can initiate as needed.

Parents should be advised not to involve children in their disputes and to realize that children probably do not understand the issues, despite their pseudosophisticated remarks. They should also be helped to perceive the unhappiness that underlies the child's surface bravado, anger, or false disinterest. Youngsters at this age are vulnerable to sexual overstimulation and should not be given the role of confidant regarding parental love affairs.

Children should be assured that the parents are working hard to stabilize the family, and they should be directed to the importance of continuing with their school work and other customary activities.

ADOLESCENTS

The incidence of disturbance among adolescents faced with their parents' divorce is higher than has been generally expected. A large number of adolescents become worried about their own future entry into young adulthood and are concerned about the fate of their own marriages and the possibility that they, too, may experience sexual and marital failure. They also become concerned with issues of morality and respond in a global way to what they experience as a need to reorganize their opinions about the world around them and to rethink values.

Of particular interest for the physician is the high potential for acute depression with accompanying suicidal preoccupation in adolescents who seemed in good psychological health prior to the family rupture. Such a response should be taken very seriously, especially if the depression is acute or lasts more than a few weeks. There is a high potential in this age group for acting out behaviors such as truancy, new sexual activity, alcohol or drug abuse, or a suicide attempt (Wallerstein and Kelly, 1974).

The physician can be very helpful in initiating contact directly with the adolescent when a prior relationship has existed. These youths are likely to welcome the opportunity for a serious, wide-ranging discussion of the impact of the divorce on their future expectations and plans. At that time the physician can ascertain whether the young person is depressed and whether he is taking appropriate or too much responsibility in the home and can respond accordingly.

The parents should be advised that although the adolescent can be very helpful to them at this critical time, the young person also needs some protection and encouragement to pursue his or her age-appropriate interests. Parents should be urged not to depend heavily on their adolescent son or daughter but to seek support for their own needs from their family therapist or friends.

PARENTS AND CHILDREN AFTER DIVORCE

The relationship between parents and children following the divorce undergoes many significant changes. The visiting parent's relationship with the children is especially likely to change. Poor—even impoverished—relationships may improve, while others that have been close and affectionate during the marriage may dwindle unexpectedly. Young children are more likely to be visited over the years than are older ones. On the other hand, some adolescent boys and girls develop and maintain a close friendship with the father—especially when the mother is depressed.

Similarly, the custodial parent and children move into new roles. Many youngsters, some very young, become closer to their mothers as proud helpers and confidants. Others move precipitously away from a closer involvement out of fear of engulfment. Altogether the divorce emerges as a nodal point of change in parent-child relationships in many of the families (Wallerstein and Kelly, 1980a and b).

There is considerable evidence that the relationship between children and both divorced parents does not lessen in emotional importance over the years. Although the mother's caretaking and psychological role becomes increasingly central in families in which the mother has custody, there is no evidence that the father's psychological significance declines correspondingly. Even in remarriage the biological father's emotional significance does not disappear or diminish markedly, and the children appear to face little conflict, creating a special slot for the step-father in addition to their relationship with the biological father. It is strikingly apparent that whether the children maintain frequent or infrequent contact with a noncustodial parent, they would consider the term "one-parent family" a misnomer. The self-image of children who have been reared in a two-parent family appears to be firmly tied to the continuing relationship with both parents, regardless of a parent's physical presence within the family.

GOOD OUTCOME

Study of factors leading to a good outcome points to the desirability of the children's continuing relationship with *both* parents in an arrangement that enables each parent to be responsible for and genuinely concerned about the well-being of the children. Children who are well adjusted and happy five years after the divorce live in families who are able to restabilize and restore the parenting after the initial dip during the breakup crisis. Improvement is also seen when divorce separates a child from a psychologically destructive or incestuous or physically abusive parent. New parental relationships within remarriages also may benefit a child. Children fare better when the friction between parents has dissipated. All in all, children who do well over the post-divorce years enjoy a sense of continuity with both parents and feel that the divorce has contributed to the quality of their lives both by removing the stress of the failing marriage and by maintaining the important gratifications.

POOR OUTCOME

At the other end of the spectrum, a considerable number of children continue to look back longingly at the failing marriage and to wish for its return. In the California Children of Divorce Proj-

ect, 37 per cent of the children were psychologically troubled and distressed at the five-year mark. The most frequent clinical finding at that time was childhood depression, seen in children who felt rejected and neglected by either the custodial or the noncustodial parent. Those who experienced repeated disappointments because of the father's infrequent or unreliable visiting or his disinterest or insensitivity to them during visits suffered grave and lasting unhappiness. Despite repeated disappointments and the passage of time, many children hold tightly to the hope that the father will eventually fulfill their expectations. Very few children are able to master their distress at being rejected by their fathers. Time does not seem to dispel their anguish, although this feeling may be carefully hidden.

Similarly, children who remain in the custody of a lonely, depressed, or emotionally disturbed mother are likely to do poorly. The additional stresses of divorce on an emotionally disturbed or physically ill parent often lead to a deterioration in parenting, no longer mitigated by the buffering presence of the healthier spouse. Children left in the custody of a chronically depressed or disinterested mother are more likely to show a decline after the divorce, although they might have held their own during the marriage with the support of the less troubled parent.

Finally, failure of the divorce to provide its intended remedy is a central cause of children's poor adjustment at the five-year mark. When parents continue to fight, or when post-divorce bitterness exceeds that of the marital conflict, children are seriously hindered in their efforts to accept the divorce. Continued parental fighting is significantly related to poor psychological adjustment among the children. Thirty per cent of the children were aware of tension and continued bitterness between their parents at the five-year mark (Wallerstein and Kelly, 1980b).

FACTORS IN OUTCOME

No single theme appears in the lives of all those children who enhance, consolidate, or continue their good developmental progress following the divorce. Nor is there a single theme among all those who deteriorate, either moderately or markedly. Rather, there are a set of complex configurations in which the available components are put together in varying combinations in the life of each individual child.

To add to the complexity, the different components of these configurations are intricately interrelated. Thus, children turn more to the noncustodial parent for support and yearn for him more intensely when the relationship with the custodial mother is conflicted or impoverished as well as when they miss him and their relationship with him on its own merits. In this way, the attitude of the child toward the father is both separate from *and* closely related to the child's feelings toward the mother and the degree and quality of the relationship within the home. Furthermore, the child's capacity to rely on friends and to turn to friends for support during the crisis also depends in some measure on the child's relationships within the family. Children with good relationships at home appear to make friends more easily and to sustain these friendships, whereas those who are unhappy at home often have trouble making friends and may try to manipulate their peers. Teachers complain about the manipulative, overbearing behavior of children who at home are frequently subdued and sad. Thus, the more fortunate children feel better both at home and in other settings while their less fortunate peers are unhappy in both places.

Briefly stated, components that seem central to the course and outcome at the five-year mark, in varying combinations of importance, include (1) the extent to which the parents have been able to resolve and put aside their conflicts and angers and to make use of the relief from conflict provided by the divorce; (2) the quality of the custodial parent's handling of the child and the resumption or improvement of parenting within the home; (3) the extent to which the child does not feel rejected in his relationship with the noncustodial or visiting parent, and the extent to which this relationship has continued on a regular basis and kept pace with the child's growth; (4) the individual assets, capacities, and deficits that the child brings to the divorce; (5) the availability of a supportive human network to the child, including siblings and extended family members, and the child's ability to make good use of it; (6) the absence of continuing anger and depression in the child; and (7) the developmental needs of the child relative to sex and age.

Although the initial breakup of the family is profoundly stressful, the eventual outcome depends in large measure not only on what has been lost but on what has been constructed to replace the failed marriage. The effect of the divorce ultimately reflects the success or failure of the parents and children to master the immediate disruption, to negotiate the transition successfully, and to create a more gratifying family to replace the family that failed.

DURATION OF THE DIVORCE PROCESS

The timetable of the divorcing process is considerably longer than most people realize. The decision to divorce leads to many changes—both intended and unintended—that affect the adults, the children, and their relationships with each other. The drama, complexity, and scope of these changes exceed the expectations of many of the participants. Often the period of disequilibrium in the lives of the family members lasts several years, and even five years after the marital separation, divorce-related issues continue to evoke strong feelings in children and adults. In the California study, the average time needed by the newly divorced woman to reestablish a sense of continuity and stability in her life was 3 to 3½ years. The men required 2 to 2½ years to establish or reestablish a sense of order in their lives. The divorce remained a live issue for half the adults at the five-year mark, especially for the women. Only rarely did both parties achieve the same degree of psychological closure, and for at least one partner, the divorce was often still a very painful issue five years later. By that time, two-thirds of the men and more than half the women viewed the divorce as beneficial. However, one-fifth of the men and one-fifth of the women deplored the divorce as a bad mistake; the rest had mixed feelings.

Similarly, as a group, the children remained strikingly aware of their family and its vicissitudes at the five-year mark. One consequence of divorce appears to be that the family becomes a focus of the conscious—even hyperalert—attention and consideration of the children. As the youngster matures, he reassesses the divorce and makes an effort to explain and master the march of family events in ways consonant with his enhanced maturity. In this way, the intellectual and emotional efforts of the youngster to cope with the family rupture appear to be reorganized at each developmental stage and extend throughout his growing up years, perhaps even into adulthood.

It is evident that many families need professional advice and guidance in negotiating this complex and tangled pathway through the divorce and the years that follow.

BEREAVEMENT AND DIVORCE COMPARED

Divorce and bereavement are examples of severe stress that is initially acute and then followed by an extended aftermath, perhaps lasting for several years. Stress of this kind engenders profound distress by demanding rapid recognition of complex major life changes and adaptation to altered circumstances. Early responses are fairly apparent at the time, either within the family or in school or on the playground. They include a wide spectrum of symptomatic behaviors that are related to age. Responses in the long aftermath emerge slowly and are less visible than those in the early phase. Ultimately, these delayed responses, which may include low self-esteem, the feeling of being unloved and unlovable, or clinical depression, may be of greater developmental significance because they are likely to be long-lasting. In fact, they may require psychotherapeutic intervention.

However, there are important differences between these two tragic experiences. First, loss due to death is final and irreversible, whereas losses associated with divorce can in reality be undone, and the child may for many years harbor a fantasy of restoring the family. Moreover, unlike death, divorce is always a voluntary decision for at least one of the partners, and the participants are keenly aware of this fact. The stress that is generated for the child carries the message that perhaps the divorce could have been avoided and that someone is to blame for the parents' unhappiness.

No matter what their age, children frequently respond with anger toward both parents for disrupting the family or toward the parent who sought the divorce. While anger is often present in bereavement — the anger of the survivor at being abandoned, or self-accusation for failure to prevent the death—there is a fundamental distinction between the anger of the bereaved child, with its roots in fantasy, and the realistic perception of the child of divorce that one or both parents failed in time of need.

Death, unlike divorce, is a universal experience. Divorce invites the anxiety-driven question, "Will divorce happen to me as well?" This nagging fear remains a perennial concern for the child of divorce and recurs as he moves toward young adulthood.

Finally, the social milieu and the available community support differ for children in each of these groups. Divorce, certainly at the outset, is a lonelier road for children and adolescents. Unlike the social network that rallies to support the bereaved immediately after the loss, there is often no support system to sustain the ruptured family when a divorce occurs. More often, friends and neighbors stay away, and extended family members may take sides in the conflict or disapprove of the divorce. Thus, divorce represents, for both children and adults, intense distress at a time of diminished support. (See also Chapter 16.)

REMARRIAGE

Most remarriage involves children from the former marriage of one or both partners. If both adults have been divorced, it is likely that their former marital partners maintain a presence in their lives as they continue to share parenting, to visit the children, and to provide economic support for the children. In addition, the former marital partners may have remarried and their partners' children or children from the new marriage may further extend the already complex kinship system. The network of relationships created by the remarried family, referred to variously as blended, reconstituted, or binuclear, has been poorly defined and insufficiently studied. As Furstenberg (1980) observes, "We have no set of beliefs, no language and no rules for a family form that has 'more than two parents' yet a substantial minority of the population of the United States will participate or already is participating in such a family system."

The remarried family begins with a past history that is often overlooked both by the adults within the family and by those who counsel remarried families. The period of single parenthood, which for women averages 2½ to 3½ years following the divorce, is a major part of the remarried families' history. During the period of single parenthood, specific patterns of parent-child interaction are likely to arise. It is not uncommon, for example, for the single parent to relate to one or more children with a special closeness and mutual dependency. These patterns are likely to change suddenly with the entry of a new parent into the family system, leaving the children with strong feelings of having been rejected by the biological parent and with intense anger at the step-parent as intruder.

There is encouraging evidence from social agencies, parent self-help groups, life education courses, adult education classes, and some preliminary research (Furstenberg, 1980; Weingarten, 1980) that adults enter the second marriage more realistically. They bring an eagerness to work at the marital relationship and a greater willingness for compromise. One Pennsylvania study reports that remarried people showed greater flexibility in the division of household tasks, more shared decision making, and a greater degree of emotional exchange between husband and wife than occurred in their first marriage (Weingarten, 1980). In general, there was less adherence to a marital bond built along strictly defined sex roles. In the California Children of Divorce Project most of the men and even more of the women expressed approval and a sense of increased contentment with the second marriage.

The changes required by the remarriage are nevertheless formidable and extend well beyond creation of the family structure and relationships. As expected, remarriage sometimes leads to geographic relocation. Such a move introduces disruption and discontinuity, including greater physical distance between the noncustodial parent and the child. As family units combine, different life styles need to be integrated. Perhaps the greatest changes occur in the new family unit with the parent-child and step-parent–step-child relationships. Indeed several myths associated with remarriage identify the step-parent–child relationship as hampering the formation of the new family unit. Three conflicting myths are (a) that step-families are essentially the same as nuclear families, (b) that step-mothers are wicked and cruel to the children, and (c) that step-parents and children will care for each other instantaneously (Visher and Visher, 1979).

STEP-PARENTS AND CHILDREN

Special attributes attach to parent-child relationships in the remarried family. Prior histories of both step-parent and step-child represent ever-present "ghosts." Children may have to deal with a range of conflicting feelings concerning their relationship with the step-parents as well as with a new and sometimes overwhelming network of new relationships with half-siblings and step-siblings of various ages.

In the California Children of Divorce Project, the step-fathers were older than their wives and had been married before. Having traveled a lonely road, they were eager for a home and a gratifying marriage. They were a sober and committed group who were supportive of their wives. With few exceptions they expected to assume the role of parent to the wives' children. Encouraged by the women, most men took this responsibility seriously and moved quickly into the role of the man of the household with the purposes, prerogatives, and authority accorded this position in traditional homes.

The precipitous entry of the step-father into the role of husband and father generated anxiety in the children. Several of the new husbands had lived in the household previously as lover and companion. During that time their relationship with the children was different and more inclined to be friendly, casual, or uninvolved. Occasionally, the men were especially pleasant or generous with the children when they were courting the mother. Perhaps in reaction to their own insecurity in the new role, a substantial number of step-fathers assumed a fairly rigid, disciplinary stance with the children, especially with the older ones. Step-fathers were variously described by complaining children as "stern" or "not affectionate"; other comments were reflective of the child's attitude of submission: "When he says something he never changes his mind;" "He has a temper," and "You don't goof when he gets mad."

Fewer children live with their step-mothers because fewer fathers have obtained custody. Nonetheless, this relationship can be as complex as that with a step-father. One major difference is that the step-mother is likely to be younger—sometimes too young to maintain appropriate distance and discipline with an adolescent child of a prior marriage. The tugs and tensions between adolescent and step-mother are formidable. Sometimes these conflicts are exacerbated by the youth's perception of the step-mother as the person who is responsible for the marital breakup. In a significant number of homes in which the adolescent lives with the father, the young person has been ejected by the mother for misbehavior and banished to the father's household for discipline the mother feels that the father is better able to provide. All these circumstances burden the new and growing relationship between step-mother and step-son or daughter. Nevertheless, there is also the potential for a close and sisterly relationship, especially between step-mother and step-daughter, for whom the relationship may remain very important over many years.

UNREALISTIC EXPECTATIONS

Not surprisingly, many of the problems in the remarried family that center on the children derive from the adults' failure to recognize that relationships are created over time with children as well as with adults. Thus, adults who fully anticipated the gradual development of the marital relationship may in contrast expect instant love, respect, and obedience from the children. Few adults appear sensitive to the need to cultivate a relationship with the child gradually and to allow for the child's suspiciousness and resistance in the initial phase. Such expectations of instant response are especially disturbing for children, who feel they are being called upon to betray their love for their parent and to substitute the step-parent in his or her place. Even very young children need to be reassured that the new adult is not being presented as a substitute for the departed parent.

CRITICAL VARIABLES IN ADJUSTMENT

Age of the child is a significant factor in the interaction between step-child and step-parent. The relationship with the younger children, mostly those below the age of 8, takes root fairly quickly and is likely to become happy and gratifying to both child and adult. Little girls and little boys alike are responsive to the affection and interest of the new step-parent.

The relationship between the step-parent and an older child tends to follow a more conflicted course. Older youngsters and adolescents may continue to resent the step-parent's presence and may ultimately fail to develop a positive attachment. These youths are likely to leave home earlier. Others gradually change their minds. In the California study some youngsters eventually sought to emulate the step-parent and placed him or her as the central figure with whom they could identify. Often the troubled relationship takes an important turn at a particular moment in time, following an incident or even a confrontation, when the child suddenly changes his attitude and decides to accept the step-parent as an authority figure and as a parent.

At the outset of the remarriage, children may be both eager and anxious. They welcome the arrival of the step-parent because of the greater security the presence provides, and they are relieved to be part of a two-parent household again. At the same time, children may resent the new adult's special place in their parent's affection and the instant authority conferred on this person. They are concerned that they might be replaced by or excluded from the new relationship. As they worry, they watch tensely for evidence of acceptance or exclusion.

Even when children feel reassured by the remarriage they may continue to worry whenever friction develops between the new parents. The new marriage evokes the memories of the early experience, and children report retiring in anxiety in their rooms or crying at night when the newly married couple quarrels. Their view of the seriousness of family friction is sometimes at odds with the adult perspective and is often surprising to the adults.

The child's relationship with the step-parent and the parent and the various ways in which this issue is resolved by the child and the adults, or continues as a source of open conflict, are of central importance in the psychological development and adjustment of the child within the remarried family. Many children are able to maintain and enjoy both relationships. Father and step-father, mother and step-mother do not occupy the same slot in the child's feelings, and the child does not confuse the relationships. These children are well able to enlarge their view of the family and to make room for all the major parental figures. All the adult relatives may be considered by the child to be of importance in providing figures for imitation and identification. Thus, the general expectation that children necessarily experience conflict as they turn from parent to step-parent as they are growing up is not borne out by observation, nor is the expectation that in the happily remarried family the biological father is likely to fade out of the children's lives. Children with step-parents whom they love and admire do not turn away from parents whom they continue to visit.

It should be noted, however, that many adults are less successful than the youngsters in defining the different roles. Rivalries between father and step-father and between mother and step-mother in the remarried families are often bitter and long-lasting. When the child experiences painful psychological conflict and feels torn between the love for the father and the love and loyalty to the step-father or between mother and step-mother, the adults are likely to be pulling in opposite directions or failing to help the child.

Unfortunately, the relationships between the step-parent and parent can all too easily become charged with the unresolved angers of the divorce and the aggravated jealousies of the remarriage. Such problems of competing parents and step-parents are especially difficult to resolve because communication is limited and there is no proper forum for discussion. Feelings may then be exacerbated as both find new grounds for accusations and counter-accusations. Often the child becomes the hapless scapegoat of adult anger and is placed in grave psychological danger by the unremitting conflict of the major adults in his life.

REMARRIAGE—CONCLUSIONS

Overall, remarriage appears to enhance the lives of many children particularly those who have not yet reached adolescence. These children are better parented by happier parents and by step-parents who take their responsibility seriously and try hard to fulfill a parental role. Such children are also content with their relationship with their visiting parent when the adults are not in conflict over the child's loyalty and affection.

For older youth, their needs often diverge from those of the remarried parents, and the remarriage that brings contentment and greater maturity to the adults may not enhance the lives of the children. In some families the relationship is not satisfactorily resolved, and the conflicts generated between step-parent and step-children can lead to rupture of the remarriage. Nevertheless, many adolescents and preadolescents, after an initial resistance and with the passage of time, are able to develop strong, close, and meaningful attachments to their step-parents, who then greatly influence their consolidation of values and choice of life style and career.

THE ROLE OF THE PHYSICIAN

The physician can be very helpful to the adults on the threshold of remarriage. Most families welcome both encouragement and competent guidance. Moreover, there is considerable evidence that adults benefit from anticipatory guidance, which enables them to discuss the impact of major events before their occurrence and to look ahead constructively at the tasks that will need to be mastered. The expectable concerns of children and adolescents can be explained by the physician to the new parent. The physician can call special attention to the differences in age and temperament among the children and to the children's hopes as well as their fears regarding the anticipated changes within their lives.

Furthermore, the physician can help family members distinguish the remarried family from the earlier family, which can never be recovered or reconstituted. Fortunately there is clear evidence from guidance of remarried families that many adults and youngsters alike can benefit significantly from even brief advice emphasizing that the relationships in the remarried family are likely to be different, less intense, and perhaps, therefore, longer lasting than those of the initial family.

Perhaps the central issue here is the honest recognition that the needs of the children and the needs of parents do not necessarily converge at all times. Different needs and wishes require careful balancing and sensitive consideration. The family may need help in learning how to provide adequate time, place, and opportunity for the continuing communication of feelings, concerns, and conflicts that inevitably arise in the course of daily living and especially during periods of great change or crisis.

JUDITH S. WALLERSTEIN

REFERENCES

Bane, M. J.: Marital disruption and the lives of children. J. Social Issues 32:103, 1976.
Bloom, B. P., White, S. W., and Asher, S. J.: Marital disruption as a stressful life event. In Levinger, G., and Moles, O. O. (eds.): Divorce and Separation. New York, Basic Books, 1979.
Blumenthal, M. D.: Mental health among the divorced. Arch. Gen. Psychiatry 16:603, 1967.
Briscoe, C. W., et al.: Divorce and psychiatric disease. Arch. Gen. Psychiatry 29:119, 1973.
Carter, H., and Glick, P. C.: Marriage and Divorce: A Social and Economic Study. Cambridge, Massachusetts, Harvard University Press, 1976.
Furstenberg, F. F., Jr.: Reflections on remarriage: Introduction to Journal of Family Issues special issue on remarriage. J. Family Issues 1:443, 1980.
Glick, P. C.: Remarriage: Some recent changes and variations. J. Family Issues 1:455, 1980.
Gove, W. R.: The relationship between sex roles, marital status and mental illness. Social Forces 51:34, 1972a.
Gove, W. R.: Sex, marital status, and suicide. J. Health Social Behavior 13:204, 1972b.

Hetherington, E. M., Cox, M., and Cox, R.: Divorced fathers. Family Coordinator 25:417, 1976.
Hetherington, E. M., Cox, M., and Cox, R.: The aftermath of divorce. In Stevens, J. H., and Mathews, M. (eds.): Mother-Child Relationships. Washington, D.C., National Association for the Education of Young Children, 1978.
Hetherington, E. M., Cox, M., and Cox, R.: Family interaction and the social, emotional and cognitive development of children following divorce. In Vaughn, V. C. III, and Brazelton, T. B. (eds.): The Family: Setting Priorities. New York, Science and Medicine, 1979a.
Hetherington, E. M., Cox, M., and Cox, R.: Play and social interaction in children following divorce. J. Social Issues 35:26, 1979b.
Kalter, N.: Children of divorce in an outpatient psychiatric population. Am. J. Orthopsychiatry 47:40, 1977.
Kaplan, D. M., Grobstein, R., and Smith, A.: Predicting the impact of severe illness in families. Health and Social Work 1:71, 1976.
Kelly, J., and Wallerstein, J.: The effects of parental divorce: Experiences of the child in early latency. Am. J. Orthopsychiatry 46:20, 1976.
Kelly, J., and Wallerstein, J.: Part-time parent, part-time child: Visiting after divorce. J. Clin. Child Psychology 6:51, 1977.
Kulka, R. A., and Weingarten, H.: The long-term effects of parental divorce in childhood on adult adjustment. J. Social Issues 35:50, 1979.
Roman, M., and Haddad, W.: The Disposable Parent: The Case for Joint Custody. New York, Holt, Rinehart and Winston, 1978.
Santrock, J. W., and Warshak, R. A.: Father custody and social development in boys and girls. J. Social Issues 35:112, 1979.
Steinman, S.: The experience of children in a joint custody arrangement. Am. J. Orthopsychiatry 51:403, 1981.
Visher, J. S., and Visher, E. B.: Stepfamilies and stepchildren. In Berlin, I. N., and Stone, L. A. (eds.): Basic Handbook of Child Psychiatry. Vol. 4. New York, Basic Books, 1979.
Wallerstein, J.: Responses of the pre-school child to divorce: Those who cope. In McMillan and Henao (eds.): Child Psychiatry: Treatment and Research. New York, Brunner-Mazel, 1977a.
Wallerstein, J.: Some observations regarding the effects of divorce on the psychological development of the pre-school girl. In Oremland, E. K., and Oremland, J. D. (eds.): The Sexual and Gender Development of Young Children. Cambridge, Massachusetts, Ballinger Press, 1977b.
Wallerstein, J.: Children and divorce. A review. Social Work 24:468, 1979.
Wallerstein, J.: Children of divorce stress and developmental tasks. In Garmezy, N., and Rutter, M. (eds.): Stress, Coping and Development. New York, McGraw-Hill Book Co., in press.
Wallerstein, J., and Kelly, J.: The effects of parental divorce: The adolescent experience. In Anthony, E. J., and Koupernik, C. (eds.): The Child in His Family. Vol. 3. New York, John Wiley and Sons, 1974.
Wallerstein, J., and Kelly, J.: The effects of parental divorce: The experiences of the preschool child. J. Am. Acad. Child Psychiatry 14:600, 1975.
Wallerstein, J., and Kelly, J.: The effects of parental divorce: The experiences of the child in later infancy. Am. J. Orthopsychiatry 46:256, 1976.
Wallerstein, J., and Kelly, J.: Effects of divorce on the father-child relationship. Am. J. Psychiatry 137:1534–1539, 1980a.
Wallerstein, J., and Kelly, J.: Surviving the Breakup: How Children and Parents Cope With Divorce. New York, Basic Books, 1980b.
Weingarten, J.: Remarriage and well-being: National survey evidence of social and psychological effects. J. Family Issues 1:533, 1980.
Weitzman, L.: The Marriage Contract: Couples, Lovers and the Law. Englewood Cliffs, New Jersey, Prentice-Hall, 1980.
Weitzman, L., and Dixon, R.: Child custody awards, legal standards, and empirical patterns for child custody, support and visitation after divorce. University of California Davis Law Review 12:473–521, 1979.
Zill, N.: Divorce, marital happiness and the mental health of children: Findings from the Foundation for Child Development National Survey of Children. Paper prepared for National Institute of Mental Health Workshop on Divorce and Children, Bethesda, Maryland, 1978.

14

Family Dysfunction: Violence, Neglect, and Sexual Misuse

Traditionally parents have been entrusted by society with the responsibility of looking after their children's best interests. Clinical experience and research dictate that optimal emotional and social development of children requires the maintenance of stable relationships with adults and that the favored context for these relationships is the family. Recently the emergence of public and professional concern about child abuse and neglect has challenged our assumptions about the sanctity of the family. We now recognize that parents vary in their ability to care for and protect their children; in fact, some families may engage in practices that interfere with or inhibit a child's development.

Which parental and family patterns are maladaptive for children and might appropriately be construed as "abusive" or "neglectful?" What is the pediatric practitioner's responsibility in such cases? This chapter reviews the history of the emergence of "child abuse" as a social problem, its definition in practice, and the three clinical entities of greatest concern: family violence, including child abuse; child neglect; and the sexual misuse of children. The impact of these problems on the child is explored, and the recognition and treatment of affected children and families are reviewed.

HISTORICAL BACKGROUND

Harm to children inflicted by their caretakers has been documented since history has been recorded (DeMause, 1974). Infanticide of unwanted infants was practiced in many civilized societies from Ancient Greece to China (DeMause, 1974; Radbill, 1968). A shift toward a more humanitarian conception of the child occurred in the seventeenth century, when the European family became a more independent and autonomous unit (Radbill, 1968). The nineteenth century witnessed the institution of foundling asylums and a greater sense of community responsibility for child welfare.

Three child welfare movements emerged in the nineteenth century in the United States. The "House of Refuge" movement in major cities enabled the state to place abandoned or neglected children in institutions in order to "save them" from becoming delinquents or criminals. The second movement began in 1875 with the case of "Mary Ellen," a girl who was cruelly abused by her foster parents. Public attention to this child's plight led to the founding of the New York Society for the Prevention of Cruelty to Children. This and other humane societies worked in conjunction with the House of Refuge movement. Placement of children in institutions became the favored approach to rescuing children in danger within their homes. The third component of the child welfare movement was the establishment of the first juvenile court in Illinois in 1899. By 1920, all but three states had juvenile courts (Pfohl, 1977).

All three movements aimed to keep children out of adult institutions, but the principal objective, according to Pfohl, was to protect society; a lesser goal was to protect the child victims themselves.

The medical discovery of child abuse is attributed to Caffey, whose 1946 paper on long bone fractures in children noted that many were of "unspecific origin." In 1957, following similar papers by other pediatric radiologists, Caffey expressed his belief that such injuries were "deliberately inflicted." In 1961, Professor C. Henry Kempe organized the first conference on "The Battered Child Syndrome," and in 1962 an article with this title by Kempe and colleagues appeared in the *Journal of the American Medical Association*, coauthored by radiologists and psychiatrists.

The syndrome involved children under 3 years of age and included "signs of neglect," diverse injuries, and a discrepancy between the injury and the explanation proffered by the caretaker. Emphasis was drawn to the presence of earlier trauma, detectable only by radiographic study.

Parents of the victims were characterized as "psychopathic, immature, impulsive and low in intellectual ability"; many had been victims of child abuse themselves. The only "treatment" that could ensure the safety of the battered child appeared to be removal from his home. The dramatic diagnostic name "battered child syndrome" attracted the attention of the medical community to a phenomenon heretofore described as "unexplained injury," and the concept of severe parental deviance was brought to public attention (Kempe, 1962).

By providing this diagnostic label, Kempe and his colleagues legitimized the attention of physicians and medical workers to a complex family and social problem. Social welfare workers and law enforcers as well as the general public soon mobilized to "do something" about child abuse. By 1966, all 50 states had drafted laws mandating certain professionals to report suspected cases of child abuse and neglect. In 1974, a National Center on Child Abuse and Neglect in the Department of Health, Education, and Welfare was mandated by Congress to gather information and disseminate knowledge about the etiology and treatment of child abuse and neglect as well as to improve state and local protective services (Public Law 93-247).

PROBLEMS OF DEFINITION

Kempe and colleagues' definition of the "battered child syndrome" was restricted to the extreme physical abuse of children. In 1971, Fontana proposed a broader definition of child maltreatment, which included malnutrition and neglect as earlier "stages" in a "maltreatment" syndrome in which physical abuse was located at the most severe end of a spectrum of injuries attributable to parental failure. In 1973 sociologist David Gil testified before the Senate Subcommittee on Children and Youth during hearings on the Child Abuse Prevention and Treatment Act and defined child abuse even more broadly as "any act of commission or omission which interferes with the optimal development of the child."

Reporting laws in all states now define child abuse and neglect broadly, including physical and emotional injury, educational and medical neglect, and sexual abuse. The working definition of child abuse and neglect is left to be defined in practice by the individual reporter, state protective systems, and the courts.

Surveys of professionals reflect considerable disagreement with respect to whether reporting laws should define child abuse broadly or narrowly.

The breadth and ambiguity of the reporting laws appear to be associated with biases in reporting, resulting in the preferential "labeling" of families among minority or disadvantaged groups as abusive, while injuries to children in white or higher income groups may be characterized as being "accidentally" incurred. Turbett and O'Toole (1980) recently studied physician's classifications of children's injuries after being presented with short case vignettes. They noted a striking preference for the diagnosis of child abuse on the same vignette when the family was black rather than white or had a low rather than high income.

Giovannoni and Becerra (1979) also used sample case vignettes to measure professional and lay consensus in evaluating the seriousness of various kinds of child maltreatment. They started with the premise, which we share, that child abuse and neglect is a form of "social deviance," the definition of which entails a value judgment as to what constitutes inadequate parental care. Their results reveal interesting patterns of judgment among four groups of professionals who must operationally define child maltreatment in their work: pediatricians, lawyers, protective service social workers, and police officers trained in family and juvenile problems. Highest interprofessional agreement was found with respect to the seriousness of physical and sexual abuse, and of parents' fostering delinquency in their children. Lawyers tended systematically to rank incidents as less serious compared with the rankings of the other three professions. There was considerable disagreement within all professional groups. Participants drawn from an ethnically diverse sample of lay people were then enlisted to evaluate these vignettes. They tended to rank *all* incidents as more serious than did the professional groups, with members of minority groups (black and Hispanic) rating the incidents as more serious than did members of the white subsample.

Giovannoni and Becerra urge greater precision in the legal and clinical definitions of child abuse and neglect in order to better inform social policy and clinical practice and avoid problems created by the diversity of values regarding child-rearing and interpretations of legal standards within and among professional groups. Some of the difficulties in understanding the etiology of child mistreatment stem from the heterogeneous nature of the cases that bear the categorical label of "child abuse." In this chapter we will examine separately three kinds of caretaking practices that have adverse consequences for children: family violence, neglect, and sexual misuse.

FAMILY VIOLENCE

The true incidence of child abuse is difficult to determine, since most cited statistics rely on figures from child protection agencies, that is, they reflect only the number of cases reported. The incidence of reported cases nationally increased by 71 per cent between 1976 and 1979 to a total of 711,142 cases in 1979 (American Humane Association, 1981). Experts agree, however, that the increase in reported cases reflects both increased public awareness of this social problem and increased agency accountability rather than an increase in true incidence of child mistreatment.

With respect to violence against children, the focus of concern has recently been expanded to include all forms of family violence. The place of violence in the management of household conflicts was studied by three sociologists, Straus, Gelles, and Steinmetz (1980), who defined physical violence as a range of specific behaviors, including hitting, biting, kicking, beating up, striking with an object, and threatening with or using a weapon. Members of this project interviewed a nationally representative sample of two-parent families. The prevalence of severe violence directed toward 3- to 17-year-old children was 3.8 per cent in the survey year. One in every 1,000 children was threatened with or subjected to a knife or gun. A projection of this ratio to the 46 million children ages 3 to 17 who lived with both parents during the survey year suggests that 1.5 to 2 million children per year are subjected to violence, of whom 46,000 are subjected to use of a weapon. In addition, eight out of 100 parents reported using one of these forms of violence against a child one or more times in the child's life. Among children subjected to lesser violence—kicks, bites, and punches—these events occurred an average of 8.6 times during the survey year. Beatings occurred an average of once every two months.

Hence, it appears that for many children, physical violence is a frequent occurrence, not a once-in-a-lifetime event. Violence was not confined to young children. When analyzed by age groups, 82 per cent of the 3- to 9-year-olds, 66 per cent of the 10- to 14-year-olds, and 34 per cent of the 15- to 17-year-olds had been subjected to some form of violence during the year (Straus et al., 1980). Although the figures are stunning, these data may suggest an underestimate of violence directed toward children by their parents for two reasons: First, the self-reported nature of data implies that many respondents may have denied or played down the use of violence in their homes, and second, two groups considered to be at greater risk of violence were omitted from the survey, i.e., single-parent households and children under 3 years of age.

Violence between spouses, a problem which has only come into prominence in the last decade, has serious implications for parent-child relationships. Present public awareness of interspousal violence has been aroused by the women's movement. As with child abuse, the true prevalence of marital violence is unknown. Since there are no laws mandating the reporting of interspousal violence, researchers have frequently had to make use of indirect measures such as calls to police departments, claims of violence in family courts, and scattered reports of battered women treated by hospital emergency rooms. Given the sources of data, estimates of incidence range from thousands of women victimized by their husbands to several million.

The national survey conducted by Straus and coworkers (1980) also measured the extent of violence between spouses. One out of every six respondents (16 per cent) reported some kind of physical violence at the hands of their spouse during the survey year. Over the course of a marriage, the chance appeared to be greater than one in four (28 per cent) that a couple would engage in an act of marital violence. Projected to the 47 million marriages in the United States, the data suggests that no fewer than 2 million women suffer severe physical violence each year. A similar number of husbands are victims of violent acts by their wives. Women who were victims of severe violence were 150 per cent more likely to inflict severe violence on their children than women who were not.

In the same study, the most frequently occurring form of family violence was between siblings. Almost 5 per cent of the children in the sample had threatened or used a knife or gun against a sibling in their lifetime. Severe sibling violence was much more frequent in families in which parents were frequently violent toward their children and toward each other, occurring in 100 per cent of such households compared to only 20 per cent of households in which parents did not use violence toward their children or toward each other.

These researchers also found that children who were victims of violence from their parents were more likely to use violence against their parents. Among those who had been hit the most by their parents, 50 per cent used violence against them, while less than one in 400 who were not hit by their parents were violent toward a parent.

UNDERSTANDING THE ORIGINS OF CHILD ABUSE AND FAMILY VIOLENCE

Initial efforts to understand child abuse focused on the *psychological problems of the parents* of the victims. For example, Kempe and colleagues (1962) described the abuser as the ''psychopathological'' member of the family, and Galdston (1965) observed the kinds of parental psychopathology in a selected, small sample. The psychodynamic approach focused on the perpetrator's personality characteristics (e.g., problems with impulse control, unfulfilled dependency needs) as the chief determinants of violence (Steele and Pollack, 1968). This intraindividual level analysis linked violence to psychopathology and to the occasionally observed concomitants of alcohol and drug abuse.

A second theoretical approach to child abuse and family violence focuses on *social factors*. Here it is assumed that violence is best understood by careful examination of the environmental realities that impinge on the family. In addition, this approach considers everyday family interactions to be precursors of violence. Within this framework many investigators have been concerned with the relationship between social stress and family violence (Newberger et al., 1977). Stress appears to be more prevalent in some families than in others owing to poverty, unemployment, single-parent family structure, social isolation, and sexual and economic inequality (Newberger and Newberger, 1981).

A third theoretical perspective on family violence builds, from social learning theory, an understanding of *violence as a learned behavior* stemming from exposure as a child (Parke and Collmer, 1975). The emphasis of this approach is on the exposure to violent models of interpersonal behavior. More than 20 articles or books support this popular perspective.

The etiology of family violence and child abuse is indeed complex. Violence is best understood as a symptom associated with the interaction of a number of factors in any given family. One must further take into account particular *vulnerabilities* in a child, parent, or family heightening their susceptibility to particular *stresses* that might in turn eventuate in violence. Bittner and Newberger (1981) have proposed a multidimensional etiological model, diagrammed in Figure 14–1, summarizing the interactions among sociocultural factors and stresses operating at the levels of society, family, parent, and child.

There has been no systematic study of what events precipitate abusive acts. Some instances are acute and self-limited; other cases are of long duration. Nonetheless, it is helpful to consider the circumstances of the family's life in the period

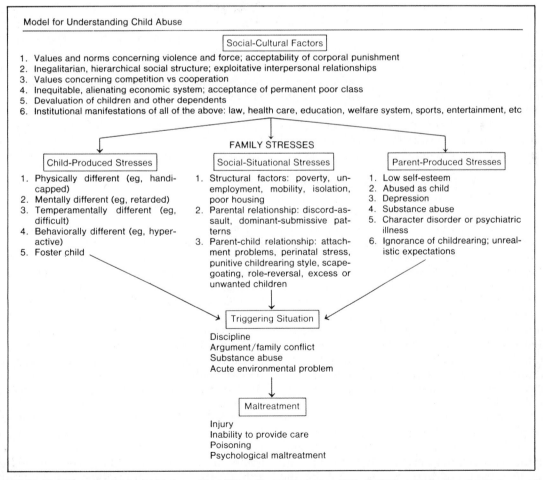

Figure 14–1. Model for understanding child abuse. (From Bittner, S., and Newberger, E. H.: Pediatric understanding of child abuse and neglect. Pediat. Rev. 2:198, 1981. Copyright American Academy of Pediatrics, 1981.)

immediately prior to the injury for which a child is brought in for care. Examples of triggering situations derived from clinical experience include a baby who, on a particular evening, would not stop crying; an alcoholic who was fired from his job; a mother who, after being beaten by her husband, could not make contact with her own mother; and the serving of an eviction notice. Any one of such stresses can become the triggering event for abuse.

It is helpful to think of the situation of an abused child being brought for care as a family crisis. A sensitive exploration of the origins of the problem and considerate attention to the recent family history by physician, nurse, and social worker can lead to clinical understanding of a given child's risk and can point the way to diminishing that risk.

Child Factors: Vulnerabilities and Stresses

The realization that children as well as their parents shape the course of family interaction is a fairly recent insight (Harper, 1975). This perspective has led to the identification of child characteristics that interfere with normal family functioning. In reviewing the literature on special characteristics of the abused child, Friedrich and Boriskin (1976) noted that behavior that makes him especially "difficult" to care for and parental perceptions of the child as "different" or "difficult" have been associated with abuse.

Included among these special characteristics of abused children are physical handicaps, congenital physical disabilities, mental retardation, schizophrenia, neurological damage, language deficits, and hyperactivity. In addition, low birth weight and prematurity have been found to be associated with abuse, whether because of early infant-mother separation or because of associated special characteristics of the infant, such as irritability. Excessive crying or fussiness is another characteristic frequent among abused children. The causal relationship between abuse and developmental disabilities may be bidirectional—developmentally disabled children appearing to be more vulnerable to abuse by caretakers, and abuse and neglect possibly resulting in developmental disabilities.

A number of researchers report that noticeable deviations in development were present in children before they were abused (Gil, 1970; Chotiner and Lehr, 1976). Statistics from these studies of incidence of preexisting handicaps or developmental problems prior to abuse range from 25 to 70 per cent. In a survey of members of Parents Anonymous, a self-help organization for parents who abuse their children, 58 per cent of the respondents reported that the abused child in their

family had developmental problems prior to the occurrence of physical abuse. Caring for a child with developmental disabilities can be very stressful emotionally, physically, and financially, thus increasing the vulnerability of these children or their siblings to physical violence.

Other characteristics of children that may make them vulnerable to abuse include age, i.e., younger children appear to be more vulnerable (Straus et al., 1980); status as a stepchild or foster child (Bittner and Newberger, 1981); and perceived similarities of the child to negatively perceived persons from the parents' past, which might include aspects of their appearance, behavior, and sex (Kempe and Kempe, 1978). It is important for the pediatrician to note when a parent describes a child as "just like" a relative or estranged husband and to question further the parent's feelings for the other person.

Separating the child's characteristics from the parents' responses to them can be challenging in the search for etiological factors in child maltreatment. While some parents are able to adjust their actions and expectations to meet the needs of their children with a minimum of difficulty, many seem to lack the resources to respond effectively to the demands of care giving. Yet another group of parents may be reasonably capable under most conditions but are nonetheless unable to cope when under severe stress or when they are confronted with a child whom they perceive as different or having very real special needs.

It is difficult to know in practice whether a child is by temperament more "difficult," irritable, provocative, active, or slow, or whether such behaviors develop in response to parental behavior and expectations.

Parental Vulnerabilities

Two factors seem to be critical in determining how vulnerable a parent is to adopting abusive behavior toward a child: (1) the parent's ability to understand and empathize with the child; and (2) the parent's own history, including the exposure to violence or deprivation in his or her family of origin.

Originally parental psychopathology was assumed to be the single reason for child mistreatment. Indeed, if present, it may adversely affect a parent's behavior toward a child. However, fewer than 10 per cent of abusive parents appear to be psychologically disturbed (Steele, 1978).

Psychologist Carolyn Newberger (1980) has developed a scale of parental awareness and reasoning about child-rearing that measures the parent's ability to think of the child as a separate individual with needs different from those of the parent, and the ability to consider the child's developmental level when making decisions about child-rearing. She found that abusive mothers scored lower than did a control group. A number of factors might interfere with a parent's level of awareness, including his or her age. For example, adolescent parents may have difficulty, given their own developmental level, in understanding and meeting the needs of another human being independently of their own needs (Belsky, 1980). It should be noted, however, that while adolescents are at high risk in this regard, reported cases of child abuse do not include great numbers of adolescent parents as perpetrators. Developmental limitations and personality characteristics might also affect the ability of the parent to empathize with their child.

Research and clinical findings indicate that parents who use violence against their children have frequently been subjected to violence as children (Newberger et al., 1977; Parke and Collmer, 1975). Straus, Gelles, and Steinmetz (1980) found that experiencing violence as a child increased the probability that a parent would be violent toward his or her child, yet many of this group of parents were *not* violent toward their children. Caution must be exercised in drawing deterministic conclusions from the association so often reported between violence experienced as a child and use of violence toward children.

The reason proffered for the observed intergenerational nature of violence is that parents provide "models" for conflict resolution and child-rearing for their children. Hence, physically violent parents may lack the learned behavioral skills to respond nonviolently under stressful conditions (Burgess et al., 1980). Exposure to noncoercive means of discipline emphasizes remorse for the consequence of one's actions and contributes to the moral reasoning of the child; children who lack this experience may later fail to engage in moral reasoning about their own child-rearing practices.

Others have argued that parents who have been physically abused as children are frequently deprived emotionally as well. Consequently, they suffer from low self-esteem, depression, and feelings of powerlessness as adults and may thus rely on coercive tactics and tend to apply them on those weaker and less powerful.

Family Stresses

A number of researchers have described an *impaired attachment relationship* between parent and abused child. A healthy attachment between parent and child requires reciprocal responsiveness to signals from each other. Factors impairing this reciprocity include perceptual handicaps, developmental disabilities, illness, or irritability on the part of either parent or child. Premature infants appear to be at greater risk for attachment difficulties (Klaus and Kennel, 1976) and for later abuse. A number of factors might contribute to this vulnerability, including greater difficulty in establishing predictable rhythms, more physical complications, and early separation of mother and child. Current neonatal practice attempts to encourage mother-child contact during hospitalization of the newborn (see Chapter 5).

The *relationship between the adults* in the household also influences the nature of the relationship between parent and child. As cited earlier, the amount of violence between spouses is related to the amount of violence directed toward children. In addition, parents who reported having observed violence between their *own* parents were more likely to direct violence toward their children.

Other family factors include the *absence of one parent* through job demands, separation, illness, divorce, or single parenthood and the *social isolation of a family* (lack of friends or relations nearby, distance from transportation, lack of a phone, noninvolvement with the community) (Newberger et al., 1977). There are many reasons for a family to be isolated, including recent moves as well as parental personality. It is important in interviewing a parent to assess their degree of social isolation and whether this is a permanent or a transient condition. Social isolation not only increases stress but also reduces the potential influence of child care practices of other adults.

Finally, *family size* appears to be a factor related to physical abuse. Straus et al. (1980) found that the likelihood of violence toward a child increases as family size increases up to five children and then drops dramatically with additional children.

Using a standardized stressful life events scale (including personal and financial losses, arrests, pregnancy, illness, and moves) Straus and colleagues also found that high numbers of *stressful life events* (eight or more) were strongly related to incidence of severe violence against children. A major stressful condition for many families in which children are abused is *poverty*. While some investigators of child and spouse abuse have claimed that socioeconomic factors were not related to the occurrence of acts of domestic violence, the same articles that have made these claims offer empirical evidence that abuse is more prevalent among those with low socioeconomic status (Gelles, 1981). Current research on family violence supports the hypothesis that domestic violence is more prevalent in low-income families (Parke and Collmer, 1975; Gil, 1970). However, this conclusion does not mean that domestic violence is confined to lower-class households (Straus et al., 1980; Gelles, 1981). Many other social stresses found to be associated with child abuse correlate with lower socioeconomic status, such as unemployment, poor housing, family size, and lack of access to child care (Newberger et al., 1977). In addition, poorer families suffer from poorer medical care, inferior education, lack of access to jobs, and residence in more stressful neighborhoods (Garbarino and Sherman, 1980). In an analysis of 58 counties in New York state, rates of child abuse were considerably higher in counties with low income and few supportive resources.

A caution must be raised regarding the association between poverty and child mistreatment, particularly when discussing reported cases of mistreatment. Child abuse occurs at all socioeconomic levels. However, as previously cited, poor and minority families are more susceptible to being labeled as child abusers, while trauma in children from families of higher status by virtue of income or race is more likely to receive the label of "accidental injury" (Newberger and Hyde, 1975; Gelles, 1975). The pediatrician must guard against assuming that because a child's parents are well dressed, well spoken, and middle class that the child's injuries could only have occurred accidentally.

Sociocultural Factors

Families live in a sociocultural context and are influenced by sociocultural norms. Gil (1975) has observed that culture in the United States values competition rather than cooperation and that physical coercion to resolve conflicts is a logical outcome. Furthermore, corporal punishment may be encouraged in some families to prepare children for adult roles in a competitive social system. There is increasing consensus on the association between the acceptance of violent means of socializing children and the occurrence of child abuse. The use of corporal punishment is widespread, and it could be argued that physical punishment of children reflects societal values expressed in a familial context. Controversy reigns regarding the legal and moral legitimacy of violence toward children. The support of corporal punishment by such institutions as the United States Supreme Court appears to sanction violent practices in the American home, some of which culminate in incidents of serious harm.

The depiction and promotion of violence in the movies and on television may also affect how adults and children approach issues of conflict. Whether media violence is associated with childhood aggressive behavior remains a subject for lively debate, but there is a developing consensus that a milieu of violence fosters actions of violence. It could be argued further that the use of violence by media heroes conveys an essential value that

violence is manly, smart, and successful (see Chapter 17). The Carnegie Council on Children asserts that the low levels of support in public welfare programs also assure a "perpetuation of exclusion" of children from the mainstream of American life. Poverty, not parental failure, is cited by Gil (1975) as the principal "abuse" of children, and its continuation is an example of "socially structured and sanctioned child abuse." Many poor children, reported as victims of child abuse and neglect, are placed in foster home care because of serious economic and familial problems and a shortage of those services in the home that enable parents to care more adequately for their offspring. Too often those foster homes and institutions are inadequate or even harmful. In following up child abuse victims compared to a group of accident victims matched on the basis of socioeconomic status, which was low, Elmer (1977) found that the two groups suffered equally from developmental and emotional problems, and concluded that "the results of child abuse are less potent for the child's development than class membership."

CHILD NEGLECT

Definition

Neglect of children can be broadly defined as a failure to provide for or to meet the child's emotional and developmental needs, including the need for adequate nutrition, clothing, safety, and shelter; for stimulation and education; and for health and dental care. Using such a broad definition, all parents might at times appear to fall short of adequately meeting a child's many needs. The question of neglect arises when lack of parental care appears to be jeopardizing the child's physical or emotional well-being or interfering with his development.

Incidence

Child neglect appears to be a more pervasive problem than physical abuse of children. When harm to a child is severe enough to require hospitalization or medical attention, it is one and a half times more likely to be due to neglect than to physical abuse (American Humane Society, 1981). Data from the National Reporting Study indicates that 63 per cent of all reported cases of child mistreatment are reported for "deprivation of necessities," while emotional maltreatment is the grounds for reporting in 14 per cent of child abuse reports. (Physical injuries constitute 15.4 per cent of all reported cases.)

Manifestations of Neglect

Physical Neglect.   Children who appear to be undernourished, suffering from malnutrition or nonorganic failure to thrive, may be victims of parental neglect. The parent may be failing to provide adequate nourishment or providing it in circumstances that make it difficult for the child to eat and thrive.

Children who appear to be dirty and poorly clothed may not suffer physical consequences of the apparent neglect but may suffer social consequences. This neglect of appearance may indicate additional emotional neglect as well.

Neglect of Medical Care.   Failure to provide for a child's routine health and dental care, as well as lack of attention to illnesses and injuries constitute medical neglect. Such absence of care may be accompanied by other forms of neglect as well.

Martin (1980) notes that children who have been physically injured by their caretakers are also more likely to have received inadequate medical care, including lack of immunizations, and their illnesses, such as ear infections, may have gone untreated. He also reports a higher incidence of undernourishment and anemia among physically abused children. Newberger et al. (1981) report that victims of physical abuse are more likely to be underweight for their age and less healthy than are control children.

Medical neglect may sometimes occur because a parent's religious beliefs may run counter to medical practice. When faced with a case of this kind, it is important for a physician to ascertain whether this is an accepted belief within a religious system or is idiosyncratic to the parent. He should also become familiar with state laws and court rulings on decision making in these cases.

Lack of Supervision.   Inadequate supervision of children may result in injury or increased risk of injury for a child. Many childhood accidents are preventable, occurring because of lapses in parental supervision. Poisonous substances, caustics, knives, or guns may be left within easy reach of an active child, or children may be left in settings that are not adequately child-proofed (for example, settings with open windows, open stairways). Sometimes siblings are left to mind other siblings when none of the children is old enough or developmentally mature enough to exercise good judgment in child care. Certainly all parents occasionally lapse in their vigilance in supervising children; however, when such lapses constitute a pattern of parental care, or when the child is left in a particularly dangerous situation, one must consider the possibility of neglect.

Educational Neglect.   Failure to meet a child's educational needs through keeping the child out of school or condoning truancy constitutes edu-

cational neglect. Education or training of a developmentally disabled child might also be neglected, as when parents refuse to take advantage of programs or services that would facilitate the child's development.

Emotional Neglect.   Emotional neglect is a broad concept, but in general it constitutes a failure to meet the child's needs for affection, attention, and emotional nurturance. It might also include fostering or lack of intervention for maladaptive behavior, e.g., delinquency, alcohol or drug abuse, suicidal gestures, overeating, or behavior problems. Parental overprotectiveness, which interferes with the emotional development of the child, might also be considered emotional neglect.

Etiology of Neglect

Very few researchers have studied the etiology of neglect separately from the etiology of physical abuse. Kadushin (1978) refers to child neglect as a "neglected topic" in the child abuse literature. The multidimensional model proposed by Bittner and Newberger (1981) as an etiological model for family violence can be applied to neglect as well (see Figure 14–1). Children in families in which many stresses are operative at the child, parent, family, and social situational levels are at increased risk of neglect by their caretakers. Instead of, or in addition to, responding to their children through violence, neglectful caretakers may respond to increased stress by withdrawal through alcohol, drugs, illness, or emotional or physical unavailability, with consequent omissions in child care. Among reported cases of child abuse, physical abuse is frequently accompanied by neglect, although neglect is also an entity in itself.

Child Vulnerabilities.   Factors that might increase a child's risk for physical abuse are also risk factors in neglect. Children who are difficult by virtue of temperament, developmental difficulties or disabilities, or age are at greater risk, as are children reminding parents of disliked or negatively perceived persons in their lives. Prematurity appears to be a contributing factor to neglect as well as abuse. Hunter et al. (1978) found a high rate of reportable abuse and neglect directed toward infants discharged from a neonatal intensive care unit. In 80 per cent of these cases, the reported condition was neglect. Premature infants or infants requiring special care may be extremely difficult to care for and not easily soothed; they may be more sensitive to stimulation and respond negatively to handling. Hence, some parents may tend to withdraw or give up when it comes to trying to meet these needs.

Parental Factors.   Descriptions of mothers who neglect their children resemble descriptions of mothers who abuse their children. They are characterized as depressed, angry, anxious, remote, helpless, desperate, and suffering from low self-esteem (Fischoff et al., 1971). In addition, they may suffer from alcoholism, drug addiction, or chronic physical illness. Hunter et al. (1978) note that many neglecting mothers were themselves severely deprived as children. Newberger and Cook (1981) have found that neglectful mothers score lower than a control group on measures of parental awareness, demonstrating less ability to understand their children as separate individuals with separate needs. Failure to supervise children appropriately may be due to a parent's inadequate understanding of child development.

Family Factors.   In one study of child neglect, families in which a child was neglected were characterized as socially isolated, large in size, and headed by a single mother (Giovannoni and Billingsley, 1970). Parent-child separations also appear to be a factor in child neglect, particularly if such separations occur during the neonatal period (Hunter et al., 1978).

Social Stresses.   As is true for families in which violence is a strategy for resolving conflict, families in which children are neglected score high on social stress inventories (Hunter et al., 1978; Giovannoni and Billingsley, 1970). Also, victims of child neglect are frequently victims of poverty as well, as is the case for abused children. Indeed, the conditions of poverty may be confused with neglect in understanding the etiology of a child's difficulties: lack of resources to provide adequate nutrition, shelter, or medical care for a child may have the same result as failures in parental care. As Elmer's follow-up study of abuse and accident victims so forcefully demonstrated, all children living in poverty are at risk of developmental impairment (Elmer, 1977).

EFFECTS OF VIOLENCE AND NEGLECT ON THE CHILD

To understand the effects of physical violence and neglect on the child one must take into account the child's age and developmental level at the time of the event, the frequency and nature of what has been experienced, and the total emotional milieu in the home. Few well-designed follow-up studies exist, and virtually no longitudinal data are available on the effects of physical abuse and neglect on children. From available clinical observations, however, it appears that physical violence and neglect affect a child at a number of levels, including the physical, cognitive, and emotional realms of development.

Physical Effects of Violence and Neglect

Injuries inflicted on children by their caretakers range from relatively minor bruises, abrasions, and lacerations to bites, burns, skull and bone fractures, and internal injuries. While many injuries heal with time, some have permanent sequelae such as neurological damage, impaired hearing, ocular damage, and physical deformities.

A number of studies have indicated that children who are injured by their caretakers are likely to be reinjured if no intervention occurs. Friedman and Morse (1976), in following up a sample of 24 abuse and neglect victims, found that in over 70 per cent of the cases, siblings had been injured as well.

In some cases, the result of physical abuse is death. It has been estimated that 1,000 children are killed by their caretakers each year (American Humane Association, 1981), and this may be an underestimate. Straus et al. (1980) found that one in 1,000 children are threatened with or subjected to potentially lethal violence through use of a knife or gun, which extrapolates to 46,000 child victims per year.

Effects of neglect may also be severe. Malnutrition may result in anemia, stunted growth and development, or death. Lack of medical care can result in permanent handicapping conditions such as hearing deficits from untreated ear infections (Martin, 1980). Failures in supervision may result in accidents with effects similar to inflicted injury (burns, fractures, neurological damage).

Cognitive Developmental Effects

Numerous studies report developmental delays in the areas of cognition, language, and motor development as well as more severe developmental disabilities among children reported to be abused or neglected (Martin, 1980; Solomons, 1979). Caffey (1972) warned that shaking infants can result in subdural hematomas and, if left untreated, mental retardation. A number of studies report that mental retardation in abused children appears to be a direct result of head trauma (Buchanan and Oliver, 1979). Others working with children who fail to thrive note that sequelae include neurological defects and intellectual debilitation.

Clearly, the developmental effects of physical violence and abuse can be severe. Developmental delays in many areas of functioning, most frequently language development, are noted to be prevalent among victims of abuse and neglect (Newberger and McAnulty, 1976). In many cases these delays appear to be environmentally induced by lack of stimulation or by emotional distress. It is important for the pediatrician faced with a case of inflicted injury or possible neglect to obtain a careful developmental assessment and to follow the child closely. Early intervention programs, day care programs, or language therapy, appropriate to the age of the child, may be important parts of an intervention plan.

Emotional Effects

Emotional effects of physical abuse and neglect are critically related to the variables mentioned earlier: the age of the child, the frequency and severity of the abusive or neglectful events, and the total emotional milieu in the home. Physical injuries may occur in the guise of corporal punishment in a strict but caring environment or may occur in the context of a rejecting, neglectful emotional environment. Neglect may occur periodically, when a parent is drinking or is ill, while at other times the child's needs may be adequately met. One parent may be abusive while the other is supportive but powerless to protect the child. In assessing the potential emotional effects of abuse and neglect on the child, it is helpful to obtain as complete a social and family history as one can, including information on the adults in the home, how each relates to the child, child care arrangements, separations, and recent traumatic events for the caretakers.

Existing follow-up studies of victims of abuse and neglect suggest that the following emotional tasks are the most impaired by aversive conditions in the home: the development of a positive self-concept; the management of aggression; and the development of social relations with others, including the ability to trust.

Effects on Self-Concept

Children who have been physically victimized or neglected by a caretaker are likely to feel that they are bad, unlovable, and unwanted. Physically abused children frequently are found to be somber and unhappy, are unable to enjoy activities, and rate themselves negatively on self-concept scales (Martin and Beezley, 1977; Kinard, 1980). Green (1978) reports frequent self-destructive behavior, such as mutilation or suicide gestures or attempts among physically abused children (seen in 41 per cent of 60 children studied); such behavior is more frequent among neglect victims than other visitors to an outpatient mental health clinic (17 per cent of 30 neglected children versus 7 per cent of 30 controls).

Aggression

Children who have been physically abused are reported to be more physically aggressive with peers than are comparison groups (Martin, 1980; Martin and Beezley, 1977; Green, 1978; Reidy, 1977). Neglected children were also rated as more aggressive than controls by their teachers. Green's data suggest that aggression is also likely to be turned against the self among abuse and neglect victims.

The frequent use of aggression by abused children has been explained as modeling after parental behavior or displacement of aggressive feelings toward peers and as a result of poor self-concept. Children who feel bad about themselves frequently "act out" and invite aggression or rejection, which confirms their poor self-image.

The association between abuse experienced as a child and later violent or delinquent behavior has been a longstanding concern. Studies reviewing the histories of delinquent boys reveal a high incidence of violence suffered as children (Carr, 1977).

Social Relations

Abused and neglected children frequently develop poor relations with peers and adults (Martin and Beezley, 1977; Kinard, 1980a). Kinard (1980a), in observing 50 physically abused children, noted frequent avoidance of peers and difficulty in giving and receiving affection in relation to parents and peers.

Attachment behavior between abused and neglected children and their parents has also been found to be aberrant, including displays of indiscriminant attachment to adults and/or avoidance of the parent.

George and Main (1980) observed that abused preschoolers approached adults less frequently than did controls and exhibited a tendency to avoid affiliative encounters with peers and adults, frequently responding to peers with approach-avoidance behaviors (e.g., walking toward a peer, but with head turned away). These investigators also noticed a tendency among abused children to harass adults (e.g., throwing a rock at a caretaker).

Other Emotional Effects

A number of investigators have reported apparent emotional precocity or pseudomaturity among abused children (Martin, 1980; George and Main, 1980). Kinard (1980) noted that physically abused children attained higher scores for mature affect and motivation on a projective test, indicating that they were able to identify feelings and motivation of characters. Martin and Beezley (1977) found that one-fifth of their sample of abused children were precocious for age in caring for parents' emotional needs and carrying out household tasks. They tended to be solicitous toward the examiners inquiring about their health and activities.

Martin and Beezley (1977) conclude that children's behavioral reactions to abuse and neglect are variable. One response pattern is the development of a cooperative pseudomature style with adults, leading to a reversal in roles with the parent. Some children become oppositional and aggressive toward adults, while others become withdrawn and avoidant.

Children suffering from depriving or abusive environments may display a number of additional behaviors that are indicative of distress: enuresis, encopresis, hyperactivity, bizarre behavior, difficulties with learning, hypervigilance, temper tantrums, sleep disturbance, and delinquent behavior (Martin, 1980).

The existing follow-up studies of abused and neglected children indicate a need for treatment of the emotional effects of abuse and neglect through psychotherapy or, in the case of very young children, therapeutic day care or, if possible, therapeutic programs for parent and child. Treatment is indicated whether or not the child remains in his original home; foster placement will not be enough to ameliorate the negative effects of abuse on a child's developing self-concept and social skills.

SEXUAL MISUSE AND ABUSE OF CHILDREN

The sexual misuse of children by adults has been labeled the "last frontier" in child maltreatment. It is the most recently discovered form of maltreatment by the pediatric community and society at large, although, as was the case with physical abuse and neglect, historians have noted its occurrence for centuries (DeMause, 1974). Sociologist David Finkelhor (1979b) notes that the "discovery" of this social problem was facilitated by the women's movement which brought the problem of rape and sexual abuse of women to public consciousness, leading in turn to an awareness of the sexual victimization of children.

Sexual abuse has been defined in a number of ways. Henry Kempe (1978) offers the following definition: "The involvement of dependent, developmentally immature children and adolescents in sexual activities that they do not fully comprehend, to which they are unable to give informed consent, or that violate the social taboos of family roles." Psychiatrists suggest that "sexual misuse" may be a more appropriate term to describe those situations in which there is "exposure of a child to sexual stimulation inappropriate for the child's role in the family" (Brant and Tisza, 1977). In the definition offered by the National Center on Child

Abuse and Neglect (1981) it is noted that "sexual abuse may also be committed by a person under the age of 18 when that person is either significantly older than the victim or when the abuser is in a position of power or control over another child."

A distillation of these definitions implies that any of the following characteristics of a sexual interaction between an adult and child or between children would qualify as abusive:

— the child is exploited for the sexual gratification of another, usually an adult
— the child is exposed to or involved in sexual activities inappropriate for his developmental level
— the child is unable to give informed consent because of age and power differences in the relationship
— the child is exposed to or involved in sexual activities inappropriate for his role in the family

As is true for other categories of abuse, the actual behaviors that are labeled abusive vary from extreme cases, on which there is consensus—actual sexual intercourse between child and adult—to less obvious and possibly incidental situations—sexual stimulation of a child through witnessing parental intercourse or seeing adult pornography. Brant and Tisza (1977) note that cultural and ethnic norms for sexual behavior must be kept in mind when applying the label of sexual abuse, and Summit and Kryso (1978) warn that "the objective distinctions between loving support and lustful intrusion are disquietingly subtle."

Incidence of Sexual Abuse

Estimates of the incidence of sexual abuse vary. Surveys of adult sexual behavior, such as that conducted by Kinsey in the 1950's, have included questions about childhood sexual experiences. Gagnon's reanalysis of Kinsey's data from 1,200 adult women indicated that 28 per cent had at least one sexual experience with an adult prior to age 13 (Gagnon, 1965). Applying this rate to the population of girls under age 13 leads to an estimated incidence of 500,000 cases of sexual abuse of girls per year. (Gagnon's definition includes exhibitionism as well as cases of physical contact.) A study by the American Humane Association of 9,000 cases of sex crimes against children in 1969 indicated that 75 per cent of the perpetrators were adults familiar to the child. In a recent survey of almost 800 college students, 19 per cent of the women and 8.6 per cent of the men had sexually victimizing experiences as children (Finkelhor, 1979a). The most common sexual experience was genital fondling. For women, half the perpetrators were family members; for men, family members constituted 17 per cent of the perpetrators.

Actual reported cases of sexual abuse per year are considerably lower in number, in 1979 numbering 7,600 cases nationally (American Humane Association, 1981).

Sexual abuse may be underreported to a greater extent than other forms of child maltreatment for a number of reasons.

1. The frequent absence of physical sequelae to the victim means that cases do not come to the attention of health professionals to the same extent as do cases of physical abuse or neglect;

2. Children are reluctant to report sexual experiences, particularly when the offender is a parent or familiar adult. (In Finkelhor's study [1979a], 63 per cent of the female victims and 73 per cent of the males had not told anyone about the experience.)

3. The denial of the problem by professionals. Rosenfeld (1979) has sensitively discussed the strong emotions engendered by these cases in health and mental health professionals.

Characteristics of Victims

Reported victims of sexual abuse are primarily girls, constituting 80 per cent of the cases reported nationally (AHA, 1981). Surveys of adult women indicate an incidence rate of sexual victimization as children ranging from 19 to 34 per cent (Gagnon, 1965; Finkelhor, 1979a). However, Finkelhor's data suggest that boys are victimized in greater frequency than had been previously thought. Based on several other surveys of adult males regarding childhood sexual experiences, Finkelhor (1982) estimates that 2.5 to 5.0 per cent of boys under the age of 13 are victimized each year, which extrapolates to a national incidence rate of 46,000 to 92,000 abused boys per year. Male victims are less likely to be physically injured than are females and therefore less likely to come to the attention of health professionals. In addition, boys are more likely to be abused by nonfamily members than are girls and are less likely to come to the attention of child protection agencies. Finkelhor (1982) also notes that when boys are abused within the family, it is more likely that there is also a female victim within the family.

Victimized children range in age from infancy to adolescence. Survey data in the aggregate indicate a vulnerable age of 8 to 12 years and that boys are abused at an older age than girls. However, reported cases indicate age of discovery of abuse, not necessarily age of onset.

Characteristics of Perpetrators

It is consistently reported that those who sexually abuse children tend to be male. This is a finding of both clinical and survey reports, with both male and female victims and in both intrafamilial and nonfamilial abuse. In Finkelhor's 1979 study, 84 per cent of the perpetrators were male; in the National Reporting Study, 86 per cent of the perpetrators involving male victims and 94 per cent of the perpetrators involving female victims were male.

Data about convicted offenders distinguish between two types of male offenders in sex crimes against children: the fixated offender and the regressed offender. The fixated offender would appropriately be labeled a pedophiliac, for whom children are the primary and exclusive sexual object. For the regressed offender, the usual sexual choice is an adult female, but stress or a crisis in family relationships may lead to a regression to a choice of a child or adolescent as a partner. There may also be a third type of offender, the indiscriminately promiscuous adult, choosing children and adults of either sex as sexual partners.

Adults who sexually misuse children are seldom psychotic and may appear to the observer as perfectly normal (Summit and Kryso, 1978).

A final important characteristic of perpetrators of sexual abuse is that they tend to be familiar to the child as family members, friends of the family, neighbors, or babysitters.

INCEST

Sexual activities among family members (exclusive of marital partners) including nonbiologically related members (step-parent, step-brother, adoptive parent) are labeled incest. The incestuous family has received considerable attention in the clinical literature in recent years, most of which has focused on father-daughter incest.

A pattern of "endogamous incest" has been described by a number of clinicians (Summit and Kryso, 1978; Weinburg, 1955; Cormier et al., 1962). The endogamous incestuous family appears on the surface to be quite normal but suffers from serious role distortion. The father-daughter (usually adolescent) relationship has become sexual as the relationship between the spouses has become bereft of sexual involvement. The involved daughter is described as taking the role of the mother in many ways, due to the mother's withdrawal through illness, depression, or emotional nonavailability. The father engaging the daughter in incest has also often victimized the mother through violence, coercing her into a passive role. Lustig et al. (1966) describe an implicit condoning of this relationship by the mother and the painful fears of separation and abandonment characterizing all members of the family. The incestuous relationship holds the family together.

Weinburg (1955) identifies another incestuous

family pattern described as less organized and very promiscuous, with greater role confusion and more blurring of boundaries than in the endogamous family. Other factors associated with incest include alcoholism, and isolation of the family (Summit and Kryso, 1978).

Less commonly reported to child protection agencies are sexual relationships among siblings or step-siblings. Survey data indicate that this may be the most common type of incest but the least harmful (Finkelhor, 1979a; Nakashima and Zakus, 1977). Incestuous sexual experiences as opposed to experiences with nonfamily members are more likely to be repeated over a long period of time (Greenburg, 1979).

SEXUAL MISUSE, ABUSE, AND VIOLENCE

To what extent are children's sexual experiences with adults coupled with violence or physical force? Finkelhor (1979a) found that 55 per cent of his respondents reported coercion through physical force or threats of physical force. Summit and Kryso (1978) describe a form of incest in which the offending male uses violence toward women and children. However, most experts concur that adults can win a child's compliance through means other than physical force, and a relatively small percentage of cases fall into this category.

EFFECTS OF SEXUAL MISUSE OF THE CHILD

Physical Effects

Depending on the type of sexual activity between adult and child, there may or may not be physical sequelae, such as venereal disease, genital irritation, abrasions, tears, bruising, other trauma to the genitals or rectum, and, in girls, pregnancy. Fewer than half the victims have physical indications of sexual abuse.

Psychological Effects

A number of factors will determine the psychological sequelae of sexual abuse to the child, including (1) the nature of the sexual activity, its frequency of occurrence, and the use of force; (2) the age and developmental status of the child; (3) the relationship between the child and the perpetrator; and (4) the family's reaction.

Short-term effects of sexual abuse will vary with the age of the child but include feelings of anxiety, mistrust, guilt, anger, fear, and depression. Behavioral symptoms may include regressive behav-

iors (enuresis, encopresis, crying, clinging); difficulties in school; withdrawal from peers; and acting out behavior that is sexual, aggressive, or self-destructive in nature (Rosenfeld, 1979; Simrel et al., 1978).

Information on the long-term effects of sexual abuse comes from clinical reports, mainly retrospective in nature. At present, there are no systematic longitudinal data on childhood victims of sexual abuse. It is assumed that children who have experienced single incidents of sexual misuse by a strange adult and are supported by their families seem to suffer fewer long-term effects, although short-term effects will occur and must be dealt with. For children abused by family members over a long period of time, however, there is usually less family support available following disclosure. Indeed, the child may be viewed as a traitor, responsible for "breaking up the family," and, if forced to testify, must bear the burden of guilt for complicity in the sexual activity and disrupting the family and possibly sending a family member to jail. Hence, in these cases, it is difficult to separate the long-term effects of sexual misuse from the disturbed family dynamics and the aftermath of disclosure of the sexual activities.

Summit and Kryso (1978) report that victims of incest tend to suffer further sexual assaults from other family members following disclosure, and tend to blame themselves, suffering from depression and impaired sexual relationships in later life. Many authors hypothesize that the older the child at the time of victimization, the worse the long-term effects. Allegedly this is because the older child is more fully able to understand the nature of the exploitation and the broken societal taboo. Finkelhor's survey supports this notion; students victimized at younger ages rated the incident as less negative than children victimized in adolescence (1979a). However, more force was also employed with the older children, which was the factor most strongly related to negative reactions to the experiences.

It has also been hypothesized that the closer and more familiar the relationship between victim and perpetrator the more traumatic the experience for the victim. Finkelhor did not find this to be the case, except when comparing abuse by a parent with abuse by all others. The former was viewed most negatively. However, the confounding of the kind of relationship by the extent of support available for the child before and after disclosure is an important consideration in exploring this relationship.

While some have argued that sexual abuse is not necessarily harmful to children, knowledge of child development and the imperfect clinical evidence to date suggest that this is not the case. Symptoms among incest victims of later disturbances in self-concept formation and relationships include depression, suicide attempts, running away, sexual acting out, and antisocial behavior (National Center on Child Abuse and Neglect, 1981).

Children thrive and develop through a nurturant relationship with parents who are able to view the child's needs separately from their own. An adult who uses children for sexual gratification is imposing his needs on the child through an activity that the child cannot comprehend fully and to which he cannot give true consent. The adult is placing the child in a role inappropriate for the child's age. The child also needs the adult to set boundaries in order to learn to control his own impulses. It is the responsibility of the adult and not the child to help the child define the boundaries of relationships and acceptable behavior. Children are sexual and their sexuality develops over time. Exploration of their own bodies and of those of their peers is to some extent normal. Interest in adult bodies and in physical contact is normal as well. However, when an adult interprets a child's behavior as sexually provocative, and uses this interpretation as a rationale for sexual relations with a child, the adult has distorted the meaning of the child's behavior, acting on his own needs instead. Children whose line of sexual development is interfered with by inappropriate adult intrusions might be expected to have later difficulty with sexual relationships and with feelings about themselves as sexual beings.

PEDIATRIC DIAGNOSIS AND MANAGEMENT OF PHYSICAL ABUSE AND NEGLECT

History

During the initial interview of a family in which child abuse or neglect is suspected, the clinician has five objectives:

1. To understand the historical and especially the medical antecedents of the child's injury and to assess the plausibility of the history.

2. To determine the dimensions of the ongoing risk to the child, so as to inform the choice of protective or family supportive interventions.

3. To gather past medical history of the child and the family members.

4. To form a relationship with the family that will foster and support their participation in subsequent diagnostic and therapeutic work with other professionals.

5. To explain to the family the case report and other aspects of the protective service process, i.e., what the pediatrician and others will be doing to protect the child and to help the parents, including an honest reckoning with the parents about the professionals' concerns.

Physical Examination

Table 14-1 summarizes the differential diagnosis of symptoms of child abuse. A complete examination including developmental assessment should be performed on any child who may be such a victim. The child's affect and his verbal and behavioral interactions with his family and other adults should also be observed and carefully noted.

In assessing the origins of the symptoms, it is wise to keep an open mind. Do not rush to conclude that a given symptom or family problem is diagnostic. Rather, the findings of the history and physical examination should guide the formation of hypotheses and a problem list. These can be more fully delineated by subsequent laboratory studies and social work and psychiatric consultations.

The cutaneous manifestations of child abuse may be ambiguous: in the first 24 hours, a bruise may be reddish, blue, or purple; from the first to the third day, the color becomes blue or blue-brown; with further metabolism of heme, the bruise acquires a greenish cast on the fifth to seventh days; and by the tenth day the bruise may appear to be yellow. Before disappearing completely after two to four more days the bruise may take on a brownish hue. The child's skin may make some of these transitional colors difficult to interpret.

Other cutaneous manifestations include bruises in the shape of a handprint, linear bruises, abrasions from whipping with a cord or rope, loop shaped marks from cords that have been folded over, crescentic bite marks, alopecia or subgaleal hematoma from pulling of the hair, and areas of abraded skin that may be caused by being bound or restrained. Impetigo in its various forms may be confused with burns or inflicted injuries. In addition to recording the distribution, shape, color, location, and approximate measurement of bruises, it may be helpful to have photographs for purposes of documentation.

The significance of obtaining informed consent for such procedures cannot be overemphasized. Parents will quickly sense any inquisitorial intentions on the part of the pediatrician, and the utility of the photographs needs to be explained openly and honestly; in some cases, it may be advisable to forego obtaining pictures in the interest of building and sustaining a helping relationship.

Four distinct patterns of inflicted burns have been described (Lenoski and Hunter, 1977). Forced immersion yields a doughnut-shaped distribution of the burn with a spared area, frequently the buttocks or back, where the body may have been in contact with a container and, thus, may have been protected from the heat of the water. A splash burn may produce nonuniform multiple noncontiguous burn areas, sometimes with "arrowhead" patterns where the water has spread laterally as it rolls off the skin. If the child's body is immersed in a maximally flexed position, the skin folds of the thorax will be spared. This will give a striped pattern. When a burn is caused by contact with a hot object, such as an iron, the object may leave a distinctive mark on the skin.

Less obvious physical signs of child abuse may reflect internal injuries. A ruptured tympanic membrane may result from a blow to the side of the head or from a basilar skull fracture. Multiple class I fractures of the teeth may result from repeated blows to the chin. A lacerated frenulum

Table 14–1. DIFFERENTIAL DIAGNOSIS OF CHILD ABUSE

Clinical Findings	Differential Diagnosis	Differentiating Tests
Cutaneous lesions		
1. Bruising	Trauma	
	Hemophilia	PT, PTT
	Von Willebrand's disease	Bleeding time
	Anaphylactoid purpura	Rule out sepsis
	Purpura fulminans	Rule out sepsis
	Ehlers-Danlos	Hyperextensibility
2. Local erythema or bullae	Burn	
	Staphylococcus impetigo	Culture, Gram stain
	Bacterial cellulitis	Culture, Gram stain
	Pyoderma gangrenosum	Culture, Gram stain
	Photosensitivity and phototoxicity reactions	History of sensitizing agent, orally or topically
	Frostbite	Clinical history and characteristics
	Herpes zoster/herpes simplex	Scraping
	Epidermolysis bullosa	Skin biopsy
	Contact dermatitis, allergic or irritant	Clinical characteristics
Ocular findings		
1. Retinal hemorrhage	Shaking or other trauma	
	Bleeding disorder	Coagulation studies
	Neoplasm	
	Resuscitation	History
2. Conjunctival hemorrhage	Trauma	
	Bacterial or viral conjunctivitis	Culture, Gram stain
	Severe coughing	History
3. Orbital swelling	Trauma	
	Orbital or periorbital cellulitis	Complete blood cell count (CBC), culture, sinus x-rays
	Metastatic disease	X-ray, CT scan; CNS examination
	Epidural hematoma	X-ray, CT scan; CNS examination
Hematuria	Trauma	Rule out other disease
	Urinary tract infection	Culture
	Acute or chronic forms of glomerular injury (e.g., glomerulonephritis)	Renal function tests; biopsy
	Hereditary or familial renal disorders (e.g., familial benign recurrent hematuria)	History
	Other (vasculitis, thrombosis, neoplasm, anomalies, stones, bacteremia, exercise, etc.)	History, cultures, IVP
Acute abdomen	Trauma	Rule out other disease
	Intrinsic gastrointestinal disease (e.g., peritonitis, obstruction, inflammatory bowel disease, Meckel's)	X-ray studies, stool tests, etc.
	Intrinsic urinary tract disease (infection, stone)	Culture, IVP
	Genital problem (torsion of spermatic cord, ovarian cyst, etc.)	History, physical examination, x-ray, laparoscopy (?)
	Vascular accident as in sickle cell crisis	Angiography, sickle prep
	Other (mesenteric adenitis, strangulated hernia, anaphylactoid purpura, pulmonary disease, pancreatitis, lead poisoning, DKA, etc.)	As appropriate

Table 14-1. DIFFERENTIAL DIAGNOSIS OF CHILD ABUSE (Continued)

Clinical Findings	Differential Diagnosis	Differentiating Tests
Osseous lesions		
1. Fractures (multiple or in various stages of healing)	Trauma	
	Osteogenesis imperfecta	X-ray and blue sclerae
	Rickets	Nutrition history
	Birth trauma	Birth history
	Hypophosphatasia	Decreased alkaline phosphatase
	Leukemia	CBC, bone marrow
	Neuroblastoma	Bone marrow, biopsy
	Status post osteomyelitis or septic arthritis	History
	Neurogenic sensory deficit	Physical examination
2. Metaphyseal or epiphyseal lesions	Trauma	X-ray and nutrition
	Scurvy	History
	Menkes syndrome	Decreased copper, decreased ceruloplasmin
	Syphilis	Serology
	Little league elbow	History
	Birth trauma	History
3. Subperiosteal ossification	Trauma	
	Osteogenic malignancy	X-ray and biopsy
	Syphilis	Serology
	Infantile cortical hyperostosis	No metaphyseal irregularity
	Osteoid osteoma	Response to aspirin
	Scurvy	Nutritional history
Sudden infant death	Unexplained	Autopsy
	Trauma	Autopsy
	Asphyxia (aspiration, nasal obstruction, laryngospasm, sleep apnea)	"Near-miss" history
	Infection—botulism?	Cultures, bacterial and viral
	Immunodeficiency?	Immunoglobulins
	Cardiac arrhythmia?	Autopsy
	Hypoadrenalism?	Electrolytes
	Metabolic abnormality—calcium?—magnesium?	Ca^{2+}, Mg^{2+}
	Hypersensitivity to cow's milk protein?	

From Bittner, S., and Newberger, E. H.: Pediatric understanding of child abuse and neglect: Pediat. Rev. 2:200, 1981. Copyright American Academy of Pediatrics, 1981.

or other intraoral trauma may result from forced feeding with a spoon.

Ocular injuries may include hyphema, corneal abrasion, subconjunctival hemorrhage, dislocation of the lens, detached retina, or retinal hemorrhages. Caffey (1972) has described the syndrome of the severely shaken infant; this syndrome may include osseous lesions such as metaphyseal avulsions or subperiosteal hematomas, intracranial hemorrhage, and retinal hemorrhage. The shaken infant may be difficult to identify, since there may be no obvious cutaneous signs. Sudden thoracic compression and consequent increased intravascular pressure may also lead to retinal hemorrhages, so-called Purtscher retinopathy.

Occult internal injuries include rupture of the pancreas and pseudocyst formation, lacerated liver or spleen, intramural hematoma of the bowel, retroperitoneal hemorrhage, renal laceration or contusion, intestinal perforation, rupture of the ureter or bladder, and chylous ascites.

The concomitant presentation of unusual chronic illness and child abuse has been noted in several recent reports. The term "Munchausen syndrome by proxy" has been applied to clinical situations in which parents make their children sick in order to attract attention to their own problems. A child with mysterious relapsing coma was found to have been given sublethal doses of chloral hydrate by his mother. Other reports include fever, bacteremia, recurrent idiopathic lesions of the skin, and sclerosed lesions of the cornea.

Laboratory Studies

Laboratory studies are helpful in delineating the nature and extent of current trauma and in defining the presence of previous trauma. Radiological findings of child abuse include multiple long bone fractures in various stages of healing, spiral fractures, epiphyseal fractures, and exaggerated periosteal reaction.

A coagulation profile (platelet count and prothrombin, partial thromboplastin, and bleeding times) will exclude endogenous disorders of bleeding. Hemoglobinuria and hematuria are known to occur with major trauma.

PEDIATRIC DIAGNOSIS AND MANAGEMENT OF SEXUAL ABUSE

Sexual abuse may present with "nonspecific" symptoms such as enuresis and encopresis, hyperactivity, fears and phobias, sleep disorders, learning problems, compulsive masturbation, sexualized play, perineal irritation, other genital injury, and distorted or pseudomature personality development (Table 14–2). Incest frequently is discovered when the child or mother reports the problem to someone outside the family.

When sexual misuse or abuse is suspected, the clinical evaluation should include a calm, careful, sensitive interview of the child alone, allowing the child to communicate with pictures, toys, and play. The parents, other close relatives, and other caretakers can be interviewed to assess risk factors in the home and to establish relationships that will endure beyond the crisis to support the family,

Table 14–2. SEXUAL ABUSE: DIAGNOSIS AND MEDICAL MANAGEMENT

Signs and Symptoms
1. Strong evidence
   Gonococcal infection: urethritis, pharyngitis, arthritis, conjunctivitis
   Trichomonas infection
   Venereal warts
   Syphilis
   Sperm or acid phosphatase present on body or clothes of victim
   Pregnancy
2. Probable evidence
   Vaginal or anal laceration
   Perineal bruises or abrasions
3. Possible evidence
   Monilial vaginitis
   Haemophilus vaginitis
   Hematuria (secondary to trauma)
   Behavioral symptoms: phobias, sexualized play, etc.

Laboratory evaluation of sexually abused child
1. Cultures: gonorrhea, monilia
2. Microscopic examination: sperm, Trichomonas, Monilia, Hemophilus
3. Blood: syphilis serology and HCG beta-subunit assay
4. Urine: routine urinalysis for blood, sperm; culture for gonorrhea; pregnancy test
5. Miscellaneous: fingernail scrapings if there was a struggle, careful search for blood, pubic hairs or semen on clothing

Treatment
1. Penicillin prophylaxis for several sexually transmitted diseases
2. Diethylstilbestrol to prevent pregnancy
3. Appropriate treatment if monilia, Trichomonas or Hemophilus vaginitis found

From Bittner, S., and Newberger, E. H.: Pediatric understanding of child abuse and neglect. Pediat. Rev. 2:205, 1981. Copyright American Academy of Pediatrics, 1981.

even in the event of ambiguous medical findings and uncertain diagnostic conclusions.

The physical examination of suspected victims of sexual abuse should first be carefully explained (Table 14–2). Foreign bodies should be sought and clothing examined for signs of semen or blood. The throat, rectum, and vagina should be cultured for gonococcus, and a serological examination for syphilis and serum assay for the beta-subunit of HCG (human chorionic gonadotropin) should be performed. If vaginal discharge is present, microscopic examination for Trichomonas and Monilia as well as culture and Gram stain will be helpful in enabling prompt treatment of infection. Vaginal contents may be gently aspirated with an eye dropper. Venereal warts may also be found on careful examination. It is well to keep in mind that the data that are gathered may subsequently be reviewed in the criminal or civil court. For girls who have passed menarche, the pregnancy test is routine. Medical management at presentation may include prophylactic antibiotic therapy and diethylstilbestrol to prevent pregnancy.

INTERDISCIPLINARY MANAGEMENT OF CHILD ABUSE

Table 14–3 summarizes the management of child abuse and divides the pediatric role into two phases, diagnosis and treatment. Principal questions are outlined, and interventions to protect the child and to help the family are summarized.

Since child abuse is regarded as a symptom of family distress and a problem with complex, multivariate origins, it should be managed by a diagnostic interdisciplinary team that includes a social worker, a pediatrician, a nurse, a psychiatrist, and an attorney. When such a diagnostic unit is not available, it may be necessary for the physician to help organize and to work with other professionals in the hospital or in the community. Management guidelines can be developed that utilize each community's resources and personnel. The protective service to which mandated reports are sent may not by itself be able to offer an adequate program of services. A social worker should be called promptly at the time of the family's presentation, both to facilitate the social assessment and to form a helping relationship.

In the initial interviews and in subsequent contacts, no direct or indirect attempt should be made to draw out a confession from the parent. Denial is a prominent ego defense among virtually all abusing parents, and their bizarre stories about how their children became injured ought not to be taken as intentional falsifications but rather as repression of profoundly distressing realities.

Table 14-3. PHASES IN MANAGEMENT OF CHILD ABUSE

Phases in Management	Primary Considerations	Interventions to Protect Child and Help Family
Diagnostic assessment Medical history Physical examination Skeletal survey	Are the physical findings at variance with the history?	Provide more comprehensive medical workup.
Laboratory tests	Is child abuse or neglect suspected?	Inform the parent of the suspicions and the physician's responsibility to protect the child.
	What is the legal responsibility regarding suspected child abuse?	Make a report to the mandated agency.
	Is the home safe for the child?	Continue the evaluation on an outpatient basis.
	Is the child at risk?	Hospitalize the child for protection and further evaluation.
Consultations for evaluation of family dynamics and child development	What is needed to make the home safe for the child's return?	Arrange for multidisciplinary conferencing for disposition planning.
Rehabilitation program	What resources will meet the needs of the child and the family?	Arrange for primary health care and appropriate treatment for the child and family.
Health needs Physical, social and environmental needs		Mobilize community resources such as child care, homemaker service, foster home placement.
Follow-up planning	Who will monitor the health and community services to the child and the family?	Provide coordination and integration of the helping resources.
Medical care Social work services Nursing Services Other services		

From Newberger, R. H., et al. *In* Hoekelman, R. A., et al. (eds.): Principles of Pediatrics. New York, McGraw-Hill Book Co., 1978. Reprinted with permission.

Other defenses such as angry outbursts against the interviewer or refusal to talk limit both the process of information gathering and the prospects for continuing a helpful professional relationship. Breakdown of the assessment process may possibly endanger the child. It is appropriate to emphasize to the parent the child's need for care—which may include admitting the child to a hospital—and the need to ensure that the child is protected from harm.

In explaining the legal obligation to report the case, the physician's compassion and honesty will go far to allay the family's anxiety. Other professionals may provide crucial aid in the evaluation process. The opportunity to observe parent-child interaction and the child's physical and psychological milestones (which might yield insight into the familial causes of a child's injury) may not be available to a physician in his office or in the emergency room. Nurses in clinical and public health settings can and do, however, make such observations, which are fundamental in case finding and evaluation. The input of these nurses contributes uniquely to diagnosis, and their perceptions should be shared appropriately with the physician and social worker attending to the family.

A home visit by a public health nurse or social worker may be an important part of all initial assessments and is made to gather data on the child's home environment that may aid in making the disposition.

A psychiatric consultation is frequently obtained in cases of child abuse and neglect. Often this consultant's perceptions lead to an understanding of what interventions can be most effective. However, this consultation should not substitute for careful diagnostic work and energetic advocacy by the social worker, pediatrician, and nurse.

Several ethical dilemmas confront pediatricians and their colleagues in the diagnosis and management of child abuse. The diagnosis itself is often impossible to make with certainty, and the physician, concerned with giving the parents the benefit of any doubt, may feel that the easiest, fairest, and most ethical approach is to send the child home without reporting. These clinical problems, once reported, may also consume substantial amounts of unremunerated time.

The reporting laws also require communication

of confidential information when a child is suspected of being at risk, and this may place the pediatrician in conflict. The Hippocratic precept *primum non nocere* is challenged when the reporting carries with it the risk of an incompetent intrusion into the life of the family by a poorly trained, inadequately supervised social worker from an overburdened and underfunded public child protection agency. The child may be separated from home or help may not materialize. It is often necessary for the interdisciplinary team to choose the least detrimental alternative.

A consensus on seven axioms of management can be found in the literature on child abuse, and these are as follows:

1. Once diagnosed, abused children—especially infants less than one year of age—are at great risk for reinjury or continued neglect.

2. In the event that the child is reinjured, it is likely that the parents will seek care at a different medical facility.

3. There is rarely any need to establish precisely who it was who injured the child and whether the injury was "intentional." The symptom itself should open the door to forming a helping alliance and planning comprehensive service for the child and the family.

4. If there is evidence that the child is at major risk, hospitalization is appropriate to allow time for interdisciplinary assessment. The complex origins of the child's injury are seldom revealed in the crisis atmosphere at the time of presentation.

5. Protection of the child must be the principal goal of intervention, but protection must go hand-in-hand with the development of a family-oriented service plan.

6. Traditional social casework alone may not adequately protect an abused child in the environment in which he received his injuries. Multidisciplinary follow-up is also necessary, and frequent contact by all those involved in the service plan may be needed to encourage the child's healthy development.

7. Problems of public social service agencies in both urban and rural agencies—specifically in numbers of adequately trained personnel and in quality of administrative and supervisory functions—militate against their effective operation in isolation from other care-providing agencies. Simply reporting a case to the public agency mandated to receive child abuse case reports may not be sufficient to protect an abused child or to help the family.

The development of programs that attend to these principles will require careful thought and planning. In the last analysis, the professionals' ability to convince patients or clients that they intend to help them depends on their ability to mobilize effective services. When case management programs and interdisciplinary cooperation improve, pediatricians and other professionals who work with children will find it easier and more rewarding to participate in comprehensive service plans.

JANE C. SNYDER
ROBERT HAMPTON
ELI H. NEWBERGER

REFERENCES

American Humane Association: National analysis of official child neglect and abuse reporting (1979). DHHS Publication No. (OHDS) 81-30232, revised 1981.

Belsky, J.: Child maltreatment: An ecological integration. Am. Psychologist 35:320, 1980.

Bittner, S., and Newberger, E. H.: Pediatric understanding of child abuse and neglect. Pediat. Rev. 2:197, 1981.

Brant, R., and Tisza, V.: The sexually misused child. Am. J. Orthopsychiatry 47:80, 1977.

Buchanan, A., and Oliver, J. E.: Abuse and neglect as a cause of mental retardation: A study of 140 children admitted to subnormality hospitals in Wiltshire. Child Abuse Neglect 3:467, 1979.

Burgess, R. L., Anderson, E. A., and Schellenbach, C. O.: A social interactional approach to the study of abusive families. *In* Vincent, J. P. (ed.): Advances in Family Interaction, Assessment and Theory. Greenwich, Ct., JAI Press, 1980.

Caffey, J.: Multiple fractures in the long bones of infants suffering from chronic subdural hematoma. Am. J. Roentgenol. 56:163, 1946.

Caffey, J.: On the theory and practice of shaking infants: Its potential residual effects of permanent brain damage and mental retardation. Am. J. Dis. Child. 124:161, 1972.

Carr, A.: Some preliminary findings on the association between child maltreatment and juvenile mistreatment in eight New York counties. Report to the Administration for Children, Youth, and Families: National Center on Child Abuse and Neglect. Processed October 20, 1977.

Chotiner, N., and Lehr, W. (eds.): Child abuse and developmental disabilities: Essays from the New England Regional Conference, 1976. DHEW Publication No. (OHDS) 79–30226.

Cormier, B., Kennedy, M., and Sangowicz, J.: Psychodynamics of father-daughter incest. Canad. Psychiat. Assoc. J. 7:203, 1962.

DeMause, L. (ed.): The History of Childhood. New York, Psychohistory Press, 1974.

Elmer, E.: Fragile Families, Troubled Children: The Aftermath of Infant Trauma. Pittsburgh, University of Pittsburgh Press, 1977.

Finkelhor, D.: Sexually Victimized Children. New York, The Free Press, 1979a.

Finkelhor, D.: Social forces in the formulation of the problem of sexual abuse. Paper delivered at the Society for the Study of Social Problems, Annual Meeting, Boston, 1979b. Family Violence Research Program, University of New Hampshire, Durham, N. H.

Finkelhor, D.: Sexual abuse of boys: The available data. *In* Groth, N. (ed.): Sexual Victimization of Males. Offenses, Offenders, and Victims. New York, Plenum Press, 1982.

Fischhoff, J., Whitten, C., and Pettit, M.: A psychiatric study of mothers of infants with growth failure secondary to maternal deprivation. *In* Cook, J., and Bowles, R. (eds.): Child Abuse: Commission and Omission. Toronto, Butterworths, 1981.

Fontana, V.: The maltreated child: The maltreatment syndrome in children. Springfield, Ill., Charles C Thomas, 1971.

Friedman, S. B., and Morse, C. W.: Child abuse: A five year follow-up of early case findings in the Emergency department. Pediatrics 54:404, 1976.

Friedrich, W. N., and Boriskin, J. A.: The role of the child in abuse: A review of the literature. Am. J. Orthopsychiatry 46:580, 1976.

Gagnon, J.: Female child victims of sex offenses. Social Probl. 13:176, 1965.

Galdston, R.: Observations of children who have been physically abused by their parents. Am. J. Psychiatry 122:440, 1965.

Garbarino, J., and Sherman, D.: High-risk families and high-risk neighborhoods: Studying the ecology of child maltreatment. Child Devel. 57:188, 1980.

Gelles, R. J.: The Violent Home. Beverly Hills, California, Sage Publications, 1974.

Gelles, R. J.: Community agencies and child abuse: Labelling and gatekeeping. Paper presented at the Conference on Recent Research on the Family, sponsored by the Society for Research in Child Development, Ann Arbor, Michigan, 1975.

Gelles, R. J.: Socioeconomic issues in child abuse. In Kerns, D. L. (ed.): Child Abuse and Neglect. To be published.

George, C., and Main, M.: Abused children: Their rejection of peers and caregivers. In Field, T. M., Goldberg, S., Stern, D., and Sostek, A. M. (eds.): High Risk Infants and Children: Adult and Peer Interactions. New York, Academic Press, 1980.

Gil, D. G.: Violence Against Children: Physical Child Abuse in the United States. Cambridge, Massachusetts, Harvard University Press, 1970.

Gil, D. G.: Unraveling child abuse. Am. J. Orthopsychiatry 45:346, 1975.

Giovannoni, J. M., and Becerra, R. M.: Defining Child Abuse. New York, The Free Press, 1979.

Giovannoni, J. M., and Billingsley, A.: Child neglect among the poor: A study of parental adequacy in families of three ethnic groups. Child Welfare 49:196, 1970.

Green, A. H.: Self-destructive behavior in battered children. Am. J. Psychiatry 135:5, 1978.

Greenberg, N. H.: The epidemiology of childhood sexual abuse. Pediat. Ann. 8:16, 1979.

Harper, L. V.: The scope of offspring effects: From caregiver to culture. Psych. Bull. 82:784, 1975.

Hunter, R., Kilstrom, N., Krayball, E., and Luda, F.: Antecedents of child abuse and neglect in premature infants: A prospective study in a newborn intensive care unit. Pediatrics 61:629, 1978.

Kadushin, A.: Neglect—Is it neglected too often? Child Abuse and Neglect: Issues on Innovation and Implementation, United States Department of Health, Education, and Welfare, 1978.

Kempe, C. H.: Sexual abuse: Another hidden pediatric problem: The 1977 C. Anderson Aldrich Lecture. Pediatrics 62:382, 1978.

Kempe, R. S., and Kempe, C. H.: Child Abuse. Cambridge, Massachusetts, Harvard University Press, 1978.

Kempe, C. H., Silverman, E. N., Steele, B. F., Dragmueller, W., and Silver, H. K.: The battered child syndrome. J.A.M.A. 181:17, 1962.

Kinard, E. M.: Emotional development in physically abused children. Am. J. Orthopsychiatry 50:686, 1980.

Klaus, M. H., and Kennell, J. H.: Mother-Infant Bonding. St. Louis, The C. V. Mosby Co., 1976.

Lenoski, E. F., and Hunter, K. A.: Patterns of inflicted burns. J. Trauma 17:842, 1977.

Lustig, N., Dresser, J. W., Spellman, S. W., and Murray, T. B.: Incest: A family group survival pattern. Arch. Gen. Psychiatry 14:31, 1966.

Martin, H. P.: The consequences of being abused and neglected: How the child fares. In Kempe, C. H., and Helfer, R. E. (eds.): The Battered Child. 3rd ed. Chicago, University of Chicago Press, 1980.

Martin, H. P., and Beezley, P.: Behavioral observations of abused children. Develop. Med. Child Neurol. 19:373, 1977.

Nakashima, I. I., and Zakus, G. E.: Incest: Review and clinical experience. Pediatrics 60:696, 1977.

National Center on Child Abuse and Neglect: Child sexual abuse: Incest, assault and sexual exploration. DHHS Publication No. (OHDS) 81-30166, 1981.

Newberger, C. M.: The cognitive structure of parenthood: Designing a descriptive measure. In Yando, S. R. (ed.): New Directions for Child Development. Vol. 7. San Francisco, Jussey-Bass, 1980.

Newberger, C. M., and Cook, S.: Parental awareness and child abuse and neglect: Studies of urban and rural parents. Paper presented at the National Conference for Family Violence Researchers, University of New Hampshire, Durham, July, 1981.

Newberger, C. M., and Newberger, E. H.: The etiology of child abuse. In Ellerstein, N. S. (ed.): Child Abuse and Neglect. New York, John Wiley and Sons, 1981.

Newberger, C. M., et al.: Child abuse and childhood accidents: A comparative analysis. Paper presented at the National Conference for Family Violence Researchers, University of New Hampshire, Durham, July, 1981.

Newberger, E. H., and Hyde, J. N., Jr.: Child abuse: Principles and implications current pediatric practice. Pediat. Clin. N. Am. 22:695, 1975.

Newberger, E. H., and McAnulty, E. H.: Family intervention in the pediatric clinic: A necessary approach to the vulnerable child. Clin. Pediat. 15:1155, 1976.

Newberger, E. H., Reed, R. B., Daniel, J. H., Hyde, J. N., Jr., and Kotelchuck, M.: Pediatric social illness: Toward an etiologic classification. Pediatrics 60:178, 1977.

Parke, R. D., and Collmer, C. W.: Child abuse: An interdisciplinary analysis. In Hetherington, N. (ed.): Review of Child Development Research. Vol. 5. Chicago, University of Chicago Press, 1975.

Pfohl, S. J.: The "discovery" of child abuse. Social Problems 24:310, 1977.

Radbill, S. X.: A history of child abuse and infanticide. In Helfer, R. E., and Kempe, C. H. (eds.): The Battered Child. 3rd ed. Chicago, University of Chicago Press, 1980.

Reidy, T. J.: The aggressive characteristics of abuse and neglected children. J. Clin. Psychol. 33:1140, 1977.

Rosenfeld, A.: The clinical management of incest and sexual abuse of children. J.A.M.A. 242:1761, 1979.

Simrel, K., Berg, R., and Thomas, J.: Crisis management of sexually abused children. Pediat. Ann. 8:59, 1979.

Solomons, G.: Child abuse and developmental disabilities. Develop. Med. Child Neurol. 21:101, 1979.

Steele, B. F.: The child abuser. In Kutash, I., Kutash, S., and Schlesinger, L. (eds.): Violence: Perspectives on Murder and Aggression. San Francisco, Jussey-Bass, 1978.

Steele, B. F., and Pollack, C. B.: A psychiatric study of parents who abuse infants and small children. In Helfer, R. E., and Kempe, C. H. (eds.): The Battered Child. 3rd ed. Chicago, University of Chicago Press, 1980.

Straus, M. A., Gelles, R. J., and Steinmetz, S.: Behind closed doors: Violence in the American family. Garden City, New York, Doubleday, 1980.

Summit, R., and Kryso, J. A.: Sexual abuse of children: A clinical spectrum. Am. J. Orthopsychiatry 48:237, 1978.

Turbett, J. P., and O'Toole, R.: Physicians' recognition and reporting of child abuse. Presented at the annual meeting, American Sociological Association, New York, 1980.

Weinburg, K. S.: Incest Behavior. New York, Citadel Press, 1955.

15

Schools as Milieux

And the whining school-boy, with his sachel and morning
face, creeping like a snail unwillingly to school.

Shakespeare

Upon enrollment in school, a child is immersed
in a new and challenging environment. With this
comes an exposure to many new adults and same-
age peers as well as to values and expectations
that are imposed directly, without the buffer of
the family (see also Chapter 8). The effects of
schooling on a child's development are profound
and complex. For many reasons, however, we are
blinded to many of these effects. On one hand,
most adults are products both of the culture that
produces schools and of the schools themselves.
Their perspectives have been shaped by their own
personal school experiences, and they have difficulty imagining a development shaped otherwise.
On the other hand, schools are only one way in
which culture influences a child's development,
and it is difficult to isolate the particular effects of
the school experience.

THE EFFECTS OF SCHOOL

Understandably, research on the effects of
school has been confusing and contradictory and
often simply reinforces prevailing social or cultural
convictions. Some investigators have concluded
that schools have little effect on children. They
argue that children have already acquired their
genetic and cultural potential and that little happens in school that can change this.

In contrast, we have learned much about rich
interaction between genetics and environment that
shapes the early years of life. It would be surprising indeed if this interaction were to end abruptly
when we reach the age of 5. Recent studies by
Michael Rutter (1980) have begun to illuminate the
relationship between the organism and the environment as the child begins the school years. His
findings are described later in this chapter.

MILESTONES AND TRANSITIONS

The nature of schooling and its specific effects
on children change dramatically as the child moves

from kindergarten or an early childhood program
through the elementary grades to the secondary
level. While there are great variations from school
to school, one can perceive general stages and
transitions that transcend these individual differences, much as individual developmental trends
may transcend some unique variations among
children. An awareness of these stages and transitions will alert the practitioner to problems
caused by transitions in school life or by a mismatch between a child's development and a
school's expectations (see later in this chapter).

Regardless of the individual school setting, five
transitions are commonly seen in a child's progression through school. These may or may not
coincide with physical transitions from one building to another, and individual school milieux may
augment or diminish the sharpness of each transition.

School entry defines the first transition for children in their academic lives. While traditionally
this comes at age 5 or 6, new laws have mandated
a downward extension of programs to the extent
that preschoolers with handicaps living in certain
states may be enrolled in school programs even at
birth. The most effective role for the school in
dealing with handicapped infants (even the wisdom of involving public schools with the child
under age 3) is still in question. Many such programs are appropriately home-based, and although administratively under the school's mandate, they do not represent an entry into the social
milieu of the school.

School entry for the 3- to 5-year-old, however,
represents a tremendous social transition. Chapter
45, on preschool screening, reviews some methods
for assessing children prior to this important transition. The wide variations between kindergarten
and early childhood programs for children in
schools makes it imperative that the practitioner
understand both the child's development and the
characteristics, goals, and expectations of the program into which the child will enter.

The next transition for most children involves

the beginning of *academic reading instruction*. For the most part, kindergarten and early childhood programs in school have a preacademic or prereading emphasis that stresses social development, peer interactions, following directions, and language skills. (In addition, early childhood programs for handicapped children usually focus on specific remediation for the child's handicaps.) Formal reading instruction, as opposed to prereading skills, traditionally begins in the first grade. In many schools, however, this transition may come in the middle of the kindergarten year or even earlier. In an apparent attempt to accelerate the cognitive development of children, many schools seem to have concluded that earlier is better when it comes to reading. To the extent that reading skills are dependent upon neurological and cognitive development (and considerable data support this view), such an approach may create problems for many children. The onset of formal reading instruction will cause specific problems for the child who is unprepared for this task. In addition, this transition in the process and pace of schooling may cause problems for the child whose adaptation is challenged by family upsets, relationship changes, the stresses of moving, a poor self-concept, or other emotional or environmental stresses.

At about the beginning of the third grade, a third, relatively unstudied transition occurs in the school life of many children. At this time, the pace of schooling again increases noticeably. The major task of school changes from "learning to read" to *"reading to learn."* No longer is the focus of instruction on the mechanics of reading but on the content of what is being read. Like all other transitions, the individual school environment may make this more or less stressful for children. However, this may be an increasingly hazardous time for the child whose reading skills have not become proficient. The practitioner should be alert to behavioral difficulties, attendance problems, or learning problems that may present at this time in the child's schooling.

Somewhat later, often at about the *start of middle school,* the child's school experiences undergo a fourth transition. Traditionally, the move to junior high school or middle school represents a change from a single-classroom/single-teacher environment to a curriculum organized by classes, in which the child moves from teacher to teacher. This particular change now often comes much earlier and may be a part of even the kindergartener's school experience. At times this may deprive the younger child of the opportunity to form a relationship with a single, caring adult. While the most secure—and perhaps even the majority of children—may develop relationships with multiple teachers, the practitioner should be alert to

this added stress for a child whose coping skills may be compromised.

Whether or not the child has experienced changing classes and multiple teachers, the middle-school transition brings with it additional stresses. Increasing emphasis has recently been focused on the role of developmental disabilities in creating "output failure" in children at this age (Levine et al., 1981), discussed elsewhere in this text (see Chapter 38). In many schools, the focus shifts from a child-centered process to a subject-oriented emphasis. A change of school buildings brings with it a complete change in the group of trusted adults in the child's environment as well as the risk that important information about a student's special needs may be lost in the transition. In addition, the child moves from an elementary school environment where he is among the oldest and most competent to one in which he is one of the youngest. This change also takes place at a time of great individual variation in physical, emotional, and sexual development (see Chapters 9 and 35). As with other transitions, most children are successful in meeting these challenges. However, an office visit at this time offers an opportunity to identify children for whom such a transition may be troublesome.

A fifth transition may take place *from junior to senior high school.* Like the previous transition, a building change may remove supportive adults, the child's records and information about planning for special needs may be lost, and the competent eighth or ninth grader becomes the young and incompetent freshman. Both the fourth and fifth transitions may be accompanied by a reduction in the number and quality of special education resources available for handicapped children. Resource programs designed to help remedy the child's handicap may not be adequate to help the child make the best use of his or her abilities in an adult, work-oriented world. At the extreme, a child with a severe reading disability may find himself in a reading program in which he is still working on primer material instead of one designed to make the best use of whatever abilities the child may have. While the move to high school may bring about increased availability of vocation-oriented programs, this may be several years too late for the child who needed such a focus during the junior high school years.

Difficulties may stem not only from periods of transition but from mismatches between the developmental rate of the child and the expectations of the school. The first grader who is somewhat late in developing the cognitive, neurological, or social skills necessary for learning to read well may experience failure in early school experiences. The "late-blooming" child who has difficulty reading throughout the first grade and then "catches up" in the second grade is a well known but little studied phenomenon. It is important to protect such a child from the emotional risks of early failure.

Developmental school mismatches may occur in children who develop earlier as well as later than other children (see Chapter 9). Studies from the National Institutes of Health indicate that early maturing boys have social, emotional, and intellectual advantages over their late maturing peers. For girls, this relationship is less strong, but the reverse may be true, with a disadvantage for early maturing girls. On the other hand, the early maturing boy who is larger and more physically competent may actually be at a disadvantage later in skilled sports. One study has suggested that such children are less likely to excel in high school sports, perhaps because their early size advantage made it unnecessary for them to develop their skills (Clark, 1968). Later, when other children catch up in size, the early maturing boy may be at a disadvantage.

SEX-ROLE SHAPING IN SCHOOL

Schools have received much criticism for reinforcing sex-role differences. However, caution is advised when one interprets the data. That school-age children voice sex-role opinions that reflect traditional viewpoints need not imply that the school itself is the major socializing factor. It is likely that the school shares this responsibility with the child's family, the media, and other influences (see Chapter 10). However, the results of this socialization process are clear. A small study in 1971 examined vocational aspirations of first and second graders. The 33 boys sampled mentioned 18 different occupations—including the frequent football player, policeman, doctor, dentist, scientist, pilot, and astronaut. In contrast, the 33 girls mentioned only eight different choices, with 25 girls choosing nurse or teacher (Wesley and Wesley, 1977). This "type" experience is repeated annually in Galveston, Texas, where medical students work in a health education program for kindergarten students. As recently as 1981, many kindergarten students believed the female medical students were "nurses"—despite a film seen immediately before the lesson that depicted a physical examination performed by a female physician.

Studies of child development and socialization suggest that learning one's own sex, and accepting that sex, is well accomplished by the time of school entry. In contrast, learning one's sex role—that is, a set of duties, rights, obligations, and expected behaviors—begins in the preschool years and continues into the school-age years and beyond. No one will dispute that boys and girls are treated differently and display differences in behavior, play activities, and choice of peer group. However, as Maccoby (1980) points out, many of these differences are differences only in degree, and studies of psychological and social characteristics show great individual variation. Wesley and Wesley (1977) suggest that relatively minor changes in school curriculum and materials might tend to exaggerate the minor differences present between girls and boys. For example, schools require more English courses and courses requiring vocabulary and memory than courses in sciences and arithmetic. Thus, boys are "forced" to perform in their "weaker" areas. In addition, boys are known to lag behind girls in neurological development by as much as a year during the early school years— a fact that may bear on the increased number of boys who experience reading failure in the early years.

Schools will most likely continue to reflect primarily societal norms, beliefs, and expectations. However, it is important to acknowledge the powerful influence teachers and school environment may have on the early solidification of sex-role expectations.

HEALTH-SEEKING BEHAVIORS IN SCHOOL

Daily in this country, thousands of school children initiate a visit to a school nurse or a health office. These visits may be triggered by minor trauma, illness, stomach ache, or headache, and most of these visits do not require medical care. What is the connection between this early child-initiated use of health services and the patterns of help- and health-seeking behaviors in adulthood? If these phenomena are truly related, an opportunity may exist in the school to influence patterns of health service utilization favorably.

To investigate this, a random sample of elementary school children was monitored over a two-year period in their visits to the school health room and to regular medical facilities (Nader and Brink, 1981; Nader et al., 1981). Over the two years, almost all those sampled (94.3 per cent) had visited the nurse at least once. One per cent of the students accounted for about one-fourth the total number of visits because of the need for medication on a regular schedule or to undergo monitoring for a significant chronic illness. When these students were excluded, it was found that girls made more visits per year than boys (5.1 versus 4.1 visits per year). Similarly, girls had a significantly greater number of specific complaints and were more likely to be included among the 10 per cent that made the most frequent visits.

This pattern is similar to that seen among adults. Likewise, there was great stability and consistency in the visiting patterns for each student. A frequent visitor and a nonvisitor in the first year were likely to be in the same category in the second year. There was clearly no reduction in the proportion of frequent visitors over the kindergarten to sixth grade range, and the single best predictor of the use of the health room by students in the second year was their use during the first year.

Many factors influence this behavior. Family learning and personal, social, and psychological factors are probably the most influential determinants of the use of health services. The school health service provides an opportunity to influence this behavior at an early stage. Although no school health program could command the resources required to alter powerful family and environmental stresses, the system may be structured to improve children's coping and problem-solving skills. For frequent visitors, the program should focus on determining the sources of stress in the child's life, offering the child alternative methods of coping with this stress while reinforcing appropriate decision-making about the use of health services.

For any child visiting the health room for a complaint, there is an opportunity to teach skills of self-care. Inquiry techniques may be employed to promote independence rather than dependency. For example, when dealing with a child who has a scraped knee, the nurse or aide can be emotionally supportive while asking the child what he thinks could be done. Reinforcement can be provided to the child for performing self-care first-aid correctly. For somatic complaints, children could be taught to assess their subjective feelings better and to match these with more objective information obtained from the nurse. Questions may be asked concerning events that preceded the child's decision to come to the school health program, and alternative methods of handling stress can be taught to children who can identify that they are using a somatic complaint and a visit to the nurse to handle a particular stressful situation. This may be an unfamiliar role for both the school nurse and other school personnel, who may view the school health service as primarily a band-aid dispensary. Such a perspective would require the active support and encouragement of child health professionals in the community.

SCHOOL AS AN EDUCATOR—RUTTER'S STUDY

In a comprehensive and carefully designed study, Michael Rutter has demonstrated that schools do affect outcomes for the children enrolled in them and that some schools do so more positively than others (Rutter, 1979, 1980). This statement would seem obvious and hardly worthy of mention were it not for the strong sentiment to the contrary that has developed over the past 20 years. The publication of the Coleman Report in 1966 and the subsequent analysis of these data have solidified the theory that there is little schools can do to reverse the strong effects of social class, early experience, family environment, and genetic predisposition. Given such potent determinants of educational outcome, some feel that what happens at school and how it happens is of little consequence. Because genetics, family environment, social class, and early experience do have profound effects on children, it is easy to understand this pessimistic outlook.

Rutter's study is important because it demonstrates that differences in schools do make a difference, and it begins to point to those dimensions of school life that make this difference. Rutter studied 1500 children just before their enrollment in 12 secondary schools in London and again three years later. The observed effects of these schools differed markedly in the rates of attendance and dropout, success on examinations, and delinquency. These differences could not be explained by existing variation among students on entry and were associated with discrete school-related factors. Some of the findings were surprising. Contrary to what might be expected, the financial resources, size of the school, age of the school, and crowding did not relate strongly to outcomes nor did the administrative status or organization.

The factors that made a difference had to do with the atmosphere created in the school, the attitudes of the staff, and the academic milieux. Schools that used more overt praise in the classroom, those which provided better conditions and a sense of respect, and those in which teachers provided positive models of organization and academic emphasis were the ones with better pupil outcomes. Schools that frequently started classes late or let classes out early tended to have poorer pupil outcomes. Finally, schools in which both curriculum and discipline were agreed upon and supported by the entire staff produced a better student outcome.

Rutter's study does not look at the large number of organizational differences that are prevalent in this country, mostly in elementary schools. One of the differences that has received much attention is the "open" versus "closed" concept. These terms are loosely applied to a variety of teaching techniques, organizational styles, and building designs, creating some confusion for the pediatrician and parent. "Open" refers to classrooms with one or more of the following characteristics: an informal environment, seating at tables or in clusters rather than in rows of desks, team teaching or multiple teachers, buildings with fewer interior walls, high noise level with multiple simultaneous activities in one class, and teaching techniques that emphasize the child's interest and "discovery" techniques.

Several points are important for the pediatrician considering the effects of "open" versus "closed" classroom. Those described as "open" may have any combination of the above characteristics; there is no standard. Consequently, research on the effects of these approaches is often inconclusive. The characteristics of the individual teacher are more important than the "openness" of the environment. For example, although open classrooms are generally somewhat more noisy and somewhat less structured, this may not hold in any individual case. A teacher may conduct a relatively quiet, very structured class in an allegedly open concept setting, while another may have an unstructured, noisy class in a traditional four-walled environment. When evaluating the effects of a classroom on an individual child, a telephone conversation with the teacher is a minimal requirement, and a visit to the school can be invaluable. Such a visit, which may take less time than hospital rounds, pays many benefits.

THE INTERACTIONS OF SCHOOLS AND FAMILIES

Much of our previous discussion has focused on the independent influence of the school on the child and the adaptations required of the child to meet the environmental demands of the institution. It would be negligent to omit from consideration two important areas of family-school interaction that provide the basis for clinically relevant problems. These problems account for a large portion of school-age children's use of health-care resources. The two areas are the family dynamics influencing the development of school phobia, and

the mismatch of expectations for a child—be they physical, intellectual, or social—held by parents, on the one hand, and by the school, on the other.

SCHOOL PHOBIA

School phobia, also called "school refusal," is not uncommon nor is it confined to a specific population or socioeconomic group. Characterized as "the great imitator," school phobia may be manifest by many somatic and psychological complaints. The diagnosis may be especially easy to miss when symptoms are expressed in place of an overt fear of school. This is particularly true because children with school phobia frequently are good students, claim to like school, and would like to return "if only the symptoms (or fears) would go away." Of primary importance in making the diagnosis is to document repeated absence from school, often for more than a week or two. Absences may even accumulate at a rate of two to three days per week over several months, although this pattern is less common. School phobia should be suspected in any child who stays home from school without a clear medical explanation and, of course, in one who claims to be afraid of school. School phobia should be distinguished from truancy, in which the child is not at school but is also not at home, since home is the comfort and refuge of the child with school phobia.

Table 15–1 lists factors within the child, family, and school that commonly contribute to school phobia. The clinician must thoroughly and efficiently investigate each of these areas. Several family characteristics and communicative patterns have been associated with the development of school phobia. Schmitt (1971) has pointed out the higher risk of school phobia in a child seen as sickly, vulnerable, and special. Likewise, parents who have been unable to separate themselves successfully from a toddler or infant may reinforce these separation difficulties later in the child's life at the time of school entry or other transitions. Both the above characteristics reflect a family-child

Table 15–1. FACTORS CONTRIBUTING TO SCHOOL PHOBIA

The Child	The Family	The School
Overdependent relationships	Overdependent patterns	Step-up in expectations (i.e., 3rd grade, junior high)
Undiagnosed learning disabilities	Child as safety valve in parents dispute	Change of important personnel (principal, teacher)
Recent absence for real illness or vacation	Chronically ill adult	Real intimidation by groups of students
	Change in job or home	Rarely, a coercive unstable teacher

dynamic portrayed beautifully by Eisenberg's phrase: "The umbilical cord pulls at both ends." In addition to this parent-child mutual dependency and difficulty in separation, several other communicative patterns have been observed. A major one is the orientation of conversation and family life around health and illnesses. Mention of somatic complaints are frequently present in interviews with these families.

When school phobia is present, there is almost always some degree of dysfunctional family communication present. Commonly, three patterns exist: In the first, one or both parents are overprotective, solicitous, and easily manipulated by the child. In another, one adult who provides the major parenting role is overprotective while another adult disagrees completely with the approach. In this situation, the child can and does take advantage of the disagreements. A final common pattern is one in which the mother is overly involved in a mutually dependent relationship with the child and the father is absent—either physically or psychologically—from active participation in family life. The task of the clinician is to determine the cause of the dysfunctional communication pattern. It may be a relatively simple lack of appropriate child-rearing information due to a lack of parenting skills or to more significant personal or family marital problems.

An early working relationship with the school is essential in both the evaluation and the management of school phobia. Although actual fear-inducing situations are rare, this possibility should be evaluated. In addition, the physician must enlist the cooperation (and a great deal of tolerance) from school personnel in order to manage school phobia successfully. Specifically, supportive resources within the school should be identified and mobilized. If needed, a gradual return to the classroom from the nurse's or counselor's office is much easier and more therapeutic than trying to effect a gradual return from the child's home.

While there are many approaches to the management of school refusal, there is strong support for a rapid return to school as the universal short-term goal for all cases. This is frequently difficult even in milder cases, and all resources of the family and the school may need to be mobilized to accomplish this goal. Specific counseling techniques and principles should be employed to support and recognize the feelings of all involved (the child, parents, and school personnel).

The difficulties often encountered in accomplishing treatment goals may be anticipated and avoided. A common pitfall for the clinician is to agree to the presumption (often advanced by the child or family) that the problem will disappear if the child is permitted to change teachers or schools. Attributing school phobia to real or imagined school or peer factors is usually a mistake.

A second pitfall is uncertainty about the diagnosis by the health-care provider. Although history and physical examination are mandatory, laboratory investigations should be limited to those absolutely indicated by the medical assessment. Parents need firm reassurance about the absence of serious physical problems that is based on sound clinical judgment. Tests that are not indicated but are performed "to reassure the parents" may have the opposite effect, reinforcing their fears about physical illness. In addition, real physical illness does not eliminate the possibility of school phobia. A child with a bona fide malady may also be missing an inordinate amount of school and have many of the same characteristics as a classic case of school refusal. Most chronically ill children want to and do attend school.

In certain cases, school phobia may be an indicator of severe psychopathology that will require intensive mental health services. Predictors of more difficult cases include (1) duration of absence greater than several months, (2) school phobia in an older adolescent, (3) older siblings who also showed school refusal, and (4) parents who steadfastly attribute the problem solely to school or physical factors. Any of these may represent sufficient reason for referral for mental health services. However, it is important to realize that the factors that inhibit school attendance may also deter the family from following through with a referral. Thus, the responsibility for follow-up and child advocacy falls squarely on the shoulders of the health-care provider. In most cases, management by the primary health-care provider is successful in interrupting a cycle of school nonattendance, which is inimical to the social and intellectual growth of the child.

MISMATCHED EXPECTATIONS

There are reciprocal interactions among the child's view of himself, the parent's perspective, and the viewpoint of the school. These expectations are influenced by real physical, cognitive, and social developmental levels or characteristics of the child. While some mismatches between the real and perceived characteristics of each are not sufficient to result in problems or situations that inhibit future growth, others can have potentially far-reaching and detrimental effects.

The major developmental task of the early and middle school-age child is an increasing sense of competence and identity. This task is complicated by mismatches in the expectations of others regarding the child: between parent and child, between parent and school, and between school and child. Depending to a great extent on the child's age and realistic assessment of his own strengths and weaknesses, it is easier for a child to cope with divergent expectation between the school and the child than with the other two situations. A notable exception to this may be in the area of achievement and cognitive development. Children who experience repeated failure—especially early in their school years—learn that they cannot learn. This may set up blocks or inhibitions to learning that are extremely difficult to overcome.

Physical Expectations

There is some evidence to support the idea that a young child's self-concept may be tied to aspects of physical growth, motor skills, and physical attributes. Judgment of one's competence to accomplish tasks is greatly influenced by previous mastery experiences of related tasks. This applies to physical tasks and a child's judgment of self-efficiency in the physical area.

Examples of common mismatches in expectations that may influence children's views of themselves include parental underevaluation of physical abilities. Parents may view a child as physically vulnerable and therefore engage in overprotection. This often imparts an unrealistic feeling of vulnerability. One clinical outcome of this may be the development of school phobia-type symptoms (see earlier discussion).

In contrast, parents may coerce their child to overachieve and encourage "superstar" level competition, reveling in their own vicarious enjoyment of previously unmet personal needs. This may result in somatic complaints or even in outright rebellion.

A common example of a mismatch of physical expectations between the school and the child is in the area of handwriting among young boys with mild fine-motor incoordination that school personnel insist can be overcome with practice.

An objective for the pediatrician is to try to have all the interested individuals (i.e., school, parents, and child) accept the wide range of normal variation when it comes to physical development and growth. Competition in early school-age should be directed more toward personal improvement and less toward the die-hard conflicts and external trappings of more senior competitive athletes. The pediatrician should advance the ideas that the physical self is a part of the total self and that a person may be better than another in something while at the same time not as proficient in something else.

Intellectual Expectations

Studies suggest that many parents tend slightly to overestimate children's intellectual abilities while many teachers tend slightly to underestimate these abilities. In a given instance, problems may result when either the parent or the teacher expects either not enough or too much from the child. One of the most difficult concepts to grasp for the parent, and often even for the teacher, is the wide range of cognitive and intellectual relative strengths and weaknesses that may exist within a given child (see Chapter 38).

The pediatrician's role is often to clarify as objectively as possible the real strengths and weaknesses and to help adults in the child's environment to understand and work together to optimize development and growth.

Social Expectations

A blatant school-age example of the mismatching of social expectations is when an adolescent male may drive a car and register for the draft but requires a signed pass to go between classes in school. Earlier in school life, differences in expectations of social competence may also be evident.

In some lower socioeconomic areas, some 8- or 9-year-old children are required to perform tasks such as shopping, housekeeping, and child care equivalent to those of a much older child. At the same time, their teacher expectations of them may seem infantile to these students.

Another example of mismatching in social expectations between parents and teachers is the case of a child who is reinforced and socialized to be creative and inquisitive (as viewed by the parent) or obnoxious and a trouble-maker (as viewed by the teacher)!

SUMMARY

Despite earlier deterministic views, it is clear that school experiences do affect the child—for better or for worse—during the middle childhood and adolescent years. Just as children continue to evolve and develop during these years, the experience of schooling can be seen in dynamic, developmental perspective. Certain common stages and transitions may be identified during which predictable developmental crises may occur.

Although the family retains its strong influence upon children and their development, the school exerts independent and important influences. Ultimately, however, the family, the school, and the youngster develop together through a dynamic interaction. Each must be considered, individually and in their interactions with one another, in order to understand fully the developing child. While physicians are comfortable including a family in their deliberations about a patient, it requires an active interest and considerable initiative to become familiar with the child's school and its effects. For physicians who may not have set foot in an elementary school since their sixth grade graduation, this presents a fascinating and worthwhile challenge.

GREGG F. WRIGHT

PHILIP R. NADER

REFERENCES

Clark, H. H.: Characteristics of young athletes. Kinesiol. Rev. 33, 1968.

Levine, M. D., Oberklaid, F., and Meltzer, L.: Developmental output failure: A study of low productivity in school-aged children. Pediatrics 67:18, 1981.

Maccoby, E. B.: Social Development: Psychological Growth and the Parent-Child Relationship. New York, Harcourt-Brace-Jovanovich, 1980.

Nader, P. R., and Brink, S. G.: Does visiting the school health room teach appropriate or inappropriate use of health services? Am. J. Public Health 71:416, 1981.

Nader, P. R., Ray, L., and Gilman, S. C.: The new morbidity: Use of school and community health care resources for behavioral, educational, and social-family problems. Pediatrics 67:53, 1981.

Rutter, M.: Fifteen Thousand Hours. Cambridge, Massachusetts, Harvard University Press, 1979.

Rutter, M.: School influences on children's behavior and development: The 1979 Kenneth Blackfan Lecture, Children's Hospital Medical Center, Boston. Pediatrics 65:208, 1980.

Schmitt, B. D.: School phobia—The great imitator. Pediatrics 48:433, 1971.

Wesley, F., and Wesley, C.: Sex Role Psychology. New York, Human Sciences Press, 1977.

16

Critical Life Events: Sibling Births, Separations, and Deaths in the Family

Among the most critical of life events for the child are those having an impact through the disruption of established patterns of interaction in the family. These include the addition of a family member (e.g., birth of a sibling or the prolonged visit of a grandparent) and a variety of temporary or permanent separations—partial or complete—from parents or siblings (e.g., because of their hospitalization, chronic illness or death) or from friends (e.g., through moves or changes in school). To a healthy child, each of these events presents the need to reorganize information, concepts, emotional ties and dependencies, and interpersonal relationships and activities, with the goal of achieving a new equilibrium in a changed world.

The cognitive and emotional resources and coping skills that normal children bring to the foregoing tasks vary with their ages, temperaments, and past experiences. These are unique to each child, and the needs of any child for help or support in coping with a critical life event can be responded to adequately only if that event can be understood in the context of these factors, which must be comprehensively assessed.

Apart from the uniqueness of each child's experience, one can identify in the fabric of each of the critical life events cited above, certain relatively common and basic threads that give some generality to a description of probable or possible effects of these events upon the child. The identification and examination of these basic elements will often indicate to the physician, nurse, psychologist, or social worker the need for a particular professional intervention (such as general health education, anticipatory guidance, specific counseling, or referral for health or social services, for psychosocial evaluation, or for psychotherapy).

Here we will consider responses of infants and children (1) to birth of siblings, (2) to certain separations, and (3) to the death of a parent or sibling. In addition to the particular features worth noting for each of these situations, some features may be considered analogous to other critical life elements.

BIRTH OF A SIBLING

The impact of the birth of a sibling is likely to be greatest upon the first-born child in the family with the arrival of the second child. The older child has previously received relatively undivided parental attention. He or she must now radically adjust personal views regarding his or her relationship to parents, their reliability, and the nature and meaning of parental priorities. The older the child at the time of birth of a younger sibling, the more likely the older child's role will undergo substantial change—a change shaped by the nature of previous relationships in the family, by the expectations of parents, and even by the sexes of the two children. Generally it will be easier, for example, for a 2- or 3-year-old girl to slip into the role of "little mother" to a baby boy or girl than for a 2- or 3-year-old boy to assume an analogous role. (See also Chapter 13 concerning sibling constellations.)

The impact of the second child is felt early by the older one, perhaps even at the time when parents begin planning for a new pregnancy and a second child. The first child may be involved to a greater or lesser extent in such planning, depending upon the age of the child, the feelings and sophistication of the parents, and the openness and general qualities of communication in the family. In any case, the status of the older child may first change significantly and perhaps dramatically with the mother's discovery of her pregnancy. From this point on, her attention is drawn toward the new baby, and her demands on and expectations of the older child may be conspicuously altered. For example, if a 2-year-old is not yet toilet-trained, the pregnant mother may intensify her concern or efforts to accomplish this, and it may become an area of conflict without either child or parent fully understanding what is happening. Feeding, sleeping, and other normal activities may be similarly entrained as areas of intensified concern and potential conflict.

Parents need to be cautioned that with the discovery of pregnancy, they should take care not to alter their attention to or their expectations of their older child (or children), especially when these changes dilute support of or impose increased demands upon the older child. It appears that children between the ages of 2 and 5 years find it most difficult to adjust to their mother's pregnancies. They do not readily understand the changes in emotional climate (between parents and between each parent and the child), in the mother's attentiveness and expectations, in the mother's shape, and the like; moreover, the readiness of parents to deal realistically with the child's needs at this time is highly variable. It is appropriate for children to be told where the baby is (to help explain the mother's changing needs and shape and why her lap is less available than before); even young children should have an opportunity to participate, at their own level of comprehension, in the arrangements being made to accommodate the new family member.

In preparation for the new family member, the arrival of the new baby should not be overdramatized, and promises should not be made that cannot be kept (such as the sex of the infant or joyousness of the event). Questions of the older child can be answered factually; those who have had experience with the birth and care of pets or animals may be better prepared to understand what is happening.

In recent years childbirth has been reexamined as a social event, with growing acceptance of the principle that its social aspects may often be more determinative of the family's future emotional and social health and functional integrity than are its medical aspects. The presence of fathers (and sometimes of godparents) in the delivery room is no longer regarded as unusual and appears often to have positive effects. Prenatal classes, instruction, and exercises have reshaped the expectations of many parents with regard to childbirth. Whether or how children should participate in this social event remains to be explored, since important questions regarding appropriate age, preparation, and expectations of the sibling during his or her mother's labor or in the delivery room have not as yet been answered. When questions about participation of older siblings arise, one must consider the immediate needs, strengths, and understanding of the individual family. (See also Chapter 5.)

Even with the best of planning one cannot fully anticipate the effect upon the family of the birth of a second child. Signs or symptoms of stress in the older child in response to his or her altered position or role will vary with each child's experiences and with the general level of emotional integrity in the family. Faced with the inevitable diversion of their mother's attention to the new baby, children may be more irritable and cry more, cling more to the mother, or exhibit regressive behavior; thumb-sucking and secondary enuresis are not unusual manifestations of a child's need to be cared for in the infantile way he or she once enjoyed. Older children, because of latent anger and frustration at their displacement, may make snide remarks or expressions about the baby; unprepared or tense parents may find these irritating or offensive. Only rarely will an older child actually attempt to harm the baby, a response that may indicate serious problems within the family.

Parents whose children are evidently emotionally upset at the presence of a new baby need to understand that at certain times the needs of the older child for their support may be more urgent than the needs of the infant. For example, given a choice between feeding the baby and meeting some legitimate need of the older child, parents may be able to delay the feeding for a few minutes with less damage to the infant's morale than to that of the older youngster. In order to keep intact the relationship with their first-born that they have enjoyed in the past, parents may need to be encouraged to set aside an uninterrupted period of time that they will spend with the older child. Also, the older child may need to be encouraged that he or she does have the right to feelings and opinions about the baby, and to produce epithets or descriptions that express these feelings but that he or she does not have the right to misbehave or to injure the baby. Often it is helpful if the older child can become involved in caring for the baby, being given—and rewarded for—helpful activities that will build self-esteem and the sense that he or she is useful and wanted.

Role of the Physician

Ideally, the physician's role with respect to reactions of the older child to the birth of a baby is preventive. The physician should be aware of the issues discussed above and should be able to make a global assessment of the strengths and weaknesses the family brings to this new experience. The physician should explore with them past difficulties and how these have been handled; what preparations have been made for pregnancy and the birth of the baby; and how, in particular, the parents have planned to handle the event with respect to the older child. If these matters are known in advance, special needs can be identified and planned for realistically.

Difficulties arising after the birth of the baby should be explored as sympathetically as possible in light of the above considerations. The *parents' own view* of the problem is crucial—what they feel the trouble may be and what strategies they may already have used to try to deal with it. It will be necessary to know whether the parents have similar or divergent views with regard to child-rearing practices and to identify any sources of conflict in the family that lie outside the child's behavior.

The outlook for problems surrounding the birth of a new baby is good, except when other more basic problems exist in the family, between the parents or between parents and the older child. With helpful interpretation and gentle encouragement and praise when due, most families easily manage the occasional abrasive exchanges between older and younger siblings or even older children and their parents over the addition of a new family member. In well-adjusted families in which siblings relatively quickly and easily establish generally positive relationships, these relationships will still be punctuated by the normal tensions and bickering that occur between siblings, but these behaviors contribute to the development of social skills as these children develop the capacity for protest, assertiveness, and the give-and-take of compromise. At its best, this all takes place in an atmosphere characterized predominantly by deep affection. In many families the arrival and integration of a new member may prove to be an event and process of virtually unalloyed pleasure for all. Indeed, this may be closer to the norm than is generally suggested by the amount of concern expressed with "sibling rivalry."

SEPARATIONS

Some critical life events result in the separation of family members from each other. These include hospitalization or chronic illness of parents or siblings or, in some instances, other types of unavoidable separations such as those imposed by military service or imprisonment of a parent. We are chiefly concerned with the effects of separation upon the otherwise well child.

Effects of separation within the family, like those of birth of a sibling, vary with the age, temperament, and previous experiences of the involved child and with the emotional climate and coping resources in the home. These effects have been most intensively studied when it has been the infant or child who has been removed from the mother or family rather than when the mother, father, or sibling has been removed from an otherwise intact family. Although there is no reason to regard these situations as wholly equivalent, clearly the child removed to a hospital or other type of care has a more wrenching experience than the child left in familiar surroundings. On the other hand these two conditions share many features (Bowlby, 1969, 1973, and 1980).

Separation in Infancy

Before the infant is 6 months old, departure of the mother seems less traumatic than later. At about six months, the normal infant begins to identify his or her mother as a particularly important person, and the appearance of separation anxiety is normal, indicating the discovery of this relationship by the infant. It has been further suggested that infants with particularly warm and deep attachments to their mothers suffer more after six months of age than do infants of the same age whose attachment has been relatively superficial; however, the situation may be more complex. For example, infants appear to be able to identify their mothers by sound, sight, and smell within the first two weeks of life. It is likely that over the greater part of the first six months of life most healthy infants actively develop a complicated signal system through which infant and mother communicate their needs and satisfactions to each other. When the mother suddenly becomes unavailable, a rapid transfer of her role in this system to another person is impossible, and the infant appears to become insecure and his or her behavior somewhat disorganized. The father and other family members (siblings and others) may be able to sustain the infant to some degree, but their emotional capacities, readiness, and skills may vary greatly. Infants whose relationship has been particularly close to their mothers rather than diffused among a number of caretakers reportedly suffer more than those who are able to maintain effective communications with some persons in the environment. Even relatively brief sharing of the care of the infant may make him or her less vulnerable when separation takes place. Whenever possible, one should take advantage of this tendency in planning separations of infants from their mothers.

When a mother must spend a short time in the hospital (a few days only), the effects on her infant under the age of six months may be negligible if she has created a sound support system. Longer separations (one month or more) are more likely to cause substantial disruption, particularly when separation occurs between six and 12 months of age, when the infant's response is likely to resemble that of his or her own hospitalization or other separation from the customary caretaking milieu.

After the first three months of life, infants separated from their mothers are increasingly at risk of *anaclitic depression*. This syndrome, first described by Spitz (1946), typically appears as a three-phase reaction: (1) angry protest, (2) resignation and depression, and (3) detachment and reorganization. Early crying may be violent or quiet; the protesting infant may lapse into lethargy, with loss of appetite and weight loss, loss of interest in surroundings, sleeping difficulties, and ultimately in some infants an apparent loss of the capacity to form close relationships with others (the "affectionless" child). Early reports and studies of anaclitic depression emphasized its frequency and serious implications, not only for readjustment but for survival itself, since mortality was high among the infants first described by Spitz. Further observations and experience with anaclitic depression have left us uncertain as to its frequency, preventability, treatability or reversibility. It is not established which infants may be most susceptible, or what aspects of temperament or family circumstances may dispose to anaclitic depression (see also Chapters 26 and 41).

Another view of the impact of separations in the first year of life was put forth by Bowlby, who found in a retrospective study of delinquent adolescents that many of them had a history of long separation from mother or family in the first year of life; Bowlby felt that these early separations contributed to the development of a personality structure that precluded warm and caring human relationships, presumably because the opportunity to form such a relationship had been aborted. Since there is no reason to doubt the potentially disastrous effect of poorly handled separations in early infancy, reasonable effort should be made to make arrangements and muster resources that will attenuate the effect of separation or provide tolerable support while the separation lasts.

The mother's return may reverse the effects of relatively short separations, but this reversal is likely to be less complete the longer the delay, owing to changes in both child and mother that make it difficult for them to recapture themes from their earlier interaction. After hospitalization, concern with her own illness or subsequent handicaps may interfere with the mother's ability to resume her earlier, primary role.

Separation After Infancy

By the age of 2 years and after, the pattern of the child's reaction to separation begins to change, a major new element being the child's attempt to understand the event intellectually and emotionally. The situation is susceptible to major and sometimes devastating misinterpretations on the part of the child, often depending on the quality of prior relationships among family members.

The infant's response to separation in the first year may be best regarded as a generalized disorientation or feeling of abandonment, with angry protest giving way to a growing sense of helplessness and despair of recapturing the comfortable and reassuring relationship with the mother. In contrast, the older child, in searching for an explanation, is more likely to regard the separation not only as abandonment but as punishment and as equivalent to loss of love. More sharply focused feelings of anger are mixed with feelings of guilt, the child believing that perhaps he or she is responsible for the parent's disappearance. Points of tension in the earlier relationship with the parent may markedly influence the child's reaction, as, for example, when disciplinary practices have been accompanied by threats of abandonment or of causing illness ("you give me a headache" or "you drive me crazy"), because the child is likely to recall these careless remarks.

Older children may react to significant separation with behavior that has some regressive features. Symptoms may include enuresis or, less commonly, soiling. Feeding disturbances, sleep disturbances, increased irritability, depression, and weeping are common, particularly among preschool children, and nightmares may occur. Sometimes the young child will engage in a persistent hunt for the absent parent, asking questions and at times leaving home to search in the neighborhood.

These problems can be mitigated to some extent if it is possible for the young child to maintain contact with the absent parent, perhaps through visits to the hospital or by telephone. Reassurances that the parent will be home after the specified period of time will not be helpful to young children, whose notion of time is poorly formed. Simply recognizing with the child that it is no fun having the parent absent and that he or she is missed, as well as giving the child permission to feel sad, will be more helpful than attempts to cheer the child up through empty diversions.

Parents may be greatly distressed at the behavior of the infant or young child when the mother returns after even a few days' absence. The mother of a 1- or 2-year-old child who expects him or her to be joyous at her return may find a sullen, suspicious child who turns away from her and seems to prefer the temporary caretaker, appearing anything but eager to reestablish an earlier warm relationship. Later the same child may scream when the mother gives any sign of going away again, clinging to her constantly for days or weeks. Parents should be prepared for the possibility of such reactions and be ready to accept the child's anger, anxiety, and need for a few days or weeks of close contact and reassurance or for extra care at bedtime.

The longer the period of separation between a mother and her infant or child, the longer the time likely to be needed to restore their original relationship, when this is possible. Recurrences of separations are particularly difficult for some children, with restoration each time being more tenuous, as the child becomes less and less willing to commit himself or herself to a trusting relationship again.

Other Separations

Prolonged, semipermanent, or permanent separations may call for special and continuing care. Emotional trauma is possible whenever an infant or child is moved from one caretaking person or setting to another. The infant or child must revise a communication system that has given stability to the earlier relationships and that, as indicated above (see Birth of a Sibling), has begun to be substantially elaborated by two months of age. Often this system is largely—and perhaps almost uniquely—invested in a single primary caretaker by six months of age. In a sense, it defines the "psychologic parent" of the child. At no age of childhood will transfer to another communication system be without stress, since it requires a period of adjustment for both the child and the new caretakers that may be prolonged and sometimes very difficult.

The importance of the notion of psychologic parent has been highlighted in recent years by the work of Goldstein, Freud, and Solnit (1973). They point out that in some cases the prolonged separations mandated by our courts and legal system do particular violence to the child and often to the psychologic parent as well. When legal conflicts over adoption and foster care arise for whatever reason, it is important that the notion of psychologic parent be understood, that the implications of separation be known, and that every effort be made to see that transfers of custody are not abrupt; rather, gradual and substantial involvement of the new caretakers in the child's communication system should be arranged, whenever possible, before the original primary caretaker moves out of the setting. (See also Chapter 13 concerning adoption and foster care.)

Analogous considerations will often be helpful to other separations, such as moves of families to new neighborhoods or cities, or the transfer of school-age children to new schools, each of which will require that young children find and adjust to new friends. Under these circumstances children may present signs and symptoms that are similar to those characterizing other separations. These problems may be particularly troublesome to adolescents and may feed the rebelliousness at that age.

The difficulties of new arrivals entering an established play group have been well described by McGrew (1972). These problems can be eased if children are able to visit their prospective new neighborhoods or schools prior to the final move and if links to persons with whom they were previously close can be maintained. Parents can often help by becoming hosts to parties or planning activities involving children in the new setting; getting to know the parents, homes, and life styles of new acquaintances can help extend the child's curiosity-at-a-distance into acceptance and friendship. Also, parents already established in the new setting can often help new arrivals by including them in neighborhood affairs.

Some studies have been made of the special problems of children in military families, including situations in which the father is absent and in danger (Noshpitz, 1979). However, little is known about the particular issues of children whose parents are in prison, which may be quite complex. Also, in these circumstances, the resources for the family may be quite meager (Sack et al., 1976). In addition to following the general principles described for coping with separations, it would appear that, for the child old enough to discuss the problems, it would be well to face them as honestly as possible, with compassionate acceptance of his or her need to know, understand, interpret, or rationalize and to be protected from inappropriate intrusion or exploitation.

Role of the Physician in Separations

The role of the physician in dealing with separations is primarily preventive, when possible. Sometimes it is possible to avoid separations or to soften their impact by skillful planning. Of primary importance is understanding the event in the parents' terms, both conceptually and practically. To plan effectively, the physician will need to know the parents' views regarding the probable impact of separation upon the child, their resources for dealing with the separation, and the nature of the support system for the child within the family. Only when these things are known can appropriate advice be given or support be found.

Some of the steps that can be taken have been described above. Even with very young children, the communication system can be kept active through frequent visits or by telephone. Time and resources spent in this maintenance are likely to be good investments.

DEATHS IN THE FAMILY

DEATH OF A PARENT

Death of a parent—the ultimate, utter separation of a caretaker from an infant or child—presents for the child many of the problems discussed in the preceding section. However, there is one difference—there is no expectation of restoration. The meaning of this event for the infant or child varies with age and with his or her previous experience with death. Probably children have a notion of the finality of death in intellectual terms at a considerably earlier age than has been thought, and the death of a child's own parent will raise questions that require sensitive reflection and reasonably prompt answers.

Human history records from the earliest times our attempts to understand death or to evade or deny it. Conceptions of the nature of death will, of course, vary depending on the cognitive and emotional status of the child as well as on familial, religious, and cultural factors. Legitimately held beliefs may vary from death as oblivion to death as a long sleep or to prompt passage to an afterlife. Often children have been introduced to notions of death through prior observations of the deaths of plants and animals, the changing of the seasons, and interpretations of their parents or other adults. Family traditions or beliefs that permit firm convictions about the meaning of death or its mystery can be appropriately shared with children, so long as this sharing is honest and authentically rooted in spiritual or religious conviction. On the other hand, attempts to protect children from their own doubts or from another view of reality by means of circumlocution, evasion, or attempts to foster the conviction in the child that the absent parent still physically lives and participates in the life of the family are inappropriate. It is best if survivors can share with bereaved children what they actually understand or feel. The situation may be particularly difficult if the dead parent is "called up" to support ineffectual controls of surviving children, such as "Your father must be turning in his grave."

The words used to explain a parent's death can be important to some children. If one persuades the child that death is sleep or that a parent "died in his (or her) sleep," the child may develop a fear of going to sleep; emphasis on irrelevant aspects of the fatal illness ("it just began as an ordinary cold") may produce anxiety or panic when the child experiences signs of trivial illness. Informing a child about the nature of the terminal illness and specifying its cause as precisely as possible should minimize the child's bewilderment or guilt.

Children can learn much about death from experiences around them, even at an early age, including the death of animals, flowers, and trees. If their early questions can be answered simply and factually, they are better prepared later on to accept and comprehend the death of a person close to them. The death of pets, in particular, provides an opportunity for children to learn a little about death and to rehearse the rituals that surround death, such as a funeral and burial. Sympathetic parents can often help children in these activities, offering interpretations and suggestions.

Parents and other adults should always be ready to share with the child what they do *not* know about death; rather than attempting to create in the child some "protective" fantasy that is not truly shared. Bowlby has suggested that effective coping of a child with the death of a parent depends on three conditions: a secure relationship with both parents prior to the loss, prompt and accurate information regarding what has happened, and the comforting presence of a surviving parent or of other adults capable of appropriate comfort and support. The needs will vary with the age of the child. For children up to one year of age, the most important effort should be to assure that the surviving parent has an adequate support system. For older infants, adequate models should be available for the development of gender roles, which may be of particular importance to preschool and early school-age children, for whom gender identity may be confused through identification or overidentification with the dead parent.

The death of a parent often comes at the end of a long illness known to be fatal. This circumstance raises questions about how the matter is to be dealt with insofar as children are concerned. Should they be told the diagnosis? Its prognosis? Should they visit a hospitalized parent? How often? Under what circumstances of medical care? How should one respond when a child asks "Is my father dying"? It is often very difficult for people suffering their own grief at events to answer questions such as these for children. Guidelines will need to be tailored to the circumstances, convictions, and resources of the persons involved, but some generalizations can be made.

First, it is appropriate that children know that a parent with a fatal illness is seriously ill, and the diagnosis may be appropriately shared at the child's level of understanding. Children's questions about whether a parent is dying may be best answered with the assurance that everything is being done to put that off as long as possible, but that everyone dies eventually. It is best if the information given children can be factual; one should not make promises that cannot be kept. On the other hand, to maintain the hope that something may be found which may be helpful is not unreasonable, while assurances are given that everything possible is being done for the patient's comfort.

Second, whether children should visit sick or dying parents will depend upon the circumstances, and most particularly upon the wishes of the dying parent. Sometimes dying parents wish to be remembered "as they were"; other parents want to share remaining time with their children and can sometimes provide much love and understanding to help their children cope with ultimate loss and attendant grief. Spouses of parents with fatal illnesses should be encouraged not to hide their own anticipatory grief from their children. When a child finds a parent weeping and asks why, the reply may simply be "I am crying because your father is sick and I wish he weren't."

In the above matters, the parents' views, feelings, and expectations are critical to management. These must be known to the physician as a basis for his or her advice or intervention and can often be learned only after gentle and patient inquiry. This inquiry can also have therapeutic value for the parent in examining his or her own needs and position relative to the child and facilitating the expression of feelings that will help in the work of grieving.

Regarding the question of whether children should view the body of a dead parent or attend the funeral, decisions depend heavily upon what the surviving parent or family is comfortable with. Attendance of the child should not be urged or forced upon an unwilling child or family by the physician. However, it is likely that the experience of the funeral will help the child learn more about how people cope with death, and the opportunity to see and share the grief of others may help the child in his or her own mourning.

The way in which children mourn varies with their ages. For young children, the process of mourning is akin to the reaction after separation from a primary caretaker, particularly in the instance of death of the mother. Older children are often able to resume normal activities fairly early and may even appear to the casual observer to be callous in their enjoyment of these activities. It is likely that children do not grieve or mourn as adults do until the time of adolescence. But children of any age will need to transfer to some other person or in some other way deal with a system of interactions and expectations that were invested in the departed parent. This process (decathexis) may have many elements: there may be anger against a departed or the surviving parent, or identification with the departed parent and an attempt to assume the parent's role in the household; he or she will need to establish an adequate support system to replace the one shattered by death (Furman, 1974).

Impediments to children's mourning include the attempts of others to shield them from what has happened or to divert them with trivial activities or concerns. The children's own concerns about the future may preoccupy them, and many children will require reassurance that their own health and material needs will be adequately met. In any case, features of denial can be expected during the grieving periods of many children. For some, it will be helpful to attend the funeral of the departed parent, to examine memorabilia with family, and to receive—again and again—answers to questions about what has happened or what seems to lie ahead. References to the departed parent should be realistic, i.e., there is no need for such canonization as might be implicit in an insistence that only the "saintly" side of the departed can be discussed.

Pathological reactions in children's mourning include extreme *denial*, in which case it may reflect parental denial or evasion. This may be particularly difficult for the surviving parent of a suicide or when the death has been particularly violent or public. Like denial, *guilt* may be a pathological reaction if unduly intense. Children sometimes feel guilty when they themselves do not experience the grief reactions seen in the adults around them. Guilt may be produced or exacerbated by unwise admonitions of surving parents or family. Under other circumstances, regression or depression may be a pathological aspect of a grief reaction. Identification with the departed parent, if overly intense, may also be pathological; in some families, children may attempt to compensate for loss of a parent by overachieving, which may ultimately deprive them of some of the normal experiences of childhood. (See also Chapter 13 for a comparison between the child's reaction to the death of a parent and to divorce.)

DEATH OF A SIBLING

Many of the considerations surrounding death of a parent also obtain in the case of death of a sibling. Many identical questions arise, such as what to tell the well sibling, whether to visit the hospital, whether surviving children should share the rituals that follow death, and how to treat the memory of the dead child. The answers to these questions involve much the same considerations as well: the skills, strength, and coping styles of parents; the advice that the questions and feelings of children be managed as factually as possible; the need to avoid canonizing the lost child; and so on. The role of the physician in these matters is also as before: to make sure that parents have a chance to know about these issues and discuss them. (See also Chapter 27.)

Death of a sibling has another element, though, of which parents may not be fully aware; it is that siblings had interactions and relationships of their own about which the parents have only incomplete knowledge. Normal rivalries, bickering, and tensions between and among siblings have often been the seeds of feelings of hostility or anger or of unconscious or conscious death fantasies, which the surviving sibling may have difficulty facing without disabling guilt or anxiety. It is important that parents give their surviving children leave to remember irritations without guilt, and especially without the feeling that any negative feelings they might have had could have contributed to the illness or death of the departed child.

Grieving parents absorbed in their own feelings will often need help in recognizing and attending to the needs of surviving children. The physician can help them become aware of these needs by sensitive inquiry as to how the parents plan to help surviving children cope with their feelings and with thoughtful suggestions.

Parents may need help in planning their own lives. Sometimes they are given the naïve advice by family, friends, or professional counselors that they should have another child soon to replace the dead one. This may be a disservice both to the parents and to surviving siblings. Parents need time to complete their mourning for a dead child before undertaking a new child-bearing venture; this usually requires a year or more. When parents have not had adequate opportunity to incorporate into their lives (and into the life of the family) the event of the death of a child, a new child may be too much regarded as if he or she were the dead child—vulnerable, in need of anxious overprotection, and having his or her identity confused with that of the dead child.

Surviving children may also need protection from their overidentification with the dead child. This overidentification may involve fears of the same illness or accidents, and defenses may involve denial, rituals, or other behavior that is not so easily interpreted. The physician's task in these matters is to help parents become sufficiently aware of these issues so that they will not become unrealistic about the hazards that may befall the surviving children who are leading normal lives.

Role of the Physician

To achieve the greatest success in helping children and their families to deal with death of a family member, the role of the physician must be primarily preventive. The physician can help families in which there is a parent with a fatal disease to understand the needs of children as well as their own needs in such a way as to plan more effectively for the care of children before and after the death of a family member. Parents can be made aware that mourning may take many forms (including the appearance of disinterest), and that it is all right to help thinly disguised or hidden feelings to be expressed rather than repressed. Parents can be encouraged to learn how to say "It's all right to cry (like me)," rather than "Don't cry!"

After the death of a parent or sibling, time will be needed for reorganizing relationships and functions within the family. This is no time for burdening young children with responsibilities beyond the ordinary or essential, and still less for asking them to assume the role of the departed. Physicians can serve families best when they can comfortably monitor all these aspects of separation or death and can make unhurried time available to parents for ventilation, discussion, and planning.

Experience has shown that the affiliation in discussion groups of parents who have shared experiences such as loss of a significant family member often helps them to see their own reactions and those of their children in a broader and healthier perspective. When such discussion groups have the guidance of professional persons with skills in group dynamics, they can be very helpful to grieving families. Such groups may be particularly helpful to families in which the loss of a significant member (parent or child) has been sudden and unexpected (as in the case of neonatal death or sudden infant death syndrome, accident, or acute illness). In such cases the family is unprepared, and the event comes as a catastrophe leaving no opportunity for anticipatory grief work. Sharing these experiences in groups can often substantially help families in the work of reestablishing their equilibrium.

VICTOR C. VAUGHAN, III

REFERENCES

Bowlby, J.: Attachment and Loss: Attachment (Vol. I). New York, Basic Books, 1969.
Bowlby, J.: Attachment and Loss: Separation (Vol. II). New York, Basic Books, 1973.

Bowlby, J.: Attachment and Loss: Loss (Vol. III). New York, Basic Books, 1980.

Forman, M. A., Hetznecker, W. H., and Dunn, J. M.: Psychosocial Dimensions in Pediatrics. *In* Vaughan, V. C., III, McKay, R. J., Jr., and Behrman, R. E. (eds.): Nelson Textbook of Pediatrics. Philadelphia, W. B. Saunders Co., 1979, pp. 67, 68, and 76–79.

Furman, E.: A Child's Parent Dies. New Haven, Yale University Press, 1974.

Goldstein, J., Freud, A., and Solnit, A. J.: Beyond the Best Interests of the Child. New York, The Free Press, 1973.

McGrew, W. C.: Aspects of social development in nursery school children with emphasis on introduction to the group. *In* Blurton Jones, N. (ed.): Ethological Studies of Child Behaviour. London, Cambridge University Press, 1972, pp. 129–156.

Noshpitz, J. D. (ed.): Basic Handbook of Child Psychiatry. (Four vols.) New York, Basic Books, 1979.

Sack, W. H., Seidler, J., and Thomas, S.: The children of imprisoned parents: A psychosocial exploration. Am. J. Orthopsychiatry 46:618, 1976.

Spitz, R. A.: Anaclitic depression. *In* Eissler, R. S. (ed.): The Psychoanalytic Study of the Child. Vol. 2. New York, International Universities Press, 1946, pp. 315–342.

Vaughan, V. C., III: The care of the child with a fatal illness. *In* Vaughan, V. C., III, McKay, R. J., Jr., and Behrman, R. E. (eds.): Nelson Textbook of Pediatrics. Philadelphia, W. B. Saunders Co., 1979, pp. 168–171.

17

Environmental Deterrents: Poverty, Affluence, Violence, and Television

Increasingly, social forces that emanate far from the immediate world of the child have come to be recognized as critical in shaping children's development. Among the most powerful of these forces are the economic and social circumstances in which the family functions and the child grows up; schools and other institutions with which the child has significant contact; the mass media, including television, radio, movies, newspapers, magazines, and books;* and the physical and social environment that surrounds the child in the neighborhood and community.

This chapter will focus on four aspects of the environment that are not covered specifically elsewhere in this volume: (1) poverty, (2) affluence, (3) neighborhood violence and danger, and (4) television. These influences on development and behavior are of particular significance for those who seek to respond to the needs of children and their families in this country today. (See also Chapter 15 concerning schools.)

In considering the role of these environmental factors in the development of children, it is essential to keep constantly in mind the critical influence of individual and family characteristics in determining the impact of outside forces. As Bronfenbrenner concludes from Elder's research on the impact of the Depression on the development of individual children, the significant determinants were not only how heavy an economic blow was dealt the family but also the age of the child when it occurred, the sex of the child, and the strength of intrafamilial linkages prior to the change in economic status (Bronfenbrenner, 1979). Similarly, the landmark longitudinal study of the children of Kauai demonstrated clearly that outcomes are determined by the interplay among the child's constitutional characteristics, stressful life events, and the attributes of the caregiving and cultural envi-

ronment (Werner and Smith, 1982). Each of the environmental factors discussed here must thus be viewed as one part of a complex set of interactions that affect a child's development.

ECONOMIC DEPRIVATION AND POVERTY

In the United States, 17 per cent of children under 18 years of age (10.7 million) were living in families classified as being below the poverty level in 1979, 7.7 million children were living in families with incomes below $5,000, and 22.5 million (one-third) were living in families whose head had not completed high school.

Poverty, however defined, is closely associated with a multitude of deterrents to healthy development. Although the mechanism of causation is not known for each of the factors involved, the association is clear and consistent.

Among the components of deprivation and poverty that may act directly, or interact with one another, to deter normal development are
— low family income, unemployment, underemployment and low educational status of adults in the family, and poor prospects of employment for adolescents
— poor living conditions, including inadequate sanitation and poor housing
— poor nutritional status and poor health status of child or family
— inadequate access to appropriate health services, social services, and social supports.

Evidence of Association Between Poverty and Deterrents to Normal Development

It is clear from a multitude of studies that poor children are more likely than the rest of the population to suffer from conditions or live in circumstances that may interfere with normal develop-

---

*Although all mass media will not be specifically discussed here, some of the principles discussed in the section on television may be extended to them.

ment (Edelman, 1980; Egbuono and Starfield, 1982; Rudov and Santangelo, 1979; Select Panel, 1981; Starfield and Pless, 1980; Starfield, 1981). Among the indicators are the following:

• From the moment of birth on, at every age, poor children are at higher risk of death than the nonpoor; disparities in the health status of poor and nonpoor children increase with age and are much greater in the teenage years than earlier.
• Children in poor families have more days of restricted activity, days lost from school, days spent in bed, and days of hospitalization as a result of acute illness and as a result of chronic conditions than nonpoor children.
• Low birth weight is more common among infants born to black than white women, to poor than nonpoor women, and to less educated than well educated women.* Poor performance on subsequent intelligence tests is associated with low birthweight, but the association is strongest for low birthweight infants who are poor.
• Elevated blood lead levels, associated with deficits in psychological and classroom performance, are nine times as high for children in families with less than $6,000 annual income than for those in families with an income of over $15,000 and are three to four times higher for blacks than for whites at each income level.
• Low socioeconomic status is associated with residual hearing impairment from otitis media, which is in turn associated with auditory processing deficits, language delays, and behavioral problems.
• Infants born to low socioeconomic status families are more likely than others to be congenitally infected with cytomegalovirus. The consequences measured in subsequent IQ tests and other behavioral evaluations are more severe for infected infants in poor families than for infected infants in nonpoor families.
• Iron deficiency anemia, which is associated with low development scores in infancy and later decreased attentiveness and conduct disorders, is three times as prevalent among the poor than nonpoor in age groups 1 to 5 and 12 to 17. Approximately 33 per cent of all black children are estimated to suffer from some kind of nutritional deficit compared with less than 15 per cent of white children.
• Births to teenagers under 16 and out-of-wedlock births—both of which are associated with greater risk of low birthweight, infant mortality, and subsequent complications—are significantly greater among nonwhites. Among 14- and 15-year-olds, the nonwhite birth rate is four times as high. The rate of out-of-wedlock births is eight times higher among nonwhites in all age groups.
• Among persons reporting to emergency rooms for drug abuse, blacks were almost three times more likely than whites to be drug-dependent.
• The median income of households in which either child abuse or child sexual abuse has been reported is considerably below the median income of all American families.
• A child in a low-income family is 15 times more likely to be diagnosed as retarded than a child from a higher income family; 50 to 80 per cent of mental retardation is estimated to be associated with economic and social inequities and injustices.
• Congenital defects were almost twice as likely to result in death among nonwhite youths 15 to 19 years of age than among whites.
• Children in low-income families are far more likely to live in housing with structural deficiencies, incomplete plumbing, and crowding—all of which contribute to accidents and to the spread of infection—than other children.
(See related chapters for fuller discussions of these problems.)

From a review of the close relationship between poverty and a wide array of deterrents to healthy development, it is reasonable to conclude that the elimination or amelioration of these deterrents requires significant changes in national policies that go far beyond the health system and the purview of health professionals. If poverty and racism were eliminated in the nation, if there were full employment and every young person could look forward to productive work, the chances for normal development would be enormously enhanced.

However, the evidence also makes clear that action short of such sweeping social changes can also be effective. Many interventions that can be provided or mobilized by health professions can

Table 17–1.   INTERVENTIONS TO AMELIORATE OR PROTECT AGAINST DETRIMENTAL EFFECTS OF POVERTY

1. Health services of special importance
   Family planning services
   Prenatal care
   Genetic services
   Nutrition
   Comprehensive care of infants and young children
   Services for adolescents
2. Extending the reach of health services and forging better links with other services
   Outreach
   Home visiting
   Coordination of health and health-related services
   Providing developmental and behavioral expertise outside the health system
3. Mobilizing community supports
   Social networks
   Day care and family-centered early interventions

---

*For some indicators, minority status and low education are used as proxies for low income.

ameliorate or protect against some of the detrimental effects of poverty and economic deprivation (Select Panel, 1981). They include (1) health services that are of special importance in protecting against the most negative effects of poverty on normal development, (2) institutional changes to make sure that appropriate services reach those who need them most, and (3) the mobilization of other sources of support for disadvantaged families (Table 17–1).

Health Services of Special Importance

A child's physical health is a primary determinant of normal development. Although many questions have been raised over the past decade suggesting that the impact of medical services on general health status may be overestimated, there is no question that the health status of children—especially poor children—is significantly affected by the receipt of appropriate health services. The health services discussed here are services to which poor families continue to have inadequate access and which have particular promise for ameliorating the detrimental effects of poverty on normal development.

Family Planning Services. Timely access to high-quality family planning services, including counseling, is an important determinant of the chances of an infant's birth being welcome and whether he or she is likely to enter a home environment that is nurturing and will encourage healthy growth and development. Appropriately available family planning services can reduce infant mortality, low birthweight, and stillbirths and increase the probability that the timing of birth, intervals between births, and family size will serve to enhance the infant's healthy development. Almost one-third of the reduction in the infant mortality rate in the United States between 1965 and 1972 resulted from shifts in the timing and spacing of births and, hence, from individual family planning decisions. In a study of low-income women in Maryland the greatest decrease in infant mortality was found to occur in areas that had made the most progress in serving women who needed family planning services (Select Panel, Vol. I, 1981).

Although unwanted and mistimed childbearing has declined substantially in recent years with increased availability of family planning services, it still presents a serious problem for many Americans. An estimated 2.8 million unplanned pregnancies occur each year, half of them terminated by abortion. Almost 16 million women at risk for unwanted pregnancy are not receiving the health care necessary for the safe and effective use of contraception. Three million of these women have low or marginal incomes and 1.8 million are teenagers who need subsidized care (Select Panel, Vol. I, 1981).

The diversity of settings in which family planning services are available seems to be a major factor in the substantially increased number of women who today use these services. For many women, being able to obtain these services as part of continuing and comprehensive care results in their more effective and appropriate use. For others, the opposite is true. The population seeking family planning services includes a large number of individuals who see their own needs as very different from those who rely on more comprehensive programs. The clients of family planning clinics are predominantly young, childless, and healthy. Some people whose primary concern is to postpone starting a family are reluctant to seek services from programs targeted to families and children and care of the sick and ailing.

For many who seek family planning services, especially teenagers, the critical needs are easy access to the facility and guarantees of confidentiality and privacy.

Prenatal Care. For reasons not yet completely understood, prenatal care is one of the most effective ways of heightening the chances of the birth of healthy infants and their subsequent optimal development. Late or no prenatal care is associated with an increased incidence of low birthweight, prematurity, and newborn mortality. Women without prenatal care are four times as likely to bear infants who subsequently die and three times as likely to bear low birthweight infants, with their increased risk of death, congenital anomalies, mental retardation, learning disabilities, and need for neonatal intensive care. Although birthweight and infant survival are strongly and consistently associated with adequacy of prenatal care, pregnant women at greatest risk are least likely to receive adequate prenatal care. Despite great improvements in the availability of prenatal care, fully one-fourth of all pregnant women still receive late, little, or no care. This percentage is significantly higher for the very young, women over 35, black women, the poor, the unmarried, the poorly educated, and those in rural populations (Select Panel, Vol. I, 1981).

Barriers to prompt and adequate prenatal care include lack of funds to pay for care out-of-pocket; lack of insurance coverage; lack of eligibility for publicly supported programs; and the absence of services that are easily accessible, appropriate, and rendered in circumstances that are acceptable to those needing care.

Prenatal care improves pregnancy outcome independently of social class. When social class, as measured by educational attainment of the mother, is held constant, infant survival reflects adequacy of prenatal care. An analysis of New

York City births conducted under the auspices of the Institute of Medicine, National Academy of Sciences, found that among college-educated mothers with adequate care, infant death rates were half as high as among college-educated mothers with inadequate care. For infants of black native-born women with one or more years of college, there was a sixfold difference in death rates between those whose mothers had adequate care (7.2 per 1000) and those whose mothers had inadequate care (42.3 per 1000) (Kessner, 1973).

The precise components of prenatal care that lead to improved reproductive outcome have not been clearly identified, but they seem to be associated with some aspects of medical and obstetrical services, especially for high-risk conditions. It is also highly likely that counseling of expectant mothers concerning potential problems for the fetus that may be caused by smoking, alcohol use, and poor nutrition—and referral, when necessary, to appropriate services—is a major contributor to the effectiveness of prenatal care.

Genetic Services.   Genetic services, which include screening, testing, counseling and treatment, can have an enormous impact on the incidence of a multitude of deterrents to normal development. This impact can be expected to increase by several orders of magnitude over the next decade or two, as our understanding of genetic diseases increases. (See Chapters 18 and 19.) However, even our present knowledge is not being universally applied and is typically least available to poor and less educated families.

Amniocentesis, which can be used to diagnose about 100 of the 2,000 identified genetic disorders during the second trimester of pregnancy, is available to only a fraction of the families that might benefit from it. For example, only one-seventh of the pregnant women over 35, who are at high risk of bearing a child afflicted with Down's syndrome, undergo amniocentesis. While there are many reasons for this, including the limited number of personnel and facilities to perform the procedure, and the controversial aspects of the subsequent course of the pregnancy, including abortion, poor families are least likely to make use of the procedure because of its expense, its inadequate coverage by third-party payers, and the lack of awareness among less educated families of its usefulness (Select Panel, Vol. I, 1981).

A similar situation exists with regard to genetic counseling. Genetic counseling is a process that attempts to help the affected individual or family (1) to comprehend the medical facts regarding a genetic disorder, including the diagnosis, probable course, and available management; (2) to appreciate the way heredity contributes to the disorder and the risk of recurrence in specified relatives; (3) to understand the alternatives for dealing with the risk of recurrence; (4) to choose the course of action that seems appropriate in view of the risk, the family's goals, and ethical and religious standards and to act in accordance with that decision; and (5) to make the best possible adjustment to the disorder in an affected family member and/or to the risk of recurrence of that disorder.

The range of problems for which counseling is provided is broad. Included are disorders resulting from simple gene abnormalities, multifactorial interactions, and chromosomal defects. Many conditions for which genetic counseling is provided will be less well defined genetically and include recurrent spontaneous abortion, multiple malformation syndromes, or unclassified mental retardation. Other issues include questions about the consequences of consanguinity, premarital counseling, genetic evaluation of children placed for adoption, and exposure to drugs or radiation before and during pregnancy (Select Panel, Vol. I, 1981).

Nutrition.   The course of pregnancy, the condition of the infant at birth, and subsequent growth and development of young children depend on adequate and balanced nutrition. Improper nutrition of a pregnant woman increases the chances of her bearing a low birthweight baby, with associated higher rates of mortality and physically and mentally handicapping conditions. Inadequate infant nutrition increases the likelihood that exposure to infection will result in disease, that resistance to disease will be weakened, that associated complications of disease will result, and that normal development will be inhibited (see Chapter 23).

In spite of these generally acknowledged facts, a sizable percentage of all pregnant women and over 90 per cent of all children born to low-income families exhibit nutritional deficiencies. Nutritional problems are especially serious for teenage mothers, a high proportion of whom are poor and who must nourish themselves during this period of their own rapid development as well as the fetus.

Publicly supported nutrition programs, including those financed by Title V (maternal and child health), the Supplemental Food Program for Women, Infants and Children (WIC), the food stamp program, and child nutrition programs in the schools and day care centers, have had a significant and measurable positive impact on child health. There is persuasive evidence of the need to expand these programs—particularly the WIC program, which provides supplemental, nutritious food to low-income pregnant and lactating women and infants and children up to age 5, who are at nutritional risk because of poor health, inadequate nutrition, or both. Studies conducted by the University of North Carolina, by the Centers for Disease Control, and by the Harvard

School of Public Health all show the association between participation in the WIC program and lower incidence of low birthweight babies (Select Panel, Vol. II, 1981).

Nevertheless, because of WIC's limited funding even before recent cutbacks, available resources have not kept up with needs, and in many communities health professionals must choose among eligible pregnant women or must decide whether to drop infants who were recently anemic to make room for new participants, risking a return of recently alleviated nutritional problems.

One area in which great strides have recently been made is in changing hospital practices to encourage breastfeeding among poor and less educated women. Hospitals are increasingly following the suggestions of the American Academy of Pediatrics intended to facilitate the decision to breastfeed and to heighten the chances of doing so successfully, including decreasing sedation; avoiding separation of mother and infant; making possible feeding on a demand schedule; discouraging routine supplementary feedings; providing rooming-in; and assuring the availability of knowledgeable, skilled, and supportive nursing personnel, both in and out of the hospital. Hospitals that have made such changes in practices and arrangements have been able to document dramatic increases in the number of young women with low incomes and little education who are successfully breastfeeding their infants.

Comprehensive Care of Infants and Young Children. The value of competent, comprehensive health care for infants and young children in providing a basis for optimal development and lifelong health is clear. Such care includes screening tests at birth to identify significant health problems that, if undetected or improperly managed, can lead to serious adverse health outcomes, including mental retardation; early attention and follow-up to families at identifiable risk of abnormal parenting practices; early detection of hearing and vision impairment and prompt attention to an infant's or child's communication needs; testing of high-risk children for lead poisoning, parasites, and sexually transmitted diseases; and careful and periodic appraisal of development and maturation and subsequent diagnostic studies to uncover sensory defects, muscular or neurological disease, or emotional deprivation.

Prompt, early treatment of illness or handicaps in young children is as essential as the availability of screening, immunization, and other preventive services. Indeed, the distinction between treatment and prevention at this stage of life is somewhat artificial. Most significant health problems that arise in this period have lifelong consequences if not effectively managed.

Yet, many low-income and minority children and children living in poor rural areas do not have full access to the services they need. There is a strong negative correlation between low family income and the timely receipt of both preventive health services and acute care services, as a result of both financial barriers and the absence of accessible providers.

Anticipatory guidance and counseling are also services that poor families could benefit from disproportionately but to which they are less likely to have access. Optimally, in the course of routine well-child care, parents can be counseled regarding accident prevention and home safety and are likely to be receptive to guidance regarding infant feeding, minor behavior problems, general growth and development, and difficulties in parent-child relations. The process of providing and exchanging information can lead to enhanced parent understanding, competence, and confidence. Such guidance is especially important to parents in dealing with temperamental differences in children, unrealistic expectations, unfounded anxieties, and the particularly onerous stresses of low income, crowded housing, poor education, and a climate of crime, violence, and scant hope for the future.

Anticipatory guidance and counseling for low-income, multiproblem families require effective links with other community resources, discussed later in this chapter, as well as special skills and sensitivities. It is important to be aware of the high proportion of factors in the lives of these families that are totally out of their control as well as the fact that the very condition of their lives militates against their being able to change those factors that might in other socioeconomic settings be amenable to change. People at the bottom of the social hierarchy see themselves at the mercy of forces quite beyond their control and understanding. The conditions in which many low-income and minority families live allow little freedom of action and scant opportunity to experience personal efficacy or control over one's fate. It is therefore a particular challenge to help such families to adopt or maintain health practices that are likely to enhance the development of their children and to strengthen and support the family's role in child development.

Among all the routine child health services, anticipatory guidance and counseling, along with effective developmental assessment, are probably most dependent on a continuing relationship with a health care provider. Yet about one-fifth of black children and one-fifth of children living in families with the lowest level of income have no regular source of health care and are therefore least likely to have access to continuing health supervision (Select Panel, Vol. I, 1981).

Even those poor families with a regular source of care often do not benefit from many of the arrangements that promote continuity and access for middle- and upper-class families. Telephone access is an important case in point. While most parents rely on their ability to get advice, information, and reassurance from pediatricians and other health professionals by phone, many clinics and outpatient departments make no provision for telephone consultation. The vast majority of poor and minority children have no effective access to telephone advice at all. In 1974, 94 per cent of the black children and 88 per cent of children in families with incomes less than $5000 had no telephone contacts with health care providers during the year (Select Panel, Vol. III, 1981).

A discussion of some of the changes needed to make continuing health care more readily available to all families with infants and young children and the required policy changes in the organization and financing of health services would be beyond the scope of this volume. Other changes that could significantly improve primary health services of poor children and that are more likely to be within the confines of decision-making in which the clinician may be influential are discussed later in this chapter.

Services for Adolescents. Despite the prominent position of adolescents in our youth-oriented society, the development and deployment of health services reflect inadequate attention to young people from ages 12 to 17, especially those in poor families.

Poor adolescents have inadequate access to most needed health services, including family planning, drug and alcohol abuse prevention, and various forms of counseling and mental health services. Moreover, to the extent that health services are available to poor adolescents, they almost always tend to be oriented more toward problems than toward normal growth and development.

Since the health problems of adolescents tend to reflect environmental, social, and psychological factors to an even greater degree than those of the rest of the population, and since many aspects of the prevailing health system deal comparatively less effectively with the psychological, social, and behavioral aspects of health, it is not surprising that adolescent health needs are often ill-served. The emphasis of third-party reimbursement for disease treatment and care rather than being on preventive services and counseling, which are highly relevant to adolescent needs, limits the effectiveness of the health system in helping adolescents, as does inadequate information about the most effective ways to design services for adolescents and a lack of consensus about the appropriate division of responsibility in this area between the schools and the health system.

Health services for adolescents should include a strong counseling component. Indeed, in areas such as family planning and the treatment of sexually transmitted diseases, the counseling components of care are of equal importance to whatever medical services are delivered. Given the many vulnerabilities of adolescents, and their many questions and concerns about a number of health matters, counseling, like education, should be a major aspect of most adolescent health services.

Privacy and confidentiality are paramount concerns of young people and must be recognized as such by the health system. As a general principle, health services—particularly family planning and other services related to sexual activity—should legally be available to adolescents without parental consent and with full assurance of privacy and confidentiality. This does not mean, of course, that parents and families should be excluded from the health care decisions of adolescents. Every effort should be made to encourage young people, and especially younger adolescents, to communicate with their families about any care they are receiving. However, requirements of parental involvement must not be permitted to become a deterrent to the receipt of needed services nor should receipt of services be contingent upon parental consent. Outreach messages must make clear that help is available on a confidential basis; information must be released only upon the young person's consent; confidentiality must also be protected in establishing fees and billing for services.

Mental health services for poor, inner-city adolescents are of particular concern and represent a nearly overwhelming unmet need. Although they constitute the fastest growing category of admissions to psychiatric hospitals, one-half to two-thirds of seriously mentally ill teenagers receive no care at all. Children with behavioral problems who become the responsibility of the state—the vast majority of whom are poor—are likely to be served in the most restrictive settings, including hospitals, jails, or residential treatment programs, although they may in fact need alternative services, such as crisis intervention, day treatment, special education, or their families may need parent training or a variety of support services. The risk of confinement is even greater for minority children, who—if they receive any treatment at all—are two times as likely as whites to be institutionalized (Select Panel, Vol. I, 1981).

The kinds of services needed by low-income disturbed children and adolescents are not those that the mental health system, as it operates in most localities, is prepared to deliver. Traditional psychotherapy seems to be comparatively ineffective with (1) children and youth from educationally disadvantaged environments where verbal communication is inefficiently used in problem-solving; (2) those whose socialization deviates sharply from mainstream expectations; and (3) those from families and neighborhoods so disorganized that the normal sources of affection, support, and discipline needed to sustain therapeutic efforts are lacking (Select Panel, Vol. I, 1981).

The narrow focus of traditional psychotherapeutic treatment can be especially disadvantageous during the adolescent years, when successful adaptation requires relationships with the broad sociocultural world beyond the family. A broad ecological perspective that encompasses the family, the school, the neighborhood, and the workplace is more likely to help a child fabricate a system of social, emotional, educational, and vocational supports that are essential to successful psychological adjustment.

Extending the Reach of Health Services and Forging Better Links with Other Services

Reducing the disparities between the poor and the nonpoor in measures of health status that are particularly relevant to developmental and behavioral factors has been most successful through changes in institutional arrangements, making services more accessible, acceptable, and easier to use (Select Panel, Vol. I, 1981). Some ways of improving access were discussed earlier relative to specific health services of special importance. More generic changes in the organization and delivery of health services and the need for better links among health services and better health and other services are described below.

When traditional public health services and personal health care have been integrated or linked and coupled with efforts to reach out into the community, the new synthesis has produced valuable benefits. A child may not have to return to the doctor for care of chronic malnutrition if a community health aide is available to help the family in obtaining food stamps and preparing an adequate diet. A health center with the capacity to mobilize action to enforce housing codes and home repair services is in a better position to respond to a child with lead poisoning than is the physician who has no contact with community agencies and cannot go beyond diagnosis and treatment of the individual. A hospital outpatient department can significantly increase its impact on underserved families through efforts to personalize care, assure continuity, and provide special counseling and outreach to adolescents and to young parents and families at special risk.

Many health institutions have changed their arrangements for rendering health services to respond better to the needs of those who have traditionally benefited least from modern medicine and have been able to document their success (Select Panel, Vol. I, 1981):

● In Cleveland, patients who are enrolled in a maternal and infant care project and who receive more patient education, nutrition counseling, social services, and appointment follow-up experience 60 per cent less perinatal mortality than a matched group of patients being seen in the same hospital who do not receive these additional services.

● In the state of Alabama, the health department established prenatal care clinics in ten of the poorest counties where none had existed. Within four years, infant mortality in Alabama was reduced from 20 per 1000 to 14.3, very close the the U.S. average—an extraordinary accomplishment in view of the fact that Alabama is 46th among the states in per capita income.

● Home visiting programs in Montreal, Denver, and rural Appalachia have been shown to improve health outcomes for pregnant women and infants and especially for children at risk of abuse.

Some of the factors that have proved most important in achieving better links among a variety of services and most effective in reaching poor families with services that are most important for enhancing normal development are discussed below.

Outreach. For many families, health services are so inaccessible—geographically or for other reasons—that they cannot make use of them without help. Some parents do not know or understand what benefits early health care would bring them and their children. This problem is acute among parents with little education and among some from other cultures. Often, health care providers must do more than open the door to needed health services. They may have to make arrangements to find, educate, and help bring in mothers and children to receive care. In recent years especially, health programs serving rural and low-income populations have found that the use of trained outreach workers can solve many problems of access.

The Council on Medical Services of the American Medical Association has identified a variety of ways in which outreach workers, particularly those who live in the area served by a health care program, can improve services. Outreach workers:

● Tend to enhance professional standards of practice because such personnel can free physicians, nurses, dentists, and other health professionals to better utilize their time and thereby extend the scope of their services to a larger patient population.

● Provide an additional source of manpower to meet community needs, especially in those areas where there is a shortage of professional health staff.

● Obviate many of the traditional problems of understanding and communication in providing health services for those in need.

● Assist the professional staff in becoming more responsive and accountable to the community served.

● Provide meaningful jobs and, as a result, benefit the community economically and socially (Schorr et al., 1976).

The effectiveness of outreach is enhanced when it involves personal contact with trained personnel who share or are sensitive to the background of the people being served and when it is conducted in a variety of places in the community.

Home Visiting.   Home visits to families before and after the birth of a baby by nurses or lay visitors under nurse supervision have been systematically undertaken in some communities in the United States and in many countries of Western Europe, often forming the cornerstone of organized efforts to improve maternal and child health. It was a more widespread phenomenon in the United States several decades ago than it is today, having fallen into disuse not because of diminishing need or lack of success but apparently because funds and professional talents are increasingly concentrated on therapeutic and more technology-intensive services.

There is now a renewed interest in home visiting services as a means of improving the delivery of preventive health services to mothers and young children at high risk and linking them to an ongoing source of care. This new interest seems to reflect a recognition of the following factors:

● Many of the most important potential gains in child health status, given the state of today's medical knowledge, cannot be made without systematic and aggressive efforts to extend the reach of health professionals beyond the walls of their own offices and institutions and out to where the people most in need of services actually are.

● A growing proportion of problems in maternal and child health are rooted in the complex life situations of families. Much of the help pregnant women, young children, and parents need and are not getting—for example, for accident prevention, nutritional problems, and child neglect and abuse—can best be provided in settings outside major health institutions.

● For many families, the traditional sources of support for pregnant women and new parents, including family and friends, are increasingly hard to come by.

● Some of the education in child-care skills, which many new mothers used to get in the hospital during the days following birth, is no longer

provided because of a shorter hospital stay after childbirth. Furthermore, some medical problems, such as neonatal jaundice, may only become apparent after the baby goes home.

● In many areas, health services may be hard to locate, and systematic efforts are required to link people most in need with requisite services.

Most programs of home visiting now in operation seem to be the product of at least some of these forces. While the programs vary widely, most have the following characteristics:

● They provide information and help with making the home a safe and nurturing environment for the infant.

● They provide information and help with nutritional problems for the pregnant woman and infant.

● They provide certain basic health services (physical and developmental assessment, immunization) or arrange for them to be provided.

● They seek to assure that the pregnant woman and family are linked with an ongoing source of health services and to social support services as necessary.

● They teach about basic health practices and provide guidance about the most effective use of professional health resources.

● They assist families who need help in integrating the infant within the family structure and life style.

● Some seek to identify families at special risk and in need of special services.

● Some seek to personalize health services in ways that are difficult to accomplish in most health care settings.

● Some seek to enhance the parents' ability to stimulate the infant's social and cognitive development.

Some programs use nurses to make the home visits, some train lay persons who work under nurse supervision, and some combine these methods. Some programs emphasize the telephone availability of the home health visitor between visits. The programs have operated under a variety of auspices, including health departments, community health centers, hospitals, medical and nursing schools, and family drop-in centers. Most begin visits during the prenatal period, some limit services to the first year of an infant's life, and some extend them until the child enters school.

Some programs are universal, offering services to all infants and pregnant women in a given area or to all where the birth occurs in a given hospital. Others offer services only to high-risk families, with risk being defined socially (low-income of family, social isolation, welfare-eligible) or medically (low birthweight infant, discharge from infant intensive care unit). Some programs combine these concepts, offering services universally but actively recruiting or providing more intensive services to high-risk families.

The reported data on the effectiveness of home visiting suggest that it can be a highly effective form of intervention, especially when begun before the birth of the baby and when it is an integral part of other efforts to improve services to pregnant women and infants, providing links to continuing care for the family and organized back-up and consultation for the home visitor (Select Panel, Vol. I, 1981).

Coordination of Services. Effective coordination of a full array of services is especially important to poor families, who often lack the resources of time and energy required to do this on their own. The family's source of primary care should have the capacity to coordinate all needed health and health related services through a working partnership among a variety of health professionals, active collaboration with patients and their families, and a strong relationship with other service systems in the community, such as the schools and the welfare system. The primary care provider should coordinate all elements of the patient's care, including that provided by other health professionals and institutions, and the participation of other agencies and providers, including those from the welfare, correctional, and educational systems. This involves referrals, the exchange of information with other providers of service, and explanations and exchanges with patients or families to achieve a greater understanding, and modifications as indicated of treatment plans or supportive services.

The degree of coordination of services that is required to make health services effective for many poor families can only be achieved through strong links between health professionals and other sources of care, services, and support in the community. Organized settings, including community health centers, health department clinics, children and youth centers, hospital outpatient departments, and group practices can most easily create these networks, but physicians in individual practice can also mobilize and coordinate the full range of services their patients need. For example, the local public health agency may provide the local physician with a part-time public health nurse or social worker; some health departments provide office space in exchange for the physician's participation in the department's health service programs. Many visiting nurse agencies have close and active working relationships with physicians in private practice. In some rural areas, the agricultural extension agent has been enlisted to provide follow-up services, and some welfare departments work closely and effectively with private practitioners to make sure that referrals are completed and to help coordinate care. Services such as food supplements and nutrition education, crisis counseling, classes in preparation for childbirth or parenting, and alcohol and drug abuse programs are frequently offered by health departments, health centers, and hospitals and are needed by some of the patients of private practitioners. Better linkages between these programs and institutions and the private practitioner can greatly enhance the latter's capacity to deal with a broad range of problems.

Providing Developmental and Behavioral Expertise Outside the Health System. Of course, the special needs of poor and deprived children for help from health professionals with particular behavioral and developmental training and skills extend beyond the health system to other settings where children live, play, and study. Children who need help—whether for a transitory crisis, serious emotional disturbance, or learning or behavioral difficulties that persist over time—are found throughout the education, correction, and social services system. Often, however, the institutions that make up these systems have been ignored or abandoned by those with the greatest relevant experience and talent. Teachers, day care workers, juvenile correctional officers, foster parents, social workers, guidance counselors, and others working with disturbed and distressed young people need far more collaboration, consultation, and support from health professionals than they are now getting in their efforts to restore the functioning effectiveness of families, neighborhoods, schools, and religious and community groups as well as to enable troubled youngsters to respond to normal sources of support, affection, instruction, and discipline.

Schools offer a particularly important opportunity for applying developmental expertise, precepts, and talents in a setting where both the immediate and long-term payoffs are likely to be great, since the precursors of serious personality disorders are so frequently apparent in the way children relate to schools (see Chapter 15).

Corrections and social service systems also need greater support from and interaction with health professionals both to improve services to individuals and to design better programs and institutional arrangements. For example, despite the evidence that regular health assessments, including developmental assessments, are particularly important for children in foster care, many such children receive no regular health services of any kind. In view of epidemiological evidence that some foster children appear to draw abuse from one foster home to another, special, in-depth attention should be provided for those who have had multiple placements or been the subjects of abuse. Others likely to be in touch with the social service system and who could benefit greatly from more psychologically informed arrangements to help them are pregnant teenagers and young mothers who have run away from home or been pushed out of the family home.

Far stronger connections are needed between child health professionals and infants, children, and adolescents who are in a setting or situation where the risks to normal development are enormous, be it a correctional institution, runaway house, welfare agency, mental hospital, or drug rehabilitation program. Currently there are numerous missed opportunities to ameliorate the effects of poverty and deprivation for children and adolescents whose special needs are signaled through contact with certain easily identifiable human service agencies or institutions.

Mobilizing Community Supports

Powerlessness has been identified as a basic problem of the poor, who receive inferior health and public services and are more likely to have their rights violated by agents of the law or other social institutions and whose options in most areas of life are severely restricted. In addition, their low educational level, restricted experience, and lack of information may make it difficult for them to understand and avail themselves of the limited resources open to them (Hetherington and Parke, 1980). In these circumstances, the role of a variety of community supports in ameliorating the negative effects of poverty and deprivation for the children of poor families can be crucial, and health professionals can make an important contribution in helping families to find and make use of such supports.

Social Networks. Both formal and informal networks can be important sources of help to families by supporting parents in their child-rearing roles, by serving as role models for both parents and children, and by providing help during crises. Social networks have been found to be a particularly important source of support for low-income black families. Community support systems of many kinds help to contribute to a sense of well-being and competence and can aid in reducing the negative consequences of stressful life events. In one study of women who had experienced high levels of stress before and during their pregnancies, it was found that women who received considerable social support had significantly fewer complications during pregnancy than women with little psychological support.

It is essential that the efforts of professionals strengthen the natural networks to which people belong and on which they depend. These may include families, friends, neighborhood groups, work relationships, religious affiliations, and a variety of voluntary associations based on principles of intimacy and mutual aid.

Day Care and Other Forms of Intervention in Early Childhood. Increasing recognition during the 1950's and 1960's of the importance of early experience for later development and achievement led to the initiation of a number of early intervention programs aimed primarily at children from disadvantaged families. The principles on which these programs were founded have been incorporated to varying degrees in a broad array of formal and informal child care arrangements that were designed primarily to make it possible for more adults to work outside the home.

Research on the effects of early intervention has generally borne out the expectations of its beneficial effects. Comparisons of various forms of early intervention have shown that programs which begin during infancy or the preschool years, have clear goals for parents, include home visits, and have high staff-to-child ratios seem to have the greatest long-term effects. Participating children were later found to be less frequently assigned to special education classes or to be held back one or more grades in school. Especially in programs that involve the parent and child in verbal interactions around a cognitively challenging task and reinforce the parent's status as the key person in the child's life, the gains include diffusion effects to younger siblings and evidence that the parents themselves showed more self-confidence and initiative (Bronfenbrenner, 1979; Schaefer, 1982).

AFFLUENCE

Our intent in including a discussion of affluence as a potential deterrent to healthy development is not to imply that the role of affluence might be analogous to that of poverty, for it is not. Poverty, as we have pointed out, almost inevitably brings with it barriers to healthy development that are difficult to overcome except through a fortunate combination of individual and family strength, appropriate and accessible health services, and effective social supports. We do not know to what extent affluence also brings with it identifiable deterrents to development. If it does, they are surely easier to overcome. Whether it does or not, one should bear in mind the enormous variations among the children of the rich as well as of the poor. As Robert Coles concluded after studying the lives of children of the wealthy, some are "boys and girls whose lives may turn out to be ethically suspect, psychologically disastrous, socially corrupt, economically privileged to the point of exploitation of others, and culturally barren. But they may also be boys and girls who, at least during childhood, show no such ruinous inclinations . . . who stand up well, and who come from quite stable, hard-working and ethically concerned families . . ." (Coles, 1977).

We include affluence in our consideration of environmental factors that may act as deterrents to normal development primarily in order to call attention to certain aspects of the affluent environment that may have negative effects—and are of interest particularly because they occur within a framework that is generally thought to provide the most propitious environment for healthy development.

Relationship Between Affluence and Development

The relationship between affluence and development, unlike that of poverty and development—for which the associations (if not the causal relationships) are amply documented—has rarely been systematically examined. There is some literature on the special problems of treating neurosis and mental illness in children of affluent families, but the possibility that growing up in an affluent family is an actual risk factor in normal development is based almost entirely on impressionistic and anecdotal evidence.

Among the negative outcomes in which growing up in affluence may be implicated are alcohol and drug abuse, motor vehicle accidents, suicide, cult membership, profound alienation, and anorexia nervosa.

If affluence is indeed a risk factor, it is unclear whether the key characteristic is great wealth or wealth just sufficient to assure that the family is free of the economic constraints that impinge on the majority of families in the society. Some have suggested that a factor more critical than the amount of wealth may be a rapid change in economic status—a sudden catapulting into wealth of a family with humble origins. Others suggest that affluence as a risk factor may be only a proxy for other more potent influences, such as the fame, power, and intense emphasis on extraordinary achievement, which may characterize a high proportion of wealthy families.

Aspects of Affluence with Possible Negative Effects

Aspects of the environment that may be present in a significant proportion of affluent families and that may be deterrents to healthy development are described below, on the assumption that health professionals may find a listing of these factors useful in alerting them to possible problem areas (Table 17–2). However, it is important to note that the suggested associations have not been subjected to widespread systematic investigation,

Table 17–2. ASPECTS OF AFFLUENCE WITH POSSIBLE NEGATIVE EFFECTS

1. Socioeconomic circumstances may lead to absence of strong, consistent parental presence and weakening of family ties.
2. The role of substitute caretakers.
3. Unrealistic or destructive parental expectations.
4. Surfeit of possessions, with diminished need for choice and relatively easy access to cars, drugs, and alcohol.

and that hard data on causal relationships are almost entirely lacking.

Parental Absence or Neglect and Weakening of Family Ties. Most clinicians who have worked with disturbed children of the rich have concluded that an inadequate parent-child relationship is the most common critical ingredient in what goes wrong for these children (Burquest, 1981; Grinker, 1978; Sorel, 1981; Wixen, 1979). A significant proportion of affluent parents live in a style that may demand, and certainly permits, long absences from home and children. Parents who find parenthood burdensome have the means to circumvent child-rearing tasks in a socially sanctioned way. Parents who are too caught up in their own pressures to achieve may not give enough time, nurturance, and attention to their children. They may underestimate the importance of establishing limits; of setting standards and goals that go beyond material success; and of a consistent, continuing, caring adult presence. Often there is a marked failure in the systematic transmission of family concerns, traditions, values, and role expectations.

There is evidence that one complicating risk factor may be the rapidity with which wealth is accumulated in some families, and the dislocations and discontinuities it may bring about in succeeding generations. The disruptions may be particularly acute for the grandchildren of wealthy, once-poor immigrants, whose children find the work ethic on which they were reared no longer relevant and become unable to transmit a coherent set of values to their own children (Grinker, 1978; Wixen, 1973 and 1979).

The Role of Substitute Caretakers. When child-rearing is left primarily to servants, a number of problems may arise. The opportunity for the creation of a strong bond between parents and children may be considerably diminished. While there are governesses, nannies, housekeepers, and other servants who provide extraordinarily consistent and profound caring and nurturance, it is also possible—especially if they are left unsupervised with young children for long periods of time—that they may be hostile and punitive, sometimes even engaging in physical or sexual abuse. There may be repeated turnover of parent figures as a result of a number of factors; servants may be dismissed when parents feel their children have become too attached to them. Servants seldom have the same expectations of children that parents do and are more likely to wait upon and indulge the children in their care than to pass on the parent's values or to set appropriate limits.

Unrealistic or Destructive Parental Expectations. When affluent parents do play a strong role in setting goals and standards for their children, they may include extraordinary pressures to measure up to the success of the parents and a narrow definition of what is acceptable. Scholastics, sports, and social achievements that are good may not be good enough; they may have to be best. The message that love and acceptance is contingent on high achievement can be particularly hard on children with average abilities.

Surfeit of Possessions and No Need for Choices. Most children in rich families are surrounded by a wealth of possessions, sometimes but not necessarily because their parents may assuage their guilt for not giving more of their time by buying expensive gifts. These children may have little experience with having to delay gratification or with the frustration that comes from having to do without a coveted possession. Furthermore, they may develop unrealistic expectations of the extent to which it is not necessary to adapt to life but to which one can make the world adapt to one's whims. Rich children's view of the world may become particularly distorted if their families use their wealth to buy their children's way out of difficulties—whether by hiring expensive lawyers in efforts to get charges of delinquency dismissed or by buying their way into exclusive schools and colleges. Easy access to automobiles, drugs, and alcohol may also be implicated in high rates of automobile accidents and substance abuse among wealthy adolescents.

Protecting Against the Detrimental Effects of Affluence

Health professionals can be especially alert to the specific signals that may characterize abnormal development in children of affluent families and can help families recognize and respond to danger signs. In addition, they can support school and community efforts to provide organized service projects and other substitutes for the challenges of fighting for economic survival, including demanding outdoor activities in which youngsters can encounter the impartial realities of the physical universe.

Most of all, health professionals can make special efforts to emphasize the importance and unique function of parenthood to those parents with the resources to leave child-rearing to others, stressing the role that parents have in providing nurturance, love, and support; in setting firm and realistic limits as to what is acceptable and unacceptable behavior; and in transmitting to their children a clear set of traditions, concerns, and values.

NEIGHBORHOOD VIOLENCE AND DANGER

Neighborhood violence and physical danger are important factors for children growing up in urban America. They are of greatest significance for children in poor and minority families, but their effects are not limited to these children (Escalona, 1975).

Neighborhoods, which should provide a sense of familiarity and protection to children, often do the opposite. In a nationwide survey, 2000 children, ages 7 to 11, were asked, "When you go outside, are you afraid someone might hurt you?" Twenty-eight per cent answered yes. Twenty per cent of American children live in neighborhoods where parents fear for their children when they go out, because there are "undesirable people in the streets, parks or playgrounds, such as drunks, drug addicts, or tough older kids." More than one child in six mentioned fighting, bullying, meanness, vandalism, or crime as the thing they would most like to change about their neighborhood (Zill, 1982).

Schools, too, are the scenes of an increasing level of vandalism and violence. In many schools security is as great a concern as education.

There has been little systematic exploration of the effects on behavior and development of children growing up in an atmosphere characterized by violence and danger. However, there is considerable anecdotal evidence of the negative impact on such factors as self-image, expectations for the future, opportunities for recreation and for interactions with friendly and supportive peers and adults as well as of the toll in terms of breeding fear and inhibiting learning. For children who live in neighborhoods where violent crime, including murder, is a common occurrence, and who go to schools that are frequently vandalized and where they themselves may be victimized, the remedies cannot lie within the health system, except to the extent that health professionals, as particularly well-informed citizens, become advocates for fundamental changes that will reduce the incidence of antisocial behavior of all kinds. Health professionals can also be influential in efforts to reduce the lethal effects of violence through control of handguns.

Families for whom the threat of violence and danger is a marginal rather than a central element of life face the challenge of raising children who will be appropriately trusting and confident but at the same time adequately cautious. The following guidelines may be helpful in advising parents in such circumstances (Viorst, 1981):

1. Children can be protected and can protect themselves from harm by using good sense and taking reasonable precautions, such as going out in a group rather than alone at risky times or in risky places, and not seeking or accepting rides from strangers.

2. Children should be taught the wisdom of avoiding fights with others who are bigger, who possess deadly weapons, or who are known for not fighting fairly.

3. Children can be taught that helping people and reaching out to them is important and can be done without taking foolish risks, such as picking up hitchhikers or letting a stranger into the house at night to use the telephone.

4. Neighborhood safety is enhanced by neighbors getting to know each other and establishing informal networks where people watch out for each other, especially where adults watch out for children.

TELEVISION

The effects of television on children have been extensively researched, especially over the past decade. Most of the research has focused on several circumscribed components of television content and their impact on certain behaviors and attitudes. In order to gain a more complete picture of the impact of television on the lives of children, it is important to examine the documented and apparent effects in relation to one another. The research findings point to the pervasive influence of television on children and suggest mutually reinforcing effects. While television is an effective conveyor of many kinds of useful information, heavy television viewing seems to have preponderantly negative effects in terms of child development (Comstock et al., 1978; Moody, 1980; Rubinstein, 1978 and 1980; Winn, 1977).

Child health professionals, in their role as advisers to children, parents, and community groups, are in an excellent position to help ameliorate television's negative influences, although there are formidable obstacles to bringing about change in children's viewing practices, in how they respond to what they see, and in what is offered on television.

Most of the early concerns about the effects of television on children centered on televised violence. The Eisenhower Commission on the Causes and Prevention of Violence, the Surgeon General's Advisory Committee, and the major academic re-searchers in the field all concluded on the basis of the evidence available in the late 1960's and early 1970's that there is a causal relationship between viewing television violence and aggressive behavior. In 1976, the AMA identified television violence as an "environmental hazard threatening the health and welfare of young Americans" (Rubinstein, 1978).

Perhaps the single most important additional theme in the burgeoning body of research that has been done since then is that television affects children in more ways and more profoundly than had been realized previously, and that television must now be considered a major agent—perhaps *the* major agent— in the socialization of children (Comstock, 1978 and 1980; Rubinstein, 1978).

Television as an Agent of Socialization

The power of television in the socialization of American children arises from its pervasiveness and its replacement of other activities that in the past served as the primary sources of information and experience about the world and about how the growing and developing child relates to it. The influence of television is weaker in those homes and among those children where family, school, and community relationships are strongest. When other influences are missing, television is most likely to fill in (Comstock, 1980 [c]; Moody, 1980).

Pervasiveness of Television

Ninety-eight per cent of American homes have at least one television set (46 per cent have more than one). Unlike most other mass media, television is always on call to most children. Access is not dependent on reading ability, mobility, or ability to pay. Thus children have greater control— and at an earlier age—over when, how much, and what television they watch than over what they read, hear from adults, or see at the movies.

Television watching became a common activity in American homes in the 1950's. Between 1950 and 1960 the number of sets in use in this country jumped from 4 million to 53 million. Television viewing has been steadily increasing in the intervening years, but children's viewing may actually have peaked in 1980.*

The average American child watches between three and four hours of television each day. When

---

*The 1981 Nielson Report on Television is the first to report a decrease in average hours of television viewed by children; a small decrease between 1979 and 1980 occurred in each age category.

At the same time, children born in the 1980's will be the first whose parents also grew up with television as a major developmental influence—with consequences as yet unknown.

the majority of American children reach high school, no other single waking activity—not school, friends, or family—has occupied as much time as does television (Rothenberg, 1975; Rubinstein, 1978; Select Panel, Vol. I, 1981).

Television may enter an infant's life very early. The mother or other caretaker may watch television while feeding the baby; as early as three months of age, infants seem to be drawn to light and movement—especially when they are in color—and sound coming from the set, and they have been observed to watch the screen in preference to the mother's face or their own hands or feet. Regular television viewing increases most rapidly during the third year of life and is greatest just before the beginning of elementary school (Slaby and Quarforth, 1980).

In 1980, the average preschooler watched over 29 hours of television per week, 6- to 11-year-olds watched 25 hours weekly, and teenagers watched for 23 hours (Nielson, 1981). These averages, of course, obscure a wide range of viewing practices among individual children and households. There are homes entirely without television (2%), and homes where the set is on constantly. One study of inner city families that included a sixth-grade child found that in 35 per cent of the homes, the television set is on virtually all the time (Medrich, 1979). In another study, 25 per cent of sixth grade children reported seeing no television on a given weekday, while an equal number watched more than 5 1/2 hours. Children from lower social status homes are found to watch more television than children from higher status homes (Slaby and Quarforth, 1980).

Replacement and Alteration of Other Activities by Television

As television has assumed its present prominent place in the lives of most American children, time devoted to other activities has diminished. Studies have shown that television watching has reduced time devoted primarily to play, sleeping, conversations among family members, care of children by adults, unstructured outdoor activities, reading by children, by adults, and by adults to children, going to the movies, listening to the radio, participating in social gatherings away from home, and reflection (Comstock, 1980[a]; Slaby and Quarforth, 1980). Underlying these changes in routine activities are three themes of major significance: An unknown amount of the learning that used to occur through direct experience now occurs vicariously, less of the information and stimuli that shape children's understandings, behavior, and attitudes are under the control of the family, and

television itself has become a major factor around which family activities revolve (Comstock, 1982).

Vicarious Learning.   Research has not focused on this aspect of television viewing, but it is reasonable to assume that some aspects of development will be affected if several hours of a preschooler's day are spent in watching television rather than in going up and down a slide, cutting and pasting colored shapes, splashing or pouring water, taking apart and putting together toys, building and demolishing block towers, digging in a sandbox, or playing peek-a-boo with a passing adult. The school-age child also is exposed to information, attitudes, relationships, and behaviors the validity and usefulness of which he or she may not be able to test against real-world experience.

Diminished Family Control.   Since television is accessible at the turn of a knob and captures children's attention for long periods of time, and because most children live in homes where there is scant supervision or restriction over what and how much television is viewed, children are exposed to messages of many kinds that differ from those their families may intend them to receive. Children used to receive a higher proportion of their information about the world through institutions (the family, school, and home) that, at least in theory, were committed to the welfare of children and the perpetuation of certain agreed-upon values. Most television, on the other hand—both programming and commercials—is designed for other purposes.

Children are exposed to commercials promoting consumption patterns and products that parents may disapprove of or cannot afford; in a recent national survey parents reported that requests from children for products they see advertised on television is one of the most common problems of child-rearing. News and entertainment programs portray criminal, sexual, or moral conduct from which children form impressions that may differ markedly from reality and from the beliefs of the caring adults in their lives. In many homes, television has become the not always welcome substitute for the family as the major transmitter of culture. An increasing numbers of families own more than one television set, more and more children look at television alone. The results seem to be a further privatization of experience and a further lessening of the chances that an adult will be around to help the child interpret what he or she has seen, heard, and felt.

Dominance of Television in the Home.   In many households, television has become the focus of activities, the organizing principle, often replacing other rituals that formerly determined meal times and gave shape to family activities. A birthday celebration may take second place to a TV special; an adult's response to a child's discovery may have to await the end of a thrilling TV drama. Family dinners may focus on the television set to the exclusion of an exchange of experience among family members.

As summarized by Professor Urie Bronfenbrenner, "The primary danger of the TV screen lies not so much in the behavior its produces as on the behavior it prevents—the talks, the games, the family activities and the arguments through which much of a child's learning takes place, and his or her character is formed" (Bronfenbrenner, 1979).

Effects of Television Viewing

1. *Increased aggressive behavior and acceptance of violence.* For more than a quarter of a century social scientists have been studying the effects of television violence on children. While there are clearly differences in how the portrayal of violence affects individual children, depending on how much of it they see, the circumstances in which they see it, and individual personality characteristics, a number of generalizations can safely be made:

Viewing of television violence by children is likely to

- result in an increase in aggressive behavior
- blunt their emotional reaction and sense of outrage at violence and wanton destruction
- reduce their willingness to intervene when fellow human beings are in danger
- increase the level of acceptability of aggressive and antisocial behavior in others (Table 17–3).

These findings emerge from studies done in the United States and elsewhere, in the laboratory and in natural surroundings. Results obtained from a wide variety of studies have been extraordinarily consistent (Comstock, 1978 and 1980; Rubinstein, 1978; NIMH, 1982).

The Eisenhower Commission on the Causes and Prevention of Violence concluded that "a constant diet of violence on television has an adverse effect on human character and attitudes. Violence on television encourages violent forms of behavior and fosters moral and social values in daily life which are unacceptable in a civilized society. We do not suggest that television is a principal cause of violence in our society. We do suggest that it is a contributing factor."

The Commission's "Task Force on Mass Media and Violence" pointed out that, as a short-range effect, those who see violent acts portrayed learn to perform them and may imitate them in a similar situation, and, as a long-term effect, exposure to media violence "socializes audiences into the norms, attitudes and values for violence."

By 1972, the Surgeon General had assembled the evidence that led him to declare "that a causal relationship has been shown between violence viewing and aggression." Since then, considerable additional confirming data have been accumulated. In a recent review of the evidence, Professor Eli Rubinstein—vice-chairman and program director of the Surgeon General's Advisory Committee from 1969 to 1972—concluded, "The original question has been answered: TV violence is harmful to the viewer." (Rubinstein, 1980[b]). An update of the Surgeon General's 1972 report, published ten years later, also found that research conducted over the last decade confirmed the causal link between televised violence and later aggressive behavior (NIMH, 1982).

The replacement of explicit violence on some TV shows, which may have occurred in response to protest and pressures from citizen and professional groups, may, ironically, have made the effects on children even more destructive. On many television programs today violence has been made to seem antiseptic, bloodless, painless, and altogether unreal. A study of 73 hours of prime-time television done for the U.S. Conference of Mayors concluded that the constant suggestion of violence, with the victims dying quietly and out of camera range, may contribute to a further desensitization of the public to actual suffering and the pain of violence (Higgins and Ray, 1978).

2. *Difficulty in distinguishing between fantasy and reality.* Closely related to the possibility that television watching contributes to desensitizing children to violence, aggression, and antisocial acts is the evidence that television watching seems to contribute to obscuring the distinction between fantasy and reality.

Even television news, existing within an entertainment environment, seeking to present facts with the tools of fantasy, ends up with a dramatized version of life, a more exciting allegory of events.

A gradual erosion of the line between fact and fantasy, news and theater, may have the effect of making people slow to react to accidents, muggings, and other forms of harm inflicted on others.

Table 17–3. EFFECTS OF HEAVY TELEVISION WATCHING

1. Increased aggressive behavior and acceptance of violence.
2. Difficulty in distinguishing between fantasy and reality.
3. Distorted perceptions of reality in relation to importance of consumption of products and services, extent of violence, and role of minorities.
4. Trivialization of sex and sexuality.
5. Increased passivity and disengagement.
6. Negative effects on cognitive learning.
7. Potential to inform, teach, and promote "prosocial" behavior not fully realized.

They may be wondering whether these things are really happening. Young children who are in the process of learning how to tell the difference between reality and make-believe are particularly vulnerable.

The possibility cannot be ruled out that an important factor in the high incidence of crime at increasingly earlier ages is the reduction in awareness of precisely what is real—to which television watching may make a significant contribution. Professionals frequently comment on this aspect of their impressions in talking with young people convicted of violent crime. In the words of one district attorney, "They don't see the person they shoot as a real live person. It's like a body they shot at, not an identifiable person" (CBS Reports, 1981).

3. *Distortions of the world as it is.* Television presents the viewer with a world quite different from the one that most Americans inhabit. This is a particular problem for children, whose knowledge of the real world and ability to test the television version against reality are still severely limited.

Little research has directly addressed the question of the effects on children of television entertainment that seems to suggest that life as it should be lived includes fast cars and boats, fancy restaurants, flashy women, iron-nerved men, lucky guesses, and powerful occult forces—especially when these images contrast with an almost total absence on the screen of people working hard and long on a difficult task, thinking carefully and rationally about complex social or intellectual problems, or caring for and being supportive of family, friends, and colleagues over long periods of time.

Two conclusions researchers have been able to draw from analyses of the content of both television advertising and programming are that:

a. Young children are easy to persuade not only of the desirability of products that may in fact run counter to their welfare, but also of the fundamental message of the estimated 20,000 commercials they see each year: that fulfillment and happiness are to be derived primarily from the consumption of products and services (Comstock, 1980[c]).

b. Such groups as women, ethnic minorities, lower-class people, foreigners, handicapped people, old people—and even children—tend to be under-represented or misrepresented in most television programming; television characterizations of people and interactions tend to be exceedingly narrow, with Hispanics (who are rapidly becoming the largest minority in America) absent or hard to find; women pictured as insignificant, dependent, and rarely working outside the home; children often failing at the things they attempt to do; and crime occurring ten times more often than it does on the streets (Rubinstein, 1978).

Social scientists have speculated that stereotyped portrayals of people and events on television may play a major role in shaping social attitudes and behaviors of children; most of the corroborating evidence for this hypothesis has come from tests of the broader hypotheses that the more television children watch, the more closely their attitudes reflect the general television stereotypes. For example, heavy viewers seem to identify more with traditional sex roles. They also see the world in a more sinister light than do those who watch less television. An extensive interview survey of children between ages 7 and 11 found that those who were heavy viewers were significantly more likely to be very fearful than was a comparable sample of children (Zill, 1982).

4. *Education about sexuality.* Sex is an important part of television, not so much because television portrays explicit sexual acts but because sexual relationships and innuendoes permeate many of the situation, comedy, and variety shows. Sexuality is often trivialized, and sex tends to be treated in a casual and humorous fashion or is shown as a weapon to be used by both men and women to gain control over others.

Studies show television to be a significant source of information about atypical sexual behavior, including homosexuality and prostitution, although it does not seem to be an important source of information on biological, health-related, and romantic aspects of sexuality. Rubinstein points out that since 1975 there has been a considerable increase in the use on TV of verbal innuendoes, portrayal or reference to casual sexual behavior, and references to socially discouraged sexual practices. With television's emphasis on youth, good looks, action, and conflict, enduring human relationships built on warmth, affection, reliability, and mutual understanding are rarely shown, and sex is rarely explored as a serious and meaningful aspect of the lives and relationships of television characters (Rubinstein, 1980).

5. *Passivity and disengagement.* A number of social scientists—and many concerned parents—have come to believe that heavy television watching, especially among young children, encourages a passive, disengaged stance toward the world (Moody, 1980; Winn, 1977).

Dr. T. Berry Brazelton has concluded that "television creates an environment that assaults and overwhelms the (child): he can respond to it only by bringing into play his shutdown mechanism, and thus becomes more passive." Many parents are concerned about the glazed eyes and near trances they see their children in as they watch television; many have witnessed what Brazelton describes as the "hooked" quiet that characterizes children watching a series of horrors on their television set (Brazelton, 1972).

Whether the passive nature of television watching has long-term effects is still a matter for speculation. Little systematic evidence is available, although there are studies which point to the association between heavy viewing and a diminished ability to play with friends, fewer hobbies, and less involvement in activities outside the home. However, cause and effect in this instance are difficult to disentangle. The comments of observers who work with children today and also did so before the television era suggest both the need for more research and the possibility of significant effects.

Many nursery school teachers believe that children's play has changed as a result of television. They note that there is not as much dramatic play as there used to be; that children show less imagination and less interest in games in which they need to take an active part. Elementary school teachers comment that children increasingly expect to be entertained at school; that they passively sit back and wait for the teacher to initiate activities; that they behave as though they will simply "turn the channel" if they are asked to do something that is not immediately gratifying and requires effort (Moody, 1980).

The physical aspects of passive viewing are also of concern. When television viewing replaces time spent in physical activity and is coupled with consumption of the candy and sugared cereals advertised on television, the detrimental consequences for children's physical well-being—including especially obesity and poor general fitness—are interactive and often serious (Richmond, 1981).

6. *Negative effects on cognitive learning.* There is some evidence, though as yet inconclusive, that heavy television watching may have negative effects on cognitive learning ability and motivation, school performance, and the acquisition of reading and other specific skills. The mechanisms by which these effects are brought about are unclear; in addition to the factors discussed previously, they may include the following:

• The effect of rapid shifts in visual images that characterize most television programming may reduce children's attention span.

• The sensory overload inherent in a high proportion of television sound and images may interfere with thoughtful processing of information and with the absorption of information in ways that make it subsequently useful.

• The fact that television viewing on normal-size screens calls for little or no eye movement has been implicated by some as a cause of reading problems in children who were heavy watchers in the preschool years.

7. *Potential to inform, teach, and promote "prosocial" behavior not fully realized.* The actual and potential capacity of television to inform, to teach, and to promote "prosocial" behavior, must be noted in order to evaluate the total effect of television on the developing child.

While there are disputes about the precise nature of learning from television, there is no question that the body of information that most children have today is vastly greater than in previous generations, largely as a result of learning from television. Pictures, sounds, and vivid impressions that were previously outside the experience of all but a very small number of children are today a part of the daily lives of a majority of children.

There is little systematic description of the effects on children of obtaining ready access through television to an almost infinite variety of literature, scientific discoveries, historical and biographical information, music, dance, theatre, and other aspects of our common heritage. There is, however, documented evidence that among adolescents a positive relationship exists between knowledge of public affairs and viewing public affairs programs on television and between attention to television news and comprehension of various aspects of government.

With regard to programs specifically designed to teach, preschool viewers of Sesame Street have shown significantly improved recognition of letters and numbers, although little acquisition of reasoning skills. Viewers of Electric Company improved in the skills specifically taught by the program, although not in a test of general reading skills. There is some evidence that Sesame Street has been more successful among children from families of higher socioeconomic status than for those from low-income and black families and that some of Sesame Street's success seems to be dependent on adult involvement and encouragement. Nevertheless, the ability of television to provide information and to teach certain skills is now undisputed.

In addition, it has been demonstrated that appropriate television portrayals can encourage socially desirable and constructive behavior. Children have been taught through television dramas to be less fearful of dogs, dentists, and other children, and programs like Mr. Rogers' Neighborhood, Sesame Street and Fat Albert and the Cosby Kids have been able to increase understanding and knowledge about positive kinds of behavior and to stimulate such behavior when the televised act had at least some similarity to a subsequent real-life opportunity (Comstock, 1980[c]).

Prospects for Change

Substantial and positive changes in how television affects children can be brought about by the efforts of parents, acting individually and jointly. Health professionals can play a critical role in assisting parents to evaluate and act on what is known about the effects of television on children.

There are many ways in which health professionals can help families to gain some control over the influence of television in their children's lives. Parents can be helped to act individually with regard to their own children; they can also be encouraged to act together, since it is easier and more effective to establish limits and patterns that are shared by other families in the neighborhood, particularly by the children's friends and peers, and that are reinforced by the institutions—especially schools—with which the children come into contact.

Most studies indicate that the majority of parents put few restrictions on their children's television viewing. In one study of first-grade children, 70 per cent of the mothers reported that they never restrict the amount of time their children may spend watching television, and 30 per cent said they never restrict the types of programs their children watch. In a study of teenagers, only 10 per cent of families report viewing rules of any kind. Furthermore, surveys indicate that children of all ages report less than half the levels of control reported by their parents (Slaby and Quarforth, 1980).

What Parents Can Do. Many families have found it possible to gain control over the influence of television in their lives, especially when parents have joined together with PTA, school, or neighborhood (Table 17–4). Some of the most promising and effective steps include the following:

1. *Limiting the time spent viewing.* These limits may involve establishing a maximum number of hours per day or week or ruling out watching at certain times of the day or week. Especially in the case of school-age children, the limiting of television watching to weekends seems to be a workable alternative for many families. Some educators suggest that available evidence warrants a complete ban on television watching by preschoolers. Families who don't wish to give up television entirely during these years may find such a step impractical and find a workable compromise in restricting watching by preschoolers to an hour a day.

2. *Restricting the specific programs that children may watch.* Some families insist that television be used

Table 17–4. FAMILY ACTION TO PROTECT AGAINST DETRIMENTAL EFFECTS OF TELEVISION

1. Establish limits on the amount of time spent watching, by day or week, and on circumstances of viewing (e.g., no viewing at mealtimes, no private set in child's room).
2. Avoid "random viewing" and supervise or consult concerning choice of programs.
3. Provide and encourage activities other than television viewing.
4. Watch television with children.
5. Teach children to watch critically.

only to watch specifically selected programs that may be jointly decided upon on a daily or weekly basis in order to avoid "random viewing." Others find it more feasible to ban specific programs that they feel may be especially damaging or unsuitable.

3. *Encouraging alternatives to television viewing.* Often active efforts to offer children attractive alternatives to television watching, especially in the preschool years, result in children's developing skills, activity patterns, and interests that they find more rewarding than watching television. Parents can encourage regular physical exercise, solitary and social play, and other activities that increase the child's direct experience with the world. Reading aloud—especially to preschoolers—is another important alternative to television watching that parents and other adults can provide.

4. *Providing adult company and commentary to children while they are watching television.* When adults watch along with children, the children are more likely to learn the cognitive material presented, to reject or not act on the aggressive behavior presented when the adult makes a disapproving interpretive comment, to understand the nature and intent of the commercials they see, and to gain additional perspective on the implicit and explicit messages offered by television.

5. *Teaching children "critical viewing skills."* When parents watch television with their children, and discuss what they see, parents can impart a sense of the discrepancies between what is shown and their own values. They can teach their children about the commercial purposes of television and an analytical posture toward the messages presented in commercials and entertainment as well as in news programs. Parents—and teachers—can teach older children something of the known effects of too much and inappropriate television viewing and enlist their own interest in avoiding the adverse effects.

Constraints on Individual Action. In working with parents in this realm, it is essential to keep in mind the constraints on their ability to exercise the kinds of control they might wish to exert. These constraints include the following:

- Lack of adult time and energy to engage in many of the recommended activities.
- Needs of adults for the respite from children's demands and from the need for adult supervision that television often represents.
- The function of television in many families as providing an escape from a generally sad and dreary existence.

These constraints are likely to represent the greatest deterrent to change among those families where television already exercises its most profound—and perhaps most negative—influence. Some students of the problem have suggested that the greatest susceptibility to the detrimental effects of television occurs in populations characterized by extreme poverty, educational limitations, or psychological disturbance within the family—precisely those families where self-regulation is likely to be most difficult.

This means that community action to improve television and to ameliorate its harmful effects on children is essential. CBS newsman Edward R. Murrow said of television that it can teach and that it can illuminate, but that "it can do so only to the extent that humans are determined to use it to those ends." This determination cannot effectively be exercised by isolated individuals.

While changes in television programming and advertising policies that could be brought about through more active governmental regulation or other forms of governmental intervention are unlikely in the near future, professionals and citizens can work together, especially through such organizations as Action for Children's Television and the National Council for Children and Television, to improve television content, discourage advertising aimed at preschool children, and bring about greater diversity of programming.

Significant changes, the impact of which is as yet difficult to predict, are certain to result from new technological developments. The advent of cable television, pay television, videodiscs, and interactive systems may mean extended viewing, more solitary viewing, less advertising, increasing differences in what the rich and the poor see on television sets and how they use them, and an abundance of choices—which may or may not mean greater diversity.

Concerned parents, educators, and health professionals can obtain more information and help in their efforts to minimize the harmful effects and maximize the positive potential of television for the children in their families and communities from the following and from the references listed at the end of this chapter, particularly those designated by an asterisk.

Action for Children's Television
46 Austin Street
Newtonville, MA 02160
  (and local affiliates of A.C.T.)

National Council for Children and Television
20 Nassau Street
Suite 215
Princeton, NJ 08540

LISBETH BAMBERGER SCHORR

REFERENCES

Brazelton, T. B.: How to tame the TV monster. Redbook, April, 1972.

Bronfenbrenner, U.: The Ecology of Human Development. Cambridge, Massachusetts, Harvard University Press, 1979.

Burquest, B.: "The Search for Structure," presented at the Central States Conference, American Society for Adolescent Psychiatry, 1981.

CBS Reports: Murder Teenage Style, September 3, 1981.

Coles, R.: Privileged Ones, The Well-Off and The Rich in America. Vol. V of Children in Crisis. Boston, Atlantic–Little, Brown Books, 1977.

Comstock, G.: Television and American Social Institutions. S.I. Newhouse School of Public Communication, Syracuse University, 1980(a).

*Comstock, G.: Television in America. Beverly Hills, California, Sage Publications, 1980(b).

Comstock, G.: Television Entertainment: Taking It Seriously. Character, Vol. 1, No. 12, October, 1980(c).

Comstock, G.: Television and American social institutions. In Pearl, D., et al. (eds.): Television and Behavior. Rockville, Maryland, National Institute of Mental Health, 1982.

Comstock, G., Chaffee, S., Katzman, N., and Roberts, D.: Television and Human Behavior. New York, Columbia University Press, 1978.

Edelman, M. W.: Portrait of Inequality: Black and White Children in America. Washington, D.C., Children's Defense Fund, 1980.

Egbuonu, L., and Starfield, B.: Child health and social status. J. Pediat. 69:350–357, 1982.

Escalona, S. K.: Children in a warring world. Am. J. Orthopsychiatry 45:765, 1975.

Grinker, R. R., Jr.: "The poor rich: The children of the super-rich." Am. J. Psychiat. 135:913, 1978.

Hetherington, E. M., and Parke, R. D.: Child Psychology: A Contemporary Viewpoint. 2nd ed. New York, McGraw-Hill Book Co., 1980.

Higgins, P., and Ray, M.: Television's Action Arsenal: Weapon Use in Prime Time. Washington, D.C., U. S. Conference of Mayors, 1978.

Kessner, D. M.: Infant Death: An Analysis by Maternal Risk and Health Care. Washington, D.C., Institute of Medicine. National Academy of Sciences, 1973.

Medrich, E. A.: Constant television: A background to daily life. J. Commun. 29:171–176, 1979.

*Moody, K.: Growing Up on Television. New York, Times Books, 1980.

*Murray, J. P., and Lonnborg, B.: Children and Television: A Primer for Parents. The Boys Town Center, Boys Town, Nebraska, 1981.

National Institute of Mental Health: Television and Behavior. Rockville, Maryland, NIMH, 1982, p. 37.

Nielson, Report on Television, 1981, quoted in ReAct, Action for Children's Television News Magazine, Spring/Summer, Vol. 10, Nos. 3 and 4, 1981.

Richmond, D. A.: Personal communication, 1981.

Rothenberg, M. B.: Effect of television violence on children and youth. J.A.M.A. 234:143–146, 1975.

Rubinstein, E. A.: Television and the young viewer. Am. Scientist 66:685–693, 1978.

Rubinstein, E. A.: Television as a sex educator. In Brown, L. (ed.): Sex Education. New York, Plenum Publishing Corporation, 1980(a).

Rubinstein, E. A.: Television violence: A historical perspective. In Palmer, E., and Dorr, A. (eds.): Children and the Faces of Television: Teaching, Violence and Selling. New York, Academic Press, 1980(b), pp. 113–127.

Rudov, M., and Santangelo, N.: Health Status of Minorities and Low-

---

*Suggested readings for parents on television viewing.

Income Groups. Bethesda, U.S. Department of Health, Education, and Welfare, DHEW Publication No. (HRA) 79-627, 1979.

Schaefer, E. S.: Professional support for family care of children. *In* Wallace, H. M., Gold, E. M., and Oglesby, A. C. (eds.): Maternal and Child Health Practices. 2nd ed. New York, John Wiley and Sons, 1982.

Schorr, L. B., Lazarus, W., and Weitz, J. H.: Doctors and Dollars are Not Enough. Washington, D.C., Children's Defense Fund, 1976.

Select Panel for the Promotion of Child Health: Better Health for Our Children: A National Strategy. Vols. I, II, and III. Washington, D.C., U.S. Department of Health and Human Services, 1981.

*Singer, D., Singer, J., and Zuckerman, D.: Teaching Television: How to Use TV to Your Child's Advantage. New York, Dial Press, 1981.

Slaby, R. G., and Quarforth, G. R.: Effects of television on the developing child. *In* Advances in Behavioral Pediatrics. Vol. 1. JAI Press, Inc., 1980, p. 225.

---

*Suggested readings for parents on television viewing.

Sorel, E.: Personal communication, 1981.

Starfield, B.: Poverty, Ill Health, and Medical Care. Presented at the Annual Meeting of the Ambulatory Pediatric Association, 1981.

Starfield, B., and Pless, I. B.: Physical health. *In* Brim, O. G., and Kagan, J. (eds.): Constancy and Change in Human Development. Cambridge, Massachusetts, Harvard University Press, 1980, pp. 272–324.

Viorst, J.: Raising children in a dangerous world. Redbook *157*:48, 1981.

Werner, E. E., and Smith, R. S.: Vulnerable but Invincible. New York, McGraw-Hill Book Co., 1982.

*Winn, M.: The Plug-in Drug. New York, Bantam Books, 1977.

Wixen, B.: Children of the Rich. New York, Crown Publishers, Inc., 1973.

Wixen, B.: Children of the Rich. *In* Noshpitz, J. D. (ed.): Basic Handbook of Child Psychiatry. New York, Basic Books, 1979.

Zill, N.: American Children: Happy, Healthy and Insecure. New York, Doubleday (Anchor Press), 1982.

IV

Biologic Influences

The 10 chapters in this section offer an overview of health related stresses that can impair development and behavioral adaptation. Included are surveys of hereditary, genetic, and dysmorphic phenomena that may alter function in childhood. Specific influences in pregnancy, labor, and delivery are considered as well as the negative effects of postnatal infections, nutritional problems, environmental toxins, and central nervous system trauma. There is general consideration of the effects of illness on behavior and development, with specific examination of some common and important prototypes of mind-body dissonance, including otitis media, bronchial asthma, gastrointestinal diseases, and chronic illness in general. The final chapter in this part explores the effects of terminal illness on children and families.

18

Heredity, Development, and Behavior*

Heredity is the process in nature by which a degree of resemblance between parents and their offspring results. The recognition of any similarity between individuals, however, also implies an appreciation of differences. The opportunity to watch and study the behavior of the developing child is a challenging privilege offered to us by nature. Behavior is defined in this chapter as the set of personality traits, intellectual and emotional qualities, and sensory responses in any social context. The concepts of differences and similarity relate more to distinct qualities or characters, such as cognitive and personality traits as well as physical features, than to whole organisms.

In human development two contrasting features can be recognized. On the one hand there is a constancy in psychomotor and mental development from infancy to childhood, and on the other hand variability and features of individuality emerge and become important early. Thus in every child's development and behavior, mankind as well as an individual can be observed. Behavior is the result of an intensive and constant interaction between heredity and environment. If this premise is accepted one should not be sidetracked by any discussion of the sterile heredity-environmental controversy still plaguing the field of behavioral science.

Because genetics is specifically concerned with heritable variation in nature, human behavior becomes one of its natural topics of interest. Moreover, the geneticist and the behavioral scientist alike have sociopolitical motivations for studying behavior, particularly when its phenotypic expression assumes the form of a learning disability, maladaptive personality, progressive neurologic disease, psychiatric abnormality, retarded psychomotor development, or mental retardation. Society itself asks them to become involved in the formulation of well founded recommendations re-

garding management, treatment, or prevention. Thus any "exposé" on the influences of heredity on mental and behavioral development in childhood should contain a didactic presentation of the relevant principles of formal genetics and clinical examples to illustrate them. In the first part particular attention will be paid to monogenic and polygenic inheritance only, because the genetics of chromosomes pertaining to behavioral traits is presented in the next chapter. Examples of how genes and genotypes affect child development are treated in the second part and principles of management and prevention in the third part of this chapter.

ELEMENTS OF FORMAL GENETICS

Mendel discovered the fundamental laws of hereditary transmission of single traits through the study of several variant characters in the garden pea. More than two decades before the discovery of chromosomes and their behavior during meiosis, he postulated the existence of particulate hereditary determinants or genes governing the appearance of each inherited character observed through several generations. Ova and spermatozoa, the products of meiotic division, form the connection between generations of people. The number of chromosomes in each gamete is only half of that in somatic cells. Genes are linearly arranged along chromosomes. This important fact in biology was established well before nucleic acids, and deoxyribonucleic acid in particular, had been identified as the chemical substance of the genetic material within chromosomes.

MONOGENIC OR MENDELIAN INHERITANCE

Principles and Definitions

Hereditary characters or phenotypic traits are determined by genes. In man, as in all higher

*Some of the laboratory results referred to were obtained with financial support of the Belgian Department of Public Health, Anthropogenetics Section, and by grant 3.0001.81 from the Fonds Geneeskundig Wetenschappelijk Onderzoek, Brussels, Belgium. ·

organisms, genes occur in pairs. At the time of gamete formation (spermatogenesis or oögenesis) the members of each gene pair, called allelic genes or alleles, segregate so that only one of them is included in a gamete. At fertilization the gene pairs are restored in the zygote. If both alleles are identical in the zygote, the latter is termed homozygous or a homozygote. If the alleles are different, the organism is called heterozygous or a heterozygote. Differences between alleles have arisen by mutation. An individual genetic make-up or constitution is called a genotype. Depending on whether the clinical expression of a mutant allele occurs in the heterozygous genotype or only in the homozygous one, the resulting phenotype or phenotypic trait is said to be dominant or recessive, respectively. The latter adjectives relate to phenotype, not to genetic constitution or genotype. The site of a gene on a chromosome is often referred to as a locus. Genes can be located on a sex chromosome or an autosome, one of the chromosomes not directly involved in sex determination. Obviously in monogenic inheritance, only alleles at a single gene locus are under consideration. The segregation of a pair of alleles is independent of any other pair during gamete formation, except when two loci are closely linked on the same chromosome.

Like Mendel, who formulated the laws of monogenic inheritance by observing the transmission of single, sharply contrasting phenotypic traits, the present day geneticist demonstrates the importance of the genotype more easily by studying behavioral traits or patterns of development unequivocally distinct from the human norm than by observations of the continuous variability of normal human behavior.

Pedigree Analysis

Autosomal Dominant Inheritance. Inadequate psychomotor development, mental deficiency, and unusual behavior, like any other phenotypic feature, can be components of autosomal dominant disorders in children and adults who by definition are heterozygotes for a mutant gene. Dominant inheritance is easily recognized in pedigrees, because affected individuals, female and male, occur in every generation (panel 1b of Figure 18–1). If one parent is affected, there is an a priori recurrence risk of one-half in the offspring, as can readily be seen from the illustration in panel 1a of Figure 18–1. Any easily recognized dominant defect, however, must be mild enough not to interfere with reproduction. Practical difficulties in identifying autosomal dominant inheritance can arise because of either new mutation (see page 319) or nonpenetrance (see page 326).

Among the better known examples of autosomal dominant disorders possibly or usually associated with learning disability, major mental deficiency, or behavioral abnormality are the following clinical syndromes: von Recklinghausen neurofibromatosis, tuberous sclerosis, Huntington chorea, myotonic dystrophy, and periodic paralysis.

Patients with neurofibromatosis may have a number of rather large, light brown spots located predominantly on nonexposed areas of the skin. These are known clinically as café au lait spots. From childhood on, any number of slowly growing neurofibromatous tumors can occur subcutaneously along the course of and connected with peripheral nerves. In the majority of patients, neurofibromas are primarily a cosmetic problem, which is sometimes compounded by the need for surgical excision. Neurofibromatosis is usually compatible with a normal life and reproduction. Unfortunately, in about 20 per cent of the patients, major clinical complications arise because virtually any part of the central and peripheral nervous systems can be involved. The possible complications are extremely varied. Among the most relevant to this discussion are seizures, and tumors of the cranial nerves, brain, or spinal cord. Often at the time medical help is solicited, the complaint relates to behavioral problems. Some patients with neurofibromatosis have only a mental handicap in addition to the cutaneous features of the disorder.

Tuberous sclerosis, or Bourneville disease, is also a multisystem disease. Among several types of skin lesions, perinasal adenoma sebaceum when present is the most helpful diagnostic feature. Seizures, often refractory to treatment, and a moderate to severe mental handicap are frequently components of the condition. Nevertheless this usually slowly progressive disorder occasionally may be compatible with normal intelligence, reproduction, and long life.

Several disorders are termed myotonia. They are characterized primarily by a difficulty in relaxing contracted muscles, noticeable in the jaw or hands. These signs of myotonia in myotonic dystrophy are soon followed by generalized muscle wasting and weakness that are apparent early in the patient's expressionless facies. Extramuscular features in this disorder include cataract, hypogonadism, and frontal balding in males. The life expectancy in myotonic dystrophy is reduced, probably because congestive heart failure and mental deterioration are often complicating elements in its natural course.

Recurrent attacks of minimal to severe flaccid paralysis, affecting either a single muscle group or all four limbs, occur in a group of disorders jointly called periodic paralysis. In one of these nosologic entities, attacks last from a few to almost 48 hours and are accompanied by a rise in the serum potassium level. In another disorder the attacks, of

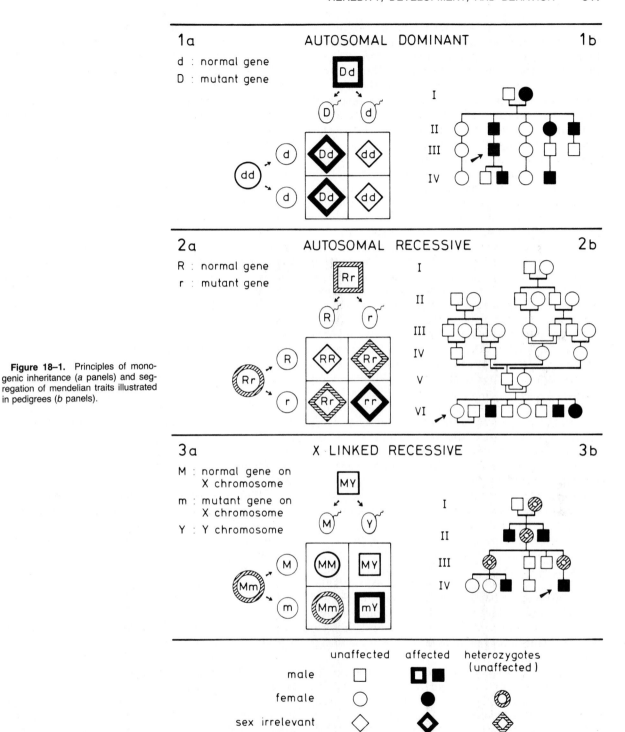

Figure 18–1. Principles of monogenic inheritance (*a* panels) and segregation of mendelian traits illustrated in pedigrees (*b* panels).

shorter duration, are accompanied by a fall in the serum potassium level. Between episodes of paralysis clinical examination reveals no physical abnormality in these patients.

Huntington chorea, an autosomal dominant disorder with a highly variable age of onset (usually during the fourth decade of life), is rarely encountered in childhood. Choreic movements and dementia are the major features in the fully devel-

oped disease, either one being of primary importance during the initial clinical stages.

Several types of hereditary blindness or deafness are also inherited as autosomal dominant traits. These conditions interfere with the usual pattern of a child's development but fortunately do not necessarily affect the final intellectual capabilities adversely.

From these examples it should be clear that

single genes can cause unusual or abnormal behavior in man, even though the pathogenesis of their phenotypic expression may remain largely unknown. They represent but a few illustrations of the fact that proper functioning of the human brain, like that of other organs, is the result of normal and coordinated activity of many genes.

Autosomal Recessive Inheritance. The recognition of autosomal recessive inheritance of a phenotypic trait in a pedigree is based on the following relevant findings (Figure 18–1, panel 2b): Both parents of the patient are clinically normal but genotypically heterozygous for the mutant gene; more than one affected individual is encountered in one sibship; males and females are affected with equal probability; consanguinity of the parents is not unusual; and the a priori recurrence risk for future affected children equals ¼ (Figure 18–1, panel 2). There are many different clinical examples of psychomotor developmental retardation due to homozygosity of a mutant gene. Not all types of intellectual disability with autosomal recessive inheritance, however, are clinically progressive. In many of them no inborn error of metabolism is demonstrable or even likely. One example is the trait segregating in the pedigree illustrated in Figure 18–1, panel 2b. Three affected siblings have oligophrenia, associated with cleft palate in only two of them. Apparently the responsible mutant gene had exerted its effect on brain as well as palatal development early in intrauterine life.

X Linked Inheritance. The X chromosome of man contains as many genes as the autosomes of comparable length. X linked recessive inheritance of a phenotypic trait can be deduced from the following features in pedigrees: If both parents are normal, only males are affected; affected nonsiblings are related to one another through the maternal line; a healthy heterozygous female has an a priori probability of ½ of having an affected son and an equal probability of having a heterozygous, though normal daughter, like herself (Fig. 18–1, panel 3a).

X linked dominant inheritance is of little practical importance in development or mental deficiency. The transmission of X linked dominant traits from affected females is identical to that in autosomal dominant inheritance. Affected males, however, can have affected daughters but no affected sons. Male to male transmission cannot be observed in X linked inheritance.

Among more than 100 nosologically defined X linked disorders, several can be cited as examples of how mutant genes affect human development and behavior (McKusick, 1978). The pedigree shown in Figure 18–1, panel 3b, illustrates a recently observed family including several people with a condition of complete androgen insensitivity known as the testicular feminization syndrome. Patients with this type of male pseudohermophroditism are phenotypic females with unmistakably female external genitalia, a blind ending vagina, and no uterus. Stature and body proportions are normal. At puberty the patients demonstrate breast development and puberal feminization. They are commonly detected because of primary amenorrhea or inguinal hernia, which sometimes contains tumor-like structures. The latter also may be found intra-abdominally and may be identified histologically as testes. Pubic and axillary hair is usually sparse. Patients with this condition are female in their psychosexual behavior, and no attempt should be made to influence their psychologic or social sex to the contrary. In genetic counseling, comments on finding a normal 46,XY karyotype should be kept to a minimum, but the monogenic nature of this disorder should be stressed in addition to the recurrence risks for siblings and children of unaffected female relatives. This androgen resistance syndrome is not due to a lack of testosterone or dihydrotestosterone but to the absence or inadequate function of an androgen receptor protein to which male hormone must bind for proper activity (Simpson 1976). This example serves to demonstrate how a single gene mutation not only can direct anatomic development but can also determine sexual differentiation of the central nervous system and the expression of gender related behavior (Naftolin, 1981).

If child development and behavior are viewed from a more purely physical standpoint, the Duchenne type of muscular dystrophy can be cited as an important example among the X linked hereditary disorders. (For discussion see page 338 in this chapter.)

Community surveys or data obtained from institutions indicate a definite excess of males among the mentally retarded. Analysis of the pedigrees of male probands with no major physical malformation regularly demonstrates a pattern of X linked recessive inheritance. Interest in this subject has recently been intensified by the discovery of an as yet unexplained association between a particular type of mental disability in males and the cytogenetic abnormality of a fragile site near the end of the long arm of the X chromosome. Mothers of these probands and other female relatives may show the same cytogenetic anomaly on one of their two X chromosomes in association with mild to moderate mental disability (see Chapter 19).

There are other types of mental retardation in males most likely due to mutant genes on the X chromosome and not associated with a cytogenetic abnormality (Herbst and Miller, 1980; Herbst et al., 1981). Some nosologic entities still in need of further clinical definition are listed in Table 18–1.

Table 18–1. TYPES OF "X LINKED" MENTAL DEFICIENCY*

	McKusick Catalog Number (McKusick, 1978)
1. Marker X chromosome syndrome† (fragile X syndrome, Martin-Bell syndrome)	30950
2. Macro-orchidism and mental deficiency without marker X chromosome	30957
3. Renpenning syndrome of mental deficiency (often small head and short stature)	30950
4. Mental deficiency and muscular atrophy type (Allan, Herndon, and Dudly)	30960

*No major physical anomalies associated. Pathogenesis unidentified.
†Sometimes mild, occasionally severe expression in heterozygous females.

Only clinical forms of X linked mental deficiency without biochemical or endocrinologic abnormalities or without any major associated physical anomalies are mentioned. Thus X linked hydrocephalus and possibly several forms of X linked microcephaly are not listed in that table.

The "Pedigree Isolated" Patient

Birth defects may represent either hereditary disorders or the effects of unidentified teratologic accidents, as may disorders of development or behavior in children. Environmental accidents of known etiologic importance in the latter instance include maternal factors, such as infection or other disease, medication, and irradiation during pregnancy; fetal factors including prematurity, complicated delivery, prolonged hypoxia, and cerebral trauma (see chapters in Part III of this book). A clinical condition due to environmental factors may mimic the phenotype of a hereditary disorder; it is then called a phenocopy.

The challenge of differentiating between the possibility of a hereditary or environmentally determined disorder is easily met if the condition occurs in several members of one family or larger kindred (familial, repeated, or multiplex occurrence). The differentiation, on the other hand, may be rather difficult when a single patient in one family or pedigree is affected (simplex occurrence). If a sufficient number of typical features of a syndrome can be recognized in such a patient, the mere establishment of the diagnosis may identify the genetic or environmental basis and permit formulation of a strong hypothesis on the risk of recurrence. The establishment of a genetic or nongenetic cause is nearly impossible in the pedigree isolated patient with a previously uncharacterized syndrome of malformations and even more diffi-

cult in the single parent with early appearing psychomotor deficiency unassociated with any physical anomaly.

The occurrence of a disorder in a sibship often but not always confirms its hereditary nature. On the other hand, the term "isolated occurrence" although it may be considered the opposite to familial occurrence, does not rule out a genetic basis for a disorder. Thus, a single involved patient in a sibship may be isolated by chance and still have a hereditary condition. Alternatively, such simplex occurrence may be due to a sporadic event of either a genetic or a nongenetic nature. In the case of a chance-isolated proband, the recurrence risk is large; in a sporadically occurring patient, it is negligibly small (Crow, 1965). A listing and a general assessment of recurrence risk in sibs and children of isolated propositi are provided in Table 18–2.

GENES AND PHENOTYPE

Basic Principle of Functional Genetics: "One Gene–One Enzyme"

Since the middle of this century it has been known that the genetic material in higher organisms is deoxyribonucleic acid, (DNA), the most important component of the nuclear chromatin in each cell. The genetic information in DNA can be considered to be written in a linear language composed of words of three nucleotides. The human genome contains many thousands of genes, only a minority of which have been identified. Genes can be recognized as separate genetic entities only after their expression or function has been altered by mutation.

How do genes govern the phenotype? The information in genes determines the linear sequence of amino acids in polypeptides in proteins, and thus their molecular structure. Most proteins function as enzymes. Each enzyme catalyzes at least one specific chemical reaction, which is usually only one in a series of reactions called a metabolic pathway. Laboratory studies using the bread mold *Neurospora crassa* experimentally demonstrated the biologic principle of "one gene–one enzyme" (Beadle and Tatum, 1941). Although modified as new knowledge has become available, this principle stands today. It applies also to man, as was first surmised by the English physician Garrod, who published his now classic work on inborn errors of human metabolism in 1909.

In Figure 18–2, part I, the principles of functional or biochemical genetics are illustrated. If a gene has been altered by mutation, one nucleotide within a particular DNA triplet, and thus within the corresponding mRNA codon, has been

Table 18-2. "SIMPLEX" OCCURRENCE OF A PROBAND

Class	Cause	Type	Recurrence Risk (First Degree Relatives) Sibs	Children
Chance-isolated propositus	Monogenic	Autosomal Dominant		1/2 × penetrance
		Autosomal Recessive	1/4	Small*
		X linked recessive (male propositus)	1/2 (males)†	Very small‡
	Chromosomal	Structural abnormality in one of parents (balanced translocation)	1/50→1/10§	Reproduction precluded
	Polygenic	"Gene complexes" suspected in both parents	1/10→1/20 (empiric risk based on population data)	
Sporadic propositus	Genetic disorder New point mutation	Autosomal dominant	Very small risk related to mutation frequency	1/2 × penetrance
		X-linked recessive (male propositus)	Very small	Very small
		Meiotic nondisjunction	Small; parental age dependent	1/2 (usually reproduction precluded)
	Chromosomal	Mitotic loss of chromosome	Very small	Small
		Chromosome mutation (de novo structural aneuploidy)	Very small	(Reproduction precluded)
	Nongenetic disorder	Unknown	Very small	Very small

*Depends on gene frequency.
†Female sibs heterozygous with probability: 1/2.
‡If reproduction: all daughters heterozygous but rarely affected.
§Average order of magnitude.

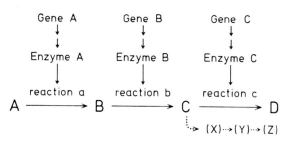

I "ONE GENE - ONE ENZYME" concept

II Point mutation : METABOLIC BLOCK

① Lack of product formation
② Accumulation of unmetabolized substrate
③ Appearance of metabolites normally undetectable

Figure 18–2. Genes as functional units and the metabolic consequences of gene mutation.

changed. In many instances this results in the substitution of one amino acid in the protein gene product. An alteration in only one amino acid among the many in a polypeptide may have severe consequences for the enzyme's catalytic role in metabolism. If it results in nearly total loss of enzyme function, a metabolic block or "inborn error" of metabolism ensues and phenotypic abnormalities are observed. The more important biochemical consequences of a metabolic block are illustrated in Figure 18–2, part II.

Inborn Errors of Metabolism

Advances in clinical and laboratory medicine, and more particularly in biochemistry, genetics, and molecular biology, have resulted in the recognition and chemical definition of an ever increasing number of inborn errors of metabolism.

Several levels of sophistication can be distinguished in the understanding of hereditary disease in general and inborn errors of metabolism in particular (O'Brien, 1969). An important first step consists of defining the clinical phenotype of a disorder, including its physical characteristics, any relevant roentgenographic findings, and its natural course. Abnormal psychomotor development often represents an important aspect of the phenotype, and if unusual behavior is broadly defined, behavioral changes are nearly always observed. Mental disability is a frequent result.

Phenotypic features of critical importance in an inborn error of metabolism are the average age of clinical onset and the character of its progression. In some disorders the course is devastatingly rapid, and barely compatible with postnatal life. In others the clinical course can be appreciated only in longitudinal studies because the disease evolves exceedingly slowly, sometimes spanning several decades.

A second level of understanding any hereditary metabolic disorder is achieved with the acquisition of morphologic or biochemical data relating to the pathogenesis. The identification of chemical compounds accumulating in the patient's tissues in this instance is of major importance in orienting the search for a primary metabolic defect.

Detection of the responsible enzyme defect constitutes the third level of scientific insight. The theoretical and practical importance of this degree of sophistication in our knowledge of an inborn error of metabolism cannot be overemphasized. It is often an absolute prerequisite for prenatal and postnatal diagnosis. It provides a scientific basis for the detection of heterozygous individuals and for sound genetic counseling. Knowing the enzyme defect helps one to understand the clinical interfamily variability between patients. Finally, without this level of understanding, molecular knowledge of the mutant gene itself could not become available.

In Table 18–3 a large list of hereditary metabolic disorders is provided. They are organized somewhat arbitrarily according to the area of metabolism most directly affected. Only those inborn errors are listed that occur in childhood and interfere with normal psychomotor or motor development or cause progressive neurologic disease or mental deficiency or disturbance of normal behavior.

The fourth level of sophistication in understanding inborn errors of metabolism consists of two different but related aspects. The first is knowledge of the complete amino acid sequence of the gene products or enzymatically active proteins, and the second consists of defining the sequence of nucleotides in the genes themselves. Such knowledge has recently become available for the structural genes of the hemoglobins, whose complete amino acid sequence has been known for several years. The increasing molecular information on gene products and genes themselves offers fascinating practical and theoretical prospects for the future, although it is unlikely that progress of this type will soon be achieved in neurobiology.

Table 18–3. INBORN ERRORS OF METABOLISM INTERFERING WITH NORMAL DEVELOPMENT OR BEHAVIOR

Name of Disorder	Heredity	Neonat. Metab. Derang.	Deficient Development			Enzyme Defect	Prog. Neur. Dis.	Seiz.	Behav. Disturb.	M.R.	Phys. Signs	Pren. Diag.	Therapy
			Ps. Mo.	Mo.	Phys.								
A: Carbohydrate Metabolism													
Glycogen storage disease, type I (von Gierke)	A.R.	+	-	-	±	Glucose-6-phosphatase	-	±	-	-	+	-	(1e)
Glycogen storage disease, type II (Pompe); juvenile forms	A.R.	-	-	+	-	α-1,4-Glucosidase	-	-	-	-	+	+	-
Glycogen storage disease, type III (Cori)	A.R.	+	-	+	±	Amylo-1,6-glucosidase	-	-	-	-	+	+	(1e)
Glycogen storage disease, type IV (Andersen)	A.R.	+	-	+	±	Amylo-1,4→1,6-transglucosidase	-	-	-	-	+	+	-
Galactosemia	A.R.	+	+	-	-	Galactose-1-phosphate uridyltransferase	-	-	-	+	+	+	1a
Galactokinase deficiency	A.R.	+	+	-	-	Galactokinase	-	+	-	+	+	+	1a
Fructose intolerance	A.R.	+	-	-	-	Fructose-1-phosphate aldolase (liver)	-	+	-	-	+	-	1a
Fructose-1,6-diphosphatase deficiency	A.R.	+	-	-	-	Fructose-1,6-diphosphatase	-	±	-	-	±	-	(1a)
B: Lipid Metabolism (Lipidoses)													
Niemann-Pick disease (sphingomyelin lipidosis)													
Type A: acute neuronopathic	A.R.	-	+	0	0	Sphingomyelinase	+	+	0	+	+	+	-
Type B: chronic nonneuronopathic	A.R.	-	-	-	-		-	-	+	-	+	+	-
Other intermediary types	A.R.												
Gaucher disease (glucosylceramide lipidosis)													
Infantile neuronopathic type	A.R.	-	+	-	+	Acid-β-D-glucosidase	+	+	0	0	+	+	-
Noninfantile non-neuronopathic type	A.R.	-	-	±	+		±	-	±	-	+	+	-
Krabbe disease (globoid cell leukodystrophy, galactosylceramide lipidosis)	A.R.	-	+	0	-	Galactocerebroside β-galactosidase	+	±	0	0	-	+	-
Metachromatic leukodystrophy (sulfatide lipidosis) infantile onset type, juvenile onset type	A.R.	-	+	+	-	Cerebroside sulfatase (arylsulfatase A)	+	-	0	+	-	+	-
Fabry disease (trihexosylceramide lipidosis)	X.R.	-	-	-	-	α-D-Galactosidase	-	-	+	-	+	+	(4b)
Wolman disease	A.R.	+	+	0	0	Acid cholesteryl ester hydrolase (acid lipase)	0	-	0	0	+	+	-
Cholesteryl ester storage disease	A.R.	-	±	-	+		-	-	-	-	+	+	-
Refsum disease	A.R.	-	-	±	-	Phytanic acid α-hydrolase	+	-	+	±	+	+	(1a) + (1b)
Tay-Sachs disease (gangliosidosis GM2, type 1)	A.R.	-	+	0	±	N-acetyl-β-D-hexosaminidase A	+	+	0	0	+	+	-
Sandhoff disease (gangliosidosis GM2, type 2)	A.R.	-	+	0	±	N-acetyl-β-D-hexosaminidase A and B	+	+	0	0	+	+	-
Other "adolescent" types of acid hexosaminidase deficiency	A.R.	-	-	-	-	N-acetyl-β-D-hexosaminidase A or A and B	+	-	+	±	+	+	-
Infantile GM1 gangliosidosis, type 1 (Norman-Landing disease)	A.R.	-	+	0	0	Acid β-D-galactosidase	+	+	0	0	+	+	-

Disorder	Inheritance	Clinical features									Enzyme deficiency	Comments
"Juvenile" GM₁ gangliosidosis, type 2	A.R.	−	+	+	+	+	+	+	+	−		(Disorders can also be classified among oligosaccharidoses)
"Chronic" type of β-galactosidase deficiency	A.R.	−	±	+	±	±	+	+	+	−		
Morquio type of β-galactosidase deficiency	A.R.	−	−	−	−	−	+	−	+	−		Can be classified as mucopolysaccharidosis IVB
C: Oligosaccharide and Glycoprotein Metabolism (Formerly Mucolipidoses)												
Dysmorphic sialidosis (sialidosis, type 1; formerly mucolipidosis I)	A.R.	−	+	+	−	±	+	+	+	−		
Cherry-red spot myoclonus syndrome (sialidosis, type 2)	A.R.	−	−	−	+	±	∓	+	+	−	Glycoprotein sialidosis	
Nephrosialidosis (infantile dysmorphic sialidosis)	A.R.	±	+	+	∓	0	+	+	+	−		
I cell disease (formerly mucolipidosis II)	A.R.	−	++	++	−	−	+	−	+	−	Glycoprotein N-acetyl-glucosaminylphosphotransferase	
Pseudo-Hurler polydystrophy (formerly mucolipidosis III)	A.R.	−	++	++	−	−	+	−	++	−		
Sialidase with β-galactosidase deficiency (Goldberg-Wenger syndrome)	A.R.	−	+	+	∓	±	+	±	+	−	Glycoprotein sialidase (?), β-D-galactosidase ?	
Infantile type	A.R.	+	+	0	−	0	0	−	+	−		
Berman syndrome (formerly mucolipidosis IV)	A.R.	−	+	+	−	−	+	+	+	−	Ganglioside sialidase?	
Mannosidosis	A.R.	−	++	++	+	+	+	+	++	−	α-D-Mannosidase	
Fucosidosis	A.R.	−	±	++	+	+	+	∓	++	−	α-L-Fucosidase	
Multiple sulfatase deficiency	A.R.	−	+	+	−	−	+	∓	+	−	Multiple sulfatases ?	
Farber's lipogranulomatous disease	A.R.	−	+	+	+	+	+	+	+	−	Acid ceramidase	
Aspartylglucosaminuria	A.R.	−	+	+	+	±	±	+	+	−	Aspartylglucosamine amido-hydrolase	
D: Mucopolysaccharidoses												
Hurler disease (MPS-IH)	A.R.	−	+	++	−	−	+	+	+	−		
Scheie disease (MPS-IS)	A.R.	−	−	++	−	−	−	+	+	−	α-L-Iduronidase	
Intermediate (MPS-IM)	A.R.	−	±	++	−	−	±	±	+	−		
Hunter disease (MPS-II: A, severe; B, mild)	X.R.	−	+	+	−	∓	+	+	+	−	Iduronate sulfatase	
Sanfilippo disease												
Type A (MPS-IIIA)	A.R.	−	+	+	−	∓	+	+	+	−	Heparan-N-sulfatase	
Type B (MPS-IIIB)	A.R.	−	+	+	−	∓	−	+	+	−	N-acetyl-α-glucosaminidase	
Type C (MPS-IIIC)	A.R.	−	∓	+	−	∓	+	+	+	−	Acetyl-coA: α-glucosaminide-N-acetyltransferase	
Type D (MPS-IIID)	A.R.	−	+	+	−	∓	+	+	+	−	N-acetylglucosamine-6-sulfate sulfatase	
Morquio disease type A (MPS-IVA) [type B (MPS-IVB): see oligosaccharidoses]	A.R.	−	−	+	−	−	−	∓	+	−	N-acetyl-galactosamine-6-sulfate sulfatase	
Maroteaux-Lamy disease (MPS-VI)	A.R.	−	∓	+	−	−	∓	∓	+	−	N-acetyl-galactosamine-4-sulfate sulfatase (arylsulfatase B)	
Sly disease (MPS-VII)	A.R.	−	+	+	−	−	+	+	+	−	β-D-Glucuronidase	
E: Disorders of Amino Acid or Organic Acid Metabolism												
Phenylketonuria (classic hyperphenylalanemia)	A.R.	−	+	+	−	+	+	+	(+)	1a	Phenylalanine hydroxylase (several qualitative variants)	
Phenylketonuria due to biopterin deficiency (at least two nonallelic forms known)	A.R.	−	+	+	−	+	+	+	(+)	(1a)	Dihydropteridine reductase Enzyme(s) for biopterin biosynthesis	

Table continued on following page

Table 18–3. INBORN ERRORS OF METABOLISM INTERFERING WITH NORMAL DEVELOPMENT OR BEHAVIOR (Continued)

Name of Disorder	Heredity	Neonat. Metab. Derang.	Deficient Development			Enzyme Defect	Prog. Neur. Dis.	Seiz.	Behav. Disturb.	M.R.	Phys. Signs	Pren. Diag.	Therapy
			Ps. Mo.	Mo.	Phys.								
Tyrosinemia (hepatorenal)	A.R.	+	±	−	−	Unknown	−	−	−	±	−	−	(1a)
Tyrosinemia (persistent)	A.R.	+	±	−	−	Unknown	−	−	−	+	−	−	(1a) + (4a)
5-Oxoprolinuria	A.R.	+	±	−	−	Glutathione synthetase	−	−	−	±	−	−	−
Hyperhydroxyprolinemia	A.R.	−	±	−	−	γ-Glutamylcysteine synthetase	+	−	−	±	−	−	−
	A.R.	−	±	−	−	Hydroxyproline oxidase	+	−	±	±	−	−	−
Primary congenital hyperammonemias Type I: carbamylphosphate synthetase deficiency (heterogeneity)	A.R.?	+	+	0	0	Carbamylphosphate synthetase	+	+	0	0	−	+	(1d)
Type II: ornithine carbamyl transferase deficiency	X.D.	+	+	0	0	Ornithine carbamylphosphatase	+	+	0	0	−	+	(1d)
Citrullinemia, type III (heterogeneity)	A.R.	+	+	0	0	Argininosuccinate synthetase	+	+	0	0	−	+	(1d)
Argininosuccinic acidemia, type IV (heterogeneity)	A.R.	+	+	0	0	Argininosuccinase	+	+	0	0	−	+	(1d)
Hyperargininemia, type V	A.R.	−	+	−	−	Arginase	−	−	±	+	−	−	(1d)
Histidinemia	A.R.	−	+	−	−	Histidase ?	−	−	±	+	−	−	−
Maple syrup urine disease (heterogeneity)	A.R.	+ or −	+	−	−	Branched brain ketoacid dehydrogenase	0 or +	+	0	0 and +	−	+	(1a)
Isovaleric acidemia	A.R.	+	+	−	−	Isovaleryl-CoA dehydrogenase	0	+	0	0	−	−	(1a)
Propionic acidemia (important ketotic hyperglycemia)	A.R.	+	+	0	0	Propionyl-CoA carboxylase	0	−	−	+	−	+	(1a)
	A.R.	+	+	−	0	Defect of coenzyme-biotin synthesis	0	−	−	+	−	+	4a
Methylmalonic acidemia(s)	A.R.	+	+	−	0	Methylmalonyl-CoA racemase	0	−	0	0	−	+	(1a)
Several types of mutant apoenzymes		+	+	−	0	Methylmalonyl-CoA mutase	0	−	0	0	−	+	
Several types of mutant enzymes in coenzyme formation	A.R.	+	+	−	0	Steps in adenosylcobalamin (AdoCbl; vit. B₁₂) synthesis	0	−	−	−	−	+	(4a)
Methylmalonic acidemia with deranged sulfur-amino-acid metabolism	A.R.	−	+	−	±	Step(s) common to both AdoCbl and methylcobalamin (MeCbl) synthesis	+	+	±	+	−	+	(4a)
Primary homocystinuria Unresponsive to vit. B₆	A.R.	−	±	−	−	Cystathionine β-synthetase (apoenzyme; coenzyme affecting enzymes)	+	±	−	+	+	+	(1a) / (4a)
Responsive to vit. B₆	A.R.	−	±	−	−		+	±	−	+	+	+	(1a) / 4a
Nonketotic hyperglycinemia	A.R.	+	+	+	−	Glycine decarboxylase	+	+	0	+	−	−	−
Serum carnosinase deficiency	A.R.?	−	+	+	−	Negatively charged serum carnosinase	+	+	0	+	−	−	−

F: Disorders Due to Malfunction of Endocrine Glands

Name of Disorder	Heredity	Neonat. Metab. Derang.	Ps. Mo.	Mo.	Phys.	Enzyme Defect	Prog. Neur. Dis.	Seiz.	Behav. Disturb.	M.R.	Phys. Signs	Pren. Diag.	Therapy
Group of entities caused by defects in homeostasis, synthesis, storage, and utilization of thyroid hormone	A.R.	−	+	+	+	Enzyme defects, unknown in most instances; rather irrelevant for clinical diagnosis and treatment	−	−	+	+	+	−	2
Pseudohypoparathyroidism	A.D.? X.D.?	−	+	−	+	Unknown	+	+	−	+	+	−	(4a)

G: Disorders in Purine and Pyrimidine Metabolism

Disease	Inheritance	Enzyme / defect										Forms of Therapy
Lesch-Nyhan disease	X.R.	Hypoxanthine guanine phosphoribosyl transferase	—	+	+	±	+	—	+	+	—	(1d)
Orotic aciduria Type I	A.R.	Orotate phosphoribosyl transferase / Orotidine-5'-phosphate decarboxylase	∓	±	±	±	±	—	+	—	+	2
Type II		Orotidine-5'-phosphate decarboxylase										

H: Miscellaneous Monogenic Metabolic Disorders

Disease	Inheritance	Enzyme / defect										Forms of Therapy
Formiminotransferase (other associated defects)	A.R.		—	+	+	+	+	±	+	—	—	—
Disorders of folate and tetrahydrofolate metabolism with central nervous system involvement	A.R.	Methylene tetrahydrofolate reductase	—	+	±	+	+	±	+	—	—	—
	A.R.	Tetrahydrofolate methyltransferase (most patients have defective cobalamin synthesis (see E))	—	+	—	+	0	±	0	—	—	— (4a)
Hartnup disease	A.R.	Unknown	±	±	—	±	—	+	—	+	—	4a
Congenital nonhemolytic jaundice; unconjugated hyperbilirubinaemia Type I (Crigler-Najjar syndrome)	A.R.	Microsomal glucuronyl transferase	—	+	—	+	+	—	+	—	±	3
Type II (milder)												
Congenital erythropoietic porphyria (Günther disease)	A.R.	Deregulation of uroporphyrinogen I (uroPgen I) synthetase and uroPgen III cosynthetase in erythroid cells	±	—	+	—	—	+	—	+	+	—
Hepatic intermittent acute porphyria (other porphyrias cause fewer or no behavioral disturbances)	A.D.	UroPgen I synthetase (general)	—	—	+	—	+	±	—	—	—	3
Hepatolenticular degeneration (Wilson disease)	A.R.	Unknown factor important in copper metabolism	—	—ms	±	+	∓	±	+	—	±	1c
Kinky hair disease (Menkes disease)	X.R.		—	+	0	+	0	0	+	+	+	—

Forms of Therapy
1. Reduction of substrate
   1a: dietary restriction
   1b: plasmapheresis
   1c: chelating of substrates, promotion of urinary excretion
   1d: administration of substances with beneficial metabolic effect
   1e: surgical procedures with beneficial metabolic effect
2. Product replacement
3. Drugs favorably influencing enzyme activity, enzyme induction
4. Gene product therapy
   4a: cofactor supplementation
   4b: allotransplantation
   4c: enzyme replacement therapy
   ( ): Treatment clinically inefficient

Abbreviations and Symbols
A.D.: autosomal dominant
A.R.: autosomal recessive
X.R.: X linked recessive
X.D.: X linked dominant
0: sign or symptom irrelevant because of rapidly fatal course
Neonat. Metab. Derang.: neonatal metabolic derangement
Ps.Mo.: psychomotor
Mo.: motor
Phys.: physical
Prog. Neur. Dis.: progressive neurologic disease
Seiz.: seizures
Behav. Disturb.: behavioral disturbance { noninfantile / patient
M.R.: mental retardation
Pren. Diag.: prenatal diagnosis

Genetic Individuality: Human Genetic Polymorphisms

The presence of monogenic disease indicates the existence of allelic variations at many gene loci in man, but random population surveys of healthy individuals also reveal differences in electrophoretic mobility, thermostability, and kinetic features of many enzyme proteins in red blood cells, serum, and several tissues. This work has shown that a small number of common variant alleles exists at about 30 per cent of all gene loci studied (Harris, 1980). These alleles are easily encountered even in small samples of a population. They are the basis of an ever increasing number of enzyme polymorphisms identified in man as in other species. Rarely observed variant alleles probably exist at nearly any gene locus. Like the more common variants, they have been generated by separate mutations that occurred in previous generations of ancestors. Variant alleles either may give rise to overt clinical consequences only in special circumstances, such as the presence of certain medications, or have no clinical implication at all (Harris, 1980). In any case they contribute to normal variation among humans.

Because of the allelic variation at many gene loci, which can be recombined in an innumerable number of combinations at fertilization, an enormous diversity exists among individuals. With the probable exception of monozygotic twins, each individual in our species is unique in his genetic and enzymatic constitution. This now acknowledged fact of human individuality may at first seem surprising. It only confirms, however, what any mother knows about her children and any teacher about his pupils. Thus, because the genotype of every human being is unique, so must be his behavior or the way he relates to and interacts with his environment.

INTERACTION OF GENES

Up to this point a gene has been defined as that quantity of DNA which contains the genetic information for a polypeptide or protein, which is usually catalytically active in a single metabolic reaction. This rigid consideration of separate gene and enzyme action is obviously incomplete and therefore incorrect. Genes act in concert with other genes. Their function is meticulously regulated by many control mechanisms, which themselves are also directly or indirectly genetically determined. In micro-organisms, several regulatory mechanisms of gene activity are known in detail. In eukaryotes in general and in higher vertebrates in particular there is still ignorance about the mechanisms of gene regulation and the very specific rules governing differentiation. It is not known which type of molecular interaction signals the activation of gene activity and which brings about its cessation. If this information is not yet known for genes completely characterized structurally, such as those involved in hemoglobin synthesis, it cannot be expected that the many genetic interactions involved in the stages of psychomotor development will be known in the near future.

For many human traits only a few types of elementary interaction of genes can be recognized and described in simple terms. Dominance and recessivity can be viewed as different modes of interaction of the products of allelic genes (see page 316). The term epistasis refers to interaction of nonallelic genes. The concept can be explained easily by considering once more the theoretical metabolic pathways in Figure 18–2. If gene A does not make a functional enzyme A, reaction a does not take place and compound B is not formed. Since the latter is the substrate for reaction b, it appears as though gene B is also not working. Gene A is said to be epistatic to gene B. Thus although genes A and B may be on different chromosomes and segregate independently upon gamete formation, they do not function independently.

Differences in Penetrance; Expressivity

Penetrance is defined as the proportion of individuals among those with the requisite genotype manifesting the expected clinical phenotype. Nonpenetrance of a hereditary disorder can explain skipped generations in pedigrees. It presents a true challenge to genetic counseling in several autosomal dominant conditions. Penetrance is probably complete in Huntington chorea from middle age on. It is not complete in tuberous sclerosis. Penetrance can also show sex differences. In the common trait, familial baldness, it approaches zero in females. It is also lower in female heterozygotes for periodic paralysis.

Expressivity is the quality of phenotypic expression of a variant gene. It can be uniform in one hereditary condition and variable in another. Dominant traits are notorious for their wide variability in severity of phenotypic expression. Sometimes expressivity cannot be detected by clinical examination or by currently available methods of examination. In such instances the hereditary trait is said to be nonpenetrant.

An influence of different nonallelic genes on the expression of a particular mutant gene is the likely explanation for variable expressivity. Nonallelic gene affects also determine whether the effect of a gene is penetrant. However, the true nature of these types of gene interactions is not known.

The Theory of Polygenic Inheritance

In contrast to the monogenically determined qualitative traits, quantitative phenotypic traits in nature are thought to be determined in part by the simultaneous and coordinated action of many genes. In its simplest form, polygenic inheritance should be considered a hypothesis that proposes that each of several or many genes has a small positive, neutral or negative effect toward the realization of a "quantitative" phenotype. Examples in nature include crop yield, plant height, weight gain in nursing animals, milk production, and many other commercially important examples in veterinary medicine and animal breeding. In humans, normal stature and blood pressure can be cited as examples. As will be discussed, behavior does not represent a merely quantitative human attribute, and it is unlikely that intelligence does either.

Quantitative traits are also called multifactorial because they result not only from the joint effects of many genes but also from more or less unknown environmental factors.

Quantitative phenotypes are said to be objectively measurable. They show a continuous frequency distribution in the population. Unlike monogenic phenotypes, they do not show mendelian segregation in families and pedigrees but tend to appear in the offspring as a blend of parental characteristics.

Differences of opinion regarding assessment of the relative importance of heredity in the etiology of phenotypic traits demonstrating continuous variation resulted in an important controversy during the early years of this century between proponents of the "new" mendelian genetic theory on one hand and the biometricians influenced by Galton's methods of quantitative analysis on the other. The former group hesitated to designate any natural phenomenon genetic unless conformance to Mendel's laws could be demonstrated. The latter maintained that continuously variable traits, to which mendelism appeared unapplicable, were nevertheless inherited and cited the finding of phenotypic correlation between parents and offspring. As will become apparent, this historic controversy, though at present satisfactorily resolved, remains pertinent in view of some present-day approaches to the genetics of human behavior.

An Abstract Model. The work on red wheat kernel color by Nilsson-Ehle in Sweden and on corn cob length by East in the United States established that both mendelists and biometricians had been partially correct in their views on heredity of continuous varying phenotypic traits (Srb et al., 1965). Even before 1910 these authors had already shown that under ideal and therefore exceptional natural circumstances, the separate small effects of several genes contributing to one quantitative trait could be defined and mendelian segregation observed.

In Figure 18–3 a theoretical example is illustrated of the inheritance of a quantitative trait numerically characterized by the phenotypic value ($\overline{w}$) averaging 0 in one and 4 in the other true-breeding plant variety in the original parental generation (P_1). Cross fertilization of P_1 individuals yields a first filial generation (F_1) exclusively composed of individuals with the average phenotypic value of 2. Upon self-fertilization of the F_1 plants, an F_2 generation is obtained, in which $\overline{w}$ also equals 2. It is composed of individuals with a discontinuous phenotypic gradation ranging from 0 to 4. Either parental phenotype is found in only one-sixteenth of the F_2 offspring. The largest single group of F_2 individuals has the intermediate phenotypic value of 2. A total of five different phenotypes can be discerned among the F_2 offspring. In the original experiments of Nilsson-Ehle a discontinuous range of seven phenotypes was recognized with regard to kernel color in wheat in F_2 individuals. The author postulated that this phenotypic trait in wheat is governed by the alleles of three loci. His several assumptions applied to our simple model are implicitly represented in Figure 18–3.

In this model only two gene loci are considered, each with two alleles. They govern the phenotypic trait in such a way that one allele (A' and B') on either locus contributes positively to the trait and the alternative allele (A and B) has no effect at all.

First, it is assumed that the alleles on either locus are codominant. This means that in the heterozygote A'A the phenotypic effect is exactly intermediate between the effects in the AA and the A'A' homozygotes. As with those of the alleles on the A locus, the effects of alleles B' and B are also simply additive. Second, there is no epistasis between the alleles of either gene locus. This assumption implies that the effect of A' equals that of B' and that the effect of A equals that of B. The overall phenotypic effect of any genotype is the simple addition of the effects exerted at each one of the loci. In Figure 18–3 all genotypes containing the same number of primed alleles have the same phenotypic value. Third, the type of frequency distribution of F_2 phenotypes could not be observed without assuming independent segregation of loci |A| and |B|. In the event of linkage, more parental allelic combinations would be expected among F_2 individuals. Fourth, in the theoretical model it is assumed that there is no differential environmental influence on any genotype. Environmental influences explain only why the phenotypes in both varieties of true-breeding parental (P_1) and F_1 populations are normally distributed around their averages (Fig. 18–4).

The broader phenotypic distribution among F_2

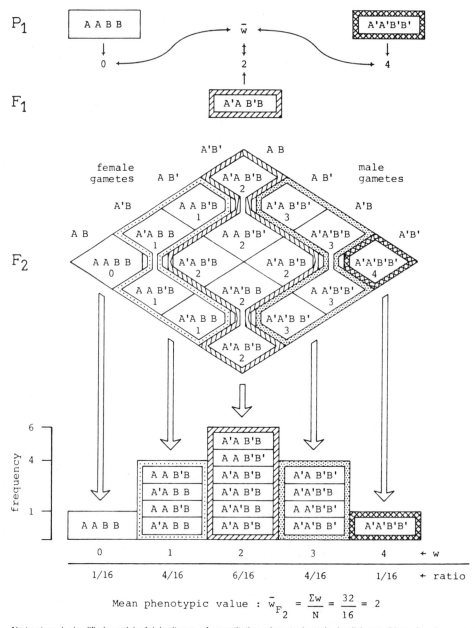

Figure 18–3. Abstract and simplified model of inheritance of quantitative phenotypic trait. Implicit conditions for demonstrating mendelian segregation are rarely if ever fulfilled in nature (see text).

individuals is accounted for by genotypic differences in addition to environmental effects. In the theoretical Figure 18–3 the distribution of F_2 phenotypes is discontinuous. Random environmental influences on each genotype cause overlap between phenotypic classes. Therefore a continuous distribution of the F_2 phenotypes forms a fairer picture of reality (Fig. 18–4). Nevertheless the phenotypic correlations between generations presented graphically can also be found only if all four conditions just cited are fulfilled. Only then can proof be obtained that quantitative traits, like qualitative ones, are governed by genes.

The Concept of Heritability.   In nature special conditions represented by the four assumptions in the abstract model are nearly never met simultaneously. Dominance is often encountered in the phenotypic effects of allelic genes. Gene interaction or epistasis is the rule rather than the exception. If more than a small number of loci are involved, linkage between some of the loci is probable. Environmental influences are not necessarily randomized in nature. These are the multiple reasons why segregation of quantitative traits cannot be observed in families, why in general any filial generation as a whole closely resembles

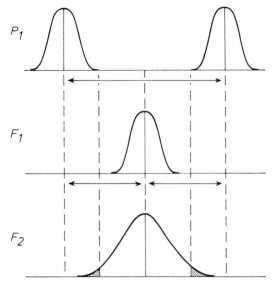

Figure 18–4. Inheritance of quantitative traits. Randomized influence of environment distributes phenotypes continuously. Abstract model of trait caused exclusively by genes with additive effect. Heritability complete (see text).

variances due to genetic causes (σ^2_G), environmental influences (σ^2_E), and interaction between genotypes and environment (σ^2_{GE}). In quantitative genetics of plants and animals the term σ^2_{GE} is often ignored. However, it is an important cause of discrepancy between prediction and actual result. More important, however, the very existence of a correlation between genotypic and environmental causes precludes the validity of applying analysis of variance to numerical data in order to estimate the genotypic and environmental contributions to a quantitative phenotype (Feldman and Lewontin, 1975). In expression (1), σ^2_G itself is conceptually partitioned as

$$\sigma^2_G = \sigma^2_a + \sigma^2_d + \sigma^2_i \qquad (2)$$

in which σ^2_a stands for variance due to genes[3] additive effect, σ^2_d for that due to dominance or any other interallelic gene effect, and σ^2_i for the variance due to interaction of nonallelic genes (epistasis). In quantitative genetics, heritability is also defined as

$$h^2_N = \sigma^2_a / \sigma^2_T$$

Finally, estimates of heritability can also be obtained from comparative phenotypic studies in relatives. If estimates of heritability are of limited value only for truly quantitative traits, their relevance becomes doubtful in the etiologic study of human development and behavior.

Threshold Traits. The model of a multifactorial etiology and polygenic inheritance has been used to explain the familial occurrence of some rather common congenital malformations (Carter, 1969; Falconer, 1960, 1965). Examples are cleft lip with and without cleft palate, clubfoot, dislocation of the hip, pyloric stenosis, several types of congenital heart defect, and most important, defects of the neural tube, some with developmental, neurologic, and behavioral consequences, such as spina bifida aperta or meningomyelocoele, anencephaly, and some forms of hydrocephaly. In each instance the congenital malformation occurs as an isolated defect in the patient and is not a component of a complex syndrome. Because there is nearly always a clear distinction between the affected and the nonaffected, each of these isolated malformations must be considered a discontinuous phenotypic trait. Falconer's multifactorial model obviously does not apply to the expression of the discontinuous trait in patients, but instead to the predisposing genetic and environmental factors. The liability for expressing the malformation is thought of as continuously distributed in the population (Carter, 1969; Falconer, 1965). Thus phenotypic values ($\overline{w}$) can theoretically be expressed in terms of two scales: the continuous underlying scale of normally distributed predis-

its parental generation, and why plant and animal breeders observe that the mean phenotype of offspring is considerably closer to the mean phenotype of the population from which parents were chosen than to that of the parents selected for breeding. Data from experimental and domestic animals relating to the latter phenomenon, called filial regression, can be used to assess the heritability of objectively measurable traits. If the difference between the population mean and that of the parents selected were entirely genetic and due to differences in additively acting genes, inheritance of the trait would be completely predictable. But quantitative traits are only partly due to additively functioning genes. Only that part of their genetic causality is defined as heritability in the narrow sense (h^2_N). Only to that extent is a quantitative trait predictably heritable and amenable to artificial selection.

Since continuously distributed quantitative traits do not lend themselves to segregation analysis, their description and assessment can be achieved only by biometric and statistical approaches and expressed in terms of the statistics, mean (μ) and variance (σ^2). A technique called analysis of variance is often applied with the purpose of statistically dissecting the relative roles of the various etiologic factors of the traits. As a statistical parameter, any total variance (σ^2_T) can be considered a composite of several partial variances. Biometricians use the expression:

$$\sigma^2_T = \sigma^2_G + \sigma^2_E + \sigma^2_{GE} \qquad (1)$$

which shows that the total variance of a phenotypic trait in the population is the sum of the posing factors as opposed to the visible scale, which is discontinuous. The application of the multifactorial model to common single malformations requires the assumption of a thresfold value of liability factors, which if exceeded in the zygote or embryo, imposes phenotypic expression. If the underlying scale is thought of as the continuous distribution of predisposing genes, phenotypic expression of the multifactorial trait is expected to occur only in those embryos with an excessive number of contributing genes, that develop in a sufficiently unfavorable environment. The model generates several predictions of incidence in relatives of the propositi. It is supported by a large number of observations, especially in the population of Great Britain, regarding primarily the incidence of cleft lip–cleft palate on one hand and neural tube defects on the other (Carter, 1969). However, the hypothesis of polygenic inheritance has not been proved by these data, which by themselves represent an empiric basis for defining the average recurrence risks useful in genetic counseling. It is significant for the discussion of behavioral traits that the proportion of the total variance in liability due to additive genetic variation—this is the heritability of the malformations—cannot be estimated from these empiric observations, since the other partial variances (e.g., σ^2_d) cannot be assessed separately. That the observations relating to cleft lip–cleft palate fit predictions made by the model of polygenic inheritance has been challenged, however (Melnick et al., 1977).

The multifactorial model has also been applied to the etiology of several diseases of adult life, such as insulin dependent diabetes mellitus, some forms of hyperlipoproteinemia, and less well defined disorders, such as some psychoses, epilepsy, and mild to moderate mental retardation. It is of interest that progress in cell biology and immunogenetics has led the way in increasing our knowledge of the etiology and nosologic definitions of such entities as insulin dependent diabetes mellitus and multifactorial hyperlipoproteinemia rather than the statistical or biometric approach (Brown et al., 1981; Craighead, 1978).

HUMAN BEHAVIOR AND THE MODEL OF POLYGENIC INHERITANCE

Abstract thinking and language are the elements of man's behavior that most evidently characterize the human brain. Because of these attributes, man has added the dimension of human cultural evolution to general biologic evolution. The brain, like any other organ, is a tissue composed of specialized cells and intercellular substances. As in other organs, brain development, gross anatomy, micro-

scopic and molecular structure, and brain function are determined by the action of many genes. This axiomatic view is supported by much evidence. The number of genes containing the information coding for brain functions must be very large. They are distributed over the entire karyotype, because even a minor structural aneuploidy involving any one of the 22 autosomes results in mental deficiency. An exceptionally large proportion (as much as 40 per cent) of all transcribable nonrepetitive DNA sequences are found represented as RNA copies in the human brain (Grouse et al., 1980).

Obviously human behavior is the result of more than mere coordinated gene action. Therefore the application of the biometric model of polygenic inheritance to the study of the relative importance of genetic and environmentally founded aspects in human behavior encounters truly insurmountable conceptual difficulties.

Conceptual Difficulties

It is generally acknowledged that the development and function of any organ are the result of gene action on one hand and environmental influences on the other, and depend on the interaction of both. Although our knowledge of the molecular basis of growth and maintenance of the human skeleton is still limited, it is generally accepted that both genetic and nongenetic factors determine adult stature, and that interaction of these two influences is probably of a rather simple qualitative and possibly additive type.

The interaction of genetic factors and environment in brain function and human behavior is entirely different in character and order of magnitude. Normal development in children requires constant interaction with the environment and more specifically with other humans. Because of his brain function and specific behavior, man shapes his own environment, which in turn shapes and modifies his behavior. Neuronal differentiation occurs concomitantly with selective gene expression and probably in part as a result of it. Moreover, studies in experimental neurobiology indicate that experience itself can modulate gene expression independently of development. Experience has a strongly positive effect on transcriptional intensity and complexity.

In cats, changes in the general environment resulting in generalized perturbation of brain action as well as specific visual changes can account for fluctuations of as much as 30 per cent in the synthesis of intraneuronal nuclear RNA (Grouse et al., 1980). In the former experiments, transcription is altered throughout the brain; in the latter, changes of RNA synthesis are restricted to the visual cortex of the animals. With either type of challenge, the complexity of the transcription products shows a positive correlation with the intensity of environmental influence.

An interesting recent suggestion is that the same electrical and chemical signals used in natural brain function may also guide and stimulate its differentiation, development, and progressive maturation. It is possible that these very signals resemble closely those that have evolved in primitive unicellular organisms as a result of constant environmental challenge and selection (Harris, 1981; McMahan, 1974; Tomkins, 1975). This phylogenetic hypothesis illustrates, and the previous arguments demonstrate, the complex and intensive interaction between genes and environment, and implies that the genetics of human behavior as a multifactorial trait cannot be approached fruitfully by applying the biometrical model of polygenic inheritance. There can be no way of separating variance due to environmental factors from that ascribed to genetic differences.

The lack of objective parameters for defining and describing normal behavior constitutes a second conceptual difficulty in studying its genetic aspects. In general, human behavior is easily discernable from that of animals. However, each human being has a unique genotype and manner of interacting with his environment and an individualized pattern of behavior. Therefore attempts to define the true norm of human behavior cannot succeed. Behavior is dependent among other factors on age, sex, social class, education, and many other ethnic and cultural influences. Hence it is impossible to study the genetics of a behavior as such (Hirsch, 1967).

A description of behavior involves the analytic distinction and evaluation of its many components, such as sensorimotor skills, personality, affect, temperament, and cognitive abilities. Sensorimotor functions can be measured with some accuracy. That these functions are governed by multiple genes is illustrated by the existence of many forms of monogenic blindness, deafness, neurogenic muscle atrophy, or muscular dystrophy. Because the genetic heterogeneity of these qualitative monogenic traits was recognized early, application of the polygenic model to assess the role of genetics in their etiology would have been redundant.

The trait called personality is rather resistant to objective measurement even though a more or less standardized approach for testing has been devised. The various normal phenotypes remain elusive and ill defined, defying analysis even by methods of quantitative genetics. Although patients with severe symptoms of psychiatric disorders are often easily differentiated from normally behaving persons, the definition of these disorders remains a challenge, and the application of the polygenic model to explain their occurrence in families has failed to reveal any real clue to their etiology.

Intelligence, the Most Thoroughly Studied Component of Human Behavior

Intelligence has often been considered an example of a continuously distributed quantitative trait. Since the early years of this century, it has been approached by various psychologic tests, with results expressed numerically by the intelligence quotient. Today's interest in intelligence has a social motivation, which in essence does not differ from that of Binet, who devised the first system of testing. Parents, teachers, and society want to know as early as possible whether a child will be able to attend regular elementary school or whether he will require special education programs. We would like the results of intelligence tests to be adequate measures of general intelligence. Intelligence quotient and intelligence are used as true synonyms in many of these studies, although everyone agrees that the IQ is but one of many possible measures of intelligence. In the tradition originating with the biometricians of last century, correlations between the IQ's in various types of relatives are studied confidently. From the correlations established, further extrapolations are made in regard to the relative roles of heredity and environment in determining intelligence.

Theories about the nature of intelligence abound in psychology (DeFries et al., 1976; Vandenberg, 1977). One group of hypotheses regards intelligence as a unitary human attribute, the single cognitive ability of man's brain. Theories in the alternate group consider intelligence to be a composite of separate and independent elements. According to one of these hypotheses the intellect is viewed as being composed of no less than 120 independent abilities identified by the use of three groups of specifically designed tests (Guilford, 1968). The unitary theories most certainly represent oversimplified abstractions of human mental function. However, they have simplified the application of intelligence, a recognized, continuously distributed variable in the population, to the model of polygenic inheritance. Moreover, positive correlations between results of intelligence tests were more easily observed in family members by assuming intelligence to be a unitary attribute.

The multiple abilities hypotheses have contributed in a major way toward a more adequate description of the full range of qualities of the human mind. It should be kept in mind, however, that all hypotheses relating to intelligent behavior are constructed to fit actual data, which consist in this instance of results from more or less specifiçally designed psychologic tests. It must not be concluded that there exists a separate genetic determinant or mechanism for either word fluency, spatial ability, or dexterity in tests with either symbolic or pictorial content. It is extremely unlikely that independent genes exist determining performance of such mental operations as convergent or divergent production as defined by two of Guilford's tests. In a review of the genetics of specific cognitive abilities, it has been shown that studies in which either monogenic inheritance or the polygenic threshold theory was tested have been inconclusive (Defries et al., 1976). Moreover, searches for linkages between so-called major genes affecting continuously variable mental traits and marker genes inevitably lead to inconsistent and conceptually incorrect conclusions (Vandenberg, 1977). Separate mental abilities discovered by psychologic testing can hardly be considered to be well defined phenotypes. However, even if the status of a phenotype is granted to spatial ability, for instance, it is readily recognized as a causally heterogeneous trait. Similar conclusions must be reached for many if not all other mental attributes distinguished in psychologic testing. Any cognitive ability is removed by many hierarchical steps from primary gene action. It results from the proper functioning of many genes and obviously from intense genotype-environment interaction. In contrast, any marker gene with which chromosomal linkage would be explored is recognized through its direct, biochemically detectable gene product.

Many psychologic studies inspired by either the unitarian or the multiple abilities theory have shown, first, that mental functioning is normally distributed in various populations and subpopulations and, second, that it is more or less positively correlated within families (Bouchard and McGue, 1981). The biometric model of polygenic inheritance has often been applied to the data. However, it should be kept in mind that heritability as a mere statistical construct is only an index of the amenability of a quantitative trait to selective breeding. Even in such instances as agricultural studies, in which correlations between genotype and environment can be fairly well randomized, the genotype-environment interaction remains a serious quandary. With regard to human intelligence, heritability is all too frequently used in a context implying at least a reasonable understanding of the relative proportion of phenotypic variation due to genetic variability. Because of the type of genetic-environmental interdependence with a primordial role in the development of the human brain and human behavior, valid statistical estimates of the genotypic or environmental contributions to the phenotypic variance are precluded. This reasoning applies not only to obser-

vations and statistics derived from pedigrees, but also to those from populations—more specifically when attempts are made to explain differences in performance between subgroups or races in intelligence testing (Feldman and Lewontin, 1975). Any data that do not deal critically with the fact that genetic as well as environmental differences can lead to rather similar distributions of human intelligence as a variable cannot be useful in establishing estimates of heritability, irrespective of whether analysis of variance or the method of path analysis is applied (Lewontin, 1975). Twin studies and observations relating to adoption are often proposed as the *par excellence* means of tackling the problems just described. (For a brief discussion, see further in this chapter.)

Mental Retardation as a Multifactorial Trait

A list of single gene defects with associated behavioral manifestations, particularly mental deficiency, is provided in Table 18–3. Too few multidisciplinary and prospective studies have been undertaken to establish distinctive behavioral features for some of the "inborn errors of metabolism." Some of the literature data have been compiled as part of a review article (Childs, 1972). In this chapter the specific behavioral attributes in some of the mucopolysaccharidoses remain to be described. In many instances of single inborn errors of metabolism however, it appears that nonspecific mental deficiency is the major phenotypic feature.

In a following chapter, mental deficiency and behavioral traits due to chromosomal aneuploidy will be discussed.

The model of polygenic inheritance has been applied also to explain the family aggregation and pattern of occurrence of some forms of mild or moderate psychomotor retardation, detected at or prior to school age in children who cannot benefit intellectually from regular elementary school training. In more than 75 per cent of the children these poor results are correlated with scores of 50 to 70 on intelligence testing offered as part of a multidisciplinary psychomedical evaluation of their development. Moderate mental disability is usually presumed or confirmed during the preschool years. In such children psychologic testing usually results in IQ scores between 35 and 50. Karyotyping and laboratory tests for detection of metabolic errors yield only normal results. No unfavorable prenatal or perinatal environmental circumstances can be identified. There is no indication of progressive neurologic disease or of decreasing mental capacity. Instead, despite a clinical history of delayed psychomotor milestones, the child's slow but steady progress can usually be documented from longitudinal observations. It is not uncommon to find one or more other children or adults with mild mental handicap in the proband's family. Obviously isolated occurrences of mental retardation also can fit this description.

Once again a conception of nonprogressive mental disability as a quantitative phenotypic trait with a multifactorial etiology does not in itself help to identify the true nature of some of the etiologic factors and define the probability of recurrence, nor does it provide insight into the relative importance of heredity and unknown environmental factors. The normal distribution of human intelligence cannot be viewed merely as a scale of occurrence in the population of a particular number of factors contributing to general mental ability, with the consequence that people having less than a minimal number of such factors fall below a threshold value and therefore are mentally deficient. Unlike common malformations in which the affected can be distinguished objectively from the nonaffected, distinction of the mentally disabled from the so-called normal is rather arbitrary and cannot be represented by a straight line or threshold value. The nature of impairment of mental abilities changes with age. In patients with mild retardation and a minor deficit in social adaptability, the mental handicap may no longer be detectable beyond school age.

A gaussian curve illustrating the distribution of normal intelligence in the human population is not more than a two dimensional reflection of biologic axioms already alluded to in previous paragraphs: a large number of genes govern normal development and function of the human brain, and a large number of variant alleles exist in the population at each of the many loci. Differences in the interaction of different alleles with the environment probably account for a major part of the variability in intelligence in the normal population. Some cases of mild mental retardation may be due to the presence of unfavorable alleles at several loci. However, it is at least as likely that single gene defects with important epistatic effects upon other genes may be the basis for many other instances of mild to moderate mental handicap, if the proper functioning of the latter genes involved in brain development depends largely on the normal activity of one of the former genes. In pedigrees with many instances of so-called pure mental retardation, a monogenic cause and a large recurrence risk must always be seriously considered.

Because mental disability is an extremely heterogeneous trait, retrospective studies in the population may be useful in establishing prevalence figures, but they are of little empiric value as a base in counseling an individual pair of parents. Thus it is not surprising that the polygenic model derived from population data must be considered

almost irrelevant in estimating recurrence risks among the patient's relatives.

Like mental retardation, epilepsy is extremely heterogeneous not only clinically but also etiologically. The multifactorial model is often invoked in order to explain the familial occurrence. Once again population data relating to prevalence must be applied with caution in individual family counseling.

INTERACTION OF GENOTYPE AND ENVIRONMENT: STUDY IN BEHAVIORAL TRAITS

Up to this point the fundamental importance of the interaction between genotype and environment in human behavior and brain development has been emphasized. Because of this interaction, the relative importance of either in generating variability in behavior and its more easily quantified component, intelligence, may never be completely known. The genetic contribution to phenotypic resemblance between relatives can be predicted in part by mendelian theory, but is certainly also dependent upon unknowns like gene interaction, allelic frequency, linkage, and assortative mating. The environmental basis for similarities between people does not rest on an a priori theory. However, it can easily be seen that environmental correlation also decreases with a decreasing degree of relationship (Lewontin, 1975). A review of all major family studies of intelligence concludes that the pattern of average correlations is consistent with that predicted by the polygenic inheritance model (Bouchard and McGue, 1981). It may be assumed, however, as some conclusions of the review confirm, that the observed results could also correspond to those predicted by the degree of environmental similarity between relatives.

Since both environmental and genetic theories predict the same pattern of correlation, the connection between "genetic relationship and environmental relationship" needs to be broken (Lewontin, 1975). Lewontin outlines six strict requirements for proper studies of adopted children, stating that "any study that does not include adopted or foster reared pairs of relatives in which the foster environments have not been randomized, will overestimate the genetic component of variance by an unknown amount." No such strictly controlled studies have yet been carried out. "Approximately designed studies do not yield approximate results." It follows that if reliable estimates of the heritability of intelligence cannot be made on the basis of currently available data (Bouchard and McGue, 1981), it could be even more inaccurate to attempt to draw conclusions about genetic differences between races and socioeconomic classes.

In examining the relative roles of genotype versus environment in multifactorial or quantitative traits, comparative twin studies are extremely useful. The results of such studies have been important in defining the mode and relative importance of heredity in the common malformations previously discussed. In studies of the etiology of variability in intelligence, they are less useful, because it cannot be assumed that intrafamilial variance in environment is the same as for various relationship groups (Lewontin, 1975). If twins, siblings, or foster siblings are raised apart, the environmental variables must also be truly randomized over the pairs under comparison. This condition is extremely difficult to fulfill and has rarely been met in reported and repeatedly quoted twin studies.

INHERITED DISEASE AND UNUSUAL CHILD BEHAVIOR OR DEVELOPMENT

It becomes apparent that the work on human behavioral genetics influenced by the concepts of quantitative genetics previously applied to animal and plant breeding programs has yielded little useful information regarding the etiology of behavioral differences in humans. Even twin and adoption studies are frequently unconsciously biased. Research on brain function, human behavior, and mental development inspired and guided by the mendelian or gene concept paradigm, however, is ultimately more likely to result in advancement of such knowledge. Biometric or galtonian approaches do not provide explanations for either genetic or environmental mechanisms of developmental and behavioral variability. Normal individual differences will ultimately be clarified through analysis of gene action and interaction in the central nervous system.

Nature itself presents an interesting opportunity to initiate an inquiry into the role of individual genes in development, in the form of the many children who present with signs of unusual behavior, arrested psychomotor development, and the loss of previously acquired developmental skills after a normal, symptom free interval of months to years. Many such instances are listed in Table 18–3. In most of the examples progressive mental impairment leading to a fatal outcome is the natural course of the disease. It is precisely this multiplicity of monogenic metabolic disorders that provides additional evidence that normal human development is governed by many genes.

In the paragraphs that follow, a few examples are given of metabolic disorders that exert a profound influence on the development and function of the central nervous system. In the first, infantile metachromatic leukodystrophy, a disorder with a rather rapid course, locomotor function initially and subsequently cognitive ability are grossly impaired. In the second clinical example, the Sanfilippo syndrome, personality and intelligence are affected first and motor function only secondarily. A great deal of the disordered chemistry of the latter condition has been elucidated. It will be used as a model to illustrate how complex and seemingly irrelevant chemical reactions may profoundly affect the proper functioning of the human brain.

PROGRESSIVE GENETIC DISORDERS

Well Defined Metabolic Disorders of the Central Nervous System

Infantile Metachromatic Leukodystrophy. This autosomal recessive disorder of childhood is characterized by a clinical onset between one and four years of age, but most typically during the second year of life after the child has already achieved milestones of psychomotor development appropriate for age. Usually a young child who has already learned to walk unassisted begins showing a gait disturbance and falls frequently, apparently because of muscular weakness. The child is found to be slightly ataxic. A neurologic examination at that stage would demonstrate transient signs of pyramidal spasticity.

Frequently ambulation is lost within a year, almost simultaneously with the ability to sit unaided. Dysphagia becomes a challenge to feeding. There is an increasing apathy and progressive loss of speech. Comprehension remains relatively less affected until well after the patient has become bedridden. Conscious contact with surroundings, however, is progressively lost and a decerebrate rigid posture adopted. Intercurrent respiratory infections are the cause of death, usually well before the tenth birthday, in an emaciated, completely unreactive patient.

Metachromatic leukodystrophy may also appear in late childhood or early adolescence. This juvenile form of the disease has a slower clinical course. Only one type of the disease is observed in any one family.

Much is known about the pathophysiology of metachromatic leukodystrophy and is applicable for diagnostic purpose. Widespread demyelination occurs throughout the central and peripheral nervous systems. Metachromatic deposits can be identified easily in the disintegrated myelin, most often enclosed by unit membranes of lysosomes. The accumulated material has been characterized chemically as cerebroside sulfate or sulfatide. Excessive quantities of this compound are stored because of deficient activity in the patients of the lysosomal acid hydrolytic enzyme, cerebroside sulfate sulfatase, also termed arylsulfatase A. Demonstrating the enzyme defect in white blood cells or cultured fibroblasts confirms a presumptive diagnosis that is usually made on clinical, neurophysiologic, and pathologic grounds. More important, the identification of the responsible enzyme defect permits diagnosis of the disorder before the onset of typical clinical symptoms or antenatally through demonstration of the enzyme defect in cultured amniotic fluid cells.

Sanfilippo Disease (Mucopolysaccharidosis Type III): A Metabolic Disorder with a Protracted Course. The mucopolysaccharidoses are a group of progressive, monogenically determined disorders, the pathologic hallmark of which is the accumulation of acid mucopolysaccharides, or glycosaminoglycans, in the lysosomes of mesenchymal and parenchymal tissues and their excessive excretion in the urine. The disorders are presented as a group in Table 18–4.

During the last 12 years major progress has been achieved in our understanding of the mucopolysaccharide storage diseases. In each a profound deficiency of a single enzyme, which normally participates in the breakdown of the mucopolysaccharides, has been demonstrated. The physiologic role of these macromolecules, often considered to be constituents of connective tissue only, remains largely unknown. Certainly the progressive accumulation in lysosomes of only partially degraded glycosaminoglycans may interfere with the proper growth and modeling of bone and exert an adverse influence on the development and function of the cardiovascular system. Mucopolysaccharide accumulation, moreover, has far reaching clinical effects on the patient's central nervous system and, consequently, on psychomotor development, behavior, and mental function.

The subject of the mucopolysaccharidoses has been adequately reviewed (Kresse et al., 1981; McKusick et al., 1978; Spranger, 1975). Unfortunately, progress in the delineation and diagnosis of these disorders has not been matched by progress in treatment (Leroy, 1979).

The Sanfilippo syndrome has been recognized as a mucopolysaccharidosis and has been aligned with the conditions known as Hurler disease (MPS-I) and Hunter disease (MPS-II) only since the discovery of excessive urinary acid mucopolysaccharide excretion in patients with the latter two disorders. The frequency of each of the mucopolysaccharidoses is largely unknown but is closer to 1.10^{-5} than to 1.10^{-4} births for the Hurler syndrome. Sanfilippo syndrome, which really comprises four different diseases, appears to be at least twice as frequent and may very well be the most frequent of the mucopolysaccharidoses (Van

De Kamp, 1981). For a summary of the clinical characteristics of the Sanfilippo syndrome, also called mucopolysaccharidosis type III, the reader may consult Table 18–4, in which the features of differential diagnostic significance are schematically presented. The descriptive terms used are applicable primarily to the characteristic findings in the fully developed clinical picture and not necessarily to either the initial or the terminal stages of the disorder.

The first parental concerns about children with Sanfilippo disease are in regard to their restlessness, nervousness, and hyperkinetic behavior, which appear between the ages of two and six years. During this period many patients wake up several times a night or go through periods of several days without sleep. This behavior is often described as anxious, unmanageable, and sometimes aggressive. Usually this behavior cannot be controlled with sedatives, which soon have adverse side effects. The more peaceful the environment, the better the chances of controlling the patient's hyperactive behavior. When restrained, such patients become desperately anxious and hyperactive. In retrospect it is frequently found that with Sanfilippo syndrome patients were somewhat delayed in speech development but had achieved the milestones of psychomotor development at appropriate ages.

At the time of the first clinical evaluation, the most consistent finding in Sanfilippo syndrome is an increased head circumference. Radiologically this macrocephaly is accompanied by thickening and sclerosis of the bony calvarium. The facial features may be slightly coarse but often do not at all resemble those in patients with MPS-I or MPS-II. The corneas are clear. Stature is usually normal, and joint mobility is not limited except at the shoulders. Moderate hepatomegaly is common, but splenomegaly is rarely found. Respiratory infections are less frequent in children with the Sanfilippo syndrome than in most of the other mucopolysaccharidoses. In about one-third of the males, repair of an inguinal hernia is necessary. Umbilical hernias are not consistently found in either sex and remain small. Conductive, perceptive, or mixed deafness is likely in many patients but is difficult to assess objectively.

Soon after the diagnosis is established, signs of retrogression of the patient's mental status become obvious. Physical changes remain minor and skeletal growth proceeds at a nearly normal rate, but psychologic contact with the family and familiar surroundings diminishes rapidly as mental capabilities become extremely limited. The patient loses intelligible speech and walks around aimlessly. Although the hyperkinesis has now diminished, spurious movements of the limbs are often noted.

At this stage, the patient with Sanfilippo syndrome behaves much like many other children

Table 18–4. MUCOPOLYSACCHARIDOSES: CLINICAL FEATURES OF DIFFERENTIAL DIAGNOSTIC IMPORTANCE*

Disorder / Symptoms	MPS-I (Hurler)	(Scheie)	(Intermediate Phenotype)	MPS-II (Hunter)	MPS-III (Sanfilippo)	MPS-IV (Morquio)	MPS-VI (Maroteaux-Lamy)
Neonatal problems	Several	None	None	Few, if any	None	None	Several
Final stature	Severely dwarfed	Nearly normal	Reduced	Severely dwarfed	Almost normal	Severely dwarfed	Severely dwarfed
Cornea	Cloudy	Cloudy	Cloudy	Clear	Clear	Cloudy	Cloudy
Kyphosis	Frequent dorso-lumbar	None	None	Rare, mild	None	Constant dorsal	Frequent, dorso-lumbar
Psychomotor retardation	Severe	None	Mild	Severe	Extreme	None	Mild
Behavior	Calm, friendly disposition	Normal	Normal	Stubborn, destructive	Restless, withdrawn	Normal	Normal
Dysostosis	Early severe	Minimal	Moderate	Severe	Minimal	Severe typical	Early severe

*Several important clinical features, though usually of consistent severity within one syndrome, are of much less or no differential diagnostic value. They are: macrocephaly, hernia, hepatosplenomegaly, hirsutism, upper respiratory infections, otitis or hearing loss, restriction of motion in all joints, and heart problems and failure.

with severely impaired mental function. Neurologic examination is difficult and often fails to demonstrate consistent findings. Convulsions are observed in some patients. Swallowing difficulties make special care necessary at feeding time, growth decelerates, bone dysplasia remains minor or absent, hepatomegaly decreases, cardiovascular symptoms are infrequent, and sexual maturation is delayed and often incomplete. Most patients do not lose independent locomotion before the age of 12, well beyond the time at which mental contact with the surroundings is lost. Some patients walk alone until early adulthood. The final stage in the disease's natural course is reached when the patient becomes bedridden. Swallowing difficulty becomes a major problem, severe pulmonary infections occur frequently, and wide irregular fluctuations in body temperature occur (probably a reflection of central nervous system dysfunction). Any of these complications may prove fatal.

Early diagnosis of Sanfilippo disease is now possible; it should be based on a demonstration of the specific enzymatic defect. Methods are available in specialized laboratories. In Figure 18–5 the enzyme defects responsible for each of the four types of Sanfilippo disease are illustrated. Arrows indicate the chemical bonds catalytically split by each enzyme. In type A (MPS-IIIA) Sanfilippo disease, sulfamate sulfatase is inactive. This lysosomal enzyme normally removes sulfate from the unique nonacetylated amino group on the second carbon atom (C_2) of glucosamine (GlcN), one of the monosaccharide components of the sulfated polysaccharide, heparan sulfate. In this biopolymer, glucosamine or N-acetylglucosamine (GlcNAc) residues at the reducing end are α-glycosidically linked to iduronic or glucuronic acid. This covalent α-glycosidic bond is enzymatically catabolized by N-acetyl-α-glucosaminidase, the lysosomal enzyme deficient in patients with type B (MPS-IIIB) Sanfilippo syndrome. This particular enzyme, however, can work effectively only if the α-glycosidically linked glucosamine is first N-acetylated at the C_2 position. This synthetic reaction is normally mediated by acetyl CoA—α-glucosaminide N-acetyltransferase, a tightly membrane bound enzyme that is deficient in patients with type C (MPS-IIIC) Sanfilippo syndrome.

Finally, a few examples of Sanfilippo D disease (MPS-IIID) have been recorded in patients in whom the breakdown of heparan sulfate is inadequate because of deficient activity of the enzyme N-acetylglucosamine-6-sulfate sulfatase (Kresse et al., 1981). In Figure 18–5 the metabolic roles of α-

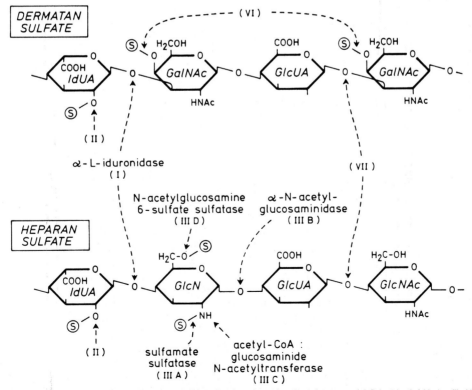

Figure 18–5. Metabolism of some acid mucopolysaccharides showing sites of action of enzymes deficient in the Hurler (I), Hunter (II), and Sly diseases (VII). Accumulation of heparan sulfate results from deficiency of any one of four enzymes (III A to D) in four genetically different types of Sanfilippo disease. Mutant genes interfere with normal child development and behavior.

L-iduronidase, iduronate-2-sulfate sulfatase, and α-glucuronidase (deficient in MPS-I, MPS-II, and MPS-VII, respectively) are also schematically indicated. Tables 18–3 and 18–4 make it clear that α-L-iduronidase deficiency, in a still unexplained manner, is involved in three clinically different disorders—the severe Hurler disease (MPS-IH), the very mild Scheie disease (MPS-IS), and the clinically intermediate entity symbolized as MPS-IM.

It is impossible to distinguish clinically among the four types of Sanfilippo diseases in an individual patient. From a study of 75 patients the impression arises that the onset of MPS-IIIA occurs earlier than that of MPS-IIIB and MPS-IIIC (Van de Kamp, 1981). Behavioral changes constitute the most frequent initial complaint in MPS-IIIA. Decreasing mental and speech development is more often the presenting feature in Sanfilippo B disease. The age of onset may be slightly later and the physical signs less pronounced in MPS-IIIB and C. Patients with Sanfilippo C disease appear to survive longer. One MPS-IIIC patient is known who started regular first grade and who survives though bedridden at the age of 22. Another such patient died at 27 years of age. In view of the few known MPS-IIID patients, any summarizing statement would be premature.

Before leaving this subject, some comments regarding the genetic aspects of human behavior are appropriate. Genetic defects of enzymes involved primarily in the hydrolysis of heparan sulfate linkages lead for the most part to severe mental and behavioral disturbances. Those defects that interfere with the catabolism of heparan sulfate and dermatan sulfate cause progressive alterations in skeletal and connective tissue in addition to central nervous system and mental disease. One can compare the features in Table 18–4.

The role of acid mucopolysaccharides clearly involves more than their being constituents of the extracellular matrix of all tissues. They are components of many subcellular organelles and likely also of the plasma membrane of most mammalian cells, including neurons. Their turnover is relatively rapid, and at some stage in metabolism they play a role in intercellular communication, tissue growth, and tissue maintenance. Whatever their exact role may be, mucopolysaccharides represent only one of the many molecules involved in normal brain function.

From this single example of Sanfilippo syndrome it should be clear that the number of genes and enzymes contributing to normal behavior must be vast indeed. As in all other inborn errors of metabolism listed in Table 18–3, mental function in Sanfilippo syndrome is adversely affected in general, not some specific mental functions at the exclusion of others. This observation supports the view that intelligence and human behavior are unitary attributes and not composed of a set of independent abilities. The latter, although distinguished by many elegant psychologic tests, probably should be viewed more as illustrating the extent of human behavioral variability rather than representing independent phenotypes governed by separate genes. Like normal hearing or vision, which can be considered schematically as end results of circuits of consecutively operating components connected in series, brain function can be viewed in such a reductionistic way. Normal behavior, however, must be the result of a complex web of many interconnecting circuits. There are a large number of ways in which one defective component can lead to abnormal function of the entire system. There are, moreover, a much greater number of ways in which the composite circuits function and relate to the environment normally, resulting in the fascinating individuality of human behavior. Mendelian heredity is basic to this individuality. Formulating hypotheses inspired by the mendelian paradigm of gene action holds the greatest promise for uncovering information about the biologic nature of the individual components in the circuit of human brain function.

The Duchenne Type of Muscular Dystrophy

Provided the physical aspects of general behavior in a child or adolescent are stressed, the Duchenne type of muscular dystrophy, an X linked recessive disease, can be cited as a tragic example, in boys, of increasing interference with mobility and locomotion beginning in childhood. This is a chronically progressive disorder, involving primarily skeletal muscle, with a fatal outcome in early adulthood. It is characterized by increasing muscle weakness and paralysis most apparent initially around the limb girdles. Histologically there is a gradual loss of skeletal muscle fibers, which are replaced by connective and adipose tissue.

Motor development in a boy with the Duchenne type of muscular dystrophy is frequently but not invariably delayed, and it is the inadequacy of gross motor function and paucity of movement in comparison with normal siblings that cause the first parental concerns. Observant parents may relate that their healthy children cry more forcefully and show more hip rocking movements while sitting on the parents' knees or standing with support than the patient. Well before the formal diagnosis of the motor handicap, the patients' difficulties in pulling themselves to standing are greater than those in unassisted walking. Early, however, affected boys assume a waddling, broad based gait. The patients soon show the tendency to walk on their toes. When walking, they typically adopt lumbar lordosis in order to achieve better support of their own body weight. They temporarily benefit from muscle and tendon retractions in maintaining these compensating mechanisms in locomotion.

Several important aspects of the natural course of Duchenne muscular dystrophy and of the multidisciplinary management of the patients will not be discussed here. The reader should consult specialized texts (Dubowitz, 1978). (See also Chapter 39G.)

It is now generally accepted that probably all patients with the Duchenne type of muscular dystrophy have a mild to moderate intellectual handicap, which is not solely related to the progressive and ultimately nearly complete motor disability. This can be demonstrated in many patients well before the age of 10 to 12 years when independent walking usually ceases and confinement to a wheelchair becomes a necessity. Formal learning ability and abstract thinking are more impaired than the patient's general emotional awareness and purposeful life. Fortunately his personal conflicts are usually adequately recognized and appropriately dealt with within the family or special schooling and management programs. Secondary orthopedic abnormalities, which become severe once ambulation is lost (such as equinovarus changes in the feet and severe scoliosis), constitute major problems of management but are rarely of any personal concern to the patients themselves. They are similarly indifferent to the obesity that is a frequent development at this stage. Anxiety and difficulty sleeping at night become increasingly significant problems in the final years and are most likely related to increasing failure of cardiac muscle.

Thus Duchenne muscular dystrophy is one of several examples of a progressive genetic disorder that does not primarily involve the central nervous system, but that both directly and indirectly affects development and behavior.

VARIANT BEHAVIOR IN NONPROGRESSIVE GENETIC DISEASE

Hemophilia A—A Treatable Disorder

Hemophilia A is the most common and most well known of the congenital disorders of coagulation. It involves a single coagulation protein called antihemophilic globulin. Its absence or malfunctioning is inherited as an X linked recessive trait and provokes easy or spontaneous bleeding, almost exclusively in boys. The clinical severity depends a great deal on the residual amount of functional antihemophilic globulin present. Patients with mild hemophilia bleed excessively only

with trauma or severe surgery. Children who have 1 to 5 per cent of the normal amount of antihemophilic globulin may bleed severely after minor injury and only occasionally have spontaneous hemorrhages. Episodes of spontaneous bleeding, involving the skin, mucous membranes, viscera, and joints, are observed in patients with virtually no circulating antihemophilic globulin and who may develop crippling hemarthroses and joint contractures.

Hemophilia A is treated with fresh frozen plasma, cryoprecipitate, or antihemophilic globulin concentrate, depending on the type and severity of bleeding or the type of prophylaxis desired.

Although the prognosis for a useful normal life is good for most patients, this treatable hereditary disorder often affects the patients' behavior, since trauma, minor accidents, and even slight physical exercise are to be avoided. The disease may affect the patient's physical development, particularly in the event of recurrent intra-articular bleeding and its orthopedic consequences. Hemophilia affects the patient and his surrounding family members mentally, emotionally, and socially. Helping children adapt favorably to the clinical implications of the disease is a multidisciplinary challenge. By itself hemophilia is a nonprogressive phenotypic trait, but the cumulative effect of its complications may progressively compound the patient's handicap.

Hereditary Short Stature

The nosology of growth deficiency is quite complex but is not the topic of this section. Many different genetic and some environmental causes of short stature have been recognized so that a classification based on etiology or specific pathogenic defects has become feasible (Rimoin and Horton, 1978). Forms of small stature associated with either intellectual handicap or physical malformations are not considered here.

Short stature can be a familial trait. It is surprising how many short parents seek expert medical advice because of slow growth or small stature in their children. Familial short stature is frequent. It often is the basis of psychologic problems in parents and adolescent children of either sex. Similar remarks may be made regarding the temporary worries in adolescents in whom there is a constitutional delay of sexual maturation and pubertal growth. Height prediction charts can be valuable aids in reassuring a boy or girl about the anticipated normal height. Shortness of stature, when objectively considered, is not and should not be a reason or explanation for unusual behavior, but shortness of stature of either a temporary or at least a nonprogressive nature may subjectively be an important source of mental or emotional problems. Fortunately such psychologic problems are not always encountered in the type of patient just discussed. They are more consistently present in patients whose adult stature is well below the normal height distribution in the population. The objective physical problems in patients with dwarfism, irrespective of its endocrinologic or direct skeletal origin, are often compounded by their own psychologic perception of themselves as human beings. Moreover, the surrounding society itself commonly adopts a subjectively special attitude toward dwarfs such that their actions and deeds are all too easily labeled as unusual.

Ultimately emotional deprivation may cause the so-called psychosocial type of growth deficiency. This is not merely a counterexample to illustrate the importance of environment instead of genotype in physical and mental development. Catch-up growth observed subsequent to removal of the child from the unfortunate psychologic environment not only is proof of the diagnosis but shows in addition that even in physical phenotypic traits, environment and genotype may interact in undefined ways.

Nonprogressive Psychomotor Retardation and Mental Disability

Atypical development and behavior are frequently the phenotypic expressions by which psychomotor retardation is first suspected in a child and may be the earliest indicators of future mental disability. Medical and other professions share an important responsibility for the early diagnosis of mental retardation followed by planning and recommendations for optimal care and management.

The many types of patients with mental handicaps can be classified according to criteria such as age, etiology, neurologic syndrome, degree of mental deficiency, behavioral features, or type of general nursing care likely to be required. The etiologic criterion for subdividing patients with mental handicap is the most ideal one with regard to prevention, assessment of clinical course and implications, management, and therapy. Unfortunately etiology-based classifications leave a considerable proportion of patients in the category of "unknown cause."

Attention is focused here exclusively on the group of patients with "pure" mental disability fulfilling the following criteria (Becker et al., 1977): Patients are physically normal, and phenotypic evidence of any syndrome of multiple congenital anomalies, including gross abnormality of the central nervous system, is lacking; the natural course is nonprogressive, and there is no clinical or chemical evidence of an underlying inborn error of metabolism; the clinical history does not suggest any environmental cause for the mental handicap.

Obviously this category of patients is clinically and etiologically very heterogeneous. It contains many severely handicapped patients living either at home or in special institutions, as well as a large group of patients with mild to moderate mental disability. Neurologic examination reveals few objective findings other than immature or primitive gross or fine motor function and inadequate coordination. Spurious movements and general hyperkinetic or hypokinetic behavior are frequent. Seizures are sometimes observed. There is usually a generalized learning disability accompanied by moderate to severe language deficit. Rarely only reading or other learning abilities are specifically deficient in mildly affected patients.

A consideration of the genetic aspects of these many forms of learning disability and mental deficiency automatically involves factors of etiology, pathogenesis, and prevention. The heterogeneous phenotypic trait of pure mental disability as already defined is more likely to have a monogenic or a multifactorial cause than to be due to chromosomal aneuploidy. Karyotypic abnormalities are usually characterized by dysmorphic features and short stature in association with mental retardation. The marker X chromosome syndrome associated with the cytogenetic feature of a fragile site located distally on the long arm of a single X chromosome is not an exception to this rule, because conceptually it can be considered to be monogenically determined (Table 18–1). Careful pedigree analysis, necessary for each proband, may provide evidence of the segregation of a mutant gene. Even the isolated patient in a pedigree does not rule out monogenic inheritance (Table 18–2). Only in the case of mild familial mental handicap may the theory of a polygenic or multifactorial etiology be difficult to disprove. Such a hypothesis is based on the assumption that low normal intellects in the "normal" members in the patients' families represent the lower extreme of the normal distribution of mental abilities in the population. Once again this hypothesis does not include any assessment of the relative roles of heredity and environment in mildly affected individuals.

The pathogenetic mechanisms involved in mental disability remain largely unknown. The many genetic defects involved must exert their unfavorable influence at some stage during embryonic or fetal life. All defects appear to involve structural elements or metabolic functions having a physiologic role limited to a particular time of cell proliferation and differentiation, organ and tissue growth, or cell migration. Therefore, at a particular time in postnatal life, what parents and medical consultants observe are signs and symptoms resulting from residual nonprogressive damage. The congenital brain damage may involve gross anatomic or histologic structural changes, ineffective cell to cell contact, failure of chemical or physical interneuronal contacts, or one of many more types of defective cerebral organization and operation. Thus there are a great number of interferences with brain development due to gene mutation. The majority have the same unfortunate phenotypic consequence—moderate to severe generalized learning disability.

A consideration of the genetics of behavioral and developmental deficiency automatically leads to the design of specific strategies for its prevention, management, or treatment.

TREATMENT, PREVENTION, AND MANAGEMENT

TREATMENT

Cure—Not in the Immediate Future

A true cure of any of the monogenic defects listed in Table 18–3, through substitution of a segment of mutant DNA by a corresponding one of a "wild-type," is unlikely to be accomplished in the near future. In the lay press, results of current recombinant DNA research are too often depicted as miraculous advances promising gene therapy in the immediate future. Detailed knowledge of the structure of limited parts of the human genome is being expanded rapidly by the application of a set of interrelated biochemical and microbiologic techniques often referred to as genetic engineering (Miller, 1981). Mammalian genes have been characterized following their insertion into plasmids within rapidly multiplying bacteria. Further maneuvers have included the insertion of genes in vitro into cultured mutant cells and the successful transfer of foreign genes into oocytes or bone marrow cells. In a few instances even phenotypic expression of the newly inserted genes has been demonstrated. Enthusiasm about these elementary successes, however, should not obviate critical thinking about future gene therapy. If it should become technically feasible, critical studies in animal models should have answered at least the following questions before any attempt is made at clinical application: Does the inserted gene remain permanently present in the cell nuclei of the target tissue? Does the gene remain unaltered? Does it have any adverse effect? Is its phenotypic expression consistent and adequately regulated (Anderson and Fletcher, 1980; Mercola and Cline, 1980)?

Future gene therapy will necessarily depend on detailed knowledge of the gene involved and of its gene products. From the scientific and the humanitarian viewpoint, the mendelian paradigm provides a sound basis for any type of treatment, since recognition of the individual nature of any variant or mutant gene and a consistent endeavour to individualize any therapeutic approach are inherent to it.

Incurable Does Not Mean Untreatable

It is not true that hereditary disorders are by definition untreatable. The minimal requisite for any type of treatment is sufficient knowledge about the primary enzyme defect. This subject has often been reviewed in a more general scope (Desnick and Grabowski, 1981, Rosenberg, 1979). The following discussion is limited to some disorders listed in Table 18–3.

There are several theoretical possibilities for exogenous manipulation of abnormal metabolism (see footnotes for Table 18–3). Dietary restriction is aimed at reducing or eliminating the adverse effects of the accumulating substrate or its unusual byproducts (Fig. 18–2, Part II). A well known example is the low phenylalanine diet prescribed for patients with classic phenylketonuria (Table 18–3E). Such dietary restriction initiated in the neonatal period prevents damage to the developing brain and the resulting severe psychomotor disability by excessive amounts of phenylalanine. The benefits of such treatment are beyond doubt, yet several questions remain regarding the modality of patient monitoring (Levy, 1979; Rosenberg, 1979). More or less effective treatment by dietary restriction is also available in other disorders of aminoacid metabolism (Table 18–3E), in some inborn errors of carbohydrate (Table 18–3A), and even in one error of lipid metabolism (Table 18–3B).

The several strategies of substrate depletion constitute an alternative way to manipulate inborn errors favorably. Plasmapheresis, one of these methods, failed to be clinically effective in Fabry disease and Gaucher disease (Table 18–3B), but when combined with dietary restriction of phytanic acid, it contributed to clinical improvement in Refsum disease (Table 18–3B; Desnick and Grabowski, 1981; Moser et al., 1979). The concentration of some unmetabolized compounds can be lowered by the administration of drugs that either chelate "substrates" or promote their urinary excretion. A well known example of this third approach is the effective treatment of patients with Wilson disease by D-penicillamine (Table 18–3H). The use of existing alternative pathways for enhanced excretion of accumulating substrates has been attempted in patients with urea cycle defects. This approach should be the subject of further scientific exploration. In acutely ill neonates with the latter disorders, peritoneal dialysis and exchange transfusion can temporarily control severe hyperammonemia but cannot be considered as long term measures. Low protein diets, sometimes supplemented with arginine, have proven to be effective in lowering the blood ammonia level. Administration of keto acids, while supporting growth as substitutes for corresponding amino acids, also provide a nitrogen sparing effect resulting in a decrease in the blood urea level (Moser et al., 1979; Shih, 1978). These forms of treatment, however, although valuable and logical, are rarely successful in severely affected infants. Portacaval shunt surgery has resulted in clinical improvement in some patients with glycogenosis, because this surgical bypass increases the availability of glucose to the peripheral tissues (Table 18–3A).

The neurologic and behavioral signs in patients with Lesch-Nyhan syndrome appear not to be related to the increased production of uric acid (Table 18–3G). Inhibitors of uric acid production, such as allopurinol, are effective only in preventing kidney stone formation, urate nephropathy, and occasionally gouty arthritis (Kelley and Wyngaarden, 1978). The recent observation of specifically decreased function of dopaminergic neuron terminals in the striatal part of the brain in patients with the Lesch-Nyhan syndrome and the absence of substantial morphologic changes indicates that hypoxanthine guanine phosphoribosyl transferase deficiency also adversely affects specific neurotransmitter production or function in an as yet unknown manner (Lloyd et al., 1981). This exemplifies once again the complexity of gene interaction in normal brain function and behavior and prompts consideration of new designs for effective therapy in this condition.

Treatment can also consist of replacement of the metabolic product not produced because of the patient's inborn error of metabolism. The effective treatment of congenital hypothyroidism by thyroxine is the best known example. It most certainly prevents inadequate growth and severe mental deficiency. Administration of uridine to patients with orotic aciduria who cannot synthesize this essential pyrimidine is usually effective. Replacement of antihemophilic globulin has often been a life saving procedure in patients with hemophilia A. The need for a fundamental understanding of the primary defect in an inborn error of metabolism before considering therapeutic approaches is further illustrated by the fact that all attempts at the administration of copper ions to patients with Menkes kinky hair syndrome, who apparently do not absorb copper normally, have been ineffective (Garnica et al., 1977, Table 18–3H).

The activity of enzymes can be altered by certain drugs. Phenobarbital, for example, is used as treatment in patients with the Crigler-Najjar syndrome, because it stimulates microsomal glucuronyl transferase. Furthermore, the hyperactivity of amino-

levulinate synthetase, which is probably responsible for intermittent hepatic porphyria, is repressed by hematin. Treatment with hematin is valuable in suppressing the acute attacks characteristic of this type of porphyria (Table 18–3H).

Inborn errors of metabolism can be the consequence of gene mutations, which interfere with binding or molecular interaction between apoenzyme and coenzyme molecules. Administration of pharmacologic doses of cofactor can have therapeutic results. Cofactors must normally be changed into active coenzymes by several enzymatic reactions. If one of the latter enzymatic steps is defective, active coenzyme must be supplied. This is an alternative explanation for the favorable effect of high doses of vitamins in some patients with a particular inborn error. The better known examples cited in Table 18–3E are propionic acidemia due to a defect in biotin biosynthesis, vitamin B_6 responsive homocystinuria and vitamin B_{12} responsive homocystinuria, and the vitamin B_{12} responsive forms of methylmalonic acidemia.

Because the substitution of defective genes is as yet not possible, gene product therapy should be the most desirable alternative. Allotransplantation of cells, tissues, or whole organs could provide functional enzyme to the mutant organism. Theoretical and technical aspects, therapeutic potential and limitations, and the variety of types of allotransplantation have recently been reviewed (Desnick and Grabowski, 1981). There are few examples with unequivocal clinical benefit. Renal transplantation in patients with Fabry disease has yielded inconsistent and at best temporary results in heterozygous women. The value of fibroblast transplantation in patients with some of the mucopolysaccharidoses remains to be determined.

Although prospects of more ample availability of some gene products through genetic engineering are brighter than ever (Hopwood, 1981), the long term goal of any type of enzyme replacement therapy aiming at the delivery of sufficient amounts of pure, highly functional and stable enzyme molecules to the proper intracellular site of action in the mutant target tissue(s) remains remote. Early clinical trials with the various types of enzyme replacement therapy have yielded no positive results. At best this type of experimentation in humans has probably not been physiologically harmful, but it may have caused psychologic damage in families, especially if unfounded hopes were raised. The failure of enzyme replacement therapy could have been predicted, because of an overwhelming lack of knowledge, which must still be overcome by future scientific work, preferably in animal models, before human trials of enzyme therapy can start anew. The half-life of the administered enzymes is too short, and they do not reach specific target organs such as the brain.

Knowledge of the physiopathology of the patient's problems does not become available automatically be mere identification of the primary enzyme defect in a monogenic disorder.

If no treatment is currently available for many enzymatically defined metabolic disorders, and none can be promised for the foreseeable future, causal treatment for less well defined genetic disorders of human development and behavior is certainly not feasible. Thus, the management of genetic disorders in general in essence becomes an object of preventive medicine.

PREVENTION

Unfortunately, because many types of inadequate child development and mental disability belong to the category of etiologically undefined disorders, preventive measures, based solely on empiric observation, cannot be applied. The more thoroughly the genetic basis of a disorder is defined, the more effective programs of prevention are likely to be.

In addition to the mendelian paradigm of genotype-phenotype correlation in human variability, ethical principles should always guide any public health application of human genetics to patients, families, or society, irrespective of the therapeutic or preventive value.

Preventive measures can be either prospective or retrospective. Genetic screening is the most important type of prospective prevention available today. Early diagnosis in the patient, genetic counseling, and prenatal diagnosis with subsequent elective interruption of pregnancy are mainly retrospective means of preventing genetic disease.

Screening Programs

Screening can be applied to entire populations or can be directed at detecting certain genetic carrier states. The subject has been adequately reviewed in recent years (Kaback, 1977; Levy, 1973). Detailed knowledge of the specific inborn error of metabolism is the scientific prerequisite in either type of screening. The goal of population screening is the detection of all presymptomatic affected individuals with the exclusion of all non-affected individuals in order to promptly initiate effective treatment. The efficacy and even feasibility of such screening depends on the availability of simple, reliable, and inexpensive tests. Phenylketonuria screening has been available for several years as a routine test for all newborns in many countries around the world. Follow-up of test results by experienced biochemists and pediatricians is of paramount importance. With equivocal test results tests should be repeated in order

to rule out false positive reactions and to insure that low phenylalanine diets are prescribed only when necessary. In blood obtained for neonatal screening, other more rarely occurring disorders of amino acid metabolism can be diagnosed, among which maple syrup urine disease, homocystinuria, tyrosinemia, and histidinemia are the best known (Table 18–3D; Levy, 1973). Galactosemia may be screened for in the same neonatal blood sample (Table 18–3A). More recently, neonatal screening for congenital hypothyroidism, an effectively treatable disorder more common than phenylketonuria, has been added to the list. As in phenylketonuria, early treatment of congenital hypothyroidism prevents severe mental disability. Cost-benefit analysis of the results of neonatal screening, an aspect of interest mainly to policy makers, is only a one dimensional assessment of its true value. An objective measurement of the alleviation of human suffering achieved can never be made.

Heterozygote screening as a measure for preventing autosomal recessive disease is aimed at identifying carrier couples. It is practical only in subpopulations with a high incidence of a particular disorder and thus a high frequency of a mutant gene. In several countries programs of this type have contributed considerably to reducing the incidence of Tay-Sachs disease in the Askhenazi Jewish population (Kaback, 1977). They have been successful for several reasons, one of which is the general availability of prenatal diagnosis of Tay-Sachs disease. Potentially important examples of heterozygote screening are sickle cell anemia and the thalassemias. Such programs, however, have not always been well received in the subpopulations at higher risk, which serves to illustrate the necessity for community education and counseling prior to initiation of any type of screening.

Genetic Counseling; Prenatal Diagnosis

There are numerous excellent texts that discuss all aspects of genetic counseling, the most specific activity of clinical geneticists. The concept encompasses availability of conclusions, as well as communication and guidance. At the time of the genetic counseling session, conclusions are available in regard to the diagnosis in previously affected individuals and the recurrence risk in siblings, children, or other family members. These conclusions are communicated by the counselor to the inquiring family or, if possible, to the patient. The communication should be sincere, objective, and didactic. It must leave time and opportunity for questions by those being counseled and for expansion and repetition by the counselor. Genetic counseling should also include information regarding clinical implications and prognosis, particularly with regard to progressive disorders. It should inform and provide nondirective advice about medical management of the patient and the social and educational assistance available. Parents and other relatives need guidance about the reproductive options open to them for dealing with the recurrence risk. They often require support during the process of deciding whether to refrain from having further children or to have future pregnancies monitored by amniocentesis and appropriate prenatal diagnostic techniques, if feasible. Thus genetic counseling has been called a prelude to prenatal diagnosis (Milunksy, 1979).

Early diagnosis in the proband provides the retrospective foundation and knowledge of the pathogenesis or intrauterine expression of the genetic disorder, the prerequisite for determining the technical strategy for effective prevention by prenatal diagnosis. In the event of a fetus at risk for any type of aneuploidy, the amniotic fluid cells obtained by transabdominal amniocentesis between 14 and 16 weeks of gestation are cultured for karyotypic analysis (see Chapter 19).

For the diagnosis of an inborn error of metabolism, the particular enzyme activity defective in a proband must be expressed in cultured amniotic cells. Furthermore, the activity in the heterozygous fetus must be distinguishable from that in the homozygous affected one. The feasibility of prenatal diagnosis by enzyme assay in amniotic fluid cells is indicated in Table 18–3 for the monogenic metabolic disorders listed. Assay of α-fetoprotein in amniotic fluid is of importance in the prenatal detection of neural tube defects such as anencephaly and meningomyelocoele or spina bifida aperta, which were previously mentioned among the multifactorial threshold traits (see page 329). Real-time ultrasonographic observation, recommended for localization of the placenta during amniocentesis, can also detect these physical anomalies of the fetus before or during the early part of the second trimester of pregnancy. Ultrasound study becomes very important in routine monitoring of pregnancies. Such diagnostic techniques are bound to contribute in a prospective way to the early intrauterine detection of major physical defects. Prenatal diagnosis of skeletal defects or dysplasias has also been achieved, but in most instances only in at-risk third trimester fetuses. Fetoscopy allows direct visualization of parts of the fetus and direct vision puncture of placental vessels for obtaining fetal blood or specimens of fetal skin. For indications, further technical aspects, and drawbacks of the procedures used in amniocentesis and prenatal diagnosis, the reader is referred to a recent comprehensive textbook by Milunski (1979). The same source can be used effectively for study of the topic of elective abortion and its medicolegal, moral, and ethical aspects.

Obviously conditions interfering with normal psychomotor and intellectual development or resulting in unusual or abnormal behavior that are not associated with a chromosomal anomaly or known metabolic defect cannot be diagnosed in utero. Prevention must then rely on early diagnosis in index patients and genetic counseling.

MANAGEMENT: THE PATIENT AND HIS FAMILY

There is one principal difference between the management of genetic traits of inadequate development and unusual behavior through measures of preventive medicine and the management of human needs, and medical problems of patients. The biologic principle of genetic individuality strongly supports the humanitarian opinion that these children and adults are merely variant human personalities, whose development, appearance, and behavior have been adversely affected by the effect of one or a small number of many genes. These people fully share the basic rights to an adequate quality of life and to a climate of general well-being and happiness. Advances in theoretical knowledge in neurobiology are laudable but should not divert attention from the basic daily needs of the patients, who may often derive direct benefit from progress in practical medicine, applied psychology, and special education. The definition of many syndromes is constantly improving, as is our insight into their natural course and clinical implications. In most areas of medicine, clinical measures for dealing with complications have become more effective. It is highly desirable that the initial diagnostic evaluation be multidisciplinary, with full consideration given to medical, psychologic, social, and educational aspects. This broad type of evaluation insures a more individualized approach to management of the patient's problems. Hospital admissions should be kept to a minimum and be as brief as possible. Parents and other family members should be informed fully and honestly. The possibility of care and management in the normal environment of the home, if feasible and practicable, must be fully explored, with options of nursery school, day care programs, and educational opportunities in the community kept open. Special stimulation projects should not be explained to parents as forms of treatment instituted to improve the patients' mental or learning ability but as ways to develop the children's strengths and as potentially interesting means for making them happier human beings. Supportive management of the patients cannot be effective without simultaneous and long term guidance and counseling for their parents and healthy siblings.

Acknowledgments

The critical reading of the manuscript by Dr. A. Garnica is gratefully acknowledged. Gratitude is expressed also to Mrs. D. Van Godtsenhoven for skillful typewriting of tables and text and to Mrs. K. Denecker for drawing the illustrations.

JULES G. LEROY

REFERENCES

Anderson, W. F., and Fletcher, J. C.: Gene therapy in human beings: when is it ethical to begin. N. Engl. J. Med., *303*:1293, 1980.

Beadle, G. W., and Tatum, E. L.: Genetic control of biochemical reactions in neurospora. Proc. Nat. Acad. Sci. U.S., 27:499, 1941.

Becker, J. M., Kaveggia, E. G., Pendleton, E., and Opitz, J. M.: A biologic and genetic study of 40 cases of severe pure mental retardation. Europ. J. Pediatr., *124*:231, 1977.

Bouchard, F. J., and McGue, M.: Familial studies of intelligence: a review. Science, *212*:1055, 1981.

Brown, M. S., Kovanen, P. T., and Goldstein, J. L.: Regulation of plasma cholesterol by lipoprotein receptors. Science, *212*:628, 1981.

Carter, C. O.: Genetics of common disorders. Br. Med. Bull., *25*:52, 1969.

Childs, B.: Genetic analysis of human behavior. Ann. Rev. Med., 23:373, 1972.

Craighead, J. E.: Current views on the etiology of insulin-dependent diabetes mellitus. N. Engl. J. Med., *299*:1439, 1978.

Crow, J. F.: Problems of ascertainment in the analysis of family data. *In* Neel, J. V., Shaw, M. W., and Schull, W. J. (Editors): Genetics and the Epidemiology of Chronic Diseases. Washington, D.C., Department of Health, Education and Welfare, 1965.

DeFries, J. C., Vandenberg, S. G., and McClearn, G. E.: Genetics of specific cognitive abilities. Ann. Rev. Gent., *10*:178, 1976.

Desnick, R. J., and Grabowski, G. A.: Advances in the treatment of inherited metabolic diseases. Adv. Hum. Genet., *11*:281, 1981.

Dubowitz, V.: Muscle Disorders in Childhood. Philadelphia, W. B. Saunders Company, 1978.

Falconer, D. S.: Introduction to Quantitative Genetics. New York, Ronald Press, 1960.

Falconer, D.S.: The inheritance of liability to certain diseases estimated from the incidence among relatives. Ann. Hum. Gent., *29*:51, 1965.

Feldman, M. W., and Lewontin, R. C.: The heritability hang-up. Science, *190*:1163, 1975.

Garnica, A. D., Frias, J. L., and Rennert, O. M.: Menkes kinky hair syndrome: is it a treatable disorder? Clin. Genet., *11*:154, 1977.

Grouse, L. D., Schrier, B. K., Letendre, C. H., and Nelson, P. G.: RNA sequence complexity in central nervous system development and plasticity. Curr. Top. Dev. Biol., *16*:381, 1980.

Guilford, J. P.: Intelligence has three facets. Science, *160*:615, 1968.

Harris, H.: The Principles of human Biochemical Genetics. Eds. Amsterdam, Elsevier/North Holland Biomedical Press, 1980.

Harris, W. A.: Neural activity and development. Ann. Rev. Physiol., *43*:689, 1981.

Herbst, D. S., Dunn, H. G., Dill, F. J., Kalousek, D. K., and Krywaniuk, L. W.: Further delineation of X-linked mental retardation. Hum. Genet., *58*:366, 1981.

Herbst, D. C., and Miller, J. R.: Nonspecific X-linked mental retardation. II. The frequency in British Columbia. Hum. Genet., *7*:461, 1980.

Hirsch, J.: Behavior-genetic, or "experimental" analysis: Am. Psychol. *22*:118, 1967.

Hopwood, D. A.: The genetic programming of industrial microorganisms. Sci. Am., *245*:67, 1981.

Kaback, M. M. (Editor): Tay-Sachs Disease: Screening and Prevention. New York, Alan R. Liss, Inc., 1977.

Kelley, W. N., and Wyngaarden, J. B.: The Lesch-Nyhan syndrome. *In* Stanbury, J. B., Wyngaarden, J. B., and Fredrickson, D. S. (Editors): The Metabolic Basis of Inherited Disease. Ed. 4. New York, McGraw-Hill Book Co., 1978.

Kresse, H., Cantz, M., Von Figura, K., Glössel, J., and Paschke, E.: The mucopolysaccharidoses: biochemistry and clinical symptoms. Klin. Wochenschr., *59*:867, 1981.

Leroy, J. G.: Management of the mucopolysaccharidoses and allied disorders. *In* Papadatos, C. J., and Bartsocas, C. S. (Editors): The Management of Genetic Disorders. New York, Alan R. Liss, Inc., 1979.

Levy, H. L.: Genetic screening. Adv. Hum. Genet., *4*:1, 1973.

Levy, H. L.: Treatment of phenylketonuria. *In* Papadatos, C.J. and Bartsocas, C. S. (Editors): The Management of Genetic Disorders. New York, Alan R. Liss, Inc., 1979.

Lewontin, R. C.: Genetic aspects of intelligence. Ann. Rev. Gent., *9*:387, 1975.

Lloyd, K. G., Hornykiewitz, O., Davidson, L., Shannah, K., Farley, I., Goldstein, M., Shibuya, M., Kelley, W., and Fox, I. H.: Biochemical evidence of dysfunction of neurotransmitters in the Lesch-Nyhan syndrome. N. Engl. J. Med. *305*:1106, 1981.

McKusick, V. A.: Mendelian Inheritance in Man. Catalogs of Autosomal Dominant, Autosomal Recessive and X-linked Phenotypes. Ed. 5. Baltimore, Johns Hopkins University Press, 1978.

McKusick, V. A., Neufeld, E. F., and Kelly, T. E.: The mucopolysaccharide storage diseases. *In* Stanbury, J. B., Wyngaarden, J. B., and Fredrickson, D. S. (Editors): The Metabolic Basis of Inherited Disease. Ed. 4. New York, McGraw-Hill Book Co., 1978.

McMahan, D.: Chemical messengers in development: a hypothesis. Science, *185*:1012, 1974.

Melnick, M., Shields, E. D., Bixler, D., and Conneally, P. M.: Facial clefting: an alternative biologic explanation for its complex etiology. Birth Defects; *13*:93, 1977.

Mercola, K. E., and Cline. M. J.: The potentials of inserting new genetic information. N. Engl. J. Med., *303*:1297, 1980.

Miller, W. L.: Recombinant DNA and the pediatrician. J. Pediatr., *99*:1, 1981.

Milunsky, A. (Editor): Genetic Disorders of the Fetus. Diagnosis, Prevention and Treatment. New York, Plenus Press, 1979.

Moser, H. W., Batshaw, M. L., Murray, C. T., Braine, H., and Brusilow, S. W.: Management of heritable disorders of the urea cycle and of Refsum's and Fabry's disease. *In* Papadatos, C. J., and Bartsocas, C. S. (Editors): The Management of Genetic Disorders. New York, Alan R. Liss, Inc., 1979.

Naftolin, F.: Understanding the bases of sex differences. Science, *211*:1263, 1981.

O'Brien, J. S.: Generalized gangliosidosis. Birth Defects, *5*:190, 1969.

Rimoin, D. L., and Horton, W. A.: Short stature. J. Pediatr., *92*:523, 697, 1978.

Rosenberg, L. E.: Therapeutic modalities for genetic diseases: an overview. *In* Papadatos, C. J., and Bartsocas, C. S. (Editors): The Management of Genetic Disorders. New York, Alan R. Liss, Inc., 1979.

Shih, V.: Urea cycle disorders and other congenital hyperammonemic syndromes. *In* Stanbury, J. B., Wyngaarden, J. B., and Fredrickson, D. S. (Editors): The Metabolic Basis of Inherited Disease. Ed. 4. New York, McGraw-Hill Book Co., 1978.

Simpson, J. L.: Disorders of Sexual Differentiation. New York, Academic Press, Inc., 1976.

Spranger, J. W.: Morphological aspects of the mucopolysaccharidoses. *In* Holter, J., and Ireland, J. T. (Editors): Inborn Errors of Skin, Hair, and Connective Tissue. Baltimore, University Park Press, 1975.

Srb, A. M., Owen, R. D., and Edgar, R. S.: General Genetics. Ed. 2. San Francisco, W. H. Freeman & Co. Publishers, 1965.

Stanbury, J. B., Wyngaarden, B. J., and Fredrickson, D. S. (Editors): The Metabolic Basis of Inherited Disease. Ed. 4. New York, McGraw-Hill Book Co., 1978.

Tomkins, G. M.: The metabolic code. Science, *189*:760, 1975.

Turner, G., and Opitz, J. M.: X-linked mental retardation. Am. J. Med. Genet., *7*:407, 1980.

Van de Kamp, J. J. P., Niermeyer, M. F., von Figura, K., and Giesberts, M. A. H.: Genetic heterogeneity and clinical variability in the Sanfilippo syndrome (types A, B and C.) Clin. Genet., *20*:152, 1981.

Vandenberg, S. G.: Hereditary abilities in man. *In* Oliverio, A. (Editor): Genetics, Environment and Intelligence. Amsterdam, Elsevier/North Holland Biomedical Press, 1977.

Vogel, F., and Motulsky, A. G.: Human Genetics. Problems and Approaches. Berlin, Springer Verlag, 1979.

19

Chromosomal Determinants

19A Chromosomal Disorders Other Than Down Syndrome

Cytogenetic abnormalities comprise perhaps the largest group of clearly defined genetic causes of developmental disability. Among individuals living within a residential institution, approximately 10 per cent have been found to have a chromosomal abnormality that can reasonably be considered to be the etiology of the mental deficiency. Since the surveys upon which these figures are based used older and less fruitful techniques than are presently available, it is reasonable to believe that these numbers actually under-represent the true proportion.

HISTORICAL ASPECTS

Down syndrome, or trisomy 21, the first chromosomal abnormality recognized in humans, was identified as such in 1959. During the next few years, a rapid succession of additional chromosomal abnormalities were reported, most of which were also associated with mental retardation. These initially included trisomy for two other autosomes (trisomy 13 and included trisomy 18). Shortly thereafter more subtle lesions, such as deletions of part of a chromosome (e.g., the cri du chat syndrome associated with deletion of chromosome 5p) and unbalanced translocations, were detected. More recently the development of methods for demonstrating very small deletions in chromosomes, by the use of what is called high resolution cytogenetics, has shown that some well known syndromes (e.g., Prader-Willi) may also result from such deletions.

Trisomies, unbalanced translocations, and deletions all represent quantitative changes in the total amount of genetic material. Despite the complexity of the cytogenetic mechanisms often involved, they can thus be considered as differing principally in degree, rather than in kind. The discovery of a new type of chromosomal change, known as a fragile site, has added an additional dimension to

the study of chromosomal abnormalities. The fragile site on the distal end of the long arm of the X chromosome is now known to be associated with X linked mental retardation. The frequency of this new disorder is believed to be second only to trisomy 21 among chromosomal abnormalities causing moderate mental retardation.

OVERVIEW AND GENERAL PRINCIPLES

In this section we review the general aspects of the types of chromosomal disorders associated with developmental disability. The major new findings regarding chromosomal abnormalities are discussed in detail in a subsequent section.

AUTOSOMAL TRISOMY AND MONOSOMY

Trisomy for an autosome occurs with high frequency among conceptuses, but the majority of these abnormalities cause fetal death. There is a significant difference among the various trisomies in the proportion of the fetuses that survive to birth. It is this difference in viability that is believed largely to determine the frequency of the various trisomies among livebirths. Monosomy for an autosome, as opposed to trisomy, is very rare. On theoretical grounds, however, monosomy should occur as often as trisomy, and thus it is presumed that the absence of monosomy among livebirths is indicative of its extreme lethality in utero. In instances in which individuals monosomic for a small part of a chromosome are born alive, it is possible to compare the clinical severity of partial monosomy with partial trisomy for the same chromosomal segment. Invariably the monosomic state is more morbid than the trisomic state. This supports the belief that monosomy for an entire autosome is nearly always lethal. It should be noted that this principal applies only to the autosomes, since monosomy for a sex chromosome occurs relatively frequently among liveborns in the form of Turner syndrome. (The individual with Turner syndrome has a single sex chromosome, an X. A Y chromosome does not occur without an X.)

Nearly every change in the amount of autosomal material, whether it be a complete trisomy, partial trisomy, or partial monosomy, is associated with developmental disability. The few quantitative changes that appear not to have clinical consequences are restricted in their location. In general, they affect regions of chromosomes that are believed not to contain critical genes. These regions include the short arms of the acrocentric chromosomes (chromosomes 13, 14, 15, 21, and 22), the secondary constrictions (on chromosomes 1, 9, and 16), and the long arm of the Y.

Notwithstanding the number of mentally retarded patients with chromosomal abnormalities that have been studied, it is still not known why a chromosomal abnormality leads to mental retardation. It is speculated that the functional and morphologic complexity of the brain must require a similar complexity in the genes governing its formation. If so, and if these genes are scattered throughout the chromosomal complement, most chromosomal changes would be likely to affect one or more "brain genes."

As noted, some chromosomal abnormalities do not cause clinical disturbances. Given that this is so, how can one be certain that the chromosomal change present in a particular child is responsible for his difficulties? This is one of the more difficult questions to answer, especially to the satisfaction of the family. If patients with similar clinical findings associated with similar chromosomal changes have been described in the literature, it is probable that the chromosomal change is the etiologic factor. Nonetheless, since the best cytogenetic technique available makes it possible to detect only relatively large changes in the amount of genetic material (containing tens to hundreds of genes), there is always some hesitancy in claiming that two patients have identical chromosomal lesions. In the majority of instances, however, this rule proves to be valid.

When there are few or no reports in the literature of patients with a similar chromosomal disorder or with similar clinical findings, the problem becomes much more complex. It is then necessary to examine other members of the family to determine whether they have a similar chromosomal change and whether they are clinically affected. When the chromosomal change is familial, the interpretation is reasonably straightforward. If familial occurrence cannot be demonstrated, the relation of chromosomal change to the patient's abnormality must remain uncertain.

MOSAICISM FOR A CHROMOSOMAL ABNORMALITY

As the experience with Down syndrome demonstrates, a small proportion of such patients may have some cells with a normal chromosomal complement. This phenomenon has also been observed with other disorders, such as trisomies 18 and 13. With a few chromosomal disorders, mosaicism (the simultaneous presence of two or more cell lines, each with a distinctive karyotype) may be found in a high proportion of patients. With trisomy 8, for example, nearly all affected individuals are mosaic. The diagnostician must be aware of this propensity, since the examination of only a limited number of cells in patients with mosaicism may fail to detect the chromosomal abnormality. Further, the clinical findings in patients with mosaicism vary markedly, depending upon the proportion of abnormal cells within each individual organ.

SEX CHROMOSOME ABNORMALITIES

Sex chromosome aneuploidy (an abnormal number of chromosomes) is also associated with developmental problems, but in general these are less severe than those associated with autosomal aneuploidy. Females with an XO karyotype (Turner syndrome) are not usually developmentally delayed. The mean I.Q. of males with an XXY karyotype (Klinefelter syndrome) is said to be slightly diminished, although many of these males function within the normal range of intelligence. It has been suggested that their verbal ability is depressed when compared with performance ability. In roughly one-third of XXY males, emotional development is delayed. Data relating to boys with an XYY karyotype and girls with an XXX karyotype are less informative. It is important to note that although there may be subtle developmental difficulties among these individuals, the majority function within the range of normal. In those situations in which there are greater numbers of sex chromosomes, however, significant developmental delay and mental retardation are likely to be present.

THE RECURRENCE RISK FOR CHROMOSOMAL ABNORMALITIES

The mechanisms underlying the various chromosomal disorders are quite heterogeneous, and the recurrence risk varies according to the individual abnormality. The information concerning recurrence of trisomy 21 appears to apply without modification to the other autosomal trisomies; that is, the rate of recurrence is believed to increase exponentially with increasing maternal age. Recurrence is probably not limited to the original trisomy, but instead includes the occurrence of any autosomal trisomy.

With familial chromosomal translocations, the risk that a carrier will have liveborn, abnormal progeny varies tremendously from translocation to translocation. It is probable that survival in utero is the primary determinant of the frequency with which abnormal progeny will be produced. Because the variability is so great, it is often necessary to use the prior experience of the individual family to determine the risk of recurrence of abnormal progeny within that family.

NEW SYNDROMES ASSOCIATED WITH CHROMOSOMAL ABNORMALITIES

THE FRAGILE X SYNDROME

Perhaps the most significant cytogenetic advance related to developmental disorders in recent years has been the identification of the fragile X chromosome. A distinct syndrome with an X linked pattern of inheritance is now associated with this chromosomal abnormality. The impact of this discovery on the field of mental retardation will be tremendous, because the fragile X disorder is a relatively common cause of mental retardation.

The history of the discovery of the fragile X chromosome is fascinating. For years it had been a puzzling observation that mental retardation is more common among males than among females. Also, for many years the medical literature has included reports of large families in which many of the male relatives were mentally retarded but in which female relatives were generally unaffected. An historically important family of this type was reported by Renpenning in 1962. The affected males in this family had no striking physical characteristics, although most were microcephalic. Subsequently male individuals with probable X linked mental retardation without striking physical characteristics were said to have the "Renpenning syndrome."

In 1969 Lubs reported his observations in regard to the chromosomes of a family with an apparently X linked form of mental retardation. He found that the X chromosome in the mentally retarded males showed a distinct structural abnormality. The distal end of the long arm frequently possessed a constriction, so that the tip appeared to be connected to the rest of the chromosome by a

thin region that frequently was broken. (The breakable region subsequently became known as a fragile site.) This work generated little immediate excitement, because other investigators were unable to identify other examples of this chromosome marker. It was not until the late 1970's that the "fragile X" chromosome was demonstrated in many other families.

Much of the credit for this development belongs to Grant Sutherland, an Australian cytogeneticist. While studying individuals with fragile sites on various chromosomes, Sutherland (1977) discovered that the ability to detect the fragile site on the X chromosome depended upon the type of culture medium used. Surprisingly, the medium popular in the 1960's was satisfactory for detecting the fragile X, whereas the supposedly "improved" and enriched media of the 1970's inhibited its demonstration. One of the principal reasons proved to be a lack of folate and thymidine in the medium. An analogous effect can be achieved by inhibiting the enzyme thymidylate synthetase by the addition to the medium of the antimetabolite fluorodeoxyuridine. It is possible that in many individuals in whom chromosome analyses have been performed in the past, the fragile X may not have been detected because the culture conditions required for its demonstration were not used.

The location of the fragile site is consistent from cell to cell, from individual to individual, and from family to family. This site is designated as fra(X) (q27) or (q28). Although fragile sites also occur on autosomes, often as a familial characteristic, the fragile site on the X chromosome is the only one known to be associated with clinical abnormality.

The clinical features present in males with the fragile X chromosome are now relatively well characterized (Table 19–1). Affected males are generally short in stature with a normal to increased head circumference. Facial features include a prominent forehead, large ears, midfacial hypoplasia, and a prominent symphysis of the mandi-

Table 19–1. CHARACTERISTICS OF MALES WITH THE FRAGILE X SYNDROME

Phenotypic features
  Normal or increased head circumference
  "Long" face
  Prominent chin
  Midfacial hypoplasia
  Macro-orchidism
Developmental features
  Mild to severe mental retardation
  Speech or language abnormality
  Autism (in some patients)
Cytogenetic finding
  Fragile site on long arm of X chromosome (in 4 to 40% of cells)
Mode of inheritance
  X linked recessive (but some females are affected)

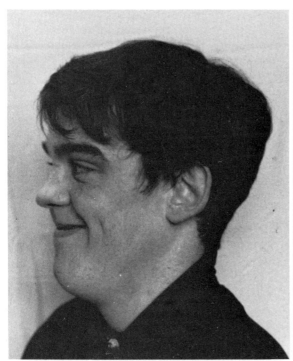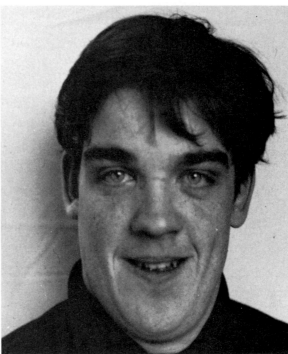

Figure 19–1. Moderately retarded 18 year old male with fragile X syndrome. Note large head, long face, and prominent mandible.

ble (Fig. 19–1). Large hands and feet are a frequent finding. Most affected postpubertal males and some children, for an as yet undetermined reason, have large testes (Sutherland and Ashforth, 1979). The presence of slightly enlarged testes actually led one examiner to suspect the diagnosis in an infant at four months of age.

Male individuals with a fragile X chromosome are typically developmentally delayed. The majority have mild to moderate mental retardation, but a spectrum from borderline intelligence to severe mental retardation has been observed. Several males with a fragile X chromosome have been reported from historical evidence to be "normal" in intelligence. The evidence was probably inadequate to exclude less severe developmental problems, such as a learning disability or a speech problem, such as apraxia. Some investigators have called attention to a characteristic speech pattern in affected individuals. There is evidence that classic infantile autism may be the presenting diagnosis in some males with fragile X chromosome (Brown et al., 1982; Meryash et al., 1982).

The expression of the fragile X in heterozygous women is extremely variable. The studies of Turner et al. (1980) indicate that as many as one-third of heterozygotes have significant cognitive impairment. The remaining two-thirds are thought not to be affected, although there are no reports detailing the results of intelligence testing among heterozygotes. As yet there is no firm explanation to account for this remarkable variation in expres-

sion. The percentage of cells exhibiting the fragile X is very low in heterozygotes (commonly less that 1 per cent), which incidentally makes demonstration of the carrier state very difficult. Heterozygotes who have an intellectual deficit, however, generally have a fragile X chromosome in a larger proportion of their cells. The presence of affected women in fragile X families has led an occasional observer to question initially the X linked nature of the disorder. Both affected women and affected men are fertile and are capable of transmitting the fragile X condition.

Not all kindreds with X linked mental retardation represent instances of the fragile X (see Chapter 18). Ironically, the family that Renpenning described has recently been found not to have the fragile X condition. It is apparent that several conditions are included within the group called nonspecific X linked mental retardation. It is estimated that 30 to 40 per cent of all kindreds with X linked mental retardation have the fragile X. It is not known how many different disorders are represented among the remaining families. It also should not be forgotten that the ability to demonstrate the fragile site in no way provides an understanding of how the fragile site relates to the clinical findings.

It is estimated that the fragile X syndrome occurs about once in several thousand live male births. Studies have shown that a minimum of 2 per cent of the male residents of schools for mentally retarded persons have the fragile X. It is the second most common cytogenetic cause of mental retardation, exceeded only by Down syndrome. Because individuals with the fragile X syndrome are usually members of the more adequately functioning segment of the mentally retarded population, it is apparent that many affected individuals—male and female—do not live in state institutions.

INDIVIDUALS WITH AN INVERTED DUPLICATED 15p

An extra small and unidentified chromosome, approximately the size of chromosome 21, is sometimes found during chromosomal analysis in developmentally delayed individuals. Both the banding pattern and the morphology distinguish this extra chromosome from a chromosome 21 or 22. This abnormality has often been referred to as a "marker" chromosome and symbolized as 47,XX (or XY), + mar. Surveys of mentally retarded populations have revealed this finding in a number of individuals.

Recently, more sophisticated staining techniques have enabled cytogeneticists to identify the extra small chromosome as one of several different entities. These different entities have been reported with sufficient frequency to permit the characterization of recognizable clinical syndromes associated with developmental disorders (Table 19–2).

Insight into the one of these disorders was provided initially by Schreck et al. (1977). They were able to show, in six of 12 developmentally disabled individuals with this marker chromosome, that the abnormal chromosome was derived from chromosome 15. This study employed antibodies specific for 5-methylcytidine, a nucleotide that occurs in high concentration in the short arm of chromosome 15, to demonstrate that the extra chromosomal material was derived from this chromosome. The abnormal chromosome was found to contain two copies of the chromosome 15 short arm. The extra chromosome has been designated an inverted duplicated 15p.

Over 30 individuals with the inverted duplicated 15p chromosome have been described. Some degree of developmental delay was present in all. Typically the patients were hypotonic and motor and speech milestones were delayed. The delay was global and nonspecific when compared with that in other developmentally disabled children. Many had a forward leaning, unstable gait and had a propensity to stumble and fall. Many developed a major motor seizure disorder, often during the second half of childhood or the second decade of life. Common physical findings included a slight downward slant of the palpebral fissures, a short philtrum, low set, posteriorly rotated ears, and minimal syndactyly of the second and third toes.

The chromosomes in the parents of these individuals have been normal. The extra chromosome in most, but not all, individuals is of maternal origin. It has been tentatively suggested that there may be a parental age effect.

INDIVIDUALS WITH AN ABNORMAL CHROMOSOME DERIVED FROM CHROMOSOME 22

The second variety of small extra chromosome was originally described by Schachenman et al. in 1965, in association with an imperforate anus and ocular coloboma. Patients with this association commonly also have preauricular tags or fistulas, congenital heart disease, and urinary tract abnormalities. This association is frequently referred to as the "cat eye syndrome," in reference to the suggestion of a vertical pupil produced by the iridial coloboma. Affected children are frequently mildly to moderately retarded, although many perform in the borderline or normal range of intelligence.

Early studies of this extra chromosome failed to determine its origin. The length of the extra chromosome varied from individual to individual, and mosaicism was frequently present. In a few instances the mosaicism has been present in multiple generations. The phenotypic expression of the syndrome varied markedly, ranging from near normality to lethal malformations. This variation may have its basis partly in the mosaicism and partly in differences in the specific genetic material present in the extra chromosome. Later studies, using recently developed banding techniques, have shown that the extra chromosome probably is derived from chromosome 22 by an undetermined mechanism.

Table 19–2. CLINICAL FEATURES IN SYNDROMES ASSOCIATED WITH DEVELOPMENTAL DISABILITY AND AN EXTRA SMALL CHROMOSOME

Inv. Dup. (15p)	"Cat Eye" Syndrome	Prader-Willi Syndrome
Upslanting eyes	Upslanting eyes	In infancy, hypotonia and failure to thrive; in childhood, hyperphagia and obesity
Low set ears	Colobomas	
Abnormal gait	Preauricular skin tags or pits	
Seizures	Congenital heart defect	
	Anal atresia with fitula	Hypogonadism
		Short stature
		Small hands and feet

Other patients with an extra small chromosome have been reported as having features both similar to and different from those of the cat eye syndrome. In some of these individuals, the extra chromosome was found to arise by inheritance from a parent with a balanced translocation between the long arm of chromosome 22 and the long arm of a chromosome 11 (Schinzel et al., 1981). (The balanced 11/22 translocation is one of the most frequently encountered reciprocal translocations in man.) Individuals with the unbalanced translocation frequently have deep set eyes, a flat nose, a prominent upper lip, a receding mandible, preauricular pits or tags, male genital hypoplasia, anal atresia or other anal anomalies, cleft palate, and congenital heart defects. Patients who have the unbalanced 11/22 translocation are all profoundly retarded, whereas those with the "cat eye syndrome" are often mildly retarded or of borderline or normal intelligence. Individuals have been described in whom the extra chromosome material has been identified as being a single band from chromosome 22. They have some features in common with individuals with the more complete cat eye syndrome. Commonly they have down-slanting palpebral fissures, preauricular malformation, and developmental delay. Anal atresia has been seen in about one-third of these individuals.

THE PRADER-WILLI SYNDROME

Prader, Labhart, and Willi in 1956 described a syndrome characterized by obesity (in childhood), short stature, cryptorchidism, and mental retardation, and a history (in infancy) of hypotonia and failure to thrive. An uncontrollable appetite in childhood is a particularly distinctive feature of this syndrome.

It is now appreciated that the history of the child with the Prader-Willi syndrome can be divided into two distinct phases. The first phase, which begins at birth, lasts from birth to about two years and is characterized by growth failure, feeding difficulty, and hypotonia. The pediatrician should consider the Prader-Willi syndrome in the differential diagnosis in a child of this age group who is failing to thrive or who has central hypotonia. The hypotonia gradually diminishes during early childhood. The second phase is heralded by the onset of hyperphagia and a decreased perception of satiety. Children in this phase rapidly become obese. A well structured program of diet control and behavior modification, however, if started early, has been shown to be helpful in controlling the obesity in some of these children.

Vanja Holm (1981) divided the symptoms of the Prader-Willi syndrome into four categories: symptoms essential for diagnosis, symptoms occurring in 50 to 90 per cent of the cases, those occurring in 10 to 50 per cent of the cases, and rare but important symptoms occurring in less than 10 per cent of the cases. The symptoms considered to be essential for diagnosis are infantile central hypotonia, hypogonadism, obesity, dysfunctional central nervous system performance, dysmorphic facial features, and short stature. A professional who is familiar with developmentally delayed children may well agree that most of these features, with the exception of hypogonadism, are general ones often found among this population. Until recently the clinician had to rely solely on clinical judgment to make the diagnosis of the Prader-Willi syndrome. The task was particularly difficult because there are several syndromes with similarities to the Prader-Willi syndrome. These include the Laurence-Moon syndrome, Fröhlich's adiposogenital dystrophy, the Bardet-Biedl syndrome, the Summitt syndrome, the Carpenter syndrome, and the Kallmann syndrome.

Now it is believed that a specific abnormality of chromosome 15 is the cause of many instances of the Prader-Willi syndrome. Ledbetter et al. (1981) have described four patients with the Prader-Willi syndrome in whom a very small segment was missing from the long arm of a chromosome 15. This discovery was the result of the use of high resolution chromosome banding. In each of these four patients the segment was missing from the region of chromosome 15 that extends from the q11 band to the q13 band. Ledbetter et al. became interested in chromosome 15 because a review of the literature revealed eight patients with the Prader-Willi syndrome in whom translocations of chromosome 15 were found. Ledbetter and his colleagues hypothesized that deletions of chromosome 15 might have occurred simultaneously with the translocations.

The report of Ledbetter et al., as well as many unpublished accounts, provides strong evidence that many (but not all) instances of the Prader-Willi syndrome are associated with deletion of a critical segment of chromosome 15. At the present time no final explanation exists for patients who clinically are diagnosable as having the Prader-Willi syndrome but who have no detectable abnormality of chromosome 15.

There have been a few reports of children with the Prader-Willi syndrome who have been found to have an extra small chromosome. (In some instances the extra chromosome was shown to be derived from chromosome 15.) The relation of this chromosomal abnormality to the deletion of chromosome 15 is unclear.

One eight year old boy has been seen with moderate mental retardation, short stature, a tendency to overeat, obesity, and undescended testicles. His parents report that he was a relatively inactive infant who gained weight poorly. The diagnosis of Prader-Willi syndrome was made, and the child was enrolled in a program for dietary management and psychotherapy along with other children with this diagnosis. (Interestingly, the parents of the other children with Prader-Willi syndrome remarked that this boy did not physically or behaviorally resemble their children.) Chromosome analysis did not reveal any abnormality in chromosome 15, but a fragile X chromosome was present in 20 per cent of his metaphase spreads. This patient dramatizes the difficulty in diagnosing the Prader-Willi syndrome on clinical grounds alone.

Among the reported children with the Prader-Willi syndrome are some who have an "atypical" form of the syndrome. In one series of 23 patients, for example, six were listed as atypical. Five of these were not significantly obese or were only transiently obese. It is probable that the Prader-Willi syndrome represents several different etiologies or several degrees of the same influential process.

OTHER DELETION SYNDROMES

A useful comparison can be made with two other clinical entities that have been found to be associated with mental retardation and with deletion of specific chromosome segments: the aniridia–Wilms tumor association, and some instances of retinoblastoma.

Congenital aniridia generally has been attributed to an autosomal dominant gene with a high penetrance. The prevalence of the defect in the general population is estimated to be 1 to 2 cases per 100,000. In the mid-1960's it became apparent that aniridia was unusually common among children with Wilms tumor. Six of 440 children with Wilms tumor in one study were found to have aniridia; another study found seven children with aniridia among 28 hospitalized for Wilms tumor. The aniridia in these patients was sporadic rather than familial. It was also noted that in children with this association there was a high frequency of nonocular abnormalities. In one series there was microcephaly in 10 of 18 children, retarded growth in 9 of 16, anomalies of the pinnae in 7 of 12 children, and genitourinary anomalies in 12 of the 22 children. Thirteen of 19 children were reported as being mentally retarded, but detailed developmental information was not provided.

Chromosome analyses in this group of patients were initially reported as demonstrating normal karyotypes. As was the case with the Prader-Willi syndrome, the abnormality was revealed only by the use of improved banding techniques. It has now been shown repeatedly that children with aniridia and a Wilms tumor have a small deletion of chromosome number 11 (Riccardi et al., 1978). The deletion always involves the 11p13 band.

Aniridia, and some instances of Wilms tumor, had been considered to result from single gene mutations. It seems likely that the deletion of chromosome 11p13 includes a number of genes and that the absence of one (or more) results in aniridia, whereas the loss of others leads to the Wilms tumor and to other abnormalities. The number of genes involved, and hence the number of abnormalities with which a given individual is afflicted, may depend on the size and location of the deleted segment. Thus, children who have growth retardation or microcephaly, as well as aniridia and Wilms tumor, presumably are missing the portion of the chromosome containing genes important for normal growth and brain development.

An analogous association exists between bilateral retinoblastoma and a deletion of the long arm of chromosome 13 (Lele et al., 1963). A distinct syndrome with retinoblastoma, growth retardation, a characteristic facies, and moderate mental retardation occurs in individuals with this chromosomal abnormality. The syndrome occurs in only a small percentage of patients with retinoblastoma. Again it is presumed that the developmental defect results from the loss of specific genes that are required for normal development.

These two examples suggest that it might be possible to explain the individual differences among Prader-Willi patients on a similar basis. It is likely that the size of the segment of chromosome 15 that has been deleted varies from patient to patient. The features in a particular child with the Prader-Willi syndrome might then depend upon which and how many genes are missing.

SUMMARY

From this discussion it should be apparent that chromosomal abnormalities are becoming increasingly important with respect to the etiology of developmental disability. As cytogenetic techniques have improved, more disorders have been found to have a chromosomal basis. Some of the newer tools now being developed, such as those involving recombinant DNA, will add a dimension to the field. It must be remembered that despite the advances we have already seen, the relationship between cytogenetic abnormalities and the pathogenesis of the developmental disability remains unknown. The physician caring for individuals with developmental disability must attempt to keep apprised of relevant cytogenetic advances.

PARK S. GERALD
DAVID L. MERYASH

REFERENCES

Brown, W. T., Friedman, E., Jenkins, E. C., Brooks, J., Wisniewski, K., Raguthu, S., and French, J. H.: Association of fragile X syndrome with autism. Lancet, 1:100, 1982.

Holm, V. A.: The diagnosis of Prader-Willi syndrome. In Holm, V. A., Sulzbacher, S. J., and Pipes, P. L. (Editors): Prader-Willi Syndrome. Baltimore, University Park Press, 1981.

Ledbetter, D. H., Riccardi, V. M., Airhart, S. D., Strobel, R. J., Keenan, B. S., and Crawford, J. D.: Deletions of chromosome 15 as a cause of the Prader-Willi syndrome. N. Eng. J. Med., 304:325, 1981.

Lele, K. P., Penrose, L. S., and Stallard, H. B.: Chromosome deletion in a case of retinoblastoma. Ann. Hum. Genet., 27:171, 1963.

Lubs, H. A.: A marker X chromosome. Am. J. Hum. Genet., 21:231, 1969.

Meryash, D. L., Szymanski, L. S., and Gerald, P. S.: Infantile autism associated with the fragile-X syndrome. J. Autism Develop. Disorders, 12:295, 1982.

Prader, A., Labhart, A., and Willi, H.: Ein syndrom von adipositas, kleinwuchs, kryptorchismus und oligophrenie nach myotonieartigem zustand im neugeborenenalter. Schweiz. Med. Wochenschr., 86:1260, 1956.

Renpenning, H., Gerrard, J. W., Zaleski, W. A., and Tabata, T.: Familial sex-linked mental retardation. Canad. Med. Assn. J., 87:954, 1962.

Riccardi, V. M., Sujansky, E., Smith, A. C., and Francke, U.: Chromosomal imbalance in the aniridia–Wilms tumor association: 11p interstitial deletion. Pediatrics, 61:604, 1978.

Schachenman, G., Schmid, W., Fraccaro, M., Mannini, A., Tiepolo, L., Perona, G. P., and Sartori, E.: Chromosomes in coloboma and anal atresia. Lancet, 2:290, 1965.

Schinzel, A., Schmid, W., Auf der Maur, P., Moser, H., Degenhardt, K. H., Geisler, M., and Grubisic, A.: Incomplete trisomy 22. I. familial 11/22 translocation with 3:1 meiotic disjunction. Delineation of a common clinical picture and report of nine new cases from six families. Hum. Genet., 56:249, 1981.

Schreck, R. R., Breg, W. R., Erlanger, B. F., and Miller, O. J.: Preferential derivation of abnormal human G-group-like chromosomes from chromosome 15. Hum. Genet., 36:1, 1977.

Sutherland, G. R.: Fragile sites on human chromosomes: demonstration of their dependence on the type of tissue culture medium. Science, 197:265, 1977.

Sutherland, G. R., and Ashforth, P. L. C.: X-linked mental retardation with macro-orchidism and the fragile site at Xq27 or 28. Hum. Genet., 48:117, 1979.

Turner, G., Brookwell, R., Daniel, A., Selikowitz, M., and Zilibowitz, M.: Heterozygous expression of X-linked mental retardation and C-chromosome marker fra(X)(q27). N. Eng. J. Med., 303:662, 1980.

19B The Child with Down Syndrome

Although John Langdon Down has been credited with first describing the condition that bears his name today, there were others, such as Esquirol in 1838 and Séguin in 1846, who prior to Down had reported children with characteristics resembling Down syndrome. In 1866, the same year that Down published his article, Duncan noted a girl "with a small round head, Chinese-looking eyes, projecting a large tongue who only knew a few words." Down's great contribution, however, was his recognition of the specific physical characteristics and his description of the condition as a distinct and separate entity: "the hair is not black as in the real Mongol, but of a brownish colour, straight and scanty. The face is flat and broad and destitute of prominence. The cheeks are rounded and extended laterally. The eyes are obliquely placed and the internal canthi more than normally distant from another. The palpebral fissure is very narrow . . . the lips are large and thick with transverse fissures. The tongue is long, thick, and is much roughened. The nose is small" (Down, 1866).

Subsequent to Down's report, numerous publications described additional phenotypic features in Down syndrome. Also, many reports dealt with speculations pertaining to the etiology, epidemiology, and treatment approaches in Down syndrome.

DEFINITION

The child with Down syndrome is a human being with a recognizable phenotype and limited intellectual endowment due to the presence of a supernumerary chromosome 21. It is thought that the excessive genetic material on a section of the lower arm of chromosome 21 and its interaction with other gene functions result in altered homeostasis leading to aberrant physical and central nervous system development.

EPIDEMIOLOGIC CONSIDERATIONS

Down syndrome is the most frequently occurring autosomal chromosome abnormality in man. It has been estimated that the present incidence rate is 1.0 to 1.2 per 1000 live births, whereas 20 years ago it reportedly was 1.6 per 1000. The decrease in the incidence over the past two decades is related to the reduction of the birth rate in women of advanced age (Hook, 1982). Although the risk for older mothers of having a chromosomally abnormal child did not change over the years, at the present time only 20 per cent of children with Down syndrome are born to mothers over 35 years of age, yet 20 years ago 50 per cent of the children with Down syndrome were offspring of older mothers.

Down syndrome has been noted to occur in all races. Recent data from the Center for Disease Control in Atlanta indicate a slight trend to a higher incidence rate in whites than in blacks (20 per cent), but this difference is not significant. It is of interest that a study in West Jerusalem revealed a high crude incidence rate (2 per 1000) for Down syndrome in non-Europeans (Harlap, 1973). With the exception of the latter study, the incidence rate in various socioeconomic classes is essentially the same.

ETIOLOGY

During the past century many hypotheses concerning the causation of Down syndrome have been reported (Warkany, 1960). After the chromosomal abnormality in Down syndrome had been discovered in 1959, investigators focused on the nondisjunctional event, and new theories of causation evolved:

1. It has been suggested that there may be a genetic predisposition to nondisjunction. Evidence in support of this theory is derived from epidemiologic studies indicating a higher risk of recurrence if there is a child with Down syndrome in the family and from patients with double aneuploidy.

2. Uchida (1981) reported that about 30 per cent of the mothers who gave birth to children with Down syndrome underwent abdominal radiation prior to conception. Alberman et al. (1972) noted a time delay between maternal radiation and the birth of a child with Down syndrome. Other investigators, however, did not observe a relationship between radiation and chromosomal aberrations (Hook, 1982).

3. Infectious diseases have been brought into the discussion as a cause of Down syndrome (Nichols, 1966; Stoller and Collmann, 1965), yet many other researchers were unable to confirm that viral diseases result in nondisjunction.

4. Another etiologic factor that has been considered is autoimmunity, particularly thyroid autoimmunity and associated thyroid disease. Fialkow's studies (1966) showed consistent differences in the presence of thyroid autoantibodies between age matched mothers of offspring with Down syndrome and control mothers.

5. Since the incidence of Down syndrome increases significantly with advancing maternal age, it has been suggested that hormonal alterations in older women might lead to nondisjunction. Endocrinologic changes, such as increased androgen secretion, decreased hydroepiandrosterone levels, a decreased systemic estradiol concentration, changes in hormone receptor concentrations, and sharply increased levels of luteinizing hormone and follicular stimulating hormone immediately before and during menopause, could increase the chance of nondisjunction.

Still other factors, such as intragametic accidents, factors relating to satellite association and nucleolar organizers, chemicals, and coital frequency, have been discussed as possible causes of Down syndrome (Crowley et al., 1982).

Although there remain many unanswered questions concerning etiologic aspects in Down syndrome, recent reports in cytogenetic and epide-

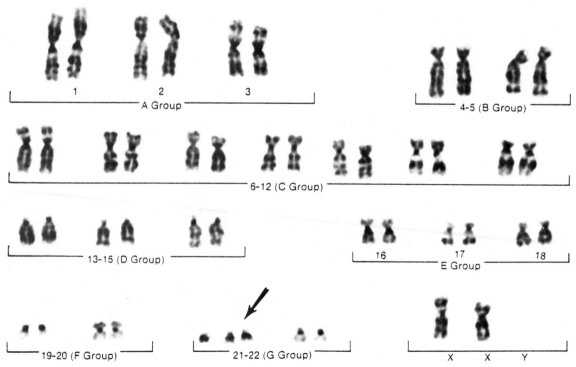

Figure 19–2. Karyotype of a female with Down syndrome, or trisomy 21 [47,XX,+21].

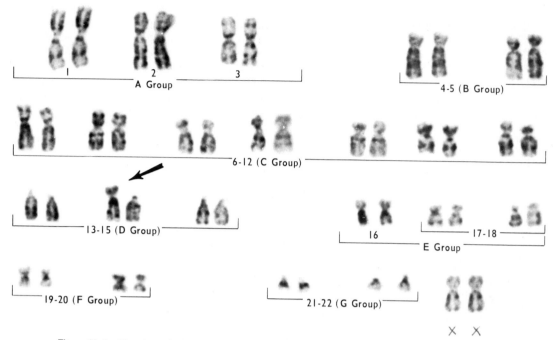

Figure 19–3.   Karyotype of a female with Down syndrome with 14/21 translocation [46,XX, − 14, + t(14q21q)].

miologic studies favor the concept of multiple causality.

CYTOGENETICS

In the early 1930's investigators already suspected that Down syndrome might be due to a chromosomal aberration. When in the mid-1950's new cytogenetic techniques allowed better visualization and more accurate studies of chromosomes, it soon was found that the child with Down syndrome had a supernumerary acrocentric chromosome (Lejeune et al., 1959). In subsequent years other chromosomal abnormalities, such as translocation and mosaicism, were reported to be associated with the phenotype of Down syndrome. Approximately 92 to 95 per cent of the children with Down syndrome have trisomy 21, as depicted in Figure 19–2, while the prevalence of translocation in various series of karyotyped individuals with Down syndrome has been reported to be between 4.8 and 6.3 per cent (Thuline, 1982). Most of the Robertsonian translocations involve attachment of the long arms of a chromosome 21 to the long arms of chromosome 14, 21, or 22 (Fig. 19–3). Translocations of chromosome 21 onto other chromosomes as well as the tandem form of translocation are rare. If a translocation is discovered in a child with Down syndrome, the parents' chromosomes are studied in order to determine whether one of the parents is a balanced carrier. This is, of course, important for genetic counseling purposes.

Mosaicism occurs in 1 to 3 per cent of the children with Down syndrome. Although in trisomy 21 and translocation the nondisjunctional event is thought to take place preconceptionally, in mosaicism nondisjunction most likely occurs during one of the first mitotic cell divisions.

An extremely rare chromosomal aberration is partial trisomy 21, when only a specific segment of the lower arm of chromosome 21 is translocated onto another chromosome (Pueschel et al., 1980).

Because of the well known maternal age effect in Down syndrome, previous investigations focused primarily on maternal aspects. In recent years it has been reported that there is also a paternal age effect, which, however, is not as highly correlated with an increased incidence of Down syndrome as advanced maternal age. In addition, cytogenetic studies of parents of children with Down syndrome revealed that in 20 to 30 per cent of the cases, the extra chromosome 21 is of paternal origin (Magenis and Chamberlin, 1981).

PHENOTYPIC CHARACTERISTICS

Since Down's first description (Down, 1866), many investigators have reported various physical stigmata observed in children with Down syndrome. Coleman (1978) mentioned that there are more than 300 abnormal features in Down syndrome. Table 19–3 presents the frequencies of the phenotypic characteristics most often noted in infants with Down syndrome (Pueschel, 1982).

Table 19–3. FREQUENCY (%) OF POSITIVE PHENOTYPIC FINDINGS IN INFANTS WITH DOWN SYNDROME

Sagittal suture separated	98
Oblique palpebral fissure	98
Wide space between first and second toes	96
False fontanel	95
Plantar crease between first and second toes	94
Hyperflexibility	91
Increased neck tissue	87
Abnormally shaped palate	85
Hypoplastic nose	83
Muscle weakness	81
Hypotonia	77
Brushfield spots	75
Mouth kept open	65
Protruding tongue	58
Epicanthal folds	57
Single palmar crease, left hand	55
Single palmar crease, right hand	52
Brachyclinodactyly, left hand	51
Brachyclinodactyly, right hand	50
Increased interpupillary distance	47
Short stubby hands	38
Flattened occiput	35
Abnormal size of ears	34
Short stubby feet	33
Abnormal structure of ears	28
Abnormal implantation of ears	16
Other hand abnormalities	13
Other abnormal eye findings	11
Syndactyly	11
Other foot abnormalities	8
Other oral abnormalities	2

Other observers' phenotypic descriptions may differ because they often examined children with Down syndrome at an older age. It is well known that certain characteristics change over time. Some stigmata such as the epicanthal folds or the initially abundant neck tissue become less prominent as the child grows, whereas others such as the fissured tongue or dental anomalies become more apparent with increasing age. Also the degree of mental deficiency and short stature are noted only in the older child.

Since some of the characteristics in the child with Down syndrome occur at a high frequency, they are considered to be "typical," leading many investigators to advocate "cardinal signs" and diagnostic indices for the clinical identification of Down syndrome. However, there are no pathognomonic signs, and physical stigmata are not identified consistently and regularly in every child with the Down syndrome. It is important to realize that there is much variation in terms of phenotypic expression in persons with Down syndrome.

When physical features are discussed with parents, it should be stressed that many of the findings observed in the child with Down syndrome do not cause any disability in the child. For example, the incurved little finger does not limit the function of the hand, nor will the slanting of the palpebral fissure interfere with the child's vision. Yet other defects, such as severe congenital heart disease, which is found in 30 to 40 per cent, and duodenal atresia, are of a serious nature and require prompt medical attention. In general, parents should be made aware of the fact that children with Down syndrome are more similar to normal children than they are different.

COUNSELING OF PARENTS

Once the diagnosis of Down syndrome has been made, the physician should communicate his message supportively with tact, compassion, and truthfulness. The initial counseling will have a vital influence on the parents' attitude and subsequent adjustment. The physician should let the parents know that the infant with Down syndrome is first and foremost a human being with the same rights as any other person with a normal karyotype. The critical role of parenthood needs emphasis—the chance for the infant to be nurtured and loved by caring parents.

It is generally accepted that if the parents are to assume full responsibility for making decisions relating to their child, they have a right to all available information. Consequently parents should be notified as soon as a definite diagnosis of Down syndrome has been made. It would seem most appropriate for both parents to be present during the initial counseling session, for they can offer support to one another. Although some physicians prefer to talk to the father, others let the mother know of the diagnosis while she is still in the delivery room. However, placing the entire burden and stress on one parent and making him or her responsible for informing the other spouse should be avoided.

The physician should know of the invariably profound emotional distress and the reactions of parents during the initial counseling session. The first counseling session might well be brief, since cognitive dysfunction and disturbance of personality integration often prevail. Hope, denial, rejection, and other defense mechanisms help the parents to sustain themselves emotionally during this initial period of adjustment. The physician's remarks should be timed to coincide with increasing parental adaptation. Follow-up sessions are necessary for review of basic considerations and communication of more details. Extra time spent by the physician in talking about various expressed concerns and issues will make the parents aware of the physician's sincere interest in helping them and their child.

The parents should be told of the meaning of the term Down syndrome, and why its use is preferred to the offending misnomers, mongoloid and mongolism. The physical characteristics should be pointed out. The anticipated developmental concerns should be discussed, and the parents should be told that motor function, mental development, and language acquisition are usually delayed in children with Down syndrome.

After the chromosomal analysis has been performed, the results should be explained in simple terms. Information should be provided concerning the risks involved in future pregnancies and the availability of amniocentesis. Questions relating to the cause of Down syndrome inevitably arise. It is important to point out that neither the father nor the mother can be considered at fault.

The impact on family life and the effects on siblings will also come up in the discussion. Parents may wish to protect brothers and sisters by not telling them about the child's problem, but they should be encouraged to be open about it. Even very young children are sensitive to parental distress and their fears may be far greater than parents realize. Although talking about the child with Down syndrome may be painful for parents, lack of openness can intensify parental isolation and promote unrealistic concerns.

Other parents who have an older child with Down syndrome might be called upon to talk to new parents. These experienced parents are often more sensitive and can help the new parents to cope more effectively than professionals can. The experienced parents are proof to the new parents that one can survive such a crisis. They also can let the new parents know that later there can be true happiness with a child with Down syndrome in the family.

DEVELOPMENTAL EXPECTATIONS

The diversity of biologic factors, functions, and accomplishments that exist in all human beings is also present in children with Down syndrome. In fact, we observe a greater variation in nearly all aspects of their life. The physical growth pattern ranges from the very short youngster to the child with above average height, from the very thin and frail infant to the heavy and overweight adolescent. Moreover, the mental abilities in children with Down syndrome span a wide range between severe retardation and normal intelligence. Likewise, the behavior and the emotional disposition vary significantly; some children may be placid and inactive, while others are aggressive and hyperactive; and there are those between these extremes.

The stereotyped picture portrayed in the past of the short, obese, unattractive individual, with open mouth and protruding tongue who is severely retarded and stubborn, is certainly not a true description of a child with Down syndrome as we know him.

It is known that the growth rate of children with Down syndrome is reduced as compared with children who do not have this chromosomal disorder. Our studies support previous reports of a reduced growth pattern (Fig. 19–4). Since hypothyroidism is known to occur at a greater frequency in Down syndrome, it is important to follow the longitudinal growth in children carefully. If the observed growth rate is less than the expected, thyroid studies should be carried out. The child with significant gastrointestinal problems or with severe heart disease also has been

Figure 19–4. *Upper curve,* Average height of normal children from birth to 12 years of age. *Lower curve,* Average height (shaded area showing the height range) of children with Down syndrome.

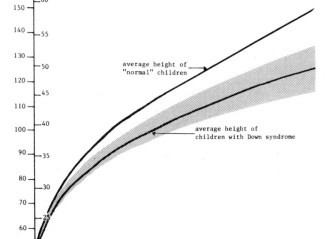

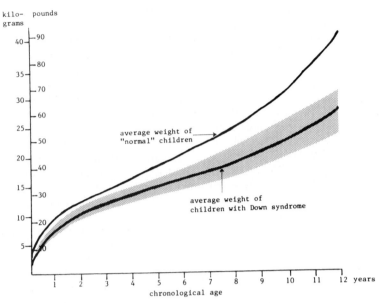

Figure 19–5. *Upper curve*, Average weight of normal children from birth to 12 years. *Lower curve*, Average weight (adjacent shaded area showing the weight range) of children with Down syndrome.

known to be smaller than children with Down syndrome who do not have these additional complications.

Since feeding problems are sometimes encountered in the young child, particularly in children with additional congenital anomalies, there might be a reduction in weight gain in early childhood. During the school years and during adolescence, however, obesity is frequently observed. Therefore, it is important to counsel parents from early childhood on, for good eating habits, a balanced diet, the avoidance of high calorie foods, and regular physical activities can prevent the child from becoming overweight. A weight chart of the young child with the Down syndrome is presented in Figure 19–5.

Parents often ask when the child will be able to sit independently or when he will finally walk. Answers to these and other questions relating to the child's motor development are provided in Table 19–4. Other developmental observations were recorded in our study of specific self-help

skills, as outlined in Table 19–5 (Pueschel, 1982). Of course, a variety of factors, such as congenital heart defects, severe hypotonia, and other interfering biologic and environmental problems, may be responsible for the marked delay in motor development and self-help skills in some children with the Down syndrome. Children who are enrolled in an early intervention program, whose parents provide environmental enrichment and stimulation, and who do not have congenital heart disease, exhibit relatively advanced development.

As in other areas of development, the intellectual abilities of the child with Down syndrome have usually been underestimated in the past. Recent reports in the literature as well as our own investigations are in disagreement with previous impressions that children with Down syndrome are severely or profoundly retarded. Figure 19–6 indicates that the majority of children with Down syndrome function in the mild to moderate range of mental retardation. Some children even have been found to have IQ's in the borderline to the

Table 19–4. DEVELOPMENTAL MILESTONES

| | Children with Down Syndrome | | "Normal" Children | |
	Average	Range	Average	Range
Smiling	2 months	1.5 to 4 months	1 month	0.5 to 3 months
Rolling over	8 months	4 to 22 months	5 months	2 to 10 months
Sitting alone	10 months	6 to 28 months	7 months	5 to 9 months
Crawling	12 months	7 to 21 months	8 months	6 to 11 months
Creeping	15 months	9 to 27 months	10 months	7 to 13 months
Standing	20 months	11 to 42 months	11 months	8 to 16 months
Walking	24 months	12 to 65 months	13 months	8 to 18 months
Talking, words	16 months	9 to 31 months	10 months	6 to 14 months
Talking, sentences	28 months	18 to 96 months	21 months	14 to 32 months

Table 19–5.   SELF-HELP SKILLS

	Children with Down Syndrome		"Normal" Children	
	Average	*Range*	*Average*	*Range*
Eating				
Finger feeding	12 months	8 to 28 months	8 months	6 to 16 months
Using spoon and fork	20 months	12 to 40 months	13 months	8 to 20 months
Toilet training				
Bladder	48 months	20 to 95 months	32 months	18 to 60 months
Bowel	42 months	28 to 90 months	29 months	16 to 48 months
Dressing				
Undressing	40 months	29 to 72 months	32 months	22 to 42 months
Putting clothes on	58 months	38 to 98 months	47 months	34 to 58 months

low average range of intellectual functioning, and only a few children are severely retarded.

While some reports mention a progressive decrease in DQ/IQ values over time (Baron, 1972; Connolly, 1978; Fishler et al., 1964; Melyn and White, 1973; Morgan, 1979), in our longitudinal study such decline in mental development was not observed (Pueschel, 1982). Melyn and White (1973) stress that the gradual decline in IQ scores in patients with Down syndrome reflects the increasingly verbal and abstract content of test materials at higher ages. Rynders et al. (1978) emphasize that the exclusive use of traditional psychometry as an index of educability for a person with Down syndrome is far too limiting. They suggest that data from studies representing a nonpsychometric point of view could serve to broaden our perspective in regard to the educability of children with Down syndrome. Studies providing a more detailed descriptive data base than that available from standardized tests could give us a richer picture of the developmental progress, strengths, and deficits of children with Down syndrome.

During the past decade research concerning behavioral systems of children with Down syndrome has been discussed in the literature (Baron, 1972; Berry et al, 1980; Cicchetti and Serafica, 1981). Baron (1972) studied temperament profiles and found that compared with normal infants, a child with Down syndrome in early life is not stereotyped in behavior. Also, Berry et al. (1980) observed that the responses of infants with Down syndrome in strange situations are qualitatively similar to those of nonretarded children. In their study, children with Down syndrome were found to be more sensitive to unfamiliar situations, separation from mother, and the presence of a new person, indicating that early social awareness of infants with Down syndrome seems to be quite well developed. Likewise, Cicchetti and Serafica (1981) noted similarities between children with Down syndrome and normal infants in their interplay among behavioral systems. Although quantitative differences in degree of responsiveness were recorded, the patterns of response were similar. Children with Down syndrome exhibited the same range of behavior as did children with normal karyotypes. These authors concluded that young children with Down syndrome are remark-

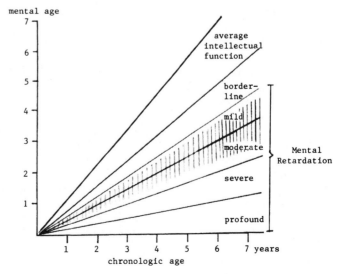

Figure 19–6. The area indicated by a vertical bar denotes the range of intellectual function of the majority of children with Down syndrome.

ably similar to their normal peers in their repertoire of social behavior and in their interactional patterns.

It is important to point out to parents that children with Down syndrome, although developmentally delayed in most areas of functioning, do make steady progress in overall development. Children with Down syndrome have definite strengths and talents, they are sensitive and aware of the feelings of others, and their overall adequate social abilities afford them participation in society as productive citizens.

MEDICAL MANAGEMENT

The child with Down syndrome is in need of the same kind of well-baby care and medical attention as any other child. Thus, the pediatrician should provide health maintenance, give immunizations, attend medical emergencies, and offer support and counsel to the family. There are, however, situations in which children with Down syndrome need special attention by the caring physician:

1. Since 70 to 80 per cent of the children with Down syndrome reportedly have hearing deficits, audiologic assessment at an early age and follow-up hearing tests are indicated. If a significant hearing impairment is identified, a referral to an otolaryngologist should be made.

2. Thirty to 40 per cent of children with Down syndrome have congential heart disease. They need long term care by a pediatric cardiologist.

3. Children with Down syndrome frequently have refractive errors, and cataracts occur more often in this population, necessitating ophthalmologic evaluations.

4. Another concern relates to nutritional aspects. Some children, in particular those with additional major congential anomalies, gain weight slowly and thrive poorly during infancy. On the other hand, obesity is often encountered during adolescence and early adulthood. The physician can prevent most of these nutritional problems by providing anticipatory dietary guidance.

5. Skeletal problems have been observed in children with Down syndrome, including patella dislocation, hip subluxation, and atlantoaxial instability. If the latter condition causes neurologic symptoms indicating spinal cord depression, or if a child holds his head in a torticollis-like position, radiologic examination of the cervical spine and neurologic consultation should be forthcoming.

6. There are other important medical aspects in Down syndrome, including immunologic concerns, metabolic dysfunctions, and biochemical derangements that may require the attention of specialists in the respective fields (Pueschel et al., 1982).

TREATMENT

Although many therapeutic modalities have been advocated druing the past century, it can be said categorically that there is no effective medical treatment available at the present time. Recent advances in molecular biology, however, make it feasible now to examine directly the genetic basis for Down syndrome (Kurnit, 1980). The challenge for the future is to further isolate, map, and characterize gene frequencies on chromosome 21 that are responsible for the physical features and the intellectual limitations observed in individuals with Down syndrome.

If one were able to identify these genes and discover their mechanisms of interference with normal developmental sequences and if one could counteract their specific actions, a rational approach to medical therapy could emerge.

EDUCATION

Significant progress has been made during the past decade in the field of behavioral sciences. Today early intervention programs, preschool nurseries, and special education strategies have demonstrated that the child with Down syndrome can participate in learning experiences that will positively influence his overall function.

Accumulating knowledge in all areas of infant research has broadened our understanding of the infant's amazing capacity to learn. There has been an attempt to apply that knowledge in programs designed for infants and young children with Down syndrome (Hayden, 1980). Recent research with both human and animal infants suggests that learning experiences made available during early life may be of crucial importance. It has been demonstrated that early intervention with infants with Down syndrome and their family results in progress that is usually not achieved by infants who have not had the learning opportunities through such programs.

During recent years an increasing number of early intervention programs have become available to guide parents in providing environmental enrichment for infants with Down syndrome. Children can benefit from early sensory stimulation, specific exercises involving gross and fine motor activities, and instruction in language acquisition. The teaching of self-help skills, including feeding, toilet training, and dressing allows the child to function more independently. It is generally agreed that it is the quality rather than the sum of total stimulation that shapes the physical and mental development of the young child. It is, therefore, the structure and content of an early stimulation program that should be emphasized rather than the indiscriminate use of nonspecific stimuli. These early intervention programs should be based on sound principles of child development and should enhance family life in general.

Just as early intervention programs foster the development of the child with Down syndrome, preschool nurseries also play an important role in the young child's life. Such a child can profit by improving gross and fine motor skills, by learning to play with others, and from social interaction with other children. Exploring the environment beyond the home enables the child to participate in a broader world.

Upon entering school, the child with Down syndrome is still very much in the process of developing and growing with his own capacity for maturation and achievement. Special education programs can help such a child to see the world as an interesting place to explore and work in. The experiences provided in the school will assist him in obtaining a feeling of personal identity, self-respect, and enjoyment. The school environment can give this child a foundation for life through the development of basic academic skills and physical as well as social abilities. School should offer an opportunity for the child to engage in sharing relationships with others and prepare him to become a productive citizen. Contrary to some reports in the literature, most children with Down syndrome are educable. Federal legislation (the Education of All Handicapped Children Act, PL 94-142) guarantees each child with a developmental disability the right to a free, appropriate education commensurate with his abilities.

Adolescence is frequently a challenging time in the life of both the young person and his family. Although this might be a troublesome time for those of normal intellectual abilities, a retarded adolescent's problems are frequently intensified. Many adolescents with Down syndrome may have physical attributes of normal youngsters, yet they often do not possess the capabilities to cope with either the demands of the environment or their own desire for independence. They are faced with the responsibility for preparing for vocational competence as well as developing social and emotional characteristics that will provide them with acceptance in society. During prevocational activities, youngsters with Down syndrome learn to develop good work habits and to engage in proper relationships with coworkers. Later, whether their employment is in private industry or in sheltered workshop situations, a person with Down syndrome can gain a feeling of self-worth and of making a contribution to society.

It is paramount that society forge attitudes that will permit persons with Down syndrome to participate in community life and to be accepted for what they are. Individuals with Down syndrome should be offered a status that allows them their rights and privileges as citizens and in a real sense preserves their human dignity. When accorded their rights and treated with dignity, persons with Down syndrome will in turn provide society with a most valuable humanizing influence.

SIEGFRIED M. PUESCHEL

REFERENCES

Alberman, E., Polani, P. E., Roberts, J. A., Spicer, C. C., Elliott, M., and Armstrong, E.: Parental exposure to x-irradiation and Down's syndrome. Ann. Hum. Genet., 36:195–208, 1972.

Baron, J.: Temperament profile of children with Down's syndrome. Dev. Med. Child Neurol., 14:640, 1972.

Berry, P., Gunn, P., and Andrews, R.: Behavior of Down syndrome infants in a strange situation. Am. J. Ment. Defic., 85:212, 1980.

Cicchetti, D., and Serafica, F. C.: Interplay among behavioral systems: illustrations from the study of attachment, affiliation and variness in young children with Down's syndrome. Dev. Psychol., 17:36, 1981.

Coleman, M.: Down's syndrome. Pediatr. Ann., 7:90, 1978.

Connolly, J. A.: Intelligence levels of Down's syndrome children. Am. J. Ment. Defic., 83:193, 1978.

Crowley, P. H., Hayden, T. L., and Gulati, D. K.: Etiology of Down syndrome. In Pueschel, S. M., and Rynders, J. E. (Editors): Down Syndrome—Advances in Biomedicine and the Behavioral Sciences. Cambridge, The Ware Press, 1982.

Down, J. L. H.: Observations on an ethnic classification of idiots. London Hosp. Clin. Lect. Rep., 3:259, 1866.

Fialkow, P. J.: Autoimmunity and chromosomal aberrations. Am. J. Hum. Genet., 18:1–5, 1966.

Fishler, K., Share, J., and Koch, R.: Adaptation of Gesell developmental scales for evaluation of development in children with Down's syndrome (mongolism). Am. J. Ment. Defic., 68:642, 1964.

Harlap, S.: Down's syndrome in West Jerusalem. Am. J. Epidemiol., 97:225, 1973.

Hayden, A. H.: Early intervention: infant learning or infant stimulation programs. Down's syndrome—Papers and Abstracts for Professionals, 3:5, 1980.

Hook, E. B.: Epidemiology of Down syndrome. In Pueschel, S. M., and Rynders, J. E. (Editors): Down Syndrome—Advances in Biomedicine and the Behavioral Sciences. Cambridge, The Ware Press, 1982.

Kurnit, D. M.: A molecular approach to Down's syndrome. Down's syndrome—Papers and Abstracts for Professionals, 3:1, 1980.

LaVeck, B., and Brehm, S. S.: Individual variability among children with Down's syndrome. Ment. Retard., 15:135, 1978.

Lejeune, J., Gauthier, M., and Turpin, R.: Études des chromosomes somatiques de neuf enfants mongoliens. C. R. Acad. Sci. Paris, 248:1721, 1959.

Magenis, R. W., and Chamberlin, J.: Parental origin of nondisjunction. In de la Cruz, F. F., and Gerald, P. S. (Editors): Trisomy 21 (Down Syndrome). Research Perspectives. Baltimore, University Park Press, 1981.

Melyn, M. A., and White, D. T.: Mental and developmental milestones of noninstitutionalized Down's syndrome children. Pediatrics, 52:542, 1973.

Morgan, S. B.: Development and distribution of intellectual and adaptive skills in Down syndrome children: implications for early intervention. Ment. Retard., 16:247, 1979.

Nichols, W. W.: The role of viruses in the etiology of chromosomal abnormalities. Am. J. Hum. Genet., 18:81, 1966.

Pueschel, S. M.: A Study of the Young Child with Down Syndrome. New York, Human Science Press, 1982.

Pueschel, S. M., Padre-Mendoza, T., and Ellenbogen, R.: Partial trisomy 21. Clin. Genet., *18*:392, 1980.

Pueschel, S. M., Sassaman, E. A., Scola, P. S., Thuline, H. C., Stark, A. M., and Horrobin, M.: Biomedical aspects in Down syndrome. *In* Pueschel, S. M., and Rynders, J. E. (Editors): Down Syndrome—Advances in Biomedicine and the Behavioral Sciences. Cambridge, The Ware Press, 1982.

Rynders, J. E., Spiker, D., and Horrobin, J. M.: Underestimating the educability of Down's syndrome children: examination of methodological problems in recent literature. Am. J. Ment. Defic., *82*:440, 1978.

Stoller, A., and Collmann, R. D.: Virus aetiology for Down's syndrome (mongolism). Nature (London), *208*:903, 1965.

Thuline, H. C.: Cytogenetics in Down syndrome. *In* Pueschel, S. M., and Rynders, J. E. (Editors): Down Syndrome—Advances in Biomedicine and the Behavioral Sciences. Cambridge, The Ware Press, 1982.

Uchida, I. A.: Down syndrome and maternal radiation. *In* de la Cruz, F. F., and Gerald, P. S. (Editors): Trisomy 21 (Down Syndrome). Baltimore, University Park Press, 1981.

Warkany, J.: Etiology of mongolism. J. Pediatr., *56*:412, 1960.

20
Congenital Anomalies

Congenital anomalies are frequently associated with variations in behavior and development. The term *congenital* simply means present at birth, and it is not synonymous with the term *genetic*. All disease processes act upon the genetic constitution of the individual, but genetic factors may or may not play a primary role in producing a particular congenital anomaly. Some malformations result from external or environmental factors, whereas others result from internal imbalances due to inherited alterations in the production of cellular enzymes or structural proteins. The latter category of defects is considered to be genetic, and this category of problems is considered in more detail in Chapter 18. The three main types of genetic problems are chromosomal, mutant gene, and multifactorial. The term *familial* refers to disorders occurring in families that may or may not have a genetic basis.

Congenital anomalies and prematurity are currently the two leading causes of infant mortality in the United States. As other kinds of problems resulting in perinatal mortality and morbidity have been controlled by improved management (e.g., infectious disease and rhesus hemolytic disease), congenital anomalies have been found to contribute to an increasing proportion of infant deaths. In fact, at the current time it can be said that children dying from congenital anomalies constitute the next major hurdle to be surmounted in reducing the perinatal mortality rate in the United States.

Anencephaly and spina bifida are the two most common major congenital anomalies in the United States, with a combined prevalence between 0.5 and 2 per 1000 live births. Both genetic and environmental factors are known to play a role in the genesis of these malformations.

Recent advances surrounding the technology of serum alpha-fetoprotein screening and diagnostic ultrasonography have made neural tube defects detectable prenatally. It is also possible to diagnose many of the genetic diseases prenatally. Better understanding of the various environmental hazards to fetal development could also lead to prevention of mortality and morbidity in this area.

One basic approach to an individual with structural defects is to identify and assemble the relevant data, interpret the findings from a developmental morphologic viewpoint, and then by induction, attempt to determine the overall diagnosis (Smith, 1982). As Sir Arthur Conan Doyle, the physician-creator of Sherlock Holmes, has emphasized:

It is a capitol mistake to theorize before one has data. Insensibly one begins to twist facts to suit theories instead of theories to suit facts.*

In assembling the relevant data, abnormalities should be validated by measurement, if possible, and near relatives should be examined for the presence of similar or related anomalies. Thereafter one should interpret the anomalies individually and collectively from a developmental morphologic viewpoint. This approach generally falls within the province of dysmorphology, the study of aberrant development of form. Once the diagnosis has been determined, one can plan the management and provide an appropriate prognosis and relevant counseling for a given child's condition.

THE DEVELOPMENTAL MORPHOLOGIC APPROACH

At the time of birth about 5 per cent of babies are observed to have structural defects. About 3 per cent have deformations, which are usually caused by late uterine constraint (Dunn, 1976). The other 2 per cent have malformations which are due to an intrinsic problem within the developing structure. A deformation is due to extrinsic forces, there being no underlying problem within the tissues of the developing structure. Of the babies who present with malformations, one-third have multiple defect disorders, some of which are caused by chromosome anomalies, mutant genes, or teratogenic exposures. These babies are considered to have malformation syndromes. The frequency of recognized malformation problems doubles by one year of age with the inclusion of defects that might not be evident in the newborn period, such as those involving the heart, kidneys or brain.

---

*Doyle, A. C.: A Scandal in Bohemia. *The Adventures of Sherlock Holmes.* New York, Harper and Bros., 1892.

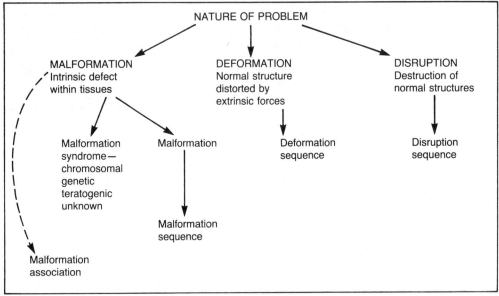

Figure 20–1.   Three categories of structural defects.

A third category of defects has recently been delineated, and it includes problems due to disruption. Infants with such problems have experienced an interruption of normal fetal development with the forceful destruction of a previously normally formed structure. Interruption of normal fetal vascular supply, through either extrinsic compression or intrinsic occlusion, is one of the more common mechanisms resulting in disruptive morphogenesis. These three categories of structural defects are summarized in Figure 20–1.

Since malformations are engendered by intrinsic problems within the developing tissue, their origin usually predates the end of the embryonic period of organogenesis (Fig. 20–2). Disruptions result from the destruction of previously formed tissues; hence, they can occur at later times during gestation. Deformations are usually due to late uterine constraint of the fetus as it begins to outgrow the confines of the uterus. Thus, such problems occur during the third trimester and are more likely in first born babies and twins and in cases of prolonged abnormal fetal lie or presentation, oligohydramnios, or uterine structural abnormality. When significant extrinsic force is brought to bear on pliable fetal parts during the early phases of gestation, it may result in disruption rather than deformation (e.g., early amnion rupture sequence).

The diagnostic process entails more than simply listing a given child's structural defects or phenotypic atypicalities. Some attempt should be made to synthesize these findings into a morphogenetic hierarchy. The interpretation of minor anomalies appropriately includes a judgment as to whether such a finding might be explained by a more primary problem in morphogenesis. Table 20–1 lists some examples of physical findings that can be interpreted as being secondary to a more primary problem in morphogenesis. Such minor anomalies can provide valuable clues to the nature and timing of the problem in development, although they usually are of no direct consequence to the patient by themselves. Figures 20–3 and 20–4 demonstrate some of these minor anomalies. A more complete description of methods of assessment and interpretation of such abnormalities and standards for measurement of some of them can be found in David W. Smith's *Recognizable Patterns of Human Malformation* (1982).

The interpretation of multiple major defects should include some consideration as to whether these defects are due to a single earlier problem or whether they are part of a broader pattern of altered morphogenesis. The term *sequence* is used to denote the initiating problem and its secondarily derived defects. Other terms (e.g., malformation, deformation, or disruption) are used to describe the underlying nature of the problem. Hence, the term cleft lip and palate malformation sequence is intended to denote an intrinsic problem in lip fusion occurring prior to day 35 of gestation, with its associated secondary failure in anterior palatal fusion (Fig. 20–5). Likewise, oligohydramnios deformation sequence describes the associated limb and craniofacial deformations that result from a prolonged deficit of amniotic fluid. The same nonspecific phenotype can be due to prolonged leakage of amniotic fluid or to a more primary renal malformation preventing fetal urine production or

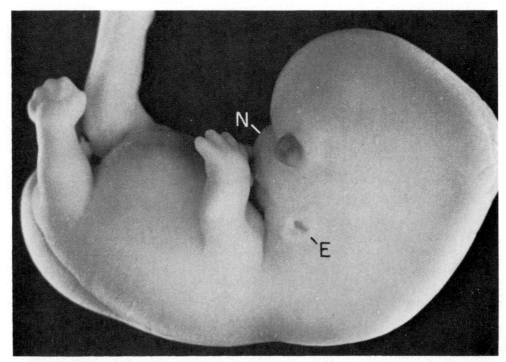

Figure 20–2. This 51 day (22 mm.) human embryo has a relatively flat nose (N) and an external ear (E) that will shift in relative position as the embryo continues to grow and develop. This is the end of the embryonic period of organogenesis and the beginning of the fetal era of histogenesis and functional maturation. The intestinal loop has largely returned to the abdominal cavity, with formation of the anterior body wall and separation of the thorax and abdomen by the transverse septum (diaphragm). The fingers are now partially separated, and the elbow is evident; the lower extremities reveal a slightly earlier phase of limb development (with a paddle-like foot plate and no obvious knee joint). The major period of cardiac morphogenesis and septation is complete. The urogenital membrane has now broken down, yielding a urethral opening. The phallus and lateral labioscrotal folds are the same for both sexes at this age, and the tail has begun to regress. Otic and ophthalmic development is nearly complete, the cerebral hemispheres are evident, and the palate is closed. (From Smith, D. W.: Recognizable Patterns of Human Malformation. 3rd ed. Philadelphia, W. B. Saunders Company, 1982.)

flow. When the deficit of amniotic fluid occurs even earlier (prior to six to eight weeks' gestation), as in early amnion rupture disruption sequence, the resultant compression may actually destroy delicate fetal parts and result in a disruption or morphogenesis.

Malformation sequences can occur as isolated anomalies, or they can be part of broader patterns of altered morphogenesis. A malformation sequence in an otherwise normal child generally carries a low magnitude recurrence risk for the same type of problem. Such recurrence risk data do not apply when the malformation sequence is one part of a broader pattern of malformation or is due to a specific teratogenic insult. Table 20–2 lists some of the common malformation sequences with their developmental timing, and Table 20–3 gives the developmental timing for some additional isolated malformations.

Malformation syndromes comprise disorders that result in multiple anomalies that are not explained on the basis of a single localized defect

Table 20–1. MINOR ANOMALIES AND THEIR DEVELOPMENTAL TIMING

Minor Anomaly	Primary Problem in Morphogenesis	Developmental Timing of Primary Problem
Syndactyly	Aberrant programmed cell death	6 weeks of fetal life
Single transverse palmar crease (four-finger line)	Aberrant flexional plane of palmar folding	11 weeks of fetal life
Aberrant scalp hair directional patterning	Problem in brain growth or structure	11 to 16 weeks of fetal life
Unusual dermal ridge patterns (dermatoglyphics)	Abnormal form in early developing hand or foot	15 to 20 weeks of fetal life
Inner canthal folds	Low nasal bridge	Variable
Upslanting palpebral fissures	Narrow bifrontal region	Variable
Downslanting palpebral fissures	Malar hypoplasia	Usually prenatal
Prominent lateral palatine ridges	Deficit of tongue thrust into hard palate	Usually prenatal
Dimples over bony prominences	Late fetal constraint	Third trimester

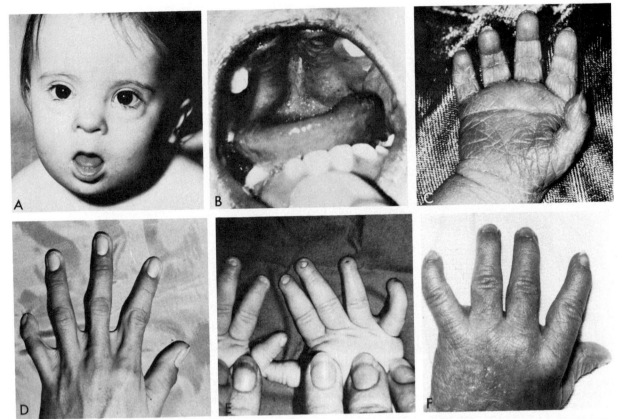

Figure 20–3. *A,* The epicanthal folds shown in this child with Down syndrome are a consequence of nasal growth deficiency, with redundant skin extending over the flattened nasal bridge to form epicanthal folds. The palpebral fissures are short owing to a smaller ocular globe size (related to the smaller size of the brain from which they are an outgrowth). They demonstrate an upward slant due to narrowing of the anterior portion of the brain. Because the latter features are part of the prenatal brain growth deficiency characteristic of Down syndrome and also of many other congenital disorders of brain development, to a certain extent they are individually nonspecific; their association with other more specific features is diagnostic. (From Smith, D. W.: Recognizable Patterns of Human Malformation. Ed. 3. Philadelphia, W. B. Saunders Company, 1982.) *B,* These prominent lateral palatine ridges in a child with Down syndrome are secondary to a deficit of tongue thrust into the hard palate, and they may be associated with a variety of disorders that result in hypotonia and diminished sucking. Their persistence in older children can be a sign of a long term deficit in neurologic function. *C,* Hand and finger creases represent flexional planes of folding. They are usually evident by 11 to 12 weeks of fetal life; hence alterations in crease patterning usually reflect an abnormality in either the form or function of the hand prior to 11 fetal weeks. This four-finger line in a child with Down syndrome is a result of the generalized shortness of the palm in this disorder, but such creases can also be seen unilaterally in about 4 per cent of normal newborns and bilaterally in about 1 per cent. *D,* Camptodactyly, or inability to extend the fingers completely, most commonly affects the fifth, fourth, and third digits in decreasing order of frequency and severity. It may be a neuromuscular sign or a sign of relatively short flexor tendons in relation to hand growth. Here it is seen in a child with the fetal alcohol syndrome, but it can also be seen with diabetes insipidus and associated microvascular disease. Also evident is partial cutaneous syndactyly as a consequence of incomplete separation of the fingers during the sixth week of fetal life. Such syndactyly most commonly occurs between the third and fourth fingers and between the second and third toes. *E and F,* Hypoplastic nails and clinodactyly of the fifth finger reflect phalangeal growth deficiency in a child with fetal hydantoin effects (*E*) and the Penta-X syndrome (*F*). Hypoplastic nails may be associated with underdeveloped distal phalanges and a predominance of low arch dermal ridge patterns. Normal individuals rarely have more than six of 10 fingertips with a low arch configuration. Such dermal ridge patterning is thought to be due to hypoplasia of the fetal fingertip pads during the sixteenth to nineteenth weeks of fetal life, when dermal ridge patterns form in relation to planes of growth stretch. Clinodactyly, or curving of the fingers, most commonly affects the fifth finger. It is the consequence of hypoplasia of the middle phalanx, normally the last digital bone to develop, and may follow a dominant pattern of inheritance in otherwise normal families.

in morphogenesis. Clinical recognition of these conditions often permits the detection of a specific mode of etiology with a well defined recurrence risk. Some of the more common etiologies include environmental teratogens, mutant gene disorders, and genetic imbalance disorders due to chromosomal abnormalities. Some of these syndromes are currently of unknown etiology, but as knowledge accumulates and diagnostic technologies improve, new etiologies can become apparent for previously recognized conditions with no known etiology. Two such examples include the recent association of some instances of the Prader-Willi syndrome

with a very small deletion from the long arm of chromosome 15 (involving a deletion of the 15q12 band), and the association of one form of X linked mental deficiency with the marker fragile X chromosome.

Occasionally multiple anomalies may occur together, on a sporadic basis, without a single underlying defect in morphogenesis or even a consistent pattern of altered morphogenesis. Such anomalies are known to be nonrandomly associated with a frequency that is much greater than their chance association would predict. The cause of such a nonrandom association is currently un-

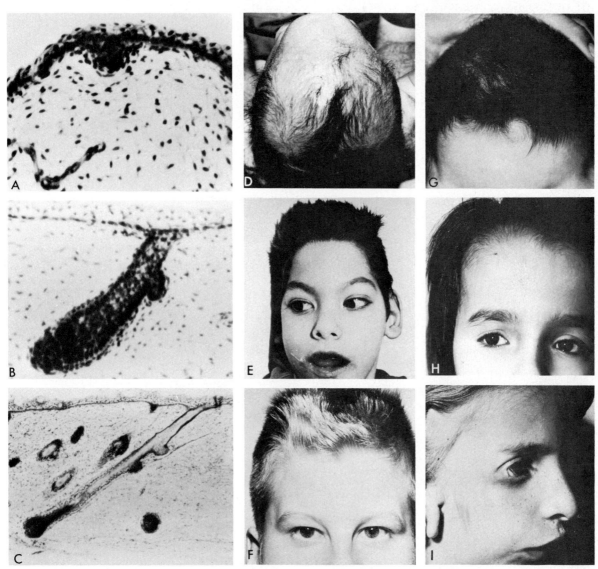

Figure 20–4. *A,* Hair follicles begin to grow down into the loose underlying mesenchyme of the scalp at 10 fetal weeks. The slope of each hair follicle, and ultimately the directional patterning of scalp hair, is determined by the direction of growth stretch exerted on the surface skin by the domelike, forward outgrowth of the early brain during the next 6 weeks. Hence, hair directional patterning reflects the growth in size and form of the developing brain during this period. *B,* At 16 weeks the hair follicle demonstrates a tangential slope in relation to brain growth. *C,* By 18 weeks the hair follicles have begun to extrude hairs onto the surface of the fetal scalp, and hair patterning is set. The parietal hair whorl normally develops at the focal point of growth-stretch tension induced by rapid brain growth between 10 and 16 weeks of fetal life. Usually it is found several centimeters anterior to the posterior fontanel. Because of the usual dominance of left hemisphere brain growth, 56 per cent of single parietal hair whorls are located to the left of the midline, 30 per cent are to the right, and 14 per cent are midline in location. In 5 per cent of normal individuals, bilateral parietal hair whorls are present (usually as a dominantly inherited trait). The anterior parietal hair stream usually sweeps forward to converge with the upsweeping frontal hair stream from the face. If this meeting occurs above the forehead, a frontal upsweep or "cowlick" may result, and this occurs in 5 per cent of normal individuals (usually as a dominantly inherited trait). *D,* Aberrant scalp and upper facial hair directional patterning may reflect an early problem in brain growth or may form during the period of hair follicle development. Here, widely spaced double parietal hair whorls in an individual with trigonocephaly suggest an early deficit in forebrain growth. *E,* With severe microcephaly the parietal hair stream may not sweep forward and a parietal hair whorl may be lacking (25 per cent of the cases). The hair follicles grow straight down into the scalp mesenchyme, and hair stands up straight with a frontal upsweep (70 per cent of cases). *F,* In individuals with a relatively small, narrow frontal area of the brain, such a frontal upsweep may imply a problem in brain development prior to 10 to 16 fetal weeks, as demonstrated by this individual with the Prader-Willi syndrome. (From Smith, D. W., and Gong, B. T.: Scalp-hair patterning. Its origin and significance relative to early brain and upper face development. Teratology *9:*17, 1974.) *G,* Note this abnormal anterior hair whorl in an individual with congenital rubella effects. *H,* At 18 weeks newly emergent hair grows over the entire face and scalp. Later, as growth of the eyebrows and scalp hair begins to predominate, hair growth over the remainder of the face is suppressed. There is usually a periocular zone of hair growth suppression. In this developmentally delayed individual with megalencephaly and widely spaced eyes on that basis, frontal hair patterning is mildly altered. With widely spaced eyes there may be a downward projection of scalp hair into the center of the forehead, known as the "widow's peak," owing to diminished periocular hair growth suppression in this area. With microphthalmos or cryptophthalmos, a projection of scalp hair may intrude into what would normally be a zone of periocular hair growth suppression. *I,* The auricle also influences hair growth in the sideburn area, and with absence or underdevelopment of the auricle, sideburn hair growth may be diminished or displaced onto the cheek, as shown in this child with the Treacher-Collins syndrome. (From Smith, D. W.: Recognizable Patterns of Human Malformation. Ed. 3. Philadelphia, W. B. Saunders Company, 1982.)

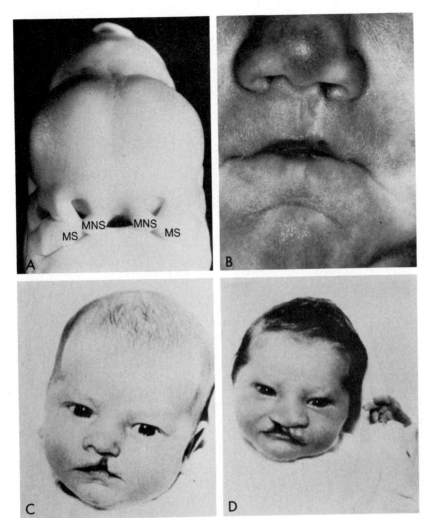

Figure 20–5. *A*, By 35 days' gestation from conception, the lip is normally fused. As shown here, the maxillary swelling (MS) fuses with the medial nasal swelling (MNS) beneath each naris. *B* to *D*, When lip fusion is incomplete, all gradations of cleft lip and its consequences can occur, ranging from a barely perceptible scarlike line at the normal site of lip closure on the left (*B*) to a widely open cleft with the secondary consequences of cleft palate, flared alae nasi, and mild ocular hypertelorism. (*D*) The risk of recurrence ranges from 3 to 5 per cent when this occurs as an isolated malformation sequence. (From Smith, D. W.: Recognizable Patterns of Human Malformation. Ed. 3. Philadelphia, W. B. Saunders Company, 1982.)

known, and all that can be done, in a diagnostic sense, is to rule out known etiologies such as chromosomal, genetic, or teratogenic causes. These disorders are termed *malformation associations,* and two examples of such conditions would include the VATER association and the CHARGE association. Such a diagnosis is generally considered to be awaiting further resolution as to its true pathogenesis and etiology.

Figure 20–6 shows a baby with the CHARGE association. Individuals with this constellation of anomalies have all or most of the following anomalies: *c*oloboma of the retina, *h*eart anomalies, *a*tresia choanae, *r*etarded growth and development, *g*enital hypoplasia in males, and *e*ar anomalies with hearing impairment. Chromosomal studies invariably have been normal and family studies have been inconclusive thus far. Most of

Table 20–2. MALFORMATION SEQUENCES AND THEIR DEVELOPMENTAL TIMING

Malformation Sequence	Primary Problem in Morphogenesis	Developmental Timing of Primary Problem
Holoprosencephaly malformation sequence	? Defect in prechordal mesoderm	21–25 days
Anencephaly-meningomyelocele malformation sequence	? Defect in neural tube closure	23–28 days
Cleft lip and palate malformation sequence	Failure of lip closure	35 days
Jugular lymphatic obstruction malformation sequence	Lag in lymphaticovenous communication	40 days
Robin malformation sequence	Early mandibular hypoplasia	8 to 9 weeks
Urethral obstruction malformation sequence	Urethral obstruction (usually prostatic urethral values)	8 to 9 weeks

Table 20–3. RELATIVE TIMING AND DEVELOPMENTAL PATHOLOGY OF CERTAIN MALFORMATIONS*

Tissue	Malformation	Defect In	Cause Prior To	Comment
Central nervous system	Anencephaly	Closure of anterior neural tube	26 days	Subsequent degeneration of forebrain
	Meningomyelocele	Closure in portion of posterior neural tube	28 days	80% lumbosacral
Face	Cleft lip	Closure of lip	36 days	42% associated with cleft palate
	Cleft maxillary palate	Fusion of maxillary palatal shelves	10 weeks	
	Branchial sinus or cyst	Resolution of branchial cleft	8 weeks	Preauricular and along line anterior to sterno-cleidomastoid
Gut	Esophageal atresia and tracheoesophageal fistula	Lateral septation of foregut into trachea and foregut	30 days	
	Rectal atresia with fistula	Lateral septation of cloaca into rectum and urogenital sinus	6 weeks	
	Duodenal atresia	Recanalization of duodenum	7 to 8 weeks	
	Malrotation of gut	Rotation of intestinal loop so that cecum lies to right	10 weeks	Associated incomplete or aberrant mesenteric attachments
	Omphalocele	Return of midgut from yolk sac to abdomen	10 weeks	
	Meckel diverticulum	Obliteration of vitelline duct	10 weeks	May contain gastric or pancreatic tissue
	Diaphragmatic hernia	Closure of pleuroperitoneal canal	6 weeks	
Genitourinary system	Extroversion of bladder	Migration of infraumbilical mesenchyme	30 days	
	Bicornuate uterus	Fusion of lower portion of müllerian ducts	10 weeks	Associated müllerian and wolffian duct defects
	Hypospadias	Fusion of urethral folds (labia minora)	12 weeks	
	Cryptorchidism	Descent of testicle into scrotum	7 to 9 months	
Heart	Transposition of great vessels	Directional development of bulbus cordis septum	34 days	
	Ventricular septal defect	Closure of ventricular septum	6 weeks	
	Patent ductus arteriosus	Closure of ductus arteriosus	9 to 10 months	
Limb	Aplasia of radius	Genesis of radial bone	38 days	Often accompanied by other defects of radial side of distal limb
	Syndactyly, severe	Separation of digital rays	6 weeks	
Complex	Cyclopia, holoprosencephaly	Prechordal mesoderm development	23 days	Secondary defects of midface and forebrain
	Sirenomelia (sympodia)	Development of posterior axis	23 days	Associated defects of cloacal development

*From Smith, D. W. (Editor): Recognizable Patterns of Human Malformation. Philadelphia, W. B. Saunders Company, 1982.

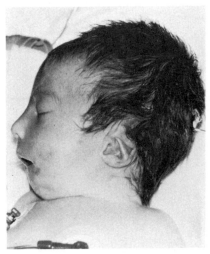

Figure 20–6. This child died during the first week of life with the complete spectrum of nonrandomly associated anomalies in CHARGE association. He had bilateral retinal colobomas, complex congenital heart disease, bilateral choanal atresia, aberrant auricles, micrognathia, micropenis, hypercalcemia, and thymic hypoplasia. The karyotype was normal, and there were no known teratogenic exposures. Other older children with this pattern of anomalies have also shown growth and mental deficiency, hearing loss, and hypogonadotropic hypogonadism. (From Smith, D. W.: Recognizable Patterns of Human Malformation. Ed. 3. Philadelphia, W. B. Saunders Company, 1982.)

these anomalies reflect an embryonic disturbance during the second month of gestation from conception, but no teratogenic exposures have been documented thus far. This is an example of a malformation association. Such a diagnostic category is useful in determining the kinds of studies to pursue when a baby presents with choanal atresia and multiple anomalies, but it should be viewed as a descriptive rather than diagnostic term.

With this brief description of the terminology currently in use within the field of dysmorphology as a preface, consideration will now be directed toward some of the more common conditions within each of these diagnostic categories. Conditions associated with the action of mutant genes or genetic imbalance due to chromosomal disorders are covered in Chapters 18 and 19. This chapter considers some of the more common disorders that result from nongenetic causes and lead to significant developmental impairments. For further details on these specific entities, the reader should consult Smith's *Recognizable Patterns of Human Malformation.*

MALFORMATIONS

Serious malformations are recognized in about 2 per cent of newborn babies, and this figure rises to about 4 per cent by one year of age as neurologic, cardiac, renal, and other disorders not apparent at birth are detected. Actually the majority of problems in morphogenesis are early lethal ones that result in unrecognized pregnancy, fetal loss,

or stillbirth. One major concern is the early detection of remediable defects of a lethal nature, such as intestinal atresia or tracheoesophageal fistula. It is also very important to recognize congenital anomalies of a potentially handicapping or deforming nature, such as dislocation of the hip or craniostenosis. Prompt recognition and initiation of therapy for such defects can prevent or vastly reduce subsequent problems. In the infant with a nonremediable handicapping problem, prompt initiation of a compassionate discussion of the nature of the problem with the parents helps to facilitate the best adaptation toward a productive course.

For life threatening malformations associated with very serious defects in brain structure and function, consultation with a dysmorphologist may be useful in the newborn nursery, in order to determine whether life perpetuating medical interference is merited. Examples of such conditions include cytogenetic anomalies that are usually lethal in the perinatal period, such as trisomy 18, trisomy 13, and triploidy. For these conditions, aspiration of bone marrow for rapid cytogenetic confirmation may facilitate such management considerations. Other examples might include holoprosencephaly, anencephaly, hydranencephaly, renal agenesis, and certain cardiac defects. For all malformation problems, management should include providing the parents with compassionate understanding and counsel, and this matter is discussed in more detail at the end of this chapter.

Obviously the initial task involves recognizing and accurately diagnosing the total nature of the malformation problem at the time of the newborn examination or during early infancy. Certain situations present an increased risk for malformation, and these are listed in Table 20–4. Such circumstances require special vigilance in the neonatal period. Malformation problems initially should be divided into two general categories: those that represent a single localized defect in morphogenesis in an otherwise normal individual, and those that result in multiple defects in one or more systems. The single primary defects are much more frequent and they are to be considered first.

ISOLATED MALFORMATIONS AND MALFORMATION SEQUENCES

Single localized defects in early morphogenesis can affect the subsequent development of other structures and result in a baby with more than one anomaly, and it is important to distinguish this kind of malformation problem from those that result in multiple malformations. The term *malformation sequence* has been utilized to distinguish such an initiating defect and its secondarily derived defects; some of the more common malfor-

Table 20–4. HIGHER RISK CATEGORIES FOR MALFORMATION PROBLEMS*

Category	Anomalies	Comment
Prematurity or postmaturity	Any	Fetal malformation may alter gestational timing
"Small for dates" babies	Any	Hypoplasia of the fetus is often accompanied by other anomalies
Breech birth presentation	Any	Certain malformed fetuses are less likely to
	Dislocation of hip deformations	assume the proper birth position
Polyhydramnios	High intestinal atresia	Defective swallowing or intestinal resorption of
	Anencephalus	amniotic fluid
	Multiple defects	
Oligohydramnios	Severe kidney defect	Lack of urine flow into amniotic space
	Urinary tract obstruction	
Twins in general	Club foot	Intrauterine crowding
Monozygotic twins	Any (2 to 3 times increased)	Probably due to same cause as monozygous twinning
Maternal diabetes mellitus	Any	Maternal metabolic aberration
	Sacral and lower limb defects	
Baby has several minor anomalies	Any, especially a known syndrome	The baby with two or more minor anomalies frequently has a serious defect as well
Parent or previous sibling has a single common malformation	Same type of malformation as in a parent or sibling	Polygenic inheritance with recurrence risk about 5%
Parent or previous offspring has a genetically determined nonpolygenic disorder	Same type of problem, with rare exceptions	Includes mendelian inheritance and certain chromosomal abnormalities
Older maternal age	Autosomal trisomy syndromes	Risk of faults in chromosome distribution increases exponentially after 35 years, especially risk for 21 trisomy
Older paternal age	Single gene disorders	The likelihood of fresh gene mutation, though rare, increases about 10-fold from the paternal age of 30 to 60 years

*From Smith, D. W. (Editor): Introduction to Clinical Pediatrics. Philadelphia, W. B. Saunders Company, 1977.

mation sequences are listed in Table 20–2. To illustrate the concept of a malformation sequence, defects in neural tube closure are to be discussed in further detail. (For discussion of other malformation sequences, the reader may wish to consult Smith's *Recognizable Patterns of Human Malformation*.)

Primary defects in neural groove closure constitute the most common single localized defects affecting the central nervous system. Figure 20–7 demonstrates how a failure in anterior neural tube closure can result in anencephaly with secondary alterations in the craniofacial region. Defects of closure at a more caudal level may result in a meningomyelocele with secondary neurologic defects resulting in club feet, hydrocephalus, and associated defects in the spinous processes. There may be multiple etiologies for defects in neural tube closure, including disruption resulting from early amnion rupture or maternal hyperthermia during the time of neural tube closure (23 to 28 days' gestation from conception). When no such etiology is appreciated, the recurrence risk is about 4 per cent for parents who have one affected child. There are some preliminary indications that this recurrence risk may be decreased in mothers who take vitamin supplements prior to conception and continue them at least through the time of normal neural tube closure. Elevation of the alpha fetoprotein level in the maternal circulation and also in the amniotic fluid, when open neural tube defects are present, facilitates prenatal diagnosis of these conditions, which are often easily detected by the ultrasonography in the second trimester.

MALFORMATION SYNDROMES

The term *syndrome* comes from the Greek word *syndromos*, which means "a running together," and it is used for clinically distinctive patterns of signs and symptoms.

A broad variety of intrinsic problems within tissues result in patterns of malformations that include the central nervous system either primarily or secondarily and thereby result in developmental or behavioral variation. In this chapter, only those that are present at birth, or congenital, will be considered in any detail. Many genetic inborn errors of metabolism result in progressive central nervous system deterioration *after birth*. Those resulting from the accumulation of metabolites within tissues, such as the mucopolysaccharidoses or lipidoses, may result in progressive coarsening of the facial features, thickening of the skin and tongue, and progressive joint contractures (Fig. 20–8). Such changes result in syndromes that may not be apparent at birth. Another example of slowly progressive postnatal dysmorphic changes would be the coarse facial features, myxedema,

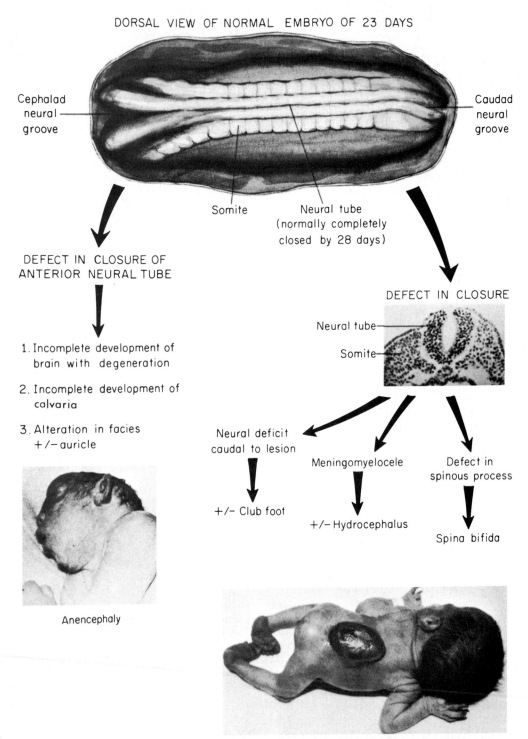

DORSAL VIEW OF NORMAL EMBRYO OF 23 DAYS

Cephalad neural groove

Caudad neural groove

Somite

Neural tube (normally completely closed by 28 days)

DEFECT IN CLOSURE OF ANTERIOR NEURAL TUBE

1. Incomplete development of brain with degeneration

2. Incomplete development of calvaria

3. Alteration in facies +/− auricle

Anencephaly

DEFECT IN CLOSURE

Neural tube

Somite

Neural deficit caudal to lesion

+/− Club foot

Meningomyelocele

+/− Hydrocephalus

Defect in spinous process

Spina bifida

Meningomyelocele with partially epithelialized sac

Figure 20–7. This figure demonstrates how a variety of different malformation sequences can result from the same type of early defect in morphogenesis, namely, a defect in proper neural tube closure. (From Smith, D. W.: Recognizable Patterns of Human Malformation. Ed. 3. Philadelphia, W. B. Saunders Company, 1982.)

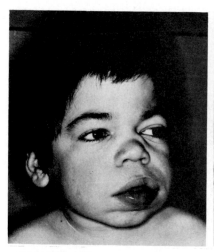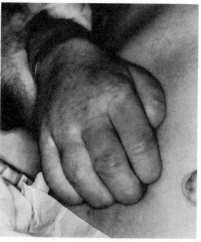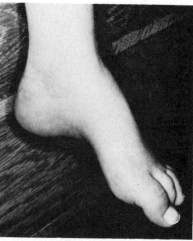

Figure 20–8. This 14 year old boy has one of the slowly progressive types of mucopolysaccharidosis, the X linked Hunter syndrome. The coarsening of his facial features and gradual onset of growth and mental deficiency began to be apparent during early childhood. He has shown evidence of neurologic deterioration, spasticity, and joint contractures. Shown here are a claw hand with thickening of the skin on the dorsum of the hand, umbilical hernia, pes cavus, and tightening of the Achilles tendon. The clarity of the corneas helps distinguish this type of mucopolysaccharidosis from other types that manifest corneal clouding.

sluggish activity and growth retardation associated with congenital hypothyroidism. Such conditions will be delineated and discussed in more detail elsewhere in this text. They are mentioned at this juncture only in the context of considering the very important question of when unusual features first became apparent in the child (Table 20–5).

It is always important to know when the child's phenotype began to vary from the normal pattern for a given family. This matter can be addressed in a nonthreatening way by asking, "Who in the family does he or she resemble?" For the child with true phenotypic similarity to what appears to be an unusual familial pattern, it is often helpful to examine as many family members as possible. Pictures of other possibly affected family members at younger ages can be most useful.

For some conditions the range of genetic expression may include malformations that are lethal in the perinatal period. Under such circumstances autopsy findings are helpful in arriving at the true diagnosis. X linked hydrocephalus is an example of one such condition, in which the pattern of genetic transmission suggests that only males are affected and there is no male to male transmission. The trait may be partially expressed by female carriers, as explained by the Lyon hypothesis. Autopsy findings suggesting aqueductal stenosis as the primary malformation leading to hydrocephalus help confirm the true diagnosis. Similarly, for a fatally malformed newborn with occipital encephalocele and polydactyly, the finding of cystic changes in the kidney and other viscera suggests a diagnosis of Meckel-Gruber syndrome, an autosomal recessive condition with a 25 per cent recurrence risk. Obviously these conditions can be diagnosed prenatally in some instances,

once families at risk for a recurrence are identified. Some of the more common congenital malformation syndromes with their distinctive findings are listed in Table 20–5.

In the case of many conditions resulting in early developmental brain anomalies we currently lack pathognomonic biochemical or cytogenetic markers that would help to confirm or reject a clinical diagnosis. Under such circumstances, consultation with specialists in dysmorphology can be useful in confirming recognition of a pattern of malformation for which there is no confirmatory laboratory test.

For many of the common disorders associated with developmental variation and having no known genetic or teratogenic cause, alterations in growth or body proportions can provide a clue to the underlying diagnosis. Prenatal growth deficiency that persists is common to Williams syndrome, the Cornelia de Lange syndrome, the Rubinstein-Taybi syndrome, and the Russell-Silver syndrome, all of which appear to be sporadic disorders with differing prognoses for cognitive development. Autosomal recessive disorders such as the Dubowitz syndrome, De Sanctis-Cacchione syndrome, Seckel syndrome, Johanson-Blizzard syndrome, Cockayne syndrome, and Smith-Lemli-Opitz syndrome can also result in rather striking, relatively proportionate growth deficiency in association with mental deficiency and other more specific features. Each of these conditions is relatively uncommon, and thus for further discussion of these entities the reader should consult other sources (e.g., Smith, 1982). The conditions listed in the former group of sporadic disorders are relatively frequent, and they are to be considered and contrasted briefly.

Table 20–5. DISTINGUISHING FEATURES IN 25 OF THE MORE COMMON SYNDROMES INVOLVING MULTIPLE DEFECTS (NOT ALL FEATURES PRESENT IN EVERY PATIENT)*

| Syndrome | Helpful Diagnostic Features | | | Mental Retardation | Short Stature | Genetics |
	Craniofacial	*Limbs*	*Other*			
1. XXY syndrome (Klinefelter)		Relatively long limbs, even in childhood	Small testes, incomplete virilization	+/–	–	XXY
2. Down syndrome	Upward slant to palpebral fissures, flat facies	Short hands with clinodactyly of fifth finger	Hypotonia	+	+/–	21 trisomy
3. 18 Trisomy syndrome	Microstomia, short palpebral fissure	Clenched hand, second finger over third; low arches on fingertips	Short sternum	+	+	18 trisomy
4. 13 Trisomy syndrome	Defects of eye, nose, lip, and forebrain of holoprosencephaly type	Polydactyly, narrow hyperconvex fingernails	Skin defects of posterior scalp	+	+	13 trisomy
5. XO syndrome (Turner)	Heart shaped facies, prominent ears, webbing of posterior neck, low posterior hairline	Congenital lymphedema or its residua	Broad chest with widely spaced nipples, hypogonadism	–	+	XO or variants
6. Turner-like syndrome (Noonan)	Webbing of posterior neck		Pectus excavatum, cryptorchidism, pulmonic stenosis	+/–	+	?
7. De Lange syndrome	Synophrys (continuous eyebrows), thin down-turning upper lip	Small or malformed hands and feet, proximal thumb	Hirsutism	+	+	?
8. Rubinstein-Taybi syndrome	Microcephaly, slanting palpebral fissures, maxillary hypoplasia	Broad thumbs and toes		+	+	?
9. Russell-Silver syndrome	Triangular hypoplastic facies with down-turning mouth	Skeletal asymmetry, clinodactyly of fifth finger	Underweight for length	–	+	?
10. Prader-Willi syndrome	+/– Upward slant of palpebral fissures	Hypotonia, especially in early infancy	Obesity from infancy, hypogenitalism	+	+	?
11. Myotonic dystrophy of Steinert	"Myopathic" facies; fine cataract		Myotonia with muscle atrophy, hypogonadism	+/–	–	Aut. dom.

No.	Syndrome	Primary features	Skeletal features	Other features			Etiology
12.	Treacher Collins syndrome (mandibulofacial dysostosis)	Malar and mandibular hypoplasia, defect in lower eyelid		Malformation of external ear	–	–	Aut. dom.
13.	Sturge-Weber malformation sequence	Flat hemangiomas of face, most commonly in trigeminal region		Hemangiomas of meninges with seizures	+/–	–	?
14.	Tuberous sclerosis (adenoma sebaceum)	Hamartomatous pink to brownish facial skin nodules	+/– Bone lesions	Seizures	+/–	–	Aut. dom.
15.	Neurofibromatosis		+/– Bone lesions	Neurofibromas, café au lait spots	–/+	–	Aut. dom.
16.	Hypohidrotic ectodermal dysplasia	Peg shaped teeth, partial anodontia, midfacial hypoplasia		Hypoplasia to aplasia of sweat glands, hyperthermia, alopecia	–	–	X linked
17.	Achondroplasia	Low nasal bridge, +/– macrocephaly	Short limbs, short hands and feet, limited elbow extension	Caudal narrowing of spinal canal, short ilium	–	+	Aut. dom.
18.	Apert syndrome (acrocephalosyndactyly)	Craniosynostosis, irregular midfacial hypoplasia and hypertelorism	Syndactyly, broad distal thumb and toe		+/–	–	Aut. dom.
19.	Crouzon syndrome (craniofacial dysostosis)	Shallow orbits, maxillary hypoplasia, craniosynostosis			–	–	Aut. dom.
20.	Hurler syndrome (MPS type I)	Coarse facies, cloudy cornea, early	Stiff joints by one year, kyphosis by 1 to 2 years	Valvular heart disease	+	+ Onset 6–18 mo.	Aut. rec.
21.	Marfan syndrome	Lens subluxation	Arachnodactyly	Aortic dilatation	–	–	Aut. dom.
22.	Osteogenesis imperfecta	Bluish scleras, odontogenesis imperfecta	Fragile bone	+/– Deafness	–	+/–	Aut. dom.
23.	Rubella syndrome	Cataract, deafness		Cardiac defect	+/–	+/–	Rubella
24.	Fetal alcohol syndrome	Short palpebral fissures, mild maxillary hypoplasia	Minor alterations	Cardiac defect	+	+	Ethanol
25.	Fetal hydantoin syndrome	Mild ocular hypertelevism, short nose, low nasal bridge	Hypoplastic distal digits, nail hypoplasia		+/–	+/–	Hydantoins (Dilantin)

*From Smith, D. W. (Editor): Introduction to Clinical Pediatrics. Philadelphia, W. B. Saunders Company, 1977.

The Williams Syndrome

Williams syndrome is a unique disorder consisting of a mild prenatal growth deficiency with relatively severe postnatal growth retardation, feeding problems, and failure to thrive. Mild microcephaly is usually present and the average IQ is 50 to 60. Such children tend to be outgoing and loquacious, with a hoarse voice and a "cocktail party manner." Cardiovascular anomalies, including supravalvular aortic stenosis, are relatively frequent findings, but infantile hypercalcemia has been observed infrequently. The facial features are more consistent and more useful for diagnostic purposes. As shown in Figure 20–9A to C, such children may show a medial eyebrow flare with a short nose, flat nasal bridge, anteverted nares, long philtrum, prominent lips with open mouth, and stellate iris patterning.

The Cornelia de Lange Syndrome

The degree of growth and mental deficiency in the Cornelia de Lange syndrome is generally more severe than that of Williams syndrome, the usual IQ being below 35. Muscle tone is generally increased, and behavior tends to show autistic qualities. Limb deficiency is characteristic: it can range from micromelia with brachydactyly (Fig. 20–9E) to oligodactyly, which characteristically affects the ulnar aspects of the hand (Fig. 20–9F).

As shown in Figure 20–9D, the facial features are characteristic. They include microbrachycephaly with marked hirsutism that involves the forehead, eyebrows, and eyelashes. Eyebrows characteristically extend across the nasion without interruption (synophrys), and the nose is short with anteverted nares. The lips and mouth are characteristic with thin lips that show a midline beaking of the upper lip and a corresponding notch in the lower lip. The angles of the mouth curve downward, and the cry tends to be low pitched, weak, and growling.

The Rubinstein-Taybi Syndrome

The Rubinstein-Taybi syndrome also includes characteristic limb malformations, with broad thumbs that show a radial angulation and broad big toes. Growth deficiency and mental deficiency are present, and the usual IQ is in the 40 to 50 range. The face usually shows a hypoplastic maxilla with down slanting palpebral fissures and a beaked nose with the nasal septum extending below the alae (Fig. 20–9G to I).

The Russell-Silver Syndrome

The Russell-Silver syndrome presents with a skeletal growth deficiency of prenatal onset that includes the facial bones. However, the cranium is generally much less affected, resulting in a relatively triangular face with an upper head that appears large. In most instances the head circumference is well within the normal range, but osseous development may be delayed in infancy, resulting in relatively large fontanels. These findings are often so striking as to result in concern over possible hydrocephalus, which these patients do not have. The skeletal growth deficiency is sometimes asymmetric in its distribution, and the fifth fingers are often incurved (Fig. 20–10). The growth rate may increase during childhood, with a tendency toward normalization of osseous maturity and an ultimate height attainment of up to 5 feet. During infancy and early childhood, such children tend to be slim and weak, often demonstrating delays in early motor progress.

During the first three years assurance of adequate glucose intake during illnesses, through frequent feedings, is imperative because the limited reserves in these patients result in a liability toward fasting hypoglycemia. After this early period, which often causes concern as to a possible risk of mental deficiency, strength improves and motor progress shows a return to normal. Intelligence is usually within the normal range. Although the disorder is often a sporadic occurrence in an otherwise normal family, autosomal dominant inheritance has been implied in several families.

Other Malformation Syndromes

It should be emphasized that most of the aforementioned disorders result in relatively proportionate growth deficiency. When growth deficiency is disproportionate, resulting in short limbed dwarfism, one of the many genetically determined osteochondrodysplasias should be considered. Achondroplasia is the most common such condition, and like some of the other rarer forms of osteochondrodystrophy, it may be associated with secondary neurologic complications. Although megacephaly is characteristic of achondroplasia, a narrow foramen magnum may also result in secondary hydrocephalus. Furthermore, spinal cord or root compression may occur as a consequence of kyphosis, spinal canal stenosis, or disc lesions in about 46 per cent of the patients (Smith, 1982). Therefore, cautious orthopedic and neurologic follow-up in such patients is very important.

Several relatively common syndromes resulting in mental deficiency are associated with unusual growth excesses. Among such syndromes are some associated with obesity. These include the Prader-Willi syndrome, Carpenter syndrome, Cohen syndrome, Laurence-Moon-Biedl syn-

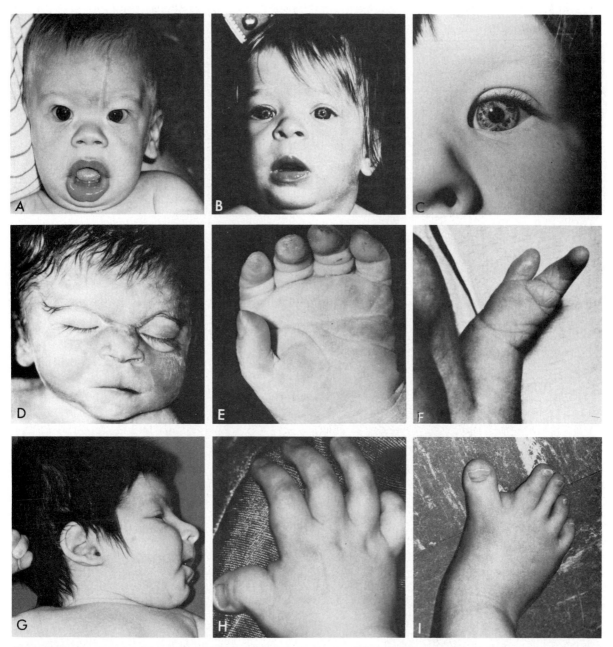

Figure 20–9. Shown here are three different sporadic syndromes of unknown etiology. Each is associated with relatively proportionate short stature and varying degrees of mental deficiency. *A* to *C*, These two young children with Williams syndrome demonstrate the facial features characteristic of this disorder. The nose is short with a flat nasal bridge and anteverted nares, giving rise to the impression of an elongated philtrum. The lips appear full, with a tendency toward an open mouth. The iris reveals a stellate pattern (*C*). *D* to *F*, The facial features shown here are typical of the Cornelia de Lange syndrome: microbrachycephaly, marked facial hirsutism, synophrys, a short nose with anteverted nares, and thin lips with a characteristic midline notch and down-turned angles of the mouth. Limb deficiency in this condition can range from brachydactyly with micromelia (*E*) to oligodactyly, which usually affects the ulnar aspects of the hand to a greater extent (*F*). *G* to *I*, This infant with the Rubinstein-Taybi syndrome shows the characteristic beaked nose with a nasal septum that extends below the alae nasi. He is relatively hirsute, with broadening and flattening of the distal thumb. The thumb (*H*) and great toe (*I*) also tend to deviate to one side, as shown here in the hands and feet of an older child with the Rubinstein-Taybi syndrome. (From Smith, D. W.: Recognizable Patterns of Human Malformation. Ed. 3. Philadelphia, W. B. Saunders Company, 1982.)

drome, and Albright hereditary osteodystrophy. Figure 20–11 compares and contrasts these obesity related conditions. Other disorders may result in more generalized macrosomia and overgrowth. Some of the more common such disorders include the Sotos syndrome, Weaver syndrome, Klippel-Trenaunay-Weber syndrome, and Beckwith-Wie-

demann syndrome (Fig. 20–12). More detailed consideration of the many syndromes that result in developmental variability is beyond the scope of this text. The foregoing discussion has attempted to highlight some of the more common conditions with unique phenotypic characteristics. The reader should consult more primary refer-

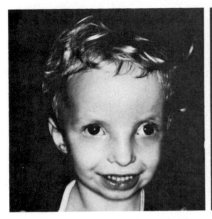

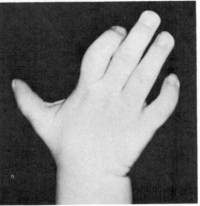

Figure 20–10. This three year old child with the Russell-Silver syndrome had delayed motor development, and his persistently small size had been a cause for concern. Here the triangular shape of his face and inturning of his fifth finger are clearly evident. In addition, he demonstrated a liability toward fasting hypoglycemia, mild skeletal asymmetry, and a deficit of adipose tissue with blue (thin) sclerae.

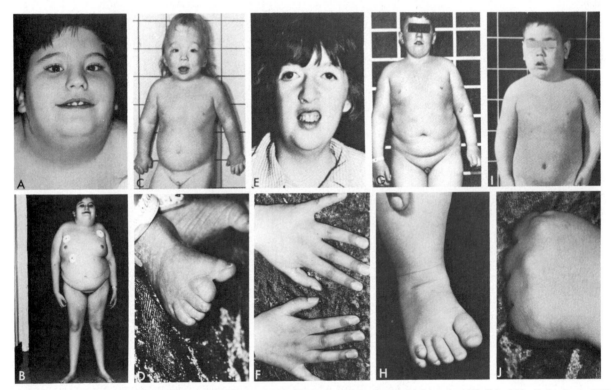

Figure 20–11. Shown here are five different disorders associated with obesity and developmental delay. *A* and *B,* This child with the Prader-Willi syndrome failed to demonstrate the specific interstitial deletion of chromosome 15 that has been found in up to half the patients with this disorder. He presented with severe hypotonia in infancy and developed obesity in early childhood due to an excessive appetite. He has relatively severe scoliosis (an occasional associated finding), which, combined with his obesity, compromises respiratory function. He has a small penis with cryptorchidism. His facial features reveal upslanted, almond shaped palpebral fissures and a frontal upsweep. (See similar features in the patient shown in Figure 20–4F). This sporadic disorder of unknown etiology carries an estimated empirical recurrence risk of 1.6 per cent; the usual IQ in this disorder ranges from 40 to 60. *C* and *D,* Carpenter syndrome is an autosomal recessive craniosynostosis syndrome associated with variable degrees of mental deficiency. The face reveals shallow supraorbital ridges with laterally displaced inner canthi. The feet usually show preaxial polydactyly with partial syndactyly. There may also be generalized aminoaciduria. *E* and *F,* Cohen syndrome is an autosomal recessive disorder associated with hypotonia and mental deficiency (IQ ranging from 30 to 70). Facial features reveal a high nasal bridge with maxillary hypoplasia and down-slanted palpebral fissures. The philtrum appears short, and some patients reveal prominent central maxillary incisors with mild micrognathia. The hands appear narrow with slim fingers, and there may be ocular problems or lumbar lordosis with mild scoliosis. *G* and *H,* The Laurence-Moon-Biedl syndrome is an autosomal recessive condition associated with mild to moderate mental deficiency, polydactyly, retinitis pigmentosa, and genital hypoplasia. Retinal degeneration usually results in impaired night vision during childhood, 15 per cent of patients showing atypical retinal pigmentation by five to 10 years of age. Central vision is lost first, followed by peripheral vision; 73 per cent of patients are blind by 20 years of age. *I* and *J,* Albright hereditary osteodystrophy syndrome shows a 2:1 female:male sex incidence, suggesting an X linked dominant mode of inheritance. Cognitive performance varies widely; the mean IQ is about 60. Stature is usually diminished, and the face tends to be rounded with a low nasal bridge and delayed dental eruption. Metacarpals and metatarsals tend to be short, especially the fourth and fifth (note fourth and fifth recessed knuckles in fisted hand), and the epiphyses are usually cone-shaped. There may be variable hypocalcemia and hyperphosphatemia with extraskeletal calcification in subcutaneous tissues and basal ganglia. (*C, E, G,* and *I* from Smith, D. W.: Recognizable Patterns of Human Malformation. Ed. 3. Philadelphia, W. B. Saunders Company, 1982.)

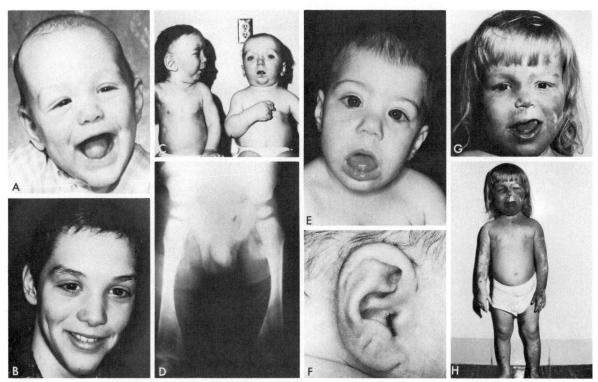

Figure 20–12. Four different sporadic disorders of unknown etiology that may be associated with macrosomia and generalized overgrowth. *A* and *B*, Sotos syndrome results in variable mental deficiency (mean IQ 72), poor coordination, and prenatal onset of excessive size, with large hands and feet. Osseous maturation is usually advanced and commensurate with height age, with phalangeal centers usually more accelerated than carpal centers. As shown in this child, as an infant and at age 10 yrs, the characteristic facial features consist of dolicocephalic macrocephaly with a prominent forehead, ocular hypertelorism, downslanted palpebral fissures, and prognathism with a narrow pointed chin. Electroencephalographic abnormalities may also be present. Although most cases are sporadic occurrences in otherwise normal families, at least five instances of presumed autosomal dominant inheritance have been reported. *C* and *D*, Weaver syndrome is also associated with large size at birth, accelerated growth, and markedly advanced skeletal maturation during infancy. There is usually mild hypertonia, developmental delay, and a hoarse, low pitched cry. These two unrelated boys, 18 and 11 months of age, manifest a wide bifrontal diameter, ocular hypertelorism, flat occiput, large ears, long philtrum, and relative micrognathia. There is limited knee and elbow extension with camptodactyly. The thumbs are broad, with prominent fingertip pads, and radiographs show broad distal ulnas and femurs (*D*). This disorder may resemble the Marshall-Smith syndrome in some cases. (From Weaver, D. D., et al.: A new overgrowth syndrome with accelerated skeletal maturation, unusual facies, and camptodactyly. J. Pediatr., *84*:547, 1974.) *E* and *F*, The Beckwith-Wiedemann syndrome consists of omphalocele, macrosomia with a large muscle mass and accelerated osseous maturation, macroglossia, and unusual linear fissures in the lobule of the external ear. Abdominal organs tend to be large, with hyperplastic and dysplastic changes (renal medullary dysplasia, pancreatic islet cell hyperplasia, fetal adrenocortical cytomegaly, and gonadal interstitial cell hyperplasia). The neonatal period may be complicated by polycythemia or hypoglycemia, and there is some undefined risk of mild to moderate mental deficiency. The estimated risk for the development of neoplasia (particularly nephroblastoma, adrenocortical carcinoma and hepatoblastoma) is 6.5 per cent. *G* and *H*, The Klippel-Trenaunay-Weber syndrome consists of asymmetrical limb hypertrophy and hemangiomas. The hypertrophy may not necessarily coincide with the area of hemangiomatous involvement, and distribution may not always be unilateral. As shown in this four year old girl with extensive involvement, when there is facial hemangiomatosis, there may be asymmetrical facial hypertrophy. When the hemangiomas follow a trigeminal facial distribution there may be overlap with the Sturge-Weber malformation sequence (facial hemangiomas, meningeal hemangiomas, and seizures). Usually only patients with facial hemangiomatosis demonstrate mental deficiency. Even without facial hemangiomatosis, there may be benign macrocephaly due to macroencephaly.

ences before attempting to make a firm diagnosis and provide counseling to affected families.

TERATOGENIC DISORDERS

Effects of Alcohol on the Fetus

The teratogenic effects of alcohol were first given widespread recognition in 1973 when Jones and Smith described a characteristic pattern of malformation in eight children born to chronically alcoholic mothers. In all these children there was prenatal and postnatal growth deficiency, with no evidence of catch-up growth; six of the children had been hospitalized for failure to thrive. The average IQ was in the mildly retarded range, and

most of the children had poor fine and gross motor coordination.

Some had limited joint mobility, altered palmar creases, and cardiac defects, and most had a characteristic face that included midface deficiency, a smooth philtrum, a flat nasal bridge, and short palpebral fissures. The nature of these defects clearly implicated the prenatal period as the time of the insult. Figure 20–13 shows the typical facial features of children with the fetal alcohol syndrome, and a typical hand is shown in Figure 20–3D.

During the past few years, hundreds of case reports from around the world have provided additional documentation of the adverse effects of alcohol exposure in utero. Table 20–6 provides a

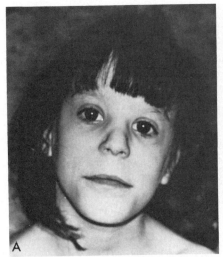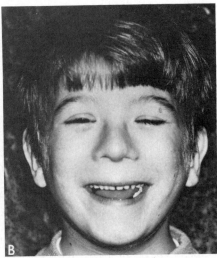

Figure 20–13. Fetal alcohol effects. *A,* This eight year old girl weighed 2.3 kg at birth after a term gestation in a gravida 4, 39 year old, chronically alcoholic woman. She is mildly retarded and growth deficient, with facial features typical of fetal alcohol syndrome (short palpebral fissures, short nose with flat nasal bridge, and thin lips with a smooth philtrum). *B,* This eight year old boy shows similar facial features and is also mildly retarded and growth deficient. His hands are shown in Figure 20–4*D.* He weighed 2 kg at birth after a term gestation in a gravida 5, para 4, abortus 1, 28 year old woman who drank one fifth of hard liquor every one to two days during the first four months of pregnancy. During this period she was hospitalized twice for detoxification and treated with disulfiram and chlordiazepoxide. She remained abstinent after the beginning of the fifth month. (From Graham, J. M., Jr.: Manual for the Assessment of Fetal Alcohol Effects. Seattle, University of Washington Press, 1982.)

list of the more commonly observed fetal alcohol effects. It has also become apparent that alcohol, like other teratogens, gives rise to a spectrum of defects, with affected children showing much individual variation in both the extent and severity of involvement. Currently the term "fetal alcohol syndrome" is reserved for the more severe end of the spectrum, in which the complete triad of growth deficiency, mental retardation, and specific facial dysmorphism is expressed. The term "fetal alcohol effects" is used to describe lesser degrees of phenotypic expression, in an attempt to answer the primary diagnostic issue, "Is the child's problem secondary to alcohol exposure in utero?"

Other studies have demonstrated dose related increases in spontaneous abortion and perinatal morbidity and mortality in drinking mothers. All these findings have been replicated in a broad variety of experimental animals, in which confounding variables can be eliminated through experimental design (Little et al., 1982). No single feature of the disorder is pathognomonic, and the fetal alcohol syndrome may mimic other recognizable patterns of malformation, especially in severely affected individuals. The differential diagnosis in fetal alcohol effects should include: the Cornelia de Lange syndrome, Noonan syndrome, Williams syndrome, Dubowitz syndrome, Stickler syndrome, X linked mental deficiency, fetal hydantoin effects, and maternal phenylketonuria.

The growth deficiency characteristic of the fetal alcohol syndrome is usually prenatal in onset and is not accompanied by postnatal catch-up growth. Such children are usually well below the third percentile for height, weight, and head circumfer-

ence, and admissions for failure to thrive have been quite common. They usually have normal levels of growth hormone, cortisol, and gonadotropins. Reductions in adipose tissue usually become more obvious in children whose mothers have consumed large amounts of alcohol throughout pregnancy. When heavily drinking alcoholic women have been able to reduce their alcohol intake during the second and third trimesters, their infants have usually shown less growth retardation than infants born to mothers who continued to drink heavily throughout pregnancy. However, even when sobriety can be achieved early in pregnancy, it does not necessarily protect infants from functional brain disturbance (Olegard et al., 1979).

Central nervous system dysfunction continues to be one of the most significant fetal alcohol effects. Studies of intellectual functioning in children with the fetal alcohol syndrome have demonstrated an average IQ slightly below 70 (in the mildly retarded range), with a wide range of individual IQ scores. An increase in the severity of fetal alcohol physical effects tends to be correlated with a decrease in intellectual performance. There also tends to be an increase in the severity of fetal alcohol effects among children born to chronic alcoholics, as opposed to those born to women in earlier phases of alcoholism.

The impact of prenatal alcohol exposure on intellectual functioning emphasizes the continuum of clinical effects that has been observed. At the milder end of this spectrum are children with hyperactivity, learning difficulties, or communicative disorders. At the other end of the scale are

Table 20–6. FETAL ALCOHOL EFFECTS*

Principal effects (seen in more than 50 per cent of the patients)
  Facial features
    Short palpebral fissures
    Short upturned nose with flat nasal bridge
    Maxillary hypoplasia
    Hypoplastic philtrum with thin upper lip
    Micrognathia or relative prognathia due to midface deficiency

  Prenatal and postnatal growth deficiency
    <2 SD below mean for both length and weight
    Absence of significant catch-up growth
    Disproportionately diminished adipose tissue

  Central nervous system dysfunction
    Microcephaly
    Altered muscle tone, especially hypotonia in infancy
    Poor fine and gross motor coordination
    Irritability in infancy
    Hyperactivity in childhood
    Mild to moderate mental deficiency

Frequent effects (seen in 26 to 50 per cent of the patients)
  Facial features
    Ptosis, strabismus, epicanthal folds
    Posteriorly rotated ears
    Prominent lateral palatine ridges

  Malformations
    Cardiac murmurs, especially septal defects
    Pectus excavatum
    Labial hypoplasia
    Aberrant palmar creases
    Hemangiomas

Occasional effects (seen in 25 per cent or fewer of the patients)
  Facial features
    Myopia, microphthalmia, blepharophimosis
    Altered helical form
    Cleft lip or palate
    Small teeth with faulty enamel

  Malformations
    Great vessel anomalies, tetralogy of Fallot
    Hypospadias, renal anomalies
    Limited joint movements, especially in fingers and elbows
    Radioulnar synostosis
    Nail hypoplasia, polydactyly
    Pectus carinatum, bifid xiphoid
    Klippel-Feil anomaly, scoliosis
    Hernias of diaphragm, umbilicus or groin
    Diastasis recti
    Hydrocephalus, meningomyelocele

*Adapted from Clarren, S. K., and Smith, D. W.: The fetal alcohol syndrome. N. Engl. J. Med., 298:1063–1067, 1978.

found children with serious developmental brain disorders and major degrees of cognitive impairment.

The mental deficiency and microcephaly associated with the fetal alcohol syndrome have been attributed to diminished brain growth. A broad variety of structural brain abnormalities have been related to prenatal ethanol exposure, the most frequent type of malformation being neuroglial heterotopias resulting from aberrant neuronal and glial migrations. Other alterations have included abnormal gyral patterning, arhinencephaly, porencephaly, absence of the corpus callosum, hydranencephaly, and hydrocephalus. Some measure of the significance of these central nervous system alterations may be derived by examining data from Sweden, which suggest a significantly increased risk of cerebral palsy in children born to alcoholic mothers; calculations indicate that every sixth case of cerebral palsy might be associated with maternal alcohol abuse during pregnancy (Olegard et al., 1979). Furthermore, a study of mildly retarded (IQ 50 to 70) school children in Götenborg, Sweden, revealed that 10 to 12 per cent of these children had the complete fetal alcohol syndrome with a confirmed history of alcohol abuse by the mother during pregnancy (Olegard et al., 1979).

A variety of studies suggesting an incidence of some degree of fetal alcohol effect in one in 300 to 400 live births implicate alcohol as the major cause of preventable mental deficiency (Olegard et al., 1979). It is impossible to calculate the cost of decreased human potential caused by fetal alcohol effects, but New York State has recently estimated that babies with some degree of fetal alcohol effect, born in that state during the course of one year, will require $155,000,000 in funds for care over the course of their lifetimes.

The role of other substances of abuse in association with alcohol remains to be clarified, as do the roles of poor nutrition and smoking, but in studies in which these effects have been evaluated statistically, alcohol emerges as the dominant teratogen, with smoking having an additive impact on intrauterine growth retardation. Of additional interest is the role of chronicity of alcoholism in the genesis of fetal alcohol effects and also the role of binge drinking in causing specific malformations. The role of genetic and metabolic factors in both the mother and her fetus remains to be clarified. However, in a large number of animal models it is clear that prenatal exposure to ethanol results in a pattern of malformation similar to that described in children born to chronically alcoholic women (Little et al., 1982).

Effects of Drugs on the Fetus

Curently a number of drugs are known to be teratogenic in the developing fetus, and an even larger number of drugs and substances are suspected of having some degree of teratogenic effect. Thalidomide and aminopterin have clearly recognized teratogenic potentials, and because of the widespread recognition of this fact, few if any fetuses are currently exposed to such drugs. As

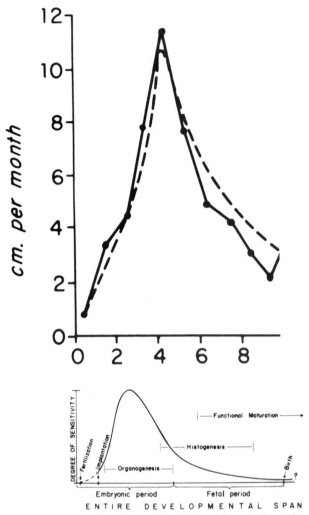

Figure 20–14. In the upper graph the fetal growth rate is shown; the lower graph indicates fetal sensitivity to teratogenic influences relative to periods of embryonic and fetal development. Periods of organogenesis and early histogenesis coincide with the most rapid period of fetal growth. For this reason prenatal growth deficiency and an increased incidence of malformations are characteristic of most known teratogens. In many instances the combination of these two factors is severe enough to result in fetal loss; hence fetal wastage is also a common effect of most known teratogens. (The lower graph was provided through the courtesy of Dr. Tom Shepard, University of Washington School of Medicine.)

mother's metabolism cannot protect her fetus from similar consequences in a subsequent pregnancy. With early malformations the severity of the defect may lead to fetal loss; hence an increased frequency of miscarriages in a mother may be a reflection of a teratogenic effect.

A second point with regard to very early exposures to teratogens within the first two weeks after conception seems worth stressing. Experience with ionizing radiation exposure during this period of gestation indicates that teratogenic exposures within this time either result in such serious fetal malformations that the pregnancy is lost (or not even recognized as a pregnancy), or a normal baby is born. This experience stems from the Hiroshima disaster, and it appears to apply to other exposures to potent teratogens within this short time.

Anticonvulsants have recently been recognized as having teratogenic potential for causing certain maternal-fetal disorders, although reliable prospective risk figures are not yet available. Following exposure to hydantoins, the risk can be as high as 10 or 11 per cent for some degree of growth and mental deficiency or specific congenital anomalies, which can include ocular hypertelorism, nasal growth deficiency, cleft lip or palate, cardiac defects, and nail hypoplasia (Figs. 20–3E, 20–15). With exposure to trimethadione the risks for growth and mental deficiency are probably even higher, and the pattern of malformations differs. Exposed children may show nasal growth deficiency with a prominent forehead, mild synophrys with an unusual upslant to the eyebrows, and a poorly developed, overlapping auricular helix.

Scattered reports suggest that other anticonvulsants, such as primidone, carbamazepine, and valproic acid, might also be teratogenic under certain circumstances. Furthermore, there is some evidence to suggest that children exposed to hydantoin may have an increased risk for neoplasia with an embryonic basis (particularly neuroblastoma and Wilms' tumor). With additional exposure to other teratogens, such as ethanol, there may very well be additive or synergistic effects.

An anticoagulant, warfarin, has also been recognized as having a teratogenic potential for causing growth and mental deficiency, seizures, marked nasal hypoplasia, and stippled epiphyses. This pattern of malformation is probably biased by the clinical nature of the sources for case assessment, and good prospective risk figures are not yet available.

Prenatal exposure to methadone or heroin has not been linked to an increased incidence of structural defects, but both result in neonatal withdrawal symptoms, and heroin may also be associated with some risk of growth deficiency and

shown in Figure 20–14, these and other teratogens have taught us that the first trimester is a critical period for embryonic organogenesis and growth. Hence this is the most important period with respect to teratogenic susceptibility. Exposures to known teratogens during this period have a high likelihood of resulting in fetal malformations and growth deficiency, depending on a number of metabolic factors that can modulate the effect of a given teratogen. In many instances only the pattern of malformations is relatively specific to a particular teratogen (e.g., fetal alcohol effects). From a practical standpoint, if a drug exposure has been found to be teratogenic during a previous pregnancy, there is a good possibility that the

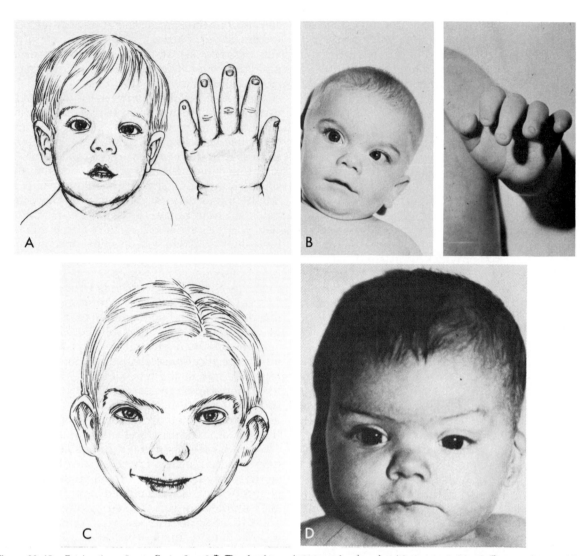

Figure 20–15. Fetal anticonvulsant effects. *A* and *B,* The drawing and photographs show facial features consistent with some degree of fetal hydantoin effects: ocular hypertelorism and nasal growth deficiency with a broad flat nasal bridge. The hands show distal phalangeal hypoplasia with small nails. *C* and *D,* This drawing and photograph show facial features consistent with fetal trimethadione effects: a short upturned nose with a broad flat nasal bridge and a poorly developed, cupped helix. Note in particular the mild synophrys with an unusual upslant to the eyebrows. (From Smith, D. W.: Mothering Your Unborn Child. Philadelphia, W. B. Saunders Company, 1979, and Smith, D. W.: Recognizable Patterns of Human Malformation. Ed. 3. Philadelphia, W. B. Saunders Company, 1982.)

possibly problems in learning and behavior. Valium may be associated with a slightly increased risk of cleft lip or cleft palate. Most data suggest that neither the phenothiazines nor the tricyclic antidepressants constitute major teratogenic hazards for the developing fetus. Lithium has been associated with congenital heart defects (particularly Ebstein's anomaly of the tricuspid valve). Amphetamines and LSD are probably not teratogenic in the human. Despite the fact that there are probably thousands of gestational exposures to marihuana and cocaine, nothing can be said about their potential for teratogenicity.

If progestogens or progestogen-estrogen combinations are teratogenic in the human, the magnitude of the teratogenic risk is extremely small. Some studies have suggested that progestogen exposures during the first trimester may be associated with an increased incidence of cardiovascular defects in the offspring, while progestin exposures between the eighth and thirteenth weeks of gestation may double the incidence of hypospadias. Other studies have failed to detect any significant increase in malformations following such exposures. Diethylstilbestrol has been associated with an increased risk of adenocarcinoma of the vagina and of genitourinary anomalies in both exposed males and females. Clomiphene, Bendectin, and adrenocorticoids are not considered to be teratogenic in the human. Streptomycin has been associated with eighth nerve damage in prenatally exposed children, and exposure to tetracycline after the fourth month of pregnancy results in yellowed deciduous teeth with enamel hypoplasia; permanent teeth are not affected. There is no indication that either penicillin, rifampin, ethambutol, or metronidazole is teratogenic in the human. Inadvertent vaccination for rubella shortly before or after contraception does not appear to constitute a significant teratogenic hazard.

Fetal Environmental Exposures

Diagnostic radiography during the first trimester usually delivers less than 5 rads to the fetus, and this is not thought to be a teratogenic dose. Serious risk to the fetus does not occur until a dose of 10 rads or more has been absorbed by the fetus. Large therapeutic doses of radiation can lead to microcephaly and mental deficiency in exposed fetuses. Table 20–7 provides estimated dosages for some of the routine diagnostic radiologic studies.

Chemical and organic solvent exposures have not been adequately evaluated, and the same is true for pesticide and herbicide exposures.

Cigarette smoking has been linked to intrauterine growth deficiency and to long term behavioral effects, which may include decreases in reading ability and other cognitive abilities. Decrements in growth appear to be related to the number of cigarettes smoked, with greater effects being related to heavier maternal smoking. There also appears to be a greater risk for perinatal morbidity and mortality in babies born to smokers as opposed to those born to nonsmokers. The influence of smoking appears to be greatest in the last four months of pregnancy; children born to women who have succeeded in stopping smoking prior to the end of the fifth lunar month do not differ in weight at birth from children born to mothers who have never smoked.

Fetal Hyperthermia Effects

Animal studies have shown heat to be a significant teratogenic factor capable of inducing a variety of defects, many of which affect the central nervous system. The defects observed in experimental animals include hernias, tooth defects, vertebral anomalies, eye defects, talipes, arthrogryposis, micrencephaly, exencephaly, and anencephaly. In the pregnant guinea pig a 1.5° C.

Table 20–7. IONIZING RADIATION DOSES FOR ROUTINE RADIOLOGIC STUDIES

Radiologic Study	Dose in Millirads
Chest x-ray	5
Cholecystography	300
Upper gastrointestinal series	330
Intravenous pyelography	585
Barium enema	485

elevation in the core temperature halts cell proliferation in the embryo, and an increase of 3° C. kills mitotic cells; this effect is most evident in the rapidly proliferating cells of the developing neuroepithelium (Edwards, 1979).

The most common abnormality in guinea pigs exposed to hyperthermia at 18 to 25 days' gestation is micrencephaly, due to death of mitotic neuroepithelial cells during a stage of development equivalent to four to six weeks in the human. The remaining stem cells are preprogramed to a finite series of divisions and are unable to compensate for the initial depopulation, resulting in progeny with micrencephaly, hypotonia, mental deficiency, occasional microphthalmia, and mild distal limb defects. Problems of spinal morphogenesis with secondary arthrogryposis are the most common consequences of hyperthermia induced at 26 to 42 days' gestation, which is equivalent to 7 to 12 weeks in the human. If exposure to hyperthermia occurs during the time of neural tube closure, neural tube defects may result (Edwards, 1979).

In the human several retrospective studies, which specifically ascertained a history of maternal hyperthermia at the time of neural tube closure, have indicated an increased incidence of exposure in patients with neural tube defects, as opposed to controls, thereby implicating maternal hyperthermia as a contributing etiologic factor in approximately 10 per cent of the babies born with neural tube defects (Pleet et al., 1981). Other studies have documented frequent occurrences of anencephaly following epidemics of influenza, but such studies are confounded by attempts to separate the role of the agents responsible for the fever and its treatment from the effects of the fever itself. An increased frequency of spontaneous abortion, stillbirths, and prematurity has also been noted after illnesses that provoke a high fever, such as influenza, typhoid fever, pneumonia and malaria. Fetal loss may very well be the most common result following sustained high fever in early pregnancy.

Among fetuses who survive a hyperthermic insult during early pregnancy, in addition to neural tube defects, the risk of other malformations appears to be increased following a febrile illness during the first trimester of pregnancy (Pleet et al., 1981). The type of malformation appears to be related to the timing of the exposure to hyperthermia (Fig. 20–16). Among 28 dysmorphic infants with a history of exposure to hyperthermia between four and 14 weeks of gestation, all survivors had mental retardation and most had altered muscle tone, including hypotonia with increased deep tendon reflexes. Eight of 11 patients exposed to hyperthermia between the fourth and seventh weeks of gestation had evidence of facial defects, including midface hypopla-

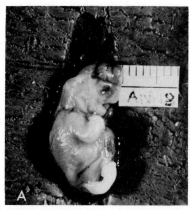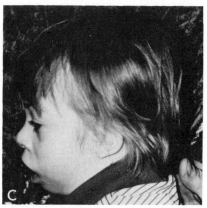

Figure 20–16. Variable effects of maternal hyperthermia during early gestation. *A,* A spontaneous abortus with anencephaly that was passed after three days of sustained high maternal fever between 24 and 27 days of gestation after conception. *B,* A microcephalic, moderately retarded 11 year old girl with hypotonia and bilateral talipes equinovarus whose mother sustained three days of high fever during the fifth week of gestation due to Hong Kong flu. Note the marked midface deficiency. (From Graham, J. M., Jr., and Edwards, M. J.: Teratogenic effects of maternal hyperthermia. *In* Op't Hof, J., and Gericke, G. S. [Editors]: Genetic Perspectives in Fetal and Neonatal Medicine. Pretoria, South Africa, Haum Publishers, 1982.) *C,* A microcephalic, profoundly retarded 12 year old girl with micrognathia, cleft palate, abnormal auricles, congenital heart disease, and bilateral talipes equinovarus whose mother had sustained a high fever between the sixth and eighth weeks of gestation due to viral laryngotracheitis.

sia, cleft lip or palate, micrognathia, and external ear anomalies. Exposures after this period were less likely to result in facial defects, but mental deficiency and altered muscle tone continued to be frequent findings (Pleet et al., 1981). Thus, in both experimental animals and humans, exposures to hyperthermia at comparable stages of gestation result in similar patterns of anomalies.

Neuronal heterotopias were seen in all four infants who died with a history of exposure to hyperthermia between the fourth and fourteenth weeks of gestation, suggesting that hyperthermia might disrupt neuronal migration as well as neuronal proliferation (Pleet et al., 1981). It is intriguing that experience in both experimental animals and man suggests that the central nervous system is the main target organ for hyperthermia, presumably because of its intense proliferative activity during stages of susceptibility. The timing of exposure, its duration, and the degree of elevation above basal core temperature are of major importance in assessing the risks of hyperthermia exposure in man.

In the aforementioned study, all instances of maternal fever were prolonged and sustained for one or more days, usually 38.9° C. or above, rather than being isolated fever spikes. Prolonged high fever during the first trimester of pregnancy, associated with viable offspring, seems to be relatively uncommon, as suggested by the findings of a prospective study of more than 50,000 pregnancies in the Collaborative Perinatal Project, which failed to reveal a single case of hyperthermia teratogenesis.

Whether environmental heat, such as might be experienced during hot summers, can influence human development similarly is unknown, but there are some indications that it might. Reports from both the northern and southern hemispheres indicate a winter-spring excess in births of children with neural tube defects, and birth dates of children in the United States admitted to an institute for mental deficiency show a similar winter-spring (February) excess. These children would have been in the first trimester of pregnancy during the summer, and in fact more admissions of mentally handicapped children followed abnormally hot, compared with abnormally cold, summers. Several surveys have shown a higher IQ in children born in summer-autumn compared with winter-spring, in both the southern and northern hemispheres.

It is unlikely that women at rest during hot summers would experience a sufficient elevation of temperature to harm the embryo, but hyperthermia could result from heavy exercise in such conditions. Moreover, a febrile illness might result in greater elevations of temperature in hot weather as the gradient between body and environmental temperatures becomes smaller or reversed. Experimental studies of possible interactions between hyperthermia and other teratogens have shown some synergistic responses. The incidence and severity of exencephaly and encephalocele in fetal hamsters were markedly increased when minimally teratogenic doses of maternal hyperthermia were combined with sodium arsenate or vitamin A, raising the possibility that hyperthermia could be an important synergistic factor in a variety of other potentially teratogenic influences. It has also been suggested that hyperthermia lowers the threshold for the expression of polygenically inherited neural tube defects.

Among retrospective studies of children with defects attributed to hyperthermia, three of 28 individuals were thought to have sustained heat exposure through prolonged sauna or hot tub bathing (Pleet et al., 1981). Healthy nonpregnant women can remain in a hot tub at 39° C. for at least 15 minutes and at 41.1° C. for at least 10 minutes without risk of reaching a core temperature of 38.9° C., the minimal elevation above core temperature associated with teratogenicity. Sauna bathing, because it allows for evaporation and convection, requires longer exposure or repeated exposures. This may explain the lack of any grossly apparent excess of hyperthermia associated malformation problems in Finland where maternal sauna bathing is usually limited to six to 12 minutes.

The results of these studies do not contraindicate sauna or hot tub use during pregnancy. They do, however, warrant caution in avoiding lengthy exposure to hyperthermia during early pregnancy. When such exposure does inadvertently occur over sufficient time to result in concern to the mother and her physician, prenatal diagnostic studies might be initiated to determine whether the fetus is affected. Because exposure to many significant hyperthermic insults may lead to fetal loss within the next six to eight weeks, invasive intrauterine studies probably should be deferred until such danger has passed, if possible. In the case of hyperthermic exposures during the third to fourth week of gestation, ultrasonographic and alpha-fetoprotein studies should be considered. Later hyperthermic exposures may be more difficult to evaluate. If the time of conception is known, repeated measurements of fetal head size or the monitoring of fetal movements may yield useful diagnostic information. Unfortunately no prospective studies of hyperthermia exposure during human gestation are available to provide reliable risk figures. For the present, until the results of such studies become available, clinical judgment and knowledge of the results of studies in experimental animals must suffice for patient counseling purposes. Obviously, further studies are urgently needed to define more clearly the risks associated with hyperthermia exposure in the human.

DEFORMATIONS

Deformations per se are rarely if ever a direct cause of serious developmental delay, but in some instances situations engendering fetal constraint can be associated with perinatal hazards that can result in central nervous system damage. In addition, failure to treat deformations appropriately during infancy can result in persistent problems of various kinds. For example, any situation that tends to overdistend the uterus may be associated with the premature onset of labor and delivery and the hazards that accompany this situation. This is a major concern when there is more than one fetus in the uterine cavity or when the uterine cavity itself is structurally abnormal. An abnormal fetal position during late fetal life may result in unusual forces of a constraining nature and also make a vaginal delivery fraught with danger to the fetus.

Breech presentation occurs in only about 4 per cent of pregnancies, but it is associated with about 32 per cent of all the extrinsic deformations noted at the time of birth (Dunn, 1976). Infants who remain in breech presentation for a prolonged period of time usually develop a characteristic head shape with marked dolichocephaly and a prominent occipital shelf (the "breech head"; Fig. 20–17). Such a deformation is more likely when the head is hyperextended in utero as a result of either a breech or transverse presentation. If the infants are delivered vaginally, birth injury due to cervical cord transection may result in marked perinatal morbidity and mortality. For this reason the current trend is toward cesarean delivery of infants with a breech or transverse presentation. In addition, when the physician in the newborn nursery examines an infant with a breech head, he may detect other deformations, such as congenital hip dislocation. There is a 17 per cent incidence of hip dislocation among all term breech

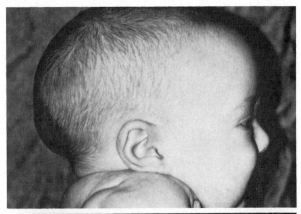

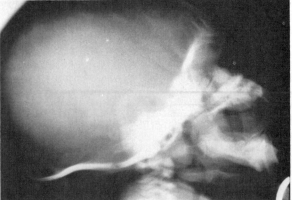

Figure 20–17. Marked dolichocephaly and prominent occipital shelf associated with hyperextension of the fetal head and prolonged breech presentation. (From Smith, D. W.: Recognizable Patterns of Human Deformation. Philadelphia, W. B. Saunders Company, 1981.)

births, and for breech presentation with extended legs in utero (frank breech) the incidence of hip dislocation is 25 per cent (Dunn, 1976). Such deformations result in lasting orthopedic problems if they are not corrected during infancy.

Recently it has been observed that craniostenosis in otherwise normal infants may be caused by fetal head constraint. If multiple sutures are synostotic as a consequence of this kind of prenatal head deformation, the rapidly growing brain may also become constrained (Fig. 20–18). In order to prevent brain damage from occurring, and to correct progressive craniofacial cosmetic deformity as the brain grows in the only direction available, an early calvariectomy must be done within the first few months of life. Before attempting such a procedure, however, the possibility of premature sutural closure due to aberrant brain growth must be ruled out.

Finally it should be noted that developmentally delayed, hypotonic infants who lie in the same position constantly after birth are at risk for the development of postnatal craniofacial deformation, particularly plagiocephaly. Periodic repositioning of such infants can help to prevent such cranial asymmetry, but if significant deformity is still present at six months of age, more active therapy may be necessary. Individually fitted plastic helmets can be fashioned to fit tightly over the parts of the infant's cranium that are prominent and loosely over the areas that are flattened. As the brain grows over the next few months, the calvarium is remodeled to fit the helmet and thus assumes a more normal shape.

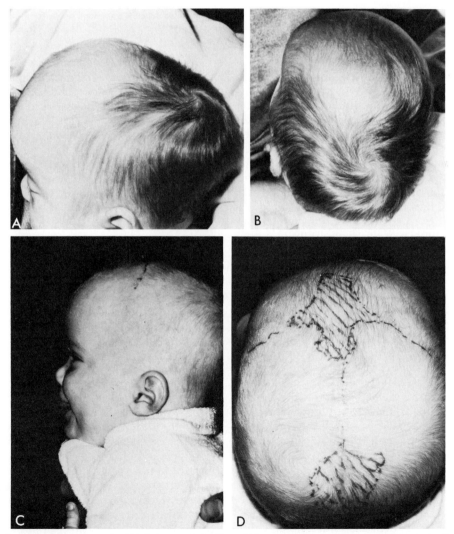

Figure 20–18. A and B, This three week old infant presented with sagittal, coronal, and lambdoidal craniosynostosis. The brain was under increased pressure, and it bulged forth through the open metopic suture, distorting the face. C and D, Following a generous calvariectomy that extended from just above the supraorbital ridges and mastoid bones to just above the foramen magnum, the calvarium was regenerated from the dura mater. Normal sutures and fontanels were formed over the next six weeks, as shown here in the same child at two months of age. (From Smith, D. W.: Recognizable Patterns of Human Deformation. Philadelphia, W. B. Saunders Company, 1981.)

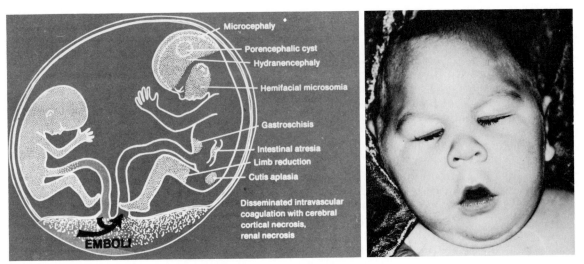

Figure 20–19. The diagram at the left demonstrates the kinds of vascular disruptive defects that have been observed in the wake of the death of one monozygotic co-twin in utero. Thromboplastin or emboli are thought to gain access to the circulation of the surviving co-twin through placental vascular anastomoses. The five month old infant on the right showed marked microcephaly, hypertonicity with opisthotonos, and hydranencephaly. At birth a macerated co-twin of the same sex was noted, and he was considered to be the cause of the disruptive brain problems in his surviving co-twin. The recurrence risk was judged to be negligible. (From Smith, D. W.: Recognizable Patterns of Human Malformation. Ed. 3. Philadelphia, W. B. Saunders Company, 1982.)

DISRUPTION

Disruptive defects of morphogenesis are a relatively newly recognized category of structural defects, and there is still much to be learned in this area. Loss of previously normally formed structures can result from vascular exchange from a dead to a living monozygotic twin through placental vascular anastomoses (Hoyme et al., 1981). Figure 20–19 shows the kinds of structural defects that can result from such a situation, which may result in the entrance of thromboplastin or emboli from the deceased co-twin into the circulation of the surviving twin where they obstruct distal circulation, resulting in areas of ischemia with subsequent loss of tissue.

Melnick (1977) has concluded from the Collaborative Perinatal Study that in about 3 per cent of near term monozygotic twins, the co-twin is deceased. Of the surviving twins of these pairs, about 1 per cent have serious brain defects as a consequence of vascular disruption. The kinds of disruptive structural defects that have been reported in survivors of such monozygotic twin pairs have included cerebellar necrosis, hydranencephaly, porencephaly, multicystic encephalomalacia, hydrocephalus, microcephaly, and spinal cord transection (Hoyme et al., 1981). It is also likely that other sources of vascular disruption (e.g., in the placenta) could lead to similar kinds of problems.

Even without the death of one monozygotic twin in utero, the placental vascular interconnections that are common in monochorionic placentas associated with monozygotic twinning may lead to difficulties. With arteriovenous placental connections, a donor twin may transfuse a recipient twin, leading to hypervolemia, an increased hematocrit, increased cardiac and renal perfusion, polyhydramnios (when the twins are diamniotic), and a larger size in the recipient twin. The elevated hematocrit may require early postnatal management because of the risk of vascular problems (e.g., cavernous sinus thrombosis or renal vein thrombosis). The donor twin tends to show the reverse situation: hypovolemia, a low hematocrit, decreased cardiac and renal blood flow, oligohydramnios, and a smaller size. The difference in size at birth tends to persist postnatally, and both twins are at risk for developmental problems as a consequence of cerebrovascular problems. The larger recipient twin usually faces a greater risk for such problems than the smaller donor twin.

COUNSELING CONSIDERATIONS

The preceding discussion has centered on a rational approach toward gaining an understanding of the nature of the primary problem that led to the presence of one or more congenital anomalies. Fortunately most children appear normal at birth. When a congenital anomaly is present, an initial thought on the minds of many parents revolves around whether such an anomaly might be associated with a risk of mental deficiency. In some instances the pattern of anomalies suggests a condition with a well established prognosis. When there is no risk of mental deficiency, such information can be vastly reassuring for the parents.

In instances of congenital anomalies that do not interfere seriously with the development of the child and that are fully correctable (e.g., a cleft lip), the parents should be told that the child is normal, that the lip did not close completely, and that a surgeon will be called in to help complete this process, which usually takes place prior to birth. Such an approach is geared toward helping the parents to accept the child as normal, rather than as being malformed, since only the lip is malformed. By explaining the defect as a normal stage in lip development prior to 35 days of fetal life, the mother can be reassured that the anomaly is not related to any environmental problems that occurred after that time. The parents should be told that closure of the lip depends on a number of genes and that the set of genes their baby derived from both parents did not allow normal lip closure to be completed. They should be told that the baby appears normal in all other respects and that the chance of a future baby's being affected with a cleft of the lip or palate is about 5 per cent.

For very serious congenital anomalies, such as cytogenetic anomalies, which are usually lethal in the perinatal period (e.g., trisomy 18, trisomy 13, or triploidy), parents should be helped to understand the difference between genes and chromosomes. For perinatal lethal chromosomal anomalies, it is often reassuring for parents to know that most individuals with this karyotype fail to survive long enough to be born and are lost as first trimester miscarriages. They should be told that the fact that their baby survived to be born may reflect the strength of their genetic background, but now that the baby lacks the protection afforded by the womb, he has a very limited capacity for survival. Most such babies die within the first weeks or months, even with optimal care, and those that do survive have a very limited potential for growth and function. It may be helpful for the counselor to assume a directive tone at this point and say, "I feel that the kindest approach for us to take is to keep the baby comfortable and fed, but not to interfere with the normal course of development by utilizing life saving medical or surgical measures. Is this the approach you would like us to take?" By presenting information in this way, the family comes to view the situation as a "late miscarriage." A similar approach can be utilized for lethal central nervous system or cardiac malformations.

This spectrum of approaches is useful only when the underlying nature of the problem is thoroughly understood. When such understanding is incomplete, counseling is much more difficult and it must remain much more tentative. For disorders that are not likely to seriously affect the child's life, the attitude should be conveyed that the child is normal. In cases of chronically handicapping disorders, the parents must realign their expectations as to what will be normal for this child and develop the capacity to take pride in optimizing development within the context of a particular condition. In seriously handicapping disorders with a high degree of lethality, medical interference toward survival should be questioned. Counseling should be provided in a compassionate, unhurried way and reinforced at frequent intervals. Every effort should be made to reinforce the strength of the family unit and to alleviate any guilt, which is not likely to be productive.

JOHN M. GRAHAM

REFERENCES

Dunn, P. M.: Congenital postural deformities. Br. Med. Bull., 32:71, 1976.

Edwards, M. J.: Is hyperthermia a human teratogen? Am. Heart J., 98:277, 1979.

Hoyme, E. H., Higginbottom, M. C., and Jones, K. L.: Vascular etiology of disruptive structural defects in monozygotic twins. Pediatrics, 67:288, 1981.

Little, R. E., Graham, J. M., Jr., and Samson, H. H.: Fetal alcohol effects in humans and animals. Adv. Alcohol Substance Abuse, 1:103, 1982.

Melnick, M.: Brain damage in survivors after death of monozygotic co-twin. Lancet, 2:1287, 1977.

Olegard, R., et al.: Effects on the child of alcohol abuse during pregnancy: retrospective and prospective studies. Acta Paediatr. Scand., 275(Suppl.):112, 1979.

Pleet, H., Graham, J. M., Jr., and Smith, D. W.: Central nervous system and facial defects associated with maternal hyperthermia at four to 14 weeks' gestation. Pediatrics, 67:785, 1981.

Smith, D. W.: Recognizable Patterns of Human Malformation. Philadelphia, W. B. Saunders Company, 1982.

Warkany, J., Lemire, R. J., and Cohen, M. M., Jr.: Mental Retardation and Congenital Malformations of the Central Nervous System. Chicago, Year Book Medical Publishers, Inc., 1981.

21
Perinatal Stresses

The association of untoward perinatal events and developmental outcome has long been studied. Little's work, published in 1862, "On the influence of abnormal parturition, difficult labors, premature birth, and asphyxia neonatorum, on the mental and physical condition of the child, especially in relation to deformities," was one of the first effective attempts to draw attention to this relationship.

Advances in perinatal care have resulted in higher survival rates and declining morbidity in small preterm infants, improved prognosis for infants who suffer perinatal asphyxia, and a better outcome for many infants malnourished in utero.

Although this has created a measure of optimism, it is only fair to note that pregnancies believed to be free of complicating influences do not always have salutary outcomes. Accordingly the subject of the developmental implications of perinatal stress must be approached with caution and humility.

Hagberg and associates (1982) have recently reviewed handicapping conditions referrable to etiologic factors in the perinatal period. They report that 45 to 65 per cent of the children with cerebral palsy and 10 to 17 per cent of those with mental retardation can be classified as having sustained perinatal insults as the presumed cause of their disabilities.

ASSESSMENT OF RISK FACTORS

A host of perinatal events come to bear on the gravida and neonate, including social, genetic, metabolic, homeostatic, infectious, nutritional, pharmacologic, and physical-environmental factors. In the past the term "high risk" pregnancy applied to both mother and baby, but improvements in obstetric care have resulted in maternal mortality rates that are so low that "high risk" pregnancy is now defined in terms of neonatal rather than maternal risk (Milligan and Shennan, 1980).

Systems for assessing the potential risks of neonatal morbidity and mortality have been devel-oped. The one most commonly employed is the "prenatal and intrapartum high risk screening assessment" developed by Hobel and others (Table 21–1). This system not only provides a listing of items considered to be potential risks to the neonate but also quantifies the degree of risk. Both prenatal and intrapartum assessments are performed, and a total score of 10 or more on either is correlated with an increased risk of compromised neonatal outcome (Hobel et al., 1973). The advantage in this assessment is that it allows for appropriate obstetric intervention and neonatal support, thus preventing or minimizing possible consequences. Major risk factors are primarily associated with prematurity and its sequelae (Philip et al., 1977); these along with asphyxia and intrauterine growth retardation constitute the most important influences on neonatal outcome. When anoxia, infection, trauma, or environmental toxins occur in the neonate already at risk due to prematurity or intrauterine growth retardation, the potential for developmental difficulties increases at least fourfold (Huff, 1979).

THE LOW BIRTH WEIGHT INFANT

Low birth weight is defined as a weight of 2500 grams or less at birth. Very low birth weight is defined as less than 1500 grams. Until the early 1970's, birth weight alone was the essential criterion for selecting preterm infants for follow-up studies. It is of interest that when singleton neonates with birth weights in the range of 700 to 1250 grams (expected gestational age, 26 to 28 weeks) are compared, a discrepancy in gestational maturity may be revealed (Vohr et al., 1979).

The increased risk of cerebral palsy in low birth weight premature infants is well recognized. Ellenberg and Nelson (1979) found that a significant number of preterm infants who developed cerebral palsy had low birth weights even for their gestational age. Accordingly analysis of outcomes in low birth weight infants should include consideration of the effects of inadequate fetal growth as well as the degree of prematurity.

390

Table 21–1. RISK FACTORS IN NEONATAL MORBIDITY AND MORTALITY*

Prenatal Factors

	Score			Score
I. Cardiovascular and renal factors		IV. Anatomic abnormalities		
1. Moderate to severe toxemia	10	1. Uterine malformation		10
2. Chronic hypertension	10	2. Incompetent cervix		10
3. Moderate to severe renal disease	10	3. Abnormal fetal position		10
4. Severe heart disease (II–IV)	10	4. Polyhydramnios		10
5. History of eclampsia	5	5. Small pelvis		5
6. History of pyelitis	5			
7. Class I heart disease	5	V. Miscellaneous factors		
8. Mild toxemia	5	1. Abnormal cervical cytologic findings		10
9. Acute pyelonephritis	5	2. Multiple pregnancy		10
10. History of cystitis	1	3. Sickle cell disease		10
11. Acute cystitis	1	4. Age $\geq$ 35 or $\leq$ 15		5
12. History of toxemia	1	5. Viral disease		5
		6. Rh sensitization only		5
II. Metabolic factors		7. Positive serology		5
1. Diabetes $\geq$ class A-II	10	8. Severe anemia (< 9 gm Hb)		5
2. Previous endocrine ablation	10	9. Excessive use of drugs		5
3. Thyroid disease	5	10. History of TB or PPD $\geq$ 10 mm.		5
4. Prediabetes (A-I)	5	11. Weight < 100 or > 200 pounds		5
5. Family history of diabetes	1	12. Pulmonary disease		5
		13. Flu syndrome (severe)		5
III. Previous histories		14. Vaginal spotting		5
1. Previous fetal transfusion for Rh	10	15. Mild anemia (9-10.9 gm Hb)		1
2. Previous stillbirth	10	16. Smoking $\geq$ 1 pack per day		1
3. Post-term > 42 weeks	10	17. Alcohol (moderate)		1
4. Previous premature infant	10	18. Emotional problem		1
5. Previous neonatal death	10			
6. Previous cesarean section	5			
7. Habitual abortion	5			
8. Infant > 10 pounds	5			
9. Multiparity > 5	5			
10. Epilepsy	5			
11. Fetal anomalies	1			

Intrapartum Factors

	Score			Score
I. Maternal factors		III. Fetal factors		
1. Moderate to severe toxemia	10	1. Abnormal presentation		10
2. Hydramnios or oligohydramnios	10	2. Multiple pregnancy		10
3. Amnionitis	10	3. Fetal bradycardia > 30 min.		10
4. Uterine rupture	10	4. Breech delivery total extraction		10
5. Mild toxemia	5	5. Prolapsed cord		10
6. Premature rupture of membrane > 12 hr.	5	6. Fetal weight < 2500 gm.		10
7. Primary dysfunctional labor	5	7. Fetal acidosis pH $\leq$ 7.25 (stage 10)		10
8. Secondary arrest of dilation	5	8. Fetal tachycardia > 30 min.		10
9. Demerol > 300 mg.	5	9. Operative forceps or vacuum extraction		5
10. MgSO$_4$ > 25 gm.	5	10. Breech delivery spontaneous or assisted		5
11. Labor > 20 hours	5	11. General anesthesia		5
12. Second stage > 2.5 hr.	5	12. Outlet forceps		1
13. Clinically small pelvis	5	13. Shoulder dystocia		1
14. Medical induction	5			
15. Precip. labor < 3 hours	5			
16. Primary cesarean section	5			
17. Repeat cesarean section	5			
18. Elective induction	1			
19. Prolonged latent phase	1			
20. Uterine tetany	1			
21. Pitocin augmentation	1			
II. Placental factors				
1. Placenta previa	10			
2. Abruptio placentae	10			
3. Post-term > 42 weeks	10			
4. Meconium stained amniotic fluid (dark)	10			
5. Meconium stained amniotic fluid (light)	5			
6. Marginal separation	1			

*Adapted from Hobel, C. J., et al.: Prenatal and intrapartum high-risk screening. Am. J. Obstet. Gynecol., *117*:1, 1973.

PREMATURITY

Prematurity is defined as birth prior to 37 weeks of gestation. Prematurity poses the greatest challenge facing neonatal and obstetric perinatologists today. It is estimated that 85 per cent of neonatal deaths not due to anomalous fetal development are associated with preterm delivery (Rush, 1976).

Newer pharmacologic approaches to the inhibition of preterm labor have been encouraging. The underlying mechanisms of premature labor are not understood, however, and there is little evidence that newer forms of treatment decrease the number of premature births (Hemminki and Sterfield, 1978). This is not to say that pharmacologic intervention may not produce a useful effect, since even a few additional days in utero may be sufficient to allow for increased pulmonary maturation (including the simultaneous use of exogenous steroids; Liggins and Howie, 1972).

Taeusch (1975) cautioned that although the benefits of steroids have been useful in preventing the respiratory distress syndrome, it is necessary to consider the possible harmful effects to mother and fetus, particularly to the developing central nervous system. However, in a recent follow-up study there were no significant deleterious effects at four years of age in children from betamethasone treated pregnancies (MacArthur et al., 1981).

Many preterm deliveries are associated with obvious causal factors, such as multiple pregnancy, fetal abnormality, or antepartum hemorrhage, but a significant proportion occur for no apparent reason (Rush et al., 1976). Naeye and Peters (1980) have suggested that amniotic fluid infection was a cause as well as a consequence of premature rupture of the fetal membranes in some cases of premature labor. Bobbit and associates (1981) added further evidence to incriminate infection as a significant etiologic factor. In their study, specimens of amniotic fluid were collected by amniocentesis from 31 patients in premature labor with intact membranes. Micro-organisms were isolated from only eight (25 per cent), yet seven of these eight patients underwent delivery within 48 hours after amniocentesis, whereas only six of the noninfected group did.

Although the actual cause of premature labor often cannot be established, obstetricians now use scoring systems to predict those gravidas at highest risk (Creasy et al., 1980; Gaziano et al., 1981). The system developed by Creasy predicts that approximately 10 per cent of the patients will be at risk. Although only one-third of these patients will deliver before term, they account for approximately two-thirds of all premature births.

The modern era of perinatal health care is characterized by vigorous obstetric management of premature labor as well as improved care for the infants. The development of intensive care for premature infants has been marred by a series of well intentioned but misdirected therapeutic interventions (Silverman, 1977). These included retrolental fibroplasia resulting from overuse of oxygen; hypoglycemia, dehydration, and jaundice enhanced by inappropriate delay in initiating feeding (Hack et al., 1979); kernicterus induced by the prophylactic use of sulfonamides and excessive vitamin K; and the use of chloramphenicol, which caused the "gray baby syndrome" characterized by circulatory collapse and death. Moreover, when oxygen restriction was imposed to reduce the incidence of retrolental fibroplasia, neonatal deaths increased, as did the incidence of neurologic damage among survivors (Bolton and Cross, 1974).

Klaus and Kennell (1976) drew attention to a major error in giving care—the imposed isolation of the premature infant from its mother and family. The effect of such isolation was the failure of parents to develop normal attachments (bonding) with their premature child, and this deficit has been causally associated with child abuse.

Prior to the mid-1960's the outlook for premature babies, especially very low birth weight infants (birth weight <1500 grams), was poor. In the 1958 British Perinatal Mortality Survey, 7 per cent of the babies weighing less than 1001 grams survived (Butler and Bonham, 1963), and reports from both sides of the Atlantic provided the gloomy figures that 50 to 60 per cent of very low birth weight survivors had a major handicap (Lubchenco et al., 1963).

With the development of greater sophistication in the management of low birth weight infants, both mortality and morbidity appear to have dramatically decreased. Pape et al. (1978) analyzed the outcome in infants born weighing less than 1001 grams before 1970 and in those admitted to their unit in 1974. They found that there had indeed been a decrease in mortality, from 75 to 53 per cent, and that the proportion of infants surviving without handicap had doubled. Davies and Tizard (1975) showed a decrease in the incidence of spastic diplegia, which is considered to be characteristic of the preterm infant, as did Hagberg and his colleagues (1975).

Hack and associates (1979) provided follow-up data from a group of very low birth weight infants born in Cleveland in the years 1975 to 1976. These children were followed prospectively for a mean of two years' conceptual age. Two hundred ninety-one babies were admitted to the neonatal intensive care unit. One hundred eighty-nine (65 per cent) survived, and of these, 160 (85 per cent) completed the follow-up. Eighty-two per cent of the babies followed appeared to be developing normally.

The sequelae of prematurity may be related to a host of events, including the degree of prematurity, the conditions leading to the onset of premature labor (e.g., infection), the circumstances of birth (e.g., trauma and asphyxia), the correlates of immaturity (e.g., respiratory distress syndrome and apnea), and the effects of fetal malnutrition.

A certain level of fetal maturation must be reached before even the most expert obstetric and neonatal care can be expected to prevail. Prior to the mid-1960's that level of maturity was approximately 28 weeks (birth weight 1000 to 1250 grams). Now the figure is closer to 26 weeks (birth weight 700 to 900 grams). Schechner (1980) has recently raised the issue, "How small is too small?" She suggests that since survival among sick infants weighing less than 750 grams is so low, and the quality of life among survivors is so poor, perhaps a less aggressive approach should be adopted for these infants. However, Britton and associates (1981) examined the outcome in infants weighing less than 801 grams who were born between 1974 and 1977. There were no survivors weighing less than 500 grams. Between 500 and 700 grams the survival rate was 17 per cent, and all the survivors had residual handicaps. Between 700 and 749 grams the survival rate was 4 per cent and between 750 and 800 grams it was 41 per cent; the incidence of handicap in these two groups was only 39 per cent. Their conclusion was that although referral to an intensive care unit may not be justified for infants weighing less than 700 grams, every effort should be made for those weighing 700 to 800 grams.

Some of the problems faced by premature infants whose weights are appropriate for their gestational age are listed in Table 21–2. Each of the conditions can potentially result in developmental problems. Low birth weight infants who suffer birth asphyxia, periventricular-intraventricular hemorrhage, or severe respiratory difficulties are the most likely to have residual problems (Hack et al., 1979). Hence, the importance of preventing

Table 21–2. CORRELATES OF IMMATURITY

Increased susceptibility to infection	Sepsis Necrotizing enterocolitis (?)
Pulmonary	Respiratory distress syndrome
Central nervous system	Apnea-bradycardia Periventricular-intraventricular hemorrhage
Cardiovascular	Patent ductus arteriosus
Gastrointestinal	Malnutrition, necrotizing enterocolitis, jaundice, rickets
Hematologic	Anemia, jaundice
Renal	Renal tubular acidosis of prematurity
Metabolic	Hypoglycemia, hypocalcemia
Ophthalmologic	Retrolental fibroplasia

asphyxia by appropriate antenatal monitoring and immediate postpartum vigilance is vital, for asphyxia has been shown to cause damage in essentially all organs, and freedom from asphyxia is probably the single most important factor that augurs for a normal outcome in a premature infant (Stewart and Reynolds, 1974).

Respiratory Distress Syndrome

The respiratory distress syndrome is a leading cause of morbidity and mortality in preterm infants. The etiologic factor is a deficiency or absence of pulmonary surfactant, which results in inadequate expansion of alveoli and an increase in the work of breathing. This, coupled with the effects of poor ventilation, results in the respiratory distress syndrome, which is characterized by hypoxemia and hypercarbia and ultimately, if uncorrected, severe respiratory and metabolic acidosis. The metabolic and hemodynamic changes that result can increase the risk of the development of intraventricular hemorrhage and its sequelae (Pape and Wigglesworth, 1979).

Most babies with the respiratory distress syndrome require some form of ventilatory support, consisting of the provision of a continuous distending airway pressure or mechanical ventilation. Saigal et al. (1982) found that the incidence of neurologic handicaps was three times greater in infants who required mechanical ventilation.

Bronchopulmonary Dysplasia

Although no single etiologic factor for bronchopulmonary dysplasia has been elucidated, the evidence points here to the combined effect of mechanical trauma secondary to assisted ventilation and the toxic effects of oxygen on the immature lung (Taghizadeh and Reynolds, 1976). The condition is suspected when a baby under treatment for the respiratory distress syndrome does not show evidence of recovery and goes on to develop chronic pulmonary changes. The pathologic picture is one of alveolar and bronchiolar necrosis and repair, with bronchial metaplasia and interstitial fibrosis. Although surviving infants eventually can be weaned from ventilatory support, they may require prolonged oxygen therapy, often in excess of six months.

The incidence varies from 5 to 28 per cent, depending on the diagnostic criteria employed, and the mortality rate has been high, ranging from 25 to 39 per cent in a report by Markestad and Fitzhardinge (1981). These authors also supplied information about a group of 26 patients with bronchopulmonary dysplasia, 21 of whom survived and were followed prospectively for two years. Lower respiratory tract infection was noted in 85 per cent of the children; 50 per cent required hospitalization during the first year and 20 per cent during the second. At two years of age only two patients had significant respiratory symptoms at rest, but in 14 there were residual radiographic changes. The average height and weight were at or below the third percentile at first, but growth occurred at an accelerated rate when there was improvement in respiratory symptoms. By two years the average weight reached the third to tenth percentile for both sexes, and height was at the tenth to twelfth percentile in boys and at the twenty-fifth percentile for girls. Growth retardation was associated with severe and prolonged respiratory dysfunction. Seventy-five per cent of the children were free of major developmental defects, and only 5 per cent had mean Bayley scores less than 85 at 18 months. One infant was found to have hydrocephalus. It was the authors' opinion that developmental outcomes were probably related to other perinatal and neonatal events rather than to the presence or absence of bronchopulmonary dysplasia per se.

Retrolental Fibroplasia

Retrolental fibroplasia was first described by Terry in 1942. It is believed to result from the toxic effect of oxygen on immature retinal blood vessels, but other factors may contribute to its development.

Although the identification of overuse of oxygen as the principal cause of retrolental fibroplasia occurred in the early 1950's, the disease continues to be a major cause of neonatal morbidity. Approximately 45 per cent of infants who weigh less than 1000 grams at birth develop acute retinal changes (Kingham, 1977). Of these, 80 to 90 per cent of the cases regress spontaneously, and when the patients are examined later in life, unexplained myopia or temporal vitreal-retinal interface changes may be the only evidence of prior disease (Yamamoto et al., 1979). The other 10 to 20 per cent develop cicatricial retinal changes that can lead to severe visual impairment (Flynn, 1979).

Periventricular-Intraventricular Hemorrhage

Periventricular-intraventricular hemorrhage is currently the most important neurologic disorder of small newborn infants (Volpe, 1981). Periventricular hemorrhage results from bleeding into the subependymal germinal matrix, which in approximately 80 per cent of the instances extends through the ependyma into the ventricular system. The germinal matrix is a highly vascular area with loose supporting tissue located in the periventricular area over the caudate nucleus. It is present from soon after neuronal migration is complete (26 to 27 weeks) until about 33 to 35 weeks' gestation, at which time it appears to undergo spontaneous dissolution. The association of periventricular-intraventricular hemorrhage with prematurity relates directly to the presence of this area of germinal matrix (Pape and Wigglesworth, 1979; Volpe, 1981).

Hambleton and Wigglesworth (1976) developed a hypothesis to explain the possible pathogenetic mechanism of intraventricular hemorrhage. They took into consideration the loose supporting tissue in the germinal matrix, the thin walled capillaries that course through this tissue, and the metabolic and hemodynamic factors that affect blood pressure and cerebral blood flow. Lou and associates (1979) determined that even slight asphyxia impairs cerebrovascular autoregulation in the newborn. This pressure-passive characteristic of cerebral blood flow results in increases in arterial blood pressure that may be transmitted directly to the capillary bed and provoke rupture of these small vessels. The various factors associated with intraventricular hemorrhage differ in nature and degree and depend on the gestational age of the infant (Pape and Wigglesworth, 1979). In the extremely premature 26 to 28 week old infant the mere process of birth may be sufficient stress to result in periventricular-intraventricular hemorrhage, whereas in older infants (those approaching 35 weeks) the occurrence of hemorrhage is more often associated with iatrogenic factors, e.g., bicarbonate administration (Wigglesworth et al., 1976).

Periventricular-intraventricular hemorrhage is a common finding in infants weighing less than 1500 grams at birth. Burstein et al. (1979) performed computed tomographic brain scans before the seventh day of life in 100 consecutive premature neonates having a birth weight less than 1500 grams who had been admitted to the neonatal intensive care unit at the University of New Mexico. Forty-four of these infants showed evidence of periventricular-intraventricular hemorrhage. The newer technique of ultrasound imaging is reported to detect smaller amounts of hemorrhage than computed tomography, and it is conceivable that information based on this technique will reveal an even greater percentage of cases of periventricular-intraventricular hemorrhage in very low birth weight infants.

Volpe (1981) has described two basic forms of clinical presentation, a catastrophic form and a saltatory or subtle form. Both varieties appear most commonly in the first two days of life. Catastrophic deterioration occurs in the infant with major hemorrhage, but smaller hemorrhages may be overlooked clinically.

With more severe hemorrhage the clinical manifestations include seizures, apnea-bradycardia,

Table 21–3. SEVERITY OF PERIVENTRICULAR-INTRAVENTRICULAR HEMORRHAGE*

Mild	Hemorrhage confined to subependymal region or with small amount of blood in the lateral ventricles
Moderate	Limited amounts of blood in the lateral ventricles
Severe	Blood fills dilated lateral ventricles

*Adapted from Volpe, J. J.: Neonatal intraventricular hemorrhage. N. Engl. J. Med., *304*:886, 1981.

respiratory distress, hypotension, a full anterior fontanelle, temperature instability, metabolic acidosis, hypoglycemia, and a falling hematocrit level. According to Volpe (1981), the most reliable sign is an unexpected fall in the hematocrit level or a failure of the hematocrit level to rise after transfusion.

The appearance of the lesion on a computed tomographic scan has been used to quantify the extent of bleeding (Table 21–3). Another system of quantification of periventricular-intraventricular hemorrhage devised by Papile and associates (1978) takes into account the extent of the hemorrhage and the presence or absence of ventricular dilation. Both models are useful in the clinical setting in terms of predicting outcome, but they may not take into consideration the pathogenetic processes.

The short term outcome is shown in Table 21–4. Infants with small hemorrhages usually survive and do not develop hydrocephalus. As the severity of the bleeding worsens, so does the outcome. In cases of severe hemorrhage, most infants die, and those who survive have a 65 to 100 per cent chance of developing hydrocephalus.

In 1970, Fedrick and Butler published an extensive study of intraventricular hemorrhage. They noted that an infant born prematurely, especially following a period of slight growth retardation and in a state of asphyxia, would be the one most

Table 21–4. SHORT TERM OUTCOME OF PERIVENTRICULAR-INTRAVENTRICULAR HEMORRHAGE*†

Severity of Hemorrhage	Deaths (% of Total‡)	Progressive Ventricular Dilation (% of Survivors)
Mild	0	0–10
Moderate	5–15	15–25
Severe	50–65	65–100

*Adapted from Volpe, J. J.: Neonatal intraventricular hemorrhage. N. Engl. J. Med., *304*:886, 1981.

†Data are based on approximately 250 cases; more than half were evaluated at St. Louis Children's Hospital, and the remainder at Emory University, the University of New Mexico, and the Massachusetts General Hospital.

‡Values are ranges that encompass 75 to 100 per cent of the cases.

likely to die as a result of intraventricular hemorrhage. They concluded by saying, "It will be interesting to see, as methods of regulating the baby's metabolism become more efficient and the relief of asphyxia at birth becomes more rapid, whether the incidence of the lesion will fall." The current incidence figure of nearly 1 in 2 infants less than 1500 grams is sufficient to justify greater devotion to the search for prevention.

Since long term follow-up data in infants with documented periventricular-intraventricular hemorrhage are still incomplete, it is not possible to estimate categorically the impact of this lesion on outcome. However, available information does suggest that in situations in which moderate to severe hemorrhage has been documented, regardless of other clinical correlates, death or severe morbidity is common, whereas mild hemorrhages appear to resolve and leave few residua (Burstein et al., 1979).

It may well be that the statistics relating to mortality and morbidity in very low birth weight premature infants relate in large part to the presence and extent of periventricular-intraventricular hemorrhage. It would therefore seem advisable to assess each newborn weighing less than 1500 grams at birth for evidence of such hemorrhage and to follow closely the neurodevelopmental outcome in proven instances for a sufficiently long interval to make possible an accurate impression of outcome.

INTRAUTERINE GROWTH RETARDATION

Deviations from normal intrauterine growth have been associated with increased perinatal mortality and morbidity (Lubchenco et al., 1972; van den Berg and Yerushalmy, 1966). Infants having a low birth weight for gestational age are termed "small for gestational age" (Jones and Battaglia, 1977).

The factors that have been associated with intrauterine growth retardation are summarized in Table 21–5. In most instances fetal malnutrition that accompanies maternal diseases is characterized by a deficient uteroplacental circulation (Crosby et al., 1977).

Hagberg and associates (1976) examined the prenatal and perinatal factors in a series of 560 children with cerebral palsy born in Sweden between 1954 and 1970 and compared these to the general population. They found a highly significant association with maternal toxemia, antepartum bleeding, and multiple births. This led them to propose the term "fetal deprivation of supply" and to suggest that these infants were rendered more vulnerable to the adverse effects of even minor degrees of hypoxia.

Table 21–5. INFLUENCES ON FETAL GROWTH[*]

Genetic	Environmental
Population differences	Maternal nutrition (e.g., anemia)
Maternal size	
Reproductive history	Drugs—cigarettes, alcohol, narcotics
Multiple gestation	
Chromosomal defect	Altitude
Fetal sex (Y-chromosome)	Placental size
	Infection
	Maternal disease (e.g., pregnancy induced hypertension)

*Obtained by combining information from Crosby, 1977; Daikoku, 1979; Jones and Battaglia, 1977; and Warshaw, 1979.

Many fetuses that are small for gestational age tolerate labor poorly. Bennet and associates (1981) compared the perinatal factors in preterm infants who developed spastic diplegia with those in infants who did not, and found a significant difference in birth weight, head circumference, and low one minute Apgar scores. They believed that these differences represent adverse intrauterine influences not only on growth but also on the ability of the infant to adapt to the extrauterine environment. The associated clinical features in infants who are small for their gestational age (e.g., hypothermia, hypoglycemia, hypocalcemia, and polycythemia) are also significant factors in influencing morbidity.

The outcome in infants who are small for their gestational age depends to some extent on the duration of the growth retardation. This was demonstrated in a study in which head growth was monitored by serial ultrasound measurements. It was found that infants whose head growth slowed before 26 weeks' gestation had significantly lower scores on cognitive and perceptual motor testing than those whose head growth slowed later (Harvey et al., 1982). Commey and Fitzhardinge (1979) reported the outcome in 71 preterm infants who were small for their gestational age. At two years of age, 24 of the 71 were at less than the third percentile for height and weight. Twenty-one per cent had neurologic sequelae, and 42 per cent had Bayley scores equal to or less than 80. All 71 of the infants had been transferred to the neonatal intensive care unit of the Hospital for Sick Children in Toronto, and it is suggested that most babies were in poor condition on arrival and suffered from hypoxemia, cold stress, and hypoglycemia as well as the effects of immaturity. Others also suggest that the transportation of small sick infants adversely affects the outcome (Kiely et al., 1981). However, Fitzhardinge and Pape (1979) have presented other information showing that infants who were small for their gestational age with a mean weight less than 1000 grams and a gestational age of 29 weeks did less

well when compared with a control group of appropriate for gestational age infants, all of whom were transported into the unit. At 18 months the infants who were small for their gestational age were significantly smaller and had significantly more neurologic problems and lower Bayley motor scores, suggesting that "the complication of intrauterine growth retardation significantly increases the risk of serious sequelae in the tiny premature infants."

Follow-up studies of term infants who are small for their gestational age have shown that they have not done as well as their appropriate for gestational age peers with reference to somatic growth, neurologic status, and developmental performance (Fitzhardinge and Stevens, 1972). A later prospective study of 96 full term infants who were small for their gestational age showed that although major neurologic defects were uncommon (1 per cent cerebral palsy and 6 per cent seizures), 25 per cent were found to have so-called "minimal cerebral dysfunction," characterized by hyperactivity, a short attention span, learning disabilities, poor fine motor coordination, and hyper-reflexia. Speech disorders were present in 32 per cent of the boys and in 25 per cent of the girls and featured immaturity of receptive and expressive language. The average IQ in the boys was 95 and in the girls, 101; however, 50 per cent of the boys and 36 per cent of the girls performed poorly in school.

As can be seen, some degree of fetal malnutrition affects term infants, but certainly not to the same extent as with preterm infants. This may be related to the findings of Harvey and his colleagues (1982), who implied that growth retardation before 26 weeks had more serious consequences.

Vohr and associates (1979) examined a group of "intermediate" infants who were small for their gestational age to determine whether the outcome would point to a more optimal time for delivery. These "intermediate" infants were born at a mean of 33.4 weeks and with a birth weight of 1220 ± 195 grams. Each such baby was paired with a birth weight match appropriate for a gestational age infant whose birth weight was 1195 ± 190 grams and whose gestation age was 29 ± 2 weeks. Both the somatic growth and the neurodevelopmental status of the small for gestational age group compared favorably with those in the appropriate for gestational age group. At two years only two small for gestational age infants (diplegia) and one appropriate for gestational age infant (hemiplegia) had problems. This encouraging study suggests that the effects of intrauterine growth retardation may be minimized by early detection of intrauterine growth retardation through careful prenatal management, effecting delivery at an optimal time (32 to 36 weeks' gestation), minimizing the potential effects of asphyxia with good neonatal care, and attention to adequate nutrition in the first two months of life. However, it should be borne in mind that some intrauterine influences that occur in the first and second trimesters of pregnancy also require closer consideration.

HYPOXIC-ISCHEMIC ENCEPHALOPATHY

The normal infant brain depends on the constant availability of oxygen, glucose, and other nutrients. These substances are delivered to the brain cells via the cerebral blood vessels. Waste products from the utilization of these substances in the cells are carried from the brain via venous channels. Any factor that interferes with this system can trigger events that jeopardize the vitality of the cerebral tissue and potentially can cause permanent damage. A relative lack of oxygen (hypoxia) can be caused by a deficit in the amount of blood perfusing the brain (ischemia) or by ventilation difficulties (asphyxia) reflected in hypoxemia and hypercarbia. Attempts to prevent asphyxia, hypoxemia, and ischemia constitute a major portion of modern obstetric care. However, the exact relationship between the presence of hypoxemia and ischemia and morbidity and mortality in the human neonate is not fully known.

Neonatal hypoxic-ischemic injuries are due to intrauterine or intrapartum asphyxia in 90 per cent of instances (Volpe, 1977) and when present clinically at birth reflect significant central nervous system insult. The newborn infant with low Apgar scores is in need of immediate attention and resuscitation. Once this has taken place, it is the neurologic picture in this critical period that will determine whether there are continuing central nervous system problems. The evolution of the clinical picture through the stages of initial shock and then irritability, lethargy, seizures, and coma has been well described by Volpe, who stressed that observation of the infant in the early hours after asphyxia is crucial for optimal management.

Utilization of the Apgar score as an index of cardiorespiratory and central nervous system wellbeing is generally accepted. It is extremely valuable for the immediate assessment and management of the newborn, but there are some questions as to its long term prognostic value. Nelson and Ellenberg (1981) reviewed Apgar scores as predictors of neurologic impairment in the National Collaborative Perinatal Project. The mortality in infants with low scores (0 to 3) was high, particularly if the score remained low. For low birth weight infants the mortality was 48 per cent with Apgar scores of 0-3 at one minute and 96 per cent

if the score remained low at 20 minutes; for full term infants the corresponding mortality rates were 6 and 59 per cent, respectively. A poor prognosis could be predicted for term infants who developed seizures four to 24 hours after birth, with low Apgar scores at 10 and 20 minutes. However, when these investigators examined the long term outcome, they found that only 27 per cent of the infants with Apgar scores less than 7 at five minutes had cerebral palsy. The remaining 73 per cent were free of handicap.

The importance of the clinical course after asphyxia cannot be overemphasized in regard to immediate management as well as for predicting long term outcome. Minkowski et al. (1977) analyzed the outcome of birth asphyxia by clinical course:

1. Infants who recover within the first week are unlikely to have abnormal sequelae.

2. Twenty to 30 per cent of those whose symptoms persist after the first week are likely to have a serious outcome.

3. Seventy per cent of those whose neurologic abnormalities and uncontrollable seizures persist will have a subsequent handicap.

In a review of the neuropathologic findings in 914 consecutive autopsies on infants under one year of age, Banker and Bruce-Gregorios (1983) found that more than 50 per cent had the lesions of hypoxic-ischemic encephalopathy. These lesions were divided into those affecting gray matter and those involving white matter. The latter represented the lesions of periventricular leukomalacia and gliosis, which could occur in isolation or in association with other lesions. Volpe has further subdivided these lesions on an anatomic basis; this classification allows for some degree of clinical correlation with specific neuropathologic findings: selected neuronal necrosis, parasagittal cerebral injury, periventricular leukomalacia, focal and multifocal ischemic brain necrosis, and status marmoratus (adapted from Volpe, 1981).

SELECTED NEURONAL NECROSIS

The term selected neuronal necrosis refers to the necrosis of specific neurons throughout the brain in a characteristic distribution, and is currently thought to be the most frequent result of hypoxic-ischemic insult in the neonate. This form of cellular injury can coexist with other types of hypoxemic-ischemic damage and has been found at autopsy in association with mental retardation, spastic quadriparesis, and seizure disorders. Clinical differentiation of this lesion from other clinical sequelae of hypoxic-ischemic lesions is often difficult.

Figure 21–1. Parasagittal cerebral injury, whose distribution is indicated by the cross hatched areas in the superomedial aspects of the cerebrum. (From Volpe, J. J.: Neurology of the Newborn. Philadelphia, W. B. Saunders Company, 1981, p. 189.)

PARASAGITTAL CEREBRAL INJURY

Parasagittal cerebral injury is the most common ischemic lesion in the full term infant. It may be unilateral but is usually bilateral. It has been suggested that the disorder occurs as a result of the strategic relationship between major blood vessels in the brain. When perfusion is reduced as a result of changes in blood pressure, the areas most likely to be compromised are those that are distal to the terminal branches. This is often referred to as the "watershed phenomenon." The most common site of vulnerability is the cortical area near the sagittal sulcus (Fig. 21–1). The clinical correlates of this lesion include hemiparesis and spastic quadriparesis, in association with significant mental retardation.

PERIVENTRICULAR LEUKOMALACIA

Periventricular leukomalacia is a lesion of the white matter, usually located in the area adjacent to the lateral ventricles, but it may be found peripherally as far as the subcortical region. These lesions have been linked to venous stasis (Schwartz, 1961), other vascular complications (Banker and Larroche, 1962), compromised cardiac function (Shuman and Selednik, 1980), and in general with perinatal illness, particularly in the infant requiring mechanical ventilation (Volpe, 1981). The association with ischemia has suggested that the lesions occur in "border zones" between major areas of arterial supply. The most important aspect of this lesion is that it occurs predominantly in premature or low birth weight infants (Banker and Bruce-Gregorios, 1983). It has been causally linked with the clinical picture of spastic diplegia.

At this point the question might appropriately be asked why premature infants tend to develop the lesions of periventricular leukomalacia whereas full term infants tend to have the more severe lesions of gray matter necrosis. An explanation is offered by the work of Myers (1979) in experimental animals. In this model the presence of glucose during the period of oxygen deprivation is strongly correlated with the development of cerebral edema. Under hypoxic conditions glucose is metabolized into lactic acid, which if present in excess of 25 micromoles per liter leads to changes in membrane physiologic properties, with a breakdown in the blood-brain barrier and shifts of fluid into the intracellular compartment, resulting in widespread tissue distruction and cerebral edema. Full term infants are more likely to have adequate blood glucose levels and glycogen stores. Therefore, if the model can be extended to human infants, they will develop the pathologic process already described. Premature and small for gestational age infants, on the other hand, are less likely to have adequate glycogen stores and tend to have problems of hypoglycemia. Therefore, the lesions they have will reflect the local effects of hypoxia and ischemia.

FOCAL ISCHEMIC NECROSIS

Focal ischemic necrosis results from the lack of perfusion of a single arterial system (Volpe, 1981). The resulting infarcts can be single or multiple and can vary in size. Cavitation is caused by dissolution of the tissue in the infarcted area. The explanation as to why cavitation results rather than the mature response of glial scar formation appears to relate to the relative lack of astrocyte response and to the myelin and water content of cerebral tissue during the later months of pregnancy and the first weeks of life. Experimental surgical compromise of major arterial vessels in preterm animals demonstrates clearly the effects of focal ischemic brain necrosis. Clinical examples of embolus and thrombus formation resulting in similar pathologic situations have been described (Cocker et al., 1965).

The clinical sequelae include a wide spectrum of disorders depending upon the location, size, and number of the focal lesions. Because many of the occlusive episodes involve the middle cerebral artery, the clinical findings will relate to functions subserved by the cortical and major corticospinal tracts perfused by that artery. Unilateral lesions appear to cause 50 per cent of the cases "congenital hemiplegia" as reviewed by computed tomographic scanning (Rothner and Cruse, 1978). This particular ischemic entity may account also for some cases of congenital porencephalic cysts and possibly some cases of hydranencephaly.

STATUS MARMORATUS

Status marmoratus is the least common of the hypoxic-ischemic encephalopathies and involves the basal ganglia, specifically the caudate nucleus, the putamen, and occasionally the globus pallidus. It occurs almost exclusively in the term infant. The name of the lesion comes from the marble-like appearance of these ganglia, which is now determined to result from hypermyelination of astrocytic processes rather than from attempts at axonal regeneration. Clinically this results in extrapyramidal or dyskinetic symptoms. Both the development of the characteristic pathologic lesion and the onset of the clinical symptoms are often delayed into the last half of the first year of life, and sometimes do not appear until the latter half of the second year of life.

BIRTH TRAUMA

There is no question that modern obstetric care has dramatically reduced the occurrence of trauma during labor and delivery, and it now occurs relatively infrequently. In this context, trauma refers to mechanical injury only, and not to the commonly associated problems of asphyxia or hypoxia.

Direct trauma resulting from excessive compressive forces causing bleeding exterior to the skull and below the skin results in caput secundum and cephalohematoma, neither of which has developmental sequelae. Skull fractures, although occurring with some frequency, usually resolve without permanent damage (Gresham 1975). Direct compression, with or without fracture, can cause brain contusion, resulting in cortical and subcortical necrosis and scarring. Rapid changes in skull shape and consequently brain shape during accommodation to the birth canal may cause tears in the brain substance. These lesions are difficult to identify and differentiate, but increased experience with computed tomographic scans should improve this situation.

INTRACRANIAL HEMORRHAGE

In a study by Volpe (1981), 70 infants in whom bloody spinal fluid was obtained by lumbar puncture were subjected to computed tomographic scanning. Primary subarachnoid hemorrhage was identified as the cause in only 29 per cent of the cases of intracranial bleeding. No bleeding site was identified in 8 per cent, and in an additional 63 per cent bleeding was due to periventricular-intraventricular hemorrhage.

Subarachnoid Hemorrhage

Primary subarachnoid hemorrhage occurs more often with trauma in the full term infant, but in the preterm infant it occurs after hypoxic events. If the infant is asymptomatic, or has seizures but is "well" during the interictal period, the outlook is usually good. For those who survive massive subarachnoid bleeding episodes the major consequence is hydrocephalus.

Subdural Hemorrhage

Subdural hemorrhages in the neonatal period occur almost exclusively as a result of traumatic lesions in the full term infant. They occur most commonly in situations in which there is cephalopelvic disproportion or abnormal delivery, e.g., breech or other presentations requiring forceps extraction.

The clinical picture following rupture of superficial vessels is determined by the extent and progression of the bleeding. The development of seizures and increased intracranial pressure are possible sequelae. Not uncommonly small or self-limiting lesions may be manifested much later as chronic subdural effusions. Fifty to 80 per cent of the infants with these lesions do well as evidenced in follow-up examinations (Volpe, 1977b).

Subdural hematoma is now found infrequently, and current clinical management is effective when the lesion occurs over the cerebral hemispheres. Posterior fossa bleeding episodes and those associated with tears of major veins and their tributaries have less favorable developmental results, and lacerations in the tentorium or falx associated with extensive bleeding present early in the neonatal period and usually have a serious and often fatal outcome.

Cerebellar Hemorrhage

Intracerebellar hemorrhage is known to occur following trauma, but it is unclear whether the injury to these tissues is direct or whether the source of the bleeding is secondary (Donat et al., 1979). Pericerebellar hemorrhages are found in premature infants and may represent an extension of periventricular-intraventricular hemorrhage.

TRAUMA TO THE SPINAL CORD AND NERVE ROOTS

Trauma occurring during the birth process may also affect the spinal cord, nerve roots, and peripheral nerves. The injuries most commonly observed involve the cervical and upper thoracic cord and follow breech extraction. Spinal cord injury and brachial plexus palsy are caused by either rotational or tractional forces applied to the spinal cord that exceed the tensile strength of the cord substance.

Spinal cord injury in the neonate produces a flaccid paralysis, anesthesia below the injury level, and bowel and bladder dysfunction; if it occurs at the C4 level or higher, it interferes with respiration.

Brachial nerve palsy is proximal (Erbs type) in 90 per cent of the cases; it includes the roots of C5-C6 in 50 per cent of the cases and C7 in the remainder (Eng, 1971). Distal paralysis (Klumpke type) involves spinal roots C8 and T1. The prognosis is excellent in proximal palsy, with a return to full function in 92 per cent of the cases by one year of age; distal paralysis has a less favorable outcome.

Facial nerve paralysis, although frequently seen, carries an excellent prognosis for complete recovery of function. Facial nerve compression during the birth process is thought to be the major etiologic event, and not the use of obstetric forceps as was previously implicated (Hepner, 1951).

PERINATAL INFECTIONS

BACTERIAL SEPTICEMIA AND MENINGITIS

Neonatal septicemia is frequently found in infants under stress from low birth weight, respiratory distress, or other conditions (Siegel and McCracken, 1981). The frequency of this septicemia appears to have been relatively stable over the past few decades, but there has been a change in the specific causative infectious organism. *E. coli* and *Staphylococcus aureus* have been the major pathogenic agents in the last three decades, and they have now been joined by the group B Streptococcus (Siegel and McCracken, 1981).

Sepsis in the neonate is known to follow infection of the amniotic fluid, urinary tract infection in the mother, or a delay in delivery after rupture of the amniotic membranes. The associated occurrence of neonatal meningitis (25 to 30 per cent), otitis media, necrotizing enterocolitis, pneumonia, and osteomyelitis has been firmly established. That septicemia often spreads to the meninges poses the greatest single threat of later neurologic involvement. The specific site of involvement, organism characteristics (virulence, amount of inoculum, presence of circulating antibodies), the general state of the infant, the type of antibiotic chosen, and the speed with which the condition is recognized and treated, all determine the type and degree of sequelae the infant develops. Motor, sensory, cognitive, and association fiber brain ab-

normalities have been described in 31 to 65 per cent of the 50 per cent who survive neonatal bacterial meningitis (Bell and McCormick, 1981). The abnormalities most often noted include hydrocephalus, mental retardation, visual and hearing deficits, and cranial nerve and long tract signs. Significant distortions of gross brain anatomy are frequently described. Group B streptococcal meningitis occurring at three to four weeks of age has a better outcome than *E. coli* or group B streptococcal meningitis that occurs during the first two weeks of life, according to Bell.

VIRAL INFECTIONS

Although viral infections occur less often than bacterial infections in the neonate, their associated morbidity and mortality can be severe (Grossman, 1980). They are usually found in two distinct clinical situations—intrauterine infections (e.g., rubella, cytomegalovirus) in which fetal infection during the first or second trimester causes severe anatomic and consequent metabolic and growth disorders, and neonatal infections (e.g., herpes virus) in which the infant is exposed to the offending virus late in pregnancy or during the birth event itself.

Intrauterine Infections

There is a wide spectrum of fetal disabilities associated with intrauterine viral infections. Many are clinically discernible at birth, but others are not manifest until the child is three to five years of age. Conditions appearing early are thrombocytopenia, low birth weight, microcephaly, cataracts, hearing loss, cardiac defects, hepatosplenomegaly, and osteolytic lesions. Possible later complications include mental retardation, motor disabilities, learning disabilities, and hearing impairment. It is now well established that the virus may continue to be isolated from body tissues for years following birth.

Neonatal Infections

Herpes virus infection is rapidly becoming a major health concern for mothers as well as their infants. This virus in the pregnant woman is a direct threat to the neonate, but appropriate prevention and treatment of the disease may reduce the incidence and sequelae. Herpes virus hominis exists in two types. Type I (oral strain) causes gingivostomatitis, recurrent herpes labialis, and herpes encephalitis in the child and adult. Type II (genital strain) causes vulvovaginitis, cervicitis, and disseminated disease in the newborn.

Although both strains have the ability to cause

severe encephalitis in children and adults, type II shows a greater tendency to cause a disseminated disease process, including encephalitis in the infant. The infant is usually infected following vaginal delivery. A cesarean section is therefore recommended, when the risk is known, to prevent potential infection. Clinical manifestations of the disease in the neonate include cutaneous herpetic lesions and nonspecific signs of poor feeding, lethargy, and weight loss. When only cutaneous lesions occur, the course is usually benign. The other symptoms mentioned have a far more significant prognosis. Only about 10 per cent of the infants affected with type II disease survive. Among the survivors, microcephaly, spasticity, convulsions, mental retardation, and blindness are seen (Bell and McCormick, 1981).

METABOLIC FACTORS

KERNICTERUS

Kernicterus involves the deposition of bilirubin in cells of the basal ganglia. The incidence of this condition in full term infants has been dramatically reduced by the prevention of Rh incompatability and by the treatment of hyperbilirubinemia with exchange transfusions. Although the lesion is now uncommon in its full expression, some deposition of bilirubin is seen in low birth weight infants at autopsy, particularly those who have had significant metabolic disturbances. The determining factors are believed to be those that increase the permeability of the blood-brain barrier to free bilirubin (Ritter et al., 1982).

HYPOGLYCEMIA

Symptomatic hypoglycemia has been associated with an adverse developmental outcome. This was a problem particularly for premature infants and infants who are small for their gestational age when delayed feeding was a common practice (Hack et al., 1979). Haworth et al. (1976), however, have shown that for infants of diabetic mothers hypoglycemia per se is not a significant factor in determining the developmental outcome.

PSYCHOSOCIAL FACTORS

Adverse psychosocial factors may operate at different levels in affecting perinatal events and hence developmental outcome. The maternal nutritional status has been suggested as influencing both the rate of intrauterine growth and the length of the gestational period (Woods et al., 1980).

Knobloch et al. (1982) evaluated the factors that affected the outcome in infants with a birth weight of less than 1500 grams. They found that significantly more mothers of infants with major handicaps had less than a twelfth grade education (40 per cent versus 18 per cent), and significantly more families were on Medicaid (75 per cent versus 29 per cent). In a follow-up study of low birth weight infants, Neligan et al. (1976) found that the outcome in the infants was strongly associated with socioeconomic status. Therefore, the importance of socioeconomic and sociocultural factors cannot be separated from the equation.

CHANGING PATTERNS IN THE INFLUENCE OF PERINATAL FACTORS

As a result of improvements in the management of pregnancy and labor and advances in neonatal intensive care, there has been a dramatic decline in perinatal mortality. Statistics in the United States show the mortality to have decreased from 1930, when the figure was 21.1 per 1000 for disorders of short gestation, low birth weight, and birth trauma, to 12.8 per 1000 in 1970 and 6.2 per 1000 in 1980 (Wegman, 1980, 1981).

The greatest impact on the reduction in neonatal mortality has come from the increased survival of low birth weight infants (less than 2500 grams) and very low birth weight infants (less than 1500 grams). Philip et al. (1981) combined data from several centers in England and North America and showed that for infants having a birth weight between 1000 and 1500 grams, the mortality improved from 44 per cent in the 1960's to 18 per cent in the 1970's; for those with birth weights between 500 and 1000 grams, there was a decrease from 91 per cent to 58 per cent.

It is of interest to compare the prevalence of cerebral palsy with the decline in neonatal and perinatal mortality. Hagberg (1979) in Sweden and Stanley (1979) in Western Australia have examined data from the period from the 1950's to the 1970's, and both have shown a change in the prevalence of cerebral palsy. The relationship with mortality statistics, however, is not a simple one; moreover, the patterns that each found differ from one another. Whereas Hagberg demonstrated a steady decline from the early 1950's to the mid-1960's and then stabilization with a slight increase in the 1970's, Stanley's data initially showed a rise with a peak in the late 1960's before falling to approximately the same level as that in the Hagberg study, and then rising again. This pattern of inconsistency is found to be more striking when results from other centers are examined as well. Kiely et al. (1981) suggest that neonatal intensive care practices have played a major role in affecting the outcome in terms of survival and handicap.

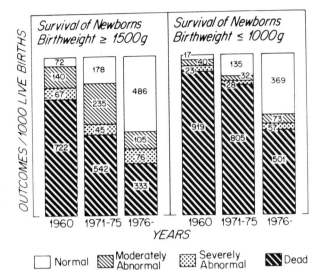

Figure 21–2. Composite data relating to the outcome in low birth-weight infants. (From Budetti, P., et al.: The costs and effectiveness of neonatal intensive care (case study #10). Washington, D.C., Office of Technology Assessment, 1981.)

Infants who have survived on low birth weight and very low birth weight status represent a significant proportion of the children who will develop cerebral palsy, and there are suggestions that this is increasing. O'Reilly and Walentynowicz (1981) examined the etiologic factors in their cerebral palsy clinic and found an increased proportion of children who had been premature, from 12 per cent before the 1940's steadily increasing to 29 per cent in the 1970's. There was a corresponding decrease in factors that were more likely to affect larger infants; e.g., dystocia decreased from 14.6 to 4.4 per cent. Dale and Stanley (1980) examined the incidence of spastic cerebral palsy between 1968–1970 and 1971–1975 and found a statistically significant reduction in the incidence in infants with birth weights greater than 2500 grams. There was no reduction for infants with birth weights less than 2500 grams between these two periods.

In a review of the literature on survival and handicap of low birth weight and very low birth weight infants, Budetti et al. (1981) demonstrated a dramatic improvement in the survival of infants without any handicap (Fig. 21–2). This, coupled with the reduction in the incidence of handicap for full term infants (Dale and Stanley, 1980), suggests that there indeed has been an overall decline in the incidence of cerebral palsy and associated handicapping conditions as a result of improved perinatal care.

ALFRED HEALY
HERMAN A. HEIN
I. LESLIE RUBIN

REFERENCES

Banker, B. Q., and Bruce-Gregorios, J.: Neuropathological alterations of the central nervous system. *In* Thompson, G. H., Rubin, I. L., and Bilenker, R. M. (Editors): The Comprehensive Management of Cerebral Palsy. New York, Grune & Stratton, Inc., 1983.

Banker, B. Q., and Larroche, J. C.: Periventricular leukomalacia of infancy. Arch. Neurol. 7:386–410, 1962.

Bell, W. E., and McCormick, W. F.: Neurologic Infections in Children. Philadelphia, W. B. Saunders Company, 1981, pp. 1–65.

Bennet, F. C., Chandler, L. S., Robinson, N. M., and Sells, C. J.: Spastic diplegia in premature infants. Am. J. Dis. Child., 135:732–737, 1981.

Bobbit, J. R., et al.: Amniotic fluid infection as determined by transabdominal amniocentesis in patients with intact membranes in premature labor. Am. J. Obstet. Gynecol., 140:947, 1981.

Bolton, D. P. G., and Cross, K. W.: Further observations on cost of preventing retrolental fibroplasia. Lancet, 1:445, 1974.

Britton, S. B., Fitzhardinge, P. M., and Ashby, S.: Is intensive care justified in infants weighing less than 801 gm. at birth? J. Pediatr., 99:937–943, 1981.

Budetti, P., Barrand, N., McManus, P., and Heinen, L. A.: The costs and effectiveness of neonatal intensive care (case study #10). Washington, D.C., Congress of the United States, Office of Technology Assessment, 1981.

Burstein, J., et al.: Intraventricular hemorrhage and hydrocephalus in premature newborns: a prospective study with CT. Am. J. Roentgenol., 132:631, 1979.

Butler, N. R., and Bonham, D. G.: Perinatal Mortality. Edinburgh, E. & S. Livingstone, 1963, p. 143.

Cocker, J., George, S. W., and Yates, P. O.: Perinatal occlusions of the middle cerebral artery. Dev. Med. Child Neurol., 7:235, 1965.

Commey, J. O. O., and Fitzhardinge, P. M.: Handicap in the preterm small-for-gestational age infant. J. Pediatr., 94:779, 1979.

Creasy, R. K., et al.: System for predicting spontaneous preterm birth. Obstet. Gynecol., 55:692, 1980.

Crosby, W., et al.: Fetal malnutrition: an appraisal of correlated factors. Am. J. Obstet. Gynecol., 128:22, 1977.

Dale, A., and Stanley, F. J.: An epidemiological study of cerebral palsy in western Australia 1956–1975. II. Spastic cerebral palsy and perinatal factors. Dev. Med. Child Neurol., 22:13–25, 1980.

Davies, P. A., and Tizard, J. P. M.: Very low birth weight and subsequent neurological defects. Dev. Med. Child. Neurol., 17:3, 1975.

Donat, J. F., Okazaki, H., and Kleinberg, F.: Cerebellar hemorrhages in newborn infants. Am. J. Dis. Child., 133:441, 1979.

Ellenberg, J. H., and Nelson, K. B.: Birthweight and gestational age in children with cerebral palsy or seizure disorders. Am. J. Dis. Child., 33:1044–1048, 1979.

Eng, G. D.: Brachial plexus palsy in newborn infants. Pediatrics, 48:18, 1971.

Fedrick, J., and Butler, N. R.: Certain causes of neonatal death—intraventricular hemorrhage. Biol. Neonate, 15:257, 1970.

Fitzhardinge, P. M., and Pape, K. B.: Intrauterine growth retardation (IUGR): an added risk to the preterm infant. Pediatr. Res., 11:562, 1979.

Fitzhardinge, P. M., and Stevens, E. M.: The small-for-dates infant. I. Later growth patterns. Pediatrics, 49:671, 1972.

Fitzhardinge, P. M., and Stevens, E. M.: The small-for-dates infant. II. Neurological and intellectual sequelae. Pediatrics, 50:50, 1972.

Flynn, J. T., et al.: Fluorescent angiography in retrolental fibroplasia: experience from 1969–1977. Ophthalmology, 86:1700, 1979.

Gaziano, E. P., et al.: Antenatal prediction of women at increased risk for infant with low birth weight. Am. J. Obstet. Gynecol., 140:99, 1981.

Gresham, E. L.: Birth trauma. Pediatr. Clin. N. Am., 22:319, 1975.

Grossman, J. H.: Perinatal viral infections. Clin. Perinatol., 7:257, 1980.

Hack, M., et al.: The low-birth-weight infant—evaluation of a changing outlook. N. Engl. J. Med., 301:1162, 1979.

Hagberg, B.: Epidemiological and preventive aspects of cerebral palsy and severe mental retardation in Sweden. Eur. J. Pediatr., 130:71–78, 1979.

Hagberg, B., et al.: The changing panorama of cerebral palsy in Sweden. Acta Pediatr. Scand., 64:187, 1975.

Hagberg, G., Hagberg, B., and Olow, I.: The changing panorama of cerebral palsy in Sweden 1954–1970. III. The importance of fetal deprivation of supply. Acta Pediatr. Scand., 65:403–408, 1976.

Hagberg, B., Hagberg, G., and Olow, I.: Gains and hazards of neonatal care: an analysis from Swedish cerebral palsy epidemiology. Dev. Med. Child Neurol., 24:13–19, 1982.

Hambleton, G., and Wigglesworth, J. S.: Origin of intraventricular haemorrhage in the preterm infant. Arch. Dis. Child., 51:651–659, 1976.

Harvey, D., Prince, J., Bunton, J., Parkinson, C., and Campbell, S.: Abilities of children who were small-for-gestational-age babies. Pediatrics, 69:296–300, 1982.

Haworth, J. C., McRae, K. N., and Dilling, L. A.: Prognosis of infants of diabetic mothers in relation to neonatal hypoglycemia. Dev. Med. Child Neurol., 18:471–479, 1976.

Hemminki, E., and Starfield, B.: Prevention and treatment of premature labor by drugs: review of controlled clinical trials. Br. J. Obstet. Gynecol., 85:411, 1978.

Hepner, W. R., Jr.: Some observations on facial paresis in the newborn infant: etiology and incidence. Pediatrics, 8:494, 1951.

Hobel, C. J., et al.: Prenatal and intrapartum high-risk screening. Am. J. Obstet. Gynecol., 117:1, 1973.

Huff, R. W., and Pauerstein, C. J.: Human Reproduction: Physiology and Pathophysiology. New York, John Wiley & Sons Inc., 1979.

Jones, M.D., and Battaglia, F. C.: Intrauterine growth retardation. Am. J. Obstet. Gynecol., 127:540, 1977.

Kiely, J. L., Paneth, N., Stein, Z., and Susser, M.: Cerebral palsy and newborn care. II. Mortality and neurological impairment in low-birth-weight infants. Dev. Med. Child Neurol., 23:650–659, 1981.

Kingham, J. D.: Acute retrolental fibroplasia. Arch. Ophthalmol., 95:39, 1977.

Klaus, M. R., and Kennel, J. H.: Maternal-Infant Bonding: The Impact of Early Separation or Loss on Family Development. St. Louis, The C. V. Mosby Co., 1976.

Knobloch, H., Malone, A., Ellison, P. H., Stevens, F., and Zdeb, M.: Considerations in evaluating changes in outcome for infants weighing less than 1,501 grams. Pediatrics, 69:285–295, 1982.

Liggins, G. C., and Howie, R. N.: A controlled trial of antepartum glucocorticoid treatment for prevention of the respiratory distress syndrome in premature infants. Pediatrics, 50:515, 1972.

Little, W. J.: On the influence of abnormal parturition, difficult labors, premature birth and asphyxia neonatorum on the mental and physical condition of the child, especially in relation to deformities. Trans. Obstet. Soc. Lond., 3:293, 1861.

Lou, H. C., et al.: Is arterial hypertension crucial for the development of cerebral hemorrhage in premature infants? Lancet, 1:1215, 1979.

Lubchenco, L. O., et al.: Sequelae of premature birth: evaluation of infants of low birth weight at 10 years of age. Am. J. Dis. Child., 106:101, 1963.

Lubchenco, L. O., et al.: Neonatal mortality rate: relationship to birth weight and gestational age. J. Pediatr., 81:814, 1972.

MacArthur, B. A., Howie, R. N., Dezoete, T. A., and Elkins, J.: Cognitive and psychosocial development of 4 year old children whose mothers were treated antenatally with Betamethasone. Pediatrics, 68:638–643, 1981.

Markestad, T., and Fitzhardinge, P. M.: Growth and development in children recovering from bronchopulmonary dysplasia. J. Pediatr., 98:597, 1981.

Milligan, M. D., and Shennan, A. T.: Perinatal management and outcome in the infant weighing 1000 to 2000 grams. Am. J. Obstet. Gynecol., 136:269, 1980.

Minkowski, A., Amiel-Tison, C., Cukier, F., Dreyfus-Brisac, J. P., Recier, J. P., and de Bethman, O.: Long term follow-up and sequelae of asphyxiated infants. In Gluck, L. (Editor): Intrauterine Asphyxia and the Developing Fetal Brain. Chicago, Year Book Medical Publishers, Inc., 1977, pp. 309–330.

Myers, R. E.: Lactic acid accumulation as cause of brain edema and cerebral necrosis resulting from oxygen deprivation. Adv. Perinat. Neurol., 1:85–114, 1979.

Naeye, R. L., and Peters, E. C.: Causes and consequences of premature rupture of fetal membranes. Lancet, 1:192, 1980.

Neligan, G. A., Kolvin, I., Scott, D. M., and Garside, R. F.: Born too young or born too small. Clin. Dev. Med., 61, 1976.

Nelson, K. B., and Ellenberg, J. H.: Apgar scores as predictors of chronic neurologic disability. Pediatrics, 68:36–44, 1981.

O'Reilly, D. E., and Walentynowicz, J. E.: Etiological factors in cerebral palsy: an historical review. Dev. Med. Child Neurol., 23:633–642, 1981.

Pape, K. E., Buncic, R. J., Ashby, S., and Fitzhardinge, P. M.: The status of 2 years of low-birth-weight infants with birth weights of less than 1001 grams. J. Pediatr., 92:253–260, 1978.

Pape, K. E., and Wigglesworth, J. S.: Haemorrhage, ischaemia and the perinatal brain. Clin. Dev. Med., 69/70, 1979.

Papile, L. A., Burstein, J., Burstein, R., and Koffler, H.: Incidence and evolution of subependymal hemorrhage: a study of infants with birth weights less than 1,500 grams. J. Pediatr., 92:529–534, 1978.

Philipp, A. G. S., Little, G. A., Polivy, D. R., and Lucey, J. F.: Neonatal mortality risks for the eighties: the importance of birth-weight-gestational age groups. Pediatrics, 68:122–130, 1981.

Philipp, E. E., et al.: Scientific Foundation of Obstetrics and Gynecology. London, William Heinemann Medical Books Ltd., 1977, Ch. 59.

Ritter, D. A., Kenny, J. D., Norton, H. J., and Rudolph, A. J.: Prospective study of free bilirubin and other risk factors in the development of kernicterus in premature infants. Pediatrics, 69:260–266, 1982.

Rothner, A. D., and Cruse, R. P.: Computed tomographic findings in childhood hemiplegia. Cleveland Clin. Quart., 45:219, 1978.

Rush, R. W., et al.: Contribution of preterm delivery to perinatal mortality. Br. Med. J., 2:965, 1976.

Saigal, S., Rosenbaum, P., Stoskopf, B., and Milner, R.: Follow-up of infants 501 to 1,500 grams birth weight delivered to residents of a geographically defined region with perinatal intensive care facilities. J. Pediatr., 4:606–613, 1982.

Schechner, S.: For the 1980's: how small is too small? Clin. Perinatol., 7:135, 1980.

Schwartz, P.: Birth Injuries of the Newborn. New York, Hafner Publications Company, 1961.

Shuman, R. M., and Selednik, L. J.: Periventricular leukomalacia: a one-year autopsy study. Arch. Neurol., 37:231, 1980.

Siegel, J. D., and McCracken, G. H.: Sepsis neonatorum. N. Eng. J. Med., 304:642, 1981.

Silverman, W. A.: The lesson of retrolental fibroplasia. Sci. Am., 236:100, 1977.

Stanley, F. J.: An epidemiological study of cerebral palsy in Western Australia, 1956–1975. I. Changes in total incidence of cerebral palsy and associated factors. Dev. Med. Child Neurol., 21:701–713, 1979.

Stewart, A. L., and Reynolds, E. O. R.: Improved prognosis in infants of very low birth weight. Pediatrics, 54:724, 1974.

Taeusch, H. W.: Glucocorticoid prophylaxis for respiratory distress syndrome: a review of potential toxicity. J. Pediatr., 87:617, 1975.

Taghizadeh, A., and Reynolds, E. O. R.: Pathogenesis of bronchopulmonary dysplasia following hyaline membrane disease. Am. J. Pathol., 82:241, 1976.

Terry, T. L.: Extreme prematurity and fibroplastic overgrowth of persistent vascular sheath behind each crystalline lens. Am. J. Ophthalmol., 25:203, 1942.

van den Berg, B. J., and Yerushalmy, J.: The relationship of the rate of intrauterine growth of infants of low birth weight to mortality, morbidity, and congenital anomalies. J. Pediatr., 69:531, 1966.

Vohr, B. R., et al.: The preterm small-for-gestational age infant: a two year follow-up study. Am. J. Obstet. Gynecol., 133:425, 1979.

Volpe, J. J.: Observing the infant in the early hours after asphyxia. In Gluck, L. (Editor): Intrauterine Asphyxia and the Developing Fetal Brain. Chicago, Year Book Medical Publishers Inc., 1977.

Volpe, J. J.: Neonatal intracranial hemorrhage. Clin. Perinatol., 4:77–102, 1977.

Volpe, J. J.: Neonatal intraventricular hemorrhage. N. Engl. J. Med., 304:886, 1981.

Volpe, J. J.: Neurology of the Newborn. Philadelphia, W. B. Saunders Company, 1981. p. 181.

Wegman, M. E.: Annual summary of vital statistics—1979: with some 1930 comparisons. Pediatrics, 66:823–833, 1980.

Wegman, M. E.: Annual summary of vital statistics—1980. Pediatrics, 68:755–762, 1981.

Wigglesworth, J. S., Keith, I. H., Girling, D. J., and Slade, S. A.: Hyaline membrane disease, alkali and intraventricular hemorrhage. Arch. Dis. Child., 51:755–762, 1976.

Woods, D. L., Malan, A. F., and Vanschalkwyk, D. J.: Maternal nutrition and the duration of pregnancy. S. Afr. Med. J., 59:374, 1980.

Yamamoto, M., et al.: A follow-up study of refractive errors in premature infants. Jpn. J. Ophthalmol., 23:435, 1979.

22

Infections of the Central Nervous System

BACTERIAL MENINGITIS

Bacterial meningitis, inflammation of the leptomeninges secondary to bacterial infection, is the most common infection of the central nervous system. Its importance stems not only from its prevalence but also from the devastation it may wreak upon its victims.

Three organisms are responsible for the majority of cases in infancy and childhood: *Hemophilus influenzae* type B, *Streptococcus pneumoniae*, and *Neisseria meningitidis*. The age of greatest risk for bacterial meningitis is between six months and one year, 90 per cent of the infections occurring in patients between one month and five years of age. This age distribution has not changed significantly for several decades, although the frequency of occurrence of various infecting agents has done so. *Hemophilus influenzae* has shown an absolute increase in incidence and has become the commonest cause of childhood bacterial meningitis, far exceeding the pneumococcal and meningococcal cases together.

Certain physical conditions carry an increased risk for the development of bacterial meningitis. In infants and children born with developmental anomalies that allow direct contact of the leptomeninges with the exterior (for example, a dermal sinus or meningocele), organisms may directly invade the cerebrospinal fluid. These children are prone to recurrent episodes of meningitis, often with an unusual infecting agent, such as Staphylococcus, a normal skin inhabitant. A similar situation may arise following head trauma, which causes a breach of the host's physical defenses. Many patients with ventricular shunts suffer repeated episodes of infection of the cerebrospinal fluid, usually in the form of ventriculitis. In this situation the organism may be unusual—again often a Staphylococcus—and since infection may be harbored in the shunt itself, eradication may be impossible without its removal.

Some children have an increased susceptibility to developing meningitis because of underlying disorders, such as pneumococcal meningitis in sickle cell disease, or because of decreased immunologic function, either congenital or acquired. Many of these children may acquire infections by unusual organisms.

The pathology of bacterial meningitis is important in considering the sequelae of the disease. The major gross finding is inflammation of the meninges with a purulent exudate. This exudate is widely distributed but accumulates particularly around the veins, in the depths of sulci, in fissures and in the basal cisterns, and over the cerebellum. It may also encase the spinal cord. Autopsy studies frequently reveal ventriculitis, and this occurrence may also be common in survivors. Histologically, signs of inflammation can be seen around cranial and spinal nerves, and also around and in the walls of blood vessels, causing thrombosis and areas of infarction in the cerebral parenchyma. Hydrocephalus, an uncommon complication of bacterial meningitis beyond the neonatal period, is most often communicating in type, secondary to adhesions around the basal cisterns or arachnoid villi, hindering normal cerebrospinal fluid circulation and resorption. Less often, stenosis of the aqueduct of Sylvius or fourth ventricle outlet foramina due to gliosis causes a noncommunicating type of hydrocephalus. Subdural effusions are common in childhood meningitis, especially that due to *H. influenzae*, but their pathogenesis remains uncertain. They may be related to transudation of protein rich fluid from subdural bridging veins and capillaries, due to inflammation.

The clinical manifestations of bacterial meningitis include symptoms and signs of systemic infection, as well as those secondary to the disease in and around the central nervous system. Classically patients complain of fever, headache, photophobia, and vomiting and on examination may be found to have nuchal rigidity, an alteration in the level of consciousness, focal neurologic signs, disturbances of vision, and rarely papilledema. Generalized or focal seizures may be part of the presentation, as may acute cerebellar ataxia. All patients suspected of having meningitis should undergo a careful neurologic examination at the time of admission, and at least daily thereafter while in the hospital.

Seizures should be promptly treated with full doses of the usual anticonvulsants given intravenously, phenobarbital and phenytoin being the usual choices. Suspicion of a subdural effusion arises in any child with focal neurologic signs, focal seizures, evidence of increasing head circumference, persistently increased intracranial pressure, or in whom there is failure of improvement following antibiotic therapy. Transillumination may yield positive findings in infants (less than one year). These effusions are frequently bilateral, and are sufficiently common that many consider them a part of the disease process rather than a complication. They are most easily identified by computed tomographic scanning. Electroencephalography may be a valuable adjunct in the detection of subdural effusions over which the voltages may be decreased, and in the identification of areas of focal damage such as may result from vascular compromise. Small effusions resolve completely and spontaneously given time. Larger or asymmetrical collections are treated by tapping. A child with a significant effusion that appears to be responsible for seizures, neurologic deficit, or a delay in recovery should be actively treated, initially by repeated tapping and, if necessary, drainage. In refractory cases, shunting or rarely surgical stripping of subdural membranes may be necessary if no resolution is occurring; such patients are likely to have significant neurologic residua, because of the underlying cerebral damage.

Computed tomography can also be used to confirm the diagnosis of hydrocephalus if suspected, and it should be possible to differentiate between non-communicating and communicating types. This complication of meningitis, which usually has its onset later in the course than do the effusions, generally requires surgical treatment by shunting.

The overall mortality rate for bacterial meningitis in childhood, although much improved from the almost 100 per cent mortality of the pre-antibiotic era, is still significant, being between 1 and 5 per cent. Focal neurologic signs, persistent or uncontrollable seizures, shock, severe obtundation, and hypothermia are all indicators of a poor prognosis. The incidence of permanent damage varies according to the etiologic agent, being highest in those surviving *H. influenzae* infection. At discharge, 20 to 30 per cent of the children can be expected to show sequelae, but a significant number improve within three to 24 months. This is most likely to occur with mild deficits such as the ataxic syndrome. Careful neurologic follow-up is important for 12 to 24 months.

The types of residual abnormality range from complete devastation with severe mental subnormality, quadriparesis, blindness, deafness, and epilepsy, through less severe degrees of deficits in the motor system, coordination, special senses, and intellectual functions. Any part of the central nervous system may be affected. Vascular damage may cause infarction in areas corresponding to large, medium, or small arteries or veins. Adhesions in the aqueduct, basal cisterns, or absorptive surfaces may cause hydrocephalus, which in itself may cause further neurologic damage as a result of increased pressure. Gliosis around the base of the brain may entrap cranial nerves, causing damage, including optic atrophy, facial palsy, and, most importantly and most commonly, deafness. Toxic effects on neurons, particularly in the cerebral cortex, may have been the cause of seizures in the acute stage, along with other factors such as electrolyte upsets, hypoxemia, and ischemia, and irreversible damage in these areas may be responsible for permanent seizure disorders. The more sophisticated the testing, the more likely are deficits to be found. Assessment of the auditory system by evoked potential recording, or of cognitive function by neuropsychologic evaluation, shows a higher incidence of abnormalities than in a control population, even in children who seem to have escaped unharmed from the disease.

NEONATAL MENINGITIS

Neonatal meningitis is considered separately because of the difference in the organisms involved, the special pathologic features, and the prognosis. The incidence varies from one institution to another, ranging from 2 to 10 cases per 1000 live births. There are two main bacteria responsible: *E. coli* and group B beta-hemolytic streptococci account for approximately 70 per cent of the infections.

The pathology of neonatal meningitis generally resembles that of meningitis in older children, but there are some important differences. The major findings are leptomeningeal inflammation, maximal at the base of the brain, and ventriculitis. The likely point of invasion of the central nervous system by bacteria is through the glycogen-rich vascular bed of the choroid plexus, where the organisms readily multiply and then enter the cerebrospinal fluid via the normal secretory mechanism. Arteritis and phlebitis are important, and commonly lead to hemorrhagic infarction. Arachnoidal fibrosis develops, especially in the basal regions, and contributes to the genesis of hydrocephalus by obstruction of the outflow pathways for cerebrospinal fluid. Long term pathologic changes include hydrocephalus, multicystic encephalomalacia, and atrophy of cortex and white matter. There may also be disruption of subsequent cerebral development.

Prognosis for recovery from neonatal bacterial meningitis depends on the speed of diagnosis and the prompt institution of effective therapy. The organism involved is important, as are various other factors, such as birth weight, asphyxia, trauma, and concurrent illness. The outlook is worst in infants infected with gram negative organisms. There the mortality is about 50 per cent, with only 50 per cent of the survivors being normal. In gram positive meningitis, 20 per cent die and 75 per cent of the survivors are without sequelae.

The types of deficit seen in these infants are variable, and the incidence depends on the sophistication of the evaluation performed. The longer the child is followed, the more likely it is that subtle deficits in higher congitive function will be found. If neurophysiologic methods, such as evoked potential studies, are used to assess special senses, a higher number of abnormal results will be found. All grades of severity are seen. Infants whose brains have undergone severe destruction typically are mentally retarded, with spastic quadriplegia, and often seizures, cortical blindness, and deafness. Failure of further cerebral growth causes microcephaly, and usually poor axial growth. Children who showed focal signs during the acute illness may have persistent hemiparesis, often with a focal seizure disorder. Hydrocephalus causes the head size to increase, long tract signs that are most marked in the lower limbs, and impairment of vision. Incoordination may occur, from minimal signs to marked cerebellar ataxia. Many children have hearing deficits caused by the disease itself or the aminoglycoside antibiotics used in treatment. As these children grow and approach school age, one must be on the lookout for possible intellectual handicaps, for the incidence of learning difficulties is increased. Hyperactivity and attentional deficits are also common. Psychologic testing is of assistance in making educational plans.

TUBERCULOUS MENINGITIS

Tuberculous meningitis remains a serious problem, despite the availability of chemotherapeutic drugs, because of its incidence in early childhood and the devastation produced in the brain. The meninges become infected by rupture of a Rich focus, a small tuberculoma in the leptomeninges, cortex, or choroid plexus. The focus discharges caseous material into the subarachnoid space, producing a thick exudate, which spreads over the pia arachnoid on the convexity, around the base of the brain and the brain stem. Involvement of the basilar regions results in hydrocephalus and cranial nerve deficits.

The clinical illness may be divided in three phases (Table 22–1).

Treatment includes specific antituberculous therapy and management of complications, such as inappropriate antidiuretic hormone secretion, as well as other abnormalities of electrolytes in the plasma, often compounded by vomiting. Seizures require prompt treatment with full doses of anticonvulsants, usually phenobarbital or phenytoin, either alone or in combination. Hydrocephalus requires relief by temporary external drainage. Corticosteroids have been recommended for use in the very ill patient, and may be of help in lowering the intracranial pressure and possibly in the prevention of adhesions. The prognosis for survival is good in patients who are treated before progression to the second phase of the illness, and for about 75 per cent of those treated in stage II. Major neurologic sequelae include spastic quadriplegia and hemiplegia, paraparesis, seizures, and optic atrophy with visual loss. Less severe problems are cranial nerve palsies, incoordination, ataxia, and mild spasticity. Some degree of intellectual deficit occurs in about 20 per cent of children, and ranges from severe retardation to subtle cognitive defects, attentional deficits, and hyperactivity. Damage to the hearing and vestibular system may be due to the disease or to therapy, and all survivors should be tested for hearing ability. Other complications include panhypopituitarism, obesity, diabetes insipidus, abnormalities of sexual development, and deficiency of growth hormone.

FOCAL PURULENT COLLECTIONS

Parameningeal infections in childhood include brain abscess, subdural empyema, and abscesses related to the spinal cord epidural space and disc spaces. All have been uncommon since the advent of antibiotics, but are of importance because of the considerable mortality and morbidity associated with them. The common underlying etiologies are listed in Table 22–2.

BRAIN ABSCESS

A brain abscess is a localized collection of pus within the cerebral parenchyma. Organisms travel

---

Table 22–1. CLINICAL STAGES OF TUBERCULOUS MENINGITIS

1. Anorexia, listlessness, fever, no neurologic signs
2. Stupor, meningeal irritation, cranial nerve palsies, abnormal neurologic signs, seizures
3. Coma, increased intracranial pressure, respiratory and temperature instability

Table 22–2.   CAUSES OF BRAIN ABSCESS

1. Cyanotic congenital heart disease (after age two years)
2. Sinusitis, mastoiditis
3. Trauma, penetrating wounds
4. Cystic fibrosis and other suppurative lung conditions
5. Immune deficiencies
6. Congenital sinus tracts

to the brain via the blood from distant foci of sepsis, or from infection in contiguous structures such as the paranasal sinuses. Frontal lobe infections are often silent for a long time, but collections in areas such as the anterior parietal or temporal lobes are likely to produce symptoms at an earlier stage. Abscesses in the brain stem are rarely diagnosed in life. The cerebellum may be affected, causing the signs of a posterior fossa mass lesion. An area of cerebral edema is usually present around the abscess, and is most marked in the early period. There may be an increase in intracranial pressure with the usual shifts in the cerebral structures that occur when the cerebrospinal fluid dynamics become deranged. Tentorial herniation may then take place, especially if a sudden change in pressure occurs, as during lumbar puncture.

The investigation of choice is the computed tomographic scan, which shows a round area of reduced density with a ring of enhancement. The treatment involves surgical drainage or aspiration.

Mortality rates remain in the region of 20 to 30 per cent despite modern management. A frequent terminal event is rupture of the abscess into a ventricle, which causes collapse with high fever, shock, and rapidly deepening coma with meningismus. The long term outlook for survivors depends on the site of the abscess, and on the speed with which the diagnosis is made and treatment instituted. Sequelae include seizures, permanent neurologic deficits, intellectual impairment, and hydrocephalus.

SUBDURAL EMPYEMA

Subdural empyema occurs by direct spread of infection from osteomyelitis of the skull, or by infection of a sterile effusion or traumatic hematoma. The collection is usually over the convexity of the brain, but may be interhemispheric. The clinical features are those of a space occupying intracranial lesion, with focal neurologic signs and signs of infection. Diagnosis is effected by use of computed tomographic scanning, and treatment is by aspiration. Craniotomy and stripping of membranes may be necessary. Long term sequelae include hemiparesis and seizures.

SPINAL EPIDURAL ABSCESS

Spinal epidural abscess is uncommon in childhood. It usually produces an acute syndrome of spinal cord compression, often with signs of systemic infection. Prompt diagnosis by myelography is essential if function is to be preserved below the level of the lesion. The child presents with an acute onset of monoparesis or paraparesis, and often with difficulty in bladder and bowel function. There may be swelling, scoliosis, and tenderness over the affected area. Demonstration of a sensory level may be difficult in the young child, although the absence of sweating may be apparent. Emergency decompression and drainage are mandatory, followed by antibiotic therapy. The prognosis depends mainly on the speed of diagnosis and treatment, delay causing permanent paralysis of the limb or limbs affected. Complications include spinal osteomyelitis, which may cause severe spinal deformities, and meningitis if organisms spread to the subarachnoid space.

Infection of the intravertebral disc space can occur in children. The symptoms are back pain, often with fever, and in young children a refusal to walk, although no weakness can be demonstrated. Radiographs show narrowing of the affected space, and there may be evidence of systemic infection, such as an increase in the erythrocyte sedimentation rate. The condition may be self-limiting, but antibiotics are generally prescribed. Deformities such as ankylosis and scoliosis may result.

VIRAL INFECTIONS OF THE CENTRAL NERVOUS SYSTEM

Viral infections of the central nervous system fall naturally into three major groups, depending on the main site of the inflammation: meningitis, encephalitis, and myelitis. Any one may be present alone, or there may be combinations. The features of each illness vary, as do the resultant morbidity and mortality. There are special epidemiologic findings related to age, season, geographic location, or social circumstances, and the disease may be acute, subacute, or chronic. Many of these potentially handicapping disorders are preventable by immunization or vector control.

VIRAL MENINGOENCEPHALITIS

"Aseptic meningitis" is generally an acute, sometimes biphasic, illness characterized by symptoms and signs of systemic viral infection, as well as those of meningeal irritation, namely, headache, photophobia, stiff neck, and vomiting. Focal neurologic signs are rare.

Treatment is entirely supportive. The course is generally short, with a steady improvement in well-being. Some patients feel tired for several weeks following the illness, a common feature of many other viral infections. There may also be a tendency to headaches and muscle pains, cramps, and tenderness for a while, but the incidence of permanent sequelae is low. Nerve deafness can occur during mumps infection. However, there can be progression to dramatic central nervous system involvement (e.g., encephalitis) with stupor, disorientation, seizures, focal signs, rostrocaudal deterioration, coma, and death, or survival with significant sequelae.

There are many agents that cause this syndrome, the most common being measles, mumps, varicella, rubella, enteroviruses, adenoviruses, arboviruses, and herpes (Table 22–3).

ENCEPHALITIS

Arboviral infections occur worldwide, and many involve the central nervous system, causing encephalitis. A brief outline is shown in Table 22–4. The essential clinical features of the commonest of these disorders are mentioned, as is the outcome in each group.

The commonest sporadic form of encephalitis is that due to herpes simplex infection. After the neonatal period this is due to herpes virus type I, and usually represents a primary infection. The virus has a predilection for the temporal and orbital-frontal lobes, where a severe hemorrhagic necrotizing inflammatory process takes place, causing destruction of cerebral tissue. The symptoms and signs are fever, vomiting, lethargy, and seizures, either generalized or focal. The neurologic examination may reveal focal signs implicating the temporal lobe as the site of the infection.

The illness is often marked by progressive deterioration, with stupor leading to coma, increased intracranial pressure, and sometimes brain stem abnormalities. A computed tomographic scan should be considered before a lumbar puncture is performed when there are signs of increased intracranial pressure; it often demonstrates focal areas of edema and necrosis. The electroencephalogram may show periodic discharges from one or both temporal regions, a diffusely abnormal background indicating the generalized encephalopathic process.

There is evidence that early treatment with antiviral drugs may improve the prognosis. Emergency brain biopsy with immediate histopathologic and immunofluorescence examination has been advocated for confirmation of the diagnosis. The outcome in herpes encephalitis is still poor,

Table 22–3. NEUROLOGIC SYNDROMES OF COMMON VIRAL INFECTIONS

Measles (rubeola)	0.1 per cent of the cases involve the central nervous system. The illness begins on the second to sixth day after the rash appears. There are the typical clinical signs and cerebrospinal fluid findings of encephalitis. Fifteen per cent die, 25 per cent of the survivors have sequelae, 60 per cent make a complete recovery.
Mumps	Meningoencephalitis occurs in 10 per cent. Over 50 per cent have pleocytosis in the cerebrospinal fluid. Hypoglycorrhacia can occur and can be prolonged; this is probably due to chorid plexus infection. Acute hydrocephalus related to aqueductal stenosis can occur. Deafness is a rare, serious complication; it is usually unilateral. This is heralded by sudden vertigo, tinnitus, ataxia, and vomiting. Eighth nerve neuritis is thought to be the cause. Facial nerve neuritis has been described with mumps, as has transverse myelitis.
Varicella (chicken pox)	Neurologic complications occur between the third and eighth days after the rash appears. Aseptic meningitis and encephalitis occur. The latter has a 27 per cent fatality rate. Other central nervous system complications include, most commonly, benign cerebellar ataxia, transverse myelitis and Reye syndrome.
Rubella (German measles)	Central nervous system involvement occurs in 1 in 5000; most commonly it is a brief and severe encephalitis with coma and seizures in 50 to 60 per cent. The mortality is 20 per cent; sequelae appear in less than 25 per cent of the survivors.
Enteroviruses Coxsackie	Herpangina, pleurodynia, myocarditis, and hand, foot, and mouth disease. Aseptic meningitis. Cerebellar ataxia. Polio-like illness.
ECHO	Summer, diarrhea, rash, aseptic meningitis; encephalitis if it occurs is benign except in infants.
Polio	Summer, enteritis, aseptic meningitis, paralysis of limbs and bulbar musculature. May be permanent. The mortality is 5 per cent.
Infectious mononucleosis (Epstein-Barr virus)	Pharyngitis, lymphadenopathy, splenomegaly, aseptic meningitis, encephalitis (can be focal), cerebellitis, myelitis, polyneuritis, Bell's palsy (cranial nerve VII), and other cranial mononeuropathies, Reye syndrome. Excellent prognosis despite dramatic illness.

Table 22–4.   ARBOVIRAL ENCEPHALITIS

Encephalitis	Vector	Reservoir	Area of Distribution	Season	Sex and Age	Clinical	CSF	Unusual Features
St. Louis	Mosquito	Birds, ? rodents	Midwest, Far West, South and Central states, rarely East	Late summer, early fall; the commonest epidemic encephalitis	Incidence in males three times greater than in females. Rare < 1 yr. Infrequent < 5 yr. Adults > children	60% aseptic meningitis → 30% severe encephalitis, ataxia, tremors, opsoclonus, myoclonus; mortality 8% in those less than 10 yr. (highest in elderly); sequelae uncommon	Polymorphs initially → mononuclear; protein, slight increase; sugar normal	Electromyograms indicate denervation
California	Mosquito	Rodents	Midwest, South, California, rural residents	July to October	Incidence in males three times greater than in females; 98% of patients less than 20 yr. of age	Aseptic meningitis → encephalitis, focal seizures, usually a mild illness with low mortality and residua	Mononuclear, occasionally hemorrhagic	Focal abnormalities on EEG, TC99 CT
Western equine	Mosquito	Birds; preceding epizootic in horses	West of Mississippi	Summer, fall	20% < 1 yr.; 30% < 5 yr.; males 60%	Severe encephalitis, seizures; course is usually 10 days; mortality 10–20%. Severe sequelae in the very young (less than one yr.)	Mononuclear	Parkinsonism as a residua; ? chronic infection
Eastern equine	Mosquito	Birds, horses will have preceding epizootic	Atlantic Coast (rare)	Summer	60% < 2 yr.	The most severe encephalitis; those less than 2 yr. old have a fulminant course — death common in less than 48 hr.; mortality 70% (less than 10 yr); severe sequelae	Polymorphs → mononuclear by 3rd day (250–1000)	—————
Venezuelan equine	Mosquito	Horses, rodents, birds will have preceding epizootic	Southern states, South America	Fall	—————	Mild aseptic meningitis with myalgia, 48 hr. illness, then 2–3 weeks of asthenia; generally full recovery	Polymorphs → mononuclear; mild increased protein	—————

but the use of antiviral therapy has reduced the mortality from about 70 to 30 per cent. Patients under 20 years of age who are noncomatose at the start of therapy do best. Survivors show a high incidence of permanent neurologic sequelae, such as motor and sensory deficits, seizures, intellectual retardation, speech, and memory deficits, and some children survive in a vegetative state.

NONBACTERIAL NEONATAL INFECTIONS

Nonbacterial infections in the neonatal period are a serious cause of mortality and morbidity, as well as a significant long term handicap. The central nervous system is frequently involved and often seems to bear the brunt of the infection. The organisms that are responsible are conveniently summarized in the mnemonic TORCH (Table 22–5). The infection may be acquired in utero, during parturition, or following birth. Subclinical infections may occur in the mother and the infant, making statistical predictions difficult.

SLOW VIRAL ILLNESSES

Neurologic infections due to the "slow" viruses comprise a small but significant group of illnesses. They have important effects on the lifespan of the affected child and on the quality of life. The most frequently seen is subacute sclerosing panencephalitis, now known to be due to measles virus. Others of note are progressive rubella panencephalitis and progressive multifocal leukoencephalopathy.

SUBACUTE SCLEROSING PANENCEPHALITIS

Subacute sclerosing panencephalitis occurs predominantly in the first decade of life. The original measles infection occurs in patients who are less than two years of age in 50 per cent of the cases, being otherwise an unremarkable example of that illness. The worldwide frequency is one case per million children. It appears to follow natural measles more often than active immunization with live attenuated measles virus, by a factor of 10 to 1. Involvement of males is three times more frequent than of females, and there is a higher incidence in rural as opposed to urban areas. The pathogenesis is still unclear. Most patients appear normal between the two illnesses, but there has been noted a coincidence of other viral infections, namely Epstein-Barr and varicella, at the time of onset of symptoms of subacute sclerosing panencephalitis. On activation, the virus spreads throughout the brain from occipital to frontal areas, with concomitant cellular destruction. The neuropathology includes encephalitis in the white and gray matter with perivascular lymphocytic and plasmocytic infiltration, as well as neuronophagia and focal gliosis. Inclusion bodies, originally described by Dawson in 1930, are seen in both neurons and glial cells. In severe cases there is diffuse gliosis and extensive demyelination.

The clinical course may be divided into four stages:

1. In the early phase there is a history of subtle changes in affect with mood swings and an intellectual decline causing deterioration in school performance. Speech and sleep may be disturbed. Examination may reveal a pigmentary macular

Table 22–5. NONBACTERIAL PRENATAL AND NEONATAL INFECTIONS

	Agents	Clinical Features
T.	Toxoplasmosis	Microcephaly, hydrocephaly, chorioretinitis deafness, parenchymal and periventricular calcifications; treatment with sulfadiazine and pyrimethamine; mild to severe psychomotor retardation
O.	Other	
	Syphilis	Meningitis, seizures, cranial nerve palsies, hydrocephalus, cerebrovascular accident, rash, snuffles, hepatosplenomegaly, bone changes
	Varicella	Congenital varicella syndrome characterized by SGA, severe scarring skin defects with associated hypotrophic limbs, chorioretinitis,microphthalmia, optic atrophy, cataracts, seizures, and mental retardation
	Enterovirus	Meningoencephalitis, myocarditis (Coxsackie B)
R.	Rubella	Microcephaly, chorioretinitis, cataracts, microphthalmia, deafness, heart defects, SGA; infectious, variable duration; mild to severe psychomotor retardation; rare late progressive panencephalitis
C.	Cytomegalovirus	Hepatosplenomegaly, thrombocytopenia, SGA, microcephaly, intracranial periventricular calcification, chorioretinitis, deafness, seizures, mild to severe psychomotor retardation; infectious, variable duration
H.	Herpes virus type II	Neonatal infection, ascending or vaginal; skin rash, severe meningoencephalitis; high mortality (80 per cent) and severe sequelae in survivors (50 per cent); antiviral agents

retinopathy. The cerebrospinal fluid is abnormal, containing both increased levels of gammaglobulin and measles antibody. This stage can last several months.

2. Motor signs develop, often manifesting as dyskinesia and dyscoordination with myoclonic jerks and signs of pyramidal tract abnormality. The electroencephalogram is abnormal, showing the typical pattern of periodic slow and sharp wave complexes, which often occur in association with the myoclonic jerks. Ocular signs such as optic atrophy and decreased visual acuity may be present. This stage can last up to one year, or the disease may appear to stabilize at this point.

3. The child lapses into stupor and coma. There is decorticate or decerebrate posturing. Autonomic dysfunction may be evident, and death may occur in this stage.

4. A phase of persistent vegetation occurs with loss of cortical function. Survival in this stage depends largely on the level of supportive care.

A small number of children have a rapidly deteriorating course, and a few may show a more slowly progressive, or relapsing and remitting, type of illness.

The presence of measles specific IgM in the cerebrospinal fluid is evidence of the active infection in the central nervous system. There is no effective or specific therapy for this devastating illness. Management is supportive and symptomatic. Special residential care is usually necessary by the time stage III is reached, or earlier, and the family needs much support both while the child is home and after.

PROGRESSIVE RUBELLA PANENCEPHALITIS

Progressive rubella panencephalitis is a late complication of both congenital and acquired rubella. It has appeared in children with "stable" congenital infections, and in a few who had a typical acquired infection with no evidence of central nervous system involvement. The pathogenesis is not clear, but seems consistent with a recrudescence of an active inflammatory process. The pathologic picture is that of chronic meningoencephalitis, often with a focal vasculopathy and areas of ischemic damage. Virus may be isolated from the brain, and the serum and cerebrospinal fluid contain high levels of rubella specific IgM antibody. The clinical features are somewhat similar to those of subacute sclerosing panencephalitis, with a progressive deterioration in mental and motor function, seizures, and prominent cerebellar signs. The duration of the illness is variable, a persistent vegetative state preceding demise in most patients. No specific therapy is available.

PROGRESSIVE MULTIFOCAL LEUKOENCEPHALOPATHY

Progressive multifocal leukoencephalopathy is a diffuse demyelinating disease, which is rare in children. It occurs almost exclusively in patients with immune deficiency, either primary or secondary, and in some chronic inflammatory diseases. The causative agent is a papovavirus. The usual clinical picture is that of dementia with multifocal neurologic deficits, often with seizures and visual disturbance. Brain biopsy of affected areas is necessary for confirmation of the diagnosis, and computed tomographic scanning is of help in identifying the sites of lesions. There is no clearly effective therapy, and the disorder is generally fatal within six months.

Chronic ECHO virus infections have been described in the central nervous system in children with agammaglobulinemia. There may also be an associated dermatomyositis-like illness.

Kuru, a disease formerly found in the Fore tribe of New Guinea and now declining in incidence, is a progressive cerebellar disorder, which is caused by an as yet unidentified agent transmitted by the ritual cannabalistic ingestion of human meat by the women and children of the tribe. Cessation of this practice has been coincident with the decrease in incidence. This condition was the prototype of "slow" virus disease.

Creutzfeldt-Jacob disease is another transmitted disease that occurs in the Western world mainly in young to middle aged adults. It is a rapidly progressive dementia, in which motor signs and myoclonic jerks are prominent. The electroencephalogram is helpful, showing periodic sharp and slow wave complexes. The disease may be transmitted by depth electrodes, neurosurgical procedures, and corneal transplants. It is invariably fatal.

MICHAEL J. BRESNAN
ELAINE M. HICKS

REFERENCES

Bell, W. E., and McCormick, W. F.: Neurologic Infections in Children. Ed. 2. Philadelphia, W. B. Saunders Company, 1981.

Feigin, R. D., and Cherry, J. D.: Textbook of Pediatric Infectious Disease. Philadelphia, W. B. Saunders Company, 1981.

Krugman, S., and Katz, S. L.: Infectious Diseases of Children. Ed. 7. St. Louis, The C. V. Mosby Co., 1981.

Volpe, J. J.: Neurology of the Newborn. Philadelphia, W. B. Saunders Company, 1981.

Weil, M. L.: Infections of the nervous system. In Menkes, J. H. (editor): Textbook of Child Neurology. Philadelphia, Lea and Febiger, 1980, Ch. 6.

23

Nutrition and Development

The emphasis in this chapter is on the role of nutrition in brain development. Because nutrition is part of a complex interaction between genetic and other social and environmental factors, it is incongruent to study the physiologic effects of malnutrition apart from the environment where the stress occurs. Yet earlier studies of malnutrition, brain growth, and learning have been devoted to generalized and gross indicators of morphologic change and have ignored other environmental influences (e.g., stimulation). Current research now explores the relationship between chemical and morphologic changes and behavior with careful consideration given to critical periods of brain growth and the multifactorial nature of malnutrition.

DEVELOPMENT AND NUTRITION OF THE CENTRAL NERVOUS SYSTEM

CRITICAL PERIODS OF CENTRAL NERVOUS SYSTEM DEVELOPMENT

The greatest percentage of brain growth (as measured by total brain weight) and differentiation of cell types and major parts of the brain occurs during prenatal and early postnatal life. It is during this period that nutrient deficiency could be anticipated to most affect development and subsequent function (Table 23–1). Major subdivisions of the brain have differing growth schedules (Fig. 23–1). The cerebellum is notable for a growth spurt that is evident at about eight months of postnatal life, accounted for by glial cell expansion. The velocity of this cell number increase in the cerebellum is more rapid than that typical of other parts of the brain. This is held to account for its enhanced sensitivity to nutritional deprivation in animal species during early postnatal life (Dobbing, 1974; Dodge et al., 1975; Winick, 1976). The total adult cell number in all parts of the brain is achieved by 15 months postnatally (Dobbing, 1974).

Much of the literature dealing with malnutrition and brain growth has assumed the existence of "critical" or "sensitive" periods of development. These are delimited intervals of rapid development in which the organism is at greater risk for per-

manent, irreversible damage if disruptive influences occur (Winick and Noble, 1966). A number of criticisms of this concept have been raised in its application to human development (Dobbing and Sands, 1981; Dodge et al., 1975). The idea of critical periods is based on behavioral and physiologic studies in animals. It is not clear that the very rapid changes in other species are characteristic of humans. In fact, no human studies have shown unequivocally that function is most susceptible to injury during its most rapid period of maturation.

According to the critical period hypothesis, final adult cell number is achieved early, and insults that limit proliferation are irreversible. It is important to note that in humans neuronal cell multiplication occurs at a time when the fetus is relatively invulnerable to nutritional influences. In addition, some investigators suggest that there is an excess in neuronal cell number and that the minimal number of neurons necessary for normal function has not been established (Dodge et al., 1975; Winick, 1976). The notion of a critical period in brain growth is probably most appropriately applied to the time during which glial cells are proliferating (Dobbing, 1974). Since glial cells retain the capacity to divide until later in life, limitations on glial cell proliferation may be reversible.

For these reasons the concept of critical periods in central nervous system development must be used in a very general way. Theoretically, exogenous factors can interfere at any time and at many sites of metabolic reactions involved in cell proliferation, migration, and differentiation (Herschkowitz and Rossi, 1972). Thus, the entire developmental period should be viewed as being "critical" for the development of the mature brain.

BIOCHEMICAL ASPECTS OF CENTRAL NERVOUS SYSTEM FUNCTION

With cell migration and the differentiation of specialized structures of specific neurons, the brain matures in biochemical function. By the second fetal month, glycolysis is present, and oxidative mechanisms are apparent during the third month. During the seventh fetal month, activity and localization of a number of enzymes occur. There appears to be a correspondence be-

Table 23–1. BRAIN DEVELOPMENT FROM THE EMBRYONIC ECTODERM

Gestational Stage	Events
1st month	Neural tube (precursor of central nervous system) has formed; closed and swellings destined to become the mid, hind and forebrain are apparent
2nd month	Paired hemispheres are formed
3rd month	Enlargement in hemispheres and thalamus; beginning of cerebellar development
4th month	Fissuration beginning with separation of temporal lobe from rest of cerebrum and development of the major divisions of cerebrum (frontal, parietal, occipital)
5th month	Development of commissures; increase in number and definition of sulci and gyri reflecting growth of cortical mantle; cell migration within the cortex largely completed by the fifth month, although may continue into early postnatal period
30 weeks to 2nd postnatal year (in certain parts of the brain)	Larger brain weight spurt involves expansion in number of neuroglial cells; these are largely oligodendroglia essential to formation of the myelin sheath; unlike neurons, these cells retain capacity to divide into adulthood; growth spurt thought to continue into 2nd postnatal year in certain parts of brain (nutritional insults affect nerve cell function because neurons surrounding satellite glial cells are metabolically interdependent)

tween the presence of certain enzymes (e.g., acetylcholinesterase) and the time when certain functions begin (Winick, 1976). The ability of the brain to synthesize protein from amino acids increases during the course of fetal life. Eight amino acids (lysine, threonine, tryptophan, methionine, phenylalanine, leucine, valine, and isoleucine) are essential to human growth, but no experimental evidence adequately documents the specific requirement of the developing central nervous system for these nutrients (Winick, 1976).

The brain is metabolically one of the most active organs of the human body. Brain oxygen utilization (50 ml. per minute) accounts for 20 per cent of the total resting utilization of oxygen. These high demands are thought to be required for the maintenance of the ionic gradients across the neuronal membrane, on which the conduction of impulses between brain neurons depends. The brain is almost exclusively restricted to glucose for energy metabolism. The billions of neurons normally depend on a continuing supply of glucose (at least 100 grams per day) to provide the energy needed. However, in starvation states the brain can use ketone bodies as a source of energy.

Neurons possess the capability to manufacture and release neurotransmitters. About 30 different chemicals are thought to be neurotransmitters. These, with the exception of acetylcholine, are amino acids or are derived from them. Only in the past 10 years have the dietary influences on the synthesis of neurotransmitters in the brain been recognized (Fernstrom and Wurtman, 1972). The rate at which neurons produce some neurotransmitters is controlled, in part, by the levels of

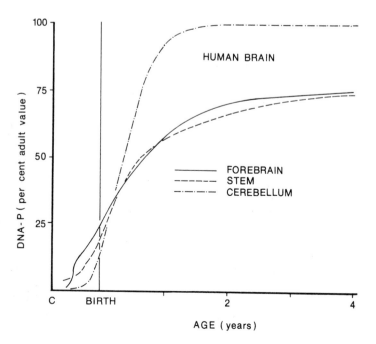

Figure 23–1. The cumulative growth schedule of different parts of the brain. Comparative values for total DNA-P, equivalent to total numbers of cells, in three human brain regions. Values have been calculated as a percentage of the adult value. (From Dobbing, J.: The later development of the brain and its vulnerability. In Davis, J. A., and Dobbing, J. [Editors]: Scientific Foundations of Paediatrics. London, William Heinemann Medical Books Ltd., 1981, pp. 744–758.)

the circulating precursors available to these neurons. The transmitters under precursor control are serotonin, the catecholamines (dopamine, epinephrine, norepinephrine), and acetylcholine (Growdon and Wurtman, 1979). Serotonin is a monoamine-5-hydroxytryptamine, which helps to induce sleep, decrease sensitivity to pain, suppress the desire for carbohydrate, and regulate temperature. Its rate of production depends on how much of its precursor, tryptophan, gets into the brain. Tryptophan competes with five other amino acids for brain entry, influenced by the effect of insulin on the levels of the other amino acids.

By a similar mechanism the availability of tyrosine affects the synthesis of the catecholamines, while choline availability in the brain directly influences the rate at which cholinergic neurons synthesize and release acetylcholine (Growdon and Wurtman, 1979). The recently acquired knowledge concerning the influence of the diet on the synthesis, storage, neuronal levels, and release of neurotransmitters is providing a new understanding of neurologic and psychiatric disorders.

It has been suggested that iron deficiency in utero or during the first two years of life can cause permanent damage through its effects on cerebral oxidative metabolism, neurotransmitter synthesis, or brain cell mitosis (Pollit and Leibel, 1976). Of importance is the question of a possible relationship between iron deficiency and alterations in behavior. Disruptive behavior in adolescents and short attention spans, along with apathy and irritability, have been reported in anemic preschool youngsters. Howell (1971) reported that three to five year olds with iron deficiency anemia displayed decreased attentiveness, narrow attention spans, and perceptual restrictions. In a study of 12 to 14 year old children from an economically deprived school, Webb and Oski (1973) noted that the anemic subjects scored lower on a standardized test of scholastic performance. From preliminary work by Pollit and coworkers (1978), it appears that adverse effects on attention and memory control processes are found, even with iron deficiency in the absence of anemia.

The hypothesis that iron deficiency may alter behavior is supported by the finding of Symes and coworkers (1969) that iron deficient rats have reduced hepatic and brain monoamine oxidase activity, and by that of Youdim et al. (1975) that platelet monoamine oxidase activity is decreased in iron deficient adults. Although Mackler and associates (1978) could not confirm the observation that the iron deficient rat has decreased brain monoamine oxidase activity, they found that the brains of these rats display decreased activity of aldehyde oxidase, a critical enzyme in the degradation of serotonin. It is evident that there are alterations in the central nervous system induced by iron deficiency. However, the exact mechanism by which iron deficiency affects central nervous system function and produces alterations in behavior is yet to be learned.

INBORN ERRORS OF AMINO ACID OR CARBOHYDRATE METABOLISM

Inborn errors of metabolism originate in the mutations of single genes that are responsible for the synthesis of a specific enzyme. Most inborn errors of metabolism share an autosomal recessive mode of transmission. This either results in a structurally altered enzyme that is incapable of normal catalytic activity or causes an inhibition of enzyme synthesis. The reduced enzymatic function produces a block in the metabolic pathway at a specific point, which in turn leads to an abnormal accumulation of substrate before the point at which the block occurs. The level of substrate, or metabolites of substrate formed by the subsidiary pathways, is increased in the blood, urine, and tissues.

The neurotransmitter system is vulnerable to impeded energy metabolism, abnormal electrolyte metabolism, and alterations of biogenic amine derivatives of amino acids and the catecholamines. Such shifts can affect neurotransmission and behavior. It is of interest that the abnormal biosynthesis of the biogenic amines can be reversed in untreated phenylketonuria by the dietary restriction of phenylalanine (Butler et al., 1981). This defect may be a partial explanation for the neuropsychiatric symptoms found in untreated phenylketonuria.

In the inborn errors of metabolism there is a wide range of clinical manifestations (Table 23–2). Treatment when available consists of controlling the substrate accumulation by the restriction or elimination of carbohydrate or protein, or in certain instances pharmacologic dosages of vitamins. Early treatment is essential in preventing the progressive loss of neurologic function, mental retardation, and other serious sequelae of metabolic disorders. For example, treatment for phenylketonuria must be started within the first month of life if normal physical and mental development is to occur (Dobson et al., 1968). There are few reports of improvement in mental development and behavior in individuals with phenylketonuria who are treated in later life (Holmgren et al., 1979). To prevent the possible toxic effects on the fetus from elevated blood phenylalanine levels in pregnant women with phenylketonuria, a phenylalanine restricted diet begun prior to conception and continued during pregnancy has been proposed (Levy et al., 1982; Pueschel et al., 1977).

Early treatment of inborn metabolic errors has been made possible by newborn screening programs, which now can be used to test for more than 20 inborn errors of metabolism. In the United States, testing for phenylketonuria is commonly mandated by state law. It has been proposed that screening tests for less common disorders may be economically justified when their cost is compared to the cost incurred when affected infants are undetected. Treatment has been improved by early identification, synthetic formulas, and special dietetic foods. Successful treatment of phenylketonuria and galactosemia are examples of the prevention of mental retardation through diet.

DEVELOPMENT AND MALNUTRITION OF THE CENTRAL NERVOUS SYSTEM

STUDY DESIGN

Because of the restriction of research design in human studies, much of our understanding of the effects of malnutrition on brain growth and development is based on animal studies. Although conclusions from animal studies can be discussed in relation to humans, concepts concerning the effects of malnutrition on specific areas of the central nervous system or specific behavior must be applied with great caution.

The anatomic and biochemical changes associated with malnutrition in an induced deficiency of a single nutrient in animals bear little resemblance to naturally occurring disorders in humans, which usually are the result of combined multiple deficiencies along with environmental influences.

Table 23–3 provides a summary of findings from various animal studies that may be useful in understanding the effects of malnutrition on the brain and behavior in humans. Instances in which parallel or analogous findings appear in the literature dealing with human malnutrition are noted.

In studying the effects of malnutrition on cognitive development in humans, the limitation of the tools for measurement and the selection of appropriate indicators of central nervous system damage cause design problems. Judgments concerning the nature and severity of the undernutrition are based on reported dietary information (e.g., 24 hour dietary recalls), biochemical measurements (e.g., hematocrit and plasma amino acid levels), and physical assessments, including anthropometric measurements and evaluation of the presence or absence of symptoms of malnutrition (edema, changes in hair texture and pigmentation, wasting). Often such judgments are imprecise or are not directly related. Easily measured indices of malnutrition are not necessarily indicative of central nervous system insult. For example, a reduction in somatic size and weight may occur

without effects on the central nervous system. Even brain size and weight and head circumference, which are often used as evidence of reduced developmental potential, do not have a one-to-one correlation with functional capacity (Dodge et al., 1975).

In human studies another design problem is that other variables associated with malnutrition are also likely to affect development and cognitive functioning. These include poverty, decreased levels of stimulation, an unstable home environment, child spacing, and infection. Because most studies have not used designs adequate for controlling these variables, many of the conclusions about the effects of malnutrition have remained uncertain (Warren, 1973). A few reports of studies using siblings as controls are available, however (e.g., Evans et al., 1980).

An additional design problem is the measurement of the behavioral sequelae of malnutrition. During periods of prolonged nutrient deprivation, abnormalities in behavior and behavioral physiology (lethargy, attention deficit, lack of motivation, and certain electroencephalographic changes) occur but do not persist after rehabilitation (Dodge et al., 1975). These time limited effects should not be confused with permanent organic brain damage caused by malnutrition.

The precision with which behavioral outcomes are measured also causes problems. Older investigations have relied upon scores from general intelligence tests. These tests are capable of detecting only the relatively large, nonspecific effects; small or focal effects are likely to be obscured in such testing. Results from recent investigations using more sophisticated measures have begun to appear (e.g., Vuori et al., 1979).

The following discussion of the relationship of malnutrition to development is divided into three parts. The first discusses evidence relating to maternal malnutrition and fetal development; the second concerns the effects of severe and moderate postnatal malnutrition; the third deals with a group of nutritional intervention studies carried out in the United States and several developing countries in the last decade.

PRENATAL MALNUTRITION

Studies of the effects of prenatal malnutrition fall into two general categories. The first includes studies in which malnutrition is evident in the mother; the second consists of inferences derived from small-for-date infants in whom malnutrition is only suspected.

The clearest example of the isolated effects of severe maternal undernutrition is the winter 1944–1945 famine in the Netherlands (Smith, 1947). At that time there was a sharp decrease in food

Table 23–2. EXAMPLES OF INBORN ERRORS OF AMINO ACID AND CARBOHYDRATE METABOLISM*

Disorder	Incidence	Biochemical Defect	Biochemical Analysis	
			Blood	Urine
Amino acids				
Classic phenylketonuria	1:10,000	Defective phenylalanine hydroxylase enzyme which prevents the conversion of phenylalanine, an essential amino acid, to tyrosine, another amino acid.	Increased phenylalanine.	Increased phenylacids: phenylpyruvic acid, phenylacetic acid, orthohydroxy-phenylacetic acid.
Classic homocystinuria	1:180,000	Defective cystathionine synthetase enzyme, which prevents the interaction between an intermediate product of the amino acid methionine with serine to form cystathionine.	Increased methionine and homocysteine.	Increased homocysteine.
Maple syrup urine disease	1:200,000	Defective oxidative decarboxylase enzymes Keto acids of the branch chain amino acids (leucine, valine, and isoleucine) are not converted to simple acids.	Increased leucine, valine, and isoleucine and their keto acids.	Increased leucine, valine, and isoleucine and their keto acids.
Hereditary tyrosinemia (hepatorenal type)	Not determined	Defect in parahydroxy-phenylpyruvic acid oxidase and perhaps other enzymes probably secondary to an as yet unknown primary defect.	Increased tyrosine and in some cases increased methionine.	Increased parahydroxy-phenylpyruvic acid; generalized aminoaciduria.
Carbohydrates				
Classic galactosemia	1:100,000	Defective galactose-1-phosphate uridyl transferase. Galactose, a monosaccharide, is not converted to glucose.	Increased galactose-1-phosphate.	Increased galactose; generalized aminoaciduria.
Hereditary fructose intolerance	Over 40 cases reported	Defect in fructose-1-phosphate aldolase. Fructose, a monosaccharide, is not converted to glucose.	Increased fructose.	Increased fructose.

*From Howard, R. B., and Herbold, N. H. (Editors): Nutrition in Clinical Care. New York. McGraw-Hill Book Co., 1982.

Table 23–2. EXAMPLES OF INBORN ERRORS OF AMINO ACID AND CARBOHYDRATE METABOLISM *(Continued)*

Clinical Symptoms	Treatment	Comment
Infant appears normal at birth, followed by hyperactivity, irritability, persistent musty odor, severe mental retardation, decreased pigmentation, and eczema, if untreated.	Phenylalanine-restricted diet: no high-quality protein foods (milk, milk products, meat, fish, poultry, eggs, nuts). Controlled amounts of fruits, vegetables, and grain products. Special formula, low in phenylalanine, provides protein and is supplemented with vitamins and minerals. Treatment monitored with blood phenylalanine levels, growth, and psychological data.	Phenylalanine must be provided in amounts sufficient to support growth. There are milder forms which may not require a diet, and there are variant forms which do not respond to treatment.
Possible mental retardation; lens dislocation; limb overgrowth; connective tissue defect leading to scoliosis, osteoporosis, vascular thrombosis; fair hair and skin.	Methionine-restricted (similar to phenylalanine-restricted) diet, supplemented with cysteine. Cysteine, a product of methionine, becomes an essential amino acid when methionine is limited. Special low-methionine formula is used, and methionine levels in the blood are monitored, along with growth and psychological functioning.	There are other forms of homocystinuria. One is responsive to high doses of pyridoxine (vitamin B_6), a coenzyme of cystathionine synthetase. Another is responsive to vitamin B_{12}, a coenzyme in the remethylation reaction of homocysteine to methionine. Prenatal diagnosis by amniocentesis is possible.
Infant appears normal at birth, with symptoms showing in the first few days. Difficulties with sucking and swallowing, irregular respiration, intermittent rigidity and flaccidity, possible grand mal seizures. Urine has the odor of maple syrup. If infant survives, mental retardation is severe if untreated.	Diet is restricted in leucine, valine, and isoleucine (similar to phenylalanine-restricted). Special formula is prepared. Blood leucine, valine and isoleucine are monitored, along with growth and psychological functioning.	There is a transient form of the disease, for which treatment is necessary only during times of illness. Prenatal diagnosis by amniocentesis is possible.
Enlargement of liver and spleen noted early in infancy; abdominal distention, liver and renal damage, vitamin D-resistant rickets.	Diet restricted in phenylalanine and tyrosine. The essential amino acid phenylalanine is a precursor of tyrosine. Methionine has also been restricted in a few individuals. A special formula is used, and blood phenylalanine, tyrosine, and methionine and growth are monitored.	Dietary treatment has not been successful in most cases. Biochemical abnormalities and renal tubular dysfunction have been corrected; however, liver disease is progressive.
Infant appears normal at birth with symptoms developing after feedings containing lactose. Symptoms include anorexia, vomiting, occasional diarrhea, lethargy, jaundice, hepatomegaly, increased susceptibility to infection. Later, cataracts and physical and mental retardation develop.	Rigid exclusion of lactose and galactose from the diet. Hydrolysis of lactose yields glucose and galactose. Diet is milk-free, free of milk products. Lactose-free formula is used. If children do not accept formula, diet will need nutrient supplementation.	Nonclassic forms include galactokinase and epimerase deficiency. The clinical features are not the same in all three forms. Prenatal diagnosis by amniocentesis is possible.
For infants, symptoms include anorexia, vomiting, failure-to-thrive, hypoglycemic convulsions, dysfunction of the liver and kidney. For older children, spontaneous hypoglycemia and vomiting occur after the ingestion of fructose.	Elimination of fructose and sucrose from the diet. Hydrolysis of sucrose yields glucose and fructose.	Differentiate from transient neonatal tyrosinemia.

Table 23–3.   A SUMMARY OF ANIMAL STUDIES ON MALNUTRITION*

Animal Studies	Analogous Human Studies	Animal Studies	Analogous Human Studies
1. Nutritional deprivation and maternal placental vascular insufficiency reduce body weight and growth of other organs more than the brain; this suggests that the brain is protected by preferential provision of nutrients	These findings resemble those reported in the literature on intrauterine growth retardation in humans (Campbell, 1976)	infection, which may, in turn, affect its development and function	
2. The timing and duration of the deprivation are the most important factors in the influence of the nutrient deficit on the brain; the earlier and the longer the period, the more profound the effects on the brain weight	None	8. The effects of malnutrition across generations (i.e., when mother and offspring are both malnourished) are greater than those of undernutrition within a single generation (Cowly and Griesel, 1963); this is presumed to be related to the effects of an inadequate intrauterine environment in fetal growth and development or to alteration of the maternal protoplasm (Dodge et al., 1975)	None
3. Cell number as measured by the total brain DNA is reduced following early malnutrition; in the majority of experiments it is the glial cells rather than the neuronal cells that are reduced; in malnutrition extending into later postnatal life, cell size is reduced	These findings are similar to those reported for marasmic infants dying of malnutrition in the first year of life (Rosso et al., 1970; Winick and Rosso, 1969)	9. Severity of the effects of malnutrition differ systematically by species; rats are the most susceptible to nutritional deprivation; in pigs, whose brain growth schedule is closest to that of humans among commonly used experimental animals (Roche, 1980), undernutrition during the fetal period has little effect; postnatal undernutrition, however, results in restriction of cellular content that is not recovered on refeeding	Primates appear to be less vulnerable to all kinds of nutritional limitations than experimental animals
4. Nutritional deprivation in juveniles (older but sexually immature animals) and adult animals results in neither cell number reduction nor histopathologic changes	None		
5. Deposition of myelin and formation of oligodendroglia is affected by nutritional deprivation in later development; myelin formation appears to be differentially affected by protein deficiency	None	10. Refeeding of pre- and postnatally malnourished animals results in catch-up growth in the brain; the degree of catch-up depends upon the severity and timing of the deprivation	Catch-up growth in brains of malnourished children is documented (Roche, 1980)
6. The cerebellum is more susceptible to nutritional insults than other areas of the brain because of its rapid growth later in development; moreover, the cerebellum appears to have the smallest capacity for recovery from nutritional deprivation upon refeeding of the animal	None	11. Augmentation of nutrients by way of reduction in litter size results in larger animals with heavier brains; functional differences between overfed and control animals have not been systemically documented	There is some evidence in human studies to indicate that supplementation results in slightly improved growth and cognitive functioning (see discussion in text)
7. Undernutrition appears to affect the development of endoneurial cells that form the brain's barrier against infectious agents; consequently restricted nutrition may render the central nervous system more susceptible to	None	12. Behavioral changes induced by undernutrition in nonprimates include delay and abnormality in the development of reflexes and motor function, abnormal electroencephalographic findings, increased emotionality and impaired learning; studies in primates have produced no systematic findings	None

*Summarized from:

Cheek, D. B.: Maternal nutritional restriction and fetal brain growth. *In* Cheek, D. B. (Editor): Fetal and Postnatal Cellular Growth: Hormone and Nutrition. New York, John Wiley & Sons, Inc., 1975.

Dodge, P. R., Prensky, A. L., and Feigin, R. D.: Nutrition and the Developing Nervous System, St. Louis, The C. V. Mosby Company, 1975.

available in certain parts of the Netherlands. At its lowest point, official food rations were only 450 calories per day per person. Most of the women included in this study had had no previous exposure to malnutrition and were living in nonpoverty circumstances. Furthermore, the famine ended as abruptly as it had begun, so that undernutrition did not extend into the postnatal period to affect the offspring. Thus, several variables usually confounded with prenatal undernutrition were fortuitously controlled. Birth records from those exposed to famine in utero were compared with those of other birth cohorts and those from the same birth cohort but from parts of the Netherlands not exposed to famine. Maternal weight, birth weight, placental weight, birth length, and birth head circumference were reduced in the famine exposed offspring. The effects on birth size were greatest if exposure occurred during the third trimester. Mental performance was assessed in a group of 19 year old men exposed to the famine in utero. There was no evidence of an effect from starvation during pregnancy on mental performance of offspring at age 19 years.

Small-for-date infants have often been assumed to be malnourished in utero, although the evidence for this is conjectural. Studies done prior to the 1960's did not usually discriminate between premature infants and small-for-date infants born at term. Thus, the neurobehavioral effects of premature birth and potential fetal malnutrition were not discriminated in these reports.

In recent years growth retarded infants determined to be of term gestational age and in whom no other etiology for growth retardation exists (e.g., infection, congenital anomaly syndromes, smoking in the mother) are suspected of suffering from malnutrition. In many of these infants a placental vascular problem plays a part in the delivery of deficient nutrients. In the remainder, no specific cause of malnutrition can be identified, although a large proportion of these small-for-date babies are offspring of mothers who are at a high "social" risk for low birth weight babies. Postmortem studies of small-for-date infants have shown reduced total brain weights and markedly diminished cerebellar weights. Head size is noted to be less affected than body length and weight in these infants, although head circumference is reduced compared to term appropriate-for-date infants (Crane and Kopta, 1979). Postnatal velocity of head growth is increased, although it is less than that typical of premature babies, and head size at three years continues to be less than the tenth percentile in a significant proportion of small-for-date infants.

Psychologic and neurologic evaluation at various ages has indicated that small-for-date babies perform less well than normal controls but better than premature babies (Dodge et al., 1975; Fan-court et al., 1976); results from some studies conflict with these findings (Babson and Kangus, 1969). In a recent investigation Lipper et al. (1981) reported that reduced neurobehavioral performance occurs only in those small-for-date infants whose head circumference is significantly reduced at birth.

POSTNATAL MALNUTRITION

In general, results concerning severe postnatal malnutrition (e.g., marasmus or kwashiorkor) are more clear-cut than those already discussed. A minority of studies have found no differences between severely malnourished and control subjects. For example, Barrera-Moncada (1963) reported the effects of kwashiorkor in a group of children between the ages of one and six years who had been nutritionally rehabilitated. He concluded that ultimate mental capacity was altered very little even though deficits in language and personal-social skills were apparent in these children. A significant number of well designed studies, however, have documented reduced intellectual performance subsequent to severe malnutrition. Stoch and Smythe (1976) found that compared with a control group matched for socioeconomic status, malnourished infants had smaller head circumferences and lower IQ scores persisting at 15 years of age. Two studies used siblings as controls in a follow-up of 37 Mexican children and 71 Jamaican children recovering from kwashiorkor (Birch et al., 1971; Hertzig et al., 1972). In both investigations a larger percentage of the index children had lower full scale WISC (Wechsler Intelligence Scale for Children) scores than their siblings. Brockman and Riciutti (1971) assessed categorization behavior in 20 marasmic and 19 control children and found persisting differences between the groups after 12 weeks of treatment. Evans et al. (1971) found significant differences on the Draw-a-Person test administered to children 10 years after the episode of severe malnutrition and to a control group of siblings born within two years of the malnourished index children. These investigators, however, did not find differences in general intelligence test scores of the malnourished children and their siblings. It is interesting that none of the reports has documented an association between the magnitude of the reduction in cognitive functioning and the timing relative to chronologic age of the nutritional insult.

Studies concerning the developmental effects of chronic moderate or marginal malnutrition have, for obvious reasons, produced less clear-cut results. Cravioto and Delacardie (1978) point out that apart from its direct effects on brain development, chronic marginal nutrition also influences mental functioning in three other ways. First, the chronically malnourished child loses learning time because of chronic or repeated illnesses. Inattentiveness, disinterest, and apathy, all associated with insufficient nutrition, may interfere with learning during periods optimal for the development of certain cognitive capacities (for example, language). Finally, the malnourished child lacks responsiveness to his environment. Personality changes may result from the pattern of ineffective interactions between the child and other persons. Differentiation of long term effects directly attributable to nutrient deficiency is thus more complicated in marginal malnutrition.

Because ascertainment is difficult, many studies have used height and weight in less than the tenth percentile as convenient measures of marginal malnutrition. Akim et al. (1956) found no differences between scores for the Gesell scales in children at a range of weight percentiles drawn from various tribes in Kampala, Africa. Using an intersensory integration test developed by Birch and Lefford (1964), Cravioto et al. (1966, 1967) and Wray (1975) showed that shorter children from rural areas in Mexico, Guatemala, and Thailand performed more poorly on tests of integrative ability than did children in higher percentiles for height. Parental size and socioeconomic status were controlled in these studies.

Botha-Antoun et al. (1968) followed a group of children who, after an uneventful early infancy, fell below the third percentile for length and weight and developed signs of malnutrition at three months. Compared to a group of matched controls from the same birth cohort, the malnourished infants walked and talked slightly though not significantly later. However, at four to five years both verbal and performance scores on the Stanford-Binet were significantly lower.

Canosa (1974) compared the performance scores in 20 malnourished children four and one-half to six and one-half years old and found differences between them and a control group of their siblings on four tests (short term memory for digits, short term memory for sentences, intentional learning, and incidental learning). Studies of mental function in children with intestinal malabsorption syndromes or other gastrointestinal abnormalities have also been used to document the effects of malnutrition. Lloyd-Still et al. (1974) tested children with various abnormalities (cystic fibrosis, ilial atresia, and protracted diarrhea) who had evidenced malnutrition but subsequently recovered. They found reduced scores on the Merrill-Palmer scale for the younger children compared with those in a control group of siblings. However, no differences were apparent in the older children tested.

NUTRITIONAL INTERVENTION

In the past two decades a series of large scale nutritional intervention studies have been carried out in New York City, Guatemala, Bogota, Columbia, and Taiwan with chronically marginally malnourished populations of pregnant women and children. The central aim in each of these programs was to improve growth, cognitive functioning, morbidity, and mortality of infants and children, and the designs for each were similar. Children and pregnant women were given calorie or protein-calorie supplements for a specified period of time. In two studies additional treatments were also evaluated. In the Guatemalan study the nutritional supplementation was accompanied by enhanced medical care. In the Bogota study education concerning child development was given to the mother.

The design measures and specific outcomes in each of the studies are summarized in Table 23–4. In all the studies the relative effects of calorie versus calorie-protein supplementation were compared. In a sense these results represent experimental evidence relating to the effect of nutrition on cognitive functioning in humans. Results from each of these studies were similar with regard to the growth parameters measured:

1. No differences in the effect of calories versus protein were noted on any growth parameter measured.

2. In each of the studies weight gain in pregnant women who received supplementation was greater than that in the control subjects.

3. The birth size of offspring of the supplemented mothers was larger in the Bogota, Guatemala, and Taiwan studies but not in the New York City study. The mean increase was 40 to 70 grams and tended to be greater in the offspring of thinner, more undernourished women.

4. A reduction in prenatal and perinatal morbidity and mortality was apparent in the Guatemala, Bogota, and Taiwan studies. A paradoxic increase in prematurity was apparent in the supplemented group in the New York City study.

From these studies of nutritional development it can be concluded that nutrient supplementation of chronically moderately malnourished infants in the prenatal and postnatal period produces modest improvements in motor, language, and attentional skills. It is likely that some of the improvement is related to amelioration of the short term effects of malnutrition. The minor improvements in cognitive functioning in the New York City and Taiwan studies suggest that the effects of supplementation may be apparent only when malnutrition is more marked.

Unfortunately the subtleties of poor nutrition are hard to measure but not without effect. For

Table 23–4. NUTRITIONAL SUPPLEMENTATION STUDIES

Study	Design Measures	Outcomes
New York City study (Rush et al., 1980)	Women weighing less than 140 pounds and with at least one other risk factor for delivering a low birth weight baby were enrolled and assigned randomly to a high calorie, high protein, or control group. At one year of age, offspring were administered the Bayley Infant Scales of Development, an object permanence, free play, and visual habituation tests.	Significant positive treatment effects for the high protein group were detected for measures of habituation, dishabituation, and length of play episodes.
Guatemala study (Freeman et al., 1980)	Four villages characterized by inadequate nutrient intake were involved in study. A high caloric supplement was made available to pregnant and lactating women and children in two villages, and a low calorie supplement was made available to the same groups in the other two. The Brazelton Assessment Scales were given to all offspring of mothers in the study, and a specially designed preschool battery with language, short term memory, perceptual, and composite subtests was administered at three, four, and five years.	The results reflect the effects of both pre- and postnatal supplementation. The nutritional supplementation had significant effects on most measures. Once an index of social status was controlled for, the greatest effect was on the language and composite measures, and the smallest was on the tests of digit memory. Children exposed to longer periods of supplementation demonstrated higher scores on some of these tests.
Bogota study (Vuori et al., 1979; Waber et al., 1981)	Pregnant women and children under seven years of age from families with evidence of malnutrition in children under five years in Bogota, Columbia, were enrolled and assigned randomly to one to six experimental groups, combining supplementation for various periods, health care, and maternal education on issues in child development. The Griffith and Einstein Scales were administered at intervals through 36 months of age. In addition, visual habituation was assessed 15 days after birth.	Nutritional supplementation effects were apparent on many subscales of the Griffiths test but particularly on motoric scales, whereas maternal education had a greater effect on language skills. Performance on the Einstein test was largely unaffected by either intervention. Supplementation effects were greatest at later ages and were contemporaneous with testing. The investigators suggested that this was evidence for indirect effects of nutritional supplementation on maturation and arousal levels. Effects, though statistically significant, were, in general, small (approximately 8 IQ points). Supplemented infants showed more mature visual habituation responses than did unsupplemented infants. For many of these outcomes, greater differences were apparent for girls.
Taiwan study (Susser, 1981)	Pregnant women with at least one male child from an area of Taiwan characterized by marginal nutritional intake were enrolled and randomly assigned to two treatment groups, one with a protein-caloric supplement and one with a low calorie supplement.	Analyses of these data are still being carried out. Item analysis of the results indicated significant supplementation effects for only one item (bring two objects together in the midline).

example, it is hypothesized that subclinical deficiences of one or more vitamins may contribute to neural tube defects. Smithells et al. (1980) studied the effects of periconceptual multivitamin supplementation in women who had previously given birth to one or more infants with neural tube defects. Folic acid was included in the preparation because of the implicated association between folate deficiency and neural tube defects. It was found that 0.6 per cent (1 of 178) of fully supplemented mothers had a child with a neural tube defect, compared with 5.0 per cent (13 of 260) in children of unsupplemented mothers. A Lancet editorial (Vitamins, Neural Tube Defects, and Ethics Committees, 1980) concluded that "Results are so shocking that they provide a strong argument for the immediate vitamin supplementation of all mothers who are at risk of having a child with a neural tube defect."

FEEDING AND DEVELOPMENT

Through feeding, the infant's biologic and emotional needs are met. The infant simultaneously experiences the comfort of a full stomach with solace and communication. Gradually the physiologic aspect of eating becomes entwined with emotions, as the feeder appeases the infant's hunger and provides interpersonal experience. These feeding interactions promote attachment and have been found to be central to the early relationship of the mother and child. The mother's behavior during feeding serves as a model for her overall behavior toward her infant (Brody, 1976). The infant's behavior during feeding also influences the attachment process. For example, the low level of responsiveness in malnourished infants elicits less response or stimulation from the mother (Pollit, 1973). Also premature or sick infants often offer misleading feeding cues that cause parents to become confused and respond inconsistently.

Feeding should be a reciprocal balance of giving and taking between the feeder and the infant. Ainsworth (1971) found that when there is a high degree of feeding synchrony between the mother and child, the subsequent attachment is strong. This synchrony develops over the first two weeks of life, as mothers learn to interpret their infant's rhythmic sucking patterns. Generally infants develop a rhythm of 10 seconds of sucking followed by a pause of four seconds. The period of sucking appears to be related to the milk flow, and the pauses are due to the mother-child interaction. Mothers who intervened during the pause period to stimulate sucking rather than allow their infant to establish their own rate of feeding were found to be intrusive (Kaye, 1977). Interestingly, the early histories of individuals with eating disturbances often fail to give evidence of gross neglect. Rather it is the subtleties of an inappropriate interaction in which the mother responds to the infant according to what she feels the infant needs, rather than the infant's cues (Bruch, 1978).

Over the years an extensive body of child development research has been devoted to the influence of the early feeding experience on development. Most theories indicate that feeding plays a role in personality development, cognition, perception, and interpersonal relationships. Although there are philosophic differences concerning the effect of feeding on development, there is agreement that the more gratifying the feeding experience, the better the effect on development. A gratifying experience involves a successful interaction between the parent and infant, synchronized with the infant's feeding ability, which goes through a marked transition during the first year of life. This transition follows the acquisition of head, trunk, gross, and fine motor control. The infant progresses from a totally dependent, somewhat passive feeder to an active participant who increasingly seeks independence. Table 23–5 summarizes the infant's feeding progression.

When parents are able to recognize their infant's feeding ability, relinquish their control, and allow them to progress accordingly, feeding problems are less likely to occur. Nevertheless this process undergoes many subtle changes and is subject to an enormous range of management issues. Usually these issues are transitory and recede as infants move on to master their next developmental level.

Infants with central nervous system dysfunction may have feeding problems from birth. Poor sucking ability, prolonged feeding time with small intakes, choking, and aspiration are common difficulties. As spoon feeding and solid foods are introduced, problems become more manifest.

The persistence of primitive reflexes and abnormal muscle tone are two characteristics of the child with central nervous system dysfunction that are commonly associated with feeding problems. The following are often identified as problem areas for which individualized solutions must be found: sitting balance; head and neck control; tongue control; sucking, swallowing, chewing, and drinking; mouth sensation (lack of mouth sensation or hypersensitivity to temperature changes or tactile stimulation); and startle reflex (a child with cerebral palsy may respond to a loud noise by a startle reflex). When observed clinically, these organic feeding problems are invariably intermixed with behavioral problems. Because these infants have lost pleasure in the function of eating, possibly because they have been gavage or force fed, there may be a neurotic superstructure around mealtimes.

Many factors influence the development of a feeding problem, and there is often difficulty in isolating the organic from the behavioral aspects. For these reasons, when there is significant difficulty, an interdisciplinary team evaluation of the infant's oral and fine motor ability is needed, along with an interactional analysis of feeding and a dietary assessment (Table 23–6).

A careful review of the diet and fluid intake should be made to determine adequacy, along with determination of the effect of the medications on the nutrient function (i.e., inhibition), taste, appetite, or feeding behavior (i.e., causing drowsiness at mealtime). Children in whom long term treatment with anticonvulsants is being carried out should be monitored in terms of the vitamin D and folate status (Hahn, 1975; Reynolds, 1974).

In the Developmental Evaluation Clinic of The Children's Hospital Medical Center, the feeding team consists of a nutritionist, physical therapist, speech pathologist, psychologist, and nurse, with a consultant pediatrician and dentist. The health care team working together, yet from different vantage points, can formulate a plan of intervention while considering the total child within the family.

Table 23–5. DEVELOPMENT OF FEEDING SKILLS

Age	Oral and Neuromuscular Development	Feeding Behavior
Birth	Rooting reflex	Turns mouth toward nipple or any object brushing cheek
	Suckle-swallow reflex	Initially sucking and swallowing are not differentiated; stimulus introduced into the mouth elicits vigorous sucking followed by a swallow if liquid is present
		Initial swallowing involves the posterior of the tongue; by nine to 12 weeks, anterior portion is increasingly involved, which facilitates ingestion of semisolid food
		Pushes food out when placed on tongue; strong the first nine weeks
	Bite	Pressure on the gums elicits a phasic bite and release
		Normal occurrence: birth to three to five months
		Retention produces biting of all objects placed in the mouth
		Interferes with mouthing activities, ingesting food, more mature biting, chewing
		By six to 10 weeks recognizes the position in which he is fed and begins mouthing and sucking when placed in this position
Three to six months	Beginning coordination between eyes and body movements	Explores world with eyes, fingers, hands, and mouth; starts reaching for objects at four months but overshoots; hands get in the way during feeding
	Learning to reach mouth with hands at four months	Finger sucking—by six months all objects go into the mouth
	Able to grasp objects voluntarily at five months	May continue to push out food placed on tongue
	Sucking reflex becomes voluntary and lateral motions of the jaw begin	Grasps objects in mitten-like fashion
		Can approximate lips to the rim of cup by five months; chewing action begins; by six months begins drinking from cup
Six to twelve months	Eyes and hands working together	Brings hand to mouth; at seven months, able to feed self biscuit
	Sits erect with support at six months	Bangs cup and objects on table at seven months
	Sits erect without support at nine months	
	Development of grasp (finger to thumb opposition)	Holds own bottle at nine to 12 months
		Pincer approach to food
		Pokes at food with index finger at 10 months
	Reaches to objects at 10 months	Reaches for food and utensils, including those beyond reach; pushes plate around with spoon; throws eating utensils; insists on holding spoon not to put in mouth but to return to plate or cup.
One to three years	Development of manual dexterity	Increased desire to feed self
		15 months—begins to use spoon but turns it before reaching mouth; may hold cup; likely to tilt the cup rather than head, causing spilling
		18 months—eats with spoon, spills frequently, turns spoon in mouth; holds glass with both hands
		Two years—inserts spoon correctly, occasionally with one hand; holds glass; plays with food; distinguishes between food and inedible materials
		Two–three years—self-feeding complete with occasional spilling; uses fork; pours from pitcher; obtains drink of water from faucet

Adapted from:
Getchel, E., and Howard, R. B.: Nutrition in development. *In* Scipien, G., and Barnard, M. (Editors): Comprehensive Pediatric Nursing. Ed. 2. New York, McGraw-Hill Book Co., 1979, p. 163.
Fetter, L.: Feeding the handicapped child. *In* Howard, R. B., and Herbold, N. H. (Editors): Nutrition in Clinical Care. New York, McGraw-Hill Book Co., 1982, p. 613.

Table 23–6. FEEDING EVALUATION OBSERVATIONS*

Oral ability
  Sucking
  Oral prehension
  Swallowing
  Breathing, swallowing coordinated
  Drooling
  Lips
  Tongue size
  Tongue thrust
  Tongue mobility
  Bite
  Munching
  Chewing
  Drinking

Oral structure
  Occlusion
  Teeth
  Caries
  Gingiva
  Oral hygiene
  Palate
  Pain on examination
  Hypersensitivity
  Teething stage

Body position
  Head control
  Sitting balance
  Placement of feet
  Usual feeding position

Hand use
  Palmar grasp
  Pincer grasp
  Opposition finger-thumb
  Hand to mouth control

Developmental feeding
Breast _____ bottle _____ weaned _____
Baby food _____ junior food _____ mashed table
  food _____ minced foods _____
Cut table foods _____ regular table foods _____

Closes hands in on bottle
Hand to mouth; sucks on fingers
Teething biscuit; holds and brings to mouth

Finger feeds
Opposes lips to rim of cup
Attempts to grasp spoon
Grasps spoon
Dips spoon in dish
Brings spoon to mouth
Holds bottle and drinks independently
Grasps cup
Raises cup to mouth
Cup-lifting, drinking, replacing
Scoops well with spoon
Feeds independently with spoon
Straw drinking
Spears with fork
Spreads with knife

Feeding environment
  Time of feedings (note how long feeding takes)
  Atmosphere of feedings (tense, pleasant, unpleasant)
  Person responsible for feeding
  Parental or caretaker's attitude toward feeding
  Past successful and unsuccessful methods
  Identify positive and negative reinforcing behaviors

Usual food intake
  24-hour recall or one or three day food diaries, depending on
  the situation

Total fluid intake

Medications
  Type
  Dosage
  Time given

Bowel concerns
  Regular
  Constipation
  Diarrhea

Anthropometric data

Clinical data

Laboratory findings

*Courtesy of Feeding Team, Developmental Evaluation Clinic, Children's Hospital Medical Center, Boston, Massachusetts.

CONCLUSION

"Nutrition is concerned directly with providing energy and ingredients necessary for cellular structure and the functioning of various metabolic systems. Indirectly, food may serve as a stimulus for behavior as well as provide a basis for social interaction" (Pal, 1974).

From this chapter it is evident that both the nutrients fed and the process of feeding are significant to development.

In general, it can be said that malnutrition has a negative influence on growth and development of the central nervous system and subsequent cognitive functioning. The critical or sensitive periods of central nervous system growth when nonrecoverable losses occur with malnutrition occur during the late prenatal life and the first one and one-half years of postnatal life.

In early postnatal life severe malnutrition has lasting effects on cognitive function. However, the magnitude of the deficit varies. The assessment of the effects of moderate chronic malnutrition is less clear. It may have effects in neurointegrative functions that need to be quantified in the future using improved testing instruments.

It is likely that disruption of glial cell division growth of dendritic trees, synaptogenesis, and deposition of the myelin sheath (all relatively late events in the development of the central nervous system) are the most important organic effects of malnutrition in humans.

ROSANNE B. HOWARD
CHRISTINE CRONK

REFERENCES

Ainsworth, M., Bell, S., and Stayton, D.: Individual differences in strange-situation behavior of one-year-olds. In Schaffer, H. R. (Editor): The Origins of Human Social Relations. New York, Academic Press, Inc., 1971, pp. 17–52.

Akim, B., McFrie, S., and Sebigajui, E.: Developmental level and nutrition (a study of young children in Uganda). J. Trio. Pediatr., 2:159, 1956.

Babson, S. G., and Kangus, J.: Preschool intelligence of undersized term infants. Am. J. Dis. Child., 117:553, 1969.

Barrera-Moncada, G.: Estudios Sobre Alteraciones del Crecimiento y del des Arrollo Psicologico del Sindrome Poluicarenceal (kwashiorkor). Caracas, Venezuela, Editora Grafa, 1963.

Birch, H. G., and Lefford, A.: Two strategies for studying perception in "brain damaged" children. In Birch, H. G. (Editor): Brain Damage in Children: Biological Aspects. Baltimore, The Williams & Wilkins Company, 1964.

Birch, H. G., Pinero, C., Alcalde, E., Toca, T., and Cravioto, J.: Relation of kwashiorkor in early childhood and intelligence at school age. Pediatr. Res., 5:579, 1971.

Botha-Antou, E., Babyan, J., and Harfouce, J. K.: Intellectual development related to nutritional status, J. Trop. Pediatr., 14:112, 1968.

Brandt, I.: Brain growth, fetal malnutrition and clinical consequences. J. Perinat. Med., 9:3, 1981.

Brockman, L. M., and Riciutti, H.: Severe protein-calorie malnutrition in infancy and childhood. Dev. Psychol., 4:312, 1971.

Brody, S.: Patterns of Mothering. New York, International Universities Press, 1976.

Bruch, H.: Eating Disorders: Obesity, Anorexia and the Person Within. New York, Basic Books, Inc., 1978.

Butler, I. S., O'Flynn, M. E., Seifert, W. E., and Howell, R. R.: Neurotransmitter defects and treatment of disorders of hyperphenylanemia. J. Ped., 98:729, 1981.

Campbell, S.: The antenatal assessment of fetal growth and development; the contribution of ultrasonic measurement. In Roberts, D. F., and Thomsen, A. M. (Editors): The Biology of Human Fetal Growth. Symposia of the Society of Human Biology. XV. New York, Taylor and Francis, Ltd., 1976.

Canosa, C.: Malnutrition and mental development in underdeveloped countries. Pädiatr. Fortbild Praxis, 37:45, 1974.

Crane, J. P., and Kopta, M. M.: Prediction of intrauterine growth retardation via ultrasonically measured head/abdominal circumference ratios. Obstet. Gynecol., 54:597, 1979.

Cowley, J. J., and Griesel, R. D.: The development of a second generation of low protein rats. J. Genet. Psychol., 103:233, 1963.

Cravioto, J., and Delacardie, E. R.: Nutrition, mental development and learning. In Falkner, F., and Tanner, J. M. (Editors): Human Growth. III. New York, Plenum Press, 1978.

Cravioto, J., Delacardie, E. R., and Birch, H. G.: Nutrition, growth and neurointegrative development. An experimental and ecologic study. Pediatrics, 38:319, 1966.

Cravioto, J., Ganoa-Espinosa, C., and Birch, H. G.: Early malnutrition and auditory visual integration in school age children. J. Spec. Educ., 2:75, 1967.

Dobbing, J.: Prenatal nutrition and neurological development. In Cravioto, J., Hombraeus, L., and Vahlquist, B. (Editors): Early Malnutrition and Mental Development. Symposium of the Swedish Nutrition Foundation. XII. Uppsala, Almqvist & Wiksell, 1974.

Dobbing, J., and Sands, J.: Vulnerability of developing brain not explained by cell number/cell size hypothesis. Early Hum. Dev., 5:227, 1981.

Dobson, J., et al.: Cognitive development and dietary therapy in phenylketonuric children. N. Engl. J. Med., 278:1142, 1968.

Dodge, P. R., Prensky, A. L., and Feigin, R. D.: Nutrition and the Developing Nervous System. St. Louis, The C. V. Mosby Co., 1975.

Evans, D., Bowie, M., Hansen, J., Moodie, A., and Vander Spuy, H.: Intellectual development and nutrition. J. Pediatr., 97:358, 1980.

Evans, D. E., Moodie, A. D., and Hansen, J. D. L.: Kwashiorkor and intellectual development. S. Afr. Med. J., 71:1413, 1971.

Fancourt, R., Campbell, S., Harvey, D., and Norman, A. P.: Follow-up study of small-for-dates babies. Br. Med. J., 1:1435, 1976.

Fernstrom, J. D., and Wurtman, R. J.: Brain serotonin content: physiological regulation by plasma neutral amino acids. Science, 178:414, 1972.

Freeman, H. E., Klein, R. E., Kagan, J., and Yarbrough, C.: Relations between nutrition and cognition in rural Guatemala. Am. J. Public Health, 67:233, 1977.

Freeman, H. E., Klein, R. E., Townsend, J. W., and Lechtig, A.: Nutrition and cognitive development among Guatemalan children. Am. J. Public Health, 70:1277, 1980.

Growdon, J. H., and Wurtman, R. J.: Dietary influences of the synthesis of neurotransmitters in the Brain. Nutr. Rev., 37:5, 129, 1979.

Hahn, T. J., et al.: Serum 25-hydroxcalciferol levels and bone mass in children on chronic anticonvulsant therapy. N. Engl. J. Med., 292:550, 1975.

Herschkowitz, N., and Rossi, E.: Critical periods in brain development. In CIBA Foundation: Lipids, Malnutrition and the Developing Brain. New York, Excerpta Medica, 1972.

Hertzig, M., Birch, H., Richardson, S., and Tizard, J.: Intellectual levels of school children severely malnourished during the first two years of life. Pediatrics, 49:814, 1972.

Holmgren, J., Blomquist, H. K., and Samuelson, G.: Positive effect of a late introduced modified diet in an 8 year old PKU child. Neuropaediatrie, 10:10, 1979.

Howard, R. B.: Nutrition in Neurological Disorders and Care of the Disabled. In Howard, R. B., and Herbold, N. H. (Editors): Nutrition in Clinical Care. New York, McGraw-Hill Book Co., 1982.

Howell, D.: Significance of iron deficiencies: consequences of mild deficiency in children: extent and meaning of iron deficiency in the United States. In Summary Proceedings of the Workshop of Food and Nutrition Board. Washington, D.C., National Academy of Science, 1971.

Kaye, K.: Toward the origin of dialogue. In Schaffer, H. R. (Editor): Studies in Mother-Infant Interaction. New York, Academic Press, Inc., 1977.

Klein, R. E., and Yarbrough, C.: Some considerations in the interpretation of psychological data as they relate to the effects of malnutrition. Sep. Arch. Latinoamericanos Nutr., 22:41, 1972.

Levy, H. L., Kaplan, G. N., and Erickson, A. M.: Comparision of treated and untreated pregnancies in a mother with phenylketonuria. J. Pediatr., 100:876, 1982.

Lipper, E., Lee, K., Bartner, L., and Grellong, B.: Determinants of neurobehavioral outcome in low birth weight infants. Pediatrics, 67:502, 1981.

Lloyd-Still, J. D., Hurwitz, I., Wolff, P. H., and Swachmann, H.: Intellectual development after severe malnutrition in infancy. Pediatrics, 54:306, 1974.

Lorenz, K.: Ker kumpan in der umwelt des Vogels. Ornithol., 83:137, 1935.

Mackler, B., et al.: Iron deficiency in the rat: biochemical studies of brain metabolism. Pediatr. Res., 12:217, 1978.

Monckeberg, F.: Effect of early marasmic malnutrition on subsequent physical and psychological development. In Scrimshaw, N. S., and Gordon, J. E. (Editors): Malnutrition, Learning and Behavior. Cambridge, The MIT Press, 1968.

Pal, B.: Malnutrition and the developing brain. J. Appl. Nutr., 26:43, 1974.

Pollit, E.: Behavior of infants in the causation of nutritional marasmus. Am. J. Clin. Nutr., 26:264, 1973.

Pollit, E., Greenfield, D., and Leibel, R.: Behavioral effects of iron deficiency among preschool children in Cambridge, Mass. Fed. Proc., 37:487, 1978 (abstr.).

Pollit, E., and Leibel, R.: Iron deficiency and behavior. J. Ped., 88:272, 1976.

Pueschel, S. M., Hum, C., and Andrews, M.: Nutritional management of the female with phenylketonuria during pregnancy. Am. J. Clin. Nutr., 30:1153, 1977.

Reynolds, E. H.: Folate metabolism and anticonvulsant therapy. Proc. R. Soc. Med., 67:6, 1974.

Roche, A. F.: Possible catch-up growth of the brain in man. Acta Med. Auxol., 12:165, 1980.

Rosso, P., Hormazabel, L., and Winick, M.: Changes in brain weight, cholesterol, phospholipid and DNA content in marasmic children. Am. J. Clinc. Nutr., 23:1275, 1970.

Rush, D., Stein, Z., and Susser, M.: A randomized controlled trial of prenatal nutritional supplementation in New York City. Pediatrics, 65:683, 1980.

Smith, C. A.: The effect of wartime starvation in Holland upon pregnancy and its product. Am. J. Obstet. Gynecol., 53:599, 1947.

Smithells, R. W., Sheppard, S., and Schorah, C. W.: Possible prevention of neural tube defects by periconceptual vitamin supplementation. Lancet, 1:339, 1980.

Stoch, M. B., and Smythe, P. M.: Fifteen year developmental study of severe undernutrition during infancy on subsequent physical growth and intellectual functioning. Arch. Dis. Child., 51:327, 1976.

Susser, M.: Prenatal nutrition, birthweight, and psychological development: an overview of experiments, quasi-experiments, and natural experiments in the past decade. Am. J. Clin. Nutr., 34:784, 1981.

Symes, A. L., et al.: Decreased monoamine oxidase activity in the liver of iron deficient rats. Can. J. Biochem., 47:999, 1969.

Vitamins, Neural Tube Defects, and Ethics Committees: Editorial. Lancet, 1:1061, May 1980.

Vuori, L., Christiansen, N., Clement, J., Mora, J. O., Wagner, M., and Herrera, M. G.: Nutritional supplementation and the outcome of pregnancy II. Visual habituation at 15 days. Am. J. Clin. Nutr., 32:463, 1979.

Waber, D. P., Vuori-Christiansen, L., Ortiz, N., Clement, J. R., Christiansen, N. E., Mora, J. O., Reed, R. B., and Herrera, M. G.: Nutritional supplementation, maternal education, and cognitive development of infants at risk of malnutrition. Am. J. Clin. Nutr., 34:807, 1981.

Warren, N.: Malnutrition and mental development. Psychol. Bull., 80:324, 1973.

Webb, T. E., and Oski, T. A.: Iron deficiency anemia and scholastic achievement in young adolescents. J. Pediatr., 82:827, 1973.

Winick, M.: Malnutrition and Brain Development. New York, Oxford University Press, 1976.

Winick, M., and Noble, A.: Cellular responses in rats during malnutrition at various ages. J. Nutr., 89:300, 1966.

Winick, M., and Rosso, P.: The effect of severe early malnutrition on cellular growth of human brain. Pediatr. Res., 3:181, 1969.

Winick, M., Rosso, P., and Brasel, J.: Malnutrition and cellular growth in the brain. CIBA Foundation: Lipids, Malnutrition, and the Developing Brain. New York, Excerpta Medica, 1972.

Wray, J.: Intersensory development in school age children at high risk of severe malnutrition during the preschool years. Unpublished manuscript, 1975.

Youdim, M. B. H., et al.: Human platelet monoamine oxidase activity in iron deficiency anemia. Clin. Sci. Mol. Med., 48:289, 1975.

24

Environmental Toxins

The concept of environmental toxicology is relatively new. An outgrowth of occupational medicine since the 1950's, it is concerned with health problems resulting from life in a technologically complex environment. The practitioner in this field necessarily must have some working knowledge of such diverse medical subjects as pathology, psychology, and epidemiology, and such nonmedical subjects as geology, ecology, economics, sociology, and political science. The pediatrician must add the disciplines of embryology and developmental medicine (Longo, 1980).

Speaking broadly, the human environment can be divided into that which is encountered as a function of normal biologic needs (e.g., drinking water, ambient air, basic foodstuffs) and that which is encountered as a function of our society (e.g., school, housing, toys, heating fuels). The former can be defined as the ecologic environment and the latter as the social environment. This distinction is of little importance in establishing the toxicity of a particular substance or the appropriate medical evaluation and management of the exposed individual, but it is of vital importance in establishing effective prevention of such encounters in the first place. It is the latter challenge that sets environmental toxicology apart from traditional medical disciplines and challenges the practitioner to reach beyond his traditional medical role.

Perhaps the most difficult responsibility of the environmental toxicologist is to weigh the potential benefits of technologic change against its potential toxic threat. Low level ionizing radiation is a case in point. There can be little doubt that the judicious use of diagnostic x-ray examination for well established medical indications far outweighs the small and, for the most part, theoretical risks to the recipient (Brent, 1980). When the use of x-ray examination at large is examined, however, such figures as an estimated 4×10^8 such examinations annually in the United States are striking (Brown et al., 1980). Skull x-ray views are widely obtained for "defensive" purposes in countless emergency rooms, and quality control in x-ray exposure of children is largely lacking. Moreover there is little transferral of previous x-ray films or communication of results from practitioner to practitioner, effectively forcing the current care giver to order further unnecessary studies. Such indiscriminate use of low level radiation poses a threat from the child's social environment. The aim of the pediatrician's intervention must lie in reducing unnecessary exposure rather than eliminating exposure in its entirety. Thus, the pediatrician must become knowledgeable about the toxic threat in its context in order to make an intelligent evaluation of its potential deleterious effects.

GENERAL PRINCIPLES

MECHANISMS OF TOXICITY

Although some environmental toxins alter the function of living tissue so directly that exposure is immediately life threatening (e.g., arsenic, organophosphates, hydrocarbons, and carbon monoxide), most produce chemical changes that may be subtle and difficult to detect without a high index of suspicion (e.g., lead, polychlorinated biphenyl compounds, asbestos, cadmium, dioxin, and low level radiation).

Mechanisms of toxicity vary widely and, depending on the degree and duration of exposure, produce clinical effects that are also disparate. Table 24–1 lists some common mechanisms of toxicity and their clinical consequences.

The interaction of toxic mechanisms with developmental changes is of particular importance to the pediatrician. This interaction may either enhance or prevent a particular clinical effect. For example, methylmercury, although generally toxic to the central nervous system, is particularly toxic to the fetal brain. Children exposed to methylmercury after birth appear to demonstrate less sensitivity to equivalent mercury levels (Amin-Zaki et al., 1978). This difference may reflect a selective distribution of methylmercury toward the fetus in placental blood. It has also been suggested that differences in fetal and adult hemoglobins may play an important role in the altered transport of mercury. In any case the stage of development and duration of exposure appear to be critical to the subsequent clinical effect, in addition to generally applicable mechanisms of toxicity.

In considering this principle, the pediatrician should be cognizant of such biologic and devel-

Table 24–1. MECHANISMS OF TOXICITY

Enzyme inhibition	Decreased growth, nerve conduction abnormality, hemolysis, nephritis
Protein precipitation	Cell necrosis, nephritis, encephalitis, pneumonitis, dermatitis
Uncoupling of oxidative phosphorylation	Malaise, weakness, hepatotoxicity
Histamine release	Edema, bronchospasm, dermatitis, hypertension
DNA breakage	Mutagenesis, carcinogenesis

opmental variables as the rate of protein synthesis, the availability of protective nutrients, the maturation of excretory mechanisms (e.g., induction of detoxifying enzymes in the liver and active transport mechanisms in the renal tubule), and the relative sensitivity of the target organ.

In general, the effects of substances that suppress protein synthetic enzymes (e.g., heavy metals) are most evident in rapidly growing tissues in which such enzymatic activity is at a high level. In more stable tissues with a slower cell turnover and lower rates of synthetic activity, such effects may be harder to detect or even clinically inconsequential.

On the other hand, substances that are toxic to transport mechanisms (hydrocarbons, metals, insecticides) or that uncouple oxidative phosphorylation (arsenic, sulfur dioxide) interfere so directly with these universal biologic activities that their effects are clinically evident almost immediately and can be assessed promptly irrespective of the age of the child or the duration of exposure.

MECHANISMS OF EXPOSURE

Biologic exposure occurs through ingestion, inhalation, and cutaneous absorption. Each of these mechanisms presents important considerations for the pediatrician.

Ingestion

Ingestion is the act of taking a substance into the gastrointestinal tract. Some substances may be excreted unabsorbed and otherwise unchanged, such as lead sulfate; others may be variably absorbed, and still others may be toxic to the gastrointestinal tract itself (iron and mercury).

Finally, some substances are chemically altered by either gastric acid or gastrointestinal flora and absorbed as different substances altogether. Ingestion tends to be the most likely mechanism of exposure to toxins in the first two years of life when oral drives are at their peak. Ecologic toxins in foodstuffs or drinking water of course can be ingested by children at any age.

Pica (the Latin word for magpie, a scavenger) is the phenomenon of ingestion of nonfood substances. Not limited to children by any means, this widespread condition contributes significantly to the risks of hazardous exposure and has been particularly implicated as a significant variable in lead poisoning. A variety of evidence has associated pica with nutritional deficiency, particularly of iron, zinc, and calcium. However, it is also observed as a cultural and emotional phenomenon. It is almost universally seen to some degree in developmentally disabled children who are institutionalized and may reflect the emotional deprivation such children suffer. Irrespective of the cause, a child with pica is an "accident looking for a place to happen," and extra care must be taken to insure a safe environment for such children.

Inhalation

Ecologic toxins do not occur commonly in forms likely to be inhaled, with the possible exception of sulfur dioxide in volcanic ash. Inhalation can occur when a substance is either volatile at room temperature (e.g., elemental mercury), is heated in a closed space without venting of fumes, or is suspended so finely in the atmosphere that it is inhaled as if it were a gas. Particulate matter laden with toxic products of combustion (e.g., sulfur dioxide, lead oxide, nitrites) can be inhaled but is more likely to be ingested on the surface of foodstuffs or, as in the case of lead, with backyard soil and household dust.

Inhalation can yield toxic effects either by irritation of the respiratory tree (e.g., formaldehyde from urea formaldehyde foam insulation) or by absorption of the toxin, generally with no chemical alteration and with relatively greater efficiency than by the gastrointestinal tract. This is particularly true in the case of heavy metals.

Important developmental variables must be considered in assessing the potential impact of an inhaled toxin. One is the diameter of the airway. If the toxin is suspended, it is possible that droplets may be sufficiently small to enter the alveoli where absorption can be highly efficient. Particulate matter, on the other hand, tends to be of such size as to be trapped in bronchi or bronchioles where absorption is less efficient. Nonetheless, if the inhaled substance is irritating (e.g., cadmium salts), local inflammation can occur with resulting absorption.

Paradoxically, if the particle size is small enough ($<1\ \mu$) particles may both enter and leave alveoli with each ventilation. In this instance the particles behave almost like a gas but without a partial pressure to produce diffusion and absorption. The net effect is to reduce the effective dose of the inhaled material.

Although airway size and tidal volume may act to reduce absorption, other variables such as body mass and increased respiratory rate, as well as smaller "dead" space, tend to have the opposite effect. The end result is that with some exceptions, inhalation dosimetry data can be applied to children and adults with equivalent accuracy.

Cutaneous Absorption

Cutaneous absorption is a less common mode of exposure. Many compounds may produce dermatitis with subsequent absorption, or may be applied as topical medication to already damaged skin. Examples are mercurous chlorides, arsenicals, and organophosphates. For a toddler with increased hand to mouth activity, the presence of a toxin on the skin simply provides another source of exposure for ingestion of the offending substance.

MECHANISMS OF EXCRETION

The urine is the usual route of excretion for most foreign substances, followed by feces (either unabsorbed or via the biliary tree), lungs, hair, sweat, and saliva. Substances may be excreted unchanged, or they may be metabolized or detoxified in the liver (usually by acetylation, methylation, oxidation, or conjugation) and the product excreted via the kidney. Rates of excretion vary widely for the same substance and may compound efforts to quantify degrees of exposure and toxicity.

In some cases normal excretory rates may be so slow that the substance in question may be undetectable in the body fluid tested despite deposition of toxic quantities in tissues. Mobilization tests are designed to demonstrate enhanced excretion under controlled conditions, allowing greater accuracy in interpretation of excretion data. Thus it is essential that the clinician know not only the route of excretion but the pattern of excretion.

Of particular importance to the pediatrician is the excretion of foreign substances in breast milk. Indeed virtually all ingested substances are excreted in breast milk to some extent (Arena, 1980). However, there are variables that have at least a qualitative influence on the final concentration of such substances. Probably the most important is the lipid solubility of the compound in question. Thus, such materials as polychlorinated diphenyl compounds, DDT, and methylmercury appear readily in breast milk because they are highly lipid soluble. Their concentration in breast milk, however, is not a direct correlate of their concentration in plasma, and the mechanisms by which specific compounds are transported into breast milk remain unclear. Other factors include the degree of

ionization at normal pH and a poorly understood relation between plasma concentration and milk flow. In general then, compounds that are lipid soluble and nonionized are most likely to appear in breast milk in significant amounts when ingested.

SPECIFIC HAZARDS

For children, life in an industrialized society includes exposure to the by-products of the technological age. Although a child's vulnerability in such exposures varies, it is fair to say that toxic effects of any specific hazard are more likely to occur in the pediatric age group. Conversely, should such toxic effects be seen, it is also probable, particularly in the case of somatic mutations resulting from radiation, that reparative mechanisms may also be more effective during childhood. Unfortunately, in cases in which that is not true, the consequence of permanent impairment makes such exposures truly tragic and, to the extent that the exposure is avoidable, negligent, or worse, criminal.

The commonest childhood hazards include the broad categories of heavy metals, air pollution, including cigarette smoke, and low level ionizing radiation. Less widespread but still common among American children is exposure to asbestos, polyhalogenated hydrocarbons, and insecticides. In addition, public attention has focused on the potential hazards from proximity to chemical "dumps." Finally, looming over this pastiche of hazards is the most ominous of all, the possibility of a nuclear accident from either a nuclear power plant or a military source.

HEAVY METALS

Lead

Lead is a metal of antiquity and, because of its ease of recovery and malleability, has been used by man for as long as recorded history. Its use is now so widespread that in North America alone an estimated nine million tons is smelted annually. Accompanying this process, an estimated 600,000 tons of lead is released into the environment, much of it carried by the prevailing winds and deposited downwind to form an everincreasing layer of leaded dust and soil, readily available for ingestion by small children (Committee on Lead in the Human Environment, 1980).

It is now common to find concentrations of lead in the soil as high as 2500 to 5000 p.p.m. (μg. per gm.). This lead "fallout," accompanied by a small but constant contribution from combusted leaded gasoline, has raised the "background" lead concentration in the population such that the recent NHANES II study by the U. S. Department of Health and Human Services demonstrated that approximately 600,000 children under the age of six have elevated blood lead levels. This should be compared to the approximately 200,000 cases of lead poisoning estimated by Dr. Jane Lin-Fu in her 1972 review of childhood lead poisoning in the United States. The difference is attributable to two factors.

First, the increasing "bombardment" from atmospheric lead has undoubtedly increased the number of exposed children many-fold. Previously the only significant routes of exposure were thought to be via the ingestion of paint or paint dust and drinking water and isolated exposures to such disparate lead sources as burned storage battery casings and lead in milk formulas from improperly soldered cans. It should be pointed out that these exposures were more than adequate to account for the incidence of lead poisoning estimated by Lin-Fu (1973) based on blood lead concentrations then thought to be acceptable.

However, the second factor accounting for the difference in incidence is that the acceptable level of lead has now been lowered from the 60 μg. per dl. level, which was associated with symptomatic plumbism, to the level of 30 μg. per dl. The latter level was recommended by the Center for Disease Control of the Department of Health, Education, and Welfare in 1978 on the basis of data correlating to levels of lead with biochemical changes in heme synthesis. These data make it apparent that enzymes in the heme synthetic pathway are sensitive to concentrations of lead at or below mean lead levels in the population at large (10 to 15 μg. per dl.; Mahaffey et al., 1982).

For example, interference with ferrochelatase (heme synthetase), the enzyme that catalyzes the insertion of ferric ion into protoporphyrin IX to produce heme, can be demonstrated at lead concentrations of 15 to 20 μg. per dl., and concomitant elevations in the protoporphyrin substrate are the result. At a blood lead level of 30 μg. per dl. or greater, there is a greater than 95 per cent chance that this elevation will be demonstrable by a finding of increased erythrocyte protoporphyrin. Thus, any child with a circulating lead level above 30 μg. per dl. can be presumed to have incurred this much interference with heme metabolism, at least.

What is most disturbing about this observation is that metalloenzymes sensitive to lead are not limited to the heme pathway. Thus, the entire cytochrome P450 system, as well as other unrelated enzyme systems such as cerebellar adenyl cyclase, has been shown to be adversely affected by lead concentrations previously believed to be acceptable (Moore et al., 1980).

Recently much attention has focused on the effects of moderately elevated lead levels on the central nervous system. No study to date has correlated sequential lead levels, at low concentrations, with neurobehavioral toxicity. Byers and Lord (1943) demonstrated that 50 per cent of the children who were neurologically symptomatic had moderate to severe sequelae. When some of these children were re-exposed to lead, severe neuropsychologic sequelae were found in 100 per cent. This study demonstrated that symptomatic children are at high risk of developing sequelae, despite adequate treatment of the acute episode. The study also pointed out the importance of preventing re-exposure at all costs. This in turn led to passage of the federal Lead Poisoning Prevention Act in 1971, which permitted large scale screening of preschool children to identify those at risk before they became symptomatic.

Over the past decade screening programs revealed areas where the incidence of asymptomatic elevated lead levels approached 30 per cent in preschool children. This in turn shifted the emphasis of behavioral research to the effects of moderately elevated levels in asymptomatic children or so-called low level lead exposure.

Although the results have not been accepted without controversy, the most important study of this question to date was performed by Needleman and his colleagues (1979) at the Children's Hospital Medical Center in Boston. Needleman obtained deciduous teeth from 2000 school children from two lower middle income communities bordering the city of Boston. After matching subjects for age, sex, and more than 30 demographic and socioeconomic variables, he performed a battery of double blind neuropsychologic tests in these youngsters and asked that their teachers respond (also double blind) to a questionnaire posing questions about the competitive behavior and adaptability of the children in school; the questionnaire is reprinted in Figure 24–1. The

Figure 24–1. Teacher questionnaire. (From Needleman, H. L., et al.: Deficits in psychologic and classroom performance of children with elevated dentine lead levels. N. Engl. J. Med., *300*:689, 1979.)

1. Is this child easily distracted during his work?
2. Can he persist with a task for a reasonable amount of time?
3. Can this child work independently and complete assigned tasks with minimal assistance?
4. Is his approach to tasks disorganized (constantly misplacing pencils, books, etc.)?
5. Do you consider this child hyperactive?
6. Is he over-excitable and impulsive?
7. Is he easily frustrated by difficulties?
8. Is he a daydreamer?
9. Can he follow simple directions?
10. Can he follow a sequence of directions?
11. In general, is this child functioning as well in the classroom as other children *his own age*?

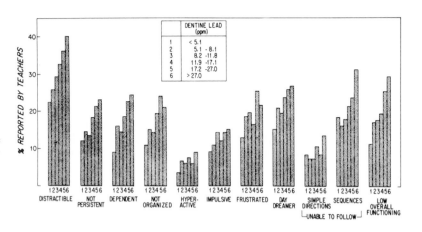

Figure 24–2. Distribution of negative ratings by teachers on 11 classroom behaviors. Dentin lead levels and teachers' ratings were completed on 2146 students. Students were classified into six groups according to dentin lead level. The relationship of negative classroom behavior to dentin lead level is seen for behavior evaluated. (From Needleman, H. L., et al.: Deficits in psychologic and classroom performance of children with elevated dentine lead levels. N. Engl. J. Med., 300:689, 1979.)

answers were then compared to levels of lead in the dentin in these children, thought to be a semiquantitative marker of lead exposure over time.

As can be seen from Figure 24–2, Needleman obtained a dose-response curve for every question asked. It has been argued that the same variables that induce children to ingest lead paint and dust may be responsible for maladaptive neurobehavioral sequelae, rather than the lead itself. Were this true, such variables would have to correlate almost precisely with the degree of lead exposure in each case to produce such a dose-response result as a coincidence. This would appear to be highly unlikely.

A more valid reservation about Needleman's data is that the exact correlation between dentin lead levels and previous blood lead levels has not yet been established. Furthermore, such important variables as age of exposure, duration of exposure, and concomitant nutritional deficiencies as iron or calcium deficiency cannot be deduced from dentin lead levels. Thus, although the impact of the asymptomatic lead burden can be acknowledged, application of these data to the individual asymptomatic child with an elevated blood lead level can only be inferred.

It would appear that studies such as Needleman's would add fuel to the demand for prompt reduction and elimination of the lead hazard throughout the country. Unfortunately, although limited funds for inspection of housing to detect lead paint hazards were provided by federal law, the funds provided only for inspection of houses in which lead poisoning cases were found (thus, in a sense, using children as a bioassay for the hazard.) Only very small amounts of money were set aside for actual abatement of the hazards, with the result that although the prevalence of childhood lead poisoning is high, the concomitant removal of the hazard is "a drop in the bucket."

For example, there are an estimated 900,000 homes in Massachusetts containing potentially toxic concentrations of lead paint (Klein, 1977). Since 1975 approximately 100,000 preschool children have been screened annually. Elevated lead levels have been found in from 8 per cent of the total in 1975 to 4.5 per cent of the total in 1980. (This drop has been attributed to the combined effect of increased awareness generated by screening programs and the reduced use of lead in gasoline.) Yet only 10,000 to 15,000 homes have been deleaded, less than 0.2 per cent of the total. Thus, to the extent that lead in paint is responsible for the incidence of childhood lead poisoning, the hazard remains relatively unabated. It requires little imagination to recognize that any reduction in surveillance will soon produce the widespread prevalence of symptomatic cases all over again.

With the increasing evidence of the adverse effects of lead even at relatively low blood concentrations has come renewed interest in the effect of lead on the fetus (Bell and Thomas, 1980). Although lead has long been thought to be abortifacient, no clear-cut epidemiologic data support this claim. There also have been reports of decreased fertility among male workers exposed to lead, but these remain to be confirmed. Animal studies suggest that lead may affect litter size and behavior as well as birth weight. Skeletal malformations have been induced with lead salts in experimental animals, but none of these data are directly applicable to humans.

In the absence of clear-cut data defining the effect of lead on the fetus, prudence suggests that pregnant women should avoid lead hazards as much as possible, particularly during the renovation of old housing. A common scenario is that of a young, possibly childless couple slowly "rehabbing" an old house, possibly rented or purchased inexpensively. Without a preschool child, they may be blissfully unaware of the hazardous concentrations of leaded paint. As they become settled, they decide to have a child, conceive, and continue their renovation work. If they know of childhood lead poisoning, they may even redouble their efforts to prepare a "lead free" environment for their expected child. However, they may be ignorant of the hazard to themselves of using such paint removal techniques as sanding or burning the old paint, thus liberating a dangerous amount of lead dust or fumes. Often such adults may become symptomatic, in which case the clinician is faced with a dilemma of whether to use chelating drugs to treat the pregnant mother. Although there is no known contraindication to chelation therapy during pregnancy, there are simply no data relating to its safety or its indications to guide the clinician. It would have been best to have avoided the exposure in the first place.

Cadmium

Cadmium is naturally found in association with zinc ores. Thus contamination of the environment has occurred largely as a function of zinc refining. Its industrial applications have included electroplating, its use in pigments in plastics and paints, and its widespread inclusion as a component metal in electronic connectors.

More widespread use of cadmium batteries increased demand for the metal to 45,000 tons annually in 1980.

Exposure usually occurs via ingestion of contaminated food stuffs or inhalation. Cigarette smoke contains relatively large amounts of cadmium (1 to 2 μg. per cigarette) with resulting inhalation of 0.1 to 0.2 μg. per cigarette. Absorption is increased in the presence of iron, calcium, or protein deficiency.

Of particular interest is the fact that, as with lead, cadmium is excreted very slowly. The half-life of cadmium in liver and kidney exceeds 10 years. Thus, at low exposures, cadmium accumulates over time. At higher exposure levels renal damage can occur, which in turn can enhance excretion.

The acute toxic effects of ingested cadmium are largely gastrointestinal, producing acute gastroenteritis and usually requiring only symptomatic management. Inhaled cadmium produces respiratory symptoms, usually after 24 hours. Inhalation of 5 mg. per cu. m. over eight hours can be lethal owing to respiratory failure caused by chemical pneumonitis or acute pulmonary edema (Friberg et al., 1974).

Chronic cadmium poisoning produces toxicity to the kidney and a syndrome characterized by osteomalacia, pathologic fractures, and bone pain (itai-itai disease). These symptoms are accompanied by disturbances in the tubular reabsorption of calcium and phosphate and alterations in vitamin D metabolism. A characteristic laboratory finding in chronic cadmium exposure is the appearance of an increased β_2 microglobulin in the urine (Piscator, 1966). Once tubular proteinuria

appears, it tends to persist even after exposure is stopped. Chronic pulmonary disease is also seen in the presence of continued inhalation.

Apparently cadmium does not cross the placenta, for newborns are virtually cadmium free at birth. Nonetheless exposure begins soon after, leading to gradual accumulation over years. Thus, in contrast to their reaction to lead exposure, small children are not likely to exhibit chronic cadmium poisoning if exposure is "low grade." However, children are subject to acute cadmium poisoning, and a number of observers have suggested that cadmium may enhance the toxicity of lead as well. In addition to the effects of chronic cadmium exposure on the kidney, an association has been supported by a number of studies between increased cadmium burden and carcinoma of the prostate. There is less convincing evidence of a similar association with respiratory cancers.

In general, the "normal" cadmium blood level for children ranges around a mean of 0.25 μg. per dl., the upper limit being 0.7 μg. per dl. Children with cadmium levels exceeding that figure should be evaluated for increased exposure to cadmium as well as to other metals.

In the absence of increased vulnerability to cadmium per se among children, it behooves the pediatrician to be aware of the long term impact of cadmium absorption via contaminated foodstuffs, and passive or active inhalation of cigarette smoke in particular. In addition, a possible role for cadmium in enhancing the toxicity of lead should add convincing weight to the argument for keeping the cadmium uptake as low as possible throughout childhood.

Mercury

Perhaps second only to lead in its ubiquity in the modern environment, mercury is a potent neurotoxin with a significant impact on man. Two fairly recent major outbreaks of congenital methylmercury poisoning with consequent neurologic deficits (Minamata Bay, Japan and Iraq) served to underscore the sensitivity of the fetal brain to organic mercury as well as the possible routes of environmental contamination, which reflect modern utilization of mercury in industry.

Man is exposed to mercury in elemental (metallic) inorganic (salts) and organic (alkyl and phenyl) compounds. The relative toxicity of these compounds and their sites of action vary depending on the form of exposure, but, as with lead, it appears that they share a common property of binding to the sulfhydryl groups of cellular proteins and consequently inhibiting cellular enzyme activity. Precise quantitation of the impact of these broad classes of mercury compounds is complicated by the fact that in vivo metabolic processes exist to convert metallic mercury to mercuric salts (oxidation) and to convert mercuric salts to methylmercury. The latter methylation is accomplished by intestinal or aquatic microbial activity and is referred to as biotransformation.

Because elemental mercury is volatile at room temperature, it is constantly being released into the biosphere from the earth's crust. Additional release occurs through burning of fossil fuels and as a by-product of industrial utilization. Approximately 10,000 tons of mercury are produced annually for use in the manufacture of chloralkalis, electrical equipment such as batteries, paints, and pigments, thermometers, fungicides and pesticides, dental amalgams, and cosmetics and pharmaceuticals. Organic mercuric compounds have been widely used as seed treatments and fungicides with consequent toxic exposures. Recent efforts to restrict this use has produced a decrease in production of alkylmercury compounds (Committee on Lead in the Human Environment, 1980).

Children can be poisoned by all three forms of mercury. Although ingested metallic mercury is generally not thought to be absorbed from the gastrointestinal tract, there is evidence that because of its volatility at body temperature, mercury vapor can be released even after ingestion of metallic mercury with consequent uptake (Berlin, 1979). In addition, small amounts of metallic mercury may be absorbed if the bowel is inflamed. Further, several recent case reports have documented release of toxic amounts of mercury vapor from infant incubators in which thermometers have broken, releasing the metallic mercury into the body of the incubator (McLaughlin et al., 1980; Waffarn and Hodgman, 1979).

Although it is generally true that exposure to the metallic form of mercury is the least likely to occur or produce damage if it does occur, its importance lies in the fact that metallic mercury is lipid soluble until it is oxidized in the red blood cell, and a portion of absorbed elemental mercury can therefore both penetrate the blood-brain barrier and cross the placenta. Thus, exposure to mercury vapor, although usually an isolated specific event, can produce both acute and chronic as well as congenital toxicity.

In contrast to the foregoing, exposure to mercuric salts is usually productive of potentially severe local inflammatory reactions. On the other hand, mercuric salts do not pass the placenta nor do they readily cross the blood-brain barrier. Probably the most important childhood manifestation of this condition is acrodynia or pink disease. A hypersensitivity reaction to cutaneously applied mercuric salts, the syndrome is characterized by a generalized rash with swelling of the extremities and face and is associated with irritability, photophobia, and profuse perspiration sometimes producing dehydration. Desquamation ensues. The syndrome responds well to reduction of mercury exposure and chelation with N-acetyl penicillamine or dimercaprol.

Because mercuric salts accumulate in the renal tubular epithelium, acute and chronic exposure can produce renal damage, with the nephrotic syndrome as a consequence.

Organic mercury exposure remains the most threatening to children because of both widespread environmental contamination and the toxicity of organic mercury to the developing brain. Because of the phenomenon of biotransformation, any source of inorganic mercury can become an indirect source of organic mercury via the food cycle, particularly seafood. In Minamata Bay, Japan, a plant utilizing mercuric salts in the manufacture of vinyl chloride discharged waste mercury into the adjoining bay. Biotransformation by aquatic protozoa and bacteria converted the inorganic mercury to methylmercury. Ingestion of large amounts of fish recovered from the bay produced chronic mercury poisoning, which had initially baffled medical authorities in the area until the mercury source was identified (Berglund et al., 1971). Unlike other forms of mercury, alkyl compounds produce a predominantly neurologic toxicity, including paresthesias, ataxia, visual disturbances including blindness, and deafness. Obtundation can follow, sometimes associated with coma. A particularly devastating feature of this episode was its impact on pregnant women and their fetuses. As a result, almost 30 per cent of the children born during the height of the epidemic had moderate to severe developmental disabilities.

Perhaps the worst episode of organic mercury poisoning occurred in the early 1970's in Iraq (Bakir et al., 1973). Rural farmers used seed grain treated with methylmercury as a fungicide to bake bread. Organic mercury poisoning struck thousands of people, and hundreds died. A high incidence of stillbirths or affected infants was observed. Further poisoning by the postnatal route of breast milk occurred but with a less profound effect. In general, the Iraq experience demonstrated the relative sensitivity of the fetus to organic mercury compared to the child and adult (Amin-Zaki et al., 1978).

The consequence of these epidemics has been to increase surveillance of methylmercury contamination of the food chain and to reduce the use of methylmercury as a fungicide and pesticide. Other widely used organic mercurial compounds, primarily phenylmercury and methoxyethyl mercury, are unstable in mammalian tissues, are rapidly degraded in the liver to inorganic mercury, and do not cross the blood-brain barrier or the placenta.

Thus, a particular set of attributes (methylation by common organisms, lipid solubility, and a long half-life of 70 to 110 days) combine to make methylmercury one of the most toxic heavy metals in man's environment. It is, unfortunately, both an ecologic and social toxin, but it seems fair to say that were it not for its industrial use, the high concentration found in Minimata Bay would be unlikely in the ecosystem.

OTHER METALS

Asbestos

A cluster of cases of mesothelioma of the pleura in 33 adults was investigated in the late 1950's in South Africa. Exposure to the mining of asbestos during childhood was a factor common to the group (Wagner et al., 1960). Some were exposed via direct contact with the mines; others were exposed because parents or older siblings who worked in the mines brought home asbestos dust on their clothing. Subsequent animal studies have shown that the size and shape of asbestos fibers are apparently the carcinogenic variables (Miller, 1978). Of importance, asbestos induced mesotheliomas require decades to develop but not decades of exposure. Indeed mesotheliomas of children and adolescents are almost certainly not due to asbestos (Grundy and Miller, 1972).

Selikoff and his colleagues (1975) have also shown that exposure to asbestos increases the probability of developing lung cancer from cigarette smoking. In fact, the combination of asbestos exposure and cigarette smoke is 100 times more carcinogenic than exposure to either substance.

Thus, as with cadmium exposure, asbestos exposure in childhood predisposes to serious illness in later life and should be rigorously avoided.

Arsenic

Acute arsenic exposure is most frequently associated with contaminated drinking water. Also a variety of forms of arsenic pesticides still abound in parts of the world and can be a source of accidental exposure.

Chronic exposure can also occur from contaminated drinking water; arsenic bearing rock strata can release arsenic into the ground water supply, thus producing a true "ecologic" toxin (Harrington et al., 1978). As expected, however, such exposures are not so severe as those caused by airborne arsenic from smelters, particularly those used to refine gold and copper ores (Baker et al., 1977; Landrigan, 1981; Milham and Strong, 1974).

Although the acute syndrome is usually manifested by severe gastrointestinal disease, chronic arsenic exposure affects the skin, peripheral nervous system, liver, blood, and lungs. The chronic skin changes associated with arsenic exposure in childhood, namely, hyperpigmentation and hyperkeratoses, predispose to squamous cell carcinoma, basal cell carcinoma, and Bowen's disease in later life.

One of the most severe episodes of subacute arsenic poisoning among children was reported from Japan where 12,000 infants drank powdered milk contaminated by arsenic (cf. Fowler et al., 1979). One hundred thirty infants died after developing a syndrome including pigmentation, liver swelling, and anemia. Of note, a number of survivors had learning disabilities and impairment of hearing in a follow-up study.

As with mercury, the toxicity of arsenic depends on the compound and the route of exposure as well as such other variables as the dose and duration of exposure.

Arsenic is a highly toxic metal. It has been shown to be carcinogenic and teratogenic in animal studies (Hood et al., 1977). There is also evidence of chromosome damage, although the clinical significance of this finding is not clear.

Fortunately arsenic poisoning can be treated with chelating agents if it is recognized in time. Nonetheless, given its widespread distribution throughout the ecologic environment as well as in man's industrial environment, both acute and chronic poisoning remain prevalent.

POLYHALOGENATED HYDROCARBONS (POLYBROMINATED AND POLYCHLORINATED BIPHENYL COMPOUNDS, DIBENZODIOXINS, HEXACHLOROPHENE)

Polyhalogenated biphenyls are fat soluble, heat stable, and nonbiodegradable. These compounds were developed for use as pesticides, flame retardants, and paint additives and as plasticizers in the electronics industry. Polyhalogenated biphenyl compounds and their highly toxic contaminants, the dibenzofurans and dibenzodioxins, have found their way into our environment by virtue of their widespread use and improper disposal. By 1972 virtually every major river in the United States had been found to be contaminated by polychlorinated biphenyl compounds. Once in water, the compounds either are ingested by fish or contaminate drinking water directly. Ingested by fish, they remain chemically and biologically intact and are eaten as part of the food chain. Thus, although manufacture of polychlorinated biphenyl compounds has been discontinued in the United States, their presence in the environment persists because they are not degraded.

Both polychlorinated and polybrominated biphenyl compounds have been shown to accumulate in fatty tissue because of their high lipid solubility. Excessive levels of polybrominated biphenyl compounds have been reported in the breast milk of women living in areas where large amounts of fish are consumed. Studies in Japan (Kodana and Ora, 1980) and in the United States (Wickizer and Brilliant, 1981) indicate that nursing infants can accumulate significant amounts of these compounds if their mothers eat large quantities of contaminated fish. Reports from Osaka prefecture in Japan suggest that breast milk may contain up to 15 times as much polychlorinated biphenyl compound as plasma (Yakushiji et al., 1979). Studies in Michigan of polybrominated biphenyl compounds in the Great Lakes confirm the significant elevation in polybrominated biphenyl compound intake when nursing mothers eat contaminated "bottom" fish more than once weekly. However, to date none of these studies has shown adverse health effects in the population studied (Barr, 1978; Kimbrough, 1980).

Significant human toxicity from these compounds has been observed in several instances, the most widely studied of which is the 1968 contamination of rice oil in Japan by the polychlorinated biphenyl, Kanechlor, as well as some polychlorinated dibenzofurans (used as organochlorine pesticides). Acute and subacute poisoning (yusho) from this episode has been characterized by marked skin change and ocular, neurologic, endocrinologic, and respiratory changes (Kuratsune, 1980). Table 24–2 shows the frequency of early symptoms seen during this episode. Almost 1600 patients were examined subsequently. As of 1973, 22 patients had died, of whom 41 per cent had malignant tumors (Urabe, 1974). Infants born to affected patients tended to be small for gestational age and showed a marked decrease in melanin in the epidermal basal cells as evidenced by an unusually dark pigmentation.

Analysis of the yusho experience is complicated by the fact that the rice oil contamination contained not only polychlorinated dibenzyl compounds but also the highly toxic polychlorinated dibenzofurans. Thus, correlation of symptoms with polychlorinated biphenyl compound levels is confounded. Recent studies of tissues from these patients have revealed some polychlorinated quaterphenyl compounds as well, further confusing the issue.

Recent public attention has been focused on one of the most toxic of this family of compounds, 2, 3, 7, 8-tetrachlorodibenzodioxin (TCDD) or dioxin. This compound is frequently found as a contaminant of the widely used herbicide, 2, 4, 5T (2, 4, 5-trichlorophenol), which, with 2, 4D (2,4-dichlorophenoxyacetic acid), is the principal ingredient of agent orange. The latter mixture was sprayed extensively as a defoliant during the Vietnam War and contained considerable amounts of TCDD, although the actual amount reaching the forest floor after spraying is not known.

Three episodes of exposure to TCDD by humans (Missouri, Seveso, Italy, and Vietnam) have been studied, but there have been only moderate gains in our understanding of the human health effects of this compound (Reggiani, 1980). The finding of chloracne appears to be a sign of toxicity common to dioxin and polychlorinated biphenyl compounds as well.

Some other toxic effects noted in humans include abnormal liver function test results, elevated cholesterol and triglyceride levels, porphyrinuria, and sensory neuropathy. Animal studies have raised questions regarding the effects of these compounds on reproduction, mutagenesis, and cell mediated immune response.

In 1971 there occurred a particularly tragic episode of accidental hexachlorophene poisoning in France, only recently reported because of legal constraints due to litigation (Martin-Bouyer et al., 1982). Two hundred four children were exposed percutaneously by use of a baby powder containing 6.3 per cent hexachlorophene. Thirty-six infants died. Table (24–3) shows the distribution of signs and symptoms in these infants. Two points are of particular interest to the pediatrician.

Table 24–2. FREQUENCY OF EARLY SYMPTOMS COMPLAINED OF BY YUSHO PATIENTS*

Symptoms	Males (n = 89)	Females (n = 100)
Dark-brown nail pigmentation	83.1	75.0
Distinctive hair follicles	64.0	56.0
Increased sweating at the palms	50.6	55.0
Acneform skin eruptions	87.6	82.0
Red plaques on the limbs	20.2	16.0
Itching	42.7	52.0
Pigmentation of the skin	75.3	72.0
Swelling of the limbs	20.2	41.0
Stiffened soles of the feet and palms of the hands	24.7	29.0
Pigmented mucous membrane	56.2	47.0
Increased eye discharge	88.8	83.0
Hyperemia of the conjunctiva	70.8	71.0
Transient visual disturbance	56.2	55.0
Jaundice	11.2	11.0
Swelling of the upper eyelids	71.9	74.0
Feeling of weakness	58.4	52.0
Numbness of the limbs	32.6	39.0
Fever	16.9	19.0
Hearing difficulties	18.0	19.0
Spasms of the limbs	7.9	8.0
Headaches	30.3	39.0
Vomiting	23.6	28.0
Diarrhea	19.1	17.0

*From Kuratsune, M.: Yusho. In Kimbrough, R. D. (Editor): Halogenated Biphenyls, Terphenyls, Naphthalenes, Dibenzodioxins and Related Products. Amsterdam, Elsevier/North Holland Biomedical Press, 1980.

Table 24–3. DISTRIBUTION OF SYMPTOMS AND SIGNS IN 224 HEXACHLOROPHENE POISONING EPISODES AMONG 204 CHILDREN*

Symptoms and Signs	No. (%)
Systemic and skin features	
Erythema of buttocks	209 (93)
Other cutaneous signs	38 (17)
Fever	99 (44)
Vomiting	77 (34)
Refusal of food	75 (33)
Diarrhea	65 (29)
Neurological features	
Drowsiness	83 (37)
Irritability	75 (33)
Coma	55 (25)
Seizures	39 (17)
Babinski signs	24 (11)
Decerebration	22 (10)
Weakness or paralysis	17 (8)
Opisthotonus	9 (4)

*From Martin-Bouyer, G., et al.: Outbreak of accidental hexachlorophene poisoning in France. Lancet, 1:91, 1982.

First, the affected batch of talcum comprised 2989 cans, of which only 199 (7 per cent) were subsequently recovered. Although the first cases of encephalopathy occurred in March 1972, the cause was not ascertained until August, during which time the talc cans were being sold and distributed. Because the first sign of hexachlorophene toxicity is a rash in the diaper area, it would have been a natural reaction to utilize more of the powder to control the rash, thus compounding the problem.

Second, there were 224 episodes among 204 individuals. Several of the fatal cases occurred among the re-exposed individuals, emphasizing the importance of factory recall procedures and the need for widespread dissemination of information regarding environmental toxins.

Finally, although TCDD is a known contaminant of hexachlorophene, the toxic events observed in this epidemic were directly due to the toxicity of hexachlorophene, not to dioxin. Long term studies of delayed and chronic health effects are underway in the affected survivors.

PESTICIDES

Among the most widespread environmental contaminants, the chlorinated hydrocarbons, including DDT, chlordane, heptachlor, and dieldrin, were extensively used until a few years ago. Two factors have altered their use. The first was the recognition of the effect of indiscriminate use of such insecticides on the ecosystem. Increasing public pressure has created a climate of concern sufficient to increase governmental surveillance and require that firms, agencies, or municipalities justify their use of any pesticide in the public domain. The second has been the finding of specific toxicity of these compounds in experimental animals and humans. Since the chlorinated hydrocarbons have been found to be carcinogenic in rodents, many of them have been removed from the market entirely. Sadly, however, this was not done until widespread ecologic damage had been done by the poisoning not only of insects but of fish, birds, and other wildlife.

More recently other compounds, namely, the organophosphates, have received increasing use. Theoretically these compounds have the advantage of being more selectively toxic to insects and of not lingering in the environment by virtue of more rapid metabolism and excretion. Certain of them, such as parathion, are highly toxic nonetheless. The common feature of this group of insecticides is inhibition of cholinesterase, which produces neurotoxicity and can be irreversible in some cases if not treated. A specific antidote (2-PAM) used with atropine can reverse the toxic effects if they are correctly diagnosed.

Compounds thought to be only mildly toxic in humans (e.g., malathion) are presently being used widely to spray crops. Although some effort has been made to restrict the "re-entry" times of workers to fields following spraying, no data specifically applicable to children exist regarding exposure and the toxicity of these compounds. Despite the clear violation of child labor laws, children, particularly of migrant workers, frequently work as field hands and may be exposed to toxic amounts of these compounds as a result of the application of re-entry restrictions based only on adult data.

Another group of insecticides are the carbamates, such as aldecor. They also inhibit cholinesterase but not irreversibly. However, some data suggest that they can be converted to the highly carcinogenic nitrosamines by soil bacteria. Others may be mutagenic in vitro.

As suggested by Kimbrough (1980), the fact that most of these compounds are petrochemicals may have an economically inhibiting effect on their use. However, the diseases they create—dermatitis, neuropathy, and even diabetes—are commonly found in the population at large and may go unrecognized as toxic effects.

EPILOGUE

This selected examination of environmental toxic threats to children merely scratches the surface. Elsewhere in this text are other examples of this phenomenon, including the potential genetic effects of cigarettes and ionizing radiation.

In many instances, such as that of atmospheric air pollution, the impact on children is difficult to separate from other population health effects. Variations in susceptibility as well as geographic and other epidemiologic factors make clinical studies more difficult. Except in patients with known respiratory disease, symptoms are difficult to quantify, making a true dose-response curve difficult to construct. Nevertheless, the continued bombardment of the environment by toxic products of combustion, such as sulfur dioxide, nitrous dioxide, and carbon monoxide, undoubtedly produces adverse health effects on some if not all of those exposed. Given the impact of respiratory illness on overall pediatric morbidity, it makes sense to encourage widespread data collection to confirm the potential effects of these known toxins under conditions of pediatric rather than adult exposure.

A special situation is the exposure to urea formaldelhyde foam insulation. Long known to be toxic to tissues as 30 per cent formalin solution, the odor of formaldehyde gas can be detected in the environment at concentrations as low as 1 to 2 p.p.m. A respiratory irritant, it can produce discomfort at these concentrations without previous sensitization. However, some individuals may respond to formaldehyde at concentrations well below 1 p.p.m. This "hypersensitivity" is not well understood. Thus, concentrations capable of producing symptoms can be liberated from the foam insulation under otherwise "acceptable" conditions. The challenge to the clinician lies in sorting out unrelated respiratory symptoms from those due to exposure from this respiratory irritant. This can be accomplished by challenging the patient under controlled conditions with minute concentrations of the potential toxin and measuring the effects on respiratory physiology. However in young children this is impractical owing to the need for their cooperation.

The Commonwealth of Massachusetts addressed this problem by placing the burden of proof on the insulation manufacturer and installer. Patients need only document the temporal presence of symptoms that could be related to formaldehyde to receive a directive from the Department of Public Health requesting removal of the insulation at the expense of the manufacturer. Recently this regulation was challenged in a state court, and a stay was granted by the Superior Court on the grounds that there was insufficient evidence supporting the toxicity of formaldehyde at low concentrations and requiring further study. In the interim, the United States Consumer Product Safety Commission has determined formaldehyde to be carcinogenic in experimental animals and has forbidden its further installation as insulation throughout the country. Litigation regarding steps requiring its removal is still in process.

The foregoing example is intended to point out the enormous complexity facing policy makers who must address the risks and benefits of potentially toxic environmental exposures.

Among the most difficult is the dilemma of nuclear power. Originally touted as a safe and inexpensive alternative to the burning of coal (air pollution) or oil (cost, limited supply), the use of nuclear power has raised increasing concern among the medical and lay community. No solution has been found for the safe disposal of radioactive waste, and the accident at the plant at Three Mile Island exposing the radioactive core to potential "meltdown" with possible release of radioactivity into the environment did little to calm skeptics about the "safety" of this form of energy. Indeed the proliferation of these plants increases the likelihood of a serious accident, raising the risk-benefit ratio beyond acceptable limits.

What role can the pediatrician play in helping patients cope with these questions? First, increased awareness—the "alert clinician" is essential to successful intervention on behalf of specific patients. Lead poisoning mimics many diseases and will escape diagnosis if the diagnosis is not entertained in the first place. Knowledge and experience can allow a pediatrician to take a short but informative environmental history on each child seen. Specific emphasis should be placed on parents' occupations and dietary history as well, to avoid missing the "fouled nest" syndrome. Toxic exposure should be thought of in any otherwise unexplained constellation of symptoms.

Second, effective advocacy for the child does not stop there. The pediatrician should inform colleagues and public health authorities of cases of toxic exposure to permit a pattern of exposure, if any, to be uncovered. Only through adequate reporting of disease can such phenomena as leukemia clusters be recognized.

Third, the pediatrician must examine the exposure from the special perspective of those who care for the developing organism. Thus levels of exposure acceptable in adults must not be assumed to be acceptable in children or infants. Special consideration must be given to altered renal and hepatic handling of substances and different dietary habits.

Fourth, the pediatrician should familiarize himself with the appropriate state or federal agency under whose jurisdiction a toxic exposure may fall. Local public health authorities may be loath to call attention to a problem for fear of spreading undue alarm. Laboratory facilities or investigators may be insufficient to the task of investigating such complex health hazards as toxic waste dump sites or polychlorinated biphenyl compounds in a polluted habor.

Agencies presently empowered to investigate environmental hazards include the Environmental

Section of the Center for Disease Control (medical evaluation), the Environmental Protection Agency (environmental evaluations), and the National Institutes of Occupational Safety and Health (NIOSH; workplace safety). There can be much overlap, and occasionally an issue such as a fungicide used in wheat may also concern the United States Department of Agriculture or the Food and Drug Administration. Finally, any publicly sold product may be under the jurisdiction of the Consumer Product Safety Commission. Thus, in a recent case examining the hazards of lead decorations on glassware intended as a promotional campaign for children, three agencies (EPA, FDA and CPSC) formed a task force to consider the issue.

Finally, the pediatrician occasionally must be a community advocate for children. It may be necessary for the pediatrician to interpret scientific data to the community to explain specific hazards. It may be useful to promote legislation when such hazards are poorly controlled. Certainly the passage of the federal Lead Poison Prevention Act was greatly assisted by documentation of the vast pediatric experience with this disease and its toxic effects.

It would appear that our society has begun to examine seriously the potential damage of ostensibly beneficial technologic change. Increased sensitivity to this issue means that our children can reap the benefits of current scientific contribution without paying the price for its excesses.

JOHN W. GRAEF

REFERENCES

Amin-Zaki, L., Majeed, M. A., Clarkson, T. W., and Greenwood, M. R.: Methylmercury poisoning in Iraque children: clinical observations over two years. Br. Med. J., 11:613, 1978.

Arena, J. M.: Drugs and chemicals excreted in breast milk. Pediatr. Ann., 9:10, 1980.

Baker, E. L., Jr., Hayes, C. G., Landrigan, P. J., Handke, J. L., Leger, R. T., Housworth, W. J., and Harrington, J. M.: A nationwide survey of heavy metal absorption in children living near primary copper, lead and zinc smelters. Am. J. Epidemiol., 106:261, 1977.

Bakir, F., Damluji, S. F., Amin-Zaki, L., Murtadha, M., Khalidi, A., Al-Rawi, N. Y., Kriti, S. T., Dhahir, H. I., Clarksen, T. W., Smith, J. C., and Doherty, R. A.: Methylmercury poisoning in Iraq. Science, 181:230, 1973.

Barr, M., Jr.: Pediatric health aspects of PBBS. Scientific aspects of polybrominated biphenyls. Environ. Health Perspect., 23:291, 1978.

Bell, J. V., and Thomas, J. A.: Effects of lead on mammalian reproduction. In Singhal, R. L., and Thomas, J. A. (Editors): Lead Toxicity. Munich, Urban und Schwarzenberg, 1980.

Berglund, F., Berlin, M., Birke, G., Cedarlof, R., von Euler, V., Friberg, L., Holmstedt, B., Jonsson, E., Luning, K. G., Ramel, C., Skerfving, S., Swensson, A., and Tejning, S.: Methylmercury in fish, a toxicologic-epidemiologic evaluation of risks. Report from an expert group. Nord. Hugien. Tids., Suppl. 4, 1971.

Berlin, M.: Mercury. In Friberg, L., Nordberg, G. F., and Vouk, V. B. (Editors): Handbook on the Toxicology of Metals. Amsterdam, Elsevier/North Holland Biomedical Press, 1979.

Brent, R. L.: X-ray microwave and ultrasound: the real and unreal hazards. Pediatr. Ann., 9:43, 1980.

Brown, R. F., Shaver, T. W., and Lamel, D. A.: Dose Levels in Diagnostic X-ray Examinations in the Selection of Patients for X-ray examinations. Washington, D.C., U.S. Department of Health, Education and Welfare, 1980.

Byers, R. K., and Lord, E. E.: Late effects of lead poisoning on mental development. Am. J. Dis. Child., 66:471, 1943.

Committee on Lead in the Human Environment: Lead in the Human Environment. Washington, D.C., National Academy of Sciences, 1980.

Fowler, B. A., Ishinishi, N., Tsuchiya, K., and Vahter, M.: Arsenic. In Friberg et al. (Editors): Handbook on the Toxicology of Metals. New York, Elsevier North-Holland, Inc., 1979, pp. 293–313.

Friberg, L., Piscator, M., Nordberg, G. F., and Kjelstrom, T.: Cadium in the Environment. III EPA 650/2–75–049 Washington, D.C., U.S. Government Printing Office, 1974.

Grundy, G. W., and Miller, R. W.: Malignant mesothelioma in childhood. Report of 13 cases. Cancer, 30:1216, 1972.

Hammond, P. B.: Metabolism of lead. In Chisolm, J. J., Jr., and O'Hara, D. M. (Editors): Lead Absorption in Children. Munich, Urban und Schwarzenberg, 1982.

Harrington, J. M., et al.: A survey of a population exposed to high concentration of arsenic in well water in Fairbanks, Alaska. Am. J. Epidemiol., 108:377, 1978.

Hood, R. D., Thacker, G. T., and Patterson, B. L.: Effects in the mouse and rat of prenatal exposure to arsenic. Environ. Health Perspect., 19:219, 1977.

Johnson, D., Kubic, P., and Levitt, C.: Accidental ingestion of vacor rodenticide; the symptoms and sequelae in a 25 month old child. Am. J. Dis. Child., 134:161, 1980.

Kimbrough, R. D.: Occupational exposure. In Kimbrough, R. D. (Editor): Halogenated Biphenyls, Terphenyls, Naphthalenes, Dibenzodioxins and Related Products. New York, Elsevier North-Holland, Inc., 1980, pp. 373–399.

Kodana, H., and Ora, H.: Transfer of polychlorinated biphenyls to infants from their mothers. Arch. Environ. Health, 35:95, 1980.

Klein, R.: Lead poisoning. In Barness, L. A. (Editor): Advances in Pediatrics: Chicago, Year Book Medical Publishers, Inc., 1977, Vol. 24.

Kuratsune, M.: Yusho. In Kimbrough, R. D. (Editor): Halogenated Biphenyls, Terphenyls, Naphthalenes, Dibenzodiozins and Related Products. Amsterdam, Elsevier/North Holland Biomedical Press, 1980.

Kuratsune, M., Yoshimura, T., Matsuzaka, J., and Yamaguchi, A.: Epidemiologic study on yusho, a poisoning caused by ingestion of rice oil contaminated with a commercial brand of polychlorinated biphenyls. Environ. Health Perspect., Experimental Issue No. 1, 119, 1972.

Landrigan, P. J.: Arsenic: state of the art. Am. J. Indust. Med., 2:5, 1981.

Lin-Fu, J. S.: Undue absorption of lead among children—a new look at an old problem. N. Engl. J. Med., 286:702, 1972.

Lin-Fu, J. S.: Vulnerability of childen to lead exposure and toxicity. N. Engl. J. Med., 289:1229, 1289, 1973.

Longo, L. D.: Environmental pollution and pregnancy: risks and uncertainties for the fetus and infant. Am. J. Obstet. Gynecol., 137:2, 162, 1980.

Mahaffey, K. R., Annest, J. L., Roberts, J., and Murphy, R. S.: National estimates of blood lead levels: United States, 1976–1980; associated with selected demographic and socioeconomic factors. N. Engl. J. Med., 307:573, 1982.

Martin-Bouyer, G., Lebreton, R., Toga, M., Stolley, P. D., and Lockhart, J.: Outbreak of accidental hexachlorophene poisoning in France. Lancet, 1:91, 1982.

McLaughlin, J. F., Telzrow, R. W., and Scott, C. M.: Neonatal mercury vapor exposure in an infant incubator. Pediatrics, 66:88, 1980.

Milham, S., Jr., and Strong, T.: Human arsenic exposure in relation to a copper smelter. Environ. Res., 7:176, 1974.

Miller, R. W.: Environmental causes of cancer in childhood. In Barness, L. A. (Editor): Advances in Pediatrics. Chicago, Year Book Medical Publishers, Inc., 1978, Vol. 25.

Moore, M. R., Meredith, P. A., and Goldberg, A.: Lead and heme biosynthesis. In Singhal, R. L.,and Thomas, J. A. (Editors): Lead Toxicity. Munich, Urban und Schwarzenberg, 1980.

Needleman, H. L., Nunnoe, C., Leviton, A., Reed, R., Peresie, H., Maher, C., and Barrett, P.: Deficits in psychologic and classroom performance of children with elevated dentine lead levels. N. Engl. J. Med., 300:689, 1979.

Piscator, M.: Proteinuria in Chronic Cadmium Poisoning. Stockholm, Beckmans Bokförlag, 1966.

Selikoff, I. J., and Hammond, E. C.: Multiple risk factors in environmental cancer. In Fraumeni, J. F., Jr. (Editor): Persons at High Risk of Cancer: An Approach to Cancer Etiology and Control. New York, Academic Press, Inc., 1975.

Reggiani, G.: Localized contamination with TCDD–Seveso, Missouri and other areas. In Kimbrough, R. (Editor): Halogenated Biphenyls, Terphenyls, Naphthalenes, Dibenzodioxins, and Related Products. Amsterdam, Elsevier/North Holland Biomedical Press, 1980.

Waffarn, F., and Hodgman, J. E.: Mercury vapor contamination of infant incubators; a potential hazard. Pediatrics, 64:640, 1979.

Wagner, J. C., Sleggs, C. A., and Marchand, P.: Diffuse plural meso-thelioma and asbestos exposure in the North-Western Cape Province. Br. J. Ind. Med., 17:260, 1960.

Wickizer, T. M., and Brilliant, L. B.: Testing for polychlorinated biphenyls in human milk. Pediatrics, 68:411, 1981.

Yakushiji, T., Waranabe, I., Kuwabra, K., Yoshida, S., Koyama, K., and Kinita, N.: Levels of polychlorinated biphenyls (PCBS) and organochlorine pesticides in human milk and blood collected in Osaka prefective from 1972 to 1977. Int. Arch. Occup. Environ. Health, 43:1, 1979.

25
Central Nervous System Trauma

INTRACRANIAL INJURIES

INCIDENCE

Each year in Great Britain there are nine deaths from head injury per 100,000 population (Jennett and Teasdale, 1981a). This accounts for 1 per cent of all deaths in the country. Trauma is the leading cause of death in the second and third decades of life, and head injury accounts for 25 per cent of these trauma deaths. Sixty per cent of all head injury fatalities are caused by road accidents. The incidence is much higher in young males. It is thought that the death rate in head injuries in the United States is at least twice that in Great Britain (Kalsbeck et al., 1980; Kraus, 1980).

It is estimated that between 200 and 300 per 100,000 of the population are admitted to a hospital each year in Great Britain and the United States because of head injuries (Kraus, 1980). These rates are higher in men than in women.

MECHANISMS INVOLVED IN BRAIN INJURIES IN CHILDREN

There are several mechanisms whereby the brain can be injured. The head can be struck by a small, rapidly moving object, which results in tearing of tissues. More commonly a moving head is suddenly decelerated or a stationary head is suddenly accelerated (Strich, 1970). In these situations the brain is damaged by striking the inner prominences of the skull, and there may be severing of blood vessels in the brain or the coverings over the brain, i.e., the meninges, skull, or scalp. In the child's brain greater shearing forces are set up within the brain substance with consequent secondary degeneration of nerve fibers that are stretched or torn (Craft, 1975; Strich, 1970).

Linear skull fractures involving the frontal or parietal bones are not as dangerous as those over the temporal and occipital bones or over the sagittal suture. In the latter cases, the sagittal sinus and its tributaries can be torn, producing a severe subdural hematoma. Temporal bone fractures can sever the middle meningeal vessels with a resultant extradural hematoma. Occipital fractures can produce injuries to the brainstem and a posterior

440

fossa hematoma, and may be accompanied by cervical spine injuries.

If a fracture line is greater than 3 mm. in width, there is likely to be a tear in the underlying dura mater. A follow-up radiograph three to four months later should be made to rule out the presence of a leptomeningeal cyst that will require surgical repair. A basal skull fracture can result in cerebrospinal fluid otorrhea or rhinorrhea or injuries to cranial nerves. Depressed skull fractures often need to be elevated surgically, and if they are comminuted, removal of bone fragments is imperative.

Brain injuries include concussion (mild closed head injury) or contusion (severe closed head injury). Cerebral laceration is usually accompanied by a compound fracture of the skull.

When there is a discrepancy between the history of the head trauma and the degree of trauma, child abuse must be suspected. In this situation other common injuries that need to be checked for are retinal hemorrhages, rib fractures, and periosteal elevation or fractures in the upper humeri.

CLINICAL APPRAISAL AND DIAGNOSTIC TESTS

With any severe head injury there is need for a full history of what occurred. Then an examination needs to be completed for clinical and medicolegal purposes. Initial efforts in management are geared toward maintaining adequate respiration, cardiac function, and blood pressure and, of course, dealing with external bleeding. A careful baseline neurologic examination is important. This should include an evaluation of the patient's mental status, cranial nerve function, and motor, sensory, and reflex systems. If the patient is in coma, a useful and simply administered scoring system could be employed. A recently devised one is the Glasgow Response Coma Scale, which yields scores for eye opening, the best motor response, and response to verbal commands. Table 25–1 gives details for this scale (Jennett and Teasdale, 1981b).

After this clinical appraisal, radiographic evaluation of the skull, spinal column, and other bones

Table 25–1. GLASGOW "COMA" SCALE*

Eye opening (E)	
Spontaneous	E–4
To speech	3
To pain	2
Nil	1

Best motor response (M)	
Obeys	M–6
Localizes	5
Withdraws	4
Abnormal flexion	3
Extensor response	2
Nil	1

Verbal response (V)	
Oriented	V–5
Confused conversation	4
Inappropriate words	3
Incomprehensible sounds	2
Nil	1

Coma score (E + M + V) = 3 to 15

*From Jennett, B., and Teasdale, G.: Assessment of impaired consciousness. *In* Management of Head Injuries. Philadelphia, F. A. Davis Co., 1981.

suspected of being involved in the injury should be made. If intracranial or intraspinal hemorrhage is suspected, a computed tomographic scan of the intracranial and intraspinal contents should be obtained (Zimmerman et al., 1978). This information is of great importance in deciding about the management of a patient with a cranial or spinal injury. Consultations with neurosurgeons, neurologists, orthopedists, and general surgeons may be needed if multiple injuries are suspected. If the intracranial pressure is increased, dehydrating agents, such as mannitol, should be used and a decision made whether an intracranial pressure monitor should be inserted. Many neurosurgeons also consider using high dose barbiturate therapy to try to reduce excessive intracranial hypertension (Shapiro et al., 1974).

CLINICAL SYNDROMES

Cerebral Concussion

A cerebral concussion is associated with reversible neuronal dysfunction with loss of consciousness for minutes or hours. There is amnesia for the period of loss of consciousness and afterward, as well as for the events preceding the injury. This is called post-traumatic amnesia. Often there is vomiting and a diffuse headache, which usually improve within one or two days. If the brain stem has been involved, unsteadiness and vertigo may be present and persist for several days. In children there may not be a loss of consciousness following a head injury, but lethargy, pallor, and vomiting can last for several hours.

Cerebral Contusion

In a cerebral contusion there is extravasation of blood into the brain tissue beneath the site of the injury or at the contrecoup site. The poles and undersurfaces of the frontal and temporal lobes are most vulnerable. Loss of consciousness is more prolonged (more than 30 minutes) and focal neurologic signs may be present. Recovery from this more severe head injury is slower. Cerebral contusion is commoner in adults than in children.

Cerebral Laceration

The cerebral laceration is associated with a very severe head injury, often a depressed skull fracture with bone fragments severing the cerebral cortex. Local bleeding may be profuse and a focal neurologic deficit is present. Often reactive brain edema complicates the management of this form of head injury, which, if severe, can result in extrusion of cerebral tissue through the skull fracture.

Intracerebral Hematoma

With severe brain injury, shearing forces can disrupt intracranial vessels with an ensuing intracerebral hematoma that, as it enlarges, will produce an increasing focal deficit. Reactive cerebral edema often aggravates this mass effect, and partial or complete herniation of the brain may result. With uncal herniation the medial part of the temporal lobe moves toward the midbrain and through the tentorial opening. Other forms of cerebral herniation include movement of the medial part of a hemisphere under the free edge of the cerebral falx or descent of the inferior part of the cerebellum and lower brainstem through the foramen magnum (tonsillar herniation). An expanding intracerebral hematoma is an extreme neurosurgical emergency requiring cerebral dehydrating agents and rapid evacuation of the hematoma.

Extradural Hematoma

In an extradural hematoma blood collects in the extradural space following damage to the middle meningeal artery. Such an injury is more likely if a linear skull fracture crosses the grooves made by this vessel in the temporoparietal bones. However, in children a skull fracture may be absent in 25 per cent of the cases. In this condition there may be a symptom-free interval of hours or occasionally days, followed by deterioration in the patient's condition, but more often signs of cerebral compression occur soon after the injury (Plum and Posner, 1980). These include lethargy, contralateral hemiparesis, and ipsilateral pupil dilation. This condition can be recognized readily on a computed tomographic scan, and a craniotomy should follow, with tying off of the middle meningeal vessels and evacuation of the extradural hematoma by the neurosurgeon.

Subdural Hygroma

In this condition there is a tear of the subarachnoid membrane over one hemisphere following a head injury, with escape of cerebrospinal fluid into the subdural space. The patient is usually an older child or young adult and may present with headache and mild contralateral hemiparesis. If the collection is large, it should be evacuated surgically.

Subdural Hematoma

Blood can collect quickly in the subdural space in the newborn infant following a difficult breech or midforceps delivery, or in the older child or young adult following a serious injury to the skull (Abroms et al., 1977). It is important to recognize that subdural hematomas can occur also in the posterior fossa (Gilles and Shillito, 1970). These are recognized in newborn infants following difficult deliveries and in older children and adults following occipital injuries. The bleeding can result from disruption of the transverse sinus or small veins traversing the space between the surface of the cerebellum and the dural sinuses.

COMPLICATIONS OF BRAIN INJURIES

Increased Intracranial Pressure

If there is prolonged impairment of consciousness following a head injury, insertion of an intracranial monitor should be considered. It is desirable to keep the intracranial pressure below 12 mm. Hg (TORR). Because the ventricles are small and because of cerebral edema, the catheter or bolt should be in the subdural space rather than in the lateral ventricle; this is then connected to a transducer and recorder. Fluid restriction, osmotic diuretics, mechanical hyperventilation, and possibly the use of barbiturates in high dosages are useful in reducing intracranial pressure. Steroids have not been as effective in producing shrinkage in head injuries as in brain tumors (Cooper et al., 1979). It is certain that prolonged increases in intracranial pressure after head injuries result in high mortality and morbidity rates.

Prolonged Post-traumatic Coma and Brain Death

With modern resuscitative and supportive measures it is possible to keep a comatose patient alive for months or years. In children it is possible for recovery to occur after weeks or months of deep obtundation. Intellectual and neurologic deficits may be present, and an extensive rehabilitation program is necessary. On the other hand, a major problem is brain death with "electrocerebral silence" on repeated electroencephalographic tracings. This is associated with an absence of pupillary and oculocephalic reflexes and an inability to sustain unassisted ventilation. The agony for the family and physicians in this situation is intense. There is need for legal consultation as well as full discussion with the family before supportive measures are withdrawn. The value of a trained counselor for the family, physicians, and nursing personnel in dealing with a child in this situation is inestimable.

Post-traumatic Epilepsy

Recent studies in Glasgow divide post-traumatic epilepsy into "early" seizures (occurring within the first week after a head injury) and "late" seizures (Jennett, 1976). The latter usually occur within six to 12 months after a head injury. It is estimated that approximately 5 per cent of all patients admitted to the hospital with nonmissile head injuries have one or more "early" seizures. These are more common if there is a depressed skull fracture, intracranial hematoma, and prolonged post-traumatic amnesia. The incidence of early post-traumatic seizures increases to 7 per cent in patients under 15 years and 11 per cent in those under one year of age. Sixty per cent of those with early epilepsy have such a seizure within 24 hours after the head injury. Two-thirds of the patients have more than one early seizure. The overall rate of status epilepticus is 10 per cent; however, in patients under age five years, this rate is doubled.

In 40 per cent of the instances focal motor seizures are noted, whereas in the remainder, generalized seizures predominate. After the occurrence of early epilepsy, the incidence of recurrent seizures can reach 15 to 30 per cent in patients with severe head injuries (Jennett, 1976).

Intracranial Infection

Post-traumatic meningitis occurs after compound skull fractures, especially with fractures of the petrous temporal bone and otorrhea, or rhinorrhea associated with a fracture of the cribriform plate. The meningitis may occur within a few days after the injury or weeks later. The commonest organism found in post-traumatic meningitis is the pneumococcus. Less frequent are *Hemophilus influenzae*, Streptococcus, Meningococcus, and Staphylococcus organisms.

Cranial Nerve and Hypothalamic Damage

Fractures of the anterior fossa floor can result in unilateral or bilateral anosmia due to injury of the olfactory nerve filaments passing through the cribriform plate. Loss of the smell sense may be present in as many as 7 per cent of the patients with head injuries who are admitted to the hospital; total recovery occurs within three months in most patients. Anterior fossa fractures or orbital fractures may injure the optic nerve and produce monocular blindness. Recovery from this injury is infrequent.

Transient diplopia due to head injury is common. If third nerve or sixth nerve palsy is present, complete recovery is the rule, but this may take several weeks or months (Roberts, 1976).

Facial nerve injury occurs with petrous temporal bone fractures; poor recovery can be expected if the fracture is in the transverse plane. With longitudinal fracture of the petrous bone, delayed facial palsy may occur two or three days after the injury. Steroid therapy with all petrous bone fractures decreases the incidence of delayed facial palsy. Hearing loss following head injury is most commonly due to damage of the organ of Corti and may be accompanied by tinnitus. Hearing loss can be unilateral or bilateral and is often permanent (Toglia and Katinsky, 1976). The lower cranial nerves are damaged more often in an extracranial location due to gunshot wounds than in head injuries. With severe head injuries, dysfunction of the hypothalamus is manifested by inappropriate antidiuretic hormone secretion, diabetes insipidus, disturbances of temperature control, and changes in appetite (Crompton, 1975). Often these manifestations of hypothalamic disturbance are short lived.

Post-traumatic Hydrocephalus

Post-traumatic hydrocephalus occurs with obstruction of cerebrospinal fluid flow due to intraventricular bleeding, and can manifest weeks or months later with headaches, vomiting, and lethargy. Large ventricles will be evident on computed tomographic scanning.

Postconcussion Syndrome

There is a relationship between the duration of post-traumatic amnesia and postconcussion symptoms (Rutherford et al., 1977). Postconcussion headache is common. It is a dull continuous frontal headache or a feeling of malaise, often associated with unsteadiness or vertigo. Associated symptoms are poor ability to concentrate, irritability, and excessive fatigue (Brooks, 1975; Merskey and Woodford, 1972). It is of interest that a migraine type of headache can be precipitated by head injuries; this can occur within hours or days after the trauma (Guthkelch, 1977).

General Issues

In the early recovery phase it is important to understand that a small child sometimes fantasizes that the injury is some form of retribution for something that he said or did. This needs to be understood and explained to the child. Furthermore, a child cannot understand why he feels discomfort and pain at a time when his doctor is saying that he is fine or is going to be fine.

The guilt of parents following an accident may give them the feeling that they are poor caregivers and that they should not try to care for their child. This should be discussed fully and the importance of their giving support to their child in the hospital setting and afterward needs to be emphasized. Also of great importance but often neglected is the emotional well-being of the professional caregivers. This is particularly true of the nursing staff in the intensive care unit setting. Group discussions about particular patients led by a trained psychologist or psychiatrist are needed regularly. This activity in the long run benefits both the patient and his family.

The treatment of complications of brain injuries include the treatment of post-traumatic hydrocephalus by shunting procedures and closure of dural tears following the discovery of cerebrospinal rhinorrhea. Nonsurgical treatment of complications of head injuries includes the treatment of post-traumatic seizures with anticonvulsants, intracranial infections with appropriate antibiotics given intravenously, and psychologic counseling for patients who have prolonged emotional reactions.

PROGNOSIS AFTER HEAD INJURY

Accurate prognosis in the acute phase of a head injury is often difficult. There are some studies that can be cited in the task of determining the prognosis (Jennett et al., 1979; Langfitt, 1978). In a follow-up study of 36 children four to 10 years after severe head injury, with unconsciousness lasting more than 24 hours, eight of 34 of the school-age children were unable to attend normal school, and a further nine showed a decline in school performance (Heiskanen and Kaste, 1974). Seventeen children were considered to be doing well at school. If a child remains unconscious for two weeks, there will be only a rare instance of adequate school performance. Seventy per cent of these children had been injured in traffic accidents, and most had suffered a diffuse cerebral injury. Twelve of the 36 had hemiparesis, and five had developed post-traumatic epilepsy.

The prognosis following severe head injury depends on the age of the patient, the presence of systemic disease, the psychosocial status of the patient, the severity of brain dysfunction on initial assessment, early complications, and the rate of recovery following the head injury (Teasdale et al., 1979b).

Jennett and others have reported the results of a computer analysis of more than 1500 patients in a data bank of serious head injuries. The latter was defined as a head injury followed by more than six hours of coma without a lucid interval. Coma in this context was defined as "the patient not opening his eyes, not obeying commands or uttering recognizable words." After analysis, clinical factors that were most predictive of the outcome included the age of the patient, the depth and duration of the coma, and the presence of an intracranial hematoma. There was a higher mortality rate in children under age five than in those aged five to 19. Compared with adults, younger patients had a greater chance for recovery after a prolonged deep coma following head injury. If an assessment of the degree of coma had been made within the first 24 hours, nonreactive pupils, the absence of oculocephalic reflexes, and extensor motor responses with stimulation were more likely to be associated with death or severe neurologic disabilities. This corresponded to Glasgow Coma Scale scores of 3 or 4.

Intracranial hematoma had a worse outcome, especially in the pediatric group (Teasdale et al., 1979a). In general, laboratory studies were not helpful in predicting outcome. However, using the data bank information, Teasdale et al. were correct in their predictions of outcome at 24 hours after the head injury in 92 to 94 per cent of the instances, based on eight clinical items: age, Glasgow Coma Scale score, best motor response score, motor abnormality pattern, pupillary responses, eye movements, the presence of apnea, and the trend toward improvement or deterioration at 24 hours after the injury.

REHABILITATION FOLLOWING HEAD INJURY

Rehabilitation involves the combined and coordinated use of medical, social, educational, and vocational measures for training or retraining the disabled individual to the highest level of functional ability.

The following is a useful classification of functional categories (Najenson et al., 1974).

Category I	Vegetative state
Category II	Independent in daily living activities
Category III	Independent and capable of sheltered employment
Category IV	Capable of simple work under normal conditions
Category V	Capable of professional work

In a comprehensive rehabilitation program for children and young adults with head injuries, early graduated mobilization is paramount (Hook, 1976). These activities are best supervised by a physical therapist and an occupational therapist. Psychologic counseling is equally important. As soon as is practicable, psychometric testing should be considered with a view to reinstituting an appropriate educational plan geared to the individual's capacity. If dysphasia or dysarthria persists following a head injury, the specialized services of a speech and language pathologist will be required. Professionals in rehabilitation medicine are often unique individuals, who through their skills and dedication to their patients can bring hope and encouragement to children and young adults with handicaps. Their work is often repetitious and the gains in individual patients may occur very slowly. The same may be true in psychologic rehabilitation in patients with handicaps. The necessary adjustment required of an active independent young person to a changed role of a dependent physically disabled person is immense. It demands the expert combined skills of a multidisciplinary group of professionals to interact with the patient as well as with one another to strive to improve the emotional and physical well-being of their patient.

There is need to consider the impact on the family of the child or young adult who has suffered a recent severe head injury. The circumstances of the accident may relate to guilt feelings on the part of the parents or a sibling. The overwhelming guilt and depression in these family members require attention and counseling in their own right. The negative effects of family members with unresolved guilt concerning the accident on the patient who is recovering from a head injury cannot be overestimated. This can be a decisive factor in the degree of motivation in the head injured patient toward his own recovery.

SPINAL CORD INJURIES

INCIDENCE

The number of new instances of spinal cord injury per year is about three to five per 100,000 population. Males outnumber females in almost all studies; this ratio can be as high as 3.5 to 1 (Kraus, 1980). The most frequent single cause of injury is motor vehicle accidents (sport injuries are also important). The overall estimate of persons in the United States sustaining spinal cord injury has been given as at least 28,000, but other estimates suggest that the number is five times greater. The number of serious injuries in automobile accidents has been reduced by the use of seat belts, and in infants by the improved design and use of car seats.

THE MECHANICS OF SPINAL CORD INJURIES

The majority of spinal cord injuries result from severe flexion of the spinal cord with or without fracture-dislocation of the spinal column. Extravasation of blood producing hematomyelia and subdural or extradural hematomas are uncommon, as are ischemic infarctions due to spinal cord injury.

CLINICAL SYNDROMES

Complete Transection of the Spinal Cord

Complete transection produces flaccid paraparesis if the spinal cord lesion is thoracolumbar in location or flaccid quadriparesis if there is involvement of the cervical segments. There is no voluntary movement, but there may be reflex movements of the limbs; sensation is preserved above the level of the lesion.

Partial Transection of the Spinal Cord

Partial transection produces a Brown-Séquard syndrome if it involves half the spinal cord. The motor tracts are severed on one side, and the crossed spinothalamic fibers are divided on the same side. This causes unilateral paralysis and a contralateral absence of pain and temperature sensation below the level of the lesion.

Spinal Cord Infarction

Infarction is difficult to diagnose and usually involves occlusion of the anterior spinal arteries. There is flaccid paraplegia or quadriplegia associated with diminished pain and temperature senses in both lower limbs, with sparing of the sacral segments. The latter occurs because these fibers are located laterally in the spinothalamic tract and the infarction in the cord is anteromedial. In addition, touch and proprioceptive senses are retained in the limbs. These sensory fibers are located in the posterior columns remote from the area of infarction. Bladder and bowel incontinence is the rule and may persist in patients in whom partial recovery from this disorder occurs. Early in the course the weakness is associated with hypotonia, but in time spasticity of the legs, hyperreflexia, and extensor plantar responses become evident.

Hematomyelia

In hematomyelia, or hemorrhage into the central canal of the spinal cord, there is a sudden onset of back pain, sometimes associated with radicular radiation of the pain and a severe sensory loss below the level of the lesion. Flaccid paralysis is accompanied by a loss of bowel and bladder control.

Epidural or Subdural Hemorrhage

Such forms of intraspinal hematomas occur with serious spinal cord injuries and present with severe back pain at the level of the lesion, followed by progressive paraplegia or quadriplegia. Occasionally signs evolve very slowly. If the trauma is minimal with these signs and symptoms, a spinal cord vascular malformation or a blood dyscrasia should be considered. The presence of progressive dysfunction of the spinal cord should prompt one to obtain an emergency myelogram and carry out surgical decompression.

COMPLICATIONS OF SPINAL CORD INJURIES

If the corticospinal motor tracts have been severed above the sacral segments, bladder incontinence will be present and soon spasticity of the bladder develops, a small contracted bladder with diminished volume and frequent reflex emptying. There is also an increase in the residual volume of urine, predisposing the patient to recurrent urinary tract infections.

Occasionally spinal cord injuries are associated with fracture-dislocations of the surrounding vertebrae. Urgent stabilization of the spinal column is necessary, often requiring a surgical procedure.

PROGNOSIS IN SPINAL CORD INJURIES

If a complete transection has occurred, the prognosis for even partial recovery is poor. If partial preservation of function is noted at the first assessment, further recovery is likely.

REHABILITATION FOLLOWING SPINAL CORD INJURIES

Rehabilitation should be planned at a regional center that deals with many such injuries. A combined multidisciplinary program is imperative, with expert supervision of physical therapy and bladder training, and the patient must receive simultaneous psychologic counseling. There is also need for reinstituting an academic program for children and young adults as soon as feasible. Issues of guilt on the part of the family, if present, need to be addressed through counseling.

ISRAEL F. ABROMS

REFERENCES

Abroms, I. F., McLennan, J. E., and Mandell, F.: Acute subdural hematoma following breech delivery. Am. J. Dis. Child., 131:192, 1977.

Brooks, D. N.: Long and short term memory in head injured patients. Cortex, 11:329–340, 1975.

Cooper, P. R., et al.: Dexamethasone and severe head injury. J. Neurosurg., 51:307–316, 1979.

Craft, A. W.: Head injury in children. In Vinken, P. J., and Bruyn, G. W. (Editors): Handbook of Clinical Neurology. Amsterdam, Elsevier North-Holland Publishing Co., 1975, Vol. 23, pp. 445–458.

Crompton, M. R.: Hypothalamic and pituitary lesions. In Vinken, P. J., and Bruyn, G. W. (Editors): Handbook of Clinical Neurology. Elsevier North-Holland Publishing Co., 1975, Vol. 23, pp. 465–469.

Gilles, F. H., and Shillito, J.: Infantile hydrocephalus: retrocerebellar subdural hematoma. J. Pediatr., 76:529, 1970.

Guthkelch, A. N.: Benign post-traumatic encephalopathy in young people and its relation to migraine. Neurosurgery, 1:101–105, 1977.

Heiskanen, O., and Kaste, M.: Late prognosis of severe brain injury in children. Dev. Med. Child Neurol., 16:11–14, 1974.

Hook, O.: Rehabilitation. In Vinken, P. J., and Bruyn, G. W. (Editors): Handbook of Clinical Neurology. Elsevier North-Holland Publishing Co., 1976, Vol. 24, pp. 683–697.

Jennett, B.: Post-traumatic epilepsy. In Vinken, P. J., and Bruyn, G. W. (Editors): Handbook of Clinical Neurology. Elsevier North-Holland Publishing Co., 1976, Vol. 24, pp. 445–454.

Jennett, B., and Teasdale, G.: Assessment of impaired consciousness. In Management of Head Injuries. Philadelphia, F. A. Davis Co., 1981a, pp. 77–93.

Jennett, B., and Teasdale, G.: Epidemiology of head injury. In Management of Head Injuries. Philadelphia, F.A. Davis Co., 1981b, pp. 1–17.

Jennett, B., Teasdale, G., Braakman, R., et al.: Prognosis in series of patients with severe head injury. Neurosurgery, 4:283, 1979.

Kalsbeck, W. D., et al.: The national head and spinal cord injury survey: Major findings. J. Neurosurg., 53:S19–S31, 1980.

Kraus, J. F.: A comparison of recent studies on the extent of the head and spinal cord injury problem in the United States. J. Neurosurg., 53:S35–S43, 1980.

Langfitt, T. W.: Measuring the outcome from head injuries. J. Neurosurg., 48:673–678, 1978.

Merskey, H., and Woodford, J. M.: Psychiatric sequelae of minor head injury. Brain, 95:521–528, 1972.

Najenson, T. L., et al.: Rehabilitation after severe head injury. Scand. J. Rehab. Med., 6:5–14, 1974.

Plum, F., and Posner, J. B.: Stupor and Coma. Ed. 3. Philadelphia, F. A. Davis Co., 1980.

Roberts, M.: Lesions of the ocular motor nerves (III, IV and VI). In Vinken, P. J., and Bruyn, G. W. (Editors): Handbook of Clinical Neurology. Amsterdam, Elsevier North-Holland Publishing Co., 1976, Vol. 24, pp. 59–72.

Rutherford, W. H., Merrett, J. D., and McDonald, J. R.: Sequelae of concussion caused by minor head injury. Lancet, 1:1, 1977.

Shapiro, H. M., Whyte, S. R., and Loeser, J.: Barbiturate-augmented hypothermia for reduction of persistent intracranial hypertension. J. Neurosurg., 40:90–100, 1974.

Strich, S. J.: Lesions in the cerebral hemispheres after blunt head injury. J. Clin. Pathol. 23(Suppl. 4):154–165, 1970.

Teasdale, G., et al.: Predicting the outcome of individual patients in the first week after severe head injury. Acta Neurochir. (Suppl.), 28:161–164, 1979a.

Teasdale, G., et al.: Age and outcome of severe head injury. Acta Neurochir. (Suppl.), 28:140–143, 1979b.

Toglia, J. U., and Katinsky, S.: Neuro-otological aspects of closed head injury. In Vinken, P. J., and Bruyn, G. W. (Editors): Handbook of Clinical Neurology. Amsterdam, Elsevier North-Holland Publishing Co., 1976, Vol. 24, pp. 119–140.

Zimmerman, R. A., et al.: Computed tomography of pediatric head trauma: Acute general cerebral swelling. Radiology, 126:403–408, 1978.

26

The Effect of General Medical Illness and Its Treatment on Development and Behavior

Up to this point, Parts III and IV of this volume have presented a wide variety of environmental and biologic influences affecting development and behavior in children. The biologic factors discussed so far have been those that are expressed primarily through the function of the central nervous system. This chapter deals with general medical illness and its treatment as a factor in development and behavior. Illness involves more of an interaction than a unidirectional effect, for development and behavior may also modify the expression of the physical illness in the individual.

This chapter is divided into the following topics: acute minor illness, hospitalization, surgery and anesthesia, early health crises and vulnerable children, chronic illness, otitis media, allergic disorders, and gastrointestinal disorders.

The first five sections deal with the varying levels of illness experience, and the last three are concerned with the specific organ systems having the most conspicuous interactions with development and behavior. The last two sections include several conditions referred to elsewhere as "psychosomatic." Terminal illness is discussed in Chapter 27.

In each section the authors describe the stresses presented by the illness, their effects on the child, and the ways in which the pediatrician or other primary care clinician can deal with the problem with the maximal benefit to the child's health, development, and behavior.

26A Acute Minor Illness

Acute minor illness is defined here as ordinary, brief health problems, such as the common respiratory and gastrointestinal infections and the familiar instances of physical trauma experienced by all children. These are the usual minor complaints that account for almost all the visits to primary health care clinicians because of illness. These illnesses and their management have not generally been considered to be significant factors in the child's development and behavior. However, there is reason to believe that this assumption of their unimportance is incorrect and is attributable to insufficient attention and research. A comprehensive critical review of the meager literature a decade ago summarized available data and called for further investigation (Carey and Sibinga, 1972). Little has been published since then (Schmitt, 1980; Sibinga and Carey, 1976). This section draws upon that review as to content and organization.

The recommendations are based on a pooling of the experience of many pediatricians rather than on formal research findings.

STRESSES ON THE CHILD AND PARENT

The child with an acute minor illness experiences some or all of the following: the discomfort of the illness and its treatment, the emotional reactions to and fantasies about the illness (such as feelings of guilt, fear, anger, depression, and apathy), the loss of normal social contacts at school and elsewhere outside the home, the restrictions (such as bed rest and diet) and the decreased or altered sensory input, and a change in the relationship with the parents, who may become more indulgent or hostile.

447

The parents endure stresses of two sorts: those that arise from the illness and its management (such as more responsibility and expense, interference with employment, and less sleep and recreation) and the extraneous factors that complicate the parents' reaction to the illness—the parents' personal problems, pre-existing physical or behavioral difficulties in the child, and situational pressures, such as unemployment or marital discord.

EFFECTS OF STRESSES

The effect of acute minor illness on children's behavior, although well known to parents and clinicians, has scarcely been studied at all. Various possible behavioral reactions have been reported, such as dependency, withdrawal, irritability, rebelliousness, and feelings of inferiority, but a comprehensive systematic delineation of these reactions and their determinants and consequences is still awaited.

The parents may also suffer from fears and anxieties, anger, guilt, fatigue, depression, and distortions and misconceptions. They are expected to master these feelings and to adjust sufficiently to meet their child's needs. Ineffective responses may consist of inappropriate feelings or lack of understanding (such as too much or too little anxiety, persistent misconceptions about the illness or its treatment or an excessive sense of personal injury), or inability to muster appropriate behavior to deal with the illness (such as difficulty in carrying out a reasonable share of the treatment process in conjunction with the doctor, as evidenced by noncompliance with the treatment plan).

It seems likely that minor illness and its management account for only a small portion of the more chronic behavioral variations in children, yet these effects are real and must be acknowledged and dealt with. Any experienced pediatrician can think of striking examples of unfavorable outcomes, such as the child who is treated by his parents for years as a semi-invalid because a physician had felt compelled to reveal the existence of a functional heart murmur without considering the impact of this information on the child and his parents, or the child who is made to feel guilty or inadequate because of recurrent respiratory or urinary infections.

NURTURING SICK CHILDREN AND SUPPORTING PARENTS

A useful way to describe principles of management of minor illness is in terms of the illness itself, the child, and the parents—how the needs of each are sometimes not met, why this happens, what the consequences may be, and finally how to avoid these complications.

TREAT THE ILLNESS

Treatment of illness can go awry in three main ways. Probably the commonest error we physicians make in the management of minor illness is in overdiagnosing and overtreating. For example, insignificant findings such as a functional heart murmur or tibial torsion may be overinterpreted, made into a "pseudodisease," and managed with greater attention than they deserve. Physicians apparently err in this direction for two principal reasons: problems within the physician himself, such as insufficient training or excessive anxiety, and pressures from the parents, such as their common urging that something decisive be done about recurrent respiratory infections. Possible consequences of overmanagement include generating fear that the child is sicker than he really is, producing harm from unnecessary procedures, and fostering excessive dependence on the physician and his treatment.

Undermanagement and mismanagement and their causes and consequences are sufficiently well known to require no further elaboration here.

Obviously the primary role of the pediatrician or other clinician is to treat the illness. This means the use of appropriate diagnostic and therapeutic measures with avoidance of the pitfalls already mentioned. Also reasonable steps should be taken to help the parents and the child prevent recurrences or spread of the illness.

NURTURE THE SICK CHILD

Undoubtedly the commonest problem in the medical management of the sick child is that the illness itself is treated but the sick child is neglected. The clinician may be inadequately sensitive and attentive to the emotional needs of the sick child. The cause is usually a narrowness of interest or a deficiency of empathy. This preoccupation with the illness itself makes it more likely that the experience will be more frightening and stressful for the child than it needs to be.

On the other hand, overattention to the child's feelings, as with the children of our friends or colleagues, may mean that proper diagnosis and treatment are in jeopardy. Third, inappropriate handling of the child, as with dishonesty or belittling of his concerns for whatever reason, must invariably make the illness experience more hazardous for the child.

Nurturing the sick child means that the physician can reduce the stress of the illness by keeping the discomfort, trauma, and restrictions of the management to a minimum; that he can promote the child's adjustment to the illness by listening to the child's concerns and responding to them honestly and supportively; and at times that the illness experience can even be an opportunity for psychological growth. By learning about the illness and his ability to tolerate discomfort and overcome a problem with the help of his family and physician, the child can gain in self-confidence and avoid developing a sense of physical inferiority or personal inadequacy.

SUPPORT THE PARENTS

Although it may be difficult to define what support of the parents should consist of, there is little problem in identifying unsupportive patterns in the physician-parent relationship.

A common form is overdomination and oversubmissiveness. The playing of too dominant a role by the physician in the care of minor illness stifles self-reliance in parents; too much submission to the parents' wishes results in abdication of the physician's proper advisory status.

Neglect of the parents entails not giving sufficient attention to their feelings, their need for information, or their help in handling the illness. Neglected parents become confused and helpless and are likely to turn elsewhere for assistance.

Inappropriate handling of the parents, giving them misinformation about the illness or its management, or doing anything that needlessly upsets them is, of course, likely to leave them with feelings of incompetence and perhaps with an attachment to the doctor through fear or misguided gratitude.

Probably the best way to be supportive to parents with a sick child is to determine their needs and expectations, to help them to deal with the illness and the sick child by providing appropriate information, encouraging self-reliance, and being available in case of further need, and to promote their general self-confidence in handling illness. Well supported, self-reliant parents are more likely to manage the illness and child well, to relieve the physician of needless repetition of instructions, and to face other aspects of child rearing with greater confidence.

The average American pediatrician spends about half his time dealing with minor illnesses, yet problems in development and behavior related to minor illness and its management have been largely ignored. It remains a major area for potential expansion of knowledge through research and for enhancement of professional skills.

WILLIAM B. CAREY

REFERENCES

Carey W. B., and Sibinga, M. S.: Avoiding pediatric pathogenesis in the management of acute minor illness. Pediatrics, 49:553, 1972.
Schmitt, B. D.: Fever phobia. Misconceptions of parents about fevers. Am. J. Dis. Child., 134:176, 1980.
Sibinga, M. S., and Carey, W. B.: Dealing with unnecessary medical trauma to children. Pediatrics, 57:800, 1976.

26B   Hospitalization

Hospitalization adds further stresses to the child's experience of physical illness, and in some cases these affect the child's behavior. This section reviews these stresses and discusses how they may affect the child and how clinicians can help parents and children to deal with these problems.

STRESSES

The stresses on the child from the experience of hospitalization must be distinguished from those of the illness precipitating the admission. The more serious illnesses requiring hospitalization usually involve the issues already discussed under the topic of acute minor illness. However, there is likely to be an increase in the extent of the physical distress and in the child's and parents' emotional reaction to the illness. Furthermore, there is of necessity a greater dependence on the physician for resolution of the problems. The potential stresses of the hospitalization itself are:

Separation. The child who is admitted to a hospital experiences varying degrees of separation from parents, siblings, home, friends, school, and the other accustomed elements of his usual life. Unlike a more pleasant separation, such as a visit to a grandmother, these experiences are likely to be sudden and poorly understood and occur at a time when the child is feeling unwell.

Unfamiliarity. No matter what the child's experience has been from reading or television viewing, admission to a hospital is still a strange and often bewildering experience for most children. The variety of unfamiliar things, people, and activities is seemingly endless.

Painful and Frightening Procedures. Not only are hospitals unfamiliar places, they may also be very unpleasant. There are painful and frightening procedures such as blood drawing, intravenous drips, physical restraints, dietary restrictions, and isolation rooms. Often their purpose is poorly understood by the child. Hospital based medical personnel know this but have to keep reminding themselves.

EFFECTS ON CHILDREN

Children who experience these stresses may respond in various ways. Most seem able to withstand brief hospitalizations without any enduring effects on their behavior; in others there are lasting negative effects from the admission. Occasionally children are able to use the challenge constructively and gain in maturity.

This variety of results appears to be related to a multitude of factors, including the nature of the illness, the pre-existing personality of the child, the attitudes of the parents, the preparation of the child prior to the hospital experience, the length of the hospitalization, the amount of stress during the hospitalization, how the stresses are handled by the professional staff, and parental visits. Research on the behavioral outcome of hospitalization has been inconclusive because these numerous variables often have not been adequately considered (Vernon et al., 1965). Furthermore, when hospital policies have shifted toward more humane treatment of hospitalized children, earlier conclusions may no longer apply.

Although the research on the effects of hospitalization on children's behavior leaves many questions unanswered, there is general agreement that children between six months and four years of age are the most vulnerable. Before six months the separation and strangeness seem not to be experienced so keenly. After four years children become increasingly able to cope with the challenge (Vernon et al., 1965).

Robertson (1970) described three phases displayed by young children during the "settling in" period in hospitals where parent visitation is minimal. The initial phase is one of protest, lasting from a few hours to several days. The child cries for a prolonged period about the separation from his mother and his fears of the unfamiliar situation. This is followed by a period of despair, in which the child feels increasingly hopeless and becomes withdrawn and apathetic. In the denial phase the child represses his feelings for his mother, shows more interest in his surroundings, but seems detached from his mother. The child who has gone through these three phases may seem well adjusted to his long term hospitalization but is likely to be lacking in deep attachment to anyone. The longer the deprivation of the child's interpersonal needs, the greater are the chances of residual behavioral disturbance.

In a massive study of hospitalization in Great Britain, Douglas (1975) found that mothers reported that 22 per cent of their children had more problems in behavior after discharge, especially more "nervous" or "difficult" behavior or sleep problems. However, 68 per cent were reported to be unchanged, and 10 per cent improved in behavior. The investigation also included an evaluation of the long range effects of childhood hospitalization on later behavior in adolescence. Douglas concluded that during childhood, single hospital admissions for up to one week carry no increased risk of later behavioral disturbance. However, one admission of over one week or repeated admissions before the age of five years were associated with an increased risk of behavior disturbance and poor reading ability in adolescence. The children most vulnerable to the adverse effects of early hospital admission were said to be highly dependent on their mothers or were under stress at home at the time of admission. It should be emphasized that the increased risk of behavior disturbance, although significant, was small and accounted for little of the variance in the adolescent's behavior.

We must not lose sight of the fact that the hospitalization experience can be a constructive one from which parents and children achieve a sense of mastery and growth. Not only can there be a sense of accomplishment from overcoming a physical problem, but a well managed admission can promote the behavioral adjustment of the child (Solnit, 1960).

HELPING PARENTS AND CHILDREN MINIMIZE THE STRESSES

There is sufficient evidence of hospitalization's being upsetting for children and possibly productive of later behavior disturbance that we should take steps to minimize these stresses and help parents and children deal with them. Expedience or the methodological flaws in existing studies should not keep us from providing the best possible management for children in our care.

Avoiding Unnecessary Admissions. The hospital admission of a child should seldom be advised simply for the convenience of the physician or parents, but should take place only when it would be unreasonable or unsafe to attempt the child's care at home. Besides the need for general anesthesia and surgery, admission should be used for intensive observation of dangerous conditions in which serious changes may occur rapidly, such as after ingestion of toxins or major head trauma, for treatment that cannot be provided by parents, such as intravenous drips, intramuscular injections, and oxygen, and in special circumstances requiring a change of the environment, such as failure to thrive, child abuse, and suicide attempts (North, 1976).

In recent years the growing popularity of day surgery has done away with many unnecessary overnight hospital stays. Brief surgical procedures, such as the repair of an inguinal hernia, can be accomplished and the child sent home after the anesthesia has worn off without having to spend the night before or after the operation away from home.

Preparation of the Child and the Parents. There is no question that preparation for the stresses of hospitalization will reduce their impact on the child. The advice of Robertson (1970) remains excellent:

The younger the child, the more difficult it is to do any worthwhile preparation. It is doubtful if under-threes can be helped in this way.

Preparation will be more effective if done by the parents and supported in the event by frequent visiting.

Preparation should be simple and truthful, but not more than the child can assimilate.

The aim should be to give an honest picture of the strange environment into which he has to go, and to give reassurance that he will return.

If there is to be an operation, the pain should be mentioned—coupled with the assurance that it will get better.

He should not be told too far ahead about going to the hospital, perhaps beginning a week in advance with a first intimation and supplementing at the pace set by his questions. But if he learns by chance some weeks ahead, for instance by overhearing parents or doctors talk about it, it is best to answer his questions as from then.

In the two-and-a-half to three-year-old age group little preparation can be done, but games of going away and coming back may help a little.

Still younger children cannot be prepared in any way. This confirms the absolute necessity for basing their care on the presence and participation of the mother.

It is also impossible to prepare for emergency admissions.

If a child has a favorite toy or a "thing" which he holds or sucks when frightened, tired, or unwell, it should always be available to him in the hospital. Even if his comforter is a "dummy" ("pacifier" in America) he should not be deprived of it.

Parents can be helped to prepare their children through the right kind of leaflet, i.e., one that is realistic about the problem without detailing incidental treatments and does not omit to deal with hurt, tears and post-separation disturbances in behavior. The ability of parents to deal with the painful task of preparation would be much improved if they could meet a suitable staff member a week or two before admission to discuss the matter.

Booklets to help children and their parents prepare for hospitalization are of two principal sorts: general ones that assist children and their parents in anticipating the feelings that they will experience, such as *Curious George Goes to the Hospital* by Rey (Boston, Houghton Mifflin, 1966; see also list at end of this section), and specific ones that contain information about the particular hospital. For example, The Children's Hospital of Philadelphia has a general information pamphlet for parents describing the process of admission, facilities and regulations while in the hospital, and discharge procedures; additional specific booklets for parents concerning the infant and pediatric intensive care units; a coloring book, *Daisy and Her Friends*, to acquaint the child with the principal events of a typical hospital experience; and other booklets for the child about specific procedures— *A Visit to the Radiology Department*, *DeeDee's Heart Test*, and *Danny's Heart Operation*. Other major hospitals offer similar publications.

Support of the Child While He Is in the Hospital. The most helpful support for the hospitalized child, especially in the six month to four year range, is to have his mother or father stay with him. This eliminates the most important part of the stress of separation and helps to fortify the child against the strangeness and discomforts of the experience. Leading children's hospitals have been allowing and encouraging this practice for years, but many institutions still regard the presence of parents as merely a disruptive nuisance. There are occasionally disadvantages to this arrangement, but they seldom outweigh the advantages. Encouraging the child to bring a few familiar and favorite books and toys further diminishes the impact of separation from home. Facilities for play and continuing education should, of course, be provided.

The next best arrangement other than having the mother or father stay with the child is to allow unlimited visiting by the parents. They can decide for themselves how much they will visit and should be allowed to take over as much as possible of the child's routine care in cooperation with the nursing staff.

The assignment of a specific nurse to the child allows for the development of feelings of attachment to the same helping and comforting member of the hospital staff.

In this supportive setting it should be possible for the child's disturbed feelings and altered behavior to be dealt with honestly and sympathetically in order to minimize the effect of the stresses of hospitalization.

Finally, procedures employed in hospitals should be constantly under review to make certain that the psychologic risks involved are justified by established benefits. Medical personnel should routinely ask themselves whether repeated lumbar punctures, isolation, or physical restraints are absolutely necessary for the child's management.

WILLIAM B. CAREY

REFERENCES

Douglas, J. W. B.: Early hospital admissions and later disturbances of behavior and learning. Dev. Med. Child Neurol., 17:456, 1975.

North, A. F., Jr.: When should a child be in the hospital? Pediatrics, 57:540, 1976.

Robertson, J.: Young Children in Hospital. Ed. 2. London, Tavistock Publications Ltd., 1970.

Solnit, A. J.: Hospitalization. An aid to physical and psychological health in childhood. Am. J. Dis. Child., 99:155, 1960.

Vernon, D. T. A., Foley, J. M., Sipowicz, R. R., and Schulman, J. L.: The Psychological Responses of Children to Hospitalization and Illness. A Review of the Literature. Springfield, Illinois, Charles C Thomas, 1965.

Recommended for Parents and Children by the Association for the Care of Children's Health

BOOKS

Fassler, J.: Helping Children Cope: Managing Stress Through Books and Stories. New York, The Free Press, 1978. Hard cover, 162 pp.

Howe, J.: The Hospital Book. New York, Crown Publishers, Inc., 1981. Soft cover, 96 pp.

Stein, S. B.: A Hospital Story. New York, Walker & Co., 1974. Hard cover, 47 pp.

PAMPHLETS*

Baznik, D.: Becky's Story: A Book to Share. Association for the Care of Children's Health, 1981. Soft cover, 31 pp.

A Child Goes to the Hospital. Association for the Care of Children's Health, 1981. Pamphlet, 15 pp.

The Chronically Ill Child and Family in the Community. Association for the Care of Children's Health, 1982. Pamphlet, 24 pp.

Preparing a Child for Repeated or Extended Hospitalization. Association for the Care of Children's Health, 1982. Pamphlet.

For Teenagers: Your Stay in the Hospital. Association for the Care of Children's Health, 1982. Pamphlet, 12 pp.

FILMS

To Prepare a Child. Media Center, Children's Hospital National Medical Center, Washington, D.C.

First Do No Harm. Media Center, Children's Hospital National Medical Center, Washington, D.C.

---

*The Association for the Care of Children's Health, formerly the Association for the Care of Children in Hospitals, is a multidisciplinary, international organization that promotes the psychosocial well-being of children and families in health care settings. Address: 3615 Wisconsin Avenue, N.W., Washington, D.C. 20016.

26C   Surgery and Anesthesia

The stresses from surgery and anesthesia are distinct from those of the child's illness and hospitalization. The nature of these stresses, their impact on children, and ways of helping children through these experiences are discussed here.

STRESSES

Physical Trauma.   Surgery can be thought of as a premeditated trauma or bodily injury, which is organized by the child's caretakers and may come to him as a surprise. Children old enough to contemplate this prospect do so with understandable apprehension.

Loss of Consciousness and Control.   The new and frightening experience of being "put to sleep" in a way that the child cannot resist or control provokes anxiety and wonder. Some children fear that it will not last long enough, others that it will be too long, and still others that they will be compromised in some way such as revealing important secrets about themselves.

EFFECTS ON CHILDREN

As with evaluating the effects of hospitalization on children, much depends on the circumstances of the surgery and anesthesia—the preparation, the duration of hospital stay, the nature of the surgery. It is impossible, therefore, to generalize about the behavioral effects of these experiences per se on children. It seems reasonable to conclude that a brief, well managed procedure, such as myringotomy and insertion of ventilating tubes, should have little or no lasting impact on the child's behavior, whatever the immediate reaction may be. Davenport and Werry (1970) found no evidence of significant posthospitalization upset in 145 American and Canadian children undergoing tonsillectomy. We may speculate further that greater gravity of the illness and the operative procedure and less skill in their management should increase the psychologic hazards of surgery and anesthesia.

HELPING PARENTS AND CHILDREN

Methodological problems in demonstrating consistent lasting behavioral effects from surgery and anesthesia have not deterred conscientious clinicians from taking steps to minimize the stresses and chances of short term emotional and behavioral disturbance. Visintainer and Wolfer (1975) demonstrated that children undergoing tonsillectomy with or without adenoidectomy and myringotomy were less upset and more cooperative during the admission if they received a combination of systematic preparation and supportive care.

Haller (1976) has stressed that the responsibility for preparation and support should be shared by the parents, the pediatric nursing staff, the anesthesiologist, and the surgeon.

The parents should obtain from the surgeon precise information about the procedure to allay their own fears and to transmit in an appropriate form to the child. Reading and play acting can help the child understand what lies ahead.

Beside providing the usual bedside care the pediatric nurse can continue the educational process concerning hospital routines, costumes, and equipment and be available as a truthful, dependable, and consistent friend.

The anesthesiologist must not only provide competent professional services but also be on hand before the procedure to explain to the child exactly what is about to happen.

The surgeon discusses with the parents and child before the admission the facts of the operative procedure. In his preoperative contact with the child in the hospital, he should be sure that the child understands why the operation is being done, what is involved, and how he will feel after the operation. After the surgery a rapid return to the child's normal routine will be welcomed by all.

WILLIAM B. CAREY

REFERENCES

Davenport, H. T., and Werry, J. S.: The effect of general anesthesia, surgery and hospitalization upon the behavior of children. Am. J. Orthopsychiatry, 40:806, 1970.

Haller, J. A., Jr.: Preparing a child for his operation. In Haller, J. A., Jr., Talbert, J. L., and Dombro, R. H. (Editors): The Hospitalized Child and His Family. Baltimore, The Johns Hopkins University Press, 1967.

Visintainer, M. A., and Wolfer, J. A.: Psychological preparation for surgical pediatric patients. The effect on children's and parents' stress responses and adjustment. Pediatrics, 56:187, 1975.

_____ 26D Early Health Crises and Vulnerable Children

Acute illness in the first months of life, even when the child recovers completely, at times may leave parents with the impression that their child is somehow defective and extraordinarily vulnerable to stress and disease. This may influence their handling of the child to the extent that his behavior is adversely affected. This phenomenon is not well defined or elaborated at present. However, steps in clinical management seem fairly clear.

STRESSES

The Child Was Ill But Recovered Completely. The original physical illness experienced by the child may range in severity from a severe life threatening one, such as sepsis, to the mere unrealized threat of it, as in physiologic jaundice of the newborn. In any case the condition is transient and the child emerges from it with no physical sequelae. It is not certain at this point, but it seems that the child's developmental or behavioral status is directly influenced little or not at all by the illness itself. This situation must be distinguished from the Damocles syndrome (see Chapter 27) in that there is no "sword" suspended over the child, i.e., no real threat of recurrent illness.

The Parents' Continuing Concern. At the time of the original illness the parents were understandably upset by the real or imagined dangers, including possible death, to their child. It is their enduring concern that is inappropriate. Even though the child has recovered fully, they may continue to regard him as unusually vulnerable to illnesses, including fatal ones, during childhood.

It should be stressed that this appears to be a defective mental image of the child, not a failure of emotional bonding. Parents may be warmly attached to their child and yet have a distorted perception of the child's health. On the other hand, parents may view their child quite accurately but with little emotional warmth. The extensive literature dealing with bonding in the last decade has generally failed to make this distinction (see Chapter 5).

These distorted parental perceptions appear to be affected by many of the same factors involved in affectional bonding, such as neonatal separation, which tends to magnify the concern that something might be wrong. The perceptions are probably also related to the mother's previous experience as a child and as a parent, by the amount and nature of family supports, and by the quality of the medical care. Parental views of the child may not be congruent with those of the attending medical personnel (Carey, 1969).

EFFECTS ON THE CHILD

The most clearly defined consequence of this continuing parental overconcern is the vulnerable child syndrome, first described by Green and Solnit (1964). They reported 25 children who had behavior problems and who had had the common feature of an illness or accident, mostly in early infancy, from which they had fully recovered although the parents had expected them to die. The principal components of the behavioral syndrome were difficulty with separation, infantilization, bodily overconcerns, and underachievement in school. They found that the same syndrome could also be brought on by the mother's displacing onto the child her concerns about herself or another family member.

Few studies since then have explored this phenomenon further. Sigal and Gagnon (1975) reported similar effects in children hospitalized early for gastroenteritis, Benjamin (1978) with a variety of illnesses, and Jeffcoate et al. (1970) with premature infants. McCormick et al. (1982) found that erroneous judgments of delayed development in their infants by 424 mothers were related primarily to early health problems

Although we may optimistically guess that most parents emerge from early health crises with appropriate attitudes and normal children, we have no estimate at this time of the incidence and prevalence of the vulnerable child syndrome. However, these children probably occupy a disproportionate amount of the physician's time. Levy (1980) determined that 25 per cent of 750 parents visiting outpatient hospital services in Boston viewed their child as "vulnerable." Further investigation revealed that in 40 per cent of these (or about 10 per cent of the total) there was no clinical basis for these parental concerns. The latter group was about evenly divided between those whose concern reflected fear of an earlier medical problem, long since resolved, and those for whom psychosocial issues seemed most important.

PREVENTING AND HELPING VULNERABLE CHILDREN

At the time of the early health crisis we must deal not only with the illness itself but also with the response of the caretakers. Parents must be fully but appropriately informed of the status of the child. In helping parents adjust to the realities of the situation, one of the most helpful measures is to give them repeated opportunities to express their concerns. Distorted perceptions can be dealt with promptly. Appropriate attitudes and plans can be approved. Family support can be mobilized and further professional assistance made available when needed. The importance of sensitive medical management at the time of the crisis cannot be overemphasized.

In the months and years that follow the health threat one should take care to stress the normality of the child and avoid labeling such children "high risk" or "at risk." One must be attentive for evidence of overconcern and inappropriate behavior in the parents. If parental worry seems greater than is justified by objective signs of illness, the disparity should be explored.

Green and Solnit (1964) described a strategy for managing an established vulnerable child reaction. A thorough physical evaluation of the child is the first step. Then the objective is to help parents understand that the child's symptoms came from his being considered special and that this attitude derives from their reaction to the acute life threatening illness. This can usually be accomplished at the primary care level, but sometimes psychiatric intervention is necessary.

WILLIAM B. CAREY

REFERENCES

Benjamin, P. Y.: Psychological problems following recovery from acute life-threatening illness. Am. J. Orthopsychiatry, 48:284, 1978.

Carey, W. B.: Psychologic sequelae of early infancy health crises. Clin. Pediatr., 8:459, 1969.

Green, M., and Solnit, A. J.: Reactions to the threatened loss of a child: a vulnerable child syndrome. Pediatrics, 34:58, 1964.

Jeffcoate, J. A., Humphrey, M. E., and Lloyd, J. K.: Disturbance in parent-child relationship following preterm delivery. Dev. Med. Child Neurol., 21:344, 1979.

Levy, J. C.: Vulnerable children: parents' perspectives and the use of medical care. Pediatrics, 65:956, 1980.

McCormick, M. C., Shapiro, S., and Starfield, B.: Factors associated with maternal opinion of infant development—clues to the vulnerable child? Pediatrics, 69:537, 1982.

Sigal, J., and Gagnon, P.: Effects of parents' and pediatricians' worry concerning severe gastroenteritis in early childhood on later disturbances in the child's behavior. J. Pediatr. 87:809, 1975.

_____ 26E   Chronic Illness

Chronic disease can no longer be considered a problem of adults and the elderly. At present no child is too young to have a chronic disease that will significantly affect future physical and psychosocial development. The treatment of chronic illness has evolved as a major challenge for modern pediatrics. This seeming paradox has arisen because of the successes of biomedical advances that have eliminated acute infectious disease as a primary concern and, at the same time, have made it possible to save and maintain lives in medical conditions in which the underlying disease cannot be cured. Even at the rapid rate of continuing advances in knowledge, the cures for most of these chronic diseases do not appear to be close at hand. It is estimated that several million children in the United States suffer from some form of chronic disease and that their number will continue to increase for many years to come. Table 26–1 shows the most recent data relating to the extent of selected chronic and handicapping illness in children as obtained from household interviews of noninstitutionalized persons.

The overall costs of this increasing national burden of chronic illness are extremely high as measured by days of hospitalization, numbers of physician visits, costs of prescribed medical regimens, reductions in the patients' functional capacity to carry on the roles of daily living, and loss of lifetime potential earnings. Equally important is the immeasurable toll in human suffering experienced by patients and their families. Understanding of the behavioral and psychosocial issues involved and use of appropriate interventions by medical professionals can do much to reduce these heavy medical, economic, and personal burdens of chronic illness. The particular challenge to pediatricians is to facilitate normal social, intellectual, and emotional development of children with chronic illness and to foster adaptations that help them maximize their functional potential throughout the life span.

The focus in this portion of the chapter is on the traditional chronic diseases such as diabetes, cystic fibrosis, rheumatoid arthritis, sickle cell anemia, spina bifida, and hemophilia. Other special categories of chronic illness are discussed elsewhere in this chapter.

The topics to be covered include the effects of chronic illness on the emotional and social development of the patient and the effects on family functioning in general and specifically in relation to family management of the illness. There is discussion of ways of coping with crises and stresses related to development that occur during the illness as well as stresses that derive from predictable crises during the medical course of the illness. For particular diseases there are unique characteristics that are specific causes of additional stress. The role of the physician's attitude and behavior as a significant factor in medical and psychosocial results in these patients is also discussed. Suggestions for therapeutic approaches to chronically ill patients and their families are given.

The medical treatment of such chronic diseases as cystic fibrosis, diabetes, and hemophilia has improved significantly in recent years. At present, with appropriate use of medications and dependable adherence to medical regimens, symptoms can be minimized and controlled with a high degree of effectiveness. Whereas death at an early age was frequent 25 years ago, survival into adulthood is now common. For these chronic disorders it is often the psychologic responses to illness rather than physical or medical factors that interfere with health status and the ability to function at highest levels of potential in daily living. Conventional wisdom informs us that individuals with comparable intrinsic medical disability can respond very differently to their disease. For some the disorder becomes the major focus of their lives and is a crippling disability. For others it is a relatively routine aspect in the background of active participation in appropriate school, social, and family roles. In other words, with comparable medical status, the patient can adopt a "sick role"

Table 26–1. SELECTED CHRONIC CONDITIONS CAUSING LIMITATION OF ACTIVITY AMONG CHILDREN AND YOUTHS UNDER 17 YEARS OF AGE: UNITED STATES, 1976*

Chronic Conditions	Percentage of Under-17 Population
Arthritis and rheumatism	1.0
Heart conditions	2.4
Hypertension without heart involvement	0.3
Diabetes	1.0
Mental and nervous conditions	6.7
Asthma	20.1
Impairments of back or spine	3.2
Impairments of lower extremities and hips	6.9
Visual impairments	3.7
Hearing impairments	5.2
Total number of patients under 17 years	2,266,695

*From National Center for Health Statistics: Health United States. DHEW Publication (PHS) 78-1232. Washington, D.C., U.S. Government Printing Office, 1978.

Editor's note: These data are based on household interviews. The section on allergy in this chapter reports that asthma affects only about 4 per cent of the population.

or a "wellness" stance that has important implications for the medical course of the illness as well as for personal growth and development. During the past decade psychiatric and behavioral science research have given us concepts, data, and practical guides that can be used for enhancing the medical, developmental, and functional well-being of children with chronic illness (Hamburg et al., 1980; Magrab, 1978).

Much of traditional medical training is ill suited for the care of patients with chronic illness. The physician is trained primarily to treat a medical problem unilaterally by medication or surgical intervention. A knowledgeable and actively cooperative patient is not necessary. In a great many cases that would be seen as inappropriate and undesirable, yet in the optimal medical care of patients with chronic illness the situation is exactly reversed. Patient and family knowledge and collaboration are not only desirable, they are absolutely necessary. The success of a long term medical regimen ultimately depends on many years of faithful and judicious implementation by patient or family of prescribed routines of self-care and self-medication on a daily basis.

Therefore, the achievement of a high degree of patient cooperation (compliance) is a critical aspect of the therapy. It has been estimated that in conditions for which efficacious treatment exists roughly half the patients fail to receive potential benefits because of noncompliance. The range of noncompliance in one study was 30 to 60 per cent, depending on the type of disorder, treatment regimen, and characteristics of the patient and medical care delivery (Becker and Maiman, 1980).

Current medical education does not systematically teach physicians the principles and practices that will significantly enhance patient compliance. As a result, many earnest physicians use fear arousal techniques with only modest degrees of success but with a high emotional cost to the patient.

Chronic disease patterns vary markedly. There are conditions for which reasonable control can be provided throughout most of the course of the disorder. Other diseases are marked by a progressive downhill course. Still other disorders are characterized by exacerbations and remissions. Nonetheless, there are some general principles that seem to govern patient and family responses to the stresses of chronic illness despite the fact of individuality of persons and diseases. The differing patterns of adaptation shown by different persons and for the same person under different developmental and situational circumstances are not random. For the most part they can be understood and, more important, often they can be predicted reliably.

The more predictable a situation, the more feasible it is to take appropriate steps. Preparations for dealing with predictable stress is called "anticipatory" coping. In living with chronic disease, especially when the illness originates in childhood, enhancing efficacy through anticipatory coping is especially important, because there is a cumulative aspect to the progression of the patient through successive stresses over extended time. The success or failure of the earlier coping efforts has great influence over the outcomes in later stages and may very likely determine the pattern of lifelong adaptations to the illness. Furthermore, the child who learns mastery of the illness not only enjoys better medical status but also may "make a virtue of necessity" and derive enhanced maturity and overall competence in other spheres as well. On the other hand, repeated failures in coping can lead to crippling effects on psychosocial development as well as a more difficult course of illness.

PREDICTABLE STRESS IN THE MEDICAL COURSE OF CHRONIC ILLNESS

Table 26–2 lists seven areas of crisis and stress that are confronted at some time during the course of chronic illness by all patients and families.

Onset and Diagnosis.   The onset of the disease and its diagnosis constitute the initial crisis. This is the critical period that includes the initial presentation of the diagnosis to the family through the stabilizing of the symptoms and that often involves a period of hospitalization. Attitudes and responses adopted at this time act powerfully in shaping the framework for future perceptions of the illness, in establishing a baseline of personal adaptation, and in defining the ways that the patient and his family will relate to the medical responsibilities. The fundamental importance of understanding the psychologic risks and opportunities of this earliest stage of the disease cannot be overestimated. For this reason, it is to be discussed in the greatest detail.

Because the diagnosis is typically an unexpected event, there is no prior psychologic preparation or

Table 26–2. PREDICTABLE CRISIS AND STRESS IN THE MEDICAL COURSE OF CHRONIC DISEASE

1. Onset and diagnosis
2. Disease specific, emotionally distressing medical symptoms
3. Hospitalization(s)
4. Response to appearance of initial major complication
5. Confrontation with significant therapeutic choices
6. Failure(s) of an expected therapeutic response
7. Threat of imminent death

possibility of anticipatory coping. As a result, most families are overwhelmed. Regardless of the category of the disease or the psychosocial attributes of the family, there is an initial reaction of shock, disbelief, and great anxiety on learning the diagnosis. Families may express initial reactions, however, in several different ways. There can be a response of hopelessness and helplessness with paralyzing immobility, or there may be anger and hostility that focuses on the physician who is the bearer of bad tidings, which may lead to medical shopping for a more favorable diagnosis. In still other cases the immediate response is one of denial. In the latter cases there is a seeming inability to grasp the true gravity of the situation. At the same time these denying persons are often very composed and initially respond to needed tasks with calm, cool efficiency. None of these responses is pathologic. All families urgently need help in coping with such a major life crisis. The factors that make diagnosis of chronic illness so stressful to families and the areas in which they need help are shown in Table 26–3.

At this time families experience overwhelming emotional disturbance that is a mix of varying emotional components and swings between anxiety, fear, anger, and depression. The distress of the family is communicated to even the youngest child and aggravates the impact of the illness on the child. Competent coping with these stresses is important not only to relieve the emotional suffering of the families but also to restore the patient's emotional equilibrium, which may be linked to significant physiologic changes directly related to his health status. This may be a serious concern in diabetes, for example. Mastery of family distress in this initial crisis is also linked to medical outcomes through the influences on subsequent efficacy of family management of the therapeutic regimen. The coping tasks confronting the family are listed in Table 26–4.

Behavioral Interventions at the Time of Diagnosis.   At the time of diagnosis it is important to involve both parents in explanations and instructions. Equality of parental involvement from the beginning is a key to supportive, successful family functioning over the long course of chronic illness; otherwise there may be marital strain and, at times, divorce. In the immediate situation, parents gain control over their disturbing emotional responses more quickly when they can share their grief and openly express negative feelings with each other as well as the medical staff and others. These sharing experiences can be thwarted when one parent, frequently the father, is much less involved than the other parent. The physician can initiate and reinforce the sharing of parental responsibility for understanding and managing the illness. Crisis research has shown that at times of

Table 26–3. DIAGNOSIS OF CHRONIC ILLNESS: FACTORS CONTRIBUTING TO STRESS

1. Anxiety about immediate medical outcome
2. Fears about future complications and about death
3. Loss of valued life goals and aspirations for the child
4. Recognition of necessity for a permanent change in living pattern due to the illness
5. Feelings of helplessness in management of the illness
6. Anxiety about planning for an uncertain future
7. Feelings of intense guilt or anger about the affliction

crisis, individuals are unusually susceptible to influence. Brief, judicious, well timed interventions at a time of crisis can have a deep and lasting impact. There are many examples of this principle in the course of treating chronic illness.

Medical Information.   The kind and amount of medical information are important issues. At times of great anxiety, information processing is seriously impaired in all persons, even the most intelligent. Patients and families therefore have great difficulty in comprehending explanations and instructions despite their intense interest and the motivation for coping. In the earliest phase there is a risk of information overload. Simple, repeated explanations that give basic information about the current episode and the general nature of the illness and that lay a realistic basis for a hopeful attitude are the most useful. When instructions are given, they should be highly specific, fairly brief, and, when possible, written out in detail. Misunderstanding and forgetting are typical.

The Family's Prior Experience.   In talking with the family, some effort should be made to assess their prior experience and competence in handling illness and other major events of life. Their prior competence in coping can serve as a rough measure of their current potential, and give a preview of preferred coping strategies. It is also important to learn whether the crisis of learning of the diagnosis of chronic illness is superimposed on other stressful events in their current life, such as job loss, recent death in the family, or contemplated divorce. If this is the case, additional family support should be mobilized or, if unavailable, referral for supportive casework or psychotherapy.

Table 26–4. PSYCHOSOCIAL TASKS IN COPING WITH CHRONIC ILLNESS

1. Maintaining emotional distress within manageable limits
2. Meeting the medical needs with appropriate knowledge, motivation, and practical skills
3. Preserving important relationships with others in a support network
4. Maintaining hope for the future
5. Achieving age-appropriate development for the patient in family, peer and school roles and responsibilities
6. Maintaining stable, equitable family functioning

Guilt Feelings. Feelings of guilt tend to be pervasive and should be addressed. If encouraged, parents will disclose nonrelevant behavior that they feel may have caused or contributed to the disease. When issues of heritability are involved in the disorder, this should be discussed. Frank discussion with the physician can do much to alleviate needless guilt. When genetic factors are involved, the groundwork can be laid for later genetic counseling.

Complications of the illness, shortened life span, and details of the longer course of illness must all be understood and honestly faced, but the initial period of diagnosis and onset of the disease is not the appropriate time to undertake definitive discussion of these matters. Pertinent questions related to these topics should be honestly, briefly, and tactfully answered. Many families seek a sense of mastery and cope by becoming information seekers. They will soon enough become experts on the particular chronic illness in their family. There are also families who are information avoiders, and they can also be recognized rather quickly. These "avoiders" may need referral for psychotherapeutic help if there is persistent continued denial and resistance to learning about the illness or its management.

The Physician's Role. The role of the physician at time of diagnosis is crucial. Especially then, but also on a continuing basis, there is a need for a calm, confident leader who will evaluate and interpret the results of procedures, laboratory studies, and reports from consultants or allied medical personnel. When there is no single reliable person to funnel information and interpret to the parents (or the adolescent patient), the ambiguities and real or apparent contradictions become a source of greatly heightened confusion and anxiety. The amount of distress thus generated is usually far out of proportion to the realities of the situation.

Patients with chronic illness often relate to a medical team. Over the course of the illness another member of the therapy team usually will discuss the family's concerns and assume the responsibility for monitoring psychosocial issues. If a team includes a social worker, this person would be a logical choice to carry out this needed function. At other times there may be a good fit of personality and temperamental style between the family and another team member, such as a nurse, dietitian, or physical therapist. These kinds of linkages should be encouraged. However, even when such supportive relationships arise between family and responsive team members, the physician, as the team leader, will be sought as the primary source of definitive information and clinical judgment unless his hurried, insensitive, or intimidating manner inhibits the patient or the

Table 26–5. PHYSICIAN ATTRIBUTES FOR EFFECTIVE CARE OF CHRONIC ILLNESS

1. "Person" rather than "illness" orientation; awareness of developmental, psychologic, sociocultural factors
2. Establishes relationship of trust, respect for family, nonauthoritarian, responsive mode of relating
3. Mutual agreement with patient on treatment goals; accommodation of medical and psychosocial needs
4. Therapeutic team that includes family, adjunct medical staff, and physician as leader
5. Emphasis on patient education, self-motivation, and personal responsibility for implementing medical regimen
6. Awareness of predictable crises in course of illness, anticipatory guidance to minimize impact; extra support during crises
7. Awareness of indications for psychiatric referral

family from communicating with the doctor about their concerns.

Discharge of the Patient from the Hospital. In the initial crisis of onset and diagnosis, the physician's manner sets the tone for future interactions. It should be recognized that taking the discharged child home from the hospital and assuming responsibility for home care are highly threatening to most families. As mentioned, salient information is often misunderstood or not retained. Families need to be able to call back and ask questions. At first they are uncertain about what is crucial and what is a trivial detail. The parents need exact information, but they also require bolstering of their sense of competence to carry out the needed health care tasks. The need for extra time, repetition of instructions, and repeated reassurance will subside and will not characterize most families unless in this initial period time and attention to launching them on home care are withheld. If parents are ignored or demeaned at this time, a secure basis for competent, confident home medical management is not formed. Table 26–5 summarizes physician characteristics that are associated with positive patient-physician interactions.

DEVELOPMENTAL ISSUES IN RELATION TO CHRONIC ILLNESS

At all ages chronic illness poses significant challenges to normal developmental processes and superimposes new tasks and challenges for the child or adolescent. These can be illustrated in terms of the varying impacts of diagnosis and onset of illness at different ages.

The Preschool Child. The preschool child responds to the immediate situation. The young child's problems include anxieties due to separation from his parents and fears due to instrusive and often painful procedures. He may chiefly need consistent, warm, reassuring support from parents and medical personnel and perhaps hospital based play therapy. At this age children often view negative or painful experiences as punishment for bad behavior. With hospital personnel, and later at home, they may try to remedy the situation by being especially well behaved. They need to know that they are not blamed and that they are still loved.

The Elementary School Child.   The elementary school age child is capable of understanding simple explanations of the illness and wants to know about the body organs involved. He is concerned about the cause and needs to know what to tell friends about the illness. Unless evoked by parents or other adults there is usually little concern about future complications. The child needs more discussion of what to do and what to expect now. Children often believe that they will be cured by a hospital stay and at the time of discharge need to understand the importance of continuing treatment at home. There is a need to come to terms with the permanence of the condition. There is usually a stage of bereavement with this realization. This grief reaction should be understood and accepted. It is important, however, that his own role in self-care be clarified and stressed. In an illness such as diabetes, when there is a "honeymoon" period of remission (to the patient, apparent cure), this is especially important. At all ages children should be encouraged to assume responsible tasks of graduated difficulty in the medical regimen to acquire a sense of mastery. In diabetes, for example, by age 10 the child should be able to administer his own shots. Responsibility for urine testing, under supervision, should have begun several years earlier and there should be some understanding of dietary needs and restrictions.

Adolescence.   Adolescence poses developmental tasks that clash directly with the characteristics and demands of chronic illness. While the individual is coming to terms with uncertainties about pubertal changes, the advent of major disease is a sharp blow to developing sexual identity and the body image. At a time when peer relations and fitting in with the crowd are paramount considerations, there is awareness of being quite different and fears about being unacceptable. When there are developmental urges toward independence, the illness can lead to wishes for security and dependence, on the one hand, or hyperindependent, rebellious, risk-taking behavior, on the other hand. Overprotectiveness of concerned parents may aggravate any or all of these conflicts. Most adolescents also are capable of understanding the uncertainty of the future. The prospects of complications and a shortened life span may cause great distress and, at times, depression.

The physician may help to avert extremes of response by honest, hopeful, and realistic discussions about the illness and the dilemmas posed as well as the possible options. Early in the course an adolescent often benefits greatly from talking with other adolescents who have successfully coped with the same illness. He can gain a sense of not being so alien, learn effective coping strategies, and benefit from a role model who concretely embodies intactness and good functioning. Adoption of mentor roles by helper adolescents is equally valuable in enhancing self-esteem and a realization that he can make important contributions to the welfare of others because of his successful mastery of difficult tasks. The role of peers in mediating adolescent intervention is so critical that many physicians find it valuable to organize adolescent groups in medical settings or to learn about peer self-help groups in the community for the continuing support and guidance of their long term adolescent patients.

PREDICTABLE STRESSES

The predictable stresses in the course of chronic illness are outlined in Table 26–2. The behavioral principles for intervention in the rest of these crises are the same as those discussed for onset and diagnosis. Although the major psychosocial tasks are initially confronted at the time of diagnosis, they also pervade the entire course of illness and recur with each new crisis. There is a need for continuing readaptation to the chronic illness at different developmental periods of life and in different stages of the illness. With each succeeding crisis, the past is revisited and layered onto the new crisis. A major difference in later crises is the possibility of anticipatory guidance. The physician can prepare the patient and his family for coping with imminent hospitalization, complications, or a lack of therapeutic response. With this type of anticipatory coping, the patient and his family do not have to try to improvise under conditions of great emotional distress and impaired problem solving ability. The problem has been defined and, with help, separated into manageable components. There is opportunity, ahead of time, to begin to work through the distressing emotions that each of these crises of loss of body function and intactness implies. Each time there is an accompanying loss of cherished life goals and a threat to close relationships. With help, ways to minimize the physical and emotional losses can be found. Alternatives for establishing bases for self-worth can be explored despite disfigurement or a new disability. In turn, when the physician has a heightened awareness of the critical nature of these events, he can be prepared to offer additional guidance and support. Principles for management of the threat of imminent death are discussed in Chapter 27.

PROBLEMS RELATED TO SPECIFIC CHRONIC ILLNESSES

Genetically Transmitted Diseases. Genetic transmission characterizes a number of chronic diseases, e.g., cystic fibrosis, hemophilia, and sickle cell anemia. For diabetes, the genetic component is also present but more ambiguous. In all these diseases there is considerable parental anguish and guilt related to the inheritance of the disease. The patient and siblings, in adolescence, should receive clear explanations of the nature of the genetic transmission of the disease and the possibilities of genetic counseling.

It is especially important that the physician be aware that these aspects of the illness are sensitive topics for the parents. Parents often have failed to communicate between themselves or with their children on these topics, and there have remained areas of silent tension. Parents also benefit from genetic counseling. Fears of pregnancy may greatly distress the mothers of children with heritable chronic diseases.

Cystic Fibrosis. Cystic fibrosis is a disease that has a progressive downhill course punctuated by repeated nearly fatal episodes. When it was first identified in 1938, over 95 per cent of the children in whom the disease was diagnosed died by the age of three years. At present a significant number of persons survive well into adulthood. Nonetheless the diagnosis means that the child faces a highly uncertain future. The most successful means of coping involves sufficient denial of the ever-hovering threat of early death so that it becomes worthwhile for patient and family to put efforts into meeting the daily medical demands and to engage in school, family, and peer activities that are growth enhancing and involve a future orientation and plans for adult roles and careers. This ideal balance of denial and reality is not easily achieved. The parents and patient often experience recurrent fear and anxiety about sudden death, even when the patient is clinically well. Except for the most advanced stages of cystic fibrosis, the emotional distress is not directly related to the medical threat of the symptoms. The symptoms are socially embarrassing; for example, repeated bouts of coughing, foul smelling stools, uncontrollable flatus, and later distorted appearance due to chest deformity, stunted growth, and diminished muscle mass. The need for pulmonary drainage three times a day may interfere with normal peer activities. In the face of these problems the patient may withdraw from school and peer activities into chronic depression and isolation. Physicians and family can do much to en-

courage the self-reliance and self-esteem that will give the patient the courage to meet the challenges. Physicians can help the family to avoid disabling overprotectiveness by stressing from the very beginning the many benefits of active participation in exercise, school, and peer activities. Hobbies, talents, and skills should be encouraged as additional sources of self-esteem.

In adolescence girls need to understand the potential hazards of pregnancy. Boys have the special problem of learning about the sterility that is part of the disease. There is often a delay in the onset of puberty that can be highly disturbing to the young adolescents of both sexes, and this delay of maturation should be discussed with them (Barbero, 1980).

Diabetes. Diabetes in children and adolescents usually means insulin dependent diabetes. For very young children there are special problems because of the denial of sweets and treats that they have learned to regard as rewards for good behavior and evidence of being loved. At the same time there are daily painful injections, which can be seen as punishments. Parents need help in being aware of and correcting these childish beliefs.

Older children feel the constraints of the demanding daily routine of urine testing, diet, and insulin shots. A balance must be struck between learning to take responsibility for the daily complex routines, and developing anxiety ridden overinvolvement. Some children learn to use the diabetes in manipulative ways to evade responsibilities, punish parents, or intimidate peers or siblings. Parents need help in finding the middle path.

Generally children can start urine testing (with help) at six or seven years and start giving their own shots at age 10. Parents benefit from guidance at this time of teaching self-responsibility to their children. Diabetic camps are often a useful transitional experience. Such a camp teaches skills, gives peer support, and provides the role modeling of other children who are carrying out the tasks. It is a time of protected separation from family. On return home, both parents and child can function on a new basis of added responsibility for the patient.

In adolescence the diabetic individual may reencounter the disease from a new vantage point. It can be a difficult time, and depression or denial may emerge from this reappraisal of the meaning of the disease. For a period of time the responsible routines that had been accepted in earlier years may be abandoned. Understanding of the emotions, clear explanations, and patience with the adolescent's exploratory needs will help to make this a transient period of disequilibrium.

The adolescent also takes more cognizance of the future and begins to worry about the long term effects of the illness and the implications for marriage and career. During adolescence the relationship with the medical staff may have to be renegotiated to reflect the growing maturity. Although it is now generally agreed that close approximation to normoglycemia is highly desirable to decrease the chances of complications, extremely tight control presents most adolescents with problems. There is a substantial risk of an insulin reaction when efforts are made to keep the blood sugar at normal levels and the urine sugar free. Insulin reactions are intrinsically threatening and in addition cause acute social distress. The individual may pass out unexpectedly and may have disturbing autonomic reactions evidenced by sweating and trembling. Perhaps the most disturbing reactions to the adolescent are the uncontrollable mood swings such as rage, silliness, or weeping for no reason. At times there can be automated behavior of which the patient has no awareness. Unless there can be some balancing of the needs to keep blood glucose levels down with enough margin of hyperglycemia to eliminate or minimize the risk of insulin reactions, the adolescent may elect to rebel and risk very poor control.

As with cystic fibrosis, adolescents need to have full and frank discussions about reproductive and family planning issues. Genetic explanations and discussions of the risks of pregnancy and the likelihood of male impotency are all matters of great concern to them. Adolescents often have problems about how to integrate the diabetes into their social life since it is an "invisible handicap." Do they "confess it" on the first date? Do they risk a friendship by informing the peer that they may have insulin reactions and instructing them about giving needed sugar? Can they risk overnight visits? What will friends think when they see them giving themselves a shot? Insulin dependent diabetes is uncommon enough that they may not know any other diabetic adolescents with whom to discuss these problems. Provision for peer groups of adolescent diabetics, either within the medical setting or by referral to groups such as the American Diabetes Association, can be very important in adolescent adaptation to diabetes.

Many adolescents are neither aware of nor prepared for the fact that the growth spurt of puberty notably increases insulin needs. They may experience this as loss of confidence in their ability to control the diabetes. Also, girls may have additional control problems due to fluctuations in insulin needs related to the menstrual cycle. Testing of the urine may be resisted by some girls during menstruation. In adolescence the "specialness" of the diet is strongly resented. It may be extremely important for the nutritionists to go over food exchanges of various items at "McDonalds" and similar fast food chains so that the diabetic adolescent can make wise choices there instead of throw-

ing caution to the wind in joining his peers. These negotiations with the adolescent can do much to prevent the impulsive abandonment of the diabetic regimen that occurs when they feel helpless, frustrated, and overly burdened by the demands of the disease.

Hemophilia. Hemophilia poses special problems because of the restrictions on physical activity, especially contact sports, in order to minimize the risk of bleeding episodes. Patients and families need specific guidance about the types and intensity of physical activities so that extremes of overprotection or rebellious, life endangering risk-taking do not occur. Most families seem capable of achieving this with some medical guidance. Another concern is the fear of taking trips or leaving the vicinity of the major medical facility that is the source of specialized treatment. Many families have been unduly restricted by these concerns and never have had a vacation. Baby sitters often are not trusted and there is no "time out" for any recreation (Salk et al., 1972).

Sickle Cell Anemia. Sickle cell anemia and thalassemia require periodic hospitalizations for crises or transfusions. These children experience a great deal of pain and realize early the imminence of death. Most of the children also miss many school days and the related educational and personal growth experiences. By the time of adolescence they are aware of the extent of their difference from others and the limitations on their vocational options. Priapism is a special problem for some adolescents. It may last for several days and require hospitalization. The patient also needs to know that the priapism may eventually lead to impotence (Nishiura and Whitten, 1980).

EFFECT OF CHRONIC ILLNESS ON FAMILIES

Malfunctioning of the family is signaled by one or more of the following responses: significant failure to adhere faithfully to important aspects of the medical regimen; criticizing, difficult, and uncooperative behavior on the part of the family; failure to keep appointments or extreme tardiness for scheduled check-ups; and evidence of poor medical progress, including frequent hospitalizations. What are the ways in which families can be helped in living with daily stress of chronic disease so as to avoid these negative outcomes?

There is now active research on the development of suitable measures to assess the impact of chronic illness on family functioning, since there is evidence that this is strongly related to the health and normal growth and development of the child, as well as being important in its own right (Stein and Reissman, 1980).

Measures of family functioning, of course, also take into account the nature of the chronic illness and the severity, duration, and fluctuations in the course of the illness as factors determining the psychologic effects of the illness. The changing developmental needs of the sick child also influence the impact on the family. For example, entry into adolescence may enormously increase family stress.

Sensitivity to the spectrum of effects on the family can enable health professionals to give needed advice, emotional support, or tangible assistance in one or all of the relevant areas that may be indicated at a given time. Review of the status of family functioning should be a part of each check-up or visit for treatment.

Families retain the coping mechanisms that they had prior to the advent of illness. Some families have more strengths than others. There are several general variables that predict impaired coping— disadvantaged or poverty status, minimal education of the mother (less than ninth grade), and minority status. When these three factors are all present, ongoing support by supplementary medical and social services should be arranged. Prior parenting style, prior marital adjustment, and previous medical stresses and experiences play a role. Finally, there is the presence or absence of additional concurrent burdens of stressful life events. Information about each of these seven factors should be assessed and included in the medical history

Economic burdens can easily be underestimated in all patients. There are significant costs for the special equipment, medications, special diets, and transportation to medical facilities. In addition, it is increasingly common for both parents to be employed. Days off from work to keep medical appointments may represent a significant loss of income, and this should encourage the scheduling of multiple procedures for a single visit when possible. The economic burdens of the illness are also often underestimated as factors in causing marital strife and in increasing parental resentment of the sick child.

Parental behavior that reflects the mutual agreement of the partners and that is consistent with age-appropriate expectations of the sick child are associated with the best medical and developmental outcomes. For reasons already discussed, when there is chronic illness, some families may tend toward indulgent overly permissive behavior because of guilt. Other parents may be overly dominant. Some may be overcontrolling by temperament, others by the desire to comply rigidly with the medical regimen for fear of dire health consequences. In many families suppressed anger or resentment of the sick child may cause one or both parents to overtly or covertly reject the child.

Any of these distortions of parenting will have negative effects on the child, which can be recognized by the physician. The guilt behavior of parents can lead to infantilized, regressed behavior in children when there is overprotection. When usual discipline is not enforced, children become manipulative, spoiled, or tyrannical. Insecure and withdrawn behavior develops when children are not given opportunities to take responsibility or when they are not encouraged to engage in the give and take of peer relations. Overcontrolling or resentful parenting can lead to many of the same responses just described. In addition there is a danger that rebellious behavior may be played out in the arena of the illness and its requirements, in order to intimidate or punish the parents, particularly during adolescence. Simple advice and counseling from the physician may help parents to feel comfortable in treating the sick child as normally as possible, including appropriate discipline. At other times it may be clear that referral for psychotherapy is needed because of parental resistance to the pediatrician's advice or helplessness in changing the distorted parenting behavior.

Sibling deprivation is likely when parents become overinvolved with the needs and routines of the sick child. The neglect of the other children is intrinsically undesirable. Furthermore it leads to anger and resentment on their part and a new burden of guilt on the part of the patient, who perceives the envy of the siblings. These additional tensions further burden the family functioning. Physicians should routinely inquire about the well-being of siblings and remind parents of the importance of their developmental needs. Signs of depression, rebellion, or undue hypochondria in siblings are warning signals of their deprivation.

The role of social support in buffering stress and contributing to improved health outcomes is now better understood (Hamburg and Killilea, 1979). Persons who tend to withdraw from social contact under stress are at high risk for coping failure and referral for psychiatric help. Inasmuch as chronic stress characterizes most families with chronic illness, there is a need for an ongoing support network to buffer the stress and bolster family functioning. This support usually comes from relatives, church affiliations, and a friendship group that includes neighbors and fellow employees. When support systems are lacking, the family or patient may frequently and inappropriately call the physician or allied medical personnel in an effort to gain needed social support. Such patients benefit greatly from learning how to find and utilize community resources. Table 26–6 summarizes the potential effects on the patient when there are failures in family functioning.

Despite the daily stress and episodic crises that characterize living with chronic disease, many

THE EFFECT OF GENERAL MEDICAL ILLNESS AND ITS TREATMENT ON DEVELOPMENT AND BEHAVIOR   **463**

Table 26–6. EFFECTS OF FAILURES IN FAMILY FUNCTIONING

Medical effects
Poor medical progress or frequent hospitalizations
Failure to maintain medical regimen
Inadequate medical supervision due to:
Broken, missed, or lateness for appointments
Uncooperativeness, frequent complaints (often unjustified)
Parenting distortions
Overprotection; infantilizing of child that stunts development
Push to pseudomaturity; leads to insecure child with shallow veneer of competence and "adultness"
Rejection; highly critical, minimal responsiveness to child; this can lead to child behavior that is:
Depressed, withdrawn
Hostile, oppositional behavior with parents
Delinquent behavior outside home
Personal distress of patient
Denial of illness or risk-taking with health damaging consequences
Withdrawal from normal peer interactions; isolation
Diminished academic motivation and skills
General sense of incompetence or helplessness
Depression about ability to control illness and loss of hope about future

ferentiated outcome variables, the physician can be sure that the multifaceted challenges of chronic illness are being met with the highest standards of modern medical care and that the patients will realize their full potential.

BEATRIX A. HAMBURG

REFERENCES

Barbero, G. J.: Cystic fibrosis in adolescence. In Shew, J. T. Y. (Editor): The Clinical Practice of Adolescent Medicine. New York, Appleton-Century-Crofts, 1980.
Becker, M. H., and Maiman, L. A.: Strategies for enhancing patient compliance. J. Community Health, 6:113, 1980.
Coogler, D. W.: Cystic fibrosis—a not so fatal disease. Pediatr. Clin. N. Am., 21:935, 1974.
Greydanus, D. E., and Hofmann, A. D.: Psychological factors in diabetes mellitus: a review of the literature with emphasis on adolescence. Am. J. Dis. Child., 133:1061, 1979.
Hamburg, B. A., and Killilea, M.: Relation of social support, stress, illness and use of health services. In Healthy People: The Surgeon-General's Report on Health Promotion and Disease Prevention. Background Papers. DHEW Publication 79-55071A, 1979.
Hamburg, B. A., Lipsett, L. F., Inoff, G. E., and Drash, A. L.: Behavioral and Psychosocial Issues in Diabetes. NIH Publication 80-1993. Washington, D.C., U.S. Government Printing Office, 1980.
Magrab, P. R.: Psychological Management of Pediatric Problems. Volume 1. Early Life Conditions and Chronic Diseases. Baltimore, University Park Press, 1978.
Moos, R. H.: Coping with Physical Illness. New York, Plenum Medical Book Co., 1977.
National Center for Health Statistics: Health United States, 1978. DHEW Publication (PHS) 78-1232, 1978.
Nishiura, E., and Whitten, C. F.: Psychosocial problems in families of children with sickle cell anemia. Urban Health, 9:32, 1980.
Salk, L., Hilgartner, M., and Granich, B.: The psychosocial impact of hemophilia on the patient and his family. Soc. Sci. Med., 6:491, 1972.
Stein, R. E. K., and Reissman, C. K.: The development of an impact-on-family scale: preliminary findings. Med. Care, 18:465, 1980.
Tavormina, J. B., Kastner, L. S., Slater, P. M., and Watt, S. L.: Chronically ill children—a psychologically and emotionally deviant population? J. Abnorm. Child Psychol., 4:99, 1976.
Zeltzer, L.: Chronic illness in the adolescent. In Shenker, I. R. (Editor): Topics in Adolescent Medicine. New York, Grune & Stratton, Inc., 1978, pp. 226–253.
Zipsook, S., and Gammon, E.: Medical noncompliance. Int. J. Psychiatry Med., 10:291, 1980–1981.

patients and families demonstrate that they are capable of good and, at times, impressive psychologic adaptation and functioning (Tavormina et al., 1976). The challenge in giving pediatric care is to foster this type of healthy psychosocial growth and development at the same time as the medical needs are being served. Appropriate care of patients requires that physiologic markers and measures of medical status be only one of the relevant outcome variables. Equal attention must be given to psychologic development, progress in school, relations with peers and siblings, and family functioning as other significant indicators of a good result or lack of progress. When there is systematic concern for and monitoring of these multiple dif-

---

26F   Otitis Media

Otitis media is a major pediatric health problem because of its relatively high incidence (particularly among preschool children), its tendency to recur, and the severe complications. Middle ear disease is probably the most frequent infectious disease of children requiring medical attention (McInerny et al., 1978). The cumulative incidence of otitis media during the first six years of life among children in the United States has been estimated to exceed 90 per cent, and 50 per cent of the children who have one episode during the first year of life experience six or more recurrences

in the next two years (Paparella and Juhn, 1979). The challenges presented by otitis media are increased by conflicting results from epidemiologic studies, controversies regarding pathogenesis and natural history, difficulties in diagnosing middle ear status, differences in opinion regarding medical versus surgical management, and uncertainty regarding long term sequelae (Paradise, 1980).

Middle ear disease is associated with a myriad of chronic disabling and even life threatening complications. Although the middle ear cleft itself is anatomically protected from external trauma by encasement deep in the temporal bone, infections can be disseminated to the mastoid process, brain, and cranial nerves. The serious acute sequelae of this dissemination range from permanent perforation of the tympanic membrane to thrombophlebitis of the sigmoid sinus, meningitis, and brain abscess (Table 26–7). In addition, during the past decade there has been increasing concern regarding the long term neurodevelopmental and behavioral effects of recurrent otitis media. Chronic middle ear disease and its associated conductive hearing loss have been linked with speech-language delays, central auditory processing dysfunctions, depressed verbal intelligence, learning disabilities, and behavioral disorders (Gottlieb et al., 1979).

This section briefly considers the pathology and clinical diagnostic assessment of the middle ear in otitis media. The major emphasis, however, is a focus on the potential health hazards associated with recurrent middle ear disease and the effects of fluctuating conductive hearing loss on a variety of neurodevelopmental skills. (See also Chapter 39 concerning hearing impairment in general.)

EPIDEMIOLOGIC AND PATHOLOGIC CONSIDERATIONS

Epidemiologic data relating to otitis media are incomplete, and the validity of the existing data is questionable because of the variability in methodology and diagnosis. However, it seems clear that ear disease is most prevalent during the first two years of life because of greater susceptibility to infection, the relatively large quantity of tonsillar and adenoidal tissue, postural factors in feeding, and the functionally immature eustachian tube.

The influence of race as an epidemiologic factor is difficult to assess. Studies in this area are influenced by socioeconomic factors and the availability of and compliance with medical care. For example, American Indians and Eskimos reportedly have a higher incidence of otitis media and the course of the disease is more severe. However, the influence of poor sanitation and hygiene, crowded living conditions, parental neglect, poor compliance with therapeutic programs, and inadequate medical care must be considered in interpreting the results of these and all epidemiologic surveys.

Sex ratio studies also require further clarification. It has been suggested that occurrence and recurrence rates, as well as the severity of the episodes, are increased in boys (Paradise, 1980). The difficulty in controlling exogenous and endogenous variables interferes with the interpretation of these epidemiologic surveys. Seasonal and climatic epidemiologic reviews reveal a higher inci-

Table 26–7. MEDICAL COMPLICATIONS ASSOCIATED WITH OTITIS MEDIA LISTED IN ORDER FROM MORE TO LEAST FREQUENT*

	Fluctuating Conductive Hearing Loss
↑ F R E Q U E N C Y	Permanent perforation of tympanic membrane†
	Osteolysis of one or more ossicles†
	Tympanosclerosis of one or more ossicles†
	Permanent occlusion of eustachian tube†
	Acute coalescent mastoiditis
	Acute petrositis (Gradenigo's syndrome)
	Facial nerve palsy
	Suppurative labyrinthitis
	Nonsuppurative labyrinthitis with sensorineural impairment
	Thrombophlebitis of sigmoid sinus
	Brain abscess
	Meningitis

*Adapted from Shambaugh, G. E., Jr.: Complications of otitis media. *In* Wiet, R. J., and Coulthard, S. W. (Editors): Proceedings of the Second National Conference on Otitis Media. Columbus, Ohio, Ross Laboratories, 1979, p. 48.
†Associated with conductive hearing loss.

dence of otitis media during the winter and spring, the peak seasons for viral upper respiratory infections. Special population studies have indicated a higher incidence of middle ear disease in children with cleft palate and Down's syndrome, reflecting possible anatomic or immunologic factors.

The classification of middle ear disease similarly reflects conceptual variations. The term otitis media encompasses a variety of middle ear disorders, including serous, acute, subacute, chronic suppurative, and mucoid otitis media. The pathogenesis of the various subtypes may be differentiated in some cases by the clinical manifestations, course, and prognosis of the disease (see Table 26–8). The role of the eustachian tube in the pathogenesis of otitis media has been defined.

The normal physiologic functions of the eustachian tube include ventilation of the middle ear to equalize air pressure on both sides of the tympanic membrane and replacement of absorbed oxygen, protection from nasopharyngeal secretions, and clearance of secretions from the middle ear by passage into the nasopharynx. Dysfunction of the eustachian tube is associated with middle ear disease. Middle ear effusion (the most common cause of conductive hearing loss in children) appears to be the result of faulty function of the eustachian tube, precipitated by either infection or allergy. Effusion associated with eustachian tube dysfunction (e.g., swelling and closure of the lining during upper respiratory infection) is usually characterized by an acute onset and a short duration. Chronic or recurrent episodes of middle ear effusions are common in children, and the condition is often bilateral. Functional obstruction of the eustachian appears to be the primary cause of otitis media with effusion, although mechanical

Table 26–8. DIFFERENTIAL DIAGNOSIS OF SEVERAL TYPES OF OTITIS MEDIA*

	Serous Otitis Media	Mucoid Otitis Media	Acute Otitis Media	Chronic Suppurative Otitis Media
Pathogenesis	Transudation due to negative pressure or local secretion Thin, watery, yellow	Absorption of watery component of serous fluid or secretory disorder of lining cells; "glue ear"	Fluid in middle ear, a culture medium for bacteria and viruses with upper respiratory tract infection Earache as main symptom	Perforation of tympanic membrane with acute otitis media Chronic tympanic membrane perforation with or without drainage
Tympanic membrane on otoscopy	Depends on amount of fluid; concave air-fluid meniscus; air bubbles trapped in fluid; fluid fills cavity	Tympanic membrane appears opaque; slight orange or blue color	Hyperemia of pars flaccida; spreads to thick, pink-red tympanic membrane; short process no longer visible	Perforation of tympanic membrane, with a drainage; yellow and foul odor
Pneumatic otoscopy	When fluid fills cavity, tympanic membrane is poorly mobile	Poorly mobile tympanic membrane	Progresses from a mobile tympanic membrane early to decreased motility as tympanic membrane becomes more inflamed	Poorly mobile to immobile
Audiogram	Mild conductive hearing loss from 5 to 20 dB. (depends on quantity of fluid)	Mild to moderate conductive loss from 10 to 50 dB.	Moderate conductive loss from 10 to 40 dB.	Moderate to severe loss depending on disease of tympanic membrane
Tympanogram	Negative pressure curve or flat curve (depends on quantity of fluid)	Usually flat	Flat or negative pressure	Flat
Mastoid x-rays	Normal to slight clouding	Normal to slight clouding	Clouding of air cells	Clouding of air cells

*Adapted from Jaffe, B. F.: Diagnosis of otitis media for the clinician. *In* Wiet, R. J., and Coulthard, S. W. (Editors): Proceedings of the Second National Conference on Otitis Media. Columbus, Ohio, Ross Laboratories, 1979, pp. 14–18.

obstruction is also common in infants and preschool children. Mechanical obstruction can result from a persistent collapse of the eustachian tube due to increased tubal compliance or an inactive opening mechanism (Bluestone, 1979).

Streptococcus pneumoniae and *Hemophilus influenzae* are among the organisms most often associated with bacterial otitis media. *Streptococcus pneumoniae* is the more common in all age groups, and the incidence of disease casued by this organism increases with advancing age. Other bacteria associated with acute otitis media include group A beta-hemolytic *Streptococcus, Neisseria catarrhalis, Staphylococcus aureus, Escherichia coli, Klebsiella pneumoniae,* and *Pseudomonas aeruginosa* (Paradise, 1980). Although acute otitis media is most often associated with viral upper respiratory infections, the virus per se is not a major epidemiologic agent in middle ear disease. The respiratory syncytial virus is recovered most often in the few cases of middle ear disease associated with primary viral infections.

EXAMINATION OF THE MIDDLE EAR

The symptoms most often associated with acute middle ear infection include earache, pulling or rubbing the ears, ear discharge, balance problems, and hearing loss. These symptoms usually prompt an otoscopic or pneumatic otoscopic examination, on which the diagnosis of otitis media is usually based. Additional methods for evaluating the middle ear include pure tone audiometry, impedance testing, tympanocentesis, and myringotomy. Hearing impairment is not a pathognomonic finding and cannot be correlated with viscosity of the middle ear fluid (Paradise, 1980).

PURE TONE AUDIOMETRY

In middle ear assessment by pure tone audiometry, pediatric audiologists usually consider a hearing loss greater than 20 dB. significant in language development (some place the threshold of concern at·10 to 15 dB.). Otologic screenings of school age children reveal that the incidence of abnormal hearing evaluations is increased in learning disabled children. The persistence of auditory acuity deficits during the school years is undoubtedly a significant factor affecting academic performance. One study reported that 38 per cent of learning disabled children had abnormal audiograms, compared to 16 per cent of the control subjects. Further, more than 50 per cent of the learning disabled children manifested abnormalities in pure tone audiometry, tympanometry, or both (Bennett et al., 1980).

TYMPANOMETRY

Tympanometry, an increasingly important adjunct or replacement for otoscopy, is an easily performed procedure that can provide valuable information regarding the compliance of the tympanic membrane in response to variations of air pressure in the external auditory canal. The air pressures involved are very small compared with those generated with the pneumatic otoscope. Maximal compliance of the tympanic membrane occurs when air pressure is the same on both sides of the membrane. Unequal pressures usually take the form of negative pressure in the middle ear space. Impaired aeration of the middle ear cavity, associated with a nonpatent eustachian tube, is ultimately manifested by absorption of the static air and a resultant negative pressure. This negative pressure may lead to transudation of fluid into the tympanic cavity. The tympanic membrane can then retract, leading to a mild conductive hearing loss, regardless of the presence of middle ear fluid. Objective evidence of tympanic membrane retraction can be visualized with tympanometry. Various disorders of the middle ear have been described in terms of the basic tympanogram pattern, the Jerger classification (see Table 26–9, Fig. 26–1). Various other methods for otologic and acoustic assessment are available (e.g., electrocochleography, auditory evoked response). Their value in diagnosis and follow-up of middle ear disease is limited, however, in office management.

NEURODEVELOPMENTAL AND BEHAVIORAL SEQUELAE OF RECURRENT MIDDLE EAR DISEASE

Physicians are generally familiar with the speech, language, educational and behavioral complications associated with deafness, particularly in the prelingual or congenitally deaf child. A relationship may similarly exist between the recurrent fluctuating conductive hearing loss associated with chronic otitis media and various neurodevelopmental disorders (Table 26–10).

NEURODEVELOPMENTAL DISORDERS

It has been proposed that repeated episodes of hearing loss during a critical developmental period (before age three years) interfere with the acquisition of central auditory processing mechanisms and speech-language skills. A speculative hypothesis suggests the following sequence of events: chronic otitis media→recurrent conductive hearing loss→disturbed central auditory processing skills→speech-language delays→impaired educational progress. The interaction is undoubtedly much more complex than this proposed linear progression, involving numerous endogenous and exogenous variables, such as genetic predisposition, inherent intellectual potential, severity and frequency of the hearing loss, and type of medical-surgical management (Fig. 26–2). However, there have been few controlled studies firmly establishing this link between transient conductive hearing impairment and subsequent neurodevelopmental disorders (Paradise, 1980; Ventry, 1980).

Research investigations suggest that if hearing

Table 26–9.   TYMPANOGRAM PATTERNS*

Type	Tympanometry Curve	Associated With	Physiologic Significance
A	Figure 26–1A	Normal middle ear function	Adequate relative compliance and normal middle ear pressure at point of maximal compliance
A$_s$	Figure 26–1B (low peaks)	Otosclerosis, thickened tympanic membrane, heavily scarred tympanic membrane, sometimes tympanosclerosis	Normal middle ear pressure and limited compliance relative to mobility of normal tympanic membrane
A$_d$	Figure 26–1B (high peaks)	Discontinuity of ossicular chain; tympanic membrane demonstrates a large monomeric membrane	Large changes in relative compliance with small changes of air pressure; extremely flaccid tympanic membrane
B	Figure 26–1C	Perforation of tympanic membrane; serous otitis media; adhesive otitis media; congenital middle ear malformation; ear canal blocked with cerumen	Function representing little or no change in compliance of middle ear as pressure is varied in external ear; no point of maximal compliance
C	Figure 26–1D Figure 26–1E	? Fluid in middle ear	Near normal compliance and middle ear pressure of −200 mm. of water or worse

*Adapted from Northern, J. L., and Downs, M. P. (Editors): Objective hearing tests. *In* Hearing in Children, Ed. 2, Baltimore, The Williams & Wilkins Co., 1979, pp. 150–158.

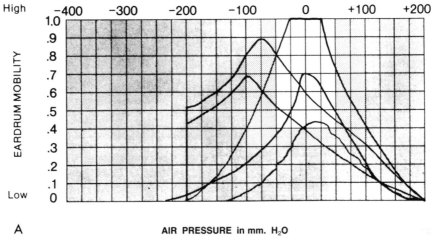

A, Normal tympanogram (type A):
        Peak: −100 mm. H$_2$O or better
Indications: Eardrum intact
                    Mobility normal
                    Middle ear pressure normal

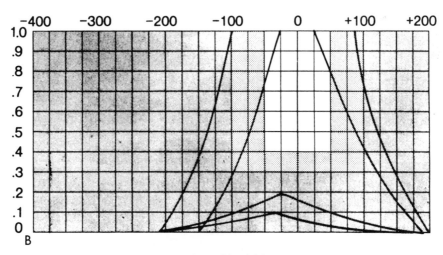

B, Abnormal tympanogram (types Ad and As):
    High peaks   (Ad)
            Peak: Normal pressure; appears "chopped off"
        Indication: Mobility excessive
        Conditions: Ossicular discontinuity or normally flaccid eardrum

    Low peaks   (As)
            Peak: Normal pressure; very shallow
        Indication: Stiff middle ear system
        Conditions: Thickened eardrum or tympanosclerosis or otosclerosis

Figure 26–1. Tympanographic patterns indicating various states of the middle ear. White areas indicate normal middle ear compliance. Gray areas indicate range of borderline function. (Courtesy of American Electromedics Corporation, Inc., Hudson, New Hampshire.)

(Illustration continued on pages 468 and 469)

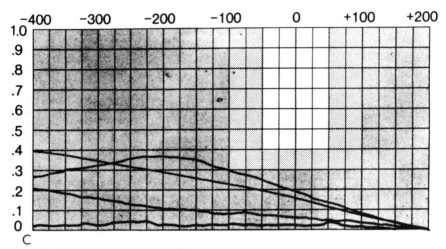

C, Abnormal tympanogram (type B):
        Peak: Flat trace, no peak
   Indications: Limited or no middle ear mobility
   Conditions: (1) PVT normal—eardrum severely retracted or otitis media
             (2) PVT large—eardrum perforated or patent ventilating tube
             (3) PVT small—complete blockage of external ear canal (wax or debris)

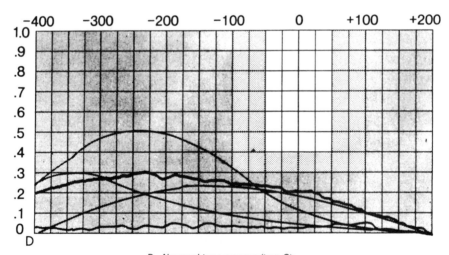

D, Abnormal tympanogram (type C):
      Rounded peaks
        Peak: No sharp peak
   Indications: Eardrum retracted
            Mobility sluggish
   Conditions: Probable otitis media

Figure 26–1 *Continued.*

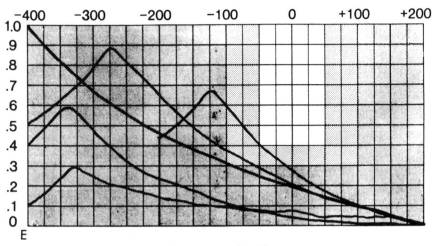

E, Abnormal tympanogram (type C):
    Sharp peaks
            Peak: -100 mm. H_2O or worse
    Indications: Eardrum retracted
            Mobility normal
    Conditions: Oncoming or resolving effusion
            Poor eustachian tube function

Figure 26–1 *Continued.*

Table 26–10. NEURODEVELOPMENTAL-BEHAVIORAL COMPLICATIONS OF HEARING LOSS

Severity of Hearing Loss	Possible Etiologic Origins	Complications			Types of Therapy
		Speech-Language	*Educational*	*Behavior*	
Slight 15–25 dB. (ASA)	Serous otitis media Perforation of tympanic membrane Sensorineural loss Tympanosclerosis	Difficulty with distant or faint speech	Possible auditory learning dysfunction May reveal a slight verbal deficit	Usually none	May require favorable class setting, speech therapy, auditory training Possible value in hearing aid
Mild 25–40 dB. (ASA)	Serous otitis media Perforation of tympanic membrane Sensorineural loss Tympanosclerosis	Difficulty with conversational speech over 3 to 5 feet May have limited vocabulary and speech disorders	May miss 50% of class discussions Auditory learning dysfunction	Psychologic problems May act inappropriately if directions are not heard well Acting-out behavior Poor self-concept	Special education resource help Hearing aid Favorable class setting Lip reading instruction Speech therapy
Moderate 40–65 dB. (ASA)	Chronic otitis media Middle ear anomaly Sensorineural loss	Conversation must be loud to understand Defective speech Deficient language use and comprehension	Learning disability Difficulty with group learning or discussion Auditory processing dysfunction Limited vocabulary	Emotional and social problems Behavioral reactions of childhood Acting out Poor self-concept	Special education resource or special class Special help in speech-language development Hearing aid and lip reading Speech therapy
Severe 65–95 dB. (ASA)	Sensorineural loss Middle ear disease	Loud voices may be heard 2 feet from ear; +/– Identification of environmental sounds; Defective speech and language If before 1 year: no spontaneous development if loss present < 1 year	Marked educational retardation Marked learning disability Limited vocabulary	Emotional and social problems that are associated with handicap Poor self-concept	Full time special education for deaf children Hearing aid, lip reading, speech therapy Auditory training Counseling
Profound 95 dB. or more (ASA)	Sensorineural or mixed loss	Relies on vision rather than hearing Defective speech and language Speech and language will not develop spontaneously if loss present < 1 year	Marked learning disability owing to lack of understanding of speech	Congenital and prelingually deaf may show severe emotional underdevelopment	As above Oral and manual communication Counseling

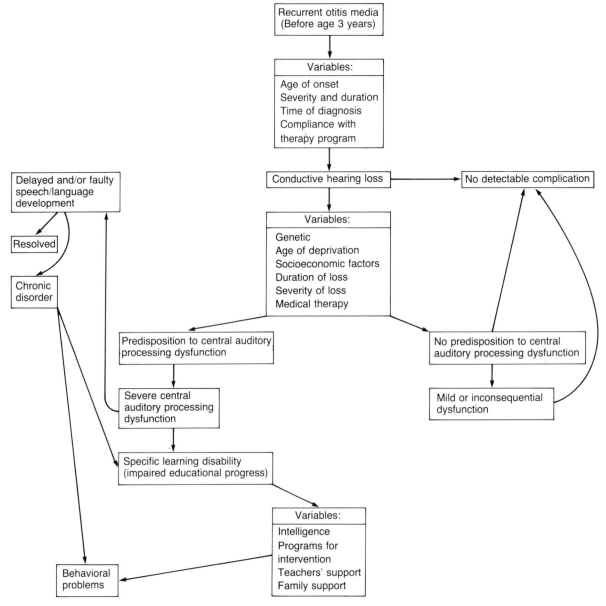

Figure 26–2. Schematic representation of a theoretical etiologic link between otitis media and neurodevelopmental disorders.

loss occurs during early developmental stages, there are neuronal changes in the central nervous system and disruptions in speech and language acquisition (Gottlieb et al., 1979). It appears that during developmental periods auditory stimuli are received and coordinated in the orderly neuro-developmental maturation of central auditory processing. The period from birth to approximately age three years is critical in speech and language development. The child progresses from the use of single words at about 12 months to meaningful word combinations by approximately two years of age. The child with profound deafness during this period will probably have difficulty in producing intelligible speech as well as in

understanding speech. Unfortunately this critical period of language development coincides with the peak incidence of chronic (recurrent) middle ear disease.

Holm and Kunze (1969) compared the language skills of two groups of children—an experimental group with histories of fluctuating hearing losses resulting from chronic otitis media and a matched control group of children with normal hearing. Evaluations of psycholinguistic processing, receptive vocabulary, and articulation suggested significant deficits in the acquisition of vocabulary, articulation skills, and ability to receive and express ideas through spoken language in the experimental group.

Kaplan et al. (1973), who reviewed data 489 Alaskan Eskimo children followed through the first 10 years of life, suggested that otitis media during the critical periods of language development was associated with impaired verbal development. The number of episodes played an important role in the development of this dysfunction.

Lewis (1976) reported that Australian aboriginal children with histories of chronic middle ear disease had inferior scores on measures of vocabulary knowledge, nonverbal intelligence, and auditory processing when compared with European and other aboriginal children. Middle ear disease was implicated as one cause of these deficiencies, acting "in concert with a variety of other determinants such as malnutrition, sociocultural inequality, reduced motivation, and indifferent states of general health."

Katz (1978) concluded that conductive hearing losses restrict the sounds of the environment from stimulating the cochleae and thereby deprive the auditory system of normal activity. The effects of this deprivation, which causes a disruption in auditory perception and language, depend on age of onset, the duration and degree of loss, unilateral as opposed to bilateral loss, and constant versus variable deficit.

Data from Cook and Teel's study (1979) of children with speech and language problems suggested that although auditory acuity was within normal limits for pure tones, negative middle ear pressure may have been associated with delayed linguistic development and competence.

The consequences of recurrent middle ear disease during the critical period of language development may be mediated by a primary interference with central auditory processing mechanisms. Language impairment may reflect underlying auditory perceptual handicaps (i.e., auditory memory, discrimination, and attention). A series of investigations suggested a possible causal association between early and severe chronic otitis media, auditory dysfunction, and language deficiencies among academic underachievers (Gottlieb et al., 1979; Zinkus and Gottlieb, 1980; Zinkus et al., 1978). One of the earliest manifestations of a central auditory processing deficit appears to be a delay in developing sequential language skills (Zinkus et al., 1978). These studies suggested that both receptive language skills and associative potentials related to symbolic language are significantly impaired among children with chronic otitis media (Zinkus and Gottlieb, 1980).

There is an increasing fund of knowledge relating disruptions in peripheral sensory organs and neurochemical imbalances in the central nervous system. It has been implied that various chemical modifiers may either enhance or impair the formation of memory (and learning) by influencing cerebral metabolism (Dunn, 1978). The effects of recurrent middle ear disease on learning and behavior may possibly reflect modifications of neuronal architecture and chemistry. The effects of visual deprivation on the neuronal integrity of the occipital lobe have been established, but the effects of auditory deprivation are not yet clear. Animal experimentation suggests that incomplete conductive hearing losses may result in abnormal central auditory neuron structures in the cochleae nuclei. These hypotheses are all based on the existence of a critical developmental period during which proper and meaningful sounds must be received in order for the central auditory processing system to mature normally (Webster and Webster, 1977).

Deaf children exhibit auditory processing dysfunctions that severely impair cognitive function. Impaired learning as a consequence of middle ear disease and fluctuating hearing loss has been attributed to numerous underlying factors. The common denominator of many studies is the implication of dysfunction in various central auditory processing mechanisms resulting from the effects of recurrent otitis media and repeated hearing loss. Assessments of intellectual ability among learning disabled children have revealed depressed verbal scores compared with performance scores (Lewis, 1976; Zinkus and Gottlieb, 1980; Zinkus et al., 1978). The deficiencies in verbal ability appear to correlate with the number of episodes of otitis media (Kaplan et al., 1973). Howie (1980) concluded that the cumulative data from medical, educational, otologic, audiologic, and psychologic sources suggest that children suffering from recurrent otitis media have temporary or permanent impairments of auditory, verbal, and intellectual skills. Needleman (1977) has further defined this relationship, suggesting that chronic otitis media affects various phonologic skills, such as auditory discrimination, auditory closure, and auditory blending. This impairment may be reversible, and children may catch up on phonologic development over time. Reading skills appear to be compromised among children with histories of recurrent middle ear disease and impaired educational progress (Masters and Marsh, 1978; Zinkus and Gottlieb, 1980; Zinkus et al., 1978).

Bennett et al. (1980) evaluated 73 students in a suburban school district who attended a full time language and learning disability program. These children were randomly matched with 73 control children who had no history of learning problems. The findings were as follows: significantly more frequent histories of otitis media among the learning disabled children, significantly more abnormal audiograms among the learning disabled children, and a tendency toward negative pressure tympanograms (Jerger's classification type c) in the learning disabled children. There was a high incidence of active but unrecognized middle ear disorders in the learning disabled group. Although a causal association between recurrent middle ear disease and learning disabilities could not be shown because the study was retrospective, the findings do suggest a relationship between the two disorders.

Brandes and Ehinger (1981) studied second and third grade students in British Columbia. They concluded that chronic middle ear disease during early childhood can produce secondary effects that persist beyond the acute episodes of conductive loss. Their data suggest that children with conductive hearing loss (compared with normal controls) perform poorly on a test battery measuring nonverbal intelligence and academic achievement. As a group, the children with hearing loss performed more poorly in specific auditory perceptual areas, but there were no significant differences between groups on the visual perception or nonverbal intelligence items. Children with conductive loss appeared to have the most difficulty in a test of selective auditory attention that measured the ability to attend to a listening task in the presence of background noise. The experimental group appeared to have greatest difficulty with tasks requiring auditory sequential memory skills. The children with hearing loss required more special services in their academic environment.

BEHAVIORAL AND EMOTIONAL PROBLEMS

The incidence of behavioral and emotional problems among deaf children is significantly higher than among hearing children, apparently the result of basic difficulty in human communication. There is less information available about the effects of fluctuating conductive hearing loss on behavioral and emotional development.

Several generalities regarding behavior can be proposed. The behavioral characteristics associated with depressed auditory acuity may be present only during the temporary episodes of hearing loss. Several well worn phrases and accusations summarize the immediate effects of the conductive hearing loss: "You aren't paying attention to me." "Why don't you listen to me?" "Your mind is a million miles away!" These descriptions suggest a child seemingly uninterested, poorly motivated, antagonistic, and disrespectful. Thus, undetected defective auditory acuity may be interpreted by peers and adults as negative behavior or an attitudinal defect. The patient may attempt to cope by constantly asking for repetitions of directions or comments. In many instances the child may act inappropriately because of poor discrimination of sounds, e.g., responding to the request, "Get me the broom," with "No, I won't go to my room."

Subtle behavioral complications may reflect underlying delays in speech and language development, auditory processing disturbances, and poor educational progress.

The final behavioral common pathway for most developmental disabilities is impaired family and peer relationships, disturbed psychoeducational interactions, and a poor self-concept. The complex neurodevelopmental complications believed to result from recurrent otitis media provide a fertile milieu for distortions of the child's self-esteem. Speech and language delays, auditory processing dysfunctions, and learning disabilities have profound effects on the child's self-confidence and self-perception. Difficulty in communications (particularly with peers), constant lack of successful experiences in social and educational encounters, and the morbidity associated with chronic middle ear disease per se may cause a deterioration of self-concept. Low self-esteem ultimately may be manifested in adolescent depression or aberrant social behavior such as juvenile delinquency.

TREATMENT

The therapeutic management for acute and suppurative otitis media is not without its confusions. It is generally agreed that routine antimicrobial treatment is required for acute infections of the middle ear, directed primarily against *S. pneumoniae* and *H. influenzae*. The use of antihistamines and decongestants is not based on established scientific investigation demonstrating definite value. Therapy for nonsuppurative otitis media is less clearly defined, depending in part upon the degree of hearing loss and residual complications. Therapeutic management often involves combinations of interventions, including antimicrobials, antihistamines, decongestants, adrenocorticosteroids, eustachian tube inflation, myringotomy, adenoidectomy, tonsillectomy, and the use of hearing aids. The relatively few studies variably support the efficacy of these interventions as routine procedures. The surgical management of middle ear disease and effusion has been the topic of considerable philosophic controversy reflecting a paucity of information regarding the natural history of this disorder (Paradise, 1980). The possibility of neurodevelopmental residua, as a consequence of recurrent middle ear disease, further underscores the need for establishing more firmly documented therapeutic approaches to the management of otitis media.

Regardless of the therapeutic plan adopted by the primary care physician, otitis media (acute or nonsuppurative) requires concerned and comprehensive management. The efficacy of the treatment plan can be monitored by periodic examination, tympanometry, and impedence testing. Accurate records should be maintained to document the frequency of middle ear problems, the degree and duration of associated hearing loss, the response to medical management, and possible neurodevelopmental complications. Children who suffer repeated episodes of middle ear disease and hearing loss may require professional assessment of speech and language skills if developmental delays are suspected. Periodic hearing evaluation, tympanometry, and impedence testing may be combined with speech-language and central auditory processing assessments for the child who is considered at risk. The possible cause and effect relationship does not mandate the use of a new or different therapy plan but underscores the responsibility for follow-up care of the immediate middle ear problem and the need for an awareness of the potential effects on speech, language, auditory processing, learning skills and behavior.

SUMMARY

Hearing loss interferes with the ability to process auditory stimuli and thereby interferes with cognitive skills. The deaf child manifests obvious difficulties with the acquisition of speech, language, and learning abilities as a result of this defective mechanism. Children with fluctuating hearing loss due to chronic middle ear disease during critical periods of development appear to exhibit similar delays in speech and language, auditory processing skills, and learning abilities. Although the hearing loss associated with recurrent middle ear disease is not of the same magnitude as that experienced by the deaf child, the complications may be significant if the loss occurs during the critical periods of language development. The disruptions of communicative, learning, and behavioral adaptations may reflect intermediary dysfunction in central auditory processing mechanisms: attention, memory, sequential memory, discrimination, synthesis, and auditory-visual integration (Rampp, 1979).

Thus, although data concerning specific relationships are not yet conclusive, it does seem that chronic middle ear disease and associated conductive hearing loss may have significant neurodevelopmental and behavioral sequelae. The available data are subject to criticisms related to retrospective design, methodologic problems, and lack of controls (Paradise, 1981). Nevertheless the accumulated evidence suggests a possible causal relationship between recurrent otitis media and

neurodevelopmental maturation, particularly mechanisms related to central auditory processing. The need for clarification of these relationships by means of well constructed studies is obvious. Such data would have profound implications for physicians who provide child health care. The diagnostic and management programs for recurrent otitis media may require modifications that provide for more active consideration of long term neurodevelopmental complications.

MARVIN I. GOTTLIEB

REFERENCES

Bennett, F. C., Ruuska, S. H., and Sherman, R.: Middle ear function and learning-disabled children. Pediatrics, 66:254–260, 1980.

Bluestone, C. D.: Eustachian tube dysfunction. In Wiet, R. J., and Coulthard, S. W. (Editors): Proceedings of the Second National Conference on Otitis Media. Columbus, Ohio, Ross Laboratories, 1979, pp. 50–58.

Brandes, P. J., and Ehinger, D. M.: The effects of early middle ear pathology on auditory perception and academic achievement. Speech Hear. Disord., 46:301–307, 1981.

Cook, R. A., and Teel, R. W., Jr: Negative middle ear pressure and language development. Clin. Pediatr., 18:296–297, 1979.

Dunn, A. J.: The neurochemistry of learning and memory. Environ. Health Perspect., 26:143–147, 1978.

Gottlieb, M. I., Zinkus, P. W., and Thompson, A.: Chronic middle ear disease and auditory perceptual deficits. Clin. Pediatr., 18:725–732, 1979.

Holm, V. A., and Kunze, L. H.: Effect of chronic otitis media on language and speech development. Pediatrics, 43:833–839, 1969.

Howie, V. M.: Developmental sequelae of chronic otitis media: A review. J. Dev. Behav. Pediatr., 1:34–38, 1980.

Kaplan, G. J., Fleshman, J. K., Bender, T. R., Barm, C., and Clark, P. S.: Long-term effects of otitis media. A ten-year cohort study of Alaskan Eskimo children. Pediatrics, 52:577–585, 1973.

Katz, J.: The effects of conductive hearing loss on auditory function. J. Am. Speech Hear. Assoc., 20:878–885, 1978.

Lewis, N.: Otitis media and linguistic incompetence. Arch. Otolaryngol., 102:387–390, 1976.

Masters, L., and Marsh, G.: Middle ear pathology as a factor in learning disabilities. J. Learning Disab., 11:54–57, 1978.

McInerny, T. K., Roghmann, K. J., and Sutherland, S. A.: Primary pediatric care in one community. Pediatrics, 61:389–397, 1978.

Needleman, H.: Effects of hearing loss from early recurrent otitis media on speech and language development. In Jaffe, B. F. (Editor): Hearing Loss In Children. Baltimore, University Park Press, 1977, pp. 640–649.

Paparella, M. M., and Juhn, S. K.: Otitis media: definitions and terminology. In Wiet, R. J., and Coulthard, S. W. (Editors): Proceedings of the Second National Conference on Otitis Media. Columbus, Ohio, Ross Laboratories, 1979, pp. 2–8.

Paradise, J. L.: Otitis media in infants and children. Pediatrics, 65:917–943, 1980.

Paradise, J. L.: Otitis media during early life: how hazardous to development? A critical review of the evidence. Pediatrics, 68:869, 1981.

Rampp, D. L.: Hearing and learning disabilities. In Bradford, L. J., and Hardy, W. G. (Editors): Hearing and Hearing Impairment. New York, Grune & Statton, Inc., 1979, pp. 381–389.

Ventry, I. M.: Effects of conductive hearing loss: fact or fiction. J. Speech Hear. Dis., 45:143–156, 1980.

Webster, D. B., and Webster, M.: Neonatal sound deprivation affects brain stem auditory nuclei. Arch. Otolaryngol., 103:392–396, 1977.

Zinkus, P. W., and Gottlieb, M. I.: Patterns of perceptual and academic deficits related to early chronic otitis media. Pediatrics, 66:246–252, 1980.

Zinkus, P. W., Gottlieb, M. I., and Schapiro, M.: Developmental and psychoeducational sequelae of chronic otitis media. Am. J. Dis. Child., 132:1100–1104, 1978.

26G   Asthma, Eczema, and Related Allergies _____

Among the medical problems intrinsically related to human development, asthma and atopic allergy have a most prominent place. In fact, age, sex, and the variable nutritional and environmental exposures associated with growing up may be more important than basic immunochemical factors in determining the presence or absence of infantile eczema, allergic rhinitis, asthma, and related clinical disorders. The answer to the many puzzles of allergy in childhood may be found not so much in the basic science laboratory as in the penetrating study of their developmental aspects.

CLASSIFICATION OF ALLERGIC REACTIONS

TYPES AND PATTERNS OF CLINICAL ALLERGY

Commonly all kinds of allergic conditions and all types of reactive airways disease are viewed as one problem, for practical purposes. It is important, however, to recognize that there exist marked clinical and etiopathologic distinctions. Clinicians must make an effort to establish a precise diagnosis in each case and to adjust their therapeutic recommendations and prognostic estimates accordingly.

From the immunopathologic point of view, four types of adverse reactions are recognized:

The type I reaction is what is called atopic allergy and is associated with reaginic antibodies, mainly of the immunoglobulin E class, which produce immediate wheal and erythema skin reactions to allergens.
Type II reactions are cytotoxic reactions, mostly involved in blood transfusions and drug hypersensitivity.
Type III reactions are often called "intermediate" or "late" to distinguish them from the immediate type I and delayed type IV reactions. They involve complement and precipitating antibodies, clinically being manifest in some cases of asthma and in hypersensitivity pneumonitis.
Type IV reactions, the delayed or cell mediated reactions, are well known, e.g., the classic 48 hour tuberculin skin reactions, and are most important in eczematous contact dermatitis, possibly also playing a role in respiratory and other forms of mucocutaneous allergy.

Clinical allergic reactions in childhood may not often be easily classified on the basis of immunopathology. More realistically, they are characterized in terms of symptom formation, their pattern (acute, intermittent, or chronic), triggering mechanisms (environmental exposures, infections, developmental stresses, psychosomatic influences), and the response to single, repeated, or combined drug therapy and especially in relation to chronic drug requirements (e.g., steroid dependence;

McNicol and Williams, 1973). Significantly, the coexistence of clinical manifestations involving more than one system—e.g., nose, lungs, skin, digestive system—is often a critical factor in permitting a final judgment regarding etiology and prognosis. A typical example is the atopic syndrome, which may start as infantile eczema and then lead to perennial allergic rhinitis or asthma, later running a variable course, and eventually disappearing in many cases (Table 26–11; Turner et al., 1980). A less well recognized syndrome consists of the concurrence of intrinsic nonatopic asthma, chronic rhinosinusitis, nasal polyps, and a secondary anaphylactoid intolerance to aspirin and related compounds. A syndrome of type III immunopathologic reactivity, which includes pulmonary infiltrates, fever, chills, dyspnea, and cough, resembling an infectious process, is diagnostic of hypersensitivity pneumonitis; although seemingly rare, it is beginning to be recognized in pediatric practice (Falliers, 1976).

Table 26–11. DEVELOPMENTAL STAGES OF ALLERGY IN CHILDHOOD

	Atopic Syndrome	Intrinsic Asthma "Triad"
Age of highest incidence	1 to 3 years	After 9 years
Sex ratio, F:M	About 1:2.5	About 3:1
Earliest manifestation	Infantile eczema	Bronchitis and asthma
Family history	Positive for atopy (5/1 of general average)	Most negative except coincidentally (1/10)*
Common symptom complex	Eczematous dermatitis Episodic urticaria Allergic rhinitis Extrinsic paroxysmal asthma	Intrinsic chronic asthma Rhinosinusitis Nasal polyposis Anaphylactoid intolerance to aspirin and related drugs
Skin tests	Many positive	Generally negative
Course and prognosis	Responds to hypoallergic management Improves with age Complete remission in about 35 per cent by puberty	Poor response to specific therapy May require continuous medication Can become progressively worse

*Random population samples show a 10 per cent or higher incidence of allergy; in atopic families this may be five times higher, up to 50 per cent. Family counseling may require some discussion of these risks.

SYMPTOM AND SYNDROME INTERACTIONS AND SHIFTS

Often, regardless of environmental influences, distinct symptoms can change markedly only in relation to the developmental stage of the growing human organism. Atopic eczematous dermatitis specifically, although often improving as a result of proper antiallergenic measures, may run a course requiring continuous medication for several months and then either subside completely, be complicated by rhinitis and asthma, or persist and even show an aggravation in the extent and severity of the skin lesions. Infants who develop atopic eczema before the third month of life have a higher than 50 per cent risk of having asthma before their sixth birthday, and if eczema appears between the third and sixth months of life, the risk is still high, estimated to be close to 35 per cent for respiratory allergy to follow. On the other hand, children who in their second decade develop chronic intrinsic asthma have a predominently negative history for atopic eczema and related problems, except for occasional episodic urticaria (Table 26–11).

It has been said that untreated allergic rhinitis may lead to asthma, but here the statistical evidence is not convincing (Broder et al., 1974), and further long term developmental surveys will be required before a definite risk can be estimated. Nasal polyps are not common in childhood. In some instances, however, they have been the first evidence of the intrinsic asthma "triad," and bronchial manifestations did not develop until weeks after polypectomy. It has been tempting to blame the surgery for this adverse sequence (which has more commonly been observed in older persons), but the possibility that this is again an unavoidable constellation of pathologic findings appears more likely (Falliers, 1976). Many underlying laboratory test abnormalities have also been correlated with the clinical problems of allergy, but a consideration of immune defects associated with an increased susceptibility to infections and of transient immunoglobulin deficiencies coinciding with the development of allergy is not within the scope of this discussion.

CLINICAL DIAGNOSIS IN RELATION TO MANAGEMENT

Unlike episodic illness, infections, trauma, and other isolated problems, allergic disease in childhood requires frequent medical consultations and extended follow-up. Therefore, whether one deals with a dermatologic, respiratory, or other manifestation, a precise etiopathologic diagnosis is a critical factor in determining what is the best immediate management and how properly to guide the family for the long term control and prevention of recurrences in each case.

Unavoidably, the correct estimate of individual risks is based not only on clinical findings but on a good personal and family history and a proper assessment of the family environment. One thing the physician treating children can certainly do is to reassure parents that the existence of a serious problem of allergy among their offspring does not necessarily mean that the entire family may be "marked" for allergy. Although there have been parents who develop allergic problems after their children, and also families with multiple allergic problems among all or most of the members, it is not often that one sees siblings with the same form of severe eczema or with intractable asthma. Studies with monozygous twins and follow-up of three generations of many families with one case of asthma indicate that a "clustering" of the same symptoms in a family pedigree is an exception and certainly not the rule (Falliers et al., 1971).

DEVELOPMENTAL CORRELATION OF ALLERGY, ENVIRONMENT, AND BEHAVIOR

MULTIFACTORIAL ETIOLOGY OF ECZEMA

Infantile eczema, the most common early manifestation of atopic allergy, usually first appears between three and six months of age. This is often a time when fruits and vegetables, cereals, and meats are introduced into the infant's diet. The cutaneous manifestations may also correlate with a transient drop in serum immunoglobulin A level and possibly with inadequate lymphocyte function. The baby also shows a rapidly developing awareness of the external environment (other than an immunologic awareness!), which curiously coincides with the clinical, immunologic, and nutritional changes. Although precise etiologic correlations (e.g., orange juice causing a skin rash) are not often possible, the significance of this coincidence cannot be ignored and will continue to require clinical study. The contribution of specific environmental allergens certainly must be examined, and efforts must always be made to modify the diet, including that of a nursing mother (Warner, 1980), to remove a blanket or a stuffed toy, and to keep hairy pets away from the infant's bedroom. Nevertheless, often the characteristic skin manifestations can progress independently and then either change themselves or become complicated by allergic rhinitis and asthma. Mucocutaneous atopy in infancy at times appears also to be associated with infantile colic and other developmental and nutritional problems, but the correlation remains tenuous. (See Chapter 28 for a discussion of colic and Chapter 33 for a discussion of the emotional factors in eczema.)

PERENNIAL ALLERGIC RHINITIS AND CHRONIC "COLDS"

Intermittent or persistent clear rhinorrhea, frequent sneezing, and evidence of nasal airway obstruction may be noted in the predisposed infant before the end of the first year of life. Asthma often may follow, with a peak incidence of the first episode at about 18 months of age.

As in the development of eczema, many factors coincide chronologically and can be etiologically important. Viral respiratory infections occur with increasing frequency at this age; the toddler begins to "get into everything," in this way increasing markedly his allergenic exposure; the diet is as varied as the family's culinary habits provide, and although emotional dependence on the key family figures may actually increase, there is a concomitant drive for independence, often requiring the placing of limits and the giving of appropriate rewards or punishments. Naturally these influences on the growing child vary widely, depending on cultural, educational, and socioeconomic determinants. It is often tempting to blame one of these factors for the development and the progression or recurrences of atopic allergy, yet the facts advocate caution in making a clinical judgment and in communicating hasty interpretations to the parents that may influence or "distort" their own perceptions and thinking.

As any physician caring for children knows, eczema may clear spontaneously in the course of human development, asthma may occur only once or twice in a lifetime, and even perennial allergic rhinitis can fluctuate markedly, with or without treatment, and can remit for years, only to reappear at times in adolescence as seasonal "hay fever." Realistically, one is inclined to attribute these otherwise unexplained transformations to genetically determined developmental pathophysiologic patterns. A more sophisticated interpretation, however, is that there is a complex interplay of inherited, environmental, and social-cultural elements, which, if chance has it that way, may favor or prevent the clinical manifestations of allergy.

ONSET AND COURSE OF ASTHMA IN EARLY CHILDHOOD

The peak incidence in the onset of asthma in childhood is close to 18 months of age. More than 70 per cent of the children with chronic asthma develop the first attack before the age of three years (Falliers, 1970). The risk of the development of asthma may be significantly higher and the age of onset earlier if the child has had atopic eczema before the age of six months, and even more so if the skin lesions were noted before the age of three months. For reasons that will be very important to investigate in depth, males outnumber females almost 2.5 to 1 in the prevalence and to a lesser extent in the severity and chronicity of asthmatic symptoms. It is not known what percentage of preschool children with a record of one or two attacks of asthma are at risk of developing persistent, chronic, and more severe symptoms.

Even when this happens, it is not possible to identify the factors that contributed to this unfavorable course. What is known, however, and what may be encouraging for the parents of a child with chronic asthma or for the patients themselves when they reach an age to have children, is that the strong genetic factors that influence the development of atopic disease (eczema, asthma, rhinitis) seem to bear no relationship to its long term course and chronicity. A survey of three generations in the families of over 1200 children with the most severe form of chronic asthma, requiring institutional management in a center in Denver, has shown that only an exceptional parent, sibling, or offspring had asthma comparable to that of the propositus (Falliers, 1970).

Often more serious than the disease itself are the adverse developmental and behavioral effects of antiasthmatic medication. The growth suppressing effects of continuous daily corticosteroid administration are known by parents as well as by physicians, but fortunately alternate day therapy and the new aerosol preparations (e.g., beclomethasone dipropionate) can obviate these. The selective beta$_2$ adrenergic bronchodilators can cause noticeable tremor and irritability when taken orally in full doses, and all theophylline preparations are apt to produce gastrointestinal problems, persistent nausea, anxiety, and insomnia. Even "old fashioned" potassium iodide has caused or aggravated adolescent acne so often that teenagers found one more reason to disregard, or discard, their medication. Oxygen therapy is not commonly required for asthma, but the lack of it has been blamed for neurologic and developmental problems in some cases.

Although many children with a tendency to wheeze may experience a temporary aggravation when they start school (the result of a combination of infectious and allergenic exposures, in addition to the physical and emotional stress of the new school routine), as a rule asthma of this type improves with age. This amelioration, or "outgrowing," of asthma may be attributed to immunologic and metabolic maturation and to the fact that after the age of five to six years the growth of the lungs includes a relative widening of the bronchial airways. However, as in the case of atopic dermatitis, the simple attainment of certain developmental turning points may be a factor. It has been reported that nearly 40 per cent of children with asthma have no further overt symptoms when they reach puberty (Johnstone, 1981).

ASTHMA, RHINOSINUSITIS, AND POLYPS IN ADOLESCENCE

One of the most serious problems in pediatric allergy, and in adolescent counseling, is the abrupt development of severe perennial intrinsic asthma, which usually has been noted to occur after the age of 10 years (Falliers, 1976). In most respects this condition resembles what more commonly appears in the fifth to sixth decades of life. It affects females significantly more than males, is triggered and aggravated by respiratory infections (often, in this group, bacterial bronchitis and not, as in early childhood, viral nasopharyngitis), and is often accompanied by chronic rhinosinusitis. Nasal polyps, which seldom complicate atopic allergy, develop within one or more years after the onset of asthma in more than one-third of the cases. When, in exceptional patients, the polyps are noted first and are removed, asthma may follow; correctly or not, it has been blamed on the surgical procedure.

For the unfortunate adolescent who suddenly has recurrent "choking" paroxysms, misses weeks of school because of asthma and bronchitis, always sounds nasal, and has a nose full of polyps, the problem can be devastating. In contrast to children who have become "used to" (if one gets used to illness!) their atopic allergy since early childhood, these otherwise healthy youngsters cannot understand and cannot "stand" what is happening to them. They blame all conceivable factors in their daily life, and their doctors do not help much, as they seem unable to identify any causes. The frequently (and for no sufficient indications) attempted allergy immunotherapy ends in failure, and this adds to the frustration. Only to complicate the problem further, there is soon evidence of anaphylactoid intolerance to aspirin, many related minor analgesics (acetaminophen almost never cross reacts with these and appears safe), and the common yellow dye (No. 5), tartrazine; the latter is found in numerous foods and drinks, including carbonated beverages ("pop"), which must be avoided.

Whether persistent or episodic, asthma and related conditions in adolescence present extraordinary problems of diagnosis and management. A truly comprehensive plan for study and treatment must consider how the dynamic biologic processes of puberty affect the clinical symptoms and how the unique social and ecologic influences in this age group modify the course of the disease and each patient's response to treatment. A compassionate physician and his team should relate to each patient on an intelligent informal level and adapt to the teenager's needs by changing accordingly. Authoritarian management plans that might have "worked" in earlier childhood, such as restrictions in diet and activity, regular hyposensitization injections, and strict schedules for medication intake, may become impractical in adolescence. Without relaxing their professional supervision, medical teams treating youngsters of this age must acquire the extra skills necessary for the understanding and management of each young person with the variable recurring problems of allergy.

MISCELLANEOUS ALLERGIES IN CHILDHOOD AND ADOLESCENCE

Generalized urticaria is seldom seen in the first one or two years of life, except possibly in the mostly papular form associated with drug eruptions. In these the "behavior" of the parent in providing or withholding medication can be important. Some parents always rush for aspirin or a cough syrup, while others refuse to give their babies any "medicines." Later in childhood, toward the end of the elementary school years, urticaria appears to be more common.

A particularly vexing problem is cyclically recurring urticaria and angioedema associated with the menarche. Monthly episodes have been attributed to autosensitivity to progesterone, or to emotional "tension," or simply to coincidence. In most cases, however, there seems to be no single cause, and the periodic rhythmic "structure" of the maturing organism must be considered accountable. Only infrequently an episode of generalized urticaria may occur following the ingestion of an analgesic containing aspirin, and this may signal the potential risk of the intrinsic asthma "triad."

THE PSYCHOPHYSIOLOGY OF CLINICAL ALLERGY

CLASSIC CONDITIONING AND AUTONOMIC "LEARNING"

As the expression of innate biologic information and as the behavioral and physical maturation of the growing child, development literally means learning. In the broadest sense, learning of course includes the immunologic recognition of antigenic substances and the development of antibodies against them. In neurophysiology, autonomic conditioning—also called operant conditioning, or instrumental learning—has attracted increasing interest as a possible mechanism for the production of disease. Understandably, the possibility has been explored that the vasodilation and the mucosal edema in respiratory allergy, and even more so the bronchoconstrictive reactions in asthma, may be conditioned (Falliers, 1969, 1983 in press). However, even though heart rate, intestinal peristalsis, peripheral blood pressure, and renal output have been effectively subjected to conditioning, no respiratory or skin disorders have been convincingly reproduced by conditioning, either in humans or in experimental animals.

Despite the present lack of "solid" experimental and clinical data, it appears still probable that autonomic conditioning may be the most important pathophysiologic process relating stages of human development and environmental experiences and events to allergic disorders. This type of operant conditioning is based on a reinforcement of periodic physiologic functions, such as acceleration of the cardiac pulse and of intestinal peristalsis. It is now recognized that bronchial airway dynamics also display well defined circadian and other rhythms and that allergic respiratory disorders represent basically a pathologic amplification of these (Falliers, 1983 in press). Adverse environmental stimuli and inordinate therapeutic intervention might indeed be the causes of this escalation, or rebound of bronchoconstriction, mucosal edema, and other aspects of allergy. The symptoms, once augmented in this way, may require further treatment and continuous attention. Early experiments in "deconditioning" or unlearning of paroxysmal cough already provide indications of the therapeutic potential of this approach (Creer, 1977).

PSYCHODYNAMIC INTERACTIONS OF PARENT, CHILD, AND ILLNESS

Much has been said and written about the pathogenic role of abnormal parent-child relationships in aggravating or perpetuating allergic diseases. Regrettably, often a major blame has been placed on the unfortunate mothers of asthmatic children for being "overprotective," "rejecting," or in other ways not handling their offspring properly and causing them to be sick. Let us recognize, from the beginning, that this "blame" often has represented a transfer of the physician's guilt for not being able to solve the problem and control the illness (Falliers, 1969).

Naturally the evolving dynamics of parent-child relationships, with their infinite nuances, complexities, and feedback processes, provide abundant chances for "abnormal" interactions. But how can any clinician define "abnormalities" when it is not often possible to determine what is the ideal protection, guidance, discipline, or independence that a growing child needs, especially a child prone to recurrent illness? Rather than becoming a judge, the provider of health care must recognize the natural variability in the "triangle" of parent-patient-illness, in which the existing symptoms or the fear of recurrences influence the persons involved and they in turn can modify the illness by their reaction to it, and the clinician's ability to change the situation by forming a closed loop, which if "healthy" can de-escalate and eliminate the symptoms but if inappropriate may further aggravate the spiral of allergic illness.

The first principle in understanding and correcting the triangular parent-patient-illness interaction is a clear conception of the medical problem, its frequency, and its severity. Isolated manifestations of asthma, paroxysmal cough, at times hysterical hyperventilation, pruritic skin disorders, prolonged school absences, and undesirable drug dependence cannot be accepted by themselves as sufficient indicators of the degree of allergic disease. Misconceptions or failure to distinguish between symptoms and disease may lead to wrong conclusions about the role of psychologic factors and to developmentally detrimental steps. If symptoms are known to be recurring frequently or to have become chronic, their place and meaning in the patient's existence and within the family constellation must be explored. Allergenic precipitants must be identified and balanced against the possible role of behavioral or emotional modifiers. Structured interview techniques have been proposed for this purpose, which fairly objectively can assess the existing level of psychopathology and its possible relationship to allergic disease (Falliers, 1969).

Many conventional psychologic test batteries have been administered in an effort to "measure" psychologic and behavioral abnormalities related to allergy and asthma. Personality inventories, such as the Minnesota Multiphasic Personality Inventory, psychomotor tests, such as the Bender-Gestalt test, and more psychiatrically oriented procedures, such as the Rorschach test, have failed to distinguish allergic children from nonallergic control subjects or to identify subgroups of children with allergic disease with clearly abnormal responses. Some tests have been adapted for special studies with asthmatic children, and one of these was the so-called PARI, the Parents' Attitude Research Instrument. Application of these questionnaires to subgroups of asthmatic children and their parents demonstrated that children with asthma who lost their symptoms when removed from the family environment showed more pathologic features in family dynamics, as compared with either healthy controls or children with chronic asthma whose symptoms persisted despite changes in interpersonal relationships. These data suggested that in some cases children with chronic allergic problems may benefit from a thorough psychologic assessment and subsequent modification of family dynamics, including possible temporary separation from their parents and siblings (Creer, 1977).

PSYCHOANALYTIC VIEWS

The function of the autonomic nervous system, which is significantly altered in allergic diseases, also bears a reciprocal relationship to altered mood states, particularly anxiety and depression. Although it has often been postulated that anxiety and anger stimulate sympathetic nervous activity and thus may be beneficial for asthma, this is not the case. Neither has there been any demonstrable proof that a parasympathetic predominance, associated with depressive states, may predispose to allergic problems. Nevertheless numerous psychiatric and psychoanalytic studies have attempted to correlate mood changes, and the often unconscious psychodynamic processes associated with these, to the clinical manifestations of allergy. Some studies have detected "more personal discomfort and unhappiness" among certain patients with respiratory allergy, while other patients were found "relatively satisfied and confident." Other investigators, however, have concluded that any degree of psychopathology noted is more likely the result than the cause of chronic and severe allergic symptoms (Falliers, 1969).

Many psychoanalytic studies have been published attributing allergic symptoms to a craving for maternal love or to a conflict between "the desire to cling and the temptation to separate." Mothers have been encouraged to show as much affection as possible for the baby with eczema, including physical closeness and hugging. Yet often, when the child develops asthma, the same mothers have been told to avoid overprotection and "let the child go." One is tempted to conclude that no psychoanalytic theory and no precise behavioral instructions can guide parental and child behavior. The professional who is called to assist a family with allergic problems must accept the fact that absolutely firm rules are impossible and unrealistic. In each case time must be spent to formulate and to modulate patterns of behavior most suitable for the child's and the parents' personal needs.

BEHAVIORAL MODIFICATION, BIOFEEDBACK, AND HYPNOSIS

Techniques derived from learning theory and from conditioning principles are relatively recent additions to psychotherapy. It is of interest to the practicing physician that one of these methods is actually called "desensitization." The goal of this type of therapy is to train the young patient to substitute a useful adaptive response, for example, relaxation, for a maladaptive one, such as anxiety or phobic withdrawal, to a stimulus associated with symptom formation. The technique is relatively simple and there have been impressive records of success in many fields (Creer, 1977). Thus this approach deserves a trial in the management of recurrent symptoms attributed to allergy. In the treatment of asthma specifically, the patients may be taken through successive steps of relaxation and reciprocal inhibition; i.e., they are taught to respond with relaxation to stimuli that usually trigger an asthmatic attack. Of course this may be an acceptable approach during quiescent periods, but when young children have serious difficulty in breathing, teaching them to relax on command not only is unrealistic but may actually be dangerous, as it may suppress their vital effort for air exchange.

Biofeedback is now a popular term, with variable and not consistently precise connotations. The use of a bodily stimulus as a signal for the enhancement or reduction of a pathologic condition has been tried for the management of intense pruritus associated with eczematous dermatitides and recurring urticaria. Only preliminary impressions are available, and these tend to support further applications of biofeedback in pediatric dermatology. More precise have been the efforts to feed back information on ventilatory function for the purpose of training children with asthma when to relax and when to adjust medication intake. Self-measurements of peak expiratory flow rates with the portable Wright peak flow meter have been instrumental in increasing each patient's awareness of the adequacy of ventilation and have prevented the two common extremes in the treatment of asthma, namely, overmedication from undue panic and undertreatment due to indifference or to the fear of side effects (Falliers, 1974).

Hypnosis has been tried in the treatment of allergic problems, on the basis of some early observations, including reports that it is capable of suppressing whealing responses to the application of allergens on the skin. Like behavior therapy, hypnosis aims at substituting new responses for old—e.g., relaxation instead of anxiety—without exploring the basic psychodynamic implications of the patient's symptoms and behavior. The difference with hypnosis is that an attempt is made to alter the allergic symptoms themselves, rather than the emotional reactions associated with them. There are still many questions concerning the specific indications for hypnosis, the method for selecting appropriate candidates, and the possible risk of symptom substitution and of adverse sequelae. Consequently this approach, as a rule, is not recommended for youngsters with allergic problems.

INDIVIDUAL AND GROUP PSYCHOTHERAPEUTIC DATA

Psychotherapy selectively directed toward the treatment of patients with allergic problems has provided some meaningful insight into the dynamics of symptom formation and the patient's attitudes toward illness. It is generally agreed, however, that referral of a child with allergic problems to a psychotherapist cannot be based on the medical symptomatology alone but on precise behavioral and psychiatric indications. Simply stated, in the absence of clearly defined psychopathology, there is no need for a child with respiratory, dermatologic, or other allergy to seek counseling. In cases in which, either experimentally or out of a sense of desperation, children have received extensive psychiatric study and treatment, the results have not justified the effort or the cost.

An exception may be group sessions, which are more intended for information exchange and discussion than for treatment, despite the commonly applied name "group therapy." Both the patients and their parents have been reported to benefit by such activities. Initially an opportunity is provided for objective discussion of what allergic problems are, the role of environment, genetic factors, and psychologic influence, and a sound approach to drug therapy. The common problems of excessive dependence on drugs—"pharmacophilia"—and its opposite, irrational fear of drugs—"pharmacophobia"—can thus be at least partially corrected. In a broader sense family treatment is directed toward the family unit and may realistically deal with problems in the family constellation related to the management of allergy. This does not necessarily require the direction of a trained psychotherapist. The whole family can be treated from the "behavioral" point of view by being encouraged to take an enjoyable vacation together, provided adequate symptom control is ensured with proper continuous medication. In many instances this type of reality therapy aimed at the family in general may require the contribution of a physician as well as a psychiatrist, social worker, school teachers, and others from the paramedical team dealing with the clinical situation (Falliers, 1969).

PSYCHOPHARMACOLOGY

Besides their specific indication for the treatment of various psychiatric disorders, psychopharmacologic agents are often considered advisable in situations in which one needs to "break the cycle" of reciprocal amplification between physical symptoms (e.g., asthma, eczema) and anxiety reactions. Some physicians have prescribed tranquilizers in order to reduce the intensity of emotional reactions precipitating an attack or aggravating these disorders. At times compounds such as hydroxyzine, meprobamate, diazepam, and the phenothiazine derivatives are tried on the assumption that if the results are good, they will prove the psychogenic origin of the allergic problem. Antidepressant drugs such as amitriptyline and imipramine also have been used. Regrettably there are no adequate studies to prove or disprove the value of this pharmacotherapeutic approach in the treatment of allergy, but logically it does not make much sense. The author has seen several young children treated by their pediatrician with methylphenidate hydrochloride for "hyperkinetic syndrome," whose coexisting chronic allergic symptoms showed no change in relation to this type of treatment. Besides, for a child prone to severe respiratory distress associated with asthma, any type of drug that might impair the ventilatory drive is contraindicated. And when some of these children end up in the hospital and need psychologic support, drugs can never be an acceptable chemical substitute for the sympathetic care, sincere affection, and continuous attention they require.

PROGNOSTIC AND PREVENTIVE EFFORTS

STATISTICS VERSUS INDIVIDUAL PROGNOSIS

Invariably, at the time the first allergic manifestations appear, parents anxiously inquire about the chances of persistence or recurrences, progression to something more serious, and similar symptoms developing in the future in a sibling. As mentioned earlier, nearly half the infants with early atopic eczema have some wheezing before they are six years of age. Statistically asthma affects nearly 4 per cent of the population in general, about 15 per cent of the members of the immediate family of an asthmatic child, and over 35 per cent of monozygous twins whose sibling already has this condition (Falliers et al., 1971), but there is no way for a clinician to establish even a remotely accurate individual prognosis. The natural course of allergic conditions in childhood and the influence of treatment are so markedly variable that no fixed developmental "milestones" can be accepted as reliable signals for significant "turning points" in allergy. This lack of definitive data, of course, must never be conveyed to the parents in the form of negative remarks, which so often cause not only discouragement but encourage poor co-operation in proper medical management.

The predominant difficulties in making an individual prognosis lead, regrettably, to a common and inexcusable mistake, and that is fatalism. Both the physician and the parents "give up," ironically by pursuing two different directions. Either they say, optimistically, that the child will "outgrow" the problem and therefore there is nothing that needs to or can be done, or, pessimistically, they accept the fate that allergy "runs in the family" and that any effort to do anything about it is doomed to failure. Actually much can be done, from basic environmental and dietary controls (the importance of which no clinician can afford to ignore) to regular prescribed treatment and prophylactic measures. The best "behavioral therapy" in allergy practice is a continuously successful medical treatment.

COMMUNICATION PATTERNS

It is precisely the impossibility of simple "yes or no" answers that in the care of children demands a comfortable doctor-patient relationship and continuous enlightened communication. Most problems of allergy cannot be solved with simple environmental manipulations and a few ordinary prescriptions. In fact, it is now recognized that more than 50 per cent of all prescriptions written are not taken as recommended or are not filled at all (Hulka et al., 1976; Rosenstock, 1975). The situation may be worse in the treatment of allergy in children. Consequently the success of treatment will depend in great part on the ability of the physician to motivate properly the parents and the patients and to ensure their compliance with his instructions. Influencing the attitudes and the behavior of the family toward the illness and its treatment must be accepted as the foundation of sound management of respiratory, dermatologic, and other forms of chronic allergic disorders in pediatric practice. Education of the parents and others who care for the children, as well as the patients themselves, about what allergy is and how to control or avoid it is a primary duty of all clinicians treating children. It is much more important—and perhaps more difficult—than making a diagnosis and writing a prescription.

Even when it comes to writing a prescription, a proper signature is not all that is needed. Almost all parents today expect explanations and a discussion: What is the "cortisone" cream going to do to the child's metabolism? What about the "nervousness" that adrenergic decongestants and theophylline are known to cause? Is the sleepiness and sedation—"doping"—from the antihistamines really unavoidable? Very often poor treatment practices and noncompliance, particularly

with theophylline and the antihistaminics, result from inadequate explanations and the parents' displeasure with the gastrointestinal and the neurologic-behavioral side effects. The successful management of allergy in children is based on communication much more than on immunologic and pharmacologic principles known to the physician alone.

Realistically one must accept the fact that the physician's time is very limited, and, beyond writing a prescription and a note in the chart, it may be difficult to talk to the parents at length. At times it is the parents—often very intelligent people with above average education—who hesitate to ask questions and "bother" the doctor! Yet many of them read much about their child's condition (it may be the "wrong" things) and need critical information, reassurance, and logical explanations and guidelines about the causes of the clinical problem and its proper management. Well trained office personnel can play a very valuable role in this respect. They can, first, provide some information about how the parent-doctor team works; they can dispense useful, well written brochures with printed or photocopied guidelines; they can correct any existing misinformation and apprehension; and they can make themselves available for continuing communication and guidance. In medical offices primarily devoted to the practice of clinical allergy, some office assistants spend 30 per cent of their time talking to people, not simply scheduling appointments but discussing clinical problems and treatment. The importance of the proper training of paramedical personnel in allergy has been recently recognized by the national specialty organizations in the United States that now offer annual instructional courses specifically for that purpose.*

PREVENTION OF PHYSICAL AND PSYCHOLOGIC IMPAIRMENT

Beyond systemic and topical therapy for allergy is the greater problem of preventing permanent physical impairment in the growing child without, at the same time, creating an overdependence on drugs and on rigid environmental restrictions. Otherwise one may be fairly successful in controlling allergy but then will be faced with the more complicated problem of psychologic abnormalities, including the inability to tolerate (psychologically, if not physically) diverse environmental changes, an unwillingness to engage in above average physical activity, and even a fear-panic reaction to the withdrawal of medication and the possibility of

*Requests for information may be addressed to the American College of Allergists, 2141 14th Street, Boulder, Colorado 80302, and to the American Association for Clinical Immunology and Allergy, P.O. Box 912, Omaha, Nebraska 68101.

recurring symptoms. Again the professional responsible for health maintenance of a growing child must be skilled enough to sense the development of any persistent physical and psychologic abnormalities and to take appropriate measures to prevent them. On each visit, and on the telephone, the child's nutrition, activity, sleep patterns, and school attendance must be discussed in relation to existing or feared symptomatology and to current or projected medication requirements.

THE "BENEFIT" OF SICKNESS AND THE ADVANTAGES OF HEALTH

The initial and ultimate goal of any therapeutic endeavor is the maintenance of physical and mental health. However, among physicians there is a somewhat naive assumption that this is also what the patient wants. The example of youngsters indulging in the self-destroying use of psychotoxic drugs, soon after medical research had been successful in protecting them against many feared contagious diseases, may be taken as an indication that the existential goal of life for them is not necessarily health. In the case of growing children with serious allergic problems particularly, the illness often becomes an asset by providing privileges not attainable otherwise. These so-called secondary "benefits" of being sick cannot be ignored and cannot be wished away with statements that may be offensive to both the patients and their parents. One of the greatest exercises in the "art" of medicine is to guide and support children who are developing in modern technologically advanced societies, to understand gradually the advantages of being healthy, not only for egocentric self-preservation but for the sake of society in general. This awareness of the value of health is something relatively new in medicine, which as a science has shown a tendency to keep a distance from the ethical questions of the philosophic disciplines. However, the time has come to recognize that all persons, young and old, and all family units, in addition to demanding a still questionable "right" to health care, share a duty and a responsibility for keeping themselves and each other well.

CONSTANTINE JOHN FALLIERS

REFERENCES

Broder, I., Higgins, W., Mathews, K. P., and Keller, J. B.: Epidemiology of asthma and allergic rhinitis in a total community, Tecumseh, Michigan. V. Natural history. J. Allergy Clin. Immunol., 54:100, 1974.

Creer, T. L.: Asthma: psychologic aspects and management. In Middleton, E., Jr., Reed, C. E., and Ellis, E. F. (Editors): Allergy Principles and Practice. St. Louis, The C. V. Mosby Co., 1977, pp. 796–811.

Falliers, C. J.: Psychosomatic study and treatment of asthmatic children. Pediatr. Clin. N. Am., 16:271, 1969.

Falliers, C. J.: Treatment of asthma in a residential center: a fifteen year study. Ann. Allergy, 28:513, 1970.

Falliers, C. J.: Self-measurements for asthma. J.A.M.A., 230:537, 1974.

Falliers, C. J.: Asthma. In Gallagher, J. R., et al. (Editors): Medical Care of the Adolescent. Ed 3. New York, Appleton-Century-Crofts, 1976, pp. 329–337.

Falliers, C. J.: The chronobiology of allergy. In Middleton, E., Jr., Reed, C. E., and Ellis, E. F. (Editors): Allergy Principles and Practice. Ed. 2. St. Louis, The C. V. Mosby Co., 1983. (In press.)

Falliers, C. J., Cardoso, R. R. deA., Bane, H., Coffey, R., and Middleton, E., Jr.: Discordant allergic manifestations in monozygotic twins: genetic identity vs. clinical, physiologic and biochemical differences. J. Allergy, 47:207, 1971.

Hulka, B. S., Cassel, J. C., Kupper, L. L., and Burdette, J. A.: Communication, compliance and concordance between physicians and patients with prescribed medications. Am. J. Pub. Health, 66:847, 1976.

Johnstone, D. E.: Current concepts in the natural history of allergic disease in children. Ann. Allergy, 46:225, 1981.

McNicol, K. N., and Williams, H. E.: Spectrum of asthma in children. Br. Med. J., 4:12, 1973.

Rosenstock, I.: Proceedings of the National Heart and Lung Institute Working Conference on Health Behavior. NIH Publication 76–868. Washington, D.C., U.S. Department of Health, Education, and Welfare, 1975, p. 135.

Turner, M. W., Brostoff, J., Mowbray, J. F., and Skelton, A.: The atopic syndrome: in vitro immunological characteristics of clinically defined subgroups of atopic subjects. Clin. Allergy, 10:575, 1980.

Warner, J. O.: Food allergy in fully breast-fed infants. Clin. Allergy, 10:133, 1980.

26H   The Gastrointestinal Tract

In recent decades several gastrointestinal diseases generally have been viewed as "psychosomatic" illnesses, the implication being that the disorders are largely a reflection of environmental stress. However, further clarification of the functioning of the autonomic nervous system has led to the recognition that the influence of the central nervous system and environment on the gut is less important than local autonomic control by peripheral nerve cells located close to the gastrointestinal tract. In fact, this more autonomous gastrointestinal tract should be acknowledged as having a considerable impact on the child's behavior and development and on his caretakers. In no way should this realization allow us to underestimate the importance of psychologic well-being and the role of stress in such disorders as peptic ulcers. However, maternal feelings and handling are now seen as having a smaller role in the etiology, and changes in them a less significant part in management, than was thought only a few years ago (Gershon and Erde, 1981).

In early infancy the interaction among behavior, development, and gastrointestinal tract function is a complex physiologic phenomenon, especially as rapid changes occur and as the gastrointestinal tract functions near or at capacity during this period of rapid growth. At times it is difficult to recognize the point at which the observed behavior or gut symptom changes from physiologic to pathologic. A one month old breast fed infant, for instance, ingests a lactose load that is larger than at any other stage of development. Similarly, if the amount is expressed in milliliters per kilogram of body weight, infants take larger swallows than children, who ingest one and one-half times more per swallow (0.33 ml. per kg. per swallow) than do adults (0.24 ml. per kg. per swallow). Feeding behavior such as burping needs to be considered in the light of our knowledge that the main pressure measured at the lower esophageal sphincter increases from about 5 mm. of mercury at birth, to about 7.5 mm. at one month, to 10 to 15 mm. at 100 days of age. This increase is related to postnatal age, not to birth weight. Moreover, the length of the sphincter increases from 1 cm. at birth to 2.5 cm. at six months of age. Thus it is obvious that the more complete our understanding of such mechanisms as sucking and swallowing, the more sensible our approach to the feeding problems in growing infants can be.

AEROPHAGIA

When young infants swallow, there is a closely correlated functioning of sucking, breathing, and swallowing mechanisms, with frequencies of sucking of two times per second, swallowing once for every five or six sucks, and breathing about every other second. At the same time, air and food cross paths. By the age of six months the child can no longer coordinate suck and swallow with breathing, and he has to stop feeding to draw a breath. At the initiation of a swallow, air is ingested into the stomach from the posterior pharynx along with fluid from the breast or bottle. With strong sucking, air enters the mouth along the side of the nipple as well. This can also occur in breast fed infants when the nipples are small or retracted.

Burping behavior of mother and child can be effective and easy, and then relatively little air stays in the stomach. With crying, however, large quantities of air can be ingested, causing abdominal distention and further discomfort, and a vicious cycle resulting in a strikingly distended, tight, tympanitic abdomen develops. Vomiting may relieve the situation, but unless the air swallowing is prevented, the abdominal distention and distress can lead to overfeeding, more crying, as well as stooling and flatus.

A logical approach would seem to lie in preventing excessive air ingestion or perhaps in substituting some of the bottle or breast feedings with spoon fed, semisolid food. Decreasing the period of time between feedings might result in less eager sucking. More effective soothing and decreased stimulation should also help (see discussion in Chapter 28 on colic).

GASTROESOPHAGEAL REFLUX

Reflux of gastric contents from the abdominal stomach into the thoracic esophagus can be a normal and natural occurrence on the benign side of a wide spectrum; on the pathologic side, vomiting can lead to aspiration, failure to thrive, or the sudden infant death syndrome (see Chapter 30). The lower esophageal sphincter is not an anatomic one but a physiologic high pressure zone relative to the proximal low pressure segment of the esophagus and the distally located stomach. Relaxation of this segment allows burping; when this relaxation is slightly longer or deeper, "wet burping" occurs. Burping ability varies, and with the development of adequate head control and shoulder girdle muscle tone, it appears easier. Gastroesophageal reflux is an episodic event that can be recorded with a variety of techniques (Gellis, 1979).

The clinical spectrum of gastroesophageal reflux has been widened considerably since manometric techniques were applied to measurement in the esophagus of infants and since acid reflux in the esophagus has been accurately measured. An additional valuable diagnostic method is the time honored barium swallow with radiographic documentation of reflux. In addition to the symptoms already mentioned, blood loss from esophagitis can lead to anemia, and a variety of recurrent or chronic respiratory problems have also been ascribed to, or are associated with, gastroesophageal reflux. Esophagitis, especially in neurologically damaged children, can lead to strictures, or to columnar epithelial replacement of the squamous epithelium of the esophagus. At present it would appear that the more infants (healthy or with a variety of symptoms) studied with more tests for gastroesophageal reflux, also including scintiscan and endoscopy, the wider the spectrum of gastroesophageal reflux will become. In contrast to the diagnosis, the conventional treatment with upright positioning, thickening of feedings, and watchful waiting for maturation is simple. It is customary to try conservative medical management for six weeks, if necessary supplemented with a floating foamy barrier (Gaviscon) or bethanechol or even cimetidine. Surgical intervention in the form of the Nissen fundoplication is popular, but objective independent data as to its real value are not yet available.

RUMINATION

Rumination is an uncommonly occurring form of intentional regurgitation of the stomach contents in infants that can lead to growth failure or dehydration. It appears to be a learned, seemingly gratifying, self-stimulating habit. Inadequate somatosensory stimulation and a defective mother-infant relationship are apparent in the history of such three to six month old infants, who are more often male than female. Characteristically the infant assists in the regurgitation with intentional moving of the jaw and mouth, quietly and seriously working at it. Sometimes fingers and fists are called into service as well. The severity and persistence of this behavioral abnormality can be so serious that behavior modification has been used. More traditional therapeutic measures relate to the surroundings and the nature of the interaction with the infant. It is hoped that evaluation of the social and emotional situation will contribute to a future for these infants that is better than their past (Wright et al., 1978).

CYCLIC VOMITING

Recurrent episodes of vomiting of more than a few days' duration, not caused by gastroenteritis or by anatomic abnormalities (such as "malrotation") and occurring beyond infancy, have become known as cyclic vomiting. It was first described by Samual Gee in 1882. Such symptoms can lead to rapid dehydration and weight loss. Terms such as cyclic, episodic, psychogenic, hysterical, and habitual vomiting have been used; rumination refers to a somewhat slower and more obviously willful vomiting in younger or retarded children. The latter condition more often leads to failure to thrive than to an acute depletion.

Cyclic vomiting occurs more frequently in girls than in boys, of elementary school age or older. Careful inquiry nearly always uncovers anxieties related to school or social or familial stresses. When alkalosis develops with decreased serum chloride and potassium levels and decreased renal function, hospital admission may be necessary for correction of water and electrolyte depletion. Often patients with such recurrent vomiting show an acceptable nutritional status between the episodes of vomiting. Although in the beginning of an episode there seems to be a partial ability to control the vomiting, once dehydration has set in and the episode is established, excessive mucus production, agitation, and inability to tolerate anything in the stomach develop. Anger, uncontrollable behavior, and lack of self-care can accompany the obvious electrolyte disturbances.

Although specific psychogenic triggering events are often obvious in the immediate past history and seem to lead to mucus production, gagging, and subsequent vomiting, a diffuse parental anxiety appears to be more significant, often related to past losses or to major traumatic family events. The child with this disorder, even when not vomiting, impresses the observer by an anxious inability to settle down for an interview and by social immaturity. Probably the "point of least resistance" is a "weak" stomach with a low vomiting threshold, as some other children have a sensitive stomach predisposing them to abdominal pain (see Chapter 28). The child is often enmeshed in family problems, has difficulty separating and escaping from parental pressures, and has secrets. Electroencephalographic abnormalities have been reported, but cyclic vomiting is not a seizure disorder, nor does it respond to anticonvulsant medication (Hoyt and Stickler, 1960; Reinhart et al., 1977).

In the last decade, behavior modification approaches have been utilized as have individual psychiatric counseling and family therapy. Encouraging and promoting psychologic maturation, environmental manipulation, and broad family support all have their place.

CHRONIC NONSPECIFIC DIARRHEA

One of the more common problems referred to the pediatric gastroenterologist is the syndrome of chronic nonspecific diarrhea. The diarrhea is of a nuisance type; weight loss usually does not occur and the diarrhea lasts for a few weeks to many months. Usually it has subsided when relative mastery of bowel control has been accomplished, by three and one-half to four years of age. By definition, malabsorption or other specific findings, such as parasites or pathogenic bacteria and immune deficiencies or sugar intolerance, are absent. Developmental and maturational factors appear to be important. For example, often the frequency of bowel movements does not really exceed the frequency of feedings, a numerical relationship acceptable for smaller infants, but unacceptable to many mothers of six to 24 month old children. Mucus and undigested food are disturbing to the parent, and bowel movements progress from loose to watery as the day wears on.

Although it is difficult to distinguish between a coincidental and a causal relationship with teething, dental eruption may represent one of the stress factors of importance. Certainly a history of overfeeding, of fat restriction, or of concurrent infections of the ear, nose, and throat or their treatment with antibiotics, or a history of acute gastroenteritis is frequently among the suggestive, but not totally etiologic, factors. Because the persistence of the syndrome can be frustrating, emotional tension and parental conflict can complicate and exacerbate the syndrome, which has been compared to the irritable bowel syndrome of adults. The frequency of "difficult children" and overly active children is increased among patients with this syndrome as is perhaps a history of "colic" at least in a referred population (Wender et al., 1976).

Obviously a thorough history including inquiry into the aforementioned factors is the first and most valuable initial approach. Dietary manipulations, and unnecessary radiographs, and other tests with low yield of positive findings should be avoided. A plea for acceptance of a less than "perfect" stooling pattern is indicated (Cohen et al., 1976).

INFLAMMATORY BOWEL DISEASE

A group of formerly enigmatic bowel diseases, usually of a chronic nature, are sometimes caused by known (e.g., Campylobacter enterocolitis) but also still unknown agents. Traditionally inflammatory bowel disease is recognized as encompassing superficial inflammation, ulcerative (hemorrhagic) colitis, and inflammation involving the complete bowel wall, Crohn's transmural colitis. The latter variety in a typical terminal ileal location in children has frequently been associated with *Yersinia enterocolitica* infection. Since inflammatory bowel disease creates vast human suffering and economic hardship, and its effect on the developing child or adolescent and his family is considerable, skillful management of the patient's developmental needs and bolstering of his support systems are of great importance.

Although one-third of the patients with ulcerative colitis respond well to appropriate simple treatment, and an additional one of four patients responds when more intensive and elaborate therapeutic measures are instituted, the remainder of these patients constitute a major challenge for the managing physician and a major stress on family functioning and development. The incidence of Crohn's disease appears to be increasing, but the response to therapy of this variety of inflammatory bowel disease at this time may be more favorable with the availability of intensive utilization of enteral and parenteral nutritional support. Nevertheless it is apparent that idiopathic ulcerative colitis and Crohn's disease cause significant major chronic intestinal, often debilitating disease in older children, adolescents, and young adults.

Ulcerative colitis involves mainly the mucosa, usually of the distal large bowel extending proximally, with an inflammatory infiltrate of polymorphonuclear cells and diagnostic crypt abscesses. The inflammation of the bowel in Crohn's disease is mucosal and also submucosal, transmural, and serosal. Regional lymphatics and lymph nodes are involved. The most typical location is in the terminal ileum, but the colon and the entire gastrointestinal tract, including the mouth, can be involved. Healthy "skipped" areas of intact bowel are free of the diagnostic noncaseating granuloma and of the transmural inflammatory process. At times the differentiation between these two forms of inflammatory bowel disease is difficult and becomes clear only as the diseases progress (Janowitz and Sachar, 1982).

ETIOLOGY

Although many causes have been advanced and later discarded, there seems to be general agreement that a genetically determined, unusual tissue response to widely present environmental agents may constitute the pathogenetic basis of the variants of inflammatory bowel disease. These environmental agents might include viral agents or variants of Pseudomonas-like micro-organisms. Many immunologic abnormalities are known in inflammatory bowel disease, but they have not been shown to be specific for inflammatory bowel disease, or universally confirmed. They may be related to the degree of progression or to many other variables, and pose major unanswered questions rather than established facts. Psychologic factors are probably not causative, but disorganized family functioning makes rational medical therapy extremely difficult and constitutes a very unfavorable prognostic factor in the author's experience (Lewkonia and McConnell, 1976).

DIAGNOSIS

Acute Salmonella enteritis, shigellosis, and *Entamoeba histolytica* infestations may mimic a few of the features of inflammatory bowel disease and therefore should be ruled out. At present Yersinia is known to cause dysentery, abdominal cramps, and also radiographically evident lesions of the small bowel suggestive of Crohn's disease. In Yersinia and Campylobacter infections, up to date, specialized bacteriologic and serologic laboratory techniques need to be utilized to identify these known bacterial causes of inflammatory bowel disease.

The diagnosis of ulcerative colitis is supported by sigmoidoscopy, which reveals friability of an inflamed mucosa and sometimes ulcerations, as well as by rectal or sigmoidal mucosal suction biopsy showing crypt abscesses and cellular infiltrates. When whitish patches are seen on a normal appearing mucosa, pseudomembranous colitis, which can occur after antibiotic therapy, should be considered.

Radiographic studies can be most helpful in

distinguishing ulcerative from granulomatous colitis (Crohn's colitis). Sometimes progression of the disease over time greatly facilitates such differentiation, for the radiologic abnormalities tend to be more typical later in the course than are endoscopic or histologic findings.

COURSE OF THE ILLNESS AND GENERAL TREATMENT

The early symptoms of inflammatory bowel disease in children are common ones indeed, and until the diagnosis is established with the institution of potentially effective therapy, many months of stress, uncertainty and doubt, multiple consultations, and variable diagnoses are the rule rather than the exception. Months of not eating, not feeling or growing well, abdominal pain, diarrhea, skin lesions, or joint problems make exploration of the coping pattern of patient and family a mandatory part of the start of therapy and follow-up, which in all likelihood will be life-long. For optimal attendance to the patient's needs the talents of members of multiple disciplines need to be coordinated, the prime responsibility lying with the gastroenterologist. Nutritionists, social workers, and psychiatrists, and at times a surgeon, radiologist, or pathologist, are essential members of such a health care team. Continued re-evaluation of parental and patient understanding and coping patterns, compliance, degree of denial, or overprotectiveness and social isolation, is most helpful.

SPECIFIC TREATMENT

Diet moderations should be instituted only when they turn out to result in fewer symptoms or a better appetite. Specifically, if a two week period of milk withholding results in significant clinical improvement, it should be sustained. Otherwise the full free nature of a liberal diet should coincide with the full free nature of liberal physical activity. The valuable effect of salicylazosulfapyridine in long term use of 2 to 8 grams a day, when tolerated without skin rashes or hematologic abnormalities such as leukopenia, makes a trial with this medication under close scrutiny one of the most helpful therapeutic approaches in inflammatory bowel disease (Azad Kahn, 1980).

We have reserved anti-inflammatory drugs such as intravenous doses of hydrocortisone or oral doses of prednisone for unresponsive or deteriorating patients in a hospital setting.

Especially for patients with Crohn's disease of the small bowel, total parenteral nutrition or enteral feeding of amino acid or peptide solutions via a nasogastric tube has been most helpful in reversing an unfavorable course of illness.

PEPTIC ULCER

An erosive defect of the gastric or more often the duodenal mucosa leads to inflammation, deformity, and scarring of the adjacent bowel wall. Frank arterial or capillary bleeding or occult blood loss can be the only sign, and sometimes typical meal related pains (but often atypical abdominal distress) complete the variable symptom complex.

Since the introduction of upper gastrointestinal endoscopy in children, it has become clear that the more vigorously the diagnosis is pursued, the higher the incidence becomes. In addition, the persistence of previously documented ulcers in patients who have become asymptomatic suggests that the true incidence is again higher than the study of symptomatic patients would indicate. This is the case especially in patients in the younger age group, up to three years; such younger patients with peptic ulcer are no longer the rarity they were until only a few years ago. Well documented peptic ulcers appear more frequently in boys than in girls and twice as often in the duodenum as in the stomach. A positive family history (25 per cent or more) suggests a genetic predisposition (Cowan, 1973).

Stress of a variable nature is well known to result in peptic ulceration in man and experimental animals (e.g., burns, trauma, treatment in intensive care units, and immobilization). It is the author's bias that the adult type of worrying is prevalent in older children with peptic ulcers (Christie and Ament, 1976).

In adults there is no convincing evidence that vigorous treatment with alkali or anticholinergic drugs changes the natural history of peptic ulcers. Thus, employment of newer therapeutic modalities, such as histamine H_2 receptor antagonists (e.g., cimetidine) and prostaglandin derivatives, deserve careful study.

Thorough attention to the patient's developing pattern of coping with stress of either an obvious or a more subtle nature (school, family, friends, economic or professional future, outlook) is important. Avoidance of other known stimuli of gastric acid secretion should be stressed (caffeine, nicotine, ethanol), especially for the adolescent or older child. The health care worker should be aware of the poorly understood, but well documented and predictable, seasonal incidence of recurrences in the spring and fall.

MAARTEN S. SIBINGA

REFERENCES

Azad Kahn, A. K., et al.: Optimum dose of sulfasalazine. Gut, 21:232, 1980.

Christie, D. L., and Ament, M. E.: Gastric acid hypersecretion in children with duodenal ulcer. Gastroenterology, 71:242, 1976.

Cohen, S. A., et al.: Chronic non-specific diarrhea. Am. J. Dis. Child., 133:490–492, 1979.

Cowan, W. K.: Genetics of duodenal and gastric ulcer. Clin. Gastroenterol., 2:539–546, 1973.

Gellis, S. S (Editor): Gastroesophageal Reflux. 76th Ross Conference on Pediatric Research, 1979, pp. 1–129.

Gershon, M. D., and Erde, S. M.: The nervous system of the gut. Gastroenterology, 80:1571, 1981.

Hoyt, C., and Stickler, G. B.: A study of 44 children with the syndrome of recurrent (cyclic) vomiting. Pediatrics, 25:775, 1960.

Janowitz, H. D., and Sachar, D. B.: Inflammatory bowel disease. Adv. Int. Med., 27:205–247, 1982.

Lewkonia, R. M., and McConnell, R. B.: Familial inflammatory bowel diseases—heredity or environment? Gut, 17:235–243, 1976.

Reinhart, J. B., Evans, S. L., and McFadden, D. L.: Cyclic vomiting in children: seen through the psychiatrist's eye. Pediatrics, 59:371–377, 1977.

Wender, E. H., Palmer, F. B., Herbst, J. J., and Wender, P. M.: Behavioral characteristics of children with chronic non-specific diarrhea. Am. J. Psychiatry, 133:20–25, 1976.

Wright, D. F., Brown, R. A., and Andrews, M. E.: Remission of chronic ruminative vomiting through reversal of social contingencies. Behav. Res. Ther., 16:134, 1978.

27
Life Threatening and Terminal Illness in Childhood

This chapter is intended as a guide to the pediatrician who will be working with the families of children confronting death. Advances in modern medical diagnosis and therapeutics have made it necessary to consider two aspects of such work. The first involves actual terminal care or intervention during an acute fatal illness. The second and more frequent aspect involves the care of the child with a chronic life threatening illness. In the following pages we shall attempt to summarize important treatment approaches arising from this distinction. We shall review the normal development of children's views of death and relate this sequence of events to the care of seriously ill children. Finally, using a family focus, we shall address the matter of home care for the dying child and aftercare for the survivors.

IMPORTANT DISTINCTIONS

The distinction between terminal illness and life threatening illness is an important one, and one that has evolved as a major issue with advances in medical therapeutics over the past three decades. Diseases that once were swiftly fatal have been converted into conditions that still threaten death, but that from a distance are difficult to gauge. Childhood cancer and cystic fibrosis are good illustrations of this phenomenon.

Children diagnosed as having acute lymphoblastic leukemia 30 years ago usually died within six months. Today such children have a 50 per cent chance of surviving five or more years in a disease-free state, and very few fail to achieve a remission of at least 18 to 24 months (Koocher and O'Malley, 1981). Half the patients diagnosed in infancy as having cystic fibrosis during the 1950's died within one year; the median duration of survival today is at least 18 years, with many patients living into their 20's and 30's.*

These new survival statistics have generated a host of psychologic stress issues. Not the least of these is the matter of long term uncertainty and

*Patient Registry Report. Cystic Fibrosis Foundation, Rockville, Maryland. March 20, 1980.

the stresses of the chronic helplessness that this induces (Koocher and O'Malley, 1981; Seligman, 1975). What of the families of such patients? Should they attempt to anticipate the child's death and accommodate to the loss, or should they attempt to deny and repress their anxieties about a potential loss while hoping for the best? Either course may lead to long term psychologic sequelae (Kemler, 1981). The uncertainty component clearly cannot be overlooked in considering the psychologic adaptation of the child patient, siblings, parents, or extended family members.

The psychologic stresses associated with the uncertainties of long term survival among children with life threatening illness have been termed the "Damocles syndrome" (Koocher and O'Malley, 1981). The quality of life among such patients and their families is often linked to their ability to ignore the uncertainty and adopt an optimistic or hopeful attitude. It would be negligent to stress only the issues of terminal care in this chapter, because the register of long term survivors grows daily and represents a major challenge in terms of preventive mental health care for the pediatrician.

BASIC TREATMENT APPROACHES

Because of the threat of death, the practitioner must be aware of the whole social ecology of the patient. If the patient does die, there will be a family of grief stricken survivors. They too should be seen as part of the "patient care" locus. A family based intervention is an absolute necessity if optimal adaptation and support of the patient are the pediatrician's goals. The costs to the patient's family are substantial in emotional and financial terms. The course of the illness is likely to sap the adaptive capacities of the parents, both as individuals and as a couple. Siblings are also likely to experience an extra burden of stress. The practitioner who does not regard the whole family as "the patient" ultimately is likely to compromise care for the ill family member.

In formulating a service plan or clinic structure for children with terminal or life threatening illness, team work is of critical importance. The most desirable strategy would be one that integrates medical and mental health care, along with home care and other ancillary services within a single system (Koocher et al., 1979). This is important because such a system enables the medical and mental health personnel to know each other well. The medical staff becomes more familiar with psychosocial issues, and the psychosocial staff learns about the diagnostic and therapeutic events their patients will be encountering. Patients benefit by virtue of the fact that all care-givers interact closely and more efficiently. It is also important for families to feel that they are not singled out as being "crazy" or in need of emotional support. Rather, they very much want mental health services to be "routine," so that no stigma accrues to those who use the service. Integrating medical and mental health services makes the use of such programs more "normal" for all concerned.

INPATIENT VERSUS OUTPATIENT CARE

Most pediatric patients with life threatening illness spend the bulk of their time outside the hospital, receiving most of their care on an outpatient basis. There is also an increasing tendency for families to provide home care for children in the terminal phase of their illness, and we shall discuss the special concerns related to that situation later in this chapter. The psychologic issues and needs of pediatric patients and their families are quite different when inpatient care is necessary, and it is important for the practitioner to recognize the distinctions.

Outpatient care is probably more reassuring for the child, who will spend more time in the company of familiar caretakers than would be the case during an inpatient stay. A degree of normality associated with sleeping in one's own home, and returning to a familiar family ecology after clinic visits can be quite supportive. In some instances, however, the stresses on family members may be greater in the sense that travel to the clinic, arranging for care of siblings, missing time at work, and other such disruptions can be quite stressful, depending on the duration of the illness and the treatments required. Some parents are reassured and feel more useful when they are able to care for their child at home, whereas others may feel insecure and tense in dealing with the side effects of treatment or other problems of care at home.

During an inpatient stay a child may require more reassurance and emotional support, as a function of the degree to which family members are able to be available and helpful. Some facilities provide for rooming-in and encourage parents to play a substantial role in caring for their child; in other circumstances this may be discouraged, or the child's condition may dictate less contact with family members. Each hospital admission may have a different meaning for the child and his family. Some may be seen as "routine" therapeutic admissions, while others may generate anxiety that the end is near (appropriately or not).

The important point for the pediatrician is that each family must be regarded in the context of the particular current treatment circumstances and the meaning of those events to them. A family who managed well under one set of circumstances a few weeks earlier may have very different reactions during a subsequent treatment phase. The distinction may well be based on a psychologic perception that is not evident to the physician in charge of the medical case management. This again underscores the importance of including mental health professionals on the treatment team.

DEVELOPMENTAL ISSUES

In order to understand the child's reaction to life threatening illness, the pediatrician must be mindful of important developmental differences among children. Age is only one variable, with emotional, social, and cognitive development each playing its own role in forming the child's reaction. Experience with loss, separation, and illness also colors the reactions of individual children. Table 27–1 highlights some of the key issues.

The ability of the child to conceptualize the nature and consequences of his illness varies dramatically as a function of cognitive development. The infant who is ill probably does not realize that circumstances could be otherwise, and makes little connection between causes and effects with respect to the disease process and the treatment program. As the toddler becomes more verbal, it is possible to communicate more directly about the disease and treatment processes; however, it is difficult for the young child to differentiate between what happens to others and what happens to himself. Magical thinking also remains quite active at this age, and the perceived causes of illness related events may differ radically from reality. By age six or seven most children have many questions, although they may not always feel free to ask them. A better sense of cause and effect in the illness-treatment process is present. Unlike the situation in younger children, death is now recognized as irreversible rather than as being akin to sleep or other personal experiences of the living. In adolescence, with the onset of hypothetical thinking, alternatives that might not have occurred to the younger child are prevalent (e.g., What will happen if the treatments do not work?).

Table 27–1.   KEY DEVELOPMENTAL TRENDS

Child's Age	Cognitive Issues	Emotional Issues	Social Issues
Birth to two years	The child functions as an egocentric being, with little sense of cause and effect. Cannot retain the concept that parents still exist if they are out of sight.	Emotional expression varies directly as a function of immediate sensation and is quite labile.	The child is totally dependent on adult care and anxious in the company of strangers or unfamiliar caretakers.
Three to six years	Although still quite egocentric, the child is able to begin to conceptualize relationships between causes and effects, especially when based on personal experience.	Less emotional lability is manifested as the child grows older, but fears and fantasies may replace separation issues as key stressors.	Early psychologic autonomy makes it easier to tolerate separation, but hospitalization and illness frequently lead to regressive needs for frequent parental reassurance.
Seven to 12 years	The child becomes aware that the experience of others may be different from one's own. Logical reasoning becomes firmly established. Many questions are asked.	Efforts at mastery and related fantasy play are used as means of improving emotional controls. Intellectual defense mechanisms come into play.	More independence from parents and competence building activity takes place, with increasing emphasis on the peer group.
Thirteen and up	With adolescence, the child becomes capable of hypothetical thought and is capable of reasoning from theoretical alternatives for the first time beyond concrete examples.	Emotional issues focus on establishing a sense of self as separate from a family membership. Future planning becomes important as the child begins to think about "growing up" in earnest.	The peer group rather than the family and the school rather than the home become the primary loci of activity and socialization.

A future-orientation becomes meaningful, and the long range consequences of illness and treatment are now salient for the first time.

Emotional expression in the infant is primitive and is directly linked to impulse and sensation. As the child gets older, mood and reactivity become more stable, but fantasies and fears become important to overall emotional reactivity and adaptation. In the preadolescent years mastery and competence building activity helps to form adaptive mechanisms in the face of emotional stress. Intellectual problem solving becomes useful as a defense mechanism. By adolescense the child is thinking seriously about growing up and all that this means, as well as developing a sense of identity distinct from membership in his family per se. The ability to cope with pain, body alterations, and similar factors related to life threatening illness (as well as recognition of the actual threat) varies dramatically as a function of the child's emotional development.

Social issues are also subject to important developmental shifts, bearing on the sorts of steps a pediatrician must take to insure the best care for the critically ill child. For the infant, separation may be akin to abandonment, producing great distress and depression. The toddler may be better able to tolerate parental separation, but may show regressive behavior such as loss of toilet training

or excessive clinging. Older children react most severely to perceived threats to their competence, expressing fears of falling behind in school or reacting with marked depression and anxiety to mobility losses. The teenager in addition suffers intense concern with the loss of peer interactions and school activities, since these are the primary delineators of evolving self-esteem in adolescence. The issues of most concern to the child in large measure are linked to the level of social functioning.

NORMATIVE ADJUSTMENT

Normal behavior in the family of the child with terminal or life threatening illness is different from that of the same family prior to the diagnosis. Although this may seem to be a truism, it is a point often overlooked by professionals immersed in the process of treating such families. For example, bright attentive parents talking retrospectively of the day they first learned their child's diagnosis frequently report, "After the doctor told us our child had leukemia, he said some other things too. . . . I don't remember what they were." The "other things" probably dealt with treatments that were to begin, but the parents, overwhelmed by the threat of a potentially terminal illness, were too stunned to absorb the information. Such reactions are indeed normal under the circumstances.

Children who are terminally ill, or who face an uncertain but potentially fatal outcome from some chronic illness, are at substantial risk for emotional disturbance as a function of stress. A host of papers, review articles, and books have documented and detailed the core stresses and common symptoms. Depending on the course and trajectory of the disease process, even children who were quite "normal" prior to becoming ill may develop increased anxiety, loss of appetite, insomnia, social isolation, emotional withdrawal, depression and apathy, and marked ambivalence toward adults who are providing primary care.

These reactions are generally best regarded as responses to acute or chronic stress, rather than as evidence of functional psychopathology. It is predictable, however, that children or families with pre-existing emotional disorders will experience an exacerbation. The primary model for the way in which this occurs is best described in the work of Seligman (1975) and other writers on the topic of learned helplessness. When a person comes to believe that the outcome he will confront (i.e., death) is independent of his own behavior, the helplessness syndrome and accompanying emotional stress are dramatic (Seligman, 1975).

Much research has focused on the child's awareness of his own fatal illness. It is clear that even young children are very aware of their medical conditions. Empirical data exist, for example, that demonstrate that anxiety levels in children with leukemia increase in parallel to increases in the frequency of outpatient clinic visits. This is the opposite of what one finds in youngsters with chronic non-life threatening illness. Other data demonstrate the increasing sense of isolation dying children tend to experience. It is clear that one cannot shield sick children from anxiety about their conditions.

Even children who otherwise seem to be coping quite well through a prolonged illness may develop specific problems, such as conditioned reflex vomiting, anxiety linked to specific medical procedures, depressive reactions to progressive loss of physical capacities, or family communication inhibitions (Koocher and Sallan, 1978). Sometimes children and their families cope quite well during periods of active treatment, even when noxious procedures are involved, only to become overwhelmed when the need for continued treatment is over (Koocher and O'Malley, 1981; Koocher and Sallen, 1978). Although this type of reaction may seem a paradoxic reaction to the cessation of a noxious experience, hospitalizations and even harsh treatment regimens may come to be imbued with some protective value.

CHILDREN'S PERCEPTIONS OF DEATH

DEVELOPMENTAL DIFFERENCES

In their review of psychologic perspectives on death, Kastenbaum and Costa (1977) note three oft-held assumptions about children's perceptions of death. First is the assumption that children do not comprehend death. Second is the assumption that adults do comprehend death. Finally, there is an assumption that even if children were able to understand death, it would be harmful for them to be concerned about it. These assumptions are clearly superficial and reflect defensiveness on the part of their proponents rather than any valid view of actual circumstances. They did, however, provide a basis on which to justify avoiding such discussions with dying children (Evans, 1968).

Early studies demonstrated that children's conceptions of death follow a developmental progression (Anthony, 1940; Nagy, 1948). Although these early works were not methodologically rigorous, they represented nearly the whole body of material on the topic for two decades. One result was that many writers simply accepted and repeated their findings without question or additional investigation (e.g., Kubler-Ross, 1968, pp. 178–79).

Anthony's study (1940) described a developmental sequence in understanding the word "dead," with a progression from initial ignorance, through personal associations, to an ultimate biologic understanding of the concept. Nagy (1948) described three biodevelopmental stages along a similar continuum. She noted that children under age five tended to regard death as a reversible process associated with separation. Her second stage involved an alleged tendency by children to "personify" death as a kind of "boogy man" or "black coachman." Finally Nagy concluded that at about age nine children begin to regard death a lawful universal process.

More recent studies have provided data-based documentation of the manner in which the death concept develops. A Piagetian framework has been applied to demonstrate how children's responses to questions about death reflect their levels of cognitive development (Koocher, 1973; 1974). Other developmental studies have documented acquisition of the universality and irrevocability of death concepts as well as highlighting children's own awareness of their potential death (Spinetta, 1974; White et al., 1978).

These studies suggest that egocentrism and magical thinking, which are a part of preoperational thought in young children, dominate concerns about death in early childhood. With the beginning of concrete operational thought at about age six or seven, the child becomes capable of taking the role of another person in the cognitive sense and thereby begins to sense the permanence of death. At this stage, however, the child may still think of death as something that occurs as a consequence of a specific illness or injury rather than as a biologic process. At the time of adolescence with accompanying abstract reasoning capability and formal operational thought, a more complete comprehension of death becomes possible. It comes as no surprise that the specific concern and fantasies expressed by children of different ages with respect to death are reflective of their cognitive understanding about it.

Bowlby's writings (1973) on the theme of separation demonstrate the importance of social relationships and the consequences of their disruption, especially during the early years of development. It is the interaction of these social relationships and the cognitive accommodation capacities, as highlighted in Table 27–1, that determine coping abilities in concert with other individual factors. A brief review of these elements from a child development standpoint may be found in a recently published book on children's conceptions of health and illness (Koocher, 1981). A detailed and well integrated review by Lonetto (1980) synthesizes the literature dealing with children's conceptions of death with heretofore unparalleled clarity.

THE DYING CHILD'S AWARENESS OF DEATH

Evan's position in regard to witholding information about the diagnosis and prognosis from the dying child was generally abandoned many years ago (Evans, 1968). The emotions of the pediatricians who must care for terminal patients remain an issue, however, at least partially because of their own intense discomfort in working with children who are so gravely ill. If there is a single article that sums the issues up best, it is the classic by Vernick and Karon (1965) entitled "Who's Afraid of Death on a Leukemia Ward?" Through the use of life-space interviews, they documented the fact that children clearly knew the seriousness of their illnesses and were eager to have someone to talk with about it. They described the communication barriers often erected by adults to "protect the child", and noted that when a child is passive with respect to discussing these concerns, it is often a reflection of the environment. A number of empirical studies have yielded supportive data (Spinetta et al., 1973).

Reviews of professional opinion and research data have consistently stressed that children as young as five or six have a very real understanding of the seriousness of their illness, and still younger children show definite reactions to increased parental stress and other effects of a terminal illness on the family. Despite this recognition by children of their serious illness, conceptualizations of death and loss issues do not really differ from the general developmental trends already noted. The predominant modes of response tend to reflect age related concerns about separation, pain, and disruption of usual life activities (Koocher, 1980).

Just as Kubler-Ross' ground breaking work in 1969 emphasized the importance of systematic psychologic intervention for the adult with terminal illness, the parallel needs of children are also being recognized. Even among normal children there are substantial elements of anxiety in regard to death (Koocher et al., 1976). It is not surprising, therefore, to find a variety of adverse psychologic symptoms and behavior problems among dying children and members of their families.

THE CRITICALLY ILL CHILD: KEY PSYCHOLOGIC ISSUES

The foregoing section of this chapter summarized some of the developmental sequences and issues related to death perceptions in childhood. The next several pages illustrate how critically ill children differ in that regard on a series of key issues.

SEPARATION

The infant who cannot see his parent cannot retain the concept that the parent still exists to provide care (Mahler et al., 1973; Piaget, 1954). The infant's reaction of acute distress to the parent's leaving is well recognized. The infant or toddler in the hospital, an unfamiliar setting with strangers as caregivers, seeks parental reassurance and comfort and finds separation even more anxiety producing than the healthy child. Arranging for parents to sleep in the hospital room with their child and to be integrally involved in their child's care is vital at this age. This includes accompanying and comforting the child during medical procedures. At times the parents' heightened anxiety, or withdrawal due to anticipatory grief, may cause them to defer basic care of their child to the nursing staff. The pediatrician should assist the parents in openly discussing their emotional reactions and in underscoring the child's need for the parents' ongoing physical care and comfort.

By approximately age three the preschooler has established a sense of psychologic autonomy and increasingly can tolerate parental absence (Mahler, 1973). The diagnosis of a life threatening illness and the need for prolonged hospitalization, however, may lead to regressive behavior and an increased need for parental reassurance. As noted earlier, Nagy (1948) and Koocher (1973) have discussed how the preschool child may perceive death as a type of separation from others. Fears and fantasies about illness and dying may be confused with concern about separation. The parents' presence and active involvement remain essential throughout the childhood years.

Separation is not as crucial an issue in adolescence as it is in childhood. The adolescent is concerned about potential family withdrawal related to anticipatory grief, the censoring of medical information, which isolates the patient, and social withdrawal of peers owing to their own anxieties about serious illness. Acceptance by peers is very inportant to the adolescent's sense of identity and self-esteem. Separation from normal peer activities poses a difficulty for many adolescent patients.

PAIN MANAGEMENT

The treatment of a life threatening illness like cancer requires many painful and noxious procedures, including surgery, bone marrow biopsy, and chemotherapy. In addition to the parents' presence, physical comforting, and reassurance, there are other strategies that the treatment team can use to deal with the anxious anticipation of aversive procedures as well as management of pain.

Every child needs preparation for procedures. An explanation of the procedure, commensurate with the child's age and ability to comprehend, helps the child to anticipate what will transpire and provides an increased sense of control. Play therapy techniques, such as providing the child with a stuffed animal and asking him to install an intravenous line using actual equipment prior to a procedure, provides an opportunity to address emotional concerns and bolster coping strategies. The young child, complaining of pain during a procedure, can learn distraction techniques like deep breathing, squeezing a parent's arm when the pain intensifies, or the use of visual imagery techniques.

Deep muscle relaxation and hypnosis can also be employed with school age children and adolescents both to aid in reducing anxiety and as a strategy to control chronic pain, such as "phantom pain" following amputation or pain due to an invasive tumor. Psychotherapy can also provide patients with a forum to discuss their emotional reactions and to better understand the emotional context of the pain they are experiencing. The following examples illustrate three approaches to pain management:

Case 1. Jill, age four, had severe anticipatory vomiting prior to chemotherapy for three months. Play therapy sessions, prior to chemotherapy, were then begun. These sessions included elaborate preparations for giving Snoopy chemotherapy. After three consultations, the incidence of vomiting had diminished significantly, and the child looked forward to treating Snoopy at the clinic. This intervention allowed Jill to express and gain some mastery over her feelings of anxiety.

Case 2. Alan, age nine, had severe phantom pain following a leg amputation for osteogenic sarcoma. He was especially fearful of physical therapy. The physical therapist was encouraged to use a Curious George doll at the beginning of each session. Alan would give physical therapy to the doll prior to receiving his own therapy. He was then able to better tolerate the sessions.

In both cases 1 and 2 the doll play allowed for an identification with the aggressor and an increased sense of control.

Case 3. George, age 15, complained of chronic pain due to a slowly healing wound. A training course in deep muscle relaxation and guided visual imagery was begun. George learned to induce relaxation and to visualize a pleasant, relaxing scene when he felt the pain becoming intolerable. George found that having this strategy as an alternative to medication made him feel that he had greater control of his pain.

CONTROL

The diagnosis of a life threatening illness creates an emotional crisis in a family. The loss of control implicit in the diagnosis is one of the most devastating aspects of the emotional crisis. Parents may experience the loss of ability to protect their child from harm and to positively influence their child's future. Their own feelings of loss of control in this situation may foster an appropriate intensification of their concern and caretaking of the child. This aids them in regaining a sense of control when they feel helpless. Often, however, this caretaking can be experienced as infantilizing by the patient.

The preadolescent child, who is gaining mastery in many new areas, may experience the loss of control to plan his life. The child may displace anger onto schedules or hospitalization that interfere with such plans. For example:

Case 4. Carl, age nine, was furious that treatment for Ewing's sarcoma interfered with his soccer team practice. He was not, however, directly angry about his diagnosis.

The preadolescent child may attempt to regain control by testing parental limits. The parents may in turn curtail discipline in order to demonstrate their special caring for their sick child. Rather than making the child feel special, however, this approach can make the child feel more out of control. Case 5 illustrates this point:

Case 5. Joey, a five year old with leukemia, was permitted to hit his parents in the face when he became angry. The parents felt that their son was naturally angry and needed to express his feelings. In addition, his parents were worried about Joey's nutritional status. They would therefore offer him multiple choices for each meal. He would make two or three choices and then eat very little.

After consultation with the oncologist and the psychologist, the parents were encouraged to set firm limits. Alternative methods for expressing anger, like hitting a pillow or yelling, were suggested. Additionally, meals were to include one choice for the entire family. Two weeks later the parents reported that Joey was less angry and was eating larger meals. The clinic staff also observed that Joey appeared calmer.

Although the loss of control is a salient issue for all patients, it has a heightened impact during adolescence because of developmental tasks specific to adolescence.

The adolescent is gaining greater autonomy in his environment and increasing independence from his parents. These gains are challenged by the hospital experience in which the patient feels deprived of the ability to make decisions while members of the medical staff make decisions that affect his health and daily life. Patients often note that they feel bombarded by intrusive medical practices. The hospital experience can encourage passivity and regressive dependence that assault the adolescent's sense of increasing mastery over his environment. The ability to experience competence through school, social experiences, and planning for the future is seriously disrupted because of prolonged treatment. This also assaults the adolescent's sense of control.

Another adolescent task is to develop a sense of mastery over one's changing body. These body image issues are directly challenged by the stress of having a life threatening illness. Physical changes due to illness can be especially trying for adolescents. For example, alopecia, a frequent side effect of chemotherapy, is a visible, inescapable reminder to the patient of his disease, as well as a possible source of embarrassment and diminution of self-esteem. Comfort and confidence in sexual attractiveness are also severely challenged by the effects of both the disease and its treatment.

The loss of control implicit in a hospitalization is also stressful. The patient deals with the disruption of his daily schedule, and the lack of familiar people and activities, and responds by feeling a loss of control. He may bring in a T-shirt, poster, or album to transmit a special interest or personal value to new hospital peers. Some adolescent patients may have a particular time of the day when they are especially uncomfortable, often coinciding with a missed special activity or lonely late evening hours.

Research, as well as clinical anecdotes, illustrates that the loss of control has unfavorable consequences for the hospitalized patient. These studies also suggest that by increasing a patient's perception of control, one can ameliorate his ability to cope with disease (Langer and Rodin, 1976;

Seligman, 1975). The maintenance of a hospital milieu sensitive to control issues is discussed later in this chapter.

HONESTY AND TRUST

As noted, a protective approach was commonly applied in discussing death and dying with pediatric cancer patients until the 1970's (Evans, 1968). Researchers and clinicians alike then began to challenge the wisdom of this standard approach. Vernick and Karon (1965) reported that children with leukemia sensed their parents' and doctors' anxiety and concern, even though the adults attempted to be cheerful. Additionally the children did not ask questions because they were apprehensive about their parents' discomfort in discussing death. Spinetta et al. (1973) also demonstrated that children as young as age six who have leukemia are aware of the meaning and possible outcome of their disease.

Older children are also well aware of their disease, its possible ramifications, and of the tendency of parents and physicians to withold or distort information. Share (1972) has discussed how the open approach to the communication of information is increasingly used with children who have life threatening illnesses. The child has many fears and apprehensions about unknown possibilities, and open communication provides an opportunity to dispel them and concentrate on the adjustment required for the realities of treatment. Kellerman (1977) has noted an inverse relationship between a child's open discussion of his illness and depression. Most important, a closed communication style leads to a sense of isolation for the child. Vernick and Karon (1965) have argued that openness allows the child to feel more secure and trusting of the medical staff and parents, noting that ". . .blows cannot always be softened, but by explanation and sharing, their impact may be made somewhat less concentrated and acute." Finally, Slavin (1981) has pointed out that by resolving debate over communication, caregivers can turn their attention to aiding children in coping with their disease. The interested reader is referred to articles on how to discuss death with children by Spinetta (1980) and Koocher (1974).

FAMILY ISSUES

The pediatrician caring for the child with a life threatening illness recognizes that providing total care entails attending to the emotional needs of the entire family. The family, not simply the patient, is the unit of care and attention. Slavin

(1981) has noted that research to date has neglected a family focus in favor of studying only the parents.

In coping with life threatening illness, the family members are faced with the problem of managing their own emotions as well as carrying on with the practical aspects of daily living. Learning to deal with chronic stress, disruption, and uncertainty makes coping a continually trying task. Slavin (1981) summarizes the burdens of coping as ". . .balancing the needs of the patient in the home with those of healthy siblings; fostering the patient's normal social and emotional development while coping with long-term uncertainty; and dealing with unresolved anticipatory grief if the child survives" (p. 30).

Increased marital stress and family financial strain are predictable phenomena. The quality of the marital relationship prior to diagnosis serves as an important predictor of the adequacy of marital coping with the crisis. Differences in parental coping styles, or being "out of synch" in the timing of emotional reactions, and differences in priorities in managing the family can also exacerbate marital tension (Sourkes, 1977). A decreased sense of parental competence resulting from the stresses can further impair coping. The survival of the marriage may be related to the survival of the patient, but professional counseling and parent support groups can play important roles in helping to manage parental and marital stress (Koocher and O'Malley, 1981).

The relationship of siblings to their parents also changes if they experience decreased parental attention and support (Sourkes, 1980). They may feel angry at both their ill sibling and their parents. They may become angry at their parents for failing to protect the patient from the disease. School and peer relationships can also be adversely affected. School performance may decline in response to family stress; in other instances siblings may plunge into academic pursuits as a means of escaping stresses at home, or to prove themselves competent in an effort to combat family feelings of hopelessness. Peer relationships can be interrupted for practical reasons, such as school absences, or because of retreat into the family. Siblings may also feel alienated from friends who do not understand their irritability and preoccupation. The presentation of somatic complaints as a means of garnering parental attention or to identify with the ill sibling has also been observed. Death and mourning for a sibling may raise many issues of vulnerability, guilt, and confusion.

Open communication between the patient's physician and the siblings, including family counseling sessions, is one very beneficial approach. Siblings have many questions and fantasies that must be addressed and clarified. Parents' efforts to spend special time with their healthy children and to aid them in maintaining a normal life are also important approaches.

IATROGENIC FACTORS

The patient may also respond to iatrogenic factors related to the disease and treatment. For example, there is now research evidence suggesting possible neuropsychologic sequelae following cranial irradiation and the intrathecal administration of drugs in the treatment of acute lymphocytic leukemia. A number of studies employing psychometric tests have revealed subtle abnormalities in the cognitive performance of treated patients (Eiser, 1978; McIntosh et al., 1976; Meadows and Evans, 1976; Soni et al., 1975). Moss and Nannis (1980) have noted significant methodologic difficulties in studies to date; however, they too observe that intellectual abilities appear to be impaired in patients receiving central nervous system irradiation, impairment being more significant in younger patients in whom the less mature central nervous system may be more vulnerable. Although Moss and Nannis do not question the benefit of prophylactic treatment of the central nervous system in terms of greatly sustained remissions in children with leukemia, they do argue for more rigorous research to ascertain the exact nature, extent, and type of risk factors for such patients.

Another possible iatrogenic by-product of treatment involves the patient's development of close ties with other patients and the pain and depression that follow the death of a clinic or hospital friend. This is especially poignant in cystic fibrosis, when adolescent patients may develop friendships lasting several years. The loss of a friend not only precipitates mourning and grief over the death, but also forces the patient to directly confront his own vulnerability and mortality.

Case 6. Ann, an 18 year old patient with cystic fibrosis, had known Will, age 19, for several years. As Will's disease progressed and hospitalizations became more frequent and more lengthy, Will became despondent and depressed. Ann, witnessing Will's sullen withdrawal, responded by becoming quite depressed. She worried about the progression of her own disease and developed refractory psychosomatic abdominal pain.

Patenaude et al. (1979) have written about the psychologic costs of bone marrow transplantation in a pediatric population. They describe the close bonding between patients who live next to each other in sterile isolation rooms for several weeks following a transplant. The death of one patient often precipitates a strong emotional reaction in another, including the potential guilt of the survivor at his own continued longevity.

Another iatrogenic factor is function loss following treatment for a life threatening illness. The patient with osteogenic sarcoma who must learn to adapt following amputation of a limb is one example of a substantial readjustment problem posed as a by-product of treatment. Even patients who are successfully treated may remain at substantial psychologic risk as they live with uncertain futures for many years. The key issue is that successful treatment must consider the whole patient and his family in the recognition that many expected but serious physical and emotional sequelae may develop as a result of treatment.

PLANNING PSYCHOLOGIC CARE FOR THE DYING CHILD

The dying child requires the support, care, and love of his family and friends. One of the interdisciplinary team's key roles is to encourage and facilitate the family's ability to offer this care. The patient's feelings that he is surrounded by a network of understanding people and is not isolated is vital in coping with end stage disease. The knowledge that he can openly discuss fears and thoughts about death without being censored is equally important. Knowing that people important to him are available to conduct "unfinished business," such as telling a rival sibling that he is loved or asking a parent to carry out a request about belongings or the funeral, can be very comforting to the patient.

THE ROLE OF THE INTERDISCIPLINARY TEAM

Parents and siblings need special support as well. Throughout the patient's treatment the family has established a network of medical providers. One typical team is composed of the inpatient physician, primary nurse, mental health professional, outpatient pediatrician, and visiting nurse. Open, frequent communication among team members can insure a uniform approach and underscore the consistent availability of the staff to the patient and family. Liaison with a school counselor, teacher, clergyman, or other community members is helpful in providing additional support. The designation of an overall team coordinator, who takes responsibility for gathering the data and presenting the synthesized information to the family, is crucial for clear communication. This prevents overly anxious parents from "splitting" team members to act out their own emotional concerns or to distort information. The patient

needs to be an involved member of the treatment team. His exclusion, rather than diminishing anxiety, exacerbates the patient's sense of loss of control and isolation. For example:

Case 7.   Susan, age nine, complained that she was treated like a baby. The doctor would always compliment her on her nightgown or ask about her stuffed animals. He would not, however, tell her about his hallway discussions with the residents. Susan complained: "I want to know if I need to have another lumbar puncture and when I can go home. I don't care if he likes my teddy bear."

THE MENTAL HEALTH PROFESSIONAL'S ROLE

The mental health professional can offer important support to the patient, family, and medical staff by consulting with other staff in the assessment of the patient's developmental level, the meaning of psychologic and psychogenic symptoms, and the care and management of the patient. This indirect consultation may lead to direct patient contact or may allow for improved care without such a referral. The mental health professional can also facilitate creation of a forum to discuss team members' emotional reactions to providing care to patients with life threatening illnesses.

The mental health professional can best serve the patient and the family when an initial evaluation is made at the time of diagnosis. At this juncture the psychologic consultant can assess the family's functioning and interactional style and make recommendations for managing the anticipated stresses. This early intervention can help manage the predictable stress reactions and prevent increased marital tension or sibling problems. It can also help the family by facilitating their communication and support for one another. This initial assessment allows for an early disposition that may include continued family or individual psychotherapy when indicated.

THE INPATIENT MILIEU: THE ISSUE OF CONTROL

There are some critical principles in establishing an environment that enhances the child's perception of control. Such an environment is ideally planned by an interdisciplinary team using a systematic approach to encourage development of the patient's sense of responsibility from the time of admission. This type of milieu is designed to help the patient continue his daily life, schedule, and habits as much as possible. Continuity of daily life activity is the key issue. Maintaining a social network through visits, phone calls, and letters is encouraged. Making the hospital room a familiar setting by bringing in important objects from home also promotes continuity and familiarity. Insuring that the patient continues to be an active, choice making, responsible person should be the central priority. This active stance is best introduced to the patient as a way of helping him cope with disease and hospitalization. Reinforcing this message by pointing out choices in daily hospital life underscores the patient's realistic options. These choices may include activity room programming, academic tutoring, developing a support network by introducing patients to each other and "veterans" to newly diagnosed patients, allowing patients to participate in decisions concerning medical procedures, and soliciting and responding to patient feedback. These components create an atmosphere in which the patient can maintain a sense of control because he understands what is happening to him and knows ways to affect and control what happens. This type of milieu should have a positive effect on the patient's ability to cope with his disease and hospitalizations.

TERMINAL CARE AT HOME

Requests to allow the child to die at home or in an alternative center such as a hospice may be posed to the pediatrician. One study of terminal care of pediatric patients at home that provides useful information about this alternative is that of Martinson (Armstrong and Martinson, 1980; Martinson et al., 1978). She studied children who died of cancer while participating in the University of Minnesota Hospital home care program. Data were reported for 32 of the children, who ranged in age from one month to 17 years. Eighty-four per cent of the participants in the program died at home, and were in the home care project for a mean of 32.7 days.

Martinson noted four essential requirements for the success of a home care program: the availability of a family member (typically the mother) to provide daily care, the availability of a nurse on 24 hour call for telephone consultation or home visits, the availability of necessary equipment such as oxygen or air mattresses, and the availability of a physician for direct consultation with the family and primary nurse and for the prescribing of medication, particularly analgesics. The nurse and physician must work together closely in order for home care to be successful. The nurse has the pivotal role in providing medical, emotional, and practical support to the patient and family.

Armstrong and Martinson (1980) reported that children under four years of age did not have a good understanding that they were dying. School age children reported feeling secure in being home with their family and were relieved that they could

avoid medical procedures associated with the hospital. Adolescents were reported to be frustrated and angry at their need to be dependent and were often depressed and in some instances accepting of imminent death. Martinson concluded that there is not sufficient demand for pediatric hospices, but that home care directed by a hospital staff is a desirable alternative for families and patients who prefer the home setting. Martinson believed that the security of the home is preferable to the hospital for the dying child. Finally, she underscored the need for postmortem follow-up and supportive services.

ALTERNATIVE TREATMENT APPROACHES

In the course of treating patients with life threatening illness, the physician is confronted with the question of how to respond to a patient's or family's request or decision to use unconventional treatment modalities. Although amygdaline (Laetrile), megavitamin therapy, and nutritional approaches are currently popular alternative treatments for cancer, there is a long history of American health treatment alternatives. Young (1967), for example, documented cancer related treatment approaches back to the colonial era in his discussion of Francis Torres' "Chinese stones" cure. Earlier in the twentieth century Harry Hoxsey's cancer cure and Krebiozen generated public attention and controversy. Young (1967) noted that the American Cancer Society listed 71 unproven methods of cancer treatment available in 1976.

The physician's response to an inquiring patient should be rooted in the context of the trust necessary in the doctor-patient relationship. Attacking the patient's intelligence or questioning his common sense injures the working alliance necessary for the mutual problem solving approach. The underlying principle is that the physician's behavior and response should help the patient adapt.

It is difficult for the physician to be familiar with all the trends and new claims for alternative treatments. There is extensive documentation of the ineffectiveness of popular approaches such as Laetrile (Braico et al., 1979; Brant and Graceffa, 1979; DiPalma, 1977; Greenberg, 1980; Herbert, 1978). The American Medical Association and the American Cancer Society provide additional information about alternative therapies. The physician who can respond in an informed and measured way in regard to potential risks and benefits of alternative therapies will be in the best position to invite his patient to re-examine such a request.

The patient who plans to proceed with the protocol recommended by the physician and to augment it with an unproven approach needs to learn the physician's appraisal and then, unless the approach is known to be deleterious, should be granted permission to proceed with this two pronged treatment strategy. The gains the patient may achieve through increased feeling of control and mastery in fighting the disease, the possible placebo effects, and the continuation of cooperation with the physician may more than compensate for the possible ineffectiveness of the patient's chosen treatment. When the patient wishes to terminate the physician's recommended treatment protocol entirely, a frank but respectful review of the empirical data relating to the two approaches must be offered. Some families will inevitably refuse treatment. In the case of the pediatric patient the physician or hospital may decide to ask for the court's intervention in determining the most suitable course of treatment (Brant and Graceffa, 1979).

Patients and their families who refuse treatment or certain procedures because of religious beliefs present more complex ethical issues. The family is often responding on the basis of deeply held beliefs, and does not wish to be in an adversarial relationship with the treatment team. It is well known, for example, that members of the Jehovah's Witness faith refuse transfusion of blood or blood products for themselves and their children on religious grounds. In cases in which a child's physical well-being is in immediate danger, courts have routinely authorized life saving transfusions for children over the objections of their Jehovah's Witness parents (Mnookin, 1978). In some instances these parents may experience emotional relief when such treatments are ordered by a secular court, since they have followed their religious beliefs precisely and have had the critical decision removed from their jurisdiction.

It is important to be mindful of the term *life saving* in the case law involving treatment of Jehovah's Witnesses. In the case of a child with terminal illness, for example, a blood transfusion might prolong life or make the patient's condition more stable without altering the terminal prognosis. In such cases courts might well sustain the religious or personal objections of family members to specific treatments.* The treatment team should give strong weight to family values in these situations, since the surviving members of the family will have to live with their memories and personal values long after the child has died. Their ability to cope can best be facilitated through respectful and supportive attention by the treatment team.

In re Storar, 433 N.Y.S. 2d 388 (N.Y., Monroe County Sup. Ct. 1980); 434 N.Y.S. 2d 46 (N.Y. App. Div. 1980).

STAGE THEORIES OF EMOTIONAL RESPONSE TO TERMINAL ILLNESS

The notion that people progress through predictable stages of emotional responses to life events has been widely discussed in recent years. Silver and Wortman (1980) have mentioned theories related to separation (Bowlby, 1960, 1973), physical disability (Gunther, 1969; Guttman, 1976), bereavement and loss (Bowlby, 1961; Engel, 1962), criminal victimization (Symonds, 1975), and terminal illness (Nighswanger, 1971). Kubler-Ross' stage theory of terminal illness is well known (Kubler-Ross, 1969). She describes a process in which the patient progresses from denial of the illness, to strong feelings of anger, to bargaining for time or improved health, to depression, and finally to acceptance of death. Silver and Wortman (1980) further record stage theories related to specific illnesses, including open heart surgery and cancer (Gullo et al., 1974).

Silver and Wortman (1980) questioned whether theories with intuitive appeal like Kubler-Ross' are indeed borne out by empirical research. They noted the relative paucity of controlled studies of stage theories in cases of response to adverse circumstances. None of the studies included pediatric patients. In reviewing the relevant primate and human studies, the authors concluded:

> . . . the limited data [do] not appear to cleanly fit a stage model of emotional response following life crises. In addition, the extreme pattern of variability that exists in response to [adverse] life events also does not support the notion of stages of response. . . . It must also be recognized that some theorists contend that people may experience more than one stage simultaneously, may move back and forth among the stages, and may skip certain stages completely (pp. 304–305).

The crucial point is to recognize the great variability in human response and the nonlinear progression of emotional responsivity, and to conceptualize in terms of reactions rather than lock-step stages. In thinking about adolescents with cancer, for example, one is likely to see a significant, direct, and continuing expression of anger beyond what Kubler-Ross has predicted. The adolescent is also much less likely to reach a stage of acceptance of death (Plumb and Holland, 1974). Case management must therefore be focused on individual patient responses and needs. The clinician should not fall into the conceptual trap of pondering why a patient has not yet reached a certain stage at an expected juncture.

Another factor impinging upon children's emotional reactivity and ability to cope with illness is their awareness and understanding of what is happening to them. For example, pediatric patients are aware of when they are feeling energetic or lethargic. A child who is ill and feels weak may be more apt to accept his prognosis than the alert anxious child, yet neither may truly accept his fate.

AFTERCARE ISSUES: ASSESSING GRIEF REACTIONS

When the patient dies, the pediatrician must consider the needs of surviving family members, and even when the patient becomes a long-term survivor, the doctor must focus special attention on the grief reactions that will normally occur. The studies by Lindemann (1944) and Schoenberg et al. (1975) are timeless accounts of human grief reactions. Although most articles about this topic have focused chiefly on adults, it is important to note that grief reactions have at least three phases that may apply to people of all ages: acute, chronic, and anticipatory grief (Kemler, 1981). The acute sadness and tearfulness that may follow immediately after a loss is probably the best recognized manifestation of grief. However, it is the long term reactions leading up to a death from chronic illness or following the loss by several months that create the more subtle management problems.

Symptoms associated with grief reactions in childhood often include tearfulness, social and emotional withdrawal, loss of interest in favorite toys or pastimes, a decreased attention span, the development of tics, loss of appetite, persistent insomnia or nightmares, decreased effectiveness in school, increases in the unfocused activity level, and expressions of guilt over past activities, especially in relation to the deceased (Bowlby, 1973; Lindemann, 1944). The natural dependency of children, along with their potential for animistic and magical thinking, makes them particularly vulnerable to prolonged adverse psychologic sequelae following an important loss (Koocher, 1973, 1974; Lonetto, 1980). At the same time the absence of symptoms of acute grief or a sharply truncated reaction may herald premature application of denial or avoidance defenses, with the potential for emergence of symptoms much later.

One key to the diagnostic assessment of childhood grief reactions is the presence or absence of anxiety. The child or adolescent who is in the process of adapting to a loss should be able to verbalize some sadness and related feelings in the course of an interview. Inability to discuss the loss, denial of effect, or anxiety and guilt relating to the deceased or surviving family members are all indicators that additional evaluation or psychotherapeutic intervention may be warranted.

Time can also be an important factor in assessing adaptation to loss, but there are no uniform guidelines to apply. Although the intensity of the depressive symptoms often abates substantially over a period of several weeks, so-called "anniversary phenomena" may trigger their return. Arrival of a birthday, holiday, or other family event may induce a return of sadness, tension, or stress, along with thoughts of the deceased person. Usually these recurrences are much less intense than the acute mourning experience. If they persist more than several days following the stimulus events or evoke a heretofore unseen intensity, a psychologic evaluation is warranted.

The clinician must also be especially sensitive to the fact that the bereaved child cannot accurately be evaluated outside the family context. Grief reactions in children are subject to both amelioration and exacerbation based on the presence or absence of emotional support within the surviving family. Behavorial contagion and social learning also play roles in determining a child's response. Religious rituals and family behavior patterns provide opportunities for observational learning and imitation that may be either facilitative or inhibitory with respect to the child's adaptation. Children may also react to mourning, depression, or anxiety in their parents or caretakers, even though they have had no personal contact with the deceased. It is therefore important for the pediatrician confronted with a patient's grief reaction to consider the parents' emotional status as well.

MANAGEMENT OF GRIEF REACTIONS

The emotional care of the bereaved child, whether a surviving sibling or another patient, should have two focal points. First is the need to help the child differentiate his fate from that of the deceased. Second is the need of the child to arrive at a sense of closure with regard to the loss. This may involve expressing feelings of guilt or responsibility for the death, as well as magical fears about what actually transpired. Both foci are important issues for any child who experiences a loss, although the need is more acute when considerable anxiety persists.

When another patient known to the child dies, the stress may be particularly intense. The need to differentiate between the recently deceased and the living is a common cognitive adaptive response among both children and adults. Adults are not immune to magical thinking, especially in times of emotional stress, but children have a particular need to distinguish between real and imagined causes of death. Investigators of cognitive development have long documented the difficulties children of different ages may have in coping with abstractions, and since death is a one-time-only, final experience for each of us, it certainly qualifies as an abstract experience when it comes to issues of individual mastery.

The questions a child may be presumed to worry about in the aftermath of a death include the following: Why (i.e., by what means) did that person die? Will that happen to me (or someone else I care about)? Did I have anything to do with it? Who will take care of me now (if the deceased was one of the child's caretakers)? Although these questions may not be specifically articulated by the child, they are almost always a part of the underlying anxiety that accompanies a prolonged grief reaction. Addressing them must involve both informational and emotional components.

In a family context, involvement in funerary rituals may also be quite helpful and supportive if such involvement is well explained and is consistent with the child's wishes. Vicarious satisfaction of adults' needs during a funeral is not a proper basis for making the decision whether to involve a child in such activities. Introduction of philosophic or religious concepts may tend to confuse and frighten young children, especially in the absence of close family support.

FOLLOW-UP VISITS

Inviting the family to return to the hospital or to meet with one or more of the professionals who helped to care for the deceased often can be helpful weeks or months after the death. Some families feel that they cannot ask for this because the patient is dead and their own needs do not seem a satisfactory basis for "imposing" on the staff.

Sometimes an invitation by the staff to discuss autopsy results provides an opportunity for surviving family members to discuss residual emotional issues. A similar opportunity is also frequently welcomed by families who are told that follow-up discussions are "routine." The family members do not want to feel "crazy," "emotionally disturbed," or different from other families, and establishing a return visit as "normal" helps them to make use of it.

Often friends, neighbors, and relatives do not understand why the immediate family has not "gotten over" the loss after several weeks or months. People outside the family may overtly or subtly give messages that they no longer wish to hear about the deceased. In such circumstances those who helped to care for the deceased child may offer the only emotional outlet available.

CHILDREN WHO SURVIVE

As noted earlier in this chapter, there are ever increasing numbers of children who become long term survivors of life threatening illness. They may live with diseases such as cancer or cystic fibrosis only to experience relapses after prolonged periods of relatively good health. These patients require special consideration and psychologic care as they struggle with the uncertainties of survival beneath a Damoclean threat (Koocher and O'Malley, 1981).

The families of such children experience similar stresses with occasional conflicting messages about whether to prepare for a death or hold realistic hope for the child's future. Encouraging such families to make use of psychologic services, while offering them appropriate medical information and sensitive follow-up care, is the best intervention strategy. The key issue is not to assume that because the disease is under control the emotions about the threat of loss are also.

A CLOSING NOTE

Many professionals are fearful of working with families of terminally ill children. Most often they cite their own helplessness or sense of inadequacy in the face of death as a basis for this fear. The patients and their families are also afraid, but want and need the support of professionals who can provide some encouragement and advice about how to cope with stress and uncertainty. The emotional rewards of working with families during this difficult time are substantial.

GERALD P. KOOCHER
STANLEY J. BERMAN

REFERENCES

Anthony, S.: The Child's Discovery of Death. New York, Harcourt, Brace and Company, 1940.

Armstrong, G. D., and Martinson, I. M.: Death, dying and terminal care: dying at home. *In* Kellerman, J. (Editor): Psychological Aspects of Childhood Cancer. Springfield, Illinois, Charles C Thomas, 1980.

Bowlby, J.: Grief and mourning in infancy and early childhood. Psychoanal. Study Child, 15:9, 1960.

Bowlby, J.: Processes of mourning. Int. J. Psychoanalysis, 42:317, 1961.

Bowlby, J.: Separation: Anxiety and Anger. New York, Basic Books, 1973.

Braico, K. T., Humbert, J. R., Terplan, K. L., and Lehotay, J. M.: Laetrile intoxication: report of a fatal case. N. Engl. J. Med., 300:238, 1979.

Brant, J., and Graceffa, J.: Rutherford, Privitera, and Chad Green: laetrile setbacks in the courts. Am. J. Law Med., 151, 1979.

DiPalma, J. R.: Laetrile: when is a drug not a drug? Clin. Pharmacol. Ther., 15:186, 1977.

Eiser, C.: Intellectual abilities among survivors of childhood leukemia as a function of CNS irradiation. Arch. Dis. Child, 53:391, 1978.

Engel, C. L.: Psychological Development in Health and Disease. Philadelphia, W. B. Saunders Company, 1962.

Evans, A. E.: If a child must die. N. Engl. J. Med., 278:138, 1968.

Greenberg, D. M.: The case against laetrile: the fraudulent cancer remedy. Cancer, 45:799, 1980.

Gullo, S. V., Cherico, D. J., and Shadick, R.: Suggested stages and response styles in life threatening illness: a focus on the cancer patient. *In* Schoenberg, B., Carr, A. C., Kutscher, A. H., Peretz, D., and Goldberg, I. K. (Editors): Anticipatory Grief. New York, Columbia University Press, 1974.

Gunther, M. S.: Emotional aspects. *In* Ruge, D. (Editor): Spinal Cord Injuries. Springfield, Illinois, Charles C Thomas, 1969.

Guttman, L.: Spinal Cord Injuries: Comprehensive Management and Research. Ed. 2. Oxford, Blackwell Scientific Publications, 1976.

Herbert, V.: Facts and fictions about megavitamin therapy. Res. Staff Phys., 43–50, December, 1978.

Kastenbaum, R., and Costa, P. T.: Psychological perspectives on death. Ann. Rev. Psychol., 28:225, 1977.

Kellerman, J., Rigler, D., and Siegal, S. E.: Psychological effects of isolation in protected environments. Am. J. Psychiatry, 134:563, 1977.

Kemler, B.: Anticipatory grief and survivorship. *In* Koocher, G. P., and O'Malley, J. E. (Editors): The Damocles Syndrome: Psychosocial Consequences of Surviving Childhood Cancer. New York, McGraw-Hill Book Co., 1981.

Koocher, G. P.: Childhood, death, and cognitive development. Dev. Psychol., 9:369, 1973.

Koocher, G. P.: Talking with children about death. Am. J. Orthopsychiatry, 44:404, 1974.

Koocher, G. P.: Initial consultations with pediatric cancer patients. *In* Kellerman, J. (Editor). Psychological Aspects of Childhood Cancer. Springfield, Illinois, Charles C Thomas, 1980.

Koocher, G. P.: Development of the death concept in childhood. *In* Bibace, R., and Walsh, M. E. (Editors): The Development of Concepts Related to Health: Future Directions in Developmental Psychology. San Francisco, Jossey-Bass, 1981.

Koocher, G. P., and O'Malley, J. E.: The Damocles Syndrome: Psychosocial Consequences of Surviving Childhood Cancer. New York, McGraw-Hill Book Co., 1981.

Koocher, G. P., O'Malley, J. E., Foster, D. J., and Gogan, J. L.: Death anxiety in normal children and adolescents. Psychiatr. Clin., 9:220, 1976.

Koocher, G. P., and Sallan, S. E.: Psychological issues in pediatric oncology. *In* Magrab, P. (Editor): Psychological Management of Pediatric Problems. College Park, Maryland, University Park Press, 1978.

Koocher, G. P., Sourkes, B. M., and Keane, W. M.: Pediatric oncology consultations: a generalizable model for medical settings. Profes. Psychol., 10:467, 1979.

Kubler-Ross, E.: On Death and Dying. New York, Macmillan Publishing Co., Inc., 1969.

Langer, E., and Rodin, J.: The effects of choice and enhanced personal responsibility for the aged: a field experiment in an institutional setting. J. Pers. Soc. Psychol., 32:951, 1976.

Lindemann, E.: Symptomatology and management of acute grief. Am. J. Psychiatry, 101:141, 1944.

Lonetto, R.: Children's Conceptions of Death. New York, Springer-Verlag New York, Inc., 1980.

Mahler, M., Pine, F., and Bergman A.: The Psychological Birth of the Human Infant. New York, Basic Books, Inc., 1973.

Martinson, I. M., Armstrong, G. D., Geis, D. P., Anglin, M. A., Gronseth, E. C., MacInnis, M., Nesbit, M. E., and Kersey, J. H.: Facilitating home care for children dying of cancer. Cancer Nurs., 1:41, 1978.

McIntosh, S., Klatskin, E. H., O'Brien, R. T., Aspres, G. T., Kammerer, B. L., Snead, C., Kalavsky, S. M., and Pearson, H. A.: Chronic neurologic disturbance in childhood leukemia. Cancer, 37:1079, 1976.

Meadows, A. T., and Evans, A. E.: Effects of chemotherapy on the central nervous system. A study of parenteral methotrexate in long-term survivors of leukemia and lymphoma in childhood. Cancer, 37:1079, 1976.

Mnookin, R. H.: Child, Family, and State. Boston, Little, Brown and Company, 1978.

Moss, H. A., and Nannis, E. D.: Psychological effects of central nervous system treatment of children with acute lymphocytic leukemia. *In* Kellerman, J. (Editor): Psychological Aspects of Childhood Cancer. Springfield, Illinois, Charles C Thomas, 1980.

Nagy, M.: The child's theories concerning death. J. Gen. Psychol., 73:3, 1948.

Nighswanger, C. A.: Ministry to the dying as a learning encounter. J. Thanatol., 1:101, 1971.

Patenaude, A. F., Szymanski, L., and Rappaport, J.: Psychological costs of bone marrow transplantation in children. Am. J. Orthopsychiatry, 49:608, 1979.

Piaget, J.: The Construction of Reality in the Child. New York, Basic Books, Inc., 1954.

Plumb, M. M., and Holland, J.: Cancer in adolescents: the symptom is the thing. *In* Carr, A. C., Kutscher, A. H., Peretz, D., and Goldbert, I. (Editors): Anticipatory Grief. New York, Columbia University Press, 1974.

Schoenberg, B., Gerber, I., Wiener, A., Kutscher, A. H., Peretz, D., and Carr, A. C.: Bereavements: Its Psychosocial Aspects. New York, Columbia University Press, 1975.

Seligman, M. E. P.: Helplessness. San Francisco, W. H. Freeman & Co., 1975.

Share, L.: Family communication in the crisis of a child's fatal illness: a literature review and analysis. Omega, 3:187, 1972.

Silver, R. L., and Wortman, C. B.: Coping with undesirable life events. *In* Garber, B., and Seligman, M. E. P. (Editors): Human Helplessness: Theory and Application. New York, Academic Press, Inc., 1980.

Slavin, L. A.: Evolving psychosocial issues in the treatment of childhood cancer: a review. *In* Koocher, G. P., and O'Malley, J. E. (Editors). The Damocles Syndrome: Psychosocial Consequences of Surviving Childhood Cancer. New York, McGraw-Hill Book Co., 1981.

Sourkes, B.: Facilitating family coping with childhood cancer. J. Pediatr. Psychol., 2:65, 1977.

Sourkes, B.: Siblings of the pediatric cancer patient. *In* Kellerman, J. (Editor): Psychological Aspects of Childhood Cancer. Springfield, Illinois, Charles C Thomas, 1980.

Spinetta, J. J.: Disease related communication: how to tell. *In* Kellerman, J. (Editor): Psychological Aspects of Childhood Cancer. Springfield, Illinois, Charles C Thomas, 1980.

Spinetta, J. J., Rigler, D., and Karon, M.: Anxiety in the dying child. Pediatrics, 52:841, 1973.

Spinetta, J. J., Rigler, D., and Karon, M.: Personal space as a measure of a dying child's sense of isolation. J. Consult. Clin. Psychol., 42:751, 1974.

Symonds, M.: Victims of violence: psychological effects and after effects. Am. J. Psychoanal., 35:19, 1975.

Vernick, J., and Karon, M.: Who's afraid of death on a leukemia ward? Am. J. Dis. Child., 109:393, 1965.

White, E., Elsom, B., and Prawat, R.: Children's conceptions of death. Child Dev., 49:307, 1978.

Young, J. H.: The Medical Messiahs: A Social History of Health Quackery in 20th Century America. Princeton, Princeton University Press, 1967.

V

Outcomes During Childhood

The 15 chapters in Part V describe common and important developmental and behavioral results that find expression in patterns of function during childhood. Often these are influenced profoundly by the multiple variables and circumstances delineated in Parts III and IV.

A wide spectrum of variation is surveyed herein, including relevant somatic functions and dysfunctions and problems of habit, sleep, learning, communication, and cognition. Represented too is a range of severity, extending from common variations, to mild or transient adjustment problems, to some of the major disorders of behavior and development that constrain the functioning of children.

28

Variations on the Theme of Pain

28A Pain Tolerance and Developmental Change in Pain Perception

The behavior of a child in pain is the public expression of an essential, private, and virtually universal human experience. It is also the final common pathway of neurologic, psychologic, and social influences that are being represented by a verbal and behavioral "language" constrained by the developmental evolution of the child. This signal is a powerful elicitor of reactions from the community in which the child lives, thereby ensuring that the total pain experience is more than just a private sensation (Craig, 1978). This reaction serves at least to shape the child's criterion of when to label an experience as painful as well as how to express it. In addition, it may act to promote functional behavior leading to the attentuation of the pain or, alternatively, promote behavior that exacerbates the pain experience and produces dysfunction.

Despite the common feeling that we know what pain means, attempts to study the phenomenon have encountered formidable conceptual and methodologic hurdles. These problems are even greater when one attempts to understand children's pain behavior because of the added variable of constantly changing developmental stages. Thus, for example, crying as a behavioral marker of pain in a three month old child has a different significance, carries a different message, and evokes a different parental-community response than crying in a school age child. It is paradoxic that despite a rich and varied clinical experience with pain in pediatric patients, there is little systematic empirical study against which to validate our assumptions concerning pain behavior in children. Nevertheless developments in pain research have helped to define concepts and approaches with potential relevance for children. An emerging theme is that psychologic (affect and cognition) and social (context and modeling) influences are integral components of the experience, both at the level of perception and at the level of expression of the pain experience (Craig, 1978; Weisenberg, 1977). In the clinical setting, when the child presents with a pain complaint, these components of the experience are relevant for both diagnosis and management (pain control and preservation of function).*

BASIC CONCEPTS

The pain experience is a complex phenomenon, as reflected in the complexity of the terms with which it is described. Although not without their critics, a number of general categories of descriptors have received wide currency, and the relationship between these levels of description are illustrated in Figure 28–1. For the individual perceiving the pain, this perception is conceived of as including a sensory component and a motivational-affective component. The sensory dimension of the experience refers to the noxious stimulus that signals actual or potential tissue damage. It is analogous to other sensations in that it can be described in terms of the stimuli that produce it (mechanical, electrical, thermal, chemical), its sensory character (pricking, burning), and "quantitative" variables such as its location, time of onset, duration, and intensity. The motivational-affective dimension of the experience refers to the emotional and aversive aspects of the perception that lead to behavior that will avoid or reduce the stimulus (Melzack and Wall, 1970). By contrast, this aspect of the pain experience is most easily described not in terms of spatial-temporal constructs, but rather in terms of qualitive psychologic notions such as anxiety provoking, distressing, or even in some instances stimulating.

Despite concern expressed by some (e.g., Liebeskind and Paul, 1977), use of these contrasting descriptive systems need not commit one to assuming the presence of a mind-body dualism or

---

*Detailed reviews may be consulted for current concepts pertaining to the neurophysiologic substrates of pain perception and their implications for pain management (e.g., Bishop, 1980; Melzack, 1973; Newburger and Sallan, 1981).

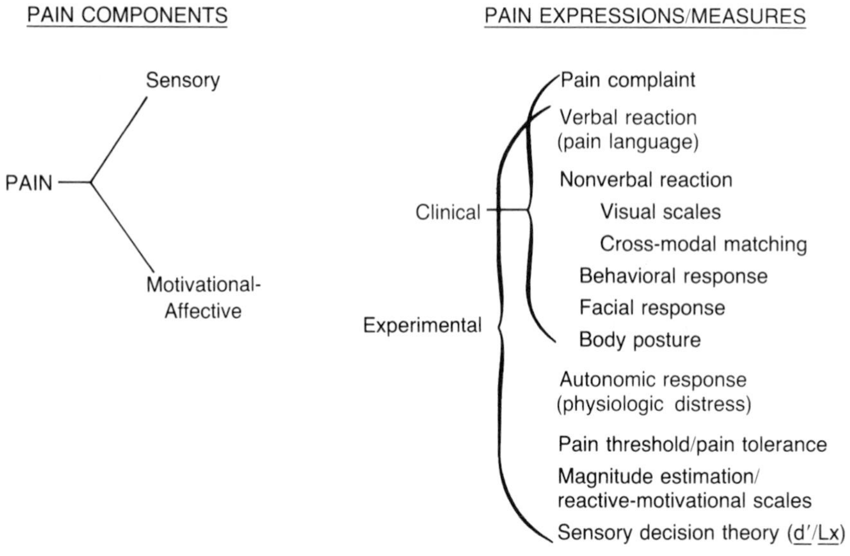

PAIN COMPONENTS   PAIN EXPRESSIONS/MEASURES

Figure 28–1. Pain perception (see text for explanation).

a differential "reality" between these two aspects. Indeed it is possible to construe each as being subserved by different neurophysiologic mechanisms (Melzack and Wall, 1970). The former may be mediated by neospinothalamic fibers projecting to the neurobasal and posterolateral thalamus and somatosensory cortex, and the latter by paramedial fibers projecting to the reticular formation, medial intralaminar thalamus, and limbic system. Both aspects are modulated by higher order cortical centers on the basis of cognitive variables and memories of past experience (Melzack and Wall, 1970; Merskey, 1975).

PAIN EXPRESSIONS AND MEASURES

EXPERIMENTAL FINDINGS

In laboratory and clinical studies of the pain experience, attempts to measure these aspects have been made in a variety of ways. In experimental exposure to an increasingly stronger stimulus, the pain threshold is defined as the point at which one first perceives the stimulus as painful. Pain tolerance refers to the point at which one is not willing to accept exposure to a higher intensity or a longer duration of the stimulus. In general, pain threshold measures have been correlated with physiologic measures and are thought to relate to the sensory component of pain perception, whereas pain tolerance has been correlated with psychologic variables and is considered to reflect affective-motivational aspects.

In the magnitude estimation procedure, the sensory component is measured by asking the subject to make a numerical comparison between a standard shock stimulus (arbitrarily rated as 10) and other stimuli above and below the standard (Tursky, 1974). The affective-motivational component is measured by asking the subject to identify the sensation threshold through to tolerance on a four point scale as the intensity is increased. This method has the advantage of permitting each subject to define his own scale for discriminating sensory component ratings, thereby producing a measure with potentially less bias for this aspect of the experience. The usefulness of this technique was illustrated in studying the question of differential pain perception in different ethnic groups. In using magnitude estimation procedures, ethnic differences were not demonstrated for sensory discrimination of shock levels, but were limited to the reactive motivational aspects of pain (Sternbach and Tursky, 1965). Consequently claims for ethnic differences based on pain threshold measures most likely relate to the bias in the particular method used for measuring pain sensation.

More recently sensory decision theory approaches have identified the sensory component with d', an index of discriminability of repeated noxious stimuli of different intensities. Lx indicates the subject's criterion for reporting pain or for labeling a particular experience as painful (Clark, 1974). Discriminability remains unaltered when attitude, expectation, and motivation are manipulated by the experimenter. By using this technique, it can be demonstrated that previously reported differences in "pain threshold" reflect differences in the subjects' willingness to label the stimulus as painful rather than real differences in pain sensitivity (Clark, 1974). On the other hand, nitrous oxide analgesia does decrease the d' value in volunteers exposed to noxious thermal stimulation (Chapman et al., 1973).

CLINICAL INVESTIGATION

In the clinical setting physicians must assess the pain complaint by relying on verbal and behavioral expressions of the patients (or their parents). In adults verbal pain descriptors can be categorized a priori into those representing sensory qualities in terms of temporal, spatial, pressure, and other characteristics (e.g., burning, aching, pounding, spreading), affective qualities using terms of tension, fear, and autonomic properties (e.g., distressing, awful, nauseating), and intensity in terms of degree (e.g., moderate, agonizing, excruciating; Melzack and Torgenson, 1971). These categories seem to have some empirical validity in both patients and experimental groups of subjects and are sensitive to differences in low back pain as compared with experimentally induced cutaneous pain (Crockett et al., 1977).

In attempts to objectify the clinical description of pain, a number of scales have been developed (Stewart, 1977). These include rating scales for intensity of pain, as well as for affective qualities. In addition to verbal response measures, cross modal comparisons to other sensory modalities (hearing, vision) have also been utilized. In the Stewart pain-color scale, for example, subjects are asked to indicate the degree of pain severity by indicating the color hue (orange to strong red) corresponding to their pain (Stewart, 1977). Although awaiting systematic study, such nonverbal measures may be particularly applicable in children for whom cross modal thinking is common (Werner, 1957) and for whom the color red appears to be preferentially matched to pain (Scott, 1978).

Pain is also expressed in terms of behavioral nonverbal reactions, and the degree of physical activity is commonly used as an index of the severity of pain. Typical examples include the immobility of the child with peritoneal irritation and appendicitis, or the child with inflamed meninges. Similarly facial expression is noted to determine whether the pain is "real" and how intense it is. However, behavioral and verbal reactions may be dissociated. The pain behavior may suggest that the stimulus is severe, whereas the severity of the pain is played down when the patient is asked to describe it verbally. Despite the clinical use made of behavioral reactions, there has been little systematic study of these aspects. An attempt to distinguish patients with severe pain from those who over-react by use of a "blink test" did not correlate with affective measures and was too variable to be discriminating (Rogers, 1971; Weisenberg, 1977).

Autonomic symptoms such as pallor, perspiration, nausea, and palpitations are commonly part of the clinical symptom complex during a pain episode. In experimental studies, autonomic signs of heart rate, blood pressure elevation, electro-dermal response, and blood volume pulse are often monitored. Although these measures are taken as objective indices of physiologic distress, they are far from being specific, since they are also sensitive to other variables, such as attention-distraction, predictability, habituation, and anticipation. Although too variable to be clinically discriminating, they are potentially useful in highly controlled experimental situations in which they may be more indicative of biologic mechanisms mediating experiences of pain and discomfort than are verbal reports, from which they may be dissociated (e.g., Craig and Neidermayer, 1974; Craig and Prkachin, 1978).

DEVELOPMENTAL CHANGES IN PAIN EXPRESSIONS

Verbal and behavioral reactions to induced pain have been studied systematically in infants and young children. In infants, crying has been used as a response to measure the reaction to a rubber band snap on the sole of the foot (Fisichelli et al., 1974; 1969). In general, the crying response is characterized by a period of relative depression at five hours of age, higher reactivity between two days and 12 weeks, and declining reactivity thereafter until one year of age. A particularly marked change occurs between eight and 16 weeks, when the percentage of no-response reactions increases from less than 10 per cent to approximately 50 per cent. Although group changes from age to age are significant, there is a lack of stability of individual responses, making the clinical value of induced crying measures somewhat tenuous. These changes are attributed to a maturing responsiveness related to central nervous system development. It is unclear whether the maturation relates more to the sensory or motivational-affective components of the pain experience, but the findings do not support a concept of pain insensitivity in infancy.

In an earlier study McGraw (1941) described sensorimotor reactions and behavioral (cognitive-conative) responses to pinprick in infants and young children. The motoric reaction develops from being an immediate diffuse unlocalized response with occasional reflex withdrawal of the stimulated body part to a focused defensive reaction by about 12 months of age. Similarly anticipatory behavioral responses to the approaching pin through the visual field of the infant develop from being nonexistent or simply exploratory to provoking fussing and crying. At a later stage infants appear to choose the posture (e.g., aggressive defense, stoic resignation) to take toward the event. In general, the appearance of anticipatory fussing, crying, and withdrawal reactions that become apparent between seven and 12 months of age is assumed to indicate the operation of learning and memory for the pain perception.

In clinical studies of reactions to inoculations, there is little evidence of "memory" (anticipatory crying or apprehensive behavior) until after six months of age (Kassowitz, 1958; Levy, 1960). At least "moderate" degrees of defensive behavior (fighting and loud crying, or tantrums) during administration were increasingly observed from six months until the second and third years of life, after which such behavior occurred less frequently (Kassowitz, 1958).

In older children, Haslam (1969) reported a generally increasing "pain threshold" to anterior tibial pressure in a relatively small number of subjects aged five to 18. In a clinical study of children with the common school aged problem of recurrent abdominal pain usually characterized by nonspecificity and diffuseness, there is an age related trend toward being able to localize the pain in children between the ages of four and 14 years (Heinild et al., 1959).

The finding of age related changes in verbal-behavioral reaction to a noxious stimulus is not surprising. Despite the variety of stimuli, settings, response measures, and conditions of testing, the overall trend is toward verbal-behavioral reactions that are increasingly localized and attenuated in terms of reported intensity. The commonly held assumption that infants are insensitive to pain is not supported by these data. Another interpretation would be that sensitivity or discriminability for the stimulus is intact, and that the infant and young child become increasingly more efficient in modulating their reaction to the stimulus. Lack of evidence for anticipatory fear reactions prior to six months of age does not imply that the stimulus was not felt or does not contribute to learning, but only that the previous experience cannot be brought to bear in the situation to initiate a defensive reaction prior to the application of the stimulus. Indeed, only one study controlled for the possible effect of earlier experience by use of a combined longitudinal and cross sectional design (Fisichelli et al., 1974). A significant effect of earlier experience was only transiently apparent for one measure (frequency of crying in response to rubber band snap pain) at about 12 weeks of age. The effect was that of increasing adaptation to the stimulus, in that those with previous experience were less likely to cry than those experiencing that stimulus for the first time. However, the increasing cognitive capacity including memory (e.g., Kagan et al., 1978) permits the development of anticipatory fear reactions to situations that include pain in the latter half of the first year (Levy, 1960).

Despite the apparent consistency of the results, the findings tell us little about the underlying pain discriminability or motivational-affective components of the pain experience. Since the level of observation is that of behavioral-verbal reactivity to the stimulus, conclusions concerning underlying sensory maturation or cognitive-affective development must of necessity be indirect and are difficult to separate. In addition, comparisons based on different pain stimuli and reaction measures introduce major obstacles to interpretation in studies with children as they do with adults (Weisenberg, 1977).

COMMUNICATIVE ASPECTS OF THE PAIN COMPLAINT

When one considers the possible meanings children expect the pain complaint to carry to observers, the complexity of the pain experience becomes even more apparent. In response to a sentence completion task, healthy 10 and 11 year old children reported pain as anxiety ("Pain is being nervous"), illness ("Pain is not growing up healthy"), helplessness ("Pain is when you scream for help and nobody comes"), and fear ("Pain is being afraid"; Schultz, 1971). In response to the question, "What does pain mean to you?" answers included fear of bodily harm and death, as well as more pragmatic contingencies such as a visit to the doctor or getting shots. Healthy younger children (five to 10 years) presented with cartoon depictions of painful events (e.g., hitting one's thumb with a hammer, getting a needle at the doctor's) responded to the question, "What's happening?" with similar though less abstract answers. Interestingly, these children also described the subject's crying as indicating anger, either self- or other-directed ("He's mad because he banged his own fingers" or "He's mad at the doctor . . . crying"; Barr and Scott, unpublished observations).

It is likely that the meaning children intend to convey in expressing their pain complaints is structured by their cognitive and emotional developmental stages. Descriptions of children's closely related understanding of illness have been seen to follow predicted conceptual stages with regard to causality in hospitalized and healthy children (Bibace and Walsh, 1980; Perrin and Gerrity, 1981; Simeonnson et al., 1979). Even in preverbal infants the typical pain expression (crying) may have different communicative functions during the early months compared with later in the first year (Bell and Ainsworth, 1972). Consequently the public expression of the pain stimulus appears to carry information to the immediate community reflecting not only perceived intensity and affect but also information and demands with changing developmental characteristics.

Pain expressions are powerful stimuli to action on the part of the community. This is particularly true in the case of infant crying in which a response on the part of parents is almost obligatory (Murray, 1979). The responses may be either beneficial for the parent-child relationship or destructive, both immediately and in the long term (Murray, 1979; Shaw, 1976). The commanding nature of witnessing painful episodes continues through to adult life (Melzack, 1973; Weisenberg, 1977). However, the overt behavioral response is itself constrained by such variables as age and sex of the respondent. Thus, for example, adolescent girls are less likely to ignore an infant's behavior than adolescent boys, even though both show the same autonomic responsivity to infant crying (Frodi and Lamb, 1978).

MODELING INFLUENCES ON PAIN EXPRESSION

In addition to the effect that pain behavior has on the community, the response to pain behavior appears to have a major role in defining the behavior and language that the child is able to use. With a sensation as uniquely private as pain, it is likely that a child would learn what reactions are appropriate or permissible by comparisons with others. Such modeling processes may encourage adaptive responses to pain situations, but may also be responsible for inappropriate pain complaints and maladaptive "sick role" behavior.

In adults the effectiveness of modeling in modifying a subject's willingness to label a stimulus as painful has been effectively demonstrated in the laboratory setting (Craig et al., 1971, 1974, 1978). Exposure to a "tolerant" model increased the current intensity that was labeled as painful, whereas exposure to an "intolerant" model decreased the level of "painful" current (Craig and Weiss, 1971). It is also possible that modeling may be effective in altering the fundamental sensory characteristics of the pain perception, since no increases were demonstrated in autonomic measures of subjective distress (Craig and Neidermayer, 1974), and a lower discriminability (d') of the shocks was found in subjects exposed to tolerant models (Craig and Prkachin, 1978). It has also been demonstrated that social manipulation producing a reduction in the facial expression of pain results in reduced autonomic and self-reported indices of pain (Kleck et al., 1976). Such findings, if true of children, would provide an empirical rationale for a social learning approach both for understanding the genesis of pain complaints and for designing therapeutic modalities.

It is probable that social learning influences affect many psychologic factors relevant to pain manifestations, including one's interpretation of internal and external events, the degree of emotional arousal associated with the pain stimulus, cognitive strategies elicited by the experience, and capacity to obtain help from the environment (Crockett et al., 1977). Such an explanation has been proposed for children with the recurrent abdominal pain syndrome on the basis of a strong association with pain complaints in family members (Apley, 1975; Oster 1972), especially when the parents had complaints at the time of evaluation (Christensen and Mortensen, 1975). The presence of such social modeling might act either to create the new behavior or to facilitate the expression of a previously learned behavior. Consequently children with the recurrent abdominal pain syndrome might express their pain complaint by facial expression differently from other children. However, no differences in facial reactivity to laboratory induced cold pressor pain were demonstrated in children with the syndrome compared with normal control subjects and frequent hospital visitors (Feuerstein et al.,1982). Nevertheless the potential for social learning approaches in clinical pain syndromes deserves wider study in the pediatric age group.

MEDIATING FACTORS AFFECTING CLINICAL PAIN BEHAVIOR

In a classic report Beecher (1956) described the reaction of men with leg wounds to demonstrate the importance of the context in which the pain experience occurs. Only 25 per cent of the men wounded in battle requested narcotics for pain relief, whereas 80 per cent of the civilians with similar surgical wounds produced under anesthesia wanted medical relief. Such findings dramatically illustrate that the degree of tissue damage is not sufficient to account for the pain reaction observed, and that factors other than the sensation of the noxious stimulus determine the pain response. Indeed the difference between patients' reactions to pain experienced in the clinical setting and patients' reactions to laboratory induced pain has been a major stumbling block in extrapolating results from the laboratory setting, since the anxiety and distress associated with the clinical disease process are seldom present in the laboratory. The gate control theory proposed by Melzack and Wall provides considerable conceptual support for such observations by stressing the role of psychologic variables in affecting the reaction to pain (Melzack, 1973; Melzack and Wall, 1970). Wall (1979) subsequently proposed that pain be considered more as a state of awareness of a need (analogous to thirst or hunger) provoked by internal events, rather than a state of awareness of sensation (analogous to hearing or seeing) provoked by an external stimulus. As a result, considerable interest in the psychologic and social correlates that mediate the clinical pain response has been generated.

Potential mediating factors include personality characteristics, social and cultural background, emotional state, and cognitive and behavioral style (e.g., field independence-dependence, copersavoiders; Sternbach, 1975; Weisenberg, 1977). Although the results have often been contradictory, a number of general principles have emerged that appear to be relevant in the context of clinical pain. First, the sensory component of pain tends to be relatively stable for individuals and does not differentiate between groups, whereas the affective-motivational component and the manner of expressing pain may be manipulated for an individual and be different between groups. Second, emotional state tends to be differentially related to clinical pain, acute pain states being associated with anxiety and chronic pain states with depression (Sternbach, 1975). Third, a number of cognitive styles appear to promote pain tolerance. Individuals who are non-neurotic, extroverted, copers, field dependent, and sensitizers tend to handle pain better and be less reactive (Weisenberg, 1977). Fourth, cultural group differences are limited to pain reactivity or tolerance when appropriate measures are used (Sternbach and Tursky, 1965), and these differences may be related to underlying (trait) anxiety group differences or attitudinal styles (denial, avoidance; Weisenberg, 1977). In addition to personality characteristics, a number of situational and contextual variables have been shown to mediate the pain response. These include the sense of control or helplessness of the patient, previous instruction, the presence of observers-models, attention-distraction, predictability of the situation, and the presence of suggestion or expectation of relief. It could be hypothesized that all these processes contribute to decreasing anxiety and therefore attenuate the perception of pain, but the exact mechanisms that are operative are not clear.

Some of these factors appear to be operative in determining children's reactions to painful experiences. Because pain episodes in children invariably implicate parents, an important variable concerns their reaction to the stimulus setting. In such situations the affective reaction of the parent may be more defining of the response than the nature of the pain stimulus (Crockett et al., 1977). Negative behavioral patterns in young children undergoing dental extraction are significantly associated with maternal anxiety, but no relationship is found with sex, age, or a history of previous unpleasant dental or medical experiences (Johnson and Baldwin, 1968). In a study of the role of modeling on behavior in hospitalized children,

early born children were more upset prior to an injection but demonstrated no differences during the injection itself (Vernon, 1974). More significantly, however, children exposed to a realistic film of other children receiving an injection demonstrated less disturbance when they themselves received an injection than did patients exposed to an unrealistic film portrayal or patients with no preparation. It is of particular interest that the patients who watched the unrealistic film demonstrated the most pain, consistent with the principle that moderate degrees of apprehension or accurate expectations are ameliorating influences. Similar results are achieved when children undergoing minor surgery and dental procedures are shown films of a peer coping with the events (Melamed et al., 1975; Melamed and Siegel, 1975). Although the specific components that are effective in these maneuvers are difficult to specify, careful attention to such situational variables can promote adaptive responses to clinical pain experiences. By the same token, situational variables may promote maladaptive clinical pain behavior, as illustrated by episodes of mass hysteria producing abdominal pain complaints (Moffatt, 1982; Smith and Eastham, 1973).

IMPLICATIONS FOR CLINICAL MANAGEMENT

Increasing clinical, experimental, and conceptual recognition of the integral role of psychologic factors in pain perception may provide guidelines to improved clinical management of pediatric pain complaints. In addition to the complexities of the pain complaint in adults, effective pediatric management must take into account both the development of the child and the parental and communal context in which the child lives. Consequently the elements that have been described as affecting pain tolerance and behavior in adults must be considered against the background of maturational, cognitive, behavioral and interpersonal changes. Thus, age related differences in publicly expressed pain behavior in children probably reflect varying combinations of developing pain sensitivity, pain tolerance, language facility, social learning, and communal reaction to the pain behavior.

Commonly pain episodes occur in acute, recurrent, or chronic situations. Typically acute pain is associated with tissue damage, but is subject to treatment, is associated with acute anxiety, and is relatively less destructive of social relationships. Acute pain may be predictable (as in dental visits or inoculations) or unpredictable (otitis, abdominal pain). Although tissue damage is common to both, anticipatory fear and anxiety are present with predictable pain. Recurrent pain is less likely to be associated with tissue damage and may or may not be subject to treatment. Its unique features are that it is seldom predictable, its duration is uncertain, and its etiology is seldom determined with certainty (e.g., colic, recurrent abdominal pains, headaches). It therefore tends to be associated with acute anxiety for a specific episode and chronic anxiety related to the attendant long term uncertainty as to its cause. With chronic pain, both the sensation of pain and the distress and suffering are prolonged and sometimes unremitting, such that the cardinal features are depression and significant functional incapacity. Paradigmatic examples include the pain of inflammatory bowel disease, cancer, and burns. A special case is the situation in which recurrent painful injections are part of a therapeutic regimen. In these cases the character of the experience is defined by whether the injections contribute immediately to relief (hemophilia) or maintenance of normal function (diabetes) or whether they represent recurrent disease (tumor chemotherapy).

Clinical assessment and therapeutic approaches should reflect these changing cardinal features affecting the pain experiences. With acute pain there is little ground for assuming insensitivity to the pain stimulus even in infants. The fact that pain tolerance increases with age provides a rationale for increasing use of techniques directed to the affective-motivational aspects of the pain experience in older children, but not for withholding appropriate analgesia from younger children and infants. Assessment of pain in younger children may be more effective if one focuses not only on the intensity of the pain but also on the affective content and communicative intent of the pain complaint. Prevention and treatment increasingly entail methods directed at anticipatory anxiety. To achieve this goal in younger children, methods for focusing parental anxiety are more likely to be beneficial. As children grow older, more opportunity for utilizing cognitive strategies aimed at increasing predictability and providing opportunities for self-control and social modeling becomes available.

With recurrent pain complaints, assessment of the affective and communicative aspects of the pain complaint has even greater importance, both in reaching a diagnosis (as in psychogenic pain) and in assessing the potential for the amount of dysfunction associated with the complaint. Measures aimed at focusing parental anxiety are of primary importance both as ends in themselves and to prevent the development of a "sick role" in the child. The role of analgesics varies with the symptom (e.g., headaches versus recurrent abdominal pain), but in all cases attention to associated situational predispositions to the complaint is critical. Although as yet undocumented, methods of self-control (relaxation, biofeedback, modeling, hypnosis, mental imagery) yielding moderate success in adults may be applicable in selected cases.

In chronic pain conditions, pharmacologic management is usually indicated. However, its effectiveness varies with the context and emotional state of the patient (Beecher, 1956; Weisenberg, 1977). In the hospital the patient's suffering is the focus of attention. It implies loss of control over the symptom or disease, and control of access to measures of relief is wholly in the hands of physicians. With chronic pain, the symptom has lost its biologic value as a warning symptom. It has itself become a disease with the expectation of future pain and suffering. Consequently the almost universal emotional state is depression. This complex may be exacerbated by undertreatment through ignorance, fear of addiction, or poor hospital practice (Marks and Sachar, 1973; Newburger and Sallan, 1981), a situation that may be even more serious in children (Eland and Anderson, 1977).

Although the principles appropriate to pain assessment and control of acute and recurrent pain are still appropriate, they interact with the relatively greater role played by pharmacologic agents. The aim of therapy is an optimal balance between function and pain control within the constraints of the disease process. Medication practices should promote psychologic health and not exacerbate dysfunctional pain behavior. In particular, the practice of giving "prn" medications requires that the patient suffer and act to reinforce pain complaints. The practical management of chronic pain has been outlined by Newburger and Sallan (1981). The principles of pharmacologic intervention include differentiation of physical dependence, which is common with narcotic analgesics, from addiction, which is not (when given on a regular schedule); adequate dosages of analgesics at predetermined intervals sufficient to maintain relief and permit optimal patient functioning in a balance between pain and sedation; combination of analgesics with other drugs to promote effective personal functioning (e.g., antidepressants, hypnotics); and avoidance of fixed combination drugs and drug side effects that compound the dysfunction and loss of control (Newburger and Sallan, 1981). In addition, combined approaches including anesthesiology, surgery, psychology, psychiatry, and specialized wards and clinics are indicated when available. For intercurrent predictable pain situations (injections, bone marrow aspirations), judicious use of relaxation procedures and methods to promote self-control deserve to be explored as adjunctive therapeutic modalities.

RONALD G. BARR

REFERENCES

Apley, J.: The Child with Abdominal Pains. London, Blackwell Scientific Publications Ltd., 1975.

Barr, R. G., and Scott, R.: Unpublished observations.

Beecher, H. K.: Relationship of significance of wound to pain experienced. J. Am. Med. Assoc., 161:1609, 1956.

Bell, S. M., and Ainsworth, D. S.: Infant crying and maternal responsiveness. Child. Dev., 43:1171, 1972.

Bibace, R., and Walsh, M. E.: Development of children's concept of illness. Pediatrics, 66:912, 1980.

Bishop, B.: Pain: its physiology and rationale for measurement. Phys. Ther., 60:13, 1980.

Chapman, C. R., Murphy, J. M., and Butler, S. H.: Analgesic strength of 33 per cent nitrous oxide: a signal detection theory evaluation. Science, 179:1246, 1973.

Christensen, M. F., and Mortensen, D.: Long-term prognosis in children with recurrent abdominal pain. Arch. Dis. Child., 50:110, 1975.

Clark, W. C.: Pain sensitivity and the report of pain: an introduction sensory decision theory. Anesthesiology, 40:272, 1974.

Craig, K. D.: Social modelling influences on pain. In Sternbach, R. A. (Editor): The Psychology of Pain. New York, Raven Press, 1978, p. 73.

Craig, K. D., and Neidermayer, H.: Autonomic correlates of pain thresholds influenced by social modelling. J. Pers. Soc. Psychol., 29:246, 1974.

Craig, K. D., and Prkachin, K. M.: Social modelling influences on sensory decision theory and psychophysiological indexes of pain. J. Pers. Soc. Psychol., 36:805, 1978.

Craig, K. D., and Weiss, S. M.: Vicarious influences on pain-threshold determinations. J. Pers. Soc. Psychol., 19:53, 1971.

Crockett, D. J., Prkachin, K. M., and Craig, K. D.: Factors of the language of pain in patient and volunteer groups. Pain, 4:175, 1977.

Eland, J. M., and Anderson, J. E.: The experience of pain in children. In Jacox, A. (Editor): Pain: A Sourcebook for Nurses and Other Health Professionals. Boston, Little, Brown and Company, 1977, p. 453.

Feuerstein, M., Barr, R. G., Francoeur, T. E., Houle, M., and Rafman, S.: Potential biobehavioural mechanisms of recurrent abdominal pain in children. Pain, 13:287, 1982.

Fisichelli, V. R., Karelitz, R. M., Fisichelli, R. M., and Cooper, J.: The course of induced crying activity in the first year of life. Pediatr. Res., 8:921, 1974.

Fisichelli, V. R., Karelitz, S., and Haber, A.: The course of induced crying activity in the neonate. J. Psychol., 73:183, 1969.

Frodi, A. M., and Lamb, M. E.: Sex differences in responsiveness to infants: a developmental study of psychophysiological and behavioral responses. Child Dev., 49:1182, 1978.

Grossberg, J. M., and Grant, B. F.: Clinical psychophysics: applications of ratio scaling and signal detection methods to research on pain, fear, drugs, and medical decision making. Psychol. Bull., 85:1154, 1978.

Haslam, D. R.: Age and the perception of pain. Psychon. Sci., 15:86, 1969.

Heinild, S., Malver, E., Roelsgaard, G., and Worning, B.: A psychosomatic approach to RAP in childhood with particular reference to the x-ray appearances of the stomach. Acta Pediatr. Scand., 48:361, 1959.

Johnson, R., and Baldwin, D. C.: Relationship of maternal anxiety to the behavior of young children undergoing dental extraction. J. Dent. Res., 47:801, 1968.

Kagan, J., Kearsley, R. B., and Zelazo, P. R.: Infancy: Its Place in Human Development. Cambridge, Harvard University Press, 1978.

Kassowitz, K. E.: Psychodynamic reactions of children to the use of hypodermic needles. A.M.A. J. Dis. Child., 95:253, 1958.

Kleck, R. E., Vaughn, R. C., Cartwright-Smith, J., Vaughn, K. B., Colby, C. Z., and Lanzetta, J. J.: Effects of being observed on expressive, subjective, and physiological responses to painful stimuli. J. Pers. Soc. Psychol., 34:1211, 1976.

Levy, D. M.: The infant's earliest memory of inoculation: a contribution to public health procedures. J. Genet. Psychol., 96:3, 1960.

Liebeskind, J. C., and Paul, L. A.: Psychological and physiological mechanisms of pain. Ann. Rev. Psychol., 28:41, 1977.

Marks, R. M., and Sachar, E. J.: Undertreatment of medical inpatients with narcotic analgesics. Ann. Intern. Med., 78:173, 1973.

McGraw, M. B.: Neural maturation as exemplified in the changing reaction of the infant to pinprick. Child Dev., 9:31, 1941.

Melamed, B. G., Hames, R. R., Heiby, E., and Glick, J.: The use of filmed modelling to reduce uncooperative behaviour of children during dental treatment. J. Dent. Res., 54:797, 1975.

Melamed, B. G., and Siegel, L. J.: Reduction of anxiety in children facing hospitalization and surgery by use of filmed modelling. J. Consult. Clin. Psychol., 43:511, 1975.

Melzack, R.: The Puzzle of Pain. New York, Basic Books Inc., 1973.

Melzack, R., and Torgerson, W. S.: On the language of pain. Anesthesiology, 34:50, 1971.

Melzack, R., and Wall, P. D.: Psychophysiology of pain. Intern. Anesthesiol. Clin., 8:3, 1970.

Merskey, H.: Pain, learning and memory. J. Psychosom. Res., 19:319, 1975.

Moffatt, M. E. K.: Epidemic hysteria in a Montreal train station. Pediatrics, 70:308, 1982.

Murray, A. D.: Infant crying as an elicitor of parental behaviour: an examination of two models. Psychol. Bull., 86:191, 1979.

Newburger, P. E., and Sallan, S. E.: Chronic pain: Principles of management. J. Pediatr., 98:180, 1981.

Oster, J.: Recurrent abdominal pain, headache, and limb pains in children and adolescents. Pediatrics, 50:429, 1972.

Perrin, E. C., and Gerrity, P. S.: There's a demon in your belly: children's understanding of illness. Pediatrics, 67:841, 1981.

Rogers, R. G.: Blink test to establish threshold for reaction to pain. Postgrad. Med., 49:108, 1971.

Schultz, N.: How children perceive pain. Nurs. Outlook, 19:670, 1971.

Scott, R.: "It hurts red": a preliminary study of children's perception of pain. Percept. Mot. Skills, 47:787, 1978.

Shaw, C.: A comparison of the patterns of mother-baby interaction for a group of crying, irritable babies and a group of more amenable babies. Child Care Health Dev., 3:1, 1977.

Simeonnson, R., Buckley, L., and Monson, L.: Conceptions of illness causality in hospitalized children. J. Pediatr. Psychol., 4:77, 1979.

Smith, H. C. J., and Eastham, E. J.: Outbreak of abdominal pain, Lancet 2:956, 1973.

Sternbach, R. A.: Psychophysiology of pain. Int. J. Psychiatry Med., 6:63, 1975.

Sternbach, R. A., and Tursky, B.: Ethnic differences among housewives in psychophysical and skin potential responses to electric shock. Psychophysiology, 1:241, 1965.

Stewart, M.: Measurement of clinical pain. In Jacox, A. (Editor): Pain: A Sourcebook for Nurses and Other Health Professionals. Boston, Little, Brown & Co., 1977, p. 453.

Tursky, B.: Physical, physiological, and psychological factors that affect pain reaction to electric shock. Psychophysiology, 11:95, 1974.

Vernon, D. T. A.: Modelling and birth order in responses to painful stimuli. J. Pers. Soc. Psychol., 29:794, 1974.

Wall, P. D.: On the relation of injury to pain: The John J. Bonica Lecture. Pain, 6:253, 1979.

Weisenberg, M.: Pain and pain control. Psychol. Bull., 84:1008, 1977.

Werner, H.: Comparative Psychology of Mental Development. New York, International Universities Press, 1957.

28B Sources of Pain

Pain in children, especially chronic pain, challenges the diagnostic acumen, therapeutic skill, and personal mettle of the physician. A subjective symptom, unseen but voiced, pain can be described only by the patient. Both psychologic and physiologic mechanisms are commonly involved.

The clinician must deal with fact, fancy, and feelings; conscious and unconscious processes; the parents as well as the child; and with psychologic as well as drug therapy. Most children are brought to the doctor seeking relief of pain due in part to fear, but others display reluctance, or at least ambivalence, about surrendering to this medical encounter. Michael Balint stated it well: "A functional illness means that the patient has a problem which he tried to solve with an illness. The illness enabled him to complain, whereas he was unable to complain about his own problem."

PSYCHOGENIC PAIN

Psychogenic pain disorder is defined in the *Diagnostic and Statistical Manual of Mental Disorders* (1980) as "a clinical picture in which the predominant feature is the complaint of pain, in the absence of adequate physical findings, and in association with evidence of the etiologic role of psychosocial factors. The disturbance is not due to any other mental disorder."

Engle (1959) and others have described a number of psychologic mechanisms to explain psychogenic pain. Derived from the Greek word meaning "to punish," pain is hypothesized in some instances to be a penitence for strong conscious or unconscious feelings of guilt resulting from aggressive, angry, or sexual feelings directed toward a relative or other significant person. Young children, of course, commonly believe that pain and illness are punishment for bad behavior or thoughts. The anxiety associated with real or imagined threats may also be experienced as pain.

Identification explains why the localization of the complaint may correspond with the actual pain experienced by a parent or other significant person, that which the child imagines they have, or with the site of previous pain in the patient. The child or parent is usually not aware of nor can he make this association. Some children, and often their parents, are hypochondriacal, misinterpreting normal physiologic sensations, like intestinal peristalsis, as evidence of a serious disease.

Although conversion disorder is not an appropriate diagnosis when the complaint is limited to pain, it may be the etiologic mechanism when there are additional manifestations, such as a gait disturbance or paresthesia. Conversion reactions lead to both primary and secondary gain, the former when unacceptable or conflictional wishes and thoughts are kept from consciousness and the latter when the child receives increased attention or is able to avoid upsetting activities, such as school. Pain may persist because it affords the child conscious secondary gain or permits the family to avoid confrontation and conflict while dealing with the child's symptoms.

THE PEDIATRIC INTERVIEW

Knowledge and experience in relation to both psychosocial and organic etiologies are prerequi-

sites to diagnostic and therapeutic effectiveness in the management of pain. A clear characterization of the pain is obviously needed, including its site, intensity, temporal relationships, frequency, duration, constancy, quality (i.e., throbbing, aching, sharp, stabbing, or tingling), and accentuating or alleviating factors. Such detail may be difficult to obtain from children under eight or nine years of age. Aware of the characteristic symptomatology and typical pain profile of clinical disorders, the physician can often recognize promptly whether the pain reflects an organic or a psychogenic etiology. This competence in dealing with biomedical as well as psychosocial disease gives the pediatrician an initial advantage over the child psychiatrist because the parents and child usually come in the belief that the pain has an organic etiology.

Psychogenic pain disorders are best approached developmentally through an understanding of the challenges and environmental stresses that a child is confronting at a given time. These are best determined in the pediatric interview, a process that allows both the collection of data and the development of a constructive relationship between the doctor, the child, and the family. The interview is the pediatrician's most effective psychotherapeutic tool.

The interview may begin with a statement somewhat as follows: "In seeing many children with abdominal (or other) pain, I've found that it's sometimes due to physical causes, sometimes to stresses at this age, and sometimes to both. But pain is pain no matter what the cause, so it's my practice to examine all possibilities thoroughly . . . physical, psychologic, whatever."

This immediate clarification precludes the parents' or child's getting the impression that the doctor, in a snap decision, has concluded that the pain is "in the child's head" or that the child is "making it up." This statement also gives the clear message that psychosocial considerations are a legitimate part of the diagnostic process. When germane, psychosocial possibilities are more easily included and accepted at the onset than after a fruitless work-up for organic disease.

The pediatric interview helps the physician understand the patient's and the parents' past and current life situations, their feelings, beliefs, and anxieties while they sense his expertise and interest in them as persons. Sharing personal facts, feelings, and problems with the physician often results in a dramatic lessening of anxiety and may alleviate pain. In addition to illuminating linkages between the child's pain and the environmental or developmental stresses being experienced, the interview provides parents, who are largely unaware of their child's problems because of their preoccupation with their personal difficulties, with the opportunity to concentrate on their child, on themselves, and on their interactions in a clarifying fashion. The physician's personal warmth, empathy, and capacity to understand how the parents and child feel helps him achieve a psychotherapeutic alliance.

The following kinds of personal, developmental, or family stresses and problems are of interest in the interview:

I. Separation experiences
  A. The anticipated, the fantasied, or the actual death of someone close to the child, e.g., a parent, sibling, grandparent, friend, or pet.
  B. Divorce or anticipated divorce or desertion.
  C. Social or vocational commitments that make the parents unavailable to the child. Such "phantom" parents are psychologically little involved with their child and are frequently absent from the home.
  D. The fear of the child, who has recovered from a critical illness or has a long term, life threatening disease, of his own premature death.
  E. Lack of communication or alienation between the patient and his family.
  F. The child placed outside his family.
II. Family illness
  A. Physical illness, e.g., cancer or myocardial infarction in a parent or a long term handicapping condition such as mental retardation or myelodysplasia in a sibling. Each parent should be asked specifically about his own health, the kind of pain experienced in the past or currently, and whether he is seeing a physician or taking medicine. The child should also be asked about complaints and illnesses in each of his parents, grandparents, and siblings, especially about the presence of pain, its location, and the child's perception of its character. Frequently pain in one of the parents serves as a model for the child's complaint.
  B. Psychologic symptoms and disorders, e.g., anxiety, depression, alcoholism, or psychosis. Helpful exploratory questions include: "Who's the nervous one in your family?" or "Who does the worrying around your house?"
  C. Parental hypochondriasis or preoccupation with illness.
III. Marital discord is obviously a major stress for children.
IV. Lack of mutuality in parent-child interactions, e.g., over-expectation, over-restriction, unfavorable comparison with a sibling, or the child's awareness that his parents are disappointed in him.
V. Difficulty with the expression of aggressive or angry feelings. The child with psychogenic pain who has difficulty in dealing with negative feelings in the belief that they are bad may deny or hide their presence. The expression of anger is especially difficult for children with chronically or seriously ill parents or in families in which the direct expression of anger is not tolerated.
VI. The child's inability to make and keep friends.
VII. The adolescent's concerns about sexuality.
VIII. School and learning problems.

The child or adolescent and the parents often spontaneously disclose problems in these areas if they are conscious of them and if the physician is a facilitative person. It is generally best for the child or parent to raise the possibility of a psychogenic etiology himself. If he poses a question such as "Could it be his nerves?" it is generally best not to make an immediate affirmative response but to sound tentative: "Well, that's an interesting idea. . . . You may have something there. I'd like to hear more of your ideas about that. . . ." Otherwise the parents may believe that they have been entrapped.

During the interview it is generally useful to learn why the parents have come now when the pain has been present for some time; what the parents and child think is wrong; how the pain has affected the child's daily life; what the symptom brings in secondary gain; whether the complaint is masking a school phobia; and what the parents and child expect the doctor to do.

A meticulous physical examination is important not only to detect abnormal findings but also for reassurance. Since parents and adolescents may talk more freely during or after the physical examination, significant information may be volunteered at that time. A thorough examination conveys the physician's interest in the patient and his complaint and in conjunction with the history may preclude or limit the need for laboratory tests, x-ray examination, or other procedures.

Such examinations, if any, should be limited to those necessary to clarify the diagnosis. They should not be performed because the parents "expect" such tests or to "reassure" them. Needless procedures do not relieve pain. If examinations mentioned by the parents are not to be done, the doctor should explain why they are unnecessary.

Once the physician is satisfied, whether on the basis of the history, the physical examination, or other evaluations, that the pain is psychogenic, the findings need to be presented in a way that is understandable and acceptable. One may begin with a statement such as, "I have good news for you. Susie's examination is completely normal, so I am pleased to say that other tests won't be necessary. As is common with children her age, her pain seems to be associated with some of the stresses we've been discussing."

One may continue: "I'm impressed that Susie is an alert, sensitive girl. We obviously wouldn't want to change that, even if we could but being bright and sensitive can be a mixed blessing. As we all know, persons who have a sensitive personality can appreciate art and music and beauty more than others, but they also seem to have more discomfort and pain. With her sensitive nervous system, it's natural that some of the things that you've been telling me about may be affecting her. We don't want to make Susie less perceptive, but more comfortable—in less pain. What we have to do here is to work on these things that are troublesome."

If there is open communication within the family, improvement often ensues without further professional help once the contributing stresses become evident. In other cases the physician may wish to be more directive, e.g., have the child return to school, refer the mother for medical or psychiatric help, recommend a tutor, or suggest that the parents decrease demands on the child or be more involved with their child. Other options include weekly appointments to see the pediatrician and, if appropriate, referral of the child for psychiatric help. When adverse circumstances such as inadequate income, death, or divorce are unalterable, the physician considers with the parent or child how they may better cope with a situation that will not change.

Another approach would be to say, "I don't know how all this started. It sounds like a viral infection, the flu, or something like mononucleosis but with the pain hanging on so long, I have a hunch that some of the stresses we've been talking about must be playing a part in all this."

Another tack might be, "I don't know whether the things we have been talking about are related to Johnny's pain or not, but as a pediatrician I know that they're important to his development and deserve our further attention."

In some situations the doctor may expect no or only limited success in the management of psychogenic pain, e.g., the patient brought reluctantly at the insistence of the school because of absenteeism, the family that is not ready to confront its problems, the child living in a chaotic situation, or a parent who for her own reasons seems to need a child with a persistent symptom.

Occasionally children or adolescents are hospitalized because of persistent psychogenic pain and a history of school absenteeism. In addition to other psychotherapeutic approaches, a structured daily hospital schedule may promote a healthier coping style. The child who has been compliant in the past, who is able to share with her doctor both positive and negative feelings, and who believes that the physician understands how she feels usually responds well to this approach. By participating in formulation of the schedule, the child identifies with both the plan and the physician.

This approach requires from the physician a sizable investment of time. Other professionals (nurse, occupational therapist, physical therapist, nutritionist, schoolteacher, and child life worker) should be involved early, individually or as a group. Weekly staff sessions encourage suggestions, observations, clarification, and consensus. The schedule should be set to the patient's estimated tolerance, fit his style (e.g., high achieving, competitive, passive, compliant, or slow to adapt), and reflect his interests, assets, and needs. Presented as the approach the physician regularly uses for children with similar symptoms, the schedule is usually not sensed as a personal punitive imposition.

A 24 hour schedule, developed and cosigned daily by the child, the physician, and the nurse, is posted prominently next to the child's bed. Bar graphs, stick figures, symbols, clock faces, and sketches may be used to make the chart more graphic and visually interesting. The child is to be fully dressed and to spend only minimal time in his hospital room during the day. Appropriate self-care is expected. All time periods are accounted for, and the term "rest period" is avoided in favor of "personal time" or "scheduled break." In children with psychogenic musculoskeletal aches, fatigue, and weakness, a full physical and occupational therapy program is prescribed.

One or two continuing projects are included each day. Group (school, child life, ward conversation, meals) and individual (physician, nurse, occupational therapist, physical therapist, peer) experiences are included to promote constructive behavior. The staff is encouraged to convey the impression that the doctor expects his orders to be followed precisely ("That's just the way things are done here"), and the doctor's personal presence must convey his firm expectation that the child will follow what has been scheduled.

An attempt is made to select a gregarious roommate who helps keep the patient busy. The staff must be encouraged to enforce the schedule even in the face of resistance, anger, delay, manipulativeness, or deception. These possible reactions should be discussed prospectively with the parents. Such challenges are met with cheerful firmness.

Nurses are asked not to reinforce the child's complaint of pain by responding to it directly. Incentives for progress are negotiated, and steps toward recovery are rewarded. The schedule is revised daily to achieve and demonstrate steady progress. Discharge from the hospital is earned by meeting previously determined levels of activity or symptom alleviation. A prompt return to school is scheduled, and a suggestion is made that the parents give a "recovery" party to which the child's friends are invited. This social event reintroduces the child to her peers and is public testimony that she is ready to resume school and other normal activities.

CHEST PAIN

Of the chief etiologies of chest pain in children listed in Table 28–1, muscle strain, fatigue, or spasm is the most common. When exertional pain with associated tachycardia or syncope suggests a cardiac basis, laboratory examinations to be considered include posteroanterior and lateral chest roentgenograms, an M mode echocardiogram, a 2-D echocardiogram, an exercise tolerance test, a 24 hour Holter monitor, a serum cholesterol determination, and possibly cardiac catheterization. Children with psychogenic precordial pain may

Table 28–1. CAUSES OF CHEST PAIN

I. Musculoskeletal
  A. Muscle strain, fatigue, and spasm
  B. Trauma, rib fractures
  C. Bone tumors or other infiltrative processes
  D. Costochondritis (Tietze's syndrome)
  E. Severe paroxysms of coughing or asthma
II. Cardiopulmonary
  A. Left ventricular outflow obstruction, e.g., idiopathic hypertrophic subaortic stenosis, subaortic stenosis, valvular aortic stenosis
  B. Pulmonary vascular obstructive syndrome or primary pulmonary hypertension, pulmonic stenosis
  C. Pericarditis
  D. Myocarditis
  E. Cardiomyopathy
  F. Aberrant left coronary artery
  G. Sickle cell anemia
  H. Mitral valve prolapse
III. Psychosocial etiology
  A. Identification with significant person
  B. Conversion symptoms
  C. Hyperventilation syndrome
IV. Esophageal
  A. Cardiospasm
  B. Hiatal hernia
  C. Gastroesophageal reflux
V. Pleura and diaphragm
  A. Pleurisy
  B. Spontaneous pneumothorax
  C. Epidemic pleurodynia
  D. Familial paroxysmal polyserositis
  E. Trichinosis
VI. Neurologic
  A. Spinal cord tumor
  B. Epidural abscess
  C. Vertebral collapse

Table 28–2. ETIOLOGIC CLASSIFICATION OF ACHES AND LIMB PAINS

I. Recurrent limb pains, growing pains
II. Foot and leg disorders
  A. Pes planus
  B. Pronated feet
  C. Genu valgum
  D. Bowlegs
  E. Tight Achilles tendon
  F. Shortened hamstring muscles
  G. Generalized ligamentous relaxation
III. Joint disorders
  A. Arthritis, including inflammatory arthropathy associated with connective tissue disease
  B. Patellofemoral pain syndrome
  C. Osteochondritis dissecans
  D. Osteochondrosis of the femoral capital epiphysis
  E. Slipped femoral epiphysis
IV. Bone tumors
  A. Ewing's tumor
  B. Osteogenic sarcoma
  C. Osteochondroma
  D. Osteoid osteoma
  E. Leukemia
  F. Metastatic neuroblastoma
  G. Primary lymphosarcoma
V. Trauma
  A. Sprains
  B. Fractures
  C. Traumatic periostitis
VI. Other bone disorders
  A. Scurvy
  B. Infantile cortical hyperostosis
  C. Vitamin A poisoning
  D. Hyperparathyroidism
  E. Osteomyelitis
  F. Gaucher's disease
  G. Disseminated fat necrosis associated with pancreatitis
VII. Muscle involvement
  A. Myositis, infectious or traumatic
  B. Rocky Mountain spotted fever
  C. Epidemic myalgia
  D. Trichinosis
  E. Dermatomyositis
  F. Guillain-Barré syndrome
  G. Leptospirosis
  H. McArdle's disease
  I. Idiopathic myoglobinuria
VIII. Diseases of the blood
  A. Severe anemia
  B. Sickle cell crisis
  C. Leukemia
IX. Rheumatic fever
X. Collagen-vascular diseases, periarteritis nodosa, serum sickness, and dermatomyositis
XI. Takayasu's disease
XII. Reflex neurovascular dystrophy
XIII. Psychogenic, including conversion reaction

have a parent or other close relative with a cardiac disorder or troubled by angina. As a rule, psychogenic pain does not awaken a child at night.

LIMB PAIN

The differential diagnosis of aches and limb pain is presented in Table 28–2. Studies by Oster (1972) and Apley (1958) documented that occasional limb pain occurs in 15 per cent of school age children. Nonarticular growing pains of no demonstrable organic etiology are common in all age groups but are most frequent in preschool children. They occur especially in the late evening or night in the thighs, calves, behind the knees, and occasionally in the upper extremities. A deep aching or discomfort, persisting from minutes to several hours, may be severe enough to cause crying. Massage, heat, or analgesics usually bring relief. The physician should also explore the possibility of contributory emotional factors in the family. Many children who experience growing pains also have headaches and abdominal pain. Nocturnal muscle cramps, more common in the gastrocnemius and soleus muscles but at times in the feet, occur

occasionally at night in older children and adolescents.

The patellofemoral pain syndrome or chondromalacia patellae, often seen in adolescents, presents with knee pain and a grinding or catching sensation. The pain, which is commonly retropatellar but occasionally anteromedial or peripatellar, is aggravated by running, squatting, kneeling, climbing stairs, and prolonged sitting. The patellofemoral compression test may cause a pain response. In this maneuver the examiner pushes the patella, held between his thumb and index finger, downward and distally against the femur while requesting the patient to tighten the quadriceps muscle or to raise his leg straight without flexion at the knee. Patellofemoral crepitus and patellar malalignment are also often noted on physical examination. Some patients have lateral patellar subluxation. Management is initially conservative with quadriceps strengthening exercises and analgesics.

Osteochondrosis of the femoral capital epiphysis (Legg-Calvé-Perthes disease) is characterized by the insidious appearance of an almost imperceptible limp. Pain, which may be referred along the distribution of the obturator nerve to the medial aspect of the thigh and knee, is usually relatively slight and may simulate that of myalgia or muscle stiffness. Pain in the knee or in the medial aspect of the thigh above the knee, referred along the course of the obturator nerve from the hip, is the earliest sign of a slipped epiphysis. A roentgenogram of the hip is indicated in all patients who have unexplained pain in the knee.

Persistent bone pain suggests a bone tumor, leukemia, or metastatic neuroblastoma. Nonosseous malignant disease usually produces more severe bone pain than would be suggested by the physical findings. Characteristically the pain caused by an osteoid osteoma is burning, most pronounced at night, and rapidly relieved by aspirin. The site of this benign tumor is usually the femur or tibia, although any bone may be involved.

Chronic or recurrent musculoskeletal pain without apparent organic cause in a child with a history of frequent school absenteeism is usually psychogenic. Chronic fatigue, weakness, anorexia, abdominal pain, and headache may also be reported. Some of these children have conversion reactions. Psychologic factors are frequently present in reflex neurovascular dystrophy, a syndrome characterized by severe limb pain, autonomic dysfunction, and vasomotor instability.

BACK PAIN

The causes of back pain are listed in Table 28–3. Infection of the intervertebral disc (discitis) occurs in young children, primarily in the lumbar region. Symptoms include irritability, refusal to sit, stand, or walk, limping, crying at night, stiffness of the back, vague back pain, abdominal pain, mild fever, and occasionally localized tenderness. Movement of the pelvis or thigh causes pain. Kernig's or Brudzinski's sign may be pres-

Table 28–3. CAUSES OF BACK PAIN

Infections of the intervertebral disc (discitis)
Calcification of intervertebral discs
Spinal cord tumor
Bone tumors of spine
Vertebral collapse and compression due to leukemia or long term corticosteroid therapy
Herniated intervertebral disc
Spondylolisthesis, spondylolysis
Ankylosing spondylitis, rheumatoid arthritis
Vertebral osteomyelitis
Tuberculous spondylitis
Psychogenic back pain, coccydynia

ent. Roentgenographic changes may not be evident for two to four weeks, but a bone scan may be positive when the child is first seen.

Spinal cord tumors may present with neck, back, or extremity pain exacerbated by coughing, sneezing, straining, or straight leg raising; motor weakness; muscle atrophy; paraspinal muscle spasm; a positive Babinski sign; and a change in bladder or bowel habits. Bone tumors of the spine such as aneurysmal bone cyst, osteoid osteoma, Ewing's sarcoma, osteogenic sarcoma, neuroblastoma, lymphomas, and leukemia may be a cause. Back pain may be due to vertebral collapse and compression with leukemia or prolonged corticosteroid therapy. Vertebral osteomyelitis may be characterized by localized back pain and fever.

Herniation of an intervertebral disc is a rare cause of low pain in older children and adolescents. The most frequent physical findings are pain on deep pressure between the fifth lumbar and the first sacral vertebrae, limited ability to elevate the extended lower extremity unilaterally or bilaterally, restriction of back movement, and flattening of the lumbar spine. Spondylolisthesis, caused usually by L5 slipping forward on S1, and spondylolysis are rare causes of low back pain in adolescents. The pain may radiate into the buttocks, legs, or groins and may be accompanied by tight hamstrings and limited ability for straight leg raising. Scoliosis may be present. The patient is less symptomatic when he leans forward.

Ankylosing spondylitis may present after the age of 15 with recurrent transient stiffness and pain involving the lumbosacral spine, sacroiliac joints, buttocks, and thighs and hips. Pain may radiate into the lower extremities. Tenderness may be present over the sacroiliac joints and lumbar spine. Anterior flexion of the lumbar spine may be limited. Peripheral arthritis may precede or accompany complaints referable to the back. Pain, stiffness, and limitation in motion of the neck may be early manifestations of juvenile rheumatoid arthritis.

MORRIS GREEN

REFERENCES

Apley, J.: A common denominator in the recurrent pains of childhood. Proc. R. Soc. Med., 51:1023, 1958.

Balint, M.: The Doctor, His Patient and The Illness. Ed. 2. New York, International Universities Press, 1964, Ch. 4.

Bernstein, B. H., et al.: Reflex neurovascular dystrophy in childhood. J. Pediatr., 93:211, 1978.

Diagnostic and Statistical Manual of Mental Disorders. Ed. 3. New York, American Psychiatric Association, 1980.

Driscoll, D. J., Glicklich, L. B., and Gallen, W. J.: Chest pain in children: a prospective study. Pediatrics, 57:648, 1976.

Dubowitz, V., and Hersov, L.: Management of children with non-organic (hysterical) disorders of motor function. Dev. Med. Child Neurol., 18:358, 1976.

Engel, G. L.: "Psychogenic" pain and the pain-prone patient. Am. J. Med., 26:899, 1959.

Engel, G. L.: Pseudoangina. Am. Heart J., 59:325, 1960.

Engel, G. L.: Pain. In MacBryde, C. M., and Blacklow, R. S. (Editors): Signs and Symptoms. Ed. 5. Philadelphia, J. B. Lippincott Company, 1970, Ch. 3, p. 44.

Green, M., and Beall, P.: Paternal deprivation—a disturbance in fathering. Pediatrics, 30:91, 1962.

Hayden, J. W.: Back pain in childhood. Pediatr. Clin. N. Am., 14:611, 1967.

Hoppenfeld, S.: Back pain. Pediat. Clin. N. Am., 24:881, 1977.

Insall, J.: "Chondromalacia patellae": patellar malalignment syndrome. Orthop. Clin. N. Am., 10:117, 1979.

Oster, J.: Recurrent abdominal pain, headache and limb pains in children and adolescents. Pediatrics, 50:429, 1972.

Oster, J., and Nielsen, A.: Growing pains. Acta Paediatr. Scand., 61:329, 1972.

Passo, M. H.: Aches and limb pain. Pediatr. Clin. N. Am., 29:209, 1982.

Peterson, H. A.: Leg aches. Pediatr. Clin. N. Am., 24:731, 1977.

Rocco, H. D., and Eyring, E. J.: Intervertebral disk infections in children. Am. J. Dis. Child., 123:448, 1972.

Schaller, J., Bitnum, S., and Wedgwood, R. J.: Ankylosing spondylitis with childhood onset. J. Pediatr., 74:505, 1969.

Schmitt, B. D.: Infants who do not sleep through the night. Dev. Behav. Pediatr., 2:20, 1981.

Tachdjian, M. O., and Matson, D. D.: Orthopaedic aspects of intraspinal tumors in infants and children. J. Bone Joint Surg., 47A:223, 1965.

28C "Colic" or Excessive Crying in Young Infants _____

"Colic" is a poorly defined and incompletely understood state of excessive crying seen in young infants who are otherwise well. It is one of the most common and most baffling of recurrent pain disorders in childhood.

Definition. There is no standard definition of colic. Pediatric texts and books of advice for parents usually (but not always) present a brief section on a phenomenon variously designated by such terms as paroxysmal fussing in infancy, infantile colic, evening colic, or three month colic.

Usual descriptions of "colic" indicate that the condition begins soon after the baby comes home from the hospital and may persist until he is three or four months of age. The crying is characterized as intense, lasting for up to several hours at a time and usually occurring in the late afternoon and evening. The affected infant is typically pictured as drawing up his knees against his abdomen and expelling much flatus. He may appear hungry but is not quieted for long by further feeding or other attempts at soothing. However, the infant eats well and grows normally.

The standard textbook description in the preceding paragraph is inadequate on several counts:

1. Without a more precise definition of the duration of the crying, it can be extended to apply to almost all young infants at one time or another.

2. Such imprecision makes most studies of the phenomenon of questionable value. Writers on the subject often express strong convictions about favorite theories of etiology and management that can neither be refuted nor verified because the infants studied are so poorly identified.

3. Figures relating to the incidence of colic are of little value.

4. Telling parents that their child has "colic" is at best a confusing message.

Probably the best definition available at this time is the one used by Wessel et al. (1954), that such a young infant is "one who, otherwise healthy and well fed, had paroxysms of irritability, fussing, or crying lasting for a total of more than three hours a day and occurring on more than three days in any one week." The excessive crying is identified in terms of intensity, duration, and frequency. The formulation achieves greater clarity if the word "paroxysm" is interpreted as crying at full force, not just any degree of fussing. Even this version, however, is applied with difficulty at times.

If a better definition is attempted, it should convey the idea that the child is otherwise well but is crying substantially more than the mean amount for infants of his age. The condition might then be better referred to as "excessive crying." The evidence to date does not justify any conclusion that the infant who is crying excessively is qualitatively different from the one who cries less. The flatus may be a consequence rather than the cause of the crying. It is not even certain that such infants are experiencing abdominal pain, as is commonly assumed. Infants flex their legs in response to a variety of noxious stimuli, such as a pinprick on the foot. The affected infant seems not to have any disease or malfunction of the bowel or any other organ but differs from the norm quantitatively as to the amount of crying.

This view is supported by the usual rapid reduction in crying within a few days by alteration of the infant's handling.

DIFFERENTIAL DIAGNOSIS

Since the phenomenon under discussion is excessive crying in otherwise well young infants, several conditions must be distinguished.

Normal Crying. Brazelton (1962) assessed crying patterns in detail in a middle class sample of 80 infants. Crying lasted about two hours a day at two weeks, increased to a peak of almost three hours by six weeks, and then gradually decreased to about one hour by three months. The upper quartile of babies were crying three and one half hours per day at six weeks. Throughout these three months the principal time for crying was in the evening.

The amount of parental complaining about crying is not necessarily proportional to the extent of crying. Some parents are upset about typical periods of fussing, while others may not seem disturbed by an excessive quantity. The first step in the differential diagnosis is to decide whether the crying is merely an ordinary amount that the parents cannot tolerate or is truly greater than average.

Faulty Feeding Technique. This includes under- and overfeeding and inadequate burping or sucking. These possibilities usually can be excluded by a routine history and physical examination.

Physical Problems in the Baby. If "colic" or excessive crying in infants is by definition found only in those who are otherwise well, various physical problems in the infant must be excluded before the diagnosis can be applied. Three kinds of problems are usually cited: acute disorders such as otitis media, intestinal cramping with diarrhea, corneal abrasion, and incarcerated hernia, all of which are relatively easily ruled out by examination of the infant; cow's milk allergy, lactose intolerance, or transmission of irritating substances such as caffeine via breast milk, all of which appear to be relatively insignificant causes of prolonged crying in otherwise well infants; and inadequately defined clinical entities, such as "immaturity" of the central nervous system or of the intestine.

CONTRIBUTORY FACTORS

If the infant is healthy but is crying substantially more than most, there is usually no single clear-cut explanation for the excess of crying. There are, however, two principal contributory factors to consider: a physiologic predisposition in the infant and inappropriate handling by the parents.

Infants vary considerably in their temperaments or emotional reactivity characteristics (see Chapter 10). Some tend to cry more than others, regardless of feeding, handling, or other factors. Perhaps this is what was meant by the old textbooks that described such babies as "hypertonic." Two temperamental predispositions have been shown so far to be associated with excessive crying. As might be expected, "difficult" infants were likely to be diagnosed as colicky, although not necessarily so. Infants with a low sensory threshold were also prone to increased crying apparently because they were more vulnerable to disorganization by an excessive sensory input from the environment (Carey, 1972).

Infants also vary in their soothability. Parents do not always know at first which methods are most effective for quieting babies in general and theirs in particular. If they are inexperienced or anxious, they may have greater trouble understanding their child's expression of needs and may respond with unsuitable manipulations. Excessive and inappropriate handling of the infant is frequently observed both as a causal factor and as a response to excessive crying.

MANAGEMENT

Most articles dealing with the management of "colic" or excessive crying are unreasonably pessimistic about the effectiveness of professional intervention. This defeatism is unjustified; appropriate management is usually successful in reducing crying to acceptable amounts.

HISTORY

Management begins, as elsewhere, with an adequate history. The first step is to define the symptom: the intensity, duration, and frequency of the crying. Parents often say that the baby is too "gassy" or too hungry, and the clinician must sift the evidence to discover that the crying is the real problem. A good way to make the parents' description of the baby more precise is to ask for a detailed narration of the baby's typical day. Information about the baby's temperamental characteristics must be based entirely on interview data and observations, since there are at present no temperament questionnaires suitable for use in the first three months of life. Having the parents demonstrate their soothing techniques may be helpful in revealing practices requiring modification. The rest of the medical history should be obtained in the usual manner but with an enrichment so as to include parental concern about the pregnancy and the child and anxieties related to their own experience as children or with rearing previous children or to inadequate family support and other stresses (Carey, 1968).

PHYSICAL EXAMINATION

The physical examination seldom reveals anything useful in regard to the management of the crying but is an absolutely necessary part of the procedure. Most parents are doubtful about reassurance that is not preceded by a careful assessment of the infant. For example, the prescription of sedative drops over the telephone is generally not successful. Laboratory tests, however, are usually not indicated.

COUNSELING

Counseling should cover several important points.

1. The clinician should reassure the parents that the examination has not revealed any problem with the infant's physical health.

2. An appropriate next step is to acknowledge, if this is so, that the infant does seem at present to be crying more than the average and that this is an unpleasant burden for most people. It may be helpful for the parents to receive some information about how much the average infant of the particular age cries. Little is to be gained, however, by awarding the child the vague diagnosis of "colic."

3. This leads to a discussion of parental anxiety and to dealing with it. The anxiety may be a factor in promoting the excessive crying, or it may be a reaction to it, or both. In any case, consideration of fears about the baby or the various pertinent psychosocial stresses often reveals the background of inappropriate parental handling of the infant (Carey, 1968).

4. Parental handling of the infant may require alteration. Parents with fussy infants usually are doing too much and need to shift their tactics. They usually will be successful if they soothe more, as by a pacifier and heating pad or hot water bottle, and stimulate less as by decreasing the picking up and feeding. A quiet environment with a minimum of unnecessary handling and correction of any faulty feeding techniques without changing the composition of the feedings seem helpful.

5. The use of medication for temporary relief of excessive crying is controversial but has a definite place in management. Most recommended are: phenobarbital elixir, ½ teaspoon (10 mg.) three times daily, or dicyclomine hydrochloride (Bentyl syrup), ½ teaspoon (5 mg.) three times daily or 1 teaspoon (10 mg.) before an anticipated evening fussy spell. Treatment for one week is usually sufficient. If excessive crying returns after that, the medication can be given for a second week. There may be some placebo effect in the administration of these substances, but it is likely that there is also a pharmacologic effect. This is probably why the use of a drug alone without other measures is only modestly effective. Medication, when indicated, should be made optional, since some parents want to try first to lessen the crying without it.

6. The expression of optimism about the immediate outcome of the foregoing measures is justified and improves chances of success. Although acknowledging with the parents that excessive crying in young infants is a poorly understood phenomenon, the clinician is on firm ground in telling parents that if the recommended steps are followed, there is an excellent chance that the crying will diminish considerably in the next two or three days. This sanguine prophecy is based on experience and is usually correct. On the contrary, telling parents that the excessive crying will go away by three months, which may be a condemnation to two or more months of further screaming, is not comforting and is likely to be counterproductive (Carey, 1968).

7. Close follow-up of the excessively fussy baby is important. A convenient way to do this is a telephone contact every two or three days until there is substantial improvement. On rare occasions it is necessary to re-evaluate the child and situation in a week or so.

8. Under very extraordinary circumstances and as a last resort, separation of the infant from the parents by a brief period of hospitalization can be dramatically effective in reducing the infant's crying. If, however, parental feelings and handling of the infant are not dealt with effectively before the parents are reunited with their infant, the old pattern of interaction and crying is likely to resume (Barbero et al., 1957).

9. Several unsuitable forms of treatment should be mentioned, if only to discourage their use. Changes in the composition of feedings, i.e., from one formula to another, are seldom appropriate solutions. Almost any formula change, in fact, almost any altered procedure done with conviction, is likely to be followed by a temporary reduction in crying because of the placebo effect.

Although the use of rectal manipulations and enemas is widely supported by tradition, there is no published evidence to establish their value.

PROGNOSIS

Standard pediatric texts and conventional wisdom repeat the notion that "colic" usually goes away by itself by three or four months and that little can be done to alter that fate. However, clinical experience and reported studies indicate that excessive crying in young infants can be sharply reduced within a few days in most instances if appropriate steps are taken. Some babies and some situations take longer, but virtually all respond to suitable management. On the other hand, with inappropriate care the fussing may even extend beyond three or four months.

The long term prognosis for individuals who as young infants cried more than average has not been sufficiently investigated. In the study already mentioned (Carey, 1972), infants diagnosed as having colic by the criteria of Wessel et al. (1954) were shown within the next few months to have a more difficult temperament and lower sensory thresholds, but these were presumably etiologic factors in the excessive crying rather than an outcome of it.

Any statement that these infants become more impatient or aggressive as children or adults is pure speculation. Since retrospective data relating to amounts of crying tend to be highly inaccurate, only prospective studies will resolve this issue.

PREVENTION

No investigation has yet proven that any particular measures will reduce the incidence of "colic" or excessive crying. Some theoretical possibilities deserve mention.

We do not know how to alter the apparently predisposing physiologic factors. We are not even fully certain what they are. It may become possible, however, by the use of such a measure as the Brazelton Neonatal Behavioral Assessment Scale to identify infants who are particularly likely to develop excessive crying. No such evidence has been published so far.

It is possible to educate parents, starting even prenatally, about infant crying and soothing. Some need to know that even completely normal infants cry some every day. Parents often are unaware that one of the commonest reasons for crying is fatigue and that an infant usually does better if not picked up at those times.

The alert clinician also deals with parental anxieties whenever they are expressed. Concerns revealed prenatally or in the newborn nursery may lead to inappropriate handling of the infant if not resolved satisfactorily.

WILLIAM B. CAREY

REFERENCES

Barbero, G. J., Rigler, D., and Rose, J. A.: Infantile gastro-intestinal disturbances: a pilot study and design for research. Am. J. Dis. Child., 94:532, 1957.

Brazelton, T. B.: Crying in infancy. Pediatrics, 29:579, 1962.

Carey, W. B.: Maternal anxiety and infantile colic. Is there a relationship? Clin. Pediatr., 7:590, 1968.

Carey, W. B.: Clinical applications of infant temperament measurements. J. Pediatr., 81:823, 1972.

Wessel, M. A., Cobb, J. C., Jackson, E. B., Harris, G. S., Jr., and Detweiler, A. C.: Paroxysmal fussing in infancy, sometimes called "colic." Pediatrics, 14:421, 1954.

28D Recurrent Abdominal Pain

DEFINITION OF THE PROBLEM

The symptom of abdominal pain is a common, troublesome, and anxiety provoking complaint of childhood. When the symptom recurs frequently over a prolonged period of time, it and its associated symptoms are referred to as the recurrent abdominal pain syndrome. The cardinal features of the syndrome are the recurrent paroxysms of unexpected pain, before and after which the child is perfectly well. To distinguish these episodes from transient gastrointestinal disturbance, most clinicians (following the example of Apley, 1975) restrict the label to pain severe enough to interfere with normal activity that recurs at least once monthly for more than three months. Usually the syndrome is described apart from the context in which the child lives—the simple recurrent abdominal pain syndrome. The extended recurrent abdominal pain syndrome can be considered to refer to the consequences of being a child with recurrent abdominal pain in the context of the child's environment.

THE SIMPLE RECURRENT ABDOMINAL PAIN SYNDROME

A number of additional clinical manifestations occur frequently enough to be considered characteristic of the syndrome. The pain is typically but not necessarily periumbilical and otherwise difficult to describe. Most commonly the pain episodes last one hour or less but occasionally last all day (Roy et al., 1975). Other "autonomic symptoms," such as nausea, perspiration, pallor, flushing, palpitations, and vomiting, are frequent, one or more being reported in up to 70 per cent of cases (Apley, 1975; Roy et al., 1975). By contrast, symptoms of organic disease (weight loss, dysuria, frequency, blood in stools, fever, jaundice, loss of consciousness following the pain) are sufficiently rare that their presence mandates search for a relevant disease process. The same children often report headaches and limb pain, although these seldom coincide with the abdominal pain episode (Oster, 1972). There are few clinical signs elicited on physical examination, the one exception being tenderness to palpation in 45 per cent of cases, often in the left lower quadrant (Dimson, 1972). Impacted stool may be found on rectal examination when recurrent abdominal pain is associated with stool retention. Significant occult blood in the stool is uncommon.

Although definitions differ slightly, the prevalence of the simple recurrent abdominal pain syndrome seems to vary between 10 and 17 per cent in a number of studies in different populations (Apley, 1975; Oster, 1972; Parcel et al., 1977; Pringle et al., 1966). It may begin in the preschool years, and Apley's series suggests that the peak time of onset is five years of age (Apley, 1975). The syndrome seems to be a problem predominantly in preadolescence, but its significance in the adolescent age group is uncertain owing to inadequate data. Although it is commonly considered a syndrome that children "grow out of," available data concerning the prognosis suggest that as many as one-third continue to experience recurrent pain episodes for many years (Apley, 1975; Christensen and Mortensen, 1975; Stickler and Murphy, 1979). In 2 to 6 per cent of the children with recurrent abdominal pain, organic disease was discovered after prolonged follow-up in selected samples (Christensen and Mortensen, 1975; Stickler and Murphy, 1979).

THE EXTENDED RECURRENT ABDOMINAL PAIN SYNDROME

The consequences of being a child with the recurrent abdominal pain syndrome tend to relate to the function or meaning of the pain experience to the child, the parent(s), the physician, and other significant people in the environment (e.g., peers, teachers). The consequences may be positive or negative depending on the appropriateness of the response. For example, the pain complaint is usually a powerful elicitor of caring behavior on the part of the child's family. However, the discovery that the pain complaint stimulates otherwise infrequent affective responses from his parents may affect the child's manner of making the complaint, or even perpetuate the complaint after the original stimulus is gone. Alternatively, the fact that the child is completely well between episodes may predispose parents or other caregivers to doubt the veracity of the complaint, resulting in caring behavior on some occasions and accusations of malingering on others. Similarly, the teacher may suspect a child who complains in the classroom of stomach pains that seem to resolve soon after reporting to the nurse, leading to attributions of school related anxiety. Unfortunately the uncertainty associated with the pain complaint may engender negative consequences for the child even from well intentioned actions on the parents' part. A typical example concerns parental fear of a life threatening disease process underlying the pain complaint. The child may be confused by the unspoken increased anxiety reflected in changed parental behavior toward him without understanding the reason for the change. In such cases disturbed parent-child relationships are the effect rather than the cause of being a child with recurrent abdominal pain.

Finally, the child is also at risk for inappropriate diagnostic and therapeutic approaches when he becomes a patient interacting with the health care system. Primarily this stems from the diagnostic dilemma facing physicians. The knowledge that a large number of occult disease processes may underlie the syndrome often stimulates extensive, usually invasive, overinvestigation, while the assumption that the pain is most often "psychogenic" may result in underinvestigation and inappropriate counseling.

Simplistically speaking, the problem of recurrent abdominal pain should probably be considered at three levels—the individual pain episode itself, the recurrence of pain episodes (the simple recurrent abdominal pain syndrome), and the syndrome understood in the context of the child's environment (extended recurrent abdominal pain syndrome). A variety of factors may be differentially associated as cause or effect at each level (e.g., interpersonal stress may cause a particular pain episode but may be an effect of being a child with recurrent abdominal pain), and appropriate management requires careful attention to these many facets of the problem. In many cases the goal of successful management is an otherwise normal child with the simple recurrent abdominal pain syndrome, while management failure is defined by worsening of the extended recurrent abdominal pain syndrome.

CONCEPTS OF PATHOPHYSIOLOGY

There is little specific information available concerning pathophysiologic mechanisms in nonspecific recurrent abdominal pain. The pain sensations are presumed to originate from nerve endings in the submucosa, musculature, or serosa of abdominal organs. No good evidence for the existence of specific pain receptors exist, and it is likely that the pain sensation is directly mediated either by the more common mechanoreceptors or indirectly via the release of humoral substances (Leek, 1977).

In the absence of a specific disease process, the inciting stimulus is likely to be increased muscle tension in the gastrointestinal tract due to distention or spasm. Studies with balloon inflation at various levels of the gastrointestinal tract have contributed to an understanding of pain symptomatology (Jones, 1938; Lipkin and Sleisinger, 1957). It appears that simple distention is not a sufficient stimulus for pain sensation, but that a change in tension in the musculature is required. This may account for the intermittent nature of the pain complaint, the pains occurring only during transient muscular contractions. A correlation between pain symptoms and motility changes at various levels of the small and large intestine has been demonstrated in selected symptomatic adult patients (Holdstock et al., 1969); unfortunately no such studies are available in children.

In general, distention from the small bowel distal to the duodenojejunal angle and from the ascending, transverse, and descending colon results in referral to the periumbilical region. Pain localized to the quadrant of origin can result from distention of the esophagus (pain referred to the sternum), the ileocecal valve, hepatic flexure, and sigmoid colon. As a rule, large bowel pain is more diffuse and less severe than that originating in the small bowel. Distention of both the small and large intestine also results in pain referred to the back, and referral to many other abdominal sites may result from large bowel distention (Swarbick et al., 1980). Consequently the clinical dictum that nonperiumbilical pain increases the likelihood of a specific organic disease process may be true in groups of patients, but it is not pathognomonic of disease in an individual case. Although pain is most commonly periumbilical in children with recurrent abdominal pain, the tendency to localize pain increases with age (Heinild et al., 1959). Whether this is due to increasing sensory specificity, language facility, or changes in underlying mechanisms causing the pain remains obscure. Although children's pain descriptions have been thought of as unhelpful and unreliable, our experience suggests that over 20 per cent of these children are able to distinguish the presenting pain complaint from others they experience in the abdominal region. Thus, although pains are difficult to describe, children do have the facility to discriminate between pains in the same area.

THE ROLE OF STRESS

Because of the paucity of evidence of underlying disease, most clinicians have considered nonorganic abdominal pain to be "psychogenic" in origin. In most descriptions, psychogencity is taken to mean that environmental or psychosocial stress factors play a role in the etiology of the syndrome. Unfortunately it is seldom indicated whether associated stress is a cause or an effect, or whether it acts at the level of particular pain episodes, the syndrome, or the extended syndrome. In principle, such factors could act to predispose children to the recurrent abdominal pain syndrome, but despite considerable clinical interest, there are no prospective studies identifying psychosocial or emotional stress as risk factors. On the other hand, stress factors could act to exacerbate particular pain episodes, or to maintain pain stimulated by other causes. Some evidence of differential environmental or individual characteristics has been reported that might account for such pain occurring in some children but not in others.

The most convincing evidence that environmental factors differ in the child with recurrent abdominal pain is the finding of an increased prevalence of abdominal pain complaints in the families of such children (Apley, 1975; Oster, 1972). This finding might support either the role of social modeling in the genesis of the syndrome or a constitutional predisposition to the complaint. However, Christensen and Mortensen (1975) reported that abdominal pain occurred more frequently only among children of parents who were complaining of pain at the time of investigation, but not among children of parents who reported recurrent abdominal pain during childhood. This finding is more consistent with the pain modeling explanation. Stressful events that could precipitate pain episodes are prevalent in children with recurrent abdominal pain (Apley, 1975; Green, 1967; Heinild et al., 1959; Roy et al., 1975; Stone and Barbero, 1970). Similarly, excessive parental anxiety, which could act as a contingent reinforcer of pain behavior, has been reported (Apley, 1975; Green, 1967; Stone and Barbero, 1970). These observations are less convincing because of their uncontrolled nature and the clinic setting in which they were obtained.

Individual characteristics differentiating the child with recurrent abdominal pain have been studied clinically and experimentally. In unreferred school children, Apley and Naish (see

Apley, 1975) reported that, as a group, children with recurrent abdominal pain were more frequently described as having "emotional disturbances" (undue fears, sleep disorders) and as more "highstrung and fussy" or "anxious, timid, and apprehensive" than control subjects, a profile also found in clinic settings (Barr and Feuerstein, 1983). However, 51 per cent of these children were considered "normal, average, good," implying that such a profile applies only to a subgroup of patients with the syndrome (Apley, 1975). Experimental studies of individual differences have described increased rectosigmoid motility following injection of prostigmine methyl sulfate (Kopel et al., 1967), implying an increased sensitivity of the colon to parasympathetic stimulation.

Attempted laboratory confirmation of the apparent clinical relationship between stress and recurrent abdominal pain suggested a deficit in general autonomic recovery (pupillary dilation) following a cold pressor stress test in two studies (Apley, 1975; Rubin et al., 1967). However, this deficit was unconfirmed when other autonomic measures were used and may not be applicable to autonomic reactivity in general (Feuerstein et al., 1982). In addition, there is no evidence of increased subjective sensitivity or differences in facial reactions to a laboratory induced pain stimulus (Apley, 1975; Feuerstein et al., 1982). In summary, although a considerable body of clinical anecdotal experience would seem to suggest a relationship between stress and abdominal pain at some level, empirical confirmation of the relationship and the psychophysiologic mechanism(s) involved is still only suggestive.

PARENTAL ANXIETY

Although not the subject of systematic study, the child with unexplained recurrent pain episodes tends to have a significant impact on the environment in which he lives. Most notably this is manifest as considerable parental anxiety about the significance of the pain for the child. Many clinicians have described the parents as "overprotective" or as demonstrating "contagious circular anxiety" in the clinic setting (Apley, 1975; Stone and Barbero, 1970). Whether this represents trait anxiety focused on this particular problem is unclear, but anxiety about the symptom in the child is virtually universal in parents of children brought for evaluation. The parental anxiety far outweighs that of the child in most cases and therefore becomes an appropriate focus for intervention. Parental anxiety varies according to whether the pain is seen more as a symptom of disease or as a transient nondisease manifestation (a "problem of living"). In some cases the pain is a signal of disturbed interpersonal relationships, which go beyond the pain symptom itself. In these cases the pain complaint is seldom the only symptom of disturbed relationships, and the parental anxiety is usually found to extend to other aspects of family function.

CLINICAL CLASSIFICATION

Although it is common to divide recurrent abdominal pain into "organic" and "psychogenic" clinical categories, a tripartite classification into organic, dysfunctional, and psychogenic is recommended as being more accurate and heuristically helpful. This takes into account the important fact that the symptom may occur in the absence of abnormal physiologic or psychologic processes and does not presume that the presence of the symptom necessarily implies the presence of disease. The classification refers to the clinical presentation at the level of the simple recurrent abdominal pain syndrome, although parental anxiety and the consequences of the extended recurrent abdominal pain syndrome are equally important in assessment and management.

The term organic abdominal pain implies that the pain sensation is generated intra-abdominally and is the result of an abnormal physiologic process such as occurs in pain secondary to distention with hydronephrosis or secondary to inflammation in Crohn's disease. In dysfunctional pain, the sensation is generated intra-abdominally but is the result of normal physiologic functioning. As subcategories of dysfunctional pain, specific syndromes are those in which the mechanism or specific pattern of pain is recognizable, whereas the designation nonspecific recurrent abdominal pain syndrome includes all those in which no mechanism or pattern is apparent.

A prototypic example of specific dysfunctional recurrent abdominal pain occurs in children with lactose intolerance following the normal decline in the concentration of small intestinal lactase enzyme during the preschool years. Insufficient hydrolysis of normal quantities of lactose in milk and milk products may predispose to gas production, distention, and pain in these children (Barr et al., 1979).

In psychogenic pain, the abdominal pain sensation may or may not be felt intra-abdominally, and disordered intra- or interpsychic emotional factors are considered to be causal of the pain complaint. Typical examples include conversion reactions and use by the child of the pain complaint simply as a means of avoiding unwanted experiences. Table 28–4 lists recognized presentations of the recurrent abdominal pain syndrome according to this classification system.

There are a number of implications for diagnosis that follow from this classification. The most ob-

Table 28–4. CLINICAL CLASSIFICATION OF CHILDREN PRESENTING WITH RECURRENT ABDOMINAL PAIN SYNDROME

Organic	Dysfunctional	Psychogenic
Gastrointestinal	Chronic stool retention	Acute reactive anxiety
Peptic ulcer	Heightened awareness of	School phobia
Gastritis	intestinal motility	Manipulation (secondary
Hiatus hernia	Lactose intolerance	gain)
Hernia	Sucrose intolerance (?)	Hysterical conversion
Volvulus, recurrent	Intestinal gas syndromes	reactions
Obstruction due to bands, adhesions	Menses, dysmenorrhea	Depression
Inflammatory bowel disease	Mittelschmerz	Complaint modeling
Crohn's disease	Reaction to normal stress	Hypochondriasis
Ulcerative colitis	and anxiety (?)	Factitious
Meckel's diverticulum	Overeating	
Neoplasms	Irritable colon	
Yersinia enterocolitis	Chilaiditi's syndrome	
Intussusception, recurrent		
Hirschsprung's disease		
Infestations (e.g., giardiasis)		
Malrotation		
Annular pancreas		
Polyps, polyposis		
Foreign body		
Mesenteric adenitis		
Malformations		
Gastric duplication		
Genitourinary		
Hydronephrosis, obstruction		
Lower tract obstruction		
Posterior urethral valves		
Atresia		
Pyelonephritis		
Renal stones		
Ovarian cyst		
Testicular or ovarian torsion		
Hematocolpos		
Endometriosis		
Neoplasms		
Hepatobiliary system		
Hepatitis		
Gallstones, cholecystitis		
Pancreatitis, especially familial		
Trauma		
Traumatic hemobilia		
Pancreatic pseudocyst		
Subserosal intestinal hemorrhage		
Abdominal wall strain		
Metabolic		
Lead poisoning		
Porphyria		
Hereditary angioedema		
Familial hyperlipidemia		
Other conditions		
Abdominal epilepsy and migraine		
Anorexia nervosa		
Sickle cell disease		
Familial Mediterranean fever		
Riley-Day syndrome		
Multiple endocrine adenomatosis		
Blood dyscrasias		
Lymphomas		
Coxsackie virus, pleurodynia		
Meconium ileus syndrome		
Brain neoplasms		
Epilepsy		
Hemolytic disease		

vious is that absence of evidence of organic disease does not imply that emotional or psychogenic factors are causal, even in the presence of recent stress events (such as death of a relative). Second, the finding of positive evidence of organic or psychologic abnormalities may be incidental to the pain complaint, as Apley (1975) has stressed. Third, constitutional and psychologic factors may interact in different ways to produce specific pain episodes. This occurs when abdominal complaints initially associated with stool retention come to be used for their "secondary gain" after the stool retention is resolved. Finally, the classification does not include the important secondary consequences of parental anxiety and the rest of the extended recurrent abdominal pain syndrome, which are equally important in management.

ASSESSMENT TECHNIQUES

The assessment of the child (and his family) presenting with recurrent abdominal pain has much in common with the assessment of other chronic and recurrent pain symptoms (see p. 513). In general, the aims of the assessment are to evaluate both the symptoms of the simple recurrent abdominal pain syndrome and the functional implications of being a child with abdominal pain. With regard to the latter, attention is paid to both the degree and the sources of parental anxiety concerning the symptom, the limitations put on activity at school or with peers, and the history of contact with the medical care system for this problem.

Usually the interview begins with a discussion of the possible types of explanations and their relative frequency even before specific information about the character of the pain is sought (Table 28–5). This tends to focus the parental anxiety from the start and to raise possibilities that may not have been considered because of concern focused only on organic disease. It also tends to prevent surprise at not finding a disease process when the investigation is complete. Patients and parents are encouraged to become coinvestigators with the physician to look for associations and clues to help explain the pain episodes. This is often facilitated by the use of patient "diaries" in which the frequency, time of occurrence, duration, and severity of the complaints can be marked, as well as other potential predisposing factors (such as milk ingestion or stool frequency). Finally the child is encouraged to be the primary respondent by discussing the symptoms with him, encouraging him to keep the diary, and leaving time for discussion with the child alone at every visit.

At the level of defining the simple recurrent abdominal pain syndrome, questions are directed

Table 28–5. PRIMARY CLINICAL DIAGNOSIS IN 50 CONSECUTIVE PATIENTS WITH RECURRENT ABDOMINAL PAIN*

		n	%
Organic			
Familial Mediterranean fever	(1)		
Hydronephrosis	(1)	3	6
Inguinal hernia	(1)		
Dysfunctional			
Specific			
Lactose intolerance		7	14
Stool retention		10	20
Increased intestinal awareness		1	2
Nonspecific			
"Spontaneous" resolution (early)†		11	22
Persistent recurrent abdominal pain, spontaneous resolution (late)‡		5	10
Persistent recurrent abdominal pain, unresolved		11	22
Psychogenic			
Manipulation (secondary gain)(1)		2	4
Acute reactive anxiety	(1)		
		50	100

*Prepared by Alissa Lipsom, M.D.

†Early spontaneous resolution means pains completely disappeared by six weeks (i.e., third clinic visit) following initial presentation without specific therapy.

‡Late spontaneous resolution means pains completely disappeared after six weeks (i.e., three clinic visits) following initial presentation without specific therapy.

at describing the phenomenon, associated symptomatology (such as autonomic symptoms), history of the first episode and associated events, factors that modify the pain, and how the symptom has evolved subsequently. While the pain description may not be helpful diagnostically (Apley, 1975), it is important as a baseline against which subsequent treatment modalities are tested.

To increase suspicion of organ disease, questions directed at "organic indicators" are asked, such as atypical descriptions of the pain episodes (e.g., constant, well localized, nonperiumbilical). Pain that wakes the patient at night is seen with organ disease, but this also occurs in dysfunctional pain due to lactose intolerance and stool retention. Symptoms of recurrent fever, jaundice, changes in stool color, appetite, or weight, and persistent vomiting and hematemesis are sought. The presence of symptoms in other organ systems raises questions of generalized disease. Questions to detect psychogenic disease have in common the attempt to understand the meaning of the pain experience for the child. The pain experience is often colored by its affective connotative qualities, which may be remembered long after the specifics of duration and severity have been forgotten. Occasionally descriptions of apparently unrelated pains may elicit emotions of implied separation, neglect, increased attention, or reward, which have implications for the role pain plays in defining the child's relationships. It should be remembered that so-called organic indicators (weight loss or disturbances in sleep or eating patterns) may also be found in psychogenic pain syndromes.

In eliciting evidence of dysfunctional pain syndromes, questions are directed at a search for normal constitutional or environmental predispositions that may contribute singly or in combination to pain episodes. An important subcategory is the recognition of patterns of normal physiologic occurrences. These may vary with the age of the child, including cramps representing a "call to stool" in children as young as two years, heightened awareness of normal movement of gastrointestinal contents in school aged children, and midmenstrual mittelschmerz in adolescent girls. A commonly associated but occult predisposition to abdominal pain is chronic stool retention. It may be overlooked because parents seldom are aware of bowel habits in their school aged children, children may be unwilling to discuss their fear of the school bathroom, and physicians may not find impacted rectal stool on physical examination (Dimson, 1972). On the basis of clinical studies, it appears that unrecognized lactose intolerance may present as recurrent abdominal pain, although the prevalence may vary with the ethnic composition of the patient population (Barr et al., 1979).

Against the background of normal and frequent stressful episodes of family life, the possibility of a constitutional sensitivity to stress in some children has been suggested (Apley, 1975). Pain episodes may not be tied to a specific stress stimulus, but rather may result from a nonspecific responsiveness. Such associations are often difficult to document but may be elicited by monitoring with a patient-completed diary.

STAGING OF EVALUATIONS

Further evaluations are generally staged according to the specific needs of the child. As a rule, the physical examination is done thoroughly during the first visit, but the rectal examination may be postponed. Findings of fullness and tenderness in the right upper and lower quadrants are important; tenderness in the left lower quadrant is not uncommon (Dimson, 1972). A negative rectal examination does not rule out stool retention, which may be limited to the colon; moreover, rectal stool may be soft in cases of long standing constipation.

Initial laboratory evaluations should include blood (complete blood count, reticulocyte count, smear, erthrocyte sedimentation rate, and blood urea nitrogen level), urine (urinalysis, culture), and stool samples (occult blood, ova, and parasites). Other noninvasive techniques that are often helpful include determination of the carmine marker transit time for evidence of stool retention and the lactose breath hydrogen test as an objective measure of lactase insufficiency (Barr et al., 1979; Dimson, 1972).

Second stage investigations should be selected on an individual basis, and generally are indicated only when some evidence other than the pain symptom itself suggests their usefulness. In the absence of other signs of organic disease, therapeutic trials directed at suspected but unproven mechanisms are often helpful, particularly with lactose intolerance and stool retention. Such trials may be difficult to carry out, and their effectiveness depends largely on the cooperation of the child. In principle, a good therapeutic trial includes a satisfactory baseline monitoring period, a treatment period, and a rechallenge. Ideally at least the child should not be aware of the treatment period he is in, although this is often not clinically feasible. The trial periods and outcome measures should be recorded at least daily. Therapeutic trials are susceptible to both type I errors (seeing an improvement that is not real) and type II errors (not demonstrating an improvement that is real) owing to measurement difficulties, other confounding influences, and lack of patient blinding. Nevertheless they are considerably more valuable than nondiscriminant use of laboratory investigations in most cases. In addition, the careful recording of symptoms often assists in bringing to light other previously unsuspected findings.

An unprepared single supine abdominal radiograph is recommended prior to performing the more invasive contrast studies (Barr et al., 1979). Ultrasound study is preferred for evaluation of gallbladder stones. Of the contrast studies, intravenous pyelography is most likely to be helpful, since half of the organic disease is in the genitourinary tract (Apley, 1975). The discriminating use of laparoscopy may be indicated in adolescent girls when endometriosis is suspected (Goldstein et al., 1979). There is still little evidence that exploratory laparotomy is useful in the majority of cases.

GUIDELINES FOR MANAGEMENT

The principles of management must be seen in the light of the natural history of the complaint and potential negative functional consequences for the child and his family. Available data suggest that at least one-third continue to experience abdominal pains for many years (Apley, 1975; Christensen and Mortensen, 1975; Stickler and Murphy, 1979). On the other hand, over 15 per cent of the children who were symptomatic for more than six months showed spontaneous resolution of their complaint within six weeks after their initial visit (see Table 28–5). The possibility that some children initially classified as having nonspecific dysfunctional abdominal pain may subsequently demonstrate pathologic disease processes must be considered (Christensen et al., 1975; Stickler and Murphy, 1979). Referral to appropriate specialists in cases of certain organic or psychogenic causes of abdominal pain syndromes can be undertaken in a primary care setting.

Traditional medical treatment by pharmacologic means (e.g., antispasmodics, anxiolytics, antidepressants) has not been demonstrated to be effective for the nonspecific dysfunctional recurrent abdominal pain syndrome. Therefore, techniques that are useful in monitoring the patient's course, focusing parental anxiety, and preventing the extended recurrent abdominal pain syndrome are indicated, and include the following:

1. The patient and parents are not told that he will grow out of it, or that organic disease has been ruled out. Rather, a "contract" is established, which includes regular scheduled visits to monitor for organic disease, including baseline laboratory tests.

2. At the initial visit the patient and parents are asked to monitor the frequency, duration, and timing of the pains in a diary. This is used to detect recurrent patterns, evaluate therapeutic trials, and seek associations with other events.

3. "Active" monitoring of the symptom is included by periodic physician initiated telephone calls between regularly scheduled visits.

4. The parent is asked to "monitor" for the possibility of organic disease by taking monthly weight measurements (and sometimes temperature measurements) during pain episodes.

5. The patient and parents are provided with a "decision tree" telling when to call earlier and what to do for specific pain episodes. (For example, "If the pain remains the same, we would like to see you in one month, but if it is accompanied by fever, vomiting, or blood in the stools, we want to see you earlier.")

6. For cases of the specific dysfunctional recurrent abdominal pain syndrome, environmental manipulation (e.g., permission to freely use a private school bathroom) may be appropriate.

7. If sources of interpersonal stress are discovered, the patient is counseled appropriately even if they are incidental to the simple recurrent abdominal pain syndrome itself. In true psychogenic pain the pain symptom may be well integrated into a broader syndrome of disordered intra- or interpersonal relationships. On the other hand, it may be quite peripheral to the underlying syndrome and may quickly disappear when attention is paid to the real focus of the patient's concerns. In such cases referral for specialized techniques of psychotherapy, play therapy, or behavioral management may be indicated.

In patients referred from primary care settings, such a "pediatric" approach can be expected to permit symptom resolution or specific therapy leading to symptom resolution in approximately 60 to 70 per cent of the cases (Table 28–5). Specific therapy for organic or psychogenic causes may be appropriate in 5 per cent each. In the persistent nonspecific recurrent abdominal pain syndrome, prevention of the extended recurrent abdominal pain syndrome is an appropriate and achievable therapeutic goal.

RONALD G. BARR

REFERENCES

Apley, J.: The Child with Abdominal Pains. London, Blackwell Scientific Publications Ltd., 1975.

Barr, R. G., and Feuerstein, M.: Recurrent abdominal pain syndrome: how appropriate are our usual clinical assumptions? In Firestone, P., and McGrath, P. (Editors): Pediatric and Adolescent Behavioural Medicine. New York, Springer-Verlag, 1983, Ch. 2.

Barr, R. G., Levine, M. D., and Watkins, J. W.: Recurrent abdominal pain of childhood due to lactose intolerance: a prospective study. N. Engl. J. Med., 300:1449, 1979.

Barr, R. G., et al.: Chronic and occult stool retention: a clinical tool for its evaluation in school-aged children. Clin. Pediatr., 18:674, 1979.

Christensen, M. F., and Mortensen, O.: Long-term prognosis in children with recurrent abdominal pain. Arch. Dis. Child., 50:110, 1975.

Dimson, S. B.: Transit time related to clinical findings in children with recurrent abdominal pain. Pediatrics, 47:666, 1972.

Feuerstein, M., Barr, R. G., Francoeur, T. E., Houle, M. M., and Rafman, S.: Potential biobehavioural mechanisms of recurrent abdominal pain in children. Pain, 13:287, 1982.

Goldstein, D. P., et al.: New insights into the old problem of chronic pelvic pain. J. Pediatr. Surg., 14:675, 1979.

Green, M.: Diagnosis and treatment: psychogenic recurrent abdominal pain. Pediatrics, 40:84, 1967.

Heinild, S., Malver, E., Roelsgaard, G., and Worning, B.: A psychosomatic approach to RAP in childhood with particular reference to the x-ray appearances of the stomach. Acta Pediatr. Scand., 48:361, 1959.

Holdstock, D. J., Misiewicz, J. J., and Waller, S. L.: Observations on the mechanism of abdominal pain. Gut, 10:19, 1969.

Jones, C. M.: Digestive Tract Pain: Diagnosis and Treatment. New York, The MacMillan Company, 1938.

Kopel, F. B., Kim, I. C., and Barbero, G. J.: Comparison of rectosigmoid motility in normal children, children with recurrent abdominal pain and children with ulcerative colitis. Pediatrics, 39:539, 1967.

Leek, B. F.: Abdominal and pelvic visceral receptors. Br. Med. Bull., 33:163, 1977.

Lipkin, M., and Sleisinger, M. H.: Studies of visceral pain: measurements of stimulus intensity and duration associated with the onset of pain in esophagus, ileum, and colon. J. Clin. Invest., 37:28, 1958.

Oster, J.: Recurrent abdominal pain, headache and limb pains in children and adolescents. Pediatrics, 50:429, 1972.

Parcel, G. S., Nader, P. R., and Meyer, M. P.: Adolescent health concerns, problems, and patterns of utilization in a triethnic urban population. Pediatrics, 60:157, 1977.

Pringle, M. L. K., Butler, N. R., and Davie, R.: 11,000 Seven Year Olds. London, Longman Group Ltd., 1966.

Roy, C. R., Silverman, A., and Cozzetto, F. J.: Psychophysiologic recurrent abdominal pain. In Roy, C. R., et al.: Pediatric Clinical Gastroenterology. Ed. 2. St. Louis, The C. V. Mosby Co., 1975, ch. 15.

Rubin, L. S., Barbero, G. J., and Sibniga, M. S.: Pupillary reactivity in children with recurrent abdominal pain. Psychosom. Med., 29:111, 1967.

Stickler, G. B., and Murphy, D. B.: Recurrent abdominal pain. Am. J. Dis. Child., 133:486, 1979.

Stone, R. J., and Barbero, G. J.: Recurrent abdominal pain in childhood. Pediatrics, 45:732, 1970.

Swarbick, E. J., Bat, L., Hegarty, J. E., Williams, C. B., and Dawson, A. M.: Site of pain from the irritable bowel. Lancet, 2:443, 1980.

28E Headache

Chronic recurrent headaches may have either an organic or a psychogenic etiology. The diagnostic possibilites are listed in Table 28–6. In some children, headache can be diagnosed and treated successfully without much difficulty; in others the diagnostic and therapeutic problems remain even after thorough exploration.

MUSCLE CONTRACTION HEADACHE

Muscle contraction, presumably secondary to tension, is the most common cause of recurrent or persistent headaches in children. Such headaches are usually generalized. In some instances the discomfort begins in the muscles of the neck, shoulders, or occiput and migrates anteriorly to the frontal region. The symptom may be described as a "tight band around the head," as "pressure from the outside," or as a persistent mild or severe aching. Although there are no prodromal symptoms, nausea, vomiting, dizziness, or anxiety may occur concurrently.

MIGRAINE HEADACHE

Migraine headaches are not uncommon in the pediatric age group; the estimated incidence is in the range of 2 to 4.5 per cent. Although children over seven years of age may have migraine headaches similar in character to those in adults, more frequently younger children present with cyclic vomiting and with generalized headaches of lesser severity. Classic migraine hemicranial headaches occur periodically in the retro-orbital, frontal, or temporal region, usually but not always on the same side. Prodromal or concurrent symptoms include photophobia, nausea, vomiting, abdominal pain, pallor, sweating, facial flushing, eyelid edema, and changes in mood. Scintillating scotoma, hemianopsia, zigzag lines, blurred vision, or paresthesias are less frequent in children than in adults. Initially unilateral and throbbing or pulsating, the headache usually assumes a generalized and constant character. Episodes generally last two to three hours but may persist for 48 hours or longer. Children commonly sleep during or after the attack. Although migraine headaches are usually infrequent, some children have several each week. Migraine and muscle contraction headaches may occur simultaneously. Eighty-five per cent of the parents of children with migraine have a history of headaches, 70 per cent of them classic migraine.

Complicated migraine is accompanied by neurologic deficits. Ophthalmoplegic migraine, usually involving the oculomotor and less frequently the abducens nerve, is rare in early childhood. Twelve to 24 hours after the onset of the headache, eye pain, nausea, vomiting, sudden (usually unilateral) ptosis, dilation of the pupil, and reduced mobility of the eye develop. These findings may persist for days or weeks. Hemiplegic migraine, at times familial, characterized by neurologic deficits, such as aphasia, paresthesias, hemiparesis, or hemisensory loss, may persist for hours to days. Alternate sides may be involved during separate episodes. Headache usually follows on the contralateral side.

Basilar artery migraine, which may occur from late infancy through adolescence, has a sudden onset with transient blindness, vertigo, dysarthria, tinnitus, transient bilateral blindness, blurred vision, oculomotor abnormalities, ataxia, loss of consciousness, paresthesias around the mouth and the distal extremities, and cranial nerve deficits. Severe headache and vomiting follow. Although some episodes last longer, most are less than three to four hours in duration. The headache may be bifrontal, bitemporal, or occipital. Acute confusional states persisting 10 minutes to 24 hours may be the initial or a later manifestation of a migraine episode in older children and adolescents; accompanying symptoms may include agitation, appre-

Table 28–6. CAUSES OF HEADACHE

I. Muscle contraction headache
II. Vascular (migraine) headaches
   A. Classic migraine
   B. Common migraine
   C. Complicated migraine
     1. Hemiplegic or hemisensory migraine
     2. Ophthalmoplegic migraine
     3. Acute confusional states
     4. Basilar migraine
     5. Alice in Wonderland syndrome
   D. Cluster headache
III. Combined headache—muscle contraction and vascular
IV. Traction headache
   A. Intracranial tumors, hematoma
   B. Brain abscess
   C. Intracranial malformation
   D. Arteriovenous malformation
   E. Pseudotumor cerebri
   F. Central nervous system leukemia
V. Inflammatory disease of the central nervous system
VI. Post-traumatic headache
VII. Ocular disorders
VIII. Sinusitis
IX. Hypertension
X. Psychogenic
   A. Conversion reaction
   B. Depression
   C. Hypochondriasis

hensiveness, and combativeness. The Alice in Wonderland syndrome, a rare presentation of migraine, is characterized by olfactory, gustatory, and auditory hallucinations, micropsia, metamorphosia, and other distortions of time and space.

Cluster headaches, characterized by bursts of one to three headaches a day over a period of weeks, are uncommon in children. In addition to severe headaches, the patient may develop unilateral conjunctional injection, rhinorrhea, and lacrimation.

Traction headaches result from traction on intracranial contents, especially vascular structures. The causes of traction headaches include intracranial tumor, hematoma, brain abscess, arteriovenous malformation, pseudotumor cerebri, and central nervous system leukemia. Headache is less commonly a symptom of intracranial tumors in children than in adults, in part because of the early separation of sutures and spontaneous decompression that occur in the young patient. It is more frequent with cerebellar tumors and less common, except for the rare meningioma, with supratentorial neoplasms. Although headaches due to a brain tumor usually occur in the morning shortly after arising and disappear after a brief time, they may be present at other times or may be precipitated by sudden exacerbations in intracranial pressure caused by coughing, sneezing, straining with a bowel movement, or a change in position. Although it is hazardous to generalize, traction headaches are usually described as dull, deep, and intermittent, though occasionally steady. That a headache may be temporarily relieved by aspirin does not rule out a traction etiology. Temporary lessening of the headache may also be due to separation of the sutures. Severe headaches that always occur in the same area, that are sudden, that fail to respond to medication, or that are accompanied by a change in personality are worrisome.

Rupture of an intracranial aneurysm may be accompanied by excruciating headache, photophobia, meningeal signs, and impaired consciousness. Arteriovenous malformations may produce a sudden headache, obtundation, focal neurologic defect, signs of meningeal irritation, and increased intracranial pressure. In addition to severe headache, patients with pseudotumor cerebri may also have blurring of vision, diplopia, nausea, vomiting, and bilateral papilledema. Results of the neurologic examination and computed axial tomography are normal. Symptoms may persist for weeks. Central nervous system leukemia causes headaches, irritability, diplopia, polyphagia, and rapid weight gain.

Post-traumatic headaches occurring in children after a severe head injury, especially one leading to unconsciousness, may persist for months or years. The patient may also complain of dizziness, nervousness, heightened sensitivity to noise, withdrawal, inability to concentrate, hyperactivity, sleep disturbance, and difficulty in controlling anger. Although the child may have experienced an unusually severe emotional reaction to the injury, the physician will want to explore the premorbid emotional adjustment and current environmental stresses. In some children a diagnosis of post-traumatic neurosis may be more accurate. Unsettled litigation related to the accident seems to contribute to persistence of the symptoms in some cases. Since this information usually is not volunteered by the parents, the physician should discreetly ask whether the accident claims have been settled. The differential diagnosis may include chronic subdural hematoma.

Headache may be caused by eyestrain secondary to uncorrected refractive errors, astigmatism, muscle imbalance, or impaired convergence; other symptoms may include burning, tearing, conjunctival hyperemia, blurring, and dizziness. Photophobia and conjunctival injection may accompany severe headache from any cause and do not necessarily indicate the presence of eye disease. Sinusitis or inflammation and engorgement of the nasal turbinates are not important causes of chronic headache in children.

Chronic hypertension is an unusual cause of chronic headache in children. Acute hypertensive encephalopathy like that associated with acute glomerulonephritis is manifested by severe headache, nausea, vomiting, confusion, seizures, and transient focal neurologic findings.

PSYCHOGENIC HEADACHE

Identification with a relative or other important person and conversion reaction may be the mechanisms in headaches of psychogenic etiology. Constant dull headaches, lasting for days, are almost always psychogenic and most frequently are due to depression.

THE DIAGNOSTIC APPROACH

THE INTERVIEW

The pediatric interview is the most helpful tool in understanding the cause of headaches. Among other details, the interviewer should review exactly what the child was thinking about and what was going on in the environment immediately before the episode. Monday morning headaches obviously suggest tension or school avoidance as etiologically important. Interpersonal interactions are of interest. Both migraine and muscle contraction headaches may be precipitated by stress, e.g., arguments between the child and someone close to him, or the experience or anticipation of disappointment, rejection, intense excitement, anxiety, or fear of failure as in school examinations, auditions, recitals, or contests. Patients with migraine headaches report numerous precipitating factors, including excessive noise, confusion, illness, diet, alcohol consumption, menstruation, orally administered contraceptives, extreme exertion, emotional stress, fasting, loss of sleep, prolonged sleep, and lengthy viewing of television or movies.

Children with either muscle contraction or migraine headaches, and at times their families, may have difficulty with feelings of aggression, anger, or resentment, especially toward parents, siblings, teachers, and close friends. This should be evaluated in the interview, recognizing that the patient or the parents, unaware of a relationship between headaches and feelings, are unlikely to volunteer this association. Since such feelings are often unconscious, the child may deny their presence. Although muscle contraction or migraine headaches may be precipitated by events that immediately evoke anger, the child who is chronically inhibited from its direct expression by strong parental disapproval, personal discomfort, or guilt may carry a burden of intense resentment. With headaches thought to be psychogenic, it is a good practice to ask the child what kinds of things make him mad, the way he deals with his anger, and how he would like to manage it. Similar information should be obtained from the parents. The presence of such affective feelings as anxiety, shame, and depression in the child should also be explored. If the child is thought to be depressed, a useful comment fronting for a question is, "You know, John, I have the feeling that things aren't going too well for you. . . ." Then, when the child looks up at the doctor, "Tell me about it. . . ."

Although personality "profiles" have no validity as being etiologically predisposing to specific symptoms or diseases, children with migraine or muscle contraction headaches often seem to be very sensitive, overly concerned with the approval of others, shy, polite, and fearful of making errors. Meticulous, serious, overconscientious, and prone to worry are other common adjectives applicable to these children.

THE PHYSICAL EXAMINATION

In addition to the usual general and neurologic assessment, the physical examination should include careful fundoscopy, examination of the visual fields, and screening for visual acuity, ability to converge, and phoria.

PROCEDURES

Since it is unusual for headaches to be the only manifestation of a seizure disorder, an electroencephalogram is but rarely needed. Computed tomography is indicated if an intracranial structural mass lesion is suspected. Skull films are rarely helpful.

TREATMENT

COUNSELING

Children and their parents are often tremendously relieved to learn that their concerns about the possibility of a brain tumor or "increased pressure" are needless. The reassurance provided by a meticulous physical examination and the physician's clarification of the etiology of the headache may in itself lead to a diminution of the complaint. Prevention and treatment of migraine and muscle contraction headaches are facilitated by attention to the child's daily schedule. In part this may include the suggestion that the child consider a less strenuous schedule, reducing the number of his extracurricular activities; avoiding excessive fatigue, excitement, or television viewing; and adopting a regular schedule for sleep and meals. The physician's periodic support may be required to reinforce such changes. Although the child may not elect to reduce his schedule, his understanding of the relation between such stress and headaches helps him accept this discomfort as an unpleasant side effect of being a perceptive, intense, and highly involved person.

Pediatric interviews are psychotherapeutic when they permit the child to talk about what he has been doing and what he has been feeling; when they help him become aware of relationships between such feelings as anger or hostility and headaches; when he comes to understand that aggressive feelings and anger are normal; and when he is encouraged to express them healthily in a socially acceptable manner without guilt. The treatment of depression in children and adolescents is discussed in Chapter 41.

DRUG THERAPY

Drug therapy for headache is difficult to evaluate because the symptom is subjective, emotional factors have a prominent contributory role, and individual responses to pain, discomfort, and medication vary.

Aspirin and acetaminophen provide highly effective symptomatic headache relief. The dosage of both aspirin and acetaminophen is 5 to 8 mg.

per kg. per dose, every four hours as necessary. Propoxyphene hydrochloride is sometimes used as a substitute for aspirin. Codeine, although highly effective in the treatment of severe headaches, is rarely indicated in children. Phenobarbital or other mild sedatives may be used with aspirin for a limited time as an adjunct in the management of severe muscle contraction headaches. Phenytoin or phenobarbital has been used in anticonvulsant doses for the prophylaxis of migraine headaches. Serum anticonvulsant drug levels should be monitored.

Although their effectiveness has not been proved in controlled studies, tranquilizing agents such as meprobamate, chlorpromazine, diazepam, and chlordiazepoxide are occasionally prescribed for symptomatic relief of muscle contraction headaches. Amitriptyline may be useful in adolescents with migraine or combined headaches, especially in the depressed teenager. In adolescents, 50 mg. at bedtime is generally the starting dose; this may be increased gradually over several weeks to a total of 200 mg. a day in divided doses.

If aspirin or a similar analgesic alone or with a sedative does not relieve migraine headaches in the adolescent with classic migraine, the combination of ergotamine tartrate and caffeine may be tried. The total dosage required for an individual patient can be determined after a few episodes. Initially the child should take one tablet of a preparation such as Cafergot, preferably during the prodromal phase or at the first sign of a headache. Few children, however, have an aura. If symptomatic relief is not obtained, the dose may be repeated every 30 minutes for four doses. Once determined, the total dosage can be taken initially. If nausea and vomiting are prominent prodromal symptoms, a rectal suppository may be used. Ergotamine should not be used in the treatment of complicated migraine. In uncontrolled studies, cyproheptadine hydrochloride (Periactin) has been thought to reduce the number and severity of migraine episodes in some children. Drug prophylaxis for migraine should be periodically discontinued to determine whether it is still indicated.

Propranolol, a highly useful drug for the prophylaxis of migraine headaches whether of the common, classic, or complex form, may be used for all age groups and is the drug of choice for the older adolescent. In children under 12 years of age the dose is 5 to 10 mg. three times daily; over that age, the dose may be increased to 10 to 20 mg. three times daily. If needed, a higher dose may be used but not in excess of 30 to 40 mg. four times daily. The drug is contraindicated in patients with asthma.

Biofeedback training and relaxation therapy are techniques that recently have been studied for the treatment of chronic muscle contraction and migraine headaches. Most of the experience has been with adults, and controlled studies in children have not been reported. Since biofeedback and relaxation therapy appear to be equally efficacious, the latter will probably have greater usefulness in the management of headaches in adolescents.

MORRIS GREEN

REFERENCES

Brady, J. P.: Behavioral medicine: scope and promise of an emerging field. Biol. Psychiatry, 16:319, 1981.

Brown, J. K.: Migraine and migraine equivalents in children. Dev. Med. Child Neurol., 19:683, 1977.

Dalessio, D. J.: Wolff's Headache and Other Head Pain. Ed. 3. Oxford, Oxford University Press, 1972.

Dillon, H., and Leopold, R. L.: Children and the postconcussion syndrome. J.A.M.A., 175:86, 1961.

Ehyai, A., and Fenichel, G. M.: The natural history of acute confusional migraine. Arch. Neurol., 35:368, 1978.

Emery, E. S.: Acute confusional state in children with migraine. Pediatrics, 60:110, 1977.

Gascon, G., and Barlow, C.: Juvenile migraine, presenting as an acute confusional state. Pediatrics, 45:628, 1970.

Golden, G. S.: The Alice in Wonderland syndrome in juvenile migraine. Pediatrics, 63:517, 1979.

Lapkin, M. L., and Golden, G. S.: Basilar artery migraine. Am. J. Dis. Child., 132:278, 1978.

Ling, W., Oftedac, G., and Weinberg, W.: Depressive illness in childhood presenting as severe headache. Am. J. Dis. Child., 120:122, 1970.

Prensky, A. L.: Migraine and migrainous variants in pediatric patients. Pediatr. Clin. N. Am., 23:461, 1976.

Prensky, A. L., and Sommer, D.: Diagnosis and treatment of migraine in children. Neurology, 29:506, 1979.

Rune, V.: Acute head injuries in children. Acta Paediatr. Scand. (Suppl.), 209:122, 1970.

Thompson, J. A.: Diagnosis and treatment of headache in the pediatric patient. Curr. Probl. Pediatr., 10:5, 1980.

Van Pelt, W., and Andermann, F.: On early onset of ophthalmoplegic migraine. Am. J. Dis. Child., 107:628, 1964.

Verret, S., and Steele, J. C.: Alternating hemiplegia in childhood: a report of eight patients with complicated migraine beginning in infancy. Pediatrics, 47:675, 1971.

Weisberg, L. A., and Chutorian, A. M.: Pseudotumor cerebri of childhood. Am. J. Dis. Child., 131:1243, 1977.

29

Variations in Feeding and Eating Behavior

_____ 29A   Common Issues in Feeding

NORMAL PARENTAL CONCERNS ABOUT FEEDING

Some of the normal parental concerns about feeding can be related to the quantity and quality, or to the pattern and regularity of feeding.

QUANTITY

A nursing mother is unable to register how many ounces or milliliters an infant ingests with each feeding, and her baby is the one who terminates the feeding. By contrast, the caretaker of a bottle fed infant is not so fortunate, being always aware of the quantity of formula ingested. Nevertheless, ideally the response of the infant, his satisfaction or satiety should provide the answer to the question, "Is he getting enough?" and not a calculation or an arbitrarily selected quantity of formula. Variation and changes in hunger and satiety should be accepted as a fact of life and not as a mistake to be corrected. The health care practitioner's responsibility is the assessment of the adequacy of nutrition and growth, of the balance between energy consumption and expenditure, and of the resulting body composition. Adequate growth velocity occurs with adequate nutrition. The use of growth percentile charts (length or height, weight, and head circumference), or of growth velocity charts, is helpful for this assessment.

On the other hand, parental concern frequently centers around overfeeding. Overfeeding, as suggested by dietary inquiry and confirmed by skinfolds in excess of the 90th percentile, should and can be discouraged, for instance, by suggesting that feeding is only one of many possible interactions between parent and child, only one of many modes of sensory input and contact. Parents should be encouraged to search for and identify their own and their infant's moods and cues and to distinguish between hunger, boredom, sadness, anger, and fatigue. (See section on obesity later in this chapter.)

The very young infant may arouse parental concern by developing "infantile colic" or excessive crying, which to some degree should be seen as a variation of normal functioning (see Chapter 28). In this situation parents and other caretakers tend to increase feedings and change associated behavior such as burping and picking up. Although reassurance and help from an experienced objective outsider can be very constructive in regard to feeding style and quantities, these guidelines are not a substitute for quiet, direct, and complete assessment of the infant and parent as seen against a background of the family situation. Changes in formulas or times or quantities of feeding should be made only for specific reasons and specific symptoms, such as documented lactose intolerance or a striking family history of cow's milk protein allergy. Changes in formulas should be maintained only if the disappearance of specific symptoms substantiates the diagnosis for which the change of formula was prescribed.

An excellent reason for decreasing the number of feedings is the persistence of the "demand" for night-time feeding by small infants. This practice may be necessary or helpful for a few weeks, but thereafter both infant and parent will benefit from discontinuation of this pattern, and from the introduction of the realistic fact that night-time is sleeping time. Separate sleeping quarters and closed doors can facilitate such maturation.

QUALITY AND TYPE OF FEEDINGS

The time of introduction of foods other than infant formula or breast milk is frequently determined by cultural and familial factors, which have nothing to do with nutrition or with the physiologic development of the infant. Some infants over-ride any imposed or suggested change in feeding practices, others adjust to changes, and some remain uncomfortable or unsatiated no matter what anybody tries. Non-nutritional aspects of eating and feeding behavior should be recognized early and put in perspective.

The mother who brings her infant with a concern such as "He won't eat vegetables for me" provides a most helpful clue and might be introduced to the thought that growing up, it is hoped, leads to independence with some unavoidable struggles for autonomy. Perhaps she will omit the "for me" the next time, as she begins to accept the infant's varying tastes and nutritional needs. She most likely has already noticed that her child has a mind and a digestive system of his own.

The mother might need help to realize that up to about four to five months of age the baby's tongue is used to push solid food out as well as back into the pharynx, and that his ability to get hold of food with his hands coincides approximately with the beginning of chewing skills at roughly six months of age. "Wet burps" should be recognized as part of "physiologic" chalasia resulting in regurgitation of small amounts of formula for two to three months at least and are usually not related to the type of feeding. Air swallowing, caused by a contraction of the pharyngeal muscles when the epiglottis has closed off the larynx, is also physiologic, but when crying or fast swallowing is present, it may cause considerable distress. It will be helpful to realize that emptying of the stomach takes between one and two hours on the average. This varies with the degree of hunger, other factors such as position, amount, and type of feeding, as well as, of course, the health of the infant. Formulas that contain more fat or are more hypertonic delay gastric emptying. Those with a high carbohydrate content or predigested protein shorten gastric emptying time.

Parental concern regarding intake of calcium, vitamins, fluoride, and other specific substances invites a thorough inquiry into the types of feeding used and the pertinent information available to or utilized by the parent. Frequently such concern represents thinking of an other than orthodox nature. The practitioner should be able to recommend or make available useful pertinent information and preferably evaluate later on the net result of his recommendation. Knowledge of the family pattern of feeding styles makes it possible to nurture sensible qualitative dietary habits.

REGULARITY OF FEEDING

Parental ability to recognize the infant's or child's hunger or desire for food varies widely and constitutes a fruitful area of clinical inquiry, pref-

erably before parental concerns in this area emerge. When it is apparent that the infant's signs are being over-read or misread, specific advice regarding frequency can be helpful. However, a better understanding of what occurs between the mother and her infant, in the rest of the family, or within the mother herself makes it possible to give additional suggestions not necessarily directed to quantity or quality or frequency of food or feeding, but to more comfortable, synchronous interaction. As far as the mother is concerned, is she learning to see things from the baby's point of view, or is she inclined to coax or force? Has the baby learned modes of communication other than hard, hungry crying? Patterns of maternal-child interaction are established during the early weeks, when the majority of interaction is related to feeding. Regularity is different from rigid control, and gratification of the baby as well as regulation of his activity-quiescence rhythm are important objectives. The mother and infant ideally might become equally active participants in the feeding interaction in terms of timing, pacing, and development of frustration tolerance. The baby and mother, it is to be hoped, will learn to influence each other's behavior positively, promoting a sense of competence in both. When feeding time ceases to be pleasurable, such a change should be recognized and not readily accepted as a fact of life.

MINOR VARIATIONS IN PARENTAL FEEDING BEHAVIOR

Maternal style and attitude, the ability to recognize the infant's signals, the time available, physical reserve and comfort are important contributors to her feeding behavior. The infant's temperament and physiologic variables such as the intensity of the gastrocolic reflex, or ease of burping and spitting up, constitute additional factors to be reckoned with when they emerge. Most often such physiologic variables are taken in stride.

PROBLEMS FROM NEONATAL COMPLICATIONS

An unexpected early rupture of the membranes and subsequent birth by cesarean section of an infant weighing 1800 grams are unpleasant and disturbing events. When, in addition, the parents have done serious planning for natural childbirth, rooming in, and breast feeding, it cannot come as a surprise that parental reserves are strained. It is the health care practitioner's task in such settings to avoid any further medical diagnostic or therapeutic intervention, unless such is unequivocally indicated. Parental concern is great for obvious reasons, most of which lie in the past, and not in the infant's wet burping or other minor feeding problems. Parents need help in adapting their feeding behavior to the altered circumstances (see also Chapter 5).

OVERFEEDING

Excesses in the frequency and quantity of feeding can lead to an exaggerated gastrocolic reflex with an increased frequency of stooling or osmotic diarrhea. This can occur when mild infectious diarrhea is treated with "clear liquids", which can result in excessive ingestion of sucrose. Overfeeding of an infant whose gastrointestinal tract is hard at work digesting his physiologically large milk intake can lead to overtaxing of his digestive and absorptive capacities. Vomiting and excessive air ingestion or inability to burp should lead parents to reconsider their thought that the infant needed more food or was hungry. Quite likely other needs were being voiced or otherwise signaled.

INCONSISTENCY DUE TO CONFLICTING ADVICE

When parental feeding behavior varies as a result of a multiplicity of suggestions and advice from others, emphasis on the individuality of the infant with support for maternal needs and self-sufficiency should accompany the health care professional's pleas for consistency in maternal handling of the infant.

PROLONGED BOTTLE FEEDING

The time of discontinuation of bottle feeding can become an issue in some families. Except for dental caries, which can result from continuous bathing of teeth in sucrose solutions, serious consequences are hardly ever caused by such prolonged use. However, the toddler who walks around carrying a bottle of milk in his hand or who goes to sleep with the bottle in his crib is likely also to have a less than adequate quantity of hemoglobin; his dietary habits and the hemoglobin level should be corrected. Other aspects of maturation and socialization will provide clues to the causes of such practices.

FOOD FADS

Sometimes parents seem to need to impose their own fads, or deviations from the norm

of eating behavior and practices, on their infant or child. There is no doubt that some atypical diets can result in excellent growth of the human infant, but specific deficiencies have been reported, resulting from the deliberate elimination of some foods for philosophic reasons.

Overconcern with the avoidance of "food additives" sometimes leads to suboptimal fluoride intake with lifelong consequences. Parental overconcern with leanness in the child can represent a subtle form of "anorexia nervosa by proxy." Exclusion of all animal products and eggs results in considerable risk of vitamin B_{12} deficiency. If, as in the diet of the Vegan cult, there is ample folic acid available, the symptoms of B_{12} deficiency are masked and spinal cord degeneration can set in. Exclusion of meats and eggs with consumption of dairy products leads to little else than iron deficiency. When dairy products are excluded as well as meat and eggs, however, the net result is a lowered intake of protein, riboflavin, and calcium. Zinc deficiency can also occur with such diets when the intake of cereal is high. The Zen macrobiotic diet can induce scurvy, hypoproteinemia, and general malnutrition. Pregnant and lactating women are quite vulnerable to these deficits, and low birth weights and poor growth in infants are well documented. A high intake of sodium and the drinking of very little fluid have obvious undesirable consequences. Occasionally children with protein calorie malnutrition identical to those seen in developing countries are reported. (See also Chapter 23 in regard to general nutrition and Chapter 9 for a discussion of adolescent fad diets.)

One can learn a great deal by taking a complete inventory of thoughts and practices related to the specific feeding philosophy encountered and by applying one's own skills to evaluate the child's growth, development, and nutrition. This is followed by sharing the evaluation with the caretakers involved.

Strong support of the "return to nature" approach, which includes breast feeding, is given in the realization of the possible risks of iron and vitamin D deficiency unless there is timely supplementation. Good maternal nutrition in terms of weight and vitamin B_{12} and ascorbic acid intake should be stressed. Soy formula supplementation is a frequently acceptable and suitable alternative to a wet nurse for parents and pediatricians alike when breast feeding is insufficient in this setting and solids or cows' milk formula is unacceptable. Sunlight or sun lamps can prevent rickets.

The use of lactase enzyme (LactAid) can be recommended for milk drinking vegetarians who are lactase deficient and therefore otherwise intolerant of milk.

If the possible benefits of avoidance of animal protein and fat ingestion can be shared between the health care practitioner and the avid food faddist, an atmosphere of cooperation may prevent antagonism and extremism.

Although the health care practitioner has to accommodate and tolerate a wide variety of nutritional beliefs of parents, his responsibility for the infant's health obviously over-rides all other obligations. At times it is exceedingly difficult to differentiate between constructive but unorthodox parental behavior and subtle forms of child abuse. Any form of neglect of the child's needs should be recognized as such. Imposed fasting, fluoride avoidance, and nonimmunization are comparable, but only nonimmunization is fairly effectively dealt with by Public Health authorities, be it very late. In rare instances, when all attempts at understanding and education, supplementation, and correction fail, it is the health care practitioner's responsibility to initiate legal action.

FOOD PHOBIAS

All children develop likes and dislikes for specific foods. Various distastes of parents in one form or another can be passed on to their children, but children also have their own opinions. Tolerance by parents of these variations is symbolic of the degree of parental control imposed and their acceptance of the child's need for independence. Thus, when parents express concern about food dislikes, it is less likely that food is the problem than an attempt to control the child.

The consistency and the "feel" of food and perhaps an exaggerated gag reflex are problems frequently seen. Minor dislikes are readily circumvented when a modicum of flexibility prevails.

Extreme persistent inability to tolerate solid food and the refusal to chew food can lead to major problems with dentition and tooth decay, unless effective psychotherapy can influence this symptom complex favorably. In two extreme cases in the author's experience with adolescents with a record of never having chewed any food, the family psychopathology was of major concern and therapy had been consistently refused. Modern treatment of major isolated phobias uses desensi-

tization and is often quite successful (see Chapter 55); for simple phobias, such treatment is hardly ever necessary.

PICA

The term pica, derived from the Latin name of the brown billed magpie, *Pica pica*, refers to a stubborn pursuit and ingestion of nonedible matters such as clay, plaster, and paint. Such dietary habits have been described since antiquity. During pregnancy and lactation they persist in many primitive societies. In the pediatric population it occurs mainly during the second year when the toddler starts getting around. Usually it does not last beyond age four or five years, and it may be as simple and innocent as the eating of sand on the beach.

When the ingested material contains lead or other toxic material (see Chapter 24), the consequences can be detrimental to development or behavior or even lethal. It would appear that less than optimal supervision, absence of better things to do, and inadequate stimulation and somatosensory input, which all occur far more frequently in social and family settings of marginal adequacy, are prerequisites for the development of problematic pica. Exploration of the housing, family, and social situation is mandatory. As with many other behavioral habits, it will persist longer and be more pronounced when the developmental process is slow or slowed down by lack of stimulation or decreased developmental potential.

MAARTEN S. SIBINGA

REFERENCES

Erhard, D.: The new vegetarians. Nutr. Today, Jan.-Feb., 1974, pp. 20–27.

Forbes, G. B.: Food fads: safe feeding of children. Pediatrics in Review, *1*:207–211, 1980.

Hegsted, D. M.: Nutrition misinformation and food faddism. Nutr. Rev., *32* (Suppl. 1):1, 1974.

Lourie, R. S., Layman, E. M., and Millican, F. K.: Why children eat things that are not food. Children, *10*:143, 1963.

29B Obesity in Childhood

The problem of obesity in infancy and childhood is multifactorial in nature and full of complexity. There are many issues to be considered with regard to etiology, pathophysiology, prevention, management, and long term implications. Cultural perspectives, maternal-child interaction, nutrition, and genetics are all interacting forces.

PREVALENCE

The use of numerous and variable criteria for the definition of obesity makes it difficult to obtain a firm grasp on its prevalence in childhood. It appears that 5 to 15 per cent of infants and preschoolers and 10 to 35 per cent of adolescents are obese. In some native American groups it may exceed 50 per cent. The prevalence is greater in females after infancy. In urban populations, particularly in inner city and disadvantaged groups, there is a steep rise in obesity in female adolescents. It is of interest that over 35 per cent of adolescent girls perceive themselves as being overweight or obese (Lloyd and Wolff, 1976). (See also Chapter 9.)

SIGNIFICANCE OF THE PROBLEM

CONSEQUENCES FOR CHILD HEALTH

It has long been assumed that childhood obesity presents no particular hazards to a child's health. This assumption has led many physicians to suggest that obese children should be left alone and that they will outgrow their baby fat. However, the psychosocial problems may be overwhelming, and obesity in childhood increases the likelihood of obesity later in life.

Mild to moderate obesity probably causes minimal morbidity. However, in infancy, obesity or rapid weight gain may be a risk factor for lower respiratory tract infection. Clinical observations by pediatricians in England revealed a higher rate of lower respiratory tract infections in infants who were 120 per cent of the weight for age standard (Hutchinson-Smith, 1971; Tracey and Harper, 1971). The explanation for this phenomenon may be mechanical, with decreased ventilatory function, or the fact that these infants are largely formula fed and are deprived of the anti-infective features of breast milk. There may be immune dysfunction; recent work has shown that diminished cellular immunity does occur in obese subjects (Chandra, 1981). Finally, obese infants may be clumsy and may be characterized by "apparent" slow motor development.

Although uncommon, the most serious and most life threatening complication of severe obesity is carbon dioxide narcosis with decreased ventilatory capacity and microatelectasis, known as the Pickwickian syndrome (Riley et al., 1976). Also lesser degrees of obesity compounded by upper airway obstruction by hypertrophied tonsils and adenoids may give rise to anoxia with poor oxygen saturation, called the "chubby puffer syndrome" (Stool et al., 1977). Because of dyspnea, even the mildest of exertion in these obese children limits their already meager activity level even further. Chronic airway obstruction with hypertrophied tonsils and adenoids may cause sleep alteration, cardiac signs, and abnormal arterial gas levels. Removal of tonsils and adenoids in this situation may be helpful.

Obese children suffer from an increase in orthopedic disorders, such as Legg-Perthes disease, genu valgum, and slipped femoral capital epiphyses. Clumsiness, shortness of breath on exertion, skin irritations due to friction, and heat discomfort are frequent complaints.

The obese adolescent may also have other nutritional problems due to erratic and unsupervised attempts at dieting. A low serum iron level and anemia are not uncommon. Menstrual irregularities may occur in obese female adolescents (Sherman and Korenman, 1974).

IMPLICATIONS OF CHILDHOOD OBESITY FOR ADULT HEALTH

Two issues are relevant and of utmost importance in the consideration of childhood obesity in relation to adult health. One important issue, to be expanded under the section on "Natural History" (p. 540), is concerned with the contribution of childhood obesity to adult obesity and the degree to which it can be prevented. The other important issue is the role of obesity as a mediator of hypertension. Childhood obesity forms a substratum for the appearance of hypertension in late childhood or adolescence.

Hypertension and hyperlipidemia are serious hazards of obesity, particularly in adolescence. These forms of morbidity link childhood obesity to some of the most serious disorders of adulthood—coronary heart disease, diabetes, and malignant hypertension. The risks of these conditions are greatest when weight gain is excessive in early adolescence (Abraham et al., 1971). Significant correlations of subcutaneous and subscapular fat folds and blood pressure were noted in more than 50 per cent of the subjects in one study, with systolic pressures over 135 mm. Hg and diastolic pressures over 90 mm. Hg (Court and Dunlop, 1974). Continuing weight gain and hypertension into adulthood constitute a particularly high risk combination.

In terms of obesity and hyperlipidemia, correlations between body weight and cholesterol and triglyceride levels are weak, as found in the Muscatine studies. However, when children with weight for height ≥ 90th percentile were examined, 17 per cent had cholesterol and 24 per cent had triglyceride levels elevated above the 90th percentile values (Lauer et al., 1975). Weight reduction consistently reduces the elevated tryglyceride levels; less so the cholesterol levels (Lauer et al., 1975).

PSYCHOSOCIAL PROBLEMS

Perhaps most devastating to a sense of well-being and self-worth are psychosocial problems facing obese children. It has been well documented that these children have a poor self-image and express feelings of inferiority and rejection. They encounter peer teasing and ridicule, are left out of games, activities, and team sports, and thus become increasingly inactive. In response they withdraw and sometimes indulge in antisocial behavior. School performance may decrease as they lose self-confidence. A vicious cycle is set into motion whereby the increasing frustration and isolation result in even greater withdrawal. As a substitute for acceptance and active involvement, eating is resorted to for gratification and solace (Bruch, 1973).

In adolescents, as in younger children, there may be a defective body image, low self-esteem, social isolation, and feelings of rejection and depression. Impaired social functioning and extreme self-consciousness have also been noted. A self-perpetuating spiral of continued feelings of rejection, isolation, depression, and boredom and inactivity results in increased eating, decreased activity, and maintenance of the obesity. Feelings of persecution also may be expressed. There is reality to this observation in the lower college admission rates and employment rates. Adolescence is normally a time of self-awareness with heightened concern with physical body changes, self-image, heterosexual relationships, and independence. These important developmental steps may be seriously distorted, delayed, or damaged in adolescent obese individuals (Bruch, 1973; Monello and Mayer, 1963; Stunkard and Mendelson, 1967).

Occasionally an overdependency syndrome, "the last bird in the nest," is seen. The child or the adolescent may be the youngest and show excessive emotional interdependence with the parents. These children may be immature and socially isolated from peers and may avoid school related activities. In short, they are overnurtured, overfed, and overprotected by the mother or father. The world is considered a dangerous place, and this message is reinforced by the overprotective family (Meyer and Neumann, 1977).

Day to day living is a constant source of frustration for the obese child and his family. The annoyance, frustration, time, and expense on the part of the child and parent, e.g., in shopping for clothing, are considerable. The teenager cannot adopt the latest fashions of his peer group.

An extreme example of the effects of severe obesity is the unfortunate obese child who cannot fit into a standard size classroom desk chair, sit at the cafeteria tables, or even use the school restroom facilities, opting to do without food and beverage at school and then gorging himself upon reaching home.

In summary, not only is self-esteem damaged, but important developmental steps on the road to adolescence and adulthood are blocked for the obese child.

DEFINITION AND MEASUREMENT OF OBESITY

Any valid and unified definition of obesity must have a high correlation with the adiposity of the individual or the estimated percentage of body weight accounted for by body fat. There are no universally accepted criteria for the diagnosis of obesity, most definitions being somewhat arbitrary. Fundamental obstacles to the accurate assessment of the prevalence of obesity, especially for infants and young children, are a lack of universally accepted definitions and variations in methods of measurement and reference data. The clinical impression, weight for age and weight for height, ponderal indices, exponential derivations of weight for height, fat folds, and combinations of these have been used. The problem with weight for age or weight for height measurements is that they do not distinguish the truly fat child from the child whose weight is increased because of increased lean body mass, as in a very muscular or a very tall child with advanced skeletal development (Forbes, 1964).

Laboratory methods are not appropriate for clinical assessment of obesity but are necessary to validate and correlate fat fold measurements with total body fat. Briefly, the laboratory methods include underwater weighing to obtain body density measurements, measurement of the absorption of fat soluble gases such as radiokrypton by fat tissue, x-ray studies to determine fat layer thickness, and measurement of total body water using potassium-40 for the estimation of the lean body mass (Bray, 1976).

Commonly used definitions are presented in Table 29–1. Weight for height is probably the simplest clinical method of identifying overweight or obese children. Commonly accepted definitions are in terms of weight for height or weight for age above standard, or greater than the 95th percentile values using the National Center for Health Statistics reference data. However, growth charts for weight for height are not useful once the adolescent growth spurt has begun, and weight for age or the use of fatfold measurements are relied upon. Fomon (1974) offers a definition based on the relationship of body weight to stature for each sex, with weight ranging 40 to 50 percentile values above height or length.

Fat fold thickness as measured by calipers has proven to be a useful and convenient tool in the clinical assessment of obesity. Subcutaneous adipose tissue, a major component of body fat in children, is readily accessible for measurement and

Table 29–1. COMMONLY USED ANTHROPOMETRIC DEFINITIONS OF OBESITY

	Overnutrition	Obesity
Wt. for ht.* (prepubertal)	110–119% std.* 90–95th percentile	≥120% std. >95th percentile
Wt. for age	110–119% std. 90–95th percentile	≥120% std. >95th percentile >2 S.D.†
Wt. and ht.	Fomon (1974)	Weight percentiles exceed height percentiles by 40 to 50 percentile points
	Age	**Obesity**
Fat folds	0 to 36 mo. (Karlberg, 1968)	>2 S.D. >90th percentile
Triceps and subscapular	0 to 18 years (Tanner and Whitehouse, 1962)	>2 S.D. >95th percentile
Triceps only	0 to 50 years (Mayer and Seltzer, 1967)	>1 S.D.

*Fiftieth percentile value of NCHS reference data (1976): 50th percentile wt. for a given ht.
†Above the mean value.

may provide information about the relative fat content of the body by measuring the thickness of a single parameter. Obesity in children has been defined in terms of thickness of fat folds for age above the 95th percentile. Triceps and subscapular skin fold thickness above 2 standard deviations, based on Karlberg's study of Swedish children aged one to three years, is useful (Karlberg et al., 1968). Standards using percentile values for American children under age five years will soon be available from the National Center for Health Statistics. The Tanner fat fold measurements developed for English children aged 0 to 18 years are very useful (Tanner, 1962). Body density measurements, when combined with skin fold measurements obtained simultaneously, allow the formation of predictive equations for the estimation of total body fat in individuals by age and sex and are based on correlations of skin fold thickness and skin density. Caliper measurements, although very convenient for clinical purposes, pose many difficulties. There is a variation in the subcutaneous fat distribution, particularly between black and Caucasian individuals. Also extensive practice is needed to obtain reproducible results.

At present there is a need for a universal definition for obesity. The fatness criteria do not define populations at risk. For example, the use of weight gain velocity would have predictive value for identifying individuals or groups at risk for developing obesity.

GROWTH AND DEVELOPMENT OF ADIPOSE TISSUE

An understanding of normal growth and development of the adipose "organ" is important in formulating a rational approach to the prevention of obesity. The hypothesis that there are "vulnerable" or "critical periods" when energy overnutrition may have long term adverse effects has practical implications for intervention.

There are periods in the life of a child when fat cells normally proliferate and grow in size (see Fig. 29–1). The first spurt actually occurs during the third trimester of fetal life, particularly the last month, and continues for the first two years of an infant's life. Fat cells normally proliferate during this time, and about 26 per cent of the body weight can be attributed to fat in the nine to 12 month old infant. The next period of hyperplasia is in

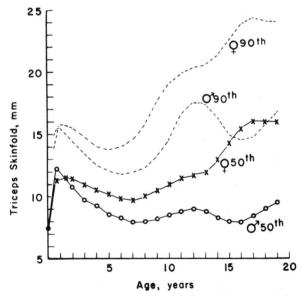

Figure 29–1. Changes in subcutaneous fat based on age. Median and 90th percentile values for triceps fatfold thickness are shown. (Adapted from Tanner, J. M., and Whitehouse, R. H.: Standards for subcutaneous fat in British children. Br. Med. J., 1:446, 1962.)

the preadolescent phase, within two years of the growth spurt. The proliferation continues into adolescence for the female, but after the prepuberal spurt the percentage of body fat actually declines in the male. In normal young male adults about 17 per cent of body weight is fat, whereas in females about 24 per cent of the body weight is fat (Knittle et al., 1981).

During these periods of normal hyperplasia, it is hypothesized that children are highly vulnerable to the effects of excessive energy intake, resulting in excessive hyperplasia. These periods have been termed critical periods, which may be appropriate times for active intervention to prevent obesity. The question that arises is whether there are inherited abnormalities of the fat cell that render them more vulnerable to hyperplasia secondary to excessive energy intake.

Controversy exists over the acceptance of fat cell hyperplasia and increased fat cell number in differentiating obese and nonobese prone individuals. There are many technical problems in assessing fat cell number having to do with the biopsy site, techniques, and validity, particularly in regard to estimating total body fat. Knittle (1981), with his carefully performed longitudinal studies and improved technique and direct measures of fat cell size, believes that he can identify the obese prone individual, tracing adipocyte development in obese and nonobese infants from under one year of age onward. In the obese child Knittle (1981) found significant increases in fat cell number and size by age two years, which persist and are increased at all ages. In the nonobese child there are no such changes in cell size and number between ages two to 10, with increments only after age 10 years, leveling off at age 14 to 16 years. Thus, after two years of age there are significant differences in adipose tissue growth and development between the obese and the nonobese child.

Are there inherent differences in the fat cells of obese and nonobese subjects, and do they respond differently from normal cells to hormonal influences? Obese cells respond less effectively to the lipolytic effects of epinephrine than the smaller cells of the nonobese child, and this is not changed by weight loss (Knittle et al., 1981).

NATURAL HISTORY AND PROGRESSION OF EARLY OBESITY TO LATER OBESITY

It is only a hypothesis that the prevention of obesity early in life can lessen the possibility of emergence of this condition in later childhood and adulthood. Our knowledge of the natural history of obesity has been based on a number of prospective and retrospective studies using a variety of different criteria for the diagnosis of obesity and yielding varied results. In a number of these studies evidence has been found for a positive correlation between early onset of obesity and continuation of obesity into later life (Knittle et al., 1981).

PREPREGNANCY AND ANTENATAL INFLUENCES

Fatness in newborns has been evaluated by weight for height and fat fold thickness measurements and is related to maternal variables. Simpson et al. (1975) were able to demonstrate a positive correlation between birth weight and maternal prepregnancy weight and between birth weight and pregnancy weight gain, in both an independent and an additive fashion. Udall (1978), who used skinfold thickness to assess neonatal and maternal obesity, found that fatter infants tended to have fatter mothers. However, it was not clear whether prepregnant maternal obesity or excessive weight gain during pregnancy correlated with the neonatal fatness. Weight gain in pregnancy in excess of 40 pounds was significantly greater in the mothers of fat infants who were large for gestational age. The infants were also longer and heavier and had a greater midarm fat area and fat fold thickness. Therefore, it was concluded that maternal obesity was strongly associated with increased subcutaneous fat in the neonate, and it was postulated that an increased transfer of free fatty acids occurred across the placenta.

NEONATAL OBESITY AND FUTURE OBESITY

The next question to be answered is whether in infants who are obese at birth there is an increased likelihood of their remaining obese. Study findings have been variable. In one study of 331 English mothers and their newborns, there was no correlation found between skin fold thickness in the mother and infant and any other variables (Gampel, 1965). Fisch et al. (1975) examined weight and length measurements in 1786 infants using growth data from the Minnesota portion of the National Institutes of Health collaborative perinatal project. Of the infants classified as obese at birth, at four years of age 26 per cent were still considered obese. When these children judged to be obese at four years of age were remeasured at seven years, 98 per cent were found to be heavier than the 90th percentile. The data suggest that the neonates judged to be obese at birth retained their heavy physique until at least age seven years. The future of such seven year olds is not known.

One of the largest retrospective studies of subsequent growth of obese and nonobese infants was reported in Sweden (Mellbin and Vuille, 1973). Weight gain in the first year was found to be the best predictor of the weight for height percentile at age seven years, with obesity in 18 per cent of the boys and 9 per cent of the girls. Charney et al. (1976) carried out the first study of infant growth data used as a predictor of adult obesity. Of the infants whose weight exceeded the 90th percentile during the first six months, 14 per cent were obese and 22 per cent were considered overweight some 30 years later. Of the children who were in the normal percentiles, only 5 per cent were obese and 9 per cent were overweight.

CHILDHOOD ONSET OBESITY

A common time for the onset of obesity is in middle childhood. Approximately one-third of adult obese patients reported onset of their weight problem in childhood. There are a variety of studies showing a high correlation of childhood obesity with future persistent obesity, and this seems to be a better predictor than infantile obesity. A nine year prospective study showed that childhood obesity was a better predictor than obesity in infancy (Abraham et al., 1971).

The Hagerstown study was especially informative because of its scope (Abraham et al., 1971). A follow-up study was carried out in 1691 adults 30 to 40 years after they participated in a study of obesity at ages nine and 13 years. Sixty per cent of the moderately overweight subjects remained moderately overweight or obese 30 years later, and 84 per cent of the markedly overweight or extremely obese subjects remained markedly obese. The odds that a child who remained obese through adolescence would ever become nonobese were found to be 28:1. In summary, there are an increasing number of studies that show a significant correlation between a childhood onset of obesity and continuation in a significantly high percentage into adolescence and adulthood.

ETIOLOGIC CONSIDERATIONS

An understanding of etiology is necessary for effective prevention and treatment of obesity. Numerous studies of families, twins, adopted and foster children, and even families and their pet animals have yielded abundant evidence of both genetic and environmental factors. The best compromise statement at this time is that there is probably an interaction of genetic make-up and adverse environmental factors leading to obesity.

On the genetics side, Knittle (1981) and others believe that there may be genetic differences in the metabolism of fat cells between obese and nonobese persons. Others suggest that there may be genetic differences in the amount of energy required for metabolism. Also it seems probable that temperamental differences, physical activity patterns, a predilection for sweets, and other factors may be under genetic control. In matings between obese and normal parents, about half the offspring become obese; when both parents were overweight, two-thirds of the offspring were obese. Although familial occurrence can imply environmental as well as genetic factors, it has been shown that the weights of adopted children do not correlate with the weights of their adoptive parents, but weights of biological children do correlate with their parents' weights. Evidence from weight correlation studies of identical and fraternal twins and siblings also tend to indicate that genetic factors in some cases play a greater role than environment in determining obesity (Lloyd and Wolff, 1976).

Garn and Clark (1976) analyzed data from the Ten-State Nutrition Survey to furnish family line information about fatness in more than 30,000 men, women, and children using fat fold data. Sibling similarities were found in fat fold measurements at a significant level of correlation throughout childhood. Spouse and father-child similarities in degree of fatness were also found.

In summary, fatness is a familial problem amply demonstrated by child-parent correlations as well as sibling correlations, husband and wife similarities, and family and pet animal similarities. Through a shared environment, families partake in similar activity patterns and eating patterns.

Age Specific Etiologic Factors

Etiologic considerations differ somewhat from age group to age group in the genesis of obesity. Therefore obesity in infancy and in the older child will be commented upon.

Infancy

Activity. Level of activity and temperament are relevant considerations in infantile obesity. One of the earlier studies to evaluate the relative effects of intake and activity on weight gain in young infants was carried out by Rose and Mayer (1968). The study demonstrated that in thin infants there was often a greater intake but much greater activity than in their overweight or obese counterparts, and that activity was the major determinant of intake. There are placid infants who sleep and eat and those who are in constant "brownian" motion, hardly sleeping at all. Activity has largely been overlooked in many evaluations of the problem of obesity.

Mode of Feeding. Some studies have shown a relationship between the mode of feeding and obesity, although there are other studies that refute such findings. In infants who are breast fed versus those who are bottle fed, it was found that breast fed infants do not gain as rapidly as bottle fed infants and that solids are usually fed earlier in bottle fed infants than in breast fed infants (Fomon, 1974). In one study it was found that in bottle fed infants the birth weight doubled earlier than in breast fed infants; moreover they received solids earlier (Neumann and Alpaugh, 1976). The bottle fed infants gained a disproportionately greater amount of weight per unit length increment than did breast fed infants whose height and weight percentiles were approximately equal.

Overfeeding is less likely to occur with breast feeding, since the infant is more able to control his intake when satisfied. Bottle fed infants through persistent coaxing have been found consistently to consume greater volumes of milk than desirable (Fomon, 1974). Long term patterns of overeating of overconcentrated formula by accident or intention may occur. Satiety mechanisms may be at work in the breast fed infant, for the fat content during a single feeding rises sharply by the end of the nursing period.

Differences in activity during feeding exist between infants who are breast fed and those who are bottle fed as observed by Bernal and Richards (1970). The breast fed infant was found to spend more time suckling, spent less time in the crib, was awake longer, and was stimulated and handled more by a given caretaker than the bottle fed infant.

Some studies have shown no relationship between breast feeding, the addition of solids, and the rate of weight gain or appearance of infantile obesity (Swiet et al., 1977). There are certainly breast fed infants whose mothers are overzealous and hyperenthusiastic about breast feeding and whose intakes far exceed what is needed.

One harmful practice is the addition of solids to a bottle of formula, particularly cereals, egg, and even strained meats. The intake of a mixture of such high caloric density may not be compensated for by a decreased volume intake in the infant under three months of age. An infant can easily be overfed by several ounces before he will refuse the feeding or vomit.

Socioeconomic and Cultural Determinants of Infantile Obesity.   Cultural ideals for the healthy infant image are often equated with a fat baby. In describing child rearing in a black ghetto, Ellis (1971) noted that the predominant attitude among mothers was to obtain maximal growth in an infant as a sign of good health. Feeding of high calorie solids and highly saturated fats such as cornbread, lard, and fatback was started early in infancy. Equating a fat baby with a healthy baby is also prevalent among Mexican-American, native American, and Samoan families in the United States (Boyd et al., 1974).

In a study in New York it was found that parental social class, introduced as a controlling variable, showed a high negative correlation with the prevalence of obesity and was a more significant predictor of overweight than other factors tested. Also obesity in infants was inversely proportional to the number of generations the subject's family had been in the United States (Stunkard et al., 1972).

Mother-Infant Interaction.   Eating behavior is easily conditioned and may be reinforced in the early months of life. Bruch (1973) suggests that owing to incorrect learning, obese infants may be incapable of distinguishing feelings of hunger and satiety. Parental attitudes can readily influence eating behavior in young infants. For example, a breast fed infant sucks until satisfied, the mother producing just enough to meet the demand. This biologic feedback system is disrupted with bottle feeding, and the mother uses a visual cue to shape the infant's eating behavior according to her own perception as to the quantity of needed feeding. She can easily then over- or underfeed while the infant learns to depend more and more on environmental cues for determination of an intake that is not based on his own needs. Later the infant is even rewarded by the mother's approval and pleasure. It is important to investigate parental attitude and behavior in this regard.

Psychologic problems in the mother may also lead to overfeeding, overeating and obesity. Maternal concern that equates the giving of sufficient food with good mothering can lead to overfeeding. Some mothers are unable to distinguish signs of hunger in their infants and feed the child with every cry. Guilt or anxiety in some mothers who have lost infants previously or have produced very small infants can lead to overfeeding. Some mothers have a need to foster an extreme overdependence of infants upon themselves, a reflection of their own lack of control. Handicapped infants are also at risk for obesity because of maternal anxiety and guilt and decreased physical activity in some. The risk factors operating in infantile obesity are presented in Table 29–2.

Etiologic Considerations in the Preschool and School Aged Child

A precipitating event or situation can often be identified in this age group. In our own clinic* and in the experience of others, obesity has been found to follow tonsillectomy or hospitalizations for serious illness and other surgery or enforced periods of prolonged bed rest. A move to a new neighborhood with loss of friends and change of school, divorce and "loss of a parent," and a family distraught over the loss of another child or

---

*UCLA Pediatric Weight Reduction Clinic.

Table 29–2. RISK FACTORS FOR INFANTILE OBESITY

Diabetic mother
Obese parent(s)
Bottle feeding
Early introduction of solids
Infant with handicapping condition or perceived as "damaged"
Reduced activity due to physical problem or child rearing practices
Mothering problems
   Anxious, insecure, inexperienced misinterpretation of infant's needs, particularly of crying
Cultural image: healthy infant = fat baby
Rapid weight gain velocity before six months
Low socioeconomic status
Recently arrived immigrant parents

close family member have been noted to be precipitating factors in childhood obesity (Neumann, 1977).

Socioeconomic status figures significantly in the etiology of obesity in school children, a factor that has been largely ignored in the past. In one large study of 3344 inner city children by Stunkard et al. (1972), there was a ninefold greater prevalence of obesity in girls of low versus high socioeconomic status. In boys, striking differences were also noted by age six years, the prevalence of obesity in boys of a lower socioeconomic status being double that in upper class boys. Not only was a greater prevalence of obesity noted by age six years among children of the poor, but the trend continued throughout childhood into adolescence. It is worth noting, therefore, that a vulnerable and high risk group for obesity is children of the poor.

Contributing factors include inferior quality of the diet with great reliance on starchy filling food, and the frequent use of fried, quick energy foods with concentrated calories, such as potato chips, soda pop, and candy rather than the more expensive protein containing foods. Meal planning, bulk shopping, and storage facilities may be lacking. Schools in depressed neighborhoods lack recreational facilities and opportunities and apparatus for adequate physical education and activity. There are few parks, playgrounds, or other recreational facilities in such areas, and even if present they may not be safe for children to play in. Parents in our own program often state that they will not let their children play in the street because of all the violence in the neighborhood and will not let them ride bicycles because of heavy automobile traffic. Child care arrangements for afterschool hours and during vacations are inadequate. Too often the television set becomes the substitute for supervised active play. The impressively higher prevalence of obesity among school age girls as compared with boys may be due in part to the availability of fewer athletic programs for girls as well as the relative de-emphasis on the importance of physical activity for girls, although this is now changing (Neumann, 1977).

Television viewing is a particular hazard with its forced inactivity, which compounds the problem for the obese child. Not only is the child exposed to abundant advertising that glorifies snack foods, soft drinks, candy, and sugared cereal, but the child can become locked into many hours of inactivity per week and more often than not believes the advertising in regard to food and medications (Lewis and Lewis, 1974). (See also Chapter 17 in regard to television.)

MANAGEMENT OF THE OBESE CHILD

Lack of firm conviction on the part of many physicians that childhood obesity may lead to adult obesity has resulted in denying intervention to many obese youngsters. This is unfortunate in light of the significant percentage of obese children who remain obese throughout their lives. For management to be successful, the physician must keep in mind the developmental stage of the child or adolescent; otherwise the program is bound to fail. This is particularly true for the adolescent. The other major pitfall is to dwell on the weight or obesity problem and ignore the needs of the child as an individual. If the child's self-esteem is completely tied to his success or failure in regard to weight loss, this is fraught with danger, for success is far from certain and loss of weight may be modest.

Self-defeating frustration often envelops the physician, the patient, and the parents when an obese child is brought for treatment. Often the physician expects little in the way of success and may be intolerant, impatient, and negative, all of which can adversely affect the outcome of treatment. The obese child's feelings of rejection and withdrawal may even be increased.

Obesity hardly fits the usual disease model, and success is painfully slow or undramatic, particularly if the goal is merely no weight gain rather than weight loss. It is particularly difficult for pediatricians in training to care for these patients when they are surrounded by "exciting and dramatic" pediatric problems. Physicians with their own unresolved weight problems often have difficulty in coping with a young obese patient. A particular danger is that the physician and family may grasp onto "magical cures" of hormones, injections, and appetite suppressants.

The parents' frustration may derive from their own negative expereience with weight loss, for many are obese and lack conviction that the child can succeed. They often sincerely wish to prevent their child from becoming increasingly obese and yet seem helpless to intervene. Diet restriction

Table 29–3. HISTORY FOR THE OBESE CHILD

Maternal weight gain during pregnancy
Birth weight—condition of infant at birth
Onset of obesity and precipitating event, if any
Early feeding history, breast versus bottle feeding, age of starting solids
Mother's philosophy about feeding and attitude toward food (reward, bribe)
Mother's or family's perception of the quantity the child should eat
Family's concept of what constitutes a healthy child
Record of weights and heights, particularly in the first year of life
Family history of obesity and any diseases aggravated or induced by obesity, particularly diabetes, hypertension, and coronary heart disease
Presence of chronic or handicapping conditions
Twenty-four hour recall for activity and food intake for a typical weekday and a weekend day
Assessment of family relationships, particularly mother-child relationship (e.g., overprotectiveness)
Degree of dependency and maturity, motivation and self-image of the child, depression
School performance and success in academic work and in physical education class
Social inter-relationships: peers, school mates
Attempts and success with previous weight reduction regimens
History of excessive sleeping during the day

Table 29–4. PHYSICAL ASSESSMENT OF THE OBESE CHILD

Measurement—weight, length/height, triceps fat fold and preferably subscapular fat folds using a fat fold caliper*
Plot of values on weight for height growth chart if prepubertal; otherwise weight/age and also height for age to note stunting
Blood pressure—cuff must cover two-thirds of the upper arm; thigh cuffs may be useful in obese teenagers
Photographs taken "before and after" weight loss are useful in older children
Distribution of fat
Presence of striae or other skin problems
Stage of puberty; Tanner staging is useful
Size of tonsils and degrees of nasal and adenoidal obstruction
Orthopedic problems, particularly of hips and knees
Dyspnea, tachycardia

*Lange-Cambridge Instrument Co., Cambridge, Maryland. Harpenden, London, England.

often becomes a battleground over control, aggression, and autonomy. A particular problem is the obese youngster who has siblings of normal weight. Continuing parent meetings of obese children have proven helpful for parents to share and express their feelings and offer realistic advice to one another about day to day problems involved in the child's diet restriction and activity program.

APPROACH TO EVALUATION

Management begins with a detailed medical history, physical examination, and psychosocial evaluation, including relevant cultural factors. Much of this information is already known when the child is under the physician's regular care. The medical and social history should include the data listed in Table 29–3.

PHYSICAL ASSESSMENT

The physical assessment should particularly include the items listed in Table 29–4.

LABORATORY STUDIES

Laboratory analyses should be kept to a minimum because the majority of patients suffer from exogenous obesity. Detection of anemia is impor-

tant because of generally poor dietary practices, particularly in adolescents. When there is a strong family history of diabetes, a glucose tolerance test should be ordered. Also a family history of elevated serum lipid levels or early cardiovascular disease is an indication for ordering serum cholesterol and fasting triglyceride levels, although such baseline measurements are valuable regardless of the family history in this high risk population.

The search for endocrinologic or other specific causes of obesity is generally unproductive. A short and sexually underdeveloped patient (height for age less than the fifth percentile), delayed bone age, or any patient with clinical findings suggestive of an endocrinologic disorder or other underlying problem is singled out for further evaluation, particularly of the thyroid or adrenal status. Thyroid studies are not indicated routinely, particularly in patients of normal height for age. The known chromosomal or other syndromes associated with obesity—Klinefelter's syndrome, the Laurence-Moon-Biedl syndrome, or the Prader-Willi syndrome—are found in only a small fraction of the cases of childhood obesity.

PSYCHOLOGIC ASSESSMENT

Psychologic assessment is accomplished through an interview with the parent and the child during the initial evaluation. Standardized personality tests may be administered, such as the Piers-Harris Children's Self-Concept Scale, Draw-a-Person, and Sentence Completion Test (Figs. 29–2, 29–3). A questionnaire sent to the school counselor requesting additional information about school performance and peer adjustment is useful. In the case of seriously disturbed children (e.g., severe depression or acting out), referral is made for psychiatric evaluation and treatment. Motiva-

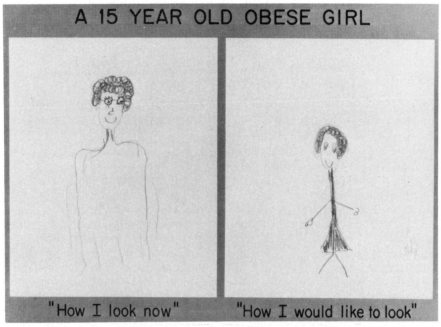

Figure 29–2. Self-portrayal by an obese adolescent girl.

Figure 29–3. Self-portrayal by an obese adolescent boy.

tion is determined by talking with the child alone and with the parents about the reasons for coming for help, who initiated the referral, and the specific weight loss or weight goal in mind. Previous attempts at weight loss and the reasons for success or failure are also discussed.

DIETARY ASSESSMENT

The dietitian takes a 24 hour recall diet history from the patient and mother, together or separately, and asks for information about the pattern of eating. Included are such matters as who does the shopping and cooking, the extent of restaurant eating, where the eating takes place and with whom, and meal or snack frequency. Food preferences and dislikes are explored, and the diet is assessed in terms of total daily caloric intake and adequacy of nutrients. It is important to discuss whether the child participates in the school breakfast or lunch program, for occasionally double meals are eaten. If no dietitian is available, a 24 hour recall using household measures to evaluate quantity and the use of a "calorie counter" for commonly eaten foods will readily establish whether excessive amounts are being ingested. Volumes of whole milk and soft drinks should be carefully noted; these are major sources of calories.

EVALUATION OF PHYSICAL ACTIVITY AND FITNESS

During the history taking, a 24 hour recall of physical activity is obtained for a school day and a weekend day. A questionnaire sent to the school physical education teacher or coach inquiring about participation, skills, and endurance may be helpful. The daily numbers of hours of television viewing and eating and snacking during this time are important in assessing activity. An estimate of the amount of chauffering the child receives in place of walking is useful in pinpointing opportunities for increased physical activity.

PREVENTION AND MANAGEMENT: PRACTICAL CONSIDERATIONS

Infancy

In the prevention and management of infantile obesity the following suggestions have proved useful (see also Table 29–5):

1. Strongly support and encourage breast feeding for infants for at least three months whenever possible. If a mother chooses to bottle feed her infant, she must be counseled not to force her infant to empty the bottle each time, and to mix and dilute the formula correctly.

2. Discourage the feeding of solids until the infant is five to six months of age, particularly if the infant is breast fed or is ingesting an iron supplemented formula. Mothers should be encourage to prepare their own baby foods, since many commercial brands have a high content of carbohydrates, particularly the desserts. If commercial foods are used, recommend the use of pure vegetables and meats rather than mixed dinners, which are lower in protein and higher in carbohydrates.

3. Explain to the mother that the child's appetite and nutritional requirements will fluctuate and

Table 29–5. MANAGEMENT OF OBESITY

| Age Group | Treatment Goal | Degree of Responsibility | | Recommended Caloric Intake* | Activity |
		Child	Parent		
Infancy	Slow weight gain velocity commensurate with length	0	+ + + +	90 kcal./kg.	Allow and encourage free movement
Toddler, preschool	Slow weight gain velocity commensurate with height	+	+ + +	60 kcal./kg.	Encourage gross physical activity in safe environment; avoid excess TV viewing
School age (prepubertal)	Keep weight level, allow child to "grow into his height" or modest weight loss if ≥120% std.; monitor height for normal growth	+ +	+ +	1200 kcal./day	Encourage group and individual physical activity; control TV viewing; organized recreation groups helpful
Adolescent (sexually mature)	Loss to goal weight (appropriate weight for height)	+ + + +	0	850 kcal./day	Encourage group and individual physical activity; encourage peer group social interaction

*Dependent on activity level.

Table 29–6. ENERGY VALUE OF INFANT STRAINED FOODS*

Type of Food	Average Energy (kcal./100 gm.)	Average % kcal. from Carbohydrate
Juices	65	96
Fruits	85	96
Vegetables		
Plain	45	80
Creamed	63	74
Meats	106	1
Egg yolk	192	3
High meat dinners (meat and vegetable)	58	56
Desserts	96	89
Cereals		
With milk	105	—
With water	52	—
Dry	360	—
Formula and cow's milk	67 (per 100 ml.)	—
Human milk	70 (per 100 ml.)	—

*Adapted from Anderson and Fomon, 1974.

diminish significantly by the age of one year as growth velocity slows. An infant should not be forced to eat more than he appears to want.

4. Obtain routine weight and length measurements, plot them on a growth chart at each visit, and share this information with the mother. Rapid weight gain and an increase in weight percentile, particularly with a discrepancy of linear growth, are to be regarded as seriously as poor weight gain.

5. Periodically assess the type and volume of milk and the amount of solids the infant consumes. At the same visit discuss with the mother the adequacy or excesses of the diet and the different caloric densities of the foods used (Table 29–6).

6. Evaluate the quality of mothering. Overfeeding may result when a mother is unable to decipher her infant's cries and offers milk in response to every crying episode.

7. Encourage parents to allow their infants free movement of the extremities and whole body. Physical activity is an important determinant of obesity, even in infants.

If an infant becomes overweight or obese, the treatment goal is not weight reduction as it is in the adult, but rather a slowing of the rate of weight gain commensurate with linear growth. Severe dietary restriction may reduce fat free body tissue and inhibit growth.

A regular diet should be given to satisfy the normal requirements for growth, with a caloric intake of about 110 kcal. per kg. per day for infants under six months, and about 90 kcal. per kg. per day for infants six to 12 months. The daily volume of milk, if excessive (over 30 ounces), should be reduced and water feedings substituted as needed. Skim milk is not recommended because of its high solute load as a result of its relatively high protein

content. In marked obesity, 2 per cent fat (low fat) milk may be given to infants over six months of age (Fomon, 1974).

Vital to the whole issue of the prevention and management of obesity in infancy is sound education of the parents, particularly the mother. Guidelines as to the nutritional values of different foods used in infant feeding, approximate amounts needed by the infant, and feeding behavior at different ages are essential. This counseling may be provided by the physician, nurse practitioner, or a specially trained nurse and occasionally in consultation with a nutritionist or dietitian if available.

Childhood

In working with obese preschool and school age children, the environment must be controlled and modified, but the child must take some degree of responsibility appropriate to his level of maturity. The child can be trained to become aware of what he eats and establish control over his food intake. Full cooperation of the family, school, and relatives is needed. Although eating and activity behavior must be changed and inner controls and proper eating behaviors developed, attention should be directed to the whole child and not just his weight problem. Being made to feel different or deprived and being excluded from social activities involving food (such as parties) can be deleterious.

As in infancy, the main goal is weight control—not weight reduction—allowing the child to "grow into his weight." The supply of calories and protein must allow for growth and development of the "lean body mass" while resulting in a decrease in the fat deposit, a difficult path to follow.

It is recommended that children under 12 years of age receive about 60 kcal. per kg. of ideal* body weight, with 20 per cent of the calories derived from protein, 40 per cent from carbohydrate, and 40 per cent from fat. The child should not be denied the basic family diet, but needs to eat smaller portions and avoid foods of a highly caloric density (Fomon, 1974). Length or height must be monitored constantly for steady linear growth. Body composition as well as body weight also should be monitored; the use of skin fold calipers can be of great help in estimating changes in the amount of body fat.

SIMPLE DIRECT COUNSELING FOR THE FAMILY

Have the mother rid the house of "junk food" and provide low calorie snacks. Cooperation of

---

*50th percentile value of National Center for Health Statistics standard.

the rest of the family in forgoing snacks in the presence of the obese child is necessary. Inappropriate use of food should be dealt with specifically: food must not be used for bribery, punishment, or reward or as a substitute for meaningful interpersonal relationships, particularly with the parents. The parents must allow the child to make his own decisions as to intake, satiety, and termination of the meal.

It is possible to modify eating behavior in this age group by furnishing appropriate size portions, avoiding the "clean plate" syndrome, limiting second servings and desserts, cutting food into small pieces and eating slowly, engaging the child in conversation while eating, avoiding the child's eating alone, allowing eating in a fixed designated place only, providing eating strategies for parties, holidays, treats, outings, and simple education about the caloric values of food, exchanges, and choices (with the help of a professional) so the child may begin to exercise some independent decision making and responsibility for his eating. Food handling jobs or duties in the school cafeteria should be avoided.

Although formal behavior modification techniques in altering eating behavior have been applied successfully to adults and to a lesser extent in adolescents, such techniques have not been generally applied to school age children. The goal of behavior modification in children is to increase adaptive behavior before maladaptive behavior becomes too resistant to change.

PHYSICAL ACTIVITY

Many obese children manage to drop out of physical education at school and out of all active sports. Making a special effort to include the child is necessary, and intervention by the physician is often needed. Unlike the extreme self-consciousness and embarrassment of teenagers who have to appear in gym suits or swim suits in public, the preteen is less sensitive in this regard. Regular daily periods of exercise (walks, hikes, dancing, skating, games) are necessary, and children should be encouraged to climb stairs and walk instead of being driven everywhere. Daily errands and walking routes can be worked out. Specific recommendations about restricting television viewing are badly needed, and also cut down on exposure to advertising. Specific referrals to the local Boys' Club, YMCA, YWCA, after school programs, park and recreation department programs, or Scouting programs are most helpful. Involving children in organized activity promotes peer contacts and relieves loneliness, boredom, and social isolation. Summer and other vacations are the most vulnerable periods for an obese youngster, and added effort is needed to involve the child in an active program or activity.

THE OBESE ADOLESCENT

Much of what has been recommended in regard to the management of the obese child applies to the adolescent. In the management of the obese adolescent two critical factors must be kept in mind. An increase in the lean body mass and fat are the major events in the growth spurt. It is essential to know where the adolescent is in terms of pubertal development and whether he is through the growth spurt. Cessation or slowing of growth can be caused by prolonged and extreme weight reduction diets. Also the fact the adolescent is struggling for independence and responsibility for himself places the burden of monitoring of food intake and activity squarely upon him, with a minimum of parental interference. Parental coercion may aggravate the normally taxed parent-adolescent relationship, with the negative outcome of a rebellious angry youngster who feels deprived.

It must be understood that weight gain in adolescence is normal and largely due to an increase in the lean body mass. The use of fat fold calipers becomes important in monitoring fat loss. Unrealistic fat loss goals, particularly in the teenager who is in early or middle adolescence, may be extremely demoralizing. A more realistic approach for the teenager who is not yet through the growth spurt is to prevent further weight gain rather than effect a sizable weight loss. The biologic and psychologic stresses on the teenager are considerable in the attempt to lose weight.

In terms of caloric restriction, if the adolescent is still in an active growth phase, a 1200 calorie diet using an exchange system will ensure adequate linear growth. If the adolescent is fully mature sexually, an 850 calorie diet is recommended. Increased daily physical activity is also prescribed.

The use of anorectic drugs, hormone injections, liquid diets, and other drugs has little place in weight reduction in the adolescent and no long term effectiveness. The goals of a long term educational change in eating behavior and an increased activity level are best accomplished through diet, an exercise program, group reinforcement and support, and practical nutrition education.

Coed "rap groups" and peer support groups are particularly valuable during this age span. Such a group often provides one of the few pleasant ongoing opportunities for socializing with peers in a mutually supportive environment. The group may be led by a social worker, a psychologist, or a skilled nurse. The UCLA weight reduction program, incorporating a multidiscipline approach and the use of groups, is described elsewhere (Meyer and Neumann, 1977).

For the morbidly obese adolescent, a protein

sparing fast under inpatient supervision and occasionally gastric stapling procedures are resorted to, with initial dramatic weight loss and no serious side effects. However, the long term outcomes are not known. Jejunoileal bypass surgery is so fraught with complications that it is seldom if ever indicated in this age group. However, despite the foregoing expedient procedures, life long modifications of eating and activity patterns need to be undertaken.

Bulimia, a recently popularized syndrome, with and without effect on body weight, probably accounts for a small percentage of the cases of obesity in teenagers. Bulimia is seen predominantly in female adolescents and in young women who are excessively concerned with their weight and who try to control their weight and intake by self-induced vomiting and even catharsis. The ingested food is often of a high caloric value and highly sweetened. Enormous amounts of food can be ingested. Eating ceases with the onset of abdominal pain, followed by self-induced vomiting and sleep, and usually accompanied by guilt and self-recrimination. Food is a constant preoccupation, and binges are often associated with times of upset. Feelings of helplessness and disturbed body image are common findings (Powers, 1980).

BEHAVIORAL TREATMENT

Traditional approaches to the treatment of obesity have yielded high drop-out rates, occasional adverse emotional reactions, poor weight loss with a high relapse rate, and poor long term success. For the past 15 years it has become apparent that in the treatment of adults behavioral treatment has been more effective than the usual dietary restriction and exercise program. There is preliminary evidence that behavioral treatment for obese youngsters may be very valuable.

A premise underlying behavioral treatment is that children will substitute beneficial eating behavior for detrimental patterns and carry these on into adult life. There are only half a dozen reports about behavioral treatment in children, but these reports thus far appear promising (Brownell and Stunkard, 1978).

Overeating is not considered a symptom of an underlying behavioral disorder but rather is a learned dysfunctional coping response to a given situation, which must be replaced by a more adaptive pattern. Careful behavioral analysis of eating behavior is part of the treatment. Eating is studied in relation to immediate antecedent events and the stimuli initiating the behavior. Behavioral programs are designed to alter the antecedent stimuli to eating and to control the eating.

In stimulus control, the events preceding the eating behavior are identified in an attempt to eliminate them. This is important in dealing with children. Such a program, for example, would eliminate visual cues such as candy, nuts and sweets, and other "forbidden foods" within sight of the child, substituting low caloric foods or no foods. Reinforcement is used to increase the likelihood of the desired behavioral response and involves tangible rewards for behavior change and weight loss, with speedy follow-up of the reward. Written contracts may be useful for older children.

The first pilot study using behavioral modification in children eight to 13 years old was carried out at the University of Pennsylvania (Jordan and Levitz, 1975). Parents were instructed in record keeping, contracts, and modeling of appropriate behavior, and the children were provided with nutritional information. Tangible rewards were given for weight loss. These rewards encouraged further adaptive behavior patterns and consisted of free passes to bowling alleys and skating rinks to increase physical activity. Weight loss was substantial, but a control group was lacking.

In another study children aged two to 11 years were randomized into a behavioral treatment group, a treatment plus reinforcement group, and a control group. This study clearly showed the advantages of behavioral modification and reinforcement in weight control (Wheeler and Hess, 1976).

Another prototype group is the deposit-refund group, which gives further evidence of the efficacy of the behavioral approach. Fifteen overweight girls and their parents paid a deposit in advance, and money was refunded for desired change and partial refunds for attendance, completion of weight records, and weight loss. They were taught stimulus control, given nutritional information, exercise, and modification of eating behavior. The group that also received reinforcement lost more weight than the group that participated only in the deposit-refund activity (Argona et al., 1975).

Slowing the rate of eating is a rational approach, because obese individuals tend to eat more quickly, although this may not always be true. The effect of slowing does reduce the amount of food eaten. A group of seven year olds were observed twice a week for six months in the school cafeteria. Rates of bites and sips were measure as well as concurrent activities while eating. One group was taught how to put their utensils down after every bite to slow down their eating rate. What emerged was that the rate of eating and sipping did slow down, as did the concurrent activity. Although there was only a slight decrease in the amount of food taken and weight loss was only slight, the study did demonstrate that the eating rate could be slowed and had potential for decreasing the amount of food ingested (Epstein et al., 1976). Other attempts at behavioral therapy have included nongroup settings but with intensive management using the same principles as in the group setting.

The general principles of all behavioral treatment are self-monitoring and the recording of body weight, food intake and physical activity, and circumstances surrounding the eating. These data are used to increase the individual's awareness of the activities he participates in, the eating he does and the circumstances under which he eats. This information serves as a useful baseline.

In terms of family intervention there has been no systematic study with or without family intervention. Family involvement in a child's weight loss program does appear to act as a reinforcer.

In summary, there is every reason to think that behavioral modification and behavior therapy will be increasingly important parts of the treatment of childhood obesity and may have much to offer. An excellent book, *Slim Chance in a Fat World*, presents a very down to earth approach to behavior therapy (Stuart and Davis, 1974).

RESOURCES FOR THE PHYSICIAN COPING WITH THE OBESE CHILD

1. Enlist the help of the family.
2. Involve help of the schools. Innovative school centered obesity treatment programs have been shown to be effective by Selzer and Mayer (1970) and by Collip (1975). Children are captives for five to six hours per day for a period of years. Advantage can be taken of all the resources of the school—the nurse, teacher, physical education teacher, and consulting dietitian. All are available in the context of everyday school activities. Health education, parent education, group interaction, and a program of vigorous daily exercise have been incorporated into these programs.
3. Make specific referrals to community recreational programs, day care, and after school programs.
4. Enlist a nutritionist or dietitian from a community hospital, intern training program, local or state health department, or government funded nutrition or food distribution programs. Some innovative physicians have banded together and hired the part-time services of a dietitian or psychologist in an office practice setting to set up groups for parent and child education or behavior modification for obese children.
5. Use a nurse practitioner or other paramedical person with an interest in working with obese children.

COMMERCIAL WEIGHT LOSS PROGRAMS

Such commercial groups as TOPS, Weight-Watchers, and Silhouette offer programs built around diet restriction, exercise, weekly weigh-ins, and some group interaction. The groups succeed at least as well if not better than regimens offered by physicians or dietitians, and about one-third of the clients lose 10 kg. of weight or more. The dropout rate is high, but the cost is nominal. These programs are primarily for adults; few are geared to the child or adolescent.

Summer weight loss camps have grown in number in recent years, many of them reputable because of their physicians, dietitians, and well trained staffs. The camps build their programs around diet restriction, exercise, group sessions, and some education. Short term results are excellent as long as the child is in the controlled camp environment. However, in the experience of many, the weight is rapidly regained when the child returns to the home and school environment. Few of these camps have year long maintenance programs, and parent education is minimal, if any. All are very expensive and reach a relatively small number of obese children and adolescents.

In assessing the success of behavioral or any other weight control programs in children, one must keep in mind the fact that the child is constantly growing and developing, with a normal tendency for weight gain and lean body mass. Unlike the situation in adults, a reasonable goal is not always weight loss but may just be a slowing down of the weight gain or maintenance of body weight while linear growth occurs. Success has to be redefined to suit the physical developmental stage of the child.

CONCLUDING REMARKS

There is a generally poor prognosis for the obese child in terms of future obesity and the poor long term success rate in maintaining normal weight and body composition. A much greater effort must be invested in the prevention of obesity. The earlier in life, the better. Every effort should be made to identify the infant or child at increased risk for obesity. Intervention also should occur as early as possible while a child is gaining in height and is under parental control and while his eating and activity patterns are still malleable, before crippling secondary psychosocial problems aggravate the obesity problems. Research is needed in the area of appropriate interventions, particularly behavioral approaches to management, because behavior change is key to the problem and ideally should involve the whole family. To maximize the chances of success in any preventive or intervention program, a sound knowledge of the physical as well as psychosocial stage of development of the child is essential.

CHARLOTTE G. NEUMANN

REFERENCES

Abraham, S., Collins, G., and Nordsieck, M.: Relationship of childhood weight status to morbidity in adults. Public Health Rep., 86:273, 1971.

Anderson, T. A., and Fomon, S. J.: In Fomon, S. J.: Infant Nutrition. Ed. 2. Philadelphia, W. B. Saunders Company, 1974.

Argona, J., Cassady, J., and Drabman, R. S.: Treating overweight children through parental training and contingency contracting. J. Appl. Behav. Anal., 8:269, 1975.

Bernal, J., and Richards, S. M. P. H.: The effects of bottle and breast feeding on infant development. J. Psychosom. Res., 14:247, 1970.

Boyd, D.: Master's Thesis: Cross-Cultural Prevalence of Infantile Obesity in Los Angeles County, 1974, U.C.L.A.

Bray, G. A.: The Obese Patient. Philadelphia, W. B. Saunders Company, 1976.

Brownell, K. D., and Stunkard, A. J.: Behavioral treatment of obesity in children. Am. J. Dis. Child., 132:405, 1978.

Bruch, H.: The Eating Disorders. New York, Basic Books, Inc., 1973.

Chandra, R. K., and Kutty, K. M.: Immunocompetence in obesity. Acta Paediatr. Scand., 69:25, 1980.

Charney, E., Goodman, H. C., McBride, M., Lyon, B., and Pratt, R.: Childhood antecedents of adult obesity. N. Engl. J. Med., 295:6, 1976.

Collip, P. J.: Obesity program in public schools. In Collip, P. J. (Editor): Childhood Obesity. Acton, Massachusetts, Publishing Sciences Group, 1975.

Court, J. M., and Dunlop, M.: In Howard, A. (Editor): Recent Advances in Obesity Research. London, Newan Publishing Co., 1974.

Ellis, E.: Obesity in children—critical approach to common pediatric problems. In Report of Second Ross Roundtable. Columbus, Ohio, Ross Laboratories, 1971.

Epstein, L. H., Parker, L., and McCoy, J. F.: Descriptive analysis of eating regulation in obese and nonobese children. J. Appl. Behav. Anal., 7:402, 1976.

Fisch, R. O., Bilek, M. F., and Olstrom, R.: Obesity and leanness at birth and their relationship to body habits in later childhood. Pediatrics, 56:521, 1975.

Fomon, S. J.: Infant Nutrition. Ed. 2. Philadelphia, W. B. Saunders Company, 1974.

Forbes, G. B.: Lean body mass and fat in obese children. Pediatrics, 34:308, 1964.

Gampel, B.: The relation of skinfold thickness in the neonate to sex, length of gestation, size at birth and maternal skinfold. Hum. Biol., 37:29, 1965.

Garn, S. M., and Clark, D. C.: Trends in fatness and the origins of obesity. Pediatrics, 57:443, 1976.

Hutchinson-Smith, B.: The relationship between the weight of an infant and lower respiratory infection. Med. Officer, 123:257, 1970.

Jordon, H., and Levitz, L.: Behavior modification in the treatment of obesity. In Report of the Second Wyeth Nutrition Symposium. Philadelphia, Wyeth Laboratories, 1975.

Karlberg, P., Engström, I., Lichenstein, H., and Svennberg, I.: The development of children in a Swedish urban community. A prospective longitudinal study. III. Physical growth during the first three years of life. Acta Paediatr. Scand. (Suppl.), 187:48, 1968.

Knittle, J. L., Merritt, R. J., Dixon-Shanies, D., Ginsberg-Fellner, F., Timmers, K. I., and Katz, D. P.: Childhood obesity. In Suskind, R. (Editor): Textbook of Pediatric Nutrition. New York, Raven Press, 1981.

Lauer, R. M., Connor, W. E., Leaverton, P. E., Reiter, M. A., and Clarke, W. R.: Coronary disease risk factors. J. Pediatr., 86:697, 1975.

Lewis, C. E., and Lewis, M. A.: The impact of television commercials on health-related beliefs and behaviors of children. Pediatrics, 53:431, 1974.

Lloyd, J. K., and Wolff, O. H.: Obesity. Recent Adv. Pediatr., 5:305, 1976.

Mellbin, T., and Vuille, J. C.: Physical development at 7 years of age in relationship to velocity of weight gain in infancy with special reference to the incidence of overweight. Br. J. Soc. Prev. Med., 27:225, 1973.

Meyer, E. U., and Neumann, C. G.: Management of the obese adolescent. Pediatr. Clin. N. Am., 24:123, 1977.

Monello, L. F., and Mayer, J.: Obese adolescent girls: an unrecognized "minority" group. Am. J. Clin. Nutr., 13:35, 1963.

National Center Health Statistics: Growth charts. Monthly Vital. Stat. Rep., 25,Suppl. 3, 1976.

Neumann, C. G.: Obesity in the preschool and school-age child. Pediatr. Clin. N. Am., 24:117, 1977.

Neumann, C. G., and Alpaugh, M.: Birthweight doubling: a fresh look. Pediatrics, 57:469, 1976.

Powers, P. S.: Obesity: The Regulation of Weight. Baltimore, The Williams & Wilkins Company, 1980, pp. 264–265.

Riley, D., Santiago, T. V., and Edelman, N. H.: Complications of obesity—hypoventilation syndrome in childhood. Am. J. Dis. Child., 130:671, 1976.

Rose, H. E., and Mayer, J.: Activity, calorie intake, fat storage and the energy balance of infants. Pediatrics, 4:18, 1968.

Selzer, C. C., and Mayer, J.: A simple criterion of obesity. Postgrad. Med., 38:A101, 1965.

Selzer, C. C., and Mayer, J.: An effective weight-control program in a public school system. Am. J. Public Health, 60:679, 1970.

Sherman, B. M., and Korenman, S. G.: Measurement of serum LH, FSH, estradiol and progesterone in disorders of the human menstrual cycle. J. Clin. Endocrinol. Metab., 39:145, 1974.

Simpson, J. W., Lawless, R. W., and Mitchell, A. C.: Responsibility of the obstetrician to the fetus: influence of pre-pregnancy weight and pregnancy weight gain in birth weight. J. Obstetr. Gynecol., 15:481, 1975.

Stool, S. E., Eavey, R. D., and Stein, N. L.: The "chubby puffer" syndrome. Upper airway obstruction and obesity, with intermittent somnolence and cardiorespiratory embarrassment. Clin. Pediatr., 16:43, 1977.

Stuart, R. B., and Davis, B.: Slim Chance in a Fat World: Behavioral Control of Obesity. Champaign, Illinois, Research Press, 1971.

Stunkard, A., d'Aquili, E., Fox, S., and Filion, R. D. L.: Influence of social class on obesity and thinness in children. .A.M.A., 221:579, 1972.

Stunkard, A., and Mendelson, M.: Obesity and body image. I. Characteristics of disturbance in the body image of some obese persons. Am. J. Psychiatry, 123:1296, 1967.

Swiet, M., Fayers, P., and Cooper, L.: Effect of feeding habit on weight in infancy. Lancet, 2:892, 1977.

Tanner, J. M., and Whitehouse, R. H.: Standards for subcutaneous fat in British children. Br. Med. J., 1:446, 1962.

Tracy, V. V., De, N. C., and Harper, J. Z.: Obesity and respiratory infections in infants and young children. Br. Med. J., 1:16, 1971.

Udall, J. N., Harrison, G. G., Vaucher, V., Walson, G., and Morm, G., III: Interaction of maternal and neonatal obesity. Pediatrics, 62:17, 1978.

Wheeler, M. E., and Hess, K. W.: Treatment of juvenile obesity by successful approximation control of eating. J. Behav. Ther. Exp. Psychiatry, 7:235, 1976.

29C   Anorexia Nervosa (Self-starvation)

Anorexia nervosa is characterized by severe weight loss due to self-inflicted starvation. However, it is a much more complex disorder than dieting out of control. Serious developmental deficits are at its roots. It is generally recognized that anorexia nervosa becomes manifest at the time of developmental transitions, most often during adolescence with its tasks of becoming self-reliant and independent and of growing beyond the immediate family. It used to be exceedingly rare, but it has been occurring with increasing frequency for the past 25 years. It is most often observed in the daughters of well-to-do and educated homes, but the social-educational background has been broadening. It has been observed in prepubertal children as young as nine years, and it now develops not infrequently in women in their 20's or 30's and also at the time of sociopsychologic tension and of decision making. It does occur in boys, but only one-tenth as often, mostly in prepuberty.

As long as anorexia nervosa was a rare condition, there was much confusion about its proper definition and the understanding of the underlying psychologic issues. Generalized conclusions were drawn from limited experience. In recent years observations of larger patient groups have resulted in the definition of a primary anorexia nervosa syndrome, as distinct from unspecific forms of psychologic undernutrition (Lucas 1981). In such atypical cases the weight loss may be secondary to loss of appetite in hysteria, depression, or schizophrenia, whereas in the primary syndrome there is no loss of appetite. The weight loss is due to deliberate abstinence from eating motivated by a fear of fatness and the relentless pursuit of thinness. It is the primary form that is on the increase, the explanation for which is only conjectural. One factor may be found in the preoccupation of the whole Western culture with slenderness. The increasing emphasis on achievement by women and on early sexual involvement seems also to play a role. Characteristic in anorexic patients is an enormous effort to be special and outstanding in every respect (Bruch 1973, 1978).

CLINICAL PICTURE

Most of the classic symptoms, both biologic and psychologic, are directly related to the malnutrition. The characteristic feature of anorexia nervosa is severe weight loss in the absence of organic disease. Its classic description is that of "a skeleton clad only with skin." The whole body, not only the fat tissue, is involved. Menses cease, the skin becomes dry with a yellowish tint, the hair is stringy, the abdomen is sunken in, and every bone shows. Remarkable are the changes in the face, with deep hollows in the cheeks, like the cachectic face of old age. As the protective layers of the body disappear, many anorexic patients find it painful to sit for any length of time.

There are many other physiologic changes due to the starvation, such as low blood pressure, slow pulse, low basal metabolism rate, anemia, and sleep disturbances (Table 29–7). These signs occur in all people who starve and do not differentiate primary anorexia nervosa from the atypical form. The differences lie in the psychologic reaction (Table 29–8). Nonanorexic starving people eat whatever is available, in contrast to the anorexic, who starves in the midst of plenty, and they will complain about the weight loss or are indifferent to it. The anorexic patient takes excessive pride in it. She defends with vigor and stubbornness her

Table 29–7. ABNORMALITIES NOTED ON PHYSICAL EXAMINATION OF 65 YOUNG ADOLESCENTS WITH ANOREXIA NERVOSA*

Abnormality	% Affected
Skin (hairiness, scaliness, dirtiness, desquamation)	88
Hypothermia (rectal temperatures <96.6° F.)	85
Bradycardia (<60 beats per minute)	80
Cachexia	72
Bradypnea (<14 breaths per minute)	66
Hypotension (systolic pressures below 70 mm. Hg)	52
Heart murmurs	38
Peripheral edema	23

*Modified from Silverman (1977).

gruesome emaciation as not being too thin. She is identified with her skeleton-like appearance and actively maintains it and claims not to "see" it as abnormal or ugly. The more pride she takes in her thinness, the stronger the assertion that she looks "just fine." In test situations anorexic patients tend to overestimate also the size of others and of abstract distances. The degree of consistent overestimation seems to be an index of the severity of the illness (Gardner and Garfinkel, 1977). The more they overestimate, the more resistant they are to treatment and the less ready to examine the faulty values and concepts with which they operate. The anorexic patient also fails to experience the body as being her own but views it as something separate from her psychologic self, something extraneous, quite often as the property of her parents. These disturbances in body image are characteristic of primary anorexia nervosa.

The symptom that arouses most concern, compassion, frustration, and rage is the anorexic patient's refusal to eat. There is no loss of appetite, as the name of the condition seems to imply, but like other starving people, anorexics are frantically preoccupied with food and eating, though they deny experiencing hunger. In advanced stages of emaciation, true loss of appetite may occur comparable to the disinterest in food during the late stages of externally induced famine. However, after recovery many confess that they actually had been tormented by hunger but were afraid to eat because they feared losing control. Many say, "I do not dare to eat. If I take just one bite, I'm afraid I will not be able to stop." This is actually what

Table 29–8. PSYCHOLOGIC TRAITS IN ANOREXIA NERVOSA

Denial of illness
Active refusal of food or binge eating and vomiting-purging
Overactivity and perfectionism
Disturbance of body image
Inaccurate perception of hunger and other bodily sensations
Deficit in sense of identity and effectiveness

occurs in many. When they give in to their hunger, they go on eating binges during which they devour huge quantities of food and fluid. Such bulimic episodes are followed by self-induced vomiting or excessive use of laxatives. Such phases of binge eating often alternate with starvation periods during which they complain about feeling "full" after a few bites of food and even after a few drops of fluid.

Characteristic of anorexia nervosa is the paradox of food refusal while being frantically preoccupied with eating. Most anorexic patients develop unusual, highly individualistic food habits. They eliminate all "fattening food," usually carbohydrates and fats. Many restrict themselves to one meal, late at night, so that they can go to sleep immediately without suffering too much guilt for having eaten at all. A few admit that if they wait until they are absolutely famished, eating gives them great pleasure, often indescribable delight. If they eat before being quite so weak and empty, they feel depressed and guilty for eating. Those who get involved with binge eating eat anything they can lay their hands on, large amounts of what they call "junk" food, mainly carbohydrates and sweets, which are not disliked at all but are deliberately excluded during the starvation phase.

In the pursuit of extreme thinness, anorexics do not rely on food restriction alone; they also engage in frantic exercise programs, such as swimming or jogging by the mile, playing tennis for hours, or doing calisthenics to the point of exhaustion. But they deny that they feel fatigue. They also spend long hours on school assignments. They usually had been good students, but now they become frantically preoccupied with making excellent grades. This overactivity stands in contrast to the apathy that accompanies malnutrition for other reasons, and it persists until the emaciation is far advanced. There is always a severe sleep disturbance. Sometimes they sleep no more than three or four hours per night.

Though they give a first impression of being alert, active, and self-assertive, anorexics suffer from a paralyzing sense of ineffectiveness, which pervades all thinking and activities. They perceive themselves as acting only in response to demands that come from others and not doing anything, except the excessive dieting, on their own initiative. This deep sense of ineffectiveness seems to stand in contrast to the reports of normal early development, which is described as having been free of difficulties and problems to an unusual degree.

FAMILY INTERACTION

It has always been a puzzle that this severe illness suddenly befalls adolescent girls who are described as having been perfect children. Evaluation of the developmental history shows that the very features that parents glowingly describe as evidence of superior behavior are indications of serious maldevelopment. The compliant and over-conforming behavior served as a camouflage for underlying serious self-doubt and a sense of inadequacy. The illness appears to be a desperate fight against feeling enslaved, not permitted or competent to lead a life of their own. This effort, instead of solving, only reinforces the difficulties. Changing the body size cannot provide what they need and search for, and it cannot correct the deficits in their overall development. They are guided by erroneous expectations when they withdraw to their own bodies and choose the road of self-starvation in a futile effort to attain selfhood and self-directed identity.

On first impression the families seem to be stable, with few broken marriages, and the parents emphasize the happiness of their homes, considering the patient's weight loss the only difficulty (Table 29–9). Extended contact reveals that underlying the parents' marital harmony is deep disillusionment with each other. The fathers are highly successful in their careers but are rather remote from their families, until the illness forces their own involvement. The mother, frustrated in her need for intimacy in the marital relationship, creates a setting of excessive closeness toward the later anorexic child and thereby dominates her. But family life is described as happy. The parents offer the best of care and educational and cultural opportunities and expect obedience and superior performance in return. For a while the child fulfills the parents' dreams of success through compliant behavior and attempts to compensate them for their own shortcomings.

Conditions change with adolescence. The child, ill prepared for the new demands for independence and self-reliance, withdraws to her own body as the realm where she can exercise control, and the illness begins. There are few conditions that provoke so much concern, but also frustration, rage, and anger, as the spectacle of a starving

Table 29–9. FAMILY INTERACTION*

Families appear stable and happy, but suffer from underlying dissatisfaction
Upwardly mobile, financially successful
Appearance-, success-, and weight-conscious
Parents relatively old
Preponderance of daughters—two-thirds of the families have no sons
Excellent child care but superimposed without regard of child's expressions of needs and desires
Later anorexic child overvalued, unrealistic expectations
Implied power struggle, which becomes manifest with onset of illness

*Less well defined since socioeconomic setting has broadened.

child refusing food. Life in the quiet, smooth running home degenerates into constant arguments and open fighting. A struggle for power, formerly concealed by the accommodating surface behavior, comes into the open. The parents' need to be in control expresses itself in their dictatorial efforts to make the patient eat and resume her former compliant behavior. A precondition for successful treatment is the resolution of the acute conflict; for long range recovery the whole pattern of family interaction needs to change so that the patient's personality can mature (Selvini, 1978).

The family constellation may be of psychologic importance. The parents are relatively old when the later-anorexic children are born, though some are firstborns. Whatever the individual characteristics of such parents, they provide well established, well functioning, and set-in-its-ways homes. Two-thirds of the families have no sons, but daughters only. In other families the anorexic girl was the youngest, with two or three older brothers, feeling under pressure to keep up with the boys. The concepts of "special achievement" or "only outstanding is good enough" are built into such families, though this is rarely openly expressed.

HUNGER AWARENESS AND INDIVIDUATION

How do such well meaning, good parents fail to transmit an adequate sense of competence and self-value to their children, who in turn strive for perfection and are quite unrealistic in their concept of what being special means? Relentless pursuit of thinness, the phobic avoidance of being fat, becomes to them a concrete way of being special and outstanding. They live in fear of not having control over their eating and also feel that they have no control over their life in general. This is expressed in their inability to make decisions and their clinging dependency on the home, while openly rejecting everything that comes from the parents.

To visualize how parental attitudes are related to a child's failure to develop a sense of autonomy, one must conceive of development as occurring in a dual mode (Bruch, 1973). Two forms of behavior need to be differentiated from birth on, that initiated in the individual and that in response to external stimuli. For healthy development, appropriate responses to clues originating in the child are as essential as stimulation from the environment. The outstanding finding in the early development of anorexic patients is the paucity of confirmation of child initiated clues. In these families growth and development are not conceived of as the child's accomplishment but as that of the parents.

The confusion in hunger awareness, the deep fear of having no inner controls, can be related directly to the early feeding experiences. Every detail of the excellent child care is done according to the mother's decision and feelings, not according to the child's clues to her own needs. A child whose mother offers food when she shows signs of nutritional distress learns to recognize "hunger" as a distinct sensation and demands food accordingly. If a mother's responses are inappropriate, be it neglectful, oversolicitous, or inhibiting, the child will fail to learn to differentiate between being hungry and other sources of discomfort, and she will grow up without discriminating awareness of her bodily sensations or a sense of control over them. This type of superimposition, or neglect to respond discriminately to a child's expressions of various needs and feelings, results also in serious psychologic disturbances, not only in confused hunger awareness. Without inner guideposts and without the experience of living her own life, such a child feels like being the property of her parents and feels helpless under the influence of internal urges and external demands. She will be deficient in her sense of self, of autonomy and decision making, and cannot differentiate her own needs and those imposed from the outside.

The pubertal growth spurt is frightening because it is associated with an increased appetite and a rapid gain in weight, and such children experience this as a loss of control. These developments coincide with growing dissatisfaction about not developing as individuals in their own right. They expect that by being slim, they will become more competent and deserving of respect.

CONCEPTUAL DEVIATIONS

Disturbed hunger awareness is only one aspect of the disturbances in conceptualization that are related to faulty transactional experiences early in life. Long before the illness becomes manifest these youngsters suffer from lack of autonomy and are overobedient and unable to make independent decisions, or even to identify their own preference. They always did what they were told to do, and never dared to test out their own capacities. This overconformity is praised by parents, and also teachers, as special goodness instead of being recognized as a sign of abnormal development. The shortcomings become dramatically apparent with adolescence. Subtle expressions had been present throughout childhood but were overlooked (Bruch, 1977; Table 29–10).

As patients progress in psychotherapy beyond their obsessive concern with food and weight, defensively claiming that everything was right in their homes, they begin to look at their early

Table 29–10. PSYCHOLOGIC ANTECEDENTS IN LATER-ANOREXIC CHILD

Excessive closeness to one parent—usually, but not always, the mother
Unresolved enmeshment with family
Deficits in hunger awareness and in control over body functions
Lack in autonomy, self assertion, and decision making, even in recognizing of preferences
Feels driven to be good and to live by the rules
Model children—academic overachievers; often praised by parents and teachers for special goodness
Overconformity in friendship patterns, usually only one friend at a time; social isolation even before onset of illness
Literal minded and inflexible in moral judgment and behavior

development in more objective terms. During this new self-evaluation it becomes apparent that they had always "marched to a different drummer" who had kept them tied to the values and convictions of early thinking.

We have learned from Piaget that conceptual development goes through definite phases, which are partly innate but also require, for appropriate development, interaction with an encouraging environment. Potentially anorexic youngsters continue to function with the morality and style of thinking of early childhood, the period of egocentricity and of preconceptual and concrete operations. The next step, the capacity for formal operations, the ability for new abstract thinking and evaluations that is characteristic of adolescent development, is deficient or absent.

The discovery of real defects in conceptualizing abilities comes as a surprise. Anorexic patients usually excel in their school performance, and this has been interpreted as indicating great giftedness and intelligence. Not uncommonly the excellent academic achievements are the result of great effort, which becomes even greater after the illness develops. Sometimes it comes as a shocking surprise that their performance on college aptitude tests, or some other evaluation of general ability, falls below what the excellent school grades had suggested. As a group they are what one might call academic overachievers.

Much more serious indications of disharmonious development are found in their everyday thinking, in their rigid interpretation of human relationships, and in their abnormal self-evaluation and self-concept. They suffer from a nearly delusional disturbance in the "body image" and are unable to "see" themselves and the severe emaciation realistically. This misperception is part of disturbed conceptualization on a much wider scale. Their abnormally low self-concept is also an expression of disturbed thinking. Each anorexic conceives of herself as inadequate, mediocre, bad, or unacceptable in some other way.

The early roots of their cognitive egocentricity have been recognized during therapeutic re-evaluation. These girls describe themselves as always having lived by the rules, driven to be good, anxiously avoiding any criticism or discontent from their parents or teachers. It has been recognized for a long time that these youngsters skip the classic period of resistance early in life; they continue to function with the morality of young children, remaining convinced of the absolute rightness of the grown-ups and of their own obligation to be obedient.

The same overcompliant adaptation is expressed in the friendship patterns. Commonly there is a series of friendships, only one friend at a time, whereby they develop different interests and a different personality with each new friend, without awareness that their own individuality has something to contribute to a friendship. Each friendship ends without any particular reason except their getting out of step. If they have one particular friend, they are invariably in the role of the follower.

There is increasing social isolation during the year preceding the illness. Some explain that they withdrew from their former friends; others feel that they were excluded. With their rigid judgmental attitude they begin to complain that the others are too childish, too superficial, too interested in boys, or in other ways are not living up to the ideal of perfection according to which they themselves function and which they also demand from others. The new ways of acting and thinking that are characteristic of adolescence are strange and frightening to them. The illness becomes manifest when they are completely out of step with their age goup.

Unsatisfying overcompliance is also apparent in their attitude toward work and school; doing what is expected remains for them the dominant theme. Even when there is some expression of independent thinking, even originality, the inner experience is that of conforming. The literal minded, concrete style of thinking expresses itself in many different ways. These youngsters cling with enormous rigidity to their early convictions and notions and to the distorted sense of reality with which they conduct their lives. They will defend with vigor and secret superiority the absolute rightness of their early convictions.

PREVENTION

The pediatrician is in the ideal position of recognizing the potential for anorexia nervosa by identifying some of the psychologic antecedents and by modifying excesses. Doing that, however, is a difficult task, since most of the deviations occur on the so-called positive side of behavior. These children rarely "give trouble," to such an extent that one might say a child who never gives trouble is already in trouble. In addition, these features are not specific for anorexia nervosa but are also observed as precursors of borderline disorders, narcissistic personality or schizophrenia. It seems that "goodness" and "obedience" have a special quality in the later anorexic child. Once the weight loss begins, the anorexia should be relatively easy to identify; characteristic are an unrealistic denial of illness, the fear of losing control, the aura of special achievement, and the feelings of ineffectiveness that underlie the defiant defense of the excessive thinness.

TREATMENT

Anorexia nervosa is a complex condition with difficult and often frustrating treatment problems (Rollins and Piazza, 1981). Treatment involves several distinct tasks, and the efforts of the pediatrician and psychiatrist need constructive integration. The pediatrician is most likely to be the first to be consulted. He faces the question of the extent to which he should handle treatment alone and when to ask for a psychiatric consultation. To a large extent this will depend on local conditions. If an experienced psychiatrist is available, early referral is advisable, at least for an evaluation, before the symptoms have become firmly established.

The psychiatric evaluation should precede extensive laboratory studies that are traditionally carried out to exclude the presence of organic illness. The exceedingly rare cases of severe weight loss for organic reasons are clinically and psychologically distinctly different from the anorexia nervosa picture. Excessive emphasis on organic factors carries the danger of family's becoming preoccupied with the frightening aspects of physical illness, and may result in neglect of the patient's emotional needs. Anorexia nervosa is a condition in which the prognosis is closely related to the pertinence of the treatment program and its early institution. Correction of the psychologic abnormalities is an essential part of this program.

Treatment involves several distinct tasks that must be integrated: restitution of normal nutrition, or at least improvement over the severe starvation; resolution of the disturbed patterns of family interaction; and individual psychotherapy to correct the deficits and distortions in psychologic functioning.

Starvation itself has a far reaching, distorting influence, biologically and psychologically. The metabolic condition may become dangerous when vomiting or abuse of laxatives or diuretics leads to disturbance in the electrolyte balance, which carries the danger of cardiac arrest. All too often, in a simplistic approach, weight increase is enforced and this is considered a cure, which, of course, cannot last. Fear of having no power in relation to others underlies the whole syndrome, and this needs to be acknowledged in the refeeding efforts. Family therapy is particularly effective with young patients soon after the onset of the illness (Minuchin et al., 1978). When the condition develops later, after the patient has moved away from home, or when it has existed for any length of time, disengagement and redirection of malfunctioning family processes are still essential, but not sufficient.

For effective and lasting treatment, the underlying misconceptions must be recognized and corrected. These patients need help in their search for autonomy and self-directed identity. Therapy must address itself to the core issues by evoking awareness in the patient of impulses, feelings, and needs originating within herself. This can be accomplished through a therapist's alert and consistent, affirming or correcting responses to any self-initiated behavior and expressions. As they gain self confidence and become more competent in living as self-directed individuals, these patients become ready to re-evaluate their overly rigid moralistic value system. This step is not easy, since they have clung with enormous tenacity to their erroneous convictions. As an anorexic patient comes to recognize and value her own abilities, she becomes capable of living as a self-directed, competent individual who can pursue goals in life other than maintaining a dangerously low body weight.

HILDE BRUCH

REFERENCES

Bruch, H.: Eating Disorders: Obesity, Anorexia Nervosa, and the Person Within. New York, Basic Books, Inc., 1973.

Bruch, H.: Psychological antecedents of anorexia nervosa. In Vigersky, R. (Editor): Anorexia Nervosa. New York, Raven Press, 1977.

Bruch, H.: The Golden Cage: The Enigma of Anorexia Nervosa. Cambridge, Harvard University Press, 1978.

Gardner, D. M., and Garfinkel, P. E.: Measurement of body image in anorexia nervosa. In Vigersky, R. (Editor): Anorexia Nervosa. New York, Raven Press, 1977.

Lucas, A. R.: Toward the understanding of anorexia nervosa as a disease entity. Mayo Clinic Proc., 56:254, 1981.

Minuchin, S., Rosman, B. L., and Baker, L.: Psychosomatic Families: Anorexia Nervosa in Context. Cambridge, Harvard University Press, 1978.

Rollins, N., and Piazza, E.: Anorexia nervosa: a quantitative approach to follow-up. J. Am. Acad. Child Psychiatry, 20:167, 1981.

Selvini, M. P.: Self Starvation. New York, Jason Aronson, 1978.

Silverman, J. A.: Clinical observations in treatment. In Vigersky, R. (Editor): Anorexia Nervosa. New York, Raven Press, 1977, p. 333.

30

Failure to Thrive

Failure to thrive (FTT) is a chronic, potentially life threatening disorder of infancy and early childhood, afflicting as many as 10 per cent of the rural outpatient population and 3 to 5 per cent of all infants under one year of age admitted to pediatric teaching hospitals (Mitchell et al., 1980). It accounts for 1 per cent of all pediatric hospitalizations and presents, in 80 per cent of cases, before the age of 18 months (Berwick, 1980; Kotelchuck and Newberger, 1978). Although there are discrepancies in the diagnostic criteria for FTT, the term is typically used to describe infants and young children whose weight is persistently below the third percentile for age on appropriate standardized growth charts or less than 85 per cent of the ideal weight for age (Barbero and Shaheen, 1967; Berwick, 1980). It may also present as acute weight loss or a failure to gain weight with the loss of two or more major percentiles on the growth curve. This common disorder has a high social cost in its consumption of medical resources and in its significant cognitive and behavioral morbidity (Berwick, 1980).

The etiology of FTT is traditionally dichotomized into organic and nonorganic categories (Hannaway, 1970). Nonorganic FTT is defined as a failure of growth without diagnosable organic disease, whereas organic FTT is a growth symptom of virtually all serious pediatric illnesses. A recent study by Homer and Ludwig (1981) suggests that three etiologic categories are necessary to adequately describe the causation of FTT: organic, mixed (organic and nonorganic), and nonorganic. In their study of a tertiary care population of 82 hospitalized children, 28 per cent of FTT cases were characterized by primarily organic contributants and 46 per cent involved primarily nonorganic contributants. Interestingly, 26 per cent of the patients presented with a "mixed" etiology (organic and nonorganic) and could not be described as organic or nonorganic alone.

Sameroff and Chandler (1975) have added the notion of a continuum of caretaking (nonorganic) casualty to the continuum of reproductive casualty in disorders such as FTT, suggesting a spectrum of risk factors responsible for the ultimate developmental outcomes in target children. Homer and Ludwig's study suggests that such a continuum

exists in FTT. It has been the authors' experience that the general approach of risk analysis, adapted from Sameroff and Chandler, is the most clinically useful and most efficient approach to questions of etiology in FTT. Rather than searching along the organic-nonorganic dichotomy, the diagnostic search is directed toward identifying positive indicators of risk, whether they be organic or nonorganic.

Whether children with FTT have organic, nonorganic, or mixed conditions, all children who fail to thrive have suffered an organic insult: their nutrition prior to diagnosis has been insufficient to sustain growth (Whitten et al., 1969). As summarized by Lloyd-Still (1976) and Pollitt and Thompson (1977), the final etiologic pathway common to all children with FTT is inadequate caloric intake, retention, or utilization. Such undernutrition has a significant and complex effect on later cognition and behavior, depending on the age at the onset of the malnutrition, its degree and its duration, as well as the socioeconomic environment of the child after nutritional rehabilitation. All etiologic categories of FTT thus are associated with a significant medical illness—malnutrition. Treatment of the malnutrition and its associated medical problems is the first and chief goal of the pediatrician. When this treatment requires the remediation of interactional, behavioral, and social patterns, the pediatrician's goal broadens to include the organizing of a treatment team adequate to the task.

In the approach to FTT, a physician must be aware that in the majority of outpatient cases, FTT is primarily nonorganic in etiology. Mitchell et al. (1980), in a study of FTT found no cases of organic growth failure in 30 patients in a primary care setting. Recent studies from tertiary medical centers report a 70 to 80 per cent nonorganic etiology for FTT (Homer and Ludwig, 1981; Sills, 1978). Given the preponderance of nonorganic causation in FTT, the physician is compelled to consider social, emotional, and environmental causes as positive diagnoses. In the face of a noncontributory physical examination and a history unsuggestive of organic disease, extensive laboratory workup and protracted hospitalization promise to contribute little to an organic diagnosis. Evaluation of organic and nonorganic etiologies in FTT should be undertaken concurrently. Medical, nutritional, social, and behavioral-interactional assessments should be obtained as part of the initial work-up. The physician can thereby reduce the cost of unnecessary diagnostic maneuvers and efficiently entertain, with the information necessary, the spectrum of etiologic possibilities that result in failure to thrive.

NONORGANIC RISK FACTORS

Failure to thrive can be assessed in terms of nonorganic and organic risk factors contributing to growth retardation. Determination of nonorganic risk requires assessment in four areas of family functioning: temperament and sickliness in the child and difficulties in the parents, feeding behavior and interactions, nonfeeding interactions, and psychosocial stressors (loss, marital stress, poverty). The gathering of information requires family interviews, observations of feeding and play, developmental assessment, and careful record review. The constellation of nonorganic risk being sought is as follows: (1) a sickly, difficult infant, an isolated overwhelmed mother, and a father emotionally or physically unavailable for support; (2) a disordered feeding situation resulting in inadequate caloric intake or retention; (3) an impoverished nonfeeding interaction; and (4) a social environment of loss, stress (especially marital), or poverty. Figure 30–1 summarizes these areas of nonorganic risk.

TEMPERAMENT AND SICKLINESS IN THE CHILD AND DIFFICULTIES IN THE PARENTS

Controlled studies of FTT populations by Pollitt (1975) and by Newberger et al. (Bithoney and Newberger, 1982; Kotelchuck and Newberger, 1978) describe temperamental and behavioral variables that identify nonorganic FTT families when compared with normal and abused control subjects. According to Newberger et al., the child at risk is perceived by his parents as behaviorally difficult and sickly. Pollitt's work (1975) elaborates on the behaviorally difficult qualities of the infant with FTT by observing the target infants to be uniquely immature as well as lethargic and passive as compared with control subjects. Gaensbauer and Sands' study (1979) of 48 abused or neglected infants describes a set of at risk infant behaviors similar to those observed by Pollitt and by Rosenn et al. (1980) in FTT infants—affective withdrawal, lack of pleasure, inconsistency and unpredictability, ambivalence-ambiguity, and negative affective communications. Clinical observations suggest that there is, in fact, a spectrum of behavioral difficulties in FTT that ultimately contribute to the perception of a "burdensome" infant. These can include problematic eating behavior, poor state control, a low threshold to overstimulation, whininess, oppositionalism, defiance, and clinging, as well as sleeping and elimination problems noted by Pollitt (1975).

Concepts such as the "difficult child" described by Chess and Thomas in Chapter 10 seem to

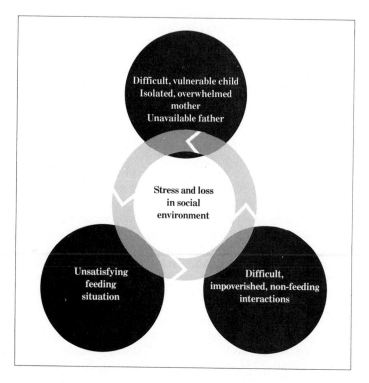

Figure 30–1. Nonorganic risk factors in failure to thrive.

apply. The perception of the FTT infant as sickly in addition to difficult only adds to the burdensomeness of the infant. Sickliness can be in the eye of the beholder: e.g., an irrational fear that the child may die may in fact be based on an early history of vulnerability (Evans et al., 1972; Sameroff and Chandler, 1975; Green and Solnit, 1964; see also Chapter 26 concerning the vulnerable child syndrome).

Mothers of FTT children, according to Bithoney and Newberger (1982), appear to be identified by their social isolation and their sense of being out of control or overwhelmed with life's situation. Fathers in the same study were present in the home as much if not more than controls, but were unavailable for emotional support or child care activities. This may be a part of or the result of marital strain. Marital dissatisfaction was a key identifier of FTT families in the Bithoney study despite the equal proportion of homes that appeared to be intact among FTT and control families. In Pollitt's controlled study, mothers of FTT children were more discipline oriented, less affectionate, and less verbally interactive with their target children. They were also identified by their reports of stressful childhoods and significant material dissatisfaction. Notably, they were not identified by a higher incidence of overt psychopathology.

FEEDING BEHAVIOR AND INTERACTIONS

Pollitt's controlled study of 19 FTT infants and families provides the best data relating to the eating disorder in FTT (Pollitt, 1975; Pollitt and Thompson, 1977). He demonstrated in his FTT sample a significantly lower total caloric intake, more feeding problems by report and by observation, skimpier, less regular meals, and poorer responses to food than in controls. Newberger's study identified FTT mothers, in contrast to controls, as uniquely reporting more feeding problems in their target children. Whitten, Pettit, and Fischoff, in their 1969 hospital simulation of inadequate mothering, also suggest inadequate caloric intake as the major causal risk factor in the nonorganic genesis of FTT.

NONFEEDING INTERACTIONS

FTT as an interactional disorder has been described by Pollitt (1975), Harper and Richmond (1979), Fraiberg (Fraiberg, 1980; Fraiberg et al., 1975), and others. The contribution of the infant as well as the caregiver to the dyadic dysfunction in FTT has been stressed. A relative poverty of interaction in FTT has been described by Pollitt et al. (1975). Identifiers of FTT interactions in this study included less verbalization, less positive affect, meeting the child's needs less often, and more punishment. Clinical observations, however, support the existence of a spectrum of interactional risk patterns including impoverished interactions and understimulation, high intensity interactions and overstimulation, and asynchronous interactions and inconsistent stimulation. The final outcome of each pattern is a relative impoverishment or neglect of the child's affective needs.

Recent studies by Gordon and Jameson (1979) and Rosenn et al. (1980) suggest that this interactional disorder is related to an attachment disorder in the FTT dyad. Gordon and Jameson (1979) observed 12 FTT dyads and 12 control dyads in a modified Ainsworth stranger situation. They found 50 per cent of the FTT infants to be "insecurely attached" to their mothers (i.e., unable to respond normally to separations) in contrast to 16 per cent of the controls. In Rosenn's controlled study of "approach-withdrawal" interactions in eight nonorganic FTT infants, 10 organic FTT infants, and seven control infants, the nonorganic FTT infants were unique in their preference for stimulation-at-a-distance versus stimulation-at-close-range, as if they preferred less intimate interactions (Rosenn et al., 1980). The observation of an insecurely attached or overly attached dyad may therefore constitute a significant nonorganic risk factor for FTT.

PSYCHOSOCIAL STRESSORS

Loss, stress, poverty, and marital strain can be causal factors in the genesis of FTT. Losses to the family may include death, illness, injury, miscarriage, separations or moves, job layoffs, major changes in self-image, or loss of the "expected child" (disappointment in the reality of the target child). Stress can come from any source that might influence family functioning, including a new pregnancy or child in the family. Pollitt (1975) describes FTT families as being unique in their high size and density index (more children in a smaller span of time than control families). Evans et al. (1972) studied family patterns of loss and stress in nonorganic FTT and were able to delineate three prognostic patterns:

1. Good prognostic pattern: acute stress or loss, strained interactions, child seen as "ill."

2. Guarded prognostic pattern: chronic stress or loss with poverty, strained interactions, child seen as "ill."

3. Poor prognostic pattern: chronic stress or loss with neglect of child, hostile interactions, child seen as "bad."

Controlled studies of FTT repeatedly point to marital dissatisfaction and distress as a prominent risk factor in the nonorganic arena. FTT families often, however, seem to be adequate on the surface, appearing as intact as control populations. Marital dissatisfaction, however, may be a family secret and may be much less discernible clinically than the child's medical presentation. Finally, poverty appears to constitute a significant stressor in FTT families, contributing through food shortage and generalized family strain to the genesis of FTT.

ORGANIC RISK FACTORS

FTT is best viewed as being caused by a combination of factors, both nonorganic and organic. The following section presents the significant organic risk factors that should be considered in the assessment of FTT: minor congenital anomalies, prenatal malnutrition, prematurity, and ongoing medical illness.

MINOR CONGENITAL ANOMALIES

Nelson's *Textbook of Pediatrics* lists multiple miscellaneous patterns of deformity that are routinely associated with short stature and decreased weight. The list is long and includes such cosmetic deformities as anteverted nostrils, microcephaly, and lateral displacement of the canthi. All have been reported as part of identifiable syndromes associated with short stature and poor weight gain. Such dysmorphic features should be looked for and catalogued because they may provide clues to appropriate management. Of particular interest is the fetal alcohol syndrome, which is documented as causing severe growth disturbance in 97 per cent of the cases both pre- and postnatally. Leonard et al. (1966) documented an increased incidence of alcohol abuse in the parents of her FTT probands but did not document the incidence of the fetal alcohol syndrome in the children, because this syndrome had not yet been fully described. Several other authors have noted a high incidence of alcoholism in FTT families (Chase and Martin, 1979; Elmer, 1960; Leonard et al., 1966; Pollitt and Thompson, 1977). Maternal cigarette smoking is also known to be a cause of intrauterine growth retardation and has been associated with later short stature.

Other forms of in utero toxin exposure have recently been shown to be associated with short stature, such as fetal hydantoin and trimethadione syndromes. Until these syndromes were recognized, children who demonstrated the associated findings—decreased height, weight, and head circumference along with developmental delay and irritability—may have been diagnosed as having nonoganic FTT (see also Chapter 20).

PRENATAL MALNUTRITION

Infants suffering from intrauterine growth retardation due to poor maternal nutrition during pregnancy, congenital infection, or inadequate placental circulation are born small for gestational age (SGA). The weight at birth is below the expected norms for gestational age. Fitzhardinge and Steven (1972), in a four year follow-up study of 96 SGA infants, found that 35 per cent were below the third percentile for both height and weight by age four. Infants with the poorest outcome in terms of growth had a lower socioeconomic status than the other study infants and were described as having parents who did not stimulate them adequately.

SGA infants are documented as suffering from an overall reduction in brain weight, cell number, and myelinization (Chase and Martin, 1979). They are well described as being disorganized in terms of motor control. They also demonstrate marked swings in emotional state and hypersensitivity to most stimuli (Brazelton, 1981). Such SGA infants are overrepresented in childhood populations suffering from pediatric social illnesses such as child abuse and neglect.

PREMATURITY

Prematurity may also predispose to FTT. The 10 to 40 per cent incidence of low birth weight (LBW) infants who subsequently fail to thrive is a significant over-representation from the 10 per cent LBW rate seen in the population at large (Bithoney and Newberger, 1982; Elmer, 1960; Kaplan et al., 1973; Riley et al., 1968). The data for appropriate gestational age (AGA) prematures are less clearcut, however, than for SGA prematures. The high incidence of FTT in LBW infants may be due to perinatal catastrophe unrelated to low birth weight and prematurity. A twofold increase in the incidence of perinatal complications has been shown in FTT children versus controls (Mitchell et al., 1980). Such perinatal complications result in neural tissue loss and are associated with subsequent effects on growth, cognition, and behavior.

ONGOING MEDICAL ILLNESS

The risk of ongoing medical illness contributing to weight loss or growth retardation requires careful evaluation. Subclinical fluctuations in chronic illness may constitute less obvious yet significant contributors to ongoing FTT; stabilization and clarification of its role in growth are crucial factors in the comprehensive treatment of FTT. For example, Mitchell et al. (1980) in a controlled study found that nonorganic FTT children have a significantly increased incidence of otitis media as compared with controls. The presenting symptoms of otitis may be irritability and poor appetite. Such symptoms can exacerbate an already difficult feeding situation.

ASSESSMENT OF FAILURE TO THRIVE

The differential diagnosis of FTT is a complex and often anxiety provoking task. On the one hand the clinician is acutely aware that virtually every serious medical disease of childhood eventually manifests itself in FTT. On the other hand the physician in a community setting sees FTT of a predominantly nonorganic origin. The physical examination and history provide the best tools to diagnose organic risk factors worthy of further pursuit (Berwick, 1980; Green and Solnit, 1964). The more comprehensive evaluation necessary to pursue nonorganic risk factors, however, may not be easy to carry out. Alliance building with FTT families is difficult. Help-rejecting styles and denial of psychologic factors are common, and loss to follow-up is a major risk in the clinical care of FTT infants. In a two year review of FTT cases at Children's Hospital in Boston, Berwick and Levy (1982) reported a 90 per cent loss to follow-up. Since that report the organization of a multidisciplinary assessment team and follow-up clinic for FTT has begun to reverse the 90 per cent figure; however, these are not easy families to work with.

Diagnostic tools described by Herzog and Harper (1981) for "psychosomatic" assessment may be particularly helpful to the clinician faced with assessing FTT: watchful waiting, judicious disregard (of equivocal biomedical data), fence sitting (between psyche and soma), and multiple diagnoses (e.g., FTT, family disarray, and maternal depression).

Given the multiplicity and chronicity of problems manifested by FTT patients on follow-up (Berwick, 1980), the difficulties and discontinuities of their medical care are of concern. Concurrent assessment of organic and nonorganic risk factors should be undertaken from the onset. A stepwise approach should follow for the nutritional, medical, social, and behavioral-interactional assessments of the FTT child and his family. The pediatrician may want to organize an assessment team for this purpose including a nutritionist, primary nurse, and social worker.

GROWTH AND NUTRITIONAL ASSESSMENT

Growth Data Analysis

Is the child failing to thrive as determined by appropriate growth charts? The NCHS charts used in this text should be used to plot all growth data (Fig. 30–2). The Boston charts currently in use in many major medical centers are not representative of the general population. The NCHS charts were obtained from children of all races, whereas the Boston charts are based on data obtained by multiple measures of 100 white middle class children of Northern European ancestry living in an urban population center.

If the child is below the third percentile for height and weight, assessment should be made of the mean midparental height (average height of both parents at any time between 25 and 45 years of age). Such assessment is done to rule out a constitutional growth deficit. Smith et al. in 1976 showed that the correlation of infant stature with mean midparental height becomes highly significant by the age of two years. The correlation was as high as 0.7 by the age of 18 years. The use of mean midparental height in diagnosing constitutional short stature is a reliable predictor after age two and may be helpful in the infant presenting with FTT. Figure 30–3 provides an example of how to use the mean midparental height charts of Tanner et al. (1970). Any child whose height falls more than 2.5 standard deviations below that expected should be evaluated for failure to thrive.

The isolated finding of depressed weight for age may not lead to a diagnosis of FTT or malnutrition. The weight for height measurement is the key in the evaluation of FTT. A child with depressed weight for age may have suffered an acute nutritional insult earlier in life, with subsequent recovery and normal growth at the time of presentation. Evidence of a depressed height age or a weight for height deficit as plotted on the NCHS charts will confirm the diagnosis of FTT. The child whose weight for height is below the fifth percentile shows evidence of an ongoing malnourished state.

Malnutrition, when severe, affects head circumference. In studying the effects of malnutrition on growth, one finds that head circumference is usually spared, decreasing only after weight and then height are affected. A head circumference that is significantly decreased may indicate malnutrition in utero or in the first year of life when brain growth, the prime determinant of head circumference, progresses rapidly.

In the growth data analysis, careful note should be taken of the time and age of the child at onset of the growth insult. History taking should be focused on this time in the child's medical, developmental, and social history in an attempt to

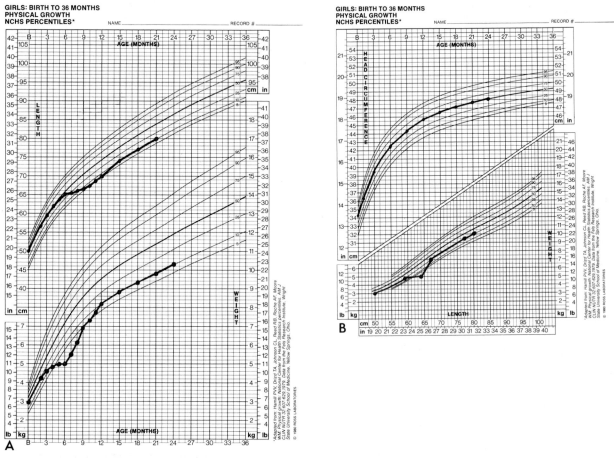

Figure 30–2. *A,* NCHS growth chart of 24 month old female with FTT from three to six months and subsequent weight recovery. *B,* NCHS head circumference and weight for height indices showing drop-off in weight for height from three to six months. (Adapted from Hamill, P. V. V., et al.: Physical growth: National Center for Health Statistics percentiles. Am. J. Clin. Nutr., *32:*607, 1979.)

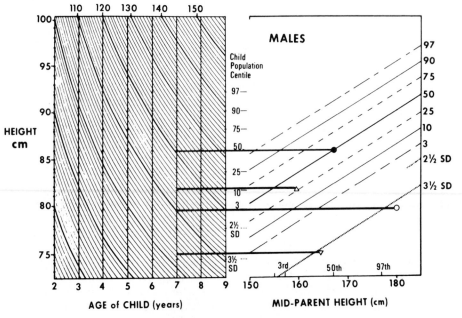

Figure 30–3. Tanner-Whitehouse chart. To use the chart, first find the child's height and then follow the curve until the child's age is reached. Next place a ruler from this point along the horizontal line across the middle of the chart to the right hand position of the chart. The point where this line crosses the vertical line of the mean parental height is noted and the percentile of the child's height, given mean parental height, is recorded. (From Tanner, J. M., et al.: Standards for children's height at ages 2 to 9 years allowing for height of parents. Arch. Dis. Child., *45:*755, 1970.)

pinpoint organic and nonorganic risks that may have been prominent (e.g., death in family, active medical illness).

Anthropometric Assessment

Careful measurement of height, weight, and head circumference should be carried out at the time of initial presentation and at all subsequent visits. Table 30–1 provides specific guidelines for obtaining the measurements necessary. After determining that the child meets the criteria for FTT, assessment can be made of the severity of the nutritional deficit. This is done by using standard anthropometric measures of triceps or subscapular skin fold thickness. The technique for obtaining the triceps measure is described in Table 30–2. The severity of the FTT as determined by these methods helps clarify for the clinician how forcefully to intervene and whether hospitalization may be necessary.

Diet History

A three to seven day diet history obtained by maternal recall and an observation of infant feed-

Table 30–1. MEASUREMENT OF WEIGHT, HEIGHT, AND HEAD CIRCUMFERENCE

Weight measurement
1. Measurement is taken using an accurately calibrated beam balance.
2. Infants are weighed in the nude; older children weighed with a minimum of clothing, i.e., no shoes, underwear only.
3. Measurements are recorded to the nearest 10 grams for infants and 100 grams for older children.
4. On follow-up weighings always weigh child on the same scale wearing the same amount and type of clothes, preferably in the nude.

Height measurement (height > 36 months of age)
1. Measurement is taken using a measuring rod or scale that is fixed either to a wall or onto the weighing scale itself.
2. Subject should remove shoes and stand on flat floor with feet parallel and torso and head erect.
3. Measurer gently lowers head piece to rest on top of head.
4. Measurement is recorded to the nearest 0.5 cm. or 1/4 inch.

Recumbent length measurement (<36 months of age)
1. Measurement is taken using a wooden length board.
2. Infant is laid on the board. The head is positioned against the fixed headboard, eyes vertical. The knees are extended and the feet are perpendicular to the lower legs.
3. The measurer moves the sliding foot piece to obtain a firm contact with the heels.
4. Measurement is recorded to the nearest 0.5 cm.

Head circumference measurement
1. Measurement is taken using a steel or fiberglass tape, not a cloth tape that may stretch.
2. With the head steady, the examiner places the tape around the head above the supraorbital ridges and the occiput so as to obtain the greatest numerical value for the head circumference.

Table 30–2. TRICEPS SKIN FOLD THICKNESS

1. Measurer tells subject to relax right arm at side and checks by gently shaking subject's arm.
2. Using the left hand, the measurer places thumb pointing downward on the medial side of the subject's arm. With thumb and index finger placed 1 cm. above the mark, the measurer grasps the skin fold parallel to the long axis of the right arm over the triceps muscle. The skin fold is lifted from the underlying muscle surface and is shaken gently to be certain that muscle has not been included.
3. The caliper is applied at the level of the horizontal mark below the thumb and index finger of the left hand. The caliper jaws are held in place while the measurer counts three seconds. Reading is taken and recorded to the nearest millimeter.
4. The entire triceps measurement is repeated two more times, starting the hand placement and entire process anew each time.

ing may be the most helpful diagnostic measures the clinician can perform. Upon obtaining the calorie count, the physician can determine whether the caloric intake (kcal. per kg.) of the infant is adequate for growth. By observing a feeding, the physician can corroborate diet history data with observed facts. The assistance and counsel of a trained nutritionist can prove very valuable in these assessments.

The average daily requirement to sustain normal growth in a healthy infant is approximately 115 kcal. per kg., although this varies somewhat for highly active, irritable, or infected infants as well as for infants with ongoing losses, e.g., vomiting and diarrhea. The average daily protein requirement for infants is 1.4 to 1.8 gm. per 100 kcal. This decreases to 1.2 to 1.4 gm. per 100 kcal. during the latter part of the first year. Caloric requirements also drop to approximately 100 kcal. per kg. per day in the same period. If the nutritional history is positive for inadequate calories and is negative for vomiting, bulky stools, or an increased metabolic- rate, the assessment should be geared to the cause of the inadequate caloric intake. Systematic observation of infant feeding behavior and infant-caregiver interactions provides critical information. In the hospitalized infant it is important to undertake such observations during a period when the medical evaluation can be suspended and family can be present. However, if the nutritional history and feeding observations show adequate caloric intake, the workup should be geared toward an evaluation of possible ongoing losses or hypermetabolism.

MEDICAL ASSESSMENT

A comprehensive medical assessment should include concurrent nutritional, social, and behavioral evaluations. Since in most cases of FTT there is no diagnosable organic cause, the initial evaluation should begin with an interdisciplinary approach. The use of a multidisciplinary team, if available, is helpful. The assistance of a nutritionist, primary nurse, social worker, and child developmentalist is invaluable and saves time and avoids needless frustration on the part of the clinician. The physician who has begun the evaluation of a child with FTT with a nutritional history and feeding observation next obtains a comprehensive medical history emphasizing known organic and nonorganic risk factors, including mean midparental height. The complexity of this assessment may make it impossible to do in the primary care setting and may require referral to a specialized center.

Review of Systems

The review of systems for FTT patients should not be different from the reviews given other patients. It should, however, emphasize organic symptoms known to be associated with FTT, such as vomiting, diarrhea, bulky or foul smelling stools, polyuria, polydipsia, and rumination as well as temperament characteristics and feeding disorders. The body systems most frequently involved in organic FTT are gastrointestinal, neurologic, and endocrinologic as shown in Table 30–3.

Medical History

The medical history should include a neonatal history. In particular, the presence of low birth weight, prematurity, perinatal complications, and toxin exposure should be explored. Finding such historical data, however, does not obviate the need for a full evaluation of the FTT. Any past hospitalizations should be noted, and a search for a history of chronic or recurrent symptoms is imperative.

Family History

The physician dealing with an FTT patient should obtain a detailed family history of organic medical disease such as gastroesophageal reflux, diabetes mellitus, or sickle cell anemia. Any history of FTT in parents or siblings should be noted. Further, any psychiatric history in the parents or immediate relatives should be catalogued, including a history of psychiatric hospitalizations and suicide attempts. A history of abuse or neglect in the parents' childhoods should also be ascertained.

Physical Examination

The physician should perform a comprehensive physical examination, including a full assessment of the child's development. Approximately 50 per cent of the children with FTT suffer from cognitive and language delays (Berwick, 1980). Other important observations include evidence of dysmorphic features or congenital anomalies. A full evaluation of the number and development of teeth is also helpful in assessing the child's musculoskeletal maturation state. Gross motor and tonal abnormalities should be evaluated. Krieger and Sargent (1967) note a characteristic "infant posture" in children suffering from FTT and deprivation, as shown in Figure 30–4. FTT infants have also been described as both significantly more "rigid" and more "flaccid" than control children (Barbero and Shaheen, 1967). Failure to identify and treat such abnormalities ultimately may limit the efficacy of other prescribed therapeutic modalities such as infant stimulation programs.

Finally, one should be especially wary for signs of neglect or abuse, bizarre or suspicious multiple skin lesions, evidence of fracture, retinal hemorrhage or detachment, severe untreated diaper

Table 30–3. STUDIES OF ULTIMATE DIAGNOSES IN FAILURE TO THRIVE*

	English (1978)	Sills (1978)	Hannaway (1970)	Riley et al. (1968)	Shaheen et al. (1968)	Ambuel and Harris (1963)
No. of cases	77†	185†	100†	83†	287‡	100‡
Diagnoses (%)						
Organic	53%	18%	49%	48%	85%	68%
Gastrointestinal§	(19)	(8)	(12)	(14)	(15)	(9)
Neurologic	(14)	(4)	(18)	(12)	(18)	(10)
Genitourinary	(4)	(0)	(5)	(5)	(5)	(4)
Endocrine	(4)	(1)	(5)	(1)	(4)	(5)
Cardiac	(7)	(1)	(4)	(2)	(13)	(31)
Other	(5)	(5)	(5)	(13)	(29)	(9)
Environmental	38%	55%	39%	31%	15%	NR‖
No cause (or constitutional)	9%	26%	12%	20%	NR‖	32%

*From Berwick, D.: Nonorganic failure-to-thrive. Pediatr. Rev., 1:265, 1980.
†Children hospitalized to diagnose FTT.
‡Survey of all children in hospital who have FTT.
§Includes cystic fibrosis.
‖Not reported.

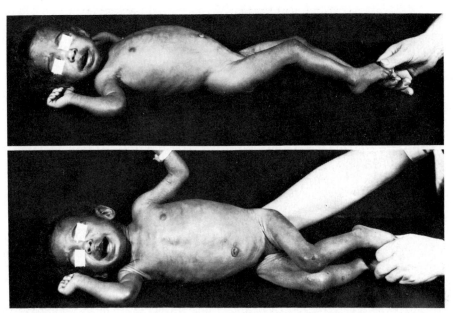

Figure 30–4. This infant demonstrates the characteristic posture seen in severely deprived infants with FTT. Note the resemblance to the decorticate posture seen with fixed neurologic disease. Unlike such neural posturing, this "deprivational" posture can be broken voluntarily if the child is offered a toy or other attractive object.

rash, or burns. The findings of FTT and evidence of physical abuse are very worrisome. Evans and her colleagues (1972) noted that the prognosis for children with FTT whose mothers viewed them as "bad," as opposed to "ill," was grave. She believed that this type of hostile parental attitude was so grave as frequently to warrant foster placement. Hufton and Oates (1977), in their six year follow-up of three FTT patients who also suffered from concomitant physical abuse, found that two of them had died. Koel (1969) reported a similar result. However, most of the medical literature dealing with FTT and physical abuse is not definitive. FTT alone cannot be considered a variant of child abuse; however, awareness of their possible association is necessary.

Laboratory Usage in FTT

The use of the laboratory should be frugal. Laboratory results are almost invariably not helpful in the differential diagnosis unless a specific indication for the tests is present. In Sills' major study of 185 patients with FTT, in every case of organic FTT there was a significant finding on the admission history or physical examination that led to the organic diagnosis (Sills, 1978). In Ambuel and Harris' study of 100 cases of FTT, 91 per cent of the cases in which an organic diagnosis was made had a suggestive finding on either the history or physical examination (Ambuel and Harris, 1963). Few laboratory tests should be done other than the baseline screens presented here without a specific medical indication on either the history or the physical examination. Tests requiring the

Table 30–4. ROUTINE LABORATORY TESTS IN FTT*

CBC
Urinalysis
Urine culture
Urine for reducing substances
BUN or creatinine
Tuberculin test
FEP/Pb
Stool for pH, reducing substances, occult blood, ova and parasites†
Sweat test
Albumin (in severe FTT only)
Chest x-ray

*Modified from Berwick, D.: Nonorganic failure-to-thrive. Pediatr. Rev., 1:265, 1980.

†Useful in recent immigrant children, e.g., Hispanics, and in children from areas where Giardia is endemic.

child to remain on nothing by mouth orders are especially confusing to the issue of weight loss or gain in the hospitalized child and should be used sparingly. Our current standardized recommendations for laboratory testing in FTT are shown in Table 30–4.

Developmental Assessment

The developmental assessment is a critical part of the work-up in FTT. Developmental morbidity in FTT is 50 to 70 per cent, with cognitive impairments (lowered IQ, speech and reading delays) most striking (Berwick, 1980). Developmental delays are predominant in the presenting picture of

FTT and often require remediation as part of the comprehensive treatment of FTT.

The initial evaluation should assess the distribution and severity of developmental delays. Clinical observations suggest that gross motor and speech delays are most severe, with relative sparing in fine motor and visual areas, and that cognitive assessment is distorted by a significantly depressed affect in the FTT infant. Development most typically appears to be uneven, affect is depressed, and the abnormalities of posture and tone already described may be prominent. With nutritional rehabilitation and developmental stimulation, "developmental catch-up" can be observed in FTT and should be followed with regular assessments.

The Denver Developmental Screening Test and the Bayley Scales of Infant Development are resources in the developmental assessment and follow-up of FTT infants (see Chapter 45). In the premature and newborn, the Brazelton Neonatal Behavioral Assessment Scale is most appropriate. Direct observation of oral motor reflexes and voluntary oral behavior should also be made to allow full assessment of feeding capacities. All developmental assessment should be undertaken in the presence of the primary caretaker to maximize performance potential. Such an assessment can provide important interactional information as well. (See Chapter 44 in regard to techniques for temperament assessment.)

INTERACTIONAL ASSESSMENT

The assessment of interaction should be made in feeding and play situations and should include observations of the infant with each primary care giver and with a stranger. In the inpatient setting the primary nurse is often most able to provide interactional observations; in the outpatient setting the clinic nurse, visiting nurse, or infant stimulation team can observe interactions in the home or, if necessary, in the clinic. In the older child with FTT, observations in a day care or nursery school setting are useful.

The interacting pair should be evaluated for richness of interaction, amount of eye and physical contact, sense of mutual pleasure, expression of warmth and affection, consistency of response, and amount and quality of verbal exchange. The individuals in the pair should be evaluated for responsivity, self-regulatory capacities, and tolerance levels. The care giver should be assessed for the appropriateness of her expectations and the infant for the effectiveness and burdensomeness of his demands. Table 30–5 reviews these assessment components.

The signs of interactive risk are the following:

Table 30–5. INTERACTIONAL ASSESSMENT*

Evaluate mother-infant pair for:
  Richness of interaction
  Amount of eye and physical contact
  Sense of mutual pleasure
  Warmth and affection
  Consistency of response
Evaluate in mother and infant:
  Responsivity
  Self-regulatory capacity
  Tolerance level
Also assess:
  Appropriateness of mother's expectations
  "Burdensomeness" of infant's demands

*Observe in feeding and nonfeeding situations.

1. Impoverished interaction (low verbalization, low play, poor eye contact, physical distance and lack of pleasure, warmth, and affection).

2. Inconsistent interaction (irregular response pattern, poor self-regulatory capacities, low tolerance level, comings and goings—physical or emotional).

3. Indiscriminate interaction (high intensity, poorly directed, or emotionally loaded overstimulation with relative neglect of infant's expressed needs).

4. Interactive mismatch (burdensome infant–poorly resilient mother; adequate mother–impossible child; normal child–inaccurate maternal expectations).

Two final categories of interactive risk that may coexist with any of the previous four are an attachment disorder in which, in the infant or the caregiver or both, reactions to separation and reunion are abnormal and result in maladaptive interaction patterns, and depression such that one or both members of the dyad are compromised in their affective ability to interact.

In-hospital observations of FTT can be geared toward problem areas in the interaction such as an attachment disorder (visitation absent or pair inseparable), a difficult infant (medically vulnerable, emotionally unavailable, passive, irritable), a feeding disorder ("walking on eggshells" during meals, poor suck or irritable feeder, lengthy or frequent feeds), a hard-to-help mother (defensive, guilt ridden, territorial with infant, inconsistent), or a father in need of mobilization (distant, frightened, alienated). Intervention can follow from the problems identified.

The special observation of feeding should be geared to diagnosing specific feeding disorders, such as infant rumination, nervous or volitional vomiting, technical problems in feeding (e.g., poor suck, tongue thrusting), forced or inadequate feeding techniques, and oppositional feeding patterns. The problem oriented approach is again useful in the feeding context and allows stepwise intervention.

Table 30–6. SOCIAL ASSESSMENT IN FTT: RISK FACTORS

Multiproblem family
   Marital stress, dissatisfaction
   Financial stress, layoffs
   Disorganized life styles
   Highly dependent relationships
   Chronic illness
Social isolation
   Mothers isolated from family, neighborhood
   Fathers unable or unwilling to help
   Help rejecting style with care givers
   Ineffective or nonuse of medical and community support
Unplanned, difficult pregnancy
   Maternal illness, depression, or loss
   Lack of birth control
   Adolescence
   Expectation of damaged child
History of loss
   Death or abandonment in family
   Loss of sense of self (e.g., adolescent pregnancy, illness)
   Loss of "expected child"

SOCIAL ASSESSMENT

A social assessment of the immediate and extended family in FTT is an invaluable source of information for the diagnostician. The risk areas detailed in Table 30–6 (multiproblem family, social isolation, unplanned difficult pregnancy, and history of loss) can be explored by a social worker or taken as part of the social history. Positive findings in the social history, especially as they coordinate in time with the onset of weight loss, constitute major pieces of etiologic information and often can explain the beginning of the nutritional deterioration.

INDICATIONS FOR HOSPITALIZATION

The following are four situations that often warrant hospitalization of the infant with FTT:

1. The seriously ill child with acute dehydration and severe protein-calorie malnutrition should never be managed on an outpatient basis. Such children are in extreme danger. Virtually all the mortality in FTT not associated with abuse is due to severe malnutrition and dehydration, with subsequent infection. Such children must be observed closely in the hospital and treated aggressively with nutritional rehabilitation appropriate to their medical condition (Viteri, 1981). The utility of invasive maneuvers such as central line placement must be weighed against the risks of infection and overmechanization of an already emotionally compromised infant.

2. Unsuccessful outpatient management for any reason is also an indication for hospitalization and more intensive evaluation.

3. Evidence of child abuse and neglect associated with FTT malnutrition should result in immediate hospitalization and protection of the infant at risk.

4. Extreme parental anxiety may also be an indication for in-hospital evaluation. The inpatient evaluation often progresses more rapidly and more efficiently, thus allaying parental fears and promoting a cooperative relationship between the physician and the parents.

TREATMENT

The treatment of FTT has initial and long-term components, which include nutritional rehabilitation, stabilization and clarification of medical problems, developmental stimulation, social intervention, parent-infant therapy, nutritional counseling, and close pediatric follow-up with frequent weighings. Table 30–7 reviews the components of treatment and therapeutic resources for the FTT child and family. The initial treatment of FTT involves nutritional rehabilitation, medical stabilization, and developmental stimulation. This phase of treatment often begins while the identification of risk factors is still going on. As contributory risks are identified, other aspects of treatment such as social intervention and parent-infant therapy can be undertaken. It sometimes proves impossible, however, to rehabilitate successfully an FTT infant nutritionally and medically without the initiation of developmental, social, and behavioral therapies. Recovery from FTT can be a long process; rapid recovery is not the rule (Berwick, 1980). The long term treatment of FTT typically involves continuing social intervention, developmental stimulation, parent-infant therapy, and nutritional counseling coordinated through close pediatric follow-up and utilizing frequent weighings as indicators of overall progress. As in the treatment of any potentially chronic disorder, the pediatrician's role is one of close surveillance and coordination of services throughout the preschool years. If the child has not been evaluated at a referral center,

Table 30–7. TREATMENT OF FTT

1. Nutritional rehabilitation 2. Stabilization of medical problems	Hospitalization, visiting nurse
3. Developmental stimulation 4. Social intervention	Infant stimulation program Physical therapy Social work
5. Parent-infant therapy	Mental health services
6. Nutritional counseling	Pediatrician Nutritional counseling (WIC)
7. Close pediatric follow-up with frequent weighings	Pediatrician Nurse

it is the pediatrician's responsibility to assemble available resources.

The treatment of FTT is most efficiently accomplished by a care giving team, typically consisting of the pediatrician as coordinator, a visiting or clinic nurse, a counseling nutritionist, and a social worker or mental health professional, often associated with an early intervention program. Members of the care giving team will benefit from regular meetings with one another and with the family to exchange information, make decisions, and coordinate care. Involvement of the parents as the fifth member of the team has pragmatic as well as therapeutic benefits.

In the case of a hospitalized child, a care team composed of a pediatrician, a primary nurse, a nutritionist, and a social worker can be assigned at admission to provide immediate and continuing information in the medical, nutritional, behavioral, and social arenas. An initial planning meeting within the first days of hospitalization is very useful in allowing the exchange of initial impressions and in assuring the success of a multidisciplinary approach. The parents should be invited to some portion of the meeting. It is often beneficial for the professional care givers to review the case among themselves first, arriving at some consensus of their own diagnostic and therapeutic impressions; the parents can then join the team to express their views, and planning can proceed with everyone present. Such meetings are especially useful in activating parents into a care giving and decision making role in the hospital, where all too frequently families experience intimidation or are rendered passive by the circumstances of medical technology and separation from their child. Concrete needs in the treatment of the FTT child, such as the mother or father rooming in as part of the hospitalization, can also be addressed if such arrangements have not been made prior to admission.

In the child seen on an outpatient basis, the care giving team may consist of the family pediatrician, the clinic nurse, an outpatient nutritionist (WIC if available*), and a social worker from the early intervention program or mental health clinic in the community. Nutritional counseling, behavioral feeding techniques, and interactional therapy may require extended office time and routine home visits. The use of visiting nurses and home intervention programs is advisable. Team coordination efforts, on an outpatient basis, often require frequent phone communications among care givers (e.g., calling in the child's weight to the parent-infant therapist or enlisting the pediatrician's aid when compliance with menal health services is a problem). Team meetings are costly in time but typically prove time efficient in terms of the overall effectiveness of the treatment efforts.

NUTRITIONAL REHABILITATION

The child with FTT requires a greater than normal caloric intake to initiate and maintain growth. Estimation of the caloric requirements for catch-up growth should be done prior to attempting nutritional rehabilitation. The Washington-Peterson formula for catch-up growth, designed by the Department of Nutrition Services at Children's Hospital Medical Center in Boston, provides a method for this calculation:

$$\text{Catch-up growth requirement} = \frac{\begin{array}{c}\text{Calories required for} \\ \text{weight age} \\ (\text{kcal./kg./day})\end{array} \times \begin{array}{c}\text{Ideal weight for} \\ \text{age (kg.)}\end{array}}{\text{Actual weight (kg.)}}$$

To use this formula:

1. Plot the child's height and weight on NCHS growth grids, noting their respective percentiles.

2. Determine the age at which the present weight would be "ideal" or 50th percentile. This is the weight age.

3. Determine the recommended calories for weight age in kcal. per kg. per day, using Table 30–8.

4. Determine the ideal weight for the child's present age (50th percentile weight for the present age).

5. Multiply the value obtained in (3) by the value obtained in (4).

6. Divide the value obtained in (5) by the actual weight.

The Washington-Peterson formula can be useful in providing general guidelines for the caloric requirements of catch-up growth in FTT. Its use should be tempered by an awareness of the factors influencing individual variations in caloric needs.

The amount of protein required per kilogram for catch-up growth can be similarly calculated, as follows:

Table 30–8. ESTIMATION OF ENERGY AND PROTEIN NEEDS: RECOMMENDED DAILY ALLOWANCES*

Weight Age	Calories (kcal./kg.)	Protein (gm./kg.)
0–6 months	115	2.2
6–12 months	105	2.0
1–3 years	100	1.8
4–6 years	85	1.5
7–10 years	80	1.2

*Federal nutrition supplementation and counseling program for *women, infants,* and *children.*

*Modified from National Academy of Sciences, Food and Nutrition Board, 1974.

$$\text{Protein requirement} \atop \text{(protein/kg./day)}} = \frac{\dfrac{\text{Protein required for}\atop \text{weight age}}{\text{(protein/kg./day)}} \times {\text{Ideal weight for}\atop \text{age (kg.)}}}{\text{Actual weight (kg.)}}$$

As nutritional rehabilitation begins, the calculated requirements for catch-up growth can be used as a guide to dietary content as well as to expected progress. In the hospitalized child the daily weight can be plotted on a graph that also shows the daily caloric intake, as demonstrated in Figure 30–5. In the nonhospitalized child such a graph can be used at less frequent but regular intervals. The graphing of weight against caloric intake provides concrete evidence of when and at what caloric level weight gain occurs. It can be used to record day and time of parental visits and rooming in, the days of parental versus nursing feedings, or episodes of vomiting, diarrhea, or intercurrent illness. What emerges is a graphic description of factors associated with weight or caloric fluctuations. Figure 30–5 shows that the mother's presence in the hospital was associated with a dramatically increased caloric intake and weight gain.

Nutritional rehabilitation can take many forms and often challenges the ingenuity of the care giving team. The use of caloric supplementation in formula preparation is often necessary, through carbohydrate additives (e.g., Polycose) or corn oil preparations. In the older child the use of high calorie milk shakes and finger foods often aids in insuring the intake of adequate calories without vastly increasing the volume of intake required. Alterations in the form of foods delivered is also helpful: e.g., returning to increased bottle feeding in the infant recalcitrant to solid food intake or introducing finger foods and independent feeding in the child caught in control struggles around "being fed." The environment of feeding also may require alteration, such as the creation of a quieter individualized feeding situation for the distractible child. Behavior management techniques may be useful in the management of rumination, vomiting, or maladaptive behavior during feeding. A general rule of thumb used in the nutritional rehabilitation of FTT children is: do not send the family away without giving them concrete help in the solutions to feeding problems. This is clearly the arena of ultimate defeat for the FTT family and one in which professionals can offer major input and ideas for change.

STABILIZATION AND CLARIFICATION OF MEDICAL PROBLEMS

The early stabilization of dehydration and attendant medical problems is the cornerstone of initial treatment in FTT. Any organic illness contributing to the FTT is an indication for standard medical treatment. The relative role of this illness in the child's growth retardation should be clarified for the family. Clarification is especially important in cases of FTT in which the extent of growth retardation is out of proportion to the effects of the organic illness identified.

More difficult and yet equally as important is the long term follow-up of the medical complications of FTT. Nutritional anemias should be treated appropriately. The increased incidence of infections in FTT children and the fact that, when infected, such children grow less well require ongoing attention and treatment.

DEVELOPMENTAL STIMULATION

The concepts of physical mobilization, cognitive stimulation, and affective reinvolvement are crucial to the recovery of the FTT infant. Services can be provided through child life programs and physical and occupational therapy departments in hospitals and through infant stimulation programs in the community. The model of working with parents and through parents to accomplish develop-

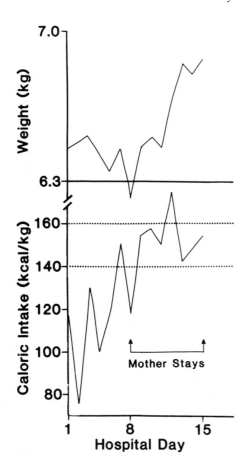

Figure 30–5. Daily weight versus caloric intake chart. Hospitalized child with FTT whose calories increased into an adequate range (140 to 160 kcal./kg./day) and whose weight began to increase coincident with the mother's rooming-in in the hospital.

mental catch-up in the target child is the model of choice.

SOCIAL INTERVENTION

Addressing the stress and loss prevalent in FTT families can be done through crisis intervention and the creation of social networks for concrete support services. The assistance of a social worker is invaluable in locating and coordinating the many agencies and resources often necessary. The difficulties in helping FTT families often can be lessened when social needs are addressed in concert with medical needs. The concrete services of visiting nurses, child care, homemakers, housing assistance, financial counseling, liaisons with job or school, and the location of mental health services are often necessary in the treatment. Coordination with child protective agencies also can be a part of social intervention in FTT, especially in families in which acceptance of services for themselves and their child must be mandated through court custody.

PARENT-INFANT THERAPY

The interactive disorder seen in some FTT families can be treated through a form of therapy developed and described by Fraiberg (1980). Parent-infant therapy attempts to mobilize the parent's personal resources in the context of the target child. In this form of therapy, typically mother-infant therapy, the mother and infant are seen together in weekly sessions. The interview involves not only the mother's description of her feelings and problems, but the active demonstration of them in relation to her infant. Different from individual psychotherapy, in which a mother might speak about her baby "who spits up all the time," the mother-infant therapist is able to observe the pain of the actual incident and attend not only to the mother's frustration and disappointment but to the adaptive handling of the very real problem as it occurs. The opportunity to deal with interactional issues as they occur in the mother-infant pair is of prime importance in FTT. The infant's health, in Fraiberg's words, "cannot wait for the resolution of the entire spectrum of a parent's personal problems." The clinician must specifically attend to the treatment of those parental issues that impede the infant's growth and development. This is done by addressing past and present feelings of pain with the parent specifically within the context of how the infant contributes to or is affected by those feelings. Parental perceptions or realities of difficult behavior or sickliness must be dealt with appropriately. Interactional

issues are clarified, and the mother receives support so that she can better meet the needs of her infant. Education regarding child development is unobtrusively introduced in each session. The goals of the therapy are to enhance the parent-infant bond, to stimulate the child's development, and to educate parents to understand and facilitate the growth of their infant. This unique treatment method includes supportive psychotherapy (through the child), developmental guidance, and environmental support with crisis intervention.

NUTRITIONAL COUNSELING

The long term nutritional follow-up of FTT infants is a necessity. FTT families benefit from concrete feeding guidelines, detailing exactly how to prepare foods and offer them to the child. Misperceptions of nutritional needs and inaccurate nutritional beliefs are common in FTT families. Feeding, as a historic battleground in the FTT home, may also be the most difficult setting in which to put abstract or general advice into action. To the extent that the battleground can be converted into a place of business, that of eating, the embattled FTT family will experience relief. Clear guidelines are often helpful in this endeavour. Nutritionally, explicit limits also allow the process of "interminable feeding" in some FTT families to stop, making way for some play and enjoyment for parents and infant in the nonfeeding situation.

However, the danger of limiting nutritional intake in any way with FTT infants and children cannot be overemphasized. Parental attention should not be focused on limits but rather on supplies, and it may be necessary to redirect attention accordingly. Changes in the child's nutritional needs must also be closely monitored. Routine recounseling with the family allows education in terms of the concrete implications of growth and development of nutritional patterns. Misperceptions that may reappear with each change in feeding also can be addressed through long term nutritional follow-up.

CLOSE PEDIATRIC FOLLOW-UP

The pediatrician is the surveyor of progress and coordinator of care in the long term follow-up of FTT. Frequent weighings provide the single most important indicator of progress in all sectors of risk. Over the long term, recovery of weight and height velocity to the child's potential is an indicator of successful treatment. Consistency, persistence, and the back-up of the care giving team are the pediatrician's most useful tools in the treatment of this complex and chronic condition.

Table 30–9. STUDIES OF SEQUELAE OF NONORGANIC FAILURE TO THRIVE*

	Hufton and Oates (1977)	Elmer et al. (1969)	Glaser et al. (1968)	Shaheen et al. (1968)
Number of cases	21	15	40	29
Average age at follow-up	7 yr. 10 mo.	3 yr. 3 mo. to 11 yr. 7 mo.	4 yr. 6 mo.	NR†
Mean interval to follow-up	76 mo.	57 mo.	41 mo.	6–40 mo.
Physical status				
Weight	24% (<10th percentile)	60% (<3rd percentile)	35% (<3rd percentile)	22% (<3rd percentile)
Height	5% (<10th percentile)	60% (<3rd percentile)	33% (<3rd percentile)	26% (<3rd percentile)
Cognitive status	48% below average on teacher ratings; 67% had reading age 1–2 yr. below chronologic age	67% mildly or moderately retarded	15% "mentally retarded"; 37% of school aged children having significant difficulty in school	NR†
Psychologic status	48% classified as "abnormal personalities" on teacher rating scale	47% with behavioral disturbance	28% with psychologic or behavioral problems	NR†

*From Berwick, D.: Nonorganic failure-to-thrive. Pediatr. Rev., 1:265, 1980.
†Not reported.

PROGNOSIS

Adequately controlled, longitudinal follow-up studies of FTT are rare. Thus, the discussion of prognosis in FTT is difficult. Only three studies in the literature to date control for mean midparental height—those by Elmer (1960), Chase and Martin (1979), and Pollitt (1975). In Elmer's controlled study, 60 per cent of the children followed for a mean of 57 months were at less than the third percentile for age in both height and weight. Only five of the 15 children in this study were within 1 cm. of predicted height based on mean midparental height. Table 30–9 succinctly reviews four major studies of the sequelae of nonorganic failure to thrive. It can be concluded from this review that the prognosis for growth in FTT is more optimistic than that for cognitive or psychologic status.

Half of all children followed for FTT have some form of cognitive impairment, with both expressive and receptive language most markedly impaired. A significant percentage of the children are described as having subnormal IQ's. In Chase's study, which controlled for parental IQ as well as socioeconomic status, the age at FTT presentation strongly affected the outcome. Children who presented before four months of age had normal developmental quotients on follow-up, whereas those who presented later had markedly depressed Yale developmental quotients. Some authors have related such findings to the relative "shallowness" of injury when FTT presents acutely, as opposed to the effects of chronic malnutrition and long-standing interactional disorders, which are less malleable to intervention.

Behavioral and affective problems are also strikingly prevalent on FTT follow-up. Estimates of the prevalence of behavioral disturbance range from 28 to 60 per cent. In Elmer's study 58 per cent of the patients were rated by blind psychiatric observers as suffering from depression and anxiety. Only two of the 15 children in the study were ultimately classified as "completely normal" in all areas measured at follow-up: growth, intellectual functioning, and behavior. This 13 per cent incidence of normality on follow-up is a sobering statistic and perhaps best conveys the long term morbidity of FTT and the factors contributing to it.

A careful reading of the literature indicates that FTT cases should be identified and treated vigorously as early as possible in the course of the disorder in order to minimize its overwhelming long term morbidity.

WILLIAM G. BITHONEY

JENNIFER M. RATHBUN

REFERENCES

Ambuel, J. P., and Harris, B.: Failure to thrive. Ohio Med. J., 59:997, 1963.

Barbero, G. J., and Shaheen, E.: Environmental failure-to-thrive: a clinical review. J. Pediatr., 71:639, 1967.

Berwick, D.: Nonorganic failure-to-thrive. Pediatr. Rev., 1:265, 1980.

Berwick, D., Levy, J. and Kleinerman, R.: Failure to thrive: diagnostic yield of hospitalisation. Arch. Dis. Child., 57:347, 1982.

Bithoney, W. G., and Newberger, E.: Non-organic FTT; developmental and familial characteristics. Pediatr. Res., 16:84A, 1982.

Brazelton, T. B.: Nutrition during early infancy. *In* Suskind, R. M. (Editor): Textbook of Pediatric Nutrition. New York, Raven Press, 1981.

Chase, H., and Martin, H.: Undernutrition and child development. N. Engl. J. Med., 282:933, 1979.

Elmer, E.: Failure to thrive: role of the mother. Pediatrics, 25:717, 1960.

Elmer, E., Gregg, G. S., and Ellison, P.: Late results of the "failure to thrive" syndrome. Clin. Pediatr., 8:584, 1969.

English, P.: Failure to thrive without organic reason. Pediatr. Ann., 7:774, 1978.

Evans, S., Reinhart, J., and Succop, R.: Failure to thrive: a study of 45 children and their families. J. Am. Acad. Child Psychiatry, 11:440, 1972.

Fitzhardinge, P., and Steven, E.: The small-for-date infant. I. Later growth patterns. Pediatrics, 49:671, 1972.

Fraiberg, S.: Clinical Studies in Infant Mental Health. New York, Basic Books, Inc., 1980.

Fraiberg, S., Adelson, E., and Shapiro, V.: Ghosts in the nursery. J. Am. Acad. Psychiatry, 14:387, 1975.

Gaensbauer, T. J., and Sands, K.: Distorted affective communications in abused/neglected infants and their potential impact on caretakers. J. Am. Acad. Child Psychiatry, 18:236, 1979.

Glaser, H. H., Heagarty, M. C., Bullard, D. M., Jr., and Pivchik, E. C.: Physical and psychological development of children with early failure to thrive. J. Pediatr., 73:690, 1968.

Gordon, A. H., and Jameson, J. C.: Infant-mother attachment in patients with nonorganic failure to thrive syndrome. J. Am. Child Psychiatry, 18:251, 1979.

Green, M., and Solnit, A. J.: Reactions to the threatened loss of a child: a vulnerable child syndrome. Pediatrics, 34:58, 1964.

Hannaway, P.: Failure to thrive: a study of 100 infants and children. Clin. Pediatr., 9:96, 1970.

Harper, G. P., and Richmond, J. B.: Normal and abnormal psychosocial development. In Barnett, H. L. (Editor): Pediatrics. New York, Appleton-Century-Crofts, 1979.

Herzog, D. B., and Harper, G. P.: Unexplained disability: diagnostic dilemmas and principles of management. Clin. Ped., 20:761, 1981.

Homer, C., and Ludwig, S.: Categorization of etiology of failure to thrive. Am. J. Dis. Child., 135:848, 1981.

Hufton, I., and Oates, R.: Nonorganic failure to thrive: a long term follow-up. Pediatrics, 59:73, 1977.

Kaplan, C. J., et al.: Long-term effects of otitis media: a ten year study of Alaskan Eskimo children. Pediatrics, 52:577, 1973.

Koel, B. S.: Failure to thrive and fatal injury as a continuum. Am. J. Dis. Child., 118:565, 1969.

Kotelchuck, M., and Newberger, E. H.: Failure to thrive: a controlled study of familial characteristics. Unpublished study, 1978.

Krieger, I., and Sargent, D. A.: A postural sign in the sensory deprivation syndrome in infants. J. Pediatr., 70:332, 1967.

Leonard, M., Rhymes, J. P., and Solnit, A. J.: Failure to thrive in infants. Am. J. Dis. Child., 111:600, 1966.

Lloyd-Still, J.: Clinical studies on the effects of malnutrition during infancy on subsequent physical and intellectual development. In Lloyd-Still, J. D. (Editor): Malnutrition and Intellectual Development. New York, John Wiley & Sons, Inc., 1976.

Mitchell, W., Gorrell, R., and Greenberg, R.: Failure to thrive: a study in a primary care setting. Pediatrics, 65:971, 1980.

Pollitt, E.: Failure to thrive: socioeconomic, dietary intake, and mother-child interaction data. Fed. Proc., 34:1593, 1975.

Pollitt, E., Eichler, A. W., and Chan, C.: Psychosocial development and behavior of mothers of failure-to-thrive children. Am. J. Orthopsychiatry, 45:525, 1975.

Pollitt, E., and Thompson, C.: Protein-calorie malnutrition and behavior: a view from psychology. In Wurtman, R. J. and Wurtman, J. J. (Editors): Nutrition and the Brain. New York, Raven Press, 1977, Vol. 2.

Riley, R. L., Landwirth, J., Kaplan, S. A., and Collipp, P. J.: Failure to thrive: an analysis of 83 cases. Calif. Med., 108:32, 1968.

Rosenn, D., Loeb, L., and Jura, M.: Differentiation of organic from nonorganic failure to thrive syndrome in infancy. Pediatrics, 66:689, 1980.

Sameroff, A. J., and Chandler, M. J.: Reproductive risk and the continuum of caretaking casualty. In Horowitz, F. D. (Editor): Review of Child Development Research. Chicago, University of Chicago Press, 1975, Vol. 4, p. 187.

Shaheen, E., Alexander, D., Truskowsky, M., and Barbero, G.: Failure to thrive—a retrospective profile. Clin. Pediatr., 7:255, 1968.

Sills, R. H.: Failure to thrive: the role of clinical and laboratory evaluation. Am. J. Dis. Child., 132:967, 1978.

Smith, D. W., Truog, W., Rogers, J. E., Greitzer, L. J., Skinner, A. L., McCann, J. J., and Harvey, M. S.: Shifting linear growth during infancy: illustration of genetic factors in growth from fetal life through infancy. J. Pediatr., 89:225, 1976.

Tanner, J. M., Goldstein, H., and Whitehouse, R.: Standards for children's height at age 2–9 years allowing for height of parents. Arch. Dis. Child., 45:755, 1970.

Viteri, F.: Primary protein-calorie malnutrition. In Suskind, R. M. (Editor): Textbook of Pediatric Nutrition. New York, Raven Press, 1981.

Whitten, C., Pettit, M., and Fischoff, J.: Evidence that growth failure from maternal deprivation is secondary to undereating. J.A.M.A., 209:1675, 1969.

31

Disordered Processes of Elimination

The term enuresis is derived from the Greek *enourein*, to void urine. There is no standardized clinical definition in the medical literature. The age at which children are expected to achieve bladder control varies among and within cultures. The frequency of wetting is usually taken into consideration, but there is no uniformly applied standard, thus making it difficult to compare published reports. The term usually implies nocturnal episodes of incontinence, and the differentiation between nocturnal and diurnal is often omitted in the literature. There is also a lack of uniformity in the distinction between primary and secondary enuresis. Primary enuresis describes the condition of a child who has never had a prolonged period during which he has been dry, and secondary enuresis describes the recurrence of wetting in a child who has been dry for a period of time. Clinicians sometimes use three months or six months, and occasionally one year, as the minimal period of dryness before labeling the onset of wetting as secondary enuresis. We would choose six months as a reasonable interval.

Some children achieve dryness at one year of age, and most children become dry between the ages of two and four. Only when bedwetting occurs beyond the age at which toilet training is customarily thought to be completed is it considered a problem. The most commonly accepted cutoff is age five, but it ranges from four to six years in various studies.

HISTORICAL BACKGROUND

The recognition of enuresis as a problem predates modern civilization. In an excellent historical review Glicklich (1951) quotes a remedy for incontinence of urine proposed in 1550 B.C., a mixture of juniper berries, cyprus, and beer. The first printed book on diseases of children was published in the Middle Ages and contained a chapter on "incontinence of urine and bedwetting." Var-

ious prescriptions for medication followed the doctrines of humors and organs: the ground cerebrum of a hare, the lung of a kid made into plaster, the pulverized bladder of a young breeding sow, the flesh of a groundhog, the inside skin of the stomach of hens. In the Renaissance an accepted theory stated that enuresis resulted from weakness of the neck of the bladder, a concept that still appears in contemporary literature. Thomas Phaer, known as the father of English pediatrics, published his *Boke of Children* in 1554 and included a paragraph entitled "of Pyssying in the bedde." He sustained the humoral and organ therapy approach, as did authors in the seventeenth century. In the eighteenth century anatomic explanations emerged, and therapy included application of blisters to the scrotum and mechanical devices to block the flow of urine.

In the nineteenth century enuresis first became viewed as an economic and social hazard, and among boys of high social class, enuretics were often denied enrollment in school. Prominent etiologic concepts included imbalance of the musculature of the bladder; primary irritability of the bladder produced by phimosis, food, or drink causing acid urine, or intestinal parasites; dreams; and penile erections. The proliferating literature also mentioned a neurogenic basis, evoking parallels with epilepsy, and included pathology of the genitourinary tract. Cures were proclaimed as a result of a wide variety of treatments.

Many of the general measures proposed in the nineteenth century persist in current practice: limiting fluid intake prior to bedtime, emptying the bladder at bedtime, avoiding deep sleep via waking the child, and various dietary restrictions. The pharmacopoeia was rapidly expanding, and several drugs were widely used, including strychnine, belladonna, and chloral hydrate. The theory of decreased bladder capacity appeared in the late 1800's. At the same time a school of thought emerged that preached the omission of treatment because various events could be relied upon to provide a curative shock to the genitourinary organs: namely, dentition, puberty, marriage, and first pregnancy. The allaying of anxiety first appeared as a therapeutic modality within this approach. At the same time various electrical stimuli were in vogue. Perhaps the first prototype of the bell and pad was devised by Nye in 1830. This apparatus relied on mild electric shock to awaken the patient. It was modified in the mid-twentieth century by the Mowrers, who replaced the electric shock with the ringing of a bell. Early in the twentieth century, following the development of psychoanalytic theory, enuresis was considered to have an emotional basis as a symptom of a personality disorder. Among the most psychoanalytically oriented clinicians, bedwetting was seen as wish fulfilling and as a regression to the early stages of infancy.

In addition to the written evidence that enuresis has been with us since the dawn of civilization, there are also contemporary descriptions of bedwetting and its treatment in cultures as diverse as those of India, West Africa, and American Indian tribes. Our current approaches to bedwetting continue to evoke an array of theories regarding etiology and therapy, all of which establish enuresis not as a disease but rather as a symptom arising from a variety of causes.

THE PHYSIOLOGY OF DRYNESS

Failure to achieve dryness cannot be understood without consideration of the mechanisms of bladder control and the behavioral stages involved in the development of daytime and nighttime dryness. Mature bladder function involves filling, desire to void, postponement, initiation of sphincter relaxation and bladder contraction, and maintenance of both of these until the bladder is empty. Afferent stimuli from the bladder are relayed to the cortex. Voluntary and involuntary inhibition of voiding are under cortical and possibly hypothalamic control. Release of this inhibition results in efferent stimuli to the bladder. When afferent stimuli arise from the bladder during sleep, the normal mature individual is awakened and becomes aware of the desire to void.

Some aspects of the mechanisms of micturition are already present at birth, with the newborn infant having periods of dryness. During approximately the first six months of life, the infant's bladder responds to distention by immediate evacuation. During the following two years there is a progressive lessening of the bladder responsiveness, presumably due to development of unconscious inhibition. Sometime between 18 and 30 months the infant begins to appreciate and express a desire to void and finally develops the conscious ability to postpone or initiate micturition. Although the development of culturally acceptable voiding behavior can be influenced by many factors, the behavior cannot be accelerated beyond the child's maturational status but can be retarded.

In a representative sample of the Baltimore metropolitan area, the percentage of all children attaining dryness rose steeply between the ages of one and three and one-half years. At each age, initial bladder control occurred in a higher percentage of females than of males, of whites than of blacks, and of infants weighing more than 2.5 kilograms at birth (Oppel, 1968a). Most children first attain bowel control, then bladder control by day, and finally bladder control at night.

PREVALENCE OF ENURESIS

Consideration of prevalence should differentiate nocturnal versus diurnal and primary versus secondary enuresis. Even when these differences are recognized, there is great variation in the reported prevalence of enuresis throughout the world.

The majority of cases of enuresis are primary and nocturnal. The recognized prevalence of secondary enuresis depends greatly on the interval chosen to categorize the child as dry. In the Baltimore survey "dry" was defined as not wetting at all for at least one month, and relapses were encountered in 25 per cent of the children (Oppel, 1968b). When the dry period is designated as six, nine, or 12 months, the reported number of secondary enuretics is greatly reduced.

Diurnal or daytime enuresis occurs much less frequently than the nocturnal variety. The overall incidence in the seven to 12 year old age group is approximately 1 per cent (Dejonge, 1973). In a large Scandinavian study 16 per cent of the enuretic sample experienced both diurnal and nocturnal wetting, and 10 per cent had diurnal enuresis alone (Hallgren, 1956). The prevalence was the same in boys and girls, although some surveys have found an overrepresentation of girls among diurnal enuretics.

Primary nocturnal enuresis is the most widely surveyed as well as the most prevalent form of bed-wetting. The problem has been extensively studied in Great Britain, the Scandinavian countries, and the United States.

Table 31–1 compares the prevalence among young children in these geographical areas. The higher prevalence among boys than among girls in all three surveys and the more frequent occurrence of bed-wetting in the United States sample are noteworthy.

The most comprehensive data for the United States were obtained in the National Health Examination Survey in the late 1960's, based on a stratified probability sample of noninstitutionalized children and youths in the entire United States. (National Center for Health Statistics, 1967,

Table 31-1. PROPORTION OF CHILDREN WITH NOCTURNAL ENURESIS

	4-4 1/2 Years		6 Years		7-7 3/4 Years	
	Males	*Females*	*Males*	*Females*	*Males*	*Females*
England*	0.14	0.11	0.12	0.08	0.08	0.06
Stockholm†	0.12	0.07	0.09	0.05	0.08	0.04
Baltimore‡	0.29	0.32	0.22	0.18	0.18	0.10

*Bloomfield and Douglas, 1956.
†Hallgren, 1956.
‡Oppel, 1968a.

1969). Eight hundred thirty-two of 3632 males were identified as enuretic, and 425 female enuretics were found among 3487 in the age group six to 11. Among adolescents aged 12 to 17, there were 193 male and 99 female enuretics, in samples of 3514 and 3196, respectively. Extensive data were gathered on these individuals by history, physical examination, objective testing, and interviews with parents and teachers. Analysis of these data revealed important associations between enuresis and certain demographic, physical, temperamental, and intellectual characteristics.*

The prevalence of enuresis at each age from six through 17 in males and females, for whites and blacks separately, is seen in Table 31-2. It is clear, both from these data and from the Baltimore study, that enuresis is encountered more frequently in the United States than in other countries where the prevalence has been reported. There is a consistent decrease in enuresis with age, and at each age there are more enuretic males than females.

The frequency of enuretic episodes was determined among the six through 11 year olds. It was found that male enuretics most commonly wet only once a month or less. By utilizing these data, it was possible to estimate a constant percentage decrement in bedwetting for each sex between ages six and 11: 11 per cent annually for males and 14 per cent for females. The number receiving any form of therapeutic intervention is not known in these samples.

*These studies are being carried out at the Center for Research in Youth Development at Stanford University by Ruth T. Gross, Sanford M. Dornbusch, J. Merrill Carlsmith, Alice Luna, and Paula M. Duke.

The National Health Examination Survey also provided the opportunity to analyze the impact of various demographic parameters on the prevalence of enuresis. Enuresis was more frequent in families of low socioeconomic status compared to middle and high, whether determined by parental education or by family income. The impact of socioeconomic status was particularly striking for females, as also noted in the British studies (Fig. 31-1). Birth order and family size were important in all socioeconomic classes for both males and females. Enuresis was more prevalent in later born children than in only and first born children, and there was a higher prevalence in households of three or more children compared to those with one or two children. An increased prevalence of enuresis in children whose birth weight was less than 2.5 kilograms was noted in males but was not consistently found among females.

CLINICAL ASSOCIATIONS

In large groups of subjects certain characteristics are found, on the average, more frequently in enuretics than in nonenuretics. The following associations with enuresis were derived by our group from analysis of the National Health Examination Survey data.

PHYSICAL CHARACTERISTICS

Both in the sample of six to 11 year olds and in the adolescent sample there was a small but con-

Table 31-2. PERCENTAGE ENURETIC BY AGE, SEX, AND RACE*

Age in Years		6	7	8	9	10	11	12	13	14	15	16	17
Males													
White	%	24.8	17.8	18.9	16.6	12.8	12.6	8.4	9.1	6.0	4.6	2.9	1.3
(Total N)	(N)	(489)	(551)	(537)	(525)	(509)	(542)	(540)	(542)	(527)	(525)	(496)	(417)
Black	%	30.0	24.8	21.3	32.0	22.9	22.0	15.9	17.6	9.2	9.3	3.5	6.2
	(N)	(86)	(81)	(81)	(78)	(67)	(86)	(101)	(80)	(88)	(84)	(57)	(69)
Females													
White	%	14.9	13.9	14.2	11.3	8.8	7.4	4.1	4.2	3.7	2.1	1.1	1.0
	(N)	(461)	(512)	(498)	(494)	(505)	(477)	(455)	(490)	(484)	(425)	(441)	(393)
Black	%	29.3	18.5	17.7	12.5	11.6	13.9	12.8	6.8	3.0	7.2	9.3	2.7
	(N)	(75)	(97)	(115)	(87)	(79)	(87)	(88)	(91)	(101)	(73)	(93)	(74)

*The National Health Examination Survey.

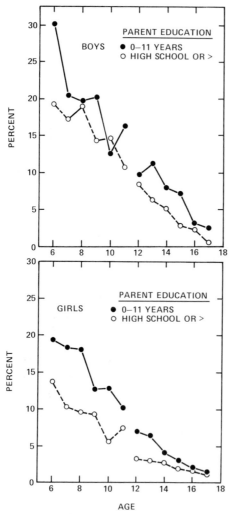

Figure 31–1. Percentage of children enuretic by age and by parent education.

two months at each age, but it was consistently present in every age group.

More striking were the differences in sexual maturation. All the individuals 12 years and older were scored for Tanner's stage of sexual maturation at the time of the physical examination. There was a higher prevalence of enuresis among late maturing males and females than among early or midmaturers. This association was most evident among 14 and 15 year old males. Table 31–3 examines the relationship of enuresis to late sexual maturation in that group of males, taking into account socioeconomic status. Among males in the low socioeconomic group at age 14, 21 per cent of the late sexual maturers were enuretic compared to 6 per cent of early or midmaturers; at age 15 the rates were 17 per cent versus 6 per cent. Late sexual maturation increases the probability of enuresis, particularly for males.

TEMPERAMENT

There were some striking differences in temperament between the enuretics and the nonenuretics. Enuretic children ages six through 11 in all social classes were more often noted by their parents to be high-strung and to lose their tempers easily. Perhaps related to these temperamental characteristics, or to the children's concern about the enuresis, enuretic boys at ages six, seven, and eight and girls at seven and eight had more difficulty in getting to sleep than did their nonenuretic peers. Additionally the enuretic children were more likely to be afraid to be alone in the dark. Enuretic children in the lower socioeconomic class, but not in the higher classes, were more likely to be described as shy. In all socioeconomic groups, enuretic children were noted by their parents to be less well liked by other children than were nonenuretics. Among the adolescents aged 12 through 15, both boys and girls with enuresis were more likely to describe themselves as tense, as having difficulty sleeping, and as having bad dreams, and their parents were more likely to describe them as nervous. Here also the direction of causality cannot be deduced from these data;

sistent decrement in average height for enuretics compared to nonenuretics. This difference was found most consistently among males, especially those of ages 14 through 16. Bone age was also determined for all individuals in the sample. For both sexes, enuretics had a lower mean bone age than did nonenuretics of the same chronologic age. This mean difference was slight, only one or

Table 31–3. PERCENTAGE OF ENURETIC MALES AGES 14 AND 15 BY EDUCATION OF PARENTS AND RATE OF SEXUAL MATURATION

| | 14 Years | | 15 Years | |
	Percentage	*(Total N)*	*Percentage*	*(Total N)*
Low parental education (0–11 years)				
Late sexual maturers	21.2	(33)	17.1	(35)
All others	6.1	(244)	5.9	(272)
Middle and high parental education (12–17 years)				
Late sexual maturers	8.9	(56)	2.9	(34)
All others	4.2	(265)	2.8	(248)

the enuresis itself may account for some of these findings.

INTELLECTUAL AND EDUCATIONAL CHARACTERISTICS

Among the six to 11 year old children of both sexes in all socioeconomic groups, the enuretics were much more likely than the nonenuretics to be described by the school as having adjustment problems and as being less attentive. The intellectual ability and the academic performance of these enuretic children, as assessed by the school, were more often below average. In all three socioeconomic classes, scores on the Wechsler Intelligence Scale for Children (WISC) and the Wide Range Achievement Test (WRAT) were strikingly lower for enuretic than for nonenuretic children. The differences between enuretics and nonenuretics in scores on both tests were even greater among adolescents than among younger children, and also greater among children and youth whose parents had not completed high school. Figure 31–2 illustrates these scores in male adolescents.

Negative characteristics of the adolescent enuretics were more often reported by the teachers, who presumably would not have knowledge of the enuresis. They noted that the enuretic youths were more often absent and more likely to need

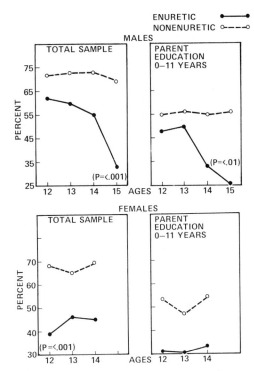

Figure 31–3. Percentage of parents who expect youth to attend college.

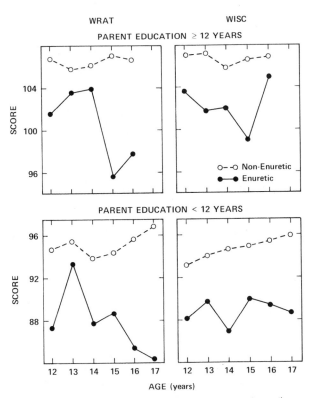

Figure 31–2. Comparison of enuretic and nonenuretic youths on achievement and intelligence test scores by age and by level of parent education.

discipline. Enuretics compared to nonenuretics were more often judged by their teachers to be in the lowest third of the class in terms of intellectual ability and academic achievement; teachers judged enuretics to be less well adjusted and less popular.

Parents of enuretic adolescents had lower educational aspirations and expectations for them. A smaller percentage of parents of enuretic than of nonenuretic males and females expected their children to attend college. These differences in parental expectations were present in all socioeconomic groups and increased with the age of the youth (Fig. 31–3). Among the enuretic youth themselves, lower educational expectations did not appear until age 14 in females and 15 in males. Prior to those ages both male and female enuretic adolescents, on the average, believed that they were at least as likely to attend college as the nonenuretic adolescents. The lowering of educational expectations for older enuretics may represent either the cumulative effect of school and parental attitudes toward the enuretic youths, or the impact upon their self-image of being enuretic at these later ages.

In the National Health Examination Survey there was a group of children who were examined twice, first at ages eight to 11 and again between the ages of 12 and 15. Eleven hundred thirty-eight males and 1039 females received these two examinations. Of this group there were 162 males who were enuretic at the earlier age. Thirty-nine per cent of these remained enuretic when examined in adolescence. Among the females, 97 were enuretic in childhood, of whom 26 per cent remained enuretic in adolescence. With these data it was possible to compare children who were enuretic at both ages to those who were enuretic early but were dry in adolescence, as well as to those who were never enuretic. If one looks at the WISC and WRAT scores for these three groups, an interesting pattern emerges. The average enuretic child had slightly lower scores for both WISC and WRAT than did the average subject who was never enuretic. However, there was no difference in the average scores of those children who were wet in childhood and dry in adolescence compared to those who were enuretic in both childhood and adolescence. The intellectual handicaps on these objective tests remained even after the enuresis was resolved.

ETIOLOGY

Enuresis as a symptom results from diverse causes, and, as mentioned earlier, the definition of the enuretic child differs considerably from one investigator to another. It is not surprising, therefore, that there are numerous theories of etiology, many of which are based on global comparisons of enuretics and nonenuretics (Table 31–4).

GENETIC FACTORS

There is a genetic component to the development of enuresis. Table 31–5 shows that monozygotic twins were concordant for enuresis about twice as frequently as dizygotic twins and also illustrates a pattern of familial incidence that is

Table 31–4. THEORIES OF ETIOLOGY: ASSOCIATED FACTORS

Genetic
Demographic
  Socioeconomic status
  Family size
  Birth order
Maturational delay
  Birth weight
  Height
  Bone age
  Motor delay
Psychosocial
  Stress, emotional disturbance
  Behavioral deviance
  Toilet training
Limited functional bladder capacity
Sleep disturbance
Psychiatric disorders
Organic
  Urinary tract infection, anomaly
  Lumbosacral anomalies
  Diabetes mellitus, diabetes insipidus, sickle cell trait

consistent with a genetic basis for enuresis. The agreement in a number of other studies in the United States, Great Britain, and Sweden—that enuresis often tends to occur in families—did not rule out the possibility that aspects of the family environment are producing results consistent with the genetic explanation. Finally, a study of young children on an Israeli kibbutz, where toilet training predominantly takes place away from the parents, continued to show a strong association of enuresis among family members. Sixty-seven per cent of enuretic children had enuretic siblings compared to 22 per cent of dry children (Kaffman and Elizur, 1977). Hallgren's Scandinavian study suggested a strong genetic base for nocturnal enuresis in both sexes. However, daytime enuresis not accompanied by noctural wetting seemed to have no genetic origin, and among those who were enuretic both day and night, there was clear indication of a family incidence only in boys.

Despite the empirical studies indicating a genetic basis for at least a subset of enuretics, the specific nature of the genetic mechanism remains unknown.

DEMOGRAPHIC FACTORS

Lower social classes show a higher rate of enuresis than others. This association has been observed in Sweden, Iceland, Great Britain, and the United States. In some of these studies, social class differences in enuresis were limited to females. In our own studies from the National Health Examination Survey, social class was found to be related in both sexes to enuresis at all ages from six through 17, and the association was greater for females than for males. In Oppel's Baltimore study, however, only the probability of relapse was associated with social class. It is difficult to assess this issue because of the diverse limits on the period of dryness in the various studies. In both Oppel's study and our own, however, enuresis was more prevalent among blacks than among whites, controlling for social class.

Even though the incidence of some types of enuresis may be higher in the lower social classes, that information is not sufficient to understand the social processes that lead to the relationship. Do the people at the bottom of the social hierarchy lack resources of time and attention to devote to the toilet training of their children? Do they use more coercive methods, thus increasing the chance of enuresis? Or is the explanation more biologically based, since the lower social classes have more children with low birth weight and other indicators of developmental delay? Finally, is a mixture of biologic and social factors influencing the prevalence of enuresis in the lower social classes? Data

Table 31–5. FAMILY INCIDENCE OF ENURESIS*

Relationship	Total No. of Children	Enuretics	
		No.	*%*
Co-twins of enuretic monozygotic twins	53	36	68
Children both of whose parents were enuretic	22	17	77
Children whose fathers were enuretic	47	20	43
Children whose mothers were enuretic	85	37	44
Co-twins of enuretic dizygotic twins	42	15	36
Siblings of enuretic twins	190	48	25
Siblings of twins, neither of whom are enuretic	388	33	9
Children neither of whose parents were enuretic	988	150	15

*From Bakwin, H.: The genetics of enuresis. Clin. Dev. Med., 48/49, 1973.

relating to the relationship of the size of the family to enuresis may shed some light on these issues. The National Health Examination Survey shows a relationship of large family size to enuresis that is strongest among the lower classes. This observation gives strong support to the possibility that limited interpersonal resources are at least one significant part of the relationship between social class and bedwetting. Given a larger family, the resources per child may be less, leading to more frequent bedwetting. The birth order of the child provides additional information. In several studies, including our own, first born and only children were less likely to be enuretic, again suggesting that increased social resources available to the child reduce the risk of enuresis.

MATURATIONAL DELAY

The possible existence of a genetic link to primary nocturnal enuresis strongly suggests a biologic factor. Some British investigators have proposed that the biologic factor may be maturational delay of neurologic mechanisms linked to bladder control (Kolvin, 1975; MacKeith, 1972). Maturational delay, whether genetic, congenital, or environmental in origin, could retard development of bladder control. The delayed maturation could then interact with social and environmental forces to cause nocturnal enuresis to persist. Almost all children have the capability to become dry by the age of five. Yet maturational delay may have caused difficulties during the early period of training so that the child subsequently would be inhibited in learning to be dry, perhaps because of the later anxiety of both child and parents.

The delayed physiologic growth and failure of enuretic children to exhibit certain learned skills at specific ages have been cited to bolster the theory of maturational delay. In several studies, particularly in England, enuretics have been found to learn to walk and talk late, to speak in a manner more difficult to understand, to be shorter, and to mature sexually at a slower rate when compared to nonenuretics. In some studies there has been

an association with low birth weight. In our own analysis of the National Health Examination Survey data we confirmed all these relationships, even when controlling for social class. As mentioned earlier, we also noted mild but consistent retardation in bone age among enuretics. These diverse associations were found not only among enuretics ages six to 11, but also in the adolescent sample, ages 12 to 17.

Not surprisingly, given the difficulties in defining enuresis and its status as a symptom of diverse conditions, there are published studies that do not confirm all the foregoing observations. Nevertheless there is enough evidence in the literature and in our own studies to suggest that maturational delay may be a major biologic factor that increases the risk of enuresis.

It is worth noting that all studies of the incidence of primary enuresis show an increased prevalence among males. Since males in the aggregate are slower in their rate of development throughout childhood and adolescence, it is possible that normal maturational delay may be the explanation for the sex difference in enuresis.

PSYCHOSOCIAL ASPECTS OF ENURESIS

Turning from the more general notion of demographic and maturational differences to more specific differences in patterns of behavior that are associated with enuresis, we can highlight only a few issues of particular concern to practicing physicians who must deal with enuretic children and their families. Well documented reviews of the literature are available (Rutter et al., 1973; Shaffer, 1973).

There is relatively consistent evidence that enuresis may be associated with stressful events in the period of childhood during which toilet training occurs, such as hospitalization, separation from parents, and break-up of a family. Emotional disturbance in the children themselves is also associated with enuresis, the relationship being stronger for females than for males. Similarly, behavioral deviance is more common among enuretics than nonenuretics, and again the difference is greater in females than in males. In the aggregate, however, the majority of cases of enuresis in both sexes show no evidence of abnormal stress or deviance, and there is no necessity for searching for deep psychogenic factors in such patients.

Toilet Training. The timing and nature of toilet training remain key issues in understanding the possible social roots of enuresis. There is no clear consensus as to whether the method of toilet training as well as its timing can either promote or retard nighttime dryness. Brazelton's success rate in an upper middle class private practice (80 per cent dry at three years; 98 per cent dry at five years) has been attributed to a relaxed individualized approach to training. One notes, however, that Brazelton defined success very liberally as a reduction in the frequency of nighttime wetting to less than once a week, and short periods of regression under unusual stress were not considered failures (Brazelton, 1962).

Many have assumed with Brazelton that early toilet training contributes to enuresis, but there is considerable evidence to the contrary. A large group of enuretic children in England were found to have been toilet trained later on the average than a control group. In a carefully monitored longitudinal study of kibbutz children in Israel, a group of children was defined as high risk on the basis of a family history of enuresis and several developmental and behavioral characteristics. For this group, toilet training was facilitated when it was begun between 12 and 15 months, and it was retarded when initiated after 20 months. On the other hand, for children not categorized as high risk, it did not matter when the training was begun (Kaffman and Elizur, 1977). The fact that enuresis is much more prevalent in the United States than in Great Britain has been considered by some to reflect the earlier and more goal directed toilet training in the British culture.

The higher prevalence of bedwetting among males, who are slower to develop than females, and among the lower classes, who may be more likely to use coercive measures in toilet training, both suggest that too early toilet training, premature for the level of physical development of the child, may create emotional difficulties. But, as already noted, some studies indicate that early training is correlated with a reduction in the probability of later bedwetting.

Some variables associated with enuresis, such as delinquency and lower levels of intellectual performance in schools, are noted long after the initial period of toilet training. They may be the product of a common underlying factor, such as maturational delay or family difficulties, that increases the probability of both enuresis and the later malfunction. It is also possible that enuresis

itself, with its attendant negative consequences for the child and the family, increases the risk of these later deviant or substandard performances.

It is likely that both directions of causation are correctly being noted by various researchers. Psychosocial problems increase the probability of enuresis, and enuresis increases the probability of various psychosocial problems (Douglas, 1964; Miller et al., 1974; Rutter, 1970).

LIMITED FUNCTIONAL BLADDER CAPACITY

It has been observed that most enuretic children void more frequently during the day than comparable nonenuretics, but that they do not void larger total volumes of urine either during the day or night. When children are asked to drink fluids freely, refrain from voiding as long as possible, and then measure the amount of urine voided, the volume measured is termed the functional bladder capacity. According to this measure, children with nocturnal enuresis have been noted to have, on the average, smaller functional bladder capacities than nonenuretic children, although there is an overlap in capacity between enuretics and nonenuretics of the same age. Proponents of the concept that limited functional bladder capacity is causally related to enuresis have noted that when urine retention training is instituted, there is a correlation between increased capacity and improvement in bedwetting (Starfield, 1972). Unfortunately, these observations have not been accompanied by cystometric studies, which would provide the most accurate estimate of bladder capacity. Finally, it cannot be determined whether limited functional bladder capacity is a causal factor or whether it is associated with the impact of enuresis on the affected child.

SLEEP DISTURBANCE

Common lore long assumed that children with nocturnal enuresis had problems with sleep arousal. Parents often described their enuretic children as unusually deep sleepers. Scientific investigation, however, comparing 100 children aged five to 15 with a matched control group, demonstrated no significant difference between the arousal time of the two groups (Bond, 1960). With the advent of the sleep polygraph, considerable interest was generated in the possibility that enuresis represents a sleep dysomia occurring in particular sleep stages. Careful investigation has failed to validate the association of enuresis with a particular stage of sleep. A large scale study involving recordings at home and in the hospital demonstrated that enuretic episodes occurred throughout the night on a random basis, with the frequency at any given sleep stage proportionate to the amount of time spent in that stage. Further, there were no differences in sleep structure or time of enuretic events between enuretic children who were considered psychiatrically disturbed and an equal group of nondisturbed enuretic children (Mikkelsen and Rapoport, 1980).

PSYCHIATRIC DISORDER

The possibility that enuresis is a psychiatric disorder has been carefully reviewed (Shaffer, 1973). On the Isle of Wight it was found that psychiatric disorders were more prevalent among enuretic than among nonenuretic children, especially in girls. Only a smaller number of enuretics of either sex, however, were rated as psychiatrically deviant; these findings, at best, could explain only a minority of enuretics. It has also been noted that when psychiatric disturbance is associated with enuresis, it is more likely to be found in children with diurnal enuresis. It is well known that secondary enuresis may occur after a disturbing event, such as hospitalization or a prolonged separation of the child from the parents. It is not known, however, whether children so affected have different patterns of achieving dryness prior to the emotionally disturbing event. Although there may be an association between bedwetting and psychiatric disorder, it remains to be determined which is causal. It is also plausible that whatever limited association does exist is the result of an underlying environmental or genetic disability that expresses itself in two different ways.

The best evidence that psychiatric disturbance is not the primary etiologic agent in enuretic children lies in the studies demonstrating no difference in treatment response, either to the bell and pad or to imipramine, between psychiatrically disturbed and nondisturbed enuretics (Mikkelsen and Rapoport, 1980).

The question has been raised whether removal of the symptom of enuresis might not be followed by the substitution of other symptoms, particularly in psychiatrically disturbed children. In the few studies that follow up treated children there is no evidence for such substitution. In fact, it appears that removal of the symptom has almost always had a positive effect on the child and the family.

ORGANIC CAUSES

Although organic causes of enuresis are relatively rare under all circumstances, they are more likely to be present in children with diurnal and secondary enuresis than in those with nocturnal and primary enuresis. When enuresis develops in a previously dry child, particularly a girl, the possibility of a urinary tract infection should be entertained. This is also true of children with nocturnal enuresis who have daytime urgency and frequency. The prevalence of enuresis in girls with urinary tract infections is as much as five times greater than in an unselected population. Furthermore, urinary tract infections are approximately five times more common in enuretic girls than in the general population. It should be noted, however, that when enuresis and urinary tract infection coexist, the bedwetting does not always cease after the infection has been cleared.

Various reports of the association between marginal obstruction of the urinary outflow tract and enuresis are unconvincing because of the inadequate criteria for both diagnosis and outcome. The association of dribbling with enuresis is rare, but dribbling should always be investigated in order to rule out the possibility of either a neurogenic bladder or an ectopic ureter.

Lumbosacral disorders that affect bladder innervation may have a causal relationship to enuresis. Of these disorders, myelomeningocele is the most obvious and most prevalent. Spina bifida occulta as an isolated finding is not associated with enuresis.

Diabetes mellitus, diabetes insipidus, and sickle cell trait may produce enuresis along with polyuria.

The total prevalence of organic causes of enuresis in some studies is as low as 1 per cent. With the exception of asymptomatic urinary tract infection in girls, the possibilities mentioned should be considered only when additional clinical findings are present.

SUGGESTIONS FOR FUTURE RESEARCH

How can the relative strength of these different theories of causation be assessed? Surveying the literature provides guidance for future research.

First, prospective longitudinal studies are needed. For too long, many researchers have relied on cross sectional and retrospective data on key variables related to enuresis. Second, such studies should carefully specify the frequency and nature of bedwetting at each age in the developing child and adolescent so that the dependent variable can be unambiguous—nocturnal or diurnal, secondary or primary. Third, indicators of maturation or of maturational delay should be recorded at each stage in the longitudinal study, so that there need be little reliance on potentially contaminated retrospective data. Fourth, the family resources available and the nature and timing of familial efforts to toilet train the child should be carefully noted. Fifth, detailed information should be collected about family structure and behavior patterns. In dealing with enuresis as a symptom, one should remember that the millions of sufferers remain a minority of all children and they come from a minority of families. Thus, for example, even though there may be a tendency for lower class families to be coercive in their child rearing, that gross generalization does not fit millions of families. Similarly, numerous middle class families may be more coercive in their child rearing than is typical in that social class. Only after gathering specific information about individual children and specific families can we move away from correlations based on aggregate relationships to a greater causal understanding of enuresis based on knowledge of individual development and corresponding social stress upon individuals. Sixth, detailed information about applications of alternative therapies should be part of the documentation for each child and family. We need to know more about differential compliance to suggested treatments, the timing of various therapies, and the response of the child and family to these interventions. Such field data may assist physicians in determining the therapy of choice for particular categories of children and families.

ASSESSMENT

Assessment of the child with enuresis should include a thorough clinical interview and physical examination, medical screening, and baseline recording of the child's enuretic behavior (Table 31–6). The interview should establish the frequency, the duration, and the time of occurrence of the enuresis, and whether it is primary or secondary. In secondary enuresis it is important to gather detailed information about events in the child's life and in the family situation at the time of relapse. It is critical to inquire into symptoms suggestive of associated urinary tract infection or metabolic disease, such as frequency, urgency, dysuria, polyuria, vomiting, unexplained fevers, and any evidence of failure to thrive.

A detailed family history should include the incidence of enuresis in parents and siblings, a careful description of the family constellation and significant life events, and the presence of relevant medical problems such as diabetes or sickle cell anemia.

A thorough developmental and psychosocial history of the child is essential in order to recognize and address possible behavioral and developmental problems.

The family and the child should be given ample opportunity to discuss their feelings and frustrations about the problem. All previous treatments and the alleged reasons for their failure should be carefully documented.

In the absence of evidence of a significant medical problem, medical screening can be very simple and should include, in addition to a thorough physical examination, a urinalysis and, in the black child, a sickle cell preparation. It is reasonable to assume that a urologic or other organic problem will manifest itself in other ways than by enuresis. There is no justification, therefore, in the absence of such evidence and with negative screening information for undertaking further medical diagnostic studies.

A baseline record of enuretic episodes should be kept in diary form by the child and his parents, for at least two weeks, before any type of intervention is attempted. It is extremely important to document the nature of the problem. It is also a frequent occurrence that the enuresis improves or disappears after the physician expresses concern and willingness to assist.

TREATMENT

See Table 31–7.

SPONTANEOUS CURE RATE

It is well known that many enuretic individuals become dry without any identifiable intervention or after several types of treatment have failed (Table 31–7). The spontaneous cure rate, however, is difficult to assess. In one large follow-up study of children with nocturnal enuresis, the annual spontaneous cure rate was determined by recall with questionnaires. The results were similar to those in the few published studies. The annual

Table 31–6. ASSESSMENT

History
    Establish frequency, duration, time of occurrence, dry
      periods
    Familial incidence, parents, siblings
    Careful documentation of developmental milestones
    Family milieux and significant life events
    Previous attempts at treatment
Physical examination
Medical screening: urinalysis
Baseline period of observation with diary, 2 to 3 weeks

Table 31–7. TYPES OF TREATMENT

Behavioral
    Counseling
    Hypnosis
    Behavior modification
      Urine alarm
      Urine retention training
      Dry bed training
Pharmacotherapy
    Imipramine and other tricyclic antidepressants

rates were as follows: between five and nine years of age, 14 per cent annually; between 10 and 19 years, 16 per cent annually. Three percent were still enuretic after the age of 20 (Forsythe and Redmond, 1974). The annual decrement in wetting in the five through nine year olds according to these data is higher than we calculated for females ages six through 11, but is the same for males. As mentioned earlier, the National Health Examination Survey unfortunately did not specify which children had received therapy for enuresis. The two sets of data are in general agreement that there is a continuing tendency toward dryness throughout development.

COUNSELING

The physician may, as a first step, choose to engage in minimal intervention. A warm supportive attitude can benefit both the child and the often distraught family. Allowing child and parents to relate their frustration and anxieties and explaining to them what we know and do not know about the problem have at times been effective, without need for other intervention. The physician may also be helpful by informing the family of those interventions that are known to be ineffectual. Among the ineffective techniques are punishment and degradation of the child, fluid restriction, and waking the child for toileting at night. Although there have been some claims for the use of various elimination diets, scientific evidence for their efficacy is lacking.

If the physician wishes to handle the problem only by counseling, he should plan to see the family on scheduled visits over a period of several months. The child should be involved as fully as possible. A system of rewards for dryness can be instituted, and it is helpful for the child to keep a calendar in which he gives himself credit for each dry night.

The physician must be mindful that other problems may coexist with enuresis. Various forms of developmental delay, including late sexual maturation, as well as academic difficulties and behavior disturbances must be investigated and addressed appropriately. In some children the behavior problems may necessitate psychotherapy or family therapy. This should be directed toward those particular problems and not toward the enuresis itself, which does not respond to a direct psychotherapeutic approach. The symptom of enuresis unattended by severe behavior problems should not be viewed as an indication for psychotherapy.

HYPNOSIS

Another form of mild intervention is the use of hypnosis, which is claimed to be successful in individuals who are able to achieve a hypnotic trance. Among 40 children who were taught the technique of self-hypnosis, excellent results were claimed by Olness for 31. In Johnson's study, in which 18 adolescent males were treated with weekly hypnotic sessions, all were enuresis-free by eight weeks, and the relapse rate over 20 months was low. These studies require replication.

BEHAVIOR MODIFICATION

The two most commonly employed and most effective methods are the urine alarm and urine retention training.

Urine Alarm. The urine alarm was popularized in 1938 by Mowrer and Mowrer, who introduced the bell and pad procedure. The method has been widely used in Great Britain but much less commonly in the United States. The purpose of this and other methods of behavior therapy is to provide the patient with the opportunity to learn techniques presumably not developed in the normal course of maturation. The alarm system, which is activated by a few drops of urine, allows the child to awaken when he urinates at night and eventually to respond to a distended bladder rather than to the alarm. With this method the cure rates in several studies were 60 to 90 per cent and the relapse rates were 20 to 45 per cent. Retreatment of the patients who relapsed was often successful (Turner, 1973).

Apparatuses less complex and simpler to use than the bell and pad have recently been introduced. One example is the Wet Stop.* A lightweight buzzer powered by a hearing aid battery is attached to the shoulder of the pajama top by a Velcro patch, and the sensor is a small plastic card that inserts into a pocket sewn to the outside of ordinary underpants (Fig. 31–4). The manufacturers claim a response rate similar to the bell and pad—about 60 per cent cured at four months.

These apparatuses can be used for children who are old enough to understand the purpose of the alarm; this certainly includes children who are old enough to enter school and can encompass some five year olds. For the successful use of these methods, the physician should plan to see the child and family and communicate with them by telephone on a regular basis until dryness is achieved.

Several studies have shown this type of conditioning to be superior to psychotherapy, nighttime awakening, and counseling. The use of drugs as adjuncts to conditioning has not proved to be a significant improvement.

Urine Retention Training. Urine retention

---

*The Wet-Stop Alarm can be ordered from Palco Laboratories, 5026 Scotts Valley Drive, Scotts Valley, California 95066.

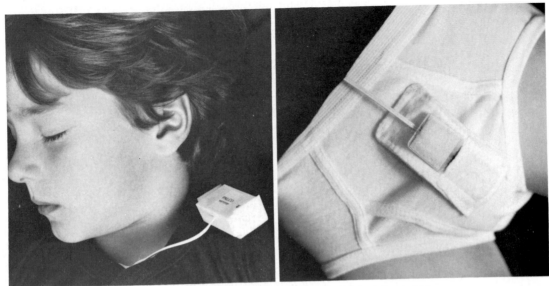

Figure 31–4. Wet-stop enuresis alarm. (Courtesy of Palco Laboratories.)

training is another type of behavior modification. In a careful study of 83 children, it was demonstrated that bladder training in the daytime resulted in enuresis improvement in two-thirds, and reduced wetting to once a month or less in one-third (Starfield, 1972). The technique, employed once daily for six months, consists of three major elements: fluids in unrestricted amounts during the daytime, voluntary retention to the point of discomfort at least once a day, and recording the amount of urine passed after maximal holding. The advantage of this method is that it does not involve expensive equipment or disruption of the family during the night. On the other hand, it does necessitate consistent compliance and the presence of a parent or involved adult on a daily basis, a potential problem for working parents.

Dry Bed Training. Other behavioral therapies combine the aforementioned methods. The dry bed training procedure is taught to the child and the parents in the course of one night by a trainer in the home. It incorporates positive reinforcement, nighttime awakening, negative response to wetting, retention training, and the use of the urine alarm. In a matched control series the remission rates were higher and more rapid than with the urine alarm alone and persisted for a six month follow-up period (Azrin et al., 1974). The paucity of research data, however, makes it difficult to evaluate and systematically compare the combined approach to other methods. An article by Schmidt reviews the various modes of therapy and provides instruction for parents (Schmidt, 1982). This review will be helpful to the physician in choosing and implementing therapy. He should be aware, however, that Schmidt is addressing only the symptom of bedwetting, and he should remain alert to the possibility of other associated problems.

PHARMACOTHERAPY

The tricyclic antidepressants form the only category of drugs that are superior to a placebo as antienuretics. The superiority of imipramine and other tricyclic compounds (trimipramine, desipramine, nortriptyline, amitriptyline) has been demonstrated in over 40 double blind studies. Among the many drugs that have been tried unsuccessfully are amphetamines, monamine oxidase inhibitors, sedative hypnotics, major tranquilizers, anticonvulsants, diuretics, pituitary snuffs, anticholinergic, and alpha adrenalytic drugs.

The antienuretic response to imipramine is rapid, occurring usually within the first one or two weeks. When the drug is stopped, relapse is likely to occur (Blackwell and Currah, 1973). The reported percentage of remissions varies from 10 to 50 per cent, with the majority of reports toward the lower end of the range.

The imipramine response has been shown to be independent of the age of the child, the sleep stage of the enuretic event, and the presence or absence of severe psychiatric problems. There appears to be a group of good responders, approximately one-half, who will have either a complete remission or a significant diminution in their wet nights. There are nonresponders (one-sixth), even at high drug dosages, and there are also transient responders (one-third) who achieve an initial good response but relapse quickly.

The plasma concentration of the drug is positively related to the antienuretic response. Doses of 75 to 125 mg. have been safely and effectively used. At these dose levels the major side effect is dryness of the mouth. Other side effects to be watched for are drowsiness, weight gain, dizziness, alteration in mental state, difficulty in concentration, sleep disturbances, and nightmares. There is a high correlation between the number of side effects and the plasma concentration of the drug. Since the laboratory determinations are not readily available in most settings, it is possible to use the side effects as a proxy measure of the blood level. If a child does not have an antienuretic response at a given dosage but has several side effects, it is likely that the child is a true nonresponder (Mikkelsen and Rapoport, 1980).

Several cases of serious accidental poisoning by an overdose of imipramine have been reported among siblings in the household, with death resulting from cardiotoxic and neurotoxic effects. For the enuretic child the minor side effects occur well in advance of any serious toxicity and at a much lower dose.

The mechanism of the antienuretic action of the tricyclic antidepressants remains unknown. The following negative observations are of interest. The antidepressant response occurs considerably later than the antienuretic response. The anticholinergic action is probably not the effective mechanism, since other anticholinergic drugs are ineffectual. Alpha-adrenalytic drugs are not effective substitutes. The drugs have no effect on the level of sleep arousal or on the time of occurrence of enuretic episodes in the sleep cycle.

Despite the lower efficacy of imipramine compared to behavior modification and the lack of knowledge of its pharmacodynamics, physicians in the United States tend to use it more frequently than the bell and pad or urine retention training. The comparative convenience of drug treatment makes it worthwhile to try this method when short term results are important, such as for summer camp, or when the family will not support the long term efforts of behavior modification.

PROPHYLAXIS

As discussed, there is no clear consensus whether the method of toilet training as well as its timing can either promote or retard nighttime dryness. It seems reasonable, however, for the physician to be alert to risk factors in the family and the child that may predispose to enuresis and, in such cases, to give the parents early and specific advice regarding toilet training. In general, the training should not proceed until the child has achieved the motor skills to walk toward and away from the toilet and to comprehend the purpose of toilet. The child should also be aware of a desire to urinate and be displeased with being wet.

Sometime during the second year this orientation can coincide with a developmental stage in which there is a desire to please the parents and master these skills, and can occur prior to a period of negativism and desire for increased autonomy. It must be remembered that there are substantial differences among children in their acquisition of these skills and their entry into various developmental phases. Parents should be sensitive to the child's reaction to the initiation of training. Providing an opportunity should not be interpreted as forcing the child, and they should be ready to postpone training if the child resists. After a lapse of time a modest attempt to train should be reinstituted, with the same willingness to withdraw if parental efforts again meet with resistance. This type of approach combines the avoidance of coercive tactics with the replicated finding that early training results in a lower incidence of enuresis.

A FRAMEWORK FOR UNDERSTANDING ENURESIS

Enuresis is a symptom with various underlying causes. Its treatment is more straightforward than our understanding of the factors that lead to one or another form of enuresis.

An organic basis for enuresis, however, is a rare phenomenon. This does not mean that enuresis is a functional disorder of exclusively psychologic origin. Indeed the research evidence leads to a strong presumption of physiologic underpinnings for the development of enuresis.

There often appears to be a genetic basis for enuresis. Maturational delay may be the mechanism by which the observed genetic basis for enuresis is expressed. Children who are slow to develop mentally and physically, even though the delay may be minimal and not apparent, are more likely to be exposed to the risk of coercive toilet training prior to the time at which they are capable of succeeding. Perhaps there is a sensitive period for toilet training, and the failure to learn appropriate behavior at that time increases dramatically the distress associated with later attempts to control micturition.

The research literature indicates that stressful situations, particularly during that sensitive period for toilet training, increases the probability of enuresis.

Although there are emotional correlates of enuresis, research does not indicate that enuresis is often a symptom of underlying psychiatric disorder. Disturbed and nondisturbed enuretics respond with equal probability of success to the more effective treatments. There is no substitution of other symptoms when enuresis is successfully treated.

Because of the complexity of underlying causal

relationships, the variability in response to different types of therapy is not surprising. Once enuresis becomes an established symptom, it is difficult to reconstruct the underlying sequence of events and thus to predict whether a specific type of behavior modification or a specific drug of undetermined pharmacologic action will be more effective. The physician must make a choice based largely on the perception of the situation, taking into account the perspectives of the child and the parent.

Enuresis is not always the final event in this chain of interactions. It may lead to a variety of negative behavioral consequences involving the self-concept of the child, the expectations and behavior of significant others, and specific decrements in performance in such areas as schooling. For these reasons the symptom of enuresis should be neither ignored nor handled with the admonition, "Don't worry; the child will outgrow it." The symptom should always be treated by one of the available effective methods, and the child should be carefully followed because of the possibility of current and future associated problems.

RUTH T. GROSS
SANFORD M. DORNBUSCH

REFERENCES

Azrin, N. H., Sneed, T. J., and Foxx, R. M.: Dry-bed training: rapid elimination of childhood enuresis. Behav. Res. Ther., 12:147, 1974.
Blackwell, B., and Currah, J.: The psychopharmacology of nocturnal enuresis. Clin. Dev. Med., 48/49:231, 1973.
Blomfield, J. M., and Douglas, J. W. B.: Bedwetting: prevalence among children aged 4–7 years. Lancet, 1:850, 1956.

Bond, M. M.: The depth of sleep in enuretic school children and in non-enuretic controls. J. Psychosom. Res., 4:274, 1960.
Brazelton, T. B.: A child-oriented approach to toilet training. Pediatrics, 29:121, 1962.
Dejonge, G. A.: Epidemiology of enuresis: a survey of the literature. Clin. Dev. Med., 48/49:39, 1973.
Douglas, J. W. B.: The Home and the School. London, MacGibbon & Kee, 1964.
Forsythe, W. I., and Redmond, A.: Enuresis and spontaneous cure rate. Arch. Dis. Child., 49:259, 1974.
Glicklich, L. B.: An historical account of enuresis. Pediatrics, 8:859, 1951.
Hallgren, B.: Enuresis: a study with reference to the morbidity risk and symptomatology. Acta Psychiatr. Neurol. Scand., 31:379, 1956.
Kaffman, M., and Elizur, E.: Infants who become enuretics: a longitudinal study of 161 kibbutz children. Monogr. Soc. Res. Child Dev., 42:2, 1977.
Kolvin, I.: Enuresis in childhood. Br. Med. J., 214:33, 1975.
MacKeith, R. C.: Is maturation delay a frequent factor in the origins of primary nocturnal enuresis? Dev. Med. Child Neurol., 14:217, 1972.
Mikkelsen, E. J., and Rapoport, J. L.: Enuresis: psychopathology, sleep stage, and drug response. Urol. Clin. N. Am., 7:361, 1980.
Miller, F. J. W., Knox, E. G., Court, S. D. M., and Brandon, S.: The School Years in Newcastle upon Tyne. London, Oxford University Press, 1974.
National Center for Health Statistics: Plan, Operation and Response Results of a Program of Childrens' Examinations. Public Health Service Publication 1000, Series 1, No. 5, 1967.
National Center for Health Statistics: Plan and Operation of a Health Examination Survey of U.S. Youths 12–17 Years of Age. Public Health Service Publication 1000, Series 1, No. 8, 1969.
Oppel, W. C., Harper, P. A., and Rider, R. V.: The age of attaining bladder control. Pediatrics, 42:614, 1968.
Oppel, W. C., Harper, P. A., and Rider, R. V.: Social, psychological, and neurological factors associated with nocturnal enuresis. Pediatrics, 42:627, 1968b.
Rutter, M., Tizard, J., and Whitmore, K.: Education, Health and Behaviour. London, Longman Group Ltd., 1970.
Rutter, M., Yule, W., and Graham, P.: Enuresis and behavioural deviance: some epidemiological considerations. Clin. Dev. Med., 48/49:137, 1973.
Schmidt, B.: Nocturnal enuresis: an update on treatment. Pediatr. Clin. N. Am., 29:21, 1982.
Shaffer, D.: The association between enuresis and emotional disorder: a review of the literature. Clin. Dev. Med., 48/49:118, 1973.
Starfield, B.: Enuresis: its pathogenesis and management. Clin. Pediatr., 11:343, 1972.
Turner, R. K.: Conditioning treatment of nocturnal enuresis: present status. Clin. Dev. Med., 48/49:195, 1973.

31B   Encopresis*

Children with encopresis lack control over their bowels, having lost or never developed this competency. Affected youngsters are apt to view themselves with shame and profound self-condemnation despite the fact that they usually are not guilty of having caused the problem. Children commonly acquire encopresis through various combinations of unforeseen events, inherited predispositions, accidental practices, inappropriate situations, and misunderstandings.

Encopresis may be defined as the passage of formed or semiformed stools in a child's underwear (or other inappropriate places), occurring regularly after age four years. Most children have most of their "accidents" late in the day, usually between 3:00 and 7:00 P.M. (Levine, 1975).

CLASSIFICATIONS

To classify children with encopresis, some authors differentiate between primary (or continuous) and secondary (or discontinuous) subtypes (Easson, 1960; Olatawura, 1973). The former condition always has been present; in other words a child never has completed training. Secondary encopresis occurs when children are completely toilet trained and subsequently regress to incontinence. Some clinicians differentiate encopresis from fecal soiling, on the basis of the degree to which full bowel movements are passed in under-

*This section was adapted from a review published in *Pediatric Clinics of North America* (April 1982).

wear. This may not stand up to clinical experience, since chronic cases commonly fluctuate, showing variations in the degree of incontinence and stool frequency. That is, the child sometimes stains or soils, at other times passes small rocklike masses, occasionally clogs the plumbing, and also has intermittently normal movements. Classifications based on onset also may be too general, as a sizable subgroup of patients may fall somewhere between the primary and secondary forms, having been partially trained.

BOWEL RELATED SYMPTOMS

Encopresis is reported to occur in 1.5 per cent of second grade youngsters (Bellman, 1966). The ratio of boys to girls varies but has been estimated at 6 to 1. There appears to be little or no social class preference and no indication that prevalence relates to family size, ordinal position, or the age of parents (Levine, 1975).

As noted, virtually all children with encopresis, at least intermittently, retain stools (Levine, 1981). This proclivity may be subtle and clinically elusive. Vague historical documentation, along with seemingly normal abdominal palpation and rectal examination, may lead the clinician to conclude that a child has "encopresis without constipation." Often in such cases a plain x-ray view of the abdomen reveals abundant retained feces. "Occult" constipation or slowly progressive stool retention with a gradual onset may be asymptomatic or may manifest itself as recurrent abdominal pain (Barr et al., 1979). Some defecate every day, but produce partial bowel movements; retention may not be gross enough to be reflected in physical findings. Some children show exclusively rectal constipation, rendering palpation of the abdomen uninformative. A rectal examination in such cases also may be ambiguous or borderline.

With increasing retention, sensory feedback from the bowel is reduced. The rectal wall stretches and diminishes in its contractile strength. There is increased water absorption from fecal material, resulting in harder and larger feces. Painful defecation may ensue, sometimes accompanied by anal fissures or hemorrhoids, promoting further toilet avoidance and more obstipation. The anal canal is apt to become further stretched and foreshortened. The external and internal sphincters then may be compromised, allowing the passage of soft feces and mucus around impactions. Paradoxical sphincteric function is associated with alteration of defecation reflexes. Physiologic pressure:volume relationships may be impaired, such that an increase in the bulk of material within the rectum is associated with reduced (rather than greater) pressure generation by voluntary and involuntary muscles. It is likely that as better tech-

niques are developed to evaluate bowel motility, a variety of physiologic motor derangements (both acquired and congenital) will be found to cause or induce vulnerability to encopresis.

OTHER SYMPTOMS

Children with encopresis manifest a wide range of clinical symptoms. In addition to incontinence, some suffer from recurrent abdominal pain. These usually are youngsters with a recent onset of stool retention and incontinence, as those with long standing encopresis seem to have greater pain tolerance to colonic distention, and rarely are aware of pain.

The prevalence of enuresis in children with encopresis varies among studies. A distended rectum may compromise bladder capacity and cause some urinary "dribbling" during the day. Treatment of the stool retention thus may alleviate such urologic problems.

LIFE STYLE AND AFFECTIVE TOLL

It is rare for children with encopresis to reveal physical symptoms in isolation (Levine et al., 1980; Pinkerton, 1958). Pre-existing, concomitant, or secondary maladaptive traits commonly occur. The lack of control over defecation is frightening to these children and dehumanizing. They are apt to fear discovery, exposure, and peer ridicule. Self-esteem commonly deteriorates, and there may ensue social withdrawal, anxiety, depression, and extra-abdominal somatic symptoms. Acting out or sociopathic behavior is not frequently encountered with encopresis; these children are more apt to isolate themselves and show signs of dependency. Many seem to have confused relationships with parents, and, in particular, they may be preoccupied with issues of autonomy and dependency.

Commonly encopresis triggers conflict, as parents, grandparents, neighbors, siblings, and professionals all promote their hypotheses about the problem. Family activities may be curtailed out of fear that the affected child could mess himself on a car trip, at a friend's home, or while dining in a restaurant. Siblings may not want friends at home because of offending indoor odors.

Children with encopresis often are bestowed unkind but relevant nicknames created by their siblings and peers. Various coping mechanisms may be mobilized to deal with this. Commonly these children are unwilling or reluctant to discuss such painful ridicule with parents or professionals.

One feature of encopresis differentiates it from most other functional disorders; namely, affected youngsters almost never have heard of any others. Parents also are apt to be unaware of its existence as a common childhood problem. They may attribute the condition to laziness or poor hygiene and are reluctant to bring it to the attention of a physician. Family shame and cultural taboos also may cause a family to delay in seeking help.

Children with encopresis tend to stand accused by the adult world. They may be admonished and told that they mess in order to win attention or else because they are lazy. They may be chastised for not coming in from play to use the toilet, despite the pathetic and entirely candid explanation, "I didn't know it was coming." The child subsequently may be charged with negligence for not changing tainted undergarments after an accident. Parents may not realize that olfactory sensation accommodates to odors that are present continuously, that people have limited perception of their personal body smells; thus a child with chronic incontinence is likely to be oblivious to his offensive aroma.

In appreciating the tragedy of encopresis, one must conceptualize a human condition in which a child is ridiculed, shamed, or blamed (by himself and others) for something he did not cause and over which he has had little, if any, actual control. Such an intolerable predicament is common among these youngsters.

PATHOGENESIS: VULNERABLE STAGES AND SYMPTOM POTENTIATION

Incontinence constitutes the end result of multiple factors that interplay. Their cumulative impact potentiates the problem. Three developmental periods are critical in the generation of encopresis (Table 31–8). Stage I ("Early Experience and Constitutional Predisposition") covers the first two years of life. Stage II ("Training and Early Autonomy") includes ages three to five years. Stage III ("Extramural Function") encompasses the early school years. As children progress through each of these spans, constitutional tendencies, the milieu of the environment, and some critical life events coincide to induce a functional bowel disorder. In all likelihood a youngster with only one potentiator is able to overcome or bypass such susceptibility. Most children with encopresis, on the other hand, appear to accumulate several potentiating factors. The existence of one or more of these seems to promote vulnerability to the others.

The timing of the onset of bowel symptoms offers some pathogenetic clues. At The Children's Hospital Medical Center, a standardized questionnaire and manual help clinicians to seek the sources and describe current manifestations of a child's encopresis (Levine and Barr, 1980). The following is an elaboration of commonly encountered potentiating factors at the three developmental stages (see also Table 31–8). A child may start to show predispositions at an early age and thereby become susceptible to subsequent potentiating factors. On the other hand, some children may not show any such tendencies until stage II or stage III.

Table 31–8. CRITICAL STAGES IN THE POTENTIATION OF ENCOPRESIS*

Stage I potentiators (infancy and toddler years)
  Simple constipation
  Early colonic inertia
  Congenital anorectal problems
  Other anorectal conditions
  Parental over-reaction
  Coercive medical interventions
Stage II potentiators (training and autonomy—three to five years)
  Psychosocial stresses during training period
  Coercive or extremely permissive training
  Idiosyncratic toilet fears
  Painful or difficult defecation
Stage III potentiators (extramural function—early school years)
  Avoidance of school bathrooms
  Prolonged or acute gastroenteritis
  Attention deficits with task impersistence
  ? Food intolerance, including lactase deficiency
  Frenetic life styles
  Psychosocial stresses

*Children who ultimately develop encopresis are likely to have accumulated multiple risk factors on this list.

STAGE I: EARLY EXPERIENCE AND CONSTITUTIONAL PREDISPOSITION

Children whose history suggests bowel dysfunction during infancy and the toddler years may be showing their constitutional or congenital predisposition to encopresis. "Early colonic inertia" has been described as an endogenous tendency toward immature or generally inefficient intestinal motility (Coekin and Gairdner, 1960). This disorder is common and in some instances may stem from genetic factors. In other cases it may mimic functional constipation resulting from other causes. Ordinarily this is a transient phenomenon, one that emerges only as a potentiator of encopresis in the presence of additional negative influences.

A few youngsters with encopresis have a history of imperforate anus or similar congenital anomalies, for which they have undergone corrective surgical procedures. The delayed onset of encopresis may result from operative effects or, alternatively, from the psychologic after-effects of early intervention in this region. More minor medical conditions also may precipitate infantile problems with defecation. For example, food intolerance may be associated with constipation. Ordinarily these problems are inconvenient but self-limited;

sometimes they may initiate a pathologic chain reaction that culminates in encopresis.

In the presence of a constitutional predisposition, potentiation of encopresis in this age group may be fostered by parental over-reaction, as extremes of grief or exaltation surrounding defecation engender excessive (or obsessive) bowel preoccupations for the developing child. Aggressive management also may predispose to dysfunction: overindulgence in suppositories and enemas, frequent digital insertions, and other coercive intrusions may create a so-called "anal stamp." As one author has observed, "The battle of the bowel seemingly won in the nursery is destined to be lost on the playing fields at school" (Anthony, 1957).

An infant or toddler may potentiate impaired defecation through voluntary withholding. The process is perceived as a negative experience, and the child is seen straining excessively during bowel movements. Such exertion occurs with the legs hyperextended, creating the impression of a struggle to defecate, but in reality the act is an effort to retain rather than expel. Fissures and perianal irritations may accompany constipation and in themselves potentiate chronic bowel difficulty, by themselves encouraging withholding.

Preventive implications are evident. In particular, consistent, low key, well informed, and nonaggressive management of simple bowel problems in infancy should minimize potentiation during this stage.

STAGE II: TRAINING AND EARLY AUTONOMY

Between the ages of three and five, children start to explore autonomy and independence. Bowel training, a newly acquired control, represents a major forward step toward independence. To some vulnerable youngsters, however, investigations of the toilet are undertaken with fear and become associated with family power struggles, at times with threatening fantasy. Children who have revealed vulnerability earlier in life (i.e., during stage I) are the most susceptible to the accretion of new potentiating factors. A child who earlier developed negative associations with defecation may dread the toilet, as that receptacle embodies still another threat surrounding the distasteful process of elimination.

Anxiety over toilets may constitute an initial potentiator in a child with no history of bowel problems. For example, when trained on a toilet seat, some youngsters have trouble performing a Valsalva maneuver while their feet are suspended in air. They may benefit from a bench, a pile of telephone directories, or some other mode of foot support. Without such leverage there may be persistent fear of falling in or of being flushed down through the bowels of the plumbing!

In some cases a child is trained to alight upon the toilet but sustains irrational fears. There may be concerns about commodes that might overflow and flood the bathroom, fantasies about sea monsters that take a nip when one sits down, or dreams about babies who get born in toilet bowels! Some apprehension is normal; in certain cases the fear is extreme and it culminates in long standing and sometimes unconscious avoidance of defecation. A child may not avoid the toilet entirely, but may devote as little time as possible thereupon, consistently accomplishing incomplete defecation. The result is obstipation followed by encopresis.

Reluctance to defecate on the toilet may occur as a result of psychosocial stresses coinciding with the training period. Inappropriate training techniques (especially extremes of coercion or permissiveness) may induce avoidance (Huschka, 1942). During stage II some relevant medical problems, such as chronic diarrhea for any reason, may contribute to negative associations with elimination and a consequent potentiation of colonic dysfunction.

Many potentiators can act synergistically with toilet phobias and improper training. For example, excessive parent-child conflict over autonomy and dependency may activate latent bowel dysfunction. In a family in which there is constant cross fire over feeding or sleeping, the toilet bowl can become another battlefield.

It should be emphasized that such conflicts are not necessarily evidence of serious family psychopathology. Frequently they are the products of misunderstandings, misplaced priorities of parenting, and conflicting or improper advice. Once again it is likely that anticipatory guidance and parent education can stave off many cases of encopresis.

STAGE III: EXTRAMURAL FUNCTION

Early in elementary school, children who have collected earlier potentiating factors may be predisposed to encopresis. New stresses come forth, the most important being school. In particular, school lavatories become a prime potentiator of encopresis. A child who used to defecate each morning at 11 A.M. at home may discover that there are no doors in front of the toilets, or that the school lavatory is a well publicized amphitheatre with a varied program of humiliating scenarios. Such a youngster may determine wisely to withhold defecation until back in the private safety of his own home. By 3 P.M., however, he has lost the urge. After several months in this holding pattern, the child may be obstipated and show paradoxical or overflow incontinence. Many other youngsters (perhaps most, in fact) decide to avoid defecation in school. Because they lack predispositions to bowel dysfunction, they can do so without risking encopresis.

It is not unusual during school for children with previous potentiation to acquire encopresis after a prolonged medical illness (such as gastroenteritis). In such cases normal bowel patterns may be deranged, and the child may not compensate.

When encopresis is potentiated during school years, a number of factors are likely to operate. A loss of rectal and anal sensation described earlier aggravates the disorder, as does a diminution in lower colonic muscle tone. A child's private anxiety about the symptom, aggravated by adult induced stresses and pressure, further worsens the condition. Dietary overindulgence, such as excessive ingestion of milk or chocolates, is common at this age and may tend to exacerbate the problem for some affected youngsters. Children with bowel dysfunction tendencies who also have attention deficits or hyperactivity are particularly prone to manifest encopresis. They are prone to be task impersistent. They seldom complete what they start—even defecation. Thus, day after day, they produce partial bowel movements, become constipated, and eventually become incontinent. A similar phenomenon may be observed in some susceptible boys and girls whose encopresis is potentiated by a frenetic life style. A child who is late getting up in the morning and skips defecation, so as to catch the school bus, chronically may withhold, accommodating to a very tight agenda with a very tight sphincter. Once again, it is likely that in such a youngster earlier or concurrent potentiating factors will be found that may encourage the development of encopresis.

PREJUDGMENTS

A multifactorial model is most likely to account for the symptom of encopresis. It is dangerously misleading and unfair to prejudge a child with this condition. Many youngsters whose life adjustment is appropriate in other respects acquire their encopresis as a result of potentiating forces well beyond their willful control and that of their parents. In other instances encopresis may be but one manifestation within a generalized life condition of maladjustment or psychosocial stress. It follows that each case must be individualized; it never should be assumed that a child with bowel incontinence is "emotionally disturbed."

THE EVALUATION PROCESS

The assessment of a child with encopresis should include careful review of the possible staged potentiators already elucidated. Moreover, it is useful to survey current clinical manifestations as well as past and current management. A questionnaire system and manual have been developed for this purpose (Levine and Barr, 1980). The following questions should be answered as a part of such an assessment:

1. To which developmental period can one trace the origins of this child's problem? One should probe carefully for the age of onset of bowel related symptoms. Were such difficulties present earlier in life? Was the onset linked in some way to training? Was this child ever fully trained? Did entry into school or some other major life event or transition coincide with the emergence of bowel dysfunction? Did this youngster have any medical treatment or condition that may have imprinted an early "anal stamp" (Anthony, 1957)? What other potentiating factors have existed?

2. What is the child's current habit of toilet utilization? Does he shun the facility altogether or even partially? Is any avoidance confined to specific territories (e.g., school)? Are bathroom visits too brief?

3. What is the current severity and frequency of this child's incontinence? Is incontinence confined to afternoons and evenings?

4. What psychosocial phenomena in the environment could be promoting or aggravating the symptom? Is there marital strife? Are there problems with a sibling? Have there been critical setbacks in the family? Are there indications of deprivation or abnormal patterns of nurturance?

5. Are there characteristics in this child that may be aggravating encopresis? Does the youngster appear to be significantly depressed? Does the child have social interactional difficulties? Is this a youngster who has significant attention deficits, such that task impersistence interferes with establishing a normal bowel routine? Are there associated learning disabilities?

6. What toll has encopresis exacted from this child? Has the youngster developed secondary maladaptive strategies? Has there been significant secondary gain from the symptom that is likely to impair resolution? What associated patterns of behavior have emerged—perhaps in response to the child's encopresis?

7. How have the child and family coped with encopresis? Have consistent therapies been tried? Has the problem provoked disagreement and strife? Are guilt and accusation associated with it? How does the child conceal the disorder?

8. What is the child's conception of the problem? Why does he think that bowel control has been lost or does not exist? How directly can the child confront and discuss the disorder?

9. Why do the parents believe that the child has encopresis? What do they believe are its underlying causes?

The answers to these questions are helpful in establishing a therapeutic alliance and in understanding the plight of a child with encopresis. Treatment and counseling plans should be derived from such background information.

The physical examination should help rule out other pathologic causes of stool retention and overflow. It should be recognized, however, that in the school aged child such conditions are extremely rare. The scepter of Hirschsprung's disease frequently overshadows encopresis. Clinicians should be aware of the great difference between the two conditions; these are summarized in Table 31–9. Children with encopresis ordinarily appear well nourished and healthy during school years. On the other hand, a child with Hirschsprung's disease is likely to look wasted and chronically ill and to have had intermittent obstructive symptoms. Children with encopresis often have a history of having produced extra large caliber stools (sometimes obstructing the plumbing); those with Hirschsprung's disease are more apt to put forth thin ribbony stools. Incontinence is the most prominent symptom in children with encopresis; in Hirschsprung's disease heavy soiling is uncommon. Children with encopresis may have acquired bowel symptoms relatively late in life, whereas those with Hirschsprung's disease generally present with serious malfunction early in infancy.

The physical examination can be used to rule out other causal conditions, such as hypothyroidism. When indicated, laboratory tests can be ordered. A sensory examination and careful neurologic assessment can uncover neurogenic factors such as spinal cord lesions. Chronic constipation also has been attributed to Crohn's disease (secondary to thrombosis and scarring of an inflamed bowel), malnutrition, and diseases that impair voluntary muscle function (e.g., amyotonia congenita, infectious polyneuritis, and cerebral palsy).

So-called ultrashort segment Hirschsprung's disease has been reported in some children (Roy et al., 1979). In these cases anal manometry is alleged to demonstrate that rectal distention fails to initiate relaxation of the internal sphincter. Such findings, however, may be liable to misinterpretation, since they may not be unique to aganglionosis. Prolonged stretching of the bowel wall in itself may compromise sphincter physiology in youngsters with a functional megacolon. In such cases a biopsy may show an apparent decrease in ganglion cells. The definitive word on ultrashort segment Hirschsprung's disease is not in yet; pediatricians therefore should be somewhat reluctant to make this diagnosis, especially if it entails major surgical intervention.

Abdominal and rectal examinations may help to estimate the degree of stool retention, but these commonly are deceptive. Many youngsters are laden with feces throughout the colon, but fail to reward the palpating hand or probing digit. A child may be packed with soft stool. A plain x-ray view of the abdomen often is helpful; it can be scored for the degree of retention (Barr et al., 1979). This is helpful in clinical evaluation, since some children have exclusively rectal constipation whereas others are encumbered with stool throughout the colon. Knowledge of the extent and degree of fecal retention is relevant in implementing and monitoring an appropriate catharsis regimen (see section on management). A barium enema seldom, if ever, is necessary.

Anal manometry is employed sometimes to evaluate sphincter function. At present its therapeutic implications are not clear-cut. It is used to diagnose Hirschsprung's disease, but children with functional encopresis also may show some manometric derangement. Abnormal findings generally will not change a medical treatment approach. A rectal biopsy may be indicated when there are signs truly suggestive of aganglionic megacolon. One might argue that a biopsy is appropriate when symptoms have not abated, despite optimal management. However, the yield, even in such cases, is low.

Girls with encopresis commonly acquire urinary tract infections as a result of ascending invasion secondary to soiling of the perineum (Shopfner, 1968). Therefore, it is wise to include a urinalysis and culture as part of the initial evaluation of any girl with fecal incontinence. Radiographic studies of the urinary tract also are indicated in long standing cases, since some girls may harbor chronic pylonephritis as a result of encopresis. Obstructive uropathies have been described in children of both sexes with chronic obstipation, these are likely to improve after restoration of normal bowel function.

Table 31–9. ENCOPRESIS AND HIRSCHSPRUNG'S DISEASE

	Encopresis	Hirschsprung's Disease
Stool incontinence	Always	Rare
Constipation	Common, may be intermittent	Always present
Symptoms as newborn	Rare	Almost always
Infant constipation	Sometimes	Common
Late onset (after age three years)	Common	Rare
Problem in bowel training	Common	Rare
Avoidance of toilet	Common	Rare
Failure to thrive	Rare	Common
Anemia	None	Common
Obstructive symptoms	Rare	Common
Stool in ampulla	Common	Rare
Loose or tight sphincter tone	Rare	Common
Large caliber stools	Common	Never
Preponderance of males	86%	90%
Incidence	1.5% at age seven to eight	1:25,000 births
Anal manometry	Sometimes abnormal	Always abnormal

Table 31–10. MANAGEMENT OF ENCOPRESIS*

Treatment Phase	Treatment Program	Comments
Initial counseling	1. Education and "demystification" of the problem 2. Removal of blame 3. Establishment and explanation of treatment plan	Include diagram, review of colonic function, shared observation and x-ray views Emphasize need for intestinal "musclebuilding"

Initial catharsis
Inpatient
	1. High normal saline enemas (750 cc. b.i.d.), three to seven days
	2. Bisacodyl (Dulcolax) suppositories b.i.d., three to seven days
	3. Use of bathroom for 15 minutes after each meal
		Patient admitted when:
		  Retention is very severe
		  Home compliance likely to be poor
		  Parents prefer admission
		  Parental administration of enemas is inadvisable psychologically

At home	1. For moderate to severe retention, 3–4 cycles as follows:
	  Day 1: hypophosphate enemas (Fleet's Adult) twice
	  Day 2: bisacodyl (Dulcolax) suppositories twice
	  Day 3: bisacodyl (Dulcolax) tablet once
	2. For mild retention, senna or danthron, one tablet daily for one to two weeks.
	Follow-up abdominal x-ray examination to confirm adequate catharsis
		1. Dosages or frequency may need alteration if child experiences excessive discomfort
		2. Admission should be considered if there is inadequate yield
		3. No lubricant during this phase

Maintenance regimen	1. Child sits on toilet twice a day at same times each day for 10 minutes each time 2. Light mineral oil (at least two tablespoons) twice a day usually for at least four to six months 3. Multiple vitamins, two a day, between mineral oil doses 4. High roughage diet: bran, cereal, vegetables, fruits 5. Use of an oral laxative (senna or danthron) for two to three weeks, then alternate days for one month (given between mineral oil doses); then laxative is discontinued; lubricant is continued.	1. A kitchen timer may be helpful 2. A chart with stars for sitting may be good for children under seven 3. Bathroom reading encouraged 4. Mineral oil may be put in juice or Coke or any other medium 5. Vitamins to compensate for alleged problems with absorption secondary to mineral oil 6. Diet should be applied but not to the point of coercion

Follow-up pattern	1. Visits every four to ten weeks, depending on severity, need for support, compliance, and associated symptoms
	2. Telephone availability to adjust doses when needed
	3. In case of relapse:
	  Check compliance
	  Use of oral laxative (e.g., Senokot) for one to two weeks
	  Adjust dosage of mineral oil
	4. Counseling or referral for associated psychosocial and developmental issues
	5. Continuing use of demystification diagram to document progress
		1. Duration of treatment program may be as long as two to three years or as short as six months
		2. Signs of relapse:
		  Excessive oil leaks
		  Large caliber stools
		  Abdominal pain
		  Decreased frequency of defecation
		  Soiling
		3. Physician should spend time alone with child
		4. In patients who are slow to respond, physician should sustain optimism; persistence cures almost all cases (eventually)

*All dosages and frequencies are for an average sized seven year old child. Appropriate adjustments should be made for smaller and larger patients.

MANAGEMENT

Various techniques have been advanced for the pediatric management of children with encopresis (Davidson et al., 1963; Levine and Bakow, 1976; Silber, 1969). Therapeutic goals include the establishment of a regular bowel habit, lessened stool retention, and restoration of neuromuscular function, together with the alleviation of emotional scars. The medical aspects usually consist of bowel retraining, with pharmacologic measures to promote complete evacuation. Table 31–10 summarizes one pediatric treatment approach.

It is likely that the medical therapy of encopresis can be facilitated through initial "demystification" and ongoing counseling. First, the youngster and parent should be reassured that many other children have this problem. A child may need to hear that there are others in his town or school who are afflicted. Such reassurance counteracts the profound isolation that these parents and children frequently feel. It is helpful for a child to be informed that "other cool kids also have a messing problem." The child should be encouraged not to feel embarrassed or "on trial" while talking to the doctor about the problem. The latter should explain that trouble with bowels is not so different from having a sore throat or a runny nose. A positive and nonaccusatory stance should be communicated. The physician might express admiration for the courage and even heroism of the child who always has to be so vigilant not to have the problem discovered by peers.

It is helpful to use drawings or diagrams to portray for the child and parents the pathogenetic mechanisms of stretched-out bowels that have lost their "feelings" and muscle tone. Figure 31–5 is an example of such illustrative material. The child is shown first the cross section of a normal colon. Its function as a pipe, carrying away and eliminating waste, is explained. The use of muscle to propel material out of the body and the role of nerves in informing the child that he needs to use the bathroom are stressed. It is then explained how some children do not empty themselves completely or often enough. At this point one might want to insert a plausible explanation for the particular child's retention. The child is then shown the cross section of an intestine in which there is a growing accumulation of "rocks" of waste. It is pointed out that when bowels get too full, there is no place for all the waste to go; consequently it stays around, it may get very hard, and it stretches the big pipe. This leads to a thinning and weakening of the muscles, so that there is not enough strength to push the rocks out, and more and more of them accumulate.

When new soft waste gets made, it tends to ooze through the spaces between the rocks and comes out in the underpants. The child should realize that a stretched colon has stretched nerves that stop working, so that he no longer gets enough warning or feeling of the need to visit the bathroom. In other words, the waste just comes out, and the first time one is aware of it is when it arrives in the underwear, i.e., much too late. It is also helpful to point out that often children get accumstomed to messing and may not be able to smell their products.

This demystification process sets the stage for a nonaccusatory, advice giving relationship with the physician. The pediatrician assumes the role of "coach." The predominant theme becomes the building up of muscles to control the bowels. The symbolism of muscle has relevance for a school aged child. It stands for growing up, acquiring autonomy, feeling effective, and gaining social control. The demystification process should be witnessed by parents, although it may create some anxiety or guilt. A parent who has been punishing a child for not coming in after an accident may feel upset at the realization that he does not feel it coming or even smell it once it is there. The physician should reassure parents that such misunderstandings are common.

It is helpful to explain that the treatment of the

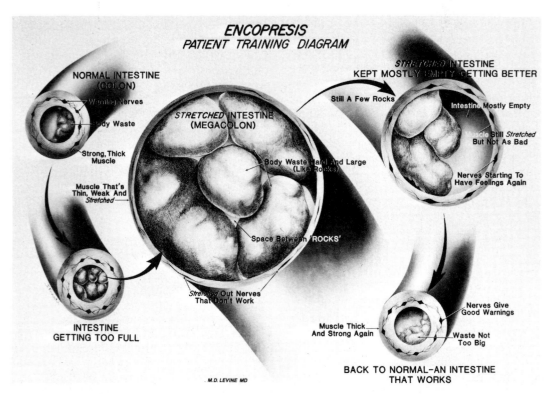

Figure 31–5. This diagram is used in the "demystification" of children with encopresis. A poster sized enlargement of this diagram serves to represent normal intestinal musculature, its distention with "rocks" of body waste, the development of a megacolon with stretched out, thin muscles and nerves, the beginning resolution of the problem, and ultimately the restoration of normal function. This teaching aid can help the clinician during initial counseling of the child and parents. An enlarged reproduction of this diagram can be obtained by writing to the author.

bowel disorder will depend upon cleaning out the rocks from the colon and then preventing new ones from forming over a long enough time for the bowel to regain its normal width, feelings, and strength. The teaching diagram shown in Figure 31–5 can be used to illustrate this. It should be emphasized that children vary markedly in the period of time needed for this restoration. Some respond within weeks; others need months or even years of treatment. Successful management usually starts with a vigorous catharsis. This is particularly true in children who have a functional megacolon with diffuse stool distribution as visualized by x-ray. Many treatment failures stem from an incomplete initial cleanout. An outpatient catharsis may include four three day cycles, beginning the first day with two adult size Fleet enemas (given together), followed on the second day by a bisacodyl (Dulcolax) suppository and on the third day by a bisacodyl tablet (Table 31–10). In many cases four or five such cycles (i.e., 12 to 15 days) may be needed before a child is fully cleansed. A follow-up film of the abdomen after two to three weeks is helpful in establishing either success or the need for an even more prolonged initial catharsis.

In a few cases hospitalization is required for the initial cleanout. This is true if the stool retention and megacolon are extremely severe, such that more vigorous expurgation is essential. Other criteria for admission include: a seriously disturbed parent-child relationship (in which case, giving enemas at home may be deleterious), parents who feel that they cannot manage this program themselves, or cases in which a period of separation would be helpful in establishing clearly for the child that incontinence is his own problem. In any case the initial catharsis is explained to the parents and the child as "a necessary evil." That is to say, one does not like to attack from below, but that approach is crucial in the short run and, it is hoped, all future therapy will enter through the ears and the mouth!

Once the initial catharsis is successful, a far less invasive, long term training routine should be established. Included are regular visits to the toilet (regardless of whether the child feels the urge). Ordinarily he is required to sit twice a day, at the same times each day, for at least 10 minutes each time. A kitchen timer may be used to document this. It is essential that the child understand that such a routine constitutes training and muscle building rather than punishment. In younger children there can be strengthening of incentive through the use of a star chart documenting toilet visitations and perhaps offering extra credit or rewards for genuine yields. The overall thrust is directed toward establishing regularity and bowel autonomy.

Defecation can be facilitated with laxatives or stool softeners. Most commonly light mineral oil is used at a starting dose of about two tablespoons twice a day (generally adequate for school aged children). The dosage may be adjusted depending upon the age and response. If mineral oil is used before or during the initial catharsis, it is likely to leak and produce a demoralizing mess. In general, mineral oil should not be used to "clean out" a child; its purpose is to help maintain good bowel function after the initial catharsis. Some children may prefer a commercial preparation of flavored mineral oil; in other cases families can use their own judgment and add enticing ingredients (such as juice or carbonated beverages). Many children tolerate mineral oil better if it is kept refrigerated. In most cases an oral laxative is also used. Senna (Senokot) or danthron (Modane) can be started at one tablet or its equivalent each day and then tapered to alternate day medication after several weeks. Usually, after one or two months, the lubricant can be given without laxative therapy. The latter may have to be reintroduced for one week during exacerbations. Increased dietary fiber may be helpful; a heaping bowl of bran cereal each morning might suffice. However, parent-child combat over food consumption must be discouraged in these families who already may be too obsessed with intakes and outputs. Thus, if the child detests bran cereal, it should not be forced upon him.

Treatment should continue for a prolonged period; training and medical management usually lasts a minimum of six months in all but the mildest cases. Partially treated encopresis commonly results in humiliating relapses. There should be follow-up visits and telephone contacts to deal with the titration of medication, with behavioral issues, and with reactions to relapses.

When pediatric efforts fail because of serious family psychopathology or actual parental sabotage, a psychiatric referral is indicated. However, the latter never should be a replacement for pediatric management. Optimally the two disciplines can work in concert to alleviate this condition.

REFRACTORY ENCOPRESIS

Some children have accidents despite optimal consistent management. At The Children's Hospital Medical Center in Boston, nearly 20 per cent of referred patients are relatively treatment resistant (Levine and Bakow, 1976). In encopresis as in other functional complaints of childhood, the remission rate is high but so is the relapse rate. The pediatrician treating encopresis must recognize that the disorder can be chronic and recurrent and that it is taxing and frequently frustrating to restore bowel function in some children. Pediatric tenacity is critical. The clinician must resist the impulse to become accusatory when musculature fails to respond promptly. The physician, parents, and child may become disenchanted with one another. Faced with the seeming futility of treatment, children often become noncompliant with recommendations. They may be accused of not taking their medications or of failing to use the toilet at the designated hours. The child may be discouraged by the apparent ineffectiveness of the intervention. He may feel understandably that it is safer not to try at all than to comply and fail. The physician should be sensitive to this all too common phenomenon. A strong alliance with the child should be forged. All must understand that sometimes encopresis requires years of treatment; no one needs to feel blamed for its chronicity.

Parents need support to accept and care for a child with encopresis. They need help in shaping reactions to exacerbations. The child should not be condemned for messing. On the other hand, he can be held accountable for noncompliance. Appropriate punishment can be administered for not taking medicine or for refusing to make regular use of the bathroom. The child should be encouraged to wash his own body, but should not be expected to launder the soiled undergarments. Mothers, fathers, sisters, brothers, and grandparents should be educated so that their comments and reactions will not be too caustic and will not add needlessly to the child's burden of humiliation.

One may want to enlist the cooperation of the school. Some children benefit from access to a private bathroom in the health room or principal's office. The teacher should be sensitive to the child's problem, allowing bathroom use whenever requested. When accidents take place during school, a change of clothing should be available (usually in the health office). The child's confidentiality and privacy are, of course, crucial.

Some children with resistant encopresis may have specific food intolerances. In particular, fatty foods and milk products may aggravate underlying bowel dysfunction. An appropriate dietary regimen may be beneficial.

In chronic cases the child must receive help in "covering up" his accidents in public. A humanitarian excuse from physical education or from taking showers in gym class may be essential. The child may need to be very careful about where he sits on the bus if the ride home commonly produces an accident. In such cases it may be important to seek another means of transportation. By demonstrating sensitivity to these intolerable human situations, the physician can strengthen his alliance with and ease the suffering of a child with encopresis. When the prolonged battle with treatment resistance ends successfully, a very grateful patient emerges triumphant and even strengthened.

THE PROTOTYPE OF ENCOPRESIS

Encopresis forms an interesting model for behavioral and developmental pediatrics. Its pathophysiology mixes diverse forces in varying degrees at predictable life stages as well as endogenous or congenital predispositions, the effects of early life experience, the impacts of nurturance, possible iatrogenic effects, developmental and maturational phenomena, training and disciplinary influences, and the impacts of school. The pediatric management of encopresis is intrinsically eclectic. It combines the role of the physician as a diagnostic formulator, a demystifier and educator, a pharmacotherapist, a counselor, an advocate and child coconspirator, and a triage officer for further referral and treatment when needed. Like other disorders of function in children, encopresis can test and strain the doctor-patient relationship as insidiously as it undermines parent-child interactions. The challenge is to sustain the helping relationship to persevere, to uncover strengths in a struggling child and family, and finally to fulfill the implied contract not to give up or abandon the cause until the damage for which the family came in has been repaired.

MELVIN D. LEVINE

REFERENCES

Anthony, E. J.: An experimental approach to the psychopathology of childhood: encopresis. Br. J. Med. Psychol., 30:146, 1957.

Barr, R. G., Levine, M. D., and Wilkinson, R. H.: Occult stool retention: a clinical tool for its evaluation in school-aged children. Clin. Pediatr., 18:674, 1979.

Bellman, M.: Studies on encopresis. Acta Paediatr. Scand., 170(Suppl):1, 1966.

Coekin, M., and Gairdner, D.: Fecal incontinence in children; the physical factor. Br. Med. J. 2:1175, 1960.

Davidson, M., Kugler, M. D., and Bauer, C. H.: Diagnosis and management in children with severe and protracted constipation and obstipation. J. Pediatr., 62:261, 1963.

Easson, W. M.: Encopresis—psychogenic soiling. Can. Med. Assoc. J., 82:624, 1960.

Huschka, M.: The child's response to coercive toilet training. Psychosom. Med. 2:301, 1942.

Levine, M. D.: Children with encopresis: a descriptive analysis. Pediatrics, 56:412, 1975.

Levine, M. D.: The schoolchild with encopresis. Pediatr. Rev., 2:285, 1981.

Levine, M. D., and Bakow, H.: Children with encopresis: a study of treatment outcome. Pediatrics, 58:845, 1976.

Levine, M. D., and Barr, R. G.: Encopresis Evaluation System. Boston, The Children's Hospital Medical Center, 1980.

Levine, M. D., Mazonson, P., and Bakow, H.: Behavioral symptom substitution in children cured of encopresis. Am. J. Dis. Child., 134:663, 1980.

Olatawura, M. O.: Encopresis, a review of 32 cases. Acta Paediatr. Scand., 62:358, 1973.

Pinkerton, P.: Psychogenic megacolon in children: the implication of bowel negativism. Arch. Dis. Child., 33:371, 1958.

Roy, C. C., Silverman, A., and Cozzetto, F. J.: Pediatric Clinical Gastroenterology. St. Louis, The C. V. Mosby Co., 1979, p. 799.

Shopfner, C. E.: Urinary tract pathology associated with constipation. Radiology, 90:865, 1968.

Silber, D. L.: Encopresis; discussion of etiology and management. Clin. Pediatr., 8:225, 1969.

32

Sleep-Wake State Development and Disorders of Sleep in Infants, Children, and Adolescents

Our modern day understanding of clinical sleep disorders dates from the serendipitous and now classic physiologic description of the rapid eye movement (REM) sleep state in the 1950's (Aserinsky and Kleitman, 1953; Dement and Kleitman, 1957). Previously sleep had been considered to be a single state of lowered central nervous system arousal. Now it is recognized as consisting of two distinct states of central nervous system activity—the REM state and the non-REM state. Both states recur cyclicly during sleep, each under the control of different neurophysiologic and biochemical regulators. Control mechanisms for the REM state reside in dopamine neurons in the pons; non-REM state control resides in serotonergic neurons in the forebrain. Physiologically the non-REM state reflects highly regulated, inhibitory activity, related to anabolic metabolism; the REM state reflects physiologic activation, important for information processing and memory storage (Denenberg and Thoman, 1981; Sterman, 1961).

Electrophysiologically the REM state is characterized by a low amplitude, fast frequency electroencephalographic pattern; binocularly synchronous rapid eye movements; inhibition of peripheral tonic muscle activity, except for phasic breakthrough of clonic twitches and body movements; and irregular accelerated cardiac and respiratory activity. In neonates the REM state is called active sleep (Fig. 32–1).

Electrophysiologically the non-REM state is characterized by tonic resting muscle activity and slowed regular cardiac and respiratory rates. Body movements are infrequent and generally limited to gross changes of position. The electroencephalogram is synchonous with wave forms of slow frequency and high voltage. In neonates the non-REM state is called quiet sleep (Fig. 32–2). During the first year of life the non-REM electroencephalographic patterns mature so that four non-REM "stages" can be scored, comparable to the four non-REM sleep stages of adults.

Research on the physiologic and biochemical mechanisms of sleep state organization and regulation and descriptions of the behavioral concomitants of sleep in humans and animals have led the way more recently to clinical investigations of sleep disorders per se. In these studies nocturnal sleep state organization is only one measure of sleep disturbance; daytime sleepiness and the impairment of waking function are equally important to assess.

In this chapter some of the developmental shifts in sleep-wake state organization from infancy through adolescence are traced, and a developmental framework for understanding sleep disorders and disturbances in infants, children, and adolescents is provided. The developmental framework is speculative but may offer directions for further research.

THE ONTOGENESIS OF SLEEP-WAKE PATTERNS

Physiologic patterns of REM activation and non-REM inhibition can be recorded in fetuses. Parmelee (1974) has established an ontogenetic map, anchored to conceptional age, for each of the physiologic variables of sleep as they develop in preterm infants. By term the REM and non-REM states are easily differentiated in polygraphic records, yet changes in their organization continue through adolescence. Four notable shifts in REM state organization after birth are:

1. A reduction in the amount of time in the REM state from eight to nine hours per 24 hours in neonates, to one to two hours in adolescents and adults (Roffwarg et al., 1966).
2. A change in the sequence of falling asleep from a waking to REM transition (REM sleep onset) in the neonate, to a waking to non-REM transition (non-REM sleep onset), beginning during the latter part of the first year of life (Emde and Walker, 1976).
3. A change in the periodicity of the REM state. As a biorhythm, REM periods recur every 50 to 60 minutes at birth;

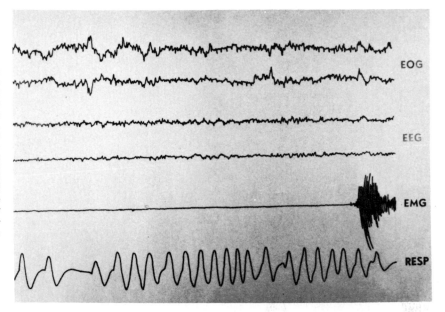

Figure 32–1. The top two tracings, left and right eyes, represent the electrooculogram (EOG) recorded from the lateral canthus referred to a neutral reference electrode. The out of phase deflections reflect REMs, or rapid eye movements. The two channels of EEG are central and occipital leads referred to a neutral reference electrode. The electromyogram (EMG) is recorded from the submental area. The respiratory tracing is obtained from a bead thermistor placed in front of the nose. The figure depicts rapid eye movements in the EOG channel, a low voltage fast EEG pattern, suppressed muscle activity, and rapid irregular respirations, characteristic of active or REM sleep.

by adolescence the REM cycle has achieved the adult periodicity of 90 to 100 minutes (Anders et al., 1980).

4. A change in temporal organization of REM and non-REM states within sleep. At birth a REM period is as long in the early part of sleep as it is at the end. By six weeks of age, a diurnal influence, leading to short REM periods during the early night and longer REM periods during the morning hours, begins to appear (Anders et al., 1982).

The non-REM state matures after birth as well:

1. The electroencephalographic pattern develops so that four non-REM sleep stages can be defined. The stage 1 non-REM electroencephalographic pattern is characterized by a low amplitude, fast tracing, not unlike waking and REM patterns; stage 2, by the presence of wave forms known as sleep spindles and K complexes superimposed upon a low voltage background; stages 3 and 4, by increased proportions of slow, high voltage, synchronous delta waves. Stage 2 non-REM sleep

begins to appear by six weeks of age; the spindles and K complexes are fully mature by two years of age. The delta waves of stage 3-4 non-REM sleep are present by two months of age.

2. As puberty progresses through Tanner stage 5, the proportionate amount of stage 3-4 non-REM sleep decreases (Carskadon et al., 1980).

3. Although stage 3-4 non-REM sleep is predominantly confined to the early night, during the preadolescent years it is not uncommon to see a second or third period of stage 3-4 non-REM sleep later in the night.

See Figure 32–3.

The diurnal organization of sleep and wakefulness changes during the first year of life. This is accomplished by the gradual consolidation of short sleep episodes into longer sustained sleep periods, and the shifting of sleep to the night-time hours.

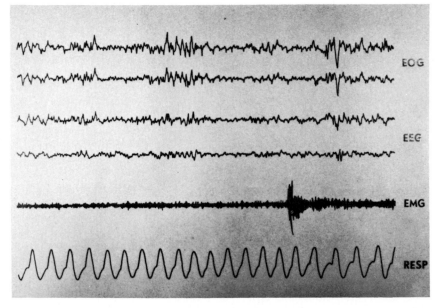

Figure 32–2. The polygraphic pattern is derived from the same electrode placements as previously. The NREM state is characterized by no eye movement activity in the EOG channel; high voltage, slow synchronous waves in the EEG tracing; the presence of muscle tone in the EMG channel; and slow regular respiratory patterns on the respiration channel.

COMPARISON BETWEEN INFANT AND ADULT
SLEEP PATTERNS

		INFANT	ADULT
1.	SLEEP STATE PROPORTIONS (%)		
	REM / NREM	50/50	20/80
2.	PERIODICITY OF SLEEP STATES	50-60 MIN REM/NREM CYCLE	90-100 MIN REM/NREM CYCLE
3.	SLEEP ONSET STATE	REM SLEEP ONSET	NREM SLEEP ONSET
4.	TEMPORAL ORGANIZATION OF SLEEP STATES	REM-NREM CYCLES EQUALLY THROUGHOUT SLEEP PERIOD	NREM-STAGES III-IV PREDOMINANT IN 1ST THIRD OF NIGHT • • • • REM STATE PREDOMINANT IN LAST THIRD OF NIGHT
5.	MATURATION OF EEG PATTERNS	LVF PATTERN HVS PATTERN 1 NREM EEG STAGE	K-COMPLEXES DELTA WAVES 4 NREM EEG STAGES
6.	CONCORDANCE OF SLEEP MEASURES (ORGANIZATION OF SLEEP STATES)	POOR	GOOD

Figure 32–3. Maturation of REM and NREM sleep parameters from infancy to adulthood.

In the neonate sleep and wakefulness are distributed evenly across the 24 hour day. Periods of sleep three to four hours in length are punctuated by one to two hour periods of wakefulness. By the end of the first month there is significantly more sleep at night. In the early months the infant's ability to sustain longer intervals of uninterrupted sleep dramatically increases from slightly less than four hours at two weeks of age to more than seven hours by five months (Anders et al., 1982). Moreover, the sleep-wake patterns of an individual infant seems to be a stable characteristic of the child (Jacklin et al., 1980).

Generally, by the child's first birthday, sleep consists of one long period at night and one to two daytime naps, an average of 14 hours daily. Interestingly, there is little change in the total amount of sleep during the first year: newborns generally sleep 14 to 16 hours per day. Beyond infancy the amount of total time spent asleep continues to decline gradually. During the second year it decreases to 12 to 13 hours daily, with one daytime nap. Between two and five years the daytime nap is given up and total sleep declines to about 11 hours. The decline continues until the adult level of seven to eight hours is reached during late adolescence.

DISORDERS AND DISTURBANCES OF SLEEP-WAKE STATES

As standardized techniques of sleep recording (polysomnography) have brought substance and rigor to an emerging new field, sleep disorder medicine, sleep disorder centers have been established to function as diagnostic laboratories in many hospitals and outpatient clinics. A recent attempt to standardize nosology has divided clinical sleep disorders into four major categories (Roffwarg, 1979):

1. Disorders of excessive somnolence, or hypersomnias.
2. Disorders of initiating and maintaining sleep, or insomnias.
3. Dysfunctions associated with sleep, sleep stages, or partial arousals, or parasomnias.
4. Disorders of sleep-wake schedule, or phase lag syndromes.

In a recent survey of medical practitioners, insomnia was reported in 5.1 per cent of the pediatricians' practices and hypersomnia in 1.3 per cent; the frequency of parasomnias ranged from 1.2 to 7.8 per cent. Nightmares occurred in 7.4 per cent of the patients. All the disorders were noted three to four times more commonly in the practices of child psychiatrists (Bixler et al., 1979). In general, however, there have been fewer studies of sleep disorders in infants, children, and adolescents than in adults (Anders et al., 1980; Dixon et al., 1981; Ferber and Rivinus, 1979).

Sleep disorder evaluations require assessment of daytime sleepiness, waking behavioral function, and nighttime sleep state organization; that is, studies of 24 hour diurnal organization and biorhythmic function. The capacity to synchronize internal rhythms with each other and with environmental cycles appears to be related to health and disease. Chronic imbalance of hemispheric function has been posited to lead to general biorhythmic disruption affecting multiple body systems, including the autonomic, sensorimotor and neuroendocrine systems (Broughton, 1975; Halberg, 1980). Sleep research thus joins chronobiology in the 24 hour investigation of sleep-wake inter-relationships. Maturational changes in sleep-wake states provide an opportunity to speculate further about sleep disorders of infants, children, and adolescents from a developmental perspective. Sleep disorders often emerge at times of physical or psychic stress and during times of rapid maturation, particularly at ages in which sleep organization is changing.

SLEEP IN BRAIN DAMAGED INFANTS

In preterm and full term infants polygraphic recordings of sleep state organization have been used to diagnose brain damage. No consistent pattern of sleep state disorganization has been noted. In some infants an absence or disturbance in cyclic organization has been noted (Monod et al., 1972). In less seriously impaired infants a failure of concordance between physiologic measures is the most common pathologic finding. The "disorganization" of physiologic patterns leads to increased proportions of "indeterminate" sleep (neither REM nor non-REM). Finally, in other groups of damaged infants excessively long single REM or non-REM periods have been reported. The results of these studies are confusing (Dreyfus-Brisac et al., 1970). Both pathologic deviations and maturational delays may present as disorganized sleep and are sometimes difficult to untangle. Definitive assessment of the functioning of the neonate's central nervous system awaits the development of complex mathematical models to analyze the simultaneous organization and integration of multiple physiologic parameters recorded over a long continuous period of sleep and waking (Prechtl, 1968).

NIGHTWAKING IN INFANCY

Clinical disturbances of sleep without apparent polygraphic abnormality are much more common in the first years of life. A developmental milestone of infancy is the achievement of diurnal regularity of sleep-wake states and periodic stability of REM–non-REM cycles. The changes in diurnal patterning bear on a common concern of parents, sleeping through the night or "settling." Moore and Ucko (1957) defined settling as sleeping without interruption from midnight to 5 A.M. They found that approximately 70 per cent of the infants settle by three months of age, 83 per cent by six months, and 90 per cent by one year. Parental reports reveal that 40 to 50 per cent of the infants experience difficulties in "nightwaking" during the second year of life after they have settled in the first.

Carey (1974, 1975) found that infants with nightwaking frequently have temperaments characterized by low sensory thresholds, and that infants breast fed beyond six months of age also awakened significantly more often than bottle fed babies. Bernal (1973) found sleep problems at 14 months of age to be associated with perinatal variables such as prolonged labor and increased latency to the first cry.

Problems of nightwaking need further clarification, however. The inability of an infant to sustain a continuous long period of sleep must be differentiated from a disturbance in the diurnal regulation of sleep. For example, an infant who sleeps from 8:00 P.M. to 3:00 A.M. at six months of age and one who falls asleep at 8:00 P.M. and wakens at 11:00 P.M. may both be labeled as nightwakers. The first infant, however, sustains a continuous sleep period of seven hours, normal for six months of age; the problem, more likely, is one of diurnal regulation. For the second infant with short sleep periods, the problem is more likely in the maintenance of continuous sleep.

Nightwaking in and of itself does not necessarily constitute a clinical problem if it does not distress the parents. In our research we reported that 44 per cent of two month old infants and 78 per cent of nine month old infants were described by their parents as sleeping through the night; however, only 15 per cent of the infants at two months and 33 per cent of the infants at nine months were actually asleep the whole night as recorded on video tape. In the majority there were short awakenings during the night that did not disrupt the parents' sleep (Anders, 1979). The latter may be bothered only when the child wakes and cries.

In the second year, problems with going to bed and falling asleep often reflect problems of separation. To delay the anxiety of separating, presleep rituals may appear. Repetitive requests for goodnights, glasses of water, or falling asleep beside the parent are common. Teddy bears or special blankets reduce separation difficulties and facilitate the transition to sleep (Paret, 1982). At this age also, children are easily excited. Fear of the dark and nightmares may result in requests for night lights. It is unlikely that polysomnographic recordings will be useful in determining electrophysiologic abnormalities in these clinical disorders. Time-lapse video recordings in the home, on the other hand, can demonstrate the disturbance (Anders and Sostek, 1976).

A short case vignette of nightwaking demonstrates the complexity of the problem:

B. was a three year old child who was referred to the Sleep Disorders Clinic by his pediatrician for an evaluation of possible "seizures" confined to sleep. A routine, waking electroencephalogram had been negative. According to B's mother, he woke regularly each night with violent shaking. These seizure-like episodes occurred three to four times each night. Time-lapse

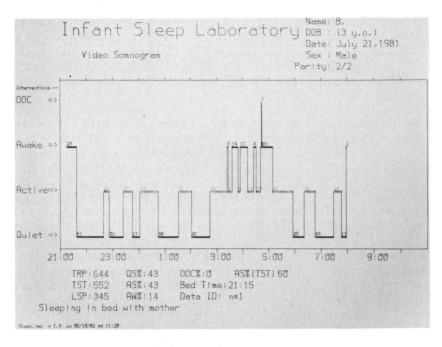

Figure 32–4. Video somnogram of infant B. The proportionate amount of active sleep (AS) during the total sleep time of 50% is significantly more than expected for a child of his age. In addition, the frequent REM periods during the early part of the night from 21:00 to 01:00 reflect an increased pressure for active sleep. TRP = total recording period, LSP = longest sleep period, QS = quiet sleep, AW = awake, OOC = out of crib.

video sleep recording on two consecutive nights demonstrated vigorous and prolonged REM periods with intense eye movements and phasic body movements throughout. Long REM periods occurred early in the night, unusual for a child of this age. Further history elicited that the symptoms worsened about six months prior to the evaluation, when B's mother heard about epilepsy in a cousin. She had been aware of B's "jerking" activity during sleep and had feared that he had epilepsy. To reduce the frequency of these attacks, she had prevented him from taking naps during the day, kept him up late in the evening, and awakened him during the "attacks" at night. The video somnogram portrayed sleep similar to that of REM-deprived adults; namely, intensification of phasic REM activities and a "pressure" for REM sleep, characterized by a short REM latency and long single REM periods. Simple clarification of the problem relieved B's mother's anxiety about epilepsy. Early bedtimes and napping were reinstituted. The symptoms disappeared rapidly and at one year follow-up, B continued to do well.

See Figure 32–4.

This vignette illustrates the need to study the sleep pattern throughout the 24 hour day and to assess wakefulness and daytime napping activity as well as night-time sleep.

A second vignette illustrates the problem of night-time separation:

R. was a two year old, younger sibling in a family of two boys with two professional parents. The two boys shared a common room until R began nightwaking at the age of six months. In order to allow his four year old brother to sleep undisturbed, R's parents moved him to another room. Over the course of the next nine months R manifested extreme difficulty in going to bed, falling asleep, and remaining asleep. Once asleep, he would wake hourly and demand the parents' presence. Night-times were grueling unless one parent would sleep the entire night next to R. When seen in the Sleep Disorders Clinic, no major nonsleep developmental or psychopathologic problems were apparent in either boy, and no extraordinary family conflicts were apparent. Video somnograms on two consecutive nights revealed frequent brief awakenings throughout the night. Treatment over several weeks, involved support for both parents in returning R to his broth-

er's room. They were encouraged to sit next to him, but not lie next to him, while he fell asleep. One year after the institution of treatment, R slept alone without undue difficulty.

Although nightwaking and sleep pattern irregularity during the first several years of life continue to plague parents and pediatricians, little more can be said at this point than that they affect 25 to 50 per cent of all infants and seem related to environmental, temperamental, and maturational factors. Diurnal regulation of REM sleep is an integral component of the sleep-wake process (Czeisler et al., 1980). From a developmental perspective, it is possible that the mechanisms responsible for the establishment of diurnal regulation and for transitions between the REM and non-REM states, which mature rapidly during this period, may be associated with these problems.

Various treatments for nightwaking have been described, ranging from the use of hypnotics, behavior therapy, chronotherapy, and supportive counseling for parents, to simply "letting the child cry." Their effectiveness has not been evaluated systematically, however. The use of hypnotics is not endorsed (Kales et al., 1980; Richman, 1981). A careful pediatric examination and a thorough developmental history, including "sleep milestones," current day-night sleep habits, bedtime rituals, the parental response and concerns, the impact of the disturbance on the family, and the presence of stress and family psychopathologic disorders, are important to obtain. At times a home visit with a time-lapse recording of the child's sleep is warranted to evaluate the maturity of the child's sleep-wake patterns, rule out sleep abnormalities, and provide insight into subtle parental behavior that may contribute to the night-

waking. In essence, nightwaking must be understood in the context of development, temperament, and family dynamics in order to prescribe an appropriate and individualized intervention.

PRESCHOOL AND SCHOOL AGE SLEEP PROBLEMS

A group of sleep disorders have been described that are particularly common in the preschool and school age years. They have been variously termed disorders of arousal (Broughton, 1968), non-REM dyssomnias (Anders and Guilleminault, 1976), and more recently parasomnias (Roffwarg, 1979). The four most common are night terrors, sleepwalking, sleeptalking, and primary nocturnal enuresis. They share certain features: The episodes occur at a particular point in the sleep cycle, namely, after a prolonged stage 3-4 non-REM period, just prior to a transition to the REM state, some 70 to 120 minutes after sleep onset; all can occur in the same child; a family history of the disorders is generally positive (Kales, A., et al, 1980a); males outnumber females 4 to 1; and there is retrograde amnesia for the episodes.

In the preschool and school age years the proportion of stage 3-4 non-REM sleep fluctuates. Both daytime stress and fatigue influence the amount of stage 3-4 non-REM sleep at night. Excessive fatigue, associated with giving up naps, and heightened stress, common in children starting school and making new friends, both may affect non-REM stage 3-4 organization. It might be predicted that if the parasomnias are associated with disturbances of stage 3-4 non-REM sleep, they should be more prevalent in this age when changes in stage 3-4 non-REM sleep state organization are prominent.

NIGHT TERRORS

Pavor nocturnus (night terrors) is a transient disorder of preschool children. It occurs frequently but irregularly and must be differentiated from the more common nightmare or anxiety dream. Nightmares consist of frightening mental experiences, which can be recalled and recounted upon awakening; they arise during REM sleep. Night terrors are not associated with recall. Instead the child suddenly sits upright in bed, screams, and appears aroused but has no recollection of the experience in the morning. The episode is brief, lasting 30 seconds to five minutes, and is associated with palpitation, sweating, and a "glassy eyed stare." The child is usually inconsolable, but at the conclusion of the episode returns rapidly to sleep. In general, night terror attacks resolve spontaneously as the child grows. Although both diazepam and imipramine are effective in controlling the episodes, their use is not recommended unless the symptoms significantly disrupt the family or interfere with the child's waking behavior (Fig. 32–5.).

Night terrors also occur in older children and adults. It is important to determine the age of onset and duration of the problem. Onset in late childhood or frequent unremitting attacks over several years suggest more serious psychopathologic disorders, and psychiatric referral is recommended. In a study of adults with persistent night terrors, Kales and her colleagues reported a mean age of onset of 12.5 years (late onset), with epi-

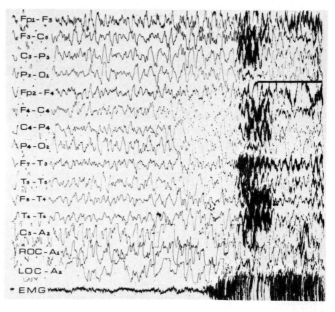

Figure 32–5. The beginning of a night terror attack in an eight year old boy. A full array of 15 EEG electrodes and an EMG channel depict stage IV sleep, characterized by high voltage, slow continuous delta activity and the onset of a night terror attack. The movement artifact reflects the child sitting upright in bed while NREM stage IV sleep continues.

sodes occurring 100 to 200 times per year. The episodes lasted about five minutes and occurred approximately 100 minutes after the onset of sleep (Kales, J., et al., 1980a).

SLEEPWALKING AND SLEEPTALKING

During school age years sleepwalking and sleeptalking commonly substitute for night terrors and, similarly, resolve spontaneously. Approximately 15 per cent of the children between the ages of five and 12 have walked at least once. Persistent sleepwalking, however, is reported to occur in only 1 to 6 per cent of children. Like night terrors, an episode begins 90 to 120 minutes after the onset of sleep. The child sits upright abruptly. Movements are clumsy, and efforts to communicate with the child usually elicit mumbled and slurred speech. Repetitive finger and hand movements are common. The child may get out of bed and walk; however, more commonly, after a period of restless movement, the child lies down and returns to sleep. Sleepwalking is not purposeful. Children frequently hurt themselves and require safeguarding.

Children in whom sleepwalking persists may have a recurrence of night terror attacks several years later. These night terrors become the prominent disorder in adulthood. Children who "outgrow" sleepwalking usually do not have significant associated psychopathologic disorders. The age of onset of somnambulism usually occurs before 10, and the episodes terminate before age 15. In contrast, adults with persistent sleepwalking or night terror attacks give a history of onset in adolescence and demonstrate significant psychopathologic disorder as adults (Kales, A., et al., 1980b).

Sleeptalking, like sleepwalking, is not purposeful. The speech is usually incomprehensible and monosyllabic. In a large sample of three to 10 year old children, parents reported that approximately 10 per cent talked in their sleep regularly (Rubens et al., 1980). In children who present with purposeful speech or directed sleepwalking without injuries, psychologic disorders are likely rather than physiologic sleep disturbances.

ENURESIS

The most prevalent parasomnia is primary nocturnal enuresis. Depending on age, sex, severity, family history, and race, 5 to 30 per cent of children between the ages of five and 15 are enuretic. Organic etiologies, including those related to genitourinary disorders, epilepsy, mental retardation, or diabetes, must be ruled out. Secondary enuresis, without organic cause, that ap-

pears after a child has had at least six months of dryness must also be excluded; a psychologic etiology is often present in secondary enuresis. The enuretic episode in primary nocturnal enuresis is characterized by autonomic arousal with tachycardia, tachypnea, erection in males, and decreased skin resistance. Immediately following micturition, children are difficult to arouse and, when awakened, indicate that they have not dreamed (Broughton, 1968). Recent studies, however, have questioned the previously reported stage 3-4 non-REM specificity of primary nocturnal enuresis (Mikkelsen and Rapoport, 1980).

The treatment of enuresis is varied and often unsuccessful. Although imipramine generally reduces the frequency of enuretic episodes, its effects are purely palliative. This medication is often demanded by parents, but its use should be restricted to short courses in severe cases with older children and adolescents, and only in conjunction with a more comprehensive treatment program. Interventions that combine a behavioral program for the child with supportive counseling and education for the parents in dealing with the child's enuresis have also been successful. Often parental misinformation, anxiety, and anger can create conflicts within the child that may precipitate, exacerbate, or prolong the disorder. Primary nighttime enuresis is a symptom, not a syndrome, and as such may include several underlying pathophysiologic etiologies. Further research is required to investigate the possibility of subclasses of enuresis and to delineate the factors related to successful treatment (Fritz and Anders, 1979).

CHILDHOOD INSOMNIA

Until recently insomnia was believed to exist only in adults. However, both adolescent and adult poor sleepers often provide clinical histories of insomnia originating in childhood. In a recent study the incidence of complaints of school age insomnia in a general pediatric population and in a child psychiatry outpatient population was compared. The characteristics of a selected sample of insomniacs and good sleepers and a comparison of child and parent reports of sleep problems in this age group were described (Dixon et al., 1981). Forty-one per cent of the parents who attended the child psychiatry clinic reported minor to severe sleep problems in their children, compared to 14 per cent of pediatric clinic parents. On the average, the duration of symptoms of insomnia was over five years. Moreover, parental reports of sleep difficulties beginning during the child's infancy were not uncommon. Although parasomnias in children are not necessarily associated with underlying psychopathologic disorders, childhood insomnia was more prevalent in the emotionally disturbed population. The majority of this sample of insomniac children had an additional diagnosis of depression, attention deficit disorder, or conduct disorder.

In another study, adults with childhood onset insomnia evidenced greater difficulty in falling asleep, less total time asleep, a lower percentage of phasic REM activity, and more neurologic "soft" signs than a group of adult onset insomniacs (Hauri and Olmstead, 1980). These results suggest that a neurophysiologic impairment may underlie persistent childhood insomnia.

ADOLESCENT SLEEP DISORDERS

Preadolescent children rarely complain of feeling sleepy. They generally awaken early each day and resist parental exhortations for naps and rests. Their weekend sleep-wake patterns do not differ from weekday patterns. In striking contrast, adolescents are renowned for their ability to sleep any time and anywhere, and to complain of tiredness. Their daytime sleepiness is related only in part to the fact that they stay up later and experience curtailed sleep at night (Anders et al., 1978; Glaubman et al., 1979). It is also related to the hormonal changes of puberty (Carskadon et al., 1980). The relationship of feeling sleepy when awake to the disorders of excessive somnolence is unclear, but the two most common, narcolepsy and the sleep apnea-hypersomnia syndrome, present with daytime sleepiness. Often symptoms are experienced first in adolescence (Navelet et al., 1976).

NARCOLEPSY

Narcolepsy is thought to be a genetic disturbance of REM sleep regulating mechanisms. Cases of familial narcolepsy have been reported, and the incidence of narcolepsy in families is approximately 60 times greater than in the general population. The peak age of onset is between 15 and 25 years of age. Thirty per cent of adult narcoleptics retrospectively report that as adolescents they had experienced excessive daytime sleepiness, which they attempted to counteract by "hyperactive" behavior (Navelet et al., 1976). A learning problem is a common presenting complaint in children with narcolepsy, and psychologic testing may suggest intellectual deficits. The full blown syndrome is characterized by the "narcoleptic tetrad":

1. REM attacks during wakefulness. The most frequent initial symptom is excessive daytime sleepiness beginning three to four hours after waking. The adolescent usually fights the sleepy feeling, but sleepiness increases so much that a frank REM nap takes place.
2. Cataplexy—characterized by episodes of inability to perform voluntary movements during wakefulness secondary to a sudden inhibition of muscle tone. The severity and extent of the attack vary from complete powerlessness to involvement of only certain groups of muscles, especially those controlling the jaw and neck.
3. Hypnagogic hallucinations—vivid visual hallucinations at sleep onset. Although occasional hypnagogic hallucinations are not uncommon in normal subjects, they rarely occur night after night. In narcoleptics they are frightening, and patients have been diagnosed as schizophrenic because of them. A clue to their existence is a reluctance to go to sleep.
4. Sleep paralysis—an inability to move voluntary muscles while falling asleep, resulting from the REM related suppression of tonic muscle activity.

Unfortunately treatment for narcolepsy remains inadequate. Psychoactive medications aid in the management of sleep attacks, and chlorimipramine provides some relief for cataplectic symptoms. The following case is illustrative:

CL was the third born in a family of five. His history revealed a normal birth after an uneventful term pregnancy. By the age of three he persisted in his tendency to sleep late in the morning and to take a short nap in the afternoon. There were no difficulties in his development. He was always considered a "long sleeper" and a quiet child during the day. By age six he began to exhibit some behavioral disturbances, which were more pronounced in midafternoon. He became "grouchy" and obviously drowsy but seemed to be fighting sleep. Similarly, at night he would refuse to go to bed, obviously frightened by the idea of sleep. He continually mentioned "fear of monsters" and over the next two years fell asleep only when someone was next to him. By age nine, CL would sleep occasionally during class at school. He was referred to the Sleep Disorders Clinic at age 12, unable to read or write and considered forgetful and poorly motivated by teachers. His referral was precipitated by an episode of falling down during an emotional outburst. The clinical interview substantiated the presence of cataplexy, sleep paralysis, and hypnagogic hallucinations. Polysomnography revealed REM sleep onsets both during daytime naps and at night. Family support, including careful explanations to the school, and Ritalin upon awakening and at noontime improved CL's symptoms.

SLEEP APNEA–HYPERSOMNIA SYNDROME

In the sleep apnea–hypersomnia syndrome, daytime sleepiness results from the multiple brief arousals at night that follow apneic episodes. The clinician can often make the diagnosis of the syndrome from the history alone. Loud inspiratory snoring followed by apneic pauses and brief arousals provide a clinical clue to the nocturnal respiratory abnormality. A cassette recorder placed by the patient's bedside easily documents the snoring.

The apneic episodes last up to one minute and are associated with hypoxemia, hypercapnea, and mild acidosis (Fig. 32–6). In a study of nine children with the sleep apnea–hypersomnia syndrome, the mean sleep apnea duration was 20.2 seconds and the mean number of obstructive apneic episodes per seven hours of night-time sleep was 319 (Guilleminault et al., 1976). Waking studies of cardiovascular and pulmonary function have

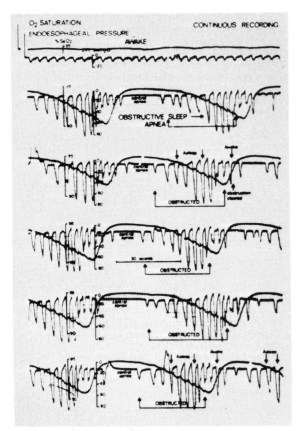

Figure 32–6. Five consecutive brief obstructive sleep apnea episodes during a continuous recording of a child with the sleep apnea–hypersomnia syndrome. Oxygen saturation is measured by means of an ear oximeter, and respiratory air flow by means of a strain gauge transducer applied to the chest. During each obstructive episode, oxygen saturation falls dramatically while thoracic pressure begins to increase in an attempt to move air through the "obstruction." A brief awakening after vigorous thoracic excursions restores oxygen saturation to its normal levels.

demonstrated no abnormalities; studies carried out during sleep, however, pinpoint the abnormality. After several years, secondary cardiovascular changes develop, which are similar to those reported in adults with the sleep apnea–hypersomnia syndrome and in children with alveolar hypoventilation. Essential hypertension, pulmonary hypertension, and cor pulmonale are the most serious cardiovascular consequences.

More than 85 per cent of the children diagnosed with the sleep apnea–hypersomnia syndrome are males. Associated symptoms include headaches, abnormal sleep behavior (i.e., excessive movement, sleepwalking, falling out of bed), enuresis, weight gain or loss, and a gradual deterioration in school performance. The syndrome appears to have an even greater impact on intellectual functioning than narcolepsy. Perhaps repetitive hypoxemia plays a role. Approximately 35 per cent of the children with the sleep apnea syndrome are identified as being borderline mentally retarded when first seen.

Treatment is not satisfactory. A careful otolaryngoscopic examination is important. Loci of airway obstruction, especially hypertrophied tonsils and adenoids, should be corrected. For adolescents without an obvious site of obstruction and with severe night-time apnea and secondary impairment of daytime functioning, a small permanent tracheostomy that functions only during the night and is closed off during the day provides instant relief and dramatic behavioral improvement. The following case vignette is illustrative:

The elder of two children, C developed normally except for intermittent loud snoring at night. When she was eight, daytime sleepiness and headaches appeared and increased over the following year. At age 10 enuresis developed and was associated with a progressive decrease in school performance. By age 12, C was physically underdeveloped and was considered borderline mentally retarded. Daytime sleepiness was dramatic. The diagnosis of "essential hypertension" resulted in a referral to the Sleep Disorders Center. Waking cardiopulmonary function studies were normal; a night-time polysomnographic recording revealed severe obstructive apnea with transient awakenings throughout the night. Otorhinolaryngologic examination revealed normal findings. Following a tracheostomy, plugged during the day and used only during sleep, C improved dramatically. She gained weight, improved in school, and again enjoyed her friends. An attempt to close off the night-time tracheostomy six months after the procedure resulted in a recurrence of the apnea episodes and C's plea to reinstate the airway. For the past five years there have been no complications in her management.

ADOLESCENT INSOMNIA

A number of surveys have suggested that 10 to 20 per cent of adolescents complain of disturbed nocturnal sleep. Stress and personality characteristics are also associated with poor sleep (Price et al., 1978; Strauch et al., 1973). In one study of 15 to 18 year olds, nearly 75 per cent of the "chronic poor sleepers" reported difficulty in falling asleep, and over 5 per cent reported feeling tired most of the time (Strauch et al., 1973). Insomnia in adolescence may be associated with depression and the symptoms are similar to those in adults. Sleep onset is normal, although occasional difficulties in falling asleep may be present. A sustained period of two to five hours of sleep with reduced amounts of stage 3-4 non-REM sleep is common. Although the proportionate amount of REM sleep is normal, a characteristic sign of depression is the shorter than normal interval from onset of sleep to the first REM period (the REM latency time). Difficulty in maintaining sleep finally leads to premature awakening. Daytime napping is present. With clinical treatment of the depression, the sleep pattern returns to normal. However, an improvement in affect usually precedes the improvement in sleep. Biorhythmic disruption is also present in endogenous depression. Temperature, neuroendocrine, and sleep-wake rhythms during the acute phase of the illness do not demonstrate normal phase relationships. After

treatment, the phase relationships revert to more normal patterns (Halberg, 1980).

Another type of insomnia that may emerge during adolescence is the delayed sleep phase syndrome. This disorder has been defined as a syndrome of sleep onset insomnia without subsequent difficulty in maintaining sleep once it has been initiated; it is associated with difficulty in morning awakening. In several studies of patients complaining of insomnia, 7 to 10 per cent fit the criteria for this disorder. Czeisler and his colleagues (1981) described resetting the circadian clocks of patients with the delayed sleep phase syndrome by a new method of treatment, called chronotherapy. In a study of five patients, treatment involved an attempt to entrain circadian rhythms to a 27 hour day by delaying sleep-dark and wake-light times by three hours each "day." All five reported a significant and lasting resolution of symptoms for at least 37 weeks after treatment.

It is possible that the availability of artificial light has contributed to the pathogenesis and maintenance of this "entrainment" disorder by enabling self-selection of light-dark cycles. Impaired processing of light-dark cues may explain the reduced capability of patients with the delayed sleep phase syndrome to reset their biologic clocks. Adolescents may be particularly susceptible to this syndrome, because the changing social demands, which result in later bedtimes, interact with the changing neuroendocrine secretion patterns of puberty, which affect sleep state relationships.

PROBLEMS AND AREAS FOR FUTURE RESEARCH

Although the number of studies investigating sleep problems in infants, children, and adolescents has increased during the past five years, the methodology of such studies is complex and the acceptance of their results still requires some skepticism. For example, sleep centers have paid little attention to the distorting effects of the laboratory setting and the polysomnograph per se in developmental studies. Laboratories commonly record sleep for more than one night consecutively in adults to assess the "first night" effect of instrumentation and sleeping in an unfamiliar setting. Such controls are rarely instituted in young children. Home recordings without instrumentation, using time-lapse video techniques, circumvent the laboratory effect, but even these studies require repeated consecutive recordings in order to substantiate the stability of sleep patterns. Also the loss of electroencephalogram and other physiologic indices of sleep in video studies limits the usefulness of video tape, especially in the investigation of the parasomnias.

The effect of environmental perturbations on sleep state organization and the influence of constitutional or temperamental differences associated with sleep state maturation require further study. Epidemiologic studies of normal and disordered sleep patterns are another area of fertile research. Longitudinal studies are needed to determine the natural course of sleep disorders in children in order to assess their long term morbidity. Finally, further research is required in the area of biorhythmic organization in order to elucidate the relationship between the temporal organization of sleep and waking and the diurnal rhythms of endocrine secretion, temperature, and motility.

Acknowledgment

Support is gratefully acknowledged from the W. T. Grant Foundation and the NIMH Psychiatric Education Branch (MH-14449).

THOMAS F. ANDERS
MARCIA A. KEENER

REFERENCES

Anders, T.: Night waking in infants during the first year of life. Pediatrics, 63: 860–864, 1979.

Anders, T., Carskadon, M., and Dement, W.: Sleep and sleepiness in children and adolescents. Pediatr. Clin. N. Am. 27: 29–43, 1980.

Anders, T., Carskadon, M., Dement, W., and Harvey, K.: Sleep habits of children and the identification of pathologically sleepy children. Child Psychiatry Hum. Dev., 9:56–63, 1978.

Anders, T., and Guilleminault, C.: The pathophysiology of sleep disorders in pediatrics. I. Sleep in infancy. Adv. Pediatr. 22:151–174, 1976.

Anders, T., Keener, M., Bowe, T., and Shoaff, B.: A longitudinal study of nighttime sleep-wake patterns in infants from birth to one year. In Call, J., and Galenson, E. (Editors); Frontiers of Infant Psychiatry. New York, Basic Books, 1982.

Anders, T., and Sostek, A.: The use of time lapse video recording of sleep-wake behavior in human infants. Psychophysiology, 13:155–158, 1976.

Aserinsky, E., and Kleitman, N.: Regularly occuring periods of eye motility and concomitant phenomena during sleep. Science, 118:243–274, 1953.

Bernal, J.: Night waking in infants during the first 14 months. Dev. Med. Child Neurol., 15:760–769, 1973.

Bixler, E., Kales, A., and Soldatos, C.: Sleep disorders encountered in medical practice: A national survey of physicians. Behav. Med. 1:1–6, 1979.

Broughton, R.: Sleep disorders: disorders of arousal? Science, 159: 1070–1077, 1968.

Broughton, R.: Biorhythmic variations in consciousness and psychological functions. Can. Psychol. Rev. 16:217–239, 1975.

Carey, W.: Night waking and temperament in infancy. J. Pediatr., 84:756–758, 1974.

Carey, W.: Breast feeding and night waking. J. Pediatr., 87: 327–329, 1975.

Carskadon, M., Harvey, K., Duke, P., Anders, T., Litt, I., and Dement, W.: Pubertal changes in daytime sleepiness. Sleep, 2:453–460, 1980.

Czeisler, C., Richardson, G., Coleman, R., Zimmerman, J., Moore-Ede, M., Dement, W., and Weitzman, E.: Chronotherapy: resetting the circadian clocks of patients with delayed sleep phase insomnia. Sleep, 4:1–21, 1981.

Czeisler, C., Weitzman, E., Moore-Ede, M., Zimmerman, J., and Knauer, R.: Human sleep: its duration and organization depend on circadian phase. Science, 210:1264–1267, 1980.

Dement W., and Kleitman, N.: Cyclic variations in EEG during sleep and their relationships to eye movements body motility and dreaming. Electroencephalogr. Clin. Neurophysiol., 9:673–190, 1957.

Denenberg, V., and Thoman, E.: Evidence for a functional role for active (REM) sleep in infancy. Sleep, 4:185–191, 1981.

Dixon, K., Monroe, L., and Jakim, S.: Insomniac children. Sleep, 4:313–318, 1981.

Dreyfus-Brisac, C., Monod, N., Parmelee, A., Prechtl, H., and Schulte, F.: For what reasons should the pediatrician follow the rapidly expanding literature on sleep? Neuropediatrie, 3:3–349, 1970.

Emde, R., and Walker, S.: Longitudinal study of infant sleep: results of 14 infants studied at monthly intervals. Psychophysiology, 13:456–461 1976.

Ferber, R., and Rivinus, T.: Practical approaches to sleep disorders of childhood. Med. Times, 107:71–80 1979.

Fritz, G., and Anders, T.: Enuresis: the application of an etiologically based classification system. Child Psychiatry Hum. Dev., 10:103–113, 1979.

Glaubman, H., Orbach, I., Gross, Y., Aviram, O., Frieder, I., Frieman, M., and Pelled, O.: REM need in adolescents as indicated by resistance to REM deprivation. Percept. Motor Skills, 48:251–254, 1979.

Guilleminault, C., Eldridge, F., Simmons, B., and Dement, W.: Sleep apnea in eight children. Pediatrics, 58:23–30, 1976.

Halberg, F.: Implications of biologic rhythms for clinical practice. In Krieger, D., and Hughes, J. (Editors): Neuroendocrinology. The Inter-relationship Between the Body's Two Major Integrative Systems. Sunderland, Massachusetts, Sinauer Associates, Inc., 1980, pp. 109–119.

Hauri, P., and Olmstead, E.: Childhood-onset insomnia. Sleep, 3:59–65, 1980.

Jacklin, C., Snow, M., Gahart, J., and Maccoby, E.: Sleep pattern development from 6 through 33 months. Pediatr. Psychol., 5:295–303, 1980.

Kales, A., Soldatos, C., Bixler, E., Ladda, R., Charney, D., Weber, G., and Schweitzer, P.: Hereditary factors in sleep walking and night terrors. Br. J. Psychiatry, 137: 111–118, 1980a

Kales, A., Soldatos, C., Caldwell, A., Kales, J., Humphrey, F., Charney, K., and Schweitzer, P.: Somnambulism. Arch. Gen. Psychiatry, 37:1406–1410, 1980b.

Kales, J., Kales, A., Soldatos, C., Caldwell, A., Charney, D., and Martin, E.: Night terrors. Arch. Gen. Psychiatry, 37:1413–1417, 1980a.

Kales, J., Soldatos, C., and Kales, A.: Childhood sleep disorders. In Gellis, S., and Kagan, B. (Editors): Current Pediatric Therapy. Philadelphia, W.B. Saunders Company, 1980b.

Mikkelsen, E., and Rapoport, J.: Enuresis: Psychopathology, sleep stage and drug response. Urol. Clin. N. Am., 7:361–377, 1980.

Monod, N., Pajot, N., and Guidasci, S.: The neonatal EEG: statistical studies and prognostic value in full term and pre-term babies. Electroencephalogr. Clin. Neurophysiol., 32:529–544, 1972.

Moore, T., and Ucko, C.: Night waking in early infancy. Arch. Dis. Child., 33:333–342, 1957.

Navelet, Y., Anders, T., and Guilleminault, C.: Narcolepsy in children. In Guilleminault, C., Dement, W., and Passouant, P. (Editors): Narcolepsy. New York, Spectrum Publications, 1976, pp. 171–177.

Paret, I.: Night waking and its relation to mother-infant interaction in nine-month old infants. In Call, J., and Galenson, E. (Editors): Frontiers of Infant Psychiatry. New York, Basic Books, Inc., 1982.

Parmelee, A.: The ontogeny of sleep patterns and associated periodicities in infants. In Falkner, F., Kretchmer, N., and Rossi, E. (Editors): Pre- and Post-natal Development of the Brain. Basle, S. Karger, 974, pp. 298–311.

Prechtl, H.: Polygraphic studies of the full term newborn. In Bax, M., and MacKeith, R. (Editors): Clinics in Developmental Medicine. London, William Heinemann, Ltd., 1968, pp. 22–40.

Price, V., Coates, T., and Thorsen, C.: Prevalence and correlates of poor sleep among adolescents. Am. J. Dis. Child., 132:583–586, 1978.

Richman, N.: Sleep problems in young children. Arch. Dis. Child., 56:491–493, 1981.

Roffwarg, H.: Diagnostic classification of sleep and arousal disorders. Sleep, 2:1–137, 1979.

Roffwarg, H., Muzio, J., and Dement, W.: Ontogenetic development of the human sleep-dream cycle. Science, 152:604–619, 1966.

Rubens, N., Reimao, M., and Lefevre, A.: Prevalence of sleep-talking in childhood. Brain Dev. 2:353–357, 1980.

Sterman, M.: Ontogeny of sleep: Implications for function. In Falkner, F., Kretchmer, N., and Rossi, E. (Editors): Pre- and Postnatal Development of the Brain. Basle, S. Karger, 1961.

Strauch, I., Dubral, I., and Struchholtz, C.: Sleep behavior in adolescents in relation to personality variables. In Jovanovic, U. (Editor): The Nature of Sleep. Stuttgart, Gustav Fischer Verlag, 1973, pp. 121–122.

33

Repetitive Behavior Patterns of Childhood

Repetitive behavior patterns, or stereotypies, are common during childhood and adolescence; they are found in association with mental retardation and abnormal or delayed development as well as among normal children. Simply defined, stereotypy is purposeless repetition with rhythmicity. This definition may be expanded to include repetitive movements, repetitive vocalizations, and even repetitive thoughts. This chapter reviews the basic hypotheses that have been proposed to explain stereotypies and provides a classification of the common as well as the more unusual types of repetitive behavior, with particular attention to epidemiology and available treatment.

Table 33–1 presents the types of specific repetitive behavior as they appear chronologically during childhood. Sections I to IV in the table are discussed in detail in this chapter. The topics under sections V to VIII are discussed in greater detail elsewhere in this textbook and thus are mentioned only briefly in the summary. Table 33–2 presents some descriptive information about types of repetitive behavior whose onset is usually during the first year of life.

ETIOLOGY

Several explanations have been proposed to explain the origin and persistence of stereotypic behavior; none, however, can be supported at the exclusion of others (Baumeister and Forehand, 1973; Kravitz and Boehm, 1971). Psychosocial issues, often given primary emphasis in the past, may serve to modify or exaggerate existing behavior rather than be its direct cause.

Neurologic explanations are founded on the observation that repetitive behavior can be elicited by specific organic intervention. Ablation of certain cortical zones in monkeys and intracerebral injection of dopamine and amphetamine into rats produce patterns of stereotypic behavior similar to those observed clinically in man after cortical damage (e.g., postencephalitic stereotypies) or overdoses of drugs (e.g., amphetamines; Kelly et al.,

1975; Kubie, 1941; Setler et al., 1978). These experiments suggest that there are specific anatomic regions in which repetitive behavior is generated and further imply that catecholamines or indoleamines may serve as the underlying neurotransmitters.

Ritvo et al. (1968), studying the rhythm of stereotypic behavior in early infantile autism, believed that "the consistency to the frequencies of these behaviors among children suggests that involuntary central nervous systems are operative." Wolff (1975) proposed that "a potential of energy, which is disposed in synchronous pulses during sleep or drowsiness, constantly builds up in the central nervous system, and periodically discharges along one of several motor channels. . . . the state of the organism determines not only the readiness for discharge but also the motor pathways by which the discharge takes place and that immediately after a spontaneous discharge the infant is relatively refractory to the 'subliminal' stimulus." Afferent input, whether visual, auditory, or kinesthetic, inhibits the discharge of such spontaneous behavior, and interruption of such inhibitory mechanisms can precipitate the release of a "discharge" behavior. An imbalance exists between this intrinsic neural energy and its inhibition at birth and during development, causing the normal overflow of stereotypic behavior in newborns, infants, and children. Persistence of these behavior patterns beyond childhood might therefore imply abnormalities in either the generation of this neural energy or the control of its expression.

Learning theorists believe that stereotypies begin as normal behavior associated with and facilitating a child's motor and ego development, which is then displaced or reinforced. Lourie (1949) and Thelen (1979) describe the appearance of rhythmic behavior in relation to the cephalocaudal progression of motor control. At two to three months a few children rock their own heads, a part of their bodies over which they then have control, whereas at six to 10 months, some children have progressed to body rocking, reflecting a caudal progression of

Table 33–1. REPETITIVE BEHAVIOR PATTERNS OF CHILDHOOD

I. The unborn child
   A. Hand sucking, thumb sucking
   B. Penile erections
II. Neonate, 0–1 month of age
   A. Rhythmic
   B. Spontaneous erections
   C. "Startle" reactions
   D. Reflex smiling
   E. "Sobbing" inspirations
   F. Myoclonic twitches
III. Childhood, onset 0–1 year
   A. Sucking behavior
      1. Thumb and finger sucking
      2. Lip sucking and biting
      3. Toe sucking
   B. Foot kicking
   C. Rocking and rolling behavior
      1. Body rocking
      2. Head rolling
      3. Head banging
   D. Bruxism
IV. Childhood, onset after 1 year
   A. Self-mutilating behaviors
      1. Nail biting
      2. Nose picking
      3. Trichotillomania
      4. Eczema
   B. Habit spasms, tics
      1. Transient tic disorder
      2. Chronic tic disorder
      3. Atypical disorder
      4. Gilles de la Tourette's syndrome
V. Stereotyped behavior associated with physical illness
   A. Drug induced: amphetamine, apomorphine, methylphenidate, ethosuccimide, pemoline, sympathomimetics (see Chapter 58)
   B. Degenerative diseases of the central nervous system, including self-destructive biting of Lesch-Nyhan syndrome, subacute sclerosing panencephalitis, postencephalitic syndrome, hepatolenticular degeneration, organic brain syndrome, and cortical atrophy (see Chapter 39)
   C. Movement and seizure disorders, including chorea, athetosis, and hemiballismus (see Chapter 39)
VI. Stereotyped behavior in children with sensory impairment (see Chapter 39)
   A. Blindism
   B. Deaf-mute subjects
VII. Stereotyped behavior in mentally retarded and multiple handicapped children (see Chapter 39)
VIII. Stereotyped behavior in children with psychiatric illness (see Chapters 39 and 41)
   A. Obsessive-compulsive disorder
   B. Infantile autism
   C. Schizophrenia

motor skills. The repetition of movement may serve as practice for a child as he learns to gain control of his body. Recent reports of infants sucking their thumbs in utero suggest that such rhythmic activities may actually begin prenatally.

These rhythmic movements may be transient, being replaced as the child develops and the attainment of new skills takes precedence. However, during the course of a child's use of such motor patterns, secondary values, such as the pleasure associated with sucking, may become prominent. The repetitive activities may then persist, serving to relieve the child's discomforts, frustrations, or unsatisfied needs (Baumeister and Forehand, 1973).

Another theory is that there exists a need or drive for movement or an optimal level of stimulation for an organism. Levy (1944) studied various instances of movement restraint in children, e.g., a playpen, a leash, or immobilization because of illness, and found that restraint eventually produced stereotypic activities. Leuba (1955) stated that "the organism tends to acquire those reactions which, when overall stimulation is low, are accompanied by increasing stimulation, and when overall stimulation is high, those which are accompanied by decreasing stimulation." Stereotyped behavior could then serve to alleviate the monotony that characterizes certain environments (e.g., children in isolation or with perceptual handicaps) or may serve to block out excessive sensory input (e.g., in autistic children; Hutt and Hutt, 1965).

A corollary of these theories is the displacement or substitution of one type of behavior for another. Wolff (1975) states that energy is displaceable and may be discharged along one or another pathway—the choice of the discharge channel being determined in part by internal or external stimulation of the nervous system. For example, a pacifier placed in the mouth might cause an infant to suck, whereas a nonspecific jar to a crib might result in a startle response. One rhythmic behavior might substitute for another. Thus, as a child responds to pressure to modify his repetitive activities, thumb sucking may yield to bruxism, nail biting, or masturbation. On the other hand, only a fragment of the original activity may persist, as when a "partialization," such as hair twirling, once associated with breast feeding during infancy, appears independently in the adolescent. Motor behavior patterns may be replaced with repetitive thoughts or ritualized compulsive behavior patterns.

CLASSIFICATION

Table 33–1 presents a classification of sterotypic behavior patterns seen during infancy and childhood. The patterns are organized chronologically on the basis of when they are likely to occur first. Many of the stereotypies associated with perceptual handicaps, systemic illnesses, drug ingestion, and psychiatric disorders are discussed more completely elsewhere in this book. Table 33–2 presents information about some characteristics of each of the habits discussed.

Table 33–2. SOME CHARACTERISTICS OF REPETITIVE BEHAVIOR

Age Group	Type of Behavior	Incidence in Term Babies	Course Onset	Loss	M:F	EEG	Development	Reference
In Utero Neonatal (0–1 month)	Thumb sucking	?	Birth	8 days				Wolff
	"Startle"		Birth	6 mo.				Wolff
	Rhythmic mouthing, sucking							
	Sobbing inspiration		Birth	2–4 wk.				Wolff
	Reflex smiling		Birth	?				Wolff
	Spontaneous erections		Birth	2–4 wk.				Wolff
	Myoclonic twitches		Birth	6 wk.				Wolff
	Hand sucking	90% at 2 hr.	Birth (median time, 54 min.)	—				Kravitz and Boehm
Infancy (1 mo.–1 yr.)	Sucking behavior							
	Thumb and finger sucking		<6 mo.	4 yr.				Gesell and Ilg, Curzon
	Lip sucking and biting	93%	Peak 18–21 mo.					Kravitz and Boehm
	Toe sucking	83.4%	Median 5.3 mo.	—				Kravitz and Boehm
	Foot kicking	99%	Median 6.7 mo.	—				Kravitz and Boehm
	Rocking and rolling behavior		Median 2.7 mo.	—				Kravitz and Boehm
	Body rocking	91%	Median 6.1 mo.		1.4:1			Sallustro and Atwell, De Lissovoy
		19–21%	Mean 6.4 mo.		1:1		Advanced motor	Bakwin and Bakwin
	Head rolling	10%	Median >12 mo. 2–3 mo.	2½–3 yr.	1:1			Kravitz and Boehm Lourie
	Head banging	5.1%	Mean 9.4 mo.	2 yr.	3:1			Sallustro and Atwell
		7.0%	Median 12 mo.		3.5:1	Normal		Kravitz and Boehm
		15.2%	—		3.4:1	—	—	De Lissovoy
	Bruxism	56%	Median 10.5 mo.	?			Normal	Kravitz and Boehm

THE UNBORN CHILD

With the advent of highly sophisticated ultrasonic techniques, information is rapidly becoming available about the activities of fetuses in utero. Hand sucking has been clearly observed, and clinically, sucking blisters are commonplace on the newborn's lips, hands, or arms. In addition, skilled observers have been able to identify spontaneous penile erections in utero. Since most prenatal ultrasonography is performed only briefly during diagnostic examinations, little information is as yet available as to the variability or frequency of in utero behavior. Since the various types of repetitive behavior noted appear so early in the newborn period, it would be fascinating to know when in development these motor patterns first appear and whether natural rhythms become entrained in utero.

THE NEONATE

Rhythmic habit patterns manifested during the first month of life have been studied extensively (Kravitz and Boehm, 1971; Thelen, 1979; Wolff, 1975; see also Chapter 5). Wolff (1975) observed the behavior of 12 healthy newborn infants over the course of the first four days of life and recorded spontaneous activities of the infants during sleep and awake states. He regularly observed startle responses, sobbing inspirations, gentle lip movements, erections, transient smiles, and twitches.

The spontaneous startle reaction consists of a brief massive jerk involving most of the infant's muscles. The shoulders are abducted, the arms are extended at the wrists and elbows, and the fingers are abducted and dorsiflexed with palmar flexion of the distal phalanges. The arms then slowly return to a partially abducted, partially flexed position. Afterward there is a brief alteration in the infant's breathing pattern. The incidence of startle responses was found to be inversely related to the frequency of other diffuse movements of the infant occurring mostly during regular sleep. Under favorable conditions of regular sleep and a relative absence of other activities, sequential spontaneous startle reactions converge at a periodicity of 30 to 90 seconds.

Other motor patterns are described by Wolff as "startle substitutes." Rhythmic mouthing or sucking occurs in bursts consisting of eight to 12 distinct lip movements at a rate of two per second, separated by intervals of total inactivity lasting four to 10 seconds; under favorable conditions, bursts of sucking come every five to 20 seconds. The occurrence of spontaneous erections during sleep approaches a grossly regular pattern in some infants and is often associated temporally with rhythmic sucking. Sobbing inspirations, "reflex"

smiling, and "myoclonic" twitches of the facial muscles and the small muscles of the hands and feet are observed more commonly during particular states; e.g., spontaneous "reflex" smiling is observed only during periodic sleep, irregular sleep, or drowsiness but never during regular sleep or when the infant is awake.

From these observations Wolff concluded that in otherwise immobile infants there are a variety of well circumscribed and apparently uninstigated behavior patterns; that in the course of a single episode of sleep or on successive days these spontaneous motor patterns are repeated in a predictable form; that whenever any type of spontaneous discharge occurs in relatively pure form, its pattern of incidence approximates a rhythm, and each type has its specific rhythm; and that different types of spontaneous behavior can substitute for each other.

Hand sucking is observed in all healthy term infants within the first hours of life. In 140 healthy infants observed by Kravitz and Boehm (1971), the median sucking time was 54 minutes; 60 per cent hand sucked within one hour, 85.8 per cent within 100 minutes, and 89.3 per cent within two hours. Whether spontaneous rhythmic mouthing and hand sucking merely precede or are developmental relatives of the thumb and finger sucking seen in older infants and children is not known.

Most of these spontaneous newborn behavior patterns do not persist. The frequency of spontaneous startle responses during regular sleep decreases from the first to the eighth day of life, sobbing inspirations and spontaneous erections decrease over the first two to four weeks of life, and myoclonic twitches persist only up to about one month. Spontaneous rhythmic mouthing, however, persists up to about six months without significant change and may reappear in older children and adults who have suffered central nervous system insults.

ONE MONTH TO ONE YEAR

Sucking Behavior

Thumb and Finger Sucking.   Various types of sucking behavior can appear during the course of a normal infancy. Thumb or finger sucking may be a natural extension of hand sucking and represents a complex coordinated act by which the infant begins to control his environment. Brazelton (1973) believes that an infant's ability to console himself through such maneuvers as sucking reflects developmental maturity of the infant's cortical functions. The choice of the thumb or finger is probably accidental, occurring in the course of a child's random movements. The thumb is inserted and is found to be pleasurable, and thus the activity persists (Bakwin and Bakwin, 1972).

The thumb is usually inserted into the mouth, lies above the tongue, and presses forward against the upper front teeth or gums and backward against the lower front teeth or gums.

Thumb and finger sucking usually appears during the early months of life and peaks between 18 and 21 months of age (Gesell and Ilg, 1937). The incidence of thumb sucking varies with race and culture (Curzon, 1974). Curzon (1974) found that 45.6 per cent of American children less than four years of age suck their thumbs, 30.7 per cent of Swedish children, 17 per cent of Indian children, and 0 per cent of Eskimo children. Curzon speculated that the Eskimo children have neither the opportunity nor the necessity for thumb sucking, since as infants, and up to three years of age, they are carried all day on their mothers' backs with a bottle of milk constantly at hand. In these studies no uniform significant sex differences or known genetic predisposition has been observed. Twenty-eight per cent of children who thumb suck continue to do so after their third birthday, but most spontaneously drop the habit by age four. A small number of children continue to suck their thumbs into their adult lives. Thumb sucking most commonly occurs when the child is restless, quiet, or about to fall asleep and may typically increase in frequency during illness. Closely related to thumb sucking are tongue sucking, pacifier sucking, and blanket sucking. Associated behavior includes pulling on one's ears or twisting one's hair while sucking.

Parents justifiably raise concerns not only about the dental implications of thumb and finger sucking but also about the effect on speech and facial structure. Dentofacial development is rapid between the ages of four and 14 years, and it is during this period that persistent thumb sucking has its most deleterious effect. In an otherwise normal infant or child, thumb sucking should be tolerated as part of the normal behavioral repertoire, and parents should be reassured that most thumb sucking resolves spontaneously by age four. Punitive measures, bitter tasting nail polish, and comments that tend to embarrass or shame a child have little effectiveness and may serve only to worsen the behavior. In a child who is not thriving, thumb sucking may be a sign of unsatisfied needs, withdrawal, and regression, and the child's total environment must be explored.

After the age of four years, persistent thumb sucking requires a dental evaluation. Many parents postpone dental consultation because they believe that no harm will occur while a child still has his baby or deciduous teeth. This is a dangerous misconception, since deciduous teeth form the pathway for the eruption of the permanent teeth. If the deciduous teeth are not properly aligned, the permanent teeth may be maloccluded. Malocclusion may be self-corrected if thumb sucking is stopped, but the longer the habit lasts, the less the self-correction. In most cases correction will not occur after 12 to 14 years of age without treatment. Malocclusion affects more than just mastication, and children may develop abnormal swallowing with tongue thrusting and distortion of facial features. Speech impediments may also result, primarily difficulty in pronouncing the consonants T and D, often with an associated lisp (Curzon, 1974).

Treatment consists primarily of oral appliances designed to concomitantly correct the malocclusion and break the sucking habit. Psychotherapy has been found to be relatively ineffective (Haryett et al., 1977).

The literature is spotted with anecdotal reports of other complications of thumb sucking. Sucking calluses on the thumb and fingers are extremely common and may become infected (Bakwin and Bakwin, 1972). Phelan et al. (1979) reported an unusual case of a severe hemorrhagic episode with shock in a five year old deaf child whose thumb sucking was so excessive that it led to a sublingual mucosal ulceration.

Lip Sucking and Biting.   Lip biting consists of the placement or thrusting of the lip between the upper and lower incisors or gums with biting. The child may simply suck his lip, usually the lower lip, or continually run his tongue around his lips. Lip sucking and biting may be so severe as to cause maceration and bleeding. A variation of this habit is the grasping of the inner aspect of the cheek, the buccal mucosa, between the teeth with biting and sucking. Kravitz and Boehm (1971) report an incidence of lip sucking and biting of approximately 94 per cent; this large percentage probably reflects their criteria of the presence of the behavior but not necessarily its persistence. Lip sucking and lip biting are initially associated with the eruption of teeth (Kravitz and Boehm, 1971). The mean age of onset is 5.3 months, corresponding to the eruption of the lower medial incisors; the behavior emerges again between six and seven years of age when children are cutting their permanent incisors. In older children the habit is particularly evident at times of stress (e.g., school examinations) and may persist into adulthood. Treatment consists of softening of the lips with creams or petrolatum to prevent maceration and to smooth rough surfaces on which the child will tear and bite. Although usually quite different clinically, the self-mutilating behavior seen in the Lesch-Nyhan syndrome must be distinguished from severe lip biting behavior.

Toe Sucking.   As the child discovers and gains motor control over his lower extremities, the toe may become a sucking object. Eighty-three per cent of normal infants manifest this activity, with a mean age of onset of 6.7 months (Kravitz and Boehm, 1971).

Foot Kicking

Almost all infants (99 per cent) kick their feet, usually during periods of excitement, whether pleasurable or uncomfortable. The median age of onset is 2.7 months (Kravitz and Boehm, 1971). Infants flex and extend the lower extremities either symmetrically or unilaterally in a rhythm that is often associated with similar movements of the hands and arms. Whether foot tapping, leg dangling, or swinging is related to this earlier habit is unclear. These habits also usually appear in older children and adults at times of stress, frustration, or impatience.

Rocking and Rolling Behavior

Rhythmic habits that involve stimulation of a child's vestibular system include body rocking, head rolling, and head banging. Epidemiologic studies have shown that differences exist among these habits with respect to age of onset, duration, and the presence or absence of sex differences, and thus suggest the importance of describing each of these behavior patterns separately rather than as a composite. The reasons repetitive, almost compulsive stimulation of the vestibular system develops and the effects of such stimulation remain speculative. It is interesting, however, that certain of these behavior patterns—body rocking and head banging—occur more frequently in developmentally precocious children (Sallustro and Atwell, 1978).

Body Rocking. Body rocking occurs with the child either in the sitting position or upon all fours, resting on the elbows and knees. It may be gently rhythmic or quite violent, occasionally being responsible for the breaking of a child's crib.

The incidence of body rocking in normal children has been reported to be between 19 and 21 per cent by Sallustro and Atwell (1978) and de Lissovoy (1961) but was reported to be as high as 91 per cent by Kravitz and Boehm (1971). This large discrepancy probably reflects the definition of body rocking by the authors. All these studies were based on retrospective reports by questionnaire by the parents. Kravitz' criterion was the presence of a specific rhythmic habit for more than two days. De Lissovoy reported the presence of "crib rocking," and Sallustro and Atwell included only children who regularly body rocked once a day or more than once a day. The median age of onset is 6.1 months according to Kravitz et al., and the mean age of onset is 6.4 months according to Sallustro. Most children (87 per cent) engaging in body rocking do so for less than 15 minutes without interruption, although some (12.2 per cent) rock for 15 to 30 minutes per episode.

Body rocking is more likely to occur when the child is listening to music, when either falling asleep or waking up, or when left alone. There have been no significant effects found relating to birth order or socioeconomic measures of the families (Sallustro and Atwell, 1978). There also are no conclusive data relating to the time when body rocking usually disappears, although in most children it is transient. In 5 per cent of the children, however, body rocking may continue for months or years and although it usually has disappeared by two to three years of age, it may persist into adolescence (Bakwin and Bakwin, 1972).

Head Rolling and Nodding. Head rolling and its equivalent, head nodding, were observed by Kravitz and Boehm (1971) in 10 per cent of the normal infants they studied. In head rolling the child lies supine in bed and rolls his head from side to side. This may be continued until the hair is completely rubbed off the back of the child's head. Head nodding occurs with the child in the sitting position and is characterized by either vigorous nodding or a lateral shaking movement of the head (Bakwin and Bakwin, 1972; Nelson, 1979). Although it has been described as early as two to three months of age (Lourie, 1949), the median age of onset in Kravitz' study was greater than 12 months, with spontaneous resolution by about two years (Kravitz and Boehm, 1971).

Head nodding must be differentiated from spasmus nutans and salaam spasms. Spasmus nutans is characterized by irregular movements of one or both eyes, deviations of the head, as well as head nodding. The onset is between four and 12 months of age with spontaneous clearing within months, although total resolution may be interrupted by temporary exacerbations (Nelson, 1979). Infantile spasms are convulsive myoclonic seizures occurring before two years of age and characterized by a sudden dropping of the head and flexion of the arms, the classic salaam posture. The electroencephalograms in such children are grossly abnormal with a pattern of high voltage slow waves.

Head Banging. Head banging is perhaps the most dramatic and most upsetting to parents of any of these maneuvers. It is characterized by repetitive hitting of the head against a solid object, usually the crib, often at rates of 60 to 80 times per minute. The most common forms of head banging have been described by de Lissovoy (1962):

1. The hands and knees position, in which the child stands on hands and knees and rocks back and forth; on the forward motion the forehead or cranial cap is struck against the crib.

2. The sitting position, in which the child is braced or sitting against the side of the crib or the head board. The knees are drawn up or the legs may be straight out; the arms and hands serve to brace the body in motion. The motion is mainly a trunk movement, or it is limited to throwing the head repeatedly to the rear, striking the crib.

3. The prone position, in which the child is lying prone; the head is raised and then dropped on the pillow or mattress or brought down with considerable force.

4. Multiple positions, in which the child kneels, stands, or sits as he holds onto the bars or the railing of the crib while striking his forehead.

5. The supine position, in which, while supine, the child rolls either his head or his whole body from side to side with the head striking the sides of the crib.

Historically head banging is associated with the presence of some type of rhythmic activity earlier in infancy, usually head rolling or body rocking (de Lissovoy, 1962; Kravitz et al., 1960). Sixty-seven per cent of one group of head bangers were also body rockers (Kravitz et al., 1960). Head banging is often accompanied by other motor rhythms. Thumb sucking or blanket holding and sucking may occur concomitantly with the banging.

Pediatricians are approached by parents because of the disruption that head banging causes in their households or even their neighbor's household and because of their concern over whether the behavior signifies serious underlying psychiatric or neurologic illness. Most distressing to the parents of head bangers is the concern over self-injury. Callus formation and, more seriously, abrasions and contusions may occur. During banging the child does not seem to experience pain or discomfort but rather can appear relaxed and happy. De Lissovoy (1961) noted the association of head banging with otitis media, and Sallustro and Atwell (1978) and Kravitz and Boehm (1971) have reported the occurrence of head banging with teething episodes, suggesting a role for head banging in pain relief or diversion.

Head banging is reported to occur in 5 to 15 per cent of normal children (de Lissovoy, 1961; Kravitz and Boehm, 1971; Sallustro and Atwell, 1978). Sallustro and Atwell (1978) reported a mean age of onset of 9.4 months and Kravitz, a median age of onset of more than 12 months. There is a definite preponderance of males over females, approximately 3 to 1. The persistence of head banging varies from less than 15 minutes to as long as three to four hours per episode and, similar to body rocking, tends to occur primarily around bedtime or awakening. Although no firm data are available relating to the duration of head banging, most children stop by four years of age, but some persist. Electroencephalograms and neurologic development are normal in these children (Bakwin and Bakwin, 1972; de Lissovoy, 1962; Kravitz et al., 1960). Head banging occurs in the other children of the family in only 20 per cent of the cases (Kravitz et al., 1960). One case report described a family in which all four children were head bangers (de Lissovoy, 1962).

Treatment. In most cases these patterns of motor behavior are transient and thus, although annoying, eventually resolve spontaneously. Parents must be reassured that no brain damage will occur and should be supported during these periods. Overconcern on their part or scolding and punishment do little and may exaggerate the habit. Knowing that his parents are concerned, the child may achieve a secondary gain through his behavior and thus the habit will be reinforced. As with sucking behavior patterns, former body rockers often develop new habits with age but return to their old habits at times of illness or stress. Follow-up of a group of head bangers revealed an exaggerated response to music in some children. When music was playing, the children could not sit still but would sway, dance, hop, or jump (de Lissovoy, 1962). Partializations of motor behavior patterns may persist. Rhythmic tapping of fingers or feet, swinging of crossed legs, or gentle rocking while immersed in thought may be reflections of former behavior patterns.

Rarely, however, such behavior becomes too annoying or becomes a danger to the child's well-being. As already discussed, these behavior patterns are more frequent in children in whom an acute or chronic illness or disability forces isolation or movement restraints. The child's emotional and physical environment must therefore be evaluated thoroughly.

In otherwise normal children several approaches have been tried. Rhythmic movements may be made purposeful through music and dancing, rocking horses, or swings in an attempt to organize or "use up" the child's motor activity (Lourie, 1949). Attempts have been made to replace the movements with rhythmic auditory stimuli, usually the metronome. Twenty-eight subjects were examined by de Lissovoy with a metronome set at the speed of the child's rhythmic rate placed beneath the crib or bed. No reaction was observed in 13 children, and seven showed curiosity or fear. Eight children, however, reacted with a modification of their rhythm or an abrupt cessation of all rhythmic activity. The modification was not always toward reduction: One child increased the force of his head banging while maintaining the same rhythm, whereas another stopped head banging at night, only to awake in the morning with more violent banging, seemingly making up for its loss during the prior evening (de Lissovoy, 1962).

In certain developmentally abnormal children, in particular autistic children, head banging may become so persistent and so forceful that actual physical injury may occur. In such cases sedatives and tranquilizers, such as diazepam, or antidepressants, such as imipramine, have been tried with variable results. The drugs of choice are phenothiazines and haloperidol, to which children may respond at times. Helmets may be necessary in such cases to protect the head.

Bruxism

Bruxism is the habit of grinding or clenching the teeth together and is seen in 56 per cent of normal infants (Kravitz and Boehm, 1971). Although gum clenching is seen in edentulous children, the median age of onset of bruxism is 10.5 months of age, after the eruption of the deciduous incisors. The incidence in adults and older children is unknown because most people are unaware of the habit; estimates of incidence have been as high as 80 to 90 per cent (Detsch, 1978). Although bruxism can occur while the child is awake, it is seen primarily during sleep. During REM sleep, muscle tension in the jaw increases while the rest of the body relaxes; the masseter muscles contract at a rate of 20.9 per hour as compared with 5.3 per hour during non-REM sleep periods (Detsch, 1978). Dental occlusal problems, temporomandibular joint syndrome, and intestinal disorders may be associated causally with this behavior or as a result of this behavior and should be evaluated. Bruxism is also observed in children with semiconscious or unconscious states due to disease, especially intracranial disease such as meningitis or encephalitis (Nelson, 1979).

ONE YEAR AND OLDER

Repetitive behavior patterns that appear after the age of one year may be divided into two categories: self-mutilating behaviors and habit spasms or tics (see Table 33–1). Self-mutilating behavior involves a process of self-injury of which the person is usually fully aware but seemingly unable to control. Tics are spasmodic involuntary movements of individual muscle groups that are not associated with a clearly definable local organic process.

Self-mutilating Behavior

Nail Biting. Nail biting is a common habit pattern during childhood and adult life and includes biting of the nail itself, biting of the cuticle or skin margins of the nail bed, and picking at the nail and surrounding tissues. It is occasionally associated with toenail biting and picking (Bakwin and Bakwin, 1972). Nail biting can result in damage to the cuticle, roughness of the nail's free edge, and bleeding or infection around the nail's margins; although malocclusion of the teeth is common in nail biters, nail biting cannot be directly related to teeth displacement (Malone and Massler, 1952).

Few children are nail biters before four to five years of age. The age incidence increases to include up to 60 per cent of 10 year olds and then gradually decreases with increasing age (Malone and Massler, 1952). Although there are no significant sex differences in incidence between five and 10 years of age, thereafter the number of girls who persist in nail biting is significantly less than the number of boys (Malone and Massler, 1952). By college age approximately 20 per cent of men and 19 per cent of women are nail biters (Coleman and McColley, 1948). After about age 30 years approximately 10 per cent of the adult male population continue to bite their nails (Malone and Massler, 1952; Pennington, 1945).

Evidence suggests a possible genetic basis for nail biting (Bakwin and Bakwin, 1972). Family studies have revealed an increased frequency of nail biting in the parents of nail biters. Monozygotic twins are concordant for the habit with approximately twice the frequency of dizygotic twins.

Malone and Massler (1952) have provided a detailed description of nail biting in childhood. Most nail biters (68 per cent) bite all 10 fingernails equally rather than selectively and all to the same degree. Nail biting is symmetric; the fingers on both hands are bitten equally with no preference for the right or left hand.

Although nail biting generates considerable antagonism between children and their parents, the nail biting itself rarely requires any treatment. Biting that results in recurrent bleeding and paronychia probably is sufficient to justify increased attention to the habit. Unfortunately many of the treatment regimens that have evolved through the efforts of parents have been generally unsatisfactory. Such methods have included contracts with their children to chew one nail only, cotton mittens or pajamas that cover both the hands and the feet, bitter nail polish, or biting substitutes such as gum chewing.

Behavioral treatment has been studied extensively and appears to be the most effective, although the relative efficacy of different treatments is less clear. An example of one such treatment is the habit reversal method for nail biting developed by Azrin et al. (1980). Nail biters were taught to engage in a competing hand grasping reaction for three minutes whenever nail biting occurred. This reaction was coupled with positive hand and nail care activities, such as filing of frayed edges and the use of hand lotions. Nail biting episodes after habit reversal were reduced more than 90 per cent through the five month follow-up period.

Negative reinforcement techniques are less successful. In a study by Azrin et al. (1980) nail biters simulated their nail biting in front of each other while telling themselves how ridiculous they appeared. Nail biting in this "negative practice" group was decreased by about 60 per cent.

The relative effectiveness of positive versus negative treatment was examined by Davidson et al. (1980). Positive treatment included relaxation training, hand and finger exercises, hand massage, nail and cuticle care, and self-reinforcement. Negative treatment included aversive imagery and self-punishment (such as burning a dollar bill or snapping a rubber band on the wrist) as well as negative self-verbalizations as in the Azrin study. Positive treatment resulted in the most durable effect. A combined treatment with both positive and negative procedures was slightly less effective than the positive treatment alone. The negative procedures were the least effective and in fact detracted from the effectiveness of placebo procedures. Crucial to all these programs was the subject's increased awareness of his nail biting. Even placebo control groups showed significant decreases in nail biting.

This study, of course, confirms what parents have repeatedly experienced in attempts to change what they consider to be maladaptive behavior in their children. Punishment and scolding, the use of bitter nail polish, and similar maneuvers are of little long lasting value and perhaps may result in new difficulties with the children. A more successful approach might include regular positive reinforcement of the child for not biting his nails through praise and recognition of his efforts as well as provision of the tools for and instruction in good nail and hand care. The child can carry a nail clipper and file at all times and be encouraged to use them as part of routine health care and whenever the urge occurs to bite the nails.

Nose Picking. Among the least acceptable of all types of repetitive behavior is nose picking. There is little systematic information about this subject, and incidence figures are not available. Observation of everyday life reveals that it is very common, involving all ages and both sexes and crossing all socioeconomic classes. As with nail biting, lip biting, and hair pulling, nose picking involves more neuromuscular coordination than such gross motor habits as head banging. Although nose grasping and "picking" are observed in the newborn infant, specific picking probably must await the development of fine coordination skills. Unlike other types of "picking" behavior, nose picking connotes personal uncleanliness and is considered socially unacceptable. Children nose pick, and even consume what is picked, without regard for social convention, much to their parents' disgust and despite multiple recriminations. Older children and adults attempt to confine this habit to occasions when they are alone or when they feel protected from public view, e.g., behind windows or within the confines of the car while driving. Nose picking may be initiated by rhinorrhea with crusting associated with allergies or infections or by local irritation and bleeding from minor injuries to the nose. Picking perpetuates the irritation and may include nasal hair plucking; a vicious cycle is created, and the picking outlasts the inciting events. Nose picking (along with other trauma) is the most common cause of recurrent nose bleeding. With the concurrence of a chronic cause of nasal irritation, as in allergic rhinitis, successful treatment of this habit may consist solely of the use of decongestants or allergic hyposensitization. Scolding and punishment are of little use in the young child. The habit usually resolves spontaneously or recedes into privacy with increasing personal modesty and social consciousness.

Trichotillomania. Hair pulling, or trichotillomania, is an uncommon habit during childhood. Although scalp hair is usually involved, hair may be removed from the eyelashes, eyebrows, and the pubic areas. Hair is either plucked out, twisted off, rubbed off in patches, or cut off with scissors. This may be performed openly, often in the presence of parents or in the company of other children. The removed hair is either quietly hidden or thrown away, or is chewed and swallowed (trichophagy). With excessive trichophagy, a trichobezoar (a massive concretion of hair in the stomach) may form and produce gastrointestinal symptoms.

The incidence of trichotillomania is not known, but most reports suggest a frequency of less than 1 per cent in children. The symptom is found in both boys and girls, although with perhaps an increased frequency among girls (DeBakey and Ochsner, 1939; Mannino and Delgado, 1969). DeBakey and Ochsner (1939) reported 172 cases of trichobezoars and found that most occurred during the teen years, with over 80 per cent of the cases occurring before the age of 30 years. Trichobezoars have been reported in children as young as one year of age and in adults as late as 56 years of age.

Patchy alopecia secondary to trichotillomania must be distinguished from alopecia areata, other trauma, tinea capitis, and the alopecia that results from systemic illness or drug therapy (Levene and Colman, 1977). The hair loss in trichotillomania results in irregular or patchy alopecia, often characterized by a "migrating" loss of hair, with an advancing border of hair loss at one side of a patch and regrowing hair along the other (Nelson, 1979). The hair is short in affected areas, but all the hair follicles are filled. In one series the average duration of hair loss was four years (Levene and Colman, 1977).

Alopecia areata is common in childhood and is characterized by single or multiple, round or oval smooth patches of complete baldness that are commonly found in the scalp or beard. At the margins of active lesions, short or "exclamation mark" hairs are found, which are of normal width at the tip and narrow at the base. Initial white regrowth of the hair is common, with a return to its former color over time. Alopecia areata may progress to complete loss of hair, alopecia totalis. Histopathologic study can help in distinguishing alopecia areata from the hair loss of trichotillomania (Muller and Winkelman, 1972).

Diffuse alopecia may occur because of drug exposure, e.g., cytotoxic therapy, anticoagulants, antithyroid medication, thallium salts, and hypervitaminosis A. Hypothyroidism, hyperthyroidism, hypoparathyroidism, hypopituitarism, poorly controlled diabetes mellitus, and even a normal pregnancy can be associated with diffuse alopecia. Iron deficiency has recently been implicated in some cases of trichotillomania. In these cases trichophagia is suspected to be a form of pica, the habit and the associated anemia resolving with iron therapy (McGhee and Buchanan, 1980). Intense dermal inflammation secondary to fungal (e.g., ringworm) and bacterial diseases (e.g., pyoderma) and immune complex disorders can cause cicatricial alopecia. Repeated traction on patches of hair with hair curlers or hot combs can result in a temporary or permanent loss of hair.

Differentiation of true trichotillomania from these other disorders may be easier than the ensuing treatment. Although it is considered benign in certain cases, especially in young children, most clinicians believe that there are usually underlying psychologic difficulties. Parents often are resistant to such implications and are usually unwilling to agree that the alopecia is self-inflicted. Several treatment regimens have been suggested, including cutting the hair short or shaving the scalp in an attempt to break the habit, as well as behavioral therapy and antidepressant medication (Bornstein and Rychtarik, 1978; Snyder, 1980). There are no controlled studies to support the efficacy of these therapies. Evaluation of possible psychologic disturbance by referral to a psychiatrist is indicated early in the course of severe and persistent trichotillomania.

Eczema. Pre-existing skin diseases may serve as a focus for persistent self-inflicted injuries (Waisman, 1965). Contact dermatitis, allergic eczema, bites, ringworm, photosensitivity eruptions, anogenital pruritus, and toxic erythema may be complicated and worsened by constant picking, itching, and digging. Acne and other nonpruritic dermatoses may also be the initial lesions in a self-perpetuating struggle between the patient and his own skin.

Eczema provides a model of a disease state self-perpetuated by habitual stereotyped behavior, in this case the "scratch that rashes." Allergic disease is thought to be the cause of eczema, but the exact mechanism in the evolution of the dermatitis is unknown. A family history of the disorder, elevated serum levels of IgE, decreased cell mediated immunity, and eosinophilia provide evidence of a biologic basis for the disease, and yet emotional stress and conflicts are known to be aggravating factors. Whereas a noneczematous child might attempt to modulate his anxiety through crying or other "acting out" behavior, an eczematous child responds with a paroxysm of scratching. Rather than lashing out, the child with eczema directs his aggression inwardly against himself. (See also Chapter 26.)

Eczema that appears in the early months of life follows a natural history, initially involving primarily the face and later the trunk and extremities. In most children the eczema either resolves completely or becomes relatively mild by five years of age. In some children, however, the disease is characterized by a relentless worsening of the rash in extent and severity, covering the child's body with a thick carpet of oozing crusted skin. He is uncomfortable and disfigured, and yet the self-destructive scratching, rubbing, and picking continue.

The traditional approach to children with eczema consists of identification of the allergic precipitants of the rash—foods, molds, perfumes, clothing—or environmental conditions that elicit worsening of symptoms—temperature change or excessive dryness. Emollients, antihistamines, topically applied steroids, and occasionally antibiotics are prescribed to break the scratch-rash cycle. Potent initiators of this cycle—emotional arousal, anxiety, or stress—are often overlooked and usually underestimated. Even when acknowledged, they are usually thought to be secondary to the skin condition rather than contributing to its persistence.

In most cases support is available from the parents not only throughout the chronic course of the disease but also during periods of exacerbation precipitated by stress. Some parents are able to cope with the complex ministrations dictated by medicine and still have enough emotional reserve to comfort and help allay the anxiety their child is feeling. The eczema in children in such families, although often chronic, is typically more limited. For others such support mechanisms do not exist and the eczema is usually more severe. For these parents the progressive disfigurement may reflect their feelings of inadequacy as caretakers and may likely engender guilt. Again a vicious cycle is established wherein the parents are overwhelmed by their child's disease; they are not fully available either medically or emotionally, and the disease worsens.

In providing effective treatment of severe eczema, the physician must thus look beyond the skin. A child's anxiety must be reduced through the use of medication or by restructuring his environment. His attention must be redirected to the outside world, and relations between the child and his parents must be re-established through shared planned activities. Time out from the disease must be provided for all concerned. For the child, sedatives may allow for a comfortable night of sleep, and for the parents, time away from the burden of constant care, at times through hospitalization of the child, may be essential.

Habit Spasms (Tics)

Disorders involving tics are divided into four categories according to age of onset, duration of symptoms, and the presence or absence of vocal tics as outlined in the American Psychiatric Association's *Diagnostic and Statistical Manual of Mental Disorders* (1980). Tics must be differentiated from other abnormal movements, including athetosis, chorea, and hemiballismus, which reflect disease of the extrapyramidal system, and dystonias, dyskinesias, myoclonic movements, and spasms. The characteristics of each of these movement disorders and their differential diagnoses are available in standard neurology textbooks (see also Chapter 39).

Motor tics are stereotyped and frequently repeated. Sniffing, swallowing, throat clearing, coughing, eye blinking, facial grimacing, and neck stretching are common. Any part of the body, however, may become involved, e.g., shrugging or jerking movements of the shoulders and shaking movements of the trunk. Although an exact incidence is not known, most people experience transient tics that last only a short time, and up to 10 per cent of the population may experience a tic lasting one month or more (Schowalter, 1980).

Transient Tic Disorder. Diagnostic criteria for the transient tic disorder include an onset during childhood or early adolescence, the presence of tics, the ability to suppress these tics for minutes to hours, a variation in the intensity of the symptoms over weeks or months, and a duration of at least one month but not more than one year (American Psychiatric Association, 1980). Although the tic may involve any body part, a facial tic, and in particular an eyeblink, is most common. Vocal tics such as grunts and barks, or clearly articulated words and phrases, may be part of this disorder, although it is rare and far more typical of Gilles de la Tourette's syndrome. After a period of remission the transient tic disorder may evolve into the Gilles de la Tourette syndrome. At least three times as many boys as girls suffer from this disorder, and there is a greater likelihood of occurrence among family members.

The respiratory system is the origin of several types of habits and tics—breath holding spells during infancy, the hyperventilation syndrome, hypoventilation or repetitive sighing respirations, and "psychogenic" or habit cough.

Breath Holding Spells. * Breath holding spells may be seen as a response to frustration or a manifestation of anger in the young infant (Nelson, 1979). The child cries violently, hyperventilates, and then suddenly stops breathing with the subsequent development of cyanosis and rigidity. Loss of consciousness, twitching movements, or frank convulsions may be seen. The spell may be followed by a few seconds of limpness. During the spell, attempts to talk to the child and to quiet him down in an attempt to abort the spell are futile. The spells end spontaneously and are physically harmless. Spells may begin as early as the first six months of life, but usually appear after one year and may persist up to five years. Because the spells are so dramatic, parents and other caretakers are usually upset and fearful of precipitating new attacks. The child may quickly learn to use this to his advantage, and the spells come to provide secondary gain. Seizure disorder is the primary differential diagnosis, but causes of tetany must also be considered. The best treatment of breath holding spells is tolerance; precipitating environmental and emotional factors should be identified and corrected. Punishments are futile and may precipitate new attacks.

Hyper- and Hypoventilation Syndromes. Hyper- and hypoventilation syndromes are usually associated with anxiety states (Wintrobe, 1974). In hyperventilation the patient usually complains of marked shortness of breath, inability to take a deep breath, and a feeling of chest tightness. In addition, numbness or "needles and pins" paresthesias of the fingers and toes, palpitations, and abdominal pain may be present. The latter symptoms are thought to be secondary to the respiratory alkalosis caused by the hyperventilation and may become so severe that tetany with carpopedal spasm may follow. Patients are usually unaware of the overbreathing and are surprised by the relief that they feel when treated at the outset with rebreathing maneuvers. Such patients often give histories of chronic fatigue and weakness and may also manifest periods of hypoventilation with repetitive sighing respirations. Treatment in these cases should be directed toward the evaluation of the underlying causes of anxiety.

Habit Cough. "Psychogenic" or habit cough, a well described entity in pediatrics, may be seen as part of the Gilles de la Tourette syndrome but usually exists independently. The exact incidence of this disorder is unknown, but reports suggest that most cases present during puberty. The clinical features of this disorder were reviewed by Berman (1966), and include "persistent violent spasms of barky, harsh, nonproductive cough, occurring almost always during the waking hours

---

Editor's note: A discussion of breath holding spells may not belong in a section on habit spasms but it is included in this chapter on repetitive behavior patterns as appropriately as elsewhere.

and unaccompanied by systemic signs and symptoms of chronic disease; a paucity or complete absence of abnormal findings in the chest; lack of response to most potent cough preparations; and a cough that remains unchanged after exertion, laughter, infection, dampness, and extremes of temperature."

The diagnosis is one of exclusion of organic etiologies. Afflicted children often undergo considerable laboratory and invasive surgical diagnostic work-ups and lose extensive classroom time. In most organic diseases, however, the cough does not disappear with sleep.

Since most tics resolve entirely with time, reassurance is probably the best approach. Identification of causes of stress and anxiety and efforts to ameliorate them may be useful in breaking the vicious cycle of tics–anxiety–more tics. Psychotherapy, relaxation, biofeedback, and other behavioral techniques have been used, but critical comparisons among these techniques or even definite evidence of effectiveness is lacking (Schowalter, 1980).

Chronic Tic Disorder. The chronic tic disorder or chronic motor tic differs from the transient form in lasting for more than one year, in the more unvarying intensity of tics during its course, and in an onset either during childhood or after the age of 40 years. Vocal tics do occur, but they are infrequent (American Psychiatric Association, 1980). Chronic motor tic occurs more often in men than women.

Atypical Tic Disorder. The category of atypical tics may be used to designate tic syndromes that do not fall neatly into the other categories because of age of onset, course, or nature of the movement involved. Thus, an atypical tic syndrome would include transient tics that originate later in life or persistent, chronic motor tics that start during the earlier years of life. Also atypical would be tics that involve unusual movements, in the absence of other tics, such as the need to kick at the floor or poke at a body part with a finger. In a child who has one tic for a few weeks and then another tic, with changes between phonic and motor tics, the diagnosis of atypical tic disorder might also be justified.

Gilles de la Tourette Syndrome. The Gilles de la Tourette syndrome is one of the most fascinating and most debilitating of the stereotypic movement disorders. Diagnostic criteria include an onset between two and 15 years of age, the presence of motor as well as multiple vocal tics, the ability to suppress these tics for minutes to hours, a variation in the intensity of symptoms over weeks or months, and a duration of symptoms for more than one year or throughout life (American Psychiatric Association, 1980).

The prevalence of the Gilles de la Tourette syndrome ranges from 0.1 to 0.5 case per thousand (American Psychiatric Association, 1980). Family studies have revealed a clear genetic contribution in multiple tic syndromes and the Gilles de la Tourette syndrome (Pauls et al., 1981). There is frequently a family history of a tic syndrome (chronic motor tics or the Gilles de la Tourette syndrome) in the parents, grandparents, aunts, uncles, and siblings of afflicted children. There appears to be a clear sex related threshold to the expression of tics and the Gilles de la Tourette syndrome. Girls with the syndrome tend to come from families with higher frequencies of tics or the Gilles de la Tourette syndrome than those of boys; in other words, they require a heavier genetic loading before the syndrome develops. The members of their families who are most likely to have the tic syndromes are male. Similarly, the offspring of women with the Gilles de la Tourette syndrome and tics are at particular risk for tic syndromes; sons are at greater risk than daughters. At present it is impossible to give precise risk estimates, nor is any type of prenatal diagnosis possible. However, available evidence can be shared with families with strong histories of tics or the Gilles de la Tourette syndrome in helping them guide their own family planning decisions.

A neurologic substrate for Gilles de la Tourette's syndrome is evidenced by a high incidence of electroencephalographic nonspecific abnormalities and "soft" signs on neurologic examination (Volkmar et al., 1982). Neurochemical studies as well as the response to neuroactive medications have suggested that the monoamines, dopamine, noradrenaline, and serotonin, may be involved in the expression of the Gilles de la Tourette syndrome (Cohen et al., 1978, 1979, 1980; Leckman et al., 1980).

Many cases have now been reported in children who have developed tic syndromes and the full blown Gilles de la Tourette syndrome while being treated for attentional problems with stimulant medications (methylphenidate, dextroamphetamine, and pemoline) (Lowe et al., 1982). Tics are one of the most common side effects of these medications and may appear in a significant percentage of all children so treated. In most cases these tics disappear soon after the withdrawal of medication. In some children, however, the motor and phonic tics may persist for months or even longer and justify the diagnosis of the Gilles de la Tourette syndrome. Among the aspects of the natural history of the Gilles de la Tourette syndrome are the attentional and "hyperactivity" problems that often precede the onset of motor and phonic symptoms. Many patients whose disorder appears to be triggered by stimulant medication may already be on the road to developing the syndrome. Others have a family history of tics or the syndrome, which may further predispose them to develop a tic syndrome. However, the capacity of stimulant medications to produce tics and stereotypic behavior in animals and man indicates that these can be produced even without genetic vulnerability. Clinical experience to date indicates that it is probably unwise for a pediatrician to prescribe stimulant medication for a "hyperactive," attentionally disturbed child with a strong family history (parent or sibling) of tics or the Gilles de la Tourette syndrome or to continue to prescribe a stimulant medication if tics appear. If a child develops tics in response to one stimulant, it is unwise to attempt a course with another.

In most children afflicted with the Gilles de la Tourette syndrome, the disease follows a typical course. As noted, attentional and behavioral problems often precede motor or phonic symptoms. Motor tics then develop, usually involving the eyes, face, neck, or shoulders. Vocal tics such as grunts, barks, yelps, throat clearing, or coughs and various other noises subsequently become associated with the motor tics. Ultimately echolalia (a repetition of one's own last words or phrases) and echokinesis (a repetition of another person's movements) may appear. Coprolalia, an irresistible urge to utter profanities, occurs in 40 per cent of the cases. Mental coprolalia, thoughts about profanities, as well as obsessive thoughts and compulsive acts or rituals may be present. Self-abusive behavior may be seen in severe cases.

As with other tic disorders, the Gilles de la Tourette syndrome imposes primarily a social disability on those affected rather than any physical limitation. When motor tics are infrequent, they are acknowledged by an observer but rarely interfere with social interaction. The motor tics in Gilles de la Tourette's syndrome, however, are frequent, often coming in rapid succession. Observers are unable to hide their recognition of the tics and often feel awkward and uncomfortable. The patient is embarrassed and often stressed, his anxiety worsening the tics. As in patients with chorea, individuals with the Gilles de la Tourette syndrome may try to disguise their movements by continuing them in purposeful activity. Vocal tics are far more pervasive in their effects on a person's ability to function socially. The least bizarre vocal tic, coughing, is interpreted by strangers as a symptom of infection. People move away, change seats, or stare at the patient with suspicion and fear. The more bizarre phonic tics, such as barking and yelping, and the most troublesome, coprolalia, create obvious difficulties. Even for people familiar with the affliction there are limits to their ability to continue to accept. Afflicted children are unable to go to school because they are disruptive in the classroom; weddings, movies, even parties are precluded. The habits may become a barrier to good relations within a family. Parents develop doubts and guilt about their ability to care for and to continue to love their child. A parent's ability

to remain supportive, understanding, and available slowly dissolves. The parents and their child move apart physically—the parents so that they are not so constantly confronted with the habits, and the child so that he does not disturb his parents. When together, frustrated parents yell at their children to stop their tics and often look at them with anger and hostility.

The pressure to start treatment quickly is difficult to resist. The treatment of Gilles de la Tourette's syndrome has consisted primarily of pharmacologic intervention. The decision to start treatment must be based on evaluation of the child's overall developmental progression and functioning. Medication is usually begun when a child is emotionally distraught or unable to function normally because of his symptoms.

Haloperidol is the drug of choice and relieves the symptoms initially in up to 80 per cent of the patients. The standard treatment, haloperidol, leads to abatement of symptoms in the majority of cases at low doses (generally 2 to 6 mg. per day). However, short term side effects, such as parkinsonian symptoms, intellectual blunting, loss of motivation, weight gain, sedation, dysphoria, school phobia, and depression, can be more disabling than motor and phonic symptoms. The long term consequences, such as tardive dyskinesia, pose serious threats. Many patients choose to endure the symptoms rather than suffer the side effects of haloperidol.

The long term roles of newer approaches to pharmacotherapy, such as clonidine and pimozide, have not yet been established in the treatment of Gilles de la Tourette's syndrome. Pimozide has a therapeutic efficacy similar to that of haloperidol, with less cognitive blunting. Clonidine hydrochloride, an alpha-adrenergic agonist that decreases noradrenergic and serotonergic activity, has been shown to ameliorate the symptoms in patients who could not tolerate or did not benefit from treatment with haloperidol. The side effects of clonidine include transient sedation and, in high doses, hypotension, but are rare at the doses prescribed. Unlike the situation with haloperidol, there is no interference with cognitive functioning.

When pharmacologic treatment is initiated, the pediatrician should assess the child's general functioning and not simply the degree of suppression of tics. If the reduction of tics is taken as the end point, the child may be overmedicated at the expense of full functioning. Perhaps the majority of children with Gilles de la Tourette's syndrome have associated learning and attentional problems; these can be exacerbated by haloperidol and often require specific intervention through tutoring or special education. Because of the likelihood of emotional problems resulting from the disruption occurring in Gilles de la Tourette's syndrome, when psychiatric symptoms or family problems are evident, referral should be made to a center skilled in the treatment of the syndrome. Such centers can offer a multidisciplinary approach that includes medical and psychiatric management of the tic disorder, family counseling, and school placement when necessary.

CONCLUSION

Repetitive behavior in children can be part of a normal child's behavioral repertoire and does not necessarily portend underlying disturbances. In fact, as shown in the foregoing discussion, certain behavior patterns are so common that their absence might be considered ominous. Kravitz and Boehm (1971) compared the age of onset of the common rhythmic habit patterns in normal children and in compromised children. The early onset of hand sucking in infants of normal birth weight with no perinatal problems (median, 54 minutes) is in marked contrast to that in infants with perinatal disorders (low birth weight, low Apgar scores, or perinatal illness; median, more than 1000 minutes). In children with Down's syndrome, there is a marked delay in the onset of the rhythmic habit patterns, and in those with cerebral palsy there is little rhythmic habit activity in the first year of life. Hand sucking and finger or hand to mouth maneuvers are considered significant indicators of mature normal cortical functioning in neonates by Brazelton (1973). These data suggest that the absence or delay in appearance of certain rhythmic behavior patterns may in fact be abnormal and of more concern than their presence.

Rare is the child who does not display one repetitive behavior pattern or another in the course of his development. Certain of these patterns, however, as well as other stereotypic movements, are seen with much greater frequency in children who suffer from certain organic diseases, perceptual handicaps, mental retardation, and psychiatric illness (Table 33–1, sections V to VIII; Sakuma, 1975).

The evaluation of children with tics, habits, or stereotypic behavior thus requires consideration of possible organic factors (such as drug exposure, organic encephalopathy, or central nervous system degenerative disorders) or associated disorders (such as mental retardation, pervasive developmental disorders, or sensory defects). These usually are easily excluded by the history and physical examination but sometimes require further diagnostic studies, such as electroencephalography, blood tests, or computed tomography. (See Chapter 39 for a more extensive discussion of these topics.)

Children with sensory handicaps commonly display stereotypies (Sakuma, 1975). Blindisms are stereotypic movements seen in blind children with normal intelligence. Movements of the upper extremities—raising, flapping, lowering, or swinging of one hand or both or pounding movements, unassociated with sounds—are most common. Movements of the hands in connection with eyes—such as pressing the eyes with the fingertips, putting fingers in the eye sockets, or rubbing of the eyelids—are performed rhythmically and are thought to be characteristic of the blind. In contrast, the upper extremities are often employed for communication by sign language in the deaf, and movements of the lower extremities—in particular, knee shaking usually with vocal sounds—represent the characteristic stereotypy of the deaf.

Stereotypy in children who suffer from both a sensory impairment and mental retardation differs from that seen in children with the sensory handicap alone. Repetitive partial body movements are the most common features in blind children with mental retardation. Swaying of the upper part of the body, flapping of hands, pulling up on the trousers, and similar maneuvers are frequently seen. In the deaf with mental retardation, whole body actions, in particular, stereotyped walking and wandering usually associated with vocal sounds, are more distinctive.

Constant repetitive behavior is seen uniformly in infantile autism. Stereotyped walking and bizarre, often intense patterns of motility, such as whirling, darting, sidewalking, lunging, hand flapping, as well as head banging and body rocking, are continued for long periods. Partial body movements also occur, perhaps less frequently, but when present often also have a bizarre quality, e.g., stretching and entwinement of fingers. Palilalia is common and is characterized by the spasmodic repetition of meaningless words (see Chapter 39).

In these abnormal populations, stereotypies are persistent, take on exaggerated form, and can consume large amounts of the child's time and energy. Whereas most repetitive behavior patterns in normal children appear during periods of isolation, boredom, and fatigue, these patterns in abnormal children may appear regardless of state or in situations that are normally of interest. Unlike normal children who may be distracted from such behavior, abnormal children seem to prefer it. It is difficult to stop stereotypic behavior in autistic children, and when stopped, they often manifest confusion. Such stereotypic behavior often interferes with therapeutic involvement with the child (Chess and Hassiki, 1978).

Central to the evaluation of a specific repetitive behavior, whether in the context of an organic illness or as an isolated "problem" in an otherwise normal child, is a review of the child's general development. Developmental assessment of a child with a tic or stereotypic behavior requires careful evaluation of the child's development, psychologic functioning, family functioning, and role in the family. The way in which the child's symptom affects, and is affected by, his environment (and particularly his parents) should be specifically evaluated by observation and discussion with the child and family. The process of thorough evaluation is, in its own right, often therapeutic; patterns of negative interaction can be muted as the child's symptom is seen in the context of developmental or family pressures. Before any available treatment is considered, the pediatrician must assess the child as a full person—the effectiveness with which he is navigating his developmental tasks, the ways in which the symptom is interfering with functioning, whom the symptom bothers (child, parent, teachers, everyone). Therapy is indicated only when development is impaired or clearly threatened, e.g., when a child's self-esteem is suffering, when he is excluded or teased, when a family cannot tolerate a symptom, or when physical harm is resulting from the symptom. Since there are no treatments that are both generally effective and benign, physicians should watch and wait and determine the need for therapy before embarking on a course of action. Many stereotypies, habits, and tics are transient, and pediatric support and guidance are all that may be required to reduce ongoing stress that may lead to their continuation or serious impact.

For each child and family, the pediatrician must define a strategy of intervention, beginning with parental and child guidance, which occurs during evaluation, and then including various other modalities, such as behavior modification, systematic guidance, psychotherapy, educational treatments, and pharmacotherapy. When the stereotypy is associated with other disorders—such as pervasive developmental disorders or sensory impairments—the intervention obviously is related to the underlying dysfunction. Further elucidation of the biologic basis of stereotypies may suggest more specific medical approaches to their treatment. Until then, pediatricians sometimes can be of most help by providing a model of patience and acceptance of the child as a person.

Acknowledgments

This research was supported in part by
Mental Health Clinical Research Center grant MH 30929,
National Institute of Child Health and Development grant HD-03008,
Developmental Psychological and Community Pediatrics grant MH 12908,
Children's Clinical Research Center grant RR00125, and
The Gateposts Foundation, Inc.

E. LAWRENCE HODER
DONALD J. COHEN

REFERENCES

American Psychiatric Association: Diagnostic and Statistical Manual of Mental Disorders. Ed. 3. Washington, D.C., American Psychiatric Association, 1980, pp. 73–79.

Azrin, N. H., Nunn, R. G., and Frantz, S. E.: Habit reversal vs. negative practice treatment of nailbiting. Behav. Res. Ther., 18:281–285, 1980.

Bakwin, H., and Bakwin, R. M.: Behavior Disorders of Children. Philadelphia, W. B. Saunders Company, 1972.

Baumeister, A. A., and Forehand, R.: Stereotyped acts. In Ellis, N. R. (Editor): International Review of Research in Mental Retardation. New York, Academic Press, Inc., 1973, pp. 55–96.

Berman, B. A.: Habit cough in adolescent children. Ann. Allergy, 24:43–46, 1966.

Bornstein, P. H., and Rychtarik, R. G.: Multicomponent behavioral treatment of trichotillomania: a case study. Behav. Res. Ther., 16:217–220, 1978.

Bornstein, P. H., Rychtarik, R. G., McFall, M. E., Winegardner, J., Winnett, R. L., and Paris, D. A.: Hypnobehavioral treatment of chronic nailbiting: a multiple baseline analysis. Int. J. Clin. Exp. Hypn., 28:208–217, 1980.

Brazelton, T. B.: Neonatal Behavioral Assessment Scale. Philadelphia, J. B. Lippincott Co., 1973.

Chess, S., and Hassiki, M.: Disorders of habit. In Principles and Practice of Child Psychiatry. New York, Plenum Press, 1978, Ch. 18, pp. 339–342.

Cohen, D. J., Caparulo, B. K., and Shaywitz, B.: Neurochemical and developmental models of childhood autism. In Serban, G. (Editor): Cognitive Defects in the Development of Mental Illness. New York, Brunner/Mazel, Inc., 1978, pp. 66–100.

Cohen, D. J., Detlor, J., Young, J. G., and Shaywitz, B. A.: Clonidine ameliorates Gilles de la Tourette syndrome. Arch. Gen. Psychiatry, 37:1350–1357, 1980.

Cohen, D. J., Nathanson, J. T., Young, J. G., Shaywitz, B. A.: Clonidine in Tourette's syndrome. Lancet, ii:pp 551–553, Sept. 15, 1979.

Cohen, D. J., Shaywitz, B. A., Caparulo, B., Young, J. G., and Bowers, M. B.: Chronic, multiple tics of Gilles de la Tourette's disease. Arch. Gen. Psychiatry, 35:245–250, 1978.

Coleman, J. C., and McColley, J. E.: Nailbiting among college students. J. Abnorm. Soc. Psychol. 43:517–525, 1948.

Curzon, M. E. J.: Dental implications of thumb sucking. Pediatrics, 54:196–200, 1974.

Davidson, A., Denney, D. R., and Elliott, C. H.: Suppression and substitution in the treatment of nailbiting. Behav. Res. Ther., 18:1–9, 1980.

DeBakey, M., and Ochsner, A.: Bezoars and concretions. Surgery, 5:132–160, 1939.

De Lissovoy, V.: Head banging in early childhood, a study of incidence. J. Pediatr., 58:803–805, 1961.

De Lissovoy, V.: Head banging in early childhood. Child Dev., 33:43–56, 1962.

De Lissovoy, V.: Head banging in early childhood. J. Genet. Psychol., 102:109–114, 1963.

Detsch, S. G.: Bruxism: emotional symptom or dental occlusal problem. U.S. Navy Med., 69:26–29, 1978.

Freiden, M. R., Jankowski, J. J., and Singer, W. D.: Nocturnal head banging as a sleep disorder: a case report. Am. J. Psychiatry, 136:1469–1470, 1979.

Gesell, A. L., and Ilg, F. L.: Feeding Behavior of Infants. Philadelphia, J. B. Lippincott Co., 1937.

Haryett, R. D., Hansen, P. C., Davidson, P. O., and Sandilands, M. L.: Chronic thumb sucking: the psychological effects and the relative effectiveness of various methods of treatment. Am. J. Orthod., 53:569–585, 1977.

Hutt, C., and Hutt, S.: Effects of environmental complexity on stereotyped behavior of children. Anim. Behav., 13:1–4, 1965.

Kelly, P. H., Seviour, P. W., and Iversen, S. D.: Amphetamine and apomorphine responses in the rat following 6-OHDA lesions of the nucleus accumbens septi and corpus striatum. Brain Res., 94:507–522, 1975.

Kravitz, H., and Boehm, J. J.: Rhythmic habit patterns in infancy: their sequence, age of onset and frequency. Child Dev., 42:399–413, 1971.

Kravitz, H., Rosenthal, V., Teplitz, Z., Murphy, J. B., and Lesser, R. E.: A study of head banging in infants and children. Dis. Nerv. Syst., 21:203–208, 1960.

Kubie, L. S.: The repetitive core of neurosis. Psychoanal. Quart., 10:23–43, 1941.

Leckman, J. F., Cohen, D. J., Detlor, J., Young, J. G., Harcherik, D., and Shaywitz, B. A.: Clonidine in the treatment of Gilles de la

Tourette syndrome: a review of data. Adv. Neurol., 35:391–401, 1982.

Leuba, C.: Toward some integration of learning theories: the concept of optimal stimulation. Psychol. Rep., 1:27–33, 1955.

Levene, G. M., and Colman, C. D.: Color Atlas of Dermatology. Chicago, Year Book Medical Publishers, Inc., 1977, pp. 206–213.

Levy, D. M.: On the problem of movement restraint, tics, stereotyped movements, hyperactivity. Am. J. Orthopsychiatry, 14:653–660, 1944.

Liley, A. W.: Studies in Physiology. Berlin, Springer Verlag, 1965.

Lourie, R. S., Role of rhythmic patterns in childhood. Am. J. Psychiatry, 105:653–660, 1949.

Lowe, T., Cohen, D. J., Detlor, J., Kremenitzer, M., and Shaywitz, B. A.: Stimulant medications precipitate Tourette's syndrome. J.A.M.A., 247:1729–1731, 1982.

Malone, A. J., and Massler, M.: Index of nailbiting in children. J. Abn. Soc. Psychol., 47:193–202, 1952.

Mannino, F. V., and Delgado, R. A.: Trichotillomania in children: a review. Am. J. Psychiatry, 126:87–93, 1969.

McGehee, F. T., and Buchanan, G. R.: Trichophagia and trichobezoar: etiologic role of iron deficiency. J. Pediatr., 97:946–948, 1980.

Muller, S. A., and Winkelman, R. K.: Trichotillomania, a clinicopathologic study of 24 cases. Arch. Derm., 105:535–539, 1972.

Nelson, W. E. (Editor): Textbook of Pediatrics. Ed. 11. Philadelphia, W. B. Saunders Company, 1979.

Pauls, D. L., Cohen, D. J., Heimbuch, R., Detlor, J., and Kidd, K. K.: Familial pattern and transmission of Gilles de la Tourette syndrome and multiple tics. Arch. Gen. Psychiat., 38:1091–1093, 1981.

Pennington, L. A.: The incidence of nailbiting among adults. Am. J. Psychiatry, 102:241–244, 1945.

Phelan, W. J., Bachara, G. H., and Satterly, A. R.: Severe hemorrhagic complication from thumb sucking. Clin. Pediatr., 18:769–770, 1979.

Ritvo, E. R., Ornitz, E. M., and LaFranchi, S.: Frequency of repetitive behaviors in early infantile autism and its variants. Arch. Gen. Psychiat., 19:341–347, 1968.

Sakuma, M.: A comparative study by the behavioral observation for stereotypy in the exceptional children. Folia Psychiatr. Neurol. Jpn., 29:371–391, 1975.

Sallustro, F., and Atwell, C. W.: Body rocking, head banging, and head rolling in normal children. J. Pediatr., 93:704–708, 1978.

Schowalter, J. E.: Tics. Pediatr. Rev., 2:55–57, 1980.

Setler, P. E., Malesky, M., McDevitt, J., and Turner, K.: Rotation produced by administration of dopamine and related substance directly into the supersensitive caudate nucleus. Life Sci., 23:1277–1284, 1978.

Snyder, S.: Trichotillomania treated with amitriptyline. J. Nerv. Ment. Dis., 168:505–507, 1980.

Thelen, E.: Rhythmical stereotypies in normal human infants. Anim. Behav., 27:699–715, 1979.

Volkmar, F., Leckman, J. F., Cohen, D. J., Detlor, J., Harcherik, D., Pritchard, J., and Shaywitz, B. A.: EEG abnormalities in Tourette's syndrome. (Submitted for publication.)

Waisman, M.: Picker, pluckers, and imposters. A panorama of cutaneous self-mutilation. Postgrad. Med., 38:620–630, 1965.

Wintrobe, M. W. (Editor): Harrison's Principles of Internal Medicine. Ed. 7. New York, McGraw-Hill Inc., 1974, p. 1302.

Wolff, P. H.: The causes, controls and organization of behavior in the neonate. Psychol. Issues, 5:1–106, 1975.

34

Body Image: Impacts and Distortions

The term body image encompasses a complex psychologic concept related to the mental representation of self and is not merely a function of objective appearance. The development of body image is an intricate evolutionary process. Dominant theories currently emphasize multideterminant factors in normal and distorted body image development.

The concept of body image development in children remains imprecise (Anthony, 1968; Clifford, 1972; Schilder, 1950). However, psychologic theory stresses early antecedents dating from infancy. Although most supporting data are anecdotal, there is a small but growing amount of empirical evidence that emphasizes the importance of understanding early body image development in children with many disorders, including genetic syndromes, congenital illnesses, and craniofacial disfigurements (Belfer et al., 1979, 1982; Green and Levitt, 1962; Kaufman, 1972; Kaufman and Hersher, 1971). The same emphasis on the understanding of body image development is necessary to appreciate certain developmental tasks for children with more subtle deficits in motor function, unobservable abnormalities, and metabolic disease.

Empirical evidence demonstrating the relevance of body image development in both normal and physically disabled or ill children is inherently limited by the nature of the construct. Components of body image include rational and irrational, conscious and unconscious forces that are built upon more or less enduring memory traces, which begin to be formed in the earliest stages of development and which are subsequently modified as the infant's world progressively expands. Schilder (1950) suggests a tridimensional unity in the development of body image, which includes not only innate perceptual factors but interpersonal, environmental, and temporal aspects as well.

Body image integrity exists on a continuum ranging from normal to pathologic. A child's body image is as multidimensional as his behavior, play, or visual productions, which are expressions of the child's sense of self. Disturbed body image development resulting from birth defects, acquired deformities (real or imagined), or physical illnesses not only affects the child but also has an impact on the child's relationship with his parents, siblings, peers, and other significant persons in his life.

The main objective in this chapter is to delineate the numerous factors that will undoubtedly confront physicians responsible for the continuing medical care of these children. Although the concluding section addresses specific psychologic and treatment considerations, recommendations for the management of children with body image disturbances are interspersed throughout. The following sections will present a four point model of body image development, a discussion of disturbances of body image development and methods of assessing body image distortions, and examples of physical illnesses, disabilities, and trauma that highlight the wide range of children's responses to these conditions and their influence on body image development and modification. The final section presents a variety of therapeutic interventions designed to aid children and parents.

MODEL OF BODY IMAGE DEVELOPMENT

Body image is defined as the aspect of the self-concept that begins to develop in the earliest stages of awareness of self, that has the physical body as a focus, and around which other aspects of self-concept are elaborated. Kaufman and Hersher (1971) make the important distinction between the body as a physical object and "the mental representation of the body which is a psychological construct."

Body image is influenced by four different yet interdependent factors: perception of body stimuli, cognitive function, response from others, and stimuli from the environment in the form of comparison with others. Although these four aspects of body image development are presented separately in the following discussion, it is important to keep in mind that these are interdependent factors, which are interpreted variously in the psychologic world of the child.

PERCEPTION OF BODY STIMULI

Normal children gain an appreciation of their own body and their environment through often delicate sensory input, and obtain feelings of mastery through an increasing sense of competence in their manipulative functions, whether it be of the hands, feet, tongue, or other bodily parts.

An infant begins to integrate multiple perceptions starting with the earliest stages of development. For example, in the feeding and sucking process the mouth is the first area to be stimulated. As development progresses, tactile impressions allow the infant to gain knowledge first of his own body and later knowledge of persons and objects in his expanding world. As the child's capacity to explore his environment grows, primary kinesthetic and tactile sensations form the foundation upon which the beginnings of self-awareness and individuality are built (Kolb, 1959).

Sensory perceptions can be altered by a congenital anomaly: Tactile stimuli are reduced in the cleft palate and in the syndactyl extremity, restricted ocular movement alters visual stimuli, and hearing is diminished in hemifacial microsomia. In the child afflicted with a congenital abnormality, a sense of bodily incompetence may develop when the abnormality is associated with a functional impairment; for example, the child with syndactyly who fails to gain tactile satisfaction or the patient with a cleft palate who has impaired oral perception and difficulty with nasal secretions (Tisza et al., 1958).

Failure to gain feelings of mastery and competence and the inability to appreciate subtle nuances of stimuli deprive children with birth defects or acquired deficiencies of a sense of wholeness and satisfaction that other children obtain. The lack of mastery of basic bodily kinesthetic and tactile sensations clearly has a negative influence on the development of a child's body image and self-concept.

With the growth of the child in size and shape and with his ever increasing motility, the body image is continuously modified through interaction with the environment. However, memory traces of early body image remain and can reappear in states of neurologic degeneration or in extreme psychologic regression.

COGNITIVE FUNCTIONS

Similar to the progressive development of a child's perceptual capacities, the development of cognitive functions influences his perception of his body and defines the limits of body image development (Goodenough, 1926; Katz and Zigler, 1967; Piaget, 1929; Shapiro and Stine, 1965).

Developmentally children learn about the body and its parts in a sequential fashion, which coincides with a progressive increase in cognitive functioning. The data of Shapiro and Stine (1965) relating to figure drawings in three and four year old children demonstrate a developmental sequential model in which, as age increases, figure drawings evolve progressively from formless scribbles to differentiated, recognizable head, body, and facial features. This research supports Schilder's (1950) concept of body image development as being "built up" over time, integrated and differentiated, and not simply as a maturational given.

Children with low intellectual potential may be limited in their perceptions of their body, and younger children, because of their immature cognition, perceive their bodies in different ways than do older children. Mental retardation can significantly limit the development of a child's body image because of the inability both to form abstract concepts of bodily functions and to relate self-perceptions to the bodies of others.

Defining the deformed child's cognitive level is complicated, because patients who look dull or retarded tend to be considered so, regardless of their true intellectual capacity. Wright (1960, 1964) suggests that normal cognitive development may be interfered with indirectly as a result of a "spread phenomenon." An illustration of the spread phenomenon applied to the facially disfigured is presented in a study by Post (cited in Macgregor et al., 1953). Photographs of disfigured adults were shown to 60 normal subjects who were asked to describe the individual in the picture. The most common characteristics mentioned were mental deficiency, physical disease, and immorality. Wright also noted that the phenomenon occurs with respect to self-perceptions as well. Cohen and Yasuna (1978) showed that facially disfigured youngsters' fears that they are mentally retarded stemmed in part from concerns about their facial structures. In the instance in which a youngster has a congenital deformity based on some definable genetic entity such as trisomy or a diffuse craniofacial abnormality, the child's appearance may contribute to the impression that the child has an intellectual deficit when in fact a normal intellectual potential may be present. It is not unlikely that children with congenital or acquired defects internalize a sense of impaired cognitive ability, which may in fact inhibit normal cognitive development.

Since the development of body image is a continuing process from infancy throughout life, it is important to recognize that a child's body image is not accurate, logical, or anatomic (Piaget, 1929).

Instead it follows prelogical modes of thinking, is under the influence of psychologic drive development, and reflects a lack of reality testing owing to the child's limited cognitive capacity as well as his limited life experiences. The child's internal sense of self does not reflect reality; rather it reflects what is psychologically important to him (Kaufman, 1972).

In spite of limited cognition, it remains possible at the earliest stages of cognitive development for a child to incorporate a sense of distorted bodily function, and in particular a perceived sense of deficit. Often this deficit is viewed by the child as a value judgment, that is, "good body" versus "bad body." This may eventually influence the child's self-esteem, as in good person versus bad person. These early self-representations have an enduring impact as the child psychologically internalizes a sense of badness.

A word of caution concerning the reliability of information given by some parents: Denial of the reality of a child's deformity and "undoing" the defect by referring to it as "cute" are common defense mechanisms used by parents whose child has a congenital or acquired defect. A sense of "specialness" may be attributed to the child that impedes reality development. In addition to these defense mechanisms, parents often respond to a child's defect through reaction formation or emotional withdrawal.

These defenses are particularly damaging to the child for two reasons. First, children tend to emulate their parents' methods of coping with anger, loss, and disappointment associated with the birth of a deformed child and tend to employ similar defensive styles. Second, these defenses can lead to an impairment of reality testing in the child such that he develops an unrealistic sense of self owing to the defect. Although some psychologic defenses are necessary in order for both parents and child to cope, the degree and intensity to which they are maintained should be taken into consideration. In the case of an older child the physician would do well to investigate the child's capacity to deal verbally with the result of the disorder without interfering too greatly with the child's adaptive defenses. One should not take away abruptly denial that facilitates coping. With a younger child the physician can gain information about the child's coping mechanisms through discussion with his parents. Healthy parents tend to speak about the child's disorder rationally and realistically. They also tend to accept and follow through on offers of help, whether it be physical or psychologic aid.

RESPONSE FROM OTHERS

The parents' perception of their child as having positive bodily attributes gives the child permission to appreciate himself and his body, to gain self-esteem, and to become comfortable with himself as he is. Early in the child's life, parental attitudes, feelings, and behavior in regard to a deformity influence the child's ability to relate positively to his defective body.

The capacity for a satisfying adaptation among children with bodily defects is dependent more upon parental, family, and cultural attitudes toward the body structure than upon the presence of a defect. A child who is quite normal but who does not fulfill expectations for athletic prowess may be viewed as defective. When family and cultural attitudes toward the defect are constructive and supportive, the child has a greater possibility for successful compensatory development without an interfering personality disorder arising. Parents can direct personality development along lines where other assets can be developed and strengthened. In this way the affected child can gain a measure of satisfaction, which to some degree compensates for an inadequate body image.

The negative effects on social interaction for a child with a congenital or acquired defect are most strongly felt in the child's encounters with his parents. During infancy the mother is usually the primary caretaker. Bowlby's (1951, 1969) ethologic theory of social attachment emphasizes the importance of mother-infant interaction as a developmental determinant. Bowlby suggests that the infant emits biologically programed responses that "trigger" nurturant behavior on the part of the mother. Two important elicitors of maternal behavior are smiling and vocalization. Children with craniofacial disfigurements, for example, may be at risk because of compromised elicitors of parental nurturance. A structural abnormality around the mouth could interfere with recognizable smile responses or vocalization.

A number of investigators have discussed parental reaction to the category of "birth defective" children. Solnit and Stark (1961) reported that parents actually experience a period of mourning following the birth of an "imperfect" child. Because the child is not the one who was expected or wished for, the mother is faced with the overwhelming task of adapting to the demands of the defective child while working through the loss. How this grief is handled is seen as a determinant of subsequent parental, particularly maternal, investment in the child. Again the "defect" may reside in a discrepancy of reality and expectation.

Greenberg (1979) places early parental reactions into the context of negative narcissistic symbiosis originally described by Lax (1971), in which the infant unconsciously represents the parent's impaired sense of self. The success with which the parents adapt to their child and productively mourn the "lost child" is dependent upon a confluence of factors, including the type of defect, its correctability, the personalities of the parents, and available support systems. Even among the most psychologically healthy parents, the birth of a defective child creates considerable stress on the parents as well as on the family system as a whole. The interaction between a mother and her craniofacially disfigured child may be affected more severely than in the case of a child with congenital heart disease because of the facially disfigured child's inability to participate in a mutually satisfying interactive process. Also of significance is the possibility that the child had a deformity that is similar to a parental deformity, and thus various parental issues are aroused, including a pathologic identity, increased guilt, and blame by others.

Many mental health professionals consider the impact so serious that early psychotherapeutic intervention is strongly advocated. The role of the physician is critical at the birth of a defective child. Parental education as to the nature and consequences of the defect may be helpful, but must be repeated more than once because of the parents' inability to deal with information at a time of acute stress. Referral to a social worker for parental or family support at the time of birth may assist in parental adjustment and minimize potential body image disturbance in the child when it is believed that the parent's response is pathologic or unusual vulnerability is noted. Preventive measures designed to aid parents in coping successfully with a child born with a birth disorder should be considered the preferred choice of treatment, and therefore referral to parent groups such as those for the parents of cleft palate children may be helpful.

The studies of Macgregor et al. (1953) indicate that the attitude of mothers whose children have facial deformities varies according to the sex of the child. The tendency for mothers to overvalue physical beauty in their daughters contributes to the female child's maladjustment to the deformity. Facial disfigurements of female children are seen as evoking feelings of guilt, hostility, humiliation, depression, or rejection in mothers. Facial deformities in male children, on the other hand, are less likely to provoke such maternal attitudes. Culturally physical beauty in females tends to be overemphasized, whereas males tend to be judged on the basis of other characteristics, such as physical stature, autonomous functioning, or success in a professional career. In the realm of functional disability, the lack of athletic prowess can have an impact on the male child as he matures equally as devastating as an esthetic deficit in a female.

The age at which the deformity occurs also has an impact on parental attitudes toward the child, thereby influencing the child's developing body image. Children born with congenital deformities appear to suffer more from social stigma as well as from a mother's sense of "punishment." Children who acquire a deformity as a result of illness, trauma, or surgical procedures are more likely to receive sympathetic acceptance from both parents and society. Moreover, children with acquired deformities have had at least the opportunity to begin the process of normal body image development prior to the trauma.

Although the mother may have been the primary caretaker for the infant during the early years, the father eventually becomes a significant figure in the child's world. Research by Mahler et al. (1975) indicates that the role of the father becomes crucial at 15 to 24 months of age. Whereas normal mothers tend toward overprotectiveness on behalf of the child, fathers psychologically represent a forward movement toward increased interaction with the broader environment. The father plays a vital role in the development of body image as the child increasingly internalizes and identifies with the father in a manner that is psychologically different from that of the mother. Edward et al. (1981) further support the importance of the father as representative of external reality.

The father also indirectly influences body image development, for he is a potential source of support for the mother of a defective child. If the father-husband is seen by the mother as sympathetic and positive, many of her attitudes of guilt, humiliation, or depression may be alleviated. The mother is often less anxious over the birth of a defective child if the father expresses more distress over the event. As with other significant persons, such as friends and relatives, the father can either support the mother's adaptation to her deformed child or increase the difficulties she may experience.

Observation of both the verbal and nonverbal communications between parents as well as those between parent and child is one source of data that permits the physician to measure the quality of the marital relationship and the degree of parent-child attachment. Parents who belittle or blame each other, be it subtle or overt, constitute a source of conflict that eventually may affect a child's sense of body image. Similarly, the mother who displays hostile or rejecting attitudes toward her child suggests a questionable bonding process. In both cases a child's body image is jeopardized. Marital conflict and a failure in attachment are diagnostic indicators for the future pathologic development not only of a child's body image but for her overall personality as well.

Siblings further contribute to the development of body image in children, both indirectly in terms of comparison and directly in terms of the nature of the responses of the siblings toward the defective child. In some cases siblings unwittingly may collude with the parents and reinforce whatever messages the child receives from the parents. However, the responses of the siblings are not necessarily those of the parents. The child may be treated with greater consideration and patience by his siblings, thus contributing to a positive self-image, or with outright hostility as a result of being perceived as receiving more attention or warmth from the parents. As the child's world expands, so does the potential for body image modification, either positive or negative.

Following the responses of parents and siblings are those of a child's peers, teachers, and the culture as a whole. Developmentally, responses from an increasing number of persons occur when the child reaches the age at which he is eligible for organized day care or begins school. Extensive research exists on the manner in which children with physical disabilities are responded to in such settings as peer groups, schools, and camps.

Concern for physical attractiveness begins early in the life of a child. Research by Dion and Berscheid (1974) demonstrates that unattractive children have less peer involvement than attractive children. Other studies indicate that the more minor physical differences young children have, the more aggressive, impulsive, or withdrawn behavior patterns they display, the less peer interaction they show, and the more negatively they are judged by their peers (Quinn and Rapoport, 1974; Rapoport and Quinn, 1975; Waldrop and Halverson, 1971). Richardson and Royce (1968) investigated children's reactions to the presence of a physical disability in handicapped and non-handicapped children. Both the disabled and the nondisabled children consistently preferred the nondisabled child.

Dion (1973) found that preschoolers three and one-half years of age reliably discriminated facial photographs of unfamiliar peers based on attractiveness. Further, when asked about friendship choice on the basis of the photographs, children demonstrated a significant preference for the better looking children and a corresponding dislike for unattractive children. Attractive children were judged to behave more prosocially while unattractive children were regarded as exhibiting more antisocial behavior. Similar findings involving the effects of attractiveness have been demonstrated with fifth graders (Cavior and Dokecki, 1970), with adolescents (Lerner and Lerner, 1977), in camp settings (Kleck et al., 1974), and cross culturally (Cavior and Dokecki, 1970).

Clifford and Walster (1973) studied the effects of attractiveness on teachers' expectations of pupil performance. The same report card was distributed to a large group of teachers with varying photographs of children attached. The attractive child was rated significantly higher on intelligence, educational attainment, educational potential, and social potential. Barocas and Black (1974), reviewing referrals for "remedial" versus "control" problems, speculated that attractive children were more likely to receive teachers' attention and help. Similarly, a number of other studies have demonstrated that attractiveness influences the frequency of interaction between teacher and student (Adams and Cohen, 1974), teachers' judgments of students' intellectual and social competence (Lerner and Lerner, 1977), and teachers' willingness to incorporate children of marginal intelligence into their classrooms (Ross and Salvia, 1975). Teachers' negative expectations for unattractive children may have important consequences. The effects of teacher expectations on student performance is well documented (Rosenthal and Jacobson, 1968). It may be that teachers' negative expectations for these children actually help to "produce" the expected intellectual and social deficits.

Physicians treating both physically normal and disordered children need to continually assess the child's internal conception of his body. Methods for assessing body image disturbance are discussed in a subsequent section of this chapter.

COMPARISON WITH OTHERS

The fourth dimension of the model for body image development relates to stimuli from the environment in the form of comparison with others. As with other dimensions of the model, a child is influenced at the earliest stages of development by factors in the environment that are recognized as being outside the child's body. Normal children compare their bodies with those of parents, siblings, and peers and later with those that are presented by the culture as a whole. During this process a child becomes aware of similarities and differences. The child with a congenital or acquired deformity confronts this growing sense of difference as cognition and self-awareness increase. All children, regardless of their physical normalcy, develop a sense of "like" and "not like." How a child reacts to inherent differences between self and others depends not only upon the objective nature of the difference but also upon the interaction of the child with significant persons in his environment as well as cultural values that are conveyed to the child in a variety of ways. Parental perceptions of their child, interactions with significant others, and cultural messages contribute to either positive or negative development of the body image. Failure to receive positive reinforcement of the body image from any of these sources may lead to the development of a negative body image, thereby contributing to the development of psychologic defenses and personality disorders. Parental attitudes and responses from other persons in the child's environment have been discussed previously. Here the concern is with the broader category of comparison with others in the larger context of society and its values toward physical appearance.

Culturally children with obvious physical defects are viewed with disapproval, revulsion, and rejection. These attitudes are evoked not only on the basis of seeing a deformed child. Such attitudes have both psychologic and social meaning. Psychologically, many people project their own inadequate body image onto those whose deficits are more visible. Projection is a defense mechanism, which allows individuals to attribute their own real or imagined unwanted characteristics onto those around them. A deformed or more subtly different child or adult provides the opportunity for individuals to place their conflicts or fears of deformity in their own body onto the body of another.

In spite of demonstrated competence in social, intellectual, and vocational skills, children with manifest defects are often stereotyped in disadvantageous ways. Children's concerns with negative stereotyping are often based on reality. These concerns should be considered as being either normal or pathologic only within the total context of a child's overall physical and psychologic adaptation to a specific defect. Typical examples of negative stereotyping are children with large noses who are assigned to a minority group and children with certain types of facial configuration who are typed as morons or as sufferers from some serious disease. Most children with body defects evaluate rejecting attitudes from familial and social sources. The child's responses to such attitudes depends upon his adaptation to the deformity. Children who refuse to look into the mirror or who refuse to go to school appear to have difficulty in integrating the deformity into a healthy sense of self. Such behavior should alert the physician to possible body image disturbances.

Cultural attitudes toward and subtly conveyed messages to children with body image disturbance are important factors contributing to body image development. Physicians responsible for the treatment of normal children and children with congenital or acquired deformities should be aware of such attitudes. Children with manifest deformities, obesity, or "differentness" are often the subject of humiliating, rejecting, and hostile responses by parents, peers, and society at large. The physician's contribution, whether directly or indirectly, can greatly facilitate a child's adjustment to a real or perceived deficit in body image by helping the child gain an enhanced understanding of the meaning of the deformity.

SOURCES OF BODY IMAGE DISTURBANCE AND METHODS OF ASSESSING BODY IMAGE DISTORTION

Body image disturbance can be found among children without congenital or acquired deformities. Many apparently normal children have body image disturbances resulting from constitutional predisposition and life experience. In fact, children without manifest body defects may suffer the effects of body image disturbance more than those with obvious body disorders. Children who are subjected to physical abuse or who were failure to thrive infants are examples of cases of possible body image disturbances in which there is no manifest deformity. Since their bodies appear intact, disturbances of body image may go unnoticed by the physician caring for such children. Cryptorchism, cerebral palsy, and hidden nevi may negatively influence a child's body image.

The complexity of body image development and disturbance can often make the assessment of such disturbances difficult. However, there are a number of reliable indicators of possible body image disturbance. Perhaps the simplest indicators are observations of the parent-child dyad, interactions between parents, and the functioning of the family as a unit. Children with manifest body defects are different, and familial and social responses to a child's differentness are good measurements of possible body image disturbance. With children who are preverbal or who lack the motor capacity for more sophisticated assessment techniques, the physician must depend solely on observation. Further, the quality of responses to any child's body may provide information concerning later body image disturbances in the context of psychosomatic conditions, such as anorexia nervosa and bulimia, or somatoform disorders, such as hypochondriasis and conversion reactions. There are a number of procedures that allow the physician to assess body image disturbance. These procedures complement the information acquired through observation of the child in his environment. As indicated, the choice of assessment technique varies according to the age of the child.

Levy (1929) suggests a modification of the physical examination of a child to include a simple psychiatric examination. Using Levy's method, the physician would follow the physical examination by asking the child to comment on what he notices about the various parts of his body, observable differences between himself and others, and preferences as to how the child would like to see his body parts at maturity. Should the child cooperate with this procedure, the physician would make further inquiry regarding ideas and feelings about the importance or lack of importance of characteristics such as height, weight, strength, and appearance.

As with any technique designed to elicit body image integration or disturbance, the child's response to such procedures is diagnostic. With Levy's method, for example, the child who spontaneously and comfortably responds to such questions indicates a degree of acceptance of his body or adaptation to deformity. Children who tend to hide their bodies or who are reluctant or embarrassed to speak about their feelings toward body parts or functioning may have body image disturbances. It should be noted, however, that the quality of the relationship between the child and the physician as well as the length of time during which a relationship with a child is developed also influence the child's responses to intimate questions.

Secord and Jourard (1953) developed a rating scale as a means of appraising body cathexis. Secord (1953) devised a word association test utilizing homonyms as a method of measuring bodily concern. Both the rating scale and the word association test would be applied most usefully to older children and adolescents. Secord's research suggests three groups of individuals relative to body concerns: the narcissistic who overvalue and overprotect the body because of its intrinsic personal value: the anxious who register bodily concerns due to physical pain, injury, or shame; and overcontrolled individuals who apparently rid themselves of bodily concerns through denial. Using the two tests, Secord found that the narcissistic group scored high for body acceptance and high on word association. The scores of those with great anxiety were low for body acceptance and high for word association. Finally overcontrolled persons scored low on both tests.

Machover (1949) designed the Draw-a-Person test, which can be used with children approximately three years old and older. The usefulness of the data obtained increases with the age of the child. The test is a projective technique devised to elicit unconscious attitudes and precepts of body image.

The usefulness of the Draw-a-Person test as an office technique is underscored by the drawings shown in Figures 34–1 and 34–2, which demonstrate the picture of a child with typical lowered self-esteem and body distortion.

Fisher and Cleveland (1957) suggested that the role of body boundaries is important to body image disturbance. Individuals may view the body as having firm protective surfaces or as open and without protection. Their research indicates a number of correlations between body image and psychologic adjustment to body damage. One correlation suggests a relationship between the boundary dimension of the body image and the development or choice of psychosomatic illness. The hypothesis is that individuals with definite body boundaries develop symptoms involving ex-

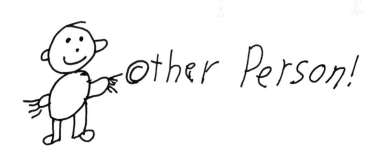

Figure 34–1. Drawing by child with low self-esteem.

Figure 34–2. Drawing by child with cystic hygroma and obesity.

saulted in both acute and chronic diseases. Healthy body image development may be disrupted by trauma or surgery.

The child with a congenital disease or a disease of early onset is faced with the task of incorporating limitations in function that affect the sense of mastery and competence that has been discussed. Depending on the organ system, the impact may vary. The child with congenital heart disease may have limited exercise tolerance, a functional limitation, or cyanosis as a physical manifestation of the disease. These features of disease, both functional and visible, serve as constant reminders of vulnerability, deficit, and limitation. The resulting sense of fragility may foster dependence, pathologic denial, reaction formation, or giving up. The pediatrician along with the parents must present to the child a balanced picture of limitations and strengths with an emphasis on the possible. When parental guilt intrudes on the parents' capacity to permit growth, intervention is indicated.

A special consideration in disease states involving vital organs is the sense of mystery, absence of control, and threat of death. The impact of the uncertainty, the unknown, and the unthinkable is to distort the body image in such a manner as to emphasize fragility. An educational approach involving the parents, the patient, siblings, teachers, and other less well informed caregivers is most important.

Although it is easier to see the impact of life threatening and chronic disease on body image, less well defined, less apparent afflictions may be equally devastating. Hemangiomas, eczema, and gross motor dysfunction impose the same burden as more life threatening illnesses. The pediatrician needs to be mindful of the impact of these types of diseases or disturbances on the development of a sense of bodily competence. Acknowledgment with the patient of the impact of eczema, for example, may provide the opportunity for the youngster to ventilate feelings of disgust, guilt, and failure. The surfacing of these issues with a helping individual may relieve the youngster and, if the concerns are significant and broached with a sense of helpfulness, may provide the caregiver with the opportunity to offer therapeutic support.

The issue of compliance is covered elsewhere in this text, but in illnesses such as diabetes a contributant to noncompliance is a failure to value one's own body. Thus, children with diabetes who sense a defective body require help to invest in themselves. That self-investment is essential to support compliance.

The child who experiences trauma, such as the traumatic amputation of a limb, and who had normal body image development may be expected to go through a grief response. There is the shock with disbelief and anger followed by an appraisal of the functional deficit and the cosmetic impact.

ternal layers of the body. Those with loose indefinite boundaries tend to develop symptoms involving interior organs. Another correlation found by Fisher and Cleveland concerns the relationship between adjustment to body handicap and crippling. These investigators viewed persons with definite body boundaries as adjusting significantly better than those having indefinite body boundaries. Definitiveness of boundary also appears to be related to a wide variety of personality characteristics in healthy individuals.

BODY IMAGE IN DISEASE STATES

The theory of body image development that has been explicated has direct relevance to children with various disease states. Body image is as-

This process of appraisal is in itself prolonged and should not be foreshortened. The initial self-assessment may either be grossly exaggerated or embody massive denial. Only after an appropriate period of mourning the loss of limb or function can the child and the parent achieve a sense of realistic deficit. Therapeutically children should be followed closely and involved actively in appropriate rehabilitation programs at the earliest time.

THERAPEUTIC CONSIDERATIONS

As in other areas, the best treatment for disturbance of body image is the prevention of body image distortion when possible. The issues related to the parental role in prevention have been discussed previously. The pediatrician, working with the parents and teachers, can further facilitate the development of a normal body image in vulnerable children through a variety of means. Alertness to specific vulnerabilities of certain children and the emphasis on compensatory strengths are vital to this strategy. Given a clumsy child, it is perfectly appropriate, through story telling or modeling, to support or encourage the child's ability in areas that de-emphasize athletic prowess. Working with the parents to help them to modify their expectations of a child who is not athletically gifted may be of great usefulness. The same approach is applicable to other areas of vulnerability in the intellectual or esthetic spheres. Regular supportive visits with a child and both parents, when available, enhance this approach. The caring relationship with the pediatrician can be most valuable and sustaining. For instance, it is rewarding to see youngsters gain self-esteem from participation in chess, cross country running, and other activities that can accentuate a child's strengths but not emphasize fine motor coordination.

Environmental manipulation, that is, the altering of the experience of a child through directed activities or changes in environment such as school or camp, is an effective therapeutic technique. Examples of such manipulations include encouraging the parents to place their child in a summer camp that will emphasize supportive activities, to enroll an artistically gifted child in art classes, or to enroll a child who has difficulty with competition in a supportive, small group athletic program. Guidance for the parents in relation to this type of manipulation can be most constructive.

Coupled with the preceding therapeutic approaches, it may be useful to systematically employ behavioral reinforcement. Essentially the therapeutic effect is to give positive reinforcement to activities, actions, and statements that imply or corroborate competence and mastery. This is in contrast to more naive comments that offer a critical appraisal of performance that may not be supportive. It is important, at times, for the pediatrician to be a parent and patient ally with the school system and school teacher to provide consistency in this type of positive behavioral reinforcement.

Unfortunate consequences of a failure to gain a sense of completeness, mastery, or competence are depression, anger, alienation, and a sense of lowered self-esteem. When the supportive interventions fail to bring about a positive change in a child as manifested by an improved sense of esteem, the pediatrician along with the family and the patient needs to consider psychotherapeutic approaches. The depression of childhood, the sense of alienation, and lowered self esteem are not self-limited; rather they can become significant personality characteristics with life-long implications for a disordered personality. Individual or group psychotherapy addressing depression and anger at an early stage can facilitate resolution of underlying conflicts and coupled with the approaches already discussed can bring a child back along a path of healthy psychologic development.

In special instances additional therapeutic approaches may be employed to help a child gain a better sense of mastery. The child who is hospitalized following trauma or surgery or with a chronic debilitating illness can benefit from child life activities based program. Because children who have body image problems may hold back from participation in these programs, the pediatrician needs to give active encouragement and support for participation.

MYRON L. BELFER

PAULINE F. LUKENS

REFERENCES

Adams. G. R., and Cohen, A. S.: A naturalistic investigation of children's physical and interpersonal characteristics as input cues that affect teacher-student interaction: some suggestive findings. J. Exp. Ed., 43:1–5, 1974.

Anthony, E. J.: The child's discovery of his body. Phys. Ther., 48:1103–1114, 1968.

Barocas, R., and Black, H.: Referral rate and physical attractiveness in third grade children. Percept. Mot. Skills, 39:731–734, 1974.

Belfer, M. L., Harrison, A. M., and Murray, J. E.: Body image and the process of reconstructive surgery. Am. J. Dis. Child., 133:532–535, 1979.

Belfer, M. L., Harrison, A. M., Pillemer, F. G., and Murray, J. E.: Appearance and the influence of reconstructive surgery on body image. Clin. Plast. Surg., 9:307–315, 1982.

Bowlby, J.: Attachment. New York, Basic Books, Inc., 1969.

Bowlby, J.: Maternal Care and Mental Health. New York, Columbia University Press, 1951.

Cavior, N., and Dokecki, R. R.: Physical attractiveness and interpersonal attraction among fifth grade boys: a replication with Mexican children. (Presented at the Southeastern Psychological Association, St. Louis, 1970.)

Clifford, E.: Psychological Aspects of Orofacial Anomalies: Speculations in Search of Data. Report 8. Washington, D.C., American Speech and Hearing Association, 1972.

Clifford, M., and Walster, E.: The effect of physical attractiveness on teacher expectations. Sociol. Ed., 46:248–258, 1973.

Cohen, F., and Yasuna, A.: Cognitive and psychological assessment of the child with craniofacial abnormalities. (Presented at Symposium on a Comprehensive Approach to Craniofacial Deformity, Boston, 1978.)

Dion, K.: Young children's stereotyping and facial attractiveness. Dev. Psychol., 9:183–188, 1973.

Dion, K., and Berscheid, E.: Physical attractiveness: perception among children. Sociometry, 37:1–12, 1974.

Edward, J., Ruskin, N., and Turrini, P.: Separation-Individuation: Theory and Application. New York, Gardner Press, 1981.

Fisher, S., and Cleveland, S.: An approach to physiological reactivity in terms of body image schema. Psychol. Rev., 64:426–437, 1957.

Goodenough, F.: Measurement of Intelligence by Drawings. Yonkers, New York, World Book Co., 1926.

Green, M., and Levitt, E. E.: Constriction of body image in children with congenital heart disease. Pediatrics, 29:438–441, 1962.

Greenberg, D. M.: Parental reactions to an infant with a birth defect: a study of five families. (Presented at the biennial meeting of the Society for Research in Child Development, San Francisco, 1979.)

Katz, P., and Zigler, E.: Self-image disparity: A developmental approach. J. Pers. Soc. Psychol., 5:186–195, 1967.

Kaufman, R. V.: Body image changes in physically ill teen-agers. J. Am. Acad. Child Psychiatry, 11:157–170, 1972.

Kaufman, R. V., and Hersher, B. A.: Body image changes in teen age diabetics. Pediatrics, 48:123–128, 1971.

Kleck, R. E., Richardson, S. A., and Ronald, L.: Physical appearance cues and interpersonal attraction in children. Chil. Dev., 45:305–310, 1974.

Kolb, L. C.: Disturbances of the body-image. In Arieti, S. (Editor): American Handbook of Psychiatry. New York, Basic Books, Inc., 1959, Vol. I.

Lax, R.: Some aspects of the interaction between mother and impaired child: mother's narcissistic trauma. Int. J. Psychiatry, 53:339–344, 1971.

Lerner, R. M., and Lerner, J.: Effects of age, sex and physical attractiveness on a child-peer relative, academic performance, and elementary school adjustment. Dev. Psychol., 13:585–590, 1977.

Levy, D. M.: Method of integrating physical and psychiatric examination with special studies of body interest, overt protection, response to growth and sex difference. Am. J. Psychiatry, 9:121–194, 1929.

Macgregor, F. C., et al.: Facial Deformities and Plastic Surgery. Springfield, Illinois, Charles C Thomas Publisher, 1953.

Machover, J. A.: Personality Projection in the Drawing of the Human Figure. Springfield, Illinois, Charles C Thomas Publisher, 1949.

Mahler, M. S., Pine, F., and Bergman, A.: The Psychological Birth of the Human Infant. New York, Basic Books, Inc., 1975.

Piaget, J.: The Child's Conception of the World. New York, Harcourt, Brace, Jovanovich, Inc., 1929.

Quinn, P. O., and Rapoport, J. L.: Minor physical anomalies and neurologic status in hyperactive boys. Pediatrics, 53:742–747, 1974.

Rapoport, J. L., and Quinn, P. O.: Minor physical anomalies (stigmata) and early developmental deviation: a major biologic subgroup of "hyperactive children." Int. J. Ment. Health, 4:29–44, 1975.

Richardson, S. A., and Royce, J.: Race and physical handicap in children's preference for other children. Child. Dev., 39:467–480, 1968.

Rosenthal, R., and Jacobson, L.: Pygmalion in the Classroom. New York, Holt, Rinehart and Winston, 1968.

Ross, M., and Salvia, J.: Attractiveness as a biasing factor in teacher judgments. Am. J. Ment. Defic., 80:96–98, 1975.

Schilder, P.: The Image and Appearance of the Human Body. New York, International Universities Press, Inc., 1950.

Secord, P.: Objectification of word-association procedures by the use of homonyms: a measure of body cathexia. J. Pers., 21:479–495, 1953.

Secord, P., and Jourard, S. M.: The appraisal of body cathexis: body cathexis and self. J. Consult. Psychol., 17:343–347, 1953.

Shapiro, T., and Stine, J.: The figure drawings of 3-year-old children. Psychoanal. Study Child, 20:298–309, 1965.

Solnit, A. J., and Stark, M. H.: Mourning and the birth of a defective child. Psychoanal. Study Child, 16:523–537, 1961.

Tisza, V. B., et al.: Psychiatric observations of children with cleft palate. Am. J. Orthopsychiatry, 28:416–423, 1958.

Waldrop, M. F., and Halverson, C. F.: Minor physical anomalies and hyperactive behavior in young children. In Hellmuth, J. (Editor): Exceptional Infant: Studies in Abnormalities. New York, Bruner/Mazel, Inc., 1971, Vol. 2.

Wright, B. A.: Physical Disability—A Psychological Approach. New York, Harper Brothers, 1960.

Wright, B. A.: Spread in adjustment to disability. Bull. Menninger Clin., 28:198–208, 1964.

35

Development of Sexuality and Its Problems

INFANCY AND EARLY CHILDHOOD

The development of sexuality is often considered a task of adolescence. The reality, however, is that sexuality begins to develop from the time of conception and often continues throughout life. Sexuality represents an amalgam of gender role, gender identity, physical characteristics, hormonal influences, society's expectations, peer and parental influences, and cognitive, psychologic and moral development superimposed on actual experiences. When it is viewed in this way, such descriptive terms as "sexually active" are rendered meaningless, as every growing child has a sex life. Development of gender is discussed in Chapter 10. This chapter concentrates on the psychosocial, biologic and behavioral aspects of the development of sexuality throughout childhood and adolescence.

FREUD'S THEORY OF SEXUAL DEVELOPMENT

Freud built a complex and convincing argument for the existence of sexuality in infants and saw the infant's orientation toward humans growing directly out of its need to discharge sexual tensions (Freud, 1964).

Freud argued that sexual impulses in the neonate are identical in kind to the adult sexual response; sexual behavior involves an interplay between excitation and satisfaction. Excitation for the newborn occurs primarily in terms of instinctual needs, the most pressing of which, according to Freud, is the need for nourishment. The satisfaction of the need for nourishment comes in feeding.

The notion of the anaclitic—which is similar to the idea of conditioned reinforcers—was invoked by Freud to explain the relationship between the satisfaction of the need for food and a more generalized notion of "pleasure." The satisfactions of feeding generalize to satisfactions associated with the process of sucking. Thus sucking for

nourishment in effect becomes sucking for pleasure. In this sense the child's first sexual object is his mother's breast, which both nourishes and gives pleasure.

Whereas Freud failed to distinguish between the nature of the satisfactions associated with nourishment and sucking for its own sake, it is clear that both forms of pleasure are part of the same phylogenetic system, one of which has specific survival value while the other serves simply to reinforce that value. The basic survival value of feeding and the taking in of nourishment is the force behind infantile sexual gratification.

Freud's major contribution to the study of sexual development, then, was his idea that sexual behavior—originally considered to be associated exclusively with adolescence and adulthood—in fact begins, although in an immature form, in newborn infants. The specific manifestations of sexuality undergo developmental transformations throughout the life cycle, like other behavior systems, but Freud argued that there are continuities between the manifestations of sexuality in infancy and adulthood. His ideas paved the way for clinicians and researchers to focus on sexual behavior in infancy and early childhood.

SEXUAL AROUSAL IN INFANCY AND EARLY CHILDHOOD

A number of investigators have observed spontaneous erection in infant males as young as two days old. Erection tends to be associated with various forms of arousal, specifically, with restlessness, stretching, limb rigidity, crying, fretting, and thumb sucking. Erection is also associated with fullness of the bladder and bowels, and frequently follows voiding; in addition, there has been some suggestion that the period between voidings of the bladder or bowels is elongated when interposed by erections. Spontaneous erections are quite common during sleep in infant males—much more common, according to some investigators, than is spontaneous erection in adult males. There is lack of agreement regarding the stage of sleep during which erection is most common. When erection occurs during REM sleep, it is typically accompanied by smiling and rhythmic mouthing or tonguing.

Stimulation of the genitals in infants of both sexes has been shown to lead to arousal. Genital stimulation of males frequently results in penile erection; in older infants such stimulation also leads to smiling and cooing. It has further been suggested that most infants, both male and female, develop orgastic capacity by age three to four years, but have few opportunities to "test" this capacity. In young males orgasm tends to occur very quickly, followed by a brief refractory period, often only several seconds in length. In females orgastic capacity also occurs very early in life but is less common than in males.

Thus, it appears that physical and psychologic response to sexual stimulation begins very early in life, and that orgasm is possible at all ages, for both males and females.

MASTURBATION IN INFANCY AND CHILDHOOD

Given that sexual arousal occurs early in life, and that there are some indications that this arousal is associated with pleasure, it comes as no surprise that masturbation is a common occurrence in both boys and girls.

Masturbation probably occurs accidentally during infancy, during the natural process of bodily exploration. When the genitals are discovered and randomly fingered, the infant discovers the relationship between this behavior and arousal and pleasure. Therefore, the infant is motivated to repeat the activity. Some writers have suggested that by age five to six years, masturbation is universal, systematic, and intentional in children of both sexes.

Alternatively, it has been suggested that masturbatory activity often begins with some physiologic discomfort in the genital area. Diaper rash, urethral infection, and bladder distention are frequently cited as drawing attention to the genital area.

The repression of masturbation is common in many cultures. In the past, Western child guidance experts suggested a variety of severe "treatments" for masturbation, ranging from the suggestion that parents refrain from allowing their infants and young children to be alone in bed while still awake, to the shackling of arms or legs and the use of tranquilizers. More recently clinicians have more commonly counseled parents that masturbation is normal and harmless.

A variety of interpretations have been made of the psychologic meaning of masturbation. Some psychoanalytic writers have described masturbation as serving as a precursor to the oral-genital shift, while others have gone so far as to argue that it involves an anxiety ridden identification with the opposite sex and a desire to become a member of the opposite sex. However, it has also been argued that masturbation in infancy and young childhood may be an entirely different phenomenon from masturbation in young and later adulthood, although the specific form of the behavior may be quite similar. Kleeman (1975), while acknowledging the frequency of masturbatory behavior among young infants, observed that emotional excitement is rarely associated with the behavior, at least through age two years. Spitz (1949) and Kinsey et al. (1948) de-emphasized the importance of childhood masturbation: "It is still to be shown that these elemental tactile experiences have anything to do with the development of the sexual behavior of the adult." Thus, although there is ample evidence that self-manipulation is common among even very young infants, and that parents and physicians are sharply aware of these episodes when they occur, there is yet some question about the significance of the experiences for the child.

SOCIOSEXUAL BEHAVIOR IN EARLY CHILDHOOD

Ample evidence exists that sociosexual experiences occur very early in a child's life, in some studies as early as age five years. The question has been raised repeatedly, by both parents and clinicians, whether such experiences have deleterious effects upon the child's development in general, or upon his development in particular. Martinson (1973) in an extensive literature review concluded that there is little or no evidence of long term consequences—either positive or negative—of early sexual experiences. The amount of arousal associated with early childhood sex play is minimal.

Regarding the origins of interpersonal sex play in children, Broderick (1968) favored a modeling explanation. In his review of cross cultural materials, he found that interpersonal sex play among children, especially oral-genital contact and copulatory attempts, were most common in cultures in which children were permitted to watch adult love making. Thus interpersonal sex play among children is common. Its form may be influenced by various social factors, but there is little evidence that such childhood experiences have any adverse developmental effects.

THE PARENTS' ROLE IN SEXUAL DEVELOPMENT

As with other forms of social development, parents exert direct socialization pressure on their children with respect to the development of sexual behavior. Both because sex is such a strong driving force in childhood behavior and because our culture has strong and rather explicit rules about "acceptable" sexual behavior, we might expect parents to be quite consistently nonpermissive about sexuality in their children. In fact, however, there appears to be considerable variability in parents' attitudes toward sexual behavior.

Research on the impact of differences in these attitudes has been striking in its consistency. In general, permissiveness toward sexual behavior is related to parental warmth, supportiveness, and nonpunitiveness, and all these appear to promote sexual adjustment later in life. The problem with this research, however, is that most of the information about parents comes from the retrospective reports of adults identified as sexually adjusted or maladjusted. Moreover, parents who are permissive about sexual issues tend to be warmer, more supportive, and less punitive than parents who are nonpermissive, so that it is difficult to know, on the basis of existing information, whether it is the permissiveness or the quality of the parent-child relation—or both—that is implicated in the development of "healthy" sexuality.

Clinicians and researchers in child development have given divergent sorts of advice to parents about various topics relevant to sexual socialization. The advice has ranged from the very general, such as Ellis' (1938) caveat that parents ought not to assist in rendering the unconscious erotic impulses of their children conscious, to the very specific, such as Finch's (1969) suggestion that in order to avoid early incestuous relations, parents ought not to let opposite sex siblings share the same bedroom. Some of the advice offered to parents seems to be in accordance with existing knowledge and theories about the effect of socialization practices on sexual development; sometimes, however, the advice has been rather counterintuitive, such as Ellis' (1938) suggestion that mothers avoid excessive tenderness and warmth (lest their sons become sexually excited). There seems to be overwhelming agreement that whatever socialization of sexual behavior does seem necessary be performed with as little fanfare as possible. Finch (1969), for example, stated: "Parents should be aware that sex play may occur and should discourage it when it is observed, but without harsh scoldings and punishment." Likewise Burch (1952) reassured parents that sex play is normal, and that "sudden scolding or a terrifying scene will do more harm than good." Ausubel

(1958) cautioned that "making a scene" can be counterproductive; he noted that children quickly learn that sexual behavior is likely to elicit a response from their parents, and thus may use their sexuality as "a weapon with which to shock adult sensibilities and express defiance of adult authority." A dramatic negative response on the part of the parent, in this context, may therefore serve to reinforce the child's behavior.

ADOLESCENCE

PSYCHOLOGIC DETERMINANTS OF ADOLESCENT SEXUALITY

Organization of the adolescent's experience of sexuality has been viewed from four major vantage points—the psychoanalytic, the psychosocial, and that of cognitive development and learning theory (Table 35–1).

Psychoanalytic theory, best articulated by Freud, sees adolescence as the time when instinctual sexuality is reawakened under the influence of surging hormones. The repression of the ego, which characterizes the antecedent latency stage, is no longer tenable, and the major task of the adolescent becomes that of fulfilling his internally driven sexual mandate within the confines imposed by the ego and the id. "In the Freudian view, the individual enters adolescence possessing a fully articulated set of erotic meanings that seek appropriate objects and behavior....Adolescence is the beginning of a quest for an appropriate interpersonal sexual script to articulate with a largely formed intrapsychic script" (Miller and Simon, 1980).

In contrast to Freud, Erikson (1950) sees adolescence as a change in social rather than biologic status. Adolescent sexuality is viewed on the continuum of identity development. Sexual identity is developed during adolescence in the context of mastering the capacity for trust, intimacy among peers, and autonomy from parents. Preparation for heterosexual selection of a mate by confirmation of a heterosexual commitment signifies successful progress in the Eriksonian view.

While concurring with Erikson in rejecting the Freudian position of sexuality as an innate drive that must be controlled, Gagnon is in disagreement with him on the matter of continuity with earlier stages. In his view early adolescence represents a break with the past, and the continuity of socialization exists not in the sexual but in the nonsexual, specifically in gender role training. The major task of the early adolescent is the integration of new definitions of a potentially sexual self with prior gender role training. He believes that the physical changes of puberty provide a signal to

Table 35–1. DEVELOPMENT OF SEXUALITY

Age	Biologic Development	Cognitive Stages (Piaget)	Social Development	Psychosexual Stages (Freud)	Psychosocial Crises (Erikson)
0	Chromosomal sex Genital dimorphism Capacity for orgasm		Gender assignment	Oral	(1) Trust vs. mistrust
1		Sensory motor	Self-exploration		
2	Growth in size		Mutual exploration	Anal	(2) Autonomy vs. shame and doubt
3		Preoperational		Phallic	(3) Initiative vs. guilt
4					
5			Genital play	Latency	(4) Industry vs. inferior
6		Concrete operational			
7					
8					
9					
10					
11					
Adolescence	Puberty Secondary sex characteristics Menarche Ejaculation	Formal	Dating Petting Coitus	Genital	(5) Identity vs. role diffusion
Young adulthood					(6) Intimacy vs. isolation

adults, who then attribute new sexual meanings to the adolescent's behavior and react to early adolescents as being more sexual than they really are (Gagnon, 1972).

The third organizing factor of adolescents' sexuality, proposed by Kohlberg and by Piaget, is cognitive development. The capacity for formal operations develops during adolescence in most but not all youths. This capacity enables the adolescent to transform erotic objects and thoughts into symbolic abstractions, which may then be explored rationally. Conflict resolution is then possible without the necessity for actually acting out the potentially conflicting scenarios (Kohlberg, 1972; Piaget, 1948).

Finally learning theory has been invoked to explain the development of adolescent sexual behavior. According to such theory, pleasurable acts will be reinforced and will tend to be repeated because they are pleasurable. Kinsey is the major proponent of this view. He argues, for example, that homosexual acts in adolescence are largely rooted in the gratification experienced during previous same-sexual experimentation (Kinsey et al., 1948). This theory addresses only the perpetuation, rather than elucidating the initiation of behavior.

It is important to remember that these sometimes disparate views of adolescent sexual development are based largely on theoretical models, which are yet to be validated. The limited body of empirical data relating to adolescent sexuality suffers from sampling bias and other methodologic limitations. The difficulty in obtaining systematic prospective data is addressed by Money (1976): ". . . childhood sexuality remains a research frontier, unopened to empirical and operational study. Any attempt to cross the frontier is subject to condemnation, as if juvenile sexology constituted a branch of pornography which, in turn, is stigmatized as illicit and immoral." A case in point is the experience of Sorensen (1973), who, in an effort to maximize parental acceptance and allow their adolescents to participate in his study of adolescent sexuality, struck from the final draft of his questionnaire exceptionally sensitive questions dealing with oral sex, anal sex, and intercourse with animals. As a result, there are no data relating to these behavior patterns.

The theories of intrapsychic determinants of adolescent sexuality thus far discussed fail to address differences between the sexes. Gagnon, however, attributes adolescent sex differences in masturbatory activity, sexual fantasies, as well as sexual intercourse to differences in earlier gender role training. For example, the central themes for female socialization involve a commitment to future marriage and to the rhetoric of romantic love, whereas for adolescent males, whose sexual activity of all types is more frequent, these acts often suggest themes of achievement and mastery. Social class differences in motivations for sexual behavior are also described, especially among young adolescent males. He ascribes the lower incidence of masturbatory behavior among working class males to a less complex fantasy life and to their view of this behavior as being unmanly; and to a lesser extent to their earlier onset of intercourse.

BIOLOGIC DETERMINANTS OF ADOLESCENT SEXUALITY

All the hypothalamic and pituitary-gonadal structures necessary for pubertal onset appear to be present from the time of birth. The reasons for their dormancy until the second decade of life remain enigmatic. It appears, however, that a feedback system exists such that gonadotropin releasing hormone stimulates gonadotropin, which in turn causes release of sex steroids from the gonads. These, in turn, inhibit further release of gonadotropin releasing hormone until sex steroid levels again fall below a threshold level, at which time the cycle repeats.

Shortly before any structural changes of puberty emerge, a shift occurs in the anterior pituitary from a consistent pattern of secretion of low levels of gonadotropins (3 to 4 mI.U. per ml.), which characterizes childhood, to one in which surges of up to 10 mI.U. per ml. of these hormones is secreted coincident with each 90 minute sleep cycle. Sex steroid concentrations fluctuate in synchrony with the gonadotropins, coincidentally for testosterone, and following a 10 to 12 hour delay for estrogens. The daytime secretory patterns remain unchanged from the prepubertal state.

The inter-relationships of the changes that characterize the development of the secondary sex characteristics of puberty are diagrammed in Figures 35–1 and 35–2 and the timing of various stages in Tables 35–2 to 35–4.

THE INTERFACE OF BIOLOGY AND BEHAVIOR

The relationship of sexual behavior to hormonal states has long been of interest to researchers. Studies of pubertal male primates have demonstrated a correlation between increasing sexual behavior and serum testosterone levels. The effect of the social setting also has been found to exert an influence, both testosterone levels and sexual behavior being decreased when adult males were present in the social group. Attempts to replicate these studies in humans thus far have been inconclusive. Even less is known about the role of estrogens in human female sexual behavior. In

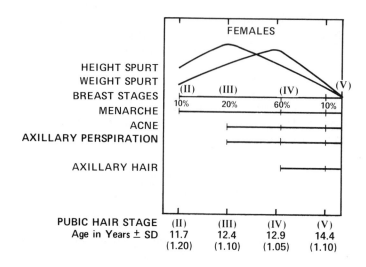

Figure 35–1. Females: Inter-relationship of secondary sex characteristics during puberty.

one of the few methodologically sound prospective studies (limited, however, by a small sample size), no relationship was found between plasma estradiol levels and sexual arousal, frequency of intercourse, or sexual gratification (Persky et al., 1978). No relationship was found between age at menarche and that at first intercourse in a study of adolescents (Morganthau and Rao, 1976).

The implications of the dramatic changes in physical appearance at puberty for the development of sexuality are myriad. It has been postulated that the primary evolutionary significance of pubic and axillary hair may have been to act as wicks for the dissemination of the odor of secretions produced by the apocrine and sebaceous glands. In lower primates, body hair and odors of sexual maturity serve as sexual attractants, territorial markers, and determinants of social hierarchy. The advanced state of socialization of the human race has reduced the sexual significance of these dermatologic appendages, but across cultures and at various periods of time within our own history, one or another secondary sex characteristic (typically the female breasts or buttocks) gains ascendance in the scale of social desirability.

Breasts are considered to be normal sequelae of puberty in females, yet our society stigmatizes gynecomastia, which commonly develops in healthy young adolescent males. Embarrassment, doubts about masculinity, and withdrawal from social interactions and often from the school locker room may result from this normal phenomenon.

The more visible of the secondary sex characteristics, particularly the breasts, are important signals to parents and peer groups that childhood has passed and that sexuality, as adults know it, becomes a possibility. The timing of appearance of these characteristics in relationship to the peer group has an impact on the young adolescent's self-image, which may indirectly affect peer interaction (see Chapter 9). The social behavior of dating may be one such example. In some circumstances, however, societal expectations may be geared more to the age than to the adolescent's stage of physical maturation. In a large study in the United States in the late 1960's, "individual

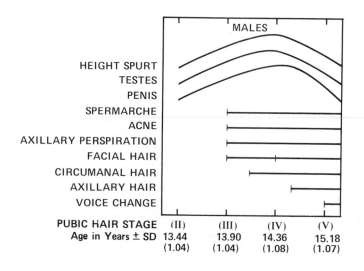

Figure 35–2. Males: Inter-relationship of secondary sex characteristics during puberty.

Table 35–2. DEVELOPMENT OF PRIMARY AND SECONDARY SEX CHARACTERISTICS, BOTH SEXES

Sex Maturity Rating	Endocrine	Pubic Hair
II (Males = 10.5–14.5 years) (Females = 10.4–12.9 years)	Gonadotropins—sleep augmentation Growth hormone—elevated	Midline Long, downy, and silky
III (Males = 11.8–14.9 years) (Females = 11.1–13.4 years)	Gonadotropin—peaks higher Growth hormone—detectable in daytime	Spreads laterally, lightly pigmented, few longer and coarser
IV (Males = 12.8–15.4 years) (Females = 11.8–14.3 years)	Continuation of above pattern Growth hormone—highest peaks	Covers entire mons, coarse and curly
V (Males = 13.8–16.3 years) (Females = 13.2–17.1 years)	Adult pattern and range No sleep augmentation	Extends to medial thighs

Table 35–3. DEVELOPMENT OF PRIMARY AND SECONDARY SEX CHARACTERISTICS, MALES

Sex Maturity Rating	Testes	Scrotum	Accessory Structures	Penis
II 10.5–14.5 years	Seminiferous tubules— ↑ size Leydig and Sertoli cells— ↑ number Greater than 3 ml. total volume	Thinning Hypervascularity	↑ Size epididymis ↑ Size seminal vesicles ↑ Size prostate	Elongation begins
III 11.8–14.9 years	Continued growth of above	Continued	Continued	Continued elongation
IV 12.8–15.4 years	Spermatogenesis	Pigmented	—	Growth of corpora cavernosa → widening of shaft
V 13.8–16.3 years	Adult size ~25 cc	—	Adult size	Adult length and width

Table 35–4. DEVELOPMENT OF PRIMARY AND SECONDARY SEX CHARACTERISTICS, FEMALES

Sex Maturity Rating	Breasts	Vaginal Mucosa	Labia Majora	Ovaries	Uterus
II 10.4–12.9 years	Budding	Thinning	Vascularization and wrinkling	Enlarges greater than 3 gm.	Corpus = cervix
III 11.1–13.4 years	Separation of areolar and breast tissue	Watery, mucoid discharge	—	Continued growth	Cervix > corpus
IV 11.8–14.3 years	—	Folds become more prominent and develop ciliated epithelial lining, pH = acid	—	Average, 6 gm.	Menarche in 90% by this stage
V 13.2–17.1 years	Adult configuration	—	—	Ovulatory cycles more frequent than anovulatory	—

differences in sexual development seem not to have a strong influence on dating, which is commonly thought of as reflecting the development of biological drives....Social pressures appear to overwhelm the individual whose rate of biological development deviates from the norm" (Dornbusch et al., 1981).

There are other sequelae of pubertal development that affect appearance and, hence, sexual attractiveness in the eyes of adolescents. The increased production of dihydrotestosterone in both sexes stimulates the size and secretions of sebaceous follicles, often resulting in the production of acne. Frequent requests for comedone extraction or lancing of cysts prior to an important social event remind us that it is of small comfort to the adolescent to know that acne signifies normal pubertal development. Another effect of the pubertal growth spurt is to cause elongation of the optic globe and, hence, myopia in the genetically predisposed. The appearance of a bespectacled *Playboy* centerfold notwithstanding, most adolescents feel that glasses detract from their sexual attractiveness. Similarly, it is socially unfortunate that completion of pubertal growth is also the time preferred by the orthodontists for affixing braces. In addition, in late adolescence, in 10 per cent of sex maturity rating males, wedge shaped indentations over each lateral frontal region of the forehead result in receding of the hairline.

SEXUAL BEHAVIOR IN ADOLESCENCE

Male Orgasm

Postpubertal orgasm in the male is accompanied by ejaculation, the discharge of semen or spermarche, which usually begins approximately one year after the onset of testicular growth. In Kinsey's sample 90 per cent experienced ejaculation between the ages of 11 and 15, with one year's difference in the mean age between the socioeconomically deprived and the advantaged (14.6 versus 13.6 years). In the majority of the sample the first nocturnal emission occurred almost one year following achievement of ejaculatory ability (Kinsey et al., 1948).

Kinsey used the phrase "total outlet" to describe the total number of orgasms achieved in a typical week through any form of sexual stimulation (masturbation, sex dreams, petting, coitus, homosexual, or animal contacts). He found the highest frequency of total outlet in the youngest group, consisting of males 15 years of age and younger (but all pubertal) with a mean of 2.9 orgasms per week. These observations run contrary to the popular notion that sexuality is awakened during adolescence, gradually reaches its peak during early adulthood, and then wanes.

In the Kinsey sample, the most frequent source of orgasm for the more highly educated youth was masturbation, then nocturnal emissions, in contrast with intercourse for the less educated group.

Female Orgasm

Kinsey found that sexual behavior less frequently culminates in orgasm for women. There were also marked differences between males and females in the age distribution of orgasmic behavior, with females 15 or younger averaging one orgasm every three weeks, and the peak of activity reached and maintained between 30 and 40 years of age. The source of orgasm for females was reported not on the basis of educational status, as with the males, but rather on the basis of marital status at the time of the survey. In the unmarried group, masturbation provided the most frequent source (84 per cent), whereas marital coitus did so for the married group.

Masturbation resulted in orgasm for 71 per cent of female adolescents surveyed by Sorensen (1973). More than half the adolescent females with intercourse experience reported that they had an orgasm during sex "rarely or never." Only one-third of this group believed it important to reach orgasm during sex, compared to more than two-thirds of those who actually experienced orgasms.

Dating and Petting

The onset of dating signifies one of the few, albeit unofficial, rites of passage in our society. It characteristically begins in the adolescent age period. Accordingly it represents an early social decision making point, which often creates tension between parents and their teenage children. As such, it may serve as a battleground for the struggle for independence. The conflict may be in response to the teenager's desire to date and the parents' opposition, or vice versa. Parental opposition is often based on their nonspoken fear that early dating may increase the risk of sexual experience and pregnancy. Rather than discussing these concerns with their offspring, parents may respond by imposing stringent restrictions about curfew, chaperones, and modes of travel. Adolescents may react by rebelling, through sexual acting out, or other mechanisms designed to test injunctions of control, rather than out of desire for the sexual act itself. Available data suggest, however, that pregnancy is unlikely, for sexual intercourse is a rare event during the early adolescent dating experience.

Unchaperoned dating was a post-World War I development, while the custom of "going steady" in high school became popular about the time of the Second World War. Dating appears not to be a single behavior but rather one that undergoes its own developmental sequence. Adolescents move from having friends of the same sex, to those of both sexes, to dating, to having the desire for having a steady date, to actual steady dating, and lastly to having sexual intercourse (Chess et al., 1976). Even within a steady dating relationship, a spectrum of behavior is experienced. Schofield (1965) describes five stages, which occurred in a fairly predictable sequence, at least among the 15 to 19 year old British youths he studied in the early 1960's: Stage 1, in which there is little heterosexual contact, is limited to kissing and the rare dates are devoid of kissing. In stage 2 sexual contact is limited to kissing and stimulation of breasts while fully clothed. Stage 3 involves "sexual intimacies which fall short of intercourse" and may include breast stimulation under clothes, genital stimulation or apposition. Stage 4 involves sexual intercourse with a single partner, and in stage 5 there may be intercourse with more than one partner.

Sorensen (1973) in studying a cross section of American youth a decade later found a similar pattern. He reported that in 22 per cent of American adolescents age 13 to 15 (20 per cent of males, 25 per cent of females), there was a total absence of any sexual contact, other than kissing, that either aimed at or resulted in pleasurable physical reactions. Another 17 per cent (14 per cent of males, 19 per cent of females) were "virgins" who had participated in beginning sexual activities (kissing, touching, exposing one's body to another for sexual pleasure) but had not yet had sexual intercourse. The duration of this phase could not be judged because of the cross sectional nature of his data. That limitation notwithstanding, the degree of satisfaction expressed by this group, both with the amount of sexual gratification as well as with peer and parental relationships, suggests that these activities may represent more than just foreplay. Petting appears to be pivotal in initiating heterosexual psychosocial encounters. In addition to providing a bridge to adult heterosexual intercourse, it teaches adolescents about each other's bodies, emotional and sexual responses, notions of masculinity and femininity, and social rules and customs of sexual behavior and thus allows for beginning consolidation of the disparate components of sexual identity.

Chilman (1979) notes that male writers are frequently puzzled by the apparent contentment of young women to restrict themselves to heavy petting without moving on to intercourse. There are a number of possible practical reasons for this behavior: It greatly reduces the chance of becoming pregnant; females may be more readily orgasmic through direct clitoral stimulation brought about by petting; and withholding full intercourse from the male until marriage is thought to increase his desire for this commitment.

In contrast to those parents who oppose their adolescents' initiation into dating, others encourage it to the point of creating intergenerational conflict over the issue. "Says 15 year old Marilyn, 'I would rather spend an evening with my girlfriends than with a boy I do not like. But my parents push me into dating. They think I don't go out enough. They are angry when I turn down a date. 'You don't have to like the boy to have fun,' they say. I feel it is dishonest to let a boy spend money on me when I have no feelings for him' " (Ginott, 1969). In Sorensen's study, 19 per cent responded in the affirmative to the statement, "Ever since I was 12 or 13 years old, my parents have encouraged me to go out on dates." Perhaps certain of these parents place a premium on the youngster's popularity, either to enhance their own social image or out of a genuine conviction that such popularity will increase the adolescent's happiness, while others may live out their sexual fantasies vicariously through their children.

The determinants of adolescent dating behavior have attracted research interest. The question posed is whether the age of onset of dating is biologically or socially determined. Data from a cross sectional study of a large national probability sample of 12 to 17 year olds were used to explore the correlates of dating. Assessing the stage of sexual maturation by Tanner staging, it was found that dating in females was more closely linked to progression through age grades than to sexual maturation (Dornbusch et al., 1981). Using a longitudinal sample and defining sexual maturation on the basis of attainment of menarche for girls and of peak velocity of growth in height for boys, another group found a dominant role for biologic factors in determining dating behavior. In the sixth and seventh grade, and again in the tenth grade, postmenarchal girls reported a higher level of dating than did those who were in later stages of development (Blyth et al., 1981).

The consequences of dating have been variously perceived. One group demonstrated a detrimental effect of dating on self-esteem among white females in the seventh grade. This finding was compounded for those who were early maturers and had entered junior high school during that grade. Moreover, they reported that girls who dated early were more likely to score low in achievement tests and to have lower grade point averages than those who had not yet dated (Simmons et al., 1979).

Sorensen's survey also suggested better school performance, as well as greater religiosity among adolescents who did not yet date. Is it that better students and those who are more religious refrain from interactions with the opposite sex or, conversely, that absence of dating allows for pursuit of scholastic and religious interests? The direction of the relationships between these variables is difficult to ascertain.

Chess et al. (1976) found physical attractiveness to affect the relations among adolescents; the less attractive the adolescent, the less likely was she or he to have a steady date. Obesity was a strong deterrent.

Gagnon (1972) believes that early dating is advantageous, especially for middle class males. "These steady dating experiences may well serve to increase the young male's investment in the rhetoric of love and emotional commitment which seem so necessary a part of the marriage pattern of this society."

Sexual Intercourse

In the past decade in the United States, the number of youth participating in sexual intercourse has increased with age throughout the adolescent years from approximately 6 to 18 per cent of females and males, respectively, at age 13 years to 60 to 79 per cent at age 19.

In Sorensen's study, 71 per cent of the males had had intercourse by age 15 years, and only 5 per cent waited until 18 years. He believed that the age of onset of intercourse may predict subsequent sexual behavior, in that 60 per cent of the sexual adventurers (those with multiple sexual partners) had had intercourse by 14 years of age, compared with 26 per cent of the monogamists (those with a single sexual partner). The Jessors, in a large sample (1,126) followed prospectively for four years after junior high school, from 1969, found 75 per cent of tenth grade and 51 per cent of twelfth grade males to be virgins (Jessor and Jessor, 1975). Over the course of the 1970's there was a 66 per cent increase in sexual activity among unmarried females 15 to 19 years of age, most of which was accounted for by whites. During this time period the average age of onset of intercourse remained constant (Alan Guttmacher Institute, 1981).

In the Sorensen study, one-quarter of the males first had intercourse with a casual acquaintance, whereas one-third of females did so with a male they intended to marry. The first intercourse occurred in the home of one of the adolescents in 40 per cent of the cases and in automobiles half as often as that.

The transition from virginity to sexual intercourse was studied by the Jessors. Characterizing nonvirginity in secondary school as a manifestation of "transition proneness," they were able to predict which students would lose their virginity in the next year. This group valued independence more and achievement less than the virgin group, and appeared to be more influenced by the views of friends than those of parents. In a cross sectional study, Sorensen reported similar differences between virginal adolescents and those who had experienced intercourse. Both studies reported greater religiosity, less tolerance for drug usage, and higher scholastic achievement among the virgins.

Consequences of Sexual Intercourse. First intercourse experiences among young adolescents are characterized by the absence of use of effective contraception. Only 32 per cent of sexually active adolescents did anything to reduce the chance of pregnancy at the time of first intercourse, condom and withdrawal (18 per cent each) being the most commonly utilized methods. In a 1971 study it was found that 50 per cent used birth control at last intercourse, whereas in 1979 the use of contraception by the same group increased to 70 per cent (Zelnik and Kantner, 1980). A careful perusal of the data, however, reveals that the increase is based almost entirely on greater dependence on totally ineffective methods, such as withdrawal, and that use of reliable methods, such as orally administered contraceptives, IUD's, or combinational barrier methods actually decreased. Earlier studies have demonstrated a lag period of approximately one year between initiation of intercourse behavior and seeking of a birth control method by adolescent females. In the 1979 study only 34 per cent of sexually active adolescents stated that they consistently practiced contraception.

Once the adolescent actually obtains an effective birth control method, there is no guarantee that he or she will continue to use it. In a study of compliance with birth control, it was found that only 45 per cent of female adolescents continued to use the prescribed method for longer than four months following its receipt. The noncompliers were the younger adolescents, those experiencing infrequent intercourse, those with a sexual relationship of short duration (less than six months), and those who had not taken the initiative for their medical visit (Litt et al., 1980). A variety of explanations has been offered for nonutilization of contraception by this age group. In a large study the majority of the two-thirds of sexually active adolescents who had never practiced contraception thought that they could not become pregnant because of the erroneous impression that it was the wrong time of the month (41 per cent). Sixteen per cent did not use contraception because they had not expected to have intercourse, and 8 per cent did not know how to obtain contraception. Only 9 per cent were actually trying to become pregnant (Alan Guttmacher Institute, 1981).

An unfortunate result of nonutilization of contraception by this age group has been an increase in pregnancy. Within the first month after initiation of intercourse, one-fifth of the first pregnancies among adolescents occur, and within the first six months 50 per cent of the first pregnancies occur. In 1978 there were one million, one hundred pregnancies among adolescents. Thirty-eight per cent of these were terminated by induced abortion and 13 per cent by spontaneous abortion. Forty-nine per cent resulted in live births, half of which were to unmarried females. In 1978 30,000 pregnancies occurred in females under the age of 15, an increase of 5 per cent from 1973. In that five year period, abortions increased by 31 per cent and live births decreased by 17 per cent (Alan Guttmacher Institute, 1981).

The sequelae of young adolescent pregnancy are myriad. They include a higher incidence of obstetrical complications, such as toxemia (15 per cent higher); anemia (92 per cent higher); postpartum hemorrhage and infection; as well as neonatal problems, such as prematurity and small-for-dates status. The infant death risk is two times greater among babies born to adolescents than to those over 20 years of age. The relative contribution of poor obstetrical care is unclear owing to delay in presentation and physiologic factors in the pathogenesis of these problems. The negative educational, social, and economic consequences to the adolescent and her baby have all been documented.

Another adverse consequence of sexual intercourse in adolescents is the possibility of contracting a sexually transmitted disease. Adolescents lead the country in the actual rate of gonorrheal infections, as well as in the rate of rise of this infection over the last 15 years. Youth and multiple sexual partners are two factors that correlate with the increased risk of pelvic inflammatory disease. The explanation for the increased risk of infection based on chronologic age is not clear.

EFFEMINATE BEHAVIOR AND HOMOSEXUALITY

A pediatrician is often consulted, directly or obliquely, by worried parents of a young boy who dresses in female clothing or enjoys playing with dolls or girls. They are concerned about the relationship of this behavior to adult homosexuality. As adolescence approaches, the gentle studious young man with a slight build may cause his former football star father to question his son's masculinity. A visit to the pediatrician for the purpose of "giving him something to help him grow" may result. The teenage male may, himself, seek counsel when faced with the appearance of gynecomastia. Alternatively, he may choose not to discuss it at all but rather to ask the doctor for an excuse from gym class for his "painful ankle" rather than suffer the embarrassment of undressing in the locker room.

To be in a position to deal with these common situations, the clinician needs information about the development of sexual preference choices and the significance of cross dressing in childhood and of same sex sexual experiences in adolescence.

Unfortunately there has been little definitive research in this vital area. The majority of studies have dealt with adults whose retrospective recall of childhood events and feelings are obviously colored by their present sexual orientation. In the few prospective studies of male children with effeminate qualities, the question of self-fulfilling prophecy must be raised as a methodologic limitation. What is the parent's or child's understanding of why they were selected for participation in the study, and how might this affect subsequent social interactions? Another major problem with the existing body of information about homosexuality is that it has been derived from studies on select samples, e.g., prisoners, psychotherapy patients, and more recently members of homophile organizations. It is reasonable to question how generalizable the information gathered from these potentially biased samples is. Finally the existing literature, until recently, has addressed itself only to male homosexuality, so that even less is understood about sexual preference among females. Keeping in mind these major limitations, we will review some of the existing body of knowledge.

Issue 1: Is homosexuality determined from early childhood, or is sexual object preference variable and not determined until late adolescence?

Bieber et al. (1962) support the position that there exists the "prehomosexual" child and that one feature distinguishing him from his peers is his early childhood sexual feelings toward males. This position is supported by Manosevitz (1972), who compared two adult samples, one homosexual and the other heterosexual, on the basis of responses to personality tests and to a specially designed life history questionnaire. Although he found no significant differences in early parental treatment in relationship to sex play, early sexual or dating experiences with girls or boys, or the age of first ejaculation between the groups, he reports other areas in which major differences were found. The homosexuals recalled more frequent and stronger sexual attractions to one or more males, including adult males, than did the heterosexual group. Fewer of the latter group admitted having had childhood sexual activity with another male. As regards masturbation, more of the homosexuals acknowledged masturbation prior to 13 years of age, and more reported fantasizing about other males while engaging in masturbatory activity. Of interest was the notation that only nine of the 25 agreed with the statement that they had been "born homosexual." On the basis of these data, Manosevitz concluded that one major distinguishing feature of the prehomosexual male child when compared to the preheterosexual is his bidirectional sexual orientation.

Whitam (1977) and Saghir and Robbins (1973) reported recall of more childhood cross dressing, doll play, and preference for female peer groups among adult homosexual males, and the latter authors, in studying adult female homosexuals, reported that 70 per cent (of 58) had been considered "tomboys" during childhood compared with only 16 per cent of heterosexuals. In adolescence, 35 per cent of the homosexuals and none of the heterosexual females recalled having been viewed as "tomboys."

Green (1979) reports one of the few prospective studies of childhood antecedents of adult homosexuality. Interpretation of his preliminary results is limited by the fact that the control group of noneffeminate boys was enrolled at a later time than the study group. As a result of the age disparity, only some of the noneffeminate group had reached adolescence at the time of the report. Preliminary data indicate that although effeminate behavior during boyhood does not consistently predict later homosexual orientation, it does appear to load in favor of such an outcome in some persons.

A different view is proposed by Glasser (1977), who is of the opinion that homosexual behavior may occur at successive developmental stages during adolescence, a period characterized by "change and flux." Accordingly its significance must be viewed against a background of other components of the developmental process. In early adolescence (puberty to 15 years), for example, the boy is involved in gradual withdrawal from dependent emotional involvement with his parents, while still acknowledging their authority. During this process the youngster becomes narcissistically self-absorbed and selects friends who possess characteristics he would like to possess, and whom he loves as he would like to be loved. In this context, mutual masturbation is viewed as a means of self-exploration and experimentation, comparison, and reassurance. Glasser notes a similar phenomenon in females, although the manifestations are more emotional than physical. He compared this "normal" homosexual activity among adolescent males with that found in those destined to be homosexuals. The former group always had a strong heterosexual interest, and the activity never occurred with an adult man.

In middle adolescence (15 to 17 or 18), the youth rejects parental ideals, values, and authority, develops his own code of ethics and morals, and turns to his peer group for support and approval. In working toward the establishment of such an identity, the adolescent, according to Glasser, is driven toward a final confrontation with his revived oedipal conflicts. How he resolves them will determine his future functioning. In his view, castration anxiety at this age may lead the young adolescent to deny the existence of people without penises or to adopt a passive and submissive posture because of fear of authority figures, either of which might drive him to homosexuality. Glasser also applies this Freudian interpretation of homosexual orientation to females and asserts that they arrive at their lesbian sexual orientation as a result of the persistence of the attempt to deny what they believe to be their anatomic inferiority. He believes that resolution of anxieties and guilt during middle adolescence will result in a heterosexual orientation, but that by late adolescence (18 to 21), homosexuality is established. This view of adolescent homosexual development remains untested, being based in psychoanalytic theory and supported by anecdotal material.

A recent study of female homosexual college students explores their perceptions of their parents. In a study of 34 lesbians and an equal number of control subjects, using the adjective check list of human behavior and the Liphe test of parent-child relationships, Neel and Martin (1981) found that the lesbians viewed their fathers as being less nurturing, more aggressive, less heterosexual, and more self-abasing, and as having less interest in and less respect for their daughters compared to the control subjects.

Issue 2: The significance of homosexual acts during adolescence. As reported by Glasser, heterosexual males may engage in sexual activity with other males during adolescence. Sorensen's study (1973) also found that 11 per cent of the males and 6 per cent of the females surveyed had had at least one homosexual experience. In the boys the first of these occurred at 11 or 12 years of age in the majority, whereas in the majority of the girls it occurred at 6 to 10 years. In approximately one-third the experience was with someone older, but in only 12 per cent of the males, and in none of the females, was the partner an adult. Continuing homosexual activity among this adolescent group was minimal. Only 2 per cent of the boys and none of the girls had a homosexual experience within the month prior to the survey. The weight of the present evidence is that homosexual experiences during adolescence are not uncommon, nor do they predict later homosexuality; conversely, however, most homosexuals report having had homosexual experiences during adolescence.

Issue 3: Is there a physical basis for homosexuality? Following the report of Kolodny et al. (1971) of low testosterone and high luteinizing hormone levels in the serum of homosexual adult males, a number of other researchers have attempted to confirm this finding. Using different techniques and sampling schedules, others have found higher but normal levels of male hormones in homosexual compared to heterosexual males, as well as elevated plasma estradiol levels in the homosexuals.

The only endocrine study of adolescents was carried out by Parks et al. (1974). In their study of six homosexual and six heterosexual males (age 17.8 ± 1.1 years) over 28 consecutive days of

Table 35–5. CATEGORIZATION OF MALE GENDER IDENTITY VARIANTS

	Trans-sexuals	Effeminate Homosexuals	Transvestites
Core gender identity	Female	Male	Male
Cross dressing	To fulfill feminine role	To attract partner	Fetishistic
Desired sex partner	Heterosexual male	Homosexual male	Heterosexual female
View of penis	Abhorrent	Source of gratification	Source of gratification

sampling, follicle stimulating hormone, luteinizing hormone, and testosterone levels were found to be comparable for the two groups and appropriate for maturational age.

From these data, sexual preference cannot be explained on the basis of hormonal differences during adolescence or adulthood. It remains possible, however, that differences may be found in prenatal or neonatal hormone levels.

Issue 4: Implications of cross dressing. During childhood, occasional cross dressing and other effeminate behavior patterns among males have no implications for later sex role choices (Bakwin, 1968).

Adolescent cross dressing, or more specifically, males dressing in female clothing, however, signifies the existence of either trans-sexualism, transvestism, or effeminate homosexuality. The differences among these three conditions, although not always clear-cut, relate to the function of the clothing, the sexual object choice, and the role of the penis in sexual gratification (Table 35–5). Trans-sexuals cross dress because they truly view themselves as females and recall that their first cross dressing experience made them feel secure and warm. In transvestites, in contrast, women's clothing is utilized fetishistically to augment or stimulate sexual arousal, whereas for effeminate homosexuals, being "in drag" serves to attract sexual partners and often has a flamboyant quality. The desired sexual partner for a trans-sexual is typically a heterosexual male; for a transvestite, a heterosexual female; and for an effeminate homosexual, a homosexual male. The penis is abhorred by the trans-sexual and is never the source of sexual gratification, in contrast to the transvestite and effeminate homosexual, both of whom derive pleasure from the genital organ.

SEXUAL LEARNING AND SEX EDUCATION

The notion that teaching about sex will stimulate children to engage in sexual experimentation cannot be supported. In fact, all existing evidence indicates that it is ignorance about sex that is more likely to result in sexual experimentation.

Given that children should be provided with sexual information, there remain a number of issues relevant to the implementation of sex education. When should it be taught? What should

be taught? Who should do the teaching, and how is it best taught? These questions are obviously inter-related, but we will attempt to review them separately.

WHEN SHOULD SEXUAL INFORMATION BE PROVIDED?

In order to decide when to initiate sex education, it is important to relate the issue to the development of cognitive function throughout childhood and adolescence. The studies by Bernstein and Cowan (1975) of children from three to 12 years of age indicate that sex information is not simply taken in but that it is transformed to the child's present cognitive level. In response to the question, "How does the baby happen to be inside the mother's body?" the youngest children (three to four years old), paralleling Piagetian levels, are preformist and believe that a baby who now exists has always existed. At level 2, children begin to attribute causality to the existence of babies, but they do so in the context of the manufacture of inanimate objects. By level 3, children in transition from preoperations to concrete operations, although aware of the three major ingredients in the creation of babies (social relationship, the external mechanics of sexual intercourse, and the fusion of biologic-genetic materials), are unable to coordinate any of the variables in a coherent system. By the stage of concrete operations (level 4), a coordinated system of biologic causality, the union of sperm and egg, is espoused without any attempt to explain why it must be so. Level 5 children envision one gamete uniting with the other and releasing the preformed embryo as a result. By 11 to 12 years of age, on the average, children have reached level 6 and recognize that the equal contribution of genetic material from both parents results in formation of the embryo.

In addition to understanding how children process information about sexuality at various ages, it is important to recognize that their social experiences will result in the acquisition of certain information regardless of whether it is taught intentionally. Kleeman (1975) has found that by the second year of life, boys have knowledge about their genitalia. Between the ages four and five and one-half 98 per cent of the 185 boys studied by Kreitler and Kreitler (1966) knew that their genitals were different from those of girls. By nine to 11 years of age more than 15 per cent had extensive knowledge of genital function during coitus, according to Conn and Kanner (1947); by the age of 12, more than 50 per cent of Ramsey's (1943) sample knew about ejaculation, the origin of babies, nocturnal emissions, contraception, masturbation, intercourse, and prostitution. By 14 years of age more than half the boys had knowledge of venereal disease. It is important to note that the Ramsey and Conn and Kanner studies were done prior to the advent of television as a major source of education, in 1940 and 1947, respectively.

Addressing the question of when to teach, Gagnon and Simon (1969) state: "If information is given too soon, it may be meaningless, or worse, anxiety provoking. If given too late, the young think we are slightly hypocritical. In the absence of highly individualized teaching in which a young person can seek information as he needs it from a responsive and caring adult, it might be best to opt for being slightly late." On the other hand, in providing information about certain topics, such as birth control to the sexually active, one does not have the luxury of being "slightly late."

WHO SHOULD TEACH ABOUT SEX?

The recent literature suggests that parents are attempting to provide some sex education for their children, but that in so doing they have a number of barriers to overcome. Most significant among these is a feeling among parents of inadequacy. This concern appears to be realistic, as in one study only 45 per cent of the mothers had correct information concerning the time of the month when pregnancy was most likely to occur (Alan Guttmacher Institute, 1981). Others worry that they must be "liberal" and comfortable with sexuality in order to be able to communicate with their children about sex. Many parents are of the belief that children do not want to talk with their parents about sex. The reality is, however, that most youngsters express the desire to do so. In Sorensen's study, for example, 50 per cent of the adolescent males and 63 per cent of the females stated that they wanted to be able to talk to their parents about sex. In surveys in Arizona and in New Zealand, parents and the schools were the youths' preferred sources of sex education.

It may be that little sex information comes from the home at least in part because many parents fail to provide unambiguous and direct answers to their children's questions. Conn and Kanner (1947) found that between the ages of four and 12, an average child asks only two questions of his parents about sex related matters. This may be, he suggests, because the parents fail to answer their children's questions adequately, and the chil-

dren therefore do not bother to ask further questions in this apparently sensitive area. Since parents seem willing to answer other sorts of questions, sex often stands out for young children as a peculiar topic, leading to heightened curiosity. Parents should be counseled that children are generally ready to accept correct information about any topic for which they seek information, and that they should be answered directly, honestly, and simply. Parents should stop wondering whether they should reveal basic information about sex to their children and ask, rather, when and how this information should be transmitted so as to be most effective and useful for the child (Broderick, 1968).

Yet, as Gagnon noted, the prevailing attitude of parents toward sexual socialization of their children seems to be that the best sexual teaching is that which teaches without provoking further questions or overt behavior. Three main types of information control seem to characterize parental attitudes toward their children's sex education. The first is called unambiguous labeling: parents call attention to some behavior and label that behavior as "wrong." Generally they do so unambiguously and without explanation. A second type of information control, called nonlabeling by Gagnon, involves an attempt to avoid the issue. A third type is called mislabeling: in this case negative sanctions are interpreted to be outside the area of sex. For example, masturbation might be warned against because of the danger of the child's "getting germs." Likewise, interpersonal sex play might be mislabeled by parents as aggression and punished as such (Gagnon, 1965). Another type of mislabeling involves avoiding labels altogether. Most commonly masturbation and the genitals are given amorphous names, if any names at all, by parents who wish to avoid sexual labels (Sears et al., 1957).

Information control of this kind may have deleterious effects on sexual development. The parental negative sanctions associated with unambiguous labeling and mislabeling may be well learned, so that the negative, "dirty" interpretations of sexual behavior may never be revised. Likewise mislabeling may result in spillage from nonsexual over to sexual areas: for example, if interpersonal sex play is mislabeled as aggression, a child's growing sense of social attitudes about aggression may translate into inferred attitudes about sexual behavior. Sears similarly suggested that mislabeling leads to misunderstandings about sex and inappropriately long persistence of the belief system that results from the mislabeling. Gagnon cautioned that punishment for mislabeled sexual behavior can cause a variety of problems for a growing child. With mislabeling it is unlikely that the child will understand why he is being punished, since it seems that the punishment is nonspecific to the behavior. Thus, such punishment is likely to have few inhibitory effects. On the other hand, punishment for mislabeled behavior can result in generalized anxiety about the behavior.

Nevertheless information control can have its payoffs (Gagnon, 1965). Nonlabeling may serve to protect a child against his parents' anxieties. Given that most adults in our culture have some anxieties about their own attitudes toward sex, it is difficult for them to be totally open with their children about sex.

Children and adolescents apparently learn early which topics might be acceptable to discuss with parents and others that are taboo. In families in which mothers and daughters communicated about sex, the daughters were more likely to initiate discussions about dating, boyfriends, and menstruation, while mothers were more apt to raise issues of sexual morality, intercourse, and birth control (Fox and Inazu, 1980). The type of issue raised by youngsters varies with their age. Preadolescents present more sexual situations to parents than adolescent children, and the former are more likely to involve the topics of genitalia, intercourse, in utero development and birth, female development and menstruation, and modesty and nudity; whereas adolescents raise issues of petting and premarital sex, dirty words and jokes, and abortion (Gilbert and Bailis, 1980).

This developmental sequence in parent-child communication about sex was confirmed by Chess et al. (1976). In the only prospective longitudinal study of the subject, they found that "by the time youths are sixteen, sex is a closed topic between themselves and their parents. The adolescents guard their privacy . . . and, for the most part, parents respect that privacy." Avoidance of discussions of sex reflects both the adolescent's effort to establish independence as well as the desire to avoid the conflict and tension that such discussions might precipitate. This notwithstanding, Fox and Inazu (1980) reported later ages of onset of coitus and more effective utilization of contraception among those who received information about sex within the family context.

The most common source of sex information seems to be same-sex peers. Peck and Wells (1923) reported that adult men identified male companions as their main source of sex information.

WHAT AND HOW TO TEACH

Against this background of limited communication between parents and their children about sexual matters came reports of the increasing rate of venereal disease and pregnancy among young people in the late 1960's. Just as our society responded to Sputnik by demanding more mathe-

matics and science in the schools, these developments placed new demands on the formal education system to provide sex education. Over the ensuing years, programs in family life education were introduced into the curriculum in most school districts; yet, in reality, by 1980 no more than 10 per cent of the students had received comprehensive sex education in school. In most programs the curriculum consists of the presentation of basic biologic factors about reproduction and less often includes attitudes, clarification of values, and decision making. Much has been written about how to set up programs, although evaluations of their efficacy are rare. When evaluations are performed, the outcome variables are typically cognitive, centering on the acquisition of factual knowledge, rather than attitudinal or behavioral outcomes. For a more complete review of the subject of program evaluation, the reader is referred to the article by Parcel and Luttman (1981).

Gordon (1975) emphasizes the importance of each school's examining its own atmosphere critically before deciding on curriculum content. If the environment is one in which student trust is lacking, attempts at teaching sexual values and attitudes will fail. In such settings he suggests a simple presentation of factual information. In recognition of the power of the peer group in the sex education process, an alternative to formal classroom teaching may be after-school programs utilizing peer counselors for sessions in the clarification of values and responsibility for sexual activity. Physicians may provide valuable help in either of these contexts. In one community, for example, a pediatrician worked effectively with the Parent-Teachers Association to organize and teach a program in pubertal development for parents and their seventh grade children.

In addition to the role of consultant, responding to requests for educational input, the physician should stimulate creation of educational opportunities within both schools and families for youngsters with handicaps. The needs of such young people for sexual information often go unrecognized by those most closely involved with their care.

The observation that children with certain handicaps, such as blindness or meningomyelocele, tend to have earlier menarche than normal suggests their need for earlier education about reproduction and the prevention of pregnancy. In one study of parents of children with meningomyelocele, however, more than one-third had the erroneous belief that their handicapped children would be incapable of reproduction. Another implication of the handicapped state is the fact that they cannot avail themselves of the informal educational opportunities for sexual exploration and transfer of information among peers available to normal children. In a study of adolescents with meningomyelocele, three-fourths had less than the average knowledge of sex, as compared with only one-quarter of intact, age matched control subjects (Hayden et al., 1979).

CONCLUSION

Every stage of development of childhood and adolescence is equally important in the development of sexuality.

Parents, as the major agents of socialization, may contribute most significantly to this process by teaching modesty, by serving as role models for mutual respect and nonexploitation between the sexes, and by providing factual information and a yardstick against which children and adolescents may measure their own behavior. Peers undoubtedly contribute heavily during adolescence by providing the most information about sex as well as an arena for sexual experimentation. Physical development, although not the primary stimulus for this experimentation, undoubtedly forces confrontation with new issues and choices during puberty. The physician, through contact with parents and the developing child throughout this period, is in a unique position to serve as a resource person to both and, in so doing, potentially to reduce the anxiety that often surrounds the issue of sexuality. Sexual development should be made a routine part of the curriculum of anticipatory guidance sessions provided with each well baby, well child, or well adolescent visit. Without waiting for questions to be asked, he should take the opportunity at each such visit to indicate the sexual behavior patterns normally engaged in at the youngster's level of development and those to be expected in the interval before the next visit. Once adolescence is reached, a similar approach should be taken with the patient and parent individually. Since the adolescent is faced with a number of options in terms of the wide range of sexual behavior patterns, it is appropriate to use these sessions to help explore the basis for decision making, as well as to discuss possible sequelae of a decision to engage in intercourse, with the goal of preventing those that are undesirable, such as pregnancy and sexually transmitted diseases.

IRIS F. LITT
JOHN A. MARTIN

REFERENCES

Alan Guttmacher Institute: Teenage Pregnancy: The Problem That Hasn't Gone Away. New York, Alan Guttmacher Institute, 1981.
Ausubel, D. P.: Theory and Problems of Child Development. New York, Grune & Stratton, Inc., 1958.

Bakwin, H.: Deviant gender-role behavior in children: relation to homosexuality. Pediatrics, 41:620–629, 1968.
Bernstein, A. C., and Cowan, P. A.: Children's concepts of how people get babies. Child Dev., 46:77–91, 1975.
Bieber, I., et al.: Homosexuality: A Psychoanalytic Study. New York, Basic Books, Inc., 1962.
Blyth, D. A., Bulcroft, R., and Simmons, R. G.: The impact of puberty on adolescents: a longitudinal study. Presented at Annual Meeting of the American Psychological Association, Los Angeles, August 26, 1981.
Broderick, C. B.: Preadolescent sexual behavior. Med. Aspects of Hum. Sexuality, 2:20, 1968.
Burch, B.: Sex and the young child. Parents Magazine, 27:36, 1952.
Chess, S., Thomas, A., and Cameron, M.: Sexual attitudes and behavior patterns in a middle-class adolescent population. Am. J. Orthopsychiatry, 46:689–701, 1976.
Chilman, C. S.: Adolescent Sexuality in a Changing American Society: Social and Psychological Perspectives. DHEW Publication (NIH) 79–1426. Washington, D.C., Department of Health, Education, and Welfare, 1979.
Conn, J. H., and Kanner, L.: Children's awareness of sex differences. J. Child Psychiatry, 1:3–57, 1947.
Dornbusch, S. H., et al.: Sexual development, age and dating: a comparison of biological and social influences upon one set of behaviors. Child Dev., 52:179–185, 1981.
Ellis, H.: Psychology of Sex. New York, Emerson Books, 1938.
Erikson, G. H.: Childhood and Society. New York, W. W. Norton & Co., Inc., 1950.
Finch, S. M.: Sex play among boys and girls. Med. Aspects of Hum. Sexuality, 3:58, 1969.
Fox, G. L., and Inazu, J. K.: Mother-daughter communication about sex. Family Relations, 29:347–352, 1980.
Freud, S.: Three essays on the theory of sexuality. In Strachey, J. (Editor and Translator): The Standard Edition of the Complete Psychological Works of Sigmund Freud. London, Hogarth Press, 1964, Vol. 7.
Freud, S.: Lecture 20 of the introductory lectures on psycho-analysis. The sexual life of man. In Strachey, J. (Editor and Translator): The Standard Edition of the Complete Psychological Works of Sigmund Freud. London, Hogarth Press, 1964, Vol. 16.
Gagnon, J. H.: Sexuality and sexual learning in the child. Psychiatry, 28:212–228, 1965.
Gagnon, J. H.: The creation of the sexual in early adolescence. In Kagan, J., and Coles, L. (Editors): 12 to 16: Early Adolescence. New York, W. W. Norton & Co., Inc., 1972, pp. 231–257.
Gagnon, J. H., and Simon, W.: Sex education and human development. In Fink, P. J., and Hammett, U. B. O. (Editors): Sexual Function and Dysfunction. Philadelphia, F. A. Davis Co., 1969.
Gilbert, F. S., and Bailis, K. L.: Sex education in the home: an empirical task analysis. J. Sex Res., 16:148–161, 1980.
Ginott, H. G.: Between Parent and Teenager. Avon Books, 1969, p. 142.
Glasser, M.: Homosexuality in adolescence. Br. J. Med. Psychol., 50:217–225, 1977.
Gordon, S.: What place does sex education have in the schools? J. School Health, 44:186–189, 1975.
Green, R.: Childhood cross-gender behavior and subsequent sexual preference. Am. J. Psychiatry, 136:106–108, 1979.
Hayden, P. W., Davenport, S. L. H., and Campbell, M. M.: Adolescents with myelodysplasia: impact of physical disability on emotional maturation. Pediatrics, 64:53–59, 1979.
Jessor, S. L., and Jessor, R.: Transition from virginity to nonvirginity among youth: a socio-psychological study over time. Dev. Psychol., 11:473–484, 1975.
Kinsey, A. C., Pomeroy, W. B., and Martin, C. E.: Sexual Behavior in the Human Male. Philadelphia, W. B. Saunders Company, 1948.
Kleeman, J. A.: Genital self-stimulation in infants and toddler girls. In Marcus, I. M., and Francis, J. J. (Editors): Masturbation: From Infancy to Senescence. New York, International Universities Press, 1975.
Kohlberg, L.: The adolescent as a philosopher. In Kagan, J., and Coles, L. (Editors): 12 to 16: Early Adolescence. New York, W. W. Norton & Co., Inc., 1972, pp. 144–175.
Kolodny, R. C., et al.: Plasma testosterone and semen analysis in male homosexuals. N. Engl. J. Med., 285:1170, 1971.
Kreitler, H., and Kreitler, S.: Children's concepts of sexuality and birth. Child Dev., 37:363–378, 1966.
Litt, I. F., Cuskey, W. R., and Rudd, S.: Identifying adolescents at risk for non-compliance with contraceptive therapy. J. Pediatr., 96:742–745, 1980.
Manosevitz, M.: The development of male homosexuality. J. Sex Res., 8:31–40, 1972.

Martinson, F. M.: Infant and Child Sexuality: A Sociological Perspective. St. Peter, Minnesota, Book Mark, 1973.

Miller, P. Y., and Simon, W.: The development of sexuality in adolescence. *In* Adelson, J. (Editor): Handbook of Adolescent Psychology. New York, John Wiley & Sons, Inc., 1980.

Money, J.: Childhood: the last frontier in sex research. The Sciences, *16*:12–15, 27, 1976.

Morganthau, J. E., and Rao, P. S. S.: Contraceptive practices in an adolescent health center. N.Y. Sci. J. Med., *76*:1311, 1976.

Neel, E., and Martin, J.: Perceptions of fathers by female homosexuals. Pediatr. Res., *15*:444, 1981 (abstract).

Parcel, G. S., and Luttmann, D.: Evaluation of a sex education course for young adolescents. Family Relations, *30*:55–60, 1980.

Parks, G. A., et al.: Variation in pituitary-gonadal function in adolescent male homosexuals and heterosexuals. J. Clin. Endocrinol. Metab., *39*:796–801, 1974.

Peck, M. W., and Wells, F. L.: On the psychosexuality of college graduate men. Ment. Hyg., *7*:697–714, 1923.

Persky, H., et al.: The relationship of plasma estradiol level to sexual behavior in young women. Psychosom. Med., *40*:523–535, 1978.

Piaget, J.: The Moral Judgment of the Child. Glencoe, Illinois, Free Press, 1948.

Ramsey, G. V.: The sex information of younger boys. Am. J. Orthopsychiatry, *13*:347–352, 1943.

Rose, R. M., et al.: Changes in testosterone and behavior during adolescence in the male rhesus monkey. Psychosom. Med., *40*:60–70, 1978.

Saghir, M., and Robins, E.: Male and Female Homosexuality. Baltimore, The Williams & Wilkins Co., 1973.

Schofield, C. B. S.: The Sexual Behavior of Young People. Boston, Little, Brown and Company, 1965.

Sears, R. R., Maccoby, E. E., and Levin, H.: Patterns of Child Rearing. Evaston, Illinois, Row, Petersort and Co., Evanston, IL, 1957.

Simmons, R., et al.: Entry into early adolescence: the impact of school structure, puberty, and early dating on self-esteem. Am. Sociol. Rev., *44*:948–967, 1979.

Sorensen, R. C.: Adolescent Sexuality in Contemporary America. Mountain View, California, World Publications, 1973.

Spitz, R. A.: Autoeroticism: some empirical findings and hypotheses on three of its manifestations in the first year of life. Psychoanal. Study Child, *3–4*:85, 1949.

Whitam, F.: Childhood indicators of male homosexuality. Arch. Sex. Behav., *6*:89–96, 1977.

Zelnik, M., and Kantner, J. F.: Sexual activity, contraceptive use and pregnancy among metropolitan area teenagers, 1971–1979. Fam. Plann. Perspect., *12*:230, 1980.

36
Socialization and Its Failure

Socialization is a major consequence of normal development in a facilitating family environment. It is the process by which the child takes on the way of life of the society and becomes a competent participant in it, developing the capacity to conform to society's rules, to enjoy and benefit from interpersonal relationships and the social milieu, and to realize capacities in a personally satisfying and socially contributory way (Clausen, 1980; Table 36–1). As with any developmental task, the process of socialization may be delayed or arrested by a variety of factors at any point along its course (Fig. 36–1). If these delays or arrests are not reversed, optimal socialization may fail to occur, leading in turn to discrete psychiatric disorders in which the failure of socialization is a major component.*

Although this outcome is the focus of this chapter, it must be put into perspective. Longitudinal studies have established that there is not a deterministic connection between infancy, childhood, and adulthood (Chess, 1979; Kagan, 1979). The processes that lead to socialization and in which the child and his environment are active participants are resilient and persistent. Missed opportunities present themselves again and new alternatives open up.

However, to acquire best the skills associated with appropriate social behavior, the developing child requires a facilitating social (family) environment and he must be also receptive to the learning experience (Winnicott, 1958b). The facilitating family is one that loves and nurtures the child, encourages and admonishes appropriately, and provides a model of mutual caring and restrained behavior. The adverse developmental consequences of isolation from these crucial social influences have been described in nonprimates, nonhuman primates, and humans (Eisenberg, 1980; Rutter, 1979a). At an intrapsychic level the child relies on the love, nurturance, and consistency in his parents to provide the firm base from which to explore his immediate family environment. Assimilating or identifying with the parental values

Table 36–1. COMPONENTS OF SOCIALIZATION

1. Conformity to society's rules
2. Pleasure from interpersonal relationships
3. Satisfaction from realization of capacities

and behavior provides the next stepping stone from which the child can confidently explore the world of the extended family—school and peers. From these experiences and interactions the adolescent gains confidence to tackle adulthood and the responsibilities of independent existence.

To avail himself optimally of a facilitating family life, and to be capable of learning social skills from the experience, the child also requires certain capacities. Principally his receptive apparatus needs to be intact. He must be able to perceive the experience his family life has to offer, to assimilate it into his own experience, to appreciate its usefulness, and to be able to recall it for future reference.

Psychosocial factors are the major sources of interference with the quality of family life, and can lead to the interruption or prevention of the provision of a psychologically facilitating environment, whereas biologic factors are most liable to interfere with the child's receptiveness to even the most facilitating environment. Although this is a useful way of conceptualizing factors that are important in socialization, it is also worthwhile to note that there are important interactional effects between the two types of factors, and that the most adverse biologic conditions tend to be associated with the most disadvantaged psychosocial environments (Rutter et al., 1970). We will review briefly the components of these factors that are considered to be crucial in the failure of the normal process of socialization.

PSYCHOSOCIAL FACTORS AND THE FACILITATING ENVIRONMENT

EARLY EXPERIENCES AND BONDING

Recent work in neonatal wards and with premature babies has suggested that early physical contact between baby and mother may nourish

---

*Editor's note: The clinical outcomes of the failure of socialization are also discussed in detail by Puig-Antich and Rabinovich in Chapter 41, "Major Child and Adolescent Psychiatric Disorders". This is not redundancy, but rather a contextual approach to the material which appropriately is handled differently in the two chapters.

Figure 36–1. Disorders of socialization.

Undersocialization ← Socialization → Oversocialization

the development of an optimal mother-child relationship, and may lead to significant gains in social competence by early childhood (Minde, 1981). Disruption of this relationship, particularly when it is not compensated for by warm consistent care, has been demonstrated to impair the capacity to form relationships; furthermore, substantial emotional and material privation in childhood has been implicated in the etiology of antisocial behavior in some children (Rutter, 1979a; see also Chapter 5).

FAMILY FACTORS

Broken homes, marital discord, large family size, and parental deviance (including criminality and alcoholism) can all reduce the quality of family life and potentially may lead to a failure in socialization of the children. However, these factors do not put children at equal risk for all sequelae of the failure of socialization, nor do children whose lives are affected by these factors invariably undergo a failure of socialization. Furthermore, the mechanisms by which these variables lead to emotional disturbance in children have not been clearly worked out (Offord, 1982).

COMMUNITY AND SOCIOECONOMIC FACTORS

Communities differ markedly in their rates of emotional disturbance among children (Rutter, 1981). In general, poor compared to middle class communities have at least twice the number of children with emotional problems (Rutter, 1981). Furthermore, particular kinds of problems appear to be increased in children in poor communities. The factors in these communities, including the characteristics of families, peer groups, schools, and even the mass media that lead to a failure of socialization resulting in increased rates of emotional disturbance, are beginning to be understood.

BIOLOGIC FACTORS AND THE RECEPTIVE CHILD

INHERITED GENETIC FACTORS

Enduring reactivity patterns or temperamental traits in the child have been noted to be extremely important in the process of development, and

there is evidence that certain of these temperamental traits may be genetically inherited (Goldsmith and Gottesman, 1981). Chess (1979) has pointed to the importance of the fit between the temperament of the developing child and his parents. A poor fit leaves the child viewed as "difficult to handle" and makes it more likely that he will not be as acceptable as a family member as other children, a process that may impede optimal socialization (see Chapter 10).

There is also evidence that adult anxiety, obsessive-compulsive disorder, and antisocial behavior may have a partial inherited basis as well, although the continuity of these back into childhood and the relationship of childhood anxiety, obsessive-compulsive disorder, and antisocial behavior to genetic variables have yet to be demonstrated.

ACQUIRED FACTORS

Frank brain damage, when manifested by cerebral palsy and epilepsy, is known to lead to a marked increase in the rate of psychiatric disturbance among children (Rutter, 1977). The types of behavioral and emotional disturbances related to brain damage, as well as the possible mechanisms whereby brain damage could lead to failures in socialization and psychiatric disturbance, are also better understood now (see Chapter 38).

FACTORS OF UNCERTAIN ETIOLOGY

Difficult temperament traits and learning problems, including educational retardation, can both play a role in the etiology of disorders in children resulting from a failure in socialization. The root causes of difficult temperament and learning problems are probably multiple in most children and can include genetic factors, brain damage, early experience, and family factors (Bates, 1980; Rutter et al., 1970).

LABELING

There is a body of literature that supports the idea that deviant behavior is promoted and maintained by society's reaction to it (Lemert, 1967; Scheff, 1966). The evidence is strongest in the case of children who have been apprehended by the police, but even here verification is lacking (Tittle, 1975).

FAILURE OF SOCIALIZATION AND PSYCHIATRIC DISORDER

Failure of socialization is usually most evident to others as a disturbance of the capacity to conform to society's rules, but the capacity for satisfying interpersonal and vocational relationships is almost invariably impaired as well. In the case of the oversocialized child, excessive conformity to society's values, in the form of rigid, self-imposed rules, effectively dampens spontaneity with peers and severely restricts the opportunities for a fulfilling life. Anxiety may develop in conforming to rules that limit opportunity by their strictness, or it may be focused on so many external objects or situations that avoiding these objects or situations may also lead to a severe restriction of social opportunities.

On the other hand, the undersocialized child's indifference to social values may parallel indifference to the feelings of others and also indifference to self-satisfying long term goals. This indifference may be frank and be accompanied by aggression and lack of anxiety or guilt, or it may be furtive and sometimes be accompanied by signs of discomfort and anxiety.

The many constituents of socialization, the range that each may encompass, and differing perceptions of what behavior is socially appropriate and acceptable have contributed to a relative lack of knowledge of the nature, prevalence, and consequence of apparent failure of socialization. To clarify these issues, we will recognize those conditions that lead to the child's being identified as deviant and in need of treatment as being examples of frank failures of socialization. Even selecting this (presumably) most affected group of children does not provide all the answers, for there is considerable debate about the limitations of the existing psychiatric and psychologic nosologic systems. We will use the recently introduced criteria of the Third Edition of the American Psychiatric Association's *Diagnostic and Statistical Manual of Mental Disorders* and refer research findings to these diagnostic groupings (American Psychiatric Association, 1980; Figs. 36–2, 36–3).

Schizoid (313.22)

Obsessive-compulsive (300.30)

Avoidant (313.21)

Separation anxiety (309.21)

Overanxious (313.00)

Figure 36–3. Oversocialization disorders.

DIAGNOSTIC FORMULATION

Since the causes of the psychiatric disorders resulting from a failure in socialization are multiple, it is necessary to carry out a comprehensive diagnostic formulation in each child suspected of having one of these disorders (see also Chapter 51). This entails the gathering of data from a variety of sources, including the parents, the child himself, the school, and, of course, any previous professional contact. The information should be organized into a scheme that lends itself to the planning of a treatment program. Figure 36–4 provides a summary of one such outline (Kline and Cameron, 1978). The etiologic factors are assigned to the categories of predisposing, precipitating, and perpetuating. Each category is subdivided into biologic and psychosocial.

Included in predisposing biologic factors are genetic or brain damage components. Psychosocial predisposing factors include any data relating to early experience or bonding and could cover, in addition, such items as family relationships and school performance. Among the biologic precipitating factors are variables such as trauma or illness. Precipitating factors in the psychosocial realm include actual or perceived loss or threat. Perpetuating factors in the biologic area might include chronic or recurrent somatic illness. In the psychosocial domain they would include an assessment of the defenses and coping mechanisms and an assessment of the liabilities of the environment.

Etiologic factors

Predisposing ⟨ → Biologic / → Psychosocial

Precipitating ⟨ → Biologic / → Psychosocial

Perpetuating ⟨ → Biologic / → Psychosocial

Strengths

Figure 36–4. Outline for diagnostic formulation.

Undersocialized
Aggressive (312.00)

Socialized
Aggressive (312.23)

Socialization

Undersocialized
Nonaggressive (312.10)

Socialized
Nonaggressive (312.21)

Figure 36–2. Undersocialization disorders.

The last category in Figure 36–4 focuses on the strength of the child (and his family). Items to be listed here range widely and could include, for instance, comments about the appealing quality of the child, his ability to put his feelings into words, his athletic gifts, and his intact self-esteem.

When the diagnostic formulation is complete, the treatment program should follow directly from it. The major task is to plan specific interventions aimed, on the one hand, at lessening the effect of particular etiologic factors and, on the other, building on characteristics that have been reported as strengths. It is hoped that the treatment plan, by lessening the impact of the etiologic factors and augmenting the strengths, will shift the balance in the favor of a more adequate adjustment for the child. Illustrative examples of the application of this scheme are provided in later sections of this chapter.

DISORDERS OF OVERSOCIALIZATION

Although the clinical manifestations differ for each disorder (Table 36–2), they have in common anxiety about normal social experiences, and all are compounded by exposure to these normal social experiences. The urge to withdraw, to become obsessive, or to panic is a powerful antagonist of intrusion into the stresses of relationships and living, and is exaggerated by the fearful perception the child has of these normal interactions. The normal anxieties of most children, which generally enhance performance by sharpening self-critical faculties, are hypertrophied to such an extent in these children that they actually prevent social exposure.

The incidence and prevalence of disorders of oversocialization can only be inferred. Even with global ratings of disturbance in which only the severity and degree of incapacity are rated without attempting categorization by diagnosis, widely differing prevalence figures have been reported. These differences can be attributed as much to differences among raters (parents, teachers, clinicians) as they can be attributed to the real effects of age and varying sociocultural conditions. Gould et al. (1981) reviewed epidemiologic studies of emotional and behavior problems in children and concluded that at least 11 per cent of children in the United States would manifest clinical maladjustment. Of the children identified in these surveys, 1 to 49 per cent had received some sort of treatment. Thus it is apparent that clinic populations are not representative of the true scope or nature of the problem.

Few rigorous epidemiologic studies have sought to categorize children by more than the simplest of diagnostic criteria. Rutter, for example, has confined his categories to conduct disorder, emotional disorder, and mixed disorder (features of both; Rutter et al., 1970). On the Isle of Wight, a rural community, he found that 3.5 per cent of 10 and 11 year old boys and girls suffered from emotional disorder as measured by teacher questionnaires (Rutter et al., 1970). However, in his study of the crowded and socially disadvantaged inner London borough, the prevalence of emotional disorder in 10 and 11 year olds was doubled to about 7 per cent in both boys and girls (Rutter, 1981). When the Isle of Wright sample was reexamined at 14 to 15 years, the subjects showed a significant increase in depression and school refusal, and emotional disorder now showed the adult pattern, for it was more prevalent in girls than in boys (Rutter et al., 1976). Similar findings have been reported in less rigorous studies (Eme, 1979).

Rutter's broad category of emotional disorders includes the disorders of oversocialization that are to be considered in this section. None of the separate diagnostic groups, however, has yet been studied on a representative population basis, nor is their representation among child psychiatric and child guidance clinic populations known with any accuracy. Indeed the existence of these separate diagnostic groups as valid clinical entities has yet to be established.

With the possible exception of schizoid disorder, disorders of oversocialization are probably less dependent on constitutional factors than are the disorders of undersocialization. Much more, they reflect a perception of the world, its rules, and its responsibilities that prevents the child from taking advantage of normal social opportunities at home, at school, and with peers. These attitudes may have been acquired because the child had some life experience that has confirmed in his own mind the dangerous or hazardous nature of the world. He may have experienced an unexpected loss or series of losses, or he may have been the victim of unusually punitive parenting.

In other children who have not been subjected to such severe influences, oversocialization has been due to the child's modeling his fearful attitudes on parents who are themselves excessively fear-ridden and socially inept and who cannot present the developing child with a realistic experience of social living. The vision these parents present of life is full of anxious apprehension and unexpected dangers; thus the socialization experience for these children is compounded by their parent's own fears. The normal vagaries and difficulties of living in a society are exaggerated and are seen as an obstacle to normal penetration of the social opportunities outside the family.

Another family constellation that can lead to oversocialization has obsessive-compulsive parents who see the world in terms of inflexible rules and extreme sanctions. They impart the view to

**Table 36–2. DISORDERS OF OVERSOCIALIZATION:
DEFINITION AND CLINICAL FEATURES***

	Schizoid Disorder (313.22)	Obsessive-Compulsive Disorder (300.30)	Avoidant Disorder (313.21)	Separation Anxiety Disorder (309.21)	Overanxious Disorder (313.00)
Essential features	Defect in capacity to form social relationships, little desire for social involvement Withdrawn, seclusive, pursues solitary interests, avoids competition	Obsessions and compulsions are central features Symptoms are ego-alien, and patient fights to control or suppress them Resistance results in mounting anxiety Patient clearly recognizes realistic senselessness of his obsessions and compulsions	Persistent, excessive shrinking from contact or involvement with strangers sufficient to interfere with peer functioning Embarrassment, timidity, withdrawal despite eagerness for participation Inhibition of motor activity or initiative Inarticulateness despite adequate communicative ability	Exaggerated distress at separation from parents, home, or other familiar surroundings Persistent morbid fears and preoccupations that danger will befall him or a close relation Phobias of animals, monsters, kidnapping, dying, accidents Nightmares and difficulty in falling asleep Difficulty in being away from home and yearning to return home	Excessive worrying and fearful behavior not related to a specific situation or object or due to a recent acute stress Much worrying about potential injury, future examinations, inclusion in peer activities, meeting expectations and deadlines Overconcern with competence and with criticism of others Difficulty in falling asleep Psychophysiologic symptoms (lump in throat, nausea, stomach aches, dizziness, nervousness)
Associated features	May show irritability when social demands are made Occasional outbursts of bizarre or aggressive behavior Mild deviation in thinking, odd preoccupations with esoteric topics Self-absorbed, excessive daydreaming No loss of capacity to recognize reality	Disorder often accompanied by traits of compulsive personality disorder Frequently shows depression and anxiety Usually chronic, occasionally episodic disorder	Tearful and anxious when pressed into social participation and competitive situations; clinging to caretakers Delay in adolescent psychosexual activity Inhibition, negativism, self-doubt Perfectionistic, self-demeaning behavior in athletic, artistic, and dramatic performances Grandiose fantasies confided to diaries and intimates Belief that great worth is unappreciated Invention of imaginary companions	Fear of the dark, demandingness, intrusiveness Demands for constant attention, complaints of being unloved Death wishes, sadness, crying, depressive symptoms Psychophysiologic symptoms (stomach aches, dizziness, palpitations) Travel symptoms, motion sickness, irritability Anxiety at change, geographic moves, nostalgia for a lost way of life Anxiety in new situations	Obsessional self-doubt, perfectionism Excessive conforming and approval seeking behavior Motor restlessness, nail biting, hair pulling, thumb sucking Hypermaturity, precocious ego development Great wish to please but rarely satisfied with performance May seem dominating to others by talkativeness and wish to demonstrate abilities Exaggerated response to pain, illness, handicap Accident proneness
Minimal duration of symptoms	None required	None required	More than 6 months	More than 2 weeks	More than 6 months

*Adapted from American Psychiatric Association: Diagnostic and Statistical Manual of Mental Disorders. Ed. 3. Washington, D.C., American Psychiatric Association, 1980.

their children that one must conform or face serious consequences, and this results in a caricature of socialization dominated by rules but devoid of pleasure; the normal constraints of society are overinterpreted and restrict rather than open up social opportunities.

Family attitudes, however, give us an incomplete understanding of the antecedents and causes of the disorders of oversocialization. Other factors that may be important can be broadly divided into those that come from within the child and those that influence the child from outside. Internal factors include biologic factors such as genetic make-up, temperament, and organic brain function and psychologic factors such as internal psychodynamics. External psychosocial factors are the family (already discussed) and the social milieu, including peers, school, and socioeconomic circumstances. However, the precise etiology of each of the disorders is not known.

The prognosis and natural history of the disorders of socialization are poorly defined, largely because of the nosologic problems to which we have previously referred. However, since all these disorders can be seen as pathologic extremes of normal traits of behavior, it is not surprising that as adults the individuals still show the vestiges of the same traits, but not necessarily to the same degree.

We do not have sufficient data to allow recommendations of effective specific management and therapy for any of these disorders. The effectiveness of a treatment is established by comparing its efficacy with that of an alternative or no treatment. These disorders have not been subjected separately to this type of evaluation; the few evaluation studies have concerned heterogeneous groups of subjects. The treatment modalities that have been most rigorously evaluated are pharmacotherapy, behavioral psychotherapy, and psychosurgery, but most treatment reports concern the use of individual or family dynamic psychotherapy, and these modalities have only rarely been subjected to any sort of systematic evaluation (Wright et al., 1976).

These disorders will now be described separately. It should be emphasized again, however, that the degree to which each represents a separate entity is presently uncertain.

SCHIZOID DISORDER

Definition of Problem and Common Manifestations

The essential features of schizoid disorder are a defect in the capacity to form social relationships, introversion, and a bland or constricted affect (see also Chapter 41). This can be differentiated from normal social reticence as early as five years of age. Children with this disorder show little desire for social involvement, prefer to be alone, and have few, if any, friends. Their friendships tend to be with younger or other poorly functioning youngsters. When placed in social situations, they are inept and awkward.

They often appear reserved, withdrawn, and seclusive and pursue solitary interests or hobbies. They may seem vague about their goals and indecisive in their activities and appear absentminded and detached from their environment. They do not appear to be distressed by their isolation. Children with this disorder generally lack the capacity for emotional display and are unable to express strong feelings. They usually appear cold, aloof, and distant, although attachment to a parent or another adult is not unusual, and they especially avoid activities in sports. They appear to resist actively normal peer and family relationships. If forced into social participation, they may become very irritable or even may be aggressive or destructive with only minimal provocation. Although this type of adjustment represents a relatively enduring characteristic, which falls within the normal range of childhood behavior, symptomatic worsening may occur during family crises, entry into school, and early adolescence. However, it is important to remember that reactive or crisis related withdrawal is frequently temporary and that the child usually returns to his normal, albeit withdrawn, pattern of behavior as conflict eases.

Prior to the recent introduction of the category of schizoid disorder, schizoid adjustment was not acknowledged until the tendency to withdrawal had become fixed and represented a personality style. In younger children the pathologic tendency to withdrawal had previously been described as a withdrawing reaction. This reaction overlapped with the anxiety disorders, and with temperamental traits of shyness and sensitivity. It should be made clear at this point that children with schizoid disorder exhibit a most marked withdrawal pattern, and there is an element of peculiarity to them. These features distinguish them from shy, withdrawn, avoidant disorder children. There are no reliable studies of the incidence and prevalence of schizoid disorders.

Antecedents and Causes

The pathogenesis of schizoid disorder is poorly understood, partly because this diagnosis has been identified only recently. As noted, oversocialization may not play a central role in its etiology. Temperament traits of reservation, withdrawal, and seclusiveness may be important in the disorder. They are common in children, although they generally do not warrant intervention. Furthermore, the same traits seem to persist from infancy through the developmental years into adult life. In the twin study by Goldsmith and Gottesman (1981), behavioral traits resembling withdrawal and seclusiveness (passivity at eight months, shyness and fearful-inhibitedness at seven years) appeared to be under genetic influence, although measures of similar behavioral characteristics at four years did not suggest genetic control. Adoption studies in schizophrenia suggest that severe schizoid withdrawal, as distinct from moderate withdrawal, may be an antecedent to either adult schizophrenia or a schizophrenic spectrum disorder (Kety et al., 1968). Thus it is speculated that schizoid disorder may be linked genetically to adult schizophrenia. The clinical observation that shy withdrawn children more often have parents with similar behavioral traits may also suggest genetic transmission, but the possibility of modeling on parental behavior traits cannot be excluded as a partial explanation.

As well as being a persistent (temperamental) pattern, acute withdrawal has also been observed as a response to certain developmental crises, such as starting school, in which it is related to separation anxiety and fear of new experiences, and at the beginning of the adolescent years, when it is theorized that an upsurge of aggressive and sexual impulses provoke a paradoxic response in some adolescents.

Early experiences and bonding, community and socioeconomic factors, and acquired biologic factors have not been implicated as causes of this disorder.

Prognosis and Natural History

For most children who are identified at some time as showing extremely withdrawn and secluded behavior, these traits represent a life-long pattern. However, although they become somewhat isolated adults, most appear to adapt fairly well to adult life. For example, in an interview follow-up study of 34 shy, withdrawn children who had been treated at a child guidance clinic, Morris et al. (1954) reported that two-thirds of the children had made a satisfactory adult adjustment. Although they continued to be quiet and retiring individuals, they were self-supporting in the community, and most had married and started families. One-third had make a marginal social adjustment, and only one individual was regarded as psychiatrically ill, but even he was living in the community.

For a smaller group of withdrawn children, extreme withdrawal is a temporary phenomenon, and following the resolution of the precipitating crisis, these children return to a much more sociable adjustment than do the first group. Finally, in a very small minority, clinically significant withdrawal and seclusiveness does herald adolescent or adult schizophrenia. However, this most serious outcome does not appear to occur more often in shy, withdrawn children generally than it does in children with any other type of disposition (Michael et al., 1957; Watt et al., 1970). It may be confined to the more peculiar and severely withdrawn children who would meet the criteria for schizoid disorder.

Guidelines for Management

There are no well controlled treatment studies of schizoid disorder, but a number of studies have evaluated social skills intervention in shy, withdrawn children who were generally drawn from noncomplainant populations such as schools (Conger and Keane, 1981). The modest success of these interventions has questionable relevance to the treatment of shyness and withdrawal of a clinical severity, and to schizoid disorder in particular, for the target behavior was generally of clinically insignificant intensity.

In spite of the absence of adequate supportive data, certain clinical principles have been recognized as important. First, the presence of a precipitating crisis or conflict should be noted, for resolution of the struggle is believed to lead to diminished anxiety and to the child's regaining enough confidence to assume his normal social relationships (Lavietes, 1980). Second, when the behavior of the child reflects similar inhibited, seclusive behavior in the parents, treatment must also focus on the parents to help them become more adaptive, flexible, and persistent in helping the child expose himself more to normal experiences. This can be a difficult task and may require a great deal of patience and understanding on the part of the therapist because the parents themselves can be fearful of the types of social experiences being advocated for their child.

These principles can be followed within a family therapy context, or they can be presented more formally as a behavior modification regimen. In both the child is given an active social role, which he performs with guidance and support and which will increase his confidence in himself, diminish anxiety in approaching social interactions and tasks, and may result in positive feedback from his peers, which further reduces anxiety.

Individual psychotherapy has also been used with these children, but its efficacy has not been evaluated. In the psychotherapeutic relationship it is important that the therapist not be as passive as the child. He can provide a model of liveliness while at the same time discussing the child's fears and anxieties openly with him without evidence of anxiety, and encouraging and supporting the child as he approaches his own fearfulness. It is important that apprehension in relation to normal events be identified and that the child be helped to understand that these feelings can be appropriate, and should not prevent active participation in life.

OBSESSIVE-COMPULSIVE DISORDER

Definition of the Problem and Common Manifestations

The essential features of obsessive-compulsive disorder are the presence of obsessions and compulsions. Obsessional thoughts are thoughts, words, or mental images that force themselves into consciousness against the individual's will. Compulsions are irrational impulses toward some form of action. These may give rise to behavior patterns that are related to the primary impulse, in that they are a reaction to it or an attempt to control it, or they appear to reflect the primary urge that underlies the compulsion. For example, a repressed sexual urge may give rise to a conscious thought about being unclean, which may in turn give rise to ritualistic hand washing behavior or an excessively careful separation of clean from unclean objects in the house.

The child recognizes the obsessions and compulsions as being foreign and as being against his will (ego alien) and generally fights to control or suppress them. This resistance results in mounting anxiety. The patient clearly recognizes the realistic senselessness of his obsessions and compulsions. The clinical disorder is often accompanied by obsessive-compulsive traits of orderliness, parsimony, and caution.

Obsessive-compulsive disorder usually follows a chronic course into adulthood, although it occasionally may be episodic. It can develop as early as four or five years of age. In most cases it develops gradually, although in about half the cases it may be precipitated into the clinical disorder by an emotional crisis, such as a death or parental separation (Werkman, 1980). By the age of 15 approximately two-thirds of adult obsessive-compulsives have already developed their condition.

The consequences for the child or adolescent can be disastrous. Apart from the constriction of personality and the general discomfort and social ineptness accompanying the disorder, rituals can become so time consuming that they seriously invade on the time available for normal functioning. The resulting social isolation can further increase anxiety, compound symptoms, and lead to depression (Hersov, 1977).

In contrast to the other disorders of oversocialization, obsessive-compulsive disorder has been recognized as a distinct psychiatric disorder in children for many years. The condition generally persists to some extent from its onset in childhood or adolescence through the entire life of the individual, and the obsessions and compulsions rarely declare themselves and then disappear entirely from the individual's behavior repertoire. Thus there is an increasing prevalence with age, which reaches a plateau in early adulthood. Nevertheless obsessive-compulsive disorder is a rare condition, although obsessional traits such as perfectionism, parsimony, and compulsive thoughts and acts are commonly found as personality traits that may even enhance the individual's success in everyday life.

Judd (1965) found obsessive-compulsive disorder to be the primary diagnosis in about 1 per cent of referrals to a large child psychiatric clinic. Hollingsworth et al. (1980) found approximately the same rate in another large clinic sample, but when he applied strict diagnostic criteria, the rate dropped to 0.2 per cent. In his Isle of Wight study, Rutter did not find any cases of obsessive-compulsive disorder among the 10 and 11 year old children he surveyed, although a number of children with emotional disorders had well developed obsessional features. Child psychiatric clinic studies also suggest that obsessive-compulsive disorder is more frequent in boys. Furthermore obsessive-compulsive traits and symptoms have been noted in more than half the families of clinically obsessive-compulsive children (Hollingsworth et al., 1980). These families tend to be middle class rather than poor, and to have an emotionally empty life style characterized by a high level of etiquette, cleanliness, and morality (Adams, 1973).

Antecedents and Causes

The cause of obsessive-compulsive disorder has long been attributed to psychodynamic factors. Freud (1959) described a characteristic configuration of defenses (reaction formation, undoing, and isolation) that the child uses to diminish anxiety about hypothesized anal aggressive impulses, and his formulation led to the notion of the "battle of the chamber pot" upon which rests the child's ability to claim autonomy for himself. It has been considered by some that losing this battle sets the stage for obsessional defenses and an obsessive-compulsive life style.

There appear to be two dynamic issues that can be identified from such formulations—the issue of how the developing child deals with aggression and rage, and the battle for autonomy between the growing child and his mother. These issues focus on the importance of the relationship of the child to its nurturing and acculturating mother who demands restraints and expects performances that may or may not be within the child's capacity.

On the other hand, behavior theorists have considered obsessions and compulsions to be conditioned responses to anxiety provoking events. Anxiety is dealt with either by association with neutral events, producing obsessional preoccupations, or by the commission of certain acts that reduce the anxiety, producing compulsive behavior.

However, some genetic studies suggest there may be an inherited component to this disorder. Although family studies indicate that there is a definite familial distribution of obsessive-compulsive traits, twin studies have been equivocal (Miner, 1973). Thus it remains possible that these traits can be attributed to either or both cultural inheritance (modeling) and genetic inheritance. That is, the child acquires similar traits to those of its parents by genetic transmission or by living with them.

Although obsessive-compulsive disorder appears to be more prevalent among middle class children, socioeconomic factors do not play a clear causative role. Furthermore, community factors, acquired biologic factors, and early experiences and bonding do not appear to be related to the etiology of this disorder.

Prognosis and Natural History

It has been shown that approximately two-thirds of the children still have a major handicap by early adult life. About one quarter show a lesser degree of handicap and relatively normal social functioning, and no more than 10 per cent appear to make a complete recovery. In Hollingsworth's study most of his subjects retained discrete obsessive-compulsive symptoms, and all had serious problems with their adult social life and peer relationships (Hollingsworth et al., 1980).

Thus the overall prognosis when obsessive-compulsive disorder develops in childhood or adolescence is poor. Generally these children go on to maintain the same pathologic traits in adulthood, often with little reduction in their severity. The prognosis may be better when there appears to be a single episode of the disorder of minor intensity, when depression is evident, and when there is a clear precipitant. Rarely these children may develop schizophrenia (Pollitt, 1975).

Diagnostic Formulation

The following case is included to illustrate the role of predisposing, precipitating, and perpetuating factors in the disorders of oversocialization. Similar mechanisms and interactive effects occur in the other disorders.

Case 1. Eighteen year old Peter was first seen at the age of 13 when he presented with extreme slowness in school and difficulty in eating. He seemed unable to make up his mind about how to proceed with his school work, and spent much of his school time neatly lining up his pens and pencils. He was indecisive about how to proceed with eating his meals and lingered over every mouthful, chewing it many times over. He had rituals about cleanliness and had to wash his hands in elaborate length until he was absolutely sure that they were clean. He procrastinated about "doing things right." Five years of treatment had not significantly improved his condition. His daily life was still dominated by lengthy rituals, which invaded almost every aspect of his functioning. Without being spoonfed, he would not consume sufficient food to properly nourish himself.

Certain predisposing events led up to this illness. Peter was the second of three boys, and he always had anxieties about going to school. He did not appear to want to leave his mother at home, and at the age of 10 he was referred to the school psychologist because of school avoidance. His mother had a metastasizing carcinoid tumor, which was first detected at about the time of the patient's birth, and her condition deteriorated suddenly just prior to her son's school avoidance. She had always been an anxious hypochondriacal woman with many fears about health, and peculiar food fads, which she had passed on to her children. For example, when Peter once vomited after a glass of milk, she believed that he had a milk allergy and milk was henceforth excluded from his diet. His father was a career scientist who was in his early sixties. He was a pedantic, extremely compulsive man whose life was dominated by strict rules and time. He had married in his forties, not before having looked after his own parents until their deaths in their eighties.

The immediate precipitant to Peter's illness was his mother's death. He was greatly distressed by it, and he recalled that no one warned him that her death was imminent, as she had been hospitalized many times in the past. His father continued to work after his mother's death, and when the patient came home for lunch, he would have to eat alone. In the past his mother had been there to feed him.

Peter's condition continued unabated for five years, and his family life may have perpetuated his condition. His father, now retired, was resigned to the burden of having to look after a psychiatrically incapacitated son. Although he met his son's material needs very well, the two of them lived in a home devoid of any life and warmth. Conversation lacked spontaneity and was confined to material needs and politics. They never shared feelings. The father was moderately depressed, but he maintained tight control over his feelings.

There may also have been biologic perpetuating factors. Because of Peter's compulsive mastication, he lost a great deal of weight, so much that his weight remained below the first or second percentile for his age until after his sixteenth birthday. It was not until his father resorted to spoonfeeding him three years after his mother's death that he began to gain weight and finally went through the physical changes normally associated with puberty.

To summarize, a number of factors contributed to the precipitation and perpetuation of illness in this predisposed boy. Already anxious and apprehensive, Peter was raised by a dying mother who seemed to have tried to control her illness by food fads and misconceptions. They had an almost symbiotic relationship from which his father, a very distant, obsessional, and bitter man, was excluded. When his mother died, the patient was just entering adolescence, and the bond with his mother was broken abruptly with no opportunity for replacement. He was unable to become overtly depressed because his father had strict sanctions against the expression of feelings. He therefore developed numerous obsessional ideas based on his mother's peculiar notions about social behavior, which led to compulsive eating and cleanliness rituals. His father had never been able to do anything more than provide for his material needs, and thus an emotional replacement for his mother had not been possible. In fact, at the present time, the original loss seems almost irrelevant, as he is relatively free of depressive symptoms of either a masked or overt type.

Guidelines for Management

There have been no controlled studies of the psychotherapeutic treatment of obsessive-compulsive disorder in childhood. Even so, the retrospective reports of psychotherapeutic interventions with these children are generally discouraging. However, Adams (1973) is not so pessimistic. He reports good success with intensive psychotherapy maintained for at least one and one-half years. He sees psychotherapy as an opportunity for these children to feel comfortable with their own feelings and impulses, and to be able to relax the inhibitions that cripple spontaneity. Thus the style of therapy he advocates is directive and open rather than being unstructured. He places emphasis on the therapist's being a very active participant, offering advice and support, and pursuing the clear communication of feelings and wishes.

Obsessional parents of obsessional children pose a particular problem. Often they have become so rigid and inflexible that they do not readily participate in treatment, and they are extremely guarded about emotional subjects. However, particularly if they can see that their child has a problem in the area of emotional expression and spontaneity, a useful strategy can be to point out how the child is crippled by the extremes of the parent's behavior. The parents can sometimes then be surprisingly sympathetic and helpful in regard to the child's suffering, although they may not be willing to change their own basic personality style.

A variety of behavior modification techniques may also be employed. The most effective appear to be those that diminish the frequency of compulsions by response prevention. In this method the compulsive ritual is deliberately triggered and then its completion is prevented, by force if necessary. Gradual approach techniques to reduce anxiety related to obsessive thoughts, such as systematic desensitization, flooding, and modeling, have not yielded as good results. However, none of these techniques has been tested in any systematic way in children.

No medication has been reported to be specific for obsessive-compulsive disorder in childhood, although the minor tranquilizers have been reported to be useful for the reduction of severe anxiety (White, 1980). Tricyclic antidepressants, which have been reported to have antiobsessive effects in adults, have only rarely been used with such children, although a recent randomized crossover trial of chlorimipramine with obsessive-compulsive adolescents by Rapoport et al. (1980) was not encourgaging. In any event, in the face of no real alternative, a six to eight week trial of chlorimipramine in full antidepressant doses is probably worthwhile.

AVOIDANT DISORDER

Definition of the Problem and Common Manifestations

The essential features of avoidant disorder include a persistent excessive shrinking from contact or involvement with strangers sufficient to interfere with peer functioning, ease of embarrassment, and excessive timidity. The child withdraws despite an eagerness for participation, shows little initiative, is inhibited in motor activities, and is inarticulate despite adequate language development.

Associated features include tearfulness and anxiety when pressed into social participation and competitive situations. These children tend to cling to caretakers, and they are inhibited and negativistic about themselves and self-critical of their performance in both athletic and artistic pursuits. They may compensate for poor self-esteem by confiding grandiose fantasies of overcoming passivity to themselves and to their close friends or believing that great worth is unappreciated, and they may invent imaginary companions to populate their lonely lives.

To be diagnosed, the disorder must be present for at least six months in a child who is older than two and one-half years. In the mild form of this disorder the child appears to be comfortable with a few close friends, but avoids the wider social sphere. The capacity to function at home and at school is not impaired. When the disorder is moderately severe, the child has difficulty participating in school work activities and appears to be alienated from peers.

In the severe clinical form of avoidant disorder, there is major interference with social participation, peer relationships, and classwork. These children, although capable of these relationships and activities, actively avoid them. They are delayed in adolescent psychosexual exploratory activity, and characteristically they are late in developing heterosexual relationships. Children with schizoid disorder, on the other hand, appear to have a basic impairment of the capacity for relationships.

Although traits of withdrawal are common in children of all ages, the fearful withdrawal required for the diagnosis of avoidant disorder is not commonly seen as an isolated symptom in child guidance clinics, and there are no data that allow a reliable estimate of its prevalence. Clinical studies suggest that avoidant disorder is more frequent in girls than in boys, but this too has yet to be verified by systematic study. Werkman (1980) believes that this reflects a socially sanctioned role of passivity and withdrawal in girls, whereas boys are generally expected to be more independent and aggressive.

Antecedents and Causes

Factors within the family may be of primary importance in the development of avoidant disorder, and certainly may compound it. First, the child may model his shy and inhibited behavior on that of a parent with similar personality traits, and there may be a transactional effect in that shy, inhibited parents may endorse and reinforce the same attitudes in their children. Second, a child may become reactively avoidant if he is overwhelmed by a dominating or overbearing parent. Third, the scapegoated child who is constantly belittled or devalued within the family may be prone to avoidant disorder.

Environmental factors also appear to be important in the development of this disorder, although there are no good studies to substantiate such a clinical impression. Werkman (1980) believes that devastating losses early in childhood or sexual trauma may predispose children to this reaction. He notes that children who are handicapped with chronic medical conditions seem prone to avoidant disorder, and he speculates that this may result from a lack of social interactions with peers.

The traits associated with avoidant disorder are generally at one end of the spectrum among the temperamental traits characterized by Chess (1979)—approach-withdrawal, adaptive-nonadaptive, and activity-passivity. Thus some dimensions of this disorder are regarded as being temperamental in nature. The evidence linking this to biologic factors is scanty. Bonding, community and socioeconomic factors, and acquired biologic factors do not appear to have a role in the etiology of this disorder.

Prognosis and Natural History

Avoidant disorder has major social consequences for the child, for its presence leads to the child's failure to develop social relationships outside the family, and it is associated with poor self-confidence and feelings of isolation and depression. The pattern may prove hard to break, particularly if the parents resist attempts at growth and development by the child, and it may not begin to be resolved until a substantial change takes place in the child's life outside his family that gives an opportunity to develop some measure of independence. Such a situation may be a geographic move or a move to a new school (Werkman, 1980).

The prognosis in this disorder is not known. The impression of clinicians is that as adults, children who had avoidant disorder have a personality that shows traits of shyness and withdrawal, but they appear to retain the capacity to function in society and do not appear to be prone to major psychiatric illness.

Guidelines for Management

The treatment of avoidant disorder traditionally has centered on individual and family psychotherapy. In individual therapy the child is encouraged to show initiative and assertiveness, and his inhibitions are explored. Assertive and appropriate behavior outside the office is encouraged and is rewarded by positive comments. In family therapy or parental counseling, the focus is on helping the child disentangle himself from a family that is supporting immature behavior. The parents are encouraged to reward initiative in the child, particularly when the child encounters situations that normally cause anxiety. The child is encouraged to actually spend time away from home, or to participate in classroom and other peer activities without the presence of the parent, as the parents' own shyness and inhibition normally make them anxious under the same conditions.

To support further the child's faltering self-esteem, it can be important to foster skill development. This generally works best when it is possible to start with one of the child's own interests and develop it into a skill; for example, the child who is interested in swimming may be encouraged to join a team, or the child who is interested in reading may be encouraged to do creative writing. A feeling of competence in skill areas reinforces the child's feelings that he can cope with other feared social interactions in his life, as well as providing new social opportunities.

Psychotropic medications are rarely indicated in avoidant disorders unless the disorder is actually a symptom of severe depression, in which case a specific antidepressant may be necessary. Minor tranquilizers should be avoided because their sedative effect may potentiate passivity and withdrawal. A night-time sedative such as diphenhydramine occasionally may be indicated, but only if the sleep disturbance is major and is interfering with the child's life (White, 1980).

SEPARATION ANXIETY DISORDER

Definition of the Problem and Common Manifestations

The essential features of separation anxiety disorder include an exaggerated distress at separation from parents, home, or other familiar surroundings, persistent morbid fears and preoccupations that danger will befall the child or a close relative, and phobias of animals, monsters, kidnapping, dying, and accidents. These children also have nightmares and difficulty in falling asleep, difficulty in being away from home, and a yearning to return home.

Associated features include fear of the dark, demandingness, and intrusiveness. Children may also insist on constant attention and complain of being unloved. They may have a death wish, be sad, cry a lot, or show depressive symptoms if a separation occurs or becomes imminent, and they may show psychophysiologic symptoms, such as stomach aches, dizziness, or palpitations, and manifest travel symptoms of motion sickness and irritability. These children are anxious in new situations, and they may also become anxious about changes in their environment, particularly geographic moves. They may long for previous residences and familiar situations.

To qualify as the clinical disorder, the features of separation anxiety disorder must be present for more than two weeks. It can affect children and adolescents of any age. Mild separation anxiety is manifested by concern about separation but not of a degree that seriously impairs the child's ability to cope with, and to function in, new situations. Moderate separation anxiety is present when there are panic reactions, but the child retains the capacity to function to some extent in new situations. Severe (or clinical) separation anxiety disorder is present when the child panics and acts on his fears, for example, to actually avoid school.

Separation anxiety disorder may be more difficult to identify in adolescents because they are more cautious about overt displays of separation anxiety. However, their behavior reveals the nature and depth of their fears. They are reluctant to be away from home, and they depend excessively on their parents or their family to do things around and outside the home that are usually normal activities for adolescents.

The clinical disorder must be distinguished from separation anxiety, which is the manifestation of symptoms of distress when the child is separated from his caretakers (mother or family). The latter separation anxiety is such a ubiquitous phenomenon that it is regarded as a normal concomitant of child development.

In fact, from the psychodynamic perspective, separation anxiety is thought to be crucial to the development of mature ego structures and to the acquisition of a sense of trust and confidence. In gaining control over their separation anxiety, children are believed to acquire valuable skills that prepare them to handle anxiety in other situations later. For example, infants play an ingenious game that may help them gain control over separation

anxiety in a symbolic way. An infant will drop an object somewhere out of sight and have the parent or somebody else retrieve it. In this way the infant probably reassures himself of the vigilance and attention of caretakers and also gains a sense of control over things (particularly parents) that disappear from view. Thus the diagnosis of separation anxiety as a disorder is made only when the behavior is not a part of an age-appropriate developmental process.

Since stringent criteria for separation anxiety disorder have been introduced only recently, there are no reliable data relating to the prevalence of this group of anxieties, with the possible exception of school phobia.

However, not all children with a school phobia have a separation anxiety disorder. Hersov (1977) found that a significant proportion of school phobic children fear some aspect of the school situation rather than having fears related to leaving home and separating from parents. Furthermore, in older children and adolescents, school phobia is more likely to be secondary to depression or schizophrenia.

The sex ratio for separation anxiety disorder is about equal, but in the case of school phobia or refusal, girls may outnumber boys. The disorder may be found in all socioeconomic classes.

Antecedents and Causes

Separation anxiety disorder may be more prevalent in families in which one or both parents have separation conflicts themselves. These parents tend to overprotect their young children from common normal social experiences that they fear may be associated with some risk, and they exaggerate current dangers and future problems. Furthermore, as well as being inculcated with fearsome implications of life outside the home, the child observes the manifest anxiety of his parents to relatively normal social situations. Thus an interactional system may develop between the parents and the child that tends to further compound these anxieties.

The psychodynamic understanding of the persistence of separation anxieties is that the child has been exposed to a separation experience in reality or in fantasy during his early years, and he retains an unconscious recollection of this experience and seeks to avoid it by protecting himself from further separations. At a deeper level it is believed that the child has not successfully resolved conflict over the original separation from the mother in infancy. Whether these dynamics can operate in the absence of the parental attitudes already described has not been explored.

Although the mutually reinforcing nature of the interactions between fearful parents and separation-anxious children are compelling and lead one to believe that this is the major determinant of separation anxieties, there is evidence from the large twin study by Goldsmith and Gottesman (1981) that dependency and separation distress may have some genetic roots as well. They found a significantly higher concordance for dependency at four years and seven years, and separation distress at seven years, in monozygotic than in dizygotic twin pairs.

Socioeconomic factors and acquired biologic factors have not been implicated in the etiology of this disorder.

Prognosis and Natural History

There are no studies of the prognosis in separation anxiety disorder. However, the subgroup of school phobia has been studied much more carefully, but because not all children with school phobia show separation anxiety, the results cannot be generalized. Nevertheless clinical reports indicate that for young children in whom separation anxiety is the major feature, the response to treatment is excellent; in some reports the response has been reported as high as 100 per cent (Werkman, 1980). Studies of older children and adolescents are not as encouraging, but this may be because most data refer to school phobic children, many of whom do not have a primary separation conflict.

Clinical evidence suggests that separation anxiety disorder does not lead to significant adult psychiatric disorder, although these children probably remain somewhat anxious and fearful individuals. Michael et al. (1957) studied a large group of shy and withdrawn children, a number of whom most likely had separation anxiety disorder, and found a very low prevalence of major psychiatric disturbance in adulthood.

In spite of the deficiencies in the objective treatment and follow-up data, the rich clinical and anecdotal reports allow some general statements to be made. Werkman (1980) regards the prognosis in these children as fair to good. Younger children appear to do particularly well in therapy, but milder separation anxieties may persist for years, or the clinical disorder may be reactivated by stresses, particularly if they involve separations in fact or in fantasy.

The prognosis is generally better when separation anxiety disorder manifests itself in younger children. This may be attributable to the availability of the mother and family, who provide the opportunity for the completion of the development task that seems to be arrested—the development of autonomy. Children who first show separation anxiety disorder in later childhood or adolescence do not have the same opportunity for parental support of the type that is developmentally appropriate, because the normal parental and societal expectations for children of this age

include a greater degree of independence, and parents and society may not be willing to provide the type of support necessary to complete the developmental task. For similar reasons, if a principal caretaking figure such as the mother or other family member is not available, the prognosis may be worse.

In the case of school phobic children, it appears that the younger ones tend to do well (Rodriquez et al., 1959). Clinical evidence suggests that these are the subgroup of school phobic children in whom separation anxiety is central to the disorder.

Guidelines for Management

Perhaps the most critical component of the treatment of separation anxiety disorder is that at some point the child must actually make the feared separation (Hersov, 1977). This principle is central to both the behavioral and the dynamic psychotherapeutic treatment of these children.

When the condition is not severe, it may resolve quickly simply with parental counseling in which the parents are encouraged, first, to allow the child to be more independent and, second, to provide the child with appropriate love and support without being overprotective. Useful ways of introducing this support include giving the child responsibility for chores and involving him, along with the parents, in outdoor or extrafamily activities from which the parents are encouraged to gradually withdraw.

When it is evident that the parents' own anxieties and conflicts are compounding those of the child, family therapy may be necessary. In addition to exploring the origins of the parental separation anxieties, structural changes within the family can also be helpful. For example, the child can be encouraged to sleep alone rather than sleeping with the parents or with siblings, and he should be given more responsibility around the house. At times, concrete limits on clinging, demanding, and overdependent behavior may be necessary. In addition, the clinician should be alert to the possibility that one of the parents (usually the mother) is clinging to the child, partly in response to her loneliness and sadness and her alienation from her spouse.

Behavior therapy simply formalizes and organizes some of these techniques. Programs generally entail gradual approach procedures in which the feared object or situation is first approached by the child in fantasy, for example, by drawing it or talking about it, and then the child begins to actually approach the real situation. Separation anxiety related school phobia in particular responds well to this type of regimen.

Individual dynamic psychotherapy has also been used for separation anxiety disorder. It is generally a more prolonged form of treatment than family or individual therapy, and this should be weighed against the apparent successes of parental counseling, family therapy, and behavioral gradual approach techniques. The task is for the child to develop a sense of independence and autonomy, and a risk in the traditional psychotherapeutic approach may be that it provides regressive opportunities. In cases in which psychotherapy has been of great value, it appears that the intervention has actually combined a behavior therapy approach with the more traditional psychotherapeutic work; thus it is difficult to be sure which of the two has been effective.

The tricyclic antidepressants, in particular, imipramine, have proved effective in small trials with preschool children showing separation anxiety related sleep disturbances and in young school phobics (Werry, 1977). However, the minor tranquilizers have a limited place in the management of separation anxiety disorder; Werry (1977) has warned that acute and situational anxiety, as is seen in school phobia, is a poor indication for their use because the minor tranquilizers may impair general functioning between periods of manifest distress.

OVERANXIOUS DISORDER

Definition of the Problem and Common Manifestations

The essential features of overanxious disorder are excessive worrying and fearful behavior that is not related to a specific situation or object or to recent stress. These children tend to worry about potential injury, future examinations, inclusion in peer activities, and meeting expectations and deadlines. They are overconcerned with competence and with the criticisms of others. They may show psychophysiologic symptoms, such as a lump in the throat, nausea, stomach aches, dizziness, and nervousness, and they may have difficulty in falling asleep because of their anxieties (initial insomnia).

Associated features include obsessional self-doubt, perfectionism, excessive conformity and approval seeking behavior, restlessness, nail biting, hair pulling, and thumb sucking. These children often appear to be overmature and show precocious ego development in some areas, particularly in their capacity to show off intellectual skills. They also may evidence a great wish to please, but they are very self-critical and are rarely satisfied with their own performance. They may try to control others by overtalkativeness and a wish to demonstrate their abilities. They may show exaggerated responses to pain, illness, and handicap, and they may be accident prone.

To meet the diagnostic criteria, the symptoms of overanxious disorder must be present in a child three years old or older for more than six months. Overanxious disorder of mild intensity does not impair functioning in school, and these children appear to be able to master life fairly effectively and without hindrance by anxiety. When they are moderately impaired, they show chronic worries that interfere with school, particularly at times of peak stress, such as examinations, but they appear to be able to function with their peers. When overanxious disorder is of clinical intensity, the children show such great anxiety that school and social functioning is manifestly impaired by the anxiety.

The prevalence of overanxious disorder is not known. Werkman (1980) cites clinical evidence that such children tend to come from small, upper socioeconomic class families and tend to be first childen and that their families show an unusual concern for performance in many areas. Furthermore, the disorder is believed to be more common in girls than in boys, although this is also an unsubstantiated clinical impression.

Antecedents and Causes

Acute or chronic environmental stress may render a child prone to overanxious disorder. Traumatic losses and other acute environmental stresses such as natural disasters or family catastrophes can lead to manifest anxiety, as can chronic stress such as marital discord or parental aggression. Their effects depend on the intensity, unexpectedness, and significance of these events in relation to current conflicts.

In a similar way to children with separation anxiety disorder and avoidant disorder, the overanxious child may model his attitudes and responses on those of chronically anxious and dependent parents (Eisenberg, 1958), and again an interactional effect can be seen. The contagiousness of parental anxieties depends on the child's age, degree of dependency, and closeness to the parent.

Psychodynamic issues in these children are thought to revolve around competitive struggles, initially with the parents in the oedipal conflict and later with sibling rivalry, and the wish to excel in comparison to peers. These conflicts are of a more mature variety than the separation conflicts seen in separation anxiety disorder and avoidant disorder. The controlling issues are similar in many ways to those seen in obsessive-compulsive children; the difference is that overanxious children are prey to their anxieties, rather than exerting rigid ritualized defenses against them.

Certain temperamental traits may set the stage for the development of overanxious disorder. Hersov believes that anxiety prone children may have been different from their siblings since early childhood in terms of their sensitivity and over-reaction to stress and their tendency to be worried over new situations. Chess (1979) sees this as a temperamental variant; however, studies of young twins have not demonstrated temperamental traits related to anxiety proneness to be genetically transmitted (Goldsmith and Gottesman, 1981). Acquired biologic factors do not appear to have an etiologic role in this disorder.

Prognosis and Natural History

The course of overanxious disorder is generally benign, and it is unusual for the disorder to be so profound that it leads to an inability to meet the minimal demands of school, home, and social life, and thus it is the least socially incapacitating of the disorders of oversocialization. However, there is suggestive evidence that it may lead into adult generalized anxiety disorder, or as adults these individuals may remain vulnerable to anxiety reactions when under stress.

The prognosis with treatment is believed to be good. These children do well in individual therapy or in family therapy. This may be because they show a more mature ego than other emotionally disordered children, and they have a greater capacity for insight. There may also be an element of overcompliance, as they wish to please the therapist (Werkman, 1980). However, it may also simply be that overanxious disorder resolves naturally, and that therapy does not impede this process.

Guidelines for Management

Hersov (1977) has suggested that children with overanxious disorder respond well to intervention that is primarily supportive, open, and structured, and he emphasizes the effectiveness of these simple techniques. In individual therapy the issues tend to focus on control. The children often try to control the therapy situation with highly organized intellectual games and discussions, but spontaneity with toys and play material can go a long way to freeing up inhibitions. The therapist must be accepting but firm and consistent. With this sort of approach such children can be helped to see that there is more fun in life than their perfectionistic, achievement driven adjustment would suggest. Similar principles should be used in the family therapy of these anxious and inhibited people.

Behavior therapy has also been used with effect in these children. Some children are sufficiently mature to be able to undergo relaxation training, but even with less mature children, components of the anxiety that appear to be situation specific can be confronted and brought under control by gradual approach techniques (Firestone et al., 1978).

Medications occasionally can be useful in these children. If they have marked initial insomnia, a hypnotic such as diphenhydramine or a short acting minor tranquilizer such as triazolam may be useful. Minor tranquilizers may be required occasionally during the day, but they should be used sparingly and only when overt anxiety is of such intensity that it seriously impairs social and academic functioning (White, 1980).

DISORDERS OF UNDERSOCIALIZATION

The major difficulty for children resulting from a failure of socialization is their willingness or inability to confine their behavior to acceptable limits as judged by adults such as parents or teachers or by their peers. The diagnostic category involved is that of conduct disorder (American Psychiatric Association, 1980).

CONDUCT DISORDER

Definition of the Problem and Common Manifestations

Overview

The essential feature of a conduct disorder is "a repetitive and persistent pattern of conduct in which either the basic rights of others or major age appropriate societal norms or rules are violated (see also Chapter 41). The conduct is more serious than the ordinary mischief and pranks of children and adolescents" (American Psychiatric Association, 1980, p. 45). The *Diagnostic and Statistical Manual of Mental Disorders* subdivides conduct disorder into four specific subtypes on the basis of the presence or absence of adequate social bonds and the presence or absence of aggressive antisocial behavior (Table 36–3). The four subtypes are undersocialized aggressive, undersocialized nonaggressive, socialized aggressive, and socialized nonaggressive.

The undersocialized subtypes show evidence of an inability to establish warm affectionate relationships with others. They exhibit a superficiality and a lack of concern for other persons, both peers and family. Appropriate guilt and remorse are generally lacking.

The socialized subtypes, on the other hand, are characterized by an ability to make social attachments to others. These children show clear evidence of family or peer group loyalty. However, they too can be callous and manipulative and show a lack of guilt with persons whom they see as being "outsiders."

The central characteristic of the aggressive subtype is a chronic repetitive pattern of aggressive

Table 36–3. DISORDERS OF UNDERSOCIALIZATION: DEFINITION AND CLINICAL FEATURES*

Conduct Disorder	Undersocialized Aggressive (312.00)	Undersocialized Nonaggressive (312.00)	Socialized Aggressive (312.23)	Socialized Nonaggressive (312.21)
Essential features	Failure to establish warm satisfactory relationships with others, including peers Repetitive and persistent pattern of antisocial behavior, which includes either physical violence against persons or property or thefts outside home involving confrontation with victim	Failure to establish warm satisfactory relationships with others, including peers Repetitive and persistent pattern of nonaggressive antisocial behavior in which societal norms or rules are broken and basic rights of others are violated	Evidence of satisfactory relationships with others, including peers Repetitive and persistent pattern of antisocial behavior, which includes either physical violence against persons or property or thefts outside home involving confrontation with victim	Evidence of satisfactory relationships with others, including peers Repetitive and persistent patterns of nonaggressive antisocial behavior in which societal norms or rules are broken and basic rights of others are violated
Associated features	Behavior difficulties at home and at school Early smoking, drinking, drug abuse, and sexual activity Poor school performance Poor self-esteem Irritability and low frustration tolerance	Behavior difficulties at home and at school, such as running away and truancy Shy or withdrawn Poor school performance Poor self-esteem Irritability and low frustration tolerance	Behavior difficulties at home and at school Early smoking, drinking, drug abuse, and sexual activity Poor school performance Involvement with police Irritability and low frustration tolerance	Behavior difficulties at home and at school, such as running away and truancy Poor school performance Poor self-esteem Irritability and low frustration tolerance
Minimal duration of symptoms	Six months	Six months	Six months	Six months

*Adapted from American Psychiatric Association: Diagnostic and Statistical Manual of Mental Disorders. Ed. 3. Washington, D.C., American Psychiatric Association, 1980.

conduct in which the rights of others are violated by either physical violence against persons or thefts outside the home involving confrontation with a victim.

The nonaggressive subtypes are characterized by both an absence of physical violence against persons and confrontation with a victim in cases of robbery. The antisocial behavior in this case centers on the violation of rules or the basic rights of others. Truancy, running away, lying in and out of the home, and stealing not involving confrontation with a victim are common symptoms.

The validity of this method of identifying subcategories of conduct disorder is controversial (American Psychiatric Association, 1980; Rutter, 1978). The one essential criterion for the establishment of the validity (accuracy) of a classification scheme of childhood psychiatric disorders is that the categories must be shown to differ on some variable or variables other than the symptoms that define them (Rutter, 1978). These variables could include ones associated with etiology, course, response to treatment, or some other aspect of the disorder.

The classification of conduct disorder just presented was first suggested following the factor analysis of clinic records. There is some evidence that the subgroups, primarily the undersocialized

and socialized subtypes, differ in family background variables and long term prognosis (Henn et al., 1980). These findings do provide some support for the validity of this subcategorization of conduct disorder. However, it should be noted that all studies addressing this classification scheme have shown considerable overlap between the groups, with many subjects unable to be placed in either category (Rutter, 1978). It is not clear yet whether subdividing conduct disordered children on the basis of personal relationships and loyalties identifies separate groups of children or whether it is identifying primarily different degrees of disturbance among children with the same disorder. Other categorization schemes for these children have been based on the presence or absence of marital discord or disharmony or the severity and variety of the antisocial symptoms (American Psychiatric Association, 1980; Rutter, 1978).

The major point is that there is no disagreement that conduct disordered children are a heterogeneous group, but there is disagreement about how the children with this disorder should be subclassified. The scheme presented in the *Diagnostic and Statistical Manual* is a promising one, perhaps the most promising one, but its validity and usefulness have not been clearly established. Therefore, the discussion of conduct disorder will not consider these subgroupings individually but rather will deal with this group of children as a single entity.

Other terms that are closely associated with the designation of conduct disorder will now be considered. They include "antisocial symptoms or behavior," "antisocial children," "aggressive children," and "delinquent." "Antisocial symptoms or behavior" refers to behavior patterns that are judged by members of the community to be outside the accepted social norms. Such socially disapproved behavior includes fighting, stealing, vandalism, and truancy. Their presence is determined by a parent, teacher, other adult, or the child himself and provides the data on which the diagnosis of conduct disorder is based.

"Antisocial children" refers in general to the same group of children as those designated by the title conduct disorder. The advantage of the *Diagnostic and Statistical Manual's* diagnostic scheme is that it spells out specific inclusion and exclusion criteria, whereas in the past, criteria for the diagnosis of antisocial or conduct disordered children have varied. For instance, Robins' (1966) criteria for the antisocial child takes into account the number of episodes of antisocial behavior and the different types of antisocial behavior. On the other hand, Rutter's (1981) diagnosis of conduct disorder has been based on a clinical interview or a validated teacher questionnaire.

"Aggressive children" is a more vague designation than either conduct disorder or antisocial children. The latter categories would include aggressive children, but many children termed aggressive would not yet have fulfilled sufficient criteria to warrant a diagnosis of antisocial or conduct disorder.

The term "delinquent" is usually reserved for conduct disordered children who are caught by the police and processed through the court system. The evidence suggests that delinquent children, thus defined, represent a more severely disturbed subgroup of children with conduct disorders. This is to say that youths who become involved with the police and the courts are the ones with the most numerous episodes of severe antisocial behavior (West and Farrington, 1977).

One last point about diagnosis. The diagnosis of children's emotional or behavioral disorders is usually based on reports of parents or teachers. The major problem here is that the teachers and parents in the majority of cases do not identify the same children as being disturbed (Rutter et al., 1970). It is not known which source of data identifies a group of children with the most valid diagnosis. Certainly work needs to be done in this area in all the childhood psychiatric diagnoses. It would be interesting, for instance, to compare children diagnosed as having conduct disorder by reliable criteria on the basis of teacher reports alone, parent reports alone, and a combination of both.

Differential Diagnosis

Conduct disordered children must be differentiated from children who exhibit sporadic antisocial symptoms. It is clear that most children display some antisocial behavior at some time (Gold, 1966). It is only when the behavior becomes frequent and severe and part of a persistent pattern that the diagnosis of conduct disorder is warranted. It is obvious then that the point at which the frequency and severity of a child's antisocial behavior result in a label of conduct disorder is somewhat arbitrary. It should be noted also that children with other psychiatric syndromes can have outbursts of temper and destructive behavior. For instance, children who are primarily depressed or primarily shy or even those with a more serious diagnosis such as borderline psychotic state may have episodes of aggressive negativistic behavior.

It should always be kept in mind that a sizable group of children referred for aggressive antisocial behavior will not qualify for a diagnosis of conduct disorder. In some the pattern of behavior may be judged not to be abnormal; in others it may be seen as being abnormal but not yet severe enough to warrant a formal diagnosis; and in still others the episodes of antisocial behavior will appear to be part of another disorder.

Overlapping Disorders

Emotional Disorders. The differentiation between children with conduct disorder and those with emotional disorder characterized by such symptoms as anxiety, sadness, fear, worry, and misery is strongly supported in the child psychiatric literature (Rutter, 1978). (The term emotional disorder would include children considered to have disorders of oversocialization.) These two broad groups of children differ in terms of sex ratio, association of the disorder with family discord and with reading difficulties, persistence of problems into adolescence, short term response to treatment, long term outcome, and degree of association in adulthood with severe psychiatric illness (Rutter, 1978; Table 36–4). It should be noted that in a substantial proportion of children there is a mixed pattern. It is common for children with conduct disorder and delinquent youths to show, in addition to antisocial behavior, symptoms such as fearfulness, worry, sadness, and misery (Offord, 1982; Robins, 1966). It is not known whether conduct disordered children with these added symptoms differ from conduct disordered children without them in important respects such as response to treatment and long term prognosis (Graham, 1974).

Table 36–4. VARIABLES DIFFERENTIATING CHILDREN WITH CONDUCT (UNDERSOCIALIZATION) DISORDER FROM CHILDREN WITH EMOTIONAL (OVERSOCIALIZATION) DISORDER*

Variable	Conduct Disorder	Emotional Disorder
Sex ratio	Boys > girls	Boys = girls
Family discord	Positive association	No association
Reading difficulties	Positive association	No association
Persistence of problems into adolescence	More	Less
Short term response to treatment	Worse	Better
Long term outcome	Worse	Better
Degree of association in adulthood with severe psychiatric illness	Greater	Less

*Adapted from Rutter, M.: Diagnostic validity in child psychiatry. Adv. Biol. Psychiatry, 2:2, 1978.

Hyperactivity (Attention Deficit Disorder with Hyperactivity).   The degree and the importance of the overlap between conduct disordered children and hyperactive children have not been satisfactorily worked out (American Psychiatric Association, 1980; see also Chapters 38 and 41). A major problem is that in many studies a clear-cut differentiation between these two syndromes has not been made (Offord, 1982; Werry, 1980). Paternite et al. (1976) have pressed for a separation in the hyperkinetic syndrome of the so-called primary symptoms such as hyperactivity, fidgetiness, and inattention from the secondary symptoms, for example, aggressive interpersonal behavior. These authors hypothesize that the primary symptomatology may occur more or less at random and that the secondary symptoms may emerge in certain children through the action of adverse environmental factors. More recent work has reinforced the value of distinguishing between, for instance, children who are exclusively hyperactive without aggression and those with aggression and hyperactivity (Langhorne et al., 1979). These two groups were found to differ in terms of important measures at initial referral, in treatment with methylphenidate, and at five year follow-up examinations (Langhorne et al., 1979). In another study delinquent boys who were also described as being hyperactive differed from nonhyperactive delinquents by having a lower birth weight than their brothers and than the nonhyperactive delinquents and possibly more delivery and postnatal complications than nonhyperactive delinquents (Offord, 1982). In addition, the onset of delinquency was earlier in the hyperactive group with many antisocial symptoms, both of which indicate a poor adult prognosis.

A useful clinical and research strategy is to separate the symptoms of hyperactivity from the symptoms of conduct disorder. The *Diagnostic and Statistical Manual* classification allows this differentiation to be made. Here hyperactive children are included in the category of attention deficit disorder with hyperactivity. The major parameters of this diagnosis are inattention, impulsivity, and hyperactivity. These parameters do not overlap with the criteria employed in the diagnosis of conduct disorder. If these more stringent criteria were to be applied strictly in the case of these two disorders, a basis would be present for determining the extent to which the resulting differentiation is a valid one.

Educational Retardation.   The school performance of antisocial or delinquent children is poor (Offord, 1982; Robins, 1966; Rutter at al., 1970; see also Chapter 38). Their difficulty in performing satisfactorily in school is not accounted for primarily by a low IQ (Offord, 1982; Rutter et al., 1970). A striking characteristic of children with conduct disorder is the increased frequency of severe reading difficulties they show. In the Isle of Wight study it was noted that one-third of the children with conduct disorder had a reading level at least 28 months behind the level that would be expected on the basis of the IQ (Rutter et al., 1970). Conversely, one-third of the children with reading retardation, as defined, exhibited clinically significant antisocial behavior. Few differences in symptomatology were found between the good and the poor readers on the Isle of Wight; the symptomatology of delinquents with and without early school failure was also remarkably similar (Offord, 1982).

Summary.   Figure 36–5 illustrates important aspects of the relationship among conduct disorder, hyperactivity, educational retardation, and emotional disorder. The precise degree of overlap among these conditions is not known and would depend, of course, on the definition used. Figure 36–5, however, illustrates in general terms most of the facets of our present understanding of this issue. It emphasizes the following points: children with conduct disorder overlap significantly with children with hyperactivity and children with educational retardation. The latter two conditions, in turn, overlap with each other. Lastly, most children with conduct disorder have "emotional" symptoms, and in some the symptom pattern is severe enough to justify the diagnosis of emotional disorder in addition to the antisocial diagnosis.

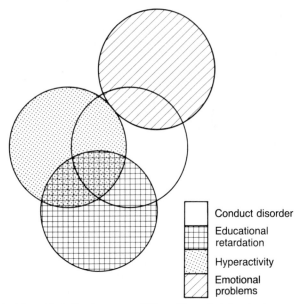

	Conduct disorder
	Educational retardation
	Hyperactivity
	Emotional problems

Figure 36–5. Overlap among children with conduct disorder, hyperactivity, educational retardation, and emotional problems.

What Figure 36–5 does not portray is the fact that children with emotional disorder can overlap with both children with educational retardation and those with hyperactivity. Again the exact degree of overlap in these instances is unknown.

Future work should focus on determining the precise relationships among these conditions using reliable diagnostic criteria. Furthermore, for each condition it will be important to learn the extent to which children with the disorder alone compared to those who have the disorder as well as another disorder or disorders differ in terms of such factors as etiology, course, and response to treatment.

Epidemiology

Children with conduct disorder compose the largest single category of emotionally disturbed youth. In the Isle of Wight study, for instance, almost three-quarters of all the boys who were considered to be psychiatrically disturbed were diagnosed as having this disorder, as were one-third of the psychiatrically disturbed girls (Rutter et al., 1970). The prevalence rates in 10 and 11 year olds on the Isle of Wight was 6.0 per cent for the boys and 1.6 per cent for the girls. Similar data from an inner London borough revealed that the prevalence rates doubled in a poor inner city area (Rutter, 1981). The frequency has always been three or four times greater in boys than in girls. This is probably the result of a combination of factors, including the tendency for boys to be more aggressive and more combative than girls even in the preschool years, their increased susceptibility

to physical stresses and to most types of psychosocial trauma, and perhaps cultural factors that result in less critical judgments of antisocial behavior in boys as compared to girls (Hutt, 1972; Rutter, 1979b). In addition, it should be noted that the onset of the disorder occurs about two or three years later in girls than in boys—ages 12 and 13 versus 9 and 10 years.

Not only does conduct disorder represent the most common group of disturbed children in community surveys, but these children are also the most frequent to come for treatment in children's mental health clinics (Robins, 1979). Many, of course, are seen in settings other than mental health ones, including school clinics (referred possibly because of academic difficulties), and by pediatricians and pediatric neurologists (referred possibly because of their hyperactivity) and lastly by the police and the courts.

In summary, no matter what else may be said about conduct disordered children, they are the most numerous of psychiatrically disturbed children, treated or untreated.

Antecedents and Causes

Psychosocial Factors

Early Experiences and Bonding. Bowlby (1951) in his landmark monograph emphasized the importance to the growth of the child of a warm continuous relationship with a caring adult, usually the mother. The absence of this relationship was termed maternal deprivation, the occurrence of which could result in long lasting deleterious effects in the developing human being.

Yarrow (1961) pointed out that the term maternal deprivation was a broad descriptive one encompassing several different patterns of maternal care, including institutionalization, separations from a mother or a mother substitute, multiple mothering in which there was no continuous person performing the major mothering functions, and, lastly, disturbances in the quality of mothering. Further work has refined to a greater degree the heterogeneous range of experiences subsumed under maternal deprivation and the mechanisms by which these separate experiences lead to quite specific outcomes in childhood and adulthood (Rutter, 1979a).

There are two major implications of this work for our understanding of the etiology of conduct disorders. First, the existing evidence is compatible with the view that when the child does not have an opportunity to form attachments or bonds with other human beings in the first three years of life, he is likely to possess a personality characterized by a lack of guilt and an inability to form lasting relationships (Rutter, 1979a). The data suggest that the bond does not have to be with a particular person such as the mother. Thus the childhood antecedent of the so-called "affectionless psychopath" appears to reflect a failure of bond formation, not a disruption of existing bonds (Bowlby, 1946).

The second implication centers on the concept of broken homes. There is a strong consistent association in the literature between broken homes and antisocial behavior and delinquency (Rutter and Madge, 1976). The association is much stronger, however, when the homes have been broken by divorce or separation rather than death of a parent (Rutter and Madge, 1976). The evidence suggests that it is not the break-up of the home itself that promotes antisocial behavior but the family discord and disharmony that precede the break (Rutter, 1979a). For instance, parental discord is firmly associated with antisocial behavior in children even when the home is unbroken. Children with conduct disorders come from families in which the parents constantly disagree and quarrel, and any break-up of the families that occurs bears only an incidental relationship to the etiology of antisocial behavior.

The mechanisms by which early failure in bonding and marital discord lead to increased rates of antisocial behavior in children have not been worked out. Winnicott (1958a) suggests, for instance, that young children who have experienced a failure in mothering, or in caretaking, if you like, in the last half of the first year of life may exhibit what he terms an "antisocial tendency." If this tendency is strong enough, it could manifest itself as antisocial behavior. Winnicott (1958a) theorizes that the child's push toward antisocial behavior arises from his perception that the caretaking environment has failed him early in his life, and he is driven, usually unconsciously, to compel the environment to make up the deficit. The antisocial behavior should be understood as a manifestation of the child's hope that if he stirs up the environment, it can make up to him the original deprivation. Obviously evidence of this theoretical construct is lacking. Its importance lies in pointing out the possibility that failures in the child's environment can result in intrapsychic disturbances, which can predispose a child to antisocial behavior.

The mechanisms by which marital discord promotes antisocial behavior in a child could include a resulting lack of supervision or inconsistent discipline because the parents are preoccupied with their own problems; a modeling effect in which the child sees and identifies with adults who are prone to fighting and violent outbursts; scapegoating of the child when the parents avoid dealing with their own problems directly by displacing them onto the child; and a possible genetic factor, which underlies both the quarreling behavior in the parents and the difficulty in impulse control in the child. Unfortunately the validity and relative importance of these mechanisms in producing antisocial behavior are not well documented. There is evidence, however, that the pattern of discipline is related to delinquency and probably acts as a causal factor. In West and Farrington's study (1977), poor parental behavior, especially harsh discipline, rejecting attitudes, and poor supervision, predicted delinquency even after taking into account the behavioral ratings of the children prior to the onset of the delinquency.

Family Factors. Broken homes, marital discord, parental deviance, and large family size have all been associated with antisocial behavior and delinquency. The role of broken homes and marital discord has been discussed in the previous section. One further point should be made in connection with broken homes. There is consistent evidence that the association between broken homes and delinquency is much stronger for female as compared to male delinquents (Offord, 1982). Why this should be so is not clear. It may be that a broken home is more devastating for a girl than for a boy. The disruption could be more painful and upsetting to the girl, who identifies more closely with the need for a settled family in which she can fulfill her role as a wife and mother. Another possibility could be that a boy from a broken home may be more readily and more quickly able to organize a life that is less dependent economically and emotionally on the family. A girl from a broken home, on the other hand, usually finds, especially if she stays with her mother, that her standard of living has been lowered and her chances of moving toward independence from a mother are not as great as those of a boy in the same circumstances.

A third possibility is based on evidence that the spontaneous rate of antisocial behavior in girls is considerably less than it is in boys (Offord, 1982). Thus, a major factor resulting in diminished parental control, such as a broken home, would be central in allowing these naturally occurring antisocial behavior patterns to flourish in girls. In boys, however, in whom these naturally occurring antisocial behavior patterns are much more common, other more minor control mechanisms such as family size may play a more important role in determining whether a boy becomes seriously antisocial. In any case, a comprehensive theory of the causes of antisocial behavior and delinquency must explain the apparently greater importance of broken homes in the etiology of female as compared to male delinquency.

Parental deviance, including particularly criminality and alcoholism, and severe psychiatric impairment are found much more commonly in the families of antisocial children compared to age matched control subjects (Offord, 1982). As with marital discord, the possible reasons that these parental characteristics are associated with increased rates of antisocial behavior can be listed theoretically, but no ranking of their importance can be identified for antisocial children in general or for particular subgroups.

Large family size has been found repeatedly to be associated with antisocial and delinquent behavior in boys but not in girls (Farrington, 1977; Offord, 1982; Rutter et al., 1970). Two major explanations for this association have been advanced. The first might be termed "inadequate family resources."Overcrowding, reduced parental supervision, and increased stress on the mother have been suggested as possible mediating mechanisms for the sibship size effect (Rutter and Madge, 1976). All these factors depend in part on a common central element: the same parents working with the same resources cannot give as much time, attention, or supervision to their children when the family is large as they could if the family were smaller.

A second explanation put forth centers on the concept of contagion (Robins et al., 1975). When one boy in a family becomes antisocial, the probability that others will be affected is increased. The more boys in a family, the more likely it is that at least one will be antisocial. Hence, boys in large families compared to boys in smaller families are at increased risk for antisocial behavior through their association with an already affected brother. The result, according to this reasoning, would be that large families would have disproportionately more antisocial boys than small families.

Recent work has revealed that the larger sibship in male delinquents compared to control families was accounted for exclusively by an excess of boys (Offord, 1982). Furthermore, among the families of delinquents, the more boys there were in families holding the number of girls constant, the greater the degree of antisocial behavior among the males. Conversely the more girls there were in the sibships, holding the number of brothers constant, the less the antisocial behavior among the boys.

These results favor a "contagion" explanation for the excess of antisocial boys in large families. However, two amendments now must be added. First, the findings suggest that boys interacting together without a strong female presence are at increased risk for antisocial behavior. This effect, the so-called male potentiation of antisocial behavior, does not depend so much on contagion, in which one boy communicates his delinquency to another, as it does on an interaction in which boys respond to each other in ways that promote antisocial behavior among them.

The second amendment to the contagion explanation centers on the role of girls. They appear to have a suppressive effect on antisocial behavior among boys. Neither of these effects, male potentiation and female suppression, is found in the siblings of female delinquents. The reason probably lies in the belief that the factors producing antisocial behavior in girls are stronger than in boys, so that relatively minor control mechanisms, such as male potentiation and female suppression, are masked.

The importance of these findings, if replicated, lies in the implications they may have for grouping children, particularly boys, in ways that will keep antisocial behavior in check. Since antisocial behavior is so widespread, especially among boys, it will be important to discover ways that can be widely employed and that are effective in keeping such behavior under reasonable control.

Community and Socioeconomic Factors

Socioeconomic Factors. Conduct disorders are known to be much more common in certain communities than in others. Why, in particular, are inner cities and low socioeconomic class associated with conduct disorder and delinquency? The reasons are not completely understood, but evidence points primarily toward the under-the-roof culture of the child's home. For instance, it has been found that children living in well-functioning homes located in high delinquency areas are about as unlikely to become delinquent as those living in low delinquency areas (Offord, 1982; West and Farrington, 1977). Similarly Rutter (1981) noted that with few exceptions the correlates of child psychiatric disturbance were the same in a poor population (an inner London borough) as in a middle class population (the Isle of Wight). In both groups, psychiatrically disturbed children lived in families with worse marriages and were more likely to have psychiatrically disturbed parents and to come from larger families. All these factors were more common in the poor areas.

Thus it is clear that antisocial and delinquent children are seen more commonly in areas of low socioeconomic class. The family and social factors associated with these disturbances are similar, regardless of the social class of the affected child. The increased prevalence of these disorders, then, in areas of poverty can be accounted for by the increased prevalence of the family circumstances from which these disorders arise. An important corollary is that when these family factors are absent, most children, irrespective of social class, are at low risk of developing a conduct disorder.

It is appropriate to mention at this point that youth, particularly adolescents, report a significant amount of antisocial behavior (Gold, 1966). The rate of such behavior is higher among boys than among girls (Gold, 1966; Offord, 1982). The contribution of this self-report literature is to emphasize the widespread occurrence of some antisocial behavior among almost all children and adolescents. It is only when these widespread and naturally occurring behavior patterns flourish and intensify that the designation of conduct disorder applies. This has implications for both the etiology and the treatment of conduct disorders in children. One does not have to center on explaining why some antisocial behavior occurs in children as much as attempting to understand why these behavior patterns flourish with such severity in some children. On the treatment side, the goal is not to eliminate antisocial behavior as much as it is to keep such behavior under reasonable control.

Lastly, with regard to the self-report literature, it should be noted that there is little evidence of a relationship between the severity of antisocial symptoms and social class. The explanation probably lies in the fact that the self-report scales emphasize trivial offenses. It is when the symptoms become more frequent and more serious that the relationship between antisocial behavior and social class emerges (Hindelang et al., 1979).

The Role of Schools. The next community factor to be discussed is the role of schools. It has been observed for some time that certain schools appear to protect children from serious antisocial behavior and delinquency much better than other schools (Rutter, 1981). The problem with these observations has been that it was never clear to what extent the differences observed among the schools were a result of the school programs themselves or were a reflection of the different kinds of students who attended these schools. In short, the schools with the better results might be attracting the better—that is, the less antisocial and superior academic—students in the first place. More recent work has again revealed marked differences in both academic achievement and rates of psychiatric disturbance in children attending secondary schools in England (Rutter, 1981). In this study, however, pupil selection as a contributing factor to these results has been ruled out more convincingly than ever before. The children had been tested extensively at age 10 before they entered secondary schools. The results, found during their secondary school careers, could not be accounted for by the intake characteristics, both academic and behavioral, of the pupils entering these schools. The factors fostering pupils' success were multiple but did not include several variables that commonly have been believed to be important, namely, resources, size of school, size of classroom, and organizational structure of the school and amount of punishment. The positive factors included ample use of rewards, praise and appreciation, a pleasant and comfortable environment, ample opportunities for children to participate and assume responsibilities, strong academic emphasis, positive models provided by teachers, effective group management in the classroom, and effective school staff organization.

Two points should be emphasized. The difference between the worst and the best schools on academic and behavioral indices were marked; second, it was not just one or two factors alone that appeared to make a difference but a complex of factors. What needs to be done now is to determine whether a "bad" school can be turned into a "good" one. There is no doubt that schools potentially can be important protective factors in limiting the antisocial behavior of children.

Television. Another community factor implicated in producing violence among children is television. There is no dispute that network television in North America is laden with violence. The 10 year average of network dramatic violence (defined as the overt expression of physical force to hurt or kill) has been almost eight incidents per hour involving more than half of all characters (Gerbner, 1978). A United States Public Health Service Report (1972) indicated that the majority of studies in young children have shown a positive relationship between exposure to filmed violence and aggressive behavior. The relationship is strongest among boys, in children with high levels of initial aggression, and in children from low socioeconomic groups. More recent survey studies indicate that for adults and children, violence laden television not only cultivates aggressive tendencies in a minority but in a more general way breeds an exaggerated sense of danger and mistrust (Gerbner, 1978). For instance, "heavy" viewers compared to "light" viewers with similar demographic characteristics overestimated the prevalence of violence and possessed a significantly greater sense of risk and suspicion. The available evidence suggests, then, that television enhances aggressive behavior among some children and, for a minority, may play a significant part in intensifying antisocial behavior (see Chapter 17).

Peer Groups. The last community factor to be mentioned is peer groups. There is evidence that for preschool children the peer group can play a part in the acquisition and maintenance of aggressive behavior (Patterson et al., 1970). Further, several studies (e.g., West and Farrington, 1977) suggest that children who have friendships with peers who themselves are delinquent may be increasing their chances of also being involved in deviant behavior.

Biologic Factors

Inherited Genetic Factors. The available twin and adoption studies suggest a genetic transmission of antisocial behavior and criminality at the adult level. In the eight twin studies they reviewed, Mednick and Hutchings (1978) reported a concordance for criminality of 60 per cent in monozygotic twins and about 30 per cent in dizygotic twins. The most rigorous study, that of Christiansen (1969), revealed a concordance rate of 36 per cent for monozygotic twins and 12.5 per cent for dizygotic twins. Christiansen also pointed out, nevertheless, the importance of a criminal subculture as another factor in the determination of criminality.

Adoption or fostering studies, however, are the stronger methodology, because they separate out the effect of environment, or cultural inheritance. The fostering studies reviewed by Mednick and Hutchings (1978) also give evidence of genetic transmission, for there is a higher rate of antisocial behavior and criminality in the biologic relatives of adoptees identified by antisocial behavior, and a higher incidence of antisocial behavior and criminality in the adopted-away offspring of male and female criminals. This cross fostering study found that the highest rate of antisocial behavior was among the offspring who not only had a biologically criminal father but were also raised by an adopted father who was criminal. Conversely the lowest rate of antisocial behavior was observed in those who had neither a biologic nor an adoptive criminal father. The other two combinations resulted in intermediate rates of antisocial behavior in offspring, and the data suggested that both environmental and genetic attributes are important determinants of criminality. This has been confirmed by Cadoret and Cain (1980). More recently, however, a large Swedish adoption study has failed to confirm the genetic transmission of criminality in their sample (Bohman, 1981).

What may be inherited then? Specific chromosomal variants have been related to aggressive and criminal behavior. The XXY configuration had been reported to be associated with criminal behavior, but recent more rigorous studies have excluded this as a significant, or even likely, cause of criminal behavior, and it is certainly not what is generally passed across generations when a liability to criminal behavior is inherited (Rainer, 1980). However, there is other evidence that genetic liability may be reflected in physiologic variables. In their elegant cross fostering study, Mednick and Hutchings (1978) reported that the criminal sons of noncriminal adoptive fathers showed the slowest electrodermal recovery times of their four groups of subjects. They reported other data suggesting that electrodermal recovery potential is heritable, and they hypothesize that a slow recovery time "might be a characteristic a criminal could pass on to a biological son which (given the proper environment circumstances) could increase the probability of the child's failing to learn adequately to inhibit asocial responses" (Mednick and Hutchings, 1978).

A number of caveats must be borne in mind in reviewing these data. First, the relationship between parental criminality and offspring antisocial behavior may be tenuous at best and also may be substantially environmentally determined. Second, the offspring data refer to adult criminals or sociopaths. Since less than half of antisocial children go on to a criminal career, these data referring to adults cannot be applied directly to children with conduct disorder. In fact, in a study of adolescent adopted-away offspring of criminal or alcoholic parents, Bohman (1981) could not correlate the social maladjustment of biologic criminal parents and their adopted-away offspring. Third, most of the data concern only the sons of criminal fathers. Only Bohman (1981) looked at the effect of the sex of the parents and offspring within the same study. However, the formulation of Mednick and Hutchings remains attractive, particularly since it attempts to account for the effect of a facilitating environment on inherited vulnerability in some cases.

The overlap between hyperactivity and antisocial behavior has already been described. There is some evidence that a small proportion of the parents of hyperactive children may have been hyperactive themselves, but this question has yet to be examined in a sufficiently rigorous manner (Bohman, 1981; Cantwell, 1978). The data relating to the inheritance of activity level as a temperamental trait are more convincing and have been supported in the largest of the numerous twin studies of temperament in which 504 same sexed twin pairs were carefully evaluated at eight months, four years, and seven years (Goldsmith and Gottesman, 1981). Monozygotic twins showed a significantly higher concordance for activity level than did dizygotic twins at the eight month and seven year examinations, but not at four years. Thus, although we know little of the heritability of deviant activity level (hyperactivity), there is evidence that temperamental trait activity may be under genetic influences. As Chess (1979) has pointed out, a high activity level may settle with age or may not present problems in the socialization process for flexible parents, but in some families there is a poor fit between parenting skills and child activity, which leads to tension between parent and child, difficulties with rearing, sometimes scapegoating, and ultimately to behavioral deviance (see also Chapter 10).

Acquired Factors. Children with evidence of frank brain damage are at increased risk for psychiatric disorder. In a population of 10 and 11 year old children, for instance, the prevalence of psychiatric disorder was increased five times in youngsters with cerebral palsy, epilepsy, or some other disorder above the brain stem (Rutter, 1977).

In addition to children with frank brain damage, there is another group who are thought to have "minimal" brain damage or dysfunction but the means are not presently available to identify them accurately. The results of their routine neurologic examinations are normal, and they have no history of definite brain injury or damage. They may display some of the so-called "soft" neurologic signs, perhaps an abnormal electroencephalogram or a history of pregnancy or birth complications. However, none of these provides definite evidence of the presence of brain damage. In spite of the limitations of the diagnostic procedures, it is generally believed that this group not only has brain damage or dysfunction but also is at increased risk for psychiatric disorder.

It was maintained at one time that children with brain damage were at risk for a specific type of psychiatric disturbance characterized by hyperactivity and impulsivity. More recent work uniformly indicates that brain damage puts children at increased risk for psychiatric disorder in general rather than for a specific type of disturbance (Rutter, 1977; Werry, 1980).

The relationship between brain damage, whether definite or not, and conduct disorder has not yet been clearly worked out. The majority of studies refer to the hyperactive child syndrome and suggest that it is associated with an increased frequency of variables that may lead to or indicate brain damage. These factors include pre- and perinatal complications, "minor" neurologic abnormalities, and abnormal electroencephalographic findings. A further problem with these studies is that the distinction between hyperactivity and conduct disorder is rarely made. Thus the extent to which the findings, uncertain as they are, apply to conduct disordered children as distinct from hyperactive children is even more uncertain (Werry, 1980).

Two major mechanisms by which brain damage could play a part in the etiology of conduct disorder can be outlined. One possibility centers on the concept of temperament, the enduring patterns of reactivity of the child. There is some evidence that brain damage, or the noxious events that may lead to it, may result in extremes of temperament (Rutter, 1977; Werry, 1980). A second mechanism could operate through the cognitive consequences of brain damage such as low IQ and reading retardation (Rutter, 1977).

In summary, then, it is thought that brain damage, whether definite or not, may play an etiologic role in conduct disorder, probably through its contribution to temperamental and cognitive abnormalities. The extent and importance of this role have not been determined.

Factors of Uncertain Etiology

Learning Problems. The school performance of antisocial children is poor (Offord, 1982; Robins, 1966). For instance, in the Isle of Wight study of 10 and 11 year old children, one-third of the children who were severely retarded in reading showed significant antisocial behavior, and one-third of those with an antisocial disorder were at least 28 months retarded in the reading level (after IQ was factored out; Rutter et al., 1970). A major issue is the nature of the relationship between school performance or learning problems on the one hand and antisocial behavior or delinquency on the other.

Figure 36–6 outlines three hypotheses concerning the nature of this relationship. With regard to hypothesis 1, although it is almost certain that some children may be held back in school because of behavioral difficulties, there is little evidence to support the idea that reading retardation develops as a consequence of antisocial behavior (Rutter et al., 1970).

Hypothesis 2 has been suggested by a number of workers (Offord, 1982; Rutter et al., 1970). Their reasoning is that early school failure leads to feelings of low self-esteem, which in turn provoke the child to engage in antisocial behavior in an effort both to raise his self-esteem and to gain a feeling of accomplishment and confidence. There is evidence, albeit indirect, supporting this idea.

Hypothesis 3 also appears to be viable. For example, the works of Sturge (1972) and Offord (1982) argue strongly for the notion that within a relatively poor urban population, educational retardation and antisocial behavior arise from common or coexisting adverse family influences and that the educational retardation itself is not causally related to the antisocial behavior. It is possible, too, that the common factors may not be limited to psychosocial ones but may include largely constitutional variables such as abnormal temperament (Rutter et al., 1970). This third hypothesis argues that learning problems and antisocial behavior occur in the same children not because one is etiologically linked to the other but because the children and the families with these disorders share common characteristics.

Studies have not yet been carried out to allow us to choose among the hypotheses, particularly

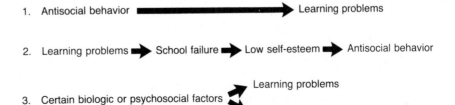

Figure 36–6. Hypothesized relationships between learning problems and antisocial behavior.

1. Antisocial behavior ⟶ Learning problems

2. Learning problems ➡ School failure ➡ Low self-esteem ➡ Antisocial behavior

3. Certain biologic or psychosocial factors ⟨ Learning problems / Antisocial behavior

between the second and the third. What are required are longitudinal studies beginning in the preschool years, or when children are just starting school, so that one can learn about the natural histories of children who show educational retardation early without behavioral problems and those who show behavioral problems without evidence of educational retardation (Murray, 1976; Rutter et al., 1970). In the meantime the existing data suggest that hypotheses 2 and 3 are both operative in accounting for the relationship between learning problems and antisocial behavior.

In concluding this section, a comment must be made about the relationship, if any, between intellectual level and conduct disorder and delinquency. In the Isle of Wight study, conduct disorder in boys, but not in girls, was significantly associated with a slightly below average IQ (Rutter et al., 1970). The results concerning IQ and delinquency are mixed with some studies reporting low IQ for delinquents as compared to controls and others reporting no difference between the IQ's of delinquents compared to their same-sexed nondelinquent siblings (Offord, 1982; West and Farrington, 1977). All studies agree on the point that the educational retardation of delinquents is more striking than their possible IQ deficits.

Difficult Temperament. The concept of the infant or child with a difficult temperament first achieved popularity in the work of Chess (1979). Ten per cent of their sample were identified as being difficult in infancy and were characterized by irregular biologic functioning, initial aversion and slow adaptability to environmental changes, high intensity in affect and expression, and negative mood.

The possible causal factors responsible for a difficult temperament are several and include genetic factors, brain damage, and the early patterns of interaction between the mother and her infant (Bates, 1980; Campbell, 1979; Goldsmith and Gottesman, 1981; Rutter, 1977; Werry, 1980). In addition, it should be made clear that the majority of work on temperament has employed parental ratings, which have shown only a modest agreement with those of external observers (Bates, 1980). Thus, the concept of a difficult temperament should not be understood as representing only the qualities of the child as rated objectively but should be recognized as consisting of the parent's perceptions of the infant's patterns of reactivity.

The issue of the predictive value of a difficult temperament will now be considered. The limited data indicate that measures of difficult temperament in infancy do not predict childhood behavior problems to a significant degree (Chess, 1979). However, by age three or four some prediction is possible. Other work reveals that elementary school aged children with a difficult temperament as measured by parental perceptions are far more

likely than the easier temperament group to show evidence of psychiatric disorder one year later (Graham et al., 1973). There is evidence too that the prediction becomes stronger if scores of difficult temperament are combined with evaluations of other aspects of family life, such as emotional relationships between the mother and child and mode of discipline (Cameron, 1978).

In summary, early evidence of a difficult temperament as perceived by the parent may put a child at increased risk for conduct disorder. Whether that risk will be realized will depend both on the inherent stability of the temperamental difficulty and the extent to which the parents and other adults in the child's life can successfully modify the perceived troublesome patterns of behavior. (See also Chapter 10.)

Labeling

In labeling theory, society's reaction to an initial or primary behavioral deviation, rather than correcting it, provokes a stronger and more enduring secondary deviation (Lemert, 1967; Scheff, 1966). Some studies suggest that the experience of apprehension by the police of children with conduct disorder escalates their antisocial behavior (Farrington, 1977). The experience itself appears to lead to increased antiauthority attitudes, perhaps a change toward more deviant peer groups, and the feeling, expressed by youth and others, that the label "antisocial" or "delinquent" applies not to a specific act but to the person himself (Becker et al., 1963; Erikson, 1966; West and Farrington, 1977). The importance of labeling and even its existence as an adverse effect are still under debate (Tittle, 1975). It is obviously an area in which much work is needed. In the meantime the clinician should be sensitive to the goal of minimizing the possiblity of labeling or stigmatization when implementing a treatment program.

Interaction Among Etiologic Factors

Figure 36–7 schematically outlines the etiologic factors potentially involved in generating conduct disorder in children. One can see immediately that this diagram does not make any attempt to spell out either what we know or think we know about causal chains or any ideas about possible interactional effects. For instance, in the case of the former, it is probable that genetic traits, brain damage, and adverse early experiences mediate some of their effects on the child through the production of an abnormal temperament. Another possible causal chain would entail poor schools leading to an increased frequency of learning problems, which in turn lead to a lowering of self-esteem in the affected child. Many other examples of suspected causal chains could be given. The

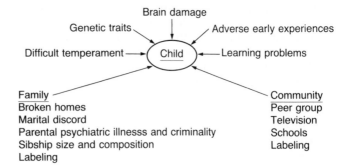

Figure 36–7. Etiologic factors in conduct disorders.

construction and validation of these causal chains, and the extent to which breaking the chains at some point with an intervention program reduces the frequency of a disorder, constitute the work of what has been termed experimental epidemiology (Robins, 1978).

Interactional effects among the etiologic factors are known to occur. Some of them appear to be additive; that is to say, the effect of the etiologic factors together is equal to the sum of each of their effects in isolation. Brain damage and an adverse environment are examples of additive factors in producing psychiatric disturbance in the child (Rutter, 1977).

Another mode of interaction among etiologic factors is the transactional effect. Here some factors increase the likelihood of experiencing another factor. For instance, there are data to suggest that a child with a difficult temperament is more likely than other children to be subjected to scapegoating in the family with resultant criticism and lowered self-esteem (Graham et al., 1973).

The third and last manner of interaction among etiologic variables is the interactive effect. In this case the effects of two etiologic factors potentiate each other, and the overall impact is greater than the sum of the individual effects. An example of this is found in the study of the relationship between psychiatric disturbance in 10 year old children and six family variables, all of which individually were strongly and significantly associated with child psychiatric disorder (Rutter, 1979b). The adverse family factors were: severe marital discord, low social status, overcrowding or large family size, paternal criminality, maternal psychiatric disorder, and admission of the child into agency care. A child with just one of these stress factors was no more likely to have a psychiatric disturbance than a child in whom there were no stress factors. However, when there were two stress factors, the risk of psychiatric disturbance increased fourfold and with more stress factors the risk climbed steeply. Thus, the stresses potentiate each other so that their overall effect is more than their individual effects when considered singly.

Work in this area is just beginning, and none of

it applies specifically to conduct disorders. Rather it focuses on pediatric psychiatric disorders in general. Nevertheless there can be no doubt that the potential etiologic factors in conduct disorder share common causal chains in some instances and interact in one or more of the ways outlined. These findings obviously have implications for assessment, management, and prevention.

Prognosis and Natural History

Children with conduct disorders have a poor prognosis as adults compared to children with emotional problems. Indeed, with the exception of childhood psychosis, such disorders have the poorest prognosis of any childhood psychiatric illness. It can be expected that at least 40 to 50 per cent of the children with persistent antisocial behavior will experience serious psychosocial difficulties in adult life (Robins, 1979).

In Robins' landmark study (1966) in which she followed up 35 years later over 500 children seen in a child guidance clinic in St. Louis in the 1920's and 100 control children, the poor long term prognosis for antisocial and delinquent children was striking (Table 36–5). Three of five of the control subjects and two of five neurotic children were normal as adults. In both these groups persons with abnormalities suffered primarily from neurotic complaints in adulthood. Thus, children referred for symptoms such as fearfulness, tics, thumb sucking, tantrums, and bedwetting had almost as good an outcome as adults as children who were never referred to the child guidance clinic.

In contrast, only one of three children with antisocial behavior, all of whom would almost

Table 36–5. RESULTS FROM ROBINS' STUDY*

1. Controls	3 of 5 normal as adults
2. Neurotics	2 of 5 normal as adults
3. Conduct disorders	1 of 3 normal as adults
4. Delinquents	1 of 4 normal as adults

*Adapted from Robins, L. N.: Deviant Children Grown Up. Baltimore, The Williams & Wilkins Co., 1966.

certainly qualify as having conduct disorders, were normal as adults. For delinquents, only one in four were normal. In both these groups the adult disturbances, when present, were serious and wide ranging, including sociopathy, criminality, alcoholism, drug abuse, marital disruption, poor work record, poor health, and financial dependency. A review of other studies makes it clear that antisocial and delinquent children have a high risk of continuing deviant behavior, although the frequency of involvement with the law falls off in the midtwenties (Robins, 1979). In addition to data supporting the continuity of deviant behavior, there is evidence that aggressiveness is a relatively stable trait. For instance, Kagan and Moss (1962) found that aggressive boys remained aggressive men, and indeed aggressivity was the most stable personality trait in their sample.

Data from longitudinal studies support the notion that antisocial behavior in childhood is most likely to persist into adulthood if it first occurs at an early age, if the behavior is serious and takes place in the community as well as at home and at school, and when the behavior leads to a court appearance and possibly institutionalization.

Two last points. It is important to note that a sizable minority of children with conduct disorder give up their deviant behavior in adolescence or adulthood. Why this is so and the factors involved in this resolution are not understood. Further work in this area, by identifying naturally occurring factors that appear to promote a good outcome, may lead to the formulation of more effective treatment programs. Last, the work of Robins and others points out that virtually all serious adult antisocial behavior begins in childhood. Thus, if effective prevention and treatment strategies could be developed for children with conduct disorders, not only would their life quality be improved in childhood, but antisocial and criminal behavior in adulthood could be drastically reduced even within one generation.

Guidelines for Management

The results of treatment of conduct disordered children have been disappointing. Few studies have shown significant, positive long term effects of treatment, and, even worse, some have shown significant adverse effects possibly through the mechanism of labeling and stigmatization of the affected child (Robins, 1979). Behavior modification techniques applied by parents in the home or by teachers at school have shown excellent short term results (Becker et al., 1967; Kent and O'Leary, 1976; Patterson et al., 1970). The training of parents in behavioral therapy techniques has been an especially promising approach (Johnson and Katz, 1973).

The following guidelines may be helpful in assessing and planning a treatment program for children, not only those with the formal diagnosis of conduct disorder or delinquency but those who display evidence of aggressiveness and some antisocial symptoms.

Assessment

The assessment should be based on data from multiple informants, including the parents, the child himself, teachers, and other professionals who have had contact with the youth. If the assessment procedure includes information from only one source, it will be both incomplete and probably distorted.

As a first step, it is important to see the entire family. This provides a forum not only for collecting needed current and historical information but also for observing the patterns of interaction among the family members. In the latter case one is particularly concerned about evidence of scapegoating of the child by the parents. That is, one should determine whether the parents focus on and complain about the child as a way to avoid dealing directly with each other's painful feelings and concerns. The factual information elicited should provide a clear understanding of the history and extent of the present problem. The actual characteristics of the antisocial behavior, including the frequency and severity of episodes, should be determined. The extent to which there may be overlapping diagnoses such as hyperactivity must be ascertained. Historical data should address the possible etiologic factors already outlined. These would include the family history, including possible inherited genetic traits, pregnancy and birth history, and other evidence of likely brain damage, early experiences and bonding, any history of a difficult temperament or learning problem, and family factors such as marital discord and parental deviance. The presence of possible mediating factors, for example, low self-esteem and inconsistent patterns of discipline, should be ascertained.

Information should be also gathered about the probable strengths of the child. Such data would include a cataloguing of his particular talents and skills, not only in academic work but in nonacademic skill areas as well, including athletic activities such as baseball and football and nonathletic skills such as music and art. The purpose is to determine the islands of achievement and competence that are present or that could be promoted.

The child usually should be seen alone. Here an opportunity is provided to observe his pattern of behavior and to gain an understanding of his view of the problem. The child's perspective may differ dramatically from that of the parents, and the reasons for this discrepancy should be explored. An excellent, easily read book about how to interview children is one by Adams (1974).

Information should be gathered from the school, ideally through an interview with the child's teacher. The teacher who deals primarily with normal children can be an excellent discriminator of normal from abnormal behavior. He can provide valuable data about the child's academic performance, including the presence or absence of learning problems, his self-esteem in the learning situation, and his behavioral patterns and their possible causes. In addition, the teacher's view of what appears to work successfully with the child can be valuable to the physician in formulating an overall treatment plan.

Data from other professionals can be helpful. Psychologic testing can provide more definitive evidence about academic and learning deficits. A physical examination, including a detailed neurologic work-up, can furnish data about the extent to which frank or suspected brain damage may be an etiologic factor.

Formulation and Treatment

When the assessment procedure is complete, the case should be formulated as outlined in Table 36–3. The treatment program should follow directly from such a formulation. Specific intervention strategies should be directed at specific etiologic factors or identified strengths. The goal is to shift the balance between the harmful effects of the etiologic factors and the curative effects of the strengths. It is not realistic nor is it probably necessary to deal effectively with all the contributing etiologic factors in order for the outcome to be successful (Rutter, 1979b).

The focus in dealing with the antisocial behavior should be not so much to attempt to determine why it has occurred as it should be to discover the reasons it has been allowed to flourish and not kept under reasonable control. Attention should be paid to the possible harmful effects of labeling. The less the treatment focuses on the antisocial behavior itself and the more it is directed to the other aspects of the child's functioning, the less likely it is that the child will internalize the view that his major characteristic is his antisocial behavior.

If the behavior of the child cannot be brought under control in the family setting, placement outside the family, even temporarily, should be considered. When the child's behavior is out of control, he is probably asking for more secure surroundings and certainly no treatment program can have much chance of success under these circumstances.

A word about pharmacotherapy. Pharmacotherapy in children has been described as merely ameliorative and not curative (White, 1980). However, it can be a valuable adjunct to other forms of treatment. None of the available psychotropic agents are specific for any particular form of undersocialization. However, they do have some specificity for certain target symptoms, particularly hyperactivity, aggression, and impulsiveness.

The symptoms of hyperactivity and impulsiveness have been demonstrated to respond best to short acting stimulants—methylphenidate and dextroamphetamine (Werry, 1977). If these drugs are ineffective, some children show behavioral improvement with a tricyclic antidepressant, such as imipramine (White, 1980). On the other hand, aggression in the absence of impulsiveness and hyperactivity is best managed with one of the neuroleptic drugs, such as thioridazine and haloperidol (White, 1980). In most instances, however, aggression that is due primarily to undersocialization rather than being associated secondarily with mental retardation or gross brain dysfunction is also associated with hyperactivity and impulsiveness, and the aggression appears to settle along with these symptoms when they are treated with the short acting stimulants. The pharmacology of these drugs is described in Chapter 58. However, there are other medication related issues that have important implications for the actual process of socialization. The side effects of medication may impede this process. Furthermore, drugs may have an important symbolic meaning to the child, and they may be perceived as helpful, or as destructive or as a sign of failure.

The side effects of the stimulants, which can exert particularly potent adverse effects on socialization, are insomnia and irritability. Insomnia extends the period in which the parents have to cope with the child, provides an opportunity for the child to see himself as additionally troublesome, and may also imply loss of control of an important function (sleep). Irritability may also make the child more difficult to handle and may compound his perception of himself as troublesome, whereas tearfulness may be thought to reflect depression.

Adverse drug related effects on learning may also be problematic, although our knowledge of the direct effects that psychotropic drugs may have on academic and social learning is inadequate. The stimulants do appear to produce favorable effects on social behavior, but their effect upon classroom learning, attention, and memory is unresolved. Werry (1981) questions "the extent to which improvement on the various tests of cognitive function represents simply an increase in performance rather than learning; that is, the child simply shows what he already knows to the full rather than in part and there is no acquisition of any really new skills." There is evidence that in certain children with attention deficit disorders with hyperactivity, the stimulants in low dosage may well set the stage for improved learning. However, methodologically adequate studies in learning disabled children have seldom demonstrated a statistically significant effect (Aman, 1980). On the other hand, they do not appear to impair scholastic learning significantly.

The side effects of the neuroleptic drugs that impede socialization most significantly are motor complications and daytime sedation. Motor complications such as akathisia and extrapyramidal reactions affect coordination in play and can demoralize the child and his parents. Daytime sedation limits the child's availability to social learning and reduces the chances for making the best use of social opportunities. Furthermore, although there is evidence that low nonsedating doses of neuroleptic drugs can enhance scholastic learning by reducing distractibility, higher doses can diminish learning, probably as a result of inattention and perceptual impairment (Werry, 1981). Since scholastic learning success is an important source of self-esteem, which reinforces social adaptability, impaired learning can impede socialization by demoralization.

Thus, in the choice of a stimulant or neuroleptic drug, the effects on scholastic and social learning and sociability need to be balanced against the degree of behavioral control obtained. However, the emotional meaning of taking a psychotropic drug also should not be underestimated. In adults, a sufficiently negative attitude toward the drug can overcome its positive therapeutic effect. Thus the drug should be presented to the child as a facilitating agent to help him control aspects of his behavior that get him into trouble. Many children understand very well how impulsiveness and aggression reduce the number of fruitful interactions and increase the number of negative interactions they have with peers and others. There is also a danger that the child may see the drug as a punishment and may be demoralized, or alternatively he may see the drug as an excuse not to be held responsible for his behavior. Lessening of responsibility is a serious consequence; the acquisition of a sense of responsibility is a crucial element of the process of socialization.

Compliance in drug therapy reflects the same issues as compliance with any sort of therapy. Werry (1981) has suggested that chaotic families are less likely than well organized families to comply with drug regimens, and they are even less likely to do so if the regimen is complicated. The same chaotic family does not provide a developmentally facilitating environment for some children, and thus careful, clear instructions about drug administration need to be given for the treatment of these children who run the risk of a double disadvantage.

The following case history illustrates some of the points made about management of children with conduct disorder.

Case 2. John was an 11 year old boy who for the preceding year had displayed increasing symptoms of verbal abuse both at home and at school and fighting with his peers. He talked back to the teachers and had been involved in several fist fights on the playground. The symptom pattern had increased significantly in the last three months.

John lived in poor, overcrowded conditions with his natural mother and stepfather. The stepfather was known to be a heavy drinker. There was constant quarreling in the home, and both the parents blamed John for many of the family's problems. In the family interview, the parents found it difficult to talk directly with each other but rather preferred to focus on John. The mother knew that he had always been a willful stubborn child. John's maternal grandfather, with whom he was close, had died suddenly three months previously. In addition, John had recently moved to a new school and found the work, especially reading, very difficult.

In the interview John displayed an ability to recognize that he was full of feelings, some of which was rage directed toward his parents and some of which was grief over the death of his grandfather. He thought that his parents never had time to listen to him, and he missed very much his talks with his grandfather. John wondered whether he was the cause of the family's quarreling and described himself as a "lousy kid." However, he did talk warmly of his music teacher at school. John's teachers recognized that the boy was full of pent-up feelings. They were concerned that he might have a specific reading disability, and they recognized that this problem had never received an adequate work-up in terms of psychologic testing. The music teacher had a good deal of affection for John and thought that he had unusual music talent with the guitar.

Figure 36–8 outlines the possible etiologic factors and the strengths of the child. The asterisks indicate the factors addressed by the treatment program. It will be noted that some of the stresses, namely, poverty, the stepfather's alcoholism, and the change in schools, were not dealt with explicitly by the management plan. However, the other factors were. Family therapy had as its goals the diminution of the scapegoating of John, the lessening of his guilt over the family discord, as well as helping the parents with behavioral techniques to deal more effectively with John's antagonistic, willful patterns of interaction. Individual interviews with John, which he invested in immediately, helped the boy deal with his unexpressed feelings, particularly the feelings of grief surrounding his grandfather's death.

In the school area, John received a complete psychologic work-up, which revealed a previously unrecognized learning problem. Specific tutoring was set up. The teachers were also helped to deal more successfully with John's manner of confrontation, and a reward system for satisfactory behavior was initiated. The music teacher arranged for John to receive guitar lessons and also spent time regularly talking with the boy about the things that were important to him.

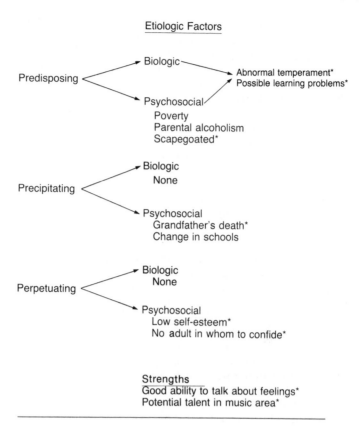

Etiologic Factors

Figure 36–8. Outline of diagnostic formulation for Case 2.

*Addressed by treatment program

Within three months, the antisocial behavior had lessened, and after six months John was not thought to be troublesome either by his parents or by the school personnel. In addition, he remarked that he felt better about himself and was much happier.

The case of John illustrates both the need for a wide ranging assessment and the willingness to provide a multipronged treatment program rather than one confined to the techniques with which the physician feels most comfortable.

INVULNERABILITY

A sizable proportion of children who are reared in families with many or all of the correlates of disorders of socialization are free of these disorders and are well adjusted. For example, one study in Britain showed that almost half the children raised in poverty, poor housing, and family adversity were well adjusted (Rutter, 1979b). In West and Farrington's longitudinal study (1977), over one-quarter of the boys from the most disadvantaged circumstances were free of evidence of serious antisocial behavior or delinquency throughout the course of the study. Further, if one studies children with persistent antisocial behavior, almost half of them are found to be without serious antisocial symptoms or criminality in adult life (Robins, 1979).

What contributes to the development of the so-called invulnerable child is not well understood (Garmezy, 1974). Beginning work in this area has suggested variables that may protect children at increased risk for psychiatric disorder from exhibiting it. They include a favorable temperament, compensating experiences outside the home, such as a superior school or opportunities for skill development, the availability of long term relationships with helping adults, fair and consistent discipline and structure, and the maintenance of an intact self-esteem (Rutter, 1979b). Much more investigation needs to be done to identify clearly the factors that prevent the emergence of psychiatric disturbance in vulnerable children and, after the disturbance has occurred, the factors that prevent it from extending into adolescence and adult life.

A helpful research design in this area is the so-called sibling design in which the index child is compared with his same sexed sibling (Offord and Jones, 1976). Here, against the background of considerable genetic and environmental similarity, factors that produce good and bad results in children within the same family can be identified. The exciting next step will be to determine the extent to which these factors can be converted into an intervention program that will be effective in either preventing the emergence of psychiatric symptoms or promoting the disappearance of symptoms once they have occurred.

PRIMARY PREVENTION

Primary prevention, that is the reduction in the number of new cases, of the disorders of the failure of socialization is attractive for several reasons. Under the present circumstances, by the time these children are diagnosed and treatment is initiated, they have suffered a good deal. The disorder has been under way for a period of time and has resulted in stress and a reduction in the quality of life. In addition, the evidence shows that existing treatment resources service only a minority of the children with these disorders and those who are seen in these facilities are not necessarily the ones most in need of treatment (Langner et al., 1974; Rutter et al., 1970). Third and last, there is no convincing evidence that our existing treatment methods are effective (Robins, 1979). Thus, if we could prevent these disorders from occurring in the first place, we would escape the three dilemmas, just outlined, associated with the treatment of established cases.

Unfortunately it is not yet clear what primary prevention programs should be implemented to prevent the sequelae of a failure in socialization. We need data about the outcome of intervention programs in this area. Each intervention program will have to be clearly described, the target groups specified, and the outcome criteria defined. The various parameters of success—for example, efficaciousness, effectiveness, and efficiency in the grading of the strength of the scientific evidence supporting conclusions in the primary prevention area—have been detailed elsewhere (Canadian Task Force on the Periodic Health Examination, 1979).

A major problem with screening programs in the behavioral area dealing with noncompliant child populations is that in order to include the children who actually will eventually have the disorder without intervention, many children must be included who are not at risk (Robins, 1974). In technical terms the sensitivity of the screening devices is high, but the specificity is unacceptably low. Thus, supposing that one had an effective intervention program for children identified by the screening process, and the money to carry it out, the good that accrued to the children correctly identified by the screening program would have to be weighed against the possible harm (through incorrect labeling, for example) that might occur to the children who were misidentified.

Obviously much work needs to be done to determine the interventions by the physician in his office (such as advice about child rearing or the distribution of pamphlets on parenting skills) carried out on a noncomplainant population of children (and their families) that do more good than harm. In the meantime, all of us are advised to tread lightly in the field of primary prevention until the data indicate the programs that definitely merit widespread implementation.

Acknowledgments

The authors would like to thank Dr. J. Beitchman, Dr. M. Jones and Dr. P. Szatmari and Mrs. M. Gieg, Mrs. S. Grainge, Miss A. Jones, and Mr. S. Offord for their help with or critical evaluations of this work.

DAVID R. OFFORD

BRENT G. H. WATERS

REFERENCES

Adams, P. L.: Obsessive Children. New York, Penguin Books, 1973.

Adams, P. L.: A Primer of Child Psychotherapy. Boston, Little, Brown and Company, 1974.

Aman, M. S.: Psychotropic drugs and learning problems: a selective review. J. Learn. Disabil., 13:87, 1980.

American Psychiatric Association: Diagnostic and Statistical Manual of Mental Disorders. Ed. 3. Washington, D.C., American Psychiatric Association, 1980.

Bates, J. E.: The concept of difficult temperament. Merrill-Palmer Quart., 26:299, 1980.

Becker, W. C., Madsen, C. H., Arnold, C. R., and Thomas, D. R.: The contingent use of teacher attention and praise in reducing classroom problems. J. Spec. Educ., 1:287, 1967.

Bohman, M.: The interaction of heredity and childhood environment: some adoptive studies. J. Child Psychol. Psychiatry, 22:195, 1981.

Bowlby, J.: Forty-four Juvenile Thieves: Their Characters and Home-Life. London, Bailliere, Tindall and Cox, 1946.

Bowlby, J.: Maternal Care and Mental Health. Geneva, World Health Organization, 1951.

Cadoret, R. J., and Cain, C.: Sex differences in predictors of antisocial behavior in adoptees. Arch. Gen. Psychiatry, 37:1171, 1980.

Cameron, J. R.: Parental treatment, children's temperament and the risk of childhood behavioral problems. 2. Initial temperament, parental attitudes and the incidence and form of behavioral problems. Am. J. Orthopsychiatry, 48:140, 1978.

Campbell, S. B. G.: Mother-infant interaction as a function of maternal ratings of temperament. Child Psychiatry Hum. Dev., 10:67, 1979.

Canadian Task Force on the Periodic Health Examination: The periodic health examination. Can. Med. Assoc. J., 121:1193, 1979.

Cantwell, D. P.: Hyperactivity and antisocial behavior. J. Am. Acad. Child Psychiatry, 17:252, 1978.

Chess, S.: Developmental theory revisited: findings of longitudinal study. Can. J. Psychiatry, 24:101, 1979.

Christiansen, K. O.: Threshold of tolerance in various population groups illustrated by results from Danish criminological twin study. In De Reuch, A., and Porter, R. (Editors): The Mentally Abnormal Offender. Boston, Little, Brown and Company, 1969.

Clausen, J. A.: Sociology and psychiatry. In Freedman, H., Caplan, A., and Sandock, B. (Editors): Comprehensive Textbook in Psychiatry. Ed. 3. Baltimore, The Williams & Wilkins Co., 1980.

Conger, J. C., and Keane, S. P.: Social skills intervention in the treatment of isolated or withdrawn children. Psychol. Bull. 90:478, 1981.

Eisenberg, L.: School phobia: a study in the communication of anxiety. Am. J. Psychiat. 114:712, 1958.

Eisenberg, L.: Normal child development. In Freedman, H., Caplan, A., and Sadock, B. (Editors): Comprehensive Textbook of Psychiatry. Ed. 3. Baltimore, The Williams & Wilkins Co., 1980.

Eme, R. F.: Sex differences in childhood psychopathology: a review. Psychol. Bull., 86:574, 1979.

Erikson, K. T.: Wayward Puritans. New York, John Wiley & Sons, Inc., 1966.

Farrington, D. P.: The effects of public labeling. Br. J. Criminol., 17:112, 1977.

Firestone, P., Waters, B., and Goodman, J.: The use of desensitisation techniques in children and adolescents: a review. J. Clin. Child Psychol., 7:142, 1978.

Freud, S. Collected Papers. London, Hogarth Press, 1959, Vol. 3.

Garmezy, N.: The study of competence in children at risk for severe psychopathology. In Anthony, E. J., and Koupernik, C. (Editors): The Child in His Family: Children at Psychiatric Risk. New York, John Wiley & Sons, Inc., 1974.

Gerbner, G.: Television's influence on values and behavior. Weekly Psychiatric Update Series, 2: Lesson 24, 1978.

Gold, M.: Undetected delinquent behavior. J. Res. Crime Delinq., 3:27, 1966.

Goldsmith, H. H., and Gottesman, I. I.: Origins of variation in behavioral style: a longitudinal study of temperament in young twins. Child Develop., 52:91, 1981.

Gould, M. S., Wunsch-Hitzig, R., and Dohrenwend, B.: Estimating the prevalence of childhood psychopathology. J. Am. Acad. Child Psychiatry, 20:462, 1981.

Graham, P.: Depression in pre-pubertal children. Dev. Med. Child Neurol., 16:340, 1974.

Graham, P., Rutter, M., and George, S.: Temperamental characteristics as predictors of behavior disorders in children. Am. J. Orthopsychiatry, 43:328, 1973.

Henn, F. A., Bardwell, R., and Jenkins, R. L.: Juvenile delinquents revisited. Arch. Gen. Psychiatry, 37:1160, 1980.

Hersov, L.: Emotional disorders. In Rutter, M., and Hersov, L. (Editors): Child Psychiatry: Modern Approaches. Oxford, Blackwell Scientific Publications, 1977.

Hindelang, M. J., Hirschi, T., and Weiss, J. G.: Correlates of delinquency: the illusion of discrepancy between self-report and official measures. Am. Soc. Rev., 44:995, 1979.

Hollingsworth, C. E., et al.: Long term outcome of obsessive compulsive disorder in childhood. J. Am. Acad. Child Psychiatry, 19:134, 1980.

Hutt, C.: Sexual differentiation in human development. In Ounsted, C., and Taylor, D. C. (Editors): Gender Differences, Their Ontogeny and Significance. London, Churchill-Livingstone, 1972.

Johnson, C. A., and Katz, R. C.: Using parents as change agents for their children: a review. J. Child Psychol. Psychiatry, 14:181, 1973.

Judd, L. S. Obsessive-compulsive neurosis in children. Arch. Gen. Psychiatry, 12:136, 1965.

Kagan, J.: The form of early development: continuity and discontinuity in emergent competence. Arch. Gen. Psychiatry, 36:1047, 1979.

Kagan, J., and Moss, H. A.: Birth to Maturity. New York, John Wiley & Sons Inc., 1962.

Kent, R. N., and O'Leary, D. K.: A controlled evaluation with conduct problem children. J. Consult. Clin. Psychol., 44:586, 1976.

Kety, S., et al.: The types of prevalence of mental illness in the biological and adoptive families of adopted schizophrenics. In Rosenthal, D., and Kety, S. (Editors): Transmission of Schizophrenia. Oxford, Pergamon Press Ltd., 1968.

Kline, S., and Cameron, P. M.: 1. Formulation. Can. Psychiat. Assoc. J., 23:39, 1978.

Langhorne, J. E., and Loney, J.: A four-fold model for subgrouping the hyperkinetic/mbd syndrome. Child Psychiat. Hum. Dev., 9:153, 1979.

Langner, T. S., Gertsen, J. C., Greene, E. L., Eisenberg, J. G., Herson, J. H., and McCarthy, E. D.: Treatment of psychological disorders among urban children. J. Consult. Clin. Psychol., 42:170, 1974.

Lavietes, R. L.: Schizoid disorders. In Freedman, H., Caplan, A., and Sadock, B. (Editors): Comprehensive Textbook of Psychiatry. Ed. 3. Baltimore, The Williams & Wilkins Co., 1980.

Lemert, E. M.: Human Deviance, Social Problems and Social Control. Englewood-Cliffs, New Jersey, Prentice-Hall, Inc., 1967.

Mednick, S. A., and Hutchings, B.: Genetic and psychophysiologic factors in asocial behavior. J. Am. Acad. Child Psychiatry, 17:209, 1978.

Michael, C., Morris, D., and Soroker, E.: Follow-up studies of shy withdrawn children. II. Relative incidence of schizophrenia. Am. J. Orthopsychiatry, 27:331, 1957.

Miner, G. D.: The evidence for genetic components in the neuroses: a review. Arch. Gen. Psychiatry, 29:111, 1973.

Minde, K. K., and Minde, R.: Psychiatric intervention in infancy: a review. J. Am. Acad. Child Psychiatry, 20:217, 1981.

Morris, D., Soroker, E., and Burrus, C.: Follow-up studies of shy withdrawn children. I. Evaluation of later adjustment. Am. J. Orthopsychiatry, 24:743, 1954.

Murray, C. A.: The Link Between Learning Disabilities and Juvenile Delinquency: Current Theory and Knowledge. Washington, D.C., U.S. Government Printing Office, 1976.

Offord, D. R.: Family backgrounds of male and female delinquents. In Gunn, J., and Farrington, D. (Editors): Advances in Forensic Psychiatry and Psychology. London, John Wiley & Sons, Ltd., 1982.

Offord, D. R., and Jones, M. B.: The proband-sibling design in psychiatry with two technical notes. Can. Psychiat. Assoc. J., 21:101, 1976.

Paternite, C. E., Loney, J., and Langhorne, J. E.: Relationships between symptomatology and sex-related factors in hyperkinetic/mbd boys. Am. J. Orthopsychiatry, 46:291, 1976.

Patterson, G. R., Cobb, J. A., and Ray, R. S.: A social engineering technology for retraining aggressive boys. In Adams, H., and Unikel, L. (Editors): Georgia Symposium in Experimental Clinical Psychology. Oxford, Permagon Press Ltd., 1970, Vol. II.

Patterson, G. R., Littman, R. A., and Bricker, W.: Assertive behavior in children: a step toward a theory of aggression. Monog. Soc. Res. Child Dev., 32:5, 1967.

Pollitt, J. D.: Obsessional states. Br. J. Psychiatry, 133:9, 1975.

Rainer, J. S.: Genetics and psychiatry. In Freedman, H., Caplan, A., and Sadock, B. (Editors): Comprehensive Textbook in Psychiatry. Ed. 3. Baltimore, The Williams & Wilkins Co., 1980.

Rapoport, J., et al.: Clinical controlled trial of chlorimipramine in adolescents with obsessive-compulsive disorder. Arch. Gen. Psychiatry, 37:1281, 1980.

Robins, L. N.: Deviant Children Grown Up. Baltimore, The Williams & Wilkins Co., 1966.

Robins, L. N.: Antisocial behavior disturbances of childhood: prevalence, prognosis and prospects. In Anthony, E. J., and Koupernik, C. (Editors): The Child in His Family: Children at Psychiatric Risk. New York, John Wiley & Sons, Inc., 1974.

Robins, L. N.: Psychiatric epidemiology. Arch. Gen. Psychiatry, 35:697, 1978.

Robins, L. N.: Longitudinal methods in the study of normal and pathological development. In Kisker, K. P., Meyer, J. E., Muller, C., and Stromgren, E. (Editors): Grundlagen und Methoden der Psychiatrie. Heidelberg, Springer-Verlag, 1979, Vol. I.

Robins, L. N., West, P. A., and Herjanic, B. L.: Arrests and delinquency in two generations: a study of black urban families and their children. J. Child Psychol. Psychiatry, 16:125, 1975.

Rodriquez, A., Rodriquez, M., and Eisenberg, L.: The outcome of school phobia: a follow-up study based on 41 cases. Am. J. Psychiatry, 116:540, 1959.

Rutter, M.: Brain damage syndromes in childhood: concepts and findings. J. Child Psychol. Psychiatry, 18:1, 1977.

Rutter, M.: Diagnostic validity in child psychiatry. Adv. Biol. Psychiatry, 2:2, 1978.

Rutter, M.: Maternal deprivation, 1972–1978: new findings, new concepts, new approaches. Child Dev., 50:283, 1979a.

Rutter, M.: Protective factors in children's responses to stress and disadvantage. In Kent, M. W., and Rolf, J. E. (Editors): Primary Prevention of Psychopathology. Volume III: Social Competence in Children. Hanover, New Hampshire, University Press of New England, 1979b.

Rutter, M.: The city and the child. Am. J. Orthopsychiatry, 51:610, 1981.

Rutter, M., et al.: Adolescent turmoil: fact or fiction. J. Child Psychol. Psychiatry, 17:35, 1976.

Rutter, M., and Madge, N.: Cycles of Disadvantage: A Review of Research. London, William Heinemann Ltd., 1976.

Rutter, M., Tizard, J., and Whitmore, K.: Education, Health and Behavior. London, Longman Group Ltd., 1970.

Sandberg, S. T., Rutter, M., and Taylor, E.: Hyperkinetic disorder in psychiatric clinic attenders. Dev. Med. Child Neurol., 20:279, 1978.

Scheff, T.: Being Mentally Ill. Hawthorne, New York, Aldine Publishing Co., 1966.

Sturge, C.: Reading retardation and antisocial behavior. Thesis, University of London, 1972.

Tittle, C. R.: Labelling and crime: an empirical evaluation. In Gove, W. R. (Editor): Labelling of Deviance. New York, John Wiley & Sons, Inc., 1975.

United States Public Health Service: Report to the Surgeon General: Television and Growing: The Impact of Televised Violence. Washington, D.C., U.S. Government Printing Office, 1972.

Watt, N., et al.: School adjustment and behavior of children hospitalized for schizophrenia as adults. Am. J. Orthopsychiatry, 40:637, 1970.

Werkman, S.: Anxiety disorders. In Freedman, H., Caplan, A., and Sadock, B. (Editors): Comprehensive Textbook in Psychiatry. Ed. 3. Baltimore, The Williams & Wilkins Co., 1980.

Werry, J. S.: The use of psychotropic drugs in children. J. Am. Acad. Child Psychiatry, 16:446, 1977.

Werry, J. S.: Organic factors. In Quay, H. C., and Werry, J. S. (Editors):

Psychopathological Disorders of Childhood. New York, John Wiley & Sons, Inc., 1980.

Werry, J. S.: Drugs and learning. J. Child Psychol. Psychiatry, 22:283, 1981.

West, D. J., and Farrington, D. P.: The Delinquent Way of Life. London, William Heinemann Ltd., 1977.

White, J. H.: Psychopharmacology in childhood: current status and future prospects. Psychiatr. Clin. N. Am. 3:443, 1980.

Winnicott, D. W.: The antisocial tendency. *In* Winnicott, D. W.: Collected Papers. New York, Basic Books, Inc., 1958a.

Winnicott, D. W.: Primary maternal occupation. *In* Winnicott, D. W.: Collected Papers. New York, Basic Books Inc., 1958b.

Wright, D. M., Moelis, I., and Pollack, L. J.: The outcome of individual child psychotherapy: increments at follow-up. J. Child Psychol. Psychiatry, 17:275, 1976.

Yarrow, L. J.: Maternal deprivation: toward an empirical and conceptual re-evaluation. Psychol. Bull., 58:459, 1961.

37

Substance Use, Abuse, and Dependence

The use of psychoactive substances to alter mood, perception, and behavior has become an integral part of coming of age in Western society. In recent years drug use by adolescents has markedly increased. Some surveys indicate that over 80 per cent of school age children in the United States have used one or more substances for nonmedical purposes. Whereas throughout history the youthful of many societies have used drugs for medical, religious, and recreational purposes, present day adolescents have access to an unprecedented array of potentially dangerous substances indigenous to many geographic areas or synthesized under laboratory conditions.

Many young people experiment with some of these drugs and do not repeat the experience. Others use the drugs intermittently or regularly in a controlled fashion and suffer few adverse consequences of this behavior. For some, early experimentation may progress to a pattern of use that is marked by compulsive and constant use and physical and psychosocial deterioration. Disability may derive from the pharmacology of the drugs, as in overdose, or the pattern of use, as in endocarditis or hepatitis from dirty needles. The leading cause of death in young people may be attributed to the physical illness, accidents, suicides, and homicides associated with drug and alcohol use (Blum et al., 1979; Hein et al., 1979). Then too, patterns may be established that will lead to severe disability in later life. Tobacco and alcohol dependence patterns established during the teen-age years are determinants of the two most important causes of premature preventable death during adulthood—lung cancer and alcohol related disabilities.

The first section of this chapter considers the determinants, epidemiology, and characteristics of drug taking in the young, followed by a discussion of general issues relating to treatment. The second section is a brief discussion of the particular drugs most liable to abuse.

GENERAL ISSUES CONCERNING SUBSTANCE USE AND MISUSE IN THE YOUNG

DEFINITIONS

Drug abuse refers to the use of any substance in a manner that deviates from the accepted medical, social, or legal patterns in a given society. There is frequently no discrete boundary that distinguishes appropriate use from abuse or misuse. Use of any drug may be intermittent or rare and not lead to adverse sequelae. In other cases the adolescent may become dependent on a drug in order to function at what he perceives to be a satisfactory level. This psychologic dependence or habituation varies in intensity and may lead to compulsive drug use, in which the acquisition and use of the drug become the primary concern of daily life. Physical dependence refers to an altered physiologic state induced by the repeated administration of a substance that requires its continued administration to prevent the appearance of a syndrome characteristic for each drug, the withdrawal or abstinence syndrome.

Addiction is a term that has been overused in the literature and the lay press to refer to both behavioral and pharmacologic events. It might more usefully be restricted to a pattern of compulsive drug use that is associated with physical dependence. In this sense of the term, neither a compulsive marijuana smoker nor a well adjusted patient being maintained on methadone would be considered an addict.

Tolerance refers to the decreased effect obtained from repeated administration of a given dose of a substance or to the need for increased amounts to obtain the effects that occurred with the first dose. Cross tolerance refers to the capacity of one drug to induce tolerance to another (Millman, 1979).

EPIDEMIOLOGY

Accurate assessment of the incidence and prevalence of drug abuse patterns is difficult, given a number of important measurement problems. Perhaps most serious is the reliance on self-report data. Since the use of most drugs is illicit or viewed as unacceptable by parents, teachers, and other adult figures, it is likely that most surveys provide conservative estimates of prevalence owing to underreporting by respondents. This occurs not only with respect to "hard" drugs such as heroin, but also with more acceptable substances such as tobacco. In fact, some recent studies suggest that the actual prevalence rates of adolescent cigarette smoking may be approximately double those reported in the most recent national surveys (Johnson et al., 1982). Another factor that contributes to the underestimation of prevalence rates is the number of tobacco, alcohol, and drug users who either drop out of school entirely or if still attending school do so irregularly. Since most surveys are conducted at schools, a significant number of substance users may not be represented.

Then too, as a result of the rapidity of change of drug abuse patterns, national survey data may be outdated by the time they are actually reported. There is also marked variation among various cultural groups and geographic locales. Problems also exist with respect to the way in which data are categorized. For example, age range or age groupings vary considerably across studies, making comparisons difficult. There is also great difficulty in distinguishing between intermittent controlled use and more compulsive and destructive use patterns. Finally data derived from overdoses, emergency room visits, or arrest records measure only those who are unsuccessful in their drug use patterns. Despite these significant limitations of the data, it is possible to highlight some of the broad trends in adolescent substance abuse over the past 20 years.

During this period there has been a rather dramatic increase in the use and abuse of all substances. With respect to drugs, this development has been associated with the emergence of a media popularized counterculture that rejected traditional values and sought to find meaning, truth, or escape in pharmacologically induced altered states of consciousness. From 1962 to 1967 the incidence of use of most drugs remained low; however, the use of marijuana began to increase during this period, particularly among metropolitan males. During the next 10 years (1967 to 1977) there was an explosive increase in the use of marijuana among youthful populations and a significant increase in the use of heroin and other opiates, cocaine, amphetamines, and psychedelics.

By 1977, over 25 per cent of 12 to 17 year olds reported having used marijuana at least once, as did 50 per cent of 18 to 25 year olds. Four per cent of 12 year olds, 15 per cent of 14 year olds, and 31 per cent of 18 to 21 year olds report current use of marijuana. However, during the period from 1974 to 1978 there was a slight leveling off of adolescents reporting current use of psychedelics, opiates, amphetamines, and depressants, whereas phencyclidine (PCP) use increased to about 6 per cent. During the last few years, at least in part because of its increased availability and purity, adolescent heroin use has increased markedly. Moreover, cocaine use is also increasing, particularly among the middle class and the affluent.

The prevalence of alcohol use in adolescents increased significantly from World War II to the mid-1960's and has remained reasonably stable since then. An estimated 70 per cent of the adolescent population has had alcohol exposures. There may be an increased prevalence of heavy drinking in today's teen-agers. Drinking to intoxication increased from 15 to 20 per cent of teenagers studied prior to 1966, to 45 to 70 per cent in the 1970's. The proportion of adolescents who reported being intoxicated at least once a month rose from 10 per cent before 1966 to 19 per cent between 1966 and 1975. Problem drinking varies widely (from 5 to 28 per cent) according to different surveys.

Several age factors become apparent when one examines the recent trends in substance abuse. First, the initiation of tobacco, alcohol, and drug use is primarily an adolescent phenomenon, and use increases throughout the adolescent years. The use of these substances rarely begins before adolescence, and if an individual has not used a given substance by the time adulthood is reached, the likelihood of becoming a regular user is significantly decreased. Finally there appears to be an increase in the abuse of all drugs in younger age groups, in which use may be even more dangerous.

There also appears to be a major shift in substance use patterns with respect to sex differences. Traditionally males were more likely to smoke, drink, or use drugs. These sex differences have decreased significantly in recent years, with many more females noted to be users.

Psychoactive drug taking, particularly of the illicit drugs, has become an integral part of the social rites of passage in the United States. Patterns of use may have changed from an epidemic situation in the late 1960's to an endemic situation at present. Knowledge about the various substances, their patterns of abuse, and their adverse consequences is widespread and has become part of the popular culture. According to this perspective, then, some adolescent substance use may be considered normative behavior within certain social contexts. Furthermore, although a significant proportion of previous substance use has led to adverse consequences, it may be argued that much current adolescent substance use is more sophisticated and controlled and frequently may not result in severe disability.

INITIATION AND DEVELOPMENT

Experimentation and Early Use. The initiation of substance use appears to be determined by the complex interaction of a variety of social, cultural, cognitive, attitudinal, personality, and developmental factors, as well as availability and legal sanctions (Blum et al., 1979; Jessor, 1975; Wechsler, 1973). Initially the use of most substances tends to be confined to social situations, solitary use being relatively infrequent. Moreover, the use of tobacco, alcohol, or drugs may provide a major focus for group interaction and identity (Becker, 1967; Jessor, 1975). Although psychosocial factors are primarily responsible for the initiation of substance abuse, psychobiologic and pharmacologic factors appear to become more important in maintaining regular use patterns.

A characteristic common to most adolescent substance users, particularly during the early stages of use, is the illusion of control. This phenomenon is in part due to inexperience, but more importantly may be determined by the unconscious mechanism of denial. Adolescents tend to exhibit a remarkable absence of concern about the potential dependency that may result from the frequent use of tobacco, alcohol, and certain drugs and overestimate their ability to avoid personally destructive use patterns. Adolescent cigarette smokers, for example, typically believe that they can quit smoking any time they want. It is not until they have actually made a serious effort to quit smoking that they gain a real appreciation of the extent to which they are both psychologically and physiologically dependent on cigarettes. Similarly, despite warning from parents, teachers, and the media, most young people believe that they are able to control their drug taking so that it will not become personally destructive for them (Millman, 1978; Zinberg et al., 1975).

Substance Use Hierarchy. Individuals appear to progress along what might be referred to as a substance use hierarchy. Initial experimentation typically begins with tobacco, beer, wine, and occasionally hard liquor. Marijuana use generally begins somewhat later. Some adolescents may go on to experiment with depressants, stimulants, and psychedelics, particularly LSD and PCP. Opiates and cocaine are usually the last substances in this decidedly nonlinear and complex progression.

However, in some sociocultural settings experimentation with opiates may begin somewhat earlier—just after the initial use of alcohol (Hamburg et al., 1975; Kandel et al., 1976).

Most adolescents stop at particular points in this sequence. Some adolescents use only one substance or one type of substance during this period. More often a variety of substances are used, depending on availability, situational factors, and the needs of the user. The pattern of use may vary from intermittent use of carefully selected substances on special occasions, to disorganized and dangerous multiple substance abuse patterns characteristic of severely disturbed adolescents. Psychologic dependence, tolerance, and then physical dependence may result in a compulsive substance abuse picture. For example, well adjusted teen-agers may use marijuana, methaqualone, or cocaine before going to a concert. In this instance use is confined to a specific situation. On the other hand, there is general disdain by this type of user for compulsive abusers of depressants or heroin who are unable to function socially or in school (Millman, 1978). Moreover, as more exotic substances are given up, there frequently is a return to substances used earlier, such as marijuana or alcohol (Hamburg et al., 1975).

Considerable controversy exists over whether the use of certain substances (e.g., marijuana) leads inexorably to other, more dangerous substances (Goode, 1974). Although many individuals do not go on to use "hard" drugs, it has been pointed out that without the prior use of tobacco, alcohol, and even coffee, there would be no progression (Blum and Richards, 1979; Hamburg et al., 1975). Moreover, the experience of altering or controlling consciousness or mood with a psychoactive substance, if perceived as positive, provides the impetus for experimentation with stronger substances. It is likely, then, that the use of tobacco, alcohol, and marijuana increases the likelihood of experimentation with heroin or depressants; use of these drugs in turn may result in the development of severe drug abuse problems in some youngsters. Nonetheless it would appear that the psychosocial and biologic predisposition of the adolescent is more critical to the development of severe drug problems than whether he has previously used tobacco or marijuana (Millman, 1978).

DETERMINANTS OF USE

A vast array of social, personality, cognitive, attitudinal, behavioral, and developmental factors have been found to be associated with tobacco, alcohol, and drug use. A multitude of studies have identified one or more variables that differentiate users and nonusers in adult and adolescent populations. Most of these studies have utilized relatively unsophisticated methodologies and have focused on only a few of the potential determinants of substance use. Thus, knowledge about the complex interaction and relative contribution of these variables is quite limited. There are few well designed prospective studies persuasively suggesting common etiologic patterns or causal pathways. None of these factors and no psychologic or other sort of examination have been shown to be predictive of the development of severe drug abuse behavior or of which people will use which drug.

Social Factors. Adolescent substance use appears to be promoted by factors relating to both the family and the larger social environment. The earliest influence to smoke, drink, or take drugs comes from the family. Adolescents growing up in families in which parents or older siblings are substance abusers tend to become drug takers themselves (Bewley et al., 1974; Borland and Rudolph, 1975; Demone, 1973; Gergen et al., 1972; Kandel, 1973; Wechsler and Thurn, 1973; Williams, 1973). Although the predominant influence coming from the family appears to be the modeling of substance use behavior of parents or siblings, parental attitudes or perceived parental attitudes can also influence the adolescent's decision to begin smoking, drinking, or taking drugs (Hunt, 1974). Other familial factors that have been found to be related to adolescent substance abuse include family instability, parental rejection, under- or overdomination by parents, and divorce (Braucht et al., 1973; Gergen et al., 1972; Seldin, 1972; Wechsler and Thurn, 1973).

Adolescents from lower socioeconomic groups are generally more likely to become substance abusers than adolescents from higher socioeconomic groups (Borland and Rudolph, 1975; Gergen et al, 1972; USPHS, 1976; Wechsler and Thurn, 1973). Although socioeconomic measures typically include income or educational components, the latter seem to be more important with respect to substance abuse. Adolescents from families in which one or both parents went to college or who themselves plan to go to college and are enrolled in college preparatory courses are less likely to smoke cigarettes or get into trouble with drugs and alcohol.

Peer influence plays a central role in the initiation, development, and maintenance of substance use (Freeland and Campbell, 1973). Smoking, drinking, and taking drugs are social activities that are rarely done alone in the early stages of use. Many studies have reported a direct relationship between individuals' substance use and that of their friends. In a 1974 HEW study, for example, 87 per cent of the teen-age smokers surveyed indicated that at least one of their best friends was a regular smoker (USPHS, 1976). The nature of peer influence is unclear. The findings in some prospective studies support the widely accepted notion that substance use by peers influences an individual's use (Jessor et al., 1973; Sadava, 1973). However, other studies suggest a process of mutual selection.

Peer influence may derive from perceived supportive attitudes of peers or from the perception that the use of a given substance is normative. Substance abusers typically overestimate the prevalence of use among their peers, and their degree of involvement with a particular substance is related to their estimate of the proportion of peers using that substance.

The media appear to be negative influences supporting substance use. Both cigarettes and alcohol are heavily advertised despite the fact that cigarette advertising no longer appears on television. Drinking is portrayed as being manly or sophisticated and has an image similar to that of cigarettes. Marijuana and other drugs, on the other hand, are not advertised; in recent years, however, marijuana and cocaine use has been increasingly portrayed in music, movies, and television as behavior that is socially acceptable, although currently illegal.

Developmental Factors. As with other child and adolescent behavior patterns, substance use and abuse must be viewed in a developmental context. The dynamics of normal psychosocial development, including changes in the adolescent's view of the world, reasoning processes, and parental and peer pressures, appear to play a role in the promotion of substance use and abuse.

Children's cognitive development undergoes significant change as they proceed through what Piaget has termed the concrete operational stage (ages eight to 11) to the stage of formal operations (ages 12 and over). During this time there is a gradual transition from thought that is rigid, literal, and grounded firmly in the here and now to thought that is more relative, abstract, and hypothetical. As a consequence the adolescent is able to envisage a wide range of possibilities and logical alternatives, to accept deviations from established rules and norms, and to recognize the frequently irrational and inconsistent nature of adult behavior.

This has direct implications for experimentation with cigarettes, alcohol, and drugs. For example, although younger children generally believe that cigarette smoking is bad and typically state that they will never smoke, as they approach adolescence and begin viewing cigarette smoking from a more relative perspective, they may formulate or consider arguments in favor of smoking (e.g., "It will help me to be more popular or attractive"). In addition, the adolescent's new cognitive orientation may permit him to discover inconsistencies or logical flaws in arguments being advanced by adults concerning the risk of substance use. Young people may use adult figures who smoke or drink as support for their own behaviors.

The differential and changing influence of peers and parents is also an important consideration in examining the initiation of substance use and abuse. During early childhood parents exert the most powerful influence on children. However, upon entry into the school environment, this influence gradually begins to decrease, while the influence of peers and other socializing agents such as teachers becomes increasingly important. Parental influence continues to decline as the child approaches adolescence, and peer influence increases dramatically. However, parents still maintain predominance over some areas of concern, such as career and educational choices. Peer influence, on the other hand, is typically predominant concerning life style matters, such as clothing and music. Because of this developmental shift in the relative importance of peer and parental influence, peers generally exert more influence with respect to cigarette smoking and the use of alcohol and other drugs than do parents.

Another important developmental phenomenon that may have an impact on experimentation and the early use of drugs concerns responsiveness to conformity pressures. As dependence on the peer group increases, there is a corresponding rise in conformity behavior. Although very young children are almost impervious to pressure to conform, the tendency to conform becomes evident during middle childhood. Conformity behavior increases rapidly during preadolescence and early adolescence and then steadily declines from middle to late adolescence.

At the same time individual susceptibility to conformity pressures varies greatly. It is likely that developmental changes in responsiveness to conformity pressures increase the adolescent's potential for yielding to the norms of the peer group with respect to smoking, drinking, or drug taking behavior.

Psychologic Factors. Despite a plethora of research data relating to the psychologic determinants of adolescent drug and alcohol abuse, controversy persists as to whether drug abuse or dependence results from specific personality patterns or psychodynamics or whether particular drug use patterns are associated with certain personality types (Millman, 1981; Zinberg, 1975). Youthful drug abusers have been described as having an external locus of control (the belief that their life is controlled by external forces such as fate or chance; Williams, 1973), lower self-esteem (Braucht et al., 1973), a higher degree of dissatisfaction and pessimism (Coan, 1973), a greater need for social approval, and less social confidence. In addition, substance users have been found to be more anxious, less assertive, more impulsive (Williams, 1973), rebellious (Jarvik et al., 1977), and more impatient to assume adult roles than nonusers.

Despite the association of these characteristics with substance use, they do not occur exclusively in users nor are they absent among nonusers. Then too, many of these data are based on retrospective formulations after people have already become heavily involved with drugs. The personality patterns or psychopathologic disorders noted may be a reaction to the drugs or use patterns in a society that stigmatizes or punishes such behavior. There are no data from prospective studies that have identified specific psychodynamic patterns or psychopathologic disorders as being predictive of drug abuse. There is no good evidence for the existence of a well defined "addictive" or "alcoholic personality" type (Millman, 1978; Zinberg, 1975).

Adolescent drug abusers vary markedly with respect to personality patterns and psychopathologic disorders; some may be normal, whereas others are significantly disabled. It is necessary to define the meaning of the alcohol or drug use in each child.

In some young people, drug taking may represent attempts at self-medication because of painful affects resulting from shame, rage, loneliness, and depression (Khantzian et al., 1974). Substance abuse may also be an attempt to satisfy or control unacceptable drives, including sexual needs, as well as primitive, sadistic, and aggressive wishes. It has been postulated that there may be an impairment in the defensive structure against these drives or feelings such that they are experienced as being overwhelming. It has been suggested that some drug abuse may be symptomatic of a masked depression and that the boredom, restlessness, apathy in school, wanderlust, philosophizing, sexual promiscuity, and frantic seeking of new activities characteristic of many youthful substance abusers may actually represent depressive states (Carlson and Kantwell, 1980; Gallemore and Wilson, 1972).

Narcissistic, borderline, and overtly psychotic adolescents also use drugs in an attempt at self-treatment of severe symptomatology (Wurmser, 1974). Intermittent or rare use of any of the drugs need not be associated with psychopathologic disorder; compulsive use patterns more often are (Khantzian et al., 1974). The more aberrant an individual's drug abuse pattern is for his social or cultural milieu, the more likely that there will be a significant degree of psychopathologic disability.

The choice of drug may also reflect personality patterns or psychiatric symptomatology. Borderline or psychotic adolescents may use opiates to control their symptomatology; depressants or alcohol may also be used (Khantzian et al., 1974;

Verebey et al., 1978). Opiates have been shown to have significant antipsychotic properties. Amphetamine use may be a form of self-medication for some young people with attentional deficits; compulsive amphetamine abusers are often unable to concentrate or eat unless they have taken the drug. Alcohol may be used by some people to suppress panic attacks or to allow the expression of long suppressed anger. Many severely disturbed young people may use only opiates or depressants and not use marijuana, hallucinogens, or stimulants, since these drugs may further weaken their hold on reality and amplify anxious or paranoid feelings or thoughts. It is interesting that some severely disabled young people do take the latter drugs. It is possible that the intense and unpleasant psychoactive effects may insulate them from their own thoughts and feelings or perhaps facilitate attempts to rationalize their "craziness" (Millman and Khuri, 1981).

In addition to the choice of drugs, abuse patterns may also reflect personality structure and psychopathologic disturbance (Zinberg, 1975). Borderline or psychotic youngsters may use a wide variety of drugs in a disorganized fashion and experience frequent adverse reactions or overdoses. Some of these people have little else of which to be proud and call themselves "garbage heads" with great pride. In Erikson's terms they depend on a "negative identity" for their feelings of worth or self-definition.

The pharmacologic effects of chronic use of particular drugs may also play a role in the development of psychopathologic disorders. For example, the continued use of opiates may result in chronic depressive states through alteration of neurochemical factors. This is particularly provocative in view of the recent work on endogenous morphine-like substances or the discovery of high affinity binding sites for benzodiazepines in the brain (Verebey, 1978).

Withdrawal of drugs from severely disturbed individuals who have been self-medicating may result in a deterioration of psychosocial functioning. It is not unusual for a well compensated heroin or depressant addict to become psychotic when detoxified. On the other hand, seemingly severe psychopathologic disorders may improve in some individuals as drugs are withdrawn.

Compulsive drug abuse patterns are also influenced by conditioned learning. The dysphoric symptoms that the drug taking behavior controlled or some of the situations attendant to drug taking behavior become, in time, the conditioned stimulus for the experience of drug craving and the associated drug seeking behavior. Long abstinent ex-addicts or recovering alcoholics experience drug craving and may even experience aspects of a withdrawal syndrome when they return to a site of former drug use or when they suffer a real or imagined loss. Then too, learning is an important determinant of the subjective drug experience. For example, marijuana and alcohol can elicit different and contrasting responses depending on the user's expectation and group pressure. Marijuana is used in some cultures as a work enhancer or appetite suppressant, in contrast to its well publicized effects in American adolescents of decreasing motivation and stimulating appetite.

Sexual Factors. As a result of lack of experience or feelings of anxiety, adolescents often have considerable difficulty with sexual performance or deriving pleasure from sexual experiences. Many young people find that low doses of sedatives, alcohol, or opiates relieve inhibitions, increase desire, and improve performance. Some young men are able to sustain an erection only when they are "stoned." The great popularity of methaqualone in recent years is a function of this phenomenon. Continued use leads to the development of tolerance and necessitates an increase in dosages. These higher doses impair sexual performance and may lead to a recurrence of the anxious, depressed feelings. Marijuana in some people heightens desire and sensitivity; in others it is associated with a disinterest in sex. The compulsive use of most drugs is associated with decreased sexual interest and performance (Millman, 1978).

Cognitive, Attitudinal, and Behavioral Factors. The knowledge and attitudes individuals have about tobacco, alcohol, and drugs also influence their initiation and early use. Logically one would expect individuals who are aware of the health and safety hazards of these substances and who have negative attitudes toward their use to be less likely to abuse them. Tobacco and drug prevention programs have been based largely on the assumption that increased knowledge about these substances and the consequences of their use would be effective in reducing use.

Some studies, however, have revealed that users frequently have more accurate knowledge about drugs than nonusers (Fejer and Smart, 1973; Goodstadt et al., 1977; Swisher et al., 1972). Moreover, studies attempting to decrease substance abuse by increasing students' knowledge either have had little or no effect on drug use or have actually increased substance abuse. It has been shown that attitudes toward the use of a particular substance were predictive of later use of that substance (Downey and O'Rourke, 1976; Jessor et al., 1973; Sadava, 1973). Strong religious attitudes have been found to be inversely correlated with substance use.

Substance abusers and nonusers have been found to differ in regard to several behavioral dimensions, suggesting a difference with respect to orientation, values, and aspirations. Substance abusers get lower grades in school and tend either to work after school hours or simply "hang out" with their friends, rather than participating in organized extracurricular activities such as sports or clubs (Demone, 1973; Jessor et al., 1973; Johnson, 1973; Kandel, 1973). Drug takers are more likely to be involved in antisocial or unacceptable behavior such as fighting, swearing, lying, cheating, stealing, gambling, and causing disciplinary problems in school.

DIAGNOSIS

The provision of appropriate treatments depends upon accurate characterization of the specific drugs of abuse and their use patterns in each adolescent, as well as the psychologic set and social situation attendant on these behavior patterns. The nature and degree of drug induced psychoactive effects and the presence of abstinence phenomena should be evaluated. This requires careful history taking, including a complete drug use history and a comprehensive physical examination.

Assessment of the mode of administration and any adverse effects of the drugs is critical to the formulation of the diagnosis. Chronic sinusitis or perforation of the nasal septum may suggest the "sniffing" of cocaine, whereby the material is insufflated and absorbed through the mucous membranes of the nasopharynx and respiratory tract. Signs of repeated intravenous injection ("tracks") suggest heroin, cocaine, or amphetamine abuse. Patterns of behavior and dress may further define the clinical picture.

Qualitative procedures for the detection in urine of the drugs of abuse are available in many laboratories. Positive results will occur if a dose sufficient to produce pharmacologic effects has been taken in the 24 hours preceding the urine sample. Since false positives occur and results are often not immediately available, these tests should be used to confirm clinical impressions.

GENERAL TREATMENT CONSIDERATIONS

In order to provide appropriate treatment, it is necessary to consider the social and psychologic characteristics of the adolescent as well as the pharmacology and patterns of abuse of the particular psychoactive substances.

Young people are often unable to appreciate the significance of their substance abuse because of denial or because these behavior patterns have not yet led to adverse consequences. It is often not until such behavior has come to the attention of family, school, or legal authorities that the adolescent is brought to the attention of treatment

personnel (Institute of Medicine, 1978). Then too, adolescents, particularly drug abusing ones, often adopt an implicitly antisocietal stance and tend to view adult authority figures, including the physician, as hostile or untrustworthy (Bernstein and Shkuda, 1974).

STAGES OF TREATMENT

Treatment should include initial and long term stages. During the initial phase, intoxication or withdrawal signs and symptoms must be dealt with (Table 37–1). Acute medical or psychologic situations should be treated. Residential or legal issues of a pressing nature must be considered as well. During this phase the decision must be made as to whether the youngster can be treated by the pediatrician in his office, referred to a psychiatrist or specialized treatment program, or admitted as an inpatient to a medical or psychiatric hospital or to a drug treatment program.

Subsequent to this initial treatment stage, which may last days to weeks, provision must be made for long term care. There is little question that it is much easier to treat an overdose or withdraw a youngster from drugs, no matter how severe the dependence, than it is to help him remain off the drugs. Long term follow-up must be assured, since these people are always at risk for relapse.

ATTITUDES OF TREATMENT PERSONNEL

The attitudes of adult authority figures may be an obstacle to case finding and to the provision of effective treatment. All too often treatment personnel consider adolescent substance abusers to be weak, lacking in will power, amoral, or even criminal. The common view that "they did it to themselves" may severely hinder treatment efforts.

Treatment of adolescent substance abusers is often extremely frustrating to the physician or the treatment team. The problems do not lend themselves to easy solutions. Regardless of the treatment modality used, progress is often slow and inconsistent. Many patients are unable to stop using drugs. Other patients, after being successfully detoxified from a given substance and after a period of sustained abstinence, may relapse to their old compulsive use patterns. Therapists often view drug taking behavior as acute illness and may have unrealistic expectations that their patients will become completely cured. In view of the chronic nature of the determinants of such behavior and the fact that the neurochemical impact of compulsive substance use may be more protracted than previously thought, a more useful

Table 37–1. COMMON DRUG POISONINGS, SIGNS OF TOXICITY, AND TREATMENT*

Drug	Mild Toxic Signs	Tissue for Diagnosis	Treatment	Severe Overdose Signs	Treatment
Opiates: Heroin Morphine Demerol Methadone	"Nodding" drowsiness, small pupils, urinary retention, slow and shallow breathing; skin scars and subcutaneous abscesses; duration 4–6 hr.; with methadone, duration to 24 hr.	Blood Urine	Naloxone (Narcan) 0.01 mg./kg. i.v., nalorphine (Nalline) 0.1 mg./kg. i.v., levallorphan (Lorfan) 0.02 mg./kg.; repeat in 10–15 min. if necessary; then repeat in 3 hr. if necessary	Coma; pinpoint pupils, slow irregular respiration or apnea, hypotension, hypothermia, pulmonary edema	Naloxone, nalorphine, levallorphan; if no response by second dose, suspect another cause; treat shock; find and detect infection
Depressants: Alcohol	Confusion, rousable drowsiness, delirium, ataxia, nystagmus, dysarthria, analgesia to stimuli	Blood, urine, breath	Alcohol excitement: diazepam or chlorpromazine	Stupor to coma; pupils reactive, usually constricted; oculovestibular response absent; motor tonus initially briefly hyperactive, then flaccid; respiration and blood pressure depressed; hypothermia; with glutethimide, pupils moderately dilated, can be fixed; with meprobamate, withdrawal seizures common; with methaqualone, coma, occasional convulsions, tachycardia, cardiac failure, bleeding tendency	Intubate, ventilate, gavage; drainage position; antimicrobials; keep mean blood pressure above 90 mm. Hg and urine output 300 ml./hr.; avoid analeptics; hemodialyze severe phenobarbital poisoning As above; diuresis of little help
Barbiturates Glutethimide (Doriden) Meprobamate (Equanil)		Blood Blood Blood	None needed for acute toxicity; withdraw drug under supervision if patient is chronic user		
Methaqualone (Quaalude, Sopor, Mandrax)	Hallicinations, agitation, motor hyperactivity, myoclonus, tonic spasms	Blood Urine			
Chlordiazepoxide (Librium)	Usually taken with another sedative if poisoning the attempt	Blood Urine			
Diazepam (Valium)					
Stimulants: Amphetamines Methylphenidate	Hyperactive, aggressive, sometimes paranoid, repetitive behavior; dilated pupils, tremor, hyperactive reflexes; hyperthermia, tachycardia, arrhythmia; acute torsion dystonia	Blood Urine	Reassurance if mild Diazepam or chlorpromazine if severe	Agitated, assaultive and paranoid excitement; occasionally convulsions; hypothermia; circulatory collapse	Chlorpromazine

conceptualization would be to view substance abuse as a chronic illness with remissions and exacerbations, more like diabetes or schizophrenia than pneumonia. To be successful, the physician and the treatment team must be prepared for a sustained effort utilizing continued enthusiasm and imagination throughout the process.

OBSTACLES TO TREATMENT

Adolescent substance abusers referred for treatment may not be willing participants. It is generally difficult to engage adolescent substance abusers in regularly scheduled formal therapy or counseling sessions, particularly at the outset of treatment. Instead it may be more efficacious to conduct short and informal sessions dealing with very practical issues of concern to the adolescent, such as legal, financial, or social problems. Discussions about such matters as music, sports, and clothing can be valuable in helping to establish a therapeutic alliance.

Adolescent substance abusers frequently know as much as, or more than, their physicians about drug related behavior. This may be quite threatening for doctors more used to dealing with patients who are relatively naive about their psychologic or physical problems. It is often extremely difficult to persuade adolescent substance users to alter their behavior with respect to substance abuse if they continue to live in an environment

Table 37–1. COMMON DRUG POISONINGS, SIGNS OF TOXICITY, AND TREATMENT *(Continued)*

Drug	Mild Toxic Signs	Tissue for Diagnosis	Treatment	Severe Overdose Signs	Treatment
Stimulants (continued):					
Cocaine	Similar but less prominent than above; less paranoid, often euphoric	Blood, clinical appraisal	Reassurance Diazepam or chlorpromazine	Twitching, irregular breathing, tachycardia	Sedation
Psychedelics (LSD, mescaline, psilocybin, STP) Phencyclidine	Confused, disoriented, perceptual distortions, distractable, withdrawn or eruptive, leading to accidents or violence; wide-eyed, dilated pupils; restless, hyper-reflexic; less often, hypertension or tachycardia		Reassure; "talk down"; do not leave alone Diazepam	Panic	Reassure; diazepam satisfactory; avoid phenothiazines Symptomatic supportive; Acidify urine with ammonium chloride
Atropine-scopolamine (Sominex)	Agitated or confused, visual hallucinations, dilated pupils, flushed and dry skin		Reassure	Toxic disoriented delirium, visual hallucination; later, amnesia, fever, dilated fixed pupils, hot flushed dry skin, urinary retention	Reassure; sedate lightly; (1) avoid phenothiazines; (2) do not leave alone
Antidepressants:					
Imipramine (Tofranil), amitriptyline (Elavil)	Restlessness, drowsiness, tachycardia, ataxia, sweating	Clinical Blood		Agitation, vomiting, hyperpyrexia, sweating, muscle dystonia, convulsions, tachycardia or arrhythmia	Symptomatic; gastric lavage
MAO inhibitors: tranylcypromine (Parnate), phenelzine (Nardil), pargyline (Eutonyl)	Hypertensive crises, agitation, drowsiness, ataxia	Clinical Blood	Withdrawal	Hypotension; headache; chest pain; agitation; coma, seizures and shock	Symptomatic; gastric lavage
Phenothiazines	Acute dystonia, somnolence, hypotension	Clinical Blood	Benadryl 0.50; withdrawal	Coma; convulsions (rare); arrhythmias; hypotension	Symptomatic; gastric lavage

*From Wyngaarden, J. B., and Smith, L. H. (Editors): Cecil Textbook of Medicine. Ed. 16. Philadelphia, W. B. Saunders Company, 1982. Used by permission.

that is conducive to their continued use of cigarettes, alcohol, or drugs. Chronic drug users may be involved with a subculture that rejects conventional values and mores. It is often imperative to attempt to move the individual to a more positive environment. Unfortunately removal to a hospital, drug treatment program, or specialized school may be the only options to effect this change.

In general, therapists should avoid assuming a critical parental role involving exhortation and threats. Optimally, the treater should strive to develop what might be characterized as a dignified supportive relationship with an accepting, interested, and knowledgeable adult.

DEALING WITH MISINFORMATION

It is not unusual for adolescents either to minimize or exaggerate the extent of their substance use. Adolescents may give much misinformation, deliberately distorting historical data, and even lying about current behavior. Adolescents frequently test limits and attempt to provoke punitive rejection. In life threatening situations strong actions may be necessary. In general, however, confrontation and coercion should be viewed as last resorts to be used only in the most compelling circumstances.

FAMILY INVOLVEMENT

It is often useful to engage the immediate family in therapeutic encounters with youthful patients, particularly when intervention in family dynamics might be expected to lead to improvement in the treatment of the patient or when particular familial behavior patterns are clearly facilitating the drug abuse behavior. In other situations, in order to encourage feelings of trust between the therapist and the patient, and to allow him to feel that he is worth the encounter, it may be more useful to see the patient alone. Family sessions might be held in addition, or with another therapist. When treating the drug abusing adolescent who may be supporting his habit by stealing or who may be involved in other antisocial acts, it may be necessary to advise family members to take appropriate measures in order to protect themselves. In some instances it may even be necessary to suggest that a family turn out a young drug user so that he will be forced to seek treatment (Kaufman and Kaufman, 1979).

USE OF MEDICATION

An underlying psychopathologic disorder may be identified when the youngster is detoxified or when the drug use is controlled. Individual treatment may then be formulated. As a general rule, most substance abusers should be treated without medication, particularly those drugs that are liable to abuse. In some instances, however, it may be necessary to administer psychotropic drugs to patients in a careful and highly controlled manner, since many compulsive substance abusers have been self-medicating and will be unable to tolerate their symptoms.

COMPREHENSIVE TREATMENT PROGRAMS

Adolescent substance abusers may require intensive and comprehensive care, including educational, vocational, legal, medical, and psychologic services. Physicians are often unable to provide this range of services for professional and economic reasons.

A variety of treatment programs are available that may provide these services. These are of differing quality and their effectiveness is generally undocumented. These include school and community based programs as well as therapeutic communities. To a large extent, comprehensive treatment programs designed specifically for adolescent populations appear to be the most efficient and most effective means of providing both care and preventive services. Many of these programs are run by graduates of the particular program, who themselves are former drug users. In addition to providing a broad range of services that might be needed by the adolescent, these programs help to reduce the sense of isolation that many youngsters feel, and the staff provides the patients with positive role models. One deficiency of these programs is that, owing to the lack of appropriate professional staff members, they are not able to handle adolescents with significant psychopathologic disorders.

One model that provides the broad range of treatment options required by adolescent substance abusers includes a coordinated inpatient and outpatient therapeutic community structure along with extensive medical and psychiatric back-up. The program cycles individuals through a series of stages. The nature and severity of their substance abuse problem determine the precise point at which they enter the program, and individuals progress at their own rate. A youngster might enter as an inpatient and eventually move to an intensive day care center, and then to a more limited outpatient situation. Thus, the early stages of the program involve rather intensive treatment, which gradually decreases as the need for services and support decreases and as the patient develops stronger ties to the larger society.

THE DANGER OF OVERTREATMENT

Some substance use may be considered reasonably normal within certain social settings; the drugs may be used in a controlled fashion and may not be dangerous. Still, the discovery that an adolescent is using one or more substances may be frightening or even incomprehensible to parents as well as authority figures. As a consequence, some adolescents who are not in need of intensive treatment are referred to therapists, treatment programs, or psychiatric facilities. Inappropriate treatment may confirm the worst fears of these adolescents concerning their abilities, potential, and even their sanity. In some cases the cure may be worse than the disease, since treatment programs may be more disruptive than the effects of the substance abuse.

Once a careful and thorough evaluation has indicated that the adolescent substance abuser is not in danger and is developing reasonably well, appropriate treatment might simply involve a few informal sessions with arrangements made for follow-up. During these sessions the therapist should attempt to provide the adolescent with reassurance concerning his sanity as well as prospects for the future.

DRUGS OF ABUSE

ALCOHOL

It is often difficult to distinguish between normal or "social" drinking and "problem" drinking in the young. Young people often drink less regularly than adults, but tend to consume a larger amount on a drinking occasion. In many adolescent groups, at a party or other social occasion, relatively heavy drinking to intoxication is the norm. This behavior has not changed significantly during the last 30 years, though younger adolescents and more females are now involved (Blane and Hewitt, 1977). It might be useful to define problem drinking as drinking that adversely affects an individual's physical or psychosocial function on a continuing basis. The term alcoholism should be reserved for a situation marked by addiction to alcohol, that is, overwhelming involvement with the acquisition and use of the drug in the presence of physical dependence.

Inner city and other socially disadvantaged young people begin drinking early, sometimes at 10 to 12 years of age. Middle class and suburban adolescents generally begin drinking at a later age. They drink in association with friends at social functions and they most often have parents who also use alcohol. Most continue to drink in a manner that is acceptable to their sociocultural milieu. As distinguished from their elders, who may be loath to use illicit substances, young people regard alcohol as one of the depressants and use it in combination with a variety of other drugs. It is used to offset the anxiety and tension induced by marijuana or the stimulants, or to prolong or potentiate the "high" obtained from other depressants or opiates. A small number come to depend on alcohol for satisfactory social functioning and begin drinking daily, often alone, for a variety of psychobiologic or social reasons. A reasonably normal or anxious adolescent who becomes a problem drinker may be reacting to an acute or chronic stressful situation, such as loss of a loved one or fear of sexual inadequacy based on several painful experiences. The depression, anxiety, and insomnia that follow a day of heavy drinking may be an important reinforcement of drinking behavior the next day. At some point this combination of habituation and early physical dependence may lead to loss of control, constant alcohol craving, and an addictive pattern, in which the use of alcohol may be a primary life concern. Concurrent dependence on other drugs is not unusual in these people (Wechsler, 1976). The use of alcohol or another drug may then lead to a deterioration of the conditions of life. Moreover, the psychopathologic pattern noted may be a reaction to the drug use rather than an etiologic factor.

At the same time it should be understood that whereas problem drinking during the adolescent years may lead to chronic alcoholism in later life, the prevalence of alcoholism during these early years remains relatively low. Those who do become alcoholic appear to have significant premorbid psychopathologic behavior patterns. Whereas illicit drug taking often appears to be attractive or romantic to many otherwise normal young people and is often a group activity, alcohol addiction has few positive associations, is often a cause of shame, and is generally associated with increasing isolation. It may represent attempts at self-medication by borderline or psychotic youngsters or may allow other severely disturbed young people to express their symptomatology, including hostile or aggressive feelings, without concomitant anxiety.

Certain subsets of young people do indulge in more excessive drinking. Children of alcoholic parents, particularly of alcoholic fathers, appear to have a genetic and familial predisposition to become alcoholic. Delinquent adolescents drink more and develop more pathologic symptoms. American Indian, black, and Hispanic children are also more likely to become problem drinkers (Blane, 1977).

Pharmacology

The psychoactive effects are similar to those of other depressants and tranquilizers. In nontolerant social drinkers the degree of intoxication depends upon the amount of alcohol consumed and correlates roughly with blood alcohol levels. The absorption of alcohol is influenced by the rate of gastric emptying and small intestine absorption. Congeners in beverage alcohol or the ingestion of food reduces the rate of absorption. The consumption of 180 ml. of distilled spirits on an empty stomach will produce a blood alcohol level of 100 mg. per 100 ml., which is associated with a state of mild to moderate sedation and inebriation. Anxiety may be reduced and individuals may perform better in social situations; sexual pleasure and performance may be enhanced. Overt signs of intoxication occur at blood levels between 100 and 200 mg. per 100 ml. At these levels there is impairment of visual-motor coordination, of the integration and evaluation of sensory information, and of sustained attention to stimuli, judgment, and sexual performance. Blood levels above 200 mg. per 100 ml. are associated with severe intoxication and marked sedation. Continued drinking of large amounts leads to the development of tolerance such that chronic alcohol abusers may appear sober and are able to function effectively even at high blood levels. The range of tolerance is narrow compared with that for the opiates;

blood alcohol levels of 450 mg. per 100 ml. are generally associated with severe somnolence or coma for both nontolerant social drinkers and alcoholics.

Cross tolerance occurs between alcohol and the general depressants, though not between the opiates and alcohol. When depressants are taken along with alcohol, synergism or potentiation of sedative effects occurs. Some synergism or potentiation may also occur with the opiates. The combination of alcohol with these drugs has proven to be lethal as a result of these additive effects and the narrow range of tolerance.

Psychoactive effects are also influenced by set and setting. If an adolescent drinks alcohol in a situation in which inebriation is anticipated, such as a rock concert or party, he will get drunk. Under other circumstances it is possible to maintain a sober demeanor despite significant drinking.

Dependence and Withdrawal Process

After a bout of acute intoxication, the anxiety, depression, feelings of guilt, headache, nausea, vomiting, diarrhea, agitation, and tremulousness, termed the "hangover," may be equivalent to the first signs and symptoms of a withdrawal syndrome. Persistent drinking of large amounts of ethanol increases the severity of this acute withdrawal syndrome, which is characterized by an increase in the intensity of these symptoms and signs, particularly tremulousness and agitation, as early as four to six hours after cessation of drinking and which generally remits or is significantly diminished by the third day. Auditory or visual hallucinations (alcohol hallucinosis) may occur in association with this syndrome. In severe cases grand mal seizures may occur 12 to 24 hours after cessation of drinking, generally followed by a postictal state. Adolescent alcoholics usually have not been drinking long enough to develop delirium tremens, though this does occur. It is marked by global confusion, disorientation, delusions and hallucinations, severe agitation, and autonomic hyperactivity. This syndrome, which has become infrequent in treated populations, peaks at 72 to 96 hours after the termination of drinking and is potentially life threatening. Hospitalization and intensive care are required.

Evidence is accumulating to indicate that following this acute withdrawal syndrome there may persist a protracted withdrawal syndrome, characterized by tremulousness, anxiety, depression, and insomnia, which may last for as long as six months after the cessation of drinking. This poorly understood though profound symptom complex may act as a physiologic and psychologic stimulus for the resumption of alcohol dependence.

Adverse Effects

The acute intoxicating effects of alcohol are more important causes of disability and death in young people than are the sequelae of chronic use. These include overdose, accidents, violent behavior, homicide, and suicide. Interestingly, a significant number of the 50 per cent of automobile fatalities involving the use of alcohol may result from chronic use and addiction rather than from episodic acute intoxication. Chronic use leads to a deterioration in social performance and to psychiatric illness, particularly depressive and anxiety states. The adverse physical effects of chronic use are less often seen in youthful populations, though they do occur. Almost every organ system may be adversely affected by this drug.

Treatment

Since many adolescent alcohol abusers deny the extent of the problem and are unable, or refuse, to alter their behavior patterns, it is often difficult to provide appropriate treatment. It may be necessary to attempt to develop a therapeutic alliance, seeing the patient often on an outpatient basis, with a view toward specific intervention at a later date. In other instances coercive measures must be employed, such as the application of familial, peer, or even financial pressure to force a young boy or girl into treatment. This problem may be compounded by the parochial attitudes of drug abuse treatment personnel and the narrow focus of many treatment programs. In general, alcohol treatment programs are focused on adult patients and are not skilled in the care of the youthful patient, particularly when there is concurrent abuse of other drugs. It is often difficult to find personnel or programs sensitive to the particular needs of this adolescent population. As with all other drug abusers, treatment must include both short and long term phases.

Alcohol Withdrawal Syndrome

If compulsive use and physical dependence are suspected, it is necessary to hospitalize the patient during the withdrawal period. Alternatives to the traditional medical or psychiatric facilities in the form of nonmedical detoxification centers or therapeutic communities have also become available. In some cases in which close supervision is possible, treatment of the withdrawal syndrome may be accomplished on an outpatient basis.

Treatment should begin with a careful search for associated surgical and medical illnesses, particularly cerebral trauma, subdural hematoma, pneumonia, and liver disease. It is good practice to give all patients multivitamin supplements; thiamin, 50 to 100 mg. intramuscularly, should be given to all severely dependent patients. Whereas paraldehyde, barbiturates, phenothiazines, and other drugs that are cross tolerant with alcohol have been used successfully to treat the withdrawal syndrome, benzodiazepines have been shown to be at least as effective and considerably less toxic than all other drugs. Chlordiazepoxide and diazepam, both long acting drugs with pharmacologically active metabolites, should be used in doses sufficient to blunt the patient's agitation and produce a state of mild drowsiness but not induce excessive sedation. Chlordiazepoxide, 25 to 100 mg., or diazepam, 5 to 20 mg., every six hours is a useful initial dosage regimen; the dosage should then be tapered gradually over the course of five to seven days or longer. In the event of extreme agitation, higher and more frequent dosages or intravenous administration may be indicated (Sellers and Kalant, 1976).

There is evidence to suggest that sensory deprivation may increase the severity of the hallucinations and agitation incident to the alcohol withdrawal syndrome. A nurse or specially trained counselor should therefore be in frequent attendance, providing reassurance and continuous cues to reality. Patients should be encouraged to remain active, eating regularly and participating in routine activities.

Long Term Treatment

A prolonged stay in an inpatient rehabilitation program or other drug free environment should be mandated for all severely dependent young people after they have been withdrawn from alcohol. Others might benefit from a day hospital treatment situation or intensive outpatient program. A variety of long term treatment approaches have been used with undocumented success; these include individual, group, and family therapy; behavior modification techniques; and pharmacotherapy of underlying psychopathologic disorders. Disulfiram (Antabuse) may be a valuable adjunct to other forms of treatment, particularly during the first few months after withdrawal. If alcohol is taken four to seven days after the administration of disulfiram, this potent aldehyde dehydrogenase inhibitor causes the build-up of toxic doses of acetaldehyde, leading to an unpleasant syndrome marked by flushing, nausea, vomiting, and cramps. Taken daily, the drug effectively prevents impulsive drinking. Patients often "forget" to take the disulfiram for the week prior to the resumption of drinking.

Alcoholics Anonymous, so valuable for many adult alcoholics, has not heretofore been particularly attractive or useful to the adolescent alcoholic or polydrug abuser. Existing groups have not been sensitive to the language or styles of young people. Recently youth oriented A.A. groups and other self-help groups based on the A.A. model have been developed and promise to be a valuable treatment resource.

CENTRAL NERVOUS SYSTEM DEPRESSANTS

The drugs in this class that adolescents most frequently abuse are the benzodiazepines, particularly diazepam (Valium) and chlordiazepoxide (Librium); the short acting barbiturates, particularly pentobarbital (Nembutal) and secobarbital (Seconal); assorted other hypnotics such as glutethimide (Doriden) and methaqualone (Quaalude); and amitriptyline (Elavil), an antidepressant with sedative properties.

Patterns of Abuse

In contradistinction to the situation with adults, young people secure these drugs illicitly and use them for recreational purposes. There is a continuum of use from the intermittent use of one or two hypnotic doses for a party or social occasion to a situation of compulsive drug use, in which 10 to 20 times the therapeutic dose may be taken daily for long periods of time. In general, drugs from other classes are taken in addition to depressants. The "high" that obtains from the use of these drugs has been characterized as that feeling of peace, numbness, and tranquility that occurs prior to sleep in normal individuals. Anxiety and inhibitions are blunted, and there may be a feeling of freedom or aggressiveness. In low doses these drugs are often used by young people to decrease sexual anxiety and are reported to enhance performance and pleasure. Higher doses lead to a diminished ability to perform sexually. Violent or aggressive behavior may be associated with use of these drugs. Compulsive users seem to be attempting to obliterate consciousness in an attempt to cope with anxiety or other symptoms, and a high incidence of psychopathologic disorder has been reported in this group (Kamali and Steer, 1976). The extent of denial in this group is noteworthy; some young people remain intoxicated for days at a time, but refuse to admit that they have ingested a general depressant. They may explain that they are tired from not sleeping or that they are sick or under the influence of a less stigmatized substance, such as an antihistamine.

Pharmacology

These drugs are general depressants of nerve tissue and of skeletal, smooth, and cardiac muscle, although at low doses the central nervous system is primarily affected. Effects vary from mild sedation to coma, depending on the drug, the dose, the route of administration, constitutional factors, and tolerance. Tolerance develops rapidly to all depressants, although, in contrast to the opiates, the range is narrow and the lethal dose is not significantly elevated from that in nontolerant individuals. Acute depressant poisoning (overdose) occurs both accidentally and incident to suicide attempts. Poisoning may occur as a result of the combination of sublethal doses of depressants and alcohol or opiates. Cross tolerance develops between all the depressants and alcohol.

Physical dependence develops to all depressants, as indicated by the development of a general depressant withdrawal syndrome. This is similar to the alcohol abstinence syndrome and varies in severity and duration depending on the particular drug, the duration of use, and the dose. In contrast to the opiate abstinence syndrome, that seen with depressants may be life threatening. Mild withdrawal signs may be limited to anxiety, insomnia, restlessness, and tremor. The more severe syndrome may be characterized by seizures occurring within the first several days, marked agitation, and a delirium that is marked by disorientation to time and place, global confusion, and auditory or visual hallucinations. The hyperthermia and tachycardia associated with the delirium can lead to exhaustion and cardiovascular collapse. In the case of barbiturates the full blown abstinence syndrome may clear by about the eighth day; with longer acting benzodiazepines seizures may not occur until the seventh or eighth day and the delirium may be similarly protracted.

Treatment

Treatment of acute depressant poisoning may require hospitalization and intensive care. If physical dependence is suspected, the patient should be placed under close observation or hospitalized and detoxification procedures instituted. Phenobarbital or a long acting diazepine such as diazepam is a suitable substitute for most of the general depressants, since the long duration of action provides for reasonably constant blood levels and protection against the development of either withdrawal symptoms or dangerous intoxication. Other general depressants may be used as well. The drug should be given after the intoxication clears but before major withdrawal symptoms have begun. Sufficient drug should be given to produce a mild intoxication, marked by inconstant, slow nystagmus on lateral gaze, slight dysarthria, and ataxia. The process should not be hurried. If increased withdrawal symptoms are noted, the process should be delayed or sus-

pended; if necessary, additional doses of the depressant may be given.

Fluid and electrolyte losses must be replaced and complicating medical and surgical conditions treated. Patients should be encouraged to be as active as possible, participating in group activities and eating regularly. Patients who are dependent on opiates as well as depressants should be withdrawn from the depressant first while being maintained on suitable doses of methadone. Withdrawal from methadone may then be effected.

After the withdrawal procedures are completed, patients often remain depressed and anxious and are unable to sleep well for prolonged periods. The danger of relapse is significant. Provision must be made for a long term rehabilitation program, which might entail an extended inpatient setting or intensive outpatient care.

CENTRAL NERVOUS SYSTEM STIMULANTS

Central nervous system stimulants are commonly referred to as "ups" or "speed." Of the various drugs in this class, the most frequently abused include amphetamines (amphetamine, dextroamphetamine, and methamphetamine), synthetic derivatives of ephedrine, and cocaine (an alkaloid of the coca plant, which grows in the Andes). Some of the other stimulants that are subject to abuse are methylphenidate (Ritalin), phenmetrazine (Preludin), and diethylpropion (Tepanil).

Stimulants are commonly used by hard driving, success oriented individuals in an attempt to improve their performance and productivity. Stimulants are generally taken orally and when used intermittently do not result in physical dependence. In some instances, with the advent of tolerance, higher and higher doses of stimulants may be required to maintain a satisfactory level of performance. Because depressants are frequently used by stimulant users in order to facilitate sleep, concurrent dependence on both drugs may occur.

The pattern of abuse manifested by polydrug abusing adolescents, using stimulants primarily for their mood elevating properties, is distinctly different from that just discussed. Among these adolescents, methamphetamine ("crystal," "meth") is the preferred drug, since it has the most powerful central effects. Although these drugs may be taken orally, the most common mode of administration involves insufflation or intravenous injection. Immediately following administration, the user experiences a "rush," which has been described as a feeling of great well-being, physical power, and intelligence. The psychoactive effects of stimulants dissipate after four to six hours and are frequently followed by an unpleasant sense of depression ("coming down" or "crashing"). Stimulants may be used compulsively for days or even weeks, followed by periods of prolonged sleep and depression.

As tolerance develops, larger and larger doses are required at increasingly more frequent intervals, to the point at which 1 gm. might be injected every two to four hours. Alcohol and other depressants are frequently used to minimize the duration and intensity of the "crashing" phase and induce sleep. During this phase there is generally a period of prolonged sleep, followed by a period during which the user experiences feelings of depression and apathy. Stimulants may also be used in combination with other drugs. One common combination, popularly referred to as a "speed ball," involves the use of heroin and either amphetamines or cocaine.

The patterns of abuse for cocaine are similar to those for amphetamines. Use may vary from occasional sniffing of cocaine in powder form by relatively well functioning adolescents to compulsive intravenous use by severely disturbed individuals. The psychoactive effects of cocaine include improvement of mood, increased energy, and a decreased need for food or sleep. However, cocaine is shorter acting and much more expensive than the amphetamines. The use of cocaine has recently increased dramatically among middle class and affluent adolescents, who use it for a variety of occasions along with alcohol and marijuana (Grinspoon and Bakalar, 1976; Petersen, 1977).

Pharmacology

When taken in low doses, stimulants produce increased alertness and increased physical and cognitive functioning, particularly when the user has not had sufficient sleep. Considerable tolerance develops to amphetamines. Cocaine produces a narrow degree of psychologic tolerance to its euphoric properties, which persists for only a short period of time. Although cross tolerance develops between the various members of the amphetamine group, it does not develop between the amphetamines and cocaine. In general, no major physiologic symptoms develop as the result of abruptly terminating the use of stimulants. However, the depression, apathy, and prolonged sleep that result are considered by some to constitute an abstinence syndrome and to provide evidence of physical dependence. Even intermittent use of these drugs may result in a protracted situation marked by generally decreased motivation, decreased performance, and irritability in some young people.

Adverse Effects

As with opiates, many of the adverse effects of stimulant use result from the intravenous administration of these drugs in nonsterile conditions. Individuals who sniff stimulants may develop nasal irritation and, on rare occasions, perforation of the nasal septum. Convulsions and strokes have been reported in connection with the use of stimulants, but severe overdose reactions or death from these drugs is rare.

The most common adverse effects resulting from stimulant use are psychologic and include paranoid ideation and stereotyped compulsive behavior. The feelings of paranoia engendered by stimulants may result in antisocial or irrational behavior. Amphetamine or cocaine psychosis may develop as a consequence of continued use. This condition may be clinically indistinguishable from acute paranoid schizophrenia. Although there is often considerable pre-existing psychopathologic disorder in compulsive stimulant abusers, psychotic episodes can develop in virtually anyone if the stimulant dose and the frequency of use are continually increased or high doses are maintained. However, the ease with which these psychotic episodes occur is probably determined by an individual's pre-existing personality structure. Psychotic breaks have been precipitated in some individuals after only one small dose of amphetamine. These psychotic reactions generally appear while the adolescent is under the influence of stimulants and usually dissipate within a few days after termination of use. Although prolonged psychotic episodes have been reported in the literature, they probably relate to pre-existing personality structure (Grinspoon and Bakalar, 1976; Petersen, 1977).

Treatment

The type of treatment indicated for adolescent stimulant users depends on the nature and frequency of use as well as the individual personality characteristics of the user. Many adolescents may require no treatment at all. Some young people who use these drugs only occasionally may profit from a combination of advice about potential adverse consequences of use, and support and encouragement concerning their ability to perform adequately without drugs. On the other hand, compulsive stimulant users should be helped or even pressured to terminate their use as quickly as possible. Hospitalization or confinement may be necessary for some adolescents. Optimally they should be placed in a safe supportive environment in order to help alleviate some of the anxiety attendant to withdrawal, as well as to help prevent relapse. Adolescent patients who experience severe paranoid ideation or overt psychotic episodes may require short term hospitalization coupled with the administration of appropriate major or minor tranquilizers. During the period following withdrawal from stimulants, patients may develop depressive symptoms severe enough to put them at risk for suicide attempts. The treatment of compulsive stimulant users or polydrug abusers is an arduous process, which generally requires long term supportive care.

OPIATES (NARCOTIC ANALGESICS)

Opiates have been used widely throughout history for medical, religious, or recreational purposes, and continue to be the most important analgesics in use today. Of these, heroin is the opiate most frequently abused in the United States. Other opiates such as morphine, codeine, and Dilaudid are subject to abuse along with several synthetic opioid substances, including Demerol, Percodan, and methadone. Initially these drugs are taken by insufflation ("snorting"), but subcutaneous injection ("skin popping") and intravenous injection ("main-lining") constitute more efficient modes of administration and are used as tolerance develops and more drug is needed to get the euphoric effects. For some adolescents, heroin use becomes the central focus of life. These adolescents may identify with the "junkie" subculture, which emphasizes street skills and disregards most scruples in a search for money for drugs. When heroin is not available, heroin-using adolescents may replace it with methadone or a variety of depressants.

As with other substance abusers, adolescent opiate users are a heterogeneous group. For some adolescents opiate use may be symptomatic of an underlying psychopathologic disturbance and may be, in effect, self-medication. In fact, opiates have been shown to possess antipsychotic properties; this effect is presently under intensive investigation. The likelihood that a psychopathologic disorder will be present in a given adolescent opiate user is related to how deviant addiction is for the particular social group of that adolescent. For example, inner city adolescent users, among whom opiate use may be quite prevalent and who can obtain the drugs easily, exhibit surprisingly little psychopathologic disturbance. However, opiate users from middle class populations are more likely to demonstrate severe personality or cognitive disorders.

Pharmacology

Tolerance develops rapidly to any opiate, and the lethal dose increases proportionately. Abstinence from heroin produces an acute abstinence syndrome, which begins in four to six hours and persists for three to five days. This syndrome is characterized by the following symptoms: drug craving, anxiety, restlessness, depression, anxiety, runny eyes and nose, irritability, yawning, perspiration, dilated pupils, sneezing, coughing, nausea, vomiting, diarrhea, abdominal cramps, and "bone" pains. Severe medical consequences such as convulsions or shock are rare. Many addicts fear that their withdrawal symptoms may be life threatening; these fears must be taken seriously, since suicide attempts or antisocial acts while in the throes of withdrawal are not uncommon.

Withdrawal from methadone has a longer time course because of its longer half-life. A protracted abstinence syndrome also occurs after withdrawal from any opioid drug and is characterized by drug craving and often by severe depression and anxiety. Psychotic episodes occasionally occur in detoxified patients; some of these patients may be using opiates as a means of self-medicating their underlying psychosis.

The discovery of structurally and sterically specific receptors for the opiates in brain, spinal cord, and other nervous tissue in all vertebrates studied, and the subsequent discovery of naturally occurring morphine-like peptides in the brain and gastrointestinal tract (enkephalins) and the pituitary gland (endorphins) in these species, may shed light on the mode of action of the opiates and on the dependence-withdrawal cycle. These endogenous morphine-like compounds elicit analgesia, produce tolerance and physical dependence, and compete with radioactive opiates for the receptor. The function of these substances is not clear, but certain of them may be neurotransmitters and may be implicated in the control of affective states, appetite drives, and pain. The opiate addiction syndrome may reflect a neurohumeral feedback mechanism similar to that observed when exogenous thyroid or adrenal corticosteroids are administered. If under resting conditions opiate receptors are exposed to a basal level of the morphine-like substances, when exogenous opiates such as heroin are administered, the overloading of the opiate receptor might suppress the synthesis or release of endogenous opioid. Termination of the exogenous opiate administration might produce a deficiency of the endogenous opioid at the receptor site and be implicated in the immediate or protracted abstinence syndrome. Moreover, there is evidence to suggest that the signs and symptoms of the opiate abstinence syndrome, including anxiety and panic, may in large part be due to noradrenergic hyperactivity.

Adverse Reactions

Serious adverse reactions from opiates are generally related to the administration of unknown quantities of unknown substances in unsterile procedures. Approximately 75 per cent of opiate deaths are due to acute heroin reactions, marked by cyanosis, pulmonary edema, respiratory depression, and coma. Although the etiology of these reactions is not fully understood, pharmacologic overdoses and allergic phenomena have been implicated. Infections of various organ systems, including skin abscesses and cellulitis, bacterial endocarditis, and viral hepatitis, account for most of the other adverse effects resulting from opiate use (Cherubin, 1968). Recently a risk of acquired immune deficiency syndrome has been linked to intravenous drug abuse.

Treatment

Acute Opiate Reaction (Overdose). The treatment of an acute opiate reaction consists of immediate nonspecific supportive resuscitative measures and the administration of the narcotic antagonist naloxone. This drug should be given in a dose of 0.1 mg. per kg. intramuscularly or intravenously every two to four hours as necessary. The antagonist should be given in all cases when an opiate overdose is suspected. A positive response, consisting of pupillary dilation, increased respiratory rate and minute volume, and increased alertness, should occur within one to two minutes after intravenous injection, and is one of the most dramatic medication responses in medicine.

Opiate Dependence. The treatment of opiate dependence must include detoxification followed by long term treatment. The most effective and most humane method of detoxification involves the use of decreasing oral doses of methadone. This can be accomplished either in specialized outpatient treatment centers or on an inpatient basis (Millman et al., 1978). Inpatient hospitalization may be warranted if the patient's condition necessitates close medical or psychologic supervision. However, among well motivated people already involved in a treatment program, outpatient detoxification is generally sufficient. Detoxification begins with 20 to 40 mg. daily, followed by a gradual reduction of the dosage to zero over the course of two weeks or longer. Clonidine (an antihypertensive drug with no narcotic properties) is currently under investigation as a detoxification agent. Owing to the chronic nature of this disease and the likelihood of relapse after detoxification, provision must be made for referral of all patients to an appropriate long term treatment program.

If the patient is dependent on a sedative hypnotic in addition to an opiate, he should be withdrawn in an inpatient setting. The optimal procedure is to first withdraw the patient from the depressant while maintaining the methadone dose at a level sufficient to prevent any opiate abstinence symptoms from emerging, yet low enough to prevent undue sedation. After the depressant withdrawal is completed, the opiate detoxification may proceed. Under some circumstances it may be necessary to detoxify the patient from both classes of drugs at the same time.

After detoxification has been successfully completed, long term outpatient treatment, including frequent contacts, as well as educational, legal, vocational, and psychologic services, is generally necessary. Some young people need prolonged rehabilitation in an inpatient setting. This may be accomplished in a "therapeutic community." The program selected should be responsive to the adolescent's psychologic and social needs.

CANNABIS (MARIJUANA AND HASHISH)

Prevalence estimates suggest that between 30 and 40 million Americans have used cannabis in one form or another. Surveys indicate that over 70 per cent of some adolescent populations have used this drug. To a large extent, cannabis has become an important part of youth culture in the United States. The two most common forms of cannabis are marijuana and hashish. Both are derived from *Cannabis sativa*, a hemp plant, which contains delta-9-tetrahydrocannabinol as its major psychoactive component. The term marijuana is used to refer to the dried mixture of leaves, vine tops, seeds, and stems of the plant. The compressed resin secreted by the flowering tops of the cannabis plant is referred to as hashish and is four to eight times more potent than marijuana. Until recently most cannabis was imported from other countries, particularly Jamaica, Colombia, and Mexico; now highly potent forms are being cultivated in the United States as well.

Patterns of Abuse

Many young people use marijuana or hashish on an intermittent basis; in these instances use is limited to social occasions, when it is smoked in a controlled and ritualistic fashion. A small number of adolescents may develop a severe habituation or compulsive use pattern, in which they smoke all day long and the factors surrounding its use become primary life concerns. These people often have severe psychopathologic disorders or social problems.

Marijuana is usually rolled into homemade cigarettes, referred to as "joints," and smoked. Hashish, on the other hand, is smoked in a wide variety of small pipes. Both forms of cannabis may be ingested in combination with a variety of foods and drinks. Although these substances vary considerably in potency, they are three to four times more potent when smoked than when ingested.

The effects of cannabis begin within minutes after inhalation, peak within one hour, and are generally dissipated within three hours. When it is ingested orally, the onset of these effects is delayed and their duration is prolonged. Some evidence suggests that metabolites of cannabis may persist for long periods of time, and that some psychoactive effects also persist. The acute physiologic effects are dose related and include an increased heart rate, dryness of the mouth and throat, fine tremors of the fingers, congestion of the vasculature of the eyes, as well as altered sleep patterns. In rare instances orthostatic hypotension and loss of consciousness may occur.

The psychoactive effects of cannabis are remarkably variable and depend on the dose, the mode of administration, the personality of the user, previous experiences with and personal expectations about the drug, and the social setting in which the drug is used. Reported effects include altered and often enhanced perception of visual, auditory, tactile, and gustatory stimuli; drowsiness; hilarious hyperactivity; altered time perception; impairment of short term memory; mood changes, including a sense of relaxed well-being or occasionally feelings of anxiety and depression; and impairment of motor performance and reaction time, particularly with respect to complex tasks such as driving in traffic. Some tolerance develops to the psychoactive and physiologic effects of the drug, such that inexperienced users demonstrate greater decrements in performance and increased physical effects compared with chronic users. A mild and variable withdrawal syndrome marked by irritability, restlessness, and sleep disturbances occurs under experimental conditions of heavy use, but has not been a clinically significant problem.

Adverse Effects

Deaths due to an overdose of cannabis have not been reported. There is considerable controversy as to whether chronic use of the drug is associated with serious adverse effects. Decreased pulmonary function, including vital capacity, and an increased incidence of bronchitis, sinusitis, and nose and throat inflammations have been reported in chronic users. Marijuana smoke contains 50 to 100 times more benzopyrene and other hydrocarbons, as well as more tar, than cigarette smoke. Some of these compounds, particularly benzopyrene, may be carcinogenic, although an increased incidence of cancer in marijuana smokers has not been reported, nor has it been well studied. The effects of cigarette and cannabis smoking may be additive. Reduced testosterone levels, altered cellular characteristics of sperm, altered menstrual cycles, and an increased incidence of spontaneous abortion have been reported, but the functional significance of these findings is unclear. The reported in vitro reduction in thymus dependent lymphocytes and inhibition of DNA, RNA, and protein synthesis require clarification.

Adverse effects associated with the use of cannabis are primarily psychologic in nature. These effects are infrequent and are similar to those seen with psychedelic drugs. Acute panic reactions, depersonalization, transient paranoid ideation, and depression are the frequently observed complications of cannabis use and usually abate in several hours. Prolonged psychotic reactions have been precipitated by cannabis, although it is probable that significant psychopathologic disorder existed in these people prior to its use. "Flashback" phenomena, similar to those seen with psychedelic drugs, have been reported, but their occurrence is rare.

An "amotivational syndrome" has been reported, in which chronic heavy users of cannabis become apathetic, are unable to pursue useful goals or master new problems, and are prone to magical thinking. This, too, is controversial and has not been confirmed by extensive field studies. It is likely, however, that in predisposed individuals the use of cannabis may contribute to an impairment of goal directivity and ambition. Through the use of this drug, some adolescents may insulate themselves from the anxieties and tasks essential for the progression from childhood to adulthood.

Treatment

Treatment of an acute cannabis reaction must include firm supportive reassurance. The adolescent should be continually reminded that he is feeling this way because of the drug he took and that the effect will wear off ("talking down"). The use of tranquilizers may be indicated for individuals manifesting aggressive or violent behavior. Treatment of the chronic user will depend on the specific psychopathologic disorder or social problems demonstrated.

PSYCHEDELIC DRUGS

During the late 1960's and early 1970's, the most frequently abused psychedelic drugs were lysergic acid diethylamide (LSD), psilocybin, mescaline (which is derived from the peyote cactus), and the substituted amphetamines such as 2,5-dimethoxy-4-methylamphetamine (DOM, "STP"). Phencyclidine ("PCP," "angel dust"), an anesthetic agent, has also been abused for its psychedelic effects. These drugs produce bizarre alterations in perception, thought, feeling, and behavior, and are sometimes classified as hallucinogens or psychotomimetics.

Patterns of Abuse

The use of psychedelics has occurred primarily among white middle class youths in this country, particularly those seeking mystical or enriching experiences. During the latter part of the 1960's, the use of psychedelics was estimated to involve 500,000 to one million people at a time when the media-popularized youth culture developed complex metaphysical and religious beliefs as well as a social style that supported the search for altered states of consciousness. Pervasive feelings of boredom and emptiness and a lack of meaning in life contributed to the popularity of these drugs. Since then, use has apparently declined, perhaps because of general disillusionment with the possibility that these drugs can provide long term answers to life's problems. Fear of the adverse consequences of these drugs may also have played a role in their decreasing popularity. More recently, increased use of phencyclidine has been noted cutting across all racial and economic lines.

Most of the psychedelic drugs available through illicit channels, whatever they were claimed to be, have actually consisted of LSD in differing doses. It is usually synthesized in illegal laboratories and made available in liquid, impregnated sugar cubes or paper, capsules, and tablets that are orally ingested. Users are generally unaware of the dose they ingest. Most psychedelic drug users do so only occasionally, with use varying from one "trip" weekly to one a month or even one a year. Psychedelic drugs are usually taken in a group setting so that support is available in the event of an acute panic reaction or "bad trip." Rarely some adolescents may take these drugs more frequently so that they may remain intoxicated for long periods ("acid heads"). These individuals often have severe psychopathologic disorders, and psychedelic drug use may be part of their delusional system. In essence some of them may be trying to attribute their strangeness or "functional difficulties" to the drug rather than to their personality make-up. Chronic phencyclidine users, in particular, have been noted to have severe social disability or psychopathologic disorders and most often take a variety of other drugs as well.

Pharmacology

LSD is more than 1000 times more potent than psilocybin and 4000 times more potent than mescaline in terms of its psychoactive effects. Central sympathomimetic stimulation is produced within 20 minutes after injection. This is characterized by mydriasis, hyperthermia, tachycardia, hypertension, piloerection, increased alertness, and facilitation of reflexes. Occasionally nausea and vomiting may occur.

The psychoactive effects of psychedelic drugs are fully manifest within one to two hours. These effects may vary markedly, depending on the personality and expectations of the user, the setting, and the dose. They produce bizarre alterations in perception, thought, feeling, and behavior. Perceptions are heightened and distorted; afterimages are prolonged and overlap with ongoing perceptions. Objects may seem to waver or melt. Illusions and synesthesias, the overflow of one sensory modality into another, are common. Thoughts may assume extraordinary importance or clarity, and body distortions are commonly perceived. Time may seem to pass very slowly. The self may be experienced as being mystically boundless and the world as possessed with unlimited and profound meaning. True hallucinations with loss of insight rarely occur, although some apparently susceptible people have them repeatedly. Mood is labile and remarkable and may range from euphoric grandiosity to paranoid reactions marked by depression and panic. The syndrome generally begins to clear after 10 to 12 hours, and distorted perceptions, tension, or fatigue may persist for an additional 24 hours.

Tolerance to psychedelics develops rapidly, so that repeated daily doses become ineffective in three to four days. Cross tolerance has been demonstrated between most of the psychedelics, implying some common mechanism of action. Physical dependence does not occur with any of the psychedelics (Wesson and Smith, 1978).

Adverse Effects

Doses of the LSD related psychedelic drugs that produced marked perceptual distortions are rarely associated with physiologic toxicity, and no toxic deaths have been reported. Pregnant women exposed to illicit LSD have an elevated incidence of spontaneous abortion. LSD may inhibit antibody formation and disrupt the body's immune system, although additional study of long term toxicity is necessary.

The most frequent complication of psychedelic use is an acute panic reaction ("bad trip," "freak out"). This syndrome is quite variable and may involve sensations of breathlessness, fear of bodily harm, paralysis, and feelings of insanity. As the drug effect wears off, the symptoms generally abate. Rarely, prolonged psychotic episodes may occur, characterized by paranoid ideation, grandiosity, bizarre delusions and hallucinations, and affectual disturbances. These psychotic episodes may be clinically indistinguishable from the functional psychoses, but generally abate more quickly or respond more readily to treatment. Some aspect of a previous "trip" may recur when the adolescent is no longer intoxicated by the drug ("flashback"). For the most part these episodes are mild, although severe episodes associated with psychotic behavior have been reported. It is clear that panic attacks and psychotic reactions occur more often in association with high doses of the drug taken by emotionally disturbed adolescents in unstable or frightening situations. It is not yet clear whether the psychedelics can precipitate prolonged psychotic reactions in reasonably healthy individuals (Wesson and Smith, 1978). It is not surprising that polydrug abusers with underlying psychopathologic disturbances generally avoid this group of drugs after one or more bad experiences.

The acute toxic effects of low doses of phencyclidine resemble those of LSD, although violent and psychotic reactions are reported to occur more frequently. At higher doses the drug is associated with severe physiologic toxicity, and deaths have been reported. The syndrome may include nystagmus, gross incoordination, hypertension, hyperreflexia which may progress to extreme muscular rigidity, arrhythmias, convulsions, and coma. A poorly documented chronic organic brain dysfunction has been described, characterized by visual and speech disturbances, impaired memory, and disorientation.

Treatment

The treatment of the acute panic reaction associated with the use of psychedelic drugs requires a warm supportive environment, someone in constant attendance, and a minimum of external stimuli. The patient should be continually reminded that the strange effects he is experiencing are a consequence of the drug that he has taken and that these effects will wear off in time. In particularly agitated patients, benzodiazepines given orally or intramuscularly may be used. "Flashbacks" may be treated with reassurance or psychotherapy when they are severe. In general, if psychedelic drug use ceases, these episodes decrease with time. The treatment of prolonged psychotic episodes is similar to the treatment of the functional psychoses (Wesson and Smith, 1978).

The treatment of the acute intoxication caused by low doses of phencyclidine is similar to that for the other psychedelics. Toxic reactions from higher doses may require hospitalization and intensive supportive care. The excretion of phencyclidine may be facilitated by acidification of the urine by the intravenous administration of ammonium chloride, 75 mg. per kilogram per day in four divided doses, or ascorbic acid, 500 mg. every four hours, with repeated monitoring of blood pH, blood gas levels, blood urea nitrogen and blood ammonia levels, and electrolyte levels. If symptoms are mild, cranberry juice and 1 or 2 grams of ascorbic acid given orally four times daily may be sufficient (Aronow and Done, 1978).

MISCELLANEOUS INHALANTS

The inhalation of amyl nitrite ("poppers," "amies") is quite widespread in some adolescent populations. Use is generally intermittent and characterized by an instantaneous feeling of flushing, dizziness, hilarity, and rapid heart rate. These effects persist for several minutes. Adverse effects are rare, although postural hypotension with loss of consciousness may occur. The inhalation of a wide range of organic solvents (e.g., toluene in glue) is quite common, though it has been accorded media attention out of proportion to the actual prevalence. The mode of administration generally involves squeezing or spraying the material into a plastic bag and inhaling the vapors. Psychoactive effects are short lived and resemble the initial intoxication and dizziness produced by alcohol. Adverse effects are rare, although suffocation due to the plastic bag has occurred. Chronic exposure, as in the case of industrial workers, may have serious adverse effects. Aerosol sprays containing fluorocarbon propellants are also inhaled for their intoxicating properties. The rare occurrence of toxicity and death may be a result of cardiac arrhythmias or upper airway obstruction and hypoxia.

The inhalation of nitrous oxide for recreational purposes occurs in some youthful propulations. The experience is marked by intoxication and hilarity. Adverse effects are rare, although forced inhalation of nitrous oxide could produce respiratory depression.

Treatment

Intermittent users of inhalants may not require specific treatment. Chronic users, particularly "glue sniffers," are reported to have severe psychopathologic disorders and require appropriate long term treatment.

TOBACCO (NICOTINE)

The initiation of cigarette smoking generally occurs during adolescence. The pattern of onset is such that a relatively small proportion of individuals begin smoking during the early teens; this proportion increases during the middle teens and levels off by the late teens. Generally speaking, if an individual has not begun to smoke by the age of 20, he is likely to remain a nonsmoker for life.

The stages in the acquisition of the smoking habit have been defined in a variety of ways. However, notwithstanding typologic differences, there is general agreement that the initiation of cigarette smoking begins with some period of experimentation followed by a period of increasingly regular smoking, which eventually results in habitual smoking. Early experimentation with the use of cigarettes tends to be confined to social situations, solitary smoking being relatively infrequent (Palmer, 1970). Later, as solitary smoking becomes more frequent, the determinants influencing smoking patterns appear to shift from primarily psychosocial factors to a mixture of psychosocial and psychopharmacologic factors. Both the act of smoking and the physiologic properties of nicotine and the other components of tobacco may become reinforcing.

The development of the smoking habit (from early experimentation to regular smoking) generally takes about three to four years. Overall, the development of the smoking habit and the general pattern of use are strikingly similar to that associated with other substances in that there is the gradual development of tolerance leading to daily use, increased dosages and a desire for repeated administrations, and ultimately psychologic and physiologic dependence (Russell, 1971).

Prevalence and Current Trends

The prevalence of cigarette smoking among men declined from 53 per cent in 1955 to 36 per cent in 1979, whereas it increased among females from 25 to 29 per cent during the same period; the overall prevalence rate is about 33 per cent. Most of the decrease seen in the smoking prevalence rates of adults is attributed to the large number of people who have quit smoking rather than a decrease in the number of people beginning to smoke. Indeed, cigarette smoking among children and adolescents continues to be a significant problem. Although cigarette smoking among youths between the ages of 12 and 18 has declined somewhat from its peak in 1974 (15 per cent), it remains at approximately the same level that it was in 1968 (12 per cent).

Traditionally smoking has been more prevalent among males than among females. Although this remains true for adults, some interesting shifts in the pattern of usage have appeared among teenagers. For example, in 1968 nearly twice as many boys (15 per cent) smoked as girls (8 per cent). By 1978 the prevalence of smoking among girls (13 per cent) exceeded that of boys (11 per cent). This was primarily accounted for by the 17 and 18 year old age group and was the result of decreases among boys; smoking among girls remained about the same as in the previous (1974) survey. Moreover, the prevalence of smoking has increased among women in the 17 to 24 year old age group and now exceeds that of males.

Another trend that has occurred with respect to smoking patterns concerns the increased use of cigarettes yielding lower amounts of "tar" and nicotine. "Lower 'tar'" cigarettes have conventionally been defined as yielding 15 mg. of "tar" or less. "Tar" is the term given to the particulate matter of cigarette smoke that remains after water and nicotine are extracted and consists primarily of polycyclic aromatic hydrocarbons. Teen-agers smoking lower "tar" cigarettes increased from 7 per cent in 1974 to 34 per cent in 1979, which is about the same as the proportion of adults currently smoking this type of cigarette. This shift has been attributed to the increased availability of "lower tar" cigarettes (59 per cent of the cigarettes manufactured in the United States) and an increased public awareness of the health hazards associated with higher "tar" cigarettes.

Pharmacology

Several thousand constitutents have been identified in tobacco and tobacco smoke; the components most likely to contribute to the health hazards of smoking are nicotine, "tar," and carbon monoxide. Of these, nicotine is believed to be the most important pharmacologic agent and the primary reinforcer of tobacco use (Jarvik, 1973). A cigarette provides an extremely efficient vehicle for the administration of nicotine; suspended on particles of "tar," nicotine is almost completely absorbed from the lung and reaches the brain within eight seconds. The smoker can regulate on a puff by puff basis critical factors such as frequency, inhalation depth, duration, and volume. Nicotine has a short half-life (40 to 80 minutes) and therefore many administrations per day are required (Russell, 1976).

The stimulant properties of nicotine appear to involve both dopaminergic pathways and specific nicotine receptors. Nicotine causes an alerting pattern in the electroencephalogram as well as decreased skeletal muscle tone, decreased amplitude in the electromyogram, and a decrease in deep tendon reflexes. The alkaloid appears to facilitate memory or attention and decreases appetite and irritability (Jaffee and Jarvik, 1978). Nicotine may cause vomiting by a complex of central and peripheral actions. The compound produces an increase in heart rate and blood pressure, increased tone and motor activity of the bowel, and increases in the concentration of growth hormone, cortisol, antidiuretic hormone, norephinephrine, epinephrine, and glycerol in the plasma (Jaffe and Jarvik, 1978).

Adverse Effects

Tobacco smoking may be the single most important preventable cause of death and disease. It is estimated that cigarette smoking is responsible for approximately 320,000 deaths annually in the United States (USPHS, 1979). Cigarette smoking among adults has been associated with coronary heart disease and arteriosclerotic peripheral vascular disease; cancer of the lung, larynx, oral cavity, esophagus, pancreas, and bladder; respiratory infections, chronic bronchitis, and emphysema; and peptic ulcers. Cigarette smoking also has been found to act synergistically with orally administered contraceptives to enhance the probability of coronary and cerebrovascular disease; with alcohol to increase the risk of cancer of the larynx, oral cavity, and esophagus; with asbestos and other occupationally encountered substances to increase the risk of lung cancer; and with other risk factors to enhance cardiovascular risk. On the average, smokers have a 70 per cent greater risk of mortality than nonsmokers.

Cigarette smoking during pregnancy has been associated with reduced birth weight, an increased incidence of spontaneous abortion and perinatal mortality, and some degree of impairment of growth and development during early childhood. Involuntary or passive inhalation of cigarette smoke can precipitate or exacerbate symptoms of existing disease states, such as asthma and cardiovascular and respiratory diseases. Pneumonia and bronchitis are more common in infants whose parents smoke. Finally smoking is a major contributor to injuries and deaths resulting from fires, burns, and other accidents.

Although the morbid or fatal consequences of smoking may not emerge until later in life, there is growing evidence that the adverse effects of smoking develop over a lifetime. Early smoking may lead to the premature development of atherosclerosis and respiratory disease. Evidence from human and animal research has indicated that smoking produces measurable lung damage in the very young. Young cigarette smokers have been found to have increased small airway dysfunction and an increased incidence of regular coughing, phlegm production, wheezing, and other respiratory symptoms.

Tolerance, Physical Dependence, and Relapse

Tolerance, primarily of the pharmacodynamic type, develops variably to many of the pharmacologic and subjective effects of nicotine (Jarvik, 1979). Cessation of tobacco use may be followed by a withdrawal syndrome that varies markedly in characteristics and severity among individuals. The syndrome develops rapidly, usually within the first 24 hours, and may persist for days or months. It often includes a craving for tobacco, nausea, diarrhea or constipation, headache, insomnia, fatigue, irritability, inability to concentrate, and increased appetite (Guilford, 1966; Russell, 1971; USPHS, 1979). Objective findings may include decreased heart rate and blood pressure (Weybrew and Stark, 1967), decreased urinary epinephrine and norepinephrine levels (Myrsten et al., 1977), increased skin temperature and hand

steadiness, a decrement in psychomotor performance on vigilance and psychomotor performance tasks (Heimstra, 1973), and weight gain in some individuals. The latter may be the result of metabolic changes, the absence of nicotine suppression of hunger contractions, affective dysphoria, or substitution of oral stimuli. Women report more withdrawal symptoms than do men. However, nearly 25 per cent of all smokers do not experience any withdrawal symptoms (Shiffman, 1979).

Many smokers adjust their smoking patterns to achieve plasma nicotine concentrations close to those to which they have become accustomed. Given low nicotine content cigarettes, they change the patterns of puffing or increase the number of cigarettes smoked to avoid declines in the plasma nicotine concentration; this behavior may be related to avoidance of early withdrawal phenomena.

There are striking similarities in the recidivism rates for opiate, alcohol, and tobacco users who have succeeded in achieving total abstinence. This is true despite differences in pharmacology and differences in their respective abstinence syndromes, which suggests the involvement of behavioral or psychologic factors in addition to pharmacologic ones (Hunt et al., 1971).

One such factor has been characterized as "craving" (i.e., the desire to smoke). It is interesting that craving can be experienced in the absence of smoking even when plasma nicotine levels are maintained by other means, such as nicotine chewing gum or intravenously injected nicotine (Kozlowski et al., 1976). This appears to operate independently of smoking deprivation and may be elicited by environmental stimuli that have become conditioned to the use of tobacco. Anecdotal evidence suggests that craving is most likely in situations strongly associated with smoking, such as coffee breaks or after meals. Relapse is most likely to occur in high risk situations, such as times of unusual stress and parties or other social functions when an alternative coping response is not available and there is the expectation that smoking a cigarette will produce a more comfortable arousal level (Gritz, 1980).

Treatment and Prevention

Since the publication of the Surgeon General's Report in 1964, millions of adults have ceased smoking. The majority of these people have stopped smoking without the aid of professional help. However, national surveys have indicated that as many as 74 per cent of current adult smokers would like to quit smoking, but have been unable to do so.

A wide range of smoking cessation techniques have been developed and tested in both clinical and research settings. These have included individual and group counseling, hypnosis, sensory deprivation, and multicomponent behavioral intervention strategies. Most of the currently available smoking cessation techniques are capable of producing initial success rates between 40 and 100 per cent. However, longer term success rates are less impressive, generally ranging from 20 to 30 per cent one year after treatment (Pechacek, 1979).

Although one might expect that it would be easier for teen-agers who have been smoking for only a few years to quit smoking than for adults who have been smoking for 10 or 20 years, it may actually be more difficult. Formal attempts to help teen-agers stop smoking appear to be even less successful than adult smoking cessation programs, particularly in terms of sustained abstinence. One reason for this may be that teen-agers are still living in the same social milieu and are subject to the same conditions and prosmoking influences as they were when they first acquired the smoking habit. The adolescent who recently quit smoking is likely to be influenced to return to smoking for precisely the same reasons he began initially. As individuals proceed out of adolescence and progress to adulthood, the factors influencing smoking initiation subside.

Then too, as with all other drugs, teen-agers are relatively naive about the extent to which they might be dependent on cigarettes. Unless they have previously attempted to quit or have known someone who attempted to quit, they tend to believe that they can stop whenever they wish. They may be quite unprepared for the difficulty of quitting smoking.

A logical alternative to the treatment of habitual cigarette smoking by means of currently available cessation techniques is to prevent its initiation and development. Traditional smoking education programs have attempted to dissuade individuals from smoking by increasing their knowledge of the potential health hazards of smoking. The fundamental assumption of this type of prevention program is that if students were cognizant of the dangers of smoking, they would simply choose not to smoke. Smoking education classes have been included in the curriculum of elementary and junior high schools, usually within the framework of science or health education. Although most traditional smoking education programs appear to be able to increase knowledge and, in some instances, alter students' attitudes about smoking, they do not have a significant impact on actual behavior (Evans et al., 1978; Thompson, 1978).

More recently researchers have developed and tested prevention strategies that focus on the social and psychologic factors promoting adolescent cigarette smoking. Two distinct types of psychosocial prevention strategies have been developed, and both have produced promising results. The first type of prevention strategy is based on a model pioneered by Evans and his colleagues (1978) in Houston. The main objective of this approach is to familiarize students with the various social pressures encouraging smoking behavior and to teach them techniques for dealing with these pressures, particularly pressures from their peers. This approach has been implemented by means of videotape and class discussion and by means of peer-led discussion and role playing (McAlister et al., 1979, 1980). In general, this type of prevention strategy has succeeded in reducing new cigarette smoking by at least 50 per cent.

A more comprehensive approach presents techniques for dealing with social pressures to smoke within the context of a program designed to improve general personal competence. This prevention program, called Life Skills Training, is based on a model of smoking onset that emphasizes the interaction of social and psychologic factors (Botvin, 1982). According to this model, cigarette smoking is promoted by prosmoking social influences, with specific psychologic factors determining the extent to which an individual is susceptible to these influences. The program includes material dealing with self-image, decision making, advertising techniques, coping with anxiety, social skills, assertiveness, and resisting peer pressure to smoke, as well as with knowledge about the prevalence and immediate consequences of cigarette smoking. This type of psychosocial prevention strategy has been found to be effective when implemented by outside health professionals, older peer leaders, and regular classroom teachers (Botvin and Eng, 1980, 1982; Botvin et al., 1980). This approach, regardless of the mode of implementation, has been found to reduce new cigarette smoking by at least 50 per cent, and when additional booster sessions are included, these reductions may be as high as 87 per cent.

CONCLUSION

Given the extent and impact of psychoactive drug taking by adolescents and preadolescents, the pediatrician should elicit a careful drug history and be prepared to recognize the physical signs and symptoms of drug use in all youngsters in this age range. The physician should be able to assess the appropriateness of the behavior for the particular social and cultural milieu of the child and whether it is presently or potentially a source of disability or danger.

When a drug or alcohol problem is uncovered, it is the responsibility of the primary care physician to treat the acute sequelae of the drug taking, including overdose, withdrawal symptomatology, or adverse physical effects. Subsequently most of these young people need ongoing comprehensive care, including frequent contacts, psychologic and social support, and perhaps vocational and legal help. It is not likely that pediatricians or other primary care physicians in private general practice will be able to provide the full range of services necessary. The economics of private practice in particular tend to prohibit pediatricians from providing optimal care. Behaviorally oriented pediatricians whose practices are organized to provide these services might be consulted. More frequently pediatricians should be prepared to effect a referral to a targeted drug treatment program. Optimally pediatricians should attempt to develop a working relationship with the particular referral resource and remain active in the treatment of their patients. Certainly follow-up should be maintained with any patients referred, since these patients tend to abandon treatment programs or therapists and relapse into alcohol or drug use. The pediatrician may be the patient's only bridge back into treatment.

The difficulty of securing specialized care for these adolescents should not be minimized. It is often difficult to assess the extent of psychopathologic disorder in these young people, and it is similarly difficult to decide whether to refer an adolescent to a psychiatrist, behaviorally oriented pediatrician, targeted drug treatment program, or hospital. To compound the problem, particular treatment programs staffed by nonprofessional counselors often bring different perspectives to bear on the drug abusing youngster and have widely differing views as to what constitutes appropriate treatment from those of pediatricians or psychiatrists. It is clear, however, that effective therapy or behavior change cannot be anticipated while an adolescent is actively abusing drugs; these behavior patterns must be interrupted.

It is the pediatrician's responsibility to follow a young person whose drug abuse behavior is intermittent or rare, remains appropriate to the youngster's social and cultural group, and has not caused and is not likely to cause medical or functional disability. Referral to a targeted drug treatment program or to a mental health professional may not be fruitful and may, in fact, be counterproductive. Drug treatment programs and many psychiatrists are sometimes overzealous in making a determination that a child has a serious problem and is in need of long term treatment. To be sure, it is often difficult to reassure parents or teachers that an adolescent is not in need of specialized care. The problem is that the disruption of the life of an adolescent mandated by certain treatment modalities, for example, a therapeutic community or special school, or the perception by the youngster that he is "sick," may be more disabling than his or her occasional alcohol or drug abuse.

It is lamentable that so few pediatricians have expertise in or are interested in these young people. A dearth of curriculum time is devoted to the subject of substance abuse in medical school and the pediatric house officer receives little training time in these areas. The same situation occurs in most medicine and psychiatry training programs. Too often the patients are perceived as unattractive, uncooperative, and the province of some other discipline. This situation should be changed so that both medical students and house officers are exposed to a well planned, comprehensive educational program in these areas. It is not sufficient to limit the exposure of trainees to hospitalized populations. As with other sorts of illness in youthful populations, hospital based trainees often come in contact with only the most severely disabled patients, those with compulsive abuse syndromes and extensive physical or psychologic disability. Exposure to these patients exclusively might stifle the enthusiasm and curiosity of physicians-to-be early in their training and lead to the well developed attitudes of negativity so often evident in their supervisors.

Whenever possible, educational experiences should be structured so that residents and medical students are exposed to drug abusing young people early in the course of their illness through school based programs, private offices, or outpatient clinics. Additionally, assignment of students and trainees to a specialized drug treatment program with the opportunity to work with nonprofessional drug and alcohol counselors may significantly broaden the physician's attitude. It may also serve to enhance the working relationship between physician and nonphysician to the benefit of the child.

These young people and their drug taking behavior patterns are fascinating to study and rewarding to treat. Although drug and alcohol abuse should be seen as chronic illness with complex determinants, remarkable change may be effected in many young peoples' lives. In others, only moderate improvement or even repeated failures may occur. As in the intensive care unit, mastery of the technical aspects of the field, including the pharmacology of the drugs and the psychobiology of the behavior, is essential before the physician is able to treat the patients with compassion, patience, continuing enthusiasm, and confidence.

ROBERT B. MILLMAN
GILBERT J. BOTVIN

REFERENCES

Aronow, R., and Done, A.: Phencyclidine overdose: an emerging concept of management. J. Am. Coll. Emerg. Phys., 7:56, 1978.

Becker, H. S.: History, culture and subjective experience: an exploration of the social basis of drug-induced experiences. J. Health Social Behav., 8:163–176, 1967.

Bernstein, B., and Shkuda, A. N.: The Young Drug User: Attitudes and Obstacles to Treatment. New York, Center for New York City Affairs, New School for Social Research, 1974.

Bewley, B., and Bland, J.: Academic performance and social factors related to cigarette smoking by school children. Br. J. Prevent. Social Med., 31:18–24, 1977.

Bewley, B., Bland, J., and Harris, R.: Factors associated with the starting of cigarette smoking by primary school children. Br. J. Prevent. Social Med., 28:37–44, 1974.

Blane, H. T., and Hewitt, L. E.: Alcohol and Youth: An Analysis of the Literature, 1960–1975. Final report prepared for National Institute on Alcohol Abuse and Alcoholism, under contract ADM 281–75–0026, March 1977.

Blum, R., and Associates: Youthful drug use. In Dupont, R. I., Goldstein, A., and O'Donnel, J. (Editors): Handbook on Drug Abuse. Washington, D.C., U.S. Department of Health, Education and Welfare and Office of Drug Abuse Policy, Executive Office of the President, National Institute on Drug Abuse, 1979, pp. 257–267.

Borland, B. L., and Rudolph, J. P.: Relative effects of low socioeconomic status, parental smoking and poor scholastic performance on smoking among high school students. Soc. Sci. Med., 9:27–30, 1975.

Botvin, G.: Broadening the focus of smoking prevention strategies. In Coates, T., Petersen, A., and Perry, C. (Editors): Promoting Adolescent Health: A Dialog on Research and Practice. New York, Academic Press, Inc., 1982.

Botvin, G., and Eng, A.: A comprehensive school-based smoking prevention program. J. School Health, 50:209–213, 1980.

Botvin, G., and Eng, A.: A multicomponent peer-leadership approach to the prevention of cigarette smoking. Prevent. Med., 11:199–211, 1982.

Botvin, G., Eng, A., and Williams, C. L.: Preventing the onset of cigarette smoking through Life Skills Training. Prevent. Med., 9:135–143, 1980.

Braucht, G., Brakarsh, D., Follingstad, D., and Berry, K.: Deviant drug use in adolescence: a review of psychosocial correlates. Psychol. Bull., 79:92–106, 1973.

Brecher, E. M.: Licit and Illicit Drugs. Mt. Vernon, New York, Consumers Union of United States, 1972.

Carlson, G. A., and Cantwell, D. P.: Unmasking masked depression in children and adolescents. Am. J. Psychiatry, 137:445–449, 1980.

Cherubin, C. D.: A review of the medical complications of narcotic addiction. Intern. J. Addict., 3:167, 1968.

Coan, R. W.: Personality variables associated with cigarette smoking. J. Person. Soc. Psychol., 26:86–104, 1973.

Demone, H. W.: The nonuse and abuse of alcohol by the male adolescent. In Chafetz, M. (Editor): Proceedings of the Second Annual Alcoholism Conference. DHEW Publication (HSM) 73–9083, 1973.

Downey, A., and O'Rourke, T.: The utilization of attitudes and beliefs as indicators of future smoking behavior. J. Drug Educ., 6:283–295, 1976.

Evans, R. I., Rozelle, R. M., Mittlemark, M. B., Hanson, W. B., Bane, A. L., and Havis, J.: Deterring the onset of smoking in children; knowledge of immediate physiological effects and coping with peer pressure, media pressure and parent modeling. J. Appl. Soc. Psychol., 8:126–135, 1978.

Evans, R., Henderson, A., Hill, P., and Raines, B.: Smoking in children and adolescents: psychosocial determinants and prevention strategies. In Smoking and Health: A Report of the Surgeon General. Washington, D.C., U.S. Department of Health, Education and Welfare, 1979, ch. 17.

Fejer, D., and Smart, R. G.: The knowledge about drugs, attitudes toward them and drug use rates among high school students. J. Drug Educ., 3:337–388, 1973.

Freeland, J. B., and Campbell, R. S.: The social context of first marijuana use. Intern. J. Addict., 8:317–324, 1973.

Gallemore, J. L., and Wilson, W. P.: Adolescent maladjustment or affective disorder? Am. J Psychiatry, 129:608–612, 1972.

Gergen, M. K., Gergen, K. J., and Morse, S. M.: Correlates of marijuana use among college students. J. Appl. Soc. Psychol., 2(1):1–16, 1972.

Goode, E.: Turning on: becoming a marijuana user. In Goode E. (Editor): The Marijuana Smokers. New York, Basic Books, Inc., 1970, pp. 122–138.

Goode, E.: Marijuana use and the progression to dangerous drugs. In Miller, L. L. (Editor): Effects on Human Behavior. New York, Academic Press, Inc., 1974, pp. 303–338.

Goodstadt, M. S., et al.: The drug attitude scale (DAS): its development and evaluation. Intern. J. Addict., 13:1307–1371, 1978.

Grinspoon, L., and Bakalar, J. B.: Cocaine: A Drug and Its Social Evolution. New York, Basic Books, Inc., 1976.

Gritz, E. R.: Smoking behavior and tobacco abuse. In Mello, N. K. (Editor): Advances in Substance Abuse. Greenwich, Connecticut, JAI Press, 1980.

Guilford, J. S.: Factors Related to Successful Abstinence from Smoking. Pittsburgh, American Institute for Research, 1966.

Hamburg, B. A., Braemer, H. C., and Jahnke, W. A.: Hierarchy of drug use in adolescence: behavioral and attitudinal correlates of substantial drug use. Am. J. Psychiatry, 132:1155–1167, 1975.

Heimstra, N. W.: The effects of smoking on mood change. In Dunn, W. L. (Editor): Smoking Behavior: Motives and Incentives. Washington, D.C., V. H. Winston and Sons, 1973, pp. 197–208.

Hein, K., Cohen, M. I., and Litt, J. F.: Illicit drug use among urban adolescents: a decade in retrospect. Am. J. Dis. Child., 133:38–40, 1979.

Hunt, W. A., Barnett, L. W., and Branch, L. G.: Relapse rates in addiction programs. J. Clin. Psychol., 27:455–456, 1971.

Institute of Medicine: A Conference Summary: Adolescent Behavior and Health. Washington, D.C., National Academy of Sciences, 1978.

Jaffe, J. H., and Jarvik, M. E.: Tobacco use and tobacco use disorder. In Lipton, M. A., DiMascio, A., and Killam, K. F. (Editors): Psychopharmacology: A Generation of Progress. New York, Raven Press, 1978, pp. 1665–1676.

Jarvik, M. E.: Further observations on nicotine as the reinforcing agent in smoking. In Dunn, W. L., Jr. (Editor): Smoking Behavior: Motives and Incentives. Washington, D.C., V. H. Winston and Sons, 1973, pp. 33–49.

Jarvik, M. E.: Tolerance to the effects of tobacco. In Krasnegor, N. A. (Editor): Cigarette Smoking as a Dependence Process. National Institute on Drug Abuse Research Monograph 23 DHEW Pub. No. (ADM)79–800. Washington, D.C., United States Government Printing Office, 1979, pp. 150–157.

Jarvik, M. E., et al.: Research on smoking behavior. National Institute on Drug Abuse Research Monograph 17, DHEW Pub. No. (ADM)78–581. Washington, D.C., United States Government Printing Office, 1977, p. 383.

Jessor, R.: Predicting time of onset of marijuana use: a developmental study of high school youth. In Lettieri, D. J. (Editor): Predicting Adolescent Drug Abuse: A Review of Issues, Methods and Correlates. Research Issues 11. Rockville, Maryland, National Institute on Drug Abuse, 1975.

Jessor, R., Jessor, S. L., and Finney, J.: A social psychology of marijuana use: longitudinal studies of high school and college youth. J. Pers. Soc. Psychol. 26:1–15, 1973.

Johnson, B. D.: Marijuana Users and Drug Subcultures. New York, John Wiley & Sons, Inc., 1973.

Johnson, C. A.: Untested and erroneous assumptions underlying antismoking programs. In Coates, T., Petersen, A., and Perry, C. (Editors): Promoting Adolescent Health: A Dialog on Research and Practice. New York, Academic Press, Inc., 1982.

Kamali, K., and Steer, R. A.: Polydrug use by high school students: involvement and correlates. Intern. J. Addict., 11:337–343, 1976.

Kandel, D.: Adolescent marijuana use: role of parents and peers. Science, 181:1067–1081, 1973.

Kandel, D., Single, E., and Kessler, R. C.: The epidemiology of drug use among New York State high school students: distribution, trends, and change in rates of use. Am. J. Publ. Health, 66:43–53, 1976.

Kaufman, E., and Kaufman, P. N.: Family Therapy of Drug and Alcohol Abuse. New York, Gardner Press, Inc., 1979.

Khantzian, E. J., Mack, J. E., and Schatzberg, A. F.: Heroin use as an attempt to cope: clinical observations. Am. J. Psychiatry, 131:160–164, 1974.

Kozlowski, L. T., Jarvik, M. E., and Gritz, E. R.: Nicotine regulation and cigarette smoking. Clin. Pharmacol. Therap. 17:93–97, 1976.

McAlister, A., Perry, C., and Maccoby, N.: Adolescent smoking: onset and prevention. Pediatrics, 63:650–658, 1979.

McAlister, A., Perry, C., Killen, J., Slinkard, L. A., and Maccoby, N.: Pilot study of smoking, alcohol and drug abuse prevention. Am. J. Publ. Health, 70:719–721, 1980.

Millman, R. B.: Drug and alcohol abuse. In Wollman, B. B., Egan J., and Ross, A. C. (Editors): Handbook of Mental Disorders in Childhood and Adolescence. Englewood Cliffs, New Jersey, Prentice-Hall, Inc., 1978, pp. 238–267.

Millman, R. B.: Drug abuse, addiction and intoxication. In Beeson, P. B., McDermott, W. and Wyngaarden, J. B. (Editors): Cecil Textbook of Medicine. Ed. 15. Philadelphia, W. B. Saunders Company, 1979, pp. 692–714.

Millman, R. B., and Khuri, E. T.: Adolescence and substance abuse. In Lowinson, J. H., and Ruiz, P. (Editors): Substance Abuse: Clinical Problems and Perspectives. Baltimore, The Williams & Wilkins Co., 1981, pp. 739–751.

Myrsten, A. L., Elgerot, A., and Edgren, B.: Effects of abstinence from tobacco smoking on physiological and psychological arousal levels in habitual smokers. Psychosom. Med., 39:24–38, 1977.

Palmer, A.: Some variables contributing to the onset of cigarette smoking among junior high school students. Soc. Sci. Med., 4:359, 1970.

Pechacek, T. F.: Modification of smoking behavior. *In* Smoking and Health: A Report of the Surgeon General. DHEW Pub. (PHS)79–50066. Washington, D.C., United States Government Printing Office, 1979, Ch. 19.

Petersen, R. C.: Cocaine: an overview. *In* Petersen, R. C., and Stillman, R. C. (Editors): Cocaine. Washington, D.C., U.S. Government Printing Office, 1977.

Russell, M. A. H.: Cigarette smoking: natural history of a dependence disorder. Br. J. Med. Psychol., *44*:1, 1971.

Russell, M. A. H.: Smoking and nicotine dependence. *In* Gibbins, R. J., Israel, Y., Kalant, H.. Popham, R. I., Schmidt, W., and Smart, R. G. (Editors): Research Advances in Alcohol and Drug Problems. New York, John Wiley & Sons, Inc., 1976, Vol. III, pp. 1–48.

Sadava, S. W.: Initiation to cannabis use: a longitudinal social psychological study of college freshmen. Canad. J. Behav. Sci., *5*:371–384, 1973.

Seldin, N. E.: The family of the addict: a review of the literature. Intern. J. Addict., *7*:97–107, 1972.

Sellers, E. M., and Kalant, H.: Alcohol intoxication and withdrawal. N. Engl. J. Med., *294*:757–762, 1976.

Shiffman, S. M.: The tobacco withdrawal syndrome. Paper presented at Conference on Cigarette Smoking as a Dependence Process, National Academy of Sciences and the National Institute on Drug Abuse, NIDA Monograph 23, 1979, pp. 158–185.

Swisher, J. D., Warner, R. W., Jr., and Herr, E. L.: Experimental comparison of four approaches to drug abuse prevention among ninth and eleventh graders. J. Couns. Psychol., *19*:328–332, 1972.

Thompson, E. L.: Smoking education programs. Am. J. Publ. Health, *68*:250–257, 1978.

U.S. Public Health Service: Teenage Smoking: National Patterns of Cigarette Smoking, Ages 12 Through 18, in 1972 and 1974. U.S. Dept. of HEW, No. (NIH)76–931. Washington, D.C., U.S. Government Printing Office, 1976.

U.S. Public Health Service: Smoking and Health. Report of the Surgeon General. U.S. Dept. of HEW, No. (PHS)79–50066. Washington, D.C., U.S. Government Printing Office, 1979.

Verebey, K., Volavka, J., and Clouet, D.: Endorphins in psychiatry: an overview and/or hypothesis. Arch. Gen. Psychiatry, *35*:877–888, 1978.

Walters, R., Marshall, W., and Shooter, J.: Anxiety, isolation and susceptibility to social influence. J. Pers., *28*:518, 1960.

Wechsler, H.: Alcohol intoxication and drug use among teenagers. Quart. J. Stud. Alcohol, *37*:1672–1679, 1976.

Wechsler, H., and Thurn, D.: Alcohol and drug use among teenagers: A questionnaire study. *In* Chafetz, M. (Editor): Proceedings of the Second Annual Alcoholism Conference. DHEW Pub. No. (HSM) 73–9083. Washington, D.C., U.S. Government Printing Office, 1973.

Wesson, D. R., and Smith, D. E.: Psychedelics. *In* Schecter, A. (Editor): Treatment Aspects of Drug Dependence. West Palm Beach, Florida, CRC Press, 1978, pp. 147–160.

Weybrew, B. B., and Stark, J. E.: Psychological and physiological changes associated with deprivation from smoking. U.S. Naval Submarine and Medical Center Report 490, 1967.

Williams, A. F.: Personality and other characteristics associated with cigarette smoking among young teenagers. J. Health Soc. Behav., *14*:374–380, 1973.

Wurmser, L.: Psychoanalytic considerations of the etiology of compulsive drug use. J. Am. Psychoanal. Assoc., *22*:820–843, 1974.

Zinberg, N. E.: Addiction and ego function. *In* Eissler, R. S., Freud, A., Kris, M., and Solnit, A. J. (Editors): The Psychoanalytic Study of the Child. New Haven, Yale University Press, 1975.

Zinberg, N. E., Jacobson, R. C., and Harding, W. M.: Social sanctions and rituals as a basis for drug abuse prevention. Am. J. Drug Alc. Abuse, *2*:165–182, 1975.

38

Developmental Dysfunction

This chapter focuses primarily on learning problems and the so-called "low severity–high prevalence" disabilities of school children. The three sections contained herein provide somewhat different perspectives on this active area of developmental-behavioral inquiry. The first section is an overview of childhood dysfunction from an empirical developmental perspective. The emphasis is on observed clinical phenomena characteristic of impaired learning and associated behavioral maladaptation. There is a relative de-emphasis on treatment strategies, as these are covered in more detail in Part VII of this book. The second section covers similar clinical phenomena but examines the issues from a neuropsychological point of view, reviewing some of the literature suggesting the relevance of cerebral localization of functions (and dysfunctions) to problems of learning and attention. The final section focuses on clinical disorders of attention and possible biochemical contributing factors. The authors review research models that have implications for clinicians and investigators. In a relatively new and rapidly progressing area of developmental and behavioral pediatrics, this chapter is offered as an exhibit of these diverse but not entirely incompatible approaches to some perplexing challenges.

38A Developmental Variations and Dysfunctions in the School Child

Within each child is a repertoire of strengths, weaknesses, and preferred styles. These emerge through a series of transactions between endogenous endowments and environmental milieux. A succession of life events, a cumulative record of successes and failures, helps to refine these functional profiles. The school years constitute a continuing provocative test for a child's evolving assets and deficits. Academic and social challenges provide continuing feedback as part of an ongoing striving to feel masterful and to buoy a healthy level of self-esteem.

In recent years there has been increasing recognition of a group of children whose functional profiles may not be well adapted to learning or social adjustment (Levine, 1982; Levine et al., 1980; Ross, 1976). They seem to suffer from one or more of the so-called "high prevalence–low severity" handicaps, a group of disabilities that occur in less severe form and in lesser number than in the conditions that limit function in children with mental retardation (low prevalence–high severity handicaps). A wide range of descriptive terms and syndromic classifications have been constructed to include such concepts as "learning disability," "hyperactivity," "dyslexia," and "minimal brain dysfunction." These labels have been difficult to define; clinicians and investigators have struggled to compile and agree upon universal diagnostic criteria, but at the heart of the matter there exists a void in the very definition of disability. With relatively subtle problems of development, how does one draw the line between a true handicap and an unusual style, a highly specialized brain, an idiosyncratic pattern, or a fundamentally normal variation? It is likely, in fact, that no brains are perfect; to varying degrees, we are all learning disabled!

To avoid some of the quandaries over criteria and definitions, this section presents an empirical, nonlabeling approach to the function and dysfunction of school age children. The focus is on specific components or elements of development and the variations and apparent deviations that are clinically observable. By adopting a view of children that shuns labels, some of the dangers of self-fulfilling prophecy may be minimized, while treatments or prescriptions can be based entirely upon the implications of individualized descriptions.

To help clarify this empirical perspective on the developmental dysfunctions of school age children, some principles can be elucidated:

• A particular weakness or disability in isolation offers little insight into a child. The same subtle handicap in two children may be quite different in its effects, depending upon each youngster's remaining profile of strengths and weaknesses, methods of coping, environmental supports, educational resources, and intrinsic resiliency. In some children, a deficit can even evolve into an advantage. A child with a language disability may become highly adept at conceptualizing the universe in spatial terms without the use of verbal descriptions. Such specialization may lead to important discoveries in engineering, nuclear science, or architecture!

• It is likely that most children who have a single area of disability of mild or moderate severity can bypass this and utilize other existing strengths. Children who experience failure in most cases have either a severe disability or multiple dysfunctions of moderate degree.

• It generally is misleading to ask whether a child's academic failure is due to emotional or neurological causes. In most cases negative outcomes represent the final common pathway of both endogenous and exogenous influences.

• Outward manifestations may be deceptive. A developmental dysfunction may present at first to clinicians, parents, and teachers as primarily a behavior problem.

• Success in childhood is, at least in part, culturally determined. Children are fortunate if their developmental profiles can satisfy the expectations of the culture in which they grow up. It has been said, for example, that middle class American culture favors children with strong language skills. Other societies might reinforce youngsters whose mechanical or visual-spatial abilities are particularly well developed (for example, so that they could be talented with a bow and arrow).

• Our expectations for children differ markedly from those for adults. We anticipate that children will be good at everything. It is assumed that adults commonly are specialized in their interests, aptitudes, occupations, and behavioral styles. There is a tendency to consider a child problematic if he displays discrete areas of academic deficiency, unusual behavior, or narrow interest. There tends to be great pressure for conformity, breadth, and uniformity among children. This is generated by adults, by peers, and by traditions.

• For the school age child there is one central mission each day: the avoidance of humiliation at all cost. In school, around the neighborhood, and at home, sustained energy and cognitive effort are mobilized to save face and sustain pride. For children with unusual or dysfunctional patterns of development, the process can be painful and arduous, and it is not unusual for them to appropriate maladaptive strategies. Sometimes their defensive techniques are mistaken for primary problems.

• School age children have minimal tolerance for failure. The success-deprived child is prone to plummet through a failure spiral that is self-perpetuating and self-destructive. It therefore becomes particularly important that clinicians, teachers, and parents learn to diagnose strengths and to help children mobilize these to avoid chronic failure in early life.

ARCS OF FUNCTION

In evaluating the performance of school children, it is helpful to apply a conceptual framework. Figure 38–1 represents such a model. It is, of course, greatly simplified and not meant as a complete model of cognition. Basic units of performance, whether they are specific tasks (such as tying shoelaces or telling time), general subject areas (such as reading or spelling), or broad areas of development (such as gross motor function or socialization) can be studied phenomenologically within this scheme. It can be seen in Figure 38–1 that there is an arc not unlike that of a simple deep tendon reflex. It includes an intake (i.e., afferent) limb, an output (i.e., efferent) limb, and central connections between the two. Selective attention is seen as the portal of entry for stimulus processing. It can be seen in the diagram that most stimuli are rejected (ignored), while some are selected to pass into the intake limb. Once this selection process has operated, sets of stimuli are grouped in one way or another so that they can be stored, understood, or used effectively. Some initial interpretation then occurs along the intake limb. Further refinement occurs as the partially assimilated information is related to previous experience, modified through higher order cognition, and used in a decision making process. Finally, the output limb is the pathway through which such integrated processed information is used to shape a response or product. It can be seen that throughout this entire process there is constant feedback, a monitoring and ongoing regulatory system that functions as a quality control process and enables the thinker or producer to continue, to modify, or to discontinue a focus or activity at any point. This model of information processing and productivity is strictly a pedagogical device to be used as the basis for our consideration of school age developmental variation and dysfunction in this section.

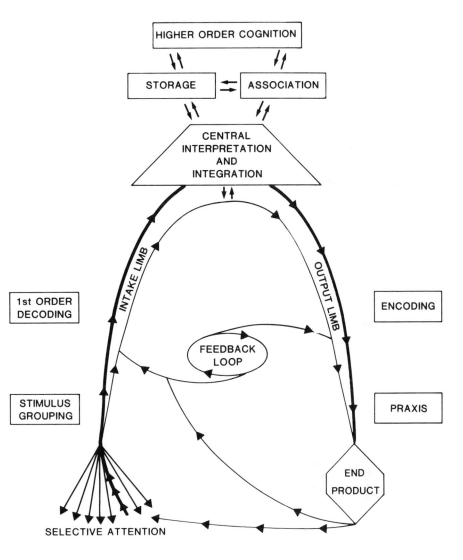

Figure 38–1. This diagram depicts the arcs of function conceptual model frequently referred to in this chapter. It portrays various tasks and areas of performance in childhood as following a predictable pathway consisting of various components of stimulus input, central storage and processing, and output with ongoing feedback that allows for persistence, modification, or discontinuation.

AT THE INTAKE LEVEL: SELECTIVE ATTENTION

The capacity to attend selectively, to focus upon appropriate sets of stimuli for adequate lengths of time, represents a critical element of development. At every instant of wakeful existence a child confronts the need to select, prioritize, and determine what is salient from amid a vast array of competing sets of stimuli. He must determine which of the many options is worthy of conscious focus, what should be rejected or relegated to the category of noise, background, or irrelevancy, which possibilities should gain entrance to the intake limb of the functional arc. Associated with this process is the perpetual decision whether to continue, to modify, or to disrupt the focus. Feedback loops provide a constant review of the appropriateness of selections.

Five standards ordinarily prevail in choosing sets of stimuli for focus and in reviewing the same

selections. Are the stimuli relevant at the moment? Are they meaningful? Are they truly informative? Are they apt to lead to satisfying or pleasurable experience? Are they likely to generate some kind of product or output in the immediate or distant future? Available stimuli that best satisfy one or more of these criteria are apt to be chosen for further processing, while others are inhibited or rejected. When stimuli deteriorate in their utility, relevance, meaningfulness, or pleasurability, concentration is diverted, as part of a continuing quest for a more information laden or satisfying input.

IMPAIRED SELECTION

A sizable group of school age children appear to have problems with selective attention (Levine and Melmed, 1982; Tarver and Hallahan, 1974). They show a marked tendency to be inappropriate in their choice of sets of stimuli; they often focus on information that is unlikely to have any beneficial yield or meet any of the aforementioned criteria. They may change foci so frequently that those that have been selected fail to be exploited for their full value. The proclivity ranges in severity from occasional conscious excursions to the frantic and purposeless meanderings of youngsters with serious attention deficits. The familiar symptom complex has been the object of continuing controversy.

Because of the common association between poor selective focus and high levels of motoric output, the term "hyperactivity" has been used to describe affected youngsters. More recently there has been widespread acceptance of the label, "attention deficit disorder" (Diagnostic and Statistical Manual of Mental Disorders, 1980). The specificity of this symptom complex has been a subject of professional disagreement (Levine and Oberklaid, 1980). As will be discussed, some children appear to have weaknesses of attention as a primary or endogenous disorder, while others may acquire the propensity.

Children with attention deficits commonly show deficiencies in one or more aspects of the selection process. Some have difficulty in distinguishing foreground from background and thus concentrate on what most would consider to be noise. Some continually stray from one focus to another, failing to attain closure or reach a state of satisfaction. Such a tendency to attend to irrelevant stimuli is called "distractibility."

There are many variations on the themes of distractibility and poor selective attention. Some youngsters reveal predominantly visual inattention; they are diverted by extraneous stimuli that they see. Frequently they have difficulty in focusing on information delivered via optic pathways. Their eyes roam aimlessly in search of novel and stimulating visual experience. Such children may show characteristic errors when they read, tending to skip words or omit entire sentences. Visual inattention to detail also may be evident in their spelling, mathematics calculations, and writing.

Other children show predominantly auditory inattention. Prolonged purposeful listening is difficult for them, as they are distracted easily by sounds most would relegate to background status. Some may complain about noise levels in the cafeteria or corridors at school. They may become fidgety and disruptive in highly verbal settings. When confronted with the demand for continuous listening, a child instead may tune out of focus. Parents may lament that the youngster seldom follows directions. The problem may masquerade as a deficit of auditory memory, but the child is only weakly attentive to language, seldom concentrating selectively and intensively enough to register words in memory.

Although visual and auditory distractibility is most commonly noted, other forms exist. Some children are distracted too easily by their own thoughts, memories, or associations. Their "free flight of ideas" promotes constant daydreaming. The child is said to be "in a world of his own." Peers may describe him as "spacy." Such a child commonly is creative and imaginative, perhaps as a result of frequent excursions to new frontiers of consciousness. A teacher or parent may mention something that reminds the child of a previous experience, which in turn engenders multiple other associations, the entire process catapulting the youngster far from classroom agendas or dinner table conversations.

Other types of distractibility are encountered. Some youngsters seem to be distracted easily by autonomic or somatic stimuli. Lowered pain thresholds or increased sensitivity to normal body sensations may promote internal distractions. Many youngsters with attention deficits have a host of somatic complaints stemming from this phenomenon.

Some children are absorbed or distracted by their own fears, anxieties, or preoccupations. They may appear to have so-called "primary attention deficits," but, in fact, their attention is nearly totally absorbed by their all-consuming emotional preoccupations (Table 38–1).

To continue to tune in selectively to the appropriate stimuli, the entire process must be reinforcing or rewarding in some way. A chosen focus should satisfy the attender. If a child has a specific learning disability or handicap that impairs information processing, a secondary form of inattention is likely to supervene (see Table 38–1). In other words, the poor quality of information input will fail to justify the expenditure of attention, so that a youngster may tune out and become distractible in highly verbal settings because of a language disability. Such a child then might stare out the window or daydream. For this reason it is helpful to have an understanding of the specific situations or conditions under which a child tends to become poorly attentive.

Some youngsters appear to have problems with attention in a wide range of settings and circumstances; they demonstrate visual and auditory distractibility, as well as a tendency toward free flight and other forms of poor concentration and impersistence. Children with attention deficits represent a heterogenous group. A broad description of their patterns of attention and inattention can be most helpful in defining their service needs.

Youngsters who have difficulty with selective attention seem also to be predisposed to a number of other traits. A symptom complex commonly is manifest, although the distribution and severity of these traits among children with attention deficits can vary considerably. The following is a description of these symptoms (see also Table 38–2):

Table 38–1.   SUBTYPES OF ATTENTION DEFICITS

Subtype	Common Characteristics
Primary	1. Often evident since early life 2. Apt to be manifest in wide range of settings and situations 3. Multiple characteristic traits in evidence, usually occurring in more than one sensory channel
Secondary to information processing deficits	1. May not appear until school age 2. Accompanied by one or more specific information processing deficits 3. May be specific to or more severe in one modality (e.g., weak auditory or diminished visual attention) and in situations that stress that area of processing 4. Likely to have fewer characteristic traits and to be less severe in manifestations
Secondary to psychosocial stresses	1. May have onset linked to a critical life event 2. Accompanied by environmental stress or deprivation 3. May be part of an identifiable psychiatric disorder (e.g., childhood depression)
Situational inattention	1. Likely to have onset linked to inappropriate matching or unrealistic expectations 2. Manifest only during certain situations or in particular settings that are not matched well with the child (such as cultural disparities between home and classroom)
Mixed forms	Any combination of some or all of foregoing types

INSATIABILITY

Many children with attention deficits have been described as "insatiable." It is difficult for them to take in and process stimuli that are fully or even moderately gratifying. It is as if the nervous system habituates too readily to novel stimuli. When a child has difficulty in achieving satisfaction, it is likely that selective attention will be an unrewarding process. Parents may comment that the child seems to want things incessantly, but that when such desires are granted, he immediately wants more or something bigger or something better. There appears to be a steady state of hunger often accompanied by an inability to be positively reinforced for good behavior or performance. In other words, feedback loops are failing to demonstrate the appetite satisfying effects of reward.

DISINHIBITION

Some children with attention deficits have trouble planning, "editing," or monitoring their behavior. They say and do things without attempting to predict their consequences. Sometimes their behavior is aggressive or disruptive; they may lose their tempers easily or transgress unknowingly with little provocation or justification. They frequently baffle themselves and are surprised by their own transgressions, most of which tend not to be premeditated but stem directly from difficulty in controlling or inhibiting the very same impulses that other children have but manage to review and reject.

The failure to predict consequences and consider actions in advance also may compromise the planning and implementation of academic work. Children with attention deficits commonly have trouble with tasks requiring planful sustained reflection. They are likely to offer the first responses that enter their minds. Some of them have a "trial and error" method of learning; they prefer to "get things over with" instead of doing them right the first time. A frenzied cognitive tempo often results in carelessly completed work. This highly impulsive style may reduce accuracy in reading, spelling, and arithmetic, despite the fact that the child shows a good grasp of underlying concepts and operations.

POORLY MODULATED ACTIVITY

There seems to be an association between poorly focused attention and high activity levels. It is critical to recognize, however, that many children show poor selective attention, impulsivity, distractibility, and disorganization without being overactive. Such "hypoactive-hyperactive" youngsters may go unrecognized because they are relatively inconspicuous and seldom disruptive. Therefore, activity levels themselves should not be considered diagnostic. In some cases normal boys and girls are observably overactive, but their mobility is purposeful, exploratory, and ultimately satisfying. As with attention, activity emerges from a process of selection as from moment to moment one chooses a pursuit in which to engage. One receives feedback regarding its appropriateness and, on the basis of such cues, either continues, modifies, or disrupts the activity cycle. A competent individual selects a high proportion of activities that have some purpose or that lead to a product or satisfaction. Many children with attention deficits choose their outputs in a manner as apparently random as their selection of stimuli for concentration. Thus, they constantly switch pursuits before attaining closure, often producing useless, rhythmic, or fidgety motor gyrations that are not at all goal directed.

It should be emphasized that with regard to both activity and attention, it is not the overall length, span, or amount that is most germane; rather it is the quality of the choices that are made from moment to moment.

IMPERSISTENCE

Many children with attention deficits tend to be impersistent; they have difficulty following through on what they start. Incomplete efforts may result from distractibility. A youngster may be diverted in the middle of a task. Insatiability may lead to impersistence, as a child perpetually feels that he would prefer to be doing something different from that in which he presently is engaged.

Some children with attention deficits appear to tire too easily when they try to concentrate. Prolonged thought processes lead to apparent fatigue, sometimes with yawning, stretching, and other outward signs of cognitive "burnout." Some observers believe that children with attention deficits sustain an imbalance between their sleep and arousal. They have difficulty in falling asleep at night and excessive fatigability during the day. As one parent recently observed, "This kid runs around like a madman, but it's as if he's doing it to try to stay awake; he keeps yawning all the time too." There is some evidence from studies of auditory evoked potentials that a subgroup of children with attention deficits may be "underaroused" during the day (Satterfield, 1973). They may not sleep adequately at night and may be less than fully alert during the day. This may account for some of those who seem to benefit from stimulant medication (see Chapter 57). At any rate, in certain children poor selective attention may relate directly to a form of fatigue, which may or may not be amenable to stimulant therapy.

INCONSISTENCY

A central feature of children with attention deficits is their inconsistency. Erratic behavior and unpredictable learning patterns are typical. Teachers may not be certain of what a particular youngster knows or has mastered, since wide fluctuations may engender disagreement and confusion. A child's ability to focus may be too dependent upon contexts. For example, it is known that children with attention deficits appear to function better in small groups or on a one to one basis and often are highly distracted by social and other stimuli in a regular classroom. Furthermore, their attention may be too dependent upon motivation.

Under highly interesting conditions a child's focus may sharpen significantly, as a favorite television program, a horror story, or some other event matched to the interests of the child seems to strengthen inordinately selective attention. This sometimes results in accusations by the adult world, as it is insisted that a child can "pay attention when he really wants to." It is important to recognize, however, that children with attention deficits have their greatest difficulty in the low and middle range of motivation, those usual levels at which much of human effort occurs. It therefore is inappropriate and misleading to evaluate a child's attention as observed during high motivational states.

Inconsistency of performance also engenders confusion when children with attention deficits undergo standardized testing at school. The results are more likely to reflect a child's momentary state of selective focus (or lack thereof) than ability, knowledge, or potential. Consequently intelligence tests, personality assessments, and achievement examinations in these children must be interpreted with great caution. Dramatically different results may be obtained from the same child on successive observations. Standardized tests were not standardized on children with attention deficits!

SOCIAL FAILURE

It has been observed commonly that children with attention deficits seem to experience difficulties in relating to their peers. Many of them consequently seek older and younger friends. Their struggle to avoid humiliation is hampered constantly by the jibes and ostracism of other youngsters. In some cases children with attention deficits are not attentive to social feedback cues. The arcs of function governing socialization may break down because youngsters are insensitive to social feedback. An affected child may say or do things that are inappropriate or ill timed but persist in this vein because negative feedback goes unnoticed.

Children with chronic disinhibition or impulsivity may acquire bad reputations. Their behavior may be interpreted as "weird" by other youngsters, who are more tactful and deliberate in their actions. Peers may feel out of control when trying to relate to a youngster who is highly impulsive. Insatiability also thwarts social integration; a child who is in a constant state of hunger commonly acts in a highly egocentric fashion, oblivious to equilibria of giving, taking, and sharing.

So it is that many youngsters with attention deficits suffer from social isolation, constant peer rejection, and verbal abuse. Potential acquaintances may perceive these children as different and fear guilt by association with them. Often children with attention deficits have little or no understanding of why they are unpopular. They have no inkling of what it is they are doing wrong or why they are not liked. Paradoxically these children may be extremely gregarious; they may make friends quickly and easily; they may lose them even faster. As a result, many retreat from peer interaction and instead develop other interests or relate to adults, younger children, or animals.

SUPERFICIALITY

Some children with attention deficits are extremely superficial in their interests and endeavors. This tendency may become increasingly evident in late elementary and junior high school. A youngster may have performed well in the early elementary grades when there was not great detail, volume, and depth of subject matter. As the quantity and complexity of academic input and output grow, the tendency to see only "the big picture" and to miss detail increasingly handicaps the student. Such a propensity may be reflected in school performance.

In some cases superficiality extends beyond formal academic performance. The child may have difficulty in developing any strong foci of involvement. There may be a history of continuing loss of interest in various hobbies, toys, or other endeavors after only a brief initial stage of novelty and romantic attraction.

ATTENTION DEFICITS AS STRENGTHS

It is inappropriate to assume that the symptom complex associated with attention deficits is inevitably maladaptive or pathological. It is probable in fact that many of these traits have a positive facet and the potential for a good prognosis. A child who is highly distractible or who drifts readily into a free flight of ideas may make interesting observations that more confined, rigid minds might never bring forth. In some cases distractibility may be an ingredient of creativity. A fast paced, cognitive tempo may result in a high degree of productivity throughout the life span (Levine et al., 1980). As a youngster ages, insatiability may evolve into ambition, and egocentricity into strong leadership. The child with an imbalance between sleep and arousal someday may be a highly productive "night person" (an alternative not offered during childhood). Thus it may be that at least some youngsters with attention deficits perform more effectively as adults than in childhood. It may be that they will be more likely to succeed when they can exploit their own strengths, create their own highly individualized agendas, and ap-

propriate a structure or organizational framework in which they can achieve greater independence.

Table 38–2 summarizes the various symptoms commonly associated with attention deficits. In attempting to understand the youngster who has many of these traits, it is useful to review the extent to which the various symptoms are manifest. It should be borne in mind that in a child who has attention deficits this propensity is likely to be acted upon by a wide range of exogenous and endogenous influences that ultimately will shape the clinical picture and determine the prog-

Table 38–2. ATTENTION DEFICITS

A. Common manifestations
  1. Poor selective attention: inability to focus on most purposeful stimulus sets for adequate lengths of time; distractibility; free flight of ideas
  2. Impulsivity: difficulty in planning and reflecting on activities in advance and during their implementation
  3. Disinhibition: problems suppressing ill advised or unneeded actions
  4. Poor modulation of activity: inappropriate nongoal directed motoric output
  5. Insatiability: difficulty in feeling and acting satisfied with current pursuits and conditions; weak re-enforceability
  6. Inconsistency: erratic quality of performance over time; emotional liability
  7. Sleep-arousal imbalance: easy fatigability at cognitive tasks; impersistence; problems in falling asleep at night
  8. Insensitivity to feedback: poor self-monitoring, carelessness
B. Possible impacts on learning
  1. Problems in following directions
  2. Poorly selective memory; tendency to retain irrelevant information while forgetting more salient data
  3. Reading, spelling, and writing errors reflecting poor visual attention to detail
  4. Illegible handwriting, possibly due to impulsivity
  5. Difficulty conforming to classroom routines
  6. Poor social adjustment in school
C. Assessment
  1. Parent and teacher questionnaires
  2. Direct observations of attention patterns during neurodevelopmental and psychological testing
  3. Direct assessment on specific tests of reflectivity, continuous performance, paired associate learning, and vigilance; these are not performed routinely for clinical purposes
D. Treatment modalities
  1. Counseling: working with child and parents to demystify problem, to develop management strategies and priorities regarding specific symptoms
  2. Regular education: teacher awareness of deficit, preferential seating in class, reduced distractions; altered length of assignments, consistent nonaccusatory feedback to child
  3. Special education: small group or one to one tutoring in nondistracting setting for part of each day; help with academic work, organizational skills, planning and attention to detail
  4. Pharmacotherapy: use of stimulant medications to enhance selective attention and reduce impulsivity (see Chapter 58)
  5. Alternative therapies: some controversial modes exist (see Chapter 59)

nosis. Environmental circumstances, reactions of adults and peers, patterns of nurturance, critical life events, and the outcomes of educational experiences can either minimize or complicate the negative impacts of attention deficits. The responses of the adult world may have a particularly significant effect on prognosis. If a child is labeled as pathological or lazy throughout the formative years, he may realize this self-fulfilling prophecy and exaggerate any deviance. On the other hand, if a youngster is helped to understand his attention problem, if the adult world can balance empathy with an insistence upon some accountability, if existing strengths are recognized and utilized, and if every effort is made to induce successful experiences and a sense of mastery, the prognosis is likely to improve, and some or all of the components of attention deficits may transform themselves into strengths.

ORIGINS AND PROGNOSIS

Just as the clinical symptomatology is varied, etiologies and pathophysiological mechanisms differ widely. A large number of investigations have been undertaken to determine the causes of attention deficits and "hyperactivity." Many etiologic offenders have been suggested. These include low level lead intoxication (Needleman et al., 1979), food additives (Swanson and Kinsbourne, 1980), other toxic substances, allergies (Rapp, 1979), perinatal medical stresses, and genetic transmission.

The prognosis of an attention deficit is difficult to determine, despite data from a number of long term follow-up studies (Weiss et al., 1979). These have suffered from methodological flaws that hamper their interpretation. In general, they have suggested that at least a subgroup of children with attention deficits experience great difficulty as young adults. Such problems as substance abuse, school dropout, unemployment, marital instability, and automobile accidents have been reported. However, it is difficult to determine which outcome measures relate specifically to attention deficits and which represent the sequelae of success deprivation, incessant adult disapproval, and ill conceived face-saving efforts. There is reason to believe that the prognosis of an attention deficit is likely to improve under optimal conditions of ongoing management. Early detection, an understanding of the problem by parents, teachers, and the child himself, and appropriate interventions may improve outcomes. Longitudinal studies of well managed children are needed.

DIFFERENTIAL DIAGNOSIS

The symptom of chronic inattention is common in childhood. It may occur either as a primary condition or as a relatively nonspecific response to a wide range of stresses. Table 38–1 is an attempt to classify various kinds of attention deficits. In children with so-called primary attention deficits dysfunction seems to be based on a fundamental weakness of selective focus. In general, the onset is early in life and the symptoms are manifest in a wide range of settings.

There are two forms of secondary attention deficits. Some youngsters become inattentive because of a specific information processing problem. If one refers to the prototype arcs of function (see Figure 38–1), these are children who obtain negative feedback when they focus. For example, if a child has difficulty in perceiving visually presented symbols, concentration on such stimuli is futile; the effort is unrewarded. The child receives negative feedback and consequently tends to reject future sets of visual stimuli. Such a youngster may become highly distractible and inattentive, especially when this weak function is stressed. Thus, inattention may be promoted by a so-called "learning disability."

Another form of secondary attention deficit is one in which the youngster's attentional strength is absorbed by emotional preoccupations. The capacity to focus selectively is overwhelmed by internal stimuli that competitively inhibit a youngster's goal directed concentration.

A third theme is that of situational inattention. This implies that a youngster is inattentive only in certain kinds of settings. In most instances there is a mismatching between the child and a specific environment. For example, a child from a bilingual family may find himself in a classroom in which very complicated English syntax is used all day. Although there is nothing wrong with the child or the classroom, the matching is inappropriate, and the child is prone to tune out.

The classification of attention deficits in Table 38–1 should not be taken to imply mutual exclusivity; a youngster may have a primary attention deficit and also a learning disability. In fact, in the School Function Program at the Children's Hospital Medical Center in Boston, approximately 70 per cent of the youngsters with attention deficits also had specific information processing handicaps. That is to say, less than one-third of the children with attentional weakness had "pure" deficits of attention. Furthermore, it is common for children with attention deficits to develop or have concurrent emotional difficulties. Their feelings of inadequacy may trigger anxiety and even childhood depression, creating a strong preoccupying focus and aggravating an underlying primary attention deficit.

In the evaluation of the child with chronic inattention, one should attempt to differentiate between possible sources of attentional weakness. It is important, in particular, to rule out or weigh the relevance of information processing problems, psychosocial stresses, and inappropriate situational settings. The existence of environmental problems should never be taken to rule out a primary attention deficit. In fact, children from difficult environments are likely to be more predisposed to attentional weakness, even on a primary basis.

Children with attention deficits often are misunderstood by the adult world. It is common for them to be condemned for the way they are. Frequently parents are accused of having caused the problem. Mothers and fathers blame each other, and parents may conspire to blame the school. The pediatrician or other clinician may be caught in the crossfire. It is important to help everyone recognize that they are all innocent victims of a perplexing and complicated clinical disorder, one for which no magic panacea or simplistic formulation exists.

PRELIMINARY ORDERING OF INFORMATION: STIMULUS GROUPING

In the previous section there was a review of the various normal and dysfunctional components of selective attention. As one pursues further the "arcs of function" model (see Figure 38–1), various areas of development come into play after sets of data pass through the screening process of selective attention.

Information upon which a child focuses needs to be sorted out and then arranged in a manner so that it can be interpreted effectively. Sounds must be grouped in the central nervous system so that they can be interpreted as words. Determinations need to be made regarding the parts of a visual image that coalesce to form an object or a symbol, or how the various components of a scene or pattern relate to each other. Such processes often are referred to as "perception." They have to do with the preliminary interpretation of information. Without such clustering, further understanding and utilization of entering stimuli would be futile. Many children with learning problems are confused, for they have trouble in grouping and studying the relationships between the parts of a data set.

Information attended to by the central nervous system generally comes in two types of "packaging"; stimuli are presented either simultaneously or successively. To interpret and use information efficiently and effectively, a child must be able to sort out data entering in these two forms. Some youngsters are adept in both, while others seem to have specific deficits in one or both. The following sections describe some of the normal developmental aspects as well as manifestations of dysfunction.

SIMULTANEOUS STIMULI

As noted, much of the information delivered to the central nervous system for processing comes in the form of simultaneously presented stimuli, an array that coheres to form a configuration or gestalt. Through central nervous system processing, the unique features of a pattern or shape are grouped and perceived as being distinct from sets of adjacent or background stimuli. In addition to such appreciation of the cohesion of a pattern, there emerges an awareness of the ways in which configurations are oriented and relate to other forms simultaneously encountered. Many of these data impart information about the world of spatial relationships. Much, but not all, gains entry to the central nervous system through visual pathways. Consequently such processing often is referred to as visual-perceptual or visual-spatial. It should be recognized, however, that simultaneously presented information can arrive through other channels. For example, stereognosis usually entails the simultaneous presentation of the salient features of a configuration so that a whole shape can be recognized by touch without visual monitoring. Certain aspects of music (e.g., chords and harmony) require the simultaneous processing of stimuli through the ears. In addition some elements of visual processing necessitate segmental integration, such as nystagmatic movements. In considering the developmental dysfunctions of children, however, it is generally believed that the intake of simultaneous stimuli through visual-spatial processing is the most relevant. It will therefore be the focus of further discussion of this area of development.

Early in life the infant absorbs a great deal of information regarding the attributes of space. Exploring the environment of a crib or a baby carriage, he assimilates the consistent attributes of configurations and their relationships to one another. Without knowing the names of various positions and juxtapositions, he becomes increasingly aware of form constancies, the permanence of objects, and such geographical relationships as in front of, behind, on top of, and underneath. As a result of such observations, rules can be generated that help to order and generalize the environment. For example, an infant learns that when an object is placed under a pillow or a blanket, it still exists and it can reappear as a consequence of appropriate goal directed actions.

As a young child grows older, he makes increasing use of the simultaneously presented information in the environment. It serves many indispensable purposes: It can help to establish consistent uses for objects and sites. It can help to automatize daily routines. Visual-spatial data can be used to program and monitor purposeful motor output. The child can generalize spatial input to learn to tie shoelaces, ride a bicycle, and dress himself. As will be described in a subsequent section of this chapter, children with visual-spatial processing problems may have difficulties in programming gross and fine motor output tasks. Finally, visual-spatial input is needed for the appreciation and storage of information laden symbols, except when intentionally bypassed (as with Braille). Reading, writing, spelling, and arithmetic depend heavily upon the mastery of visually presented symbols. The abilities to perceive and remember the salient features of these symbols, to distinguish them from one another, and to reproduce them effectively (as in writing or spelling) are key elements of academic mastery.

PERCEPTUAL DYSFUNCTION

Some children have inordinate difficulty in interpreting, storing, and applying spatial data (Table 38–3). Early in life they may encounter problems with the recognition of consistencies of shape, with integrating the collective attributes of a configuration and seeing its relationship to its surroundings and backgrounds. Concurrently they may have difficulty in understanding spatial concepts, and they are prone to confusion over directionality, which may lead to trouble in distinguishing left from right. They may have trouble in remembering the side to which a small "b" points in contrast to a small "d." They may be slow to learn to dress independently, with particular confusion over left and right shoes or the fronts and backs of articles of attire. During toddler and preschool years such children may refrain from activities that entail a heavy visual-spatial input. Certain sports or crafts especially may be frustrating.

When children with visual-spatial disorientation enter early elementary school, they confront the frustrating challenge of the visual symbol system. Written letters, numbers, and finally words may elude them, tending to look all too similar and therefore difficult to recognize and even harder to reproduce. Characteristic letter reversals may be seen when an affected child starts to write, although it should be borne in mind that such errors are not necessarily diagnostic of visual-spatial problems. They also may be seen with other developmental dysfunctions (e.g., impulsivity or deficiencies of memory).

Because they have so much difficulty in mastering the attributes of the visual-spatial world, such children often have to struggle to appreciate even verbally presented spatial concepts. Prepositions like "near," "away from," "over," and "beside" may create confusion. When given verbal direc-

Table 38–3. VISUAL-SPATIAL DISORIENTATION

A. Common manifestations
   1. Whole-part dissociation: difficulty in linking the parts to a whole in a visual configuration
   2. Foreground-background confusion: problems in distinguishing foregrounds and backgrounds; trouble in discerning information embedded within a complex visual configuration
   3. Confusion over position and juxtaposition: difficulty in appreciating relative positions of multiple configurations, including such attributes as "below, above, between, and behind"
   4. Problems with laterality and directionality: confusion over left and right, difficulty in distinguishing and retaining unique relationships relating to a vertical midline (e.g., distinguishing b from d)
   5. Weaknesses in visual discrimination: trouble in distinguishing between similar but not identical shapes or patterns
   6. Deficits of visual attention and retention: difficulties in focusing upon and retaining visual detail, possibly secondary to weaknesses of processing

B. Possible impacts on learning
   1. Trouble in learning to associate visual symbols (i.e., letters, words, and numbers) with specific sounds and meanings
   2. Difficulty in distinguishing between visually similar symbols
   3. Delay in establishing a strong sight vocabulary for reading and spelling
   4. Trouble in revisualizing words, which results in characteristic spelling errors (e.g., "lite" for "light")
   5. Confusion when attempting to write letters, numbers, and words
   6. Possible difficulties with athletic activities requiring visual-spatial discriminations (e.g., catching or throwing a baseball)
   7. Possible associated problems with eye-hand coordination affecting handwriting and craft activities of various types

C. Assessment
   1. Observation of characteristic visual errors in reading and spelling
   2. Results of subtests of intelligence assessments: especially WISC-R subtests of block design and object assembly
   3. Specific visual-perceptual test batteries, such as the Frostig tests of visual perception
   4. Visual perceptual motor assessments (usually involving the copying of geometrical forms), such as the Beery VMI and Bender Gestalt tests (these particular tests have a strong element of integration and output, in addition to visual-spatial orientation)
   5. Motor free tests of visual perception: assessments of embedded figures, progressive matrices, and matching of designs

D. Treatment modalities
   1. Regular education: multisensory approaches to reading and spelling; use of strong phonetic or linguistic reading materials; emphasis on verbal rather than visual instructions; simplification of visually presented materials
   2. Special education: individualized help to enhance perceptual processes (effectiveness controversial); teaching of verbal remediation strategies (e.g., subvocalization, the use of word cues and rules, the bypass of visual processing)

tions that have a strong spatial component, the child may need to struggle to comply.

As is the case with other developmental dysfunctions, the clinical manifestations of visual-spatial difficulties change with age. Ultimately an affected child is likely to master the symbols of the alphabet as well as the visual configurations of numbers and to associate these with appropriate verbal labels. However, more complicated configurations constructed from visual symbolic elements may compose the next stumbling blocks. The child who experiences difficulty in appreciating whole word configurations may be delayed in acquiring a sight vocabulary for reading. The latter consists of words whose meanings and pronunciations are recognizable nearly instantly. They are so familiar that they can be processed with only minimal allocation of selective attention and higher order cognition. Their recognition is practically automatic. For a child with significant spatial processing difficulties, such automaticity may be tardy in arriving. The affected youngster approaches each word as an unfamiliar, seemingly previously unmet configuration. He must subject the word to the processes applied to an unfamiliar pattern; namely, he must break it down into its component parts, identify these, and then reassemble them. Such word analysis is of course time consuming. When a great deal of attention and conscious effort is diverted toward active decoding, the expenditure to sound out each component may be so great as to interfere with comprehension. Some children may find that they can comprehend better than they can decode. They may therefore perform better when reading silently than aloud. A child may engage in what is called "word by word reading." This highly mechanical style is likely to be slow, laborious, and generally unrewarding. It should be acknowledged that some children with strong word analytic and language abilities may be able to develop an efficient phonetic reading style and so compensate for weak visual-spatial processing skills.

As children with visually based word decoding problems progress through school, they often improve. With the proper instruction and, it is hoped, with the moral support they need to sustain self-esteem, most ultimately master reading, although they may remain a bit delayed in their skills compared with peers. A relatively small subgroup, however, persist as nonreaders.

Children with visual-spatial problems slowly develop their sight vocabulary. The recognition of familiar word configurations and their consistent associations with verbal utterances arrives eventually but is too slow to become automatic. Many such children persist in having to struggle with writing and spelling. Although they may learn to recognize previously encountered words, they may have trouble in retrieving exact configurations so as to reproduce them accurately on paper. Thus, their continuing confusion over spatial attributes may create problems for written output. Planning spatial utilization on the page, for example, may be particularly devastating. Such issues will be covered in a later section of this chapter (see page 729).

Sometimes it can be difficult to differentiate between deficiencies of visual perception and weak visual attention to detail. However, these two phenomena frequently are found in association. A child may become visually inattentive as a result of repeated frustrations with visual-spatial processing. In other words, visual attention goes unrewarded, and negative feedback occurs. When trying to process spatial information, the child experiences frustration and negative reinforcement. This may be associated with a lifelong tendency toward superficial and poorly sustained visual selective attention. The latter in turn may result in a lessening of visually oriented experience. The child who seldom is alert to visual configurations certainly will be delayed in mastering them, whereas one whose processing constantly leads to rewarding insights will gain continually in this area of development, as he seeks increasingly complex and challenging visual-spatial experiences.

It was once thought that most or all of the specific learning problems of school age children derived from visual perceptual handicaps. At present there is increasing evidence that such dysfunctions have their major impact during the earliest grades in school. Moreover, it is likely that most children with isolated visual-spatial dysfunctions in the presence of intact or strong abilities in other areas of development will overcome or bypass their weakness, ultimately experiencing improved academic achievement. In particular, it is likely that youngsters with advanced language and higher order conceptual abilities often will succeed in circumventing a handicap in visual-spatial processing.

ASSESSMENT

The evaluation of simultaneous processing or visual-spatial function is fairly well established, although results can be misleading. First, one seeks a history consistent with this dysfunction. Sometimes, but not always, there is evidence of early clumsiness, a reluctance to participate in visual-spatial oriented activities, and problems in learning to trace, to draw, and to distinguish left from right.

A variety of standardized tests have been developed to evaluate visual-spatial orientation at various ages. Most commonly children are asked to copy geometrical forms. It is assumed that those with visual-spatial disorientation will have difficulty with this. Although this generally is the case, some precautions are judicious. If one refers to the prototype of the arcs of function, it can be seen that copying a geometrical form involves some developmental capacities in addition to visual-spatial orientation. First, a child may need to attend selectively to the visually presented stimulus set. A child who is highly impulsive or inattentive to detail may copy geometrical forms poorly as a result. Second, a child's ability to remember the various attributes of a configuration may affect performance. A youngster who needs to keep looking back at the original stimulus may lose track of form copying. Finally, it should be recognized that such tasks involve a strong motor component. A child who has problems with fine motor output may do poorly not because of visual-spatial processing difficulties but because of the troubles he has in executing motor tasks with a pencil (see section on output, p. 730).

It also should be stressed that some children who have visual-spatial problems may copy forms fairly well. A child may have developed a strategy in which a visual form can be broken down into its component subsegments and these reassembled on paper. A child may be able to sketch or put together fragments of a configuration without actually having an appreciation of its overall gestalt or shape.

There are motor-free tests of visual perception. These frequently require a child to match similar configurations from among contrasting ones or to find certain shapes embedded within others. It should be emphasized that all such assessments have a strong component of sustained visual attention to detail.

Certain sections of intelligence tests may provide clues regarding visual-spatial processing. On the Wechsler Intelligence Scale for Children—Revised (WISC-R), deficiencies may underlie relatively low scores in object assembly, picture completion, and block design, although it is important to recognize that other developmental dysfunctions can contribute to problems on these subtests.

A careful analysis of a child's academic performance may be strongly suggestive of visual-spatial problems. Delays in acquiring letter naming abilities and characteristic confusion over similar letters or words may be encountered. Thus, as with the assessment of other areas of development, multiple interlocking sources of diagnostic agreement are needed to establish that a child may have delays in the processing of simultaneous data and, in particular, information relating to visual configurations.

SUCCESSIVE STIMULI: TIME AND SEQUENCE

In the previous section it was stated that stimuli may be grouped in configurations requiring simultaneous appreciation to extract their meaning and uniqueness. A second possibility is that bits of information may be grouped in a particular sequence or serial order. As with simultaneous stimuli, full appreciation depends upon a child's early experimentation and experience with the grouping, ungrouping, and regrouping of stimuli. Just as space is a medium within which simultaneous stimuli cluster in configurations, time is the dimension in which successive stimuli are arranged in patterns. Sequences are meaningful according to the order in which stimuli are arranged to form communicative units. Verbal language may be viewed as one such sequence. Comprehension obviously depends upon an awareness of the order of syllables in a spoken word. There also is a need to aggregate and segregate groups of sounds (i.e., phonemes) to construct meaningful word units. Visual sequences require the same kinds of analysis. A light flashing the Morse code must be interpreted with a clear appreciation of which dots and dashes go together to form visual sequential units of information.

It is clear that time and sequence are fundamental organizing media in human experience. Meaningful successive data may be transported through different sensory modalities and find expression in a wide range of motor and linguistic outputs. Visual sequences, auditory and verbal sequences, motor rhythms, and multistep processes are commonly encountered in daily life, and all share a fundamental temporal orientation.

Very early in life, infants begin to acquire information about time and serial order. They may learn that sucking upon a nipple in a particular rhythmic sequence yields plentiful nutriment. They may observe continuously that certain routines in the day follow each other predictably in a temporal sequence. Causal relationships depend upon consistent time sequences. Thus, an infant may learn that hunger generally is followed by feeding, which in turn leads to satiation. Such replicable processes help to reinforce temporal-sequential organization. As children grow older, their experiences with sequences and time relationships become increasingly extensive and complex. As we shall see, a well developed orientation in this area is critical for their academic achievement and adjustment to adult expectations.

DEFICIENCIES OF SUCCESSIVE PROCESSING: TEMPORAL-SEQUENTIAL DISORGANIZATION

Some children seem to have inordinate difficulty with the appreciation, storage, and utilization of time and sequences (Table 38–4). Such dysfunction can be clinically elusive and yet severely disabling to the growing child. Initially a youngster may demonstrate marked confusion over temporal relationships. Time oriented prepositions, such as "before" and "after," may be a source of confusion even though the child seems to acquire language easily. Other time sequences, such as the days of the week and the months of the year, may frustrate the youngster. In preschool or kindergarten the child may put forth such jumbled utterances as, "Are we gonna have lunch before, or did we eat it after?" Alternatively he may plead with his parents, insisting, "I want to go to bed late tonight so I can stay up until before the movie is over."

As children with temporal-sequential disorganization progress through school, they are likely to encounter a succession of obstacles. Time related functions, as might be expected, are particularly hazardous. They have trouble in mastering the days of the week and the months of the year and may be delayed in learning to tell time. They have real problems following multistep instructions. The teacher may ask the class to carry out an exercise entailing several steps in a particular order (e.g., "Find your pencil, open your book to page 8, and copy down the new words on the list"). The delayed sequencer may become confused and anxious, and the child may be able to retain only one step of an instruction, since the processing of serial data is a vain struggle.

The acquisition of new academic skills can be particularly taxing in the presence of sequencing problems. Just as the youngster who fails or falters with simultaneous processing may have trouble in appreciating whole word configurations, the child with temporal-sequential disorganization may encounter difficulty in breaking words down into their component sequential parts and binding them together again. Such deficiencies of word analysis may interfere with the decoding of unfamiliar words. On the other hand, if a child has strong language and visual-spatial skills, sight vocabulary may not be impaired.

The child with a sequencing problem may struggle inordinately to establish the correct order of letters of the alphabet, despite the fact that their unique visual configurations are discernible. Spelling may be affected, as the serial order of letters in a word eludes the child. Some children with sequencing problems experience difficulty with mathematics. In particular, they appear to be prone to encounter delays in mastering the multiplication tables and solving multiple step computations. As children with problems of sequential organization progress toward late elementary school, their greatest lags occur in writing, spelling, and organization. The latter may be particularly pervasive and disconcerting (see section on output, p. 729), resulting in difficulty in following

Table 38–4. TEMPORAL-SEQUENTIAL DISORGANIZATION

A. Common manifestations
1. Confusion over time relationships: difficulty in mastering temporal prepositions (before, after); delay in learning days of the week, months of the year; delay in learning to tell time
2. Confusion over multistep directions: difficulty in appreciating, storing, and retrieving data presented in a specific serial order; sometimes deficit is modality specific (i.e., either visual or auditory)
3. Trouble with multistep processes: difficulties in assimilating and implementing operations requiring multisteps in a particular order
B. Possible impacts on learning
1. Difficulty in following multistep directions in the classroom
2. Trouble in mastering sequences of symbols (as in the order of letters in a word or particular sequences of numbers)
3. Difficulty in remembering and implementing complex multistep processes, as in long division and complicated multiplication problems
4. Organizational problems, especially in late elementary and junior high school: poor study habits; difficulty in organizing written narrative; problems in planning and staging long term assignments; difficulty in adhering to schedules and generally poor time orientation
C. Assessment
1. Characteristic history of time confusion and difficulty with multistep orientations as reported on parent and teacher questionnaires
2. Possible low scores on digit span and picture arrangements subtests of the WISC (N.B.: normal scores on these subtests do not rule out a sequencing deficit, since alternative strategies may have been used by the child)
3. Various tests of sequential memory, such as visual and auditory sequential memory subtests of the Illinois Test of Psycholinguistic Abilities and the various visual and auditory attention span subtests of the Detroit Tests of Learning Ability
4. Failure on neurodevelopmental examination items, such as imitative block tapping, object spans, finger tapping exercises, sequential finger opposition tasks, and complex serial verbal commands
D. Treatment modalities
1. Regular education; presentation of succinct instructions and information (i.e., small "chunks" of input); need for repetition; avoidance of tasks entailing multistep operations or presentation with visual demonstration models
2. Special education: specific help with the mastery of increasingly long modality specific sequences; organizational help (outlining, scheduling skills, staging of narratives, notetaking); teaching of subvocalization skills

a schedule in school, problems in planning and organizing projects, trouble in arranging paragraphs or sentences in a report, and in knowing how to start, stage, and finish multiple step operations.

It is common for disabilities of temporal-sequential organization to masquerade as primary behavioral or emotional problems. An affected child may be accused of not really trying, of not wanting to follow directions, or of not attempting to concentrate. As they grow older, such youngsters commonly are thought to be lazy or poorly motivated, because it is so difficult for them to organize and complete assignments. Some children with sequencing deficits are in a perpetual state of fear and panic in a classroom. They develop profound feelings of inadequacy, as they struggle to deal with the constant onslaught of data delivered in sequences. Despite good concentration, they may be able to extract only a portion of each sequential set. Their anxiety is heightened as they gaze about the classroom and observe other youngsters having no trouble whatsoever focusing on a problem, or understanding and applying sequentially presented information. As part of a campaign to avoid humiliation and save face, an affected youngster may become a class clown, or act defiant, or tune out totally because of repeated frustrations and negative reinforcement when trying to attend to sequences.

Some children with problems of sequential organization shy away from certain athletic activities that place a heavy emphasis on multistep processes. A child may be well coordinated and have no difficulty whatsoever in catching or throwing a ball, but may nevertheless withdraw from complex athletic events or dance lessons because of their emphasis on motor sequential organization. He may have trouble in mastering the steps or rules of the game. Complicated motor outputs necessary for certain sports, for dancing, and for various handicrafts may inhibit the youngster who is wary of displaying such inadequacy.

ASSESSMENT

Although problems with temporal-sequential organization are common, they often pass undiagnosed. Parents and teachers may be more sensitive to simultaneous or visual-spatial processing problems than they are to difficulties with serial order. Because of their persistent nature, it is particularly important that these dysfunctions be detected at an early age. As might be surmised, there are rich historical cues that include long standing difficulties with temporal prepositions, trouble in completing multistep tasks, a poor record of following serially presented verbal directions, a delay in learning how to tell time, as well

as a variety of other problems in mastering time or sequences. All may constitute evidence of this dysfunction.

Some children with sequencing problems manifest their difficulties more in one modality or activity than in others. Compensatory strengths in other developmental areas may modify the clinical picture of the child with a sequencing problem. Thus, if a youngster has very strong auditory processing abilities, difficulties with the appreciation and storage of verbally presented sequences may be less severe, while, in the same youngster, visual data in serial order may be harder to handle. A child with particularly well developed gross motor skills may overcome the potential for motor sequencing problems; he is so agile that motor function in general is facilitated. For this reason, in the evaluation of children with possible sequencing problems, it is important to include a variety of tasks, utilizing specific input and output modalities of the arcs of function.

Visual sequencing can be assessed by having a child imitate tapping in a particular order. Pointing to a series of objects or squares in a certain sequence is one example. Alternatively the child could be asked to imitate directly certain patterns in which the examiner opposes one of his fingers to another. Fine motor function would be an important contaminating variable in that task.

Auditory sequencing frequently is tested using a digit span or a series of numbers, which the child is expected to repeat in the correct order. Apprehension, anxiety, familiarity with numbers, and the rate of auditory processing in general can obscure the pure assessment of sequencing using a digit span. Some of these concerns can be overcome, for example, by trying word spans also, by relaxing the child to alleviate apprehension, by altering the rate of presentation of stimuli, and by asking the child to make eye contact and to concentrate assiduously.

Other assessment techniques for sequencing include having the child carry out increasingly complex multistep serial commands, observing the youngster imitating certain motor rhythms, and asking a child to arrange various objects in a particular order. A youngster's mastery of sequences from daily experience also can be revealing; for this the child might recite the days of the week (either forward or backward, depending on the age) or the months of the year. Asking a child to count backward or tell time also is helpful.

A wide range of sequencing tasks are included in various intelligence tests as well as in examinations for learning disabilities. The picture arrangement subtest of the WISC-R is one such example. It should be emphasized, however, that some children with sequencing difficulties may perform well on this subtest, since it also entails a high level of social awareness and experience (fortified by reading comic books or watching a great amount of television). The tests of visual and auditory sequential memory in the Illinois Test of Psycholinguistic Abilities are likely to tap into sequential organization. The Detroit Tests of Learning Aptitude contain relevant sequencing tasks. In all cases, however, it should be recognized that a child's associated strengths and weaknesses may alter performance and either mask a sequencing problem or create the semblance of one.

Thus, it can be seen that, as with visual-spatial orientation, temporal-sequential organization can be clinically elusive. A firm diagnosis depends upon confirmatory evidence from parents and teachers, direct observations of difficulty with sequencing tasks, and a history of the kinds of academic problems that could be at least partially explained by delays in this area.

COMBINED VISUAL-SPATIAL AND TEMPORAL-SEQUENTIAL DYSFUNCTIONS

It should not be assumed that problems with temporal-sequential organization and difficulties with visual-spatial orientation are mutually exclusive. Some youngsters may be afflicted with both types of dysfunction. One can imagine the plight of a youngster with combined temporal and spatial disorientation. Yet such a child may have compensatory strengths. It is possible, for example, to have weaknesses in both these areas of development and still show strengths in language and cognition in general. However, those with combined deficiencies are likely to show a slow adjustment to school. They may present primarily as behavior problems. Commonly they are noted to have temper tantrums and show a great deal of emotional turmoil in an academic setting. In many cases these children show their greatest difficulties during the earliest elementary school grades. As they enter mid to late elementary school, compensatory strategies, especially strong conceptual abilities, may supervene.

LANGUAGE PROCESSING

The effects of language are felt over all segments of the arcs of function. To begin with, auditory attention and feedback are critical components of academic and social function. The more meaningful words are to a child, the more intently that youngster is likely to focus upon them. Language can facilitate function in other developmental areas. Visual-spatial processing, for example, can be aided via language. A child can give himself verbal clues and understand language classification systems that simplify and generalize the visual environment. Language is durable; ideas, experiences, and techniques can be stored in the form of their verbal descriptions. Much controversy exists, but there clearly is a relationship between language and higher order conceptual abilities (e.g., reasoning, inferring, and thinking abstractly). Finally, language is an important product, a means of encoding, a medium of output and communication.

So-called receptive language is critical for learning. The rapid conversion of spoken words into meanings is critical for knowledge and skill acquisition. Within the arcs of function, language becomes a particularly important step at the level of "first order decoding" (see Figure 38–1). This is a stage at which stimulus sets (either simultaneous or successive) are imparted with some basic meaning according to previously assimilated rules. Although at a later stage these will undergo further integration with memory stores and more intensive scrutiny during higher order conceptual processes, at this preliminary level symbols start to be translated into ideas. As a child develops, an increasing proportion of meaning extraction occurs at this level, at which language interpretation is rapid and relatively automatic. For example, if a child is given a word problem in arithmetic, after processing the sequence of information presented, a meaning to the question emerges rapidly. The actual solution of the problem depends upon reference to memory and higher order conceptual functions. To the extent that a child is not adept at language processing, much more higher order conscious effort will have to be diverted into even understanding the word problem.

Not all such preliminary decoding is mediated through language. A child may have an instant appreciation of the significance of certain visual configurations (such as a ball heading toward his bat) without naming or conceptualizing the phenomenon in words. A child may observe that a particular mathematics problem involves addition or subtraction without actually using language cues to establish this. Some children, on the other hand, may require a considerable degree of first order decoding with language. This is particularly true if there are weaknesses in visual processing or other areas. A child may tend to subvocalize or give himself phonetic clues, establishing a strong preference for language as a means of decoding almost everything.

Since language is probably the most academically relevant form of preliminary decoding, the remainder of this section will deal with the phenomenology of receptive language.

Multiple steps are involved in a child's efficient and accurate processing of language. (These issues are covered in more detail in Chapter 40.) The following is a brief summary of some of the components of receptive language function:

Auditory Acuity. Sensorineural hearing loss, ear canal problems, middle ear disease, or abnormalities of anatomical structures can interfere with the passage of sounds to higher cortical centers, thereby causing symptoms that may mimic a problem with central auditory processing. Although there is considerable controversy, there exists some evidence that chronic serous otitis media may be an antecedent of hearing deficits and later language disabilities (see Chapter 26).

Auditory Attention. Selective attention to human speech sounds is critical for optimal language function. Children with weak auditory attention may have difficulties in suppressing extraneous background noises. Such an auditory figure-ground problem may present clinically as a behavior disorder or a global attention deficit. Some children with problems of this type may dislike noisy surroundings (such as cafeterias and playgrounds). As with other weaknesses of attention, poor auditory focus may be either a result or a cause of language impairment; in some cases it represents both. Children with auditory inattention may manifest impulsivity when they have to perform in highly verbal settings. They may appear far more reflective and attentive when confronted with visually presented instructions. It is likely that some youngsters who are labeled "hyperactive" have exclusively auditory attentional weaknesses. Since school constitutes a highly verbal setting, problems with selective focus, activity control, and reflective behavior may be evident only in the academic arena. Discrete weaknesses of auditory attention therefore should be part of the differential diagnosis in a youngster who has trouble in concentrating in school.

Auditory Discrimination. The ability to differentiate between similar auditory sound units is critical for optimal language processing. Confusion between discrete syllables can result in problems when a child attempts to read, an act that necessitates the firm establishment of sound-symbol associations. Obviously when the sounds are indistinct, such associations weaken.

Segmentation and Blending of Words. In previous sections it was emphasized that the understanding of sequences and configurations is based upon initial processes of grouping, ungrouping, and regrouping. Such analyses and syntheses enable a child to understand which basic bits of information are attached to which other ones to create the most meaningful and useful patterns. The same processes may be applied to language. A child must have an innate sense that words consist of combinations of sounds that indeed can be broken down and reassembled. Sequential organization and the ability to analyze the sound fragments of a word without losing their serial order therefore are necessary steps in the preliminary decoding of words, especially in early stages of mastery. Word analysis skills in reading are especially dependent upon this aspect of language.

Appreciation of Syntax. An appreciation of the rules of syntax is a fundamental component of language processing. The order of words and the grammatical construction of sentences provide major clues to meaning. The capacity to decode and utilize such sentences in an increasingly complex and efficient manner is a major academic facilitator. Children who have difficulty with the rapid and effective processing of syntax may become confused about instructions in the classroom. They tune out of selective focus and become highly distractible, for sustained auditory attention leads only to frustration amid the complexity of syntactical structure. In other words, on the arcs of function there is consistently negative feedback during periods of auditory attention, so that the child tends to focus on more extraneous or irrelevant stimuli.

Receptive Vocabulary. The school age child expands his receptive vocabulary (i.e., the store of words he comprehends) exponentially with age. The rate of this growth is determined by sociocultural factors as well as developmental and neurological variables. An impoverished receptive vocabulary may be one indication of a generalized developmental delay, or deprivation or lack of experience, or an isolated language disability. A child with a reduced receptive vocabulary is likely to show similar limitations in expressive language. The rapid automatic understanding of conversation and of classroom verbal instructions may elude a youngster with a limited store of vocabulary. The appreciation of syntax and the acquisition of a rich receptive vocabulary are two major components of auditory comprehension.

Metalinguistic Awareness. The ability to infer meaning from language clues, intonations, and contexts is an important aspect of receptive language. Ordinarily children do not need to attend to every individual phoneme in a sentence to understand its literal or its implied meaning. An "inner knowledge" of how language works has been called "metalinguistic awareness." Meaningful clues are inherent in the rhythm, the prosody, and the intonation of spoken language. Some children are particularly expert at extracting meanings from these hints. They may not even need to attend very carefully to understand what is being said. This is likely to be a significant facilitator of academic progress.

Rate of Language Processing. Rate or efficiency is a critical and frequently neglected aspect of language processing. A child may understand verbally presented information but may process it at a rate slower than that of his peers. Such a youngster may chronically straggle, forever striving to catch up with the rapid onslaught of words. Prolonged latencies of response to language input constitute serious impediments in the classroom. This becomes particularly germane in the late elementary school grades, when the complexity, quantity, and rate of verbal output from teacher are likely to increase markedly. Again, impaired youngsters may become secondarily inattentive.

RECEPTIVE LANGUAGE DYSFUNCTION

Developmental language disabilities may be elusive and subtle (Table 38–5). Children with such handicaps may experience considerable maladjustment, behavior problems, and social failure.

Table 38–5. DEVELOPMENTAL LANGUAGE DISABILITIES

A. Common manifestations
  1. Difficulty in understanding verbally presented instructions or information
  2. Trouble in discriminating between similar sounds
  3. Diminished vocabulary—either receptive, expressive, or both
  4. Trouble in finding words quickly enough (dysphasias)
  5. Problems with auditory attention to detail
  6. Difficulty in remembering verbally presented information
  7. Behavioral deterioration in highly verbal settings
  8. Articulation problems
  9. Problems with socialization
  10. Difficulty in keeping up with rapidly presented verbal inputs (i.e., relatively slow rate of verbal processing)
  11. Trouble in distinguishing auditory foreground from background noises
B. Impacts on learning
  1. Delayed acquisition of reading skills
  2. Difficulty in establishing firm sound symbol associations (phonetic equivalence of visual symbols)
  3. Trouble in following verbal directions in the classroom
  4. Delays in spelling, with characteristic errors that are good visual but poor auditory approximations of words (e.g., "laght" for "light")
  5. Problems with written language
  6. Difficulty in participating in classroom discussions
  7. Difficulty in solving word problems in mathematics
C. Assessment
  1. Possible historical evidence of delayed acquisition of language (especially the use of complete sentences and syntax)
  2. Possible history of recurrent otitis media during infant or toddler years
  3. Low verbal subtest scores on the WISC-R, especially comprehension, vocabulary, and similarities
  4. Relatively low scores on specific tests for language disabilities, such as the Illinois Test of Psycholinguistic Abilities
  5. Symptoms suggestive of language disability on parent and teacher questionnaire forms
  6. Reading and spelling errors characteristic of language disabilities
  7. Immature patterns of speech (e.g., relatively impoverished vocabulary, overly simple syntax, and verbal economy in general)

Difficulties in processing verbal input may present initially in some (but not all) cases as delayed acquisition of spoken language. A child who does not process efficiently may have difficulty in encoding ideas in his own words and sentences. Other children give an early history of difficulty in following directions or of apparent auditory inattention. Poor word pronunciation or articulation also may constitute an early indicator of receptive language processing problems, although it should be emphasized that such pronunciation difficulties commonly occur in isolation.

As children with language processing problems enter school, they are vulnerable to academic failure (Wiig and Semel, 1976). A child with a receptive language dysfunction may have trouble in acquiring basic reading skills. The establishment of sound-symbol associations may be delayed because the child has poor auditory discrimination or a weak sense of the actual sounds of words and the way they are differentiated from each other. Such a youngster may prefer a sight method or predominantly visual approach to reading. Language difficulties also may engender problems with spelling. An affected youngster may write words that are good visual approximations but make no sense phonetically (e.g., "laght" for "light").

Processing instructions, participating in class discussions, and even interacting with friends may be problems for youngsters who have to struggle to interpret language efficiently and accurately. Ultimately many such youngsters have trouble with written expression, a skill often described as the highest form of language. They may be weak at expressing themselves in writing because they have not yet mastered the fundamentals of language processing.

ASSESSMENT

A typical history of delayed language acquisition, articulation problems, characteristic spelling and reading errors, and difficulty in concentrating on verbal input may or may not be present. A number of standardized language tests exist. These include the Illinois Test of Psycholinguistic Abilities and certain parts of the Detroit Tests of Learning Aptitude. Various verbal subtests of the WISC-R also are thought to be related to language abilities. These include vocabulary and similarities tests. It is important to add, however, that cultural factors, experience, and higher order conceptual abilities can influence the results of these subtests.

In assembling a battery to screen for language processing difficulties, the clinician should include some tests of receptive vocabulary (perhaps the Peabody Picture Vocabulary Test), an assessment of the comprehension of syntax, a sentence repetition test, and some assessment of the ability to analyze and resynthesize words. Tests of overall language development also would include some assessment of auditory memory, word finding and naming abilities, and spontaneous speech. In general, when there is a strong suspicion of a significant language processing disability, the services of a specialized speech and language therapist should be sought.

MEMORY

It is nearly impossible to separate memory from learning. The storage of relevant information and skill is a critical component of development and academic achievement (Ring, 1975). In fact, the processes involved in retaining, recognizing, and retrieving data are so pervasive that there is good reason to question whether this should be considered a separate area of development in childhood. The word "memory" is vague. One needs always to ask, "Memory for what?" The following generalizations about memory function show its close association with other areas of development:

• The capacity to store and retrieve data is proportional to the meaningfulness of incoming information. Previous experience and exposure, the perceived relevance of the data, and the frequency and duration of earlier exposures to it all relate to the durability of new sets of stimuli. Thus, it is easier to store and remember five words in one's native tongue than it is to perform the same task with five words in an unfamiliar foreign language. The more intensively and the more diversely one can relate newly presented information to that which has been previously encountered and stored, the more easily one can retain the novel input. One might anticipate therefore that children who confront new knowledge and skills in a way that enables them to see their relevance to what they already know are more likely to retain what is novel. Strong conceptual and integrative abilities can serve to enhance memory.

• The effectiveness with which information is processed relates to its inclusion in memory stores. If the reception is poor, if a child has difficulty in perceiving or grouping the components of the stimulus set, the storage of such information is likely to be poorly established. Subsequent attempts at retrieval may be labored or futile. This means that a youngster with auditory processing problems may have concomitant difficulties with the storage of verbally presented information. A child with visual-spatial disorientation is particularly susceptible to difficulty with visual memory, while a youngster who has temporal-sequential disorganization may have real problems in remembering things in their correct order. In some cases

a child's processing problems may improve with age, while the child is left with relatively intransigent memory deficits in that modality. For example, a youngster with visual-spatial disorientation ultimately may improve and show enhanced abilities in perceiving configurations and spatial relationships. The same child, however, may continue to have difficulty with revisualization or visual retrieval memory. Ultimately he may have a fairly good sight vocabulary and adequate reading abilities but reveal persistent problems with spelling and written output because of trouble in retrieving the visual configurations of words.

• Memory is in a dynamic equilibrium with selective attention. If a child tunes out or attends only superficially to the stimulus set, although he or she momentarily may appreciate and interpret it, storage may only be transient, although some data may be retained subliminally (i.e., "incidental learning"). Selective attention of sufficient intensity is needed to register salient or central stimuli firmly in short and long term memory stores. Some children seem to have problems in what might be called the "attention-retention dimension." These are youngsters in whom it is difficult to decide whether there is a problem in focusing or one in retaining and retrieving information. In many such cases the difficulty rests in both areas. That is to say, chronically weak attention seems to result in deficiencies of memory, and difficulties in storing information render selective attention somewhat futile and thereby depress the latter, as data pass "in one ear and out the other." It is not unusual for a child with attention deficits to be referred for help because of apparent problems with memory.

• Another important parameter links memory and attention, and this has to do with processes of selection and inhibition. Just as a child needs to select the most purposeful stimuli to focus, he also is expected to choose the most important bits of information storage. In both instances much (in fact, most) is rejected or discarded. An efficient memory is one that stores primarily those data that are apt to be of relevant information or pleasure at a later time. Thus, the capacity to forget selectively is critical. Some youngsters store and therefore retrieve a great deal of irrelevant trivia while having only the most transitory grasp of information that could be more beneficial to them. A child may be the only one in the family who can remember what color shoes Uncle George wore four years ago at Christmas. In fact, everyone may be astounded by the child's extraordinary memory. Incidentally they also may have noted that he has trouble in taking a telephone message or that he can never remember to take home the right books or complete his assignments. Such an indiscriminate storage of information may masquerade deceptively as a miraculous memory!

• Aside from their relationship to attention and their possible modality specificity, individual memories may be further subspecialized. There are wide differences in the capacity to remember faces, to recall jokes, to store and retrieve motor patterns for dancing, to retain specific skills, and to retrieve words quickly from auditory memory stores.

• Traditionally, developmental psychologists studying memory have divided this function according to the duration of retention. Thus, one speaks of iconic memory (the most instantaneous "photographic" image), instant recall (such as that of a parrot), short term memory, and long term retention. These, of course, represent rather artificial categories, since the duration of memory storage is likely to be on a continuum. Selective decay of memory traces after they have served their purpose is also of great importance. If one has an appointment this afternoon at 2 P.M. in room 703, it is critical to remember this now, but, on the other hand, the time and place of that meeting will have little relevance and, it is hoped, will have been expurgated from memory three or four weeks hence. Long term memory stores may need to be the most immediately accessible. Day to day information needed for practical living is likely to be on file therein.

• Another important differentiation involves the distinction between recognition and retrieval memories. This is particularly relevant for learning. A child, for example, may learn to recognize certain words but have difficulty in retrieving their precise configurations in order to spell them aloud or write them on paper. Alternatively a youngster may have a good receptive vocabulary but be unable to find words quickly enough (i.e., retrieve them).

MEMORY DYSFUNCTIONS

In establishing a clinical profile of a child, specific components of memory should be accounted for. In general, these tend to be part of the evaluation of other areas of development. The following are some of the dimensions of memory that are likely to be particularly relevant to academic function in childhood, although their place in cognitive psychology is a matter of debate: visual-spatial recognition and retrieval, visual and auditory sequential memory, memory for language, and memory for motor patterns (see p. 731). Children with serious difficulties in one or more of these areas are likely to experience delays in acquiring academic skills and knowledge. It should be emphasized, however, that weaknesses in these areas are unlikely to occur in isolation; they are almost always accompanied by current or ear-

lier problems with information processing or selective attention.

Many children learn to compensate for memory deficits. They preferentially choose modalities of input that promote better storage. For example, youngsters with problems in visual memory may prefer to transform visual stimuli into descriptive language that is more readily registered in memory and retrieved. Some children who have difficulties on the "attention-retention dimension" subvocalize frequently. They may constantly repeat instructions verbally to themselves so that they can have a second chance at attending to their details. It is likely that in many cases such strategies are beneficial and should be encouraged. It should be added that all developing children become increasingly skilled memory strategizers. Various mnemonic devices and rehearsal techniques become important parts of their aptitude for learning.

ASSESSMENT

It follows from what has been stated that the evaluation of various aspects of memory function must be tied closely to the specific processing modalities in question. Thus, the evaluation of verbal memory should be part of a more complete assessment of language. Visual recognition and retrieval should be examined at the same time that visual processing and visual-perceptual and motor functions are assessed. For example, a youngster may be given some geometrical forms to copy as part of an evaluation of visual-motor abilities. He then should be asked to analyze geometrical forms and remember their details, subsequently copying them from memory. Motor-free visual perception tasks can also have a memory component. A child can be shown various geometrical forms and later be asked to identify the configuration to which he was previously exposed from among a series of similar alternatives.

Often it is difficult to separate out sequential memory, since so many of the assessment tasks require a child to repeat a sequence after the examiner has done so. To differentiate the memory component from a pure appreciation of sequences, some assessment techniques should involve direct imitation of a particular serial order pattern.

INTEGRATED FUNCTIONS

With the steady influx of information through sensory channels, from memory stores, and via other internal sources, there clearly is a need for orchestration or integration, so that incoming data can be used in decision making. It may be, for example, that a particular stimulus set will arrive for processing through two sensory channels at the same time. In watching television or attending a play, integrated verbal and visual stimuli are, of course, the media of communication. Under such circumstances the capacity to integrate visual and auditory input is fundamental and in fact fairly primitive developmentally. As children progress through the educational system, the demands for such intersensory integration increase. There is a requirement to integrate and associate visual stimuli (such as letters or words) with specific vocalizations. Memory for discrete sounds must be linked tightly to constant visual configurations.

It is not surprising, therefore, that studies of certain tests of educational readiness have demonstrated that tasks involving so-called "intersensory integration" have a high level of predictive validity with regard to later school function. An example of such a task might entail having a child match the shape of an object that he is holding in his hand behind his back with an identical one arranged along with objects of alternative shapes on a table in front of him. Successful completion of this matching exercise would require the simultaneous integration of visual and haptic (i.e., touch) or stereognostic input and feedback systems.

Higher order integrated functions serve other purposes also. There must be integration between inputs and outputs. The successful completion of a task may involve the simultaneous integration of multiple functions. A player attempting to hit a baseball, for example, must process a visual-spatial input (i.e., the trajectory of the ball), mobilize previously acquired knowledge of an appropriate stance, judge the timing of the speed of the ball, and retrieve the appropriate motor pattern for swinging effectively at the approaching sphere. Similarly, a gymnast or dancer must integrate inputs of body position sense, movement, and external spatial coordinates to program and implement a motor pattern. It is critical that these various subcomponents be coordinated to achieve a smooth and successful effort.

Some children may have more difficulty with specific integrative linkages. A particular youngster may be adept at processing visual-spatial information, and visual memory also may be intact. The youngster may have no difficulty with basic motor coordination. However, when the child needs to integrate a visual input with a fine motor response, there is a breakdown. Other youngsters may have trouble with the integration of verbal inputs with motor responses, so that it may be difficult for them to write from dictation. One might infer thereby that as part of the evaluation of a child who is having school problems, there should be some consideration of specific integrative functions between discrete intake pathways or modalities on the one hand and forms of output on the other.

HIGHER ORDER CONCEPTUALIZATION

Atop the arc of function illustrated in Figure 38–1 there is an amalgam of processes that usually are thought of as the highest of the higher cortical functions of man. As children develop, their capacities to conceptualize, to generate and apply rules governing the order of the universe, to think on a more abstract level, to reason, to infer, to extrapolate, and to generalize from contexts coalesce and serve as crucial facilitators and integrators of experience.

Higher order conceptual abilities can redeem a child who has specific information processing deficits or gaps anywhere along the intake limb of the arcs of function. Thus, a youngster who has difficulty remembering the visual configuration of words may compensate by being good at applying rules, such as "'i' before 'e' except after 'c.'" A blossoming of higher order conceptual abilities frequently occurs in early adolescence during the period that Piaget referred to as "formal operations" (see Chapters 3 and 9). Often a youngster who earlier had to struggle with learning may suddenly improve dramatically because of superior conceptual functions. On the other hand, in some cases a child who has fared well with perceptual and retentive abilities may decline in ability because of difficulties in dealing on a more abstract conceptual level.

ASSESSMENT

There are few, if any, well standardized tools to evaluate the higher order conceptual abilities of a child. A variety of tasks requiring inferential reasoning have been used. For example, a youngster may be shown a series of visual configurations and asked to predict on the basis of the pattern which series or individual configuration will come next. Tasks involving the formation of analogies, or the classification of words or objects by their properties, may help to elucidate a child's relative strengths or weaknesses in working with abstract concepts. Various assessments have utilized the work of Piaget to document passage into specific cognitive developmental stages. Many instruments have suffered because it has been difficult to separate conceptual abilities from other components of function, such as language processing. Visual inferential reasoning may be quite dependent upon visual attention to detail, earlier successful experience in this modality, good visual perception, and well developed visual retentive abilities. It also can be difficult to discriminate between verbal reasoning and language skill in general. Nevertheless the clinician needs to keep in mind that higher conceptual abilities are important for learning and for rendering experience increasingly meaningful and enriching in its associations for the developing child.

OUTPUT FUNCTIONS AND THEIR ANTECEDENTS

In recent years there has been a great deal of emphasis on learning disabilities and, specifically, on the dysfunctions that interfere with the early mastery of academic skills. Chief among these foci has been reading. This process of recognition and decoding of visual symbols is, of course, a critical element in school success. Equally relevant, however, are a series of performance areas in which the stress is on encoding rather than decoding. This entails the organization and communication of one's own findings or thoughts, usually in response to specific problem solving or product oriented demands. Writing, spelling, complex mathematical computations, and long term work projects all involve high levels of productivity and efficient, accurate output. As children reach the late elementary school grades, there is a pronounced shift in emphasis from a predominantly decoding experience to a requirement for high volumes of output (especially written). Such increased volume expectations, frequently accompanied by time constraints, are met fairly readily by most children.

Returning to the conceptual model of functional arcs (see Figure 38–1), we see that the output limb is dependent upon antecedent segments of the arc. Like a simple spinal reflex, the efferent action is stimulated and selected as a response to an afferent input. Similarly, a child's generation of a product as well as his long range productivity is dependent upon various contributions from the intake limb. Thus, to be productive a child must be able to attend selectively to stimuli in the environment that will help shape output. For example, if the product is to be a paragraph copied from a blackboard, the child will need to select the written material while filtering out extraneous stimuli on the slate or in other parts of the room. To pursue this example further, the student will need to perceive and group the visual stimuli on the blackboard appropriately, to have some understanding of their meaning (although this may not be crucial), to be able to associate that which is being copied with some previous experience, to have associations with this and perhaps to draw inferences that will make the paragraph more meaningful. If one or more of these antecedent steps breaks down or is ineffective, the entire process may be overly laborious and possibly unsuccessful. Thus, it is important to recognize that every output is dependent to varying degrees upon selective attention, information processing, memory, and higher order conceptualization.

Once a child has decided upon a response that has been ordered and interpreted, various steps involved in output directly transpire. First the child selects a modality of response. Three output channels are most commonly employed: expressive language, fine motor activity, and gross motor activity. These also may be combined in the development of a single product. For example, written output would entail a fusion of expressive language and fine motor functions. Once the modality for response is selected, a blueprint needs to be "drawn" to code the child's response. To facilitate this process, vast numbers of such "blueprints" are stored in long term memory. A previously used plan can be appropriated to encode the spelling of a word, the visual configuration of a letter, or the pattern of one's own signature. Once the plan has been created or selected, it must be put into effect. This entails knowing what combinations of muscular activities (sequential facilitations and inhibitions) are required for the moment to moment execution of the plan. For example, if the blueprint concerns the spelling of the word "dog," the praxis consists of the muscular movements with a pencil in hand to represent the word on paper.

Just as the various feedback loops are critical for the monitoring of stimulus grouping and decoding, they are important as perpetuating forces in output. While writing a word or while steering a car, there is a steady, often unconscious awareness of how the task is going. Such constant feedback can enable one to continue, disrupt, or modify what is occurring on the output limb.

It should be emphasized that the arcs of function can be applied to a single product or performance over a week or month or an entire academic career. It is likely that the most effective continuous performance occurs when there is ongoing positive feedback, when end products elicit pride, sustain self-esteem, and are comparable or superior in quality to those of peers. On the other hand, when striving toward productivity elicits shame, guilt, or embarrassment, when the effort required is excessive and the rewards too scanty, output is likely to deteriorate in both its quality and its amount. It is also important to recognize that output is energized by two other factors, namely, motivation and practice. A child who is chronically unsuccessful is unlikely to be motivated. A child who faces chronic failure is apt to become inhibited with regard to output, to avoid certain cycles of productivity whenever possible, and consequently not to gain from the lubricant effects of practice in that aspect of output.

DEVELOPMENTAL OUTPUT FAILURE

A significant number of school age children experience serious problems with output (Levine et al., 1981). Many never have difficulty with the aspects of education that emphasize exclusively the intake limbs. Often reading skills are acquired easily, whereas increasing demands for written output lead to serious academic deterioration. For this reason the phenomenon of developmental output failure is encountered most commonly in late elementary and junior high school. It is found among children who are slow to automatize and integrate the various segments of the arc needed for efficient output. In some cases they show no evidence of any "learning disabilities" but appear rather to have "working disabilities." Because they may perform well on multiple choice tests (predominantly emphasizing decoding and recognition abilities), such youngsters may be misunderstood and falsely accused of laziness. They may be serenaded continuously by a background chanting the familiar refrain, "You can do better." For most of these children, increased volumes of written output constitute the major source of anguish and failure. For this reason the following description of the developmental components of output failure focuses largely upon their contribution to impairment of written output. At the same time it will be noted that the specific areas of development under discussion are germane to other output modes of expression and performance. Much of this phenomenology is summarized in Table 38–6.

FINE MOTOR OUTPUT

So-called manual dexterity constitutes a major output modality. The effective use of writing utensils, forks and knives, or artistic and craft materials is among the many operations benefiting from well controlled use of one's fingers.

In a recent study it was found that nearly two-thirds of the children with developmental output failure had difficulties with one or another aspect of fine motor output (Levine et al., 1981). The precise nature of their fine motor difficulties showed considerable variation, however. As already stated, the analysis of any output function requires an understanding of the various input components upon which it is based. In the fine motor area there are three major sources of input and feedback that are of critical importance: visual, propriokinesthetic, and motor memory.

Many fine motor activities require what is commonly referred to as "eye-hand coordination." That is to say, on the arcs of function, the motor output is based upon visual input data. If one is threading a needle, putting together the parts of a model airplane, or repairing a motor, it is likely that much of what is implemented is based on a predominantly visual input. It is important to see the hole in the head of the needle and to aim one's

finger movements accordingly. The initial input thus is a visual one. Ongoing feedback (i.e., how close the thread is coming to the needle) also is visual. Thus the overall process might be described as a visual–fine motor arc. It is likely to be conditioned by previous experience (i.e., practice) and perhaps facilitated by certain conceptualized rules (such as where to keep one's elbows and how to twist the threat to sharpen its end point). Some children with problems of eye-hand coordination have difficulty in copying a geometrical form, even though they may have fewer problems with pure visual perceptual tasks (i.e., those unattached to a motor response). They may have trouble in integrating a visual input with a motor input (see section on integration). Some youngsters with illegible handwriting may demonstrate these weaknesses of visual-motor integration. One would expect that they also would have problems with other tasks involving manual dexterity, since so many of these fine motor demands require a visual input.

Some children have problems in manipulating a pencil for writing but show no apparent fine motor deficits in other areas of performance. Thus a child may be said to have a slow or very sloppy handwriting, despite the fact that he is superb at assembling model airplanes, fixing engines, and connecting tiny transistors in a radio. Some such children are imperceptive to proprioceptive and kinesthetic feedback from their digits. Those fine motor activities in which there is a need for constant somesthetic feedback from the fingers—in particular those entailing steady and complex rhythmic movements (such as are necessary for writing)—are the ones most likely to be impaired in such instances. On a neuromaturational examination such a child is likely to show finger agnosia or have trouble in localizing his fingers in space without close visual monitoring (Kinsbourne and Warrington, 1963; see Chapter 45B). One can observe such a youngster writing and find that he keeps his eyes extremely close to the page. Some children with finger agnosia try to compensate for their poor proprioceptive and kinesthetic feedback by developing a very awkward and perhaps painful pencil grasp. Their discomfort constitutes feedback as they write! It is important to stress that much of the feedback loop that helps to guide written output is not visual but depends on an internal dynamic awareness of moment to moment finger and pencil location.

The combined input and feedback from visual and proprioceptive-kinesthetic pathways guides the written hand. It cannot do so without one other important component, namely, motor memory. As noted previously, a virtual library of specific motor plans, often called "engrams" or "kinetic melodies," is stored in memory. In many respects they are not unlike the rolls on an old

Table 38–6. DEVELOPMENTAL OUTPUT FAILURE

Component	Subcomponent	Common Manifestations
Motor	Weak eye-hand coordination	Poor pencil control; slow writing; difficulty with nonpencil fine motor tasks
	Finger agnosia (weak propriokinesthetic feedback)	Awkward, slow pencil movement; abnormal grasp; close visual monitoring while writing; little (if any) trouble with nonpencil fine motor tasks
	Apraxia (problem with motor planning)	Poor spacing and utilization of page, inconsistent letter formations, hesitancy; possible associated oral dyspraxia
Memory	Motor memory deficit (poor formation or retrieval of engrams)	Hesitancy, false starts, poor mastery of finger movement patterns, inconsistent handwriting
	Impaired revisualization and visual memory	Poor spelling (often phonetically correct errors), hesitancy, awkward letter formation, slow copying from blackboard
	Sequential memory deficit	Narrative organizational problems, sequencing errors in spelling, problems in scheduling work, mathematics weaknesses
	Other memory weaknesses	Trouble with rapid retrieval of vocabulary and rules (of grammar, punctuation, capitalization); difficulty with simultaneous access to multiple memory stores
Language	Vocabulary and word finding weaknesses	Immature vocabulary in writing; reluctance to participate in class discussions, problems with creative and expository writing as well as oral reports
	Problems with syntax formation and grammatical construction	Use of primitive, vague, simple sentences; excessive written economy; similar findings with spoken language; grammatical errors
Attention	Impulsivity	Poor planning and organization; poorly legible but highly inconsistent handwriting; careless errors in spelling, mathematics despite good knowledge
	Problems with persistence and delay of gratification	Easy cognitive fatigability, extreme brevity of output, incomplete assignments, distractibility
	Organizational deficits	Trouble in starting, difficulty in bringing together needed materials at right time, forgetfulness
	Inattention to detail	Problems with mathematics, self-monitoring, proofreading with frequent careless errors
Secondary effects (possible complications)	Chronic success deprivation	Inhibition, fear of failure, nonrisk taking strategy, lack of work incentive, anxiety, depression
	Social dysfunction	Excessive responsiveness to peer pressure, alienation from adults, antisocial activities
	Parent-child tension	Poor communication at home, perceived negative comparisons with siblings, alienation, withdrawal from family life
	Overexposure to criticism in school	Misunderstandings, feelings of humiliation, refusal of help, superimposition of working and learning inhibitions

fashioned player piano! They contain the visual patterns and coded finger movements needed to reproduce letters or words on paper. The more complex they become, the more sequential they are in their orientation. Thus, a youngster's sequential memory may be related to such stereotyped motor output. There is considerable variation in the accuracy, speed, and efficiency with which children are capable of forming, storing, and retrieving these "kinetic melodies." In late elementary and junior high school, there may be some youngsters who are slow to retrieve them.

Sometimes this is just a matter of maturation, and in several years the entire process will be more completely automatized. In the meantime, however, they may experience considerable difficulty, accusation, and failure, with a subsequent loss of motivation and incentive.

Another source of fine motor difficulty for some children is an apraxia. Children with this kind of disorder may have little or no difficulty in responding to visual information or proprioceptive kinesthetic feedback. Motor memory also may be intact. However, they seem not to know which muscles to facilitate and which to inhibit during the motor act of writing. Some of these children complain that although they know how to do things, they never can get their fingers to do them right! Many also have an oral apraxia or dyspraxia and may present with a history of having had speech therapy or of continuing to have articulation problems.

Some youngsters who have a motor apraxia also demonstrate difficulties with motor planning in general. They may have problems in utilizing the space on the page effectively. Their writing appears to be almost randomly placed and spacing within and between words may be highly erratic. This can be seen particularly when a youngster has evidence of both an apraxia and a high level of impulsivity.

Thus it can be seen that a number of subcomponents of developmental dysfunction in the fine motor area may account for graphomotor weaknesses. As with other aspects of performance, problems with selective attention also can interfere. For example, a youngster who is chronically impulsive may present with highly inconsistent but generally atrocious handwriting. It is noteworthy that many studies of the effects of stimulant medication on children with attention deficits have documented improvement in handwriting. These youngsters may operate at such an impulsive, frenetic cognitive tempo that there is little time and only a minimally systematic approach to scanning the memory stores for the appropriate engrams.

REVISUALIZATION

Closely linked and yet distinct from motor memory is the process of revisualization. This is an operation in which one searches one's visual memory to bring forth the configuration of a word, a letter, or some other symbol. Children with revisualization problems may forget what a particular word or letter looks like. In some cases they have no difficulty whatsoever in recognizing the correct spelling of a word, but when they are required to recall that configuration without any visual cues, they fail consistently. Thus, on a multiple choice test in which a youngster is expected to select the properly spelled word from among three or four alternatives, there may be no problem. However, on a dictated word list the same child performs dismally, often substituting phonetic spellings (e.g., "lite" for "light"). Problems with revisualization should be suspected when a child has writing difficulties and tends to make phonetic spelling errors.

In some instances the child's writing seems to reflect a "flickering" memory. Both revisualization and the retrieval of motor patterns for letter and word formation seem to be highly inconsistent.

There may be small elements of words or letters that are executed perfectly well, followed almost always capriciously by errant excursions of the pencil. One may note that the same letters are made differently through a passage. Some of these youngsters also appear to have problems with selective attention. The weak visual attention to detail is reflected in omissions and other careless mistakes in both letters and word execution.

EXPRESSIVE LANGUAGE

Commonly it is stated that writing is the highest form of language. If a child has not mastered the spoken word, he is likely to encounter serious problems with its written representation. Some children who have had chronic word finding problems (i.e., aphasias or dyphasias) may be particularly vulnerable to writing failure. They may have to struggle inordinately to find the words to express their ideas during writing. The tremendous effort at word finding may compete or interfere seriously with the other retrieval memory components of writing. The intense search for words may eclipse memory for spelling, punctuation, and capitalization. The whole effort can become inordinately difficult, stressful, and unsuccessful.

Some youngsters may have problems in writing owing to their tenuous grasp of syntax and the rules of grammar. Linguistically unsophisticated youngsters are likely to experience trouble with written expression. This condition can exist because of underlying neurological dysfunction or sociocultural factors in the background of the child.

Further elaboration of clinical disorders associated with poor expressive language can be found in Chapter 40.

OTHER DEVELOPMENTAL FACTORS ASSOCIATED WITH OUTPUT FAILURE

Because effective inputs are required to program efficient outputs, varying combinations of information processing problems can be encountered in youngsters with writing difficulties. For example, if a child has deficiencies of temporal-sequential organization, there may be significant impairment in the capacity to orchestrate a narrative flow. Arranging paragraphs in the correct order and establishing a logical sequence of ideas may be very difficult for such a youngster. In an earlier section of this chapter the monumental organizational problems of children with sequencing deficits in early adolescence were stressed. Poor study habits, absentmindedness, and an inability to schedule the various work stages of long range projects may seriously impede output. Other organizational problems also may thwart effective output. These include problems in integrating data from multiple sources, trouble with the retelling or resynthesis of ideas in a reading assignment, and difficulty in summarizing.

MULTIPLICITY OF DEFICITS

Children who have only one reason for output failure often can overcome their disadvantage and demonstrate a reasonably high level of productivity. Thus a child with a fine motor apraxia but a high level of competence in memory, language, and organization may well bypass his motor weaknesses and find a way to become a good writer. On the other hand, children with two or more dysfunctions are far more susceptible. In a recent study of students with developmental output failure presenting at a diagnostic clinic, the majority had more than one developmental dysfunction contributing to their lack of productivity (Levine et al., 1981). Almost every conceivable combination of deficits was represented within this sample.

Particularly common were clusters of dysfunctions that included poor selective attention, word finding problems, and one or more of the fine motor weaknesses. The combination of impulsivity and poor revisualization also was prevalent. It should be stressed, however, that no distinct consistent syndromes emerged. This would suggest strongly that children with manifestations of poor productivity should be evaluated in terms of a broad range of parameters to identify deficits and strengths. The overlapping areas of function contributing to output are depicted in Figure 38–2.

COMPLICATIONS OF DEVELOPMENTAL OUTPUT FAILURE

A number of developmental and environmental factors complicate the picture of output failure (see Table 38–6). First, impaired productivity tends to manifest itself at a very inappropriate time of life, namely, late latency and early puberty. Youngsters at this age are apt to be especially self-conscious. They want very much to resemble their peers.

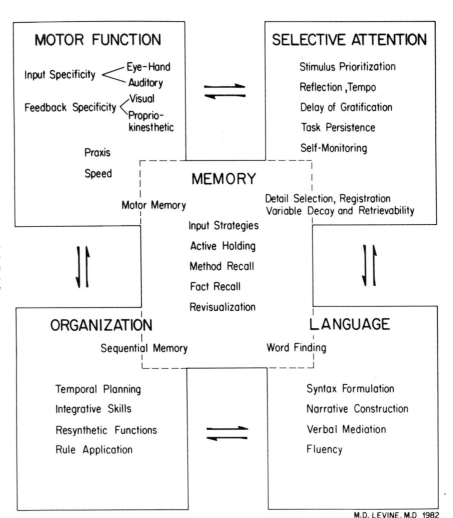

Figure 38–2. Diagrammatic representation of various developmental components of productivity in childhood. It can be seen that these areas overlap, particularly with regard to their dependency upon rapid retrieval memory.

M.D. LEVINE, M.D 1982

Sloppy, skimpy, written reports besmirched with the teacher's red marks may symbolize permanent documentations of their differentness, of their abnormalcy. This may lead to shame and academic inhibition. Rather than be reproached for such products, the youngster may tend to rationalize, pretending that he has no homework. Ultimately he may refuse to do written assignments, preferring the macho of bold defiance to the humiliation of poor grades.

It is not unusual for children in this age group to resist help. They shun resource rooms or learning centers. They are embarrassed at being tutored. They may insist that they do not need any help. They may be uncooperative or belligerent when offered assistance. Once again, an obsessive desire to be perceived as normal may override all other needs.

Entry into the sixth or seventh grade can be tumultuous. Middle and junior high schools have a tendency to have fewer resources. The transition to a multiple teacher format may be particularly traumatic for a child who feels academically inept and vulnerable. He may feel a loss of privacy, a broadened exposure, and susceptibility to humiliation. At this stage of education there may be fewer specialized services available. Classroom teachers may not be aware of developmental problems with output. A child who reads well but has output failure may be in particular jeopardy. His disability may lack credibility, his plight worsened by a barrage of accusations of laziness.

It also is possible that the problems of a youngster with developmental output failure will not be understood by his parents. They too may feel that he is primarily lazy or unmotivated. The predicament can be aggravated by the presence of one or more siblings who are highly productive and successful. A child with output failure may have to work for two or three hours to produce a two or three page report. The product may be messy and disorganized. It may fail to conform to rules of spelling or punctuation. Despite earnest effort, the youngster may be required to copy it over (a form of torture for affected students). It may be that the best he can receive is a C− or a D, while a sibling "breezes through" a similar assignment in 20 minutes, achieving at worst a B+ or an A−. Even unspoken or implicit comparisons are painful. The affected youngster may be depleted of all initiative and incentive to work.

Output failure may be complicated further by environmental problems. A child who leaves an unhappy home to face day long criticism because of impaired productivity may be particularly vulnerable to sociopathy and other emotional complications of combined output and environmental failure. It is possible that a student may encounter difficulties with productivity at the same time that his parents are working through their own midlife crises. A wide communication gap may exist and aggravate the youngster's personal feelings of anguish and worthlessness. The only salvation may lie in antisocial acts, in the formation of social subcultures consisting of peers who face similar failures.

It should be re-emphasized that although output failure manifests itself primarily with writing, many affected children also have problems with arithmetic. Some have particular difficulties with reading comprehension. They may be able to answer questions about a paragraph (such as one found on a multiple choice test), but they are poor at resynthesizing or encoding the details in their own words and with their own organizational schemes. In particular, they may be unable to apply what they have read to inform their own writing. Over and above their academic weaknesses, many such youngsters seem to have generalized organizational problems, poor study habits, and a clinical picture suggestive of chronic success deprivation.

GROSS MOTOR FUNCTION

The area of gross motor function represents another important output channel, but one that is not directly related to academic achievement. Nevertheless children seem to take seriously their own physical mastery over space. While not every child needs to become a great athlete, it is clear that feelings of effectiveness in this area can be helpful in the attainment of appropriate levels of self-esteem, personality development, and social skill.

GROSS MOTOR DYSFUNCTION

Children who have delays in the development of gross motor function may develop profound feelings of inadequacy. On the other hand, youngsters who have strength in this area of development often can mobilize such assets to overcome any negative feelings about academic performance and to sustain self-esteem.

Deficiencies of gross motor function are quite analogous to those described in the fine motor area. Some children seem to have trouble predominantly with those gross motor acts based on visual input and ongoing visual monitoring. The game of baseball, for example, stresses visual-spatial processing and appropriate gross motor responses. Catching or hitting a baseball requires a well developed appreciation of spatial attributes, a keen ability to judge trajectories, and a firm linkage between visual inputs and well timed motor responses. Children who have difficulty with visual-spatial processing may thus encounter problems with games that require judgments about small spheres flying through space. The same youngsters may be adept at sports in which the major input is proprioceptive and kinesthetic. For example, a child with visual-spatial problems but intact somesthetic inputs and monitoring may become an excellent swimmer, a very good gymnast, or an accomplished skier.

Some children seem to have weaknesses of gross motor memory. They have trouble in retrieving the plans for previously implemented complex motor acts. They may prefer sports in which there is a more straightforward or simple physical component. Various track and field events may appeal to such youngsters. They may avoid events in which too many different complicated motor sequences need to be retrieved. A child may want to participate in ballet or other forms of dance but may be limited by poor retrieval of motor patterns.

It should be stressed that some youngsters with gross motor weaknesses can learn to play a wide range of sports. In some cases their deficits are of such a degree that a great deal of extra practice will be required. Often children fear such practice, because while they are in the process of learning, they are liable to be ridiculed by their peers. The answer may be privacy, i.e., helping the child to learn to play initially without an audience. Certain sports may be more gratifying to children with gross motor problems. For example, soccer is an activity that offers a reasonable amount of privacy. There tends to be a crowd where the ball is! This may tend to diminish accountability somewhat, thus enabling a youngster with motor deficits to feel like a real team member and not be ridiculed by his colleagues.

ASSESSMENT

Because gross motor function is important from the point of view of young children, its assessment through a history and perhaps direct observation can contribute to a comprehensive understanding of a child's functional profile. Standardized gross motor tasks can be used as part of a neurodevelopmental examination (see Chapter 45). A historical review of the child's various athletic endeavors (including triumphs and failures) also may be helpful. This can be included in standardized questionnaires.

RELEVANCE OF SOCIAL DEVELOPMENT

Children with developmental dysfunctions commonly (but by no means universally) are prone to have difficulties in social interaction, with peers in particular (Bryan, 1977). Many of them experience confusion and conflict in attempting to form and sustain relationships with others of their own age group. In some instances they are rejected and become isolated, or else they withdraw voluntarily out of fear of not being able to compete and compare successfully. Some children with gross motor delays may withdraw, anticipating that association with groups of other youngsters could lead to humiliation on a playing field.

It is not unusual for youngsters with developmental dysfunctions to seek the company of adults or younger neighbors. They may demonstrate remarkable social poise in all circumstances except when they find themselves in a milieu of others of their own age.

Many reasons can be advanced for the association between social failure and developmental dysfunction. Sometimes children seem to harbor disabilities of social judgment or perception. Their shortcomings in peer interaction may parallel closely their cognitive deficits. In fact, it is possible to conceptualize a model of social productivity that utilizes the arcs of function already described. For example, a child with attention deficits may be highly impulsive and disinhibited in his behavior. Such a pattern may be found offensive by peers. They may reject and ostracize the affected youngster. Few children who are highly impulsive also are popular. Some children with chronic impulsivity seldom are reflective enough to predict the social consequences of their actions. They commit one faux pas after another. They rapidly acquire a bad reputation, one from which recovery is quite difficult.

Other examples of a close relationship between social failure and developmental dysfunction can be cited. Children with language disabilities may have difficulty in controlling social interactions with words. They may also be relatively imperceptive to verbal feedback clues during social intercourse. Some children with visual processing problems may be imperceptive in "reading" faces and acquiring social feedback in that manner. Some children with developmental dysfunctions appear to be socially "illiterate." They seem not to understand or be able to carry out basic principles of relating. They have difficulty in sharing; they may not understand or be able to implement the fundamental give and take dynamics of relationships. Their peers may resent them for this.

In many respects school is as much a social as an academic test. Clinicians evaluating the social competence of children need to inquire carefully about certain "hot spots," locations and moments in the school day at which social pressures reach a peak. These include the bus stop, the bus, the hallways in front of lockers, the bathrooms, the cafeteria, and the gymnasium. One can learn a great deal about a child's social experiences and stresses by trying to recreate a picture of what it is like in any of these scenes. It is useful, first of all, to have a child depict a typical day in that location and then to have the youngster talk about how he or she fits within that context. It can be helpful to have a sense of what names the youngster is being called by others. Children who are low on the social ladder often endure an unfortunate set of peer labels. Such barbed appelations as "retard," "mental," "redneck," or "faggot" are not unusual. Often youngsters will not talk about being called these names, but they are most anxious to share their anguish if appropriate enquiries are made in a supportive and private setting.

BEHAVIORAL CORRELATES OF DEVELOPMENTAL DYSFUNCTIONS

Just as it is difficult to separate social interaction from cognitive development, it is equally arbitrary to divorce patterns of behavior and learning disorders (Levine et al., 1980). When confronting a child whose learning and behavior are a problem in school or at home, it is helpful to identify possible linkages between the two. The following classification of behavior patterns in the presence of developmental dysfunction may be helpful:

Direct Behavioral Manifestations of Dysfunction. Some behavioral traits may be actual symptoms of a child's disability. For example, a child with attention deficits may be destructive and disinhibited, as described earlier in this chapter. Aggressive, poorly fettered outbursts may be part of a broader picture of attention deficits. In other words the behavior may be one component of an underlying neurodevelopmental problem.

Behavioral "Metastases." Sometimes a child's developmental dysfunctions lead indirectly to what appears on the surface to be a separate behavioral issue. For example, some children develop encopresis (fecal incontinence) because they sit on the toilet for only very brief periods of time (see Chapter 31). They are as impersistent at defecation as they are at most other activities in their lives! Ultimately they become constipated, develop megacolon, and manifest incontinence. Thus, in such cases the encopresis represents a kind of "metastasis" of the developmental dysfunction to the large intestine!

Secondary Behavior Patterns. In some cases a child may develop an emotional problem because of chronic failure stemming from a developmental dysfunction. The child may become increasingly anxious and depressed because of long standing frustration and failure in school. Some or all of the symptoms of childhood depression may appear, including self-deprecatory statements, loss of appetite, somatic symptoms of various sorts,

excessive fatigue, and overall sadness (see also Chapter 41). In such instances the symptoms clearly are reactive, traits that develop as secondary manifestations of a learning disorder.

Compensatory Behavior. Sometimes a child develops an undesirable behavior pattern in an effort to compensate for an apparent weakness. For example, some youngsters with the kinds of social failure we have described may feel out of control in social settings. They may react by becoming increasingly aggressive and controlling as a way of dealing with their own feelings of social impotence. A child with gross motor dysfunction may become overly aggressive and "physical" to make up for deep seated feelings of weakness and motor ineffectiveness.

Face Saving Strategies. As has been noted repeatedly, it is often the case that youngsters with developmental dysfunction are embarrassed about their problems. In some instances they try to cover up. Sometimes maladaptive strategies are employed. These might include excessive clowning in class, defiant behavior, efforts to bribe or buy off friends, a gamut of charismatic guises, or a veneer of extreme indifference. Sometimes such strategies can engender more trouble than the dysfunctions they were designed to mask. It is important, however, for teachers and clinicians to recognize such strategic maneuvers, to separate them from primary psychopathological disorders, and to realize that a child cannot be stripped bare of his strategies until he is able to substitute better ones or alter the conditions that necessitated them in the first place.

Psychosocially Induced Behavior. In some instances children with developmental dysfunction may manifest behavioral traits that are unrelated etiologically to their learning problems. One may sometimes advance the argument that underlying psychosocial difficulties are causing or promoting poor adaptation to academic settings. It certainly is the case that children with serious psychopathological states are likely to show impairment in school performance (see Chapter 41).

In most cases one creates a false dichotomy by asking whether a youngster's school problems are "emotional" or "organic." It is likely that the two are closely bound, and that there is little value in the struggle to determine which came first or which is primary and which is secondary. Often definitive answers are not forthcoming. It is indeed unfortunate if a child has developmental dysfunctions and is labeled as having a "pure emotional" problem. It may be equally catastrophic if in a child with developmental weaknesses the emotional needs are neglected. In most cases learning disorders represent a final common pathway in which both environmental influences and constitutional predispositions have coalesced.

ACADEMIC IMPACTS

Many factors come together to determine the ultimate academic impacts of a child's profile of strengths and weaknesses. The developmental dysfunctions outlined earlier in this chapter have a significant bearing on the learning process. However, it is important to recognize that clear-cut one-to-one relationships between specific disabilities and deficiencies in skill acquisitions do not exist. A particular child may have visual-spatial processing difficulties and yet learn to read with ease, whereas another youngster with the same degree of dysfunction may experience a major delay. What then are the variables that seem to influence the outcome? The following is a summary of possible forces that can determine the impact on learning of an underlying developmental dysfunction:

The Presence or Absence of Other Disabilities. As was mentioned earlier, a single dysfunction is likely to have less of an effect on the learning process. In fact, single disabilities are quite common in the community; many high achieving students have isolated areas of developmental dysfunction. Two or more areas of deficit are more apt to precipitate failure. If a child has an attention deficit but is very much intact in every other area, the prognosis for learning is far better than if he has the same problem accompanied by significant language disabilities.

Socioeconomic Status. Socioeconomic status is known to be an influential variable. Many studies have demonstrated that children from higher socioeconomic brackets are better able to overcome disabilities (see Chapter 11). There are likely to be many reasons for this phenomenon.

Early Educational Experience. If a child with developmental dysfunction has been fortunate enough to have had positive and encouraging exposures early in his education, resiliency may be maximized. On the other hand, a youngster who has been overexposed to criticism and humiliation as a result of his developmental problems may develop inhibitions about school and learning. His developmental dysfunctions thereby are potentiated and have a greater impact upon learning.

Motivation. For a variety of reasons some children may be more motivated than others to overcome their developmental dysfunctions. Strong role models in the home environment may induce such motivation. A desire to overcome adversity and achieve success is an important and little understood variable.

Self-esteem. It is likely that success is a self-perpetuating process. If a youngster feels good about himself, if he does not develop a negative self-image, the capacity to overcome inherent deficiencies will be enhanced. It is likely that children who are chronically deprived of success, who seldom if ever feel masterful, who are unable to taste achievement, may continue to decline academically. This is why it is important for the adults in his world to program successful experiences for a child. Whenever possible, compensatory strengths need to be found, and these must be mobilized to help a child develop a sense of mastery.

Virtually any of the developmental dysfunctions enumerated in this chapter can interact with other endogenous and exogenous factors to create problems in reading, spelling, writing, mathematics, or organization in general. For this reason, terms like "reading disability," "dyslexia," and "dyscalculia" are not particularly helpful in most cases. In attempting to formulate the reasons for a child's failure, a simplistic term like "minimal brain dysfunction" is of little value. A more comprehensive approach can have far richer therapeutic implications (for further elaboration see Chapter 52 on Diagnostic Formulation).

NEUROPSYCHOLOGICAL APPROACHES

Chapter 38A has adopted an organizational schema that is based on a developmental information processing and output model. There are many other defensible approaches to this subject area. They may have significant implications for both diagnosis and treatment.

The field of neuropsychology has a long tradition of research and service. Neuropsychologists attempt to relate functional impairments to specific localized lesions of the central nervous system. Recently there has been increasing study of neuropsychological bases of learning disorders.

Denckla (1978) has described certain typical hemisyndromes. Children with a right hemisyndrome manifest three or more signs involving the right side of the body, such as distal weakness, hypotonia, and increased reflexes. These are said to imply that the left cerebral hemisphere may be dysfunctional or dysgenetic. Such children are said to have a "left hemisphere style." According to Denckla, they "behave to a striking degree like mild or subtle versions of adult aphasia and/or related disorders." She adds: "They respond in the holistic, big configuration, simultaneous-sensitive manner of the 'strong right brain,' relatively speaking, and have difficulty balancing this style with the detail-attentive sequential, analytic, linguistic style of the left brain that is so important academically."

Conversely, a left hemisyndrome is said to be

characterized by left sided distal weakness, hypotonia, and increased reflexes. Affected youngsters are described as having problems with "higher order verbal skills, spatial orientation skills, mathematical understanding, and normal give and take of social-emotional expressions." Neuropsychologists attempt to find functions of the brain that are represented geographically close to each other. It is thought that lesions in certain locations thus may produce clusters of symptoms or manifestations that are predictable and consistent.

There have been neuropsychological approaches to children with attention deficits. Most research has tended to focus on the function of the reticular activating system. This area of the brain stem has a major role in sustaining arousal. It was noted earlier in this chapter that many children with attention deficits are said to be underaroused. The possibility of a "lesion" in the reticular activating system has been suggested. Some investigations of attention deficits have concentrated on frontal lobe function. The frontal lobes are active in selectively heightening cortical tone in certain areas of the brain during particular activities. Thus, if someone is proofreading or looking carefully at a picture, the highest cortical tone should be in the occipital regions of the brain where most visual processing occurs. At the same time other parts of the brain are reduced in their tone, so as not to distract from the major operation at hand. One might surmise that some children with attention deficits have impairments of appropriate parts of the brain. This may account for some distractibility and diffuseness in their performance.

Neuropsychological approaches are based to some extent on the hypothesis that there is consistence of localization of function from child to child. There is still the need for further investigation to document this and to show the extent to which generalizations about adults with brain damage can be extrapolated to account for learning disorders in childhood.

Some promising new computerized radiological techniques should help to refine further our concepts of cerebral localization and the anatomical correlates of learning disorders. It may be, however, that there is more plasticity than one suspects. Newer radiological techniques may demonstrate localizable lesions only in a subsample of children with learning disorders. Further considerations of a neurophyschological model are found in Chapter 38B.

REMEDIATION

Only through a comprehensive multidisciplinary approach can most children with developmental dysfunctions show optimal levels of resiliency. Appropriate sensitivity and support must come from parents, teachers, siblings, mental health specialists, and pediatricians. The specific details of intervention are covered in other chapters in Part VII of this book. The following is a list of some general remediation strategies that must be thought about in dealing with children with learning problems.

Modified Regular Education. The regular classroom teacher needs to have a good understanding of a child's developmental dysfunctions. Certain "bypass strategies" can be used in the classroom. For example, a child with developmental output failure might be allowed to type assignments, to write shorter reports, to submit a recorded cassette rather than a written paper (see Chapter 56). A child with attention deficits may need to sit close to the front of a room to reduce distractions and to allow for more direct feedback. The classroom teacher may need to select an appropriate set of curriculum materials to match a child's learning style. For example, a youngster with both language and visual perceptual difficulties may require a strong multisensory approach to reading, one that allows for multiple channels of intake in order to reinforce memory.

Special Education. A strong special educational program can help to remediate areas of developmental weakness (see Chapter 56). Within a learning center or resource room a youngster can receive individualized attention. A child with a sequential organizational problem can be helped to gain a better appreciation of time, to store and retrieve increasingly complex data delivered in sequence, and to carry out multistep operations in a systematic way. A student with visual-spatial problems can be helped to appreciate, store, and retrieve increasingly complex visual configurations. The special educator can also help a child acquire basic skills. He can discover teaching methods that seem to work best with the youngster. The special educator also can counsel the child about his problems and advise the regular classroom teacher about daily management.

Special Services. More specialized intervention is sometimes helpful (see Chapter 57). For example, a speech pathologist may be needed to work with a child who has a language disability. An occupational therapist may participate in the care of a youngster with motor deficits.

Counseling. The following are among the indications for therapeutic counseling: the development of serious maladaptive strategies; the existence of significant anxiety and depression; the presence of serious complicating psychosocial factors at home; and the clinical recognition of a serious psychopathological condition.

Some children may not require referral to a mental health specialist but still need counseling (see Chapter 53). In particular, they need to have their developmental dysfunctions explained to them. Such a process of "demystification" can be undertaken by a teacher, a pediatrician, or another professional. In many instances the parents require the same form of education. It also can be useful to include siblings, since the latter are forced to endure much of the impact on family created by a handicapped child.

Medications. Some children, particularly those with attention deficits, may benefit from psychopharmacological drugs (see Chapter 58). There are specific indications for these, and their administration must be closely monitored.

It is likely that most children with developmental dysfunctions will require more than one of the foregoing therapeutic approaches. Close coordination of services can be critically important. In many instances the pediatrician can play an important role in promoting this (see Chapter 60).

PROGNOSIS

Because children with developmental dysfunctions differ so markedly from each other, prognostication is hazardous. All those working with parents and children should strive to orient everyone directly involved and should set short term rather than long range goals. These goals should be attainable. They should be given priorities. There should be a recognition that results cannot be achieved overnight. There is a paucity of data relating to the natural history of developmental dysfunctions. There is even less information about the outcome for children with developmental dysfunctions who are well managed throughout childhood. At present, the most secure position is one that insures a child optimal services and support from year to year. Long term predictions represent an open invitation to self-fulfilling prophecies and constraints of opportunity.

DEVELOPMENTAL VARIATIONS AND LEARNING PROBLEMS AMONG ADOLESCENTS

The developmental dysfunctions and learning problems of secondary school students pose a formidable challenge to physicians and other professionals. Far less research and theoretical consideration has been directed toward this age group than toward the school problems of younger children. A number of methodological barriers have slowed progress in this area. There also are issues that constitute important considerations for those dealing with adolescents who are failing in school. They include the following:

● Older youngsters may have superimposed over their developmental dysfunctions a series of protective (and sometimes maladaptive) behavioral shrouds to save face. These may tend to obscure the underlying developmental problems. In addition, such defense mechanisms may constitute greater management problems than the disabilities themselves. An adolescent who gets into trouble, is rebellious, or seems to adopt an indifferent attitude may appear exclusively to have an emotional disorder rather than an end stage learning problem. The clinician needs to be aware that in many cases both forces operate. That is to say, a failing adolescent may show signs of significant emotional disturbance along with constitutional predispositions to school failure. The picture can be complicated further by the addition of family stresses, the absence of academically oriented role models, the presence of poverty, and the existence of inadequate secondary schools.

● In addition to secondary maladaptive strategies, children with chronic success deprivation due to learning problems may have secondary affective symptoms. In particular, depression is a common sequela of chronic failure in adolescence. Moreover, disturbed family relationships may be the result as often as the cause of poor learning in secondary school. A child who chronically disappoints his parents or one who is jealous of siblings whose academic performance is superior may exert a potent negative impact on family function.

● Many of the traditional tests and other assessment techniques for learning problems tend to "ceiling out" at or around puberty. There are very few assessments of language ability, visual processing, and sequential organization that are commonly applied in the diagnosis of adolescents with learning problems. The relative unavailability of such tools may create a sense that developmental dysfunctions need not be part of the differential diagnosis of the floundering teenager. The state of the art of assessment in this age group remains primitive, but there is every reason to believe that learning problems do not evanesce at age 13 or 14. The use of existing tools (such as the Detroit Test of Learning Abilities), careful history taking, and neurodevelopmental examinations may reveal specific weaknesses that are relevant to a child's academic difficulties.

● There may be a tendency for school personnel, parents, and students themselves to be fatalistic about learning problems in adolescents. It is common to encounter the attitude that somehow a teenager has lost all developmental resiliency. There may be more of an effort to bypass, to ignore, or to resign oneself to dysfunction in this age group. It is ironic that great resources are allocated to the rehabilitation of a 75 year old stroke victim while there is a common feeling that a 16 year old child with a language disability may be too old to be helped with language! Such intervention commonly is replaced by vocational training, placement in low achieving groups, or various constraints on opportunity. Although such alternative pathways often are appropriate for a youngster, there should be, at the very least, serious consideration of direct intervention for a learning problem in this age group.

• Adolescents may have undergone attitudinal changes that make it difficult to help them with their developmental dysfunctions (Handel, 1975). Some are embarrassed at getting help in school. Some develop antiacademic attitudes as a response to potent peer pressure (Carter et al., 1975). Others have become preoccupied with nonacademic sources of gratification (such as the opposite sex, sports, motorcycles, or drugs) such that their learning problems rank low on the tally of priorities. Thus, one may find oneself diagnosing and working with youngsters who do not particularly want or value the assistance offered.

The foregoing impediments should not engender feelings of hopelessness on the part of clinicians. There also is a positive side to working with adolescents who have learning problems. It is possible to retrace the historical course of their academic careers and thereby develop good hypotheses about their learning styles and patterns. An adolescent can be an available collaborator in the diagnostic process. As children proceed through the stage of formal operations, they are better able to understand and talk about their own cognitive styles. It is much easier to educate them about their learning problems and strengths than it is to clarify such issues for younger children. Using such an approach, a clinician can develop a strong alliance with an adolescent by co-conspiring to develop strategies to overcome academic failure.

The learning disorders of adolescents are likely to manifest themselves in a manner that is different from what is encountered during elementary school. There may be considerably less difficulty attributable to perceptual problems and specific language disabilities. Problems with rapid retrieval memory become especially common and troublesome (Landau and Hagen, 1974). Many of the components of output disorders described in this chapter become increasingly prominent as youngsters are challenged with greater volumes of homework, increased demands for writing, and the necessity for greater efficiency and attention to detail. Typical chief complaints in this age group include poor study habits, forgetfulness, an inability to complete assignments, trouble in taking tests, and general disorganization. In addition to these problems of productivity, some adolescents begin to show problems with higher order conceptualization. These cognitive difficulties in some cases may reflect delays in arriving at the stage of formal operations (Inhelder and Piaget, 1958). Adolescents with such learning problems are said to be too "concrete." They may have trouble in

dealing with abstract concepts, employing generalizations, drawing inferences, understanding irony and metaphor, and dealing with certain symbolic processes for mathematics (especially algebra).

In the adolescent age group, mathematics disabilities become accentuated further. This is one form of dysfunction that is more common in girls than in boys. Some students with "math phobia" have been discovered to be so humiliated by their inability to grasp and utilize higher order mathematics concepts that the very sight of such challenges creates inordinate anxiety and inhibition, further compromising their ability to succeed in this area.

Another common learning disorder of adolescence relates to an inability to master foreign languages. The reasons for this kind of dysfunction are poorly understood. It represents a disability that can occur in isolation, or alternatively it may be a logical extension of concomitant or earlier learning difficulties. It should not be assumed that youngsters with good language abilities in English inevitably succeed in a foreign language. It is obvious that the acquisition of the latter depends upon multiple other forms of linguistic, cognitive, and memory functions.

The capacity of a youngster to take tests may increasingly become a problem in high school. Parents may express concern because a child who succeeds academically does poorly on standardized achievement tests or the multiple choice admission tests for colleges. Some youngsters appear to be unable to cope with the format of such examinations. They may have trouble in deciding among multiple choices. Their learning styles may be such that they are far better at filling in blanks or writing essays than using other people's terminologies and conceptual frameworks to answer questions. The latter, of course, is required to succeed on machine-scored multiple choice examinations. There are some data to suggest that such youngsters can benefit from specific tutoring in test-taking abilities. Since they lack a sense of the hidden logic or structure of multiple choice examinations, they can be helped to master this. Such a "survival skill" in our society may be important and may prevent undue humiliation in an otherwise motivated and successful student.

The clinician seeing an adolescent with poor test-taking ability should encourage a great deal of practice with these kinds of examinations, beginning well in advance of the time that such tasks will become major determinants of a student's future. A youngster who appears to have difficulty with multiple choice test-taking should begin systematic practice early in his secondary school career. It is also important for the clinician to differentiate between youngsters whose cognitive styles are such that these kinds of tests are almost inappropriate for them and those who display a high level of test-taking anxiety. In some cases, of course, both these factors may be operating synergistically. Those whose anxieties seriously interfere with test performance may need to have some counseling from an experienced guidance counselor or mental health professional who can help a youngster understand the sources of this tension and develop some more adaptive mechanisms to deal with it.

Another common problem for adolescents has to do with intense academic pressure emanating from parents. A high school student may come to feel that no matter what he does in school, he can never please his parents. Their expectations may be far too great or even unrealistic. In some cases parents do not openly coerce their children but instead set an example or espouse values perceived by an adolescent as well beyond attainment possibilities. Sometimes youngsters need to understand that little can be done about the pressure—except to acknowledge and try to contain it. Unspoken academic pressures can be almost impossible to eradicate.

It is, of course, preposterous to ask high achieving parents to perform less well themselves. On the other hand, certain kinds of counseling strategies can be helpful. The precise nature of the pressures needs to be identified. Some effort needs to be made to rechannel them. Parents have to be helped to see their children's own strengths and to acknowledge and exploit these appropriately. It is important to detect and depict for all concerned major discrepancies of expectation. Parents and children need to understand that many different kinds of minds make up a culture. A youngster who is good with his hands can make as important a contribution and have dreams that are as lofty as one who is adept with words. Parents may need to understand that a youngster who builds and fixes motorcycle engines or one who designs and sews dresses may be more influential, happy, and wealthy than one who becomes a literary critic, a lawyer, or a college professor. It may be that the most important consideration is the extent to which an adolescent can commit himself to various interests and career inclinations rather than the precise identity of those orientations.

Recently an additional potential problem area has been identified. Males who are late sexual developers in adolescence may engender negative educational expectations from their parents. For example, their parents may be less likely to expect them to go to college (see Chapter 9B). Pediatricians can identify the late developers and should be prepared to monitor their scholastic performance and support their educational aspirations.

The clinician dealing with an adolescent who has learning problems or school maladjustment has many therapeutic tools at his disposal (Duke, 1980). These include the following:

Developmental Counseling. The pediatrician or family physician can spend a great deal of time discussing a youngster's learning difficulties with him. A process of "demystification" can be undertaken in a manner similar to that suggested for younger children. Advice can be given regarding the pursuit of strengths without dowsing the hope for improvement in weak areas. In some cases referral to a mental health specialist may be needed when a youngster is particularly depressed or has developed a range of maladaptive strategies to deal with failure. Moreover, when a serious family psychopathological disorder complicates an underlying learning disorder, such therapeutic intervention can be critical.

Special Educational Programs. Junior and senior high schools continue to offer resource room or learning center help in most communities. An adolescent may need some urging to take advantage of this. As noted, there may be some shame associated with receiving special educational assistance. This is particularly prominent during junior high school years. Nevertheless the clinician can be very helpful in identifying services that are available in the school and in advocating that a youngster receive these. Help in specific academic areas often can be rendered. In addition, the clinician can encourage the special educator to work on specific areas of processing and organization that seem to be impaired (Mann et al., 1978). Many youngsters need targeted help with organizational skills and study habits. Such techniques as underlining, outlining, taking notes, scheduling work in advance, and self-monitoring may need to be taught explicitly to secondary school students.

Tutorial Help. Some youngsters may benefit from tutorial help outside school. They may feel stigmatized within the school setting and prefer to have a private tutor. When parents can afford it, this can be quite helpful to a youngster. Once again, basic academic areas can be worked on. The clinician or diagnostic team can help by identifying specific areas of weakness and strength.

Vocational-Educational Counseling. Many adolescents can benefit from seeing a counselor who can help them with their vocational plans (Brolin and D'Alonzo, 1979). Such an individual also can assist in identifying specific areas of strength and interest. Standardized tests that characterize aptitudes or inclinations are commonly employed.

Medications. The use of stimulant medication for adolescents with attention deficits has been called into question repeatedly. It once was said that these pharmacological agents are ineffective in the adolescent age group. Moreover, there has been concern about giving drugs to teenagers in view of the high prevalence of substance abuse in our times. Obviously, therefore, some caution is advised. One probably ought to have a higher threshold for using stimulant medications in this age group. On the other hand, youngsters with significant attention deficits during the teenage years may benefit from stimulant medication, and its use should not be ruled out entirely.

Bypassing Weaknesses. Adolescents need to be helped to circumvent some of their learning problems while continuing to pursue a sophisticated education. Society is demonstrating a greater recognition of these difficulties in this age group. For example, some youngsters are permitted to take college entrance examinations untimed. A number of colleges throughout the United States are accepting and making provisions for youngsters who have "learning disabilities." Some college students with learning problems have been able to select courses especially carefully. Those with writing difficulties can be told to approach various professors before signing up for a course, in order to find out which ones would allow for less written output or substitute assignments with less writing.

It is clear that the developmental dysfunctions and learning disorders of adolescents represent a great challenge to the fields of developmental and behavioral pediatrics as well as those of education, psychology, and psychiatry. Further research is needed to elucidate the phenomenology and management of academic underachievement in this age group (Miller, 1981). In the meantime there needs to be, at the very least, a sensitivity to the existence of such morbidity and a professional's willingness to listen, to understand the plight, and to advocate.

MELVIN D. LEVINE

REFERENCES

Brolin, D. E., and D'Alonzo, B. J.: Critical issues in career education for the handicapped student. Except. Child., 45:246, 1979.

Bryan, T. H.: Social relationships and verbal interactions of learning disabled children. J. Learn Dis., 10:501, 1977.

Carter, D. E., DeTine, S. L., Spero, J., and Benson, F. W.: Peer acceptance and school-related variables in an integrated junior high school. J. Educ. Psychol., 67:267, 1975.

Denckla, M.: Minimal brain dysfunction. In Education and the Brain. Chicago, The National Society for the Study of Education, 1978.

Diagnostic and Statistical Manual of Mental Disorders (DSM III). Washington, D.C., American Psychiatric Association, 1980.

Duke, P.: The role of the pediatrician in the adolescent's school. Pediatr. Clin. N. Am., 27:163, 1980.

Handel, D.: Attitudinal orientation and cognitive functioning among adolescents. Dev. Psychol., 11:667, 1975.

Inhelder, B., and Piaget, J.: The Growth of Logical Thinking from Childhood to Adolescence. New York, Basic Books, Inc., 1958.

Kinsbourne, M., and Warrington, E.: The development of finger differentiation. Q. J. Exp. Psychol., 15:132, 1963.

Laudau, B. L., and Hagen, J. W.: Acquisition and retention in normal and educable retarded children. Child Dev., 45:643, 1974.

Levine, M. D.: The low severity-high prevalence disabilities of childhood. In Barness, L. (Editor): Advances in Pediatrics. Chicago, Year Book Medical Publishers, Inc. 1982, p. 529.

Levine, M. D., Brooks, R., and Shonkoff, J.: A Pediatric Approach to Learning Disorders. New York, John Wiley & Sons, Inc., 1980.

Levine, M. D., and Melmed, R.: The unhappy wanderers: children with attention deficits. Pediatr. Clin. N. Am., 29:105, 1982.

Levine, M. D., and Oberklaid, F.: Hyperactivity: symptom complex or complex symptom? Am. J. Dis. Child., 134:409, 1980.

Levine, M. D., Oberklaid, F., and Meltzer, L.: Developmental output failure: a study of low productivity in school-aged children. Pediatrics, 67:18, 1981.

Mann, L., Goodman, L., and Wriderholt, J. L.: Teaching the Learning Disabled Adolescent. Boston, Houghton-Mifflin Co., 1978.

Miller, S. R.: A crisis in appropriate education: the dearth of data on programs for secondary handicapped adolescents. J. Spec. Educ., 15:351, 1981.

Needleman, H. L., Gunnoe, C., Leviton, A., Reed, R., Peresie, H., Maher, C., and Barrett, P.: Deficits in psychological and classroom performance in children with elevated dentine lead levels. N. Engl. J. Med., 300:689, 1979.

Rapp, D. J.: Food allergy treatment for hyperkinesis. J. Learn. Dis., 12:608, 1979.

Ring, B.: Memory problems and children with learning problems. Acad. Ther., 11:111, 1975.

Ross, A. O.: Psychological Aspects of Learning Disabilities and Reading Disorders. New York, McGraw-Hill Book Co., 1976.

Satterfield, J.: EEG issues in children with minimal cerebral dysfunction. In Walzer, S., and Wolff, P. (Editors): Minimal Cerebral Dysfunction in Children. New York, Grune & Stratton, Inc., 1973.

Swanson, J. M., and Kinsbourne, M.: Food dyes impair performance of hyperactive children on a laboratory learning task. Science, 207:1485, 1980.

Tarver, S., and Hallahan, D.: Attention deficits in children with learning disabilities: a review. J. Learn. Dis., 7:560, 1974.

Weiss, G., et al.: Hyperactive as young adults: A controlled prospective ten-year follow-up of 75 children. Arch. Gen. Psychiatry, 38:657, 1979.

Wiig, E., and Semel, E.: Language Disabilities in Children and Adolescents. Columbus, Ohio, Charles E. Merrill, 1976.

38B   Hemispheric Specialization: Pragmatic Applications to Developmental and Behavioral Pediatrics

Data emanating from research on hemispheric specialization have already yielded a promising theoretical framework by which to appreciate the brain's orchestration of mood, cognition, and activity. This research has also suggested new approaches to common and uncommon pediatric problems, such as learning disability. In the future this research may reveal ways to enhance our creativity, teach messy children to be neat, and even improve our sense of humor.

BASIC THEORY

Research on hemispheric specialization began in the 1960's, when it was restricted to a handful of epileptic adults with intractable seizures who had undergone a surgical procedure that severed the corpus callosum and thereby the connections between the hemispheres. These "split brain" patients revealed subtle but consistent changes: they possessed two separate, conscious minds with differing functions. Experiments were designed to tease out and further delineate the advantages of each. Recently research activities have escalated as new techniques have permitted the study of intact subjects, including children. This has shifted the focus from disconnectedness to connectedness, from simple functions to complex behavior.

Each hemisphere is thought to have its own area of expertise. In the typical right handed person, the left brain handles language processes, verbal images, and mathematical calculations, while the right side is specialized for music and spatial relations, such as geometrical shapes and patterns on which mechanical skills depend. There is some overlap: For example, the right hemisphere can do simple addition up to 10 and has the use of a few words with syntax at about the level of a two year old. Women seem less specialized than men in both verbal and spatial skills, a finding consistent with the clinical observation that three times as many men with left hemisphere lesions become aphasic as do women with comparable damage. Women are better able to focus on one particular task, whether it be following a recipe or talking to a friend; men are better at endeavors that require the simultaneous integration of two separate cognitive approaches at the same time. Women may be more easily able to suppress the activity of one hemisphere when the other is active. Although in the past left handed individuals were thought to have mirror image brains or a reversal of cerebral dominance, we now recognize that most left handed persons are left dominant; handedness per se is a poor index of hemispheric specialization (Devel, 1980).

The most fundamental difference between the hemispheres is not that each is specialized to work with different materials such as words or patterns, but that each is suspected to have a separate cognitive style or system for processing information. The left is apt to be analytic, logical, and deductive: it appreciates each tree but may miss the forest. The right hemisphere employs a holistic, synthetic, or inductive approach, which immediately perceives the "gestalt" of the forest but misses the trees. Both modes must be not only present but integrated for adequate or optimal functioning. In practice both hemispheres perceive the same data and search for an answer; the left hemisphere is thought to make tentative propositions based upon logical constructs, while the right hemisphere simultaneously supplies intuitive cues and a strong sense of affirmation, which the left needs to validate its original assumption (Sperry et al., 1979). The two modes are somewhat antagonistic: if the left hemisphere is to process written material, then the right must be partially inhibited (Levy et al., 1972; Witelson, 1977b).

The difference between the two hemispheres in conceptual style has prompted some to refer to the left as "masculine" and the right as "feminine." The left is said to be plodding and pedantic; it pays no-nonsense attention to detail. The right is thought to be intuitive, to live by shifting metaphor, symbol, or pun, to arrive at an inspirational hunch. The left is considered instrumental, the right expressive. The left renders a sound executive decision; the right adds spice to a cocktail party. However pleasant or unpleasant the analogy, it cannot survive when we consider that girls are superior in verbal skill, a left hemisphere function, and that boys demonstrate an increasing advantage in spatial abilities from early adolescence on (Maccoby and Jacklin, 1974).

The left and right sides of the brain may differ in yet another respect; namely, the left brain is involved in intended, conscious activity; the right contains programed, automatic functions. This dichotomy necessarily involves an internal transposition: A child struggles to learn how to tie his shoelace; later he does it without thinking. A beginner typist hunts for the correct key whereas the expert floats through 80 words a minute.

The frontal lobe is not as clearly specialized as the remainder of the cerebral cortex. The diversity of frontal lobe connections—with the cortex, the limbic system, and the basal ganglia—suggests a higher level, integrative function. The frontal lobe serves to inhibit less socialized expressions of eroticism, irritability, and anger, which allows considered functions such as initiative, planning ahead, social behavior, and creativity to come into existence. The memory and intelligence of patients who have undergone frontal lobotomy remain reasonably intact, but most have become listless and dependent, with diminished drive and little appreciation of pleasure, pain, or social nuances (Heilman and Valenstein, 1979). The frontal lobes appear far less specialized than other areas of the cortex, perhaps because their function is largely to supervise and collate input from both hemispheres. Nevertheless lesions of the dominant frontal lobe do tend to interfere more with verbal fluency and spontaneity (Benton, 1968; Milner, 1964), whereas nondominant hemispheric lesions present with accentuated visuospacial and emotional changes (Benton, 1968; Heilman and Valenstein, 1979; Milner, 1971).

BRAIN DEVELOPMENT

Anatomic asymmetry is present in fetuses. Newborn infants already demonstrate left cerebral specialization for speech sounds, and 90 per cent of infants in the first week of life will turn their heads preferentially to the right (Turkewitz, 1977). During childhood there is progressive myelinization of the corpus callosum, and commissural transmission increases in conjunction with the myelinization process. There is no demonstrable evidence of transmission across the commissure until after age three and one-half years (Salamy, 1978), an observation that has prompted Galin's speculative suggestion that young children may operate much like split brain patients (Galin, 1977). For example, a pregnant mother might painstakingly explain her imminent departure for the hospital to her five year old daughter. The child would understand completely, be able to repeat exactly, yet simultaneously experience anger and rejection. This could occur habitually if a caretaker were in the habit of coupling contradictory verbal and nonverbal messages, such as the statement, "I love you," accompanied by a scowl, which connotes hatred. In effect, this delivers highly charged but opposing input simultaneously to each hemisphere; to avoid continuing conflict, the child could learn to disconnect the hemispheres. Presumably a conditioned, functional disconnection such as this would impede the child's ability to understand emotional issues, to attach emotional significance to intentional acts, or to remember dreams.

Brain maturation also may explain why children who become bilingual before age five rely more on the left hemisphere for language processing whereas the right brain is involved when language is acquired later, near age 11.

As development proceeds, there is necessarily a loss in plasticity. Removal of one hemisphere in a young child alters the developmental sequence so that both verbal and spatial functions are acquired by the remaining hemisphere; no permanent dysphasia develops (Basser, 1962). However, there is an intellectual price to pay for such plasticity: When cognitive abilities are compressed into a single hemisphere, general intelligence is likely to be lower (Lansdell, 1969; Milner, 1974). There also seems to be a succinct relationship between the acquisition of cerebral laterality and the rate of physical maturation (Waber, 1976), perhaps explaining why so many hyperactive children with confused laterality and dominance are both physically and emotionally immature (Serafetinides, 1981). In any case there is an internal sequential development and stabilization of function that matches physical and behavioral gains. This process continues up to or beyond age 13.

RELATIONSHIP TO DISEASE

Clinical disorders secondary to cerebral dysfunction may be related to a deficit in one hemisphere, consequent disinhibition of the other side, disordered interhemispheric interaction, and the success or failure of compensatory mechanisms (Wender, 1971; Wexler, 1980). In reviewing the literature, Flor-Henry (1973) associates the adult schizophrenic syndrome with left hemispheric abnormalities. Other studies indicate left hemispheric instability and impaired interhemispheric transfer (Beaumont and Dimond, 1973; Green, 1978; Wexler, 1980). Manic-depressive patients are thought to have primary right hemispheric dysfunction (Flor-Henry, 1973), and lithium exerts its effect upon the nondominant hemisphere (Small et al., 1973). Because both these conditions are, in part, hereditary, such research eventually may enable us to detect children who are genetically predisposed so that we may attempt to prevent expression of the disease process. There also are studies of functionally rather than severely impaired patients. University of Michigan researchers suggest that the perfectionistic, conscientious, obsessive-compulsive person depends heavily on the left hemisphere, whereas the labile, attention seeking hysteric relies on the right hemisphere (Smokler and Shervin, 1979).

Investigations that associate disorders of cerebral dominance with childhood disease are concerned primarily with autism and dyslexia.

DEVELOPMENTAL DYSLEXIA

The majority of children classified as being "learning disabled" by the schools have reading problems. In our culture an adolescent who can read only at the second grade level is severely handicapped; he can understand simple traffic signs but may be unable to comprehend or fill out job applications. Because verbal symbols are processed in the left brain, we might assume that all "dyslexics" suffer from a left hemispheric impairment. Unfortunately the case is not that simple.

First, there is some evidence that some asymmetry of the cerebrum or establishment of either left or right dominance may be necessary for the acquisition of reading skills. Children with poorly established, confused, or fluctuating lateralization of ear, hand, and eye preference may evidence multiple severe problems in learning to read. Any established dominance is said by some investigators to be better than none at all (Kershner, 1975). Once dominance is defined, both sides participate in the reading process: The left commonly provides linguistic skills such as word recognition and association; the right orients words in space, preventing rotations and left-right reversals. One group of poor readers may demonstrate difficulty with linguistic processing, labeling, and syntax (Wiig, 1976); another reveals symbol rotation, reversal, and distortion (Keney and Keney, 1968). Thus, such children can present a hemispheric deficit that is left, right, or mixed (Keefe and Swinney, 1979; Kershner, 1977; Naylor, 1980). Presumably a fourth, transfer deficit may also exist, one symptom of which could be an impediment in converting intentioned reading to automatic reading.

Once again, hemispheric specialization theory seems to jibe with clinical impressions. Can it also suggest avenues of therapy? To consider this question, we must assume that reading disorders, whether due to maturational lags (Gorden, 1980) or not, are amenable to remediation. An alternative concept would be that such disabilities are genetically programed, i.e., the product of immutable internal timetables that determine whether improvement occurs. There does seem to be a strong genetic component; in one study 90 per cent of first degree relatives of "dyslexics" with a right dominant profile demonstrated the same profile; however, most of these relatives claimed never to have had reading difficulties (Van Den Honert, 1977). Reading problems derive from more than one variable: presumably the process could be mitigated through early intervention aimed at ameliorating the imbalance. Demonstrable increases in performance secondary to intensive training have been reported (Van Den Honert, 1977), although this issue remains highly controversial.

Specialization theory holds that youngsters with reading disorders with combined or predominantly left hemispheric deficits, who are usually boys, should respond best to an individualized remedial program that stresses language skills, including sequencing and auditory perceptual discrimination. At the same time, right brain processing can facilitate left brain learning through techniques such as the "whole word" rather than through the phonics approach to reading (see Chapter 56). Similar methods include the pairing of words with pictures and the teaching of sign language or Chinese logographics, which can then be paired with English words (Van Den Honert, 1977; Witelson, 1977a). Children with a primary deficit in the right hemisphere often are girls (Witelson, 1977c); they might benefit from an individualized program with a holistic approach such as tracing letters on each other's back, analyzing pictures, painting a picture and telling a story, or reading and constructing maps. Some

additional suggestions are contained in a review by Ferry et al. (1979).

Pharmacologic and other agents have been shown to alter the ratio of electrical activity between the hemispheres; these include stimulants, which decrease interhemispheric differences. Hyperactive children who demonstrate clinical improvement after taking stimulants are reported to reveal a modulated cortical response to stimuli (Rosenthal and Allen, 1978). Meditation, relaxation techniques, biofeedback, yoga, hypnosis, and dreaming all reportedly increase right brain activity (Ray et al., 1977). These may well become ancillary aides to education in the future.

FUTURE DIRECTIONS

Pragmatic contributions stemming from brain specialization research are thought by some to be promising; the potential for future developments awaits further critical scrutiny. An art teacher in California tried a technique presumably to enhance creativity by inhibiting left brain input. She asked her students to copy upside-down drawings; apparently because the pictures were disoriented in space, the left brain was said to relinquish its dominance and artistic ability might then flourish (Edwards, 1977). There is hope also for parents frustrated by children who leave a trail of toys, clothing, and dirty dishes. Mills ties neatness to left dominance and sloppiness to right; neat people are alleged to be visually oriented, verbal, and sequentially organized, whereas messy ones are thought to be kinesthetically oriented and are said to think by pictorial images rather than words (Mills, unpublished study). If this proves true, the same remedial principles used in the classroom could be applied at home; if not, at least parents could accept sloppy children with greater equanimity by recognizing messiness as but one aspect of a cognitive style! Last, but not least, a sense of humor is thought to depend upon the right hemisphere. Patients with right cerebral injury crack off-target jokes with inappropriate punch lines. They miss the point of a story and are deaf to subtle shades of meaning (Gardner, 1981). Perhaps our sense of humor could be enhanced through techniques that facilitate a relative increase in right brain activity (Ray et al., 1977). Hemispheric specialization research has indeed stimulated fresh approaches to old problems. Further research will be needed to confirm and extend the very tempting preliminary findings and informed hypotheses in this field.

ALAYNE YATES

REFERENCES

Basser, L. S.: Hemiplegia of early onset and the faculty of speech with special reference to the effects of hemispherectomy. Brain, 85:427, 1962.

Beaumont, J. G., and Dimond, S. J.: Brain disconnection and schizophrenia. Br. J. Psychiatry, 123:661, 1973.

Benton, A. L.: Differential behavioral effects in frontal lobe disease. Neuropsychologica, 6:53, 1968.

Devel, R. K.: Cerebral dominance and hemispheric asymmetries on the computer tomogram in children. Neurology, 30:934, 1980.

Edwards, B.: Communication. Brain-Mind Bulletin, March 20, 1977.

Ferry, P. C., Culbertson, J. L., Fitzgibbons, P. M., and Netsky, M. G.: Brain function and language disabilities. Int. J. Ped. Otorhinolaryngol., 1:13, 1979.

Flor-Henry, P.: Psychiatric syndromes considered as manifestations of lateralized temporal-limbic dysfunction. In Laitinen, L. V., and Livingston, K. E. (Editors): Surgical Approaches in Psychiatry, Baltimore, University Park Press, 1973.

Galin, D.: Presentation, University of California at Berkeley Symposium, Educating Both Halves of the Brain, 1977.

Gardner, H.: How the split brain gets a joke. Psychol. Today, 2/81:74, 1981.

Gorden, H. W.: Cognitive asymmetry in dyslexic families. Neuropsychologica, 18:645, 1980.

Green, P.: Defective interhemispheric transfer in schizophrenia. J. Abnorm. Psychol., 87:472, 1978.

Heilman, K. M., and Valenstein, E.: Clinical Neuropsychology. New York, Oxford University Press, 1979.

Keefe, B., and Swinney, D.: On the relationship of hemispheric specialization and developmental dyslexia. Cortex, 15:471, 1979.

Keney, A., and Keney, V.: Dyslexia. St. Louis, The C. V. Mosby Co., 1968.

Kershner, J. R.: Reading and laterality revisited. J. Spec. Ed., 9:269, 1975.

Kershner, J. R.: Cerebral dominance in disabled readers, good readers and gifted children. Child. Dev., 48:61, 1977.

Lansdell, H.: Verbal and non-verbal factors in right-hemisphere speech. J. Comp. Physiol. Psychol., 69:734, 1969.

Levy, J. Trevarthen, C., and Sperry, R. W.: Perception of bilateral chimeric figures following hemispheric deconnexion. Brain, 95:61, 1972.

Maccoby, E., and Jacklin, C.: The Psychology of Sex Differences. Stanford, Stanford University Press, 1974.

Mills, S.: Odd couple syndrome. Unpublished study.

Milner, B.: Some effects of frontal lobectomy in man. In Warren, J. M., and Akert, K. (Editors): The Frontal Granular Cortex and Behavior. New York, McGraw-Hill Book Co., 1964.

Milner, B.: Interhemispheric differences in the localisation of psychological processes in man. Br. Med. Bull., 27:272, 1971.

Milner, B.: Hemispheric specialization: scope and limits. In Schmitt, F. O., and Worden, F. G. (Editors): The Neurosciences: Third Study Program. Cambridge, MIT Press, 1974.

Naylor, H.: Reading disability and lateral asymmetry: an information processing analogy. Psychol. Bull., 87:531, 1980.

Prior, M. R., and Bradshaw, J. L.: Hemisphere functioning in autistic children. Cortex, 15:73, 1979.

Ray, W., Frediana, A. W., and Harman, D.: Self regulation of hemispheric asymmetry. Biofeedback Self Regul., 2:195, 1977.

Roemer, R. A., Shagass, C., and Straumanis, J. J.: Pattern evoked potential measurements suggesting lateralized hemispheric dysfunction in chronic schizophrenics. Biol. Psychiatry, 13:185, 1978.

Rosenthal, R., and Allen, T.: An examination of the attention, arousal and left hemispheric dysfunction of hyperactive children. Psychol. Bull., 85:689, 1978.

Salamy, A.: Commissural transmission: Maturational changes in humans. Science, 200:1409, 1978.

Serafetinides, E. A.: Psychopathology and the cerebral hemispheres. In Serafetinides, E. A. (Editor): Psychiatric Research in Practice: Biobehavioral Themes. New York, Grune & Stratton, Inc., 1981.

Small, I. F., Small, J. C., and Milstein, V.: Interhemispheric relationships with somatic therapy. Dis. Nerv. Syst., 34:170, 1973.

Smokler, I. A., and Shervin, H.: Cerebral lateralization and personality style. Arch. Gen. Psychiatry, 36:949, 1979.

Sperry, R., Zaidel, E., and Zaidel, D.: Self recognition and social awareness in the deconnected minor hemisphere. Neuropsychologica, 17:153, 1979.

Turkewitz, G.: The development of lateral differentiation in the human infant. Ann. N.Y. Acad. Sci., 299:309, 1977.

Van Den Honert, D.: A neuropsychological technique for training dyslexics. J. Learn. Disab., 1:15, 1977.

Waber, D. P.: Sex differences in cognition: a function of maturation rate? Science, 192:572, 1976.

Wender, P. H.: Minimal Brain Dysfunction in Children. New York, John Wiley & Sons, Inc., 1971.

Wexler, B. E.: Cerebral laterality and psychiatry: a review of the literature. Am. J. Psychiatry, 137:279, 1980.

Wiig, E.: Language disabilities of adolescents: implications for diagnosis and remediation. Br. J. Disord. Commun., 13:3, 1976.

Witelson, S. F.: Developmental dyslexia: two hemispheres and none left. Science, 195:309, 1977a.

Witelson, S. F.: Early hemispheric specialization and interhemispheric plasticity: an empirical and theoretical review. In Segalowitz, S. J., and Gruber, F. A. (Editors): Language Development and Neurological Theory. New York, Academic Press, Inc., 1977b.

Witelson, S. F.: Neural and cognitive correlates of developmental dyslexia: age and sex differences. In Shagass, C., Gershon, S., and Friedhoff, A. J. (Editors): Psychopathology and Brain Dysfunction. New York, Raven Press, 1977c.

38C　Biologic Influences in Attentional Disorders* _____

At perhaps the most elemental level, attentional deficits in children may be conceptualized as the vector of two primary forces: primary biologic disturbances, which interact with environmental conditions, the resultant yielding a symptom complex characterized by attentional deficits, impulsive behavior, and often, though not always, increased motor activity. Depending upon the circumstances, one or the other of these primary factors may be more important in the expression of the child's symptoms, but in this review we have focused upon an examination of one of these vectors—the evidence supporting the notion of a role for biologic influences in the genesis of attentional deficits in children. This evidence is derived from a number of diverse lines of investigation incorporating the results of epidemiologic and psychopharmacologic studies as well as data derived from experimental animal models of the human disorder.

In a sense, our current views of the biologic processes in attentional disorders have been shaped by historical influences that have evolved since the early part of this century. More recent concepts differ from the older views in that importance is placed on inherited or intrinsic factors as primary etiologic agents, and that no matter what the primary etiology (traumatic, infectious,

*Study supported by USPHS grants NS 12384, AA 03599, Mental Health Clinical Research Center grant 1 P 50 MH-30929, Clinical Center grant RR 00125, the State of Connecticut, and the Thrasher Research Foundation.

teratologic), the pathogenesis or final common pathway involves biochemical changes. There is no question in the mind of any serious investigator or experienced clinician that environmental factors significantly influence the expression of attentional deficits regardless of the etiology or severity of disorder. This notion does not negate the concept of attentional deficits as an entity but rather regards the expression of attentional deficits as reflecting environmental factors acting upon a vulnerable biologic substrate with a lowered threshold for being influenced by environmental distractors. Such a concept posits that the insult or inherited defect produces a lowered threshold and that, depending on the particular environment, the symptoms may or may not be expressed. Attentional deficits are viewed not as all or none phenomena, but rather as occurring along a spectrum, a perspective consistent with the clinical observation of how differently affected children may appear in the one to one relationship in the physician's office as compared to that in a large classroom. This sensitivity to environmental influences is utilized by special educators in providing structured settings with low pupil to teacher ratios.

This review includes information about both etiologic agents that have been suggested as predisposing to attentional deficit syndromes and the biochemical mechanisms influenced by these etiologic factors. Special emphasis has been placed on details of research areas that have not been recently reviewed, and in assessing strategies that have proven helpful in attempts to relate biochemical processes to behavior.

The discussion necessarily represents a selected view, reflecting our own interests (and biases), and references have been chosen that illustrate these positions. Certain other areas, although supporting a biologic influence, have been discussed by us elsewhere and are not included in this chapter (S. Shaywitz et al., 1978). (See also Chapter 38A.)

ETIOLOGIC FACTORS IN THE GENESIS OF ATTENTIONAL DISORDERS

EXOGENOUS INFLUENCES

Background: The Concept of "Brain Damage"

Reports dating back to the turn of the century describe a characteristic behavioral pattern in adults following brain injury. A recurrent thread of impulsivity, distractability, low frustration tolerance, and memory difficulties highlights these observations. The pandemic of 1917 produced a new series of descriptions of behavioral disturbances, this time following encephalitis in children. However, perhaps the most influential and enduring support for the notion of a specific behavioral syndrome in children arising from brain damage is that provided by the work of Strauss and associates. In a now classic report Strauss and Lehtinen (1947) formulated the conceptual entity of the "brain injured (damaged) child." In a survey of a population of mentally defective institutionalized children that utilized the history and neurologic examination as support for the diagnosis, children were classified into two groups—an exogenous group in whom the neurologic examination or history was thought to provide evidence for a cerebral insult, and an endogenous group for whom such information was lacking. Although behavioral ratings provided by teachers and caretakers indicated much overlap in the cognitive and emotional styles of both groups, there appeared to be an excess of hyperactive, distractable, impulsive, emotionally labile, and perseverative behavior patterns in the exogenous or brain damaged group. As if by sleight of hand, these vaguely defined behavioral traits based upon nonvalidated indicators of brain injury were now to be considered, in and of themselves, as indicators of brain injury or damage! Often forgetting or being unaware of the tenuous foundation that served as the rationale for diagnosing brain damage, physicians embraced the concept of the brain damaged child as both an investigational tool and a diagnostic category.

Since its promulgation, the notion that "all brain lesions, wherever localized, are followed by a similar kind of disordered behavior" has been criticized, for it has never been established that children designated as "brain damaged" on the basis of the behavioral pattern described by Strauss and Lehtinen have sustained central nervous system "damage," although it is not for want of trying. The development of each new diagnostic technique has invariably generated the expectation that this procedure would at last provide documentation of the hypothesized brain abnormalities. As might be expected, the advent of computed tomography raised hopes that this new and powerful diagnostic technique would prove more helpful than earlier, less sophisticated procedures in elucidating "brain damage" in this group of children. However, when quantitative techniques, "blind" analysis, and inclusion of a contrast group are incorporated into the experimental design, computed tomographic scans of children with attentional deficits cannot be considered abnormal. If anatomic evidence of "brain damage" does exist in this group, its documentation must await the application of still newer diagnostic techniques, such as positron emission tomography or nuclear magnetic resonance.

Specific Exogenous Etiologic Agents

Infectious and Metabolic Factors. The preceding discussion, however, cannot ignore the clinical impression that attentional difficulties may occur as the sequelae of a number of traumatic and metabolic encephalopathies. In a recent study of the long term consequences of the Reye syndrome, the most common metabolic encephalopathy in childhood, attentional difficulties were noted in the majority of survivors of the Reye syndrome but not in sibling controls (S. Shaywitz et al., 1982). The findings are particularly intriguing in light of the evidence from several lines of investigation, which demonstrates alterations in central dopaminergic mechanisms during the acute stages of the Reye syndrome and fits with the notion (see following discussion) that attentional difficulties may be related to disturbances in brain monoaminergic functioning (B. Shaywitz et al., 1979).

Toxic Agents: Lead. Lead poisoning represents a more controversial etiology of attentional difficulties. Thus it has long been recognized that children who survive the effects of acute lead intoxication are frequently left with significant neurologic handicaps and intellectual sequelae. Less well appreciated are studies that indicate that persistently elevated lead levels without clinical evidence of encephalopathy may also be associated with cognitive and behavioral difficulties. A major limitation in the interpretation of such studies has been the confounding effects of the body lead burden with other dependent variables, such as social class. Yule et al. (1981) have shown that even when socioeconomic status was taken into consideration, significant associations still emerged between what many would consider to be only minimally elevated lead levels and scores on tests of reading and spelling as well as on general intelligence tests. Whether such slight elevations in blood lead concentration are associated with the spectrum of attentional difficulties and whether these problems persist are questions that have not yet been resolved (Ernhart et al., 1981; Needleman et al., 1981).

Head Trauma. Head trauma represents still another insult often implicated in the genesis of attentional difficulties. As was the case for lead intoxication, controversy exists over the causal relationship between head trauma and the development of disorders of attention, activity, and cognition. Thus, although clinical lore suggests that even minimal degrees of head trauma may result in behavior similar to patterns described as sequelae of the Reye syndrome, recent evidence indicates that deficits after head trauma are observed only with injuries severe enough to result in at least one week of post-traumatic amnesia. Rutter (1981) argues that in cases of minimal degrees of head trauma it is difficult to decide whether the behavioral and cognitive deficits observed are the consequences of the injury or simply represent the fact that children with attentional difficulties and hyperactivity are more likely to behave in a manner that may lead to accidents.

ENDOGENOUS INFLUENCES: GENETIC FACTORS

Evidence suggesting an inherited or genetic influence represents a cogent argument for the belief that endogenous factors play a role in the expression of attentional disorders. The first suggestion of this relationship emerged from investigations demonstrating an increased incidence of a history of hyperactivity in the parents of hyperactive children as compared to controls.

Although they implicate a familial transmission for certain symptoms, such studies do not resolve the relative contribution of genetic as compared to environmental factors, and thus an alternate experimental strategy, a comparison between monozygotic and dizygotic twins, has been employed. Such studies have found a high concordance for hyperactivity in monozygotic as compared to dizygotic twins.

In the hierarchy of research strategies to test for genetic influences, the adoptive studies represent a more critical attempt to link attentional deficits to genetic influences. These studies indicate that the incidence of attentional difficulties in nonbiologic relatives of adopted children with attentional deficits does not differ from that observed in a control group. However, no study to date has examined the incidence of attentional difficulties in the biologic families of adopted children who have attentional deficits.

PRENATAL FACTORS: MINOR CONGENITAL ANOMALIES

Further support for a biologic basis for the expression of attentional deficits comes from evidence indicating that children with attentional problems and hyperactive-impulsive behavior patterns demonstrate an increased number of minor congenital anomalies.

Such congenital anomalies are considered separately from the exogenous or endogenous etiologies, since they may be the result of aberrant prenatal development reflecting either genetically determined altered embryogenesis or an insult early in pregnancy. The presence of an increased number of anomalies in children with disorders of activity and attention regulation would suggest that such children are born with a biologic susceptibility to these difficulties that precedes postnatal environmental experiences. Characteristically, minor anomalies are of no medical or cosmetic consequence to the patient and would be overlooked on a general physical examination.

Historically, the interpretation of the significance of minor anomalies dates back to observations that in the examination of children from birth to adolescence it is rare to detect three or more anomalies in a child otherwise free of indicators of major system disturbance. Furthermore, more than 40 per cent of the children with idiopathic mental retardation demonstrate at least three of these minor anatomic defects. Such findings suggest that the presence of a high number of associated anomalies is indirect evidence of a significant impairment in embryonic development that not only influences the occurrence of the anomaly but is also associated with disordered central nervous system development.

Waldrop and associates have reported a series of investigations that indicate the stability and reliability of anomaly scores during childhood (Waldrop and Goering, 1971; Waldrop et al., 1968). Quinn and Rapoport (1974) have related an increased number of anomalies to a history of both obstetric complications and paternal hyperactivity, although other investigations applying the same scoring system have not been able to replicate these findings. Of particular interest has been the recurring association of high anomaly scores with difficulties of modulation of activity, impulsivity, and aggression. Significant relationships have now also been reported between anomaly scores and hyperactivity at ages two and seven, negative peer judgments, early onset of hyperkinesis, and increased levels of dopamine beta-hydroxylase.

The finding of minor congenital anomalies in children with attentional disorders assumes particular relevance in light of accumulating evidence that suggests that prenatal exposure to ethanol may produce both physical and behavioral manifestations. Physical findings in the fetal alcohol syndrome center on indications of pre- and postnatal growth deficiency—dysmorphogenesis manifesting primarily as short palpebral fissures, a hypoplastic philtrum, a thinner upper vermilion, and midfacial hypoplasia; and central nervous system dysfunction presenting as mental deficiency (Clarren and Smith, 1978). We have recently reported 15 children referred to our Learning Disorders Unit whose mothers were found to have a history of heavy drinking during pregnancy (S. Shaywitz et al., 1980). Most exhibited evidence of microcephaly and postnatal growth deficiency, and all had a continuum of dysmorphic features. Although all had intelligence in the average range, all were experiencing academic failure due to problems of activity and attention regulation. In a more recent report we described two additional children who presented with severe language disabilities

in addition to the dysmorphic features (S. Shaywitz et al., 1981).

Such findings provide support for the belief that milder degrees of central nervous system dysfunction frequently may be encountered in the offspring of alcoholic women, and suggest consideration of an expansion of the concept of the fetal alcohol syndrome to include behavioral and learning deficits as manifestations of central nervous system involvement. From a public health perspective, alcohol exposure in utero may be an important, preventable determinant of attentional disorders in childhood.

THE RELATIONSHIP BETWEEN CENTRAL NERVOUS SYSTEM DYSFUNCTION AND BEHAVIOR

Even in cases in which brain damage is assumed and there is a behavioral disturbance, the nature of this association remains problematic. Rather than a causal relationship, it may be coincidental or a result of a nervous system injury causing impaired cognitive function, which in turn leads to a secondary behavioral disturbance (Rutter, 1981; Rutter et al., 1970). Others have postulated altered environmental reactions to the limitations produced by structural damage to the nervous system, which in turn provokes abnormal behavior (Sameroff and Chandler, 1975). In this view, the catenation of developmental events originating with a constitutional defect leading to behavioral deviation as the inevitable outcome is discarded in favor of a more dynamic, transactional model acknowledging the bidirectionality of the effect of cerebral insult on both the child and his environment.

PATHOGENESIS OF ATTENTIONAL DISORDERS

PHARMACOLOGY

Central to our belief in a biologic influence in attentional problems is the rapidly accumulating evidence suggesting a relationship between brain monoaminergic mechanisms and the expression of particular behavior patterns, evidence more readily appreciated in light of an understanding of the pharmacology of central catecholamines (Cooper et al., 1978; Fig. 38–3). The catecholamines dopamine and norepinephrine are derived from the amino acid tyrosine. The first, and rate limiting, step in their synthesis is hydroxylation via tyrosine hydroxylase, to form l-dihydroxyphenylalanine (l-DOPA); aromatic amino acid decarboxylase then catalyzes the formation of dopa-

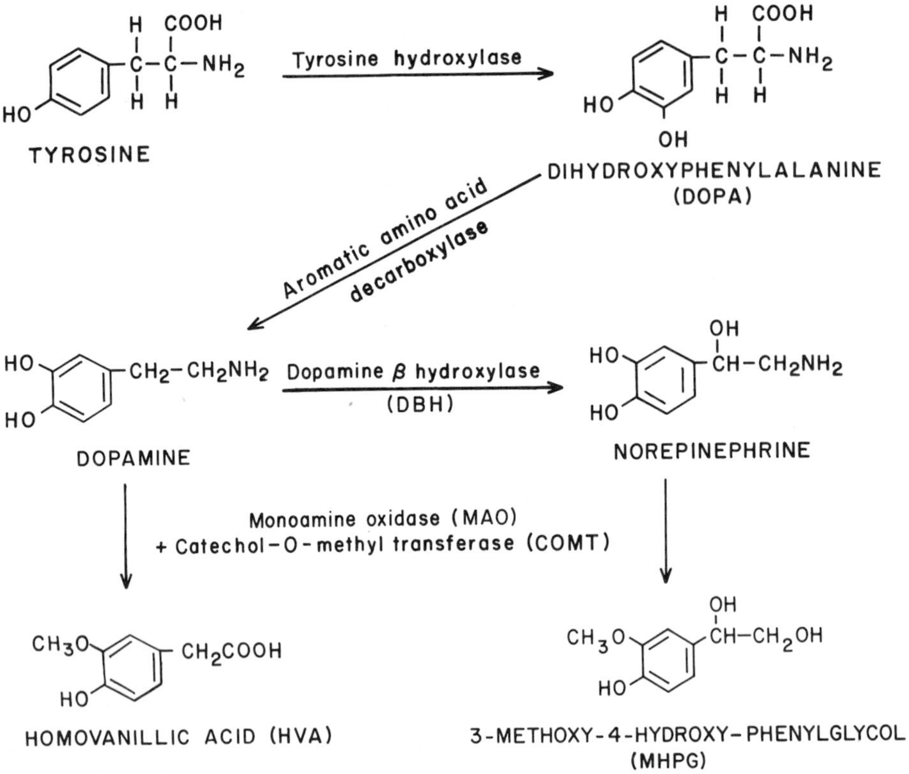

Figure 38–3.

mine from l-DOPA, and the formation of norepinephrine from dopamine proceeds via the enzyme dopamine-beta-hydroxylase.

Catabolism of dopamine proceeds by one of two routes: deamination via monoamine oxidase resulting in the formation of dihydroxyphenylacetic acid (DOPAC), or O-methylation by catechol-O-methyltransferase (COMT) to 3-methoxytyramine (MTA). DOPAC is subsequently O-methylated (via COMT) and MTA is deaminated (via monoamine oxidase) to yield homovanillic acid. For norepinephrine the combined actions of COMT and monoamine oxidase result in an O-methylated deaminated glycol, 3-methoxy-4-hydroxyphenyl glycol (MHPG).

The indoleamine serotonin is derived from the amino acid tryptophan, and as in the synthesis of catecholamines, the rate limiting step is hydroxylation to 5-hydroxytryptophan, followed by decarboxylation to 5-hydroxytryptamine (serotonin). Degradation proceeds via monoamine oxidase to 5-hydroxyindoleacetic acid (5-HIAA; Fig. 38–4).

BACKGROUND

The observations of the effects of stimulant drugs on hyperactive children provided some of the earliest evidence linking brain catecholamines to attentional difficulties. For almost 50 years it has been known that administration of amphetamines to children with hyperactivity, attentional

disorders, and impulsivity often produces a remarkable ameliorative effect, improving attention and reducing activity without producing drowsiness. Numerous studies over the past two decades have confirmed these effects (Barkley, 1977; Whelan and Henker, 1976), although the long term effects of stimulants remain controversial. Pharmacologic studies in both animals and man have documented that the stimulants amphetamine and methylphenidate exert their actions via central catecholaminergic mechanisms. Because amphetamine or methylphenidate often ameliorates the symptoms of hyperactivity and attentional difficulties, and since both agents exert their effects via central catecholaminergic pathways, this commonality between the symptoms and the stimulants prompted Wender (1971) to suggest that brain catecholamines are influential in the genesis of attentional difficulties.

However, it is obvious that the response to a drug does not necessarily indicate the pathogenesis of a disease. Thus children with rheumatic fever respond well to aspirin, yet clearly rheumatic fever is not due to lack of aspirin! Furthermore, as noted by Rapoport and Ferguson (1981) although most children respond to drugs that act as agonists on dopaminergic pathways, some children have a positive response to dopamine antagonists as well. Clinical studies of children with attentional difficulties could reconcile some of these issues and would provide considerably more

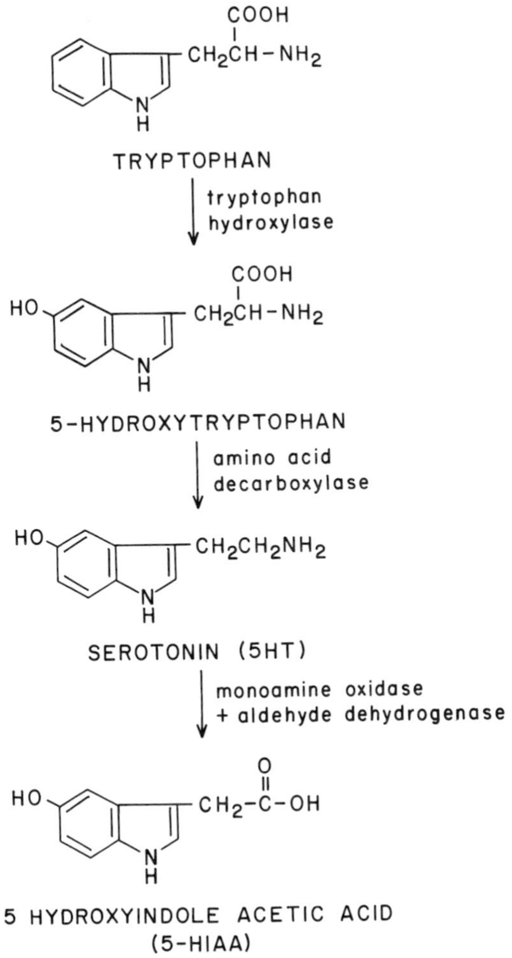

TRYPTOPHAN

tryptophan hydroxylase

5-HYDROXYTRYPTOPHAN

amino acid decarboxylase

SEROTONIN (5HT)

monoamine oxidase + aldehyde dehydrogenase

5 HYDROXYINDOLE ACETIC ACID (5-HIAA)

Figure 38–4.

convincing evidence of a relationship between central monoaminergic mechanisms and the behavior patterns exhibited in the clinical disorder.

CLINICAL STRATEGIES

Attempts to relate a clinical disorder to a biochemical dysfunction must rely on both the availability of measurable and accessible samples of the suspected biochemical marker compound and a well defined clinical cohort. In this instance clinical confirmation of a biochemical influence in attentional deficits requires the availability of the principal monoamine metabolites and the enzymes concerned with their synthesis and degradation (Table 38–7).

"Baseline" Concentrations of Amines and Metabolites

Recent evidence indicates that urinary concentrations of MHPG might be a useful index of brain norepinephrine metabolism (Young et al., 1981). Using MHPG as an index of central noradrenergic

activity, Shekim et al. (1979) reported reduced concentrations in hyperactive boys, concentrations further reduced by the administration of amphetamine, although these findings have not been supported by other investigators. Investigations of amines and their metabolites in the cerebrospinal fluid have provided a window on brain metabolism that may more accurately reflect monoaminergic function in the central nervous system than measures in urine or plasma. Not only have recent techniques made possible the examination of nanogram concentrations of the amine metabolites in the cerebrospinal fluid, but considerable evidence indicates that the concentrations of these metabolites reflect the activity of the parent amines in the brain (Cohen et al., 1980; B. Shaywitz et al., 1980). Shetty and Chase (1976) reported that although baseline cerebrospinal fluid concentrations of each metabolite were similar in both hyperactive and control groups, administration of amphetamine resulted in a significant reduction in homovanillic acid, the metabolite of dopamine, but not 5-HIAA, the metabolite of 5HT, in the hyperactive children. Utilizing the technique of probenecid loading, we have found that concentrations of homovanillic acid relative to probenecid were significantly reduced in the lumbar cerebrospinal fluid of hyperactive children as compared to control subjects; levels of 5-HIAA were not altered (B. Shaywitz et al., 1980). Our findings are consistent with the notion that brain dopamine systems, but not 5HT systems, may be abnormal in some children with attentional difficulties.

Pharmacologic "Probes"

Although the strategies just described continue to be employed, most recently an entirely new kind of investigation has emerged that potentially may provide far more information about the functional characteristics of brain monoaminergic systems in attentional disorders. The basis of this new research strategy is the utilization of phar-

Table 38–7. MONOAMINERGIC RELATED COMPOUNDS EXAMINED IN ATTENTIONAL DISORDERS

Compound*	Body Fluid Sampled		
	Blood	Urine	Cerebrospinal Fluid
DA	x		
NE	x		
5HT	x		
HVA	x	x	x
5-HIAA		x	x
MHPG		x	x
MAO	x		
DBH	x		

*DA = dopamine. NE = norepinephrine. 5HT = serotonin. HVA = homovanillic acid. 5-HIAA = 5-hydroxyindoleacetic acid. MHPG = 3-methoxy-4-hydroxyphenyl glycol. MAO = monoamine oxidase. DBH = dopamine β hydroxylase

macologic "probes" that act on particular mono-aminergic systems. The stimulant methylpheni-date is an example of such a probe, since good evidence from animal investigations suggests that this agent acts to stimulate the impulse mediated release of a reserpine sensitive pool of dopamine. In continuing studies of the effects of methylphen-idate (S. Shaywitz et al., 1981), we found that peak concentrations occurred within two to three hours and that the plasma half-life averaged two and one-half hours, findings consistent with the "behavioral half-life" of two to four hours reported by Swanson et al. (1978). Methylphenidate also produced a significant and consistent elevation in the serum growth hormone level and a reduction in the serum prolactin concentration, effects most reasonably understood as the actions of methyl-phenidate on dopamine pathways, presumably within the tuberoinfundibular dopamine system, and therefore providing an index of dopamine function within this pathway. By comparing these effects with those observed after a probe believed to affect primarily norepinephrine mechanisms, such as clonidine (Cohen et al., 1980), we may be able to discriminate particular subtypes of mono-aminergic dysfunction in children with attentional difficulties.

ALTERNATE STRATEGIES: EXPERIMENTAL ANIMAL MODELS

Rationale

As a research strategy, the relevance and appli-cability of any animal model may be justified on the basis of both ethical and scientific considera-tions. Thus, it is permissible to perform certain types of experiments on animals that almost cer-tainly would be considered dangerous if applied to humans. Scientific considerations provide an even more cogent rationale for the use of an animal model. Not only may animals be reared under controlled conditions, but their behavior is simpler and, it is hoped, less variable than that of man. Furthermore, their behavior can be measured by utilizing more precise techniques than are possible in human investigations. Still another advantage of animal models, and of particular relevance to the investigation of the developing nervous sys-tem, is the ease with which the entire course of maturation from prenatal life through adulthood may be examined using a time frame that is far shorter than a similar span in humans, and thus more convenient and accessible to quantitative measurement techniques.

Dopamine Depletion

Such considerations led investigators to work toward the development of an animal model that would prove helpful in the examination of atten-tional difficulties in children. Although a number of animal models have been suggested (S. Shay-witz et al., 1978), the model that most closely parallels the clinical disorder is that described initially in 1976 and now confirmed by investiga-tors throughout the world (B. Shaywitz et al., 1976). It is produced in the developing rat pup by the combined effects of desmethylimipramine (DMI) and the intracisternal administration of the neurotoxin 6-hydroxydopamine (6-OHDA), a pro-cedure designed to selectively ablate central do-pamine systems. Such treatment results in rapid and permanent reduction of brain dopamine to concentrations 10 to 25 per cent of those in con-trols, while brain norepinephrine and 5HT remain unaffected.

The developmental pattern in dopamine de-pleted rat pups has many parallels with the clini-cal disorder observed in children. For example, DMI/6-OHDA treated animals are significantly more active than their littermate controls during the period of behavioral arousal that occurs be-tween two and three weeks of age. However, with maturation, the hyperactivity disappears, a find-ing that corresponds to the clinical syndrome in which hyperactivity is often pronounced until 10 to 12 years of age but then abates. Furthermore, pups treated with DMI/6-OHDA remain active throughout the experimental observation period; i.e., they fail to habituate, a finding that bears striking similarity to the frequent inability of the child with attentional difficulties to adjust appro-priately and easily to a change in his environment. Deficits in cognitive performance represent a sig-nificant clinical problem in the child with atten-tional disorders, and these behavior patterns are demonstrated in appetitive, escape, and avoidance tasks in the DMI/6-OHDA treated rat pups.

Still another parallel with the clinical syndrome is demonstrated in the response of the DMI/6-OHDA pups to pharmacologic drugs used in the management of children with attentional difficul-ties. As discussed previously, administration of stimulants such as amphetamine or methylpheni-date often produces a remarkable ameliorative effect on both the hyperactivity and attentional deficits in affected children, a response termed "paradoxical," although whether such an effect should indeed be considered unusual has recently been questioned (Rapoport et al., 1980). However, the observation that such stimulants improve the attention of affected children (at least for the short term) has been confirmed on numerous occasions, and similar effects in the animal model would not be unexpected. In contrast to the hyperactivity, which all investigative groups employing the DMI/6-OHDA model have uniformly reported, it has been more difficult to arrive at a consensus in regard to the response to stimulants. In our hands, administration of amphetamine and methylphenidate increases activity in normal pups but decreases activity in DMI/6-OHDA animals.

The DMI/6-OHDA model has been particularly helpful in allowing us to examine the interaction between biologic factors (represented by preferential depletion of brain dopamine) and environmental influences (embodied in alterations in litter composition) upon locomotor activity and avoidance performance. For this study DMI/6-OHDA pups were reared from five days of age through weaning with other comparably treated animals or with littermates that were sham treated. Both the hyperactivity and cognitive deficits observed in the DMI/6-OHDA pups were significantly improved by rearing the dopamine depleted pups with normal littermates, rather than solely with other damaged animals, findings in a way comparable to clinical reports suggesting that particular modifications of the environment may be therapeutic in children with attentional difficulties.

Norepinephrine Depletion

All the experiments discussed have focused primarily upon the hyperactive motor behavior and cognitive deficits rather than attentional deficits. In experiments performed in adult animals, Mason and Iversen (1978) have shown that interruption of the dorsal noradrenergic pathways ascending to the forebrain (by lesions of the dorsal noradrenergic bundle) exerts little effect on motor activity or acquisition of avoidance learning but significantly influences the ability of the damaged animal to extinguish a previously learned response. They equated this so-called dorsal bundle extinction effect with attention, implying that lesions affecting noradrenergic pathways influence attentional processes. We and others have shown that selective perturbation of noradrenergic systems in the developing rat pup has little effect on activity, although in some experiments such treatment does affect cognitive performance. Although studies similar to those performed by Mason and Iversen have not yet been reported in young animals, it is intriguing to speculate that the dopaminergic systems mediate the hyperactivity and some cognitive deficits, while noradrenergic pathways are responsible for effects classified as "attentional."

Cerebrospinal Fluid: The Interface Between Human and Animal Studies

We have recently developed a technique that has permitted us to examine the relationship between the cerebrospinal fluid monoamine metabolites, homovanillic acid and 5-HIAA, and brain concentrations of their parent amines, dopamine and serotonin, and we have studied these parameters in both normal developing rats and pups treated with DMI/6-OHDA. In normal rat pups both brain and cerebrospinal fluid concentrations of homovanillic acid and 5-HIAA decrease with age, a finding similar to that observed in children. Pups treated with DMI/6-OHDA demonstrated not only the significant reduction of brain dopamine but a similar reduction in brain levels of homovanillic acid, which is paralleled by a significant reduction in homovanillic acid levels in the cerebrospinal fluid. These animal studies, for the first time, permit us to examine simultaneously both brain and cerebrospinal fluid. As such, they offer a unique opportunity to experimentally validate the hypothesized relationship between cerebrospinal fluid metabolites and brain monoamines and, it is hoped, will allow us to evaluate more critically the results obtained in human cerebrospinal fluid.

Relevance of Animal Studies

The extrapolation from what appears to be a very compatible and overlapping developmental profile in an experimental animal to a common clinical disorder of childhood raises the question of the appropriateness of animal models to the investigations of behavioral disorders in children. Although we appreciate the great caution that must necessarily be taken in the interpretation of such investigations, we also believe that animal studies derive their justification not only from the possibility of their link to a particular clinical syndrome but from the insights they may provide in understanding certain behavioral phenomena occurring in the developing organism in general.

It must be emphasized that the scientific methodology itself tends to foster an oversimplified view of behavior, a view in which behavioral regularities are abstracted as general principles of behavior while any apparent aberration in expected outcome is ignored. Furthermore, it is obvious that behavior has multiple determinants, thus necessitating a cautious approach in predicting or generalizing about the effects of a particular treatment or lesion on behavior. The resultant pathologic syndrome depends upon a complex interaction of external events as well as the independent variable under investigation. Furthermore, the specific behavior induced must be considered within the relevant pattern of behavior in the particular species.

Despite these considerations, most biologically oriented scientists believe that generalization from nonhuman behavioral investigations to man is justifiable, but differ widely over the degree to which they are willing to accept such extrapolation. Clearly there are areas in which the use of animal models would be inappropriate. Such limitations and reservations include investigations of unique human behavior such as speech. Other areas, however, are more readily and acceptably subject to extrapolation and generalization, particularly behavior patterns in the neonatal and developing organism that appear less complex than adult patterns. It seems likely that immature non-human behavior more nearly approximates that in the immature human in terms of physiologic-behavioral interactions and development. Because maturational differentiation progressively separates different species, it is probable that any generalizations made across species would be more relevant between the developing animal and the child than between mature animals and adult humans. Notwithstanding the difficulties involved and the cautions that must be exercised in extrapolating from animal models to human disease states, many investigators believe that the intelligent utilization of relevant animal models of neuropsychiatric disorders may presage significant advances in our understanding of these processes.

However, the real value of such experiments does not depend solely upon the slavish mimicry of a human disorder in another species. It is obvious that no thoughtful investigator really believes that rat pups are simply small human beings with fur coats and long tails. Rather such animal models have a heuristic value in generating principles and generalizations about human behavior. The critical investigator who employs animal models must sail the narrow straits, scrupulously avoiding the twin Scyllas of anthropomorphic interpretation and reductionist simplification in the generalization from animal experiments to human disorders, and the Charybdis of being so paralyzed and rigid in the interpretation of animal investigations that generalizations are impossible and a potentially valuable research tool is indiscriminately and prematurely abandoned.

In our view the DMI/6-OHDA model system has significant potential in elucidating the neurochemical mechanisms underlying particular behavior patterns occurring in the developing organism, patterns that may have their counterpart in attentional disturbances in children. Such an approach is not limited by the methodologic restrictions imposed by human investigations and provides a unique opportunity to explore the relationship between the cardinal symptoms of the disorder and an alteration in a specific catecholaminergic system. If we are able to discover something of the mechanisms involved in the evolution of a specific behavioral parameter in our model system, we will have taken a significant step in understanding the mechanisms involved in the production of this symptom in other model systems as well. Whether this strategy will ultimately enable us to unravel and understand attentional disorders awaits future investigations.

CONCLUSION

In this discussion we have reviewed selected biologic factors that have been implicated in the etiology of attentional disorders, and focused upon the role of brain monoaminergic mechanisms as, in a sense, the final common pathway for the expression of these behavior patterns. Although we have emphasized the biologic factors, the importance of environmental influences cannot be overestimated. Any reasonable perspective on neuropsychiatric disorders recognizes the transactional nature of the relationship between underlying biologic factors, be they exogenous or endogenous, and environmental influences. The question for present day investigators is not "nature *or* nurture" but rather the relative contribution of each to the development of attentional disorders.

SALLY E. SHAYWITZ
BENNETT A. SHAYWITZ

REFERENCES

Barkley, R. A.: A review of stimulant drug research with hyperactive children. J. Child. Psychol. Psychiatry, *18*:137–165, 1977.

Clarren, S. K., and Smith, D. W.: The fetal alcohol syndrome. N. Engl. J. Med., *298*:1063–1067, 1978.

Cohen, D. J., Detlor, J., Young, J. G., and Shaywitz, B. A.: Clonidine ameliorates Gilles de la Tourette's syndrome. Arch. Gen. Psychiatry, *37*:1350–1357, 1980.

Cohen, D. J., Shaywitz, B. A., Young, J. G., and Bowers, M. B., Jr.: Cerebrospinal fluid monoamine metabolites in neuropsychiatric disorders of childhood. In Wood, J., (Editor): Neurobiology of Cerebrospinal Fluid. New York, Plenum Publishing Corp., 1980, Ch. 46.

Cooper, J. R., Bloom, F. E., and Roth, R. H.: The Biochemical Basis of Neuropharmacology. Ed. 3. New York, Oxford University Press, 1978.

Ernhart, C. B., Landa, B., and Schell, N. B.: Subclinical levels of lead and developmental deficit—a multivariate follow-up reassessment. Pediatrics, *67*:911–919, 1981.

Mason, S. T., and Iversen, D. S.: Reward, attention, and the dorsal noradrenergic bundle. Brain Res., *150*:135–148, 1978.

Needleman, H. L., Bellinger, D., and Levitan, A.: Does lead at low dose affect intelligence in children? Pediatrics, *68*:894–896, 1981.

Quinn, P., and Rapoport, J.: Minor physical anomalies and neurological status in hyperactive boys. Pediatrics, *53*:742–747, 1974.

Rapoport, J. L., Buchsbaum, M. S., Weingartner, H., Zahn, T. P., Ludlow, C., and Mikkelsen, E. J.: Dextroamphetamine: its cognitive and behavioral effects in normal and hyperactive boys and normal men. Arch. Gen. Psychiatry, *37*:933–943, 1980.

Rapoport, J. L., and Ferguson, H. B.: Biological validation of the hyperkinetic syndrome. Dev. Med. Child Neurol., *23*:667–682, 1981.

Rutter, M.: Psychological sequelae of brain damage in children. Am. J. Psychiatry, *138*:1533–1544, 1981.

Rutter, M., Graham, P., and Yule, W.: A neuropsychiatric study in childhood. Clinics in Developmental Medicine, No. 35/36 S.I.M.P. London, William Heinemann Medical Books Ltd., 1970.

Sameroff, A. J., and Chandler, M. J.: Reproductive risk and the continuum of caretaking casualty. In Horowitz, F. D., Hetherington, M., Scarr-Salapatek, S., and Siegel, G. (Editors): Review of Child Development Research. Chicago, University of Chicago Press, 1975, Vol. 4, pp. 187–244.

Shaywitz, B. A., Cohen, D. J., and Bowers, M. B., Jr.: Cerebrospinal fluid monoamine metabolites in neurological disorders of childhood. In Wood, J. H. (Editor): Neurobiology of Cerebrospinal Fluid. New York, Plenum Publishing Corp., 1980, Ch. 17.

Shaywitz, B. A., Venes, J., Cohen, D. J., and Bowers, M. B., Jr.: Reye syndrome: monoamine metabolites in ventricular fluid. Neurology, 29:467–472, 1979.

Shaywitz, B. A., Yager, R. D., and Klopper, J. H.: Selective brain dopamine depletion in developing rats: an experimental model of minimal brain dysfunction. Science, 191:305–308, 1976.

Shaywitz, S. E., Caparulo, B. K., and Hodgson, E. S.: Developmental language disability as a consequence of prenatal exposure to ethanol. Pediatrics, 68:850–851, 1981.

Shaywitz, S. E., Cohen, P. M., Cohen, D. J., Mikkelson, E., Morowitz, G., and Shaywitz, B. A.: Long-term consequences of Reye syndrome. A sibling matched controlled study of neurologic, cognitive, academic and psychiatric function in Reye syndrome. J. Pediatr., 100:41–46, 1982.

Shaywitz, S. E., Cohen, D. J., and Shaywitz, B. A.: The biochemical basis of minimal brain dysfunction. J. Pediatr., 29:179, 1978.

Shaywitz, S. E., Cohen, D. J., and Shaywitz, B. A.: Behavior and learning difficulties in children of normal intelligence born to alcoholic mothers. J. Pediatr., 96:978–982, 1980.

Shaywitz, S. E., Hunt, R. J., Jatlow, P., Cohen, D. J., Young, J. G., Pierce, R. N., Anderson, G. M., and Shaywitz, B. A.: Psychopharmacology of attention deficit disorder: pharmacokinetic, neuroendocrine and behavioral measures following acute and chronic treatment with methylphenidate. Pediatrics, 69:688–694, 1982.

Shaywitz, S. E., Shaywitz, B. A., and Cohen, D. J.: Classification of attention deficit disorder—newer diagnostic schema and assessment scales. Schizophr. Bull., 8:360–424, 1982.

Shekim, W. O., Dekirmenjian, H., Chapel, J. L., Javaid, J., and Davis, J. M.: Norepinephrine metabolism and clinical response to dextroamphetamine in hyperactive boys. J. Pediatr., 95:389–394, 1979.

Shetty, T., and Chase, T. N.: Central monoamines and hyperkinesis of childhood. Neurology, 26:1000–1006, 1976.

Strauss, A. A., and Lehtinen, L. E.: Psychopathology and Education of the Brain-Injured Child. New York, Grune & Stratton, Inc., 1947.

Swanson, J., Kinsbourne, M., Roberts, W., and Zucker, K.: Timeresponse analysis of the effect of stimulant medication on the learning ability of children referred for hyperactivity. Pediatrics, 61:21–29, 1978.

Waldrop, M. F., and Goering, J. D.: Hyperactivity and minor physical anomalies in elementary school children. Am. J. Orthopsychiatry, 41:602–607, 1971.

Waldrop, M. F., Pederson, F. A., and Bell, R. Q.: Minor physical anomalies and behavior in preschool children. Child Dev., 39:391–400, 1968.

Wender, P. M.: Minimal Brain Dysfunction in Children. New York, John Wiley & Sons, Inc., 1971, pp. 12–30.

Whelan, C. K., and Henker, B.: Psychostimulants and children: a review and analysis. Psychol. Bull., 83:1113–1130, 1976.

Young, J. G., Cohen, D. J., Shaywitz, B. A., Anderson, G. M., and Maas, J. W.: Clinical studies of MHPG in childhood and adolescence. In Maas, J. (Editor): MHPG in Psychopathology. New York, Academic Press, Inc., 1981.

Yule, W., Landsdown, R., Millar, I. B., and Urbanowicz, M.: The relationship between blood lead concentrations, intelligence and attainment in a school population: a pilot study. Dev. Med. Child Neurol., 23:567–576, 1981.

39

Major Handicapping Conditions

It is appropriate to consider in a unified fashion the special circumstances of certain major dilemmas in childhood that have in common a serious impingement on development. These are the so-called "developmental disabilities." They are presented here in a functional classification, which cuts across considerations of mode of origin and acknowledges the correlated toll of human resources in each instance. Mental retardation, the sensory handicaps, and cerebral palsy share similar elements of challenge in educational planning,

supports to families, and defense of human rights. It is encouraging that clinical facilities and human service agencies in recent years have come to acknowledge the obligation for more creative study, and design for care in these high severity situations. The reader will note here the concern for accuracy in identification, preoccupation with the dyamics of the disability, examination of available interventions, and acknowledgment of professional responsibility.

39A Mental Retardation*

DEFINITION OF MENTAL RETARDATION

Mental retardation is a symptom, a test result, an educational dilemma, a personal and social anxiety, and a public signal of concern. Obviously it refers specifically to exceptionality of cognitive function, but it tends immediately to incorporate a complex of insinuations regarding origin, inscrutability, personal traits, segregation, and outcome. There is a necessity to acknowledge but challenge the lore of retardation, and to look afresh at the settings of this individual manifestation, knowledge about its natural history, approaches to supportive intervention, and its cultural importance. One hastens to admit that categorical presentations are of use principally to provide a studied atmosphere in which one can then move on quickly to a more personalized description of the background and needs for each child.

The most widely employed definition of mental retardation is that promulgated about 20 years ago by the American Association on Mental Deficiency: "Mental retardation refers to significantly subaverage general intellectual functioning existing concurrently with deficits in adaptive behavior, and manifested during the developmental period" (Grossman, 1977). All three elements of this proposal are important—a measured impair-

ment of intelligence, functional implications for personal adaptation, and an expression of the process during the first 18 years of life. The first two features are to be discussed. The latter component reserves the term mental retardation for the dynamics of primary human development and maturation—a huge territory of its own and separate in many conceptual ways from complications of mentation that can affect the adult.

Measurement of intelligence is considered in Chapters 46 and 50, with both the utility and limitations presented there in relation to the concept of the "intelligence quotient." If one admits these constraints, the basic definition of retardation uses ranges of IQ for the classification of degree of handicap. The accepted convention now is to view intelligence on a gaussian distribution curve, with retardation commencing at two standard deviations below the median, and with subsequent levels charted at 3, 4, and beyond. In applying Stanford-Binet results, these groups would be:

Mild retardation	IQ's 67–52
Moderate retardation	51–36
Severe retardation	35–20
Profound retardation	Below 20

As essential as these designations may be for statistical and educational studies, the stereotypes that have emerged in association with them have often confounded the planning for individual children. "Mild retardation" was traditionally referred to as "educable mental retardation" in schools,

*Preparation of this material was supported in part by the U.S. Department of Health and Human Services, Maternal and Child Health Service (Project 928), and Administration on Developmental Disabilities (Project 59-P-05163).

and such a child was considered to be eligible for partial integration and compromised achievement. "Moderate retardation" was viewed as "trainable mental retardation", a condition usually handled in segregated classes and thought to be inconsistent with the learning of useful reading or independence. "Severe" and "profound" retardation became associated with a defeatist prospect, below the level of justification for energetic educational investment (see p. 768).

Between 2 and 3 per cent of the population test at the "significantly subaverage" level, with the majority (85 to 90 per cent) falling in the area of "mild" handicap. A special quandary exists regarding the classification of young people whose intelligence measures below 1 but above 2 standard deviations from the median—IQ 85 to 68, the so-called "borderline" range. The latter can be considered as "borderline normal" or "borderline retarded," but in 1973 the American Association on Mental Deficiency elected to omit this group from the terminology of mental retardation. This step was motivated by a requirement for fairness, since involved here are many persons who are in special phases of transition or under environmental pressure and whose functional outcome is appropriately considered to be much more dynamic. There is some contrast between "borderline" children as seen in public schools or community programs and those who are referred to hospital clinics for more complex collections of personal difficulties. The latter often require a cluster of services similar to those for more retarded children.

The "adaptive behavior" concerns in the definition of mental retardation acknowledge that "retardation" has a personal and cultural aspect as well. Considered here are the individual strengths and adjustments characterizing the person or the deficits therein. In typical situations adaptive and social skills parallel rather closely the intellectual capacity of the young person. In atypical situations (with unusually creative or unusually desultory support, or in some idiosyncratic settings) there may be notable deviation. Relevant instruments exist for the measurement of adaptive behavior (e.g., the Vineland Social Maturity Scale and the American Association on Mental Deficiency Adaptive Behavior Scale), but often it is a "gestalt" impression that speaks to these issues.

In the young child, adaptation implies achievement of useful coordinated psychomotor skills, then communication capacity, and finally gains in "activities of daily living" (self-help and independence in eating, dressing, toileting). These obviously are related to basic neurological issues as well. Later one looks to social skills, accommodation to the educational environment, achievement of prevocational and vocational mastery, and then ability to manage outside living requirements. In the latter sequence experiential and mental health factors are important.

Persons who are retarded, those who are related to retarded persons, or those who are involved in professional or voluntary efforts with retarded persons come to realize that there is a "world" of "mental retardation." For all these people the categories and definitions are incomplete and often inadequate. There is a subculture associated with "significantly subaverage intellectual functioning" among humans. It has its own special warmth and special desperation. The awakening of human rights efforts in the past two decades has brought a particular grace to the field of mental retardation. One can only hope that the new atmosphere of individual consideration, energetic support, and freedom from prejudicial generalizations will prevail in our culture.

ORIGINS OF MENTAL RETARDATION

When mental retardation is identified in a child, there is a sense of urgency to determine the causative factor or factors. This derives primarily from a need to counsel the family accurately, but also pertains to the eventual ability to guide appropriate interventions and training. In the broader view the settings for retardation must be understood as a basis for public health and preventive activities.

It is often stated that the etiology of mental retardation is usually unknown. Such an assertion is inadequate. It is true that in the majority of children with very mild handicaps one can make only somewhat vague conclusions about background issues of inferred relevance, the so-called "cultural-familial" factors. Discrete elements of importance become more apparent as the degree of handicap increases. These two populations are on a continuum, however, and a thoughtful search for contributing conditions is justified in all instances. In young children, in those in whom direct contact is possible with the parents, in those for whom full historical information is available, and for those in whom modern biomedical pursuit can be undertaken, the yield of accurate surmise about causation is high. Retrospective studies of older patients, such as those in state residential facilities, are of restricted value. In any instance one may be limited to declarations about the probable setting, and some uncertainty may remain regarding the precise pathogenesis. Such information is still helpful to the family, as is also their ability to dismiss some of their own fears about other possible causes in which they may be accusing themselves of having an operational role.

Impressions about the mechanisms of mental retardation are influenced by the circumstances in which children are seen. Professionals who work in the child study sections of school systems see large numbers with mild, more dynamic handicaps of apparent cultural and environmental origin. Workers in family health clinics or mental health clinics are impressed by the frequency of troubled support systems for families and of polygenic inheritance. Pediatricians and hospital clinics, on the other hand, see a disproportionate number of children with organic difficulties and complications of illness. Finally, child development centers and mental retardation institutes tend to draw the very complex child, with mixed biological and social liability, who has bewildered the educational and health care systems. It is in the latter setting that some of the most analytical studies are undertaken, and the lessons from this work are to be emphasized here. Table 39–1, for

Table 39–1. SETTINGS IN WHICH MENTAL RETARDATION OCCURS: EXPERIENCE OF THE DEVELOPMENTAL EVALUATION CLINIC, CHILDREN'S HOSPITAL MEDICAL CENTER, BOSTON (2200 RETARDED CHILDREN)*

	Percentage of Total Group
I. HEREDITARY DISORDERS (preconceptual origin; variable expression, multiple somatic effects, frequently a progressive course)	5
A. Inborn errors of metabolism (e.g., Tay-Sachs disease, Hurler disease, phenylketonuria)	
B. Other single gene abnormalities (e.g., muscular dystrophy, neurofibromatosis, tuberous sclerosis)	
C. Chromosomal aberrations, including translocation	
D. Polygenic familial syndromes	
II. EARLY ALTERATIONS OF EMBRYONIC DEVELOPMENT (sporadic events affecting embryogenesis; phenotypic changes, usually a stable developmental handicap)	32
A. Chromosomal changes, including trisomy (e.g., Down syndrome)	10
B. Prenatal influence syndromes (e.g., intrauterine infections, drugs, unknown forces)	22
III. OTHER PREGNANCY PROBLEMS AND PERINATAL MORBIDITY (impingement on progress of fetus during last two trimesters or newborn; neurologic abnormalities frequent, handicap stable or occasionally with increasing problems)	11
A. Fetal malnutrition—placental insufficiency	1
B. Perinatal difficulties (e.g., prematurity, hypoxia, trauma)	10

IV. ACQUIRED CHILDHOOD DISEASES (acute modification of developmental status; variable potential for functional recovery)	4
A. Infection (e.g., encephalitis, meningitis)	
B. Cranial trauma	
C. Other (e.g., cardiac arrest, intoxications)	
V. ENVIRONMENTAL AND BEHAVIORAL PROBLEMS (dynamic influences, operational throughout development; commonly combined with other handicaps)	17
A. Deprivation	6
B. Parental neurosis, psychosis	3
C. Childhood neurosis	4
D. Childhood psychosis, including autism	4
VI. UNKNOWN CAUSES (no definite hereditary, gestational, perinatal, acquired, or environmental issues; or else multiple elements present)	31

*Adapted, with permission, from Crocker, A. C. *In* Thompson, G. H., et al. (Editors): Comprehensive Management of Cerebral Palsy. New York, Grune & Stratton, Inc., 1983.

example, reflects the diagnostic experience of a university affiliated facility for mental retardation, as it carefully evaluated children referred for developmental review from community, school, and residential programs.

Any schema that outlines the backgrounds for the occurrence of mental retardation inevitably becomes a check list for disordered human development in general. As such, it chronicles all the steps in the developmental sequence at which pernicious influences can intrude. The clinical outcome may well be predominantly mental retardation, but it may just as well be cerebral palsy, epilepsy, blindness, deafness, physical handicap, or even emotional disturbance or learning disability (with any combinations thereof). Reference is made to sections B, C, D, E, and F of this chapter for further analysis of the special concerns about causation of related handicaps, and to section G for a discussion of complex disabilities.

Hereditary Disorders. Hereditary disorders as a background for significant mental retardation are much less frequent than popular belief would indicate. Single gene aberrations with biochemical markers, the so-called "inborn errors of metabolism" (such as Tay-Sachs disease, Hurler disease, phenylketonuria, and galactosemia), can produce a devastating cerebral handicap but are of very low incidence. The phakomatoses, such as neurofibromatosis and tuberous sclerosis, are more frequent but have a wide variation in developmental effect. Occasional families are found who demonstrate unique patterns of cortical disorder with recessive or dominant inheritance, and certain notable pedigrees show apparent polygenic or variable expression genetic liability. (See Chapter 18 for a full review of hereditary concerns.)

Chromosomal Aberrations. Chromosomal aberrations are probably underestimated as a source of serious developmental handicap. (See Chapter 19.) Down syndrome has long been identified as one of the major discrete biomedical origins of retardation. Technologic advances in the past five years have provided evidence that other chromosomal rearrangements may lie behind the problems of many troubled young people. The incidence of the fragile X syndrome appears to equal that of the Down syndrome in males; the Prader-Willi syndrome is commonly generated by chromosomal change; and other "minor" chromosomal variations are now being identified when carefully sought in retarded children.

Congenital Anomaly Syndromes. "Birth defects" or prenatal influence syndromes—the outcomes of embryodysgenesis— constitute an enormously important background for developmental handicap, as discussed in Chapter 20. In some instances these involve a cluster of dysmorphic features of familiar nosology, one of the named syndromes. More commonly, however, one encounters the small child with various mild phenotypic variations who also has developmental delay, with the assumption implicit that the central nervous system has shared in the malformation complex. Topographical indicators include changes in the position or configuration of the eyes or ears, midface deficiencies, an unusual palatal shape, a small mandible, changes in the digits, palmar creases, or dermatoglyphics, hernias, genital variation, or unusual feet. The child may have a normal birth weight or be small for dates, body growth is commonly slowed, and head size is often small (but sometimes proportionate to small bodily growth). Computed tomographic scanning studies of the central nervous system usually reveal normal findings unless there is an odd cranial configuration, and the brain changes can be assumed to be in cortical layering or at other subtle levels. It is a rare experience when one is able to provide the family with a presumed mechanism for these first (or second) trimester influences.

Fetal Malnutrition. Fetal malnutrition refers to diminished support for fetal growth as pregnancy proceeds, especially regarding factors in placental integrity or vascular configuration. Reduced size of the infant or the early onset of labor may result, sometimes with untoward developmental consequences directly or as listed in the following paragraph.

Perinatal Stress. Perinatal stresses, considered in detail in Chapter 21, affect particularly the vulnerable infant. This refers to premature birth or obstetrical complications in the full term infant. There is a potentially compromising complex of troubling events—trauma, central nervous system hemorrhage, pulmonary difficulties, acidosis, hypoglycemia, and sometimes infection—that can operate negatively on the extrauterine adjustment of the immature infant brain. These important issues can produce immediately recognizable complications for development or place the child in an uncertain "at risk" status.

Acquired Conditions. Specific conditions acquired in childhood cause mental retardation in relatively few instances. Most significant are complications of central nervous system infections (encephalitis, meningitis; see Chapter 22) and cranial trauma (household and motor vehicle accidents; see Chapter 25). The role of toxins is often less discrete but unquestionably important (see Chapter 24).

Deprivation. Deprivation in children constitutes a vast area of varying characteristics, which include psychosocial disadvantage and disordered parenting. Specific issues of inadequate stimulation, deficient interpersonal nurturance, physical abuse, and malnutrition may be operative. Family chaos, cultural maladjustment, poverty, and inept support systems are common. (See Chapters 14 and 23 for specific analyses of these potential developmental deterrents.) Such situations may complicate the course in children who already have specific handicaps.

Psychiatric Disorders. Various psychiatric disorders in child, parents, or both can also lead to a modified developmental course and ultimate mental retardation. Maternal schizophrenia can be such a setting, as can also a serious intrinsic psychic atypicality in the child. Childhood autism is a compelling example of distracted development, presumably of potentially heterogeneous origin. (See Chapters 39F and 41 for further discussion.)

A detailed search for the factors just described may yield negative or inconclusive results in an appreciable number of children with significant mental retardation, and one is required to list the apparent causation as unknown. In the experience reported in Table 39–1 this represented almost one-third of the total. In some of these children multiple factors may be present, but none of ascendant importance, and no final conclusions can be drawn. In others historical review and current study provide no viable hypothesis regarding the route of developmental handicap. Here, as mentioned, one can nonetheless discount certain parental fears. Otherwise such children stand as testimony to the incomplete knowledge that presently exists about the vulnerabilities and liabilities of the small maturing human.

DIAGNOSTIC FORMULATION

SCREENING FOR MENTAL RETARDATION

The early recognition of cognitive handicap is the primary objective of developmental screening. By using screening instruments devised in recent years (Thorpe and Werner, 1974), it is now possible for the practitioner to detect significant cognitive handicap in infancy. (See Chapter 45.) This clinical capability makes unacceptable the formerly common tendency to defer diagnosis until a child was old enough to attend school. Furthermore the emphasis on intervention in the preschool years provides treatment options unavailable until recent years. The primary care physician or other health care provider responsible for infants and young children cannot neglect the need to appropriately identify children requiring special attention owing to mental retardation or related problems.

THE CONCEPT OF BEING "AT RISK"

Concurrent with the intent to recognize the early signs of mental retardation, many studies have sought subpopulations of children most likely to have cognitive difficulty. In the nineteenth century, recognized authorities of the time postulated a primary familial or genetic cause for mental retardation. Therefore it was logical to assume that mentally retarded parents, or families whose adult members were known to have a mental handicap, might be most likely, or at greatest risk, to have mentally retarded progeny. Ignorance regarding the mechanisms of inheritance, the imprecision of diagnosis, and frequent eugenic motivations discredit many such theories and their implications for medical practice. For a time early in this century, the etiology of retardation in fact suffered from a lack of scientific scrutiny. The concept that biological or environmental factors might predispose to retardation was submerged beneath the intense emphasis on classification.

Advances in molecular biology, submicroscopic anatomy, and the understanding of diverse physiological feedback loops have altered these perceptions. We now appreciate that a multitude of genetic factors, untoward events in intrauterine life, and postnatal influences impinge on the developing central nervous system. When the capacity of that system to compensate is exceeded, there is inevitable deficit. Infants and children known or suspected of having been subjected to potentially damaging effects are at risk for developmental compromise.

PARENTAL APPREHENSION

Of course the suspicion of risk does not confirm any diagnosis. Knowing that an infant was born

Table 39–2. EARLY INFANT BEHAVIOR PATTERNS THAT SUGGEST COGNITIVE HANDICAP

Nonresponsiveness to contact
Poor eye contact during feeding
Diminished spontaneous activity
Decreased alertness to voice or movement
Irritability
Slow feeding

prematurely, had acidosis in the first day of life, and subsequently had a grand mal seizure does not establish a development diagnosis. Those clinical events define that child to be at risk for a developmental disorder. Therefore the challenge to the practitioner to predict effect from cause creates considerable apprehension. This apprehension is not the sole province of the professional.

Parents naturally expect their child to be normally developed at birth, to thrive in infancy, and to progress through successive stages of maturation. Even a serious maternal problem during pregnancy, or a severe neonatal illness, does not necessarily alert parents to a potential developmental disorder. Indeed the fortunate human tendency is to expect resolution of any problem. The infant is clearly resilient despite many untoward events. The sick child can become well. Medical technology can offer notable supports.

The infant who eventually will have the diagnosis of mental retardation may early exhibit developmental compromise and send warning signals to parents. Despite the basic parental expectation of normal development, some infant behavior patterns will clearly precipitate disquiet. Such patterns are indicated in Table 39–2. None of these is specifically diagnostic for cognitive handicap. Their appearance in an infant who has been developing normally is ominous and could also suggest meningitis, encephalitis, or other central processes. When the behavior is typical and consistent from birth, few parents will conclude that their infant is entirely normal.

CONCERN IN THE PEDIATRIC OFFICE

Parental sensitivity obligates the practitioner to assess carefully the infant's neurological and developmental status. A practical approach in obtaining the necessary information and structuring it for evaluation should be reflected in the office medical record. There the pertinent prenatal and perinatal history may suggest a risk factor. Confirmation of negative metabolic screening results for phenylketonuria, hypothyroidism, and galactosemia should be sought if not available.

Parental concern about infant development, even though not observed during an office examination of the infant, should never be dismissed as unimportant. If the physician ignores evidence of developmental delay or suggests no specific course of action, the parent's confidence in the physician's judgment will pale. Discussion about parental concern is always appropriate.

COMMUNITY CASE FINDING PROGRAMS

The concern that mentally retarded children should be identified as early as possible and then be provided services has prompted the organization of formal screening programs. Generally these programs are administered under the auspices of local public health, social service, or education agencies. The first systematic attempt to identify children with mental retardation in a large child population occurred in the federally supported Head Start Program. This group of children from low income households has provided a "captive audience" for follow-through services after screening.

An even more extensive effort occurred with the initiation of the Early and Periodic Screening, Diagnosis, and Treatment (EPSDT) program under Title XIX of the Social Security Act. Children receiving medical assistance benefits in this program may have annual comprehensive screening by participating physicians. Components of the screening include a physical examination, dental inspection, hearing and vision testing, blood and urine testing, and assessment of the developmental status.

Some states and localities organize screening programs sponsored by public health agencies. These programs are targeted on children who do not have a regular source of primary care. Generally protocols for these programs indicate the criteria for referral to a physician, dentist, or other professional. Even if a family does not have a personal physician, recognition of a problem in an infant or parental concern usually leads to preliminary assessment in a screening program. There is persistent criticism in some segments of the health care community that there is too much duplication of effort in screening. Nevertheless it is unlikely that a five year old child with mental retardation can now enter kindergarten in most communities without some previous evaluation. That situation remains possible in areas or communities where a family may be isolated for cultural reasons or because of extreme poverty. Population density probably is not a determinant if there is a basic system of screening services.

The nature of the system will continue to evolve. A dynamic interaction between office or clinic based primary medical care and publicly supported programs that have broad accessibility will guarantee the refinement of an efficient, capable screening network.

SPECIFIC ROLE OF GOVERNMENT

Since 1935 the federal government in the United States has exercised a major role in the design and delivery of child health services. Title V of the Social Security Act established a federal-state partnership to insure an intensive effort to lower the infant mortality rate, provide primary care to children from low income families, and finally prevent and treat handicapping conditions through the prompt identification and management of disabled children. State based Maternal and Child Health (MCH) Programs and Crippled Children's Services (CCS) programs have had authority for these responsibilities. MCH activities have concentrated on prevention relating to mental retardation. In most states, neonatal metabolic screening programs, chromosome diagnostic laboratories for prenatal and postnatal testing, and developmental screening programs are either administered or funded by MCH federal and state funds. CCS programs emphasize case finding and treatment of handicapped children. The service model varies extensively from state to state. Some programs have organized extensive clinical service programs, built around permanent or itinerant clinics. Other states contract with major medical centers to provide specialized services at university hospitals or other facilities. In most CCS programs limited medical, surgical, or rehabilitative care can be purchased for financially eligible families.

Owing to the historical priority on serving physically handicapped children, many CCS programs provide few services to children with mental retardation after diagnosis, unless there is a concomitant physical or medical problem. However, in some states the CCS agency may directly fund a child development program that offers tertiary level referral services to a large population of children.

IDENTIFICATION IN THE SCHOOL SYSTEM

The design of screening services to identify children with mental retardation changed substantially with the passage of the Education for All Handicapped Children Act (PL 94–142) in 1975. (See Chapter 61.) In order to receive federal special education funds, states must make an effort to locate all children requiring special services. Every child is affirmed to have the right to an appropriate education. Although mechanisms for locating children vary, now every school district in the nation must have a procedure to identify, assess, and plan an educational program. Within two years after implementation of the Act the federal government reported that almost 8 per cent of the children attending school throughout the nation were receiving special education services, including almost 770,000 children with mental retardation (more than 1.5 per cent of the student population).

Identification efforts commonly are focused on four year olds by means of child find programs or preschool screening. In selected states or regions a specific referral mechanism has been created to encourage physicians to refer younger children or even infants. The logic of referral is dependent on the school's offering an appropriate program or service for the identified child. Preschool programs are most frequently designed for children with hearing loss who consequently have delays in speech or language development. Other programs concentrate on multiply handicapped children, including young children with cognitive handicaps.

Children who have entered school without any recognition of their cognitive problem may not be appropriately evaluated until sometime in the early grades. Recognition and referral are then dependent on the classroom teacher. Placement in an educational program is usually determined by the urgency of the child's needs and the options within the school system. A considerable number of classification problems persist.

SCREENING INSTRUMENTS

A positive advance in the recognition of developmental delay has been the utilization of standardized screening protocols. The neurological examination has been the traditional mode by which physicians assess development, in conjunction with a history of the infant's developmental "milestones." When a child has the benefit of longitudinal primary care from a single physician or clinic, subtle developmental abnormalities can be followed in consecutive well child visits. A frustration for many practitioners is the lack of a firm or objective evaluation base on which to make decisions. Mental recall of the quality or onset of developmental gains is inadequate without excellent medical records. Structuring the record to provide a comprehensive, consistent data base involves the use of developmental screening instruments.

Observations in the 1930's by Gesell et al. (1940) and the theoretical approaches to child development advocated by Piaget (1952) have validated a systematic assessment of developmental function. Using the knowledge of normal patterns of task achievement, investigators have constructed observational tools to follow an infant's developmental progress. One tool, the Denver Developmental Screening Test, has come to be used widely by pediatric clinicians. The test enables the examiner to evaluate quickly a child's developmental

status in five areas: gross motor, fine motor, adaptive, language, and personal-social. Brief questions are required of the parent or caretaker, and only simple objects are necessary to elicit the infant's capabilities. The instrument has been standardized to assure reliability and consistency among trained examiners. (See Chapter 45.) The recording of data is uncomplicated when the scoring sheet is used. The Denver Developmental Screening Test represents one of several similar efforts to provide a practical instrument for the clinician.

The infant or child who appears to be delayed in development according to such an instrument then deserves careful monitoring, usually accompanied by additional medical and cognitive assessment. Caution, however, is warranted lest the examiner equate the observed developmental lag with a prediction of a specific mental handicap. Developmental screening instruments are not designed as predictive tests; they are used to assess developmental function at a point in time. The examiner should consider the infant's or child's status in the context of the history of predisposing biological factors, the developmental history, the character of family nurturance, and the current physical examination. If the accumulated knowledge regarding the child points to a serious multifaceted lag, major concern and prompt assessment are desirable.

Scattered lags in selected developmental areas indicate the need for a carefully focused approach. The infant who is delayed in language, for example, deserves a thorough otological and hearing evaluation.

THE UTILITY OF EARLY IDENTIFICATION

The clear rationale for screening children and identifying those with disability derives from the thought that the developmental handicap is not static, that the developmental process is in fact dynamic and subject to modification. This notion has gained widespread support from studies demonstrating the environmental influences on child development.

The transposition of developmental theory regarding children without known disability to children with disability represents a novel event in medical services. Prior to 1960 there was little published information about the effect of special services for preschool retarded children. As institutionalization of young children became uncommon, interest and motivation expanded in many communities to serve these children. Studies of the impact of early services began to appear, concentrating on children with a diagnosis such as the Down syndrome (Hayden and Haring, 1976). The concomitant proliferation of health and social services spawned a new generic service, variously termed infant stimulation or early intervention.

The concept of intervention as it relates to mental retardation is diagrammed in Figure 39–1. The premise is that intervention changes outcome, in this case developmental status. Greater amounts of intervention, in both quality of therapy and frequency of service, alters outcome favorably (toward "A" on the curve). However, as the severity of the biological or constitutional deficit increases, the outcome can be less favorably altered (toward "B" on the curve). An infant with a developmental delay secondary to deprivation might respond to intervention with a favorable outcome ("A"), whereas another infant with severe encephalopathy secondary to hypoxia would not ("B"). The corollary to this premise is that for any specific child, continued intervention will improve the outcome in comparison to a child with a similar disability who has not benefited from this intervention (represented in Figure 39–2).

These premises bolstered the widespread appearance of community programs, sponsored by a variety of public and voluntary agencies and facilities. Early evidence of program effects validated the conviction that families gained extensive knowledge regarding caring for their retarded or handicapped child, especially in feeding, play, and techniques of stimulation. A milieu of community caring, some promise of continuity of service, and a chance to develop a positive parental role reinforced families attempting to cope with their responsibilities.

Less convincing has been the specific impact of the intervention program on developmental outcome in the child. There are multiple reasons. Considerable effort is required to carefully select children for controlled studies. Children with heterogenous diagnoses, social or cultural backgrounds, and "uncontrolled" interventions in other settings do not provide useful information

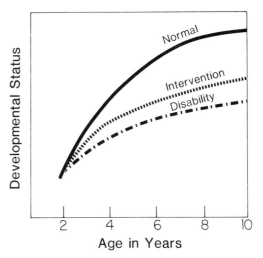

Figure 39–2. The outcome after intervention.

about outcome (Denhoff, 1981). Even early in the new intervention era it was perceived as ethically unacceptable to deprive a control cohort from substantial intervention. Therefore, relatively small variables had to be compared. The major difficulty, however, rested with a lack of sustained observation over many years to assess ultimate outcome while maintaining defined intervention criteria. (See Chapter 59.)

Currently these frustrations have only partially altered the enthusiasm for early identification. Inasmuch as parent oriented intervention encourages effective family coping and a chance for the child with mental retardation or other disability to remain in the home, most programs continue to be well utilized. The need continues to define standards for programs, use highly trained personnel wisely, and integrate the components of intervention with maximal efficiency.

THE DIAGNOSIS OF MENTAL RETARDATION

OFFICE DIAGNOSIS AND REFERRAL

The meaning of mental retardation is ambiguous among the general public. Therefore clinicians have become appropriately sensitive to the understanding of families when the diagnosis is initially made. Parents who expect their young infant to be normal, to achieve advanced academic and vocational skills, and to be socially independent may be devastated by application of the term to their child, even if some term such as mild, borderline, or nearly normal is used.

Since it is now accepted that cognitive function is by definition relative (to the status of peers), it is probably most important to discuss the concepts of intelligence, emotion, and living skills with the

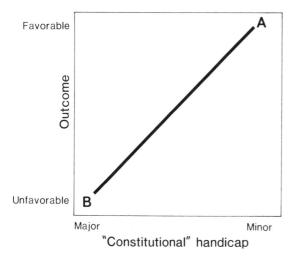

Figure 39–1. The concept of intervention.

parents before concluding for them that their child's function will be less than normal. These discussions are clearly most difficult when the child is very young, when the impairment may be subtle to the inexperienced observer, or when a parent has mental retardation or an emotional disorder.

Therefore the clinician should be confident that the diagnosis is valid. In most cases a primary care physician or other health professional is well advised to confirm clinical impressions. Screening tests are insufficient to substantiate a diagnosis. Therefore the administration of standardized psychometric tests is necessary, usually requiring a referral to another diagnostic specialist, or a team. A prompt referral, with maintenance of communication with the family, will in most cases preserve an effective primary care relationship and reinforce the appropriateness of the physician's judgment in the mind of the family.

SPECIALIZED DIAGNOSTIC SETTINGS: INFANT FOLLOW-UP

There have been extensive changes in patterns of practice in evaluating children for possible mental retardation. Primary has been the motivation to address the needs of specific populations of children.

Neonatal intensive care ushered in a new generation of infants with unknown outcomes. In fact early studies of the outcomes in very low birth weight infants from the 1940's reported severe handicaps in 80 per cent of the survivors. Rapid reductions in morbidity have followed so that currently groups of infants at greatest risk for mental retardation include only those of extreme prematurity and low birth weight (less than 28 weeks' gestation, or 1000 grams), infants with prolonged neonatal illness secondary to respiratory insufficiency, acidosis, or other metabolic problems, and finally newborns who have sustained a specific central nervous system insult, such as intracranial hemorrhage or meningitis.

Initially neonatal intensive care physicians organized follow-up clinics to document outcomes in children discharged from their units. These clinics have not only a medical emphasis but also a developmental orientation. In a generally systematic fashion it is possible to evaluate children early who may have sustained damage. In most current series, about 15 to 20 per cent of surviving infants with birth weights under 1500 grams are found to have a substantial neuromotor handicap.

Since most intensive care nurseries draw referrals from a large number of primary care settings and at times extensive geographical areas, it is logical to organize a focused follow-up program. Developmental assessment in these programs provides critical information about the effect of neonatal care practices, skilled consultation for the primary physician, and prompt recognition of the special needs of potentially retarded or handicapped infants.

CHILD DEVELOPMENT CLINICS AND TEAMS

The need for comprehensive assessment has stimulated the formation of teams of specialists working in child development clinics. In most clinics these teams were multidisciplinary at onset; that is, professionals from the disciplines of pediatrics, neurology, psychology, audiology, physical therapy, and others would each examine the child. Their diagnostic observations and conclusions would be summarized in a collective report. Recently the function of many teams has become more interdisciplinary, implying a greater degree of collaborative thought about the child's status and needs, with members of each discipline represented attempting to complement the contributions of other team members by virtue of his professional knowledge and skill. (See Chapter 49A.)

Child development clinics typically receive referrals from a variety of sources, including physicians. Concern about the duration and cost of evaluation discourages a monolithic approach to the assessment of every child referred. An expanded team of as many as 10 to 12 professionals may be necessary to understand properly the disability of some children. Others can be appropriately evaluated by smaller teams composed of the professionals specifically required for certain developmental problems.

The staffs of child development clinics in most communities have broader responsibilities of teaching and consultation to programs and facilities serving individuals with mental retardation. They therefore often have an excellent overview of the strengths and deficiencies of the service system in their area.

TECHNICAL AIDS IN DIAGNOSIS

The physical examination, with emphasis on neuromotor function, is at the core of the medical diagnostic process. A careful description of physical anomalies is essential in order to recognize patterns of malformation associated with mental retardation. In infants the measurement of body proportions and head circumference may help eliminate dwarfism with associated anomalies if the findings are normal. Transillumination of the skull may suggest incipient hydrocephalus, hydranencephalus, or porencephaly in an infant with major developmental compromise.

Clusters of malformations generally indicate the need for chromosomal analysis to rule out trisomies and deletions. Developmental lag with anomalies can be pursued by confirming that neonatal screens for phenylketonuria and hypothyroidism are negative. Plasma amino acid studies and urine screens for mucopolysaccharides may also be useful. Further metabolic studies generally should be only performed if the history, clinical examination, or other factors suggest a specific disorder. This is particularly the case when an infant has evidence of progressive disease.

Computed tomographic scans make it possible to locate central nervous system malformations without an invasive procedure. A scan is indicated if there is a focal neuromotor deficit, an abnormal cranial vault, expanding head size, or progressive deficit with seizures.

Laboratory tests or other diagnostic procedures are never a substitute for a thorough assessment of the clinical history and examination.

THE MEANING OF RETARDATION FOR THE AFFECTED CHILD AND FAMILY

THE PSYCHOLOGY OF EXCEPTIONALITY

Children and adults with mental retardation share an attribute with all other minority groups: they are different. The difference may be real or perceived, but the effect is substantial. Difference begins with the altered expectations of parents when they learn about retardation. The newborn infant with the Down syndrome may lack none of the capabilities of other neonates, yet that infant is perceived as different. The difference derives from common parental expectations that child development leads to maturity and independence in adult life. Since their infant is not predicted to achieve that status, the child is seen to be exceptional.

Exceptionality has had negative meaning for the child with mental retardation, largely because of the expectation that the child with mental retardation will learn poorly, will have diminished social experiences, and will remain dependent on adults indefinitely. For many retarded children the perceptions of parents and others constitute a significant part of their handicap. They are protected from experiencing many of the successes and failures of all other children. The development of a positive self-image may be limited. The retarded youth may never emerge through a preadolescent bewilderment about self-identity, appropriate social relationships, and future expectations.

In the interim, parents may have to confront a potentially nonsympathetic extended family and community. This reaction does not necessarily stem from an uncaring or ignorant society. Instead their child is viewed as exceptional because he cannot compete with, or "match," the peer group of children. Sometimes implied is a global sense of family inferiority. A family can become withdrawn and isolated. Only the most fully coping parents may emerge from this milieu to advocate effectively for their specific child's needs or those of similar children.

PERSONAL ADAPTATION AND PROGRESS

The observation that many young adults with substantial mental handicaps are achieving accommodation with themselves and the community suggests that new outcomes are possible. Antisocial behavior is not to be expected. Appropriate behavior and socialization constitute the goal for mentally retarded persons in the community.

These radically different expectations have resulted from the complex process termed normalization. A common conviction among observers of behavior of residents in state residential facilities was that the institutional environment led to most inappropriate behavior patterns. Self-stimulation, self-abuse, aggression, and poor basic skill development often could be traced to a living situation in which there was no expectation of any other behavior. Repressive control using medication or physical restraint was employed. Extracting residents from these dehumanizing settings into more normal human circumstances has been a societal goal. Through creation of alternative community residential programs and major programmatic improvements in public facilities, mentally retarded persons can be given the opportunity to participate in the life of the community in a much more constructive and positive manner.

Progress in the adaptation of these persons is dependent on a commitment from society to permit normal adjustment. That potential adjustment is not simply a function of intelligence or physical ability.

FAMILY ADJUSTMENTS

Children with mental retardation are the future beneficiaries of these massive changes. A new sequence of opportunities exists, beginning with early identification and the prompt initiation of services, continuing with special education, and then developing into vocational training and residential options.

This changing outlook has created realistic potential for effective family coping. If raising a retarded child is an acceptable parental responsibility, it is possible for families to continue their community life without embarrassment or feelings of inadequacy. Community school programs, options for respite care, and more developed social experiences for retarded children make it possible for a family to attend to the needs of brothers and sisters and the marital relationship. All the energies of a parent, traditionally the mother, are not expended in the physical care of the retarded child. Families can grow and develop in the context of their changing needs and opportunities.

VALUE SYSTEMS AND ADVOCACY

A recognition of the special needs of retarded children and the community's response to those needs ultimately confirms the value system by which children and families are served. In the era of institutionalization the accepted value of the cognitively handicapped child was low. At the solitary recommendation of the physician a child could be removed from the family and admitted to a large impersonal institution where the only expectation was humane custodial care.

The acceptance of such children and the attempt to provide adequate nurturing environments in the community have created a much different system of values. The question has become not whether services should be provided, but how they can best be designed to meet the needs of children. The extension of this basic philosophy pervades all human service areas, including health care. No longer can an individual physician make isolated judgments about the necessity or desirability of medical intervention for children with mental retardation. Whenever there are therapeutic options, the needs of the child, the prognosis of the condition, and the family's understanding of benefits of treatment can be considered to reach a decision.

When there is disagreement regarding an appropriate course, organized advocacy may be necessary. A major force for change has been the use of the courts to compel governments and institutions to take new directions and facilitate the development of community resources. Class action suits have led to consent decrees in many states, which are presently altering the mix of services available for substantially retarded children and adults.

CONSUMER MOVEMENT AND ADAPTATION

Since need invariably creates impetus for change, a hallmark in the development of services for retarded children and youth has been continued effort of family organizations and coalitions.

In many communities parents have performed the necessary work to begin a group home, including arrangement for property acquisition, long term financing, staffing, and the initiation of day care programs. Parents of retarded children have been instrumental in advocating nationwide special education legislation to insure appropriate education for their children.

Perhaps the most remarkable change in the family support of retarded children has been a willingness of families to adopt children with major developmental problems and to bring those children into the mainstream of community living. This development, a strikingly new phenomenon in recent years, now permits placement of severely handicapped children who in the past could have been served only in a revolving succession of foster homes or in long term residential facilities. The adoption of retarded children validates the dignity of these children and provides credibility to their needs in the family's community.

SERVICES FOR THE CHILD WITH MENTAL RETARDATION

SPECIALIZED HEALTH CARE

All children with cognitive handicaps require the coordinated services of a variety of health professionals. This need has led to the evolution of specialized health care capabilities. The selection of specific physicians and other professionals is determined by the nature of the child's disability. The child who has mental retardation without concomitant specialized health care needs may require only coordinated primary care. In this case the physician may be a liaison to schools and other agencies providing specialized services to the child. The management of growth disorders, orthopedic problems, nutritional deficiency, seizures, severe behavioral difficulties, reconstructive surgery, dental care, and other special needs must be developed in the context of the child's particular disorder.

The pediatrician may have an ambivalent role in this situation. It is difficult to coordinate specialized services from an office setting that is physically distant from the source of those services. Children receiving special services at major centers frequently do not have a coordinating physician. This situation places a burden on the family to resolve gaps in service or to reconcile contradictory recommendations.

Physicians practicing in communities that are geographically remote from specialized resources generally endeavor to maintain a relationship with the family and an awareness of the child's ongoing care. This facilitates an appropriate response to acute illness and referral when there are new developments. The pediatrician or family physician can fulfill a supportive role for the parents, facilitate access to specialty services, assist in the location of appropriate respite care facilities, and be responsive to the comprehensive needs of a child with mental retardation in the community.

THE EDUCATIONAL SEQUENCE

Children require an evolving educational program as they progress through developmental stages. In infancy and early childhood, intervention stresses the acquisition of normal developmental skills or compensation when a physical or cognitive handicap prevents the acquisition of such skills. A major emphasis is on working closely with parents so that they understand the abilities and needs of their child. Parents should be incorporated into the intervention approach. Most programs in fact emphasize relatively limited contact between the professionals and the infant but stress the teaching of parents.

As very young children progress and become capable of concentrating on learning tasks, more conventional teaching methods become possible. Preschool programs emphasize a low ratio of students to teacher in order to provide individualized attention for children. This is particularly important for children with mental retardation who may have limited attention span, who may require patient assistance with tasks requiring fine motor coordination, and who may have concomitant physical problems.

During middle childhood the emphasis on specific educational strategies for each child continues. Federal legislation under the Education for All Handicapped Children Act requires an individual educational plan that is implemented in the least restrictive environment. This encourages the placement of children with mild mental retardation in schools, and even classrooms, with other children who have less extensive handicaps. Some children are able to participate in the range of nonacademic activities with their chronological peers and then undertake intensive work in specific learning areas with other children who have special needs.

A specific need is for thoughtful preparation of children for life long vocational pursuits. This means that the youngster with mental retardation should not be required to continue with tasks requiring early academic skills and then be expected to master some type of meaningful activity without other prevocational training. Even in the preadolescent period some attention can be given to the strengths of the individual in pursuing skill development. This may require the combined resources of the school and local industries, whose representatives work in concert in preparing the mentally retarded youth for community living.

A general requirement is careful continuing evaluation of the child's potential and accomplishments, both to support and acknowledge progress in school and also to establish realistic positive expectations for future performance. Children with mental retardation should not be perceived as being in static programs that are primarily caretaking in nature. Instead the real change that occurs with development and increasing ability should be documented and the academic and other abilities of the child emphasized. Supportive teachers, counselors, and other school personnel can assist families in recognizing their child's progress and in modifying school programs as the child grows. (See Chapter 56.)

LIFELONG SUPPORTS

For the young person who is mentally retarded there now are a substantial number of programs available in the preschool and school years, generally incorporating the sharing of goals and components with the family. Securing continuity and coordination becomes a greater challenge as the teen years and young adult life approach. The "individual educational plan" strives to consider the whole person, his traits, and his needs, but it is understandably designed to feature educational issues primarily. The maturing person has more complex requirements, and nonschool factors assume a larger role. (Section H of this chapter considers specifically the adjustments implicit in the adolescent years.)

There is no single modus coordinating the guidance of retarded youths. The physician who knows the young person and the family can have a strategic role, particularly in correlation with counseling in the school system. Child development centers can also provide wise advice. In most states the "mental health" or "mental retardation" agencies become partners in programming responsibility as school functions decline, but there are often gaps. Families who have maintained diligent surveillance of the educational efforts may be fatigued at this point. Community agencies—including health centers and consumer groups (such as the Associations for Retarded Citizens)—often provide entry into social and recreational programs of major value as well as individual counseling.

Fears about what is to happen as the adult years proceed are universal. The last decade and a half has seen the development of extensive programs for retarded adults. Adult day activity centers are now found in most communities, usually with a vocational orientation. Depending on the skills of the person involved, the program may incorporate relatively simple prevocational projects, it may be a full "workshop" providing moderate reimbursement to the client for contract work, or it may grade projects into competitive employment opportunities (see Chapter 57G). The availability of guided out-of-home residential facilities is now widespread. Here again a spectrum of independence is possible, varying from highly supervised settings with assistance in health care issues, to community group homes of many types, and on to supervised apartment living. These domiciliary arrangements are managed by social service or mental health–mental retardation agencies or, more commonly, by private vendors who are licensed by local governments. Family participation in the affairs of the day programs and the living facilities is encouraged, with different degrees of success. At any rate, the former specter of the older retarded person remaining idle and disillusioned in the home of aging parents is now no longer valid. Our culture assuredly has considered the needs of retarded adults, and in many ways has made provisions for them, but it would be unfair to claim that the system is fully adequate.

SEVERE AND PROFOUND RETARDATION

In the total spectrum of mental retardation there exists a minority within the minority whose handicap is of unusually serious nature—children (and adults) who have severe or profound retardation. This group—at most, 5 per cent of all cognitively disabled persons—raises special issues because of the magnitude of their quantitative and qualitative atypicality. It can be fairly said that their need for us to teach them is extraordinary, and that their ability to teach us is equally remarkable. The general public, and many professionals as well, are bewildered by individuals with IQ's below 35 (or, particularly, below 20) and lack frames of reference for interaction with them. Some of their notable features are the following:

1. Lack of self-care and even survival skills. These are persons with a truly pervasive handicap who often are not able to dress, feed, or toilet themselves even in adult life, and who have a compelling dependency on others throughout their lives.

2. Communication blockade. Inadequate language is invariable, and often there is no successful verbal communication whatever. This can lead to social failure, complicated by the inability of others to interpret the person's feelings.

3. Deviant behavior. With blunted exchange and reward on other levels, there is often a resort to bizarre repetitive or stereotypic behavior, sometimes self-stimulating. This can include rocking, twirling, and posturing, with alarm induced when actions become self-injurious.

4. Serious organic handicaps. Although severe or profound retardation is found in all segments of the causation schema (see Table 39–1), it is more frequent in those with hereditary or malformation problems. Usually there are combined disabilities (including seizures and sensory problems) and many times serious health issues as well. Motor function difficulties or complications often reduce the person's potential for ambulation.

5. Greater commitment for intervention. Incidental learning (such as that from ambient experiences and social exchange) is reduced, so that efforts for progress require more discrete programming.

6. Teacher confusion regarding potential. Standard test instruments are less relevant at this far end of the spectrum, and personal progress and its documentation are slow and atypical. Gestures of simple "enrichment" compete with actual structured "training" in the investment of care providers.

When a child has no body control at all, or a young adult has no language or is unable to ambulate, the challenge to parents, teachers, and other professionals is enormous. It is axiomatic in human development that continued learning and progress are a fundamental response to adequate stimulation for all except those who are in coma, but in these situations the gains can be painfully slow and the feedback next to nil. Most clinical psychologists restrict the use of the term "profound mental retardation" to individuals in whom the adaptive components are massively deviant, rather than utilizing the IQ definition primarily. Dybwad (1981) has pointed out that "profound mental retardation" is in itself probably not a useful term; "profound disability" is more reasonable, since in these persons complexity of handicap is the rule.

It is in persons with these major degrees of mental retardation and other disabilities that human rights are most vulnerable. Prior to the social revolution of the 1960's and 1970's, such handicaps were invariably the basis for exclusion and segregation. Since that time, the promulgation of "no reject" programs for education, family support, and general activities has allowed many quiet miracles of personal triumph to occur. The latter refers to fulfillment for the individuals with special needs, as well as for those who relate to them or work with them.

The ultimate role of the large public residential institutions—the "state schools," "training schools," or, in the modern idiom, "developmental centers"—is unsettled. These troubled facilities were initially established, beginning at the turn of the century, with the alleged purpose of providing specialized treatment (Nelson and Crocker, 1978). It soon became apparent, however, that programs were severely compromised by the many hundreds (even thousands) of needy humans, with the attendant management complications. When enrollment reached almost a quarter million persons 20 years ago, a reaction developed to further admission and the philosophic revision of "deinstitutionalization" began. As the population in the state residential facilities declined (now nationwide to about half of the former number), those remaining were increasingly the complex older individuals with severe or profound retardation. Most people believe that even these clients are best served in smaller, more normalized environments within the community. There is currently a firm resistance to admitting children to state schools; the same basic reasons apply to adults as well.

PREVENTION OF MENTAL RETARDATION

As has been mentioned, the causes and settings for mental retardation are heterogeneous, and thus it can be assumed that programs aimed at a reduction of the incidence of mental retardation will draw on many bases. In general, it can be said that the feasibility of intervening for prevention is predicated on an improved understanding of the biomedical and social substrates for child development as well as the roots of their deviation, creation of special technologies that give support to normal progression, and a societal commitment that assigns high value to an improved outcome for children, with the attendant assignment of public resources. The last 15 years have seen substantial gains in this regard, with the achievement of methodology now that can "prevent" the occurrence of perhaps one-third of the traditional situations that lead to significant mental retardation (Crocker, 1982).

It is simpler to account for the effects of primary prevention procedures, which actually eliminate the occurrence of the condition leading to the developmental handicap, and secondary activities, which intervene through early identification and special support to avert retardation per se. More difficult to document are the effects of tertiary prevention measures in which assistance to children and families with problems may minimize the long term disability or prevent complications. Some of the most successful interventions accomplished in recent years relate to primary or secondary limitations of diseases with a serious morbidity but a low incidence; many of the more prevalent situations have not yet yielded to public health approaches. A current scorecard follows:

Nearly Total Elimination:
Congenital rubella, by early immunization and antibody screening.
Retardation in phenylketonuria, galactosemia, and congenital hypothyroidism, by newborn screening followed by dietary management or replacement therapy.

Kernicterus, by reduction of sensitization through the use of globulin therapy.
Major Reduction:
Tay-Sachs disease, by carrier screening and prenatal diagnosis in persons with increased risk.
Morbidity from prematurity, through newborn intensive care nurseries.
Measles encephalitis, by early vaccination.
Significant Current Efforts Underway:
Neural tube defects, by maternal serum alpha-fetoprotein screening and prenatal diagnosis.
The Down syndrome, through counseling of older pregnant women and prenatal diagnosis.
Lead intoxication, by environmental improvement, screening of lead levels, and chelation when necessary.
Fetal alcohol effects and syndrome, by public education.
Morbidity from head injury, via the use of child restraints in automobiles.
Special Assistance and Relief:
Early identification, followed by stimulation and training, for children with the Down syndrome, multiple handicaps, and hearing impairment.
Support to families with handicapped children, to provide guidance and resources.
Genetic counseling when special risk is involved.
Improved management for difficult pregnancies.

Regrettably there are many children whose initial retardation handicap could not have been prevented by usual means. These include the majority with congenital anomalies from unknown prenatal influences, most nonfamilial chromosomal disorders (including the children with the Down syndrome born of younger mothers—75 per cent or more of the total), most serious childhood neuroses and psychoses, which interfere with development, and the very substantial number of children in whom the basis of retardation cannot be identified at all, even on careful study.

The outlook is good for continuing improvement in the prevention of mental retardation syndromes. Child care professionals can assist in this movement by the promotion of immunization, newborn screening, guidance during pregnancy, and the use of child safety measures. Gains will also be made through the wider employment of early developmental screening, comprehensive assessment of children with known handicaps, and thoughtful genetic counseling. Public support should be marshaled for programs for the infant and young child. Basic and applied research regarding the nature of cortical handicap must not lapse in times of economic pressure.

ALLEN C. CROCKER
RICHARD P. NELSON

REFERENCES

Crocker, A. C.: The involvement of siblings of children with handicaps. *In* Milunsky, A. (Editor): Coping with Crisis and Handicap. New York, Plenum Publishing Corp., 1981.
Crocker, A. C.: Current strategies in prevention of mental retardation. Pediatr. Ann., 11:450, 1982.
Crocker, A. C.: Cerebral palsy in a spectrum of developmental disabil-

ities. *In* Thompson, G. H., Rubin, I. L., and Bilenker, R. M. (Editors): Comprehensive Management of Cerebral Palsy. New York, Grune & Stratton, Inc., 1983.

Denhoff, E.: Current status of infant stimulation or enrichment programs for children with developmental disabilities. Pediatrics, *67*:32, 1981.

Dybwad, G.: Personal communication, 1981.

Featherstone, H.: A Difference in the Family: Life with a Disabled Child. New York, Basic Books, Inc., 1980.

Gesell, A., Halverson, H. M., Thompson, H., Ilg, F. L., Castner, R. B., Ames, L. B., and Amatruda, C. S.: The First Five Years of Life. New York, Harper & Brothers, 1940.

Grossman, H. J.: Manual on Terminology and Classification in Mental Retardation. Washington, D.C., American Association on Mental Deficiency, 1977.

Hayden, A. H., and Haring, N. G.: Early intervention for high-risk infants and young children: Programs for Down syndrome children. *In* Tjossem, T. D. (Editor): Intervention Strategies for High Risk Infants and Children. Baltimore, University Park Press, 1976.

Milunsky, A.: The Prevention of Genetic Disease and Mental Retardation. Philadelphia, W. B. Saunders Company, 1975.

Nelson, R. P., and Crocker A. C.: The medical care of mentally retarded persons in public residential facilities. N. Engl. J. Med., *299*:1039, 1978.

Piaget, J.: The Origins of Intelligence in Children. New York, International Universities Press, Inc., 1952.

Scheiner, A. P., and Abroms, I. F. (Editors): The Practical Management of the Developmentally Disabled Child. St. Louis, The C. V. Mosby Co., 1980.

Thorpe, H. F., and Werner, E. E.: Developmental screening of preschool children: A critical review of inventories used in health and educational programs. Pediatrics, *53*:362, 1974.

39B The Child with Hearing Impairment

Hearing impairment is one of the most common and important disorders affecting children and their families, yet it remains poorly understood by many pediatricians. The population of hearing impaired children in the United States is a heterogeneous one, including many with such mild hearing loss that they might have gone unnoticed except for routine hearing screening programs, referral for inattentiveness at school, or articulation disorder. These children can look forward to essentially normal lives. At the other end of the spectrum are children with profound hearing loss, which if present from the time of birth will require special education services in childhood and will affect their vocation, social relationships, and most other aspects of their life experience. This section will explore the issues having an impact on the children in the latter category especially.

The past 100 years has seen a rapid evolution in society's methods of responding to the needs of deaf children. Until the time of the Industrial Revolution, persons with serious hearing impairment could be reasonably well integrated. With increasing urbanization and reliance on an industrial base at the turn of the century, deaf persons found such integration increasingly difficult, and they were often sent to special schools or other residential settings where services could be provided in a centralized fashion. The period following the end of World War II has seen another shift, to greater decentralization of services and consideration for the hearing impaired person's individual rights and needs. This is now evidenced by successful ventures on the Broadway stage about deaf persons by deaf persons, provision of interpreters for the deaf in regular classrooms, and deaf lawyers arguing cases before the Supreme Court. These developments are confronting deaf persons as well as the rest of society at an ever increasing tempo, stressing the deaf person's ability to cope with change in the same fashion that changes in the modern world exert stress on the person with normal hearing. The challenge to the "normal" deaf person now is to decide whether to emerge from the relatively isolated world known only to people of similar disabilities and to compete for financial and social recognition with their hearing peers. Such a choice is not yet generally available to multihandicapped deaf persons—those whose hearing loss is complicated by serious deficits in vision, cognition, motoric ability, or other functions, as has been seen so commonly in the child with congenital rubella. No doubt, as this movement by developmentally disabled persons toward acquiring abilities for communication is facilitated by modern electronics and computer technology, we will see more seriously disabled deaf persons entering the competitive arena to a greater extent than is now the case. For in the end, the major problem for the hearing impaired person to overcome is, in fact, communication. Most deaf persons are not mentally retarded, but may have some functional delay in development because of an inability to communicate information with a hearing world. The ability to breach this communication gap through modern technology and education techniques has the potential for rescuing many thousands of deaf persons from their imposed exile. It is the role of pediatricians to participate in this rescue activity on behalf of affected patients, as well as to engage vigorously in all primary, secondary, and tertiary prevention modalities to reduce the numbers of hearing impaired persons.

ETIOLOGY OF SERIOUS HEARING IMPAIRMENT

The causes of deafness in childhood are diverse and complex. Three categories are included in most surveys:

1. Acquired deafness, in which a clear biological insult to the mechanism of hearing in an otherwise normal individual is shown to be the cause of hearing impairment. Examples include congenital rubella, meningitis, and asphyxia in prematurity.

2. Genetic deafness, in which inherited factors play an etiologic role, for example, in which a parent (dominant) or sibling (recessive) had evidence of similar hearing impairment.

3. Unknown, in which no clear or compelling evidence existed of acquired or genetic factors.

Published reports of the prevalence of each of these three factors among groups of deaf children must be carefully interpreted because of the selection process performed by the group carrying out the survey. For example, in a series of 348 deaf children seen over a period of two years in a large otology clinic in New York City, 39 per cent were characterized as having acquired deafness, 40 per cent unknown, and 21 per cent genetic (Ruben and Rozycki, 1971). On the other hand, a large number of deaf children surveyed in Great Britain by a genetics service found 54 per cent with genetically determined deafness, and 46 per cent with an acquired form of hearing loss (Fraser, 1964).

Statistical reports of the burden of each etiologic factor among deaf children today do not appear to be available at the present time. All pediatricians would acknowledge that it is of great value to be informed of the most likely diagnostic considerations when confronted with a deaf child as a patient for the first time. The pediatrician's role as family advisor would be greatly enhanced if such information were made available in regard to the probability of a certain diagnosis at a given age of the child—whether the patient is a deaf infant, preschooler, or teenager. That such information is not now readily available is lamentable. It is hoped that a well planned and coordinated survey using present day, state of the art diagnostic techniques may be undertaken soon to determine the true incidence of the various causes of deafness in children. The etiologic background for the hearing impairment of children attending a hospital clinic (for general audiologic service, not just diagnostic) is shown in Table 39–3.

Among children attending residential schools

**Table 39–3. CURRENT EXPERIENCE OF A HOSPITAL AUDIOLOGY SERVICE IN ETIOLOGY OF HEARING IMPAIRMENT (ALL DEGREES), 200 PATIENTS, OCTOBER 1, 1980, TO SEPTEMBER 30, 1981*

	Age (yr.) at Clinical Contact					
	0–2	3–5	6–11	12–15	16–20	Total
Sensorineural						
Prenatal and perinatal issues						
Prematurity			2		1	3
Newborn anoxia		2	2	2		6
Rubella	1		5	3	6	15
Cytomegalovirus			1			1
Hyperbilirubinemia			3		2	5
Familial conditions	1	4	9	4	2	20
Total						50
Childhood acquired						
Meningitis	1	2	5	1		9
Viral, unknown		1		1		2
Mastoidectomy				1		1
Anoxia—cardiac arrest				1		1
Brain tumor					1	1
Head trauma				1	1	2
Noise induced				2		2
Total						18
Unknown	4	9	16	14	4	47
Conductive						
Prenatal and perinatal issues						
Atresia of ear canal				1		1
Treacher-Collins syndrome					1	1
Total						2
Childhood acquired						
Middle ear effusion	8	21	24	6	1	60
Eustachian tube dysfunction		2	4	4	2	12
Tympanic membrane perforation		1				1
Impacted cerumen			1			1
Ossicular disarticulation					1	1
Ear canal stenosis			1			1
Total						76
Unknown		1	4	1	1	7

*Data courtesy of Richard S. Sweitzer, Ed.D., Director of Audiology, Kennedy Memorial Hospital for Children, Brighton, Massachusetts.

for the deaf, the largest cause of hearing loss is still genetic. Although some of these genetic disorders lend themselves to early diagnosis because of the presence of certain dysmorphic features (e.g., white forelock and hypertelorism of Waardenberg's syndrome), most others are probably multifactorial in origin and are based on the presence of genes recessive for deafness and silent in each parent that may be expressed clinically in their children. Such children deserve careful evaluation to insure that the etiology of the deafness is indeed most probably on a genetic basis. The diagnosis of "genetic deafness" is often based on the exclusion of other known causes, and it should be followed by genetic counseling so that parents and siblings can be informed about the risk of recurrence. In general, the birth of a "genetically deaf" child to hearing parents serves to identify those parents as being far more likely to have similarly affected children than families in the general population (approximately 2 per cent versus 0.3 per cent). The element of proper medical assessment of all deaf children to determine etiology must be stressed. In a review of a large number of children attending a school for the deaf recently, approximately 10 of them were rediagnosed as having congenital rubella (because of the presence of some subtle features such as congenital rubella retinopathy) instead of genetic deafness. This new diagnosis was greeted with some consternation by the parents after accepting the latter diagnosis and its implications for many years.

Another major group of children with serious hearing impairment are affected because of congenital infection. (See Chapter 22.) For many years the prototype of this infection has been rubella. Some of the features of rubella and congenital rubella are worth reviewing here because they may be generalized to other congenital infections as well.

The rubella epidemic of 1963–1964 in the United States was the largest outbreak of the disease that had been seen in the preceding 30 years. It is estimated that 20,000 infants were affected and that 1 per cent of all pregnancies during this period were affected. It was also estimated that by the time this group of children reached the age of 21 years, the services they required in health care, education, and other interventions would cost approximately two billion dollars! The impact of this epidemic was such that it lent great impetus to the development of a live attenuated rubella virus vaccine, which was licensed in 1969. Since that time, more than 70 million doses have been administered (primarily to infants and preschool age children) with a consequent reduction of dramatic proportions in the incidence of rubella and congenital rubella. At the present time only sporadic cases of rubella are reported, and these usually in unimmunized individuals in the late teenage years or in their twenties. Strong consideration is now being given to immunizing this population, the last major reservoir of susceptible individuals in this country, in an effort to further reduce the incidence of rubella induced birth defects. There are, however, theoretical hazards in immunizing a young woman with live rubella vaccine (a potential teratogenic agent) at a time when she may unknowingly be in the early stages of pregnancy.

Rubella, like many other viral illnesses, can occur in a spectrum, from being totally subclinical to one in which clinical disease is present with many complications. However, the clinical severity of the mother's illness in the first trimester of pregnancy has no bearing on the expression of clinical disease in the fetus. In fact, in a large group of patients followed longitudinally in New York City since 1965, 20 per cent of the children with rubella induced birth defects had mothers who had no clinical history of rash or illness during the pregnancy. Among those children with congenital rubella related hearing impairment, the hearing loss was usually bilateral, and hearing occasionally was more seriously affected in one ear than in the other. The hearing loss was always sensorineural in type (the virus either destroyed or impaired the development of the cells in the organ of Corti) and ranged from mild through moderately severe or profound. There were a number of children who appeared to have a speech impairment that was out of proportion to the degree of hearing loss noted on audiometric testing. Since rubella virus infection in utero is seldom limited to the inner ear but is often recovered from the central nervous system, heart, liver, bone, and other organs, it has been proposed that such children may have a "central language" impairment complicating the peripheral hearing loss, a language impairment of the expressive or receptive type or some combination of the two. These children commonly have difficulty with short term memory, auditory sequencing, and other "soft signs" of neurological dysfunction. It is important for the clinician to recall the generalized nature of such congenital infections when caring for a deaf child with congenital rubella, and to search for a possible occult urological abnormality or endocrine deficiency when clinically warranted. Similarly it is now accepted that the virus may never be totally eliminated from the child, but it may take up permanent silent residence in a number of organs, and be responsible for clinical disease in the second or third decade of life (e.g., diabetes, thyroid disease, degenerative neurological disorders such as progressive rubella panencephalitis) when the body's defenses are no longer able to keep it quiescent. (See Chapter 39G.)

Congenital rubella accounts for approximately

20 per cent of the children attending schools for the deaf and for a greater percentage enrolled in programs for the deaf and blind. By and large these are children whose mothers had rubella during the great epidemic in 1963–1964 and who are now late teenagers. It has been the vigilance on the part of pediatricians and public health agencies in the use of rubella vaccine that has virtually ended rubella as a major cause of hearing impairment in children. Prudence would dictate that the same devotion to immunization practices continue to be exercised by clinicians if they are to prevent this cause of deafness in children.

Unfortunately there are other congenital viral infections that can cause hearing impairment in children. Much less is known about the natural history and prevention of these infections, and preventive measures of the type brought about with rubella vaccine are still awaited. Particularly perplexing is the problem of cytomegalovirus infection as a cause of deafness in children. It is known that approximately 1 per cent of all newborns actively shed cytomegalovirus in the urine and are presumed to be congenitally infected. When followed carefully for a period of years, many of these children are found to be developmentally delayed or hearing impaired. Research on the nature of this disorder is continuing, but prevention is complicated because cytomegalovirus, a member of the herpesvirus family, may not lend itself to the prevention of infection through vaccination because of theoretical concerns regarding the virus' oncogenic potential.

It is probable that other congenital infections may lead to hearing impairment. Progress in this area has been hampered because hearing loss may not be clinically apparent for years after the birth of the child and there are problems in undertaking retrospective research of this type. It is hoped that closer collaboration between obstetricians and pediatricians in the immunological and virological monitoring of pregnant women and newborn infants may bring more light to this field.

A third major group of children have serious hearing impairment caused by diverse etiologies, predominantly resulting from problems encountered during the perinatal period. These include premature infants with asphyxia and acidosis, intracranial bleeding, sepsis, and meningitis; infants with hyperbilirubinemia and kernicterus; infants exposed to toxic agents such as aminoglycoside antibiotics; and older children with infections of the central nervous system, tumors, or other degenerative neurological disorders. For present purposes, it must be pointed out that otitis media should always be considered in a child with sensorineural hearing loss who appears to undergo further deterioration in speech and hearing. It is imperative that such children be identified and treated promptly before additional developmental impairment supervenes. Children in whom otitis media is the sole cause of the hearing impairment, of course, have a conductive hearing handicap, which generally causes no more than moderate to severe loss in the affected ear. Management decisions in this group of children are far different from those in children with severe sensorineural hearing loss. (Children with recurrent otitis media are discussed in Chapter 26.)

The pediatrician's role in the prevention of this diverse group of disorders is variable, but there is a constant need for the physician to be attentive to the possible association of hearing loss with the disorder in question, regardless of whether it is prematurity, meningitis, or hyperbilirubinemia. Such children must be identified by their pediatricians as being at high risk for having or developing serious hearing impairment and referred routinely and early for diagnostic audiological evaluation by persons experienced in testing young infants. There is no question that the pediatrician plays a major role here in secondary prevention by diagnosing sensorineural hearing loss in a young infant and recommending prompt referral for amplification with hearing aids and auditory training. Such deaf children often can develop useful speech and have an entirely different life to look forward to from that of another child with "prelingual deafness" not so identified and managed.

In the area of tertiary prevention the pediatrician should have an established relationship with the child, family, and community so that the deaf child can have access to all services consistent with the child's developmental needs. This is necessary regardless of whether such access relates to proper educational placement, psychological counseling, or advice to deaf teenagers in regard to birth control and the potential for having developmentally disabled children of their own.

DEVELOPMENTAL SERVICES

It is prudent at this juncture to consider hearing impairment in children from a developmental point of view. Many pediatricians are justifiably concerned with conditions in the prenatal or perinatal period that predispose the child to permanent sensorineural hearing loss. Infants and children considered to be at high risk for serious hearing impairment should undergo audiological evaluation as soon as the diagnosis of hearing loss is suspected. Objective measurements of hearing can be made in the newborn nursery, and newer electronic methods to assist in establishing the diagnosis are being developed. In the past it was common experience for parents and teachers to lament the fact that they had been advised that despite their concern that the child did not appear to respond to sound appropriately, their physician would not refer the child for evaluation because "he was too young to test." It is important to recognize now that no child is too young or too multihandicapped to receive a thorough professional evaluation. The information derived from the hearing evaluation must be shared fully with the family and referring physician, and management decisions must be based on this information. Early diagnosis and referral of the developmentally disabled deaf child do make a difference in the ultimate outcome.

As the child grows older, the pediatrician will continue to be called on by the family, and later by the deaf child as an adolescent, for support. He will need to devote extra time to such families because of their special needs. Depending on the etiology of the hearing impairment and its possible association with central nervous system dysfunction, the child may present with specific learning disabilities or behavior disorders. The pediatrician must be prepared to "listen" to the deaf child, which sometimes means that communication must be effected through an intermediary who is familiar with manual communication. Although this is important in the pediatrician's office, it takes on even greater importance in a hospital emergency room or inpatient service. Taking a history, obtaining informed consent, and complying with nurses' instructions is difficult enough when dealing with patients with normal hearing. In a hospital where hearing aids and glasses may not be readily available for the deaf, visually impaired child, such procedures may represent formidable obstacles to good patient care, and they are important considerations for clinicians to be aware of. It must always be kept in mind when dealing with deaf persons that they are very dependent on their vision in order to "lip" or "speech read" and upon their hands to communicate through "sign language" when their speech is not intelligible. In such cases it is helpful to communicate through the use of pictures, drawings, or the written word. In most urban areas trained interpreters for the deaf are readily available for use when family members or other knowledgeable individuals are not available. When hospitalization of a deaf child is being considered for elective reasons, it is worthwhile for the child and family to visit the inpatient unit prior to admission so that issues surrounding the need to maintain communications can be carefully reviewed by the nursing staff and others, and plans developed to deal with the impairment.

When deaf children grow into deaf adolescents, new developmental crises must be overcome by the child and family. The children, infantilized by some parents because of the disability for many years, go through puberty and present the family with incontrovertible evidence that they are children no longer. Conflicts frequently arise between parents and children at this time that may be more strident than in families in which the children have no hearing impairment. Sexual acting out may be common among deaf teenagers, with probably a higher than average pregnancy rate because of their somewhat reduced access to counseling and birth control devices. Pediatricians are often sought out by parents at this time for advice regarding the availability of sterilization procedures because of their fear that they would become responsible for raising the child who is the product of such sexual experiences. This is especially the case when their child is "multihandicapped deaf" and the parents express fear of the child's sexual exploitation. Such fears have probably been heightened by the changing societal attitudes toward placement of developmentally disabled persons in "the least restrictive environment." The days of placement of a multihandicapped deaf person in a large public residential facility have in all probability passed into history, and with it the assurance of parents of developmentally disabled children that there would always be a place supported by the state where their child would be offered all basic services and protected from harm. We now know all too well that such large public facilities did not often live up to this expectation; in fact many developmentally disabled persons showed deterioration in their abilities following admission. It remains at least in part the responsibility of pediatricians to promote and advocate better community based alternatives for deaf children and their families and to be aware of the network of educational, vocational, and residential arrangements that are in existence to support this effort.

AUDIOLOGICAL SERVICES

It is important to recognize the special set of relationships that must exist between the audiologist, speech-language pathologist, otologist, and pediatrician in order for optimal care to be rendered to the deaf child. The pediatrician, responsible for primary care of the patient, must seek out professionals who are sensitive to the special needs of the child and family and who can work in harmony on the child's behalf. The pediatrician must be comfortable with sharing responsibility for decision making with other members of the team, each of whom may be required to play the role of case manager from time to time. Conflicts between professionals regarding who may or may not be more expert in a certain area of diagnosis or management only delays proper remediation and causes more anguish to the family. The pediatrician may be called on to mediate in such disputes, as in discussions with people in school systems, vocational programs, or the legal system.

His role as an advocate for the "whole child" and not as proponent for any particular discipline's point of view is of great importance to the family.

Before discussing the various audiological procedures available, it is well to keep in mind the following general principles of hearing testing in young children: Test results may not always be reliable and reproducible from one observer to another; they may vary from time to time depending on the child's state of health, hunger, distractibility, mood, and effect of medication. There is no single test that is best for all children, and numerous assessments are often required to get a reasonably accurate picture of the state of the child's hearing in each ear. Amplification in a child with proven hearing loss may be worth a therapeutic trial, but may not be successful because of the age of intervention and other conditions.

Evaluation of a child's hearing must take into consideration the capacity to perceive various frequencies of sound (pitch) as well as the loudness of sound (expressed in intensity as decibels). The frequency range of the normal human ear is 20 to 20,000 Hertz (Hz), though the frequency range for speech ranges from 100 to 10,000 Hz.

The normal adult can hear the softest sound at 0 decibels (threshold), and the loudest sound at the threshold of pain is approximately 115 decibels. Conversational speech is perceived at 50 decibels; a whispered voice can be perceived at 10 to 20 decibels by persons with normal hearing.

Hearing impaired children often develop speech only with difficulty because speech sounds are complex in frequency and intensity. For example, a hearing loss occurring primarily in the high frequency range would lead to great difficulty with consonants, especially the "s" and "t" sounds (which also are of low intensity), which the deaf child is unable to hear himself uttering. It is this lack of effective auditory feedback in the deaf child that is usually responsible for the distortions and omissions in his own speech, as well as for his inability to comprehend the speech of another person speaking to him. These speech and language deficits generally carry over into the realm of education; reading scores of deaf children are often far below those of their age matched hearing peers.

The audiogram developed in the evaluation of the child is a graphic representation of the characteristics of the child's hearing with each ear. The audiogram enables the audiologist to assign a category of hearing loss, although it must be recognized that some children with other handicapping conditions clinically may appear to have functional hearing not reflected in the audiogram. The pediatrician must be wary of the designation of children with certain disorders, such as autism, as being "deaf" because of certain test results. Such children have been mistakenly fitted with

hearing aids in the past, a practice that is potentially harmful because of the damage that may be caused to normal hearing by excessive amplification of sound. The audiogram is also useful in determining the type of hearing aid that may be of most value to the deaf child and in tailoring the frequency and intensity characteristics of the hearing aid to the specific characteristics of the child's hearing loss.

In young infants and children who are unable to cooperate voluntarily with the audiologist, hearing can be measured by using certain behavioral measures (startle response, localizing toward origin of sound, cessation of activity), first in a sound field and later through the use of earphones to measure the hearing in each ear separately. Other children may respond to conditioning and reinforcement techniques in the testing environment (conditioned orientation reflex audiometry, visual reinforcement audiometry, tangible reinforcement operant conditioning audiometry, play audiometry) in an effort to obtain reliable and reproducible results. Experienced audiologists try a number of these techniques to assess young infants and children to obtain the most valid data possible.

Objective testing of hearing is also becoming more prevalent with assessment of acoustic reflex thresholds through tympanometry. This technique, though widely used, is subject to some error because of many false positive and false negative results, and such data must be interpreted with great care. More encouraging perhaps is the increasing reliability of brain stem evoked response audiometry, which measures electrical signals from the brain following an auditory stimulus. This technique is being studied increasingly as a hearing testing device for seriously multihandicapped children and those with significant behavioral impairment in whom measurements can be made after sleep is induced with sedation.

The evaluation of the hearing impaired child by audiologists is only part of the evaluation process. Generally the young deaf child should undergo a speech and language evaluation and an evaluation by a special educator who should have experience with a broad range of deaf children. The speech and language therapist attempts to determine the child's ability to understand language in a broad sense—the ability to understand gestures and symbols in addition to spoken language. Once the deaf child is fitted with hearing aids, it is the responsibility of the speech therapist to determine the child's ability to develop meaningful speech, to teach the child to "speech read," to determine the appropriate role to be played by "total" communication and how the family and other persons in the child's environment may work together in promoting the child's awareness and development of language.

The pediatrician may be asked by the family for

assistance during this period in a number of different ways. Otitis media in a child with moderate to severe sensorineural hearing loss can cause further deterioration in residual hearing, which may be manifested by the child's inattentiveness despite amplification, irritability and other mood changes, or deterioration of speech. Prompt and vigorous medical treatment of otitis media in such children can be an important adjunct to the acquisition of language skills. Consultation with an otolaryngologist regarding the management of a deaf child with chronic serous otitis media can also be helpful. Myringotomy and insertion of tympanostomy tubes have often proved to be of great value in management of the deaf child with mixed hearing loss (combined sensorineural and conductive hearing loss). The pediatrician should also have some familiarity with the operation of hearing aids, including insuring that the child has a well fitted ear mold that does not traumatize the skin of the external ear or ear canal, that the hearing aid is equipped with fresh batteries and has no broken parts or cords, and that the family or staff of the habilitation agency encourages the child to wear it.

EDUCATIONAL SERVICES

The pediatrician may find himself in the middle of the long standing controversy between persons favoring the "oral" approach to habilitation and those in favor of the "manual" approach. This dispute has polarized many in the field of special education for the deaf and represents two reasonable points of view. The "oralists" hold that deaf children will have to compete as adults in a hearing world where the ability to communicate effectively through speech is crucial to any successful endeavor and any reliance on the use of manual communication has generally been frowned upon. Those favoring "manual" communication (use of the manual alphabet and sign language) point out that deaf children may develop a broader appreciation for "language" at an earlier age through manual communication, and that this enables them to communicate with other deaf persons, a group with whom they may well identify themselves to a considerable degree through much of their lives.

The outcome of this controversy appears to be a compromise between the two, which has been referred to as "total communication," a system that relies on speech reading, manual communication, and the development of oral speech when possible. "Total communication" has proven to be very helpful in multihandicapped deaf children who, despite having only a moderate hearing loss, have been unable to develop speech because of a central language disorder. The addition of manual communication to the habilitation program of these children has often led to the use of more spontaneous "language," a lessening of frustration because of the inability to communicate, and frequently a facilitation of the development of oral speech itself. The "total communication" approach is widely practiced now in classes for deaf children in regular schools, in special residential schools for the deaf, and in programs of higher education.

The advent of Public Law 94–142, the Education for All Handicapped Children Act, has made a major impact on the management of deaf children in our society. Its emphasis on insuring that education takes place in the "least restrictive environment" is an important step in guaranteeing civil rights for deaf persons and has brought about fundamental changes in the character of educational services for deaf children. Communities are now required to provide free public education for all deaf children. This has led to one noted court case in which a local school board was requested to provide an interpreter for the deaf for a deaf child in a class of children with normal hearing so that the deaf child could keep up with her hearing peers. In the past such "bright" deaf children were educated in special schools for the deaf, often on a residential basis. Parents of these children now have a legal basis on which to have their children educated at the local neighborhood school, closer to their homes and families.

Increasingly the role of special state and private schools for the deaf is becoming one of providing educational services for children with more complex needs—those who have other handicapping conditions in addition to deafness, including mental retardation, neuromuscular disorders such as cerebral palsy, and serious behavioral impairments. These special schools in turn have been able to provide a new level of service to these multihandicapped children through an interdisciplinary team, which often includes physical and occupational therapists as well as nurses and pediatricians. Now many of these children can be "mainstreamed" into special classes in the regular public schools. It can be seen that a new set of dynamic relationships has been established among the public schools, special schools for the deaf, residential facilities for mentally retarded persons, and other agencies, which have been oriented to the special needs of individual deaf children, rather than the needs and characteristics of their own programs, by PL 94–142 and similar legislative and judicial mandates. It remains to be seen what effect recent conservative trends and a reduction in federal funds for education and social programs will have on the revolutionary changes in habilitation services for deaf children that were so prominent in the 1960's and 1970's.

THE ROLE OF THE PEDIATRICIAN

The pediatrician as a physician concerned with promoting the development of all children may play a number of roles in regard to services for deaf children, whether the pediatrician is in general private practice, has a subspecialty interest (such as pediatric neurology), has a fulltime position in a teaching hospital, or has an administrative position in government.

As a practitioner in general practice, the pediatrician is often the first to identify the child as being deaf and to refer him for evaluation and diagnosis. He follows the child through various developmental phases over a prolonged period of time. It is the pediatrician who intervenes with antibiotics to treat otitis media and gives mumps vaccine to prevent mumps induced hearing loss. He recognizes that deaf children may have the same predilection to develop learning disabilities as children with normal hearing and that they may deserve a trial of a stimulant drug (amphetamine or similar agent) as an adjunct to management when there is an attentional deficit disorder. In the management of the brain injured, multihandicapped deaf child, the pediatrician may observe gratifying improvement in learning in the classroom following appropriate doses of thioridazine or similar compounds. Such drugs may actually make the highly distractible child available for learning from his teacher for the first time. Since such drugs also may be associated with serious side effects, open communication must be maintained between physician, school, and family to insure that such side effects are recognized. Drug holidays on weekends or vacations may be useful in the management of children who are receiving such medication.

Of course, when the deaf child grows past adolescence, it may no longer be appropriate for the pediatrician to remain the primary care physician. At this time it would be of great value for the pediatrician to summarize the child's medical "experience" for the internist or family practitioner so that some continuity of health services can be maintained.

The pediatrician who is based primarily at a teaching hospital has a responsibility to offer training related to the management of deaf, language impaired children to young physicians in a residency training program as well as to attending pediatricians in private practice. This can be arranged through conferences at the hospital, in speech and hearing clinics, and at special education programs in the community. This pediatrician, or a counterpart in community practice, is often a consultant to special schools for deaf children or regular schools where deaf children may be integrated in regular or special classes. These pediatricians have a special responsibility to offer inservice training to school nurses, special education teachers, and administrators regarding the medically related issues affecting deaf children, such as genetics, management of infectious diseases, behavioral counseling, and so forth. Fortunately many school systems are now moving toward fuller utilization of pediatricians' expertise in such areas.

Pediatricians with administrative or clinical positions in government agencies have special opportunities to serve seriously impaired deaf children. For example, pediatricians in state departments of health can develop and promote programs that prevent deafness in children, such as childhood immunizations, special care for high risk newborns, and genetic counseling. They may also participate in the development of new systems of health care, such as regionalized health programs for the evaluation and treatment of children with serious hearing impairment. In their special relationships with state legislators and federal agencies, they can advocate the support of services for the prevention and treatment of deafness in children.

PHILIP R. ZIRING

REFERENCES

Fraser, G. R.: Profound childhood deafness. J. Med. Genet., 1:118–154, 1964.

Gerber, S. E., and Mencher, G. T.: Early Diagnosis of Hearing Loss. New York, Grune & Stratton, Inc., 1978.

Jaffe, R. F.: Hearing Loss in Children. Baltimore, University Park Press, 1977.

Konigsmark, B. W.: Hereditary deafness in man. N. Engl. J. Med., 281:713–720, 1969.

Kukla, D., and Thomas, T.: Assessment of Auditory Functioning of Deaf/Blind/Multihandicapped Children. Dallas, South Central Regional Center for Deaf-Blind Children, 1978.

Miller, M., and Rabinowitz, M.: Audiological problems associated with pre-natal rubella. Int. Audiol. 8:90–98, 1969.

Myklebust, H.: The Psychology of Deafness. New York, Grune & Stratton, Inc., 1960.

Northern, J., and Downs, M.: Hearing in Children. Baltimore, The Williams & Wilkins Co., 1975.

Rubin, M.: Hearing Aids: Current Developments and Concepts. Baltimore, University Park Press, 1976.

Ruben, R. J., and Rozycki, D. L.: Clinical aspects of genetic deafness. Ann. Oto. Rhinol. Laryngol., 80:255, 1971.

Sanders, D.: Aural Rehabilitation. Englewood Cliffs, New Jersey, Prentice-Hall, Inc., 1971.

Schein, J. D., and Delk, M. T.: The deaf population of the United States. Silver Springs, Maryland, National Association for the Deaf, 1974.

Ziring, P. R.: Congenital rubella: the teenage years. Pediatr. Ann., 6:11, 1977.

Ziring, P. R., Florman, E. E., and Cooper, L. Z.: The diagnosis of rubella. Pediatr. Clin. N. Am., 18:87–91, 1971.

39C Visual Impairment and Blindness*

In the United States it is estimated that the prevalence of blindness and serious visual impairment in the pediatric population is approximately 64 children per 100,000 population. Another 100 children per 100,000 have less serious visual impairment. Hence it is quite likely that every pediatrician or family medicine specialist in a primary care setting will have at least several blind children in his practice. The blind child, like any other youngster with a life-long disability, presents special social, educational, and psychological needs, which if identified and met can reduce the degree of handicap associated with the disability.

DEFINITIONS, INCIDENCE, AND PREVALENCE

No common world-wide definition of blindness exists. In fact, in 1970 the World Health Organization (WHO) surveyed its member countries and found 65 different definitions in use. Hence, estimating the incidence and prevalence across regions of the world is difficult. The member countries of WHO have consequently agreed to the following definition promulgated by that organization (always with reference to the better of two eyes):

1. Visual impairment: Snellen acuity no better than 6/18 m (corrected) or visual field no better than 20 degrees.
2. Social blindness: Snellen acuity no better than 6/60 m (corrected) or a visual field no better than 20 degrees.
3. Virtual blindness: Snellen acuity no better than 1/60 m or a visual field of less than 10 degrees.
4. Total blindness: no light perception.

The United States Public Health Service defines blindness as the best corrected acuity in both eyes of 6/60 m or visual fields of less than 20 degrees bilaterally. Using this definition, Goldstein (1980) compiled incidence and prevalence rates for the United States from a number of primary demographic resources. Table 39–4 shows these data

*Preparation of this material was supported in part by USPHS Research Grant EY-01790 from the National Eye Institute.

for 1970 as a function of age group. About 0.5 per cent of all children are significantly visually impaired, and only about 10 per cent of all expected cases of blindness in the total population occur in children and adolescents. Since retrolental fibroplasia is no longer of frequent occurrence, one can expect that there will be a steady decline in blindness prevalence among children.

Table 39–5 lists the usual causes of visual impairment and blindness in persons of all ages, classified by onset and site of the disorder, as reported in England and Wales between 1955 and 1960 (Sorsby, 1966). Table 39–6 shows estimates of the incidence of blindness in children by age and cause, summarized from data gathered by the National Eye Institute in 1969 and 1970 from 16 states participating in a standardized coordinated reporting effort (Kahn and Moorhead, 1973). Table 39–6 indicates that only one of five cases of childhood blindness is reported during the child's first five years of life. The leading causes of childhood blindness are congenital cataract, prenatal retinopathy, and optic nerve disease.

The National Eye Institute has estimated that as of 1970 the leading causes of new cases of blindness in all age groups in the United States were glaucoma (1.5 per 100,000 population), senile cataract (1.7 per 100,000), and retinal diseases other than prenatal disease but including diabetes (3.4 per 100,000). Since continued progress is being made in the prevention of glaucoma and in the treatment of retinal diseases and cataract, we might anticipate that there will be a continued reduction in the incidence of blindness in the United States over the next 20 years.

About 20,000 school age children require braille (or another tactual system) as a reading medium and must be taught as nonseeing children. Most of these children are now receiving and will in the future receive their education in the public educational system as "mainstreamed" students, as most states are reserving their schools for the blind for only a very small number of blind, multiply handicapped children. Hence, for the foreseeable future, significant numbers of blind and visually

Table 39–4. AGE SPECIFIC INCIDENCE AND PREVALENCE RATE FOR BLINDNESS PER 100,000 POPULATION IN THE UNITED STATES FOR 1970*

| Age (Years) | Rate per 100,000 Population | | Estimated Census Total | |
	Incidence	Prevalence	Population	Blind
Under 20	7.1	63.8	76,970,000	49,119
20–64	17.7	206.2	106,176,000	218,930
65 and Over	97.6	995.5	20,066,000	199,751
All ages	21.6	230.2	203,212,000	467,800

*From Goldstein, H.: The reported demography and causes of blindness throughout the world. Adv. Ophthalmol., 40:1–99, 1980.

Table 39–5. CLASSIFICATION OF MAJOR CAUSES OF BLINDNESS*

I. Congenital defects
   A. Globe as a whole
   B. Lens
       Cataract
       Dislocation
   C. Uveal tract
       Malformations
       Congenital syphilis
       Toxoplasmosis
   D. Retina
       Retrolental fibroplasia
       Aplasia
       Retinoblastoma
       Congenital detachment
       Optic atrophy
II. Abiotrophic defects
   A. Cornea
       Dystrophies
       Keratoconus
   B. Choroidal dystrophies
   C. Retina
       Retinitis pigmentosa
       Macular dystrophy
       Genetic detachment
III. Glaucoma
IV. Interstitial keratitis
V. Primary cataract (senile)
VI. Iris and iridocyclitis
VII. Choroiditis
VIII. Myopia
       Choreoretinal atrophy
       Retinal detachment
IX. Other macular lesions
X. Idiopathic retinal detachment
XI. Diabetic retinopathy
XII. Vascular and hypertensive retinopathy
XIII. Optic atrophy other than congenital

*Adapted from Sorsby, A.: The Incidence and Causes of Blindness in England and Wales, 1948–1962. Ministry of Health Reports on Public Health and Medical Subjects, No. 114. London, Her Majesty's Stationary Office, 1966.

impaired children will live in the community and require specialized services.

DEVELOPMENTAL ISSUES RELATING TO VISUAL IMPAIRMENT

The relationship of visual impairment and blindness to child development has been reviewed in depth by Warren (1976). Only some of the key issues will be considered here.

The deprivation of vision has important implications for development, on both theoretical and empirical grounds. It is known from studies of normal child development that visual experience facilitates (and probably underlies) many important concepts of space and form that are in turn important to the development of perception and perhaps even intelligence itself. When a child is forced to interact with the world without vision,

he is doing so without the sensory system most adapted for spatial and shape information gathering. How well these functions can be subsumed by other perceptual modalities such as touch and audition seems to depend on the nature of the task, the adaptability of the child, and the way in which the residual modalities are enriched by experience.

Research in the nineteenth and early twentieth centuries was directed at studying what epistemologists have referred to as sensory compensation, loosely defined as the adaptation of one modality to perform tasks usually accomplished by another, dysfunctional, modality. Little evidence has emerged to demonstrate any consistent changes in fingertip sensitivity, for example, in persons blinded so severely as to necessitate use of the hands as the primary modality for perceiving shape. Some investigators have demonstrated cortical electrophysiological deviations in blind persons, such as diminished occipital and increased somatosensory activity, as measured by the electroencephalogram. However, no specific links have been made between such changes and sensory capacity behaviorally. The bulk of the data relating to the comprehension issue suggests that a considerable amount of perceptual learning can occur with increased usage of a perceptual modality. This learning tends to increase the user's attention to attributes of a stimulus that differentiate one shape from another and improve the information pick-up process in that modality. Most researchers now agree that this learning explanation accounts for observations of compensation.

Recently it has been learned that interactions between infant and caregiver during the immedi-

Table 39–6. INCIDENCE OF BLINDNESS IN CHILDREN BY AGE AND CAUSE (1969–1970)*

| | Incidence (Cases/100,000) | | |
| | All | Age of Children | |
Cause	Ages	Under 5	5–19
All causes	5328	147	585
Glaucoma	544	0	0
Cataract, total	788	28	80
Prenatal	145	28	72
Other (senile)	633	0	8
Retinal disease, total	1585	13	104
Prenatal	302	9	78
Diabetic	464	0	1
Other	819	4	25
Retrolental fibroplasia	52	11	34
Myopia	118	2	35
Corneal or scleral disease	136	2	9
Uveitis	154	1	13
Optic nerve disease	352	19	84
Multiple affections	504	0	3
Other	443	48	163
Unknown	662	23	60

*Adapted from Kalen, H., and Moorhead, H.: Statistics on Blindness in the Model Reporting Area. Washington, D.C., National Eye Institute, 1973.

ate postpartum period and thereafter form the basis for attachment, a very important cornerstone of social development. A significant element of this interaction is visually mediated via face to face contacts. The consequences of a lack of visual mediation in the case of the blind infant are unclear, but recent data indicate substantial delays, which if not corrected may have as yet unknown effects on social and emotional development.

In much of the literature relating to child development, the concept of early experience is set apart as a special kind of learning, a building block for later development. A strong component of this concept is the notion of so-called "critical periods" for the emergence of certain behavior patterns if specific eliciting experiences are made available to the infant. Many critical periods occur during the first days, weeks, and months of life. If a baby has vision available during these periods and later loses it, will he be better off developmentally than a congenitally totally blind infant?

Children who are blinded after one to two years of age are generally referred to as being adventitiously blinded. A considerable period of the early development of these children is accomplished with vision available. A sizable number of studies have compared the developmental course in such children with that of those who are born blind. In general, adventitiously blinded children perform most tasks of spatial and form perception more accurately than congenitally blind children and often do not differ in performance from blindfolded sighted children (Hatwell, 1978; Worchel, 1951). The finding is not universal, and the outcome of specific experiments depends upon certain task and subject variables, such as the amount of specific visual experience with the stimuli being used and the particular attribute of shape or space being perceived. Worchel (1951) has argued that early visual experience permits concept learning in the visual modality early in life that facilitates "visualization" of tactual and auditory material once vision is no longer available. The "visualization" hypothesis has obvious heuristic appeal but has never been directly tested and in fact has been challenged by some as an explanatory construct (Juurma, 1965). Particularly important to explaining the effects of early visual experience are the data indicating negative instances: sometimes greater early tactile experience outweighs early visual experience and sometimes it makes no difference. Ultimately the explanation probably relates to the efficiency of the principal perceptual system being used by the subjects to prehend the stimulus or perform the task (Davidson, 1976). Hence there is a need to compare the capacities of perceptual systems.

HAPTIC PERCEPTION COMPARED WITH VISUAL PERCEPTION (PERCEPTION OF SHAPE AND SPACE)

The term "haptic" refers to the process of actively exploring an object or perceptual space with the hands. The haptic perceptual system combines input from tactile sensations and kinesthesis. Many experts on haptic perception have emphasized the importance of active exploratory movements in haptic information gathering. The active movement enables the individual to identify distinctive characteristics of the stimulus, such as relationships between parts of a form, that are not perceptible without movement. Most theorists agree that exploratory perceptual movement is an important feature of the information gathering process, and considerable research has been reported on both eye and hand movement functions.

Direct comparisons of eye and hand have not been reported. However, some perceptual consequences of anatomical and operating-characteristic differences between the two systems are known.

Simultaneous Versus Successive Pick-up of Information. The eye is neuroanatomically developed to gather information from disparate points in a visual array simultaneously (e.g., without scanning). Even when movement is made necessary by a large visual display, the eye is capable of prehending a larger space per unit of time than are the hands and fingers. The hand, because of this limitation, is required to gather information from disparate points in a haptic array successively. The perceptual process is much slower by hand than by eye. Also successive prehension of different parts of a shape by the hand places an extra burden on short term haptic memory to integrate the parts into a whole percept.

Developmental Differences. Only scanty data are available that compare developmental patterns of vision and haptic perception, and there is some controversy about this in the existing literature. It does appear that seeing persons develop visual perceptual proficiency at a faster rate than nonseeing persons develop haptic proficiency, a finding of considerable importance to understanding developmental delays in blind children. Also there is a hint that the haptic system may develop differently when vision is available compared with when it is not. The hand is used more for orienting stimuli for the eye when both are present but is used for active primary information gathering when the eye is absent. Of some theoretical and practical significance is the finding by some investigators that perceptual development probably proceeds in a similar sequence and follows the same rules of information processing for the eye and the hand; hence enrichment strategies for visual learners (such as educationally disadvantaged or developmentally delayed seeing children) and for haptic learners may be qualitatively similar.

AUDITORY PERCEPTION COMPARED WITH VISUAL PERCEPTION (LOCALIZATION AND DISTANCE PERCEPTION)

Haptic perception is adapted only for proximal stimulus prehension. In the absence of vision, the auditory system is the only remaining means a blind person has of gathering information about distal space. However, the ideal stimulus for specifying spatial information is light, not sound. Few distinctive aspects of space, such as distance, depth, symmetry, and motion, are detectable ordinarily by auditory cues alone. The blind person must learn to use the auditory cues available to facilitate the localization and tracking of spatial attributes and, when none are available, to supply his own.

One of the best examples of an adequate substitution of auditory for visual cues in determining spatial properties is the phenomenon of echo location. In avoiding collisions with obstacles during locomotion, blind persons skilled in mobility utilize the reflections of self-produced sounds to judge distances of objects in their paths. The skill, thoroughly investigated by Dallenbach and his colleagues, is similar to echo location behavior in bats and is learned through attention to differential loudness of the reflected sounds. It can be acquired surprisingly quickly by naive blindfolded sighted subjects.

Unfortunately not all spatial features can be detected by audition or touch, and this situation may be responsible for some developmental delays in early infancy and childhood for blind youngsters.

THE DEVELOPMENT OF THE BLIND CHILD

INFANCY

A number of comparisons have been made of various developmental milestones in blind and sighted infants. In general, significant delays have been found in blind infants in the development of certain gross motor behavior patterns related to the development of locomotion, in prehension skills, and in the development of attachment behavior. The blind infant eventually develops all these functions during the second and third years of life. However, many links probably exist between such early behavior and the normal development of important cognitive, perceptual, and personality functions, suggesting that the blind infant ordinarily may be considered at risk for later developmental delays without adequate parental education and specific intervention in infancy.

Gross Motor Development

Table 39–7 shows a comparison of gross motor milestones in blind and sighted infants as reported by Fraiberg (1971) and her colleagues. There is a uniform delay of gross motor functions in this series of 10 blind infants. The greatest delays occur in mobility and locomotion related behavior in-

Table 39–7. GROSS MOTOR ITEMS AND AGE ACHIEVED BY BLIND (CHILD DEVELOPMENT PROJECT) AND SIGHTED (Bayley)*

Item	Age Range†		Median Age		Difference in Median Age
	Sighted	Blind	Sighted	Blind	
Elevates self by arms, prone[a]	0.7–5.0	4.5–9.5	2.1	8.75	6.65
Sits alone momentarily	4.0–8.0	5.0–8.5	5.3	6.75	1.45
Rolls from back to stomach[a]	4.0–10.0	4.5–9.5	6.4	7.25	.85
Sits alone steadily	5.0–9.0	6.5–9.5	6.6	8.00	1.40
Raises self to sitting position[a]	6.0–11.0	9.5–15.5	8.3	11.00	2.70
Stands up by furniture (pulls up to stand)[a]	6.0–12.0	9.5–15.0	8.6	13.00	4.40
Stepping movements (walks hands held)[ab]	6.0–12.0	8.0–11.5	8.8	10.75	1.95
Stands alone[a]	9.0–16.0	9.0–15.5	11.0	13.00	2.00
Walks alone, three steps[a]	9.0–17.0	11.5–19.0	11.7	15.25	3.55
[Walks alone, across room][a]	[11.3–14.3]	12.0–20.5	[12.11]	19.25	7.15

*Adapted from Adelson, E., and Fraiberg, S.: Insights from the Blind. New York, Basic Books, Inc., 1977, p. 204.
†All ages given in months. Ages rounded to nearest half month. Three cases corrected for three months' prematurity. Age range includes 5–95% of Bayley sample, 25–90% of Denver sample, and 10–90% of Child Development Project sample.
[a]One child had not achieved by two years.
[b]Not observed for one child prior to walking alone.
[]: Item from Denver Developmental Screening Test.

cluding precrawling, sitting, pulling to standing, and walking. Ninety per cent of these blind babies were substantially delayed in the onset of walking and independent walking, which did not appear until a median age of almost 20 months.

Prehension of objects by the hands is also delayed in the blind infant. This response appears in its completed form by about age six months in nonblinded infants; major milestones in its achievement include bringing the hands to the midline, hand in hand play, grasping and transfer from hand to hand, gross swiping, and finally reaching accurately to a target with coordinated prehension. The essential ingredient in the normal emergence of this sequence is coordination of eye and hand, an impossible task for the blind child. As a result, such a child generally does not show a completed prehension response until about nine months of age, and some year-old blind babies still do not show the response. Of considerable importance is the potential effect of these prehension delays on the development of haptic exploratory skills. The blind infant paradoxically may show poorer haptic skill than sighted infants.

Fraiberg (1971) and her associates attribute these gross motor and guidance response delays to two factors: the forced substitution of auditory for visual cues to identify distal shapes and spatial properties and the corresponding difficulties this substitution presents for establishing the "world out there" (to borrow Fraiberg's term), and the dependency on an adequately developed object concept prior to the onset of effective use of auditory cues by the blind infant to identify the object's location in space. Touch alone or touch and audition together are effective substitutes for vision in establishing the concept that an object in space has permanence despite its temporary disappearance from view. But audition alone is not. Therefore distal objects not prehensible by nonmobile blind infants probably are not recognized as features of a permanent world until much later

in development. Sighted babies, on the other hand, do not depend on object permanence as a precursor to reaching and locomotion, because objects can be located and their movement monitored visually on a continuous basis.

Developmental interventions in infancy that are designed to enhance the development of aurally based object permanence have been successful. Figure 39–3 shows data from one such intervention program designed and implemented by Fraiberg in her well known Michigan Child Development Project. The 10 blind babies in Fraiberg's apparently successful program are compared with another series of 66 blind infants who did not receive interventions, reported by Norris et al. (1957). The essence of the intervention program was the introduction of paired auditory-tactual cues to sustain the infant's interest in and contact with the distal world, and encouragement of physical activity. The data show a substantial acceleration of gross motor functions in Fraiberg's series, but all milestones are still delayed in relation to sighted babies. Unfortunately because no follow-up data for Fraiberg's babies are available, one cannot say anything about durability of the intervention effects.

The Development of Attachment

With sighted infants the infant and caregiver develop a pattern of reciprocal reactions that lead to the emergence of attachment. The milestones include smiling at the familiar face (six months), stranger avoidance (seven to 15 months), and distress at separation from the caregiver (six to nine months). The blind infant, however, shows a different pattern. Smiling (developed as a response to a familiar voice) occurs only inconsistently, even in 12 month old blind infants. Exploration of the caregiver's face and smiling in response to familiar tactile-kinesthetic handling occur, however, at the same time as does smiling

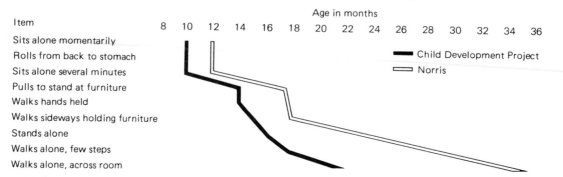

Figure 39–3. Data from Fraiberg's intervention. The subjects were 10 blind infants (dark line) who were compared to Norris' nonintervened sample of 66 infants (light line). (From Adelson, E., and Fraiberg, S.: Insights from the Blind. New York, Basic Books, Inc., 1977, p. 210.)

at a caregiver's face in the sighted child. Anxiety toward strangers appears at one year of age in the blind child, but a reaction to separation is not apparent until 11 to 20 months of age (a six month delay compared to the typical sighted infant). These delays are no doubt related to the inability of the blind infant and caregiver to "communicate" by visual face to face contact. The typical blind infant may make its needs for attachment and comforting known through a sign system of hand gestures, which include "tactile seeking," as described in the following excerpt from Fraiberg's protocols:

> Toni is seven months old. Her mother (a very experienced mother, remember, with five older children) tells us, "She's not really interested in her toys."
> We assemble a group of Toni's crib toys, stuffed animals and dolls, and invite the mother to present them to Toni, one by one. As each of the toys is placed in her hands, Toni's face is immobile. She gives the impression of "staring off into remote space." Naturally the totally blind child does not orient his face toward the toy in his hands. Since visual inspection is the sign that we read as "interest," and averted eyes and staring are read as the sign of "uninterest," Toni "looks bored."
> Now we watch Toni's hands. While her face "looks bored," her fingers scan each of the toys. One stuffed doll is dropped after brief manual scanning. A second doll is scanned, brought to the mouth, tongued, mouthed, removed, scanned again. Now we remove doll number 2 and place doll number 1 in Toni's hand. A quick scanning of fingers and she drops it again. She makes fretful sounds, eyes staring off into space. We return doll number 2 to her hands. She quiets instantly, clutches it, brings it to her mouth, and explores its contours (1977, p. 104).

It is not always apparent to caregivers that their blind baby is communicating through tactile seeking, yet this link may be important in establishing attachment behavior.

Language Development

The literature relating to the development of receptive and expressive language in blind infants and young children indicates a parity with sighted children. There are some exceptions, however. Infants blinded as a result of retrolental fibroplasia may show a more limited vocabulary development than other blind children. The important variable here may be an increased anticipation by the caregiver of the congenitally blinded child's needs, thus decreasing the demand for language production. Word meaning also may be less well established in many young blind children, regardless of the etiology of blindness, signaled by the occurrence of idiosyncratic "verbalisms." Such usage may stem from the lack of a sensory referent or analogue for words dependent upon visual imagery for meaning. Finally, minor articulation problems may develop in the blind infant who does not have the benefit of observing facial and oral positioning correlated with specific utterances.

Intervention

As pointed out earlier, most of the delays in early sensorimotor development in the blind infant occur as a result of a deprivation of visual information. Substitution systems can be effectively taught to the infant and the expected defects can be reduced. Fraiberg (1971) has developed a comprehensive intervention protocol. Although limited evidence is available to document its effectiveness, the data relating to outcome are encouraging and the protocol is in wide use in blind infant programs in the United States. Fraiberg (1977) has also published a popular text readable by parents and other caregivers that highlights areas of concern and suggests some remediation techniques. Of considerable importance to the primary care provider is the counseling of parents regarding special differences between blind and sighted infants, such as the use of a tactile sign system discussed earlier. Without this basic information, delays in social, cognitive, and perceptual development may be exacerbated.

CHILDHOOD

The period between age one and six years in the development of the blind child has rarely been studied. Therefore we have little in the way of a map to help us understand the consequences of early infancy on later development. A considerable amount of literature is available on social, cognitive, and perceptual development in the school aged blinded youngster (Warren, 1976). Only selected topics will be reviewed.

Cognitive and Perception Development

In general, patterns of cognitive and perceptional development in otherwise unaffected blind and nonblind children run a parallel course for auditory-based functions. Blinded children appear to have somewhat better auditory attention, but tasks involving abstract reasoning with either tactual or auditory verbal material are equivalently performed. If any discrepancies do exist, they usually can be explained by the sometimes limited experience of blind children with stimulus materials being used.

Tests of abilities, including IQ tests as well as functional batteries, have been used to estimate the pattern of global intellectual development in blind children. Several conclusions are of importance. First, as will be seen in a later section, few IQ tests are available with normative data relating to blind individuals, or that are "culture fair." Hence, comparing blind and sighted persons on the basis of IQ is very difficult and probably inappropriate. Second, certain etiologies for blindness lead to either increased or decreased overall cognitive potential. For example, a sizable number of studies have shown that children blinded as a result of retinoblastoma may have a significantly higher level of ability than children blinded by other conditions. In several studies the mean Hayes-Binet IQ for patients who were bilaterally enucleated because of retinoblastoma was about 120, or 21 points above that of nonretinoblastoma control group patients as well as that of a unilateral retinoblastoma control group. The advantage probably reflects a combination of genetic and environmental factors as yet only partially understood. A second example is the incidence of lowered IQ scores in children blinded by retrolental fibroplasia. In several series of such children, as many as 35 per cent showed neurological abnormalities, and more than 40 per cent had IQ scores below the average range. Some controlled series have also indicated that the probable reason for lowered IQ is not the retrolental fibroplasia per se but possibly the neurological sequelae of prematurity.

The most significant deficits in cognition and perception in blind children occur in tasks that are aided by vision or visual experience. The best studied example of this phenomenon is the development of conservation. Conservation refers to a cognitive concept or rule of perceptual transformation: a physical property (such as the amount of solid or the volume) remains invariant even though its perceptible shape or appearance is transformed. A block of clay loses no mass if it is rolled into a wiener shape even though its physical dimensions have been altered and it looks different. The seeing child makes this judgment correctly at about eight to nine years of age, coincident with the onset of the Piagetian cognitive stage of concrete operational thought. The blind child in repeated studies has been shown to lag behind in the onset of this response by two to three years. The reason for the lag is unclear, but it probably lies in the blind child's lack of vision, which would permit experience in the continuous monitoring of perceptual transformations.

Other concepts in the blind child are delayed, such as seriation and certain classification abilities that are significantly dependent on visual experience. The delays are much less significant, or nonexistent, in adventitiously blinded youngsters, reinforcing the experiential hypothesis. It is likely that these deficits are responsive to remediation, but no data exist to support this contention. Of some concern is the possible long term negative consequence for spatial learning in blind children caused by such early delays in concept formation, but again few data are yet available.

Functions of Haptic Exploratory Search

It is important to examine how blind children and adolescents employ their hands as instruments for information pick-up and processing. We know that regardless of the presence or absence of vision, experience influences the selection of styles in the haptic examination of an object, and the particular style used by a perceiver can influence not only what is learned about the shape but also what is stored and remembered for later use.

Sighted preschoolers tend to be very passive when asked to use their hands to explore a shape, pressing down or squeezing it and paying little attention to critical details that distinguish the shape from others. Between about five and eight years of age, sighted children increase their active exploration, using styles that globally search the shape in a comparative fashion. By age 12 years, haptic search ability is quite well developed.

Despite the obvious importance of the question, we know almost nothing of the young blind child's development of haptic search. What little we do know suggests no particular advantage to the blind five to eight year old over sighted comparison subjects in the use of efficient search. In fact, some of the author's studies have led to the conclusion that it may take the blind child longer than the sighted child to develop coordinated haptic exploration. A hypothesis to explain the findings implicates the carryover from preschool years of some deficits in organizing spatial dimensions in the absence of vision. By age 10 or 12, however, the blind child shows haptic search ability that appears to be more active and more efficient than sighted children's efforts at isolating features and discriminating different shapes. The proficient blind haptic explorer also tends to remember what was felt more accurately than the sighted child.

Because of the importance of exploratory search by hand to the blind child, monitoring the child's opportunities to use the hands and the relative progress of search development is a key component of any educational assessment.

Some Educational Issues

The primary care provider must be familiar with several special needs of the blind child in his education in order to be an effective advocate.

Reading. Many legally blind children can read normal newspaper print with some additional magnification, but others require large print texts and newsprint. Some but not all reading materials are available in large print.

About 50,000 persons in the United States have visual impairment that is severe enough to require an alternative to print in order to read. About

20,000 are school age children, according to the American Printing House for the Blind. Several alternatives are available. The most widely used system, of course, is braille. Despite the universality of braille (it has been in use since the middle of the nineteenth century), the relative ease in obtaining braille materials, and the generally few restrictions placed on the reader when using it (books are portable and can be read anywhere), braille suffers from some serious systematic deficiencies. First, the reading rate for even the best braille readers is about one-third as fast as that in the visual reading of print. Second, not all materials are available in braille and transcription. The American Printing House uses computerized transcription, but there is a delay in obtaining this service. Third, the space required for storage of braille books is considerable (one 250 page print book may require 12 or more volumes of braille pages), preventing most individuals from maintaining large libraries at home. Finally, there is a very poor understanding of the basic perceptual and linguistic mechanisms underlying the acquisition of braille reading skill. As a result the teaching of braille reading is often idiosyncratic from school to school and teacher to teacher, and not all readers ever obtain maximum efficiency.

Alternatives to braille have recently been developed. The talking book program has existed in most locations for some time and is available in most libraries. Recent research on rate controlled speech (redundant aspects of speech are subtracted from taped recordings of voice, accelerating the speed of some speech while losing no intelligibility) suggests that this medium may be viable for mass use in the future.

The Stanford Research Institute has developed a device called OPTACON (optical to tactile converter), which receives optical input from a standard printed text and converts it to a tactual dot figuration, which can be distinguished by the finger. The device is portable but expensive (about $1000). Also, since it senses only one character at a time, it leads to very slow reading rates (about 35 wpm). It does not do well with print that is not clear and sharp; hence one can read a newspaper only with difficulty. This device is being studied and may hold some promise for wider use if its cost can be reduced and the reading rate accelerated.

Grunwald has developed a reading machine that reproduces braille from material coded on tape and passes the resulting signals across the reader's fingertips passively. This device is not in commercial production. It is inflexible insofar as it does not allow the finger to search a text; the finger can "read" in only one direction, one cell at a time, as the materials come off the tape.

Print-to-speech converters are now available. The best known is the Kurzweil device, present in libraries in many larger metropolitan areas. This device accepts a print format and "reads aloud" for the user. The cost of the device is about $20,000; it requires a large area for storage and is relatively inaccessible.

Despite new developments, it is very likely that braille will continue as the principal medium for nonvisual reading for some time to come.

Mobility. A key milestone leading to achieving independence by the blind child and adolescent is the acquisition of orientation and mobility skills. Although guide dogs are in wide use among adult blind persons, most blind school aged youngsters utilize the long cane as an aide.

Orientation and mobility instruction can be started in most public school settings very early in the child's educational career and is provided in most states through cooperative arrangements between schools and agencies for blind persons.

Alternatives to the conventional mobility aide are being developed and will probably be in general use within the next 10 to 20 years. The most promising such device is a computer, which converts visual images, sensed by a microcomponent video camera mounted in a pair of glasses, to tactual impressions on the back. More recently a system has been developed in which the input from the camera is transmitted directly to the central nervous system by means of optic nerve electrode implants.

Residential Versus Day School. Most states operate residential schools for the blind. Prior to 1974 the majority of significantly visually handicapped children were educated in these facilities. In that year the United States Congress passed PL 94-142 guaranteeing an appropriate public education for all handicapped persons. Along with the general movement toward "mainstreaming" handicapped children in public educational facilities, which occurred after 1974, more and more blind children have been returned to their local districts for their schooling. At present, few schools for the blind accept children whose handicaps are limited to only blindness. On the contrary, the state school populations are now made up of blind, multiply handicapped children. Most regional deaf-blind educational programs are located at state schools, and in a number of states the residential units have been refurbished and reclassified as intermediate care facilities for mentally retarded persons.

The public educational sector has been only somewhat successful in displacing the specialty programs once offered at the state schools. Braille and orientation and mobility instruction are usually provided by itinerant teachers; few districts have self-contained classes for visually impaired children. In many places accessibility to barrier-free facilities is a problem. Nearly all visually impaired children are provided for through the local district's Committee or Commission for the Blind (or Handicapped), and parents often need help in seeing that their child receives the proper support for an appropriate education.

SOME PERSPECTIVES ON THE BLIND ADULT

Long term planning for children requires a look to the adult community into which they will become assimilated. Despite the existence of educational and vocational programs for blind persons in the United States since the early nineteenth century, the blind adult, like other developmentally disabled persons, has not been fully accepted by society. In fact, blindness is a physical handicap easily detectable by anyone, often leading to stigmatization. Few efforts have been made nationally or locally to make the public environment barrier free for visually impaired persons (such as providing braille numbering of elevator control buttons or braille identification of rest rooms). Only one-fourth of the adult blind population is involved in agencies serving the blind; the remaining three-fourths are screened out because they are unemployable or multiply handicapped (Scott, 1969). Employment opportunities for seriously visually impaired adults are limited, and considerable discrimination is still prevalent.

On the other hand, things have improved dramatically over the past 15 years and continue to improve with advances in technologies for teaching reading and developing visual prostheses. Also early intervention with seriously visually impaired children should reduce the number of adult blind persons with significant handicaps.

DEVELOPMENTAL AND BEHAVIORAL DISORDERS AND BLINDNESS

EMOTIONAL ADJUSTMENT

The work on emotional adjustment of blind children is scanty. Overall the data suggest that the blind child is more likely to have adjustment problems than the sighted child. As with other adjustment reactions of childhood, there seems to be a link between the degree of the disturbance and the attitudes of the parents toward the child and the disability condition. There is also a hint that partially sighted children may have more difficulty in adjusting to visual impairment than totally blind children, although the data are insufficient to date.

SIGNIFICANT EMOTIONAL DISTURBANCES

Several types of severe behavioral and emotional disturbances occur with a higher incidence in blind than in sighted populations.

Autism and Retrolental Fibroplasia. At least three significant series of patients with retrolental fibroplasia have been reported in which about 15 per cent showed autistic behavior, including withdrawal, noncommunication, sterotypy, and self-injurious activity. The origin of this correlation is unknown, but some interplay between central nervous system damage, parent-child relationship, and sensory deprivation is suspected. The retrolental fibroplasia per se may have less to do with the behavioral problems than originally thought. Moreover, serious emotional disorders can result with nonadaptive confluence of these three factors in other etiologies of blindness, such as retinoblastoma and cataracts. Early diagnosis and intensive treatment are keys to anything other than a guarded prognosis for these children.

Sterotypy. In many blind children there is a relatively high prevalence of repetitive head and body, limb, and hand movements, actions sometimes referred to collectively as "blindisms." Such socially inappropriate behavior also occurs in other disabled and handicapped children but usually only in those who are the most severely mentally retarded; therefore its occurrence in the blind child is of some concern. The patterns of sterotypy occur in blind infants and continue through childhood and adolescence unless treated. Their origin is unclear, but most of the literature points to a compensatory response to increased sensory stimulation. There is no indication that blindisms are necessary to the overall functioning of the blind child; these behavior patterns respond favorably to straightforward operant behavior therapy and should be treated.

BLINDNESS AND MENTAL RETARDATION

Between 3 and 15 per cent of the blind population are thought to function in the range of mild mental retardation or lower. As noted earlier, several of the etiologies for blindness involve a higher incidence of mental deficiency than others, with special reference to retrolental fibroplasia and deaf-blindness due to rubella. Other reasons for this higher than usual prevalence may include the much more frequent institutionalization in the past of young blind children whose IQ's were low, but not low enough to have led to institutionalization if they were not blind. Also IQ tests may selectively discriminate against visually impaired children.

Few studies exist regarding the capabilities of the blind, mentally retarded client. Most data come from studies of the deaf-blind population and therefore represent instances of the most severe multiple handicaps. There may be qualitative differences between children who are blind and mentally retarded and those who have one disorder or the other. For example, blind children who are also mentally retarded seem more capable of efficient haptic search of objects than sighted, mentally retarded control subjects, thus reflecting the effects of added haptic experience. In general, however, the blind, mentally retarded person is probably far more handicapped than someone with either one or the other disorder, and is at high risk for severe educational and vocational impairment unless specialized services are provided.

DEVELOPMENTAL EVALUATION OF THE VISUALLY IMPAIRED CHILD

Assessment of the child's developmental status is of great importance in proper educational and habilitative planning. Like most developmental evaluations, that of the blind child is best done by an interdisciplinary team.

Since few causes of visual impairment result in total loss of vision, most patients have some useful vision. Assessment of functional vision in the infant, however, is difficult. Obviously early and careful assessment can lead to maximization of early partial sight. In addition to ophthalmologic studies, a number of visually guided behavior patterns are useful in estimating infant visual status. For example, visually guided reaching and its precursors, such as swiping, cannot occur on time without vision. Infants with partial vision develop these responses normally; hence their presence implies functional vision. Often infants with field defects are identifiable by visually guided responses that occur only in functioning fields, or only when targets are at specific distances. Finally the presence of a gaze disorder in infancy should not be confused with functional blindness.

As has been noted, blind infants are at risk for significant developmental delays; therefore some attempt should be made within the first three to six months of life to obtain a baseline measurement of the infant's gross motor development. The Denver Developmental Screening Test or other similar developmental checklists can be useful for in-office screening, but physical therapy and occupational therapy consultations provide more information to the parents in regard to intervention. Parents should be counseled about their child's possible developmental delays and what they can do about it. Finally, an assessment of cognitive

and emotional functioning should be obtained within the first year of life from a pediatric psychologist or other developmental specialist. Not every such specialist has training or experience with visually impaired infants and young children; expertise in this area should be carefully considered.

Follow-up with most blind children should be periodic, from preschool years through the early grades, because developmental problems may persist. Most school districts provide specialized teams of allied health and educational professionals who can accomplish the evaluations and help in the planning process for the child. However, few persons on such teams have extensive experience with visually impaired children, and one might seek additional expertise with the help of the local Association for the Blind. This agency also can usually provide assessments of orientation and mobility, braille reading readiness, and other haptic skills, all of significance to the thorough evaluation of blind children.

Psychological and educational testing of blind and visually impaired children often includes instruments standardized on sighted populations. Many professionals have pointed out the inadequacy of such tools for assessing blind children. However, very few tests exist that are appropriate for blind individuals at this age. A good assessment therefore includes multiple measurements of cognitive and educational potential and emotional status, some of which may be standardized on blind populations. One should place little reliance on evaluations that use only one measurement technique.

COMMUNITY RESOURCES FOR VISUALLY IMPAIRED PERSONS

As should be evident from the discussion to this point, the blind child and his family need a variety of nonmedical services, including education, social and recreational, vocational, orientation and mobility, and financial. As is true for almost every developmental disability, the provision of services in these areas has evolved from separate and often noncommunicating sectors. Included in the mix are agencies of federal, state, or local government and private, not-for-profit providers. Some agencies offer comprehensive services in a particular area of specialization, such as state education departments, which operate state schools for the blind. Other agencies provide only specialized service (the Commission for the Blind in some states offers only advocacy assistance). There is overlap at times, and obtaining services can be a confusing and time consuming activity for the family of a blind child. The primary care provider can facilitate the process by assisting the family in obtaining accurate information about services and helping them to contact relevant agency counselors.

As is true of any other developmental disorder, education of blind children between ages five and 21 in all states is the responsibility of the local school district. Many states have extended the age at which public education is first available to below five years, and almost all states have a legal mechanism to provide educational services for preschool handicapped children. Any educational needs of a visually impaired child should be referred to the chairman of the local school district's committee on the handicapped (or its equivalent). Social, recreational, and sometimes special educational and vocational needs of visually impaired children are usually met by either state or local governmental or private agencies, such as the Lighthouse or other local Associations for the Blind. Each state and each community may have very different such agencies, and the primary care provider should not assume that a given agency can meet a given need. These facilities historically have served primarily the adult blind population, but many have programs for children. Most of these agencies also provide braille transcription and orientation and mobility instruction, and some operate recreational programs for adolescents.

Vocational needs, although they are the responsibility of the Office of Vocational Rehabilitation, may also be met in part by the Association for the Blind. Financial aid via the federal Disabled Children's Program, Physically Handicapped Children's Program, Medicaid, and Supplemental Security Income are administered by the state Departments of Social Services and Health.

Advocacy agencies have been developed in many states over the past few years. Operated by state government, they may offer services specifically for visually impaired persons or in general to all handicapped individuals. Advocates are often helpful in assisting a client in obtaining services and can take on responsibilities that otherwise would fall to the primary care provider. They can be particularly helpful when the child has multiple needs or when multiple handicaps exist, requiring transactions with numerous agencies. The advocacy agencies are accessible through the Association for the Blind.

ASSOCIATED ISSUES

Prevention of developmental delay secondary to blindness is still an evolving field. The most effective tool is early habilitation. Few communities have home-based early intervention programs specialized for visually impaired infants and preschoolers. Without such programs, infants and their families may not obtain educational-habilitative services until the child can begin a day program at age two or three years, and valuable time is lost. Before such programs can become a preventive force, the health care sector must become active in the early detection of blindness related developmental delays and interact with the educational sector to facilitate the development of community resources if none exist.

Another important secondary and tertiary preventative mechanism is parent education. The blind infant and young child, as indicated earlier, has special needs, which if not met can lead to developmental disability. The primary care provider must take time to counsel parents about such needs and act as a continuing resource for developmental consultation if necessary. Such advocacy may be needed through the early elementary school years for blind children, and requires more continuous patient contact than would be dictated by well child care schedules.

PHILIP W. DAVIDSON

REFERENCES

Davidson, P. W.: Some functions of active handling: studies with blinded humans. New Outlook for the Blind, 196–202, 1976.

Fraiberg, S.: Intervention in infancy: a program for blind infants. J. Am. Acad. Child Psychiatry, 10:381–405, 1976.

Fraiberg, S.: Insights from the Blind: Comparative Studies of Blind and Sighted Infants. New York, Basic Books, Inc., 1977.

Goldstein, H.: The reported demography and causes of blindness throughout the world. Adv. Ophthalmol, 40:1–99, 1980.

Hatwell, Y.: Form perception and related issues in blind humans. In Held, R., Leibowitz, H., and Teuber, H. L. (Editors): Handbook of Sensory Physiology. (Vol. 8). Perception. New York, Springer-Verlag New York, Inc., 1978.

Juurma, J.: An Analysis of the Components of Orientation Ability and Mental Manipulation of Spatial Relationships. Report 28. Helsinki, Institute of Occupational Health, 1965.

Kahn, H., and Moorhead, W.: Statistics on Blindness in the Model Reporting Area. DHEW Publication (NIH) 73–427. Washington, D.C., National Eye Institute, 1973.

Norris, M., Spaulding, P., and Brodie, F.: Blindness in Children. Chicago, University of Chicago Press, 1957.

Scott, R.: The Making of Blind Men: A Study of Adult Socialization. New York, Russell Sage Foundation, 1969.

Sorsby, A.: The Incidence and Causes of Blindness in England and Wales, 1948–1962. Ministry of Health Reports on Public Health and Medical Subjects, No. 114. London, Her Majesty's Stationery Office, 1966.

Sorsby, A.: Blindness: statistics. In Sorsby, A. (Editor): Modern Ophthalmology. Ed. 2. Philadelphia, J. B. Lippincott Company, 1972, Vol. 4, pp. 1171–1182.

Warren, D. H.: Blindness and Early Childhood Development. New York, American Foundation for the Blind, 1976.

Worchel, P.: Space perception and orientation in the blind. Psychol. Monogr., 65, 1951.

39D Cerebral Palsy

Cerebral palsy is a disorder of movement and posture secondary to a static encephalopathy, with the insult to the brain occurring prenatally, perinatally, or in early childhood. The specific label of "cerebral palsy" is confined to the designation of motor problems of a nonprogressive nature. The detection of a cerebral lesion of a static nature as early as possible can prevent a delay in recognizing either a surgically treatable lesion or a heredofamilial neurodegenerative disease and guards against the psychological devastation resulting from the revision of a diagnosis to a more severe and more progressive disease.

With the changing picture seen in infants with cerebral palsy, it may not be until the preschool age that the type of disorder can be determined. Because of frequent uncertainties about the exact etiology and the poor clinicopathological correlations that exist, the classification of cerebral palsy must be based on the clinical assessment of the type of neuromotor dysfunction (Table 39–8). Because cerebral palsy usually results from diffuse cortical damage, it is not infrequent that the motor disability is of the mixed type, spastic ataxia being the most common.

ETIOLOGY

The causes of cerebral palsy are diverse. Consequently the manifestations of the disease can vary extensively, and the sites involved can be either diffusely spaced over the cortex and brain stem or limited to one specific area. The anatomical findings depend on the timing and the type of insult. Chromosomal aberrations or teratogenic factors acting during the first eight weeks of pregnancy affect embryogenesis and result in gross malformations. Teratogenic insults after the first trimester affect brain maturation. Fetal infections at this stage of fetal development result in destructive lesions. Hypoxic-ischemic incidents may cause microanatomical abnormalities secondary to abnormal patterns of migration of neural crest neurons. Perinatal complications of the hypoxic or ischemic type result in ischemic or hemorrhagic brain infarcts. Premature infants are especially susceptible to these types of cerebral disease. Postnatal causes may be infections, meningoencephalitis, head trauma, and environmental toxins, such as lead and carbon monoxide. The most common cause of cerebral palsy used to be prematurity, but with the advent of modern neonatal intensive care units, full term babies may be proportionately more at risk than premature infants.

If a child has a motor disability secondary to an upper motor neuron lesion, the physician must attempt to clarify the etiology. In the search for the cause of static encephalopathy, the history can prove very helpful, particularly if high risk factors known to be associated with cerebral palsy are identified (Table 39–9). The knowledge that a high risk incident occurred offers only circumstantial evidence of a cause and effect relationship. All that can be deduced if an embryo, fetus, infant, or child is exposed to a cerebral insult is that he has a statistically greater chance of being damaged as compared to those not similarly exposed (Table 39–9). Low Apgar scores at one, five, or 10 minutes

Table 39–8. CLASSIFICATION OF CEREBRAL PALSY

Spastic
Monoparesis
Hemiparesis
Diplegia (legs>arms)
Triplegia
Quadriparesis (arms>legs)
Athetoid
Rigid
Ataxic
Tremor
Atonic
Mixed
Spastic-athetoid
Rigid-spastic
Spastic-ataxic

Table 39–9. HIGH RISK FACTORS

Prenatal
Hyperemesis gravidarum
Toxemia
Teratogenic drugs
Placenta previa
Placenta abruptio
Intrauterine viral or bacterial infection (toxoplasmosis, rubella, cytomegalovirus, herpes, and syphilis)
Chromosomal abnormality
Maternal malnutrition
Positive family history
Perinatal
Prematurity
Breech or face delivery
Intrauterine asphyxia
Asphyxia neonatorum
Low Apgar score—especially at 5, 10, and 20 minutes
Seizures
Respiratory distress syndrome
Hyaline membrane disease
Hyperbilirubinemia
Postnatal
Head trauma
Intracranial infections (encephalitis, meningitis)
Toxic encephalopathies (e.g., lead)
Cerebrovascular accident

after birth are examples of high risk factors. Seizures in the newborn period, especially during the first 48 hours of life, are exceptionally "high risk" factors.

The findings on physical examination contribute to the determination of the etiology and diagnosis. The presence of many congenital stigmata would indicate a maldevelopment of the brain. A specific congenital syndrome might be obvious from the clinical findings. A chromosome karyotype would aid in confirming the presence of a maldevelopment and might establish an etiology. Evidence of chorioretinitis on examination of the eye grounds would indicate an intrauterine infection as the cause.

Progressive encephalopathy must be considered if any one of the following is evident by history or by examination: a family history of a neurodegenerative disease, loss of any previously achieved milestones, macrocephaly, papilledema or optic atrophy, pigmentary degeneration of the retina, macular degeneration, hepatomegaly, an unusual smell to the skin or urine, delayed eruption of primary teeth, and chronic diarrhea or constipation.

Absence of these signs, however, does not rule out an advancing lesion. For example, not all progressive encephalopathies present initially as a regression in development. If the pathological process is slow in evolving, a delay in the achievement of milestones will be evident in the presenting complaint. It may not be until several years later that a loss of a previously achieved function will be manifested. Although this sequence of events would ultimately lead to the suspicion of a progressive central nervous system lesion, another caveat exists. In the normal course of events, a cerebral palsied child with a static central nervous system abnormality frequently evidences a change in the clinical manifestation as he grows older. The variations in the manifestations of fixed encephalopathy with regard to tone and involuntary movements can be very misleading.

In determining the type of investigations to pursue when searching for an etiology, the history and clinical findings offer leads. For example, the presence of microcephaly or chorioretinitis would necessitate an investigation of intrauterine infection such as toxoplasmosis, rubella, cytomegovirus, herpes, and syphilis).

When clues to the etiology are absent, a screening examination should include a urine examination for metabolic screening and amino acid levels as well as a skull x-ray view to discover evidence of calcification. A computed tomographic scan is not done routinely, but only when the possibility of a mass lesion, hydrocephalus, or an anomaly of the brain is suspected.

CLINICAL PRESENTATION

INFANCY

Clinical Manifestations

Cerebral palsy in infancy usually presents with a variation in tone, delayed achievement of motor landmarks or asymmetry in the functional use of the extremities, and hyperreflexia.

Infants may show increased tone of the antigravity muscles or, by contrast, initially may show hypotonia, which often changes to hypertonia of either the rigid or the spastic type at one to one and one-half years of age.

Hypertonia is manifested by a tendency toward extensor posturing, which is exaggerated by the supine position (Fig 39–4). Normally the tonic labyrinth reflex increases the impulses to the extensor muscles of the neck and back when the infant is supine. However, if this extensor pattern is especially strong, extensor posturing may occur when the infant is prone, a position that under normal circumstances augments the flexor pattern. The extensor posture may result in the infant's precociously being able to elevate his head and neck. By placing the baby in ventral suspension (supporting the baby in the prone position with the hands under the abdomen), this deceptive extensor tone will emerge, as evidenced by the head and neck's achieving a level above that of the horizontally level trunk. In normal babies who are less than three months of age, the pattern in ventral suspension is characterized by the head's being slightly flexed and the trunk, slightly convex.

Hypertonicity of the antigravity muscles of the

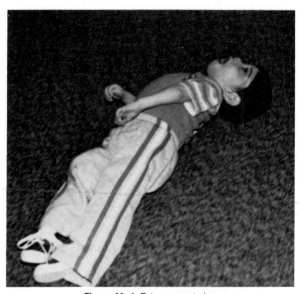

Figure 39–4. Extensor posturing.

lower extremities can be demonstrated by pulling the arms of a supine baby as if to achieve a sitting position. Normal infants will come to a sitting position with the hips and knees flexed. The baby with increased extensor tone tends to come up to a standing position with the hips and knees extended and the feet in an equinus position. The same infant, when placed in vertical suspension by providing support under the axilla, will tend to scissor, because of increased tone of the adductors and internal rotators of the hips.

Hypotonia will be noted by the increased range of motion of passively moved proximal and distal joints. The parameters for this are ill defined. Estimation of hypotonia is often based on a clinical judgment. The use of the anterior "scarf sign" for estimating increased range of motion of the shoulder joint and the "hip sign," which measures the extent of abduction of the hips, may be of assistance to the examiner in judging tone. To demonstrate the anterior scarf sign, one shoulder of the supine infant is held against the examining table while that arm is pulled across the chest immediately below the neck. With the chin in the midline, the elbow normally cannot be drawn past the chin. The ability to do so indicates increased range of motion of the shoulder joint. The range of motion of the hips is gauged by passively abducting the hips of the supine infant and keeping the knees extended. If more than 160 degrees' abduction across both hips is achieved, hypermobility exists.

Increased range of motion of joints does not necessarily reflect "cerebral hypotonia." If it is associated with decreased reflexes or their absence, lower motor neuron disease, of either anterior horn cell, peripheral nerve, myoneural junction, or muscle origin, is more likely.

Although hypotonia is seen as a nonspecific manifestation of many mental retardation syndromes and metabolic abnormalities, such as chronic malnutrition, it may also indicate cerebral palsy, especially if it is associated with hyperreflexia.

Evaluating the status of a few of the primitive reflexes is an additional aid available in judging the possible presence of a motor dysfunction resulting from an upper motor neuron lesion. For example, an asymmetric tonic neck reflex that is judged to be "obligatory" is an abnormal finding at any age and places the infant at risk for having cerebral palsy (Fig. 39–5). Although such a reflex may be seen in normal babies under six months of age, it is rarely, if ever, of the obligatory type. Most infants, when in the quiet awake state, go into a fencing position when the head is turned to one side. However, with crying or agitated activity, the normal response is for the extremities

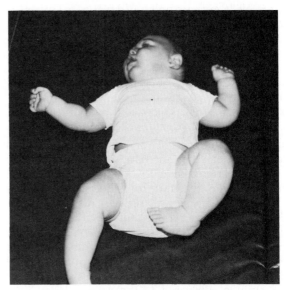

Figure 39–5. Asymmetric tonic neck reflex.

to move out of the fencing position in spite of the head's being maintained toward the side. This is not true when an obligatory tonic neck reflex is present. Although the baby may be very agitated and remain so for approximately 30 seconds, the extremities will remain in the position imposed by the reflex. A nonobligatory tonic neck reflex after six months of age is also an indicator of cerebral palsy. Although an obligatory tonic neck reflex and the delayed disappearance of a nonobligatory asymmetric tonic neck reflex are occasionally seen in primarily retarded, nonmotor handicapped infants, they are more indicative of a motor deficit of the cerebral palsy type.

Involuntary movements of the athetoid, choreic, and ataxic varieties are rarely manifested prior to one year of age. For example, characteristically there is a delay in the appearance of dyskinetic movements in infants who have suffered bilirubin encephalopathy. After the first month of life they become hypotonic and normoreflexic and remain so for the first year of life. There is a delay in the achievement of motor milestones as a result of immature motor patterns. It is not until between one and two years of age that there is a gradual change from the hypotonic to a rigid state, which is then associated with the first appearance of slow athetoid movements. Prior to the development of the involuntary movements, a history of marked neonatal hyperbilirubinemia, the finding of an obligatory asymmetric tonic neck reflex, and paresis of upward gaze may offer clues to the presence of cerebral palsy.

The classic symptoms of spastic hemiparetic cerebral palsy are also delayed in evolving. If the insult to the brain occurred perinatally, no asymmetry in the movements of the upper or lower extremities is seen during the first four months of life. In fact, the presence at this stage of an asymmetry of active movement suggests lower motor neuron disease (e.g., brachial palsy) rather than a cerebral lesion. Not only is paresis of the upper and lower extremities in the first four months of life rare, but tone and reflex asymmetry are also uncommon. However, as an infant approaches four months of age, the mother may begin to notice that one hand remains fisted while the other hand tends to open up and be solely involved with reaching for objects. At this stage asymmetry of tone and reflexes may be evident. However, these findings may be difficult to elicit, especially when the infant is not relaxed. Asymmetry of tone may be assessed more easily by observing resistance to supination of the wrist, in addition to the limited "flapping" of the hemiparetic side as compared to the normal side. Nonetheless even these signs may prove difficult to evaluate, and it then becomes essential to observe for a "functional" discrepancy. For an infant younger than one year of age to exhibit consistent use of one extremity at the exclusion of the other is considered abnormal. If the infant is uncooperative and will not perform voluntarily, there are other tests that do not require cooperation and that may demonstrate abnormal asymmetry of hand function as early as five months of age. If a cloth is placed over an infant's face, the hemiparetic baby usually pulls the cover off with only one hand, in contrast to normal babies who are more likely to use both hands.

When the infant reaches the age of six to seven months, the lateral propping reactions can be assessed when the baby is tilted in the direction of the involved extremity. This asymmetry of response is more dramatically demonstrable at seven to nine months of age when anterior propping reactions (parachuting) have evolved. The hemiparetic arm fails to extend as if in a protective response. Instead it remains rather immobile, with flexion at the elbow.

Signs or symptoms of asymmetry of the lower extremities do not become apparent until the baby is eight or nine months of age. When the hemiparetic infant first begins to crawl, he will do so with an asymmetric pattern, propelling forward with only one arm and leg. When the baby is held in a standing position, there may be a tendency for the involved lower extremity to go into internal rotation, with an equinovarus posturing of the foot. These findings antedate tone and reflex asymmetries, which may not become manifested until one to one and one-half years of age. The final degree of spasticity may not be reached until two to two and one-half years of age.

Associated Handicaps. The associated handicaps that are frequently found in infants with cerebral palsy are strabismus, nearsightedness, and hearing deficits (Table 39–10). It has been estimated that over 75 per cent of cerebral palsied infants have phorias or tropias. If noticed in an infant who is younger than four months of age, this should be reassessed during subsequent visits. If it persists after four months of age, referral to an ophthalmologist is essential. An attempt should be made to examine the eye grounds at six months of age to see whether there is a refractive error. This can be judged by the number of diopter changes in the ophthalmoscope lens that are nec-

Table 39–10. ASSOCIATED DISABILITIES AND COMPLICATIONS IN CEREBRAL PALSY*

Condition	Relative Frequency†	Clinical Types with Highest Relative Frequency*
Mental retardation	+ + +	Atonic, rigid, spastic quadriparesis
Epilepsy	+ +	Acquired and congenital hemiplegia, spastic quadriparesis
Visual handicaps		
Strabismus	+ + +	Spastic diplegia and quadriparesis
Refractory errors	+ +	Spastic: athetoid 2:1
Hemianopsia		Hemiplegia (+)
Other handicaps	+	
Hearing impairment	+	Postkernicterus (+ + +)
		Other (+)
Dysarthria	+	Athetosis (+ + +)
		Spastic quadriparesis (+)
Cortical sensory deficit		Hemiplegia (+ +)
Unequal extremity growth		Hemiplegia (+ +)
Scoliosis	+	Severe spastic and spastic athetoid (+ + +)
Dental dysmorphogenesis	+	
Joint contractures	+ + +	Spastic types
Perceptual deficit	+ +	

*Adapted from Molnar, G. E., and Taft, L. T.: Pediatric rehabilitation. Curr. Probl. Pediatr., 7:28, 1977.
† + = 25%. + + = 25 to 50%. + + + = 75%.

essary to obtain a sharper picture of the retinal vessels. If more than 2 diopters is required, a referral to an ophthalmologist is indicated. At six months, one year, and two years of age, visual acuity can be assessed. This can be done by the use of the Stycar test in which white balls of different sizes are rolled at a distance of 10 feet from the child. The visual angle at a distance of 10 feet plus and the size of the balls that succeed in capturing the infant's attention indicates the true visual acuity.

A sensorineural hearing loss is often associated with cerebral palsy. Infants with postbilirubin encephalopathy are especially at risk, an estimated 66 per cent having partial to profound deafness. Early detection of a hearing loss is essential in order to facilitate language development by offering early enhancement of speech sounds. Routine hearing assessment begins with testing of the newborn. From birth to four months of age the sound of a rattle or the crackling of tissue paper should result in a quieting or an alerting response. To simply use a wooden clapper or a loud hand clap to elicit the auditory blink reflex is not a sensitive enough test, since it takes a 60 to 70 decibel sound to produce the blink; therefore, this method can rule out only a severe hearing deficit. The orientation reflex toward sounds develops in an infant of normal intellect at approximately four to six months of age. If there is any historical or objective evidence to suggest a hearing impairment, a formal audiometric examination is necessary. If this proves impractical because of the infant's aberrant behavior or cognitive level, a brain stem or cortical auditory evoked response can be pursued.

Cerebral palsy is associated with mental retardation in over 50 per cent of individuals. The level of intellectual functioning may be difficult to judge in infancy, because many of the developmental assessment tests are based on motor accomplishments. The motor handicap resulting from cerebral palsy can affect fine motor, gross motor, adaptive, and social behavior and result in a low developmental quotient and falsely suggest poor cognitive abilities. Language development should not be interfered with by the motor handicap and in a normal hearing cerebral palsied infant can offer a clue to the intellectual status. Nonetheless, one caution with the assessment is to avoid making any conclusive statements as to the intellectual capabilities of the cerebral palsied infant.

Potential Complications. Spastic infants have the potential for developing tightness and ultimately contractures of the joints. All joints should routinely be tested to see whether there is tightness or limitation of movement. Special attention should be given to hip mobility. The spastic adductors of diplegic and quadriplegic infants may result in subluxation of the femoral head. This is so frequent a complication that it has been advised to routinely obtain x-ray views of the hips at two years and five years of age if the adductors are spastic, regardless of whether classic clinical findings of subluxation are present.

Intervention Strategies

Informing Interview. The informing interview can be regarded as an intervention strategy during which the physician can witness the parents enter and move through the crisis process. By sensitively, but frankly, stating the diagnosis and its implications, he introduces the "impact stage" when the parents come to recognize that they are now in a crisis. The physician next can anticipate a fledgling "response stage" during which the parents may try to deal with the crisis by tried and true coping mechanisms of the past, only to find that they are of little use in dealing with a matter of such severity. The physician and other staff members will then observe a turning point that, it is hoped, will represent the beginnings of a resolution of the parents' dilemma. Alternatively the trend may unfortunately be toward a worsening of the situation as a result of denial, fear, or anger.

To insure that the turning point occurs in an adaptive rather than a maladaptive direction, the physician can follow certain guidelines from the beginning. First, the most facilitative approach would involve taking the time to establish rapport with the parents. Second, he can provide a focus or structure for the events occurring in the parents' lives. By providing basic information about the child's diagnosis, by gradually unfolding the specific details of their child's particular condition, by admitting to uncertainty about the specific outcome or prognosis, and by setting out alternative courses of action (which actually need not be considered until sometime later), the physician can at least plant the seed for the parents' later understanding of their own dilemma, from which they could eventually make some responsible decisions.

Third, the approach would involve acknowledging their feelings. By speaking calmly and repetitively, by using their names frequently, by enlisting eye to eye contact, and by mirroring their feelings to convey an acceptance of them, the physician can convey the message that they are in the presence of an advocate for their concerns. Fourth, the physician can help the parents mobilize their resources by exploring their coping methods with them and can assist them in identifying sources of support or of stress that can aid or abet their efforts to deal effectively with their situation. Last, he can review with them the alternative courses of action, and either participate in or refer them to others who can help in the implementation of their plans, both immediate and long term.

One of the first questions that parents ask after learning that their child has cerebral palsy is, "Why us?" Reflecting on the history taking, when the course of the pregnancy, delivery, and period of early infancy are reviewed, the parents might generate a number of hypotheses of their own about the etiology of their child's handicap. If the parents succeed in identifying (or misidentifying) possible etiological factors, such as the mother's having smoked during pregnancy, or the father's having dismissed the urgency of the mother's request to hurry along to the hospital, they might blame themselves or one another for the outcome. Both parents, but especially mothers, characteristically feel that in some way they were responsible for causing their offspring's problem. When we consider that no etiology can be ascertained in approximately one-third of the children with cerebral palsy, there is considerable room for doubt and speculation on the part of the parents and, therefore, guilt. Parental feelings related to etiology need to be explored. Sometimes reassurance that, however unrealistic, guilt feelings are common among parents of cerebral palsied children, can prove ameliorative.

The search for a known "cause" for the cerebral palsy is an activity in which many parents of a newly diagnosed child engage. They may be concerned about whether the problem with their infant can recur with future offspring. This topic warrants discussion with the physician. If the etiology was related to a reproductive complication, the parents should be advised to consult the obstetrician. If a genetically dependent syndrome is suspected, the physician can recommend genetic counseling. Although the matter of etiology is critical, the physician must guard against the possible preoccupation of the parents with the question, "What happened?" and redirect them instead to the more relevant questions, "What can I expect from my child?" and "What can I do for my child?"

Treatment of Motor Handicaps. An early referral to a center that has an occupational-physical therapy program can be extremely helpful. Although there is no physical therapy technique that has been proved to change the natural course of the motor dysfunction, there are a number of techniques that permit the involved infant to achieve better functional use.

Most therapists are using the neurodevelopmental approach as devised by the Bobaths. One of the advantages of this technique is that the babies are placed in positions that minimize the spasticity and suppress some of the primitive reflexes that may interfere with voluntary function. For example, a baby with an exaggerated symmetric tonic neck reflex extends his arms reflexively when there is extensor posturing of the spine and neck (Fig.

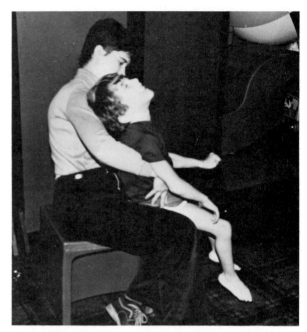

Figure 39–6. Symmetric tonic neck reflex (extensor).

39–6). The strength of this symmetric tonic neck reflex pattern may make it impossible for the infant to voluntarily flex his arms in order to execute hand-mouth activities. By placing the baby in a sitting position with the trunk slightly flexed, the symmetric tonic neck reflex will now result in flexion of the upper extremities, making hand-mouth and eye-hand activities easier for the baby to accomplish (Fig. 39–7). Mothers are taught these techniques and incorporate them into the everyday management of their infants.

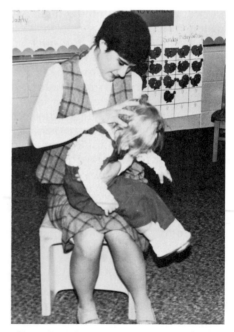

Figure 39–7. Symmetric tonic neck reflex (flexor).

Prevention of Complications. At the first sign of tightness of joint movement, passive stretching through the range of motion is essential in order to prevent fixed contractures. This can be taught to the parents by the physical therapist. Recognition of phorias, tropias, or cataracts requires an early referral to an ophthalmologist. A suspicion of a hearing loss requires an evaluation by an audiologist.

Psychosocial Aspects

Infant. Even in infancy, the child with cerebral palsy must deal with special problems in psychosocial adjustment. For the normal child the first weeks of life are characterized by intricate visual, tactile, and auditory exchanges with the care giver. These activities unite the mother and infant in a social interaction involving reciprocity, rhythmicity, and rudimentary turn-taking that is increasingly being recognized as being basic for later social and linguistic communication. To the extent that the cerebral palsied child appears normal and healthy in the neonatal period, a bond identical to that of the "normal" infant-mother dyad can be established. However, if low birth weight, prematurity, or other difficulties mark the infant's history, the child will be at risk not only for the physical handicapping condition of cerebral palsy, but for a potential social handicap as well, incurred by subtle disruptions in the mother-infant interaction. For example, the reduced responsivity often found in premature infants can interfere with and decrease the mother's contact with her infant.

Parents. On the one hand, the parents of cerebral palsied children often have an advantage over parents of children with other handicapping conditions, since the earliest parent-infant interactions might have occurred in the context of normalcy. An initial period of synchrony and acceptance in the relationship among family members would provide a solid foundation upon which to resolve the new dilemma brought on by the diagnosis.

On the other hand, the early months preceding a diagnosis may well have included building an expectation for the child's performance both within and outside the family, and may even have involved the mental structuring of the child's future. With the onslaught of a diagnosis, the parents must deal with feelings of being cheated out of the future as they had envisioned it and of a loss of the child as they had come to think of him. They may enter into a process of mourning the loss of their normal child or of resenting the child for failing to conform to their now well established construct of the future.

Parents of cerebral palsied infants may project their feelings of frustration onto the infant and to some extent may delude themselves about the infant's coping skills, perceiving the infant to be worse off psychologically than he is. Seeing the youngster struggle to reach for an attractive toy might strike them as "pathetic" and an affront to the child's self-esteem, but at this age the infant has not yet succeeded in differentiating "self" from "other" and therefore in a cognitive sense has a limited self-concept to be offended. His attempt to reach out, however unsuccessful, is an end in itself and part of the human organism's inevitable striving toward mastery. If focus were placed on the positive aspect of the infant's struggles rather than on the recurrent failures, the infant's evolving sense of self would have a better chance of including self-esteem and confidence.

In this period of development, of potential consequence is a lack of sensorimotor experiences required for cognitive growth. If an infant is dependent upon a care giver to place objects in his hand or to make things move, skills such as concept formation and the understanding of cause and effect relationships are in jeopardy. Parents can be encouraged to permit independent exploration by their infants and need not be preoccupied in providing constantly for their child's needs. Oversolicitude, even toward the infant, can stifle development.

Equally delusory would be a denial of the child's handicaps such that demands would be made upon him that exceeded his abilities.

Apart from the infant, the parents have their own problems to deal with. With a handicapped child, their own dependency needs might be unmet and heightened. More than ever they will want to be nurtured and supported themselves. With their own needs so pronounced, turning inwardly toward themselves, one parent might be negligent of the other and in turn resent the neglect of his spouse. Perceiving the spouse's reaction as a loss of support, and feeling hostile about this additional loss, the parent will be ill prepared to offer comfort. Thus a vicious cycle of isolation can evolve.

Siblings. Siblings of handicapped infants might react to the news of the infant brother's or sister's handicap with confusion. Within the family constellation they might find the strain between the parents puzzling and feel like helpless bystanders, or they might see the parental preoccupation with and concern for the infant threatening and react with feelings of anger and hostility toward the parents and child alike. They might adopt a rivalrous stance and attempt to compete for the parents' attention and concern through acting-out behavior. Often associated with their possibly maladaptive response are feelings of guilt over putting the parents through additional worry and unhappiness.

Outside the family, siblings of the infant with cerebral palsy might confront the notion of stigma from the perspective of the stigmatized and feel ashamed of their association with a handicapped child.

Many of the brothers' and sisters' reactions are dependent upon the messages they receive from their parents. Optimal adjustment to the arrival of a cerebral palsied infant in the household requires that the parents communicate to their other children feelings of acceptance of the cerebral palsied child as he is. The child's defects can be viewed with sympathetic tolerance. And insofar as the child is able, the expectations will be set forward that he will eventually assume responsibility for conforming to the general household routine and for participating in household activities.

Anticipatory Guidance

Very early the parents should be informed of the difficulty in judging cognitive and other outcomes. They should be told that the clinical picture may not evolve fully for a period of a year or two. The primary care physician should offer his assistance in guiding the family and advising them of the different therapeutic interventions that they might hear about. Attention should be paid to controversial therapies. The physician should have available published position statements about the various therapeutic intervention techniques used in cerebral palsy. This information can be shared with the parents if there is any question about the therapy they might be considering.

The possibility of the future use of casts, bracing, or surgery should be mentioned, with the explanation that they may serve to prevent or minimize contractures and offer stability to an extremity or to the trunk.

Parents have an early concern as to whether the child is going to be ambulatory. If the cerebral palsy is of the hemispastic type, the parents can be reassured that almost all children walk, although at a delayed age. However, one cannot comment so early on the cosmesis of the gait. Most ataxic children, if this is a primary difficulty, achieve improved balance as they get older, and most of them walk independently, although possibly at a very late age.

It is important for the physician not to focus counseling on future independent ambulation of the infant. This can become a preoccupation with most parents.

PRESCHOOL PERIOD

Clinical Manifestations

By the time the cerebral palsied youngster reaches the age of two to five years, the type and extent of the motor disability are well defined. The hypertonus and dyskinetic movements reach a maximal intensity by the age of three to four years. Some hypotonic infants, in spite of having exaggerated deep tendon reflexes, show no increase in tone but remain hypotonic throughout life. Such a child may have cerebellar dysfunction, which counteracts the hypertonicity resulting from pyramidal tract involvement and maintains the child in a hypotonic state. The hypotonic cerebral palsied youngster usually exhibits severe motor and intellectual impairment. The prognosis is very guarded for independent functioning with this type of cerebral palsy.

The physician must routinely monitor the deep tendon reflexes of hypotonic infants. If they were once normal to exaggerated and change to a diminished or absent status, a neurodegenerative disease is possible.

Associated Handicaps. It is during the toddler age that the cognitive abilities of the youngster with cerebral palsy are easier to determine. If the child has developed language, especially receptive language, the Peabody Picture Vocabulary Test can be used to estimate the extent of the child's receptive language ability. This test requires minimal motor skills in order for the child to respond. To indicate his understanding of the spoken word, the child need only point to the correct picture among four on a page.

In spastic children, tightness and limitation of joint movement may have developed because of contractures. In the hemiplegic child, shortening of the hemiparetic extremities may become more apparent.

The onset of seizures in a cerebral palsied patient can occur at any time. However, in the majority of cases they begin in the preschool years. They can range in type from classic grand mal to the more difficult to control myoclonic seizures. Psychomotor seizures can mimic petit mal or a behavior disorder. Episodes of sudden immobility or unusual behavior patterns that are stereotyped and are unprovoked may prove to be manifestations of psychomotor seizures.

Psychosocial Aspects

Child. An over-riding developmental issue that the preschool age child has to contend with is his developing sense of self. The world of the average preschool child is no longer confined to the family unit but reaches out to the wider circle of the neighborhood and community at large. As the child becomes acquainted with others, he gradually recognizes that he is somehow "different." Whatever the handicap, it can be gleaned as evidence of some kind of damage, which, for all the preschooler knows, may mean impending death or disfigurement. Constantly identifying with monsters or ineffectual characters could lead to the development of unfavorable feelings about oneself.

The preschool period is one of active imagining. Fantasy play represents one of the most fulfilling, as well as one of the most revealing, modes of self-expression among children of this age. The therapeutic use of this common behavioral phenomenon could prove very beneficial to the child's developing resolution of his problem. Via fantasies that are guided by sensitive and skilled adults, cerebral palsied children can come to relinquish the magical thinking and fearful ideation that might have plagued them and acquire instead more adaptive ways of viewing themselves and their condition. Through behavioral rehearsal in fantasy play too, the children could develop strategies for dealing with mundane problems that might confront them.

Parents. Parents may be struggling with the child's new need for autonomy, wanting to protect the child from the outside world and its insensitivity.

At the age when sex role socialization is at its height, parents often react differently to their cerebral palsied sons and daughters. For example, it may be that parents of boys have a stronger career or achievement orientation with respect to their sons than to their daughters. In a study of cerebral palsied children Zisserman (1978) concluded that her finding of greater knowledge about cerebral palsy in the parents of boys "may be one manifestation of the greater impact of the son's disabilities upon the family integration and of the parents' concern about the future careers of their handicapped *male* offspring" (italics in original).

The time following the diagnosis is the time when the bulk of intervention should properly occur, with subsequent intervention aimed at maintaining support. Needed too, however, is a review of the past and the successful and not so successful strategies that were used for dealing with various problems.

A review also of what lies ahead is in order, now that the parents are moving closer to the next developmental stages. Present concerns can also be addressed. Among them might well be the parents' reaction to their changing child. His growing need for independence might conflict sharply with the parents' established routine of care and protection. Their own fears for the child's limitations, or for their loss of a now accustomed and appreciated feeling of being needed, can stand in the way of granting autonomy. Or it could be that the parents regard dependence and a lack of initiative as good behavior or manageable behavior, and are reluctant to see the advent of autonomy and assertiveness.

Siblings. The brothers and sisters of a cerebral palsied child could by now be reacting to the child's newly evolving sense of self, either negatively by playing up to his self-derogations or positively by encouraging their afflicted sibling to abandon such thoughts and to value himself more highly.

SCHOOL AGE AND ADOLESCENCE

Clinical Manifestations

The period of transition between the latency period and adolescence is not one of significant change in the clinical manifestations of the motor dysfunction. If anything, by the age of nine or 10 the youngster will probably reach his maximal functional ability, at least for gross tasks needed for the activities of daily living. However, it is a period in a youngster's life of heightened anxiety over his handicapped condition and the prospects for future independent functioning.

It is usually during the early school years that a decision is made about the promise or futility of continued attempts at ambulation. Parents as well as professionals are resistant to "give up," holding onto the idea that further physical therapeutic intervention may permit some ambulation or improve existing ambulation skills. One must consider, in the decision making process, the limited benefits, relative to the risks, that might accrue from continuation of intensive physical therapy. The child comes to see himself as a failure if he is unable to ambulate. Energies that might be used for more meaningful pursuits at this time are drained. At times, accepting a wheelchair existence can be a relief to the youngster who can then proceed to accommodate to a nonambulatory life situation. Many youngsters at this age actually lose some of their ambulatory skills and in fact can become wheelchair bound. It has been suggested that the growth spurt that occurs in adolescence causes a further stretch on the muscles, making them "more spastic." Also progressive deformities may occur at this age and add to difficulties in mobility. Last, there is some evidence that at the time of adolescence there is an increase in the amount of energy (as measured by maximal oxygen consumption) required to do the same motor task as compared to an earlier age. Therefore, it becomes more difficult for the individual to perform a motor task, resulting in easy fatigability.

Associated Handicaps. The progression of scoliosis must be watched for during the rapid growth spurt in adolescence. There is also a tendency for leg length discrepancy to increase during this age.

Many of the operative repairs necessary for improving function of cerebral palsy are now undertaken. Missing school at this early age and the isolation by hospitalization can have serious emotional consequences.

Psychosocial Aspects

The successful resolution on the part of the child and the parents alike of the autonomy dependency conflict of the earlier stage will enhance the child's adjustment to the separation involved in entering into a school or remedial program. In a study of the development of physically handicapped children, Minde and his associates (1972) found that children with strongly dependent relationships with their mothers, who were inclined to think of their mothers as "omnipotent, all-curing beings," had the most difficulty in adapting to the school routine. He noted, "When these . . . children first came to school they quickly refused to do almost anything for themselves, such as going to the bathroom, helping with dressing or undressing, or eating lunch." With the support of their teachers and parents, the children gradually overcame the period of regression and eased into the school routine.

Between approximately five and nine years of age, children with normal intelligence pass through stages of cognitive development that permit them to reflect on their condition and assess it in an altered way. Although previously they might have recognized their "differentness," their "magical thinking" may have transformed it into a temporary disability associated with an otherwise healthy and well functioning body. As Minde explains, they might "claim to have okay legs' and yet not be able to walk." Just as preschoolers can regard their sex, their clothing, and other situations in their lives as changeable, they can envision their handicap as reversible as well. However, with maturity and with exposure to other handicapped children in school, they realize that their handicap is permanent. This realization is often difficult to accept. Sometimes they will admit to a part of it (such as a paralysis of one limb) while denying the rest of it. Sometimes they will attempt to deny all of it and strenuously set out to prove their "normality," even to the point of aspiring to the local Little League, for example. As evidence of the reality of their limitations gathers, they frequently enter a severe depressive period during which they must come to terms with the implications of the handicap and must somehow incorporate this knowledge into their own life plans and expectations. Until resolution, they exhibit withdrawal from others, a refusal to do school work, a wish to die, a sense of futility, or a belief that they are being punished.

An added dilemma at this stage involves the growing awareness by their healthy peers of the handicapped child's condition and limitations. When the healthy child interacts with a handicapped child, for example, one confined to a wheelchair, he responds to the child's condition with discomfort or avoidance. The stigmatizing effects of a child's wheelchair and handicap interfere with the flow of the interaction. Healthy children are more inclined to play cooperatively with and talk with and move toward a healthy peer than a handicapped peer, and to prefer the nonhandicapped child as a future playmate. In the Minde study mentioned earlier, a loss of peer companionship was noted: "Although each child was said to be fully accepted by his peer group when admitted to the school, [his] previously tolerant friends at home had grown up from sandbox toddlers to baseball and hockey players and now wanted little or nothing to do with [him]. Thus . . . 22 of the 31 children over eight years old had no constant outside friends during the summer and only four had contact with normal peers during the winter months."

With the onset of adolescence, the psychosocial adjustment of the cerebral palsied child is threatened by a number of changes that occur at this time, such as a deterioration in his physical status, a realization of the unlikelihood of a cure, the abandonment by the peer group, a restriction on the activities from which teenagers typically derive a feeling of competence (e.g., driving and athletics), newly evolved sexual feelings, a confrontation with one's future in terms of marriage and vocation, and a shift in services from a pediatric to adult setting, with its concomitant loss of stable support personnel (Freeman, 1970).

A growing recognition of the chronicity of his illness and its implications for every aspect of both current and future daily functioning is harsh. The cerebral palsied child's fears and anxieties in this regard may be based on a realistic assessment of the possibility or on his own construction of the possibility in such a way that they become reality, as in a self-fulfilling prophecy.

When the social and vocational outcomes of cerebral palsy in young adults were studied, it was shown that more than half were unemployed, and, of those who were employed, most were working in unskilled positions, frequently in sheltered workshops rather than in competitive employment. The cerebral palsied adults were to a great degree economically dependent, even though most were independent in a physical sense as well as in terms of the educational level they had achieved. At odds with their potential was not just their economic status but their social functioning, nearly one-third being described as "social isolates." The variable most responsible for the young cerebral palsied person's inability to fulfill his potential, apart from serious retardation or physical disability, was social stigma. This suggests that optimal development may result more from the cerebral palsied patient's ability to manipulate positively the attitudes of others in his environment than from an exclusive focus on physical remediation.

In adolescence, sexuality arises as an issue that will continue to require attention throughout the lifespan. In the teenage years the individual with cerebral palsy may remain physically dependent on others and be unable to engage in sexual exploration or sexual activity and even to initiate the most casual interactions with members of the opposite sex. When the opportunity for cross sex interaction exists, the stigmatization of the cerebral palsied adolescent often prevents the development of intimacy. And in circumstances in which an intimate relationship is successfully formed, physical as well as psychological barriers may exist. For example, in a study of a young cerebral palsied woman with leg spasms and limited voluntary movement of the lower extremities, sex counseling was required to deal with her specific physical handicaps as they related to sexual intercourse. Counselors reviewed the feasibility of various positions as well as the use of props and the assistance in positioning of her legs by her sex partner. In addition, gynecologic, genetic, and orthopedic consultations were used to deal with the possibility of her eventually bearing and rearing children.

A number of guidelines exist for the treatment of the cerebral palsied adolescent, which take into account the adolescent's newly emerged sensitivities. From the patient's and the family's standpoint, the handicap is "different" once adolescence is reached and should be treated as such. This involves freshly reviewing the diagnosis and its implications in order to provide both the child and the family with enough information to begin considering the social, sexual, marital, and vocational possibilities that lie in the future. Until adolescence, these issues seem so distant that concerned individuals seldom attempt to consolidate the medical data with the nonmedical consequences of the handicap.

A second suggestion concerns the inclusion of the adolescents in conferences and inviting their participation in decisions directly related to their daily life. Treating the adolescent more like an adult would have the effect of fostering independence and mastery and discouraging regressive tendencies.

Third, the physician must be encouraged to accentuate the young person's assets and to shift the emphasis away from the limitations and disease and toward his potentialities.

Fourth, a new frankness or openness about the patient's diagnosis, management, and prognosis is in order so that a reciprocal openness can be promoted in the patient. The physician who is honest with and receptive to the patient is most likely to facilitate the patient's expression of his fears and concerns. When these are brought to light, the physician can respond with reassurance or with clarification, as appropriate.

Fifth, the use of psychiatric or psychologic counseling should be considered at the very first sign of concern or puzzlement in the patient rather than as a last resort. Equipping the teenager with positive methods of coping early will prevent complications later and permit a longer period of adjustment to the future.

One of the most important methods that a cerebral palsied individual develops concerns the manipulation of the attitudes of others. At a meeting of the American Academy for Cerebral Palsy, a number of handicapped individuals shared their thoughts on the matter. For example, a young cerebral palsied adult recalled her own adolescence and the difficulties she had with the expectations that people had about her limitations. When she announced her plans to go to college, many acquaintances expressed concern that she would have no one to take care of her. She related, however, how she dealt with the situations in which she had to rely on the help of others: she simply stated, "I need this kind of help," giving clear and matter of fact requests or instructions to individuals, most of whom she found eager to provide assistance.

CONCLUSION

Cerebral palsy is a lifetime problem. Different developmental stages bring new sets of problems, challenges, and accomplishments for the child and the care givers alike.

In spite of strenuous efforts and creative attempts to ameliorate the conditions associated with cerebral palsy, however, the outlook has troubling aspects. It has been learned that the potential for independent functioning, educational success, and motor ability will remain unfilled if the individual for whom care is being provided has failed to achieve psychological well-being within his social environment. Throughout the child's development and our care of him, attention must continually focus on feelings of self-worth and satisfaction with self-mastery.

LAWRENCE T. TAFT
WENDY S. MATTHEWS

REFERENCES

Ballard, R.: Early management of a handicap: the parent's needs. In Oppe, T. E., and Wookford, F. P. (Editors): Early Management of Handicapping Disorders. New York, Associated Scientific Publishers, 1976.

Bobath, K., and Bobath, B.: Cerebral palsy. In Pearson, P. M., and Williams, C. E. (Editors): Physical Therapy Services in the Developmental Disabilities. Springfield, Illinois, Charles C Thomas, 1972.

Finnie, N. R.: Handling the Young Cerebral Palsied Child at Home. New York, E. P. Dalton and Co., 1975.

Freeman, R. D.: Psychiatric problems in adolescents with cerebral palsy. Dev. Med. Child Neurol., 12:64–70, 1970.

Klapper, Z., and Birch, H.: The relation of childhood characteristics to outcome in young adults with cerebral palsy. Dev. Med. Child Neurol., 4:645–656, 1966.

Minde, K., et al.: How they grow up: 41 handicapped children and their families. Am. J. Psychiatry, 128:1554–1560, 1972.

Molnar, G. E., and Gorden, S. W.: Cerebral palsy: predictive value of selected clinical signs for early prognostication of motor function. Arch. Phys. Med. Rehabil., 57:153, 1976.

Molnar, G. E., and Taft, L. T.: Pediatric rehabilitation. Part 1. Cerebral palsy and spinal cord injuries. Curr. Probl. Pediatr., 7:28, 1977.

Paine, R. S.: The evolution of infantile postural reflexes in the presence of clinical brain syndromes. Dev. Med. Child Neurol., 6:4, 1966.

Perlman, J., and Roth, D.: Stigmatizing effects of a child's wheelchair in successive and simultaneous interactions. J. Pediatr. Psychol., 5:43–55, 1980.

Zisserman, L.: Sex of a parent and knowledge about cerebral palsy. Am. J. Occup. Ther., 32:500–504, 1978.

39E Seizure Disorders

An epileptic seizure is a clinical event caused by a sudden excessive electrical discharge of the neurons. Seizures are a common problem in pediatric practice and are variable in their manifestations, being at one extreme acutely life threatening and at the other scarcely noticeable. The word "epilepsy" implies a tendency to recurrent seizures. There are multiple causes of seizures, although in many instances in childhood no specific etiology is identifiable (i.e., the idiopathic or primary epilepsies). A single seizure may occur during a disorder that temporarily lowers the seizure threshold, an example being a metabolic disorder such as hypoglycemia. This does not necessarily imply that the child has epilepsy. The clinician faced with a child who has had a seizure must therefore decide whether a tendency to recurrent seizures—epilepsy—is present.

The evaluation of a child with possible seizures should identify the type of episode that occurred. A simplified outline of seizure types in childhood and adolescence by age of onset is given in Table 39–11.

The importance of obtaining accurate historical information cannot be overemphasized. Further, careful neurodevelopmental examination is essential in every child. Most are found to be normal, but because of the cerebral origin of seizures and the frequent association with other diseases, a degree of physical or intellectual impairment may be present.

The electroencephalogram is the most useful test in the evaluation of seizures, and any child suspected of having had a seizure should undergo electroencephalography. This should include a period of wakefulness with hyperventilation and intermittent photic stimulation, and a period of light sleep. In interpreting the results of the electroencephalogram it is important to realize that normal variations occur that may resemble epileptiform activity. Examples are ctenoids (14 and 6 per second positive spikes), small sharp spikes (benign epileptiform transients of sleep), and hypnogogic hypersynchrony in young children. There are some epileptic persons in whom the electroencephalogram is normal even in sleep; this is most common in partial or focal epilepsy of temporal or frontal lobe origin.

Further testing is dictated by the clinical situation. A computed tomographic scan is not necessary in every child, but is most likely to be positive when a structural cerebral lesion is suspected. A lumbar puncture is indicated if there is any possibility of a central nervous system infection. In a few children a search for inborn errors of metabolism may be indicated.

There are several nonepileptic paroxysmal disorders that are common in childhood and that therefore enter the differential diagnosis of seizures (Table 39–12).

Breath Holding Spells. Breath holding spells occur in children between the ages of six months and six years and are of two types—"red-blue"

Table 39–11. TYPES OF SEIZURES IN CHILDHOOD BY AGE OF ONSET

Type	Age
Neonatal	Birth–1 month
Infantile spasms	6 to 12 months
Myoclonic-atonic	1 to 4 years
Febrile seizures	6 months–6 years
Primary generalized	
Absence (petit mal)	5 to 15 years
Tonic-clonic (grand mal)	5 to 20 years
Partial (focal)	
Simple, e.g., jacksonian	Any age
Complex, e.g., psychomotor	Any age
Secondary (symptomatic)	Any age

Table 39–12. NONEPILEPTIC PAROXYSMAL DISORDERS IN CHILDHOOD

Breath holding spells
  Red-blue (cyanotic)
  White (pallid) = infantile syncope
Sleep disorders
  Hypnogogic myoclonus
  Night terrors
  Sleep apnea
Syncope
Benign paroxysmal vertigo
Migraine

and "white." Red-blue spells are episodes of apnea that follow a bout of crying. White spells are due to vasovagal syncope and are usually precipitated by pain or fear. Crying is often minimal or "silent"; the child is pale and rigid and often arches. A careful history can differentiate these from each other and from seizures, but either type may terminate in tonic clonic movements resulting from transient cerebral hypoxia-ischemia. The electroencephalogram is normal in both, but ocular compression performed under controlled conditions with simultaneous electroencephalographic and electrocardiographic recording may precipitate a "white" attack with brief cardiac asystole and confirm that diagnosis.

Sleep Disorders. Various disorders of sleep and arousal may mimic epilepsy. Hypnogogic myoclonus, the limb jerking that many people experience on falling asleep, is benign, as opposed to myoclonus on arousal, which is usually epileptic. Night terrors are arousals that occur in slow wave sleep (stage III or IV, nonrapid eye movement sleep), usually about an hour after falling asleep. The child may cry out and appear awake and terrified for five to 10 minutes, although he is still asleep. Sleep apnea is secondary to upper airway obstruction and is manifested as noisy irregular respirations with snoring, nocturnal restlessness, and daytime somnolence.

Syncope. Syncope is common in childhood and often runs in families. A careful history may suggest this diagnosis, although as in breath holding attacks, tonic or clonic movements may occur at the end of an attack, especially if the head has been kept upright.

Benign Paroxysmal Vertigo. Benign paroxysmal vertigo is a disorder of young children who have acute attacks of ataxia, vertigo, and occasionally vomiting, which may be very alarming but are short lived. Nystagmus is present in attacks, but between spells the child is normal.

Migraine. Migraine is notably common in children, in whom the manifestations are often rather atypical. Transient focal neurological signs or confusional states may occur, and suggest epilepsy. A history of migraine is generally present in the family. The possible coexistence of migraine and definite seizures, however, warrants a careful evaluation, because a structural lesion, such as an arteriovenous malformation, may be present.

APPROACH TO MANAGEMENT

In considering the management of the child with true epilepsy, it is essential to define the aims and principles involved in such an endeavor. The entire life and social milieu of the person should be considered. Seizure control must be obtained at the least cost to the patient in terms of cerebral damage, drug toxicity, and expense, both financial and emotional.

The causes of both seizures and epilepsy are legion. Any condition that can lead to neuronal instability or damage, causing discharges, should be considered for prompt intervention, and this includes perinatal problems, metabolic disorders, infections of the central nervous system, febrile illnesses in the young child, trauma to the head, and some vascular events. Many disorders are not amenable to specific therapy (for example, inherited diseases like phakomatosis), but genetic counseling may play a significant part in reducing their incidence. Seizures themselves can be damaging to the brain, especially prolonged status epilepticus, and probably also some of the minor motor seizures in early childhood, although there is no clear evidence that short focal or generalized seizures cause cerebral damage. These potentially damaging situations should therefore be promptly treated both with specific anticonvulsants and with full supportive measures to preserve the *milieu intérieur*. It seems clear that in many of the chronic epilepsies, the longer the seizures elude control, the more intractable they become, possibly as a result of the "kindling" phenomenon.

The consequences of seizures are many, varied, and unfortunately common. The possible damage to the brain, already mentioned, is one factor that is usually uppermost in the minds of those dealing with a patient with seizures, both physician and parents. However, there are others of equal long term importance, namely, the psychosocial complications. If possible, one should try to see the child and the parents each alone on at least one occasion, because children are often unwilling to voice fears in front of their parents, and the latter usually feel more comfortable if they are able to discuss the situation in private. Parents may wish to express anxieties that they do not want to communicate to their child. One must attempt to educate, often through the parents, any others likely to be involved with the seizures, including teachers, classmates, friends, and the extended family. Only if these individuals have some knowledge of the child's disorder can they be fully sympathetic and supportive to the child and family. In older children and teenagers, the social aspects loom large as the person becomes more aware of his dependency and loss of independence. These adolescents often feel very angry about their lack of control of their lives, and as a result compliance to therapy can be a major problem. At times psychiatric consultation can be of special value to the patient and the family. Voluntary organizations (and their local chapters) can also be a source of great support for the child and the family by providing a means whereby they can meet in groups with other affected children and families to learn at first hand how best to cope with the various problems that arise.

PRINCIPLES OF DRUG THERAPY

The drug treatment of epilepsy is an area in which advances in pharmacology and pharmacokinetics have made an enormous impact on clinical practice by increasing knowledge of the metabolism of the drugs used and their interactions. Probably the most significant factor has been the ability to measure serum levels of anticonvulsants and to utilize this information in management.

Several features of drug metabolism are important for the efficient management of these medications. The half-life of a drug is the time it takes for the blood level to fall by 50 per cent and is related to the biotransformation of the substance to other compounds and to excretion. It also determines the length of time before a steady state is reached in the body; in general this occurs in about 4 to 5 half-lives. Thus, the half-life of a drug will determine how long it will take for the drug to reach a constant level on a particular maintenance dose. If a loading dose is given, the time taken to reach a steady state will be shortened and is dependent on the time taken for distribution of the drug in the body. There may, however, be complications in giving a loading dose, such as oversedation in the case of phenobarbitone. The half-life also determines the necessary frequency of dosage, those whose half-life is shorter being given relatively more frequently. In order to achieve a fluctuation of less than 50 per cent in the blood level, a drug should generally be given at intervals of 1 half-life or less. A once or twice daily schedule is preferable for all patients because this increases the convenience and therefore compliance. The half-life is also important when overdosage occurs; the longer the half-life, the longer the period required for the level to fall to a nontoxic range, and the longer the drug should be withheld.

For many anticonvulsants, a close relationship exists between the levels of drug in the blood and in the brain (and thus for seizure control), and this has led to the recognition of certain ranges of blood levels that are considered to give a useful treatment effect without toxicity, the "therapeutic range." There is, however, a wide range of variability from patient to patient, and thus these levels must be used with caution and must always be considered only as a guideline for finding the optimal dose for the individual patient.

The determination of the blood level of an anticonvulsant is not a procedure that should simply be routine. In children whose seizures are well controlled on nontoxic doses of medication, frequent determinations of levels are not necessary.

In difficult situations, however, they are of particular utility. This is especially true in situations, such as status epilepticus, in which it is essential to know the exact medication level in order to plan a safe and effective regimen of treatment of the patient and also in the neonate, in whom drug metabolism is unpredictable. Another factor of importance in considering blood levels of drugs is the reliability of the laboratory determination; there can be significant fluctuations in the results given by different laboratories.

The basic principle of anticonvulsant therapy is to make an accurate diagnosis of the type of seizure the child is having, decide which drug is likely to be the most effective, and then prescribe it in appropriate amounts. The pharmacokinetic characteristics of the drug in question should be considered in deciding about the timing of doses. There should be a plan for building up the dose, or for loading, and the blood levels should be checked when the correct dose for weight and age has been reached or signs of toxicity appear. Further action depends on the response of the seizures. If the seizures are uncontrolled, and either the drug level is therapeutic or toxic effects occur at higher doses, an additional drug should be chosen. When the level of the second drug reaches a therapeutic level, and the seizures are controlled then the first drug is withdrawn. Rarely is it necessary to continue with two drugs at this stage. One should aim for control with a single medication (monotherapy). Only when one has run out of alternatives should combinations (polypharmacy) be tried.

A patient who does not effectively absorb a drug is rarely encountered. In such instances the dose should be increased slowly until therapeutic levels are obtained or until the seizures stop. Once control is achieved, the need for frequent checking of blood levels drops; this should probably be done every six to 12 months in children thereafter. In adults, once a year is enough. In the event of toxicity, the appropriate action depends on the situation, and one should always remember that drug-drug interactions may be important.

INDIVIDUAL ANTICONVULSANTS (DISCUSSED IN ORDER OF INTRODUCTION)

Phenobarbitone. Phenobarbitone is the oldest anticonvulsant in use and still one of the best. It is safe, effective, cheap, and well tolerated, although with some notable exceptions. Its major uses in childhood are in the control of neonatal seizures, prophylaxis of febrile seizures, and in some children with primary or secondary generalized tonic-clonic seizures. The major problem with this medication is the incidence of side effects involving behavior and changes in alertness. In about 25 per cent of young children taking therapeutic doses of the drug, there are intolerable alterations in behavior, mainly hyperactivity, insomnia, and outbursts of temper. The parents often describe their offspring as having become a "different child." There is therefore often poor drug compliance. In a few children, especially in the older age groups, the reverse occurs, with sedation and a decrease in responsiveness to the environment. This is obviously a worrying feature in a school age child who may be functioning at a level below his potential.

Other toxic and idiosyncratic side effects are rare. Rashes and blood dyscrasias are very uncommon. A state of physical dependence develops with prolonged treatment, and these children are at risk of developing status epilepticus should the drug be discontinued suddenly. Interactions with other drugs are common, for phenobarbitone is an inducer of hepatic microsomal enzymes and may increase the breakdown of other medications. Its exact mode of action is still uncertain, but it appears to act by inhibiting membrane depolarization shifts, which occur in seizure foci, and by affecting neurotransmitters.

Mephobarbital. Mephobarbital (Mebaral) is a barbiturate that seems empirically to have some advantage over phenobarbitone in that the behavioral effects appear to be less marked with this drug. The dosage is 60 to 100 per cent greater than that of phenobarbitone.

Phenytoin. Phenytoin (Dilantin) is the next in order of introduction and has had extensive use. It appears to have a different mode of action from phenobarbitone in that it inhibits post-tetanic potentiation of nerve impulse transmission and limits the spread of epileptogenic activity from a seizure focus. Oral absorption may be very erratic, especially in the neonate, and occasionally an older patient is found who does not absorb it in the usual way. It is effective in generalized tonic-clonic seizures, either primary or secondary, but care must be taken in children who may have absence attacks because these may be exacerbated by this drug. Focal epilepsies generally respond well, and also the tonic seizures often seen in the group of children with minor motor seizures, although phenytoin does not help the other types of seizures in these patients.

The common side effects of note are those on skin and gums. There is a high incidence of hirsutism and acne in patients given long term phenytoin therapy, and significant gingival hypertrophy occurs in many patients. Coarsening of facial features also occurs. The gum problem can be minimized by scrupulous dental and gingival hygiene, but these side effects are sufficient to make one consider alternative forms of therapy, particularly in teenage girls. Other problems resulting from phenytoin therapy include rashes, a common occurrence that must be taken seriously because some cases progress to a full blown Stevens-Johnson syndrome and rarely blood dyscrasias. Long term phenytoin therapy has been associated with abnormalities of vitamin D metabolism, producing rickets or osteomalacia by interfering with hepatic metabolism of the vitamin. This is most likely to occur in children on a poor diet, or in those with little exposure to sunlight. Thus the populations most likely to be affected are deprived groups and retarded individuals in institutions. These children probably should be given supplementary vitamins.

There are several effects of phenytoin—both acute and chronic—on the central nervous system. Acute overdosage causes a cerebellar syndrome with nystagmus and ataxia. At high levels, there will be sedation and encephalopathy. Chronic toxicity is considered to cause slowing of the waking background of the electroencephalogram and produces a drop-out of Purkinje cells from the cerebellum, with a resultant permanent deficit.

Primidone. Primidone (Mysoline) is a relative of phenobarbitone and is metabolized to the latter compound. It is of limited use in childhood, having little to add to phenobarbitone and being potentially more sedative.

Ethosuximide. Ethosuximide (Zarontin) has been the drug of first choice in petit mal or absence seizures for 20 years. This is its only indication. It is generally well tolerated, although intractable hiccoughs are a recognized side effect and occasional blood dyscrasias may occur. It is ineffective against focal or generalized motor seizures. Occasionally it may be useful as an adjunct to other medications in patients with myoclonic-atonic epilepsy who have atypical absence attacks and 2 to 4 Hz. atypical spike and wave discharges on the electroencephalogram.

During the 1970's three valuable new drugs were introduced, and these have become increasingly widely used in clinical practice. In order of introduction, they are:

Carbamazepine. Carbamazepine (Tegretol) is chemically related to the tricyclic group of antidepressants. It is potentially the drug of first choice in the treatment of partial seizures, either simple or complex, and is also highly effective in the management of generalized tonic-clonic seizures. It has no use in petit mal and only limited effectiveness in the minor motor seizures. Its mode of action is unclear. Symptoms of overdosage include dizziness, ataxia, blurred vision, and sedation, and these may be experienced in the early phase of treatment before tolerance develops. Some patients are very sensitive to this drug and can take it only in small doses, although it may often be effective even at low levels. Gradual introduction with a slow increase in dosage may obviate these

Table 39–13. CHARACTERISTICS OF THE COMMON ANTICONVULSANTS IN ORDER OF INTRODUCTION

	Indications	Daily Dose	Elimination Half-Life in Hours	Therapeutic Range	Toxic Effects and Comments
Phenobarbitone	Generalized tonic-clonic, focal, myoclonic-akinetic (atypical petit mal), psychomotor	5 mg./kg./day until age 5 yr.; then decreasing until reaches 1.5–2.0 mg./kg. in adult	A and C:* 40–140 N: 60–170	10–50 μg./ml.	Drowsiness to which tolerance quickly develops; rash; rare hematological difficulties; paradoxical excitement-hyperkinesis; this effect may be offset empirically by mephobarbital (Mebaral)
Diphenylhydantoin (phenytoin, Dilantin, DPH)	Generalized tonic-clonic, psychomotor, focal, myoclonic-akinetic (atypical petit mal)	6 mg./kg./day until reaches 50 kg.	A: 10–40 C: 5–15 N: 10–30 P: 10–40	5–20 μg./ml.	Gum hypertrophy, hirsutism: nystagmus (20 μg./ml.), ataxia (30 μg./ml.), pseudo-dementia (40 μg./ml.); maculopapular rash, erythema multiforme, exfoliative dermatitis; hepatitis; blood dyscrasias, megaloblastic anemia; lymphadenopathy, pseudolymphoma; lupus; interference with vit. D, hypocalcemia; teratogenicity in pregnancy

Drug	Indications	Dosage	Age*	Blood level	Side effects
Primidone (Mysoline)	Generalized tonic-clonic, psychomotor, myoclonic-akinetic (atypical petit mal)	10–20 mg./kg./day	A: 6–18 C: 5–11	4–12 μg./ml.; also measure phenobarbital level	Drowsiness, ataxia; behavioral disturbances, paranoia, psychoses, megaloblastic anemia, folate deficiency; rash, usually remits and allows continuation; seems to provide little advantage over phenobarbital, its main effective metabolite; introduce slowly
Ethosuximide (Zarontin)	Absences (petit mal)	20–30 mg./kg./day	A and C: 20–60	40–100 μg./ml.	Gastric distress, hiccoughs, vomiting; confusion; rare blood dyscrasias, rashes, lupus.
Carbamazepine (Tegretol)	Generalized tonic-clonic, psychomotor, focal	10–20 mg./kg./day	A and C: 10–30	3–12 μg./ml.	Drowsiness, ataxia; rash, blood dyscrasias, leukopenia, thrombocytopenia, hepatic toxicity; introduce slowly
Clonazepam (Clonopin)	Myoclonic-akinetic (atypical petit mal)	0.05 mg./kg./day; increase slowly to 0.2 mg./kg./day	A and C: 20–40	5–50 μg./ml.	Sedation, ataxia, hyperkinesis, anorexia, increased salivation; preferably not given simultaneously with valproic acid (petit mal status); introduce slowly
Sodium valproate (V.P.A., Depakene)	Absences (petit mal), myoclonic-akinetic (atypical petit mal), generalized	15–50 mg./kg./day	A and C: 6–15	50–100 μg./ml. (uncertain)	Sedation, nausea, vomiting are usually transient; hepatotoxic—most significant; hematologic, nephrotoxicity and alopecia rarely; introduce slowly

*C = children. A = adults. N = neonates. P = prematures.

effects. Toxic effects include rashes, leukopenia, thrombocytopenia, and hepatic dysfunction, but these are not common. These parameters, however, especially the leukocyte and platelet counts, must be checked regularly in the first six months of therapy. Occasional patients may become water overloaded because of the drug's effect on osmoreceptors, increasing the effect of antidiuretic hormone.

Clonazepam. Clonazepam (Clonopin) is the latest and apparently most effective of the benzodiazepines to be introduced for oral anticonvulsant use. It acts by limiting spread of epileptogenic activity from focal lesions through increasing polysynaptic inhibition in the central nervous system by enhancing gamma-aminobutyric acid. Its major use is in the myoclonic seizures of childhood, and also in infantile spasms. It is also effective in focal and generalized motor epilepsies. The major side effect is sedation, which mandates very slow introduction of the drug. Because therapeutic levels are very variable, determination of blood levels is of little use. In some children a behavior change similar to that seen with phenobarbitone occurs, although tolerance may develop and this effect decreases. Petit mal (absence) status or stupor may occur when this drug is given with valproic acid, and hence this combination is to be avoided. Some infants and children have troublesome hypersalivation and an increase in bronchopulmonary secretions when taking clonazepam, and this effect may be severe enough to require cessation of therapy.

Valproic Acid. Valproic acid (sodium valproate, Depakene) seems to work by potentiating inhibitory neurotransmitters (gamma-amino butyric acid). Its major use is in absence epilepsy, in which it is highly effective in the majority of patients. However, its expense and the risk, although low, of potentially fatal toxic reactions keep it in second place to ethosuximide for most patients. Other types of seizures in which valproic acid is of help include the minor motor group and tonic-clonic seizures. It has a relatively short half-life, and thus its daily administration must be divided into at least three doses. It has a number of side effects, the most common being gastrointestinal, with many patients experiencing nausea, vomiting, and gastric discomfort after ingestion. Paradoxically some patients, especially females, gain weight excessively. There is little sedative effect. The major problem is the effect on the liver. There have been over 40 deaths from hepatic failure associated with use of this drug, although many of these patients were also taking other medications. Those at apparently greatest risk are infants and children with severe seizure diorders accompanied by mental retardation. In a number of patients taking valproic acid there is a minor elevation of the liver enzymes levels, but they are asymptomatic, and no progression is seen in most cases. It is essential, however, to check these parameters regularly, at least in the first six months of treatment, and to stop the drug if the liver function test results become progressively abnormal. A lowered dosage may allow a return to normal levels. Pancreatitis has also been associated with this medication. Thrombocytopenia and alterations in platelet function can occur, and should be checked. This may be important in patients requiring surgical procedures while taking Valproic acid.

Valproate has important interactions with other anticonvulsants in common usage. It impairs phenobarbital excretion and therefore elevates phenobarbital blood levels, necessitating a decrease of 30 to 50 per cent in the phenobarbital or primidone (Mysoline) dosage. Valproate is highly protein bound and displaces phenytoin, causing a drop in the total plasma concentration. However, because of a concomitant increase in free phenytoin, its anticonvulsant effect is basically unchanged. Clonazepam and valproate in combination may precipitate absence status, but can be used together if indicated.

STATUS EPILEPTICUS

The treatment of status epilepticus (i.e., recurrent seizures without recovery of consciousness) is a medical emergency. In childhood the majority of instances occur before the age of three years, and these are often associated with fever. There is a potential for cerebral damage as a result, and therefore prompt and effective action is essential (Table 39–14).

It is essential to remember the other aspects of the child's care. One must try to find out the underlying reason for the seizure, which may be a first complex febrile seizure or a severe relapse of a chronic seizure disorder perhaps precipitated by noncompliance or a medication change, or it may be the presentation of a new central nervous system infection such as meningitis or encephalitis. In addition to the tests performed in the emergency room, a lumbar puncture may be indicated, although this is preferably performed when the seizures are controlled. Investigations such as electroencephalography and computed tomography are not likely to be of help in the acute stage unless there are focal features indicated by the seizures or the neurological examination. The general support of the child is important, including care of the comatose patient and the management of fluids and airway. Overhydration is to be avoided, for it may exacerbate potential cerebral edema. Good oxygenation is essential to avoid hypoxia of the metabolically very active discharging neurons. Once the seizures have been controlled, a treatment plan should be devised for

Table 39–14. MANAGEMENT OF STATUS EPILEPTICUS

1. Clear and maintain airway and oxygenation.
2. Stat intravenous infusion having drawn blood for metabolic studies (glucose, calcium, electrolytes), blood levels, and toxic screen. Give glucose I.V., 50% 1–2 cc./kg.
3. Obtain history.
4. Anticonvulsants.

Plan A: If no previous medications (in particular, phenobarbital) and respirations are not compromised:
1. Diazepam, 1 mg./yr. of age to *maximum* dose of 10 mg. I.V., 1 mg./minute (i.e., slowly). Respiratory arrest may occur; do not use unless ready to support; use plan B.
2. Phenytoin 10–15 mg./kg. I.V. slowly over 30 minutes, not to exceed 25 mg./min., diluted in saline. Monitor EKG. If a known epileptic individual has been taking Dilantin previously, use 5 mg./kg. I.V. Do not give I.M. as it crystallizes and is poorly absorbed.
3. Repeat the Diazepam in 30 minutes if necessary. If seizures persist, give paraldehyde, 0.3 cc./kg. per rectum in oil. Can be repeated in one hour.
4. Phenobarbital, 5 mg./kg. I.V. slowly, can be added in intractable situations. Monitor respirations. Can be repeated in 30 to 60 minutes.

Plan B:
1. Dilantin as above.
2. Phenobarbital 10 mg./kg. I.V. to maximum dose of 200 mg.
3. Paraldehyde as above.

Rarely general anesthesia using pentobarbital or halothane may be necessary.

Maintenance: Dilantin and phenobarbital I.V. or P.O.; maintain levels.

---

the gradual return to maintenance oral medications. Any underlying disease should be appropriately treated.

Status epilepticus is a potentially fatal complication and should therefore be viewed with a high degree of concern. The overvigorous administration of large doses of anticonvulsants by the intravenous route is not without risk, however, and the treating physician must be able to fully support the patient if complications arise.

THE MAJOR CLINICAL SEIZURE SYNDROMES

SEIZURES IN THE NEONATAL PERIOD

Seizures are the most common neurological abnormality in the neonatal period. Their incidence may be impossible to assess accurately, because many infants have such subtle clinical manifestations of seizure activity that they pass unnoticed in the nursery. The great importance lies in the high frequency of associated disorders, both of the central nervous system and of other systems, which may seriously prejudice future neurological development. Some of these are amenable to prompt treatment. Seizures at this age are always secondary to underlying abnormalities, even though an exact etiology may not be identified in

as many as one quarter of the instances. It is essential to approach the convulsing newborn with an attitude of inquiry as to the basic cause of the seizures, and not merely treat symptomatically.

The patterns of seizures in the newborn period are varied, and a full description of all possible manifestations is not appropriate here (Table 39–15). It is of great importance to know that a large number of infants may be having seizures that are unrecognized because of the subtlety of their features.

Many newborn infants are "jittery." This is not a convulsive phenomenon, although some of the predisposing conditions are also potential causes of seizures—such as the metabolic derangements of hypoglycemia, hypocalcemia, and hypomagnesemia, hypoxic-ischemic insult, and maternal drug addiction. This situation should therefore be carefully evaluated in order to correct any underlying abnormalities and prevent convulsions from occurring, since the latter would significantly affect the infant's prognosis for full recovery. Jitteriness is characterized by tremulous movements, sometimes appearing to be clonic, which are extremely stimulus sensitive and can be inhibited by restraint (unlike seizure jerking, which continues regardless). Another distinguishing feature is that jitteriness is not accompanied by gaze abnormalities or jerking eye movements.

Table 39–15. TYPES OF SEIZURE ACTIVITY IN THE NEWBORN

Minimal or subtle	
Oculomotor signs	Nystagmus
	Eye deviation
Orofacial signs	Chewing
	Non-nutritive sucking
	Drooling
Limb movements	Stiffening
	"Bicycling"
	"Swimming"
Cardiorespiratory	Apnea
	Bradycardia
Autonomic	Pallor
	Flushing
Clonic	
Multifocal	Jacksonian (rare)
	Migratory
Focal	
Lateralized	
Tonic	
Focal	
Lateralized	
Generalized	
Myoclonic (rare)	
Generalized	
Focal	
Multifocal	
Lateralized	

The electroencephalogram is useful in the management of neonatal seizures. There are important modifications in the technique to be employed, compared to that for older children, if results are to be meaningful. The criteria for normality are dependent on the gestational age of the infant; significant variations in the electroencephalogram occur in premature infants. The electroencephalogram may be used for prediction of outcome in some full term infants, and this may be helpful to clinicians and parents.

Despite the many causes of neonatal seizures, in approximately 25 per cent of infants the origin remains unknown (Table 39–16). Etiologic considerations are of importance in that some disorders (e.g., hypoglycemia) should be treated rapidly in order to prevent damage. The investigation of these infants therefore should seek common correctable diseases first and be designed to cover all the likely causes whenever possible. A full history (including pregnancy, labor, delivery, and the family) is essential, and a careful neurological and developmental examination should be performed if the baby's condition permits. Blood is drawn for metabolic and septic studies, and urine, skin, and cerebrospinal fluid are cultured for bacteria and viruses. An interictal electroencephalogram may be helpful, and an emergency computed tomographic scan may make the diagnosis in some situations, especially when an intracranial injury or malformation is suspected. Treatment should initially be directed toward correction of the causes; then consideration of anticonvulsants is appropriate (Table 39–17). If seizures are prolonged or life threatening, full doses first of phenobarbitone and then of phenytoin may be given intravenously. Diazepam (Valium) has the advantage of rapid effect, but there are serious draw-

Table 39–16. ETIOLOGIES AND CLINICAL CONSIDERATIONS IN NEONATAL SEIZURES

Causes	Age at Seizure (0–3 vs. >3 days)	Predisposing Factors	Diagnostic Tests	Treatment	Prognosis
Hypoxic ischemic insult	0–3	Low birth weight, maternal disease or drugs, difficult delivery	History, fetal monitoring, Apgar testing, resuscitation	Symptomatic	Poor
Metabolic derangements Hypoglycemia	0–3	Low birth weight, maternal diabetes mellitus, hypoxia-ischemia, infection	Blood glucose	I.V. glucose	Variable
Hypocalcemia (± Hypomagnesemia)	0–3	Low birth weight, hypoxia-ischemia, infection	Blood calcium	I.V. calcium	Variable
	>3	High phosphate, maternal hyperparathyroidism		I.V. calcium	Good
Electrolytes	0–3	Infection, hypoxia-ischemia (SIADH)*	Electrolytes	Fluid balance	Variable
Intracranial hemorrhage Intraventricular hemorrhage	<3 days	Prematurity	CT scan, ultrasound	Symptomatic L.P.'s	Poor
Subarachnoid	<3 days	Full term, traumatic, ischemia-hypoxia	CT scan, L.P.	Symptomatic	Good
Subdural Supratentorial Infratentorial	Anytime	Trauma	CT scan	Neurosurgical	Good
Infection Acquired (bacterial/viral)	>3	Low birth weight, PROM*, vaginal, hypoxia-ischemia, handling	Septic work-up	Antibiotics Antiviral	Poor
Congenital (TORCH*)	Anytime	Maternal infection	Antibody titers	Symptomatic	Poor
Trauma	0–3	Low or high birth weight, instrumental delivery	History, skull x-ray, CT scan	Possible neurosurgical	Variable
Malformations	Anytime	Unknown	CT scan, chromosome studies	Symptomatic	Poor
Drug withdrawal	0–3	Maternal drug abuse	History, toxic screen, mother and child	Temporary phenobarbital	Good for seizures; may be other effects, e.g., fetal alcohol syndrome

*SIADH = Syndrome of inappropriate antidiuretic hormone.
PROM = Premature rupture of membranes.
TORCH = Toxoplasmosis, rubella, cytomegalovirus, herpes infection.

Table 39–17. THERAPY OF NEONATAL SEIZURES

1. Correct underlying metabolic disorders
   a. Hypoglycemia: glucose, 25 per cent, 2–4 ml./kg.
   b. Hypocalcemia: calcium gluconate, 5 per cent, 4 ml./kg. I.V. (EKG monitor)
   c. Hypomagnesemia: Magnesium sulfate, 50 per cent, 0.2 ml./kg. I.M.
2. Anticonvulsants
   a. Phenobarbital, 10–20 mg./kg. I.V. stat; maintenance, 3–4 mg./kg./day I.V./I.M./P.O.
   b. Phenytoin, 10–20 mg./kg. I.V. stat; maintenance, 3–4 mg./kg./day I.V. (not I.M. or P.O., as not well absorbed).
   c. Paraldehyde, 0.3 cc./kg. P.R.
   d. Pyridoxine hydrochloride, 50 mg. I.V., with EEG monitor in resistant cases.

Caveats: Do not overtreat. The prognosis is related to etiology, not the seizures per se.

backs to its use in the neonate. The decision about the duration of maintenance therapy is one for which there are no clear guidelines. Each instance must be considered in light of the persistence of seizure activity or electroencephalographic abnormality. In many infants it is possible to discontinue medication before discharge from the hospital.

The outlook for neonates with seizures has changed in the last few years; mortality has decreased from over 50 per cent to about 20 per cent. The incidence of sequelae in survivors, however, seems to have changed little and remains at about 30 per cent. Thus, half the infants recover completely. The best predictors of outcome are the nature of the underlying disorder and the electroencephalogram, neurological examination in the newborn being a rather poor prognostic indicator. Infants who have a severe lesion, such as a cerebral dysmorphism, or a serious hypoxic-ischemic insult may be predicted to fare badly. Those whose electroencephalogram is relatively well organized, and whose seizures are well defined, are likely to do relatively well. Of full term infants with seizures who have a normal electroencephalogram in the neonatal period, 80 per cent are normal at follow-up. Following discharge, all such infants deserve careful follow-up for at least one year, for it is only in this fashion that evolving early deficits will become apparent and can be appropriately managed. The types of deficits found in follow-up of survivors include combinations of motor deficits, spastic diplegia, hemiplegia, quadriplegia, choreoathetosis, cerebellar ataxia, microcephaly, mental retardation, deafness, varying types of chronic seizure disorders, and learning disabilities.

INFANTILE SPASMS

Between the ages of one and six months, seizures are uncommon, owing to the developmental stage of the brain and in particular the relative paucity of connections within it, rendering the convulsive threshold higher than in later life. When seizures occur, however, they generally carry a gloomy prognosis, since only severely abnormal brains can produce seizures at this age. At about three to six months an otherwise unusual type of seizure begins to appear, "infantile spasms." These are part of a triad known as West's syndrome (after the English physician who first described it in his own son), the other features being an electroencephalographic pattern termed "hypsarrhythmia" and concomitant cessation or regression of intellectual development.

The spasms are short generalized seizures, usually involving the trunk musculature, which produce sudden flexion movements, causing the infant to double over at the waist—the "salaam" attack. There may be associated flexion of the neck, and the limbs may be outstretched and stiff. Less frequently extension movements occur. There may be a bifacial spasm or "smile" or a cry, or some infants may laugh inappropriately. The spasms typically occur in flurries and most often are associated with either the drowsy or arousal states. They are commonly mistaken for colic. Less dramatic forms such as head nodding are seen, but these are often undetected for a period. A cessation of development may have already been noticed by the parents.

The two groups into which patients with infantile spasms may be divided are the symptomatic or secondary group, in whom there is an identifiable cause, and the cryptogenic group, in whom no such cause can be found (Table 39–18). These two groups differ in features at the onset, in prognosis, and possibly in the recommended treatment. The evaluation of these children must include a thorough search for a cause in every infant, even though only a few cases have therapeutic implications (e.g., some inborn errors of metabo-

Table 39–18. CAUSES OF INFANTILE SPASMS

Cryptogenic
Symptomatic
　Cerebral dysmorphism, including
　　Tuberous sclerosis
　　Aicardi's syndrome, i.e., females, agenesis of corpus collosum and retinal colobomas
　　Chromosomal disorders, e.g., trisomy 21

　Cerebral damage
　　Prenatal infection: cytomegalovirus, toxoplasmosis, rubella
　　Postnatal infection: meningitis, encephalitis
　　Perinatal encephalopathies: hypoxia, ischemia, intracranial hemorrhage
　　Trauma: peri- and postnatal
　　Metabolic: phenylketonuria, pyridoxine dependency, pyridoxine deficiency

　Postvaccinal: pertussis

lism). Others have a genetic basis, and counseling should be available for these families. The electroencephalogram shows severe disruption with an abnormal background, multifocal discharges, and a "burst suppression" pattern, which may be most obvious in sleep. Other studies include a computed tomographic scan, an amino acid screen, screening by Wood's light for the hypopigmented ash leaf spots of tuberous sclerosis, and a careful funduscopic examination.

The treatment is somewhat controversial, although since the introduction of steroid therapy in the 1950's there has been agreement that in all cryptogenic cases treatment should be undertaken with either corticosteroids or ACTH. The majority of infants with no identified cause for spasms are given ACTH by daily intramuscular injection in doses ranging from 60 to 80 units. This regimen is continued for two weeks or until toxicity or side effects occur; then decreasing doses are used for a total of six to eight weeks. On this schedule good initial control of seizures is achieved in many children, but the long term results are far from encouraging and the majority of children have recurrent seizures and significant neurodevelopmental deficits. The management of the symptomatic situations is not so clear, since their outcome is predictably poor regardless of therapy. Alternative medications to ACTH or steroids include clonazepam (or nitrazepam outside the United States), sodium valproate, and phenobarbital. Many children have chronic seizure disorders that evolve into the minor motor or myoclonic type and defy good control no matter what is tried.

In addition to specific treatment of the seizures, these patients and their families require much support throughout their lives. The handicaps they demonstrate are varied in both their type and severity, depending on the underlying lesion. Many infants show a degree of physical as well as mental retardation, being of short stature, and often with a strikingly small head size. There may be deficits of vision and hearing, as well as various neurological deficits involving the motor, sensory, extrapyramidal, and cerebellar systems. Many children are seriously handicapped and require extensive support. It is essential that these problems be addressed promptly in order to minimize the degree of handicap and to institute the forms of support therapy that can be of greatest assistance.

MYOCLONIC SEIZURES

The group of seizure disorders in early childhood variably termed minor motor or myoclonic-atonic (astatic) epilepsy, is a significant problem in the practice of the pediatric neurologist. They are anything but "petit" and are very "mal." This is generally a poorly understood group of seizure problems, often with devastating effects on the children concerned and difficult to control. They comprise a mixture of seizure types, the full discussion of which is beyond the scope of this presentation, but their main manifestations are various motor phenomena, with additional features.

Myoclonic seizures are sudden flexion contractions of muscle groups. They may be generalized or focal. In the generalized type the child is usually thrown off his feet or may jackknife to the ground. In the focal type the degree of involvement is variable, from that of a small distal muscle to a proximal muscle or group thereof. Seizures may be unilateral or bilateral. Atonic seizures produce a sudden loss of muscle tone and thus cause a loss of posture. The child may sink or crumple to the ground, slump over, or merely nod his head, depending on the number of the muscle groups affected. In myoclonic and atonic seizures there is no aura, and little or no loss of consciousness. The seizure lasts only a fraction of a second, and the child picks himself up again. The suddenness of the attack makes injury a very real possibility. These children are very likely to hit the head and face as they fall.

Tonic seizures are intermixed with both myoclonic and atonic seizures and may also be generalized or focal; in many children they occur most often at night, when they happen during slow wave sleep and often cause an arousal. They may occur with such great frequency that the child may experience a severe interference with the normal sleep cycle, causing chronic sleep deprivation and daytime drowsiness. This in itself predisposes to seizures, and thus a vicious cycle is set up. Tonic seizures produce a stiffening of the affected parts, interfere with consciousness if they are bilateral, and have the potential for producing respiratory embarrassment, although most of these seizures are very shortlived.

Another type of associated seizure is the "atypical petit mal attack." In this type there is an absence seizure, often with some additional features such as automatisms or clonic movements. The onset is not so sudden as in the classic petit mal seizure and the return to the resting state is not so prompt. Typically all these types of seizures coexist in the one patient, the relative frequency of one or another type varying from time to time. The children with these types of seizures can be classified into two groups as noted in Table 39–19.

The evaluation of the child with "minor motor seizures" involves taking a careful history and performing neurological, ophthalmological, and developmental examinations. The typical electroencephalogram shows a slow spike wave with polyspikes, a poorly organized background, and

Table 39–19. MINOR MOTOR EPILEPSY

1. Cryptogenic, i.e., previously normal*
2. Symptomatic
   Previous infantile spasms (see Table 39–18)
   Perinatal encephalopathies
   Infectious—meningitis, encephalitis, slow virus (e.g., subacute sclerosing panencephalitis)
   Metabolic—degenerative (e.g., ceroid lipofuscinosis)
   Cerebral dysgenesis (e.g., tuberous sclerosis)

*The Lennox-Gastaut syndrome includes children in this group who have a typical slow spike wave EEG and whose mental development is slowed without a degenerative disorder being recognized.

discharges consisting of diffuse fast (beta) activity, which may be correlated with tonic seizures. If there is evidence of a progressive disorder, such as cerebral degeneration (ceroid lipofuscinosis), or of slow virus disease (subacute sclerosing panencephalitis), the cerebrospinal fluid should be examined for an increased protein level or measles virus antibodies. In many cerebral degenerative conditions, the cerebrospinal fluid protein level is increased. Confirmation of the diagnosis requires further studies, e.g., skin biopsy for ceroid lipofuscinosis. A computed tomographic scan will show cerebral atrophy in some instances of static encephalopathy, regardless of the cause.

The management of these disorders is one of the most difficult problems in childhood epilepsy. The older anticonvulsants do not have much effect, apart from phenytoin, which is useful for tonic seizures, and phenobarbitone for the major tonic-clonic seizures, which some of these children have in addition to the other attacks. More effective drugs are those of the benzodiazepine group, namely, clonazepam and niatrazepam (not available in the United States). Sodium valproate is also effective, but if it is given with clonazepam, it may cause absence status. Usually these children are taking at least two drugs, and control is often less than satisfactory. Because of the limited drug choices, "cycling" of medication is inevitable. The spike wave status (stupor) these patients occasionally develop may be aborted by the intravenous administration of diazepam. Methsuximide (Celontin) and ethosuximide (Zarontin) may be useful adjuncts in some patients. Others gain surprising, but often only temporary, relief when acetazolamide is added to the regimen. In some situations ACTH seems to arrest a progressive course by stopping the seizures and improving mentation, although unfortunately not permanently.

It is also in this condition that the ketogenic diet has most application. This entails the adjustment of the child's intake of fat and carbohydrate so that ketosis develops and is persistent. This often has a dramatic effect on the seizure frequency, allowing a reduction in the dosage of anticonvul-

sants. It is a difficult diet from the practical point of view, requiring a diligent parent to implement it successfully. The ease with which it can be organized and tolerated has been increased by the use of medium chain triglycerides as the source of fat. Vitamin and mineral supplements are necessary. In very difficult situations one may be forced to resort to some of the toxic anticonvulsants less used nowadays, such as trimethadione (Tridione) or phenacemide (Phenurone).

All these children require careful assessment of intellectual capabilities and planning for their education. Some may be educable in a normal class, especially if the teacher is sympathetic, but a child who has frequent seizures, especially of the atonic type, is not easy to deal with, even in a very protective atmosphere. Parents need much support in all aspects of their child's care. Often respite admissions to the hospital or another facility are necessary. Sleep disturbance due to frequent nocturnal seizures may be successfully treated by bedtime sedation to deepen sleep and allow normal sleep cycles to occur. This measure alone may dramatically improve the child's well-being. Children who have frequent atonic spells (drop attacks) should wear a helmet for protection.

The prognosis in childhood myoclonic and minor motor seizures is generally not good, for the reasons outlined above, namely associated neurological and developmental deficits, difficulties of seizure control, and the incidence of progressive disorders.

There is a group of children who have only myoclonic seizures. These children are usually older than four years of age at the onset and have a normal history and family history, and the neurological and developmental examinations are normal. No underlying disease is apparent. The electroencephalogram may show spikes and waves or bursts of polyspikes, usually in a normally organized background for age. These children have a good prognosis.

FEBRILE SEIZURES

Febrile seizures are the commonest type of seizure in childhood, occurring in 3 to 5 per cent of the pediatric population. They occur in a well defined age group and have distinguishing features that clearly separate them from other types of seizures, but there are often complicating factors that cloud the issue. The features of "simple" febrile seizures are as follows:

1. The age range is six months to six years.
2. The seizure is short in duration (less than 15 minutes).
3. The seizure is generalized but may be either tonic, clonic, or atonic (rarely fully developed tonic-clonic).
4. There are no focal phenomena.

5. Seizures occur as a single episode during an episode of fever, or at least within a 24 hour period.

6. The findings on neurological examination are normal.

7. The fever is due to extracerebral causes and classically is in the initial stage of rapid ascent when the seizure occurs.

8. There is no history of any neurological problem, including seizures other than febrile, or of perinatal difficulty.

9. There is no family history of seizures other than those of the febrile type.

Additionally, the interictal electroencephalogram is normal. Any divergence from these features may have different implications for the outcome, and the seizure is termed a "complex" or atypical febrile seizure.

The tendency for seizures with fever to occur appears to be an inherited trait, apparently of autosomal dominant inheritance with incomplete penetrance, so that a positive family history in siblings and parents of affected infants is frequently elicited and reassuring. All children must be carefully examined in a search for the cause of the fever. Further investigations will depend on the clinical circumstances, but in most infants (under 18 months), and especially in those with their first episode, a lumbar puncture should be undertaken to eliminate the diagnosis of meningitis. Usually blood is drawn for the routine metabolic studies such as glucose, calcium, and electrolyte determinations, but these are unrewarding in this group of patients, as are skull x-ray views. Of greater importance may be a blood culture. In the acute stage, treatment is confined to that required for the cause of the fever and the administration of antipyretic drugs. In the typical single simple febrile seizure, no anticonvulsant is indicated.

The child who has a "complex" or atypical febrile seizure, however, requires further investigation and therapy. A prolonged seizure often is considered significant enough to warrant immediate institution of anticonvulsant therapy, and, obviously, the child in status epilepticus should be treated as an emergency. Some children have a focal convulsion, and especially if there is transient hemiparesis (Todd's paresis), a computed tomographic scan to rule out a structural lesion may be deemed necessary.

The place of electroencephalography in the evaluation of children with "simple" febrile seizures is somewhat controversial in that to place a child firmly in the "simple" category requires a normal interictal (two weeks after the seizure) electroencephalogram. However, if the situation is clinically typical, the findings on the electroencephalogram are unlikely to alter one's management; thus there is little point in obtaining one. This observation is enhanced by the fact that 15 per cent of the electroencephalograms show nonspecific abnormalities anyway, and this only causes confusion and worry. In recurrent cases, or in those with atypical features, however, an electroencephalogram should be obtained in order to better predict the outcome and the need for treatment.

The question of when to recommend continuous prophylactic anticonvulsant medication is still the subject of much discussion. Children who have "simple" febrile seizures have an increased incidence of later nonfebrile seizures as compared to control subjects, but this incidence (1 to 2 per cent) is still very small. Children with atypical features are more likely to have epilepsy in the future, the rate being increased almost 15-fold.

Currently there is no evidence that phenobarbital prophylaxis begun following a simple or complicated febrile seizure prevents epilepsy, but it does reduce the recurrence rate and perhaps the incidence of complicated recurrences when it is given in adequate doses. The blood level of phenobarbitone should be in the region of 15 mcg. per ml. in order for the drug to afford protection. Phenobarbitone given at the onset of fever is ineffective (unless given intravenously) because of the time taken to reach therapeutic levels.

The major reason for medicating a child with febrile seizures is to prevent further seizures, particularly prolonged ones, for these protracted convulsions may cause cerebral damage. Unfortunately there is no way of predicting which children are going to have prolonged seizures, and the majority of prolonged seizures occur on the first occasion. Other factors of importance are the specific drug to be used and its possible and actual effects on the child and the family. Although febrile seizures are stressful events for parents, few wish to chronically medicate their otherwise healthy young child, especially if there are possible side effects. The drug most frequently used, phenobarbitone, is associated with intolerable behavioral effects (hyperkinesis and insomnia) in 20 per cent of the children, and it also may have ill defined effects on cognitive skills and learning ability.

PRIMARY GENERALIZED SEIZURES

Primary generalized epilepsy is the term used to describe both absence or classic "petit mal" and tonic-clonic or "grand mal" seizures when either or both occur in a typical, fairly well defined clinical setting. The other types of seizure classified as generalized (namely, myoclonic, tonic, and atonic) do not fit into the same pattern, and are dealt with elsewhere.

The mechanisms underlying the primary generalized epilepsies are still uncertain. There are two major theories—first, that there is a central "pacemaker" situated somewhere in the upper brainstem or deep gray matter that is responsible for firing electrical discharges to all areas of the cerebral cortex at the same time (the "centrencephalic" theory) and second, that the structures in the upper brain stem and deep gray matter are firing volleys of electrical activity to the cortex, but that the cortex itself is "over-reactive" and in response to these signals produces synchronous discharges.

Absence or Petit Mal Attacks

In the absence or pure petit mal type of generalized epilepsy, the onset and subsequent disappearance almost always occur in the first two decades of life, usually between the ages of four and 14 years of age. Girls are more commonly affected than boys, and there is a positive family history of this type in some instances. The typical history is of frequent short episodes of lack of attention or loss of awareness, usually in the order of five to 20 seconds in duration, rarely more than 30 seconds long. During this time the child is motionless, but usually does not lose tone or posture, and stares vacantly. In some cases there are subtle motor accompaniments, usually of the eyelids (flickers), 3 cycles per second nystagmus, and occasionally clonic movements of the hands; an object held in the hand may be dropped. There may be automatisms, especially in the attacks that are somewhat longer in duration. A rare child will fall or be incontinent of urine. After the attack is over, the child rapidly returns to normal consciousness. On questioning the child, one may find an awareness of the attack, but no history of an aura is elicitable. Many episodes may be occurring daily, which seriously curtail learning ability and often cause a deterioration in school work. These children are neurologically normal and have a normal IQ. An attack may be precipitated by hyperventilation, a maneuver that may be carried out easily and safely in the office as a diagnostic procedure. The electroencephalogram is abnormal in almost 100 per cent of the children prior to treatment, showing the classic pattern of 3 per second bilaterally synchronous spike and wave discharges either at rest, on hyperventilation, in drowsiness, or on intermittent photic stimulation.

The treatment of absence seizures is straightforward in most children, the drug of choice being ethosuximide. Valproic acid may be equally effective but its relative toxicity, the need for careful monitoring, and the expense keep it in second place. Acetazolamide is an excellent adjunct. The duration of therapy depends on the response, both clinical and electroencephalographic. Continued therapy is recommended for two to four years following the remission of clinical seizures in association with a normal electroencephalogram. The prognosis in typical situations is good.

There is an increased risk of major tonic-clonic seizures. This concern depends on the prior occurrence of such attacks, the age of onset of the absence attacks, (those under five and over 10 having an increased risk), and a family history of major convulsions. The presence of neurological deficits or intellectual impairment also increases the chances of tonic-clonic seizures. The rate of occurrence of major seizures is about 25 per cent. It is recommended that a child with absence seizures in whom there are atypical electroencephalographic features or who has had tonic-clonic seizures be treated with both a drug for absence attacks and a drug for tonic-clonic seizures. Valproic acid rather than ethosuximide may be the drug of choice in such situations, because the latter drug does not protect against major seizures. Some neurologists prefer to use a combination of ethosuximide and a barbiturate. The combination of ethosuximide and phenytoin occasionally may aggravate absence attacks.

Generalized or Grand Mal Attacks

The generalized tonic-clonic seizure is the epileptic convulsion well known to the general public, and the phenomenon pictured by most people when the word epilepsy or grand mal is used. As in the absence seizure group, tonic-clonic seizures that are primary occur in a fairly well defined population of young people. The onset usually occurs in childhood, between the ages of five and 15 years, in a child who is neurologically normal and of normal intelligence. There may be a positive family history. There may be a brief but nonspecific aura of impending trouble. There is a sudden tonic phase in which there may be a cry, and in which the whole body stiffens; the child usually falls to the ground, sometimes injuring himself. In this phase there is interference with respiration and cyanosis, and the tongue may be lacerated. Eventually relaxation occurs, and clonic jerking of all four limbs appears. There is then some respiratory excursion, and the color may return to normal. The clonic phase may last for just a few seconds or persist for many minutes before relaxation of the muscles occurs and normal breathing resumes. During the seizure there may be incontinence of urine or feces, the pupils are usually dilated, and the pulse and blood pressure are increased. The plantar responses are usually of the extensor type for a while postictally. After a period of depressed consciousness the child awakens, often confused, feeling tired, and with a headache. Subsequently sleep occurs, following which the child awakens refreshed and returns to normal. Recurrent tonic-clonic episodes without regaining consciousness constitute status epilepticus; this occurs in less than 5 per cent of epileptics and is a medical emergency.

The interictal electroencephalogram in children with primary generalized tonic-clonic seizures shows irregular spike and wave discharges or polyspikes, which are generalized and sometimes are seen only in sleep. A normal awake and asleep interictal electroencephalogram is seen most often in the child with a "single" seizure and in cases with metabolic causes. The drugs of choice for treatment are phenytoin, carbemazepine, and phenobarbitone, in that order. Generally there is a trend away from the use of barbiturates in children of school age because of the potentially sedative effects of these drugs and, hence, the placing of phenytoin and carbemazepine at the top of the list. It is usual to allow at least two and often four years of freedom from seizures to elapse, and to have a normal electroencephalogram in waking and sleep before tapering medications.

PARTIAL (FOCAL) SEIZURES

Partial seizures, which are always secondary to a gross or microscopic cerebral lesion, are classified according to their clinical features. Careful analysis of the patient's subjective complaints and objective evidence gained by observers may allow appropriate subclassification (Table 39–20).

This classification of partial seizures includes so-called psychomotor seizures and temporal lobe epilepsy, but emphasizes the multiple sites of origin in the brain. Such seizures may present at any age, although complex partial seizures are rarely diagnosed in very young children. This may be because of the necessary reliance on a clear history of sensory and psychosensory features, which may not be forthcoming from a small child. However, it is surprising how often the youngster may give a good description of the aura, and therefore it is worthwhile inquiring about this feature. In many instances the focal part of the seizure is brief, and therefore missed, and only the generalized tonic-clonic convulsion is described. This may lead to an erroneous diagnosis of generalized epilepsy.

Causes of focal seizures in childhood include any localized lesion in the cerebral cortex. This may be traumatic, inflammatory, ischemic-hypoxic, vascular, or neoplastic. The pathological change most commonly responsible for temporal lobe partial seizures in childhood is scarring of the mesial portion of the temporal lobe, incisural sclerosis, which may be the result of hypoxia and ischemia either at the time of birth or during a prolonged, usually febrile, seizure later in infancy or early childhood. In partial epilepsy the electroencephalogram usually shows a focal discharge.

Table 39–20. CLASSIFICATION OF PARTIAL (FOCAL, LOCAL) SEIZURES

A. Simple partial: no impairment of consciousness
  1. With motor signs
    a. Focal motor without march
    b. Focal motor with march—Jacksonian
    c. Versive—head and eye turning
    d. Postural—arm flexion
    e. Phonatory—vocalization or speech arrest
  2. With somatosensory or special sensory signs (simple hallucinations)
    a. Somatosensory (e.g., tingling)
    b. Visual (e.g., light flashing)
    c. Auditory (e.g., buzzing)
    d. Olfactory—smells, usually unpleasant (rule out tumor)
    e. Gustatory—tastes, usually unpleasant (rule out tumor)
    f. Vertiginous
  3. With autonomic symptoms or signs—flushing, piloerection
  4. With psychic symptoms (rare without impairment of consciousness)
    a. Dysphasia (e.g., jargon speech)
    b. Dysmnesia (e.g., déjà vu)
    c. Cognitive (e.g., dreamy states, distortion of time sense)
    d. Affective (e.g., fear, anger)
    e. Illusions (e.g., macropsia)
    f. Structured hallucinations (e.g., music, scenes)
B. Complex partial (psychomotor)
  1. Simple partial onset followed by impairment of consciousness
    a. Simple partial onset (A 1 to 4) followed by impaired consciousness
    b. With automatisms
  2. With impairment of consciousness at onset
    a. With impairment of consciousness only
    b. With automatisms
C. Partial seizures evolving to generalized tonic-clonic convulsions
  1. Simple partial seizures A, evolving to generalized tonic-clonic convulsions
  2. Complex partial seizures B, evolving to generalized tonic-clonic convulsions
  3. Simple partial seizures evolving to complex partial seizures evolving to generalized tonic-clonic convulsions

In temporal lobe epilepsy a sleep tracing is often necessary to activate discharges. Additional leads such as nasopharyngeal or sphenoidal electrodes increase the yield of abnormalities. The search for an etiology is made relatively easy by the availability of the computed tomographic scan, which reveals scars, abscesses, or tumors. However, small tumors may not show up initially, and the examination may have to be repeated at intervals determined by the clinical status of the patient. Additional neuroradiological studies such as arteriography and pneumoencephalography are seldom indicated nowadays, but occasionally may give further necessary information in difficult situations. When secondary generalization is very rapid, further electroencephalographic studies may be helpful, especially a record during repetitive injection of small doses of short acting barbiturates, a method that will distinguish between primary and secondary generalization. The assessment of special senses is of importance when seizures arise in areas subserving vision and hearing. Routine clinical testing should be carried out, including charting of visual fields, and recording of evoked potentials may give more information about the extent of the lesions in the relevant areas of cortex.

The management of focal seizures includes that of the underlying lesion (e.g., excision of a tumor) and anticonvulsant medication. In most children seizures can be brought under control with either phenytoin or carbamazepine. Primidone and less often phenobarbitone may be effective. Sodium valproate may be helpful.

It is important to remember that children with partial seizures may have additional problems, such as focal neurological deficits, which must be addressed in the overall management of the disorder. Depending on the site and extent of the underlying lesion, the child may otherwise be completely normal, may have hemiparesis or hemiplegia, with or without a speech disorder, may have hemianopsia, and may be intellectually impaired to a variable degree. Associated deficits should receive appropriate attention, and educational plans should be made on the basis of thorough psychological assessment, care being taken to assist in specific learning disabilities. Psychiatric consultation is necessary in some children with temporal lobe epilepsy, because concomitant and resultant behavior disorders, and more rarely organic mental disorders (such as psychosis), may be present. Rarely the person with psychomotor seizures may be physically or verbally abusive to others, but this usually occurs only if an attempt is made to restrict activity during the seizure. True rage outbursts involving physical violence are not a common feature of seizures, being seen more frequently as an interictal phenomenon in persons who have epilepsy, and then with clear precipitants.

A particular type of focal seizure seen in children is the so-called sylvian seizure (rolandic seizures, benign midtemporal spike). In this seizure there is often an initial sensory symptom of numbness or tingling in the tongue or face on one side, followed by facial spasms or twitching with speech arrest and drooling of saliva. They are particularly likely to occur at night or in the morning. There may be a progression to a focal clonic seizure involving the arm and leg on the side of the facial twitching and, especially in sleep, generalization to a tonic-clonic convulsion. The attack is usually short lived, and the child rapidly returns to normal. There is typically no history of other convulsive phenomena in the child or the family, and the findings on neurological examination are normal with no abnormality of intellectual develop-

ment. The electroencephalogram shows active central and midtemporal spikes, often in waking but with a marked enhancement in sleep. The record is otherwise normal. These children, who usually present at about six to 10 years of age, are easily controlled by moderate doses of either barbiturates or phenytoin, and when the electroencephalogram normalizes—generally in the early teen years—the dosage of the drug may be decreased and medication stopped. The prognosis is excellent.

It is particularly important to follow a child with focal epilepsy when the etiology is unclear in order to permit prompt recognition of the occasional situation in which a progressive lesion, most often a tumor, is present despite initial negative investigations. Features that should alert the physician to the possibility of such a lesion include a deterioration in previously satisfactory seizure control despite adequate medication levels, a deterioration in school performance, a change in the findings on neurological examination, a change in the electroencephalogram, and the appearance of calcification on skull x-ray views or on the computed tomographic scan.

In a few children with chronic focal epilepsy (epilepsia partialis continua), a chronic inflammatory process may be responsible, the so-called chronic encephalitis of Rasmussen, a diagnosis that can be confirmed only by pathological examination of brain tissue, often obtained at the time of surgical attack in a case of a focal epilepsy. In situations in which a tumor is found, it is usually a slow growing glioma, although some children have hamartomatous malformations.

EPILEPSY SURGERY

The place of surgery in the management of children with epilepsy is well established. This is particularly true of temporal lobe epilepsy. Such measures are not to be lightly undertaken, however, since the results can be devastatingly bad if the situation is not carefully evaluated from many perspectives (not just that of the seizure disorder). The indications for removal of a section of brain in order to cure epilepsy include at least the following:

1. Electroencephalographic demonstration of an electrical focus that is responsible for the patient's seizures (at least most of them, or the type that are the major problem).

2. Seizures refractory to medical management (including prolonged trials of all possibly effective drugs in therapeutic dosage).

3. Absence of serious additional neurological and psychiatric abnormalities.

4. An intelligence level in the normal range.

Those situations with the best results not only

fit these criteria but also are those in which a structural lesion is demonstrable in the area of the electroencephalographic focus. Intensive psychological evaluation is necessary before surgery, the identification of the side of dominance for speech being essential and the assessment of representation of memory being equally important. Intracarotid injection of a short acting barbiturate (Amytal) with testing of speech and memory functions (Wada test) will help with this assessment. At surgery the brain surface recording of the electroencephalogram, corticography, is an essential part of the operation in defining the limits of the focus prior to excision and in testing the remaining cortex. The potential benefits and risks must be clearly stated to the child and the parents, and sufficient time must be allowed for them to consider the situation carefully.

GENETIC COUNSELING

The risk of inheriting idiopathic or primary generalized epilepsy (absence or tonic-clonic) is less than 10 per cent if one parent is affected but can be as high as 25 per cent if both parents are affected. In the case of symptomatic (secondary) epilepsy the risk of inheritance is not increased unless the cause in the parent is one of the specific inherited diseases, such as tuberous sclerosis or neurofibromatosis, in which the risk could be as high as 50 per cent.

EPILEPSY IN PREGNANCY

Seizure frequency occasionally can be increased during pregnancy, and this may be related to poor drug absorption and a hypermetabolic state causing lowering of previously adequate blood levels. The monitoring of blood levels is therefore mandatory with appropriate adjustment of dosage. Status epilepticus occurs rarely and is an indication for rapid control, preferably with diazepam. Because of the possibility of bleeding related to deficient coagulation factors, treatment with vitamin K is recommended for the mother and infant. Withdrawal symptoms may be seen in infants of mothers who are taking barbiturates. Breast feeding is not contraindicated.

There appears to be a slightly increased risk of teratogenic effects, but whether this is medication, seizure, or "epilepsy" related is uncertain. The babies may have a lower birth weight and a higher perinatal morbidity rate.

DRIVER'S LICENSE

Epileptic drivers appear statistically to have poorer driving records than nonepileptic drivers. Well controlled epileptic individuals should be allowed to drive under controlled circumstances, which vary from state to state. The patient should be given careful advice concerning compliance with medication taking, avoiding alcohol and other drugs, and adequate sleep.

MICHAEL J. BRESNAN
ELAINE M. HICKS

REFERENCES

Berg, B.: Prognosis of childhood epilepsy—another look (editorial). N. Engl. J. Med., 306:861, 1982.
Committee on Drugs: Valproic acid: benefits and risks. Pediatrics, 70:316–319, 1982.
Delgado-Escueta, A. V., et al.: Current concepts in neurology—management of status epilepticus. N. Eng. J. Med., 306:1337–1340, 1982.
Hauser, W. A., et al.: Seizure recurrence after a first unprovoked seizure. N. Engl. J. Med., 307:522–528, 1982.
Holowach Thurston, J., et al.: Prognosis in childhood epilepsy: additional follow-up of 148 children 15 to 23 years after withdrawal of anticonvulsant therapy. N. Engl. J. Med., 306:831, 1982.
Kurokawa, T., et al.: West syndrome and Lennox-Gastaut syndrome: a survey of natural history. Pediatrics, 65:81, 1980.
Masland, R.: The physician's responsibility for epileptic drivers. Ann. Neurol., 4:485–486, 1978.
Matsumoto, A., et al.: Long-term prognosis after infantile spasms: a statistical study of prognostic factors in 200 cases. Dev. Med. Child Neurol., 23:51–65, 1981.
Nelson, K. B., and Ellenberg, J. H.: Febrile Seizures, New York, Raven Press, 1980.
O'Donohoe, N. V.: Epilepsies of Childhood. In Postgraduate Pediatric Series. London, Butterworth & Co. (Publishers) Ltd., 1979.
Penry, J. K., and Daly, D. D.: Advances in Neurology. Vol. XI. Complex Partial Seizures and Their Treatment. New York, Raven Press, 1975.
Tyrer, J. H. (editor): The Treatment of Epilepsy. Current Status of Medical Therapy. Philadelphia, J. B. Lippincott Company, 1980, Vol. V.
Volpe, J. J.: Neurology of the Newborn. Philadelphia, W. B. Saunders Company, 1981.
Wilder, B. J., and Bruni, J.: Seizure Disorders: A Pharmacological Approach to Treatment. New York, Raven Press, 1981.
Woodbury, D. M., Penry, J. K., and Pippenger, C. E. (Editors): Antiepileptic Drugs. Ed. 2. New York, Raven Press, 1982.

39F Childhood Autism

Since Leo Kanner's description of the condition in 1943, autism has been the subject of a great deal of theory, treatment effort, and research. Nevertheless an accurate diagnosis and effective treatment of autism still pose a serious challenge

to even the most competent practitioner. Autistic children often present the symptoms characteristic of a number of other diagnostic categories (e.g., mental retardation, schizophrenia). In addition, autism has been resistant to many medical and psychotherapeutic practices that have been successful with other syndromes (e.g., chemotherapy for the treatment of manic-depressive disorders, psychotherapy for the treatment of phobic disorders).

Before attempting a diagnosis or treatment of autism, therefore, one must understand autism and the behavioral excesses and deficits characteristic of autistic children, understand the approaches to treatment that have been shown to be most effective when working with the autistic child, and learn how to work cooperatively with others in providing a comprehensive program of services for the autistic child, adolescent, and adult.

DEFINITION

Kanner's definition of autism was the result of his systematic observation of 11 children with a previously undiagnosed syndrome. Kanner noted that each of the children displayed a variety of behavior patterns that distinguished them from children with other psychiatric disorders. The features included: social withdrawal, which he described as "an extreme autistic aloneness" evident from the early months of life; severe communication deficits, such as mutism or the inability to "convey meaning" through whatever speech may exist; a desire for the "maintenance of sameness" in the environment; and a severe limitation in the occurrence and variety of "spontaneous activity" as evidenced by a preoccupation with manipulating objects and stereotyped play habits. Kanner labeled this syndrome "early childhood autism."

Since Kanner's early work, much has been learned about autism. This increase in interest and knowledge is reflected in a number of elaborations of Kanner's (1943) definition as well as in the publication of several new definitions for the condition. For example, Ritvo and Freeman's (1977) definition of the condition was published under the auspices of the National Society for Autistic Children and has become the Society's official definition of autism. An abstract of this definition and description follows:

Autism is a severely incapacitating, life long developmental disability, which typically appears during the first three years of life. It occurs in approximately five of every 10,000 births and is four times more common in boys than in girls. It has been found throughout the world in families of all racial, ethnic, and social backgrounds. No known factors in the psychological environment of a child have been shown to cause autism. The symptoms are caused by physical disorders of the brain. They must be documented by the history or be present on examination. They include the following:

1. There are disturbances in the rate of appearance of physical, social, and language skills.
2. Responses to sensations are abnormal. Any one or a combination of sight, hearing, touch, pain, balance, smell, taste, and the way a child holds his body is affected.

3. Speech and language are absent or delayed in developing, although specific thinking capabilities may be present. Immature rhythms of speech, a limited understanding of ideas, and the use of words without attaching the usual meaning to them are common.
4. There are abnormal ways of relating to people, objects, and events.

Typically these children do not respond appropriately to adults and other children. Objects and toys are not used as normally intended. Autism occurs by itself or in association with other disorders that affect the function of the brain, such as viral infections, metabolic disturbances, and epilepsy. On IQ testing, approximately 60 per cent have scores below 50, 20 per cent between 50 and 70, and only 20 per cent greater than 70. Most show wide variations in performance on different tests and at different times. Autistic people live a normal life span. Since symptoms change, and some may disappear with age, periodic re-evaluations are necessary to respond to changing needs. The severe form of the syndrome may include the most extreme forms of self-injurious, repetitive, highly unusual, and aggressive behavior. Such behavior may be persistent and highly resistant to change, often requiring unique management, treatment, or teaching strategies. Special educational programs using behavioral methods and designed for specific individuals have proven most helpful. Supportive counseling may be helpful for families with autistic members, as it is for families who have members with other severe life long disabilities. Medication to decrease specific symptoms may help certain autistic people live more satisfactory lives.

BEHAVIORAL CHARACTERISTICS

An understanding of the behavioral features characteristic of autistic children is essential to accurate diagnosis and effective treatment. Newsom et al. (1979) and Ritvo and Freeman (1977) have provided useful summaries of these autistic behavior patterns.

SENSORY PROCESSES

"Blind while seeing, deaf while hearing" is a statement frequently used to describe the autistic child's apparent insensitivity to the sights and sounds of his environment. Although examination of autistic individuals may rule out blindness or deafness, their behavior is very similar to that of children suffering from severe losses of hearing and vision.

Vision. There may be lack of eye contact with others in the environment, frequent use of peripheral vision, an appearance of "looking through" rather than at others, and a preoccupation with visual details of objects and sources of illumination. However, Newsom and Simon (1977) using a simultaneous discrimination procedure have found that only two of the 12 autistic children they tested had subnormal vision. Their results support the widely held assumption that the majority of autistic children are not visually impaired.

Hearing. One may find a lack of responsiveness or an exaggerated responsiveness to variations in sound, a failure to respond to sounds, such as speech from others, and a preoccupation with self-induced sounds. Newsom et al. (1979) described the use of operant conditioning procedures to measure hearing thresholds in two severely retarded, nonverbal autistic children. Both children were found to have "slight bilateral hearing losses" (25 to 40 db.).

Tactile, Olfactory, and Gustatory Sensitivity. Included are diminished or exaggerated pain and temperature sensitivity, specific food preferences and exaggerated sensitivity to food textures, a preoccupation with the texture of objects in the environment (e.g., shoes, clothing), and sniffing, licking, or ingesting inedible objects.

PERCEPTION

Many types of autistic behavior have been attributed to perceptual abnormalities. For example, autistic children may display an anticipatory response to gestures, a delayed capacity to appropriately utilize objects or to assign meaning to them, and an insistence on "sameness" characterized by an acute awareness of a sequence of events and a difficulty in accepting a change in sequence.

Autistic children perform poorly relative to nonautistic children of the same age in visual and auditory discrimination tasks, and they have difficulties when they are expected to learn stimuli presented in more than one modality. It has also been demonstrated that in the presence of multiple stimuli, autistic children generally "overselect" and learn to respond to a very limited number of stimuli. There is also evidence that autistic children "overgeneralize" in learning situations; i.e., the child responds to irrelevant stimuli in the situation. Research on overselectivity and overgeneralization has helped to explain some of the reasons it is difficult to teach new forms of discrimination and new concepts to autistic children.

SPEECH AND LANGUAGE

Communication problems constitute the most debilitative behavioral deficit in autistic children. It has been estimated that approximately 50 per cent of autistic children remain essentially nonverbal throughout their lives; i.e., despite the apparent absence of a physiological basis for their mutism, they have not learned to speak. If an autistic child has not acquired communicative speech by the age of five, his prognosis for ever achieving it is generally poor. As a result the major focus of therapy for autistic children is the development of speech and language. Observations of autistic children make reference to a number of speech and language disorders.

Echolalia. Immediate echolalia refers to the autistic child's repeating all or part of what has just been said to him. For example, a question to the child, such as "How are you today?" may elicit the response, "How are you today?" Delayed echolalia refers to the autistic child's repeating what is spoken to him but only after a period of time has passed. For example, the child may suddenly repeat some part of a parent's disciplinary command ("Don't play in the living room!") several hours or even days after the interaction has occurred.

It has been estimated that immediate or delayed echolalia is characteristic of up to 75 per cent of the autistic children who eventually gain speech. However, although the imitation of speech sounds is an important aspect of normal speech development, echolalia can pose a serious obstacle to speech training in autistic children, so much so that the elimination of echolalia is considered to be an indicator of success in the treatment of autistic children.

According to the results summarized by Newsom et al. (1979), research on immediate and delayed echolalia has several implications for teaching language to autistic children: Delayed and immediate echolalia appear to be separate phenomena: delayed echolalia appears to be maintained by "intensive reinforcement," whereas immediate echolalia apparently occurs as a result of the child's failure to adequately comprehend what is said to him. Therefore, these separate problems should be treated differently. Delayed echolalia, like immediate echolalia, may represent a developmental stage in language acquisition, a stage that autistic children apparently fail to outgrow.

Affirmation by Repetition. Affirmation by repetition refers to the autistic child's echolalic response to questions for which a "yes" answer is expected. For example, the child who is asked "Do you feel good?" responds "Do you feel good?" Newsom et al. (1979) suggest that this behavior develops as a result of its being reinforced by others and probably has no more meaning or purpose than any other echolalic response.

Elective Mutism. Elective mutism refers to a disorder in which there is no language retardation but rather a selectivity in the use of spoken language. The disorder generally develops at about age three to five years, after a period of normal speech development. The usual pattern is that the child is able to talk but does not or cannot in certain circumstances or to certain people (e.g., talking at home but not at school). At least two distinct subgroups of electively nonspeaking children have been described: immature, unresponsive, or oppositional children who use mutism in an attention seeking or evasive manner, and anxious, fearful children who use mutism as a fear reducing mechanism. Whatever the type or cause of this withholding of language, there appears to be no abnormality in language comprehension or production; when these children do speak, their sentence structure, vocabulary, and articulation are normal.

Pronominal Reversal.   Pronominal reversal refers to the tendency of the autistic child to confuse pronouns, especially the pronouns "I" and "you." The result is that the child rarely uses the pronoun "I" in his speech. Although a number of theories have been presented to explain this phenomenon (e.g., Bruno Bettelheim's contention that pronoun reversal is an attempted "denial of self"), it is best understood as another example of echolalia. For example, the child may respond with a "you" answer to a "you" question (e.g., "Do you like candy?") and with an "I" answer to an "I" question (e.g., "Do I like candy?").

Other Speech and Language Disorders. Rutter (1972) has pointed out that autistic children are typically slow in producing meaningful words and are significantly delayed in their use of phrases. Sentences, when they are produced, are grammatically incomplete, with articles, prepositions, and conjunctions omitted. In about 20 per cent of the children, Rutter reports a sequence of normal language development followed by a regression between 18 and 30 months of age, with subsequent impaired language development.

According to Rutter, even the autistic child with language competence uses speech for communication far less frequently than does the nonautistic child. The older autistic child with good verbal fluency is rarely able to join spontaneously in conversation with others. In addition, there may be abnormalities in the speech delivery of the older autistic child (e.g., flat delivery with little emotion or inflection and use of a special voice, such as a whisper).

Research in this area has resulted in a number of procedures used to train speech in nonverbal autistic children, especially the teaching of sign language.

SOCIAL SKILLS

The autistic child may be socially withdrawn to the extent that he is indifferent or actively resistant to interaction with others. The child may display a lack of eye contact, a lack of appropriate responsivity to people and physical contact, an absence of cooperative play and friendships, superficial immature interactions with peers and adults, anxiety toward strangers, and absence or delay of an appropriate smiling response. The autistic child's most obvious social skill deficit has been described by Schreibman and Koegel (1981): "They may react to people more as if they were objects than as if they were people. For example, a child may climb into his mother's lap, not for affection, but in order to reach a cookie jar."

BEHAVIORAL EXCESSES

Autistic children are likely to prefer self-stimulation to interaction with or stimulation from others. Self-stimulatory behavior appears to have little utility other than to provide sensory input for the child and may include rhythmic body rocking; arm, hand, or object spinning; finger flapping; facial grimaces; bizarre postures and body movements; gazing at hands and light sources; and making bizarre sounds and noises. Self-stimulation is of importance because the bizarre conspicuous nature of the behavior makes community integration of the child difficult and further may actually interfere with the child's learning new skills.

Recent research on the reduction of self-stimulatory behavior in autistic children has been focused on the sensory reinforcement assumed to be maintaining such behavior. Rincover (1978) successfully reduced or eliminated self-stimulatory behavior such as object spinning by avoiding the visual or auditory elements produced by such behavior (e.g., by placing a pad under the object).

Autistic children may engage in severe forms of oppositional behavior. Tantrums may include screaming, crying, kicking, scratching, throwing or destroying objects, and so forth. Severe tantrums may include acts of self-injury (i.e., the child may mutilate or inflict injury on himself or herself by hitting, biting, or hair pulling) or aggression toward others in the environment (e.g., parents, siblings, therapists, teachers).

Research on self-injury and aggression in autistic children shows that demands appear to evoke the behavior, and it is apparently reinforced by a termination of demands. It is also thought that these behavior patterns must be controlled or eliminated if effective training and education of the autistic child are to be possible.

DIFFERENTIAL DIAGNOSIS

The differential diagnosis of autism is often difficult, since many of the types of behavior identified as being characteristic of autistic children are also exhibited by other types of children. Moderate to severe social withdrawal, a great frequency of self-stimulation, and pronounced communication and academic deficits are seen in children with a number of other diagnoses, including serious mental retardation, specific sensory deficits, childhood schizophrenia, physical or psychological trauma, progressive organic brain disease and congenital, developmental, and acquired disorders in the mechanisms associated with language processes. Similarly, self-stimulation and mild levels of social withdrawal and language delay are not unusual in otherwise normal children.

What may prove to be the most difficult task for the practitioner and diagnostician is to distinguish autism from mental retardation and childhood schizophrenia. However, as described by Ritvo and Freeman (1977), Schreibman and Koegel (1980), and others, there are several important distinctions. For example, although both autistic and mentally retarded children typically score low on IQ tests, retarded children tend to do more poorly on a wider variety of tasks than autistic children. Autistic children may show isolated areas of exceptionally high levels of functioning ("islands of ability"), such as the "autistic savants" described by Rimland (1978). For example, Rimland describes how one eight year old child was capable of the rapid accurate multiplication of four digit numbers.

Relative to schizophrenic children, autistic children show a much earlier age of onset (less than 30 months) and more significant language deficits. Also schizophrenia is characterized by the presence of a thought disorder (according to the American Psychiatric Association's *Diagnostic and Statistical Manual*), and the presence of such a disorder is difficult to assess in a nonverbal child.

The Diagnosis of Autism.   Another obstacle to differential diagnosis concerns the diagnosis of autism itself. As previously described, there are a number of definitions for the condition, and they are not in complete agreement. In addition, the definitions for autism do not describe a specific disease or condition but rather a syndrome characterized by a large number of symptoms. Furthermore, not all the symptoms need to be present for a diagnosis of autism, making it possible for children with very different symptomatologies to be diagnosed as being autistic. The diagnosis, therefore, becomes a question of "How autistic is the child?", that is, how many of the symptoms does he exhibit and to what extent does he exhibit them? As Rutter (1978) has noted: "Kanner's use of the term *autism* was more than a simple label, and that is . . . the trouble. . . . It was also a hypothesis . . . that behind the behavioral description lay a disease entity" (p. 141).

Koegel and Schreibman (1981) have noted a number of problems resulting from this heterogeneity among autistic children. Specifically the diagnosis does not facilitate communication, because professionals often disagree in their definitions of autism and because the diagnosis does not specify the behavioral excesses and deficits of the individual child. Moreover, the diagnosis does not suggest a treatment, since most treatment methods are symptom specific (e.g., language training) as opposed to syndrome specific. Finally, the "autism" diagnosis does not indicate a prognosis: Although most autistic children do not improve without intensive treatment, some do, and the degree of improvement in response to treatment may be highly variable.

Given these problems, Christian (1981), Schreibman and Koegel (1981), and others recommend a careful assessment of the behavior patterns characteristic of each child rather than an attempt to define the syndrome. The result is a more "functional diagnosis" of autism, one that is child specific, prognostic, and conducive to more effective communication among parents, physicians, psychologists, and teachers. Functional diagnosis has been facilitated by the publication of rating scales, surveys, and lists of the types of behavior characteristic of autistic children (as described by Christian, 1981) and detailed descriptions of autistic behavior, such as that included in the first section of this chapter (see also Newsom et al., 1979).

ETIOLOGY

The psychogenic hypothesis considers autism to be the result of an internal disturbance produced by environmental factors, with particular attention given to the mental health of the parents. Kanner offered one such hypothesis, referring to parents of autistic children as "refrigerator type." Bettelheim, a major proponent of the psychogenic theory, has described autism as being due to "the parents' wish that the child should not exist." These concepts are not widely supported in current times.

In contrast, the biological or "biogenic" theory of causation considers autism to be the result of physical disorders of the brain, biochemical deficiencies, or metabolic disturbances. The biological view is supported by evidence that autism is seen in the first months of life and is associated with signs of neurological dysfunction (e.g., disturbances of developmental rates, responses to stimuli, and speech and language capacities). However, as noted by Schreibman and Koegel (1981):

At this point in time there are many . . . theories relating to causation. However, there is no solid evidence to support any one of these theories. The general consensus now among professionals is that the disorder is probably organic in nature, and most likely does not have an environmental cause. However, to attempt to be more specific than that at this time is to move completely into the realm of speculation (p. 502).

Several instances have now been recorded of a presentation as autism by individuals with the fragile X syndrome. (See Chapter 19.)

PROGNOSIS

As previously described, the heterogeneity autistic children makes it difficult to indicate a general prognosis for the condition. However, it is possible to determine a prognosis for a specific child who has been diagnosed as being autistic. Rutter (1972) suggests a consideration of the following factors in determining a prognosis:

1. A critical prognostic factor in autism is the measured level of intelligence. Children with performance IQ's of less than 50 or 60 have a very poor prognosis and are likely to require long term care. However, this prognosis can be improved with early intensive treatment.

2. The degree of language impairment is the critical prognostic factor in autistic children of nearly normal intelligence. If severe deficits in receptive and expressive language are still evident at about five years of age, the child is likely to have a poor prognosis for nearly normal adjustment. However, autistic children of average intelligence may have a good prognosis for at least a fair level of ultimate adjustment despite significant language deficits at age five.

3. The prognosis is also a function of the treatment that the child receives. The work by Lovaas and his colleagues suggests that intensive behavioral training at a very early age (three years) and a very structured preschool program are likely to be most beneficial.

APPROACHES TO TREATMENT

PARENT CENTERED PSYCHOTHERAPY

The major approaches to the treatment of childhood autism have included parent centered psychotherapy, chemotherapy, and behavior modification. Parent centered psychotherapy was the first procedure used to treat autism and was based on the assumption that parent behavior was responsible for the child's problems. Parents of autistic children have been portrayed as being "aloof," "lacking real warmth," and "unstimulating to their children." Kanner described these parents as being obsessive, perfectionistic, humorless individuals who were reported to be "cold," "formal," and "extremely anxious" in social situations. The goal of therapy therefore was to "cure" the parents in the hope that a remediation of the child's problems would follow.

However, the theoretical underpinnings for this approach to treatment have been refuted. McAdoo and DeMyer (1977) reviewed the available research on the parents of autistic children and reached the following conclusions:

1. Parents of autistic children show no more signs of mental or emotional illness than parents of children with nonpsychotic organic disorders or nonpsychotic emotional disorders. Parents of autistic children demonstrate significantly less psychopathology than do parents who are patients in a psychiatric outpatient clinic.

2. Parents of autistic children as a group do not have extreme personality traits such as coldness, obsessiveness, social anxiety, or extreme rage. No specific defects in acceptance, nurturing, warmth, feeding, and tactile and general stimulation of their infants have been identified.

3. The uncertainty and confusion demonstrated by most parents in relation to rearing their autistic child may result from the stress of trying unsuccessfully to understand and influence a child with a neurobiologically based defect in both verbal and non-verbal communication. Such a situation would constitute a severe and continuing stress which may result in personality changes as well as pose a potential threat to the marital bond (p. 165).

In addition, there are findings indicating that parents of autistic children are typically well adjusted and that many have normal children as well. Nevertheless, countless autistic children in desperate need of services have been left in waiting rooms while their parents have been the sole focus of expensive, time consuming, and often unnecessary psychotherapy.

CHEMOTHERAPY

Chemotherapy for autistic behavior is suggested by the biological or "biogenic" theory, i.e., that autism results from physical disorders to the brain, biochemical deficiencies, or metabolic disturbances and that these problems can be remedied by pharmacological agents. However, although the biological approach has generated a wealth of promising research, as yet limited insight has been gained into new technologies of treatment. Results of numerous studies in the areas of biochemistry, chemotherapy, and cognitive dysfunction have remained ambiguous. In summarizing the present status of this research, Ritvo (1976) has concluded:

Follow-up studies are just beginning to appear in the medical literature. They will eventually answer the question as to the natural history of the disease. Since we do not know the specific cause or neuroanatomical or neurobiochemical pathology involved, specific etiologically based therapy is unavailable (p. 5).

Therefore, although chemotherapy may be utilized to decrease specific symptoms, it is not now considered a viable treatment for autism. As noted by Campbell et al. (1977): "Clearly, more well controlled and well thought out studies are needed to ascertain the utility of drugs and therapeutic adjuncts in the treatment of infantile autism; or whether their role is in management only" (p. 159).

BEHAVIOR MODIFICATION

The treatment of choice for autistic children as recognized by the National Society for Autistic

Children consists of "special education programs using behavioral methods and designed for specific individuals" (Ritvo and Freeman, 1977). The use of behavioral methods is a child centered approach to treatment that is intensive and highly structured. The approach is based on the realization that the autistic child is severely withdrawn and concerned only with maintaining sameness in his environment, and the determination that such a situation must not be allowed to continue (i.e., the child must attend, comply, and respond before he can be helped).

Intrusion into the world of the autistic child may be met with resistance that resembles a life or death struggle (e.g., severe tantrums of long duration, aggression toward others, destruction of property, and even self-mutilation). Oppenheim (1974) has described this type of resistance in her nonverbal autistic son, Ethan:

> Ethan, then four and one-half years old, did not *look* when I tried to show him something; he gave no indication that he was *listening* to what I was telling him; and he made not the slightest effort to *imitate* the procedures I was trying to teach him. . . . I realized, simply on a common sense basis, that there was one fundamental prerequisite which had to be accomplished before teaching of any kind could begin. . . . I had to require Ethan to *attend*. The ensuing battle of wits between Ethan and myself was a shattering experience for me. . . . But the resultant change in Ethan taught me a lesson I have never forgotten: the establishment of control is the crucial step in any educational program for autistic children (p. 34).

Rationale. As described by Christian (1981) and Luce and Christian (1981), the behavioral approach is concerned primarily with the child's observable behavior rather than some inferred pathogenic state or internal disturbance. Attention is focused on the environment—on the environmental conditions that act as antecedents and consequences for the child's "autistic" behavior. Treatment is educational, i.e., structuring the environment in such a way to increase the probability of adaptive behavior while decreasing the probability of maladaptive behavior.

Specifically, a desirable consequence ("reinforcement") of a behavior increases the likelihood that that behavior will occur again; an undesirable consequence ("punishment") decreases the likelihood that the behavior will recur. Antecedent conditions associated with desirable consequences of a behavior tend to occasion the occurrence of the behavior in the future; antecedent conditions associated with undesirable consequences of the behavior tend to discourage the future occurrence of the behavior. For example, if a child receives more reinforcement (praise, eye contact, interest) from social interaction with the mother than from the father, the mother's presence will occasion more social interaction from the child than will the father's.

These operations are powerful in their effects on behavior regardless of whether the behavior being affected is appropriate or inappropriate. Therefore, if attention is reinforcing to the child and tantrum behavior brings attention from the mother, there is an increased probability that tantrums will occur again in similar circumstances, i.e., when the child seeks attention and the mother is present.

The fundamental components of the behavioral approach include assessing the special needs of the child, developing plans and goals for training based on the child's needs, analyzing the behavior as a function of its environmental context, applying procedures of proven effectiveness in children displaying severe behavior problems, and evaluating the effectiveness of each training effort.

Since the behavioral approach has been identified as the treatment of choice for educating and habilitating autistic children, the actual procedures involved in this approach will be presented in the following discussion. The basic components of this approach applied to the special needs of autistic children have been described in a number of publications. The following discussion is based largely on the procedural models presented by Christian (1981), Schreibman and Koegel (1980), and Rincover et al. (1978).

Determining the Child's Special Needs

As previously explained, it would be a mistake to expect that all autistic children will display the same types of behavior and thus present the same needs. Those with experience in working with autistic children are constantly amazed at the diversity of skills and deficits they exhibit. Some children display "special abilities," such as the "autistic savants" previously described, whereas others may display severe deficits with no apparent special skills. Some autistic children may be self-injurious to the extent that they must be protected from themselves. In some instances, children labeled "autistic" actually display only a few of the characteristic behavior patterns typical of autism.

"Baseline" Observation. Therefore, the needs of the individual child must determine the problems targeted for intervention, the goals set, and the strategies selected for implementation. For example, a great frequency of stereotyped, self-stimulatory behavior is considered by many to be a "classic" characteristic of autistic behavior (e.g., flapping the hands or fingers, rocking, facial grimacing). One may observe a child engaging in self-stimulation and decide to "change" the behavior. However, if one first takes a preliminary or "baseline" record of how frequently the child engages in the behavior, one finds that he actually spends little time in self-stimulation. Such observation often reveals other needs that are more deserving of therapeutic intervention. One also learns when and in what context problem behavior is displayed, information that is important in determining the strategy to employ.

This preliminary "baseline" assessment also provides an important index for use in evaluating the effectiveness of one's intervention. For example, if a child makes no spontaneous sounds during the baseline observation and is capable of five spontaneous sounds after two weeks of intervention, it is obvious that progress is being made. This type of reinforcing feedback is important to the parent or professional, since the child may be resisting even the most effective intervention, thus providing one with little encouragement and, indeed, increasing the difficulty of one's task.

Standardized Assessment.   The use of a standardized assessment instrument can provide information concerning a child's adaptive behaviors, e.g., skills necessary for independent day to day living. For example, the Adaptive Behavior Scale assesses the levels of the child's skills in a large number of categories, including those of particular relevance to the autistic child, i.e., adaptive behavior, such as self-care, socialization, and communication, and maladaptive behavior, such as aggression and a high rate of self-stimulation. Such an instrument provides one with a large number of specific baseline records against which the effectiveness of behavior change and education efforts can be determined.

Measures of adaptive behavior are much preferred over instruments yielding an intellectual quotient (IQ scores) when one is working with autistic children. Investigators have noted that IQ scores, although of value in assessing the potential of academic performance for average, middle class children, often fail to provide the most useful (i.e., prescriptive) information in the case of the handicapped child. It has also been observed that handicapped children typically are unable to perform tasks required in standardized psychological assessment.

In summary, baseline assessment is possible only when one uses objective terms to describe the target behavior. The objective description of the problem behavior includes a determination of such parameters as the frequency, duration, and context of the behavior. For example, suppose that a child engaged in 10 tantrums (defined as crying, screaming, or head slaps); each tantrum had an average duration of 85 seconds; two occurred in the living room and eight during mealtime. This wording is quite different from the more subjective terms one typically uses to describe a behavior problem; e.g., "My child has tantrums all the time" or "He was a bad boy today."

Procedural Strategies for Behavior Change

Functional analysis of the child's behavior thus becomes a critical component of effective behavioral change. Such an analysis enables one to understand the operation of the environment in occasioning and maintaining the child's behavior. This understanding makes it possible for one to learn how to use the reinforcing or negative aspects of the environment in establishing structure and control, motivating the child, and giving the child feedback concerning his behavior.

Christian (1981) has provided examples of these environmental operations:

1. Operations that increase the frequency of a behavior
   a. Providing desirable consequences for the behavior (e.g., a food item or praise for good work)
   b. Removing an undesirable consequence for the behavior (e.g., letting the child come out of his room when screaming ceases)
2. Operations that decrease the frequency of a behavior
   a. Presenting an undesirable consequence for the behavior (e.g., a scolding, sending the child to his room for a brief "time out" period)
   b. Withholding a desirable consequence that the child is accustomed to receiving (e.g., ignoring rather than attending to temper tantrums)
   c. Removing a desirable consequence (e.g., temporary, 10 second removal of a child's plate until screaming at the table ceases)
   d. Requiring some effort or change in the child's behavior as an undesirable consequence (e.g., requiring that the child engage in a brief period of calisthenic exercise contingent upon the inappropriate behavior)

However, knowing the operations that bring about behavioral change is only one aspect of effective intervention with autistic children. Researchers have found that education and rehabilitation of autistic children are possible only when a number of critical issues and procedural guidelines are adequately addressed (Rincover et al., 1978; Schreibman and Koegel, 1981). These issues and guidelines have been summarized by Christian (1981).

Early Intervention.   Identification of handicaps and intervention should take place as early as possible with autistic children, preferably before three years of age and always before five years of age. The most important reasons to begin work with the child at an early age are that language plays a critical role in the development of other skills and should be taught as soon as possible, and the longer the history of deviant behavior, the more difficult it is to change.

Adequate Training Environment.   As previously discussed, instructional control must first be established; i.e., the child must comply with simple requests to "Sit down" or "Look at me." Training sessions should be kept brief (five to 10 minutes) at first and gradually extended. An attempt should be made to minimize extraneous and potentially distracting stimuli in the environment where training is to take place. As previously described, autistic children tend to "overselect" certain stimuli in their environment for attention.

The Proper Use of Instructions. Instructions must be brief, clear, consistent, and to the point. One must be sure that the child is attending to the instruction; requiring eye contact from the child as the first step in giving instructions is a way to ensure that he is attending and thus is more likely to follow instructions. The child must be required to attend and to perform if training is to be effective. Having the child "practice" the same response or behavior a number of times can help to ensure that he is attending to and complying with instructions.

The Effective Use of Reinforcers. A first step in the effective use of reinforcers is to determine what is reinforcing for a particular child by "sampling," e.g., presenting several items and seeing which one he prefers. Although food reinforcers are typically very effective, one must choose a type and amount of food that does not lead to rapid satiation (e.g., sunflower seeds, raisins, dry cereals).

One immediately reinforces each occurrence of the desired behavior when teaching a new skill. One may go to an intermittent schedule of reinforcement later in the teaching process, i.e., reinforcing every other behavior, or perhaps reinforcing the child for every five minutes of the desired behavior.

The Use of Attention as a Reinforcer. One must realize the power of one's attention as a reinforcer. If close observation of a pattern of behavior such as tantrums indicates that it is consistently followed by attention, systematic withholding of attention contingent upon the behavior becomes the treatment of choice. It is also important concurrently to make a special effort to attend to the child when he is not engaging in the undesirable behavior, i.e., the use of attention to reinforce the child when he is not engaged in a tantrum.

One must be prepared for the child to actively resist this procedure, e.g., the frequent report by parents that "ignoring it just made it get worse." Many parents and helping professionals may abandon the procedure in response to heavy resistance from the child, out of fear that the problem will continue to escalate. However, discontinuing the procedure will no doubt be perceived as a desirable consequence by the child, thus reinforcing his tantrum behavior. The point to remember is that resistance by the child is a predictable result of withholding reinforcement and that the target behavior will decrease if one is patient and willing to stick to the procedure. However, this may not be possible if the child resorts to unacceptable levels of self-injury, aggression, or destruction of property.

Positive types of social interaction, such as the giving of attention (e.g., a look, a nod, an answer to a question), affection (e.g., smiling, a touch, a hug), and verbal praise (e.g., "good work," "O.K."), are also powerful ways of maintaining appropriate behavior. Attention and affection can also increase the effectiveness of another reinforcer (e.g., food) when the two are used together. In using praise in combination with food reinforcement, attention often becomes effective in maintaining the behavior when food reinforcement is faded out.

Successive Approximation of New Behaviors. The teaching of new skills or types of behavior to autistic children often requires "shaping" or the reinforcement of successive approximations of a desired behavior. One begins with a response already in the child's repertoire. A number of behavioral steps or approximations are then identified between this initial behavior and the terminal behavior (or what one wants the child to be able to do). Each step becomes a target behavior, and only after stable performance is achieved at each step should the child be taken to the next step. If the child fails to master a step along the way, one returns to the previous step, regains stable performance, and then proceeds in approximating the terminal behavior. The shaping procedure is complete when stable performance of the terminal behavior is achieved.

However, autistic children often require patience and perseverance through many trials. Even the simplest of skills may require the patient training of a large number of approximations. Keeping the goal in mind and learning not to spend too much or too little time at any one step are important elements of successful shaping.

Shaping is useful in remedying all the major skill deficits of autistic children—deficits in the areas of language, socialization, self-help, and motor and academic-preacademic skills. For example, shoe tying is easily broken down to a number of distinct steps, each approximating the terminal behavior of being able to tie one's shoelaces without assistance. The teaching of social interaction may begin with reinforcing fleeting face to face contact, moving to eye to eye contact, and so forth. Similarly, teaching the word "ma ma" begins with reinforcing the child when he emits any sound and proceeding with reinforcing approximations of "mm," then "ma," and then "ma ma."

After a new behavior has been shaped, it can be linked with others to form a behavioral "chain." Establishing chains or routines ("chaining") that are under the control of a single instruction is important in working with autistic children. For example, "Get dressed" can come to occasion a number of specific types of behavior (finding clothes, putting on shirt, pulling on pants). "Time to get up" initiates a long morning routine of behavior patterns that are essential to independent functioning.

The Use of Prompts and Cues. Prompts and cues are stimuli that may occasion a desired response prior to training or with little training. Prompts ensure that the learning situation is successful, i.e., increasing appropriate responding, thereby maximizing the reinforcement that the child receives.

Knowing when to employ prompts and when to fade them out is of critical importance in working with the autistic child, since prompting is an essential ingredient of every training effort. With a prompt, one is attempting to ensure that the child engages in the desired response in the presence of the relevant stimulus or antecedent event. When prompts are no longer needed, they are "faded" out so that only the training stimulus remains.

The use of prompts may include one's physically assisting the child in the performance of the desired response (e.g., holding a child's lips in the appropriate position to produce an "oo" sound), giving the child the correct answer (e.g., in a discrimination task), or modeling the desired response for the child (e.g., producing speech sounds).

As noted by Christian (1981), the most effective training programs for autistic children employ a combination of shaping, chaining, prompting, and fading procedures. For example, Lovaas (1977) and his colleagues have made effective use of these procedures in their development of a language acquisition training protocol for autistic children. Lovaas' approach consists of a number of programmatic steps, which include building verbal responses, labeling discrete events, establishing relationship between events, using abstract terms and conversation, giving and seeking information, developing grammatical skills, using recalls, and encouraging spontaneity. This highly effective program represents both the critical elements of the behavioral approach and the success possible when these elements are correctly utilized.

Consistency of Application. The implementation of procedural strategies must be consistent across home, school, and treatment environments if the behavioral approach is to be maximally effective. Consistency of the types of target behavior, goals, and procedures is essential if skills learned in one environment are to "generalize" or be maintained in another environment. Consistency is particularly important with autistic children, because the child's difficulty in accepting environmental change and other characteristic problems such as "overselectivity" to certain environmental stimuli often interact to hinder generalization of skills and control.

Close communication between parents and helping professionals and continuing education for parents, teachers, and other helping professionals are essential in promoting consistency and generalization. Communication can be facilitated by developing a system of written correspondence between the teacher and parents and between other helping professionals and the teacher. For example, daily school reports, hospital behavior checklists, and home behavior checklists can be used to improve communication, thus fostering a "team approach" and ensuring consistency of effort.

Parent Education

In recent years parent training has emerged as a major component in the treatment of autistic and severely disturbed children and adolescents. The familiar picture of parents bringing their child to a professional for treatment is changing to a more practical model of parents being trained as paraprofessionals to administer treatment to the child in the home.

The rationales for training parents as therapists have included the following: Since parents spend the most time with the child, they are in the best position to provide training for him. In addition, trained parents are more likely to keep a difficult-to-manage child in the home environment, thus minimizing the need for restrictive, possibly long term institutional treatment. Finally there is the issue of practicality. There are a greater number of autistic children in need of treatment than there are professional therapists with expertise in their treatment, and there is a growing demand for child and family intervention.

Another issue in training parents as therapists concerns the nature of the training that they should receive. For example, one rationale for teaching parents behavior modification methods has already been mentioned: these procedures have been identified as the "treatment of choice" for autistic children. Other advantages in teaching parents behavioral training methods include the following:

1. Principles of behavioral intervention can be easily carried out in the home.

2. Behavioral methodology is concerned with easily observed phenomena and thus is readily acceptable by parent trainees.

3. Parents can be taught simultaneously in groups, with relatively short training periods required.

4. Professional staff in some cases may exert

more impact on treatment by training parents than by working in one to one treatment sessions.

5. Parents more readily accept treatment models that do not imply parental disorders.

6. Successful behavioral intervention in the home increases the probability that the autistic child can remain at home (O'Dell, 1974).

When the child's problem behavior is of enough severity (e.g., aggression, self-injury, severe social and self-help skill deficits), or other variables make it necessary that the child be temporarily removed from the home environment, parent training takes on special significance. If the child is to be returned to the home, the skills learned in the residential treatment setting must be transferred or generalized to the home environment.

Parent training helps to ensure that the structure and procedure in other settings, such as school and day treatment, are consistent in or "generalized" to the home environment. Extensive research has shown that this generalization will simply not occur without active parent participation and ongoing parent training.

Lovaas et al. (1973) have provided a graphic illustration of the importance of parent training as a component of residential treatment for autistic children. Their findings indicated that autistic children whose parents were trained to apply behavioral methods continued to improve after discharge, whereas children who were sent to institutions or foster homes with untrained parents regressed. For the latter group, troubling behavior increased in frequency (e.g., self-stimulation and echolalia), and children appeared to lose all that they had gained of social nonverbal behavior, and much of what they had gained in appropriate verbal and play behavior. However, it is significant that Lovaas and his colleagues found that a reinstatement of behavior therapy for several of the institutionalized (no parent training) children temporarily re-established some of the therapeutic gains that had been made while in training.

TREATMENT SETTINGS

The trend of treatment for autistic youth is moving progressively in the direction of community based programs. The ideal of autistic children and adolescents living at home and attending structured day treatment-education programs is becoming a reality with the development of these programs in many communities across the country. It is in these types of settings that autistic individuals are most likely to receive the least restrictive, most progressive services.

However, extensive parent support is essential for the success of these programs (Lovaas et al., 1973), and such support may not be available to

many autistic children. In addition, the severity of the autistic child's behavior problems may make living at home impossible. In these instances the least restrictive alternative would appear to be the community based group home program with well trained, live-in, teaching "parents" and a structured day treatment-education program. This type of program has been shown to be successful in the remediation of serious behavioral obstacles to home living while helping parents or guardians provide home support to facilitate the child's return to a less restrictive environment.

The application of this technology with autistic children has been described by Lovaas (1978):

To avoid institutionalization, which is the most common trap and frightening future for autistic children, we have begun exploring training professional parents to take care of the autistic child in family residences ("teaching homes") within the child's own community, near his parents, and schools and medical facilities that are already available. Four to six children per home constitute a heterogeneous group, so that the more regressed autistics can model after nonpsychotic peers. The children are cared for by "teaching parents," who are college graduates enrolled in master's programs (education, social welfare, or psychology), and whom we explicitly train in behavior therapy techniques. At the same time we take explicit measures on the child's progress within those teaching homes, and explicitly arrange for surveillance and input by the child's natural parents to protect the child's welfare and improve upon such a program. The efficient, short-term education of parents to become professional parents is probably the critical variable in assuring the success of this operation. Autistic children are difficult to manage, and without an efficient training procedure which can turn out a large number of (para) professionals to successfully care for them, they all too often end up in the large traditional and deadly (state) institutions.

Unfortunately group home programs are not yet available for all persons who need them. In addition, the programs that are available are likely to have selection criteria (e.g., age, degree of parental involvement, severity of disturbance) that may leave many autistic persons with no alternative aside from the residential treatment center.

Therefore, autistic children require a continuum of services—from in-home living to the residential treatment center. However, for the residential treatment program to become an effective component of this service continuum, it must serve as a "springboard" to less restrictive components of the continuum (e.g., community based group homes). This can be accomplished to the extent to which the residential program approximates the philosophy, goals, and procedures of less restrictive programs.

For example, the criteria (e.g., the skill level of the individual) for successful transition from a residential treatment program must be identical to those for successful acceptance in a community based program. This requires a degree of synchronization of programs along the service continuum so that placement in one program is the first step in the child's preparation for the next, less restrictive program.

Specifically the training procedures and curricula employed at each level must be based upon the "criterion of ultimate functioning." One must constantly consider where the autistic child will eventually be living and what skills can be taught to aid in that adjustment. Similarly attention must be paid to teaching community living skills and to the utilization of incidental teaching strategies whenever possible, as this type of programming serves to increase the probability that skills acquired will also be used and maintained. Finally institutional environments should be designed to be rich in opportunities for learning adaptive behaviors rather than designed to withstand maladaptive behavior.

Consistent with the criterion of ultimate functioning, the residential treatment program could develop on-grounds approximations to community based living so that children are prepared for successful transition. For example, an on-grounds "group home" would be valuable in helping a child bridge what may be a considerable gap between institutional and community living. At a more basic level the child's living area must be designed and furnished to the extent possible to approximate his "ultimate" living environment. "Step level" programming is particularly useful in this regard. Similarly a vocational training program and an on-grounds simulated workshop may be developed to increase the child's exposure to recreational and educational activities in the program's local community as well as in his home community.

THE FUTURE

Solving the child's immediate needs is not the only challenge facing those working with autistic children. Parents and helping professionals must continue to meet the needs of the child as he enters adolescence and adulthood. Parents of children who require residential, day treatment, and vocational-educational services are finding it difficult to obtain these services effectively. For example, Sullivan (1976) has observed that "There is a sudden drop in the number of available programs, both day and residential, once the child reaches puberty or late adolescence" and that "Suitable programs for adult autistic persons are so rare as to be practically nonexistent."

However, parents of autistic persons are becoming more militant, and there is the prospect of parents and helping professionals working together to bring about an increase in the quantity and quality of services available to autistic persons. In addition, the increased attention being given to the rights of clients and the rights of students will no doubt benefit the autistic individual (McClannahan and Krantz, 1981). For example, with the requirements of new legislation (P.L. 94–142), local education agencies are more likely to join parents and professionals in the development of a continuum of services for autistic children—particularly in terms of special education programs through adolescence.

SUMMARY

Childhood autism is a developmental disability characterized by an onset before five years of age; a disturbance in the rate of appearance of physical, social, and communication skills; an abnormal response to sensation; absence or delayed development of speech and language; and abnormal ways of relating to people, objects, and events. Autism is distinguished from mental retardation by the fact that retarded children do poorly on a wider variety of intelligence tests than do autistic children; autism is distinguished from childhood schizophrenia especially by its early age of onset. There is no known cause for autism, although the disorder is probably organic in nature and apparently does not have an environmental cause. Autistic children require a behaviorally oriented educational approach to treatment if they are to make adequate gains in communication, social, self-help, and academic skills. This approach to treatment requires that one develop an understanding of the child's needs, the relationship between the child's behavior and his environment, and the way in which aspects of the environment can be used in educating and rehabilitating the child. Parent education is an essential component of this approach. Autistic children appear to require a continuum of services, from in-home living to the residential treatment center; each service program in the continuum must prepare the child for the next, less restrictive component. The prognosis for the autistic child depends upon his level of measured intelligence, the degree of language impairment, and involvement in treatment. Children who are involved in a program of intensive behavioral training at an early age (three to four years) appear to have the best prognosis.

WALTER P. CHRISTIAN

REFERENCES

Campbell, M., Geller, B., and Cohen, I. L.: Current status of drug research and treatment with autistic children. J. Pediatr. Psychol., 2:153, 1977.

Christian, W. P.: Reaching autistic children: strategies for parents and helping professionals. In Milunsky, A. (Editor): Coping with Crisis and Handicap. New York, Plenum Press, 1981.

Kanner, L.: Autistic disturbances of affective contact. Nerv. Child, 2:217, 1943.

Koegel, R. L., and Schreibman, L.: How to Teach Autistic and Other

Severely Handicapped Children. Lawrence, Kansas, H & H Enterprises, Inc., 1981.

Lovaas, O. I.: The Autistic Child: Language Development Through Behavior Modification. New York, Irvington Publishers, 1977.

Lovaas, O. I.: Parents as therapists. *In* Rutter, M., and Schopler, E. (Editors): Autism: A Reappraisal of Concepts and Treatment. New York, Plenum Press, 1978.

Lovaas, O. I., Koegel, R. L., Simmons, J. Q., and Long, J. S.: Some generalization and follow-up measures on autistic children in behavior therapy. J. Appl. Behav. Anal., *6*:131, 1973.

Luce, S. C., and Christian, W. P. (Editors): How to Work with Autistic and Severely Handicapped Youth: A Series of Eight Training Manuals. Lawrence, Kansas, H & H Enterprises, 1981.

McAdoo, W. B., and DeMyer, M. K.: Research related to family factors in autism. J. Pediatr. Psychol., *2*:162, 1977.

McClannahan, L. E., and Krantz, P. J.: Accountability systems for protection of the rights of autistic children and youth. *In* Hannah, G. T., Christian, W. P., and Clark, H. B. (Editors): Preservation of Client Rights: A Handbook for Practitioners Providing Therapeutic, Educational, and Rehabilitative Services. New York, Free Press, 1981.

Newsom, C. D., Carr, E. G., and Lovaas, O. I.: The experimental analysis and modification of autistic behavior. *In* Davidson, R. S. (Editor): Modification of Pathological Behavior. New York, Gardner Press, 1979.

Newsom, C. D., and Simon, K. M.: A simultaneous discrimination procedure for the measurement of vision in nonverbal children. J. Appl. Behav. Anal., *10*:633–644, 1977.

O'Dell, S.: Training parents in behavior modification. Psychol. Bull., *81*:418–433, 1974.

Oppenheim, R. C.: Effective Teaching Methods for Autistic Children. Springfield, Illinois, Charles C Thomas, 1974.

Rimland, B.: Inside the mind of the autistic savant. Psychol. Today, August 1978.

Rincover, A.: Sensory extinction: a procedure for eliminating self-stimulatory behavior in psychotic children. J. Abnorm. Child Psychol., *6*:299, 1978.

Rincover, A., Koegel, R. L., and Russo, D. C.: Some recent behavioral research on the education of autistic children. Educ. Treat. Child., *1*:31, 1978.

Ritvo, E. R.: Autism: from adjective to noun. *In* Ritvo, E. R., Freeman, B. J., Ornitz, E. M., and Tanquay, P. M. (Editors): Autism: Diagnosis, Current Research, and Management. New York, Spectrum Publications, 1976.

Ritvo, E. R., and Freeman, B. J.: National Society for Autistic Children definition of the syndrome of autism. J. Pediatr. Psychol., *2*:146, 1977.

Rutter, M.: Psychiatric causes of language retardation. *In* Rutter, M., and Martin, J. A. M. (Editors): The Child with Delayed Speech. Philadelphia, J. B. Lippincott Co., 1972.

Rutter, M.: Diagnosis and definition of childhood autism. J. Autism Child. Schiz., *8*:139, 1978.

Schreibman, L., and Koegel, R. L.: A guideline for planning behavior modification programs for autistic children. *In* Turner, S. M., Calhoun, K. S., and Adams, H. E. (Editors): Handbook of Clinical Behavior Therapy. New York, John Wiley & Sons, Inc., 1980.

Sullivan, R. C.: Autism: Current trends in services. *In* Ritvo, E. R., Freeman, B. J., Ornitz, E. M., and Tanquay, P. M. (Editors): Autism: Diagnosis, Current Research, and Management. New York, Spectrum Publications, 1976.

39G The Child with Multiple Handicaps

DEVELOPMENTAL IMPLICATIONS OF MULTIPLE PROBLEMS

THE BACKGROUND FOR MULTIPLICITY OF DISABILITY

In Chapter 39 up to this point we have considered, in sequence, particular components of central nervous system handicap: limitation in intelligence, or mental retardation; difficulties in sensory function, leading to deafness or blindness; impairment of motor function, presenting as cerebral palsy; disturbance of rhythms in the neuronal network, with seizure disorders; and troubled integrative function in communication and interpersonal relations, as seen in autism. The common thread in these circumstances is an interference with the integrity of the central nervous system (and associated end organs). It follows that the troubling force in a "major handicapping condition" is likely to have sufficient magnitude to produce diffuse effects. One can take as axiomatic the expectation that "pure" forms of the foregoing conditions are seldom found when the disability is major—multiplicity appears either by virtue of a common origin or as a result of evolving complications.

One used to hear reference to the "common" or "simple" type of mental retardation, whatever the supposed meaning. Modern workers, especially in dealing with children who have a serious cognitive handicap, are accustomed to finding more complicated situations, because of associated disabilities, simultaneous health care problems, or accompanying emotional or adjustment challenges. The statistics reflecting associated handicaps in individuals with mental retardation are heavily influenced by the nature of the population sampled and the threshold used for defining the difficulty. If one looks at abnormalities in motor function, for example, the incidences are found to vary. Accardo and Capute (1979) cite an incidence of 10 per cent for cerebral palsy generally among retarded children. A serious physical handicap (difficulties in ambulation, orthopedic complications) is found in one-third of the clients in a state residential facility for retardation (Nelson and Crocker, 1978). And some degree of motor function abnormality can be identified in over half the patients in a typical hospital referral clinic for mental retardation, including aberrations in tone, strength, balance, coordination, motor planning, and visual motor perception (Crocker, 1983). The incidence of seizures in these three groups was 4, 34, and 10 per cent, respectively. Sensory handicaps are always seen more frequently among retarded individuals, with significant hearing impairment in about 10 to 20 per cent in a typical series and serious visual impairment in about 5 to 10 per cent.

Study of the settings for cortical handicap presented in Table 39–1 in this chapter reveals that the origins for mental retardation coincide compellingly with those for cerebral palsy and epilepsy, particularly in regard to prenatal and perinatal phenomena. Hospital clinics for children with cerebral palsy usually report simultaneous mental retardation in one-half to two-thirds of the children (Accardo and Capute, 1979; Crocker, 1983). The incidence of retardation in seizure clinics is significant but lower—25 per cent according to Accardo and Capute (1979).

SPECIAL HEALTH CARE REQUIREMENTS

The support system for a population of children with developmental disabilities must provide assistance in a predictable group of continuing health needs. Mental retardation programs, for example, make an important contribution in the treatment of emotional disturbances, motor function and orthopedic problems, and seizures. The most widespread of these is emotional difficulty, of significance in half or more of the patients, and varying from mild behavioral disturbance to depression, acting out, self-abuse, or even overwhelming thought disorder. These issues are discussed further in Chapter 39H.

Other complications commonly encountered with handicapped children, influenced in part by their special life style, include nutritional difficulties (both feeding problems and obesity), chronic ear infections, pulmonary infection, dental or periodontal disease, injury, and pica. Some special syndromes have obvious liabilities regarding cardiac disease, hydrocephalus, urinary infections, growth failure, or reduced fertility. Strabismus, undescended testes, and scoliosis are frequent anomalies.

FUNCTIONAL EFFECTS OF MULTIPLE HANDICAPS

The specific or actual additive constraint to human development caused by the presence of more than one disability is obvious. When a deaf child cannot utilize manual language because of visual impairment or a motor function handicap, he is doubly penalized. The complication of a serious mobility limitation intrudes upon a retarded child's potential for access to learning experiences. Far more significant, than these concrete elements, however, have been the societal, political, and even professional reactions to children with a multiplicity of handicaps. For many generations, in the United States and elsewhere, a child with retardation, deafness, blindness, or cerebral palsy would suffer almost certain exclusion from the usual educational and recreational systems if he had other troubles in addition, particularly the less savory attributes of uncontrolled seizures, acting out behavior, a bizarre physical appearance, or an interfering somatic illness. A time of discovery began in the 1970's with the securing of a "no reject" principle inherent in the various "Education for All Handicapped Children" Acts. Three prototypic situations will illustrate the traditional dilemmas of compounded handicaps:

Deafness with Other Problems

In earlier times the children with early onset "pure" deafness were educated for the most part in segregated and highly structured "schools for the deaf"; the minority who had other difficulties in addition generally melded with the seriously mentally retarded population by default. The huge rubella epidemic of 1963–1964 created a sudden and dramatic increase in the numbers of sensorineural deaf children, many of whom had visual handicaps, central processing problems, behavioral atypicality, and sometimes mental retardation. The flexibility of the deaf school criteria was critically challenged, and many small new educational facilities were created to develop experimental curricula that could reach out and meet the needs of multihandicapped deaf children. In Massachusetts, their names (Little People's School, Learning Center for Deaf Children, and Willie Ross School—named after an involved child) reflected a more expeditious approach to child needs than had the earlier institutions (e.g., Boston School for the Deaf). Beyond this a federal program reflected specific Congressional support for the education of deaf-blind children. When an inventory was established of these young people with double sensory handicaps, it was found that many were already grouped with the severely and profoundly mentally retarded clients in state residential facilities, often without cognitive handicaps having been proven.

Serious Physical Handicap (Including Cerebral Palsy) and Mental Retardation

The difficulty in making an accurate determination of intellectual function in a child with a massive physical handicap often led to educational nihilism if some evidence of mental retardation was detected. Many young people with borderline or mild retardation who had striking cerebral palsy were allowed to languish in state residential facilities, to have their true learning potential uncovered only when the movement of "deinstitutionalization" got under way. Further, until the advent of P.L. 94–142, it was usual to exclude from early educational efforts children who had major bodily handicaps that could lead to fragility of health or a fatal outcome, such as those with the Cornelia de Lange syndrome or Hurler disease. This deprived the child and family of coordinated supports and reinforcement of personal progress within the years available (Crocker and Cushna, 1972). Finally, until the recent revision of federal guidelines, it was difficult to obtain vocational rehabilitation services for a physically handicapped young adult if any degree of retardation was also present.

Mental Retardation Accompanied by Behavioral Disturbance

The classic instance of programmatic confoundment, and often rejection, has been that seen in the presence of the "dual diagnosis" of mental retardation and mental illness. Psychiatric treatment facilities have been skittish about accepting retarded clients, and classes for mentally retarded children have been bewildered when atypical behavior is prominent. The usual cause in current times for referral of retarded children to segregated residential schooling is the inability of community programs to deal with serious emotional disturbance. The urgent need for coordinated effort between mental health and retardation resources is demonstrated as well by serious childhood psychosis (and autism) in which retardation is often a sequela of a troubled early course.

The ultimate additive handicap for a disabled child, of course, is social and personal. For a child from a minority group, the vulnerability is yet greater. If the family is in chaos, support systems are ineffective, and the child experiences devaluation and social isolation, a critically compromising group of complications can ensue. A loss of individual compensation and personality integration will place serious limits on his adaptive outcome.

CHILDREN WITH MULTIPLE HANDICAPPING CONDITIONS

There are a vast number of diseases and syndromes that cause multiple handicaps. The conditions discussed in this section are representative. They include static conditions amenable to habilitation and a productive life (such as the myelodysplasias and congenital musculoskeletal anomalies) in addition to other fixed disorders necessitating lifelong comprehensive services (e.g., congenital rubella, Prader-Willi syndrome). Progressive conditions (e.g., mucopolysaccharidoses) create a different set of dynamics in health care. Young persons with these conditions are depicted in Figure 39–8.

MYELODYSPLASIA

Disturbances in the development of the neural tube create the largest group of major malformations known in man. The spectrum of anomalies ranges from defects of the cephalad neural tube incompatible with extrauterine life (such as anencephaly) to occult spina bifida, usually a serendipitous finding on pelvic x-ray examination. Many infants are born with serious malformations that are treatable by surgical and medical management. Collectively these defects, variously referred to as meningocele, myelomeningocele, or spina bifida or by other specific terms, are called the myelodysplasias.

Neural tube defects occur at a rate of 1 to 2 per 1000 births. Anencephaly or other multiple malformation syndromes, which result in early neonatal death, account for almost 70 per cent of the instances of neural tube defects. The survival rate in infants with encephalocele or myelomeningocele is about 80 per cent.

Characteristics

Population studies now demonstrate an increase in long term survival. A majority of children with myelomeningocele have physical handicaps necessitating assistance with ambulation. A smaller number have documented mental retardation.

The neurological level and associated medical problems are the primary determinants of functional outcome. Neurological difficulties involve both motor and sensory problems corresponding to the denervated segmental levels. The motor function is characterized by a flaccid paralysis secondary to the muscle involvement. A combination of spastic and flaccid paralysis is present in 10 to 40 per cent of the cases. Abnormal limb positions result from muscle imbalance. Children with disorders at the lumbar and sacral-spinal levels may require surgical and orthotic management to permit functional ambulation. Children with lesions at spinal levels above the lumbar region usually are not effective ambulators and require a wheelchair or other adaptive means of locomotion.

Children with spina bifida have a greater incidence of congenital malformations of other neuroectodermal and mesodermal tissues. Seventy per cent of the children with myelomeningocele are likely to have hydrocephalus, 50 to 80 per cent requiring shunting. A revision of the shunt is frequently required, especially before age four. Hydrocephalus and its treatment complications are the major cause of death in infancy and the preschool years.

The major causes of morbidity and mortality among children with myelomeningocele beyond

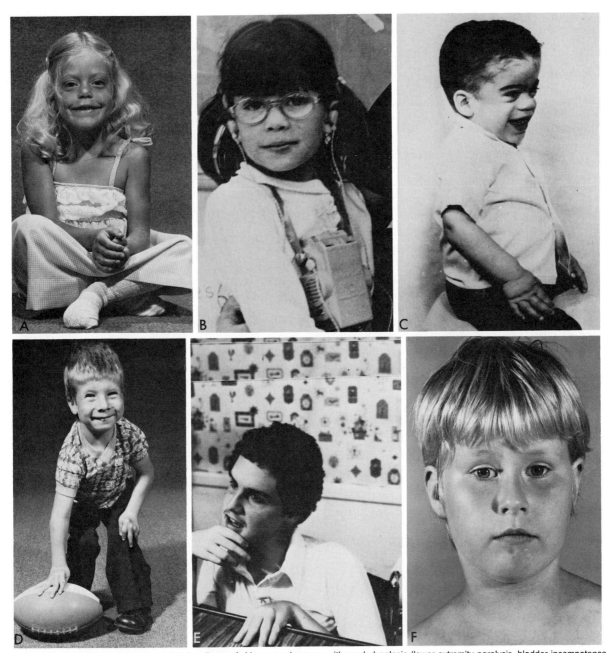

Figure 39–8. Children who have multiple handicaps. *A,* Lisa, age six years, with myelodysplasia (lower extremity paralysis, bladder incompetence, shunt for hydrocephalus). She attends first grade, with resource room. *B,* Linda, when four years old, with congenital rubella (profound deafness, serious visual impairment). *C,* Roland, three and one-half years with the Hurler syndrome (skeletal changes, developmental delay, hepatosplenomegaly). He died a few months later from congestive heart failure. *D,* Mark, age four years, with arthrogryposis (severe joint impairment, multiple fractures), doing well in preschool. *E,* Stewart, 17 years of age, with muscular dystrophy, oldest of three seriously handicapped brothers. *F,* Heidi, seven years old, with the Prader-Willi syndrome (mild retardation, doing fairly well on diet for weight control).

three years of age are pyelonephritis and renal failure. Urological evaluation and management should be begun early in the neonatal period. The management of the neurogenic bowel and bladder remains a lifelong process. Depending on the level of lesion and the associated denervation of bowel and bladder, specific techniques and surgical procedures have been devised to optimize continence and prevent renal deterioration.

The neonatal survival rates in infants with all types of neural tube defects and the incidence of major handicapping conditions in long term survivors in recent years have stimulated the search for ways of making a reliable prenatal diagnosis of these disorders. The capability for making a prenatal diagnosis now hinges on the observation that an open neural tube defect leaks large quantities of alpha-fetoprotein into the amniotic fluid

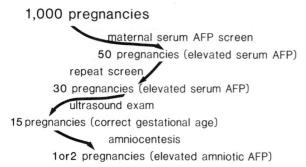

Figure 39–9. Prenatal diagnosis of neural tube defects: diagnostic sequence in a cohort of low risk pregnancies.

and across the placenta; elevated amounts are measurable in the maternal serum late in the first trimester. The radioimmunoassay of alpha-feto-protein permits a diagnostic sequence summarized in Figure 39–9.

The concept of treatment selection of newborns with neural tube defects derives from the convictions of clinicians that certain factors predict the eventual handicap of the child. Such factors include a paucity of cerebral tissue on a computed tomographic scan, gross hydrocephalus at birth (the occiputofrontal circumference exceeding 2 S.D. above the mean), associated major anomalies of other organ systems, and concomitant birth injury. There is no consensus among those in major centers caring for children with neural tube defects that treatment selection currently can be based on any existing criteria.

Aggressive neonatal management of liveborn infants with myelomeningocele is usually desirable. Management includes repair of the myelo-meningocele and shunting of hydrocephalus if the condition is severe or progressive in the first weeks or months.

Treatment

The goal of comprehensive habilitation of the child with myelodysplasia is to provide an environment facilitating the achievement of maximal levels of motor, intellectual, and social functioning. The immediate and long term functional goals are established through an understanding of the neurologic dysfunction, the associated medical problems, the level of cognitive function, and psychosocial adjustment (Akins et al., 1980).

The main concerns of management in the functional habilitation of the child with myelodysplasia are to prevent or reduce deformities, to train the child in adaptive functioning and self-care skills, to achieve some means of independent locomotion, to control excretion of urine and feces, to foster the best possible personal adjustment, and to provide proper education and vocational rehabilitation.

Experience has shown that the management of the child with myelodysplasia is frequently best done through a team approach. The team works with the child, family, and primary care physician to develop a management strategy for each child. The clinic team typically includes specialists in pediatrics, physical medicine and rehabilitation, neurosurgery, orthopedics, urology, genetics, nursing, psychology, and social services. Depending on the age of the child, level of the neurological lesion, the acute or chronic nature of the problem, and the individual needs, the child and family may see any or all of these specialists during a single clinic visit. Recommendations for short term and long term management are given to the referring physician and the parents.

CONGENITAL RUBELLA SYNDROME

The consequences of prenatal rubella infection serve as a prototype for understanding many of the teratogenetic influences on the developing fetus. In 1941 the Australian ophthalmologist Gregg first reported the sequelae of fetal rubella acquired during the first trimester of gestation via the placental circulation. Successive reports of the systemic nature of these infections appeared in the literature documenting the pervasive destruction of developing tissue by viral replicative infection. The full consequences of this process did not become apparent until the epidemic of 1963–1964 when the congenital rubella syndrome was widely reported as a result of infections of thousands of children (see Chapter 39B). Fetal infection occurs in approximately half the mothers who contract rubella during the first trimester, and the risk of further damage persists during subsequent months of fetal development. The extent of long term residua probably relates to the virulence of a specific epidemic. This may account for the less than complete clinical awareness of the syndrome prior to the major epidemic.

The availability of immunization against rubella in the late 1960's provided an opportunity to eradicate the serious consequences of maternal rubella infection. The primary emphasis in rubella vaccination is to create immunity in girls entering adolescence and the child bearing years, to avoid natural viral infection. Rubella is generally of little consequence to the young female adult but obviously is the cause of major handicaps in the susceptible offspring. Establishing the immune status of the adolescent girl, especially in the premarital years, is now a common emphasis in medical care of adolescents. Several states have a requirement that rubella hemagglutination-inhibition antibody titers be determined as part of the required premarital serology studies.

Characteristics

The rubella virus has a predilection for a variety of tissues. The affinity for tissues of ectodermal origin, however, creates the most serious consequences. Central nervous system involvement centers chiefly on diffuse gray matter disease resulting in microcephaly and impairment of mental function, which can range from retardation to more subtle behavioral and learning dysfunctions that become evident only in middle childhood. The cochlea is damaged in the majority of instances (as high as 80 per cent in some series), leading to serious sensorineural hearing loss. The anterior structures of the eye can also be damaged, resulting in congenital cataracts of the lens, glaucoma, and corneal opacities. Chorioretinitis, microphthalmia, and strabismus are also common.

The other major organ system infected by the virus is the cardiovascular system, with patent ductus arteriosus, peripheral pulmonic stenosis, and fibromuscular proliferation in the medium and large size arteries. The heart itself can show a variety of septal defects and myocardial inflammatory disease. Disseminated active viral infections in other soft tissues, including the liver and lung, can result in devastating neonatal disease. When not fatal, such infection can leave all the chronic effects associated with inflammation in those organs. A variety of other chronic conditions have been reported to occur more frequently in children with congenital rubella syndrome, including diabetes mellitus, chronic renal disease, hemolytic anemias, and other structural defects, such as meningocele, hypospadias, and dermatoglyphic abnormalities.

Perhaps of greater interest has been the increased understanding of more subtle influences on neurodevelopment as a result of longitudinal study of children with documented congenital rubella encephalitis (Hanshaw and Dudgeon, 1978). A variety of minor motor abnormalities, including abnormal tone and reflexes, feeding difficulties, and delays in developmental milestones, may be common in these children during the preschool years, even without mental retardation. By early adolescence many youths have specific learning deficits and behavioral disturbances. In situations in which it is no longer possible to document the occurrence of congenital rubella infection, there are undoubtedly children who present with sensorineural hearing loss without the other major stigmata of the congenital rubella syndrome.

Diagnosis and Treatment

In the newborn with lethargy, petechiae, and hepatosplenomepaly one should suspect an intrauterine viral infection. Generally rubella virus can be cultured from a variety of sites, including the urine and stool. Antibodies acquired from the mother are also detectable in the serum. Such a classic appearance, however, is exceptional, and therefore clinical suspicion must be based on less obvious symptoms. The prenatal history may reveal a brief maternal illness characterized by fever and malaise and perhaps accompanied by a rash. There is no direct treatment for the fetus during intrauterine life. If maternal rubella can be documented during the first trimester or early in the second trimester, the mother can be counseled regarding termination of the pregnancy. Persistence of the replicating virus in the newborn suggests that appropriate antiviral chemotherapy might diminish some of the continuing damage to tissues.

Implications for Life Adaptation

The outcome for a child with the congenital rubella syndrome clearly is dependent on the extent of multisystem involvement due to the primary infection. The spectrum of outcomes therefore ranges from fulminant intrauterine and neonatal disease leading to death in infancy (an uncommon event) to chronic manifestations. Some children may have combined sensory handicaps, including profound deafness and blindness. Others may have a single major sensory deficit combined with a remediable condition, such as congenital heart disease. The presence of mental retardation influences the child's capabilities in terms of education, vocational training, and life adjustment.

As is mentioned earlier in this chapter, congenital rubella is a fascinating example of a condition that induced the development of the whole range of specialized services for handicapped children because of the unique needs of many children with the syndrome. Approximately 20,000 children were identified with the syndrome in 1964. As these children grew through infancy, the preschool years, middle childhood, and into adolescence, a whole new series of family coping and educational strategies were conceived. Almost none of the traditional schools for children with single sensory handicaps (blindness or deafness) were prepared to provide an adequate structural curriculum for children with multiple handicaps, either the combination of these two sensory handicaps or a combination of retardation with one or more sensory deficits. Fortunately the assumption that the deaf blind child was so compromised as to prevent any meaningful environmental interaction has been successfully refuted by the experience of some specialized facilities. Total communication strategies, electronic communication devices, tactile training, and a variety of other educational techniques have now opened new vistas for these children. Fortunately their younger counterparts, now estimated at 20 to 40 new cases per year in the United States, will benefit from the extensive experience accumulated during the past two decades.

THE HURLER SYNDROME

The Hurler syndrome is produced by a rare inborn error of metabolism involving the mucopolysaccharides. These are large carbohydrate based polymers with important structural functions in connective tissue, cartilage, the cornea, the heart, and cerebral cortex. The eponym derives from the name of a Swiss pediatrician, Gertrud Hurler, who described a brother and sister with this disease in 1919. Involved children have an unusual and characteristic appearance, which denotes serious bodily handicaps. Death occurs during early or middle childhood. As with many inborn errors, there is autosomal recessive hereditary transmission. The parents (heterozygotes) are not clinically affected, but in the child with the double (homozygous) abnormality there is an absence of "iduronidase" activity. It is the critical deficiency of this enzyme that leads to all the subsequent complications. (See Chapter 18.)

Characteristics

The child with the Hurler syndrome is an extraordinary individual, engaging, receptive, and altogether lovable. His appearance in the newborn period is nearly normal, but in the months that follow there is a gradual change. The head is relatively large, with a prominent forehead, a broad nose, and a rather flat facial contour. Enlargement of the liver and spleen gives an appearance of fullness to the abdomen. There may be inguinal or umbilical hernias. A limitation in all the joints causes flexion contractures, preventing normal extension of the fingers, elbows, hips, and knees. There is an unusual configuration of certain vertebral bodies, with a "kyphos" in the midback. The corneas are cloudy, although this does not seriously handicap vision. Linear growth is enhanced at first, with all children being well above the ninetieth percentile at their first birthday. Because of the cartilage disturbance at the epiphyses, there follows a gradual growth arrest (by about three years). This may be the only condition in which a child can be both a giant and a dwarf in the same lifetime. There is hirsutism of the limbs and trunk, chronic nasal discharge is usual, hearing handicaps are frequent, and seizures may occur. Cardiac abnormalities are produced by changes in the structure of the valves, subintimal deposits in the coronaries and aorta, and mucopolysaccharide accumulations within the myocar-

dial cells. Repeated respiratory infections are common.

The early developmental progress is nearly normal, but the children come to follow the usual decelerating "metabolic" developmental course. Although independent walking may be achieved by the age of one to one and one-half years, motor skills generally remain limited. Some words will be learned at a later age, but language development is severely constrained. Toileting independence is almost never acquired. After about three years, further learning is only sporadic, and one begins to see the loss of previously learned skills. The ability to walk is gradually lost, and facility with language is lost as well, the child showing increasing lack of interest in the environment. This encroaching dementia represents the effect of neuronal destruction and loss and is followed by instability in homeostasis. Death occurs from refractory congestive heart failure or from the complications of decerebration.

Diagnosis

The Hurler syndrome is usually diagnosed in the second half of the first year of life, at which time the pediatrician and others become concerned by the remarkable appearance of the child (as well as the enlargement of the liver and spleen). By this time radiological study of the skeleton reveals many changes (particularly in the metacarpals, spine, and hips). A screening test is available for the detection of increased mucopolysaccharide excretion in the urine (Berry spot test), but precise identification of the syndrome requires documentation of an iduronidase deficiency in the white blood cells or cultured skin fibroblasts. Not uncommonly the child may have been treated for hernia or respiratory infections in the early months of life, the unwary observer not having suspected the Hurler diagnosis. Iduronidase analysis can be performed by an experienced laboratory technician on cultured amniotic cells as well, providing the capacity for prenatal diagnosis in subsequent pregnancies.

Treatment

The child with the Hurler syndrome has continuing health care needs, often best guided by those who are familiar with this special situation. Respiratory infections and middle ear infections require diligent treatment. Hernia repair may be required, but incarceration virtually never occurs and anesthesia represents a significant risk. The nasal discharge is relieved only by adenoidectomy, an approach not generally justified. Heart failure responds at first to the use of digitalis preparations. Subarachnoid cysts may produce increased intracranial pressure and hydrocephalus, sometimes requiring shunting operations. The ultimate hope for therapy would be the prompt and effective provision of enzyme replacement. This is not at present possible, although courageous efforts (e.g., bone marrow transplantation) have been pursued on an experimental basis.

In addition to care for the involved child, the family should benefit from a program of informed support. They are bewildered by the patient's unusual situation and fragile outlook and require guidance in their own adjustment. The security of knowing that a partnership exists for the child's care, and that he is valued by workers in the health system, will do much to assist them (Crocker and Cullinane, 1972). There is a need for accurate genetic counseling, provision of respite care, contact with other families, and encouragement for the brothers and sisters.

Implications for Life Adaptation

The very young child with the Hurler syndrome benefits substantially from a stimulation and training program. He will warm to the social interaction and give back a buoyant fellowship. There will be gains in language, motor, and self-help skills and the acquisition of limited preacademic competence. These are precious accomplishments and a store against the coming times when learning is diminished. Medical problems (infections, cardiac disease) may be intrusive, and coordination is needed between the clinical and educational resources. Because of the progressive cerebral disease there should be a realistic tailoring of goals. The limitation in joint motion cannot be effectively modified by physical therapy. It is appropriate to keep the child's world from shrinking as his skills diminish, but eventually the activities possible become very limited. Although unusual in appearance, the child with the Hurler syndrome has a special nobility, and it is common for his continued enrollment in a preschool program to be highly valued by the local workers and other children. Finally the inscrutable effects of the biological handicap exert their inevitable outcome.

ARTHROGRYPOSIS

Despite the presence of functioning internal organ systems, a constellation of seriously deforming congenital anomalies places an extraordinary burden on the young child. Within this vast array of diagnoses, which include a variety of malformation patterns having a chromosomal or metabolic basis, are disorders collectively referred to as the arthrogryposes (Drennan, 1978). In general these disorders are the result of the prenatal onset of joint contractures. The specific etiology of the contractures is unknown, but neurological deficits,

Table 39–21. PRIMARY ETIOLOGIES OF ARTHROGRYPOSIS

Neurological deficits
    Anterior horn cell diseases
    Lower motor neuron lesions
    Prenatal central nervous system damage
Muscle and tissue disorders
    Primary muscle aplasia
    Fibrous or fatty degeneration
    Abnormal tendinous insertions
    Connective tissue disorders
Skeletal defects
    Defective limb bud rotation
    Vascular anomalies
    Fetal arthropathy
Intrauterine fetal crowding
    Malposition of fetus
    Uterine deformity
    Oligohydramnios
    Fetal hypotonia

muscle disorders, connective tissue or skeletal defects, or intrauterine fetal crowding may contribute either in isolation or in combination. Table 39–21 indicates several specific diagnoses that may be responsible for the joint abnormality within the broader classifications mentioned. Syndromes that include maldevelopment of other organ systems or that have chromosomal or metabolic markers can be specifically recognized. However, in other situations one can only speculate that an unknown teratogenic effect during the early weeks of fetal development caused the malformations observed.

The severity of extremity involvement clearly determines the extent of the handicap. Although extremity involvement does not necessarily create multiple handicaps, in severe cases the extent of the physical abnormality so fundamentally changes a child's pattern of early mobility, acquisition of fine motor skills, self-image, and activities of daily living that the consequences are quite substantial. The atypicality of such children is obvious to even the most casual observer. Therefore the consequence of what might be viewed as a strictly musculoskeletal deficit are much more fundamental to the life and development of the child.

Diagnosis

The evaluation of the newborn with multiple extremity anomalies should proceed in a logical pattern. The history of onset and the character of fetal movement, the mode of delivery, and an estimate of the amount of the amniotic fluid may give clues to the specific etiology. Diminished intrauterine movement suggests a neuromuscular etiology. A breech delivery suggests some difficulty in fetal rotation, which may be a consequence of the shape of the uterus or the quantity of the amniotic fluid. Fetal crowding at critical stages of extremity development may be secondary to oligohydramnios. When other organ systems are involved so that the child has multisystem defects, chromosomal disorders may be suspected, including a variety of trisomies and deletion syndromes. Radiological evaluation of the skeletal system aids in determining the precise nature of the anatomical deficit, including abnormal joint structure, synostosis, or connective tissue webs. Associated skeletal problems, such as congenital scoliosis due to malformed vertebrae and hip dislocation, are also important to identify. Examination of palmar finger creases, and sometimes sole creases, may suggest the timing of the developmental abnormality and provide therapeutic insight into rehabilitation of the affected extremity. Skin dimples suggest very early malformation, since the contact of skin and osseous structures without interposing adipose or other soft tissue frequently results in an attachment of the skin to the underlying bony structure. Generalized neurological and muscular function are also important, since the presence of systemic disease may well establish a specific diagnosis and predict the long term development and outcome.

In a majority of the cases of arthrogryposis there are no specific neurological, muscular, or chromosomal features. Therefore the defect in prenatal development can be judged to be static rather than progressive, and the approach to treatment relies upon the capability of utilizing surrounding tissues as well as adaptive equipment.

Treatment

The general approach in therapy is to maintain and, when possible, restore function in the involved extremity or extremities. Since most commonly multiple contractures involve the lower extremities to a greater extent than the upper, early effort should be devoted to stabilizing the major joints to promote weight bearing. Thereafter surgery may be helpful in utilizing functional muscles and tendons in an effort to compensate for undeveloped or malformed structures. The need to continually improve function has historically prompted the use of multiple procedures for each affected joint. Aggressive surgical treatment is generally used if mobility or extremity utility will predictably improve after a procedure. In recent studies children with extensive arthrogryposis commonly have undergone more than 10 significant orthopedic procedures during early and middle childhood. Radical approaches, including amputation, may well be indicated, even in early life, to encourage ambulation when the nonfunctional extremity is an impediment to walking. Such decisions are difficult, since parents generally react to recommendation for amputation with great fear.

They are understandably apprehensive about a procedure that their child might subsequently judge to be mutilating, when he is an adolescent or young adult. By late infancy evaluation will reasonably predict the long term capability of the extremity, and the surgeon may be quite confident in suggesting the outcome with or without a major procedure.

Treatment for associated anomalies should be assessed on the basis of specific requirements for the child. By definition this implies the need for multiple specialists and the coordination of care, especially when major operative procedures are to be undertaken.

Implications for Life Adaptation

Only in the most severe cases, in which there is a serious anomaly of an essential organ system, such as the cardiovascular system, will death occur prematurely. Otherwise the child may progress quite normally in cognitive development and have the same general needs as any other child.

THE DUCHENNE TYPE OF MUSCULAR DYSTROPHY

The Duchenne type of muscular dystrophy is a progressive disorder that results in loss of ambulation, eventual total physical incapacitation, and death in late adolescence or early adulthood. In contrast to most of the other multiply handicapping conditions discussed in this section, this disorder disrupts the apparently normal development of the young child by superimposing a chronic loss of function.

The disease is evidenced initially by the onset of weakness, generally most obvious in the proximal muscles of the legs. The characteristic "waddling gait" is another manifestation of weakness in the hip girdle area and is a consequence of alternating weight transfer from hip to hip. The specific pathophysiology of the disease has not been delineated. In the Duchenne type of dystrophy the myopathy appears to be a primary disease of striated muscle. Variations of the disease or other dystrophies may be secondary to other biochemical or neurophysiological changes and have different clinical manifestations.

Characteristics

The progressive loss of function in children with the Duchenne type of dystrophy generates an increasing handicap. At diagnosis the child's gait and ability to climb stairs may appear to be normal to the casual observer. Eventually because of muscle weakness the child will use a railing or other assistance in negotiating a stairway. The rate of climb slows as weakness progresses. Walking continues to be accomplished unassisted, as are other gross motor maneuvers such as rising from a chair, but eventually the child loses the ability to lift his legs and climb steps. As the disease continues, the child is not able to rise from the sitting position but is able to walk. Thereafter walking depends upon stabilization in the lower legs through the use of a brace and assistance in balance using a walker. Then the child will be able to stand in bracing support but will be unable to walk even with assistance. Finally the child is confined to a wheelchair and useful motor activity is restricted to the upper extremities.

There is some controversy about the possible primary intellectual deficit of this progressive dystrophy (Marsh and Munsat, 1974). Progressive physical incapacity can easily lead to withdrawal from social interaction, the fostering of dependence on parents and siblings, and the onset of inappropriate interpersonal relationships that may reflect the child's preoccupation with his disease process. Children followed over time have a slower rate of learning than their peers, and there may be a loss of intellectual capability as measured by standard psychometric tests. The basic performance of a specific child relates most directly to the family circumstances and the child's effectiveness in coping with his disease process. With progressive disability the likelihood of a severe respiratory compromise or the onset of characteristic cardiomyopathy determines the child's longevity. When the child intellectually understands the course of his disease, there may be chronic depression. Family counseling and unified strategy in working with the child become essential to maintaining an equilibrium of emotions at home and in preventing unfavorable reactions from brothers and sisters as well as between parents.

Diagnosis and Treatment

The toddler or young preschool child who clinically is suspected of having dystrophy should be the subject of intensive scrutiny to specify the type of dystrophy, which establishes the prognosis. Careful physical examination that centers on the neuromuscular status of the child is indicated. Absence of reflexes or pathological neurological signs are not characteristic of the Duchenne type of dystrophy in the early stages of this disease. Gowers' maneuver may be used to demonstrate the pattern of proximal weakness. In this maneuver the child must use his hands and arms to push off the floor from a sitting position to rise to standing in order to compensate for the loss of proximal muscle strength in the thighs. The hypertrophic appearance of the calf muscles is the classic presenting physical sign. The creatine phosphokinase level should be measured in the serum. Intermediate levels of this enzyme may indicate another type of dystrophy or another myopathy. The disease should be confirmed by muscle biopsy, which in the Duchenne type of dystrophy shows a variety of degenerative changes.

Once the diagnosis is established, the general therapeutic course should be discussed at length with the family. Eventually the child himself will ask direct questions deserving a response. In the early years the overall goal of treatment is to maintain ambulation and the child's normal daily activities. This includes school attendance and participation in community activities. With progressive weakness the habilitative measures necessary for the maintenance of ambulation may be undertaken. At this point the child clearly will recognize the presence of disability and must become an active participant in treatment. Physical therapy to prevent the onset of lower extremity contractures and the maintenance of an active life pattern are overall goals.

When the child becomes wheelchair bound, prevention of deforming scoliosis and attention to respiratory infection will be important. The eventual management of a life threatening respiratory or cardiac complication may well be determined by active discussion among the responsible physicians and the family. Heroic measures that cannot change the ultimate course of the disease process may well be avoided if in fact they result in extreme emotional stress to the family and child.

Implications for Life Adaptation

The specter of terminal disease makes it difficult to consider the future in the usual sense. Since the child will function well intellectually until the final stages of the disease, it is important to maintain an active family life for as long as possible. The future may be defined in achievable increments of weeks or months giving the child expectations for specific events and activities. Many families consider it important to emphasize the needs of brothers and sisters. If there is more than one affected child in the family, professional counseling and support may be critical to maintenance of the the family. Denial of reality is at all times inappropriate and should be confronted in a supportive yet determined manner.

THE PRADER-WILLI SYNDROME

One of the distinct benefits of recent developments in the understanding of metabolic disease has been the elucidation of clinical entities that probably share a common physiological pathway.

Until recently, as an example, there was no recognition that children who presented with a history of infantile hypotonia, obesity, short stature, hypogonadism, and mental retardation share a similar disease process, now described as the Prader-Willi syndrome. Especially when mental retardation was the predominant clinical cause of poor adaptation and learning, the other physical features appeared to be secondary. An irony in the examination of many children with serious mental retardation who have concomitant physical findings is that these findings are assumed to be part of a pattern of retardation when in fact their presence may indicate a specific complex mechanism. In the 30 years since the first description of children now considered as having the Prader-Willi syndrome, there has been considerable interest focused on the origins of uncontrolled weight gain and calorie ingestion in these children.

Characteristics

Owing to the nonspecific nature of the syndrome it is still common for individual cases not to be diagnosed until adolescence or early adulthood. With the current emphasis on careful developmental evaluation of young children, this will not as likely be a pattern in future years. The infant who is hypotonic because of a muscular disorder and who shows evidence of delay in other areas of development may well fall into a course consistent with the syndrome. It has now been recognized that many of these children show peculiar food related behavior patterns that can be documented prior to the onset of obesity, including the sneaking of food, gorging at meals, consumption of food products not normally considered appetizing to children, and a general preoccupation with food, food storage, and eating. These behavior patterns become particularly obvious prior to the onset of rapid weight gain in many patients.

Diagnosis and Treatment

Concomitant delays in general development are typical in children with mild to moderate mental retardation. These require that the clinician be aware of the overall pattern of development and issues surrounding eating and weight gain in order to be suspicious of the diagnosis. There are no pathognomonic diagnostic features, although recently deletion of chromosome 15 has been reported with great frequency in individuals with the Prader-Willi syndrome (Ledbetter et al., 1981). (See Chapter 19.)

The treatment for individual children centers on their specific cognitive and nutritional needs. The approach to special learning needs resulting from the mental retardation is similar to that in other children with the same degree of handicap. Strenuous approaches are advocated to control weight gain. Such measures include the temporary placement of the child in a controlled environment, generally a specialized residential program, to monitor food intake and institute appropriate dietary change reinforced by behavior modification techniques. A protein sparing, modified fast has been reported to yield at least temporary success in some individuals who normally have not experienced satiety in their typical eating environment. The complications of extreme obesity, including poor skin hygiene, clinical diabetes mellitus, and stress to the cardiovascular system, are all positively affected by weight control and weight loss when such approaches have been attempted. Radical changes in the home environment, including the locking of refrigerators or the elimination of food from the house, generally place great stress on other family members and result in hostility to the patient. Maintaining steady weight over a long period of time has not been clearly demonstrated in any longitudinal series.

Implications for Life Adaptation

The successful control of weight and motivation of the young person with the Prader-Willi syndrome may make possible much more comprehensive integration into community life, including sheltered employment and semi-independent residential living. Without weight control, severe obesity can readily shorten life expectancy and create major problems in ambulation or other problems that affect the ability of a family to care for a young person. A comprehensive approach to behavior management and activities of daily living is beneficial.

RICHARD P. NELSON
ALLEN C. CROCKER

REFERENCES

Accardo, P. J., and Capute, A. J.: The Pediatrician and the Developmentally Delayed Child. Baltimore, University Park Press, 1979.

Akins, C., Davidson, R., and Hopkins, T.: The child with myelodysplasia. In Scheiner, A. P., and Abroms, I. F. (Editors): The Practical Management of the Developmentally Disabled Child. St. Louis, The C. V. Mosby Co., 1980.

Crocker, A. C.: Cerebral palsy in a spectrum of developmental disabilities. In Thompson, G. H., Rubin, I. L., and Bilenker, R. M. (Editors): Comprehensive Management of Cerebral Palsy. New York, Grune & Stratton, Inc., 1983.

Crocker, A. C., and Cullinane, M. C.: Families under stress: the diagnosis of Hurler's syndrome. Postgrad. Med., 51:223, 1972.

Crocker, A. C., and Cushna, B.: Pediatric decisions in children with serious mental retardation. Pediatr. Clin. N. Am., *19*:413, 1972.

Drennan, J. C.: Arthrogryposis multiplex congenita. *In* Lovell, W. W., and Winter, R. B. (Editors): Pediatric Orthopedics. Philadelphia, J. B. Lippincott Co., 1978.

Hanshaw, J. B., and Dudgeon, J. A.: Viral Diseases of the Fetus and Newborn. Philadelphia, W. B. Saunders Company, 1978.

Ledbetter, D. H., Riccardi, V. M., Airhart, S. D., Strobel, R. J., Keenan,

B. S., and Crawford, J. D.: Deletions of chromosome 15 as a cause of the Prader-Willi syndrome. N. Engl. J. Med., *304*:325, 1981.

Marsh, G. G., and Munsat, T. L.: Evidence for early impairment of verbal intelligence in Duchenne muscular dystrophy. Arch. Dis. Child., *49*:118, 1974.

Nelson, R. P., and Crocker, A. C.: The medical care of mentally retarded persons in public residential facilities. N. Engl. J. Med., *229*:1039, 1978.

39H Emotional Problems in a Child with Serious Developmental Handicap

The importance of emotional development and disorders in a severely developmentally handicapped child is frequently overlooked or even denied. Some professionals may believe that below a certain cognitive level, emotional reactions and disorders are not possible. For instance, in the past it was believed that retarded persons are unaware of society's values and that therefore they do not know that they do not meet these values. Thus they would not feel unworthy and cannot develop depression (Gardner, 1967). Others tend to see only the external behavior of these individuals, as learned and without affective content. Both these approaches have in common the tendency to regard developmentally disabled persons, particularly those who are mentally retarded, as being unable to experience human emotions and therefore essentially as less than fully human.

On the other hand, a recent resurgence of interest of mental health professionals in mental retardation has increased our knowledge about the emotional development and psychopathology of retarded persons (Cushna et al., 1980; Donaldson and Menolascino, 1977). We know that they may experience richness of emotions even if they are not able to communicate them in the way we are accustomed to; they may react with deep emotions (rather than with automatic behaviors) to events in their lives; and they may develop the same emotional disorders as "normal" persons, which may be diagnosed and treated by the same, albeit modified, techniques.

RETARDED PERSONS AS INDIVIDUALS

Developmentally disabled persons form a more heterogeneous group than nonretarded ones. In fact, a mildly retarded adult living in community residence and partly self-supporting has much more in common with his nonretarded neighbor than with a profoundly retarded, nonverbal dormitory mate in an institution where he had grown up—yet technically both are classified as retarded.

It is not an accident that the title of this chapter is "Emotional Problems in a Child with Serious Developmental Handicap." The singular noun, "child," emphasizes the need to individualize each child as a unique person, rather than sterotype him by virtue of belonging to a group with a common denominator of a specific handicap. This applies particularly to mental retardation, which is not a specific disorder but rather a behavioral-functional syndrome. In fact, "deindividualization" has been pointed out by Wolfensberger (1972) as being a hallmark of an institution, where retarded persons are treated as part of a group only, rather than as individuals.

The heterogeneity of developmentally handicapped children from the point of view of their emotional development and psychopathology can be understood if we consider that multiple factors are operational here. Garrard and Richmond's schema (1965) illustrates well how psychosocial adaptation depends on a complicated interaction of many biological and environmental factors. The resulting combinations of these factors may well be idiosyncratic for specific children. It has been pointed out that the behavior of a retarded person is not an immutable result of low intelligence and that these individuals display a variety of behavioral patterns (Zigler and Balla, 1977).

Thus, one cannot speak about the "specific" personality of a retarded child. At the most one can talk about common denominators, or personality patterns, that frequently are encountered in these children, but by no means unique to them. Retarded persons have been described as being passive, impulsive, overactive, dependent, immature, and self-centered. Obviously all these characteristics are encountered in nonretarded persons as well. Although some traits might be statistically more frequent in retarded persons taken as a group, one cannot expect them to occur in every retarded individual.

In more recent years studies utilizing scientific research techniques have documented the more frequent occurrence of some personality features in groups of children classified as being retarded, in comparison with nonretarded ones. Krupski (1979), observing classroom behavior of educable retarded children, found that they were more distractible than nonretarded peers. Overdependency is another characteristic noticed in many retarded persons (Zigler and Balla, 1977). Social deprivation, negative experiences with others, and the experience of repetitive failing may influence the personality development of the retarded persons more than the mental retardation per se (Zigler and Balla, 1977). Bernstein (1970) has pointed out that low self-esteem of a retarded child is linked to the reactions of those in his environment. Krupski (1979) has shown that the ability of retarded students to attend decreased when the task was more difficult for them. Zigler and Balla (1977) also pointed out the retarded child's high degree of expectancy of failure as a consequence of the experience of failing.

Although it would be a mistake to ascribe the personality and psychopathology in a retarded child to a fixed neurological defect (Philips, 1966), the latter should not be ignored either. As a group, retarded persons (particularly those who are severely and profoundly retarded) have a higher prevalence of neurological dysfunction. The presence of organic brain disorder does predispose the child to the development of a behavioral disorder (Rutter, 1981; Rutter et al., 1970). Although the type of behavior disorder manifested may not be specific, it has been pointed out that perseveration and socially disinhibited behavior may be more common in this group (Chess, 1972; Rutter, 1981). Furthermore, other physical handicaps may be associated with mental retardation and may have an indirect but important bearing on the child's behavior. Sensory handicaps are important in affecting the development of communication skills and thus may contribute to the social isolation of the child. Multiple handicaps are particularly pernicious (as is described later). Motor handicaps may prevent a cognitively delayed child from compensating through gross motor activities. Other physical stigmata may "mark" the child as looking "retarded" and subject him to stereotyped reactions in society.

TRANSITIONAL CRISES AND THE EMOTIONAL DEVELOPMENT OF SEVERELY DEVELOPMENTALLY HANDICAPPED INDIVIDUALS

INFANCY AND EARLY CHILDHOOD

A severely developmentally disabled infant is surrounded by confusion and puzzlement. The parents are puzzled and anxious because of the delayed development, often accompanied by diminished responsiveness. They may be preoccupied with guilt feelings, denial of the child's problems, and conflicting advice and opinion from others. They may respond by overstimulating the child or by pulling away from him, to which the child may react with symptoms such as feeding and sleep disturbances, irritability, withdrawal, or failure to thrive. A vicious circle of parental confusion-anger-depression and the child's disturbed behavior–abnormal responsivity may ensue.

The pediatrician, often untrained and inexperienced in the early recognition of developmental disabilities, is puzzled by child's unusual presentation if it is not accompanied by obvious signs of causative physical disorder. Hoping that the disorder will "go away," he may advise that the child will "grow out of it" or that the mother is overanxious. Worse, when retardation is diagnosed early, an inappropriate and destructive opinion that "nothing can be done" may be voiced.

The child's reaction can be only speculated upon. He may find the outside world confusing and overwhelming; he may respond with increased withdrawal and unresponsiveness, not unlike an abused child or with irritability and vegetative symptoms (Galdston, 1968). In any case one can see here the potential beginnings of a distorted view of the world around him and of his own inferior position in it.

Both parents and professionals are often under the mistaken impression that young children do not develop emotional disorders, and that if they do, they are too young to be seen by a psychiatrist. As the result, referrals of these children for psychiatric consultation are rare. Actually this is one of most important ages in the genesis of emotional disorders and one that offers the best chances for prevention through specialized intervention. In fact, a subspecialty of infant psychiatry has been recognized in recent years. Pediatricians in their capacity of well baby care providers are in the best position to recognize these disorders and initiate intervention.

Maternal-infant bonding is one of the most significant processes of this early period of life, and it may be quite disturbed in the case of a handicapped infant (Emde and Brown, 1978; Hagamen, 1980; Klaus and Kennel, 1976). Stone and Chesney (1978) followed for one year 15 infants with a variety of handicaps: Down syndrome, brain injury, blindness, and multiple handicaps. Disturbances were present in one or more of the attachment types of behaviors, such as social smiling, vocalizing, or eye contact. Limpness while held and not demanding the caretaker's attention were also noticed. Such behavior is obviously crucial in securing the mother's emotional response and interaction with the child (Emde et al., 1976; Klaus et al., 1972). The mother who receives little, or abnormal, feedback from an infant, especially if she is depressed after learning that he is handicapped, may reach out less to such child than to one who responds normally. In the past such maternal withdrawal from the child was often erroneously seen as a cause of the child's disturbance, and these mothers were subjected to prolonged case work focused on their presumed rejection of the child. Prospectively, already at this stage the family may start seeing the child as being unrewarding and a burden.

These problems are well illustrated in blind infants, who were extensively studied by Fraiberg, whose work was summarized in the book, *Insights from the Blind* (1977). A blind infant's lack of eye contact, smiling in response to the human face, and facial expression of emotion may all interfere with the development of human attachments unless early intervention is provided. In fact researchers noticed that many blind children appear to be disturbed, to the point that they resemble children with infantile autism (Fraiberg, 1977; Fraiberg and Freedman, 1964; Keeler, 1958; Norris et al., 1957).

Children with multiple handicaps, especially sensory handicaps, are particularly vulnerable to the development of emotional disorders. These handicaps may exert a synergistic effect, since closing of several communication channels prevents the development of compensatory mechanisms utilizing intact channels, which would occur in the case of a single handicap. This was well demonstrated in a study of 41 hearing impaired young children with congenital rubella, conducted in our clinic. It was found that the adjustment was best in children with one disability and worst in those with four disabilities (deafness, blindness, mental retardation, motor impairment). All the latter children were, in fact, "autistic."

Handicapped young children manifest the same emotional disorders at this age as nonhandicapped ones. Early diagnosis of a pervasive developmental disorder, of which infantile autism is best known, is possible, although not always easy, since very young children with multiple handicaps, including retardation, may appear to be "autistic-like" by virtue of superficially poor relatedness. The presence of other characteristic symptoms will help to establish the diagnosis. These disorders are described in detail in Chapter 39F.

A disturbance defined recently as "reactive attachment disorder of infancy," encompassing situations previously called "nonorganic failure to thrive," may be diagnosed at this age period (American Psychiatric Association, 1980). By definition the onset occurs before age eight months in the presence of a lack of adequate caretaking, and its manifestations include poor social responsivity, weak cry, hypomotility and hypotonia, weight loss, or failure to gain weight. The diagnosis may be particularly difficult if the child is also retarded or autistic, both of which conditions may include some similar symptoms. Children with uncomplicated retardation, although developing slowly, do not usually show poor relatedness and failure to thrive. The relating difficulty of autistic children is usually much more pervasive and persistent. Children with a reactive attachment disorder may improve quite rapidly when proper emotional and physical care is given to them. Related to the foregoing is the syndrome of child abuse-neglect, to which developmentally disabled children are particularly vulnerable; this disorder is described in detail in Chapter 14.

MIDDLE CHILDHOOD

A school age child faces a new and different set of challenges, the most important of which is adaptation to the school and to the peer group. Both require an increased degree of separation from the parents, independence, and confidence, as well as frustration tolerance and self-control. Developmentally disabled and retarded children may be handicapped in these respects. Conflicts develop owing to a discrepancy between their abilities and the expectations and demands of the environment—school, parents, and peers. Children who are aware of their handicaps also develop a conflict between what they feel they should achieve and what they actually can achieve. As a result a state of chronic stress and anxiety may evolve, against which a variety of compensatory defense mechanisms may be employed, which may result in behavior patterns that further reduce the child's ability to adapt. These defenses may be more immature than in a nonretarded child of the same age (Bernstein, 1970). Thus, a retarded child may withdraw or resort to fantasy solutions (seen by others as lying) or aggression (direct, or an unstructured temper tantrum). Their common denominator will be despair, confusion, anger, a sense of unworthiness, and rejection. Other symptoms may include overactivity, anxiety, school phobia, or distractibility. Most important, the poor image of the self, passivity, dependence, and inappropriate attention getting, which result from repeated experiences of failure, progressively become fixed features of the child's personality.

The child's caregivers often convey to him a double message. Overprotection and infantilization are frequent. A child may not be expected to develop even the self-care skills of which he may be quite capable, and may not be taught to perform useful chores, which would give him a sense of belonging and contributing to family life. Yet he may be pushed and drilled in academic tasks beyond the possessed cognitive capacity. The sum total will be a further deepening sense of low self-image, as well as parental and child frustration.

ADOLESCENCE AND YOUNG ADULTHOOD

Many professionals and laymen, including the parents, tend to deny that developmentally handicapped and retarded persons ever become adolescents. This view in a sense denies that these persons become sexual beings and expresses the belief that they remain eternal children. Others overstress the issue of sexuality of a retarded adolescent, seeing him as an uninhibited person, who should be sterilized. In our experience adolescence is the most challenging developmental crisis, for both the family and the individual are confronted with the fact that the developmental handicap is permanent and that another evaluation or a different education will not make it go away. In the majority of retarded persons, physical development is normal. Thus these youngsters face the same developmental challenge as "normal" ones to adapt to the pubertal changes of their bodies, but because of their lowered intellectual capacities, they are more perplexed and stressed by what is happening to them than are their nonretarded peers.

Psychological functioning in retarded adolescents, frequently arrested in earlier developmental stages, is often characterized by immaturity, concreteness, rigidity, passivity-dependency, and compulsivity. Because of his egocentricity or inability to perceive others' points of view, especially concerning abstraction, a retarded child may stand out as "childish" or "stupid" (Bernstein, 1970). He may be unable to keep up with the average adolescent's fast flowing conceptual thinking, preoccupation with ideas, and plans for the future. As a result, he becomes increasingly isolated socially. The most common concern of a parent of a retarded adolescent was expressed by the mother of one of our patients: "He is always, always, alone."

Eisenberg (1980) has characterized an average adolescent as being "no longer a child, still not an adult." The retarded adolescent's dilemma is even more serious. Owing to the reality of delayed cognitive-emotional development, past experiences of failure, and a low self-image, he commonly sees himself as a dependent child. Indeed such children are often still trying to master earlier developmental tasks, such as learning simple concrete skills. Their search for autonomy and individuation is frustrated, if not made impossible, by real personal deficiencies requiring continued supervision and dependence on their caregivers. This is particularly true if the latter are overpro-

tective and ambivalent and focus on the child's limitations rather than on promoting his strengths. More often than not, pediatricians contribute to this attitude by treating their retarded adolescent patients as young children. It is indeed rare that a physician will talk to a verbal, mildly retarded adolescent in privacy and in a respectful, nonpaternalistic fashion. Mildly and moderately retarded adolescents (which means the majority) are aware of their deficiencies yet are confused about their nature and cause, as well as the abilities they may have. Yet commonly parents and physicians maintain a conspiracy of silence and rarely give them an understandable and acceptable explanation. As a result of well meaning overprotection, retarded adolescents may not be given an opportunity to learn through the taking of risk, even in such simple activities as using a stove, hair dryer, or public transportation. As pointed out by Perske (1972), this "dignity of risk" is very important for human development.

Sexuality, although an issue earlier in life, becomes a major source of confusion and conflict at this stage of a retarded person's development. It is a subject by itself, beyond the scope of this chapter. Recent reviews of the topic by this author may be consulted by those interested (Szymanski and Jansen, 1980; Szymanski, 1981). For many years retarded persons had been stereotyped, together with other "deviants" as being sexually uninhibited "moral morons," to be restrained by outside controls, for their own and society's protection. On the other hand, another view has held that they are asexual eternal children who do not need sexual expression. Retarded adolescents and young adults have been often victims of sexual exploitation. An even more frequent form of sexual abuse has been to deny them rights to sexuality, ranging from forced sterilization to forced living in sexual segregation. Even in current times it is not infrequent that a parent requests a pediatrician to arrange for an involuntary sterilization for a retarded adolescent child (which besides being morally wrong is legally virtually impossible to obtain). It is now recognized more and more that retarded persons are sexual beings, capable of sexual expression (which does not necessarily lead to conception) and who need early and appropriate sexuality education.

In summary, severely developmentally handicapped adolescents are subjected to external and internal stresses, against which they may develop a variety of defensive reactions. Some resort to regression, an increase in dependency, or an avoidance of challenge. Others become withdrawn, and depressed. Still others attempt to imitate superficially their nonhandicapped peers in external behavior, become victims of sexual and other types of exploitation, become aggressive, and are led to delinquency. All are vulnerable to developing emotional disorders, especially depression.

EMOTIONAL DISORDERS IN RETARDED CHILDREN: PREVALENCE AND DIAGNOSIS

Prevalence of emotional disorders in retarded children has been the subject of a number of studies (reviewed by Webster, 1970, and Szymanski, 1980b). In spite of design deficiencies, these studies agree in finding a high prevalence of these disorders, generally within the 30 to 60 per cent range. The Diagnostic and Statistical Manual of Mental Disorders (DSM-III), recently introduced, facilitates the diagnosis, with features such as the use of specific diagnostic criteria and a multiaxial approach (American Psychiatric Association, 1980). In our Developmental Evaluation Clinic there is currently a study on the adaptation of the DSM-III for use with developmentally handicapped children referred for comprehensive assessment. Preliminary results indicate that in a sample of 126 retarded children, symptoms warranting formal psychiatric diagnosis were present in 70 per cent. The most common diagnoses were pervasive developmental disorders (in 16 per cent of the children), adjustment disorders (in 15 per cent, affective disorders (in 13 per cent), and attention deficit disorders (in 10 per cent). Thus diagnostic criteria of mental disorders can be utilized with developmentally delayed children, provided that they are modified according to the developmental level, as exemplified in the discussion that follows.

Psychiatric diagnostic techniques that are appropriate for use with retarded persons have been described (Bernstein, 1970; Szymanski, 1977, 1980b), and only highlights will be pointed out here. Because of the multiple handicaps that are often involved, as well as the child's dependency on and involvement with multiple caretakers, the psychiatric diagnosis has to be established within a comprehensive context. Past and present development, a behavioral history, medical assessments, a family assessment, and school information have to be integrated with information gathered when examining the child. Nonverbal and verbal techniques are used, adapted to the child's communication level. An initial period of noninterventive observation of the child's play by a friendly interviewer may help break the ice and reduce the child's anxiety. Information should be collected concerning the child's functioning in all important areas, such as the ability to relate to others, the range and appropriateness of behavior and affect, attention span; activity level, impulse control, presence or absence of thought disorder, quality of organization and major themes of play, the child's self-image and understanding of his own handicap, his capacity for symbolic and conceptual thinking and play, the level and quality of communication, the relationship with the parents, and parental management of the child. This information has to be evaluated in reference to the child's developmental level, educational experience, and the family values. For instance, in the absence of other evidence of delusion, a child's belief that a cartoon character is real may be a function of his cognitive level and information received from others. "Talking to himself", a commonly heard complaint, in a moderately retarded child who otherwise relates well to others may be a variation of the "imaginary friend" theme linked to loneliness, rather than evidence of psychosis.

IMPORTANCE OF PSYCHIATRIC DISORDERS

The psychiatric diagnosis is not a mere "label," nor it should be treated as such. Proper diagnosis may resolve doubts and anxiety, lead to appropriate treatment, or prevent an inappropriate one. For instance, bizarre acting youngster who is in a behavioral program because of "attention getting" may actually be psychotic and in need of appropriate treatment, whereas another, with superficially similar symptoms and taking high doses of phenothiazines, may be nonpsychotic, reacting to institutional boredom and deprivation.

PROBLEMS IN DIAGNOSIS

In the following section, three of the specific classes of mental disorders are briefly reviewed as examples. The focus is on adaptation of the diagnostic criteria to the developmentally disabled population, since the general features of these disorders are discussed elsewhere in this book.

Pervasive Developmental Disorders

This class, introduced in the DSM-III, groups disorders that were described previously by variety of terms—autism, childhood schizophrenia, symbiotic psychosis, and atypical child. Its current major subdivisions are infantile autism and childhood onset pervasive developmental disorders, each of which may occur as full syndrome and as a residual state. According to the DSM-III, these disorders are "characterized by distortions in the development of multiple basic psychological functions that are involved in the development of social skills and language, such as attention, perception, reality testing, and motor movement." Specific criteria are set, that have to be met in order to justify the diagnosis. This is particularly helpful in the case of infantile autism, which in the past was frequently diagnosed according to clinician's school of thought, rather than according to clinical manifestations. One of the frequent misconceptions was that autistic children were very intelligent and thus that autism and mental retardation were mutually exclusive. A recent review of the evolution of the concepts of autism and its relationship to mental retardation has been made by Tanguay (1980), who points out that 75 per cent of autistic children function in the retarded range. Thus these disorders frequently coexist, and they may be "caused" by similar conditions (or rather associated with them), such as congenital rubella. Another misconception was the view that autism was a specific disease entity. As pointed out by Tanguay (1979), autism should be seen as a behavioral syndrome, which probably consists of many subsets, each with a specific causal relationship to an etiological factor. With further progress in research on autism, particularly its biological aspects, we may be able to "chip away" these specific subsets from the heterogeneous group of "autism," perhaps to the point that the nonspecific general diagnosis of autism will be less justified. (Further details of the diagnosis of autism can be found in Chapter 39F.)

Organic Mental Disorders

Woodward et al. (1970) have pointed out the loose use of the term "organic" diagnosis as well as the physician's belief that one had to diagnose mental disorders as either organic or nonorganic. In fact one wonders whether such a general diagnostic statement should be made at all, since essentially all mental disorders are to a degree "organic," even if it meant the encoding of a certain traumatic memory in a neuron. As pointed out by Lipowski (1980), the DSM-III provides a significant step toward clarifying this diagnosis. It makes it clear that differentiating organic mental disorders as a separate class does not imply that the "nonorganic" disorders do not depend on brain processes. Demonstration of the existence of a specific organic factor or recognition of a specific brain syndrome is required in order to make an "organic" diagnosis.

Depressive Disorders

There is virtually no literature dealing with nonpsychotic depression in retarded persons. Some of earlier literature (reviewed by Gardner, 1967) questioned whether they were less vulnerable to depression because they were unaware of societal values and thus felt less unworthiness and guilt because of their own deficits. Other studies suggested that retarded persons' experiences of chronic rejection made them more vulnerable to depression. In fact, a similar controversy existed as to whether nonretarded children could suffer depression, yet many current studies indicate that this disorder is not uncommon among children (Carlson and Cantwell, 1980; Kashani et al., 1981). The actual symptoms of depression may be modified by the presence of retardation. Persons with poor verbal and conceptual capacities may not be able to complain of feeling depressed, guilty, or unworthy. On the other hand, somatic complaints and behavioral symptoms such as apathy and regression may assume a major importance.

PREVENTION OF EMOTIONAL MALADJUSTMENT IN HANDICAPPED CHILD: THE ROLE OF THE PEDIATRICIAN

The knowledge that developmentally handicapped persons are at risk of developing emotional disorders makes it mandatory to provide them with early preventive help. Pediatricians are especially important here as the primary care providers who see the handicapped child early and frequently are the first ones to make, or at least to suspect, the diagnosis of retardation. Retrospective studies indicate that in recent years the parents of children with the Down syndrome have tended to ascribe less importance to pediatricians' early advice about their later concrete decisions concerning their children, such as institutionalization or school placement (Pueschel and Murphy, 1977; Springer and Steele, 1980). Repeated clinical experience has demonstrated, however, that pediatricians may have much more importance.

Physicians often believe that they fail if they do not diagnose and "cure" the patient's illness. Facing a child with "incurable" retardation, especially if of unknown origin, may result in emotional stress for the physician, which in the past was even linked with a tendency to recommend institutionalization (Goodman, 1964). Experience has shown that the physician has considerable importance for the family, not as "healer" of the child's retardation but of the parents' emotional wounds; as a source of support and guidance; and as a teacher of a proper attitude toward the child. On the other hand, an improper attitude on the part of the physician is well remembered by the family for many years. Many parents of the patients we have seen recollected how painful to them were pediatricians' statements that "There is nothing you can do" which (besides being untrue) conveyed to them an attitude of helplessness and hopelessness, intensifying their guilt feelings.

What the family needs, first of all, is to learn from the pediatrician to regard the child as an individual who has human worth as does any other child, rather than as one of "those children," undeserving of professional time. The physician's inquiring about and pointing out the child's personal characteristics, talking to him, and responding with understanding to parental questions convey the message of "individualization." Providing the parents with up to date information about child's condition, referring them for a second opinion to a specialized clinic and for support to a parents' group, advising on services, and giving "permission" to set limits and proper expectations for the child are important direct services. In the 1980's there can be no excuse that one is not trained in this field; continued medical education is every physician's responsibility. Behaviorally trained pediatricians can assess the parents' psychosocial background, the influence of the child's handicap on the family, and the need for periodic respite. They can direct the family to set up an appropriate stimulation program, to set the limits, to promote the child's independence and to focus on the child's strengths rather his disabilities. The pediatrician should also be able to distinguish early the transient behavior disorders of the child related to an environmental situation from a more pervasive, functionally limiting, emotional disturbance, especially if it involves dysphoric moods. In the case of the latter, referral for psychiatric consultation may be indicated. One, however, should avoid the common tendency to refer only patients who are disturbing to others, while neglecting those who are otherwise disturbed.

UTILIZATION OF PSYCHIATRIC CONSULTATIONS

Many child psychiatrists and clinics have not been responsive to the needs of retarded patients, nor have they had experience in working with them, but this does not relieve them of the responsibility to acquire such expertise and provide services to all disturbed children who need them, regardless of the level of intelligence.

A satisfactory consultation depends to a considerable degree on good communication between the consultee and the consultant. Thus a clear statement of the reasons for consultation is necessary. Particularly important is the "hidden agenda," which frequently goes unsaid (e.g., "Medicate this child, since nothing else will work!"). Conversely, prompt and preferably detailed feedback by the consultant to the consultee is important as well. The consulting psychiatrist, jointly with the pediatrician, should "synthesize" the available information about the child into a comprehensive assessment, rather than limit himself to psychological interpretations only. A statement regarding the psychiatric diagnosis, if warranted, in addition to the one relating to retardation, must be included. The pediatrician and the consultant should be able to work as an interdisciplinary team, which includes also the child's parents, teachers, and other caregivers.

MENTAL HEALTH INTERVENTIONS

Emphasis in the following brief paragraphs is placed on modifications applicable to retarded patients. Detailed description of various treatment techniques can be found in textbooks edited by Menolascino (1970) and Szymanski and Tanguay (1980).

Psychotherapy. In spite of the available literature on effectiveness of psychotherapy with retarded persons (reviews by Jakab, 1970, and Szymanski, 1980a), some professionals may still not believe (or rather, not know) that these patients do respond well, both to individual and to group psychotherapy. The treatment techniques need to be adapted to the patient's level, particularly his communication ability. The approach often needs to be reality oriented, concrete, repetitive, and structured. Nonverbal techniques need to be utilized as well. Group therapy is particularly effective with adolescents and young adults who need to learn to relate to their peers.

Psychoactive Medication. An excellent review of current knowledge of the use of these drugs with retarded patients has been written by Rivinus (1980). In using these drugs, one has to follow the same principles of sound medical practice as with nonretarded patients. Unfortunately the psychotropic drugs have often been misused, particularly in institutions, under pressure from the administration and the nonmedical staff to have the clients rendered cooperative and docile. These drugs should be prescribed (as with any other drug) by a physician who is thoroughly familiar with them. They should be a part of a comprehensive treatment program, not a substitute for it. One should remember that components such as psychotherapy, milieu therapy, and behavior modification have a synergistic effect with the medications. These drugs should be utilized for their specific action in specific disorders. For example, thioridazine should be used as an antipsychotic drug and not as a nonspecific "major transquilizer." When used in this way and in proper dosage, these drugs are often very effective and permit the retarded patients to participate meaningfully in a variety of treatment and educational activities. Thus, whenever possible, a definite psychiatric diagnosis should be established prior to prescribing the drug, which should be used to treat a disorder, rather than a symptom disturbing to the caregiver.

LUDWIK S. SZYMANSKI

REFERENCES

American Psychiatric Association: Diagnostic and Statistical Manual of Mental Disorders. Ed. 3. Washington, D.C., American Psychiatric Association, 1980.

Bernstein, N. R.: Intellectual defect and personality development. In Bernstein, N. R. (Editor): Diminished People. Boston, Little, Brown and Company, 1970.

Carlson, G. A., and Cantwell, D. P.: Unmasking masked depression in children and adolescents. Am. J. Psychiatry, 137:445, 1980.

Chess, S.: Neurological dysfunction and childhood behavioral pathology. J. Autism Childhood Schizophrenia, 2:299, 1972.

Cushna, B., Szymanski, L. S., and Tanguay, P. E.: Professional roles and unmet manpower needs. In Szymanski, L. S., and Tanguay, P. E. (Editors): Emotional Disorders of Mentally Retarded Persons. Baltimore, University Park Press, 1980.

Donaldson, J. Y., and Menolascino, F. J.: Past, current and future roles of child psychiatry in mental retardation. J. Am. Acad. Child Psychiatry, 16:38, 1977.

Eisenberg, L.: Normal child development. In Kaplan, H. T., Freedman, A. M., and Sadock, B. J. (Editors): Comprehensive Textbook of Psychiatry. Ed. 3. Baltimore, The Williams & Wilkins Company, 1980.

Emde, R. N., and Brown, C.: Adaptation to the birth of a Down's syndrome infant: grieving and maternal attachment. J. Am. Acad. Child Psychiatry, 17:229, 1978.

Emde, R. N., Gauensbauer, J., and Harmon, R. J.: Emotional Expression in Infancy: A Biobehavioral Study. New York, International Universities Press, 1976.

Fraiberg, S.: Insights from the Blind. New York, Basic Books, Inc., 1977.

Fraiberg, S., and Freedman, D.: Studies in the ego development of the congenitally blind child. Psychoanal. Study Child, 19:113, 1964.

Galdston, R.: Dysfunctions of parenting: the battered child, the neglected child, the exploited child. In Howells, J. G. (Editor): Modern Perspectives of International Child Psychiatry. Edinburgh, Oliver and Boyd, 1968.

Gardner, W. T.: Occurrence of severe depressive reactions in the mentally retarded. Am. J. Psychiatry, 124:142, 1967.

Garrard, S. D., and Richmond, J. B.: Diagnosis in mental retardation, and mental retardation without biological manifestations. In Carter, C. H. (Editor): Medical Aspects of Mental Retardation. Springfield, Illinois, Charles C Thomas, 1965.

Goodman, L.: Continuing treatment of parents with congenitally defective infants. Soc. Casework, 9:92, 1964.

Hagamen, M. B.: Family adaptation to the diagnosis of mental retardation in a child and strategies of intervention. In Szymanski, L. S., and Tanguay, P. E. (Editors): Emotional Disorders of Mentally Retarded Persons. Baltimore, University Park Press, 1980.

Jakab, I.: Psychotherapy of the mentally retarded child. In Bernstein, N. R. (Editor): Diminished People. Boston, Little, Brown and Company, 1970.

Kashani, J. H., Husain, A., Shekim, W., Hodges, K. K., Cytryn, L., and McKnew, D. H.: Current perspectives on childhood depression: an overview. Am. J. Psychiatry, 138:143, 1981.

Keeler, W. R.: Autistic patterns and defective communication in blind children with retrolental fibroplasia. In Hoch, P. H., and Zubin, J. (Editors): Psychopathology of Communication. New York, Grune & Stratton, Inc., 1958.

Klaus, M. H., Jerauld, R., Kreger, N. C., McAlpine, W., Steffa, M., and Kennell, J. H.: Maternal attachment; importance of the first post-partum days. N. Engl. J. Med., 286:460, 1972.

Klaus, M. H., and Kennell, J. H.: Maternal-Infant Bonding. St. Louis, The C. V. Mosby Company, 1976.

Krupski, A.: Are retarded children more distractible? Observational analysis of retarded and nonretarded children's classroom behavior. Am. J. Ment. Defic., 84:1, 1979.

Lipowski, Z. J.: A new look at organic brain syndromes. Am. J. Psychiatry, 137:674, 1980.

Menolascino, F. J.: Psychiatric Approaches to Mental Retardation. New York, Basic Books, Inc., 1970.

Norris, M., Spaulding, P., and Brodie, F.: Blindness in Children. Chicago, University of Chicago Press, 1957.

Perske, R.: The dignity of risk. In Wolfensberger, W. (Editor): The Principle of Normalization in Human Services. Toronto, National Institute on Mental Retardation, 1972.

Philips, I.: Children, mental retardation and emotional disorder. In Philips, I. (Editor): Prevention and Treatment of Mental Retardation. New York, Basic Books, Inc., 1966.

Pueschel, S. M., and Murphy, A.: Assessment of counseling practices at the birth of a child with Down's syndrome. Am. J. Ment. Defic., 81:325, 1977.

Rivinus, T. M.: Psychopharmacology and the mentally retarded patient. In Szymanski, L. S., and Tanguay, P. E. (Editors): Emotional Disorders of Mentally Retarded Persons. Baltimore, University Park Press, 1980.

Rutter, M.: Psychological sequelae of brain damage in children. Am. J. Psychiatry, 138:1533, 1981.

Rutter, M., Graham, P. J., and Yule, W.: A neuropsychiatric study in childhood. Clin. Devel. Med., No. 35–36, 1970.

Springer, A., and Steele, M. W.: Effects of physicians' early parental counseling on rearing of Down syndrome children. Am. J. Ment. Defic., 85:1, 1980.

Stone, N. W., and Chesney, B. H.: Attachment behaviors in handicapped infants. Ment. Retard., 16:8, 1978.

Szymanski, L. S.: Psychiatric diagnostic evaluation of mentally retarded individuals. J. Am. Acad. Child Psychiatry, 16:67, 1977.

Szymanski, L. S. Individual psychotherapy with retarded persons. In Szymanski, L. S., and Tanguay, P. E. (Editors): Emotional Disorders of Mentally Retarded Persons. Baltimore, University Park Press, 1980a.

Szymanski, L. S.: Psychiatric diagnosis of retarded persons. In Szymanski, L. S. and Tanguay, P. E. (Editors): Emotional Disorders of Mentally Retarded Persons. Baltimore, University Park Press, 1980b.

Szymanski, L. S.: Coping with sexuality and sexual vulnerability in developmentally disabled individuals. In Milunsky, A. (Editor): Coping with Crisis and Handicap. New York, Plenum Press, 1981.

Szymanski, L. S., and Jansen, P. E.: Assessment of sexuality and sexual vulnerability of retarded persons. In Szymanski, L. S. and Tanguay, P. E. (Editors): Emotional Disorders of Mentally Retarded Persons. Baltimore, University Park Press, 1980.

Szymanski, L. S., and Tanguay, P. E.: Emotional Disorders of Mentally Retarded Persons. Baltimore, University Park Press, 1980.

Tanguay, P. E.: Early infantile autism: New advances in research. Presented at the Annual Meeting, American Academy of Child Psychiatry, Chicago, 1979.

Tanguay, P. E.: Early infantile autism and mental retardation: differential diagnosis. In Szymanski, L. S., and Tanguary, P. E. (Editors): Emotional Disorders of Mentally Retarded Persons. Baltimore, University Park Press, 1980.

Webster, T. G.: Unique aspects of emotional development in mentally retarded children. In Menolascino, F. J. (Editor): Psychiatric Approaches to Mental Retardation. Basic Books, New York, 1970.

Wolfensberger, W.: The Principle of Normalization in Human Services. Toronto, National Institute on Mental Retardation, 1972.

Woodward, K. F., Jaffe, N., and Brown, D.: Early psychiatric intervention for young mentally retarded children. In Menolascino, F. J. (Editor): Psychiatric Approaches to Mental Retardation. New York, Basic Books, Inc., 1970.

Zigler, E., and Balla, D.: Personality factors in the performance of the retarded: Implications for clinical assessment. J. Am. Acad. Child Psychiatry, 16:19, 1977.

40
Communicative Disorders

Mastery of basic communicative skills is one of the major developmental achievements of early childhood. When a child fails to speak by age two years, the family initially brings their concern to the attention of the child's pediatrician or family physician. These concerns should never be dismissed lightly, since the ability to communicate is central to human relationships and learning. Communication is the result of a complex interaction process. Because of the central role of communication in human existence, communicative disorders have long been of concern to many different professionals, and the dynamics of professional relations are well known to specialists in this area. This chapter emphasizes the need for interprofessional cooperation in the assessment and treatment of disordered communication during childhood.

Language is a system of symbolic representation used by human beings to communicate information. Writing, finger spelling, and sign language are alternative systems for the representation of thought. Speech is the planned execution of the oral movements necessary to articulate language.

Disorders of communication result from a wide variety of causes and are associated with an even greater diversity of acute or chronic medical or developmental problems. Little information is available about socioeconomic variation or incidences in various chronic medical or developmental conditions. Deviance in communication is present when a person's communication interferes with the speaker's ability to be understood, when a listener is distracted from the interest of the message, or when the communication pattern deviates from what is accepted by the social group. The presence of a significant communicative disorder has an important effect upon the child's social-emotional interaction within the family setting, the peer group, and the learning environment.

The prevalence of communicative disorders is difficult to estimate because of variability in the ways such information is collected. Seven to 10 per cent of the general population is considered to be functioning below the norm in one or more aspects of communicative function (Bax and Hart, 1976). Speech and language disorders are the commonest developmental problems affecting pre-

school children. The prevalence of such disorders does appear to diminish with age, attesting to the developmental nature of many of these conditions and providing some factual basis for reassuring the family that the child will "grow out of it."

One study reported a 15 per cent prevalence of speech problems in three year old children, decreasing to 5 per cent for five and one-half year old children living in an area of central London (Bax and Hart, 1976). In the National Child Development Study of England it was reported that 1.4 per cent of children aged seven years had largely unintelligible speech, while a larger group revealed some speech disorder (Peckham, 1973). Impaired development of speech and language is a common indicator of cerebral dysfunction that may lead to later difficulties in reading and writing, and children with reading disability may have difficulties throughout their school career and possibly later. Males generally are more often affected by developmental language and reading disorders than females, and girls appear to manifest a developmental advantage for certain verbal functions until adolescence. Such sex differences and a variety of family studies suggest that genetic factors may play a role in developmental language and reading disorders.

NORMAL DEVELOPMENT OF LANGUAGE AND COMMUNICATION

Normally children learn to speak fluently, to express themselves adequately in most situations, and to understand what is said to them in everyday contexts by the time they are five years of age. Language comprehension and production continue to develop for another few years, but the linguistic achievements of the five year old are still quite remarkable. There may be considerable variability among normal children with respect to the rate and extent of language acquisition in these first five years. In our society parents may become unduly anxious about the speech and language status of their children if it compares unfavorably with that of other children in their community. It may be very important for the child's physician, to whom these concerns are first addressed, to determine whether the preschool child is indeed at risk for significant speech and language impairment or whether he simply lags behind his peers temporarily.

The problem is a particularly difficult one, because the likelihood is great that the child who is not talking by age two years, or who is still unintelligible at age three years, will eventually catch up with his peers. Yet for the small number of children who will show persistent speech and language deficits, early identification and intervention are probably very important for amelioration of their condition. In the case of hearing impairment, the earlier the child receives appropriate amplification and language stimulation, the better the prognosis for language development will be.

Significant delays in language production or perception are not necessarily predictive of later language or learning disability (just as accelerated speech and language development is not necessarily predictive of superior academic performance in the school years). Yet such delays may indicate serious underlying deficits. It is therefore important for the pediatrician to take the parents' concerns seriously and to employ a set of guidelines for referral of such children to other professionals for their evaluation and recommendations.

Before presenting such guidelines, it may be helpful to describe normal language acquisition in some detail and to indicate major language milestones, together with approximate ages for their attainment, for the normal child. Such milestones are most often documented in terms of language production, but it is also important to be aware of the child's level of language comprehension and his ability to communicate with others. If the child uses spoken language in a stereotyped manner without apparent meaning or understanding of what he is saying, his productive capacity may lead the examiner to overestimate his potential for language and for learning. It is also important in reviewing language milestones to remember that although a child may have an inborn capacity for learning language, this capacity will not function in the absence of a supportive environment in which the child is exposed to natural language stimulation.

It is convenient to consider language milestones in terms of a prespeech period (usually the first 12 to 18 months of life) and a language acquisition period (18 months and beyond), although such periods are not, in fact, clearly separated from one another developmentally. During the so-called prespeech period, the infant acquires social interactive skills, sound production, and speech perception skills. These abilities may be relatively separate in the first months of life, but increasingly they interact as the child grows older and each makes an important contribution to language development.

PRESPEECH PERIOD

Linguistic and auditory milestones for the purpose of screening children in the prespeech period are shown in Table 40–1. A second more detailed developmental sequence shown in Table 40–2 addresses speech sound production, speech and nonspeech auditory perception, and communication milestones for the first 15 months of life. These are important in deciding whether intervention should be offered and, if so, the level of functioning at which intervention should begin.

The development of sound production reflects increasing speech motor control, i.e., the ability to differentiate movements of tongue, lip, palate, and jaw and to coordinate such movements with respiratory and laryngeal movements so that the prosodic features of pitch, loudness, and timing

Table 40–1. LINGUISTIC AND AUDITORY MILESTONES

Language Milestone	Months of Age	Language Milestone	Months of Age
1. Alerting	1	17. Three words	14
2. Social smile	1½	18. One step command (without gestures)	15
3. Cooing	3	19. Four to six words	15
4. Orient to voice	4	20. Immature jargoning	15
5. Orient to bell (I)	5	21. Seven to 20 words	18
6. "Ah-goo"	5	22. Mature jargoning	18
7. Razzing	5	23. One body part	18
8. Babbling	6	24. Three body parts	21
9. Orient to bell (II)	7	25. Two word combinations	21
10. "Dada/mama" (inappropriately)	8	26. Five body parts	23
11. Gesture	9	27. Fifty words	24
12. Orient to bell (III)	10	28. Two word sentences (noun-pronoun inappropriately and verb)	24
13. "Dada/mama" (appropriately)	10		
14. One word	11	29. Pronouns (I, me, you, inappropriately)	24
15. One-step command (with gesture)	12		
16. Two words	12		

DURING THE FIRST 15 MONTHS OF LIFE

Month	Sound Production	Auditory Perception	Communication
0–1	Reflexive sound production: cry, discomfort sounds, vegetative sounds. These sounds are either vocalic or consonantal. Pain and hunger cries are differentiated in first week.	Responses to instrumental music and speech are differentiated. Responses show preference for human speech. Ability to turn head in horizontal plane to sound source in immediate vicinity is present.	Infant links mother's voice with her face and is comforted when crying by her voice. Infant responds differently to infant crying and to adult speech. Infant stares intently at faces.
2–3	Consonantal sounds are produced at the back of the mouth. Brief consonantal elements are superimposed upon vocalic sounds.	Sucking and heartrate responses indicate discrimination of different CV syllables (ba vs. da; da vs. ta).	Social smiling and cooing vocalizations appear. Infant is attentive to nodding, smiling adult and is likely to vocalize in response to adult. Turn-taking is managed chiefly by adults; mother and infant may vocalize in chorus.
4–6	Reflexive sound production declines. Infant begins to experiment with and to prolong noncry sounds. Examples are the use of extreme high and low pitches, extreme pitch glides, and "raspberries." Greater variety of vocalic sounds appear. Longer duration vocalic consonantal elements may be combined in "syllables" toward the end of this period; also longer series of vocalic + consonantal sounds are produced.	Head turn responses are made to change in background stimulation. Responses indicate discrimination of vowels and of CV syllables and also of pitch contours. However, if vowel contrast and pitch contour contrasts are presented simultaneously, the vowel contrasts are more salient for the infant and pitch is ignored. Infant responds differently to different tones of voice.	Laughter appears and is elicited by tickling, blowing on belly, and so forth. Infant is preoccupied with objects and may appear to take people for granted. Mother tends to follow infant's gaze rather than engaging infant directly in mutual gaze.
7–9	Consonants and vowels are combined in consonant-vowel (CV) syllables, and these syllables are produced in series (e.g., da da). Intonation contour is added to series of syllables. High front vowel sounds are produced, including /i/, and also more "difficult" consonants, e.g., /s/ and /ts/.	Infants may be able to discriminate final syllables of two contrasted series, e.g., pataka from patapa, if final syllable is of sufficient duration. They may discriminate falling and rising intonation contour of phrase, e.g., "see the cat," if final syllable is stressed. Infants recognize sounds that go with familiar toys and people. Sound localization improves; infants can locate sound source in vertical plane. Infants display interest in music.	Imitation of adults takes form of primitive sound making responses. Turn-taking in these and other exchanges is well developed. Infants show a great deal of interest in adult's mouth when adult is speaking. Infants may produce different sound combinations when playing with objects and when interacting with adults.
10–15	Rounded vowels emerge, e.g., /u/. Protowords are produced with phonetically consistent forms, i.e., with some free variation but with certain components such as nasal consonant + vowel or stop + vowel, always present in a given protoword. Expressive jargon is produced to greater extent by some infants, i.e., long series of CV, VC, or CVC syllables with natural intonation contour and intensity variation, which sounds like "a foreign language" to adults but in which phonetically consistent forms may be embedded. Protowords are produced to a greater extent by others.	Infant begins to show preference for novel utterances. Responds with gesture to patty cake, peek-a-boo, and other games. May respond with changes in gaze to questions such as "Where's the _____?" Infant knows his own name and responds to "No!" commands. Sound localization indicated by head movement in all places.	Infant begins to express wishes and desires ("protoimperatives") by combining gestures and sounds in protowords. He also shares with adults, by means of gestures and sounds or protowords ("protodeclaratives"), his interests in toys and play activities. Gaze at adult begins to accompany the "protodeclaratives" and "protoimperatives." Exchanges are well timed and infant initiates interaction more often. Infant imitates adult speech patterns more clearly.

may also be controlled in relation to articulatory movement.

In the development of speech perception, the infant becomes aware of different aspects of speech production by adults (rhythm, pitch, intensity, syllabic content) and is increasingly able to discriminate vowel and consonant phonemes (the smallest units of speech), pitch contours, duration, and intensity levels.

The development of communicative behavior reflects an increasing responsiveness to others on the part of the infant, an awareness of the responses of others, and an interest in influencing the course of interaction. Children learn that their own behavior (smiling, vocalizing) may have a powerful effect upon the behavior of others. Later they learn to use gesture, facial expression, and vocal behavior as they engage in joint activities with adults and make demands of them.

By paying attention to these aspects of language development, it becomes possible to estimate how far behind a given infant may be in any one aspect of development and also to examine his developmental profile. For example, a child who is not producing words but who uses a great deal of expressive jargon, who is visually alert, who communicates by means of gesture, and who actively explores his environment may be considered to be much more advanced than a child of the same age who is not attentive to others, who plays very little with toys, and who is producing few consonantal sounds. The child's profile may also provide clues to the nature of his language disorder. However, there is no substitute for careful psychologic, audiologic, and speech and language evaluation and follow-up in the child who is found to be significantly delayed in prespeech language milestones.

LANGUAGE ACQUISITION PERIOD

Expression-Production

During the prespeech period the sounds produced by infants from different cultures and different language backgrounds are similar. Within the second year of life, however, infants may begin to incorporate speech sounds from their own language environments into their wordlike productions and into words used referentially, i.e., to make demands ("up," "mine") or to refer to familiar objects and people. At first only a few words are acquired, and they may not be clearly differentiated from one another. For example, "bubba" may be used to refer to "baby" and "bottle," and "da" for "daddy" and "that." It may be quite clear, from the child's gaze or his grasping behavior, what he intends to name. Early words

may be idiosyncratic, used for only a short time, and then dropped.

In the second half of the second year, children frequently show a marked vocabulary spurt—from five to 10 to a total of 50 words. Words are no longer dropped from this vocabulary. In the course of this semantic development, the words used by the child may at first be overextended in meaning; e.g., the word "cow" may be used to refer to a dog or a horse as well as a cow. Children at this age may appear to be so anxious to talk about what interests them that they press words into service that are only metaphorically related to referents. At the same time the child may not use category words, such as food or plant, because these category words are not relevant to their interests. It should also be noted that the child may be able to recognize many more words than he can produce at this time.

The one word (or "holophrastic") stage of language development is followed by growth in the use of multiword utterances and, thus, of increased mean utterance length. As he progresses to longer utterances, the young child first may put together a series of single words, each followed by a pause; e.g., "See . . . truck . . . vroom," or a multiword utterance that represents one long word (one that the child cannot yet analyze into its component parts); e.g., "Where'd it go?" Soon, however, he combines different content and function words with one another in new ways, as when he says, "See truck go" or "Daddy go bye bye."

Children differ from one another considerably at this point with respect to their strategies for acquiring the language that they adopt. Some children are highly imitative and may benefit directly from adult speech models in known contexts. Others are more expressive; they are especially able to convey their attitudes and wishes so as to manipulate others, but they may use less intelligible speech in combination with gestures to accomplish this. A third group is more analytic in its approach to language acquisition. Such children learn by analyzing and resynthesizing spoken materials, especially those that are referential. They may produce single words earlier than children in the first or second groups. It may happen that a child will adopt a strategy that is not compatible with his parents' style of communication. If so, it is not profitable to reinforce the parents' approach, for example, by suggesting that they require the child to name each food or toy that he wants before he is allowed to have it.

Whatever his initial strategy, the child eventually combines content words (such as "car," "baby," "dog") and function words (such as "go," "put," "this," "more") to express different relations; e.g., "my book" (possession), "more duckie" (recurrence), "daddy car" (location), or "daddy throw ball" (agent-action-object). At first the actions and objects referred to by the child must be present and ongoing; only later does he become able to talk about absent objects or past and future events.

Unlike the series of single words or memorized sequences that the child produced earlier, these true multiword utterances express meanings that are more complex than the sum of the single word parts. This complexity is achieved by means of the grammatical form of the utterance, i.e., its syntax. Grammatical (syntactic) forms, in English, make use of word order and word endings (e.g., plurals, possessive forms, and verb tense endings). The use of these forms is subject to rules, and it is clear from the novel utterances that young children produce ("I goed to school" or "I see sheeps") that they generate hypotheses about these rules. Sometimes these hypotheses lead to overextension of the rules, as in the examples already given in which the irregular word forms "went" and "sheep" would be correct. Standardized tests have been developed for the assessment of language expression in children three to eight years of age.

Reception-Comprehension

Comprehension also shows marked development in the second and third years of life. However, the development of comprehension and production may be asymmetrical in relation to one another. It has been observed recently that mothers frequently overestimate the comprehension of young preschool children, not realizing how much they themselves contribute to the child's responses by gesture, gaze, and the use of situational support. This overestimation benefits the child by giving him the opportunity to practice interactive sequences with his mother, without excessive demands being made upon comprehension. More specifically, from 10 to 21 months of age, it has been shown that children understand what is said to them on the basis of single word understanding, taken together with the support of natural gestures on the part of the mother and of the familiar everyday contexts of the speech addressed to them. For example, the child may "understand" the command, "Go get a clean diaper," because he knows the word "diaper" and knows that a clean diaper is required at changing time. Also he knows where the clean ones are kept and his mother is likely, in addition, to look or gesture in that direction.

Similarly the child at that time might be capable of saying the word "diaper" and indicating, by gesture, that someone needs to lift him up in order that he may reach a clean diaper. In producing speech the young child may use words to indicate those elements of the communication that are not immediately obvious from the situational context or, at older ages, that are not immediately present and visible. In comprehension, however, the child is highly dependent upon the situational context in the early stages of language development and will understand words only if they relate clearly to that context. References to absent objects, to objects not clearly indicated by the context, or to past and future events are understood only at later ages.

It should also be noted that parents, or at least middle class American English parents, talk much more simply and clearly to young children than to other adults and that they may also offer frequent repetitions of parts of what they have said. "Go get your red truck. The red one. The red truck." In this manner they aid their children's comprehension as well as offering them good models for production.

Later in the development of language production, children learn to express more complex relations. They produce longer and more complex utterances, question forms, a variety of negative forms, and passive constructions (e.g., "The boy was hit"). They learn the use of conjunctions and relative pronouns and are able to introduce a variety of noun and verb clauses into their utterances. Standardized tests are available for the assessment of language reception in children 18 months to eight years of age.

Articulation

Articulatory competence begins to be mastered from the time of the young child's first vocabulary spurt (18 to 24 months). As he acquires additional words, it becomes necessary for him to distinguish them from one another, either by intonation or by the use of additional vowel or consonant contrasts. When multiword utterances appear, the child in addition must begin to learn the phonologic rules of language, i.e., rules for permitted sound combinations and for the formation of plurals and other word endings in different language contexts. He may also develop some rules of his own, e.g., rules for reducing difficult consonant clusters, such as /str/, so that "string" becomes "ting."

Children must use the muscles controlling respiration and phonation and the movements of the tongue, lips, and jaw in a highly coordinated manner, in order to produce speech. They must learn the optimal target positions for these structures in order to produce difficult consonant phonemes and vowels, and they must learn how to time speech movements in relation to one another in order to produce intelligible speech. The order in which the phonemes of a given language are learned (e.g., 46 phonemes for the English language: nine vowels and 37 consonants) depends on the degree of difficulty (both in perception and production) and the frequency with which they are used in the child's language environment. Templin (1966) indicates that five year old children produce 88 per cent of the English phonemes correctly. In general, the phonemes /h/, /w/, /m/, /n/, /ng/, /f/, /p/, and /t/ are the first to be produced correctly, 75 per cent of children producing these sounds correctly by age three years. Conversely production of the phonemes /th/, /v/, /z/, /zh/, and /dz/ does not achieve this accuracy rate until age six or seven years. By age eight years, phonologic development is usually completed. A number of standardized tests are available for the assessment of articulation and in addition speech-language pathologists and developmental psycholinguists have developed procedures for documenting the phonologic rule systems used by children with aberrant patterns of speech.

Interaction Communication

With respect to the development of social communicative aspects of language, or its pragmatic aspects, the child must acquire the ability to establish and maintain social exchanges, to influence others to act in a desired way, to label and indicate, and to make reference to objects and events. These functions are critical if the child is to use language effectively in different social contexts.

In addition, the child must learn to modify messages to suit different speech contexts, i.e., to address others in more and less formal contexts, to use polite forms with adults, and to introduce emphasis. Finally the child must learn how to link his utterances with those of a conversational partner in order to sustain a topic and maintain the flow of meaning in a conversational exchange. These linkages enhance the child's development of turn-taking skills in discourse, e.g., by requesting clarification or repetition of a misunderstood word or acknowledging a previous speaker's remark.

Procedures for the assessment of the social communicative aspects of language are still in early phases. They may make an important contribution to our understanding of language development and language disorders in children. It is important to document language delay and disturbance in relation to the child's nonverbal intelligence. The relationship between language development and cognitive development is still being debated. Although not yet fully understood, this relationship is an important one in the language-delayed child as well as the normal child. If the child presents with an overall delay and his linguistic and nonverbal cognitive abilities are commensurate with one another, different strategies will be adopted

in intervention than would be applied in the case of specific language delay.

SUMMARY

A young child cannot acquire language in an environment that is not supportive. He must hear speech that relates directly to his own everyday experience and to his knowledge of the world. He must be responded to as he attempts to talk about what interests him. If he learns to express himself and to relate his experiences to others, his understanding of objects and events may be enhanced. Thus language development and cognitive development are closely related in the normal child. It is necessary for the child to have normal hearing, to perceive elements of speech, and to be able to discriminate and contrast them in the speech of others. The child must have normal speech motor abilities and must develop auditory-motor linkages and the ability to monitor his own speech output. Finally, it is necessary for him to be able to interact meaningfully with others and to adapt to different social contexts as he internalizes the model of the language spoken in his environment and produces novel utterances based upon that model. Guidelines for the referral of children whose development of language and communication deviates from normal are shown in Table 40–3.

THE NEUROLOGIC BASIS FOR CHILDHOOD COMMUNICATIVE DISORDERS

The primary etiologies of most communicative disorders remain largely unknown. It is clear that normal language development is dependent upon the proper functioning of those sensory, perceptual, motor, and cognitive mechanisms that are utilized in the reception and production of speech. When one or more of these mechanisms does not develop properly, impaired language function usually results. Various research strategies have been utilized in attempting to understand the exact nature of specific types of language impairment.

Research approaches emphasizing the cognitive-linguistic aspects of language disability have suggested that the grammar of language impaired children is qualitatively different from that of normal children. Others have suggested that language impaired children fail to use the base aspects of grammar in expressing and expanding grammatical relations because of an underlying cognitive deficit in symbolization. Cognitive deficits in learning higher order concepts affect the acquisition of syntactic rules, as well as other aspects of the comprehension and use of language. The etiologic relationship of auditory perceptual problems to

Table 40–3. GUIDELINES FOR REFERRAL OF AN INFANT FOR AUDIOLOGIC SPEECH-LANGUAGE OR PSYCHOLOGIC EVALUATION

Interaction-Communication

1. Excessive crying after three months of age.
2. Lack of crying, or crying that is perceived as abnormal in infancy.
3. Lack of eye contact or of smiling after three months of age.
4. Lack of cooing vocalizations or response to smiling adults from three to six months of age.
5. Expressions of dislike at being held (squirming, crying, consistent tenseness that is relieved by placing child in infant seat) in first six months.
6. Failure of appearance of laughter, or lack of laughter in interactive situations, by six months of age.
7. Failure to respond in interactive peek-a-boo and patty cake games by one year of age.
8. Failure to indicate communicative intention nonverbally by one year of age.

Reception-Comprehension

1. Failure to respond to environmental sounds.
2. Failure to quiet to mother's voice when infant's fussing or crying and when mother is out of immediate line of sight and not in contact with infant.
3. Difficulty in localizing a sound source correctly after nine to 12 months.
4. Failure to respond to voices of family members in the second six months of life, when these persons return after an absence and are still out of immediate line of sight and not in contact with infant.
5. Failure to understand common words or commands by age 18 months.
6. Failure to indicate one or two familiar objects or people when these are named with gesture in the second year of life.
7. Failure to indicate one or two familiar objects or people when these are named (without accompanying gesture toward or gaze at object-person) early in the third year of life.
8. Failure to understand simple discussions of past or future events by age three.
9. Report that a child does not understand what is said to him, "takes no notice" of what is said to him, or "takes a long time to catch on" to what is said in the third year of life.

Expression-Production

1. Failure to produce any consonantal sounds (raspberries, nasals, and stops) in the first year of life.
2. Failure to produce consonantal sounds toward front of mouth in second six months of life.
3. Failure to produce high front /v/, back /a/, and rounded vowels /u/ by 15 months of age.
4. Failure to produce prolonged vocalic or consonantal sounds, to combine primitive consonantal and vocalic elements in a single segment, or to produce series of segments or syllables in noncry in the first 12 months of life.
5. Failure to produce reduplicated babbling by nine to 10 months of age.
6. Lack of or use of an excessive amount of expressive jargon after 18 months of age.
7. Failure to produce recognizable words by age two.
8. Failure to use several recognizable two word combinations that combine two ideas by age two years, six months; e.g., "more juice."
9. Lack of multiword utterances (phrases, sentences) by age three.
10. Lack of intelligible speech by age three.
11. Many initial consonants omitted at age three. Lack of final consonants by age four.
12. Continuing substitution of easy sounds for more difficult sounds after age five.
13. Persisting faults of speech articulation after age seven.
14. Decrease in amount of speech produced, instead of steady increase, at any age from three to seven years.
15. Sentences that are poorly formed, confused, marked by word reversals or telegraphic style after age four (dialectal variations not to be considered in this category).
16. Noticeable stuttering or other types of abnormality of rhythm or rate (rapid speech, cluttering) after age four.
17. Monotonous, unusually loud, hoarse, harsh, or inaudible voice.
18. Pitch that is not appropriate to child's age and sex.
19. Noticeable hypernasality or lack of normal resonance.
20. Embarrassment or disturbed feelings about his speech on the part of the child at any age.

---

linguistic difficulties in language impaired children is still unclear.

As audiologic procedures have improved, it has become apparent that simple peripheral hearing loss in language impaired children is usually not sufficient to explain the severe language delay. Some language (and reading) impaired children demonstrate auditory perceptual deficits, and many studies have emphasized a predominance of auditory temporal processing difficulties in children with delayed language development.

The ability to analyze and process rapidly changing acoustic information is basic to phonetic analysis of the speech code. However, recent studies have demonstrated that even young infants and subhuman primates have the ability to accomplish such perceptual processing of vocal sounds. The ability to recognize different vowels or consonants as belonging in the same category usually improves significantly with age in normal children. Such improvement with age may not be seen in language impaired children. These observations suggest that the normal capacity to make speech-sound discriminations in infancy is not fully developed, and that progressive tuning of the auditory processing system occurs during development. The conclusion that auditory temporal processing may be a developmental phenomenon that is disturbed in some children with developmental language disorders is supported by the finding that those speech sounds that incorporate rapidly changing acoustic spectra (and hence are the most difficult for these children to perceive) are also the most difficult ones for them to produce correctly.

Perceptual processing deficits in language impaired children are not specific to the auditory system but may be present in other sensory modalities as well, suggesting a generalized problem in the management of sensory information, rather than a problem relating to a specific sensory area within the central nervous system. A large proportion of children referred for primary reading impairment also have associated language impairment, suggesting that both deficits may be secondary to a more general communicative disorder. The close relationship between oral and written communicative skills should be fairly obvious when it is realized that the basic skill in reading involves decoding a visual symbol to an already acquired auditory-verbal symbol.

Thus, auditory perceptual impairment in the detection of acoustic features that normally cue recognition of certain phonemes has been suggested as one underlying deficit in some childhood language disorders. Language impaired children with such a deficit may also manifest a general disability in sequencing within visual and motor modalities as well (Stark and Tallal, 1980). Children with reading disability may also demonstrate impaired sequential processing abilities, and these sequencing deficits may be most apparent when the items to be sequenced are presented rapidly. It is possible that sequencing deficits and impaired auditory memory for speechlike stimuli are secondary to underlying problems in auditory rate processing (Stark and Tallal, 1980). Adults with focal left hemisphere brain damage may have impaired auditory rate processing abilities, and these impairments may correspond to the degree of language comprehension impairment in these subjects. Furthermore, recent evidence suggests that the superiority of the left hemisphere for linguistic processing actually may be a reflection of the dominance of the left hemisphere for processing rapidly changing acoustic events (Schwartz and Tallal, 1980).

Studies of childhood language disorders acquired as a consequence of brain damage suggest that hemispheric specialization for language occurs early in life and that there may be a continual leftward lateralization of these functions with time. The end result of such hemispheric specialization for language function has been studied in patients who have undergone cerebral commissurotomy for the relief of intractable epilepsy (Zaidel, 1978). These studies suggest that the disconnected right hemisphere understands oral and written language in terms of larger patterns related to meaning (vocabulary recognition) rather than by phonetic analysis, which appears to be a function for which the left hemisphere is specialized. In split-brain studies there have been no reports of speech sound production by a disconnected right hemisphere, nor has the disconnected right hemisphere been shown to utilize the syntactic aspects of language. Both these functions (phonetic analysis and syntactic analysis) appear to be left hemisphere functions. The disconnected right hemisphere also appears to have a limited short term auditory memory for verbal material. Zaidel (1978) has shown that responses from the disconnected right hemisphere are poorest for material that requires good short term memory but is not complex in syntactic structure, implying greater difficulty with auditory and verbal memory than with grammatical complexity.

Work by Galaburda et al. (1978) has suggested that structural differences between the hemispheres may underlie cerebral dominance. These investigators noted structural asymmetries in the auditory cortex and in the sylvian fissure that are present even in the human fetus. The sylvian asymmetries also have been hypothesized, on the basis of endocranial contours, to have been present in Neanderthal man and in some of the great apes. Galaburda et al. (1978) suggested that these structural differences relate to functional differences between right and left hemisphere function.

The notion that structure and function are integrally related is central to our understanding of the pathogenesis of many conditions. Specifically Galaburda et al. (1978) suggested that morphologic asymmetries favoring the auditory cortex of the brain may be important for left hemisphere dominance for language. They summarize previous work indicating that the posterior portion of the temporal lobe (the planum temporale) is larger on the left than on the right in most human brains. They suggest that a reversal of the usual left-right asymmetry of the brain near the posterior language zone might leave the left hemisphere anatomically ill suited to perform certain functions related to language.

A reversal in the usual left-right morphologic asymmetry of the parieto-occipital region has been noted in some children with developmental dyslexia, and such children are more likely to have lower verbal IQ's and to experience delayed speech acquisition than dyslexic children with the more usual pattern of cerebral asymmetry. A similar reversal of cerebral asymmetry has recently been noted for autistic children, and severe expressive language delay is also characteristic of that disorder. Similarly, some studies of electrophysiologic data from dyslexic children have suggested aberrant physiology in much of the cortical region ordinarily involved in reading and speech.

Thus, altered structure or function within the cerebral cortex has been demonstrated in some children with dyslexia, and impaired auditory temporal processing has also been demonstrated in some dyslexic children with underlying developmental language disability. Language learning develops predominantly in the left hemisphere, which is thought to be specialized for the temporal and sequential analysis of verbal and nonverbal information and for the analysis of configurations into component parts. Psychologic study of adults with surgically severed, interhemispheric corpus callosal connections suggests that speech sound discrimination is restricted to the left hemisphere, possibly because this hemisphere usually is innately specialized for processing rapidly changing acoustic events (Schwartz and Tallal, 1980; Zaidel, 1978). The right hemisphere is specialized to synthesize a percept of the whole from fragmentary information and to process the configurational features of perceptions that cannot be analyzed into component parts; it also appears to have some capacity for semantic representation in the language learning process (Zaidel, 1978).

Sex differences in hemispheric maturation and specialization may underlie the developmental advantage demonstrated by girls for certain verbal functions; boys demonstrate a similar developmental advantage in tasks requiring skill in spatial visualization (a predominantly right hemisphere task). Such innate differences may underlie the greater incidence of developmental language and reading disorders in boys than in girls. The hypothesis that the left hemisphere in the male is more vulnerable to developmental or environmental disturbances remains to be proven.

The role of auditory processing deficits in the genesis of communicative disorders continues to be an area of active investigation. Certainly interactions between the developing phonologic, syntactic, and semantic spheres play important roles in the development of communicative skills. The totality of functions performed by the intact cerebral cortex clearly exceeds the sum of functions performed by either hemisphere alone, and the development of various cognitive functions is heavily dependent on interhemispheric cooperation and competition. In the absence of definitive neurocognitive screening tests to demonstrate underlying deficits, clinicians must rely on knowledge of normal communicative development to screen certain conditions that place the child at higher risk for impaired communication. These high risk conditions are set forth in the next section, followed by a delineation of specific types of communicative impairment.

FACTORS ADVERSELY INFLUENCING THE DEVELOPMENT OF COMMUNICATIVE SKILLS

The previous sections of this chapter have emphasized the complex developmental interactions that occur within the pragmatic, phonologic, syntactic, and semantic spheres as communicative skills evolve and emerge in the young infant and child. In discussing these developmental phenomena, certain social, anatomic, physiologic, and cognitive prerequisites have been stressed. Obviously deficiencies within any of these important areas will have a negative impact on the development of communicative skills.

Ineffective social interaction or frank deprivation can have a deleterious influence on language learning skills. This important variable is not to be discussed further, since the nature of its pervasive influence has been discussed in Chapter 14. The following discussion will present a few common clinical examples of how communicative skills may become disordered when certain prerequisites are deficient.

HEARING IMPAIRMENT

Intact hearing is important in the early development of auditory-motor feedback loops during infancy. Deafness is always associated with impaired communicative skills, and recently data have been accumulating to suggest that even mild to moderate hearing loss associated with chronic otitis media may be a cause for concern (Chapter 26).

Children with chronic otitis media and fluctuating hearing loss have been found to be delayed in receptive and expressive language skills. Conversely, children with cleft palate who are vigorously managed from early infancy to control middle ear infusion have significantly better language development than those who are managed more conservatively, with long periods of untreated effusion. Such children may also demonstrate auditory processing deficits and problems in learning to read and spell (Bennett et al., 1980).

Several factors can enhance the child's susceptibility to chronic otitis media. Some of the more common factors include allergy, ethnic background, parental neglect, anatomic midface deficiency due to craniofacial malformation syndromes that compromise eustachian tube function, and specific malformations of the oropharynx, such as cleft palate or choanal atresia (which may or may not be part of a broader pattern of malformation). Examples of some of these disorders are shown in Table 40–4.

In addition there is a long list of genetic and metabolic disorders associated with primary hearing loss due to malformations in the ossicular chain (e.g., Treacher Collins syndrome), otosclerosis (e.g., osteogenesis imperfecta), and sensorineural deafness (e.g., Waardenburg syndrome). These have been well catalogued in books by Konigsmark and Gorlin (1976) and by Fraser (1976). Clearly fetal teratogenic insults can also result in malformations of the peripheral hearing mechanism. Congenital syphilis and rubella infection previously have been very common causes of

Table 40–4. SYNDROMES ASSOCIATED WITH A PREDISPOSITION TOWARD CHRONIC OTITIS MEDIA

Genetically determined acrocephalosyndactyly
  Apert syndrome
  Crouzon syndrome
  Pfeiffer syndrome
  Saethre-Chotzen syndrome
 	
Genetically determined craniofacial syndromes
  Treacher Collins syndrome
  Stickler syndrome
  Cleidocranial dysostosis
  Frontometaphyseal dysplasia
  Mohr syndrome
  Otopalatodigital syndrome
  Velocardiofacial syndrome
	Genetically determined dwarfism (limiting anterior cranial base growth)
	  Kniest syndrome
	  Spondyloepiphyseal dysplasia congenita
	  Diastrophic dwarfism
	
	Chromosomal disorders
	  Trisomy 21
	  18q syndrome
	  4p syndrome
	
	Teratogenic disorders
	  Fetal alcohol effects
	  Fetal hydantoin effects
	  Fetal hyperthermia effects

sensorineural deafness, and other fetal infections may result in a similar outcome (e.g., cytomegalovirus and toxoplasmosis; Fraser, 1976).

Finally, external ear malformations are known to be associated with conductive hearing loss due to middle ear anomalies (Jaffe, 1976). Such malformations may derive from teratogenic insults during the sixth week of gestation, when the pinna is being formed from the six branchial arch derived auricular hillocks and the ossicular chain is also being formed from the first and second branchial arches. The early development of this part of the face is nourished by a transient vessel derived from the first aortic arch, the stapedial artery. Poswillo (1973) has suggested that some of the sporadic first and second branchial arch syndromes resulting in malformed ears and conductive deafness may stem from vascular disruption of this artery. Anomalies have been produced in experimental animals by administering teratogens that cause rupture, hemorrhage, and tissue necrosis at the site where the stapedial artery branches from the early carotid artery. The end results are strikingly similar to those in patients with branchial arch syndromes such as the facioauriculovertebral malformation sequence (Poswillo, 1973). Teratogenic insults due to maternal hyperthermia at this time of gestation have resulted in similar ear malformations.

Some indication of the significance of the association between conductive hearing loss and isolated ear malformations can be derived from examining the data of the Collaborative Perinatal Project. Among 53,257 children born to over 50,000 women, 600 children were determined to have a malformation of the external ear or branchial cleft or both. Forty-two of these children appeared to have a broader pattern of altered morphogenesis. Among the 558 individuals with nonsyndromic external ear malformations, 523 had an isolated ear malformation, which was associated with hearing loss in 6 per cent of the cases and with

defective speech articulation in 12 per cent of the cases (Melnick, 1980).

Perinatal and postnatal insults may also result in varying degrees of hearing loss that can interfere with the development of communicative functions. Etiologic considerations should include prematurity or low birth weight, birth asphyxia or subsequent anoxic insults, kernicterus, central nervous system infections, ototoxic medications (e.g., aminoglycoside antibiotics, diuretics), central nervous system trauma, acoustic neuroma (e.g., neurofibromatosis), external auditory canal obstruction (e.g., foreign body, cerumen), and otosclerosis. In some of these conditions the central nervous system may be affected in addition to the peripheral hearing mechanism. (See Chapter 39B.)

ADVERSE GENETIC FACTORS AND TERATOGENIC-TOXIC EXPOSURES

The kinds of prenatal conditions that may result in significant mental deficiency are covered in Chapter 21; communicative impairment is usually a part of these conditions as a secondary consequence. There are other disorders resulting in mild central nervous system impairment that may present as communicative disorders in otherwise apparently normal children.

One such etiologic category would include children with sex chromosome anomalies. A history of delayed speech and language development is characteristic of boys with an extra X chromosome, and many of them develop language and learning disorders. Girls with an extra X chromosome also demonstrate delayed speech and language development in association with later learning disabilities. The incidence of XXY individuals at birth is 1.3 per 1000 live-born male newborns, making it the most common disorder of sex chromosomes in the human. The incidence of female newborns with an extra X chromosome is about one per 1000 live-born female newborns.

Inborn errors of metabolism, in contrast to chromosome disorders, are found much less frequently, but some of them are more likely to be diagnosed because of newborn screening programs. Phenylketonuria is one of the more common conditions discovered in such programs, with an incidence of about one in 10,000 births. It is usually treated promptly with dietary therapy, thereby preventing the serious degrees of mental deficiency associated with this disorder in the past. Such treatment, however, may not prevent the development of speech and language problems in children who have been treated for phenylketonuria and who have normal intelligence. In the past histidinemia, another inborn error of metabolism with an incidence of about one in 14,000, has been linked to speech and language disturbances. As more knowledge has accumulated, it has become clear that most histidinemic children with such difficulties also had subnormal intelligence, whereas those with normal intelligence did not have these problems. This may imply that mildly altered neurochemical factors play a significant role in the genesis of some language and learning disorders.

In addition to chromosomal aneuploidy involving the sex chromosomes and genetic alterations in metabolism, toxic environmental exposures can also result in varying degrees of central nervous system impairment that adversely affect the development of communicative skills. Examples of such exposures might include prenatal exposure to alcohol or anticonvulsants, perinatal hypoxia or hypoglycemia, or postnatal exposure to subacute doses of lead. In each of these instances long term exposure to large doses can result in serious degrees of mental deficiency; lower level exposures approximate a dose-response curve resulting in more subtle central nervous system alterations. The latter may express themselves through impaired development of communicative skills. Chil-

dren exposed to such insults should be screened for these kinds of disorders.

ALTERATIONS IN OROFACIAL ANATOMY

It has already been mentioned that palatal clefting results in a secondary predisposition toward otitis media and thereby may result in conductive hearing impairment, which affects speech and language development. Enlarged tonsils and adenoids can have a similar consequence, and both these anatomic deviations can affect velopharyngeal competence and result in predictable patterns of misarticulation and deviant resonance (to be discussed more fully in the next section). Before removing enlarged adenoids, one should carefully evaluate their role in establishing velopharyngeal closure. Examples of some of the kinds of deviations in orofacial anatomy that might be associated with impaired communication are shown in Figure 40–1.

Malformations of the tongue are a rare cause of speech difficulties (Table 40–5). If the tongue is grooved in the middle of the tip by a tight frenulum when it is protruded slightly beyond the lower incisors, such a tongue-tie may have to be clipped surgically before adequate articulation can be established. In cases of severe ankylosis of the tongue, the performance of frenulectomy does not guarantee normal acquisition of speech sounds. It does provide the child with an improved range of motion, thereby permitting the child access to the learning of appropriate motor patterns for speech sound production. Milder instances of tongue-tie are not known to be associated with significant articulatory disturbances. Similarly, alterations in the size of the tongue (either deficiency or excess) can have an adverse impact on articulation.

Neuromuscular disorders of the tongue, such as the dysarthria that may be seen in some individuals with cerebral palsy, can also have a deleteri-

Table 40–5. SYNDROMES ASSOCIATED WITH TONGUE MALFORMATIONS

Macroglossia	Oral
Beckwith-Wiedemann syndrome	Orofaciodigital syndrome
Congenital hypothyroidism	Ellis–van Creveld syndrome
Hurler syndrome	Hypoglossia-hypodactyly syndrome
Generalized gangliosidosis	Frontonasal dysplasia malformation sequence
Maroteaux-Lamy mucopolysaccharidosis	Mohr syndrome
Robinow syndrome	Opitz-Frias syndrome
Scheie syndrome	Popliteal web syndrome
	Meckel syndrome
Cleft or irregular tongue	
Hypoglossia-hypodactyly syndrome	
Cerebrocostomandibular syndrome	
Facioauriculovertebral malformation sequence	
Mohr syndrome	
Multiple neuroma syndrome	
Orofaciodigital syndrome	
Meckel syndrome	

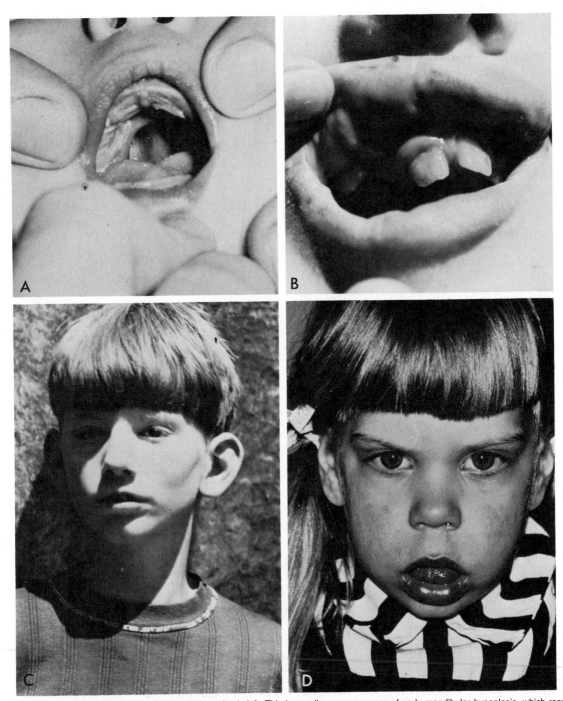

Figure 40–1. *A,* An example of a U shaped posterior palatal cleft. This is usually a consequence of early mandibular hypoplasia, which results in upward and posterior displacement of the tongue, with secondary obstruction of the posterior palatal closure. Such defects can be isolated anomalies or part of a broader pattern of altered morphogenesis. *B,* The upper lip shows a lingual frenula and cleft upper alveolar ridge in a girl with orofaciodigital syndrome, type 1. This can affect articulation adversely. *C,* This boy with the Moebius syndrome has congenital facial nerve paralysis, resulting in an expressionless face. This can also result in articulatory difficulties. In some instances other cranial nerves may be involved, leading to further difficulties. (From Smith, D.W. *Recognizable Patterns of Human Malformation,* Philadelphia, W. B. Saunders Company, 1982.) *D,* This girl with the Beckwith-Wiedemann syndrome has macroglossia. Such enlargement of the tongue can lead to difficulties with articulation. Wedge resection of part of the anterior tip of the tongue can improve both articulation and cosmetic appearance. (Photo courtesy of D. W. Smith.)

ous effect on speech production. Finally, maladaptive tongue habits, e.g., tongue thrust or reverse swallowing, are said to be associated with articulation deficits and dental malocclusion. Tongue thrust is manifested when the tip of the tongue is displaced anteriorly against the upper incisors during speech and swallowing acts. The relationship of tongue thrust to speech disorders and dental malocclusion is unclear at this point.

TYPES OF COMMUNICATIVE IMPAIRMENT

DISORDERS OF RESONANCE

Resonance disorders result from a disruption in the normal oronasal sound balance. These disorders are heard as hypernasality or hyponasality.

Normal speech sound production is dependent on the separation or coupling of the oral and nasal cavities. The relationship between the oral and nasal cavities is regulated through the action of the velopharyngeal mechanism. With the exception of the phonemes /m/, /n/, and /ng/, varying degrees of velopharyngeal closure, ensuring effective separation of the oral and nasal cavities, are required for the production of consonants and vowels. The velopharyngeal seal results from elevation and posterior movement of the velum and simultaneous reduction in the dimension of the medial and superior pharyngeal space through muscular contraction. When activated, the seal allows for the direction and shaping of the intraoral air stream, permitting the production of all nonnasal consonants and vowels.

Disturbances in resonance are of two principal kinds: hypernasality and hyponasality. Hypernasality is heard when the velopharyngeal seal is incompetent. When significant obstruction of the nasal passageways or nasopharyngeal space occurs, hyponasality will be heard. Since hyponasality is associated with nasal or nasopharyngeal obstructions, significant enlargement of adenoid tissue, or edema of the nasal mucosal lining due to allergy or chronic infection, referral to an otolaryngologist is indicated for medical or surgical management. Treatment of hyponasality by a speech-language pathologist is not indicated.

However, the language specialist has an important role in the diagnosis and planning of treatment for individuals with hypernasality. Hypernasality is due to dysfunction or incompetence of the velopharyngeal mechanism. The three primary causes are:

1. Abnormal function of the velum or hard palate, as seen in individuals with complete or partial clefts, submucous clefts, or foreshortening of the velum.

2. Adventitious injury to the velum from projectile injury or postsurgical complications.

3. Significant incoordination, weakness, or paralysis of velopharyngeal functioning due to neurologic insult, for example, pseudobulbar dysarthria, acute myopathies, and peripheral neuropathies.

Varying degrees of audible symptoms may be associated with impairment in the velopharyngeal seal. There may be an increase in nasal tone, audible nasal emission of air, decreased ability to direct the air stream into the oral cavity, and problems in the building and maintenance of intraoral breath pressure. Vowels may become nasalized, with distortion or omission of consonants. Because velopharyngeal inadequacy is a major component of oral and craniofacial anomalies, management is best accomplished through the use of an interdisciplinary team of surgeons, otolaryngologists, dentists, nurses, speech-language pathologists, audiologists, and psychologists. Management must consider the cognitive and psychologic state, as well as anatomic and sensory factors. Decisions include the kind of reconstructive surgery required, the coordination of surgical and orthodontic correction with speech and language therapy, meeting the child's psychoeducational needs, and utilizing audiologic management as indicated.

VOICE DISORDERS

Voice disorders are heard as deviations in the quality, pitch, or loudness of the voice. Voice disorders most often result from alteration in the vibration or movement of the vocal folds. Some deficits reflect altered respiratory support. The most common causes in children are vocal abuse leading to vocal nodules, vocal misuse (e.g., excessive vocal loudness or inappropriate modal pitch), allergy resulting in laryngeal tissue edema, nonmalignant growths (e.g., juvenile papillomas), neurogenic conditions at either a central or a peripheral level, trauma to the larynx, and psychogenic factors (e.g., severe anxiety or sexual identity issues).

The diagnosis and treatment of voice disorders require a careful analysis of medical, neurologic, and psychologic factors, including a history of vocal use, careful delineation of alterations in quality, pitch, and loudness of the voice, and an examination of laryngeal anatomy and function through direct or indirect laryngoscopy. Psychologic factors can also contribute to voice disorders and therefore need to be considered.

On the basis of these evaluations, decisions are made concerning the management of the problem. Conservative management consists of the use of medication or voice therapy. However, in cases of laryngeal web or juvenile papilloma, surgical intervention is indicated. This necessitates voice rehabilitation to prevent recurrence. For disorders arising from edema associated with allergy, the use of medication with voice therapy consultation is indicated. For children whose voice disorders have a psychologic basis, the combined use of psychotherapy and vocal re-education may be indicated. Lastly, in children with histories of vocal abuse or misuse and in whom vocal nodules may be present, conservative noninvasive management is indicated. The treatment centers on voice therapy—not surgery—to bring about changes in the underlying laryngeal tissue.

Vocal therapy includes the provision of information and insight into the causes of the presenting disorder, the use of methods and techniques to modifying vocal behavior, the use of self-monitoring techniques in various speaking situations, and prevention of further laryngeal damage or recurrence of the original problem.

Too often it is assumed that only surgical alteration of the vocal folds will be needed. Contending with postsurgical laryngeal changes, such as residual scarring, necessitates continued follow-up in such children and the provision of voice therapy support services.

DISORDERS OF FLUENCY

Disorders of fluency are marked by problems of rate and rhythm. The most common disorder of fluency encountered in early childhood is "stuttering." Various theories have been advanced to explain the onset and course of stuttering; these include psychoanalytic formulations, biochemical and neurologic hypotheses, and explanations based upon learning theory. Current theories of stuttering emphasize the importance of the development of speech fluency, the role of oral language formulation deficits in the genesis and maintenance of dysfluent speech, and the relationship of emotional and environmental factors. Dysfluent speech (stuttering) is viewed as having diverse etiologies, courses, and prognoses.

Dysfluent speech is usually brought to the attention of the pediatrician or speech-language pathologist in the preschool years, i.e., between the ages of two and four years. Although late onset dysfluency may occur in older children, little is known of this disorder. Its genesis may be related principally to a stress reaction or performance anxiety, with dysfluent speech as one of the cardinal symptoms.

Superficially three factors operate together. The speaker exhibits a set of speech behavior patterns, usually consisting of sound or whole word repetition, prolongation of sounds, or stoppage of speech production (blocks). The reason may not be evident, but such dysfluency is judged by a listener (family, peers) to be deviant, leading to a reaction toward the child. For some families, guilt and fear redirect the interaction with the child. The speaker, through awareness of his speech pattern and the listener's response, begins to incorporate the negative judgment and to anticipate the undesired speech pattern, and attempts to avoid the dysfluency. When he is unable to do so, the dysfluent pattern and associated behavior and the speaker's negative self-image become entrenched.

The importance of early referral of the child and family to a speech-language pathologist is therefore evident. Care must be taken not to give false reassurance to the family. The goals in seeing both the child and family are to determine the nature of the dysfluency, determine the social and environmental patterns influencing the dysfluency, determine the kind of treatment necessary for the child and family, provide the family with insight into the problem and methods for interim management, and provide the child and family with continuing consultation and follow-up.

Specific treatment for dysfluent speech is frequently recommended, but the type and nature of treatment should vary with the age of the client. Treatment for the preschool child often involves the parents in a counseling program. As the child increases in age, more direct therapeutic strategies are used, including careful examination of speech disturbing situations, provision of insight into the factors that maintain the dysfluent speech pattern, and instruction in direct means by which the dysfluencies may be modified or controlled.

DISORDERS OF ARTICULATION

Disorders of articulation are heard as deviations in the production of speech sounds, i.e., consonants and vowels. Speech sound production is a learned auditory-motor act. The child generates a set of hypotheses about what speech sounds are and how they are produced and combined, and these are revised over time. The child learns the essential acoustic features of what constitutes speech sounds and how, through a coordinated series of oral, nasal, and laryngeal adjustments, these sounds are produced. In addition, the child learns a set of production strategies that change the production of a sound to accommodate and mark essential grammatical features, for example, the variable production of /s/ in plural forms, as in ba*t*s, be*d*s, and ba*tch*es.

The acquisition of all speech sounds is expected to be completed by the eighth year of life, and variability in acquisition for individual children is to be anticipated. The evaluation of a true articulation disorder requires careful consideration of the bases and the factors maintaining various types of errors, the speech sounds involved, and the speaker's effectiveness and degree of frustration in conveying information to the listener.

Disorders of articulation are characterized by four types of errors: substitution errors (e.g., *wab*bit for *rabbit*); omission errors (e.g., boo for boo*r*); addition errors in which an extra sound is placed inappropriately in a word; and distortion errors of consonant and vowel sounds. The degree of severity depends on the perceived interaction of these types of error and the frequency with which a sound is used in the language. Disturbances of intonation and accent may also be present, as in speech disorders due to neurologic disorders; these may also negatively affect the speaker's intelligibility.

The causes of articulation disorders are heterogeneous and in some instances unknown. An interacting set of variables may be present and explain the person's inability to produce speech, as in the child with cleft lip-palate, the child with dysarthria associated with cerebral palsy, or the child with a specific language disorder. Care must be taken to assess the interaction of variables if appropriate diagnosis and treatment are to be accomplished.

Among the factors that can cause or contribute to disorders of articulation are the following:

1. Impairment of hearing, such as bilateral sensorineural hearing loss and chronic middle ear disease (serous otitis media, ossicular chain deformity) during the early years of infancy and childhood.

2. Structural disorders of the oral cavity such as cleft lip-palate, severe malocclusion, macroglossia, and significant ankylosis of the tongue.

3. Central and peripheral nervous system disorders, e.g., the dysarthrias, a group of speech disorders due to severe incoordination, weakness, or paralysis of the oral musculature. Such disorders also affect palatal, laryngeal, and respiratory actions in speech sound production.

4. Presumed central nervous system based problems in volitional oral motor patterning for the purposes of speech sound production. Children with this problem show a normal range of motion, tone, and strength for oral motor functions. Although there is an absence of weakness or paralysis, the child seems to be unable to use the capacity present in the oral musculature for the proper production or sequencing of speech sounds.

For children with mild to moderate articulation deficits, traditional therapeutic methods afford a good prognosis for normal speech production. However, for the child with a severe communicative disorder related to limited cognitive ability, or severe neuromotor speech disturbance (e.g, dys-

arthria), supplemented or augmentative communication systems are available. Such systems can take the form of aided communication devices—communication boards, charts, computer assisted communication systems, or unaided procedures (e.g., manual sign language or gestural systems). Determination of candidacy should include a careful assessment of cognitive level, neuromotor status, and knowledge of linguistic rule systems. Selection of the specific augmentative system is determined by the user and environmental needs.

LANGUAGE DISABILITIES

Language disability is a general term that refers to a heterogenous group of disorders characterized by deficits in the comprehension, production, and use of language. (See also Chapter 38.) Such disabilities are diverse and are associated with an even more diverse set of etiologies. Disturbances in language, for example, occur with a high degree of frequency in children with hearing impairment, mental retardation, or pervasive developmental disorders (e.g., autism). This section considers only those children with specific language disabilities, historically referred to as aphasic children or children with developmental aphasia.

Children with specific language disabilities also manifest etiologic heterogeneity, e.g., teratogenic consequences, perinatal difficulties, metabolic and genetic factors, and possibly postnatal toxic exposures. However, such children have normal hearing sensitivity bilaterally, normal intelligence, a normal capacity to relate to others and enter into reciprocal activities, and no evidence of primary emotional disturbance. In the majority of instances, children with specific language disabilities are identified in the preschool years. Parents of these children present to the pediatrician with a variety of complaints, e.g., "My child is not talking like other kids," or "My child is two and a half and only has 50 words," or "I don't understand Timmy when he talks," or "Chuck doesn't make sense when he speaks."

When should a pediatrician be concerned and refer a child to a speech and language pathologist for assessment? Specific guidelines for referral are included in Table 40–3. In all instances children whose speech and language status is questioned should undergo a hearing assessment, preferably by a certified audiologist with specific pediatric experience and training.

Language disabilities may be chronic in children, initially appearing in the preschool years and persisting through the school years into adulthood. Because the education system assumes that the child has a knowledge of language before entering school, and because the formulation of curricula for reading and writing assumes language to be available to the learner, children with language impairments are at high risk for academic failure of varying degrees and types. Indeed recent experience indicates that children with language disorders and concomitant problems in the processing of auditory-verbal information have been identified as a major subgroup within the group of children with learning disabilities.

The realization that childhood language disorders are diverse in their aspects reduces the likelihood that professionals will adopt a single process etiology model for both assessment and treatment. The child with language impairment may then demonstrate problems not only in language knowledge and use, but also in self-control and regulation, problems in learning style, and difficulties in the acquisition and development of reading, writing, and mathematical abilities. The remainder of this section deals with the clinical manifestations of language disorders in children.

ISSUES IN THE PSYCHOLOGICAL ASSESSMENT OF CHILDREN WITH LANGUAGE DISABILITIES

The diagnosis of specific language disabilities is determined by a number of factors, among which is the assumption of normal intelligence. It is in the area of the psychologic assessment of language impaired children that special problems are encountered. The first problem arises from the intelligence measures to be used. Most tests of intelligence, e.g., the Stanford-Binet Intelligence Scale, rely extensively on the use of verbal skills to determine the level of intellectual functioning. Clearly the language impaired child will be disadvantaged by reliance on such a scale. Consequently other scales that involve nonverbal methods of determining intellectual performance should be used, e.g., the Leiter International Scale or the Hiskey-Nebraska Test of Learning Aptitude. Children with primary language problems function within normal intellectual limits on nonverbal scales. This does not preclude the possibility that the same children might also demonstrate problems in visual-motor integration.

Similarly children with language disability may demonstrate problems in self-regulation, appearing to be easily frustrated, excitable, impulsive, and inattentive, especially in activities that involve language. In addition, in their early years children with language problems may appear to be overly dependent upon their families. This can result from their lack of communicative effectiveness within their environments and from the difficulties they encounter in being understood. Some children show problems in the appropriate management of their aggression. Because of this and because of difficulty in communicative effective-

ness, many of these children develop difficulties in peer relationships. The evaluation of these factors is important in understanding the language impaired child, and appropriate referral should be made to a psychologist familiar with the developmental assessment of children with language disabilities.

Hearing Assessment

All children suspected of having communication impairment should undergo a hearing assessment conducted by a qualified audiologist. For some children with specific disorders of receptive language, for example, repeated sessions with an audiologist may be needed. This is necessary because the child may have difficulty in learning a response set, and preteaching and follow-up will be necessary to be certain of the child's responses. In addition, careful audiologic follow-up is necessary in children with a history of chronic middle ear disease. Although the precise effects of middle ear disease on the development of competency in language are still unclear, the presence of persistent, even mild conductive hearing loss may adversely affect the development of language.

Problems in the Comprehension of Language

Understanding or comprehension refers to the individual's ability to derive meaning from the spoken message. This is achieved through reference to words, the relationships specified by words, and sentence structure. In addition, the individual comprehends against a background of shared experiences, presuppositions about the speaker's message, and ability to understand the context in which the message is heard. For adequate understanding, reasoning ability, as well as judgments concerning the speaker's intent, will be used.

Children with specific disorders of comprehension (receptive language) demonstrate a number of clinical signs. These children have varying problems in the understanding of single words and of relationships represented by words. They also have reduced efficiency in comprehending sentences with different grammatical structures and lengths, as well as problems in understanding social conversation, idioms, and humor. Such disorders are seen in varying degrees and in varying aspects, depending upon the age expectations for the child.

Simply because the parent reports that the child "understands everything," or because the child can follow simple commands in the pediatrician's office or turns to the speaker when spoken to or called by name, there is no assurance that the child's understanding skills are intact. Whenever the parent or pediatrician has concern over the child's ability to understand language, formal assessment of hearing and of the child's breadth of understanding should be undertaken. In all instances care should be exercised in assuming the intactness of comprehension abilities.

Assessment of comprehension by the speech and language pathologist of necessity includes measurements of the child's ability to understand single words, to understand sentences containing various grammatical sequences (e.g., response to questions), to comprehend negative sentences and sentences of differing tenses, and to follow commands of increasing length and complexity. In addition, the child should be engaged in conversation so that the clinician can judge the child's ability to follow a conversation, change topics, and respond appropriately to implied information. These assessments should be consistent with expectations for the child's chronologic age and mental status.

There are different types of comprehension problems. For some children the understanding of single words, simple questions, and certain kinds of directions will be age appropriate. However, these children may have difficulty in understanding complex language, e.g., "Before the door was opened, the boy put his coat on. . . . When did the boy put his coat on?" or "Match the appropriate prefix with the root word." On the other hand, children are seen who demonstrate categorical reductions in their comprehension abilities despite intelligence that is within normal limits. Children with comprehension problems may also evidence problems in oral language formulation (expressive language); these problems also show individual variation.

Problems in Expressive Language Production

Expressive language is the principal means by which an individual represents thought and self in a social interactive context. Deficits in expressive language are manifested by age appropriate discrepancies in the use of syntax, morphology, and word finding abilities and in the child's ability to comment on various topics in story form or conversation. In assessing expressive language, reference is made to the analysis of topic, the context of communication, the child's intent, the principal means by which the child communicates, the relationships specified by the child, the child's overall sentence building skills, his word finding abilities, and the appropriateness and organization of narrative in conversation.

The parents of children with language problems complain that their children have difficulty expressing their ideas in words. They complain about not being able to follow the child—"If I

didn't know what he was talking about, I wouldn't understand him." Parents are usually sensitive to the fact that their child is not speaking like other children or that this child is developing language ability quite differently from the ways used by his older siblings. The parents notice that the child becomes frustrated and, when not understood, refuses to repeat. In addition, they note that the child may not speak a great deal and that he may rely on getting something for himself or use gestures or simple pointing in order to communicate. Especially in the early preschool years, tantrums and problems in control may be evident.

The following examples of children's expressive language are offered to demonstrate the problems these children have in the production of syntax and morphology, word finding and the maintenance and cohesiveness of narrative organization.

Child W. H. Normal hearing, normal intelligence, problems in the understanding of complete sentence structures; chronologic age, six years.
"I'm put it over for you."
"That a Christmas carol sing," stated in an attempt to name a wreath.
"That's cartoon show. . . about bad guys in a ship and on the Argo (pause), and they got away motion gun (pause). This thing (gesturing ray gun), this thing here goes clicked up really fact. . .when they fire it."
Child K. K. Normal hearing, normal intelligence, no measurable problems in understanding for chronologic age; chronologic age, four years, three months.
"I didn't be here."
"After that I being there many more."
". . . cuz he got a shoe on his hook."
"Look it (pointing) all paper under there have buy some more."
Child M. J. Normal hearing, normal intelligence, no measurable problems in understanding for her chronologic age; chronologic age, five years, four months.
"What we gonna do?"
"Michael have a big monster, can knock me down."
"Then the boy getting a cookie; then he gonna fall down hurt hisself."

Thus, children with expressive problems demonstrate an array of disturbances in oral language formulation that span specification of meaning, production of age appropriate grammar, difficulties in word finding during confrontational speaking situations, and reduced ability to form cohesive narratives.

Children with oral language formulation deficits evidence disruption in their ability to use the rules of syntax in a manner commensurate with mental age. Receptive language skills are often within normal limits, although dependent upon the tests used, but specific areas of understanding may be found to be reduced, e.g., the language of time and space, or comprehension of various grammatical constructions. A gestural representational system may be noted and may be used alone or during attempts to communicate through words or phrases. The language deficit may extend to the use of only single words, the use of telegraphic syntax (e.g., "Boy push girl truck"), problems with the use of tense and number, difficulty with the formation of questions that begin with "what" or "why," or the use of irregular forms of the grammar. Examples of such children's utterances have already been provided. In psychologic studies employing nonverbal measurements of intellectual levels, they evidence normal functioning; but problems in visual motor-spatial integration may be present. These children may demonstrate an inability to use language in diverse social contexts. Problems may also be present in word finding skills and in narrative organization and development.

Word finding problems represent a momentary inability on the part of the child to recall the name of an object or event of which the child has knowledge. It is most commonly observed under confrontation situations or naming, in answering questions, or in prolonged explanations. Parents report that "He can't say what he wants," "I can't follow his stories," or "It seems as if it's on the tip of his tongue and he just can't get it out." During confrontation naming of pictures or objects, the child may evidence increased latencies of response time. The child may attempt to represent thought by using gestures that conform to the object's shape, features, or uses. Some children speak in definitions, descriptions, or illusions. They may use associated labels, e.g., "rain" for "umbrella," "fish scooper" for "pelican," or "smoke" for "pipe." Sometimes a word that sounds like the one sought after will be used, e.g., "slow" for "low," "tornado" for "volcano." They may mix the phonemic sequences within a word, e.g., "donimo" for "domino," "nirosus" for "rhinoceros." They may communicate in indefinite narratives in which only illusion is present. Many of these children become dysfluent and need careful management before true stuttering is established.

Narrative deficits may be manifested in a number of ways. Some children are unable to comment on content or demonstrate reduced story telling skills. Some have difficulty in elaborating on a story or may speak only in the most concrete and basic way about events. Others have difficulty in maintaining organizational features and seem to ramble and build incoherent narratives (see the example of Child W. H. just given). Problems in word finding and problems in serial order maintenance compound the problems of building narrative.

The therapy for the child with language impairment depends upon the severity and type of disability. For some children, thereapeutic approaches that emphasize the cognitive prerequisites of language are used to assist him in acquiring the cognitive forms that will be represented in language. For other children, at both preschool and school age levels, placement in a substantially separate program is necessary to ensure the use of an adaptive modified curriculum in instruction and to accommodate to individual needs for explicit teaching, assistance in organization, and provision of language modification. For other children, direct treatment is chosen to assist the child in the acquisition of language rules. These therapies in the school years must be carefully based upon the functional requirements of language in the child's school subject areas. Therapy begins with what the child knows and proceeds developmentally, assisting him in acquiring new and appropriate strategies to facilitate the production of language consistent with his age and with social needs.

In summary, children with specific language disorders have no other primary sensory, emotional, or cognitive impairment to account for their delayed language development. Prompt recognition of deviation from normal linguistic milestones is most important, so that early therapy for language disorders can be instituted. Without such therapy, persistent language disability can disrupt social growth and development, limiting the child's expression of feelings and severely compromising the development of his self-concept and adaptive skills.

JOHN M. GRAHAM, Jr.
ANTHONY S. BASHIR
RACHEL E. STARK

REFERENCES

Bax, M., and Hart, H.: Health needs of preschool children. Arch. Dis. Child., 51:848, 1976.

Bennett, F. C., Ruvska, S. H., and Sherman, R.: Middle ear function in learning-disabled children. Pediatrics, 66:254, 1980.

Fraser, G. R.: The Causes of Profound Deafness in Childhood. Baltimore, Johns Hopkins University Press, 1976.

Galaburda, A. M., LeMay, M., Kemper, T. L., and Geschwind, N.: Right-left asymmetries in the brain: structural differences between the hemispheres may underlie cerebral dominance. Science, 199:852, 1978.

Jaffe, B. F.: Pinna anomalies associated with congenital conductive hearing loss. Pediatrics, 57:332, 1976.

Konigsmark, B. W., and Gorlin, R. J.: Genetic and Metabolic Deafness. Philadelphia, W. B. Saunders Company, 1976.

Melnick, M.: The etiology of external ear malformations and its relation to abnormalities of the middle ear, inner ear, and other organ systems. Birth Defects, 16:303, 1980.

Peckham, C. S.: Speech defect in a national sample of children aged seven years. Br. J. Dis. Comm., 8:2, 1973.

Poswillo, D.: The pathogenesis of the first and second branchial arch syndrome. Oral Surg., 35:302, 1973.

Schwartz, J., and Tallal, P.: Rate of acoustic change may underlie hemispheric specialization for speech perception. Science, 207:1380, 1980.

Stark, R. E., and Tallal, P.: Perceptual and motor deficits in language-impaired children. In Keith, R. W. (Editor): Central Auditory and Language Disorders in Children. Houston, College Hill Press, 1980.

Templin, M.: The study of articulation and language development during the early school years. In Smith, F., and Miller, G. A. (Editors): The Genesis of Language. Cambridge, M.I.T. Press, 1966.

Zaidel, E.: The split and half brains as models of congenital language disability. Paper presented at the Symposium on the Neurological Bases of Language Disorders in Children: Methods and Directions for Research, National Institute of Communicative Diseases and Stroke, Bethesda, Maryland, January 16-17. 1978.

41

Major Child and Adolescent Psychiatric Disorders*

This chapter offers a broad overview of major psychiatric disorders. As such it overlaps in some respects other sections of this book. Such minimal redundancy has been preserved to allow for an intact comprehensive classification system in this critical area of developmental-behavioral morbidity.

DEFINITIONS AND PREVALENCE OF CHILD PSYCHIATRIC DISORDER

Child or adolescent psychiatric disorder is defined in this chapter as any constellation of symptoms expressed as disturbances of thought, emotions, or behavior that is accompanied by impairment in academic or social functioning, or by personal suffering. The functional impairment should be attributable to the disorder or its symptoms in the clinician's best judgment.

Epidemiological studies using similar definitions of child psychiatric disorder have found relatively stable rates in the general populations of preschoolers, school age children, and adolescents. In all three age groups, rates of child psychiatric disorder are between 5 and 15 per cent. Uniformly urban areas show substantially higher rates than rural areas.

Any attempt to define psychiatric symptoms and disorders in children must take cognizance of the fact that what are normal thoughts, emotions, and behavior for a particular chronological age and developmental level may be deviant for another. The norms for psychological functioning at any chronological age include a range of functional abilities. For example, although the anxiety that keeps the eight month old infant clinging to its mother when confronted by an adult stranger is very common in the general population of infants and thus, it is considered normal; the same is not true of an eight year old who is too frightened to greet an unknown adult and clings to his mother. Temper tantrums in a two year old are a common sight, which no one would consider particularly

pathological behavior. If a tantrum were to occur in the same child four years later, it would strike us as unusual and we would begin to think in terms of pathological conditions. Most six year olds can be expected to express anger and displeasure without the physical manifestations of a tantrum. It is our view, however, that developmental changes in the symptomatic expression of psychopathology have been overrated, and that this notion has contributed to the perpetuation of nosologic confusion in child psychiatry. In fact, only in preschoolers do developmental factors strongly change psychopathological expressivity, and these effects decrease quickly with increasing age.

Isolated problems such as those just mentioned are frequently seen in children at certain developmental levels. With the addition of other behavioral symptoms and functional difficulties, the same problems can become nonspecific indicators or even key elements of severe psychopathological conditions. Thus the same behavior pattern that exists in a normal child at a certain age can also be part of the symptom picture of a severe psychopathological state at another age. Psychological states can be arranged in a spectrum, which ranges from normal to severely pathological. Unfortunately there are no sharp divisions between normality and pathology; rather, varying degrees of inner and outer behavioral deviance and functional impairment. This point bears on the question of continuity or discontinuity of normality and psychopathology.

Although certain diagnostic entities like separation anxiety disorder may be put on a continuum with normality, others like childhood schizophrenia, early adolescent mania, and endogenous major depressive disorder show evidence of qualitative differences, which would be difficult to explain within the continuity hypothesis. Thus it can be difficult to distinguish normality from pathology if one restricts the diagnostic view to symptom existence and severity and does not take into account functional impairment. Furthermore, in children, psychological functioning can also be examined by determining the rate and success of different developmental processes. Any aspect of

*The writing of this chapter was partially supported by grants MH 30838 and MH 30839 from the National Institute of Mental Health.

865

maturation and development can provide the substrate for symptom formation or functional impairment. Therefore, we have to examine physical, emotional, social, and intellectual functioning in order to be well informed when making a diagnosis.

Functional impairment should be assessed independently of symptoms. The three main areas to be inquired about are academic performance, peer relationships in and out of school, and social functioning at home and at school, although other aspects of the child's functioning may also be relevant in specific disorders. Some authors have suggested that functional impairment should not be assessed as present unless the patient fails in at least one aspect of his functioning. In our opinion this is too extreme a position, which can result in unwarranted treatment delays. A child with a recent (one month) onset of a major depressive episode may compensate for a marked decrease in concentration ability by increased effort. Thus he will not fail in academic tasks, but he may need to spend double the amount of time and effort to complete work he could have accomplished as well but much faster just one month previously.

In episodic disorders (major depression, separation anxiety disorder) functional impairment should be assessed in relation to the preonset baseline, with appropriate developmental corrections if the episode has been of long duration. In chronic, long standing disorders it is much more useful and more practical to compare the child's current functioning to the peer group's average function, or to age norms if they exist. Obviously, IQ at both tails of the gaussian distribution should be taken into account when a clinical judgment of chronic functional impairment is made.

ASSESSMENT OF SYMPTOMS OF CHILD AND ADOLESCENT PSYCHIATRIC DISORDER

Although the format of interviewing the parent(s) and the child and obtaining information from school has changed little since the beginnings of child psychiatry, the techniques of clinical assessment have changed considerably during the last 20 years. This evolution has occurred in close connection with progress in the development of diagnostic classification systems that have influenced the assessment process, and vice versa. Such advances have been largely fueled by similar advances in adult psychiatry and by the increasing importance of child psychiatric diagnosis as a determinant of treatment choice, along with the availability of a widening scope of therapeutic techniques.

Initially a substantial proportion of child psychiatrists totally disregarded classification systems not in consonance with the psychodynamic-psychoanalytic unitary view of mental illness in childhood and focused instead on the degree of severity of the clinical picture. Unconscious mental conflict was seen as the root of mental disorder, and the task of the child psychiatrist was to find out what conflicts were underlying every abnormal behavior, in order to make the child aware of them through interpretive work. It was thought that making the unconscious conscious would produce therapeutic benefit in all mental disorders, regardless of diagnosis or severity. Although lip service to "constitutional factors" was paid by a few prominent child psychoanalysts interested in problems of diagnosis, this unitary etiological model of childhood mental disorder prevailed, and the only treatment method remained psychotherapy.

Consistent with this model, the clinician was interested in assessing worries, preoccupations, and symbolic meanings in order to infer unconscious psychic conflict. Symptoms were viewed as ephemeral surface manifestations of conflictual, arrested, or abnormal development and would disappear once conflict was worked through and psychological development was "moving again" in the proper direction. In the interview with the parents, the clinician was especially interested in a chronicle of the child's social relationships in early childhood with the parents, and with parental reaction to and expectations from the child, as these were thought to be extremely important determinants (and, later, reflections) of pathogenic psychic conflict.

This unitary model came to be seriously challenged during the 1960's and 1970's by other models of child psychiatric disorder. The evidence supporting such models came from findings in three main research areas: behavior therapy, epidemiology, and pediatric psychopharmacology. Findings in these three areas have resulted in major shifts in both assessment methods and classification systems.

The basic tenet of the theoretical underpinning of behavior therapy is that behavior is dependent on immediate antecedents and immediate contingencies, and that by changing these, the therapist can reliably alter behavior away from maladaptation. This functional analysis therefore shifted the focus to behavioral phenomenology, to observation, to "microanalysis" of behavior, and to careful description and identification of strictly defined single behavior patterns, which could be reliably assessed.

Epidemiology is a powerful tool to generate and test hypotheses regarding etiology of behavioral disorders as well as to arrive at prevalence rates for different disorders, traits, or characteristics in the general population. Epidemiology has been mostly used for data collected from sources other than the child himself, namely, teachers and parents. Therefore, it is more likely to identify observable behavior than "inner" psychic phenomena like mood, obsessions, or hallucinations, unless they are accompanied by persistent overt specific behavioral correlates.

The Isle of Wight study was one of the first child psychiatric studies to use standardized structured interviews with the child himself and well defined diagnostic categories. As part of that study, Rutter and Graham (1968) demonstrated that 10 year olds can be interviewed, and that the information obtained is not only reliable but also valid. This was a major advance, which, with a few exceptions, unfortunately did not have much of an impact on child psychiatric practice in this country.

During the late 1960's and 1970's, pediatric psychopharmacology developed substantially. Several properly controlled double blind studies demonstrated the effectiveness of stimulant drugs in the treatment of the hyperkinetic syndrome (attention deficit disorder with hyperactivity). The need for reliable rating scales to diagnose this disorder and measure the effects of stimulant drugs led to the development of the Conners Parent and Teacher Questionnaires, the Behavior Problem Checklist, and others. The symptoms of attention deficit disorder are mainly observable and can be assessed reliably by teachers and by independent observers. Thus these rating scales focus on strictly defined individual behavior patterns. Another major advance in pediatric psychopharmacology was the evidence produced by Gittelman-Klein and Klein (1973) supporting the effectiveness of imipramine in a specific type of school phobia (separation anxiety disorder). In that study both observational measures and the child's self-report were used to assess drug and placebo effects. This disorder is characterized by both behavioral and emotional symptoms.

Historically the most recent impetus for the further development of child psychiatric assessment, and for emphasizing the importance of diagnostic criteria, has been provided by researchers studying major depression in children and adolescents. The existence of major depression before puberty has been controversial. Nevertheless, during the 1970's, child psychiatrists became increasingly interested in this disorder. At present there is a substantial consensus about its existence, as well as active research. When our group entered this field, our first question was, "Can major depression in children be diagnosed using the same symptom criteria that have been established for adult patients—the Research Diagnostic Criteria?" Among the clinical instruments then available for children there was none that assessed all symptoms of major depression as defined by the Research Diagnostic Criteria. By using a crude interview instrument, it was possible to diagnose prepubertal depression, on the basis of the same symptom criteria as for adult depression. Weinberg et al. (1973) had also modified criteria derived from adult depression to diagnose depressed children. Thus the interest of the clinician shifted from worries and preoccupations to symptoms, which in turn led to more emphasis on talking with the child as opposed to playing with the child as a means of psychiatric assessment.

Although the assessment of psychodynamic constellations and symbolic meanings through play is an important part of the general psychiatric assessment of children, it has only a minor, if any, role to play in the clinical process of arriving at a diagnosis of most child psychiatric disorders. Play interviews by and large do not assess symptoms, and may be distracting for both the child and the interviewer. In assessing symptoms like depressive mood, anhedonia, or hallucinations, the child must pay attention to relatively new concepts and questions; he must think and answer, and he is more likely to do that in an interview situation without the distraction of toys.

Most clinical researchers interested in prepubertal or adolescent depression strongly emphasize the diagnostic importance of the interview with the child. This represents a major shift from the past, when most child psychiatrists used information from parents and teachers to reach a diagnosis (or a treatment recommendation), paying little attention to what the child said in a usually nonstructured play interview. This is historically surprising, because Rutter and Graham (1968) and also Herjanic et al. (1975) had demonstrated that children are reliable reporters. The importance of the child interview is enhanced because the most crucial symptoms of depression are "inner" symptoms, which may completely escape observation by others and therefore can be accurately reported only by the child. Sensitive parents are usually the best informants in regard to observable behavior, including verbal and nonverbal behavior, and are best able to provide a chronological framework to evaluate the episode.

Once it became clear that unmodified symptom criteria for major depressive disorder in adults could be used to identify depressed children, the next task was to develop an interview schedule that could reliably measure all symptoms of the depressive syndrome as well as other psychiatric disorders in children and adolescents.

At present there are three diagnostic instruments available to measure the depressive syndrome in subjects younger than the age of 18 years. All three instruments are in an advanced stage of development. These instruments are:

1. Diagnostic Interview for Children and Adolescents (DICA), developed by Herjanic. This

schedule records information obtained in the interview with the child. It does not record the interview with the parent. It covers the child's lifetime, not the present episode.

2. Interview Schedule for Children (ISC), developed by Kovacs. This instrument also records the interview with the child. It also covers the child's lifetime, and includes all the symptoms that have been reported in the literature to be part of depression in childhood. A separate, differently structured form covers the interview with the parent.

3. Schedule for Affective Disorders and Schizophrenia for School Age Children (Present Episode; Kiddie-SADS-P, K-SADS-P), developed by Puig-Antich and Chambers. This instrument records information from the interview with the child and with the parent. Both interviews follow the same structure, and summary ratings are achieved item by item. The K-SADS-P measures psychopathological behavior only during the current episode of disorders. A second schedule (K-SADS-E) is used for the past history.

Although there are important differences in scope and method among these schedules, a final critique and comparison of these instruments should await development of their final form and are beyond the scope of this chapter.

DIAGNOSTIC CLASSIFICATION

The bulk of child psychiatric disorders can be classified into two broad categories: emotional and conduct disorders. Emotional disorders are also referred to as neurotic disorders. The term neurotic disorder is used here without any etiological implications, but rather in a purely phenomenological sense to refer to a mental disorder in which the predominant disturbance is a symptom or group of symptoms that is distressing to the child and is recognized as alien. The ability to know what is real and what is not is basically unimpaired, and the child's behavior does not violate the rights of others. Nevertheless, functioning can be profoundly impaired. These disturbances are relatively enduring or recurrent without treatment and are not transitory stress reactions. The distressing symptoms are: self-consciousness, social withdrawal, shyness, separation anxiety, generalized anxiety, phobias, rituals, obsessions, crying, hypersensitivity, depression, and chronic sadness. Epidemiological studies show that these disorders occur in 2.5 per cent of children living in rural areas. The prevalence is greater in urban areas and increases in adolescence. In early and middle childhood sex ratios are equal, and these shift in adolescence to the adult pattern of greater frequency in women than in men (Rutter et al., 1970).

In the *Diagnostic and Statistical Manual of Mental Disorders,* the recent official American Psychiatric Association (1980) classification of mental disorders, the categories of emotional disorder that apply to children include three that are specific to children—overanxious disorder, avoidant disorder of childhood, and separation anxiety disorder (many cases of school phobia fall into this category); and a number from the adult nosology—obsessive compulsive disorders, phobic disorders, somatoform disorders (conversion hysteria), and depressive disorders. Excessive states of anxiety either alone or in combination with depressive affect are present in virtually all emotional disorders. The symptoms of emotional disorders in children are frequently not as well differentiated as they are in adults. As a result it is sometimes difficult to categorize an emotional disorder in a child because there is no predominant type of symptom that colors the clinical picture or appears to be specifically related to functional impairment. Many children with emotional disorders grow up to be healthy adults, with the likely exceptions of those with definite childhood diagnoses of obsessive-compulsive, separation anxiety, and severe depressive disorders.

The essential feature of the conduct disorder is a repetitive and consistent pattern of behavior that violates age appropriate developmental and societal norms. The conduct disturbance goes well beyond mischief and pranks. Conduct disorders are characterized by marked disobedience, disruptiveness, destructiveness, and in some cases delinquent behavior. The diagnosis of conduct disorder is therefore dependent on the ascertainment of a behavioral pattern that is persistent and that includes the repetitive occurrence of antisocial acts. Delinquency is not a psychiatric diagnosis. It is a legal term, and it only means that the child was caught, reported, and dealt with legally by societal agencies. Conduct disorder is a psychiatric diagnosis, for which being or not being caught is irrelevant. Apparently, low IQ children with conduct disorders are more likely to become labeled as delinquent than high IQ children with the same diagnosis.

Some workers in this area believe that a distinction should be made between the undersocialized child with a conduct disorder and children who are better socialized. Children with undersocialized conduct disorders are characterized by a failure to establish normal relationships with others. The relationships they have lack affection and empathy. Peer relationships are largely missing, and the relationships that do exist are based on immediate advantage. Egocentrism is evidenced by a readiness to manipulate others for advantage without any thought of reciprocation. Feelings of guilt or remorse are missing. Such a child would as soon inform on companions as assist them, strictly depending on his own short term advantage. By contrast, children who have socialized conduct disorders usually show evidence of social attachment to peers or family members, but they can be callous and manipulative toward others whom they consider "outsiders." They are similar to the children with unsocialized conduct disorders in that they do not experience guilt when the "outsiders" are made to suffer.

Another important subcategorization of conduct disorders centers along the axis of presence or absence of aggressive behavior. Unaggressive conduct disorder behavior patterns include stealing without confrontation with the victim and truancy.

The difference between conduct disorders and emotional disorders goes beyond differences in the symptom composition of the clinical picture. Children with conduct disorders have been found to differ from children with emotional disorders in sex distribution, presence of reading disability, family functioning and relationships, size of sibship and ordinal position in the sibship, parental criminality, personality disorders or parental neuroses, short term response to treatment, and long term outcome.

Although conduct disorders and emotional disorders are distinguished by the symptom picture and other variables, in most clinical studies of psychiatrically disturbed children a substantial proportion of the children have symptoms of both. Rutter et al. (1968) in the Isle of Wight Study classified those children with symptoms of both conduct and emotional disorders as having "mixed disorders." The children with mixed disorders appeared more like those with pure conduct disorders than those with emotional disorders. The symptom patterns on parental interview ratings, sex ratio, size of sibship, and prevalence of reading retardation were all quite similar to those of the "pure" conduct disorder group. In follow-up studies at age 14 years a majority of the children originally classified as having mixed disorders were found to have either pure or mixed conduct disorders, and only a small minority had emotional disorders.

Traditionally, depressive disorders have been classified as emotional disorders. A group of prepubertal boys have recently been identified who meet criteria for major depressive disorders and also have symptoms of conduct disorders (Puig-Antich, 1982). These cases originally would have been classified as "mixed" disorders. In the majority of these cases the onset of conduct disorder followed the onset of depression. When these children were treated with imipramine, it was found that the conduct disorder symptoms disappeared along with the depressive symptoms. This raises the important question of whether depressions really should be classified as emotional disorders.

Although it is true that many cases of depression in children fit all the criteria for an emotional disorder, some do not. For example, there are psychotic depressions in children in which reality testing is not maintained, mainly because of the appearance of hallucinations. Rare cases of bipolar affective disorder exist in prepubertal children and become quite frequent in adolescents. Although many depressive children do meet emotional disorders criteria, not all of them do. An alternate approach might be to divide emotional disorders into depressive disorders and nondepressed emotional disorders, or to separate major depressives completely as affective disorders, as in the adult classification. As we will see later, clear biological differences have been demonstrated between those with major depressions and those with nondepressed emotional disorders.

The number of children whose diagnosis falls outside the categories of emotional disorder and conduct disorder is relatively small, and the etiology of the remaining disorders is heavily weighted toward biological processes. Such categories include infantile autism, childhood schizophrenia, pervasive developmental disorders, attention deficit disorders, and Gilles de la Tourette's syndrome.

THE MAJOR CLINICAL PICTURES

In the description of the major diagnostic entities, it will be unavoidable to borrow at times from DSM III, which remains the most commonly agreed upon consensual psychiatric nosological document to date. Nevertheless substantial refocusing to pediatric samples, treatment methods, and new research will be added.

MAJOR DEPRESSIVE DISORDER

The main symptoms of a major depressive disorder in childhood and adolescence are either a persistent depressive mood or pervasive loss of interest or pleasure in all or many usual activities (Table 41–1). Depressed mood in adolescents and in children, as well as in adults, is not always referred to as depression or sadness. In fact, we have made it mandatory in our own work always to inquire about eight different labels youngsters may use: sad, depressed, low, down, down in the dumps, empty, blue, very unhappy, or a "bad feeling inside" he cannot get rid of. The minimum persistence of a depressive mood is three periods per week, each lasting at least three continuous hours. For the assessment of length of each daily period, it is helpful to use time milestones—getting up, breakfast, getting to school, lunch, afternoon classes, the favorite TV program's

Table 41–1. DIAGNOSTIC CRITERIA FOR MAJOR DEPRESSIVE EPISODE*

A. Dysphoric mood or loss of interest or pleasure in all or almost all usual activities and pastimes. The dysphoric mood is characterized by symptoms such as the following: depressed, sad, blue, hopeless, low, down in the dumps, irritable. The mood disturbance must be prominent and relatively persistent, but not necessarily the most dominant symptom, and does not include momentary shifts from one dysphoric mood to another dysphoric mood, e.g., anxiety to depression to anger, such as are seen in states of acute psychotic turmoil. (For children under six, dysphoric mood may have to be inferred from a persistently sad facial expression.)

B. At least four of the following symptoms have each been present nearly every day for a period of at least two weeks (in children under six, at least three of the first four).
1. Poor appetite or significant weight loss (when not dieting) or increased appetite or significant weight gain (in children under six, consider failure to make expected weight gains).
2. Insomnia or hypersomnia.
3. Psychomotor agitation or retardation (but not merely subjective feelings of restlessness or being slowed down) (in children under six, hypoactivity).
4. Loss of interest or pleasure in usual activities, or decrease in sexual drive not limited to a period when delusional or hallucinating (in children under six, signs of apathy).
5. Loss of energy; fatigue.
6. Feelings of worthlessness, self-reproach, or excessive or inappropriate guilt (either may be delusional).
7. Complaints or evidence of diminished ability to think or concentrate, such as slowed thinking, or indecisiveness not associated with marked loosening of associations or incoherence.
8. Recurrent thoughts of death, suicidal ideation, wishes to be dead, or suicide attempt.

C. Neither of the following dominate the clinical picture when an affective syndrome (i.e., criteria A and B above) is not present, that is, before it developed or after it has remitted:
1. Preoccupation with a mood-incongruent delusion or hallucination (see definition below).
2. Bizarre behavior.

D. Not superimposed on either schizophrenia, schizophreniform disorder, or a paranoid disorder.

E. Not due to any organic mental disorder or uncomplicated bereavement.

*Reprinted from Diagnostic and Statistical Manual of Mental Disorders, Third Edition, 1980, with permission of the American Psychiatric Association.

length, and so on. Careful interviewing is crucial. Some children report short periods of sadness, but on further questioning their "mood baseline" is constant unhappiness, always considerably lower than their peers'. It is also important to make sure that periods of depressive mood are not just secondary to the absence of the main attachment figures or to being away from home. Some children with separation anxiety disorder report sadness, but such depressive mood is completely relieved by the presence of the mother or by being at home. We do not count such an exquisitely reactive depressive mood as indicative of a major depression, although in some major depressives, the diagnosis of separation anxiety can also be made. In such cases, although mood is worsened by physical separations (and their anticipation), the children continue to experience dysphoria at home, in the presence of their mothers. Diurnal variation and lack of reactivity of depressive mood may be present, but are not necessary for a positive finding of the symptom of depressive mood.

The assessment of anhedonia and lack of interest is not particularly difficult in children. The key is to compare how much fun they can have at present as compared to before the onset of the episode, or as compared to their peers, and to make sure that the number of opportunities is about the same for both elements of the comparison. Persistent boredom, in the presence of available stimulation, is an excellent indicator of anhedonia and lack of interest in children and adolescents.

The presence of depressive mood or pervasive anhedonia or lack of interest is not sufficient for the diagnosis of major depression to be made. The mood disturbance should be associated with at least four of eight symptoms that are characteristic of the depressive syndrome. These symptoms include appetite disturbance; sleep disturbance; loss of energy, fatigability, or tiredness; psychomotor agitation or retardation; feelings of excessive or inappropriate guilt; loss of interest or pleasure in usual activities; difficulty in concentrating or thinking; and thoughts of death or suicide or any suicidal behavior. A decrease in appetite may be accompanied by significant weight loss or the failure to make the expected weight gain. On the other hand, appetite may be increased, with consequent weight gain. The sleep disturbance may involve difficulty in falling asleep (initial insomnia), waking up during sleep (except for urination; middle insomnia), early morning awakening (terminal insomnia), or sleeping too much, including napping (hypersomnia). In adolescents, difficulty in falling asleep can be extreme, resulting in a 180 degree circadian inversion of the sleep-wake cycle. The patient sleeps during the day and is awake all night and cannot reverse the pattern without treatment. All forms of sleep disturbance are accompanied by a feeling of not having slept well (nonrestorative sleep).

Psychomotor agitation takes the form of an inability to sit still, pacing, handwringing, pulling or rubbing hair, skin, clothing, or other objects, outbursts of complaining or shouting (temper tantrums), and nonstop talking. Psychomotor retardation is a visible generalized slowing down of physical movements, reactions, and speech. There may be long speech latencies. Some children may show generalized hypoactivity and sluggishness. Complaints of chronic fatigue and need to rest (not sleep) are frequent. The child may reproach himself for minor failings that are exaggerated, and negative environmental events are seen as confirming the negative self-evaluation. Inordinate guilt becomes bound to current or past, real or imagined failings and can reach delusional proportions. Frequent manifestations of excessive guilt are feeling responsible for accidents or other persons' acts, when objectively the youngster has made no contribution to them, or he begins to feel guilty again about a past misdeed that had already been forgotten. Difficulty in concentration, slowed thinking, indecision, memory difficulties and easy distractability are common.

Features commonly associated with depressive episodes include depressed appearance, low self-esteem, tearfulness, feelings of anxiety, irritability, fear, brooding, excessive concern with physical health, phobias, and, in adolescents, panic attacks. In depressed youngsters, separation anxiety and conduct disorder behavior may accompany the depressive episode. In adolescents, negativistic behavior may appear. Feelings of not being understood, restlessness, grouchiness, aggression, and the desire to leave home are common. Sulkiness, a reluctance to cooperate in family activities, withdrawal from social and school activities, inattention to personal appearance, and increased sensitivity to rejection in relationships are also common. Substance abuse may develop and is usually an attempt by the adolescent to self-administer psychopharmacological "treatment."

Adult major depressive disorders have been divided into various subtypes. Thus, the more severe subgroup, with many vegetative signs of depression, is called the endogenous subtype. Although the etymology of *endogenous* indicates that it is internally produced, the current use of the word in psychiatry indicates only a particular symptom pattern, without etiological assumptions. Endogenous depressions have been found to be common both in midchildhood and in adolescence. They can be identified by using the same symptom criteria as in adults. As we will see later, they also share psychobiological characteristics with adult endogenous depression.

Endogenous depression is distinguished from other depressive disorders by a pervasive loss of interest or pleasure, a lack of reactivity to positive environmental changes, a worsening of mood in the morning, excessive guilt, psychomotor changes, early morning awakening, loss of interest, poor appetite and weight loss, and a distinct quality of the depressed mood. What is meant by a distinct quality of the depressed mood is that the feelings of depression are clearly distinct from the feelings experienced after a personal loss. It is sometimes difficult to question children about quality of mood if they have not suffered the loss of an adult or pet. Also young children cannot understand an abstract concept of this magnitude.

A second subtype of major depressive disorder that occurs in children is psychotic major depressive disorder. When depressive hallucinations or delusions are present along with the depressive symptoms, the individual is diagnosed as having a psychotic major depressive disorder. In prepuberty most psychotic depressives are not delusional, but present with depressive hallucinations. In a recent series, slightly more than one-third of prepubertal major depressives showed depressive hallucinations (Chambers et al., 1982). Adolescents can present both depressive delusions and hallucinations. Hallucinations are defined as perceptions in the absence of an identifiable external stimulus. They occur in clear consciousness and should be distinguished from such hallucination-like phenomena as illusions, elaborated fantasies, eidetic imagery, and imaginary companions.

Depressive hallucinations are characterized by the following criteria: They are temporally and thematically consistent with the depressive mood, and their form has certain typical characteristics. Temporal consistency with depressive mood indicates that the child hallucinated only while he was clinically depressed. Thematic consistency is demonstrated when the content of the hallucinations is congruent with the depressive mood. Thus, voices tell the child to kill himself or that he is no good. In "pure" major depressions, conversing or commenting voices are not a feature. Instead the child typically hears one voice talking directly to him, most frequently with suicidal commands.

In assessing hallucinations, we have applied strictly the same assessment methodology used in adults, with some language modifications tending toward concreteness. Frequently child psychiatric clinicians tend to discuss such phenomena as a "normal product of the child's fantasy life." We withhold judgment on this point, but we point to the evidence in our own studies that the plasma level of imipramine (and its main metabolite, desmethylimipramine) needed to induce a clinical response in the depressive syndrome in prepubertal major depressives was significantly higher among the psychotic subtypes than in the rest of the sample.

Delusions or false beliefs are difficult to assess and can be confused with elaborated fantasies, overvalued ideas, and cultural beliefs. In our experience four types of delusions occur in children with major depressions—delusions of guilt, sin or poverty, somatic illness, persecution, and nihilistic expectations. The common link is that the child may say that these are happening because he is such a terrible person. If a child has the symptoms of a depressive syndrome and hallucinations or delusions, which are not thematically and temporally congruent with the depressive syndrome, we consider this a schizoaffective disorder in keeping with the adult nosology. Schizoaffective disorders are not frequent, but we have made such a diagnosis in a few children and adolescents.

Definitive evidence of hereditary factors in the etiology of major depression in adults has not been found, but very suggestive evidence does exist. Family aggregation studies have yielded a substantial body of data consistent with the presence of hereditary factors in the etiology of major depression in adults. Several twin studies have convincingly demonstrated that monozygotic twins have a significantly higher concordance rate for affective disorder than dizygotic twins, results that constitute another source of evidence strongly supporting the role of genetic factors.

Little research has been done to establish the genetic basis of major depressive disorders in children. Several authors have indicated the presence of depressed parents in the families of children diagnosed as depressed, but the diagnostic evaluation of the parents was incomplete. We have reported the frequent occurrence of pedigrees similar to Winokur's depressive spectrum disease in a group of 13 carefully diagnosed prepubertal major depressive children (Winokur, 1979). "Depressive spectrum disease" pedigrees contain first degree relatives with diagnoses of depression, sociopathy, and alcoholism. Further, studies in both children and adolescent major depressives show a high morbidity risk for major depressive disorders in first degree biological relatives over 16 years of age. Thus, there appear to be more similarities than differences in family histories of prepubertal, adolescent, and adult major depressive disorders.

In some depressed children we find evidence of a depressed mood or loss of interest or pleasure but not many of the associated symptoms of the depressive syndrome. The lower symptomatic limit of severe depression in children has not been established other than arbitrarily at the present time.

Biological Correlates of Prepubertal Major Depressive Disorders

The validation of most psychiatric disorders has been traditionally based on data gathered mainly from family studies, epidemiologic studies, and follow-up studies. In the case of major depression in adults, a variety of biological correlates have been found, which in turn provide further validation of the disorder. These include, among others:

1. Neuroendocrine correlates: cortisol hypersecretion, lack of inhibition of cortisol during dexamethasone suppression testing and inhibition of the cortisol response to d-amphetamine, growth hormone hyporesponsivity to insulin induced hypoglycemia and to desmethylimipramine, and blunting of the thyrotropin response to the infusion of thyrotropin releasing hormone.

2. Polysomnographic correlates: increased REM density, shortened first REMP latency, decreased delta sleep, decreased sleep efficiency, and abnormal temporal distribution of REM sleep.

Therefore, the investigator in the field of childhood major depression can attempt validation of the syndrome using a much wider set of measures than those usually available to investigators of other psychiatric diagnoses at all ages.

We have been carrying out a controlled study whose objective is to test the validity of the diagnosis of major depressive disorder in prepuberty by comparing children fitting unmodified research diagnostic criteria for this diagnosis with two groups of controls (normal subjects and nondepressed children with emotional disorders) along several psychobiological parameters that have been shown to be associated with adult major depressive disorders.

Cortisol hypersecretion measured by frequent plasma sampling through an indwelling cannula for 24 hours or shorter periods has been shown to be associated with endogenous depressive episodes in about 40 per cent of adult patients. The likelihood that an adult with this diagnosis will actually be a cortisol hypersecretor increases with age. In prepuberty only 10 to 15 per cent of endogenous depressive children present this pattern of hypersecretion of cortisol when compared to themselves upon recovery. No data are available in adolescent endogenous depressives. From the available studies it appears that the relationship between age and cortisol hypersecretion is present throughout the lifespan.

Another method of detecting abnormalities in cortisol secretion is the dexamethasone suppression test. Following administration of 1 mg. of dexamethasone, the plasma cortisol level does not fall below 5 mcg. per dl. in the next 24 hours in approximately 60 per cent of adult endogenous depressives. In a large sample of adult depressives between ages 15 and 85 years, no age effects were found in the sensitivity of the test. In a recent pilot study, similar sensitivity (71 per cent) was found in prepubertal endogenous depressives. Thus, it appears that the results of studies of cortisol secretion in prepubertal endogenous depressives, although not identical, are highly consistent with similar data in adults with the same diagnosis.

The growth hormone response to insulin induced hypoglycemia has been found to be markedly blunted in approximately 40 per cent of endogenous adult depressives. Three-quarters of prepubertal children with the same diagnosis hyposecrete growth hormone in response to induced hypoglycemia. In addition, the low responsivity of growth hormone to induced hypoglycemia probably persists after recovery from the depressive syndrome, at least in severely depressed prepubertal children. These neuroendocrine results, as well as the family history studies alluded to earlier, point again to the similarity between prepubertal and adult major depressive disorders. Other neuroendocrine studies are in progress in various centers including the dexamethasone suppression test, the thyrotropin response to the injection of thyrotropin releasing hormone, and the growth hormone response to desmethylimipramine.

The original purpose of the "neuroendocrine window" research strategy in depression was to establish the potential chain of inference from endocrine abnormalities associated with affective disorders, to the neurotransmitter disregulations responsible for both mood disorders and their endocrine correlates. Currently available data indicate that a functional deficit in norepinephrine systems may be present in prepubertal endogenous depression and may also persist after recovery from the depressive episode. A review of neurochemical regulation of mood, pituitary function, and affective disorders is beyond the scope of this chapter.

In contradistinction to adult major depressive disorders, prepubertal children with this diagnosis have shown no differences from normal and pathological controls in a variety of polysomnographic variables that are known to be characteristically altered in depressive adults: first REMP latency, REM density, delta sleep time, and sleep efficiency. The neuroendocrine results reviewed so far, as well as the high familial morbidity risks for affective disorders, make it unlikely that this discrepancy in sleep findings reflects a basic difference in the nature or pathophysiology of depression at different ages.

Another possible interpretation is that age effects are responsible for the differences in the sleep correlates of depression at different ages. Thus the lack of sleep correlates in prepubertal major depressives is consistent with strong age effects on electroencephalographically recorded sleep across the lifespan, and the sleep correlates of depression may actually be secondary to an interaction between depression and age, rather than a reflection of intrinsic pathophysiological mechanisms common to depressive illness at all ages.

Treatment

Data are available in regard to the effectiveness of tricyclic antidepressant treatment of major depressive disorders in youngsters. In a recently completed study a highly significant relationship was found between maintenance plasma levels of imipramine and its main metabolite, desmethylimipramine, and the clinical response of the depressive syndrome at five weeks in prepubertal children (Puig-Antich et al., 1979). Interestingly, children who had presented with depressive hallucinations during their depressive episode needed higher maintenance plasma levels in order to induce a clinical response than children with nonpsychotic major depression. Thus, from logistic regression equations, it was possible to predict that with an imipramine-desmethylimipramine maintenance plasma level of 200 ng. per ml., 89 per cent of nonpsychotics would fully respond, in contrast to only 38 per cent of psychotic subtypes. In both subgroups, the plasma levels and response rates were positively and linearly correlated.

In this study the effectiveness of imipramine in prepubertal depression was also investigated by means of a double blind placebo controlled five week trial. Both groups, imipramine and placebo treated, were significantly improved at the end of the trial, but no significant differences in clinical response were found between them. The clinical response rate for both groups was nearly 60 per cent. Nevertheless, when the imipramine group was split into patients with "high" and with "low" maintenance plasma levels, clinical response rates were 100 and 33 per cent, respectively. The overall conclusion from both studies was that imipramine is probably effective in prepubertal major depressives and that for optimal results imipramine administration should be monitored by plasma level determinations. The aim of imipramine dose titration should be not a particular dose but a plasma level targeted on the basis of the severity of depressive syndrome and the presence of psychotic depressive symptomatology. Such a procedure is likely to result in a higher mean imipramine dose than that used in previous studies.

It is well known that the half-life of imipramine and desmethylimipramine is shorter in prepubertal subjects than in adults, reflecting a variety of factors, including a higher liver mass per kilogram of body weight and a higher metabolic rate. As a result, compared to adults, higher dosages per kilogram of body weight are needed in children to produce equivalent plasma levels.

The issue of effectiveness of imipramine and other tricyclic antidepressants in adolescent major depressive disorder is not at all settled. In our experience the metabolism of imipramine in adolescents is similar to that in children. Thus, relatively high dosages (3.5 mg. per day and over) are commonly necessary. For proper studies of effectiveness to be carried out, pharmacokinetic variability and the heterogeneity of the clinical picture should be taken into account.

If one considers only therapeutic effects, it appears that simply raising the imipramine dosages would solve the problem. The existence of toxic side effects and their potential seriousness add a new level of complexity to the psychopharmacological treatment of major depression in youngsters. The main side effects of imipramine in children and adolescents are:

1. Excessive lengthening of the PR interval (an upper limit of 0.21 second is the maximum tolerable PR length between ages six and 16 years).

2. Excessive widening of the QRS complex (over 130 per cent of baseline).

3. An excessive increase in the resting heart rate (over 130 beats per minute).

4. An excessive increase in systolic or diastolic arterial blood pressure (over 140/90 mm. Hg).

5. Orthostatic hypotension leading to persistent and marked unsteadiness or falling following a postural change.

6. Persistent drowsiness and somnolence during school hours.

7. Chest pains for which no organic cause is found.

8. A rare syndrome of behavioral toxicity consisting of poor concentration, perplexity, forgetfulness, and difficulty in learning.

9. A persistent increase in irritability and aggressive behavior, which appears to be dose related.

10. A syndrome of anticholinergic blockade.

11. Any other persistent side effect that produces marked functional impairment or personal distress and suffering.

In most cases such side effects, when they occur, can be dealt with by dose adjustment or changes in the prescribed daily pattern of imipramine intake. Except for electrocardiographic changes, the other side effects listed tend to be relatively uncommon. When imipramine has to be discontinued, nortriptyline and desmethylimipramine can be quite useful. Nevertheless, in the rare instance in which a child develops an allergic reaction to imipramine, he is highly likely to be allergic to desmethylimipramine also.

Assessment of the effectiveness of an individual therapeutic trial of imipramine (or other antidepressants) should be carried out for five weeks. The measures of outcome should be the symptoms of the depressive syndrome, not other psycho-pathological aspects of the child's clinical picture. If a clinical response is absent, incomplete, or partial, plasma level adjustments or diagnostic re-evaluation may be necessary. Psychotic depressives may benefit from the addition of a neuroleptic drug.

When a clinical response of the depressive syndrome has occurred, the treatment is not ended. This is the appropriate time to re-evaluate the psychosocial functioning of the child at school, with peers, and within the family and to consider the need for psychotherapeutic intervention.

Scientifically acceptable evidence supporting the effectiveness or lack of effectiveness of various child psychotherapies, parental counseling, and family therapy in prepubertal or adolescent major depression is lacking; it is also lacking in most other childhood disorders in which psychosocial interventions are thought to be the treatment of choice by many clinicians. In adult depression, recent studies have shown that interpersonal psychotherapy is helpful in improving the psychosocial functioning of patients in whom tricyclic antidepressants have induced and maintained improvement of depressive symptoms.

Mother-child relationships are markedly impaired in depressed children during illness as compared to those in normals and nondepressed neurotics. Similarly, depressed children, while ill, have markedly restricted peer relations. Compared to normals, nondepressed, emotionally disturbed children are also impaired in their interaction with their mothers and in peer relations, but to a lesser degree than depressed children.

Impaired social functioning and depression in children may each result from the other, or both may be caused by common antecedents. Fourteen initially depressed children were re-examined three to four months after clinical recovery from the depressive syndrome, and after discontinuation of imipramine without relapse for one month. Their peer relations after recovery were almost as impaired as during the illness. Mother-child relationships, on the other hand, had improved moderately, although the depth of communication was still significantly worse than that of normals, and of children with emotional disorders during their illness. These findings suggest that the effect of depression on interpersonal relations is a long term one or else that there is an independent psychopathological process, perhaps with a common etiology.

Our clinical impression is that adolescents present with a very similar clinical course. It is only after depression has been successfully treated, and the direct effects of mood disorder on interpersonal relations have been eliminated, that the clinician can identify maladaptive patterns of interaction that are intrinsic to the youngster's personality. It is also at this point that the child is ready to participate in psychosocial treatment with possible benefit. We have been impressed with the pragmatic advantages of short term group treatment with youngsters, both from the point of view of cost effectiveness and because of the ease with which social skill problems appear "in vivo" in front of the therapist in this once a week "social laboratory."

MANIA AND HYPOMANIA

Reports of affective disorders in prepubertal children that resemble adult mania are extremely rare. A composite picture of these reports would be that of a prepubertal child who begins to present with extreme mood swings, with frequent but relatively short elated and depressive periods. The manic and depressive episodes fully meet diagnostic criteria, and there may be euthymic intervals (Table 41–2). The length of the cycle period is short, and the child frequently changes from mania to depression. As the child enters puberty, the cycle period and the duration of the episode lengthen, and by midadolescence the symptoms conform to the classic picture of manic-depressive illness of the bipolar type. An adolescent onset of bipolar illness is not by any means rare, and mania should be considered in the differential diagnosis of any agitated state in an adolescent.

An important sign for the confirmation of the diagnosis of bipolar illness in adults is a history of mania in close biological relatives. A substantial portion of children and adolescents with clear-cut manic episodes in whom family data are reported present with a positive family history of manic and depressive disorders, or at least depressive disorders.

The essential feature of a manic episode is a distinct period when the predominant mood is either elevated, expansive, or irritable, accompanied by hyperactivity, pressured speech, flight of ideas, inflated self-esteem, a decreased need for sleep, distractability, and involvement in dangerous activities. The elevated mood may be described as euphoric, unusually good, cheerful, or high. The euphoric mood has an infectious quality for the unfamiliar observer but is clearly recognized as excessive by the familiar observer. Although elevated mood is considered the prototypical symptom, the predominant symptom may be irritability. This becomes apparent when the youngster is thwarted. The behavior of manic adolescents often has a flamboyant or bizarre quality; for example, dressing in strange or colorful clothes, distributing candy or money, or giving advice to strangers. Speech is typically loud, rapid, and difficult to interpret. Frequently there is a flight of ideas, i.e., a nearly continuous flow of accelerated speech with frequent, although logically intact changes in topic. The changes in topic are based on understandable associations, distracting stimuli, or plays on words. When the flight of ideas is severe, the speech may become disorganized and incoherent. Grandiose delusions involving a special relationship to God or some well known figure are common. There is a decreased need for sleep; the child wakes several hours before the usual time, full of energy. When the sleep disturbance is severe, the child may go for several days without any sleep at all and yet not feel tired.

Treatment

To our knowledge, systematic studies of the psychobiology of mania in adolescents (or in chil-

---

Table 41–2. DIAGNOSTIC CRITERIA FOR A MANIC EPISODE*

A. One or more distinct periods with a predominantly elevated, expansive, or irritable mood. The elevated or irritable mood must be a prominent part of the illness and relatively persistent, although it may alternate or intermingle with depressive mood.

B. Duration of at least one week (or any duration if hospitalization is necessary), during which, for most of the time, at least three of the following symptoms have persisted (four if the mood is only irritable) and have been present to a significant degree:
  1. Increase in activity (either socially, at work, or sexually) or physical restlessness.
  2. More talkative than usual or pressure to keep talking.
  3. Flight of ideas or subjective experience that thoughts are racing.
  4. Inflated self-esteem (grandiosity, which may be delusional).
  5. Decreased need for sleep.
  6. Distractibility; i.e., attention is too easily drawn to unimportant or irrelevant external stimuli.
  7. Excessive involvement in activities that have a high potential for painful consequences which are not recognized, e.g., buying sprees, sexual indiscretions, foolish business investments, reckless driving.

C. Neither of the following dominate the clinical picture when an affective syndrome (i.e., criteria A and B above) is not present, that is, before it developed or after it has remitted:
  1. Preoccupation with a mood-incongruent delusion or hallucination (see definition below).
  2. Bizarre behavior.

D. Not superimposed on either schizophrenia, schizophreniform disorder, or a paranoid disorder.

E. Not due to any organic mental disorder, such as substance intoxication.

*Reprinted from Diagnostic and Statistical Manual of Mental Disorders, Third Edition, 1980, with permission of the American Psychiatric Association.

dren) are lacking. The same is true for psycho-pharmacological studies. Nevertheless the consensus among the few pediatric psychiatrists who have had a research interest in adolescent mania is that lithium carbonate is at least as effective in the prophylaxis against future affective episodes as in adult bipolar patients. Children and adolescents tolerate lithium very well. Severe toxicity is quite rare. As in adults, it is advisable to start treatment of the acute episode with a neuroleptic drug for rapid control of manic symptomatology while lithium is titrated to a plasma level between 0.7 and 1.0 mEq per l. It is advisable not to expect a full prophylactic effect of well regulated lithium administration until the sixth week of treatment.

Among adult bipolar patients, "rapid cyclers" (patients with at least four major affective episodes per year) are considered poor responders to lithium carbonate treatment. Most young adolescent bipolar patients have short cycle periods and would qualify as "rapid cyclers." In our experience, however, short cycles in adolescents have not as much bearing on the effectiveness of lithium.

Severe mania can easily be mistaken for delirium. Highly accelerated thought processes can prevent the patient from properly answering questions on orientation. Electroencephalographic tracings can be helpful in the differential diagnosis, for manics do not show any of the profound electroencephalographic abnormalities found in delirious patients.

Finally, the extreme rarity of prepubertal mania is a very interesting phenomenon. It is consistent with the lack of euphoric response to d-amphetamine in both prepubertal hyperkinetics and normal subjects (Rapoport et al., 1980), and it suggests that puberty has an effect on brain function without which euphoria, elation, and full fledged mania are almost impossible.

SEPARATION ANXIETY DISORDER

Separation anxiety disorder is characterized by excessive anxiety following separation from major attachment figures, such as parents or other caretakers, or from home or other familiar surroundings. (See also Chapter 36.) Excessive anxiety is triggered both by actual separation experiences and also by the anticipation of such experiences. The severity of anxiety may be maximal and involve actual physical clinging and attacks of panic. These children are uncomfortable whenever they travel independently or without their parents or siblings away from their homes or other familiar areas. They avoid or refuse to visit or sleep at a friend's home, go out of the house on errands, or attend camp or school. In severe cases they may be unable to stay in a room by themselves and instead cling to or "shadow" the parent around the house. When a separation is anticipated or occurs, prepubertal children frequently present physical complaints such as stomach aches, headaches, nausea, and vomiting. In contrast, adolescents are more likely to develop cardiovascular symptoms, such as palpitations, dizziness, and faintness. Frank panic attacks are also more common in adolescents.

A frequent concomitant of excessive separation anxiety is morbid fears and preoccupations that accidents or illness will befall their parents or themselves. Fears of getting kidnapped, lost, or other mishaps are also common. Younger children have less specific, more vague concerns. As the child gets older, the fears usually become crystallized around identified potential dangers, although for many older children and adolescents, fears may remain ill defined so that they are left with pervasive anxiety. Young children do not usually show anticipatory anxiety, probably because of cognitive immaturity.

Children with this disorder often have fears of animals, monsters, and situations that are perceived as presenting dangers to themselves and their families. They may have exaggerated fears of muggers, burglars, kidnappers, car accidents, or plane travel. They frequently have difficulty in falling asleep and may request that someone stay in the bedroom with them until they fall asleep. They may make their way to their parents' bed or sleep outside the parents door. Morbid fears are often expressed in nightmares in these children. On separation from home, they feel acute homesickness and feel uncomfortable to the point of misery and even panic. They cannot wait to return home and are preoccupied with reunion fantasies until they do. On occasion a child may become violent toward an individual who is forcing separation. Some may prefer to avoid friends and relatives rather than give an explanation for their discomfort.

A depressive mood that is completely relieved by returning home or to the major attachment figures is frequently part of the disorder. If a depressive mood is reported even at home with the parents when no separation experience is impending, the diagnosis of major depression should be considered, for the two syndromes frequently coexist. Many children with separation anxiety disorder are among the group of children who used to be diagnosed as having school phobia. In fact this is not a true phobia of school. Any situation involving separation, as elementary as going to sleep, will bring on the anxiety. Thus school refusal is a better name for this particular symptom of separation anxiety disorder. (See also Chapter 15.) Separation anxiety disorder, of course, is not the only cause of school refusal. Children may refuse to go to school for a wide variety of reasons, from true physical illness or justified fears of an adolescent gang to repeated lack of success in academic tasks or overly harsh teachers or separation anxiety disorder.

Diagnostically it is important to differentiate truancy from the school refusal of separation anxious children. Truancy, a behavior pattern indicative of conduct disorder, involves school absences in which parents are frequently lied to and deceived, and the child spends the school time out of the home and out of school. Children with separation anxiety rarely deceive their parents. They simply want to stay home or remain at the parent's side so as to avoid the unpleasant feelings of anxiety.

Children with excessive anxiety of any type may report visual illusions in the dark and voices calling only their names. These episodes should be differentiated from true hallucinations, for such anxiety related phenomena do not carry the diagnostic implication of psychosis.

Separation anxiety is a developmentally appropriate response during the preschool years, and its normal resolution has wide chronological age limits. Extreme caution should be exercised in making this diagnosis in the preschool years, or even during the first months of school attendance. Normal developmental variations of separation anxiety during the late preschool years usually do not require more active intervention than parental counseling.

The extreme form of separation anxiety disorder involving school refusal may begin in preschool years but most often occurs between ages 11 and 12. Exacerbations and remissions are frequent. In extreme cases anticipatory anxiety and avoidance of separations persist for many years, and the child can neither attend school nor function independently in many different areas. The disorder usually develops following a stressful life event, for example, a loss, the death of a relative, friend, or pet, an illness in the family, or a move or change of schools. Children with this disorder tend to come from families that are close knit and well functioning. The etiological significance of this familial pattern is not clear. The disorder is equally frequent in both sexes and is more common in family members than in the general population. Separation anxiety disorder is distinguished from the other anxiety disorders of childhood because the anxiety is focused solely on separation.

Imipramine has been shown to be effective in the treatment of separation anxiety disorder when used in dosages similar to the ones prescribed for patients with major depression. No plasma level–clinical response studies have been carried out. A study of clomipramine in "school phobic" children carried out in Great Britain revealed no differences between responses to placebo and drug, but the drug dosage was well below those used in the successful imipramine study. The effects of imipramine are on separation anxiety proper. Once separation anxiety is blocked, a return to school can be spontaneous or the child may need a behavioral program of slow increments of separation experiences.

Until the advent of evidence of the effectiveness of imipramine in this disorder, school phobias in adolescents were thought to be associated with a poor prognosis in regard to return to school. At that time the aim of treatment was immediate return to school. In retrospect it is not clear that this was the best strategy. Children who return to school under severe parental pressure while still highly separation anxious are miserable there, and it is likely they cannot concentrate or perform scholastically under such conditions. Responsivity to imipramine does not correlate with age; thus, it may be more rational to block separation anxiety first by pharmacological methods, and then to desensitize the child "in vivo" relatively rapidly if such a behavioral program proves necessary.

As in other child psychiatric disorders, assessment of interpersonal relationships and determination of an indication for specific psychosocial treatment may be better left until the time the acute episode is under control.

OVERANXIOUS DISORDER

In children with overanxious disorder, the major symptom is excessive worrying and fearful behavior that is not focused on a specific situation or person. This should be an established symptom picture lasting for several months, not transient and due to a recent psychosocial stress. Normal childhood worries about future events, examinations, the possibility of injuries, relationships with peers, and meeting expectations are greatly exaggerated. An inordinate amount of time may be spent asking about the discomforts or dangers of a situation, such as a routine visit to the pediatrician. The anxiety may be expressed as concern about his own competence and the evaluations by others. The child may need frequent reassurance about his competencies and past, present, and future behavior. There may be preoccupations with adults who seem to be "mean" or critical. As the child grows older, the preoccupations become more definite and apparent as excessive concern about conventional forms of judgment, such as that of peers, social or athletic acceptance, school grades, and behavior of family members that might embarrass the child. In some cases physical concomitants of anxiety are prominent—shortness of breath, a lump in the throat, nausea, dizziness, headache, gastrointestinal distress, or other somatic discomfort. Difficulty in falling asleep is common. Some children complain about feeling "nervous" or "scared" or of having "butterflies in the stomach" or being "unable to relax." Episodes of depersonalization or derealization may also occur.

It is important to differentiate this syndrome from simple "shyness." In fact, shy children have an excellent long term outcome, whereas overanxious children are more likely to continue to present similar clinical pictures of anxiety disorder. In severe cases this disorder can be incapacitating and may result in difficulty in meeting realistic demands at home and in school. Poor school performance and failure to engage in age appropriate activities are frequent complications. The disorder is more common in boys than in girls.

It is common for pervasive anxiety to be accompanied by obsessional, phobic, or hysterical symptoms. The distinction between anxiety, obsessive-compulsive, phobic, and hysterical disorders is a matter of symptom patterns. Anxiety disorders often arise in children with a pre-existing tendency to overreact to ordinary stresses. These anxiety prone children have been distinct from their siblings in terms of their sensitivity and reactivity to stress since early childhood. It is probably most useful to view this as a temperamental variation that has its roots in biological and genetic influences. This temperamental attribute is neither fixed nor immutable but can be actualized by environmental stresses. Anxiety disorders can be precipitated by frightening experiences, such as an operative procedure, death of a friend or relative, or an accident. In other cases there is a transmission of anxiety from chronically anxious, fearful parents. It is not just that the child picks up parental problems; chronic parental anxiety may introduce stress in the parent-child interaction, and insecure parents may prolong and compound a child's anxieties by their uncertain and indecisive response to them.

Treatment

There is little knowledge about the effectiveness of different treatments in this disorder. Psychotherapy is routinely used by most practitioners. There is little knowledge regarding benzodiazepines and other minor tranquilizers in youngsters with chronic anxiety. In situations in which parental anxieties interact with the child's, family therapy and direct treatment of parental anxiety would seem to be indicated on purely common sense grounds.

OBSESSIVE-COMPULSIVE DISORDER

The main symptoms of this disorder are recurrent obsessions or compulsions. (See also Chapter 36.) Obsessions are recurrent and persistent ideas, thoughts, images, or impulses that are not experienced as being under voluntary control but rather as thoughts that intrude in one's consciousness and are perceived as foreign, senseless, or repugnant. A feeling of resistance to the obsession is considered necessary in adults, but this may be difficult to assess in children. It has yet to be demonstrated that the presence or absence of the feeling of resistance is crucial.

Compulsions are repetitive, seemingly purposeful behavior patterns that are performed according to set rules or in a stereotyped manner. The behavior is performed in order to produce or prevent some future event or situation. The activity is not connected in a realistic way to the desired effect, or, if related, it is excessive or illogical in some other way. The child feels compelled to perform the act, and there may be a desire to resist the compulsion. It is important to differentiate true compulsions from games or habits. It is helpful to ask the child to imagine that he is prevented from being able to carry out the ritual or to think of occasions on which this has happened. In true compulsions the youngster's response is either anxiety or more commonly anger; this is not so with games or habits. The most common obsessions are repetitive thoughts of violence, contamination, and doubt. The most common compulsions involve hand washing, counting, checking, touching, and collecting. Hand washing is repeated, homework is checked and rechecked, and the child is unable to inhibit this behavior even though he appreciates its futility.

Mild rituals and obsessions are a common part of normal development. Bedtime and dressing rituals in toddlers and preschoolers have been well described. Children often seem compelled to touch certain objects, either walk on or step over the cracks in the sidewalk, or make certain signs. Ritual elements are widespread in the songs and games of children. But in such cases, there is no accompanying functional impairment. The compulsive behavior that is part of normal development is not experienced as being alien or incongruous, and there is no internal need to resist the behavior. It is experienced as pleasant, is eagerly performed and short lived, and is easily stopped under pressure without appreciable anxiety.

The obsessional child frequently does not keep his thoughts or compulsions to himself. Rather he tries to get others, usually his parents, to participate in the rituals, insisting that they answer repetitive questions or assist him in compulsive acts. An obsessive-compulsive child often can control his family with his symptoms. Frequently parents become frustrated by their child's behavior because all attempts to deal with the rituals by permissiveness, reasoning, or punishment fail. Adolescents with obsessive-compulsive disorders often lose or lack social skills and become isolated from their peers and their social activities. Obsessional thoughts and compulsive behavior may develop as secondary features in depressive disorders, but the obsessions and compulsions tend to disappear when the depressive episode ends.

This disorder is equally common in males and females. The disorder is not seen very frequently, and the onset is often sudden. These children have been considered normal before the appearance of symptoms. Their parents often have compulsive traits. The diagnosis is usually made after several years of symptoms. The disorder is frequently resistant to psychosocial therapies and the diagnosis frequently does not change at long term follow-up. Obsessional preoccupations are sometimes an important part of the prodrome of schizophrenia in adolescents. However, only a very small number of adolescents with long standing obsessional symptoms become schizophrenic. Obsessions and compulsive symptoms are also seen in the Gilles de la Tourette syndrome, which is distinguished by recurrent, involuntary, repetitive, rapid movements, including multiple vocal tics. The vocal and motor tics clarify the diagnosis.

In a recent study of childhood obsessive-compulsive disorder, Rapoport et al. (1981) found that depressive symptoms were common to all the subjects studied. The depressive symptoms appeared months or years after the obsessive-compulsive symptoms and were most severe in children in whom there was an early onset of the obsessive-compulsive disorder. All the children studied met criteria for the diagnosis of major depressive disorder and had had suicidal thoughts at some time since the onset of their illness. Sleep studies showed findings similar in some respects to those found in adult depressive patients with primary affective disorder—a short first REMP latency and total sleep time and strong trends toward increased sleep latency and reduced sleep efficiency. Dichotic listening tests demonstrated a significant lack of the usual laterality effect in the speech perception of these patients, but the significance of such findings is unclear.

Treatment

The treatment of obsessive-compulsive disorder in children and adolescents is again rather uncharted territory. As in other neurotic disorders, individual psychotherapy has been widely used, but no studies of effectiveness have been carried out. Patients with moderate to severe obsessive-compulsive disorders do not appear to be very responsive to psychosocial therapy.

The first task of the clinician confronting a case with this diagnosis is to determine whether there is a strong affective component, such as associated major depression or separation anxiety disorder.

If present, these should be treated pharmacologically and the patient's obsessive-compulsive symptomatology re-evaluated after a clinical response to antidepressant medication.

If the clinical picture is that of a "pure" obsessive-compulsive disorder, there are no established guidelines for treatment in children. In adults with obsessive-compulsive disorder, clomipramine has been shown to be effective in decreasing rituals and obsessive symptoms, especially when associated with "in vivo" implosion treatment. Similar studies have not been published in children.

PHOBIC DISORDERS

Phobic disorders are characterized by a persistent, abnormally intense fear of a specific object, activity, or situation that results in a desire to avoid such object, activity, or situation (Table 41–3). The reaction to the phobic stimulus usually has some significant maladaptive effect on life adjustment and social or role functioning suffers. Irrational fears are commonplace in children, but they usually disappear during the course of normal development and usually do not have a significant effect on life adjustment. Thus, phobic disorders have to be differentiated from the normal fears of children. Childhood fears and phobias involve similar subjective feelings and behavioral changes. However, in phobias the excessive persistent nature of the response to the phobic stimulus leads to a maladaptive response. Phobic disorders in children commonly involve fears of animals, death, insects, the dark, noise, and school. In prepubertal children, fears of specific situations and animals are most common, and social anxiety and agoraphobia are rare. Different types of phobias differ in their usual age of onset. Animal and insect phobias are distinct from all other types of phobia because they start by age five and almost none begin in adulthood. This type of phobia is also distinguished by a good response to behavior modification treatment. Agoraphobias start at any age from late childhood to midlife, with peaks of onset in late adolescence and about the age of 30. Social phobias begin at or after puberty, and specific situational phobias can begin at any age.

Simple phobia is sometimes referred to as a specific phobia, and the key symptom is a persistent, abnormally intense fear of and desire to avoid some object, activity, or situation other than being alone or in public places away from home (agoraphobia) and extreme embarrassment or humiliation in certain social situations (social phobia). Simple phobias often involve animals, fear of closed spaces (claustrophobia), and fear of heights (acrophobia). Most simple phobias that start in childhood disappear without treatment. The impairment is minimal if the phobia stimulus is rarely encountered.

Table 41–3. DIAGNOSTIC CRITERIA FOR PHOBIC DISORDERS*

Diagnostic Criteria for Simple Phobia
  A. A persistent, irrational fear of, and compelling desire to avoid, an object or a situation other than being alone, or in public places away from home (agoraphobia), or of humiliation or embarrassment in certain social situations (social phobia). Phobic objects are often animals, and phobic situations frequently involve heights or closed spaces.
  B. Significant distress from the disturbance and recognition by the individual that his or her fear is excessive or unreasonable.
  C. Not due to another mental disorder, such as schizophrenia or obsessive compulsive disorder.
Diagnostic Criteria for Agoraphobia
  A. The individual has marked fear of and thus avoids being alone or in public places from which escape might be difficult or help not available in case of sudden incapacitation, e.g., crowds, tunnels, bridges, public transportation.
  B. There is increasing constriction of normal activities until the fears or avoidance behavior dominate the individual's life.
  C. Not due to a major depressive episode, obsessive compulsive disorder, paranoid personality disorder, or schizophrenia.
Diagnostic Criteria for Social Phobia
  A. A persistent, irrational fear of, and compelling desire to avoid, a situation in which the individual is exposed to possible scrutiny by others and fears that he or she may act in a way that will be humiliating or embarrassing.
  B. Significant distress because of the disturbance and recognition by the individual that his or her fear is excessive or unreasonable.
  C. Not due to another mental disorder, such as major depression or avoidant personality disorder.

*Reprinted from Diagnostic and Statistical Manual of Mental Disorders, Third Edition, 1980, with permission of the American Psychiatric Association.

Agoraphobia is a marked fear of being alone, being in public places from which egress might be difficult, or being in a place where assistance is unavailable in case of incapacitation. It is distinct from separation anxiety in that individuals other than major attachment figures can allay the fear. Daily activities become increasingly constricted as fears and avoidance behavior take up more and more of the individual's time and energy. Situations that prove difficult for adolescent agoraphobics include being in crowds and being in tunnels, on bridges, in elevators, or on public transportation. Often a family member or friend must accompany these adolescents if they are to leave their home.

The initial phase of the disorder frequently consists of recurrent panic attacks, which are very frightening. The individual becomes reluctant or refuses to enter a variety of situations that are associated with the attacks. Even after panic attacks subside, the anticipatory anxiety and the reluctance to enter situations associated with panic attack remain. Not all agoraphobias are associated with panic attacks. The severity of the disturbance waxes and wanes, and periods of complete remission are possible. During exacerbations of the illness an adolescent may be housebound. Some adolescents may attempt to alleviate their anxieties with alcohol, marijuana, and other depressants. Many adolescents who develop agoraphobia with panic disorder have a history of separation anxiety disorder in childhood. This is a relatively common disorder in adults and is seen more frequently in females than in males. Panic attacks respond to imipramine treatment. Behavior modification may be needed after cessation of the panic attacks to return the youngster to normal mobility.

A social phobia is a persistent fear of and compelling desire to avoid situations in which the adolescent may be exposed to the scrutiny of others. The adolescent also fears that he will behave in a manner that will draw attention to himself so as to be embarrassed or humiliated. Marked anticipatory anxiety attends the entering of such a situation, and the adolescent attempts to avoid it. This disturbance causes a significant amount of distress, and adolescents are well aware of their discomfort. Examples of social phobias are fears of speaking in public, using public lavatories, eating in public, and writing in the presence of others. Generally an individual has only one social phobia but agoraphobia or simple phobia may coexist with social phobia. The disorder begins in early adolescence and is usually chronic and may undergo exacerbation when the anxiety impairs performance of the feared activity. The disorder is relatively rare. Anticipatory anxiety does not respond to any known pharmacological intervention. Psychosocial treatments are widely used.

CONVERSION SYMPTOMS

Every pediatrician is likely to be confronted at some point in his career with a patient who presents one or more physical symptoms, for which after thorough clinical investigation there is no known explanatory physiological mechanism (e.g., areas of anesthesia or paralysis of one or more limbs that cannot be attributed to any specific lesion site or likely combination of sites) and no demonstrable physical or neuroradiological findings. From the days of Charcot it has been known that a high level of suggestibility (and hypnotizability) tends to be present in many such patients, and that under hypnosis a considerable proportion of such symptoms are reversible.

Although theoretically, in dealing with such patients, willful malingering should be suspected, it is a well accepted clinical consensus that most are not malingerers but suffer from symptoms that are completely beyond conscious voluntary control and that frequently are accompanied by substantial disability. Hysteria was the diagnostic label chosen for this somewhat heterogeneous group of patients. Freud's psychological investigations in hysterics enhanced the suspicion that psychological factors were paramount in the etiology of these disorders, although Freud's theoretical scheme could never solve the problem of symptom choice, i.e., why, given a particular psychological conflict, one patient would just be anxious, another would develop a phobia and still another, paralysis.

Consistent with what he believed to be true. Freud coined the term "conversion" to name the hypothesized mechanism by which "psychological energy" becomes "physical energy." It should be readily observable that if the foregoing construct is accepted, conversion symptoms could not be diagnosed without strong evidence of specific psychological causation. This is impractical, however, because specific evidence may not be forthcoming, even after months or years of intensive psychotherapy, and the diagnosis might only be made retrospectively. Nevertheless, a diagnosis is important before planning treatment. At present, the American Psychiatric Association classification accepts evidence of the etiological involvement of psychological factors if there is a temporal relationship between a particular event and the onset or exacerbation of the symptom, or if there is evidence of secondary gain derived from the symptom. Unfortunately this line of investigation with the historically accepted assumptions leads to an untenable logical circle.

Other approaches using a phenomenological classification and follow-up have been more useful, both in adult and in child psychiatry. Thus, adults with "conversion" and other unexplainable symptoms with no demonstrable physical basis can be classified in several groups (following the American Psychiatric Association nomenclature).

1. Somatization disorder. In this disorder, recurrent and multiple somatic complaints of several years duration dominate the clinical picture. Medical attention is sought repeatedly for a variety of clinical problems, including neurological symptoms, gastrointestinal symptoms, pain, psychosexual dysfunction, dizziness, and cardiopulmonary symptoms, but no physical disorder is found. The course is chronic and fluctuating, with practically no spontaneous remissions. The onset is usually insidious. The family history is rich in somatization disorder and antisocial personality.

2. Conversion disorder. In this disorder a single conversion symptom appears abruptly and has a short course and an abrupt recovery. There is usually a strong relationship to stressful events for the patient.

3. Psychogenic pain disorder. In this disorder persistent pain not explained by any physical disorder dominates the clinical picture, is not under voluntary control, and may possibly be related to stressful events for the patient. The course may be chronic or short lived. In other words, the conversion symptom here is pain.

Of relevance to this discussion is that the peak onset of all these disorders is during adolescence, and that all three types of disorders can be seen during prepuberty. Nevertheless a note of caution is in order in view of Caplan's study (1970). Caplan carried out a follow-up study of all the children and adolescents diagnosed over a period of 22 years as having conversion symptoms. On follow-up four to 11 years later, he found that 13 of the 28 patients (46 per cent) had a nonpsychiatric medical diagnosis, which in retrospect may have accounted for the initial symptoms, which in fact were mistakenly identified as conversion symptoms. Such an outcome was frequent in patients whose initial "conversion symptoms" had been visual loss. Moreover, Rivinus et al. (1975) found poor school performance, visual loss, and gait and postural disturbances to be frequent in children with organic neurological disease whose initial presentation resembled a psychiatric disorder. Finally, conversion symptoms with an insidious onset and chronic course are also more likely to be due to neurological disorder.

True acute onset, short lived conversion symptoms are not infrequent in children. It is important to realize that conversion symptoms remain a diagnosis of exclusion, and that their presence is not infrequently accompanied by the presence of organic, especially neurological, disease. Thus true epileptics frequently also have pseudoseizures.

Although the presence of a conversion symptom may seem to some to be fertile ground for in depth psychological investigation, it is important that the clinician recognize that such an "unexplainable" symptom may actually be the first manifestation of a pathological neurological process; appropriate work-up and close follow-up should be routine.

Finally it should be pointed out that hypnosis can transiently reverse subjective symptoms, even when they are clearly organically based. Therefore, hypnosis effects should not be taken as incontrovertible evidence of conversion.

The prevalence of conversion symptoms has been decreasing steadily during the twentieth century (except during periods of war). We do not thoroughly understand this phenomenon. At present these symptoms are mostly limited to relatively backward subcultural groups.

It is important to differentiate somatization disorder from hypochondria. In the latter the predominant disturbance is an unrealistic interpretation of physical signs or sensations as being abnormal, leading to a preoccupation (or even delusion) with the fear or belief of having a serious disease. Negative medical work-up and consequent professional reassurance do not influence the patient's preoccupation with having a serious disease. This preoccupation is accompanied by significant social and occupational impairment. By contrast, in somatization disorder the patient presents with symptoms, not with the fear of having a specific disease.

The acute conversion symptoms require little treatment and will most likely disappear spontaneously once the psychosocial stressors are at least temporarily removed. During this time it is important to work with the family, the patient, and other significant persons in his life, with the aim of avoiding repetition of the same environmental precipitants.

In contrast, when these symptoms show a chronic course, as in somatization disorder and many cases of psychogenic pain, treatment is arduous and frequently unsuccessful. Psychotherapy, hypnosis, and parental counseling have all been used with mixed results. The possible coexistence of affective disorders, especially in psychogenic pain, should be explored and, if they are present, the disorder should be treated pharmacologically.

ATTENTION DEFICIT DISORDER

Attention deficit disorder was originally described as consisting of developmentally inappropriate inattention, impulsivity, and hyperactivity (see also Chapters 38 and 57). It is now recognized that individuals exist who have all the stigmata of the syndrome without hyperactivity. In addition, inattention has been suggested as the key feature of the syndrome underlying other manifestations like impulsivity and hyperactivity. In the past a variety of names have been used to describe this disorder, including minimal cerebral dysfunction and minimal brain damage. The choice of these terms to describe the syndrome has been particularly unfortunate because they imply that brain damage or brain dysfunction is present in children with this syndrome.

If what is meant by brain damage is demonstrable structural damage to the brain, the term minimal brain damage is misleading. Although some children with attention deficit disorder may suffer from brain damage, the majority do not. Although soft neurological signs may be associated with the syndrome, as well as minor congenital anomalies, such associated features do not imply brain damage and are not present in all or even a majority of cases. Minimal brain dysfunction is also an inappropriate diagnostic label for a group of children whose difficulties are manifested by behavioral abnormalities and in whom no specific brain dysfunction, central to the disorder, has been demonstrated.

Typically the child with attention deficit disorder appears deviant to the school personnel early in his elementary school years. However, a careful history is likely to reveal that differences were present from early childhood, although they were not as troublesome then, or were accepted by the parents as part of the child's personality. Children with attention deficit disorder with hyperactivity seem to have more energy than others, even as infants. Parental descriptions like "always on the go" or "seems driven by a motor" are frequent. They sleep less than siblings and are able to wear out clothes, shoes, toys, and baby sitters much more quickly than other children. Objective measures of gross motor activity have not demonstrated hyperactivity as such in these children during their school years. This has raised the question whether they actually have a greater amount of motor activity. The difference may not be in the amount of activity but rather in its quality. In younger children, hyperactivity is manifested by gross motor activity, and as the child gets older, "hyperactivity" is more likely to reveal itself in poorly organized motoric patterns, difficulty in sitting still, fidgeting, wiggling, restlessness, tapping, or drumming.

Distractability and short attention span are more noticeable at school than at home probably owing to the greater demand for sustained concentration in the first setting. These children often give the impression that they are not listening or have not heard directions. Their work is sloppy and is performed in an impulsive fashion. Oversights, omissions, insertions, or misinterpretations of easy items characterize these children's work, even in situations that hold a great deal of intrinsic interest for them. At home, attentional problems are shown by a failure to follow parental requests and instructions. Play situations that require attention and concentration easily demonstrate the disturbance. The impulsivity of the hyperactive child is manifested by temper tantrums, fights over trivialities, low frustration tolerance, and a tendency to act before thinking and to become overexcited and uncontrollable in stimulating situations like parties and family gatherings. In adolescence, hyperactivity, impulsivity, and excitability tend to diminish while attentional difficulties tend to persist. As a result it is difficult to make a correct diagnosis in adolescence without a good prepubertal history.

Originally it was thought that antisocial behavior was a component of this syndrome. Careful clinical studies have shown that only a small minority of children with this syndrome present with conduct disorder when initially seen. Family studies suggest aggregation of alcoholism, antisocial personality, and hysteria in families of children with attention deficit disorders. The mechanisms underlying such familial aggregation are unknown. Learning difficulties are frequently associated with attention deficit disorder. Several studies have found that a majority of hyperactive children have lower educational achievement when compared with a matched group of normal controls. In part this may result from difficulties in attention and lack of persistence. Nevertheless, associated specific reading or arithmetic disabilities are rather frequent.

It is likely that the diagnosis of attention deficit disorder includes a heterogeneous group of children with disorders of different etiologies. In some cases the disorder may be secondary to a structural abnormality in the brain; in others the syndrome may result from developmental delays, from disorders of arousal, or from disorders having a genetic basis. Nevertheless, evidence for these etiological mechanisms is not strong, and in most cases the etiology remains unknown. Prevalence figures vary from 5 to 20 per cent in school age populations; the boy:girl ratio varies from 4:1 to 9:1. There is little question that regardless of etiology, attention deficit disorder represents a substantial handicap for many children vis-à-vis school performance. Furthermore functional impairment frequently extends to little success in peer relations and poor adjustment in the family. Persistent negative feedback from the environment in a multiplicity of settings may be a substantial factor in the development of secondary complications like low self-esteem and maybe conduct disorder. In fact cross situational behavioral deviances and functional impairment may be important indices of minimum severity for the diagnosis to be made. Follow-up studies of attention deficit children in young adulthood have shown residual symptoms of inattention and some degree of social maladjustment but not the predicted increase in antisocial behavior and antisocial personality disorders.

There is little question that stimulants are very effective in improving attention span and reducing impulsivity and hyperactivity in children with attention deficit disorder. It has not been conclusively shown that the long term outcome is changed by long term treatment. Nevertheless, given the few risks entailed by the use of these drugs in prepuberty, and the clear therapeutic effects achieved, it appears reasonable to treat these children with stimulants, if for no other reason than to break the cycle of persistent negative feedback due to the disturbing effects of their behavior on others. Although behavior therapies have been strongly recommended, little evidence exists for their effectiveness, especially in studies in which the results with stimulants and behavior modification are compared. Studies of the effects of food additives on attention deficit disorder have not supported the early sensationalistic claims. Although in a very few children food additives have been shown to worsen the symptomatology, this is definitely not true in the majority of children with this diagnosis.

It stands to reason that a major role of treatment should be to avoid the development of chronically faltering self-esteem. Pharmacological measures may be necessary to enable the child to be socially or academically successful to a reasonable degree. In severe cases neuroleptic drugs may be necessary, especially in the evening, to enable the child to function in the family situation, or in instances in which associated conduct disorder is not controlled by stimulants or when stimulants are simply not effective.

Hasty diagnostic evaluations may lead to the misdiagnosis of depression as "hyperactivity" in prepubertal children. In such cases stimulant medication usually worsens the clinical picture by having an adverse effect on depressive mood and increasing irritability. Discontinuation of the drug and careful psychiatric re-evaluation of the child, including a semistructured interview with the youngster, usually lead to the correct diagnosis.

Adolescents and adults with a history of attention deficit disorder and residual inattention or impulsivity may also benefit from the use of stimulants. However, careful attention to the possibility of substance abuse or psychological dependence should be paid by the physician. In contrast, in prepuberty, substance abuse during stimulant treatment of attention deficit disorder is extremely rare.

Narcolepsy in children may be misdiagnosed as attention deficit disorder. It is clear that basing the diagnosis of attention deficit disorder simply on a behavioral rating scale is poor practice and should be discouraged. Parental rating scales are helpful only as an initial screening device or guide for a proper diagnostic evaluation. Once the diagnosis is made, the treatment response is best measured by a teacher rating scale.

Validation of the diagnosis of attention deficit disorder with or without hyperactivity has not been very successful. At one point the response to d-amphetamine was thought to be "paradoxical" (calming) and thus was thought to validate the diagnosis indirectly. Rapoport's studies on d-amphetamine responsivity of normal prepubertal children disproved this view (Rapoport et al., 1980). Such "paradoxical" responses are not specific to attention deficit disorders but are generalized in prepubertal children. Until new evidence comes to light, the basic fact remains that children fitting the diagnostic criteria for attention deficit disorder are functionally impaired, sometimes to a high degree, and that stimulants have been shown to help them considerably.

(For further coverage of this subject, the reader is referred to Chapter 38.)

CONDUCT DISORDERS

Conduct disorder refers to a cluster of persistent patterns of behavior that violate the rights of others or age appropriate societal norms. (See also Chapter 36.) The syndrome has been divided into four subtypes: undersocialized-aggressive; undersocialized-nonaggressive; socialized-aggressive; and socialized-nonaggressive. These subtypes are based on the presence or absence of adequate social bonds and the absence of a pattern of aggressive antisocial behavior, and were derived largely from large studies in which similar symptom groupings resulted from factor analyses. Nevertheless it should be stated that not all studies reproduced the same subgroups and that they have limited usefulness in predicting the course and outcome of the disorder. Some investigators believe that a more useful distinction could be made on the basis of the variety, frequency, and seriousness of antisocial behavior rather than the type of disorder, since these variables have been shown to have usefulness as a predictor of outcome in this disorder.

Clinical judgments on socialization are based on reports of the existence of special friendships, appropriate guilt, and selfless mutual concern and loyalty within the peer group. Nonaggressive conduct disorder behavior patterns include substance abuse, persistent truancy, running away from home overnight, frequent lying in a variety of social settings, theft not involving confrontation with a victim (in and out of the house), and chronic rule breaking. Aggressive behavior refers to persistent physical fighting and bullying producing serious harm, vandalism, rape, physical cruelty to persons, breaking and entering, setting fires, assault, and other physical violence against persons or property, including armed robbery, purse snatching, and extortion. Nonsocially sanctioned cruelty to animals is frequently associated with aggressive conduct disorders.

It is important to emphasize the differences between conduct disorder and juvenile delinquency. Conduct disorder is a psychiatric diagnosis based on the presence of repetitive patterns of specific behavior that violate the rights of others or age appropriate societal norms. Most of this behavior happens to involve legally punishable offenses, but the diagnosis is based only on the presence of such behavior patterns (not on an isolated incident), regardless of whether the

youngster has or has not been caught, arrested, convicted, or sentenced. Delinquency, on the other hand, is a legal term, which indicates only that the youngster has been declared guilty in a court of law of at least one punishable offense. Although there is substantial overlap, a youngster diagnosed as having conduct disorder may or may not be delinquent, and a delinquent youngster may or may not fit the criteria for conduct disorder (of any type). There is some evidence that given similar antisocial behavior patterns, youngsters with conduct disorder are less likely to be delinquents, the higher their IQ and the higher their social class.

Functional impairment is greatest in the undersocialized aggressive subtype and lowest in the opposite subtype, but it is usually generalized to different social settings. Academic underachievement is commonly associated with the disorder, but not necessarily secondary to conduct disorder. In fact, one-third of prepubertal onset conduct disorders are associated with developmental reading disability, suggesting that repeated academic failure may increase the likelihood for the eruption of conduct disorder behavior. This association may create a vicious circle ending in school suspension and expulsion. Precocious sexual activity and early substance use and abuse frequently lead to such severe complications as venereal disease, unwanted pregnancy, physical injury, medical disease, illicit drug dependence or addiction, and legal problems.

Antecedents of conduct disorder include many factors reflecting psychosocial disadvantage, of which the most frequent are parental rejection, neglect, and abuse, early institutional living, frequent shifting of major attachment figures, illegitimacy, and inconsistent child rearing techniques, including harsh or cruel disciplinary methods, parental diagnoses of antisocial personality, alcoholism and drug addiction, parental criminality, large family size, and absence of the father. Although it is clear that the environment in which future conduct disorder children are raised tends to be far below "average expectable," and could be characterized as low in terms of a love/hate ratio, it is important to withhold judgment regarding the mode of transmission of these disorders. Thus it may be tempting to some to assume a purely environmental transmission mechanism, but evidence indicating the potential importance of hereditary factors is available. Adolescents who were adopted away as infants are more likely to develop antisocial behavior if their biological parents carried a diagnosis of alcoholism or antisocial personality. Studies of adoptees also indicate that environmental variables (already described) in the adoptive parents are important, especially in the development of antisocial behavior in boys. Therefore, the etiopathogenesis of conduct disorder and antisocial behavior is beginning to reveal itself as a rather complex multidetermined set of interactions, the unraveling of which carries a possibility of great benefit to the individual and to society.

Follow-up studies of children and adolescents diagnosed as having conduct disorders or as being juvenile delinquents are greatly in agreement regarding the strong predictive power of conduct disorder for future psychiatric disorders and social adjustment problems, antisocial personality, alcoholism, and adult criminality.

Recently Robins (1978) reviewed her own findings from four different longitudinal studies of male conduct disorder–antisocial personality and concluded the following:

1. The diagnosis of adult antisocial personality virtually requires the past presence of persistent childhood antisocial behavior (conduct disorder), or, stated differently, the absence of a conduct disorder diagnosis up to age 16 years virtually guarantees that the child will not be an antisocial personality as an adult.

2. Only about half of those with conduct disorders in childhood and adolescence become antisocial persons as adults.

3. The severity and number of instances of antisocial behavior in childhood are the best predictors of adult functioning and adjustment among all childhood behavior patterns.

4. Childhood behavior predicts adult behavior better than any other variable in the child's environment (social class, family background).

It should be added that alcoholism and other drug use disorders are found quite frequently as one of the adult outcomes of childhood conduct disorder. The more aggressive and more unsocialized the child's behavior has been, the greater the likelihood of a poor outcome.

Treatment

To complete a rather gloomy picture, it should be stated that the treatment of moderate and severe conduct disorders, juvenile delinquency, and adult antisocial personalities is a relatively hopeless undertaking. Although different modalities of insight oriented and interpersonal psychotherapy have been used for many years, evidence of their effectiveness is lacking and there is some evidence for their being noxious. A. Freud (1970) has specifically warned against their use in the treatment of conduct disorder. Moreover, "treatment" through juvenile court has been shown to be highly deleterious. It has been repeatedly demonstrated that court appearances and detention in correctional institutions for juveniles are good predictors of future deviance when compared with the results in children who committed similar offenses that were undetected.

Behavioral modification techniques using the teachers or the parents as cotherapists have shown some promise, but the samples are constituted mostly of mild nondelinquent conduct disorders, and parents of antisocial children may not be the most trainable and most consistent "therapists." In a series of 27 consecutive referrals of aggressive children, Patterson (1973) after eight weeks showed a 60 per cent reduction in undesirable behavior. However, Reid and Hendricks (1973) reanalyzed the data and were able to show that in the subgroup of 14 children who in addition to aggressivity also had been referred for stealing, only six responded, and their families had much lower rates for friendly positive behavior than the families of aggressive children who did not steal. Thus a response tended to occur in children with less severe conduct disorders and with lower familial risk factors.

One factor that has not received a great deal of attention in the development of conduct disorder is the presence of another child psychiatric disorder. Maletzsky (1974) has addressed this issue partially in delinquent adolescents. He carried out a three month, placebo controlled, double blind, random assignment trial of d-amphetamine in delinquent adolescents using a sequential design. After 28 patients had been treated, the differences in drug-placebo outcome reached significance. He also reported that in the drug group, the adolescents with higher pretreatment hyperactivity scores were the ones who improved the most; a history of hyperactivity before puberty also correlated with the clinical response to d-amphetamine. Maletzsky concluded that childhood hyperactivity and adolescent delinquency may be differing manifestations of a basic neurochemical process. Later findings by Rapoport et al. (1980) regarding the lack of specificity of the clinical response to d-amphetamine in hyperactive children casts some doubt on Maletzsky's hypothesis. Nevertheless his data suggest that successful treatment of attention deficit disorder in adolescents with a concomitant diagnosis of conduct disorder may have a secondary beneficial effect on conduct disorder if a direct effect of stimulant treatment on the conduct disorder can be ruled out.

Maletzsky (1974) addressed indirectly the issue of subgrouping some conduct disorders according to the presence of other psychiatric diagnoses and treating those as specifically as possible. This is a new strategy that is fully justified in view of the dearth of efficacious treatments available for moderate and severe conduct disorders. Maletzsky's conceptual and psychopharmacological strategy bears a strong resemblance to our own observations in prepubertal children fitting criteria for both major depression and conduct disorder. Briefly stated, it has been observed that in prepubertal children fitting the criteria for both diagnoses, in whom the onset of major depression preceded the onset of conduct disorder, successful psychopharmacological treatment of the depressive syndrome with tricyclic antidepressants was followed by the persistent abatement of conduct disorder behavior. In addition, discontinuation of drug treatment three months after recovery did not precipitate a relapse of conduct disorder patterns in the subjects who were studied unless a relapse of the depressive syndrome also occurred. A relapse or a recurrence of the depressive syndrome in these children was always followed shortly by the reappearance of the conduct disorder behavior pattern. The appearance of conduct disorder never preceded the onset of a dysphoric mood and the depressive syndrome.

It appears then that improvement in conduct disorder behavior as a result of improvement of the depressive syndrome is possible. The clinical course in a small sample of these children assessed both retrospectively and prospectively suggests a second hypothesis—that the major depressive syndrome may trigger the emergence and persistence of conduct disorder behavior patterns; it may possibly lower the threshold for the emergence of conduct disorder in children whose environmental and genetic predisposition may be great but not sufficient to permit the emergence of such behavior patterns in the absence of a depressive disorder.

It is reasonable to expect that prepubertal boys fitting criteria for major depression and conduct disorder may be subject to the same long term outcome risks as "pure" conduct disorders. This hypothesis receives further support from family aggregation studies in early onset unipolar depressive adult probands in whom alcoholism and, in some studies, antisocial personality appear frequently among first degree biological relatives. Family histories of prepubertal major depressive disorders present a similar pattern.

On the basis of these considerations it may be hypothesized that antisocial personalities in depressive spectrum disease families are likely to begin in childhood or adolescence as conduct disorders, as with any other kind of antisocial personality. It is highly unlikely that they arise "de novo" in adult life. Boys who secondarily develop conduct disorder during a depressive episode may be the same subjects who will appear as antisocial personalities or alcoholics in the pedigrees of their "early onset" unipolar depressed sisters in adulthood.

If these hypotheses are true, long term management of these children that focuses on the very early treatment of every major depressive episode throughout childhood and adolescence may radically change the long term outcome.

SCHIZOPHRENIA AND RELATED PSYCHOTIC CONDITIONS OF CHILDHOOD AND ADOLESCENCE

The concepts of "childhood psychosis" and "childhood schizophrenia" have evolved over the years (Table 41–4). Originally childhood psychosis was used to designate a very heterogeneous group of children and adolescents who had frank psychotic symptoms (hallucinations, delusions, thought disorder) or whose behavioral deviance was so massive that they could not be thought of as "neurotic" in any way.

From within this heterogeneous group two rather clear-cut syndromes have been isolated. One, infantile autism, is characterized by an onset before three years of age, a pervasive lack of responsiveness to other people, gross deficits in language development (including lack of speech, or echolalia), pronominal reversal and noncommunicative speech, insistence on sameness, stereotyped movements, lack of imaginative play, and absence of delusions, hallucinations, and thought

---

Table 41–4.   DIAGNOSTIC CRITERIA FOR SCHIZOPHRENIC DISORDER*

A. At least one of the following during a phase of the illness:
   1. Bizarre delusions (content is patently absurd and has *no* possible basis in fact), such as delusions of being controlled, thought broadcasting, thought insertion, or thought withdrawal.
   2. Somatic, grandiose, religious, nihilistic, or other delusions without persecutory or jealous content.
   3. Delusions with persecutory or jealous content if accompanied by hallucinations of any type.
   4. Auditory hallucinations in which either a voice keeps up a running commentary on the individual's behavior or thoughts, or two or more voices converse with each other.
   5. Auditory hallucinations on several occasions with content of more than one or two words, having no apparent relation to depression or elation.
   6. Incoherence, marked loosening of associations, markedly illogical thinking, or marked poverty of content of speech if associated with at least one of the following:
      a. Blunted, flat, or inappropriate affect.
      b. Delusions or hallucinations.
      c. Catatonic or other grossly disorganized behavior.
B. Deterioration from a previous level of functioning in such areas as work, social relations, and self-care.
C. Duration: continuous signs of the illness for at least six months at some time during the person's life, with some signs of the illness at present. The six-month period must include an active phase during which there were symptoms from A.

*Reprinted from Diagnostic and Statistical Manual of Mental Disorders, Third Edition, 1980, with permission of the American Psychiatric Association.

disorder (see also Chapter 39). If the disorder is not associated with mental retardation, if speech develops, and if no seizures or other neurological complications appear, autistic children grow into odd, socially awkward adults who are not psychotic (no hallucinations, no delusions). Family histories indicate no schizophrenia, occasional diagnoses of autism, and speech delay in biological relatives. Autism may also be associated with organic conditions, such as congenital rubella or phenylketonuria.

A second group that has emerged is classic adult type schizophrenia. This is most clear in adolescents and late prepubertal children. The clinical picture in older children and adolescents diagnosed as being schizophrenic resembles the adult syndrome so closely that there is little controversy about the validity of this diagnosis and considerable agreement about the symptomatology of schizophrenia in these age groups. The essential features for the diagnosis of schizophrenia are the presence of delusions or nonaffective auditory hallucinations or thought disorder during the most severe phase of the illness, accompanied by deterioration from the previous level of functioning. The disorder should last for at least six months (including prodromal and residual phases). It should not be associated with affective episodes and should not be attributable to organic brain disease. During the prodromal or residual phases, functional deterioration is associated with at least two of the following symptoms: social isolation, poor scholastic performance, bizarre behavior, poor personal hygiene and grooming, flat or inappropriate affect, odd or bizarre ideas, unusual perceptual experiences, and digressive overelaborate speech.

The delusions or false beliefs are characteristically bizarre with no possible basis in fact. Persecutory delusions and delusions of reference, in which events, objects, or other people are given special significance, are frequently seen. Certain delusions are far more common in this disorder than in other psychotic disorders. These include the belief that one's thoughts are broadcast from one's head to the world (thought broadcasting), that foreign thoughts are being inserted into one's mind (thought insertion), that thoughts have been removed from one's head (thought removal), or that one's feelings, impulses, thoughts, or actions are being imposed by some external force (delusions of being controlled). Less frequently somatic, grandiose, religious, and nihilistic delusions are seen.

The clinician should be very careful in diagnosing delusions in a child. To be reasonably sure, it is not sufficient that the belief be false. What is important is the logic of the false belief. After the main types of delusions have been inquired for directly in the child interview, it is helpful to ask the child how he knows that what he is saying is actually true. Truly delusional youngsters forcefully assert that they know it without a doubt, although no reasons will be given for such certainty even when specifically asked for. Another way to check the correctness of a rating of delusions is to ask the youngster whether he thinks that it is probably true (but it may not be) or whether he knows for sure that it is true. If the youngster knows for sure but cannot adduce evidence for the correctness of his belief, he is likely to be delusional, especially if his belief is patently false and he is intelligent enough to realize it. True delusions should be differentiated from overvalued ideas (of which the patient is not fully certain) and from elaborated fantasies (like those concerning an imaginary friend).

A disturbance in the form of thought is often present and is referred to as a "formal thought disorder." The most common example is loosening of associations, in which ideas shift from one topic to another without any apparent connection and without the speaker's being aware that the topics are unconnected. Statements are juxtaposed without any connection, or the individual may shift from one frame of reference to another without comment. If loosening of association is severe enough, speech becomes incomprehensible. Other disturbances of form of thought include poverty of content of speech, neologisms, perseveration, clang associations, and blocking.

The major disturbances of perception are various forms of hallucinations. The most frequent are auditory hallucinations, perceived as voices coming from outside the head. The voices can be single or multiple, and it is characteristic of schizophrenia that the voices carry on a conversation among themselves that the patient is listening to, or they comment on his behavior. A command hallucination may be obeyed and result in danger for the child or adolescent or others. Tactile hallucinations if present typically involve electrical, tingling, burning, or crawling sensations; visual, gustatory, and olfactory hallucinations occur less frequently. Other perceptual abnormalities include sensations of bodily change, perceptual distortions, and hypersensitivity to smell, sight, and sound. The affect disturbance involves blunting, flattening, or inappropriate affect. Blunted affect is characterized by a marked reduction in the intensity of affective expression. Flat affect involves the absence of affect expression, the face is blank, and the voice is expressionless. Inappropriate affect indicates that the affect expressed is not congruent with the content of the child's or adolescent's speech or ideas. Frequently there is a tendency for the individual to withdraw from involvement with the external world and to become engrossed in his fantasies and ideas no matter how distorted or unreal.

Various disturbances in psychomotor behavior occur, particularly in acute and severe chronic forms. These include mannerisms, echopraxia, and catatonia. In catatonic states the child or adolescent is mute and does not interact with the environment. He may maintain a rigid posture, resisting efforts to be moved (catatonic rigidity); he may make purposeless stereotypic motor movements (catatonic excitement) or may voluntarily assume inappropriate or bizarre postures (catatonic posturing).

In spite of the lack of interaction with the environment, catatonic patients are well oriented and know what is going on around them. Because of their muteness and immobility, a psychiatric diagnosis is difficult to make without a sodium amytal interview. When sodium amytal is given slowly intravenously, a full mental status appraisal can be obtained and a diagnosis can be made. Catatonia is a relatively good prognostic sign, and may be related to affective disorders or to mixed schizoaffective disorders.

The development of the active phase of the illness is usually preceded by a prodromal phase in which there is a clear deterioration in functioning. Social withdrawal, peculiar behavior, impairment in personal hygiene, inappropriate affect, disturbances in communication, bizarre ideation, and unusual perceptual experiences are characteristic. A change in personality is often noted by friends, teachers, and family. The prognosis is poorer as the prodromal phase lengthens. Unfortunately the prepsychotic abnormalities do not form a sufficiently distinctive picture to be useful for prediction.

The family history in schizophrenic children and adolescents is frequently positive for schizophrenia. Besides clear-cut schizophrenia, a variety of related diagnoses can be found in adolescents and older children. Among them are schizophreniform disorders (disorders fitting all the criteria for schizophrenia except for duration, which is less than six months and more than two weeks), brief reactive psychoses (acute psychotic episodes of sudden onset, lasting for one day to two weeks without a prodromal phase and with eventual return to a prior level of functioning, that follow recognizably severe stressful life events), and schizoaffective disorders.

The earliest age of onset of schizophrenia in prepuberty had been thought to be seven years. During the early school years before puberty, most children with this diagnosis show evidence of very severe chronic undifferentiated subtypes. Evidence of neurological signs is frequently present, and some have expressed the opinion that such early schizophrenia may be secondary to organic brain disease. It is equally possible that the frequent "soft" neurological signs in these children are indicative of the severity of the schizophrenic

disorder. Recent studies have reported schizophrenia in children as early as ages four and five years. In these children a variety of minor congenital anomalies and muscle biopsy findings have been described. The validity of such a diagnosis has not been proven so far, but work is proceeding. It should not be forgotten that these children may have the most severe form of the disorder.

A variety of other children in the prepubertal age range present with a puzzling clinical picture that includes gross and sustained impairment in social relationships, as well as a variety of severe psychological manifestations, including severe sudden unexplained anxiety or rage attacks, resistance to environmental change, odd psychomotor and speech behavior, abnormal sensitivity to sensory stimuli, or self-mutilation. These children usually have low IQ's, and the onset of illness is between three and 12 years of age. The findings in these children do not fit criteria for schizophrenia, and little else is known about them.

Treatment

The treatment of schizophrenia in children and adolescents is similar to the current treatment of adult schizophrenics. It is usually more successful in adolescents than in prepuberty, but this probably reflects the frequency with which chronic undifferentiated subtypes appear in the preadolescent group. In adolescents it is crucial to inquire about hallucinogenic drug use. "Flashbacks" should be differentiated from schizophrenia, and the use of PCP ("angel dust") should be investigated in any youngster with psychotic symptoms.

SUICIDE AND SELF-DESTRUCTIVE BEHAVIOR

Suicidal behavior is not a disorder. It is a symptom that can be a part of a variety of disorders. Suicidal behavior refers to any behavior that carries the potential for serious self-injury or death and that is preceded by thoughts or ruminations about one's own death and distinct death wishes. Suicide gestures, which also occur in children, are always accompanied by death wishes, although the determination to kill oneself is definitely weaker than in suicide attempts. In a suicide gesture there is also the desire to manipulate some other person or persons, and this is usually attempted by a dramatic act or actions that precede or are part of the self-destructive behavior. Not all self-destructive behavior is suicidal. The self-destructive finger and lip biting of the Lesch-Nyhan syndrome is without ideation and must be prevented by physical restraint. Impulsive behavior is a spur of the moment manifestation. It occurs before thinking and in the absence of a wish to harm or kill oneself. Impulsive behavior is frequent in children with conduct disorder or attention deficit disorder, especially those with lower IQ's. Delusional children have been known to threaten or engage in self-destructive behavior. The prepubertal child who stands on the window sill with a towel on his back threatening to jump is convinced that he can fly. It is his distorted judgment produced by his delusions that has put him at risk, not any wish to die.

The assessment is likely to be incomplete unless certain facts about suicidal children are understood. Usually problems other than suicidal behavior have brought the child to the attention of a clinician. Parents and adult relatives are frequently unaware of suicidal thoughts, plans, and even behavior in their children. The only way the clinician can obtain this information is to ask the child directly. Some clinicians may feel uncomfortable in asking a child about suicidal ideation or planning, but deleterious effects in questioning nonsuicidal children have never been demonstrated. On the other hand, suicidal children feel quite relieved at being able to talk openly about suicidal thoughts, plans, and acts, which they had kept to themselves.

Once basic information has been gathered, it becomes necessary to make the differential diagnosis between impulsive and suicidal behavior. The child must be questioned directly, and the examining clinician should present the questions in as gradual and neutral a manner as possible. Frequently a series of questions is used: "Have you ever thought about killing yourself?" "Have you thought about how you would do it?" "Have you ever tried it?" "What is the closest you have come to doing it?" "What did you do?" "Did you really want to die?" "How many times have you tried it?" At this point the examining clinician should have enough information to distinguish a suicide attempt from impulsive behavior and to make some estimate of the potential danger of the situation.

Dangerous suicidal behavior must also be distinguished from suicide gestures. A suicide gesture usually can be distinguished from a suicide attempt by examining the context and environment in which the self-destructive behavior occurred. In what kind of atmosphere did the self-destruction occur? Were other people around? Was there an attempt to communicate via the self-destructive actions or through a suicide note left in view? Was the method of suicide relatively nonlethal, such as pills or wrist cutting? Truly suicidal children make their attempts when they are alone so they will not be found. They leave no hints or clues as to their plans, and suicide notes are not left so that they are not found prematurely. By asking the questions just presented, it is frequently possible to distinguish a gesture from a suicide attempt, although at times the distinction is difficult. Al-

though the treatment plan may vary depending upon whether self-destructive behavior was a suicide attempt or a suicide gesture, the first step in any plan should be to protect the child, if necessary through hospitalization.

Epidemiological studies reveal additional information. In Shaffer's study of all the suicidal deaths in children ages 10 to 14 in the years 1962 to 1968 in England and Wales, most of the children showed either antisocial or mixed antisocial and emotional disturbance before their death (Shaffer, 1974). Suicide was often precipitated by a disciplinary crisis. The most common situation before suicide was one in which the child knew that his parents were to be told of some type of antisocial behavior. Thus, children with conduct disorder who have a major depressive disorder and are facing a disciplinary crisis are the most vulnerable to their own self-destructiveness and must be closely observed and possibly protected.

Studies of adolescent suicide victims and suicide attempts reveal that female adolescents are most likely to make suicide gestures and infrequently commit suicide; male adolescents make few suicide gestures, but they are more likely to commit suicide than female adolescents. Although the number of prepubertal children who commit suicide successfully is very small, it is our impression that many children who attempt suicide fail because they lack the cognitive skill to design and execute a successful attempt. However, their wish to die is just as intense as in adolescence or adulthood. For example, an eight year old boy attempted to hang himself from the shower curtain rod using a wet towel. He failed because his grasp of geometry and physics was developmentally limited. Whenever it is difficult to make an accurate assessment of the degree of a child's self-destructiveness, it is wise to err on the side of caution and hospitalize a child protectively. If hospitalization is not called for, follow-up must be rapid and certain, for many of the situations that precipitate suicidal behavior are extremely fluid and require constant monitoring and frequent reevaluation. It should be remembered that even suicide gestures should be taken very seriously. In fact, the choice of this method to attract attention to oneself is already quite pathological. Unless the psychiatric diagnosis leading to a suicide gesture is addressed (most frequently affective disorder), it is likely that a second instance of suicidal behavior will occur. That the first (or last) behavior was gestural does not guarantee that the next one will not be a serious attempt. In fact most adolescents who kill themselves have "gestured" or said something about it before hand. In addition, true suicide gestures may produce death out of miscalculation. Therefore, it is good clinical judgment to treat any suicidal behavior as dangerous and act protectively even if it appears to be a gesture.

The rate of successful suicide increases substantially with puberty and continues to rise throughout adolescence. In addition, there appears to have been a steady increase during the last decade, the reasons for which are not understood. Recent studies of adult depressives have identified a biological marker of suicide—a low 5-hydroxy-indole acetic acid level in the cerebrospinal fluid. No similar work has been carried out in adolescents.

SUMMARY

The diagnosis and treatment of major psychiatric disorders of childhood and adolescence have evolved considerably during the past decade. This progress has been based on significant improvements in methods of assessment, systematic research on the validity of the adult types of diagnostic categories, considerable expansion of work on familial aggregation and other techniques bearing on the question of familial transmission of the disorders, marked progress in psychobiological strategies, the development of methods of behavioral modification and pediatric psychopharmacology, and the continued work on child psychiatric epidemiology. Besides the clear advantages for patient care, this increase in the body of scientific knowledge of child psychiatry is further defining the discipline; it brings it into the mainstream of modern adult psychiatry, and it makes it much more relevant to other medical specialties, especially pediatrics. This rapid evolution is continuing at an increased pace. Different treatment modalities now have more known indications, contraindications, and interactions. Thus at present the relevant treatment questions have become the following: For this disorder, in this age group, which treatment modalities are indicated? And for each one, which part of the clinical picture will be improved and in which sequence in relation to other treatments should it be administered, and for how long?

Acknowledgment

We gratefully thank Miss M. Quattrock for editorial assistance and for the typing of this manuscript.

JOAQUIM PUIG-ANTICH
HARRIS RABINOVICH

REFERENCES

American Psychiatric Association: Diagnostic and Statistical Manual of Mental Disorders. Ed. 3. Washington D.C., American Psychiatric Association, 1980.

Caplan, H. L.: Hysterical "conversion" symptoms in childhood. M. Phil. dissertation, University of London, 1970.

Chambers, W. J., Puig-Antich, J., Tabrizi, M. A., and Davies, M.: Psychotic symptoms in prepubertal major depressive disorder. Arch. Gen. Psychiatry, 39:921–927, 1982.

Freud, A.: The symptomatology of childhood: a preliminary attempt at classification. Psychoanal. Study Child., 25:19–41, 1970.

Gittelman-Klein, R., and Klein, D. F.: School phobia; diagnostic considerations in the light of imipramine effects. J. Nerv. Ment. Dis., 156:199–215, 1973.

Herjanic, B., Herjanic, M., Brown, F., and Wheatt, T.: Are children reliable reporters? J. Assoc. Child. Psychol., 3:41–48, 1975.

Maletzky, P. B.: D-amphetamine and delinquency: hyperkinesis persisting? Dis. Nerv. Syst., 35:543–547, 1974.

Patterson, G. R.: Reprogramming the families of aggressive boys. *In* Thoresen, C. E. (Editor): Behavior Modification in Education. Chicago, National Society for the Study of Education, 1973.

Puig-Antich, J., et al.: Plasma levels of imipramine (IMI) and desmethylimipramine (DMI) and clinical response in prepubertal major depressive disorder: a preliminary report. J. Am. Acad. Child Psychiatry, 18:616–627, 1979.

Puig-Antich, J.: Major deficiencies and conduct disorders in prepuberty. J. Am. Acad. Child Psychiatry, 21:118–128, 1982.

Quitkin, F., et al.: Atypical depressives: a preliminary report of antidepressant response, sleep patterns, and cortisol secretion. Presented at the meeting of the American Psychopathological Association, New York City, May 1981.

Rapoport, J. L., Buchsbaum, M. S., Weingartner, H., Zahn, J. P., Ludlow, C., and Mikkelsen, E. J.: Dextroamphetamine: its cognitive and behavioral effects in normal and hyperactive boys and normal men. Arch. Gen. Psychiatry, 37:933–943, 1980.

Rapoport, J., et al.: Childhood obsessive-compulsive disorder. Am. J. Psychiatry, 138:1545–1554, 1981.

Reid, J. B., and Hendricks, A. F.: A preliminary analysis of the effectiveness of direct, home interventions in treatment of predelinquent boys who steal. *In* Clark, I. N., and Hamerlynck, L. A. (Editors): Critical Issues in Research and Practice. Champaign, Illinois, Research Press, 1973.

Rivinus, T. M., Jamison, D. L., and Graham, P. J.: Childhood organic neurological disease presenting as psychiatric disorder. Arch. Dis. Child., 50:115–119, 1975.

Robins, L. N.: Sturdy childhood predictors of adult antisocial behavior: Replications from longitudinal studies. Psychol. Med., 8:611–622, 1978.

Rutter, M., and Graham, P.: The reliability and validity of the psychiatric assessment of the child: the interview with the child. Br. J. Psychiatry, 114:563–579, 1968.

Rutter, M., Tizard, J., and Whitmore, K.: Education, Health and Behavior. London, Longman, 1970.

Shaffer, D.: Suicide in childhood and early adolescence. J. Child. Psychol. Psychiatry, 15:275–291, 1974.

Weinberg, W. A., Rutman, J., Sullivan, L., Pencik, E. C., and Dietz, S. G.: Depression in children referred to an educational diagnostic center. J. Pediatr., 83:1065–1072, 1973.

Winokur, G.: Unipolar depression. Arch. Gen. Psychiatry, 36:47–53, 1979.

42

The Gifted Child

Giftedness—the mere mention of the word, with its otherworldly connotation, evokes confusion in the minds of most people. What is giftedness? How do we know whether someone is gifted? What does it mean to that person's life and the lives of those around him?

The answers to these questions are important to physicians caring for children, because increasingly parents have come to request developmental advice from health care providers. Although developmental and intellectual sophistication is not a disease or an illness, it represents a significant area of concern for parents, encompasses approximately 3 per cent of the population, and has associated with it a large number of misconceptions.

Unfortunately the literature yields few clear answers. Most of the research is old and plagued by significant methodological flaws. The field itself has been subject to enormous fluctuations of interest with periods of intense research alternating with periods of almost total dormancy. Controversies continue to rage, and there remains limited consensus among educators regarding such fundamental issues as the criteria for being considered gifted or the most appropriate academic programming for gifted children.

Our purpose in this chapter is to analyze the critical issues concerning developmentally advanced or academically superior children and to offer practical approaches to their problems for the physician who provides their care.

THE EVOLUTION OF THE CONCEPT OF GIFTEDNESS

The concept of giftedness has been undergoing continuous evolution for at least 4000 years. In each society giftedness has received its own unique identification, depending upon individualized cultural assumptions. The seminal figure in the saga of giftedness in America is Lewis Madison Terman (1877–1956), a psychologist who initially standardized intelligence testing and then systematically studied a cohort of individuals who performed in the superior range. For our purposes the development of the concept of giftedness can be looked at in three distinct stages—the pre-

Terman era, the Terman era, and the post-Terman era.

The Pre-Terman Era. The notion that certain individuals have superior intellectual abilities is an ancient one. The Chinese imperial civil service as early as 2200 B.C. had developed a series of elaborate examinations of intelligence and proficiency that determined promotion. The Palace School in the Ottoman Empire was another example of attempts to select those most intellectually able to rule. The empire was searched for those who appeared most academically and physically competent. These individuals were screened rigorously and, if they were determined worthy, were sent to the Palace School for a broad education and preparation for eventual leadership.

In this country, despite statements such as the one by Jefferson ("We must dream of an aristocracy of achievement arising out of a democracy of opportunity"), identification and programming for those academically abler have been intermittent. In 1868 W. T. Harris instituted flexible promotion as a way of providing for abler students in St. Louis. Rapid achievement classes sprung up sporadically in school systems across the country and were designed to hasten the progress of our most academically competent children. It was clear in this country in the beginning of the twentieth century that giftedness was defined by high academic achievement. Some research had taken place,[*] but there was limited understanding of gofted children and even more limited programming.

The Terman Era. Lewis Terman began his *Genetic Studies of Genius* in 1922, after completing his standardization of the Stanford-Binet test, an individual intelligence test. It was his goal to discover characteristics of gifted children, and he designed a study to gather data regarding their medical, psychological, and social attributes. In Terman's study, teachers were asked to nominate children they believed to be the brightest in their classes. This group was then screened with either the Stanford-Binet test or an equivalent intelligence measure. A score of 140 was selected as a cut-off point on this examination so that subjects

---

*The Twenty-third Yearbook of the National Society for the Study of Education, published in 1924, listed 453 papers in its annotated bibliography on giftedness (Passow, 1981).

would have "a degree of brightness that would rate with the top one percent of the school population" (Terman, 1925). Eventually 1568 California school children were entered into the study. Although children were selected from all socioeconomic strata, Terman's group was overrepresented with children of well educated, white, professional parents. Terman's research design used group averages to compare his gifted children with controls. He followed his children (known as "Termites") for over 30 years, until his death in 1956, and published five volumes describing the averaged details of their lives at various stages.

Terman's initial study and all the subsequent follow-up studies suggested that his cohort of intellectually gifted children were and remained physically, morally, and intellectually superior to control subjects. His work dealt a death blow to the myth that gifted children were "undersized, sickly, hollow-chested, stoop-shouldered, clumsy, nervously tense, and bespectacled" (Terman and Oden, 1959), and created a new concept of intellectual superiority correlating closely with superiority in all areas of their lives. Terman's children were larger and more precocious academically, earned substantially more graduate degrees, had predominantly professional and managerial careers, and by age 40 had written 67 books, 1400 professional or scientific articles, and 400 short stories and plays. This record was 10 to 30 times as large as that for his control subjects. In addition, evaluations of their mental health and general life contentment suggested that they were superior to the control group in these areas as well.

Terman's work revolutionized thinking about this population. It served as a model for other longitudinal studies of gifted children and set forth clearly the notion that intellectually superior children were a generally homogeneous group who displayed "superior" attributes in other areas as well as intelligence. Terman's work entrenched the intelligence test as the sole measure of giftedness and implied most strongly that such an instrument could predict success in other areas of life.

Major criticism has been leveled at Terman's research in recent years. The focus of the criticism has been primarily on Terman's methodology. Terman's sample was biased heavily in favor of socioeconomically advantaged children. His screening mechanism (having teachers nominate candidates) resulted in the selection of only well rounded high achievers. He controlled his research, using group averages and not matched controls, and socioeconomic factors were not taken into account. Accordingly, for example, when Terman tells us that his gifted group is physically larger than controls, he does so using a control

population that for the most part belongs to a different socioeconomic stratum from that of his experimental group. Later researchers have found that when socioeconomic factors are held constant, there is little difference in stature among gifted and nongifted children.

Regardless of these criticisms, however, Terman's work exploded destructive myths about gifted children and remains the most comprehensive longitudinal look at this population of children who score in the superior range on intelligence measures.

The Post-Terman Era. In the decades following Terman's initial publication and the spate of corroborating investigations, gradual changes began to occur in the conceptualization of giftedness. The notion that high academic achievement or superior performance on an intelligence test was adequate to define giftedness was replaced by the notion that mere academic potential should not be considered a sign of giftedness unless it yielded socially productive contributions (gifted is as gifted does). The demand for evidence of gifted behavior as a criterion for giftedness deflated the idea that giftedness was an all or none phenomenon and allowed consideration of a continuum of giftedness (Renzulli, 1978); that is, one could be gifted in certain areas and not others at certain times and not at others.

This development was an outgrowth of the studies of eminence, which looked retrospectively at individuals who had been successful in their fields of endeavor. It was discovered that high intelligence scores and great academic achievement were far from the sole predictors of success. Multiple reviews of the research dealing with the relationship between academic aptitude (standardized test scores, high school grades, college grades) and professional achievement found a minimal correlation between those indicators and a wide variety of measures of success in the adult (world peer recognition and independent assessment of an individual's contribution to his field). Hoyt (1965) concluded from a review of 46 studies in this area that "There is good reason to believe that academic achievement and other types of educational growth and development are relatively independent of each other."

In addition to the shift from potential to product as the criterion for giftedness, a liberalization of the areas of achievement acceptable to demonstrate giftedness also occurred. Areas other than solely academic ones were suggested, and talent (usually thought of as a more mechanical or technical isolated gift) became integrated into the notion of giftedness.

The intrusion of talent into gifted criteria and the beginning of the breakdown of IQ as the sole indicator of giftedness led to a widely touted definition of giftedness by the United States Office of Education. The Marland Report defined giftedness in 1972 in the following way:

Gifted and talented children are those identified by professionally qualified persons who, by virtue of outstanding abilities, are capable of high performance. These are children who require differentiated educational programs and/or services beyond those normally provided by the regular school program in order to realize their contribution to self and society.

Children capable of high performance include those with demonstrated achievement and/or potential ability in any of the following areas, singly or in combination: 1) general intellectual ability; 2) specific academic aptitude; 3) creative or productive thinking; 4) leadership ability; 5) visual and performing arts; 6) psychomotor ability.

Although this definition allows for a more liberal reinterpretation of what constitutes giftedness, it appears to have some major flaws. Renzulli (1978) suggests that motivational issues (task commitment) are critical aspects of productivity and are not accounted for in the Marland definition. He reviewed a later Terman study that analyzed the 150 most successful and 150 least successful men in his group. Success was determined by the extent to which a subject had made use of his superior academic ability and was individually defined for each field of endeavor. If an academic career was chosen, for example, recognition in the field was heavily weighted. In business, earned income was more important. Although the most and the least successful men did not differ in terms of IQ criteria, personality factors appeared to differentiate between them: "The four traits on which the most and least successful groups differed most widely were persistence in the accomplishment of ends, integration towards goals, self-confidence, and freedom from inferiority feelings. In the total picture, the greatest contrast between the two groups was in all around emotional and social adjustment and drive to achieve" (Renzulli, 1978).

In addition to task commitment, other investigators suggest that creativity is a critical aspect of giftedness. Guilford (1977) and Torrance (1974) have developed instruments that attempt to measure creativity. They assume that divergent thinking, that is, the tolerance of ambiguity, allows one to be creative. To date, most tests of creativity have not been correlated with subsequent development of creative work. The subjectiveness implicit in the concept of creativity has made this area difficult to measure. Regardless, most investigators believe that it is a vital part of the concept of giftedness.

Renzulli (1978) has attempted to integrate these diffuse criteria into a single conceptualization of giftedness. He states that giftedness "consists of an interaction among three basic clusters of human traits, these clusters being above average general ability, high levels of task commitment, and high levels of creativity. Gifted and talented children are those possessing or capable of developing this composite set of traits and applying them to any potentially valuable area of human performance. Children who manifest or are capable of developing an interaction among these three clusters require a wide variety of educational opportunities and services that are not ordinarily provided through regular instructional programming." (See Figure 42–1.)

In summary, thinking in regard to gifted children changed dramatically from an emphasis solely on academic prowess and intellectual potential to a demand for the demonstration of gifted behavior that may eventually yield social productivity. It does not appear that any single instrument or cluster of instruments is predictive of who will eventually make valuable contributions to society. Instead Passow (1981) suggests that "Identification of gifted and talented is related not only to systematic observation and interpretation of test and observation data, but to the creation of the right kinds of opportunities which facilitate self-identification—identification by performance and product which results in the manifestation of gifted and talented behaviors." It appears as though giftedness, similar to learning disabilities, is the product of individual human variation and the interaction of an individual's strengths and weaknesses with his environment.

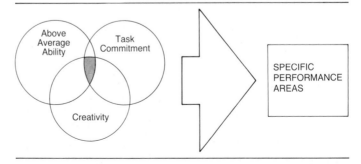

Figure 42–1. Graphic representation of the definition of giftedness.

ATTRIBUTES AND DEVELOPMENT IN GIFTED CHILDREN

As suggested previously, most current theoreticians believe that gifted children are a heterogeneous group who display infinite variation in their development. Much previous research, however, has attempted to characterize the development of one subgroup of gifted children, those who achieve in the superior range on IQ measures. Almost all this research suffers by comparison with current social science research in its lack of controlling for socioeconomic status and in its primarily retrospective analysis of early development. These studies appear to reflect primarily the development of upper class children who tend to do most well on IQ measures (30 to 40 per cent score in the gifted range, as opposed to a predicted 10 to 12 per cent). The following research should be viewed with that important caveat.

Physical Development. Terman's work, which used an IQ cut-off to incorporate the top 1 per cent of the population, suggested that as a group, the gifted children he studied tended to be more mature physiologically when compared with his control subjects. On 37 anthropometric measures, Terman's gifted group proved to be superior to children of comparable age in measures of height, weight, lung capacity, and muscular strength (Terman, 1925). Other investigators reported similar data.

In a large retrospective study of children with superior intelligence at age seven, Fisch et al. (1976) found that favorable parental, social, and educational backgrounds were maximally correlated with superior intelligence, whereas perinatal factors, such as Apgar scores, neurological or physical abnormalities, infections, anoxia, or trauma, were identical for high IQ, average IQ, and low IQ groups. They found that larger head size at one year of age was an early finding associated with superior intelligence, as was greater height and weight at four years.

All these studies suffer from a significant flaw that was previously mentioned in regard to Terman's work. Recent investigators have demonstrated that if socioeconomic factors are held constant, the differences that have been identified in these studies tend to disappear. It is clear, for example, that children of professional parents are physically superior to and score more highly on IQ tests of mental abilities than children of lower socioeconomic group parents (Tanner, 1978).

Terman also examined motor milestones. He found that his sample tended to be ambulatory one month earlier on the average than his control population. Freeman (1979), in the Gulbekenian study, another longitudinal examination of gifted children, attempted to hold social class constant and found that walking was not correlated with IQ.

Cognitive Development. Investigations of the cognitive development of gifted children have taken two pathways—the search for qualitative and quantitative differences in the thinking of these children.

Attempts to find qualitative differences in the thinking of gifted children, that is, different approaches to problem solving, have yielded little. Examination of the rate of acquisition of skills, however, has been far more fruitful, and it is the rate of knowledge acquisition that is primarily measured when one is attempting to determine giftedness in a young population.

Early language development and early reading ability have both been assessed as possible markers of later superior cognitive abilities. Terman (1925) and Freeman (1979) found that their gifted children spoke approximately three and one-half months earlier than their control populations, and they identified a clear relationship between language complexity and later high IQ. Early reading was also more frequent in Terman's group, and nearly half of his group of gifted children had learned to read before entering into first grade and 20 per cent had done so before age five. Robinson (1981) reported that gifted preschoolers in his study achieved Piagetian conservation and the understanding of gender, an aspect of social cognition that has been shown to be related to general intellectual development, many months before children of average ability.

Personality Development. In the area of personality development, a discrepancy exists between the highly gifted (IQ greater than 180, or development two times the chronological age) and the more average gifted (IQ 130 to 150). The highly gifted (geniuses) do not fare as well as those with more moderate "gifts." Hollingsworth's study (1942) of the highly gifted found them to display underachievement, alienation, and often suicide. She reported that they had "great difficulty in finding playmates who were congenial, both in size and mental ability. Thus, they are thrown back upon themselves to work out forms of solitary intellectual play" (Robinson, 1981, p. 75).

The research on the personality of more average gifted children is much more positive. They appear to rate significantly higher than peers on measures of earnestness, trustworthiness, honesty, and emotional stability, as well as the capacity for objective self-appraisal. Their moral judgment appears to be more mature in regard to distinctions between the intention and outcome of an action and right and wrong behavior.

Other investigators attempted to assess the popularity of gifted children. Gallagher (1958) summarized this work and found that the academically most able were generally well liked and often among the most popular children in the class. A clear positive correlation between group acceptance and academic performance was found.

As regards friends, most studies suggest that gifted children select friends more frequently commensurate with their mental age; accordingly, in one study, 25 per cent of gifted children expressed preferences for older playmates versus 9 per cent of average children.

The classic stereotype regarding the personality of gifted children is that they are social isolates. With the exception of the extremely gifted, who have been found to have an increased number of social and psychological problems, this does not appear to be so in the majority of bright children. As one might expect, they run the gamut of personality styles, from outgoing to retiring, from self-confident and assertive to feeling inferior.

Longitudinal Studies. The monumental study of Terman, which has been cited frequently throughout this chapter, continues to represent the major repository of information about the lives of certain individuals with superior IQ's. It suggests, as has been previously discussed, that superior intelligence, defined during childhood by performance on a standard intelligence test, is associated with a high degree of personal and social adjustment, which is equal to or better than that of the population at large.

In addition to physical and occupational data, Terman looked at mental health, productivity, and life satisfaction. In all areas his gifted population was equal to or superior to the general population. He had effectively destroyed the notion that gifted children were meek, frail, and inadequate in areas other than intellectual endeavor. Whether he was merely describing the characteristics of our upper and upper middle classes continues to be disputed.

All the longitudinal studies mentioned were performed before the liberalization of the concept of giftedness and the shift to an emphasis on potential for productiveness as a measure of giftedness, as opposed to merely high academic functioning. Looked at with today's eyes, these studies are not particularly useful in attempting to unravel the complex hereditary and environmental issues that continue to plague this field. They are useful, however, in dispelling the myths of giftedness suggesting that children who are identified as being gifted will inevitably lead lives of eccentricity and isolation. Regardless of the reasons one may perform in a superior manner on a specific intelligence test, enormous variability and little predictability are possible about the life one will live.

Robinson (1979) summarized the research on the development of gifted children as follows:

> While some studies have reported that the average levels of personal and social adjustment, physical health, and the like are slightly higher for gifted children . . . the mean differences favoring the gifted group tend to be small and sometimes disappear when the comparison group is appropriately matched for variables such as social class. Much more striking than the mean difference is the variability within the gifted group. In fact, intellectually advanced children are about as heterogeneous as any other population on measures other than those directly related to the instruments used to identify them. There is . . . no such thing as a "typical gifted child."

THE EDUCATION OF GIFTED CHILDREN

Gifted education in America has been more susceptible to shifts in national sentiment than any other area in education. Although there was some early interest in gifted education following the publication of Terman's studies, serious programming for abler children was begun primarily in the years immediately following the launching of Sputnik. The technological vulnerability perceived following the launching of the Russian satellite triggered an enormous outpouring of public and private funds to assist in the development of enrichment and accelerated programs for intellectually talented students, particularly in the sciences.

In the 1960's, however, the focus on gifted children disappeared almost as dramatically as it had occurred. The civil rights movement brought with it a sense of egalitarianism and a desire to help the socially disadvantaged. Technology was viewed with increasing derision as an instrument of war and was disdained by many individuals. Programs for gifted children, with their technological flavor and primarily middle class student population, were not, accordingly, an area of interest or research for young teachers or academicians.

The 1970's and early 1980's have witnessed a resurgence of interest in this population. Broader definitions of giftedness have allowed the inclusion of children into gifted programs who might not have been able to meet the IQ criteria. School systems have been able to reconcile a concern for the socially disadvantaged child with an interest in gifted children. In addition, most states now have laws that mandate the identification and evaluation of gifted children, and many states have statutes mandating programming.

Rationale for Gifted Education. Why should we have a different or enriched educational experience for some children? The gifted education movement has grappled for many years with this most fundamental question. Does gifted education foster elitism in our public schools, whose primary goal should theoretically be the development of an egalitarian democratic society? If we are to believe Terman's work, his cohort of children, without the aid of special programming, led happy, productive lives. What, then, is the rationale behind programming for gifted children?

Explanations in support of gifted education use as their primary assumption the hypothesis that intellectual potential unchallenged will not blossom maximally. This assumption is based on a variety of diverse studies. In one study, intellectually superior students who were perceived by their teachers to have average intelligence declined in performance, as compared with students with similar capabilities who were recognized by their teachers as being superior. Other studies have found that children who are identified as gifted and placed in special programs achieve at a higher level. For example, a group of investigators reported that when a cohort of gifted children was placed in a special class and a matched cohort was placed in a regular classroom, the gifted class moved at a rate twice that of the regular class. It is the argument that lack of nurturance can blunt potential that is the underlying basis for gifted programming.

Current Selection Criteria. Once we have decided on the appropriateness of gifted programming, another fundamental question emerges. How do we identify who is gifted and will be admitted to our program?

As previously suggested, the definition of giftedness is in flux at this time, and this is reflected in the criteria used for admission into most gifted programs. Despite the stated recognition of the fact that multiple criteria are to be used for decisions about who is or is not gifted, most school systems at the present time use performance on intelligence or achievement tests as the major criterion. More creative approaches use some combination of parent, peer, and teacher nomination and an evaluation of creativity and perseverance instead of or in addition to standardized test data. Unfortunately, however, the latter methods are extremely subjective.

Renzulli (1981), in keeping with his notion that giftedness is not an absolute concept, implores us to think in terms of gifted behavior that children may demonstrate at one time and not another, depending upon a variety of factors. He suggests that giftedness is not an all or none phenomenon, and that many more children display gifted behavior (produce products that might be construed as gifted, or act in a way that displays particular abilities) than can be identified on intelligence measures. To accommodate this framework, he has developed the revolving door identification method in which children can earn their way into an enrichment class by a number of pathways. In this classroom they focus only on particular areas in which they have displayed gifted behavior. If they continue to display gifted behavior, such as creativity or perseverance, they may earn the right to continue in the resource classroom and enjoy the freedom implicit in such a room. If they have displayed little task perseverance, creativity, or sophisticated understanding in the area that was presumed to be their gifted one, they are returned to their classroom once their project is completed. Simultaneously other children who have displayed ifted behavior in the mainstream may now have the opportunity to pursue their areas of interest in the enrichment class. This model provides for the continuous recognition of talents in children and allows more children to benefit from this type of educational experience. Such a method of selection also demands continuous performance from children and does not reward those who are solely gifted test takers.

Programming for Gifted Children. The issue of how to program for gifted children is another controversial area. Educators appear to be divided into two distinct groups, one group favoring enrichment (children remain with other children of their chronological age but receive special programming) and the other group favoring acceleration (placing younger children in programs that were intended for older children).

Acceleration was once the more common practice, but because of concerns that accelerated children would not develop appropriately in psychological and social areas when removed from same age classmates, this method has been declining in favor among educational practitioners. Recent research, however, suggests that the expected social problems for these intellectually advanced but chronologically younger children may have been overestimated, that acceleration might be a more useful strategy than is presently assumed. In the area of mathematics, for example, Stanley's study of mathematically precocious youths at Johns Hopkins University has demonstrated significant advantages in accelerating students into college and graduate study without significant negative consequences (Stanley et al., 1977). In addition, acceleration continues to be the most common practice among children who are presumed to be highly gifted.

The enrichment model is the predominant one used in American gifted education at this time. Children identified as being gifted are placed in special programs either part of the day or all day. In these resource rooms the focus is often on expanding upon factual material that is presented in other settings, with the teacher acting as a catalyst or resource person. The use merely of additional work or busy work runs counter to the concept of the enrichment model.

Another controversy rages. Should gifted children be kept in one self-contained classroom, or should they primarily be "mainstreamed" and receive only occasional resource room enrichment? Implicit in this question are corollary questions. Do gifted children in the mainstream make adequate progress and do they serve as valuable models for chronological peers, or is their development in some way blunted and their relationship with peers complicated by elitism and stigmatism? The literature attempting to answer these complicated questions is preliminary and poorly controlled. At the present time, however, it appears to suggest that removal of advanced students from the mainstream might have beneficial effects upon their own development and does not appear to alter significantly the behavior of their mainstreamed peers, who no longer have this group as intellectual models.

What we can say most clearly at the present time is that multiple variables must be considered in the decision to provide special education for a child who is thought to be gifted, regardless of the type of program (acceleration or enrichment, whole day versus part-time). The decision should be made on an individual basis, taking into account the child's chronological age, physical status, motor coordination, emotional maturity, family constellation, personality style, and areas and degree of giftedness.

SPECIAL PROBLEMS IN GIFTED EDUCATION

Two special subsets of gifted children deserve particular attention—the underachieving gifted child and the minority gifted child.

Underachieving Gifted. The concept of "underachieving gifted" presents a paradox if we are to accept the definition of giftedness proposed earlier, which suggests that being gifted requires the demonstration of gifted behavior and not merely the displaying of potential on intelligence tests. Despite this seeming contradiction, most investigators believe that there exists a substantial population of children with great ability who for a variety of reasons are not able to demonstrate their capabilities in clearly observable ways (production of superior academic work, demonstration of artistic talents, demonstration of leadership potential). Implicit in the assumption of gifted underachievement is the notion that if we change certain conditions, achievement commensurate with ability will occur.

Studies of gifted underachievement began in the late 1940's, with Terman's work analyzing the differences between successful gifted children and those who were less successful. This work reached its peak in the 1960's, when many investigators focused their energies on attempts to improve the educational status of disadvantaged and handicapped children. This body of literature attempted to define the characteristics of the underachieving gifted child. What seems to be most clear, at least in middle class underachieving gifted children, is a pervasive self-perception of inadequacy in these children. They believe that they are not capable of performing adequately academically, and this often correlates with their generally low self-esteem. This low self-concept is reflected in two classic behavioral styles, aggression and withdrawal. Such styles often perpetuate poor academic performance, which further aggravates the feelings of inadequacy, which in turn perpetuates such problem behavior.

The reasons for not realizing one's potential are as numerous as there are individuals. They would appear to fall into some broad categories, however—developmental, psychological, and social. Some underachievers appear to have demonstrated developmental lags or attentional difficulties early in their academic careers. Despite the subsequent maturation in regard to these problems, they are left with the perception that they are less able than others to perform, and this residual feeling acts as a damper on their achievement, despite testing evidence to the contrary. In other children, complex family dynamics have left them without the self-esteem necessary to achieve maximally. Some children emerge from environments in which academic success is not a high priority, and there is limited reinforcement for academic effort. Finally, many children would appear to have some combination of these factors as an explanation for their academic difficulties despite adequate potential.

A variety of innovative projects have been designed to help underachieving gifted children. The Cupertino experience described by Whitmore (1980) is of interest. Children discovered to have "gifted" potential without displaying gifted behavior were identified. Many were extremely shy and nonassertive and others acted out aggressively. They were often being considered for placement in classrooms for emotionally disturbed children. Instead they were placed in a special extended learning program whose goals were to decrease self-degrading comparisons with high achievers, to develop acceptance of self through acceptance of others with similar problems, to enjoy intellectual stimulation with a curriculum centered on strengths, to achieve genuine success, and to develop more adequate social skills. The Cupertino project was extremely successful through the use of a student centered individualized curriculum, a close pupil-teacher relationship, and the active involvement of parents.

It is important to note that poor motivation and low self-esteem can counteract ability and create a state of chronic underachievement. Students in this category often develop inappropriate behavioral styles. It is important to consider the cognitive abilities and talents of all children who present behavior problems before special programming takes place.

Minority Gifted. Traditional criteria for identifying gifted children using standardized intelligence tests effectively excluded a larger percentage of minority children then would be predicted on the basis of a normal distribution curve. The reasons for this IQ discrepancy are complex and controversial and are discussed by other authors in this volume. Regardless of the origins of this discrepancy, whether they be test artifact, environmental, or biological, minority children perform on the average 1 standard deviation below white children on most standardized intelligence measures. Because these tests reflect knowledge that is valued by the dominant culture, these tests are somewhat predictive of subsequent school performance. As we have mentioned, however, they are not predictive of subsequent societal contributions.

With the broadening of criteria used for eligibility into gifted programming to include isolated talents, creativity, and leadership, an increased number of minority children have become eligible. Standardized testing has been used with local instead of national norms and has been found to be helpful in selecting the children within a given community who are clearly most academically able. With the recognition that identification should not be absolute but relative, depending upon community standards, gifted programming now can be developed in schools where, according to previous criteria, few children would have been eligible.

Many authors have attempted to define the characteristics of the "culturally different" gifted child. The suggestion has been made that these children are more frequently visual than auditory learners, that they are more problem centered than abstract centered, that they require somewhat more structure than do dominant culture gifted children, and that multimedia teaching is most successful in this group of children. Gallagher (1975) notes quite correctly, however, that extreme caution must be used in generalizing about such a diverse group as the "culturally different." It would appear that the sine qua non of all gifted education should apply to this group as well: identify individual strengths and teach to them.

As with other children, strong liaison with the parents of this group is vital to the success of gifted programming. Perhaps more important for this group than for others, community individuals who might offer an occupational role model for the child should be brought to the classroom to offer alternative possibilities about careers and to expand the horizons of a child who might not have considered such lofty possibilities for himself. The curriculum, of course, should be tailored to the student population, with particular emphasis on cultural heroes, for the same analytic thinking can be fostered on culturally relevant material as on more obtuse, less related material.

The child from a nondominant culture clearly has a unique set of educational problems. Economic disadvantage, lack of successful community role models, bilingualism, and a seeming lack of educational relevance have all acted to depress the academic performance of these children. Expanding gifted criteria to allow for the incorporation of the unique attributes of this group of children can have positive effects on them as individuals and on the nation as a whole. This should not be interpreted as suggesting a lowering of standards for this group of children. As has been suggested, however, motivation is vital for continued academic progress, and some alteration of traditional curricula is often critical to kindle or rekindle the desire for academic success in culturally different children.

PEDIATRIC APPROACH TO THE DEVELOPMENTALLY ADVANCED CHILD

Although giftedness clearly cannot be construed as being a medical problem, pediatricians in their newly expanded role concerned with quality of life issues may have a part to play in this arena. The family of a developmentally advanced child, especially before school age, uses the pediatrician as their major child health and development professional. Accordingly the pediatrician may be called upon to help the family in sorting out questions concerning giftedness or potential giftedness.

The pediatric role in the area of gifted and talented children resembles that for other uncommon children: (1) Rule out illness as a cause of the exceptionality (with the exception of the few conditions listed in the next section, this is rarely an issue in gifted children). (2) Help the child and the family in procuring the most appropriate diagnostic evaluation and educational program, including appropriate referrals when necessary. (3) Support the family and the child in coping with the psychosocial problems that are often part of being different or living with one who is different.

If parents suggest to the pediatrician that they believe their child is gifted and are uncertain how to proceed, the pediatrician should discuss their reasons for believing their child is gifted. What is the child doing that led them to this conclusion? This necessitates some understanding of the margins of normal development. If the child's development appears to fall within that framework and the pediatrician believes that the child is not exceptional, this opinion should be shared with the family. If parents still desire a more complete evaluation, or if the physician believes that there is justification in their claim,* the next step is the initiation of the evaluation process. At present, evaluation of potentially gifted children is the responsibility of the local public school system, as determined by individual state law.

Evaluation of younger children will most likely

---

*Parents were found to be accurate predictors of giftedness in the Seattle Project (Robinson, 1978).

Table 42–1. CLUES TO GIFTEDNESS

Toddler and Preschooler	School Aged Child
Early onset of language	Intellectual curiosity (enjoys learning)
Vocabulary and syntax greater than age expectation	Large fund of general information
Long attention span	Interested in solving problems, often finding unique solutions
Self-taught reading skills	Nonconformity
Large fund of general knowledge	Extreme persistence at tasks of interest
Asks frequent questions	Fluency in verbal and nonverbal communication
Demonstrates unusual talent (musical, artistic)	Good retention of learned material
	Able to generalize learned material to other areas
	Demonstrates leadership
	Demonstrates unusual talents

consist of standardized intelligence tests that attempt to elucidate academic potential (see Chapter 46 on intelligence testing). In older children, in addition to intellectual testing, achievement tests may be administered and an evaluation of creativity may be performed. If a child has written poems or stories, has done creative work in other fields, or has produced any creative products, these might be considered part of the process to determine giftedness. As previously stated, there continues to be an enormous controversy over what constitutes giftedness, and in certain communities this may lead to confusion and contradictory statements among professionals and parents involved with the child. If controversy develops within a community, the National Association for Gifted Children or the Association for the Gifted might offer clarification and support.

Once the testing has been completed and the child has been categorized as being gifted, the pediatrician should have the child and family visit the office to discuss the implications of this pronouncement. Demystification of the concept is vital. Parents and children have enormous misconceptions about what it means to be gifted. Removal of the otherworldly connotation of the word "gifted" should be attempted, and the child should be told that he is like other children, except in having special abilities in certain areas that may allow him to achieve at a different rate from his peers and allow him to see things in a way that might be slightly different. The child's and family's myth of the development of gifted children should be explored and discussed. The presumed burdens of giftedness should also be aired. Counseling at this point should cover three specific areas—home, school, and community.

Home. Advice about dealing with gifted children at home is often sought from the pediatrician because few other professionals are in a position to discuss home life intimately with the family.

First and foremost, it is important to support the parents in their child rearing practices. Either prior to or as a result of the labeling of their child, parents often develop difficulties in relating to their child. They are fearful that their child is brighter than they are and that they are not adequate to provide the needed stimulation to allow

their child's "gift" to blossom. They appear to lose balance as parents. The pediatrician should reassure them that for their child to have been identified as having special abilities, they must have certainly done something right to the present, and it is important that they be encouraged about their own parenting skills.

Parents should be told to treat their child as they do his other siblings. It is important that they not focus on the label "gifted" in the presence of the child or in their dealings with him or his siblings. They should be comfortable realizing that there might in fact be areas in which their child's abilities surpass their own and that the child might often ask questions that they cannot answer. The parents need to develop the self-confidence to say "I don't know, but let's find out together" instead of the more defensive "Because I said so."

As regards stimulation, parents should be instructed not to push their children. Stimulation should be child centered, and the child's lead as regards the type of material and depth of the exposure should be followed. Materials should be flexible and allow for the generation of creative products. Museums, books, and other educational material should be provided in the child's expressed area of interest, and not all areas of endeavor should move be foisted upon the child. Because many children who have great intellectual ability are extremely persistent, they will let parents know what they want if the parents are able to listen.

The family life of gifted children often becomes complicated. Parents often forget natural instincts and treat their children differently, a tendency analogous to that in the "vulnerable child syndrome" described by Green and Solnit in 1964 in children with chronic disease. Siblings often feel inferior because of the special attention focused on a gifted brother or sister. This may be magnified if they are close in age. If the gifted child is surpassing a chronologically older sibling in academic work, it might be particularly painful to the older child, and parents should attempt to find ways to continue to assert that older brother or sister still has the special privileges and responsibilities of age (staying up later, different chores). It is important to allow and cultivate each sibling's uniqueness and specialness. Special time with parents and fostering of musical talents or athletic skills of siblings may help them to feel more adequate and more involved in family life. As previously mentioned, children with increased intellectual abilities often play with children who are on their developmental level and, therefore, older. If the sibling of a gifted child is slightly older, his friends might well become friends of the gifted child and this might create further tension, which should clearly be addressed by the family.

Parents might wish to read one or more of the periodicals and reference texts that have been developed for parents and teachers of gifted children (e.g., *G/C/T,* or *Everyday Enrichment* by Herbert Kanigher) to allow them further insight into dealing with their child.

School. The pediatrician has a more limited role as the gifted child enters into school. (See also Chapter 15.) Parents receive more of their advice from teachers and parents of other gifted children at this time. Certain issues merit discussion, however.

The argument over acceleration versus enrichment continues to rage, as has been discussed in the education section of this chapter. The pendulum at this writing has swung toward enrichment classes, but major investigators in the field continue to support acceleration. This of course must be individualized, depending upon the child's physical size, chronological age, personality, family constellation, and degree of advancement. One is more inclined to urge the acceleration of a large, emotionally mature child than a shy, retiring, small child who is more likely to be overwhelmed by a new set of very different peers and circumstances.

Educators of gifted children, as all professionals, are variable in their enthusiasm, creativity, humor, and intelligence. The emphasis in gifted education should not be on busy work or a greater quantity of work, but instead should focus on projects that expand the child's perspective in a given area. A creative teacher can act as a catalyst and resource person for the child, helping the child to generate and prove hypotheses. If specific gifted programming does not exist, the classroom teacher can function in this capacity. It is hard for parents to judge the competence of their child's teacher or program. Excessive homework, however, is not to be expected or tolerated. Such work will only further the intellectual aspect of the child's development at the expense of appropriate play and social interactive time.

Boredom with school work, school avoidance behavior such as complaining of headaches or abdominal pains that occur only on school mornings, a developing sense of elitism, or the development of peer animosity should make a parent concerned about the quality of his child's program.

Close parent-teacher liaison is vital in working with gifted children, and this should be fostered if it has not developed.

Community. Parents should be urged to join or begin a local chapter of the National Association for Gifted Children, or other organization lobbying for gifted children. In such an organization they can receive suggestions and support from others who are experiencing the excitement and frustration of life with a gifted child. Parent organizations can apply pressure to the school board to create and develop more and better programs for gifted children. A parent group can generate new ideas for extracurricular activities (for example, a "fellowship" program after school for gifted children in areas of interest, working at local companies using computers, working after school at the zoo with the veterinarian). Noncompetitive interest groups can be developed and staffed. Another possibility is to bring local experts "in residence" to the school. A doctor might visit one morning a week for a month, a lawyer similarly, and children with specific interest in these areas could develop projects in conjunction with these local experts.

The physician might want to be involved with these parents' groups to provide guidance and to learn more about these children and the problems they encounter.

MEDICAL PROBLEMS ASSOCIATED WITH SUPERIOR INTELLIGENCE

Although increased intellectual ability, unlike decreased intellectual ability, is rarely associated with disease, Bakwin and Bakwin (1972) report five conditions that have been correlated with increased intelligence on standardized measures: infantile autism, in which certain subgroups supposedly have very specialized forms of intelligence; idiopathic precocious puberty; adrenogenital syndrome; retinoblastoma; and anoxic episodes at birth in certain subgroups of children.

Whether the superior intelligence of these children is biologically based or whether it is the result of increased stimulation through increased adult contact, either in the hospital or at home, is unclear.

SUMMARY

Increasingly, issues surrounding cognitive aspects of development, particularly in the preschool years, have fallen under the purview of the physician caring for children. Although the medical role with developmentally sophisticated children is less clear than with developmentally delayed children, some understanding of the gifted child is important.

As has been proposed, major theoretical shifts have occurred in conceptualizations about gifted children. Terman's work successfully deflated the myth that gifted and talented children were physically weak and frail, had poor eyesight, and were doomed to lead sickly lives culminating in nervous exhaustion. This research denigrated the "early ripe, early rot" phenomenon and found that in such children intellectual skills seemed to predict a plethora of abilities outside the intellectual realm.

Terman substituted a new myth of superchildren for the old neurasthenic myth. Only recently in this era of individual differences have we begun to look at giftedness as the interaction of intelligence, creativity, and persistence revealed through a continuum of behavior patterns in a very heterogeneous group of children.

Terman's work dominated educational programming and planning for gifted children as well. IQ testing has been and remains the major criterion for entry into programs for the gifted. Decisions regarding enrichment versus acceleration must be individualized at this time because current knowledge does not support any one position.

The physician's role in this area is to support the families of the children in the inevitable conflicts that they undergo and to child-center the child's academic stimulation. As an independent, knowledgeable, and sensitive individual outside the educational system with a long standing relationship with the family, the physician is in an ideal position to positively influence the academic development of the child, the stability of the family, and the conceptualization of gifted children within the community.

Organizations for Gifted Children

National Association for Gifted Children (NAGC)
217 Gregory Drive
Hot Springs, Arkansas 71901

The Association for the Gifted (TAG)
c/o Council for Exceptional Children
1920 Association Drive
Reston, Virginia 22091

Publications Concerning Gifted Children

Gifted Child Quarterly
G/C/T
Roeper Review
Journal of the Education of the Gifted
Journal of Creative Behavior

NEIL L. SCHECHTER

REFERENCES

Bakwin, H., and Bakwin, R.: Behavior Disorders in Children. Philadelphia, W. B. Saunders Company, 1972.

Fisch, R. O., et al.: Children with superior intelligence at seven years of age: a prospective study of the influence of perinatal, medical, and socioeconomic factors. Am. J. Dis. Child., 130:481–82, 1976.

Freeman, J.: Gifted Children. Baltimore, University Park Press, 1979.

Gallagher, J. J.: Social status of children related to intelligence, propinquity, and social perception. Elemen. School J., 58:225–31, 1958.

Gallagher, J. J.: Teaching the Gifted Child. Boston, Allyn and Bacon, 1975.

Green, M., and Solnit, A. J.: Reactions to the threatened loss of a child: a vulnerable child syndrome. Pediatrics, 34:58–66, 1964.

Guilford, J. P.: SPS Catalog. Orange, California, Sheridan Psychological Services, 1977.

Hildreth, G. H.: Introduction to the Gifted. New York, McGraw-Hill Book Company, 1966.

Hollingsworth, L. S.: Children Above 180 IQ. Yonkers, World Book Company, 1942.

Hoyt, D. P.: The Relationship Between College Grades and Adult Achievement: A Review of the Literature. Research Report No. 7. Iowa City, American College Testing Program, 1965.

Marland, S. P.: Education of the Gifted and Talented. Report to the Congress of the United States by the U.S. Commissioner of Education. Washington, D.C., U.S. Government Printing Office, 1972, Vol. 1.

Newland, T. E.: The Gifted in Socioeducational Perspective. Englewood Cliffs, New Jersey, Prentice-Hall, Inc., 1976.

Passow, A. H.: The nature of giftedness and talent. Gifted Child Quar. 25:5–10, 1981.

Renzulli, J. S.: What makes giftedness: re-examining a definition. Phi Kappa Deltan, 60:180–4, 1978.

Renzulli, J. S., Reis, S. M., and Smith, L. H.: The Revolving Door Identification Method. Mansfield, Connecticut, Creative Learning Press, 1981.

Robinson, H. B.: The uncommonly bright child. In Lewis, M. B., and Rosenblum, L. A. (Editors): The Uncommon Child. New York, Plenum Press, 1981.

Robinson, H. B., Roedell, W. C., and Jackson, N. E.: Early identification and intervention. In Passow, A. H. (Editor): The Gifted and the Talented: Their Education and Development. Seventy-Eighth Yearbook of the National Society for the Study of Education. Chicago, University of Chicago Press, 1979.

Stanley, J. C., George, W. L., and Solano, C. H. (Editors): The Gifted and Creative: A Fifty Year Perspective. Baltimore, Johns Hopkins University Press, 1977.

Tanner, J. M.: Education and Physical Growth: Implications of the Study of Children's Growth for Educational Theory and Practice. London, Hodder and Stoughton, 1978.

Terman, L. M.: Genetic Studies of Genius. Vol. 1: Mental and Physical Traits of a Thousand Gifted Children. Stanford, Stanford University Press, 1925.

Terman, L. M., and Oden, M. H.: Genetic Studies of Genius. Vol.V: The Gifted Group at Mid-Life. Stanford, Stanford University Press, 1959.

Torrance, E. P.: Torrance Tests of Creative Thinking. New York, Personnel Press, 1974.

Whitmore, J. R.: Giftedness, Conflict, and Underachievement. Boston, Allyn and Bacon, 1980.

VI

Assessing and Describing Variation

The chapters contained in this part explore the range of processes of assessment in developmental and behavioral pediatrics. In the first three chapters specific techniques commonly employed in health care and other settings are reviewed. The following four chapters survey specific target parameters of assessment (e.g., intelligence, personality, and educational skill). Then follows a series of discussions of the diagnostic role of various collaborating professions. The concluding chapter proposes one model for integrating assessment data.

43

Interviewing

Behavioral pediatrics emphasizes the interaction between psychological and physical aspects of health and illness. Since psychological issues involve emotions that are often difficult both to perceive and to express, special skill is required to reveal this aspect of experience. Components of that skill include careful observation, sensitivity to one's own behavior (including how one is perceived by others), and a knowledge of the common psychological factors that affect communication. The tool of assessment and therapy in the behavioral field is the interview, a conversation between doctor and patient aimed at obtaining or imparting information. Interviewing takes place in many different contexts of pediatric practice. The history taking session is the most obvious interviewing occasion. Explaining the diagnosis or treatment, obtaining follow-up information, and counseling are also examples of situations requiring interviewing skills.

The content of these interviews—the questions that should be asked and the information that is imparted—varies depending upon the nature of the medical or psychosocial problem. The content of the interview is reviewed in this and other pediatric texts, and will not be discussed here (Friedman and Hoekelman, 1980; Green and Haggerty, 1977; Hoekelman et al., 1978). This chapter deals with the communication process, the art of asking questions so that the most useful information is obtained and imparting information in a manner that is most helpful to the patient.

In the first section techniques are described that enhance the possibility of effective communication. These techniques apply primarily to the beginning of the interview and the setting in which it is conducted. The next section deals with techniques that affect the interaction of a conversation. These skills require one to detect and manipulate the psychological factors that affect communication. The final section provides suggestions for handling the most commonly encountered difficult interviews. One caveat is that many of these comments are particularly aimed at the pediatric trainee in a university hospital setting. The private

practitioner will recognize when suggestions are not appropriate for his setting or level of skill.

TECHNIQUES THAT ENHANCE INTERVIEWS

BREAKING THE ICE

The initial few moments before the interview begins set the tone for the rest of the session. First impressions are formed on the basis of this encounter. In doctor-patient interactions in which there are often educational and socioeconomic differences between the parties, these first impressions may crucially affect the interview. This introduction should serve to make the patient feel important and reveal the doctor's human qualities. Both factors help bridge the gap between doctor and patient and may increase the latter's willingness to divulge emotion laden information. These opening few minutes are handled best by the physician who takes a genuine interest in the patient and his family. Thus the skilled interviewer will introduce himself, address the patient by name, and notice the patient's appearance and the subtle (or obvious) signs of apprehension or irritation that may accompany this initial encounter. He should be curious about who this person is and the events that immediately preceded the interview, such as a long trip to the hospital, a long wait before seeing the doctor, or the possible frustrations of parking or finding the interview location. Children should be noticed and acknowledged individually, and the father's presence should be acknowledged and encouraged.

The doctor may also reveal enough about himself to help put the patient at ease. It is part of the art of communication to know just how much personal information will help the patient feel comfortable without revealing so much that the appropriate differences in roles are blurred. It is usually safe to comment on what you know about where the patient lives and the colorful and interesting clothes that he is wearing and to share frustrations such as parking and having to wait or enduring cumbersome procedures such as obtaining clinic cards or completing admission information.

Following these introductions the interview begins. The physician should ask the initial question or make a statement that indicates this transition.

THE INTERVIEW SETTING

Clinic rooms and hospital rooms are designed to facilitate physical care, often with little regard to the needs of good interviewing. Space and the arrangement of furniture within it can powerfully affect the interview, especially when psychological information is sought or emotionally laden material is being explained. A few moments should always be taken to obtain chairs and rearrange them so that direct eye contact is possible at an optimal conversational distance. The patient and doctor should either both be seated or both be standing. It is emotionally difficult for a sitting patient to talk to a standing doctor, but this combination often occurs in health care settings. If chairs are placed beside each other in a clinic room, they should be moved so that direct eye contact is made. These arrangements are more difficult on a hospital ward, but it is usually possible for the physician to put a chair at the bedside and lower the bed until patient and doctor are at the same eye level. If a desk or table is provided, it should not be between the doctor and the patient. Such a physical barrier between the two participants is often perceived by the patient as increasing the emotional distance as well as promoting the feeling that the doctor is in a position of authority.

NOTE TAKING

Particularly when the exchange contains emotionally laden information, note taking can produce problems that inhibit the interview. The patient may be sensitive to what the doctor considers important enough to note. Such behavior as straining to see what is written or stopping in midsentence when the doctor begins writing indicates this sensitivity, and in this situation, note taking should be abandoned. Note taking should not prevent good eye contact. If notes are brief, eye contact need not be significantly disrupted. If jotting down a quotation or a particularly long notation is necessary, it helps to explain the need for interruption to the patient, who is then likely to tolerate the disruption more easily. For example, the physician might say, "Excuse me a moment while I write this down so I won't forget it later."

DISTRACTIONS

The most common distraction in the pediatric setting is the behavior of young children when the parent is being interviewed. Children are normally irritable and demanding of the parent's attention when they are very young or ill. The developmentally immature child is also typically distracting in the interview setting. Providing age appropriate toys that engage the child should always be tried first. If there is a sink in the clinic room used for the interview, the child may be engrossed by water play that can usually be kept within bounds by supplying paper cups and other receptacles and specifying rules for their use. If this fails, it may be necessary to remove the irritable child to the care of another person for the duration of the history taking. If the child presents a discipline problem in the interviewing setting, it helps to put forbidden objects out of reach or to place the parent's or the doctor's chair as a barrier in front of off-limits territory, e.g., the set of drawers, the door leading out of the room, or the lamp that can be broken. It is often wise to allow only one child at a time in the interview setting, and parents should be encouraged to leave siblings at home unless their presence is important in the interview.

Another common distraction comes from interruptions by health care personnel. In one clinic this distraction was nicely controlled following the installation of small "seeing eyes" on each clinic room door that allowed nursing personnel to see whether the room was occupied and by whom without disrupting the interview within.

PRIVACY

The patient's willingness to talk is seriously compromised when privacy is ignored, yet in many medical settings little attention is given to assuring privacy. In the clinic setting additional health personnel such as nurses, pharmacists, and psychology or social work students may be present. If so, they should be introduced and their reason for participation explained. If video or audio taping or one way mirrors are used, one should always explain their presence and indicate when they are in use, obtaining the patient's permission. If these acts of common courtesy are adhered to strictly, most patients can adjust to the presence of other people with minimal loss of the trust required to produce good communication.

Both privacy and the need for a distraction-free setting necessitate interviewing rooms on the hospital wards. It is a reflection of the low priority given to psychosocial issues that such rooms are usually not available. Interviewing rooms should be free of telephones and have a sign on the outside of the door indicating when they are in use.

INTERACTIONAL COMMUNICATION SKILLS

The techniques described in this section enhance communication by helping the patient to feel more comfortable in revealing emotionally laden information. It is natural to hide thoughts that provoke shame, embarrassment, guilt, or other emotional responses, yet such feelings usually accompany the important information that leads to an understanding of the patient's behavior. Therefore, these techniques are vital to any interview that ranges beyond simple yes-no answers to direct questions. These techniques are best learned through supervised experience; however, the descriptions and the clinical vignettes presented here will provide a framework for further practice.

EMPATHY

Empathy is the capacity for understanding, with emotion, another's feelings or ideas. When one party in a conversation comes to believe that the other truly understands his own experience—particularly the emotional quality of that experience—the bond between the two is enhanced and much more information is likely to be revealed. For example, the mother of a very sick child explains that her husband was out of town when her child became ill. You note that her eyes begin to tear and respond, "You must have felt awfully alone." Such a statement, if it accurately reflects the mother's feelings, may bring a rush of emotion and a wish to tell this understanding doctor much about the illness that until now she was afraid to reveal. The physician heard the mother's words, saw the signs of emotion behind the words, and responded from a sense of what must have produced the feeling. This is empathy.

One requirement for the skillful use of empathy is that it be genuinely felt by the interviewer. To "understand" someone's predicament but not feel it emotionally is more akin to pity than to empathy. The interviewer must possess the ability to put himself "in the patient's shoes" emotionally. This is difficult to do unless the physician appreciates and respects the patient's feelings. This

means that the physician must like the patient. At this point the reader may ask, "Am I supposed to like all my patients? I find many of them irritating and difficult." This is so true of human interaction, but most skilled interviewers find that negative feelings toward patients decrease as they gain more respect for the variety of human experience. A good interviewer comes to like a greater percentage of the patients he encounters. One empathic young doctor related that she never felt comfortable managing child abuse cases until, being a parent herself, she came to understand how feelings that lead to abuse could develop even in a loving, caring parent. As with this physician, empathy often increases as one gains life experiences. It is not necessary, however, to share the experience in order to gain empathy. A genuine curiosity about people and a willingness to listen to their thoughts and feelings also promote empathic understanding.

There are situations in everyday medical practice in which empathy could be expressed but is not. This is often the result of the physician's reluctance to abandon the formal and factual communication style learned in medical history taking, when an empathic style means uncovering emotions that are difficult to handle. Physicians often feel very uncomfortable with tears and avoid situations that might produce them. The skilled interviewer must learn to manage emotional situations rather than avoid them. Frequently the best response to an emotional reaction is to remain silent, while maintaining eye contact, until the patient regains control. This behavior conveys the unspoken message that such emotions are not out of place and will not impair the patient-doctor communication. In fact, once the emotion that is close to the surface has been expressed, communication is improved because the patient no longer has to strain to keep such feelings under control.

REPETITION AND REVIEW

Communication is greatly enhanced if each party in a conversation truly listens to what the other is saying. Whenever there is direct indication that one has been understood, there is often a feeling of increased contact with the person who is listening. Such understanding is effectively communicated by accurately repeating what the other person has said. It is helpful, therefore, for the physician to periodically review the history by repeating a summarized version. The patient experiences the positive feeling that accompanies being accurately heard and also has the chance to correct misinterpretations. The physician also can use this technique to enhance the quality of the history by lending a degree of coherence to what may have been a poorly organized account.

In the following example of a common medical history, the physician's review clarifies the narrative and helps the mother fill in missing information. One should try to sense also the feeling of satisfaction the mother must experience in knowing that the doctor has understood her account:

Doctor: Tell me about Johnny's illness.
Mother: He started a fever last night. He was up most of the night crying. Actually I wondered if he was sick before that. On Friday he didn't seem hungry and he vomited after his lunch. His sister, Liz, has been sick all week. Last night he didn't want anything to eat at all. He only drank a little juice. He kept pulling on his ear. Finally he got to sleep in the early morning, but today he seems sleepy and he has no energy.
Doctor: Let me repeat what I've heard. You first noticed something Friday when he seemed less hungry and vomited once. Then yesterday he developed a fever in the evening and was up most of the night crying. He was also pulling on his ears and wouldn't eat, though he drank a little. How was he during the day yesterday?
Mother: He had a fever, but it wasn't very high. Also he wasn't as active as usual, but I wasn't worried. I thought he just had a beginning cold.

THE LANGUAGE OF THE INTERVIEW

The skillful interviewer learns to adapt his language to a style that is familiar to the patient. Developing this skill requires close observation of the patient's words as well as sensitivity to one's own vocabulary and expressive style. Less skilled interviewers are often unaware of the words they choose to explain an illness or treatment and are surprised at what they said when video or audio tapes are reviewed later. The interviewer should also be aware that the patient might not have understood him. Exit interviews have shown that patients often understand much less than the doctor thinks (Korsch and Aley, 1973). It is useful, when time is available, to practice asking the patient to repeat what he understood of your explanation. This should be done in a nondemanding way; otherwise the patient may feel that he is being quizzed. One might say, for example, "I know Johnny's problem is complicated and the words I use may be unfamiliar to you. Would you be willing to tell me what you heard so I can be sure I explained it clearly?" This wording makes the physician, rather than the patient, appear to be responsible for the clarity of communication.

When explaining medical conditions, the physician should learn to replace technical terms with common language. When discussing behavioral-emotional issues, he should be aware that the nontechnical terms he uses may mean something quite different to the patient. The skilled interviewer needs to remain continually alert to be certain that he understands the patient's use of particular words. For example, diagnostic terms such as retardation, hyperactivity, and depression often mean something quite different to parents and to children. The interviewer should determine what the patient understands by terms such as "hyperactivity" or "slow learner." One should avoid, however, asking such questions as, "What do you think hyperactivity means?" because the patient may feel that he is being put on the spot. Some empathetic encouragement is helpful, such as: "Parents often have some idea of what 'slow learner' means from conversations with friends or things that they have read. It helps me to know what thoughts you have about this problem."

Patients also lend personal meanings to abstract descriptors, such as "lazy," "sensitive," "spoiled," or "angry." The interviewer should remain constantly alert to the use of these general terms and query the patient as to their special meaning. Such inquiry might be phrased: "You said she is sensitive. What does she do that shows her sensitivity?" Or "Lazy? In what sense do you mean that?"

CONTENT AND PROCESS

It is very helpful during an interview to be able to switch from a focus on what is being said (content) to notice and comment instead on how it is being communicated (process). The following vignette illustrates this shift. This segment of the interview is not intended to illustrate skillful interaction. It takes a while for this physician to realize that he is missing the mother's primary concerns. This type of interaction, though awkward, occurs quite frequently.

Mother: I think Bobbie is just too skinny.
Doctor: On my physical exam he looked just fine.
Mother: But his ribs stick out so much.
Doctor: It's O.K. for his ribs to show that much.
Mother: My friend Jane's boy is just his age and he weighs 5 pounds more.
Doctor: Her son's build may be different. On the growth chart Bobbie's weight is just right for his build.
Mother: But he eats hardly anything. For breakfast I can't get him to finish even one piece of toast.
Doctor: It's normal for children this age to eat very little at one meal. He may eat more at another. In any case, whatever he does eat is fine for him since his weight is in the normal range for his age.
Mother: I get worried that he will get sick if he doesn't put on more weight. He has to eat something.
Doctor: You know, Mrs. Smith, each time I say that Bobbie seems fine, you bring up another concern. I get the feeling that nothing I say could make you stop worrying about him.

Until this last statement, the physician had been responding each time to what the mother just said (content). A pattern of interaction, however, is being established, and the last statement acknowledges and describes that pattern (process). Such a comment at this point in the interchange suddenly shifts the focus of communication and usually helps to reveal the real concerns. The following illustrates what may have transpired as the result of this shift:

Mother: Last winter he had such a terrible time. He was sick all of the time. Now that winter's coming, I'm so afraid that's going to happen again. My friend Jane got some vitamins from her doctor and I thought something like that might help Bobbie.

Until now the mother and the physician have focused on the child's growth and eating habits. This response indicates the mother's real concern that he is ill more frequently than normal and that this is a serious problem.

On the other hand the mother might respond as follows:

Mother: I just can't get anyone to find out what's the matter with Bobbie. I know he's not acting like a boy of his age should and nobody knows what's the matter. I guess I'm just going to have to keep trying until somebody is able to figure it out.

With this response the mother clearly indicates her perception that something is wrong with her child. The next step for the physician should be to explore some of the psychological reasons for the mother's distorted perception of her child's health.

Whenever he senses mounting frustration in an interview situation, the physician should consider not just what is being said (content) but the way it is conveyed (process), which includes his own emotional reactions. Focusing on the quality rather than the content of communication, however, can be seen as critical and the patient may become defensive. Therefore, timing and the quality of the relationship are important factors to consider. In the next sequence a fairly straightforward initial history session seems to call for a comment on the style rather than the content of the mother's replies, yet a process statement in this situation would be premature.

Doctor: How long has he had asthma?
Mother: Three or four years. His other doctor said he shouldn't play any sports because he starts wheezing whenever he runs around.
Doctor: Is he on any medicines?
Mother: Yes. He takes an antihistamine, and we keep theophylline at home to use whenever he starts to wheeze. He's missed a lot of school. His other doctor said I should keep him home whenever the weather is bad.
Doctor: How many attacks did he have, say, in the last year?
Mother: Just one or two. He's supposed to stay in bed at the first sign of a cold because if he gets an infection you can be sure he'll have a bad attack.

The pattern of the mother's response indicates considerable anxiety, perhaps to the point of overprotection. At this early stage of a new patient-doctor relationship it would be premature to com-

ment, for example, that "Every time I ask a question you are anxious to tell me how easily he gets sick." Instead, an empathic statement like "I can see that his asthma has been a real worry" would be more appropriate. Later, as the relationship becomes better established, a statement commenting on the pattern of communication might be very helpful.

In general, shifting the focus of attention from what is being said to how it is communicated helps to reveal hidden issues and feelings, which can then be handled more directly. This is one of the most useful techniques in managing the difficult interview, as will be discussed later in this chapter.

ACTIVE LISTENING

Active listening occurs when one indicates by an active response that he is both hearing the other person's words and understanding the feeling behind those words (Ginnott, 1967; Gordon, 1975). A passive response, by contrast, is usually a statement or phrase that constitutes a reaction to only the patient's words. In the following segment an active listening response is helpful but not often considered:

Doctor: How are you feeling today?
Patient: (10 year old hospitalized with cystic fibrosis): I don't like that new intern.
Doctor: Oh? He seems pretty nice to me.
Patient: He poked me five times to get my IV started.
Doctor: I'm sure he tried his best.
Patient: They shouldn't let him start my IV's. He doesn't know how to do it.
Doctor: I bet it really feels good when your IV gets started right the very first time.
Patient: Yeah!

This child responds to the pain through anger directed at the person he views as responsible. In fact, the intern may have been clumsy, but the more mature response expected of the older patient is based upon an awareness that procedures will sometimes go well and at other times will be painful despite everyone's desire to minimize pain. The doctor's comment to this child acknowledges the underlying wish that procedures would go smoothly. In this, the child and the doctor can agree, and the empathic agreement serves to increase the closeness of the contact and make the child feel less alone with his pain. The "active listening" component of this interchange is to hear beyond the attack on the intern to the underlying frustration with procedures. Another example will help illustrate this principle:

Patient: (An eight year old girl with osteomyelitis, being given antibiotics intravenously): My brother's birthday party is tomorrow.
Doctor: How old will he be?

Patient: He's gonna be six. They're going to a movie. I sure wish I could go.
Doctor: Maybe both of you can go to a movie with your Mom when you get out of the hospital.
Patient: There won't be any good movies then.
Doctor: I'm sure there will be some good movies coming up by then.
Patient: Yeah, but I won't get to see him open his presents.
Doctor: It's kind of lonely being here in the hospital, isn't it?
Patient: (tears in eyes): I miss everybody at home. (Pause, followed by brightening.) "Judy (the child life worker) is going to help me make a yarn picture for his present.

Here the active listening consists of hearing the loneliness behind the complaints and apparent refusal to be comforted. When the underlying emotion is acknowledged, one frequently sees the phenomenon illustrated here. The child leaves the complaints behind and involves herself with present activities.

Whenever the interviewer experiences frustration from failing to comfort the patient emotionally, an active listening response should be considered. The technique is also useful whenever one suspects that an important statement is being made indirectly. Another example will illustrate the latter point:

Child: (A five year old, being evaluated for seizures): This hospital is ugly.
Doctor: What's ugly about it?
Child: The nurses are all mean.
Doctor: Don't you like Susan, your nurse?
Child: They don't have anything for kids to do.
Doctor: I bet you wish you could go home.
Child: (with sadness): Can I go today?

This child wants to go home but expresses herself by complaining about the hospital. Children typically communicate in this indirect way. A natural response is to defend the hospital against her attack or persuade her to adopt a more positive attitude. Either approach will likely only further entrench her negative style. Active listening means hearing the homesick emotions behind the angry words. The interviewer knows when he has accurately identified the underlying emotion by the quality of the child's response to the active listening statement.

KEEPING THE INTERVIEW INTERESTING

Often a history taking session seems factual and dry, and the sensitive interviewer becomes aware of feeling bored. By contrast, a lively interview is interesting, and both interviewer and patient are alert. Lively is not the same as histrionic. Liveliness stems from the patient's emotional involvement with what he is saying. This involvement is particularly important when the goal of history taking is exploration of the interaction between the psychological and physical aspects of the illness. In such interviews information related in a dry factual way is likely to be safe material that is easy for the patient to discuss but not very useful. Emotion-containing information is more difficult to reveal but of greater value in understanding the problem. All the techniques described serve to liven the interview. In order to change the atmosphere of a dry interview, the physician must be observant and prepared to respond to subtle cues indicating the feelings that underly the patient's words. The skilled interviewer will notice the patient's use of unusual words or phrases or subtle changes in facial expression. A useful technique is to repeat unexpected words with a questioning lilt. The following vignette illustrates this approach:

(Mother of an 11 year old boy with chronic rheumatoid arthritis. The teacher has reported to the physician that the child seems depressed.)

Doctor: How does he do in school?
Mother: He loves school. He gets almost straight A's. The teacher says he's one of the hardest working boys in his class.
Doctor: Does he have any difficulty in handling gym?
Mother: He tries hard, but most things he just can't do. The teacher lets him do other things like keeping score whenever the games are too hard for him.
Doctor: Does he seem to you to be happy?
Mother: Oh, yes! We don't let his illness get him down.
Doctor: We?

This mother is giving an unrealistically optimistic picture of the boy's school adjustment, particularly since the physician has been told by the teacher that the child seems depressed. Until the physician's last comment, the interview has been unproductive and somewhat dull, since questions have been answered in clichés. The mother's last statement, however, contains a surprising word in her use of the plural pronoun when the singular would be more appropriate. The alert physician will pick up this incongruity and echo it back as a question. This response may help the mother reveal, for example, that she has been worried about her son. This change will be felt as increasing both the mother's and the physician's interest in the interaction.

Besides attending closely to the words, the interviewer needs to remain alert to changes in facial expression or body language that suggest that something of interest is happening within the patient's thoughts. The following vignette illustrates a situation that calls for this technique. In this instance the presence of both parents is helpful, because at any point in the interview, one parent can react both to the doctor's question and to the other parent's response.

(Parents of a 12 year old girl with chronic abdominal pain.)

Doctor: Do you know of anything that might be bothering Mary?

Mother: That's what we can't figure out. She seems so well adjusted.

Father: She plays in the school band. She has lots of friends and she gets good grades.

Mother: She's really helpful around the house. I can always count on Mary when I need a job done.

Father: She's always been the easy one to discipline. She never seems to complain when we ask things of her. Sometimes I come down kind of hard on the kids, but never with Mary. I don't have to.

Mother: (One corner of her mouth turning up to a slight smile.)

Doctor: Did you think of something amusing?

Mother: I was remembering the kids last week joking around, mimicking their father being heavy-handed. Mary was joking most of all. Maybe being strict bothers her more than we think.

During this segment of the interview the parents are painting an idyllic picture of Mary. The physician just listens and observes closely. The slight change in the mother's expression gives a clue about an intruding thought. Commenting on the changed expression is often all one needs to do to encourage further communication that begins to reveal emotionally meaningful material.

REMAINING PATIENT CENTERED

Physicians usually cast themselves in the role of advice givers; medical training reinforces the importance of that function. The advice giving role appears to be successful when the patient is accepting and compliant. Two problems develop, however, when the physician-patient interaction is based primarily on the giving and accepting of advice. First, the patient may question or reject the advice but remains reluctant to do so openly because the physician occupies a position of superiority. Instead the advice is rejected covertly and is manifested in noncompliance. Second, the patient adopts a passive role in the management of his health problems and fails to contribute to his own care, expecting everything to be done for him. For these reasons it is wise always to encourage the resourcefulness and independence of the patient. The best way to accomplish this goal is to diminish the advice giving role as much as possible and to remain patient centered in the interview situation.

One technique that fosters the patient's independence is to ask for his thoughts about the nature of the illness—what caused it and how it should be managed. Two useful questions are: "What did you think might be the matter?" and "What did you hope this visit (to the office or hospital) would accomplish?" (Korsch and Aley, 1973). The patient may come back with a response like, "That's what I came to you for, doctor, to find out what the problem is." In this case further inquiry should help bolster the patient's confidence that his ideas are valuable: "It is my job to find out what's wrong, but sometimes parents have a notion about what might be the problem, and those notions give me valuable information and clues." These questions may reveal what a parent fantasizes about his child's illness, as illustrated in the following interchange:

Mother (of a two year old with fever): He became ill so suddenly!

Doctor: He has a fever of 103°, but I don't find any signs of bacterial infection, so it is probably a virus.

Mother: He seems to be breathing funny.

Doctor: I've checked his lungs and heart and both are O.K.

Mother: Why would he get sick so quickly?

Doctor: Viruses often affect a child that way.

Mother: Are you sure there isn't any infection?

Doctor: Yes. (Pause.) What do you think might be wrong?

Mother: He was premature and had trouble breathing in the nursery. He was so sick! They had him on a respirator for several days. I worry so that his lungs are weak because of that.

The reader should notice that this vignette illustrates a typical manifestation of the vulnerable child syndrome (Green and Solnit, 1964). Because of the patient's earlier problem as a baby, this mother is now very sensitive that any illness may be a sequel of that episode. The doctor's question helps to bring out the reason for her unrealistic worries. The physician knows, now that the mother's concerns have been revealed, that whenever this child becomes ill, she may overreact; and if she does, he can respond effectively by specifically reassuring her about the consequences of the problem in infancy.

Another helpful technique is to encourage and support the patient's attempt to find solutions to his problem. This approach is illustrated in the following exchange:

Father: I've been concerned about Mark's wheezing whenever he runs hard. I'm afraid he's not going to learn to play and interact with the other boys.

Doctor: Does he stay away from other kids' fun?

Father: Well, yes; he hasn't gone out for soccer or football.

Doctor: Does he do anything else with other children?

Father: Yeah. He wrestles around and he rides his bike. He asked us the other day if he could take karate or tennis.

Doctor: What did you think about that?

Father: You know, I think he's figuring out how to pick sports that don't mean a lot of running. Maybe we should let him try those and see if his asthma will be O.K.

Doctor: That sounds like a good idea.

By remaining noncommittal and encouraging the father to continue his own internal conversation, the physician has helped him come to his own conclusions. The father's own solutions are more likely to be acted upon than any ideas the doctor might generate.

It is also helpful to encourage expression of suggestions that go against the interviewer's advice. The physician should state his own opinion clearly but then simply acknowledge the disagreement, thus encouraging the patient to take responsibility for his own ideas. In the following example the mother has previously established a pattern of chronic complaints, and the physician has investigated the quality of care given:

Mother: My husband and I don't like this hospital. Susie just isn't getting good care.
Doctor: What problems do you see?
Mother: Her IV infiltrated twice now. If the nurses had been watching, that wouldn't have happened.
Doctor: It's hard to keep IV's in Susie. Her veins are very fragile. I've known about your concerns, and I have inquired about her care. I feel sure that proper supervision was given.
Mother: Yesterday she came down with that fever. The other girl in her room had a temperature, too. I don't think they're being careful enough about spreading germs.
Doctor: I know it's hard when your child is in the hospital and then she gets sick with a virus too. I'm afraid this frequently happens on a children's ward. There's not much we can do to prevent it. These are usually mild illnesses. We would have to put Susie into isolation to prevent this, and in my judgment, that would not be worth the strain on you and Susie.
Mother: We think she should be somewhere else.
Doctor: I feel comfortable with having her at this hospital. However, I will help you carry out the decision you make and will give a recommendation if that's what you wish.
Mother: Well, maybe we shouldn't rush to decide. You think she's getting good care here?
Doctor: Yes, I do.

The important technique here is for the physician to acknowledge the parent's right to make his own decision. This does not mean that the doctor relinquishes his opinion, which he continues to state clearly. Patients who chronically complain are forced by this technique to act upon their concerns, which in turn often helps them adopt a more mature response.

THE DIFFICULT INTERVIEW

In this section the techniques already described are brought to bear on three of the most commonly encountered difficulties in the interview situation—the patient who talks too much, the one who talks too little, and the angry patient. With the patient who talks too much the physician feels frustrated, because he cannot control the interview and precious time is used in unproductive ways. Patients who talk too little provoke the physician into overly dominant participation in the interview, asking more and more questions and even suggesting responses. However, most physicians experience the greatest difficulty in handling the angry, demanding patient. The angry patient pro-

vokes in the unskilled interviewer a defensive stance accompanied by feelings of irritation and anger. Once this cycle of angry demands from the patient followed by defensiveness in the physician is established, the interview will be completely unproductive. The techniques described here are aimed at avoiding these unproductive reactions. They are just guidelines, which need to be practiced through supervised experience.

THE OVERLY TALKATIVE PATIENT

When excessive talking is encountered, one must determine whether this pattern is due to the patient's style, which means that it is typical of that person's verbal interaction in all situations, or whether there are emotional reasons for overtalkativeness in this particular situation. The person with an overly talkative style may also be tangential and show poor verbal organization. Another excessively verbal style is accompanied by an obsessive quality characterized by a need to overqualify everything.

These excessively verbal patterns are characterological, meaning that they are seen in all conversation situations and have been typical of that person for many years. Often the patient is aware of his style but has difficulty in modifying it. In the case of the loosely organized and disjointed account, it is often useful to interrupt, using the technique of repetition and review to do so tactfully. The following illustrates this approach:

Doctor: How long has she been sick?
Mother: About two days. No, maybe a week or so. Last week she didn't look good on Tuesday. I thought she might be coming down with something then. Dr. Jones told me that I should watch if she ever got diarrhea. You know she had a really bad bout of that a couple of years ago.
Doctor (interrupting): You're uncertain, then, whether her illness began two days ago or really started last week?
Mother: No. The fever began two days ago.
Doctor: After you noted the fever, what happened next?

By repeating the mother's answer up to the point at which she began to wander, the physician has put her back on track while giving her the positive feedback of his own willingness to listen. When handled in this manner, the overtalkative patient can usually accept interruptions gracefully. Such patients are usually aware that they easily wander off track and even appreciate the added structure. The use of this technique will probably have to continue throughout the interview, since disorganized communication is difficult for this patient to change.

The patient who is obsessive in verbal style may also be helped by interruptions, particularly ones that repeat the question and give permission to omit detail. The following vignette illustrates a helpful response:

Doctor: When did you first notice his school problem?
Father: He's always had trouble in school. Well, actually, it's only been since kindergarten, about the middle part of kindergarten. Let's see—was it before Christmas? No, he still had Mrs. Brown as a teacher then. No, Mrs. Brown didn't start until February some time. It was before Mrs. Brown started teaching.
Doctor (interrupting): I wondered when you first noticed his problem. I only need to know in a general way.
Father: Oh. (Pause.) Somewhere in the middle of kindergarten.

Sometimes excessive talking reflects the patient's need to ventilate. This is more likely to be the case in a first interview or when an emotional reaction is described as part of the account. A statement that acknowledges the need to ventilate (a process statement) can sometimes shorten this period while still satisfying the parent's emotional need. The following interchange illustrates these points:

(Parents of a seven year old boy who is having problems in school. This is the first diagnostic interview.)

Doctor: Tell me about Adam's problems.
Mother: We were really shocked. The teacher told us last week that they want to hold him back. Up until then we had no idea he was having difficulty.
Father: His kindergarten teacher said he was a little immature but would probably grow out of it. But they didn't do anything else. We asked if he should be tested, but they said no.
Mother: I think the kindergarten teacher just let him play with blocks while the others did their school work. We were really mad when we found that out.
Father: We think he might have been O.K. if they had helped him. Instead they just let him do what he wanted.
Mother: This year we wanted to be sure they didn't ignore him, so we asked the teacher to keep in touch with us. She kept saying not to worry. Then they dropped this bombshell on us. We're about to take this whole thing to the superintendent.
Doctor: I can see that you are very upset with the school. I would like to hear more about that, but first I wonder if you think Adam has any kind of problem. Did you come mainly because of the school's concerns?
Father: Oh, no. We've been concerned about how hard it seems for him to learn his letters.
Mother: And sometimes he seems so much younger than the others in his grade.

These parents reveal their underlying feelings and need to ventilate early in the interview. The physician's statement acknowledges in a nonjudgmental way the parents' need to discuss the school management issue. In this example the parents indicate their willingness to shift the focus to their child. At times the need to ventilate is so strong that the patient will not shift attention elsewhere. In this case the interviewer may need to be more direct. For example, the doctor might say, "I can see that you would like to talk about this. However, we have only 15 minutes. Should we use that time to discuss this subject?" This is an effective way of handling such a situation. The physician establishes the limits and gives the patient the choice of how to use the time.

THE RETICENT PATIENT

Talking very little may also stem from the patient's long established style or reflect psychological factors operating in the particular interview setting.

The patient who is stylistically undertalkative has a long history of poor verbal skill. When tested, such people often reveal a limited vocabulary and difficulties with grammar and syntax. These problems may be accentuated by low self-esteem in the presence of the more verbal physician. Such patients are often unfairly judged to be intellectually limited, since many people intuitively judge intelligence on the basis of verbal skill alone. If style is the basis of the poor verbal response, the physician should resist the natural tendency to ask more questions and instead learn to cultivate silences punctuated by facilitative phrases, such as, "Tell me more about that." "What's that like for you?" Or "What's going on right now?"—again followed by silence. Impatience often makes the situation worse. Empathic statements may be helpful, such as, "I know this must be hard to talk about." Or, "It's difficult to find the right words."

When it is due to psychological factors, the undertalkative response is usually based upon lack of trust, especially when psychosocial information is being sought. It is also more likely to be seen in adolescents who as part of normal development are experiencing mistrust and suspicion of people in authority, such as doctors. Patients with a history of problems with authorities are often particularly reticent in an interview. Any intervention that increases trust will improve communication, but, realistically, increasing trust is a long range goal and the interview is happening now. A more immediately useful technique is to comment on the style of the interview (process) as it begins to emerge out of the content. The following interview segment illustrates this technique:

(A 16 year old boy living in a drug rehabilitation unit, referred because of stomach pain.)

Doctor: Tell me about your stomach pain.
Patient: What do you want to know?
Doctor: When did it start? What was it like?
Patient: I dunno. Just a pain.
Doctor: Could you describe it for me?
Patient: Nothing special.

Doctor: Does anything make it better or worse?
Patient: Nope.
Doctor: I've asked several questions and you haven't said very much. I get the feeling this pain hasn't bothered you.
Patient: That's right. They made me come.

Once it is clear to the physician that very little information will be offered, it is helpful, as illustrated here, to comment on the style of the interaction. This boy's response confirms the hunch that his reticence is due to psychological factors. He feels that he has been pushed to come to the doctor's office. It may not be true that he is unconcerned about his pain; but this is the stance he has chosen to take, and the history will continue to be unproductive unless his feelings change. Sometimes the confrontation of a process statement will improve communication, because the patient, once he has asserted his rebellion, may experience less need to maintain it.

THE ANGRY PATIENT

Although they appear to be angry at the physician, these patients are usually frustrated by their inability to affect what is happening to them. They may be upset, for example, with the hospital, the medical system, the disease that has so disrupted their lives, and the low income that fails to meet their needs. Interviews with such patients are difficult because the anger appears to be directed at the interviewer and this provokes a defensive response. It is important to avoid this defensive reaction. The most useful technique is to recognize and then identify the demand that is implicit in the expression of anger. While the patient is talking, it helps to ask yourself, "What does this person want?" and "What does he want from me?" Then either state what you think is the demand or ask the question, "What would you like me to do?" The following is an illustrative interaction between an angry parent and the attending physician.

(Father of a child brought to the emergency room with an injured arm.)

Father: This hospital is run by a bunch of incompetents.
Doctor: What's the problem?
Father: We've been waiting an hour. My boy is in pain, and no one seems to care. They haven't even given him any pain medicine, though I begged them to.
Doctor: I can see he is in pain. The staff have instructions not to give medicines before the doctor examines him, and everyone has been tied up with a very bad accident.
Father: I've seen several people just standing around. Someone could have called another doctor. I asked them to.
Doctor: I came as quickly as I could.
Father: Why didn't they call someone else?
Doctor: I don't know. (Pause.) What would you like me to do?

Father: I know it's not your fault. I've been sitting around here fuming and I had to get it off my chest.

The father seems to be demanding that this physician do something about his problem. The skilled interviewer recognizes that his anger is more diffuse. Usually the angry person needs a few moments to ventilate. Then, by asking "What would you like me to do?" the interviewer narrows the focus to himself; this usually helps the patient to identify his frustration more specifically. Often both the patient and the doctor can agree about the source of frustration, and this increases the contact between them. For example, in the foregoing vignette, the father may have replied, "I just want this emergency room to be run more efficiently." In this case the physician could respond, "I sense how frustrating this has been. I wish I could help, but I had no control over what happened this afternoon." The source of the frustration is identified by an empathic statement, the doctor establishes the limits of his influence, and the patient feels that he has been heard.

It is important to recognize when anger is directed at the physician's behavior. Because of the physician's status, it is difficult for most patients to express such feelings, and the wise physician remains sensitive to the subtle indications of such anger. Examples of statements that are clearly directed at the doctor include: "I'm upset that you didn't return my call." "I think this charge is too high," or "You said you would check Susie this afternoon but you didn't." In these instances it is entirely appropriate for the physician to defend his behavior if he feels it was justified or to apologize if he feels it was not.

Some patients are continually frustrated by their inability to affect what happens to them and respond habitually by making frequent demands of others. Such people are usually ineffectual and have difficulty in accepting responsibility for their own choices. Instead they find it easier to blame their frustrations on others. A phrase that describes this style is "help-rejecting complainer." The following interview segment illustrates a typical interview situation with this type of person:

(Mother of a six year old boy with school problems. Attention deficit disorder has been diagnosed, and treatment with stimulant medication has begun.)

Mother: What am I going to do when the teacher calls and wants me to take him home because she can't handle him?
Doctor: Hopefully the medication will help, and that won't happen anymore.
Mother: It happened last week. Sometimes the medicine doesn't help enough.
Doctor: We've talked to the school about different methods of handling him. They will put him with the resource teacher when he is difficult.
Mother: Yeah, but she's not any good at managing him.

Doctor: Is there anyone at the school who you think can handle him well?
Mother: No.
Doctor: Well, then, maybe he's better off coming home when that happens.
Mother: But I can't be around all the time just in case they can't manage.
Doctor: What would you like to see happen?
Mother: I don't know.
Doctor: Do you see any alternatives other than those we've discussed?
Mother: I guess I could let the school try their plan.

In the middle of this discussion the doctor should sense that he is responding with suggestions that the mother repeatedly rejects. This style is best handled by encouraging the parent to take responsibility for developing approaches to the problem. Such patients typically find it difficult to generate their own solutions. They are accustomed to responding repeatedly to other people's ideas. It is important that the interviewer refrain from encouraging this pattern.

SUMMARY

Interviewing is a skill that is vital to the successful practice of pediatrics. It requires sensitivity to the psychological aspects of communication, including skilled observation of the other person and sensitivity to oneself as an important part of the interaction. Skilled interviewing is facilitated by establishing an appropriate interview setting and by sensitively managing the few moments before the interview begins. This lays the groundwork for the interaction that constitutes the substance of the interview. Acquiring communication skills means learning techniques that facilitate the flow of information. These skills include employing empathy, learning to listen actively, observing and then appropriately identifying the style of communication as well as content, becoming sensitive to the use of language, and promoting the independence and resourcefulness of the patient. The reward for these efforts is nothing less than enormously improving one's ability to help patients cope with the physical and the psychological aspects of illness.

ESTHER H. WENDER

REFERENCES

Friedman, S. B., and Hoekelman, R. A. (Editors): Behavioral Pediatrics: Psychosocial Aspects of Child Health Care. New York, McGraw-Hill Book Company, 1980.
Ginott, H. G.: Between Parent and Child. New York, Macmillan, Inc., 1967.
Gordon, T.: Parent Effectiveness Training. New York, The New American Library, Inc., 1975.
Green, M., and Solnit, A. J.: Reactions to the threatened loss of a child: the vulnerable child syndrome. Pediatrics, 34:58–66, 1964.
Green, M., and Haggerty, R. J. (Editors): Ambulatory Pediatrics II: Personal Health Care of Children in the Office. Philadelphia, W. B. Saunders Company, 1977.
Hoekelman, R., et al. (Editors): Principles of Pediatrics: Health Care of the Young. New York, McGraw-Hill Inc., 1978.
Korsch, B. M., and Aley, E. F.: Pediatric interviewing techniques. In Current Problems in Pediatrics. Chicago, Year Book, 1973.

44

Behavioral Assessment

44A Clinical Assessment of Behavioral Performance or Adjustment _____

Observe, record, tabulate, communicate: use your five senses.
See, then reason, compare and control.

Osler

For the clinician concerned with the social and emotional health of the child, valid and reliable judgments about the child's behavior and his environment are essential. The first part of this chapter discusses the use of observations and questionnaires in the clinical setting to enhance the understanding of children's behavioral performance or adjustment. It is not intended that all possible procedures or tests be presented, since several reference texts provide that information (Buros, 1980; Johnson, 1976; Johnson and Kopp, 1980). (See Chapter 43 on the assessment of behavior via the interview and Chapter 50 concerning criteria for selection of tests.) The second part of this chapter covers clinical assessment of behavioral style or temperament.

OBSERVATIONS

ASSESSMENT OF THE CHILD AND HIS INTERACTIONS BY OBSERVATIONS

Direct observation of a child and his family in the office or hospital ward can provide valuable information in the assessment and management of the child as long as the observer is aware of the setting specificity of behavior. For example, the interaction between a child and a parent suspected of abuse may provide better clinical insight than a history obtained from that parent. The clinician who provides continuity of care also has the advantage of being able to conduct multiple observations and can note changes in behavior and interaction over time. Observations need not be limited to the office or hospital; direct observation of the child in the school and of the family at home may yield insights that would be unobtainable from any other source. However, the average clinician, who may be poorly trained to observe interactions in the office or hospital, is probably even less well trained to observe behavior in other settings.

Observations may simply record what happens spontaneously or they may be structured. For example, if a child misbehaves in the office and the parent does not intervene, the clinician might encourage the parent by saying "Please feel free to handle her the way you would at home." The nature of the interaction can then be recorded and intervention, if necessary, can be guided by the information so obtained. More formal structuring such as role playing and staging can also be used to create specific situations that might not normally occur in the setting. Although such structured observations have been used in a variety of circumstances, for example, with behavior modification programs, no formal reliability or validity studies have been performed on them. The behavior seen may thus lead to inaccurate assessments.

The clinician may record the data collected from observations informally or may make use of a standardized form. These forms ensure that all data considered essential are collected and retrievable, and they facilitate the use of the information for research, peer review, program planning and evaluation, and teaching purposes. Examples of standardized forms based on observations made by clinicians in clinical settings include the Mother-Infant Form of Gray et al. (1979), the Brazelton Neonatal Behavior Assessment Scale (1973), the Attachment Indicators During Stress (AIDS) form of Massie and Campbell (1980), and the Approach-Withdrawal Scale of Rosenn et al. (1980).

The Mother-Infant Form of Gray et al. lists parental reactions that can occur immediately after the delivery of a newborn, such as "no eye contact." This information and the results of questions about the prenatal and postpartum periods and about the family circumstances have led to an index of risk for child abuse.

The Brazelton assessment scale is discussed in Chapter 5.

The AIDS scale is used with infants from birth to 18 months to detect troublesome mother-infant interactions during the stress of an ordinary physical examination. The clinician records the behavior of the infant and the mother in several categories, including gazing, vocalizing, touching, holding, affect, and proximity. These observations are made when the infant is separated from the mother during the examination and then reunited after it. The separation and reunion components are reminiscent of the Strange Situation Test of Ainsworth and Bell (1974), and both procedures are based on the same attachment theory. This scale is new, and its predictive validity is not known. However, the scale has been standardized on a large number of interactions, and reliability information is available.

The Approach-Withdrawal Scale developed by Rosenn et al. (1980) is a way of scoring a structured social interaction between an infant and the clinician. The staging of the interactions is a condensation of the usual sequence of parent-infant interaction seen in the hospital and includes an initial approach, presentation of a toy, picking up of the infant, and termination of contact. During each of the nine stages assessed, the child is rated on a seven point "approach-withdrawal" scale. Rosenn et al. found that this scale could be used to differentiate between children with nonorganic failure to thrive and those with organic failure to thrive or other organic disease. A similar but less structured scale has been used to assess the behavior of adolescents in a hospital setting (Kaplan et al., 1981).

Although observing behavior is a common daily activity of most people, knowing what to observe and how to interpret it requires training and experience. Because behavior reflects developmental level, a knowledge of normal development is essential. For example, the constant holding by a mother of a four month old would be interpreted as normal, but the same action with a two year old might be interpreted as failure to allow the child to explore.

The office or hospital is a setting foreign to the child, and differences in behavior between two children may be related as much to their social backgrounds and past experiences as they are to differences in prevailing behavior patterns. The presence of the observer has a major effect on the behavior displayed, and the office setting may have little relevance to the conditions of everyday life.

The scales just discussed take observations that customarily would be made by the experienced clinician, formalize them into a checklist, and then compress them into a score (or scores). However, because behavior is so wondrously complex and multifaceted, to assign it a single score is to deprive it of much of its meaning. Variation in behavior cannot be ranked in such a linear fashion. Thus, although direct observation of behavior can provide useful information, the clinician must be aware of the hazards involved in moving from observation to interpretation.

ENVIRONMENTAL ASSESSMENT

Since child behavior and functioning are significantly related to the social environment in which they occur, some way of measuring the latter is also needed. The most carefully researched assessment of the home environment is the HOME Scale developed by Caldwell and her colleagues (Bradley and Caldwell, 1981). However, because it requires an interview and observations in the home, it is not practical for the usual office setting. Coons et al. (1977, 1982) have attempted to derive from it a questionnaire version, which is discussed in the section on developmental screening. (See Chapter 45.) Rutter (1980) and his colleagues have developed an observation measure to assess the school environment. However, this also is not practical for most clinicians.

QUESTIONNAIRES

Because the direct observation of behavior outside the practice setting requires the investment of so much time, most busy clinicians must rely on less direct information to ascertain what a child does at home and in school. The most commonly used means of obtaining this information is the interview, which is discussed in Chapter 43. However, questionnaires have been developed to assess the same information. Like the observation forms already discussed, these questionnaires provide a standardized format to insure that all the information considered essential is collected. In addition they allow the clinician to collect information more efficiently by requiring that the parent, teacher, or child fill out the form outside the clinical setting.

Before using any of these questionnaires, the clinician must first answer these questions: Why is a questionnaire being used? Are available tests psychometrically sound? Which available form is most appropriate to use to collect the desired information, and what limitations of reliability and validity are inherent in the data obtained from the questionnaire?

CLINICAL USES OF QUESTIONNAIRES

There are a variety of uses for questionnaires. These include:

1. Opening up channels of communication. Perhaps the most important use of questionnaires is to create a dialogue between the clinician and the person filling it out. Many parents will not raise questions or express concern about their child's behavior to the physician because they think he is not interested in them (Chamberlin, 1975). Giving the mother an opportunity to identify concerns on a questionnaire conveys to her that these are legitimate areas for discussion. In a similar way, some persons working with teenagers have found this a useful way to locate areas of concern that might not be brought up by more direct questioning (Levine, 1970). Finally, the use of a brief teacher check list as part of the annual health assessment is a useful way to improve teacher-physician communication.

2. Helping the parents recognize their child's individuality. Questionnaires dealing with individual differences such as temperament have been found useful in helping parents realize that all children are not alike and can differ behaviorally in some significant ways (see the second part of this chapter).

3. Recognizing inadequacies in the child's environment. Using data gathered from home and school questionnaires, the clinician can determine the factors thought to be important for normal child development, such as the amount of positive contact between the parent and the child or the structure of the home's physical environment.

4. Identifying behavior and emotional problems. Before a test is adopted to screen for behavior problems, a number of criteria need to be fulfilled including assurance that intervention is possible, effective, and not harmful, that the condition being screened is important, and that the method of screening has good predictive value. Because much of the rest of this book deals with the first two problems, only the method of screening is discussed in this section.

Researchers interested in the prevention of behavioral problems have hoped to be able to identify early patterns of behavior sufficiently predictive of later disturbances to warrant the establishment of extensive screening and early treatment programs. Unfortunately the situational specificity and poor predictability of behavior patterns have made it clear that if a behavior screening program were established on a widespread basis, it would identify large numbers of children who, if followed over time, would be found to have transient situational difficulties that resolve spontaneously. If all these children were referred for more extensive diagnosis, they would soon overwhelm any type of treatment facility, and much needless parental anxiety would be created (Chamberlin, 1981). This poor predictive power of currently available screening tests diminishes the value of questionnaires for this purpose. A practical alternative then is to use questionnaires to identify problems that cause enough current pain in the family to justify intervention. Describing a typical day, for instance, may identify "vicious circle" interactions that often respond to brief intervention techniques (Chamberlin, 1974).

Once the clinician has decided why he wants to assess behavior, he must determine the specific aspects to measure. Because no single unifying theory of child development exists on which to base data collection, the clinician must find the areas that he believes to be most helpful for understanding and treating children and their families. There is the danger, however, that the information gathered will be viewed in a narrow disease model approach. Rather than assessing the child's needs and assets, the clinician might simply determine what is "wrong" with the child and attempt therapy on the basis of a diagnostic label or score. Such an approach often leads to ineffective therapy. Thus the information generated from questionnaires should always be integrated into the total evaluation of the child.

CRITERIA FOR SELECTING QUESTIONNAIRES

A number of criteria based in part on the standards recommended for psychological tests by the American Psychological Association can be used for judging questionnaires. (See also Chapter 50.) These criteria include the following:

1. Intended population. For what population is the questionnaire designed? Is it appropriate to use in another group of children, e.g., the disabled?

2. Purpose and rationale. Are the purpose (e.g., screening, diagnosis, providing a behavior profile) and the rationale (results of studies demonstrating its utility) clearly stated?

3. Administration and scoring. Are instructions comprehensive and clear?

4. Standardization. Was the measure standardized on a sufficiently large and varied sample?

5. Reliability. Are test-retest, interobserver, and internal consistency reliabilities adequate?

6. Validity. Does the test really measure what it is designed to? How well do its findings agree with those from other accepted measures?

7. Theoretical rationale. Are the questions based on a reasonable theory of development, or are they simply chosen for convenience?

Whenever a clinician selects a questionnaire, he should use the foregoing criteria in the process of evaluation. He should be wary if these standards are not met.

SPECIFIC QUESTIONNAIRES

Parent Questionnaires

One of the major advances in the past decade in understanding the way behavioral problems develop has been the recognition of the effect of individual differences in children on parents and parenting (Thomas et al., 1968). The assessment of temperament by questionnaire is discussed in the final section of this chapter.

A host of questionnaires are available to assess behavior problems and adjustment. Most include behavioral descriptions of the child related to problems of daily routine (food refusal, resistance to bedtime or night awakening, resistance to toilet training, or stool withholding), aggressive-resistant behavior (negativism, temper tantrums, aggressive responses to siblings or peers), overdependent or withdrawing behavior (demands to be waited on, separation upset, clinging, fears, shyness), "hyperactivity" or excessive restlessness, and undesirable habits (thumb sucking, nail biting, throat clearing, playing with genitals; Chamberlin, 1975).

The Child Behavior Checklist of Achenbach and Edelbrock (1981) is an example of such a questionnaire. It consists of 20 social competence items and 118 behavior problems intended for children four to 16 years of age. This scale has also been standardized on several large populations, and several determinations of reliability and validity have been performed. The Personality Inventory for Children developed by Wirt et al. (1977) has also been used in clinical settings, but its 600 questions have discouraged extensive use.

Third Party Questionnaires

Because of the poor predictability of problem behavior, increasing attention is being focused on the positive aspects of child behavior and functioning, and questionnaires are being developed to tap this area as well. Since peer relations and learning tasks are central to this assessment, most of these are filled out by teachers in day care or classroom settings. A typical example is the one

devised for preschool aged children by Kohn and Rosmann (1972).

Recent studies showing relationships between early observation measures of mother-child attachment (12 to 18 months) and later measures of social and cognitive development suggest the possibility of developing a questionnaire measure for this as well, but the authors are not aware that any have been tested for clinical use (Chamberlin and Keller, 1982).

One of the most widely used questionnaires to assess the classroom behavior of children is that developed by Connors (1969). Like other questionnaires, this has been shown to be a good approximation of the child's current behavior in that setting and it is sensitive enough to show the effects of drugs such as methylphenidate that are used to increase the attention span of children with some types of learning problems. However, its psychometric accuracy is diminished by its use of impressionistic questions.

Rutter (1967) and Graham and Rutter (1968) have developed behavioral questionnaires for both parents and teachers that have been used in the initial screening of large populations of school aged children for psychologic problems. However, these must be followed by a careful clinical assessment, since only about half the children identified with each instrument turn out to have psychologic problems severe enough to need the attention of a mental health professional. The parent and school data collection system of Levine et al. (1980a,b) provide similar information about the child's behavior in both settings and also about the resources available to that child. (See Chapter 45B.)

Because functioning in any setting is related to the characteristics of the setting as well as to the characteristics of the child, it is important to get an independent assessment of the classroom environment before relying entirely on the teacher's report. It is not uncommon for a child's "hyperactivity" to disappear with a change in teachers. There have also been a number of reports of the successful treatment of "hyperactivity" by changing the teacher's response pattern to the child. Thus, at least some school problems arise out of a lack of fit between a particular type of child and his classroom environment. Moreover, an assessment such as "hyperactivity" must be clearly defined in operational terms before it can have any useful meaning for therapeutic intervention.

Self-assessment Questionnaires

To assess feelings and self-perceptions, and for the assessment of behavior in older children (especially those above age 16 for whom the value of parental reports declines), self-administered ques-

Table 44–1. SUMMARY OF INSTRUMENTS PRESENTED

	Test	Authors	Purpose	Age of Intended Population	Reliability		Validity	
					Test-retest	Inter-observer	Concurrent	Predictive
Scales based on observations of behavior	Mother-Infant Form	Gray et al.	Screen for child abuse potential	Newborn	N*	A*	A	A
	Attachment Indicators During Stress	Massie and Campbell	Assess attachment during routine physical examination	0–18 mo.	N	A	A	N
	Approach-Withdrawal	Rosenn, Loeb et al.	Differentiate organic from nonorganic failure to thrive	6–16 mo.	A	A	A	N
Environmental assessment questionnaire	Home Screening Inventory	Coons et al.	Assess current home environment	Infant to preschool	N	A	A	A
Parent questionnaires	Child Behavior Check	Achenbach and Edelbrock	Assess current home behavior	4–16 yr.	N	A	A	A
	Parent Questionnaire	Graham and Rutter	Assess current home behavior	10–11 yr.	A	A	A	N
Third party questionnaires	Social Competence and Symptom Scales	Kohn and Rosmann	Assess current behavior in day care or home	Preschool	N	A	A	A
	Teacher Rating Scale	Conners	Assess and monitor "hyperactivity" in classroom	School age	A	N	A	A
	Behavior Questionnaire for Teachers	Rutter	Assess current behavior in school	7–13 yr.	A	A	A	N

*A, information available. N, information not available.

tionnaires are used. These questionnaires have been used with adolescents to help identify concerns that might not be brought up spontaneously (Levine, 1970). Although tests like the California Personality Inventory have been used for research (Lawlor and Davis, 1981), no single self-assessment questionnaire has been accepted as the best for use with teenagers.

An interesting potential use of questionnaires was the approach taken by Gleser et al. (1980) in which adolescents and their parents were given the same 40 item behavioral questionnaire (the Adolescent Life Assessment Checklist) to assess not only the occurrence of problems but also the degree of agreement between parent and child. Although these authors found a fair amount of agreement between the two, there were also striking differences. The clinical significance of these differences and the use of such a technique in a nonresearch setting should be explored.

RELIABILITY AND VALIDITY OF QUESTIONNAIRE DATA

Information based on parent, teacher, or patient reports may be misleading because the information often involves selective recall or other distortions especially in the directon of social desirability. However, a number of investigators have found that parents and teachers can be accurate observers and reporters of a child if they are asked to describe specific behaviors rather than interpret them and to limit their observations to the present or recent past (within a few weeks). When these two constraints are followed, independent observers in the home and classroom have corroborated the accuracy of both parent and teacher reports (Chamberlin, 1976; Graham and Rutter, 1968; Rutter, 1967; Thomas et al., 1968). There are, however, two major limitations to the use of such data that the practitioner should be aware of.

1. Measures of current behavior and functioning are not very good predictors of future behavior and functioning. Many persons developing check lists have assumed the validity of the "trait" theory of personality. That is, children develop certain personality traits, such as being "aggressive" or "inhibited," and these traits are consistently displayed in a variety of different settings. Although such a conception seems to have a certain amount of "face" validity, it is not well supported by research. What one finds instead is that although there is enough continuity of behavior and functioning to make modest predictions about the future for groups of children, there is enough variation so that predictions about individuals are often quite inaccurate (Chamberlin, 1981). Longitudinal follow-up studies of children and adults

have repeatedly demonstrated this even when dealing with fairly extreme groups of well or maladjusted individuals. This general lack of consistency over time is at least in part related to the second major limitation of questionnaire data:

2. There is a considerable situational component to a person's functioning at any time. This has been demonstrated repeatedly in the low correlations found between home and school behavior patterns in numerous reports and from the studies showing how changes in caretaker behavior (teachers, group leaders) can have rather marked effects on the behavior of the persons experiencing them (Chamberlin, 1981; Sandberg et al., 1980).

CONCLUSION

Informal and standardized observations and questionnaires have been found to be reasonably accurate ways of assessing a child's functioning in a particular setting at a particular time, but are not very accurate for predicting behavior in the future or in other settings. The information obtained can be used to guide the management of the child and can help open channels of communication with parents, patients, and teachers. However, the clinician must know why he is collecting the data and what information he should collect and must choose the method of collection with great care. The data obtained must be interpreted with the knowledge that measures of current behavior are not good predictors of future behavior and that behavior is not always stable over different situations. All these decisions require an insightful and well trained clinician. No paper test can be a substitute for mature judgment.

GREGORY S. LIPTAK
ROBERT W. CHAMBERLIN

REFERENCES

Achenbach, T., and Edelbrock, C. S.: Behavioral problems and competencies reported by parents of normal and disturbed children aged four through sixteen. Monogr. Soc. Res. Child Devel., 46:1–81, 1981.

Ainsworth, M., and Bell, S.: Mother-infant interaction and the growth of competence. In Connally, K. J., and Bruner, J. (Editors): The Growth of Competence. London, Academic Press, 1974.

Bradley, R., and Caldwell, B.: Pediatric usefulness of home assessment. In Camp B. (Editor): Advances in Behavioral Pediatrics. Greenwich, Connecticut, JAI Press, Inc., 1981, Vol. 2, pp. 61–80.

Brazelton, T. B.: Neonatal behavioral assessment scale. Clinics in Developmental Medicine. No. 50. Philadelphia, J. B. Lippincott Company, 1973.

Buros, O. K. (Editor): The Eighth Mental Measurement Yearbook. Highland Park, New Jersey, Gryphon Press, 1980.

Chamberlin, R. W.: Management of behavior problems in preschool children. Pediatr. Clin. N. Am., 21:33–47, 1974.

Chamberlin, R. W.: Behavior problems of preschoolers. In Haggerty, R., Roghmann, K., and Pless, I. B. (Editors): Child Health and the Community. New York, John Wiley & Sons, Inc., 1975, Ch. 3D.

Chamberlin, R. W.: The use of teacher checklists to identify children at risk for later behavioral and emotional problems. Am. J. Dis. Child., 130:141–145, 1976.

Chamberlin, R. W.: Relationship of preschool behavior and learning patterns to later school functioning. In Advances in Behavioral Pediatrics. Greenwich, Connecticut, JAI Press, Inc., 1981, Vol. 2, pp. 11–127.

Chamberlin, R., and Keller, B.: Research on parenting: implications for primary health care providers. In Advances in Behavioral Pediatrics. Greenwich, Connecticut, JAI Press, Inc., 1982, Vol. 3.

Connors, C. K.: A teacher rating scale for use in drug studies with children. Am. J. Psychiatry, 126:152, 1969.

Coons, C., et al.: A home screening questionnaire. In Frankenburg, W. (Editor): Developmental Screening. Denver, University of Colorado Medical Center, 1977.

Coons, C. E., et al.: Preliminary results of a combined developmental and environmental screening project. In Anastasiow, N. J., Frankenburg, W. K., and Fandall, A. W. (Editors): Identifying the Developmentally Delayed Child. Baltimore, University Park Press, 1982.

Gleser, G. C., Seligman, R., Winget, C., and Rauh, J. L.: Parents view their adolescents' mental health. J. Adoles. Health Care, 1:30–36, 1980.

Graham, P., and Rutter, M.: The reliability and validity of the psychiatric assessment of the child. II. Interview with the parent. Br. J. Psychiatry, 114:581–592, 1968.

Gray, J. D., Cutler, C. A., Dean, J. G., and Kempe, C. H.: Prediction and prevention of child abuse. Sem. Perinatol., 3:85–90, 1979.

Johnson, K. L., and Kopp, C. B.: A Bibliography of Screening and Assessment Measures for Infants. Los Angeles, University of California, 1980.

Johnson, O. G.: Tests and Measurement in Child Development: Handbook II (2 vols.) San Francisco, Jossey-Bass, 1976.

Kaplan, S., Rosenstein, J., Skomovowsky, P., Shenker, I. R., and

Ramsey, P.: The hospitalized adolescent interaction scale: an instrument to measure patient behavior on an adolescent medicine ward. J. Adoles. Health Care, 2:101–6, 1981.

Kohn, M., and Rosman, B.: A social competence scale and symptom check list for the preschool child. Devel. Psychol., 6:430–444, 1972.

Lawlor, C. L., and Davis, A. M.: Primary dysmenorrhea: relationship to personality and attitudes in adolescent females. J. Adolesc. Health Care., 1:208–212, 1981.

Levine, C.: Doctor-patient communication with the inner city adolescent. N. Engl. J. Med., 282:494–495, 1970.

Levine, M. D.: The ANSER System. Cambridge, Massachusetts, Educator's Publishing Service, 1980.

Levine, M. D., Brooks, R., and Shonkoff, J.: A Pediatric Approach to Learning Disorders. New York, John Wiley & Sons, Inc., 1980.

Massie, H. N., and Campbell, K.: The Massie-Campbell Scale of Mother-Infant Attachment Indicators During Stress. Detroit, Wayne State University Press, 1980.

Rosenn, D. W., Loeb, L. S., and Jura, M. B.: Differentiation of organic from non-organic failure to thrive syndrome in infancy. Pediatrics, 66:698–704, 1980.

Rutter, M.: A children's behavior questionnaire for completion by teachers: preliminary findings. J. Child Psychol. Psychiatry, 8:1–11, 1967.

Rutter, M.: School influence on children's behavior and development. The 1979 Kenneth Blackfan lecture. Pediatrics, 65:208–220, 1980.

Sandberg, S. T., Wieselberg, M., and Shaffer, D.: Hyperkinetic and conduct problem children in a primary school population: some epidemiological considerations. J. Child Psychol. Psychiatry, 21:293–311, 1980.

Thomas, A., Chess, S., and Birch, H. G.: Temperament and Behavior Disorders in Children. New York, New York University Press, 1968.

Wirt, R. D., Feat, P. D., and Broen, W. E.: Personality Inventory for Children. Los Angeles, Western Psychological Services, 1977.

44B Clinical Assessment of Behavioral Style or Temperament

The first part of this chapter has discussed in general terms the clinical assessment of behavior through observations and questionnaires and in particular the ways of appraising behavioral performance or adjustment. This concluding section takes up in greater detail the clinical assessment of behavioral style or temperament.

Since Part II of this volume has already described the phenomenon of temperament (see especially Chapter 10), there is no need to do so again here. Topics covered in this section include the clinical usefulness of temperament data, an evaluation of the techniques available for clinical determinations, and a listing of available questionnaires and their psychometric characteristics (see also Carey, 1981a, 1982a).

CLINICAL USES OF TEMPERAMENT DATA

Temperament data are useful to the clinician in three principal ways (Carey, 1981b, 1982b). First, general discussions of the phenomenon of temperament between the clinician and the parents help to provide them with background informa-

tion against which they may see their child in better perspective. For example, parents should find enlightening the concept of apparently inborn behavioral variables, such as activity, that permits conclusions such as the one that a high degree of activity may be an extreme of normal rather than an abnormality.

Second, it is sometimes valuable to determine the specific temperament profile of a particular child. This is especially true in the case of a difficult infant, for it allows the clinician and the parents to understand that the child's negativity and low degree of adaptability are probably not the fault of the parents' handling or feeding.

Third, when temperament-environment conflicts have resulted in secondary symptoms of behavioral maladjustment, it is helpful to determine the contribution of the child's temperament to the clinical problem. Such information may provide an explanation of the magnitude and direction of the child's symptoms and sets realistic goals for therapeutic intervention. With appropriate alterations of parental management, the reactive symptoms in the child should disappear. Meanwhile parents, teachers, and other caretakers must learn to live in a more tolerant and more flexible manner with the child's temperament, which is evidently less changeable.

There is no clearly established and recommended frequency at which clinicians should routinely obtain temperament data for the children under their care. One possible scheme would be to do a determination at about six months routinely but thereafter only as indicated by problems in interactions or the child's adjustment.

TECHNIQUES FOR TEMPERAMENT DETERMINATIONS

As mentioned in the first section of this chapter, there are three principal methods for obtaining data relating to behavior—interviews, observations, and questionnaires. This is true for both behavioral performance and style.

INTERVIEWS

The best known interview technique for obtaining temperament data is the one devised about 25 years ago by the New York Longitudinal Study of Thomas, Chess, and colleagues. (See Chapter 10.) Although their interview was sufficient for the needs of their study, neither it nor any adaptation of it has found wide usage in research or clinical practice. Its flexibility makes it more sensitive to varying situations, but it also is less capable of standardization. Its length of one to two hours allows great richness of detail to be developed in behavioral descriptions but renders it impractical in clinical settings. Nevertheless clinicians can informally make use of the concepts and specific content areas of the New York Longitudinal Study interview in clinical situations as long as they do not yield to the temptation to generalize about a characteristic from a few bits of data.

OBSERVATIONS

Primary care clinicians have an opportunity to observe the behavior of the children under their supervision. However, as has been stressed elsewhere in this volume, the samples of behavior supplied in offices and clinics are usually too brief and are often atypical. There may someday be a standardized test for clinical measurement of temperament, but there is none today. Furthermore, there is no standardized comprehensive technique for research either. Various studies have devised methods for use with a particular investigation, but these are not easily applicable elsewhere. Gregg (1972) devised a Clinical Behavioral Observation Instrument for data collection in the pedia-

trician's office, but it employed separate observers, rated only four of the nine New York Longitudinal Study characteristics, and has not been further refined since its original publication. Recently Matheny (1980) and colleagues have developed an elaboration of the behavioral items on the Bayley development scale for their ratings of temperament in their study of twins. This too requires trained observers and has not been applied clinically.

The Brazelton Neonatal Behavioral Assessment Scale (1973) is regarded by some as an appropriate way to determine "constitutional temperament." However, despite its considerable value for studying neonatal problems and for helping parents to understand their newborns, one must recognize that newborn behavior is affected by nongenetic prenatal and perinatal factors, is not very stable from one day to the next, and does not provide an adequate view of the temperamental characteristics thought to be influenced by genetic factors. Comparison between newborn behavior, as measured by the Brazelton scale, and later temperamental traits has been hindered by these factors and by the fact that some of the newborn measures, such as muscle tone, are not temperament variables, and none of the scale items are the same in both content and dimensions as the temperament characteristics of the New York Longitudinal Study.

QUESTIONNAIRES

With the increased recognition of the theoretical and clinical importance of temperament and the lack of practical ways of measuring it by interview or observations, a series of questionnaires has been developed in the last dozen years. Table 44–2 lists currently available scales, and Table 44–3 presents samples of items from two of them. (See also Hubert et al., 1982.)

Questionnaires have the advantage of being standardized as to norms for characteristics at various ages, are much briefer than comparable interview and observation techniques (since questionnaires require only 20 to 30 minutes of the mother's time and about 10 to 15 minutes for the clinician to score), and are low in cost and high in acceptability. Two problems in their use are the question of their appropriateness with less literate parents of below high school graduation level and their external validity.

The issue of validity is not easily resolved (Carey, 1982c; in press [b]). Behavioral scientists presently tend to speak of any parental judgments of temperament as "perceptions" and of their own data, no matter how brief and unrepresentative, as scientific "observations." It would be more

Table 44–2.   QUESTIONNAIRES FOR MEASURING TEMPERAMENT
I. Principal questionnaires in English using New York Longitudinal Study categories

Age Span	Name of Test	Authors	No. Items	Retest Rel.†	Alpha Rel.†	Reference and Comments
0–4 months	None					
4–8 months	Infant Temperament Quest. (1970)	Carey	70	0.84m	0.47m	Use of ITQ 0–4 months not recommended J. Pediatr., 77:188, 1970; replaced by revised version
	Infant Temperament Quest. (1978) (ITQ)	Carey and McDevitt	95	0.86t	0.83t	Pediatrics, 61:735, 1978;* revised version
8–12 months	None					Use of ITQ may be possible if standardized for period
1–3 years	Toddler Temperament Scale (TTS)	Fullard, McDevitt, and Carey	97	0.88t	0.85t	J. Pediat. Psychol. (in press)*
3–7 years	Parent Temperament Quest.	Thomas, Chess, and Korn	72	na	na	Thomas and Chess: Temperament and Development, Brunner/Mazel, 1977
	Teacher Temperament Quest.	Thomas, Chess, and Korn	64	na	na	Ibid.
	Teacher Temperament Quest. (short form)	Keogh, Pullis, and Cadwell	23	na	na	J. Educat. Meas. 19:323, 1982
	Behavioral Style Quest. (BSQ)	McDevitt and Carey	100	0.89t	0.84t	J. Child Psychol. Psychiatry, 19:245, 1978*
8–12 years	Middle Childhood Temp. Quest. (MCTQ)	Hegvik, McDevitt, and Carey	99	0.87m	0.82m	J. Dev. Behav. Pediatr., 3:197, 1982*
12–18 years	None					
18–21 years	Early Adult Temp. Quest.	Thomas et al.	140	na	0.82m	Educat. Psychol. Meas. 42:593, 1982

*Addresses from which some of these questionnaires may be obtained. $10.00 per scale prepaid requested to cover expenses. Forms may be photocopied from sample received.

ITQ: William B. Carey, M.D.
319 W. Front Street
Media, Pa. 19063

TTS: William Fullard, Ph.D.
Department of Educational Psychology
Temple University
Philadelphia, Pa. 19122

BSQ: Sean C. McDevitt, Ph.D
Devereux Center
6436 E. Sweetwater
Scottsdale, Arizona, 85254

MCTQ: Mrs. Robin L. Hegvik
307 North Wayne Avenue
Wayne, Pa. 19087

†Retest reliability and alpha reliability (internal consistency) figures are given as median category (m) or total (t) values. na indicates that data are not available.

II. Other research questionnaires in order of publication

Buss, A. H., and Plomin, R.: A Temperament Theory of Personality Development. New York, John Wiley & Sons, Inc., 1975.
Garside, R. F., et al.: Dimensions of temperament in infant school children. J. Child Psychol. Psychiatry, 16:219, 1975.
Bates, J., Freeland, C., and Lounsbury, M.: Measurement of infant difficultness. Child Devel., 50:794, 1979.
Persson-Blennow, I., and McNeil, T.: A questionnaire for measurement of temperament in six-month-old infants: development and standardization. J. Child Psychol. Psychiatry, 20:1, 1979.
Persson-Blennow, I., and McNeil, T.: Questionnaires for measurement of temperament in one- and two-year-old children. J. Child Psychol. Psychiatry, 21:37, 1980.
Bohlin, G., Hagekull, B., and Lindhagen, K.: Dimensions of infant behavior. Infant Behav. Devel., 4:83, 1981.
Hagekull, B., and Bohlin, G.: Individual stability in dimensions of infant behavior. Infant Behav. Devel., 4:97, 1981.
Rothbart, M. K.: Measurement of temperament in infancy. Child Devel. 52:569, 1981.
Lerner, R. M., Palermo M., Spiro, A., III, and Nesselroade, J. R.: Assessing the dimensions of temperamental individuality across the life span: The Dimensions of Temperament Survey (DOTS). Child Devel., 53:149. 1982.
Strelau, J.: The temperament inventory: a pavlovian typology approach. University of Warsaw Institute of Psychology. Unpublished manuscript, 1980.

Table 44–3. SAMPLE ITEMS FROM TEMPERAMENT QUESTIONNAIRES

Infant Temperament Questionnaire (Carey and McDevitt)

Using the following scale, please circle the number that indicates how often the infant's recent and current behavior has been like that described by each item:

Almost Never (1)	Rarely (2)	Variable; Usually Does Not (3)	Variable; Usually Does (4)		Frequently (5)				Almost Always (6)
1. The infant eats about the same amount of solid food (within 1 oz.) from day to day.			Almost never	1 2 3	4 5 6				Almost always
2. The infant is fussy on waking up and going to sleep (frowns, cries).			Almost never	1 2 3	4 5 6				Almost always
3. The infant plays with a toy for under a minute and then looks for another toy or activity.			Almost never	1 2 3	4 5 6				Almost always
4. The infant sits still while watching TV or other nearby activity.			Almost never	1 2 3	4 5 6				Almost always
5. The infant accepts right away any change in place or position of feeding or person giving it.			Almost never	1 2 3	4 5 6				Almost always
6. The infant accepts nail cutting without protest.			Almost never	1 2 3	4 5 6				Almost always

Middle Childhood Temperament Questionnaire (Hegvik, McDevitt, and Carey)

Using the scale below, please mark an "X" in the space that tells how often the child's recent and current behavior has been like the behavior described by each item:

Almost Never (1)	Rarely (2)	Variable; Usually Does Not (3)	Variable; Usually Does (4)		Frequently (5)		Almost Always (6)
1. Runs to get where he/she wants to go.			Almost never	__:__:__:__:__:__ 1 2 3 4 5 6			Almost always
2. Avoids (stays away from, doesn't talk to) a new sitter on first meeting.			Almost never	__:__:__:__:__:__ 1 2 3 4 5 6			Almost always
3. Easily excited by praise (laughs, claps, yells, etc.).			Almost never	__:__:__:__:__:__ 1 2 3 4 5 6			Almost always
4. Frowns or complains when asked by the parent to do a chore.			Almost never	__:__:__:__:__:__ 1 2 3 4 5 6			Almost always
5. Notices (looks toward) minor changes in lighting (changes in shadows, turning on lights, etc.).			Almost never	__:__:__:__:__:__ 1 2 3 4 5 6			Almost always
6. Loses interest in a new toy or game the same day she/he gets it.			Almost never	__:__:__:__:__:__ 1 2 3 4 5 6			Almost always

appropriate to say that perceptions are general or hasty impressions, and that ratings are multiple scored judgments of certain behavior patterns in specific settings. Both parents and professionals can have perceptions and make ratings.

The ideal test of the validity of the parental ratings on a temperament questionnaire would be a comparison with a comprehensive standardized professional rating scheme. As already mentioned, there is none in existence now. However, every adequately designed test so far has demonstrated at least moderate validity of parental reports. Data from parents must be contemporaneous and relate to specific behavior patterns rather than general impressions. Comparisons of parental and professional ratings must involve the same content and dimensions of behavior, a requirement overlooked in the only two published reports claiming to discredit the validity of parental ratings (Carey, 1982c; in press [a]). Clinical users of temperament questionnaires may therefore be reassured of at least a moderate degree of validity. Although distortions may occur, these can be minimized by the interviewing and observations that should always accompany the use of a questionnaire.

TRANSLATIONS

Some of the questionnaires listed in the tables have been translated into various foreign languages for clinical and research use here and abroad. Unfortunately no directory of these translations exists. The formation of such catalogues would be facilitated if information on them were submitted to the first authors of the various scales. Assembled lists then could be shared with other potential users.

WILLIAM B. CAREY

REFERENCES

Brazelton, T. B.: Neonatal Behavioral Assessment Scale. Philadelphia, J. B. Lippincott Co., 1973.

Carey, W. B.: The importance of temperament-environment interaction for child health and development. *In* Lewis, M., and Rosenblum, L. (Editors): The Uncommon Child. New York, Plenum Press, 1981a.

Carey, W. B.: Intervention strategies using temperament data. *In* Brown, C. C. (Editor): Infants at Risk: Assessment and Intervention. Piscataway, New Jersey, Johnson & Johnson, 1981b.

Carey, W. B.: Clinical appraisal of temperament. *In* Lewis, M., and Taft, L. (Editors): Developmental Disabilities: Theory, Assessment and Intervention. Jamaica, New York, S. P. Medical & Scientific Books, 1982a.

Carey, W. B.: Clinical use of temperament data in pediatrics. *In* Porter, R., and Collins, G. M. (Editors): Temperamental Differences in Infants and Young Children. Ciba Foundation Symposium No. 89. London, Pitman Books, 1982b.

Carey, W. B.: Validity of parental assessments of development and behavior. Am. J. Dis. Child., *136*:97, 1982c.

Carey, W. B.: Some pitfalls in infant temperament research. Infant Behavior and Development. (In press [a].)

Carey, W. B.: The validity of temperament assessments. *In* Brazelton, T. B., and Als, H.: Behavioral Assessment in Newborn and Older Infants. Hillsdale, New Jersey, Erlbaum Assoc. (In press [b].)

Gregg, G. S.: Clinical experience with efforts to define individual differences in temperament. *In* Westman, J. C. (Editor): Individual Differences in Children. New York, John Wiley & Sons, Inc., 1973.

Hubert, N. C., Wachs, T. D., Peters-Martin P., and Gandour, M. J.: The study of early temperament: measurement and conceptual issues. Child Devel., *53*:571, 1982.

Matheny, A. P., Jr.: Bayley's Infant Behavior Record: behavioral components and twin analyses. Child Devel., *51*:1157, 1980.

45

Developmental Assessment

_____ 45A Infant and Preschool Developmental Screening

In his textbook Nelson (1969) stated the goal of pediatric care: "The goal in the medical management of the child is to permit him to come into adulthood at his optimal state of development, physically, mentally and socially, so that he can compete at his most effective level."

It is the physician's role both to provide anticipatory guidance to the parent and to monitor the growth and development of the child. When deviancy arises, the physician should identify it promptly and take appropriate action. This activity is a major component of health maintenance.

One way of identifying deviancy is through screening. Since the term screening is commonly used in a variety of ways, it is essential that it be defined for the purpose of this discussion. Screening, as defined by the World Health Organization, is the application of quick, simple, and accurate tests or procedures to an asymptomatic population to separate individuals who have a high likelihood of harboring the problem in question from those who do not. Thus, suspect or positive screening results should be followed with thorough diagnostic evaluations to determine whether children who are suspect on screening actually have the problem in question.

The purpose of screening is to facilitate the earlier-than-usual diagnosis and treatment of a disease or handicapping condition at the time when treatment is most effective (Fig. 45–1). Through screening and early treatment of disease, the disease process may be either reversed or ameliorated. Although the goal of primary prevention is to prevent a disease, screening is a secondary preventive measure. Thus, screening is intended to identify and remediate the process early in order to improve the outcome. With its emphasis on prevention and the promotion of develop-

ment and health, screening is an essential aspect of health maintenance.

As more and better screening procedures have been developed, some common misconceptions have developed:

1. The assumption has been made by some that suspect screening results are the same as a diagnosis, so that individuals who are not diseased or handicapped have been treated.

2. The assumption has also been made that all disease conditions lend themselves to screening. For many conditions, such as childhood diabetes mellitus and cystic fibrosis, it remains to be shown that treatment instituted during the presymptomatic period improves the outcome more than when treatment is begun after the patient becomes symptomatic.

3. The further assumption has been made that any screening test will do. Pediatricians are generally not trained to select the most appropriate screening test, and thus often select the first test with which they become acquainted. Since criteria in screening test selection are well established, it behooves the clinician to apply these criteria to select the test that is most appropriate.

These misconceptions are particularly apparent when one focuses attention upon the development of children. As a part of health maintenance, pediatricians are continually required to assess the development of children. This comes about not only because parents pose questions regarding the development of their children, but also because physicians are concerned about the development of children who were subject to adverse circumstances during the mother's pregnancy or delivery. Since many physicians lack specific training in screening, there is a common tendency on the part of health clinicians to misuse developmental screening procedures.

In light of the misuse of developmental screening tests, it is the purpose of this chapter to review the case for routine and periodic developmental screening as part of health maintenance, to review a few procedures designed for this purpose, and

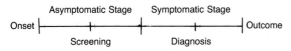

Figure 45–1. Progression of handicapping process.

to suggest a practical scheme that can be adapted to fit into most pediatric and general medical practices.

DEVELOPMENTAL PROCESS

Before considering the rationale for early identification and treatment of developmentally handicapped children, it is important to review briefly three major theories of child development as described by Sameroff (1975). These are the main effect model, the interactional model, and the transactional model. (See also Chapters 3 and 10.)

MAIN EFFECT MODEL

Gesell proposed that genetic factors, as well as constitutional defects caused by pregnancy and delivery complications, exert such strong influences that an understanding of these factors makes it possible to predict a child's later developmental status (Gesell and Armatruda, 1941; Fig. 45–2). Although this theory is still very popular among physicians, longitudinal studies have not given it much credence. For instance, although a study of several hundred newborn infants in St. Louis, Missouri, demonstrated that anoxic infants scored more poorly in the neonatal period and at three years of age, it is also showed that the anoxic infants showed minimal impairment of function by seven years of age (Graham et al., 1956). Other studies undertaken to predict later developmental status on the basis of biological factors also proved most prediction to be unreliable. Similarly the classic studies of prematurity by Drillien (1963) demonstrated that later IQs corresponded to birth weights, but there was still a major degree of variation within each birth weight group.

Effect ⟶ Outcome

Figure 45–2. Main effect model.

INTERACTIONAL MODEL

Although the previous model proposes that nature primarily determines later developmental status, the interactional model proposes that nature and nurture interact to produce a predictable outcome (Fig. 45–3). Thus, while children with constitutional problems raised in deviant environments have poor outcomes, those raised in nurturing environments may well have a good outcome. Drillien (1963) lent support to this theory by demonstrating that low birth weight infants raised in high socioeconomic status families later

Figure 45–3. Interactional model.

had IQs that were higher than those of identical infants reared in low socioeconomic status families.

TRANSACTIONAL MODEL

The transactional model stresses the plasticity of developing children and their environments (Fig. 45–4). This theory emphasizes that the child and the environment continuously interact upon each other and produce changes in each other. Thus accurate later predictions are most difficult.

The experience of some clinicians suggests that there is an element of truth in each of these theories. Some genetic or biological traits and some environmental factors yield fairly predictable outcomes, and thus support the main effect model. As the geneticist Dubzhanski stated, infants are born with a "norm of reaction" or a "reaction range." The clinician often does not know the breadth of that range, since individuals differ greatly. The environment sets limits of variability and interacts with the biological factors, yielding some predictable outcomes. This supports the interactional model. What is not known at this time is the range or plasticity of the biological and environmental factors. Therefore, predictions of later developmental outcome in the individual child using the transactional model is subject to a moderate degree of error.

Although it is generally agreed that most clinicians are limited in their ability to modify an individual's genetic or biological make-up, there are nevertheless medical treatments that mitigate these handicapping conditions, such as drugs for epilepsy, physical therapy or surgery for limb contractures, and hearing aids for the hearing impaired. In light of the foregoing theories of child development, the clinician must assess both the biological status and the home environment of a child to determine the degree to which each may be fostering or hindering the child's development. Although some aspects of the child's biological

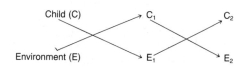

Figure 45–4. Transactional model.

make-up and environment may have little plasticity, for the individual case the degree of plasticity available is difficult to determine if no effort is made to intervene.

RATIONALE FOR PERIODIC DEVELOPMENTAL ASSESSMENT OF ALL CHILDREN

When considering how best to foster the optimal development of all children, one might ask what the range of plasticity of development is at different ages. To state it another way, one might ask, What is the rationale for early intervention? This is a complex issue, since the development of a given child is dependent upon a myriad of biological and environmental factors. Certainly some of these conditions are more plastic or amenable to treatment than others. In most cases it is impossible to predict the outcome of early intervention solely on the basis of a presenting symptom (such as a developmental deviation) without evaluating a child to determine the causative factors. Even when one knows the etiology of a particular child's developmental retardation (which does not prove to be the case in about 50 per cent of children evaluated in most developmental clinics), it is very difficult to predict the degree to which one can reverse the biological or environmental factors deterring a child's development. As in the case of many chronic handicapping pediatric conditions (such as cystic fibrosis, childhood rheumatoid arthritis, and diabetes), the physician caring for the developmentally retarded child is left to design a specific intervention program even when there is uncertainty about the program's efficacy in effecting a cure.

It is not the purpose of this chapter to provide a lengthy review of the efficacy of early intervention, but the subject must be briefly addressed, since the most common reason physicians give for not undertaking periodic and routine developmental assessment of all children is their view that early intervention in cases of mental retardation is of little if any value. Fortunately, 75 per cent of the mentally retarded population is of the cultural-familial type, in which IQs range from 50 to 75. One researcher has stated that more than half of these individuals could achieve IQ scores above the retarded range if a more stimulating environment were provided. Early intervention is even efficacious in the case of organic mental retardation, as in the Down syndrome. For instance, of the children with the Down syndrome who received early intervention at one state hospital, 90 per cent had vocabularies of 250 words at eight years of age (Hayden and Haring, 1974). The

University of Minnesota's program for early treatment of children with the Down syndrome demonstrated greater increases in cognition and communication skills among the experimental than among the control group (Rynders and Harrobin, 1972). Other studies of children with cerebral palsy, hearing and visual impairment, and emotional disturbances have also demonstrated the efficacy of early intervention.

In summary, developmental problems may be due to one or more of a large variety of biological and environmental factors. Some are more amenable to treatment (more plastic) than others. It is the role of each pediatric health care provider to determine the factors that may be retarding a given child's innate developmental potential, and to insure that each child who is found to be functioning at a subnormal level obtains early and appropriate habilitation and education. Since developmental problems may arise at any time in a child's life, as is the case for medical conditions such as tuberculosis, it is the responsibility of the clinician to reassess periodically the development of all children if early treatment is to be effected.

ROUTINE DEVELOPMENTAL SCREENING AS AN ESSENTIAL PART OF HEALTH MAINTENANCE

What case can be made for the routine developmental screening of all children? Since physicians caring for infants and preschool age children have frequent contact with the children and their families during the formative years from birth to six years of age, those physicians are in a strategic position to identify handicaps in the early stages. Although most pediatricians assume that they are astute in identifying children with significant developmental delays, studies and experience have shown this not to be true.

How do most primary care physicians currently attempt to identify such handicapping conditions? Surveys of pediatricians from 1964 to 1974 in Great Britain and the United States have shown a very consistent pattern. The physicians rely upon a history of developmental landmarks; clinical judgment, which includes a so-called response to the clinical situation; a few motor developmental landmarks, and, later, speech and language ability. Various studies have consistently demonstrated these methods to be insufficient, since recall of developmental landmarks is grossly inaccurate, and major milestones of achievement such as sitting, standing, and walking may be normal in 50 per cent of all retarded children. In a study by Bierman et al. (1964) of 681 two year olds, pediatricians correctly identified only three of 11 retarded children. In another study by Korsch et al. (1961) it was noted that pediatricians consistently over-rated the IQs of retarded children. In other words, the physicians who were studied failed to identify the retarded adequately.

The findings in these studies are more than 10 years old. It might be expected that the situation has changed today, since developmental screening is taught in most pediatric training programs. In view of this, it is disturbing to find a recent study indicating that only 10 per cent of pediatricians surveyed actually screened the development of all children, even though early identification of developmental deviations through routine screening is one of the most productive activities a physician can undertake. Such screening is productive because developmental problems are some of the most common conditions for which screening can be done, and because early treatment is generally more effective than treatment instituted at the usual time. Developmental disabilities in children are far more numerous than such problems as renal disease, diabetes, tuberculosis, and heart disease, for which most physicians screen all children during the first few years of life.

The main reason cited by physicians for failing to conduct routine developmental screening was "lack of time." This situation is alarming, since professionals and the lay public rely upon physicians to identify the majority of developmentally delayed children as early as possible. It is not uncommon to hear a parent say, "If there is anything wrong with my child, the doctor will catch it."

Since physicians often fail to identify delayed children during the first few years of life, it is often not until these children reach school age that their deviations are revealed. It is this failure of health care providers to identify and refer such children to the schools that has resulted in legislation mandating that each school district identify all the deviant children within their district as early as possible. In other words, it is the health profession's failure to identify these children early that has prompted the schools to step in and play this role. The schools' decision to do so has already created a conflict as to whose responsibility it is to play this role. If pediatricians and family physicians continue to abdicate their role in early identification and remediation of handicapping conditions, they are no longer fulfilling their primary role as put forth by Nelson (1969).

DEVELOPMENTAL SCREENING

One can make a strong case for developmental screening to facilitate early treatment and education if the problems are treatable, if earlier than

Table 45-1. CHARACTERISTICS OF AN IDEAL SCREENING TEST

Reliability. Results should be consistent from one testing time to the next (test-retest stability) and between two persons observing or taking the same measurements (inter-rater reliability).

Validity. Accuracy should be achieved in separating diseased from nondiseased subjects, and results should be confirmed by the diagnosis. The test should have predictive and concurrent validity, ferreting out the majority of individuals who will become handicapped if not treated.

Suitability for Mass Testing. To facilitate the screening of large numbers of persons, procedures should be acceptable, simple, quick, and economical.

—Acceptability suggests that the test should not be painful, embarrassing, or discomforting to those undergoing screening. Also the test should be acceptable to professional persons who will administer follow-up procedures.

—Simplicity refers to the ease with which the test can be administered. Complex procedures and equipment are likely to be cumbersome and often are unavailable in sparsely populated areas.

—Quickness is desirable because the faster the procedure, the more persons who can be screened in a unit of time, and the lower will be the cost of administering the test.

—Economical costs of collecting the specimen and administering and interpreting the test should be considered and kept at a minimum.

Appropriateness for the Screening Population. The test may be appropriate for one population but inappropriate for another. For instance, telebinocular vision screening instruments are inappropriate for preschool children, who are unable to focus into the eyepieces, whereas they are appropriate for older children.

usual treatment improves the outcome, and if the condition is relatively prevalent. Yet the final criterion—whether it is ethical to screen for a disease or handicap—rests with the availability of an appropriate screening test or procedure. The ideal screening test is quick, simple, economical, appropriate to the population, reliable (reproducible), and above all accurate (Table 45-1). At present no single developmental screening test meets all these criteria. Procedures that are quick, simple, and economical have not been demonstrated to be highly reliable or valid. Those that are valid are generally not quick, simple, or economical. (See also Chapter 50.)

The importance of the foregoing criteria becomes apparent when one considers several facts. There is the need to rescreen periodically the development of all children; the prevalence of developmental problems is about 10 per cent; and early identification and treatment are essential for

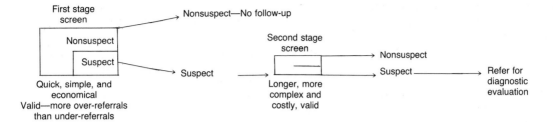

Figure 45–5. Two stage screening process.

many developmental delays. In addition, the importance of validity is underscored because false negative screening results will keep children with problems from obtaining prompt appropriate treatment, and over-referrals of children who do not have problems will lead to the use of costly diagnostic procedures, which often are only available on a limited basis.

TWO STAGE SCREENING

To achieve a screening process that meets all the criteria of the ideal screening test, a two stage developmental screening process has been recommended. This process is illustrated in Figure 45–5. In such a two stage process the first stage is quick and simple, making it economical; it is also valid or accurate in identifying all the developmentally handicapped children screened. It is essential that such a first stage screen err in terms of over-referrals (false positives) rather than under-referrals. This is important because under-referrals would not be identified until a later re-screening, when treatment may not be as efficacious.

In such a two stage screening process all the children who are suspect on the first stage screen are subjected to the more lengthy, more complex, and therefore more costly second stage screen. Such a second stage screen may take the form of a test that must be administered by a trained assistant, who in turn requires space and time to perform such an evaluation. The main purpose of the second screen is to decrease the number of false positive findings generated by the first stage screen while not increasing the number of false negatives.

In such a two stage process, as many as 75 per cent of the subjects may be nonsuspect as a result of the first stage screen; this means that only 25 per cent would require the more lengthy second screen. In the case of developmental screening, only about 10 per cent of the children would have positive results on the second screen, and therefore only 2.5 per cent of the total population

screened would require referral for a diagnostic evaluation.

A similar two stage screening process was employed by McDonald et al. (1963) for screening a large number of government workers for diabetes mellitus. The first stage consisted of a two hour postprandial blood sugar test with the criteria for positive set low enough to generate few under-referrals at the expense of over-referrals. Individuals who were suspect on the first stage were rescreened with the longer, more costly but more valid glucose tolerance test.

Another and more common example is the screening for airplane hijackers, in which the first stage requires walking through a large metal detector, and the second stage requires a more expensive individual examination by a person using a hand-held metal detector, which is "brushed" over the entire body to identify metal. For such a two stage process to work, and to minimize the chance of a plane hijacker going through the screening process without being identified, each stage of the screen must be highly sensitive; i.e., each stage must miss few hijackers.

Two Stage Developmental Screening for Infants and Preschool Aged Children

The recommended two stage developmental screening processes have generally required that the first stage consist of either a parent-answered questionnaire or a very brief prescreening test, while the second stage consists of a test applied by a specially trained assistant. The following presents an overview of some of the more commonly used first and second stage screening procedures.

First Stage Screen: Use of Parent Questionnaire

Revised Parent Developmental Questionnaire. One developmental questionnaire, covering the age span from four weeks to 36 months, was devised by Knobloch et al. (1980). This parent-answered questionnaire is designed to assess five areas of behavior (adaptive, gross motor, fine motor, language, and personal-social) as well as parental reports of hearing, vision, seizures, and other concerns. When a child is 28 weeks of age, 20 to 30 minutes are required to complete the questionnaire. Results are scored as normal, questionable, or abnormal. The questionnaire was validated at only one age—28 weeks— by comparing parental reports with the results of a 40-week Gesell Developmental and Neurological Examination. False negative results were obtained in 12.6 per cent (2.6 per cent for major anomalies and 10 per cent for minor anomalies). False positive results were obtained in 6 per cent. If one employs this questionnaire, it is important that the second stage or "back-up" be Knobloch's Revised Developmental Screening Inventory.

Denver Prescreening Developmental Questionnaire. The Prescreening Developmental Questionnaire (PDQ) was created to identify children who require a more lengthy and thorough screening with the Denver Developmental Screening Test (DDST; Frankenburg et al., 1976).

To devise the PDQ, 97 of 105 standard DDST items were formulated into questions. To administer the questionnaire, parents are asked to answer 10 age appropriate questions, a process that takes only five minutes. The PDQ was validated against the DDST for the entire age span from three months to six years. In a study of 1155 cases, parental responses agreed 93.3 per cent with DDST items. First stage screening revealed 31.2 per cent suspects who required a second stage screen.

In a subsequent cross validation study among parents having at least a high school education, the PDQ was found to serve as an excellent first stage or prescreen, since it identified all the children who were mentally retarded (as identified with the Revised Bayley Scales of Infant Development and the Stanford-Binet scale). Results among parents having less than a high school education were less accurate, suggesting that the PDQ should not be used with these parents.

Minnesota Infant Development Inventory. This scale, consisting of 75 items and covering the age range from one to 36 months, holds promise, since its authors, Ireton and Thwing, are developing and validating it very thoroughly. Currently only a research form of the inventory is available.

First Stage Screen: Use of Brief Tests

Abbreviated Denver Developmental Screening Test (DDST). Although the Denver Developmental Screening Test (DDST) has enjoyed worldwide popularity because of its simplicity and validity, its chief drawback for mass screening

purposes is the time required to administer it (15 to 20 minutes). To overcome this disadvantage, a two stage screening process—a first stage consisting of an abbreviated version of the DDST and the second stage consisting of the full DDST—was devised.

The abbreviated-to-full DDST process allows immediate follow-up on suspect first stage screening results by the clinician before the child leaves the test setting. This is especially important when it is difficult to guarantee that a child will return for a second stage screening appointment.

To administer the abbreviated DDST, one administers only 12 items instead of the usual 20 to 25 items. This process is highly sensitive, since it identifies all the full-DDST suspects in half the usual time. In a random screening sample only 25 per cent of the subjects would be suspect on the first stage; thus, only 25 per cent of the children screened would require the completion of the full DDST (about 10 more minutes of time).

Second Stage Screen: Use of Developmental Questionnaire

Denver Prescreening Developmental Questionnaire. Although a large variety of developmental questionnaires exists, almost none have been designed as a second stage screen. The Denver Prescreening Developmental Questionnaire (PDQ), although originally designed as a first stage screen, has (in a subsequent study involving 10,000 children) proved to be effective as both first and second stage screens. Persons wishing to use the PDQ as both first and second stage screens should review the description of the PDQ as a first stage screen in the foregoing discussion. To use the PDQ as a second stage screen, one readministers the PDQ two to three weeks later and uses the same scoring criteria as for stage 1. Children who are suspect on the second stage PDQ should be referred for a full diagnostic evaluation. Use of the DDST as a follow-up of the second PDQ has proven to be less cost efficient, and therefore is no longer recommended.

Second Stage Screen: Use of Observational Tests

Revised Developmental Screening Inventory. The oldest of the screening measures discussed here is the Revised Developmental Screening Inventory (RDSI) developed by Knobloch et al. (1980). It encompasses the age range from four weeks to 36 months and is designed to assess five fields of behavior. For each area of behavior the examiner assigns a maturity level (in weeks or months) and one of three diagnostic categories—abnormal, borderline or questionable, and normal or advanced. The test has not been standardized, but has been compared with other methods of evaluation to establish reliability and validity. A comparison of individual RDSI results with complete developmental evaluations (consisting of the Gesell Developmental and Neurological Examination) has shown substantial agreement. The RDSI's validity as a predictor of later abnormality was 0.81.

The inventory requires a greater level of sophistication among evaluators than other screening procedures, and was designed primarily for physicians and nurses. It can be integrated into regular evaluations by neurologists, pediatricians, and child psychiatrists. However, the lack of standardization data, the limited number of well designed reliability and validity studies, the limited age span, and the 20 to 30 minutes required to evaluate each child limit its usefulness in mass screening endeavors.

Denver Developmental Screening Test. The Denver Developmental Screening Test (DDST) is a device for the detection of developmental delays during infancy and the preschool years. The test yields an overall developmental profile with special emphasis on gross motor, language, fine motor-adaptive, and personal-social skills. It has been well constructed and evaluated, and the few attractive materials and clear pictorial charts make it simple to administer, score, and interpret. Results of the evaluation are classified as abnormal, borderline or questionable, untestable, and normal (Frankenburg and Dodds, 1967).

The 105 test items were drawn from 12 developmental and preschool intelligence tests. The items were administered to 1036 infants and children ranging in age from two weeks to 6.4 years. Each item is depicted by a bar, which is located under a reference line to clearly depict the ages at which 25, 50, 75, and 90 per cent of the children passed each item. Approximately 20 age appropriate items are administered to determine whether a child can perform the majority of tasks passed by 90 per cent of the normative sample.

Test-retest reliability (stability) over a one week interval was 95.8 per cent for the age span from two months to 5.5 years. Interobserver agreement was 90 per cent. DDST results of abnormal, questionable, and normal correspond with mean developmental quotients and intelligence quotients of 69.13, 83.75, and 95.74, respectively. In a study of the test's predictive validity, it was determined that infants and preschool aged children who scored abnormally on the DDST had an 89 per cent probability of failing in school five to six years later. The test was demonstrated to show very minor developmental differences when holding social class constant and comparing Anglo-American, Hispanic, and black children. Thus the test is perceived not to yield ethnic biases.

The proper administration of the test can be learned by anyone who has completed grade school. It takes about 10 to 15 minutes to screen each child. The training time varies from a few hours for physicians and nurses to a few days, depending upon the level of formal education of the trainee and the availability of children for practicing test administration.

Test users should remember that the DDST is not a diagnostic instrument, and thus the results should not be used to plan treatment. The test's ease of administration and interpretation, as well as its high degree of validity, perhaps explains why it is used to screen several million children annually throughout the world.

Other Developmental Screening Tests. Although it is beyond the scope of this presentation to describe all the available developmental screening tests, a few of them are summarized in Table 45–2.

SCREENING OF THE HOME ENVIRONMENT

Since the child's home environment is a major determinant of future development, it is appropriate to use a combined developmental screen of the child and a screen of the home environment. The rationale for this approach is discussed earlier in this chapter under "Developmental Process." In a combined developmental-environmental screen the former ascertains a child's developmental status at a given time and therefore reflects the child's biological integrity and past experiences. The environmental screen increases the predictive accuracy of later development by determining whether a child's given environment is likely to promote or retard future development. Through this scheme the child who is abnormal on a developmental screen and who resides in a home that does not promote development is highly suspect, and is unlikely to manifest normal development when he or she reaches school age. Similarly, the child who is nonsuspect on the developmental screen and has a supportive home environment is highly likely to develop normally in the future.

Assessment of a child's home environment is still an evolving process. To date, the best known screening instrument to determine the home environment's potential for fostering cognitive development is Caldwell's HOME Scale (Bradley and Caldwell, 1977). Although the authors have recommended it as a screen, the requirement of one to three hours' total time to make a home visit preclude it from being used to routinely screen

Table 45–2. PARTIAL LISTING OF AVAILABLE SCREENING TESTS

Name	Kind	Standardization Population	Time	Administration	Age	Source
The Denver Developmental Screening Test (DDST)	Developmental	Children representative of Denver population	10–25 min.	Individual	Birth to 6 years	LADOCA Pub. Found., E. 51st Ave. & Lincoln St., Denver, CO 80216*
The Developmental Screening Inventory (DSI)	Developmental	None for this test; items selected from Gesell developmental schedules	20–30 min.	Individual	4 weeks to 18 months	Department of Peds., Albany Medical College, Albany, N.Y. 12208†
The Goodenough-Harris Drawing Test	Developmental	Private preschool and public school children	15–30 min.	Individual or group	3 to 12 years	The Psychological Corp., 1372 Peachtree St. NE, Atlanta, GA 30309
The Peabody Picture Vocabulary Test (PPVT)	Intelligence	Private preschool and public school children	10–15 min.	Individual	2 years, 6 mo. to adulthood	American Guidance Svc., Publishing Building, Circle Pines, MN 55014
The Preschool Attainment Record (PAR)	Developmental	None as such; item placement based on other tests	10–15 min.	Individual	6 mo. to 7 years	American Guidance Svc., Publishing Building, Circle Pines, MN 55014
The Quick Screening Scale (QSS)	Developmental	None as such; item placement based on other tests	30–45 min.	Individual	1 to 9 years	The Psychometric Affil., Chicago Plaza, Brockport, IL 62910
The Quick Test (QT)	Intelligence	Caucasian children and adults without geographical representatives	10–15 min.	Individual	1.5 years to adult	Psych. Test Specialists, Box 1441, Missoula, MT 59801
The Slossen Intelligence Test (SIT)	Intelligence	Urban and rural New York children	15–30 min.	Individual	1 mo. to adult	Slossen Education Pub., P.O. Box 280, East Aurora, NY 14052
The Raven Progressive Matrices	Intelligence	Scottish children	Up to 30 min.	Individual or group	5 to 11 years	The Psychological Corp., 757 Third Avenue, New York, N.Y. 10017
The Vane Kindergarten Test (VKT)	Intelligence	Urban and rural children from NE U.S.	24–40 min.	Individual or group	4 to 6½ years	Clin. Psych. Pub. Co. Inc., 4 Conant Sq., Brandon, VT 05733

*PDQ and HSQ also available from this source.
†Revised Parent Developmental Questionnaire also available from this source.

How many children's books does your child have of his *own*?

_____ 0—too young
_____ 1 or 2
_____ 3 to 9
_____ 10 or more

About how often do you take your child to the doctor?

Do you have any friends with children about the same age as your child?

Yes _____
No _____

Figure 45–6. Sample questions from the Home Screening Questionnaire.

large numbers of children. To overcome this disadvantage, Coons et al. (1981) developed the Home Screening Questionnaire (HSQ), which comes with two age forms. It must be acknowledged that there is no published report of the use of either the HOME Scale or the HSQ in a primary health care setting.

Home Screening Questionnaire (HSQ). The two versions of the Home Screening Questionnaire (HSQ) are those for birth to three years and three to six years; these correspond to the HOME Scales developed by Caldwell. Interview items from the HOME Inventory Scales were selected and were reconstructed into questions (30 for 0–3 HSQ and 34 for 3–6 HSQ). The questions were revised to obtain the highest possible degree of agreement between a parent answered HSQ and a HOME Inventory completed by a trained interviewer. Sample questions from the HSQ are shown in Figure 45–6. Both forms of the HSQ are parent answered and require approximately 15 minutes to complete. The questions range from the third to sixth grade reading level and are appropriate for use with children who reside in poverty circumstances. Although prediction of later school problems can be improved through the additional assessment of the home environment, one must also recognize the limitations suggested by Sameroff (1975) in discussing his transactional model of development. He wrote that "One must monitor both the child and environment continuously throughout development" to make relatively accurate predictions, since the child and the environment act upon each other to produce changes, and these changes continue to affect both the child and the environment.

RECOMMENDATIONS FOR DEVELOPMENTAL-ENVIRONMENTAL SCREENING OF INFANTS AND PRESCHOOL AGED CHILDREN

Many approaches to developmental screening have been advocated. The reasons for these many approaches are the variety of definitions of what is meant by "development." For instance, for some, development may encompass temperament, neurological status, cognitive states, emotional status, and social status, as well as other areas. Another reason for the broad diversity of approaches is that in the psychosocial field there has been a tendency to develop a variety of screening tests, which generally are gleaned from diagnostic tests without regard to the accuracy of the screening test. The approach suggested here is to screen for cognitive problems and other problems that cause school failure. The reason for selecting school failure is that, in contrast with most other aspects of developmental screening, it meets most of the criteria in selecting conditions for screening. That is to say, the problem is highly important to the individual and to society; early treatment improves outcome; a firm diagnosis is possible; treatment before the usual time is most efficacious; adequate resources are available to diagnose and treat the problem; the financial savings achieved through early treatment far outweigh the total cost of screening, diagnosis, and treatment; and valid screening procedures are available.

Although there are a number of viable alternatives as to methods of screening, the following represents one approach that has evolved over the past 14 years. Various parts of this approach have been employed throughout the world. The general format is to use a two phase screen involving the abbreviated DDST followed by the full DDST, and the HSQ followed by a home visit using an interview (such as the HOME Scale). Since the prevalence of school failure is far higher among children residing in homes of low socioeconomic status where the parent(s) is likely to have less than a high school education, a selective screening approach is proposed. This could involve a more intensive screening using the HSQ for high risk, low socioeconomic status families. Screening among the higher socioeconomic status families, in which the prevalence of school failure is far less, is more selective and is confined to use of the DDST or PDQ.

SCREENING POPULATIONS IN WHICH ONE OR BOTH PARENTS HAVE LESS THAN A HIGH SCHOOL EDUCATION

The process described consists of a first stage screen with the routine administration of the abbreviated DDST at the following seven age groupings: three to six months, nine to 12 months, 18 to 24 months, and three, four, five, and six years of age (Fig. 45–7). This first stage screen can be incorporated readily into the medical evaluation at recommended ages, and requires only about

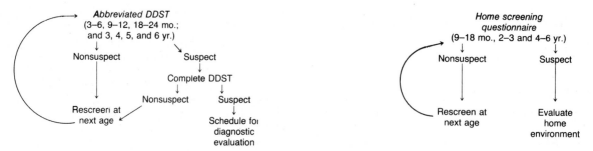

Figure 45–7. Recommended two stage screen for children of families of low socioeconomic status (one parent has less than a high school education).

five minutes of additional time. Children who are suspect on the abbreviated DDST should receive the full DDST immediately on the same occasion. Children who are still suspect on the second stage, full DDST are scheduled for a complete diagnostic evaluation. Children who are nonsuspect on the full DDST are scheduled for the next periodic rescreen with the abbreviated DDST.

The second routine screen for all children is that of the home environment with the HSQ, which is scored by a nonprofessional assistant within a period of about five to eight minutes.

Just as all children should undergo routine and periodic rescreening with the abbreviated DDST, they should also undergo periodic rescreening with the HSQ or their equivalents. Between birth and six years it is recommended that the HSQ be administered three times (at nine to 18 months, two to three years, and four to six years of age). Nonsuspect scores should be noted and rescreening should be performed at the next recommended age. Suspect scores warrant making a home visit to obtain a firsthand view of the home environ-

ment. A procedure that may be used as part of such a home evaluation is the HOME Scale (Bradley and Caldwell, 1977). The findings during the home visit should be considered in further planning (such as planning for parent education, parent counseling, or other procedures). The logistics of such home visits have yet to be worked out.

SCREENING CHILDHOOD POPULATIONS WHEREIN BOTH PARENTS HAVE AT LEAST A HIGH SCHOOL EDUCATION

Although some clinicians may prefer to use the Denver PDQ, the abbreviated DDST followed by the full DDST is both an economical process and one that can be easily incorporated into the routine and periodic medical evaluations.

The procedure is identical to that for children whose parent(s) has less than a high school education, except that the HSQ is used on a selective basis (Fig. 45–8). That is to say, the HSQ is used

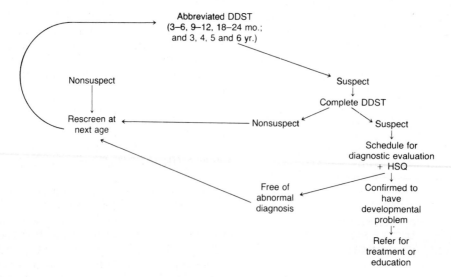

Figure 45–8. Recommended two stage screen for children of families of middle and high socioeconomic status (both parents have graduated from high school).

here routinely only as part of the diagnostic evaluation.

DIAGNOSTIC EVALUATION

Since developmental problems may be due to one or more of a variety of causes, it is necessary to consider the various possibilities. These may include sensory or neurological impairment, speech and language problems, metabolic and genetic problems, abuse and neglect, and impaired intellectual-developmental function. The tests used to confirm the presence of disability generally consist of one or more of a variety of measures such as the Bayley Scales of Infant Development, the Gesell test, the Stanford-Binet scale, and others. These tests must not be confused with screening measures, since they are diagnostic procedures that require extensive time (generally at least an hour) as well as administration and interpretation by a specialist who has had special training in this field.

It is recommended that the home environment be evaluated routinely in the diagnostic stage in an attempt to rule out neglect as a factor that may partially be responsible for the child's slow development. Such an evaluation can be done with the HSQ or a home visit (if the home evaluation has not been performed recently) even if no one suspects the home environment to be contributing to the child's developmental problem.

CONCLUSION

Formal developmental screening should be incorporated into all routine health maintenance protocols for a number of reasons:

1. Early identification and treatment of developmental problems are more efficacious and less costly.

2. The common practice of relying upon a developmental history and uncritical clinical observation fails to identify the majority of children with mild but significant developmental delays.

3. Developmental problems in the infant and preschool age population are far more common than the combined prevalences of phenylketonuria, hypothyroidism and other genetic disorders, diabetes, and tuberculosis—for which formal and routine screening have been incorporated into most pediatric practices.

4. The use of a two stage developmental screening program in pediatric practice is appropriate, since it requires a minimum of time and space while still providing accurate results.

5. The Home Screening Questionnaire is ideally suited for selective and routine screening of children whose parents have less than a high school education. It is also ideally suited for use as part of the diagnostic evaluation if the home environment has not been evaluated already.

6. All children who are developmentally suspect require a thorough diagnostic evaluation, which generally can be provided by the child's physician together with two or three consultants.

The incorporation of these developmental screening methods into routine health maintenance can be one of the most rewarding child health activities undertaken by physicians, since screening enables the physician to play one of the major roles of a pediatrician, permitting children to reach adulthood at their optimal state of development.

WILLIAM K. FRANKENBURG

REFERENCES

Bierman, J. M., Connor, A., Vaage, M., and Honzik, M. P.: Pediatricians' assessment of intelligence of two year olds and their mental test scores. Pediatrics, *34*:680, 1964.

Bradley, R., and Caldwell, B.: Home observation for measurement of the environment: a validation study of screening efficiency. Am. J. Ment. Defic., *81*:417–420, 1977.

Coons, C. E., et al.: Home Screening Questionnaire Reference Manual. Denver, Colorado, J.F.K. Child Development Center, 1981.

Drillien, C. M.: The Growth and Development of the Prematurely Born Infant. Edinburgh, E. & S. Livingstone, 1963.

Frankenburg, W. K., and Dodds, J. B.: The Denver Developmental Screening Test. J. Pediatr., *71*:181, 1967.

Frankenburg, W. K., et al.: The Denver Prescreening Developmental Questionnaire. Pediatrics, *57*:744–753, 1976.

Gesell, A., and Armatruda, C.: Developmental Diagnosis. New York, Paul B. Hoeber, 1941.

Graham, F. K., Matarazzo, R. G., and Caldwell, B. M.: Behavioral differences between normal and traumatized newborns. II. Standardization, reliability and validity. Psychological Monographs, *70* (21, whole no. 428), 1956.

Hayden, A., and Haring, N.: Programs for Down's syndrome children. Paper presented at Conference for Early Intervention with High Risk Infants and Young Children, University of North Carolina, Chapel Hill, May 5–8, 1974.

Knobloch, H., Stevens, F., and Malone, A. F.: Manual of Developmental Diagnosis, Hagerstown, Harper & Row, 1980.

Korsch, B., Cobb, K., and Ashe, B.: Pediatricians' appraisals of patients' intelligence. Pediatrics, *27*:990, 1961.

McDonald, G. W., et al.: Large scale diabetes screening program for federal employees. Public Health Rep., *78*:553, 1963.

Nelson, W. E., Vaughan, V. C., III, and McKay, R. J.: Textbook of Pediatrics. Philadelphia, W.B. Saunders Company, 1969, p. 1.

Rynders, J., and Harrobin, M.: Enhancement of communication skill development in Down's Syndrome children through early intervention. Annual Report: Research Development and Demonstration Center in Education of Handicapped Children. Minneapolis, University of Minnesota Press, 1972.

Sameroff, A. J.: Early influences on development: fact or fancy? Merrill-Palmer Quart., *21*:267–294, 1975.

45B The Developmental Assessment of the School Age Child

It is common in pediatric practice to assess development in very young children. Direct observations have been standardized and organized into screening and assessment instruments (as described earlier in this chapter). Developmental examinations permit the clinician to observe specific areas of function in a systematic manner. Age appropriate tasks detect weaknesses or delays as well as domains of strength or accelerated attainment. A developmental examination need not generate a specific score or percentile ranking such as one might derive from psychometric or achievement tests. Within health care settings, developmental testing, as such, has been limited largely to children under the age of six. This section explores the extension of developmental assessment up through the school years.

JUSTIFICATIONS

As children enter school, there may be a tendency to believe that development is assessed adequately through school performance and, when indicated, by tests of intelligence, such as the WISC-R or the Stanford-Binet scale. Although these methods of evaluation have major implications regarding development, relevant data about a child's function can be assembled by complementing such findings with other standardized observations. As part of the developmental model, a clinician attempts to account for specific components of function (i.e., strengths, weaknesses, and preferred styles). Some of these may be measured only indirectly by intelligence tests and school performance. In evaluating development in a school child, one starts with certain areas that one wishes to account for and then determines the most appropriate or practical methods of data collection or direct observation. Thus, a determination may be made that gross motor proficiency is of importance to school age children and therefore should be included in a developmental assessment process. Parent interviews or questionnaires can be used to cull some evidence about the adequacy of performance in various gross motor activities. In addition, age appropriate gross motor items can be part of a direct developmental assessment in a clinical setting. To strengthen such an evaluation, one would use several tasks that might yield information about different components of gross motor performance; this is elaborated upon later in this section.

For the physician or nurse, the developmental assessment of a school child can be part of an expanded neurological examination involving sophisticated observation of higher cortical function and motor output. Often this is combined with an assessment of the maturation of the central nervous system or a search for minor neurological indicators. The entire "package" is known as a neurodevelopmental examination. To be optimally useful in the evaluation of a child with possible developmental delays, the neurodevelopmental examination should fulfill the following criteria:

1. It should be appropriately standardized for age. In any developmental examination it is important that the tasks used to elicit strengths and weaknesses be at the right level of difficulty for the chronological age of the child being assessed. When this is not the case, high false negative and false positive rates are likely to be found. The standardization of such observations can be difficult. Pilot testing of any neurodevelopmental assessment can require large numbers of subjects and a great deal of time. Moreover, the characteristics of the sample population in whom the instrument is standardized can have a major influence on the normative data collected. Cultural and socioeconomic biases, in particular, can bear heavily on the established norms. This is especially true with regard to tests of language ability, in which normal standards can vary widely from community to community, cultural group to cultural group, and even between neighborhoods in the same town. In some instances an examiner must attempt to establish his own "local norms."

2. The neurodevelopmental examination should have demonstrable reliability. In particular, two observers administering or rating the performance in the same child should be likely to record the same or very similar findings. So-called test-retest reliability or the establishment of consistency of performance over time also is helpful.

3. The validity of the examination ought to be documented. A neurodevelopmental examination should have some concurrent validity with other fully standardized instruments. Because it may be measuring somewhat different functions, it need not show perfect agreement with other tools. One also might want to establish predictive validity, especially if the examination is intended to offer any prognostic implications (such as educational readiness).

4. The neurodevelopmental examination should not be designed for use in isolation. It should be

1.0	PERFORMANCE AREA (continued)	TYPICAL PERFORMANCE				VARIABILITY OF PERFORMANCE			
		Strong for age	Appropriate for age	Delayed less than 1 year	Delayed more than 1 year	Consistent performance	Somewhat variable	Highly unpredictable	Cannot say
1.18	Word Pronunciation								
1.19	Comprehension of Verbal Instructions								
1.20	Oral Sentence Structure and Fluency								
1.21	Vocabulary								
1.22	Writing from Dictation								
1.23	Copying Written Material								
1.24	Orienting Letters (b - d, etc.)								
1.25	Keeping Place in Reading								
1.26	Knowing Left from Right								
1.27	Discriminating Similar Words/Letters								
1.28	Retaining Recent Instructions								
1.29	Retaining Yesterday's Lessons								
1.30	Retaining Skills following Vacations								
1.31	Visual Memory								
1.32	Remembering Routines								
1.33	Using Numbers								
1.34	Performing Tasks in Correct Order								
1.35	Getting Letters in Correct Order								
1.36	Getting Words in Correct Order								
1.37	Understanding Time								
1.38	Motivation								
1.39	Imagination								
1.40	Creativity								
1.41	Sense of humor								
1.42	Enthusiasm								

Figure 45–9. A page from a school questionnaire, part of the ANSER System (Levine, 1980). Items are clustered according to specific categories: 1:18–1:22 tap language skills; 1:23–1.27, visual-spatial orientation; 1:28–1.32, memory; 1.33–1.37, temporal-sequential organization; and 1.38–1.42, other relevant attributes. Each child is rated in comparison with peers of comparable age. Consistency levels are included, since many youngsters (especially those with attention deficits) are apt to perform erratically in one or more areas.

melded to a systematic history or the use of standardized parent or teacher questionnaires. These data collection tools should probe the same areas of development that are being observed during the neurodevelopmental examination. An example of this type of instrument is found in the ANSER system, a series of standardized health and developmental questionnaires that have been developed and used at the Children's Hospital Medical Center in Boston (Levine, 1980). These questionnaires also include a Self-Administered Student Profile, in which the child himself is asked to rate specific components of development by identifying with certain quotations from other children who have experienced lags in these areas (Levine et al., 1981a). Some examples of neuro-

developmental items on parent and teacher questionnaires are shown in Figures 45–9 and 45–10. In addition to interview and questionnaire data, the neurodevelopmental examination can be supplemented by standardized testing performed in school or by other professionals in the community. No single set of observations—be it a neurodevelopmental examination, an intelligence test, or parent or teacher reports—should be taken as the ultimate word. Instead it is the compilation and integration of data from multiple sources that is likely to provide the most accurate assessment of development.

5. The neurodevelopmental examination is neither a screening test nor a definitive evaluation. In general, such assessments represent second

7.0	SKILLS AND INTERESTS	Following is a checklist of skills and interests. Please put an X in the box in the column which best describes how well this child performs each skill. In considering each of these skills, it is useful to compare the child to other children his or her own age.			
7.1	Balancing	Falls easily	Seldom loses balance	Has excellent balance	Cannot say
7.2	Throwing or catching a ball	Has difficulty	Can do this	Very good for age	Cannot say
7.3	Carrying things	Often spills things	Sometimes spills things	Rarely spills things	Cannot say
7.4	Running	Slow or awkward	Average	Fast and excellent	Cannot say
7.5	Playing sports	Poor	Average	Excellent	Cannot say
7.6	Using a pencil	Poor	Average	Excellent	Cannot say
7.7	Tying shoelaces	Poor or unable	Adequate	Does easily	Cannot say
7.8	Dressing self	Poor or unable	Adequate	Fast and easy	Cannot say
7.9	Putting things together	Poor	Average	Excellent	Cannot say
7.10	Understanding spoken instructions	Often confused	Sometimes confused	Understands well	Cannot say
7.11	Telling a story	Often confused	Sometimes confused	Does it easily	Cannot say
7.12	Finding the right words for things	Trouble finding words	No problems	Exceptionally good	Cannot say
7.13	Pronouncing words	Often mispronounces	Rarely mispronounces	Never mispronounces	Cannot say
7.14	Remembering spoken instructions	Often forgets	Sometimes forgets	Rarely forgets	Cannot say
7.15	Remembering phone numbers	Often forgets	Sometimes forgets	Rarely forgets	Cannot say
7.16	Remembering familiar places	Often forgets	Usually remembers	Remembers very well	Cannot say
7.17	Remembering things in the right order	Has difficulty	Does adequately	Does very well	Cannot say
7.18	Telling time	Has difficulty	Does adequately	Does very well	Cannot say
7.19	Counting money	Has difficulty	Does adequately	Does very well	Cannot say
7.20	Reading	Has difficulty	Does adequately	Does very well	Cannot say
7.21	Spelling	Has difficulty	Does adequately	Does very well	Cannot say
7.22	Telling left from right	Has difficulty	Does adequately	Does very well	Cannot say

Figure 45–10. A page from a parent questionnaire, part of the ANSER System (Levine, 1980). Parents are asked to rate a range of skills and interests that are grouped as follows: 7.1–7.5 refer to gross motor skills; 7.6–7.9, fine motor function; 7.10–7.13, language abilities; 7.14–7.16, memory; 7.17–7.19, temporal-sequential organization; and 7.20–7.22, other. Such parental perceptions can be used in conjunction with direct assessments as well as a school questionnaire and a self-administered student profile completed by the child.

level diagnostic procedures. That is to say, they are too detailed for screening purposes and too cursory to constitute the ultimate word in any one area of function. Usually the neurodevelopmental examination of a school age child requires nearly an hour. This makes it unlikely that such instruments can be used for community wide screening purposes. A five or 10 minute neurodevelopmental examination in this age group might be dangerous because of the high likelihood of false negative findings in seeking some of the many low severity disabilities.

A history can be used as a prescreening tool to identify the youngsters who could benefit from a neurodevelopmental examination. The latter can then aid in further triage. After completing a neurodevelopmental examination and integrating findings with data from questionnaires, a pediatrician can determine whether a youngster needs a more thorough evaluation of one or more areas, such as motor function, language, or general psychological state. The neurodevelopmental examination can offer the physician an opportunity to observe a child's behavior and coping skills and strategies during the performance of age appropriate tasks (Levine et al., 1980a). In particular, observations of selective attention can be relevant and useful. A child may come to a physician's office and appear to be perfectly attentive and well organized during the history and routine physical examination. However, the neurodevelopmental assessment, by "provoking" the youngster with age appropriate tasks, may elicit the attentional weaknesses that have been described to the clinician by parents and teachers. The direct observations of the physician can help to clarify clinical complaints without attempting the total diagnostic formulation.

7. A neurodevelopmental examination should not be used to generate a label for a child. Because most extended neurological examinations are not summarized as a score or rank, they are largely qualitative assessments. This makes it hazardous to use a neurodevelopmental examination to derive a specific diagnosis, such as "minimal brain damage" or "dyslexia." Instead findings from such an assessment are best summarized as a profile of strengths and weaknesses that work either for or against the acquisition of competence in various life pursuits.

AREAS FOR ASSESSMENT

There are multiple conceptual frameworks that can underlie the design of a neurodevelopmental examination. In this section one particular model familiar to the author is explored and elaborated upon. Other systems of neurodevelopmental assessment also might be defensible. For the purposes of this discussion, eight areas of neurodevelopmental assessment are considered: neuromaturation, gross motor function, fine motor function, visual-spatial orientation, temporal-sequential organization, language, memory, and behavioral and stylistic observations.

This discussion is not intended to constitute a manual of administration for a neurodevelopmental examination. Instead each of the foregoing areas is reviewed with some discussion of the types of observations that might be relevant on an examination. Further information about each of these areas of development and function is contained in Chapter 38.

MINOR NEUROLOGICAL INDICATORS

The documentation of the existence of certain neurological markers commonly found in association with learning problems has become an important part of the pediatric neurodevelopmental assessment. The particular findings sought in such an investigation sometimes are referred to as "soft" neurological signs. Others call them indicators of neuromaturational delay. Two basic types of minor neurological indicators have been described—signs whose clinical significance is dependent upon the child's chronological age and signs that constitute relatively mild versions of "hard" neurological findings that are abnormal at any age. Although the literature has not made this distinction consistently, the latter would appear to be more reflective of some localized neurological deficit, whereas the former often are said to reflect "immaturity."

Most minor neurological indicators that change with age are common in young children and become progressively more unusual during the later elementary school years. The persistence of these neuromaturational indicators beyond the chronological ages at which they usually disappear has been associated with learning disorders, behavioral problems, and other symptoms of developmental dysfunction.

The most commonly elicited minor neurological indicators to be described are summarized in Table 45–3.

SYNKINETIC ("MIRROR") MOVEMENTS

When a toddler or a preschool child is asked to perform a particular act with one hand, it is not unusual to see an almost perfect mimicry of the activity on the contralateral side. For example, when such a youngster is asked to oppose the thumb and forefinger repeatedly, the opposite hand may mirror the action (Grant et al., 1973). As youngsters progress through the elementary school years, synkinetic movements become increasingly rare. At first they are replaced by imitative postures but not dynamic representations of the contralateral movement. To elicit synkinesias in older school children and adolescents one may need to use particularly difficult or complex motor tasks. Persistence of true mirror movements beyond the age of about eight years is unusual and is thought to occur commonly in children with learning and behavioral dysfunctions (Cohen et al., 1967).

In addition to synkinetic or mirror movements, children are likely to show other forms of associated ("overflow") movement—especially when they are young or have learning or behavioral dysfunctions. Thus, a youngster consistently may show rhythmic mouth movements, head bobbing, or back and forth motions of the tongue while using his hands and feet to execute a task. These associated movements most commonly flow in a caudad to craniad direction.

While performing fine and gross motor testing, the physician can be alert to these associated movements of various types.

Dysdiadochokinesis. The ability to perform rapid pronation and supination can be tested by having a child imitate the turning back and forth of a door knob or a light bulb. Such rapid alternating movements elicit the sign of dysdiadochokinesis. Some children are unable to suppress activity in proximal muscle groups while performing a more distal motor act. They then display excessive flailing of the limbs. This is a phenomenon that is very common in preschool children but becomes increasingly associated with dysfunction in older youngsters (Toumen and Prechtl, 1970).

Stimulus Extinction. Young children have difficulty in perceiving simultaneously presented sensory stimuli. In some instances more proximal stimuli are noted over distal ones. Two point discrimination in general may be poor. Thus, a child with eyes closed may be touched simultaneously on the hand and face but report only the stimulus on the face (sometimes called rostral dominance). This response pattern is common in young children but generally is not encountered after the age of seven (Kraft, 1968). Its persistence may be associated with developmental dysfunction. Recent investigations, however, have indicated that poor performance sometimes may be related to expectations or "mind sets" (Nolan and Kagan, 1978). That is to say, conceptual presumptions (concerning interpretation of the instructions) rather than neuromaturational delay may affect the results.

Finger Agnosia. A youngster's ability to locate

Table 45–3. INDICATORS ASSOCIATED WITH NEUROLOGICAL IMMATURITY*

Task	Immature Response	Norms (References)
Rapid alternating hand movements (pronation-supination)	Dysdiadochokinesis	Mature by 7 years (progresses over 4–8 years; Grant et al., 1973)
	Synkinesia	Markedly decreased after 9 years (Cohen et al., 1967)
Repeated finger-to-thumb apposition	Awkward execution	Mature after 8 years (Grant et al., 1973)
	Synkinesia	Markedly decreased after 9 years (Cohen et al., 1967)
Sequential apposition of thumb to each of the fingers, forward and backward	Synkinesia	Markedly decreased after 9 years (Cohen et al., 1967)
Alternating squeezing and relaxing single hand grip of rubber toy	Synkinesia	Markedly decreased after 9 years (Cohen et al., 1967)
Finger localization (identification of stimulated fingers)	Inability to respond correctly	Mature level reached at 11–12 years (progressive development beginning before 6 years; Benton, 1959)
Finger differentiation (discrimination of simultaneous stimulation of two fingers of one hand)	Inability to discriminate	Mature response by 95% at 7 years (Kinsbourne and Warrington, 1963)
Double simultaneous stimulation of face and hand	Extinction of distal stimulus (rostral dominance)	5 years—14% mature 6 years—54% mature 7 years—88% mature (Kraft, 1968)

Identification of right and left on self (laterality) — Incorrect response

N	Age (Years)	Percentage Correct
82	4	32
92	5	59
65	6	73
25	7	92
20	8	95

Berges and Lezine, 1965.

Identification of right and left on examiner — Incorrect response

N	Age (Years)	Percentage Correct
82	4	7
92	5	15
65	6	27
25	7	48
20	8	70

Berges and Lezine, 1965.

Execution of crossed commands on self (e.g., "Touch your left knee with your right hand") — Incorrect response

N	Age (Years)	Percentage Correct
82	4	7
92	5	32
65	6	61
25	7	84
20	8	90

Berges and Lezine, 1965.

Task	Immature Response	Norms (References)
Consistent hand preference (hand dominance)	Inconsistent persistence	Established by 7 years
Stabilization of ipsilateral eye and hand preference	Mixed preference ("mixed dominance")	Established by 8 years

Indicators Associated with Classic Abnormalities ("Equivocal Hard Signs")

Motor impersistence (Garfield, 1964)	Mild ataxia
Choreiform syndrome (Prechtl and Stemmer, 1962)	Nystagmus
Slight asymmetry of deep tendon reflexes	Strabismus
"Hyperactive" deep tendon reflexes	Borderline abnormalities or asymmetry of muscle tone
Awkward postures or gait	Equivocal abnormal reflexes (e.g., Babinski)
Mild tremors	

*Adapted from Levine, M. D., et al. A Pediatric Approach to Learning Disorders. New York, John Wiley & Sons, Inc., 1980.

his fingers in the absence of visual cues can be tested by asking him with eyes closed how many digits the examiner is touching (Kinsbourne and Warrington, 1963). Alternatively one may ask the child how many of the digits are held between two of the examiner's fingers. Another method of eliciting this is to have the youngster imitate the examiner's finger movements. The latter should oppose specific fingers with his hands close to his ears. The child is then asked to imitate. Some youngsters are unable to do so or must look at their own fingers to accomplish the task. There is some indication that difficulties with finger localization may reflect relatively poor feedback from digits and consequently may be associated with inefficient writing and other fine motor deficits (see Chapter 38).

Motor Impersistence. The inability to maintain a fixed motor stance with the arms extended, mouth open, and tongue protruding is a sign originally described in association with known "brain damage" (Garfield, 1964). However, it has been found to occur commonly in youngsters with primary attention deficits and other learning problems.

Choreiform Movements. Involuntary, rotatory, and rhythmic movements when present are mostly commonly seen in the outstretched fingers or tongue when a child is asked to close his eyes, extend both arms, spread his fingers, open his mouth, protrude his tongue, and sustain this posture for at least 30 seconds. A number of studies in children have correlated these involuntary movements with school failure and behavior problems. Prechtl and Stemmer (1962) first described "the choreiform syndrome" to characterize children with this finding and a history of hyperactivity, impulsivity, poor frustration tolerance, emotional lability, and problems with reading and socialization. Since such symptoms are common in children with learning disorders of all types, the existence of this particular symptom complex as a syndrome has come into question.

Poorly Established Lateral Preference. The tendency toward preferential use of one side of the body is likely to reflect the development of one hemisphere of the brain for a particular set of functions. Hand preference is usually well established between four and six years of age. Eye preference is established by the third year of life. Ear and foot preference can also be evaluated, but less is known about their clinical utility. Children who show delays in establishing lateral preferences may have problems in other developmental areas. This has been a controversial subject in the literature. Mixed preference (e.g., a tendency to be right-handed and left-eyed) has been associated in some studies with an increased incidence of reading problems. However, contradictory find-

ings have been reported. There have been no consistent documented associations between left-handedness and learning problems.

Deficient Left-Right Discrimination. As children grow older, they become increasingly competent in identifying left and right and make various discriminations based upon these. In some cases, however, left-right differentiation may be a delayed acquisition, one that is part of a generalized developmental lag (Benton, 1959). By age six, most youngsters can tell left from right on their own bodies. Before their eighth birthdays they usually are able to cross the midline (e.g., "Touch your right knee with your left hand"). By age nine or 10, generally they can identify left and right on the examiner (e.g., "Touch my left hand with your right hand"). By early adolescence, children are competent at distinguishing readily left and right starting from new bases and space (e.g., "right face, left face," as in marching commands). Problems with left-right discrimination may result from complex mixtures of maturational, developmental, or basic processing problems. Some youngsters have difficulty in carrying out left-right commands not because they are confused about laterality but because they have language processing problems or deficits of rapid retrieval memory. Left-right reversals of letters and consequent reading retardation were described in the early and influential studies of Orton (1937). Such confusion was attributed to incomplete cerebral lateralization (i.e., dominance). It is now known that there can be many other causes of such reversals.

USE AND MISUSE OF MINOR NEUROLOGICAL INDICATORS

The indicators described in this section can become an important part of a neurodevelopmental examination. Most do not have direct implications for intervention, but may help in suggesting a strong constitutional input when a child is failing. In some instances neuromaturational delays will be accompanied by other forms of maturational lag, reflected in delayed skeletal age, lags in dentition, or social immaturity. In a recent study the composite rating of associated movements was found to correlate with the presence or absence of attention deficits and various kinds of motor dysfunction. There was little correlation between associated movements and language disabilities.

There is a danger of some misinterpretation and misuse of minor neurological signs. For example, one single indicator, in isolation, may not be meaningful. Highly successful children may show one or more of these signs, whereas other youngsters with significant developmental dysfunctions may have no evidence of neuromaturational delay.

Clusters appear to be more accurate discriminators than any single isolated sign. It is inappropriate to interpret neuromaturational indicators as evidence that a child will catch up eventually. The word "maturation" is misinterpreted widely in children with learning disorders. A so-called "maturational lag" should not be taken to imply that a child will catch up or remit spontaneously. Thus, evidence of immaturity of the nervous system should not be used as justification for therapeutic inaction or nihilism.

SPECIFIC AREAS OF DEVELOPMENT

The neurodevelopmental examination can be used to tap directly certain specific areas of a child's development. Although there are many different formats and classifications upon which to base neurodevelopmental examinations, in general it is likely that the following areas should be covered when evaluating the neurodevelopmental status of a school age child:

Gross Motor Function. A youngster's ability to coordinate muscles and move effectively in space seems to relate closely to self-esteem and socialization during the school years (Shaw et al., in press). For this reason it is important to include some direct measures of this functional area. Gross motor items should include those that test eye–upper limb coordination (such as catching and throwing a ball, body position sense (such as balancing with eyes closed), and various stressed gaits and motor sequential tasks.

Fine Motor Function. The child's ability to manipulate his fingers effectively is important for written output and various craft activities (Levine et al., 1981b). Direct observations of fine motor ability should include tasks of eye-hand coordination (such as stringing beads) as well as some tests of neuromotor speed (for example, rapid opposition of the thumb and forefinger) and the execution of various motor engrams with the fingers (repetitive patterns of sequential finger opposition). The latter tests should be performed both with eyes open and with eyes closed.

Visual-Spatial Orientation. As part of a neurodevelopmental examination, it is appropriate to evaluate a youngster's capacity to interpret visually presented information. The ability to appreciate inter-relationships in space is thought to be relevant to the acquisition of reading and other symbol recognition skills. Most commonly this is tested by having a child copy geometrical forms that grow in complexity with the youngster's age. It should be recognized, however, that such tasks involve a great deal of motor planning, sustained attention, and organization. Thus, having a child copy geometrical configuration also taps into functions that have little to do with visual perception. A careful neurodevelopmental examination should supplement such observations with some motor-free visual perceptual tasks (such as matching designs). In making observations about a child's visual perceptual abilities, it is useful to differentiate between the youngster's capacity to appreciate and analyze patterns, on the one hand, and his ability to attend to visual detail, on the other. Thus a youngster may "get the big picture" or understand the general attributes of a gestalt but ignore details (perhaps because of attentional difficulties). The implications for treatment differ considerably depending upon whether the child is having difficulty with visual configurational analysis or visual attention to detail.

Temporal-Sequential Organization. As is discussed in Chapter 38, a youngster's appreciation of time and sequence is critically important for the acquisition of reading, spelling, and arithmetic skills. It also has a direct impact on a child's capacity to follow directions and organize output. For this reason the neurodevelopmental examination should include a series of sequencing items. Tasks involving the retention of auditory sequences are important. The most common of these is the digit span. Forward and reverse spans can be useful, although results may be "contaminated" by a youngster's familiarity or lack of comfort with numbers, by anxiety, by more generalized problems with memory, and by deficiencies of attention. It is sometimes helpful to supplement digit spans with word spans. A neurodevelopmental examination also should include some visual sequences. Tapping objects in a particular order after the examiner does so is an example of this. Another aspect of sequential organization entails the youngster's ability to follow multistep commands. Standardized serial commands can be used: these should become increasingly lengthy and complex according to the child's chronological age. Finally, a series of motor sequencing tasks also can be revealing. Various forms of rapid alternating movements in a particular order can be used for this.

Memory. Short and long term memory can be sampled during the neurodevelopmental examination. Often this is done in conjunction with the examination of other areas of development. Thus, an analysis of sequencing also entails some insight into sequential memory. While having the child copy geometric forms, as part of the evaluation of visual-spatial orientation, it is useful to have him do some of these from memory. A language assessment should include auditory memory. As noted in Chapter 38, a youngster's memory deficits are likely to relate to other processing problems in a particular area of development. In assessing memory, one needs to bear in mind constantly its close association with attention. If a child is not focused properly on a task, he is unlikely to retain very much.

Language. A neurodevelopmental examination should include some direct sampling of a child's linguistic abilities. Included should be an evaluation of receptive language, comprising such areas as auditory attention, auditory discrimination, and verbal comprehension. Some assessment of both receptive and expressive vocabulary is needed. Having a child identify pictures is one way to sample expressive vocabulary. Asking him to choose among several optional pictures when given a word is a common method to evaluate receptive vocabulary; the Peabody Picture Vocabulary Test is an example of the latter. While evaluating a child's expressive language abilities, informal assessments of the length and complexity of sentences, use of grammar, and sophistication of vocabulary can be undertaken. The clinician also needs to be alert for problems of articulation.

Behavior and Style. The physician can use a neurodevelopmental examination to make behavioral observations that can contribute to diagnostic formulations. While observing the neurodevelopmental assessment the physician can judge how the child deals with positive and negative reinforcement, how he copes with frustration, whether he develops good cognitive strategies, and what patterns of selective attention and activity emerge during the performance of developmentally appropriate tasks. One can look for changes in the child's attention and impulse control induced by different types of tasks. One can look for distractibility and a tendency for performance to either deteriorate or improve over time. Levels of anxiety and defensiveness associated with cognitive challenges can be discerned.

The examination also can be helpful in assessing the extent to which a child can form an alliance with the physician. Is he "slow to warm up"? Is he able to make eye contact? Is he trusting? If the parents observe the neurodevelopmental examination (usually a good idea), one may make some observations about the nature of parent involvement with the child. One can note how they support or exert pressure. One can watch for parental reactions to the child's successes and failures. One can assay the child's capacity to become involved in tasks and gauge the amount of feedback he wants or demands. Throughout developmental testing one can discern leitmotifs that have stylistic implications regarding the child's function.

Training. It is imperative that the clinician administering a neurodevelopmental examination have adequate training to do so. Once one is certain of the reliability of the instrument being used, there is the need for assurance that the examiner is reliable also. Some interobserver reliability testing should be part of the background of every neurodevelopmental assessor. Specific training programs for this purpose are being developed throughout the country. (See Chapter 63.) Physicians interested in becoming more involved in neurodevelopmental assessment should undergo direct supervisory training. It is unlikely that one will be able to read a manual and thereby effectively administer and interpret a neurodevelopmental examination. Training to use such an instrument must be supplemented by knowledge of the development of the school age child. Obviously it is impossible to separate interpretive findings from familiarity with the subject matter.

THE PEEX: AN EXAMPLE

At the Children's Hospital Medical Center in Boston, the Pediatric Early Elementary Examination (PEEX) is used as a neurodevelopmental examination for first, second, and third grade children. Table 45–4 summarizes the items on this neurodevelopmental examination. It can be seen that the PEEX permits judgments about the child's developmental attainment in various areas. In addition, there are simultaneous observations of attention activity and behavior during the examination. A composite rating of minor neurological signs also is included. In field studies this instrument has been shown to correlate strongly with findings on the WISC-R and with parent and teacher ratings of behavior and learning.

A series of videotapes have been developed to train physicians in the use of the PEEX. The examination is used mainly in private practices, clinics for children with learning disorders, and school based consultations by pediatricians and nurses. It is designed to complement (but not replace) various psychometric tests while expanding upon standard neurological examinations.

INTEGRATION AND FORMULATION

In comparing questionnaire data with direct observations and other standardized testing, the physician becomes an integrator of findings. Formulating a child's problem is very much like preparing a legal case for argument in court. One needs to seek multiple interlocking bits of evidence that make a sound argument to explain a child's difficulties. Any one fragment of evidence is likely to be circumstantial or a so-called "red herring." The observations of a single observer may be invalid. For example, a child with attention deficits may proceed through an entire neurodevelopmental examination and display no problems with

Table 45–4. AREAS AND TEST ITEMS ON A SCHOOL AGE NEURODEVELOPMENTAL EXAMINATION (PEEX)

Area	General Description	Items*
Minor neurological indicators	Various signs thought to be associated with developmental dysfunction and learning problems	Visual tracking Stimulus extinction Left-to-right discrimination Rapid pronation-supination Imitative finger movement Finger differentiation Associatiated movements
Temporal-sequential organization	Tasks that assess a child's knowledge of everyday sequences and time relationships as well as ability to understand and remember new data in the correct order	Digit spans Word spans Object spans Imitative block tapping Count forward and backward Days of week, telling time Gross and fine motor sequences Understand temporal prepositions
Visual-spatial orientation	Tasks that reveal a child's ability to appreciatiate spatial relationships and visual detail, to integrate such inputs with motor responses, and to remember visual configurations	Connect dots to imitate designs Copy designs (direct and from memory) Matching designs Visual recognition Visual retrieval Object spans
Auditory-language function	Exercises tapping a child's understanding and use of language	Comprehension of oral directions Comprehension of complex sentences Confrontation naming test (word finding and expressive vocabulary)
Fine motor function	Activities entailing eye-hand coordination, finger localization, neuromotor speed, pencil control, and fine motor pattern mastery	Finger opposition (simple and sequential) Sentence copying (legibility and speed) Pencil control Motor speed test Imitative finger movements Finger differentiation
Gross motor function	Exercises requiring motor praxis, inhibition, rhythm, body position sense, and eye–upper limb coordination	Hopping in rhythm Tandem gait Catch a ball Sustained stance (eyes closed) Associated movements
Memory	Ability to recall at end of PEEX certain stimuli presented near the beginning	Object recall, other visual recall Form copying (revisualization) Word recall

*Certain items are listed in more than one area because they entail several developmental components.

distractibility, impulsivity, or attentional weakness, yet the same youngster may be rated as having terrible attention by parents and teachers. It is quite possible that on a one to one basis, in a physician's office, the manifestations will not be evident. In such a case the clinician may have to discard his own findings and accept those that are more reflective of day-in and day-out function. In other instances weaknesses may be misinterpreted or unnoticed by a teacher or a parent. Direct observations of function may elicit these.

EDUCATIONAL READINESS

The neurodevelopmental examination also can be useful in determining educational readiness. Children during the preschool years may raise various issues about their preparedness for kindergarten or first grade. The very same neurodevelopmental areas described for school age children are relevant in youngsters between the ages of three and five. The Pediatric Examination of Educational Readiness (PEER) has been validated for this purpose (Levine, 1982; Levine et al., 1980a). Like the PEEX, which is used in older children, the PEER can detect relative strengths and weaknesses that may be predictive of later school failure. It can allow for further evaluation and for early intervention. In order to permit a fairly broad look at various components of function, it is difficult to offer a very rapid 10 or 15 minute screening test for learning disabilities. Examinations such as the PEER examination require more time (as much as one hour). As a result, in many cases it is practical to prescreen and only use a neurodevelopmental examination in these youngsters who are at risk for failure. Questionnaires can be used for this prescreening process. Children who have had a great deal of illness, who display developmental lags, who have behavioral dysfunctions, or who come from difficult home situations may benefit most from a neurodevelopmental examination prior to school entry.

NEURODEVELOPMENTAL EXAMINATION OF THE ADOLESCENT

Neurodevelopmental problems in adolescence constitute a serious area of morbidity, one that is poorly understood but is receiving increasing attention in clinics and research settings. The evaluation of a failing adolescent can be difficult because of a lack of sophisticated standardized instruments. Many tests for specific learning disabilities tend to "ceiling out" at an earlier age. It may be difficult therefore to detect a problem with sequential organization or visual-spatial processing in a 16 year old. On the other hand, careful history taking often can suggest the existence of such a disorder. The ANSER system described previously includes a special questionnaire for the adolescent age group. Included is an inventory of earlier academic problems (i.e., during preschool and the elementary years). This can provide important clues to the existence of underlying disorders. In school, careful analyses of the kinds of problems the child is having and specific error patterns (in reading, spelling, and arithmetic) can be useful in pinpointing a discrete area of developmental dysfunction. At the present time, efforts are under way to develop appropriate neurodevelopmental examinations for the adolescent age group.

THE FUNCTIONAL PROFILE

The ultimate aim of a neurodevelopmental examination is to help in elucidating a child's strengths and weaknesses and their relevance to his performance in life. Systematic observations are not designed to derive a specific quantitative score or introduce a label for a child. Instead, empirical observations of strengths and weaknesses can have direct implications for intervention and future monitoring. The physician can compile such observations in a narrative description of a child, helping thereby to mobilize specific services for the patient and providing a helpful source of informed advocacy. (See also Chapters 51 and 52.)

MELVIN D. LEVINE

REFERENCES

Benton, A.: Right-Left Discrimination and Finger Localization Development and Pathology. New York, Hoeber Medical Division, Harper and Row, 1959.

Berges, J., and Lezine, T.: The imitation of gestures. Clinics in Developmental Medicine, No. 18. London, William Heinemann, 1965.

Cohen, H., Taft, L., Mahadeviah, M., and Birch, H.: Developmental changes in overflow in normal and aberrantly functioning children. J. Pediatr., 71:39, 1967.

Garfield, J.: Motor impersistence in normal and brain-damaged children. Neurology, 14:623, 1964.

Grant, W., Boelsche, A., and Zin, D.: Developmental patterns of two motor functions. Dev. Med. Child Neurol., 15:171, 1973.

Kinsbourne, M., and Warrington, E.: The development of finger differentiation. Quart. J. Exp. Psychol., 15:132, 1963.

Kraft, M.: The face-hand test. Dev. Med. Child Neurol., 10:214, 1968.

Levine, M. D.: The ANSER System. Cambridge, Massachusetts, Educator's Publishing Service, 1980.

Levine, M. D.: The Pediatric Examination of Educational Readiness. Cambridge, Massachusetts, Educators Publishing Service, 1982.

Levine, M. D., Brooks, R., and Shonkoff, J.: A Pediatric Approach to Learning Disorders. New York, John Wiley & Sons, Inc., 1980a.

Levine, M. D., Clarke, S., and Ferb, T.: The child as a diagnostic participant: Helping students describe their learning disorders. J. Learn. Disab., 14:527, 1981.

Levine, M. D., Oberklaid, F., Ferb, T. E., Hanson, M. A., Palfrey, J., and Aufseesen, C. L.: The pediatric examination of educational readiness. Validation of an extended observation procedure. Pediatrics, 66:341, 1980b.

Levine, M. D., Oberklaid, F., and Meltzer, L.: Developmental output failure—a study of low productivity in school age children. Pediatrics, 67:18, 1981b.

Nolan, E., and Kagan, J.: Psychological factors in the face-hand test. Arch. Neurol., 35:41, 1978.

Orton, S.: Reading, Writing and Speech Problems in Children. New York, W. W. Norton & Co., Inc., 1937.

Prechtl, H., and Stemmer, C.: The choreiform syndrome in children. Dev. Med. Child Neurol., 4:119, 1962.

Shaw, L., Levine, M. D., and Belfer, M.: Developmental double jeopardy: a study of clumsiness and self-esteem in learning disabled children. Dev. Behav. Ped. (In press.)

Toumen, B., and Prechtl, H.: The Neurological Examination of the Child with Minor Neurological Dysfunction. Little Club Clinics in Developmental Medicine, No. 38. London, Spastics Society, 1970.

46

Intelligence and Its Measurement

The measurement of intelligence stems from the idea of individual differences. Individuals approach problem solving with greater or lesser facility. There appears to be in certain individuals across the life span an identifiable attribute that Eysenck (1962) viewed as "mental speed." For others it is broader comprehension, or mental grasp, in explaining problematic situations or in understanding sequences in time. There are conceptual skills—the use of comparatives, or distinctions, of groupings or sets, of numbers or logic, all of which differ among individuals. These differences characterize some as being faster or slower learners, as being more or less rewarding to teach, or as being better or poorer candidates for more intensive teaching or training.

PRACTICAL APPLICATIONS

These differences are probably most evident in the field of mental retardation. Binet and Simon (1905) said: "We are of the opinion that the most valuable use of our scale will not be its application to normal pupils, but rather to those of inferior grades of intelligence." With mentally retarded individuals, the deviation from age graded expectations, the compromises of problem solving and mental grasp, and the flaws in the actual building blocks of thought can be directly and individually observed. It is probably in this area that we have some of the best examples to help us to appreciate the accomplishments of the maturing human mind. In observation of negation and compromise as well as more lengthy acquisitions over time, human mental maturation and its varying aberrant forms reveal some clearly outlined contrasts. However, there are also pitfalls in following this procedure.

To use an example, one might take the instance of a 17 year old girl. This girl is an extremely labile individual, subject to outbursts of resentment and offensive language. Intelligence tests have described her as having a six year mental age. She understands simple functions but cannot explain natural consequences, such as the fact that the

wind and not the water moves a sailboat, that it is proper to return a lost object to its owner. Her understanding of sequence is so poor that she does not always perceive the consequences of certain causes culminating in an effect. She does not know the days of the week in order, nor does she really understand well such concepts as more than or less than. When asked why objects are similar or alike, she instead concentrates on their differences. Yet as an individual, this 17 year old has also been labeled as emotionally disturbed. She has never been taught to read or write. Her negative behavior patterns have driven teachers and proposed advocates from assisting her.

When she is contrasted with a similar 17 year old, perhaps in the same classroom, who falls into an almost identical pattern of test performance, one notes that there is a great difference. The second young woman has learned to read to about a third or fourth grade level, has mastered basic addition and subtraction, and has been trained to use a hand calculator so well that she even assists her mother with minor grocery purchases. This second pupil has a warm and charming smile, which elicits adult and peer support appropriately. These two persons, of course, cannot be treated or related to as similar merely because each derives a test score with a six year, five month mentality.

This issue, however, does not even approach another important dimension of the injustice of disregarding individuals and reacting to them as if each were a dehumanized and impersonal numeric IQ 42. The concern here really centers upon the recognized weakness of mental age comparisons, which dominated discussions of intelligence for almost half a century. Of course, a retarded 17 year old woman is not a six year old child. There are years of accumulated experience; there are tempered and tried problem solving approaches, which have produced different frustrations and different senses of accomplishment. There is also the issue of competence. In a very promising approach Scarr (1981) views competence and the motivational aspects of social adjustment as being responsible for making mentally retarded individuals with similar conceptual building blocks into very differently functioning individuals. She points out well that it is not only the basic cognitive skills that an individual employs that determines what that person will be able to master or what he will fail in.

In the concrete application to the two 17 year olds instituted, one clearly sees that the concept of mentation cannot be separated from its application, just as there is no removal of mind from body, or of problem solving or its speed from the social environment, which modifies motivation and opportunity for social support or reinforcement.

MENTAL AGE

Alfred Binet (1905) provided the contribution of basic mental age levels with derived expectations for children's accomplishments at particular chronological ages. His was a discovery brought about through social necessity. Binet was commissioned by the French Ministry of Education to settle the question of the capability for training (or "educability") of a large number of urchins who were creating a social problem in the streets of Paris at that time. Sorting them for various educational, training, or institutionalized placements on the basis of their level of functioning was a reasonable solution. In the United States in the early years of this century Goddard (1920) also found it advantageous to sort or group residents of the Vineland Training School following Binet's procedures. Intelligence tests as we know them work very well to designate levels of retardation. In other words, they efficiently sort out the lowest percentiles of retarded individuals within respective age cohorts. This leads to expedient pedagogic or instructional groupings, which may be expected to advance together more uniformly in learning or training programs.

Louis Terman (1916) was able to influence his group of Stanford psychologists in aligning the Binet items in a modified style with the expectations of a white, middle class, American elementary school curriculum. Later statistical modifications, including the IQ score, factor analysis, and the Spearman g factor, supported an idealism that was tied into a cultural value of achievement according to individual ability. Along with this concept came a rather conformist or uniformist impression of American culture, which prevailed throughout the first half of this century.

Wechsler's arguments against mental age and his use of deviation measures are tied into a deeper appreciation for the plurality of humankind. Wechsler's earlier (1958) definition of intelligence as allowing individuals "to deal effectively with environment" reflected the need to individualize

and to consider variation from an otherwise oversimplified uniform IQ concept. His later (1975) definition is a lasting contribution—"the capacity of an individual to understand the world about him and his resourcefulness to cope with its challenges." His second greatest contribution was the provision of a multiplicity of subtest scores that appropriately allows the opportunity to identify the manner in which a child acquires this thing called intelligence.

BASIC ACHIEVEMENT

It should be evident that only a small portion of what a child learns is acquired through being taught. What is learned is actually provided through more natural interactions between the child and what he encounters in his environment. There is the very natural relationship between curiosity and exploration. The infant learns to roll over with the rewards of an immediate different visual perspective, or different body pressures and effects upon limb extension, with the ever increasing possibilities of locomotion. Motor accomplishments, like sitting, standing, or walking, are gross indications of how rapidly an individual is mastering the challenges of his environment. Gesell (1947) used early motor accomplishments to objectify developmental levels, in many ways a refinement of the pediatric developmental evaluation, but beyond gross motor indicators come finer qualifications. Probably most significant among these is the opposition of the thumb to the other fingers. Man's greatest physical capacity is first employed in the use of the pincer grasp. Then each hand becomes independent. The child can hold not only two but three objects. The hands combine in their work. They can assist in both procurement and locomotion, and they are employed by different individuals at different rates of accomplishment. However, the acquisition of either gross or fine motor accomplishments does not measure alertness or even explorative quality. As the infant advances, new episodes involving the problem solving approach seem to be more clearly related to what otherwise might be termed "mentation." These are the simple puzzles, the heaping of objects into the container, the use of a stick as an extensor for procuring a more distal object, the finger or the peg in the hole, and learning to reach around a piece of clear glass to procure an object.

These accomplishments were also aligned into developmental expectations. Early infant tests merely grouped such problem solving accomplishments by age level expectations. Later infant tests were refined by applying more detailed statistical procedures, such as factor analysis, to these alignments. Using the California Growth Studies, Hofstaetter (1954) was able to demonstrate the advancement of motor intelligence to spatial problem solving and eventually to the more centralized language mastery predictions. His staging of these early factors is probably the most important contribution in understanding succeeding mental processes in the preschool child. It leads from basic motor accomplishments into the predominance of language mastery with its very important role in understanding and predicting later intellectual accomplishments.

MEASUREMENT ISSUES

In consideration of the foregoing, it is not surprising that the mental measurement movement became involved in a Holy Grail search for the "g factor." The concept that some all encompassing and unifying explanation would lead to more nearly perfect predictions is an understandable, human faulted, intellectual wish, the likeness of which can be found in other sciences. In mental measurement pursuits it took the form of leading some researchers to pursue correlations among various forms of tests to see which types of response or performance would be the strongest or most consistent predictors of future test measurements. This led to the remarkable discovery that vocabulary or word mastery had the highest correlation with all other mental trait measurements and also with composite scores. It was later discovered that not only was word mastery the most powerful predictor of all other scores, but it also was the most contaminated by correlation with socioeconomic status. This placed mental measurement specialists in a troubling ethical bind. They could rely upon word mastery as the strongest predictor of achievement, school success, and performance in other test areas, but resorting to this more efficient predictor biased their results by giving subjects from higher socioeconomic backgrounds the advantage of appearing superior. In terms of test application, this accounted for unfairness in competition for admission to more prestigious academic institutions and also constituted the means of avoiding exclusion (or perhaps even banishment) to undesirable "special" education settings.

The judicial system continues to struggle with this issue at present, and various "due process" procedures have been initiated in several states to prevent this abuse of special class placements. However, the awareness that Boston had loaded its classes intended for retarded children with emotionally unstable, minority poor, bilingual, and other culturally deviant troublemakers was an exposé that shaped much of the current special

education legislation in this country (Jones, 1970). In the pre-1970 Boston school situation there was no room for the long waiting lists of retarded children, since their resources were being misemployed by the school system to deflect other troubling or disturbing influences from the school mainstream. In this manner, measurement procedures were being misused to the detriment of the children they were intended to help. The due process review system by which measurements were placed in a more appropriate perspective was imperative. Measurements thereby became evidence upon which an appointed panel of reviewers could make a better informed decision about placements that affect life opportunities or educational privilege. The school psychologists, in turn, were also elevated from the perfunctory role of a lowly test technician (who often performed under conditions that in themselves invalidated resultant test scores) to recognition as being contributory professional team members. In this legally recognized position, psychologists found themselves capable of guiding a collective decision making process in their more appropriate role of measurement specialists.

Psychologists had learned to account for deviation in language acquisition in assessing the ability of deaf children. Marshall Hiskey (1941) had produced a remarkable instrument that avoided spoken language but provided norms for both deaf and hearing children beginning at age three. Some of the differences between these two groups was slight, but the normative data provided good evidence of the differences in advancement in various skill areas.

This accentuates the fact that a sensory handicap in itself does not uniformly depress intellectual development. The provision of two sets of norms also allows professionals responsible for guiding children with hearing handicaps the advantage of realizing the magnitude of deviation at various age levels in comparison with nonimpaired children. In the current climate of educational mainstreaming, it is this measurement that is most informative in implementing special education programs. There is undeniable value in obtaining a measurement of performance of a handicapped child that may be interpreted in relation to his rank among others who share the same handicap. However, the critical issue is the amount of difference this makes in our intent to educate the child within the community in a mainstreamed program. Informed professionals must know how the child ranks within broader populations. For deaf children Hiskey painstakingly provided both sets of norms. With other handicaps it may be necessary to use two sets of instruments (with understandable slippage in the comparisons) in order to gain the same comparison.

Table 46–1. LEXICON OF ABILITY TERMS

Constitutional (innate)

Aptitude: A natural ability. A tendency or capacity for learning. A talent. A combination of characteristics indicative of an individual's ability to learn or develop proficiencies.

Capability: Traits conducive to efficiency and ability. The capacity for specific use or development.

Capacity: Power to receive, hold, or store. Power to accommodate problems.

Ideation: The capacity to formulate abstractions or mental images.

Individual differences: Variation of potential among members of the human species.

Intellect: The power of knowing. The capacity for rational knowledge.

Potential: An existing possibility. The power to develop into actuality.

Experiential (accomplished)

Achievement: An accomplishment. The result of an endeavor. That which an individual has learned or mastered.

Competence: Demonstrated ability, experience, or training necessary for adequate performance. Fitness.

Knowledge: The range of an individual's acquired information and intellectual understanding.

Mastery: Possession of thorough understanding. Accomplished skill or technique.

Proficiency: Thorough competence derived through training and practice.

Qualification: Fitness for employment or engagement. Meeting with standards.

Understanding: A grasp of the meaning of, the nature of, the significance of, or an explanation of something.

Mixed

Ability: Power, skill, or resource to accomplish an objective.

Adaptive behavior: Effectiveness of degree to which an individual meets standards of personal independence and social responsibility.

Attention: Alerting to or orienting toward. Focus on relevant information. Ability to sustain selective focus.

Cognition: The process of knowing, involving both awareness and judgment.

Comparison: Identifying relative features and values.

Comprehension: Mental grasp, apprehension, or full understanding.

Conceptualization: Mental operations of processing. Formulations or organization of observations or relationships.

Intelligence: The general capacity to use or exercise the intellect. "The capacity of an individual to understand the world about him and his resourcefulness to cope with its challenges" (D. Wechsler, 1975).

Problem-solving: Means of resolving issues or seeking solution to a presented task.

Reasoning: The process of consecutive logical thinking. The drawing of inferences and conclusions.

Skill: Effective performance or execution of a task.

Thinking: Rational ordering. The processing of ideas. The internal conversation.

Physical and gross motor handicaps provide another challenging deviation, especially in early childhood. It has been noted that much of infant testing depends upon physical mastery and hand control. Children with cerebral palsy and other major physical impairments do acquire receptive language capacities, often without the possibility of experiencing the motor mastery of normal children. Clinically the child's capabilities are often discovered with surprise. Every clinician has a number of anecdotes of the mental prowess of some very physically incapacitated children. The derivation of the inner language of these children and their communicative processes await further study and better documentation. It is probably sufficient to say that scientists and educators have much to learn from these potential studies.

Similarly children with learning disabilities provide many "surprises" concerning their abilities. These may take the form of either sudden demonstrations of abilities that were unanticipated or of disappointments that are engendered by false hopes never supported by solid measurements and subsequently failing to materialize into actual achievements. Learning disabilities are often diagnosed on the basis of widely disparate scores within one or in comparison of several instruments. Wide discrepancies between the verbal and performance measures of the Wechsler, or among subtests, are considered as evidence of learning disabilities. The presumption is that because a child is deficient in only certain areas of functioning, his "strength" scores reveal that he should be able to compensate for these deficiencies. The "false hopes" complicate planning for such children when there is denial of the fact that composite scores still provide evidence of the extent of efforts that will be required to support this child in the educational aims of advancing with his peers.

Among other unresolved measurement issues are questions such as effects of sibling ordering. It seems clear from the studies of academic primogeniture that first born siblings perform better on verbal tests (Altus, 1965; Schacter, 1963; Sutton-Smith and Rosenberg, 1970). The issue of whether intelligence or composite scores increase or decrease by sibling rank has never been resolved. Fifty years ago the renowned written debate between Thurstone and Jenkins (1929), taking the increase-with-rank position, and Terman (1925), taking the opposite view, could never be settled. Probably the resolution does lie in the types of instruments used by the different researchers, but to this time no encompassing explanation has been generally accepted to account for the differences between the two sets of data and their interpretations.

VARIABILITY OF MEASUREMENT

At the foundation of many of the controversies surrounding measurement issues was the belief that one numerical score could represent an individual or could predict future accomplishments.

This was clearly allied with the attempts in the early development of infant tests to develop "final" IQ predictors at as young an age as possible. Many of the classic longitudinal studies, such as the Fell's research (Sontag et al., 1956) and the California Growth Studies (Bayley, 1955), revealed powerful predictive measures in infancy. They also revealed the variability of measurements among young children, and the statistical problems necessitating that variance be taken into account in order that predictive measurement could be validly applied.

An example of the difficulty in accounting for variance would be the fact that the Wechsler Preschool Scale could be applied to children only as young as four years of age. The variance in the types of tasks employed by this test is too great below that age. This could be expected, since the tasks in the children's form of the Wechsler scale are mainly adaptations of the original tasks chosen by the designer for the adult scale. The "downgrading" or the attempts to apply these to younger groups of children not surprisingly meets limits beyond which predictive validity is lost because the variability of performance on these tasks is so great for children that young.

A similar problem was met in standardizing tests for blind children. Several adaptations have been developed, but fail to meet tests of predictive validity owing to the extreme variance among the samples of blind subjects. This variance, quite logically, also increases among younger subjects. In part this may be due to the degree of central nervous system involvement commonly associated with the impairment of a major sensory system. The point is that early childhood and types of impairment produce variance that limits predictive validity. The converse of this point is similarly important although not as definitely established. Intellectual assessment in patients with many other types of handicaps or impairments should be approached cautiously.

THE DEBUNKING OF THE CONSTANT IQ

Nancy Bayley (1955) probably said this most effectively:

It becomes evident that the intellectual growth of any given child is a resultant of varied and complex factors. These will include his inherent capacities for growth, both in amount and in rate of progress. They will include the emotional climate in which he grows, whether he is encouraged or discouraged, whether his drive (ego-involvement) is strong in intellectual thought process, or is directed toward other aspects of his life-field (p. 813).

Mental growth is no exception to the rule that growth occurs when the organism meets most appropriate circumstances. Harsher conditions inhibit it. It is understandable then that intellectual stimulation should produce changes in measurement scores.

This fact produced shock waves 40 years ago when the "Iowa Preschool Studies" were published (Wellman, 1938; Wellman et al., 1940). Great skepticism was expressed regarding the techniques, conditions, and legitimacy of approach. At that time in history the IQ was not supposed to change. It was considered to reflect a certain immutable influence of which measurements were merely a glimpse of reality. Variability was to be considered measurement error or as imperfections of the technology in documenting a more unchanging existence.

However, the Iowa studies showed that preschoolers in stimulation programs left those programs with higher IQ's, and also that they fared better in later education. Skodak and Skeels (1949) went a step further. They showed that children removed from the detrimental influences of a depression engendered orphanage could also demonstrate IQ gains by being cared for and concomitantly nurtured by adult retarded women on wards of a fairly well programmed (at that time) state institution for retarded persons. Skeels (1966) later went on to show that the improvements in early test scores were related to other far reaching inherent gains in terms of life adjustment, emotional satisfaction, and attainment of goals commonly valued as indices of personal fulfillment.

Similar arguments have been produced over the success of the stimulation of children from deprived environments who had attended Project Head Start. Early reports confirmed the achievement of IQ increments. Later reports became embroiled in controversy over their lasting effects, giving the impression that these gains became negligible after the subjects had progressed two or three years further into their educational course. It seems definite that IQ scores do not maintain their gains over this period of time for the Head Start pupils. On the other hand, it took Gray et al. (1981) to apply other indices of successful adjustment or accomplishment, in such areas as achievement, teacher ratings, and high school graduation. They showed that the stimulated subjects maintained certain advantages attributed to their early educational experiences, perhaps accomplishments that might have been predicted by higher scores on ability instruments.

CONCLUSION

These findings and discoveries teach that intelligence is not a unitary concept. There is no one all-embracing ability factor. There are multifaceted components within human differences that allow certain individuals to be more clever, more alert to events around them, and more competent at deriving conclusions that guide their future interactions with their environment. Just as mind cannot exist apart from body, intelligence cannot exist apart from life accomplishments. But within each individual's capacity to respond, there are clear differentiations, which demonstrate the advantages of some individuals in comparison with others. These change with time and with environmental conditions. They also are complex and require clear and objective interpretation by the measurement specialist regarding influencing and interacting conditions.

BRUCE CUSHNA

REFERENCES

Altus, W. D.: Birth order and academic primogeniture. J. Pers. Soc. Psychol. 2:872–876, 1965.

Binet, A., and Simon, T.: The development of the Binet-Simon Scale. L'Anee Psychologique, 11:191–244, 1905. (Also translated by Elizabeth S. Kite: The Training School at Vineland, 1916.) In Rosenblat, R., and Allinsworth, W.: The Causes of Behavior. II. Boston, Allyn and Bacon, 1966.

Bayley, N.: On the growth of intelligence. Am. Psychol., 10:805–810, 1955.

Boring, E. G.: Intelligence as the tests test it. New Republic, 35:35–37, 1923.

Eells, K., Davis, A., Havighurst, R. J., and Tyler, L.: Intelligence and Cultural Differences. Chicago, University of Chicago Press, 1951.

Eysenck, H. J.: Know Your Own IQ. Baltimore, Penguin Books, 1962.

Gesell, A., and Amatruda, D. S.: Developmental Diagnosis. Ed. 2. New York, Hoeber-Harper, 1947.

Goddard, H. H.: Human Efficiency and Levels of Intelligence. Princeton, Princeton University Press, 1920.

Gray, S. A., Ramsey, B. K., and Klaus, R. A.: From 3 to 20: The Early Training Project. Baltimore, University Park Press, 1981.

Hiskey, M.: A new performance test for young deaf children. Educ. Psychol. Meas., 1:77–84, 1941.

Hofstaetter, P. R.: The changing composition of "intelligence": a study of the t-technique. J. Genet. Psychol., 85:159–164, 1954.

Jensen, A.: How much can we boost IQ and scholastic achievement? Harvard Educ. Rev., 39:1–123, 1969.

Jones, H. (Editor): The Way We Go to School. The Exclusion of Children in Boston. Boston, Beacon Press, 1970.

Scarr, S.: Testing for children. Assessment and the many determinants of intellectual competence. Am. Psychol., 35:1159–1166, 1981.

Schacter, S.: Birth order, eminence and higher education. J. Pers. Soc. Psychol., 28:757–768, 1963.

Skeels, H.: Adult status of children with contrasting early life experiences. Monogr. Soc. Res. Child Devel., 105:31, 1966.

Skodak, M., and Skeels, H.: A follow-up study of one hundred adoptive children. J. Genet. Psychol., 75:85–125, 1949.

Sontag, L. W., Baker, C. T., and Nelson, V. L.: Mental growth and personality development. Monog. Soc. Res. Child Devel., 68:23, 1959.

Sutton-Smith, B., and Rosenberg, B. G.: The Sibling. New York, Holt, Rinehart and Winston, 1970.

Terman, L. M.: The Measurement of Intelligence. Boston, Houghton Mifflin, 1916.

Terman, L. M.: Genetic Studies of Genius. Stanford, Stanford University Press, 1925.

Thurstone, L. L., and Jenkins, R.: Birth order and intelligence. J. Educ. Psychol., 20:641–651, 1929.

Wechsler, D.: The Measurement and Appraisal of Adult Intelligence. Ed. 4. Baltimore, The Williams & Wilkins Co., 1958.

Wechsler, D.: Intelligence defined and undefined: a relativistic appraisal. Am. Psychol., 30:135–139, 1975.

Wellman, B. L.: Our changing concept of intelligence. J. Consult. Psychol., 2:97–107, 1938.

Wellman, B. L., Skeels, H. M., and Skodak, M.: Review of McNemar's critical examination of the Iowa studies. Psychol. Bull., 37:93–111, 1940.

47
Educational Assessment

The past few decades have heralded a clearer understanding of the complexity of learning difficulties, and we have witnessed major advances in the development of specialized educational techniques. Even with the advent of a variety of measurement procedures, the successful diagnosis of educational problems still rests on the achievement of a fine balance between the science and the art of the evaluation process. The science of assessment involves a continual testing of the hypotheses postulated by the diagnostician to explain the probable etiology of the child's special learning characteristics. In contrast, the art of assessment rests not only on professional expertise but also on the quality of the interpersonal relationship between the clinician and the child. Finally the synthesis and interpretation of the accumulated diagnostic information constitute an exacting and specialized task, which requires a unique blend of creativity and skill.

Historically educational assessment has been regarded as a process whereby achievement tests were applied and then interpreted primarily in terms of grade level attainment. The major concern was to examine what the child had learned and at what level the educator could program instruction. Currently there is greater emphasis on the evaluation of learning capacity and cognitive style and their inter-relationship with skill attainment. In other words, the focus now is not only on what the child has learned, but on how the child learns and why learning may be delayed or advanced. Present practice therefore stresses the importance of a holistic picture of each child, particularly when developmental delays exist that may influence educational performance in a variety of ways.

Unfortunately, as occurs in any developing field, the proliferation of novel assessment procedures has at times outstripped the professional expertise for application of these techniques. The result has been widespread and often differing interpretations of test findings, and disagreements between parents, professionals, and school systems have frequently emerged. In addition, test scores have often been used for the classification and labeling of children rather than for the provision of appropriate intervention. It is evident, then, that there is a need for a better understanding of the results and implications of diagnostic assessments, the limitations of specific tests, and the relevance of these findings for the delivery of appropriate instruction at school. During the past decade these issues have assumed increasing importance for the primary care pediatrician, who is often involved in the early guidance of children with various handicapping conditions, developmental delays, language problems, and behavioral difficulties.

The approach supported in this chapter emphasizes the developmental, cognitive, social, emotional, and instructional factors that are relevant for adequate educational functioning. Consequently the terms "educational assessment" and "psychoeducational assessment" are used interchangeably. The methods and techniques described are applicable in most settings, including schools and clinics, and are generally appropriate for the evaluation of all individuals, regardless of their specific needs. Specific modifications of assessment procedures are addressed in the section beginning on page 968. An approach is utilized that emphasizes each child's unique learning style, and use of the labels "minimally brain damaged," "dyslexic," and "learning disabled" is not supported.

CURRENT CONCEPTS AND MODELS APPLIED TO EDUCATIONAL ASSESSMENT

"Educational assessment" generally refers to that procedure by which data are assembled from a variety of sources in order to provide guidelines about the child's learning style as well as his strengths and weaknesses in specific learning areas. This continuing process evolves as a collaborative effort between the child and his parents, teachers, special educators, psychologists, physicians, and other specialized professionals. There is continuous review and refinement of diagnostic hypotheses as information is gathered from these various disciplines, culminating in a formulation that emphasizes the child's personal characteristics and provides suggestions for appropriate instruction. Assessment could therefore incorporate a number of measures. These might include tests of auditory and visual perception, psycholinguistic abilities, cognitive skills, and basic academic skills, such as reading, written language, and arithmetic. Findings might also be integrated with observations about the child's adaptation to the requirements of the school curriculum. In sum, educational assessment is broader than testing and has the purpose of seeking a resolution among the presenting symptomatology, the probable etiology, and the most appropriate intervention strategies. This process is facilitated by interdisciplinary action and cooperation, and the integration of findings obtained by these professionals.

PURPOSES OF EDUCATIONAL ASSESSMENT

The more important aims of educational assessment include determination of individual variation and exceptionality, measurement of educational change and progress, and formulation of an appropriate individualized educational plan for the child with special needs.

Assessment of Individual Variation and Exceptionality

The child's specific needs as well as his strengths and weaknesses are identified and matched as closely as possible with an appropriate potential teaching program. In the large-group situation that exists in the regular classroom, this is often difficult because instruction is generally developed for the average student. Nevertheless some programmatic changes can always be introduced in order to accommodate to the learning styles of different children. To this end, evaluation procedures help to determine whether particular teaching techniques are well suited to a child's learning style and the areas in which that child would benefit from intervention.

When a learning problem seems to exist, differential diagnosis may be undertaken. During this process, assessment techniques are applied to determine and contrast the relative contributions and manifestations of a child's learning difficulty. In addition, a profile of the child's underlying abilities and acquired skills is matched with specific instructional methods and curriculum content.

Measurement of Educational Change and Progress

An assessment model incorporating a continuing measurement of progress has different implications for successful intervention than does an evaluation utilizing a single measure of educational status. The difference between the two can be likened to the diagnosis and management of a chronic as opposed to an acute medical problem. Few instruments currently exist for the objective evaluation of educational progress, and although there is an abundance of teacher-constructed materials, these are usually not applied in clinical diagnosis. A number of standardized tests have alternate forms so that one version can be applied in the first testing and a second matched version can be administered in a follow-up session. These alternate forms are not as common as one would expect and thus tests are often reapplied on different occasions, a procedure that with the correct interpretation and error analysis can provide useful results. However, the practice effect resulting from repeated exposure to the same test may suggest a false display of progress on any particular measure. Therefore, a continuing challenge to professionals involved in educational assessment is to develop techniques that are sensitive to subtle improvements in a child's overall functioning and that can differentiate the effects of development from those changes that may result from intervention. The value of a one time evaluation is often insignificant when contrasted with the impact of a program in which the child's progress is regularly audited by parents as well as professionals. In summary, an effective follow-up system should incorporate continuous modification of management plans through assessment of educational progress, monitoring of the appropriateness of recommendations, and a review of management plans.

Formulating an Individualized Educational Plan for the Child with Special Needs

When a child has been referred for an evaluation because parents or teachers suspect that a learning problem exists, the diagnosis frequently culminates in the development of an individualized educational plan (see Chapter 56 for discussion). In brief, such a plan incorporates the recommendations generated from an assessment, and translates these suggestions into an educational prescriptive plan appropriate for that child's strengths and weaknesses. When the differential diagnosis has indicated that a mismatch may exist between the child's learning style and the type of instruction provided, the individualized educational plan should specify the objectives and techniques for improving the learning situation.

EDUCATIONAL ASSESSMENT AS PART OF THE COMPREHENSIVE CLINICAL EVALUATION

Traditionally educational assessment has occurred in the school setting. Recently, however, this has been more frequently incorporated in interdisciplinary evaluation procedures as part of a comprehensive clinical assessment. Professionals from fields such as medicine, education, psychology, speech and language, and occupational therapy evaluate the child and coordinate their findings in the development of an appropriate diagnosis. These evaluations are particularly helpful when the parents or the school system is seeking an independent opinion and a multi-professional assessment is needed within a single setting. When learning problems or specific school difficulties are suspected, the psychoeducational component of the evaluation is critical. The team member who performs this function is generally the special educator, educational psychologist, developmental psychologist, or the learning disabilities specialist, depending on the composition of the clinic team and the nature of the presenting problem. This component of the assessment usually emphasizes the child's perceptual and conceptual ability, educational skills, and overall performance in relation to other children of the same chronological age and grade level. School reports and findings from previous evaluations are also reviewed in conjunction with results obtained from the clinic assessment. This usually culminates in recommendations relating to classroom management, provision of specialized services, and specific instructional techniques (see Chapter 55).

PATTERN PROFILE ANALYSIS AS AN APPROACH TO ASSESSMENT

Test scores frequently have little meaning for the child, the teacher, the parents, or the health care professional. Reports that focus only on age and grade equivalents and include a limited interpretation of findings are not very helpful. Test scores should be evaluated in relation to the performance of other children in the same chronological age group who are subject to comparable environmental and social experiences and are from similar cultural backgrounds. In addition, the child's profile of strengths and weaknesses in multiple educational and academic skills must be assessed in relation to performance in various developmental, cognitive, and behavioral areas, using pattern profile analysis. For instance, a child

with above average intelligence whose reading and writing performance fall within the average or slightly below average range may be underachieving. It is particularly important in these instances that the diagnostic report incorporate behavioral observations, error analysis, and an interpretation of the child's learning style.

TECHNIQUES AND METHODS IN EDUCATIONAL ASSESSMENT

FORMAL STANDARDIZED MEASURES

Commercially prepared, standardized (formal) tests are frequently applied in school settings in accordance with exact procedures for administration, scoring, and interpretation specified in the test manual. These formal tests generally provide two levels of information: They determine the child's performance as it relates to his age group and they provide diagnostic information about the child's functioning within specific skill areas. Currently there is considerable controversy between and among researchers and practitioners concerning the technical adequacy of specific tests that are used frequently for educational assessment and decision making. It is advised that multiple sources of data be used and corroborated for diagnostic assessment, and that clinicians always avoid generalizing from isolated test scores.

A few specific applications of standardized tests will be discussed here as part of educational assessment; for more specific information, the reader is referred to Chapter 50.

Paper and Pencil Group Tests

In most school districts students are tested at regular intervals with paper and pencil group achievement tests. Examples of the more frequently cited tests are the Stanford Achievement Test, the California Achievement Test, and the SRA Achievement Test. These tests serve a variety of functions: They provide regional comparisons of achievement, they are useful for determining placement decisions for specific children, and they may be utilized to evaluate program effectiveness and to suggest weaknesses in the curriculum. Paper and pencil tests are also helpful for providing a general screening system and alerting teachers to the difficulties that specific children may be experiencing. Findings need to be corroborated by the child's daily classroom performance and by other informal teacher assessments. In such instances the results may be reviewed with the student, the parents, and possibly other teachers and might result in a referral to the school psychologist or a clinic for further evaluation.

Awareness of the limitations of these group

administered paper and pencil tests is important when decisions are based on the test scores obtained. If an extended evaluation is sought for a particular child, previous reports are usually requested, and health care professionals may be faced with discrepant scores, which result from the application of different tests. A common reason for these divergent results relates to the application of group tests in one situation and individual measures in another. Cautious interpretation of findings is critical because group scores are frequently depressed when compared with individual scores, particularly for children with special needs. Because of their specific reading and writing difficulties, children with learning problems or more severe handicapping conditions are unfairly penalized by group time limits, machine scorable answer sheets, and written instructions.

Diagnostic Tests

Most clinical diagnostic measures have been developed as individual tests for administration on a one to one basis. Comprehensive psychoeducational batteries assess a range of cognitive, language, and achievement areas and include the Wechsler Intelligence Scales, the Detroit Tests of Learning Aptitude, the Peabody Picture Vocabulary Test (PPVT), the Peabody Individual Achievement Test (PIAT), the Woodcock-Johnson Psycho-Educational Battery, and the Wide Range Achievement Test (WRAT).

In most instances appropriate assessment involves the administration of a variety of diagnostic tests so as to elucidate the possible symptomatology and etiology of the learning problem. Combinations of subtle difficulties are noted as they occur across specific tests, and provide the basis for the differential diagnosis of particular developmental delays and learning difficulties (Table 47–1). For example, visual-spatial perception weaknesses can be diagnosed from delays on certain Wechsler subtests in combination with particular processing difficulties, such as directional confusion, spatial disorganization, and frequent rotations when drawing. Associated educational difficulties may include reversals of letters, numbers, and words in reading and writing, disorganized writing, and difficulty with vertical alignment of mathematics items.

Screening Techniques

To ensure that sensory, physical, and developmental problems are identified at an early age, screening of large numbers of preschool children is a fairly common procedure (see Chapter 45). In particular, reading readiness and school readiness tests are regularly administered. A variety of instruments have been published to assist in this task; these include such well known tests as the Denver Developmental Screening Survey and the Metropolitan Readiness Test (see Chapter 50). Administration of these measures usually occurs in combination with questionnaires, informal observations, and parent interviews. As was indicated previously, the object of screening usually is to determine which children require further evaluation or need close monitoring of their learning progress. Because this process is expensive and time consuming, ideal screening instruments should not overidentify too many children (false positives), nor should they miss children with genuine problems (false negatives). Currently numerous early childhood programs are evolving new techniques for more precise identification of preschool developmental difficulties. One example is the Brookline Early Education Project (BEEP), an interdisciplinary longitudinal, diagnostic and educational program in Brookline, Massachusetts (Pierson, 1977). Within this project, developmental measures such as the Pediatric Examination of Educational Readiness (PEER) are currently being refined for early identification of subtle learning problems (Meltzer et al., 1981).

Controversy still exists about the potential benefits versus the possible harm resulting from the screening and labeling of children for the purpose of providing early intervention. One reason for this concern is the lack of consensus in the research literature regarding the efficacy of such programs (Hodges and Cooper, 1981; Palmer and Anderson, 1979).

In particular, a child's performance in later learning can never be definitely attributed to the effects of intervention. The question of how much evaluation to provide for children identified as exceptional in screening programs is also a difficult one. A number of alternatives exist. The first is to monitor the child's progress closely with the expectation that at some stage there will be clarification as to whether further assessment is required. An alternative solution is to offer every preschooler a complete interdisciplinary work-up. An intermediate type of assessment procedure may be most appropriate for children who have moderate cognitive and learning difficulties that are clearly recognized by the various professionals working with them. In these instances the pediatrician or neighborhood health center staff could confer with school authorities to determine which specialists should be involved in the diagnosis and program planning. When children present subtle or unusual combinations of disabilities, or more serious handicaps, extensive interdisciplinary evaluations may be necessary. For instance, a child with an attention deficit might need an extended

Table 47–1. DIAGNOSTIC TESTS FOR EVALUATING SPECIFIC DEVELOPMENTAL DELAYS AND LEARNING DIFFICULTIES

Diagnostic Formulations	Standardized Tests and Subtests	Examples of Processing Delays	Associated Educational Difficulties
Visual discrimination and spatial perception delays	Wechsler Scales (picture completion, block design, object assembly) Bender Visual-Motor Gestalt Test Beery-Buktenica Test of Visual-Motor Integration Primary Mental Abilities Test (spatial relations, perceptual speed)	Directional confusion (left/right; up/down; before/behind) Spatial disorganization Figure-ground confusion Rotations, reversals, and inversions	1. Reading Reversals of letters, numbers, words (p/q, b/d, was/saw, on/no) 2. Written expression Unequal sizing and spacing of letters and words Disorganized Slow writing speed Words cramped together on a line Difficulty with copying tasks 3. Mathematics Reversals of numbers (6/9, 31/13) Difficulty with vertical alignment of math items Confusion of operations ($\times$/+) Problems with place value, equations, geometry
Delays in processing of sequential information	Wechsler Scales (coding, digit span, picture arrangement, arithmetic) Illinois Test of Psycholinguistic Abilities (sequential memory subtests)	Difficulty in telling time Problems in remembering everyday sequences (e.g., days, months, seasons) Often forgets alphabetical sequence	1. Reading Confused letter and word order Weak blending skills Disorganized reading comprehension

Detroit Tests of Learning Aptitude (subtests measuring visual and auditory attention)	Poor understanding of daily routines Difficulty in remembering spoken instructions, messages, telephone numbers Confusion of narrative organization Delayed mastery of hopping, sports	2. Written expression and spelling Order of letters and words confused when writing from dictation Recognition memory better than retrieval 3. Mathematics Difficulty in counting, multiplying Inaccurate memory for math facts
Language difficulty (receptive or expressive)	Wechsler Scales (information, similarities, arithmetic, vocabulary, comprehension, digit span) Peabody Picture Vocabulary Test Detroit Tests of Learning Aptitude (verbal opposites, verbal absurdities, oral commissions, auditory attention span, oral directions) Illinois Test of Psycholinguistic abilities (grammatical closure, sound blending, auditory reception)	Problems discriminating speech sounds Frequent mispronunciations of words Limited verbalizations Hesitancy following verbal instructions Limited vocabulary Difficulty with word finding Poor comprehension of syntax	1. Reading Reading comprehension lower than word recognition scores Weak blending skills Confusion of prepositions, relational words, pronouns Reading often unintelligible 2. Written expression Confusion of vowel sounds when spelling (rid/read; leud/lead) Writing unintelligible 3. Mathematics Paper and pencil computation better than mental computation

Note: Procedures similar to those listed here are used when diagnosing other developmental and processing delays, e.g., selective attention, memory, higher-order conceptualization. This table provides only a few examples of standardized tests; more refined assessment methods differentiate processing delays in analytical, verbal, and nonverbal areas. The reader is referred to relevant tables in Chapter 38 for additional information.

assessment involving numerous professionals because of the enigmatic quality of the symptoms.

The following example illustrates a situation in which the combination of difficulties was misdiagnosed because a complete interdisciplinary evaluation was not used initially:

Jose, a very active five year old with limited language skills, was screened for kindergarten entry and referred to a hospital for further evaluation. Although he received a full pediatric work-up, he was given only brief screening in the areas of language, fine motor function, and cognition. Because his mother did not speak English, the social service interview was omitted. Jose's discharge diagnosis read: "Healthy male with developmental immaturity in attention and language." He was placed in a bilingual program for immature five year olds.

At the end of the year he was again referred to a hospital for an expanded work-up. Assessment indicated a moderate sensorineural hearing loss, severe communication disorder, and serious adjustment problems. His nonverbal intelligence was fully normal. It was recommended that he receive a total communication program, with intensive language therapy, which could include sign language and the use of a portable communication board. It was also suggested that a bilingual home coordinator work with the child's mother.

As in Jose's situation, considerable confusion still pervades clinics and schools regarding the similarities and differences between screening and assessment. Serious problems such as hearing loss and emotional disorders often remain untreated because observed behavior patterns were ignored at the screening interview and further testing was not recommended. In other instances children have been placed in remedial programs that were not suited to their needs. In order to ensure appropriate utilization of screening and assessment procedures, it is critical that health care professionals remain familiar with current research regarding the purposes and efficacy of such programs.

INFORMAL EDUCATIONAL ASSESSMENT

Informal educational measures are generally criterion referenced rather than norm referenced and can be administered over a period of time rather than in a single session. They are designed to provide flexible test procedures and consequently facilitate the assessment of each child's unique profile of strengths and weaknesses. In the educational sphere numerous commercial or teacher designed inventories exist for the informal measurement of reading, arithmetic, and writing skills; these are discussed in the section beginning on page 1015.

Traditionally informal educational procedures have been utilized in the regular classroom setting or the special education setting. Recently an increasing number of psychoeducational specialists and special educators in clinic settings have incorporated these techniques into their diagnostic batteries. Children referred to clinics often have been tested previously on most of the common standardized measures; therefore informal inventories and procedures are especially helpful. A few well known procedures are described in the following section.

Classroom Observations and Anecdotal Records

Within the large group setting of the classroom, the competent teacher constantly evaluates each child's progress toward defined goals and maintains anecdotal notes, work samples, and results of informal teacher designed tests. Classroom record keeping frequently takes different forms depending on the style and orientation of the specific teacher. For instance, checklists and inventories, both commercially designed and teacher made, are often used to document students' progress at regular intervals. Many teachers chart the changes in students' behavior and performance, particularly when improvements are small or when behavioral techniques require analysis of the antecedents and frequency of occurrence of such behavior. These records usually form the basis for discussion with parents and other professionals. They also constitute a critical component of the assessment procedure, because they provide the most accurate picture of the child's daily performance in a group learning situation.

Diagnostic Teaching

Diagnostic teaching or "clinical teaching" generally refers to that procedure whereby assessment and intervention are refined in a continuing alternating process until suitable educational progress has occurred and remediation can be terminated (Fig. 47–1).

By using this procedure, the results of previous tests can be reviewed in conjunction with background information and questionnaires; these allow the clinician to formulate hypotheses about the child's specific learning problems. These ideas are then translated into a miniteaching plan, which is applied, modified, and adjusted according to the child's success or failure in the test situation. On the basis of the success of these teaching techniques, management strategies are recommended to the school for incorporation into the daily teaching and remedial program. The advantages and benefits of clinical diagnostic teaching include some of the following: First, the diagnosis does not end when treatment begins; there is continuous feedback among diagnosis, remediation, and educational change, and there is also constant modification of intervention strategies.

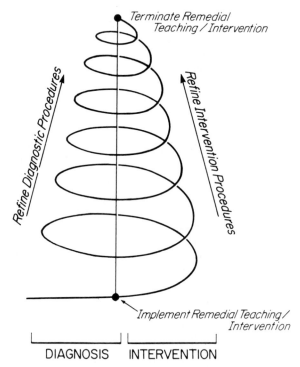

Figure 47–1. The diagnostic-teaching process. Diagnostic teaching is a continuous process whereby diagnosis and intervention are refined in an ongoing, alternating fashion. Both processes are modified and improved until intervention can be terminated.

Second, because monitoring is a formal component of the diagnostic teaching plan, remediation can be designed to meet the specific needs of the individual child. In other words, group oriented instruction may be avoided and a more individualized program introduced. Third, flexibility of assessment procedure is assured. Finally, over a period of time the clinician evaluates a child's individual reactions to specific task demands and accordingly modifies recommendations for remediation.

Piagetian Techniques

Over the past decade Piagetian principles and techniques have been applied increasingly to assess children's readiness for instruction and acquisition of particular cognitive and developmental processes. Although there is still controversy regarding the relationship of these tasks to distinct academic skills (Elkind, 1976; Meltzer, 1978), there has been widespread acceptance of the philosophy and theoretical rationale underlying Piaget's system. Consequently there has been increasing emphasis on modifying standardized tests to suit the actual needs and cognitive level of each child, rather than adhering only to the objective but sometimes static standards of formal tests. Table 47–2 lists some common Piagetian tasks that are applied to children aged six to

eight years to evaluate the level of cognitive development and abstract reasoning. For further information, readers are referred to the book by Ginsburg and Opper (1981).

An innovative diagnostic procedure that uses a Piagetian orientation while transcending the limitations of the specific tasks has been recently developed by Feuerstein and his associates (1981). Their "learning potential assessment device" emphasizes the process rather than the product of learning by teaching the child specific concepts in the test situation and evaluating his ability to apply these principles. Their assessment procedure, which questions the utility of formal standardized tests, holds much promise for the future.

FOCUS OF EDUCATIONAL ASSESSMENT

This section provides a synopsis of the major skills that are evaluated as part of an educational assessment procedure, namely, reading, spelling, written expression, and mathematics. In keeping with the approach just outlined, an integrated diagnostic-prescriptive model is emphasized throughout.

READING

Current epidemiological studies suggest that approximately 15 per cent of school children exhibit significant reading disabilities. In the United States alone this accounts for more than eight million children, with similar rates estimated for Canada, the United Kingdom, France, and Denmark. Follow-up studies suggest that reading disorders are a major cause of school drop-out and delinquency, and that persistent reading problems often result in secondary emotional or behavioral disturbances (see Benton and Pearl, 1978, for recent reviews). A vast body of literature has emerged that deals with reading readiness, prevention of reading problems, and identification of reading disabilities; nevertheless there are still no absolute criteria for designating reading as "advanced" or "delayed." This definitional confusion can be traced to the early characterizations of reading problems as "word blindness," "strephosymbolia," or twisted symbols, and to the more recent controversy about the term "dyslexia."

Recent reviews (e.g., Resnick, 1979) suggest two predominant themes in reading research. The first approach considers reading as a decoding process, which comprises identification of printed symbols and their translation into an approximation of oral language. The second viewpoint emphasizes an autonomous language process, so that compre-

Table 47–2.   SAMPLE BATTERY OF PIAGETIAN TASKS FOR COGNITIVE ASSESSMENT

Task	Description	Area of Cognition Assessment	Indicators of Possible Delays
Conservation of number (six years)	a. Comparison of the number of blocks in two rows—one row bunched b. Comparison of the number of blocks in 2 rows—one row spread out	These tasks evaluate the child's cognition that the property of number remains unchanged when its perceptual appearance is altered.	Overdependence on visual appearance of the rows of blocks; child cannot understand that the number of blocks remains the same even though the shapes of the rows differ.
Conservation of continuous quantity (seven years)	a. Comparison of equivalent quantities of liquid in standard tumbler and tall, narrow cylinder b. Comparison of equivalent quantities of liquid in standard tumbler and three small wine glasses	These tasks evaluate the child's understanding that the property of quantity remains invariant (is conserved) when it is transformed in shape and perceptual appearance	Overdependence on perceptual appearance; child cannot grasp that liquid quantities remain the same even though they look different and that: a. Variations in the length and width of the container are reciprocal (compensation) b. No liquid has been added or taken away (invariant quantity) c. The liquid can be poured back into the original container and returned to its original state (reversibility)
Seriation (seven years)	a. Serial ordering of 10 strips of wood b. Deducing certain size estimates (e.g., biggest, middle sized, smallest) from a set of 10 strips of wood	These tasks measure the child's ability to order a series mentally in two directions simultaneously and to coordinate relationships (e.g., 3<4<5).	Judgments of size are based on only one perceptual cue, e.g., the top of the wooden strips, not the bottom
Classification (eight years)	a. Comparison of a class of wooden beads with its subclasses of yellow and green beads b. Deducing part-whole relationships	Classification refers to the ability to form logical groups and to understand inclusive relationships	Child cannot understand that subgroups share some characteristics with the overall category—e.g., green beads can differ from yellow beads in terms of color, but both are part of the overall class of "woodenness"

hension of the written word is distinct from the understanding of spoken language. The latter view suggests the importance of evaluating numerous developmental processes as part of a reading assessment, namely, visual and auditory perception, receptive and expressive language, cognition, and attention.

Because of the developmental and instructional precursors of reading efficiency, assessment of early reading skills often differs significantly from evaluation of more advanced reading. Word recognition and word analysis skills compose the most critical features of beginning reading, so that assessment generally emphasizes reading accuracy and rate. Adequate diagnostic inventories not only elicit the grade level obtained but, more importantly, include an analysis of the reading style and specific errors (Table 47–3).

As summarized in Table 47–3, stylistic weaknesses to be noted are associated with dysfluencies, slow reading rhythm, and possible overdependence on a particular reading method. The latter is reflected in a limited sight vocabulary or inefficient word attack skills. Omissions, substitutions, and reversals of letters, syllables, and words are also noted. Cautious interpretation of these error patterns is important, for recent research suggests that reversals in reading do not necessarily reflect underlying reading disabilities. Interpretation of these errors in conjunction with developmental findings generally yields valuable diagnostic information.

At the higher grade levels, reading for meaning assumes greater importance, and the manner in which reading comprehension is assessed often is critical to the final diagnostic formulation. For example, most comprehension tests examine

Table 47–4. COMPONENTS OF READING COMPREHENSION INVENTORIES

Reading Comprehension Measure	Observations
A. Answers to structured questions (open ended or multiple choice)	Recall of main idea Memory for factual details Vocabulary Drawing inferences
B. Retelling story orally (as in essay writing, science, history)	Formulating main idea in relation to supportive factual details Vocabulary Drawing inferences Organizing information for retelling story Sequencing facts in correct order
C. Retelling story in writing	Formulating main idea in relation to supportive factual details Vocabulary Drawing inferences Organizing information for retelling story Sequencing facts in correct order Use of syntax and grammar

Note: All three measures of reading comprehension are compared. Oral and silent reading comprehension are compared.

memory for facts rather than the skill with which information is organized or inferences drawn. Since everyday performance at school and at home demands adequate formulation and sequential organization, the diagnostic assessment should incorporate oral and written measures of the ability to retell a story. As is evident from Table 47–4, comparison of the child's skills across different task formats provides useful diagnostic information. Frequently youngsters with reading difficulties succeed when they are prompted with specific structured questions, but flounder when no cues are provided and they are required to integrate, synthesize, and cognitively restructure the material.

SPELLING

Spelling is a skill that relies heavily on memorization for the correct sequential ordering of specific letters within words. Because of the inconsistencies that characterize the English language, a one to one correspondence does not exist between each symbol and a specific sound. Instead a number of possibilities exist, which depend on other features of the words. Consequently spelling, or encoding, is often a more difficult task than reading or decoding; in fact, fluent readers often display spelling deficits. During the reading process, context cues help the reader to recognize specific words, whereas these clues are minimized

Table 47–3. SOME COMMON INDICATORS OF DIFFICULTY WITH ORAL READING

Reading Style	Reading Errors
Dysfluencies Word-by-word reading Over-reliance on a particular reading method to sound out each word Consistent use of finger pointing Frequent subvocalizing-re-auditorizing Numerous self-corrections Often loses place on page—omits words or sentences	Omissions of letters, syllables, words ("sog"/song) Substitutions of letters, syllables, words ("hen"/den) Additions of letters, syllables, words ("hitted"/hit) Reversals of letters, words ("big"/pig; "was"/saw) Sequencing confusions ("fatsen"/fasten) Problems decoding beginning sounds in words Difficulty in decoding word endings Confusions of various vowel combinations

Note: These parameters are observed and compared for single-word reading and oral reading of a paragraph. Errors are identified after brief testing; more specific diagnostic information is obtained from in-depth evaluation.

during spelling. Spelling difficulties may result from deficits in one or a number of areas, including reading, analysis of the structural features of words, blending or synthesis of sounds, generalizing across sound patterns, remembering the letters in the correct sequential order, integrating visual, auditory, and motor processes, automatizing all these skills, and "revisualizing" the whole word correctly.

Comprehensive evaluations of achievement generally include formal spelling tests, which yield grade level scores. These measures are useful for screening rather than for diagnosis, for they reflect the child's level of functioning rather than the nature or type of deficiency. Informal spelling inventories often provide more appropriate diagnostic information and incorporate specific analyses of spelling errors (Boder, 1973; Camp and Dolcourt, 1977). The particular method utilized for assessing specific spelling skills is often critical because this can influence the diagnostic formulation significantly. Techniques that are frequently used require that the child spell single dictated words, write dictated sentences, select specific words from a multiple choice array, or proofread for misspellings.

Specific spelling errors that occur most frequently include phonetic errors, in which the pronunciation of the error resembles the actual word ("dus"/does, "pese"/peace, "luvs"/loves). Visual errors are identified when the correct letters are included in the word but their sequential order is incorrect and they do not make sense phonetically ("lghit"/light). Finally, mixed errors generally do not make sense visually or phonetically and include reversals ("was"/saw), omissions ("hep"/help), and semantic substitutions. Figure 47–2 shows a spelling sample from a second grader with fairly severe visual-spatial difficulties. His reliance on auditory cues to spell produced numerous phonetic errors ("lite"/light, "wach"/watch, "nacher"/nature). Reversals (b/d) occurred when he was presented with more difficult word patterns ("bress"/dress, "orber"/order) and not with basic consonant-vowel-consonant patterns, which he had already consolidated in his sight-word vocabulary (boy, and).

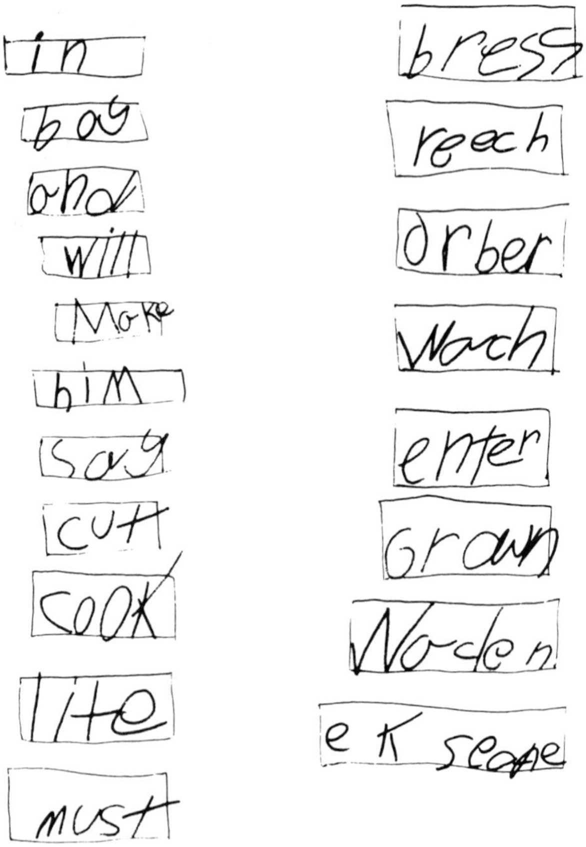

WRITTEN EXPRESSION

Current research and the clinical literature highlight the need for evaluating written language more specifically to assess the zones of commonality and difference that characterize reading and writing deficits. Because diagnostic assessment of writing problems is not as advanced as the evaluation of reading difficulties, many children with writing deficits are frequently mislabeled and are

Figure 47–2. Spelling sample from a seven year old second grader with fairly severe visual-spatial difficulties. The words he has attempted to spell are as follows: "in, boy, and, will, make, him, say, cut, cook, light, must, dress, reach, order, watch, enter, grown, nature, explain."

not provided with appropriate remedial services, particularly in junior high school.

A number of explanations and terms have been proposed to describe deficient written output. Graphomotor problems, or difficulties with pencil control, have been emphasized and are referred to as "dysgraphia" or "agraphia." Fine motor function deficits, visual-perceptual-motor difficulties, and finger agnosia have also been implicated. Finally adequate memory and language skills are considered critical for appropriate written language. In fact, proficiency in writing is dependent on the integration and automatization of motor, perceptual, language, and cognitive processes. A few components of writing include planning and executing the correct graphic form for each letter while revisualizing the letter formations, ordering the letters and words correctly to produce an appropriate syntactic sequence, and keeping a single idea in mind while formulating this in the correct word sequence.

Assessment of written language skills differs in relative emphasis according to the grade level of the child. In the early grades diagnostic procedures generally focus on the motor and visual-motor components of writing. Writing problems

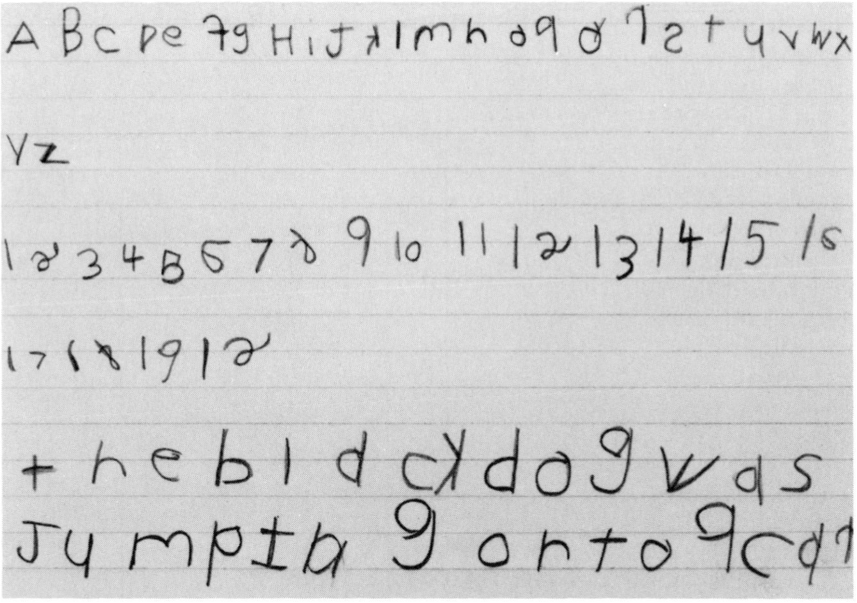

Figure 47–3. Writing sample from a seven and one-half year old repeating first grade. The writing screen reflects reversals of letters and words, poor letter alignment and spacing, and frequent errors on a copying task. The sentence for copying reads as follows: "The black dog was jumping on top of the cat."

at this age level may in fact arise from difficulties in various areas, including fine motor function, pencil grasp, pencil pressure, hand position, visual and auditory perception, memory processes, tactile-kinesthetic input, and sustained attention. Common manifestations of such difficulties include frequent reversals and substitutions of letters and words, poor spatial organization, extremely slow writing speed, and general "illegibility." Figure 47–3 illustrates these deficits in a seven and one-half year old who was repeating the first grade. He was unable to reproduce the alphabetic sequence without frequent letter reversals and rotations; copying a sentence was even more of a problem.

At the late elementary and junior high school levels the content of writing assumes greater importance. Adequate diagnosis and remediation emphasize the memory and language components of writing, as reflected in the organization of ideas, formulation, and the fluency of written expression. Reduced written productivity may reflect possible word-finding problems, limited familiarity with the format of written language, deficient expressive language, mixed syntax and formula-

tion, or poor sequential memory. In addition, difficulties with penmanship, spatial organization, graphomotor planning, and finger differentiation may have an impact on the overall speed, volume, and quality of the written product. The specific developmental and cognitive deficits that seem to cluster with these writing difficulties have been characterized clinically as "developmental output failure" (Levine et al., 1981). This problem is usually manifested in terms of reduced written output or productivity, frequent refusals to complete work, failure to submit assignments, a tendency to "forget homework," and chronic underachievement. Figure 47–4 presents a writing sample from an 11 year old with a superior verbal IQ and severe spatial organization deficits who had experienced written language difficulties since the first grade. Even though his reading skills and oral language level were three years above his grade level, he was unable to produce even a few sentences without erasures, retracings, poor spacing of letters and words, and many spelling errors. Writing speed and volume were always reduced.

Few tests exist for the adequate differential diagnosis of written language, primarily because of

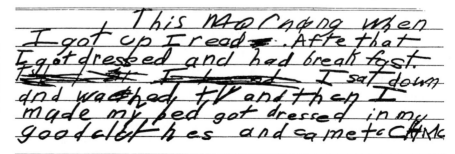

Figure 47–4. Writing sample from an 11 year old fifth grader with a superior verbal IQ (140) whose reading achievement was three years above grade level. Errors are associated with severe deficits in perception, visual-spatial organization, and motor planning components of writing.

Table 47–5. COMPONENTS OF A DIAGNOSTIC INVENTORY OF WRITTEN LANGUAGE

Motor Planning	Spatial Organization	Letter Production	Mechanics	Language Usage	Speed of Writing
Overall quality of letter forms	Letter spacing	Erasures and retracings	Spelling	Vocabulary level	Time taken to complete various writing tasks
Motor rhythm–motor movement for writing	Word spacing	False starts	Punctuation	Sentence structure	
Writing fluency	Ability to stay on the line	Mixture of print and cursive	Capitalization	Subject-verb agreement	
Pencil pressure	Reversals of letters and words	Mixture of upper and lower case		Grammatical usage	
	Rotations or inversions of letters and words			Overall thematic quality	

Note: Each of these parameters is observed across a number of tasks, namely, writing the alphabet, copying, writing from dictation, reproducing a sentence from memory, and writing a paragraph.

the difficulty in evaluating this skill. Two of the more popular inventories include the Myklebust Picture Story Language Test (PSLT) and the Test of Written Language (TOWL). At the Children's Hospital Medical Center in Boston a diagnostic inventory of written expression is currently being developed and evaluated. This is designed to satisfy certain criteria: brief administration time, simplicity of items, complexity of diagnostic information, an integrated paradigm for writing assessment that yields information about the child's possible developmental and cognitive weaknesses, and provision of suggestions for specific remedial strategies. Table 47–5 summarizes some of the elements of this inventory.

This diagnostic writing inventory compares various processes in a number of simple tasks requiring different types of input and output, namely, the ability to reproduce the alphabet automatically, a task that emphasizes revisualization and sequential memory; the ability to copy appropriately when motor planning is stressed and memory is not required; efficient writing from dictation, which involves auditory memory and the translation of a blueprint from memory into writing; the ability to reproduce from memory a sentence that the child has read previously (this task involves visual memory in conjunction with integration, organization, and motor output); and the ability to formulate a story about a picture and to produce this in writing with no input from the examiner (this is the most complex of written tasks and involves organization and integration of all the skills and processes mentioned).

A timely referral for diagnostic assessment of written language difficulties may avert much anxiety, lowering of self-concept, and embarrassment, and the early introduction of remediation often can prevent the onset of difficulties in curriculum areas such as geography, social studies, and history.

ARITHMETIC-MATHEMATICS DISABILITIES

Arithmetic disabilities, often referred to as "dyscalculia," are characterized by a wide variety of perceptual and cognitive disorders. These include spatial difficulties, perceptual weaknesses, language problems, reading problems, deficient acquisition of the number concept, disturbances of visual-perceptual-motor output, and attention disorders. Differential diagnosis of arithmetic problems is not yet as refined as is the assessment of reading delays. Consequently difficulties with arithmetic operations and computation often are attributed simply to poor instruction in the early grades. In reality the student may exhibit specific processing weaknesses, which contribute to delays in the early acquisition of the number concept.

In recent years a considerable body of research has related the acquisition of the number concept, assessed primarily by means of Piagetian tests, to arithmetic skills. The tasks, which measure concepts such as conservation, one to one correspondence, seriation, and classification have been adapted from Piaget's original techniques and have been applied in the classroom and clinical setting. As yet, however, there has been no conclusive evidence to indicate that arithmetic disabilities are necessarily a function of delayed acquisition of the number concept. Rather, development of the number concept and "readiness" for learning arithmetic seem to be necessary but not sufficient conditions for adequate performance on arithmetic measures.

Appropriate techniques for assessing arithmetic difficulties are scarce. Paper and pencil tests include the Wide Range Achievement Test and the Stanford Arithmetic Test: more extended diagnosis is possible through application of the Keymath Diagnostic Arithmetic Test. Ideally diagnostic assessment provides an index of the child's performance relative to his current grade level, in addition to analysis of style and types of errors. It is also important to measure the understanding of concepts related to time and money, because these often provide clues to the existence of possible sequencing delays. It is critical to differentiate between arithmetic disorders, which are associated primarily with computation weaknesses, and those resulting from limited understanding of the symbolic operations, which provide the "language of arithmetic" (e.g., $+$, $-$, $\times$, $\div$). In particular, direct comparisons of the child's computational skills, conceptual understanding, and ability to apply operations are helpful. The child's approach to paper and pencil calculations is frequently evaluated in terms of part-whole perception difficulties, reversals of digits, spatial organization difficulties, and confusion of operational signs (e.g., $+$, $\times$).

CAUTIONS REGARDING TEST INTERPRETATION

Much has been written about the negative effects that can result from misinterpretation of test data and misuse of scores. Recently, widespread concern has been expressed about the application of standardized tests to members of minority groups who may not be adequately represented in the standardization sample. In fact, states such as California have imposed a moratorium on IQ tests for the placement of children in classes for the "retarded" as a result of a 1964 class action suit. Unfortunately, alternative assessment procedures are frequently inadequate, and informal observations and reports may often be influenced by spurious factors such as physical appearance, verbal precocity, family influence, or personality.

A few critical issues should always be taken into account to ensure appropriate educational testing practices:

1. Tests should be selected so as to be compatible with the child's specific cultural and socioeconomic status.

2. Tests should be administered by qualified professionals under conditions specified in the test manual.

3. Results should be integrated with other assessments and reports concerning the child.

4. The test format is critical. For instance, a spelling test could be dictated orally or presented in a multiple choice format; each of these involves different processing modalities.

5. Group and individual tests frequently yield different results.

Within the boundaries of these constraints, important issues include the following:

1. Children differ in their test taking ability; for instance, some children excel under test conditions and others are immobilized by anxiety.

2. The context of the assessment may influence test performance. For example, a school setting is very different from a clinic or hospital setting.

3. Consideration must be given to factors that influence the child's attention, motivation, and fatigue.

4. Certain children react to the specific expectations and cues of the examiner (e.g., smiles, eye contact).

5. Family stress may reduce the child's attention and effort.

6. Fatigue may affect performance detrimentally.

7. Test interpretation is dependent on the skill of the specific clinician.

8. Cultural factors often influence the value placed on achievement orientation; consequently minority groups are frequently misplaced in classes for the mentally retarded despite average intelligence.

Finally, repeated assessment using the same test often reduces the reliability of findings. Consequently, when a second opinion is elicited because a family or professional is not satisfied with an assessment, all previous test reports should be made available. In these instances review of previous evaluations, in conjunction with questionnaires, rating scales, and informal observations by relevant professionals, may provide the information needed for adequate diagnosis.

EDUCATIONAL ASSESSMENT OF SPECIAL POPULATIONS

The educational assessment of exceptional individual is a clinical specialty that provides a unique challenge to diagnosticians. The puzzle of diagnosis is particularly subtle and complex, and discrete portions of information are often difficult to integrate in order to provide a coherent picture of the child's needs. Once again the critical features of assessment are the individualization of test administration and the flexibility of clinical interpretation.

The approach presented in this section is based on a categorization of exceptionality along several dimensions: area of exceptionality (e.g., intellectual, sensory, motor, physical, personality, medical, social), degree or extent of exceptionality (the individual's difference from the norm), and breadth or range of exceptionality (the number of skills and functional areas affected in one individual). In general, the area of exceptionality is first determined, after which the latter two dimensions are assessed. For instance, once it has been established that a child is hearing impaired, it is important to evaluate whether he can read and write and how much difficulty he exhibits in these areas.

This section provides a brief discussion of the major issues concerned with the assessment of special populations. The models and techniques described in the first three sections provide the basic principles for this discussion. It is assumed that the reader is familiar with the issues regarding definition and classification of various special populations (see Chapter 39). For further information concerning assessment in each of the exceptional groups, readers are referred to the books by Dunn (1973) and Meyer (1978). Finally, terms such as "learning disabled" and "gifted" are used in this section to provide a point of reference and to help clarify communication; they do not refer to distinct syndromes.

ASSESSMENT OF "LEARNING DISABLED" CHILDREN

The term "learning disability" has evolved as a categorical entity that is currently not well defined and is not consistently diagnosable. The rapid growth of this area of special education has been characterized by definitional confusion and controversy, and by an abundance of different labels such as "perceptual impairment," "brain injury," "dyslexia," "language disability," and "minimal brain damage." The most recent official definition proposed (Public Law 94-142) considers a learning disability to imply a disorder in language usage, which is manifested in imperfect listening, think-

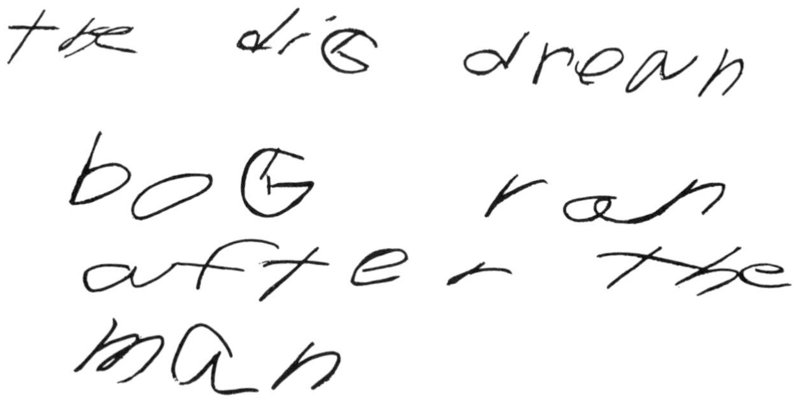

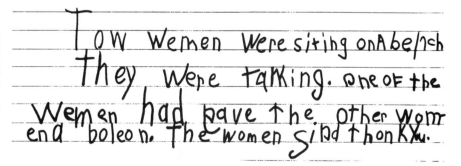

Figure 47–5. Writing sample from a seven and one-half year old second grader with severe visual-spatial difficulties. He had attempted to write the dictated sentence, "The big brown dog ran after the man."

ing, speaking, reading, writing, spelling, or mathematics ability. Nevertheless recent literature reviews indicate that there has been limited success in the various attempts to apply this definition in research and clinical domains (Hallahan and Bryan, 1981; Sabatino, 1980). The currently accepted viewpoint is that the term "learning disability" is more useful as a concept than as a unique category of exceptionality.

A nonlabeling approach to this group of children is proposed here. Furthermore a diagnostic procedure is advocated that determines the child's profile of relative strengths and weaknesses, the multiplicity of weaknesses, and the ability to compensate for specific difficulties (Levine et al., 1980; Satz and Morris, 1980). In other words, assessment emphasizes cognitive and academic skills and typically includes measures of perception, information processing or thinking, selective attention, memory, language, social interaction skills, and basic skills such as reading, writing, spelling, and arithmetic. The child with specific reading problems who has been labeled as "dyslexic" can be identified in terms of this diagnostic system, and assessment can be focused on the existence of delays in various processing areas, such as visual-spatial or temporal-sequential processing. Figure 47–5 provides an example of visual-spatial difficulties that influenced the writing of a seven and one-half year old second grader. Reversals of letters and words (b/d, p/q, u/v), in conjunction with

poor organization and spacing of writing, are fairly common in children with severe visual-spatial deficits.

Specific writing deficits are often associated with severe auditory sequential memory difficulties and possible language delays (Fig. 47–6). Characteristic errors include frequent spelling errors, confusions of vowel sounds in words, and some unintelligible writing.

ASSESSMENT OF MENTALLY RETARDED CHILDREN

The educational assessment of mentally retarded individuals is frequently very similar to the evaluation of children with subtle learning problems. In particular, children with mild to borderline mental retardation may develop primary reading skills at a relatively normal age; many are able to identify upper and lower case letters by kindergarten age and to proceed fairly easily with the learning of sound-symbol relationships. In this population reading achievement becomes more problematic at the higher grade levels when reading comprehension is emphasized. In fact, almost by definition, mildly retarded persons do not exceed a fifth to sixth grade level of reading comprehension; most do not exceed a fourth grade level. In general, the reduced likelihood of educational success in this group is attributable to the higher

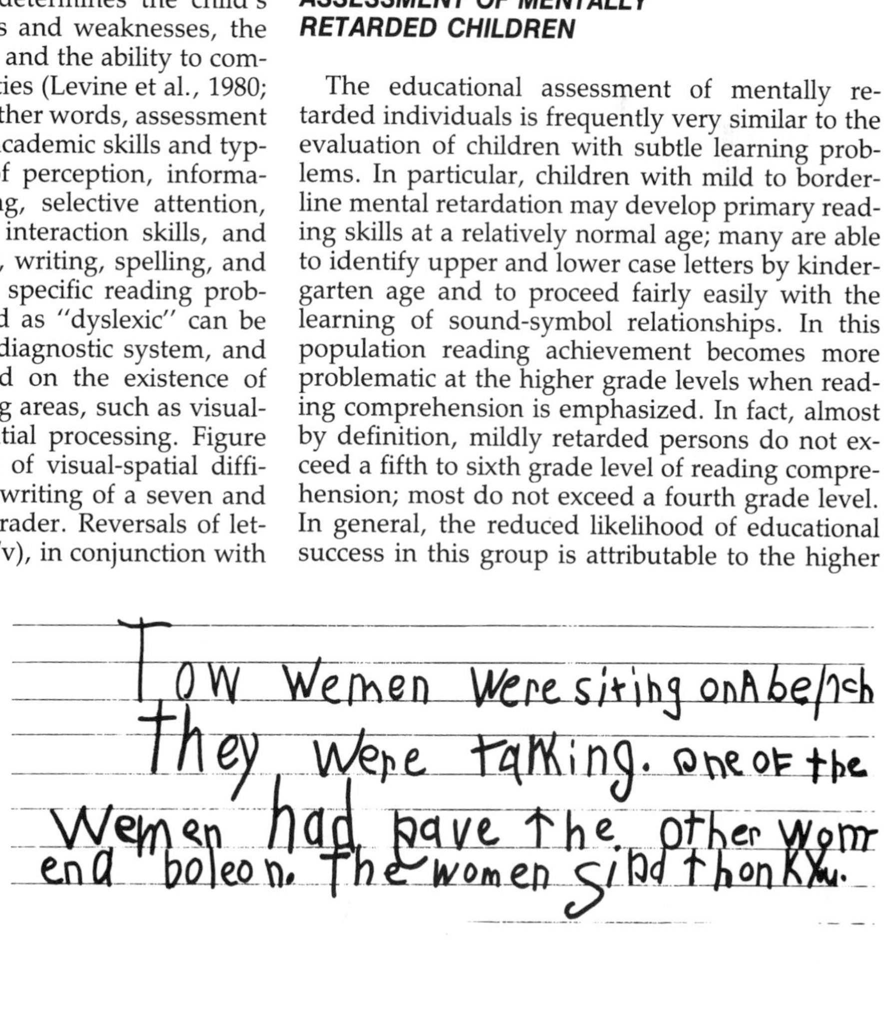

Figure 47–6. Writing sample from an eight and one-half year old third grader with severe auditory-sequential memory difficulties and evidence of a language disability. His description of a particular picture reads as follows: "Two women were sitting on a bench. They were talking. One of the women had gave the other woman balloon. The woman said thank you."

than normal incidence of multiple delays in perceptual and cognitive areas (e.g., auditory perception, visual perception, attention, memory).

Specific adaptations of test formats for this group include paper and pencil tests and multiple choice tasks, both of which reduce the requirements for oral language. Such modifications are particularly helpful at the higher grade levels for assessing reading comprehension. For instance, the child with limited expressive language skills may answer, "I don't know," to any question requiring more than a one word response and consequently benefits from a multiple choice format to demonstrate his reading comprehension. Other format changes may be necessary for children who have learned to read from color coded texts or from materials that use an augmented teaching alphabet. Diagnostic teaching is particularly useful in conjunction with informal educational tests, as is illustrated in the following case study:

Angie, a 20 year old nonverbal resident of a state school for mental retardation, was being evaluated by use of a pictorial communication board when the speech therapist noticed that she recognized a few printed words. Consequently an educational evaluation was scheduled to determine her reading ability and potential for further progress. This was a complex process because she had never been exposed to schoolwork because of her presumed total deficiencies. To attempt to assess Angie's capacities in educational areas, the special educator developed multiple choice formats for all testing. Angie proved adept at recognizing the names of people and objects in her environment and could even complete verbal analogies at the seven year level. After several sessions of diagnostic teaching, she could respond to written directions, such as, "Put the book in the desk. Give me the red pen." As a result of this diagnostic teaching, Angie's entire program was re-evaluated to consider transfer from the state school to a community setting.

As is clear in Angie's situation, learning potential should always be assessed, no matter how severe the problem may seem. In all instances reading instruction is a life skill that is critical for daily survival and for the enhancement of leisure enjoyment.

Individuals who are severely and profoundly retarded are evaluated especially by history and direct observation. Ideally assessment should occur in a setting familiar to the client. Standardized tests are of limited use, although developmental and behavioral checklists, parent and teacher questionnaires, and subtests dealing with specific skills or abilities may be helpful. Probably more than in any other exceptional population, the focus of assessment is on the child's strengths and weaknesses in developmental areas, his investment in various tasks and situations, and his ability to communicate needs and feelings. As is the case with all educational assessment, parents and school staff frequently raise a variety of issues, such as, Are special services needed? Why is behavior deteriorating? Does the student have the ability to learn a particular new skill? What methods might be used to effect change? Once again, these queries can be optimally addressed in the context of an interdisciplinary evaluation with the cooperation of the involved school staff and personnel, who work on a daily basis with the child (see Chapter 39A, p. 768).

ASSESSMENT OF CHILDREN WITH SEVERE SENSORY AND MOTOR HANDICAPS

Publications that deal with the educational evaluation of children with sensory impairment often assume the existence of adequate cognitive abilities and normal development of remaining senses. In some instances these assumptions may not be valid; in fact, the existence of multiple handicaps frequently has been documented in youngsters with severe hearing or visual impairment. Assessment techniques and intervention strategies need to be modified accordingly; for instance, reduced tactile sensitivity in "legally blind" children excludes fingertip learning of Braille. For blind children who are also retarded it may be impossible to master Braille because this is a difficult alphabetic code—in fact, more difficult to learn than normal print. In the case of the hearing impaired population, when a child has been given amplification and optimal early stimulation yet fails to develop interactive language, other problems should be suspected.

Educational assessment of such a child is critically important at the earliest possible age so that the child's potential for learning can be identified and an appropriate educational program developed to maximize the outcome. Specialized evaluation within an interdisciplinary or transdisciplinary team model is often critical for assessment of this population, so that the expertise of specific professionals (e.g., ophthalmologists and audiologists) is available.

Educational Evaluation of Children with Significant Hearing Impairment

Educational assessment of children wih significant hearing impairment requires a specialist who has been specifically trained to work with such children, who is comfortable with sign language, and who is acquainted with the school program the child attends (see Chapter 39B, p. 776). The selection and interpretation of educational tests are often difficult in this population; nonverbal tasks are frequently necessary. When regular classroom instruction is anticipated, standardized measures can be useful for assessing comparability between the hearing impaired child and the average population. Paper and pencil tests, such as selections from the Primary Mental Abilities Battery, may be used to evaluate visual, spatial, and perceptual abilities. Subtests from standardized intelligence tests, human figure drawing, form copying, puzzles, and connect-the-dot tasks may also yield helpful information about perceptual skills.

The concerns expressed by parents of hearing impaired children are usually similar to those expressed by parents of other exceptional groups. Issues pertain to the child's academic potential, his likelihood for achieving literacy, and the need for specific educational programs, support services, teaching methods, and materials. At the preschool level the child's social maturity and readiness for formal education are usually the major concern.

Assessment of Children with Visual Impairment

Assessment of children with severe visual impairment is also highly specialized, and the interdisciplinary team often includes a wide range of professionals, such as an ophthalmologist, a psychologist, the child's teacher, an educational specialist, and a pediatrician (see Chapter 39C, p. 787). Educational assessment frequently addresses issues pertaining to the child's capacity to adapt to the specialized school environment, the possibility of regular classroom instruction, and the specific materials needed for optimal learning to occur. Educational evaluation determines the effectiveness of alternative materials for children who cannot see printed matter. For instance, consideration is given to Braille instruction, optical scanners, speech compressors, and other aids. Many legally blind children are often able to read with the aid of a magnifying device or with large print books. The current trend is to teach the child to utilize this vision most effectively.

When the clinical educational specialist is asked to evaluate the visually impaired child's readiness for regular classroom instruction, assessment focuses upon general maturity, social and interaction skills, knowledge of the world, physical mobility, and cognitive and academic development. To facilitate regular classroom instruction, public school teachers who are inexperienced with blind students may require support in dealing with their own reactions to blindness. School districts and parents may disagree about the child's need for regular instruction or for specialized private school placement; in these instances an independent clinical evaluation may be sought. The interdisciplinary team will play a critical role in the assessment of the child's needs, as is illustrated in the following case history:

Sandy, a 10 year old legally blind child with atypical emotional responses, attended a private school for the blind. Academically her progress was poor, although both print and Braille teaching had been introduced. Socially, however, she was improving dramatically; she was becoming more verbal, and there was a reduction in her often bizarre behavior. Because of her limited academic gains, the local school district advised the parents that they would not continue to pay for private schooling. The parents sought an outside clinical evaluation to assist in program planning. After a meeting with Sandy and an assessment of the various options, a plan was worked out whereby the school would reimburse the tuition and state funding would pay the remaining charges. The school authorities agreed that they did not have an appropriate program for Sandy, and the private school prepared an individualized educational plan that was less academically oriented.

Programs for blind children such as Sandy customarily incorporate plans for regular assessment of progress by staff members who are knowledgeable about the specific program content and who have the highly specialized skills required to determine Braille mastery and the special numeric system used. When a child in such a system is referred to a clinic, major concerns are frequently associated with the child's motivation, mental health, or other nonacademic issues. In such a situation the educational specialist ideally should observe the child in both the clinic and the school setting and assess the child's attitudes toward school and particular subject areas and his openness to services. The question of program modification, or even a total change of placement, may need to be considered.

ASSESSMENT OF CHILDREN WITH SEVERE COMMUNICATION DISORDERS

The general guidelines we have discussed are equally applicable to children with severe communication disorders. Once again an interdisciplinary evaluation is most helpful, in collaboration with the professional staff members who are working with the child on a daily basis. Modification of testing techniques is critical for assessment of this group.

In children with physical limitations, such as cerebral palsy, communication may occur through subtle movements of eyes, facial muscles, and mouth; careful interpretation is needed in all instances. Educational assessment is used to determine the augmented communication system best suited to the child's needs by evaluating the child's sensory intactness, intellectual capacity, manual skills, and attention, taking into consideration the child's chronological age. Within this population the child's specific learning potential is frequently underestimated because he is nonverbal. The following study illustrates this issue:

Jon was a 10 year old boy with cerebral palsy who had spent four years in a class for nonacademic multihandicapped children. Because several of the children were nonverbal, the teacher used paired signing and speech. Jon's physical handicap precluded good use of signing; consequently he had no means of expressing himself orally. During the clinic evaluation he demonstrated a sign vocabulary of 20 words and names to which he was frequently exposed. When he was given a headlight that he could use as a pointer, he was able to pick out words in ¼ inch print. An intensive reading program was recommended because this would be extremely useful as a communication medium.

Children who are more cognitively impaired than Jon, or whose visual problems make print reading difficult, may still learn to use a communication board on which pictures represent concepts. Planning must involve parents, teachers, and communication specialists.

Assessment of the child with features that conventionally have been labeled "autistic" is most heavily dependent on a careful review of previous records, questionnaires, parent and school reports, and informal observation in a setting in which the child feels comfortable. Frequently the cognitive abilities of autistic children are underestimated, and their strengths may not be identified because of their severe communication disorders and difficult behavior in the school or clinic setting (see Chapter 39). Children with serious emotional disturbance are often uncooperative for assessment, and indeed for instruction; academic skills are usually absent or delayed. However, they may exhibit competence in one or more subjects that are of particular interest for them. The following illustration represents an example of a child whose academic skills were ignored because of his severe social difficulties.

Rob, a 14 year old autistic boy, had taught himself to read, although he lacked many basic social skills and seldom used language interactively. His poor cooperation at the clinic precluded standardized testing, but he was persuaded to respond using an IBM test response form. Eighth grade vocabulary and reading skills and fourth grade math reasoning skills were documented; appropriate educational materials were consequently recommended for his individualized program.

EDUCATIONAL ASSESSMENT OF GIFTED CHILDREN

Over the past few decades conceptualization of the terms "gifted" and "advanced" has changed gradually from an exclusive emphasis on superior intelligence as defined by an IQ higher than 140, to a multifaceted concept encompassing above average performance in intellectual areas, creativity, leadership, and exceptional talent in the sciences and the arts. There is little consensus about the most appropriate battery of tests for identifying children with advanced or superior abilities. Renzulli and Smith (1977) and Newland (1976) have specified a number of guidelines for assessment of the gifted, which include: simultaneous use of data derived from intelligence tests, creativity tests, teacher reports, parent information, peer nomination, school grades, and achievement tests; modified interpretation of IQ tests for identifying gifted children and replacement of the traditional IQ cutoff with evaluation of cognitive processes such as analysis, synthesis, and lateral thinking; case study data, including individualized methods of identification; direct observation by professionals in multidisciplinary or interdisciplinary teams; follow-up and refinement of assessment; and use of criterion-referenced measures developed specifically for the gifted population.

Currently, popular assessment procedures utilize an "identification matrix" (Callahan, 1981); data obtained from different sources are weighted according to their relative importance in the context of the child's school and home environment. The use of standardized achievement tests for the identification of gifted children is frequently limited. Usually such children are able to succeed beyond the highest level on these tests, and consequently their advanced understanding and conceptual development cannot be evaluated. Nevertheless these tests provide useful information about the child's patterns of strengths, weaknesses, and levels of mastery; these form a basis for curriculum planning.

SUMMARY

The concepts, principles, and approaches that provide the fundamental basis for assessment in this group are generally applicable to all populations. It is to be hoped that a shift will occur from standardized instruments toward more individualized case study approaches, and from assessment of exceptional children toward evaluation of achievement in "normal" children who do not display any overt problems. In situations in which there is individualization of all instruction, labels such as "gifted" or "learning disabled" become irrelevant. A noncategorical nonlabeling approach would result in individualized educational programs specifically tailored toward the unique learning styles of every child, no matter how severe their difficulties or how advanced their skills.

GENERAL CONSIDERATIONS AND CONCLUSIONS

This chapter has considered many aspects of educational assessment, but has so far omitted one of the most critical components of the evaluation procedure—the process of demystification. Once the evaluation has been completed and a diagnostic formulation determined, the findings need to be communicated to the parents and the child in a manner devoid of complicated terminology and professional jargon. When a number of professionals have been involved in the assessment, this responsibility usually falls on the coordinating member of the team, and it is this person who is required to address the critical issues with the family. In this regard specific diagnostic categories often require clarification. In some instances parents or professionals may express relief when the child's problem is assigned a particular label, because this facilitates the delivery of appropriate services. In other instances labeling can stigmatize a child and may result in a self-fulfilling prophecy whereby parents, teachers, and peers expect the child to be different and therefore treat the child as being somewhat deviant. The use of specific descriptive labels is dependent not only on the type of problem or its severity but also on the needs of the child, family, and school system. Pediatricians are often in a good position to explore these issues with the family, to counsel parents about appropriate intervention, and to serve as a liaison for the child, the parents, the school, and community personnel.

It is hoped that future research on the efficacy of educational assessment will provide additional insights into the ramifications of the medical-diagnostic and educational-prescriptive models of evaluation. Further integration of these two approaches could result in a reduction of school failure, improvement of educational services, and ultimate alleviation of the problems of school dropout and delinquency.

LYNN J. MELTZER

JEAN M. ZADIG

REFERENCES

Benton, A. L., and Pearl, D. (Editors): Dyslexia; An Appraisal of Current Knowledge. New York, Oxford University Press, 1978.

Boder, E.: Developmental dyslexia: a diagnostic approach based on three atypical reading-spelling patterns. Dev. Med. Child. Neurol. 15:661, 1973.

Callahan, C. M.: Superior abilities. In Kauffman, J. M., and Hallahan, D. P. (Editors): Handbook of Special Education. Englewood Cliffs, New Jersey, Prentice-Hall, Inc., 1981, pp. 141–164.

Camp, B., and Dolcourt, J.: Reading and spelling in good and poor readers. J. Learn. Disab., 10:300, 1977.

Dunn, L. M.: Exceptional Children in the Schools. New York, Holt, Rinehart, & Winston, 1973.

Elkind, D.: Child Development and Education: A Piagetian Perspective. New York, Oxford University Press, 1976.

Feuerstein, R., Miller, R., and Jensen, M. R.: Can evolving techniques better measure cognitive change? J. Spec. Educ., 15:201–270, 1981.

Ginsburg, H., and Opper, S.: Piaget's Theory of Intellectual Development: An Introduction. Ed. 2. Englewood Cliffs, New Jersey, Prentice-Hall, Inc., 1981.

Hallahan, D. P., and Bryan, T. H.: Learning Disabilities. In Kauffman, J. M., and Hallahan, D. P. (Editors): Handbook of Special Education. Englewood Cliffs, New Jersey, Prentice-Hall, Inc., 1981, pp. 141–164.

Hodges, W., and Cooper, W.: Head Start and follow through: influences on intellectual development. J. Spec. Educ., 15:221, 1981.

Lerner, J.: Learning Disabilities: Theories, Diagnosis, and Teaching Strategies. Ed. 3. Boston, Houghton Mifflin Co., 1981.

Levine, M. D., Brooks, R., and Shonkoff, J.: A Pediatric Approach to Learning Disorders. New York, John Wiley & Sons, Inc., 1980.

Levine, M. D., Oberklaid, F., and Meltzer, L. J.: Developmental output failure: a study of low productivity in school-aged children. Pediatrics, 67:18, 1981.

Meltzer, L. J.: Abstract reasoning in a specific group of perceptually-impaired children: namely, the learning-disabled. J. Genet. Psychol., 132:185, 1978.

Meltzer, L. J., Levine, M. D., Palfrey, J. S., Aufseeser, C. L., and Oberklaid, F.: Evaluation of a multidimensional assessment procedure for preschool children. Dev. Behav. Ped., 2:67, 1981.

Meyer, E. L.: Exceptional Children and Youth. Denver, Love Publishing Co., 1978.

Myklebust, H. R.: Development and Disorders of Written Language. New York, Grune & Stratton, Inc., 1973, Vol. 2.

Newland, T. E.: The Gifted in Socioeducational Perspective. Englewood Cliffs, New Jersey, Prentice-Hall, Inc., 1976.

Palmer, F. H., and Anderson, L. W.: Long-term gains from early intervention: findings from longitudinal studies. In Zigler, F., and Valentine, J. (Editors): Project Head Start: A Legacy of the War on Poverty. New York, The Free Press, 1979, pp. 433–466.

Pierson, D. E., and Nicol, E. H.: The Fourth Year of the Brookline Early Education Project: A Report of Progress and Plans. Newton, Massachusetts, MASBO Cooperative Corporation, 1977.

Renzulli, J. S., and Smith, L. H.: Two approaches to the identification of gifted children. Except. Child., 43:512, 1977.

Resnick, L. B.: Theories and prescriptions for early reading instruction. In Resnick, L. B., and Weaver, P. (Editors): Theory and Practice in Early Reading, Hillsdale, New Jersey, Lawrence Erlbaum Associates, Inc., 1979, pp. 321–338.

Sabatino, D. A., and Miller, T. L.: The dilemma of diagnosis in learning disabilities: problems and potential directions. Psychol. Schools, 17:76, 1980.

Satz, P., and Morris, R.: Learning disability subtypes: a review. In Pirrozolio, F., and Wittrock, J. (Editors): Neuropsychological and Cognitive Processes in Reading. New York, Academic Press, Inc., 1980.

Wiederholt, J. L.: Adolescents with learning disabilities: the problem in perspective. In Mann, L., Goodman, L., and Wiederholt, J. (Editors): Teaching the Learning-Disabled Adolescent. Boston, Houghton Mifflin, 1978.

Ysseldyke, J. E., and Shinn, M. R.: Psychoeducational evaluation. In Kauffman, J. M., and Hallahan, D. P. (Editors): Handbook of Special Education. Englewood Cliffs, New Jersey, Prentice-Hall, Inc., 1981, pp. 418–440.

48

Projective Techniques in Personality Assessment

Recently there was a diagnostic case conference about a child who had received a thorough evaluation at a psychiatric facility. A plethora of data was presented about the child's past and present functioning, information that had been gathered through interviews with the child, the child's parents, the child's pediatrician, and the child's school. It was then time for the psychologist who had tested the child to present his findings. The senior staff member who was chairman of the conference introduced the psychologist by saying, "Now we will hear whether our impressions about the child are accurate and we will get a truly in-depth picture of the child's problems." Projective testing was placed on a pedestal, judged to provide the most accurate and most significant assessment data.

More recently a psychologist proclaimed that he never uses projective tests. "I feel that projective tests are worthless; I can get more out of a half-hour interview that I can out of several hours of giving the Rorschach, TAT, or Sentence Completion." Similarly, Gittelman (1980) in a comprehensive review of the research related to the diagnostic efficacy of projective testing of children wrote: "The current status of projective tests in children can be summarized succinctly: sometimes they tell us poorly something we already know. . . . Resources in the mental health field are limited. On the basis of empirical data, it is clear that the use of projective testing in children does not deserve a high priority" (p. 434).

The controversy concerning the use of projective testing in the personality assessment of both children and adults continues unabated; some psychologists have written of its decline (Lewandowski and Saccuzzo, 1976), while others have described its merit (Levy and Fox, 1975). Many psychologists regularly carry out projective testing even in the face of research that questions the validity and reliability of projective test procedures and data, especially for arriving at a diagnosis (Thelen et al., 1968; Wade and Baker, 1977). Thus,

it seems legitimate to ask whether projective testing is nothing more than a mirage, created and sustained by psychologists thirsting for instruments to diagnose emotional health and illness. Are Rorschach and TAT cards and similar material the present day counterparts of the "Emperor's New Clothes," illusory garments possessed by psychologists? Or are they similar to fabrics with much sartorial promise, requiring the sensitive hands of a tailor to shape and fit them into a worthwhile product? If the latter, is there a standard pattern that guides the work of each tailor?

These and related questions will be addressed in this chapter. One should note, first of all, that projective tests are not the only techniques used in personality assessment, but given their popularity or notoriety (depending on one's theoretical persuasion), they will compose the major subject of this chapter. Also, it is not the author's intent to present in detail those arguments and research findings that have been advanced both for and against projective testing. Rather, it may be helpful to state from the outset that as a clinician who has used projective techniques with hundreds of children and adolescents, this contributor believes that such testing can play an important part in clinical practice. However, practitioners must be aware of the strengths and limitations of different projective procedures and be prepared to employ them in a creative manner. To use projective techniques in a routine, stereotyped fashion without careful consideration given to diagnostic questions renders these procedures sterile and justifies calls for their demise. Projectives are not simply instruments for obtaining a differential diagnosis (a practice that research calls into question) or for labeling a child; rather they can provide stimuli and strategies for understanding a child's cognitive and affective functioning in order to assist the child's development.

This chapter will elucidate the concept of projection and its application to assessment procedures, the major projective techniques and the kinds of responses that each requires, the significant influence of the examiner-child relationship and the testing situation on a child's responses, and possible limitations, abuses, and misuses of projective testing. In addition, a phenomenological approach to projective testing that has been found very useful by the author will be outlined; test administration procedures will be described with a focus on the assessment of cognitive organization, affects, beliefs, emotional conflicts, and coping strategies. The relevance of projective data for planning a treatment program and the use of projective techniques by pediatricians will be considered.

THE CONCEPT OF PROJECTION AND PROJECTIVE TECHNIQUES

The concept of projection has a long, rich, and sometimes confusing history in psychological theory (Abt and Bellak, 1950; Rabin, 1960; Rapaport, 1952). Many clinicians immediately think of projection as a pathological process, as a defense mechanism in which a person's unacceptable thoughts and feelings are attributed to other people or things. An example of this view would be a child who has difficulty in accepting his angry feelings so that these feelings are unconsciously projected onto an older child, who is then perceived to be the one who is angry and threatening. However, as Rabin (1960) has indicated, this pathological concept of projection is too narrow and too misleading when we refer to projective techniques. The concept of projection when applied to assessment procedures implies a process by which a person's feelings, thoughts, and emotional conflicts are elicted by and represented through different test stimuli.

Given this broader concept of projection, many psychologists who defend the use of projective tests typically see them as potent vehicles for uncovering hidden or defended-against thoughts and feelings, vehicles immune from the conscious manipulation of patients. These psychologists often judge the data derived from projectives as more valid and more accurate indicators of the patient's personality make-up than data obtained from what they see to be more structured or objective assessment procedures, including direct interview material. On the other side, their more disbelieving colleagues, armed with selected research findings, have long since discarded their Rorschach and TAT cards.

One need not "throw out the baby with the bath water." Rather, if the diagnostic task is seen to entail the understanding and articulation of a child's emotional and cognitive functioning, it becomes evident that each test procedure or interview question, regardless of how structured or unstructured, offers unique opportunities for insight about the child. Any test, even a seemingly neutral measure of intelligence, is capable of yielding data that are projective in nature (Waite, 1961).

In essence, it is not that a Rorschach card contains more of that ingredient called projection or elicits more significant material than a direct question such as "How do you feel?" The child's response to each procedure as well as a comparison of these responses in light of the different test requirements produces important information about a child's cognitive abilities, feelings, thoughts, conflicts, and coping strategies when he is confronted with diverse test demands. Arguments about the merits of one projective procedure over another without consideration given to the situation and the diagnostic questions are often meaningless exercises in futility. It is most critical to ask: "For this patient at this point of time, what test or series of tests or evaluation procedures will best answer the questions that have been raised?"

DIMENSIONS OF PROJECTIVE TECHNIQUES

The kinds of tests and assessment procedures that fall under the category of projectives are diverse. Some are more structured than others, demanding different levels of organizational ability. Some require verbal responses while others require written responses. Some are administered individually while others may be given in a group setting. Some demand the active participation of the examiner in eliciting and elaborating responses while others require minimal inquiry or involvement by the examiner.

There are many ways of classifying projective techniques. As noted by Lindzey (1959) and Rabin (1960), one of the most useful approaches for categorizing projectives is based upon the kinds of responses required of the patient. Five categories that have been described include association, construction, completion, choice or ordering, and expression. Before giving a brief description of each of the five categories, it is important to note that some projective techniques do not fall neatly into any single category. However, it is useful, when administering projective tests, to consider the types of responses demanded of the child.

1. Associative tests require the child to respond to stimuli with the first word, or image, or idea that comes to mind. The Rorschach and Early Memories tests fall within this category.

2. Construction procedures ask the patient to construct or create something, typically in response to a visual stimulus. The thematic tests,

which require a child to create stories in response to pictures that he is shown, are examples of construction techniques.

3. Completion tests include assessment procedures in which the child is administered an incomplete product, which he is requested to finish. Perhaps the test used most frequently in this regard is the Sentence Completion Test. The initial requirement of Winnicott's "Squiggle Game" would also be classified within the completion category.

4. As the name implies, choice or ordering responses present stimuli that the child is asked to order in terms of some preference. Santostefano's innovative Miniature Situations Test, in which a child is requested to select the order in which he would like to engage in certain tasks, represents this category of test response.

5. Expressive techniques contain some of the elements found in constructive methods but do not rely on test stimuli. Expressive projective strategies include free play, puppet play, and free art work as well as tests that require one to draw a person or a family.

When evaluating a child through the use of projective techniques, it often is useful to administer several different assessment procedures varying in terms of the dimensions of how structured they are, the organizational skills they require of the child, and the kinds of responses involved. Such a cross section of dimensions yields a comprehensive picture of the child's cognitive and affective functioning.

MAJOR PROJECTIVE TECHNIQUES USED WTIH CHILDREN AND ADOLESCENTS

It is well beyond the scope of this chapter to describe in detail each of the projective techniques

Table 48–1. MAJOR PROJECTIVE TECHNIQUES USED WITH CHILDREN AND ADOLESCENTS

Rorschach
Thematic cards
  1. Children's Apperception Test (CAT)
  2. Thematic Apperception Test (TAT)
  3. Tasks of Emotional Development Test (TED)
Sentence completion
Drawing techniques
  1. Draw-A-Person (D-A-P)
  2. House-Tree-Person (H-T-P)
  3. Kinetic Family Drawings (K-F-D)
Squiggle Game
Three Wishes
Cole Animals
Early Memories
Miniature Situations Test (MST)
Projective Play

listed in Table 48–1 and the particular measures that clinicians have used to order and understand the data that are derived. Entire chapters and books have been devoted to just one assessment procedure, such as the Rorschach Test or the Thematic Apperception Test (TAT), so that the following brief descriptions will not be able to capture the full richness and complexity of these procedures.

RORSCHACH TEST

This is perhaps the best known and most widely used projective test. It consists of 10 inkblots, which are shown to the child, and he is requested to say what each resembles, reminds him of, or could be. Five of the inkblots are achromatic (black, white, and gray), whereas the others contain colors. Illustrations of two of the cards are found in Figure 48–1.

The stimuli of the Rorschach Test are relatively unstructured and ambiguous, therefore requiring the child to use many organizational skills in producing articulated responses. As Santostefano (1963) has noted, the Rorschach Test yields valuable information not only about a child's feelings and conflicts, but also about perceptual and cognitive skills. In order to assess a child's inner life with its complexity of thoughts, feelings, and coping styles, different dimensions of the Rorschach Test have been studied with a number of scoring systems. The reader is referred to books and articles by Ames et al. (1952, 1959), Exner (1974), Halpern (1953, 1960), and Hertz (1960) for discussions of the application of Rorschach scoring and interpretation to children and adolescents and to those by Lerner (1975), Rapaport et al. (1968), Rickers-Ovsiankina (1960), and Schachtel (1966) for additional information about the features of the Rorschach Test.

The dimensions of the Rorschach scoring system that have received special attention include the following:

Area

A child's response may include the entire (whole) inkblot, a large detail of the stimulus, a small detail, or the white space. Younger children typically give whole responses that are diffuse and not very well articulated; with development, more detailed responses appear. The use of white space is often seen to be linked to oppositional personality qualities.

Determinants

This dimension refers to the qualities of the areas chosen that determine the nature and con-

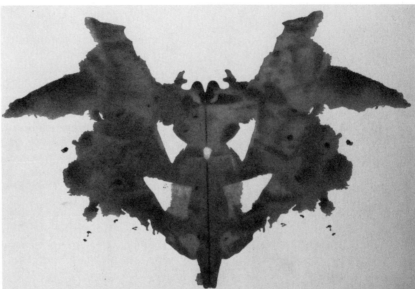

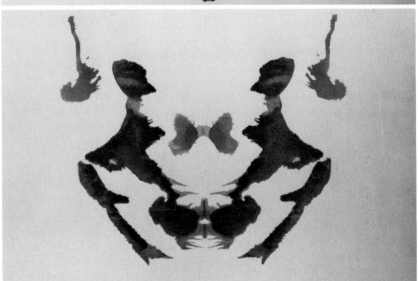

Figure 48–1. Examples from the Rorschach Test.

tent of the response. Possible determinants include the following:

Form. Pure form responses are predicated entirely on the outline or contour of the area, such as "It's a bat; it's shaped just like a bat." Different scoring systems also take into consideration the level of accuracy of the form response: that is, the extent to which the response resembles the stimulus form found on the card. Typically the less the resemblance between the content and the form, the more pathological the response.

Color. A response is given a color score when the card's color is used, at least in part, in the creation of the percept. In scoring color, consideration is also given to how large a role form plays, so that "a colorful butterfly" involves a more articulated form than "a splotch of blood." Theoretically, color is seen to be linked to an expression of feelings or affects, to the emotional interaction between a child and his world. The manner in which the child uses color on the Rorschach Test is believed to reflect how well feelings are organized, represented, and expressed.

Movement. This score is given when the child attributes movement to figures or objects in his response. Movement may be seen in humans or animals as well as in inanimate objects (e.g., "a top spinning," "a rocket blasting off"). Since the Rorschach card is static, the attribution of movement is thought to reflect creativity, capacity for empathy, capacity for rich interpersonal relationships, capacity for delay, and fantasy production. However, inanimate movement is associated with inner tension and conflict.

Clinicians are sensitive to the nature of the

movement response. For example, "a man running in a race" and "a man sitting on a rock" represent vastly different degrees of movement, suggesting differences in how freely a person uses fantasy and action.

Shading. A shading score is given when the response is based, in part, on the lightness and darkness of the card and is perceived to be shading by the child. Variations in shading include responses involving the experience of texture and vista (three dimensional appearance). Shading responses are hypothesized to be an indication of the child's anxiety. If shading occurs in a response with good form level, the child is believed to handle anxiety in a more mature, adaptive fashion; if the shading response is diffuse and poorly articulated, it is believed to be a sign that anxiety is not very well organized and that it is not dealt with very effectively.

Content

As the name implies, this determinant refers to the specific content of the response. Generic categories are used in scoring content and include Humans, Parts of Humans, Animals, Parts of Animals, Anatomy, Mythological-like Animals and Humans, Food, Nature Concepts, Geographical Concepts, Plants, and Man-made Objects. At all ages animal responses are the most frequent. Human content responses are believed to be associated with interest in other people and are found to increase with age. The nature of the interest in other people—for example, warm or distant—would be based on other indicators, such as the use of color and the actions, if any, that the person is perceived to be engaged in.

Pathological Thinking

It has been this author's experience that the Rorschach Test offers a significant lens through which to assess pathological or deviant thinking. Data pertaining to cognitive disturbance and disordered thinking are critical in understanding the world of the child and in planning intervention programs. Frequent examples of pathological thinking in order of probable disturbance (from least to most) include the following:

1. Fabulized responses, in which more affect is attributed to a percept than is typical (e.g., "a mean bear").

2. Fabulized combinations, which involve the impossible combination of two details appearing in close physical proximity on the card (e.g., "two elephants with tuxedos on").

3. Confabulations, which are fabulized responses taken to an extreme without any justification (e.g., "an evil witch who is filled with hatred and wants to destroy the world").

4. Contaminations, in which two separate concepts are fused together without any regard to how independent they are (e.g., in response to a red area on one of the cards, the child sees both blood and fire and then combines the two into "bloody fire").

Before leaving this brief description of the Rorschach Test, it is important to note that a number of clinicians have attempted to apply a developmental scoring system to Rorschach responses (Becker, 1956; Exner, 1974; Friedman, 1953; Hemmendinger, 1953; Lerner, 1975; Phillips, et al., 1959), often guided by the developmental theory advanced by Werner (1948). Such a scoring system provides a measure by which to assess a patient's level of cognitive functioning.

THEMATIC CARDS

There are several projective tests that require the child to create a story in response to a picture that he is shown. These tests provide more structured stimuli than the Rorschach Test and include such procedures as the Children's Apperception Test, the Thematic Apperception Test, and the Tasks of Emotional Development Test.

Children's Apperception Test (CAT)

The CAT consists of 10 cards depicting different scenes and themes. There are two comparable forms, one with animal figures and the other with human figures. Illustrations of two of the cards are found in Figure 48–2. The 10 cards tap themes related to sibling rivalry, aggression, achievement, sexuality, parents, and fears. The CAT was originally developed for use with children between the ages of three and 10. The pictures depicting animals are more commonly used for younger children.

Thematic Apperception Test (TAT)

This test consists of 31 cards including one that is blank. One of the TAT cards is depicted in Figure 48–3. Typically only about 10 of the cards are administered to children and adolescents. As with the CAT, the different TAT cards are seen to tap various themes, including achievement, aggression, parental punishment, nurturance, and attitudes about parents and adults. Kagan (1960) believes that the cards are most appropriate for people who are at least 11 or 12 years old. It is the author's experience that many child clinicians use the cards with younger children.

Figure 48–2. Examples from the Children's Apperception Test.

Tasks of Emotional Development Test (TED)

The TED consists of four sets of photographs, one set each for boys and girls ages six to 11 and one set each for adolescent boys and girls ages 12 to 18. The sets for the younger children contain 12 cards, whereas those for the adolescents have 13 cards. The photographs represent different tasks of emotional development, such as socialization with peers, relationships with adults, attitudes toward learning, sibling relationships, separation from parents, and establishment of a positive self-image.

The stories a child creates in response to the

CAT, TAT, or TED pictures have been analyzed in different ways. In terms of the child's emotional life, clinicians have been interested in how the main hero in the story is presented—that is, the hero's self-image, the particular strengths, weaknesses, conflicts, needs, anxieties, and defenses of the hero, and how the hero views others in his world. As Kagan (1960) has noted, a basic assumption of thematic tests is that there exists a strong relationship between how the characters are portrayed in the story and the story-teller's psychological make-up and view of his environment; this assumption is supported by some research and has been called into question by other research (Gittelman, 1980; Sigel, 1960).

Figure 48–3. Example from the Thematic Apperception Test.

Additional information is obtained from thematic cards when a story seems to bear little resemblance to the picture shown the child. For example, if a child is shown a picture of an obviously angry child and creates a story about a happy youngster, the clinician is provided with evidence that the child has difficulty in handling angry feelings and has to deny what is the obvious content of the card.

In addition to providing a possible glimpse into the child's emotional world, these thematic tests have been used by clinicians to gather information about a child's cognitive functioning (Kagan, 1960). The elaboration of the characters and stories, the originality, sophistication, and coherence of the plot, and the complexity of the language used are several dimensions related to cognitive and language skills.

SENTENCE COMPLETION TEST

A number of clinicians have developed tests in which a child is given a sentence stem and asked to complete it. These tests typically include questions related to affects, aggression, ego ideal and self-image, and views of parents, siblings, peers, and school. As with other projective instruments, careful attention is paid not only to the specific content of the response, but to the ease or diffi-

culty with which a response is offered. Samples of stems from a Sentence Completion Test developed by the author with two colleagues, Doctors Kalman Heller and Ethan Pollack, are listed in Table 48–2.

DRAWING TECHNIQUES

The use of drawings as a diagnostic projective instrument has a long history in child psychology. Three of the most popular drawing tests are Draw-

Table 48–2. SAMPLE STEMS FROM A SENTENCE COMPLETION TEST DEVELOPED BY HELLER, POLLACK, AND BROOKS

The thing I do best is
My teachers think I
The worst thing about me is
I feel sad when
I feel happy when
Once I tried to do this very hard thing and
My mother is okay but
My father is okay but
I have trouble with my schoolwork because
My parents think I
When I have to go to the doctor
I hate my brother/sister when
If I were the youngest/oldest child in my family, then
I cry when

A-Person, House-Tree-Person, and Kinetic Family Drawings. Another test, the Squiggle Game, also relies on drawings, but is conceptually different from these three and will be discussed separately.

Draw-A-Person (D-A-P)

In this test, which was originally used by Goodenough (1926) as a measure of intelligence, the child is requested to draw a picture of a person. Typically the product is done in pencil, although many clinicians give the child the opportunity to use crayons or magic markers. After the child completes the drawing, a request is frequently made to draw a person of the opposite sex from that in the first picture. Clinicians often gather additional information by asking the child questions about the figures he has drawn, such as what the figure is doing, feeling, or thinking. The drawings are seen to reveal a child's self-image or the image a child would like to have, or a view of significant others. DiLeo (1970, 1973), Koppitz (1968), and Machover (1949, 1953) have written of the D-A-P as an instrument for diagnosing personality and cognitive functioning in children and adolescents.

House-Tree-Person (H-T-P)

This test requires a child to draw a picture of a house, a tree, and a person, each on separate sheets of paper. The test was introduced by Buck (1948). Hammer (1958) has written that the house represents feelings about the home environment, including relationships with parents and siblings, whereas the tree is believed to symbolize the patient's more unconscious feeling about himself. The person is hypothesized to reflect closer to conscious images of self or significant others.

Kinetic Family Drawings (K-F-D)

This test was introduced by Burns and Kaufman (1970, 1972) and requires the child to draw a picture of all members of the family, including the child, doing something. Burns and Kaufman believe that having the figures engaged in movement or activity, unlike what is required in other drawing techniques, helps to mobilize a child's feelings about himself as well as relationships within the family.

SQUIGGLE GAME

This technique was developed by Winnicott (1971) and can be used for both diagnostic and therapeutic purposes (as can other projective strategies). The game requires that the clinician draw a squiggle and request the child to "make a picture out of the squiggle." When the picture is finished, the child is asked to draw a squiggle, and the clinician completes a picture from the child's squiggle. The clinician's own drawings and the questions he asks the child serve to articulate and elaborate different emotional themes confronting the child. Berger (1980) has described the use of the Squiggle Game in the general practice of pediatrics.

THREE WISHES

As the name implies, in this projective technique the child is asked what three things he would wish for if any three wishes could be granted. The child's responses as well as the stated reasons for these responses offer insight into the child's interests, needs, and aspirations as well as areas in which the child feels vulnerable. For instance, a child's wish for all the candy in the world may represent a deep-felt need, whereas the wish to be a famous surgeon would most likely reflect the ideal toward which the child is striving. A wish to be bionic and indestructible is frequently associated with underlying feelings of being weak and small.

COLE ANIMALS

Many child clinicians use this test, which asks the child the animals he would most like to be and least like to be. Once the child has selected the animals, the clinician probes further into why the child selected those particular animals. A child wishing to be a small kitten because "small kittens are fed and taken care of" is expressing different needs from those of a child who wishes to be a cheetah because "cheetahs would come in first in a race."

EARLY MEMORIES

In the past 15 years there has been an increased interest in the study of early memories as a projective tool. Monahan (1981) has developed an early memories test in which the child or adolescent is asked for the following set of five early memories: earliest, earliest of mother, earliest of father, a birthday, and school. Memories are scored along such dimensions as clarity of the memory, activity versus passivity of the child in the memories, and degree of interaction with another person. Monahan has used his procedure as a diagnostic instrument for differentiating suicidal from nonsuicidal adolescents hospitalized at a psychiatric facility.

MINIATURE SITUATIONS TEST (MST)

In an attempt to have the tasks in the evaluation reflect as closely as possible situations or issues the child faces in the "real" world, Santostefano (1968, 1970, 1971, 1977) has developed test procedures that require some choice and action on the part of the individual. For instance, in one MST procedure designed to assess the expression of aggression, the child is presented with three games and asked to do them in the order he would most like to do them in. In one item the child is asked to tie, knock down, and stab an enemy toy soldier; each of these actions requires a different degree of directness. The order that is selected by the child and the manner in which the actions are performed reveal different features of the child's personality and coping styles, thereby providing critical data in planning an intervention program.

Another MST procedure concerns parent-child interactions. A parent and child are asked to act upon each other in some way; the test taps motives related to dependence, aggression, nurturance, and dominance. The parent and child take turns in deciding on one of two choices, such as the parent giving the child a drink from a cup or from a baby bottle. The clinician is active in asking questions about why particular choices are made.

PROJECTIVE PLAY

Projective play encompasses a wide spectrum of activities and is applicable to both diagnostic and therapeutic endeavors. Under this category of diagnostic strategies are included free play involving various toys such as clay, darts, army men, cars, trucks, dolls (Murphy and Krall, 1960), the use of puppets (Woltmann, 1951), finger painting (Napoli, 1951), and engagement in different board and card games (Beiser, 1979). Each of these forms of projective play varies along different dimensions, including the activity level of the clinician, the structure provided by the task and the clinician, and the control assumed by the child. The common thread running through all these diagnostic strategies is that they serve to articulate the major emotional issues facing the child and the ways in which the child manages these issues. For instance, one child may reveal the difficulty he is having with aggression by making "dumb" moves in checkers that keep him from beating the clinician; another child struggling with similar problems may demonstrate his fear of aggression by placing a lion puppet in a cage so that "it cannot get out and kill everyone in the world." Given the relative lack of structure of projective play, the clinician must be alert to the symbolic messages that are being communicated by the child and then offer input to facilitate the diagnostic process.

POSSIBLE LIMITATIONS AND ABUSES OF DATA FROM PROJECTIVE TECHNIQUES

As noted earlier, some clinicians place projective techniques in an exalted position, while others perceive them as being unscientific and of little value. It is misleading to set up straw men and depict psychologists as simply divided into two polarized enemy camps, since there are many psychologists who subscribe to a more balanced picture of the worth of projectives. The latter group recognizes that there are certain test variables that place some limits on the conclusions that can be drawn from projective data, but that these are not so great as to render projective testing a useless exercise. As a matter of fact, an awareness of these test variables helps the clinician to gain a more realistic view of what projective techniques can and cannot do, so as to guard against misusing or misinterpreting the data that are derived from such techniques.

The test variables being alluded to may be divided into two large and, at times, overlapping groups, one related to the influence of the testing situation on a patient's responses and the other concerning how interpretations are formulated from the assessment data.

Test Situation Variables

Some clinicians administer, score, and interpret projective tests with veritable blinders on. Little if any consideration is given to such important test variables as what the testing means to the child, the degree of rapport between clinician and child, the cooperation of the child, the particular coping strategies recruited by the child, the attention strength of the child, and the testing style of the clinician. Whether they realize it or not, clinicians who pay only lip service to these variables are working under the assumption that the child's responses remain invariant across all situations; it is as if the child's personality is believed to be so fixed that external situational factors have little influence on the responses that are elicited by the test material.

Although a child's personality may be a relatively constant entity, that constancy should not be taken to mean that the complexity and content of test responses and behavior are not influenced by situational variables. Situational factors have an impact on an individual's personality, and behavior varies as a function of the interaction between the individual and the environment that he is in at that time (Mischel, 1969, 1977). The assessment process is not immune to this impact (Bersoff, 1973; Brooks, 1979; Korchin, 1976; Laosa, 1979). Projective test responses may change from one session to the next depending on such factors as the child's gaining a better understanding of the purpose of the testing (e.g., "These tests are not to prove I'm crazy, but to help me"), the structure and sensitivity provided by the clinician, and an increase in rapport between the clinician and the child.

Some clinicians may argue that these caveats offer ample evidence for doing away with projective tests, since responses are so susceptible to situational factors that they become meaningless. However, the question of susceptibility applies to any assessment procedure (even a multiple choice paper and pencil test), so that the important issue is how well one understands the influence of situational factors on test responses, especially in terms of interpreting projective test data. This position means that to formulate interpretations from test responses, one must recognize that some of these responses may have been different under other testing conditions. Situational variables challenge the clinician to be creative in the diagnostic process by selecting and modifying particular test variables in order to observe the impact this has on the child. In a similar vein, Bersoff (1973) has recommended "psychosituational testing" in which the individual is evaluated as he interacts and is affected by the environment.

POSSIBLE ABUSES IN INTERPRETATION OF TEST DATA

The issue of abuse in interpreting projective test data touches not only upon understanding the influence of test situational variables but several other related questions as well. For example: What is the purpose of the testing? What is the relationship between the fantasies expressed on test responses and a child's "real-life" behavior? Do the tests predict future behavior? An entire book replete with hundreds of studies could be written in an attempt to answer these queries. The conclusions would be tenuous at best. One thing that is clear, however, is that clinicians must constantly guard against falling prey to abuses in test interpretation, abuses that may appear in the following forms:

1. The "sign approach" occurs when one particular kind of test response is inferred to symbolize major personality characteristics. For example,

although many clinicians believe that "eyes" on the Rorschach are a sign of a paranoid personality, there is no evidence for this. Chapman and Chapman (1969) demonstrated that the Rorschach signs that clinicians inferred for homosexuality had no basis in reality. Interpretations predicated on the sign approach are irresponsible, going far beyond the data and assuming that complex human behavior can be represented by one or two isolated test responses.

2. The struggle to define the relationship between a child's verbalizations and fantasies in testing and the child's real-life behavior has long plagued psychologists, and the results of numerous studies assessing this relationship have been contradictory (Gittelman, 1980; Sigel, 1960). The innovative test strategies developed by Santostefano (1971, 1980) represent one attempt to address this complicated matter. There are no easy solutions in understanding the correlation between fantasy and overt behavior, yet there are clinicians who consciously or unconsciously assume that high correlations exist; for example, they believe that aggressive content on the Rorschach or TAT test is isomorphic with the child's behavior in his environment. Clinicians must use great caution when extrapolating from test responses to behavior outside the testing session. Findings should not be interpreted in a vacuum, but in conjunction with other data, such as parental observations and school reports.

3. Closely related to the problem of correlating test responses with the child's real-life behavior is the question of prediction; that is, can a child's behavior be predicted from his test responses? As Mundy (1972) noted in her review of projective testing with children, in most instances test performance cannot predict the future with a secure level of accuracy. Rather, testing can portray the child's inner world at the time of testing. A high probability of prediction is difficult not only because of the limitations of correlating test responses with other behavior, but because future behavior is a function of many personal and environmental variables that cannot be captured in testing.

4. Another possible abuse of projective testing concerns its use in attaching a label to a child, a label that all too often becomes part of a damaging prophecy. Attaching a diagnostic label to a child (a label frequently predicated on insufficient data) may result in one's responding to the child as if he possessed the malady or disease captured in the label, thereby reinforcing the behavior associated with the label. Ellis (1967) and Sarbin (1967) have addressed the general issue of diagnostic labeling and its dangers in the field of mental health.

PROJECTIVE TESTING: A PHENOMENOLOGICAL APPROACH

The reservations about the efficacy, reliability, and validity of projective testing are many, but as noted earlier, if clinicians do not attempt to go beyond the data and are cognizant of the strengths and limitations of projective procedures, they will find such techniques extremely useful in understanding the inner world of the child and in planning intervention programs.

In the phenomenological approach advocated by the author, projective techniques are used figuratively to put oneself "inside the child's head," that is, to understand and describe the child's perception and experience of himself, his inner needs, affects, and conflicts, other people, and challenges and stresses in the environment. During an evaluation of a child and his parents, the parents might be asked to describe a typical day in their child's life, but to do so as if they were the child. How easily and accurately parents describe their child through the child's mind, as well as the kinds of descriptions they offer, reveals much about their level of empathy and understanding. After the evaluation the clinician should attempt to answer the same question about the child that was posed to the parents.

Such a construction of a picture of the inner world of a child is guided by certain principles that provide a sense of direction. These include the following:

DEVELOPMENTAL PRINCIPLES

The meaning or significance of a child's test performance is best understood within a developmental framework, such as the biodevelopmental approach proposed by Santostefano (Santostefano, 1978; Santostefano and Baker, 1972). Developmental principles offer blueprints for assessing the level of a child's cognitive and affective functioning, especially in comparison with other children. These principles help the clinician to evaluate the severity of particular problems and to develop appropriate intervention programs.

ADAPTATION

The "fitting together" or adaptation between a child and his environment (Hartmann, 1958; Laosa, 1979) is considered to a great extent in the portrait one paints of the child by asking such questions as: How successfully is the child adapting to his world, both the inner world of thoughts and feelings and the external world of people and tasks? What are the stresses and challenges confronting the child? What are the coping strategies that the child has recruited to manage these stresses and challenges? How adaptive or maladaptive are these strategies?

FOCUS ON STRENGTH AND COMPETENCE

Too often, psychological evaluations focus on disorder rather than strength. This focus is reinforced by such diagnostic frameworks as the American Psychiatric Association's *Diagnostic and Statistical Manual*, in which mental health is literally downplayed by diagnosing emotional strength primarily in terms of the absence of signs of disease. During an evaluation the clinician must actively look for the child's strengths and assess the kinds of experiences that foster the child's sense of competence and mastery (White, 1959). A clinician who searches primarily for disorder is wearing blinders that distort and add a certain destructive quality to the evaluation.

The evaluation process, guided by principles of development, adaptation, and mastery, contains the following steps that increase the clinician's understanding of the phenomenal world of the child:

Articulation of Diagnostic Questions

The evaluation should be directed by several well articulated diagnostic questions, not simply for the sake of obtaining a diagnostic label but rather to answer key questions about the child, especially questions raised by the child's parents, school, or physician. For example, if a child is school phobic, the testing should address such questions as how the child perceives school and school personnel, how he perceives leaving home to go to school, and what the child believes will happen in school or at home if he attends. If diagnostic questions are poorly defined, the assessment work-up is often a chaotic exercise in futility and waste.

Selection of Assessment Procedures

Once the diagnostic questions have been delineated, and even before the clinician sees the child, the former should decide which projective techniques would prove most useful in answering the questions. This does not mean that modifications in the choice of projective tests and strategies cannot be made once the clinician starts to evaluate the child. Rather, an early plan of action ensures that the clinician will not administer the same projective tests to all children regardless of the diagnostic questions. The diagnostic process must be extricated from an assembly line mentality.

Development of an Alliance with the Child

The concept of alliance implies trust and cooperation between the patient and the clinician, without which the task in which they are engaged would be compromised (Brooks, 1979; Levine et al., 1980). To nurture this crucial aspect of the evaluation, the clinician must take time to help the child comprehend the reasons for the testing so that the process is demystified and made less frightening. Difficulties in forming an alliance, as well as the kinds of coping strategies used by children to avoid taking the tests (e.g., the child's saying, "The tests are stupid" or "I'll give you the tests"), are important factors to be observed and managed by the clinician. There is more to administering a test than simply handing a card to a child and asking for a response.

Use of a Flexible Approach to Testing

An evaluation using projective techniques should be conceived of as a very active process in which the clinician constantly observes and generates hypotheses about the child's performance, keeping in mind the original diagnostic questions, but not hesitating to raise new questions. As new hypotheses and questions are generated, the clinician must feel comfortable in employing projective tests in a flexible and creative way to answer these questions.

Ironically, flexibility and creativity may sometimes run counter to standardized procedures for administering tests and thereby threaten to some extent comparisons with norms that are available for other children. In all likelihood, however, one should not feel too concerned (or guilty) about wandering from the instructions in the testing manual if such a transgression facilitates the opportunity to gain valuable information. For example, in scoring the Rorschach test, many clinicians are interested in the number of movement responses a child offers without prompting from the clinician. However, if a child verbalizes a response that can be used to gain entry into his inner world, the clinician need not hesitate to ask additional questions, including those that encourage the child to think in terms of movement images. This should be done even though the action may lead to movement responses on subsequent cards that might not have been given if the additional questions had not been asked.

As an illustration of this kind of modification of test administration, the example can be cited of a child who saw "a bat with holes in it" on the first Rorschach card. Given certain diagnostic questions about the child concerning self-esteem, relationship with peers and adults, and coping style, the child was asked such questions as: "What is the bat doing?" "How did the bat get the holes?" "How does the bat feel having holes?" "Do the holes stop the bat from doing certain things?" "Do the holes help in any way?" "What do other bats feel about this bat?" "Can anyone help the bat?"

Some professionals, while not denying that significant information can be obtained by modifying standard administration procedures, have raised an important question, namely, "Is it not advantageous to administer tests in a standardized way first and only then return to particular responses for elaboration?" Appelbaum (1959) has found this a useful procedure with the Rorschach test. However, sometimes if the clinician does not ask certain questions immediately following the child's response, the feelings and thoughts surrounding the response are lost or their richness is reduced.

The author is not advocating a thoughtless, random, chaotic approach to testing, or suggesting that norms are unimportant, even though many of the research findings with projective tests question the relevance of available norms. Rather, as Korchin (1976) has written: ". . . the clinician must be sensitive, ingenious, and inventive. It is sometimes necessary to improvise new procedures or alter standard tests even on the spur of the moment. . . . Variation from standard procedure means, of course, that available norms may be inapplicable. However, the gain in personologically useful information can offset the loss of standardization and norms" (p. 210).

Areas to Assess

Although different diagnostic questions exist for each child, there are certain general areas that should be assessed in every psychological evaluation. These include the following:

Cognitive Skills. Too often projective techniques are construed as providing knowledge only about a child's emotional functioning. They also are invaluable in eliciting data about how a child organizes and structures information, especially when the information is presented in an ambiguous form, such as on the Rorschach cards. Critical questions about a child's cognition may be answered via projective testing. For instance, how clearly is the child able to articulate and express thoughts and feelings? When the child is asked to construct stories from pictures, do the stories make sense; that is, does one idea flow smoothly from another or is the story line difficult to follow? Does the child go off on tangents and, if so, when? On the Rorschach (or related tests), are the child's images well articulated and reality related, or are they diffuse, bearing little resemblance to the stimulus card? If the child's thinking becomes disorganized, is there particular affective content that prompts this disorganization? Does the child show disordered thinking characteristic of a significant emotional problem?

The ways in which children organize information about their inner world of thoughts and feelings and their external environment reveal much about the severity of their emotional problems and guide the clinician in determining the kinds of interventions that are desirable. For example, if a child has difficulty in describing feelings and is prone to impulsively act out rather than reflect upon feelings, a treatment program must focus initially on strengthening the child's cognitive functioning. If a child becomes confused when confronted with particular themes, such as loss and separation, these themes should be addressed in treatment to help the child understand and master them.

The spheres of cognitive and emotional functioning are inextricably interwoven, and projective test data can provide information about both.

Affects. The main affects or emotions in a child's life are important data obtained from projective testing. The clinician should be especially sensitive to how easily and comfortably the child is able to express feelings. Some children are scared by feelings and resort to different coping strategies to avoid them; some children acknowledge certain feelings such as anger, but keep in check other feelings such as love.

It is useful when looking at the affects expressed or defended against on projective techniques, to do so in terms of the feelings children have about themselves, about their sense of competence, and about significant people and events in their environment. Do they like themselves? Do they feel good about their achievements? Do they feel worthwhile? Do they primarily feel sad or angry or happy or scared, and what triggers these feelings? Do other people make them happy or sad and in what way? Are their lives permeated by hope or despair? A clinician can answer many of these questions with projective procedures, opening a door to the inner world of the child.

Beliefs. Closely related to affects are the beliefs that a child holds. Not only does the clinician ask what children feel but also what they think about themselves and their world. Do they believe that they are competent or incompetent and in what areas? What do they believe is valuable and what worthless? What do they believe leads to success as opposed to failure? How do they view other people, such as their parents or teachers or peers?

Do they believe that these other people can help them, or do they see them as destructive?

Very importantly, projective instruments can also provide information about how rigidly or flexibly beliefs are maintained. For example, if a child creates a story in which the main character is portrayed as believing that adults are not to be trusted, the clinician can explore this belief and determine how modifiable it is. The clinician might ask the child if there is anything that could be done to change the story character's view about adults. A child who says no is demonstrating greater rigidity (and possibly greater resistance to help) than a child who offers some possibilities for increased trust.

Belief systems are major determinants of behavior (Beck, 1976), and projective techniques help to highlight their sources and contents.

Major Conflict Areas. Given the affects and beliefs that dominate a child's life, the clinician should delineate the major conflict areas that are associated with these affects and beliefs and assess the severity of these conflicts. This information is critical in determining whether treatment is indicated and what the focus of that treatment should be.

For example, one five year old boy believed that he had caused his mother's miscarriage because he accidentally fell against her several days before the event. This led him to believe that any action or aggression could lead to murder, a belief that was graphically illustrated on one of his projective responses involving an image of statues; he said that sometimes it might be better to be a statue than a person since statues could not move and hurt things. The thought of being like a statue made him sad. The clinician was able to talk with him about this during the evaluation and in so doing became even more aware of his distress and his willingness to work with a therapist. In this example, the major conflict area concerned the expression of aggression, and it was rooted in a distorted belief ("I caused the miscarriage and I can cause more harm if I show angry feelings") and associated with feelings of sadness and helplessness. The extent to which this boy wanted to go to avoid any action (i.e., to become a statue) revealed the intensity of the conflict.

Although some children face major conflicts related to such themes as aggression, independence, dependence, separation, and competence, they do not always experience a great deal of distress and anxiety, since the conflict figuratively has become woven tightly into their personality. Projective data help to determine not only the particular conflict issues, but the impact of these issues on the child.

Coping Strategies. When children experience conflict and anxiety and when they feel that their self-esteem is eroding, they will use coping strategies or defenses to lessen their distress and sense of incompetence. Earlier in this chapter emphasis was placed on the importance of the clinician's being sensitive to the particular coping strategies recruited by children to manage the testing situation. For example, children who are anxious about the testing because they believe that they will look dumb or that they will reveal things they wish to hide might use such coping maneuvers as refusing to answer questions or saying that the test is stupid or asking the clinician for an inordinate amount of help or taking control and saying that they want to administer the test to the clinician.

In addition to the child's overt behavior during the evaluation session, coping strategies are also manifest in the test content, especially through the figures that are created in the test responses. As an illustration, one child saw a smiling face on the Rorschach test. In response to several questions about why the face was smiling, he said that the person was unhappy and angry because no one loved or cared for him but did not want anyone to know how he really felt, for fear that they would become angry with him. The child's response indicated that he was coping with sadness and conflicts pertaining to anger and nurturance by placing a mask of happiness over his face—a coping strategy that unfortunately was not very adaptive, since it precluded anyone's being of help to the child.

Different kinds of coping strategies are revealed on projective tests, some more adaptive and flexible than others. A child's particular coping style and the occasions when this style is used give clues about a child's areas of vulnerabilities and suggest strategies the clinician can employ to modify or make more adaptive the child's coping maneuvers.

Synthesizing the Data

The data obtained from projective tests should be synthesized with all other data (e.g., parent, school, pediatrician reports) to help answer diagnostic questions. Given the phenomenological approach advocated in this chapter, at the conclusion of a diagnostic work-up, the clinician should attempt to describe the child's inner world—feelings, thoughts, conflicts, images of self and others, openness to change, and coping strategies. The data should be understood within a developmental model to determine the nature and severity of the child's problems and whether any intervention program is necessary. The data should also offer guidelines in selecting the form of the intervention program. It is beyond the scope of this chapter to specify the signs that suggest the need

for treatment, but the reader is referred to the book by Bush (1980) for a discussion of this matter.

PROJECTIVE DATA TO HELP PLAN A TREATMENT PROGRAM

Treatment approaches with children typically involve working with the child, the parents, and the child's school. One example of the importance of projective test data in planning treatment strategies is found in a structured therapeutic technique developed by the author (Brooks, 1981). In this technique, which is called Creative Characters, the therapist selects the major emotional issues confronting the child and then develops characters to represent the child and significant people in the child's world; the stories that are created with these characters capture the problems facing the child and provide a format for communicating solutions to these problems.

A case that illustrates the Creative Characters strategy and its reliance on projective data involves a 10 year old boy, Joey, who was referred for numerous problems, including a short attention span, low frustration tolerance, the bullying of other children, poor school performance, and refusal to do many tasks at home and school. Joey had a history of early medical problems, undergoing surgery at four weeks of age for pyloric stenosis and an adenoidectomy when five months old. He had had recurrent upper respiratory infections and bouts of otitis media. When two and one-half years old, he underwent another adenoidectomy, also including a tonsillectomy. Although he had problems in learning at school, he was found to have at least average intelligence.

The projective testing was very revealing. What emerged was a picture of a boy who saw himself as very defective and incompetent but attempted to hide this sense of incompetence behind a tough-guy stance. For instance, on the Rorschach test he saw images of monsters that acted tough but were really scared, of bugs that kept fighting to determine which was most powerful, and of broken, defective trees. On the TAT and Sentence Completion procedures, several of his responses concerned not being able to do something, getting frustrated, and quitting.

Joey created a story about a boy who was born to quit and in response to queries stated that he thought he was born to quit and that God had made him that way. Although he added that he did not think he could change, other stories were more encouraging, including one about a doctor who helped a boy and another about a lonely, poor child who received help to escape from the slums. On the basis of his respones, it appeared that Joey was not always certain that traditional helping figures such as policemen could be trusted, suggesting that patience and time would be needed to develop an alliance with him. His lack of trust seemed to have its roots in the frequent surgery he faced as a young child.

Given the data from the projective tests as well as the other information obtained from the parents, school, and pediatrician, it was decided that the issues of low self-esteem, defectiveness, and lack of trust in others as well as his self-defeating coping strategies of quitting and bullying had to be addressed and highlighted in therapy. Because his interest in policemen was evident on several projective responses, characters and stories were created having to do with children learning to trust policemen. To confront the issue of bullying and quitting, a police chief character was "interviewed"; the chief explained to the newscaster why some police recruits act like bullies or quit their training and what could be done to remedy this situation. During the course of treatment, Joey (as the police chief) said that a police doctor could assist the police recruits to recognize that when they acted too tough or when they quit, they were really scared and felt that they could not handle their job. The police doctor taught them strategies for facing rather than avoiding stress. Joey eventually moved from the policeman theme and wrote a trilogy about a dog named Quitter that thought that he was born to quit but learned not to quit and then helped another dog that faced a similar predicament. By the end of the first story Quitter's name was changed to Try. Most importantly, Joey's improvement in therapy was paralleled by significant improvement in his daily life, including less quitting and a greater ease in accepting the help of others.

The data obtained from the projective testing about Joey's phenomenal world served as beacons of light throughout the therapy work with him. The projective tests were not used to label Joey but rather to provide a picture of his feelings, thoughts, cognition, and conflicts, and of the coping strategies he employed. Possessing this information about Joey's inner world, one could develop characters and stories to articulate, elaborate, and resolve Joey's problems. The case of Joey is an example of how projective responses may be invaluable in helping to understand a child and in directing treatment efforts.

USE OF PROJECTIVE TECHNIQUES BY PEDIATRICIANS AND REFERRAL TO A MENTAL HEALTH PROFESSIONAL

Some pediatricians have used projective procedures with their patients (Berger, 1980). At this point, however, most pediatricians have neither the time nor the training to go beyond a few structured, direct interview questions in an attempt to understand the inner world of the child. Detailed projective tests should be administered only by professionals who have received closely supervised training in their administration and who have taken courses describing when to use these procedures and their reliability and validity. As argued earlier in this chapter, projective techniques can be important instruments for use in the assessment of cognitive and emotional functioning in children, but if used by untrained individuals of any profession, they can pose a danger in contributing to a view of the child that is inaccurate and biased.

Close collaboration between a pediatrician and mental health professional is optimal. If a pediatrician has done an interview screening with a child with emotional problems and still has questions about the nature and severity of these problems, a referral to a mental health professional for additional assessment is warranted. Levine et al. (1980) have discussed when and how pediatricians should make such referrals, stressing that pediatricians should refer to mental health professionals who will work closely with the pediatrician in diagnosing the child's problem and developing a treatment program.

ROBERT B. BROOKS

REFERENCES

Abt, L., and Bellak, L. (Editors): Projective Psychology. New York, Alfred A. Knopf, Inc., 1950.

Ames, L. B., Learned, J., Metraux, R. W., and Walker, R. N.: Child Rorschach Responses: Developmental Trends from Two to Ten Years. New York, Paul B. Hoeber, 1952.

Ames, L. B., Metraux, R. W., and Walker, R. N.: Adolescent Rorschach Responses. New York, Paul B. Hoeber, 1959.

Appelbaum, S. A.: The effect of altered psychological atmosphere on Rorschach responses: a new supplementary procedure. Bull. Menninger Clin., 23:179, 1959.

Beck, A. T.: Cognitive Therapy and the Emotional Disorders. New York, International Universities Press, 1976.

Becker, W. C.: A genetic approach to the interpretation and evaluation of the process-reactive distinction in schizophrenia. J. Abnorm. Soc. Psychol., 53:299, 1956.

Beiser, H. R.: Formal games in diagnosis and therapy. J. Am. Acad. Child Psychiatry, 18:480, 1979.

Berger, L. R.: The Winnicott Squiggle Game: a vehicle for communicating with the school-aged child. Pediatrics, 66:921, 1980.

Bersoff, D. N.: Silk purses into sows' ears: the decline of psychological testing and a suggestion for its redemption. Am. Psychol. 28:892, 1973.

Brooks, R.: Psychoeducational assessment: a broader perspective. Prof. Psychol., 10:708, 1979.

Brooks, R.: Creative Characters: a technique in child therapy. Psychother. Theory Res. Prac., 18:131, 1981.

Buck, J. N.: The H-T-P technique: a qualitative and quantitative scoring manual. J. Clin. Psychol., 4:317, 1948.

Burns, R. C., and Kaufman, S. H.: Kinetic Family Drawings (K-F-D). New York, Brunner/Mazel, Inc., 1970.

Burns, R. C., and Kaufman, S. H.: Actions, Styles and Symbols in Kinetic Family Drawings (K-F-D). New York, Brunner/Mazel, Inc., 1972.

Bush, R.: A Parents' Guide to Child Therapy. New York, Delacorte Press, 1980.

Chapman, L. J., and Chapman, J. P.: Illusory correlation as an obstacle to the use of valid psychodiagnostic signs. J. Abnorm. Psychol., 74:271, 1969.

DiLeo, J. H.: Young Children and Their Drawings. New York, Brunner/Mazel, Inc., 1970.

DiLeo, J. H.: Children's Drawings as Diagnostic Aids. New York, Brunner/Mazel, Inc., 1973.

Ellis, A.: Should some people be labeled mentally ill? J. Consult. Psychol., 31:435, 1967.

Exner, J.: The Rorschach: A Comprehensive System. New York, John Wiley & Sons, Inc., 1974.

Friedman, H.: Perceptual regression in schizophrenia: an hypothesis suggested by use of the Rorschach test. J. Proj. Tech., 17:171, 1953.

Gittelman, R.: The role of psychological tests for differential diagnosis in child psychiatry. J. Am. Acad. Child Psychiatry, 19:413, 1980.

Goodenough, F. L.: Measurement of Intelligence by Drawings. New York, World Book Co., 1926.

Halpern, F.: A Clinical Approach to Children's Rorschachs. New York, Grune & Stratton, Inc., 1953.

Halpern, F.: The Rorschach test with children. In Rabin, A. I., and Haworth, M. R. (Editors): Projective Techniques with Children. New York, Grune & Stratton, Inc., 1960.

Hammer, E. F.: The Clinical Application of Projective Drawings. Springfield, Illinois, Charles C Thomas, 1958.

Hartmann, H.: Ego Psychology and the Problem of Adaptation. New York, International Universities Press, 1958.

Hemmendinger, L.: Perceptual organization and development as reflected in the structure of the Rorschach test responses. J. Proj. Tech., 17:162, 1953.

Hertz, M. R.: The Rorschach in adolescence. In Rabin, A. I., and Haworth, M. R. (Editors): Projective Techniques with Children. New York, Grune & Stratton, Inc., 1960.

Kagan, J.: Thematic apperceptive techniques with children. In Rabin, A. I., and Haworth, M. R. (Editors): Projective Techniques with Children. New York, Grune & Stratton, Inc., 1960.

Koppitz, E. M.: Psychological Evaluation of Children's Human Figure Drawings. New York, Grune & Stratton, Inc., 1968.

Korchin, S. J.: Modern Clinical Psychology. New York, Basic Books, Inc., 1976.

Laosa, L. M.: Social competence in childhood: Toward a developmental, socioculturally relativistic paradigm. In Kent, M. W., and Rolf, J. E. (Editors): Primary Prevention of Psychopathology. Vol. III: Social Competence in Children. Hanover, New Hampshire, University Press of New England, 1979.

Lerner, P. M. (Editor): Handbook of Rorschach Scales. New York, International Universities Press, 1975.

Lerner, P. M.: The genetic level score: a review. In Lerner, P. M. (Editor): Handbook of Rorschach Scales. New York, International Universities Press, 1975.

Levine, M. D., Brooks, R., and Shonkoff, J. D.: A Pediatric Approach to Learning Disorders. New York, John Wiley & Sons, Inc., 1980.

Levy, M. R., and Fox, H. M.: Psychological testing is alive and well. Prof. Psychol., 6:420, 1975.

Lewandowski, D. G., and Saccuzzo, D. P.: The decline of psychological testing. Prof. Psychol., 7:177, 1976.

Lindzey, G.: On the classification of projective techniques. Psychol. Bull., 56:158, 1959.

Machover, K.: Personality Projection in the Drawing of the Human Figure. Springfield, Illinois, Charles C Thomas, 1949.

Machover, K.: Human figure drawings of children. J. Proj. Tech., 17:85, 1953.

Mischel, W.: Continuity and change in personality. Am. Psychol., 24:1012, 1969.

Mischel, W.: On the future of personality measurement. Am. Psychol., 32:246, 1977.

Monahan, R. T.: Suicidal children's and adolescents' responses to Early Memories test. Paper presented at conference of The Society for Personality Assessment, 1981.

Mundy, J.: The use of projective techniques with children. In Wolman, B. B. (Editor): Manual of Child Psychopathology. New York, McGraw-Hill, Inc., 1972.

Murphy, L. B., and Krall, V.: Free play as a projective tool. In Rabin, A. I., and Haworth, M. R. (Editors): Projective Techniques with Children. New York, Grune & Stratton, Inc., 1960.

Napoli, P. J.: Finger painting. In Anderson, H. H., and Anderson, G. L. (Editors): An Introduction to Projective Techniques. New York, Prentice-Hall, Inc., 1951.

Phillips, L., Kaden, S., and Waldman, M.: Rorschach indices of developmental level. J. Genet. Psychol., 94:267, 1959.

Rabin, A. I.: Projective methods and projection in children. In Rabin, A. I., and Haworth, M. R. (Editors): Projective Techniques with Children. New York, Grune & Stratton, Inc., 1960.

Rapaport, D.: Projective techniques and the theory of thinking. J. Proj. Tech., 16:269, 1952.

Rapaport, D., Gill, M. M., and Schafer, R. (revision edited by Holt, R. R.): Diagnostic Psychological Testing. New York, International Universities Press, 1968.

Rickers-Ovsiankina, M. A. (Editor): Rorschach Psychology. New York, John Wiley & Sons, Inc., 1960.

Santostefano, S.: Psychological testing in evaluating and understanding organic brain damage and the effects of drugs in children. J. Pediatr., 62:766, 1963.

Santostefano, S.: Miniature situations and methodological problems in parent-child interaction research. Merrill-Palmer Quart. Behav. Develp., 14:285, 1968.

Santostefano, S.: Assessment of motives in children. Psychol. Rep., 26:639, 1970.

Santostefano, S.: Beyond nosology: diagnosis from the viewpoint of development. In Rie, H. E. (Editor): Perspectives in Child Psychopathology. New York, Aldine-Atherton Co., 1971.

Santostefano, S.: Action, fantasy and language: developmental levels of ego organization in communicating drives and affects. In Freedman, N., and Grand, S. (Editors): Communicative Structures and Psychic Structures. New York, Plenum Publishing Corp., 1977.

Santostefano, S.: A Biodevelopmental Approach to Clinical Child Psychology: Cognitive Controls and Cognitive Control Therapy. New York, John Wiley & Sons, Inc., 1978.

Santostefano, S.: Cognition in personality and the treatment process: A psychoanalytic view. In Solnit, A. J., Eissler, R. S., Freud, A., Kris, M., and Neubauer, P. B. (Editors): The Psychoanalytic Study of the Child. New Haven, Yale University Press, 1980, Vol. 35.

Santostefano, S., and Baker, A. H.: The contribution of developmental psychology. In Wolman, B. B. (Editor): Manual of Child Psychopathology. New York, McGraw-Hill, Inc., 1972.

Sarbin, T. R.: On the futility of the proposition that some people be labeled "mentally ill." J. Consult. Psychol., 31:447, 1967.

Schachtel, E. G.: Experiential Foundations of Rorschach's Test. New York, Basic Books, Inc., 1966.

Sigel, I.: The application of projective techniques in research with children. In Rabin, A. I., and Haworth, M. R. (Editors): Projective Techniques with Children. New York, Grune & Stratton, Inc., 1960.

Thelen, M. H., Varble, D. L., and Johnson, J.: Attitudes of academic clinical psychologists toward projective techniques. Am. Psychol., 23:517, 1968.

Wade, T. C., and Baker, T. B.: Opinions and use of psychological tests: a survey of clinical psychologists. Am. Psychol., 32:874, 1977.

Waite, R.: The intelligence test as a psychodiagnostic instrument. J. Proj. Tech., 25:90, 1961.

Werner, H.: Comparative Psychology of Mental Development. Chicago, Follett Publishing Co., 1948.

White, R.: Motivation reconsidered: the concept of competence. Psychol. Rev., 66:297, 1959.

Winnicott, D. W.: Therapeutic Consultations in Child Psychiatry. New York, Basic Books, Inc., 1971.

Woltmann, A. G.: The use of puppetry as a projective method in therapy. In Anderson, H. H., and Anderson, G. L. (Editors): An Introduction to Projective Techniques. New York, Prentice-Hall, Inc., 1951.

49

Interdisciplinary Assessment

49A   The Function of Teams

The terms "team" and "teamwork" encompass a flexible group of concepts in human services. Several general principles are basic to all real teamwork. Brill (1976) has defined a team as ". . .a group of people each of whom possesses particular expertise, each of whom is responsible for making individual decisions, who together hold a common purpose, and who meet together to communicate, collaborate, and consolidate knowledge, from which plans are made, actions determined, and future decisions influenced."

Teamwork in clinical pediatrics is essentially a task oriented, problem solving approach to diagnosis and treatment as well as to prevention of illness and promotion of optimal development in the child. It is a progressive process, beginning with definition of the problem and purpose and proceeding to the detailing of goals, tasks, roles, and methods of intervention. The team then evaluates and revises when necessary. Successful use of the team model depends upon the capabilities of individual team members and their ability to meet together to communicate information, to synthesize that information into a sound understanding of the issues before the team, and to arrive at collaborative decisions regarding actions to be taken. This final step is a most crucial and challenging one, which involves reaching agreement about treatment priorities and formulation of a plan that is both feasible and acceptable to the involved family.

THE STRUCTURE OF TEAMS

For children or families in whom there are a multiplicity of functional issues, it follows that problem solving, whether diagnostic or treatment oriented, will proceed most effectively when the actions of diverse professionals can be melded into a team approach. In current practice, team formats are common in many clinical situations, such as that for the child with psychosomatic illness, the child in a cerebral palsy center, assessment in a

mental retardation study clinic, educational planning in the special services department of a school system or collaborative agency, and program development for a child or family in a complex social dilemma. Teams may be based in a central facility or may be part of outreach activities, which work directly in community settings. How a team utilizes the strengths and contributions of its various professional members is a test of the level of mutual respect that exists in that agency among different disciplines. This, in turn, is influenced by local biases, the maturity of the workers involved, and the requirements of funding and reimbursement patterns.

The dynamics of professional interaction in assessment or treatment circumstances are diagramed in Figure 49–1. In this illustration the ultimate communicating and interpreting individual is indicated by the darker ring; ideally this person would also assume responsibility for continuing coordination.

Solo Action or Practice.   In this model a single individual accepts the responsibility for identifying the cogent issues and designing relevant management (e.g., physician, social worker, community health nurse). Services for clients with multifactorial problems are unlikely to be well guided in this practical but rather assumptive mode.

Consultation Practice.   Here the resources of an outside expert (e.g., neurologist) are obtained, to enrich understanding of the basic problems, but limited continuing contact from this direction occurs with the child or family.

Partnership Practice.   If the primary clinician establishes a partly shared responsibility with another analogous worker, the scope of interaction with the child and family inevitably is enhanced.

Multidisciplinary Services.   This very expedient system, popularized in the 1950's, brings into focus the reports, if not the continuing guidance, of a varied group of professionals in regard to a given situation. The various workers (e.g., psychology, speech pathology, psychiatry) may or may not be located in the same facility and often

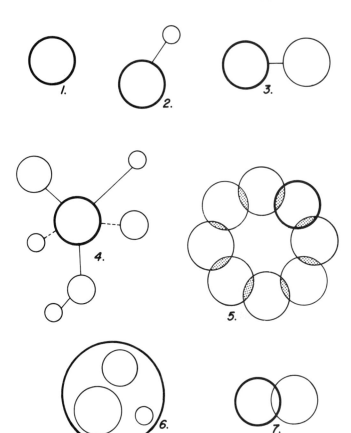

Figure 49-1. The dynamics of interaction in assessment or treatment. (See text.) 1, Solo practice. 2, Consultation. 3, Partnership. 4, The multidisciplinary team. 5, The interdisciplinary team. 6, The transdisciplinary team. 7, Collaborative practice.

do not have the opportunity to confer together directly in behalf of the client. The degree of equilibration of understanding of the child's problem and the balance of design for services depend on the personalities and working styles of the contributors and the universality of view of the coordinator.

Interdisciplinary Services. This rich and fulfilling model for the study and management of multiproblem children and families was a creation of the revolution in human services that characterized the late 1960's and the 1970's. When fully implemented, the interdisciplinary team allows a true sharing in the consideration of all issues. The team works in substantially one location, meets as a group to establish common convictions about the nature and best solutions for the clinical and social challenges, and remains in place to follow through on the results of programming. One worker must be designated to serve as coordinator for a given client, but in the developing discussions, the views of all professionals are equally valued and family contact responsibilities are assigned as appropriate. Some overlap of observations and ideas is intrinsic, but this serves as a control on arbitrary decisions or pursuit of vested interests. Examples of situations in which group action of this sort is especially appropriate would be the child with major feeding problems, the

nonverbal child, the child in an abuse or neglect dilemma, or the multihandicapped child in whom there have been conflicting interpretations about his capacities. The major difficulty with this model, of course, is that it incurs high costs in terms of professional resources and work arrangements. Slippage in potential reimbursement occurs because of the time required for complimentary observations, conferences, preparation of coordinated reports, and follow-through. In the end, this type of activity requires subsidization—either because of its value as a training milieu or for the accomplishment of a special project assignment.

Transdisciplinary Services. In certain circumstances in which expedited communication with clients is of value, special arrangements have been established whereby one professional takes on the responsibility for translating the work of a group. Here the understanding of other disciplinary roles is at a sophisticated level, so that the special educator may transmit the view of the physical therapist or speech clinician, for example. For this method to be effective, there must be a stable team and a modulated level of client handicap.

Collaborative Practice. The truly mutual sharing of client handling by two (or possibly three) professionals of different disciplines has emerged as a relatively new concept. In this model a pair of workers, who maintain a high degree of feedback and are of equal rank, divide the task of working with the child and family in a common setting. Examples include pediatrician and nurse practitioner, orthopedist and physical therapist, and physician and psychologist. The family can gain from a smoothed professional program, while the assignment utilizes each worker in his area of most useful offering. This break with traditionally authoritarian unilateral management requires training and experience, but can be extremely rewarding.

TEAM DYNAMICS

As could be anticipated, there is a life to teams and a requirement to attend to their sustenance. Team members make a personal contribution to the process, often fairly costly in time and resources, and there must be returned a product of sufficient value to justify this investment. As Magrab et al. (1982) have pointed out, three qualifying elements determine the functional tone of a team: leadership style, role clarification, and group atmosphere. Leadership, or coordination, may be laissez faire, democratic, or autocratic. Each conceivably could be successful as long as an unswerving adherence is maintained to accountability regarding goals. The leader must be sensitive to interactions within the group, as well as to the personal traits and needs of the members. In return for the trust and opportunity being given to him, he must offer additional input and preservation of standards. Role clarification for the members of a team is often neglected, with resultant professional dissatisfaction. The functional responsibilities actually may be best defined in "role agreements," determined in advance. In the interdisciplinary team model, as mentioned, one element of the role may be acceptance of rotation through the leader or coordinator position. The group atmosphere for a team is a magical synthesis, with significant effect on the effectiveness of decision making. Optimally there are trust, openness, respect, and mutual reliance among the members. If team loyalty is sustained, one can hope that nonfunctional behavior (such as lateness or procrastination in reports) will be reduced.

THE USE OF TEAMS AND THEIR LIMITATIONS

Personal and practical considerations often determine the ultimate format in which various mini- or maxiteams assemble and function. It is clear, however, that in current times the concept of the physician as a lone problem solver must be modified, especially when life situation elements impinge on potential clinical outcomes or the problems have complex functional components. Many pediatricians now in training have come to appreciate the professionally expanding and fulfilling aspects of practice shared with other therapists, behaviorists, and social scientists. Similarly educators and child care workers have found physicians to be sympathetic team members when a good exchange of professional insights can be assured. In some fashion the special richness of the interdisciplinary team spirit should be sought whenever a multifactorial problem is at hand; the size of the working group, or its physical proximity, may need to be modified, but this cooperative sharing of contributions remains a desired goal.

In the real world it must be admitted that there are significant limitations (and liabilities) to the promulgation of teams. Formal group structures may require time and resources that simply are not available. Many professionals report experiences in which the team process proved to be wasteful and clumsy. With some anger they equate teams with committees, and they consider the term "interdisciplinarianism" to be symbolic of unwieldiness. One could note that team actions can be very demanding of the time of clients and families as well, especially in settings in which compliance is socially or psychologically difficult. Moreover, teams can become distorted by "group think," whereby purposes are being served other than those of effective client intervention. Some of these issues represent failures of the system; others are practical matters. In the ideal circumstance, shared professional action stands to offer the best comprehensive services to the child and family with complex difficulties.

In the remainder of this chapter the special contributions of social work, physical therapy, occupational therapy, and nursing are presented as these relate to the assessment of children with special needs. Interdisciplinary activities are emphasized. In Chapter 57 the therapeutic services of these professionals are further described, as well as other therapies and vocational counseling. The work of additional team members appears in other chapters—such as speech pathology (Chapter 40), nutrition (Chapters 23 and 29), psychology (Chapters 46, 50, and 55), and education (Chapters 47 and 56).

ALLEN C. CROCKER
MARIE M. CULLINANE

REFERENCES

Brill, N. I.: Teamwork: Working Together in the Human Services. Philadelphia, J. B. Lippincott Co., 1976.
Ducanis, A. J., and Golin, A. K.: The Interdisciplinary Health Care

Team: a Handbook. Germantown, Maryland, Aspen System Corporation, 1979.

Dyer, W. G.: Team Building: Issues and Alternatives. Reading, Massachusetts, Addison-Wesley Publishing Co., Inc., 1977.

Elder, J. O., and Magrab, P. R.: Coordinating Services to Handicapped Children. Baltimore, Paul H. Brookes Publishing Co., 1980.

Holm, M. D., and McCartin, R.: Early Intervention: A Team Approach. Baltimore, University Park Press, 1978.

Magrab, P. R., Elder, J. O., Kazuk, E., Pelosi, J., and Wiegerink, R.: Developing a Community Team. Washington, D.C., American Association of University Affiliated Programs, 1982.

49B   Social Work

THE ROLE OF THE SOCIAL WORKER

The role of the social worker on the team is to evaluate the psychosocial aspects of the child's problem as a basis for developing an accurate diagnosis and a realistic treatment plan. Because children are emotionally and physically dependent on their parents, developmental and behavioral problems can be adequately understood or planned for only with detailed information about their environment and with the active participation of their parents or caretakers. Commonly community resources are a necessary part of the treatment program and thus information about what is actually available must be secured.

A major function of the social worker is to interview the parents in order to obtain pertinent information about the child's role in the family, as well as information about family relationships and particular stresses or strengths that might compromise or enhance the parents' ability to implement a treatment program. The social worker also has as a major goal the development of a good working relationship with the family by focusing on their concerns and expectations from the evaluation, as well as by offering information and emotional support regarding the evaluation procedures. The social worker also uses this information to help other team members to be sensitive to the family's concerns and limitations. The social worker can contact appropriate community agencies and facilitate communication with the clinical team. Making recommendations for community services without sensitivity to their impact can be alienating and can create obstacles to the development of the kind of rapport necessary for working together to improve services. It is sometimes helpful if the involved community agencies can be invited to planning conferences so that there can be personal dialogue. Conversely, professionals should make occasional visits to community centers to get a better sense of how community programs function. Because families are sometimes resistant to utilizing recommended services, the social worker can play a role in exploring the underlying reasons and in assisting them in making the necessary contacts.

BEHAVIOR PROBLEMS

Many families who seek help in altering a child's behavior have little knowledge of the diagnostic or treatment approaches that are usually involved. They may expect that the discussion will be restricted to the area of their concern, that it have an organic focus, or that it may involve disciplinary measures for the child. Sometimes they may be seeking immediate action or resolution. Often they have only a limited awareness of their own role in the problem. They may be unprepared for the extensive history taking about family relationships that is required to understand fully the child's needs and to develop an appropriate treatment plan. Particularly if a crisis has precipitated the request for help, they may see little relationship between their needs and what is planned. Often they are not expecting that they as well as the child may have to change.

In gathering information about the child's functioning, the social worker begins with the immediate concerns of the parent, clarifying the specific nature of the problems, the event that precipitated the request for help, and the previous efforts the parents have made to cope with the identified problems. The parents' responses to these questions provide insights into their values, attitudes, and behavior, as well as the child's. Information is solicited about other aspects of the child's adjustment that may result in identification of additional problem areas. The parents may have found it difficult to discuss some of these. Possibly because of lack of knowledge about appropriate behavior, they may not have perceived these as problems. However, there may be behavior patterns from which the parents obtain gratification, and they may be resistant to plans aimed at change. The social worker obtains a detailed history of the child's earlier development with considerable emphasis on how the parents dealt with the child's needs for nurturance, discipline, and independence. Some information is sought about the functioning of other family members, as well as the family's relationship with the extended family and the community.

Usually parents who are seeking help for a

child's behavior are themselves under considerable stress. An important component of the interviews is responding to expressed feelings of anger, anxiety, or depression by illuminating them and offering concern and support. For many parents this discussion demonstrates that they are valued and that talking about a problem can bring about improvement. Often the history taking process can help the parents to clarify their thinking about the nature and scope of the child's problem and enhance their awareness of contributing factors. This experience can be used as the basis for engaging the parents' interest in a plan of help that will take time and in which they must participate. In some instances this increased understanding can result in immediate positive changes in the way they relate to the child.

The social worker analyzes the available information to identify for the team the major factors that might be contributing to the child's problems. Recommendations are also made as to the kinds of treatment programs the family would be able to support. Information about available resources in the child's community is a prerequisite for making realistic referrals for services.

The range of services utilized will vary greatly. Young children are totally dependent upon their parents to take them for services, and their behavior is so enmeshed with that of the parent that the prognosis for change is poor unless the parent can be motivated to make significant changes in the child's life situation. Nursery schools, extended school days, and involvement of the child in activities outside the home can help supplement counseling for the parents. Adolescents require less support from their parents in using therapeutic services, provided the parents give permission. In some instances the child is responding to major family problems, and referrals may be made for family counseling. The treatment of choice may involve concurrent counseling for parents, with the social worker often assuming this responsibility. In exceptional instances, separation of child and family is indicated because of the nature of problem and the family's inability to participate in a home based treatment program. The social worker can play a major role in assisting the family in following through on recommendations.

NEGLECTED OR ABUSED CHILDREN

Children who present with indications of neglect or inflicted injury always reflect disturbed family relationships. (See Chapter 14.) They should be referred to a social worker. In many states children with these symptoms must be reported to a designated child welfare agency. The parents feel threatened by such a referral, for they may be unsure of the implications. It is important that a social worker be part of the clinical team assessing such situations. The social worker can contribute by exploring the circumstances to determine whether such a conclusion is appropriate, to assess the urgency of the situation in terms of child's immediate safety in the home, and to help the family identify the areas of stress for which they must seek help. The family will be concerned about the referral to a social worker, possibly fearing retaliation or loss of the child. They should be informed that the primary goal of the social worker is to learn about the family's needs in order to develop a plan of care.

The social worker usually begins the contact by exploring the family's concerns and attempting to alleviate those that are unrealistic. However, the family must understand that evaluation of the total situation is mandatory and that their cooperation is essential. Most families respond to this structure and often are relieved that someone is intervening, even though they are unable to ask for help. Information about the circumstances surrounding the "accident" or their view of the child's condition is obtained as well as the whereabouts of significant family members and their role in the situation. The focus is not on fixing blame. One learns about the relationship between parents and child, and the aspects of caretaking that are gratifying and those that are stressful. It is important to learn how age-appropriate the parents' expectations of the child's behavior are. The parents are encouraged to talk about any personal stresses they are experiencing, how they handle stress, and what supportive relationships are available. A part of the history taking usually involves learning about the parents' early history and their readiness for assuming the responsibilities involved in child care. Information is also obtained about how the parents have handled other obligations.

The social worker attempts to demonstrate to the parents a concern for their welfare and protection, as well as for that of the child. It is important to be honest and direct in response to the parents' questions and to avoid precipitating confrontations or battles for control. From this information the social worker formulates a psychosocial diagnosis, including assessments of the parents' capacity to be sensitive to the child's needs and to provide for them, their ability to inhibit impulsive behavior, the supports available to them at times of stress, and their willingness to ask for help when they feel overwhelmed and out of control. The age and needs of the child are also important in deciding whether the parents are able to care for the child pending the development of a treatment program (which could involve mobilizing community resources).

Follow-up care almost invariably involves a referral to a child welfare agency for counseling and supervision of the child's care in the home. Often supplementary services are appropriate; these can help improve the parents' confidence and skills. Such services may include improved health services, parent education programs and support groups, respite care, and homemaker services. Other forms of help that are directed more to the parents' personal needs may be part of the rehabilitation program. These may include psychiatric care and vocational or employment counseling. It is evident that these families need to be followed closely by the clinical team to ensure that contact with the necessary programs is made and maintained.

DEVELOPMENTAL OR SIGNIFICANT LEARNING PROBLEMS

The identification of a developmental handicap in a child is invariably a source of great stress for the family. (See Chapter 39.) One should anticipate that the parents will experience a grief reaction as they become aware that the child's delays will require significant modifications throughout his life. The social worker's assessment will give a more individualized picture of what the child represents to his family and the unique way in which the parents will respond to the information. It is important to remember that the diagnosis of a handicap in a child is a crisis for the parents and that appropriate supports provided at that time will make a significant difference in their ultimate adaptation to the child's needs.

Parents of handicapped children have to assume a number of extra responsibilities if their child is to develop his abilities to the fullest. These include following through on special treatments, providing supplementary stimulation, making informed decisions about appropriate school programs, helping the child make social contacts, and advocating the development of needed community resources. A prerequisite for carrying out these functions is the obtaining of accurate information, the acceptance of the handicap in the child, and confidence in dealing with professionals and the network of agencies. Since it is the family members who bear the primary responsibility for implementing the treatment program, it is important to design a program that is compatible with the family's resources and motivations. Without an advocate the parents may have difficulty in bringing their own concerns and needs into the discussion. The social worker can provide support and guidance for the parents in assuming these roles.

The social worker, with the assistance of members of appropriate specialties from the team, can be helpful in interpreting the child's needs to community agencies and substantiating the child's right to services. Families may find it difficult to use certain services that are designated for handicapped children. They may have preconceived ideas about these programs and may have difficulty in accepting that their child's needs will be appropriately served in this context. Conversely, community agencies that customarily serve the normal population have anxieties about serving handicapped clients.

Families may seek evaluations at critical periods in the child's life to review progress, assess needs, and obtain guidance. Entering school and emerging adolescence are frequent precipitants. The social worker then focuses more on learning about the family's present management of the child and future aspirations, and less on the impact of the initial diagnosis. However, for some families unresolved issues continuing from this early period may directly affect their relationship with the child and their response to the current evaluation. The social worker's assessment should attempt to reconstruct the sources of such behavior. Does the focus of the problem lie in a psychopathologic disorder of the parent, inadequate support from other key relationships in their lives, poor management of the initial diagnosis, or a lack of adequate response from community agencies? The longer the pattern is established, the more difficult it is to effect change. Sometimes the emergence of a new developmental stage in the child or family can create disequilibrium in established patterns of behavior, rendering them less effective and satisfying, and the family may be more open to considering recommendations aimed at change.

ANN P. MURPHY

REFERENCES

Gurney, W.: Building a collaborative network. Soc. Work Health Care, 1:185, 1975.

Horwitz, J.: Interprofessional team work. Social Worker, 38:5, 1970.

Kane, R.: Inter-professional education and social work. Soc. Work Health Care, 2:229, 1976.

49C   Physical Therapy

ROLE DEFINITION: PHYSICAL THERAPY

The physical therapist is increasingly called upon to be a part of interdisciplinary assessment teams in both clinical and educational settings. The type of physical therapy study is guided by both the age of the child and the reasons for assessment. The goals may include one or more of the following:

1. To identify aspects of movement or posture that may benefit from a physical therapy program, and to make suggestions regarding the content of that program, e.g., a child with spina bifida who may require increased shoulder and arm strength in order to perform transfers from chair to bed.

2. To ascertain the level of development of both gross and fine motor skills, as well as the quality of their performance.

3. To clarify the relationship between the child's movement, in both its qualitative and developmental aspects, and his function in daily living activities and play, e.g., the child with congenital heart disease who experiences delay and difficulty in the achievement of gross motor skills.

4. To explain the relationship (if any exists) of movement and motor development to the problem(s) for which the child was referred, e.g., the infant who is referred with delayed overall development and who also has hypotonia.

The goals of the therapist must take into account the specific concerns of the parents or caretakers, since these questions are important sources of information about the child. Such discussion also provides insights regarding the parents' present state of knowledge and consideration of the child's abilities.

CONTENT OF THE ASSESSMENT

The physical therapist may carry out detailed testing in one or more areas and screen others, depending upon the age of the child and the problem for which he was referred. The main emphasis for the severely physically handicapped child, for example, may be on evaluating the need for adaptations of his living environment, whereas for the developmentally delayed infant the emphasis probably would be on aspects of his motor development.

HISTORY

Information about early motor achievements of a very young child is, of course, important, as is information about his preferred play activities and functioning in daily living skills. This knowledge can be helpful both in structuring the evaluation and in determining areas of emphasis.

POSTURE AND SPONTANEOUS MOVEMENT

Observation of the child's posture and spontaneous movements is usually done while the history is being taken. The infant or young child who may be apprehensive can then be observed while being handled by his parent; the older child, if there is sufficient space, can demonstrate his mobility without the sense of being under surveillance. The child's habitual postures and the amount of self-initiated movement that occurs can be noted.

STRENGTH

The child's strength may be examined reliably using the techniques of manual muscle examination described by Daniels et al. (1956). Modifications of these tests for young children and for those who are not able to understand verbal instruction have been made by Zausmer (1953). Mechanical and electronic devices have been utilized for testing strength in some muscles in adults and older children, but they have not yet been adapted for infants and young children.

RANGE OF JOINT MOTION

The active and passive ranges of motion of the joints of the extremities, neck, and trunk are measured to determine whether any restriction of movement is affecting function. This type of measurement is most commonly used with children who have neuromuscular and musculoskeletal problems. In children who have central nervous system disorders, the range of joint motion may be related to aspects of muscle tone, habitual postures, or asymmetries of movement.

MUSCLE TONE

Muscle tone is evaluated while taking into consideration the state of the child, using the methodology of André-Thomas et al. (1960) and others. Muscle tone has been found to be a useful parameter in examining the movement characteristics of infants, with the qualification that serial studies are necessary for making any type of prognostication from this assessment.

PRIMITIVE REFLEXES AND POSTURAL REACTIONS

These are measured using the techniques developed by Milani-Comparetti and Gidoni (1967), and more recently, in the case of primitive reflexes, the tests of Capute et al. (1978) for qualitative aspects as well as their timing. This type of testing has been utilized prognostically as well as in planning treatment for children with cerebral palsy or mental retardation. Observations of this type also can be used as indicators of problems in movement in children with learning disabilities.

DEVELOPMENTAL TESTS

The development of movement in infants can be examined by use of the tests developed by Gesell (Knobloch et al., 1980) and Bayley (1969). The achievements recorded using these tests are supplemented by recording the quality of movement patterns as well as any deviant motor characteristics that are important. There may be, for example, a sitting or walking pattern that is abnormal even though independence in the activity has been achieved. Recently there has been developed a scale for measurement of movement in the first year of life that combines the observation of motor milestones with that of other aspects of movement, such as muscle tone and the degree of active movement (Chandler et al., 1980). This test is not yet standardized, but reliability and validity studies have been done. For the preschool child and older child, the maturation of motor patterns can be examined by using the tests developed by the Dutch neurologist Touwen (1979). The achievement of more advanced gross and fine motor skills in the learning disabled or older retarded child can be evaluated with the test developed by Stott et al. (1971).

ACTIVITIES OF DAILY LIVING AND ADAPTIVE EQUIPMENT

These areas may be the most important focus for the physically handicapped child. Components of the evaluation can range from feeding (both oral-motor components and self-feeding) to adaptations of seating and living environments. This area is usually evaluated together with the nurse, orthetist, occupational therapist, speech pathologist, and others because specific expertise is needed.

ANALYSIS AND INTEGRATION OF FINDINGS

The physical therapist, following the assessment, examines the relationships of the findings in the various areas in order to describe their meaning in regard to the child's function in an integrated way, with a minimum of isolated statements. It is not useful, for example, to have a description of individual muscle strength unless this is related to everyday functions.

RESULTS OF THE ASSESSMENT

The results are usually stated in terms of the child's activities and function, with the questions of the parents providing the initial frame of reference. The findings are related to those of other disciplines, such as discussion of sitting balance as it affects performance in some aspects of psychological testing, or the relationship between control of the trunk and development of postural reactions to respiratory control for speech. The rationale for treatment is discussed in terms of specific goals, but less emphasis is placed on specific methods of implementation, in the event that these are not readily available to the family in their community. The rationale for not recommending treatment or for suggesting consultation by the physical therapist with other caregivers, is stated. The latter may occur, for example, in the situation of the multihandicapped child when the evaluation has brought out other more pressing treatment needs than physical therapy. The relationship of therapy to play or sports activities should be explained, such as when supervised recreation can be considered a substitute or an adjunct to the therapeutic program and when it cannot.

The assessment by the physical therapist, as with other assessments, describes a single point in the development of movement of the child and should be regarded as such. The dynamics of this development and the child's potential for change are considerations. The need for re-evaluations at reasonable intervals to examine the effects of treatment as well as to establish new goals should be explained to the parents.

ALICE M. SHEA

REFERENCES

André-Thomas. Chesni, Y., and Saint-Anne Dargassies, S.: The Neurological Examination of the Infant. Clinics in Developmental Medicine, No. 1. London, National Spastic Society, 1960.
Bayley, N.: Bayley Scales of Infant Development. New York, The Psychological Corporation, 1969.

Capute, A., Accardo, P. J., Vinning, E., Rubenstein, J., and Harryman, S.: Primitive Reflex Profile. Baltimore, University Park Press, 1978.

Chandler, L., Andrews, M. and Swanson, M.: Movement Assessment of Infants, A Manual. Rolling Bay, Washington, 1980.

Daniels, L., Williams, M., and Worthingham, C.: Muscle Testing Techniques of Manual Examination. Ed. 2. Philadelphia, W. B. Saunders Company, 1956.

Knobloch, H., Stevens, F., and Malone, A. F.: Manual of Developmental Diagnosis: The Administration and Interpretation of the Revised Gesell and Amatruda Developmental and Neurological Examination. Philadelphia, J. B. Lippincott Co., 1980.

Milani-Comparetti, A., and Gidoni, E. A.: Routine developmental examination in normal and retarded children. Dev. Med. Child Neurol., 9:631, 1967.

Stott, D. H., Moyes, F. A., and Henderson, S. F.: Tests of Motor Impairment. Guelph, Ontario, Brook Educational Publishing Ltd., 1971.

Touwen, B. C. L.: Examination of the Child with Minor Neurological Dsyfunction. Ed. 2. Clinics in Developmental Medicine, No. 71. London, William Heinemann Medical Books Ltd., 1979.

Zausmer, E.: Evaluation of strength and motor development in infants. Phys. Ther. Rev., 33:575, 1953.

49D Occupational Therapy

The occupational therapist assesses a child's fine motor and perceptual skills as they relate to the performance of everyday activities. Strength and problem areas in activities of daily living, play and school-work skills are determined, with or without special adaptations or appliances, and in relationship both to age expectations and the demands of a given environment. The purpose of the evaluation is to determine what the child can do, and why he can do some things and not others (Pearson and Williams, 1972). To discover this, the therapist investigates the neuromuscular, sensory integrative, and psychosocial functions that can foster or hinder performance.

The scope of an occupational therapy evaluation depends upon the environment in which a child performs (home, school, institution) as well as on his age and disability and is influenced by the functions of an occupational therapy center and its goals of service. In screening programs, such as those found in schools or early intervention programs, the therapist determines whether a follow-up evaluation or service is needed. In comprehensive evaluation centers, assessment of the actual and potential functional capacity assists parents and teachers in developing realistic expectations for the child and helps in determining program placement or the need for treatment in occupational therapy. When a child has been referred for direct treatment in occupational therapy, the evaluation first provides information with which to plan and carry out a therapeutic program and, second, provides a baseline by which to measure progress. The content of and approaches to evaluation, to be discussed, have been developed within treatment settings and are most effective when used in continuing assessment closely related to treatment.

DEVELOPMENTAL ASSESSMENT OF PERFORMANCE

Developmental assessment in occupational therapy is in the fine motor area and includes activities of daily living, play activities, and school-work tasks, as well as the motor and perceptual skills needed to accomplish such tasks. If a formal developmental evaluation is not otherwise available, the occupational therapist uses a standardized developmental test to determine the overall level at which a child is functioning and to provide a context for interpreting fine motor development. However, the core of the developmental assessment in occupational therapy is the performance check list, which is constructed to meet the needs of a particular therapeutic setting. These lists are traditional in occupational therapy and physical therapy, especially in testing for performance of the activities of daily living and motor skills. In pediatrics the reviews are ordered in developmental sequences within behavior categories. Approximate chronological ages at which individual items are accomplished are taken from research literature or standardized scales and are included on test forms to serve as guides to normative levels of behavior. These developmental surveys contribute valuable information to the overall assessment of a child's actual and potential development, but they are not developmental tests and differ markedly in purpose and procedure from standardized instruments. Performance check lists are designed to provide detailed information about a child's performance in a circumscribed area and a description of current performance rather than serving as indexes of specific developmental level.

According to Wolf (1969), the first known check list of self-care activities was published in 1935. The evaluation of daily life activities came into widespread use in occupational therapy a few years later. Areas of performance usually assessed in occupational therapy include feeding (sucking, chewing, swallowing, self-feeding, using appropriate utensils, table manners); dressing (putting on and removing clothing, fastenings, adjusting clothing, choosing appropriate clothing, neatness and attractiveness); grooming and hygiene (personal health needs, bathing, toileting, hair care, make-up); homemaking (cooking, cleaning, sewing); play skills (manipulation of toys, games, pencils, crayons, scissors, recreational equipment); and school and work skills (writing, desk activities, prevocational activities).

In addition to these specific skills, the therapist assesses basic upper extremity skills, such as reach, grasp, release, pincer grasp, supinate grasp, and bilateral hand use. As appropriate, standardized tests are used to evaluate the more advanced manipulative skills. A special form of upper extremity evaluation is the use of prosthetics or orthotics. The therapist evaluates the developmental readiness of a limb-deficient child and the need for an orthotic device as well as the use of such devices.

The selection of items for a performance check list is individualized to a treatment setting, but standardized administration of individual items is used to allow assessment of developmental change in task accomplishment. Once the baseline measure has been obtained, the item is often readministered in a variety of ways to determine whether factors such as altered position, adaptive equipment, extra demonstration, nonverbal presentation, or time of day will enhance the child's performance. For example, the therapist may determine whether a child with hemiplegia who is unable to tie his shoes could learn to tie a one handed bow or whether elastic shoelaces and a shoe horn would foster earlier dressing independence.

In addition to trying out adapted equipment and methods during the developmental assessment, the therapist observes the quality of performance to determine to what degree perceptual or motor disabilities interfere with performance. Physical factors such as poor coordination and timing of movement, joint limitation, maladaptive postures, evidence of weakness, and undue stress or fatigue are indications of the need for treatment or for further evaluation. Failure in fine motor tasks despite apparently adequate intellectual and motor abilities suggests a need to explore sensory and perceptual functions. For example, in assessing a child's ability to write, the quality of sitting posture and postural background movements are assessed, since the quality of pencil use is influenced by the stability and ease of sitting without leaning on the table and by the ability to shift posture with movements of the arm. The way the child holds the pencil is noted, as well as whether the pencil grasp is mature and whether the pencil point can be seen by the child without tilting the body laterally. The child's ability to use his two hands together, one to hold and shift his paper, and his ability to start, stop, and change directions at will are evaluated. Finally, eye-hand coordination and visual-spatial abilities such as part-whole relationships and organization on a page are determined.

The check list for the activities of daily living has been and continues to be designed for use in a particular center and is tailored to meet the goals of that center. However, a number of check lists have been published that can be used as devised, or as sources for items in a center produced instrument.

An excellent pediatric assessment of the activities of daily living, developed in the Occupational Therapy Department in the Childrens' Hospital at Stanford, California, has been published by Coley (1978). Coley provides developmentally ordered subskills for many of the items and discusses gross motor, perceptual, and psychological functions that influence performance in such activities. She also includes items appropriate for older children and adolescents. The Erhardt (1979) Developmental Prehension Assessment assesses the development of basic hand and approach skills from birth to 15 months and the development of pencil grasp. Hopkins and Smith (1978) have published a check list for oral-motor assessment and have discussed the normal development of sucking, swallowing, tongue placement, and lip closure as they relate to feeding. A check list developed for use in early intervention programs has been published by Schafer and Moersch (1977). This evaluation tool was developed collaboratively by physical therapists, occupational therapists, and speech therapists and was designed specifically for use in planning and implementing therapy in a transdisciplinary setting.

ASSESSMENT OF DISABILITY AREAS

The developmental performance check list identifies the age appropriate skills that the child has and begins the determination of why he can do some skills but not others. This information together with an understanding of the child's diagnosis identifies the need for further assessment of neuromuscular, sensory integrative-perceptual, or psychosocial functions.

The evaluation of neuromuscular status is done as it affects upper extremity performance. The areas assessed include joint range of motion, strength of individual muscles or muscle groups, hand and pinch strength, reflex status and postural stability, and background body movement for reaching. A description of these evaluation procedures can be found in the books by Hopkins and Smith (1978) and Trombly and Scott (1977).

Two approaches to the evaluation of perceptual abilities are used in occupational therapy either separately or in combination.

The Perceptual-Motor Approach. The perceptual-motor approach breaks down performance into developmentally sequenced perceptual skills.

Body awareness, spatial relationships, visual and tactile form perception, motor planning, and constructive praxis are areas frequently included in the assessment.

The Sensory Integration Assessment. The sensory integration assessment examines perceptual disabilities in relationship to underlying disorders of tactile, proprioceptive, and vestibular functions. The tests used to assess sensory integration were developed for use in learning disability to describe and delineate a child's disorder as a basis for planning treatment. The battery includes 18 standardized tests, which examine somatosensory perception, balance and praxis, visual spatial functions, and postrotary nystagmus (Ayres, 1980).

Although most of this discussion has focused on the evaluation of perceptual and motor disabilities, the occupational therapist is also concerned with the psychological basis of nonperformance. Functioning in everyday activities is not just a matter of ability to apply a skill; it also depends on the motivation to perform and the establishment of habits of performance. It is important to know the child's interests, how he spends his time, as well as his coping behavior patterns, such as frustration tolerance, relationship to authority, impulse control, and adaptation to change. Play requires both developmental social play skills (i.e., dyadic and group interaction) and independence in and self-regulation of play behavior. The assessment of play behavior includes an investigation of what, with whom, and when a child engages in play. Florey (1981) discusses the areas and functions of play and describes instruments that have been developed to investigate play behavior in children with delays in development. These instruments include the play history, guides

to the observation of play, and check lists of play development.

Finally, the occupational therapist is accustomed to working on interdisciplinary teams and depends on team members as sources of the information needed both for interpreting the occupational therapy findings and for planning treatment. The occupational therapist seeks information about the child's diagnosis and medical condition and his gross motor, cognitive, language, and personal-social development, and in turn contributes information about the child's fine motor and perceptual abilities as they relate to common skills.

ANNE HENDERSON

REFERENCES

Ayres, A. J.: Southern California Sensory Integration Tests Manual. Los Angeles, Western Psychological Services, 1980.

Coley, I. L.: Pediatric Assessment of Self Care Activities. St. Louis, The C. V. Mosby Company, 1978.

Erhardt, R. P.: Erhardt Developmental Prehension Assessment, 1979. (Available from R. P. Erhardt, 2109 3rd St. N., Fargo, North Dakota 58102.)

Florey, L. L.: Studies of play: implications for growth, development and clinical practice. Am. J. Occup. Ther., 35:519, 1981.

Hopkins, H. L., and Smith, H. D.: Willard and Spackman's Occupational Therapy. Philadelphia, J. B. Lippincott Company, 1978.

Pearson, P. H., and Williams, C. E.: Physical Therapy Services in the Developmental Disabilities. Springfield, Illinois, Charles C Thomas, 1972.

Schafer, D. S., and Moersch, M. S. (Editors): Developmental Programming for Infants and Young Children. Ann Arbor, University of Michigan Press, 1977.

Trombly, C. A., and Scott, A. D.: Occupational Therapy for Physical Dysfunction. Baltimore, The Williams & Wilkins Co., 1977.

Wolf, J. M.: The Results of Treatment in Cerebral Palsy. Springfield, Illinois, Charles C Thomas, 1969.

49E   Nursing

Nursing is a profession requiring expertise in areas of health and child care, with the ability to assess needs and to respond by practical assistance. The professional nurse is a valued member of the interdisciplinary assessment team. Her specific skills and specialized knowledge are applicable to a broad scope of problems, enabling her to function in numerous roles appropriate to developmental issues.

The nurse can perform an independent assessment of the child's health and developmental status. She is prepared to evaluate environmental factors in the home, school, and community setting. She can function as a case manager in assisting the family and the team to set program priorities, and is able to work with multiple agencies

in coordinating and expediting services. She has particular ability in advising other professionals in aspects of home management.

The nature of the situation dictates the methods and services chosen to obtain information for a nursing assessment. The repertoire of assessment methods includes the interview, observation, physical examination, interaction with the child and parent, and the use of a variety of standardized nursing assessment tools.

THE INTERVIEW

An in-depth interview with the parents provides information about the child's health, the developmental level of the child, the family's functioning, and the nature of his environment and his interactions within it. It should include a detailed review of the birth and health history, including dental care and immunizations. Attention should be given to interventive services received by the child and family.

The goal in assessing family functioning is to identify strengths for coping and managing, preferences, and needs. A family profile includes the age, ordinal position, health history, and education and occupation of the mother, father, siblings, and other persons living in the household. Information regarding the religious orientation, cultural background, economic status, interaction with the community, and daily living patterns is also relevant. The family organization and the goals and values of its members are identified.

It is important to become aware of events that are happening in the family, for these have an impact on the behavior of family members. Questions are designed to gain information about child rearing practices and expectations of the child's performance. Patterns of dealing with problems, approaches that have been tried, and their successes and failures are explored. Through a guided discussion of topics, data are obtained about the child's behavior in the activities of daily living— sleep patterns, abilities in performing self-help skills (feeding, toileting, dressing, tasks of hygiene), and household responsibilities. Information about the personality or temperament of the child is essential in evaluating the management of his daily care. A review of the routine of a day's activities often enables the parent to recall both pleasurable and stressful situations.

ASSESSMENT OF FAMILY INTERACTION

Another focus of the assessment involves interactional elements within the family, with attention primarily centered on evaluation of the parent-child relations. These assessments are often made by observing family members in informal settings or performing specific care tasks, such as the mealtime situation. This may be done informally or more precisely by use of a tool such as the Nursing Child Assessment of Feeding.

OBSERVATION

The nurse's observation and interaction with the child provides opportunity for evaluation of health and functional levels. In addition, observation affords a chance to study the issues identified by the parent during the interview. A physical examination may include appraisal of general appearance, posture, and body movements. Height, weight, and head circumference are measured and graphed and vital signs are recorded. These measurements are taken periodically to denote rates of growth and changes. A complete physical examination may be performed by nurses trained in such assessment.

DEVELOPMENTAL ASSESSMENT

In evaluating development and functional levels, the nurse relies on direct observation of behavior or activities. She often watches the child eating, toileting, dressing, undressing, or carrying out tasks using fine or gross motor skills. She may engage the child in a play session to learn about his interactions and ways of handling situations. She looks at the manner in which the child approaches the task, his social behavior, his use of his body, his problem solving ability, and his attention and perseverance at accomplishing a task.

Assessment tools are utilized to observe behavior patterns and functional skills in an objective, systematic manner (see Table 49–1). Most instruments arrange items in an orderly way, from simple to more complex tasks.

HOME VISIT

Assessment carried out in the home setting provides the advantage of observation of the child and family in familiar surroundings. It assists in understanding the concerns engendered by the parent's or child's frame of reference. Data are gathered in regard to the general neighborhood of the home, and the adequacy and provision of the home setting for the child's comfort, safety, and needed activities. It also provides opportunity for appraisal of family interactions and the environment as a setting for nurturance of the child's learning and development. A standardized tool, The Caldwell Inventory of Home Stimulation enables the nurse to sample certain aspects of the quantity and quality of social, emotional, and cognitive support available to the young child within the home. Comparison of behavior patterns at home and in the clinic shows how situational factors affect performance. Often the nurse is the only team member who has seen the entire family as well as the home setting. In some instances assessment may include the school or daycare environment of the child. This knowledge is necessary for realistic planning of intervention strategies.

Table 49-1. ASSESSMENT TOOLS FREQUENTLY UTILIZED BY NURSES

Instrument	Ages	Description	Where to Obtain
Denver Developmental Screening Test	1 month–6 years	Concise, easy to administer, systematic approach to assessing preschool child; tests areas of personal-social skills, fine-motor adaptive skills, language skills, and gross-motor skills	LADOCA, Project & Publishing Foundation, Inc., E. 51st Avenue and Lincoln Street, Denver, Colorado 80216
Washington Guide to Promoting Development in the Young Child	1–52 months	Assists in systematically observing the child in areas of feeding, sleep, play, language, discipline, toileting, dressing; expected tasks are presented, with activities suggested to provide for enhancement of growth and development	In Teaching Children with Developmental Problems: A Family Care Approach, by K. Barnard and M. Erickson. St. Louis, The C. V. Mosby Co., 1976
Developmental Profile (Alpern-Boll)	Birth to preadolescence	Screens a child's development in physical, self-help, social, academic, and communication areas; a questionnaire administered to parent or teacher	Psychological Development Publications, 7150 Lakeside Drive, Indianapolis, Indiana 46273
Neonatal Behavioral Assessment Scale (Brazelton)	First month	Behavioral assessment scale and psychological scale for the newborn infant; valuable in the detection of abnormality as well as developing insight about infant's repertoire of behavior	T. Berry Brazelton: The National Behavioral Assessment Scale. London, William Heinemann Ltd. 1973 (Philadelphia, J. B. Lippincott Co., 1972).
Nursing Child Assessment Feeding Scales	Birth to 1 year	Assesses parent-child interactions during feeding; subscales address parent characteristics and infant characteristics; requires observation of the entire feeding	NCAST, T-436 Health Sciences Building, SC-74, University of Washington, Seattle, Washington 98195
Nursing Child Assessment Teaching Scales	Birth to 3 years	Assesses parent-child interactions during teaching; subscales address parent characteristics and infant characteristics	NCAST, T-436 Health Sciences Building, SC-74, University of Washington, Seattle, Washington 98195
Caldwell Home Inventory	1. Birth to 3 years 2. 3–6 years	Home observation for measurement of environment in areas of social, emotional, and cognitive supports	Center for Early Development and Education, University of Arkansas, 814 Sherman, Little Rock, Arkansas 72202

ASSESSMENT IN SPECIAL AREAS

Nurses working in maternity settings provide immediate assessment of infant physiology at birth by observation and use of the Apgar system for scoring heart rate, respiratory effort, muscle tone, and reflex responses. Infant assessment continues as a nurse cares directly for the baby and assists or instructs the mother and father. She is a critical observer of infant appearance, posture, tone, body movement, and temperament. Measurement of specific factors in infant behavior and other attributes significant to the early detection of potential difficulties is approached through use of the Brazelton Neonatal Behavioral Assessment Scale. Throughout infancy she appraises the quality and development of the parent-child relationship.

In primary pediatric care, nurses continue periodic assessment of child health and development, as well as parental child care practices. The Denver Developmental Screening Test and the Washington Guide to Promoting Development in the Young Child are useful for this purpose.

Preschool screening programs involve nurses in evaluation of child health and developmental status regarding readiness for school entry, as well as screening for deficits in vision, hearing, and other areas of function. Because immunization against infectious diseases is a major preschool concern, the nurse is essential in helping the family to achieve compliance.

School health nurses provide health services, screening programs, and health education for the student population. As health professionals they are equipped to offer valuable consultation to educators in regard to special needs of students with chronic health problems and disabilities. The capabilities of the nurse are also utilized in curriculum areas of human genetics, sexuality, and preparation for parenthood.

Nursing assessment of the child and family is seen as part of a dynamic process in which information is obtained and integrated into a plan for care throughout the child's development. Its focus is on wellness, individual strengths, and promotion of the best circumstances possible for growth and development of the child within the family.

EUNICE SHISHMANIAN
CHRISTINA BROWNE

REFERENCES

Haynes, U.: A Developmental Approach to Case Findings. DHEW Publication No. (HSA) 79–5210. Washington, D.C., Department of Health, Education and Welfare.

Krajicek, M., and Tearney, A.: Detection of Developmental Problems in Children. Baltimore, University Park Press, 1979.

Powell, M. L.: Assessment and Management of Developmental Changes and Problems in Children. St. Louis, The C. V. Mosby Co., 1981.

50

Standardized Psychological Testing

Standardized psychological tests are frequently used to measure aspects of human behavior. In recent years these measures have been characterized as either a boon or a bane to mankind (Hernnstein, 1973; Kamin, 1974). This dispute has arisen in part because standardized psychological and educational tests deal with differences between individuals, always an area of controversy, and are not like measures of the physical world:

A standardized test is a task or set of tasks given under standardized conditions and designed to assess some aspects of a person's knowledge, skill, or personality. A test provides a scale of measurement for consistent individual differences regarding some psychological concept and serves to line up people according to that concept (Green, 1981).

Standard conditions must be observed if comparison with others is desired or performance changes over time are to be described. The more staunchly the procedures are adhered to, the more comparable the results obtained by different examiners. Test norms cannot be appropriately used unless the testing procedures and conditions are clearly stated and faithfully followed.

Lack of agreement about what tests measure is at the heart of the controversy. Measuring behavior or concepts theoretically underlying behavior is fraught with more difficulty than applying a yardstick to the features of a more concrete nature. In developing a psychological test there are no ultimate measures of the behavior or concept to calibrate against. The Bureau of Standards has no criterion against which to compare, for example, a new intelligence test. Intelligence is an artificial construct that is used to aid in understanding observable behavior, particularly behavior in school. What intelligence tests measure continues to be debated by social scientists, legislators, and now members of the judiciary. The changed social context of testing appears to have promoted such widespread concern.

Of greater importance than the current rankling is the usefulness of standardized tests. Tests must be of value in determining a child's present level of functioning and capacity so that his development can be monitored as he prepares to take his place in this world. Do better instruments or techniques than tests exist? None have been brought forward. The focus has often been on the negative aspects of testing. The issue remains, How well do the current tests help us to understand and predict behavior? The chapter will attempt to provide some answers to the question.

The discussion is intended as an introduction to standardized psychological and educational tests. Those who wish a more extensive treatment of assessment issues are referred to other sources. A good summary of current assessment issues can be found in a special issue of the *American Psychologist* (October 1981). For a comprehensive discussion of assessment, the texts by Cronbach (1970) and Anastasi (1982) are suggested. A text by Sattler (1982) provides a thorough review of the assessment of children's intelligence and special abilities.

TYPES OF TESTS AND ISSUES

A wide variety of instruments exist to measure the behavior of children. Although both group and individually administered measures are widely used, this chapter favors individually administered instruments. Individualized testing is often required for children because of the reading and compliance skills necessary for performing most group measures. Individual testing allows direct, intensive, and complete observation of a child's test behavior. The expense is greater with individual testing but is necessary when the quality of data collection in a child is critical.

Meier (1976) has developed a three stage model for the screening of children that covers primary screening through intervention. At stage I, brief tests and interviews can be used by paraprofessionals for a rough sorting of children who may have or be prone to acquire developmental problems. Group measures are often used at this level. Stage II involves a professional or team of professionals who make individual assessments of the child to decide whether the child has significant problems. Individually administered standardized tests provide much of the data. Children found to have significant problems are referred to appropriate programs and follow-up (stage III).

The definition of a "significant problem" depends upon the criteria of the professionals doing the assessment. Mass screening has been done in the name of early intervention or to demonstrate the need for establishing intervention services for children. Children who appear to be "at risk" for developmental problems are referred to appropriate educational and treatment programs. Identification of problems depends upon whether programs and treatment exist or whether a need must be demonstrated to establish services: economic issues often are a major consideration. What becomes of the costly and elaborate findings of a testing program if there are no ways to develop treatment or to refer the child to appropriate programs? Assessment procedures should focus on data that will lead to practical help for the family and child.

NORM REFERENCED AND CRITERION REFERENCED TESTS

The major types of standardized tests are norm or criterion referenced measures. Norm referenced measures are the tests most frequently encountered. The scores on such tests are interpreted by use of the test norms, i.e., the performance of a specified population of persons on the same test. The criterion referenced test, rather than using the scores of other individuals, employs as its standard a specific content domain. Minimum competency tests are such measures. The child taking a minimum competency test must meet or exceed a certain arbitrary standard in various subject areas in order to pass the test. The child's performance may not necessarily be compared to that of others. Some tests, however, can be both norm and criterion referenced, such as certain educational achievement tests.

The scores on norm referenced tests have no significance in themselves but derive their meaning from the performance of others. The ascertainment of a child's developmental milestones has been a useful norm referenced pediatric technique for determining a child's approximate developmental maturity. The acquisition of milestones, however, may vary across different minority (ethnic and cultural) populations of children within our society, and this must be taken into account in interpreting a specific child's development. The choice of the norms is of critical importance.

The criterion referenced test contrasts the person with the test content. This approach is often used in educational achievement testing to determine a child's level of sophistication and to define problem areas in a subject area such as mathematics. Computer generated teaching programs also use it, with the regular feedback on perform-

ance inherent in this approach serving as a source of continued motivation. The developmental realm also can be surveyed with a criterion referenced approach. Infant sensorimotor development can be described by using a criterion referenced measure like the Ordinal Scales of Infant Development (Uzgiris and Hunt, 1975). The test was developed using Piaget's stage theory of sensorimotor development. The test results depict the stage or stages of development of an infant and make possible an understanding and explanation of his present behavior and help predict the behavior patterns he will acquire next.

Another frequent use of criterion referenced techniques is in "behavioral assessment." Professionals with a strong background in learning theory often see problem behavior not as a sign of other underlying problems but as the problem itself. They are less concerned with the etiology of the behavior or how it compares with the problems of others. The focus of study is when, where, and under what circumstances the behavior occurs. The interest in assessment is in establishing a baseline for judging the problematic behavior patterns and the frequency with which they occur, so that the future frequencies of the behavior can be contrasted to discover whether a prescribed treatment is effective.

MEASURING DEVELOPMENT

Developmental assessment is a means of ascertaining the functional maturity of the child's central nervous system by assessing his behavior. In measuring the development of children, the assumption is made that behavior develops in regular patterns and is therefore predictable. Children, however, do not develop at the same rate. Normal differences in learning rates and styles have caused some confusion, because ascertainment of developmental age is not necessarily a direct measure of intellectual potential or disability. Children who are greatly delayed in developmental age usually have significantly lower scores on later tests of intelligence and achievement, whereas children who are advanced in development may not eventually prove to have above average intellectual skills. By the age of one year, moderate or greater mental retardation can be detected, whereas mild mental retardation is often discovered by two years of age and confirmed by three years of age. (See also Chapter 45.)

The Gesell Scales and Denver Developmental Screening Tests are frequently used measures of development. They essentially are standard developmental series of observations similar to those routinely made by a pediatrician. These schedules are not numerically scored but are interpreted more impressionistically with a view to determining the maturity of the child. They would seem best used as criterion referenced measures in describing the present abilities of a child or as a rough screening technique for detecting developmental delay rather than as norm referenced measures in making judgments as to the significance of specific developmental delays.

TEST ADMINISTRATION

That only qualified professionals should select, administer, score, and interpret standardized tests is evident. Distributors of tests restrict sales to qualified professionals much as medication sales are controlled. Testing is not done casually; important reasons for testing must exist. If used by people who are less than adequately trained, tests can be used to do great harm. Examples of the misuse of tests abound. The assessment of children with significant motor or sensory handicaps and children from minority populations in our society requires additional special training and experience.

The administration of a test must follow standard procedures in most situations. The procedures of many tests are deceptively simple, and they are easily subject to misapplication. Extensive professional education, training, and supervision are necessary prerequisites to appropriate use of standardized tests.

RAPPORT

Rapport with the child being tested is important in order to be able to obtain valid and replicable results. The aim of the assessor is to secure the confidence and cooperation of the child so that he will participate as fully as possible in the procedure. If the child is not motivated to perform well on intellectual and achievement measures, the results will not reflect what the child really can do. The examiner must be able to relate well to the child, not under- or over-rating the child because of the child's personal characteristics.

Tests are not administered in an automatic rote manner; the process must be attuned to the child without deviating from the test procedures. It is a time intensive endeavor which requires that the examiner have a thorough knowledge of test materials and procedures as well as wide experience with the manifold problems that occur during assessment. Most of all, the competent examiner must have developed skills in observing, recording, and interpreting behavior, as well as being able to modify his own behavior to appropriately promote the performance of the child being tested.

In scoring and interpreting the test results, the characteristics of the examiner as well as the individual characteristics of the child often need to be taken into account.

COACHING AND PRACTICE

The effects of coaching and practice on test scores constitute another controversial issue. Practice on specific items of an intelligence or achievement test can increase scores, but practice on similar but not identical items of a test produces its greatest impact on scores when the test is poorly constructed and does not affect scores on well constructed tests. Individuals with deficient educational backgrounds are more likely to benefit from coaching prior to taking achievement tests if that coaching is in the form of tutoring to improve deficient skills. The score is then changed by virtue of improved skills, not special test tricks.

Long term programs to improve the general intellectual skills, and thus test scores, of handicapped and educationally disadvantaged young children have demonstrated a degree of success. Intelligence test scores improve, but not usually in a dramatic fashion unless the program also makes extensive efforts to change the child's circumstances as well as stimulating the child. The characteristic of these improvements in test scores of being stable over time has not yet been well studied, and the features of a program that contribute to score gains are also not clear.

SOCIAL AND ETHICAL ISSUES

The test results obtained by children are considered private and confidential data that are available to others only with parental permission (Buckley Amendment). The parents, however, have direct access to whatever is recorded in their child's record, and at the age of 18 years the young adult has the same access to his records that the parents had.

The use of tests by schools to place large numbers of minority children in special education programs has led to court battles. The norms used with many current tests rate the performance of many minority children below those of other children. Efforts to produce tests that are less discriminatory have been forthcoming. One of these tests is the System of Multicultural Pluralistic Assessment, a test for black, white, and Hispanic children aged five to 12 years. This test is a battery of scales that purports to measure cognitive, perceptual-motor, and adaptive behavior by including medical and social data and by using multiple norms to estimate a child's learning potential.

However, it was standardized only on children in California, and no data have been presented to show whether its use will lead to educational decisions that are not racially or culturally discriminatory.

In the past a number of criterion referenced as well as norm referenced tests have been described as being culture-free, i.e., with cultural experiences not playing a part in test performance. There do not appear to be any truly culture-free tests, however, because culture appears to be invariably intertwined with what and how quickly we learn. This is particularly true of intelligence, as Scarr (1978) has noted: "Intelligence tests are not tests of intelligence in some abstract, culture-free way. They are measures of the ability to function intellectually by virtue of knowledge and skills in the culture . . . which they sample."

There is no simple solution to assessing minority children except for a greater concern with and sensitivity to individual differences among members of minority groups and an awareness of the shortcomings and hazards of using standardized tests, so that assessment results truly can be used to help children rather than harm them.

COMMUNICATING TEST RESULTS

The efforts that have gone into the testing of a child can be seriously diluted if the evaluation results cannot be effectively communicated to the child's parents, to the child, and to agencies that may provide assistance. Informing parents that the assessment indicates that their child has a significant problem can be a most difficult task. Not only must the test results be clearly and carefully explained, but the parent's expectations and characteristics must be taken into account. Reporting test results not only is often an information giving session, but can be one of counseling as well, with therapeutic issues coming to the fore. Written communications of test results must also be clearly written with specific recommendations so that both professionals and parents can understand them. A more exhaustive review of issues in communicating assessment results to parents is presented by the author in a test describing the assessment of infants and young children (Schnell, 1982).

ASSESSMENT CONCEPTS

Different standardized tests have different psychometric characteristics making them better for some situations than for others. Major psychometric characteristics to be taken into account are the reliability, validity, and method of test score interpretation. Adequate reliability and validity are essential for a test to produce useful scores, which in turn depend upon norms or appropriate content for particular interpretation and use.

RELIABILITY

A test must be consistent if useful results are to be obtained. The psychometric properties of reliability are usually measured by the stability of the test scores over time as well as the internal consistency of the items of the test. The most frequently used statistical procedures that estimate reliability are correlational techniques. Correlation coefficients thus indicate the degree of reliability; coefficients of .80 or higher are usually needed in order for most intelligence and achievement tests to be considered reliably stable.

Stability over time is a form of reliability that refers to the capacity of a test to generate the same or similar results on two different occasions. If the test is measuring something that is expected to change over time, as on achievement measures, the intervals between test administrations to assess stability should be rather short. If the test measures something, such as intelligence, that is expected to be relatively stable over time, longer test intervals are used. This test-retest reliability is reduced by error resulting from factors that promote change over time, such as variations in the administration of the test on the two different occasions or nonsystematic variations in the behavior of the child taking the test due to changes in health or motivation.

The internal consistency of a test reflects the consistency of response to the items throughout the test during one administration of the test. The usual method of measuring internal consistency is to split the test in half by separating the odd and even numbered items and correlating responses to them. Generally the longer the test, the more internally consistent it will be. Timed tests, however, do not lend themselves well to split-half analysis.

No tests can be perfectly reliable, for error always plays a role. A measure of how error may have influenced the specific score of an individual is provided by the standard error of measurement. This is the standard deviation of the distribution of error scores around the true score, and can be computed by use of the stability coefficient and the standard deviation of the test. As the standard error of measurement grows larger, the uncertainty about the score increases. To obtain a more useful estimate of an individual's performance, a confidence interval can be set up around the obtained test score using the standard error of measurement. For example, if a child obtained an

IQ score of 110 on an intelligence test and the standard error of measurement of the test for a child of that age was 3, the child's score can be represented as 110 ± 3, or as a range of 107 to 113. The use of one standard error of measurement allows the statement that the chances are that the child's score 68 times out of 100 should fall within the 107 to 113 range. By using more than one standard error of measurement and thereby establishing a greater range, the level of confidence can be increased from 68 per cent to even higher levels. Many groups tests consistently report scores in ranges using the standard error of measurement. The results of individual tests if portrayed in ranges, rather than as single numbers, might lead to less misunderstanding.

VALIDITY

Test validity indicates what the test measures. The validity of a test never can be any better than its reliabilty. If test results are not stable and consistent, they cannot measure anything very well. Because tests are used for many different purposes, different estimates of validity are used. There are two major types of validity, one that concerns itself directly with the test makeup and another that relates the test results to other independently recorded criteria.

The kinds of items a test contains are referred to as its content validity. Content validity is not measured statistically but rather is determined by judging whether the test items are representative of the area being measured. It is especially useful in evaluating achievement tests. Another way of evaluating a test's makeup is through evaluating its construct validity. This kind of validity clarifies the way in which the scores can be explained psychologically, i.e., whether the test measures a psychological construct or trait. Construct validity can be determined by examining items to relate them to the appropriate psychological constructs. Statistical techniques also can be used, such as factor analysis of item content or correlation of the test results with those of another test that is a demonstrated measure of the trait or construct. Both the lack of relationship with certain tests as well as the strong relationship with others can help to explain what a test is measuring.

Criterion related validity examines the relationship between test scores and other measures or outcomes. These other criteria are independently measured. If the criterion measures are obtained at about the same time that the test of interest is administered, one has obtained what is described as concurrent validity. The results of the test are correlated with the results of recognized tests of the area of interest, just as might have been done

in establishing construct validity. If the test results are highly correlated, this may be taken as evidence that they are measures of the same thing.

A more stringent test of criterion related validity is to establish predictive validity, the correlation between scores on the test and performance after, an interval of time, on the basis of other criteria. Relating scores on an intelligence test to later performance in school is an example of this strategy. Because of the difficulty in carrying out the longer term research that predictive validity requires, many test manuals do not report predictive validity measures but rely on measures of concurrent validity.

SCORES AND NORMS

Raw scores from a test rely for their interpretation on comparison with something. A raw score by itself is meaningless. Norms are the standard comparison on norm referenced tests. They are developed by establishing how a representative group of individuals perform on the test. For norms to be appropriately used, they must be representative of children who are similar to the child taking the test. If the norm group is not appropriate, the score from such a comparison will provide misleading information. If a child lives in a rural area and only urban children are represented in the norms, the norms will be of less help in telling us how that child compares to other rural children. The norms would match the child only against city children. If no more appropriate norms were available, the test might not be worth administering.

Two major kinds of norms are utilized in interpreting test scores—national and local norms. National norms are the ones most frequently available because of their general usefulness. They have usually been developed from a national sample of children who are selected on the basis of how well they represent various segments of the national population of children as derived from census data. As our population changes, norms need to be continually updated in order to be representative. One of the major criticisms of national norms for specific tests is that instead of being representative, they are biased to favor or discredit one group or another.

For many purposes national norms are not an appropriate comparison group. If one wants to know how well a child may perform in a given school system, comparison with other children in the system (local norms), who may not be representative of the national population, would be most relevant. Similarly, if one wanted to know how well a profoundly deaf child might perform in a program for the deaf, comparison with similarly hearing impaired children would be more useful than comparison with a normal population of children. These local or more specific norms are needed to facilitate better test interpretation.

To improve interpretation, raw scores can be converted to a number of different derived scores, such as age equivalents, grade equivalents, percentiles, and standard scores. The normal course of development is frequently used as a comparison. The six year old who receives the same raw score on an intelligence test as a typical 11 year old is said to have a mental age of 11. Similarly, this first grade child may earn a score at the fourth grade level on reading skills. These scores do not mean that this six year old is like a normal 11 year old or a fourth grader, but only that his raw score is similar. The way the score was obtained and the performance pattern on the tests may be very different from that of an 11 year old or a fourth grader. Because a mental age of 11 was obtained does not mean that the six year old is in any other way like an 11 year old. The score has to be carefully and specifically interpreted.

Percentiles have been a popular way to evaluate a score. The score is related to a point in a distribution of scores at or below which a certain percentage of children fall. When 85 per cent of the children fall below a specific score, the score is at the 85th percentile. Other percentile techniques involve dividing the percentile distribution in parts: 4 for quartiles or 10 for deciles. Although percentiles appear to be easy to interpret, they do present certain pitfalls. The percentile is essentially a ranking procedure, and the differences from one percentile to the next at different places in the distribution do not represent equal units. Differences in the middle of the distribution tend to be very small as compared with much larger differences toward the end of the distribution. Raw score differences between the 50th and 60th percentiles are smaller than those between the 85th and 95th percentiles. It would be easier for a child's score to move from the 50th to the 60th percentile than from the 85th to the 95th. Children at the 50th and 60th percentiles tend to be more similar in performance than other children at the 85th and 95th percentiles.

A common practice on many tests is to convert raw scores to standard scores. Standard scores describe how far from the mean of the distribution a score falls relative to the standard deviation and can have any mean and standard deviation. Stanines have a mean of 5 and a standard deviation of 2, whereas T scores have a mean of 50 and a standard deviation of 10. The deviation IQ is the most frequently used standard score on intelligence tests, with a mean of 100 and usually a standard deviation of 15 or 16 points, depending upon the test. When IQ's were first used, they were calculated as a ratio score comparing mental age (MA) with chronological age (CA; MA/CA × 100 = IQ). When a ratio IQ is used, the standard deviation differs with age, changing the interpretation; thus deviation IQ's (standard scores) have replaced them.

DEVELOPMENT AND EVALUATION OF TESTS

Tests are developed for a multitude of purposes. Often, however, a test designed for a specific purpose or group receives wider application in other situations and with other populations. The adequacy of these tests in other circumstances has to be carefully scrutinized; the reliability, validity, and norms, or content of the test for the purpose at hand must be carefully checked. The test manual is usually the first place to begin in evaluating a test.

The *Mental Measurement Yearbook* provides a comprehensive list and critique of individual standardized tests (Buros, 1978). *Tests and Measurements in Child Development: A Handbook* provides information about unpublished tests suitable for children from birth to 12 years of age (Johnson and Bommarito, 1971). The *Journal of Educational Measurement* and the *Journal of Counseling Psychology* present current research data and reviews of tests.

Many home grown tests are put together from items on well used and standardized tests. A potentially erroneous assumption is that by using items from other tests, the "new test" will assume all the properties of the well standardized tests. Norms may not be developed beyond using the age placement of the borrowed items, and reliability and validity data may be sparse. The standard procedures and conditions for test administration may be vague. These collections of items are usually assembled to meet a local need, but they may not have wider application unless extensive work is undertaken to determine their psychometric properties. The usefulness of such instruments, without adequate research data to support them, needs careful consideration.

USING TESTS

PLANNING AN EVALUATION

The most crucial step in planning the assessment of a young child is defining what should be assessed. This decision should be based in large part on the referral question, which unfortunately can be rather vague. A frequent referral question is, "Why is Johnny doing so poorly in preschool?"

This question must be made specific by a more detailed description of Johnny's behavior. The foregoing question gives no clue to whether Johnny is physically ill, mentally retarded, emotionally disturbed, learning disabled, sensorially handicapped, or motorically handicapped, whether he does not speak English, or whether he comes from a different cultural background. Another possibility may be that the standards of the preschool program are inappropriate. Starting an assessment with such a vague question would require a very expensive and extensive evaluation, calling upon the use of many tests and the efforts of numerous disciplines. To make the child's assessment most effective, a detailed description of the types of behavior of concern and the circumstances in which they occur is of great help. The information is gathered in several ways. The referral agent is usually asked to provide further information. The gathering of data from parent and teacher questionnaires as well as information from interviews is a further step. If data from these sources do not significantly delineate the question, observation of the child in the problematic situation is often called for.

After clarification of the referral question, the next issue involves a determination of the resources that will be available to help the child after testing. If the assessment can lead to no positive action, there is often little need for it. For standardized tests to be of use, there should be a means to do something for the child and family. Alternatively there may be a purpose in compiling documentation that a need exists for certain resources and thus may help a number of children and families.

The nature of resources potentially available helps in determining the extent of testing. How much of the assessment information can be utilized to assist the child and his family? If the sophistication of the resource programs available is not equal to the detail of the test data, some of the data might serve to confuse rather than lead to positive action.

There is always the temptation to display diagnostic acumen by going further than necessary. The use of standardized tests is not an academic exercise, involving the squeezing of all possible data from the situation. Instead the test data should lead to practical help for the family and child. The test data should be used in the best interests of the child.

Standardized tests are chosen for the assessment that provides the most needed information. When a child has complex problems, often a battery of tests is needed to provide complete data. Common test batteries include at least one intelligence test, achievement measures, a test of visual motor skills, and one or more tests of personality. A combination of standardized tests and informal procedures is often used. The contrasting data from the tests then can lead to a differential diagnosis and appropriate recommendations.

ADAPTING TESTS

In planning an evaluation, testing procedures occasionally must be modified to fit the situation and the child, e.g., in extreme situations, as in testing a child with a motor and a language handicap or a child who is autistic. If the test items or directions are changed because a different mode of presentation or response is required, the norms of the test can no longer be used. Observations of the child's behavior may be more relevant to understanding the child than administering the test according to the directions and watching the child not respond. Observations outside the standardized testing situation can also provide information about a child's strengths and weaknesses, particularly about how the child has adapted to his environment and the impact of his behavior on those around him.

Some changes in standard test procedures are permissible. For example, if the order of presentation of items on the Stanford-Binet Intelligence Scale is changed, minority group children may receive higher scores while scores of other children remain unaffected. When more difficult items are paired with easier items, the motivation of the minority child to succeed is increased and performance is improved. Whenever there is doubt about the reliability or validity of test results, they are interpreted in a cautious manner in the context of other behavioral and background data.

Minority group children present difficult assessment issues. "Minority" can refer not only to children who come from heterogeneous racial, ethnic, and socioeconomic backgrounds but also to children who have significant disabilities that interact with their behavior. There is no clear direction to take in using standardized tests to assess minority group children. In using tests with such children, the examiner must be familiar with the cultural patterns of families and with the ways in which major motor and sensory deficits influence learning and behavior. When the professional does not have specialized experiences and training, referral to another who does is required. The ability of the examiner to establish good rapport with a child can transcend racial and ethnic issues.

When a child and his family speak a language other than English, the assessment should be conducted in the language that is most often used with the child. To be maximally useful, the testing ought to be done by someone who not only speaks the language but also has knowledge of the cultural subgroup. It should be noted that translating a test from English into another language also directly affects the use of the norms of the test unless those norms have been developed from such a translation.

TEST PERFORMANCE

The child's behavior is never "wrong," and reliance solely on test scores from individually administered tests can be insufficient. Systematic observation of how and when a child responds and copes with testing procedures can be used to understand how the child functions. The difficulty encountered in attempting to engage the child on standardized test items yields important data, such as information about antecedents and consequences that are maintaining the child's behavior. Observation of the "untestable" child generally reveals severe learning or behavior problems, as well as coping strategies that are potential strengths. The child who is difficult to engage can be assessed, although standardized tests may have to be abandoned.

Reliance on tests alone in assessing children may obscure other information needed for planning relevant programing. Behavior that is elicited by assessment procedures other than standard tests is often important. The language disabled child may appear "hyperactive" when tasks requiring expressive speech are administered, but the same child may attend well on nonverbal performance items. Behavior is best understood in the context in which it occurs. When only behavior patterns that are intentionally elicited are considered, valuable information is lost. Observations from caregivers and teachers also can be used to put test performance in perspective.

LABELING

One of the most frequent by-products of testing is a label or classification of the child. Unfortunately labels are fraught with hazards for the child. As Hobbs (1975) indicates in his excellent book on the labeling of children:

Classification can profoundly affect what happens to a child. It can open doors to service and experiences the child needs in order to grow in competence, to become a person sure of his worth and appreciative of the worth of others, to live with zest and to know joy. On the other hand, classification, and the consequences that ensue, can blight the life of a child, reducing opportunity, diminishing his competence and self-esteem, alienating him from others, nurturing a meanness of spirit, and making him less a person than he would become.

Although labeling often can lead to negative consequences, it is essential in a number of situations. Labeling may enable services to be obtained

for children, allow for the planning and organizing of programs for them, and assist in describing the outcomes of intervention efforts.

Whether a child is described as normal or deviant is culturally determined by a particular society at a particular time and for a particular purpose. The behavior of children who are described as being mentally retarded or emotionally disturbed or as having physical disabilities can be placed on a continuum, from very deviant to within normal expectations, depending upon the predominant views of the society. In our society a child is thought to be deviant if he is not trying to surpass his peers. In a culture like that of the Zuni Indians of the Southwest, children are taught not to excel at the expense of their peers. A Zuni child who consistently betters his peers on a test would be labeled as deviant in his cultural group.

Test results often lead to deficits being described by labels. A more fruitful approach is to describe a child's needs rather than coding deficits. This approach leads to developing programs to meet needs, as is implicit in Public Law 94–142. (See Chapter 61B.)

REVIEW OF TESTS

CLASSIFICATION OF TESTS

Standardized tests may be classified according to many different criteria. Cronbach (1970) divides tests into two groups. One group includes tests used to assess how well a child can perform at his best, to earn the best score he can. Two major subgroups are tests of intelligence and achievement tests. Intelligence tests attempt to measure general abilities that are important in thinking and problem solving, whereas achievement tests measure specific accomplishments, as in the outcome of school instruction. The second group of tests measures typical rather than best performance. The tests in this group are measures of personality and interest. Tests of typical behavior provide the best data to characterize personality.

INTELLECTUAL ASSESSMENT

Intelligence tests sample only part of the wide domain of abilities most laymen label as intelligence. Intelligence tests usually emphasize language development, problem solving, and abstract abilities at the exclusion of many specific abilities, motivation, and creativity. The results of intelligence tests are not all encompassing. With all their faults, intelligence tests results continue to be critical in understanding and predicting behavior, particularly school performance. They have assisted in describing children who function outside the average range of intelligence, those who are depicted as being mentally retarded or intellectually gifted. The patterns of strengths and weaknesses displayed on intelligence tests also have been used to describe learning disabilities. (See also Chapter 46.)

Intelligence tests are used for various age levels. Those used for infants have many limitations. Their predictive powers for normal children are not great, but they have proven to be relatively good discriminators of children with significant developmental delays. Inconsistencies in the behavior of infants and preschool children can easily compromise test reliability, making it essential to interpret tests results cautiously. Although acceptable levels of reliability are more difficult to achieve with younger children, there are reliable tests available to assess handicapped infants and preschoolers.

The validity of tests used with young handicapped children must be carefully considered. Research demonstrates that the younger the disabled child, the less predictive intelligence test scores will be of later school age test scores and academic performance. Many of the indicators of later specific learning difficulties and behavior problems are not easily measurable with current techniques before the age of six years. With normal children, scores on infant tests do not correlate with later measures before 18 to 24 months of age. As a child's age increases, the correlations of his scores with later scores increases, so that by four years of age there is a very solid relationship.

The infant test is a good measure of present developmental status, and the correlations between scores that show definite developmental delays and later scores are significant. As a general rule, the more severe a child's handicap, the more predictive tests are of later learning. Infant tests have predictive validity when a child scores 2 or 3 standard deviations below the mean. Children with severe motor, sensory, and cognitive deficits can be identified. Although the tests are more predictive of future problems when the child is severely handicapped, assessment of such a child is difficult and requires specialized training.

Bloom (1964), in his review of longitudinal studies of intelligence, has concluded that intelligence at age 17 can be accounted for by the following developmental pattern: 20 per cent is developed by age one year, 50 per cent by age four, 80 per cent by age eight, and 92 per cent by age 13.

Although correlations become increasingly greater with age, there are IQ fluctuations for individual children. Some of these fluctuations reflect the fact that the age of acquiring various abilities sampled by the tests can be different for different children; changes as great as 50 IQ points

have been reported over time. The effects of extreme environments can change the IQ score by at least 20 points. No decisions should be made about a child on the basis of a single test administered during the first years of life. "A single early measure of general intelligence cannot be the basis for a long term decision about an individual. When the results of several test administrations are combined, the reliability of these scores is enhanced and their predictability increased" (Bloom, 1964).

The many limitations of the IQ score need to be taken in account in interpreting it. The IQ score is a gross summary score of intellectual ability. As a score it is not perfectly stable over time; some children's scores may change significantly. Patterns of success and failure in children receiving the same score can also vary, indicating different intellectual strengths and limitations, despite the same overall score. The IQ is not a pure measure of intellectual ability, because no matter what intelligence test is used, only certain aspects of intelligence are sampled. In attempting to depict potential, past experience is reflected in the IQ as well. Some useful purposes are served, however, because IQ scores can assist in understanding the present behavior of a child and in predicting his future behavior, particularly school performance. In making decisions about children, comprehensive information along with the IQ, when appropriate, should be used.

At any age, intellectual handicaps are suggested when the IQ score is more than 1 standard deviation below the mean of the test. The nature of the handicap cannot be described by the IQ alone. The pattern of skills and limitations, educational achievements, motivation, and resources available must be taken into account.

Most importantly, an intellectual handicap cannot be adequately described without taking adaptive behavior into consideration. Adaptive behavior refers to how individuals operate at daily living tasks and how they meet the expectations of their environment. Children who eventually are designated as mentally retarded are as individually distinctive and statistically variable as children whose intelligence is within the average range.

Some changes in IQ have been demonstrated to be related to different personality patterns of children and to differences in environmental factors. Gains in IQ have been found in children who are self-assured, competitive, and not easily frustrated, while losses in IQ have been demonstrated with dependent, passive, and easily frustrated children. Environmental factors like socioeconomic status and constancy of emotional adjustment have also been linked to IQ changes. Suffice to say that scores must be considered in the context of all the available information about the child and family.

INFANT SCALES

Infant scales are often considered to be tests of development rather than intelligence, because they rely so heavily on sensory and motor development with fewer language items and because in normal populations they correlate poorly with later scores on intelligence tests. They are useful in the identification of developmentally delayed infants in the birth to two to two and one-half year range. The level of development of the child is depicted either in normative terms or with regard to stages of development. Test items require adequate sensory and motor development as well as appropriate cognitive development. The presence of motor or sensory handicaps significantly decreases performance on these scales, particularly at the earlier age levels. In the later stages of the tests, at or above the 18 month level, expressive language items begin to play a more important role.

Toys of high interest value, such as bells, formboards, brightly colored boxes, and blocks, are employed as test items on infant scales. The examiner presents to the infant the toy to be played with in order to induce the infant to imitate an action with the toy or to follow the examiner's direction in playing with the toy. Activities with the same blocks can be scored across a range, from the first few months of life through two and one-half years of age. Infant scales create structured play situations, the examiner having to move the infant from one situation to another while keeping him alert and interested.

Individual Infant Scales

Bayley Scales of Infant Development (Psychological Corp., 1969).   The Bayley Scales are currently the most popular of all the infant tests. Nancy Bayley and her coworkers spent almost 30 years in developing and standardizing them. The test consists of three major scales designed to evaluate mental skills (Mental Scale), motor skills (Psychomotor Scale), and socioemotional maturity (Behavioral Record) in babies from birth to 30 months of age. Each of the scales is scored separately, developmental indices being obtained from the norms for the Mental and Psychomotor Scales. The Behavioral Record is a series of ratings of different behavior patterns and can be treated as a criterion referenced scale. Many of the items are adopted from earlier tests, such as the Gesell Developmental Schedules and the Cattell Infant Intelligence Scale. None of the scales correlates highly with later measures of intelligence for a normal population, but they do have utility in identifying developmentally delayed infants.

Ordinal Scales of Infant Psychological Development (1972).   These scales are not sold as a formal test but rather are presented in a book by Ina Uzgiris and J. McV. Hunt, *Assessment in Infancy: Ordinal Scales of Psychological Development.* The Ordinal Scales of Infant Development are a criterion referenced test based on Piaget's stage theory of cognitive development that assesses infants during the sensorimotor period of development, from birth through a mental age of 24 months. The six subscales assess the infant's progress in becoming aware of his environment (sensory) and his subsequent actions (motor). Two of the most useful subscales are the Object Permanence Scale (pursuing an object when it is not visable) and the Understanding of Means-End Relationships Scale (learning that various activities produce specific results). The major advantage of the scales is that they permit one to focus on what the child is able to do with an explanation of why the child is doing it rather than concentrating upon deficits. The test data are helpful to parents in interpreting their child's behavior and are also useful in planning appropriate play activities for the child. Other sensorimotor scales such as the Casati-Lezine and the Decarie Scales, based on Piaget's theory, are also broadly used.

PRESCHOOL INTELLIGENCE TESTS

Preschool intelligence tests are used in the age range of about two and one-half through six years of age. They are frequently used to describe children with developmental delays and handicaps who may have difficulty when they enter regular public school programs. Preschool intelligence tests examine language development and the integration of perceptual and motor skills. Difficulties occur when one attempts to distinguish between delays in development due to differences in experience and learning opportunities and those attributed to biological handicaps.

Preschool children can exhibit great variability in performance. Large differences often exist among them in how well socialized they are; e.g., some children easily separate from their parents and readily follow the directions of the examiner, whereas others may be negativistic and oppositional, having little experience with adults besides their parents. There are many differences in experience and thinking processes between the preschooler and the school age child. Preschool children are rather egocentric in their thinking, often lacking interest in whether the examiner understands what they are talking about and being unable to take the perspective of another. These characteristics often make the preschooler's normal behavior appear unusual when contrasted with that of older children.

Besides the major tests to be presented there are other tests that can be used in specific situations to better understand the intellectual functioning of the preschooler. Because language delay in the preschooler is the most frequent reason for referral for diagnostic evaluation, tests are needed that do not require verbal responses to assess intellectual development. There are several major nonverbal tests. The Merrill-Palmer Scale of Mental Tests is one such scale for preschool children. Although its standardization is poor, its performance items are very attractive to children, who may become interested in them even though they refuse to participate in more rigorously structured tests. The newer and more adequately standardized Extended Merrill-Palmer Scale, unlike the original, requires extensive verbal mediation and thus cannot serve the same purpose. Other nonverbal tests like the Leiter International Performance Scale and the Columbia Mental Maturity Scale are appropriate in the preschool range and into the school years, or beyond.

Another area of importance that aids in the interpretation of intelligence test results is the development of adaptive skills. The measurement of adaptive behavior concerns itself with social maturity and the level of self-help skills. These skills are essential for normal living regardless of the IQ. Information for the scales is not obtained directly from the child but from observation of the parents, caretakers, and teachers. The Vineland Social Maturity Scale is often used to study communication, locomotion, eating, dressing, general self-help, occupation, self-direction, and socialization skills from birth into adulthood. Adaptions of it, like the Preschool Attainment Record for only the infant and preschool period, are also available. The Adaptive Behavior Rating Scale is the most widely used measure of adaptive behavior. It is produced by the American Association on Mental Deficiency for use with mentally retarded people from infancy into adulthood. The scale has 23 subscales and measures both adaptive and maladaptive behavior.

Stanford-Binet Intelligence Scale (Houghton-Mifflin, 1972). The Stanford-Binet Scale is the oldest and most widely used test for preschool children. It begins at the two year level and continues with items up into adulthood. Although consisting heavily of verbal items to the exclusion of many performance items, it is a very reliable and valid intelligence test. Because of its reliance on verbal items it is not useful for children with speech and language deficits and may not serve well in characterizing children with learning disabilities. The only scores obtained from it are IQ and mental age scores. Several independent schemes exist to analyze item content in terms of a child's strengths and limitations, but the unequal distribution of items of various types throughout the test makes this a difficult task. Adaptions of

the Binet Scale exist for use with children with visual deficits. (Hayes-Binet and the Perkins-Binet Tests of Intelligence for the Blind)

McCarthy Scales of Children's Abilities (Psychological Corporation, 1978). The McCarthy is a well standardized measure of the mental and motor functioning of children two and one-half to eight and one-half years old. It was developed as a systematic assessment of the strengths and weaknesses within specific cognitive areas of development. As such, it is useful in assessing preschool children who have learning deficits but are not mentally retarded. Children with major learning disabilities sometimes may obtain overall scores on this test in the retarded range. The test consists of 18 separate subtests for which attractive toylike materials are provided. The 18 separate subtests are summarized by an overall general cognitive score and five major area scores: verbal, perceptual-performance, quantitative, memory, and motor. The test requires a great deal of clerical work to complete and can require a lengthy administration time. A shorter screening version has been developed, which consists of six of the 18 subtests. That version is more weighted than the original toward perceptual performance and motor items and less to verbal and quantitative items. No general score is obtained, but the six individual scores are used to establish a risk classification for learning problems.

Wechsler Primary and Preschool Test of Intelligence (Psychological Corporation, 1967). The Wechsler Primary and Preschool Test of Intelligence (WPPSI) was modeled after David Wechsler's widely used tests for older children and adults. It was developed as a preschool measure, the original intent being for it to extend downward below its present starting age of four years. The test, however, did not work well in children younger than five years, and the effort was abandoned. Thus it is presently used for children from age four to six and one-half years. The WPPSI has 11 subtests, similar to the other Wechsler Scales except for the replacement of three subtests to make them more attractive to preschoolers. The 11 subtests are divided into verbal and performance scales, with a full scale IQ representing the total test performance. The test format may not appeal to some young preschool children, and its truncated age range limits its usefulness. Although it too has a long administration time, it provides useful diagnostic information for the assessment of children who are learning disabled or mentally retarded.

SCHOOL AGE INTELLIGENCE TESTS

In contrast to preschool tests, the content of tests for school age children is strongly related to academic achievement. Much intelligence testing in schools is done with group rather than individual measures. Individual intelligence testing is done for the same diagnostic purposes as preschool assessment except that the results for school age children are often clearer. The Wechsler Intelligence Scale for Children-Revised is the dominant individually administered test. The Verbal Scales, with performance scale omitted, are often used in the evaluation of visually handicapped children. In addition to an individually administered intelligence test, other supplementary measures are often used in cognitive assessment. A test of visual motor skills such as the Bender Visual Motor Gestalt Test, which requires the reproduction of geometric forms, is often used. Other tests of particular skill areas or emotional functioning are frequently called for. Achievement test results can be contrasted with intelligence test results to determine the level at which a child is working in relation to his potential.

Individual School Age Intelligence Tests

The Wechsler Intelligence Scale for Children-Revised (Psychological Corporation, 1974).   The Wechsler Intelligence Scale for Children-Revised (WISC-R) is the major test used for school aged children, ages six to 16 years, 11 months. It is well standardized and has demonstrated excellent reliability and validity. The Wechsler Adult Intelligence Scale-Revised (WAIS-R), restandardized in 1981, covers the ages from 16 through adulthood. The WISC-R is divided into a verbal scale assessing language related skills and a performance scale assessing perceptual-motor skills. A full scale IQ score summarizes the total test performance. Each of the verbal and performance scales are made up of five regular and one optional subtest. Significant differences between scale scores or subscale scores have been used diagnostically for describing learning disabilities, emotional problems, and mental retardation.

ACHIEVEMENT TESTING

Achievement tests are generally group tests that are used more often by schools than by other agencies. They are used for the periodic evaluation of a child's general progress in various subject areas, the test content being closely related to the content of the instruction. Much informal achievement testing is done by individual teachers to assess the progress of their students in learning segments of curriculum material. Readiness tests have been devised for preschool children to determine whether learning problems exist that may interfere with learning in a regular school curriculum or to assess prerequisite skills for instruction in school reading, math, and writing programs. Many of these tests have been developed by particular school systems to meet their own needs. At present there are few readiness tests that can be used easily in many different situations, because the results are so closely tied to the educational techniques, goals, and resources of individual elementary schools. (See also Chapter 47.)

Educational achievement tests should be both norm and criterion referenced to be most useful, because they measure various school subject areas. The child's performance on them can then be described with regard to a reference group and in terms of a criterion as to his strengths and weaknesses in specific segments of material. A child may have good skills in punctuation but use adverbs and adjectives poorly. The criterion referenced information gives data relating to the specific areas in which the child needs help, whereas the normative data describe performance in terms of a reference group who have received similar instruction and are of the same age or grade. Unlike intelligence tests, these tests are measures of present behavior rather than predicting future behavior.

Both group and individual tests of educational achievement are widely used, and group tests can also be administered individually. Some of the tests are multiple skill batteries, which incorporate various subject areas; others measure a specific content area, such as reading or math. A number of achievement tests use a special format or are provided in alternate forms for children with special needs, such as those with sensory, motor, or language impairment.

Selected Educational Achievement Tests*

Individually Administered Multiple Skill Batteries

Peabody Individual Achievement Test (American Guidance Service, 1970).   This norm referenced screening test provides scores for reading recognition, reading comprehension, math, spelling, and general information. For kindergarten through the twelfth grade, it has age as well as grade norms. Since all but the last subtest require only a pointing response from the student, this test is useful with speech or language impaired and physically handicapped children.

Brigance Diagnostic Inventories (Curriculum Associates, Inc. 1977, 1978, 1980).   This criterion referenced instrument consists of three batteries covering the preschool, elementary, and secon-

---

*Jean M. Zadig, Ph.D., and Lynn Meltzer, Ph.D., developed the majority of the material used in this section on selected educational achievement tests.

dary levels. It provides information about skills in both cognitive and social areas and is particularly useful to educators in developing individual educational plans.

Wide Range Achievement Test (Guidance Associates, 1978). This test is often used as a quick screen for spelling, word recognition, and math skills. The standardization sample has not been adequate, and its use with children having special needs may be inappropriate.

Woodcock-Johnson Psychoeducational Battery (Teaching Resources, 1978). This multiple skill battery purports to assess cognitive ability, academic achievement, and interests of children aged three through adulthood, using 27 separate scores. The administration time is lengthy, and scoring and interpretation are highly complex. The test is relatively new, and its usefulness is still being evaluated.

Group Administered Multiple Skill Batteries

California Achievement Test (California Test Bureau–McGraw-Hill, 1977, 1978). This norm and criterion referenced test is intended for children in kindergarten through grade 12. The test covers prereading skills, reading, spelling, language, math, and the use of references. The battery is well respected by educators as providing useful data about students.

Stanford Achievement Test (Harcourt Brace Jovanovich, 1973). This comprehensive battery is both norm and criterion referenced. It assesses skill development in reading, language, spelling, math, social studies, science, and listening comprehension. Three forms of the test cover grades 1.5 through 9.5, with extensions to complete the range of kindergarten through grade 12. The tests are well standardized, and there are special forms for students with visual or auditory impairment.

Iowa Tests of Basic Skills (Riverside Publishing Co., 1978). This battery is similar to the California and Stanford Achievement Tests. It provides a comprehensive assessment of a number of subject areas, with a range from kindergarten through grade 12.

Individually Administered Reading Tests

Diagnostic Reading Scales (California Test Bureau–McGraw-Hill, 1972). This popular screening test measures skills in oral and silent reading and paragraph comprehension for grades 1 through 8. Error analysis yields information about instructional needs. Standardization, reliability, and validity data are vague.

Woodcock Reading Mastery Tests (American Guidance Service, 1973). This norm and criterion referenced battery assesses letter and word identification, word attack, word comprehension, and paragraph comprehension in pupils from kindergarten through grade 12. In addition to traditional scores, it yields scores of reading proficiency at different levels of difficulty, the so-called Mastery Score.

Durrell Analysis of Reading Difficulty (Psychological Corporation, 1955). This test is used for assessing skills in word recognition and comprehension for grades 1 through 6.

Gates-McKillop Reading Diagnostic Tests (Teachers College Press, 1962). This test, despite limitations, is widely used for the diagnosis of reading problems in grades 2 through 6. The battery contains 17 subtests; two different forms of the test are available.

Gilmore Oral Reading Test (Psychological Corporation, 1968). This test for grades 1 through 8 is broadly utilized in spite of some technical limitations. It can provide useful information about oral reading, accuracy, comprehension, and rate of reading when used diagnostically by an experienced reading specialist.

Group Administered Reading Tests

The Stanford Diagnostic Reading Test (Harcourt Brace Jovanovich, 1974, 1977). This well standardized test is norm and criterion referenced. Four overlapping levels cover grades 1 through 12. It is most useful in determining a student's strengths and weaknesses.

The Prescriptive Reading Inventory (California Test Bureau–McGraw-Hill, 1972). This is a criterion referenced test for pupils in kindergarten through grade 6. It provides information relative to individual skill mastery and grouping for teaching purposes.

Individually Administered Math Tests

Key Math Diagnostic Arithmetic Test (American Guidance Services, 1971). This math skills development measure is intended for kindergarten through grade 8. Subtests are grouped in the areas of content, operations, and applications. The test is useful in screening students for strengths and weaknesses in math.

Group Administered Math Tests

Diagnostic Math Inventory (California Test Bureau–McGraw-Hill, 1977). This criterion referenced test assesses skills in relation to 325 identified objectives. The range covered is grades 1.5 to 8.5.

Stanford Diagnostic Mathematics Test (Harcourt Brace Jovanovich, 1976). This test is both norm and criterion referenced. It is well standardized but is used primarily in the diagnosis of problems and the planning of individualized math programs.

PERSONALITY TESTS

Personality tests are concerned with intrapersonal, interpersonal, motivational, and attitudinal aspects of the child, the relatively stable emotional qualities underlying a child's behavior. The goal of personality testing is not to obtain an optimal assessment of abilities or skills but rather to effect an assessment of typical performance. Frequently personality testing is utilized to determine the emotional adjustment or maladjustment of a child. Younger children may be very forthright in discussing their emotional concerns, negating the need for elaborate personality testing. (See also Chapter 48.)

Considerable disparity exists among various personality theories, no one theory being able to account for all of a child's behavior. This diversity of views has led to numerous different approaches to the measurement of personality. The field of personality assessment abounds with measurement devices of many types. Personality assessment, however, can be summarized under three major techniques—interview-observational procedures, projective techniques, and self-report inventories. Interview and observational data can be collected directly from the child or others who know the child well, such as parents or teachers. Projective techniques provide procedures allowing the child to provide structure for relatively unstructured material, whereas self-report inventories are usually paper and pencil questionnaires that call for yes-no responses or a choice between items.

INTERVIEW-OBSERVATIONAL PROCEDURES

The interview or observation employs the observer as part of the test vehicle. In collecting direct observational data on a child, observers base their judgments more on what the child does than on what the child does not do—amount and kinds of activity, postures, facial expressions, quality, style, and content of speech. The reliability obtained from the interview or observation is increased when the length and number of the sessions are increased. Research has also demonstrated significant differences in skills among different interviewers and observers. (See also Chapter 43.)

By observing children, analyzing one's own be-

havior, and a variety of educational experiences, each person over time develops an integral set of theories of human behavior that allows him to explain why children behave the way they do in various situations. These theories can resemble formal personality theories but often are a mixture of formal and informal influences. This internal theorizing usually leads to stereotyping, in which a child is seen as a representative of a certain group of children and all the characteristics of that group are attributed to the child. Stereotyping provides a method of organizing perceptions into finite categories so that behavior can be readily understood. Stereotypes are essential to bring coherence to the myriad number of interpretations possible, but can lead to distortions of perception with the intrusion of personal needs and misunderstanding as a result of oversimplification. Stereotypes can serve as a starting point in understanding a child, not as an end. The richness and variety of a child's behavior patterns can characterize his uniqueness as well as his similarity to a specific group of children.

Whether a child's behavior is being directly observed or information from others is being gathered, the reliability and validity of the data have to be critically evaluated. Are these reliable samples of a child's typical behavior? The accuracy of the data can be improved when they are obtained under controlled standardized conditions. Many interview formats and behavior rating scales have been developed, with none presently achieving the widespread popularity that some of the projective techniques and self-report inventories enjoy. To be discussed are several examples of rating scales.

BEHAVIOR RATING SCALES

Nursing Child Assessment Project Scales (Barnard, 1978). This series of four scales is used as an early predictor of child health and development; they formed a part of the Nursing Child Assessment Project. A well organized program has been developed for training professionals to a high level of inter-rate reliability in their use so that the scales can be maximally effective. The Sleep/Activity Record is used to document the newborn and young infant's unique state related behavior patterns in eight areas. The Feeding and Teaching Scales have six similar subscale formats and assess the infant's and parent's adaptive behavior from observations of their interactions in feeding and teaching situations, during the infant's first year of life. With the final scale, the Home Observation for Measurement of the Environment (HOME) developed by Caldwell in 1978, information is gathered for the six subscales directly by observing the child's environment as well as by interviewing the mother. It may be used during the first three years of the child's life. Each of the four scales exhibits good reliability, but only preliminary validity data are available. (See Chapter 49E.)

Temperament Questionnaires. Carey has developed a series of temperament questionnaires to be used with children from four months of age through 12 years. Each questionnaire requires a parent to respond to questions about their perceptions of their child's behavioral characteristics. Unlike some of the other frequently used parental questionnaires, such as the Behavior Problem Checklist (Quay and Peterson, 1979) and the Child Behavior Checklist (Achenbach and Edelbroch, 1979), instruments describing temperament examine relatively stable behavioral patterns of a child rather than focusing on specific behavioral deviancies. Temperament scales generally demonstrate adequate reliability, but more data on validity are needed. (See also Chapter 44.)

Behavior and Attitude Check List (Sattler, 1981). This checklist consists of seven point scales for each of 28 items. It is used to record observations in 11 areas during a testing session. Reliability and validity data are lacking.

Conners Rating Scales (Conners, 1969, 1970, 1973). These three symptom rating scales were developed by Conners to assess hyperactivity. They consist of a 93 item parent questionnaire, a 39 item teacher questionnaire, and a parent-teacher questionnaire with 10 items in common. The three scales were adopted by the National Institute of Mental Health Early Clinical Drug Evaluation Unit to assess changes in hyperactivity resulting from medication.

PROJECTIVE TECHNIQUES

In projective assessment the test data are determined by the characteristic course of the child's spontaneous thought processes, his reactions to and structure of unstructured material. Projective tests elicit responses to more or less complex ambiguous situations. When one is confronted by ambiguous stimuli, the natural tendency is to project one's own needs, experiences, and view of the world on them. The child is not usually aware of how his responses will be interpreted and responds more openly and directly. There are no right or wrong answers, for responses are viewed as fantasy material to be interpreted. A wide variety of materials are used in projective testing—incomplete sentences, pictures, inkblots, drawings, play materials, for example. (See also Chapter 48.)

Varying characteristics of the thought process are elicited by different projective techniques. For example, tests that require the child to make up a story, like the Thematic Apperception Test (TAT), which presents pictures of different people in a variety of vague situations, often evoke themes representative of the child's needs through study of relationships between people. Conversely, the Rorschach Test, which presents pictures of inkblot designs and asks the child what he sees, often elicits responses that characterize psychological defenses. The degree of structure varies from measure to measure, from tests like the TAT, which require complex verbal responses, to the completion of incomplete sentences with a word or two. Most of the tests are given individually, although a few have been adapted for group use. There are some tests with which several different scoring and interpretative systems can be used. Projective interpretations can also be made from observations of how a child plays with various symbolic toys like a doll house and dolls, puppets, and a toy telephone, but no formal scoring systems are available.

Projective Tests*

Drawing Techniques. One of the most straightforward tests of personality involves asking the young child who has the requisite perceptual motor skills to draw a picture of a person (Draw-A-Person Test). Similar tests call for drawing a picture of the family (Kinetic Family Drawings) or other objects as well (House-Tree-Person Test). The way the picture is organized, the details included or omitted, the way the features are drawn, and the things the child says about the drawing provide the test data that are usually needed. Often the picture represents the child who is doing the drawing and can provide information about his self-concept. As the child gets older, the results begin to reflect artistic ability more than emotional issues. The amount and organization of detail in the drawing of a person can also be scored as a rough assessment of the child's intellectual development.

Sentence Completion Techniques. Incomplete sentences can constitute the most easily administered and most flexible of the projective tests. The main ingredients are sentence stems, which the child is required to complete in written or verbal form: "The best thing that ever happened to me was. . . ." There are many different individual sentence completion tests. The stems can range from the innocuous to the emotionally loaded. The themes consistently chosen and the descriptions used form the test data.

---

*See Chapter 48.

Thematic Techniques. The Thematic Apperception Test (TAT) is the prototype thematic projective test, which was designed to reveal the dominant needs or motivations of the individual. It consists of 30 black and white pictures and one blank card, with different cards for men and women and boys and girls (four to 14 years of age). Although the original directions call for presenting 10 cards (one at a time) in two different sessions, many examiners prefer to use only one session and present a total of 10 cards. The subject is asked to make up a story for each picture, describing what is happening, what will happen, and what the characters are feeling and thinking. A large amount of normative data has been gathered regarding the responses to each card. The child frequently identifies with a figure (hero) in his stories and often presents events from his own past. Impulses representing needs are symbolically presented in the stories. Some theorists speculate that the technique taps unconscious, preconscious, and conscious material. The TAT has been utilized extensively in personality research. McClelland, Atkinson, and their colleagues have employed some of the TAT cards in extensive research on the achievement need.

Tests such as the Children's Apperception Test use pictures of animals instead of people on the assumption that young children respond more easily to pictures of animals. This assumption is not always borne out in practice. Another test, the Tasks of Emotional Development Test, uses pictures of children in typical problematic situations, e.g., a child standing next to a cookie jar. The more concrete nature of the pictures does not appear to produce more pertinent diagnostic material than those that are more ambiguous.

Inkblot Techniques. The Rorschach Test is the original inkblot test introduced in 1921 by Hermann Rorschach. It is extensively used but frequently criticized because of issues over validity. This standard psychodiagnostic test is said to be diagnostic of personality as a whole and has been used frequently to determine the degree of maladjustment. A major assumption is that it yields unconscious material. Ten cards, printed with bilaterally symmetrical inkblots, are presented, one at a time. As each card is presented, the person is asked to "Tell what the blot looks like, reminds you of, what it might be; tell everything you see in the blot." The examiner records the responses, and after all 10 cards have been gone through the examiner inquires about where on the blots the percepts were seen and what about the blot made the individual think of the response. The chief scores concern the location, determinants, and content of the percept. The popularity of the percepts is also considered. Six major individual scoring systems have been devised to represent the responses, and different examiners may use different systems. Another inkblot test designed to improve the psychometric properties of the Rorschach Test is the Holtzman Inkblot Technique, which uses 45 cards instead of 10 but allows only one response per card. This test can be used readily in a group administration.

SELF-REPORT PERSONALITY INVENTORIES

Self-report inventories are group tests that can also be administered individually. Most are computer scored, and the results are analyzed by various computer programs. They are used principally in clinical and counseling contexts, although they can be employed for screening purposes. Using a questionnaire format, they require the test taker to indicate whether certain kinds of behavior describe him or not, and how the test taker would respond to certain situations. A review of the field indicates several theoretical positions around which the most commonly used paper and pencil inventories have been constructed. Three procedures have generally been followed in developing personality inventories—the criterion group procedure, the construct approach, and the factor analytic method.

The criterion group procedure is illustrated by the Minnesota Multiphasic Personality Inventory (MMPI) and the California Psychological Inventory. Criterion groups are defined on some reasonable empirical basis; one is a normal group and the other is chosen to represent a particular trait of interest. The two groups are then given a number of items and their responses are recorded. The responses that differentiate between the two groups are then utilized as the basis for constructing a scale on the test. The Schizophrenia Scale of the MMPI, for example, was constructed this way as items were selected that differentiated between a group of normal individuals and a group of schizophrenic patients. The reason the items discriminate between the groups may not be readily apparent. The major problem in this approach lies in the representativeness of the original criterion and control group, since the adequacy of their definition and selection is related directly to the accuracy of a prediction obtained from these inventories.

The construct approach requires that the test constructor have a definite idea of each variable that he desires to test. The Edwards Personal Preference Schedule typifies the construct approach; it was developed to measure the strength of major needs assessed by the TAT. Items are selected or devised that will sample the variables of interest. The items are analyzed by administering them to a heterogeneous group of people, frequently employing correlational and factor analytic techniques to select the items that correlate significantly with a construct for inclusion in a scale. With this method the test constructor knows what he has tried to put into his items from the very beginning, as opposed to other approaches that use a pool of heterogeneous items. A major limitation lies in the fact that different test constructors initially may include different bits of behavior under the same construct. Although both may use the same label, the end result may be that the scale developed by each is different.

Guilford and Zimmerman in their Temperament Survey, Cattell in the Sixteen Personality Factor Questionnaire, and Eysenck in the Maudsley Personality Inventory have employed factor analytic methods. In personality inventory development, factor analysis subjects the correlations within a pool of personality inventory items to a close scrutiny from which hypothetical factors emerge. These represent clusters of items, each of which presumes to measure a different and unique aspect of personality. This is not, however, a technique to use in detecting all possible personality parameters, because it cannot transcend the limitations of the original test data. The application of a label to a factor is often a source of difficulty because of terminological confusion, which in some cases has led to the proliferation of innumerable labels for similar or identical concepts. In an effort to avoid labeling, some test developers have used numbers or letters or have made up new words to designate their factors.

Factor analytic research has illustrated that the reliable variance in the majority of personality inventories, regardless of their theoretical bases, can be accounted for by two basic personality dimensions. The bulk of the common factor variance has been subsumed under a general factor of neuroticism and a bipolar factor of introversion-extroversion. Correlates of the neuroticism factor suggest that it measures a lack of general psychological maturity and a willingness to admit to psychological concerns. The introversion-extroversion factor is related to a desire for environmental and social predictability and structure, along with concern about the reactions of others to one's behavior.

California Psychological Inventories (Consulting Psychologist Press, 1957). This test is composed of 18 scales. About half of its true-false items are taken from the MMPI. The test was developed for use with normal populations, beginning at 13 years of age. Unlike the MMPI, which focuses on clinical interpretation, the California Psychological Inventory endeavors to predict such criteria as potential delinquency and probability of high school dropout.

The original MMPI was developed to discriminate between normal and specific diagnostic populations. It is the most widely used personality inventory. Although designed primarily for adults, it has been used for adolescents as well. Many new scales have been added to the standard 13. Numerous methods of profile analysis have been developed, including computer generated analysis, which gives summary descriptions of the test taker.

The Edwards Personal Preference Schedule (Psychological Corporation, 1953). The Edwards Scale is an inventory purporting to measure the strength of 15 needs also tapped by the TAT. It consists of pairs of statements from which test takers must choose one that is most characteristic of themselves. The test is designed for the older adolescent and the adult.

Cattell Inventories (Institute for Personality and Ability Testing). Cattell and his associates developed a series of personality inventories based upon their factorial research, the most well known of which is the Sixteen Factor Questionnaire designed for ages 16 years and up. Similar inventories have been developed for children from the age of six years.

Missouri Children's Picture Series (Sines, Paulker, and Sines, 1974). This experimental scale for children five to 12 years old does not require the child to be able to read or give verbal responses. The format is unique in presenting cards with line drawings of common events involving children. The drawings represent the questions on the usual self-report inventory. The response called for instead of a yes or no is for the child to place the card in one of two piles, a fun pile or a not-fun pile. The responses are then scored on six empirically derived scales: conformity, male-female, maturity, aggression inhibition, and hyperactivity.

INTEREST MEASUREMENT

Another aspect of the child's personality is his interests. Interests have been investigated mainly as an adjunct to educational and vocational counseling. Interests are not clearly elicited by direct interview because responses often can be unreliable. These less than accurate responses result in part from the child's insufficient information about different occupations, school programs, and activities. The interests of young children through elementary school can be quite ephemeral or can be more related to personality structure. In junior high school, impressions of adult occupations become more realistic and perceptions of individual differences become more acute. After age 17, interests become moderately stable but never totally fixed. General lines of interest can remain unchanged while specific interests continue to vary. Interest test results can modulate what a child claims are his interests. Interest tests are also a form of self-report inventory that give cues regarding personality development and possible maladjustment. Most interest tests can be administered individually or in a group format. Two widely used occupational interest inventories are described in the following section; they are among the most psychometrically intricate tests.

INTEREST INVENTORIES

Kuder General Interest Survey (Science Research Associates, 1964). There are a number of different forms of the Kuder interest inventories. The Kuder General Interest Survey is designed for grades 6 to 12 and is a revision of the Kuder Vocational Preference Record. A forced choice format is used in which the child must indicate for groups of three items which item he most likes and which he likes least. Scores are computed in percentiles standing in 10 interest categories. This interest profile is then compared with those of people in different occupational groups for interpretation. A verification scale is also used to detect those who are not appropriately following the instructions or are attempting to slant the results. Adequate reliability and validity data are available.

Strong-Campbell Interest Inventory (Stanford University Press, 1974). The Strong-Campbell Inventory is an extremely sophisticated, computer scored interest inventory for high school students and adults. It consists of 325 items grouped into seven parts with results reported in standard score format. The first five parts require the person to respond by marking like, dislike, or indifferent to items in the following categories: occupation, school subjects, activities, amusement, and day to day contacts with various kinds of people. In the final sections a choice is required between sets of two statements as to which statement is liked better; then a yes, no, or question is called for in response to a number of self-descriptive statements. The last two sections result in scores on two special scales, one of which assesses the tendency to continue one's education; the other is an introversion-extroversion scale indicating interest in working with other people. Administrative indices characterize the degree of carelessness in responding and test taking response sets. The main results of the test are scored on six general occupational themes describing the type of person and congenial working environment, which subdivide into 23 basic interest groups. The individual's performance is compared to that in 124 different occupational criterion groups. Scores on the six general occupational themes and the 23 basic interest groups serve as a reference to aid in explaining the similarities with certain occupations. Impressive reliability and validity data have been developed.

RICHARD R. SCHNELL

REFERENCES

Achenbach, T. M., and Edelbroch, C. S.: The child behavior profile. II. Boys aged 12–16 and girls aged 6–11 and 12–16. J. Consult. Clin. Psychol., 47:223–233, 1979.

Anastasi, A.: Psychological Testing. Ed. 5. New York, Macmillan Publishing Co., Inc., 1982.

Bloom, B. S.: Stability and Change in Human Characteristics. New York, John Wiley & Sons, Inc., 1964.

Buros, O. K. (Editor): The Eighth Mental Measurements Yearbook. Highland Park, New Jersey, Gryphon Press, 1978.

Cronbach, L. J.: Essentials of Psychological Testing. Ed. 3. New York, Harper and Row, 1970.

Green, B. F.: A primer of testing. Am. Psychol., 36:1001–1011, 1981.

Hernnstein, R. J.: IQ in the Meritocracy. Boston, Atlantic, Little, Brown, 1973.

Hobbs, N.: The Futures of Children. San Francisco, Jossey Bass, 1975.

Johnson, O. G., and Bommarito, J. W.: Tests and Measurements in Child Development: A Handbook. San Francisco, Jossey Bass, 1971.

Kamin, L. J.: The Science and Politics of IQ. Potomac, Maryland, Lawrence Erlbaum Associates, Inc., 1974.

Meier, J. H.: Developmental and Learning Disabilities. Baltimore, University Park Press, 1976.

Quay, H. C., and Peterson, D. R.: Behavior Problem Checklist. Miami, 1979.

Sattler, J. M.: Assessment of Children's Intelligence and Special Abilities. Ed. 2. Boston, Allyn & Bacon, Inc., 1982.

Scarr, S.: From evolution to Larry P., or what shall we do about IQ tests? Intelligence, 2:235–342, 1978.

Schnell, R. R.: The psychologist's role in the family conference. In Ulrey, G., and Rogers, S. (Editors): Psychological Assessment of Handicapped Infants and Young Children. New York, Thieme-Stratton, 1982.

Uzgiris, I. C., and Hunt, J. M.: Assessment in Infancy: Ordinal Scales of Psychological Development. Urbana, University of Illinois Press, 1975.

51
Pediatric Role Definition

DIFFICULTIES WITH ROLE DEFINITION

Any attempt to define a special role for the pediatric caretaker—whether pediatrician, family physician, nurse practitioner, or other professional—within the broad area of developmental and behavioral pediatrics is fraught with hazards. Review of the pediatric literature from the past few decades clearly documents the mandate for significant pediatric involvement with children with developmental and behavioral concerns. Yet despite a generally accepted belief that such pediatric involvement is indeed "an idea whose time has arrived" (Richmond, 1975), no specific pediatric roles have been generally agreed upon. Difficulties with role definition reflect, in part, the broad range of problems and conditions considered within the limits of behavioral and developmental pediatrics. In addition, the primary care provider is faced with a wide array of options for involvement within this area, further complicating the process of role definition.

VARIETY AND NATURE OF PROBLEMS CONSIDERED

The unique nature of developmental and behavioral disorders as compared with other areas within the field of pediatrics contributes to the difficulties with role definition. Several features shared by many of these problems determine their uniqueness:

Frequent Uncertainties Regarding Etiology and Prognosis. For many developmental and behavioral problems, our understanding of etiology is limited and the prognosis is unknown. Specific etiologies are rarely apparent in children with, for example, attention deficits and information processing problems or learning disabilities. The ability to predict future functioning for these children is extremely limited. Even for relatively discrete and easily recognized disorders, such as neural tube defects, at least six different etiologies have been suggested—both genetic and nongenetic—and the prognostic importance of certain criteria is highly controversial.

Composite of Health Maintenance, Chronic Care, and Acute Care. Children with developmental and behavioral disorders present a unique challenge to the caretaker, in that issues of health maintenance are complicated by the need to address additional problems relating to the chronic nature of handicapping conditions as well as the "crises" precipitated by acute illness. For the child with Down's syndrome, for example, the caretaker must deal with varied issues, such as developmental delay, an ill understood increased susceptibility to infections, potential catastrophic complications relating to cardiac, hematologic, or gastrointestinal disorders and defects, in addition to health maintenance.

Frequent Need for Multiple Assessments. Developmental and behavioral disorders present unique diagnostic challenges, frequently defying methods of simple assessment and rapid confirmation by laboratory studies. Indeed, for the child suspected of developmental delay or a disorder of attention or presenting with a behavioral problem, multiple observations and perspectives at different times are critically important in increasing the reliability of assessments.

Involvement of Multiple Professions in an Interdisciplinary Approach. Experience in a consultative clinic evaluating preschool children has documented the need for comprehensive evaluations involving multiple disciplines for behavioral or developmental dysfunction (Oberklaid et al., 1979). Assessment and intervention for children with school related problems may necessitate contributions from medical (pediatrics, nursing, neurology, psychiatry, otolaryngology, ophthalmology), educational (teacher, principal, learning disability specialist), mental health (psychology, guidance counselor), and social service (social worker) personnel as well as those in other fields (speech-language therapy, physical therapy, occupational therapy). Any description of pediatric roles must include guidelines for interdisciplinary cooperation in assessment and management.

Multiplicity of Problems in a Given Child. Children with disorders of development and behavior infrequently manifest discrete, isolated problems. For the learning disabled child, for example, frequent accompanying concerns may include diminished self-esteem, problems with peer interactions, and acting-out behavior in the classroom. Myelodysplasia may be complicated by numerous related problems, including paraplegia, neurogenic bladder, bowel incontinence, kyphoscoliosis, hydrocephalus, and mental retardation. Thus, any definition of pediatric roles must include functioning as a coordinator or ombudsman for handicapped children.

Complex Effects on Families. Parental reactions to being informed of their child's handicapping condition are exceedingly varied, with possible responses including shock, anger, helplessness, panic, devastation, withdrawal, silence, denial, mourning, self-blame, and unrealistic guilt as well as overprotection.

The effects upon siblings are similarly variable and may involve support and affection, teasing or disgust, guilt, jealousy, regression, or acting-out behavior. Rearing a child with developmental-behavioral dysfunction may have either a positive or a negative effect upon parental relationships. A marriage may be strengthened by the opportunity to share the responsibility for caring for a handicapped child, providing mutual emotional support during stressful times and involving the extended family in a helpful manner in child care. However, the parents' relationship will be weakened when they are confronted by continuing conflict as to "who is to blame," exclusive parental communication through or about the child, overinvolvement by one parent (perhaps to the exclusion of the other), or a family life that revolves totally around the child while neglecting both parents' needs. Any description of roles must thus include a significant counseling component.

Despite common features shared by many developmental and behavioral problems, any description of pediatric roles must also consider their remarkable heterogeneity. Quite obviously, no single approach can be outlined that will be equally appropriate for such diverse conditions as learning disabilities, cerebral palsy, myelomeningocele, and sleep disorders. Furthermore, each problem requires of the caretaker differing expertise and knowledge. Thus, any description of pediatric roles must be specific enough to consider the unique features of problems of development and behavior, and flexible enough to take into account their heterogeneity.

DIFFICULTIES WITH EARLY IDENTIFICATION AND ASSESSMENT

A further problem in defining a specific pediatric role is the difficulty in identifying and evaluating many children with evidence of developmental or behavioral dysfunction. With regard to identification, few guidelines are available to aid the pediatric caretaker in determining the optimal frequency and developmental stages at which to screen children and obtain the maximal yield at minimal expense. Furthermore, there as yet exists no definitive evidence that screening of children for many developmental problems is worthwhile.

Certain screening practices during routine health maintenance are well accepted. (See Chapter 45.) For example, the cost effectiveness of screening for phenylketonuria is well demonstrated despite its low frequency of 1 in 15,000, for the screening test utilized is both sensitive and specific. Other screening practices are well accepted despite a lack of documentation of efficacy. For example, all pediatric caretakers know the importance of routinely charting height, weight, and head circumference, yet no studies have documented the usefulness of such measurements in identifying children with otherwise unsuspected health problems. Even screening for sensory impairment—certainly considered routine and sound medical practice—is somewhat controversial in that little information is available concerning the reliability or validity of the techniques utilized in screening for visual acuity, and methods of hearing screening are least well developed for the first year of life, when identification is most crucial (North, 1974).

For children with physical, sensory, and gross developmental problems, the reasons for early identification and diagnosis are obvious. These handicaps, however, differ significantly from a problem such as potential learning failure. In screening for the latter, the caretaker is attempting to identify a condition not yet manifest in the child! In addition to difficulties with screening, evaluation of developmental dysfunction is problematic in that there exists no single developmental assessment tool suited for use in the medical setting that is both comprehensive in scope and free from significant limitations (Thorpe and Werner, 1974). Thus, difficulties with identification and assessment of developmental and behavioral problems further complicate the process of role definition.

INTERVENTION STRATEGIES UNKNOWN FOR MANY PROBLEMS

Delineation of appropriate guidelines for pediatric involvement with intervention for handicapping conditions is further hampered by the lack of documented efficacy for many treatment approaches to such problems. Unlike other areas within pediatrics, recognition of a given condition may well not lead to initiation of a specific treatment plan. For example, even following the identification of factors predisposing to school failure, the optimal intervention is frequently unknown. Critics of early identification and intervention cite a "similarity of remedial recommendations." Some argue that in the absence of clearly documented effective treatment, identification of some problems may do more harm than good by generating parental anxiety. Despite the unpopularity of such criticisms among pediatric caretakers in general, advocates of substantial involvement with developmental and behavioral dysfunction lack the documentation to refute effectively those skeptical of such involvement. Longitudinal studies demonstrating the benefits of pediatric participation in the identification, assessment, and management of a variety of developmental and behavioral problems must precede acceptance of clearly defined pediatric roles.

THE VARIETY OF POSSIBLE PEDIATRIC ROLES

Many opinions have been expressed concerning the pediatric caretaker's diagnostic role in dealing with developmental and behavioral problems. This variety of opinions and the wide array of options for involvement available to the caretaker further hinder the process of role definition. For no condition is this variety more apparent than with school related problems of learning and behavior. A wide variety of possible roles have been offered including the identification of any physical or psychological handicaps that might impair learning; searching for evidence of neuromaturational delay or "soft neurological signs," providing a neurodevelopmental assessment at the time of entry into school, communicating with the teacher regarding specific concerns, providing guidance regarding acceptable treatment approaches, prescribing psychoactive medication, acting as an advocate for the child and parents, dealing with parental anxiety, supporting educational plans, serving as school health physician, supporting alternatives to academic achievement—and even no involvement at all unless "organic" problems are considered to contribute to the child's dysfunction. Given the number of possibilities, the lack of agreement concerning specific functions is hardly surprising.

GOALS OF THIS CHAPTER

Because of the difficulties inherent in attempting to define specific pediatric roles in assessment, the remainder of this chapter will attempt to identify the factors that determine the individual caretaker's functioning within this broad area. In addition, various roles possible for the pediatric caretaker will be described. Finally, the levels of competency required for these various roles will be identified.

FACTORS MOLDING PEDIATRIC ROLES

Multiple factors determine the level of involvement of the pediatric caretaker in the case of children with problems of development and behavior (Fig. 51–1). Certain of these factors are discipline related, inherent characteristics of the field of pediatrics. Others reflect the setting within which the caretaker functions, e.g., practice related variables, present day technology and knowledge, and patterns of morbidity. Also highly significant are physician related factors, such as training, experience, attitudes, and interests.

DISCIPLINE RELATED FACTORS

Goal. The primary goal of the pediatrician, as defined by the American Academy of Pediatrics, is the "promotion of optimal growth and development in children." Pediatric care is "devoted to the attainment by all children of their maximum potential for physical, emotional and social health." Furthermore, the Academy directs that "children with chronic handicaps should be able to function at their optimal level" (American Academy of Pediatrics, 1972). Hence, the basic goal of the pediatric caretaker as defined by the Academy mandates specific involvement with the developmentally dysfunctioning child as well as the promotion and monitoring of a child's development and behavior.

Child Advocacy. Other aspects of the field of pediatrics influence the role of the caretaker. The pediatrician's role as an advocate for children is well established; the American Academy of Pediatrics has been supportive of child health programs and children's rights for the past five decades. Such advocacy and involvement logically extends to children with chronic handicapping conditions of development and behavior—acting as a voice for those who might not otherwise be heard.

Access and Continuity. The pediatric caretaker's unique early access to infants and young children is a further factor in role determination. As a result of his continuing relationships with families as the general physician for young children, the pediatrician is well suited to monitor growth and development and to intervene when appropriate. Pediatric caretakers are the only professionals having routine, regular contact with infants and families. Furthermore, familiarity with a child's family and the community affords the pediatrician the opportunity to view the child in

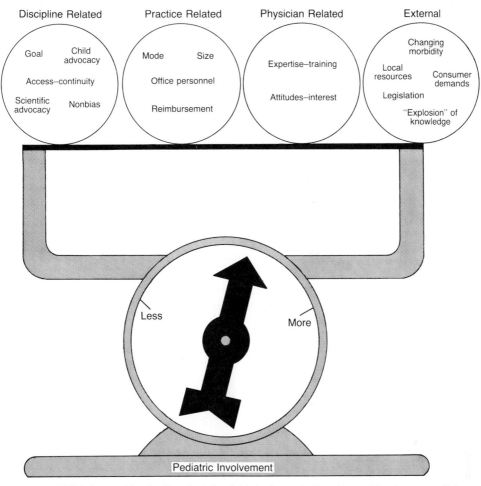

Figure 51–1. Multiple factors determine the extent of pediatric involvement with problems of development and behavior.

the context of his own world. By providing continuing care, the opportunity also exists for multiple observations and longitudinal perspectives to increase the reliability of individual assessments.

Scientific Advocacy. Physicians and other health care providers are traditionally perceived as being knowledgeable about a broad range of childhood issues. Hence, they are often cast in the role of experts in the assessment and management of developmental disabilities. The pediatric caretaker is not infrequently called upon to provide guidance regarding acceptable treatment approaches to a variety of problems.

Nonbias. Other discipline related factors that distinguish the field of pediatrics from the mental health professions further influence roles in assessment and management. The pediatric caretaker is a generalist who can integrate contributions from multiple disciplines and consultants and who can operate relatively free from any strong disciplinary biases. As the traditional etiologic paradigm of "organic versus emotional" is increasingly challenged, pediatricians can comfortably accept a "transactional" model of devel-

opment, and view developmental and behavioral outcomes as the result of an ongoing reciprocal interaction between a child and the environment, as well as view most performance failures in children as "dysfunction"—attributed exclusively to neither specific organic nor psychogenic etiologies. Such an approach affords pediatricians the opportunity to be nonaccusatory, as children and parents are supported as "innocent victims of dysfunction." Counseling is descriptive and advice giving rather than analytical, striving toward the demystification of symptoms for parents, children, and teachers. Hence, certain aspects of the field of pediatrics afford the caretaker the opportunity to counsel without the perceived stigma of the mental health professions.

PRACTICE RELATED FACTORS

Obvious differences among pediatric practices limit the generalizability of statements regarding the proper role of the child health care provider with developmental concerns.

Mode. The mode of pediatric practice varies,

with solo, partnership, groups of three or more pediatricians, and multidisciplinary groups all significantly represented. Surprisingly, a recent analysis of primary care approaches to developmental disabilities revealed that customary procedures were not affected by the mode of practice (solo versus group), although intuition might well suggest otherwise (Shonkoff et al., 1979). Clearly, the busy solo practitioner may require an approach to developmental assessment different from that of the group practitioner encouraged by peers to pursue an area of interest.

Size. The majority of pediatricians assume responsibility for 5000 to 10,000 patient visits per year. The diversity of practice size is further illustrated by the observation that approximately 25 per cent of pediatricians see 2000 to 5000 patients and almost 10 per cent see over 10,000 patients per year. Logic would dictate that the time spent per patient be more limited within an extremely busy practice setting. However, caretakers have devised numerous methods in order to devote time to developmental and behavioral issues, such as after-hours school and parent conferences as well as adjusting the content of health maintenance visits to be more developmentally oriented while reducing the time spent with the physical examination. In fact, the study just discussed also failed to demonstrate any correlation between customary approaches to developmental problems and size of the practice.

Office Personnel. To the extent that tasks and responsibilities are delegated to midlevel practitioners, such as nurse practitioners and physician assistants, the productivity of physicians is increased. At the present time physician's assistants are utilized in 10 per cent of private pediatrics practices, while almost 15 per cent employ pediatric nurse practitioners or associates. Developmental screening within the practice setting is frequently the responsibility of these members of the health care team. The use of a standardized instrument (such as the Denver Developmental Screening Test) is reported more frequently by pediatricians who employ nurse practitioners. Thus, the presence within a practice of certain personnel may significantly affect pediatric roles in assessing development.

Reimbursement. As noted, pediatric involvement with developmental and behavioral problems is time consuming. Multiple visits are frequently required for assessment, with additional sessions necessary to share information and counsel families (see Chapter 53). Unfortunately, as noted by the Task Force on Pediatric Education, fee-for-service reimbursement policies currently encourage caretakers to treat as many patients per day as possible, while actually discouraging the time consuming approach needed to effectively diagnose and treat biosocial and developmental problems (Task Force on Pediatric Education, 1979). In addition, reimbursement is not generally provided by insurance companies for the counseling required in the care of such children. Hence, current fee-for-service reimbursement practices as well as the procedure oriented compensation schedules adopted by most third party payers actually serve as a financial disincentive to recognize and treat these problems. The Committee on Children with Handicaps of the American Academy of Pediatrics has mandated that physicians must receive third party compensation for coordination, staffing, and follow-up efforts on behalf of the developmentally disabled, as well as compensation for counseling and school visits on behalf of these patients. Significant modification of current reimbursement practices is mandatory prior to defining a more active pediatric role in this area.

Evidence does exist that pediatric caretakers are willing to approach developmental and behavioral problems despite these difficulties with compensation. In the study already noted, the socioeconomic status of the patient population within a given practice (as defined by the percentage of Medicaid eligible patients) did not affect the approaches utilized within the practice. Thus, the likelihood of families assuming responsibility for charges (as opposed to insufficient third party reimbursement) did not appear to affect the caretaker's approach to a developmental or behavioral problem. Furthermore, experience within a primary care clinic offering comprehensive assessments for school related problems has demonstrated that parents will accept charges based upon time requirements for such evaluations as appropriate compensation for services provided (Dworkin et al., 1981). Pediatric caretakers must be comfortable with charging fees appropriate for the time spent in dealing with problems. This may prove philosophically difficult for physicians accustomed to relatively lower fee schedules as compared with peers in other disciplines of medicine. Indeed, for a more significant involvement with issues of development and behavior, the caretaker must receive—and expect to receive—adequate compensation.

EXTERNAL FACTORS

Numerous variables relating to the world within which the pediatric caretaker functions influence the specific role he assumes in the area of developmental assessment and management. In addition to factors relating to the practitioner's practice setting already discussed, others relate to the characteristics of contemporary medical practice, such as changing patterns of morbidity, consumer demands and community expectations, availability of consultants and local resources, legislative trends, as well as the state of knowledge in the field of child development and behavior.

Changing Morbidity. An evolution in pediatric practice has added to the pressures on pediatricians to assume more responsibility for developmental concerns. Scientific advances such as improved infant feeding practices, immunizations, and antibiotics have made more caretaker time available for dealing with what has been referred to as the "new morbidity." Numerous studies have documented this changing content of pediatric practice. A significant portion of this morbidity includes problems such as learning disabilities, attentional problems, mental retardation and language or speech impairment.

Consumer Demands. Pediatric caretakers increasingly are being called upon to serve as consultants to Head Start programs, nurseries, and school systems. Parents, school systems, and government agencies are now seeking medical assistance in evaluating and managing a wide range of developmental dysfunctions, such as language disorders, learning disabilities, behavioral problems, and attentional disorders, in addition to the more severe cognitive, sensory, and motor handicaps. Expectations facing the pediatrician vary from one community to another. For example, for the school system offering comprehensive screening for school readiness prior to school entry, pediatric based assessment need not necessarily include hearing screening. For the community in which such screening is not routine, the physician may well be expected to identify those children at risk for school problems.

Social movements such as the movement for independent living, consumerism, self-help, demedicalization, and self-care have significant implications in determining pediatric roles. Pediatric caretakers are often called upon to facilitate optimal functioning of the disabled child with minimal dependence upon medical personnel and procedures. Aiding the child with myelodysplasia in accomplishing self-catheterization to provide continence as opposed to ileal loop diversion or a life in diapers is but one example of services that may be expected of the physician by parents and children. Trends toward deinstitutionalization and returning children to the community have an impact upon the composition of pediatric practice. As children with a variety of handicapping conditions return to their homes and communities to live, families will appropriately look to their traditional sources of health care to provide medical supervision for these children.

Local Resources. As discussed previously, optimal care for children with developmental disabilities requires a multidisciplinary approach with maximal utilization of community resources. The availability of consultants within a given area significantly affects the role assumed by the pediatric caretaker.

With adequate resources and available consultants, the pediatrician may more appropriately assume the role of ombudsman, coordinating services while providing primary care. When confronted with a lack of available experts, the caretaker may need to assume a broader role out of necessity—a role for which he may feel unprepared. Local resources vary significantly. For example, the almost universal availability of consultants in a wide array of disciplines within the New England states contrasts with the experience of rural physicians in upstate New York who indicated a need for increased availability of mental health services, neurologists, physical and occupational therapists, and speech therapists (Dworkin et al., 1979).

Legislation. At both state and federal levels, legislation influences the extent of physician involvement with children with handicapping conditions. On the national level, Public Law 94–142 (The Education for All Handicapped Children Act) does not specifically define a role for physicians (see Chapter 61A). However, physician involvement is certainly implied, with minimal involvement including the counseling of parents whose children are undergoing evaluation. State laws vary significantly. In some, no specific physician role is identified. In others, such as Massachusetts and New York, the physician assumes a clearly defined role in each child's multidisciplinary assessment. Thus, with respect to legislative factors, the extent of the physician's involvement with the assessment and management of developmental and behavioral problems is quite varied. However, at the very least, a supportive, counseling function is within the spirit of federal legislation.

In addition to directly defining specific physician involvement, legislation also has an indirect impact upon pediatric roles. For example, the pediatric caretaker may well be called upon to ensure that the educational setting for a child with a seizure disorder and learning disability is in the least restrictive environment, as mandated by PL 94–142. Alternatively, physician involvement may be sought by a school system planning a new building to ensure access for handicapped children as required by the Architectural Barriers Act of 1968 (PL 90–480 and as amended in PL 94–541).

"Explosion" of Knowledge. Factual information relating to child development has increased tremendously in recent years. The availability of a wide assortment of screening and assessment instruments, such as the Neonatal Behavioral Assessment Scale of Brazelton, the Infant Temperament Questionnaire of Carey and McDevitt, and the Pediatric Examination of Educational Readiness from the Brookline (Massachusetts) Early Education Project, provides the pediatric caretaker with the tools to assess systematically different areas of child functioning. Longitudinal studies are beginning to document the natural history of perplexing problems of learning and attention. Recent advances have significantly changed the outlook for children with conditions such as spina bifida. In addition, knowledge of recent advances in the field of developmental and behavioral pediatrics has been facilitated by the appearance of new textbooks, journals, and educational programs, such as the Pediatrics Review and Education Program of the American Academy of Pediatrics. With the growth of the field of developmental pediatrics, the role of the child health professional will continually undergo change and refinement. As knowledge escalates, so will expectations.

PHYSICIAN RELATED FACTORS

Expertise and Training. Numerous recent studies have evaluated the adequacy of contemporary pediatric education in relation to the need for knowledge in pediatric practice. (See Chapter 63.) The results of these studies clearly indicate that the majority of practitioners believe that pediatric education is not properly oriented toward the needs and realities of a general practice. In addition to more "traditional" medical areas, such as allergy and orthopedics, subjects stressed as requiring more emphasis in training include behavior problems, mental retardation, health maintenance issues, normal growth and development, emotional problems, and child guidance. The Task Force on Pediatric Education has labeled as "inadequate" the current educational content of many programs in the biosocial and developmental aspects of pediatrics. Among pediatricians discussing such issues, 79 per cent rated their formal training in development as inadequate! Almost 50 per cent regarded medical school as being of no value as a source of knowledge in developmental pediatrics. Residency training was viewed with only slightly higher regard, with 30 per cent viewing it as highly valuable (Dworkin et al., 1979). These negative feelings regarding competency and training certainly influence the willingness of the physician to assume a greater role in assessment and management.

Attitudes and Interest. Knowledge and expertise, rather than favorable attitudes and interest, may be the limiting factors in greater physician involvement in the area of developmental-behavioral pediatrics. The previously mentioned survey

of New England pediatricians documented "positive" attitudes regarding issues in developmental pediatrics. For example, 84 per cent of those surveyed agreed with the statement that screening for potential learning disabilities in the preschool child should be a routine part of primary pediatric care. The majority of pediatricians believed in the ultimate value of early identification of a variety of handicapping conditions, such as learning disabilities, behavior problems, language impairment, hearing impairment, and blindness. Furthermore, two-thirds of those surveyed disagreed with the opinion that management of developmental problems is not feasible within the context of the practice setting (Shonkoff et al., 1979). Attitudes thus do not appear to hinder the involvement of pediatricians with problems of behavior and development.

ROLE DESCRIPTIONS

Now that we have identified the many difficulties with role definition and the factors molding pediatric roles, descriptions of such roles will follow. Of necessity, such descriptions must be general in nature. For a given practitioner, the manner in which the multiple factors merge to determine specific functioning in the area of developmental-behavioral pediatrics will be unique. A general classification of pediatric roles is feasible, however (Table 51–1). For the primary care physician with no special expertise or interest in this area, responsibilities include a screening role in health care maintenance, the facilitation of appropriate assessments via referrals, and a limited role in clinical problem solving. For the pediatrician with a special interest (and perhaps special training) in child development, a more significant role in clinical problem solving is appropriate, as is a commitment to more extended developmental consultation, participation in assessments as a member of a multidisciplinary team, and participation in medical education. On a different level is the pediatrician with specific interest, training, and expertise in developmental pediatrics who functions as a full-time specialist.

THE PRIMARY CARE PHYSICIAN WITHOUT SPECIAL INTEREST

The Committee on Standards of Child Health Care of the American Academy of Pediatrics has identified a basic role for the pediatrician in screening for developmental and behavioral problems:

. . .the pediatrician must seek to identify more subtle evidence of dysfunction that may impair the growing child's acquisition of language or other skills necessary for school

Table 51–1. A GENERAL CLASSIFICATION OF PEDIATRIC ROLES WITH DEVELOPMENTAL-BEHAVIORAL CONCERNS

Description	Functions	Competency Requirements
Primary care physician without special interest	1. Periodic evaluation of child's health status 2. Provision of appropriate anticipatory guidance and counseling 3. Provision of care to handicapped children during acute illnesses 4. Facilitation of appropriate assessments via referrals 5. Coordination of required medical care 6. Limited role in clinical problem solving	1. Knowledge of basic principles, processes, and milestones of child development 2. Knowledge of medical implications of chronic handicapping conditions 3. Knowledge and understanding of agenda items for anticipatory guidance 4. Ability to communicate effectively with the family of a handicapped child 5. Familiarity with community resources 6. Ability to act as "ombudsman" 7. Understanding of developmental screening, diagnostic and assessment procedures 8. Development of a plan for personal continuing education
Primary care physician with special interest and expertise	As above, in addition to: 1. More extensive role in clinical problem solving 2. Consultant role as member of a multidisciplinary team 3. Participation in medical education	As above, in addition to: 1. Knowledge of assessment and management of specific behavior and developmental problems 2. Ability to function effectively within a multidisciplinary team 3. Familiarity with the organization and functioning of schools 4. Ability to teach the principles of developmental-behavioral pediatrics
Full-time specialist in developmental-behavioral pediatrics	As above, in addition to: 1. Provision of detailed definitive consultation and clinical problem solving 2. Active participation and leadership in multidisciplinary assessments 3. Major role in medical education, including curriculum development and evaluation 4. Design, implementation, and evaluation of developmental research	As above, in addition to: 1. Skills in curriculum development and educational strategies 2. Ability to plan, execute, and evaluate research

success. The pediatrician must also recognize early signs of behavior problems that may be the foundation for more serious abnormalities in later childhood.

Although specific criteria for optimal pediatric performance in the practice setting are difficult to determine, six general areas of functioning for the generalist providing child health care can be identified:

Periodic Evaluation of the Child's Health Status. The pediatric caretaker assumes an obvious screening role for developmental-behavioral dysfunction during child health maintenance. For the child who is apparently growing and developing satisfactorily, routine procedures during well-child visits include obtaining an appropriate interval history, measurements (including height, weight and head circumference during the first two years of life) to assess physical growth and maturation, sensory screening unless provided by adequate community resources, developmental screening (see next paragraph), and physical examination. The Committee on Standards of Child Health Care broadly defines the goal of the latter as "not only a thorough search for the detection of abnormalities, but also a recognition and acknowledgement of positive attributes such as stature, muscular development, intellectual achievement and emotional growth."

Developmental screening issues are discussed in full in Chapter 45. The infrequent use of developmental screening tests by practicing pediatricians has been documented by several studies. At the present time, physicians most frequently screen for developmental delay using subjective (observation, discussion with parents) rather than objective means. Some caretakers screen only selected areas of development, such as speech or drawing abilities. The limitations of relying upon "clinical judgment" have been well demonstrated. Although specific criteria for optimal pediatric performance in this area have been difficult to determine, and routine developmental screening using standardized tools cannot be definitively advocated on the basis of benefits documented by long range studies, a more objective approach to such screening appears empirically useful. This may be accomplished less formally (and probably less reliably) through the use of flow sheets incorporating age appropriate tasks selected from developmental schedules, such as that of Gesell. The use of a standardized developmental screening test, such as the Denver Developmental Screening Test, is preferred by pediatricians who are recent graduates from medical school and more likely to employ nurse practitioners (Shonkoff et al., 1979). Any child suspected on the basis of subjective evaluation of manifesting developmental delay should be screened via a standardized format.

Provision of Appropriate Anticipatory Guidance and Counseling. The process of anticipatory guidance involves the provision of counseling and information to parents to assist them in creating the optimal home environment for the enhancement of their child's physical, emotional, and cognitive development. (See Chapter 53.) Such counseling is best provided within a clinical setting emphasizing the continuity of care during health maintenance visits. Several studies have suggested that altering the content of well-child visits to include a discussion with parents of ideas and opinions regarding normal child development may foster more optimal child development—cognitive, social, emotional, as well as physical (Casey et al., 1979).

The primary caretaker, by virtue of a unique relationship with families and children, also assumes an important role in counseling the family of a handicapped child: to help the family achieve the best adjustment possible. Despite the possible involvement of multiple specialists from a variety of disciplines, the family (and child) is most likely to turn to the primary caretaker for support and counseling. Given the pediatrician's familiarity with the child, family, and community resources, he is uniquely suited to understand the handicapped child within the larger setting of the family and community and thus more effectively can provide comprehensive care and counsel. Counseling functions may include informing parents that their child is handicapped, dealing with reactions of parents and family members (including the child) to such information, monitoring the impact of a handicapped child upon parental relationships and on siblings, as well as attempting to facilitate positive support for the family from the community.

Provision of Care to Handicapped Children During Acute Illnesses. When asked to describe their subsequent involvement with families of children who are mentally retarded, 94 per cent of pediatricians indicated that they would continue to provide routine medical care (Shonkoff et al.,

1979). As a result of deinstitutionalization, children with chronic handicapping conditions are returning to the community for health care. Most pediatric practices include a significant number of children with a variety of handicapping conditions, such as mental retardation, language and speech impairment, learning disabilities, disorders of attention, cerebral palsy, hearing impairment, and serious emotional disturbance. Despite the frequent relationship between an acute illness and chronic handicapping condition (e.g., urinary tract infection in the child with myelodysplasia, or upper respiratory infection or sinusitis in the child with Down's syndrome), few caretakers would regard these problems as not being within their area of responsibility. (See also Chapter 26 concerning needs and care of all children with general physical illness.)

Facilitation of Appropriate Assessments Via Referrals. The American Academy of Pediatrics is quite specific in assigning to the primary health caretaker the responsibility for being aware of community resources to ensure that appropriate referral is carried out for the child suspected of developmental or behavioral dysfunction. Familiarity with and availability of local resources are necessary for the care of the child with a chronic handicapping condition. (See Chapter 60.) Although the New England states may be unique, in that section of the country the location of practice in a rural area does not necessarily imply isolation from specific consultants. In some cases, this has meant reliance on a child development unit in an accessible, if not convenient, major medical center. In other instances practitioners utilized community based developmental programs with a full range of services (Dworkin et al., 1979).

Coordination of Required Medical Care. As previously discussed, children with handicapping conditions tend to manifest multiple problems requiring the involvement of numerous specialists. Because such specialists tend to view a child within the isolated context of their area of expertise, the risk of fragmentation of care is great. Thus, the primary care provider must assume a role in coordinating required medical services, avoiding redundancy while ensuring that the ultimate goal of assessment and management is to enable the child to function optimally in society. Experts have suggested that the pediatrician-generalist act as "ombudsman" for the handicapped child and his family, reviewing all aspects of the child's functioning with specialists as well as the child's family, school, and community.

Limited Role in Clinical Problem Solving. Despite the wide variety of problems encompassed by the field of developmental and behavioral pediatrics, caretakers without any special expertise or interest in this area many nonetheless feel comfortable in the assessment and management of a limited number of common conditions. Most pediatricians, for example, may view problems such as enuresis, sleep disorders, encopresis, and colic as manageable within the context of their practice. Even with a problem of traditionally less pediatric concern, such as school failure, successful pediatric involvement has been demonstrated. In regard to the child who is failing in school, over 50 per cent of pediatricians interviewed indicated that they would initiate contact with the school to obtain additional information about the child's problem (Shonkoff et al., 1979).

THE PRIMARY CARE PHYSICIAN WITH SPECIAL INTEREST AND EXPERTISE

For the caretaker with special interest or training in the area of development and behavior, additional functions are possible beyond those already discussed.

More Significant Role in Clinical Problem Solving. Parents and educators are demanding increasing pediatric attention to the needs of children with developmental and behavioral dysfunction. (See Chapter 38A.) With the appropriate interest, expertise, and pediatric setting, a more significant role in problem solving can be identified.

One example of a more significant role in clinical problem solving is a pediatric based approach to the school age child with problems of learning and behavior. Such an approach to the child with school problems appropriate for the primary care setting has recently been described and evaluated (Dworkin et al., 1981). Within this approach, two office visits are utilized for assessment (Fig. 51–2). At the initial visit relating to the problem, data collection is performed via history taking and a complete physical examination including a standard neurological examination and vision and hearing screening. At the conclusion of this visit, which requires approximately 30 minutes, extensive parent and teacher questionnaires are distributed. Subsequently a second 45 minute office visit is scheduled, usually within one week, to perform neurodevelopmental assessment employing an age appropriate examination. Within the model described, the Pediatric Elementary Examination (PEEX) and the Pediatric Evaluation of Educational Readiness (PEER) are assessment tools utilized. (See Chapter 45B.)

As a result of pediatric based assessment, the need for referral for psychoeducational evaluation is determined. Criteria for referral include delays

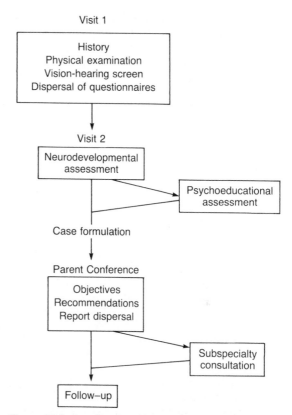

Figure 51–2. An office based approach to the child with school problems. (Reproduced with permission from Dworkin, P. H., et al.: Pediatric-based assessment. Children with school problems. J. Sch. Health, 51:325, 1981. Copyright 1981, American School Health Association, Kent, Ohio 44240.)

in discrete areas of function suggesting the possibility of learning disabilities, overall delays indicative of possible cognitive impairment, or academic delays for which no apparent etiology is identified. Following completion of the evaluation, a follow-up appointment of 30 minutes duration is scheduled at the physician's office with the parents and the child. Discussion includes case formulation with a description of the child's strengths and weaknesses, as well as recommendations. The need for subspecialty consultation or further testing is also discussed. The pediatrician assumes responsibility for compiling a report including the results of assessment and recommendations for management and follow-up. An addendum from the educator describes the results of educational assessment and recommendations for specific educational intervention. Reports are distributed to schools and appropriate consultants with the written consent of a responsible family member.

Pediatric based approaches to a variety of other developmental and behavioral problems might similarly be devised. The mechanism of evaluation already detailed requires only two office visits for assessment and one visit for a follow-up conference, with a total time requirement of approximately two hours. The use of standardized questionnaires significantly facilitates the gathering of data. Furthermore, communication with schools via written reports has the advantage of precision and availability for later reference. Pediatrician attendance at school based conferences is infrequently required, only for children with complex diagnostic formulations. The charge for pediatric based assessment within this model (approximately $140) is prorated from a fee of $20 for a routine office or clinic visit typically requiring 15 to 20 minutes. Parents uniformly accept charges as appropriate compensation for services provided.

Consultant Role as a Member of a Multidisciplinary Team. The opportunity for greater involvement by the physician with special interest and expertise is present at various levels. Consultation is often requested by a variety of organizations such as Head Start, nurseries, and special schools. Community service programs such as mental health and adoption agencies may also seek pediatric support and guidance. Questions may relate to difficulties encountered with specific children within such programs whereby the pediatric caretaker refers the patient for assessment and management. Alternatively, pediatric advice may be sought in the establishment of screening practices and the identification of program candidates or other policies. The pediatrician may also be called upon to perform such screening.

For the pediatrician with adequate skills, training, and interest, involvement with schools is possible beyond performing physical examinations, updating immunization status, and approving children for competitive sports. In certain states, for example, multidisciplinary assessments of children with special needs must include medical evaluation. Thus, the pediatric caretaker serves as a member of the team developing a child's educational plan. Pediatric evaluation may then include neurodevelopmental assessment and behavioral observation in addition to the traditional medical history and physical examination. Limited examples are now available of the practicing pediatrician who participates as a consultant in the area of school dysfunction as a member of a multidisciplinary team (McCormick, 1977).

Participation in Medical Education. As attempts are made to correct well documented deficiencies in the areas of development and behavior within medical training, the need for individuals with expertise in this area to provide instruction is apparent. An important objective of residency training, for example, is the presentation of an optimal model of the pediatric generalist who can deal effectively with behavioral and developmental dysfunction and interact effectively with agencies and professionals from other disciplines. The physician dealing with such problems in the office setting can serve as an effective role model for trainees observing within the practice, or can serve as a part-time attending physician within ambulatory care departments or appropriate subspecialty clinics within teaching institutions. For training programs including an experience in school health, opportunities for on-site training outside the hospital setting also exist.

THE FULL TIME SPECIALIST IN DEVELOPMENTAL PEDIATRICS

More controversial is the role of the physician with special interest and expertise in the area of developmental-behavioral pediatrics who pursues this area full time as a "specialist." There has been an intense and emotional debate over the appropriateness of developmental pediatrics as a true subspecialty. Somewhat simplistically, two general roles may be described:

The Private Practice of Developmental Pediatrics. At the present time a limited number of individuals engage in the subspecialty practice of developmental and behavioral pediatrics. One major problem is the fiscal viability of such specialization. An additional difficulty is identifying an appropriate patient population and avoiding overlap with other traditional specialties, such as neurology, psychiatry, and psychology. Indeed, as the pediatric caretaker assumes the role of a subspecialist, risks of fragmentation of care and biased approaches to problems re-emerge. These are the very attributes of pediatric care utilized to defend increased pediatric involvement within this area! Thus, the feasibility of full time pediatric practice in the area of development and behavior cannot be determined at present. One might envision for such a subspecialist more extensive roles in clinical problem solving and neurodevelopmental assessment, multidisciplinary team evaluations, and medical education than that described for the caretaker with some special interest and expertise.

The Developmental Pediatrician in an Academic Center. Developmental and behavioral pediatrics is becoming increasingly recognized as a specialized area of interest within medical centers. In some institutions this area is contained within an ambulatory pediatrics program. In others, the developmental pediatrician functions within a self-contained unit for handicapped children. Given the variety of problems considered under the heading of developmental-behavioral pediatrics, one could not expect one individual to serve as the ultimate referral source. The developmental pediatrician, however, may possess expertise in selected areas for detailed definitive consultation and problem solving. Examples of such areas include neonatal behavioral assessment, parent-infant interaction, the child with multiple handicaps, birth defects, mental retardation, behavioral problems, school related problems, and adolescence. In addition to a significant role in diagnosis and management of select problems, major responsibilities of the academic developmentalist include training and research.

IMPLICATIONS FOR COMPETENCY

There is as yet no uniform agreement concerning the expertise required of pediatric caretakers to deal competently with issues of behavior and development. However, in considering possible roles and functions of the physician in this area, implications concerning knowledge and expertise are apparent. One might therefore suggest a minimal level of competency appropriate for the practitioner with no special interest, as well as more sophisticated skills and understanding for those more actively involved with the handicapped child. (See also Chapter 63 for a discussion of training.)

BASIC SKILLS AND UNDERSTANDING

In order for the primary care physician with no special interest or expertise in developmental and behavioral problems to perform the various functions described, certain necessary skills and knowledge can be described:

Knowledge of Basic Principles, Processes, and Milestones of Child Development. Without an understanding of normal growth and development, the caretaker will obviously be unable to fulfill a monitoring role and identify deviations from normal requiring assessment and intervention. Such knowledge is the "basic science" for pediatrics. These topics are covered in detail in Part II of this volume. Also necessary is some knowledge of the principal environmental influences described in Part III.

Knowledge of Medical Implications of Chronic Handicapping Conditions. Appropriate health monitoring, provision of care during acute illness, and coordinating the care of the handicapped child require an understanding of commonly associated medical problems and conditions as well as a basic familiarity with those conditions themselves. (See Parts IV and V.) Examples include the possibility

of atresia of the gastrointestinal tract in the neonate with Down's syndrome and the 20 per cent incidence of cardiac anomalies in the female with Turner's syndrome.

Knowledge and Understanding of Agenda Items for Anticipatory Guidance. A basic premise of anticipatory guidance is that the optimal time for the introduction of information relating to common problems that arise during child rearing is prior to the age at which such problems are likely to appear. Thus, knowledge of anticipatory guidance requires familiarity with affective, cognitive, and physical development.

Ability to Communicate Effectively with the Family of a Handicapped Child. In order to counsel families properly, the caretaker must be aware of typical reactions of parents and siblings to the news that a child is handicapped and recognize the impact of a handicapped child on familial relationships. Effective communication in this area is based on a general competence in interviewing and counseling. (See Chapters 43 and 53.)

Familiarity with Community Resources. In order to facilitate appropriate referrals and insure optimal care, the practitioner must be cognizant of resources available in different communities—within both the health and educational systems.

Awareness of Legal and Legislative Aspects of Handicapping Conditions. Informing parents of their rights under legislation such as PL 94–142 is a critical pediatric role. In order for the pediatrician to be an effective advocate for handicapped children, his familiarity with their rights as defined by relevant legislation is imperative.

Ability to Act as Ombudsman. In order to coordinate care effectively for the handicapped child, one must possess an understanding of the roles and competencies of other professionals involved with services for such children. Such an understanding facilitates communication between the primary caretaker and other specialists.

Understanding of Developmental Screening, Diagnostic, and Assessment Procedures. Practicing pediatricians tend to employ developmental screening tests infrequently. Often such tests are utilized only after delay has been established by other criteria, thereby making screening irrelevant, or inappropriately viewed as a mechanism to establish a diagnosis. Many experts advocate that pediatric caretakers possess the ability to administer and correctly interpret a standardized screening instrument, such as the Denver Developmental Screening Test, as well as be familiar with the indications, limitations, and benefits of commonly employed assessment procedures. Another basic skill is effective screening for behavioral variation and deviation (discussed in Part V) by some combination of available assessment techniques. (See Part VI.)

Development of a Plan for Personal Continuing Education. Numerous options exist for the practitioner to keep abreast of the significant advances in knowledge within the field of developmental-behavioral pediatrics (see following discussion). Without the development of a personal strategy, information and approaches will rapidly become outdated.

BEYOND BASIC COMPETENCY

For the physician who is more involved with handicapped children on either a part or full time basis, other necessary skills can be identified:

Detailed Knowledge of Assessment and Management of Specific Behavioral-Developmental Problems. As discussed previously, expertise across the broad realm of such problems in a given individual is unlikely. Examples of special areas of interest include the pediatrician involved with school related problems of learning and behavior, the consultant with expertise in developmental screening and preschool issues who advises a community Head Start program, as well as the caretaker who provides medical supervision to an Easter Seals program for children with cerebral palsy. (See Parts VI and VII of this volume for details on techniques.)

Ability to Function Effectively Within a Multidisciplinary Team. Even for the pediatrician without special involvement with the handicapped child, frequent communication with specialists from other disciplines is not unusual. With greater pediatric involvement, more significant interaction is mandatory and inevitable. Thus, the caretaker must be comfortable with the process of multidisciplinary assessment, in addition to possessing a sophisticated understanding of what other disciplines can and cannot provide. (See Chapter 49.)

Familiarity with the Organization and Functioning of Schools. For the physician who assumes a greater role in the assessment and management of school related problems, a working knowledge of the school environment is essential. Implications concerning a child's strengths and weaknesses cannot be drawn without considering the context within which the child must function. (See Chapter 15.)

Ability to Teach the Principles of Developmental-Behavioral Pediatrics. Because of the precarious nature of training in this area and the lack of faculty with sufficient expertise to provide instruction, the caretaker with a special interest and expertise will be called upon to participate in medical education as well as to provide instruction to professional and lay groups. The quality of such instruction must be commensurate with that of the better established subspecialties. For the full time specialist in an academic setting, skills in curriculum development and knowledge of educational strategies are also important. (See Chapter 63.)

Ability to Plan, Execute, and Evaluate Research. Difficulties inherent in performing research in the behavioral sciences do not lessen the importance of performing such studies. Rather, research strategies and protocols must be well designed to deal with the infinite variables confronting the investigator. (See Chapter 62.) The full time specialist in an academic center must be skilled in research methodology in order that significant advances may continue within this field.

MECHANISMS TO IMPROVE COMPETENCY

Ideally residency training affords the physician the opportunity to acquire the basic skills necessary to achieve competency in dealing with behavioral and developmental issues, as well as to avail himself of training (elective rotations during residency training, fellowships) to increase his skills beyond the basic level. Numerous studies have documented the general inadequacy of present training in this regard. As a result of increasing awareness of these deficiencies, as cited by the Task Force on Pediatric Education and others, significant and important changes in pediatric training on all levels are occurring. (See Chapter 63.)

Numerous opportunities exist for the practitioner to improve his competency in this area. Most physicians view interdisciplinary professional contacts as a valuable continuing source of knowledge. Although clinical experience is also highly regarded, few pediatric caretakers view such experience as an adequate substitute for more formal training.

The physician now has available a plethora of new texts and journals from which to select for ongoing education. Postgraduate courses are continually offered and are viewed as a valuable source of knowledge. Comprehensive programs such as that on "Pediatric Developmental Diagnosis" developed by the John F. Kennedy Child Development Center of the University of Colorado Health Sciences Center, and the American Academy of Pediatrics' "New Directions in Care of the Handicapped Child" are available nationwide. Most preferred by pediatricians is a part time longitudinal clinical experience. Many medical centers offer practitioners "minifellowships," in which for one-half or one day per week, or alternatively several consecutive weeks, the practitioner participates in activities under supervision and receives instruction in assessment and management.

tency and function at a higher level of involvement.

PAUL H. DWORKIN

SUMMARY

There are numerous difficulties in attempting to define the role of the pediatric caretaker with developmental and behavioral concerns. However, multiple factors can be identified that determine the level of involvement of a physician with such children. These include discipline related factors that are inherent characteristics of the field of pediatrics, practice related factors, external factors, and important physician related factors. For a given physician, these factors merge to produce a particular level of involvement. A general classification of pediatric roles includes the primary care physician with no special interest or expertise in development and behavior, the primary caretaker with special interest, and the full time specialist in the area of developmental and behavioral pediatrics. For each role, specific functions can be identified, as can the competency required to perform such functions. Numerous options exist for the caretaker who desires to improve his competency and function at a higher level of involvement.

REFERENCES

American Academy of Pediatrics Committee on Standards of Child Health Care: Standards of Child Health Care. Ed. 3. Evanston, Illinois, American Academy of Pediatrics, 1977.

Casey, P., Sharp, M., and Loda, F.: Child-health supervision for children under two years of age: A review of its content and effectiveness. J. Pediatr., 95:1, 1979.

Dworkin, P. H., Shonkoff, J. P., Leviton, A., and Levine, M. D.: Training in developmental pediatrics. How practitioners perceive the gap. Am. J. Dis. Child., 133:709, 1979.

Dworkin, P. H., Woodrum, D. T., Brooks, K. S., and Marshall, R.: Pediatric-based assessment. Children with school problems. J. Sch. Health, 51:325, 1981.

McCormick, D. P.: Pediatric evaluation of children with school problems. Am. J. Dis. Child., 131:318, 1977.

North, A. F.: Screening in child health care: Where are we and where are we going? Pediatrics, 54:631, 1974.

Oberklaid, F., Dworkin, P. H., and Levine, M. D.: Developmental-behavioral dysfunction in preschool children. Am. J. Dis. Child., 133:1126, 1979.

Richmond, J. B.: An idea whose time has arrived. Pediatr. Clin. North Am., 22:517, 1975.

Shonkoff, J. P., Dworkin, P. H., Leviton, A., and Levine, M. D.: Primary care approaches to developmental disabilities. Pediatrics, 64:506, 1979.

The Task Force on Pediatric Education: The Future of Pediatric Education. Evanston, Illinois, American Academy of Pediatrics, 1978.

Thorpe, H., and Werner, E.: Developmental screening of preschool children. A critical review of inventories used in health and educational programs. Pediatrics, 53:362, 1974.

52

Comprehensive Diagnostic Formulation

As the process of assessment nears completion, the clinician arranges and weighs evidence from various sources to compile a diagnostic formulation, one that will evolve into a therapeutic plan. The preceding chapters in this book have explored a wide range of biological and psychosocial factors influencing children's development and behavior. There has been extensive review of the many possible symptomatic manifestations of these factors, along with consideration of assessment techniques by which they may be evaluated. Only a comprehensive diagnostic formulation that integrates pertinent data from these multiple sources will effectively coordinate the clinician's plan of management, while facilitating discussions about the child with the family and allowing for effective communication with colleagues and referral resources. This chapter delineates one approach to such a formulation.

DEFICIENCIES COMMONLY ENCOUNTERED IN PRESENT DIAGNOSTIC PRACTICE

Diagnostic reasoning commonly employed in clinical practice today may be susceptible to problems of oversimplification of various sorts and to a tendency to view the child too narrowly. Perhaps the commonest weakness in current diagnostic practice is the use of the child's worst problem as the main or only diagnosis. To refer to a child as a "c.p.," "asthmatic," or "drug abuser" may identify the most troublesome focus of parental and professional concern and even may be a useful form of mental shorthand for the clinician. However, such labels fail to consider the important array of relevant strengths and weaknesses of the child and his milieu. Since all children with a specific condition, such as asthma, are not the

same, a false sense of homogeneity may be conveyed through inappropriate use of labels. Furthermore, certain particularly meaningless and perhaps misleading labels are often used as summary statements about children. "Hyperactivity" is a prime example of this practice; the term is poorly defined and means different things to different people (Carey and McDevitt, 1980). "Emotionally disturbed" is another diagnosis that is too vague to convey a specific meaning and is potentially harmful to parents and children.

Another diagnostic distortion occurs when the examiner puts his own main interest or area of expertise first and gives little or no attention to other aspects of the child. To the allergist, the child's hypersensitivities may be taken as his most pressing or only problem. To the family therapist, family dynamics are of paramount and sometimes exclusive significance. Although various aspects of the child and his situation may contribute to a comprehensive diagnostic formulation, no single facet should be mistaken as constituting an adequate account of the total child.

The "problem oriented" approach has its supporters, who maintain that documenting specific clinical concerns clearly insures that they will be remembered and dealt with adequately. One cannot quarrel with that goal. However, the message of this chapter is that unless all pertinent weaknesses and strengths of the child and his situation are assembled into a single formulation, there is a real danger that some complication or critical redeeming aspects of a child will be overlooked.

ELEMENTS OF COMPREHENSIVE DIAGNOSTIC FORMULATION

Having strongly urged a comprehensive diagnostic formulation, one must acknowledge that it

Table 52–1. COMPREHENSIVE DIAGNOSTIC FORMULATION

		Relevant Findings (Strengths and Deficits)	Service Needs
CHILD			

Physical health	General somatic state
	  Organic and functional
	  Nutrition, growth, and physical maturation
	Neurological status
	  Sensory, reflex, motor, coordination		

Development	Capacities
	  Motor
	  Language
	  Information processing
	  Attention and organization
	  Social skills
	  Intelligence		

Behavior	Temperament (style)
	  Clusters (e.g., difficult, easy)
	  Traits (e.g., adaptability, attention)
	Performance (adjustment)
	  Social competence
	  Task performance (especially school)
	  Self-direction, care, esteem
	  Coping style
	  General mental and emotional state (e.g., anxiety, depression)		

INTERACTION WITH ENVIRONMENT			
Input	Parental care (e.g., attitudes and expectations, feelings, management) Sociocultural situation Nonhuman environment		
Outcome	Effect of child on parents and other caretakers Complaints by caretakers		

SUMMARY OF FINDINGS:

PLANS:
  To meet needs

  For follow-up

is difficult, and perhaps imposssible, to present a single scheme upon which all potential users can agree. The following are some basic elements of one articulated diagnostic profile (Table 52–1).

STATUS OF THE CHILD

For purposes of diagnostic formulation it seems reasonable to suggest a three part separation of the status of the child into physical health, development, and behavior. Two of these can be subdivided, making five categories to account for in describing the status and needs of a child.

Physical Health

General Somatic State. An appraisal of the general somatic state describes the child's organic and functional status and includes organic illness, malfunction or handicaps of various organ systems (such as the skin and respiratory and cardiovascular systems), as well as the nutritional status, growth and physical maturation, and such problems as malnourishment, obesity, and disturbances of growth or bodily development.

Neurological Status. The neurological status subsumes sensory and motor function, reflexes, and coordination. Problems in this area might be sensory loss, including vision and hearing, "cerebral palsy," convulsive disorders, "soft signs" (minor neurological indicators), or incoordination. These biologic influences are considered extensively in Parts IV and V of this volume.

Development

The section of the diagnostic formulation on capacities includes the various elements of development and their current level and degree of appropriateness for age. Included would be a child's gross motor function, fine motor skill, language development, visual-spatial orientation, temporal-sequential organization, higher order conceptual abilities, and various aspects of social perception and skills. Also included within this category would be the current status of the child's selective attention and organizational ability. The latter characteristics are hard to classify and, as such, are also found in this comprehensive formulation under temperament. Determinations of intelligence should be included here. Assessment of these various capacities has been discussed elsewhere, particularly in Chapters 45, 46, and 47.

Behavior

The child's temperament or behavioral style should be considered and evaluated independently of his behavioral performance or adjustment. The various dimensions of temperament and the clinical clusters derived from them have been described fully in Chapter 10, and their assessment is discussed in Chapter 44.

As already noted, a quandary arises in regard to the placement of attention, considered to be an aspect both of information processing and of temperament. There may be similarities or differences in the various components of attention required for specific learning tasks and those involved in a child's overall interaction with his environment. For the present, the characteristic can tentatively appear under both headings.

Behavioral performance or adjustment deals with the child's social relations or competence, his task performance and the degree of achievement and satisfaction (especially in school), and his relations with himself, or his self-direction, self-care, and self-esteem. Findings of concern in these three areas might be, respectively, aggressiveness or inhibition, low achievement or satisfaction, and dependency, self-neglect, or self-abuse or low self-esteem. (See especially Chapters 43, 44, and 48.)

Also included at this point are the child's coping or adaptive style (see Chapter 10) and any further observations as to his general mental or emotional state; e.g., evidence of anxiety, depression, or poor reality testing.

Motivation is an important dimension of behavior, but since it is estimated only with great difficulty by the primary care clinician, it is not included as a component of this comprehensive profile.

THE CHILD'S INTERACTION WITH HIS ENVIRONMENT

The interaction of the child with his environment should reflect the ways in which the environment is affecting the child and how the child is altering his milieu. These environmental factors are covered in Part III of this volume.

1. Parental care consists of parental attitudes including expectations (How realistic and how supportive are they?), parental feelings (What is the amount of attachment or detachment and of affection or anger and rejection?), and the actual management of the child (What is the amount and quality of physical care, stimulation, and organization of the child?).

2. The sociocultural situation describes the impact of siblings, other family members, neighborhood influences, television, school, and medical care. Which are helpful and which are not?

3. An appraisal of the nonhuman environment rates the degree of suitability or hazard in housing, pets, environmental substances and so forth.

The impact of the child on his surroundings is a highly important part of the interaction that is frequently either ignored or not explicitly included in diagnostic formulations. One rates here the degree of pleasure or displeasure and satisfaction or dissatisfaction experienced by parents, teachers, and others who regularly encounter the child. It should include a statement about the aspects of the child that are most bothersome and that have led to clinical attention.

At the end of the comprehensive diagnostic formulation it is appropriate to summarize the various details in a single statement or two. One example might be as follows:

> This eight month old male infant is physically, neurologically, and developmentally normal and has a relatively easy temperament, but he has become very demanding of his mother's attention. The reason seems to be that the authoritarian grandmother persuaded the mother that the baby should not be allowed to cry because of his umbilical hernia. The mother has become angry about the baby's demands and thinks that there is something wrong with him.

Sometimes clinical syndromes emerge from certain combinations of findings. For example, the so-called "vulnerable child syndrome" (see Chapter 26) describes the child who is physically and developmentally normal but has a particular pattern of behavioral maladjustment related to the continuation of inappropriate parental concern and handling following recovery from a worrisome early illness. This summary is the section where such syndromes could be mentioned.

PLANS FOR SERVICE NEEDS

Finally it is desirable to select from the list of findings in the comprehensive diagnostic formulation those areas calling for action on the part of the clinician. Not all of the suspected or definite problems need be dealt with. For example, if parents are coping well with a child with difficult temperament, intervention is not indicated as it would be if there were parental-child conflict because of it. Similarly, a pediatrician generally should not attempt to influence the course of a parental divorce but should help the family understand and cope with its impact on the child. The service needs for the demanding infant in the previous section would include sufficient examination to reassure the clinician and the family that there is no physicial problem with the child other than the umbilical hernia, suggestions to the mother about revision of her handling of the baby, and help for the mother in evaluating more critically the advice received from her own mother.

Having thus defined appropriate service needs, the clinician can proceed to implement them, which is the process of management described in the next several chapters in this book. Plans for follow-up complete the formulation.

ADVANTAGES OF A COMPREHENSIVE DIAGNOSTIC FORMULATION

The advantages in the use of the comprehensive diagnostic formulation can be found in practice, research and education.

In practice one gains the assurance in making a complex diagnosis that a broad range of pertinent factors are considered so that relevant issues are unlikely to be omitted. This wide view of the child enhances the clinician's diagnostic reasoning, his discussions with the patient and family, and his communications with other professionals. One might argue that such a comprehensive evaluation is not necessary in the immediate management of acute minor illnesses such as otitis media or gastroenteritis. However, if professional contacts extend into well child care or involvement with chronic physical problems, a broader evaluation becomes very helpful. With concerns in the area of development and behavior, this is a necessity, not a luxury. Plans for management of problems in the latter areas stand a far better chance of meeting the child's needs if they are based on a truly comprehensive, empirical assessment rather than on incomplete data or stereotypic diagnostic labels.

In carrying out research the use of this model of formulation encourages more precise definition of subjects, thereby allowing studies to become more interpretable and more significant. For example, as already mentioned, investigations referring to patients simply as "hyperactive" without any further clarification of their overall function are of little value. In whose view is the child excessively active and how was that determined? Will not the study outcomes be affected by how a child rates in each of the five areas of the comprehensive formulation (i.e., neurological status, temperament, behavioral adjustment, and so forth)? There are serious dangers in attempting to study a child or a cohort identified only by a single symptom.

In medical education the use of this approach to diagnosis would encourage both teachers and students to think of children in terms of their true complexity and avoid overly facile diagnoses based on inadequate information or narrow observer bias.

PROBLEMS IN FORMULATION

One must acknowledge that, although primary care clinicians need to reason comprehensively, this admirable goal is not easily achieved. In the first place, the various professional persons dealing with the children's development and behavior may not agree that the five part division of the child's status is an acceptable one. Various settings and points of view may argue for modification of contents and subdivisions. For example, the advocates of neurology and psychiatry may plead for an expansion of their spheres of interest. One should not object to this as long as the other elements of the formulation are retained and considered in the final diagnosis and service plan.

Another problem is the lack of standardized criteria for diagnostic ratings in some areas, particularly behavioral adjustment. One can agree that this topic deals with the child's relationship to others, to tasks, and to himself, but the dividing line between normal and abnormal is broad and variable. Also, although attention as a temperament characteristic can be given a total numerical score on a questionnaire, it can vary markedly in the information processing area, depending upon the setting in which it is appraised.

How is the clinician to arrive at a comprehensive diagnostic formulation if there is a major area of missing data, as with the pediatrician evaluating a problem of school adjustment without specific data about information processing skills? Clearly he must refrain from proposing a final diagnosis until such assessments are available. On the other hand, the same information may be largely superfluous in other situations, as in the case of helping the child and surviving parent to deal with the death of the other parent. All five areas of the formulation should be borne in mind, but clinical data in each are sought only to the extent appropriate for competent management of the child.

Finally, a major problem in the use of this sort of diagnostic profile is its implementation, that is, in persuading oneself and others to give up old habits of abbreviated and distorted conceptualizations and to think comprehensively.

WILLIAM B. CAREY

MELVIN D. LEVINE

REFERENCES

Carey, W. B., and McDevitt, S. C.: Minimal brain dysfunction and hyperkinesis. A clinical viewpoint. Am. J. Dis. Child., *134*:926, 1980.

Levine, M. D., Brooks, R., and Shonkoff, J. P.: A Pediatric Approach to Learning Disorders. New York, John Wiley & Sons, Inc., 1980.

Rose, J. A.: Pediatric interview manual. Unpublished manuscript, 1962.

Talbot, N. B., and Howell, M. C.: Social and behavioral causes and consequences of disease among children. *In* Talbot, N. B., Kagan, J., and Eisenberg, L. (Editors): Behavioral Science in Pediatric Medicine. Philadelphia, W. B. Saunders Company, 1971.

VII

The Enhancement of Development and Adaptation

Part VII of *Developmental-Behavioral Pediatrics* examines the wide array of therapeutic options. As in Part VI, there is broad representation of professional disciplines. The modes of intervention most commonly applied and considered are elucidated by the individual professions that usually offer them. Such a framework, however, is not meant to imply the existence of rigid boundaries between disciplines. On the contrary, it is expected that the techniques and principles underlying them will have direct relevance to the therapeutic styles and practices of readers from all disciplines.

53

Pediatric Developmental-Behavioral Counseling

To heal sometimes, to relieve often and to comfort always

Edward Trudeau

Counseling in a pediatric context can be defined as a process of helping adequate parents and children make positive changes in their behavior and mental health. Behavior problems are very common in childhood. Some 10 to 15 per cent of all children have behavioral problems that interfere with life adjustment. Physicians caring for children are confronted with behavioral problems virtually every day (Prazar and Charney, 1980; Starfield et al., 1980).

Although counseling styles vary, some common characteristics of counseling in the pediatric setting can be elucidated. Frequently encountered differences between pediatric counseling and psychotherapy are summarized in Table 53–1. The fact that the pediatrician works with basically healthy families has a significant impact on the nature of counseling. In many cases his role consists of delineating and clarifying problem behavior patterns and trying to change them through active advice. Frequently this requires only one or two visits. Usually such efforts are successful and

Table 53–1. THE PEDIATRIC COUNSELOR'S ORIENTATION (RELATIVE TO THE PSYCHOTHERAPIST'S ORIENTATION)

Works mainly with stable children and parents
Focuses more on the present
Focuses more on behavior than on thoughts or feelings
Focuses more on normal development
Requires less extensive evaluations
Leads the interview more (provides less total listening time)
Uses more action oriented, direct, specific approaches
Uses more empirical approaches (if an approach works, one does not need to know the theory behind it)
Uses more behavioral modification
Relys more on education, reassurance, specific advice and environmental intervention
Provides briefer follow-up visits (20 or 30 minutes)
Provides fewer visits (two or three for most problems; six maximum)
Sets a shorter time frame (usually three months)

highly efficient. The primary care physician has an advantage over many other counselors, namely, in knowing how the family operates and having their trust because of his previously established efficacy with physical illnesses. Families who do not respond to pediatric counseling can be referred for more formal psychotherapy at a later time. The sensitive pediatrician usually can detect these seriously disturbed families and refer them early.

This chapter reviews several types of pediatric counseling that fall within the primary care domain. The behavioral problems selected as examples are commonplace ones, and the counseling methods discussed can be integrated into the practicing pediatrician's time frame. Levels of intervention are covered in approximate order of increasing complexity and time requirements. Generally the pediatrician acts at the lowest level of intervention that is effective for the issue with which he is dealing. Pediatricians are in a unique position to be eclectic. Most problems require a combination of treatment approaches (e.g., education, reassurance, advice, and advocacy). Although every pediatrician provides some counseling, individual interest and training vary greatly. Each physician should participate in this aspect of health care only to the degree to which he feels comfortable.

ESTABLISHING RAPPORT AND TRUST

The forging of a strong physician-parent-child alliance is critical. This alliance can be strengthened at every office visit. A number of factors are likely to facilitate the establishment of rapport. These are summarized below:

A family's confidence in the physician's ability. As noted, this frequently derives from his expertise in more general medical matters. Accumulated experience surrounding illnesses is likely to fortify the family's respect for the clinician.

A sense on the part of parents that the physician respects them. Periodic praise for the job they are doing and the conveyance of admiration for them are likely to be helpful.

A sense that the physician really is interested in their problems. This can be communicated through appropriate open ended questions. (See Chapter 43 on Interviewing.)

A sense that the physician really is available to them with regard to behavioral issues. The physician can communicate an interest by asking about behavioral concerns in a relaxed unhurried manner. For certain vulnerable families, longer appointments may need to be scheduled when sufficient time is available.

A sense that the physician has some personal involvement. It is critical that the physician demonstrate that he knows the family as individuals rather than just another constellation of "clients." This approach entails an ability to communicate to the family that the physician has acquired (and remembers) unique information about them. This might include the approximate ages of children, any chronic diseases the children might have, and various demographic details.

A sense of comfort. This can be fostered through certain kinds of courtesies. It is helpful to show the parents respect, calling each by an actual name rather than "mother" or "father." A relaxed atmosphere, perhaps laced with some humor, is essential. The physician can help to create the atmosphere that, even in the face of behavioral difficulties, all is not gloomy and grave.

A sense that the physician really understands the family's struggles. This consists mainly of becoming a good listener and offering undivided attention. It also includes a high degree of empathy. In listening, the physician must project a feeling that he takes seriously the worries of parents and children. Such statements as "That must have been frightening to you" can be helpful. In this regard it is important to avoid suggesting that the parents (or the child) are over-reacting. It is essential that families and children not feel criticized for their concerns. This does not preclude helping them reorder their priorities in a sensitive and delicate manner. In general, whatever causes apprehension should be construed as a legitimate concern.

A guarantee of confidentiality. This implies that the parent (or child) can speak freely in the physician's office about personal matters or sensitive issues. This requires the cooperation of the office staff as well. Sometimes the physician needs to be explicit about confidentiality, particularly when speaking with adolescents. (See Chapter 9.) In that instance an explicit "contract" must be established early in the relationship. Parents may have to agree to this also.

Table 53–2. TYPES OF PEDIATRIC COUNSELING

1. Releasing painful feelings with ventilation—first deal with pressing emotional issues
2. Education—supplying needed general information
3. Reassurance—specific information that counteracts fears
4. Listening and clarifying the problem—providing parents with a clearer perspective about the child's problem
5. Approval of the parents' approach—helping parents to use their own resources
6. Specific advice—suggestions about altered parental handling of specific problems
7. Environmental intervention—suggestions about other changes in the child's environment
8. Extended counseling—more visits for more complicated problems

A meeting of the family's expectations. Substantial insight may be needed for a physician to discern what it is that families actually want. Some seek reassurance. Others desire highly specific day to day management advice. Still others are hoping for a referral. An understanding of what people want and why they seem to want it can be an important element in establishing rapport. The capacity to meet needs and fulfill wants simultaneously can facilitate a helping relationship.

Once a meaningful alliance has been established, specific techniques of counseling can be employed. Most pediatricians tend not to adopt any stereotyped or consistent approach. The nature of the problems, the family's coping style, and the likelihood of a child or parent's benefiting from various approaches are among the factors that need to be taken into consideration. Some general modalities of counseling are covered in the following sections. They are summarized in Table 53–2.

RELEASING PAINFUL FEELINGS*

Some parents and patients are in acute emotional distress when they visit their physician. They are preoccupied with painful issues. Until these painful feelings find an outlet, the parent probably will not be able to relate an accurate medical history or interpret medical advice. Also any counseling that requires thinking (e.g., behavior modification) may be less successful until compelling emotional issues are dealt with.

The process of releasing painful feelings may be called "ventilation," a term that has numerous implications. The angry parent may need to ventilate grievances. The frightened patient may need to express fears (e.g., a hyperventilation attack). The mourning patient may need to grieve about

---

*See also Chapter 43.

the loss (e.g., sudden death of a parent). A patient who has been attacked (e.g., a rape victim) may need to pour out her feelings about what has happened.

The Technique of Ventilation. The setting must be relaxed and private. The patient or parent is encouraged to talk. Usually ventilation begins spontaneously. The process can also be initiated by openers such as "Something's bothering you. Why don't you tell me about it?" More specifically, the physician can state, "You look angry (worried or sad). Why don't you tell me what's bothering you?" The repetition of emotionally laden words that the parent has used will help to continue the process (e.g., "You felt put-down"). The essential response to ventilation is noncritical listening. Censure conveys a lack of license to talk freely. Even if the parent's feelings seem excessive, the physician must express agreement that the situation is "unfortunate." The parent or child does not want to hear that "It could be worse" or "Be grateful that (such and such) didn't happen." The success of ventilation depends on the patient's perception that he has expressed personal anguish to someone who really understands. As a distressed parent or child expresses deep feelings, there may be a recovery of composure and emotional equilibrium.

EDUCATION

Education involves the presentation of facts or medical opinions to a parent or child. Education is undertaken mainly to impart information, but it also serves a critical role in reducing anxiety, dispelling misconception, and fostering feelings of effectiveness on the part of a parent or child.

Requested Education. Education may be particularly effective when it has been asked for. In these instances the timing of the education is optimal. Adolescents commonly ask questions about acne, venereal disease, the prevention of pregnancy, and smoking. These topics obviously deserve thoughtful answers. A family may ask their physician about the pros and cons of getting a dog. The physician can remind them that most children under age three cannot be taught to treat a dog appropriately and risk being bitten. If parents wish to have a dog during this developmental period, they can be advised not to leave their child alone with the dog at any time.

A breast feeding mother commonly asks when she should add solids to the child's diet. A reasonable response would be that solids should be introduced when the child is between four and eight months of age. Since he should actively participate in the feeding process, the child who is younger than four months of age is unable to

sit with enough support to turn his head voluntarily and refuse the spoon when he has had enough. By eight months of age, breast milk usually no longer fully meets the child's nutritional needs.

Some parents ask about the effect of television on their child and seek guidelines. (See Chapter 17.)

Anticipatory Guidance. Anticipatory guidance represents the pediatrician's nearly unique access to preventive counseling. Topics such as nutrition, accident prevention, behavior management, developmental stimulation, sex education, and general health education all may be covered during every visit. Most expectant parents have many questions that can be discussed with their pediatrician several weeks prior to delivery. The most frequent concerns include arguments for and against breast feeding and circumcision, preparation of the breasts if breast feeding is to be used, hospital policies about rooming in and parent-infant contact in the delivery room, separation problems with siblings during the mother's confinement, ways of decreasing sibling rivalry, and essential baby equipment.

Preventive Information about Discipline. The subject of discipline can be brought up and covered during almost every well child visit. The following are typical issues:

Birth to Four Months. Ask about crying. Find out what the parents do when the child cries and be certain that they are not physically punishing the infant. Explain that a baby cannot be "spoiled" during the first four months and may need to be comforted if he is crying.

Four to Eight Months. Remind the parents that punishment is generally unnecessary and ineffective before the child begins to crawl. Encourage them to give the baby the most attention when he is being good, not when he is crying.

Nine Months. Advise the parents to use a negative voice and eye contact rather than physical punishment. Remove the child from dangers. If the child does not respond to disapproval for safety issues, use a single slap on the hand.

Twelve Months. Warn the parents not to discipline the child for normal exploratory behavior, unless it is dangerous. Discuss the importance of positive reinforcement for good behavior.

Eighteen Months. Warn the parents not to discipline the child for normal negativism or toilet training accidents. Review with the parents how to respond more effectively to these two difficult developmental phases.

Two Years. Review with the parents the type of discipline they are using and the importance of consistency. Encourage temporary social isolation using the time-out chair or time-out room. With parents who desire to use physical punishment as one of their discipline techniques, encourage them to hit the child only with their hands, to strike only the buttocks or hands, to hit only once for each infraction, and not to use this technique exclusively.

Three to Six Years. Make sure that the child minds adults properly. Ask the parents whether the child exhibits any behavior that they would like to change. Again discuss the importance of positive reinforcement for good behavior.

Six Through 12 Years. Make sure that the child minds his teacher appropriately. Ask the parents whether there is any behavior pattern that they would like to change. If these issues are dealt with during health supervision visits, such children rarely will be brought in at a later date with discipline problems.

Printed or Audiovisual Approaches. Comprehensive education of the parents can be very time consuming for the physician. More efficient methods are available. Printed materials include information sheets written by the physician, health pamphlets, or books (e.g., Brazelton, 1969). Not only do information sheets save the physician's time, they also give the father (or other family members who were not present during the office visit) the opportunity to read what the physician recommends. In addition, these handouts can provide more information than most physicians have the time to give, and they help to prevent recall problems for the mother. Some offices also have 10 minute audio cassettes or video cassettes that are available in the waiting room to impart information. These aids may be directed at particular age groups or cover specific chronic disease or management issues.

REASSURANCE

Reassurance can be defined as a special kind of education that counteracts fears. Reassurance relieves or removes unnecessary anxiety, especially regarding one's physical or emotional health. Reassurance is the physician's most commonly used type of counseling. Parents need some reassurance during almost every office visit. Reassurance is more likely to be effective if the following guidelines are followed. (See also Chapter 26 in regard to the management of minor illness.)

Properly Timed Reassurance. Reassurance should be preceded by data collection. It should never be too hasty or offered too early. In patients with emotional concerns a careful history should be elicited. Reassurance based on meager data is apt to be hollow and unconvincing to the parent. Only after the parent or child feels that the physician has explored the problem fully and understands it will the reassurance be acceptable.

Specific Reassurance. The most effective reassurance is specific and focused. The targeted concern or worry is identified by listening carefully. For example, a parent may be afraid primarily of a brain tumor in a child with recurrent headaches, appendicitis in a child with recurrent abdominal pains, or a heart attack in a child with chest pains. Once the precise over-riding fear is identified, the physician can carefully investigate that specific concern and offer reassurance when it is unfounded. Blanket reassurance (e.g., "There's nothing to worry about," "Everything will be just fine," or other extravagant promises) leads the parent to suspect the physician of being somewhat dishonest or insensitive and dilutes the value of any specific advice. Pronouncing children with psychosomatic complaints well is a common and often helpful pediatric role. For example, "Your child is in excellent physical health. His pains are due to worrying."

Honest Reassurance. What the physician tells the parents must be honest. If the physician is caught in one lie, the balance of his reassurance will be thrown into question. On the other hand, the physician need not reveal everything he is thinking. Any nonessential data that would be anxiety producing can be withheld (e.g., the differential diagnosis).

Reassurance of Universality. Physicians can offer great comfort to parents and children by commenting on the universality of their problems (when appropriate). A statement such as "That argument goes on in homes everywhere where there's a 16 year old" can alleviate much anxiety.

Nonverbal Reassurance. Nonverbal messages often communicate more to the parent than the physician's words. The physician can show concern for his patient without expressing alarm. He can appear relaxed without being remote. For example, a physician can examine a patient's heart without wearing a worried facial expression. If the parent relates a history of symptoms that have frightened him to a physician who remains calm, he will often conclude, "If this doesn't upset my doctor, I guess everything is going to be all right." In many ways the parent or child takes a reading of the physician's nonverbal cues and determines for himself whether he should feel assured.

Examples of Reassurance. Parents of infants need reassurance that their child's red face and grunting with bowel movements do not mean that the child is constipated. Head rolling and body rocking in infants do not represent an emotional problem but are the baby's way of making a transition into sleep. Thumb sucking in young children is comforting but does not mean that insecurity is present. This example is also a reminder that reassurance is age dependent and that after age five or six, thumb sucking should be discouraged because it can cause malocclusion of the permanent teeth. Although the parents of young infants should be reassured about postural abnormalities of the legs and feet (e.g., toeing in or bowlegs), one must be careful not to raise false hopes about the rapidity of the self-correction. One can reassure the parents that the correction will be complete, but that it will not begin until the child starts to walk and then will take approximately 12 months or longer of walking before the child's legs and feet will begin to look straight.

School age children who are reacting to a divorce need reassurance that visitation with both parents will continue, that their parents still love them, and that home life eventually will return to normal. Children with a history of retentive soiling need reassurance that their bowel movements will be pain free if they take a stool softener and that they do not need to hold back the bowel movements to protect themselves from pain. School phobic children need a "clean bill of health" before their parents will be able to return them to school on a daily basis. Adolescent patients are concerned often about rapid growth and body change. They need reassurance that their particular somatotype, genital size, breast size, and other body parts are normal.

PROBLEM IDENTIFICATION: LISTENING AND CLARIFICATION

Listening. To be effective at counseling, one needs a complete and accurate picture of a problem. A common error in giving advice is offering it too quickly. The parents or child should be listened to if one wishes to understand their world. Any conclusions about whether parents are reasonable or unreasonable should be delayed until they have been allowed to describe their unique situation. Listening in itself is therapeutic; it conveys respect and encourages independent decision making. However, listening cannot be totally open ended. The pediatrician may need to lead the interview because of time constraints on his evaluation. (See also Chapter 43 on interviewing.)

Minimal Psychosocial Data Base. Children with one or two behavior symptoms (e.g., thumb sucking or nightmares) can be treated by offering direct advice if the physician observes a happy child and a positive parent-child interaction. An expeditious approach usually can also be taken with families whom the physician knows from long experience to be stable. In these cases only four additional questions need to be asked. First, "Does he have any other behavioral problems?" If not, "Why do you think she is acting this way?" Third, "What have you already tried?" In this way the physician will not prescribe advice that has

Table 53–3. MINIMAL PSYCHOSOCIAL DATA BASE

1. Number of behavioral symptoms (one or two)
2. Parents' theory of cause
3. Parents' previous approach to problem
4. Stresses on family

already failed. Finally, "Are there any unusual stresses on your family?" This abbreviated evaluation (summarized in Table 53–3) is based on an assumption (supported by a long acquaintance) that the family generally functions well.

Complete Psychosocial Data Base. In children with multiple symptoms (e.g., multiple discipline problems) or complex problems (e.g., encopresis), a complete psychosocial data base should be collected before advice is offered. One must carefully document rather than assume that these parents and children are basically healthy. The content of this data base is outlined in Table 53–4. The same approach should be used with children functioning poorly in more than one setting (i.e., home, school, neighborhood). Some physicians may wish to substitute a questionnaire that the parents can complete at their leisure at home prior to the appointment. If this evaluation is done totally by interview, considerable time may be required. Any child who does not respond to advice offered following a brief evaluation should undergo a complete evaluation.

Clarification of Problems. Clarification involves identification of the problem and an explanation of its possible causes and effects. The objective of clarification is to make possible the parents' understanding of their child's behavior. The physician must carefully define with the parents' help the behavior patterns they want changed. The parents must have the final word

Table 53–4. COMPLETE PSYCHOSOCIAL DATA BASE

1. Chief complaint
   Parents' main concern
   Parents' theory of cause
   Parents' previous approaches to the problem
2. The child
   Other types of misbehavior
   Frequency of criticism or punishment (times per day)
   Child's good points
   Peer status
   School status
3. Child rearing style
   Strict vs. democratic vs. lenient
   Number of child rearers
   Consistency vs. inconsistency
4. The family
   Constellation
   Marital status
   Family stresses
   Support systems

about the selection of target behavior. The physician can state, "If I understand you correctly, you are most concerned about. . . ." Pediatricians, however, may not have enough time to allow parents to work out their own understanding of the cause of their child's problems. Once the physician understands the situation, he may explain it in understandable terms. In some cases the parents are either too strict or too lenient. Sometimes the central issue is a vicious cycle or power struggle (e.g., pressure brings resistance; constant criticism leads to giving up and depression). The parents should be given credit if their analysis of the problem seems correct. Once the physician has presented an interpretation to the parents, he can ask whether it makes sense to them.

Reducing Parental Guilt. The parents of children with emotional problems usually feel somewhat responsible and guilty. Once the parent-child relationship has been examined and parents have been advised to change what they are doing, this guilt is inevitable.

Guilt can be reduced in several ways. The guilt can be universalized (e.g., "Everyone tries that"). The physician can personally absolve the parents (e.g., "I can easily understand why you tried that"). The blame can be shared with schools, relatives, siblings, and other etiological factors (e.g., "Your actions were just one of the reasons behind this problem"). The parents can be reminded that the harm was not intentional. Also the parents' errors can be relegated to the past (e.g., "That was long ago and much has happened since then"). Mainly the physician can show empathy and emphasize "All parents make some mistakes, and that is part of being an involved parent." Positive aspects of the parent-child relationship can be underscored. Sometimes problems stem from parental leniency and overindulgence, and the physician can state, "You love him too much" or "You tried too hard." One can end with the viewpoint that "The main need now is to look ahead rather than behind."

APPROVAL OF THE PARENTS' APPROACH

A definite trend in the delivery of health services is self-care. Parents are being encouraged to become active participants in their family's health service. Just as parents learn how to manage coughs and colds by themselves, the common sense they possess about human behavior should also be supported. The physician is in a position to foster independent decision making. Not only is this approach sound economically, it also enhances parental self-esteem and instills a sense of competency. Inexperienced parents are often overanxious and insecure. They need to be temporarily dependent on their physician. Bringing such parents to a level of independent problem solving and self-care is a gradual but achievable process.

Reinforcement of Parents' Strategies. After clarification of a problem, parents often formulate their own custom made treatment plan. Others can be encouraged to propose a solution. One can ask, "Do you have any idea how you might want to treat this?" One can draw out the parents' deliberations over possible approaches. The physician can approve of their previous actions if they were at all reasonable. He may encourage them to proceed with a change they had considered. He can endorse their plans and encourage them to adhere to them. Often they seek the physician's approval to do what they wanted to do anyway (e.g., use a pacifier). In this way independent thought is encouraged and the parents' self-confidence is strengthened. The physician constantly operates on the premise that a wide variation of workable approaches exists for most problems and that the selection of a strategy must take into account the parents' culture and value system. Since the parents will have to live with the consequences of the plan, they should be encouraged to arrive at final management decisions themselves.

Approval of Parenting. Parents can be complimented regarding their parenting skills during every visit. The fact that they love and care for their children can be mentioned. Parents of children with emotional problems are usually on the defensive and need to have their self-esteem bolstered.

Avoidance of Criticism. Criticism of parents has several unfortunate side effects. First, it engenders guilt. Many parents normally blame themselves for causing their child's sickness (e.g., by something they fed the child), and the physician should alleviate rather than accentuate such self-accusation. Second, parents who are criticized may become angry at the physician, and his medical advice then may be followed poorly. Even harmful approaches usually can be changed without causing anger by stating, "Recently we have found that a different approach may be even more helpful."

SPECIFIC ADVICE

The physicians should make specific recommendations for the relief of symptoms. Advice is indicated whenever a simple behavioral problem exists for which the parents have not devised an approach. The direct giving of advice is the mainstay of brief counseling. (See also Chapter 54.) Suggestions about child rearing are among the most common types of advice offered. Standard advice can be given for symptoms without a differential diagnosis. More individualized advice must be prescribed for problems with several variations or etiological subtypes. Practical, clear-cut instructions are more likely to be successful. Pediatricians should have treatment packages (consisting of one to 10 pieces of advice) for all common parental complaints. In emotionally healthy individuals one does not need to worry about symptom substitution. The physician should restrict advice to his areas of expertise. The physician should avoid giving speculative advice in areas in which he is not trained no matter what they may be. Examples of specific advice giving geared to the age of the child are contained in Chapters 5 to 9.

The following is one more specific and very familiar advice giving scenario: Negativism is a normal healthy phase seen in most children between two and three years of age. The perspective that this phase is important for self-determination and identity needs to be shared with parents. To the child, "No" means "Do I have to?" It should not be confused with disrespect. If the parent can keep a sense of perspective or even humor about this phase, it will only last six to 12 months. Second, the child should not be punished for saying "No." Third, the parents should try to minimize their directives and rules; they should avoid unnecessary demands and keep safety as their main priority during this time. Fourth, they should give the child extra choices and alternatives to increase his sense of freedom. Examples are: letting him choose the book he wants to read, the toys that go into his bathtub, and the fruit he wants for a snack. The physician can ask him which ear he wants looked at first. The more quickly the child gains a feeling that he is a decision maker, the more quickly this phase will be over. Fifth, the child should not be given choices when no choice exists. Taking a bath and going to bed are not negotiable. Sixth, when a request must be made, it should be presented in as positive a manner as possible (e.g., "Let's do this or that"). Confrontation should be avoided. The parent must avoid the two extremes of punishing the child or giving in to all the child's "no's."

Gain the Parents' Acceptance of the Advice. The physician needs feedback from the parents about the advice that has been suggested. To avoid confusion, he can ask the parents to repeat back the substance of what he has said. He can say, "Please tell me what I have gotten across." If misunderstandings are present, they can be ironed out before action is taken. To avoid noncompliance with advice, he must also ask whether this particular advice is acceptable to the parents. He can ask, "Does that seem reasonable to you?" or "How do you feel about that approach?" If the parents seem unconvinced, he must decide whether to persuade them to accept this particular advice or to suggest another option.

Write Down Advice for the Parents. The physician should jot down the main suggestions that the parents have agreed upon and give it to them as they leave. He might use carbon paper, so that his records also contain a copy of the treatment plan. In this way he can be assured that the plan will not be undermined by forgetfulness. Parents usually appreciate this added demonstration of concern.

Follow-up Visits. If advice is given, the results of the advice should be learned. Advice should be followed by at least one visit or phone call. This is in contrast to prior techniques of reassurance and education in which follow-up may be optional. If more than two follow-up visits are needed, probably the physician is entertaining the need for extended counseling and needs to stop to acquire a more complete psychosocial data base and a more precise concept of the etiology. The focus of the follow-up visit is on how the problem has been responding to the advice. In some cases the parents can keep a written record that can be reviewed during the visit (e.g., a diary of triggering events in a child with breath holding spells).

Pitfalls. A common pitfall in giving advice is rigidity on the part of the physician. The gap between the physician's request of the parents and the behavior that the parents are willing or able to provide should be kept to a minimum. If the physician's expectations are too high, he will lose the family to follow-up. Advice should always be broached as a consideration rather than a dogmatic order.

Two examples of areas in which physicians commonly give advice that is in conflict with parents' inclinations are the child's sleeping in the parents' room, and weaning. Some mothers (especially those who are breast feeding) prefer to have their babies sleep in a cradle in the parents' bedroom until they reach an age at which nighttime feedings are unnecessary (i.e., three or four months of age). No proven harm comes from this approach. Although many physicians expect the baby to be weaned from breast or bottle by one year of age, many parents wish to continue breast or bottle feeding to two years of age. If the infant feeds only from the breast or bottle three or four times a day, also eats solid foods, does not carry a bottle around with him during the day or go to sleep with it at night, and is otherwise developing normally, the physician need not challenge the parents' position in this matter.

ENVIRONMENTAL INTERVENTION

Environmental intervention consists of recommendations for specific changes in the patient's environment. These recommendations attempt to reduce factors that are contributing to the patient's problems or to mobilize people outside the family unit who can help. Environmental intervention is a part of many treatment plans. The physician becomes effective in this sphere after he acquires a thorough knowledge of the community's resources. Often a social worker can advise him when he is uncertain about available help for a specific problem. Usually the school system and other agencies respond positively to the physician's suggestions.

In simple problems, environmental manipulation may be curative (e.g., a night light for the child who fears the dark). In children with multifactorial problems, it may offer temporary improvement while counseling paves the way to more permanent solutions (e.g., school phobia). In complex, long standing problems, environmental manipulation should not be substituted for referral because it is seldom a panacea or shortcut (e.g., multiproblem families). Environmental interventions on behalf of the patient at home, in the school, and in the community are best illustrated with several specific examples:

Home Recommendations. Home recommendations can be used to change the home environment. For discipline problems a time-out room can be designated and prepared. A quiet place can be provided for study. For sleep problems the baby can be moved to a separate room, the older child can be given a bed of his own, or the adolescent can be permitted to decide about his own bedtime. Once a child climbs over a crib railing, the railing should be left down or the crib mattress placed on the floor until a regular bed can be obtained. Chores or allowances can be increased or reduced, depending on circumstances. The television set might be disconnected temporarily to encourage studying or conversation. Pets provide an on call, 24 hour per day, affectionate, cuddly relationship. Pets can be acquired for shy or anxious children, as well as for handicapped ones. They may also be helpful during times of loss or loneliness, as following a move, death, or divorce.

School Recommendations. The following recommendations can be implemented to improve the child's school environment. Nursery school or Head Start school may be indicated for the child who is overprotected or understimulated. The child with a physical handicap may require that his classes be on the ground floor. Children with learning disorders require remedial classes or tutoring after school. (See Chapter 38.) High school students may be enlisted as tutors for younger children.

In the case of the child who has missed considerable school because of a prolonged illness or school phobia, the physician needs the school's help in returning the patient to full school attendance. (See Chapter 15.) The homeroom teacher can serve temporarily as a surrogate mother and make the return as nontraumatic as possible. Someone representing the school may have to call for and accompany the child to school for a week. For the student who develops physical symptoms of anxiety while at school (e.g., abdominal pain), the physician may need to request that the school nurse permit the child to rest periodically in her office for 15 minutes rather than sending him home. For some anxious children the physician may need to request a decrease in school stress (e.g., a change in teacher, a shower excuse, or a gym excuse). For the child whose parents are going through a divorce the physician can alert the child's teacher that he needs extra understanding, a temporary reduction in criticism, and possibly the assignment of a popular older student to help him. Most children with problems can receive considerable support from their teachers if the pediatrician keeps them informed as to the child's special needs. When a child is seriously disturbed, the physician may request the school's assistance in recommending that the parent take the child for psychotherapy.

Community Recommendations. The most helpful general advice that can be given is "full activity on doctor's orders." This advice is especially beneficial for the depressed child or the overprotected child. Children with socialization or peer avoidance problems need more peer contact time. The possibility of joining clubs, teams, or other recreational outlets should be explored. Shy, anxious children need the reassurance of adult supervision and structured activities. A summer camp program serves a similar purpose, but the camp counselors must be prepared to deal with homesickness. Special camps exist for many children with chronic diseases or handicaps. Infants with developmental delays due to environmental deprivation might be enrolled in stimulation programs. Some children are helped by adult friend programs (e.g., Big Brother or Child Companion Programs).

The physician can interact with people from the patient's neighborhood and community in other ways. He may help the adolescent to obtain his first job by placing a call on his behalf. He may need to encourage the adolescent to change jobs when he is locked into one that is dissatisfying. Lessons in athletics or the arts can be encouraged to help the success-deprived patient. If a patient is overextended and under considerable stress, he can be encouraged to decrease his extracurricular activities or even discontinue a job.

Mobilizing a Support System. Physicians understand the value of support systems and can help mothers mobilize these. Taking care of a newborn during the first three months of life often requires at least two adults. The extended family may need to be enlisted if the mother has not done so. It is crucial that a relative or friend help care for siblings and assist the mother in obtaining naps, so that she will not be excessively fatigued. Sometimes a support system exists but needs to be consolidated. The father should be invited to come to a health supervision visit during the first year of life (as should a grandmother if one lives in the home), so that he will know that his child's pediatrician values his input and also so that he will be more accepting of the physician's telephone advice in future acute illnesses.

Twenty per cent of children live with single parents and these parents need special support. (See Chapter 13D.) Often two such parents can live together and share the responsibilities for child rearing. Cooperative sitting and shopping often can be arranged. Mothers who work can be reassured that there is no evidence of short or long term harmful effects of daycare. When no support group is available, the physician and public health nurse may temporarily provide a support system for the mother. Volunteers also may be helpful, especially mothers who have successfully managed a similar problem in their own child (e.g., colic, breast feeding, breath holding, or attention deficits). A physician may decide to keep a card file of the names and the phone numbers of successful mothers who are willing to provide such support and teaching. For the family in a serious crisis, temporary placement of the children with a relative, friend, or even foster home may need to be considered.

Implementation of Environmental Intervention. In order of increasing time commitment, environmental change can be initiated by having the parents do everything, making a telephone call oneself, writing a letter, or attending a conference. Having the parents explore the possibilities in their neighborhood and then coming up with a plan constitutes the easiest approach (e.g., finding an extracurricular activity for their child). If a parent-teacher conference is the recommendation, clearly the parents can carry out this plan without the physician's further input. Telephone calls by the physician to other agencies or professionals can have an important impact (e.g., calling the clergyman when the family has suffered a tragedy). More commonly the physician makes phone calls to relatives (e.g., calling grandparents for

support or calling the father if he is unreasonable about child custody or when the disciplinary approaches of the father and mother are highly inconsistent). A brief letter takes little more time than a telephone call. Often the physician writes the teacher, principal, school nurse, counselor, social worker, or several of these people regarding school recommendations. He may need to phone a camp director to gain special permission to allow a handicapped child to attend camp. Scheduling a special conference in his office may be the only way of dealing with an alcoholic father who is having a devastating impact on his son. Occasionally the physician will need to attend a school staff meeting (e.g., when one of his patients has frequent seizures in school).

EXTENDED COUNSELING

In brief counseling, specific advice or options are offered for one or two isolated behavioral symptoms. Good results are expected with one or two follow-up visits. Direct advice can be given after a minimal psychosocial data base has been obtained. Every pediatrician should provide brief counseling and advice. By contrast, extended counseling requires longer visits and more extensive contacts. Extended counseling is needed for children with multiple or complicated symptoms and should be preceded by the obtaining of a complete psychosocial data base. (See Table 53–4.) Extended pediatric counseling may require four to six visits (or more).

Two examples of extended counseling that are common in the practice of pediatrics are psychosomatic counseling and discipline counseling. Pediatricians must be fully trained to evaluate and treat children who have any symptom that might stem from organic causes as well as psychological ones. No other profession has the background to assess these complaints efficiently and completely. Skills in discipline counseling or child management counseling are a prerequisite to the enjoyment of pediatric practice. The 1978 Report by the Task Force on Pediatric Education documented that 49 per cent of parents needed help from their pediatrician with discipline problems. Child rearing problems are mentioned during at least one-quarter of office visits.

PSYCHOSOMATIC COUNSELING

Physicians are essential to the evaluation of acute and chronic psychosomatic complaints (e.g., acute hyperventilation or recurrent abdominal pain). In managing these children, the most difficult step is to get the parents to change their focus from organic to emotional issues. The pediatrician can best negotiate this transition by performing a thorough physical evaluation, obtaining reasonable laboratory studies, and then discussing his formulation of the problem with both parents. The family will demonstrate less resistance to a psychogenic explanation of the symptom if this diagnosis is reached by the pediatrician rather than a psychiatrist. Psychosomatic counseling includes emphasizing that the symptom is real, explaining its temporal relationship to stress, discussing target organs, and offering a treatment plan that reduces the symptom.

DISCIPLINE COUNSELING

Discipline counseling or child management counseling can be based on principles of behavioral modification. The main advantage of these principles is that they can be easily taught to the parents and can be used to effect change in the child without direct involvement of the physician or other professionals. These principles can also be applied to a wide variety of behavior problems or situations. Once learned, this approach can be transferred to the rearing of future siblings. These advantages are summarized beautifully in a Chinese proverb that states, "Give a man a fish and you feed him for a day; teach him to fish and you feed him for a lifetime." Behavioral modification works best, however, if it is preceded by listening, problem clarification, and other previously discussed parts of counseling. The eight steps in discipline counseling are discussed in the following sections (see Table 53–5).

Teach the Basic Principles of Behavior Modification in the Office. Most parents lack a system for disciplining their children. The following four principles of behavior modification can be reviewed. (See Chapter 55 for details.) (1) All behavior (appropriate or inappropriate) is learned. (2) Behavior is shaped predominantly by its consequences. (3) If the consequence is pleasant (a reward), the behavior is more likely to be repeated. (4) If the consequence is unpleasant (a punishment), the behavior is less likely to be repeated.

Table 53–5. STEPS IN DISCIPLINE COUNSELING

1. Teach the basic principles of behavioral modification
2. List the types of problem behavior
3. Assign priorities to the problems
4. Devise a treatment plan for each problem
5. Demonstrate appropriate responses in the office
6. Praise the child for adaptive behavior
7. Write down the treatment plan
8. Provide follow-up visits

(See Chapter 10 in regard to temperament differences, especially adaptability, in children.)

The three components of a behavior modification system (rules, punishment techniques, and rewards) should be discussed. Rules are the parents' statements regarding the desired and undesired types of behavior. Punishment techniques include ignoring the behavior (extinction), verbal disapproval, temporary isolation ("time out"), physical punishment, and temporary removal of a privilege. Rewards or positive reinforcements include praise, attention, special activities with the parent, star charts, candy, and money. Punishments (i.e., actions) are useful mainly in stopping the child from doing something inappropriate. Rewards or words are more helpful for encouraging a child to do something he is not doing (e.g., for improving school performance; Drabman and Jarvie, 1977).

List the Types of Problem Behavior. The parents can be asked to describe the behavior patterns that they and the child's teacher object to. The parent should also describe the punishment or consequences used for each of these kinds of misbehavior. Preferably this information is collected in advance by the parents, who bring in a list or circle items on a behavioral check list. Vague descriptions of the misbehavior (e.g., hyperactive, irresponsible, mean) are not helpful. Rules must be built around specific misbehavior patterns, such as "has trouble concentrating," "doesn't complete chores," or "hits his brother."

Give the Problems Priorities. The parents will be more successful if they concentrate on the two or three most important rules initially. The highest priority is given issues of safety in which the child may harm himself (e.g., running into the street). Of equal importance is the prevention of harm to others, such as the parents, other children, or animals. Destructive behavior toward property is of the next importance. Finally, any behavior that infringes on the rights of others can be considered. Types of misbehavior that should not be dealt with during the first visits are those that involve a body part and that therefore are uncontrollable by the parent if the child decides to continue the power struggle (e.g., soiling or wetting). If the family has serious social problems, specific crisis and safety issues can be brought up during the first visit, but child management counseling should be deferred. Once the types of misbehaviors are given priorities, the physician himself should tell the child that he will be corrected for infractions of certain rules and what the desired adaptive behavior is. The list of new rules should be posted in a conspicuous place in the home, and the child should be informed of the purpose of this placard.

Devise a Treatment Plan for Each Problem. An

appropriate parental response for each of the child's types of misbehavior should be defined. The two most helpful techniques are ignoring the behavior and temporary isolation. Any behavior that is harmless can be successfully removed by ignoring it (e.g., temper tantrums and whining). Temporary isolation is the mainstay of discipline counseling, especially for the child who has received too much spanking. Time out can be accomplished in a crib, playpen, chair, or bedroom. Time out must be quiet time and should not start until the child stops crying and protesting. The duration of the quiet time is brief (e.g., one minute per year of age). Time out is so effective that it is taught to all parents. In addition, parents are asked to discontinue all physical punishment for two months, since most children with disciplinary problems are already too aggressive. The physician must also gain the parents' acceptance of this treatment plan. Most parents can have time out explained, be helped with the designing of approaches to two problems, and then sent home with encouragement to come up with reasonable treatment plans for the remaining problems before the next visit.

Demonstrate Appropriate Responses in the Office. The physician can use the office visit to teach the parents the child management responses he wishes them to imitate. He can help the parent ignore the child who is having a simple temper tantrum. If the child hits the physician, he can state, "You must be angry with me. We don't hit people. Next time tell me you are angry instead." If the child is out of control in the office and the mother is unable to administer discipline, the physician can take over. Obviously he should never spank a child in his office. He can use a time out chair facing a corner effectively. Even when the child is not misbehaving, the physician may wish to rehearse the time out chair regimen in the office if the mother seems hesitant about accepting its value. One can tell the child, "We're going to pretend you just pushed your sister. That kind of behavior is not allowed. You are going to have to go to the time out chair for four minutes of quiet time. I'll tell you when you can get up." The physician can quickly guide the child into the chair, keep a serious demeanor, and start the time out routine over if the child does not comply. He may then wish to transfer some of his power to the parent by having her also rehearse the time out procedure in his office setting.

Praise the Child for Adaptive Behavior. For every type of maladaptive behavior there is an adaptive behavior pattern that the parents desire their child to manifest (e.g., "Don't push Paul" versus "Play with Paul"). Parents commonly forget to give positive reinforcement to their children when they are being good. Parents can be re-minded that the positive feedback a child receives each day must outweigh the criticism and other negative feedback. Parents can also be reminded that any positive attitude or behavior that they wish to teach their child must be modeled in their own lives (e.g., hard work, kindness, and impulse control).

Write Down the Treatment Plan. (See foregoing section on specific advice.)

Follow-up Visits. The physician can ask the parent to keep a written record between visits. This diary can be used to record the frequency and duration of the instances of misbehavior as well as the result of punishment. This provides material for discussion. The parents also can be requested to continue to discuss with each other at home at least once a week the issues that were raised during the counseling session. If the mother seems to lack confidence or commitment, the physician can provide telephone support approximately 48 hours after the initial visit. The second visit should be scheduled approximately one week after the first one. The problem identified should remain the focus of follow-up visits. One can assess progress on the basis of symptom elimination, symptom improvement, a lack of change, worsening of symptoms, or the occurrence of a new symptom. The physician can then refine and recalibrate the treatment plan with the parents' contribution. The parents should be congratulated about any success they have had. At this point the physician usually can set a time limit on the number of sessions he plans to set aside for this problem (usually two to six). The family is next seen two to six weeks later, depending on the progress that has already been made. If the parent on the next visit states that progress has been good, the physician can ask, "What does he still do that wears you down?" If the treatment plan fails after several visits, the family should be referred to a mental health resource. (See Chapter 54.)

Other Options. One efficient way to provide discipline counseling is to have a brief first visit and send the parents home to read a book that teaches the principles of behavior modification. The physician can guide the parents to preferred books such as Christopherson's *Little People* (1977) or Patterson's *Living with Children* (1975). The parents can then make a list of target behavior patterns, attempt to apply what they have learned, and come up with a plan for each problem. They can then return to the physician's office if there are specific questions. The parents might also wish to enroll in a child rearing discussion group, if available (e.g., a Systematic Training in Effective Parenting program; Dinkmeyer and McKay, 1982). The benefits of the group approach are a lower fee for the session, a reduction in the parents' guilt about causing the problem, a sense of helping other people, and an increased likelihood that the parents will accept observations and recommendations made by their equals. Some physicians may consider leading or co-leading child rearing group sessions in their office as long as the problems dealt with are relatively minor. The total number of parents should probably be kept to less than 15. One format is to have two evening sessions geared to a particular age group and offered every other month. Infants, preschoolers, school age children, and adolescents can be discussed on a rotating basis.

Pitfalls. Some parents expect shortcuts and quick cures for discipline problems. They return on the second visit and state that the treatment plan did not help. They plead with the physician for a completely new treatment plan. If the physician is already using a sound approach, he should modify it rather than replace it. A good example is temporary isolation, which usually is effective for discipline problems if used properly. Sometimes parents are frustrated because of unrealistic expectations. They should be warned at the first visit that there is often an early upsurge of maladaptive behavior when a new discipline system is initiated. This initial "response burst" is especially common in children with strong willed, difficult temperaments. (See Chapter 10.)

Treatment failures are not always due to excessively lenient or inconsistent parents. Some result from negative parent-child relationships or situations in which an intense power struggle is present. Discipline counseling as discussed in this section often is not effective until the relationship improves. A child tends not to respond to disapproval until he has known approval and love. Special attention must be paid to having the positive surpass the negative feedback for the child. Also the parents should spend a regular period, perhaps a minimum of 15 to 30 minutes daily, doing something the child enjoys. Some children are truly rejected or neglected and need referral for family therapy. (See Chapter 54.)

OPTIONAL TYPES OF EXTENDED COUNSELING

Three additional areas in which counseling is most commonly requested center around divorce, school problems, and adolescence. Involvement in this additional counseling should be considered optional for the busy practicing pediatrician. The physician who elects to engage in these areas of expanded counseling must set attainable therapeutic goals regarding what will and will not be attempted. A time limit of perhaps four to six visits should be discussed with the family on the initial encounter. Although the physician will use some behavior modification and advice, most extended counseling entails active listening, clarification, and support. Each type of counseling will be discussed briefly.

Divorce Counseling. Over one million children are affected by divorce for the first time each year. The main goal of counseling is to minimize the harm to the children. The approach is to convince parents that children need a secure relationship with both of them and that they must keep their sons and daughters out of their battles. Together the parents must agree not to argue about issues such as visits and support in the children's presence. Parents should carefully elucidate the good points as well as the bad points of the spouse. The children should be told that they need not take sides, that both parents love them, and that they will never need to choose between them. Visits should be scheduled, predictable, and remembered. When the children are told of the impending divorce, they need to know that the parents are to blame, not they. They also must hear that divorce is going to be final and that there is nothing they can do to reunite the parents. They need to be reassured that both parents will continue to care for them and see them and, if possible, that their school and home will not change. The acute symptoms of anxiety, anger, and sadness often require that opportunities be provided for ventilation. A pediatrician can be very helpful to families at the time of marital discord, separation, or divorce. (See also Chapter 13D.)

School Problem Counseling. The main goal of this type of counseling is to insure that the child's educational needs are assessed and adequately met. Parents sometimes consult the physician because they are unhappy with the school's approach to their child but are unable to clarify what is actually happening within the school. The parents can be told that they have a right to know what their child's problem is and what the school proposes to do about it.

Following the school staff's assessment, the parents should receive a full report concerning the child's problem. They should also be told who the school contact person is for their child. Usually meetings with the special education teacher, the school psychologist, or both will assure the parent that the appropriate actions are being taken on behalf of their child.

If parents remain dissatisfied about the school's handling of their child's learning problems, the physician can attempt to review the school problem for the family. For children who have been evaluated by the school, the physician can have the parents sign release forms and obtain copies of evaluation summaries from the school. The physician should also request a copy of the child's individualized educational plan and current behavioral objectives. Pediatricians vary in the kinds and degrees of evaluation and counseling they then offer. (See also Chapter 47.)

Adolescent Counseling.   Adolescents have difficulty in expressing verbally what is happening to them. A goal of counseling is usually to improve self-esteem and increase their freedom of communication. Their main need is for an objective, active listener who can respond with understanding statements, a paraphrasing of what they are trying to say, and a clear interpretation of reality. Listening proves to adolescents that someone really cares about them. In a sense they need an adult-adult relationship. The physician can fulfill this requirement. If the adolescent can put his strong feelings into words, they are less likely to be expressed in regretted actions later. For the adolescent who is unable to deal with a particular issue, the physician can help him talk by using Rothenberg's third person technique: "Most girls your age are concerned about . . . ." If a difficult question is raised during the interview, the physician can resist answering it with direct advice and instead request the adolescent go home and take an hour out of the next week to think about it. Usually the adolescent will return with an appropriate resolution.

Some adolescents have strong, repeated disagreements with their parents, and alienation counseling is in order. The physician's goal is to negotiate a feasible compromise between the adolescent and his parents by meeting with all of them. Often, if given the responsibility, the adolescent can arrive at rules for dating, driving, dress, and family chores that are acceptable to the parents. If not, the physician can arbitrate the differences. There are several books that can help parents improve their communication skills (e.g., Ginott, 1969). (See also Chapter 9.)

Self-help Counseling.   Counseling that is intended to increase the patient's commitment to a treatment goal can be called self-help, self-control, or motivational counseling. The key to overcoming problems like obesity is helping the child or adolescent take responsibility for the symptom. The patient must become an active participant in the entire treatment process. The adolescent must understand that the problem is his and that the physician and parents can only offer suggestions and support. The physician can also provide some measure of accountability (e.g., weekly office weights). Self-help counseling is also useful in dealing with cigarette smoking, a poor exercise program, environmental stresses, deficient study habits, and other maladaptive tendencies that are difficult to change. Usually children under age eight are not sufficiently motivated to participate

in such a program. A treatment plan can be developed by the physician and contracted with the older child or adolescent. Another approach is to ask the youngster to develop a treatment plan and bring it in for discussion with the physician. Such patients often benefit from cue cards or reminders placed in conspicuous locations throughout the house (e.g., mirrors and doors). (See Chapter 29 for a detailed discussion of obesity.)

Group Counseling Options.   In terms of training and cost effectiveness, the physician is not the best person to manage many of the health needs reviewed in the previous paragraph. Sometimes a psychologist, social worker, or former patient may offer group sessions for these conditions. Courses may be provided at a local church or college. Sometimes a good book or an audio cassette series is available on the topic. Some junior high schools, for example, offer free group counseling for children failing in school or those reacting to a divorce. If the pediatrician decides to lead a group by himself, it should not include seriously disturbed parents or children.

Pitfalls.   One error in extended counseling is taking on a patient who clearly needs long term psychotherapy. A variation on this error is to remain involved with a case despite a lack of progress. Children with serious emotional problems should be referred to a mental health setting. (See Chapter 60.) Major education problems require the collaboration of an educational specialist. Multiproblem families should be referred to a social worker. Some patients can be referred on the initial visit (Table 53–6). If progress has not been accomplished by the fifth or sixth session of extended counseling, referral often is indicated. Referral should not mean abandonment. Continuing contact is important for the patient's emo-

Table 53–6.   CRITERIA FOR REFERRAL TO MENTAL HEALTH RESOURCE

1. Life threatening behavior (e.g., suicide attempt, homicide attempt, drug abuse, poisoning in children over 10 years old)
2. Chronic aggressive or antisocial behavior (e.g., juvenile delinquency, runaway, firesetting, repeated stealing, extreme cruelty to animals)
3. Bizarre behavior (e.g., autism, thought disorder)
4. Chronic withdrawn behavior (e.g., chronic depression)
5. Academic failure on emotional basis
6. Victim of repeated physical or sexual abuse
7. Rejected child (i.e., parent can describe no good points)
8. Previous psychiatric hospitalization
9. Previous unsuccessful psychotherapy
10. Multiproblem, dysfunctional families
11. Inadequate progress after six or more sessions of pediatric counseling

tional health and also increases the prospects of compliance with the referral. Some patients also have physical symptoms that require ongoing review by the general physician. The physician also can be helpful in monitoring the quality and effectiveness of the therapy. In addition, he can assess over time the appropriateness of the match between a therapist and a family. Moreover, the physician can support and reinforce compliance with therapeutic recommendations. In helping to select a therapist or agency, the physician should insist upon one that provides him periodically with information about the family's progress.

THE LOGISTICS AND ECONOMICS OF COUNSELING

Although some may feel that it is unrealistic for the pediatrician to become involved with time consuming behavioral problems, he is very well suited for this role. Most primary physicians are efficient people. If he is the regular doctor, he has two additional advantages. He knows the family well and the evaluation can be done in much less time. In addition, the parents already trust his advice. He can attain the same results that it would take an unknown counselor much longer to achieve. This section reviews some aspects of office organization that may improve the physician's efficiency in counseling.

Special Issues with Adolescents. In general, the physician should try not to split the family unit either during an evaluation or treatment. However, adolescents sometimes need access to a personal advocate and protected confidentiality. (See Chapter 9.) In most offices patients over age 12 are seen separately from their parents. This approach is essential for the older adolescent and somewhat optional for the younger adolescent. The private interview helps to build independence and allows the adolescent to disclose his troubles more fully. To prevent the overlooking of a major concern of the parents, the office nurse can ask the parent after she has placed the teenager in the examination room: "Is there anything special that you want the doctor to know before he sees your boy?" If the teenager is younger than 15 and the parent comes in with him, they can be seen together for the interview if the parent implies that they prefer that. The main error is not in seeing them together but in seeing the parent alone, which causes the adolescent to suspect that the adults may be conspiring against him. The parent then can leave during the physical examination, and the physician can defer more sensitive questions to this time.

Counseling: Whom to Include. The counseling time spent with the child compared with that with the parents increases with age. Children less than six years of age often can be treated by working exclusively with the parents (ideally both of them). The child can be left with a sitter while the parents meet with the physician. Benefits of this approach are that once the parents have learned behavioral modification principles, they can become the primary therapists for their child around the clock. The youngster needs to come in only if he needs to overcome a problem that requires special motivation on his part (e.g., thumbsucking or daytime wetting). By school age the parents and children often are seen together and therefore share the counseling time equally. If the adolescent has a personal problem, the parents may not be seen at all. If the difficulty is largely a family communication problem, the parents and adolescent come together during part of the visit, leaving some private time for the adolescent to meet with the physician. For parents who need individualized counseling, the presence of a part-time social worker in the physician's office is very helpful (e.g., for marital problems).

Data in Advance. The initial evaluation visit proceeds much more efficiently if the family or adolescent has completed a behavioral screening questionnaire in advance. (See Chapter 44.) Parents of a child with a bedwetting problem may be asked for several bladder capacity measurements. If the child has been referred by an outside agency or physician, a referral letter containing a summary of the problem and the specific questions that the referring professional wishes answers to should be provided. Also the results of previous laboratory studies should be known. If time is short, this information may be gained by phone. Of more importance, the consulting physician should send a report to the referring professional or agency following his evaluation so that communication is optimal and environmental intervention is maximized.

Scheduling Appointments. The initial evaluation commonly requires 45 to 60 minutes. A common error is to set aside inadequate time or to try to carry out an abbreviated evaluation during a visit for another purpose. If a child with complicated psychosocial problems is detected during a health supervision visit, the patient should be rescheduled for a longer visit at a later time. Follow-up visits are usually 20 to 30 minutes long, depending on the problem. Parents should be given an exact date and time for follow-up. Telling them simply to come back if advice does not work out is not sufficient. If physicians agree that acute otitis media requires follow-up, they should readily see that a treatment plan for discipline problems or encopresis also needs to be monitored. Some physicians prefer to use the 5 to 6 P.M. time for initial evaluations because their office staff has left and their overhead is thereby reduced. However, others find themselves more tired and less sensitive at that hour.

Fees for Counseling.   Many physicians charge inadequately for the counseling they provide, and this may be one of the reasons they become disillusioned about dealing with psychosocial issues. The pediatrician must keep in mind that his productivity with counseling may be higher than that of any mental health professional. He should charge a reasonable amount for this time, for example, $1.00 per minute. This amounts to $60.00 for the initial evaluation and $30.00 for follow-up visits. The fees and the estimated total number of visits should be discussed with the family before the initial evaluation is scheduled. This can be done by the physician if the subject comes up during a health supervision visit or by the office assistant who schedules long appointments. If the parents cannot afford to pay the physician for the amount of time he spends with them, they might be referred to a mental health clinic or another center with a sliding fee scale. If families are being seen in a prepaid health maintenance organization, such counseling fees can more readily be absorbed.

CONCLUSION

Counseling is an unavoidable, intrinsic part of pediatric care (Task Force on Pediatric Education, 1978). Parents often seek out physicians who feel comfortable confronting both physical and emotional issues. Optimal pediatric care requires competent counseling skills. Full enjoyment of practice is enhanced through a knowledge of behavior modification principles and child rearing counseling techniques. Without such skills, physicians may turn excessively to psychotropic drugs and psychiatric referrals. Through experience, course work, reading, and discussion groups, pediatricians can upgrade their counseling skills to match their competency in treating physical illness.

Acknowledgments

The author was fortunate to have helpful reviews of this manuscript by Dr. Norman Scott, Dr. Melvin Weiner, and Jean Vail. Little, Brown and Company has kindly allowed prepublication of material from *The Pediatrician's Role in Behavioral Problems*.

BARTON D. SCHMITT

REFERENCES

Brazelton, T. B.: Infants and Mothers: Differences in Development. New York, Delacorte Press, 1969.
Christopherson, E. R.: Little People: Guidelines for Common Sense Child Rearing. Lawrence, Kansas, H. and H. Enterprises, Inc., 1977.
Drabman, R. S., and Jarvie, G.: Counseling parents of children with behavior problems: the use of extinction and timeout techniques. Pediatrics, 59:78, 1977.
Dinkmeyer, D., and McKay, G. D.: Systematic Training for Effective Parenting: The Parent's Handbook. Circle Pines, Minnesota, American Guidance Service Inc., 1982.
Ginott, H. G.: Between Parent and Teenager. New York, Avon Books, 1969.
Patterson, G. R.: Living with Children. Champaign, Illinois, Research Press, 1975.
Prazar, G., and Charney, E.: Behavioral pediatrics in office practice. Pediatr. Ann., 9:220, 1980.
Starfield, B., et al.: Psychosocial and psychosomatic diagnoses in primary care of children. Pediatrics, 66:159, 1980.
Task Force on Pediatric Education: The Future of Pediatric Education. Evanston, Illinois, American Academy of Pediatrics, 1978.

54

Psychotherapy with Children

All too often, physicians refer patients for psychotherapy without a clear understanding of what such treatment entails, or for which psychological disturbances it is best suited. In the past 20 years there has been a bewildering proliferation of therapies for children and in the variety of professionals who treat behavior disorders. Respectable approaches to treatment all have certain features in common: they attempt to relieve distress and psychological pain; they attempt to change maladaptive or destructive behavior and interaction patterns; and they attempt to enhance self-esteem. In addition, all the respectable therapies derive from theories of personality development, which influence the therapeutic techniques employed.

The success of treatment is dependent to some extent on its theoretical orientation, the nature and severity of the symptoms and the context in which the symptoms occur. The primary physician plays an important role in making an appropriate referral and, subsequently, in maintaining continued contact with the patient, the family, and the therapist through the course of therapy. It is important for physicians to understand the theoretical principles underlying a particular therapy and to appreciate the indications for its prescription. Several modalities of treatment are described for purposes of comparison, but the intent of this chapter is to define psychotherapy, to describe some of its theoretical foundations and methods, and to highlight issues that enhance the likelihood of a successful outcome. Of all the psychological therapies, psychotherapy is the most complex in terms of theory, method, and the demands it makes on the patient and family (Ornstein, 1976). As general physicians become better informed about the indications for psychotherapy and its advantages and disadvantages, referrals for and expectations of treatment should become more realistic.

Psychotherapy, conducted by a specially trained professional, is a method of psychological treatment based in psychodynamic theory. Although the term psychodynamic psychotherapy might be more accurate, it is redundant. Psychotherapists use the patient's fantasies and dreams, the child's symbolic play, and aspects of interpersonal relationships as indirect expressions of "unconscious" processes. Through an understanding of unconscious feelings and memories, the therapist and patient come to understand the origins of the psychological conflicts and the symbolic meanings of the symptomatic behavior.

THE THEORY OF PSYCHOTHERAPY

Psychodynamic theory traces its origins to Sigmund Freud's observations that many symptoms of psychopathological illness, especially those of a nonpsychotic nature, could be understood as symbolic representations of unconscious mental processes. Freud's discoveries led to a theory of psychological functioning and personality development, and to a method of treatment now called psychoanalysis. Psychodynamic theory is derived from psychoanalytic theory, and psychotherapy stems from psychoanalysis. An important aspect of psychodynamic theory is its emphasis on emotions and on the role that both conscious and unconscious affect plays in normal development, in the pathogenesis of behavioral symptoms, and in everyday interactions (Emde, 1980).

The theoretical and epiphenomenal construct of the unconscious, as a psychological domain of the mind, is critical to the understanding of psychotherapy. The unconscious for Freud implied more than events that were forgotten or out of awareness. Freud believed that the unconscious represented functions of the mind involving feelings, memories, and experiences that had been perceived as psychologically painful or "traumatic," and had been excluded from conscious recall by a "defense mechanism" called repression. Other defense mechanisms aided repression in this process, and in toto the defenses of the mind served the purpose of maintaining mental health. Failures of defenses, associated with stressful experiences or experiences reminiscent of prior psychological trauma, potentiated the possibility of a reappearance of the unconscious feeling or memory, and a breakthrough of unconscious anxiety. To avoid the recurrence of painful feelings into consciousness, psychological symptoms emerged. Freud viewed psychological symptoms as a second line of defense (second to defense mechanisms) in maintaining adaptation.

Freud's treatment method involved a technique called free association, which leads the patient and therapist to an understanding of the unconscious derivatives of symptoms and defenses. During free association the therapist focuses on repetitive themes in the patient's behavior and interactions, on hidden meanings behind slips of the tongue and other parapraxes, on the symbolic meanings of behavior such as compulsive behavior and obsessive thoughts, and on the latent meanings of dreams and fantasies.

Psychodynamic theory, therefore, distinguishes between memories and feelings that have been forgotten but can be recalled easily, and those that are part of the repressed unconscious reservoir of the mind. In his later writings and in the theoretical models of many psychoanalysts who followed, Freud emphasized defensive "style" or character as well as unconscious conflict. This shift in emphasis to "ego psychology" has broadened the scope of psychoanalytic theory to include environmental variables that shape personality. The clinical applications of ego psychology have expanded psychotherapy's focus to include the interactive character style of the patient.

Freud's hypotheses, although difficult to substantiate by modern biologic technology, were rooted in the biological and neurophysiological theories of his day. However one conceives the unconscious, the construct has proven to be clinically useful, especially in the treatment of a large group of symptoms, broadly defined as neurotic.

Were he alive today, Freud undoubtedly would have reformulated his theory to accommodate modern biology. It is interesting to speculate, for example, that the unconscious might be viewed as a unique sector of the dendritic tree, arborized after birth as a result of specific experiences, or it might represent states of central nervous system arousal responsive to specific conditions of stress. Either the dendritic branch pattern or the central nervous system response pattern would most likely have been set during the preverbal period of infant development. The alterations nevertheless would continue to exert an influence on subsequent development and behavior, possibly through altered neurochemical substrate characteristics. Such early influences are not easily related at a later age. Psychological trauma that occurred after an age when verbal memories were possible might be "repressed" by storage in a

biological form that would make them equally unavailable for subsequent recall.

Although in recent years controversy has developed about the efficacy of formal psychoanalysis as a treatment for psychological disorders, there is less controversy about the contribution of psychoanalysis to the theory of mental development and mental functioning (Emde, 1980). For example, Bowlby's theories of attachment and Klaus and Kennells' theories of bonding are modern elaborations of Freud's theories of psychological development (Bowlby, 1980; Klaus and Kennell, 1976).

Psychodynamic theory holds, therefore, that nonpsychotic psychological symptoms, manifested as disturbances of thinking, feeling, or behavior (such as unbidden repetitive thoughts, absence of or excessive intensity of feelings, and repetitive nonadaptive behavior patterns), serve the purpose of maintaining some psychological equilibrium. The symptoms can be understood in a symbolic and dynamic way if the unconscious conflicts they reflect can be unraveled. In the exploration of symptoms, three overlapping components are examined: the real life pressures on the individual, that is, the stresses that have triggered the symptom, such as a child's hospitalization, failure at school, or loss of a loved one; the preverbal or nonverbal unconscious memories of real or imagined painful experiences of the past that are surmised from symbolic and repetitive themes in free association or play; and the special character of the individual's relationships with others, including the therapist, which are endowed with attributes of early childhood relationships. For example:

A six year old boy's dog is killed in an accident. Shortly thereafter the child develops nightmares that progress to incapacitating panic attacks at school. In the physician's office, the boy is excessively frightened of the examination. The history reveals that at 18 months of age the child had been hospitalized for a hernia repair. The child's play reveals a recurrent theme of abandonment. The feelings of being frightened and helpless in the hospital at 18 months and of being angry at his parents for seeming not to protect him at that time are unconscious. The death of the pet reawakens them, and the symptoms that develop attempt to keep them from consciousness.

Psychodynamic theory maintains that all interpersonal behavior, including symptomatic behavior, is motivated and is designed to achieve a goal, though the goal of psychopathologic behavior is usually maladaptive. One purpose of symptoms is to reduce anxiety, a signal that the painful unconscious feelings, memories, or experiences are threatening to emerge. The motivation for symptomatic behavior is unconscious.

The construct of "imagined" versus "real" events is also central to psychodynamic theory.

Certain important real experiences may have been "misperceived" by a developmentally immature infant, yet it is this misperceived or imagined event that persists as a source of conflict and anxiety. Similarly, when a young child wishes something that is forbidden, whether forbidden by a real parental prohibition or by an imagined (internalized) parental prohibition, the simultaneous presence of the wish and prohibition results in a conflict that engenders anxiety. Thus, certain wishes, prohibitions, and memories—real or imagined—experienced during particularly vulnerable times in early development, become unconscious. They may arouse unconscious anxiety when subsequent situations reminiscent of the original event are encountered. The deployment of successful defenses may alleviate the anxiety, or the emergence of psychological symptoms develops to circumvent the re-experience of the unconscious feeling or memory. Children are less able than adults to understand and tolerate anxiety or to allay it through constructive action. Therefore, they are more likely to develop behavioral disturbances. For example:

A 14 year old high school freshman was evaluated because of a learning problem in school. Physical examination and psychological testing revealed no basis for the symptom. Both parents were successful professionals with high expectations of their children. The patient could not explain why he procrastinated and had difficulty in achieving his potential. His therapy led him to understand his conflict about growing up and being successful. On the one hand, he would lose the protection of his family and the security of his dependency. On the other, he feared that in his success, he would outdo his father and incur his wrath and revenge, which he imagined to be present as a younger child. His symptom was designed to maintain his status as little boy forever.

Psychological symptoms can be understood as inhibitions to optimal functioning. They are the least adaptive way of expressing unconscious anxiety. A more appropriate method of behavioral expression is through the development of characterological defenses, or personality style. Such personality characteristics (defense mechanisms) may be manifested as "habits" or characteristic patterns of responding by thoughts or feelings in stressful situations. Characterological defenses can be differentiated from coping mechanisms in that they are more likely to protect from imagined trauma and are not always successful. Coping mechanisms, on the other hand, are designed to respond to real stresses and handicaps. Thus, the caution of a child who cannot sit down at the dinner table without first washing his hands until they are spotless represents a defense against symbolic dirt and contamination, whereas the caution of the child with juvenile rheumatoid arthritis who avoids undue exposure to cold climates represents good judgment and successful coping. When hand washing becomes ritualized and in-

hibits normal functioning, the defense has become a symptom.

PSYCHOTHERAPY AND OTHER THERAPIES

The general practitioner who encounters a child with a psychological problem or functional symptom first needs to decide whether to manage the child within the confines of a busy office practice, or whether to refer the child to another specialist, a therapist. If the physician elects to treat the child, therapy may involve giving advice to or counseling the parents. Occasionally family therapy or behavioral therapy or a combination of these modalities is attempted, but treatment of the individual child is rarely provided by the general physician. Children who require child centered therapy are referred to child therapists, usually social workers, psychologists, or child psychiatrists. The therapist's professional affiliation does not necessarily determine her method of therapy, her theoretical orientation, or her skill level. The following brief descriptions provide a frame of reference by which to compare and contrast psychotherapy with a few of the other more common therapeutic techniques.

GUIDANCE THERAPIES

Giving Advice. Advice giving is a method of treatment in which the experience, the wisdom, and the "omnipotence" of the professional are used to tell a child and parent "what to do" about a problem. This is the most common type of interaction between physician and patient. "Take this medicine" or "That's not good for you" are examples. Research dealing with compliance has suggested, however, that patients do not always follow the advice of their physicians (Francis et al., 1969; Litt and Cusky, 1980). Patients who do, generally have had a trusting, long term relationship with the physician, and their physician's advice does not conflict too sharply with their own beliefs and values. In the management of acute illnesses or injuries, including behavioral reactions to acute stress, advice giving is appropriate. However, in the medical and psychological management of chronic illness, which includes most of the behavior disorders of childhood and physical handicaps, the giving of advice generally does not suffice.

Counseling. Counseling is a special form of guidance therapy in which the professional listens sensitively to the family's problem and attempts to guide them to their own solution. In the extreme, counseling is totally nondirective. The counselor responds minimally and only to facilitate continued discourse. In a medical practice, however, counseling more often involves listening to the family's problems in a nonjudgmental way, exploring available alternatives and helping the family come to some reasonable solution or decision, using all the psychological and social supports that are available, including the patience and empathy of the physician.

Goal Directed Therapy. Brief goal directed therapy shares certain features with counseling. Goal directed therapy specifies a set of objectives and the number of sessions required to achieve them. The sessions then are used "to work" on particular aspects of the problem in a predetermined sequence. Crisis intervention is an example of a specialized kind of brief, goal directed therapy.

In a 10 session set of brief therapy hours, for example, a child's misbehavior at school is explored in detail with the parents and teachers during the first three sessions; after each subsequent session, goals for improved behavior are set and monitored while the reasons for success and failure are explored. The final sessions review progress and delineate further areas of work for the child and family alone (Rosenthal, 1979). The focus of both goal directed therapy and counseling is on the current problem and on the environment in which it exists.

The guidance therapies are most useful when the behavioral problem is acute and of recent onset, and when the precipitating event is apparent. It may be necessary, for example, for some children to have an opportunity to express to a professional their feelings of grief about the death of a loved one (including a pet) or the loss of a parent through divorce. The physician listens sensitively to the child's expression of feelings, and provides the child with an opportunity to share the emotional experience of the loss, including the feelings of helplessness and anger. Acute behavioral symptoms, such as nightmares or bed wetting, which are frequently associated with acute crises, may be relieved by such individual or family counseling as more constructive alternatives to symptomatic grieving are explored.

BEHAVIOR THERAPY

Behavior therapies have become increasingly popular over the past five to 10 years in managing a range of psychological problems and disorders, from infantile autism to single symptoms such as enuresis and encopresis. An essential principle of behavior therapy states that positive and negative reinforcers alter the frequency of occurrence of a particular behavior. In other words, rewards or punishments that are consistently applied in relation to a desired or nondesired behavior alter its frequency. Behavior therapy is derived from social learning theory, which has contributed significantly to our understanding of normal and psychopathological behavioral development. In group settings, applications of behavior therapy involve a token economy or a system of peer governance to shape responses (Liberman, 1971). (For a more comprehensive review of behavior therapy, the reader is referred to Chapter 55.)

According to psychodynamic theory, the removal of a symptom alone, without resolution of the underlying unconscious conflict, should result in another symptom to avoid the emergence of the persisting unconscious traumatic affect or memory. It has been shown, however, that such psychological symptom substitution does not occur very often (Miller et al., 1972). The clinical success of behavior therapy in symptom removal, on the other hand, does not negate the psychodynamic idea that symptoms reflect unconscious conflict. Symptom removal presumably affects the patient's psychological and social adaptation positively and, in turn, improves his overall sense of well-being. Success breeds further success, and the context in which the conflict originally surfaced changes. Furthermore, behavior therapies do not work in all situations. Disorders amenable to psychotherapy need to be more carefully distinguished from disturbances responsive to behavior therapy. Even in the latter disorders, the effectiveness of behavioral treatment is optimized when the motivational (dynamic) characteristics of a child and his family are taken into consideration. For example, the star chart, a common behavioral reward paradigm, is not equally effective for all children with encopresis. Some children may not be sufficiently motivated by gold stars. Similarly, "grounding" as a form of punishment for an "acting out" adolescent may paradoxically serve as a positive reward for the adolescent who wants to stay at home because of low self-esteem and difficulty with peer relationships. Behavioral treatments are most successful when they are individualized and are integrated with the child's social context in a psychodynamically meaningful way.

Many behavior therapists now believe that the benefits of behavior therapy are enhanced by simultaneous supportive psychotherapy. The reduction of parental and patient anxiety, the restoration of parental authority and control, and the positive feelings that the child experiences from a nonpunitive therapist enhance the child's self-esteem and the family's compliance with the behavioral contract.

As in brief therapy, the objectives, the time course, and the rewards and punishments of a behavior therapy contract are worked out early in the course of treatment. Breakdowns in the contract are often as instructive as successes. The skillful behavior therapist learns about family communication patterns and interaction styles from such failures, and uses the information to construct new, more appropriate behavioral tasks.

A question about behavior therapy is the degree to which successful symptom reduction generalizes to new and different settings. A child with a specific fear may demonstrate reduced anxiety after desensitization in a laboratory setting, but the fear may reappear outside the laboratory. An autistic child may learn to make eye contact in school for a reward of candy, but when the reinforcer is absent or the setting is changed, the desired behavior disappears. Failure in the generalizability of a response is usually related to the severity of the psychopathological disorder and the degree of cognitive impairment, rather than the nature of the target symptom. The child with infantile autism is less likely to generalize than the child with a specific phobia.

Behavior therapists have had their greatest success with monosymptom disorders. Children with one symptom, such as enuresis, encopresis, or a specific fear, generally respond well to behavior therapy. Success has also been notable in the return to school of children with the school avoidance syndrome and in weight restoration in adolescents with anorexia nervosa, although the underlying symptoms of distorted body image, low self-esteem, and fear of loss of control may not be affected. Well circumscribed behavior patterns, like weight in pounds and days of school attendance, are more readily and more reliably quantified. Base-line rates and rates of change can be charted, and the success of the behavioral intervention quickly determined.

PHARMACOTHERAPY

The indications for and the details of pharmacotherapy are reviewed in Chapter 58 of this volume. Suffice it to say that in the pharmacological management of childhood behavior disorders, pediatricians and primary care physicians are more likely to use sedatives, minor tranquilizers, and stimulants; child psychiatrists are more likely to use neuroleptics and antidepressants. In general, medication should be used sparingly and cautiously in children, not only because of the potential danger of side effects and the long term complications for growth and development, but also because their use implies to the child that behavior is controlled from without and does not require self-control. In circumstances when the child is unable to exert self-control over symptomatic behavior, the judicious use of an appropriate

medication is enormously helpful. Unfortunately all too often medications are prescribed because of a need for action, a need "to do something," or to relieve the anxiety of parents, teachers, or the physician.

COGNITIVE THERAPIES

Cognitive therapy is now being used with children (Coates and Thorensen, 1980). Instead of a behavioral contract, a cognitive strategy is designed that anticipates problems in advance and rehearses nonpathological alternative responses. Instead of a positive reinforcer for an obese child who loses weight, the cognitive therapist and patient design alternatives to eating excessively. The child becomes sensitive to feelings of hunger, when previously he had been unaware of such feelings, and practices avoiding the usual response more each day, i.e., delays eating when hungry. He learns to anticipate the feeling and to substitute behavior other than eating, such as exercising or taking a drive. Planning a menu in advance and shopping with prerehearsed meals in mind are other strategies. The reward comes from the enhanced self-esteem derived from accomplishing the task as planned. Cognitive therapy has been especially successful in adults and adolescents with depressive disorders in which the feelings of helplessness, low self-esteem, and negative self-image are targeted specifically, and exercises to regain psychological strength are rehearsed and enacted (Beck, 1970).

PSYCHOTHERAPIES

Finally, the psychotherapies are those therapies that rely on a psychodynamic understanding of the psychological conflicts of individuals in families and groups. The psychological symptoms interfere with current relationships and inhibit the individual's ability to function adaptively. The original conflicts that underlie the symptoms are re-experienced and understood during the course of therapy. The success of psychotherapy therefore depends on the establishment of an emotional alliance between the patient and the therapist. In the alliance, the patient, through free association and dynamically meaningful interactions, recollects traumatic psychological events of the past. The therapist's responsibility is to interpret the psychodynamically meaningful material. The affective restructuring that takes place through the alliance leads to a decrease in the symptoms.

The psychotherapist's task is to untangle symptoms and defense and coping mechanisms in a child's behavior patterns and to attempt to understand the sources of conflict that produce the psychopathological aspects of the behavior. Understanding the conflict, and therefore the meaning of the behavior per se, does not lead necessarily to a change; however, understanding the conflict often helps the child and family to reduce unnecessary pressures, which in turn may lead to the relief of symptoms.

Family therapy has become popular with pediatricians. It must be distinguished from parental advice giving or counseling. Family psychotherapy is based on the principle that the child's symptoms serve as an unconscious expression of a more pervasive family conflict. The child's symptomatic behavior maintains an equilibrium in a dysfunctional family. For example, covert marital conflict is kept hidden for fear the marriage would not survive, or an emotionally disturbed parent is able to maintain a shaky equilibrium because the family focuses attention on the problem child. In family treatment the entire family system is the "patient." Interviews with an entire family can be helpful to the physician in clarifying the nature of family interactions surrounding the symptomatic behavior of a child, even when formal family therapy is not undertaken.

Group psychotherapy theory holds that members interact with each other as an extended family during therapy (Yalom, 1970). The therapist attempts to understand relationships between individuals in terms of each member's underlying psychological conflicts, as remembered or re-experienced in the group. Group therapy needs to be differentiated from the aid provided by self-help groups, such as Alcoholics Anonymous and Weight Watchers, and self-actualizing groups, such as EST, which are more advice-giving and supportive in their orientation.

SOME PRINCIPLES OF PSYCHOTHERAPY

The techniques for establishing a psychotherapeutic relationship with a child and fostering its growth through the course of treatment are learned as part of psychotherapy training (Ornstein, 1976). For physicians who do not intend to practice psychotherapy, the skills of the psychotherapist are not as important as an understanding of their rationale. The psychotherapist attempts to impart symbolic meaning to a behavior. Thus, the task is similar to that of an art critic who attempts to interpret the meaning of a creative work. For example, in understanding a poem or the undercurrents of a play, the trained listener searches for verbal and nonverbal clues that reveal the creator's intentions. The listener's past experiences, intuition, training, and empathy are all called upon in the response and interpretation. Similarly, the psychotherapist listens to the content and observes the process of psychotherapy sessions. Content and process are both important. Clues to process are provided by nonverbal messages, feeling tones of the session, and the intuitive impressions derived from clinical experience. Certain clinical data are particularly likely to yield clues to the symbolic meanings of content: dreams, fantasies, fleeting thoughts, and, with children, recurrent play themes. Content alone, however, is not sufficient for adequate understanding. A child whose play is continually characterized by violence and victimization is not necessarily interested in fighting. The play theme instead may represent unconscious feelings of helplessness and attempts at mastery.

In the therapy of a child, play is analogous to the dreams and fantasies of the adolescent and adult. Through play, children express their conflicts. When a theme recurs often enough in therapy, the therapist comes to understand a portion of the conflict, its symbolic meaning or painful affect, and makes an interpretation. If the interpretation is accurate, the child generally feels the correctness and comes to some understanding of the symptom or troublesome behavior.

Psychotherapy therefore is a form of intervention derived from a particular set of principles. A basic objective is to develop a meaningful relationship, or therapeutic alliance, with the child. The relationship implies strong positive and negative feelings that are experienced by both the child and the therapist. These feelings emerge only if the relationship is continuous and regular, and if feelings of trust and a sense of attachment have developed. The strength of this bond should be anticipated and supported by the referring physician, and must be understood by the family. It is through the process of developing trust and attachment that the work of therapy gets done. Too often, parents, and sometimes the physician, feel excluded from the relationship and feel jealous or competitive. Unwittingly they attempt to sabotage the work of therapy. But just as during normal development, attachment to parents is followed by separation and individuation, so too, as therapy progresses, the attachment to the psychotherapist lessens as the child develops more autonomy and strength to cope with his own stresses.

The term "therapeutic alliance" in psychotherapy refers to both the conscious and the unconscious aspects of the relationship between the child and the therapist and serves as the vehicle whereby the process of therapy takes place. The conscious elements of the therapeutic alliance are reflected in shared therapeutic objectives of the child and the therapist.

Transference reactions represent the unconscious elements of the therapeutic alliance. Transferences may be positive or negative. They reflect the unrealistic expectations and irrational beliefs about the therapy—the unconscious misreading of the therapeutic relationship as a reliving of earlier attachments. Positive transferences are generally expressed as the child's feelings and beliefs that the therapist has the attributes of the all-good and powerful, all-knowing and all-protective, fantasy parent. Negative transferences are frequently expressed as "resistances." The child views the therapist as being uninterested or punitive. Such resistances are an important aspect of the therapeutic process and help the therapist understand the child's early development and current behavior.

Transference feelings are bidirectional; that is, both the child and therapist experience transference reactions. The therapist's transference reactions are termed countertransference reactions, and the competent therapist insures that they do not interfere with the therapeutic process. Because they are remembrances of things past, transference feelings need to be "worked through" during the course of therapy. Through an understanding of these transference and countertransference reactions, the therapist comes to define the psychological conflicts that burden the child.

Psychotherapy attempts to understand the emotional components of all the patient's or family's relationships. The relationship with the therapist, although critical, is not a substitute for relationships with other significant adults and friends. However, the relationship with the therapist must be protected if the child is to develop the trust necessary for therapy to proceed. It is for this reason that the child's therapy needs to be confidential. Often parents, and sometimes the physician, feel angered by the psychotherapist's insistence on privacy. Sometimes, of course, psychotherapists carry their insistence on confidentiality to unnecessary extremes. Children do not mind sharing some information. Certainly children in danger of harming themselves or others, or children whose behavior might jeopardize their future, need protection by their parents or an appropriate authority. The therapist's role can never be that of substitute parent. The therapist must not lose sight of the family's values, and must remain aware that the ultimate responsibility for the child's welfare lies with the parent or legal guardian.

If psychotherapy with children is to be successful, the therapist must establish a therapeutic alliance with the parents as well as with the child. A positive relationship with the child's physician is also helpful. Too often psychotherapy breaks down when one side of the triad becomes alienated. Occasionally, significant other figures in the child's life, such as probation officers, school personnel, or extended family, also need to be involved.

The frequency of sessions, the length of treatment, and the cost of psychotherapy are often criticized. Regular contact over an extended period of time is essential, however, to establish a therapeutic alliance. Since continuity of a psychological theme from one session to another is important, the child should be seen at least once a week; for some children twice-weekly sessions may be necessary. It is difficult to establish a therapeutic relationship in less than six months, and it may require nine months to a year. The frequency and duration of psychotherapy sessions require an extensive commitment of energy, time, and resources. Although difficult to predict in advance, the factors that influence frequency and duration include the severity of the problem, the ability of the child to form a therapeutic relationship, the course of the symptoms during the early phases of treatment, and the psychological support and flexibility of the parents. The family's financial circumstances must also be considered. A good working alliance with the family indicates that both the family and the therapist recognize the special demands of the therapeutic situation, and that the therapist recognizes the special needs of the family.

Just as regular contact with a therapist over a period of time enhances the development of a therapeutic relationship, so too does constancy of the psychotherapy setting. The therapist's office should be comfortable, and therapeutic interactions must be age-appropriate. Commonly, if psychotherapy involves play, as it does with younger children, the toys the child uses should become familiar. Some psychotherapists have special toy sets for individual patients; others tend to use a general set of toys. It is important that whatever the child customarily plays with be present from session to session. It is equally important that the child's play productions, such as drawings or clay models, be preserved and kept confidential until the child wishes to share them.

INDICATIONS AND CONTRAINDICATIONS FOR PSYCHOTHERAPY

A number of factors need to be considered before the physician refers a child for psychotherapy. In some cases the indications are clear-cut, and the major issue is the successful preparation of the child and family for the referral. The referring physician must have some understanding of the process of psychotherapy in order to accomplish this. If the physician is ambivalent about psychotherapy, his concerns will be transmitted to the patient and the likelihood of a successful referral will be reduced. In addition to the physician's conviction, however, the family's motivation for psychotherapy must be developed and supported. The methods, the limitations, and the expected outcome of therapy should be discussed. There should not be a disparity between the patient's, the family's, the referring physician's, and the therapist's expectations. The family must realize and be able to afford the financial burden and the commitment of time. The parents must become partners in the therapeutic alliance.

The child must be able to form an emotional attachment. Psychotherapy is not indicated for children with early infantile autism and other psychotic disorders of childhood in which attachment capacities and social interactions are impaired. Psychotherapy also is not indicated for children of chaotic families who cannot make a commitment to long term, regular treatment. Similarly, children with serious disturbances of conduct who tend to act out their conflicts, rather than talk about them or symbolize them through play, are generally poor candidates for psychotherapy.

In some situations a definitive diagnosis or the most appropriate method of treatment is not clear. Likewise, some children who may ultimately become amenable to successful treatment by psychotherapy, or parents who may ultimately accept the requirements for such treatment, may need a period of longer initial diagnostic exploration and preparation either prior to referral to the mental health professional or prior to the onset of psychotherapy.

Children and their parents must be able to acknowledge the psychological components of the disturbance and accept that a "talking" or a "play" approach to the problems will have a chance of success. In other words, psychotherapy, like other medical therapies, works best when the patients' expectations of outcome are optimistic. The placebo effect of all medical treatments continues to be important. Psychotherapy works least well in patients whose doubts are great. In general, most nonpsychotic and nonconduct disordered youngsters are suitable for psychotherapy, given adequate financial resources, commitment of time and energy, and proper motivation. Referral for psychotherapy should be made in such cases, especially when behavioral programs or attempts at brief counseling or short term, goal oriented therapy have been unsuccessful. A child psychiatrist or other mental health professional with special training in psychotherapy should be selected. It is helpful if the physician knows the psychotherapist personally. Such knowledge is best gained from previous collaborative work in which there was successful management of behavioral disturbances in other patients. A social acquaintance is not necessarily useful in making a decision for referral. The physician should call the psychotherapist, whether he knows him or not, and explain the problem, the reasons for the referral, and the extent to which the family has been prepared for psychotherapy. The referring physician and the psychotherapist should then remain in regular contact during the course of treatment in order to share information and to provide reassurance that therapy is progressing satisfactorily.

PSYCHOPATHOLOGY IN PARENTS

Not infrequently a practitioner interviewing parents about a symptomatic child notes that the parent may need psychiatric evaluation or psychotherapy. The physician may even consider, with good reason, that the child's problem might be ameliorated if the parent alone were treated, or the parent's marital problems relieved through some kind of counseling. As many practitioners have regretfully discovered, these are sensitive situations. Direct and uninvited reference to parents' problems often produces angry or frightened responses and flight from the physician. The same conditions apply to the child psychiatrist who must initially accept the child's problem as presented by the parent and focus treatment in the area of parental concern. Later, after a therapeutic relationship has developed, the parent may be helped to see connections between the parents' problem and the child's behavior. Sometimes parents need to see what can be done with the child's problem before recognizing their own contribution and need, as in the following example:

> The parents brought their fearful five year old girl to a child psychiatrist for several interviews. After some brief work with the child and counseling with the parents, the fearfulness greatly diminished. The father then could ask for an individual interview and requested psychotherapy for his own neurotic conflicts, saying "I didn't think there was anything to this psychotherapy until I saw how much it helped my daughter."

Sometimes parents are literally unable to see any connection between their feelings or behavior and the child's symptoms. When parents do not ask directly for help with their own problems, it is best to make a referral based only on the parents' concern about the child's symptoms.

THE EVALUATION OF PSYCHOTHERAPY OUTCOME

The effectiveness of psychotherapy for both adults and children has been difficult to establish (Andrews and Harvey, 1981; Smith et al., 1980).

It is a complex treatment, which defies standardization and experimental design. Control or comparison groups of untreated patients have been difficult to assemble. Similarly, comparisons between treatment methods, and between individual psychotherapists, have been unsatisfactory (Luborsky et al., 1975). Even comparable diagnoses and indices of symptom severity have been problematic (D'Angelo and Walsh, 1967). Finally, the benefits of short term versus long term outcomes have been hard to establish (Heinicke, 1969; Heinicke and Goldman, 1960). Studies that have compared improvement among untreated patients on a "waiting list" with matched patients who received time-limited psychotherapy or behavior therapy have suggested that patients who received psychotherapy improved more in the immediate follow-up period than the waiting list control subjects, but all groups improved (Miller et al., 1972).

In the classic studies of long term outcome, children treated in a child guidance clinic with a variety of therapy techniques and by a variety of therapists all did well as adults if their presenting diagnoses did not involve psychosis or serious antisocial behavior (Robbins, 1966). They were no more likely to need psychiatric help as adults than the general population of adults if their childhood psychopathological disorders had been diagnosed as neurotic. The form of treatment was not as important as the presenting diagnosis. In a recent study of the long term effects of multiple modality treatment of hyperactive children, Cantwell and his colleagues concluded that psychotherapy, combined with pharmacotherapy and supportive therapy at school and home, was superior to any single mode of treatment, or no treatment at all (Satterfield et al., 1980).

There is no doubt that children's behavioral symptoms, their psychological adjustment, and their sense of self-esteem improve in association with psychotherapy, at least over the short term (Levitt, 1971; Wright et al., 1976). There is also no doubt that such improvement alters their environment's reactions to them. Such effects, in and of themselves, warrant a course of psychotherapy. A child who is able to change from functioning in a disapproving environment to one whose interactions are appreciated and welcomed optimizes his potential for normal psychosocial growth and development. Yet psychotherapy is not the only way to produce change, and occasionally improvement occurs with no therapy at all. Shifts in environmental support, in levels of stress, and in maturational competence are all powerful forces affecting behavior and psychological development. A prescription for psychotherapy should nevertheless be given when symptoms and maladaptive functioning are diagnosed that are of a variety that is responsive to this form of treatment.

SUMMARY AND CONCLUSIONS

This chapter has attempted to highlight the theory and practice of psychotherapy with children. The focus has been on the psychodynamic understanding of behavior and the importance of process in psychotherapeutic work. Dynamic psychotherapy has been contrasted with other forms of psychological therapy suitable for children. The indications for each have been presented. More knowledge about effectiveness and outcome of each of the treatment modalities is required, but psychodynamic psychotherapy remains an important treatment method for many emotionally and behaviorally disturbed children. Psychotherapy will have an optimal effect only if its benefits and limitations are understood by the families and the referring physicians.

Acknowledgment

Support is gratefully acknowledged from the W. T. Grant Foundation and the NIMH Psychiatric Education Branch (MH-14449).

THOMAS F. ANDERS

CHARLES WALTON

REFERENCES

Andrews, G., and Harvey, R.: Does psychotherapy benefit neurotic patients? A reanalysis of the Smith Glass and Miller data. Arch. General Psychiatry, *38*:1203–1208, 1981.

Beck, A.: Depression: Causes and Treatment. Philadelphia, University of Pennsylvania Press, 1970.

Bowlby, J.: Attachment and Loss. New York, Basic Books, Inc., 1980, Volumes 1 to 3.

Coates, T., and Thoresen, C.: Behavioral self-control and educational practice. In Berliner, D. (Editor): Review of Research in Education. New York, Praeger Publishers, 1980.

D'Angelo, R., and Walsh, J.: An evaluation of various therapy approaches with lower socio-economic-group children. J. Psychol., *67*:59–64, 1967.

Emde, R.: Toward a psychoanalytic theory of affect: I. The organizational model and its propositions; II. Emerging models of emotional development in infancy. In Greenspan, S., and Pollock, G. (Editors): The Course of Life: Psychoanalytic Contributions Toward Understanding Personality Development. Bethesda, Maryland, National Institute of Mental Health, 1980.

Francis, V., Korsch, B., and Morris, M.: Gaps in doctor-patient communication: Patients' response to medical advice. New Engl. J. Med., *280*:535–540, 1969.

Heinicke, C.: Frequency of psychotherapeutic sessions as a factor affecting outcome: analysis of clinical ratings and test results. J. Abnorm. Psychol., *74*:553–560, 1969.

Heinicke, C., and Goldman, A.: Research on psychotherapy with children: A review and suggestions for further study. Am. J. Orthopsychiatry, *30*:483–494, 1960.

Klaus, M., and Kennell, J.: Maternal-Infant Bonding: The Impact of Early Separation or Loss on Family Development. St. Louis, The C. V. Mosby Company, 1976.

Levitt, E.: The results of psychotherapy with children: an evaluation. J. Consult. Psychol., *21*:181–196, 1971.

Liberman, R.: Behavioral group therapy: a controlled study. Br. J. Psychiatry, *119*:535–544, 1971.

Litt, I., and Cusky, W.: Compliance with medical regimens during adolescence. Ped. Clin. N. Am., *27*:3–15, 1980.

Luborsky, L., Singer, B., and Luborsky, L.: Comparative studies of psychotherapy: is it true that "everyone has one and all must have prizes"? Arch. Gen. Psychiatry, *32*:995–1008, 1975.

Miller, L., Barret, C., Hampe, E., and Noble, H.: Comparison of reciprocal inhibition, psychotherapy and waiting list control for phobic children. J. Abnorm. Psychol., 79:269–279, 1972.

Ornstein, A.: Toward a theory of psychoanalytic psychotherapy with children. Compr. Psychiatry, 17:3–36, 1976.

Robins, L.: Deviant Children Grown Up: A Sociological and Psychiatric Study of Personality. Baltimore, The Williams & Wilkins Company, 1966.

Rosenthal, A.: Brief focused psychotherapy. In Harrison, S. (Editor): Basic Handbook of Child Psychiatry. III. Therapeutic Interventions. New York, Basic Books, Inc., 1979.

Satterfield, J., Satterfield, B., and Cantwell, D.: Multimodality treatment: a two-year evaluation of 61 hyperactive boys. Arch. Gen. Psychiatry, 37:915–919, 1980.

Smith, M., Glass, G., and Miller, T.: Benefits of Psychotherapy. Baltimore, Johns Hopkins University Press, 1980.

Wright, D., Moelis, I., and Pollock, L.: The outcome of individual child psychotherapy: increments at follow-up. J. Child Psychol. Psychiatry, 17:275–285, 1976.

Yalom, I.: The Theory and Practice of Group Psychotherapy. New York, Basic Books, Inc., 1970.

55
Behavior Management

With the rapid growth of developmental and behavioral pediatrics, the pediatrician is now, more than ever, called upon to identify, evaluate, and treat children experiencing behavioral difficulties. To meet this growing demand, the pediatrician requires a sophisticated understanding of the many factors contributing to the development of maladaptive behavior and a current knowledge of the complex factors pertaining to its remediation (Russo and Varni, 1982).

In the past, evaluation and treatment of pediatric behavior problems involved long term psychotherapy with a primary focus on analytic assessment and dynamic therapy. Psychotherapy of this sort required a period of years and did not fit well into the context of pediatric practice. Furthermore, many pediatric behavior problems do not warrant such radical psychological intervention. Recently a behavioral technology for modifying maladaptive behavior has emerged as an extension of basic research on animal learning and conditioning to problems in human behavior. This technology provides procedures that can be used at home, at school, in the hospital—in fact, anywhere there exists a need to alleviate behavior problems and establish more desirable behavior. These procedures offer an effective approach to the assessment and treatment of a host of behavioral difficulties encountered by pediatricians every day. This chapter is therefore intended to provide the pediatrician with a working knowledge of the behavioral approach to the treatment of pediatric behavior problems.

The behavioral approach is most clearly understood by distinguishing it from more traditional psychological approaches. Traditionally psychology has been influenced by what is generally called "the medical model." In medicine, internal conditions of the individual (e.g., bacteria, virus, lesions, organ dysfunction) account for symptoms (e.g., fever, infection, discomfort). The "real" problem is not the symptom but the internal condition or disease. It is not sufficient to treat the symptom; the underlying cause must be remedied. An extension of this medical model to psychology has led to a concern with internal, underlying causes of behavior rather than with behavior itself. Although the medical model is of obvious value in medicine, its extension to the treatment of behavior problems has been disappointing.

Behavioral scientists view behavior primarily as a product of external environmental determinants. They argue that there is little to be gained from conceptualizing maladaptive behavior as a symptom of some internal condition. Regardless of the cause, environmental factors influence behavior in lawful and predictable ways. The principles that describe the relationship between behavior and environmental events (antecedents and consequences) are used to modify maladaptive behavior, to facilitate coping, and to develop adaptive behavior.

Pediatric behavior management involves the application of experimentally derived principles of learning and conditioning to behavior problems in children. These principles fall into two general categories corresponding to two learning processes: respondent conditioning and operant conditioning. The initial sections of this chapter explain the fundamental principles of respondent and operant conditioning and provide guidelines for their application in pediatrics. Then strategies for incorporating these principles into the design and implementation of behavior and stress management programs are discussed in detail. Examples have been selected to represent the variety of problems that pediatricians encounter on a daily basis. (The child in most examples is denoted merely by "he" rather than by "he/she." This convention is adopted for simplicity and in no way reflects a gender preference in the use of behavior management procedures.) Special attention is paid to the role of learning and behavior in chronic disease.

It is expected that ready access to procedural suggestions and guidelines may be helpful to the pediatrician designing specific assessment and behavior management programs. For convenience, Table 55–1 outlines the techniques that we discuss with page numbers referenced. We have included

Table 55–1. SUMMARY OF BEHAVIOR MANAGEMENT PROCEDURES

Procedures based on respondent conditioning
    A. Systematic desensitization
    B. Assertive training
    C. Other procedures
       1. For constipation
       2. For enuresis
Procedures based on operant conditioning
    A. Procedures to increase behavior
       1. Positive reinforcement
       2. Token economies
       3. Negative reinforcement
    B. Procedures to decrease behavior
       1. Extinction
       2. Punishment
    C. Procedures to develop new behavior: shaping
    D. Procedures to generalize behavior from one setting to another
    E. Other procedures
       1. Stress management
       2. Biofeedback

it here as a general overview and introduction to the principles and procedures presented in the chapter.

PRINCIPLES OF BEHAVIOR MANAGEMENT AND THEIR APPLICATION

RESPONDENT CONDITIONING

Principles and Procedures

Respondent conditioning (also called Pavlovian conditioning or classical conditioning) depends on the fact that certain stimuli automatically elicit certain responses without any prior learning or conditioning. For example, food in the mouth automatically elicits salivation; dust in the eye automatically elicits the formation of tears. Such "automatic" stimulus-response relationships are called unconditioned reflexes; the stimulus is an unconditioned stimulus (UCS) and the response is an unconditioned response (UCR). An unconditioned reflex is symbolized as UCS-UCR.

Respondent conditioning occurs when a neutral stimulus (i.e., one that does not initially elicit the response being conditioned) is followed closely in time by an unconditioned stimulus that does elicit a particular response. As a result of this temporal pairing, the formerly neutral stimulus comes to elicit the same response elicited by the unconditioned stimulus. At this point the neutral stimulus has become a conditioned stimulus (CS) and the response it elicits is a conditioned response (CR). This respondent conditioning paradigm is presented diagrammatically in Figure 55–1.

In the first and most famous experiment on respondent conditioning, Ivan Pavlov demonstrated that a dog's salivary response could be conditioned to occur at the sound of a bell. In this experiment, the bell (initially a neutral stimulus) was paired with meat powder (UCS) that automatically elicited salivation (UCR). As a result of this stimulus pairing, the bell alone (now CS) came to elicit salivation (CR).

Respondent conditioning has been used to explain phobic behavior. Phobias are assumed to be conditioned fears of formerly neutral objects. Consider, for example, how a small child might develop a phobia for water. Suppose the child is playfully following a floating beachball in a lake near shore. Never having had any reason to fear water, he is not afraid now. Suddenly he trips and falls so that his head slips under water. Quite naturally the child may begin to choke and will become frightened. The next time he is near a lake he will probably cry and show other types of fear behavior. Thus, a stimulus (a lake) that previously did not elicit fear behavior comes to do so because it was paired with a stimulus (falling, choking) that did elicit fear.

In this example, the child's experience in the lake will have two important effects on his behavior. First, as mentioned, he will have a fear reaction whenever he visits a lake (a reaction to lakes that has been respondently conditioned). Second, he will tend to avoid lakes whenever possible. He will be reluctant to go to the beach, or to ride in boats since these situations are similar to the original fear producing event. The hypothetical child will have acquired a phobia for water.

Watson used a respondent conditioning procedure to condition a fear response in an 11 month old infant, "Little Albert" (Watson and Rayner, 1920). Watson first demonstrated that Albert showed no fear of a variety of furry things, including a white rat, a rabbit, a dog, and a cotton ball. Then, when Albert was playing with the rat, Watson banged a steel bar with a hammer just behind Albert's head. The loud noise elicited a fear response (e.g., crying) from Albert. After a total of six pairings of the loud noise with the white rat, Albert showed a strong fear response

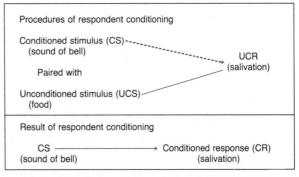

Figure 55–1. Diagram of respondent conditioning paradigm.

to the rat. Moreover, the rabbit, the dog, and the cotton ball also elicited the fear response.

Subsequent experiments by Jones (1924) clearly demonstrated that fear reactions can be eliminated through the use of respondent conditioning procedures. Indeed, effective procedures for the treatment of phobias have been the most important contribution of respondent conditioning to the field of behavior therapy.

Applications of Respondent Conditioning

Systematic Desensitization

Indications for the Use of Systematic Desensitization. Many children have fears that are so intense that they are virtually incapacitated by them. For example, a child might have such an intense fear of going to school that he feels sick each morning before school or is unable to stay in class. Another child might be so terrified of doctors that he cannot tolerate a physical examination. Trying to convince these children that their fears are irrational is rarely successful. Systematic desensitization is a technique based on the principles of respondent conditioning that is often quite effective in the treatment of such irrational incapacitating fears or phobias.

Wolpe (1958) developed systematic desensitization based on the notion that opposite responses to a given stimulus would inhibit each other. That is, if a response incompatible with fear could be made to occur in the presence of a stimulus that elicited fear, the incompatible response would inhibit the occurrence of fear on subsequent presentations of that stimulus. In systematic desensitization, phobic patients learn to inhibit fear by relaxation.

Carrying Out Systematic Desensitization. Systematic desensitization involves three general steps: progressive muscle relaxation training, the creation of the anxiety hierarchy, and the actual desensitization.

Progressive Muscle Relaxation Training. The clinician begins the process of systematic desensitization by first training the child to produce a relaxation response that is antagonistic to anxiety. The essential feature of progressive muscle relaxation is the alternate tensing and relaxing of different muscles in the body (Jacobson, 1938). In a typical session of relaxation training, the child either sits comfortably in a chair or reclines on a sofa or bed. The clinician instructs the child to close his eyes, take three deep breaths, and then alternately tense and relax muscles according to the following sequence:

1. Left hand—by clenching the left fist.
2. Right hand—by clenching the right fist.
3. Left arm—by clenching the left fist and bending the arm inward toward the body.
4. Right arm—by clenching the right fist and bending the arm inward toward the body.
5. Both arms together—(3) and (4) at the same time.
6. Shoulders and neck—by shrugging the shoulders and raising them toward the ears.
7. Forehead—by gently raising the eyebrows.
8. Eyes and nose—by wrinkling the eyes and nose at the same time.
9. Inside of mouth—by clenching the teeth and pressing the tongue to the roof of the mouth.
10. Outside of mouth—by squeezing the lips tightly closed.
11. Upper body—by leaning forward and trying to touch the elbows behind the back.
12. Stomach—by taking a deep breath and holding it.
13. Buttocks—by squeezing them together, raising the body slightly.
14. Right leg—by lifting the straightened leg and pointing the toes toward the head.
15. Left leg—repeating (14) with the left leg.

At each step of this sequence the clinician instructs the child to tense the particular muscle group while paying close attention to the internal activities and sensations associated with the tension. After the child has maintained the tension for three to five seconds, he is told to release it slowly, noticing the difference in how he feels as the muscle relaxes. By using this procedure, the child is taught to relax in the clinician's office (about 20 to 30 minutes of actual training time) and is then instructed to practice twice a day between weekly office visits. Initially the child will take 15 to 20 minutes to practice relaxation; but as his skill improves, he will become able to achieve the relaxed state within a period of two or three minutes (usually after four to six weeks of consistent practice).

Creating the Anxiety Hierarchy. While the child masters progressive muscle relaxation, the clinician may begin to establish a hierarchy of anxiety provoking stimuli and events. The clinician must obtain a detailed description of all the stimuli and situations related to the child's debilitating fear. The child is then asked to rate these stimuli on a 100 point scale in which a rating of 0 represents the stimulus that elicits the least amount of anxiety and 100 represents the stimulus that elicits the greatest amount of anxiety. The anxiety eliciting stimuli or situations are then arranged in a hierarchy, with those that elicit the least anxiety at the bottom and those that elicit the most anxiety at the top. (*Note*: For a child who is too young to use the rating scale, the hierarchy may be established by interviewing him.) The following hierarchy might be established for a child with school phobia:

17. Answering a question incorrectly in front of the class.
16. Being called upon by the teacher to stand up in front of the class.
15. Sitting at the desk, working.
14. Entering the classroom.
13. Walking down the hall to the classroom.

12. Walking into the school.

11. Playing with other children in the school playground before class.

10. Walking to school.

9. Saying good-bye to family at home.

8. Putting coat on to leave for school.

7. Getting school books ready for school.

6. Eating breakfast before school.

5. Getting washed and dressed for school.

4. Getting out of bed.

3. Being awakened by mother to get ready for school.

2. Going to bed the night before, anticipating school the next day.

1. Laying out clothes the night before to wear to school the next day.

Note that the child in this hypothetical example is most afraid of making a mistake in front of his peers. Other children may find working at their desks most anxiety provoking; still others may rank "playing in the playground with other children" as most anxiety provoking. It is therefore important to construct a hierarchy that is valid for the particular child. A hierarchy that is valid for one child may not be valid for another, even though both may have the same general type of phobia.

The nature of the hierarchy established for a particular child may indicate other treatment strategies in addition to, or even instead of, systematic desensitization. For example, a child who is most afraid of working at his desk may have an academic or learning deficit requiring remediation; a child who is most afraid of playing in the playground with peers may benefit from social skills training to facilitate peer interactions. The anxiety hierarchy not only is a necessary component of systematic desensitization but is also, in many cases, an important diagnostic tool.

Desensitization. After the clinician has taught the child to induce self-relaxation and has constructed a hierarchy of anxiety provoking stimuli, desensitization begins. While in the relaxed state, the child is instructed to imagine the least anxiety provoking stimulus in the hierarchy. If the child experiences any anxiety whatsoever while visualizing the scene, he signals the clinician by raising one finger (a minimal response selected to avoid disruption of the relaxed state). If no anxiety occurs after seven to 10 seconds of visualizing the scene, the clinician instructs the child to stop imagining the scene and to relax. (*Note*: It is not advisable to praise the child for visualizing a scene without anxiety, as this might discourage him from reporting anxiety when he experiences it.) Following 15 to 20 seconds of relaxation, the child is instructed to imagine the scene again. After he successfully imagines the scene on two consecutive trials, he is instructed to move to the next scene in the hierarchy. As soon as the child experiences anxiety, he is instructed to cease imagining the scene and to induce the relaxed state.

When he is completely relaxed, he is instructed to imagine the previous scene in the hierarchy. If no difficulty is encountered, the next scene in the hierarchy is attempted again. In this way the child gradually moves through the hierarchy from stimuli that are least anxiety provoking to those that are most anxiety provoking. At each step, relaxation counteracts the anxiety usually elicited by the scene.

There has been little empirical evaluation of the efficacy of systematic desensitization in the treatment of childhood phobia. Findings with adults suggest that upon completion of the last (most anxiety provoking) scene in the hierarchy, some adults are able to encounter the actual feared situation with little difficulty while others require further in vivo desensitization. Our experience in treating phobic children suggests that it is often advisable, particularly when children are between five and eight years old, to supplement traditional systematic desensitization as follows:

1. When the child has completed 75 to 85 per cent of the hierarchy, in vivo desensitization should be incorporated into the treatment, so that he begins to confront the actual feared stimuli. In vivo desensitization should be structured in a gradual hierarchy and the child should be instructed to use relaxation to control his anxiety. (*Note*: Recent findings with phobic adults question the necessity of desensitizing the patient to imagined stimuli. Indeed, in many cases of pediatric phobia we have had great success with progressive muscle relaxation training followed by carefully planned in vivo desensitization alone.)

2. The clinician should arrange for positive reinforcement to occur on at least the first several occasions when the child interacts with the actual stimuli to which desensitization was directed. (Positive reinforcement is discussed in detail later in this chapter. For now, suffice it to say that by positive reinforcement we mean praise, attention, rewards).

3. Regular follow-up sessions should be conducted to ensure that there is no recurrence of phobic symptoms. If a relapse occurs, booster sessions are in order.

Assertive Training

Lange and Jakubowski (1976, p. 38) define assertion as follows: "Assertion involves standing up for personal rights and expressing thoughts, feelings, and beliefs in direct, honest, and appropriate ways which respect the rights of other people." Wolpe (1958) originally described assertion as an anxiety inhibiting response much like relaxation, and considered assertive training to be an application of respondent principles much like systematic desensitization.

Indications for the Use of Assertive Training. Assertive training is particularly appropriate for children who experience anxiety in social situations. Examples include situations in which a child shows shy, withdrawn behavior when a more assertive response is appropriate, or awkward behavior when a certain amount of social skill is required. Children who are "picked on" or "scapegoated" often can benefit from assertive training; similarly, children who are obviously uncomfortable with their peers or who tend to avoid peer interactions are often good candidates for assertive training.

Carrying out Assertive Training. Rimm and Masters (1974, p. 73) report that "by far the most commonly used assertive training technique is behavioral rehearsal. This technique requires that the client and therapist act out relevant interpersonal interactions." This role playing helps the clinician to identify those situations that are most problematic for the child and allows him to determine types of behavior that should be changed in the child to facilitate social interactions. Also role playing enables the child to practice more appropriate assertive responses to difficult social situations in relatively nonanxiety provoking simulations of those situations. The child is first taught to use assertion in small ways, and is later taught to use it in more difficult situations.

Other Applications of Respondent Conditioning

Chronic Constipation. Defecation, the desired response in cases of constipation, can often be elicited by a laxative. Since reliance on a laxative may not be desirable, Quarti and Renaud (1974) used a respondent conditioning procedure to wean patients with chronic constipation off the drug. Quarti and Renaud instructed patients to administer a distinctive electrical stimulus—a mild nonpainful electrical current—to themselves immediately prior to defecating. Defecation was initially elicited by a laxative, but the dosage of the laxative was gradually reduced until defecation was elicited by the electrical stimulus alone. Some patients were able to eliminate the electrical stimulus by presenting it at the same time every day until environmental stimuli naturally present at that time came to elicit defecation. Thus, these patients achieved regularity without the continued use of artificial stimulation.

Enuresis (Bed Wetting). A common explanation for enuresis is that pressure in the child's bladder when he has to urinate while asleep is inadequate stimulation to awaken him. A device that has been used effectively to treat enuresis, particularly primary enuresis in young children, consists of a bell connected to a conductive pad placed under the bottom sheet on the child's bed. This "bell and pad" device is wired so that the bell will ring and awaken the child as soon as a drop of urine touches the pad. Eventually the child will begin to awaken before urinating; apparently the awakening response becomes respondently conditioned to the stimulus of pressure in the bladder. It is highly recommended that the bell and pad be used together with a program of positive reinforcement for urinating in the toilet during the night. (Guidelines for the use of positive reinforcement are provided later in this chapter.)

Concluding Remarks

In general, when a stimulus fails to elicit an appropriate response, the problem usually can be corrected by therapeutic strategies derived from the principles of respondent conditioning. Pediatricians are faced with problems of respondent conditioning—such as phobia and enuresis—every day. Nevertheless such problems represent only a small portion of the behavioral difficulties encountered in pediatric populations. Much of the remainder of this chapter is devoted to a discussion of the principles and procedures important for understanding and treating the wide variety of behavior problems not subsumed within the respondent conditioning framework.

OPERANT CONDITIONING

THE OPERANT APPROACH

Many behavior patterns are not automatic or elicited by stimuli and cannot be explained in terms of respondent conditioning. Indeed most behavior patterns are considered to be emitted and are thought to be controlled by their consequences. Such types of behavior are called operants because they operate (or have some effect) on the environment (by generating environmental consequences). The role assigned to antecedent stimuli in the operant paradigm differs from that in the respondent paradigm. Whereas antecedent stimuli elicit respondent behavior, they influence operant behavior by providing a context conducive to its occurrence (or nonoccurrence). The principles of operant conditioning describe the relationship between behavior and the antecedents and consequences that influence or control it. Interventions based upon the principles of operant conditioning modify behavior by manipulating its antecedents and consequences.

It is important to note that a response may be

elicited (respondent behavior) yet be controlled by its consequences (operant behavior). For example, crying may be elicited in a child when he falls and scrapes his knee on a sidewalk. Once the crying begins, it may be controlled by its consequences. If the crying child receives attention and sympathy from his parents, he may continue to cry; if he is teased as a "crybaby" by his peers, he will probably stop. Regardless of how a behavior begins, it can be maintained or eliminated by its consequences.

By far, most of the behavior problems encountered in pediatric settings are operant in nature; that is, they are modifiable by alterations in the environmental antecedents and consequences. Pediatric behavior problems fall into three general categories: behavioral deficiencies (e.g., not eating enough at mealtime, not doing homework assignments), behavioral excesses (e.g., having a tantrum at bedtime, spitting up food at mealtime), and behavioral inappropriateness (e.g., eating paint chips and other inedible items instead of food). The reader may find it helpful to keep this classification in mind when selecting the most appropriate intervention for a behavior problem, because some procedures help to increase the frequency of a behavior pattern or develop a new behavior, others help to decrease the frequency of the behavior, and still others help to change the time or place of occurrence of the behavior. Procedures in the first two categories focus upon the consequences contingent upon behavior (see Figure 55–2); procedures in the latter category focus upon the antecedent stimuli controlling behavior. Principles pertaining to procedures in each category are to be described along with guidelines for their use in pediatric behavior management programs.

	Positive Reinforcer	Unpleasant Stimulus
	Positive reinforcement*	Punishment†
	Time out†	Negative reinforcement*
	Extinction†	

*Strengthens or increases behavior.
†Weakens or decreases behavior.

Figure 55–2. Summary of procedures that strengthen or weaken behavior.

PROCEDURES TO INCREASE BEHAVIOR

The principle of reinforcement describes an increase in the frequency of a response when it is immediately followed by certain consequences. Such consequences are called reinforcers. A reinforcer, then, is an event that increases the frequency of the behavior that precedes it. There are two general categories of reinforcement: positive reinforcement and negative reinforcement.

Positive Reinforcement

Principles and Procedures

Positive reinforcement occurs when there is an increase in the frequency of a response that is followed by a favorable event. As described by Martin and Pear (1978, p. 18), the principle of positive reinforcement has two parts: "(1) In a given situation somebody does something that is followed immediately by a certain consequence; then (2) that person is more likely to do the same again when he next encounters a similar situation. The consequence is a positive reinforcer."

The reader may note a similarity between the technical term, positive reinforcer, and the everyday word, reward. It is useful, however, to distinguish between these two terms. A positive reinforcer is defined by its "effect on behavior: if an event follows the behavior and the frequency of the behavior increases, the event is a positive reinforcer. Rewards are defined as things given or received in return for service, merit, or achievement. Although rewards may be highly valued, at least subjectively, they do not necessarily increase the frequency of the behavior they follow. Whether a reward is also a positive reinforcer can be determined only by the effect it has upon the behavior it follows.

Positive reinforcement is probably involved in situations such as the following: (1) A child who cries at bedtime is allowed to stay up later; (2) a teenager who mows the neighbor's lawn receives money for his work; (3) a young patient who must take bad tasting medicine receives a piece of candy when he takes it without complaining; (4) a student who does his homework earns special privileges in class. As a result of positive reinforcement, the child probably will not cry the next time he is put to bed; the teenager will probably mow his neighbor's lawn the next time he is asked; the young patient will probably take his medicine willingly; and the student will continue to do his homework assignments.

In some situations, positive reinforcement may be working inadvertently. For example, a child is reprimanded by his mother (e.g., she may say,

"Stop that!") when he misbehaves. Since a reprimand is a form of attention for the child, it can reinforce his misbehavior so that he may be more likely to get into trouble in the future. Reprimands are particularly reinforcing for children who have little opportunity to get attention otherwise. It is important to remember that events that are positively reinforcing for one child may not be so for another.

Indications for the Use of Positive Reinforcement

Positive reinforcement is an appropriate procedure when:

1. Low frequency behavior has to be increased. For example, a child who is a "picky eater" must learn to eat a variety of foods; a diabetic child who occasionally complies with dietary restrictions must learn to adhere to a strict dietary regimen; a child who is disruptive in the classroom must learn to spend more time working quietly at his desk.

2. New behavior patterns have to be established. For example, a diabetic child whose disease is newly diagnosed must learn to administer insulin to himself; a child who has refused to attend school for the last several months must return to his class; a child who is shy and withdrawn must learn how to interact with peers.

3. Behavior patterns have to be developed in new situations. For example, a child who is encopretic must learn to defecate in the toilet; a retarded child who is overly affectionate must learn when affectionate behavior is appropriate.

Guidelines for the Effective Use of Positive Reinforcement

Selecting the Target Behavior to Be Increased

1. The behavior should be a specific behavior (e.g., putting toys away in the toy box before supper) rather than a general category of behavior (e.g., cooperativeness). If a vague category of behavior is selected, a consistent program of positive reinforcement will be impossible.

2. The response requirement for reinforcement should be relatively lenient at the beginning of the program. That is, the target behavior should be one with which the child already responds at least occasionally. As the child begins to respond with this behavior more frequently, the program may be expanded to increase the frequency of occurrence of more difficult behavior patterns or to develop new behavior patterns not yet in the child's behavioral repertoire (procedures to develop new behaviors are discussed separately under the heading of "Shaping").

Assessment. Assessment of the target behavior should begin several days before instituting a program of positive reinforcement and should continue throughout the program in order to make it possible to evaluate the program's efficacy. An increase in response frequency during the program compared with baseline (i.e., before the program began) would indicate an effective program of positive reinforcement.

Choosing Reinforcers. Different children are often reinforced by different things. For example, some children will work to earn preferred foods; others may not be at all motivated by food. Some children will work very hard to earn time with their friends; others prefer to spend time alone watching television or listening to music. Furthermore, a stimulus or event may be a reinforcer for a child at one time and not at another time. It is therefore important to select a reinforcer that is effective for the child with whom one is working.

There is often a considerable amount of trial and error involved in the selection of an effective reinforcer for a particular child. One method that facilitates this process is simply to observe the child in his everyday activities. The activities he engages in most often, the people he interacts with most frequently, and the foods and material items he seeks out most actively are all potential reinforcers. Another method to facilitate the selection of effective reinforcers is a reinforcer survey—a series of questions about the child's preferred activities or foods that can be administered to the child or his parents either during an interview or as a written questionnaire. An example of a reinforcer survey is shown in Figure 55–3.

The only "sure-fire" way to learn whether a stimulus is reinforcing is by observing its effect on the child's behavior. If the frequency of a behavior increases when it is followed by a certain stimulus, that stimulus is a positive reinforcer. In other words, a stimulus is defined as a positive reinforcer only by its effect on behavior.

Delivering Reinforcement. Reinforcement should be delivered in a positive and conspicuous manner immediately after occurrence of the target behavior. The child should be told why he is receiving reinforcement and praised for his accomplishment. Pairing praise with established reinforcers can enhance the reinforcing effects of praise and facilitate later attempts to wean the child from the program (see following discussion).

Reinforcer "Menus." It is sometimes a good idea to allow the child to choose from a number of available reinforcers, that is, from a "reinforcer menu." For example, when reinforcing a young child for putting his toys away before supper, his mother may allow him to choose from the following reinforcer menu: staying up 15 minutes past bedtime, having a special dessert at supper, reading a story with her after supper. The advantage of this is that at least one reinforcer among the selection is likely to be strong.

1. Foods: What are your favorite things to eat and drink?
   A. Breakfast:
      1. What are your favorite foods for breakfast?

      _____
      _____
      _____

      2. Would you like to have these things for breakfast more often than you do now?

      _____
      _____

   B. Lunch:
      1. What are your favorite foods for lunch?

      _____
      _____
      _____

      2. Would you like to have these things for lunch more often than you do now?

      _____
      _____

   C. Supper:
      1. What are your favorite foods for supper?

      _____
      _____
      _____

      2. Would you like to have these things for supper more often than you do now?

      _____

   D. Desserts and snacks:
      1. What are your favorite things for dessert or snacks?

      _____
      _____
      _____

      2. Would you like to have these things more often than you do now?

      _____

   E. What things do you like to drink most?

      _____
      _____
      _____

2. Activities: What things do you like to do the most?
   A. Around the house:
      1. What are your favorite things to do when you are at home (like having friends over, playing games, going outside to play, staying up late, cooking, helping "Mom" with chores, doing jobs, hobbies, watching TV, reading, listening to music)? Be sure to list all the things you like to do.

      _____
      _____
      _____

      2. Would you like to be able to do some of these things more? _____ If so, which ones?

      _____
      _____
      _____

   B. When you go out:
      1. What are your favorite things to do when you go out (like play in the backyard, go to a friend's house, go to a movie, go shopping)? Be sure to list all the things that you like to do.

      _____
      _____
      _____

      2. Would you like to be able to do some of these things more often? _____ If so, which ones?

      _____
      _____
      _____

   C. If you had some extra money, what would you like to buy?

      _____
      _____
      _____

   D. Are there people whom you especially enjoy being with? _____ Whom do you like to be with most?

      _____
      _____
      _____

Figure 55–3. A reinforcer survey to help identify reinforcers.

Satiation and Deprivation. Satiation refers to the condition in which a child has been exposed to a reinforcer to such an extent that it is no longer reinforcing. Most reinforcers are not effective unless the child has been deprived of them for some time prior to their use. A child who gets his favorite dessert after every meal will probably not be reinforced by a special serving of that dessert. It is a good idea, therefore, to use as reinforcers preferred items that are not readily available to the child.

Immediacy. Delayed reinforcement is much less effective than immediate reinforcement. To be maximally effective, a reinforcer should be given immediately after the desired response. However, some kinds of reinforcers (e.g., a movie, a party, a car trip, a picnic) cannot conveniently be given immediately after the desired response. A solution

to this problem is discussed in the section on token economies.

Frequency of Reinforcement. At the beginning of a positive reinforcement program it is best to reinforce the desired behavior each time it occurs (i.e., continuous reinforcement). In practice it is virtually impossible to constantly monitor a child's behavior to ensure continuous reinforcement of the target behavior. However, as continuous a schedule as possible should be used to establish the behavior. Once learned, a behavior can be maintained by intermittent reinforcement, that is, by reinforcing the behavior occasionally rather than every time it occurs. It is important to remember, however, that the frequency of reinforcement should be reduced in a gradual manner.

Instructions. In order for a program of positive reinforcement to work, it is not necessary for the child to be able to talk about, or even understand, the program. Programs are effective even for very young children with a limited capacity to verbally grasp the "intricacies" of the program. Nevertheless instructions can facilitate the effectiveness of a program for children who can understand them.

Weaning the Child from the Program. In learning to use positive reinforcement to change their child's behavior, parents often ask, "Will I always have to reward my child when he does something I want him to do?" The answer to this question is both yes and no: Reinforcement will probably always be necessary to maintain the child's desirable behavior; but at the end of a carefully planned program, reinforcement need not be as contrived or as frequent as in the initial stages. For example, suppose that a mother wishes to teach her child to follow instructions. At first she may have to reinforce the child with a special treat each time he does what is asked of him. Eventually, however, she may be able to maintain this behavior by merely praising the child. That is, the child may be weaned from the program to the point at which more "natural" reinforcers (e.g., praise) can maintain the desired behavior.

The following considerations are important for weaning a child from a program of positive reinforcement:

1. Initially praise and other potentially reinforcing stimuli normally available in a child's natural environment may not be effective. The reinforcement program should be designed to establish these stimuli as conditioned reinforcers by ensuring that they are presented to the child together with existing reinforcers. When natural events or stimuli (e.g., praise) acquire reinforcing properties (i.e., become conditioned reinforcers), they may effectively maintain behavior patterns developed during the course of the program.

2. Since desirable types of behavior are not necessarily reinforced every time they occur in the child's natural environment, it is important to introduce intermittent reinforcement as soon as possible. The more similar the program's reinforcement schedule is to that of the natural environment, the more easily the child may be weaned from the program.

When a Program of Positive Reinforcement Fails. There may be problems with a program if there are lengthy periods of time during which the child does not earn the positive reinforcer or there is no increase in the reinforced behavior soon after the program is implemented. Some of the most common problems and their resolution include the following:

1. The response requirement for reinforcement is too stringent. Resolution: Change the response requirement to one that is easier for the child.

2. The events or stimuli programed as reinforcers are ineffective, perhaps because of an error in reinforcer selection or because of overexposure to the stimulus (causing satiation). Resolution: Try other reinforcers or increase the schedule of reinforcement so that the child receives a smaller quantity of the reinforcer.

3. Reinforcers are not being delivered immediately after the desired response. Resolution: Change the program as necessary to ensure prompt delivery of reinforcers.

4. The schedule of reinforcement is increased too quickly. Resolution: Return to a richer (perhaps, continuous) reinforcement schedule.

Token Economies: A Special Application of Positive Reinforcement

Tokens, such as poker chips, coins, tickets, stars, stickers, points, or checkmarks, are conditioned reinforcers. Thus they do not reinforce in their own right, but acquire reinforcing properties through association with stimuli and events that are reinforcing. Tokens are, in fact, generalized conditioned reinforcers because they can be exchanged for a wide variety of reinforcers, called back-up reinforcers.

A reinforcement system based on tokens is referred to as a token economy. The tokens are earned and used to purchase back-up reinforcers such as food, activities, privileges, and material possessions. The token value of back-up reinforcers must be specific so that it is clear how many tokens are required to purchase various reinforcers. Also the target behavior must be specified (as in any program) along with the number of tokens earned by its performance.

We have used token reinforcement systems for many patients, both in the hospital and at home. Inpatient token systems are typically administered by nursing staff; home token systems are administered by the child's parents. A hospitalized anorectic patient who had refused to take anything by mouth (she was fed via a nasogastric tube) earned points for eating increased quantities and varieties of foods. She exchanged her points for shopping trips, records, and individual time with selected members of the nursing staff (a valuable commodity for most patients housed on a busy inpatient unit). Over a period of 12 weeks she progressed from drinking small sips of water to eating three 500 calorie meals per day. In a continuing research project we are currently evaluating the efficacy of a home token system in reducing harmful scratching in children with atopic dermatitis. One child earns stickers for periods of nonscratching. Stickers are awarded by the child's parents and can be exchanged for the opportunity to have a party, go on a picnic, see a movie, and so on. Preliminary findings suggest a decrease in scratching as a result of this token program. For other examples of token economies, the reader is referred to the books by Kazdin (1975) and Martin and Pear (1978).

Advantages and Disadvantages of Token Economies

Tokens offers certain advantages over other systems of positive reinforcement:

1. Tokens are powerful reinforcers and often can maintain higher levels of behavior than other conditioned reinforcers such as praise and approval.

2. Tokens nearly always can be delivered immediately after the target response. Thus, they can bridge the delay between the desired behavior and back-up reinforcers when back-up reinforcement cannot conveniently be delivered immediately.

3. Since tokens can be used to purchase a variety of back-up reinforcers, they are less subject to satiation than other reinforcers. If a child loses interest in one or two of the back-up reinforcers, there are usually several others that have not lost their reinforcing value.

4. Tokens are very useful for inpatient settings where there are several children with different reinforcer preferences. Tokens conveniently permit the administration of a single reinforcer (i.e., tokens) to all children. Individual preferences can be exercised in the exchange of tokens for back-up reinforcers. Token exchange can be facilitated by a central "store," stocked with a variety of items, where all children can spend their tokens.

5. Tokens permit the parceling out of reinforcers (e.g., activities, material possessions), which might have to be earned in an all or none fashion. The tokens can be saved toward the purchase of the back-up reinforcer.

A potential problem with any token system is removing it after behavior gains have been made. Since tokens constitute a set of reinforcing events not available in most settings (except tokens such as money or grades), specific procedures must be used to withdraw the token system without a decrement in performance. These procedures are discussed along with other guidelines for designing a token economy in the next section.

Some Guidelines for Designing a Token Economy

The reader is encouraged to review the guidelines for developing a program of positive reinforcement as they apply to designing a token economy. In addition, the following considerations are important:

Selecting the Type of Token. The main considerations in selecting the type of token are the type of child for whom the token system is designed and the setting in which the system will be implemented. It is a good idea, whenever possible, to include the child in token selection. For very small children, stickers pasted on a chart that is readily accessible to the child usually work well. The number of tokens earned is in full view of the child, and they cannot be misplaced. For older children, marks or points tabulated on a record sheet are usually adequate.

Making Tokens Reinforcing. Tokens need to be established as conditioned reinforcers because they have little or no reinforcing value in their own right. For some children it is sufficient to explain that tokens can be exchanged for back-up reinforcers. After the explanation, tokens immediately take on a reinforcing value that is maintained by the actual exchange process.

For other children (e.g., very young or developmentally delayed children), tokens do not acquire their reinforcing value until they are specifically paired with back-up reinforcers in a special "token training" sequence. Tokens are given noncontingently several times. Immediately after receiving the tokens, the child is taught to trade them for a known reinforcer. For example, a retarded child may be given a token just prior to supper. A few seconds later, as the child sits down to eat, he is taught to return the token. Thus, the token is exchanged for access to food. By several repetitions of this sequence with a variety of back-up reinforcers, the token becomes a conditioned reinforcer (Kazdin, 1975).

Delivering Token Reinforcement. Tokens should be delivered in a positive and conspicuous manner immediately after the target behavior occurs. The child should be told why he is receiving the token and praised for his accomplishment.

How Many Tokens Should Be Given? Tokens should be paid frequently during the early stages of the program. The number of tokens that a child should receive for a given behavior depends upon:

1. The baseline frequency of the behavior (i.e., the frequency of the behavior before instituting the token system). The less frequent a behavior, the more tokens that should follow its occurrence.

2. The number of opportunities the child has to earn tokens. The more opportunities available, the fewer the tokens that should be paid each time.

3. The cost (in tokens) of back-up reinforcers. The more expensive (in tokens) the back-up reinforcers, the more tokens that should be paid per response, at least initially.

In general, the child should be earning enough

tokens to allow him access to an ample selection of back-up reinforcers. However, he should not be allowed to become "so rich that he doesn't have to work." The number of tokens paid for a given behavior may be adjusted (usually decreased) during the course of the program as the child's behavior improves.

Assigning Token Values to Back-up Reinforcers. Three factors should be considered when assigning token values to back-up reinforcers.

1. The token value of each back-up reinforcer should be directly related to its monetary value. The more an item costs in money, the higher the token value that should be assigned to it.

2. The token value of each back-up reinforcer should be directly related to its reinforcing value. That is, more powerful reinforcers should cost the child more tokens.

3. Back-up reinforcers that are therapeutically beneficial for the child should be less expensive (in tokens). The token value of back-up reinforcers may be adjusted (usually increased) during the course of the program.

Exchanging the Tokens for Back-up Reinforcers. In the beginning of the token program, a child should be allowed to spend his tokens frequently (once a day, at least). As the child progresses and becomes accustomed to saving tokens, token exchanges may occur less often.

Weaning the Child from the Token System. Since social reinforcement (e.g., praise) rather than the giving of tokens is the prevalent form of reinforcement for children in the natural environment, a token economy should be designed such that social reinforcement gradually takes the place of tokens. This may be done in two ways: Tokens may be eliminated gradually by making the schedule of token delivery more and more intermittent, by decreasing the number of behaviors reinforced with tokens, or by increasing the delay between the target response and token presentation. Or tokens may be decreased in value gradually by increasing the cost of back-up reinforcers, or by increasing the delay between token acquisition and token exchange for back-up reinforcers. At present the optimal combination of these procedures is not known (Martin and Pear, 1978).

When a Token System Fails. A token economy may be in trouble if there are long periods of time during which the child earns no or very few tokens, or if there is no increase in the reinforced behavior soon after the system is implemented. Also there may be a problem if the child does not exchange the tokens that he earns for the back-up reinforcers. Should any of these situations arise, the token system will merit careful scrutiny with specific attention to the areas of potential difficulty outlined for positive reinforcement programs.

Negative Reinforcement

Principles and Procedures

Negative reinforcement refers to the process of effecting an increase in the frequency of a response by removing an unpleasant stimulus immediately after the response is performed (Kazdin, 1975). Note that reinforcement (positive or negative) refers to an increase in the frequency of behavior. Just as a stimulus may be positively reinforcing to some children and not to others, an unpleasant stimulus may be negatively reinforcing to some children and not to others. Also a stimulus may be a negative reinforcer for a child at one time and not at another time. A negative reinforcer, like a positive reinforcer, is defined by its effect on behavior.

Negative reinforcement is probably involved in situations such as the following: An individual standing at a bus stop on a cold day puts on his coat and feels warmer, or a patient with a headache takes an aspirin to relieve the pain. In each case an unpleasant stimulus (either cold or pain) is removed after a specific response is performed (either putting on a coat or taking medicine).

There is an interesting combination of positive and negative types of reinforcement that frequently occurs in interactions between parents and their children and that may foster socially undesirable behavior. In such interactions the parents' behavior is negatively reinforced because it terminates undesirable behavior initiated by the child. At the same time the child's undesirable behavior is positively reinforced. For example, parents may hug a child who is whining. Whining is an unpleasant event for the parents that terminates when they hug the child. Hugging the child is thus negatively reinforced. At the same time the child's whining is positively reinforced by attention from the parents and is likely to occur more frequently in the future. This pattern, characteristic of numerous parent-child interactions, can explain how many behavior problems develop.

Escape and Avoidance Conditioning

A behavior pattern is increased in frequency through negative reinforcement when it results in escape from or avoidance of an unpleasant event. Escape behavior allows the child to terminate a continuing unpleasant event. Examples of negative reinforcement through escape have already been cited: the person who puts on his coat when he feels cold and the patient who takes an aspirin when he has a headache are both negatively reinforced by escape from an unpleasant stimulus (cold or pain).

Avoidance behavior allows the child to prevent

or indefinitely postpone contact with an unpleasant event. Examples of negative reinforcement through avoidance include drinking alcohol sparingly to avoid a driving mishap and doing homework to avoid a penalty. Avoidance behavior may develop after a child has learned to escape from an unpleasant event and is facilitated when a signal warns of the forthcoming unpleasant event. For example, when a child misbehaves, his mother may frown and say "no." If the child persists, his mother may take more aggressive action (e.g., send the child to his room) and the child will probably stop misbehaving to escape the unpleasant situation. The next time the child misbehaves, he may stop what he is doing when his mother frowns and says "no." He has learned to recognize a frown and the word "no" as a warning and acts to avoid more severe consequences. In some cases, of course, avoidance behavior is acquired without direct experience of the unpleasant event. A child learns not to touch hot stoves and electrical outlets without experiencing the direct consequences of doing so. In such cases verbal cues from others (usually parents) are sufficient to teach the child that certain things are to be avoided.

Escape and avoidance conditioning is often involved inadvertently in the development or exacerbation of many problems encountered in pediatric settings. For example, escape and avoidance contingencies, at least partially, may maintain the soiling of encopretic children who experience difficulty with peer interactions. An encopretic child may learn that when he soils in a difficult social situation, such as recess at school, he is excused from the situation (to change his clothes). Soiling is negatively reinforced by escape from a difficult situation. The child may also learn that he can avoid recess entirely by soiling just before it begins. Similarly a child who suffers from migraine headache may learn that he can escape from or avoid unpleasant tasks and situations when he has a headache. Problems of this sort are considered further under the heading, "Learning and Behavior in Chronic Disease."

Indications for the Use of Negative Reinforcement

Negative reinforcement is not often used in behavioral programs. Escape and avoidance conditioning involves unpleasant stimuli that can produce undesirable emotional behavior such as crying, aggression, and general fearfulness. Moreover, any stimulus associated with aversive stimulation tends to become aversive itself. This means that there is a danger that the child's parents, as administrators of a program of negative reinforcement, may become conditioned aversive stimuli for the child. In view of these considerations, positive reinforcement is by far the treatment of choice when the goal is to increase behavior. Only when a carefully designed program of positive reinforcement has proven ineffective should negative reinforcement be considered.

Guidelines for the Effective Use of Negative Reinforcement

When the use of negative reinforcement is clearly warranted, the reader is encouraged to review the guidelines for the effective use of positive reinforcement, for these apply also to negative reinforcement. In addition, the following guidelines may be helpful:

Escape Versus Avoidance.   Given a choice between maintaining behavior by use of an escape or avoidance procedure, the latter is preferable for two reasons. First, in escape conditioning the unpleasant stimulus must be present for the target response to occur; in avoidance conditioning, the unpleasant stimulus is presented only if the target response does not occur. Second, avoidance behavior is more durable than escape behavior. Note, however, that it may be necessary to establish behavior by escape conditioning before it can be maintained by an avoidance procedure.

Using a Warning Signal.   During avoidance conditioning, a signal should warn the child of the impending unpleasant event. The signal improves performance by warning that failure to respond will result in aversive stimulation.

Positive Reinforcement.   It is a good idea to use positive reinforcement for the target response in conjunction with negative reinforcement. That is, the target response should be followed not only by escape from or avoidance of an unpleasant event but also by a positive consequence. This will help to strengthen the desired behavior and to counteract the harmful effects of the aversive stimulation.

Instructions.   As with positive reinforcement, instructions are not necessary for negative reinforcement to work. Nevertheless they can facilitate the effectiveness of a program for children who can understand them.

Discontinuing the Program.   If there is little or no increase in the target behavior soon after negative reinforcement is implemented, the procedure cannot be ethically justified and should be discontinued.

PROCEDURES TO DECREASE BEHAVIOR

Extinction

Principles and Procedures

Extinction refers to withholding reinforcement from a previously reinforced response and results in a decrease in the frequency of that response. As described by Martin and Pear (1978, p. 40), the principle of extinction has two parts: "(1) If, in a given situation, somebody emits a previously reinforced response and the response is not followed by the usual reinforcing consequence, then (2) that person is less likely to do the same thing again when he next encounters a similar situation." Stated simply, if a response that used to be reinforced is no longer reinforced, it will, over time, decrease in frequency. In a general sense, if a behavior is ignored, it will go away. Numerous examples of extinction are evident in everyday life. For example, a child stops trying to initiate interactions with a friend if his efforts are never successful; a student stops raising his hand in class if he is never called upon by the teacher; a child stops nagging his mother for a cookie just before supper if his efforts go unreinforced (i.e., if he is not given the cookie).

Many undesirable types of behavior with which children respond are maintained by attention from parents and other adults in their environment. Withholding parental attention following the occurrence of undesirable behavior is an extinction procedure used frequently in pediatric settings. Suppose, for example, that parents report that their child cries continuously at bedtime unless someone stays with him until he falls asleep. Extinction may be a useful procedure to eliminate the problem. It appears that the child's crying is positively reinforced by attention from the adult who stays with him. Under extinction, this reinforcement would be discontinued; that is, crying would no longer result in someone staying with the child at bedtime. The child's crying would simply be ignored. During the first several nights of extinction, there may be an increase in crying (see the next section on "Characteristics of the Extinction Process"). However, with consistent application of the procedure, the child's crying at bedtime will gradually decrease and disappear. Extinction has also been used successfully to eliminate tantrums in young children, to decrease complaints and other types of behavior in response to pain (e.g., refusal to participate in therapeutic exercises and activities) in patients with chronic recalcitrant pain (e.g., Fordyce, 1976), to reduce scratching in patients with dermatitis, and so on.

Extinction is most effective when combined with positive reinforcement for some desirable alternative behavior. Thus, extinction of bedtime crying might be combined with positive reinforcement for a quiet night; extinction of tantrums might be combined with positive reinforcement for desirable behavior that is incompatible with tantrums; extinction of pain behavior might be combined with positive reinforcement for "well behavior"

(e.g., participation in activities); extinction of scratching might be combined with positive reinforcement of no scratching. The combination of the two procedures will decrease the frequency of the undesirable behavior much more effectively than extinction alone.

Extinction must not be used indiscriminately to eliminate undesirable behavior. In some cases apparently undesirable behavior may be communicating an important need that should not be ignored. Crying, for example, may indicate injury, illness, fear, or other forms of discomfort. Prior to implementing extinction to reduce the frequency of its occurrence, the behavior must be examined closely in terms of the desirability of decreasing it.

Characteristics of the Extinction Process

Gradual Reduction in the Frequency of the Behavior. The process of extinction is usually a gradual one. That is, extinction typically does not show an immediate reduction (or elimination) of the extinguished response, but rather a gradual reduction over time. Several unreinforced responses occur before the effect of extinction becomes apparent. In clinical applications the delayed effects of extinction are particularly problematic when the undesirable types of behavior are dangerous to the child or to those around him. Although extinction reduces the frequency of a behavior pattern, dangerous behavior usually requires an intervention that yields more rapid results than extinction.

Extinction Burst—"Things Get Worse Before They Get Better." When extinction first begins, the frequency of the undesired response may increase before it decreases. The increase in responding that occurs early in extinction is called the extinction burst. For example, when parents first begin to extinguish a child's tantrums, the child may scream louder and longer than ever before; similarly, when a mother begins to ignore her child's nagging for a snack just before supper, the child may become more persistent than ever before. Thus, under extinction it is usually necessary to endure increased levels of the undesired behavior before a decrease occurs. This feature of the extinction process has three implications for the use of extinction as a treatment strategy:

1. A burst of responses is especially serious when the behavior is physically harmful to the child or to others. Although extinction can eliminate the behavior, the increased risk of physical harm during the burst period argues against its use for dangerous behavior.

2. When extinction is used as a treatment strategy, it is important to prepare the parents (and anyone else who must ignore the behavior) for a

possible burst of responses. If unprepared for a worsening of behavior, parents may conclude that the procedure has failed and abandon its use.

3. Some parents may find it very difficult to tolerate the undesirable behavior as it intensifies at the beginning of extinction. Thus, during the extinction burst the likelihood that reinforcement will occur is increased. For example, a tantrum may become worse when parents ignore the behavior. When the tantrum is worse, parents will have increased difficulty in tolerating the behavior and may give in to the child by providing attention and comfort. Parental reinforcement at this time will have the disastrous effect of increasing the probability of very intense tantrums because reinforcement is provided when the behavior is worse than usual. The success of an extinction procedure depends upon its consistent implementation, especially during the difficult period of the extinction burst. For some parents, who may experience particular difficulty during the burst, extinction may not be the treatment of choice.

Spontaneous Recovery. After extinction has progressed, the response may temporarily reappear even though it has not been reinforced. This phenomenon is referred to as spontaneous recovery. As with the extinction burst, a major concern with spontaneous recovery is that the response will be reinforced. If reinforcement occurs, it follows a long series of unreinforced responses—not unlike a highly intermittent schedule of reinforcement—and can seriously disrupt the extinction process. If extinction continues and no reinforcement occurs, the spontaneously recovered response will quickly decrease. It is important to prepare parents for the possibility of spontaneous recovery and to advise continued application of the extinction procedure.

Possible Side Effects. The cessation of reinforcement sometimes can produce emotional responses, such as aggression, agitation, or rage, particularly when extinction first begins. Parents should be prepared for the occurrence of such behavior.

Factors Affecting the Extinction Process

Frequency of Reinforcement. Behavior patterns maintained by intermittent reinforcement prior to extinction will take longer to extinguish than those previously maintained by continuous reinforcement. Many behavior problems encountered in pediatric settings are maintained by intermittent reinforcement. Consider, for example, a child who cries and has tantrums when he wants to have "his way." Sometimes tantrums might have the desired effect and the child might get "his way"; other times he might not. Such intermittent reinforcement produces behavior that can

persist for some time, even in the absence of reinforcement (i.e., extinction). This means that in extinguishing behavior that has been reinforced intermittently, parents must be prepared for extinction to take considerable time.

Avoidance Conditioning. Research has shown that avoidance types of behavior are highly resistant to extinction. That is, behavior that is negatively reinforced by avoiding unpleasant stimulation tends to persist long after the unpleasant stimulation is removed from the situation. It is the nature of the avoidance procedure that when the behavior occurs regularly, the unpleasant stimulus is never presented. Presumably, when the procedure no longer applies (i.e., when the unpleasant stimuli are removed from the situation), it takes quite some time for the child to discover that performing the behavior is not what keeps the unpleasant stimulus away. Therefore, extinction of avoidance behavior (by discontinuing the threat of unpleasant stimulation) is a lengthy process.

Indications for the Use of Extinction

Extinction is an appropriate procedure whenever it is necessary to reduce or eliminate undesirable behavior as long as the behavior does not place the child or those around him in physical danger and the circumstances are such that the procedure can be implemented consistently.

Guidelines for the Effective Use of Extinction

Selecting the Behavior to Be Decreased. As with positive reinforcement, the behavior selected for extinction should be a specific behavior. It is not a good idea to try to extinguish many types of behavior at once. Rather than planning a major character improvement, select one (or, at most, two) of the child's troublesome behavior patterns to extinguish. The program may later be expanded to include additional problematic behaviors.

Assessment. The frequency of the undesirable behavior should be monitored for several days prior to extinction and throughout the extinction process. This information will be important for the later evaluation of the effects of extinction. (Assessment issues are discussed more fully in a separate section.)

Identifying the Reinforcer. Extinction requires that the reinforcer(s) maintaining the undesirable behavior be identified and then withheld when the behavior occurs. This sounds quite simple, but in practice it is sometimes difficult to identify the reinforcer. For example, a child may behave aggressively with peers. There may be a variety of reinforcers maintaining this behavior, including the control aggression exerts over peers, a submissive response from the victim, admiration from friends, or attention from a teacher or parent. An extinction program that merely minimizes parental attention to the aggressive behavior is likely to be ineffective. Extinction requires that all sources of reinforcement for the undesirable behavior be identified and eliminated.

Controlling the Source of Reinforcement. Once the reinforcer maintaining an undesirable behavior has been identified, the extinction procedure requires that the reinforcer be withheld following each and every occurrence of the behavior. Any accidental reinforcement may reinstate the undesirable behavior and prolong the extinction process.

Unfortunately it is not always easy to control the source of reinforcement for undesirable behavior. Even when the child's parents implement extinction consistently, other sources of reinforcement may frustrate their efforts. For example, visiting friends or relatives may pay attention to a behavior that parents have been conscientiously ignoring; siblings or peers often reinforce behavior that the parents would like to extinguish. It is very difficult to enlist the cooperation of these individuals in a program of extinction, and, if possible, it is often best to implement extinction in their absence (Martin and Pear, 1978).

Things Get Worse Before They Get Better. The early stages of the extinction process are characterized by an increased frequency of the undesirable behavior and by the occasional outburst of aggression. Therefore, it is important to provide parents with much support and encouragement when they first embark on extinction to ensure their consistent implementation of the program even when it may appear to them to be failing.

Choosing the Setting to Carry Out Extinction. Since the early stages of extinction can be particularly trying for parents who are ignoring disruptive behavior, it is important to consider the setting in which extinction will first be carried out. In general, the setting should be selected to minimize the influence of alternative reinforcers (i.e., from other people) on the behavior to be extinguished and to maximize the chances of the parents' persisting with the program. For example, it would be inadvisable for a mother to initiate extinction of her child's temper tantrums at a family get-together where alternative sources of reinforcement abound. Similarly, it would be unwise for her to initiate the program in a busy department store where nasty looks from other shoppers would decrease the chances of her carrying through effectively. Extinction should begin in the setting where it has the greatest chance of success.

Combining Extinction with Positive Reinforcement. Extinction is a procedure that teaches a child what not to do. Used in isolation, extinction can create a "behavioral vacuum" by eliminating one behavior without replacing it with another more desirable behavior. When a behavioral vacuum exists, there is a risk that it will be filled with another undesirable behavior. Rather than merely teaching a child what not to do with extinction, it is important also to teach him what to do instead. Whenever one source of reinforcement is eliminated (through extinction), another, more desirable, way of earning reinforcement should be provided. Thus, extinction works effectively when used in combination with positive reinforcement for some desirable alternative behavior. Whenever possible, the alternative behavior to be reinforced should be one that is incompatible with (i.e., cannot occur at the same time as) the undesirable behavior to be extinguished. For example, a mother extinguishing her child's tantrums might also positively reinforce the desirable behavior of playing quietly. The guidelines outlined previously for implementing a program of positive reinforcement should be followed.

Instructions. As with any program, the effects of extinction do not depend upon the child's understanding of the procedure. However, whenever possible, it is a good idea to tell the child about the plan in advance.

When a Program of Extinction Fails. If the undesirable behavior does not begin to decrease after extinction has been applied consistently for several days, there may be some difficulty with the program. There are three common reasons for the failure of an extinction procedure:

1. The attention (or other events) withheld following the undesirable behavior is not the reinforcer maintaining the behavior.

2. The undesirable behavior is being reinforced (perhaps intermittently) from another source.

3. The desirable alternative behavior is not being strengthened appropriately (refer to the guidelines for positive reinforcement).

If extinction does not appear to be working, these reasons should be examined carefully.

Punishment

Principles and Procedures

Punishment refers to a decrease in response rate when the response is followed by an unpleasant consequence. As described by Martin and Pear (1978, p. 178), the principle of punishment has two parts: "(1) If in a given situation somebody does something that is followed immediately by a certain consequence, then (2) that person is less likely to do the same thing again when he next encounters a similar situation. The consequence is called a punisher." This definition is somewhat different from the everyday usage of the term. Punishment usually refers to a penalty imposed for engaging in a certain behavior. The technical definition includes an additional requirement: the frequency of the behavior must decrease. Thus, punishment, like reinforcement, is defined by its effect on behavior.

It is important to note that punishment, in the behavioral sense, does not necessarily involve physical pain. It is not used as a form of retribution or payment for the misbehavior. Also, what acts as a punisher for one child may not be a punisher for another child; and what acts as a punisher for a child at one time may not be a punisher for that child at another time. Only if the frequency of a response is reduced can punishment be operative.

Punishment is often confused with negative reinforcement, and it is important to distinguish between these two terms. Reinforcement, of course, refers to procedures that increase the frequency of behavior; punishment refers to procedures that decrease the frequency of behavior. In negative reinforcement an unpleasant event is removed following a response; in punishment an unpleasant event is presented following a response.

Types of Punishment

Punishment may take one of two forms: the presentation of unpleasant events or the removal of positive events after a response.

Presentation of Unpleasant Events. Punishment by presenting an unpleasant event contingent upon an undesirable behavior is often employed by parents to decrease a child's undesirable behavior. Reprimands, threats, and spankings are common consequences when a child misbehaves. Reprimands that are delivered quietly and privately to the child have been found to be more effective in suppressing behavior than those delivered loudly and publicly. Threats do not suppress behavior unless they are consistently followed by the threatened unpleasant consequences. The efficacy of reprimands, threats, or spankings as punishers varies greatly from child to child. In fact, for some children, the "negative attention" of these consequences serves to reinforce their undesirable behavior. For this reason it may be better to simply ignore a child's undesirable behavior rather than to provide any form of attention (positive or negative) when it occurs.

Removal of Positive Events. Time out from positive reinforcement is the most commonly used form of punishment in behavior management programs. Time out refers to the removal of all positive reinforcers for a specified period of time following the occurrence of undesirable behavior. This usually involves isolating the child for a specified time period by sending him to his room or placing him in a chair located in a quiet area

(e.g., "the quiet corner"). The time-out period need not be lengthy (five or 10 minutes is usually sufficient). However, it is important not to terminate time out while the child is engaged in undesirable behavior (e.g., screaming, crying). It is sometimes necessary to extend the time-out period beyond the specified time period to ensure at least one or two minutes of appropriate behavior before terminating time out.

Examples of situations involving time out abound in pediatric settings. Parents use time out when they isolate a child in his room for 10 minutes contingent upon misbehavior at home; teachers use time out when they "send a child to the corner" for five minutes contingent upon misbehavior in class. A less dramatic use of time out is exemplified by a parent who removes a child's dinner plate for five minutes contingent upon misbehavior at the dinner table. We have taught parents to use time out to eliminate head banging in toddlers. Specifically, whenever the child begins to bang his head, his parents are taught to isolate him in an environment carefully designed to ensure that he can do no harm by further head banging. Usually a playpen with ample padding on hard surfaces works well. The child is placed alone in the playpen (without toys) as soon as head banging occurs, and must remain there until no head banging (or other undesirable behavior) occurs for approximately one or two minutes.

It is sometimes difficult to eliminate all sources of reinforcement during the time-out period. For example, the child who is isolated in his room may have access to a number of reinforcing activities (e.g., listening to music, playing with toys); the child who is sent to the corner may receive attention and social reinforcement from classmates; the child in the playpen may receive attention from a sibling. It is important to ensure that the child spends the time-out period in a setting that minimizes the availability of all reinforcers.

Characteristics of the Punishment Process

Immediacy of Effects. A reduction in response rate usually occurs soon after a punishment procedure is implemented. The greater the intensity of the unpleasant event, the more immediate the response suppression. In general, if there is no immediate response suppression, it is probably not advantageous to continue the punishment procedure.

Specificity of Effects. Punishment often leads to effects that are specific to the situation in which the response is punished. Thus, the punished behavior may be suppressed only in the presence of the person who administered the punishment and not in the presence of other adults. Also, response suppression may occur only in the setting where punishment occurred and not in other settings. This means that in order to completely suppress undesirable behavior with punishment, it may be necessary for a variety of individuals to administer punishment in a variety of settings.

Response Recovery After Punishment Withdrawal. The effects of punishment are often short lived so that when the punishment contingency is withdrawn, the response rate increases or returns to its baseline (i.e., prepunishment) level. One way to minimize response recovery when punishment is withdrawn involves positive reinforcement for desirable behavior that is incompatible with the punished behavior. For example, time out for head banging in a small child might be used in combination with positive reinforcement for appropriate play behavior. When the punishment procedure is discontinued, the reinforced response can replace the previously punished response and can be maintained with continued reinforcement.

Possible Side Effects of Punishment—A Cautionary Note. Punishment must be approached with caution because there may be several undesirable side effects associated with its use. Even though the target behavior may be eliminated, other consequences resulting directly from punishment may be worse than the original behavior or at least as problematic in their own right. Some of the potential side effects of punishment are as follows:

1. Punishment can produce undesirable emotional reactions, such as crying, general fearfulness, or aggression. These emotional states may be temporarily disruptive to the child and may interfere with his performance of desirable behavior. For example, when a child receives a spanking, crying, anger, and other similar emotional states will probably occur. The child may be temporarily unresponsive to his social environment until he is no longer upset.

2. Any stimulus associated with the punishing stimulus may become a punishing stimulus (i.e., a conditioned punisher) itself. For example, if a parent punishes a child for fighting with a sibling over a toy, anything associated with the situation—such as the toy, the sibling, the home environment, the parent—will tend to be punishing to the child. The child may attempt to avoid or escape these stimuli. Thus, instead of helping the child to interact with his sibling, punishment may drive him away from his family and home.

3. Punishment does not establish any new behavior; it merely suppresses old behavior. That is, punishment cannot teach a child what to do; it can only teach him what not to do. The goal of any behavior management program should be the development of appropriate behavior. Punish-

ment, at least by itself, cannot accomplish this goal.

4. Children can readily learn aggressive behavior patterns from parental models. If a parent applies punishment to a child, the child is likely to do the same to others. Thus, by punishing children, parents inadvertently may be providing models of aggressive behavior for them to follow.

5. Parents who use punishment are reinforced for punishing. Because punishment results in quick suppression of undesirable behavior, a parent who uses it will be negatively reinforced (by removal of the undesirable behavior). Hence, the parent is likely to rely increasingly on punishment, neglecting the use of positive reinforcement for desirable behavior and running the risk of undesirable side effects.

Indications for the Use of Punishment. It follows from the previous discussion that punishment is a procedure to be used with extreme caution because of possible undesirable side effects. Furthermore, the use of unpleasant stimuli to control behavior is questionable on ethical grounds. Therefore, punishment should be used only as a last resort after other procedures have proven ineffective. In most cases undesirable behavior can be controlled without punishment. Behavior incompatible with the undesirable behavior may be reinforced, while the undesirable response is extinguished. Reinforcement combined with extinction can alter behavior effectively while minimizing the risk of side effects.

Guidelines for the Effective Use of Punishment. Punishment is obviously unpleasant for the child being punished and is generally unpleasant for the parent doing the punishing. Furthermore, the effects of punishment are setting specific, often temporary, and usually associated with problematic side effects. Therefore, punishment should be used only as a last resort when other procedures have failed or when the target behavior is particularly dangerous. The guidelines outlined in the following paragraphs should be followed carefully:

Selecting the Target Behavior. Punishment is most effective when a specific behavior (e.g., head banging) is identified, rather than a general category of behavior (e.g., self-injurious behavior).

Assessment. Like all behavior management programs, punishment requires a thorough assessment of the target behavior (to be fully discussed later in this chapter). The frequency of the target behavior should be monitored for several days prior to implementing punishment and throughout the punishment procedure. If no decrease in responding occurs soon after punishment begins (one to two days at most), the procedure should be discontinued. *Note:* When the target behavior is extremely dangerous, the baseline period must be shortened or omitted. It is still important to monitor the frequency of the behavior from the beginning of the punishment program in order to evaluate its efficacy.

Delivering the Punisher.   Punishment most effectively suppresses a behavior if it is delivered immediately after every occurrence of that behavior. Delayed punishment is less effective than immediate punishment, and occasional or intermittent punishment is less effective than punishment that occurs after every instance of the undersirable behavior. The person administering the punisher should do so in a calm, matter of fact manner.

Eliminating Reinforcement for the Target Behavior.   Since the undesirable behavior is occurring, something is probably reinforcing it. Prior to implementing punishment, it is important to identify the reinforcers for the target behavior and eliminate them. Punishment and positive reinforcement should never be contingent on the same behavior.

Reinforcing Alternative Appropriate Behavior.   To decrease a child's undesirable behavior with punishment, it is maximally effective to concurrently increase an alternative desirable response with reinforcement. Suppression of the punished response can be facilitated by reinforcing incompatible behavior. Programing positive reinforcement in this way has another advantage. It gives the person doing the punishing an opportunity to be associated with positive reinforcement (as well as with the punisher), thereby minimizing the risk of that person's becoming a conditioned punisher.

When Punishment Fails.   If the punished behavior does not decrease almost immediately after punishment is introduced, the procedure cannot be ethically justified and should be discontinued.

PROCEDURES TO DEVELOP NEW BEHAVIOR: SHAPING

Principles and Procedures

Shaping is a procedure used to establish a new behavior pattern not presently performed by the child. The frequency of a new behavior pattern cannot be increased simply by waiting until it occurs and then reinforcing it. The behavior may never occur. In shaping, a new behavior pattern is developed by reinforcing small steps or approximations toward the desired behavior. Shaping begins by reinforcing a response that occurs at least occasionally and that at least remotely resembles the final desired response. When this initial response is occurring with great frequency, it is no longer reinforced and a slightly closer approx-

imation to the final response is reinforced. The final desired response is finally established by reinforcing successive approximations of it.

Parents use shaping (along with other procedures) to develop talking in their children. When an infant first begins to babble, some of the sounds he makes remotely approximate words. When this happens, parents usually reinforce the behavior with lavish attention, hugs, and smiles. The child then enters a stage in which "baby talk" (i.e., closer approximations to actual words) occurs and is reinforced. Finally, the child is required to use actual words, then short sentences, and then more complex sentences before reinforcement is given. Although oversimplified, this description illustrates the importance of shaping in the process by which children gradually progress from babbling, to baby talk, and finally to complex speech.

Misapplications of shaping are often involved in the development of behavior problems encountered in pediatric settings. One misuse of shaping, commonly observed in retarded children, leads to self-injurious behavior (Martin and Pear, 1978). Suppose, for example, that a small child receives very little attention from his family when he engages in socially appropriate behavior. However, if the child accidentally falls and lightly bumps his head, his parents are likely to run over to him, assess the extent of the injury, and provide much attention. Because of this reinforcement, and because anything else the child does seldom evokes attention, he is likely to repeat the behavior of falling and bumping his head. After the first few times, his parents will realize that the child is not really hurt and will probably stop reinforcing him when he falls. Since the behavior is being extinguished, the child may begin to exhibit it with increased intensity; that is, he may begin to hit his head with greater force. Alarmed by the louder "thud," the parents may again run over to the child. If this shaping process were to continue, the child would eventually hit his head with sufficient force to cause physical injury. Self-injurious behavior is very difficult to eliminate, and it is a good idea to intervene in this shaping process as early as possible.

Properly used, shaping provides an important component of any behavior management program designed to develop new behavior patterns. For example, the first author incorporated shaping into a program of positive reinforcement used to reinstate eating in a child who had refused all oral intake of food (mentioned previously). The child was initially reinforced for raising a glass of water to her lip, then for placing a drop of water on her tongue, then for swallowing a small sip of water. In a carefully planned sequence of steps, reinforcement was provided for drinking gradually increased amounts of water, drinking gradually increased amounts of a dietary supplement (Ensure), drinking gradually increased amounts of other beverages (e.g., juice, tea), eating a small bite of a preferred food, eating gradually increased quantities of preferred foods, eating gradually increased varieties of foods, and eating balanced meals. This shaping program effectively re-established a more normal eating pattern in a child who had refused all oral intake of food.

Indications for the Use of Shaping

Shaping is the procedure of choice to develop new types of behavior (i.e., behavior patterns not yet in the child's behavioral repertoire).

Guidelines for the Effective Use of Shaping

Selecting the Target Behavior. The first step in shaping is to precisely identify the final desired behavior, referred to as the target or terminal behavior. All the characteristics of the target behavior (its frequency, intensity, and so on) should be specified. Also, the conditions under which the behavior is or is not to occur should be identified. Thus, a specific behavior (e.g., eating three 500 calorie meals per day) rather than a general category of behavior (e.g., oral food intake) should be selected as the target.

Selecting the Initial Behavior. In a shaping program it is crucial to know not only what the program is intended to accomplish (the target behavior) but also where it is starting. Since the target behavior does not occur initially, a starting point must be determined. That is, it is necessary to select an appropriate behavior to reinforce at the beginning of the program. This behavior, referred to as the initial behavior, should be one that the child uses at least occasionally (so that its occurrence can be reinforced) and that at least remotely approximates or resembles the target behavior. It may be necessary to observe the child's behavior for several days before starting a shaping program in order to ensure appropriate selection of the initial behavior.

Selecting the Shaping Steps. Before beginning the shaping program, it is a good idea to outline the successive approximations (steps) through which the child will be moved to approximate the target behavior. Note that the successive approximations mapped out in advance are, at best, merely "educated guesses." During the program it is important to be flexible so that the steps outlined beforehand can be modified according to the child's performance.

Moving from One Step to the Next. Unfortunately there is no hard and fast rule that specifies the number of reinforcements that should be given for any one successive approximation before moving to the next. In general, it is important not to move too quickly from one approximation to the next. Moving to a new approximation before the previous one has been well established can result in losing the previous approximation without establishing the new one. If a behavior is lost because of moving through the program too quickly, it is best to resume reinforcement of an earlier approximation and repeat the steps more slowly. It is also important not to progress too slowly. If one approximation is reinforced for so long that it becomes extremely strong, new approximations are less likely to appear. These guidelines, in fact, say little more than "Do not shape too quickly or too slowly." Regrettably the research necessary to provide more concrete guidelines has not yet been done.

Positive Reinforcement. Positive reinforcement is an integral component of any shaping program. Thus, the guidelines discussed earlier for the effective use of positive reinforcement are relevant here (with the exception of guidelines pertaining to the frequency of reinforcement, since the rules for reinforcement delivery are specified by the shaping procedure).

When Shaping Fails. There may be problems with a shaping program if the child stops earning reinforcement or does not progress smoothly from one step to the next. Some of the most common problems and their resolution include:

1. The events or stimuli programed as reinforcers are ineffective. This possibility should be investigated first. Resolution: Switch to an effective reinforcer.

2. The child was moved too quickly from one step to the next. Resolution: Return to the previous step. When the child has received additional reinforcement at this earlier step, try the next step again. If the child continues to have difficulty, despite "retraining" at previous steps, add more steps at the point of difficulty.

3. Sometimes poor progress occurs because the steps are too small (resulting in boredom or inattentiveness). Resolution: Whether poor performance is the result of steps that are too large or too small is difficult to determine. However, the former problem is much more common than the latter. Nevertheless, when difficulties are thought to be the result of steps being too easy, it may be appropriate to explore the effects of increased step size on the child's performance.

PROCEDURES TO GENERALIZE BEHAVIOR FROM ONE SETTING TO ANOTHER

When a behavior management program is used to teach a child an appropriate behavior in one setting (the training setting), it is usually desirable for the child to perform that behavior in other settings as well. That is, it is desirable for the behavior to generalize from the setting in which it was taught to other settings. Generalization occurs when a behavior becomes more (less) probable in one setting as a result of having been strengthened (weakened) in another setting. For example, when an encopretic child learns to use the toilet appropriately at home, it is hoped that he will also use the toilet at school, at his friend's house, and so on. Similarly, when one parent has successfully extinguished a child's temper tantrums, it is hoped that nontantrum behavior will generalize to the other parent, relatives, friends, and strangers. Unfortunately generalization does not readily occur. It is not sufficient to develop behavior in one setting and hope for it to occur spontaneously in other settings. Rather, procedures to ensure the generalization of behavioral gains must be carefully programed. Two factors are especially important:

1. The more similar the training setting is to other settings, the more generalization will occur. For example, to generalize an encopretic child's toilet usage from home to school, it may help to simulate certain aspects of school at home (and vice versa).

2. Train the target behavior in as many settings as possible. Generalization is greatly enhanced when the desirable (undesirable) behavior is reinforced (extinguished) by as many people in as many situations as possible. Thus, generalization of toilet usage in the encopretic child may be enhanced if it is reinforced not only by his mother at home but also by his father at home, his teacher at school, and so on.

OTHER BEHAVIOR MANAGEMENT PROCEDURES

Stress Management

Two approaches to stress management have already been described: systematic desensitization for phobic behavior and assertive training for socially withdrawn behavior. Other problems encountered in pediatric settings also stem from exceedingly high levels of anxiety. Anxiety provoking situations (or stimuli) are many and varied. For example, some children report anxiety when they write examinations, when they argue with a friend, or when their parents "fight." The problem may be neither phobia nor withdrawn behavior and still be anxiety related.

Anxiety may be associated with a variety of somatic and behavioral concomitants, such as nausea, vomiting, dizziness, headache, tics, rapid heart and respiratory rates, asthmatic episodes,

reduced or increased appetite, nail biting, or hair pulling (trichotillomania). When a child presents with such difficulties in the absence of organic dysfunction, it may be worthwhile to evaluate his strategies for coping with stress. When a deficit exists, training in stress management may be appropriate. There exists no universally accepted approach to stress management training. Described next is a procedure with which we have had some clinical success with children. Further research is necessary to identify the procedure whereby children can most efficiently learn stress-coping skills.

Principles and Procedures

The goal of stress management training is the development of skills to cope with stressful situations. Once the anxiety provoking situation(s) and the child's response to it have been specified, training focuses on two areas: relaxation training and reinforcement of nonanxiety responses.

Relaxation Training. Like systematic desensitization, stress management training involves teaching the child relaxation as a response incompatible with anxiety. Relaxation may be of either the progressive muscle type (described earlier), the autogenic type (Shultz and Luthe, 1959), or the meditative type (Benson, 1975). In autogenic relaxation the child is taught to imagine tension draining from his body and to focus upon sensations of heaviness and warmth as his body relaxes. In meditative relaxation the child is taught to focus his attention on the rate and volume of his breathing.

Once the child has learned a relaxation procedure, he should practice it two or three times per day. When he has mastered relaxation (usually after two or three weeks of practice), he may begin to supplement the daily practice by using the procedure to control anxiety in stressful situations.

Reinforcement of Nonanxiety Responses. Whenever possible, stress management training involves both extinction of anxiety responses and reinforcement for adaptive coping responses to stressful situations. Thus, there is an attempt to minimize attention to the anxiety from parents and others and to limit the possibility of escape from the stressful situation contingent upon anxious behavior; simultaneously reinforcement is maximized for behavior appropriate to the situation.

Many anxiety responses are not easily monitored. That is, it is difficult for a parent to keep track of how often a child bites his nails, pulls his hair, feels nauseous, and so on. In such cases a program of reinforcement may be facilitated by teaching the child to monitor and record the frequency of his own behavior.

Indications for the Use of Stress Management

As already described, stress management training is appropriate for a child who lacks the skills necessary to cope with stressful situations.

Guidelines for the Effective Use of Stress Management

Choosing the Type of Relaxation. There has been no systematic evaluation of the different types of relaxation training to determine whether one is more efficacious than another. The selection of the relaxation procedures depends very much upon the subjective judgment and preferences of the clinician.

Practicing the Relaxation Procedure. The relaxation procedure should be carefully taught to the child in the clinician's office. The child and the clinician should then agree upon two specific times for the child to practice relaxation every day (usually once in the morning and once in the evening). The importance of consistent practice should be stressed to the child.

Using Relaxation in Stressful Situations. When the child learns to induce relaxation with ease, he should be instructed to use the skill to control anxiety in stressful situations. (*Note*: This is in addition to, not instead of, the twice-daily practice.) The child should be taught to induce relaxation just prior to a stressful situation (whenever possible) before the anxiety response occurs and during the stressful situation when early signs of mounting anxiety are detected. For example, a student who has experienced anxiety during examinations may benefit from relaxation prior to the examination and during the examination to prevent the escalation of the anxiety response. (*Note*: It is generally inadvisable to encourage the use of relaxation to abort a full blown "anxiety attack," since such efforts are often unsuccessful.)

Reinforcing Nonanxiety Behavior. It is a good idea to specify appropriate responses that are incompatible with the anxiety response. Positive reinforcement should be programed to increase the occurrence of these responses in stressful situations (following the guidelines for the effective use of positive reinforcement). For example, a child who bites his nails while studying might earn reinforcement for clasping his hands in his lap except to write or turn pages. In such cases a parent cannot easily monitor the child's behavior, and it may be necessary to teach the child to monitor and record his own behavior while he studies. In fact, the child might even administer his own reinforcement. He might, for instance, allow himself a short break contingent upon 15 minutes of studying without nail biting.

Other Coping Skills. Once relaxation has been mastered, it may be useful to add other behavioral and nonbehavioral interventions (e.g., assertive training, family therapy) to address the issues underlying the child's anxiety. Not only must the child learn to relax, he must also learn to reshape his coping responses.

When Stress Management Training Fails. Stress management training may not eliminate anxious behavior entirely, particularly if the anxiety response pervades many aspects of the child's life. However, most children experience some relief within the first few weeks of stress management training. If not, a revision of the training program may be indicated.

Three problems are commonly encountered in stress management programs: Factors contributing to the child's anxiety are overlooked, reinforcement for anxious behavior is not eliminated, and the child is noncompliant with the relaxation regimen. When the program fails, it may be necessary to resolve one or more of these difficulties.

Biofeedback

Biofeedback is a relatively new field of scientific endeavor, yet it represents a major advance in both basic research and clinical practice: it has not only widened the scope and capabilities of behavioral research on physiological functioning, but has also shed new light on the etiology and treatment of psychophysiological disorders. Biofeedback is defined as "the application of operant conditioning methods to the control of visceral, somatomotor, and central nervous system activities. The individual is fed back information about his own biological responses. The information is a sensory analog (usually auditory or visual) of the actual simultaneously occurring responses. Hence, the term biofeedback—"bio" meaning that the information is biological and "feedback" meaning that the information is precisely coupled in time with the ongoing biological events (Shapiro and Surwit, 1981, pp. 45, 51).

In less than 20 years biofeedback has spawned hundreds of basic and clinical research papers, numerous books, and several specialized clinics and training programs. Biofeedback research has produced techniques that have rendered physiological activities amenable to control by the same environmental contingencies that control other forms of behavior. It thus offers a behavioral strategy to modify physical symptoms of psychophysiological and other disorders.

Because biofeedback involves the treatment of disease, medical participation is a necessary component of any biofeedback program. However, it is unlikely for a physician to have the behavioral expertise necessary to conduct a successful program of biofeedback. Therefore, the use of biofeedback in therapy for physiological disorders is typically a collaborative effort involving both medical and behavioral specialists.

Rather than detailing the "do's and don'ts" of implementing a biofeedback program, the following discussion offers an overview of the field focusing upon the fundamental principles and procedures of biofeedback, some basic research on selected physiological responses and the clinical applications of this research, and issues relevant to the use of biofeedback in the treatment of physiological disorders. Biofeedback research involving nonhuman subjects is beyond the scope of this chapter, but has been reviewed by Shapiro and Surwit (1976). The reader may note that the following discussion is of research involving adult subjects and patients. This reflects that fact that biofeedback has yet to be systematically evaluated in the treatment of pediatric disorders.

Principles and Procedures

The basic biofeedback procedure can be outlined as follows. The patient is seated comfortably in a quiet, controlled environment. Physiological responses are measured and recorded by sensors (electrodes), which are connected by lead wires to electronic equipment located nearby. Whenever a response of interest occurs (e.g., an increase in heart rate, a decrease in systolic blood pressure, a decrease in muscle activity, an increase in digital skin temperature), a signal is presented to the patient. The signal may be a discrete stimulus, such as a flash of light or a brief tone, that signals the presence or absence of changes in the appropriate direction in a "yes-no" fashion, or continuous feedback, such as a varying light or tone, meter readings, or a continuously changing graphic display, that fluctuates in direct correspondence with fluctuations in the physiological response.

Basic Research and Clinical Applications

This section reviews empirical research on blood pressure control with biofeedback in which there has been substantial investigation as well as related clinical application. Other physiological responses that have been controlled by biofeedback will be mentioned briefly. This review, by no means exhaustive, is intended to present the methodology of basic biofeedback research and to raise issues important for its clinical application.

Blood Pressure.   Most of the human studies on blood pressure control with biofeedback involve the "constant cuff" method described by Shapiro and Surwit (1981, p. 52) as follows:

In the constant cuff method, a blood pressure cuff is wrapped around the upper arm, and a crystal microphone is placed over the brachial artery under the distal end of the cuff. The cuff is inflated to about average systolic pressure and held constant at that level. Whenever the systolic pressure rises and exceeds the occluding cuff pressure, a Korotkoff sound is detected from the microphone. When the systolic pressure is less than the occluding pressure, no Korotkoff sound is detected. [With] a regulated low-pressure source and programming apparatus, it is possible to find a constant cuff pressure at which 50% of the heart beats yield Korotkoff sounds. This pressure is by definition median systolic pressure. Inasmuch as the time between the R-wave in the electrocardiogram and the occurrence of the Korotkoff sound is approximately 300 msec., it is possible to detect either the presence of the Korotkoff sound (high systolic pressure relative to the median) or its absence (low systolic pressure relative to the median) on each heart beat. In this way, the system provides information about directional changes in pressure relative to the median on each successive heart beat, and this information can be used in a biofeedback procedure. Subjects are provided with binary (yes-no) feedback of either relatively high or low pressure on each heart beat. After a prescribed number of feedback stimuli or a change in median pressure, rewarding slides or other incentives are presented.

Basic research using the constant cuff method suggests that blood pressure can be self-regulated by normal subjects with a fair degree of consistency and specificity (Shapiro et al., 1970, 1972). Subjects taught to increase or decrease systolic blood pressure showed changes (in the appropriate direction) varying from 3 to 10 per cent of the baseline value at the end of a single session of training. More significant results were reported for diastolic pressure, with subjects showing increases up to 25 per cent and decreases up to 15 per cent of the baseline value (Shapiro et al., 1972). In general, blood pressure increases were easier to obtain in normal subjects than decreases. Most of this research consists of one session experiments. More research is needed to determine whether greater changes can be brought about with longer training.

Interestingly, changes in the heart rate did not consistently correlate with changes in blood pressure. Shapiro et al. (1970) reported that the heart rate was not associated with learned changes in systolic pressure. However, Fey and Lindholm (1975) reported that the heart rate increased or decreased in groups receiving feedback for increasing or decreasing systolic blood pressure, respectively. Brener (1974) reported learned changes in diastolic blood pressure to be unassociated with changes in the heart rate. However, Shapiro et al. found that the heart rate was not independent of learned changes in the diastolic pressure. Further research is needed on the specificity of control of blood pressure and other cardiovascular functions.

The basic research on blood pressure control in normal subjects has provided a foundation for the clinical application of biofeedback to hypertension. Benson et al. (1971) used biofeedback to lower the systolic blood pressure in five of seven patients they treated for hypertension. (Of the two who did not respond, one did not have an elevated systolic blood pressure and the other had renal artery stenosis.) Medication dosage and diet were kept constant throughout the study. The five patients responding positively showed decreases in systolic pressure of 34, 29, 16, 16, and 17 mm. Hg within 33, 22, 34, 31, and 12 training sessions, respectively. Other investigators have reported similar success in the treatment of hypertension by biofeedback to lower the systolic pressure. Abnormally high diastolic levels have been more resistant than high systolic levels to treatment by biofeedback, although Goldman et al. (1975) reported reductions in both systolic and diastolic levels for seven patients with hypertension. These and other studies have been reviewed in detail by Shapiro and Surwit (1981). It is important to note that there is considerable variability in the response of hypertensive patients to biofeedback. The factors contributing to this variability have yet to be identified.

Surwit et al. (1978) compared two types of biofeedback training and meditative relaxation in the treatment of 24 borderline hypertensive adults. The first group received binary feedback for simultaneous decreases in blood pressure and heart rate; the second group received feedback for reduced muscle tension in the frontalis and forearm; the third group received training in meditative relaxation. All three treatment groups showed a significant reduction in blood pressure, suggesting that each of the behavioral procedures was equally effective as a clinical intervention. On the basis of evidence of this sort, Reeves and Shapiro (1978) have made an important statement concerning the use of biofeedback in the treatment of hypertension:

Since biofeedback and relaxation have by and large resulted in equivalent [blood pressure] reductions, it would seem that less costly and simpler relaxation procedures would be the treatment of choice. However, the combination of biofeedback and relaxation has produced the most substantial pressure reductions. Unfortunately, studies combining biofeedback and relaxation have employed designs which make it impossible to assess the relative contribution of biofeedback and relaxation to such reductions. At any rate, research addressing whether biofeedback itself adds anything unique, over and above simpler relaxation techniques, is needed.

Also comparative studies of biofeedback and other behavioral methods with medical management of hypertension are needed to fully assess biofeedback as a viable approach to the treatment of hypertension.

Other Responses. Various physiological responses have been brought under control using biofeedback, with important implications for the clinical application of biofeedback. For example, biofeedback of the heart rate has been used to train adult subjects to regulate the heart rate in the treatment of cardiac arrhythmias (e.g., paroxysmal atrial tachycardia, sinus tachycardia, ventricular arrhythmias); electromyographic feedback has been used in the treatment of several neuromuscular disorders and to facilitate relaxation training; control of digital skin temperature with biofeedback has been used in the treatment of migraine headache and Raynaud's disease; electroencephalographic feedback to increase sensorimotor rhythm has been used in the treatment of seizure disorders; biofeedback for sphincteric response has been used for the treatment of fecal incontinence. Research on these responses has been reviewed by Shapiro and Surwit (1981).

Limitations on the Clinical Applications of Biofeedback

We have seen that, using biofeedback, subjects can learn control of physiological responses that may help in the treatment of physical problems previously thought to be treatable only by somatic therapies. However, further research is needed, and several issues must be resolved before biofeedback can be advocated as a standard treatment for any disorder. Questions of efficacy and the comparative effects of biofeedback and other behavioral methods (e.g., relaxation training) have already been raised. In addition, there are numerous practical and therapeutic factors that may limit the clinical utility of biofeedback.

Unresolved Technical Issues. There exists no specific standarized form or structure for biofeedback training. There is insufficient empirical evidence to resolve such technical issues as the following: Is it better to use visual, auditory, or tactile feedback? Is continuous, moment to moment feedback preferable to discrete, more intermittent feedback? How many sessions of biofeedback training are necessary to know whether the patient will benefit from it? How often should training sessions occur? Is it advisable to recommend specific aides (e.g., visualizations, specific thoughts) to the patient during training? What types of individuals are most likely to respond to biofeedback? At this time the clinician must select a specific procedure for a specific patient in a trial and error fashion on the basis of the individual characteristics of the patient, the physiological symptom in question, the physiological system for which feedback is to be given, and so on.

Economy. Biofeedback training often requires a great deal of time and effort on the part of both the patient and the clinician to yield a clinically useful result. A cost-benefit analysis might militate against the use of biofeedback, particularly if an equal therapeutic effect could be obtained from medication that is simple and painless to use. Unless medication is ineffective or its side effects are dangerous, it is unlikely that biofeedback will be the treatment of choice.

Motivation. Shapiro and Surwit (1981) pointed out that "It is not sufficient to assume that feedback indicating therapeutic improvement will, in and of itself, act as a reinforcer and maintain the persistent practice required to gain therapeutic benefit" (p. 64). Three motivational problems are encountered in clinical applications of biofeedback:

1. Many of the disorders to which biofeedback has been applied have no immediate adverse consequences. Hypertension, for example, usually causes minimal discomfort to the patient, until the disorder becomes very severe. Therefore, there is no immediate reinforcement for the hypertensive patient who practices biofeedback, and long term compliance with a biofeedback program is difficult to maintain. One might expect biofeedback to be most appropriate for the treatment of such disorders as migraine headache and Raynaud's disease in which training can lead to immediate relief of pain (although there are no data on relative efficacy of biofeedback treatment for disorders such as hypertension as opposed to migraine headache).

2. The symptom itself may be reinforcing for the patient. That is, the disorder may have secondary gain. For example, the disorder may elicit sympathy and support from caring relatives and friends and may provide the patient with a socially acceptable reason to avoid unpleasant tasks and responsibilities. Biofeedback does not take into account the more subtle contingencies of secondary gain implicit in many disorders. As a result, many clinical applications of biofeedback have been undermined by these contingencies. "A behavioral therapy designed to treat a disorder supported by secondary gain would therefore have to include techniques aimed at making up any social deficit left by the removal of the symptom" (Shapiro and Surwit, 1981, p. 64).

3. The patient may engage in behavior that conflicts with the aims of biofeedback therapy. For example, a patient with Raynaud's disease who is learning to increase the digital skin temperature may enjoy ice skating even though lengthy periods in cold places may be countertherapeutic. It seems futile to attempt to treat a disorder by biofeedback without simultaneously intervening to deal with factors that may be exacerbating the problem.

Transfer of Training. There is no reason that biofeedback, like a course of radiation therapy, can always be expected to produce sustained effects. It is possible for a patient to show perfect control over his problem during a feedback session and have no control at home. During a biofeedback session the patient acquires control over a physiological response in a restful, laboratory setting using precise information about the response fed back to him by specially designed electronic apparatus; at home he must attempt physiological control in difficult and stressful situations without feedback. Research has not adequately addressed the problems of generalization encountered in clinical applications of biofeedback.

Patient Characteristics. Most clinical and experimental work relating to biofeedback has been done with highly educated, motivated adults. It is presently unclear how other individuals, especially children, will respond to biofeedback. Research is continuing in our laboratory and at other sites to assess the utility of biofeedback in the treatment of pediatric disorders.

Concluding Remarks

It is clear that biofeedback is in the early stages of its scientific development and that numerous issues concerning its clinical utility are unresolved. Yet there is sufficient evidence to warrant the consideration of biofeedback as a treatment modality when medical management is ineffective or its side effects are dangerous. In particular, "the selectivity of physiological control often achieved by biofeedback methods would suggest that the methods would have a unique advantage in disorders in which the symptom is quite specific, for example, cardiac arrhythmias, seizure disorders, and various neuromuscular disorders" (Shapiro and Surwit, 1981, p. 61).

BEHAVIOR MANAGEMENT IN THE PRACTICE OF PEDIATRICS

HOW TO DESIGN, IMPLEMENT, AND EVALUATE A BEHAVIOR MANAGEMENT PROGRAM

So far we have described a variety of principles and procedures for overcoming pediatric behavior problems (including behavioral deficits, excesses, and inappropriateness). It is probably obvious by now that most pediatric behavior management programs involve combinations of these various principles and procedures. The design and implementation of a behavior management program involve decisions and practical steps that are not really a part of the scientific procedures themselves. These decisions and steps include identifying the target behavior, assessing the target behavior and the antecedent and consequent events associated with it, selecting the intervention strategy, and implementing the program.

These components of behavioral programing are discussed separately in the sections that follow. The goal is to elucidate the role of the clinician (e.g., pediatrician, psychologist) in the process of developing, implementing, and evaluating a behavior management program. The discussion focuses upon home programs administered by parents but is readily extended to school or hospital ward programs administered by teachers or nurses. The decisions to be made and the steps to be executed are exemplified with reference to a common pediatric behavior problem—failure to comply with parental requests (i.e., noncompliance).

---

1. Would you describe the problem to me?
2. What is it that the child actually does (or does not do) to cause a problem?
3. Can you give me a few examples of situations in which the problem occurs?
4. Are there other things that the child does that seem to be related to the problem behavior? (For example, if the problem is that the child does not follow instructions, does he also create a tantrum or yell when he is asked to do something?)
5. Can you state the problem in positive terms? (For example, if the goal is to reduce tantrums when the child is left alone, positively stated the goal may be to increase independent play when the child is left alone.)
6. Now that you have had a little time to think about the problem, could you be more precise in specifying the problem in terms of the actual behavior patterns of the child?
7. How often does the behavior problem occur? Does it seem to happen more in some situations than others?

Figure 55–4. Some questions to ask a patient to help specify the goals of a behavior management program.

IDENTIFYING THE TARGET BEHAVIOR

The goals of the behavior management program are usually identified during the first or second interview the clinician has with the parents and child. Three guidelines for selecting the types of target behavior are probably obvious by now:

1. Regardless of whether the goal is to increase or decrease behavior, the target behavior must be clearly specified.

2. The response requirement at the beginning of the program should be relatively lenient to afford the child the maximal opportunity to achieve early success (reinforcement).

3. When the goal is to decrease an undesirable type of behavior, it is important to select an alternative form of desirable behavior to be increased (ideally, one that is incompatible with the undesirable behavior pattern). Both the desirable and the undesirable behavior patterns must be clearly specified.

Identification of the behavior pattern to be changed may appear to be a relatively simple task. Very often it is not. In many cases parents provide general or global statements regarding behavior problems that are insufficient for actually beginning a program. For example, rather than stating that their child does not follow instructions, the parents of a noncompliant child may report that the child "is always bad," "never listens," "always argues," and so on. Such summary statements are too general to be of much use in formulating the goals of a behavior management program. In selecting target behavior patterns, the clinician should provide careful guidance for the parents to ensure that the types of target behavior are defined explicitly so that they can be observed, measured, and agreed upon by everyone (including the child) involved in the program. The questions in Figure 55–4 might help the clinician guide the parents and child to a clear specification of the behavior to be changed by a behavior management program.

ASSESSMENT

When the target behavior problem has been identified in precise terms, assessment can begin. The objectives of behavioral assessment are to determine the frequency of the target behavior, the conditions under which it occurs (i.e., the antecedent stimuli), the consequences that maintain it, and reinforcers that might be used in a behavioral intervention.

Assessment of behavior is essential for at least three reasons. First, assessment provides the information necessary for the clinician to evaluate the desirability and feasibility of behavioral intervention. A behavior management program may be unnecessary or inappropriate if the behavior problem occurs infrequently, the source of reinforcement maintaining the problem is difficult to identify or hard to control, or reinforcers cannot be found. Careful assessment can prevent the clinician from attempting a behavioral intervention that is either unwarranted or doomed to failure. Second, behavioral assessment helps the clinician to identify the best strategy for intervention. This point will be discussed further under the separate heading, "Selecting the Intervention Strategy." Finally, assessment is essential to reflect behavior change after the program is begun in order to evaluate its efficacy.

Strategies of Assessment

Behavioral assessment relies upon objective information obtained by direct observation of the behavior to be changed. Reliance upon more subjective sources of information (such as parental reports) in the absence of objective assessment may distort the extent to which the behavior actually occurs. For example, a child's tantrums may be so intense that parents recall them as occurring more often than they actually do. In contrast, some children may have tantrums so often that parents become somewhat accustomed to a high rate and perceive them as occurring less often than they actually do. Human judgment does not always correspond to actual data obtained by observing behavior. For this reason, behavioral assessment emphasizes the objective measurement of behavior through direct observation.

Pediatric behavior problems generally occur at home (or perhaps in school) and are usually brought to the clinician's attention by the child's parents. The clinician rarely has the opportunity to directly observe the child engaging in the problem behavior. Therefore, the clinician must enlist the parent's cooperation in behavioral assessment. In most pediatric behavior management programs, the parents begin the behavioral assessment one or two weeks prior to intervention. A measure of either the frequency or the duration of the target behavior is typically used.

Frequency Measures

Frequency counts involve simply tallying the number of times the behavior occurs in a given period of time. This system of recording behavior is called continuous recording, and the measure obtained is the response rate. While recording response rate, it is convenient to record also the

antecedents and consequences associated with each occurrence of the behavior, thereby providing a complete record of the behavior and the factors controlling it.

Measures of response rate are particularly useful when the target behavior is discrete and when the behavior takes a relatively constant amount of time each time it is performed. Because a discrete response has a clearly delineated beginning and end, separate instances of the response can be counted. The performance of the behavior should take a relatively constant amount of time so that the responses that are counted are relatively equal. For example, if a child has tantrums for 15 minutes on one occasion and for 45 minutes on another, these might be counted as two instances of tantrums. However, a great deal of information would be lost by simply counting the number of tantrums, since they differ in duration.

With continuous recording, parents should be instructed to record each occurrence of the target behavior along with a brief description of what happened just before and just after the behavior. When teaching parents to do continuous recording, the clinician may find the following guidelines helpful:

1. Parents may be unable to maintain accurate records throughout the day. It is helpful to designate a specific segment of time (e.g., two hours after school, at mealtime, one hour after supper) and request that parents record each instance of the behavior during this time interval.

2. The clinician can facilitate the task of record keeping by providing parents with a data sheet that specifies the information to be collected and how it is to be recorded. Figure 55–5 presents a

Date/Time	Request	Did the child comply?	What happened then?
Sept. 1 7:30 A.M.	Make your bed	No	I (mother) asked him several times, and the bed was still unmade when he left for school so I made it myself.
Sept. 1 2:30 P.M.	Hang up your coat	No	I asked him several times and then finally brought the coat to him and waited there until he went to hang it up.
Sept. 1 6:15 P.M.	Clear the table before dessert	No	I (father) asked him several times, and then my wife partially cleared it herself and served dessert.
Sept. 1 6:45 P.M.	Finish clearing the table	Yes	—
Sept. 10 7:45 P.M.	Walk the dog	No	I (father) asked him several times and then did it myself.

Figure 55–5. A sample data sheet for recording a child's compliance with parental requests. The data are hypothetical to exemplify the interactions that typically occur between parents and a noncompliant child (before intervention).

hypothetical example of a data sheet that parents might use in the behavioral assessment of their child's noncompliance.

Measures of Duration

Measures of duration are used to assess how long a behavioral episode lasts. The parent simply starts and stops a stopwatch or notes the time when the response begins and ends. The onset and termination of the response must be carefully defined. For example, in recording the duration of a tantrum, a child may cry continuously for several minutes, whimper for short periods, stop all noise for a few seconds, and begin intense crying again. In recording the duration, a decision must be made as to what constitutes the beginning and the end of a tantrum. (In most cases this decision is arbitrary and should be made by the clinician and parents jointly.)

The use of response duration is generally restricted to situations in which the length of time a behavior pattern occurs is of major concern. In most behavior management programs the goal is to change the frequency of a response rather than its duration. There are notable exceptions, of course. For example, in dealing with tantrums, parents are often more concerned about duration than frequency; sometimes in dealing with noncompliance, parents are concerned not only about whether the child complies but also about how long he takes to do so. In such cases parents may be taught to record the response duration (or latency) in a format much like that illustrated in Figure 55–5.

SELECTING THE INTERVENTION STRATEGY

Once the initial behavioral assessment is complete (usually one or two weeks), the clinician meets with the parents and child to select an appropriate intervention strategy. Early in this meeting the clinician must identify the individuals who will conduct the program, usually the child's parents. Then the clinician must consider the alternative combinations of principles and procedures appropriate for the goals of the particular program (summarized in Table 55–1) and review the guidelines for their effective application. Parental compliance with program implementation can be enhanced by involving parents in this early stage of program planning.

The behavioral assessment may greatly facilitate the selection of an appropriate intervention strategy by elucidating the environmental factors maintaining the behavior problem. Consider, for example, the behavioral assessment of noncompliant behavior in Figure 55–5. Several types of parental behavior related to the child's noncompliance are apparent: The parents repeat instructions excessively; they physically intervene to procure compliance; and they do not positively reinforce compliance. To put it simply, the parents reinforce noncompliance with attention and extinguish compliance. These observations suggest two avenues for behavioral intervention: Withhold parental attention contingent upon noncompliance (perhaps a brief time out is indicated), and reinforce compliant behavior with parental praise and attention and with primary reinforcers such as edibles or activities (especially early in the program).

IMPLEMENTING THE PROGRAM

Training Parents in Behavior Management

Once a behavior management program has been designed, it is important to ensure that parents are well trained in its implementation. If a program is to be effective, the contingencies need to be applied to the child in a precise and consistent manner. Program implementation can be greatly facilitated by providing parents with a written list of the steps they are to follow.

Review and Revision

The clinician usually meets with the parents and child weekly or biweekly in the early stages of program implementation. During these meetings the clinician reviews parental records of the child's behavior to ensure that the program has been properly implemented and that behavior change is occurring in the desired direction. The parents in our hypothetical example might provide the records in Figure 55–6 when first implementing a behavioral program for noncompliance. In comparing the records before intervention (Figure 55–5) and during intervention (Figure 55–6), it is apparent that the parents are using the program with some success.

The behavioral improvements observed at the beginning of a behavioral intervention are often small and seemingly insignificant (especially to the child's parents). In our hypothetical example, compliance increased from 20 per cent (the child complied with one of five requests on Figure 55–5) to 40 per cent (the child complied with two of five requests in Figure 55–6) at the beginning of the behavioral intervention. But the parents must still contend with a child who is noncompliant more than half the time. When behavioral gains are relatively minimal, parents require a great deal of

Date/Time	Request	Did the child comply?	What happened then?
Sept. 10 7:30 A.M.	Make your bed	No	The child spent five minutes alone in his room, but didn't make his bed after that, so he spent another five minutes alone in his room. Then he made the bed, and I (mother) thanked him and told him how well he made it.
Sept. 10 2:30 P.M.	Put your football away	No	He spent five minutes alone in his room—then he did it. I (mother) praised him and gave him his favorite snack.
Sept. 10 6:15 P.M.	Clear the table	Yes	His father and I were both amazed—the child got an extra dessert.
Sept. 10 6:45 P.M.	Do the dishes	Yes	I (mother) praised him and helped him dry the dishes.
Sept. 10 7:45 P.M.	Walk the dog	No	He spent five minutes alone in his room—then walked the dog and went up to his room to do his homework.

Figure 55–6. A sample data sheet for recording a child's compliance with parental requests (hypothetical data that might be recorded during intervention).

support and encouragement to maintain the program consistently. Indeed the astute clinician may notice a procedural violation in Figure 55–6 when the parents failed to reinforce the child for his eventual compliance with the request to walk the dog. It is important for the clinician to meet frequently with the parents and child early in intervention to ensure immediate detection and correction of such procedural deviations that might undermine the program's efficacy.

Once a program has been implemented consistently for a period of one or two weeks, significant behavioral gains should become apparent. If not, the clinician must examine the program design for possible flaws and weaknesses (it may help to review the guidelines for effective application of the procedures) and revise it accordingly. When a program has proven effective, the clinician may consider either expanding it to address additional behavior problems or weaning the child from it. As the child's behavior steadily improves, the clinician may gradually decrease the frequency of contacts with the family until only infrequent follow-up visits are necessary.

CONCLUDING REMARKS

The behavioral approach emphasizes the treatment of problem behavior by modification of the environment in which the behavior occurs. To the extent that pediatric behavior problems occur at home or in school, their remediation is not possible within the context of the pediatric office visit. Thus, the role of the clinician in behavioral intervention is not as a therapist working directly with the child, but as an advisor consulting with the parents (or teachers) and the child as they embark together on a program of behavior management.

For problems that are identified early, the pediatrician may effect significant improvement simply by offering treatment suggestions to the parents during routine office visits. For problems that are more long standing, intensive behavioral work is indicated. Although many of the principles and procedures of behavior management are described in this chapter, the reader is cautioned that there is "no substitute for experience." The pediatrician contemplating behavior management for a patient would be wise to seek the assistance of a behaviorally trained psychologist. In fact, when faced with a particularly recalcitrant behavior problem, the pediatrician may be well advised to refer the patient for treatment by a behavioral psychologist. Whether the pediatrician chooses to follow a child in behavioral treatment or to refer him to a behavioral psychologist, a strong background in behavior management is a vital asset. A behaviorally trained pediatrician is in a unique position to provide empirical assessment of pediatric behavior problems, behavioral intervention to avert developing behavior problems, and, most importantly, early detection of problems requiring referral to a behavioral psychologist.

LEARNING AND BEHAVIOR IN CHRONIC DISEASE

A BEHAVIORAL PERSPECTIVE ON CHRONIC DISEASE

The extent to which learning is a factor in disease has yet to be fully elucidated, but there is evidence that it is considerable. The results of a recent study by Ader and Cohen (1975) are striking. Using a respondent conditioning paradigm in rats, they demonstrated conditioned suppression of the immune response to a foreign protein (injected sheep erythrocytes). With cyclophosphamide, an immunosuppressant, as the unconditioned stimulus, they were able to elicit immunosuppression to the previously neutral stimulus, saccharine (conditioned stimulus). Appropriate control groups did not exhibit this phenomenon. Data such as these indicate the power of conditioning procedures and implicate learning and conditioning as being important in the disease process.

Learning and behavior play an especially important role in chronic disease. Symptoms of chronic disease are observable events or types of behavior (e.g., wheezing in an asthmatic child, convulsing in an epileptic child), which, because of their recurrent nature, are susceptible to the influence of environmental contingencies. Although there is generally a definite pathophysiologic process involved, symptoms of chronic disease can affect the environment in ways that can maintain and exacerbate the problem. A comprehensive approach to the treatment of chronic disease must include not only medical management of the physiological disorder but also behavioral management of the environmental contingencies involved in maintaining or exacerbating disease symptomatology.

In the last five years a new field of scientific endeavor devoted to the study of behavior and disease has arisen out of the collaborative efforts of behavioral and medical scientists. This field is behavioral medicine. Pomerleau and Brady (1981, p. xii) define behavioral medicine as: "(a) the clinical use of techniques derived from the experimental analysis of behavior—behavior therapy and behavior modification—for the evaluation, prevention, management, or treatment of physical disease or physiological dysfunction; and (b) the conduct of research contributing to the functional analysis and understanding of behavior associated with medical disorders and problems in health care." In other words, the term behavioral medicine is used to refer to clinical treatment or research involving the application of the principles of respondent and operant conditioning to medically related behavior problems. Although a number of other definitions detail more comprehensively the medical and behavioral contributions to assessment and treatment, this definition underscores the need for behavioral service in pediatric practice.

Pomerleau and Brady (1981) identify four principal lines of development in behavioral medicine at present: intervention to modify a behavior pattern (i.e., a disease symptom) that in itself constitutes a problem; intervention to modify the behavior of health care providers (e.g., physicians, nurses) in order to improve delivery of health services to the patient; intervention to improve the patient's adherence to a prescribed treatment regimen; and intervention to promote prevention of disease. Although a thorough review of the field of behavioral medicine will not be attempted, the next section reviews the behavioral medicine approach to the treatment of two medical disorders—asthma and seizure disorders. This review is intended to provide the reader with a working knowledge and understanding of current developments in behavioral medicine as they relate to the practice of pediatrics. (More comprehensive reviews are available elsewhere, e.g., Pomerleau and Brady, 1981; Russo and Varni, 1982.)

BEHAVIORAL PROCEDURES IN THE TREATMENT OF CHRONIC DISEASE

Asthma

Asthma has been defined as ". . . a diffuse obstructive disease of the airways characterized by a high degree of reversibility with appropriate therapy" (Ellis, 1975, p. 504). There are several etiologies that are important to varying degrees in children. These include biochemical abnormalities, infections, and immunological and endocrine factors. Although psychological factors are rarely considered to be of primary importance, they may precipitate asthmatic symptoms (Ellis, 1975). Turnbill (1962) has even suggested that asthmatic symptoms may be acquired through a process of respondent and operant conditioning. As stated by Christophersen and Rapoff (1981, p. 112), "It is more likely that under certain conditions, with specific individuals and at specific points in time, asthmatic attacks may be triggered . . . or maintained by emotional and behavioral factors." Although the role of psychological factors in asthma is not well understood, there is a growing body of research evaluating behavioral treatment strategies designed to provide relief from symptoms and improve medical treatment of asthmatic patients. A representative selection of this research will be reviewed.

Neisworth and Moore (1972) suggested that a child's asthmatic symptoms may be maintained by parental reinforcement. They described behavioral treatment of a seven year old asthmatic male who exhibited prolonged wheezing and coughing at bedtime. The mother frequently cautioned the child to limit his activities and to take his medication. She was particularly attentive to the child at bedtime. The authors speculated that the mother was reinforcing "sick" behavior and taught her to use an extinction procedure whereby she discontinued all special attention and medications at bedtime. In addition, the child was reinforced if he coughed less than usual during the night (i.e., he earned lunch money instead of having to take his lunch the next day). Using appropriate experimental controls, the authors demonstrated a reduction in asthmatic symptoms, which was maintained at an 11 month follow-up. Neisworth and Moore argued that environmental contingencies can exacerbate or remediate symptoms, but emphasized that their study ". . . does not purport to obviate 'organic' factors in the etiology or maintenance of asthmatic responses" (p. 98).

In an uncontrolled case study, Creer et al. (1974) used behavioral procedures to treat a 10 year old asthmatic child who was spending an inordinate amount of time in the hospital. A behavioral analysis suggested that asthmatic symptoms culminating in hospitalization were maintained by avoidance of school and peer interactions. A time-out procedure, implemented during hospitalization, required the child to remain alone in his room with only his schoolbooks available for the duration of his hospital stay. In addition, the child was reinforced for school attendance. The authors reported not only a decrease in the frequency and duration of hospital stays, but also an increase in school performance.

Although effective medical treatment is available for the symptomatic relief of asthma, the benefits of this treatment are minimal if the child is uncooperative or is unable to follow treatment regimens. Renne and Creer (1976) used behavioral procedures to teach four asthmatic children to use an intermittent positive pressure breathing apparatus, a device to administer bronchodilator medication to the child's airways. They identified three types of behavior (eye fixation, facial orientation, and diaphragmatic breathing) requisite for the appropriate use of this apparatus. Tickets that could be traded for gifts were awarded contingent upon improved performance of the target behavior. The authors demonstrated increased effectiveness of intermittent positive pressure breathing treatment as a function of the reinforcement procedure.

Numerous studies have examined the impact of progressive muscle relaxation on objective (e.g., maximal peak flow values) and subjective (e.g., frequency of wheezing attacks as recorded by the patient) measures of respiratory function in asthmatic patients (e.g., Moore, 1965). Relaxation is sometimes incorporated into a program of electromyographic feedback (e.g., Kostes et al., 1976) or reciprocal inhibition (relaxation paired with stimuli related to asthmatic symptoms; e.g., Moore, 1965). The results of these studies must generally be considered tentative because of weaknesses in experimental design. However, it appears that some asthmatic children obtain symptomatic relief when they are taught relaxation. Further research is needed to evaluate the importance of biofeedback and reciprocal inhibition.

Several conclusions can be drawn concerning the behavioral management of asthmatic children. Behavioral treatment should be considered adjunctive to the medical treatment of asthma. Relaxation, biofeedback, and reciprocal inhibition are promising but should be used cautiously under close medical supervision. Most importantly, children with asthma (or any chronic disease) can derive a special status by using their symptoms to manipulate parents. Within reasonable limits, asthmatics should be treated like other children. Parents of asthmatic children often can benefit from behavioral management training to enhance the social and behavioral development of their children (Christophersen and Rapoff, 1981, p. 115).

Seizure Disorders

The term seizure disorder is used to refer to a ". . . variable symptom complex characterized by recurrent, paroxysmal attacks of unconsciousness or impaired consciousness, usually with a succession of tonic or clonic muscular spasms or other abnormal behavior" (Baird, 1975, p. 1380). Seizure disorders are thought to have a genetic or hereditary basis. However, it has been suggested that some seizures may be maintained in part by environmental factors (Baird, 1975). A number of studies have evaluated the efficacy of respondent and operant conditioning procedures in controlling seizures. A representative sample of these studies is to be reviewed.

Efron (1957) suggested that uncinate seizures could be inhibited by properly timed administration of olfactory stimuli. One patient aborted seizures by smelling jasmine scent and other powerful odors when she felt a seizure coming on. The olfactory stimuli were then paired with a visual stimulus (a bracelet) until the visual stimulus alone effectively aborted seizures. Presenting the stimuli before seizure climax yielded the best results. Although the patient learned to abort seizures, the overall frequency of seizures did not decrease.

Ince (1976) used systematic desensitization to eliminate petit mal seizures in a 12 year old boy who experienced numerous daily seizures at school and was fearful of attending school. Progressive muscle relaxation was taught to the child and hierarchies were constructed around anxiety provoking situations (e.g., having a seizure in class or on the baseball field). The patient was systematically desensitized to each element in each hierarchy. Then desensitization was directly applied to seizures. At the onset of a seizure the child was taught to initiate relaxation by saying "Relax" to himself 10 times. Subjective reports from the child, his parents, and teachers indicated that the child was seizure free, even nine months after treatment was terminated.

Sterman et al. (1974) taught four epileptic patients to control their sensorimotor rhythm using electroencephalographic feedback. A biofeedback apparatus provided visual and auditory feedback for appropriate electroencephalographic signals, and patients were instructed to "relax and think positive thoughts." These authors reported reductions in electroencephalographic and clinical variables indicative of seizure activity. They indicated that tonic-clonic and myoclonic seizures were significantly reduced in frequency.

Several studies have suggested that the frequency of seizures can be reduced by operant conditioning procedures. Balaschak (1976) described a teacher administered behavioral program that was effective in reducing the seizures in an 11 year old child. The teacher reinforced the child with positive attention and other rewards for seizure free periods during the day. In addition, the teacher paid as little attention as possible to the child during a seizure. Subjective reports by the teacher and records of the school nurse indicated a substantial reduction in seizure frequency.

Zlutnick et al. (1975) conceived of seizures as the climax of a chain of preseizure symptoms and behavior patterns. They reasoned that seizures may be prevented by interrupting this chain before the seizure climax occurs. They tested this hypothesis with five epileptic children, all of whom had seizures at least once a day. Treatment for four children involved shouting "no" and shaking the child at the onset of preseizure behavior (e.g., staring, lowered activity level). This interruption procedure yielded varying results ranging from total elimination of seizures in some children to transitory effects in others. Preseizure behavior in the fifth child was treated by a procedure that involved placing his arms by his side and delivering reinforcers (praise and primary reinforcers) after five seconds of "hands down." At a nine month follow-up examination this child's seizure frequency was close to zero. The authors suggested that seizures can be modified by manipulation of environmental events before the seizure climax and emphasized the need for close communication between medical and behavioral clini-

cians in the treatment of seizure disorders. The study by Zlutnick et al. is particularly noteworthy because it was one of the few studies in this area that employed an adequate experimental design and method of data collection.

In summary, the available research dealing with behavioral control of seizures is suggestive and exploratory. When positive results have been obtained, it often has been difficult to attribute them to behavioral treatment because of inadequacies in experimental design. Nevertheless, when seizures are recalcitrant to medical management, behavioral intervention may be a viable alternative. Note, however, that any attempt at behavior management of seizures should occur under controlled medical conditions and in conjunction with established anticonvulsant medication regimens.

Concluding Remarks

The foregoing review suggests the feasibility of behavioral medicine in the practice of pediatrics. Behavioral medicine procedures have been used effectively in the management of a variety of chronic pediatric disorders in addition to asthma and seizure disorders. These include migraine headache and other forms of chronic pain, diabetes, enuresis, encopresis, obesity, neuromuscular disorders, and dermatological disorders. Although much of behavioral medicine research is preliminary, it signifies a substantial improvement in the management of problems encountered in pediatrics. As stated by Christophersen and Rapoff (1981, p. 120) in a recent review of behavioral pediatrics, ". . . there [are] sufficient data in behavioral pediatrics, so that now . . . a marriage between pediatrics and behavioral sciences . . . is not only possible but is well founded. Over the next decade, we expect a tremendous outpouring of data which will be of great benefit to the parents and children who have come to rely on the pediatric health care delivery system."

SUMMARY AND CONCLUSIONS

Out of the experimental analysis of behavior has emerged a scientific technology for behavior change based on the principles of respondent and operant conditioning. Procedures exist to increase and decrease the frequency of behavior and to generalize behavior from one setting to another. The application of these procedures to the practice of pediatrics has greatly enhanced the clinical management of pediatric behavior problems. The very recent application of behavioral procedures to medically related behavior problems has been especially important. The field of behavioral medicine represents a collaborative effort of medical and behavioral scientists and practitioners that can only continue to advance the practice of pediatric medicine.

Acknowledgment

Preparation of this chapter was supported in part by a Postdoctoral Fellowship awarded to Dr. Hirsch by the Medical Research Council of Canada.

DEBRA L. OLENICK HIRSCH

DENNIS C. RUSSO

REFERENCES

Ader, R., and Cohen, N.: Behaviorally conditioned immunosuppression. Psychosom. Med., 37:333–340, 1975.

Baird, H. W.: Convulsive disorders. *In* Vaughan, V. C., McKay, R. J., and Nelson, W. E. (Editors): Nelson Textbook of Pediatrics. Ed. 10. Philadelphia, W. B. Saunders Company, 1975.

Balaschak, B. A.: Teacher-implemented behavior modification in a case of organically based epilepsy. J. Consult. Clin. Psychol., 44:218, 1976.

Benson, H.: The Relaxation Response. New York, William Morrow & Co., Inc., 1975.

Benson, H., Shapiro, D., Tursky, B., and Schwartz, G. E.: Decreased systolic blood pressure through operant conditioning techniques in patients with essential hypertension. Science, 173:740, 1971.

Brener, J. A.: A general model of voluntary control applied to the phenomena of cardiovascular change. *In* Obrist, P. A., Black, A. H., Brener, J., and DiCara, L. V. (Editors): Cardiovascular Psychophysiology. Chicago, Aldine, 1974.

Christophersen, E. R., and Rapoff, M. A.: Behavioral pediatrics. *In* Pomerleau, O. F., and Brady, J. P. (Editors): Behavioral Medicine: Theory and Practice. Baltimore, The Williams & Wilkins Company, 1981.

Creer, T. L., Weinberg, E., and Molk, L.: Managing a hospital behavior problem: malingering. J. Behav. Ther. Exp. Psychiatry, 5:259, 1974.

Efron, R.: The conditioned inhibition of uncinate fits. Brain, 80:251, 1957.

Ellis, E. F.: Asthma. *In* Vaughan, V. C., McKay, R. J., and Nelson, E. W. (Editors): Nelson Textbook of Pediatrics. Ed. 10. Philadelphia, W. B. Saunders Company, 1975.

Fey, S. G., and Lindholm, E.: Systolic blood pressure and heart rate changes during three sessions involving biofeedback or no biofeedback. Psychophysiology, 12:513, 1975.

Fordyce, W. E.: Behavioral Methods for Chronic Pain and Illness. St. Louis, The C. V. Mosby Company, 1976.

Goldman, H., Kleinman, K. M., Snow, M. Y., Bidus, D. R., and Korol, B.: Relationship between essential hypertension and cognitive functioning: effects of biofeedback. Psychophysiology, 12:569, 1975.

Ince, L. P.: The use of relaxation training and a conditioned stimulus in the elimination of epileptic seizures in a child: a case study. J. Behav. Ther. Exp. Psychiatry, 7:39, 1976.

Jacobson, E.: Progressive Relaxation. Chicago, University of Chicago Press, 1938.

Jones, M. C.: The elimination of children's fears. J. Exp. Psychol., 7:383, 1924.

Kazdin, A. E.: Behavior Modification in Applied Settings. Homewood, Illinois, The Dorsey Press, 1975.

Kostes, H., Claus, K. D., Crawford, P. L., Edwards, J. E., and Scherr, M. S.: Operant reduction of frontalis EMG activity in the treatment of asthma in children. J. Psychosom. Res., 20:453, 1976.

Lange, A. J., and Jakubowski, P.: Responsible Assertive Behavior. Champaign, Illinois, Research Press, 1976.

Martin, G., and Pear, J.: Behavior Modification: What It Is and How to Do It. Englewood Cliffs, New Jersey, Prentice-Hall, Inc., 1978.

Moore, N.: Behavioral therapy in bronchial asthma: a controlled study. J. Psychosom. Res., 9:257, 1965.

Neisworth, J. T., and Moore, F.: Operant treatment of asthmatic responding with the parent as therapist. Behav. Ther., 3:95, 1972.

Pomerleau, O. F., and Brady, J. P.: Introduction: The scope and promise of behavioral medicine. *In* Pomerleau, O. F., and Brady, J. P. (Editors): Behavioral Medicine: Theory and Practice. Baltimore, The Williams & Wilkins Company, 1981.

Quarti, C., and Renaud, J.: A new treatment of constipation by conditioning: A preliminary report. *In* Franks, C. M. (Editor): Conditioning Techniques in Clinical Practice and Research. New York, Springer-Verlag, New York, Inc., 1974.

Reeves, J. L., and Shapiro, D.: Biofeedback and relaxation in essential hypertension. Int. Rev. Appl. Psychol., 1978.

Renne, C. M., and Creer, T. L.: Training children with asthma to use inhalation therapy equipment. J. Appl. Behav. Anal, 9:1, 1976.

Rimm, D. C., and Masters, J. C.: Behavior Therapy: Techniques and Empirical Findings. New York, Academic Press, Inc., 1979.

Russo, D. C., and Varni, J.: Behavioral Pediatrics: Research and Practice. New York, Plenum Press, 1982.

Schultz, J. H., and Luthe, W.: Autogenic Training: A Psychophysiological Approach in Psychotherapy. New York, Grune & Stratton, Inc., 1959.

Shapiro, D., Schwartz, G. E., and Tursky, B.: Control of diastolic blood pressure in man by feedback and reinforcement. Psychophysiology, 9:292, 1972.

Shapiro, D., and Surwit, R. S.: Learned control of physiological function and disease. *In* Leitenberg, H. (Editor): Handbook of Behavior Modification and Behavior Therapy. Englewood Cliffs, New Jersey, Prentice-Hall, Inc., 1976.

Shapiro, D., and Surwit, R. S.: Biofeedback. *In* Pomerleau, O. F., and Brady, J. P. (Editors): Behavioral Medicine: Theory and Practice. Baltimore, The Williams & Wilkins Company, 1981.

Shapiro, D., Tursky, B., and Schwartz, G. E.: Differentiation of heart rate and systolic blood pressure in man by operant conditioning. Psychosom. Med., 32:417, 1970.

Sterman, M. B., Macdonald, L. R., and Stone, R. K.: Biofeedback training of the sensorimotor electroencephalogram rhythm in man: effects on epilepsy. Epilepsia, 15:395, 1974.

Surwit, R. S., Shapiro, D., and Good, M. I.: A comparison of cardiovascular biofeedback, neuromuscular biofeedback, and meditation in the treatment of borderline essential hypertension. J. Consult. Clin. Psychol., 46:252, 1978.

Turnbill, J. W.: Asthma conceived as learned response. J. Psychosom. Res., 6:59, 1962.

Watson, J. B., and Rayner, R.: Conditioned emotional reactions. J. Exp. Psychol., 3:1, 1920.

Wolpe, J.: Psychotherapy by Reciprocal Inhibition. Stanford, Stanford University Press, 1958.

Zlutnick, S., Mayville, W. J., and Moffat, S.: Modification of seizure disorders: the interruption of behavioral chains. J. Appl. Behav. Anal., 8:1, 1975.

56
Special Education

The preparation of a chapter on special education for a book such as this requires considerable selectivity and ingenuity with regard to inclusions and exclusions. Even brief coverage of the field is obviously impossible. However, much relevant material will be found in other chapters of the book, most notably Chapter 47, Educational Assessment; Chapter 57, Specialized Services; Chapter 61, Laws and Legal Issues; and Chapter 39, Major Handicapping Conditions. The reference list at the end of this chapter also includes a few general references to which the reader may turn for broader coverage.

The inclusion of some historical material seems essential to an understanding of current policies and practices. However, most developments in the field have come during the last quarter century, as the right to educational opportunity has been extended to all children, regardless of the degree of their disability. Thus, the chief focus of the chapter is upon current program alternatives, curriculum modifications, and strategies for working with problem learners.

An attempt has been made to avoid educational jargon, but inevitably a few terms bear explanation. The term handicapped children, as defined by PL 94–142 (Federal Register, August 1977) includes 11 categories for which services are mandated: deaf, deaf-blind, hard of hearing, mentally retarded, multihandicapped, orthopedically impaired, other health impaired, seriously emotionally disturbed, students with specific learning disability, speech impaired, and visually handicapped. Logically children are considered to be handicapped at school only if they differ from other pupils in one or more educationally significant characteristics and consequently require modification of the standard curriculum or learning environment for optimal progress or adjustment.

The pediatrician or other noneducator may be disconcerted by the use of the terms *pupil* and *student* rather than *child* or *patient*. They are employed here frequently, not only because they are part of the language of the school, but because they apply equally to young children, older children, adolescents, and young adults who participate in educational programs.

THE HISTORICAL CONTEXT

1900–1950

Until modern times, formal education in the Western World was available chiefly to affluent and academically talented males. The nineteenth century saw a gradual acceptance of the idea that governments had a responsibility for offering some type of free education to the poor, in the interests of having an informed citizenry. By 1900, the right to attend public school was extended to American children of both sexes. Nevertheless the student had to accept whatever was offered in the way of programs and facilities, and these could be very limited indeed. With the opening of free schools came the observation that some children whose phenotype was not remarkable could not progress normally in the classroom. The need to identify such children and to exclude them, or make other educational arrangements for them, prompted the Paris school authorities to commission development of the first scale of mental development, the Binet-Simon Scale of 1905. The first day classes in the United States for children with learning difficulties were founded between 1928 and 1935; these classes enrolled children who would currently be described as slow learners, learning disabled, and mildly retarded. The children of immigrants, with their cultural and linguistic differences, were also featured prominently in the classes.

Conspicuously absent from the public school scene were children with severe sensory or motor handicaps, seizure disorders, serious emotional or behavioral problems, and significant mental retardation. The earliest European efforts at educating such children were privately sponsored, and usually involved religious communities or physicians whose primary motivations were charitable or scientific. The nineteenth century American counterparts, both private and state supported, included residential programs for blind, deaf, and mentally retarded children. By the turn of the century, most states had one or more of these institutions, but many children with severe handicaps remained at home without any formal education. The goal of the first schools for retarded young people was the return of "defective" children to normalcy; when this proved unrealistic, the goal shifted to "moral improvement" and the protection of society.

MORE RECENT DEVELOPMENTS IN AMERICAN SPECIAL EDUCATION

As the twentieth century progressed, there was increasing pressure from parents and some professionals to broaden the scope of public education. Multiple categories of disability had been recognized, and efforts to educate different populations began to proliferate. The 1950's saw the establishment of "trainable" classes for children not considered "educable" in the academic sense. It should be noted that the presence of any retarded children in the public schools was only gradually accepted; typically the educable class was housed in the school basement or an annex. Stairs and other physical barriers in most buildings partially precluded their use by physically handicapped or blind children, and it was generally considered that the presence of children with obvious handicaps would make normal children (and their teachers) uncomfortable. Parochial schools, which have played a significant part in American education, also concentrated on normal children, although a number of private programs for handicapped children were sponsored by religious organizations.

The civil rights movement of the 1960's led to the disclosure that numbers of poor and minority students were being shunted into special classes, where they failed to acquire literacy skills. To the embarrassment of special educators, no well designed studies existed to document the efficacy of special education programs; the children apparently had not gained significantly from their treatment, and the negative aspects of their placement were obvious to everyone. In retrospect it appears that lack of teacher accountability for student progress, low expectations shared by teachers and pupils alike, and the use of a single curriculum for a heterogeneous group of problem learners virtually ensured mediocrity. The overuse and misuse of psychological tests as the basis for classifying children was eventually recognized, leading to a drastic reduction in their role. Experience also showed that many children did not fit neatly into the categories for which teachers were trained and classrooms established. Hence, the trend of the 1970's was to break down barriers, get rid of labels, and attempt to design individual educational programs that would meet the unique needs of each child.

As medical science eradicated some disabling disease conditions and corrected many birth defects, state and private residential schools began to see a change in their clientele. The long overdue closing of most state residential institutions to young children, and the rapid expansion of community based programs, left such institutions with the most challenging, older pupils whose multiple needs could not be met easily in the schools or at home. Meanwhile the increasing cost of education and the concomitant reluctance of school districts to fund private programs that duplicated the services they offered had a powerful impact on private schools. The inexorable trend for all children, regardless of the degree of handicap, was toward community placement, with education provided by the school district. When parents could not care for the child at home, a foster placement or group home might be sought. This forced school personnel to develop programs, often in collaboration with adjacent school districts, for every type of handicapped pupil.

EARLY CHILDHOOD PROGRAMS

Although public kindergartens had become available in most states by midcentury, nursery schools remained essentially the preserve of the middle class until the Head Start movement of the 1960's. In serving disadvantaged preschoolers, Head Start uncovered many who had mild deficits or who were at risk of developing educational handicaps. The subsequent mandate to have 10 per cent of the enrollment be handicapped children, implemented in 1976, emphasized the commitment to these children. (The efficacy of the programs has been explored by Zigler and Valentine [1979].) For young children with more serious handicaps, federal and state agencies and private foundations continued to support clinical nurseries.

Unfortunately many other children with mild difficulties were not identified until age five or six, when their parents presented them for public school enrollment. Often they were sent home to "mature" for a year, or spent two years in kindergarten where they received two hours a day of activities geared to the normal child's needs. Most school districts offered special education only after a child had failed at school for two or three years. In part this reflected the limited ability of educators to recognize subtle developmental problems through evaluation procedures. The federally sponsored University Affiliated Facilities Programs and other interdisciplinary projects helped greatly to fill this void so that screening, evaluation, and planning could run smoothly. Great strides were made in training personnel for early childhood special needs programs and in developing curricula and materials. An often stated goal of early identification and intervention is a reduction in the number of children subsequently requiring special education services.

COORDINATED PLANNING AND DECISION MAKING IN CURRENT PRACTICE

THE INDIVIDUALIZED EDUCATIONAL PLAN

A requirement for the identification of educational special needs is the interdisciplinary evaluation, as described in Chapters 47 and 61. Although the composition of the team varies with the complexity of the student's problems, there is assurance that more than one professional will be involved in the decision to provide special education services. The findings of the team are presented at a meeting to which parents and involved school personnel are invited. Pediatricians and other professional advocates for the child may also be present. The conclusions reached by the group are summarized in the individualized educational plan. The salient points addressed in this plan are the nature of the student's problems, the educational services that will be delivered (including frequency, duration, and location of service delivery), the short and long term program objectives, the class setting(s) in which the child will be educated, a plan for evaluating the student's progress, and the time interval until formal review. When several professionals are involved in providing services, the plan should also specifiy how they will coordinate their efforts.

The individualized educational plan is a binding document that parents must accept or reject within a specified time period after their receipt of the plan. If they reject all or part of the plan, various levels of appeal are allowed, depending upon applicable state laws. Occasionally parents who believe that their child or his special education needs are poorly understood request further evaluation, after which the school team will reconsider the specific provisions of the plan. This individualized approach to educational programming is helpful even for "normal" students, and in this respect special education has profoundly influenced regular education.

INTERDISCIPLINARY SERVICES

Since the current model employed in special education for the assessment of exceptional children and the delivery of services is an interdisciplinary one, the need for interdependence among professionals has increased as formats for educational management have proliferated. The composition of a particular child's service team depends, of course, upon his needs. One child may be serviced by a large group, including a regular class teacher, resource room staff, speech-language therapist, clinical psychologist, and occupational therapist. Another child may require the support of only two or three school staff members, but need the services of independent professionals, such as the pediatrician, physical therapist, or the staff of a developmental clinic.

When there is confusion or disagreement about the nature of the child's difficulties or the means by which these should be addressed, the opinion of an independent clinic team is often sought. In such cases these professionals describe the child and his needs, answer the specific questions posed, and advocate for the child while facilitating communication between the parents and school personnel. It is always helpful if representatives from the child's school service team can be present for a conference at the clinic or if members of the clinic team can attend a meeting held at the school. Under some circumstances parents may wish the clinic to monitor their child's progress on a long-term basis.

"MAINSTREAMING"—PROBLEMS AND ISSUES

Current federal and state laws require that children with special needs be placed in the least restrictive environment that can meet their educational needs. This practice is known as "mainstreaming" and provides for maximal integration of children with special needs into the regular classroom setting. The child may be taught in the large group setting for the entire school day or may be removed for brief or extended periods so that developmental and educational difficulties can be addressed by specialists. Mainstreaming is therefore heavily dependent on the determination of the child's individual learning style and educational needs (see Chapter 47) and the development of an educational program that matches these needs. Mainstreaming reduces the isolation and stigmatization that frequently result when the child is separated from peers for special education. Furthermore, the policy allows for the introduction of a coordinated approach towards individualized and group teaching.

The practical implementation of the mainstreaming imperative differs according to the nature and severity of a child's special needs: a child with mild learning problems will be serviced in a different fashion from one with more serious handicaps. Districts vary considerably in their policies regarding mainstreaming, however. School administrators play a critical role in setting the climate, planning in-service programs, and actively encouraging acceptance of children with special needs by other staff members. As might be expected, regular teachers differ in their capacity and willingness to participate in mainstreaming efforts. The teacher forced to accept an exceptional child may transmit resentment or discomfort to all members of the class. Conversely the teacher known for success with problem learners may be assigned so many of these children that effective individualization becomes impossible. Depending on some of these practical limitations, a child with severe handicaps may be mainstreamed in one district, whereas a child with less severe deficits may be placed in a self-contained classroom in another district. The following histories illustrate the problem:

> Dick is a quadriplegic, nonverbal boy of 14 who, despite all odds, has been successfully mainstreamed in a junior high school. An aide attends class with him to interpret Dick's communication, which occurs primarily through a communication board, activated with a headlight beam. The school psychologist, regular and special education teachers, and peer/tutors have invested considerable time and effort in Dick's program.
>
> Alex is a 14 year old boy with significant reading delays, high borderline intelligence, nearly normal social and communication skills, and no physical handicaps. He is mainstreamed only for lunch period and nonacademic activities. Although over age, he has remained in the elementary school "because his reading level is only at grade 5." The school administration has not developed a normalized program for Alex and he feels cut off from his adolescent peer group.

Obviously these examples suggest considerable variation in the interpretation of the mainstreaming mandate. It may be pointed out that many small school districts have always practiced integration because there were no alternatives. Students who showed limited academic progress were usually retained several times before they left school at age 16. The analogous current practice of placing older pupils, such as Alex, in primary classes for reading instruction is unfortunate. If the intent of mainstreaming is normalization, there should not be an age difference of more than one or two years between the special student and the pupils with whom he is mainstreamed.

Research has not yet demonstrated significant advantages or disadvantages to the mainstreaming policy, and consequently a frequently heard argument is that mainstreaming is still an "experiment." Many regular classroom teachers continue to believe that they are untrained and lacking in the necessary support services when children with special needs are placed in their classrooms. In addition the attitudes of normal pupils toward those with special needs do not necessarily promote acceptance and social integration. However, it does seem that when districts have made efforts

in good faith to mainstream appropriate students, there has been reasonable success.

RELATIONSHIP OF REGULAR EDUCATION AND SPECIAL EDUCATION

Effective integration of children with special needs depends upon careful planning, coordination, integration, and team effort, as well as the provision of appropriate services. The informed support of parents is also critical, since their negative attitudes transmitted to the child can undermine the best of programs. It is also important to recognize the added time and effort involved when a regular classroom teacher must individualize instruction for one or more pupils with special needs; the task requires knowledge, ability to utilize consultation, and a high level of motivation and insight. Frequently the teacher must monitor the performance of the exceptional child while attending to the needs of the larger group. In-service training programs for these teachers can be valuable for increasing their understanding and skills and supporting their morale. When health professionals and other clinicians can be involved, the dialogue is mutually beneficial and close professional relationships are forged.

When regular class teachers have specific problems with special needs students in their classrooms, helpful suggestions may be offered by the pediatrician or another involved professional who feels comfortable with school related issues. Examples of suggestions that may be given include helping the child to develop regular routines, developing a written contract for completing activities so that the child has a measure of control over assignments, discussing in advance any necessary changes in the schedule, and in the case of junior high school students or those in team teaching situations, recognizing the many transitions required in moving from one classroom to another.

The sensitive teacher is often able to develop a relationship with the special needs student that provides the understanding and encouragement needed for continued effort.

THE SPECIFIC LEARNING DISABILITIES FUROR

The identification of specific learning disabilities as a separate, hitherto unidentified, entity in the field of special education and the strong parent and professional advocacy for this group of children led to the founding of private and public school programs designed for them. A somewhat artificial dichotomy was set up between children with specific learning disabilities (typically white and middle class) and those viewed as being globally slow. The former were perceived as children with normal potential, realizable through specialized instructional techniques, whereas the latter were relegated to "educable" or "social adjustment" classes. The facts that there was considerable overlap between these groups and that within each group were children who required similar specialized educational resources were ignored. In some states the legislation of an IQ cutoff of 90 for determining eligibility for learning disabilities services resulted in many children being deprived of needed services. These included disadvantaged and minority children as well as those with mild handicaps. The interested reader is referred to Cruickshank's excellent discussion (Cruickshank, 1980).

Recent refinement and improvement of techniques for assessment and remediation of learning difficulties have resulted in more appropriate delivery of educational services to children, regardless of their specific IQ scores. In addition, regular and special education teachers have become more sophisticated in the individualization of instruction and more accepting of children with uneven maturation and varying cognitive styles.

PROVIDING SPECIAL EDUCATION SERVICES

PLACEMENT OPTIONS

Public Law 94–142 mandates that schools provide a continuum of alternative placements for children with special needs. As mentioned under "Mainstreaming," the least restrictive placement consistent with the child's need is preferred. Dunn's inverted pyramid model displays 11 major administrative plans in special education, ranging from the most integrated programs, which serve the greatest number of children, to the most segregated, which serve the fewest (Dunn, 1973). The selection of the service format most suitable for a particular child is both complex and challenging and involves the entire school team as well as the parents. The pediatrician and other health care professionals may serve an important role in helping parents to understand and assess the relative merits of the various alternatives. In particular, attendance at the team meeting at which the individualized educational plan is generated composes a critical component of this decision making process; the pediatric advocacy role can often be extremely useful at this meeting. When helping parents to decide educational placement for their child, professionals should be aware of the provisions of the law and the alternatives available

locally. A summary of the program and service formats available for children with special needs is provided in Table 56–1. For further discussion of these programs, the reader is referred to the books by Dunn et al. (1973) and Lerner (1981).

In evaluating the program options summarized in Table 56–1, it should be remembered that the optimal program must address not only the child's cognitive-instructional needs but other considerations as well—e.g., social maturity, the ability to tolerate the stress of particular school environments, the need for specific educational or behavioral strategies, and physical stamina. Frequently there is considerable fragmentation of a child's program so that he is sent from one classroom or specialist to another throughout the day. Such a child may experience frustration and isolation from a consistent peer group.

When resources for particular special needs groups are unavailable in a district, regional arrangements such as collaboratives are often formed between districts. This allows students to receive appropriate services without resorting to residential placement. The major disadvantages include the amount of time lost in daily travel and the separation of students from peers in their own communities.

PROFESSIONAL SERVICES FOUND IN SCHOOLS

Within the school setting a variety of professional services should be available; these differ somewhat from one state to another, although greater homogeneity has been achieved in recent years. Typically a school district offers a variety of psychological services: evaluation, referral for mental health therapy, and adjustment and guidance counseling. In addition, remedial reading, tutoring or small group instruction for learning disabilities and speech and language therapy are provided. Recently many districts have begun to offer occupational therapy for pupils with visual-spatial, fine motor, and sensory integration problems, and physical therapy or adaptive physical education for those whose motor coordination and gross motor skills are deficient. Others providing assistance in the schools include teacher aides, school volunteers, and peer tutors. Such helpers can be valuable in providing individualized teaching and supervision, particularly with severely handicapped children. When staff-pupil ratios are optimal, the staff may also use behavioral management principles in their structuring of children with hyperactivity or attention disorders.

Extended day programing (i.e., 9 to 5) and 12 month programs are enormously helpful when handicapped children present challenging care problems at home. Some districts offer these services to autistic and other severely disabled children on the basis that they could not otherwise be maintained in the community.

ALTERNATIVES FOR CHILDREN WITH GRADE DEFICIENT SKILLS

The practice of retaining a child in a grade because of his failure to master the curriculum content of that grade is usually not an appropriate alternative, except at the preschool or primary level or during the transition from an ungraded class to a regular grade. The social stigma of "staying back" and the paucity of evidence that the practice was beneficial to the pupil were early reasons given for discontinuing retention. The trend toward adapting the curriculum to the child and the whole mainstreaming philosophy argue against separating the problem learner from his age peers.

It is still appropriate, when a child's birthday falls late in the year or when he is physically immature, to consider delaying entrance into first grade. If continuation in a kindergarten program or placement in a transitional class seems best, however, it is essential that careful assessment be carried out to determine whether special services, such as speech-language therapy, are needed during that year. If the child's social development will be furthered by participation in a kindergarten program, traditionally two and one-half hours daily, it may be important to schedule special services at another time during the school day. The following example illustrates how such a program can be arranged:

Adam, a six year old, was delayed in developing fine motor and visual perceptual skills and had poor expressive language skills; he was also emotionally immature. He participated in a regular kindergarten class each morning, and then reported to the primary resource class for the lunch period. In the afternoon he received language therapy twice weekly, occupational therapy once a week, and a special gym class once a week. Adam also spent approximately eight hours weekly in the resource room where he received intensive help with reading and math readiness skills. It was planned that he would enter first grade at age seven, with minimal special education needs.

The provision of reduced size classrooms, ungraded classes, and similar specialized options is most often found at the primary level. There is an expectation, sometimes unfounded, that by fourth grade most children will be able to succeed in regular class. Resource rooms at the intermediate level and tutoring when indicated can serve some of the same needs. However, it is common to find that the resource room staff must cope with a large and heterogeneous group of students who arrive and depart at different times. The constant shifting of pupils may be distracting to children who have attentional or behavioral problems, and the pupils in fact may receive relatively little individual attention.

SPECIAL EDUCATION SERVICES FOR OLDER STUDENTS

Middle school or junior high school programs for special needs pupils also vary from integrated arrangements to self-contained classrooms. This age is a difficult one for "normal" boys and girls in our culture, and most parents are anxious about adjustment issues as well as school progress. In many instances young people with significant developmental delays remain in the elementary school until age 13 or 14. Extending this retention for a longer period is unfortunate, since it increases their social isolation and impedes the appropriate growth toward maturity and independence.

Parents may raise the question of vocational education for students who have not progressed far in the academic areas or who complain of boredom and frustration at school. However, the days of apprenticing a 12 year old to the blacksmith or the cobbler are past; and even normal students are not admitted to technical programs until age 14 or 15. The fact that many states provide special education through the twenty-first year has made early enrollment in vocational courses even less appropriate for handicapped youth. Obviously pupils with both normal and special needs can benefit from shop courses, home economics courses, and arts and crafts activities, but the primary focus of education in early adolescence should be upon the development of oral (and written, if appropriate) communication skills, mastery of quantitative skills to the extent possible, and knowledge of practical areas, such as nutrition, hygiene, budgeting, and the responsibilities of citizenship. These can be adapted to the abilities and limitations of the student and presented in ways that support his interest and best efforts.

In the midadolescent years many pupils with special needs explore a variety of work areas, often set up as job stations at the school. These experiences develop special interests and realistic aspirations for employment. The final years of education (usually between ages 18 and 21) find the student ready for supervised work in the community, as part of the school program.

In some respects young people who continue to require special education at the secondary level represent the core of the special needs school

Table 56–1. SERVICE FORMATS FOR PUPILS WITH SPECIAL NEEDS

Service Format	Rationale and Possible Modifications	Advantages	Disadvantages
Regular classroom	Student benefits from and contributes to normal school experience Individualized materials may be used with minor modification of instructional style and increased tolerance for shorter attention span and lower work output	1. Does not label the student with special needs 2. Enhances social skills 3. Develops in all pupils an awareness and appreciation of human differences	1. Student may require disproportionate amount of teacher's time and attention 2. Student may not have all his special needs met 3. May be disruptive for regular class 4. Other pupils may ostracize child
Regular classroom with minimal supportive assistance	Student benefits from regular class but receives individual or small group instruction in problem area, e.g., language development, reading	1. Student is not segregated but receives attention to his special needs in a one-to-one or small group setting 2. Special teacher can provide consultation with regular class teacher 3. Regular teacher has more time for working with other pupils	1. Tutoring may be poorly coordinated with regular classroom work 2. Attention is drawn to the different pupil 3. In some cases tutor may be a school volunteer or other person with minimal qualifications 4. Brief (10 to 15 minute) periods may not benefit child
Regular classroom with resource room placement for a significant portion of the school day	Student enjoys the advantages of both placements Teachers and other staff work together to deliver individualized educational plan Length of time out of classroom may be reduced as student progresses	1. Specialized personnel deliver services in key areas 2. Small group ensures closer monitoring and individualization of instruction 3. Pupil still benefits from contact with normal pupils	1. Program may be fragmented, with staff not working together 2. Special student must cope with frequent moves, two different classes, and many peers

Placement	Description	Advantages	Disadvantages
		4. Easier for regular class teacher 5. There is less stigmatization than if pupil received full time special education	3. Resource room may be a disruptive environment, with pupils constantly shifting
Self-contained special class in regular school, special center operated by public school district or collaborative	Intensive special education delivered in an integrated fashion Classroom and other physical resources adapted Students may have contact with "normal" peers for selected nonacademic activities Staff are well trained and administration is simplified	1. Pupils with severe special needs are accommodated in or near their own community 2. Class size, curriculum goals, and materials are appropriate to students' needs 3. Students learn social skills with others of similar maturity	1. Pupils are stigmatized and isolated from normal peers 2. Some pupils spend considerable time being transported to class 3. "Normal" pupils cannot benefit from contacts with handicapped children
Private day school or residential school (state or private)	Pupils having severe special needs and for whom no appropriate public program is available can be served in a private program, which agrees to implement the individualized educational plan Parents may prefer to send child to private school at personal expense, or child may be homeless	1. Pupil receives an appropriate education, usually with other pupils who have similar special needs 2. Total program may be better integrated, with all needed services in one place 3. Family usually has more respite	1. Student is deprived of all school contact with peers and vice versa 2. Student is often placed out of his own community 3. Special school may enroll a heterogenous group 4. In residential school, student is deprived of a normal home life and receives skilled care for only 25 to 30 hours weekly 5. Education is very costly
Hospital or home tutoring	Short term alternative for pupils who cannot participate in a group or are waiting for placement	Provides some limited skill training and cognitive stimulation	1. Student has no contact with peers 2. Student receives only a few hours of education weekly

population—those with serious mental retardation or multiple handicaps and those who represent the failures of an imperfect system. Students with mild mental retardation or significant reading handicaps due to other factors are most often managed in a team teaching format. Some are able to succeed in nonacademic courses at a vocational-technical high school, or they may participate in a special program there. In such cases remedial academic instruction continues, with increasing emphasis upon what is most practical for each student.

Unfortunately many school districts fail to provide adequately for adolescent pupils with special needs, thus encouraging them to drop out of school early. The entry of these handicapped youths (including those with poor educational skills) into the labor force is particularly unfortunate in urban areas where high unemployment pits them against nonhandicapped adults in the search for unskilled work. The existence of a strong relationship between learning disabilities and school failure, on the one hand, and juvenile delinquency, on the other, has long been the focus of extensive research (Rutter, 1975).

No discussion of special education at the secondary level would be complete without acknowledgment that tutorial or small group instruction for those with subtle learning impairments is often unavailable in public high schools. For the most part, students with mildly deficient reading skills, difficulties in written language production, spelling disorders, or reduced auditory processing skills are left unsupported. Many have been unable to pass the recently instituted competency tests and thus leave high school without a diploma. The weaknesses and uneven distribution of services for such pupils constitute a serious problem, which must be confronted in the years to come; it represents one of special education's major frontiers.

SUMMER SCHOOL PROGRAMS

In many school districts summer school programs are offered for selected pupils, most commonly those needing remedial or make-up instruction. Programing for children with major special needs often involves a day camp program under broader sponsorship, putting less emphasis on skill acquisition. Private programs, ranging from advance credit courses to residential schools with intensive behavioral programs, are also available in some localities, and school districts occasionally pay for a child to attend such a program if they agree that it is educationally necessary. Most often, however, parents must defray the expenses or find community sponsorship to underwrite the cost.

COMPUTER ASSISTED INSTRUCTION

A word should be said about the growing availability and use of computers in education as this affects children with special needs. The decrease in the size and cost of computers and the increase in types of minicomputers on the market have led to the expectation that before the turn of the century the field of education will be profoundly changed by the applications of computer technology. It will be possible for computers to select and present appropriate learning material to the student, based upon all that is known about his cognitive abilities and instructional needs, monitor performance, and make constant program corrections to insure progress. No one anticipates that the teacher will become obsolete, but the teacher's role will be greatly changed, removal of the more tedious tasks freeing him for the more creative aspects of teaching. The implications for mainstreaming are enormous, since the individualizing of curriculum content and the appropriate modification of materials will no longer burden the teacher. The reader interested in pursuing this fascinating subject is referred to the periodical, *Classroom Computer News.*

CURRICULUM MODIFICATIONS FOR SPECIAL POPULATIONS

SUBTLE AND MILD LEARNING PROBLEMS

The majority of school children with special educational needs are those with subtle or mild learning problems. Much of what is said about the learning characteristics and educational needs of these students is equally applicable to other exceptional students. Although they represent a broad range of intellectual functioning, from superior to mildly mentally retarded, a disproportionate number fall in the low average to borderline categories, represented by IQ scores between 100 and 70. In looking at a particular child's learning problems, local norms must always be considered along with other information that might influence scores. A slow learning child in an upper middle class district where the mean IQ is 115 is relatively more handicapped at school than a child of similar ability in an area where the mean is 85 or 90. The reason behind a child's poor showing on tests is obviously of great significance; some represent the low end of the normal distribution, others have been compromised by prenatal or other early developmental factors, while still others (as has been well documented) score poorly because of cultural or linguistic differences, specific learning disabilities, or deficient academic skills. The prognosis for relative success in school (and adjustment in adulthood) varies greatly with the etiology of the problem, the quality of education, family and community support, and the personal strengths of the individual.

The past 20 years have seen a proliferation of specific remedial approaches and techniques for children with subtle learning problems. In the 1960's perceptual and motor training methods were popularized by Frostig and Kephart, and it was claimed that these methods improved the reading skills of children with certain learning problems (Mann, 1979). Concurrently, multisensory training was incorporated into the various Head Start programs, and was reported to address and prevent learning difficulties in the culturally different child (see the review by Zigler and Valentine, 1979). However, over the last few years these training procedures have been scrutinized and frequently criticized, because their efficacy has not been demonstrated. Nevertheless the principles of process training that were generally espoused and practiced by the perceptual training movement are now integrated into a broader approach toward remediation.

Current instructional techniques are separated according to two major schools of thought, namely, basic ability approaches and direct skill approaches. Basic ability approaches emphasize remediation of the abilities and processes underlying a particular problem and reflect aspects of the perceptual training movement. They stress the determination of each child's cognitive style, perceptual skills, and language abilities and attempt to match teaching strategies with the child's processing and learning profile. In contrast, the direct skills approaches generally analyze the subject matter or content to be learned, regardless of the child's learning profile. A number of different techniques and teaching procedures may be applied, irrespective of the child's learning characteristics, developmental levels, and chronological age. In addition, teaching materials that employ a task analysis approach have recently burgeoned.

Although the merits of the basic ability and the direct skills approaches are each widely espoused, research studies have not established the efficacy of either, when applied in isolation (Arter and Jenkins, 1978; Mann, 1979). One major criticism of the basic ability approach is that remediation of processing abilities as isolated and discrete functions does not necessarily generalize to the improvement of academic skills. Thus, training a child to trace shapes and discriminate objects will not necessarily result in improved writing skills unless the child is taught to trace letters and to differentiate alphabetic symbols with different spatial orientations (e.g., b/d, p/q). At the opposite end of the spectrum it is often suggested that materials developed for children without deficits are not appropriate for those who do manifest delays in specific areas. When the two discrete approaches are combined, however, they provide an integrated remedial system and teaching is individualized according to the student's needs. A few specific techniques pertinent to this holistic method are outlined in the following section.

Remediating Basic Abilities in Relation to Learning Style

For simplicity of presentation, the subject matter in this section is subdivided into discrete areas. In all instances an integrated remedial model is suggested so that a variety of teaching options are applied to combine the remediation of processes with the teaching of specific skills. To illustrate, a child with reading problems and temporal-sequential weaknesses should have training in temporal-sequential processing, as well as the application of specific reading approaches. For greater detail, the reader is referred to the texts by Wiig and Semel (1976) and by Mann et al. (1978); the latter deals specifically with remediation for adolescents.

Visual-spatial difficulties generally result from a limited understanding of left-right orientation, body position, or background and foreground. Consequently reading and spelling performance may be characterized by confusion in the recognition and reproduction of the correct direction in which letters and words are oriented. The resulting reversals of letters (b/d, p/d, b/p) or words (was/saw) are often remediated using mnemonic cues, which help the child to remember directional orientation, and through multisensory techniques, which emphasize tactile, visual, and auditory components of letters. In the earlier grades, useful techniques include design matching, pegboards, formboards, and mazes, which stress left-right orientation and orientation in space.

Sequencing (Auditory, Visual, and Cross Modal)

Children whose learning difficulties are associated with processing weaknesses in sequential organization usually experience major difficulty with any tasks requiring the serial ordering of information. The resulting delays in reading accuracy and spelling mastery often can be remediated by exposing the child to activities that emphasize sequential ordering of letters, words, and pictures and that stress analysis as well as synthesis. Multiplication tables will not be easily learned, so that activities that emphasize multisensory learning of the sequences are helpful. Because understanding of time related concepts is poor, helpful activities include work with calendars and clocks and practice with automatization of known sequences, such as the days of the week, months of the year, and alphabetic and number sequences.

Fine Motor Coordination

Children with fine motor coordination weaknesses often have difficulty with written expression because of their awkward manipulation of the pencil and poor integration of the visual-perceptual-motor demands of tasks. Useful remedial activities for improving eye-motor coordination include tracing, copying, and chalkboard games and the use of scissors, puzzles, formboards, connect-the-dot worksheets, and coloring games. More controversial techniques relate to the remediation of pencil grasp. Often children develop awkward pencil grasps that translate into writing problems at a later stage. In the younger age groups, retraining the child to develop a well positioned tripod grasp can often be helpful in preventing future penmanship problems.

Auditory Perception and Receptive and Expressive Language

Learning difficulties associated with subtle auditory processing weaknesses and receptive and expressive language disabilities frequently result in poor understanding of instructions, problems narrating a coherent story, and limited verbalization. The consequent delays in reading comprehension, written language, and performance on all verbal tasks generally are remediated by providing activities that enhance listening skills, vocabulary, syntax formation, and narrative organization. Frequently the special educator collaborates with the speech-language therapist to provide integrated remediation of these deficits.

Conceptualization

Difficulties in conceptualization are often evidenced in a limited ability to categorize and classify experiences and information. Children who display problems in this area show delays on reading comprehension tasks and demonstrate poor integration of information for oral or written expression. Further, their techniques for problem solving are often limited, and they tend to associate events in a fragmented fashion rather than

to formulate overall themes or organizing principles. Numerous activities for improving logical thinking may be extrapolated from Piagetian approaches, and include logic and classification games, grouping and categorizing, and probability games. Frequently materials for teaching higher order mental functions are incorporated into the regular reading and science curriculum, although they may not be recognized as "cognitive."

Attention Problems

The child with a major attention deficit usually requires modification of the learning environment rather than a specific curriculum. An optimal setting can offer a highly structured environment, reduction of distracting stimuli, and a considerable amount of one-to-one or small group instruction. Strategies for improving attention may include visual and auditory vigilance tasks, which encourage attention to detail, and highly structured worksheets, which reduce the amount of extraneous information.

Because children with attention problems experience significant organizational difficulties, management techniques frequently emphasize the establishment of regular daily routines in the classroom and at home, short concentrated work periods, which are usually more productive than longer time spans, and the use of calendars on a regular basis at home and at school. Such techniques allow for short bursts of physical activity between tasks to counter restlessness, provide for discussion and clarification of changes in the youngster's schedule to ensure some measure of self-control, and encourage the development of a private communication system between the teacher and the child so that they can cooperate to monitor episodes of inattention. The child can signal the teacher privately when instructions need to be repeated. Accommodations such as these can be used in the regular or special education setting and provide effective methods for improving behavior and attention regardless of whether medication (e.g., Ritalin) has been prescribed.

Remediating Direct Skills and Content Areas

Reading and Spelling Skills. Over the years, techniques for teaching reading have proliferated from a variety of theoretical viewpoints and have resulted in numerous reading programs, which differ in their orientations and methods. Nevertheless applied research suggests a fairly consistent theme: Programs that emphasize analysis of the alphabetic code are most effective for most children in the early grades, and language based programs stressing comprehension are most rele-

Table 56–2. COMMON APPROACHES TO BEGINNING READING

Reading Approach	Description	Children for Whom These Programs Are Well Suited	Children Whose Learning Styles Are Not Well Matched with These Programs
Basal readers (e.g., Scott-Foresman, Houghton-Mifflin)	Use sequenced readers, which emphasize reading for meaning rather than word analysis	Children with good visual perception strategies and possible sequencing deficits benefit from the sight-word or look-say approach	May be inappropriate for children with attention deficits or visual perceptual problems because of their inability to attend to all the word features simultaneously
Phonics programs (e.g., Hay-Wingo, Orton-Gillingham)	Stress sound-symbol associations and blending these sounds into words	Helpful for children with multiple processing weaknesses who need a structured multistep reading program and who learn best when tasks are broken down into their component parts	Not appropriate for children with weaknesses in sequential processing and auditory memory because of their difficulty in blending the sounds into whole words
Linguistic programs (e.g., Merrill Linguistic Readers, SRA)	Also emphasize decoding skills, but select vocabulary from phonemically regular words or word families, so that child learns the rhyming patterns and associates the sound with a particular pattern of letters	Appropriate for use with children who display sequential memory problems and good oral language skills	Inappropriate for children with weaknesses in "gestalt" processing and strengths in sequencing (auditory or visual)
Language-experience approaches (e.g., Open Court Carden)	Presented as total language arts programs, these incorporate the child's own language experience and vocabulary, and teach listening, speaking, reading, and writing in an integrated manner	Most suitable for children with global language disabilities or children whose spoken English differs significantly from that contained in the basal readers	Not well suited for children with major attention problems, who benefit from a highly structured, sequenced approach
Multisensory approaches (e.g., Fernald, V-A-K-T)	Emphasize the simultaneous presentation of auditory, visual, and tactile stimuli (e.g., letters are learned through tracing sandpaper letters, saying the sound, and seeing the symbol)	Most young children benefit from a multisensory approach, as long as this is structured and organized in a sequential fashion, as in Montessori programs	Children who show major strengths in one sensory area and significant deficits in another may learn optimally through a program that capitalizes on their strengths
Modifications of traditional alphabet (e.g., i.t.a., Distar)	Incorporate an expanded alphabet, which simplifies written language by matching each sound with a different alphabetic symbol	Helpful for children whose language and learning problems cause difficulty with the rules and exceptions of English phonics	Because children must shift to a regular alphabetic system in third grade, these programs are problematic for children whose cognitive style prohibits easy transitions from one learning approach to another

vant for later grades. Furthermore, approaches to teaching that stress structured reading curricula and the direct teaching of reading skills have demonstrated greater efficacy (e.g., Chall, 1967). Despite the current debate over whether most reading problems are a consequence of poor teaching or derive from underlying neurological and developmental deficits (Resnick and Weaver, 1979), a number of remedial techniques have been designed for children with specific learning problems.

Table 56–2 provides a summary description of a few well known approaches for teaching beginning reading. It is evident from this table that certain techniques are more appropriate for children with specific developmental profiles (e.g., visual-spatial deficits), whereas others are better suited to children with different learning styles (e.g., language processing difficulties). An awareness of the advantages and disadvantages of these techniques is often important for the pediatrician, since these methods are frequently major components of the individualized educational plan.

It should be noted that each of these approaches is usually accompanied by workbooks and teacher-made materials, which may emphasize particular language arts skills. Many teachers employ an eclectic approach, ordering materials from other publishers to supplement the reading program.

Written Expression: Focus on Bypass Strategies. As emphasized in Chapter 47, proficiency in written language is dependent on the integration and automatization of a number of motor, perceptual, language, and cognitive processes. Consequently appropriate remediation can be provided only when the source of difficulty is located through diagnostic assessment. Furthermore, decisions about remediating weaknesses or teaching bypass strategies are generally dependent on the grade level of the child, his developmental strengths and weaknesses, and the locus of the problem. Remediation generally addresses writing deficits at a number of levels, including readiness to write, pencil grasp, penmanship or writing fluency, organization and spacing, punctuation and capitalization, and writing content. Writing readiness activities may include visual-perceptual-motor training through tracing activities, dot-to-dot games, cutting, and left-to-right drawing. From the beginning of first grade, writing remediation often stresses the development of an adequate pencil grasp and pressure, as well as the improvement of visual memory for specific letters and alphabetic sequences ("revisualization"). Educational materials may include print of a larger size or a different color to strengthen visual memory and to enhance recall. In particular, memory for alphabetic symbols that are often reversed owing to their similar visual appearance (e.g., b/d,

p/q, p/b) is generally enhanced through multisensory materials so that learning can be consolidated through all modalities.

At the upper elementary and secondary levels, written expression is a daily requirement and writing content assumes critical importance, with speed and volume of output an expectation. Overall productivity is frequently reduced as a result of numerous possible combinations of perceptual, language, and cognitive deficits. Bypass strategies should be considered to reduce pressure on the student; these may include some of the following: modified expectations for volume of written output, ungraded writing assignments and different grading systems for contents and "mechanics," increased time for completing written work, untimed standardized achievement tests, and the use of a typewriter or tape recorder to encourage expression of ideas.

Recently greater interest has been shown in the assessment and remediation of specific writing problems (Hammill, 1975). It is to be hoped that future research and clinical studies will result in exploration of this area.

Arithmetic and Other Mathematics Skills. A child's understanding of quantity is heavily dependent upon his level of cognitive development. Early programs for teaching arithmetic readiness and for remediating difficulties emphasize conceptual understanding, one-to-one correspondence, seriation, and number conservation. More recently they have incorporated strategies based on Piagetian tasks (Furth and Wacks, 1976). Mastery of the ordinal system for counting is often difficult for children with sequencing problems; thus remedial activities stress the ordering of number sequences and the understanding of temporal concepts such as "first" and "last." Recognition of the visual symbols is emphasized for children with visual-spatial difficulties and the use of concrete manipulatives; for example, fraction bars, Dienes blocks, and geoboards may help children to note part-whole relationships. Depending on a child's age and special needs, a calculator may be used to capitalize on problem solving skills. (Computation instruction can be provided in a remedial setting such as a resource room.)

SENSORY AND MOTOR HANDICAPS

Physically handicapped children and those with single sensory handicaps are often ideal candidates for mainstreaming with normal children. They may require adaptive equipment, special materials, and resource room support, but the educational goals for them are essentially the same as those for their age peers. Suitable low vision aids and specialized training, such as Braille reading and speech-language therapy, are supplied by the ancillary school staff or itinerant consultants, who also assist the classroom teachers. Of course, there are parents who for personal reasons prefer a private, usually residential school; in sparsely populated areas with few resources, this alternative may be necessary. However, given their choice, most children would elect to remain in the community.

Because of the eradication of several causes of sensory and motor handicaps, there have been changes in the types of students presenting at school with these problems; a number are multihandicapped and require self-contained special classes. Unfortunately the classes to which they are assigned are often so heterogeneous that individual pupils are not well served.

Deaf children, as opposed to those with mild to moderate hearing impairment, are also more likely to require placement in special class. The development of language skills in these children is given the highest priority. Nevertheless some do not become effective oral communicators. Sign language, finger spelling, written communication, and other alternatives are employed. In the past, deaf pupils were sent to state or private schools at very early ages. Studies of graduates found them to have academic performance only at fourth to fifth grade levels. Public school programs in the last decade have appeared to be having somewhat greater success in bringing out the potential of these young people and improving their possibilities for social and occupational integration into the community.

SEVERE COMMUNICATION DISORDERS

Only recently has attention been focused upon the educational needs of pupils with severe communication disorders, including nonverbal children. An interdisciplinary team, composed of professionals with expertise in alternative communication systems, technologic aids, and evaluation of cognitive skills in nonverbal persons, must cooperate in planning, implementing, and monitoring the educational program. Because school personnel feel inadequate to deal with these students, they may prefer to enroll them in private programs or to provide home tutors. However, it is usually possible to obtain consultation from a medical center or university clinic and to arrange for appropriate staff to be hired. This alternative, although costly, is less expensive than a residential school and offers greater benefit to the student. The selection of program options is heavily influenced by the student's cognitive ability, since the mentally retarded child is less able to use written language and some other alternatives to speech.

EMOTIONAL DISTURBANCE

Education for seriously emotionally disturbed children has lagged behind that of most other special groups. The child whose behavior was considered unacceptable or distressing at school was usually excluded; it was common until recently to find such children receiving instruction four or five hours weekly from a home tutor sent by the school district. In private residential schools for disturbed children, the goal of treatment was to improve the child's adjustment and behavior; academic progress was of low priority. Often the treatment centers did not even employ qualified teachers but used mental health staff in the classrooms. A number of factors led to change in this situation: the availability of mental health services in the community, the trend away from confining children with special needs to residential facilities, and recognition of the role education can play in normalizing the life of the disturbed child. Currently most school districts operate or collaborate in the sponsorship of programs for disturbed children. Except that these pupils have a higher than expected incidence of specific learning disabilities, their instructional needs are similar to those of other children. The learning environment and relationships with staff and peers are the two areas that most often require special consideration. The student is gradually integrated with peers as his emotional gains and academic progress allow. Obviously disturbed children who have serious learning handicaps, sensory impairment, or other multihandicapping conditions present a special challenge in terms of programing; they require the services of a number of school and outside resources.

The majority of children identified as autistic or schizophrenic fall into this category, and the best arrangements for them appear to be intensive behaviorally oriented programs that operate for extended days and provide support and training to families.

MILD MENTAL RETARDATION

Mildly retarded children, or those scoring two to three standard deviations below the mean, are a very heterogeneous group. Unfortunately the traditional educable class tended to lump them together without regard for their differences. Several factors have led to a significant reduction in the number of educable pupils served in self-contained classes: the reassignment to regular classes of poor and minority students who were inappropriately placed in special classes, the mainstreaming of mildly retarded students whose learning and behavioral characteristics make this feasible, the availability of an individualized educational plan for each child with special needs, and the presence in the school of a variety of resource rooms and special services, including consultants for classroom teachers.

Mildly retarded children who do not have specific cognitive and perceptual learning disabilities, as discussed earlier, have academic instructional needs similar to those of normal children. Their rate of learning of course will be slower, and the skill levels at high school graduation will usually be those of an average fifth or sixth grader. (Parents sometimes mistakenly interpret this to mean that their child will stop learning or reach a plateau in early adolescence.) As a mildly retarded child grows older, it is critical that his reading materials in all subjects be suited to his age and interests. A book designed to appeal to a seven year old youngster seldom motivates the mildly retarded 10 year old; but ungraded "high interest, low vocabulary" materials are readily available.

The large number of mildly retarded pupils who exhibit specific learning problems will benefit from the same approaches discussed under "Subtle and Mild Learning Problems." In the past such children often failed to achieve literacy (fourth grade reading level) because their cognitive-perceptual difficulties were dismissed as part of their global retardation.

The issue of reading instruction for mentally retarded pupils is discussed in Savage and Mooney's *Teaching Reading to Children with Special Needs* (1979). They point out that although intelligence and reading ability are positively correlated, especially if a verbal measure of intelligence has been used to classify the child, the correlation is lower at beginning reading levels at which "decoding" is emphasized and higher at more advanced levels at which the student must demonstrate inferential comprehension, often by answering open ended questions or by integrating new material with previously learned material. This has obvious implications for mainstreaming, particularly after the third grade, when higher order cognitive skills are increasingly required.

The primary focus of education for mildly retarded pupils is academic, until they reach mid-adolescence and begin to be oriented toward vocational preparation. Reading, oral and written communication, basic quantitative skills, and the acquisition of science and social studies concepts are stressed, although students are also involved in other activities that support the development of a healthy self-concept, good work habits, and the constructive use of leisure time.

A small percentage of mildly retarded pupils are not successful in an academic program, usually because of a second handicap that interferes with learning to read. The educational goals for these children must be reviewed frequently and every effort made to capitalize on their strengths. Projects involving television, computers, tape recorders, and various functional tasks replace most academic subjects, and students learn compensatory strategies for independent living.

MODERATE AND SEVERE MENTAL RETARDATION

Although combined here in the interests of brevity, moderately and severely retarded individuals compose two distinct groups testing, respectively, three to four and four to five standard deviations below the mean on individual intelligence tests. Sensory or motor impairments are common in both groups, and an assumption is made that most individuals will not achieve full independence at adulthood. Moderately retarded pupils were formerly called trainable, to distinguish them from educable. Although full literacy is still not considered a realistic goal for most of them, it is now recognized that the majority can make some progress in acquiring primary reading skills. These are extremely useful for shopping, public transportation, banking, and leisure activities (i.e. reading the *TV Guide*), to say nothing of the enormous personal satisfaction reading brings. Other educational priorities include developing oral communication and social skills and such quantitative concepts as simple counting, recognizing coins, and using a digital watch. Health and sex education are important needs also.

Higher functioning, severely retarded children are often grouped with moderately retarded children in educational programs, partly because both are low incidence conditions and partly because instruction must be individualized in any case. Neither group can be mainstreamed for a significant part of the school day. Objectives for the severely retarded child include socialization and personal care (e.g., independent use of the toilet) and simple communication. In selecting materials for older, severely retarded pupils, teachers avoid those that are obviously intended for preschool children. Commercial materials are available, but teachers of these students must also be creative in adapting tasks. Color, form, and size discrimination, matching, sorting, and simple assembly line tasks prepare the young adult for transition to a workshop or day activity center.

The postschool adjustment of moderately and severely retarded individuals usually hinges upon interpersonal skills, attention, and dependability, rather than a slight edge in IQ. Teachers therefore make a concerted effort to inculcate these attributes from preschool through secondary education. Parents of the students often look to the school for guidance in setting limits and deciding upon objectives for home and school training. They may also be helped by meeting with school personnel and other parents to share their concerns and discuss common problems. The long term involvement of clinic based mental retardation professionals can also be invaluable when parents have questions about program changes or family crises involving the severely retarded young person. Because of the unique needs of each severely retarded student, the formulation of the individualized educational plan requires the contributions of many professionals. Considerable individual attention may be needed, including one-on-one teaching, if progress is to be made.

PROFOUND MENTAL RETARDATION

Profoundly retarded young people, who may function below the one year developmental level, have only recently received educational attention. Their scores on standardized tests are meaningless as indicators of educational or habilitative potential, and the child's teacher often finds careful observation the best tool. The student may be called blind or deaf and still prove to have useful function when he is motivated by familiar (pleasant) sights and sounds. He may register awareness of a favorite teacher upon hearing that person's voice or may respond with anticipation to the sight and smell of food. Although some profoundly retarded children show a limited ability to form human relationships, most eventually respond to a friendly and consistent teacher or care worker. The medical problems of these children, many of whom are nonambulatory, sometimes lead health professionals to suggest that they belong in institutions or at home. Parents also may worry that their child will be stressed or fatigued by school attendance; they may question the ability of school staff to manage feeding, toileting, and other needs. However, when profoundly retarded children do attend school, they generally show no ill effects. Teachers and other staff provide them with social and sensory stimulation (music, finger painting, swimming, rocking, touching various textures). The teacher often becomes a strong support to the family; it is not uncommon to find them providing occasional respite care.

The usefulness of educational programs for profoundly retarded children is often questioned on the basis of "cost effectiveness," given the minute potential for improvement. In fact the value of the programs may never be empirically demonstrated; however, parents and involved teachers do not question the improvement in the quality of life represented by school programs.

ISSUES OF POLITICS AND PUPIL RIGHTS

As this chapter is being prepared, the future of PL 94-142 is uncertain. Faced with the loss of promised funds, many states are moving to limit special education services. Professionals involved with handicapped children and their families need to remain informed about the status of special education at the local, state, and federal levels. Parents often ask for support in securing educational services that the school has been reluctant or unable to provide. These include special therapies, social or recreational programs after school hours, summer school, camp, or private school. In some cases parental requests may be unreasonable, but often the service is necessary. Health care professionals may provide valuable assistance to families in obtaining appropriate services.

Insurance companies, public and private agencies, community organizations, and even the parents themselves may share in the funding of services. Sometimes parents need clarification of the fact that PL 94-142 does not promise public funding for the best program that can be located but, rather, an appropriate program. Of course, if parents choose to pay their child's educational expenses themselves, their freedom to select among options is unquestioned.

When a child has severe handicaps or is difficult to manage at home, parents may question whether he would learn more in a 24 hour placement or would have better care from skilled workers. Experience suggests that neither of these is probable, but the parents may be saying that their family's needs can be met best by removing the handicapped member. If the pediatrician or other professional is drawn into a controversy about who should bear the cost of residential placement, it is important for him to be straightforward about the rationale for the placement. When family stresses are the issue rather than the quality of the school program, cost sharing or alternative sources of funding are appropriate. When one considers the cost of many residential programs (i.e., $40,000 or more per year), it is not surprising that the expense is often a controversial point. Nevertheless financial constraints should not be the reason for denying students an education. The alternative for some would be a return to the custodial institutions, which ironically are even more expensive to maintain than quality programs in the community.

When other professionals deal with school personnel, they should be aware that criticism, or even well meant suggestions, may be quite threatening. Despite their obvious bias, educators regard themselves as the best judges of educational services, just as pediatricians regard themselves as the best arbiters in medical situations. A collaborative relationship in which all parties work for the child's best interests is the ideal. Obviously the quality of services and the effectiveness of personnel vary; this is true for regular as well as special education. The advocate recognizing a limited teacher or an inadequate individualized educational plan should, of course, press tactfully for modification of the plan.

SUMMARY

The third quarter of this century has brought impressive expansion and improvement in the quality of special education services. For most students with developmental disabilities, appropriate community based programs have become a reality. However, as funding priorities shift, the specter of service reductions is an immediate threat. Parents and professionals must work together to ensure effective services for every child and young person.

JEAN M. ZADIG

LYNN J. MELTZER

REFERENCES

Arter, J. A., and Jenkins, J.: Differential Diagnosis–Prescriptive Teaching: A Critical Appraisal. Arlington, Virginia, Educational Resources Information Center Report 150578, 1978.

Auckerman, R.: Approaches to Beginning Reading. New York, John Wiley & Sons, Inc., 1971.

Chall, J. S.: Learning to Read: The Great Debate. New York, McGraw-Hill, Inc., 1967.

Classroom Computer News. Cambridge, Massachusetts.

Cruickshank, W.: An interview with William Cruickshank by G. Markel. Directive Teacher, Summer/Fall, 1980.

Dunn, L.: Exceptional Children in the Schools. Ed. 2. New York, Holt, Rinehart, & Winston, 1973.

Furth, H., and Wacks, H.: Thinking Goes to School: Piaget's Theory in Practice. New York, Oxford University Press, 1976.

Hammill, D. D., and Bartel, N. R.: Teaching Children with Learning and Behavior Problems. Boston, Allyn & Bacon, Inc., 1975.

Lerner, J.: Learning Disabilities: Theories, Diagnosis, and Teaching Strategies. Ed. 3. Boston, Houghton Mifflin Co., 1981.

Journal of Special Education (whole summer issue). New York, Grune & Stratton, Inc., 1981.

Mann, L.: On the Trail of Progress: A Historical Perspective on Cognitive Processes and Their Training. New York, Grune & Stratton, Inc., 1979.

Mann, L., Goodman, L., and Wiederholt, J. (Editors): Teaching the Learning Disabled Adolescent. Boston, Houghton Mifflin Co., 1978.

Meyen, E.: Exceptional Children and Youth. Denver, Love Publishing Co., 1978.

Resnick, L. B., and Weaver, P. (Editors): Theory and Practice in Early Reading. Hillsdale, New Jersey, Lawrence Erlbaum Associates, Inc., 1979.

Rutter, M.: Helping Troubled Children. New York, Plenum Press, 1975.

Savage, J., and Mooney, J.: Teaching Reading to Children with Special Needs. Boston, Allyn & Bacon, 1979.

Wiig, E., and Semel, E.: Language Disabilities in Children and Adolescents. Columbus, Ohio, Charles E. Merrill Publishing Co., 1976.

Zigler, E., and Valentine, J. (Editors): Project Head Start: A Legacy of the War on Poverty. New York, The Free Press, 1979.

57

Other Specialized Services

57A Coordination of Services

Many professionals involved with child study and care stand ready to dispense relevant services, although their recruitment and coordination may prove challenging. The alignment of diverse professional groups in the instance of assessment is discussed in the section on "The Function of Teams" in Chapter 49A. Consideration will now be given to the continuing or longitudinal management of services (see also Chapter 39A, page 766).

A model of progression can be formulated that focuses on six major life periods and their varying requirements, as in Figure 57–1. The individual exemplified is multiply handicapped and needs significant developmental support. The relative importance of the various services evolves through a predictable series of changes:

In the perinatal period the disabled child may require substantial clinical intervention, with various types of medical assistance (e.g., pediatrics, nursing, physical therapy). Direct support for the parents in accommodating to the situation can be provided by social service and other counselors and assistants.

In infancy and the preschool years the health care related services should begin to modulate and a sequence of developmental supports is undertaken ("early intervention," "stimulation programs"), contributed by specialists in education, physical therapy, occupational therapy, and lan-

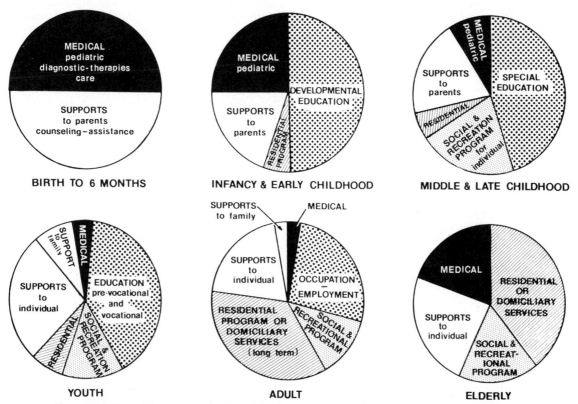

Figure 57–1. Sequential program needs of an individual with mental retardation and other developmental disabilities.

guage therapy, for example. This program is first promoted in home visits and later in a child development center or school. A potential exists for respite care assistance to the family in the home or out of the home.

In the school years the resources of special education bring a wide range of learning experiences, often supported by particular therapies as well (e.g., speech therapy, physical therapy, occupational therapy, music). The other elements are present but less prominent. A new factor is the provision of direct recreational and social fulfillment for the individual, acknowledging his broader personal needs.

In the young adult years the educational offerings include vocational training, as appropriate. Counseling and guidance are now more important for the individual himself, with correlated assistance to the family as well. The residential possibilities may include community homes or projects, particularly if there are pressures in the family or special behavioral management needs.

An adult completes schooling in the usual sense and spends his day activities in a vocational setting (workshop, competitive employment). He also characteristically leaves home, assuming residence at the level of independence his disabilities allow. Recreation and leisure activities are important, as is appropriate guidance in life projects.

For the elderly person the emphasis on vocation decreases, although creative daytime programs are essential. Health care needs may become a significant element in the service package.

In an open society there is obviously no single means of insuring a smooth course through such a diversity of life requirements. For the troubled small infant, a physician or medical facility may serve in a directing role, with the help of social workers. In the early preschool years, community or state agencies often coordinate the services for the young child.

Currently the responsibility of the school district for the child begins at age three to five years. Although an attempt is made to incorporate the major elements in the individualized educational plan, the school does not ordinarily assume responsibility for the whole year, budgets for personnel are tight, and turnover may hamper continuity. State health departments have varying levels of commitment for the guidance of handicapped children (until age 21). In the young adult years other state agencies attempt to provide support, including those devoted to mental health, mental retardation, developmental disabilities, vocational rehabilitation, social services, and welfare. In some individuals with sensory handicaps (blindness, deafness), special state offices may offer critically valuable services. Likewise, for the senior citizen, state and voluntary agencies play a role. Throughout the process, various protection and advocacy groups can be contacted for assistance.

All these factors constitute a complicated network, and individual issues have a proclivity for falling through the cracks. In our culture there is no master system to guide a person with sustained developmental or behavioral special needs. At various periods, from various sources, there may arise an appropriate "administrator," "case manager," "program monitor," "service coordinator," or "advocate," but organizational realities allow regrettable lapses. In the final analysis, it falls to the family of a troubled person to provide the major initiative. In this regard they often become truly expert (although they did not expect to be) in the coordination of resources. Their activism and promotional abilities may be very skilled. It is to be hoped that they can make good professional and consumer group friends along the way who can be supportive. In the later years the involved individual also comes to develop special insights about the world of services; he also needs friends.

ALLEN C. CROCKER
MARIE M. CULLINANE

57B   Social Work

SCOPE OF SOCIAL WORK PROFESSION

The goal of social work is the improvement of the social functioning of individuals, families, and groups. The social worker views the person in the situation and assesses the factors both within the person and within the situation that produce stress or dysfunction and suggests the resources that might contribute to resolution. Historically social work has dealt especially with populations that are economically and socially deprived, but over time its concern has expanded to include all people who experience problems in their interpersonal relationships and in carrying out their social roles and responsibilities. These may be individuals whose functioning is compromised by personality problems or physical and intellectual handicaps or those who are encountering external stresses beyond their ability to manage. Increasingly social work has focused on identifying populations at risk in order to plan how supports could be made available prior to actual breakdown or social dysfunction.

The social worker can intervene through planning, developing, and administering the policies, programs, and services that meet identified human needs. The majority of social workers are involved in clinical activities, providing a range of direct services to individuals, families, and groups with psychosocial problems. Such intervention can be short term, focusing on specific issues of decision making or the resolution of a crisis, or long term, with the goal of helping people work through maladaptive behavior patterns or attitudes that impede functioning. Some individuals and families require continuing support and assistance in dealing with all aspects of their life situation.

PHILOSOPHY AND STRATEGIES OF HELPING

The social work profession places high value on respecting individual differences and a person's right to self-determination. In most situations the prospective client has to indicate a desire for help. Exceptions are protective situations in which the client's behavior is jeopardizing the welfare of a child who is dependent upon him, or in which he is by age or developmental level deemed unable to make decisions for himself. Individuals or families may seek help from a social agency out of concern for themselves or other significant people in their lives.

The social worker clarifies the nature of the request, solicits data about relevant current and historical factors, and obtains some general information about other aspects of the person's functioning and life situation. As a part of this process the social worker may help the client to redefine the scope of the problem. It is also important to find out the kind of help the person anticipates and how congruent this is with resolving the identified problem and what the agency can offer. The social worker makes an assessment of the factors contributing to the person's concerns and recommends a plan that will help. The plan may include a series of interviews to help the person explore his own role in the problem. There may be recruitment of others in the helping process, including additional community agencies, and the provision of information or advice. A helping contract is then negotiated between client and worker as to the nature of the problems to be addressed and the means to be utilized.

The amount of responsibility clients are expected to take depends upon their abilities and the kinds of problems under discussion. An important value of social work is helping individuals develop their personal skills and level of social functioning. Therefore it is important that the individual concerned take as much responsibility as possible in dealing with his problem, leading to acquisition of skills in problem solving and an improved self-concept. Often the problem that brings the person for help is a prototype of other issues he will have to face in the future. In these situations the social worker's role is to provide emotional support, opportunities for ventilation of feelings, help in clarification of the pertinent issues, facilitation of the development of self-awareness, and validation of constructive coping mechanisms. The social worker may be directive or may take a more educational approach in the case of persons who are limited in social awareness, social judgment, or cognitive abilities.

With the client's permission the social worker may intervene directly in the person's life situation by interpreting the needs to others of significance to the patient. Even very capable individuals may require such assistance in fulfilling their special needs or in substantiating the community's responsibilities to develop appropriate services. Advocacy is an aspect of every social worker's functioning, but some agencies and individual workers have this as a specialty. Advocacy involves helping client groups become aware of their rights and helping them negotiate complex or resistant access systems.

In addition to work with individuals, social workers often see family members as a unit, in order to work on specific issues that affect all members. They may also organize and lead group therapy sessions for individuals with common or complementary problems. The particular approach chosen depends upon the needs of the client but also varies according to the opinion of the agency about the origin of the problems and the most effective strategy to resolve them.

EDUCATIONAL PREPARATION AND REGULATION

The typical preparation of the social worker is a master's degree in social work, usually encompassing two years of study with a significant proportion of time spent in supervised practice. Increasingly, different levels of preparation have come into existence, beginning at the junior college and baccalaureate levels. These programs prepare professionals to work in situations in which there is less need for sophisticated assessment skills and more emphasis on offering support and concrete services. Doctoral programs in social work prepare individuals for careers in teaching, research, or more complex clinical roles.

Many states require certification or licensing of social workers. The National Association of Social Workers is also involved in setting standards of professional practice.

ORGANIZATION OF SERVICES

Social workers usually function in the context of an agency program, but increasingly many are engaged in part time or full time private practice. Social agencies can be public or private in their sponsorship. Most private agencies depend to some extent on fees from clients to support their operating costs. National standard setting organizations influence the quality of services provided by agencies.

Services in different parts of the country are organized differently and vary greatly in the commitment they make to clients with psychosocial problems. Often the federal government may influence local programs by setting minimal standards if the community is to receive federal funds to subsidize services. Social agencies usually have areas of specialization in terms of the age range and types of problems they are set up to serve. Prospective clients often have to live within a certain geographical area, usually defined by the sources of financial support that fund the agency. Agencies vary greatly in the kind of commitment they expect from the client in initiating contact and keeping scheduled appointments. There has been experimentation with different strategies to make services more accessible to those who might not be willing or able to make this kind of commitment. Neighborhood based service programs, as well as walk-in or telephone crisis centers, have been developed and publicized. They are prepared to meet immediate needs and also try to serve those who will need more extended help in seeking services.

TYPICAL SOCIAL AGENCIES

Many agencies serve the needs of children and adolescents either directly or indirectly. Most communities of any size have a family agency. This type of agency serves individuals with personal problems, offers marital counseling, and assists those with concerns about parenting. They may also organize support groups and educational programs devoted to common life crises or social roles. Large communities may have a number of agencies of this type, sometimes sponsored by different religious denominations.

Another agency common to most communities is a child welfare agency. If there is none at the local level, there is invariably a publicly sponsored child welfare program at the state level. These agencies tend to specialize in protective services for children who are neglected or abused. They also work with families who request placement of their children because they believe that they cannot cope with their children's needs. The goal of the agency program is to help the parents work out the problems that are interfering with satisfactory parenting, so that the family can remain together as a unit. These agencies often recruit and supervise foster homes or alternative care facilities.

Placement of children, however, is used as a last resort. Once children and their parents are separated, reconciliation becomes more difficult. Indefinite foster care has proved to have detrimental effects on the welfare of children. Public policy now puts time limits on the use of this as an option, and if the parents are unable to reunite the family, steps are taken to release the child for adoption. Although social workers have a social and quasilegal sanction to offer unrequested services to families who are presumed to be neglecting or abusing their children, they have no authority to separate children involuntarily from their parents without involving court action.

Child welfare agencies also recruit adoptive homes for children who are free for permanent placement. Once this service was restricted to normal infants who were easy to place, but now that social values and practice have changed, agencies are being more creative in developing other needed services. They often recruit permanent homes for handicapped children and those who used to be considered impossible to place. They also subsidize adoptive placements when finances are the barrier to a permanent arrangement. They are engaged in developing more creative programs for adolescents who are pregnant and need counseling about their future.

SOCIAL WORKERS AS TEAM MEMBERS

In addition to staffing social agencies, social workers are employed in other organizations and institutions whose clients have, or are prone to have, psychosocial problems. Mental health clinics, residential treatment centers, juvenile correction facilities, services for developmentally disabled individuals, health centers, and hospitals usually have social workers as an integral part of the program. The extent of the social work component depends on the funding of the program and its philosophy of responsibility for meeting the psychosocial needs of its clientele.

The particular role of the social worker also varies. In most instances the social worker works with the family or the affected individual or with the community whence he comes. The social worker's function may be restricted to routine arrangements centered on application for services to meet the client's needs. More often the social worker has responsibility for making individualized assessments of the family and community resources and for contributing to the treatment program by working with the identified problems. In some programs social workers provide the basic treatment services for the affected individuals.

Many community systems that serve normal children have social workers available as consultants or special resources. There are social workers in most public school systems. Frequently the social worker is responsible for family assessments of children who are presenting behavior problems or exhibiting academic or learning problems at school. The social worker is then responsible for helping the family locate and use appropriate community services. In some school systems the social worker may have continuing therapeutic responsibilities directly with the identified children or their families. Nursery schools and day care centers, particularly those that specialize in low income or high risk populations, usually have social workers on the staff. More often than not they may interview all families at the point of intake to determine the soundness of the family's functioning, and may offer support groups dealings with parenting as well as individual counseling when it is needed. In these situations social services are made available to people who might never be motivated to seek help at social agencies.

ANN P. MURPHY

REFERENCES

Klenk, R., and Ryan, R.: The Practice of Social Work. Belmont, California, Wadsworth Publishing Company, 1970.
Meyer, C.: Social Work Practice. New York, The Free Press, 1976.
Siporin, M.: Introduction to Social Work Practice. New York, Macmillan Publishing Co. Inc., 1975.

57C   Physical Therapy

Since the beginnings of the profession, in the early years of this century, physical therapists have been involved in the care of children, including those with developmental disorders. This work has become more widespread in the past two decades, and an increasing number of physical therapists are now exclusively engaged in work with children. Practice settings range from hospitals, clinics, and institutions for retarded persons to rehabilitation centers, homes, and public schools.

The entry level educational requirement for the physical therapist is presently the baccalaureate degree, but this will move to postbaccalaureate level by 1990. In addition to the basic science courses and studies of the effects of disease and trauma, the physical therapist's preparation also includes course work in human development and its disorders, particularly in relation to the neurophysiological and developmental aspects of movement. Recently there have been a number of graduate programs established at the master's degree level that offer study in specific areas, including pediatrics, to broaden the therapist's preparation for clinical, educational, and research functions.

The American Physical Therapy Association has included a Section on Pediatrics since 1974, which as grown to more than 2000 members. The Association in 1981 approved a Specialization in Pediatrics for clinical practitioners, which will be further defined and regulated by a council, which was also formed in 1981. Competencies in a variety of treatment areas in pediatrics, including musculoskeletal, cardiopulmonary, neuromuscular, and developmental disabilities, have been formulated, and these will form the basis for the council's deliberations.

The physical therapist traditionally has evaluated and treated patients only following referral from a physician or other licensed medical practitioner. This policy is presently being reviewed by the Association with the purpose of developing guidelines for having evaluation by the physical therapist be an entry point into the health care system. These guidelines will continue to emphasize the communication between the therapist and the physician, which has been a very important part of the referral process.

The role of the physical therapist will be described by utilizing examples of practice in a variety of situations.

SCREENING AND ASSESSMENT BY THE PHYSICAL THERAPIST

The physical therapist for several years has participated in posture screening programs, usually in cooperation with orthopedic surgeons, nurses, and physical educators in public school settings, for the purpose of early detection of deformities such as scoliosis. Physical therapists and occupational therapists are now participating in early screening of motor development in public schools, as well as in high risk infant follow-up programs.

Assessment is a part of every treatment program, but may also be a part of an interdisciplinary evaluation to determine the child's abilities and disabilities, as described in Chapter 49C.

TREATMENT

The greatest proportion of the practice of physical therapy with children is devoted to the treatment of acute or chronic conditions with the overall purposes of gaining optimal function in posture and movement and preventing, to the degree that this is possible, development of deviant postures and movements.

Treatment programs may include exercise and other movement techniques that utilize sensory cues (tactile, kinesthetic, proprioceptive, visual) to facilitate movement. The utilization of sensory factors, as well as behavioral methodology, has made it possible to treat the child without his active cooperation and therefore to extend treatment to very young children or to those who are significantly retarded. Play and developmentally appropriate toys and games are important components of treatment with all young children, particularly those who are new to the process of therapy and those who are receiving therapy for an extended period.

Treatment for children with cerebral palsy has evolved over time as knowledge of the functions of the central nervous system has expanded. The most widely utilized treatment approach has been that of Karel and Berta Bobath (1980), developed in the past 35 years and based upon a hierarchical concept of central nervous system function. This method involves the inhibition of abnormal movement patterns and the facilitation of postural reactions and normal movement, with emphasis on the function of the child in daily living activities. Other aspects of methodology are described in the volume edited by Pearson and Williams (1972) and in the chapter by Harryman in the book by Johnston and Magrab (1976).

Treatment during infancy ("early intervention") is being implemented increasingly. It draws on the concept of plasticity of the central nervous system, as well as the belief that development of deviant postures and movements can be alleviated best during that period (Connolly et al., 1980).

Other types of treatment programs include teaching the use of adaptive equipment and assistive devices, gait training for severely retarded children using behavior modification, and training in the use of biofeedback devices in gait or manipulation when the child has specific movement deficiencies.

The approach to the individual child is based upon the findings of an evaluation as well as his response to treatment. This response is evaluated periodically. The importance of re-evaluation cannot be overemphasized, particularly in the care of children who may require long term treatment. There may be periods in the child's life when physical therapy should be emphasized, and other times when it can be given a relatively minor role so that other aspects of the child's development and learning can take precedence. The concurrent provisions of more than one physical therapy program with essentially the same goals is to be discouraged.

HOME TREATMENT PROGRAMS

The optimal management of motor disorders involves the transfer of some aspects of treatment to the home. Careful consideration must be given to providing a clear understanding of the content and goals of the program to the family, and to the child himself if he is able to understand. The time commitment necessary on the part of the child and family and their motivation to provide the necessary emotional and time resources to the program must be considered. Home programs should be discussed and revised at regular intervals to meet the changing requirements of the child over time as well as to discuss their current feasibility. This can be provided in a supervising clinical center, or by therapists who visit in the home from a base such as a Visiting Nurses Association.

CONSULTATION BY THE PHYSICAL THERAPIST

The physical therapist increasingly is being called upon to provide consultation in a variety of situations. The integration of handicapped children in public schools has required instruction to school staffs regarding functional capabilities of individual children, positioning and transfer techniques for physically handicapped persons, and adaptations of furniture and equipment. In many practice environments there is a need for consultation with other professionals, such as occupational therapists, regarding joint programing for children with learning disabilities or feeding problems, pediatricians regarding atypical motor development of infants, or prosthetists and orthopedic surgeons regarding utilization of prosthetic devices.

RESEARCH IN PHYSICAL THERAPY

Research and postgraduate education are being fostered in the profession under the auspices of the Foundation for Physical Therapy. The Foundation awards traineeships for graduate education at the doctoral level and research grants to individual investigators. Research emphasis has been on exploration of treatment methods, as well as kinesiology. There are studies in progress that will evaluate the effects of treatment in developmental disorders. There is, in addition, a need for information about the natural history of movement disorders in children. This type of longitudinal data will be very important in the evaluation of treatment procedures.

FUTURE GOALS

Physical therapy, although a relatively young profession, has taken some important steps in its self-definition in the past two decades. This definition, it is hoped, will retain the best of its original components but will also call on the flexibility that has been characteristic of the profession in the past to respond to the changing needs of children and families and the health care system itself.

ALICE M. SHEA

REFERENCES

Bobath, K.: A Neurophysiological Basis for the Treatment of Cerebral Palsy. Clinics in Developmental Medicine, No. 75. London, William Heinemann Medical Books Ltd., 1980.

Connolly, B., Morgan, S., Russell, F., and Richardson, B.: Intervention with Down syndrome children: follow-up report. Phys. Ther., 60:1405, 1980.

Harryman, S. H.: Physical therapy. In Johnston, R. B., and Magrab, P. R. (Editors): Developmental Disorders: Assessment, Treatment, Education. Baltimore, University Park Press, 1976.

Pearson, P. H., and Williams, C. E. (Editors): Physical Therapy Services in the Developmental Disabilities. Springfield, Illinois, Charles C Thomas, 1972.

57D   Occupational Therapy

Occupational therapy offers a service designed to overcome or reduce the limitation that disability places on an individual's independence in everyday activity. "Occupation" in occupational therapy means "being productively occupied," not "employment," and refers to roles, habits, and skills associated with the performance of activities of daily living, play-leisure, and school-work activities. Occupational therapy was founded in the belief that occupation had healing qualities for individuals with debilitating physical or mental illnesses. An early definition stated that occupational therapy was "a method of treatment by means of instruction and employment in productive occupation," the objectives of which were "to arouse interest, courage, and confidence; to exercise the mind and body in health activities; to overcome functional disability; and to reestablish a capacity for industrial and social usefulness" (Dunton, 1940). The scope of occupational therapy has greatly expanded over the last 60 years, but the philosophical base remains the same. "The idea that, by engaging in occupation designed as therapy, man can restore, increase and maintain his ability as an occupational creature is the foundation of occupational therapy" (Keilhofner et al., 1980).

The medium of occupational therapy is purposeful activity. Activities used with children include the activities of daily living, play and school activities, and, with older children, manual arts and crafts and prevocational tasks. An activity may be an end in itself, e.g., the development of a specific daily living skill, such as making a bed from a wheelchair or cutting meat with one hand, but more often is a means of restoring or improving functions, such as strength or range of movement, eye-hand coordination, ability to learn, work habits, and ability to get along with others. Activities used in occupational therapy are goal directed, involve active participation, and generally require a purposeful motor response.

Treatment with children follows a developmental progression, which considers the development of the child's healthy area of performance as well as treatment of dysfunctional performance. Whenever possible, activities and their settings are planned in accord with the child's chronological age or highest functioning level as well as the developmental level of the more dysfunctional components of performance.

Therapy may be administered in one or more forms, each based upon a period of evaluation. Direct therapy may be given by the occupational therapist or by an occupational therapy aide under the direction of the therapist. Planning and supervising home treatment programs for parents or other caretakers can include teaching parents specific therapeutic techniques, adapting home environments, and planning ways in which developmental activities can be incorporated into routine family life. Finally, consultation may be provided to other professionals, e.g., teachers in public schools or other therapists in transdisciplinary programs.

The disorders affecting children commonly referred to occupational therapy programs are cerebral palsy, learning disability, congenital orthopedic and neuromuscular conditions, developmental delay, mental retardation, juvenile rheumatoid arthritis, and primary behavior disorders.

There are two approaches to treatment in occupational therapy—first, training in specific skills of everyday living, and second, building better foundations for later skill development. These approaches are used singly or in combination, depending upon the age of the child and the course of the disability. For example, in younger children with neurological impairment, the emphasis in treatment is on the development of perceptual or motor foundation abilities, whereas with orthopedically handicapped children of all ages, the emphasis is on specific developmental skills.

PROFESSIONAL TRAINING

Entry into the profession of occupational therapy requires the successful completion of an accredited educational program and the passage of a national certification examination. Fifteen states also require licensure for practice, and licensure bills are pending in other states. Educational programs are accredited by the American Occupational Therapy Association (AOTA), on the basis of standards developed by the AOTA Commission on Education in collaboration with the American Medical Association, and include both four or five year baccalaureate programs and two year master's degree programs. The majority of occupational therapists graduate with bachelor's degrees (89 per cent in 1981). The advisability of requiring that the entry level of education for all occupational therapists be at the master's degree level has been under consideration by the Association, but concerns about the continuing national shortage of therapists and the costs of service delivery and of education led to the decision that now was not the time for such a change.

Occupational therapy services are also provided by two other categories of personnel, the certified occupational therapy assistant (COTA) and the occupational therapy aide. The occupational therapy aide receives on the job training, but the standards of practice and of the two year training program of the occupational therapy assistant is maintained by the AOTA. Both, however, work under the supervision of occupational therapists.

Postprofessional graduate programs offer advanced professional content in specialized areas of practice. Over 10 per cent of the occupational therapists certified at the bachelor's level return for master's degrees in occupational therapy or in related fields, but the percentage is considerably higher in pediatric occupational therapy, because a master's degree is often required for practice with children. More than half the 20 programs offering postprofessional master's degrees and one of the two established doctoral programs offer specialized training related to the treatment of handicapped children.

A master's degree is almost universally required for academic teaching, and universities are increasingly requiring a doctoral degree. A master's degree is also commonly required for administrative and clinical teaching positions and for specialized areas of practice. Therapists with doctoral degrees are primarily in academic positions, but the numbers in program administration, research, and private practice are rapidly increasing.

DEVELOPMENT OF SPECIFIC SKILLS

The occupational therapist can foster the earlier development of specific skills by designing and providing adapted equipment, teaching special techniques, and providing programs for developmental training. E. C. High has compiled an excellent resource guide to habilitative techniques and equipment for positioning and seating, feeding, dressing, and hygiene. The guide is intended for cerebral palsied persons but can provide information useful for other handicapping conditions as well. Other useful references include a manual of techniques and aids for the activities of daily living in mentally retarded children (Copeland et al., 1976), activities for children whose developmental level is from 0 to 36 months (Schafer and Moersch, 1977), and the valuable guide to caretaking for the young cerebral palsied child by Finnie (1975).

Before beginning skill training, the therapist determines whether the child has the necessary skills or whether the skills can be learned without eliciting maladaptive postures or movements. For example, to begin self-feeding with a spoon, a multiply handicapped child must have the following motor abilities as well as to be able to anticipate the spoon when being fed: he must be able to suck from a spoon and swallow, flex his head with good stability, and be able to lift his hand to his mouth without eliciting primitive tonic neck reflexes. If the child does not have these prerequisite skills, the occupational therapist works with the speech therapist, the physical therapist, and the child's caretaker to develop a program for the development of oral motor and postural abilities. Other prerequisite motor abilities, however, can be compensated for by the use of adapted equipment. Specially designed chairs can support the child who does not have trunk stability, and a swivel spoon with a built up handle can compensate for immature grasp and poor wrist and forearm control. With aids such as these, self-feeding with a spoon can be initiated following the usual developmental sequence (Coley, 1978). Training in the activities of daily living is carried out in collaboration with the child's primary caretaker so that practice of the skills can be incorporated into the daily routine.

DEVELOPMENT OF FOUNDATIONS FOR SKILL

To build foundations for later skill development, the occupational therapist plans an activity program designed to improve basic fine motor skills, basic whole body skills, perceptual abilities, and sensory integration. Treatment of disability in these areas generally follows a developmental sequence.

Development of Basic Fine Motor Skills. Games, toys, and manual arts and crafts are used by the occupational therapist in the development of manipulative abilities. A developmental sequence is usually followed, both in general patterns of motor development (i.e., gross to fine, proximal to distal) and in specific basic hand skills (i.e., reaching, grasping, releasing, pincer grasp, and supinate grasp). Developmental training in bilateral use of the hands begins with reaching to the midline, progresses to symmetrical and then asymmetrical hand use, and includes the difficult ability to use the hands together behind the neck or the back. Treatment to improve hand skills may include increasing muscle strength or range of motion and the development of speed and accuracy in the use of the upper extremities.

A special area of manipulation training is in the use of prosthetic and orthotic devices. Both pre-prosthetic and prosthetic types of training are provided for limb-deficient infants and children (Clark and Patton, 1980). Orthotics, such as free and friction feeders, and devices that support and control movement at the elbow and shoulder require protracted training for their use. The occupational therapist also designs or makes hand and wrist splints to improve function.

Development of Basic Whole Body Skills. Head control and trunk stability are prerequisites to the optimal use of the upper extremities, and mobility greatly increases the number of developmental play experiences a child can have. These functions are therefore of concern to the occupational therapist as well as to the physical therapist, and they work together in planning adapted equipment for stability in sitting, standing or lying, and wheelchairs and scooters for mobility, as well as in providing activities that will foster development (Howison et al., 1978; Williamson, 1981).

Development of Perceptual Functions. Both the use of the hands in manipulating objects or tools and the movement of the body in the spatial environment can be impaired by perceptual deficits. The occupational therapist is concerned with the organization of visual, tactile, and proprioceptive percepts in the guidance of motor activity. Treatment may be planned to improve eye-hand coordination and the ability to plan and sequence movements, or to deal with the spatial relationships of body parts and the body to objects. The therapist works with the teacher in the development of paper and pencil constructive and visual spatial abilities.

Sensory Integration. A special approach to the development of perceptual organization is sensory integration therapy. Sensory integration therapy was developed for use with children with learning disabilities but more recently has been adapted for the treatment of autistic, mentally retarded, emotionally disturbed, and some types of language disturbed children. Sensory integration is defined as the process of structuring sensory experience to permit an adaptive response. Emphasis is on input from the tactile, proprioceptive, and vestibular systems and on the gross motor activities that stimulate these systems. Sensory integration theory holds that sensory stimulation accompanied by the demand for an adaptive motor response leads to sensory integration, and that sensory integration in turn influences the maturation of postural responses, motor planning, visual spatial skills, and the ability to learn academic skills. The relationship of sensory integration therapy to academic improvement is still controversial, but research indicates at least an indirect influence on learning in some children (Ayres, 1978).

Development of Psychosocial Skills. The occupational therapist is also concerned with psychosocial development and works with other professionals to insure that habits as well as skills of self-care are developed. In program planning, the development of self-esteem, feeling of competence, peer relationships, and coping behavior is incorporated into the overall occupational therapy program for developing skills.

For a child whose primary problem is psychiatric, the goals are similar but the process differs. The occupational therapist combines the traditional theoretical frameworks of psychiatry with the use of activity and the interactional processes that occur during an activity. In normal development many interactional and ego skills are learned in conjunction with age appropriate activities. The occupational therapist structures constructive and creative activities first to meet the child's level of self-identity and interpersonal skills and then to demand a higher level skill.

ANNE HENDERSON

REFERENCES

Ayres, A. J.: Learning disabilities and the vestibular system. J. Learn. Disab., 11:18–29, 1978.

Clark, S. D., and Patton, J. G.: Occupational therapy for a limb deficient child: a developmental approach for treatment planning and selec-tion of prostheses for infants and young children with unilateral upper extremity limb deficiencies. Clin. Orthop., 148:47–54, 1980.

Coley, I. L.: Pediatric Assessment of Self Care Activities. St. Louis, The C. V. Mosby Company, 1978.

Copeland, M., Ford, L., and Solon, N.: Occupational Therapy for the Mentally Retarded. Baltimore, University Park Press, 1976.

Dunton, W. R., Sr.: Occupational therapy. In Barr, D. P.: Barr's Modern Medical Therapy in General Practice. Baltimore, The Williams & Wilkins Company, Vol. 1, 1940.

Finnie, N. R.: Handling the Young Cerebral Palsied Child at Home. New York, E. P. Dutton & Co., 1970.

Howison, M. V., Perella, J. A., and Gordon, D.: Cerebral palsy. In Hopkins, H. L., and Smith, H. D. (Editors): Willard and Spackman's Occupational Therapy. Philadelphia, J. B. Lippincott Company, 1978.

High, E. C.: A Resource Guide to Habilitative Techniques and Aids for Cerebral Palsied Persons of All Ages. (Available from Job Development Laboratory, Rehabilitation Medicine, George Washington University, Room 420, 2300 Eye Street, N.W., Washington, D.C. 20037.)

Keilhofner, G., Burke, J., and Igi, C.: A model of human occupation. Part 4. Assessment and intervention. Am. J. Occup. Ther., 34:777–788, 1980.

Schafer, D. S., and Moersch, M. S. (Editors): Developmental Programming for Infants and Young Children. Ann Arbor, University of Michigan Press, 1977.

Williamson, G. G.: Pediatric overview. In Abreu, B. C. (Editor): Physical Disabilities Manual. New York, Raven Press, 1981.

57E Nursing

MODELS OF PRACTICE

Variability in academic programs prepares nurses for different levels of function. Nationally there is a lack of uniformity in the terminology used to define a nurse. There is also a lack of clarity in expectations for nursing practice roles. For the purpose of this discussion a registered nurse is understood to be licensed to practice professional nursing, holds ultimate responsibility for direct and indirect nursing care, and is a graduate of an approved school of professional nursing. Responsibilities include nursing care, health maintenance, teaching, and counseling.

All registered nurses are prepared as generalists at the basic or baccalaureate level, and utilize a theoretical base to guide their practice.

SPECIALIZATION IN NURSING

Beyond basic preparation the nurse may develop skill and knowledge in a particular area of practice or take further professional study. These nurses are recognized by specific titles.

The pediatric clinician, by virtue of extended experience and continuing education, has a high degree of expertise in a specific area of pediatric nursing. Examples are nurses working in surgical intensive care or dialysis units.

A pediatric nurse practitioner is a registered nurse with preparation in a specialized educational program. The duration of programs varies from three months to one year. This preparation at present may be in the context of a formal continuing education program, a baccalaureate nursing program, or a master's degree program. Certification is now required to practice in this role. The pediatric nurse practitioner has advanced skills in assessment of physical and psychosocial health-illness status. Preparation enables her to function in primary pediatric care and other child health areas. Emphasis is placed upon preventive aspects of care, health maintenance, treatment of routine pediatric illnesses, and health screening. These

Table 57–1. CATEGORIES IN LICENSURE

Title	License	Educational Preparation	Usual Length of Training
Nursing assistant (aide)	None	Job training	Months
Licensed practical nurse	LPN	Training program	1 year
Registered nurse	RN	Diploma program	2–3 years
		Associate degree program	2 years
		Baccalaureate degree program	4 years

nurses have made a major contribution in the delivery of well child care and in increasing health care to underserved groups in inner city and rural areas and in school health programs. The demonstration of safe, high quality health care by the nurse practitioner, as well as strong public acceptance, has contributed to the continued use of the pediatric nurse practitioner role in a variety of practice settings.

The clinical nurse specialist holds a master's degree with a concentration in a clinical area of nursing. She has advanced knowledge, skill, and competence in a specialized area. She has the background for making clinical judgments and providing consultation. She frequently teaches skills to other nurses while giving direct care to clients. She may also carry out clinical research to evaluate the effects of nursing methods and interventions for the improvement of care.

NURSING ROLES IN VARIOUS SETTINGS

Hospital Inpatient Setting. The nurse directs and manages the total care plan for the child during hospitalization. Activities range from carrying out highly technological treatments to the broadly supportive aspects of physical and emotional care. She monitors the patient's condition and adjusts care at all times throughout hospitalization. She is the consistently available contact person for the family, aiding their understanding of their child's needs and progress. In preparation for discharge from the hospital she coordinates with the family and community services for continuing care of the child at home.

Other Settings. As outlined in Table 57–2, the nurse participates in a wide variety of environ-

Table 57–2. NURSING ROLES

Setting	Function
Primary care	
Physicians' offices	Periodic health assessment
Neighborhood health clinics	Immunizations
Health maintenance organizations	Collaborates in management of care during illness
Well baby clinics	
Specialty clinic	Monitors and manages programs of specialized care
Home	Assesses the child's condition and family situation
	Provides direct care
	Provides direction of care
School and camp	Conducts screening programs
	Provides health services
	Provides health education
	Consults with educators regarding student health issues

mental settings outside the hospital—primary care settings, specialty clinics, homes, schools, and camps. In these locations the nurse may be one of the few consistent care givers to provide continuity over time and thus increase compliance with medical and other treatment regimens.

NURSING FUNCTIONS

Wherever health services are delivered, the first personal contact of the consumer may well be with a nurse. She is relied upon to make the initial and continuing assessment of the child and family, independently or in collaboration with members of other disciplines. She provides direct care, teaches family members to carry out treatment programs, and provides supportive supervision. She assists in times of stress, promotes positive family adaptation and self-reliance, and is a counselor regarding health issues. The scope of her consideration encompasses concern for all the children in the family.

In order to ensure continuity of care and to aid the family in dealing with systems, the nurse coordinates services within her own facility and with personnel of other community agencies. She is an advocate for the child and family at both the individual and the community level.

The nurse has broad opportunities and capabilities as an educator. She is constantly teaching the family or the child in such areas as hygiene, nutrition, sexuality, developmental stimulation, and behavior management. Teaching may involve interpretation of the health plan in understandable terms or mapping out complex medical regimens in ways that are least traumatic and most developmentally appropriate. In small group teaching of such topics as family planning, substance abuse, and parenting concerns, she assists family members to become knowledgeable so that they may make choices in health behavior in their own situations. The nurse is active in community-wide planning for improvement of health services for families and children.

New and expanded roles have broadened the scope of pediatric nursing and have led to a collaborative practice model of health care delivery for children. In all settings and activities related to child health, nursing plays a special and critical role.

EUNICE SHISHMANIAN
CHRISTINA BROWNE

REFERENCE

A statement on the Scope of Maternal and Child Health Nursing Practice. American Nurses Association Publication MCH-10, 1980.

57F   Recreation, Art, Music, and Dance Therapy and Adapted Physical Education

Each of these fields provides a broad based program for healthy individuals as well as a specialized therapeutic modality for those with special needs. Each modality has an important role to play in the humanistic approach to assessment and therapeutic intervention. Each field requires professional preparation at an accredited institution of higher education and offers either voluntary or mandatory registration or certification of personnel; in many cases certification is a criterion for employment in the chosen profession.

The following sections provide an introduction to the inter-relationship and value of these specialized areas. A listing at the end provides a representative sample of professional organizations and national service programs from which more definitive information may be obtained.

TERMINOLOGY AND PHILOSOPHICAL FOUNDATIONS

Frequently there are as many definitions for a term as there are individuals who care to define it, and so it is with the fields of specialization contained in this section.

RECREATION AND THERAPEUTIC RECREATION

Joseph Lee, the recognized "Father of Recreation," in *Play in Education* (1917) indicates that "play is growth," "play trains for life," and "play is purposeful." Play, although usually self-directed, is but one of the many activities within the realm of the diversified field of recreation.

The *American Heritage Dictionary* provides the following definitions: "recreate—to impart fresh life to, refresh mentally or physically"; "recreation—refreshment of one's mind and body after labor through diverting activity, play"; "therapeutic—having healing or curative powers gradually or methodically ameliorative"; "therapy—the treatment of illness or disability." To state it simply, recreation is the worthy use of leisure time.

Many professionals consider the following criteria essential for an activity to be considered recreational: the activity must be engaged in during leisure time; it must be worthwhile, provide a diversion, be enjoyable, and be voluntary. Recreation is essential for a balanced life.

Although many definitions of therapeutic recreation have been cited in texts over the years, the most commonly promulgated statement, which continues to appear even in the most recent texts

(Chubb and Chubb, 1981), comes from an unpublished article prepared by a task force and presented at the Ninth Regional Institute on Therapeutic Recreation at the University of North Carolina in 1969. Therapeutic recreation is defined as ". . . a process which utilizes recreation services for purposive intervention in some physical, emotional and/or social behavior and to promote the growth and development of the individual." Frye and Peters (1972) further emphasize recreation and therapeutic recreation's dual concepts and usages: "The word 'therapeutic' refers to a healing or curative process or again, the term may be used in a general, broad sense to describe anything of a positive or beneficial nature. Understood in this light, all desirable recreation may be looked upon as therapeutic, and all recreation assumed to be dedicated to therapeutic recreation as the basic professional commitment."

Recreation therapy is considered a specific clinical service:

This service is a specific evaluation and treatment process carried out by a trained clinician with a high level of clinical knowledge and expertise. . . . It should be noted that the object of the activities used is not necessarily to teach skills, but rather to use the cognitive, affective, and sensory-motor qualities inherent in each activity. This phase of the treatment program is best characterized by the word "therapy," not the word "recreation." Activities of this nature are the therapist's disability management tools. . . . Recreational therapy is a treatment modality which does not duplicate the traditional therapies of the medical model, but increases the depth of patient treatment service. . . . Therapeutic recreation, the second component of the Recreation Therapy Department's service continuum, is a service which has several different goals. These objectives include the promotion of appropriate patient adjustment to the hospital setting, the provision of opportunities for creativity, the development of self-esteem, self-expression, feelings of accomplishment, emotional well-being, social opportunities, and other more general, but equally significant, objectives. It is a process which provides for individual as well as group leisure experiences (Parker and Downing, 1981).

Rehabilitation, education, and recreation are identified by Gunn and Peterson (1978) as the three domains within therapeutic recreation services. The range of services of therapeutic recreation, according to O'Morrow (1976), include activity programing, leadership and instruction, counseling (recreation or leisure), administration, facilities, equipment and supplies, consultation, education and training (leisure education), and research.

ADAPTED PHYSICAL EDUCATION

Adapted Physical Education is a program of developmental physical activities specifically de-

Figure 57–2. Children in supportive therapeutic and leisure activities. *A,* A music session in a special education program combines proper listening and rhythmic skills needed in a child's development. *B,* Gymnastics (here with a trampoline) helps the special student develop coordination and muscle tone. *C,* Through a qualified sailing program, a student is able to learn the proper techniques of boating, while at the same time having a worthwhile leisure activity. *D,* Camping can be used as a problem solving activity, developing self-confidence and individual enjoyment. (Illustrations provided by Richard Crisafulli, Director of Physical Education and Recreation, St. Coletta Day School, Braintree, Massachusetts.)

signed for each individual, contributing to the cognitive, affective, and psychomotor aspects of behavior. The individual educational plan developed for the child with special needs must include regular physical education, and adapted physical education when appropriate. In order to provide additional lifetime skills and enhancement of the total development of the child, therapeutic recreation often should also be prescribed. Today the most widely accepted definition for adapted physical education comes from the Committee on Adapted Physical Education of the American Association for Health, Physical Education, and Recreation:

Adapted physical education is a diversified program of development activities, games, sports, and rhythms suited to the interests, capacities, and limitations of students with disabilities who may not safely or successfully engage in unrestricted activities of the general physical education program (Crowe et al., 1981).

Recent state and federal laws have mandated education for all handicapped persons, accessibility by removal of architectural barriers, and the right of handicapped individuals to participate in and benefit from programs. Federal legislation (PL 94–142) clearly defines physical education (the only curriculum area specifically identified) as an essential program and mandates its inclusion in the educational curriculum. The prescription for adapted physical education and recreation therapy must clearly meet the needs of each individual. Implementation of the program must be under the leadership, direction, and supervision of qualified adapted physical education instructors and recreation therapists. Assignment to specialized physical education programs or recreation therapy activities does not remove the necessity for the early physical education program (Crowe et al., 1981).

ART THERAPY, DANCE THERAPY, AND MUSIC THERAPY

The American Art Therapy Association, Inc., defines art therapy as follows:

Art therapy is a human service profession. Art therapy offers an opportunity to explore personal problems and potentials through verbal and non-verbal expression and to develop physical, emotional and/or learning skills through therapeutic art experiences. . . . Therapy through art recognizes art processes, forms, content, and associations as reflections of an individual's development.

The use of art as therapy implies that the creative process can be a means both of reconciling emotional conflicts and of fostering self-awareness and personal growth. The benefits of art therapy experiences are applicable to populations with special needs. Art therapy, like the other creative arts, therapies, and therapeutic recreation, may be primary, parallel, or adjunctive therapy in psychiatric centers, clinics, community centers, nursing homes, drug and alcohol treatment clinics, schools, half-way houses, prisons, developmental centers, residential treatment centers, general hospitals, and other clinical, educational, and rehabilitative settings.

The American Dance Therapy Association defines dance therapy as "the psychotherapeutic use of movement as a process which furthers the emotional and physical integration of the individual." The philosophy of dance therapy is further clarified by the Association as follows:

It espouses a holistic view of the individual, recognizing the complex interaction of psyche and soma in sickness and health. . . . dance is the most fundamental of the arts, involving as it does one's body—one's self. It is an especially intimate and powerful medium for therapy, and one that is universally applicable. It engages the total person and provides a vehicle for self-expression, understanding and growth for individuals of diverse needs. . . . Its benefits can be applied to persons in many different treatment settings.

The American Association for Music Therapy promotes the following definition:

Music therapy is the use of music as a tool to effect behavioral, physical, cognitive, and emotional changes through treatment, rehabilitation, education and training of adults (including the specialty of geriatrics) and children with physical, mental and emotional disabilities.

The National Association for Music Therapy, Inc., states that:

Music therapy . . . is the use of music in the accomplishment of therapeutic aims: the restoration, maintenance, and improvement of mental and physical health. It is the systematic application of music, as directed by the music therapist in a therapeutic environment, to bring about desirable changes in behavior. . . .

CORRELATION AND INTEGRATION

Dudley Allen Sargent, a man well ahead of his time, led the way for preventive health care and modern rehabilitation through physical training based on individual and group physical education and gymnastic activities. In his text, *Physical Education* (1906), Sargent enumerated the "aims of physical training under four general heads: hygienic, educative, recreative and remedial." A review of the literature provides a striking similarity of the current philosophy of adapted physical education and therapeutic recreation to the pioneering efforts of Sargent. These goals are also basic to the "activities of daily living" programs

of the more traditional therapies. Today there are more commonalities than differences in the roles of the recreation therapist and the adapted physical educator, since the delivery of service is no longer restricted to the traditional environments of hospitals or schools.

Multidisciplinary recreation programs, designed specifically for the child, not only enrich the individual's life and increase emotional growth, but also enhance his creativity and appreciation for the cultural and performing arts. Venting pent-up anger, fear, or anxiety provides a healthy outlet when it is accomplished through recreational programs. Participation in these activities in the out-of-doors provides additional ways of correlating and integrating worthwhile and enriching activities with nature. The natural surroundings provide an ideal environment for developing an appreciation of the world. The impressionable young person may develop a sense of values, a conservation consciousness, and an appreciation for the creation and beauty of the earth. Through these expressive programs, children, most particularly those who are developmentally disabled, can achieve a higher level of functioning and move closer to and be in tune with nature. The need for these values and appreciations is more critical today than ever before.

Music and dance provide such a diversity of programs that, whether in active participation or passive involvement, they have tremendous appeal to all people. When combined, they provide a dynamic positive force for wellness (Kraus, 1977). Free expression and creative movement provide wholesome outlets for all people, but have a more significant value as a form of nonverbal expression for individuals with emotional stress and for psychiatric patients. Music, dance, and other activities in the broader field of recreation provide unique opportunities for the enhancement of desirable interpersonal relationships and social dynamics (Kraus, 1977). In addition to providing beneficial programs, the therapists also furnish psychodiagnostic tools for the assessment of an individual's status and well-being.

THE TEAM APPROACH: INTEGRATED SERVICES—HOLISTIC ISSUES

Throughout recorded history philosophers and leaders of the various cultures have proclaimed the importance of dealing with the whole person. One can find innumerable quotations from famous individuals that emphasize the importance of recreation for a balanced life. Juvenal's observation that "a sound mind can only exist in a sound body" has been carried down through the ages (Jenny, 1955). It was also recognized that recreational activities, when integrated with other activities, provide unlimited opportunities for the development of full potential, of the total person. During the Golden Age of Greece, Plato wrote in *The Republic*, "Music and gymnastics and other recreational activities will minister to better natures, giving health to soul and body." The mind, body, and spirit, as held so dearly in the Greek tradition, is today the focal point of many rehabilitation centers, educational, and recreational agencies.

Sargent set the pace for sound health through preventive medicine. He was truly an advocate and pioneer in the humanistic and holistic concepts endorsed today (Makechnie, 1979):

> Our chief province should be to make the weak strong, the crooked straight, the timid courageous. . . . Let us give courage to the timid, strength to the feeble, grace to the awkward, and hope to the despondent (Sargent, 1906).

Most adults are physically unfit and, because of the work ethic, have never had an opportunity to develop leisure attitudes, an understanding of leisure education, or the development of additive life skills. A critical need in today's society is to provide leisure education for all individuals, starting literally at birth and continuing throughout life. This is even more essential for individuals with developmental disabilities.

Qualified leisure educators and qualified leisure counselors can serve as facilitators as they bring each individual to the realization and appreciation of the worthwhile use of leisure time. The services of these highly specialized facilitators aid the individual client, but in the broader context they also make a valuable contribution to integrated services, to "the team," and to society in general. The leisure educator and leisure counselor must be a member of the team of specialists in order to assess, prescribe, oversee, and evaluate the needs of each client. This is important not only during illness or rehabilitation but also during the transition into the community and the mainstream of society.

The total team approach is critical to preventive as well as rehabilitative and restorative services for all individuals, but most particularly for those with physical, intellectual, learning and behavioral disabilities. These services must be provided to all, particularly to children during the most impressionable years. Recreation providers, leisure counselors, therapists in all modalities, educators, representatives of the medical profession, the clergy, the family, and all other professionals who deal on a day-by-day basis with the individual must work harmoniously in order to achieve the most beneficial results for the total well-being of each client. Regardless of modality or academic degrees, all professionals must be equal members of the team.

VALUES OF RECREATION, ART, MUSIC, DANCE, AND RECREATION THERAPY

An ancient Chinese proverb succinctly sets the mold for recreation: "Life is not a vessel to be drained, but a cup to be filled." Recreation through its many dimensions and diversified forms helps to fill that cup. Healthful living and attainment of the fullness of life can be maximized through participation in wholesome leisure time activities. The greatest value of recreation is its power to enrich the lives of individuals. Recreation also contributes to the physical well-being of the participant, provides a healthy happy outlook on life, increases self-worth and feelings of satisfaction and accomplishment, increases growth of social attitudes, improves interpersonal relationships, increases skills and knowledge in a breadth of diversified fields, improves morale and morality, contributes to the reduction of crime and delinquency, and increases opportunities for democratic behavior, cooperation, and community solidarity. It is important to emphasize that the values derived from recreational experiences are related in large measure to those characterizing leadership. It is critical that recreation and therapeutic recreation be incorporated as a dynamic service within the framework of the community, schools, general and rehabilitative services of hospitals, rehabilitation centers, and other programs and facilities serving "normal" individuals as well as those with special needs. Programs must be directed by responsible qualified role models.

The neuromuscular and physiological involvement of a client is the fundamental common denominator for the integration and correlation of the fields of recreation, therapeutic recreation, adapted physical education, corrective therapy, and art, music, and dance therapies.

From the earliest cultures, music has served as a universal language. It continues to break down barriers in communication, provides positive group dynamics, and even has been known to improve international relations.

Music can calm or stimulate the listener. As a therapeutic tool it can stimulate self-expression and creativity in movement, providing emotional release, while at other times it may be effectively used for sedative purposes through its calming effect (Shivers and Fait, 1975). Mitchell and Zanker (1948) found that "music had a definite value in strengthening group cohesion and interpersonal relationships, in facilitating emotional release, and in enabling the patient to experience better personality integration."

The use of music has no age limitations; it spans the life of an individual from birth to death. The professionally prepared, certified music therapist selects the appropriate musical instrument, electronic device, or the musical selection that can be most beneficial to an individual, but the services and programs are particularly valuable to those with developmental-behavioral deficiencies.

The dance therapist also employs musical instruments, electronic devices, and a wide range of music to elicit responses in the form of physical movement. Dance also is used as a medium of communication. It has a universal appeal because of its wide diversity of forms, from classical and traditional ethnic dances to contemporary fad dances. Therefore, it provides an excellent program, once tailored to individual needs, for all people, with particular value for those with developmental or behavior immaturity.

Under the direction of a certified dance therapist, individuals with disabilities benefit from dance therapy in many different ways.

For some the dance becomes a means of expressing deep-seated feelings which cannot be expressed in any other way.... For others, the participation in a rhythmic movement offers an opportunity for learning kinesthetic sensitivity, balance, and spatial relationships. To still others it presents a social challenge and permits heterosexual interaction (Bunzel, 1948).

The importance of movement in the child's total development has been readily recognized and is supported by research. There are indications that movement may influence the child's ability to learn, to conceptualize, or to help him cope with behavioral problems. For some children movement can be the chief medium for communication through which ideas and emotions are revealed. For others, dance and movement can help to deal with issues and problems that they might find difficult to talk about or to discuss. In any case, it is fair to assume that movement-dance therapy in the school has the potential to help the young person to better understand himself and others around him (Balazs, 1977).

Art, in graphic or pictorial form, provides individuals with a medium of expression, a diagnostic instrument, and a therapeutic outlet. "Art activities have shown themselves to be useful for achieving emotional catharsis, especially among patients who have difficulties in releasing tensions or in translating their ideas into words" (Naumberg, 1966).

Although "play" is a vital part of recreation, play therapy is a method that psychologists and psychotherapists use to assist clients whose psychologic disorder responds to catharsis. Play is a valid form of self-expression that appears under no other circumstances (Shivers and Fait, 1975). Swimming and "water play" provide enjoyment, but also have significant therapeutic values. Physical therapists, recreation leaders, recreation therapists, and physical educators have long recognized the value of water for the enhancement of developmental skills in a relaxing buoyant environment.

For many decades, world-renowned contemporary physicians and pioneers in rehabilitation programs for psychiatric and medically ill patients proudly enunciated the value of recreation, not only for the so-called "normal" individual, but as a vital link to the prevention of illness, and the restoration and long-term rehabilitation of individuals with physical and/or mental problems. One of the most outspoken advocates for providing recreation for the ill and disabled, Dr. Howard Rusk, Director of the Institute of Physical Medicine of New York City, writes: "I firmly believe that both individual and group recreation for patients have a direct relationship upon their recovery; these, in my opinion are definitely adjunctive therapy." Rusk also firmly endorsed the need for recreation or leisure counseling as a means of preparing a psychiatric patient for permanent return to the community. "Recreation counseling can help a discharged psychiatric patient avoid the solitary ways that set the stage for re-admission to a psychiatric institution" (O'Morrow, 1976).

William Menninger described the use and values of recreation therapy for mentally ill patients, noting:

It has been the privilege of many of us practicing medicine in psychiatry to have some very rewarding experiences in the use of recreation as an adjunctive method of treatment.... Recreation has not only played an important part in the treatment program of many mental illnesses, but it has been a considerable factor in enabling former patients to remain well (O'Morrow, 1976).

HUMANISTIC-HOLISTIC VALUE

Through its humanistic philosophy, recreation contributes to the well-being of the whole person. The environment is another factor that can enhance the value of programs. The out-of-doors, although not practical for some programs, lends itself to an aesthetic and wholesome setting and a natural environment for programs related to recreation and the adjunctive therapies. Special programs involving democratic decision making, such as camping, outdoor education, and those with a potential high risk factor like Project Adventure and Outward Bound, provide opportunities for the development of awareness, appreciation, trust, personal challenge, and individual and team cooperation. The physical challenge of such programs, although great, is but a small segment of the total involvement. Individuals with special needs gain immeasurably by the holistic approach and values of these types of programs. Through the multisensory, three dimensional experiences provided in outdoor education programs, a golden opportunity is afforded for a total educational experience for well or disabled individuals, simultaneously providing them with essential individualized therapeutic services.

INTERDISCIPLINARY USE

Physical therapy, occupational therapy, play therapy, hydrotherapy, and recreation therapy all provide activities and services somewhat similar to and in some cases the same as, those provided by the adapted physical educator. In 1953 George Stevenson, the medical director of the National Association for Mental Health, wrote:

Recreation really re-creates. It affects the individual deeply, is somewhat akin to therapy but is more positive than therapy. Recreation is pointed less toward the correction of a disorder than toward the elevation of the quality of living (Frye and Peters, 1972).

INTERGENERATIONAL INTEGRATION AND "MAINSTREAMING"

One of the major assets of recreational programing is that participation in most activities can effectively break the traditional barriers, such as age, race, sex, and handicapping conditions. Regardless of the potential for intergenerational programing and integration of able with disabled persons, some administrators and service personnel either integrate with great reluctance and skepticism or in some cases, despite federal and state laws, refuse to accept the challenge. In essence they are depriving each individual of an invaluable experience in sharing, growing, and caring. Folk dancing is an excellent recreational activity that combines music and dance, making it appealing to all. Folk dancing also provides unexcelled opportunities for children, parents, and grandparents to enjoy active participation simultaneously in a multidimensional recreational program. The color, the form, the music, and the synchronized flow of body movement graphically depict art in the fluid form. The demanding vigorous physical movement of some dances, in contrast to the delicate and intricate grace of other dances, not only contributes to physical fitness through gross and fine motor skill development but challenges the intellect as well. A well planned program implemented by a good leader improves social interaction, strengthens emotional stability, and enhances cultural and spiritual growth. Whether one is a spectator or an active participant, folk dancing provides an enjoyable experience while promoting a better understanding and appreciation of other people.

Pediatricians and child care workers may find programs in therapeutic recreation, adapted physical education, art therapy, music therapy, and dance therapy at a local education agency. In addition to public school departments and municipal recreation departments, many social service agencies provide specialized programs in these fields encompassing all age groups.

Advocacy groups for the various types of disabilities also provide a resource network, and in some cases programs or funding for disabled individuals, or bona fide groups serving clients with special needs.

POSTSCRIPT

It is clear that although each modality discussed in this section may be conducted independently with a great deal of success, a coordinated and integrated program can maximize the potential value of each element and contribute toward the client's attainment of the fullness of life.

The Credo for the Recreation Profession succinctly sums up the contribution the field of recreation makes toward improving the quality of life of the whole person:

That because of the recreation movement, more men have a song in their hearts and sing it out, sense more of the drama of the world, see beauty more clearly in all about them, feel the poetry of the world, like to be among the trees, find joy in watching plants grow, have pride in their bodies, want to be more skillful in the use of their hands and all their powers, are happy to use their minds for just the fun of it, enjoy people more, and find satisfaction in serving their neighborhoods and their cities (American Recreation Society, 1962).

EDITH G. DeANGELIS

RESOURCES

Adaptive Environments Center
Massachusetts College of Art
26 Overland Street
Boston, Massachusetts 02215

American Alliance for Health, Physical Education, Recreation, and Dance
Unit on Programs for the Handicapped
Information and Research Utilization Center
1900 Association Drive
Reston, Virginia 22070

American Art Therapy Association
428 East Preston Street
Baltimore, Maryland 21202

American Association for Music Therapy
35 West Fourth Street (777 Education Building)
New York, New York 10003

American Dance Therapy Association
Suite 230
2000 Century Plaza
Columbia, Maryland 21044

American Society of Group Psychotherapy and Psychodrama
39 East 20th Street
New York, New York 10003

Art with the Handicapped Artists
  Foundation of Canada
Box 322 Station 1
Toronto, Ontario
Canada M5S 2S8

Association for Poetry Therapy
Castriel Institute
47 East 51st Street
New York, New York 10022

Hospital Audiences, Inc.
1540 Broadway
New York, New York 10036

National Arts and the Handicapped Information Service
Office for Special Constituencies, Room 1200
National Endowment for the Arts
2401 East Street, N.W.
Washington, D.C. 20506

National Association for Drama Therapy
c/o Barbara Sandberg
Theater Department
William Patterson College
Wayne, New Jersey 07470

National Association for Music Therapy, Inc.
P.O. Box 610
Lawrence, Kansas 66044

National Center on Educational Media and Materials for
the Handicapped
The Ohio State University
Columbus, Ohio 43210

National Recreation and Park Association
1601 North Kent Street
Arlington, Virginia 22209

National Therapeutic Recreation Society
c/o National Recreation & Park Association
1601 North Kent Street
Arlington, Virginia 22209

Rehabilitation Services Administration
Service Projects Branch
Division of Special Projects
Room 3329, Switzer Building
330 C Street, S.W.
Washington, D.C. 20202

The Association of Handicapped Artists, Inc.
503 Brisbane Building
Buffalo, New York 14203

The National Access Center
1419 27th Street, N.W.
Washington, D.C. 20007

The National Center for the Arts and Aging
The Ohio State University
Columbus, Ohio 43210

The National Committee Arts for the Handicapped
Suite 905
1701 K. Street, N.W.
Washington, D.C. 20006

REFERENCES

American Art Therapy Association: Art Therapy. Baltimore, American Art Therapy Association, Inc. (brochure).
American Association for Music Therapy: Policies and Procedures for Granting Certification. New York, American Association for Music Therapy.
American Dance Therapy Association: What Is Dance Therapy? Columbia, Maryland, American Dance Therapy Association (brochure).
American Recreation Society: Credo for the Recreation Profession. San Francisco, Recreation Center for the Handicapped, 1967.
Balazs, E.: Dance Therapy in the Classroom. Waldwick, New Jersey, Hoctor Products for Education, 1977.
Best, G. A.: Individuals with Physical Disabilities. St. Louis, The C. V. Mosby Company, 1978.
Bunzel, G. G.: Psychokinetics and dance. J. Health Phys. Ed. Recr., 19:180, 1948.
Chubb, M., and Chubb, J. R.: One Third of Our Time? New York, John Wiley & Sons, Inc., 1981.
Clarke, H. H., and Clarke, D. H.: Developmental and Adapted Physical Education. Ed. 2. Englewood Cliffs, New Jersey, Prentice-Hall, Inc., 1978.
Cratty, B. J.: Adapted Physical Education for Handicapped Children and Youth. Denver, Love Publishing Company, 1980.
Crowe, W. C., Auxter, D., and Pyfer, J.: Principles and Methods of Adapted Physical Education and Recreation. Ed. 4. St. Louis, The C. V. Mosby Company, 1981.
Frye, V., and Peters, M.: Therapeutic Recreation: Its Theory, Philosophy, and Practice. Harrisburg, The Stackpole Co., 1972.
Geddes, D.: Adapted Physical Activity: an Individualized Approach. Boston, Allyn & Bacon, 1980.
Gunn, S. L., and Peterson, C. A.: Therapeutic Recreation: Program Design, Principles, and Procedures. Englewood Cliffs, New Jersey, Prentice-Hall, Inc., 1978.
Jenny, J. H.: Introduction to Recreation Education. Philadelphia, W. B. Saunders Company, 1955.
Kraus, R.: Recreation Today: Program Planning and Leadership. Ed. 2. Santa Monica, Goodyear Publishing Co., Inc., 1977.
Lee, J.: Play in Education. New York, The Macmillan Co., 1917.
Makechnie, G. K.: Optimal Health: The Quest; A History of Boston University's Sargent College of Allied Health Professions. Boston, Boston University, 1979.
Mitchell, S. D., and Zanker, A.: The use of music in group therapy. J. Ment. Sci., 44, 1948.
National Association for Music Therapy: A Career in Music Therapy. Lawrence, Kansas, National Association for Music Therapy, Inc. (brochure).
Naumberg, M.: Dynamically Oriented Art Therapy: Its Principles and Practices. New York, Grune & Stratton, Inc., 1966.
O'Morrow, G. S.: Therapeutic Recreation: A Helping Profession. Reston, Virginia, Reston Publishing Company, Inc., 1976.
Parker, R. A., and Downing, R.: Recreation therapy: a model for consideration. Ther. Recr. J., 15:2, 1981.
Pomeroy, J.: Recreation for the Physically Handicapped. New York, The Macmillan Company, 1964.
Sargent, D. A.: Physical Education. Boston, Ginn & Co., 1906.
Sherrill, C.: Adapted Physical Education and Recreation: a Multidisciplinary Approach. Ed. 2. Dubuque, Iowa, Wm. C. Brown Company, 1976.
Shivers, J. S., and Fait, H. F.: Therapeutic and Adapted Recreational Services. Philadelphia, Lea & Febiger, 1975.
Stein, T. A., and Sessoms, H. D.: Recreation and Special Populations. Ed. 2. Boston, Holbrook Press, Inc., 1977.
Stein, T. A., Sessoms, H. D., and Murphy, J. F.: Recreation and the Economically Deprived. Boston, Holbrook Press, Inc., 1977.
Wiseman, D. C.: A Practical Approach to Adapted Physical Education. Reading, Massachusetts, Addison Wesley Publishing Company, 1982.

57G   Vocational Rehabilitation

In the early teen years, as the child with special needs approaches the middle and high school grades, the academic focus begins to move toward a life preparation curriculum. Family supports correspondingly must shift toward increasing the child's capacity to interact with the community socially, emotionally, and vocationally. The school program should begin to provide the troubled student with an opportunity to experience different career clusters, develop realistic vocational goals, and acquire the skills necessary to achieve these goals. The high school curriculum focuses upon enhanced career awareness and specific skill mastery so that upon graduation the disabled student will be prepared to move into the employment world.

For some students during these transitional years, the resources of the local vocational rehabilitation agency will be necessary. Traditionally these services are introduced during the later school years when the rehabilitation counselor begins the eligibility determination process for the disabled student. The services available through vocational rehabilitation are directed toward employment and training. Other needs such as residential and social needs may require the assistance of public agencies such as departments of mental health, social service, or welfare.

VOCATIONAL REHABILITATION: FROM PAST TO PRESENT

Developmentally, vocational rehabilitation has gone through a number of major changes during the past six decades. With the passage of the Smith Fess Act (PL 66-236), the retraining of injured workers became a national priority. The Act fostered the development of a vocational rehabilitation system at the federal level and, subsequently, the establishment of bureaus, departments, or offices of vocational rehabilitation in all the states. Although this development started in the 1920's at the national level, it was not until some two decades later that each state responded to the federal initiative.

The initial focus of vocational rehabilitation had been directed toward aiding workers who were injured on the job. Such individuals would be eligible for training so that they could re-enter the competitive employment market. This focus remained the exclusive mission of the vocational rehabilitation network for more than two and one-half decades. It was not until the mid-1940's that the emphasis shifted from retraining to training. This change recognized the needs of disabled individuals who through specific training services would be able to enter the employment market for the first time. Individuals with mental retardation and other developmental disabilities as well as disabled individuals who had never been employed could now be eligible for vocational rehabilitation services.

During the early 1970's more subtle changes occurred. Although there was still a focus upon employment, the ultimate goal moved from competitive employment to gainful employment. Gainful employment was further defined to include not only competitive employment but sheltered and home bound employment as well. This change reflected a change in philosophy for vocational rehabilitation, indicating that more substantially disabled individuals would be considered eligible for services. A symbolic change in terminology also occurred with the development of the Rehabilitation Amendments of 1973, when for the first time the term vocational was not included as part of the title of this federal bill.

Along with the program changes during this time, there was also a shift toward the provision of services to more severely disabled individuals. The law, however, still required that such individuals reasonably could be expected to benefit "in terms of employability from vocational rehabilitation services" in order to be considered an eligible client. In an attempt to clarify the term severe disability, a modified categorical approach was developed. The term severely disabled by definition indicates a "disability which requires multiple service over an extended period of time and results from amputation, blindness, cancer, cerebral palsy, cystic fibrosis, deafness, heart disease, hemiplegia, mental retardation, mental illness, multiple sclerosis, muscular dystrophy, neurological disorders (including stroke and epilepsy), paraplegia, quadriplegia and other spinal cord conditions, renal failure, respiratory or pulmonary dysfunction, and any other disabilities specified by the Secretary in regulations he shall prescribe."

The change in focus to gainful employment, as well as the added emphasis upon the provision of services to severely disabled persons, was symptomatic of the evolutionary changes the vocational rehabilitation system was experiencing. The passage of the Rehabilitation, Comprehensive Services, and Developmental Disabilities Act of 1978 (PL 95–602) was an attempt to coordinate all these changes as well as other initiatives for disabled persons, such as the Developmental Disabilities Act, into a comprehensive service delivery system. This Act provides a continuation of the traditional vocational rehabilitation services and research, develops a National Council on the Handicapped, special employment opportunities for disabled individuals, and comprehensive services for independent living. The purpose of the Act is to develop and implement, through research, training, service, and the guarantee of equal opportunity, comprehensive and coordinative programs of vocational rehabilitation and independent living.

SERVICES OFFERED

A state division of vocational rehabilitation has as its principal focus the provision of rehabilitation services to disabled adults of all ages. These services are directed toward increasing the employability of the individual. The vocational rehabilitation agency may provide a number of services including evaluation of rehabilitation potential; counseling, guidance, referral, and placement; vocational and other training services for handicapped individuals (personal and social adjustment training, books, other training materials); physical and mental restoration services; limited maintenance services; interpretive services for deaf individuals; mobility services for blind persons; occupational licenses, schools, equipment, and initial stock and supplies; transportation in connection with the rendering of vocational and rehabilitation services; and telecommunications, sensory, and other technological aids and devices.

Although a wide range of services may be offered through vocational rehabilitation, the initial effort is directed toward the determination of eligibility and whether there is a reasonable expectation that the individual will be able to engage in employment (competitive, sheltered, or home bound). The determination of eligibility is made through the evaluation of the rehabilitation potential of each client or applicant. This evaluation is a preliminary diagnostic study to determine that the individual has a substantial handicap in employment and that rehabilitation services are needed.

The diagnostic study consists of an evaluation of pertinent medical, psychological, psychiatric, vocational, educational, cultural, social, and environmental factors that bear on the individual's handicap in employment. To the degree necessary, the rehabilitation agency may conduct an evaluation of the individual's personality, cognitive level, educational achievement, work experience, vocational aptitude and interest, personal and social adjustment, employment opportunities, and other data helpful in determining the nature and scope of the services needed. In certain instances the agency may provide evaluation services for up to 18 months for individuals whose vocational potential may be unclear or difficult to assess.

The core of the rehabilitation process is the development of the individual written rehabilitation plan (IWRP). This plan, developed jointly by the vocational rehabilitation counselor and the disabled individual (or, in appropriate cases, his parent or guardian), sets forth the "terms and conditions as well as the rights and remedies, under which goods and services are provided to the individuals." The plan is reviewed annually, at which time the individual may modify the plan as stated. The plan must include a statement of long range rehabilitation goals for the individual and intermediate rehabilitation objectives related to the attainment of these goals; a statement of specific vocational rehabilitation services to be provided; a projected date for implementation and the anticipated duration of each such service, and an evaluation procedure and schedule for determining whether such objectives and goals are being achieved; and, when appropriate, a detailed explanation of the availability of a client assistance project.

In addition to the development plan and the possibility of an extended evaluation (up to 18 months), all state vocational rehabilitation agencies are required to have a due process procedure. This procedure is designed to allow individuals who have been denied services or who do not agree with the goals and objectives stated in the rehabilitation plan to appeal the agency action or decision. The role of the disabled individual or parent or guardian is crucial in both the development and the implementation of the rehabilitation plan.

HOW TO GAIN ACCESS TO THE SYSTEM

Within each state there is a bureau, division, or office of vocational rehabilitation. Generally the agency utilizes a field or area office structure so that potential clients can have access to the counseling and placement services offered. Referrals can be made by the individual himself, a parent or friend, or a professional or other interested individual. In most instances the referring individual is encouraged to have the potential client contact the counselor directly and set up an initial meeting if possible. At this point the eligibility determination process can be initated. If recent evaluation data are available, this information will be requested. If this is not the case, the counselor may request that such data be obtained. All disabled individuals are entitled to apply for services and to have a complete review of their application by the agency. Any costs associated with the determination of eligibility are assumed by the agency. Once an applicant is determined to be an appropriate client for the vocational rehabilitation system, a rehabilitation plan is developed.

In most instances the counselor who has worked with the client in the development of the rehabilitation plan continues to provide direct services. The vocational rehabilitation counselor, who usually has a master's degree in rehabilitation counseling, provides regular and continuing counseling services to his clients. In addition to this specific training in counseling methodologies, the rehabilitation counselor has a broad knowledge of handicapping conditions, medically, socially, and emotionally. He must be able to deal with client issues, concerns, and needs, as well as be responsive to the industrial trends. Traditionally training programs at the master's level provide the counselor with a knowledge of individual and group counseling technique, vocational and other evaluation procedures, and a knowledge of placement strategies. The counselor serves as a resource to the client, a coordinator of services, and a monitor of client progress. Although states may use a system in which the rehabilitation counselor has a specialized case load (e.g., mental retardation, epilepsy, deafness), in most instances the counselor carries a general case load that is responsive to a geographic area rather than to diagnostic categories.

SUMMARY

The scope of the services offered by the vocational rehabilitation system is limited to vocationally related activities. The needs of the disabled adult are often complex and involve not only vocational but social, emotional, and residential supports as well. It is imperative that the vocational rehabilitation agency work closely with other agencies, such as departments of mental health or hygiene, the division of employment security, the department of education, and the

private sector, to insure that a comprehensive plan is developed for each client. The rehabilitation counselor can serve as a facilitator in some instances, and in others his role will be as an active member of an interdisciplinary team that deals with the disabled young adult. Comprehensive service planning is essential if a total rehabilitation program is to be implemented.

WILLIAM E. KIERNAN

REFERENCES

Bellamy, G. T., O'Connor, G., and Karan, O. C. (Editors): Vocational Rehabilitation of Severely Handicapped Persons: Contemporary Service Strategies. Baltimore, University Park Press, 1979.

Lynch, K. P., Kiernan, W. E., and Stark, J. A. (Editors): Prevocational and Vocational Education for Special Needs Youth. Baltimore, Paul H. Brooks Publishing Co., 1982.

Magrab, P. R., and Elder, J. O. (Editors): Planning for Services to Handicapped Persons. Baltimore, Paul H. Brooks Publishing Co., 1979.

58

Child and Adolescent Psychopharmacology

Although it has been almost 50 years since the introduction of the first specialized psychopharmacological treatment of children, little new research on behalf of children has been contributed until the last few years. Amphetamines were first used to treat children with behavior disorders in 1937, and that treatment has developed into the use of psychostimulants for treating hyperactivity and attention deficit disorders in children. In the 1950's, new drugs for treating depression and psychosis revolutionized the psychiatric treatment of adults with major mental disorders. Psychotropic medications were applied to a wide variety of forms of emotional distress, and entered the public consciousness as an example of advanced technology in affluent countries (Baldessarini, 1977; Shader, 1975). Mind altering street drugs became commonly available for recreational use, and have become ingrained as a part of the youth and poverty cultures. Throughout this period, children were protected from the risks and benefits of psychotherapeutic drugs: New knowledge about children and pharmacology was not being integrated into medicine, even on an experimental basis for the most severely afflicted children.

Over the last few years, there has been a renewed interest in refining indications for the use of psychostimulants in children. The nature of childhood depression has been clarified by findings that some patients are more treatable when antidepressant medication is a part of the treatment program. In university settings, lithium has been used for treatment of certain children with a particular type of severe psychopathological disorder, suggesting the possible appearance of manic-depressive illness in childhood. Antipsychotic drugs are being employed in the treatment of certain childhood forms of psychosis, agitation, and impulsivity.

The use of drugs to alter the behavior and feelings of children is a highly sensitive matter, and is often not viewed as being comparable to other forms of medical treatment by children, parents, or the community. Such use of drugs to alter the mind raises ethical issues, which are still only partially understood by physicians and the public, and their role in the psychiatric treatment of adults is still being debated.

At the present time, clinicians will find that psychostimulants are an established and commonly used treatment for certain disorders in children. Other drugs are now being introduced in child psychiatry, and these newer treatments will remain, for a time, in the specialized repertoire of the child psychiatrist (Biederman, 1982; Kalogerakis, 1982; Klein et al., 1980; Werry, 1978, 1982; White, 1977; Wiener, 1977). The pediatrician may be expected to preside over the established psychopharmacological treatments: psychostimulants for hyperactivity and attention deficit disorder, antidepressants for bed wetting, and neuroleptics for certain forms of psychosis, agitation, and verbal-motor tic.

This chapter will review the established psychopharmacological treatments of children, as well as provide some general comments concerning the newer treatments that may be prescribed for children under the specialized care of a child psychiatrist. The formally established indications for psychotropic drug therapy will be reviewed first (Table 58–1). Then possible indications for drug therapy in children will be considered, including a survey of newer and experimental treatments. Finally, the management of the child receiving prescribed drugs, as well as family and community issues, will be discussed.

FORMALLY ESTABLISHED INDICATIONS FOR PSYCHOTROPIC DRUG THERAPY IN CHILDREN

ATTENTION DEFICIT DISORDERS AND HYPERACTIVITY

Criteria for the use of psychostimulants in treating behavior disorders in children have been only partially defined, even after a half century of their

Table 58–1. INDICATIONS FOR PSYCHOTROPIC DRUG THERAPY IN CHILDREN

Formally established indications
    Attention deficit disorder and hyperactivity
    Enuresis
    Verbal-motor tic (Tourette's) disorder
    Extreme agitation
    Overt psychosis
Possible indications
    Affective spectrum
        Childhood depression
        Childhood mania
        Psychotic depression
        Phobic disorders
        Separation anxiety disorder
        Panic disorders
        Obsessive-compulsive disorder
    Adult-form schizophrenia
    Atypical development (childhood schizophrenia, infantile autism)
    Behavioral impulsivity
    Anxiety and sleep induction disorders
    Sleep-arousal disorders
        Sleep terror disorder
        Sleepwalking
        Narcolepsy

clinical use. Chapter 38 of this book outlines current thinking about the nature of motoric overactivity, impulsivity, and distractibility.

In general, impulsivity in some neurologically based functions (motoric overactivity, emotional changeability, or cognitive distractibility) may be seen in stimulant-responsive children. However, such criteria are only loosely defined, and it remains difficult to state clear and specific "indications" for psychostimulant therapy. In practice, clinicians often try psychostimulants in an empirical trial and infer a diagnosis from a positive drug response: Such a diagnostic method is fallacious and unscientific, since the pharmacological response may be found some day in a variety of disorders. At present, correct diagnosis is often viewed as less important than correct treatment.

Conditions that can mimic the clinical presentation of chronic motoric overactivity, emotional changeability, and cognitive impulsivity include lead toxicity, anxiety, boredom, constipation, hunger, hypoglycemia, thyroid and other metabolic diseases, childhood mania, depression, and psychosis. For some children, a psychiatric evaluation may be needed to rule out certain factors before employing psychostimulant treatment.

Follow-up studies in children with stimulant-responsive impulsivity suggest a high risk for impaired functioning in later life. Although many children are able to develop good adjustment in adulthood, there appears to be an over-representation of delinquency, sociopathy, impulsivity, alcoholism, and depressive disorders. Although the motoric impulsivity appears to improve with age, it appears that the cognitive and emotional aspects of the impulsivity are more likely to persist into adulthood and become integrated in the adult personality.

Pharmacological Treatment of Attention Deficit Disorders and Hyperactivity

The quieting effect of stimulants, often called "paradoxical," on overly active or impulsive children remains poorly understood. Recent research has questioned classic beliefs about the quieting effect of stimulants, proposing that this "paradoxical" effect is actually nonparadoxical (instead suggesting that the quieting is an ordinary and expected result of a "tuning" or "sharpening" of underactive attention or arousal mechanisms) and that the "paradoxical" effect may not be specific to hyperactive children (but may be seen either as a normal developmental stage or even in normal children under some circumstances).

However, clinicians are generally aware that not all children respond therapeutically to stimulants. The studies that suggest that normal children have a quieting response to stimulants are done in particular laboratory settings, and their relevance in normal "real life" settings is unclear. At a clinical level, it seems clear that many children neither are calmed by stimulants nor receive any therapeutic benefit, so that claims that all children respond "paradoxically" in a laboratory experiment are difficult to understand clinically.

Some researchers even question the clinical effectiveness of psychostimulants in any children, suggesting that the diagnosis is so imprecise that practicing clinicians are nonspecifically treating a group of more or less undiagnosed children (Weiss and Hechtman, 1979). There is probably considerable truth in the claim that many and perhaps most empirical trials of stimulants are made without an adequate evaluation or clinical diagnosis. Whatever the standards of practice may be, there does appear to be a group of children for whom stimulants can be correctly used, in whom attention deficit disorder can be adequately diagnosed, for whom empirical trials can be performed with reasonable precision and care, and whose target symptoms can be helped by medications.

Target symptoms for psychostimulant treatment include attention deficit, distractibility, hypermotoric behavior, aggressivity, emotional lability, and quick temper.

Symptom improvement may be clear-cut and dramatic in cases of "pure" attention deficit disorder, but only partial in children in whom there is a mixed picture of attention deficit disorder associated with other characterological or developmental problems. For instance, a child may show some motoric and cognitive improvement after taking stimulants but continue to have impulsive behavior based on ego underdevelopment or delinquent behavior based on malsocialization. In such cases, the stimulant medication should be continued because of its value in treating a part of the symptom complex, and other treatment modalities should be instituted to influence the other symptoms.

The following specific clinical findings may increase the likelihood of a therapeutic response to psychostimulants when associated with symptoms of impulsivity (motoric overactivity, cognitive distractibility, or emotional changeability):

1. Organic brain damage, such as intrauterine damage, obstetrical difficulty at the time of birth, or postnatal neurological trauma.

2. A family psychiatric history in which the males were overactive and distractible during childhood.

3. Congenital body (somatic) damage, such as neurological "soft signs" or minor physical abnormalities (anomalies).

4. A history of "speeding up" or agitation in response to common medical sedatives (such as barbiturates or diazepam).

5. A history of calming or quieting in response to ordinary medical stimulants (such as pseudoephedrine or caffeine).

6. An intrauterine history of overactivity, with an unusual amount of "kicking" described by the mother in comparison to that with her other children.

Other associated clinical features—electroencephalographic changes, intelligence differences, specific learning or language disorders, and other medical conditions—have not been found to predict a helpful psychostimulant response.

Short acting stimulant medications are generally considered the first line of drug treatment. D-amphetamine (Dexedrine) and methylphenidate (Ritalin) are the two short acting psychostimulants currently in use in the United States. Clinically there is little difference between the two drugs, except a slightly longer half-life for methylphenidate. D-amphetamine is generally preferable, since generic and trade forms of methylphenidate are more expensive. Only in the case of the diabetic child with attention deficit disorder should methylphenidate be used first, owing to the greater effect of D-amphetamine on blood glucose.

Recommended regimens (Table 58–2) and common side effects (Table 58–3) are listed at the end of this chapter. The clinical effects appear within 30 minutes and last for four to six hours. Thus the therapeutic effects of a short acting stimulant at a given dosage can be evaluated in one or two days.

If administration of multiple doses during the daytime presents logistical difficulties, the slow release forms of d-amphetamine and methylphenidate can be given once a day in the morning before school. If a dose is ineffective, it should be increased in a stepwise fashion up to the indicated dose limits.

If one of the short acting stimulants is found to be ineffective, it is good clinical practice to proceed with a trial of the other short acting stimulant: about one-quarter of stimulant-responsive children respond to one but not the other drug. If both drugs fail (and diagnosis is still considered secure), it is then reasonable to try magnesium pemoline (Cylert), which is available only in an expensive brand, but which may be useful in certain cases when the short acting stimulants fail. (Tricyclic antidepressant medications may also be helpful in some cases, though many children appear to develop a tolerance to the therapeutic effect after several months, so that their long term effectiveness for most children is minimal. However, antidepressants may be preferred if depressive symptoms are seen concurrently; the impulsivity as well as depression may then be persistently improved by antidepressants.)

All the stimulants have the potential disadvantage of causing initial insomnia and nightmares in one-quarter of the children who receive them. This is an easy side effect to treat: By shifting the time of administration into the early part of the day (before 2:00 P.M.), these problems during nighttime sleep generally can be avoided. For children with persistent insomnia, chloral hydrate or diphenhydramine (Benadryl) may be used to help initiate sleep. Benzodiazepines (such as Valium or Librium) and barbiturates should always be avoided, since they tend to produce paradoxical excitation in these children, which could worsen their impulsivity. In general, it is usually best to avoid evening doses of stimulants for all children.

There appears to be some retarding effect of psychostimulants on body growth and height development, although the extent and frequency of this side effect are still being documented (Roche et al., 1979). There may be a complete return to normal body size, without any residual stunting, by way of a "growth rebound" after the medication is stopped. However, if stimulants are maintained through the period of adolescence when bone epiphyses are closing, it is likely that the growth retardation will become permanent. In general, the extent of growth slowing appears to be 1 to 3 centimeters over the course of several years: this loss, even if permanent, would be less significant for most children than the developmental effects of an unmedicated attention deficit disorder.

Weight loss and decreased appetite may be found in some children. If there is a weight loss, it is self-limited and stops in a few weeks.

In certain children, stimulants may cause an alteration of mood, including crying, uncomfortable feelings, or a sense of unhappiness. In most cases, a lower dose of the psychostimulant is sufficient to provide good control of the symptomatic impulsivity without these side effects of mood change. Often these children have a family history of affective illness and may be better treated with an antidepressant such as imipramine.

Cardiac and cardiovascular effects should be monitored, but typically do not produce any clinically significant problems.

A more significant potential problem is the aggravation or even initiation of a tic disorder, including Tourette's syndrome (Lowe et al., 1982). The exact frequency of this problem is unclear, but it appears to be relatively rare. A neurological examination that documents possible movement abnormalities prior to medication can help delineate the presence or degree of any such motor side effects that may be attributable to the medication.

The most significant risk is the precipitation of psychosis in children who have an associated or underlying psychotic disorder. Careful psychiatric screening can identify such children at risk, and the use of stimulants in these individuals should be avoided or employed only under the supervision of a specialist familiar with the pharmacological management of psychotic disorders in children. In general, most children are not at risk for developing psychotic reactions in the ordinary usage of this medication.

Psychostimulants are frequently used for several years, with a full course of treatment typically ceasing about the time of puberty when there is often a reduction in motoric hyperactivity. However, since the impulsive, emotional, and cognitive aspects of the syndrome may persist beyond the period of motoric hyperactivity, it is reasonable to consider the use of these medications in adolescence or adulthood. The use of stimulants during adolescence is troublesome, partly because of the incidence of drug abuse and partly because of the potentially permanent effect on height.

When stimulants are used over the course of several years, it is advisable to make attempts at dose changes about once or twice a year. It is worth trying both slightly lower doses and slightly higher doses, since the optimal dose may drift either way, depending on whether "outgrowing" the syndrome or increasing body weight is more dominant at that time.

Sometimes it appears that a psychostimulant has "lost effect," and impulsive symptoms emerge on a previously satisfactory dose. Generally the pharmacological effect is not lost, but increased environmental stress causes the return of the symptom. Maintaining the medication and dealing with the reactive stress are recommended at those times.

When discontinuing medication, it is often said that the medication may be merely stopped without tapering the dose. However, mild withdrawal symptoms (subjectively noticeable fatigue and mood change) are often present and can be minimized by tapering over several days.

Nonpharmacological Treatment of Attention Deficit

Surprisingly pharmacological treatment with psychostimulants is not sufficient to produce significant improvement in school grades in most children (Aman, 1980). This is true even for many children in whom there is a good therapeutic effect on behavior and attention. Contrary to traditional practice, nonpharmacological modalities of treatment should be considered at the same time as psychostimulant medication is initiated.

Many of these children have associated specific learning disorders or other educational disabilities. Individualized educational plans, which permit the teaching of these children with special needs, are frequently an important part of the overall treatment program. Sometimes environmental manipulations involving reduction of environmental sensory stimulation may be useful: Using quieting carpets, dull lights, subdued colors, allowing only a few toys out of the box at one time, limiting the number of friends visiting at once, and avoiding parties and supermarkets may be helpful. At school, special arrangements can include self-contained classrooms and avoidance of overstimulating places (school bus, cafeteria, hallways).

In some children the avoidance of artificial food dyes may provide a clinically beneficial result (Swanson and Kinsbourne, 1980). These children cannot be identified except by clinical trials of diets, and they probably reflect only a small proportion of the cases.

Given the burden of this condition on other areas of development, and the relatively poor progress that many of these children show on follow-up studies, it is probably advisable for many of them to be evaluated for psychotherapy—particularly if grades do not improve in response to other aspects of the treatment program. Psychotherapy should be considered if there remains a significant problem in school grades, interpersonal interactions, emotional stability, or cognitive effectiveness. A good partial response to one or several treatment modalities should not be taken as evidence that psychotherapy is not needed: This evaluation is probably best done after a period of weeks with psychostimulant treatment.

BED WETTING (ENURESIS)

The antienuretic effectiveness of the antidepressant imipramine has been formally accepted by the United States Food and Drug Administration, but its clinical usefulness for treating bed wetting is very limited (Mikkelsen and Rapoport, 1980).

Under all circumstances a medical work-up for correctable organic causes should precede any consideration of pharmacological intervention. (See also Chapter 31.) If bed wetting occurs only during the night, it is probably preferable to employ behavioral therapy methods (such as bladder control exercises, the use of devices that produce arousing sounds when a blanket is moistened, or midsleep awakenings for trips to the bathroom at regularly scheduled intervals). If urinary incontinence also occurs during the day, medication might be considered, but bladder control exercises may still be preferred. In general, a low dose of imipramine (10 to 75 mg. nightly) may be sufficient. Symptom control may appear after two to five nights but is often only transitory. Most children return to enuresis after stopping the medication; only 15 per cent of children receive permanent sustained relief after a course of imipramine.

The mechanism of action of imipramine in treating enuresis is unclear. Other antidepressants (amitriptyline and the monoamine oxidase inhibitors, but not pure anticholinergic drugs) have an antienuretic effect. It is hypothesized that the antidepressants may act by correcting a developmental lag of neural control mechanisms governing bladder sphincter or body tone, by stimulating "attention" directed toward visceral functions, or by correcting a pathological form of "deep sleep" associated with depression or a neurodevelopmental abnormality in sleep physiology.

VERBAL-MOTOR TIC (TOURETTE'S) DISORDER

Certain types of verbal and motor involuntary movements (the Gilles de la Tourette syndrome) may respond to low doses of antipsychotic drugs; for example, haloperidol in low doses (2 to 5 mg. daily) may be sufficient to control a dramatic and severe presentation. Other neuroleptic drugs can be effective, and minor tranquilizers may also be adequate for control in some children with this disorder. Over the course of years, a slowly increasing dose of an antipsychotic drug may be required, but the use of drugs to control this potentially crippling and socially disfiguring illness may be more or less essential. (See also Chapter 33.) For children who do not respond to neuroleptic drugs, clonidine has been found to be a clinically effective alternative in experimental protocols (Cohen et al., 1980).

EXTREME AGITATION

Acute or chronic intervals of extreme nervousness, unmanageable behavior, unpredictable destructiveness, sudden assaultiveness, or psychotic disorganization may be seen in children presenting in emergency rooms or inpatient hospital settings. Such factors as situational stress (family), environmental stimulation (noise, novelty, movement), biological cycling (monthly, diurnal, menstrual), drug intoxication (recreational abuse, poisoning), medical-psychiatric disease states, or psychological vulnerability may contribute to the extreme agitation. Sometimes such conditions cannot be adequately diagnosed (or even tolerated) in the diagnostic setting without physical or chemical management to provide control and restraint.

Such extreme agitation may be treated nonspecifically with medication while evaluating for a more specific treatment. Table 58–2 describes various chemical restraints that may be used in these settings. Individual doses of diphenhydramine or chlorpromazine can provide "chemical restraint" of undiagnosed disruptive behavior.

It is important to emphasize that chemical restraints must be considered a nonspecific means of control, similar to tying a patient down with ropes: restraints should be used only for the sake of protection and should not substitute for continuing evaluation. It is a matter of debate whether this use of medication should be considered treatment rather than protective management.

OVERT PSYCHOSIS

Sometimes extreme agitation or overtly psychotic symptoms of delusions and hallucinations persist beyond the first few hours of antipsychotic treatment. In these cases, the treatment of presenting agitation merges into the treatment of psychosis.

Psychosis is a final common pathway for a number of different medical and psychiatric conditions, as well as for drug intoxication and withdrawal. (See Chapter 41.)

Physical safety is the highest priority in management. Then medical evaluation is paramount and needs to be fully coordinated with psychiatric evaluation. These evaluations are best performed with the minimal feasible medication in order to evaluate the natural presentation.

Once treatment begins, neuroleptic medications may be given in repeated doses every two to four hours until target symptoms are controlled or until side effects are limiting. Table 58–2 gives recommended regimens for achieving symptom control, and a moderate degree of judgment and experience is helpful in determining when and how to withdraw medication over the course of time.

For children who need to be maintained on full antipsychotic doses of neuroleptic drugs, there are generally numerous side effects that must be anticipated and managed in "exchange" for the tremendous advantage of a chemical control of psychosis.

Sedation, postural hypotension, and other anticholinergic and antihistaminic side effects are very standard, but these side effects are often transient (lasting several days or weeks) or can be minimized by switching to a drug with a different side effect profile.

Of sensitive concern to many adolescents are some potentially embarrassing and persistent side effects, including weight gain and easy sunburning. Adolescent males sometimes have difficulty in maintaining penile erections, or may experience "retrograde" ejaculation of semen into the urinary bladder, which may be surprising and cause concern regarding personal integrity. Also a worsening of acne may be seen, and has a special importance for adolescents. Although sunburning and acne are easily managed by sun blocking lotions and dermatological treatment, all these symptoms tend to remain—without progressively improving—until the medication is stopped.

These neuroleptic drugs often cause motor side effects. For the first several weeks of treatment, acute dystonic reactions, restlessness, and Parkinson-like movements may appear. These effects are subjectively upsetting but are not dangerous. By lowering the neuroleptic dosage or by use of a supplementary anticholinergic drug, such as benztropine (Cogentin), 0.5 to 1.0 mg. daily, or diphenhydramine (Benadryl), 10 to 25 mg., three times daily, these motor symptoms can be controlled completely in most cases.

Of major concern for a patient who is maintained on the extended neuroleptic treatment of psychosis is a late appearing motor side effect—involuntary dyskinetic movements of the face and mouth, which may begin to appear after six months of neuroleptic treatment. This "tardive dyskinesia" may worsen as the dose of the neuroleptic drug is lowered, and has been irreversible in some adults even after several months without the medication. This is a serious side effect, which may be seen in children, but it is relatively unusual since most children do not need the very high

dosages and lengthy treatment that some psychotic adults require. Nonetheless children undergoing lengthy institutional treatment are clearly at risk, and all children who receive even low doses of antipsychotic drugs should be examined periodically during neuroleptic drug treatment for dyskinetic movements (Gaultieri et al., 1980).

Although the late appearing motor changes have attracted particular attention, it is conceivable that long term–high dose neuroleptic drug treatment may also produce tardive cognitive and emotional changes (Gaultieri and Guimond, 1981).

Despite these significant problems, the neuroleptic drug treatment of psychosis may be preferred to undertreatment of the psychosis itself, and can proceed with thoughtful selection of doses and duration as well as careful monitoring of effects. At doses below the antipsychotic range (e.g., chlorpromazine below 3.0 mg. per kg.), neuroleptic drugs have relatively few side effects. The treatment of tic disorders or impulsivity syndromes may entail only mild transient sedation. However, when lengthy treatment and high doses are required to manage psychosis or chronic psychotic agitation, neuroleptic drug treatment can become costly in personal terms. Such treatment should be reserved for children requiring special intervention for assuring physical safety or for initiating effective involvement in developmentally oriented treatment programs.

POSSIBLE INDICATIONS FOR DRUG THERAPY

Recent psychopharmacological research is expanding the number of childhood conditions that someday will be considered to be formal indications for drug treatment. These new treatment modalities remain somewhat experimental in nature and should not be considered for routine use in the community, although information about their effectiveness and appropriate application will be available in the coming years.

CHILDHOOD DEPRESSION

Depression as a drug treatable syndrome has only recently been defined in children (see Chapter 41.) In the last few years, there has been general acceptance of the notion that children can have endogenous depression similar to the serious depressions seen in the adult psychiatric population. Previously children were seen as having sadness but not drug treatable depressive illness, or "masked depression" in which other symptoms were more prominent, or "depressive equiva-

Table 58–2.   RECOMMENDED TREATMENT REGIMENS FOR PSYCHOTROPIC MEDICATIONS IN CHILDREN

I.   Stimulants

Drug	Trade Name	Dose Equivalency (mg.)	Minimum Age for Advertising
Dextroamphetamine	Dexedrine	5	3 years
Methylphenidate	Ritalin	10	6 years
Magnesium pemoline	Cylert	18.75	6 years

Treatment of attention deficit and hyperactivity
  Dextroamphetamine
    Initial dose: 5 mg. at 8 A.M. for two days
    Dose progression: 5 mg. at 8 A.M. and noon for two days; then increase by 5 mg. per day every two to four days
    Daily dose range: 10 to 40 mg.
    Halve these doses for children under six years old
    Available in oral elixir and long acting form (Dexedrine Spansule)
  Methylphenidate
    Initial dose: 10 mg. at 8 A.M. for two days
    Dose progression: 10 mg. at 8 A.M. and noon for two days; then increase by 10 mg. per day every two to four days
    Daily dose range: 20 to 80 mg.
    Halve these doses for children under six years old
    Available in long acting form (Ritalin-SR, 20 mg.)
  Magnesium pemoline
    Initial dose: 37.5 mg. in the morning for seven days
    Dose progression: Increase by 18.75 mg. per day weekly
    Daily dose range: 37.5 to 112.5 mg.

II.   Antidepressants

Drug	Trade Name	Dose Equivalency (mg.)	Minimum Age for Advertising
Imipramine	Tofranil	150	Depression: 12 years Enuresis: 6 years
Desipramine	Norpramin	150	12 years
Amitriptyline	Elavil	150	12 years
Doxepin	Sinequan	150	12 years
Trimipramine	Surmontil	150	—
Nortriptyline	Aventyl	50	12 years
Protriptyline	Vivactil	25	—
Maprotiline	Ludiomil	150	18 years
Amoxapine	Asendin	250	16 years
Trazodone	Desyrel	300	18 years

Treatment regimens: Doses given for imipramine: adjust to dose equivalent for other antidepressants
  Treatment of depression:
    Initial dose: 0.5 mg. per kg. nightly for two days
    Dose progression: Increase by 0.5 mg. per kg. per day every three days
    Daily dose range: 4.0–5.0 mg. per kg. (for older adolescents, 3.5 mg. per kg)
  Treatment of enuresis:
    Initial dose: 0.5 mg. per kg. nightly for two days
    Dose progression: Increase by 0.5 mg. per kg. per day every one to two weeks
    Daily dose range: 0.5 to 2.0 mg. per kg.
  Treatment of attention deficit disorder and hyperactivity:
    Initial dose: 0.3 mg. per kg. nightly for one week
    Dose progression: Increase by 0.3 mg. per kg. nightly every seven days
    Daily dose range: 0.3 to 2.0 mg. per kg.

lents" in which particular symptoms were believed to reflect an underlying depression. Children were said to lack the psychological development required to sustain a depression of the adult type. Now the antidepressants originally used in the treatment of endogenous depression in adults have been found to be effective in treating certain depressive children, aiding research in the identification of drug treatable disorders in children.

The indications for the use of tricyclic antidepressants remain somewhat unclear at this time.

The *Diagnostic and Statistical Manual* criteria for major depression define a population of children who respond to antidepressant medication (Petti and Law, 1982; Puig-Antich et al., 1978), but the rigid application of those adult derived criteria to children is not a satisfactory means of identifying depressed children. Use of these criteria selects only the most severe cases of childhood depression and does not tap the much larger pool of depressed children who might also benefit from antidepressant treatment. Furthermore, these chil-

Table 58–2. RECOMMENDED TREATMENT REGIMENS FOR PSYCHOTROPIC MEDICATIONS IN CHILDREN *(Continued)*

III. Antipsychotics

Drug	Trade Name	Dose Equivalency (mg.)	Minimum Age for Advertising
Chlorpromazine	Thorazine	100	6 months
Thioridazine	Mellaril	100	2 years
Haloperidol	Haldol	3	3 years
Fluphenazine	Prolixin	2	12 years
Perphenazine	Trilafon	10	12 years
Trifluoperazine	Stelazine	5	12 years
Thiothixene	Navane	5	12 years
Loxapine	Loxitane	15	12 years
Molindone	Moban	10	16 years

Treatment of severe agitation:
  Chlorpromazine regimen (oral doses):
    Initial dose: 0.5 mg. per kg.
    Dose progression: Adm inister 0.5 to 2.0 mg. per kg. every two hours until controlled (or side effects are limiting)
    Daily dose range: 5.0 to 15.0 mg. per kg.
  Haloperidol regimen (oral doses; for intramuscular route administer half the oral dose):
    Initial dose: 0.5 mg.
    Dose progression: Administer 0.01 to 0.10 mg. per kg. every two hours until controlled (or side effects are limiting)
    Daily dose range: 0.1 to 0.5 mg. per kg.
Treatment of acute psychosis:
  Follow regimen for treatment of severe agitation
  Once symptoms are controlled, maintain on once-daily dose regimen, with daily (oral) dose range of:
    Chlorpromazine: 4.0 to 10.0 mg. per kg.
    Haloperidol: 0.1 to 0.3 mg. per kg.
  Taper as symptoms allow
Treatment of impulsivity:
  Chlorpromazine regimen (oral dose):
    Initial dose: 0.2 mg. per kg. test dose, then 0.6 mg. per kg. daily, divided into four doses
    Dose progression: Increase by 0.6 mg. per kg. per day (divided into four doses) every three to seven days
    Daily dose range: 0.6 to 4.0 mg. per kg. divided into four doses a day
  The less sedating antipsychotic drugs appear less effective for treating impulsivity
Treatment of verbal-motor (Tourette's) syndrome:
  Haloperidol regimen (oral dose):
    Initial dose: 0.5 mg. daily for two to three days
    Dose progression: Increase by 0.5 mg. daily every two to three days
    Daily dose range: 2 to 10 mg.

IV. Lithium

Minimum age for advertising: 12 years
Treatment regimen (halve these doses for children under age 12):
    Initial dose: 300 mg. twice daily
    Dose progression: Increase by 300 mg. every five to seven days, obtaining serum levels after five days on a given dose
                    Maximize the number of divided doses each day, up to four daily doses
  Therapeutic range for serum lithium level: 0.5 to 1.5 mEq. per liter
  Slow release tablet (Lithobid) may be advantageous to reduce side effects for some children. Also available in oral liquid (lithium citrate) form.

dren often have other prominent clinical features, including separation anxiety and age specific vegetative symptoms, which the criteria do not recognize. At the present time, while such alternative criteria are being developed, the *Diagnostic and Statistical Manual* criteria are probably as good as any method currently available for identifying antidepressant responsiveness in children.

A family psychiatric history of affective disorders (as well as affective spectrum disorders and responsiveness to antidepressants) is an important means of identifying children who may be vulnerable to developing affective illness. This vulnerability is usually viewed as a life-long characteristic, and does not specify that an individual has an illness at a particular time. Thus, a child may have a significant family psychiatric history of affective illness, but not have a drug responsive illness.

Also the dexamethasone suppression test for affective illness, a biochemical test of hypothalamic-pituitary adrenal overactivity in major affective disorders, has some diagnostic validity in children, and its clinical usefulness is currently under exploration (Poznanski et al., 1982).

A finding of major clinical significance concerns the very high percentage of parents of tricyclic responsive depressed children who themselves have major depressive and other affective illnesses. Although the data at this time are preliminary, several researchers are finding that over 50 per cent of the children with major depression have a parent who has had major depression or an affective illness. Since these parents, whose affective character disorders influence their parenting styles, are generally a part of the caretaking environment of the child, it is likely that not only biochemical genetics but also interpersonal factors associated with parental affective illness—both early developmental and current situational—may contribute to the emergence of affective illness in the youngsters. An interacting psychodynamic feature, in which both parent and child covertly experience separation anxiety relieved by the presence of the other, may lead to a setting in which both parent and child psychodynamically support each other's depression. In such cases, which may be the majority, it may be advisable to treat both parent and child with psychopharmacological (and psychotherapeutic) treatments simultaneously.

The technical use of tricyclic antidepressants is different in children and in adults. Higher doses are required to achieve therapeutic blood levels in children: Adults are generally treated adequately with imipramine, 3.5 mg. per kg., but children below the age of 14 years usually need 5.0 mg. per kg. Furthermore, children often require four to six weeks of treatment in order to achieve a peak therapeutic response, whereas adults generally show a full response in two to four weeks. Thus children need higher doses and longer treatment before improvement is seen.

Adjustment of antidepressant dosage by monitoring blood levels is unreliable. Although preliminary reports suggest that the therapeutic blood range for tricyclic antidepressants in children is similar to that in adults, currently available commercial reports of tricyclic levels do not include the hydroxylated metabolites of imipramine, which constitute almost 50 per cent of the total active metabolite mass in children (Potter et al., 1982). At present it is more advisable to use clinical side effects and electrocardiographic conduction slowing as determinants of maximal tolerable doses for children and adolescents.

Traditionally imipramine has been the antidepressant used predominantly in the treatment of children. Other tricyclic antidepressants with different side effect profiles are also effective, including amitriptyline (which, like imipramine, is available in the generic form), desipramine, and nortriptyline. Doxepin, protriptyline, trimipramine, and the newer "second generation" antidepressants have not yet been extensively studied in children. The side effects of the monoamine oxidase inhibitor antidepressants have discouraged recent investigations.

The side effects of the tricyclic antidepressants are listed in Table 58–2. Generally the antihistaminic and anticholinergic side effects (sedation, dry mouth, constipation, orthostatic hypertension, blurred vision) may limit the maximal daily dosage of a given drug, but sometimes cardiac conduction slowing is most significant.

The cardiac effects of antidepressants are usually less problematic in children than in adults, but it is good practice to be more conservative and careful in the cardiac monitoring of tricyclic antidepressants in children (Saraf et al., 1978). Since there has been one case of a child who died suddenly while receiving antidepressant medication (though at an unreasonably high dose of 15 mg. per kg.), it is good practice to obtain a baseline electrocardiogram in all children, to raise the dose gradually to 3.5 mg. per kg. (comparable to the adult oral dose), and then to obtain sequential electrocardiograms (perhaps every two weeks) while raising the dose from 3.5 to 5.0 mg. per kg. Electrocardiograms should be obtained not sooner than 48 hours after a dose increase. Once the electrocardiogram has been obtained on 5 mg. per kg. dose load, repeat electrocardiograms are unnecessary. The finding of a significant conduction slowing (PR > 0.20, QRS > 0.12) may require lowering of the dose, though higher doses may be considered under surveillance of cardiac consultation.

Children have a higher incidence of seizures with tricyclic antidepressant medications than adults. This may be due to the high doses of antidepressants used in children, their lower degree of neurobiological development, or the high prevalence of brain damaged children in the child psychiatric population. These drug induced seizures are generally found only in children who have abnormalities on neurological examination, clearly abnormal electroencephalograms, or a history of clinical seizures. It is advisable to have a full neurological examination with electroencephalography, as well as a general pediatric examination, prior to starting antidepressant treatment. In the event that a seizure occurs while antidepressant medication is being given, it is not necessary to stop the tricyclic antidepressant; instead phenytoin, added to the antidepressant, is generally sufficient to prevent further seizures. As with other types of drug induced seizures, the prophylactic use of phenytoin is not recommended (Itil and Soldatos, 1980).

Abdominal cramps are frequently reported by children (and often by adults as well). Weight loss is more frequent in children than in adults, although this tends to occur in children who are either overweight or overeating as a part of their depressive symptomatology. Weight loss usually stabilizes after about six weeks and is not itself a reason to alter the drug regimen.

One effect of antidepressants in children, only rarely seen in adults, is anger. The significance of this effect is unclear, but it may be a treatment emergent effect (similar to antidepressant induced mania in adults), or it may be a side effect. If overt psychosis emerges during antidepressant drug treatment, an antipsychotic drug (neuroleptic or lithium) should be employed while the antidepressant dose is lowered or discontinued.

CHILDHOOD MANIA

Similar to childhood depression, childhood mania has only recently been recognized as a possible clinical entity, and it appears in children with a different clinical picture from that in the adult form of the illness. (See Chapter 41.) At present, the symptoms of childhood mania are not well defined, and lithium is not formally indicated for any childhood disorder.

In general, the appearance of lithium-responsive illness in childhood is not associated with overt psychotic symptoms. Neither hallucinations nor delusions are part of the clinical picture in childhood. Instead a variety of behavior disorders (impulsivity, aggressivity, obliviousness, and temper tantrums) may be found to occur with a cyclic variation over time. Angry interpersonal interactions, low self-esteem, and intervals of "depression" are typical.

In certain cases, some features of attention deficit disorder may be found, including overactivity, distractibility, and impaired school performance (Davis, 1979; DeLong, 1978). Other associated symptoms, which may be too nonspecific to be helpful in the differential diagnosis, are anxiety, muscle tension, lying or stealing, and an extraordinary degree of social provocativeness and intrusiveness. This clinical description, however, is somewhat characteristic of many children in psychiatric treatment, who are often labeled as having a "borderline" or "tension discharge" disorder. Elucidation of more specific criteria for the identification of "manic" or lithium-responsive children awaits further research.

Different observers of lithium-responsive children have stressed the prominence of affective features (Youngerman, 1978), similarity to thought disorder (Carlson and Strober, 1978), behavioral lability ("emotionally unstable character disorder"; Klein et al., 1980), temper tantrums (Davis, 1979; DeLong, 1978), and physiological findings (McKnew et al., 1981; Popper, 1982).

Owing to the strong biological underpinnings of manic-depressive illness in adults, current research is being undertaken to distinguish lithium-responsive children from other children by using biological signs in addition to behavioral and psychological symptoms.

In a study of six children of manic-depressive parents, the two children who fulfilled formal *Diagnostic and Statistical Manual-3* criteria for bipolar affective illness were found to respond to lithium, and were also found to have neurophysiological responses (augmentation on average evoked potential measurements) that are characteristic of adults with bipolar illness (McKnew et al., 1981).

Lithium-responsive children have been observed to show a cyclic alternation between two physiological (vegetative) states, with periodic intervals of a "low" period, in which there was decreased dream recall, unusually sound sleep, overly lengthy sleep, dysphoric affect on awakening, and an improvement in this dysphoria over the course of one-half to four hours, and a "high" period, in which there was exceedingly violent imagery in nightmares, restless sleep with multiple awakenings, and difficulty in falling asleep (taking over two hours), with motor restlessness and excessive appetite and thirst. These two states appeared to correspond to "depressive" and "manic" periods, and the cyclic alternation was reported as early as two years of age (Popper, 1982). This cyclic variation in vegetative physiology may possibly be seen, then, many years prior to the appearance of overt psychotic and depressive symptoms.

Psychodynamically these children are particularly fearful of abandonment but defend energetically against dependence. This counterdependency is prominent in their attitudes toward medication, psychotherapy, and help in general, and typically complicates their process of receiving treatment.

Lithium-responsive children often have a family history of affective illness (particularly bipolar disorder), alcoholism and substance abuse, violence, and actual abandonments.

These children respond to lithium by improvements in their behavioral impulsivity, affective and cognitive symptoms, and vegetative signs.

Lithium may help to identify children who are at risk for developing adult forms of depression or psychosis, and who may present with a "low grade psychosis" or "prepsychosis." Alternatively some lithium-responsive children may grow into psychiatrically unimpaired adults. Follow-up studies are required before the long term prognosis in these children can be inferred.

At present diagnostic and prognostic descriptions of lithium-responsive children can be offered only on the basis of tentative findings, and the

Table 58–3.   ADVERSE EFFECTS AND MEDICAL MANAGEMENT OF PSYCHOTROPIC MEDICATIONS IN CHILDREN

I. Psychostimulants
 A. Work-up
  1. Physical and neurological examination, including height and weight, blood pressure, pulse, and dyskinetic movements
  2. Routine laboratory tests (complete blood count with differential, blood chemistry profile, and urinalysis)
 B. Common adverse effects
  1. Difficulty in falling asleep
  2. Mild elevation of pulse and blood pressure
 C. Less frequent adverse effects
  1. Decreased appetite
  2. Crying and dysphoria
  3. Growth retardation
  4. Drowsiness
  5. Nervousness and irritability
  6. Sweating
 D. Unusual but serious side effects and precautions
  1. Potential for medication abuse
  2. Hypertension
  3. Lowering seizure threshold
  4. Worsening of tic disorder or dyskinesia
  5. Psychotic breakthroughs
 E. Contraindications
  1. Marked anxiety, agitation, or psychosis
  2. Glaucoma
  3. Verbal-motor tic (Tourette's) disorder
 F. Toxicity and overdose
  1. Irritability, restlessness, agitation, nausea, diarrhea
  2. High fever, sweating, pallor, flushing
  3. Arrhythmias, tachycardia, significant hypertension
  4. Delirium, tremor, convulsions, coma
 G. Drug interactions
  1. Increases blood level of tricyclic antidepressants
  2. Increases metabolism of phenytoin
  3. Opposes effect of antihypertensives
  4. Acetazolamide increases renal absorption of amphetamines
 H. Follow-up
  1. Record height and weight every four weeks
  2. Follow blood pressure, pulse, and dyskinetic movements
  3. Yearly physical examination and routine laboratory tests, including liver function tests

II. Tricyclic antidepressants
 A. Work-up
  1. Physical and neurological examination, including weight, blood pressure, pulse, and dyskinetic movements
  2. Routine laboratory tests
  3. Baseline electrocardiogram for children under 18 years
  4. Baseline electroencephalogram for children under 18 years
 B. Common adverse effects
  1. Sedation
  2. Dry mouth, blurred vision, constipation, nasal stuffiness
  3. Orthostatic hypotension
 C. Less frequent adverse effects
  1. Cardiac conduction slowing, asymptomatic
  2. Insomnia and nightmares
  3. Abdominal cramps
  4. Mild tachycardia and elevated blood pressure
  5. Weight loss
  6. Urinary retention
 D. Unusual but serious side effects and precautions
  1. Cardiac conduction slowing with heart block
  2. Seizures (if pre-existing neurological abnormality)
  3. Hypertension
  4. Anger
  5. Aggravation of glaucoma and cystic fibrosis
 E. Contraindications
  1. Major cardiovascular disease or recovery from heart surgery
 F. Toxicity and overdose
  1. Atropine-like (dry mouth, blurred vision, mydriasis, paralytic ileus, bladder paralysis)
  2. Cardiovascular (hyper- and hypotension, arrhythmia, tachycardia, rapid weak pulse, cyanosis, shock)
  3. Neurological (respiratory depression, hyperpyrexia, convulsions, confusion, delirium)
 G. Drug interactions
  1. Increases effects of sympathomimetic amines
  2. Interacts additively with anticholinergic agents
  3. Increases sedation of other medications
  4. Potentiates effects of alcohol
  5. Psychostimulants increase blood pressure and blood tricyclic antidepressant levels
  6. Possibly increases cardiac arrhythmias under anesthesia
 H. Follow-up
  1. Record weight every four weeks
  2. Follow blood pressure, pulse, and dyskinetic movements
  3. Electrocardiograms are needed repetitively (every two weeks) to monitor cardiac conduction slowing while raising doses beyond imipramine, 3.5 mg. per kg.
  4. Warnings regarding machinery in shop class, driving, and alcohol
  5. Yearly physical examination and routine laboratory tests, including liver and thyroid function tests

III. Antipsychotics
 A. Work-up
  1. Physical and neurological examination, including weight, blood pressure, pulse, and dyskinetic movements
  2. Routine laboratory tests (complete blood count with differential, blood chemistry profile, and urinalysis)
 B. Common adverse effects
  1. Sedation
  2. Orthostatic hypotension
  3. Dry mouth, blurred vision, constipation, nasal stuffiness
  4. Motor restlessness (akathisia)
  5. Acute dystonias
  6. Parkinsonian motor symptoms
  7. Weight gain
 C. Less frequent adverse effects
  1. Sunburning (mainly with chlorpromazine and thioridazine; use sun-blocking lotions)
  2. Acne-like rash (mainly chlorpromazine)
  3. Retrograde ejaculation

Table 58–3.   ADVERSE EFFECTS AND MEDICAL MANAGEMENT OF PSYCHOTROPIC MEDICATIONS IN CHILDREN *(Continued)*

4. Difficulty in maintaining penile erections
5. Paradoxical agitation
D. Unusual but serious side effects and precautions
  1. Tardive dyskinesia: involuntary movements, especially of face and mouth, which may worsen on lowering dose of antipsychotic; neurological effect is generally related to dose and duration of antipsychotic treatment; particularly serious, since some cases may be irreversible
  2. Eye changes: retinitis pigmentosa (thioridazine only); reversible cloudy reflex (chlorpromazine)
  3. Obstructive jaundice and hepatitis
  4. Leukopenia (if sore throat, get complete blood count with differential immediately)
  5. Systemic lupus erythematosus-like syndrome
  6. Aggravation of glaucoma and cystic fibrosis
  7. Amenorrhea and galactorrhea; gynecomastia
  8. Possible lowering of seizure threshold
E. Contraindications
  1. Liver damage
  2. Severe central nervous system dysfunction, obtundation, coma
  3. Respiratory depression
  4. Blood dyscrasia
F. Toxicity and overdose
  1. Atropine-like (dry mouth, blurred vision, mydriasis, paralytic ileus, bladder paralysis)
  2. Severe extrapyramidal symptoms
  3. Disturbances of cardiac conduction
  4. Temperature dysregulation
  5. Central nervous system depression, seizures, disorientation
  6. Respiratory depression
G. Drug interactions
  1. Interacts additively with anticholinergic drugs
  2. Potentiates sedation of other medications
  3. Potentiates effects of alcohol
  4. Probably enhances morphine analgesia
  5. Antacids may decrease absorption of antipsychotics
H. Follow-up
  1. Record weight every four weeks
  2. Follow extrapyramidal signs, blood pressure, pulse, and possibly late developing dyskinesia
  3. Repeated warnings regarding:
    a. Sunburning
    b. Susceptibility in extremely hot or cold weather
    c. Sickness and colds (?hepatitis, ?leukopenia)
    d. Machinery in shop class, driving, and alcohol
    e. For adolescent males, retrograde ejaculation and difficulty in maintaining penile erections
  4. Yearly physical examinations and routine laboratory tests, including liver function tests
IV. Lithium carbonate
  A. Work-up
    1. Physical and neurological examination, including weight, blood pressure, and pulse
    2. Routine laboratory tests (complete blood count with differential, blood chemistry profile, and urinalysis)
    3. Baseline electrocardiogram for children under 18 years
    4. Baseline electroencephalogram for children under 18 years
    5. Electrolytes
    6. T_4, T_3 resin uptake, thyroid stimulating hormone, and thyroglobulin antibodies
    7. Creatinine blood levels—three determinations on different days; then average to get baseline value
    8. Optional: baseline creatinine clearance and urine concentrating capacity
  B. Common adverse effects
    1. Mild gastrointestinal irritation (take medications with meals)
  C. Less frequent adverse effects
    1. Lithium induced dehydration: especially with exertion or hot weather; dehydration may result in high and toxic lithium levels; push liquids liberally when dehydration is anticipated
    2. Polyuria and polydipsia
    3. Tremor (helped by diazepam or propranolol)
    4. Acne-like rash
  D. Unusual but serious side effects and precautions
    1. Possible renal damage—see text (potentially significant)
    2. Goiter (especially in patients with thyroid dysfunction)
    3. Drowsiness, dizziness, ataxia, slurred speech, and muscle weakness may be present in therapeutic range
    4. Leukocytosis
    5. Elevated blood sugar—may be significant for diabetics
  E. Contraindications
    1. Significant renal, cardiovascular, or thyroid disease
    2. Pregnancy (teratogenic cardiac abnormalities)
    3. Dehydration syndromes
  F. Toxicity and overdose
    1. Ataxia, confusion, tremor, muscle weakness, drowsiness, slurred speech, tinnitus
    2. Increasing neuromuscular irritability
    3. Lethargy, obtundation, coma, seizures
  G. Drug interactions
    1. Potentiates antithyroid activity of potassium iodide
    2. Decreases pressor effect of norepinephrine
    3. Diuretics increase lithium blood levels
    4. Marijuana may increase lithium blood levels
  H. Follow-up
    1. Record weight every four weeks
    2. Monthly checks of blood lithium levels
    3. Every three months, check blood creatinine levels and morning urine specific gravity
    4. Follow blood pressure, pulse, and dyskinetic movements
    5. Repeated warnings regarding:
      a. Dehydration toxicity—encourage adequate fluid intake before exertion and on hot days
      b. Sudden dietary intake of salt can lower blood lithium level abruptly and cause symptom return—encourage consistent salt intake
      c. Machinery in shop class, driving, alcohol, and marijuana
    6. Yearly physical examinations and routine laboratory tests, including thyroid stimulating hormone

clinical application of lithium treatment in the community should probably be delayed or reserved for only the most severely disturbed children.

When lithium is employed, a special work-up is required (Table 58–3). Therapeutic blood levels for children and adults are probably approximately comparable, and levels up to 1.5 mEq. per L. may be required for full symptom control during certain phases of the illness. Owing to the cyclic nature of this illness, a level of 0.5 mEq. per L. may be adequate at some periods, but 1.5 mEq. per L. may be required at other times in the disease cycle. Many children attain better results with the slow release capsule than with standard preparations of lithium.

Once a stable dose of lithium has been determined, blood lithium levels (a burden for some children who are afraid of hypodermic needles) and blood pressure checks should be obtained every four weeks, and a serum creatinine level (and possibly a test of urine concentrating ability) should be obtained every three months to appraise kidney function. Thyroid stimulating hormone levels should be obtained every 12 months along with the routine physical examination.

Side effects of lithium are similar in children and adults, including the common dehydration toxicity: shifts in salt and water balance (during summer heat or exertion) may cause elevations of blood lithium levels into the toxic range (Table 58–3).

The risk of lithium induced kidney damage has been raised as a potential problem with long term use in adults and may limit its clinical applicability. At the present time the significance of the kidney changes in adults is unclear—it may be drug induced, or it may be disease related; it may be serious, or it may be functionally insignificant in most individuals. It is not known whether children are more or less prone than adults to develop these renal changes. For now, lithium should be considered only for children whose conditions are severe enough that their overall development may be damaged by withholding the treatment.

If lithium is found to be therapeutic for a child, it should not be assumed that life-long pharmacological treatment is needed. These children are uniformly troubled with a variety of psychodynamic issues, and lengthy intensive psychotherapy and psychoeducational treatments are frequently required. Progress may be unworkably slow if medication is not employed. Their counterdependence toward any help—including psychological and medical treatment—typically complicates their course. Management of this counterdependence often *is* their psychotherapy: For some, successful limit setting may constitute the beginnings of productive treatment. Several months or years of combined pharmacological and psychodynamic treatment may be sufficient to allow the child to attain an adequate degree of psychological development so that continued use of medication might not be required.

Although long term follow-up studies of these children will be many years in the making, the use of lithium in children may contribute to our current thinking about the developmental precursors of the major adult psychiatric illnesses.

PSYCHOTIC DEPRESSION

Major depression can appear at times with overt psychotic symptoms, such as delusions and hallucinations. (See Chapter 41.) The joint appearance of depression and psychosis is not well understood in adults or in children. Such conditions may be labeled psychotic depression, delusional depression, depressive psychosis, or paranoid depression. They may be viewed as representing a particularly severe sort of depression, a different category (not necessarily more severe) of depression, or the combined occurrence of two separate conditions (psychosis with depression).

Since children are developing their sense of reality, "reality testing" is incompletely developed, and this may result in the exaggerated appearance of "psychotic symptoms."

Owing to the suicidal and unpredictable nature of many of these children, intensive treatment programs should be made available.

For the most rapid and complete pharmacological treatment, it is recommended that both antipsychotic and antidepressant drugs be used in combination. If delusional depression is treated with antipsychotic drugs alone, there is a rapid control of overt psychotic symptoms and agitation, but affective symptoms become dominant over the course of time. If treatment is instituted with antidepressants alone, improvement in the affective disorder is found after a delay of several weeks, with later improvement in the psychotic symptoms secondary to the lifting of the depression. A response of both affective and psychotic symptoms is most rapid with the initial use of both drugs simultaneously.

Similarly the psychodynamic treatments that work best are directed to the treatment of depression as well as to treatment of psychosis.

PHOBIC DISORDERS

Phobias (fears) are common in children, particularly at ages three to five years when specific fears may be seen in developmentally normal children. Sometimes such "phobias" persist or reappear, reflecting conflicts based on oedipal dynamics (with a specific sexual and aggressive significance of the fears). These phobias are probably not drug treatable.

There is also a different group of phobias—which improve with antidepressant medication—based on separation anxiety (rather than oedipal anxiety). A fear of being away from mother, out of the house, at school, in a new place, or in crowds can sometimes be a phobia based on such (preoedipal) separation anxiety.

GENERAL COMMENTS ON SCHOOL ABSENCE

School "phobia" can be due to many causes, and a careful differential diagnosis is required in order to determine whether there is an underlying separation anxiety (true "school phobia"). Some cases of school absence result from physical intimidation and require social interventions. More often truancy, and occasionally psychotic paranoia, may account for school absence without any true phobic dynamics. (See Chapter 15.)

Truancy is a type of school absence associated with counterdependent behavior and is seen in active and socializing children who often are developing character disorders; it is best responded to by limit setting, and antidepressants are usually ineffective. In contrast, phobic children tend to be dependent, clinging, and passive and may reveal little or no sense of social involvement with peers and the peer groups. Paranoia may be the presenting symptom of a psychotic disorder and may be made worse both by antidepressants and by limit setting. When truancy and paranoia can be ruled out, lengthy school absence is often treatable with an antidepressant, even when other manifestations of depression are not evident.

In general, if school absence continues for more than two weeks, and truancy and paranoia are both ruled out, an antidepressant is likely to be helpful. The diagnosis is usually a depressive, phobic, or panic disorder.

SEPARATION ANXIETY DISORDER

Certain children become anxious when separated from parents or from some other personally significant individual. They often cling to people and avoid situations in which they act independently. Sometimes their anxiety is accompanied by fantasies of harm befalling their parents while away. These children—many of whom have excessive school absences—often have a depressive, phobic, or panic disorder, and may respond to antidepressants.

PANIC DISORDERS

Some children present with anxiety attacks, which include many somatic symptoms, such as hyperventilation, restlessness, nausea, dizziness, and trembling, and with a mixture of fear, anger, anxiety, and confusion. Such episodes of "panic" tend to last five to 20 minutes, almost never exceeding 30 minutes, and may occur several times a day or week. They may occur episodically for several weeks or months at a time, and in adults may be experimentally induced by the infusion of sodium lactate. Such panic attacks often are seen in children who have phobic or depressive disorders, but sometimes the panic anxiety is the only symptom observed. As with such panic disorders in adults, these conditions are often seen in families with affective or phobic disorders and may be successfully treated with antidepressants.

Superimposed on the panic attacks there may develop a fear of having a recurrence of the panic attacks. This "anticipatory anxiety" may become quite generalized and may persist long after the panic attacks have been stopped by antidepressants. The anticipatory anxiety does not respond to antidepressants, can respond to minor tranquilizers, but is probably best managed by supportive and behavioral measures. It is helpful to advise such patients that the anticipatory anxiety may not improve substantially until after they have seen for themselves that they can trust the medication to control the panic attacks.

OBSESSIVE-COMPULSIVE DISORDER

Overt ritualistic types of behavior are sometimes observed in children with obsessional thinking. (See Chapters 36 and 41.) This disorder has been recently described, and its distinctive clinical features may interfere with a child's ordinary functioning (Rapoport et al., 1981). Particularly when associated with a family history of depression, such obsessional symptoms may be successfully treated with antidepressants.

GENERAL COMMENTS ON THESE "AFFECTIVE" DISORDERS

The foregoing disorders—childhood depression, childhood mania, psychotic depression, phobic disorder, panic disorder, and obsessive-compulsive disorder—may be viewed as composing a series of conditions associated with depression, or as constituting variations of a more basic disorder. All these conditions are responsive to antidepressants, although they may not appear to be depressions. All are associated with cognitive as well as affective disturbances, and thus the label of "affective illness" may be too limiting.

In addition to the common characteristic of pharmacological responsiveness to "antidepressants," all these conditions present intermittently with episodes triggered by loss, are psychodynamically associated with an underlying separation anxiety, and generally occur in families with other affective illnesses. Furthermore, children with these conditions generally have a parent who has an affective disorder that is treatable with antidepressant medication.

This group of disorders is distinguished by relatively specific pharmacological, psychodynamic, phenomenological, and familial-genetic characteristics.

In treatment, these children may show only limited improvement until the parent's affective disorder is treated pharmacologically and psychotherapeutically.

ADULT–FORM SCHIZOPHRENIA

Schizophrenia may appear before the age of 16 with overt hallucinations, delusions, and disordered reality testing similar to that in schizophrenic adults. The juvenile appearance of the adult form of schizophrenia needs to be distinguished from childhood schizophrenia (atypical development, infantile autism), to be described. When an overtly disturbed child has an early developmental period with generally adequate functioning, except for prominent social isolation (the "loner") and a subsequent (after age eight) acute or subacute onset of psychotic disorganization, the early appearance of adult schizophrenia should be considered. This illness may be treated with antipsychotic drugs; however, frequently such medications are not very successful, and may offer relatively little improvement over no medication at all. Such cases are generally resistant to other forms of therapy as well, and the prognosis may be very poor. Since misdiagnosis is common for this particular condition, aggressive empirical trials of psychotherapy and pharmacotherapy may be considered.

ATYPICAL DEVELOPMENT (CHILDHOOD SCHIZOPHRENIA, INFANTILE AUTISM)

Certain psychotic disorders of childhood are based on abnormal primary development rather than regression from a higher level of functioning. (See Chapters 39F and 41.) These conditions are not directly related to psychosis in adulthood and

later childhood, and instead are examples of arrested or severely delayed development, probably secondary to as yet undefined organic disease (Ornitz and Ritvo, 1976). Children with "atypical development" have never developed functionally beyond a primitive (psychotic) level of understanding of reality. They have little balanced emotional contact with things and people in the world, and use much of their mental power to comprehend ordinary events in the world. When a child's psychosis is based on such abnormal development rather than regression from a previously high level of functioning, medications are usually of little benefit. (Antipsychotic medications may be helpful in an early onset of the adult form of schizophrenia or other psychotic conditions in which the medications serve to re-establish the previously attained higher functions.) It is not surprising that medications cannot restore an education, understanding, or development that the child had never himself attained. Instead a comprehensive program entailing a controlled milieu, social interactions commensurate with the child's abilities, and some cognitive, sensory, and motor training can have a powerful impact on the child's development and future.

For many children with atypical development, neurological or electroencephalographic abnormalities may be found, and the use of anticonvulsants may then be considered.

These children may show periods of extreme agitation or anxiety during which their behavior becomes disorganized and unmanageable. They may experience such agitation intermittently for several weeks or months, particularly during stressful periods of development (such as puberty). At these times individual doses of sedative drugs for the purpose of calming the agitation may be helpful (see foregoing section on "Extreme Agitation").

In general, the long term use of psychotropic medications in children with atypical development does not contribute significantly to their overall development, and other treatment modalities may be employed more effectively.

Certain children with atypical development, such as those with high blood levels of serotonin, may some day be treatable by specific biological intervention (Geller et al., 1982).

BEHAVIORAL IMPULSIVITY

Impulsive behavior, angry outbursts, recklessness, assaultiveness (to self or others), destructiveness, or repeated incidents of extreme agitation may be seen as presenting symptoms of "tension discharge" or "borderline" disorders in children.

(See Chapter 36.) Such diagnoses are poorly delineated and probably represent different conditions with superficially similar symptomatology.

A pattern of repetitive impulsivity often may be adequately treated with low divided doses of sedating antipsychotic drugs, for instance, chlorpromazine (25 to 50 mg.), thioridazine (25 to 50 mg.), or loxapine (5 to 10 mg.) given four times a day.

Nonsedating drugs are often preferred by clinicians, but are generally less effective for treating such impulsivity. Also divided doses are more effective than single larger doses for treating this behavioral impulsivity, even though neuroleptic drugs can be used once daily for treating psychosis. It seems that the management of neuroleptic drugs for treating this diagnostically mixed group of behavioral impulsivities is different from the management of neuroleptic drugs for treating adult psychosis.

Some of these children appear to be on their way to developing overt psychotic conditions in later years, and the use of low and divided doses of neuroleptic drugs is recommended for the putatively "prepsychotic" or "low grade psychotic" phase. Again, the developmental precursors of major adult psychosis cannot yet be identified, but differential drug responses may help identify such children at risk.

Management with low doses of sedating antipsychotic agents should be considered a nonspecific means of controlling behavioral impulsivity, to be used while proceeding with diagnostic evaluation for alternative and more specific treatment with stimulants, antidepressants, or lithium.

In general, minor tranquilizers are difficult to use in this group because of the risk of drug abuse.

ANXIETY AND SLEEP INDUCTION DISORDERS

Apart from anxiety related to specific disorders, children may experience anxiety due to situational stress. (See Chapter 32.) Clinicians sometimes employ sedatives to treat anxiety or difficulty in falling asleep. In hospitals such medication also may be used to control anxiety associated with (or preventing the administration of) medical and surgical procedures.

In children benzodiazepines (such as diazepam, 2 to 5 mg.) should be used cautiously when the neurological status is not known. In some cases anxiety may be "covered" by diphenhydramine (Benadryl), 25 to 50 mg., or chloral hydrate, 500 to 1000 mg. In more severe cases it may be necessary to use chlorpromazine, 25 to 50 mg., or

haloperidol, 2 to 5 mg. If needed, these doses may be repeated once. For children under six years old, these doses may be halved.

Most clinicians prefer to avoid such treatment when possible. These treatments may be considered for occasional short term use in special situations, but persistent anxiety or difficulty in falling asleep (more than four weeks) should lead to further evaluation.

SLEEP-AROUSAL DISORDERS

Sleep disorders are common in children, and should always be evaluated in the context of searching for both associated medical and other psychiatric disorders. When appropriate evaluation has ruled out other medical and psychiatric diagnoses, certain sleep symptoms or disorders may be treated with medication. Sleep terror disorder (an acute disruption of sleep in which a child shouts or screams "during sleep" without awakening for minutes, quiets gradually in several minutes, and does not recall the incident in the morning) may be treated by benzodiazepines or antidepressants. Sleepwalking and sleep-talking may be treated with antidepressants. Narcolepsy may be treated by stimulants. Sleep that is "too deep" may also be treated by imipramine and should be evaluated for the possibility of associated affective disorder.

ADMINISTRATION OF PSYCHOTROPIC MEDICATION IN CHILDREN

ETHICAL ISSUES IN CHILD PSYCHOPHARMACOLOGY

The practitioner in child psychopharmacology is faced with the common problem of applying new research to clinical practice. New information is becoming available concerning the usefulness of drugs, and there is a temptation to employ such new information and new treatments even before full knowledge of the appropriateness of such treatments is obtained. Alternatively, and equally dangerously, there is a tendency to be overly careful and "protective" of children—a clinical attitude that may similarly deprive children of good medical care.

The use or nonuse of psychotropic medications in children requires especially thoughtful judgment by clinicians about the individual case. The Food and Drug Administration (FDA) provides guidelines for the use of particular drugs in children of certain ages, but there is also legal recognition of the physician's and patient's rights to make decisions based on individual clinical observations. That is, the FDA guidelines are not legally binding, and the clinician's judgment is legally recognized. If a clinician believes that a particular drug would be appropriate for a child, that judgment takes precedence over FDA guidelines. The FDA itself has emphasized this point, asserting that "Under the Federal Food, Drug, and Cosmetic Act, a drug approved for marketing may be labeled, promoted, and advertised by the manufacturer only for those uses for which the drug's safety and effectiveness have been established and which the FDA has approved . . . . The [Food, Drug, and Cosmetic] Act does not, however, limit the manner in which a physician may use an approved drug. Once a product has been approved for marketing, a physician may prescribe it, for uses or in treatment regimens or patient populations that are not included in approved labeling. . . . Accepted medical practice often includes drug use that is not reflected in approved drug labeling" (Department of Health and Human Services, 1982).

There are important practical matters that a clinician will want to consider in prescribing psychotropic drugs in children:

1. There is a lack of well defined and well established indications for many of the uses of psychotropic medications in children.

2. At present there is uncertainty about the short term as well as long term side effects of such drugs in children.

3. Furthermore, the effect of these drugs in developing children cannot be extrapolated from animal research or adult psychopharmacology: possible drug effects on the developing brain, endocrine systems, and somatic system can be determined only by direct application and observation in children.

4. There are special problems of informed consent in children and in psychiatric patients: the nature of informed consent, how children participate in the process, and the validity of consent given by psychiatric patients are all topics of continuing discussion. The nature of informed consent for individuals who are both children and psychiatric patients has never been formally addressed.

5. The clinician must consider the risks of medication abuse by the individual patient, as well as by peers, the family members, and the community at large.

Such issues will undoubtedly give both clinicians and parents reason to pause before rushing to use psychotropic medications.

Nonetheless the risks of withholding treatment or withholding research from children also has great consequences.

The ethical dilemma concerning the use or non-use of psychotropic medications in child psychiatry will become increasingly important as more information about the usefulness of drugs in children is accumulated.

CONSENT FOR MEDICATION TREATMENT IN CHILDREN

"Informed consent" must be obtained prior to starting a treatment, with special considerations if such a treatment is a nonestablished treatment. (See Chapter 61.) For children, full consent contains several components:

"Informed Consent" of the Parents

Since children are unable to make judgments legally in their own behalf, the parents become responsible for making judgments for them. Written release forms may be used when nonestablished therapy is suggested, describing the nonestablished nature of the treatment for the child's disorder, the expected risks and side effects, and an explicit statement that there are unknown risks in using the drug in young people. Such consent forms do not relieve the physician of legal responsibility, but can be used to demonstrate that the parents contributed to the decision making process regarding the use of medication for their child. Such forms are frequently misunderstood by parents, and do not in any way substitute for good verbal explanations and for meetings in which the parents can ask questions of the prescribing physician.

"Emotional Consent" of the Parents

Aside from whatever information may be provided, the parents or guardians who are responsible for the decision to use the drug should "wrestle" with the decision. There should be nothing automatic or rapid about a parent's decision to use an established or a nonestablished treatment for his child. The parent who says, "The doctor knows best," and turns the decision over to the judgment of the doctor, has avoided taking emotional responsibility for the decision. It is suggested that nonestablished (and generally established) treatments be delayed until parents are able to do their best in coming to grips emotionally with the fact that pharmacological treatment is being recommended for their child.

"Assent" of the Child

Since children are unable to provide "consent," the notion of assent has been developed: they can contribute something to the consent process, though it may not be the legal equivalent of informed understanding. The use of "assent" is a sensible step, which helps the clinicians and parents to be aware of whatever hesitations or concerns the child may have. What constitutes appropriate "assent" depends on the child's capacity for understanding and is best approached in a manner that respects the individuality of each child.

Optional "Concurrent" Consent of a Guardian

For cases in which the competence of the parents' consent may be questioned (for some psychotic parents, or for parents whose sadistic impulses may not be well controlled), it may be advisable to obtain the additional concurring consent of an independent guardian.

Documentation of Consent

The documentation in a medical chart should contain an explicit statement of the consent, a clinical description of the parents' and the child's responsiveness in the consent process, including a statement of their degree of understanding and emotional involvement in the consent process, diagnosis, specific target symptoms for the medication treatment, drug dose and regimen, special neurological or medical conditions that may interact with the drug, and the possible risks of medication abuse.

MEDICATION ABUSE

All psychotropic medications may be abused, since any mind altering effect may be pleasurably experienced as being "different" or as "a trip." The risks of medication abuse are particularly significant in adolescents. Adolescents or children may themselves have an inclination to abuse, may socialize with drug abusing peers, or may feel tempted to sell the pills for money. Children who are not inclined to these forms of drug abuse may nonetheless experience "strong arming" in the schoolyard, where other children may exert coercion to obtain pills from them.

Specific precautions may be exercised to minimize the chances of medication abuse. It is recommended that children and adolescents be advised not to carry pills around in their pockets, in order to decrease the temptation of selling or being strong armed. For medications that can be administered before or after school, it may be possible for all doses to be taken under the parents' supervision. If medications need to be administered during the daytime at school, the drugs may be kept by the school nurse in a special location, and the child may go to see the nurse for the medication. Many schools are requiring this procedure as a routine. This arrangement has the additional advantage of allowing a professional to monitor the child on a daily basis for both therapeutic and side effects. The prescribing doctor can maintain open communication with the school nurse as a means of obtaining clinically useful information.

Medication abuse can sometimes be avoided by prescribing the less abusable drugs. Instead of using stimulant medications, tricyclic antidepressants sometimes can be employed. Instead of minor tranquilizers, such as barbiturates or benzodiazepines, a physician may prescribe a low dose of a major tranquilizer.

Another potential source of medication abuse is the family. Any member—parent or child—may be considered at risk for substance abuse, suicidal overdose, or (especially for younger siblings) accidental ingestion. It is not unusual for family members, including parents, to try to get a therapeutic effect from the child's medication. It is suggested that parents be specifically advised to tell the physician if they feel an inclination to try the medication for themselves.

Because of the risks of accidental or suicidal overdoses, it is advisable to prescribe a limited number of pills at a time. Antidepressants are particularly dangerous when taken in overdoses and constitute a very high fraction of accidental lethal overdoses in children (generally children who ingest their parents' antidepressants). Antidepressants prescribed for adults or children should be stored carefully in places where such accidents are unlikely.

There are a number of settings in which medication should not be prescribed:

1. When the risks of medication abuse cannot be managed. In situations in which these risks cannot be adequately controlled, it may be necessary to forego drug therapy—even when it is indicated.

2. When proper administration of the drugs (or observation of the therapeutic and side effects) cannot be obtained. Unfortunately certain underfunded public facilities may be unable to use these treatments safely.

3. In place of psychotherapy.

4. In settings in which drugs are the only available treatment, that is, in the absence of psychotherapy.

5. When the child's parents disapprove of the use of the medication. It is unreasonable to expect a child to take a medication when even one parent disapproves. Sometimes a parent's reluctance may be based on reasonable opinion, but sometimes it may reflect a resistance to obtaining (or accepting) help, treatment disruption, or oppositional split-

ting with the other parent. In these situations it is advised that discussions with the parents continue to focus on the treatment resistance, while postponing treatment with medications.

6. When the child reasonably refuses to use the medication, despite support and approval for treatment by both parents. This situation is uncommon. (More often, when the parents support and the child refuses treatment, the child is oppositional or unreasonable.)

MANAGEMENT ISSUES IN PRESCRIBING MEDICATION FOR CHILDREN

In presenting the idea of medication to children, one focuses on a symptom that the child recognizes and wants to change. Needless to say, it is unhelpful to pick a symptom that may be prominent but that the child does not recognize, or that he observes but is not troubled by.

Children are frequently willing to look for help in treating their difficulty in falling asleep, anxiety, nightmares, "getting into trouble," or fear of being alone. It is worth time to allow the child to state his wish for help in terms that are acceptable to him. A child who is reluctant to admit personal problems is unlikely to accept medication or psychotherapy.

Pills are a concrete form of help and can help clarify the dynamics of getting and using help for the child and the family.

In allowing a child to express his feelings about the medication, one may hear about the child's fears of personal weakness, dependence, life-long use, "being crazy," having one's mind "controlled," hangovers, sedation induced passivity, muscular weakness, genital dysfunction, sterility, overexcitation, or being poisoned. The expression of such fantasies is not a waste of time for the pediatrician or psychopharmacologist. Exposing and working with such fantasies about the pills can promote the individual's effective use of medication. The child may have covert but identical fears of receiving help, or of receiving psychotherapy as well.

It is also worth time to allow the child to admit reluctance to taking medication or, alternatively, to express an excessive, magically based interest in taking the medication—and to deal with both as psychological resistance to the treatment.

Often a child resists accepting help—any help—because of a history of being coerced by a parent. To coerce such a child to take medications is to bypass the appropriate management of his resistances, and places him in a passive and nonresponsible position.

Often a drug treatment is best delayed while such resistances are worked with psychothera-

peutically. It then can be helpful to retain focus around the child's refusal to accept medication, as a clear and concrete way of working with the child's refusal to accept help.

If the child's medication refusal is viewed as unreasonable and as preventing necessary treatment, the child is sometimes forced to take the medication. The legality of such an action is uncertain, but clinically the treatment must then deal with the child's anger at the coercion. Therapeutic coercion is often a part of the psychiatric management of children, but may be inadvisable for nonestablished treatments.

Frequently the parents can be brought usefully into this aspect of management. Any psychological resistance that either parent may have to medications will probably show up in the child's medication-taking behavior. That is, if even one parent is hesitant about the use of medication, the child is likely to resist its usage. This is true even when the parents are divorced and the resistant parent is living far away; the child is likely to talk with the distant parent, and the parent's doubts can be communicated and expressed in the child's attitudes and behavior. Only when both parents are solidly behind the treatment and able to help a child deal with feelings about the treatment will a child be able to work consistently with the treatment, whether pharmacological or psychodynamic. The management of parental resistances to medication usage replicates and parallels the management of the parents in any other aspect of the child's therapy.

Again, getting "quick" consent loses the opportunity to strengthen the alliance around understanding the problems and the need for special intervention—and ultimately loses effective support for the treatment. This process takes time.

In general, it is clinically far better to delay starting a treatment than to try to force a treatment over resistance, since the child or his family will stop the medication sooner or later. The best time for dealing with the resistance to medication is at the start of the treatment, not after the passive "compliance" breaks down.

In addition to the psychological "resistances," the administration of medications may be modified by "reality" factors and "family ego." If a family lacks the money to buy the pills, or no one is at home with a child during the day, or the parents do not remember to give the pills, the medications will probably be administered inconsistently. The ability of the family to organize itself for a medication treatment provides the prescribing doctor with a sense of the reality of the child's home life and the capacity of the parents and child for personal self-care and organization.

It is essential to explain to each parent the nature of the treatment. Often a parent has feelings and fears about medications, which may take some attention to manage. (If one parent is away, arrangements should be made to communicate with that parent, in order to achieve joint parental support for the child.)

Many parents need to have specific explanations concerning their responsibility in making the decision in regard to the treatment, and for supporting the treatment of the child. Furthermore it is often necessary to tell parents specifically to keep the pills in their own control at home, and to be responsible themselves for administering the pills to the child or adolescent at home. Asking a young child or midadolescent to keep such powerful medicines (fantasies of "poisons") in their bedrooms may be unduly stressful. Parents can usefully maintain such responsibility as a part of appropriate parenting. This is not recommended as an area in which to promote the child's "autonomy": not until late adolescence is it usually advisable to ask youngsters to be responsible for keeping the bottles with the pills.

It is always appropriate to discuss the possibility of medication abuse with parents, including specific discussion of abuse by family members. Parents should be given a clear explanation of the side effects and drug interactions (including alcohol) and be urged to telephone the physician if unexpected side effects appear. Also intervening colds or medical problems, and unanticipated questions, should result in direct telephone contact between parent and doctor. (Usually in multidisciplinary practices in which the psychologist or the social worker is the main contact person with the parents, direct contact with the physician about such problems works better than attempting to filter all such communications through the therapist.) Furthermore it is good practice to advise a parent that if a dose is accidentally missed, the next dose should not be doubled. Permission may be requested to inform school officials about the treatment and to maintain open discussion with the school nurse about therapeutic and side effects. In writing, the doctor may inform the school nurse of the drug, dose, target symptoms, and side effects. For legal reasons the diagnosis should probably be withheld, but target symptoms may be specified. A separate labeled medication container can be sent to the school, by way of a separate prescription to the family. The school nurse can administer medication during the school day, follow side effects and therapeutic effects, and provide feedback to the prescribing physician.

Given the new developments in childhood psychopharmacology, one should expect considerable variation in the knowledge and understanding that parents and school officials have concerning psychotropic medications. This is an opportunity for public education regarding these treatments.

Some schools are not equipped to provide adequate medical coverage, and the staff may be particularly uncomfortable if some of their students are receiving these medications.

PSYCHIATRIC DISORDERS OF PARENTS

It is useful to remember that the parents of many child psychiatric patients themselves have a significant psychiatric disorder. In reaching out to these parents, it is well to remember that many of them have phobias (including resistances to travel to appointments), separation anxiety disorders, psychoses, and depressions. One cannot automatically assume that the parents are able to act in their child's (or their own) best interest, though their intentions may be good. When parents troubled by psychosis or depression are living at home and exerting a major influence on the child and the child's milieu, it may be essential to consider reaching out to the parents and help them accept psychiatric and possible psychopharmacological treatment for themselves. The treatment of major parental psychopathological illness is sometimes an essential feature in the treatment of the child. To treat a child's separation anxiety but then to send him home to a depressed and clinging mother may be self-defeating—that is, defeating the long term development of the child.

THE ROLE OF DRUG THERAPY IN PSYCHIATRIC TREATMENT AND PSYCHOLOGICAL DEVELOPMENT OF CHILDREN

When a drug is prescribed to a child, the parents need to be given a statement of the significance of this intervention, the role of the psychotherapy in a drug treatable illness, and a long term view of the child's condition. The nature of the illness is often viewed by parents in terms of how long the drug will be used and whether symptoms might return after administration of the drug is stopped.

For most conditions these questions are fairly straightforward. However, it may be more difficult to explain how psychotherapy can help in a condition that is viewed as drug treatable, biochemically based, and genetic. It may be easier to explain how psychotherapy may help the child's adjustment to the illness and general development, but to explain how the disease manifestations may be improved by psychotherapy as well as medications requires providing a "model" of how mind and body may interact.

In the case of depression, for example, several months of drug treatment may be sufficient to treat a particular depressive episode; however, the clinician may choose to inform the parent that such episodes may occur intermittently throughout life. Psychotherapy might then alter the psychological vulnerability to certain events and decrease the child's over-responding to anxiety or depressive affects, and thereby diminish the likelihood of depressive psychosomatic reactions in the future. That is, depressive or anxiety episodes may be induced by particular environmental events, and psychotherapy may decrease the child's psychological vulnerability to such situations. Because psychotherapy provides increasingly more effective ego defenses to such affects, the need for medication over the course of years may decrease. Drugs such as antidepressants or lithium may be used for several months or years, until the other treatments (individual psychotherapy, family modalities, situational interventions, intensive residential treatments) provide sufficient support for the child to proceed without medication. Drugs may be useful during the lengthy period of waiting until treatment facilitated ego development takes hold.

There is evidence that both psychotherapy and psychopharmacotherapy provide different and noninterchangeable aspects of a psychiatric treatment (Rounsaville et al., 1981).

It is still possible to find clinicians who believe that one treatment is "superior" to the other or who view psychiatric treatment in the polarized terms of treating just the mind or the body. These physicians will have difficulty in providing effective integration of treatment modalities, since their conceptions of their patients will be unscientifically split.

SUMMARY

It is appropriate to hesitate before using psychotropic medications with children. In addition to the known short term and long term effects of these drugs, there are unknown possibilities for specific developmental effects that may be observed only in children. However, these risks may be outweighed for an individual child by the risks of chronically impaired development. That is, the decision to withhold medication may result in the continuation of disordered psychological development, and the risks of such a complicated psychological development may be considered worse than the unknown risks of using medication.

For children with illnesses that are sufficiently severe to justify the use of a nonestablished treatment, it would be expected that medication would not be the sole treatment modality.

Even in cases in which a drug is clearly therapeutic, it is frequently acceptable practice to withhold a medication in the psychiatric treatment of a child. The clinician can always avoid the ethical dilemmas associated with the use of medications in children by just not using these treatments. However, increasing knowledge about the potential usefulness of psychotropic medications in children increases the dilemma of the physician who needs to weigh increasing knowledge against remaining uncertainties.

Decisions about the potential usefulness of medications may now be made early in the diagnostic process. Often therapists choose to withhold medications until after a "trial" of psychotherapy. However, in individual cases it may be less acceptable to "wait until psychotherapy fails" when a therapeutic effect of a drug may be predictable quite a bit earlier.

A common logical error made in psychological evaluations is to employ a drug when certain symptoms cannot be explained psychologically. However, having a good environmental or psychological explanation of why a child is depressed (or anxious) does not give information about whether the child's body has responded with a psychosomatic reaction to the depressive or anxious affect. Knowledge of the dynamics of the case does not mean that one knows how the body has reacted to those dynamic events. To defer a medication treatment because a psychological explanation is available is to misunderstand how the body interacts with emotions.

At the level of the family, the caretaking environment provided by the parents is often influenced by their own psychopharmacologically treatable illnesses. Parenting and caretaking styles are certainly influenced by genetic factors, since several "character disorders" may be genetically modified, including sociopathy, depression, somatization disorder, alcohol abuse, and probably certain types of violence associated with these disorders. Knowing the psychiatric functioning of the parents—and of the family—provides a sense of both the biogenetic predisposition and the familial interactions that influence the child's early developmental period and the current caretaking milieu. The psychiatric treatment of a child may entail the psychopharmacological and psychodevelopmental treatment of the parents.

At the present time, child psychopharmacology itself remains open to future development. The potential risks of withholding research from children cannot be underestimated.

CHARLES W. POPPER

RICHARD FAMULARO

REFERENCES

Aman, M. G.: Psychotropic drugs and learning problems: a selective review. J. Learn. Disab., 13:87–89, 1980.

Baldessarini, R. J.: Chemotherapy in Psychiatry. Cambridge, Harvard University Press, 1977.

Biederman, J.: New directions in pediatric psychopharmacology. Drug Ther., 12:147–170, 1982.

Carlson, G. A., and Strober, M.: Manic-depressive illness in early adolescence. J. Am. Acad. Child Psychiatry, 17:138–153, 1978.

Cohen, D. J., Detlor, J., Young, J. G., and Shaywitz, B. A.: Clonidine ameliorates Gilles de la Tourette syndrome. Arch. Gen. Psychiatry, 37:1350–1357, 1980.

Davis, R. E.: Manic-depressive variant syndrome of childhood: a preliminary report. Am. J. Psychiatry, 136:702–706, 1979.

DeLong, G. R.: Lithium carbonate treatment of select behavior disorders in children suggesting manic-depressive illness. J. Pediatr., 93:689–694, 1978.

Department of Health and Human Services, United States Government: Use of approved drugs for unlabeled indications. FDA Drug Bull., 12:4–5, 1982.

Gaultieri, C. T., Barnhill, J., McGimsey, J., and Schell, D.: Tardive dyskinesia and other movement disorders in children treated with psychotropic drugs. J. Am. Acad. Child Psychiatry, 19:491–510, 1980.

Gaultieri, C. T., and Guimond, M.: Tardive dyskinesia and the behavioral consequences of chronic neuroleptic treatment. Dev. Med. Child Neurol., 23:255–258, 1981.

Geller, E., Ritvo, E. R., Freeman, B. J., and Yuwiler, A.: Preliminary observations on the effect of fenfluramine on blood serotonin and symptoms in three autistic boys. N. Engl. J. Med., 307:165–168, 1982.

Gittelman-Klein, R., and Klein, D. F.: School phobia: diagnostic considerations in the light of imipramine effects. J. Nerv. Ment. Dis., 156:199–215, 1973.

Itil, T. M., and Soldatos, C.: Epileptogenic side effects of psychotropic drugs. J.A.M.A., 244:1460–1463, 1980.

Kalogerakis, M. G.: Pharmacotherapy in adolescent psychiatry. In Esman, A. H. (Editor): The Psychiatric Treatment of Adolescents. New York, International Universities Press, 1982.

Klein, D. F., Gittelman, R., Quitkin, F., and Rifkin, A.: Diagnosis and Drug Treatment of Psychiatric Disorders: Adults and Children. Baltimore, The Williams & Wilkins Co., 1980.

Lowe, T. L., Cohen, D. J., Detlor, J., Kremenitzer, M. W., and Shaywitz, B. A.: Stimulant medications precipitate Tourette's syndrome. J.A.M.A., 247:1168–1169, 1982.

McKnew, D. H., Cytryn, L., Buchsbaum, M. S., Hamovit, J., Lamour, M., Rapoport, J. L., and Gershon, E. S.: Lithium in children of lithium-responding parents. Psychiatr. Res., 4:171–180, 1981.

Mikkelsen, E. J., and Rapoport, J. L.: Enuresis: psychopathology, sleep stage, and drug response. Urol. Clin. N. Am., 7:361–377, 1980.

Ornitz, E. M., and Ritvo, E. R.: The syndrome of autism: a critical review. Am. J. Psychiatry, 133:609–621, 1976.

Petti, T. A., and Law, W., III: Imipramine treatment of depressed children: a double-blind pilot study. J. Clin. Psychopharmacol., 2:107–110, 1982.

Popper, C.: The use of lithium in children and adolescents. Syllabus and Scientific Proceedings, 1982 Annual Meeting, American Psychiatric Association. Washington, D.C., American Psychiatric Association, 1982, p. 153.

Potter, W. Z., Calil, H. M., Sutfin, T. A., Zavadil, A. P., Jusko, W. J., Rapoport, J. L., and Goodwin, F. K.: Active metabolites of imipramine and desipramine in man. Clin. Pharmacol. Ther., 31:393–401, 1982.

Poznanski, E. O., Carroll, B. J., Banegas, M., Cook, S. C., and Grossman, J. A.: The dexamethasone suppression test in prepubertal depressed children. Am. J. Psychiatry, 139:321–324, 1982.

Puig-Antich, J., Blau, S., Marx, N., Greenhill, L. L., and Chambers, W.: Prepubertal major depressive disorder: a pilot study. J. Am. Acad. Child Psychiatry, 17:695–707, 1978.

Rapoport, J., Elkins, R., Langer, D. H., Sceery, W., Buchsbaum, M. S., Gillin, J. C., Murphy, D. L., Zahn, T. P., Lake, R., Ludlow, C., and Mendelson, W.: Childhood obsessive-compulsive disorder. Am. J. Psychiatry, 138:1545–1554, 1981.

Roche, A. F., Lipman, R. S., Overall, J. E., and Hung, W.: The effects of stimulant medication on the growth of hyperkinetic children. Pediatrics, 63:847–850, 1979.

Rounsaville, B. J., Klerman, G. L., and Weissman, M. M.: Do psychotherapy and pharmacotherapy for depression conflict? Arch. Gen. Psychiatry, 38:24–29, 1981.

Saraf, K. R., Klein, D. F., Gittelman-Klein, R., and Greenhill, P.: EKG effects of imipramine treatment in children. J. Am. Acad. Child Psychiatry, 17:60–69, 1978.

Shader, R. I. (Editor): Manual of Psychiatric Therapeutics. Boston, Little, Brown & Co., 1975.

Swanson, J. M., and Kinsbourne, M.: Food dyes impair performance of hyperactive children on a laboratory learning test. Science, 207:1485–1487, 1980.

Weiss, G., and Hechtman, L.: The hyperactive child syndrome. Science, 205:1348–1354, 1979.

Werry, J. S. (Editor): Pediatric Psychopharmacology: The Use of Behavior Modifying Drugs in Children. New York, Brunner/Mazel, Inc., 1978.

Werry, J. S. (Editor): Advances in pediatric psychopharmacology. J. Am. Acad. Child Psychiatry, 21:1–37, 105–152, 1982.

White, J. H.: Pediatric Psychopharmacology. Baltimore, The Williams & Wilkins Co., 1977.

Wiener, J. M. (Editor): Psychopharmacology in Childhood and Adolescence. New York, Basic Books, Inc., 1977.

Youngerman, J., and Canino, I.: Lithium carbonate use in children and adolescents. Arch. Gen. Psychiatry, 35:216–244, 1978.

59

Alternative Therapies

"I know something interesting is sure to happen" she said to herself, "whenever I eat or drink anything: so I'll just see what this bottle does."

Alice's Adventures in Wonderland, Chapter IV

Alternative therapies for developmental disabilities are those apart from the traditional (respectable) programs usually endorsed by organized medicine because of their survival after critical scrutiny and controlled experimental studies. For some alternative therapies, proven benefit beyond that of placebo effect, itself significant, may be shown in the future; in others, no benefit and even potential harm are likely. A review of some of these common therapies as well as an awareness of the dynamics leading to choices of unproven therapies by parents should be useful in working with and understanding families' attempts to navigate among a bewildering array of options in ways that may appear to be irrational.

The fact that parents of developmentally handicapped children often progress through a chronic grieving process marked by guilt, frustration, and failed expectations provides a great degree of understandable motivation to try any treatment program promising success. Often confused by diagnostic interpretations resulting from their child's evaluation because of their own emotional state, which can impair information processing, parents may pursue opinions and unproven interventions endlessly. Faulty family-professional communications at the outset with insufficient emphasis on the necessity of both parents' presence at interpretive conferences, lack of patient repetitive explanation, failure to use simple language and to check that recommendations have been understood ("What is your understanding of what I have been saying about Paul's problem?"), as well as poor follow-up, set the stage for the familiar parental quest "to get to the bottom of it" by shopping elsewhere. Lack of honest validation for a particular therapy, great cost, time, and associated emotional bankruptcy attending the pursuit of unproven remedies do not reduce a family's fervor that is usually rationalized by the statement; "It can't do any harm . . . and it might help,"

especially if their therapist seems to be sincere and caring.

A number of complex and interrelated factors appear to influence a family's particular choice of therapy in addition to the advice of their physician—the shared belief that a given therapy works, influential family and friends, cultural assumptions concerning various therapies, or the unconscious preference to focus upon complicated (often scientific sounding) procedures in order to displace frustration and anxiety about a child. Also the exotic and often unpronounceable words for medical technology and frequently impersonal delivery styles can serve to increase feelings of helplessness, worthlessness, and loss of self-determination as parents and child become bewildered by the disability, sensing their lives to be out of control. As helplessness increases, feelings of personal responsibility decrease. Often exposed to vague and technical explanations, outnumbered in team-parent conferences, frustrated by long waiting room intervals, uncertain follow-up, physician failure to include both parents, and few opportunities to share realistic treatment expectations, parents exhibit "noncompliance" with the doctor's orders, and attempt uncritical self-selection of alternative therapies. In this context family behavior can be viewed as quite understandable and rational. Having made their own choices for the first time by choosing "their" therapy, the parents may feel that they are doing something themselves and that they have some measure of control and, by logical extension, some favorable influence over outcome. One need only discuss (listen to) the rituals of motor patterning or special diets with involved parents to sense the psychological importance of heightened parent participation these programs demand.

In addition to these forces, the public is deluged with usually responsible information about new medical advances provided by the 3500 scientific writers associated with the National Association of Scientific Writers and the American Medical Writers Association. Boasting much larger readership in the millions, irresponsible tabloids sold through supermarkets feature unproven "amazing" and "incredible" cures, often implying that the medical profession maintains a conspiracy to suppress health information or fails to accept the "true healing" knowledge provided by a self-styled expert. Enthusiasm is directly proportional to the conspicuous absence of controlled data. In 1977 the Quackery Committee of the Pennsylvania Medical Society found that one-quarter of 500 nationally circulated magazines accepted "flagrantly misleading" ads. The combined efforts of the Food and Drug Administration, Federal Trade Commission, and Postal Service are generally unsuccessful in controlling the advertising of unproven health remedies whose proponents can dismantle their quickly formed corporations and businesses, creating new ones virtually overnight before legal machinery can enforce the law.

Analysis of various naturopathic, chiropractic, religious, magical, "established," or "unestablished" medical practices credits a substantial portion of the healers' success in all these models with an intuitive use of psychotherapeutic principles combined with the ability to arouse expectant trust and hope in the sufferer. This occurs in a context providing a way of conceptualizing and clarifying previously mysterious symptoms or behavior for the family, thus bringing relief of anxiety. The therapists' personal qualities of deep interest, warmth, acceptance of the patient, dedication, and reasonable informed optimism, as well as training and experience promote the patient's positive perception. The illness itself, because of attendant anxiety, creates heightened dependency needs in the family, thus usually facilitating the healers' role. New treatments seductively suggest excitement and hope, as does any novel experience in our daily lives. All new forms of treatment seem to work best immediately after their introduction, presumably because of this favorable emotional response to novelty. Lest one forget the powerful interplay between the state of mind and healing, it is important to recognize that most medical treatment until relatively recently consisted of placebo effect.

In previous times, when physicians were concerned primarily with acute illness, family and physician attitudes concerning their roles were more clear and unilateral than the focus on developmental disability now permits. That is, the knowledgeable doctor was superordinate, commanding blind obedience in his patients. The patient and his ill informed and frightened family were subordinate, passive, respectful, and even deferential.

Times and goals have changed with the shift from acute illness to chronic disability. One is less able to provide cure than give management. The latter role is more difficult for the traditionally trained physician and now requires that the patient and the family be drawn into the relationship as cooperative participants.

HYPNOSIS

Although the hypnotic state can be defined only in operational terms, hypnotherapy is steadily proving its usefulness as a clinical tool in medicine. Well aware of the long affiliation with religion and magic over the centuries, most pediatricians remain skeptical about its efficacy, having had no supervised training and being uncertain as to how best to locate qualified hypnotherapists. Helpful applications of hypnotherapy in children have included such problems as enuresis, stammering, tics, obesity, anorexia nervosa, various habit disorders, asthma, warts, anesthesia, hemophilia, learning difficulties, encopresis, cyclic vomiting, preoperative anxiety, and psychogenic epilepsy. Significant reports have related the successful adjunctive use of hypnosis in patients with burns or terminal disease, medical illnesses in which psychological factors aggravate already difficult management situations. Yet misconceptions fostered by fictional literature and Hollywood plots are widespread and are based upon unsubstantiated fears that a child's will may be controlled, that he may become gullible and less able to judge right from wrong, and that he may never awaken from the trance or may incur addiction and emotional or intellectual damage. For these reasons many physicians unfortunately relegate hypnosis to that of a treatment of last choice despite the statement in 1956 by the British and American Medical Associations that it is "valuable as a therapeutic adjunct." Opportunities for supervised training are increasingly available in medical schools and postgraduate courses.

Therapeutic hypnosis can be considered as "an altered state of consciousness, usually involving relaxation, in which a person develops heightened concentration on a particular idea or image for the purpose of maximizing potential in one or more areas" (Olness and Gardner, 1978). Relaxation, which does have objectively measurable physiologic correlates, is used to diminish fear. A state of suggestibility is produced through a variety of induction techniques encouraging acceptance and subsequent implementation of suggestions for a patient's benefit. Those who distort perceptions easily, are imaginative, highly motivated, and trusting and who relate well with the operator are more likely to exhibit "trance capacity" and can be guided successfully by suggestions to experience reality differently (e.g., an altered perception of pain). Hypnotic responsivity remains a relatively stable attribute for individuals, and therapeutic results are not particularly dependent upon the depth of hypnosis reached. The respect for and resonance with a child's own experience producing a successful interaction between subject and operator are of paramount importance. Children are good subjects because they are less inhibited than adults, are interested in trying a new pleasant technique, and have a more accessible imagination, easily becoming totally absorbed and readily intertwining fantasy and reality. Other factors accounting for increased hypnotic susceptibility in children have included help-seeking behavior, responsiveness to authority, trust, willingness to experience a regressive (passive) state, desire for mastery of a new skill, and limited reality testing. Techniques of induction involving image and relaxation are provided in the reading references at the end of this chapter but in conjunction with references, the interested reader is encouraged to contact reputable organizations for lists of pediatric workshops or the location of qualified therapists.*

As with any therapy, there are risks. A carpenter whose only tool is a hammer is likely to perceive everything as a nail. Hypnotherapy is not a panacea and must be used in the context of a child's illness with a responsible understanding of the particular disease process and of the relevant dynamics operating. To simply remove a symptom without evaluating its differential diagnosis (e.g., headache) or to attempt to remove unwanted behavior symptoms, which may provide strong secondary gain or be stabilizing a significant psychiatric disorder, is to court failure and the development of complications.

Guided in constructive ways through relaxation and hypnotic suggestion, children are especially able to use their imaginative capacity for their own benefit to modify pain perception, attitudes toward illness, and some body functions not usually considered to be voluntary. As our awareness of stress related children's diseases increases, the improved understanding of this adjunctive therapy provides not only an alternative but also a therapeutic sense of mastery for the children who are benefited.

As with other therapies, parents' attitudes must be respected and their cooperation recruited for success. Memories of humiliating nightclub tricks may constitute a parent's only experience with

hypnosis and can account for understandable skepticism. So prevalent are misconceptions that some experienced hypnotherapists find themselves using terms such as "suggestive therapy," "relaxation," or "relaxation–mental imagery" to circumvent the widespread negative connotation of hypnosis. The parents' acceptance and enthusiasm for hypnosis increase a child's likelihood of success. Approached thoughtfully, parents can be taught to assist children in self-hypnosis after having their own questions answered and, in some instances, having observed a hypnotic session once an alliance has been established between the child and the hypnotherapist. Enhancement of trust and confidence through participation is reported by parents, who then are likely to perceive the hypnotic process and relief of distress as a related extension of their parenting role.

DIET

The overuse of stimulant medication without proper evaluation in children with attention deficits, as well as the fears that medication will addict, stunt growth, or be used as a pharmacologic solution for social problems, has heightened popular excitement to find a "more natural" intervention for these disorders. Such children are characterized by inattentive, impulsive, and distractible behavior leading to conflict and failure in school, often with social correlates. (See Chapter 38A.)

Rivers of ink have been spilled during public debate since Dr. Benjamin Feingold hypothesized in 1973 that "hyperactivity" was due primarily to ingestion of low molecular weight chemicals such as salicylates and common food additives. His book, *Why Your Child Is Hyperactive* (1975), claiming 30 to 50 per cent success in 25 children in his clinical practice, provided anecdotal confirmation reporting the disappearance of symptoms after four weeks without salicylates, artificial foods, and flavors. His theory postulates a toxin and a genetically determined defect in the central nervous system of vulnerable children and also implies a need for life long abstinence from the harmful agents. These hypotheses have spread without controlled studies through national television talk shows, magazine articles, and parent groups with unrestrained enthusiasm. Criticism of the thesis has included lack of controls and failure to select cases by any consistent validated process. In fact, patients with attention deficit disorders and hyperactivity do not represent a homogeneous group of children but rather a common symptom pathway of organic brain injury, language processing deficits, psychosocial adversity, chaotic child rearing environments, and deficits of sequential or-

---

*Two such organizations are:

1. The Society for Clinical and Experimental Hypnosis, 129A Kings Park Drive, Liverpool, New York 13088, publishers of the *International Journal of Clinical and Experimental Hypnosis* (quarterly).

2. The American Society of Clinical Hypnosis, Suite 218, 2400 East Devon Avenue, Des Plaines, Illinois 60018, publishers of the *American Journal of Clinical Hypnosis* (quarterly).

ganization, to name a few etiologies. For this reason it is highly unlikely that any single treatment modality will show effectiveness when applied to such a heterogeneous population of disorders. Whether restrictive diets may prove worthwhile to a specific subgroup of hyperactive and attention disordered children remains to be seen and awaits more exacting patient selection as well as controlled double blind studies. The National Advisory Committee on Hyperkinesis and Food Additives (1975) and the United States Food and Drug Administration (Interagency Collaborative Group on Hyperkinesia, 1976) concluded after detailed review that no controlled studies had demonstrated that hyperkinesis is related to food additives, and no confirmation exists for the claim that hyperactive children improve on diets without salicylates and food additives. Further, long term nutritional needs of children may not be adequately met.

Despite the minor role dietary factors appear to play, there is no serious argument against the systematic trial of a diagnostic elimination diet provided unrealistic promises and unreasonable protocols are not placed on the family. Medical supervision is recommended, and attention to vitamin C supplementation is desirable in view of the exclusion of many fruits thought to be potentially harmful because they contain "natural salicylates." As a result of the antitechnological desire of many to avoid medicines in favor of "natural" remedies, food product labeling, public pressure to remove additives, and increased awareness of quality dietary requirements have been constructive consequences of the diet therapy movement. As is true of studies relating to vegetarianism, the attitudes of parents pursuing special diets can become sufficiently complex to require first a nonjudgmental inquiry about philosophical beliefs relating to diet and life style before the physician can make effective suggestions for any additional therapy.

More recent enthusiasm for treating problem behavior by allergy strategies has been expressed in the popular press, favoring diagnostic use of sublingual antigens to identify food induced respiratory, gastrointestinal, and behavioral symptoms. The technique involves placing three drops of 1:100 aqueous extracted and glycerinated allergenic extract under the tongue of a child and waiting for symptoms, which are then "treated" by administering a "neutralizing dose" consisting of a 1:300,000 solution of the same ingredient. Unfortunately, after review of available data, the Executive Committee of the American Academy of Allergy (1981) found no persuasive controlled studies showing any diagnostic or therapeutic effect of sublingual antigen administration, nor any known mechanisms accounting for claimed neutralizing effects of diluted allergenic extracts. Similar statements rejecting subcutaneous provocation and "neutralization" as a method for the diagnosis and treatment of allergic disease were also made.

HYPOGLYCEMIA

Hypoglycemia has been invoked as a cause of fidgeting and inattention, which contribute to reduced learning. Presumably diet therapy is then prescribed. However, there has been no evidence to support this contention. Silver, in 1975, recommended that a five to seven hour glucose tolerance test be employed before accepting this diagnosis.

VITAMINS

More than 1.3 million dollars per year in the United States are spent on vitamins. Vitamins are not synthesized by the body, are essential for maintenance of normal metabolism, and when provided in chemically pure form must be treated with the same responsibility and respect due any drug. A potentially dangerous aura of benevolent security surrounds vitamin administration, since it is tempting to believe that their natural occurrence in normal diets guarantees their safety when ingested in unlimited quantities. Excessive vitamin dosing can cause toxicity (particularly true of the fat soluble group). Fortunately the Food and Drug Administration has already identified vitamins A and D as worthy of special precaution. New regulations prohibit sale without a prescription of preparations of vitamin A with more than 10,000 International Units per dose unit (twice the recommended daily allowance for adults) or preparations of vitamin D with more than 400 International Units per dose unit (the recommended daily allowance).

Given the ready availability of most vitamins without legal restriction, the power of advertising, and the remarkable benefit in certain defined clinical disorders, it is understandable that vitamins have come to enjoy a magical status and therefore overuse.

Clinical entities justifying increased vitamin intake include the steatorrheas (vitamins A, D, E, K); vitamin B_{12} malabsorption, diphenylhydantoin therapy (folic acid and vitamin D), and inborn errors of metabolism that affect apoenzyme at the cofactor binding site, disturbing the metabolism of the biologically active derivative from the vitamin. When recommended daily allowances are exceeded by a factor of 10 (or five times in the case of vitamins A and D), megadosing occurs by definition of the National Academy of Sciences, risking possible toxicity. Megavitamin doses were especially applied to nicotinic acid and nicotinamide therapy for the treatment of schizophrenia, which was found not only to be useless by the American Psychiatric Association Task Force on Vitamin Therapy in Psychiatry, but also the cause of serious toxicity—persisting skin erythema, pruritus, tachycardia, liver damage, hyperglycemia, and hyperuricemia.

Megavitamins have been characterized as the "ideal" antipatient drug with no proven benefits, great expense to the patient, and capacity for harm. Advocates have emphasized a vastly increased dosage of B complex vitamins and folic acid with no documented benefits in behavior, learning, or autistic disorders. Despite the fact that most Americans receive vitamins and minerals in their regular diet, manufacturers insinuate that most diets need supplementation to correct some implied imbalance.

In 1968 Pauling used the term "orthomolecular medicine" to mean the "treatment of mental disorders by the provision of the optimum molecular environment for the mind, especially the optimum concentration of substances normally present in the human body." Soon after, reports appeared encouraging massive vitamin doses in the treatment of mental retardation, psychoses, and autism without any peer review or appropriate scrutiny to justify the claims. In instances in which no biochemically tested quantitated deficiency is present and the specific clinical special situations do not exist, vitamin therapy is not justified. "Megavitamin treatment therapy as a treatment for learning disabilities and psychoses in children, including autism, is not justified on the basis of documented clinical results" (American Academy of Pediatrics Committee on Nutrition).

Trace elements also have received some attention in the management of developmental-behavioral disabilities. Copper, zinc, magnesium, chromium, calcium, potassium, sodium, and iron are necessary for normal nutrition in small amounts, and in some areas subnormal levels of minerals are detected by hair analysis, the results of which are used to justify subsequent "replacement" therapy in children with a variety of disorders. No responsible study has been reported thus far to support either a causative or a therapeutic role for trace minerals in these children.

PATTERNING AND RELATED THERAPIES

Patterning was developed by Doman et al. (1960) at the Institutes for the Achievement of Human Potential (Philadelphia) and is a form of treatment based on the premise that a variety of sensory and motor experiences can facilitate "neurological organization." The basic assumption unsupported by any human or animal studies is that passive manipulation of the limbs and head (patterning) will affect the brain. Programing may include visual, mobility, auditory, tactile, language, and manual categories. It has been argued without substantiation by proponents that lower brain structures must be developed before organization of cortical structures occurs and that, in order to improve the neurological development of brain or help retarded children, lower level functions require therapeutic emphasis. In these methods the advocates argue that one reaches "the brain itself by pouring into the afferent sensory system. . . . all of the stimuli normally provided by his environment, but with such an intensity and frequency as to draw ultimately a response from the corresponding motor systems."

In addition to patterned manipulations, other techniques may include sensory stimulation, rebreathing of expired air, and restriction of fluid intake. The suggestion was made that by increasing the carbon dioxide content, one could help relieve some types of muscular rigidity. Although it is correct that carbon dioxide is an effective cerebral arteriovasodilator and can increase cerebral blood flow, there is no evidence that this mode of therapy influences an established or fixed brain lesion. There is no evidence that fluid or diet restrictions affect rigidity or spasticity. The American Academy of Pediatrics, American Academy of Cerebral Palsy and Developmental Medicine, and Canadian Association for Retarded Children have reviewed the available literature, concluding that there are no data supporting these neurophysiologic "retraining" methods or the underlying hypothesis that massive inputs of afferent stimuli can reflux to the brain or stimulate regeneration of damaged neurons.

SENSORY INTEGRATION THERAPY

Sensory integration therapy was developed by Ayers, an occupational therapist with long experience in treating cerebral palsied and learning disabled children. Her hypothesis included the following syndromes of sensory motor dysfunction based on an analysis of perceptual, motor, and psycholinguistic test scores as well as basic neurophysiologic data: postural, ocular, and bilateral integration (including problems with right-left discrimination); praxis (including finger agnosia); functions of the left side of the body; form and space perception; and auditory language function.

Later revisions described "tactile defensiveness," characterized by a withdrawal response to certain kinds of tactile stimuli combined with deficits and tactile perception, distractibility, and over-activity. Many now standardized test items are contained in the Southern California Sensory Integration Test Battery. As a result of studies on learning disabled children, Ayers described those with inadequate sensory integration at the brain stem level as having problems with distractibility, extraocular control, visual orientation to environmental space, sound processing, and immature postural reactions. Individualized occupational therapy treatment programs often with school endorsement and scheduling include such modalities as controlled vestibular and somatosensory stimulation (passive and active) with spinning, riding a scooter board down a ramp, electrical vibratory brushes, and manipulative puzzles. These are thought possibly to influence and facilitate the integration of primitive reflexes (e.g., asymmetrical tonic neck reflex) and that equilibrium reactions will subsequently affect midbrain organization favorably, thus aiding interhemisphere cortical connections, with a consequent improvement of reading skills. Tactile stimulation is applied in order to "lower the level of excitation of the reticular arousal system." Formal eye muscle exercises are not used, but it is implied that visual processing improvement favorably influences brain stem relationships to postural and ocular mechanisms, thus improving eye tracking. Every component of the program is directed at a specific goal based upon the hypothetical model. Research attempts to show that enhanced high cortical functions depend upon strategies to improve brain stem organization, but methodologic weaknesses cloud their proper clinical analysis.

Given current knowledge, the physician would do well to embrace the following attitudes: Younger children (under seven or eight) seem to gain increased perceptual performance from perceptually oriented activities. There is thus far no good reason to assume that the type of sensorimotor training suggested as a prerequisite to symbolic learning has any influence on neurologic maturation. An increasing amount of research now suggests that remedial measures for reading readiness derived from theories of visual deficits have little usefulness in correcting reading problems. Perceptual motor training has little relationship to improved academic achievement. The commonly used tests to identify "disorders" in sensoricognitive processes (Frostig Developmental Test of Visual Perception and Illinois Test of Psycholinguistic Abilities) contain such highly intercorrelated subtests that measurements of deficiencies in functions cannot be performed accurately. Basing therapy upon differential subtest results may be unjustified. A powerful argument can be made in favor of direct instructional methods for academic problems with deliberate instructional techniques, trial remediation, pointed information, and key concepts in a context of individualized support, warmth, and enthusiasm. (See also Chapter 56.) There is a questionable need to divide variables of school learning into basic sensory and intersensory processes. It is more effective to consider the learner's task in units that closely approximate the skill he needs to learn. Certainly the clumsy child can benefit from noncompetitive supportive athletic-gym programing, but subjugation to special programs of creeping, spinning, and posture integration can be humiliating and regressive to children, eroding any incentive to try other more useful programs when the adults' enthusiasm is not borne out by the results.

With programs designed to improve central nervous system function, the physician must remember that the usefulness of any program depends upon the specific quality of the skill a child is expected to acquire. The acquisition of a skill needed daily, or even the rebirth of enthusiasm, justifies therapeutic intervention. Neither parents nor physician should expect this benefit to be generalized beyond the "splinter" skill acquired in managing pegs and discriminating shapes. As always, nonspecific benefits may accrue, as Kinsbourne and Caplan (1979) have emphasized, when a child is desperately in need of information, time, and understanding and given these ingredients by a therapist who departs from the "theoretical clutter of an irrational remediation technique." (For further consideration of the roles of occupational therapy, see Chapter 49D.)

OPTOMETRIC TRAINING

Controversy continues over school and private programs utilizing "visual" training to improve visual tracking, binocular functions, and visual-perceptual motor skills. The following conclusions reached by the American Academy of Pediatrics, the American Academy of Ophthalmology and Otolaryngology, and the American Association of Ophthalmology were provoked by the flourishing programs for learning disabled children started by developmental optometrists:

Children with learning disabilities have the same incidence of ocular abnormalities as those who achieve normally and read at grade level. When present, such abnormalities require therapy. Learning problems require multidisciplinary approaches in which eye care has no use in isolation. Peripheral eye defects do not produce reading disorders and reversals of letters, words, or numbers. Special glasses have no value in the specific therapy of reading disorders, excluding correctible ocular defects. Claims that academic abilities of learning disabled children can be improved by visual training, muscle exercises, ocular pursuit, or glasses are without scientific basis.

The only confirmed indication for visual training is convergence insufficiency when accommodation at near point produces diplopia. Unless valid documentation of the value of optometric training is forthcoming, it seems prudent to rely on special educators to use educational strategies for remediation or bypass of specific visual-perceptual or visual-motor problems that are possibly related to learning difficulties.

THERAPEUTIC DECISION MAKING

Culturally shared beliefs, the conviction of the therapist, and the lack of obvious specific therapy all influence the choice of treatment for disabled children. In our culture there is a tendency to be more permissive with aggressive treatment measures because both physician and family are more comfortable with an optimistically presented plan. Aggressive optimism is valued; passivity and skepticism are constrained. There is a strong bias toward therapy of some sort, particularly among humans who consistently demonstrate powerful altruism in order to defend their offspring. There is a potential for unlimited exploitation in parents who "cannot do enough" for their child. "Doing nothing" can always be criticized, but judicious restraint and family developmental counseling may be preferable to bankruptcy and regressive therapies. The physician must resist temptations to follow each treatment fad, realizing that simplistic new theories can be put forth in a matter of minutes, yet take years to refute. Widespread promotion of hypotheses without careful scrutiny can generate enormous hope and energy, but in fact can be a disservice without appropriate controlled studies. Promoters of hypotheses may even claim conspiracy when required by the establishment to test their own theories by traditional methods. A challenging role remains for the developmental pediatrician as an informed advocate, a community resource for assessment of scientific information, a monitor of progress and family ally, continuously helping them to accept reasonable goals, noting that their child, no matter what his handicap, is still acceptable, still loved, and still valued. Serious questions concerning the adequacy of current periodic well care visits as a vehicle for providing this support have been raised. Improvements in service delivery must focus upon improved cost-benefit models to better reconcile the time needs of these families with current high practice overheads.

GEORGE STORM

REFERENCES

Ambrose, G.: Hypnotherapy with Children. London, Staples Press, 1961.

American Academy of Allergy: Position statements: controversial techniques. J. Allergy Clin. Immunol., 67:333–338, 1981.

American Academy of Pediatrics Committee on Nutrition: Megavitamin therapy for childhood psychoses and learning disabilities. Pediatrics 58:910, 1976.

Ayres, A. J.: Sensory integration and learning disorders. Los Angeles Western Psychological Services, 1972.

Bax, M., and MacKeith, R.: Treatment of cerebral palsy. Dev. Med. Child Neurol., 15:1, 1973.

Bierman, C. W., and Furukawa, C. T.: Food additives and hyperkinesis. Pediatrics, 61:932, 1978.

Benson, H., Arns, P. A., and Hoffman, J. W.: The relaxation response and hypnosis. Int. J. Clin. Exper. Hypnosis, 29:259, 1981.

Benton, C.: Comment: the eye and learning disabilities. J. Learn. Disab., 6:334, 1974.

Chapanis, N. P.: The patterning method therapy: a critique. In Black, P. (Editor): Brain Dysfunction in Children: Etiology, Diagnosis. New York, Raven Press, 1981, p. 265.

Cott, A.: Orthomolecular approach to the treatment of learning disabilities. Schizophrenia, 3:95, 1971.

Cousins, N.: Anatomy of an Illness. New York, Bantam Books, 1981.

Dikel, W., and Wolness, K.: Self-hypnosis, biofeedback and voluntary peripheral temperature control in children. Pediatrics, 66:335, 1980.

Doman, R., Spitz, E., Zuckman, E., Delacato, C., and Doman, G.: Children with severe brain injuries: neurological organization in terms of mobility. J.A.M.A. 174:257, 1960.

Duff, R. S.: Counseling families and deciding care of severely defective children: a way of coping with "medical Vietnam." Pediatrics, 67:315, 1981.

Feingold, B.: Why Your Child Is Hyperactive. New York, Random House, 1975.

Frank, J. D.: Persuasion and Healing. New York, Schocken Books, 1963.

Frankel, F. H.: Hypnosis: Trance as a Mechanism. New York, Plenum Publishing Corp., 1976.

Gardner, G. G.: Hypnosis with children. Int. J. Clin. Exper. Hypnosis, 22:20–38, 1974.

Gardner, G. G.: Parents: obstacles or allies in child hypnotherapy? Am. J. Clin. Hypnosis, 17:7, 1974.

Gardner, G. G.: Hypnotherapy in the management of childhood habit disorders. J. pediatrics, 92:838–840, 1978.

Gardner, G. G.: Teaching self-hypnosis to children. Int. J. Clin. Exp. Hypnosis, 29:300, 1981.

Gardner, G. G., and Olness, K.: Hypnosis and Hypnotherapy with Children. New York, Grune & Stratton, Inc., 1981.

Gonick-Barris, S.: Art for children with minimal brain dysfunction. A. J. Art Ther., 15:67, 1976.

Harley, J., Matthew, C. G., and Eichman, P.: Synthetic food colors and hyperactivity in children: a double-blind challenge experiment. Pediatrics, 62:975, 1978.

Harley, J., et al.: Hyperkinesis and food additives: testing the Feingold hypothesis. Pediatrics, 61:818, 1978.

Harris, S.: Effects of neurodevelopmental therapy on motor performance of infants with Down's syndrome. Dev. Med. Child Neurol., 23:477, 1981.

Hockelman, R. A.: Got a minute? Pediatrics, 66:1013, 1980.

Interagency Collaborative Group on Hyperkinesis: First report of Preliminary Findings and Recommendations to the Assistant Secretary for Health. Washington, D.C., Department of Health, Education and Welfare, 1976.

Kaffman, M.: Hypnosis as an adjunct to psychotherapy in child psychiatry. Arch. Gen. Psychiatry, 18:725–738, 1968.

Keogh, B.: Optometric vision training programs for children with learning disabilities; review of issues and research. J. Learn. Disab., 7:36, 1974.

Kinsbourne, M., and Caplan, P. J.: Irrational methods for remediation of learning disability. *In* Children's Attention Learning Problems. Boston, Little, Brown and Company, 1979, p. 196.

Klopp, K. K.: Production of local anesthesia using waking suggestion with the child patient. Int. J. Clin. Exp. Hypnosis, *9*:59–62, 1961.

Levine, M., Brooks, R., and Shonkoff, J.: Medical therapies and interventions. *In* A Pediatric Approach to Learning Disorders. New York, John Wiley & Sons, Inc., 1980.

Lewin, R.: Child Alive. New York, Anchor Press/Doubleday, 1975.

Lipton, M., and Wender, E.: Statement Summarizing Research Findings on the Issue of the Relationship Between Food-additive-free Diets and Hyperkinesis in Children. New York, The Nutrition Foundation, 1977.

Lyon, R.: Auditory-perceptual training: the state of the art. J. Learn. Disab. *10*:564, 1977.

MacKeith, R.: Do disorders of perception occur? Dev. Med. Child Neurol., *19*:822, 1977.

McKinlay, I.: Strategies for clumsy children. Dev. Med. Child Neurol., *20*:494, 1978.

Merrill, R. E., and Adams, R. C.: Recreation therapy. *In* Brain Dysfunction in Children. New York, Raven Press, 1981, p. 239.

National Advisory Committee on Hyperkinesis and Food Additives: Report to the Nutrition Foundation. New York, The Nutrition Foundation, 1975.

Olness K., and Gardner, G. G.: Hypnotherapy in pediatrics. Pediatrics, *62*:228, 1978.

Pless, I. B.: On doubting and uncertainty. Pediatrics, *58*:7, 1976.

Reisinger, K. S., and Bires, J. A.; Anticipating guidance in pediatric practice. Pediatrics, *66*:889, 1980.

Silver, L. B.: Acceptable and controversial approaches to treating the child with learning disabilities. Pediatrics, *55*:406, 1975.

Sparrows, Z.: Patterning treatment for retarded children. Pediatrics, *62*:137, 1978.

Vellutino, F. R., Steger, B. M., Moyer, S. C., Harding, C. J., and Niles, J. A.: Has the perceptual deficit hypothesis led us astray? J. Learn. Disab., *10*:375–385, 1977.

Waitzkin, H., and Waterman, B.: The Exploitation of Illness in Capitalist Society. New York, Bobbs-Merrill, 1974.

Williams, J. I., Cram, D. M., Tansig, F. T., and Webster, E.: Relative effects of drugs and diet on hyperactive behaviors: an experimental study. Pediatrics, *61*:811, 1978.

Yule, W: Issues and problems in remedial education. Review article. Dev. Med. Child Neurol., *18*:675, 1976.

60
Referral Processes

The last 20 years have brought significant changes in the services to children with developmental disabilities. These changes will require that the primary care physician spend an increasing amount of time providing ongoing care to this population. The management concept of deinstitutionalization has appropriately closed residential institutions to all but those children who require extensive nursing care. Children who have been cared for previously out of the home have now assumed their rightful place in the community practice of the primary care physician. Another factor that adds to the physician's involvement is the passage of PL 94-142. This law, the Education for All Handicapped Children's Act, requires comprehensive evaluation and individual program planning and implementation for all children identified as being developmentally disabled (Palfrey et al., 1978). Although the law does not stipulate a specific role for the physician, it would seem appropriate that he would actively participate in the screening, evaluation, and program design for these youngsters.

In light of this increased time commitment, the milieu of the busy office practice makes it untenable for the primary care physician to be the sole provider of care. Moreover, the complexity and diversity of the problems these children present require the involvement of many professionals. The need for current knowledge and specialized contact that a consultant or an interdisciplinary team can provide has to be weighed against the management skills and continuity provided by the primary care physician (Table 60-1). The means for utilizing the technical expertise of the

tertiary care center or the consultant in alliance with the much needed guidance of the individualized management by the primary care physician constitute the topic of this chapter.

DETERRENTS TO PRIMARY CARE FOR THE CHILD WITH A DEVELOPMENTAL DISABILITY

The decision to utilize consultative or referral services is based on the knowledge of the physician about behavioral and developmental disabilities, the nature of the youngster's disability in terms of severity and complexity, the physical, emotional, and financial resources of the family, the administrative organization of the physician's office in terms of the availability of time and of allied personnel, the attitudes and personal philosophies of the physician, and the accessibility of consultative and treatment resources. These factors are so intimately entwined that it is difficult to dicsuss any one of them separately.

The role of the primary care physician in the care of the child who is developmentally disabled is presented in Chapter 51, and has been discussed elsewhere in the literature (American Board of Pediatrics, 1975; Battle, 1980; McInerny, 1975; Table 60-1). The ability to function effectively as a coordinator and general manager of services for a child with a developmental or behavioral disability is a sophisticated task, which requires basic knowledge in child development, neurology, education, and rehabilitation. This basic body of knowledge is not presented in most pediatric training programs. This deficiency not only hinders the physician's ability to effectively manage his patients, but also functions as a major barrier to referral and consultation. Chamberlin (1980) indicated that the physician who does not count on his allied professionals for support "is likely to be unaware of what it might contribute to the family." This lack of knowledge obviously leads to either late referral or no referral at all.

Several efforts have been made to rectify this deficit. The Task Force on Pediatric Education,

Table 60–1. THE ROLE OF THE PRIMARY CARE PHYSICIAN

1. Screening and early identification
2. Coordination and participation in the diagnostic and evaluative process, program planning, and habilitation
3. The interpretations of the findings presented in item 2
4. Continuing counseling and support
5. Day to day health maintenance
6. Advocacy
7. Monitoring of the habilitative process

(1978) suggested that residency training programs should result in pediatricians who are competent in both the biomedical and biosocial aspects of pediatrics. The Task Force stated that pediatricians should also have the ability to relate to a variety of allied professionals. The American Academy of Pediatrics (1981) has made a concerted effort to address this issue with the practicing pediatrician. They offer intensive courses that emphasize the role of the primary care physician in the treatment of handicapped children.

The lack of training is not the only deterrent to providing services to children with developmental disabilities. Even if physicians were adequately trained in this area, the traditional office routine and the personnel and scheduling patterns would require significant alteration if the primary care physician were to play a major role. Pless (1980) has made several suggestions about how to overcome these practice barriers. He suggested the "red tagging" of children who are at high risk for developmental disabilities in order to make possible more extensive evaluation and early identification, the introduction of the nurse practitioner to assume responsibility for well child care using the time gained to provide services to youngsters with more complex problems, the sharing of space with such allied professionals as psychologists and social workers, the initiation of an appropriate hourly rate of compensation for services provided, and the development of an adequate record keeping system, such as the goal oriented record.

Attitudes and personal philosophies may also be a deterrent to seeking consultation. The physician's disenchantment with the quality of life of the individual with severe mental retardation may result in a conclusion that further remediation is not worthwhile. Although in almost all instances children and families benefit from habilitation, in certain remarkable circumstances such decisions may be appropriate. They should be extensively discussed with the child's family.

THE RELATIONSHIP BETWEEN THE PRIMARY CARE PHYSICIAN AND THE FAMILY

Despite these obstacles, many physicians in busy office practices have been successful in the care and management of children with complex problems. The physician looks upon himself as the child's advocate who knows the family best and maintains his role as the primary caretaker to the limit of his expertise. He utilizes the referral and consultation process with discretion and supports the philosophy that "pediatricians' understanding of the family provides the best care

available for the child with chronic illness" (McInerny, 1975).

There is little question that the pediatrician is in the best position to document the developmental process. Careful observation and screening of the child and the family monitor the child's growth while simultaneously providing the family with the necessary guidance to aid in an orderly developmental evolution. Often the relationship with the family is established during a prenatal visit or through the care of other siblings. This asset, coupled with a knowledge of the community in which the family lives, makes the primary care physician invaluable to the consultant and the family.

THE CONSULTATION AND REFERRAL PROCESS

A distinction should be made between consultation and referral. Although this separation is somewhat arbitrary, referral generally implies diagnosis and evaluation followed by long or short term therapy or services. Consultation, on the other hand, implies an advisory process, which is more short term and finite. The need for longitudinal observation in the interpretation of developmental behaviors makes a single consultative visit unlikely, and follow-up is generally the rule. As noted previously, the complex nature of developmental and behavioral problems rarely utilizes the knowledge and services of a single discipline and generally requires multiple visits. Similarly, almost all families who have children with developmental disabilities require continuing counseling, as well as educational or genetic guidance. The presentation of information following extensive diagnostic study frequently requires multiple visits before the family understands the nature of the child's disability. This is less true when the referral is for a specific medical problem, such as constipation, visual defects, or cardiopulmonary difficulties.

THE BENEFITS OF CONSULTATION AND REFERRAL

The benefits of consultation are multiple. They include the accessibility to the most current information and treatment and at a minimum should provide confirmation and support of pre-existing impressions and diagnoses, evaluation and planning of treatment and programs by an individual or interdisciplinary team, periodic monitoring of the effectiveness of the treatment programs, and the assurance that all is being done that can be done.

THE DISADVANTAGES OF CONSULTATION AND REFERRAL

The disadvantages are that the consultation may be expensive, time consuming, and anxiety producing. The family may be confused by the introduction of additional professionals who give conflicting recommendations. In general, consultants do not have the advantage of a long term relationship witht the family and are not as accessible in time of crisis as the primary care physician. Although these shortcomings can lead to confusion and fragmentation of service, with careful monitoring and communication by the primary care physician these shortcomings can be easily overcome.

THE DECISION TO REQUEST CONSULTATION

Consultation can be obtained from individuals or an interdisciplinary team that is community or hospital based. The decision to consult is often dependent upon the nature of the youngster's disability. In general, the more complex and severe disabilities require consultation.

The circumstances of serious behavior disorders, psychoses, complex psychoneuroses and personality disorders, and severe family disorder almost always require referral to a psychiatrist, psychologist, or mental health center. However, the primary care physician should be in a position to skillfully manage less complex behavioral disorders such as encopresis, night fears, school phobias, loss reactions, and oppositional behaviors. The latter situations may be complicated by a family's inadequate emotional resources. In these instances the help of a competent social worker will be valuable. Children with other types of developmental disabilities, such as mental retardation or learning disabilities, require the services of many professionals. They need the expertise of orthopedists, ophthalmologists, audiologists, occupational and physical therapists, and speech pathologists.

THE ROLES AND RELATIONSHIP OF THE PARENTS, PRIMARY CARE PHYSICIAN, AND CONSULTANT

Understanding the roles, responsibilities, and relationships of individual specialists is important if their services are to be optimally utilized. Figure 60–1 describes the transaction of the consultative process. It stresses the inter-relation among the parents, the primary care physician, and the consultants. A breakdown in the relationships among the members of this complex interaction leads to confusion and ultimately fragmentation of care. A lack of understanding on the part of the parents may be interpreted as noncompliance, whereas in reality it may be due to a lack of communication by and between professionals. Levine and Kliebhan (1981) recently described the relationship between the physician and physical and occupational therapists. Although they focus on a particular group of therapists, what they have to say has general application. They stress the need for improved communication between physicians and allied professionals. There is little question that the physician must have an understanding and knowledge of terminology, observational techniques, and treatment of neurodevelopmental disorders if optimal communication is to occur among the specialist, the family, and the physician. Table 60–2 presents factors that potentially can diminish the effectiveness of the consultative process. All these pitfalls can be overcome by continuing communication between the members of the consultative partnership.

The process of consultation and referral should be carefully planned. Although the suggestion for consultation may seem to be a simple, routine procedure for the physician, this often represents

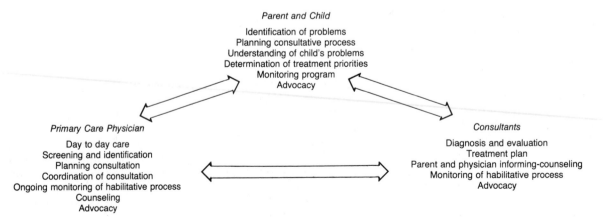

Figure 60–1. The roles, responsibilities, and relationships of parents and professionals in the consultative process.

Table 60–2. FACTORS THAT DIMINISH THE EFFECTIVENESS OF THE CONSULTATIVE PROCESS

Deterrents Presented by the Primary Care Provider	Deterrents Presented by the Consultative or Referral Source

1. Lack of recognition of significant problems
2. Misdiagnosis and improper referral
3. Objectives of consultation not clear to family or consultant
4. Inadequate support and coordination of the consultative process
   a. Poor communication with consultant
   b. Lines of responsibility not clear
5. Inadequate discussion with the family of the results of the consultation
	1. Referral process too cumbersome, and
	   a. Application too complex, unclear, and too lengthy
	   b. Appointment times inconvenient
	2. Poor "fit" between consultant and family
	3. Misdiagnosis
	4. Treatment plan inappropriate
	   a. Poor communication with primary care physician
	   b. Plan impractical and too complex
	   c. Not well explained to family
	   d. Lines of responsibility not clear
	Infection
	5. Excessive use of jargon with the family and primary care provider

a crisis for the family. At a minimum the family finds the process anxiety producing. The request for consultation frequently occurs when an existing problem has worsened, or when a physician's diagnostic suspicions are substantiated. In the former instance it may be a time when a behavior has become intolerable, and in the latter case it may occur when the possibility of a developmental delay becomes apparent to the parents or the physician. The fears and worries that the consultative process provokes should be discussed carefully with the parents and the consultant. Although the parents have had pre-existing concerns, the consultative process may raise some feelings and questions for the first time. The responsibility for answering these questions should not be abdicated to the consultant. Although not all answers are available prior to consultation, these issues and feelings should be addressed by the primary care physician by use of existing information. The adaptation of a family to a child with a developmental disability is an evolving process. Continuing discussion promotes understanding of the disability and helps resolve existing conflicts and concerns.

Every effort should be made to have both parents present at the time of consultation. A study by Tarran (1981) indicated that parents of children with cerebral palsy preferred to have the diagnosis of the child's handicap discussed when they were together. They preferred to be told as early as possible and wanted this to be followed by opportunities for further discussion.

The physician should know as much as possible about the consultant's approach to the problems for which the child is being referred. The reasons and purpose of the consultation should be reviewed with the consultant and parents, and the questions to be answered should represent the needs of both the family and the physician. If the purpose is continuing treatment, the potential goals should represent the priorities of the family-physician alliance. If the consultative team or person is to assume major responsibility for the management of the diagnostic or habilitative process, this should be made known to the family and consultant. Selection of the one to be responsible for the day to day care of the patient should be discussed. A parent, for example, should not be confused as to whom to call in the event that his youngster has a seizure. In almost all instances this primary care responsibility is best handled by the family physician with consultative support from the specialist. As noted previously, the degree of involvement of the consultant is dependent upon the complexity of the child's problem and the resources and attitudes of the primary care physician. In general, the more complex the problem and the fewer the resources of the primary care physician and the family, the greater the involvement of the consultant or the interdisciplinary team.

These guidelines also apply to children being referred for behavioral or psychosomatic disorders. As noted previously, the less complex behavioral problems can be managed by the primary care physician. The fact that they can be managed without consultation should not detract from the potential long term significance of these developmental issues. Minor problems that are poorly managed can take on crisis proportions or result in nagging, neurotic disabilities. One must also guard against the denial that certain behavior patterns or symptoms are significant. Such subversion can result in the delay or the neglect of referral for appropriate psychiatric care or treatment. The stigma associated with mental illness continues to remain an obstacle to treatment.

The ease of acquiring services is also an important factor in determining the utilization and ultimate effectiveness of treatment. The concept that if the family is in sufficient need they will overcome all obstacles to psychiatric care has not been substantiated. Adolescents feel uncomfortable about leaving school to see a therapist, and not all employers are understanding about the need for family therapy. A convenient time for the first several visits is helpful in establishing a relationship with the family.

The process and goals of psychiatric consultation should be carefully discussed with the child, family, and consultant. All referrals for consultation require a "proper fit" between the consultant and the family. Maximal benefit is achieved when the family can feel comfortable with the theoretical model proposed by the therapist. The psychoanalytic or nondirective approach may be suitable for some families and irritating and nonproductive for others. The family may also prefer consultation with a psychologist or social worker rather than a psychiatrist. The decision about whom to send the family to should be determined by the expertise of the professional and the needs of the family.

The consultative process can occur in one of several ways. The referring physician may decide to proceed with the consultation in a stepwise fashion with the return of the family to the primary care physician after each consultation. This is generally time consuming and may be confusing. The matter is best resolved by referring the family to an interdisciplinary team and having the team decide which consultations to draw upon. Hospitalization for diagnosis and evaluation of developmental disabilities is generally not necessary unless the family has to travel long distances for multiple appointments or there is a need for siezure control, extended observation, or behavioral management.

All consultations should be initiated by a telephone contact with the consultant and should be followed by a letter, which accompanies all the necessary records. Nothing is more disheartening to a family than a consultant who is unaware of their child's problems. Similarly, it is distressing to the consultant when a family does not have an inkling as to why they have been referred.

Upon completion of the initial evaluation, the consultant should provide information to the parent and the primary physician in a concise understandable report. Lengthy narratives may not be read by the primary care physician. A copy of the report should be sent to the referring physician and optimally to the parents. At this point the primary care physician should review the evaluation with the family and the need for follow-up should be established. The continuing roles of the consultants should be made clear, and if an interdisciplinary team evaluated the child, the individual responsibilities of the team members should be identified for the parents. Subsequent visits will depend upon the need for additional observation and treatment or the occurrence of new problems or questions. A family's perception of their child may change with age and school entry. Questions regarding pubescence or residential placement may also indicate the need for reassurance, help, and a return visit to the consultant. The consultant may want to monitor the child's progress or defer diagnosis until additional observations are made.

The effective implementation of the treatment program will contribute to the success of the consultative process. Some consultants provide advocacy for the child and family, but this responsibility generally falls to the primary care physician. The physician must understand the child and the adaptive phases experienced by the family. All the available technical knowledge cannot substitute for empathy, understanding, and a willingness to give the necessary time and energy that the care of developmentally disabled children requires. Parents continually need advice, direction, and a discussion of specific methods for dealing with difficult problems. Indifference on the part of the physician will lead to a sense of resentment and boredom. Support and interest in the child and family provide dignity for the child and satisfaction for the physician and make a difficult life situation easier for all.

ALBERT P. SCHEINER

REFERENCES

American Academy of Pediatrics: New Directions in Care for the Handicapped Child. Evanston, Illinois, American Academy of Pediatrics, 1981.

American Board of Pediatrics: Foundations for Evaluating the Competency of Pediatricians. Chicago, American Board of Pediatrics, Inc., 1975.

Battle, CU: The role of the primary care physician. In Scheiner, A. P., and Abroms, I. F. (Editors): The Practical Management of the Developmentally Disabled Child. St. Louis, The C. V. Mosby Co., 1980, Ch. 1.

Chamberlin, HR: The interdisciplinary team: contributions by allied and non-medical disciplines. In Gabel, S., and Erickson, M. T. (Editors): Child Development and Developmental Disabilities. Boston, Little, Brown and Company, 1980, Ch. 21.

Levine, M. S., and Kliebhan, L.: Communication between physician and physical and occupational therapists: a neurodevelopmentally based prescription. Pediatrics, 68:208–214, 1981.

McInerny, T.: The role of the primary physician in group practice. In Moore, T. D. (Editor): The Care of Children with Chronic Illness. Report of the 67th Ross Conference on Pediatric Research. Columbus, Ohio, Ross Laboratories, 1975, pp. 72–74.

Palfrey, J. S., Mervis, R. C., and Butler, J. A.: New directions in the evaluation and education of handicapped children. N. Engl. J. Med., 298:819, 1978.

Pless, I. B.: Practical problems and their management. In Scheiner, A. P., and Abroms, I. F. (Editors): The Practical Managment of the Developmentally Disabled Child. St. Louis, The C. V. Mosby Co., 1980, Ch. 17.

Tarran, E. C.: Parents' views of medical and social work services for families with young cerebral palsied children. Dev. Med. Child Neurol., 23:173–182, 1981.

Task Force on Pediatric Education: The Future of Pediatric Education Report. Evanston, Illinois, American Academy of Pediatrics, 1978.

VIII
Professionals and Children

The concluding set of chapters surveys several subject areas that influence the present and future delivery of care oriented toward developmental and behavioral health. Included are discussions of legal provisions and legislative entitlements for the young, guidelines for valid research, and considerations of the means for preparing physicians to assume meaningful roles in this aspect of general pediatrics. Part VIII concludes with a statement of some of the moral and ethical issues that emerge while caring for and caring about the developing child.

61

Laws and Legal Issues

61A   Legal Issues Relevant to Child Development and Behavior

Pediatricians and other professionals involved in fostering the health and life adjustment of children need to be aware of the range of legal issues relating to such care. This chapter surveys some of the most pertinent of these, including laws concerning privacy, confidentiality, informed consent, and the control of behavior. Other chapters in this book refer to legal rights of children in such areas as abuse, neglect, and family disruption (Chapters 13 and 14) and special education entitlement (Chapter 61B). In this chapter legal precedents are cited as they illustrate principles. Comparisons are made between laws as they apply to children and adults.

In recent years, health care providers have been confronted with a great deal of legal policy pertaining to the protection and adequate treatment of persons in their care. Martin (1975) has identified four sources of legal policy concerning health and human services: the due process clause of the Constitution, the equal protection clause of the Constitution, the application of administrative guidelines, and judicial precedent.

1. Due process requires that when services involve deprivation of an individual's liberty or property (e.g., administering a drug that limits the patient's control of his own behavior, withholding mail or personal belongings from institutionalized clients), there must be due process of law. Due process is understood to require a hearing before an impartial entity, the opportunity to hear witnesses, and the opportunity to appeal the decision.

2. Equal protection of the law requires that a human service agency or practitioner treat each patient group substantially the same as other patients or patient groups entitled to the same treatment. Violations of the equal protection clause can be seen in the discriminatory assignment of patients to service programs and research projects on the basis of race; the refusal of a public school to educate a child on the basis of his race, religion, or handicapping condition; and the often arbitrary

and capricious treatment of the institutionalized mentally ill and developmentally disabled.

3. The third source of legal policy concerns the application of administrative guidelines by the courts. This involves the government's use of funding contingencies to ensure that patients' rights are protected and the standards for human service delivery are met. Examples of these guidelines can be seen in the requirements of PL 94-142 (e.g., individualized education plans for handicapped children) and Titles XVIII and XIX of the Social Security Amendments of 1965 (e.g., regulations concerning the use of seclusion and restraint in institutions eligible for funding under Titles XVIII and XIX).

4. The fourth and final source of legal policy is the subject of much of the remainder of this chapter—court cases that build legal doctrine by precedent. These court decisions have not only determined "standards" for human service delivery and specified the rights of clients and patients, they have provided a valuable source of information for predicting the nature of legal policy in the future (see footnote 1 following references).

LEGAL POLICY AND THE PEDIATRIC PATIENT

These forms of legal policy have only recently been applied to children. At present the child is not considered to be the "holder of privilege" with regard to the rights of privacy, confidentiality, and informed consent. Nevertheless an understanding of current legal policy is important for several reasons. First, when the patient is not considered legally competent to be the holder of privilege, it is important to understand the law as it applies to the individual or individuals acting in the patient's behalf (e.g., an adult acting on behalf of a pediatric patient). Second, the courts have been known to interpret legal precedent rather broadly. When there exists no legal precedent concerning the pediatric patient, the courts may turn to precedent concerning adult patients in rendering decisions. Third, in some cases a child may be judged to be "competent" and therefore entitled to the rights of an adult (e.g., *Planned Parenthood of Missouri v. Danforth*, 1976). Finally, it is simply good practice to treat the pediatric patient with the same respect that one affords an adult. This view has received the support of the courts and is certain to influence any decision regarding pediatric patients. Indeed, the Danforth court determined that "constitutional rights do not mature and come into being magically only when one attains . . . age of majority."

Therefore, understanding the laws concerning privacy, confidentiality, and informed consent as they apply to the pediatric patient requires an understanding of how they apply to the adult patient. In the remainder of this section, legal precedent will be discussed first, as it was established (typically with adults), and second, as it applies to the pediatric patient.

PRIVACY

Privacy refers to the right of the individual to keep information about himself out of the hands of other individuals, including agencies and officials of the government. These private facts include identifying characteristics of the individual, test scores, disabilities or handicaps, and medical or psychological services. The legal basis for the individual's right to privacy was stated by Mr. Justice Brandeis in *Olmstead v. United States:*

> The makers of our Constitution undertook to secure conditions favorable to the pursuit of happiness. . . . They sought to protect Americans in their beliefs, their thoughts, their emotions, and their sensations. They conferred, as against the Government, the right to be left alone—the most comprehensive of rights and the right most valued by civilized men.

In addition to the right to privacy guaranteed by the Constitution, further definition has been established through legislation. The statutory rights to privacy include the concept of privilege as well as rights provided via statutes such as the Education for All Handicapped Children Act (PL 94-142), the Developmentally Disabled Assistance and Bill of Rights Act (PL 94-103), the Privacy Act (5 U.S.C. 552a [1974]), and state statutes concerning the delivery of medical and psychological services.

Therefore, privacy is a right concerned with maintaining one's individuality. This was further evident in *Griswold v. Connecticut* (1965). In this case, employees of a Connecticut Planned Parenthood program were charged with violating state law that prohibited anyone from aiding others in preventing conception. The state law was struck down by the Supreme Court, which argued that

personal affairs fall "within the zone of privacy created by several fundamental constitutional guarantees" and quoted the Ninth Amendment, which suggests unnamed rights retained by the people. The Ninth Amendment is generally construed to contain implicitly a right of privacy or the general right to be left alone, an interpretation that has been supported in other legal cases (e.g., *Roe v. Wade*, 1973). In the Griswold decision, Mr. Justice Goldberg maintained that for the state to encroach upon this right to be left alone, there must be a compelling state interest that is directly related to the accomplishment of a permissible state policy.

THE PEDIATRIC PATIENT

The minor's right to privacy was supported in *Merriken v. Cressman* (1973). This case concerned the efforts of a public school program to identify potential drug users in its student body. Students were asked to complete questionnaires that contained intimate questions about family relationships. They were also asked to list the names of students who made unusual or odd remarks. Although the results of the questionnaire were to be confidential, information about potential drug abusers was to be distributed to school personnel. The Court noted that although the students were children (eighth graders), they still maintained the right to privacy.

Further support for the child's right to privacy was provided in *Planned Parenthood of Missouri v. Danforth* (1976) and *Bellotti v. Baird* (1979). In both these cases the issue was whether parental consent should be required for a minor to receive an abortion; i.e., a minor's right to an abortion depends on her right to privacy, which is in conflict with the constitutional right of her parents to supervise the upbringing of their children. In these cases the Court determined that the state had no "compelling" interest in safeguarding the authority of the family relationship and that the state could no longer impose a blanket provision requiring consent of a parent for a minor to obtain an abortion.

In the Danforth case the Court maintained that "the family will not be strengthened by providing a parent with absolute power to over-rule a girl and her doctor's decision." The Court further determined that "constitutional rights do not mature and come into being magically only when one attains state defined age of majority." The Court went on to emphasize the need to ascertain whether the minor understands the consequences of the act and gives her "informed consent."

In the case of *In re Smith* (1972), the Maryland Court of Special Appeals considered the question of whether a parent has the authority to require a child to submit to an abortion when the child wants to keep her baby. Here, too, the rights of the minor were confirmed, the Court stating that ". . . not only do minors have the right to consent to their own abortions, but they also have the right to withhold their consent over the objections of their natural guardians." Similarly, in *Carey v. Population Services, Inc.* (1977) the Supreme Court ruled that a New York provision prohibiting the distribution of contraceptives to persons under 16 was not justified because the right to privacy in connection with decisions affecting procreation extends to minors as well as adults.

INVASION OF PRIVACY

For conduct to constitute an invasion of privacy, there must be an intentional intrusion upon the private affairs or concerns of another that would be viewed as offensive or objectionable to the reasonable person of ordinary sensibilities. However, as suggested by Schwitzgebel and Schwitzgebel (1980), Rosoff (1981), and others, health care professionals might be expected to maintain a higher standard of conduct than the typical citizen in avoiding an invasion of a patient's privacy.

Invasion of privacy might consist of a public disclosure of private facts (e.g., *Merriken v. Cressman*, 1973), publicly placing a person in a false light, psychological testing without adequate consent or testing conducted by persons other than well trained practitioners (*Merriken v. Cressman*, 1973), and various forms of commercial exploitation. However, public disclosure of private facts is not a necessary condition for an invasion of privacy to occur. Other forms of invasion include opening one's mail, examining private bank accounts, and the utilization of electronic recording devices.

CONFIDENTIALITY

Whereas privacy is concerned with the disclosure of information, the patient's right to confidentiality is concerned with what happens to private or privileged information after it has been disclosed. Schwitzgebel and Schwitzgebel (1980) have distinguished between privileged communication and confidentiality. Privileged communication refers to "a legal right existing by statute which protects the client (patient) from having his or her confidences revealed publicly, without permission, during legal proceedings. Generally, the right belongs to the client, not to the therapist" (p. 200). The principle of the patient as "holder of privilege" has been supported by most state laws

(e.g., California Evidence Code 911-920 and 1010-1020) and through case law (*In re Lifschutz*, 1970).

Confidentiality, on the other hand, is not limited to situations involving evidence and legal proceedings and is not a legal right authorized by statute. Rather it is an ethical practice that is addressed in the codes of conduct of various professional organizations intended to safeguard the patient's confidentiality and to provide sanctions for professional misconduct. Confidentiality therefore is understood to ". . . protect a client (patient) from unauthorized disclosures of information given in confidence, without the consent of the client" (Schwitzgebel and Schwitzgebel, 1980, p. 200).

Wigmore (1961) has suggested four criteria for the establishment of privileged communication between practitioner and patient. Specifically, the patient must offer communications in the confidence that they will not be disclosed; confidentiality must be considered essential to the satisfactory maintenance of the therapeutic relationship; the relationship is one that the community feels should be fostered; and there would be greater injury than benefit resulting from disclosure of the communication.

The amount of confidentiality privilege and responsibility afforded treatment personnel varies according to professional role and to local and federal statutes. In general, when there is more than one statute that can be applied for the protection of privileged communication, the most rigorous will be used. Traditionally physicians (especially psychiatrists) have been accorded the most protection, followed by psychologists and the nonpsychological therapists.

However, the proposed *Rules of Evidence for U.S. District Courts and Magistrates* abolishes the physician-patient privilege while retaining the psychotherapist-patient privilege. According to Schwitzgebel and Schwitzgebel (1980):

> The rationale of this policy was that in actual practice so many exceptions to the physician-patient privilege have been found necessary, in order to prevent fraud or to promote the public interest, that the privilege is meaningless. Psychotherapists are protected because, presumably, the matters dealt with are less understandable or more socially sensitive. Therefore, these matters should not be accessible to the public or to the patient for the patient's own welfare (p. 202).

THE PEDIATRIC PATIENT

In the case of the pediatric patient, the parent or legal guardian has generally been considered to be the legal holder of privilege. However, as Rosoff (1981) has noted:

> It is difficult to imagine a provider being held liable for respecting the confidence of a minor patient except in cases where it appears likely that the minor or others may be harmed and the provider fails to take reasonable action to prevent this (p. 207).

Nevertheless, in some states, health care providers forfeit their right to obtain reimbursement from parents (or guardians) unless the parents are informed of the care being provided the child and consent to it.

Pediatric practitioners should therefore communicate to their patients the fact that parents are the holders of privilege and that, by law, parents are entitled to have access to the child's communications. The practitioner should also explain other circumstances in which disclosure is mandatory.

CONFIDENTIALITY OF PATIENT RECORDS

The issue of confidentiality of patient records is complicated by the increasing emphasis now being placed upon the ready accessibility of fiscal and treatment-related information. Requests for information from patient records may come from state and federal health agencies, employers curious about the treatment history and present ability of their employees, police departments anxious to learn the identity of individuals being treated for alcoholism or drug addiction, life insurance companies interested in potential suicide risk, and professional standards organizations requiring reviewers' access to medical and clinical records (Gobert, 1976).

The patient's rights to privacy and privileged communication may be violated unless he gives written consent for the release of information in response to requests such as these. In short, the patient's right to privacy includes the right to control the dissemination and use of information concerning his psychological or mental state (*Olmstead v. United States*, 1928).

Another issue concerns the patient's ability to gain access to his own medical record. The Freedom of Information Act (5 U.S.C. 552) provides that anyone has the right of access to (and to receive copies of) any record, document, or file in the possession of the federal government. The revised Family Educational Rights and Privacy Act of 1974 gives students 18 or older and their parents the right to inspect relevant school records. However, those records "created or maintained by a physician, psychiatrist, psychologist, or other recognized professional or paraprofessional acting in his or her professional or paraprofessional capacity or assisting in that capacity" and those records "created, maintained, or used only in connection with the provision of treatment to the student" are not to be "disclosed to anyone other than individuals providing the treatment" (sp. U.S.C. 1232g).

Similarly, state and private organizations may not be required by the Privacy Act or the Freedom of Information Act to disclose information from a patient's medical record (Schwitzgebel and Schwitzgebel, 1980). Furthermore, there is little case and statutory law in such situations. In *Gaertner v. Michigan* (1971) and *Bush v. Kallen* (1973), the courts decided that hospital records are typically hospital property but that parties with a legitimate interest could inspect the record provided it resulted in no danger or harm to the patient. In *Whalen v. Roe* (1977), the Supreme Court ruled that the patient has no constitutional protection of privacy interests in his own medical records. In *Gotkin v. Miller* (1975), the court found no constitutionally protected property right in direct and unrestricted access to one's own medical records. However, lower courts have enjoined the collection of records of sensitive information that invaded individual's privacy (*Phoenix Place v. Michigan Department of Mental Health*, 1978; *Merriken v. Cressman*, 1973).

MANDATORY DISCLOSURE AND LOSS OF PRIVILEGE

A major exception to the accepted right of privilege concerns the situation in which there is a likelihood that the patient's behavior will represent a "clear and imminent danger" to other members of society. In *Tarasoff v. Regents of the University of California* (1974), the California state supreme court held that "once a therapist does in fact determine, or under applicable professional standards reasonably should have determined, that a patient poses a serious danger of violence to others, he (she) bears a duty to exercise reasonable care to protect the foreseeable victims of that danger." The decision also included the statements that "If a doctor or psychotherapist knows, or ought to know, that his patient might inflict harm upon a third person, the doctor has a legal obligation to warn the intended victim or those who reasonably could be expected to notify the person of the peril" and ". . . the public policy favoring protection of confidential character of patient-psychotherapist communications must yield to the extent to which disclosure is essential to avert dangers to others. The protective privilege ends where the public peril begins."

In addition, in some jurisdictions, statutes require mandatory disclosure of a specific type of crime, such as child abuse or drug abuse, when it becomes known to a health care practitioner or professional school employee. These statutes typically provide civil immunity for practitioners from lawsuits by patients alleging harm.

In certain situations the patient may lose his privilege of confidential communication. In some cases conditions for loss of privilege are specified in state statutes. Shah (1970) has described situations in which loss of privilege is probable: a dispute between individuals in a privileged relationship such as a practitioner's being sued by his patient for malpractice; a situation in which a practitioner's services are solicited to aid in some illegal activity; a case in which, during the course of diagnostic assessment or treatment, the practitioner determines that the patient is in urgent need of hospitalization; a patient's involvement in workmen's compensation litigation or in prosecution for homicide; a contest regarding the validity of a document such as the patient's will; and a situation in which loss of privilege is permitted and ordered by judicial discretion.

There are other situations in which the patient is generally considered to have automatically lost his privilege of confidential communication. These situations include child-custody cases in which a party's mental condition is used as part of a claim or a defense or when the practitioner suspects child abuse by either party; civil proceedings in which a patient uses his mental condition as an element of his claim or defense—however, the patient must actually introduce his mental condition as evidence (i.e., it cannot be assumed or speculated from an examination of his or her medical records) (*Roberts v. Supreme Court of Butte County*, 1973); and when the patient makes communications to a psychiatrist in a court ordered examination after having been informed that his communications would not be privileged (*Gibson v. Virginia*, 1975) (Schwitzgebel and Schwitzgebel, 1980).

BREACH OF CONFIDENTIALITY

Brooks (1980) has summarized the legal basis of liability concerning an unjustified invasion of privacy or breach of confidentiality. These are the sources of law that support the patient's right to privacy and that may provide a legal remedy for an unwarranted loss of privilege. Remedy in each case may be in the form of damages as compensation for the injury to the patient or an injunction to prevent further disclosure of private facts. Although it is unlikely that a pediatric patient will be a litigant in such a proceeding, it is important to understand the options available to those acting on his behalf.

The first basis of law is the common law contract; i.e., the physician enters into an agreement with a patient to provide medical treatment and is expected to keep in confidence all disclosures made by the patient. This contract is further defined by Section 9 of the Principles of Medical Ethics, which states that a physician "may not reveal the confidences entrusted to him (or her) in the course of medical attendance, or the deficiencies he (or she) may observe in the character of patients, unless he (or she) is required to do so by law or unless it becomes necessary to protect the welfare of the individual or of the community." A breach of this implied covenant of secrecy, whether deliberate or negligent, can result in professional liability for the practitioner.

A second legal basis for liability is that of tort, i.e., a legal wrong or infliction of injury that conflicts with the professional's duty not to injure the patient. There may be punitive damages sought by the patient if the injury is deliberate. Brooks (1980) has observed that:

> . . . a tort may be based on the breach by a professional of a statute specifically addressed to confidentiality, or it may be based on a breach of public policy as evidenced in a privilege statute. While the prime function of a privilege statute is to prevent the revealing of confidential information as evidence in court, the public policy expressed in these statutes is widely regarded as applicable to all revelations of confidential information (p. 365).

A third basis for liability is action under a federal statute for a violation of the individual's constitutional right to privacy, although this interpretation of the Constitution has never been addressed by the Supreme Court. A final basis of liability would be a violation of a specific state or federal statute that identifies types of information that must be kept confidential. These include civil rights law, state licensing statutes for professionals, and federal alcohol and drug abuse statutes (*Federal Register*, 1975).

Breaches of confidentiality may be in the form of verbal disclosure of privileged information, disclosure via film or other recordings, and disclosure through written material. For example, in *Doe v. Roe* (1977), a therapist was found liable for publishing a book about a patient that included information sufficient to reveal the patient's identity to a number of readers.

INFORMED CONSENT

The legal basis for the patient's right to informed consent was stated by Mr. Justice Cardozo in *Canterbury v. Spence* (1972):

> Every human being of adult years and sound mind has a right to determine what shall be done with his own body, and a surgeon who performs an operation without his patient's consent commits an assault for which he is liable in damages.

Informed consent refers to the patient's right not to be touched or treated without his authorization. According to Rosoff (1981), it requires disclosure by the practitioner of the patient's condition or problem, the nature and purpose of the proposed treatment, the risks and benefits of the proposed treatment, the probability that the proposed treatment will be successful, possible alternatives to the proposed treatment, and prognosis if the proposed treatment is not given.

In the light of these disclosures, the patient must give his "express" or "implied" consent. Express consent is that which is spoken or written by the patient, although the patient's written consent should always be secured (*Kelly v. Gershkoff*, 1973). Implied consent is that which is reasonably implied by the conduct of the patient, although the patient must be aware of what he is implying in a particular context (*O'Brien v. Cunnard Steamship Company*, 1891).

The legal issues involved in informed consent have been summarized by Cataldo and Ventura (1981):

1. The fountainhead of the informed consent doctrine is the patient's right to exercise control over his own body (*Schloendorff v. Society of New York Hospital*, 1914).
2. The doctrine takes full account of the fact that the patient depends completely on the physician for information with which he (the patient) makes decisions (*Cobbs v. Grant*, 1972).
3. A physician cannot properly undertake to perform therapy without the prior consent of his patient (*Mohr v. Williams*, 1905; *McClees v. Cohen*, 1930, regarding consent to operative procedures).
4. In order for the patient's consent to be effective, it must have been "informed" in the sense that the patient received a fair and reasonable explanation (*Kenny v. Lockwood*, 1931).
5. The doctrine of informed consent imposes on the physician the duty to explain the procedure and to warn the patient of risks so that the patient can make an intelligent and informed choice (*Salgo v. Stanford University Board of Trustees*, 1957).
6. The doctrine requires the physician to reveal the nature of the ailment, the nature of the proposed treatment, and the probability of success, alternatives, and risks (*Natanson v. Kline*, 1960; *Scaria v. St. Paul Fire & Marine Ins. Co.*, 1975, regarding medical malpractice).
7. The law does not allow a physician to substitute his judgment for that of the patient in the matter of consent to treatment (*Collins v. Itoh*, 1972).
8. With regard to the use of a standard consent form, unless a person has been adequately apprised of risks and alternatives, any consent given is necessarily ineffectual (*Pegram v. Sisco*, 1976). Authorization by such an inadequate consent form is simply one additional piece of evidence (*Garone v. Roberts' Technical and Trade School*, 1975).
9. The informed consent doctrine is properly cast as a tort action for negligence, as opposed to battery or assault (*Cobbs v. Grant*, 1972; *Downer v. Veilleux*, 1974; *Trogun v. Fruchtman*, 1973), except that surgery performed without a patient's informed consent is a technical battery (*Shetter v. Rochelle*, 1965, 1966).
10. Beyond these general principles, there is considerable disagreement about whether the duty to warn should be based on a professional or a general reasonableness standard of care, whether expert testimony is required to prove standard of care, the appropriate test for proving causal connection between failure to disclose and any resulting injury or damages, and how one determines that consent is truly informed (Cataldo and Ventura, 1981, pp. 303–304).

In addition to case law, the patient's right to informed consent is addressed in codes of ethics for health care professionals as well as in the accreditation standards of the Joint Commission on the Accreditation of Hospitals:

The patient has the right to reasonably informed participation in decisions involving his health care. To the degree possible, this should be based on a clear, concise explanation of his condition and of all proposed technical procedures, including the possibilities of risk of mortality or serious side effects, problems not related to recuperation, and probability of success. The patient should not be subjected to any procedure without his voluntary, competent and understanding consent, or that of his legally authorized representative. Where medically significant alternatives for care or treatment exist, the patient shall be so informed.

The patient has the right to know who is responsible for authorizing and performing the procedures or treatment.

The patient shall be informed if the hospital proposes to engage in or perform human experimentation or other research/education projects affecting his care or treatment, and the patient has the right to refuse to participate in any such activity."

Therefore, consent includes at least three elements: competence, knowledge, and voluntariness (Friedman, 1975; Martin, 1975). Competence refers to the individual's ability to make a well reasoned decision, to understand the nature of the choice presented, and to meaningfully give consent. Knowledge refers to the individual's understanding of the nature of treatment, the alternatives available, and the potential benefits and risks involved. Voluntariness refers to the individual's voluntary agreement to participate in treatment.

THE PEDIATRIC PATIENT

However, as noted by Rosoff (1981):

As a general proposition, the consent of a minor to treatment is ineffective. Thus, the health care provider must secure the consent of the minor's parent or other person standing *in loco parentis* or risk liability to the minor and/or the parent (p. 188).

Exceptions to this proposition include emergencies in which immediate treatment is required to preserve the life of or prevent the serious impairment to the health of a child. If parents or persons standing *in loco parentis* cannot be located within the time available in these situations, the courts have usually held that the existence of an emergency obviates the need for informed consent. Courts have also recognized the consent of a minor for simple treatment of a nonurgent nature if he can demonstrate the capacity to understand and appreciate the risks and benefits of the treatment.

INFORMED CONSENT LITIGATION

Until recently, court decisions about informed consent litigation applied the professional standard-of-care consideration. This general philosophy was defined by the court in *Natanson v. Kline* (1960):

> The duty of the physician to disclose, however, is limited to those disclosures which a reasonable medical practitioner would make under the same or similar circumstances. How the physician may best discharge his obligation to the patient in this difficult situation involves primarily a question of medical judgment. So long as the disclosure is sufficient to assure an informed consent, the physician's choice of plausible courses should not be called into question if it appears, all circumstances considered, that the physician was motivated only by the patient's best therapeutic interests and he proceeded as competent medical men would have done in a similar situation.

Whereas this professional standard of care is physician based, the current rule on informed consent is patient based: It focuses on the informational needs of the average patient rather than on professional established standards of disclosure. In *Canterbury v. Spence*, the patient's right to informed consent is based on his right to self-determination:

> The root premise is the concept, fundamental in American jurisprudence, that "every human being of adult years and sound mind has a right to determine what shall be done with his own body. . . . " True consent to what happens to one's self is the informed exercise of a choice, and that entails an opportunity to evaluate knowledgeably the options available and the risks attendant upon each.

According to Cataldo and Ventura (1981), this new standard means that "expert testimony by medical experts is not necessary for settling the issue of negligence in obtaining informed consent—rather, only evidence as to whether the patient was truly informed." According to this doctrine of informed consent, patients are protected against the possibility of a "conspiracy of silence" by physicians and cases are easier to try. In addition, this patient based standard means that correctly performed medical procedures that result in harm to the patient can be just cause for compensation if the physician was negligent in informing the patient of the reasonable risks and alternatives (*Sard v. Hardy*, 1977).

SUMMARY

At all levels of our judicial system, litigation on behalf of patients and clients is increasing. At issue are the rights of these citizens to adequate care, privacy and confidentiality, informed consent, due process, and protection from any and all practice and procedure that threatens their dignity as human beings.

The challenge facing the health care practitioner is to provide a high quality of service based on the patient's special needs, but to do so only within the limits of the patient's rights under the law. This requires that the practitioner understand the law and its implications for health care as well as the various procedures that can be employed to protect patients' rights (see footnote 2 following references).

The challenge for the pediatric practitioner is even more demanding, since the rights of pediatric patients are only just emerging and vary greatly from state to state. Continuing education in the areas of patients' rights, children's rights, and procedures effective in protecting the rights of the pediatric patient is therefore essential for the present day pediatric practitioner.

LAWS REGARDING BEHAVIOR CONTROL

Health care essentially consists of a practitioner's attempts to control and change a patient's behavior in order to promote improvement in the patient's state of health. Specifically the patient comes to treatment with a goal, namely, to develop some understanding or to alleviate a problem that interferes with effective living. In providing the means by which the patient's goal can be attained, the practitioner often assumes a degree of behavioral control over the individual. This control may be exerted via medication; surgical procedures involving deprivation, restraint, isolation, or aversive stimulation; or hospitalization.

By cooperating with the practitioner, the patient assumes that his loss of behavioral control is a reasonable risk given the potential benefit of improved health. Thus, the obese patient is willing to be deprived of food in response to the recommendation of his physician. Furthermore, he may consent to medication and even surgery in attempts to alleviate the condition.

However, not every patient is capable of understanding the risks and benefits of a health care procedure and of reaching an informed decision to cooperate. These individuals—the mentally ill, the developmentally disabled, children, and students—represent the bottom scale of our society in terms of power and are most likely to be deprived of their rights. For example, parents may decide to restrict the food intake of their overweight child. Institutional staff may administer weight control medication to their overweight, mentally retarded residents or limit them to only two meals per day.

In addition, practitioners are not always sufficiently competent in their area of expertise or properly respectful of their patient's rights. It is not surprising, therefore, that the interaction of pediatric patients, willing to sacrifice control of their behavior or powerless to prevent it, and practitioners, with the means to control behavior, may result in abuses of the pediatric patient's rights. As described by Martin (1981):

> It is not unusual to find instances in which individuals have been deprived of food and water in order to force changes in their behavior or to punish them . . . the practitioners have seen virtue in such deprivation because they considered it effective in bringing about a desired change [in patient behavior] (p. 3).

Intervening in someone's life to change his behavior may represent a deprivation of liberty sufficient to raise constitutional issues. Behavior control procedures may be unnecessarily restrictive, and constitute cruel and unusual punishment (in violation of the Eighth Amendment) or curtail personal freedom without due process (in violation of the Fourteenth Amendment). In addition, recent court decisions suggest that any type of medical or psychological treatment and even the mere assignment of a patient to a special treatment program may constitute a deprivation of liberty, i.e., the patient's freedom to associate with others. In *Goss v. Lopez* (1975), the Supreme Court argued that a student charged with misconduct would receive a label that could damage his standing with fellow students and his teachers, thus depriving him of an essential element of liberty.

Cases such as *Goss v. Lopez* provide a good picture of how the courts are beginning to consider the rights of children in situations in which their behavior is under the control of adults. Indeed it is apparent that the current trend in legal policy is to extend rights such as autonomy, due process, protection against harm, least restrictive alternative, freedom from involuntary servitude, and minimum standards of care to children. An examination of legal issues concerning behavioral control must therefore address each of these areas.

It follows that if one is to understand laws concerning behavior control as they apply to pediatric patients with behavior, learning, or developmental problems he must understand these laws as they apply to the adult patient. Therefore, the remainder of this section describes legal precedent concerning behavior control, first, as it was established (typically with adults) and, second, as it applies to the pediatric patient.

AUTONOMY

The constitutional right to privacy is actually the right of the individual to autonomy, a right to make significant decisions about what happens to his mind or body. Martin (1981) has observed that "the first amendment to the U.S. Constitution has been interpreted as protecting the right to generate ideas. If speech is to be free, then the thought process must not be tinkered with" (p. 6).

For example, in *Kaimowitz v. Michigan Department of Mental Health* (1973), the court ruled that the First Amendment "protects the generation and free flow of ideas from unwanted interference with one's mental processes." Similarly, in *Mackey v. Procunier* (1973), the court ruled that the involuntary administration of a drug raised "serious constitutional questions respecting cruel and unusual punishment or unpermissible tinkering with the mental processes."

However, these decisions are only the most recent manifestations of legal interest in preserving the individual's right to autonomy. In *Union Pacific Railroad v. Botsford* (1891), the court ruled that:

> No right is held more sacred, or is more carefully guarded by the common law, than the right of every individual to the possession and control of his own person, free from all restraint or interference of others, unless by clear and unquestionable authority of law.

This decision was supported by *Griswold v. Connecticut* (1965) in which the Supreme Court maintained that for the state to encroach on the individual's right to be left alone, there must be a compelling state interest, which is directly related to the accomplishment of a permissible state policy. In *Stanley v. Georgia* (1969), the United States Supreme Court stated that "our whole constitutional heritage rebels at the thought of giving government the power to control men's minds."

Behavioral control procedures that have been identified by the courts as posing a threat to the individual's right to autonomy include chemotherapy, surgery, electroconvulsive therapy, and behavior modification. The courts have shown particular interest in protecting incompetent individuals from the involuntary administration of these forms of behavioral control. The procedures that are most likely to be used with the pediatric patient are chemotherapy and behavior modification.

Chemotherapy

The very purpose of medication is to attain some degree of therapeutic control or alteration of body chemistry and, ultimately, of the patient's behavior. In *Wyatt v. Stickney* (1972), the court ruled that no medication may be administered without a written order of a physician. In addition, the use of medication must be consistent with the patient's program of treatment, may not be administered in excessive quantities, and may not be used as a punishment procedure or as a substitute for treatment.

Court decisions concerning the use of medication have been determined largely on the basis of whether the treatment in question is a recognized medical practice, i.e., whether it is experimental, the extent to which it interferes with the patient's thought processes, and whether the drug regimen is therapeutic or is being used primarily to control the patient's behavior. In addition, a number of legal cases have been concerned with the right of the patient to consent to behavior control medication.

Schwitzgebel and Schwitzgebel (1980) and Rosoff (1981) have summarized a number of legal cases involving the use of medication for behavioral control. For example, in *Scott v. Plante* (1976), a patient was given psychotropic medication without his consent, despite the fact that he had apparently not been found to be legally incompetent to either consent to or refuse treatment. The Court ruled that in the absence of emergency, a patient or someone standing *in loco parentis* has the right to a hearing before medication is administered without his consent. In *In re Cleo Lundquist* (1976), the Court insisted that there be consent of the patient or guardian or a court order before a psychotropic drug could be administered, and that less restrictive treatment procedures be exhausted before considering medication; whereas in *In re Paul Fussa* (1976), a court allowed forced administration of the same drug (Prolixin) on the grounds that it is a commonly employed medical practice.

Behavior Modification

Briefly defined, behavior modification refers to the management of environmental antecedents and consequences to influence the behavior upon which they are contingent. (See Chapter 54.) As Martin (1975) has observed, "The only acceptable goal for any . . . program of behavior change should be to change overt behavior. It should not be a goal to change a person's mind or attitude" (p. 57).

Behavior modification is distinguished from drug therapy and surgical procedures, since it is a form of treatment based primarily upon psychological principles of behavior and social or environmental interventions rather than upon biological principles and physiological intervention (Martin, 1975; Stolz et al., 1975). In addition, as evident from the preceding discussion, issues concerning behavior control do not apply uniquely to behavior modification, but extend to any procedure that influences behavior.

The basic concern with behavior modification procedures has always been that they will somehow decrease the extent to which an individual can control his environment or can feel free to make his own choices. However, as noted by Ball (1968), behavior modification may actually result in the individual's becoming freer to select from alternatives that were previously unavailable to him. This increased freedom is made possible by

overcoming debilitating behavior patterns that restrict the individual's opportunities for a more normal adjustment.

Nevertheless, despite the potential of behavior modification to increase an individual's autonomy, there are cases in which individual rights have been abused. Much of this abuse has taken place in prisons and in public institutions for the mentally retarded and mentally ill. In the case of *Clonce v. Richardson* (1974), the court decided that a behavior modification project that begins by depriving the person of something represents a sufficient change in status that the individual must be accorded full due process guarantees of notice, hearing, and opportunity to contest his inclusion in the project.

The *Kaimowitz v. Michigan Department of Mental Health* decision, discussed previously, might also be applied to behavior modification programs. Martin (1975) has noted that "if the Kaimowitz reasoning is widely adopted, it might mean that behavioral approaches which are experimental, and thus require consent, may not be attempted with prisoners or mental patients who are involuntarily confined" (p. 172).

The Pediatric Patient

In *Morales v. Turman* (1973), a federal district court ruled that:

> It is not sufficient for defendants to contend that merely removing a child from his environment and placing him in a "structured" situation constitutes constitutionally adequate treatment . . . nor [do] sporadic attempts at . . . "behavior modification" through the use of point systems rise to the dignity of professional treatment programs geared to individual juveniles.

The court also suggested that although certain behavior modification procedures were sufficient deprivations of liberty that they required due process before being attempted, other procedures (e.g., "time out" for less than one hour) are slight enough deprivations that they do not require due process consideration.

In addition to those provided by legal precedent, guidelines for the use of behavior modification procedures and the protection of patient rights have been developed by various states and professional organizations (Association for the Advancement of Behavior Therapy, 1978; Stolz, 1976; Wexler, 1975). Similarly, Griffith (1980) has suggested a number of procedures that can be used to create a legally safe environment, and Risley and Sheldon-Wildgen (1980) have described the development and function of Human Rights Committees—necessary safeguards in institutional settings in which behavior modification and other behavior control procedures are employed.

DUE PROCESS

The Fifth Amendment to the Constitution provides, in part, that "no person shall be . . . deprived of life, liberty, or property, without due process of law." The Fourteenth Amendment applies to the states and holds that "nor shall any state deprive any person of life, liberty, or property without due process of law." The legal protections afforded under these due process provisions include, among others, the right to a hearing, to counsel, to be present at proceedings, to cross-examine witnesses, to call witnesses, and to receive independent evaluation or a "second opinion."

The importance of this law of behavior control can be seen in the substantial number of legal cases that have dealt with the rights of patients and students to procedural due process of law. For example, a number of cases have concerned the due process rights of the institutionalized patient, since commitment, even civil commitment, involves a loss of liberty (e.g., *Fhagen v. Miller*, 1972; *In re Barnard*, 1971; *Lessard v. Schmidt*, 1972; *People v. Bailey*, 1966; *Specht v. Patterson*, 1967). In addition, the case of *In re Buttonow* (1968) addressed the due process right of patients admitted to a health care institution on a voluntary basis. Finally, in the case of *Clonce v. Richardson*, the Court addressed the patient's right to due process concerning involvement in a special treatment program within an institution. It ruled that a treatment program that begins by depriving the person of something represents a sufficient change in status that the individuals must be awarded full due process guarantees of notice, hearing, and opportunity to contest their involvement in the program.

The Pediatric Patient

In the case of *In re Gault* (1967), due process rights were extended to institutionalized juveniles. This case involved a 15 year old boy who was sent to a state industrial school for making obscene phone calls. He complained that the juvenile hearing did not accord him full due process rights. In its decision, the Supreme Court held that the impact of the hearing—a potential loss of liberty—required that the youth receive full due process rights. The Court required that the child receive notice of the charges against him, the right to counsel at a hearing in which he could confront accusers and cross examine them, the right to refuse to testify against himself, the right to a transcript of the hearing, and the right to appeal. In a similar ruling (*Morales v. Turman*), a federal district court suggested that many of the treatment techniques used in a juvenile detention center

constituted sufficient deprivation of liberty that they required due process before being attempted.

More recent litigation has addressed the due process rights of students, rights that were provided them in the Education for All Handicapped Children Act (PL 94–142), which was signed into law in 1975. *Goss v. Lopez* involved the suspension of several high school students suspected of being involved in a lunchroom disturbance. The Supreme Court held that such an action influenced the students' freedom of association and the way other students and teachers viewed them. This was considered a sufficient deprivation of liberty that the constitutional requirement of due process was raised. The Court ruled that due process in such cases requires at least a notice and a hearing.

In *Mills v. Board of Education of District of Columbia* (1972), a federal district court ruled that all school age children are entitled to public education and that they should be assigned to regular public school classrooms. It further held that students can be reassigned to other educational services only if the alternative is more suited to their educational needs and only after notice and a hearing to challenge the reassignment. Similarly, in *Pennsylvania Association for Retarded Children v. Commonwealth of Pennsylvania* (1971), the court ruled that educational assignments had to follow due process and must be re-evaluated every two years with notice and a hearing.

PROTECTION AGAINST HARM

Behavior control procedures that expose the patient to harm or danger may be found to be in violation of the Eighth Amendment's prohibition against cruel and unusual punishment. According to Schwitzgebel and Schwitzgebel (1980), "a procedure may constitute cruel and unusual punishment if it violates minimal standards of decency, is wholly disproportionate to the alleged offense, or goes beyond what is necessary" (p. 84).

The Pediatric Patient

Cruel and unusual punishment as defined in case law has included, among other matters, deprivation of basic amenities (bed, clothing) as a disciplinary action (*Wright v. McMann*, 1972), the forceful injection of drug into a juvenile as a disciplinary procedure (*Nelson v. Heyne*, 1974), and the performance of degrading, unnecessary acts as a disciplinary procedure (*Morales v. Turman*). In addition, a number of cases have addressed the use of corporal punishment, restraint, and isolation in the treatment of children and adults with behavior disorders and developmental disabilities.

Corporal Punishment

The courts have been inconsistent in their rulings concerning the use of corporal punishment. For example, in *Ingraham v. Wright* (1977), the Supreme Court ruled that spanking, as a disciplinary procedure, does not violate the Eighth Amendment. In earlier cases—*Gary W. v. State of Louisiana* (1976), *Morales v. Turman*, and *Nelson v. Heyne*—courts ruled that corporal punishment as well as any actions generally described as causing physical pain or discomfort was unacceptable.

In *Ingraham v. Wright*, the court reasoned that local civil and criminal laws could adequately regulate the disciplinary procedures employed by teachers. As noted by Schwitzgebel and Schwitzgebel (1981), Massachusetts and New Jersey are the only states prohibiting the use of corporal punishment as a disciplinary procedure.

Restraint and Isolation

Case law has generally indicated that restraint and isolation cannot be used as punishment to control patient behavior. The only legally acceptable occasion for using these behavior control procedures is when their use is based on substantial evidence that the patient's behavior constitutes an immediate threat to his physical well being or that of others (*Wyatt v. Stickney*).

Physical Restraint. Physical restraint may involve the use of a mechanical apparatus or the physical holding of the patient by a staff member. Guidelines for the use of physical restraints have been provided in case law, in the standards of regulatory agencies, and in federal regulations pertinent to human services. Specifically, *Wyatt v. Stickney* prohibited the physical restraint of mental patients except for emergency situations. The court also stipulated that the restrained patient must be given bathroom privileges every hour and bathed every 12 hours.

More demanding guidelines concerning the use of physical restraint have been provided by the Joint Commission for the Accreditation of Hospitals and the United States Department of Health, Education, and Welfare. The Joint Commission's guidelines specify that restraint may not be used as a punishment procedure, for staff convenience, or as a substitute for a treatment program. In addition, the Commission sets a 12 hour time limit on the use of restraint and requires that the patient be given 10 minutes of exercise for every two hours of restraint.

In 1974 the Department of Health, Education, and Welfare developed a set of patient rights to be enforced in all skilled nursing facilities that participate in the Medicare or Medicaid programs (1978a). In 1976 a similar set of rights was developed for patients served in intermediate care facilities (Department of Health, Education, and Welfare, 1978b). These guidelines identify "the right to be free from mental and physical abuse and chemical and physical restraint" and include the following policies and procedures governing the use of restraints:

a. Orders indicate the specific reasons for the use of restraint.

b. Their use is temporary and the resident will not be restrained for an indefinite amount of time.

c. Orders for restraints shall not be enforced for longer than 12 hours, unless the resident's condition warrants.

d. A resident placed in the restraint shall be checked at least every 30 minutes by appropriately trained staff and an account is kept of this surveillance.

e. Reorders are issued *only* after a review of the resident's condition.

f. Their use is not employed as punishment, for the convenience of the staff, or as a substitute for supervision.

g. Mechanical restraints avoid physical injury to the resident and provide a minimum of discomfort.

h. The opportunity for motion and exercise is provided for a period of not less than 10 minutes during each two hours in which restraints are employed, except at night.

i. The practice of locking residents in their rooms or using locked restraints also constitutes physical restraint and must be in conformance with the requirements contained in this standard ("Interpretative Guidelines," Department of Health, Education, and Welfare, 1978b).

Isolation. Isolation may include the long term or brief "time out" seclusion of a patient in a locked room. Long term isolation of mental patients has been ruled a violation of the Eighth Amendment, i.e., that an extended period of solitary confinement is cruel and unusual punishment (*Wright v. McMann; Sinclair v. Henderson*, 1971). The potential harm of long term isolation was described by the court in *Morales v. Turman*:

> Experts were unanimous in their opinion that solitary confinement of a child in a small cell is an extreme measure that should be used only in emergency situations to calm uncontrollably violent behavior, and should not last longer than necessary to calm the child. . . . The child should not be left entirely alone for long periods. . . . Prolonged confinement of a child to a single building can be harmful unless the child is receiving a great deal of attention during the time of confinement. Experiments in sensory deprivation have shown that the absence of many and varied stimuli may have a serious detrimental effect upon the mental health of a child.

Case law regarding the use of isolation has generally supported the view expressed in *Morales v. Turman*. In *Negron v. Ward* (1976), the court ruled that isolation could only be used to prevent injury and only for a maximal duration of two hours. In *Welsch v. Likins* (1974), it was held that the patient in isolation must be checked every half hour (in *Morales v. Turman*, the court required that juveniles be placed in isolation for treatment purposes for a period not exceeding one hour); *Wyatt v. Stickney* established the right of patients to be free from isolation and specified one hour as the time limit on therapeutically justifiable isolation.

"Time out" refers to a behavior control procedure in which the patient is placed in a locked room for a brief period of time (five to 10 minutes) contingent upon the occurrence of some serious maladaptive behavior. However, as noted by Martin (1975), "It is hard to monitor length of time in time out. It is supposed to last only a few minutes, but I have seen mental institution inmates taken to their room for time out and then forgotten for hours by an overworked aide" (p. 86).

The courts have apparently shared Martin's concern. In *Morales v. Turman,* the court declared that isolation for disciplinary reasons was a sufficiently severe deprivation of liberty that it required due process procedures, i.e., in advance of isolation there must be notice of intent to discipline, a period of time to allow the patient (or his parent or guardian) to prepare a defense, and a hearing. In addition, the court ruled that during isolation the patient must be visited at least once each hour. It is important to note, however, that the *Morales* decision recognized that brief time out (one to five minutes) was mild enough that it did not warrant full due process procedures.

Conditions for time out were also provided by *Wyatt v. Stickney.* The Joint Commission's standards permit the use of time out, but only if the room is not locked. However, it was in the case of *Gary W. v. State of Louisiana* that the most restrictive guidelines for the use of time out were delineated:

No child shall be placed alone in a locked room either as punishment or for any other purpose. Legitimate "time out" procedures may be used under close and direct professional supervision. These standards shall apply to "time out" procedures:

They are to be imposed only when less restrictive measures are not feasible.

Placement shall be in an unlocked room with a staff member constantly nearby in a place where the staff member can supervise the child.

The child shall have access to bathroom facilities as needed.

The period of isolation or segregation shall not exceed 12 hours unless renewed by a qualified professional.

Except in an emergency situation in which it is likely that a child would harm himself or others, the decision to place a child in "time out" shall be made pursuant to a written order by a qualified professional. . . . Any such order must specify the terms and conditions of "time out" and the rationale for the decision.

Emergency use of "time out" shall be authorized only by the superintendent . . . and shall be limited to a period of not more than one hour.

LEAST RESTRICTIVE ALTERNATIVE

The doctrine of the "least restrictive alternative" was first cited in the case of *Lake v. Cameron* (1966). In this case the Federal Court of Appeals for the District of Columbia ruled that an involuntarily committed psychiatric patient was entitled to the least restrictive alternative available and that the

burden was on the state to explore and exhaust all less restrictive alternatives before deciding on confinement. The majority opinion stated that "Deprivations of liberty . . . should not go beyond what is necessary for [the patient's] protection." The *Lake v. Cameron* decision therefore was similar to that of *Shelton v. Tucker* (1960) in which the Supreme Court ruled that "the breadth of legislative abridgement must be viewed in light of less drastic means for achieving the same basic purpose."

In addition, the doctrine has been extended in other cases that noted that the confinement of a patient to maximal security or even full time hospitalization is justified only when less restrictive alternatives have proven ineffective (*Covington v. Harris,* 1969; *Dixon v. Weinberger,* 1975; *Lessard v. Schmidt*). In *Lessard v. Schmidt,* for example, the court required that before a specific treatment procedure is implemented, a practitioner must first show "(1) what alternatives are available; (2) what alternatives were investigated; and (3) why the investigated alternatives were not deemed feasible."

In *Wyatt v. Stickney,* the court required that each patient have an individualized plan of treatment and that the plan contain "a statement of the least restrictive treatment conditions necessary to achieve the purposes of commitment." More specifically, the Wyatt decision applied the least restrictive alternative doctrine to the actual procedures employed within the institutional setting and not just to the question of a patient's confinement. This extension of the doctrine was supported by the court's decision in *New York Association for Retarded Children v. Carey* (1975).

The Pediatric Patient

The use of the least restrictive alternative to treatment as well as to confinement is also required by some state statutes and by federal laws such as PL 94–142. This kind of legislation is covered in detail in Chapter 61B.

INVOLUNTARY SERVITUDE

When labor is performed that would be compensated if the laborer were not a patient, courts and regulatory agencies are finding noncompensation for the patient to be a case of "involuntary servitude," which constitutes a violation of the Thirteenth Amendment. In *Jobsen v. Henne* (1966) the court determined that work programs that supply institution maintenance needs as well as provide therapy for the patient are constitutional. *Weidenfeller v. Kidulis* (1974) ruled that work programs in which no therapeutic purpose was served and in which there was no compensation for the patient were unconstitutional. In *Dale v. State* (1974) the court ruled that voluntary labor, including hospital operation and maintenance, is permitted by patients competent to give informed consent.

A broader limitation on involuntary servitude was provided by *Wyatt v. Stickney*, in which the court found that no patient "shall be required to perform labor which involves the operation or maintenance of the hospital or for which the hospital is under contract with an outside organization." The Wyatt court further ruled that in situations in which hospital operation and maintenance are not involved, involuntary labor can be required only when it has been properly approved as a therapeutic activity and when it is adequately supervised by an appropriate staff member.

Specifically, *Souder v. Brennan* (1973) established that when the work assignment is clearly more in the nature of institutional maintenance than therapy, and when the institution receives the economic benefit from the work, the minimal wage must be paid. However, as noted by Schwitzgebel and Schwitzgebel (1980), the Souder opinion relied on the earlier case of *Maryland v. Wirtz* (1968), which was subsequently overruled by the Supreme Court, making "the applicability of the federal minimum wage standard . . . dubious, and state statutes and regulations, if any, . . . applicable" (p. 47).

Freedom from involuntary servitude is also addressed in federal regulations regarding human services (Department of Health, Education, and Welfare, 1978a, 1978b). The Department of Health, Education, and Welfare describes "the right not to be required to perform services for the facility that are not included for therapeutic purposes in one's plan of care."

The Pediatric Patient

Morales v. Turman found the use of make-work for children—in which nothing is accomplished by the work (e.g., moving rocks from one place to another)—to be cruel and unusual punishment. However, *Gary W. v. State of Louisiana* held that:

A child may be required to perform without compensation such housekeeping tasks as would be performed by a child in a natural home, foster home, or group home, provided that nothing in the child's individual treatment plan forbids such work.

MINIMUM STANDARDS

A number of court cases have established standards and guidelines that serve to regulate the nature of services that patients receive. First, courts have established that each patient (child and adult) has a right to a program of treatment based upon his special needs. The right to treatment was established in *Rouse v. Cameron* (1966), *Wyatt v. Stickney*, and *O'Connor v. Donaldson* (1975).

In the Wyatt case, it was held that involuntarily committed patients "unquestionably have a constitutional right to receive such individual treatment as well as give each of them a realistic opportunity to be cured or to improve his or her mental condition." The Court in *Rouse v. Cameron* held that the patient has a right to treatment or release and stated that "involuntary confinement without treatment is shocking." In the O'Connor case it was held that there is a constitutional right to individual treatment, i.e., when the justification for commitment is treatment, it violates due process if the treatment is not provided. Similarly, if the justification for treatment is dangerousness to self or others, treatment is the *quid quo pro* society must pay as the price of the extra safety it derives from the denial of an individual's liberty.

However, the patient's *right to refuse treatment* has also been established by the courts, although the right of the pediatric patient to refuse treatment has not been generally acknowledged. In a modification of the *Wyatt v. Stickney* decree, a federal court substantially restricted the freedom of hospital based psychiatrists to use certain potentially hazardous forms of treatment (e.g., psychosurgery). In effect, the court viewed the patient's right to refuse treatment as being basically synonymous with his right to give informed consent prior to treatment, i.e., refusal of consent as a refusal of treatment. This rationale was supported in *Kaimowitz v. Michigan Department of Mental Health* in which it was held that an involuntarily confined patient could not give truly informed and noncoercive consent to psychosurgery. Similarly, in *In re Cleo Lundquist* and *Rennie v. Klein*, courts ruled that hospitalized mental patients have a qualified constitutional right to refuse medications.

Secondly, case law has addressed the nature of the environment in which patients are served in regard to whether those environments provide for the patient's basic rights. Perhaps the most significant case in this area was *Wyatt v. Stickney*. In this case a federal district court, affirmed by the Fifth Circuit Court of Appeals, held that there was a constitutional right to adequate care in state institutions and that it could determine whether care was adequate and could formulate workable standards for institutional treatment.

The Wyatt court focused on three fundamental conditions for effective and adequate treatment: a humane physical and psychological environment, a sufficient number of qualified staff members to administer adequate treatment, and individualized treatment plans. The court recognized the rights of patients to the least restrictive conditions necessary for treatment; to be free from isolation (with one hour specified as the time limit for therapeutically justifiable isolation); not to be subjected to experimental research without consent; not to be subjected to psychosurgery, electroconvulsive treatment, aversive conditioning, or other unusual or hazardous treatment procedures without expressed and informed consent after consultation with counsel; to keep and use personal possessions; not to be required to perform institutional maintenance work, and to receive minimum wage if such work is voluntarily performed; to a comfortable bed and privacy; to send sealed mail to anyone and to receive sealed mail from public officials, private physicians, and attorneys; to have interactions with the opposite sex; to have access to a day room with television and other recreational facilities; to adequate meals; to an adequate staff; and to an individualized plan of treatment with a projected timetable for meeting the specific goals for treatment, criteria for transition to less restrictive treatment conditions, and criteria for discharge (Martin, 1975).

The Pediatric Patient

Whereas the Wyatt case established minimal standards for the treatment of psychiatric and developmentally disabled patients, *Morales v. Turman* enumerated standards of care for juveniles. Similarly, in *Inmates of Boys' Training School v. Affleck* (1972), it was held that juveniles cannot be held in any facility without their being provided the following:

A room equipped with lighting sufficient for reading until 10:00 P.M.; sufficient clothing to meet seasonal needs; mattresses, pillows, and bedding—which must be changed once a week—including blankets, sheets, and pillowcases; personal hygiene supplies, including soap, toothpaste, towels, toilet paper, and a toothbrush; a daily change of undergarments and socks; minimum writing materials—i.e., pen, pencil, paper, and envelopes; prescription eyeglasses, if needed; access to books, periodicals, and other reading materials; daily showers; access to medical facilities and nursing services; and general correspondence privileges.

In addition, courts have recognized the patient's right to exercise and to be outdoors on a regular basis (*Bartley v. Kremens*, 1975; *J.L. and J.R. v. Parham*, 1977) and to free exercise of religion (*Wyatt v. Stickney*, *Winters v. Miller*). Standards for the physical environments in which patients are housed have been specified in case law (*Wyatt v. Stickney*) as well as in the standards of regulatory agencies (Joint Commission regulations). Minimal standards for educational services have been provided by the *Education for All Handicapped Children Act* (PL 94-142).

SUMMARY

Controlling human behavior either in the context of a treatment facility or in society at large has tremendous legal and ethical implications. Medical and psychological techniques have the potential to produce dramatic and irreversible changes in the individual, and the patient must be protected against their misuse. This is particularly true for the pediatric patient, who is not always capable of understanding the risks and benefits of a health care procedure and of reaching an informed decision to cooperate. In short, pediatric patients are essentially powerless to resist attempts by others to control their behavior.

Protection for the rights of patients against unwanted or unwarranted behavior control has been the subject of a large amount of case law in recent years. Cases have focused on patients' rights to autonomy, due process, protection from harm, least restrictive treatment, involuntary servitude, and minimal standards of treatment, and have many implications concerning the treatment of the pediatric patient. In addition, these basic rights have been addressed in federal laws and the guidelines of agencies concerned with the accreditation and regulation of human service programs.

It is imperative, therefore, that the pediatric practitioner understand these laws and regulations concerning the control of human behavior and to ensure that the basic rights of his patients are protected. The individual practitioner must be guided by an understanding of the purpose for which his patient's behavior is to be controlled and the extent to which control over the behavior interferes with the patient's freedom.

WALTER P. CHRISTIAN

REFERENCES

Association for the Advancement of Behavior Therapy: Ethical Issues for Human Services. New York, Association for Advancement of Behavior Therapy, 1978.

Ball, T. S.: Issues and implications of operant conditioning: the reestablishment of social behavior. Hosp. Commun. Psychiatry, 19:230, 1968.

Bartley v. Kremens, 402 F. Supp. 1039 (E.D. Pa. 1975), rev'd. 431 U.S. 119 (1977).

Bellotti v. Baird, 443 U.S. 662, 61 L.Ed. 2d 797 (1979).

Brooks, A. D.: Mental health law. *In* Feldman, S. (Editor): The Administration of Mental Health Services Ed. 2. Springfield, Illinois, Charles C Thomas, 1980.

Bush v. Kallen, 201 A. 2d 142 (App. Div. N.J. 1973).

Canterbury v. Spence, 464 F. 2d 772 (D.C. Cir. 1972).

Carey v. Population Services, Inc., 431 U.S. 678 (1977).

Cataldo, M. F., and Ventura, M. G.: Patient rights in acute care hospitals. *In* Hannah, G. T., Christian, W. P., and Clark, H. B. (Editors): Preservation of Client Rights: A Handbook for Practitioners Providing Therapeutic, Educational, and Rehabilitative Services. New York, The Free Press, 1981.

Clonce v. Richardson, 379 F. Supp. 338 (W.D. Mo. 1974).

Cobbs v. Grant, 8 Cal. 3d 229, 104 Cal Rpter. 505, 502 P.2d 1, 9 (1972).

Collins v. Itoh, 160 Mont. 461, 503 P2d 36, 40 (1972).

Covington v. Harris, 419 F. 2d 617 (D.C. Cir. 1969).

Dale v. State, 355 N.Y.S. 2d 485 (Sup. Ct. N.Y., App. Div., 1974).

Department of Health, Education and Welfare: 42 Code of Federal Regulations, 405.1121(k): (1), (6), (9), (11), and (12) (1978a).

Department of Health, Education and Welfare: 42 Code of Federal Regulations, 442.311 (1978b).

Dixon v. Weinberger, 405 F. Supp. 974 (D.D.C. 1975).

Doe v. Roe, 400 N.Y.S. 2d 668 (1977).

Downer v. Veilleux, 322 A.2d 82, 89–90 (Me. 1974).

Federal Register (40 FR 27802): Title 42, Code of Federal Regulations, Part 2 (42 CFR Part 2). Confidentiality of alcohol and drug abuse patient records. FR, July 1, 1975, Washington, D.C., Department of Health, Education, and Welfare.

Fhagen v. Miller, 29 N.Y. 2d 348, 278 N.E. 2d 615 (1972).

Friedman, P. R.: Legal regulations of applied behavior analysis in mental institutions and prisons. Ariz. Law Rev., 17:39, 1975.

Gaertner v. Michigan, 187 N.W. 2d 429 (Mich. 1971).

Garone v. Roberts' Technical & Trade School, 47 App. Div.2d 305, 366 N.Y.S.2d 129, 133 (1975).

Gary W. v. State of Louisiana, 437 F. Supp. 1209 (1976).

Gibson v. Virginia, 219S.E. 2d 845 (1975).

Gobert J. J.: Accommodating patient rights and computerized mental health systems. North Carol. Law Rev., 54:154, 1976.

Goss v. Lopez, 95 S. Ct. 729 (1975).

Gotkin v. Miller, 514 F. 2d 125 (2nd Cir. 1975).

Griffith, R. D.: An administrative perspective on guidelines for behavior modification: the creation of a legally safe environment. Behav. Therap., 3:5, 1980.

Griswold v. Connecticut, 381 U.S. 479 (1965).

In re Barnard, 455 F. 2d 1370 (D.C. Cir. 1971).

In re Buttonow, 244 N.E. 2d 677 (N.Y.Ct. App. 1968).

In re Fussa, No. 46912 (Hennepin City Probate Court, June, 1976).

In re Gault, 387 U.S. 1 (1967).

In re Lifschutz, 467 P. 2d 557 (Cal. 1970).

In re Cleo Lundquist, No. 140 151 (Ramsey City Probate Court, April, 1976).

In re Smith, 295 A 2d 238 (Md. 1972).

Ingraham v. Wright, 525 F. 2d 909 (5th Cir. 1976), aff'd 97 S. Ct. 1401 (1977).

Inmates of Boys Training School v. Affleck, 364 F. Supp. 1354 (D.R.I. 1972).

J.L. & J.R. v. Parham, 412 F. Supp. 112 M.D. (Ga. 1976), 431 U.S. 936 (1977).

Jobsen v. Henne, 355 F. 2d 129 (2d Cir. 1966).

Kaimowitz v. Michigan Department of Mental Health. Cir. No 1 73–19434– AW (Cir. Ct. July 10, 1973).

Kelly v. Gershkoff, 112 R.I. 507, 312 A 2d 211 (s. Ct. 1973).

Kenny v. Lockwood, (1932) 1 D.L.R. 507, 520 (Ont. 1931).

Lake v. Cameron, 364 F. 2d 657 (D.C. Cir. 1966).

Lessard v. Schmidt, 349 F. Supp. 1078 (E.D. Wisc. 1972).

Mackey v. Procunier, 447 F. 2d 877 (9th Cir. 1973).

Martin, R.: Legal Challenges to Behavior Modification: Trends in Schools, Corrections and Mental Health. Champaign, Illinois, Research Press, 1975.

Martin, R.: Legal issues in preserving client rights. *In* Hannah, G. T., Christian, W. P., and Clark, H. B. (Editors): Preservation of Clients' Rights: A Handbook for Practitioners Providing Therapeutic, Educational, and Rehabilitative Services. New York, The Free Press, 1981.

Maryland v. Wirtz, 392 U.S. 183 (1968).

McClees v. Cohen, 158 Md. 60, 62–63, 148 A 124 (1930).

Merriken v. Cressman, 364 F. Supp. 913 (E.D. Pa. 1973).

Mills v. District of Columbia Board of Education, 348 F. Supp. 866 (D.D.C., 1972).

Mohr v. Williams, 95 Minn. 261, 104 N.W. 12, 15 (1905).

Morales v. Turman, 364 F. Supp. 166 (E.D. Tex. 1973).

Natanson v. Kline, 186 Kan. 393, 350 P. 2d 1093, 1106; 187 Kan. 186, 354 P.2d 670 (1960).

Negron v. Ward, 74 Civ. 1480 (S.D. N.Y., 1976).

Nelson v. Heyne, 491 F. 2d 352 (1974).

New York Association for Retarded Children v. Carey, 393 F. Supp. 715 (E.D. N.Y. 1975).

O'Brien v. Cunard Steamship Co., 154 Mass. 272, 28 N.E. 266 (1891).

O'Connor v. Donaldson, 422 U.S. 562, 572 (1975).

Olmstead v. United States, 277 U.S. 438 (1928).

Pegram v. Sisco, 406 F. Supp. 776, 779 (W.D.Ark.), aff'd, 547 F.2d 1172 (8th Cir. 1976).

Pennsylvania Association for Retarded Children v. Commonwealth of Pennsylvania, 343 F. Supp. 279 (E.D. Pa. 1972).

People v. Bailey, 21 N.Y. 2d 588, 237 N.E. 2d 204, 289 N.Y. 2d 943 (1968).

Phoenix Place v. Michigan Department of Mental Health, No. 77–737–260CZ (Cir. Ct. Mich., 1978).

Planned Parenthood of Central Missouri v. Danforth, 428 U.S. 52 (1976).

Rennie v. Klein, 462 F. Supp. 1131 (D.N.J. 1978).

Risley, T. R., and Sheldon-Wildgen, J.: Suggested procedures for Human Rights Committees of potentially controversial treatment programs. Behav. Therap., 3:9, 1980.

Roberts v. Supreme Court of Butte County, 9 Cal. 3rd 330, 2d. 309 (1973).

Roe v. Wade, 410 U.S. 113, 152 (1973).

Rosoff, A. J.: Informed Consent: A Guide for Health Care Providers. Rockville, Maryland, Aspen Systems Corporation, 1981.

Rouse v. Cameron, 373 F. 2d 451 (D.C. Cir. 1966).

Salgo v. Stanford University Board of Trustees, 154 Cal. App.2d 560, 317 P.2d 170, 181 (1957).

Sard v. Hardy, 34 Md. App. 217, 367 A.2d 525 (1976); 34 Md. App. 231 & 235 (1977).

Scaria v. St. Paul Fire & Marine Ins. Co. 68 Wis.2d I, 227 N.W. 2d 647, 654 (1975).

Schloendorff v. Society of New York Hospital, 211 N.Y. 125, 105 N.E. 92, 93 (1914).

Schwitzgebel, R. L., and Schwitzgebel, R. K.: Law and Psychological Practice. New York, Wiley, 1980.

Scott v. Plante, 532 F. 2d 939 (3rd Cir. 1976).

Shah, S. A.: Privileged communications, confidentiality, and privacy: privileged communications. Pro. Psychol., 1:159, 1970.

Shelton v. Tucker, 364 U.S. 479 (1960).

Shetter v. Rochelle, 2 Ariz. App. 358, 409 P.2d 74, 82 (1965); 2 Ariz. App. 607, 411 P.2d 45 (1966).

Sinclair v. Henderson, 331 F. Supp. 1123 (E.D. La. 1971).

Souder v. Brennan, 367 F. Supp. 808 (D.D.C. 1973).

Specht v. Patterson, 386 U.S. 605 (1967).

Stanley v. Georgia, 397 U.S. 557 (1969).

Stolz, S. B.: Ethical issues in behavior modification. *In* Bermant, B., and Kelman, H. (Editors): Ethics of Social Intervention. Washington, D.C., Hemisphere Publishing Corp., 1976.

Stolz, S. B., Wienckowski, L. A., and Brown, B. S.: Behavior modifications: a perspective on critical issues. Am. Psychol., 11:1027, 1975.

Tarasoff v. Regents of the University of California, 17 Cal. 3d 425, 131 Cal. Rptr. 14, 551 p2d 334, 83 ALR 3d 1166 (1976).

Trogun v. Fruchtman, 58 Wis.2d 596, 207 N.W.2d 297, 311–13, (1973).

Union Pacific Railroad v. Botsford, 141 U.S. 250, 251 (1891).

Weidenfeller v. Kidulis, 380 F. Supp. 445 (E. D. Wisc. 1974).

Welsch v. Likins, 373 F. Supp. 487 (D. Minn. 1974).

Wexler, D. B.: Behavior modification and other behavior change procedures: the emerging law and the proposed Florida Guidelines. Criminal Law Bull., 11:600, 1975.

Whalen v. Roe, 429 U.S. 589 (1977).

Wigmore, J. H.: Evidence in Trials at Common Law. Boston, Little, Brown & Co., 1961.

Winters v. Miller, 446 F. 2d 65, 71 (2d Cir. 1971).

Wright v. McMann, 387 F. 2d 519 (2d Cir. 1967).

Wyatt v. Stickney, 325 F. Supp. 781 aff'd on rehearing, 334 F. Supp. 1341 (M.D. Ala. 1971); aff'd on rehearing, 344 F. Supp. 363; aff'd in separate decision, 344 F. Supp. 387 (M.D. Ala. 1972); aff'd sub nom, *Wyatt v. Aderholt*, 503 F 2d 1305 (5th Cir. 1974).

FOOTNOTES

1. For additional information concerning legal decisions and legal policy related to health and human services, the reader is referred to the following:

Brooks, A. D.: Mental health law. *In* Feldman, S. (Editor): The Administration of Mental Health Services Ed. 2. Springfield, Illinois, Charles C Thomas, 1980.

Hannah, G. T., Christian, W. P. and Clark, H. B. (Editors): Preservation of Clients' Rights: A Handbook for Practitioners Providing Therapeutic, Educational, and Rehabilitative Services. New York, The Free Press, 1981.

Martin, R.: Legal Challenges to Behavior Modification: Trends in Schools, Corrections and Mental Health. Champaign, Illinois, Research Press, 1975.

Martin, R.: Legal issues in preserving client rights. *In* Hannah, G. T., Christian, W. P., and Clark, H. B. (Editors): Preservation of Clients' Rights: A Handbook for Practitioners Providing Therapeutic, Educational, and Rehabilitative Services. New York, The Free Press, 1981.

Miller, H., Brodsky, S., and Blecchmore, J.: Patient's rights: Who's wrong? The changing role of mental health professionals. Pro. Psychol., 7:274, 1976.

Rosoff, A. J.: Informed Consent: A Guide for Health Care Providers. Rockville, Maryland, Aspen Systems Corporation, 1981.

Schwitzgebel, R. L., and Schwitzgebel, R. K.: Law and Psychological Practice. New York, John Wiley & Sons, Inc., 1980.

Simon, G.: Psychology and the "treatment rights movement." Pro. Psychol., *6*:243, 1975.

United States Department of Health, Education, and Welfare: Litigation and Mental Health Services. DHEW Publication (ADM) 76–261. Washington, D.C., U.S. Government Printing Office, 1975.

2. Procedures for the protection of patients' rights of privacy, confidentiality, and informed consent are discussed in the following publications:

Cataldo, M. F., and Ventura, M. G.: Patient rights in acute care hospitals. *In* Hannah, G. T., Christian, W. P., and Clark, H. B. (Editors): Preservation of Clients' Rights: A Handbook for Practi-

tioners Providing Therapeutic, Educational, and Rehabilitative Services. New York, The Free Press, 1981.

Hannah, G. T., Christian, W. P., and Clark, H. B. (Editors): Preservation of Clients' Rights: A Handbook for Practitioners Providing Therapeutic, Educational, and Rehabilitative Services. New York, The Free Press, 1981.

Rosoff, A. J.: Informed Consent: A Guide for Health Care Providers. Rockville, Maryland, Aspen Systems Corporation, 1981.

Schwitzgebel, R. L., and Schwitzgebel, R. K.: Law and Psychological Practice. New York, John Wiley & Sons, Inc., 1980.

61B   Legislation for the Education of Children with Handicaps

It sometimes occurred to me that she was like a person alone and helpless in a deep, dark, still pit, and that I was letting down a cord and dangling it about in hopes she might find it; and that finally she would seize it by chance and, clinging to it, be drawn up by it into the light of day, and into human society.

Samuel Gridley Howe, describing his work with Laura Bridgman

In the mid-1970's, pediatricians joined other child development specialists in heralding the revolutionary legislative advances that assured the right of a full appropriate public education to all handicapped children in the United States. The establishment of this right came as the result of a long progression of evolving beliefs about handicap, as well as a century's development of special educational philosophy and practice, and a number of complex shifts in public attitudes toward responsibility for handicapped children. Although there are likely to be several more chapters in the story before the end of the century, the formulation of new principles about the education of handicapped children is substantial and is likely to have an enduring impact on pediatric practice and pediatric patients.

HISTORICAL BACKGROUND

In early American times, when communities faced dire hardship in everyday living, when infant and child mortality was exceedingly high and epidemics were commonplace, there were few resources for health care or education of any kind. Little was known of the cause(s) of disease and affliction, and the lack of information led to deep seated uneasiness about handicap. The belief was even articulated that handicaps were punishments or signs of displeasure from God (Cone, 1979). In this climate it is not surprising that there was only minimal public attention to the specific needs of handicapped children.

A noticeable shift in attitude about handicap occurred in the late eighteenth and early nineteenth centuries as medical discoveries supplied rational explanations for many illnesses and disabling conditions. Gradually the notion developed that some individuals were more fortunate than others and thus had an obligation to serve their less fortunate neighbors. Moreover, acute awareness of the misfortune of handicapped children was paralleled by scientific interest in the process of education itself. Dr. Samuel Gridley Howe, Thomas Hopkins Gallaudet, and Hervey Backus Wilbur studied and then applied the teaching methods originated by Braille, Michel, Roche-Ambroise, and Seguin in France (Bremner, 1970). Using individual, sequential teaching, and special materials, they were able to demonstrate the efficacy of education for the blind, deaf, and retarded children with whom they worked. Their successes led to the establishment of small schools for the education of handicapped children.

By the 1840's, in a number of progressive states, including Massachusetts, Connecticut, New York, and Ohio, reformers began to call for public commitment to educational institutions for handicapped children. Samuel Gridley Howe articulated the American ideal of education when he explained, "It is a link in the chain of common schools—the last indeed, but still a necessary link in order to embrace all the children of the State."*

Initially there was tremendous excitement and optimism based on the results from the small schools and institutions. As a result a number of

---

*Howe, S. G. as quoted in Bremner, R. H. (Editor): Children and Youth in America. A Documentary History. Vol. I: 1600–1865. Cambridge, Harvard University Press, 1970, p. 782.

states opened institutions for larger numbers of children. It became the norm for pediatricians and other physicians to recommend placement of children with certain handicapping conditions in institutions. In short order these institutions became overcrowded, and the principles of individualization, close observation, and painstaking care, which had characterized the early work of Wilbur and Howe, was lost. The enthusiasm generated in the 1840's began to disappear in the second half of the century, and new attitudes toward special care for handicapped children began to emerge. Perhaps the most devastating was the appearance of an association of handicap and moral turpitude (Rosen et al., 1976). Undoubtedly the philosophy of Social Darwinism allowed a decline in the care of many handicapped individuals.

After the turn of the century, segregation of handicapped children from the community became more and more common. State laws were written specifically to exclude children with certain handicapping conditions from the provisions of compulsory education statutes. Pediatricians were trained that it was "best for the parents" to alleviate their suffering by removing their handicapped children to special schools and institutions. Often it was on the basis of a newborn's physical examination that a child would be placed in the care of the state indefinitely.

The treatment of children in institutions and the availability of special educational programing varied enormously. In many instances inadequate facilities and staff made it impossible for states to provide more than subsistence custodial care. Many children suffered for lack of pediatric, diagnostic, and educational intervention, and their problems were exacerbated by what Spitz termed "hospitalism." In other cases, however, significant progress was achieved. Great advances were made in the education of deaf and blind children. Furthermore, in the early part of the twentieth century two follow-up studies documented the efficacy of educational programs for a substantial number of retarded youngsters. The first was a study showing positive outcomes for youngsters in ungraded public school classes in New York City. The second study, that of Walter Fernald, was all the more powerful because it disproved his own hypothesis that the youngsters who left Waverly State School (largely against protest) would never be able to adjust to the noninstitutional community (Rosen et al., 1976).

Subsequent studies made it increasingly clear that formal programing and specific stimulation were as essential for handicapped children as for nonhandicapped children. But it became increasingly evident that even in the presence of the most zealous professional involvement, the negative consequence of divorcing handicapped children from their families and their natural communities was taking its toll.

By the 1950's and early 1960's the ideas that had been evolving came to full articulation. On the theoretical front, the adoption by educational specialists, pediatricians, and psychologists of the developmental model was a major breakthrough. The developmental model derived from the systematic observations of psychometricians following Binet, and pediatricians following Gesell, held that given sufficient opportunity, most children would develop—albeit at their own rate. This notion was in many ways a gauntlet thrown out to pediatrics, psychology, and special education, challenging the creation of diagnostic tools precise enough to define a child's level of function and educational materials refined enough to take advantage of "teachable moments." Diagnosis, documentation, and evaluation began to be viewed as essential components of education. Professionals began to call for educational opportunities for handicapped children within the public schools—for more individualized programing and for greater emphasis on the full extent of an individual child's functioning.

As important as the consensus of the professional community was, it was the dissatisfaction of parents about their children's exclusion from society that undoubtedly had the greater impact. Invigorated by the civil rights victories of other minority groups, parents became increasingly vocal about the segregation of their children into institutions outside the mainstream of American life. They were concerned that their children were not receiving the most appropriate education. The conditions in the large state institutions were far from ideal, and in some cases shockingly bad. Many communities lacked alternative public educational opportunities. As a result, several outspoken parent groups launched lawsuits, charging discrimination against their children. The two landmark cases were the *Pennsylvania Association for Retarded Citizens v. Pennsylvania*,* in which the right to a free, appropriate education for mentally retarded children was upheld, and *Mills v. The Board of Education*,† in which the Federal District Court in the District of Columbia concluded that all handicapped children have a right to education, even when funds available for education are limited.

The theoretical shifts in understanding handicapped children, the civil rights movement, and recognition that children with handicaps had been "handled" outside society for too long led many state legislators and Congress to consider the ed-

---

*Pennsylvania Association for Retarded Children v. Commonwealth of Pennsylvania. 343 F. Suppl. 279, 1972.

†Mills v. Board of Education of the District of Columbia, 348 F. Suppl. 866, 1972.

ucational needs of handicapped children. Ultimately the states and Congress moved toward the creation of statutes that assured the right of free public education for all handicapped children.

PL 94-142: THE EDUCATION FOR ALL HANDICAPPED CHILDREN ACT

In 1975 both houses of Congress passed PL 94-142, The Education for All Handicapped Children Act, by large majorities. The law represented the culmination of a series of legislative enactments, including PL 89-750, which established the Bureau of Education for the Handicapped in 1966, PL 91-517 the Developmental Disabilities legislation of 1971, which established the University Affiliated Facilities for training, and PL 94-585, the Rehabilitation Act of 1975.

At the time of passage of PL 94-142, it was recognized that the legislation would need to be broad enough to accomplish three major roles. First, uniform standards for appropriate educational services were seriously wanted. Second, a major shift toward community and school system responsibility was essential if handicapped children were to receive the most appropriate services. And, third, it was recognized that the legislation represented an entitlement that systematically had been withheld from children with handicaps. To ensure the attainment of these three goals, the authors of the federal legislation created a program that is comprehensive in scope and specific in detail.

Under the regulations passed in 1977, states are required to provide "a full appropriate public education" for all handicapped children. The law applies to children who are mentally retarded, hard of hearing, deaf, speech impaired, visually handicapped, seriously emotionally impaired, other health impaired, or multihandicapped as well as children with specific learning disabilities. Priority is given to the most severely impaired youngsters.

The state education agencies are specifically required to identify, locate, and evaluate all children with handicaps, and to prepare and implement individualized education plans for these children. They are further required to insure placement of children in the least restrictive environment possible and to uphold procedural safeguards for parents ("due process"). Children placed by the state in private schools are to enjoy the same rights and safeguards as children in public schools. "Related services" needed by students to benefit from special education are also to be provided. These include transportation, counseling, physical therapy, occupational therapy, speech and hearing, school health services, and diagnostic health services. Finally, states must provide in-service training for special and regular education teachers.

PL 94-142 is unique among federal education grant programs in that it guarantees rights to handicapped children and their parents without regard to the level of federal expenditure. Under the law, some federal funds are authorized for the program, based on a formula whereby the number of handicapped children ages three to 21 who are receiving a public education in a state is multiplied by a percentage of the nationwide average per pupil expenditure. The majority of the funds for the program, however, must come from the states and localities. To insure compliance with its provisions, the federal legislation combined with Section 504 (the "civil rights" act for the handicapped) stipulates that all federal funds for the handicapped can be withheld if a state does not carry out the provisions of PL 94-142.

PEDIATRIC ROLE

Because PL 94-142 is fundamentally a special education law, the pediatric role is implied rather than specified. Nonetheless it is clear that pediatricians are expected to serve as consultants in the educational process for children with handicaps. Furthermore, the majority of states, in changing their education laws to conform to the federal mandate, made dramatic changes with regard to physician responsibility. Prior to the 1970's pediatricians were often asked to exclude children from school by certifying that the children's physical or medical condition prevented or "rendered inadvisable" attendance at school. The new special education laws, in contrast, include physicians among the specialists qualified to designate children as "exceptional" and therefore in need of special services (Bolick, 1974). With this change in language come significant new responsibilities. It is no longer sufficient for pediatricians and other specialists to define the manifest disabilities of children. Now they must define the children's potential abilities and outline the individual services that are likely to be of the greatest benefit. To do this well, pediatricians need to engage in every aspect of the education of all handicapped children's program, including identification, evaluation, service provision, and parent participation (Palfrey et al., 1978).

IDENTIFICATION

Pediatricians are in a unique position to act as case finders for handicapped children's educational programs. Aware of the history of the children and of "at risk" indicators, pediatricians may be able to identify childhood handicaps early in their course. Previously this early identification often led to the disheartening situation of informing parents about handicaps without being able to offer constructive suggestions about intervention. There was therefore a tendency to protect parents from this information as long as possible. With the advent of early educational programs for young children with handicaps, pediatricians can temper the news of disability with recommendations for intervention and support.

Early identification efforts have been most successful for conditions that are severe and relatively obvious (e.g., sensorineural deafness, retrolental fibroplasia, cerebral palsy, and mental retardation associated with phenotypic stigmata). The identification of children with less clearly defined or less serious conditions poses a complicated set of issues, including the predictive implications of findings in the toddler and preschool period, and the diagnostic acumen of current pediatric practice.

A recognized hazard of the movement toward early identification, and particularly the early identification of the subtle developmental handicaps and behavioral deviation, is early overidentification. Experience with developmental variation and the transient nature of many common behavioral manifestations of young children are likely to protect most pediatricians from false positive identification. On the other hand, the general practice of reassuring parents concerned about developmental delays that "he'll outgrow it" may preclude an investigation that would have great value to the child and family. The challenge to pediatricians is to judge which parental concerns to pursue. The use of specially designed questionnaires and observation techniques may help pediatricians to make these decisions in an informed, stepwise fashion. An active period of waiting, watching, and gathering of information ultimately can be reassuring to parents and often will identify specific problems that can be referred for educational or other intervention. (Keough and Becker, 1973; Oberklaid and Levine, 1980).

EVALUATION

Individual diagnostic evaluation is the pivotal component of the education for all handicapped children program. Under PL 94-142, individual evaluations are required for every child receiving special education.

The basic evaluation team consists of the child's teacher, a representative of special education, one or both parents, and other individuals at the discretion of parents or agencies, as well as the child when appropriate.

The individualized education plan must contain four components: a description of the child's present level of educational performance, a statement of instructional goals for the year, a statement of the educational services that will be provided and the amount of time that will be spent in ordinary classrooms, and a specification of the duration of special services and how they will be evaluated.

Although PL 94-142 does not specifically require physician input in regard to the individualized education plan, many states do. There is considerable variation, however, in the extent of the involvement required. In some states all that is necessary is a statement regarding a recent physical examination. In others specific questionnaires are used to pinpoint areas of educational relevance. In still others the physician or the physician's representative is asked to be present at the individualized education plan meeting (Twarog et al., 1979).

There is also variation from state to state regarding which children need medical evaluations. In a few states there must be medical consultation for every child who is to receive special educational services. In other states certain categories of handicap necessitate certain specialist referrals (for instance, ophthalmological evaluation might be required for all blind children, but the same children might not have a general physical or neurodevelopmental assessment), and, finally, in some states a panel of consultants decides on a case by case basis what the most relevant evaluation should be.

Because pediatricians are being asked to participate in the evaluation of children with the whole range of handicaps from severe retardation and physical disability to emotional disorder to sensory problems and learning disabilities, it becomes increasingly important for physicians to recognize the issues and questions that are foremost in the minds of the educators and parents.

When children are referred to pediatricians for the "medical" portion of the individualized education plan, it is likely that the evaluation team wants a number of issues to be addressed. First, there is the lingering hope that somehow the doctor, the modern-day shaman, can determine the cause of the child's problem and that then the child can be cured. It is thus extremely important that the physician document the efforts that have been (and are being) made to establish a cause or cure for the disorder and spend some time explaining what is and is not understood about the child's problem, treatment, and prognosis. Second, educators wish to know the behavioral consequences of the child's disorder. Will the child be experiencing seizures in the classroom; is he likely or unlikely to interact with other children; are there any safety considerations? Third, the team is likely to want as full an exploration of the child's history, current neurodevelopmental status, and attention-activity modulation as possible.

Although assessment of some of these areas will undoubtedly be covered by other specialists, the pediatrician is often in a good position to synthesize many issues, adding the longitudinal perspective derived from history taking or (ideally) from a long term relationship with the child. In order to have a better personal understanding of the developmental, functional, and behavioral issues under team consideration, the physician may wish to obtain some observational data (Levine et al., 1980). This process may help the pediatrician to concentrate the medical reports on issues of highest relevance for the evaluation team.

SERVICES

The major service to be provided under PL 94-142 and state laws is special education. However, the law also calls for the provision of "related services" needed by the student to benefit from special education.

The "related services" aspect of the law challenges school systems to work with community health and mental health agencies to insure that no barriers to special education remain unaddressed.* A variety of systems have been established by school districts across the country to meet the related services section of the law. Although many of these have worked well, there have been persistent problems in a number of areas, including counseling, supervision, and the extent to which school systems should be involved in the diagnosis and therapy of certain mental and physical conditions. Beyond the professional issues, there are also serious reimbursement problems. The most devastating has been the denial of proprietary insurance benefits to families when medical services were ordered by schools and were thus construed as educational rather than health needs. Although the unexpected consequences of the related services aspect of the law are probably balanced by the increased opportunities for children, effective coordination of services remains an unfulfilled goal.

PARENT PARTICIPATION

The education for all handicapped children program creates a major contact point for physicians with the education system regarding the families for which they provide care. PL 94-142 is in many ways a "consumer law," and the position of parents vis-à-vis the educational system is protected

by the due process clauses within the law. Physicians, knowing families, are in a position to help parents share actively in the decision making process. This is a role to which parents are unaccustomed, and they often want to have a helping professional available for consultation and advice.

Parents also frequently want a second opinion, and PL 94-142 allows them to seek this from their physicians. When this happens, physicians should consider the request as a call for mediation by the parents between themselves and the school. They should therefore obtain as much educational information as possible while maintaining their distance from the school. In addition, physicians should try to avoid preconceived biases toward one or another specialty. They should try to interpret the specialist's reports within the context of the whole child.

As advocates for children, physicians can join with parents to monitor their state education laws to see that these conform to PL 94-142 (Jacobs and Walker, 1979). This is particularly important in times of federal and state funding cutbacks, when compromise of quality and standards may be the consequence.

EXPERIENCE WITH THE EDUCATION FOR ALL HANDICAPPED CHILDREN ACT

PL 94-142 has been in effect since 1977. The overall experience with the law has been extremely favorable. The Office of Special Education reports that as a result of the law, almost 75 per cent of the nation's handicapped children are receiving special education and related services as compared to less than half prior to PL 94-142's enactment, that 84 per cent of the states and territories have reported annual increases in the number of handicapped children served, that the number of preschool children receiving special education has increased by 20,000 in three years (a growth rate of more than 10 per cent), and that the number of previously institutionalized children being served by local education agencies has increased by almost 40 per cent. (U.S. Department of Health, Education, and Welfare, Office of Education, August 1979).

The program has, of course, suffered a number of birth pangs. There is some concern that the wide variation among states in terms of services to handicapped children represents a lack of access for a significant number of children. Furthermore, there continue to be difficulties relating to issues of labeling and categorizing youngsters, in providing services to multiply handicapped and behaviorally disturbed children, in expanding the program to high school and preschool children, and

---

*U.S. Department of Health and Human Services, Office of the Assistant Secretary for Health and the Surgeon General, 1981.

in arranging related services (U.S. Department of Health, Education, and Welfare, Office of Education, January 1979).

Despite these problems, great progress has been made and substantial changes are apparent in schools and communities. Although it remains to be seen how the program will fare in periods of economic hardship and political conservatism, the principles upon which the legislation is based are sound, and the right of handicapped children to free, appropriate public education has been unequivocally established. It is now up to all involved to maintain and improve the programs that are provided within the community.

FUTURE DIRECTIONS FOR PEDIATRICS

To meet the needs of the ideal multidisciplinary program outlined in PL 94-142, pediatrics as a specialty needs to fulfill obligations in four areas. These are training, research concerning assessment techniques for all handicapping conditions, more active participation in the evaluation and management of children involved in the program, and exploration of early intervention.

Pediatricians at all levels need more training in developmental assessment, in the care and management of handicapped children, and in the implications of the special education legislation. A number of studies have indicated the widespread dissatisfaction with current training in developmental pediatrics (Dworkin et al., 1979), and at least one study has indicated that less than half the practicing pediatricians in a large community were conversant with the provisions of PL 94-142 or their own state special education law (Jacobs and Walker, 1979).

Research dealing with assessment techniques in pediatrics is still sorely needed. There are still many areas of children's development that are only partially understood and for which the assessment batteries are primitive. Although it is possible to assign a psychometric number to most children, that number does little to explain style, focus, motivation, attention, underlying emotional status, self-concept, or developmental status. Without a grasp of these parameters it may be impossible to design an appropriate program. It is not enough to determine what a child can do; it is also necessary to understand how he does it and what modifications are likely to be necessary to allow him to do it better. For severely neurologically impaired children this may mean extensive evaluation of different postures and wheelchair placements; for a child with activity problems it may mean a two to three month trial with medications; for a child with learning problems it may

mean collaboration with educational specialists to determine the best teaching strategies to meet the child's own learning style.

In order to meet teachers' and parents' requests for physician input into the education for all handicapped children act, more active participation of physicians in evaluation and strategy sessions may be needed. This may mean a sacrifice of time for physicians. Work is needed by specialists in education and pediatrics to determine equitable reimbursement for that time, since it frequently can be an essential component of the process for the child and family.

When physicians are called into evaluation teams, it is important that they realize the limitations of their role. It is often the tendency of physicians to assume the position of chairman of the board and to feel responsible for knowing everything about a given child. This is not necessary. It is sufficient to provide the observations gleaned in the office and to submit them as evidence to be corroborated by the assessments of other members of the team.

Finally, since pediatricians are involved with very young children, there is a major need for pediatric investigation of early intervention and support for the programs that have been shown effective. There is a general belief that "the earlier, the better," but more data are needed to define what the components of that "earlier" should be. Pediatricians have often served as the evaluators of new techniques, including antibiotics and artificial formulas. They can now help to monitor the effectiveness of early educational intervention.

SUMMARY AND CONCLUSIONS

Major innovations have occurred in the education for all handicapped children. Attitudes toward the rights of handicapped children have grown in a fundamental manner, as has scientific understanding of developmental deviation. In many ways society has begun to tolerate individual differences and to accept individual variation. The acceptance in pediatrics of the developmental model has had a major impact on this societal thinking.

Pediatricians have long been involved in the care and treatment of handicapped children, but frequently have felt stymied when asked by parents about the functional and developmental aspects of their care. With special education legislation in place, there are now educational resources for all handicapped children, and pediatricians can join other professionals in planning the full range of services for children with handicaps who are under their care.

JUDITH S. PALFREY

REFERENCES

Bolick, N.: Digest of state and federal laws: Education of handicapped children. Reston, Virginia Council for Exceptional Children, 1974.

Bremner, R. H. (Editor): Children and Youth in America, A Documentary History (two vols.). Cambridge, Harvard University Press, 1970.

Cone, T. E.: History of American Pediatrics. Boston, Little, Brown and Company, 1979.

Dworkin, P. H., et al.: Training in developmental pediatrics. Am. J. Dis. Child., 133:709, 1979.

Jacobs, F. H.: Identification of Preschool Handicapped Children: A Community Approach, Community Health Studies. Boston, Harvard School of Public Health, 1979.

Jacobs, F. H., and Walker, D. K.: Pediatricians and the Education for All Handicapped Children Act of 1975 (Public Law 94-142). Pediatrics, 61:135, 1978.

Keough, B. K., and Becker, L. D.: Early detection of learning problems: questions, cautions, guidelines. Except. Child, 40:15, 1973.

Levine, M. D., Brooks, R., and Shonkoff, J.: A Pediatric Approach to Learning Disorders. New York, John Wiley & Sons, Inc., 1980.

Oberklaid, F., and Levine, M. D.: Precursors of school dysfunction. Pediatr. Rev., 2:1, 1980.

Palfrey, J. S., Mervis, R. C., and Butler, J. A.: New directions in the evaluation and education of handicapped children. N. Engl. J. Med., 298:819, 1978.

Rosen, M., Clark, G. R., and Kivitz, M. S. (Editors): The History of Mental Retardation. Collected Papers (two vols.). Baltimore, University Park Press, 1976.

Twarog, W. T., Levine, M. D., and Berkeley, T. R.: Patterns of Physician Participation in the Evaluation of Handicapped Children for Special Education Programs. A Report on State Regulations. Report prepared for the Bureau of the Education for the Handicapped, U.S. Office of Education under grant PR 451BH70139, 1979.

U.S. Department of Health and Human Services, Office of the Assistant Secretary for Health and Surgeon General: Better Health for Our Children: A National Strategy, Vol. II. Washington, D.C., U.S. Government Printing Office, 1981.

U.S. Department of Health, Education, and Welfare, Office of Education: Progress Toward a Free Appropriate Public Education. Washington, D.C., January 1979.

U.S. Department of Health, Education and Welfare, Office of Education: Progress Toward a Free Appropriate Public Education: Semi-Annual Update on the Implementation of Public Law 94-142: The Education for All Handicapped Children Act, August 1979.

62

Methodological Issues in Behavioral and Developmental Pediatrics*

Although the specific details vary between settings, topic area, and investigator, the issue of methodological rigor hangs constantly over the heads of clinicians and researchers alike. Most providers gradually formulate their private ideas about what works and what does not work, frequently independently of what may be happening in the field at large. Every practicing pediatrician† develops a style of practice that includes the number of patients he sees each day, the amount of time he spends with each patient, how he spends that time, and what he charges for that time. When faced with decisions about what medicine to prescribe and in what quantities and how often, he frequently resorts to a relatively small number of preferred medications, at dosage levels that may or may not correspond to the little black book of recommendations that he carried in his pocket as a resident. Ultimately all these decisions are based on whatever input the individual practitioner considers to be important at the time. Every professional makes these types of decisions, although many decisions are made in the absence of any conscious effort. The important point is that methodological rigor is an issue that is faced by every pediatric health care provider; some face it directly, perhaps even scheduling time specifically to address the issue, and some face it as part of their day to day activities.

Methodological rigor is not something that comes easy to some and hard to others. For those who choose to value it highly, it is hard; to those who are less concerned, it is easy. Thus, to some clinicians and to some researchers, it is of paramount concern; to others it is unimportant. Yet when we want to be confident about our results, either as individuals or in holding our work up for public or professional scrutiny, we must face the issue of methodological rigor. Thus, the discussion that follows is intended as a positive

discussion of an issue that is important to "us," not to "them."

Behavioral and developmental pediatrics includes a wide range of clinical and research activities that are demonstrated by the breadth of chapters in this book. As with any new, emerging field, much of the early work can be characterized as "pilot studies," which lack the necessary rigor to qualify as "hard science."

Within the general rubric of behavioral and developmental pediatrics lie several separate, but related, areas of inquiry. Although much of this chapter is devoted to intervention studies, it does not exclude such prototypes as epidemiological investigations, longitudinal research, field testing of developmental assessment tools, or studies that focus on the natural history of a disorder or a disease.

GENERAL ISSUES

EXPERIMENTAL DESIGN

The experimental design in a particular study allows the author and the reader to determine how well the investigator handled threats to internal (extraneous variables that confound effects) and external (how generalizable are the results?) validity. The goal of any study is to provide a believable demonstration that does not raise any serious doubts in the mind of the reader.

Case Studies. In clinical practice, many of the nonroutine patients who are seen are treated as case studies. That is, without any real concern for experimental controls, the clinician arrives at a tentative diagnosis and introduces a treatment regimen, relying upon "treatment outcome" as his measure of the accuracy of his diagnosis and the confirmation of the appropriate selection of the treatment regimen. A strong tendency exists to conclude, with a favorable outcome, that the physician's treatment was responsible for the improvement in the patient's condition. Often, over a period of time, the individual clinician accumu-

*Preparation of this manuscript was partially supported by a grant from the National Institute for Child Health and Human Development (HD 03144) to the Bureau of Child Research, University of Kansas.
†The generic term "pediatrician" is used throughout this chapter to refer to any pediatric health care provider.

lates a collection of case studies, wherein he treats what he thinks is the same condition with the same regimen, hoping to achieve uniformly positive outcomes. These collections of case studies frequently are interpreted as "proof" that the diagnoses and corresponding treatments were "correct." Since these ministudies constitute such a large part of clinical practice, they are a valuable first step in the development of effective management protocols.

Case studies frequently play an important role in the development of protocols by researchers. When one is faced with a disease entity that is not adequately covered in the literature, and with a patient who sincerely desires alleviation of his symptoms, there is little other choice.

Some of the most poignant case studies have been published in the area of child abuse. In Helfer and Kempe's *The Battered Child* (1974), case studies were used very effectively to illustrate findings that were encountered time and time again by the authors. In attempting to reconstruct the histories of parents who had battered their children, the authors could not control for threats to internal and external validity, nor could they neglect their responsibilities as physicians and caregivers to obtain the most thorough histories possible. Many new areas of investigation are approached in such a fashion. The investigators conduct as thorough a search as possible, attempting to identify virtually any variable that may have been a part of the etiology of the presenting condition. Consistencies usually become increasingly apparent as the search progresses.

As any field of inquiry matures, or as more becomes known about a particular disease entity, the need for controls to internal and external validity must be addressed. Perhaps the greatest drawback or handicap that stems from the publication or discussion of case studies is that they set a precedent for a level of inquiry that, if not advanced, can serve as a deterrent to further scientific inquiry. In spite of the fact that a substantial body of literature now exists in the areas of behavioral and developmental pediatrics, the argument that the measurement problems in the psychosocial domain are insurmountable is still all too often offered as an excuse for the lack of actual data collection. In dealing with the idiosyncrasies involved in the psychosocial interactions of parents and their children, the need for some type of an experimental or quasiexperimental design is great. Fortunately the technology for single subject research has advanced significantly over recent years, particularly within the behavioral domain. These designs offer a reasonable compromise for the investigator who is interested in conducting legitimate inquiry into an area within behavioral and developmental pediatrics, but who simply is not in a position to conduct large scale group research, either because the subjects are not available in sufficient numbers or because the cost of conducting such studies would be prohibitive.

Single Subject Designs.   Recently investigators have reported a variety of single subject designs that at least partially control for threats to internal validity without necessitating the addition of more subjects. The oldest of these (the first known discussion was by Claude Bernard in 1865) is the reversal design. With this model the investigator begins recording relevant symptoms during a baseline or preintervention phase. Then the treatment protocol is introduced. After the patient's condition stabilizes under the latter, and provided there is no inherent danger to the patient, the investigator removes his intervention and observes whether the patient's condition returns to the baseline or preintervention level. An example of the use of such a design would be an institutionalized retarded male who has been taking chlorpromazine for 12 years for the control of his ward behavior. After gathering data relating to the percentage of time that the patient was in bed, inappropriately, during the baseline period, a placebo was substituted for the drug. For a period of 12 days the patient was not observed in bed during the normal day activities, suggesting that the chlorpromazine may have resulted in substantially lower activity levels. The managing physician then reinstituted the chlorpromazine for an additional 18 days, during which the patient spent the majority of his daytime in bed. In this fashion the physician was able, with confidence, to make the statement that the patient's daytime behavior was inappropriately suppressed. That is, when chlorpromazine was taken, the patient spent the majority of time in bed. When the medication was not taken, he spent his time out of bed (Fig. 62–1). Obviously many conditions exist in which the physician cannot or would not choose to effect such a reversal, such as when a return to baseline

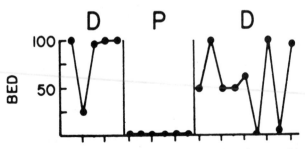

Figure 62–1. Percentage of intervals during which patient stayed in bed occurred across drug (D) and placebo (P) conditions. Each point summarizes two days data. (From Marholin, D., Touchette, P. E., and Malcolm, R. M., et al.: Withdrawal of chronic chlorpromazine medication: an experimental analysis. J. Appl. Behav. Anal., *12*:159, 1979.)

or pretreatment levels would be hazardous to the health of the patient. Such conditions are frequently encountered and can be adequately managed using one of a variety of "multiple baseline" designs.

Multiple baseline designs involve the collection of information or data on several baselines simultaneously. The three most commonly used multiple baseline designs are multiple baseline across subjects, across responses, and across situations.

1. In a study employing a multiple baseline across subjects, preintervention measures are taken simultaneously in preferably at least three different patients. One of the three patients is then selected for the intervention, while the other two remain untreated, with their condition being monitored. Later the protocol is introduced for a second patient (so that now two patients are on the protocol and one is not). After more time has passed with all three patients' conditions being measured, the protocol is introduced to the third patient. In this way the investigator is able to make the statement that "When I introduce my protocol, the patient's condition improves, and when I do not introduce my protocol, the patient's condition does not improve."

An example of the use of such a design is reflected in Christophersen's (1977) study of the effects of parents' use of child auto restraint seats on their child's behavior (Fig. 62–2). For the first three children, an immediate and dramatic im-

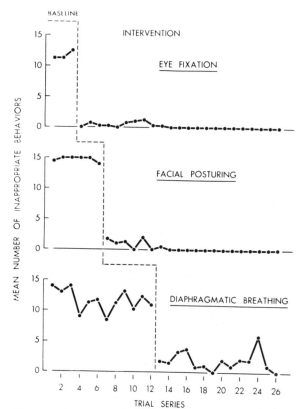

Figure 62–3. The mean number of inappropriate events recorded by the experimenters over a 26 trial period for four subjects on three target responses. The maximal number of inappropriate responses per trial was 15 for each behavior. (From Renne, C. M., and Creer, T. L.: Training children with asthma to use inhalation therapy equipment. J. Appl. Behav. Anal., 9:1, 1976.)

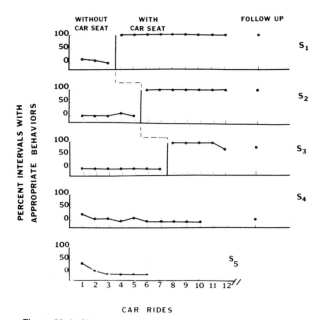

Figure 62–2. Child behavior during 15-minute automobile rides before and after sequential introduction of car seats, including three-month follow-up. Subjects 4 and 5 refused to use car seats. (From Christophersen, E. R.: Children's behavior during automobile rides: do car seats make a difference? Pediatrics, 60:69, 1977.)

provement occurred in the children's behavior when, and only when, the parents used the restraint device. Since riding in an automobile without a restraint device is unsafe, a reversal design was deemed to be inappropriate, yet the use of a multiple baseline across subjects design provided an adequate demonstration of the effects of intervention.

2. In a multiple baseline across responses design, several responses are simultaneously assessed for their baseline or preintervention levels. Then the experimental procedures are introduced for one of the responses, while the remaining responses are assessed. Later the intervention procedures are introduced for a second response and then a third. In the example found in Figure 62–3, from Renne and Creer (1976), the intervention was introduced for "eye fixation" first, resulting in a dramatic decrease in that behavior, with no corresponding decrease or change in the other two behaviors. Later the intervention was introduced for the "facial posturing" and then later for the "diaphragmatic breathing." In each instance the

experimenters demonstrated that when, and only when, they introduced their intervention, the subject's behavior evidenced a change.

3. In a multiple baseline across situations design, the experimental conditions would be introduced only in one situation, for example, at school, and nothing would be changed in other settings (e.g., the home). Later the intervention could be done at the other settings.

Some conditions dictate the use of single subject designs. When only a single subject (or a few) is available for a particular condition or disease (e.g., the single patient with Tourette's symptoms), it is simply not possible to wait until enough patients are available to form two groups. When a new disease or behavior problem or an entirely new treatment protocol is being developed, the researcher inevitably begins with one subject and studies that subject in painstaking detail, prior to introducing the procedure to yet another subject. Only after the investigator has carefully detailed the results and side effects that can be anticipated can the procedures be attempted with a large group of subjects.

In actual practice, single subject designs typically precede the use of group designs. For example, early investigations of the effects of biofeedback procedures for the control of anal sphincter pressure involved only one or a few subjects using single subject experimental designs (Kohlenberg, 1973). When the effectiveness of the procedures was evident with a number of single cases, larger scale investigations were conducted to further demonstrate the generality of the procedures. For a detailed discussion of single subject designs the reader is referred to the book by Hersen and Barlow (1976).

Group Designs.   For all the elaborate designs that fall within this basic category, there is a single common feature—at least two groups, who are treated differently in at least one respect, are compared. If the two groups are identical in all but one way, and one group does significantly better than the other, the experimenter has shown some support that a certain treatment is effective, or that one treatment is "better" than another. Rarely is a single experiment sufficient to establish, beyond reasonable doubt, that the effective or better treatment is also the treatment of choice, for many other factors must be considered before such a sweeping generalization can be justified. Although many investigators accept this point with chagrin, it remains true. With the literal explosion in the number of articles published in the medical and behavioral sciences, it is all too easy to overlook the fact that unless the study was performed correctly (that is, unless numerous scientific criteria are met), the article contributes little, if anything, to medical science and introduces the risk of harm when new and unusual procedures are recommended without experimental support.

There are several conditions that require group designs:

1. When epidemiological data are needed (e.g., the occurrence and prevalence of disease in medicine).
2. When actuarial data are needed (e.g., the percentage of patients who can be expected to react favorably to a drug; the number and side effects that are associated with a particular drug).
3. When comparing treatments in which serial or order effects are known or suspected to occur.
4. When the intent is to draw generalizations to a much larger population.
5. When comparing two or more treatment strategies for the same disease entity or problem area.
6. When assessing the effects of a treatment for different age groups.
7. When sex differences are being examined.

In the traditional pretest–post-test control group design, one group usually is subjected to the procedures under investigation (i.e., the experimental group) while the other group is simply monitored (i.e., the no treatment control group). Another design involves alternative treatment groups, in which two or more groups are subjected to different treatments and the outcomes are compared.

Conversely, group designs have the disadvantage that they do not allow the prediction of the response or side effects in an individual patient to a treatment protocol—these designs are based on averaged responses, which can be misleading at times. Likewise, although developmental norms can tell us what the average child of a particular age is like, it tells us little about a single individual. Also, in the pilot stages of development of entirely new interventions, it may be difficult to justify subjecting an entire group of subjects to an untried procedure, since very little is known about it. For a detailed discussion of group experimental designs, the reader is referred to the book by Campbell and Stanley (1963).

Group designs have the disadvantage of taking the focus away from the individual patient and placing the focus on the "average" patient from the group. The individual's response to a treatment is frequently obscured when an analysis of group design data is conducted.

The use of group designs is generally accompanied by measures of central tendency (e.g., mean, median, mode) that, even when accompanied by indicators of variability or deviance (e.g., standard deviation, range, variance), simply do not detail how any one subject reacted. The best way of dealing with this shortcoming is to include, in tabular form when feasible, a listing of the scores (or behaviors) of each subject at each measurement point. A secondary advantage to the inclusion of such individual subject data is that it allows the interested reader to scrutinize the data carefully, verify the analysis, and reanalyze it using different procedures.

Important side effects—both beneficial and harmful—often go undetected when new procedures are evaluated using group designs. In the pilot stage of development of new treatment procedures, these unanticipated side effects may prove to be the most important facts learned about the procedure. It may be difficult and unwarranted to justify subjecting an entire group of subjects to an untried procedure.

Each set of experimental designs plays an important role in behavioral and developmental pediatrics. The advantage of first investigating a number of single cases intensively is to assess for subtle—or not so subtle—effects associated with treatment procedures that would not be obvious within a group design. Later studies using group designs, with random selection or assignment, are necessary to determine generality to a larger sample. Selection procedures for research subjects is an important component of any experimental or quasiexperimental design.

SELECTION OF SUBJECTS

The issue of subject selection refers to how the participating patients in a study were selected. The substance of this issue lies in the fact that subject selection can and does play a major role in the generalizability of the results of one study to the general population. In the preceding discussion regarding experimental design, the point was made that many investigations actually begin with an "interesting patient" and gradually expand into a full scale investigation. With reference to a case study, the subject was usually selected because "he was there," either in the sense that he was the only patient who had the condition that warranted further investigation or in the sense that he was the first patient who fit the experimental protocol after the study was initiated. In either case the issue of subject selection usually cannot be addressed with a case study, since there are simply too many uncontrolled factors involved. A case study is precisely that—an intensive study of one subject. Frequently little generalization is indicated or warranted. However, this fact does not obviate the need on the part of the experimenter to describe as many relevant details about the subject as feasible.

Written reports of investigative studies often provide too little detail regarding how the patients in those studies were selected. Ideally, if a pediatric group is reporting on a topic like enuresis, either every enuretic patient in the practice will be studied or a sample of enuretic patients will be selected randomly from the total number who meet the entry criteria for the study. The entry criteria can be misleading, depending upon the site in which the study takes place. If the study is conducted by selecting enuretic children from the local public school system, there is a high probability that it is a representative sample, since, by law, every child must attend school. However, such a study would omit any children enrolled in a nonpublic school, and may influence the generalizability of the results to this population owing to some unknown or unacknowledged selection factors.

If all the enuretic children seen in a pediatric clinic were eligible for participation in a study, some description should be provided of the demographic characteristics of the patient population served. Some university based pediatric clinics serve almost exclusively low income families, whereas some private practice groups, by virtue of their office location and fee schedule, serve middle or upper middle income families. Obviously a university hospital based psychiatry department that serves indigent families or families on public assistance could hardly be called a representative sample of "average" enuretics that a pediatrician could be expected to encounter.

Since most investigators are quite limited in their selection of sites at which research can be conducted, the description of the locations where the study was done becomes mandatory. If a single case, several cases, or a "representative sample" is presented, the investigator needs to identify clearly how these cases were chosen. A similar responsibility must be carried by the reader or reviewer who is seeking "incidence" estimates for an individual disease or problem. For example, in Wright, Schaefer, and Solomons' (1979) discussion of the incidence of "encopresis," they distinguish incidence reports from psychiatry clinics, institutionalized children, hospital based pediatric groups, and the general population.

The more restrictive that the selection is for a population under study, the less generality the results have for other populations. However, if the population is clearly defined, as in the case of the study by Azrin et al. (1973) of enuresis treatment with institutionalized retardates, there is an increased likelihood that other professionals dealing with identical or very similar populations could implement the procedures with their populations. The central issue is that investigators need to describe clearly how the subjects were selected, i.e., by what criteria and from what population, for inclusion in a particular study. Similarly, the reader of an article must ascertain whether the findings in a particular study are applicable to his own setting.

The issue of selection of subjects is particularly important in epidemiological studies and in articles that report the development of a new screening tool or assessment device. A pediatrician whose primary practice is limited to middle and upper middle class families who reside in single family dwellings built after 1970 will obviously take less interest in the problem of lead ingestion than the pediatrician who serves poor families who reside in housing built prior to 1955.

Another example is the fluoridation of water supplies. The Kansas City, Missouri, metropolitan area is served by several different water districts. For years the pediatrician had to determine which water district a child resided in in order to determine whether a child lived in an area with a fluoridated water supply. The American Academy of Pediatrics has recommended that if there is any doubt about the fluoridation of the water supply, the family is asked to take a sample of their water to the County Health Department for analysis of the fluoride level. Even a massive epidemiological study of the incidence of dental caries in children of varying ages must be able to identify and separate populations exposed to differing levels of natural or artificially fluoridated water supplies.

In the original publication of the Denver Developmental Screening Test the authors carefully detailed the demographic characteristics of the 1036 presumably normal children who were used for creating the test norms. They also cautioned the reader about interpreting the results of the testing with children of lower income families. In subsequent publications the authors continued to carry messages of caution and addressed the issues of reliability and validity that will be discussed later in this chapter.

Whether the Denver Developmental Screening Test is used as a research or an assessment tool in pediatric practice, the more the individual child varies from the standardization sample, the more difficulty there is in interpreting the scoring. If the test is administered to a child from a low income, blue collar, single parent family who immigrated recently from Laos, the validity of the interpretation of the resulting score would be difficult to determine using existing norms.

Brazelton and his colleagues have published a variety of studies that have examined such issues as the comparability of the scores on the Brazelton Neonatal Behavioral Assessment Scale in several different countries and in infants from different socioeconomic classes. It is only through such laborious and time consuming investigation that the influence issue of subject selection can be determined.

STUDY SETTING

Frequently articles appear in the literature without concern for the setting in which they were conducted. For example, many studies have been conducted on the use of "time out" as a disciplinary procedure for young children (e.g., Bernal, 1969; Forehand et al., 1979). These generally were conducted by behavioral psychologists in psychological clinics. The question of whether these procedures would be as effective when implemented, for example, by pediatricians in a general pediatric practice is left unanswered (cf. Drabman and Jarvie, 1977). Similarly, in Olness' (1975) description of the use of hypnosis in the management of childhood enuresis, the procedures were implemented by a pediatrician (well trained in the therapeutic use of hypnosis) in a general pediatric practice. The question of whether similar hypnosis procedures would be as effective when implemented by other pediatricians remains unanswered.

Ideally, since many problems are usually detected by the patient's primary care physician (either the pediatrician or the family practitioner), procedures for the management of the specific problems should be piloted by these practitioners in their normal (usual) practice settings. To the extent that these studies are properly designed and conducted by the investigator, the practitioner can probably be confident that the treatment recommended would be useful in a practice setting.

Azrin and Foxx's (1974) *Toilet Training in Less Than a Day* provides an illustrative example. The pilot work and the research on the "dry pants training" procedure described by Azrin and Foxx were implemented by professionals well trained in the dry pants procedures. Yet the book was marketed for use by parents in the privacy of their own homes. As one subsequent study has shown (Matson and Ollendick, 1977), the dry pants training procedures were not as effective when implemented by parents. Parents who both read the book and had some professional supervision had much better results than parents who just read the book. These findings emphasize the importance of piloting treatment programs in settings similar to those in which the program will ultimately be used.

The most important setting characteristics relate to the type of services offered and the training and practice of the providers. These two characteristics alone can account for an enormous difference in the populations that are treated. For example, a private pediatrician's office staffed by ambulatory care pediatricians in a well to do suburban area obviously serves a different clientele than a children's and youth clinic in a poor urban area or a behavioral pediatrics clinic that is staffed, in part, by residents from a child psychiatry training program. Any publication or training workshop that was sponsored by one of these three settings would probably reflect the biases of the setting in which the original work was conducted.

As with other issues raised in this chapter, the point here is that when the original work was conducted in a setting different from the one in which it is being introduced, the clinician or researcher needs to be cognizant of the difference in setting and, perhaps, take appropriate precautionary measures that may not have been included in the original work.

MEDICAL COMPLIANCE

The issue of medical compliance has been called one of the best documented but least understood phenomena in medicine (Becker and Maiman, 1975). When a study is conducted in a clinical research unit with inpatients, or with inpatients supervised by a well trained and well motivated medical and nursing staff, some caution must be exercised in attempts to generalize the results to general outpatient practices. Although much of the basic research on disease control of such entities as diabetes and asthma is usually conducted in well staffed hospital settings, the general implementation of the findings with outpatients usually occurs under less than ideal circumstances with less than optimal results. In fact, the problem of medical compliance has been an unknown influence that undoubtedly has biased drug studies as well as behavioral pediatrics treatment programs.

Ideally investigators will include some reliable measure of patient compliance when they report the results of their intervention efforts. The most prevalent measure—patient report or physician estimate—is usually no better than chance and cannot be depended upon (Gordis, 1979). Unfortunately much of the literature in behavioral pediatrics depends upon parents for either the collection of "data" or reports of cures or outcome. For example, Olness (1975) with enuretics and Levine and Bakow (1976) and Wright (1973) with encopretics depended entirely on parent reports as did Foxx and Azrin (1973a) for their data on the outcome of toilet training.

More objective measures of compliance include pill counts (Sackett, 1979), actual observation of parent implementation of procedures, and urine, serum, or saliva biochemical assays. To date, one of the most sophisticated measures of compliance is the long term assay, best represented by the hemoglobin A_{1c} determination to measure the control of diabetes. Since this measure actually assesses the patient's state during the three month period prior to the assay, there is virtually nothing that the patient can do, other than comply with his treatment regimen, to affect appreciably the HbA_{1c} measure.

Unfortunately objective measures sometimes have important confounding variables that must be addressed. With pill counts, if the patient is aware that the pill count is to be done, the patient can destroy some of the pills so as to appear more compliant. Also if the patient starts the medication regimen one day late but thereafter complies perfectly, he will be adjudged partially noncompliant when in fact the regimen was followed very carefully. With most of the biochemical assays, e.g., blood sugar levels or urine sugar levels, compliance in the 24 hour period immediately preceding the assay usually results in an assay with normal or nearly normal values, which may be misleading.

For a complete review of the factors affecting compliance, the reader is referred to the book by Sackett and Haynes (1976).

OBSERVATION AND MEASUREMENT PROCEDURES

Whatever the form of measurement used, from biochemical assays to observation of the rate of occurrence of behavior problems, interobserver agreement checks are not a luxury—they are absolutely essential! In the behavioral literature, the general rule of thumb is that, on a random basis, at least 20 per cent of the observations should have independent interobserver agreement checks, preferably with the checks evenly distributed over subjects and over experimental conditions.

The purpose of interobserver agreement checks is to reduce the likelihood that any of the experimenter's biases might have inadvertently influenced the data collection. For example, if hired observers are being paid to conduct observations of the effects of a particular medication on the activity levels of preschoolers, the observers unknowingly may allow their definitions or procedures to change just enough to demonstrate an improvement over baseline, when the child is started on medication. A commonly accepted control procedure in such cases is the double blind crossover design. With this design, neither the patient nor the investigator knows whether any individual coded packet contains a placebo or one of the drugs under investigation. At a preselected time the patient is changed from one set of packets to another, with the patient and physician still "blind" as to the contents. In this way neither the physician nor the parent can alter his observations in any systematic way, since there is no knowledge of what treatment is in effect. Although the double blind crossover design is a highly sophisticated procedure, its use does not obviate the need for interobserver agreement checks. Nor is it always possible to keep the parent or physician blind to the treatment condition, since commercially prepared placebo drugs are frequently a slightly different color to avoid the inadvertent substitution of a placebo for the real drug.

The major factors that are documented to affect the accuracy of the recording of human observers are bias, reactivity, and observer drift.

Observer Bias. Bias is defined as "having observers respond systematically to some variable other than the behavior at hand" (Bailey, 1977, p. 115). Probably the best documented example of observer bias would be some of the early reports of attempts to persuade parents to purchase and use child restraint devices for automobile travel. Kanthor (1976) reported, entirely on the basis of parent reports, that 69 per cent of parents correctly restrained their children during automobile rides. However, in another study by Williams (1976) that has been replicated by other authors, the actual observed use of child restraint devices is 7 per cent. Undoubtedly the parents in the Kanthor study overestimated their use of child restraint devices.

Observer Reactivity. Observer reactivity has been defined as "interference or the intrusiveness of the observer himself upon the behavior being observed" (Johnson and Bolstad, 1973, p. 38). The factors that have been identified as contributing to reactivity are the conspicuousness of the observer, observer attributes such as sex, age, race, and socioeconomic status, and observer interaction or participation in the study.

Observer reactivity could and probably does play a part in studies such as the completeness of the physical examination during well child visits (Leake et al., 1978) or the use of a written protocol to guide house officers in the work-up and management of dysuria (Dworin and Stross, 1979). In these examples the residents probably would perform at their absolute best, since they were aware that their performance was being recorded. In interpreting the results of such studies, the reader needs to keep this motivational factor in mind, assuming that the residents' performance under more natural circumstances would probably be less than optimal.

There are, of course, cases in which reactivity plays an important and additive role. For example, in studies like that of Frankenburg and Dodds (1967) the participating mothers probably tried to maximize their child's performance. In fact, in the Brazelton Neonatal Behavioral Assessment Scale (1973), the observers are specifically instructed to score only the infant's best performance—ignoring those trials with less than optimal performance.

Observer Drift. Observer drift refers "to the tendency of observers to change the manner in which they apply the definitions of behavior over time" (Kazdin, 1973, p. 143). Johnson and Bolstad (1973) discussed instrument decay or observer drift as a problem that increases the longer the research lasts. "In the case of human observers, the decay may result from processes of forgetting, new learning, [and] fatigue" (Johnson and Bolstad, 1973, p. 18). Hersen and Barlow (1976) assumed drift to be a universal problem in observational measurement.

Inadequacy of behavioral definitions and codes may contribute to observer drift. Complexity of behavior observed may lead to drift from the original definition. Studies consistently have found a decline of up to 25 per cent from training reliability figures to actual data collection conditions. Thus, it is apparent that agreement scores during training may not be an adequate estimate of interobserver agreement during actual data collection; hence, the need for checks throughout an entire investigation.

Consensual Observer Drift. Consensual observer drift refers to the tendency for pairs of observers to change their application of a behavioral definition consistent with each other but drifting from the criterion measurement (Johnson and Bolstad, 1973). This bias can be addressed periodically by incorporating a third observer to assess the pair's data.

The majority of research in pediatrics, from epidemiological to developmental and intervention studies, has not adequately addressed the issue of accuracy of recordings. Most research reporting incidence rates of problems has relied on parental accounts. Much epidemiological work either has relied upon clinic records of unknown (but usually questionable) reliability, or has utilized surveys and questionnaires that have inherent problems with reliability. Even studies on intervention procedures for such clinical problems as enuresis or encopresis have relied totally upon parental reports.

Pediatric researchers must control for the problems that are inherent in either accepting parental reports or relying on one or two observers to record accurately throughout an entire study. In the medical laboratory not only are overt reliability tests performed (under the auspices of "quality control checks"), but many laboratories even conduct covert checks when known samples are sent through to again check on the quality control. Although interobserver agreement procedures are cumbersome and expensive, the outcome, i.e., more reliable data, probably justifies the added expense of time and effort.

Reliability of observational measurement in behavioral and developmental pediatrics research is essential for establishing the quality and credibility of research findings. Continuing efforts to reduce sources of unreliability in observational measurement must be made to assure progress in the field.

Several concerns regarding reliability occur when the primary instrument is a written test or some form of developmental assessment. These concerns were adequately addressed in a recent article by Robinson et al. (1980) on the standardization of an inventory for children with conduct disorders. The authors addressed both the individual items on the inventory and the overall scale.

Item Analyses. The reliability of each test item can be examined in terms of its relationship to the total test and its stability over time. The relationship of each item to the total test is assessed by calculating the correlation between each item score and the total inventory score. The issue of stability over time is addressed by calculating the correlation of the response to each item on the first administration with a later retest response.

The validity of each item is evaluated by examining the difference between the mean scores obtained by normal children and children known to have a single disorder in common. The minimal criterion established for item validity would be a significant difference between affected children and normal children.

Scale Analyses. Reliability of the inventory is assessed by using three commonly accepted methods: split-half coefficients, test-retest reliability, and internal consistency. The split-half coefficient is calculated by randomly dividing the inventory test items in half and calculating the scores on the two halves. The test-retest reliability is assessed by correlating the scores obtained on the first administration with the scores obtained from a later second administration, usually administered within six months of the first testing (Anastasia, 1976).

Validity. The validity of a test concerns what the test measures and how well it does so (Anastasia, 1976). For this reason the validity of a test must be determined with reference to the particular use for which the test is being considered. There are three principal categories of validity: content, criterion-related, and construct validity.

Content validity is the degree to which the measurement instrument items represent the universe of the behavior being measured as judged by content experts. For example, an achievement test in elementary school arithmetic must be able to answer the question, "How much has a child learned in the past?" As such, the test outcome is based entirely on how accurately the total of the individual test items reflects what a child should be expected to know at the time of the testing.

Criterion related validity assesses the relationship between the measurement tool and a crite-

rion. As such, it indicates the effectiveness of a test in predicting an individual's behavior in specified situations. For example, an aptitude test used to predict performance in high school mathematics must be able to answer the question, "How well will a given child learn in the future?" This refers to the "predictive validity" of a test. In a test for diagnosing learning disability, the question would be, "Does a child's performance reflect a specific disability?" This refers to the "concurrent validity" of a test.

Construct validity is the degree of relationship between the measurement and the construct being evaluated. For example, a test of logical reasoning must be able to answer the question, "How can we describe a child's psychological functioning?" Examples of such constructs are IQ, depression, verbal skills, and mechanical comprehension.

Perhaps the most important point regarding the validity of a test is that validity refers to observable behavioral events and, for that reason, can be tested. For a comprehensive discussion of tests and test construction, the interested reader is referred to the book by Anastasia (1976).

GENERALITY OF TREATMENT EFFECTS

"A behavioral change may be said to have generality if it proves durable over time, if it appears in a wide variety of possible environments, or if it spreads to a wide variety of related behaviors" (Baer et al., 1968, p. 94). Thus, if an intervention procedure produces both immediate and long term improvement, that procedure is said to have generality. Clearly the majority of research studies report only the immediate effects of the procedures under investigation. Occasionally the author provides anecdotal reports from family members that a behavior change was maintained for an extended period of time. Rarely do authors provide long term follow-up data (i.e., for a period of years) using the same data acquisition procedures that were used in the initial intervention, including interobserver reliability checks. As an example, one area of pediatric research has dealt with getting parents to use safety restraint seats every time they transport their infants or children in an automobile. Most of the literature published to date has reported only the short term effects of procedures—usually no more than four to six weeks after intervention. Yet to be fully protected, children must be transported in safety restraint seats throughout childhood. Christophersen and Gyulay (1981) reported a procedure for getting parents to comply with the health care provider's instructions on the use of car seats and furnished data immediately after intervention, three months later, six months later, and one full year later. Although this type of follow-up data is difficult to obtain and delays publication of research findings, it provides the best assurance that a procedure will stand the test of time.

Stillman et al. (1977), in a study on the effectiveness of a set of procedures for training medical students to interview mothers, provided a follow-up assessment one year later. This procedure showed that not only were the training procedures effective at the time, but the improvement was still present at a later point in time. However, without the use of either an untrained control group or a group trained using an alternative set of procedures, it was not possible to state positively that the original interviewing training protocol, and not the clinical experience in the interim, were responsible for the effects seen one year later.

Hutter et al. (1977) used another type of control group in their study of interviewing skills. Fifteen months after their training program, they assessed the students' interviewing skills, as well as that of the students who had graduated one year earlier and who accordingly had a full year more of clinical experience. This specially trained group scored significantly better than the previous year's class.

This type of a control group represents a sensitivity on the part of the authors to the problem of assessing the durability of training procedures over time. Unfortunately, when such control procedures are not included, there is no substantive way to compensate for the omission. The reader is left to wonder whether the training effects were durable.

Similarly, much of the published literature is concerned with treatment procedures that were evaluated originally in a teaching hospital or on a special research grant. However, in some cases, unless the results generalize to the natural environment, many of these educational techniques have limited utility. Many practitioners are all too familiar with the lack of generalization of compliance with chronic disease regimens from the inpatient setting to the natural home. Numerous articles have recently appeared that address this issue. For example, a juvenile onset, insulin dependent diabetic is admitted to an endocrinology service for the purpose of regulating his insulin requirements. Under these nearly ideal circumstances, in which meals are planned and prepared under the direction of a registered dietitian and served by a registered or licensed practical nurse, and when insulin injections and urine testing are supervised by the nursing staff, all but the most recalcitrant juvenile diabetic can be well regulated. However, not long after discharge from the hospital, all too often the results of the inpatient hospitalization are not found to generalize to the home.

One confounding factor that is often discussed privately but rarely appears in the literature is related to the enthusiasm of the investigators who conducted the original work. A dedicated team of researchers operating in a well funded university setting, which includes research fellows and postdoctoral students, may very well be in a position to obtain results that are hard to get in a real-life office setting. Clearly, when the staffing on a given project far exceeds what is customary and usual, this needs to be stated in any reports describing the outcome.

Obviously, in the aforedescribed situations, one alternative that needs to be considered is that short term and long term results, or hospital versus home results, may represent two different but related problems. Providing parents with the impetus for initial change may have to be approached with entirely different procedures from those necessary for long term maintenance. Likewise, the hospitalized juvenile diabetic probably lives under a totally different set of circumstances than those of the same diabetic in his own home. The delineation of the parameters of generalization is an important but often neglected task in behavioral and developmental pediatrics.

SIGNIFICANCE

The issue of whether a finding is significant can be addressed in two distinctly different ways. From the statistical standpoint, the question of significance means whether the observed change was due to a chance variation or to the intervention efforts of the investigator. Elaborate mathematical computations are currently available for estimating the statistical significance of the change. Usually the researchers, with the help of a statistician, are able to arrive at a figure, like the 0.01 level, which means that only one time in 100 will a change of that magnitude occur as a result of chance variation. Thus, although the possibility exists that the change was due to chance, it is not very likely.

A totally different but equally important question is whether the change was socially significant (cf. Baer et al., 1968). That is, assuming that during the baseline or preintervention period the patient's behavior was unacceptable because of the type of behavior (e.g., head banging or other self-injurious behavior), the situation in which the behavior occurs (e.g., toilet training), or the rate at which the behavior occurs (e.g., certain types of self-care behavior), the researchers would not consider the change to be significant unless the post-treatment behavior was within normal limits.

Thus, three distinct possibilities emerge. One is that the investigator can produce a statistically significant change that is also socially significant (e.g., teaching enough self-help skills so that a child can take care of himself). The second is a change that is not statistically significant but is socially significant (e.g., in toilet training a child, reducing bowel accidents from one per day to none would not be statistically significant but would certainly please the parents). The third is a change that is statistically significant but not socially significant (e.g., reducing the incidence of self-injurious head banging from 200 times per day to only 20 times per day is still unacceptable to most parents).

Ideally most of the treatment programs described in the literature on behavioral and developmental pediatrics should be both socially significant and, within appropriate experimental designs, statistically significant. In actuality, many do not meet these two criteria.

ADEQUATE DESCRIPTION OF PROCEDURES

Baer et al. (1968) categorized the technological description of treatment or intervention procedures (i.e., the independent variable) as an important dimension for behavioral research. By "technological" they meant simply that the techniques were completely and adequately described.

Foxx and Azrin's (1973b) book, *Toilet Training the Retarded*, provides an eloquent illustration of a technological description. They provide step by step directions for implementing procedures to train self-initiated toileting. They include definitions of most terms that might not be generally understood and provide flow charts detailing exactly what to do and when. Although research papers published in journals cannot be expected to detail procedures as clearly as a book, more concern for the inclusion of detailed descriptions would benefit many articles.

The medical literature frequently has an advantage in that there is a long history of using clearly described, highly reliable procedures. For example, when a researcher describes a decrease in the white blood cell count per high power field after administration of a commercially available drug, most researchers reading or hearing this description would be able to administer the same drug under the same conditions and note the corresponding changes in the white blood cell count. However, a behavioral scientist may describe the use of a "time out" procedure for discipline, but there is no commonly accepted standard of how to use "time out." The easiest way to rectify this problem is to provide detailed descriptions of the procedure (or reference to an earlier published description) and conduct and report the results of interobserver reliability checks on the actual use of a procedure like "time out." Although such a requirement initially may appear to be too expensive or too time consuming, the overall savings in terms of the ability of researchers to communicate accurately what they did and what happened will probably more than offset the initial added expense. Although there has been a tendency for some behavioral scientists to try to simplify protocols by dropping out previously required reliability checks, few medical laboratories would consent to dropping their use of "quality control checks." The concern for technological rigor exemplified by the medical technologist might very profitably be carried over into the behavioral scenes.

In Frankenburg and Dodds' instruction manual for administering the Denver Developmental Screening Test, they not only state the conditions under which the test should be administered, they also provide both a check-list and a self-test. With a description that is this precise, the reader can administer the test at another location, with reasonable assurance that his results will be comparable to the original work.

OVERVIEW

In order to meet each of the dimensions of an intervention study that have been discussed, the study must:

1. Utilize appropriate experimental and control procedures to minimize threats to internal and external validity.

2. Specify exactly how the subjects for the study were selected and provide detailed demographic characteristics for those subjects.

3. Specify precisely the setting and circumstances under which the study was conducted.

4. Include objective measures of the patient's compliance with intervention procedures and, when appropriate, measures that document the correct introduction of the intervention procedures.

5. Include routine interobserver checks in order to establish the reliability and validity of the measurement procedures.

6. Include repeated measures, over time, beginning at the preintervention level and continuing to collect follow-up data for an extended period of time after introduction of the intervention procedures.

7. Obtain both statistically and socially significant outcomes.

8. Provide enough technological description of the intervention and measurement procedures

that a typically trained reader could replicate the study.

Some of these dimensions can be, and frequently are, compromised or not addressed at all. In the case of the experimental design, the omission of the patient compliance and the reliability of the measurement procedures can render the results uninterpretable; hence, they represent fatal research flaws. In the case of the omission of the selection of the subjects and the setting, the generalization of treatment effects, the adequacy of the technological description of the procedures, or the social and statistical significance of the study, the research flaw is not fatal, but depending on the magnitude of the omission, the results become more difficult to interpret.

To maximize the utility of the research or clinical studies, the study conditions must closely approximate the circumstances under which the procedures will ultimately be used. For example, Brazelton (1975) reported that he spent approximately 85 per cent of his office time providing anticipatory guidance. Yet Reisinger and Bires (1980) reported that the pediatricians in their study averaged only 97 seconds of anticipatory guidance with mothers of children under five months of age, and seven seconds when the children were 13 to 17 years old. Clearly the level of anticipatory guidance practiced by Brazelton would not be found in the practices studied by Reisinger and Bires. Yet the utility of both these studies is not compromised. For those pediatricians who devote a significant amount of their clinical efforts to anticipatory guidance, the writings of Brazelton will be very appropriate. Likewise the pediatrician who schedules many brief office visits can readily identify with the findings of Reisinger and Bires. The only time that problems develop is when the author or the presentor does not provide enough of a description of the practice setting to allow the reader or listener to determine this applicability of a set of procedures to his practice.

DISCUSSION

The collection of hard scientific data in any area of medicine is a difficult task. The composition of the articles, reviews, and professional addresses that describe such work is an equally demanding task. Yet without such endeavors the data base of medicine will not grow, and the frustration of unresolved conditions, disease processes that we do not understand, and patients who suffer or die will continue unabated. Many medical scientists have devoted their entire professional careers to the conduct and dissemination of the scientific enterprise.

Whenever a new area is investigated, the entry and the initial work are difficult and the rewards of solving a new problem illusive. Perhaps the last remaining frontier in medicine, as a scientific enterprise, is the behavioral or psychosocial domain. The researcher initially is frustrated because the cultures, stains, and assays of medical science do not readily transfer—new ones must be developed. When simple tests like the Denver Developmental Screening Test are introduced (Frankenburg and Dodds, 1967), they come to enjoy widespread adoption, the scientific enterprise has been advanced, and since the road is smoother, the way is easier for those who follow.

When Lee Robins (*Deviant Children Grown Up*, 1966) publishes perhaps discouraging or unwelcome data on the therapeutic process, the work can be interpreted as a slap on the face, and the field suffers. Or the work can be considered a new challenge, with a renewed spirit, and the field prospers.

There are no simple answers in scientific medicine—only simple solutions. The process of developing procedures for assessment, description, comparison, or intervention that stand up to professional scrutiny is a demanding challenge. The work described in this book represents some of the best work in the field of behavioral and developmental pediatrics, but it is faulty. The responsibility of the authors and their readers is to extend this work, polish it, and improve it to the point at which children will reap long term benefits from it.

EDWARD R. CHRISTOPHERSEN

REFERENCES

Anastasia, A.: Psychological Testing. Ed. 4. New York, Macmillan Publishing Co., Inc., 1976.

Azrin, N. H., and Foxx, R. M.: Toilet Training in Less Than a Day. New York, Simon & Schuster, 1974.

Azrin, N. H., Sneed, T. J., and Foxx, R. M.: Dry bed: a rapid method of eliminating bedwetting (enuresis) of the retarded. Behav. Res. Ther., 11:427, 1973.

Baer, D. M., Wolf, M. M., and Risley, T. R.: Some current dimensions of applied behavior analysis. J. Appl. Behav. Anal., 1:91, 1968.

Bailey, J. S.: A Handbook of Research Methods in Applied Behavior Analysis. Tallahassee, Florida State University, 1977.

Becker, M. H., and Maiman, L. A.: Sociobehavioral determinants of compliance with health and medical care recommendations. Med. Care, 13:10, 1975.

Bernal, M. E.: Behavioral feedback in the modification of brat behaviors. J. Nerv. Ment. Dis., 148:375, 1969.

Bernard, C.: An Introduction to the Study of Experimental Medicine. New York, Dover Publications, 1957. (Originally published in 1865.)

Brazelton, T. B.: Anticipatory guidance. Pediatr. Clin. N. Am., 22:533, 1975.

Brazelton, T. B.: Neonatal Behavioral Assessment Scale. Philadelphia, J. B. Lippincott Co., 1973.

Campbell, D. T., and Stanley, J. C.: Experimental and Quasi-experimental Designs for Research. Chicago, Rand McNally, 1963.

Christophersen, E. R.: Children's behavior during automobile rides: do car seats make a difference? Pediatrics, 60:69, 1977.

Christophersen, E. R., and Gyulay, J.: Parental compliance with car seat usage: a positive approach with long-term follow-up. J. Pediatr. Psychol., 6:301, 1981.

Drabman, R. S., and Jarvie, G.: Counseling parents of children with behavior problems: the use of extinction and time-out techniques. Pediatrics, 51:78, 1977.

Dworin, A. M., and Stross, J. K.: The use of protocols as educational tools for house officers. J. Med. Educ., 54:954, 1979.

Forehand, R., Flanagan, S., and Adams, H. E.: A comparison for four instructional techniques for teaching parents the use of time-out. Behav. Ther., 10:94, 1979.

Foxx, R. M., and Azrin, N. H.: Dry pants: a rapid method of toilet training children. Behav. Res. Ther., 11:435, 1973a.

Foxx, R. M., and Azrin, N. H.: Toilet Training the Retarded. Champaign, Illinois, Research Press, 1973b.

Frankenburg, W. K., and Dodds, J. B.: The Denver Developmental Screening Test. J. Pediatr., 71:181, 1967.

Gordis, L.: Conceptual and methodological problems in measuring patient compliance. In Haynes, R. B., Taylor, D. W., and Sackett, D. L. (Editors): Compliance in Health Care. Baltimore, Johns Hopkins University Press, 1979.

Helfer, R. E., and Kempe, C. H. (Editors): The Battered Child. Ed. 2. Chicago, The University of Chicago Press, 1974.

Hersen, M., and Barlow, D. H.: Single-case Experimental Designs: Strategies for Studying Behavior Change. New York, Pergamon Press, 1976.

Hutter, M. J., Dungy, C. I., Zakus, G. E., Moore, V. J., Ott, J. E., and Favret, A. C.: Interviewing skills: a comprehensive approach to teaching and evaluation. J. Med. Educ., 52:328, 1977.

Johnson, S. M., and Bolstad, O. D.: Methodological issues in naturalistic observation: Some problems and solutions for field research. In Hamerlynck, L. A., Handy, L. C., and Mash, E. J. (Editors): Behavior Change: Methodology, Concepts, and Practice. Champaign, Illinois, Research Press, 1973.

Kanthor, H. A.: Care safety for infants: Effectiveness of prenatal counseling. Pediatrics, 58:320, 1976.

Kazdin, A. E.: Methodological and assessment considerations in evaluating reinforcement programs in applied settings. J. Appl. Behav. Anal., 6:517, 1973.

Kohlenberg, R. J.: Operant conditioning of human and sphincter pressure. J. Appl. Behav. Anal., 6:201, 1973.

Leake, H. C., Barnard, J. D., and Christophersen, E. R.: Evaluation of pediatric resident's performance during the well-child visit. J. Med. Educ., 53:361, 1978.

Levine, M. D., and Bakow, H.: Children with encopresis: a treatment outcome study. Pediatrics, 58:845, 1976.

Marholin, D., Touchette, P. E., and Stewart, R. M.: Withdrawal of chronic chlorpromazine medication: an experimental analysis. J. Appl. Behav. Anal., 12:159, 1979.

Matson, J. L., and Ollendick, T. H.: Issues in toilet training normal children. Behav. Ther., 8:549, 1977.

Olness, K.: The use of self-hypnosis in the treatment of childhood nocturnal enuresis. Clin. Pediatr., 14:273, 1975.

Renne, C. M., and Creer, T. L.: Training children with asthma to use inhalation therapy equipment. J. Appl. Behav. Anal., 9:1, 1976.

Reisinger, K. S., and Bires, J. A.: Anticipatory guidance in pediatric practice. Pediatrics, 66:889, 1980.

Robins, L. N.: Deviant Children Grown Up. Baltimore, The Williams & Wilkins Co., 1966.

Robinson, E. A., Eyberg, S. M., and Ross, A. W.: The standardization of an inventory of child conduct problem behaviors. J. Clin. Child. Psychol., 9:22, 1980.

Sackett, D. L.: Methods for compliance research. In Haynes, R. B., Taylor, D. W., and Sackett, D. L. (Editors): Compliance in Health Care. Baltimore, Johns Hopkins University Press, 1979.

Sackett, D. L., and Haynes, R. B. (Editors): Compliance with Therapeutic Regimens. Baltimore, Johns Hopkins University Press, 1976.

Stillman, P. L., Sabers, D. L., and Redfield, D. L.: Use of trained mothers to teach interviewing skills to first-year medical students: a follow-up study. Pediatrics, 60:165, 1977.

Williams, A. F.: Observed child restraint use in automobiles. Am. J. Dis. Child., 130:1311, 1976.

Wright, L.: Handling the encopretic child. Profess. Psychol., 4:137, 1973.

Wright, L., Schaeffer, A. B., and Solomons, G. (Editors): Encyclopedia of Pediatric Psychology. Baltimore, University Park Press, 1979.

63

Physician Education in Developmental-Behavioral Pediatrics*

The body of knowledge encompassed by the preceding chapters of *Developmental-Behavioral Pediatrics* forms the basis for effective decision making and clinical judgments in the practice of pediatrics and several other medical specialties. How can this information be taught to physicians? What are the differing needs of all physicians, of pediatricians, of subspecialists in developmental-behavioral pediatrics, of specialists in other medical disciplines? In this chapter we attempt to answer these questions by discussing issues relating to the planning, content, and evaluation of a curriculum in developmental-behavioral pediatrics.

APPROACHES TO CURRICULUM PLANNING

Although the majority of the approximately 240 pediatric residency programs in the United States and Canada offer specific training in developmental or behavioral pediatrics, much of the training appears to be fragmented and unsystematic. Moreover, fewer than half the medical school pediatric departments offered any form of training in this field to their students in 1979.† In a study carried out by Dworkin and his colleagues (1979), 50 per cent of practicing pediatricians indicated that medical school was of no value as a source of knowledge in developmental pediatrics, and 20 per cent indicated that residency was of no value. In a more extensive survey of 7000 recent graduates of pediatric residency programs, over 50 per cent indicated that their residency experience was insufficient in the care of patients with psychosocial or behavioral problems, and over 40 per cent had similar concerns about the care of patients

with chronic cerebral dysfunctions (The Task Force on Pediatric Education, 1978). As pediatric educators plan to initiate or expand a curriculum in developmental-behavioral pediatrics, they must consider the purpose of such training, its most appropriate timing, the resources they have available within their programs, and models that have been tried by others. We will review each of these elements prior to considering the content of a developmental-behavioral pediatric curriculum and its evaluation.

PURPOSE

The underlying purpose of teaching developmental-behavioral pediatrics is to enhance the lives and the opportunities for learning and fulfillment of all children with the developmental or behavioral variations described in the preceding chapters. The devastating impact of a severe handicap on a family, let alone the child, as well as the frequency with which less severe but still damaging variations occur makes it crucial that physicians handle such problems with a broad base of knowledge and practiced skill. With the possible exception of infectious disease, there is no more frequent problem for the practicing physician than developmental or behavioral variation. The tools we use for assessment and intervention are still somewhat crude, but in the hands of a well trained and experienced physician they can be effective.

The majority of medical school pediatric department chairmen and pediatric residency training program directors need more practical reasons for introducing a new curriculum or expanding the current teaching efforts in developmental-behavioral pediatrics. They and others might say, "Let the schools deal with developmental and behavioral variations." But there are other purposes for such training at the medical student, resident, fellow, and practitioner level: to meet the demands

*Preparation of this chapter was supported by Maternal and Child Health Services Training Grant 922 and by U.S. Office of Special Education, Department of Education, Grant G007903056.

†Details of the surveys of medical schools and residency programs carried out by the Pediatric Education Project, The Nisonger Center, The Ohio State University, can be obtained by writing the authors.

of primary care pediatricians; to respond to the pleas of advocacy groups, particularly parents and professionals from other disciplines, that physicians have at least a basic knowledge of child development and conditions reflecting developmental and behavioral variation; to introduce and demonstrate concepts and skills that will be valuable in all aspects of medical practice, e.g., the interdisciplinary team approach; and to build a base for scientific research into one of society's most challenging problems: How can human intelligence and competence be fostered in each child? Each of these issues is to be discussed.

The dissatisfaction of primary care practitioners with their medical school and residency training in developmental-behavioral pediatrics was discussed earlier. Pediatricians in general practice often have an area of special interest, and an even larger number indicate that they would like to develop such an interest, either through an additional year or more of fellowship or through continuing education. Aside from allergy, the areas of special interest most frequently cited by practitioners are developmental and behavioral pediatrics (Delaney et al., 1980). Advances in prevention and disease control, population shifts, and technological advances are also likely to increase this interest (Haggerty, 1974; Richmond, 1975).

Parents have probably been most vocal in expressing their dissatisfaction with pediatricians' knowledge, skills, and attitudes toward handicapped children (Gorham et al., 1975; Guralnick et al., 1980). They cannot understand why physicians are so uncomfortable in dealing with their questions and in interacting with their child, or why they make statements that conflict with what every parent, other professional, and knowledgeable physician tells them. It is rare now, fortunately, for physicians to recommend early institutionalization for a developmentally disabled child, but less rare for the pediatrician to say, "He'll grow out of it" or "Keep him out of school for another year until he matures."

In part, these statements reflect a belief that development proceeds along some predetermined course, uninfluenced by anything save the passage of time. Such ignorance or nonacceptance of the well described dynamic and interactive nature of development not only leaves the pediatrician who is untrained in developmental-behavioral pediatrics out of step with other professional colleagues, but leaves infants in his practice at risk of failure to develop fully because of his not looking at the infant's learning environment. A number of studies indicate that the quality of parental care and of environmental intervention programs are significant factors influencing the child's development and subsequent school success (Haskins et al., 1978; Sameroff and Chandler, 1975). Further,

pediatric intervention with the support of a home health visitor can influence parenting practices (Gray et al., 1977). Pediatricians and other physicians need this knowledge to utilize correctly as much as they need knowledge of emerging antibiotic resistance to micro-organisms.

In addition to meeting the needs of practitioners, parents, and other professionals for improved knowledge and skills in developmental-behavioral pediatrics, implementation or refinement of such a curriculum may serve the purpose of teaching skills useful in other areas of medicine but best taught by the developmental-behavioral pediatric faculty. A prime example is the interdisciplinary process. For a physician to participate effectively in a truly interdisciplinary team, skills are required that are taught infrequently by departments other than those devoted to developmental-behavioral pediatrics and psychiatry. Modern medicine, however, demands smoothly functioning teams in oncology, cardiology, neonatology, and rehabilitation, to mention but a few. Group problem solving of this type is one of the most difficult processes to carry out successfully. To be effective as a member of such a group, the physician must recognize the dynamics of the team's functioning and must have the interpersonal skills to share both leadership and membership roles. The emotional climate, particularly such factors as trust, openness, respect, and interdependence, must be favorable for effective decision-making to occur (Johnston and Magrab, 1976).

The final purpose in upgrading the teaching of developmental-behavioral pediatrics is its increasing importance as a field for research efforts. Solutions to society's most pressing problems—violence, injustice, and apathy—are related in ways as yet unclear to child rearing and educational practices. Further, there are still opportunities for making astute clinical observations with major preventive ramifications (for instance, the rediscovery of the fetal alcohol syndrome within the past decade). Excellent teaching attracts the best students. They then maintain their interest, strengthen residency and fellowship programs, and become the future faculty for research and teaching. The goal of enhancing a child's intelligence and competence underlies the efforts of everyone from perinatologist to parents. The scientific basis of that goal's achievement is still in its infancy and is the area of expertise of the developmental-behavioral pediatrician.

TIMING

In planning when to teach the various aspects of developmental-behavioral pediatrics, each medical school pediatric department must consider the minimal knowledge, skills, and attitudes necessary for all graduates of their school and what is best taught in pediatric or other residencies. Residency training program directors initially need to review or supplement the diverse training received in different medical schools before proceeding with the more advanced curriculum at the resident level. Decisions regarding the content of fellowship training depend on the resources of the particular location and are not an issue for most department chairmen or general pediatric residency training program directors. The content of continuing education for practitioners is variable, consisting of basic issues similar to those presented to residents, with a particular emphasis on recent advances.

During the first year of medical school, the basic sciences occupy nearly the entire curriculum. Developmental psychology deserves its place as a basic science along with anatomy, physiology, and biochemistry. In addition, the interest in people with which medical students enter their freshman year is often stifled rather than fostered by the overwhelming number of facts they are expected to learn. This can be countered to some extent, however. For example, a freshman course in Galveston exposes the students to normal children in public elementary schools, focusing their learning on variations in cognitive, social, and physical development between and within age groups. Ninety-eight per cent of the medical students indicated that the experience made them aware of the need to adapt their communication to the child's level, and 74 per cent agreed that the experience increased their skills in talking to children (Vanderpool and Parcel, 1979).

During the second year of medical school the courses on history taking and physical diagnosis offer an excellent opportunity to teach normal growth and development of infants and children. Further, behavioral and physical deviations (such as encopresis or dysmorphic syndromes) can be integrated into the pathology curriculum. The developmental-behavioral pediatric faculty members should take every opportunity to teach freshman and sophomore students as well as those in the clinical years so that their interest in children's development can be fostered.

The clinical years, particularly the third year clerkship, provide an opportunity to teach all future physicians about the developmental-behavioral functioning of each child they see, regardless of the setting (outpatient clinic or ward) or the condition (mild cold or terminal leukemia). The student clerk has more time to relate to both the child and the parents. With appropriate supervision, the broad realities of the child's life (play, friends, family) and the parents' fears and aspirations can be recognized by the student. Every child can be a developmental-behavioral case study, and every student will thus become a better physician dealing with children regardless of the specialty chosen.

The fourth year often offers an opportunity for students who have decided on their future specialty to choose an elective. The multidisciplinary needs of children with developmental-behavioral problems allows an integrated approach for students considering a career in neurology, orthopedics, psychiatry, ophthalmology, otolaryngology, and physical medicine and rehabilitation as well as pediatrics or family practice. Physicians entering any of these specialties as well as those entering medicine and surgery would benefit from exposure to a number of handicapped children and young adults with a variety of conditions. Since 10 to 15 per cent of the population have developmental or behavioral difficulties, most physicians have frequent contact with these individuals regardless of their chosen field of practice.

The timing of experiences during the years of pediatric residency requires review of basic developmental-behavioral knowledge and interviewing skills during the PL-1 year and introduction of a core residency curriculum when most appropriate in the individual program. There is a common belief that developmental-behavioral pediatrics should not (some say "cannot") be taught in the first year when residents are learning to deal with life and death issues. However, in four of 11 residency programs described by Friedman et al. (1981) in which a mandatory block behavioral pediatrics rotation was used in the PL-1 year, residents and faculty agreed that that was when it should be taught, and the residents did relate successfully to behavioral topics. Even when the core block rotation in developmental or behavioral pediatrics is postponed until the PL-2 or PL-3 year, child development issues and other introductory knowledge and skills should be integrated into the outpatient, ward, and continuity clinic experiences of the PL-1 year.

The majority of programs teaching developmental-behavioral pediatrics choose a block rotation, stressing that, as in other core areas of pediatrics, the residents need to be free of other major responsibilities in order to learn (Friedman et al., 1981). There are others, however, who believe strongly that developmental-behavioral issues are so totally interwoven into the care of every child that the experience is best integrated into the regular rotations throughout the residency years. It is more difficult to draw the resident's attention away from the pressures of acute medical management while on ward or outpatient services (as is well known by the difficulty in establishing a smoothly operating and well attended continuity clinic experience each week throughout the residency), but there is no doubt that the appropriate patients are available in the intensive care nursery, general outpatient clinics, inpatient services (endocrinology, cardiology, intensive care), as well as outpatient specialty services such as newborn high risk follow-up, school behavior, or birth defects clinics. The attraction in the majority of programs, however, of one or two month block rotations is the opportunity for the resident to follow complex cases from intake through interdisciplinary assessment and staffing to the initiation of management. This is difficult to schedule even with a block rotation, but much more difficult in the face of routine ward or outpatient clinic obligations.

In the initial one or two month block, there is time for only a limited experience with even the most common of the major handicapping conditions, e.g., developmental dysfunctions affecting school learning, mental retardation, cerebral palsy, communication disorders, or major psychopathological states. Therefore, the PL-2 or PL-3 resident with a particular interest in any aspect of developmental-behavioral pediatrics, after having completed the core experience, should have available elective months during which thorough exposure to a single area can be obtained. Often this may mean a rotation in another institution that offers a fellowship experience, but it may require only an additional month for expansion of clinical experience and skills.

Pediatric training programs that offer developmental-behavioral pediatric rotations need to consider the training of nonpediatric residents as well. With little or no developmental-behavioral pediatrics in most medical school curricula, it is clear that specialists in other fields will obtain training only if it is provided during their residency. The most urgent need is for training of family practice residents who will be providing a significant portion of primary medical care to children, but residents in neurology and psychiatry also need a high level of expertise in these areas. There is an assumption on the part of schools and the public that any neurologist or any psychiatrist is an expert in the management of school learning or behavior problems. This is true, of course, only to the degree that such expertise is taught through exposure to developmental-behavioral pediatrics and the related disciplines. Of the other specialties, residents in orthopedics, neurosurgery, physical medicine and rehabilitation, otolaryngology, and ophthalmology should have some formal exposure to the management of children with developmental disabilities. Particularly the orthopedist and neurosurgeon need to be taught the full range of clinical attitudes and communication skills because of frequent contact with children with cerebral palsy and central nervous system malformations, and their families.

RESOURCES

In planning a developmental-behavioral pediatric curriculum, pediatric department chairmen and pediatric residency training program directors must consider their resources in terms of faculty, students, clinical opportunities (patients and the organization of services), and supplementary teaching opportunities (lectures, seminars, readings, audiovisual programs, and visits to schools, day care settings, and institutions). The approximately 240 North American pediatric residencies, and to a lesser extent the approximately 120 medical schools, vary widely in number of faculty and students and in breadth of clinical services available. All, however, have some opportunities to teach developmental-behavioral pediatrics and should assign a member of the pediatric faculty to oversee the design and implementation of the curriculum. Whenever possible, the faculty member should be a pediatrician with fellowship training in developmental or behavioral pediatrics who has a commitment to cover all aspects in the curriculum. The temptation to turn the training responsibilities over to a neurologist or a psychiatrist should be avoided, since developmental-behavioral pediatrics is increasingly a pediatric subspecialty.

MODELS

Two efforts of national scope focusing on residency curricula have resulted in models or guidelines available to pediatric residency training program directors. The results of the W. T. Grant Foundation's encouragement of behavioral pediatrics teaching in 11 residency programs have been summarized by Friedman et al. (1981). The report is of interest in its description of a wide variety of approaches to behavioral pediatric teaching. A detailed curriculum in developmental pediatrics has been developed by a national task force supported by the Office of Special Education, United States Department of Education, beginning in 1979 (Guralnick et al., 1982). The curriculum focuses on the development of clinical skills required for the comprehensive diagnosis, assessment, and medical management of children with a variety of handicapping conditions. In addition to protocols describing specific clinical activities involved in screening, diagnosis, assessment, interdisciplinary planning, parent informing, and long term management, goals, objectives, content outlines, and core and supplementary reading lists are provided. The curriculum has been implemented in 20 residency programs.

CURRICULUM CONTENT

The design of the curriculum content requires the identification of goals and objectives for each level of participation. Here we consider primarily the content for the medical student, the pediatric resident, and the fellow. The content for the continuing education of practitioners will bridge the same body of skills and knowledge taught to the resident and fellow, depending on the degree of interest and previous training or experience of the practitioner. In keeping with the abilities cited by the American Board of Pediatrics, Inc. (1974), objectives will be considered in terms of attitudes, factual knowledge, interpersonal skills, technical skills, and clinical judgment.

MEDICAL STUDENT CURRICULUM

The developmental-behavioral pediatric curriculum for the medical student aims at providing the minimal level of competence for any physician, regardless of the field of practice (Table 63–1). Attitudes that must be conveyed include an acceptance that a difference in behavior or development be viewed as a variation, not necessarily a

Table 63–1. CONTENT OUTLINE OF MEDICAL STUDENT CURRICULUM

Goals for attitudes
- Accept differences in behavior or development as variations, not necessarily disorders
- Recognize the limits and adverse consequences of labeling children
- Develop an attitude of advocacy for the rights of children with behavioral-developmental variation
- Display empathy and support for the parents and other family members of a child with a behavioral-developmental variation

Goals for factual knowledge
- Understand the competence of infants and the relationship between biological endowment and environmental experience in learning and development
- Be aware of the range of normal development in motor, language, cognitive, and social skills and the impact of dysfunction on the development of these skills
- Cite the common prenatal, perinatal, and postnatal factors that lead to dysfunction and describe the current common preventive efforts and their effectiveness

Goals for interpersonal and technical skills
- Obtain accurate information from parents about a child's past and current behavior
- Appraise the learning environment at home and in school
- Elicit feelings from parents
- Use a developmental screening procedure reliably
- Observe and describe accurately common dysmorphic features
- Recognize common dysmorphic syndromes

Goal for clinical judgment
- Recommend consultation or therapeutic intervention at the appropriate time

disorder. Related to this, the medical student should become aware of the inadequacy of labels in describing a child's development or behavior and the pejorative nature and limiting consequences of most labels. Curricula should foster an attitude of advocacy for the rights of the child with a behavioral variation or a developmental handicap. Finally they should develop an attitude of empathy and support for the child's parents and other family members.

The factual knowledge that medical students should obtain, as already discussed, includes recent advances in developmental psychology, such as environmental influences on the competence of the infant. The normal processes of developing motor, language, cognitive, and social skills as well as the degree of variation and the impact of dysfunction must be appreciated. Medical students should also learn the more common genetic and prenatal, perinatal, and postnatal factors that lead to developmental-behavioral deviation and be aware of current efforts at prevention and their effectiveness.

Interviewing skills unique to developmental-behavioral pediatrics that should be learned by medical students include an accurate elicitation of information about the child's past and current behavior and skills and an appraisal of the learning environment at home and in school. Further, the student must retain the ability to elicit feelings of the parents. Helfer (1970) noted that, in contrast to learning the latter skill, seniors were less effective than freshman medical students in obtaining such information. He provided the following example:

> One programed mother had a three year old retarded child with meningomyelocele and hydrocephalus. She was three months pregnant and was most concerned about the outcome of this pregnancy. No senior discovered this fact, but in his interview the first freshman who interviewed her asked, "Are you going to have any more children?" After learning of her pregnancy he replied, "You must be worried that you'll have another baby with the same problem" (pp. 625–626).

Technical skills at the medical student level include direct appraisal of development through use of a screening instrument, recognition of the more common syndromes, and accurate observation and description of common dysmorphic features. Clinical judgment includes learning a sense of timing for recommending a consultation or therapeutic intervention.

PEDIATRIC RESIDENT CURRICULUM

The content of the developmental-behavioral pediatric resident curriculum includes the core attitudes, knowledge, and skills necessary for the competent functioning of a pediatrician in general

Table 63–2. CONTENT OUTLINE OF CORE PEDIATRIC RESIDENT CURRICULUM*

Goals for attitudes
- Display the attitudes expected at the medical student level (see Table 63–1)
- Demonstrate an awareness of and interest in the status of public acceptance of children with developmental-behavioral variations, including recent changes in children's rights and commonly held misconceptions about such children
- Display a positive accepting attitude toward children with developmental-behavioral variations by sensitively and appropriately interacting with the children and their families
- Demonstrate an understanding of ethical issues regarding children with developmental-behavioral variation, such as therapeutic abortion and passive euthanasia, and formulate his own position on these issues

Goals for factual knowledge
- Display the factual knowledge expected at the medical student level (see Table 63–1)
- State the common definitions and classification systems of developmental-behavioral variation, such as mental retardation, learning and communication disorders, cerebral palsy, hearing or visual impairments, major psychopathological disorders, and multiple handicaps
- Describe the presentation, natural history, and associated developmental problems of the conditions already cited
- State the incidence, clinical manifestations, and prognosis of the major etiologies of the conditions already cited
- Describe in detail the approach to prevention of the conditions already cited in terms of prenatal, perinatal, and postnatal preventive efforts
- Define the roles and contributions of other disciplines in the development of a management plan for a child with developmental-behavioral variation
- State the educational rights of children with developmental-behavioral variation and describe access to the community resources available to meet their educational and other needs
- Display familiarity with classic and contemporary developmental, psychological, and educational research and demonstrate awareness of particular research design constraints in these areas

Goals for interpersonal skills
- Display the interpersonal skills expected at the medical student level (see Table 63–1)

- Advise parents appropriately regarding contemporary issues in child development, such as both parents working outside the home, the selection of a preschool, and the impact of television
- Participate effectively as a member of a multidisciplinary or interdisciplinary diagnostic team and as a member of a community or school based planning team
- Demonstrate awareness of common immediate and continuing parental reactions to a child with a significant developmental-behavioral variation
- Maintain a continuing relationship with parents and other professionals participating in the long term management of children with developmental-behavioral variation, serving either as team member or as case manager

Goals for technical skills
- Display the technical skills expected at the medical student level (see Table 63–1)
- Demonstrate the regular use of developmental screening in a continuity practice
- Identify and organize potential contributing or etiological factors by obtaining a complete medical history, performing a full physical examination, and taking other initial medical steps for each child suspected of having a significant developmental-behavioral variation
- Assess, in a preliminary manner, the areas of hearing, vision, motor, language, and socioemotional development and function
- Apply specific management techniques in medical interventions, behavior modification, genetic counseling, and family counseling

Goals for clinical judgment
- Display the clinical judgment expected at the medical student level (see Table 63–1)
- Translate the theories and processes of child development into clinical judgments regarding normal children and those with developmental-behavioral variations
- Evaluate parent-child interactions and child rearing practices and determine the implications of the findings for the child's development or behavior
- Integrate the clinical findings from a preliminary assessment into a written report or an oral summary
- Respond sensitively and appropriately to ways in which a child with a developmental-behavioral variation can influence family dynamics
- Make appropriate decisions regarding the long term management of a child with a developmental-behavioral variation

*Adapted from Bennett, F. C., et al.: Curriculum in developmental pediatrics. (In preparation.)

practice or in a subspecialty such as neonatology or cardiology. Beyond the core requirements, elective opportunities for general pediatric residents merge into the content of subspecialty training at the fellowship level. As outlined in Chapter 51, the practicing pediatrician without a special interest in developmental-behavioral pediatrics requires less expertise than the pediatrician who assumes a more extensive management role because of a special interest in the field.

The core content at the pediatric residency level is summarized in Table 63–2. The goals included in the table are adapted from the curriculum developed by Bennett et al. (in preparation). For

examples of specific objectives within each of the ability areas, see the section on Curriculum Evaluation on page 1217.

Because of variability in the clinical resources in pediatric training sites, the objectives of the residency curriculum can be met in a number of different ways. For instance factual knowledge can be taught by the resident's reading selected chapters or articles, attending lectures or seminars, observing films or slide-tapes, or discussing patients with faculty or fellow residents. Regardless of the method of learning, the resident's knowledge should find application in clinical settings. In developmental-behavioral pediatrics, as in other disciplines, it is clinical responsibility for the diagnosis and management of patients that provides the best learning opportunity.

Supervised resident experience should include thorough developmental-behavioral evaluations of children with the more common dysfunctions. To the extent possible, the resident should take a major role in the management of children with varying degrees of mental retardation, with cerebral palsy, with learning or communication disorders but average intelligence, and with common behavioral deviations such as recurrent abdominal pain, enuresis, tics, and obesity.

In addition to supervised exposure to a wide range of clinical conditions, residents should be afforded supervised experiences with developmental-behavioral problems of children of different ages. The attitudes, knowledge, and skills necessary for the competent management of an infant with Down's syndrome are totally different from those needed for the care of an adolescent with the same condition. The resident who enters general pediatric practice will encounter this condition as often as diabetes mellitus. (Encopresis will be encountered five times more often than either.) Therefore adequate training necessitates direct, supervised clinical experience with a number of children of different ages with at least the common developmental-behavioral variations. Further, several should be fully assessed in the context of an interdisciplinary team, and as many as possible should be managed over time, for instance, in the resident's continuity clinic.

FELLOWSHIP CURRICULUM

In many regards, fellowship training in developmental-behavioral pediatrics is an opportunity to broaden and deepen the knowledge and skills introduced at the pediatric resident level. Extensive clinical experience with a wide spectrum of children, longer follow-up, and continuing work with members of related disciplines form the basis of training in this subspecialty. There are, however, three additional measures unique to the fellowship level that are crucial for the further development of knowledge and clinical services in developmental-behavioral pediatrics. These are the improvement of teaching skills, an introduction to developmental-behavioral research, and the development of skills in program planning and administration, including a working knowledge of governmental procedures, public health policies, and the relationships among community resources. A further skill emphasized in some fellowship programs is the "transdisciplinary" approach in which the tools, techniques, and methodologies of several medical and nonmedical disciplines are used in patient management (Capute and Accardo, 1980).

CURRICULUM EVALUATION

The evaluation of a developmental-behavioral pediatric curriculum can be approached along a spectrum of sophistication. As a general rule, the more sophisticated, complex, or comprehensive the evaluation of a curriculum, the more resources, particularly in terms of faculty and student time, are used in the evaluation of learning rather than the learning itself. It is clear, however, that without any evaluation, ineffective or insufficient training efforts may continue unmodified, thus totally wasting faculty and student time. An estimated 5 to 10 per cent of the time (or resources) devoted to evaluation of the individual student, resident, or fellow and the evaluation of the overall impact of the curriculum would seem reasonable. Specifying goals and objectives for attitudes, knowledge, skills, and clinical judgment in terms that are measurable is essential for the accurate evaluation of a curriculum's effectiveness.

EVALUATION OF STUDENT OR TRAINEE

The basic measure of effectiveness of any course or training rotation is learning on the part of each student or trainee. A variety of observations can be made to assess such learning. Attitudes and skills can be assessed by written or oral checklists filled out by the student or the instructor. Knowl-

Table 63–3. MATRIX ASSOCIATING TASKS AND ABILITIES*

Abilities	Tasks		
	1: Gathering, Organizing, and Recording Data	2: Assessing Data	3: Managing Problems and Maintaining Health
A. Attitudes	A-1	A-2	A-3
B. Factual knowledge	B-1	B-2	B-3
C. Interpersonal skills	C-1	C-2	C-3
D. Technical skills	D-1	D-2	D-3
E. Clinical judgment	E-1	E-2	E-3

*From The American Board of Pediatrics, Inc.: Foundations for Evaluating the Competency of Pediatricians. Chicago, American Board of Pediatrics, Inc., 1974.

Table 63–4. EXAMPLES OF EDUCATIONAL OBJECTIVES FOR EACH CELL OF TASKS-BY-ABILITIES MATRIX*

Clinical problem: A 26 year old mother has a two year old daughter who has Down's syndrome with mild to moderate developmental delay. The mother is 12 weeks pregnant.

- A-1 Attitudes: data gathering
  The resident displays sensitivity in interacting directly with the child in the process of assessing her current adaptive developmental level.
- A-2 Attitudes: assessment of data
  The resident chooses to take extra time to explain to the mother the definition of "mild to moderate developmental delay" and the implications for future function on discovering that she believes that it means that her child will never progress beyond the current 12 to 15 month developmental level.
- A-3 Attitudes: management
  The resident discusses the relationship between amniocentesis and abortion with the mother regarding her current pregnancy.
- B-1 Factual knowledge: data gathering
  The resident recognizes the chromosome karyotype of 46, xx, t (14q21q).
- B-2 Factual knowledge: assessment of data
  The resident states three theories explaining the increased frequency of Down's syndrome with advanced maternal age.
- B-3 Factual knowledge: management
  The resident describes how the public school will involve the mother in planning for future appropriate classroom placement.
- C-1 Interpersonal skills: data gathering
  The resident uses vocabulary that the mother is able to understand when obtaining perinatal and developmental history from her.
- C-2 Interpersonal skills: assessment of data
  The resident listens to the prognosis of the two year old as stated by other professionals at the staffing conference.
- C-3 Interpersonal skills: management
  The resident discusses general health care needs (diet, exercise, sleep, play) with the mother.
- D-1 Technical skills: data gathering
  The resident reliably carries out a brief preliminary assessment of hearing, vision, motor, language, and socioemotional development of the two year old.
- D-2 Technical skills: assessment of data
  The resident accurately interprets the data obtained in the preliminary developmental assessment.
- D-3 Technical skills: management
  The resident correctly advises the mother regarding the recurrence risk of Down's syndrome in discussing with her the current pregnancy.
- E-1 Clinical judgment: data gathering
  The resident observes the interaction between the mother and child during the interview and relates the observation to subsequent recommendations for behavior management.
- E-2 Clinical judgment: assessment of data
  The resident asks to confer with both parents when it is discovered that the parents differ in their beliefs regarding abortion.
- E-3 Clinical judgment: management
  The resident inquires about the professional training and developmental perspective of staff members at the preschool before referring the child there for placement.

*Adapted from Richardson, H. B., et al. In Guralnick, M. J., and Richardson, H. B. (Editors): Pediatric Education and the Needs of Exceptional Children. Baltimore, University Park Press, 1980.

edge can be measured more objectively by written or oral tests administered formally (e.g., items from standardized multiple choice examinations, such as that of the American Board of Pediatrics) or informally (the instructor's probing the student's knowledge with questions during case discussions). Interpersonal and technical skills can be evaluated by direct observation or rating of videotapes in actual or simulated clinical settings. Clinical judgment also can be observed in clinical situations or through written or oral patient management problems.

Developmental-behavioral pediatric teaching can be evaluated following the same organizational approach as other aspects of pediatric practice. For instance, the American Board of Pediatrics divides the clinical problem-solving process into abilities and tasks according to the matrix presented in Table 63–3 (American Board of Pediatrics, Inc., 1974). Examples of objectives amenable to evaluation within each cell of the matrix are presented in Table 63–4 (also see Vaughan, 1980).

EVALUATION OF THE OVERALL CURRICULUM

Evaluation of the overall acceptance and effectiveness of a course or training rotation is done constantly by medical school pediatric department chairmen and pediatric residency training program directors. However, the usual approach to such evaluation is not designed to yield information that can specifically attribute learning to participation in the rotation itself. As noted, attitudes, knowledge, skills, and judgment can be assessed through a combination of objective and subjective measures. In addition, students or residents provide feedback regarding what they like or dislike and what they find valuable. Faculty members indicate by their participation, or lack of it, what they find of practical value in teaching students or residents. Important as these types of evaluation are for appropriate decision making in curriculum planning, a more systematic and comparative approach to measuring the impact of a course or rotation should be attempted when possible. Two recently reported designs for overall curriculum evaluation of developmental-behavioral pediatric rotations are described briefly as examples among the various approaches that might be undertaken.

Phillips et al. (1981) described the evaluation of a mandatory two month behavioral pediatrics rotation for first year pediatric residents emphasizing normal growth and development, common behavior problems, and the psychosocial needs of the hospitalized child or adolescent. The residents were asked to report on their perceived compe-

EXPERIMENTAL DESIGN

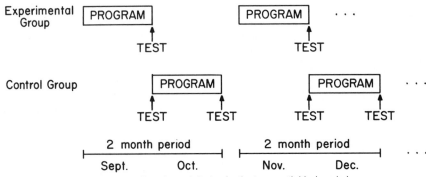

Figure 63–1. Experimental design for the two month block period.

tence on a five point scale in diagnosis and management of a number of physical (e.g., streptococcal pharyngitis), behavioral (e.g., school phobia), and mixed (e.g., Down's syndrome) clinical entities at each of four points during the year (beginning of the year, just before behavioral pediatrics rotation, just after behavioral pediatrics rotation, and at the end of the year). Data were obtained from 37 residents over a three year period. The results showed a gradual increase in perceived competence over the course of the year in the diagnosis and management of physical conditions. However, a marked increase in the scores for the behavioral conditions occurred, which correlated directly with the beginning and end of the behavioral pediatric rotation. The mixed conditions revealed an intermediate pattern of increase in perceived competence by the residents. Although the outcome measure (perceived competence) is subjective, the study design establishes that the increase in relation to behavioral and mixed entities results from the specific behavioral pediatrics rotation, not from gradual growth over the course of the year.

In a study using an objective measure of resident knowledge about handicapping conditions and their management, skills in describing specific types of behavior of handicapped children, and attitudes toward handicapped children as judged by how positively the resident described the children's behavior, Richardson and Guralnick (1978) noted significant changes that could be attributed to a brief developmental pediatrics rotation. Specifically, a post-test only with control group design was used by randomly assigning residents to the first or second months of five sequential two month experimental periods during the school year (September through June). In this manner, for example, two residents in the experimental group would receive a post-test on completion of their rotation. At the same time or the following week, the next group of two residents (control

group) would receive the same test at the beginning of their rotation (Fig. 63–1). By repeating the sequence of experimental and control residents every two months, an adequate total subject population in each group was obtained. In addition, although their post-test scores were not part of the analysis, control group residents would then participate in the rotation.

The results of the comparison between the experimental and control groups (nine residents in each group) revealed that residents in the experimental group were significantly more knowledgeable and held more positive attitudes toward handicapped children than residents in the control group. The curriculum was not effective, however, in increasing the residents' skills in describing specific behavior patterns in the children observed.

The ultimate test of a curriculum's effectiveness is an objective measure of its long term impact on pediatric behavior outside the experimental setting. Such tests are generally beyond the scope and budgets of most medical school or residency training programs. Nonetheless the faculty members responsible for components of a developmental-behavioral pediatric curriculum should carry out whatever evaluation is possible in a carefully designed and objective manner. Only in this way will pediatric educators be able to clarify what is effective and revise their training programs accordingly.

H. BURTT RICHARDSON, JR.
MICHAEL J. GURALNICK

REFERENCES

American Board of Pediatrics, Inc.: Foundations for Evaluating the Competency of Pediatricians. Chicago, American Board of Pediatrics, Inc., 1974.
Bennett, F. C., Heiser, K., Richardson, H. B., and Guralnick, M. J.: Curriculum in developmental pediatrics. (In preparation.)

Capute, A. M., and Accardo, P. J.: A fellowship program on the needs of exceptional children. *In* Guralnick, M. J., and Richardson, H. B. (Editors): Pediatric Education and the Needs of Exceptional Children. Baltimore, University Park Press, 1980.

Delaney, D. W., Bartram, J. B., Olmsted, R. W., and Copps, S. C.: Curriculum in handicapping conditions: implications for residency, fellowship, and continuing education of pediatricians. *In* Guralnick, M. J., and Richardson, H. B. (Editors): Pediatric Education and the Needs of Exceptional Children. Baltimore; University Park Press, 1980.

Dworkin, P. H., Shonkoff, J. P., Leviton, A., and Levine, M. D.: Training in developmental pediatrics. Am. J. Dis. Child., 133:709–712, 1979.

Friedman, S. B., Phillips, S., and Parrish, J.: Current status of behavioral pediatric training for general pediatric residents: a study of 11 funded programs. Pediatrics. (In press.)

Gorham, K. A., Des Jardins, C., Page, R., Pettis, E., and Schreiber, B.: Effect on parents. *In* Hobbs, N. (Editor): Issues in the Classification of Children. San Francisco, Jossey-Bass, Inc., 1975, Vol. 2.

Gray, J. D., Cutler, C. A., Dean, J. G., and Kempe, C. H.: Prediction and prevention of child abuse and neglect. Child Abuse Neglect Int. J., 1:1–14, 1977.

Guralnick, M. J., Richardson, H. B., and Kutner, D. R.: Pediatric education and the development of exceptional children. *In* Guralnick, M. J., and Richardson, H. B. (Editors): Pediatric Education and the Needs of Exceptional Children. Baltimore, University Park Press, 1980.

Guralnick, M. J., Richardson, H. B., and Heiser, K. E.: A curriculum in handicapping conditions for pediatric residents. Except. Child., 48:338–346, 1982.

Haggerty, R. J.: The changing role of the pediatrician in child health care. Am. J. Dis. Child., 127:545–549, 1974.

Haskins, R., Finkelstein, N. W., and Stedman, D. J.: Infant-stimulation programs and their effects. Pediatr. Ann., 7:124–144, 1978.

Helfer, R. E.: An objective comparison of the pediatric interviewing skills of freshmen and senior medical students. Pediatrics, 45:623–627, 1970.

Johnston, R. B., and Magrab, P. R.: Introduction to developmental disorders and the interdisciplinary process. *In* Johnston, R. B., and Magrab, P. R. (Editors): Developmental Disorders: Assessment, Treatment, Education. Baltimore, University Park Press, 1976.

Phillips, S., Friedman, S. B., Smith, J. C., and Felice, M.: Evaluation of a residency training program in behavioral pediatrics. Pediatrics. (In press.)

Richmond, J. B,: An idea whose time has arrived. Pediatr. Clin. N. Am., 22:517–523, 1975.

Richardson, H. B., and Guralnick, M. J.: Pediatric residents and young handicapped children: curriculum evaluation. J. Med. Ed., 53:487–492, 1978.

Richardson, H. B., Guralnick, M. J., Taft, L. T., and Levine, M. D.: A comprehensive curriculum in child development and handicapping conditions: Prospects for design, implementation, and evaluation. *In* Guralnick, M. J., and Richardson, H. B. (Editors): Pediatric Education and the Needs of Exceptional Children. Baltimore; University Park Press, 1980.

Sameroff, A. J., and Chandler, M. J.: Reproductive risk and the continuum of caretaking casualty. *In* Horowitz, F. D., Hetherington, M., Scarr-Salapatek, S., and Siegel, G. (Editors): Review of Child Development Research. Chicago, University of Chicago Press, 1975, Vol. 5.

The Task Force on Pediatric Education: The Future of Pediatric Education. Evanston, Illinois, American Academy of Pediatrics, 1978.

Vanderpool, N. A., and Parcel, G. S.: Interacting with children in elementary schools: An effective approach to teaching child development. J. Med. Ed., 54:418–420, 1979.

Vaughan, V. C., III: Evaluating competencies: Relationship to content of a developmental disabilities curriculum for physicians. *In* Guralnick, M. J., and Richardson, H. B. (Editors): Pediatric Education and the Needs of Exceptional Children. Baltimore, University Park Press, 1980.

64

The Right to Be Different

This book has explored variation in the developing human. It has presented a view of "normal children," but it also has told of the lives of some who are small, some who are fearful, and some who are developmentally delayed, gifted, atypical in their behavior, unusual in appearance, early or late or uncertain sexually, deaf, economically impoverished, terminally ill, or otherwise variant. The features have been described. It is now appropriate to consider some premises regarding variation and deviation—and the stakes.

It is predictable that in a culture that strives for uniformity, substantial personal cost is associated with differentness. Equality and conformity therein run the risk of becoming confounded, engendering the implicit danger of reduced opportunity for fulfillment in the areas of education and treatment, life activities, and social acceptance.

HUMAN RIGHTS

A gratifying byproduct of the social revolution that took place in the 1960's and 1970's has been a codification of the fundamental liberties and prerogatives of all humans. These affirmations were promulgated in organizational assertions or activist tracts, sometimes with legislative (or other governmental) approbation. The landmark occasions for handicapped persons are represented by Public Law 94-142 (see Chapter 61), Title XIX, the Developmentally Disabled Assistance and Bill of Rights Act, the Section 504 amendment to the Vocational Rehabilitation Act, and the endorsement by the United Nations of the Declaration of General and Special Rights of the Mentally Retarded (developed by The International League of Societies for the Mentally Retarded).

Most listings of human rights as they pertain to exceptional individuals deal with a common basic inventory (Crocker and Cushna, 1976). They invoke a defense of "normal rights," such as family living, educational opportunities, treatment and habilitation services, employment opportunities, establishment of contracts, and confidentiality in personal records. It is then essential to state as well certain "special rights," which acknowledge particular vulnerability, including issues of guardianship, protection in drug or behavioral treatments, consent in experimental procedures, counseling regarding reproduction, and intelligent exposure to life situations involving risk. In recent times an increasing emphasis has been placed as well on securing a setting of specifically "normalized" experiences and circumstances (Wolfensberger, 1972) and freedom from architectural and environmental barriers.

It is reasonable to state that a fair amount of success has been achieved regarding the rights to education and treatment, the legalistic aspects of protection, and architectural barriers, although vigilance is continually needed. A substantial lag pertains in the social realm.

THE VERY STANDARD DEVIATION

What is meant by different? This has quantitative and qualitative elements. Not uncommonly an instrument-dependent norm is established (e.g., regarding intelligence or visual acuity) by studies of an accessible population. In the same human group, one standard deviation of measurement away from the norm then represents a "borderline" situation, with troubling unusualness, but redemption if personal factors are favorable. It is a reality that two standard deviations of "downward" variation indicate true exceptionality (e.g., mild mental retardation; see Chapter 39A). Four or five such standard deviations produce an abnormality so assertive as to be culturally startling, and a sequence of potential exclusion begins. These difficulties are multiplied when behavior or social style is simultaneously aberrant.

The relativity of exceptionality is further demonstrated by consideration of special human circumstances. Groce (1980) has described the responses on the island of Martha's Vineyard in the eighteenth and nineteenth centuries to an extraordinary incidence of hereditary deafness, potentiated by environmentally conditioned consanguineous mating. Serious deafness existed in some remote villages at a rate from 4 to 25 per cent, producing a situation in which this usually exceptional handicap became a "normal" element of rural life. As a consequence young people grew up learning sign language as a natural aid to local relations; church services were automatically conducted in sign and verbal communication simultaneously, and no prejudicial implications existed for deafness.

Other instances of acceptance of pervasive handicap are known in certain large pedigrees in which "the disease" in the family has become a fact of life (such as the unusual reticuloendotheliosis described by Omenn [1965]). This is particularly true in X linked syndromes, in which there can be a quiet tolerance of the fatal involvement of many male children. A special population of individuals with Hunter disease exists in the Catskill area of New York State, in which death from this handicapping mucopolysaccharidosis occurs typically only in the fourth or fifth decade (Bebee and Formel, 1954). The involved males have been well integrated into the rural setting, although they have compelling personal special needs. One can also note the varied meaning of Down syndrome in differing circumstances. In the open population this constitutional disability can be critically exclusionary, but in the exceptional culture of a state residential facility for mental retardation, persons with Down syndrome often have a notably sanctioned status.

It has been appropriately championed that virtually all humans who have some degree of differentness resemble "normals" in many more ways than they differ from them. Moreover, they may differ from each other in more ways than they differ from normalcy. They harbor the same emotional feelings, yearnings for rewarding social relations, potential for growth, and liability for injury, suffering, and disease. Brightman (1975), in a small book for the instruction of children, has warmly expressed this universality across diversity by voicing, "I hope you can see, just how much you're like me. . . ."

THE ENABLING OF LABELING

Issues relating to the effects of assigning diagnostic "labels" to variant children have been vigorously discussed in the past decade, with a predictable ambiguity in the conclusions. In the educational or clinical milieu there is a tempting utility in appropriating an accurate descriptive term linking a child to others with the same or similar characteristics, so that specialized services can be mobilized. Further, it is a reality that consumer group achievements in the promotion and defense of programmatic rights have rallied around common convictions solidified by categorical labeling. At the same time, fears of elaborating self-fulfilling prophesies and biased service design are well founded. Mandell and Fiscus (1981) have summarized some risks of diagnostic labels in the educational setting:

- Labeling detracts from the development of appropriate individualized programs.
- Labels once assigned are difficult to remove.
- Labeling and subsequent categorical placement limit the opportunities for normal children to become familiar with pupils with handicaps.
- Labeling tends to establish conditions for mistakes in assessment of minority group children.
- Labeling minimizes the significance of societal and environmental factors.

Unquestionably the act of labeling courts the risk of mislabeling. This can occur in a variety of ways. It is perhaps best illustrated in considering children with behavioral and relatively low severity learning problems. In responding to the compulsion to find a diagnostic tag for a child, a clinician may seek to identify and feature only one aspect of a child's complex plight. There may be endless discussions about whether his failure to learn is primarily "emotional" or a "learning disability." In reality it is likely that such students suffer from an amalgam of factors that may include neurological predispositions, negative educational experiences, unresolved stresses at home, maladaptive coping strategies, difficulties with attention, or inadequate or overly coercive role models. To reduce such factors to one diagnosis may represent a real injustice and, more important, may thwart the mobilization of a much needed multifaceted intervention program.

Diagnostic labels also run the risk of reflecting the training and disciplinary biases of the labeler. A professional who is most comfortable in dealing with family problems may brand a child who is having school problems as "emotionally disturbed." One who is more neurologically biased may relate the same observable phenomena in school to "minimal brain damage." Although disciplinary inclinations never can be totally eliminated, it is likely that labels will encourage these propensities.

Diagnostic labels may become as permanent as a tattoo. That is to say, they may penetrate so deeply as to be nearly impossible to excise from the identity of a person. This may affect how others view the individual and also how that human perceives himself. In short, children grow and they change; their labels may grow and not change. Labels may artificially exceed the state of the art of assessment. In the field of developmental and behavioral pediatrics, for many clinical conditions diagnostic criteria are far from clear-cut. Features of some disorders tend to overlap. For example, many of the traits described as being typical of children with depression also are applicable to those with deficits of attention. In both groups, self-deprecatory comments, problems with sleep, dysphoria, somatic complaints, agitated behavior, and low self-esteem are common. Whether a child is labeled as having an attention deficit, a depression, or both may be a fairly arbitrary decision.

Closely related to the issue of labeling is the broader matter that might be called assessment semantics. The selection of words to describe a child's behavior, a perceived delay, or an individual difference can have major implications in terms of the way in which the world views and treats that child. For example, a particular youngster may be called "immature" by his teacher. That person may be referring to a pattern of behavior that includes impulsivity, emotional lability, overactivity, and an inability to persist at tasks. The entire complex of traits, in fact, may be secondary to some underlying learning problems that are impairing his adjustment to school. However, because the word "immaturity" was selected to describe the phenomenon, the implication is that he will "outgrow it," that no service is needed, and perhaps that he should be retained in first grade! The more basic decision about whether a child really has a problem can be influenced strongly by choices of words to describe phenomena. A youngster may be described as "highly independent," or, alternatively, the same child may be thought of as "a disciplinary problem" or "a loner." The connotations differ considerably. A child who has unusual interests tending toward noncomformity might be described by some observers as "socially maladjusted," whereas others might commend his willingness to "do his own thing." It is of interest that the term "eccentric" seldom is used in describing children. This may reflect a tendency to equate and label highly unusual behavior as pathological.

RESPONSES TO VARIATION: PUBLIC AND PRIVATE

Bewilderment is a common reaction to the unfamiliar. Social ostracism by peers and adults may be practiced against a child with unusual physical features or behavior, generally as a reflection of observer uncertainty. In earlier times this often led to a lasting residential segregation (see review by Cushna, 1976). Now, armed with right-to-least-restrictive-education legislative support, the child with special needs poses as well a perceived economic threat to the mainline citizenry. Development of "individualized educational plans" at the time of this writing is undertaken in 10 to 20 per cent of school children, varying in proportion to the basic affluence and resources of the school district. The fact is that these support systems—even when they include psychotherapy, other special services, recreational components, and 12 month coverage—are still less costly than the outcomes of neglect (or doing it by the old institutional system).

Bewilderment is assuredly also shared by the unusual child himself, awash in the phenomena of a world he never made. Being out of synchrony with one's peers robs one of needed positive feedback and is destructive of self-image (see Chapter 39H, section on adolescence). In recent times, supported by the Association for Retarded Citizens groups, young adults with mental retardation have begun to meet to discuss issues of self-concept and rights.

Differentness in their children exacts a toll from parents as well. Featherstone (1980) has documented eloquently the universal responses of guilt, fear, loneliness, and anger felt by mothers and fathers of disabled children. "This type of child should not be in a public place," a mother was told by another shopper when she took her nonacting-out but unusual appearing, profoundly retarded daughter to a supermarket (reported in a recent clinic experience). Effects on brothers and sisters are significant, but usually manageable (Crocker, 1981).

In recent years there has been a gratifying new awareness of stylistic difference. Variations in styles of being parents present a wide range of workable and interesting modes of nurturance. At the same time enhanced appreciation of infant and child temperament has promulgated a greater clinical tolerance for behavioral variation or the expression of unique styles during childhood. With this has come a new awareness of "the match," the nearly fortuitous but critical encounter between a child's unique stylistic and cognitive repertoire and the personal needs, expectations, and values of an important adult, such as a parent or a teacher.

GETTING CHILDREN CHANGED

Attempts to remedy variation can result from a complex of motivations. Pressing to change a child into a more conforming presentation, to relieve the stress of unusualness per se, must be thoughtfully reviewed. The use of psychoactive medications, for example, as a quick lift of the burden in behaviorally atypical children, may be a substitute for a more penetrating, child focused program of primary assistance in behavioral adaptation or, alternatively, tolerance by adults. A provocative project is now under way in Europe and Israel to ameliorate the image of children with Down syndrome by extensive reconstructive surgery on the eyelids, tongue, nose, and lips.

Counseling and guidance also raise significant ethical questions. Is there a danger that the therapist or counselor will superimpose his own values on a naive child? To what extent do various forms of therapeutic counseling edge the child toward uniformity and conformity? Should one press the development of social skills for a youngster who prefers to be alone? If a child appears to have gross motor delays and shows little or no interest in sports, is the adult world privileged to insist upon physical education for the child? It may be that such a child would prefer not to feature motor pursuits in his repertoire. Children, in fact, may have a right to "specialize." They may need help in resisting the over-riding drive of adults to make them good at almost everything. Specialization appears to be a right that is reserved for grown-ups. In certain instances it may need to be afforded to children also. At the very least, one should pause when a youngster about to undergo therapy protests, "You know, I like the way I am." One must modulate one's zeal to promulgate a standard form of adjustment. There may be some youngsters who are not fitting in particularly comfortably during childhood but are destined to be more effective adults than they are children. Altering their childhood conformation, in fact, may worsen their prognosis.

A FINAL WORD

If there is a commitment to the vitality of a pluralistic society, with its enrichment by the contributions of diverse citizens, there is indeed "the right to be different." There are increasing examples of this in our culture:

• The new movement for assistance in adoption of "hard to place" children (handicapped, behaviorally atypical, biracial) has been strikingly successful. In effect this has transformed the rejected child into the avidly chosen child. Several hundred adoptions of children with Down syndrome have now taken place, and in some cities there is a waiting list of couples for adoption of infants and young children with Down syndrome.

• The reform of "deinstitutionalization," bringing seriously disabled persons into circumstances of community living, has led to substantial victories. Major functional gains are commonplace when individualized programs are provided.

• In the "new" special education, the excitement of learning has been successfully incorporated into a vastly broader scope. The nonverbal child who "speaks" via a communication board,

the autistic child who goes home in two years after child and parent training, the atypical youth who receives reimbursement for graded accomplishments in a vocational training program— these are victories of the social revolution.

• When parent-professional communication is thoughtfully developed, the quality of parenthood in unusual circumstances can be enormously reinforced, to the gain of all parties. Self-help groups have been important as well in improving parental understanding of differentness. As one mother reported, "First I realized what she would never be, then I learned what she did not have to be, and finally I think I have come to terms with what she is and can be" (Pueschel et al., 1978).

• The orientation of young adults to the helping professions developed a momentum in the 1960's and 1970's and has been sustained. "Working with children" remains a major preoccupation of thoughtful youth, and assuredly youthful enthusiasm and insights are needed in developmental and behavioral programs. Acceptances of differences in human style and expression have characterized the recent generation.

The variations in human presentation delineated in this book can be considered the products of phenotypic, genotypic, sociocultural, and individual circumstances that characterize our species. The concept of "normal" or "average" is statistically perceivable, but often subject to political inducement, and assuredly humanistically irrelevant on many occasions. We have much to learn from a far broader view. And much to celebrate.

THE EDITORS

REFERENCES

Beebe, R. T., and Formel, P. F.: Gargoylism: sex-linked transmission in nine males. Trans. Am. Clin. Climatol. Assoc., 66:199, 1954.

Brightman, A. J.: Like Me. Cambridge, Behavioral Education Projects, Inc., 1975.

Crocker, A. C.: The involvement of siblings of children with handicaps. In Milunsky, A. (Editor): Coping with Crisis and Handicap. New York, Plenum Publishing Corp., 1981.

Crocker, A. C., and Cushna, B.: Ethical considerations and attitudes in the field of developmental disorders. In Johnston, R. B., and Magrab, P. R. (Editors): Developmental Disorders; Evaluation, Treatment, and Education. Baltimore, University Park Press, 1976.

Cushna, B.: They'll be happier with their own kind. In Koocher, G. P. (Editor): Children's Rights and the Mental Health Professions. New York, John Wiley & Sons, Inc., 1976.

Featherstone, H.: A Difference in the Family; Living with a Disabled Child. New York, Basic Books, Inc., 1980.

Groce, N.: Everyone here spoke sign language. Nat. Hist., 89:10, 1980.

Mandell, C. J., and Fiscus, E.: Understanding Exceptional People. St. Paul, West Publishing Co., 1981.

Omenn, G. S.: Familial reticuloendotheliosis with eosinophilia. N. Engl. J. Med., 273:427, 1965.

Pueschel, S. M. (Editor): Down Syndrome; Growing and Learning. Kanas City, Sheed Andrews & McMeel, 1978.

Wolfensberger, W.: The Principle of Normalization in Human Services. Toronto, National Institute on Mental Retardation, 1972.

INDEX

Note: Page numbers in *italics* refer to illustrations. Page numbers followed by (t) refer to tables.